LDA	lithium diisopropylamide
LDMAN	lithium 1-(dimethylamino)naphthalenide
LHMDS	=LiHMDS
LICA	lithium isopropylcyclohexylamide
LiHMDS	lithium hexamethyldisilazide
LiTMP	lithium 2,2,6,6-tetramethylpiperidide
LTA	lead tetraacetate
LTMP	=LiTMP
lut	lutidine
m-CPBA	*m*-chloroperbenzoic acid
MA	maleic anhydride
MAD	methylaluminum bis(2,6-di-*t*-butyl-4-methylphenoxide)
MAT	methylaluminum bis(2,4,6-tri-*t*-butylphenoxide)
Me	methyl
MEK	methyl ethyl ketone
MEM	(2-methoxyethoxy)methyl
MIC	methyl isocyanate
MMPP	magnesium monoperoxyphthalate
MOM	methoxymethyl
MoOPH	oxodiperoxomolybdenum(pyridine)-(hexamethylphosphoric triamide)
mp	melting point
MPM	=PMB
Ms	mesyl (methanesulfonyl)
MS	mass spectrometry; molecular sieves
MTEE	methyl *t*-butyl ether
MTM	methylthiomethyl
MVK	methyl vinyl ketone
n	refractive index
NaHDMS	sodium hexamethyldisilazide
Naph	naphthyl
NBA	*N*-bromoacetamide
nbd	norbornadiene (bicyclo[2.2.1]hepta-2,5-diene)
NBS	*N*-bromosuccinimide
NCS	*N*-chlorosuccinimide
NIS	*N*-iodosuccinimide
NMO	*N*-methylmorpholine *N*-oxide
NMP	*N*-methyl-2-pyrrolidinone
NMR	nuclear magnetic resonance
NORPHOS	bis(diphenylphosphino)bicyclo[2.2.1]-hept-5-ene
Np	=Naph
PCC	pyridinium chlorochromate
PDC	pyridinium dichromate
Pent	*n*-pentyl
Ph	phenyl
phen	1,10-phenanthroline
Phth	phthaloyl
Piv	pivaloyl
PMB	*p*-methoxybenzyl
PMDTA	*N,N,N′,N″,N″*-penta[...] triamine
PPA	polyphosphoric acid
PPE	polyphosphate ester
PPTS	pyridinium *p*-toluenesulfonate
Pr	*n*-propyl
PTC	phase transfer catalyst/catalysis
PTSA	*p*-toluenesulfonic acid
py	pyridine
RAMP	(*R*)-1-amino-2-(methoxymethyl)pyrrolidine
rt	room temperature
salen	bis(salicylidene)ethylenediamine
SAMP	(*S*)-1-amino-2-(methoxymethyl)pyrrolidine
SET	single electron transfer
Sia	siamyl (3-methyl-2-butyl)
TASF	tris(diethylamino)sulfonium difluorotrimethylsilicate
TBAB	tetrabutylammonium bromide
TBAD	=DBAD
TBAF	tetrabutylammonium fluoride
TBAI	tetrabutylammonium iodide
TBAP	tetrabutylammonium perruthenate
TBDMS	*t*-butyldimethylsilyl
TBDPS	*t*-butyldiphenylsilyl
TBHP	*t*-butyl hydroperoxide
TBS	=TBDMS
TCNE	tetracyanoethylene
TCNQ	7,7,8,8-tetracyanoquinodimethane
TEA	triethylamine
TEBA	triethylbenzylammonium chloride
TEBAC	=TEBA
TEMPO	2,2,6,6-tetramethylpiperidinoxyl
TES	triethylsilyl
Tf	triflyl (trifluoromethanesulfonyl)
TFA	trifluoroacetic acid
TFAA	trifluoroacetic anhydride
THF	tetrahydrofuran
THP	tetrahydropyran; tetrahydropyranyl
Thx	thexyl (2,3-dimethyl-2-butyl)
TIPS	triisopropylsilyl
TMANO	trimethylamine *N*-oxide
TMEDA	*N,N,N′,N′*-tetramethylethylenediamine
TMG	1,1,3,3-tetramethylguanidine
TMS	trimethylsilyl
Tol	*p*-tolyl
TPAP	tetrapropylammonium perruthenate
TBHP	*t*-butyl hydroperoxide
TPP	tetraphenylporphyrin
Tr	trityl (triphenylmethyl)
Ts	tosyl (*p*-toluenesulfonyl)
TTN	thallium(III) nitrate
UHP	urea–hydrogen peroxide complex
Z	=Cbz

Handbook of Reagents for Organic Synthesis

Acidic and Basic Reagents

OTHER TITLES IN THIS COLLECTION

Reagents, Auxiliaries and Catalysts for C–C Bond Formation
Edited by Robert M. Coates and Scott E. Denmark
ISBN 0 471 97924 4

Oxidizing and Reducing Agents
Edited by Steven D. Burke and Rick L. Danheiser
ISBN 0 471 97926 0

Activating Agents and Protecting Groups
Edited by Anthony J. Pearson and William R. Roush
ISBN 0 471 97927 9

Handbook of Reagents for Organic Synthesis

Acidic and Basic Reagents

Edited by

Hans J. Reich
The University of Wisconsin at Madison

and

James H. Rigby
Wayne State University

JOHN WILEY & SONS

Chichester • New York • Weinheim • Brisbane • Toronto • Singapore

Telephone (+44) 1243 779777

Email (for orders and customer service enquiries): cs-books@wiley.co.uk
Visit our Home Page on www.wileyeurope.com or www.wiley.com

Reprinted February 2000, May 2001, November 2003

Other Wiley Editorial Offices

John Wiley & Sons Inc., 111 River Street, Hoboken, NJ 07030, USA

Jossey-Bass, 989 Market Street, San Francisco, CA 94103-1741, USA

Wiley-VCH Verlag GmbH, Boschstr. 12, D-69469 Weinheim, Germany

John Wiley & Sons Australia Ltd, 33 Park Road, Milton, Queensland 4064, Australia

John Wiley & Sons (Asia) Pte Ltd, 2 Clementi Loop #02-01, Jin Xing Distripark, Singapore 129809

John Wiley & Sons Canada Ltd, 22 Worcester Road, Etobicoke, Ontario, Canada M9W 1L1

British Library Cataloguing in Publication Data

A catalogue record for this book is available from the British Library

ISBN 0 471 97925 2

Typeset by Thomson Press (India) Ltd., New Delhi
Printed and bound in Great Britain by Antony Rowe Ltd, Chippenham, Wiltshire
This book is printed on acid-free paper responsibly manufactured from sustainable forestry in which at least two trees are planted for each one used for paper production.

Contents

Preface

As stated in its Preface, the major motivation for our undertaking publication of the *Encyclopedia of Reagents for Organic Synthesis* was "to incorporate into a single work a genuinely authoritative and systematic description of the utility of all reagents used in organic chemistry." By all accounts, this reference compendium has succeeded admirably in attaining this objective. Experts from around the globe contributed many relevant facts that define the various uses characteristic of each reagent. The choice of a masthead format for providing relevant information about each entry, the highlighting of key transformations with illustrative equations, and the incorporation of detailed indexes serve in tandem to facilitate the retrieval of desired information.

Notwithstanding these accomplishments, the editors have since recognized that the large size of this eight-volume work and its cost of purchase have often served to deter the placement of copies of the *Encyclopedia* in or near laboratories where the need for this type of insight is most critically needed. In an effort to meet this demand in a cost-effective manner, the decision was made to cull from the major work that information having the highest probability for repeated consultation and to incorporate same into a set of handbooks. The latter would also be purchasable on a single unit basis.

The ultimate result of these deliberations is the publication of the *Handbook of Reagents for Organic Synthesis* consisting of the following four volumes:

Reagents, Auxiliaries and Catalysts for C–C Bond Formation
Edited by Robert M. Coates and Scott E. Denmark

Oxidizing and Reducing Agents
Edited by Steven D. Burke and Rick L. Danheiser

Acidic and Basic Reagents
Edited by Hans J. Reich and James H. Rigby

Activating Agents and Protecting Groups
Edited by Anthony J. Pearson and William R. Roush

Each of the volumes contains a complete compilation of those entries from the original *Encyclopedia* that bear on the specific topic. Ample listings can be found to functionally related reagents contained in the original work. For the sake of current awareness, references to recent reviews and monographs have been included, as have relevant new procedures from *Organic Syntheses*.

The end product of this effort by eight of the original editors of the *Encyclopedia* is an affordable, enlightening set of books that should find their way into the laboratories of all practicing synthetic chemists. Every attempt has been made to be of the broadest synthetic relevance and our expectation is that our colleagues will share this opinion.

Leo A. Paquette
Columbus, Ohio USA

Introduction

Recognizing the critical need for bringing a cost effective as well as handy reference work dealing with the most popular reagents in synthesis to the widest possible audience of practicing organic chemists, the editors of *The Encyclopedia of Reagents for Organic Synthesis (EROS)* have developed a list of the most important and useful reagents employed in contemporary organic synthesis. The result of this effort is a collection entitled "Handbook of Reagents for Organic Synthesis" that contains over 500 reagent entries from the original *Encyclopedia* that run the gamut of functions from oxidation and reduction to activation and protection. To assist the reader in quickly locating a reagent of interest, the "Handbook" has been divided into four volumes, each of which is devoted to a set of closely related reagents, these include: "Reagents, Auxiliaries and Catalysts for C–C Bond Formation"; "Oxidizing and Reducing Agents"; "Acidic and Basic Reagents"; Activating Agents and Protecting Groups".

Compiled in the present volume, entitled "Acidic and Basic Reagents," is a group of articles on the most useful and important acidic and basic agents that were originally included in the very popular *Encyclopedia of Reagents for Organic Synthesis*. Each article contains all of the information found in the original version as well as extensive listings of functionally related reagents that are located in the full *Encyclopedia*. Following this **Introduction** are listings of **Recent Review Articles and Monographs** on subjects related in a general sense to the topic of acids and bases as well as relevant **Organic Syntheses** that deal with either the preparations or reactions of reagents featured in this volume. To make this work as up to date as possible, particular emphasis was placed on including references appearing since the original publication date of EROS in 1995. It is hoped that by including these listings, the reader will be able to quickly access a broad range of information of interest that is beyond the scope of the reagent entries themselves.

Acids and bases are among the most fundamental and versatile reagents for effecting organic transformations, and in selecting candidate entries for inclusion in this particular collection, the editors adopted a fairly broad set of criteria for defining what exactly constitutes an acidic or basic reagent. Therefore, not only are the usual acids and bases such as hydrochloric acid, aluminum chloride, potassium *t*-butoxide and so forth included, so are compounds that behave like acids or bases but that are normally classified in other ways. For instance, articles on ligands or complexing agents such as 18-crown-6, 1,1′-bi-2,2′-naphthol and triethylphosphite can also be found in this volume along with the more traditional Bronsted/Lewis acids and bases. Furthermore, in recognition of the growing importance of biocatalysts in the field of organic synthesis, articles on esterases and lipases are also included in this volume since, in the broadest sense, the net conversions effected by these enzymes can be viewed as being of an acidic or basic nature.

In compiling the master list of reagents to be included in the "Handbook" the editors endeavored to consider not only species of the broadest current synthetic importance and accepted utility, but also reagents that may not yet be fully integrated into mainstream synthetic practice, but will certainly become so in the future. A good example of this is the entry on the very potent, but little used, phosphazene base P_4-*t*-Bu, one of the so-called Schwesinger bases, which the editors believe will become, in the fullness of time, a more broadly used reagent. A number of reagents present in this four volume series were identified as exhibiting multiple applications that could have simultaneously placed them appropriately in several different volumes. To avoid unnecessary duplication, articles describing these types of reagents were assigned to a single volume and only cross references to these particular entries are included in the other relevant volumes. For example, the entry for ethylaluminum dichloride is found in the volume entitled "Activating Agents and Protecting Groups" with only a "boiler plate" cross reference listing in this volume. Similarly, hexamethyldisilazane can serve as a base in some instances, but it can also be employed to transfer Me_3Si and, as such, will be found in the volume focusing on "Activating Agents and Protecting Gruops".

The editors of *The Encyclopedia of Reagents for Organic Synthesis* hope that you find this compilation of the best and most important acidic and basic reagents to be a valuable addition to your chemical library.

Hans J. Reich
The University of Wisconsin at Madison
James H. Rigby
Wayne State University

Organic Synthesis Examples

Metalation

"Erythro-Directed Reduction of a β-Keto Amide: *erythro*-1-(3-Hydroxy-2-methyl-3-phenylpropanoyl)piperidine," Fujita, M.; Hiyama, T. *Org. Synth.* **1990**, *69*, 44.

1. LDA
2. PhCOCl

"Asymmetric Synthesis of 4,4-Dialkylcyclohexenones from Chiral Bicyclic Lactams: (*R*)-4-Ethyl-4-allyl-2-cyclohexen-1-one," Meyers, A. I.; Berney, D. *Org. Synth.* **1990**, *69*, 55.

LDA/EtI
LDA/allyl bromide

RED-Al
$Bu_4NH_2PO_4$

"Intramolecular Oxidative Coupling of a Bisenolate: 4-Methyltricyclo[2.2.2.0^{3,5}]octane-2,6-dione," Poupart, M.-A.; Lassalle, G.; Paquette, L. A. *Org. Synth.* **1990**, *69*, 173.

PPA
HOAc, 100°C

1. LDA (2.2 equiv.)
2. $FeCl_3 \cdot DMF$ (inverse addition)

"2-Substituted Pyrroles from *N-tert*-Butoxycarbonyl-2-bromopyrrole: *N-tert*-Butoxy-2-trimethylsilylpyrrole," Chen, W.; Stephenson, E. K.; Cava, M. P.; Jackson, Y. A. *Org. Synth.* **1991**, *70*, 151.

BuLi, THF; Me_3SiCl

"Preparation and Use of (Methoxymethoxy) Methyl Lithium: 1-(Hydroxymethyl)cycloheptanol," Johnson, C. R.; Medich, J. R.; Danheiser, R. L.; Romines, K. R.; Koyama, H.; Gee, S. K. *Org. Synth.* **1992**, *71*, 140.

$Bu_3SnCH_2OCH_2OCH_3$ → 1) BuLi, THF, −78°C; 2) cycloheptanone

HCl / MeOH

"Ethyl 1-Naphthylacetate: Ester Homologation via Ynolate Anions," Reddy, R. E.; Kowalski, C. J. *Org. Synth.* **1992**, *71*, 146.

+ CH_2Br_2
1) LiTMP, −78°C
2) LiHMDS + LiOEt −78°C → −20°C
3) *sec*-BuLi, −78°C
4) BuLi, −20°C
5) EtOH/AcCl

"9-Bromo-9-phyenylfluorene," Jamison, T. F.; Lubell, W. D.; Dener, J. M.; Krisché, M. J.; Rapoport, H. *Org. Synth.* **1992**, *71*, 220.

1. C_4H_9Li/ether
2. fluorenone/THF
3. H_2O

48% HBr/H_2O
Toluene

"Stereoselective Aldol Reaction of Doubly Deprotonated (*R*)-(+)-2-hydroxy-1,2,2-triphenyl Ethyl Acetate (HYTRA): (*R*)-3-hydroxy-4-methylpentanoic Acid," Braun, M.; Gräf, S. *Org. Synth.* **1993**, *72*, 38.

2 LDA, THF/0°C; isobutyraldehyde, −126°C

KOH/H_2O/MeOH, reflux

"Synthesis of (*S*)-2-Methylproline: A General Method for the Preparation of α-Branched Amino Acids," Beck, A. K.; Blank, S.; Job, K.; Seebach, D.; Sommerfeld, Th. *Org. Synth.* **1993**, *72*, 62.

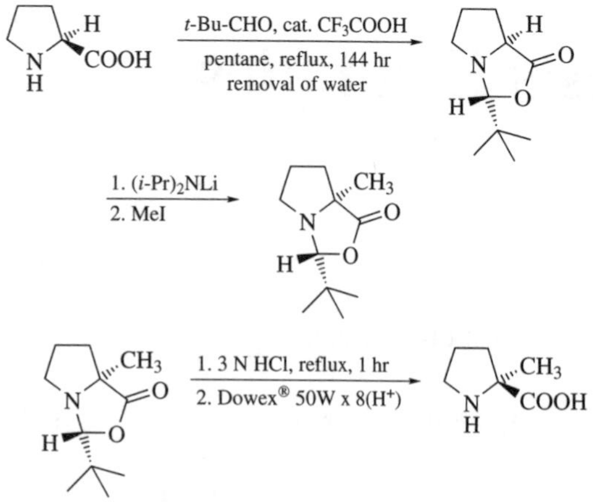

"7-Methoxyphthalide," Wang, X.; de Silva, S. O.; Reed, J. N.; Billadeau, R.; Griffen, E. J.; Chan, A.; Snieckus, V. *Org. Synth.* **1993**, *72*, 163.

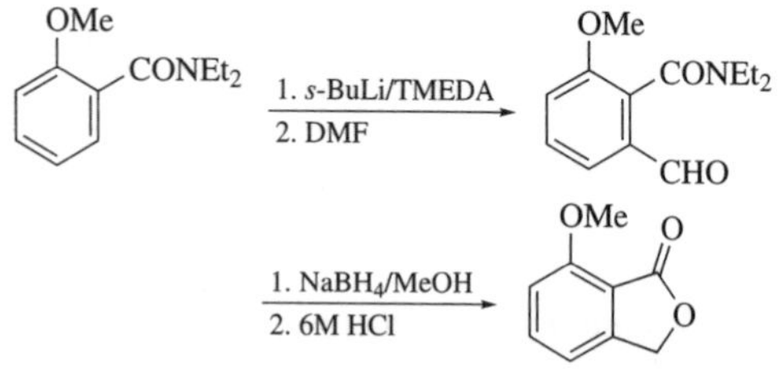

"Synthesis of 7-Substituted Indolines via Directed Lithiation of 1-(*tert*-Butoxycarbonyl)indoline: 7-Indoline Carboxaldehyde," Iwao, M.; Kuraishi, T. *Org. Synth.* **1995**, *73*, 85.

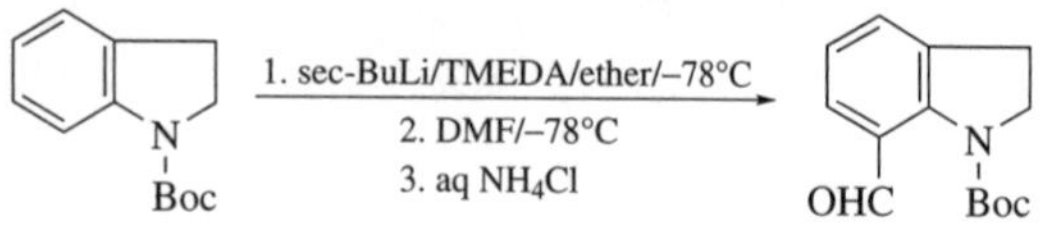

"Regio- and Stereoselective Intramolecular Hydrosilation of α-Hydroxy Enol Ethers: 2,3-*syn*-2-Methoxymethoxy-1,3-nonanediol," Tamao, K.; Nakagawa, Y.; Ito, Y. *Org. Synth.* **1995**, *73*, 94.

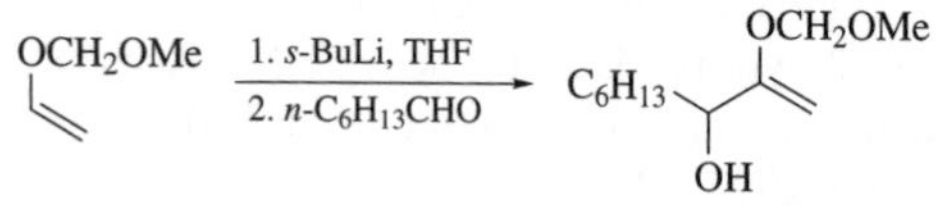

"Detrifluoracetylative Diazo Group Transfer: (*E*)-1-Diazo-4-phenyl-3-buten-2-one," Danheiser, R. L.; Miller, R. F.; Brisbois, R. G. *Org. Synth.* **1995**, *73*, 134.

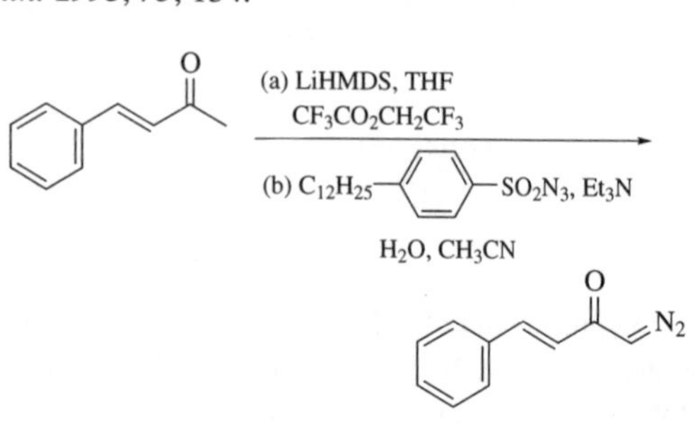

"Bis(trifluoroethyl) (carboethoxymethyl)phosphenate," (Patois, C.; Savignac, P.; About-Jaudet, E.; Collignon, N. *Org. Synth.* **1995**, *73*, 152.

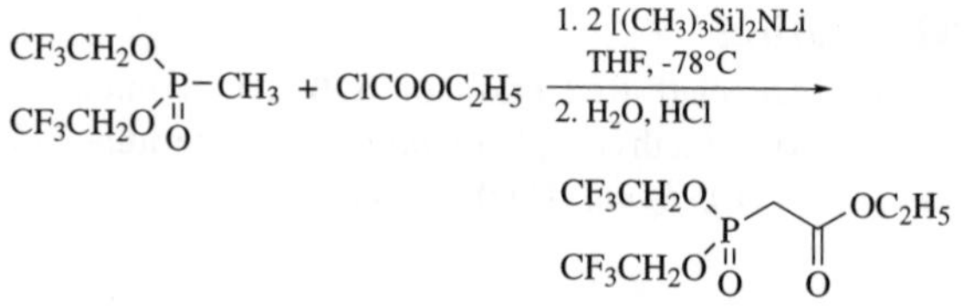

"4-Ketoundecanoic Acid," Tschantz, M. A.; Burgess, L. E.; Meyers, A. I. *Org. Synth.* **1995**, *73*, 215.

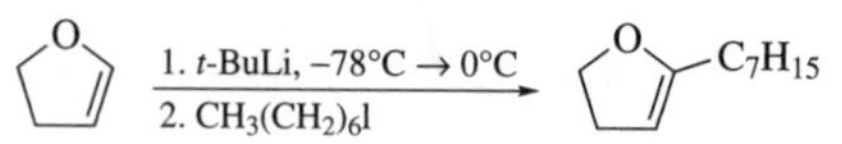

"Cyclopentanone Annulation via Cyclopropanone Derivatives. (3aβ,9bβ)-1,2,3a,4,5,9b-Hexahydro-9b-hydroxy-3a-methyl-3*H*-benz[e]inden-3-one," Bradlee, M. J.; Helquist, P. *Org. Synth.* **1996**, *74*, 137.

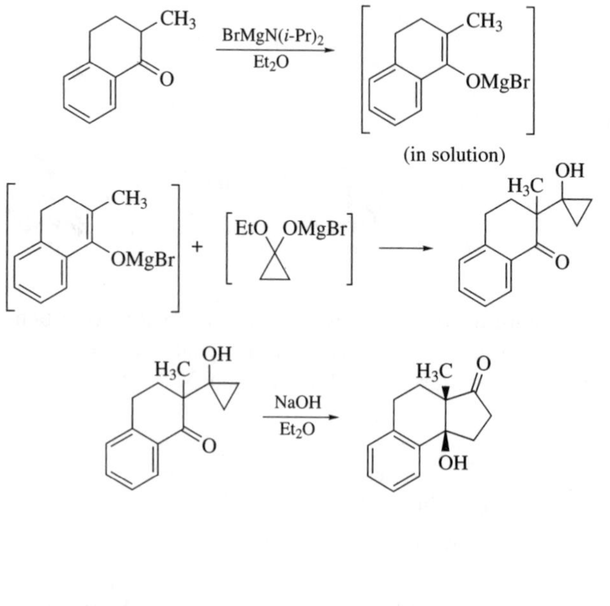

"Regioselective Synthesis of 3-Substituted Indoles: 3-Ethylindole," Amat, M.; Hadida, S.; Sathyanarayana, S.; Bosch, J. *Org. Synth.* **1996**, *74*, 248.

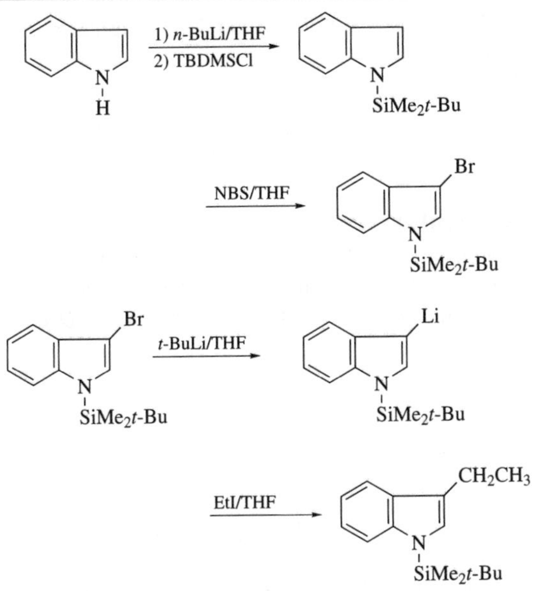

1) n-BuLi/THF 2) TBDMSCl

NBS/THF

t-BuLi/THF

EtI/THF

"(S)-(−)- and (R)-(+)-1,1′-Bi-2-Naphthol," Kazlauskas, R. J. *Org. Synth.* **1991**, *70*, 60.

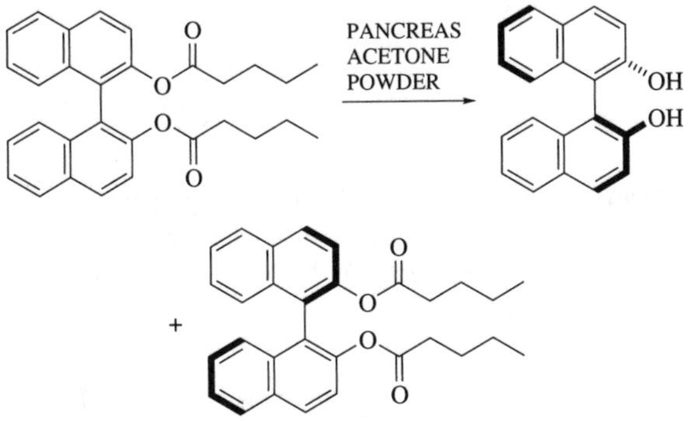

PANCREAS ACETONE POWDER

"3-3-Dimethylpyrrole and 2,3,7,8,12,13,18-octaethylporphyrin," Sessler, J. L.; Mozaffari, A.; Johnson, M. R. *Org. Synth.* **1991**, *70*, 68.

KF/IPA

"A Hydroxymethyl Anion Equivalent: Tributyl[(methoxymethoxy)methyl]stannane," Danheiser, R. L.; Romines, K. R.; Koyama, H.; Gee, S. K.; Johnson, C. R.; Medich, J. R. *Org. Synth.* **1992**, *71*, 133.

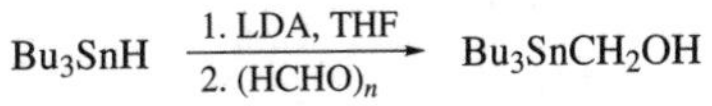

$$Bu_3SnH \xrightarrow[2.\ (HCHO)_n]{1.\ LDA,\ THF} Bu_3SnCH_2OH$$

"Asymmetric Catalytic Glyoxalate-ene Reaction: Methyl (2R)-2-Hydroxy-4-phenyl-4-pentenoate," Mikami, K.; Terada, M.; Narisawa, S.; Nakai, T. *Org. Synth.* **1992**, *71*, 14.

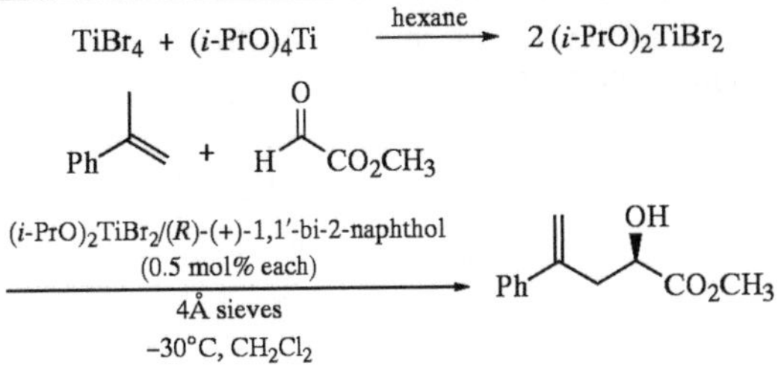

$$TiBr_4 + (i\text{-}PrO)_4Ti \xrightarrow{hexane} 2\ (i\text{-}PrO)_2TiBr_2$$

$(i\text{-}PrO)_2TiBr_2$/(R)-(+)-1,1′-bi-2-naphthol (0.5 mol% each), 4Å sieves, −30°C, CH_2Cl_2

"Asymmetric Hydrogenation of 3-Oxo Carboxylates Using BINAP-Ruthenium Complexes: (R)-(−)-Methyl 3-Hydroxybutyrate," Kitamura, M.; Tokunaga, M.; Ohkuma, T.; Noyori, R. *Org. Synth.* **1992**, *71*, 1.

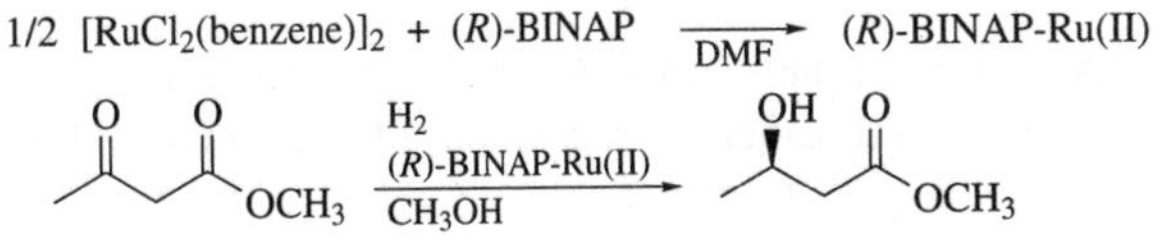

$$1/2\ [RuCl_2(benzene)]_2 + (R)\text{-BINAP} \xrightarrow{DMF} (R)\text{-BINAP-Ru(II)}$$

H_2, (R)-BINAP-Ru(II), CH_3OH

"(3,3-Difluoroallyl)trimethylsilane," Gonzalez, J.; Foti, M. J.; Elsheimer, S. *Org. Synth.* **1993**, *72*, 225.

DBU

"Phenylthioacetylene," Magriotis, P. A.; Brown, J. T. *Org. Synth.* **1993**, *72*, 252.

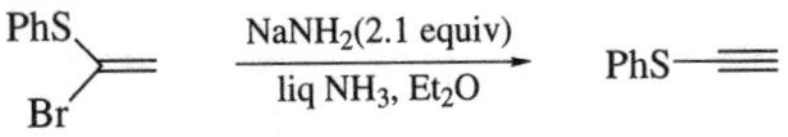

$NaNH_2$(2.1 equiv), liq NH_3, Et_2O

"Stereoselective Alkene Synthesis via 1-chloro-1-[(dimethyl)phenylsilyl] alkanes and α-(dimethyl)phenylsilyl ketones: 6-methyl-6-dodecene," Barrett, A. G. M.; Flygare, J. A.; Hill, J. M.; Wallace, E. M. *Org. Synth.* **1995**, *73*, 50.

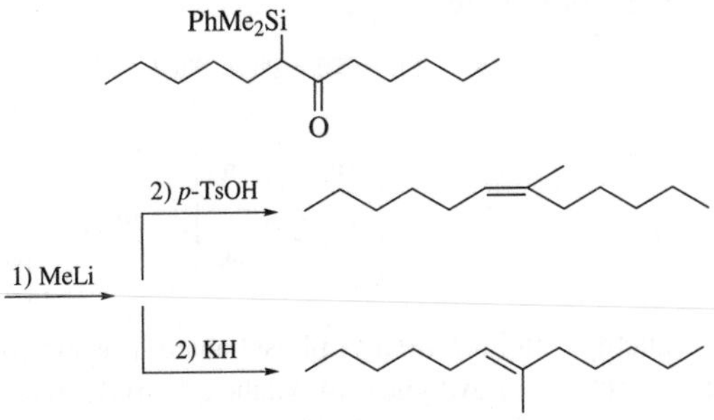

1) MeLi; 2) p-TsOH / 2) KH

"An Improved Preparation of 3-Bromo-2-(H)-pyran-2-one. An Ambiphilic Diene for Diels-Alder Cycloadditions," Posner, G. H.; Afarinkia, K.; Dai, H. *Org. Synth.* **1995**, *73*, 231.

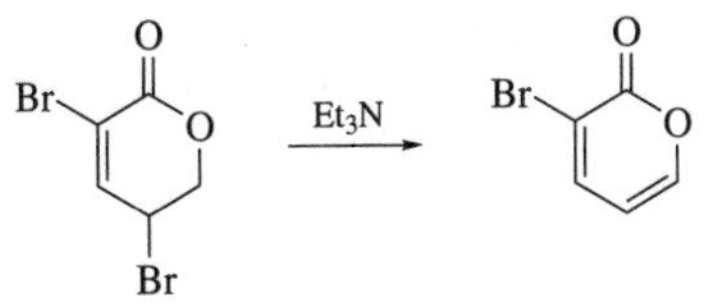

Et_3N

Avoid Skin Contact with All Reagents

"1,3,5-Cyclooctatriene," Oda, M.; Kawase, T.; Kurata, H. *Org. Synth.* **1995**, *73*, 240.

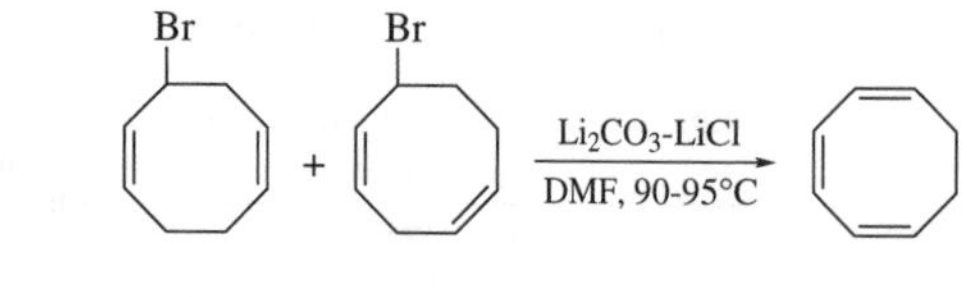

Br Br + Li_2CO_3-LiCl DMF, 90-95°C

"3-Pyrroline," Meyers, A. I.; Warmus, J. S.; Dilley, G. J. *Org. Synth.* **1995**, *73*, 246.

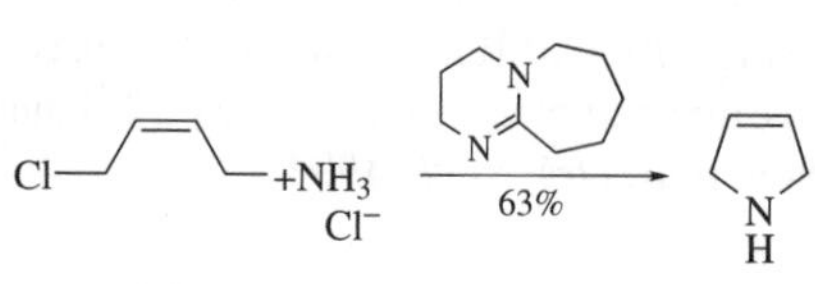

Cl– +NH_3 Cl^- 63% N H

"Acetylenic Ethers from Alcohols and Their Reduction to *Z*- and *E*-enol Ethers: Preparation of 1-Menthoxy-1-butyne from Menthol and Converstion to (*Z*)- and (*E*)-1-Menthoxy-1-butene," Kann, N.; Bernardes, V.; Greene, A. E. *Org. Synth.* **1996**, *74*, 13.

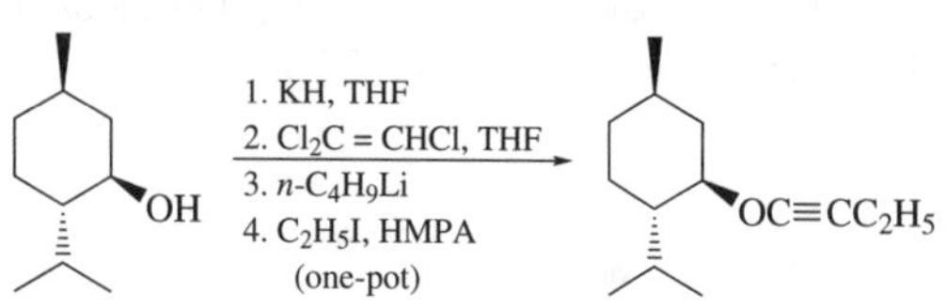

OH 1. KH, THF 2. $Cl_2C = CHCl$, THF 3. *n*-C_4H_9Li 4. C_2H_5I, HMPA (one-pot) $OC{\equiv}CC_2H_5$

"(*R*)-(+)-2-(Diphenylhydroxylmethyl)pyrrolidine," Nikolic, N. A.; Beak. P. *Org. Synth.* **1996**, *74*, 23.

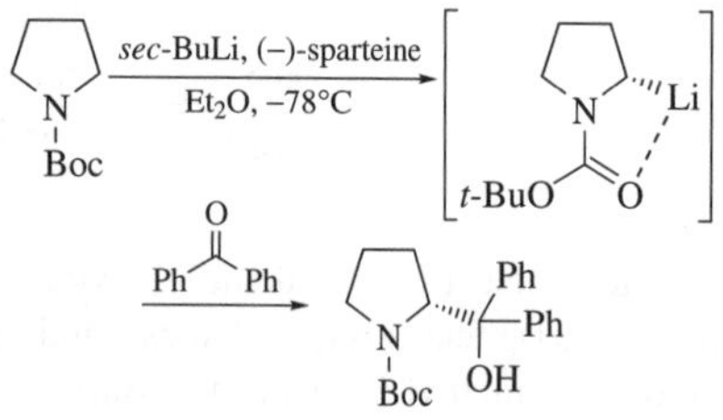

N Boc *sec*-BuLi, (–)-sparteine Et_2O, –78°C N Li *t*-BuO O
O Ph Ph N Boc Ph Ph OH

"1,2,3-Triphenylcyclopropenium Bromide," Xu, R.; Breslow, R. *Org. Synth.* **1996**, *74*, 72.

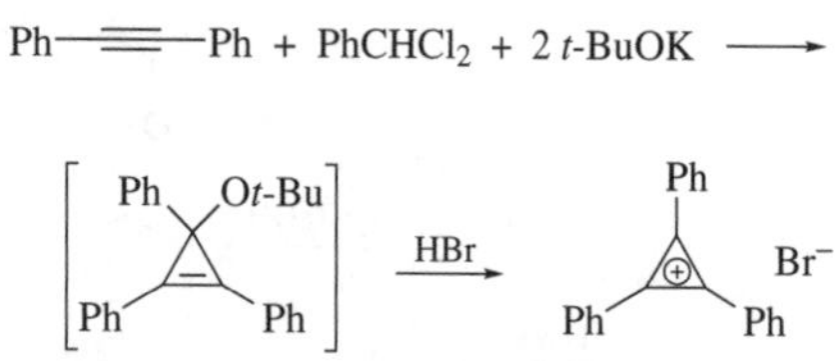

Ph—≡—Ph + $PhCHCl_2$ + 2 *t*-BuOK ⟶

Ph O*t*-Bu Ph Ph HBr Ph Ph Ph Br^-

"Diethyl (dichloromethyl)phosphonate. Preparation and use in the Synthesis of Alkynes: (4-Methoxyhenyl)ethyne," Marinetti, A.; Savignac, P. *Org. Synth.* **1996**, *74*, 108.

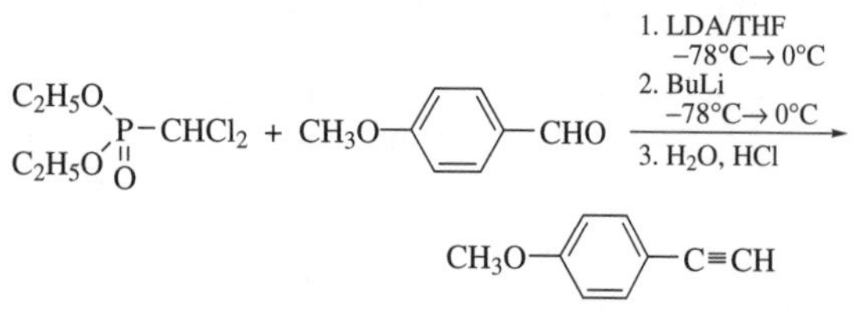

C_2H_5O P–CCl_3 C_2H_5O O 1. *i*-PrMgCl THF/ether, –78°C 2. EtOH, H_2O, HCl C_2H_5O P–$CHCl_2$ C_2H_5O O

C_2H_5O P–$CHCl_2$ C_2H_5O O + CH_3O–C6H4–CHO 1. LDA/THF –78°C→ 0°C 2. BuLi –78°C→ 0°C 3. H_2O, HCl

CH_3O–C6H4–C≡CH

"Phenyl Vinyl Sulfide," Reno, D. S.; Pariza, R. J. *Org. Synth.* **1996**, *74*, 124.

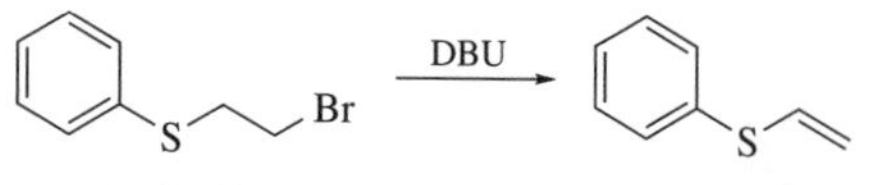

S Br DBU S

Acid Catalysts

"(*E*)-1-Benzyl-3-(1-iodoethylidene)piperidine: Nucleophile-promoted Alkyne-iminium Ion Cyclizations," Arnold, H.; Overman, L. E.; Sharp, M. J.; Witschel, M. C. *Org. Synth.* **1991**, *70*, 111.

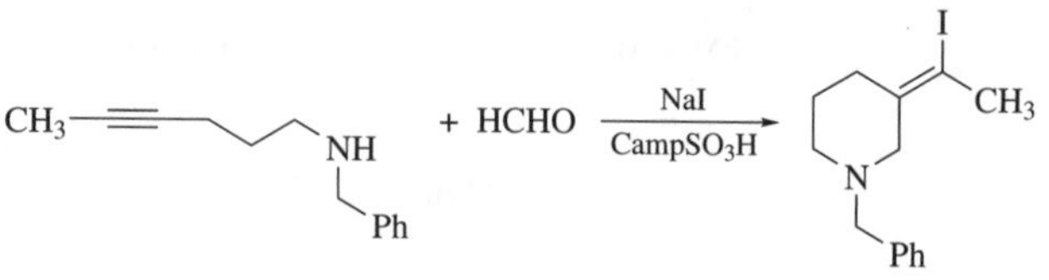

CH_3 NH Ph + HCHO NaI $CampSO_3H$ I CH_3 N Ph

"Tetrahydro-3-benzazepin-2-ones: Lead tetracetate oxidation of isoquinoline enamides," Lenz, G. R.; Lessor, R. A. *Org. Synth.* **1991**, *70*, 139.

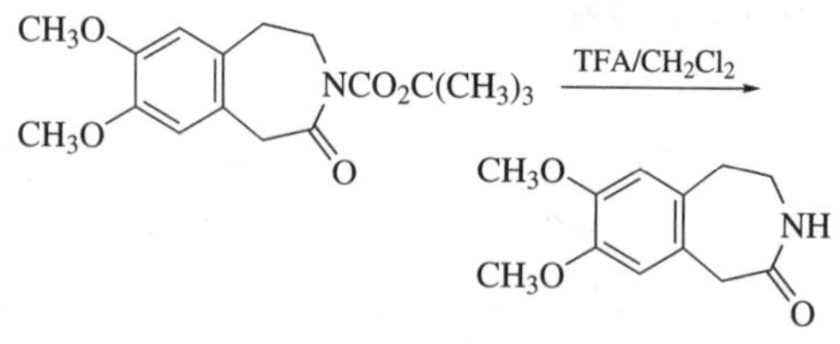

CH_3O CH_3O $NCO_2C(CH_3)_3$ O TFA/CH_2Cl_2 CH_3O CH_3O NH O

"Direct Degradation of the Biopolymer Poly[(*R*)-3-hydroxybutyric acid] to (*R*)-3-hydroxybutanoic Acid and its Methyl Ester," Seebach, D.; Beck, A. K.; Breitschuh, R.; Job, K. *Org. Synth.* **1992**, *71*, 39.

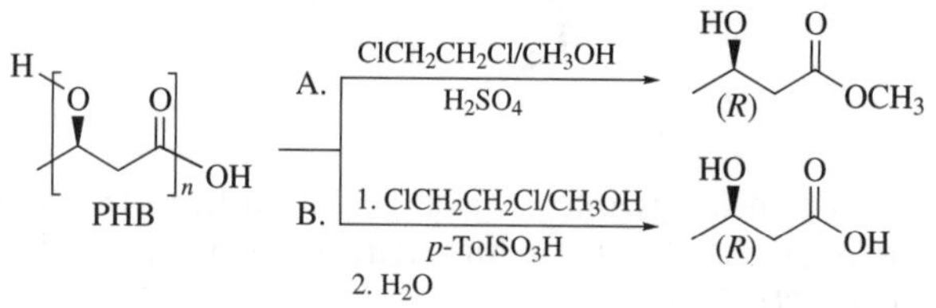

H O O OH *n* PHB A. $ClCH_2CH_2Cl/CH_3OH$ H_2SO_4 HO O (*R*) OCH_3
B. 1. $ClCH_2CH_2Cl/CH_3OH$ *p*-$TolSO_3H$ 2. H_2O HO O (*R*) OH

"Preparation and Reactions of Alkenylchromium Reagents: 2-Hexyl-5-phenyl-1-penten-3-ol," Takai, K.; Sakogawa, K.; Kataoka, Y.; Oshima, K.; Utimoto, K. *Org. Synth.* **1993**, *72*, 180.

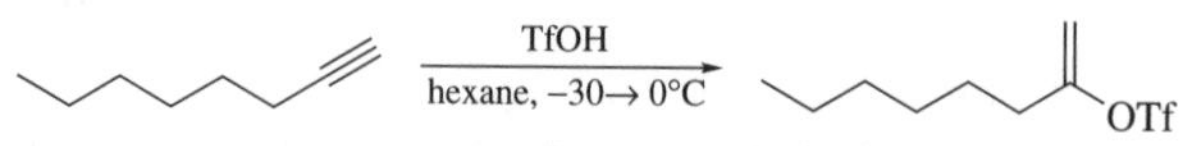

TfOH hexane, –30→ 0°C OTf

"Spiroannelation of Enol Silanes-2-oxo-5-methoxy-spiro[5.4]decane," Lee, T. V.; Porter, J. R. *Org. Synth.* **1993**, *72*, 189.

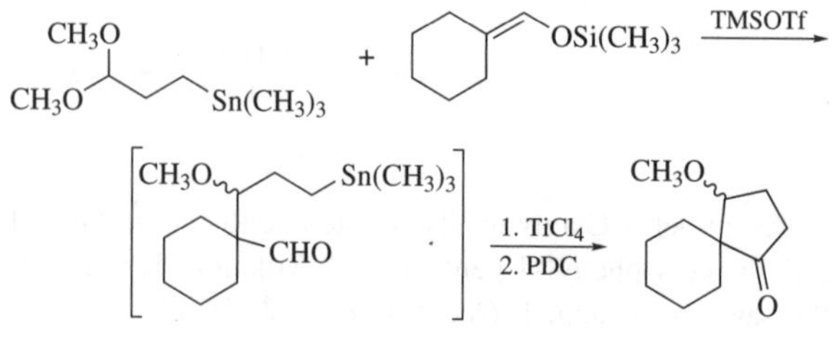

CH_3O CH_3O $Sn(CH_3)_3$ + $OSi(CH_3)_3$ TMSOTf

CH_3O $Sn(CH_3)_3$ CHO 1. $TiCl_4$ 2. PDC CH_3O O

Lewis Acids

"Substitution Reactions of 2-Benzenesulfonyl Cyclic Ethers: Tetrahydro-2-(phenylethynyl)-2*H*-pyran," Brown, D. S.; Ley, S. V. *Org. Synth.* **1991**, *70*, 157.

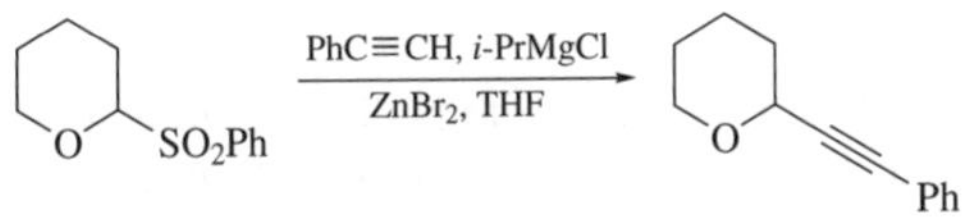

"2-Methyl-1-3,-cyclopentanedione," Meister, P. G.; Sivik, M. R.; Paquette, L. A. *Org. Synth.* **1991**, *70*, 226.

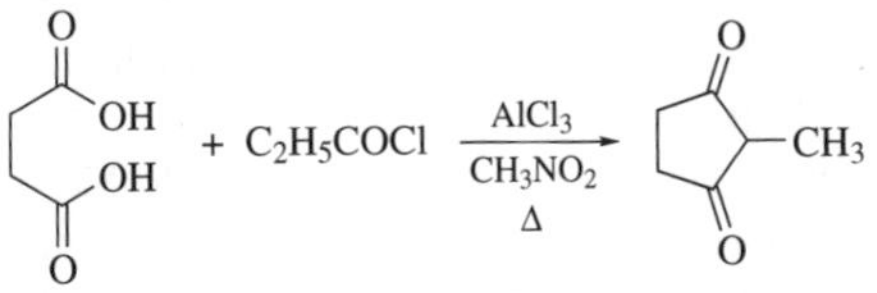

"Asymmetric Catalytic Glyoxylate-ene-Reaction: Methyl (2*R*)-2-Hydroxy-4-phenyl-4-pentenoate," Mikami, K.; Terada, M.; Narisawa, S.; Nakai, T. *Org. Synth.* **1992**, *71*, 14.

$$TiBr_4 + (i\text{-PrO})_4Ti \xrightarrow{\text{hexane}} 2\,(i\text{-PrO})_2TiBr_2$$

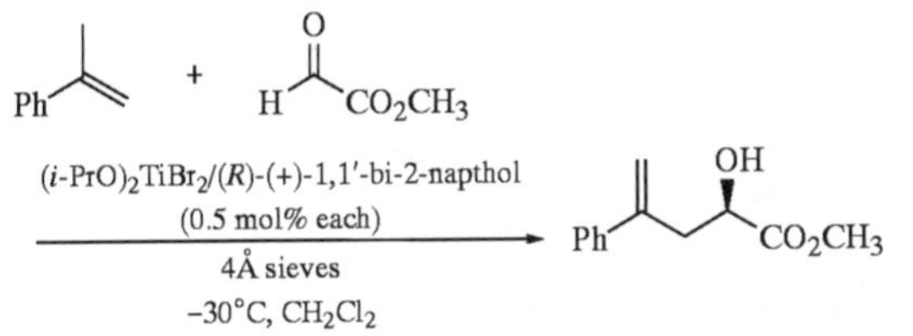

"Ubiquinone-1," Naruta, Y.; Maruyama, K. *Org. Synth.* **1992**, *71*, 125.

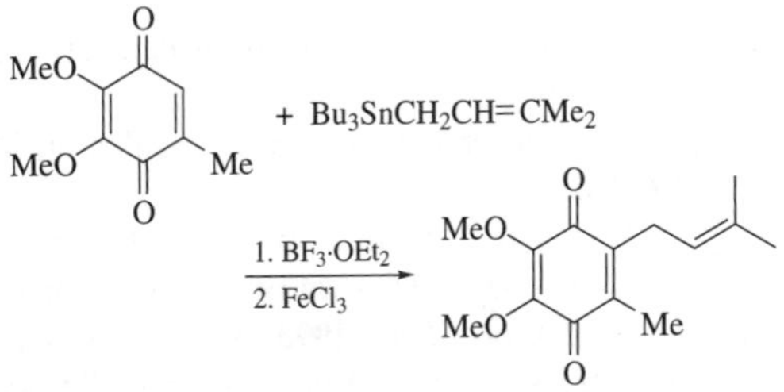

"(*R*)-(−)-2,2-Diphenylcyclopentanol," Denmark, S. E.; Marcin, L. R.; Schnute, M. E.; Thorarensen, A. *Org. Synth.* **1996**, *74*, 33.

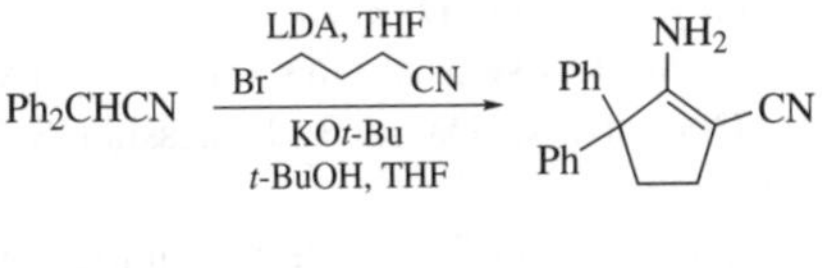

"Nitroacetaldehyde Diethyl Acetal," Jäger, V.; Poggendorf, P. *Org. Synth.* **1996**, *74*, 130.

$$O_2N\text{-}CH_3 + HC(OC_2H_5)_3 \xrightarrow[-90°C]{ZnCl_2} O_2NCH_2CH(OC_2H_5)_2$$

Hydrolyses-Enzymatic

"Lipase-Catalyzed Kinetic Resolution of Alcohols via Chloroacetate Esters: (−)-(1*R*,2*S*)-*trans*-2-Phenylcyclohexanol and (+)-(1*S*,2*R*)-*trans*-2-phenylcyclohexanol," Schwartz, A.; Madan, P.; Whitesell, J. K.; Lawrence, R. M. *Org. Synth.* **1990**, *69*, 1.

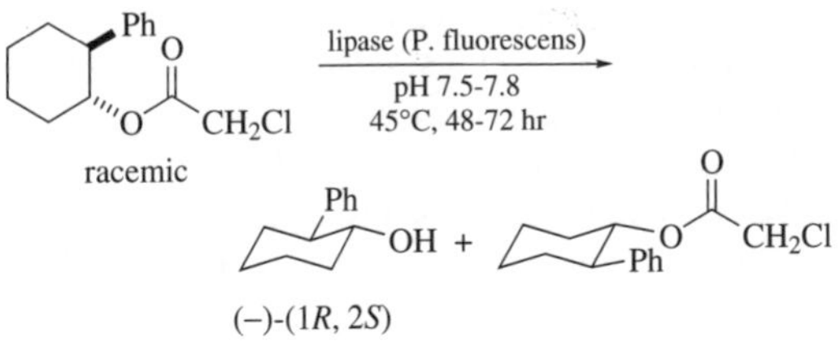

"Enantioselective Saponification with Pig Liver Esterase (PLE): (1*S*,2*S*,3*R*)-3-Hydroxy-2-nitrocyclohexyl Acetate," Eberle, M.; Missbach, M.; Seebach, D. *Org. Synth.* **1990**, *69*, 19.

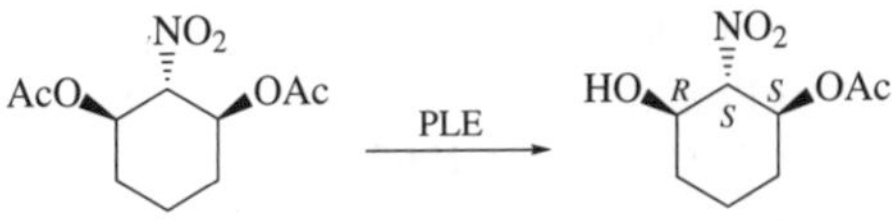

"Enantiomerically Pure Ethyl (*R*)- and (*S*)-2-Fluorohexanoate by Enzyme-Catalyzed Kinetic Resolution," Kalaritis, P.; Regenye, R. W. *Org. Synth.* **1990**, *69*, 10.

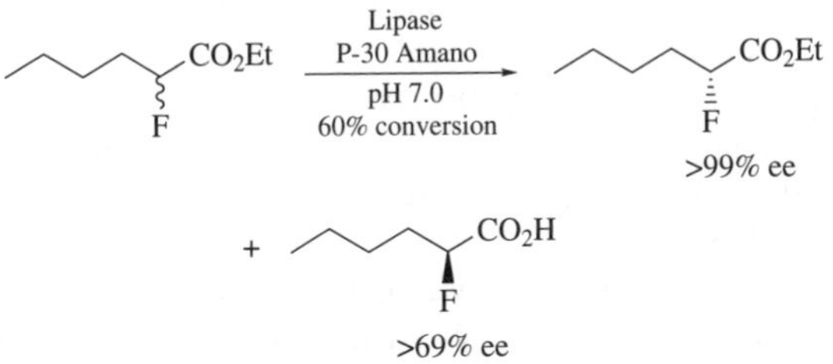

"Enantioselective Hydrolysis of *cis*-3-5,-Diacetoxycyclopentene: (1*R*,4*S*)-4-hydroxy-2-cyclopentenyl acetate," Deardorff, D. R.; Windham, C. Q.; Craney, C. L., *Org. Synth.* **1995**, *73*, 25.

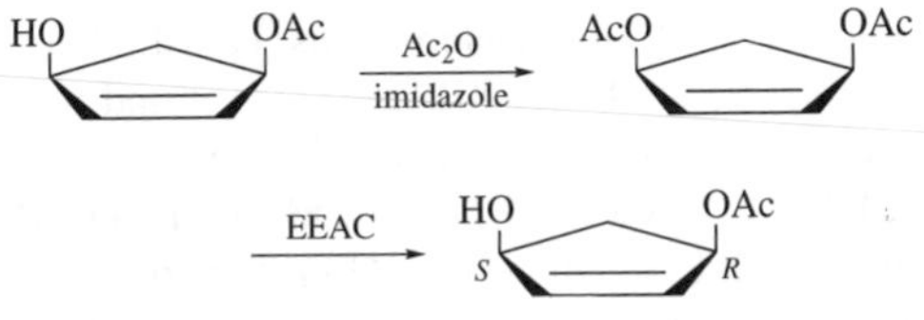

Recent Review Articles and Monographs

Metalation

"Metalations by Organolithium Compounds," Mallan, J. M.; Bebb, R. L. *Chem. Rev.* **1969**, *69*, 693.

Preparative Polar Organometallic Chemistry, Brandsma, L.; Verkruijsse, H. Springer-Verlag: Berlin, 1987.

The Chemistry of Organolithium Compounds, Wakefield, B. J.; Pergamon: Oxford, 1974. *Organolithium Methods*, Wakefield, B. J. Academic: London, 1988.

"Selective Carbanion Chemistry and Anion-cation Interactions in Solution: A Survey," Seyden-Penne, J. *New J. Chem.* **1992**, *16*, 251.

"Organometallics in Synthesis," Schlosser, M., Ed. Wiley: Chichester, U.K., 1994.

"Lithium Chemistry: A Theoretical and Experimental Overview," Sapse, A.-M.; Schleyer, P. v. R. eds. NY, Wiley, 1995.

"Heteroatom-Faciliated Lithiations," Gschwend, H. W.; Rodriguez, H. R. *Org. React.* **1979**, *26*, 1. "Lateral Lithiation Reactions Promoted by Heteroatomic Substituents," Clark, R.D.; Jahangir, A. *Org. React.* **1995**, *47*, 1.

"Lewis Acid Complexation of Tertiary Amines and Related Compounds: A Strategy for α Deprotonation and Stereocontrol," Kessar, S.V.; Singh, P. *Chem. Rev.* **1997**, *97*, 721.

"Directed Lithiation of Aromatic Tertiary Amides: An Evolving Synthetic Methodology for Polysubstituted Aromatics," Beak, P.; Snieckus, V. *Acc. Chem. Res.* **1982**, *15*, 306. "Dipole- Stabilized Carbanions: Novel and Useful Intermediates," Beak, P.; Reitz, D. B. *Chem. Rev.* **1978**, *78*, 275. "Aromatic Organolithium Reagents Bearing Electrophilic Groups. Preparation by Halogen-Lithium Exchange," Parham, W. E.; Bradsher,C. K. *Acc. Chem. Res.* **1982**, *15*, 300.

"The Directed Ortho Metalation Reaction. Methodology, Applications, Synthetic Links, and a Non-aromatic Ramification," Snieckus, V. *Pure Appl. Chem.* **1990**, *62*, 2047. "Directed Ortho Metalation. Tertiary Amide and O-Carbamate Directors in Synthetic Strategies for Polysubstituted Aromatics," Snieckus, V. *Chem. Rev.* **1990**, *90*, 879.

"Synthesis and Reactions of Lithiated Monocyclic Azoles Containing Two or More Heteroatoms. Part II: Oxazoles," Iddon, B. *Heterocycles* **1994** *37*, 1321. "Synthesis and Reactions of Lithiated Monocyclic Azoles Containing Two or more Hetero-Atoms. Part III: Pyrazoles," Grimmett, M. R.; Iddon, B. *Heterocycles*, **1994**, *37*, 2087. "Synthesis and Reactions of Lithiated Monocyclic Azoles Containing 2 or more Hetero-Atoms. Part IV: Imidazoles," Iddon, B.; Ngochindo, R. I. *Heterocycles*, **1994**, *38*, 2487. "Synthesis and Reactions of Lithiated Monocyclic Azoles Containing Two or More Hetero-atoms. Part V. Isothiazoles and Thiazoles," Iddon, B. *Heterocycles* **1995**, *41*, 533. "Metalation of Diazines," Turck, A.; Plè, N.; Quèguiner, G. *Heterocycles*, **1994**, *37*, 2149. "Synthesis and Reactions of Lithiated Monocyclic Azoles Containing Two or More Hetero-atoms. Part VI. Triazoles, Tetrazoles, Oxadiazoles, and Thiadiazoles," Grimmett, M.R.; Iddon, B. *Heterocycles*, **1995**, *41*, 1525.

"Metalation and Electrophilic Substitution of Amine Derivatives Adjacent to Nitrogen: α-Metallo Amine Synthetic Equivalents," Beak, P.; Zadjel, W. J.; Reitz, D. B. *Chem. Rev.* **1984**, *84*, 471.

"Generation and Reactions of sp^2-Carbanionic Centers in the Vicinity of Heterocyclic Nitrogen Atoms," Rewcastle, G.W.; Katritzky, A.R. Adv. *Heterocyclic Chem.* **1993**, *56*, 155.

"Lithioalkenes from Arylsulphonylhydrazones," Chamberlin, A. R.; Bloom, S. H. *Org. React.* **1990**, *39*, 1. "Recent Applications of the Shapiro Reaction," Adlington, R. M.; Barrett, A. G. M. *Acc. Chem. Res.* **1983**, *16*, 55.

"Oxiranyl Anions and Aziridinyl Anions," Satoh, T. *Chem. Rev.* **1996**, *96*, 3303.

"Synthetic Uses of the 1,3-Dithiane Grouping from 1977-1988," Page, P. C. B.; van Niel, M. B.; Prodger, J. C. *Tetrahedron* **1989**, *45*, 7643.

"The Synthetic Utility of α-Amino Alkoxides," Comins, D. L. *Synlett* **1992**, 615.

"Potassium Hydride in Organic Synthesis," Pinnick, H. W. *Org. Prep. Proc. Int.* **1983**, *15*, 199.

"Arene-catalysed Lithiation Reactions," Yus, M. *Chem. Soc. Rev.* **1996**, *25*, 155.

Bases and Ligands

"Hydroxide Ion Initiated Reactions Under Phase Transfer Catalysis Conditions: Mechanism and Implications" Rabinovitz, M.; Cohen, Y.; Halpern, M. *Angew. Chem. Int. Ed. Engl.* **1986**, *25*, 960.

The Chemistry of the Hydroxyl Group, Part 1; Fyfe, C. A. Wiley: New York, 1971.

"Superbases for Organic Synthesis" Schlosser, M. *Pure Appl. Chem.* **1988**, *60*, 1627.

"Specific Transition State Stabilization by Metal Ions in Reactions of Functionalized Crown Ethers," Cacciapaglia, R.; Mandolini, L. *Pure Appl. Chem.* **1993**, *65*, 533. "Catalysis by Metal Ions in Reactions of Crown Ether Substrates," Cacciapaglia, R.; Mandolini, L. *Chem. Soc. Rev.* **1993**, *22*, 221.

Hydrides of the Elements of the Main Groups, Wiberg, E.; Amberger, E. Elsevier: New York, 1971.

"Is *N,N,N',N'*-Tetramethylethylenediamine, Good Ligand for Lithium?" Collum, D. B. *Acc. Chem. Res.* **1992**, *25*, 448.

"Regioselective Manipulation of Hydroxyl Groups via Organotin Derivatives," David, S.; Hanessian, S. *Tetrahedron* **1985**, *41*, 643.

"Esterifications, Transesterifications, and Deesterifications Mediated by Organotin Oxides, Hydroxides, and Alkoxides," Mascaretti, O.A.; Furlan, R.L.E. *Aldrichimica Acta* **1997**, *30*, 55.

"4-Dialkylaminopyridines as Highly Active Acylation Catalysts," Höfle, G.; Steglich, W.; Vorbrüggen, H. *Angew. Chem. Int. Ed. Engl.* **1978**, *17*, 569.

"2,6-Di-*tert*-butylpyridine - An Unusual Base," Kanner, B. *Heterocycles* **1982**, *18*, 411.

"Bicyclic Amidines as Reagents in Organic Synthesis," Oediger, H.; Möller, F.; Eiter, K. *Synthesis* **1972**, 591.

"The Baylis-Hillman Reaction: A Novel Carbon-Carbon Bond Forming Reaction," Basavaiah, D.; Rao, P.D.; Hyma, R.S. *Tetrahedron* **1996**, *52*, 8001.

Phase Transfer Catalysis Dehmlow, E. V. ed.; 2nd ed., Verlag Chemie: Deerfield Beach, FL, **1983**.

Salt Effects in Organic and Organometallic Chemistry Loupy, A.; Tchoubar, B.; VCH: Weinheim, 1992.

Chiral Bases and Ligands

"Enantioselective Synthesis with Lithium/(−)-Sparteine Carbanion Pairs," Hoppe, D.; Hense, T. *Angew. Chem. Intl. Ed. Engl.* **1997**, *36*, 2282.

"Metallated 2-Alkenyl Carbamates: Chiral Homoenolate Reagents for Asymmetric Synthesis," Hoppe, D.; Krämer, T.; Schwark, J. -R.; Zschage, O. *Pure Appl. Chem.* **1990**, *62*, 1999.

"Regioselective, Diastereoselective, and Enantioselective Lithiation-Substitution Sequences: Reaction Pathways and Synthetic Applications," Beak, P.; Basu, A.; Gallagher, D.J.; Park, Y.S.; Thayumanavan, S. *Acc. Chem. Res.* **1996**, *29*, 552.

"Some Stereochemical Aspects of Bisquinolizidine Alkaloids Sparteine Type," Boczoń, W. *Heterocycles* **1992**, *33*, 1101.

"Asymmetric Carbon-carbon Bond Formation Using Sulfoxide-stabilized Carbanions," Walker, A.J. *Tetrahedron: Asymmetry* **1992**, *3*, 961.

"Asymmetric Synthesis Using Homochiral Lithium Amide Bases," Cox, P.J.; Simpkins, N.S. *Tetrahedron: Asymm.* **1991**, *2*, 1.

Protic Acid Catalysts

"Polyphosphoric Acid as a Reagents in Organic Chemistry" Popp, F. D.; McEwen, W. E. *Chem. Rev.* **1958**, *58*, 321.

Friedel-Crafts Alkylation Chemistry, Roberts, R. M.; Khalaf, A. A.; Markel Dekker: New York, **1984.**

Superacids, Olah, G. A.; Prakash, G. K. S.; Sommer, J.; Wiley, New York, **1985**.

"The Chemistry of Formic Acids and Its Simple Derivatives Gibson," H. W. *Chem. Rev.* **1969**, *69*, 673.

"The Hydrogen Halides" Downs, A. J.; Adams, C. J. In *Comprehensive Inorganic Chemistry*; Bailar, J. C., Ed.; Pergamon: Oxford, **1973**; Vol. 2, p 1280.

"The Combination of Hydrogen Fluoride with Organic Bases as Fluorination Agents," Yoneda, N. *Tetrahedron* **1991**, *47*, 5329.

"Cleavage of Ethers," Bhatt, M. V.; Kulkarni, S. U. Synthesis, **1983**, 249.

Synthetic Reagents, Pizey, J. S. Ellis Horwood: Chichester, 1985; Vol. 6.

"Trifluoromethanesulfonic Acid and Derivatives," Howells, R. D.; McCown, J. D. *Chem. Rev.* **1977**, *77*, 69.

"Acid/Base-Induced Selectivity of Molecular Sieves in Catalytic Conversion of Polar Molecules," Eder-Mirth, G.; Lercher, J. A. *Recl. Trav. Chim. Pays-Bas* **1996**, *115*, 157.

"Enantioselective Protonation of Enolates and Enols," Fehr, C. *Angew. Chem. Int. Ed. Engl.* **1996**, *35*, 2566.

Lewis Acids

Lewis Acids and Selectivity in Organic Synthesis Pons, J.-M.; Santelli, M., Eds. CRC: Boca Raton, FL, 1995.

Boron Trifluoride and Its Derivatives Booth, H. S.; Martin, D. R.; Wiley: New York, 1949.

"Reactions of Boron Trichloride with Organic Compounds" Gerrard, W.; Lappert, M. F. *Chem. Rev.* **1958**, *58*, 1081.

Boron Fluoride and Its Compounds as Catalysts in Organic Chemistry Topchiev, A. V.; Zavgorodnii, S. V.; Paushkin, Ya. M.; Pergamon: New York, 1959.

"Boron Halides" Greenwood, N. N.; Thomas, B. S. In *Comprehensive Inorganic Chemistry*; Trotman-Dickenson, A. F., Ed,; Pergamon: New York, **1973**; Vol. 1, pp 956.

"New Synthetic Applications of Dialkylboron Halide Reagents," Guindon, Y.; Anderson, P. C.; Yoakim, C.; Girard, Y.; Berthiaume, S.; Morton, H. E. *Pure Appl. Chem.* **1988**, *60*, 1705.

"Lewis Acids and Selectivity in Organic Synthesis," Pons, J.-M.; Santelli, M., CRC Press, 1995.

"$LiClO_4$ in Ether–an Unusual Solvent," Waldmann, H. *Angew. Chem. Int. Ed. Eng.* **1991**, *30*, 1306.

"Recent Developments in Preparative Sulfonation and Sulfation," Gilbert, E. E. *Synthesis* **1969**, 3.

"Novel Lewis Acid Catalysis in Organic Synthesis," Suzuki, K. *Pure App. Chem.* **1994**, *66*, 1557.

"Iodotrimethylsilane–A Versatile Synthetic Reagent," Olah, G. A.; Narang, S. C. *Tetrahedron* **1982**, *38*, 2225.

"Trialkylsilyl Perfluoroalkanesulfonates: Highly Reactive Silylating Agents and Lewis Acids in Organic Synthesis," Emde, H.; Domsch, D.; Feger, H.; Frick, H.; Götz, A.; Hergott, H. H.; Hofmann, K.; Kober, W.; Krägeloh, K.; Oesterle, T.; Steppan, W.; West, W.; Simchen, G. *Synthesis* **1982**, 1.

"Mechanisms of Epoxide Reactions," Parker, R. E.; Isaacs, N. S. *Chem. Rev.* **1959**, *59*, 737.

"Carbonyl Addition Reactions Promoted by Cerium Reagents," Imamoto, T. *Pure & Appl. Chem.* **1990**, *62*, 747.

"Rare Earth Metal Trifluoromethanesulfonates as Water-Tolerant Lewis Acid Catalysts in Organic Synthesis," Kobayashi, S. *Synlett*, **1994**, 689.

"Lewis-acid Catalysis of Carbon Carbon Bond Forming Reactions in Water," Engberts, J.B.F.N.; Feringa, B.L.; Keller, E.; Otto, S. *Rec. Trav. Chim. Pays-Bas* **1996**, *115*, 457.

"Lanthanides in Organic Synthesis," Kagan, H. B.; Namy, J. L. *Tetrahedron* **1986**, *42*, 6573.

"Lanthanides in Organic Synthesis," Imamoto, T. Academic Press: London, 1994.

"Application of Lanthanide Reagents in Organic Synthesis," Molander, G. A. *Chem. Rev.* **1992**, *92*, 29.

"Organomercury Compounds in Organic Synthesis," Larock, R. C. *Angew. Chem. Int. Ed. Engl.* **1978**, *17*, 27.

"Organomercurials in Organic Synthesis," Larock, R. C. *Tetrahedron* **1982**, *38*, 1713.

"Organomercury Compounds in Organic Synthesis," Larock, R. C.; Springer: Berlin, **1985**.

"Solvomercuration/Demercuration Reactions in Organic Synthesis," Larock, R. C.; Springer: Berlin, 1986.

Chiral Lewis Acids

"Chiral Lewis Acids in Catalytic Asymmetric Reactions," Narasaka, K. *Synthesis* **1991**, 1.

"Synthesis and Applications of Binaphthylic C_2-Symmetry Derivatives as Chiral Auxiliaries in Enantioselective Reactions"

Rosini, C.; Franzini, L.; Raffaelli, A.; Salvadori, P. *Synthesis* **1992**, 503.

"Asymmetric Boron-Catalyzed Reactions" Deloux, L.; Srebnik, M. *Chem. Rev.* **1993**, *93*, 763.

"Chiral Titanium Complexes for Enantioselective Addition of Nucleophiles to Carbonyl Groups," Duthaler, R.O.; Hafner, A. *Chem. Rev.* **1992**, *92*, 807.

"Practical and Useful Methods for the Enantioselective Reduction of Unsymmetrical Ketones" Singh, V. K. *Synthesis* **1992**, 605.

"Asymmetric Syntheses with Chiral Oxazaborolidines" Wallbaum, S.; Martens, J. *Tetrahedron: Asymmetry* **1992**, *3*, 1475.

"Oxazaborolidines and Dioxaborolidines in Enantioselective Catalysis," Lohray, B.B.; Bhushan, V. *Angew. Chem. Int. Ed. Eng.* **1992**, *31*, 729.

"Catalytic Enantioselective C-C coupling – Allyl Transfer and Mukaiyama Aldol Reaction," Bach, T. *Angew. Chem. I. E. Engl.* **1994**, *33*, 417.

"Chiral Lewis Acid Catalysts in Diels-Alder Cycloadditions: Mechanistic Aspects and Synthetic Applications of Recent Systems" Dias, L. C. *J. Brazil Chem. Soc.* **1997**, *8*, 289.

"Enantioselective Synthesis with Optically Active Transition-metal Catalysts," H. Brunner *Synthesis* **1988**, 645.

"Natural Products by Enantioselective Catalysis with Transition Metal Compounds," Brunner, H. *Pure App. Chem.* **1994**, *66*, 2033.

"Transition Metal or Lewis Acid-Catalyzed Asymmetric Reactions with Chiral Organosulfur Functionality," Hiroi, K. *Rev. Heteroatom Chem.* **1996**, *14*, 21.

"Asymmetric Ene Reactions in Organic Synthesis" Mikami, K.; Shimizu, M. *Chem. Rev.* **1992**, *92*, 1021.

"Asymmetric Catalysis for Carbonyl-Ene Reaction" Mikami, K.; Terada, M.; Narisawa, S.; Nakai, T. *Synlett* **1992**, 255.

Solid Phase Catalysts

Molecular Sieve Catalysts, Michiels, P.; De Herdt, O. C. E., Eds.; Pergamon: Oxford, 1987.

An Introduction to Zeolite Molecular Sieves, Dyer, A. Wiley: New York, 1988.

"Organic Reactions at Alumina Surfaces," Posner, G. H. *Angew. Chem. Int. Ed. Engl.* **1978**, *17*, 487.

"Organic Reactions on Alumina," Kabalka, G. W.; Pagni, R. M. *Tetrahedron* **1997**, *53*, 7999.

Biocatalytic Hydrolysis

"Biocatalysis as a New Powerful Tool for the Synthesis of Enantiomerically Pure Chiral Building Blocks," Santaniello, E.; Ferraboschi, P. *in Advances in Asymmetric Synthesis*; Hassner, A., Ed., SAI: Greenwich, CT; Vol. 2; 1997.

"Biocatalytic Deracemization Techniques. Dynamic Resolutions and Stereoinversions," Stecher, H.; Faber, K. *Synthesis* **1997**, 1.

"Enzymes in Organic Synthesis" Jones, J. B. *Tetrahedron* **1986**, *42*, 3351.

"General Aspects and Optimization of Enantioselective Biocatalysis in Organic Solvents: The Use of Lipases," Chen, C.-S.; Sih, C. J. *Angew. Chem. Int. Ed. Engl.* **1989**, *28*, 695.

"Asymmetric Transformations Catalyzed by Enzymes in Organic Solvents," Klibanov, A. M. *Acc. Chem. Res.* **1990**, *23*, 114.

"Pseudomonas fluorescens Lipase in Asymmetric Synthesis," Xie, Z.-F. *Tetrahedron: Asymmetry* **1991**, *2*, 733.

"Esterolytic and Lipolytic Enzymes in Organic Synthesis," Boland, W.; Frößl, C.; Lorenz, M. *Synthesis*, **1991**, 1049.

Alumina[1]

Al2O3

[1344-28-1] Al_2O_3 (MW 101.96)

(a mildly acidic, basic, or neutral support for chromatographic separations; a reagent for catalyzing dehydration, elimination, addition, condensation, epoxide opening, oxidation, and reduction reactions)

Alternate Name: γ-alumina.
Physical Data: mp 2015 °C; bp 2980 °C; *d* 3.97 g cm^{-3}.
Solubility: slightly sol acid and alkaline solution.
Form Supplied in: fine white powder, widely available in varying particle size (50–200 μm; 70–290 mesh), in acidic (pH 4), basic (pH 10), and neutral (pH 7) forms.
Drying: the activity of alumina has been classified by the Brockmann scale into five grades. The most active form, grade I, is obtained by heating alumina to 200 °C while passing an inert gas through the system, or heating to ~400 °C in an open vessel, followed by cooling in a dessicator. Addition of 3–4% (w/w) water and mixing for several hours converts grade I alumina to grade II. Other grades are similarly obtained (grade III, 5–7%; grade IV, 9–11%; grade V, 15–19% water).[2,3]
Handling, Storage, and Precautions: inhalation of fine mesh alumina can cause respiratory difficulties. Alumina is best handled under a fume hood and stored under dry, inert conditions.

Introduction. Alumina is one of the most widely used packing materials for adsorption chromatography and is available in acidic, basic, and neutral forms. Use of the correct type is important to avoid unwanted reactions of the substrate being purified.[1,3] Possessing both Lewis acidic and basic sites, alumina has been found to catalyze a wide range of reactions, generally under conditions that are milder and more selective than comparable homogeneous reactions.[1]

Dehydration and Eliminations. One of the earliest uses of alumina as a catalyst was for the dehydration of alcohols.[4,5] These reactions generally require high temperature and yield primarily non-Saytzeff products. Complex terpenes have been dehydrated with ***Pyridine*** or ***Quinoline*** doped alumina (eq 1).[6b] Numerous other groups can be eliminated in the presence of alumina, including OR, OAc, O_3SR, O_2SR, and halides.[1,7] Some of these eliminations proceed under mild conditions,[1] often during chromatographic purification (eq 2).[7d] Sulfonates can be eliminated in the presence of acid and base sensitive groups, without skeletal rearrangements. However, a large excess of properly activated alumina is required, and poor stereo- and regiocontrol are observed.[7e] Dehydrohalogenations, particularly dehydrofluorinations, occur readily over alumina (eq 3).[8] Stereoselective syntheses of vinyl halides have been developed that take advantage of desilicohalogenation[9] or deborohalogenation[10] of vinylsilane or vinylboronic acid derived dihalides. Benzol[*c*]thiophene has been synthesized by dehydration of a sulfoxide precursor.[11] The oxidation of selenides to selenoxides and their elimination to alkenes can be accomplished in one step using basic alumina and ***t-Butyl Hydroperoxide*** in THF.[12]

alumina, pyridine, 220–230 °C, 44% (1)

alumina activity I, 65–80% (2)

alumina, CCl_4, 25 °C, 24 h; 60% + <4% (3)

Alumina has been used for various dehydration reactions, including those leading to piperidines,[13] pyrroles (eq 4) and pyrazoles,[14] and other heterocycles.[15] It is also an effective catalyst for the selective protection of aldehydes in the presence of ketones.[16]

$EtNH_2$, alumina, 0 to 20 °C, 96% (4)

Addition and Condensation Reactions. Alumina promotes the addition of various heteroatom species, whether by electrophilic or nucleophilic processes. In contrast to the elimination reactions described earlier, alumina also promotes the intramolecular addition of OH and OR groups to isolated (eq 5)[6c] and carbonyl-activated alkenes.[17] It is also reported to catalyze the conjugate addition of other nucleophiles, such as amines.[18] In the presence of alumina, ***Iodine*** can be used to iodinate aromatics, hydroiodinate alkenes, and diiodinate alkynes (eq 6).[19] Hydrochlorinations and hydrobrominations of alkenes and alkynes give the Markovnikov products, with good stereoselectivity.[20]

basic alumina, 84%; acidic alumina, PhH, 75 °C, 72% (5)

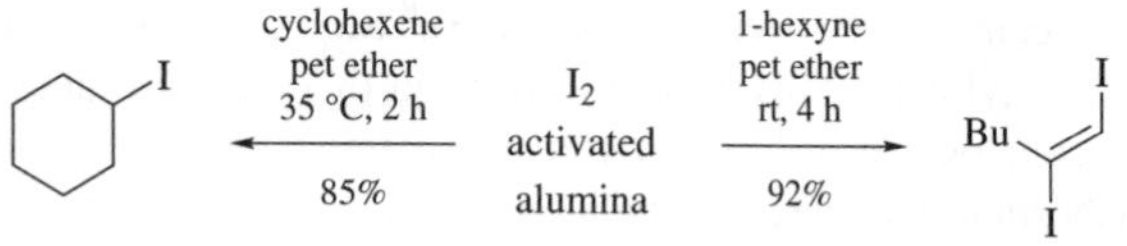

(6)

Aldol-type condensations between aldehydes and various active methylene compounds,[21] Michael reactions (eq 7),[22] as well as Wittig-type reactions[23] can be carried out on alumina under mild conditions, often without a solvent. An interesting nitroaldol reaction–cyclization sequence gives 2-isoxazoline 2-oxides with good diastereoselectivity (eq 8).[24]

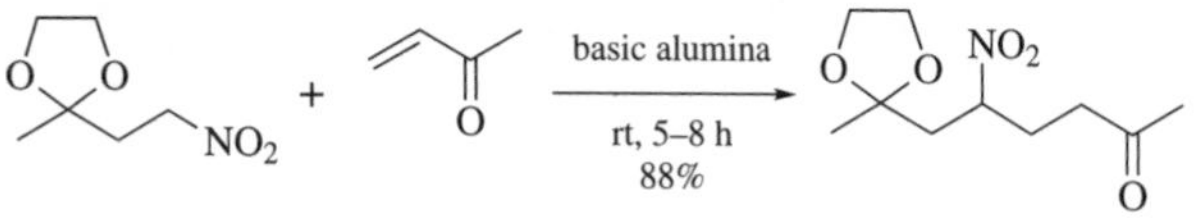

(7)

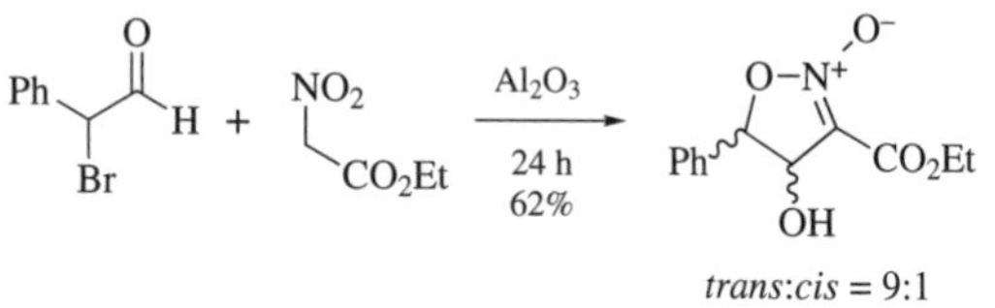

(8)

trans:*cis* = 9:1

Orbital symmetry controlled reactions that have been promoted by alumina include the Diels–Alder,[25] the ene,[26] and the Carroll rearrangement.[27] These reactions proceeded under milder conditions and with greater stereoselectivity. In a spectacular example, chromatographic purification promoted a diastereoselective intramolecular Diels–Alder that produced the verrucarol skeleton (eq 9).[25b]

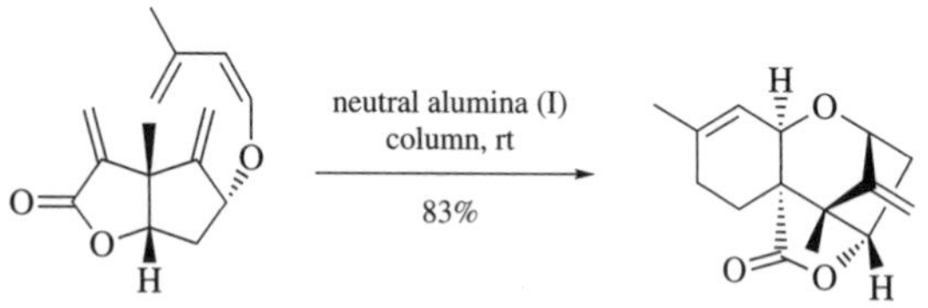

(9)

Alkylation reactions that have been induced by alumina include per-*C*-methylation of phenol,[28] intramolecular alkylation to yield a spiro-fused cyclopropane,[29] and *S*-[30] and *O*-alkylations (eq 10).[31] The activation of ***Diazomethane*** by alumina has provided methods for the conversion of ketones to epoxides[32] and for the selective monomethylation of dicarboxylic acids.[33] Basic alumina has been used for the generation and trapping of dichlorocarbene.[34]

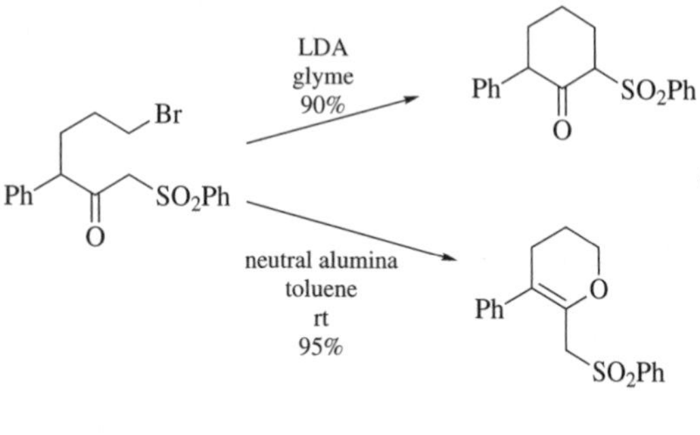

(10)

Epoxides. Epoxides can be opened under mild, selective conditions using alumina impregnated with a variety of nucleophiles, such as alcohols, thiols, selenols, amines, carboxylic acids (eq 11),[35] and peroxides.[36] Use is made of this process in a route to (*Z*)-enamines (eq 12).[37] Formation of C–C bonds by intramolecular opening of epoxides has been reported (eq 13),[38] as have alumina catalyzed epoxide formations[23,39] and rearrangements.[40]

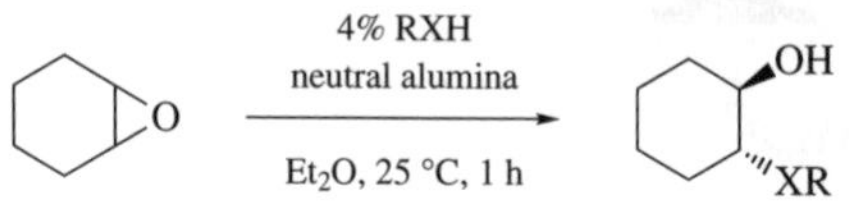

(11)

RX = MeO, 66%; PhS, 70%, PhSe, 95%; *n*-BuNH, 73%

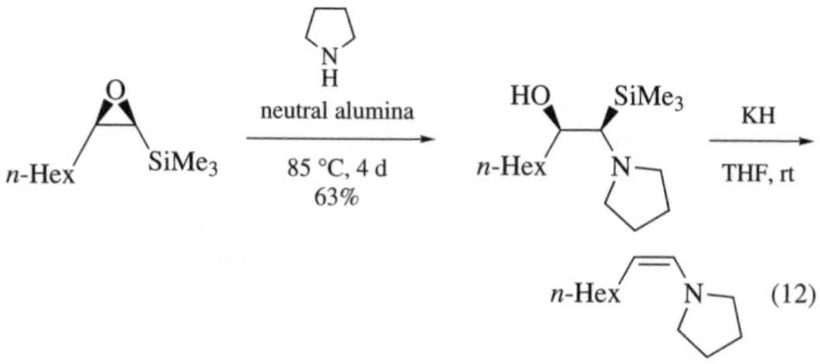

(12)

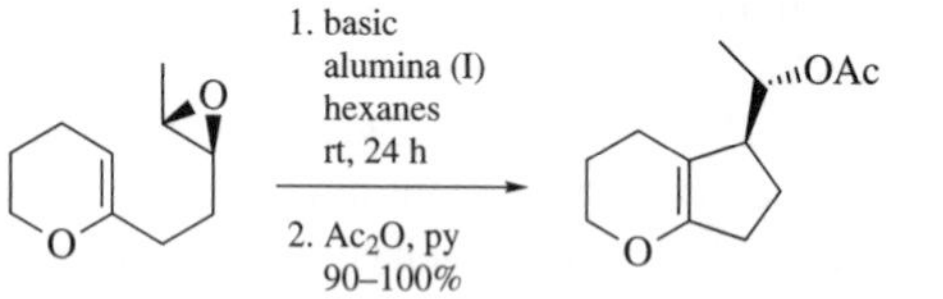

(13)

Oxidations and Reductions. Posner has shown that Oppenauer oxidations, with Cl_3CCHO or PhCHO as the hydrogen acceptors, are greatly accelerated in the presence of activated alumina.[41] Secondary alcohols are oxidized selectively over primary alcohols (eq 14) and groups susceptible to other oxidants (sulfides, selenides, and alkenes) are unaffected. Even cyclobutanol, which is prone to fragmentation with one-electron oxidants, can be oxidized to cyclobutanone in 92% yield.

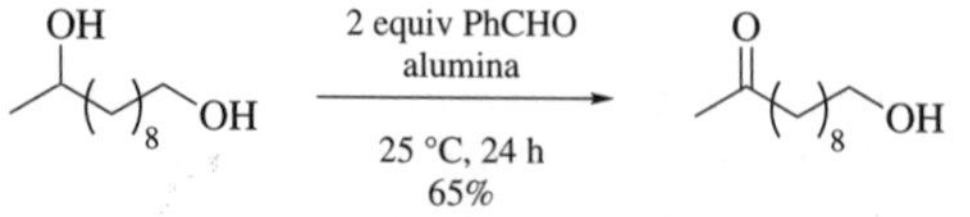

(14)

The complementary reduction reaction (Meerwein–Ponndorf–Verley), using isopropanol as the hydride donor, is also facilitated by alumina and allows the selective reduction of aldehydes over ketones.[42] Functional groups that survive these conditions include alkene, nitro, ester, amide, nitrile, primary and secondary iodides, and benzylic bromide.

Air oxidation of a fluoren-9-ol to the fluoren-9-one and thiols to disulfides are accelerated on the alumina surface.[43] Alumina has also been used as a solid support for a variety of inorganic reagents,[44] and for immobilizing chiral catalysts.[45]

Miscellaneous Reactions. Many rearrangements are catalyzed by alumina.[1] The Beckmann rearrangement[46] of the *O*-sulfonyloxime shown gives the expected amide with activated alumina, and the corresponding oxazoline with basic alumina (eq 15).[46d] Alumina has long been used for isomerization of β,γ-unsaturated ketones to the conjugated ketones.[47] Isomerizations of alkynes to allenes,[48] and allenes to conjugated dienoates[49] have also been reported (eq 16).

activated alumina 80% (15)

alumina PhH, reflux, 5 h 82% (16)

Alumina promotes the hydrolysis of acetates of primary alcohols,[50] the deacylation of imides,[51] the hydrolysis of sulfonylimines,[52] and the decarbalkoxylation of β-keto esters and carbamates.[53] It can also be used for acylations and esterifications, with high selectivity for primary alcohols over secondary alcohols.[54]

Related Reagents. Molecular Sieves.

---

1. Posner, G. H. *AG(E)* **1978**, *17*, 487.
2. Perrin, D. D.; Armarego, W. L. F. *Purification of Laboratory Chemicals*; Pergamon: New York, 1988; pp 20, 310.
3. (a) Furniss, B. S.; Hannaford, A. J.; Smith, P. W. G.; Tatchell, A. R. *Vogel's Textbook of Practical Organic Chemistry*; Longman-Wiley: New York, 1989; p 212. (b) *FF* **1967**, *1*, 19.
4. Knözinger, H. *AG(E)* **1968**, *7*, 791.
5. (a) Hershberg, E. B.; Ruhoff, J. R. *OS* **1937**, *17*, 25. (b) Newton, L. W.; Coburn, E. R. *OSC* **1955**, *3*, 312. (c) Sawyer, R. L.; Andrus, D. W. *OSC* **1955**, *3*, 276.
6. (a) von Rudloff, E. *CJC* **1961**, *39*, 1860. (b) Corey, E. J.; Hortmann, A. G. *JACS* **1965**, *87*, 5736. (c) Barrett, H. C.; Büchi, G. *JACS* **1967**, *89*, 5665.
7. (a) Kobayashi, S.; Shinya, M.; Taniguchi, H. *TL* **1971**, 71. (b) Ishii, H.; Tozyo, T.; Nakamura, M.; Funke, E. *CPB* **1972**, *20*, 203. (c) Gotthardt, H.; Hammond, G. S. *CB* **1974**, *107*, 3922. (d) Mayr, H.; Huisgen, R. *AG(E)* **1975**, *14*, 499. (e) Posner. G. H.; Gurria, G. M.; Babiak, K. A. *JOC* **1977**, *42*, 3173. (f) Vidal, J.; Huet, F. *TL* **1986**, *27*, 3733.
8. (a) Strobach, D. R.; Boswell, G. A., Jr. *JOC* **1971**, *36*, 818. (b) Boswell, G. A., Jr. *JOC* **1966**, *31*, 991.
9. (a) Miller, R. B.; McGarvey, G. *JOC* **1978**, *43*, 4424. (b) Miller, R. B.; McGarvey, G. *SC* **1977**, *7*, 475.
10. Sponholtz, W. R., III; Pagni, R. M.; Kabalka, G. W.; Green, J. F.; Tan, L. C. *JOC* **1991**, *56*, 5700.
11. Cava, M. P.; Pollack, N. M.; Mamer, O. A.; Mitchell, M. J. *JOC* **1971**, *36*, 3932.
12. Labar, D.; Hevesi, L.; Dumont, W.; Krief, A. *TL* **1978**, 1141.
13. (a) Bourns, A. N.; Embleton, H. W.; Hansuld, M. K. *OSC*, **1963**, *4*, 795. (b) Glacet, C.; Adrian, G. *CR(C)* **1969**, *269*, 1322.
14. (a) Texier,-Boullet, F.; Klein, B.; Hamelin, J. *S* **1986**, 409. (b) Tolstikov, G. A.; Galin, F. Z.; Makaev, F. Z. *ZOR* **1989**, *25*, 875.
15. (a) LeBlanc, R. J.; Vaughan, K. *CJC* **1972**, *50*, 2544. (b) Higashino, T.; Suzuki, K.; Hayashi, E. *CPB* **1978**, *26*, 3485. (c) Bladé-Font, A. *TL* **1980**, *21*, 2443. (d) Hooper, D. L.; Manning, H. W.; LaFrance, R. J.; Vaughan, K. *CJC* **1986**, *65*, 250. (e) Hull, J. W., Jr.; Otterson, K.; Rhubright, D. *JOC* **1993**, *58*, 520.
16. Kamitori, Y.; Hojo, M.; Masuda, R.; Yoshida, T. *TL* **1985**, *26*, 4767.
17. McPhail, A. T.; Onan, K. D. *TL* **1973**, 4641.
18. (a) Pelletier, S. W.; Venkov, A. P.; Finer-Moore, J.; Mody, N. V. *TL* **1980**, *21*, 809. (b) Pelletier, S. W.; Gebeyehu, G.; Mody, N. V. *H* **1982**, *19*, 235. (c) Dzurilla, M.; Kutschy, P.; Kristian, P. *S* **1985**, 933.
19. Pagni, R.; Kabalka, G. W.; Boothe, R.; Gaetano, K.; Stewart, L. J.; Conaway, R.; Dial, C.; Gray, D.; Larson, S.; Luidhardt, T. *JOC* **1988**, *53*, 4477.
20. Kropp, P. J.; Daus, K. A.; Tubergen, M. W.; Kepler, K. D.; Wilson, V. P.; Craig, S. L.; Baillargeon, M. M.; Breton, G. W. *JACS* **1993**, *115*, 3071, and references cited therein. Addition of HN_3: Breton, G. W.; Daus, K. A.; Kropp, P. J. *JOC* **1992**, *57*, 6646.
21. (a) Rosan, A.; Rosenblum, M. *JOC* **1975**, *40*, 3621. (b) Texier-Boullet, F.; Foucaud, A. *TL* **1982**, *23*, 4927. (c) Rosini, G.; Ballini, R.; Sorrenti, P. *S* **1983**, 1014. (d) Varma, R. S.; Kabalka, G. W.; Evans, L. T.; Pagni, R. M. *SC* **1985**, *15*, 279. (e) Nesi, R.; Stefano, C.; Piero, S.-F. *H* **1985**, *23*, 1465. (f) Rosini, G.; Ballini, R.; Petrini, M.; Sorrenti, P. *S* **1985**, 515. (g) Foucaud, A.; Bakouetila, M. *S* **1987**, 854. (h) Moison, H.; Texier-Boullet, F.; Foucaud, A. *T* **1987**, *43*, 537.
22. (a) Rosini, G.; Marotta, E.; Ballini, R.; Petrini, M. *S* **1986**, 237. (b) Ballini, R.; Petrini, M.; Marcantoni, E.; Rosini, G. *S* **1988**, 231.
23. Texier-Boullet, F.; Villemin, D.; Ricard, M.; Moison, H.; Foucaud, A. *T* **1985**, *41*, 1259.
24. Isoxazoline: Rosini, G.; Galarini, R.; Marotta, E.; Righi, P. *JOC* **1990**, *55*, 781. Rosini, G.; Marotta, E.; Righi, E.; Seerden, J. P. *JOC* **1991**, *56*, 6258.
25. (a) Parlar, H.; Baumann, R. *AG(E)* **1981**, *20*, 1014 (b) Koreeda, M.; Ricca, D. J.; Luengo, J. I. *JOC* **1988**, *53*, 5586.
26. (a) Tietze, L. F.; Beifuss, U.; Ruther, M. *JOC* **1989**, *54*, 3120. (b) Tietze, L. F.; Beifuss, U. *S* **1988**, 359.
27. Pogrebnoi, S. I.; Kalyan, Y. B.; Krimer, M. Z.; Smit, W. A. *TL* **1987**, *28*, 4893.
28. Cullinane, N. M.; Chard, S. J.; Dawkins, C. W. C. *OSC* **1963**, *4*, 520.
29. Baird, R.; Winstein, S. *JACS* **1963**, *85*, 567.
30. Villemin, D. *CC* **1985**, 870.
31. (a) Ogawa, H.; Chihara, T.; Teratani, S.; Taya, K. *BCJ* **1986**, *59*, 2481. (b) Cooke, F.; Magnus, P. *CC* **1976**, 519.
32. Hart, P. A.; Sandmann, R. A. *TL* **1969**, 305.
33. Ogawa, H.; Chihara, T.; Taya, K. *JACS* **1985**, *107*, 1365.
34. Sarratosa, F. *J. Chem. Educ.* **1964**, *41*, 564.
35. (a) Posner, G. H.; Rogers, D. Z. *JACS* **1977**, *99*, 8208. (b) Posner, G. H.; Rogers, D. Z. *JACS* **1977**, *99*, 8214. (c) Evans, D. A.; Golob, A. M.; Mandel, N. S.; Mandel, G. S. *JACS* **1978**, *100*, 8170.
36. Kropf, H.; Amirabadi, H. M.; Mosebach, M.; Torkler, A.; von Wallis, H. *S* **1983**, 587.
37. Hudrlik, P. F.; Hudrlik, A. M.; Kulkarni, A. K. *TL* **1985**, *26*, 139.
38. (a) Boeckman, R. K., Jr.; Bruza, K. J.; Heinrich, G. R. *JACS* **1978**, *100*, 7101. (b) Niwa, M.; Iguchi, M.; Yamamura, S. *TL* **1979**, 4291.
39. (a) Dhillon, R. S.; Chhabra, B. R.; Wadia, M. S.; Kalsi, P. S. *TL* **1974**, 401. (b) Antonioletti, R.; D'Auria, M.; De Mico, A.; Piancatelli, G.; Scettri, A. *T* **1983**, *39*, 1765.
40. (a) Tsuboi, S.; Furutani, H.; Takeda, A. *S* **1987**, 292. (b) Harigaya, Y.; Yotsumoto, K.; Takamatsu, S.; Yamaguchi, H.; Onda, M. *CPB* **1981**, *29*, 2557.
41. (a) Posner, G. H.; Perfetti, R. B.; Runquist, A. W. *TL* **1976**, 3499. (b) Posner, G. H.; Chapdelaine, M. J. *S* **1977**, 555. (c) Posner, G. H.; Chapdelaine, M. J. *TL* **1977**, 3227.
42. Posner, G. H.; Runquist, A. W.; Chapdelaine, M. J. *JOC* **1977**, *42*, 1202. Also see: Suginome, H.; Kato, K. *TL* **1973**, 4143.
43. (a) Pan, H.-L.; Cole, C.-A.; Fletcher, T. L. *S* **1975**, 716. (b) Liu, K.-T.; Tong, Y.-C. *S* **1978**, 669.
44. Review: Laszlo, P. *COS* **1991**, *7*, 839. Recent examples: (a) Singh, S.; Dev, S. *T* **1993**, *49*, 10959. (b) Lee, D. G.; Chen, T.; Wang, Z. *JOC* **1993**, *58*, 2918. (c) Morimoto, T.; Hirano, M.; Iwasaki, K.; Ishikawa, T. *CL* **1994**, 53. (d) Santaniello, E.; Ponti, F.; Manzocchi, A. *S* **1978**, 891.
45. Soai, K.; Watanabe, M.; Yamamoto, A. *JOC* **1990**, *55*, 4832.
46. (a) Craig, J. C.; Naik, A. R. *JACS* **1962**, *84*, 3410. (b) Gonzalez, A.; Galvez, C. *S* **1982**, 946. (c) Luh, T.-Y.; Chow, H.-F.; Leung, W. Y.; Tam, S. W. *T* **1985**, *41*, 519. (d) Nagano, H.; Masunaga, Y.; Matsuo, Y.; Shiota, M. *BCJ* **1987**, *60*, 707. See also: (e) Métayer, A.; Barbier, M. *BSF* **1972**, 3625.

47. (a) Marshall, J. A.; Roebke, H. *JOC* **1966**, *31*, 3109. (b) Hudlicky, T.; Srnak, T. *TL* **1981**, *22*, 3351. (c) Reetz, M. T.; Wenderoth, B.; Urz, R. *CB* **1985**, *118*, 348. (d) Hatzigrigoriou, E.; Roux-Schmitt, M.-C.; Wartski, L. *T* **1988**, *44*, 4457. Also see: (e) Scettri, A.; Piancatelli, G.; D'Auria, M.; David, G. *T* **1979**, *35*, 135.
48. (a) Larock, R. C.; Chow, M.-S.; Smith, S. J. *JOC* **1986**, *51*, 2623. (b) Manning, D. T.; Coleman, H. A. J. *JOC* **1969**, *34*, 3248.
49. Tsuboi, S.; Matsuda, T.; Mimura, S.; Takeda, A. *OSC* **1993**, *8*, 251.
50. Johns, W. F.; Jerina, D. M. *JOC* **1963**, *28*, 2922.
51. Boar, R. B.; McGhie, J. F.; Robinson, M.; Barton, D. H. R.; Horwell, D. C.; Stick, R. V. *JCS(P1)* **1975**, 1237.
52. Coutts, I. G. C.; Culbert, N. J.; Edward, M.; Hadfield, J. A.; Musto, D. R.; Pavlidis, V. H.; Richards, D. J. *JCS(P1)* **1985**, 1829.
53. (a) Greene, A. E.; Cruz, A.; Crabbé, P. *TL* **1976**, 2707. (b) van Leusen, A. M.; Strating, J. *OSC* **1988**, *6*, 981.
54. (a) Posner, G. H.; Oda, M. *TL* **1981**, *22*, 5003. (b) Rana, S. S.; Barlow, J. J.; Matta, K. L. *TL* **1981**, *22*, 5007. (c) Posner, G. A.; Okada, S. S.; Babiak, K. A.; Miura, K.; Rose, R. K. *S* **1981**, 789. (d) Nagasawa, K.; Yoshitake, S.; Amiya, T.; Ito, K. *SC* **1990**, *20*, 2033.

Viresh H. Rawal, Seiji Iwasa,
Alan S. Florjancic, & Agnes Fabre
The Ohio State University, Columbus, OH, USA

Aluminum Chloride[1]

[7446-70-0] $AlCl_3$ (MW 133.34)

(Lewis acid catalyst for Friedel–Crafts, Diels–Alder, [2 + 2] cycloadditions, ene reactions, rearrangements, and other reactions)

Physical Data: mp 190 °C (193–194 °C sealed tube); sublimes at 180 °C; *d* 2.44 g cm^{-3}.
Solubility: sol many organic solvents, e.g. benzene, nitrobenzene, carbon tetrachloride, chloroform, methylene chloride, nitromethane, and 1,2-dichloroethane; insol carbon disulfide.
Form Supplied in: colorless solid when pure, typically a gray or yellow-green solid; also available as a 1.0 M nitrobenzene solution.
Handling, Storage, and Precautions: fumes in air with a strong odor of HCl. $AlCl_3$ reacts violently with H_2O. All containers should be kept tightly closed and protected from moisture.[1c] Use in a fume hood.

Friedel–Crafts Chemistry.[1,2] $AlCl_3$ has traditionally been used in stoichiometric or catalytic[3] amounts to mediate Friedel–Crafts alkylations and acylations of aromatic systems (eq 1).

RCl / $AlCl_3$; RCOCl / $AlCl_3$ (1)

This is a result of the Lewis acidity of $AlCl_3$ which complexes strongly with carbonyl groups.[4] Adaptations of these basic reactions have been reported.[5] In chiral systems, inter- and intramolecular acylations have been achieved without the loss of optical activity (eq 2).[6]

$NHCO_2Me$, Cl → ($AlCl_3$, benzene, 50–60%) → MeO_2CHN
$NHCO_2Me$, Cl → ($AlCl_3$, 55–75%, >98% ee) → $NHCO_2Me$ (2)

Friedel–Crafts chemistry at an asymmetric center generally proceeds with racemization, but the use of mesylates or chlorosulfonates as leaving groups has resulted in alkylations with excellent control of stereochemistry.[7] The reactions proceed with inversion of configuration (eq 3). Cyclopropane derivatives have been used as three-carbon units in acylation reactions (eq 4).[8] In conjunction with triethylsilane, a net alkylation is possible under acylation conditions (eq 5).[9] These conditions are compatible with halogen atoms present elsewhere in the molecule. Acylation reactions of phenolic compounds with heteroaromatic systems have also been accomplished (eq 6).[10]

CO_2Me, X (*S*) → ($AlCl_3$, PhH, 50–80%, >97% ee) → CO_2Me, Ph (*S*) (3)
X = OSO_2Me, OSO_2Cl

CO_2Et → ($AlCl_3$, PhH, 93%) → (4)

R' → (RCOCl, $AlCl_3$, Et_3SiH, 43–94%) → R, R' (5)

OH, OH + Cl, Cl, N, N, Cl → ($AlCl_3$, 63%) → HO, OH, Cl, N, N, Cl (6)

Treatment of aryl azides with $AlCl_3$ has been reported to give polycyclic aromatic compounds (eq 7),[11] or aziridines when the reactions are run in the presence of alkenes (eq 8).[12]

N_3 → ($AlCl_3$, 89%) → N H (7)

N_3, R + ( )$_n$, n = 1, 2 → ($AlCl_3$, 63–93%) → ( )$_n$ N Ar (8)

The scope of Friedel–Crafts chemistry has been expanded beyond aromatic systems to nonaromatic systems, such as alkenes and alkynes and the mechanistic details have been investigated.[13] The Friedel–Crafts alkylation[14] and acylation[15] of alkenes provide access to a variety of organic systems (eq 9). The acylation of alkynes provides access to cyclopentenone derivatives (eq 10).[16] In addition, one can use this chemistry to access indenyl systems[17] and vinyl chlorides.[18] Allylic sulfones can undergo allylation chemistry (eq 11).[19]

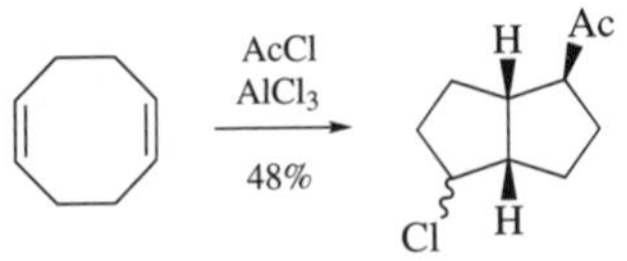

AcCl, $AlCl_3$, 48% (9)

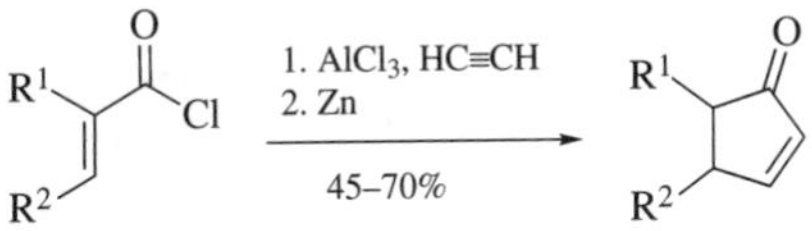

1. $AlCl_3$, HC≡CH; 2. Zn; 45–70% (10)

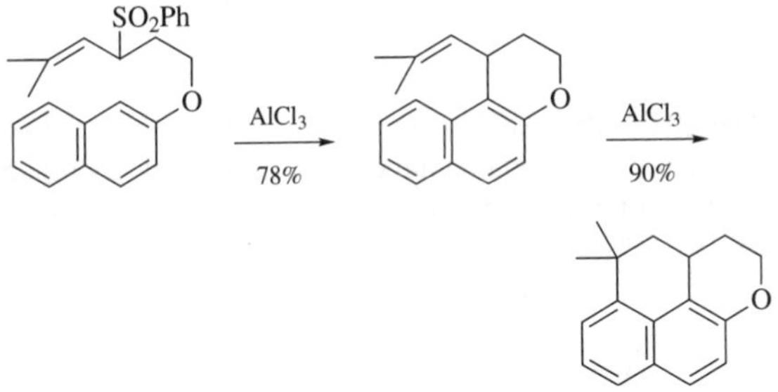

$AlCl_3$, 78%; $AlCl_3$, 90% (11)

The use of silyl derivatives in Friedel–Crafts chemistry has not only improved the regioselectivity but extended the scope of these reactions. Substitution at the *ipso* position occurs with aryl silanes (eq 12).[20] The ability of silyl groups to stablize β-carbenium ions (β-effect) affords acylated products with complete control of regiochemistry (eq 13).[21]

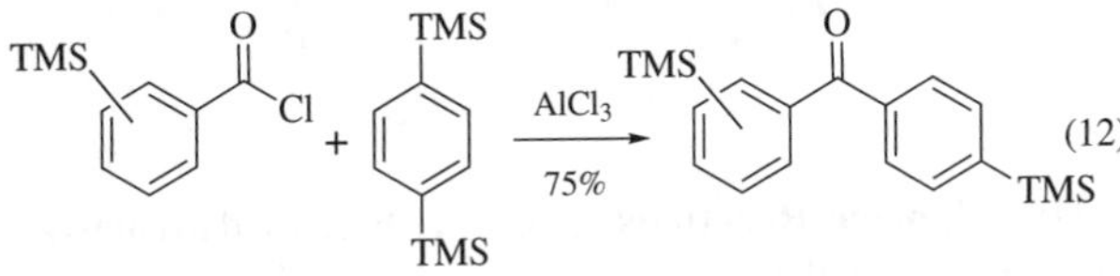

$AlCl_3$, 75% (12)

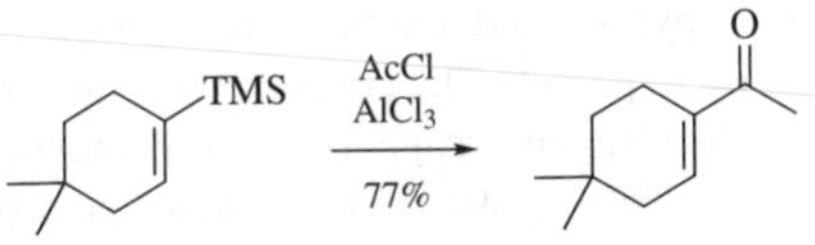

AcCl, $AlCl_3$, 77% (13)

The use of silylacetylenes gives ynones (eq 14),[22] cyclopentenone derivatives (eq 15),[23] and α-amino acid derivatives (eq 16).[24]

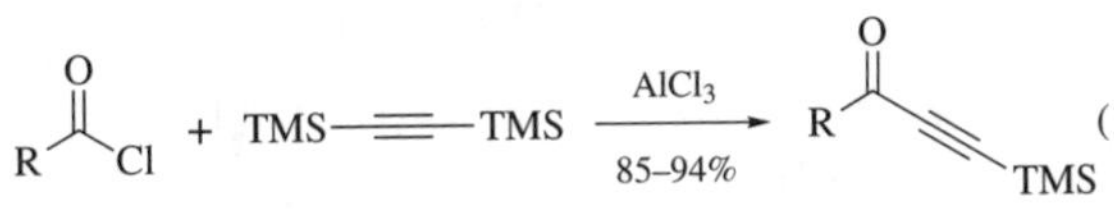

$AlCl_3$, 85–94% (14)

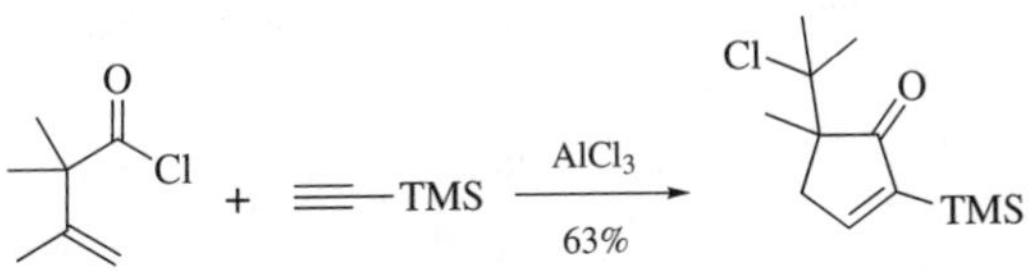

$AlCl_3$, 63% (15)

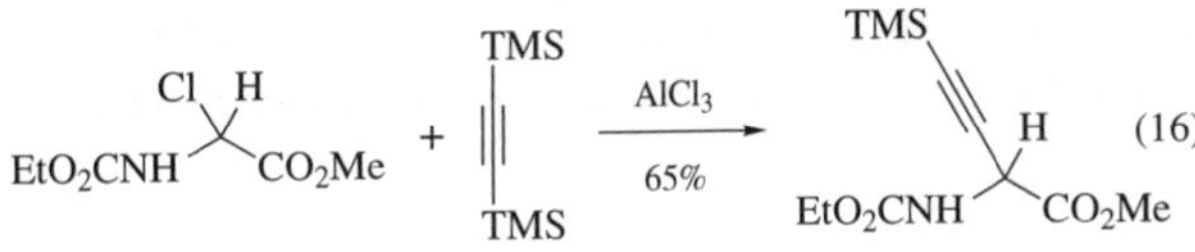

$AlCl_3$, 65% (16)

Propargylic silanes undergo acylation to generate allenyl ketones (eq 17),[25] while alkylsilanes afford cycloalkanones (eq 18).[26]

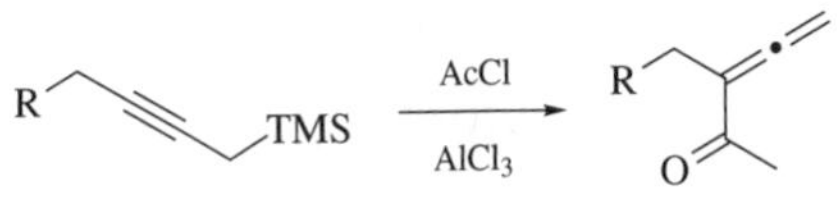

AcCl, $AlCl_3$ (17)

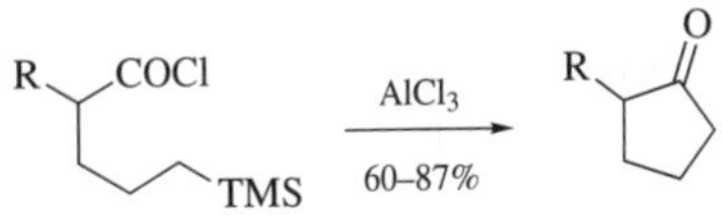

$AlCl_3$, 60–87% (18)

Several name reactions are promoted by $AlCl_3$. For example, the Darzens–Nenitzescu reaction is simply the acylation of alkenes. The Ferrario reaction generates phenoxathiins from diphenyl ethers (eq 19).[27] The rearrangement of acyloxy aromatic systems is known as the Fries rearrangement (eq 20).[28] Aryl aldehydes are produced by the Gatterman aldehyde synthesis (eq 21).[29] The initial step of the Haworth phenanthrene synthesis makes use of a Friedel–Crafts acylation.[30] The acylation of phenolic compounds is called the Houben–Hoesch reaction (eq 22).[31] The Leuckart amide synthesis generates aryl amides from isocyanates (eq 23).[32]

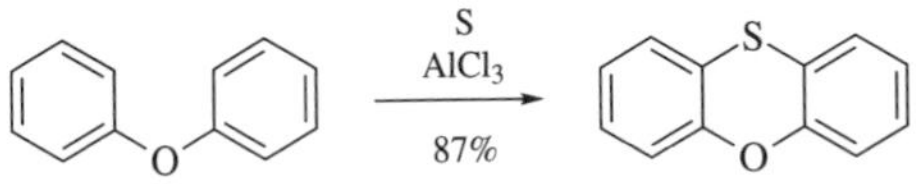

S, $AlCl_3$, 87% (19)

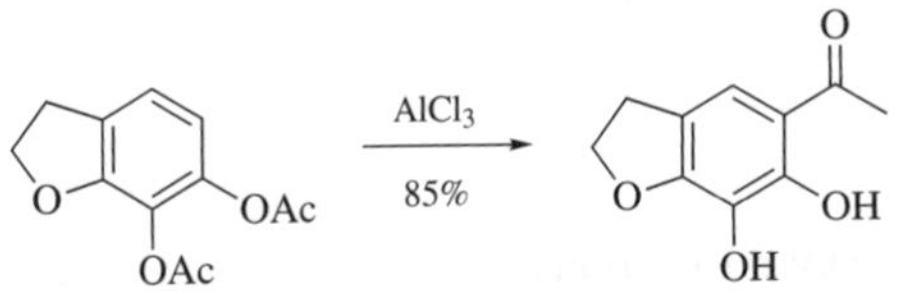

$AlCl_3$, 85% (20)

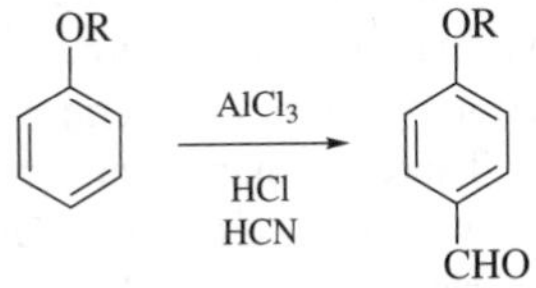

$AlCl_3$, HCl, HCN (21)

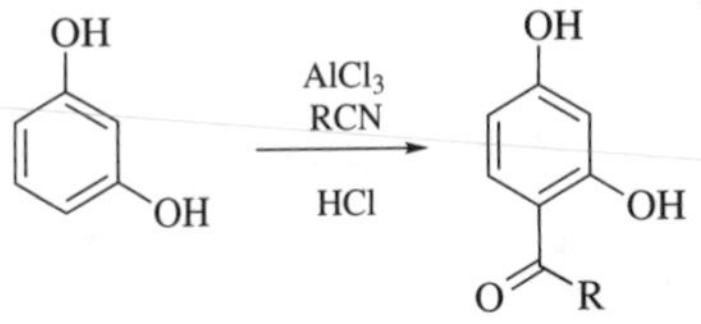

$AlCl_3$, RCN, HCl (22)

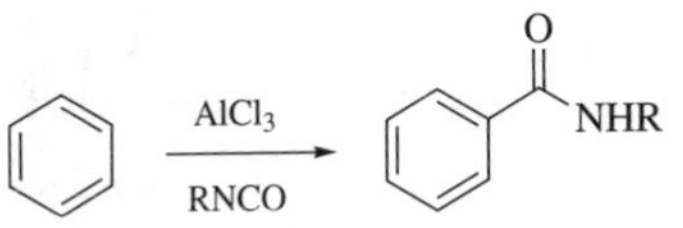

$AlCl_3$, RNCO (23)

Amides can also be obtained by $AlCl_3$ catalyzed ester amine exchange which proceeds primarily without racemization of chiral centers (eq 24).[33] The reaction of phenols with β-keto esters is known as the Pechmann condensation (eq 25).[34] Aryl amines are used in the Riehm quinoline synthesis (eq 26).[35] Aromatic systems may be coupled via the Scholl reaction (eq 27)[36] and indole derivatives are prepared in the Stolle synthesis (eq 28).[37] In the Zincke-Suhl reaction, phenols are converted to dienones (eq 29).[38]

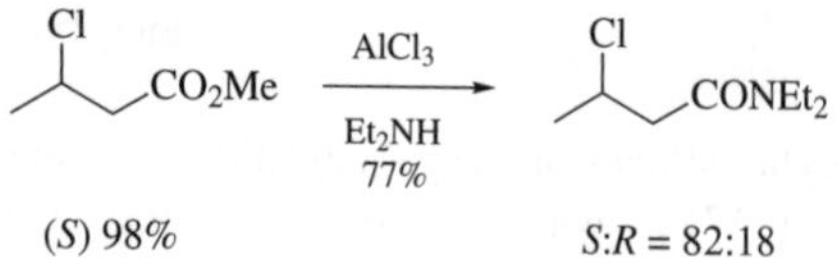

Cl, CO_2Me → ($AlCl_3$, Et_2NH, 77%) Cl, $CONEt_2$ (24)

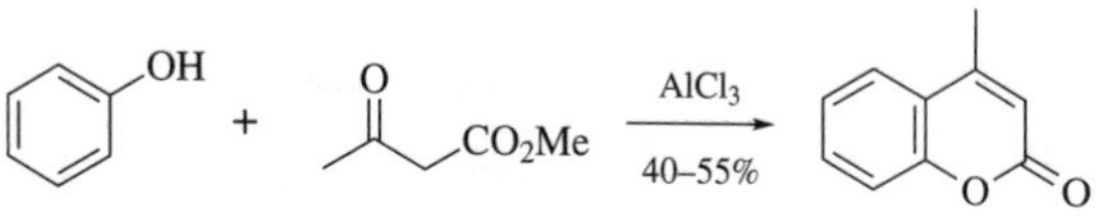

(*S*) 98% *S*:*R* = 82:18

OH + O, CO_2Me → ($AlCl_3$, 40–55%) O, O (25)

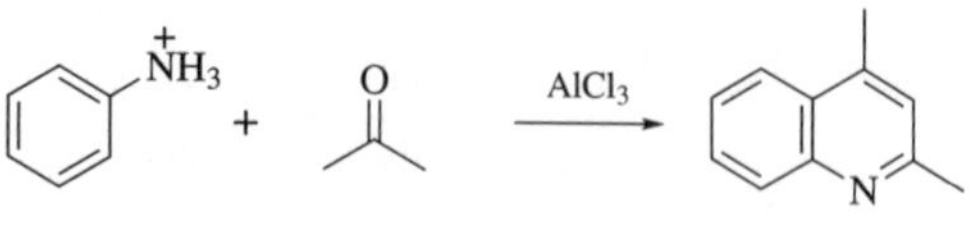

$\overset{+}{N}H_3$ + O → ($AlCl_3$) N (26)

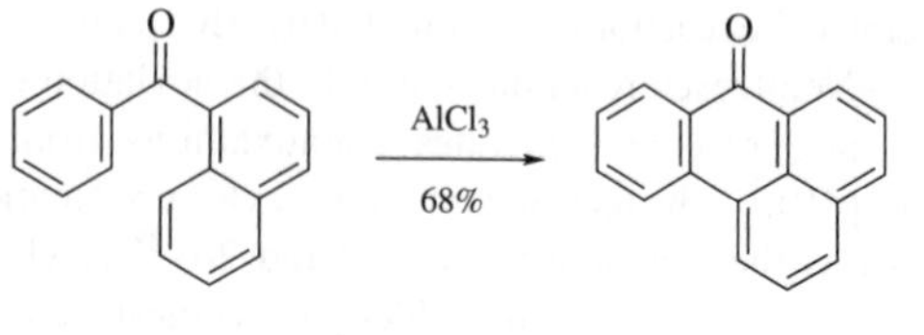

O → ($AlCl_3$, 68%) O (27)

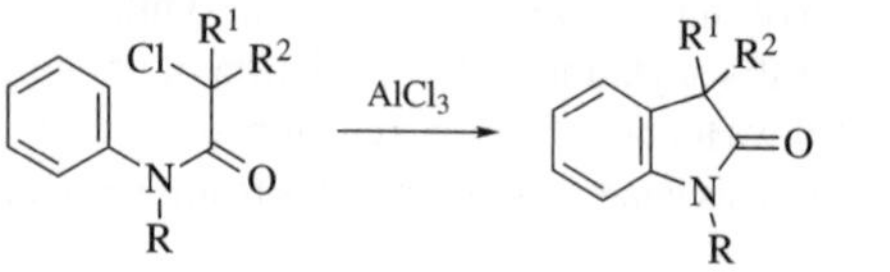

Cl, R^1, R^2, N, O, R → ($AlCl_3$) R^1, R^2, O, N, R (28)

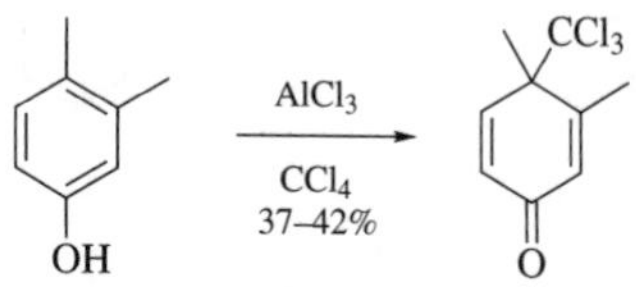

OH → ($AlCl_3$, CCl_4, 37–42%) CCl_3, O (29)

Diels–Alder Reactions. There is some evidence that $AlCl_3$ catalysis of Diels–Alder reactions changes the transition state from a synchronous to an asynchronous one.[39] This also enhances asymmetric induction by increasing steric interactions at one end of the dienophile. There are many examples of $AlCl_3$ promoted Diels–Alder reactions (eq 30).[40] Hetero-Diels–Alder reactions can be used to generate oxygen (eq 31)[41] and nitrogen (eq 32)[42] containing heterocycles.

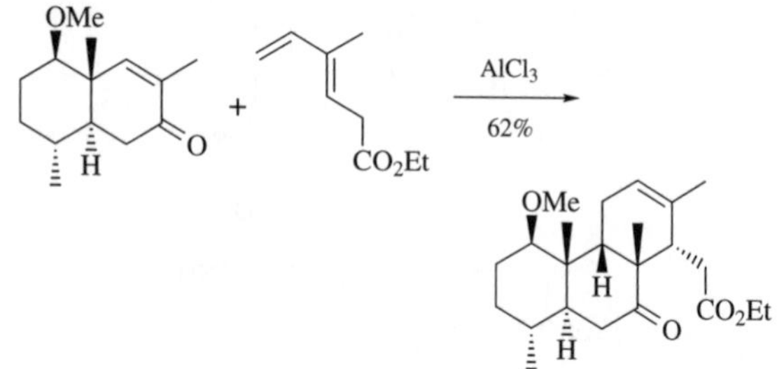

OMe, H, O + CO_2Et → ($AlCl_3$, 62%) OMe, H, H, O, CO_2Et (30)

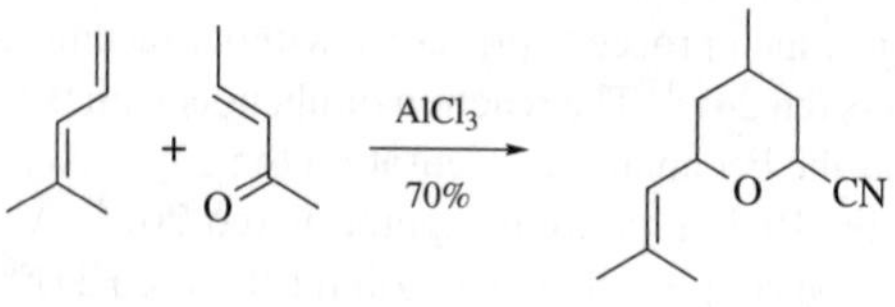

+ O → ($AlCl_3$, 70%) O, CN (31)

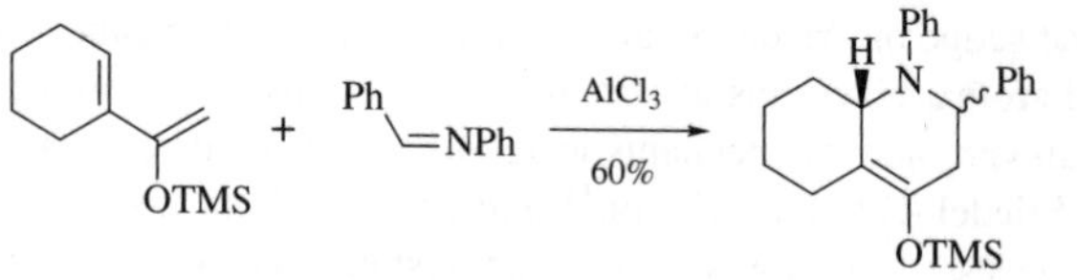

OTMS + Ph, =NPh → ($AlCl_3$, 60%) H, Ph, N, Ph, OTMS (32)

$AlCl_3$ can also be used to catalyze [2 + 2] cycloaddition reactions (eq 33)[43] and ene reactions (eq 34).[44]

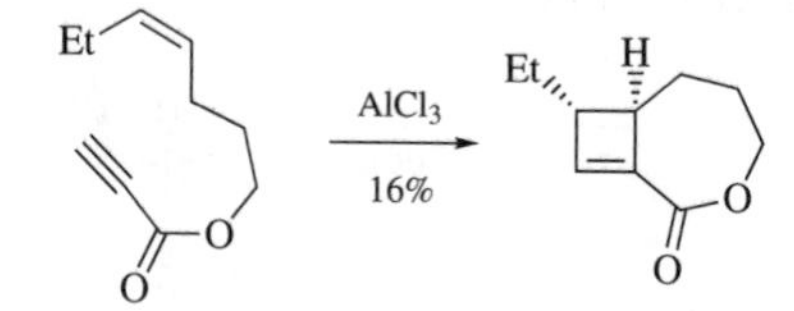

Et, O, O → ($AlCl_3$, 16%) Et, H, O, O (33)

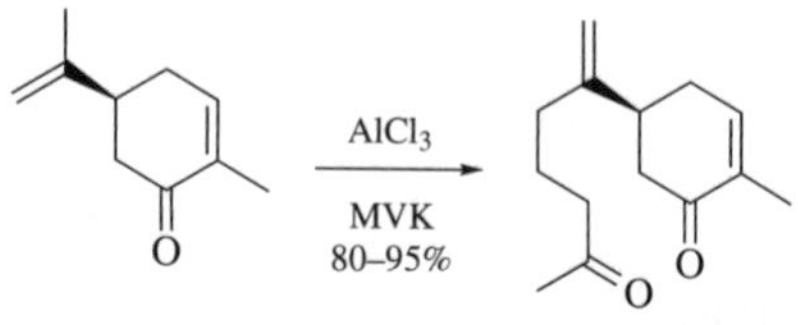

O → ($AlCl_3$, MVK, 80–95%) O, O (34)

Rearrangements. $AlCl_3$ catalyzed rearrangement of hydrocarbon derivatives to adamantanes has been well documented (eq 35).[45] Other rearrangements have been used in triquinane synthesis (eq 36).[46]

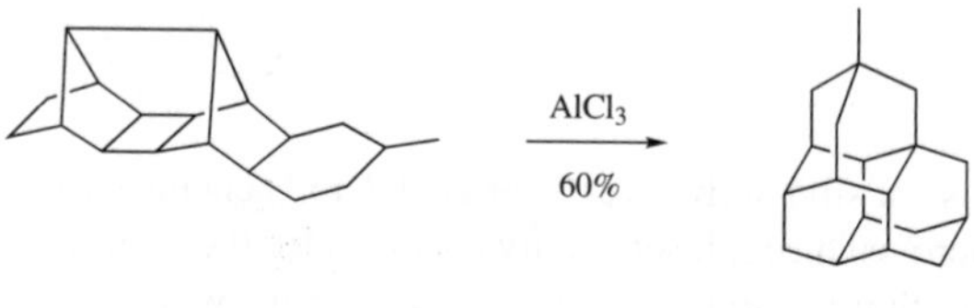

→ ($AlCl_3$, 60%) (35)

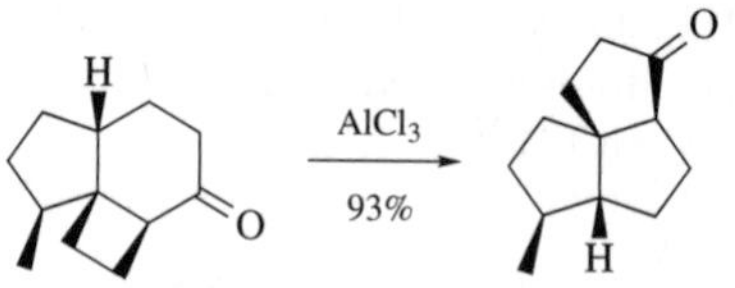

H, O → ($AlCl_3$, 93%) O, H (36)

Miscellaneous Reactions. $AlCl_3$ has been used to catalyze the addition of allylsilanes to aldehydes and acid chlorides (eq 37).[47] Cyclic ethers (pyrans and oxepins) have been prepared with hydroxyalkenes (eq 38).[48] The course of reactions between aldehydes and allylic Grignard reagents can be completely diverted to α-allylation by $AlCl_3$ (eq 39).[49] The normal course of the reaction gives γ-allylation products.

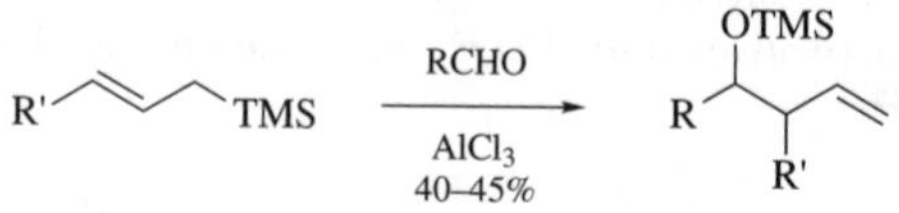

R', TMS → (RCHO, $AlCl_3$, 40–45%) OTMS, R, R' (37)

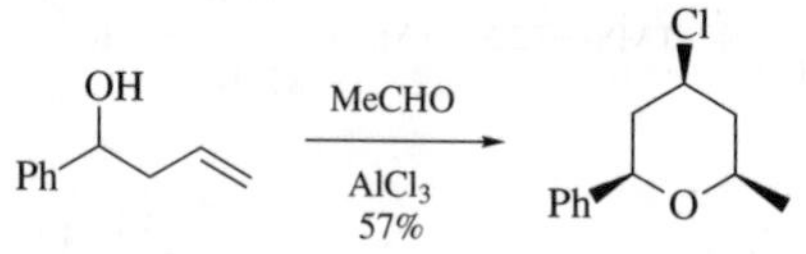

OH, Ph → (MeCHO, $AlCl_3$, 57%) Cl, Ph, O (38)

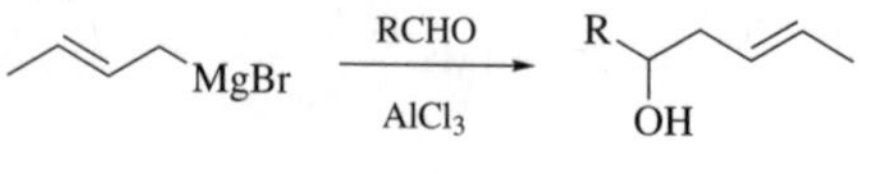

MgBr → (RCHO, $AlCl_3$) R, OH (39)

$AlCl_3$ can be used to remove *t*-butyl groups from aromatic rings (eq 40),[50] thereby using this group as a protecting element for a ring position. $AlCl_3$ has also been used to remove *p*-nitrobenzyl (PNB) and benzhydryl protecting groups (eq 41).[51] The combination of $AlCl_3$ and ***Ethanethiol*** has formed the basis of a push–pull mechanism for the cleavage of many types of bonds including C–X,[52] C–NO_2,[53] C=C,[54] and C–O.[55] Furthermore, $AlCl_3$ has been used to catalyze chlorination of aromatic rings,[56] open epoxides,[57] and mediate addition of dichlorophosphoryl groups to alkanes.[58]

OH, *t*-Bu, *t*-Bu, *t*-Bu → ($AlCl_3$, 80%) → OH (40)

R, $NHCO_2PNB$, S, N, O, CO_2PNB → ($AlCl_3$, 24–82%) → R, NH_2, S, N, O, CO_2H (41)

Related Reagents. Antimony(V) Fluoride; Ethylaluminum Dichloride; Hydrogen Fluoride; Iron(III) Chloride; Iron(III) Chloride–Alumina; Tin(IV) Chloride; Titanium(IV) Chloride; Zinc Chloride.

1. (a) Thomas, C. A. *Anhydrous Aluminum Chloride in Organic Chemistry*; ACS Monograph Series; Reinholdt: New York, 1941. (b) Shine, H. J. *Aromatic Rearrangements*; Elsevier: Amsterdam, 1967. (c) *FF* **1967**, *1*, 24. (d) Olah, G. A. *Friedel–Crafts Chemistry*; Wiley: New York, 1973. (e) Roberts, R. M.; Khalaf, A. A. *Friedel–Crafts Alkylation Chemistry*; Marcel Dekker: New York, 1984.
2. Gore, P. H. *CR* **1955**, *55*, 229.
3. Pearson, D. E.; Buehler, C. A. *S* **1972**, 533.
4. (a) Tan, L. K.; Brownstein, S. *JOC* **1982**, *47*, 4737. (b) Tan, L. K.; Brownstein, S. *JOC* **1983**, *48*, 3389.
5. Drago, R. S.; Getty, E. E. *JACS* **1988**, *110*, 3311.
6. McClure, D. E.; Arison, B. H.; Jones, J. H.; Baldwin, J. J. *JOC* **1981**, *46*, 2431.
7. Piccolo, O.; Spreafico, F.; Visentin, G.; Valoti, E. *JOC* **1985**, *50*, 3945.
8. Pinnick, H. W.; Brown, S. P.; McLean, E. A.; Zoller, L. W. *JOC* **1981**, *46*, 3758.
9. Jaxa-Chamiel, A.; Shah, V. P.; Kruse, L. I. *JCS(P1)* **1989**, 1705.
10. (a) Pollak, A.; Stanovnik, B.; Tisler, M. *JOC* **1966**, *31*, 4297. (b) Coates, W. J.; McKillop, A. *JOC* **1990**, *55*, 5418.
11. Takeuchi, H.; Maeda, M.; Mitani, M.; Koyama, K. *CC* **1985**, 287.
12. Takeuchi, H.; Shiobara, Y.; Kawamoto, H.; Koyama, K. *JCS(P1)* **1990**, 321.
13. (a) Puck, R.; Mayr, H.; Rubow, M.; Wilhelm, E. *JACS* **1986**, *108*, 7767. (b) Brownstein, S.; Morrison, A.; Tan, L. K. *JOC* **1985**, *50*, 2796.
14. Mayr, H.; Striepe, W. *JOC* **1983**, *48*, 1159.
15. (a) Ansell, M. F.; Ducker, J. W. *JCS* **1960**, 5219. (b) Cantrell, T. S. *JOC* **1967**, *32*, 1669. (c) Groves, JK. *CSR* **1972**, *1*, 73. (d) House, H. O. *Modern Synthetic Reactions*; Benjamin-Cummings: Menlo Park, CA, 1972; pp 786–816.
16. (a) Martin, G. J.; Rabiller, C.; Mabon, G. *TL* **1970**, 3131. (b) Rizzo, C. J.; Dunlap, N. A.; Smith, A. B. *JOC* **1987**, *52*, 5280.
17. Maroni, R.; Melloni, G.; Modena, G. *JCS(P1)* **1974**, 353.
18. Maroni, R.; Melloni, G.; Modena, G. *JCS(P1)* **1973**, 2491.
19. Trost, B. M.; Ghadiri, M. R. *JACS* **1984**, *106*, 7260.
20. (a) Eaborn, C. *JCS* **1956**, 4858. (b) Habich, D.; Effenberger, F. *S* **1979**, 841.
21. (a) Fleming, I.; Pearce, A. *CC* **1975**, 633. (b) Fristad, W. E.; Dime, D. S.; Bailey, T. R.; Paquette, L. A. *TL* **1979**, 1999.
22. (a) Walton, D. R. M.; Waugh, F. *JOM* **1972**, *37*, 45. (b) Newman, H. *JOC* **1973**, *38*, 2254.
23. Karpf, M. *TL* **1982**, *23*, 4923.
24. Casara, P.; Metcalf, B. W. *TL* **1978**, 1581.
25. Flood, T.; Peterson, P. E. *JOC* **1980**, *45*, 5006.
26. Urabe, H.; Kuwajima, I. *JOC* **1984**, *49*, 1140.
27. Ferrario, E. *BSF* **1911**, *9*, 536.
28. (a) Blatt, A. H. *OR* **1942**, *1*, 342. (b) Gammill, R. B. *TL* **1985**, *26*, 1385.
29. Truce, W. E. *OR* **1957**, *9*, 37.
30. Berliner, E. *OR* **1949**, *5*, 229.
31. Spoerri, P. E.; Dubois, A. S. *OR* **1949**, *5*, 387.
32. Effenberger, F.; Gleiter, R. *CB* **1964**, *97*, 472.
33. Gless, R. D. *SC* **1986**, *16*, 633.
34. Sethna, S.; Phadke, R. *OR* **1953**, *7*, 1.
35. Elderfield, R. C.; McCarthy, J. R. *JACS* **1951**, *73*, 975.
36. Clowes, G. A. *JCS(P1)* **1968**, 2519.
37. Sumpter, W. C. *CR* **1944**, *34*, 393.
38. Newman, M. S.; Wood, L. L. *JACS* **1959**, *81*, 6450.
39. Tolbert, L. M.; Ali, M. B. *JACS* **1984**, *106*, 3806.
40. (a) Cohen, N.; Banner, B. L.; Eichel, W. F. *SC* **1978**, *8*, 427. (b) Fringuelli, F.; Pizzo, F.; Taticchi, A.; Wenkert, E. *SC* **1979**, *9*, 391. (c) Ismail, Z. M.; Hoffmann, H. M. R. *JOC* **1981**, *46*, 3549. (d) Vidari, G.; Ferrino, S.; Grieco, P. A. *JACS* **1984**, *106*, 3539. (e) Angell, E. C.; Fringuelli, F.; Guo, M.; Minuti, L.; Taticchi, A.; Wenkert, E. *JOC* **1988**, *53*, 4325.
41. Ismail, Z. M.; Hoffmann, H. M. R. *AG(E)* **1982**, *21*, 859.
42. LeCoz, L.; Wartski, L.; Seyden-Penne, J.; Chardin, P.; Nierlich, M. *TL* **1989**, *30*, 2795.
43. Jung, M. E.; Haleweg, K. M. *TL* **1981**, *22*, 2735.
44. (a) Snider, B. B.; Rodini, D. J.; Conn, R. S. E.; Sealfon, S. *JACS* **1979**, *101*, 5283. (b) Mehta, G.; Reddy, A. V. *TL* **1979**, 2625. (c) Snider, B. B. *ACR* **1980**, *13*, 426.
45. (a) Bingham, R. C.; Schleyer, P. R. *Top. Curr. Chem.* **1971**, *18*, 1. (b) McKervey, M. A. *CSR* **1974**, *3*, 479. (c) McKervey, M. A. *T* **1980**, *36*, 971.
46. Kakiuchi, K.; Ue, M.; Tsukahara, H.; Shimizu, T.; Miyao, T.; Tobe, Y.; Odaira, Y.; Yasuda, M.; Shima, K. *JACS* **1989**, *111*, 3707.
47. (a) Deleris, G.; Donogues, J.; Calas, R. *TL* **1976**, 2449. (b) Pillot, J.-P.; Donogues, J.; Calas, R. *TL* **1976**, 1871.
48. Coppi, L.; Ricci, A.; Taddai, M. *JOC* **1988**, *53*, 911.
49. Yamamoto, Y.; Maruyama, K. *JOC* **1983**, *48*, 1564.
50. Lewis, N.; Morgan, I. *SC* **1988**, *18*, 1783.
51. Ohtani, M.; Watanabe, F.; Narisada, M. *JOC* **1984**, *49*, 5271.
52. Node, M.; Kawabata, T.; Ohta, K.; Fujimoto, M.; Fujita, E.; Fuji, K. *JOC* **1984**, *49*, 3641.
53. Node, M.; Kawabata, T.; Ueda, M.; Fujimoto, M.; Fuji, K.; Fujita, E. *TL* **1982**, *23*, 4047.
54. Fuji, K.; Kawabata, T.; Node, M.; Fujita, E. *JOC* **1984**, *49*, 3214.
55. Node, M.; Nishide, K.; Ochiai, M.; Fuji, K.; Fujita, E. *JOC* **1981**, *46*, 5163.
56. Watson, W. D. *JOC* **1985**, *50*, 2145.
57. Eisch, J. J.; Liu, Z.-R.; Ma, X.; Zheng, G.-X. *JOC* **1992**, *57*, 5140.
58. Olah, G. A.; Farooq, O.; Wang, Q.; Wu, A.-H. *JOC* **1990**, *55*, 1224.

Paul Galatsis
University of Guelph, Ontario, Canada

Aluminum Isopropoxide[1]

Al(O-*i*-Pr)$_3$

[555-31-7] $C_9H_{21}AlO_3$ (MW 204.25)

(mild reagent for Meerwein–Ponndorf–Verley reduction;[1] Oppenauer oxidation;[13] hydrolysis of oximes;[16] rearrangement of epoxides to allylic alcohols;[17] regio- and chemoselective ring opening of epoxides;[20] preparation of ethers[21])

Alternate Name: triisopropoxyaluminum.
Physical Data: mp 138–142 °C (99.99+%), 118 °C (98+%); bp 140.5 °C; *d* 1.035 g cm^{-3}.
Solubility: sol benzene; less sol alcohols.
Form Supplied in: white solid (99.99+% or 98+% purity based on metals analysis).
Preparative Methods: see example below.
Handling, Storage, and Precautions: the dry solid is corrosive, moisture sensitive, flammable, and an irritant. Use in a fume hood.

NMR Analysis of Aluminum Isopropoxide. Evidence from molecular weight determinations indicating that aluminum isopropoxide aged in benzene solution consists largely of the tetramer (**1**), whereas freshly distilled molten material is trimeric (**2**),[2] is fully confirmed by NMR spectroscopy.[3]

(1) (2)

Meerwein–Ponndorf–Verley Reduction. One use of the reagent is for the reduction of carbonyl compounds, particularly of unsaturated aldehydes and ketones, for the reagent attacks only carbonyl compounds. An example is the reduction of crotonaldehyde to crotyl alcohol (eq 1).[1] A mixture of 27 g of cleaned ***Aluminum*** foil, 300 mL of isopropanol, and 0.5 g of ***Mercury(II) Chloride*** is heated to boiling, 2 mL of carbon tetrachloride is added as catalyst, and heating is continued. The mixture turns gray, and vigorous evolution of hydrogen begins. Refluxing is continued until gas evolution has largely subsided (6–12 h). The solution, which is black from the presence of suspended solid, can be concentrated and the aluminum isopropoxide distilled in vacuum (colorless liquid) or used as such. Thus the undistilled solution prepared as described from 1.74 mol of aluminum and 500 mL of isopropanol is treated with 3 mol of crotonaldehyde and 1 L of isopropanol. On reflux at a bath temperature of 110 °C, acetone slowly distills at 60–70 °C. After 8–9 h, when the distillate no longer gives a test for acetone, most of the remaining isopropanol is distilled at reduced pressure and the residue is cooled and hydrolyzed with 6 N sulfuric acid to liberate crotyl alcohol from its aluminum derivative.

Al(O-*i*-Pr)$_3$, 60% (1)

The Meerwein–Ponndorf–Verley reduction of the ketone (**3**) involves formation of a cyclic coordination complex (**4**) which, by hydrogen transfer, affords the mixed alkoxide (**5**), hydrolyzed to the alcohol (**6**) (eq 2).[4] Further reflection suggests that under forcing conditions it might be possible to effect repetition of the hydrogen transfer and produce the hydrocarbon (**7**). Trial indeed shows that reduction of diaryl ketones can be effected efficiently by heating with excess reagent at 250 °C (eq 3).[5]

(3) (4) (5) (6) (7) (2)

Al(O-*i*-Pr)$_3$, 250 °C, 95% (3)

anthraquinone → anthracene 75%

anthrone → anthracene 92%

A study[6] of this reduction of mono- and bicyclic ketones shows that, contrary to commonly held views, the reduction proceeds at a relatively high rate. The reduction of cyclohexanone and of 2-methylcyclohexanone is immeasurably rapid. Even menthone is reduced almost completely in 2 h. The stereochemistry of the reduction of 3-isothujone (**8**) and of 3-thujone (**11**) has been examined (eqs 4 and 5). The ketone (**8**) produces a preponderance of the *cis*-alcohol (**9**). The stereoselectivity is less pronounced in the case of 3-thujone (**11**), although again the *cis*-alcohol (**12**) predominates. The preponderance of the *cis*-alcohols can be increased by decreasing the concentration of ketone and alkoxide.

(8) → (9) + (10) 7:1 (4)

(11) → (12) + (13) 3.2:1 (5)

This reducing agent is the reagent of choice for reduction of enones of type (**14**) to the α,β-unsaturated alcohols (**15**) (eq 6). Usual reducing agents favor 1,4-reduction to the saturated alcohol.[7]

(**14**) BPC = biphenylcarbonyl; $Al(O\text{-}i\text{-}Pr)_3$ → (**15**) 20%, each isomer (6)

The Meerwein–Ponndorf–Verley reduction of pyrimidin-2(1*H*)-ones using ***Zirconium Tetraisopropoxide*** or aluminum isopropoxide leads to exclusive formation of the 3,4-dihydro isomer (eq 7).[8] The former reducing agent is found to be more effective.

$Zr(O\text{-}i\text{-}Pr)_4$ or $Al(O\text{-}i\text{-}Pr)_3$, *i*-PrOH, 90 °C, 2 days, 20–91% (7)

R = H, halide

Reductions with Chiral Aluminum Alkoxides. The reduction of cyclohexyl methyl ketone with catalytic amounts of aluminum alkoxide and excess chiral alcohol gives (*S*)-1-cyclohexylethanol in 22% ee (eq 8).[9]

$Al(OR)_3$ → 22% ee (8)

Isobornyloxyaluminum dichloride is a good reagent for reducing ketones to alcohols. The reduction is irreversible and subject to marked steric approach control (eq 9).[10]

$OAlCl_2$ → 70% ee (9)

Diastereoselective Reductions of Chiral Acetals. Recently, it has been reported that ***Pentafluorophenol*** is an effective accelerator for Meerwein–Ponndorf–Verley reduction.[11] Reduction of 4-*t*-butylcyclohexanone with aluminum isopropoxide (3 equiv) in dichloromethane, for example, is very slow at 0 °C (<5% yield for 5 h), but in the presence of pentafluorophenol (1 equiv), the reduction is cleanly completed within 4 h at 0 °C (eq 10). The question of why this reagent retains sufficient nucleophilicity is still open. It is possible that the *o*-halo substituents of the phenoxide ligand may coordinate with the aluminum atom, thus increasing the nucleophilicity of the reagent.

$Al(O\text{-}i\text{-}Pr)_3$ (3 equiv), C_6F_5OH (1 equiv), CH_2Cl_2, 0 °C (10)

Chiral acetals derived from (−)-***(2R,4R)-2,4-Pentanediol*** and ketone are reductively cleaved with high diastereoselectivity by a 1:2 mixture of diethylaluminum fluoride and pentafluorophenol.[11] Furthermore, aluminum pentafluorophenoxide is a very powerful Lewis acid catalyst for the present reaction.[12] The reductive cleavage in the presence of 5 mol % of $Al(OC_6F_5)_3$ affords stereoselectively retentive reduced β-alkoxy ketones. The reaction is an intramolecular Meerwein–Ponndorf–Verley reductive and Oppenauer oxidative reaction on an acetal template (eq 11).

$Et_2AlF\text{–}2C_6F_5OH$ (1.2 equiv) or $Al(OC_6F_5)_3$ (5 mol %), CH_2Cl_2 or toluene; $R^1 > R^2$; K_2CO_3 (11)

The direct formation of α,β-alkoxy ketones is quite useful. Removal of the chiral auxiliary, followed by base-catalyzed β-elimination of the resulting β-alkoxy ketone, easily gives an optically pure alcohol in good yield. Several examples of the reaction are summarized in Table 1.

Table 1 Reductive Cleavages of Acetals Using $Al(OC_6F_5)_3$ Catalyst

$Al(OC_6F_5)_3$ (5 mol %), CH_2Cl_2

R^1	R^2	Yield (%)	Ratio (*S*:*R*)
C_5H_{11}	Me	83	82:18
i-Bu	Me	61	73:27
i-Pr	Me	90	94:6
Ph	Me	71	>99:1
Ph	Et	78	92:8
c-Hex	Me	89	95:5
CH_2 CH_2 (*t*-Bu)		67	81:19 (*trans*:*cis*)

Although the detailed mechanism is not yet clear, it is assumed that an energetically stable tight ion-paired intermediate is generated by stereoselective coordination of $Al(OC_6F_5)_3$ to one of the oxygens of the acetal; the hydrogen atom of the alkoxide is then transferred as a hydride from the retentive direction to this departing oxygen, which leads to the (*S*) configuration at the resulting ether carbon, as described (eq 12).

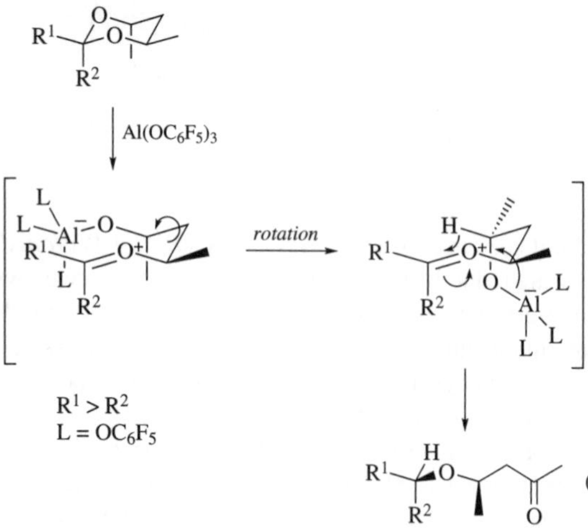

$R^1 > R^2$
$L = OC_6F_5$

(12)

Oppenauer Oxidation.[13] Cholestenone is prepared by oxidation of cholesterol in toluene solution with aluminum isopropoxide as catalyst and cyclohexanone as hydrogen acceptor (eq 13).[14]

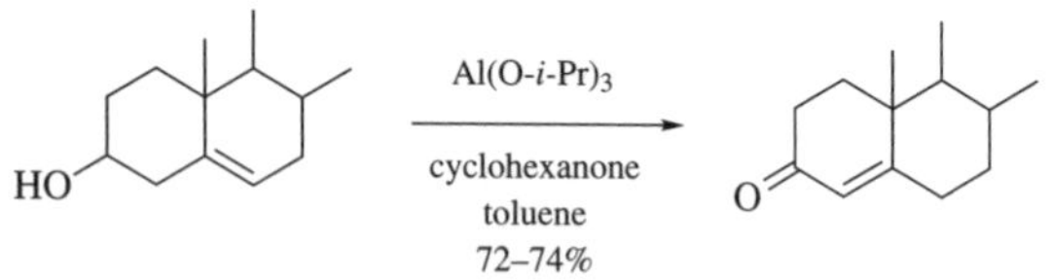

(13)

A formate, unlike an acetate, is easily oxidized and gives the same product as the free alcohol.[15] For oxidation of (**16**) to (**17**) the combination of cyclohexanone and aluminum isopropoxide and a hydrocarbon solvent is used: xylene (bp 140 °C at 760 mmHg) or toluene (bp 111 °C at 760 mmHg) (eq 14).

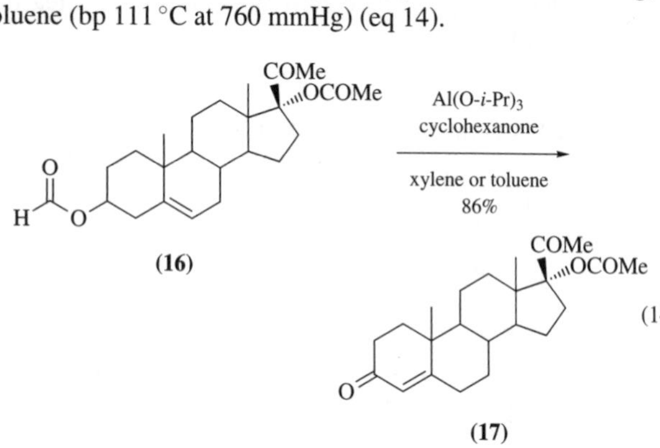

(14)

Hydrolysis of Oximes.[16] Oximes can be converted into parent carbonyl compounds by aluminum isopropoxide followed by acid hydrolysis (2N HCl) (eq 15). Yields are generally high in the case of ketones, but are lower for regeneration of aldehydes.

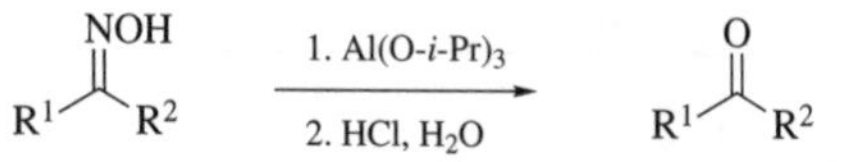

(15)

Rearrangement of Epoxides to Allylic Alcohols. The key step in the synthesis of the sesquiterpene lactone saussurea lactone (**21**) involved fragmentation of the epoxymesylate (**18**), obtained from α-santonin by several steps (eq 16).[17] When treated with aluminum isopropoxide in boiling toluene (N_2, 72 h), (**18**) is converted mainly into (**20**). The minor product (**19**) is the only product when the fragmentation is quenched after 12 h. Other bases such as potassium *t*-butoxide, LDA, and lithium diethylamide cannot be used. Aluminum isopropoxide is effective probably because aluminum has a marked affinity for oxygen and effects cleavage of the epoxide ring. Meerwein–Ponndorf–Verley reduction is probably involved in one step.

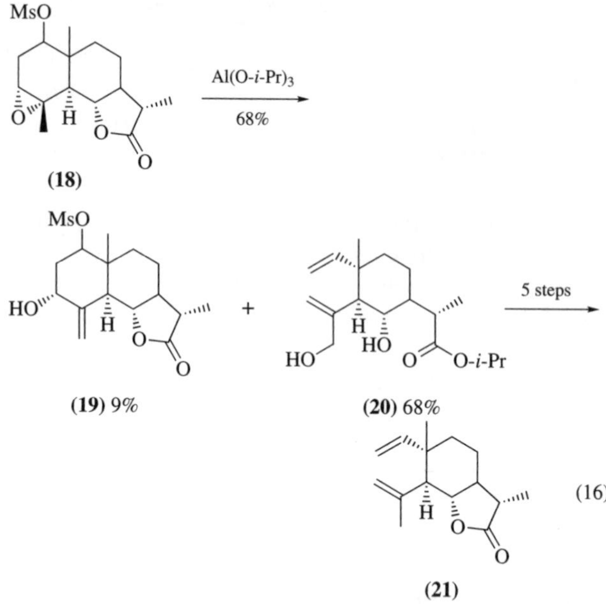

(16)

α-Pinene oxide (**22**) rearranges to pinocarvenol (**23**) in the presence of 1 mol % of aluminum isopropoxide at 100–120 °C for 1 h.[18] The oxide (**22**) rearranges to pinanone (**24**) in the presence of 5 mol % of the alkoxide at 140–170 °C for 2 h. Aluminum isopropoxide has been used to rearrange (**23**) to (**24**) (200 °C, 3 h, 80% yield) (eq 17).[19]

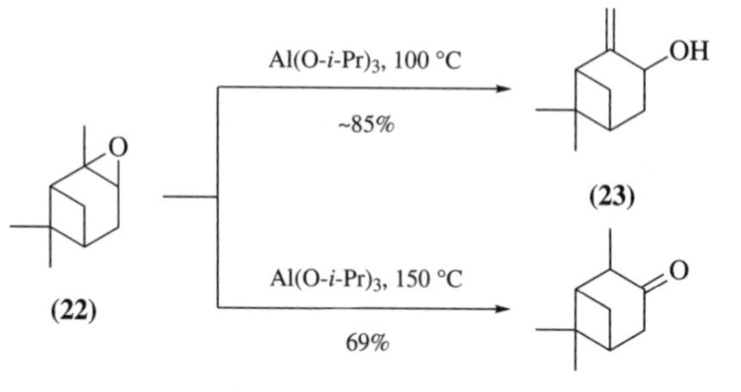

(17)

Regio- and Chemoselective Ring Opening of Epoxides. Functionalized epoxides are regioselectively opened using trimethylsilyl azide/aluminum isopropoxide, giving 2-trimethylsiloxy azides by attack on the less substituted carbon (eq 18).[20]

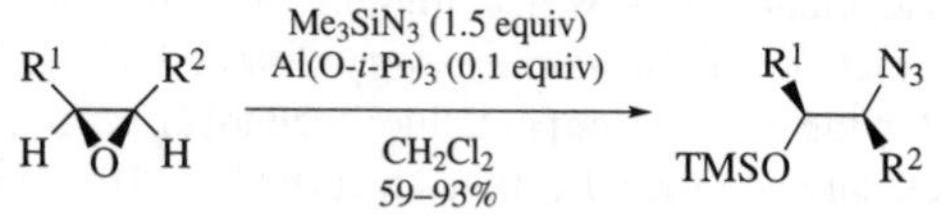

(18)

Preparation of Ethers. Ethers ROR′ are prepared from aluminum alkoxides, $Al(OR)_3$, and alkyl halides, R′X. Thus EtCH-

MeOH is treated with Al, $HgBr_2$, and MeI in DMF to give EtCH-MeOMe (eq 19).[21]

$$Al(OR)_3 + R^1X \xrightarrow[20\text{–}80\%]{\text{DMF, reflux, 2 days}} ROR^1 \quad (19)$$

R, R^1 = alkyl; X = halide

Related Reagents. Titanium Tetraisopropoxide.

1. Wilds, A. L. *OR* **1944**, *2*, 178.
2. Shiner, V. J.; Whittaker, D.; Fernandez, V. P. *JACS* **1963**, *85*, 2318.
3. Worrall, I. J. *J. Chem. Educ.* **1969**, *46*, 510.
4. Woodward, R. B.; Wendler, N. L.; Brutschy, F. J. *JACS* **1945**, *67*, 1425.
5. Hoffsommer, R. D.; Taub, D.; Wendler, N. L. *CI(L)* **1964**, 482.
6. Hach, V. *JOC* **1973**, *38*, 293.
7. Picker, D. H.; Andersen, N. H.; Leovey, E. M. K. *SC* **1975**, *5*, 451.
8. Høseggen, T.; Rise, F.; Undheim, K. *JCS(P1)* **1986**, 849.
9. Doering, W. von E.; Young, R. W. *JACS* **1950**, *72*, 631.
10. Nasipuri, D.; Sarker, G. *JIC* **1967**, *44*, 165.
11. Ishihara, K.; Hanaki, N.; Yamamoto, H. *JACS* **1991**, *113*, 7074.
12. Ishihara, K.; Hanaki, N.; Yamamoto, H. *SL* **1993**, 127; *JACS*, **1993**, *115*, 10695.
13. Djerassi, C. *OR* **1951**, *6*, 207.
14. Eastham, J. F.; Teranishi, R. *OSC* **1963**, *4*, 192.
15. Ringold, H. J.; Löken, B.; Rosenkranz, G.; Sondheimer, F. *JACS* **1956**, *78*, 816.
16. Sugden, J. K. *CI(L)* **1972**, 680.
17. Ando, M.; Tajima, K.; Takase, K. *CL* **1978**, 617.
18. Scheidl, F. *S* **1982**, 728.
19. Schmidt, H. *CB* **1929**, *62*, 104.
20. Emziane, M.; Lhoste, P.; Sinou, D. *S* **1988**, 541.
21. Lompa-Krzymien, L.; Leitch, L. C. *Pol. J. Chem.* **1983**, *57*, 629.

Kazuaki Ishihara & Hisashi Yamamoto
Nagoya University, Japan

Antimony(V) Fluoride[1]

SbF_5

[7783-70-2] F_5Sb (MW 216.74)

(one of the strongest Lewis acids,[1] used as catalyst for Friedel–Crafts reactions, isomerization, and other acid related chemistry;[1] an efficient acid system for preparation of carbocations and onium ions as well as their salts;[1] a fluorinating agent, and a strong oxidant)

Alternate Name: antimony pentafluoride.
Physical Data: bp 149.5 °C.
Solubility: SbF_5 reacts with most organic solvents, forming solids with ether, acetone, carbon disulfide, and petroleum ether. SbF_5 is soluble in SO_2 and SO_2ClF.
Form Supplied in: viscous liquid; commercially available.
Handling, Storage, and Precautions: SbF_5 is extremely corrosive, toxic, and moisture sensitive. It can be purified by distillation. SbF_5 fumes when exposed to atmosphere. It should be stored under anhydrous conditions in a Teflon bottle and handled using proper gloves in a well ventilated hood.

Antimony pentafluoride is one of the strongest Lewis acids reported and is capable of forming stable conjugate superacid systems with HF and FSO_3H. Its complex with ***Trifluoromethanesulfonic Acid*** is, however, less stable and cannot be stored for extended periods of time. Nevertheless, CF_3SO_3H/SbF_5 is also a useful acid system when prepared in situ.[2] The most important properties of SbF_5 include its high acidity, strong oxidative ability, and great tendency to form stable anions.[3] The chemistry of SbF_5 is mainly characterized by these properties. The major applications of SbF_5 in organic synthesis include oxidation, fluorination, and as a catalyst for Friedel–Crafts type reactions and other acid related chemistry, and as a medium for preparation of carbocations and onium ions.[1]

Friedel–Crafts and Related Chemistry. The use of SbF_5 as a Friedel–Crafts catalyst has significantly expanded the scope of these reactions.[3] This is made possible through the strong Lewis acidity and oxidative ability of SbF_5. Influenced by the strong acidity, compounds of otherwise very weak nucleophilicity such as perfluorocarbons can react readily with aromatics. Furthermore, the strong oxidative ability of SbF_5 enables unconventional substrates such as methane to be oxidized to form positively charged species, which in turn can be applied as electrophilic reagents. On the other hand, these factors also considerably restrict the use of SbF_5 in organic synthesis since the high reactivity makes it a less selective Lewis acid for many reactions.

Alkylation of arenes proceeds readily under SbF_5 catalysis.[3] In addition to alkyl halides, alkyl esters and haloesters have also been applied to alkylate arenes under the reaction conditions.[4] Perfluoro or perchlorofluoro compounds have been similarly used as alkylating agents in the presence of SbF_5.[5] For example, perfluorotoluene reacts with pentafluorobenzene to form perfluorodiphenylmethane in 68% yield after the reaction mixture is quenched with HF. When H_2O is used in the quenching step, perfluorobenzophenone is obtained in 93% yield (eq 1).[5a]

$C_6F_5CF_3$ → 1. C_6F_5H, SbF_5; 2. HF (68%) → $C_6F_5CF_2C_6F_5$
$C_6F_5CF_3$ → 1. C_6F_5H, SbF_5; 2. H_2O (93%) → $C_6F_5C(O)C_6F_5$ (1)

Acylation of pentafluorobenzene to form ketones either with acid halides or with anhydrides has been achieved with an excess of SbF_5.[6] In the case of dicarboxylic acid dichlorides or anhydrides as acylating agents, diketones are obtained as the reaction products (eq 2). The reaction of the anhydrides of dicarboxylic acids with aromatic compounds does not stop at the keto acid stage, as is the case when other Lewis acids are used. Using phosgene in place of acid chlorides results in the formation of pentafluorobenzoic acid in good yield.[3]

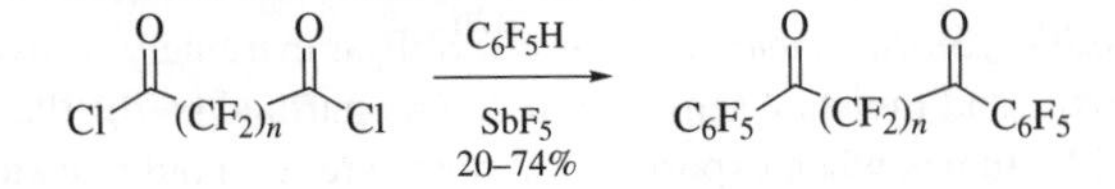

$$\mathrm{ClC(O)(CF_2)_nC(O)Cl} \xrightarrow[\text{20–74\%}]{\mathrm{C_6F_5H},\ \mathrm{SbF_5}} \mathrm{C_6F_5C(O)(CF_2)_nC(O)C_6F_5} \quad (2)$$

Perfluorobenzenium salt can be conveniently prepared from the reaction of 1,4-perfluorocyclohexadiene with SbF_5.[7a–c] In the presence of SbF_5, this salt is able to react with three equivalents of pentafluorobenzene to yield perfluoro-1,3,5-triphenylbenzene (eq 3).[7a–c] When 2,2′-di-*H*-octafluorobiphenyl is used as the substrate, perfluorotriphenylene is obtained in 50% yield (eq 4). Perfluoronaphthlenenium ion also reacts with polyfluorinated arenes in a similar fashion.[7a–c] The above reaction offers a facile approach for the preparation of these perfluorinated polynuclear aromatic compounds. Oxidation of perchlorobenzene with SbF_5 in the presence of pentafluorobenzene leads to the formation of coupling products.[7d]

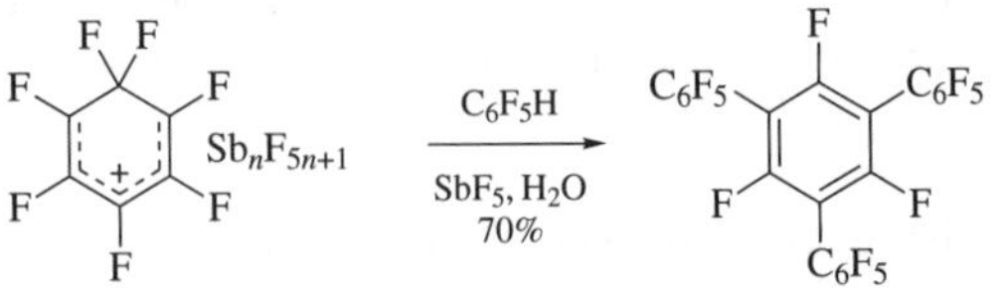

$$\mathrm{C_6F_7^+\ Sb_nF_{5n+1}^-} \xrightarrow[\text{70\%}]{\mathrm{C_6F_5H},\ \mathrm{SbF_5, H_2O}} \text{1,3,5-}\mathrm{(C_6F_5)_3C_6F_3} \quad (3)$$

$$\mathrm{C_6F_7^+\ Sb_nF_{5n+1}^-} + \text{2,2}'\text{-di-}H\text{-octafluorobiphenyl} \xrightarrow[\text{50\%}]{\mathrm{SbF_5}} \text{perfluorotriphenylene} \quad (4)$$

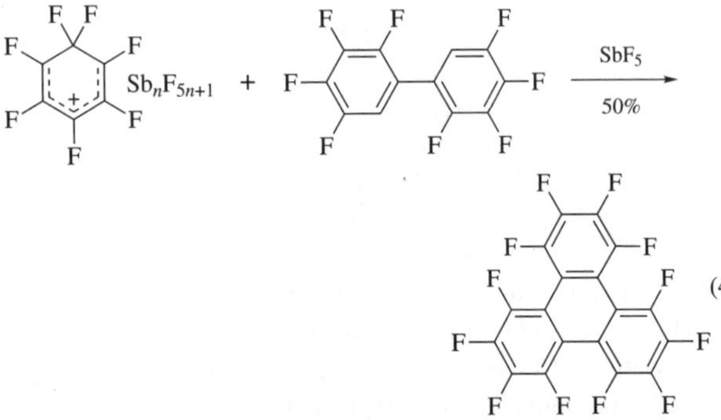

Friedel–Crafts sulfonylation of aromatics with alkane- and arenesulfonyl halides and anhydrides has been studied.[8] Good yields of sulfones are generally obtained (eq 5). In the case of pentafluorobenzenesulfonyl fluoride with pentafluorobenzene, decafluorodiphenyl sulfone is formed along with decafluorodiphenyl.[8c] A convenient approach for synthesizing symmetrical aryl sulfones is to react aromatics with ***Fluorosulfuric Acid*** in the presence of SbF_5 (eq 6).[9] Certain phenylacetylenes react with SO_2 and benzene in the presence of SbF_5 to form benzothiophene *S*-oxides[10] (eq 7). In some cases, 1,1-diphenylvinylsulfinic acids were also obtained as side products of the reaction. Sulfinyl fluoride reacts with arenes similarly under the catalysis of SbF_5 to give sulfoxides (eq 8).[8c]

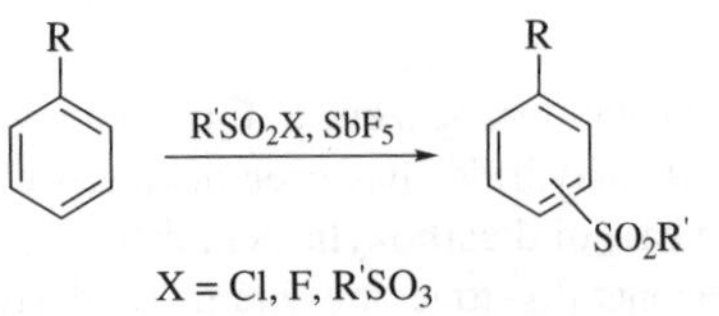

$$\mathrm{RC_6H_5} \xrightarrow{\mathrm{R'SO_2X,\ SbF_5}} \mathrm{RC_6H_4SO_2R'} \quad (5)$$

X = Cl, F, R′SO$_3$

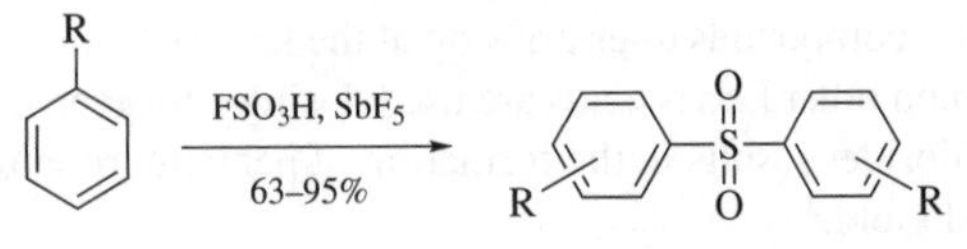

$$\mathrm{RC_6H_5} \xrightarrow[\text{63–95\%}]{\mathrm{FSO_3H,\ SbF_5}} \mathrm{RC_6H_4S(O)_2C_6H_4R} \quad (6)$$

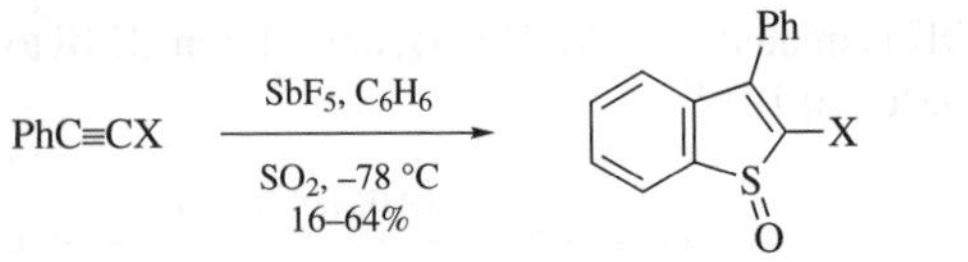

$$\mathrm{PhC{\equiv}CX} \xrightarrow[\mathrm{SO_2,\ -78\,^{\circ}C},\ \text{16–64\%}]{\mathrm{SbF_5,\ C_6H_6}} \text{3-phenyl-2-X-benzothiophene } S\text{-oxide} \quad (7)$$

X = Cl, Br, Ph

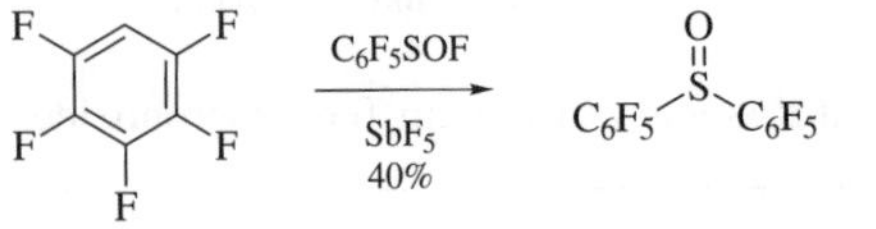

$$\mathrm{C_6F_5H} \xrightarrow[\mathrm{SbF_5},\ 40\%]{\mathrm{C_6F_5SOF}} \mathrm{C_6F_5S(O)C_6F_5} \quad (8)$$

Oxidation of elemental sulfur and selenium with SbF_5 leads to the formation of doubly charged polyatomic cations.[11] These cations are able to react with polyfluorinated arenes to form diaryl sulfides or selenides (eq 9).[12a–c] Under similar treatment, polyfluorodiaryl disulfides or diselenides also react with aromatics to form diaryl sulfides and selenides, respectively (eq 10).[12a,12b]

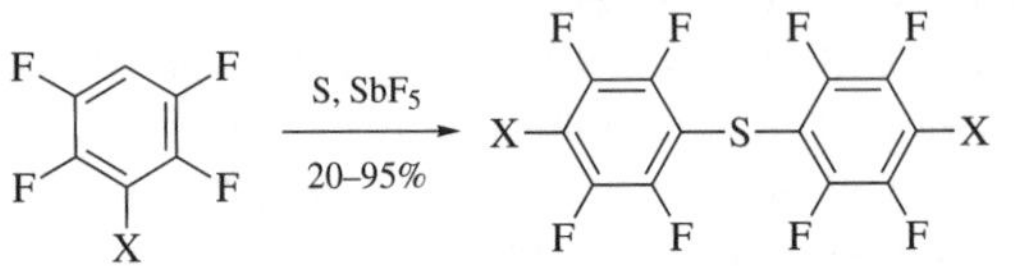

$$\mathrm{C_6F_4HX} \xrightarrow[\text{20–95\%}]{\mathrm{S,\ SbF_5}} \mathrm{X{-}C_6F_4{-}S{-}C_6F_4{-}X} \quad (9)$$

X = H, F, Br, SMe

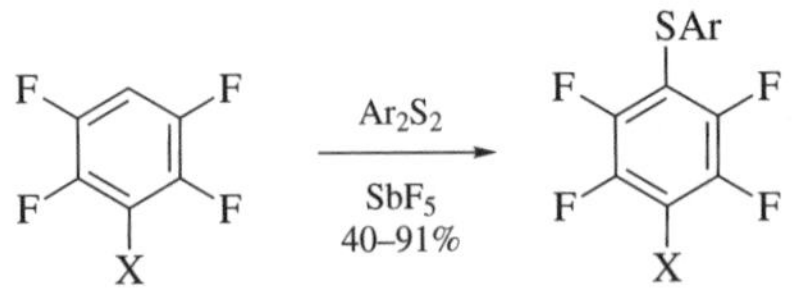

$$\mathrm{C_6F_4HX} \xrightarrow[\mathrm{SbF_5},\ \text{40–91\%}]{\mathrm{Ar_2S_2}} \mathrm{ArS{-}C_6F_4{-}X} \quad (10)$$

X = F, Cl; Ar = polyfluoroaryl

In the presence of SbF_5, inorganic halides such as NaCl and NaBr can serve as electrophilic halogenating agents.[5b,13] Even deactivated arenes such as 5*H*-nonafluoroindan can be brominated under the reaction conditions (eq 11).[13b]

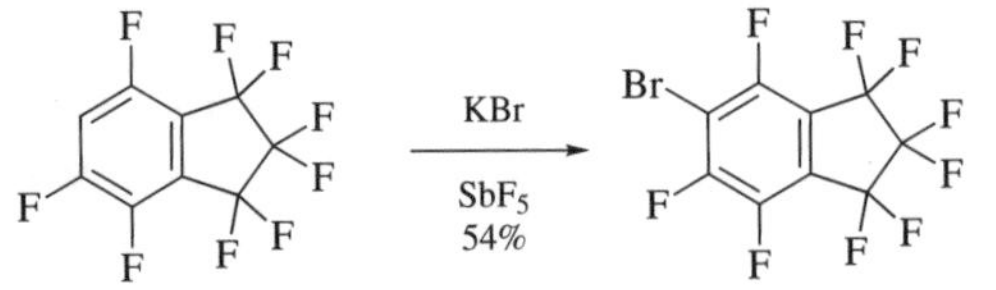

$$5H\text{-nonafluoroindan} \xrightarrow[\mathrm{SbF_5},\ 54\%]{\mathrm{KBr}} 5\text{-bromononafluoroindan} \quad (11)$$

Functionalization of alkanes has been achieved under SbF_5 catalysis.[3] Halogenation of alkanes is one of the widely studied reactions in this field.[14] A valuable synthetic procedure is the reaction of alkanes with methylene chloride (or bromide) in the presence of SbF_5 (eq 12).[14d] In these reactions, halonium ions are initially formed, which in turn abstract hydride from hydrocarbons. Quenching of the resulting carbocations with halides leads to the desired haloalkanes.

$$\mathrm{RH} \xrightarrow[\text{64–88\%}]{\mathrm{CH_2X_2,\ SbF_5}} \mathrm{RX} \quad (12)$$

RH = secondary, tertiary alkanes
X = Cl, Br

In the presence of SbF_5, alkyl chlorides are ionized to form carbocations, which can be trapped with CO.[15] Quenching the reaction intermediates with water or alcohols yields the corresponding carboxylic acids or esters. For instance, halogenated trishomobarrelene reacts with CO in SbF_5/SO_2ClF to give, after treatment with alcohol and water, a mixture of acid and ether (eq 13).[15a]

(13)

53% 24%

Studies have been carried out on the alkylation of alkenes in the presence of SbF_5, especially fluoroalkenes.[16] For example, treatment of 1,1,1-trifluoroethane with SbF_5 in the presence of tetrafluoroethylene yields 90% of 1,1,1,2,2,3,3-heptafluorobutane.[16c] Perfluoroallyl or -benzoyl compounds with varying structures have also been utilized as alkylating agents (eqs 14 and 15).[16b,e–h] At elevated temperatures, intramolecular alkylation was observed in certain cases with perfluorodienes, forming perfluorocyclopentenes or perfluorocyclobutenes (eq 16).[17]

(14)

X = F, CF_3

(15)

33% 39%

(16)

Acylation of alkenes can also be similarly effected with SbF_5.[18] Reaction of acetyl fluoride with trifluoroethylene produces 1,1,1,2-tetrafluorohexane-3,5-dione in 40% yield.[18a] When antimony pentafluoride is used in excess (more than 6 molar equivalents), 1,1,1,2-tetrafluorobutan-3-one is formed as a side product. Benzoyl fluoride reacts with difluoroethylene to form the expected ketone in 39% yield.[18a] α,β-Unsaturated carboxylic acid fluorides have also been used in reactions with perfluoroalkenes.[18b] In these cases, α,β-unsaturated ketones are obtained (eq 17). Enol acetates of perfluoroisopropyl methyl ketone and perfluoro-*t*-butyl methyl ketone react with acetyl fluoride to provide the corresponding β-diketones (eq 18).[18c]

(17)

(18)

32–35% 32–40%

Oxygenation of dienes with molecular oxygen to form Diels–Alder-type adducts can be effected by Lewis acids and some salts of stable carbenium ions.[19] In the case of ergosteryl acetate, SbF_5 is by far the most active catalyst (eq 19).[19a]

(19)

Isomerization and Rearrangement. Isomerization of perfluoroalkenes can be realized with SbF_5 catalysis.[20] The terminal carbon–carbon bonds of these alkenes are usually moved to the 2-position under the influence of this catalyst (eq 20). A further inward shift generally occurs only if H or Cl atoms are present in the 4-position of the alkenes. As a rule, the isomerization leads to the predominant formation of the *trans* isomers. Terminal fluorodienes also isomerize exothermally into dienes containing internal double bonds in the presence of SbF_5. With a catalytic amount of SbF_5, perfluoro-1,4-cyclohexadiene disproportionates to hexafluorobenzene and perfluorocyclohexene. The disproportionation proceeds intramolecularly in the case of perfluoro-1,4,5,8-tetrahydronaphthalene. The starting material is completely converted to perfluorotetralin (eq 21).[20f]

$$R_FCF_2FC{=}CF_2 \xrightarrow{SbF_5} R_FFC{=}CFCF_3 \quad (20)$$

(21)

Like most Lewis acids, SbF_5 promotes the rearrangement of epoxides to carbonyl compounds,[21] and SbF_5 is an efficient catalyst for this reaction. However, the migratory aptitude of substituent groups in the reaction under SbF_5 catalysis is much less selective compared to that promoted by weak Lewis acids such as ***Methylaluminum Bis(4-bromo-2,6-di-t-butylphenoxide)***.[21a] Nevertheless, SbF_5 is well suited for the rearrangement of perfluoroepoxides.[21b–e] Excellent yields of ketones are obtained from the reaction. When diepoxides are used, diketones are obtained as the reaction products (eq 22).[21c]

$$\xrightarrow{SbF_5} CF_3CO(CF_2)_2COCF_3 \quad (22)$$

Fluorination and Transformation of Fluorinated Compounds. SbF_5 is a strong fluorinating agent. However, its use is largely limited to the preparation of perfluoro- or polyfluoroorganic compounds.[22] SbF_5 has also been used for transformation of perfluoro compounds.[22]

When hexachlorobenzene is subjected to SbF_5 treatment, 44% of 1,2-dichloro-3,3,4,4,5,5,6,6-octafluorocyclohex-1-ene is obtained as the major product of the reaction (eq 23).[23] Using $SbCl_5$ as reaction solvent leads to better reaction control, resulting in an increase in the product yield.[23c] In the case of polyfluoronaphthalenes, polyfluorinated tetralins are obtained.[23d]

Treatment of hexafluoro-2-trichloromethyl-2-propanol with SbF_5 gives perfluoro-*t*-butyl alcohol in 92% yield (eq 24).[24]

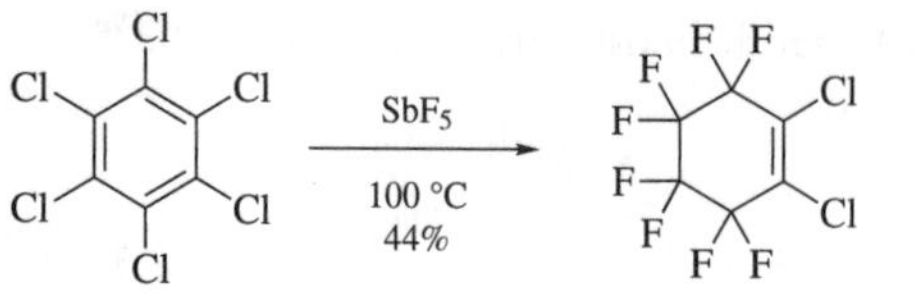

(23)

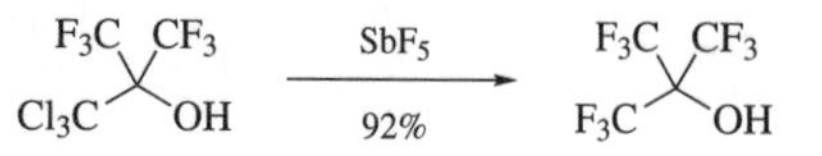

(24)

Hydrolysis of CF_3 groups to CO_2H can be induced by SbF_5.[25a] In this reaction, the CF_3-bearing compounds are first treated with SbF_5 and the resulting reaction mixtures are subsequently quenched with water to form the desired products. For example, perfluorotoluene was converted to pentafluorobenzoic acid in 86% yield by this procedure. When perfluoroxylenes and perfluoromesitylene go through the same treatment, the corresponding di- or triacids are obtained in high yields. It is also possible to partially hydrolyze perfluoroxylenes and perfluoromesitylene through a stepwise procedure and to isolate intermediate products. The hydrolysis procedure is also applicable to other halogen-containing compounds.[25b]

Under SbF_5 catalysis, ***Trifluoromethanesulfonic Anhydride*** is readily decomposed to trifluoromethyl triflate in high yield (eq 25).[26] This method is the most convenient procedure reported for preparation of the triflate.

$$(CF_3SO_2)_2O \xrightarrow[94\%]{SbF_5,\ 25\ ^\circ C} CF_3SO_3CF_3 \qquad (25)$$

Preparation of Carbocations, Onium Ions, and Their Salts. SbF_5 is a preferred medium for the preparation of carbocations and onium ions.[1] In fact, the first observation of stable carbocations was achieved in this medium.[1,27] By dissolving *t*-butyl fluoride in an excess of SbF_5, the *t*-butyl cation was obtained (eq 26). Subsequently, many alkyl cations have been obtained in SbF_5 (either neat or diluted with SO_2, SO_2ClF, or SO_2F_2). The 2-norbornyl cation, one of the most controversial ions in the history of physical organic chemistry,[28] was prepared from *exo*-2-fluoronorbornane in SbF_5/SO_2 (or SO_2ClF) solution (eq 27).[1] Bridgehead cations such as 1-adamantyl, 1-trishomobarrelyl, and 1-trishomobullvalyl cations have also been similarly prepared with the use of SbF_5.[15a,29]

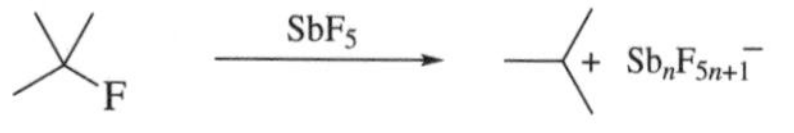

(26)

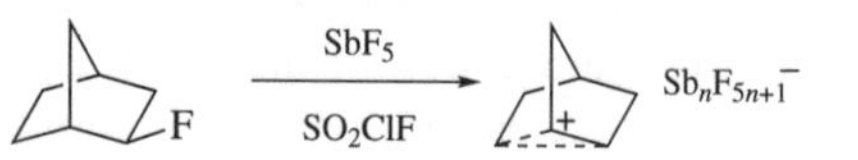

(27)

Carbodications have also been studied.[1] One convenient way of preparing alkyl dications is to ionize dihalides with SbF_5 in SO_2ClF (eq 28). In these systems, separation of the two cation centers by at least two methylene groups is necessary for the ions to be observable.[1] Aromatic dications are usually prepared by oxidizing the corresponding aromatic compounds with SbF_5 (eq 29).[1,30] In the case of pagodane containing a planar cyclobutane ring, oxidation leads to the formation of cyclobutane dication which was characterized as a frozen two-electron Woodward–Hoffmann transition state model (eq 30).[31]

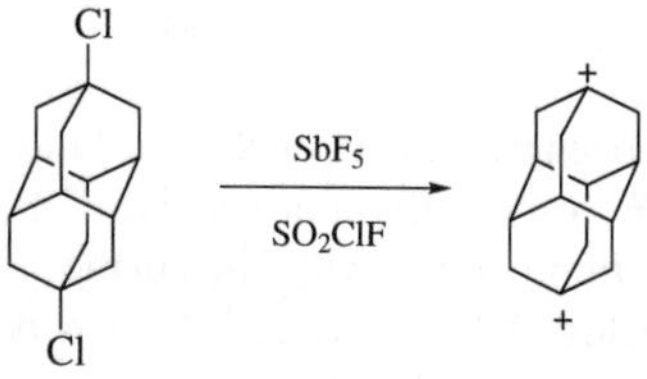

(28)

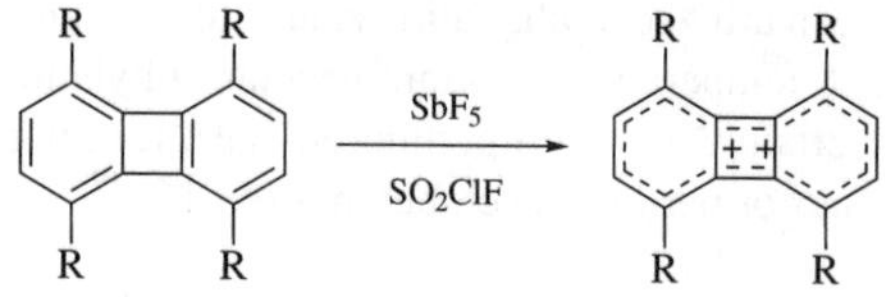

(29)

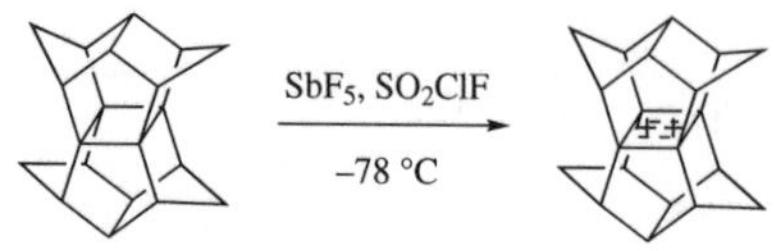

(30)

SbF_5 has also been found useful in the preparation of homoaromatic cations.[1,32] For example, the simplest 2π monohomoaromatic cations, the homocyclopropenyl cations, can be prepared from corresponding 3-halocyclobutenes in SbF_5 (eq 31).[32a]

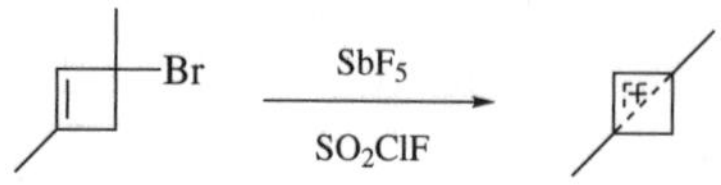

(31)

Perfluorocarbocations are generally prepared with the aid of SbF_5.[33] Perfluorobenzyl cations and perfluorinated allyl cations are among the most well studied perfluorinated carbocations. It was reported that in the perfluoroallyl cations containing a pentafluorophenyl group at the 1- or 2-position, the phenyl groups were partially removed from the plane of the allyl triad (eq 32).[33b]

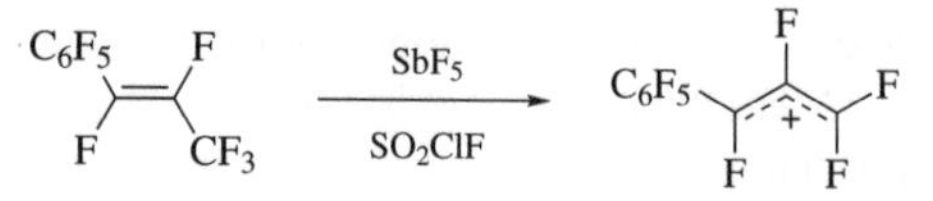

(32)

Use of SbF_5 as ionizing agent offers a convenient route to acyclic and cyclic halonium ions from alkyl halides.[1] Three-membered ring cyclic halonium ions are important intermediates in the electrophilic halogenations of carbon–carbon double bonds.[1] More recently, a stable 1,4-bridged bicyclic bromonium ion, 7-bromoniabicyclo[2.2.1]heptane, has been prepared through the use of SbF_5 involving an unprecedented transannular participation in a six-membered ring (eq 33).[34]

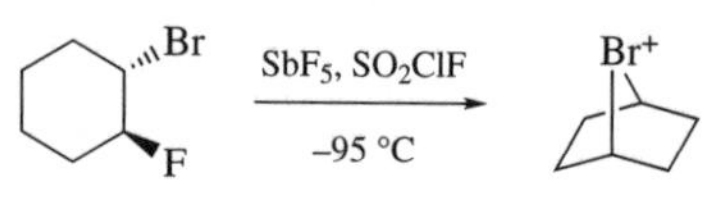

(33)

Other Reactions. Reaction of α,β-unsaturated carbonyl compounds with diazo compounds generally gives low yields of cyclopropyl compounds. The rapid formation of 1-pyrazolines and their subsequent rearrangement products 2-pyrazolines is the reason for the inefficiency of cyclopropanation of these substrates.

However, in the presense of SbF_5, the cyclopropanation of α,β-unsaturated carbonyl compounds with diazocarbonyl compounds proceeds very well to produce the desired products in good yields (eq 34).[35]

$$R^1R^1C{=}CH{-}C(O)R + R^2C(O)CHN_2 \xrightarrow[11\text{–}80\%]{SbF_5} R^2OC\text{-}c\text{-}C_3H_2(R^1)_2\text{-}COR \quad (34)$$

$R = H, OMe; R^1 = H, Me; R^2 = Ph, OEt$

Related Reagents. Aluminum Chloride; Fluorosulfuric Acid–Antimony(V) Fluoride; Hydrogen Fluoride–Antimony(V) Fluoride; Methyl Fluoride–Antimony(V) Fluoride.

1. Olah, G. A.; Prakash, G. K. S.; Sommer, J. *Superacids*; Wiley: New York, 1985.
2. For example, see (a) Ohwada, T.; Yamagata, N.; Shudo, K. *JACS* **1991**, *113*, 1364. (b) Farooq, O.; Farnia, S. M. F.; Stephenson, M.; Olah, G. A. *JOC* **1988**, *53*, 2840. (c) Carre, B.; Devynck, J. *Anal. Chim. Acta.* **1984**, *159*, 149. (d) Choukroun, H.; Germain, A.; Brunel, D.; Commeyras, A. *NJC* **1983**, *7*, 83.
3. Yakobson, G. G.; Furin, G. G. *S* **1980**, 345.
4. (a) Olah, G. A.; Nishimura, J. *JACS* **1974**, *96*, 2214. (b) Booth, B. L.; Haszeldine, R. N.; Laali, K. *JCS(P1)* **1980**, 2887.
5. (a) Pozdnyakovich, Y. V.; Shteingarts, V. D. *JFC* **1974**, *4*, 317. (b) Brovko, V. V.; Sokolenko, V. A.; Yakobson, G. G. *JOU* **1974**, *10*, 300.
6. Furin, G. G.; Yakobson, G. G.; *Izv. Sib. Otd. Akad. Nauk SSSR, Ser. Khim. Nauk* **1974**, 78 (*CA* **1974**, *80*, 120 457c.)
7. (a) Shteingarts, V. D. In *Synthetic Fluorine Chemistry*, Olah, G. A.; Chambers, R. D.; Prakash, G. K. S., Eds.; Wiley: New York, 1992; Chapter 12. (b) Pozdnyakovich, Y. V.; Shteingarts, V. D. *JOU* **1977**, *13*, 1772. (c) Pozdnyakovich, Y. V.; Chuikova, T. V.; Bardin, V. V.; Shteingarts, V. D. *JOU* **1976**, *12*, 687. (d) Bardin, V. V.; Yakobson, G. G. *IZV* **1976**, 2350.
8. Olah, G. A.; Kobayashi, S.; Nishimura, J. *JACS* **1973**, *95*, 564. (b) Olah, G. A.; Lin, H. C. *S* **1974**, 342. (c) Furin, G. G.; Yakobson, G. G.; *Izv. Sib. Otd. Akad. Nauk SSSR, Ser. Khim. Nauk* **1976**, 120 (*CA* **1976**, *85*, 77 801z.)
9. Tanaka, M.; Souma, Y. *JOC* **1992**, *57*, 3738.
10. (a) Fan, R.-L.; Dickstein, J. I.; Miller, S. I. *JOC* **1982**, *47*, 2466. (b) Miller, S. I.; Dickstein, J. I. *ACR* **1976**, *9*, 358.
11. Gillespie, R. J.; Peel, T. E. In *Advances in Physical Organic Chemistry*, Gold, V., Ed.; Academic Press: London, 1971; Vol. 9, p 1.
12. (a) Furin, G. G.; Terent'eva, T. V.; Yakobson, G. G. *JOU* **1973**, *9*, 2221. (b) Yakobson, G. G.; Furrin, G. G.; Terent'eva, T. V. *IZV* **1972,** 2128. (c) Yakobson, G. G.; Furin, G. G.; Terent'eva, T. V. *JOU* **1974**, *10*, 802. (d) Furin, G. G.; Schegoleva, L. N.; Yakobson, G. G. *JOU* **1975**, *11*, 1275.
13. (a) Lukmanov, V. G.; Alekseeva, L. A.; Yagupol'skii, L. M. *JOU* **1977**, *13*, 1979. (b) Furin, G. G.; Malyuta, N. G.; Platonov, V. E.; Yakobson, G. G. *JOU* **1974**, *10*, 832.
14. (a) Olah, G. A.; Mo, Y. K. *JACS* **1972**, *94*, 6864. (b) Olah, G. A.; Renner, R.; Schilling, P.; Mo, Y. K. *JACS* **1973**, *95*, 7686. (c) Halpern, Y. *Isr. J. Chem.* **1975**, *13*, 99. (d) Olah, G. A.; Wu, A.; Farooq, O. *JOC* **1989**, *54*, 1463.
15. (a) de Meijere, A.; Schallner, O. *AG(E)* **1973**, *12*, 399. (b) Farcasiu, D. *CC* **1977**, 394.
16. (a) Belen'kii, G. G.; Savicheva, G. I.; German, L. S. *IZV* **1978**, 1433. (b) Belen'kii, G. G.; Lur'e, E. P.; German, L. S. *IZV* **1976**, 2365. (c) Petrov, V. A.; Belen'kii, G. G.; German, L. S. *IZV* **1980**, 2117. (d) Karpov, V. M.; Mezhenkova, T. V.; Platonov, V. E.; Yakobson, G. G. *JOU* **1984**, *20*, 1220. (e) Petrov, V. A.; Belen'kii, G. G.; German, L. S.; Kurbakova, A. P.; Leites, L. A. *IZV* **1982**, 170. (f) Karpov, V. M.; Mezhenkova, T. V.; Platonov, V. E.; Yakobson, G. G. *IZV* **1985**, 2315. (g) Petrov, V. A.; Belen'kii, G. G.; German, L. S.; Mysov, E. I. *IZV* **1981**, 2098. (h) Petrov, V. A.; Belen'kii, G. G.; German, L. S. *IZV* **1981**, 1920.
17. (a) Petrov, V. A.; Belen'kii, G. G.; German, L. S. *IZV* **1982**, 2411. (b) Petrov, V. A.; Belen'kii, G. G.; German, L. S. *IZV* **1989**, 385. (c) Petrov, V. A.; German, L. S.; Belen'kii, G. G. *IZV* **1989**, 391.
18. (a) Belen'kii, G. G.; German, L. S.; *IZV* **1974**, 942. (b) Chepik, S. D.; Belen'kii, G. G.; Cherstkov, V. F.; Sterlin, S. R.; German, L. S. *IZV* **1991**, 513. (c) Knunyants, I. L.; Igumnov, S. M. *IZV* **1982**, 204.
19. (a) Barton, D. H. R.; Haynes, R. K.; Magnus, P. D.; Menzies, I. D. *CC* **1974**, 511. (b) Barton, D. H. R.; Haynes, R. K.; Leclerc, G.; Magnus, P. D.; Menzies, I. D. *JCS(P1)* **1975**, 2055. (c) Haynes, R. K. *AJC* **1978**, *31*, 131.
20. (a) Filyakova, T. I.; Belen'kii, G. G.; Lur'e, E. P.; Zapevalov, A. Y.; Kolenko, I. P.; German, L. S. *IZV* **1979**, 681. (b) Belen'kii, G. G.; Savicheva, G. I.; Lur'e, E. P.; German, L. S. *IZV* **1978**, 1640. (c) Filyakova, T. I.; Zapevalov, A. Y. *JOU* **1991**, *27*, 1605. (d) Petrov, V. A.; Belen'kii, G. G.; German, L. S. *IZV* **1989**, 385. (e) Chepik, S. D.; Petrov, V. A.; Galakhov, M. V.; Belen'kii, G. G.; Mysov, E. I.; German, L. S. *IZV* **1990**, 1844. (f) Avramenko, A. A.; Bardin, V. V.; Furin, G. G.; Karelin, A. I.; Krasil'nikov, V. A.; Tushin, P. P. *JOU* **1988**, *24*, 1298.
21. (a) Maruoka, K.; Ooi, T.; Yamamoto, H. *T* **1992**, *48*, 3303. (b) Zapevalov, A. Y.; Filyakova, T. I.; Kolenko, I. P.; Kodess, M. I. *JOU* **1986**, *22*, 80. (c) Filyakova, T. I.; Ilatovskii, R. E.; Zapevalov, A. Y. *JOU* **1991**, *27*, 1818. (d) Filyakova, T. I.; Matochkina, E. G.; Peschanskii, N. V.; Kodess, M. I.; Zapevalov, A. Y. *JOU* **1992**, *28*, 20. (e) Filyakova, T. I.; Matochkina, E. G.; Peschanskii, N. V.; Kodess, M. I.; Zapevalov, A. Y. *JOU* **1991**, *27*, 1423.
22. Yakobson, G. G.; Vlasov, V. M. *S* **1976**, 652.
23. (a) Leffler, A. J. *JOC* **1959**, *24*, 1132. (b) McBee, E. T.; Wiseman, P. A.; Bachman, G. B. *Ind. Eng. Chem.* **1947**, *39*, 415. (c) Christe, K. O.; Pavlath, A. E. *JCS* **1963**, 5549. (d) Pozdnyakovich, Y. V.; Shteingarts, V. D. *JOU* **1978**, *14*, 2069.
24. Dear, R. E. A. *S* **1970**, 361.
25. (a) Karpov, V. M.; Panteleev, I. V.; Platonov, V. E. *JOU* **1991**, *27*, 1932. (b) Karpov, V. M.; Platonov, V. E.; Yakoboson, G. G. *IZV* **1979**, 2082.
26. Taylor, S. L.; Martin, J. C. *JOC* **1987**, *52*, 4147.
27. Olah, G. A.; Tolgyesi, W. S.; Kuhn, S. J.; Moffatt, M. E.; Bastien, I. J.; Baker, E. B. *JACS* **1963**, *85*, 1328.
28. For reviews, see (a) Olah, G. A.; Prakash, G. K. S.; Saunders, M. *ACR* **1983**, *16*, 440. (b) Brown, H. C. *ACR* **1983**, *16*, 432. (c) Walling, C. *ACR* **1983**, *16*, 448.
29. (a) Schleyer, P. v. R.; Fort, R. C. Jr.; Watts, W. E.; Comisarow, M. B.; Olah, G. A. *JACS* **1964**, *86*, 4195. (b) Olah, G. A.; Prakash, G. K. S.; Shih, J. G.; Krishnamurthy, V. V.; Mateescu, G. D.; Liang, G.; Sipos, G.; Buss, V.; Gund, T. M.; Schleyer, P. v. R. *JACS* **1985**, *107*, 2764.
30. (a) Mills, N. S. *JOC* **1992**, *57*, 1899. (b) Krusic, P. J.; Wasserman, E. *JACS* **1991**, *113*, 2322. (c) Mullen, K.; Meul, T.; Schade, P.; Schmickler, H.; Vogel, E. *JACS* **1987**, *109*, 4992. (d) Lammertsma, K.; Olah, G. A.; Berke, C. M.; Streitwieser, A. Jr. *JACS* **1979**, *101*, 6658. (e) Olah, G. A.; Lang, G.; *JACS* **1977**, *99*, 6045.
31. Prakash, G. K. S.; Krishnamurthy, V. V.; Herges, R.; Bau, R.; Yuan, H.; Olah, G. A.; Fessner, W-D.; Prinzbach, H. *JACS* **1988**, *110*, 7764.
32. (a) Olah, G. A.; Staral, J. S.; Spear, R. J.; Liang, G. *JACS* **1975**, *97*, 5489. (b) Olah, G. A.; Staral, J. S.; Paquette, L. A. *JACS* **1976**, *98*, 1267.
33. (a) Pozdnyakovich, Y. V.; Shteingarts, V. D. *JFC* **1974**, *4*, 283. (b) Galakhov, M. V.; Petrov, V. A.; Chepik, S. D.; Belen'kii, G. G.; Bakhmutov, V. I.; German, L. S. *IZV* **1989**, 1773. (c) Petrov, V. A.; Belen'kii, G. G.; German, L. S. *IZV* **1984**, 438.
34. Prakash, G. K. S.; Aniszfeld, R.; Hashimoto, T.; Bausch, J. W.; Olah, G. A. *JACS* **1989**, *111*, 8726.
35. Doyle, M. P.; Buhro, W. E.; Dellaria, J. F. Jr. *TL* **1979**, 4429.

George A. Olah, G. K. Surya Prakash, Qi Wang & Xing-ya Li
University of Southern California, Los Angeles, CA, USA

B

Barium Hydroxide

$Ba(OH)_2$

[17194-00-2]	BaH_2O_2	(MW 171.35)
($\cdot H_2O$)		
[22326-55-2]	BaH_4O_3	(MW 189.37)
($\cdot 8H_2O$)		
[12230-71-6]	$BaH_{18}O_{10}$	(MW 315.48)

(used as a base catalyst in a variety of organic reactions such as decarboxylations, aldol or aldol-type reactions, Claisen–Schmidt reactions, Michael additions, and Wittig–Horner reactions)

Physical Data: $Ba(OH)_2 \cdot H_2O$, *d* 3.743 g cm^{-3}. $Ba(OH)_2 \cdot 8H_2O$, mp 78 °C; *d* 2.180 g cm^{-3}.

Solubility: monohydrate is slightly sol water. Octahydrate is freely sol water and methanol, slightly sol ethanol; practically insol acetone.

Form Supplied in: monohydrate, white solid; octahydrate, transparent crystals or white masses.

Handling, Storage, and Precautions: poison; corrosive. May be fatal if swallowed, inhaled, or absorbed through skin. Material is extremely destructive to tissue of the mucous membranes and upper respiratory tract, eyes, and skin.[1] Incompatible with acids. Absorbs moisture and CO_2 from air. Use only in a chemical fume hood. Avoid contact with eyes, skin, and clothing. Wash thoroughly after handling. Keep container tightly closed.

Introduction. The octahydrate, $Ba(OH)_2 \cdot 8H_2O$, is the most common form. Upon dehydration at elevated temperatures (200–500 °C) it is converted to the anhydrous form. On standing at rt, anhydrous $Ba(OH)_2$ is partially rehydrated and the resulting equilibrium water content is related to the dehydration temperature. The solids resulting from such a treatment are referred to in the literature as 'activated barium hydroxides'. An activated $Ba(OH)_2$ is a mixture of anhydrous $Ba(OH)_2$ and the monohydrate in variable ratios. In the crystal structure, the monohydrate is the main component over the solid surface.[1] The most popular activated $Ba(OH)_2$ is sometimes referred to as C-200 (i.e. dehydrated at 200 °C) and was determined to be $Ba(OH)_2 \cdot 0.8H_2O$.[2]

Partial Hydrolysis of Dimethyl Dicarboxylates. The monomethyl ester of undecanedioic acid was prepared by partial hydrolysis of the dimethyl ester with methanolic $Ba(OH)_2$ (eq 1). The precipitation of the barium salt of the monoester prevents further hydrolysis to the diacid. The procedure has an advantage over the fractional distillation of partially esterified diacids, particularly for the high-boiling half esters which may disproportionate at elevated temperatures. This method was not satisfactory for low molecular weight half esters because of the higher solubility of their barium salts in methanol.[3]

MeO O O OMe (9) → 1. $Ba(OH)_2$, MeOH; 2. H^+; 60–64% → MeO O O OH (9) (1)

Ketocarboxylic Acids from Ketodicarboxylic Esters. The ketodicarboxylate derivative (**1**) was hydrolyzed and decarboxylated when treated with $Ba(OH)_2 \cdot 8H_2O$ in refluxing ethanol for 20 h to give the crystalline γ-keto acid (**2**) in 98% yield (eq 2).[4]

(**1**) O, CO_2Me, CO_2Me → $Ba(OH)_2$, EtOH, reflux, 20 h; 98% → (**2**) O, CO_2H (2)

Monocarboxylic Acids from Dicarboxylic Acid Anhydrides. In the total synthesis of (±)-pentalenolactone, the intermediate dicarboxylic anhydride (**3**) was converted to carboxyenone (**4**) upon treatment with 1.2 equiv of $Ba(OH)_2 \cdot 8H_2O$ in water and heating at reflux temperature for 5 h (eq 3). The cleavage of the anhydride, the formation of the enone from the β-methoxy silyl enol ether, and the decarboxylation of the vinylogous β-keto barium carboxylate were all achieved in one operation giving the desired product (**4**) in 99% yield.[5]

(**3**) OMe, Me_3SiO, O, O, O, O → $Ba(OH)_2 \cdot 8H_2O$, water, reflux, 5 h → (**4**) O, O, O, H, O, OH (3)

Reaction with α-Chlorolactams: a Stereoselective Favorskii-Type Ring Contraction. Treatment of the bicyclic α-chlorovalerolactams (**5**), (**6**), and (**7**) with aqueous $Ba(OH)_2$ at reflux temperature promoted a Favorskii-type ring contraction to give the octahydroindoles and octahydroisoindoles (**8**), (**9**), and (**10**) respectively. While chlorolactams (**5**) and (**6**) rearrange in a diastereoselective fashion (eqs 4 and 5), (**7**) gave a 1:1 mixture of (**10a**) and (**10b**) (eq 6). The observed diastereoselectivity was attributed to *cis*-1,3 and *cis*-1,2 steric interactions in (**5**) and (**6**), respectively, and the lack of such interactions in (**7**).[6]

Preparation of Diacetone Alcohol. Heating acetone in a flask fitted with a Soxhlet extractor containing $Ba(OH)_2$ for 95–120 h gave 71% yield of distilled diacetone alcohol (eq 7).[7]

(5)

1. $Ba(OH)_2$, H_2O reflux, 2 h
2. H_2SO_4 or CO_2 94%

(8a) 9:1 (8b) (4)

(6)

1. $Ba(OH)_2$, H_2O reflux, 2 h
2. H_2SO_4 or CO_2 96%

(9) only product (5)

(7)

1. $Ba(OH)_2$, H_2O reflux, 2 h
2. H_2SO_4 or CO_2

(10a) 1:1 (10b) (6)

$Ba(OH)_2$, reflux, 71% (7)

Heterogeneous Claisen–Schmidt Condensation. Activated $Ba(OH)_2$ (C-200) was used as a heterogeneous catalyst in Claisen–Schmidt condensations. The method was applied to reactions of various aromatic aldehydes with methyl ketones to give the corresponding styryl ketones. Generally, the aldehyde is mixed with the methyl ketone (2 equiv) and the C-200 catalyst (12%) in ethanol and the mixture is heated at reflux for 1 h. The yields ranged from 25 to 100% (eq 8).[8] The procedure was extended to the synthesis of *o*-hydroxychalcones, e.g. the reaction of *o*-hydroxyacetophenone with benzaldehyde in 96% ethanol and a catalytic amount of C-200 gave *o*-hydroxychalcone in 89% yield (eq 9).[9]

$Ba(OH)_2$ (C-200), EtOH, reflux (8)

R^1 = H, R^2 = Ph, 94–98%; R^1 = Me, R^2 = *n*-Pr, 99–100%; R^1 = NO_2, R^2 = *t*-Bu, 24–25%

+ PhCHO, $Ba(OH)_2$, EtOH, reflux, 4 h, 89% (9)

The yields from these reactions were as good as or better than other methods. Secondary reactions such as the Cannizzaro reaction of aldehydes and aldol additions of ketones were not observed. Unlike other methods, protection of the phenol was not necessary and the product did not cyclize to the flavenone. The absence of cyclized products as well as the (*E*) configuration of the alkenic bond were explained by a rigid transition state in which the carbanion and aldehyde are adsorbed on the insoluble catalyst surface.[9]

Michael Addition to Chalcones. Partially dehydrated $Ba(OH)_2$ efficiently catalyzed the Michael additions of active methylene compounds to chalcones. The reaction temperature and amount of catalyst determined the isolated product (eq 10). Thus the reaction of chalcone with ethyl acetoacetate (25 mmol each) at rt in EtOH with 5 mg catalyst gave primarily the Michael addition product (**11**) in 90% yield. By increasing the amount of catalyst to 50 mg at rt, the reaction proceeded further to give the cyclic product (**12**) in 95% yield, resulting from Michael addition followed by intramolecular aldol reaction. The dehydrated cyclic adduct (**13**) was the major product (80%) at reflux temperature.[10]

25 mmol; $Ba(OH)_2$, catalyst, EtOH (10)

(**11**) rt, 5 mg catalyst; 90%

or (**12**) rt, 50 mg catalyst; 95%

or (**13**) reflux, 50 mg catalyst; 80%

Addition of Diethyl Malonate to Coumarin. An attempted Michael addition of diethyl malonate to coumarin (**14**) in the presence of solid $Ba(OH)_2$ (C-200) in EtOH gave the unusual 1,2-addition–elimination product (**15**) in 30% yield (eq 11). The result was explained in terms of the structure of the catalyst and the chelation of coumarin to the barium ion of the lattice.[11]

(**14**) $CH_2(CO_2Et)_2$, EtOH, $Ba(OH)_2$ (**15**) (11)

Synthesis of Δ^2-Isoxazolines and Δ^2-Pyrazolines. The addition of NH_2OH and $PhNHNH_2$ to chalcone and related enones (**16**) was catalyzed by $Ba(OH)_2$ (C-200) to give good yields of Δ^2-isoxazoline (**17**) and Δ^2-pyrazoline (**18**) derivatives respectively (eq 12). The products were obtained in yields between 65 and 75% and with complete regioselectivity.[12]

Oxidation of Benzyl Halides to Benzaldehydes. Benzyl halides were oxidized to benzaldehydes by ***Dimethyl Sulfoxide*** in the presence of activated $Ba(OH)_2$ (C-200) in 1,4-dioxane. The optimum conditions include the use of a 5 M solution of the aryl halide in 1,4-dioxane with 3 equiv of DMSO and about 0.2 equiv of the $Ba(OH)_2$ catalyst, then heating the mixture at 130 °C for 3 h. The reaction does not take place in the absence of the C-200 catalyst. The effect of each variable in the reaction on the yield was studied, but no specific yields were given for isolated products except for $PhCH_2Cl$, which was reported as 100%. The proposed mechanism of the reaction involves an initial displacement of the halide by DMSO followed by elimination of Me_2S with the aid of the basic OH groups on the catalyst surface (eq 13).[13]

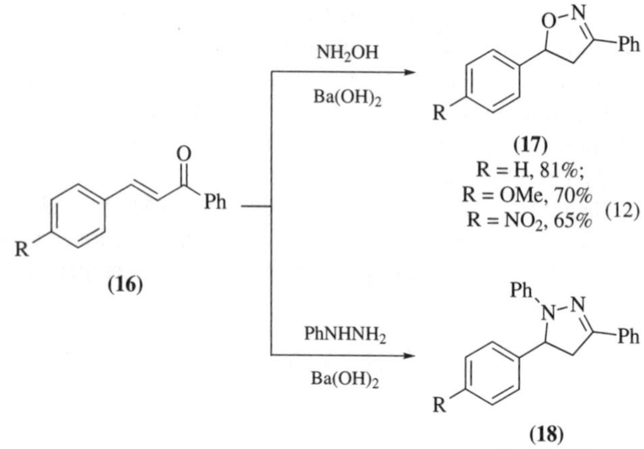

(16) → NH_2OH, $Ba(OH)_2$ → (17) R = H, 81%; R = OMe, 70%; R = NO_2, 65% (12)

(16) → $PhNHNH_2$, $Ba(OH)_2$ → (18) R = H, 65%; R = OMe, 70%

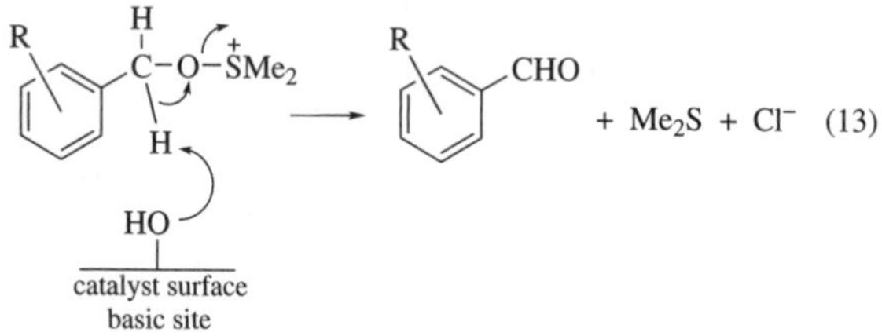

catalyst surface basic site → CHO + Me_2S + Cl^- (13)

Hydrolysis of 2-Alkoxyimidazolines. In a procedure for the stereoselective synthesis of vicinal diamines, several alkenes were converted into 2-alkoxyimidazolines (**19**) via consecutive reactions with cyanamide/NBS, HCl/EtOH, and Et_3N/EtOH. The intermediates (**19**) were hydrolyzed with $Ba(OH)_2$ at 120 °C for 18 h (eq 14) to give the diamines (**20**) in 79–99% yields for this step and in 61–71% overall yields.[14]

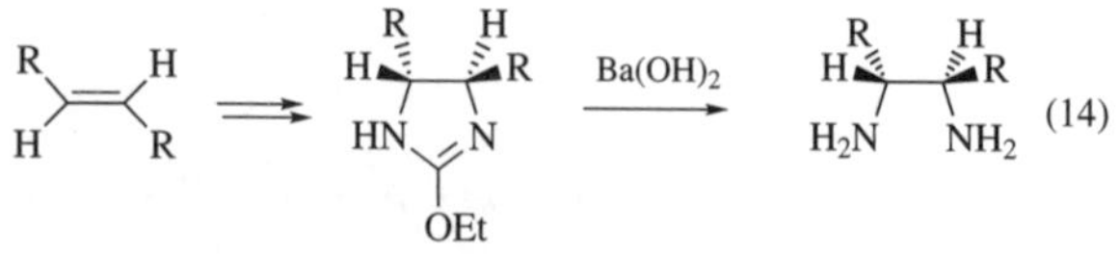

(19) OEt → $Ba(OH)_2$ → (20) H_2N NH_2 (14)

A Skeletal Rearrangement in a 6β,19-Oxido Steroid. When 6β,19-oxido-2,17-dihydroxyandrosta-1,4-dien-3-one (**21**) was heated at reflux with $Ba(OH)_2 \cdot 8H_2O$ in pyridine for 22 h, and then worked up under acidic conditions, it rearranged to the furopyranone (**23**) in 80–85% yield (eq 15). The rearrangement included a ring B contraction, a double bond isomerization and a benzylic acid rearrangement. The acidic work-up caused the carboxy intermediate (**22**) to cyclize with the vinyl ether to give (**23**).[15]

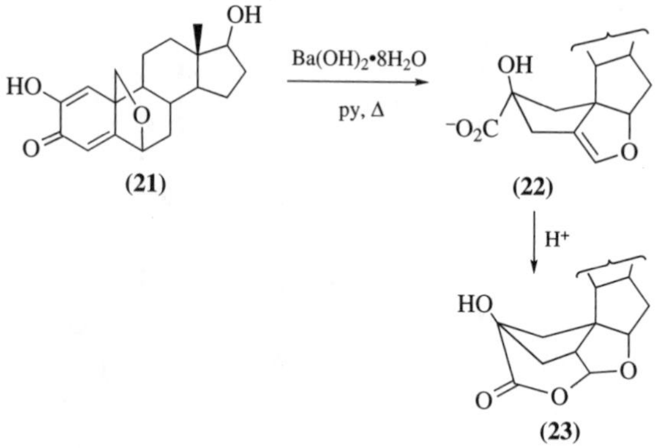

(21) → $Ba(OH)_2 \cdot 8H_2O$, py, Δ → (22) → H^+ → (23) (15)

Wittig–Horner Reaction. Activated $Ba(OH)_2$ (C-200) catalyzed the reactions of aldehydes with triethylphosphonoacetates (**24**) in 1,4-dioxane in the presence of a small amount of water at 70 °C to give the corresponding 3-substituted ethyl acrylates (**25**) (eq 16). The yields of the products from aromatic aldehydes ranged from 0% for indole-3-carbaldehyde to 100% for furfural and *m*-nitrobenzaldehyde. The method was applied to hindered aldehydes such as pyrene-1-carbaldehyde and to unsaturated aldehydes such as *cis*- or *trans*-citral.[16] The reaction was also applied to various aliphatic aldehydes using 2-oxoalkanephosphonates to give the corresponding (*E*)-α,β-unsaturated ketones in high yields and stereoselectivities.[17]

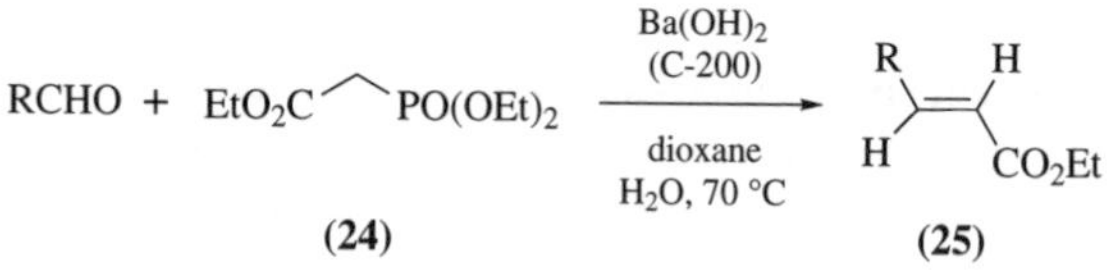

RCHO + EtO_2C–$PO(OEt)_2$ (24) → $Ba(OH)_2$ (C-200), dioxane, H_2O, 70 °C → R(H)C=C(H)CO_2Et (25) (16)

Deacylation of 2-Alkyl-2-halo-1,3-dicarbonyl Compounds. Treatment of various α-alkyl-α-bromoacetoacetic esters (**26**) with a suspension of anhydrous $Ba(OH)_2$ in absolute ethanol at 0 °C for 30 min effected their deacetylation and gave the corresponding α-bromo esters (**27**) in very good yields without any competing saponification or Favorskii rearrangement (eq 17).[18] The starting α-alkyl-α-bromoacetoacetic esters were prepared from the corresponding acetoacetic esters in two steps (NaH/RX then NaH/Br_2) and used directly in the above reaction. The overall yields for the three-step sequence ranged between 70 and 85%. A similar deacylation of the substituted 4-alkyl-4-chloroheptane-3,5-dione (**28**) was reported in the synthesis of racemic juvenile hormone using $Ba(OH)_2$ in ethanol at 0 °C for 25 min (eq 18) to give the corresponding α-chloro ketone (**29**).[19]

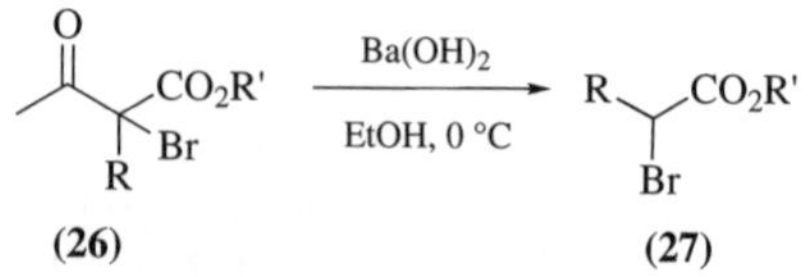

(26) CO_2R', Br, R → $Ba(OH)_2$, EtOH, 0 °C → R–CH(Br)–CO_2R' (27) (17)

(28) → (29): $Ba(OH)_2$, EtOH, 0 °C (18)

R = [structure with CO_2Me]

Removal of the *N*-Protecting Benzyloxycarbonyl Group. The alkynyl amine (**31**) was obtained in 75% yield by removal of the *N*-protecting benzyloxycarbonyl group from the benzyl carbamate (**30**) using 0.15 M $Ba(OH)_2$ in 3:2 glyme–H_2O at reflux temperature for 40 h (eq 19). Other reagents (Me_3SiI, BBr_3, Me_2BBr, BF_3/EtSH, $AlCl_3$/EtSH, MeLi/LiBr, or KOH/EtOH) caused partial destruction of the alkyne.[20]

(30) → (31): 0.15M $Ba(OH)_2$, 3:2 glyme–H_2O, reflux, 40 h, 75% (19)

Related Reagents. Lithium Hydroxide; Potassium Hydroxide; Sodium Hydroxide.

1. For complete safety data on $Ba(OH)_2 \cdot 8H_2O$, see *The Sigma-Aldrich Library of Chemical Safety Data*, 2nd ed.; Lenga, R. E., Ed.; Sigma-Aldrich: Milwaukee, 1988; Vol. 1, p 335C.
2. Barrios, J.; Marinas, J. M.; Sinisterra, J. V. *BSB* **1986**, *95*, 107.
3. Durham, L. J.; McLeod, D. J.; Cason, J. *OSC* **1963**, *4*, 635.
4. Miller, R. B.; Nash, R. D. *T* **1974**, *30*, 2961.
5. Danishefsky, S.; Hirama, M.; Gombatz, K.; Harayama, T.; Berman, E.; Schuda, P. F. *JACS* **1979**, *101*, 7020.
6. Henning, R.; Urbach, H. *TL* **1983**, *24*, 5339.
7. Conant, J. B.; Tuttle, N. *OSC* **1944**, *1*, 199.
8. Sinisterra, J. V.; Garcia-Raso, A.; Cabello, J. A.; Marinas, J. M. *S* **1984**, 502.
9. Alcantara, A. R.; Marinas, J. M.; Sinisterra, J. V. *TL* **1987**, *28*, 1515.
10. (a) Garcia-Raso, A.; Garcia-Raso, J.; Campaner, B.; Mestres, R.; Sinisterra, J. V. *S* **1982**, 1037. For a discussion on the mechanism of the Michael addition of active methylene compounds to chalcone, see (b) Iglesias, M.; Marinas, J. M.; Sinisterra, J. V. *T* **1987**, *43*, 2335.
11. Sinisterra, J. V.; Marinas, J. M. *M* **1986**, *117*, 111.
12. Sinisterra, J. V. *React. Kinet. Catal. Lett.* **1986**, *30*, 93.
13. Climent, M. S.; Marinas, J. M.; Sinisterra, J. V. *React. Kinet. Catal. Lett.* **1987**, *34*, 201.
14. Kohn, H.; Jung, S.-H. *JACS* **1983**, *105*, 4106.
15. Chorvat, R. J.; Bible, Jr., R. H.; Swenton, L. *T* **1975**, *31*, 1353.
16. Sinisterra, J. V.; Mouloungui, Z.; Delmas, M.; Gaset, A. *S* **1985**, 1097.
17. Alvarez-Ibarra, C.; Arias, S.; Banon, G.; Fernandez, M. J.; Rodriguez, M.; Sinisterra, V. *CC* **1987**, 1509.
18. Stotter, P. L.; Hill, K. A. *TL* **1972**, 4067.
19. Johnson, W. S.; Li, T.-t.; Faulkner, D. J.; Campbell, S. F. *JACS* **1968**, *90*, 6225.
20. Overman, L. E.; Sharp, M. J. *TL* **1988**, *29*, 901.

Ahmed F. Abdel-Magid
The R. W. Johnson Pharmaceutical Research Institute, Spring House, PA, USA

Benzyltrimethylammonium Hydroxide

$Ph\frown\overset{+}{N}Me_3\ ^{-}OH$

[100-85-6] $C_{10}H_{17}NO$ (MW 167.25)

(quaternary ammonium salt; strong base; phase transfer catalyst[1])

Alternate Name: Triton® B.
Physical Data: none available (generally obtained in solution).
Solubility: sol water, alcohols, hydrocarbons, aromatic hydrocarbons, halogenated solvents.
Form Supplied in: 40 wt % in H_2O; 35–40 wt % in MeOH (Triton B). Typical impurities are amines or benzyl alcohol.
Purification: [2] methanol solution can be decolorized with charcoal, concentrated to a syrup, and dried under vacuum at 75 °C and 1 mmHg pressure. Anhydrous reagent is obtained by drying over P_2O_5 in a vacuum desiccator.
Handling, Storage, and Precautions: hygroscopic solutions; highly toxic; decomposition can occur on heating.[3]

Epoxidations. Benzyltrimethylammonium hydroxide (**1**) is a common reagent in organic synthesis. It finds use as a catalytic base in the epoxidation of α,β-unsaturated ketones with ***t-Butyl Hydroperoxide***,[4] as demonstrated in a step toward the total synthesis of a complex anthracyclinone (eq 1).[5] The combination of Triton B and either *t*-BuOOH or ***Hydrogen Peroxide*** has been used to advantage in situations where conventional epoxidation conditions have proven either destructive[6] or ineffective.[7] Epoxidations of other α,β-unsaturated systems, such as sulfonates,[8] have also been reported.

t-BuOOH, Triton B, CH_2Cl_2, 90% (1)

Oxidations at Activated Methylenes. Benzyltrimethylammonium hydroxide has been used as a catalyst for the oxidation of methylene and/or methine groups. The oxidation of a tricyclic ketone to an anthraquinone was accomplished in good yield using this approach (eq 2).[9] Similarly, benzylic anions have been generated and efficiently trapped with air in DMSO.[10]

Dehydrohalogenation. A variety of 1-bromoalkynes have been prepared through the action of Triton B on 1,1-

dibromoalkenes.[11] Both aryl and alkyl substituents were examined, with recoveries ranging from 35 to 87%, and generally above 60% yield (eq 3).

In an extension of this reaction, a synthesis of propargyl amines using Triton B was developed in which 1,1-dibromoalkenes were dehydrohalogenated, isomerized in situ to bromoallenes, and subjected to nucleophilic attack by primary or secondary amines (eq 4).[12] Yields ranged from 5 to 88%. Attack by methanol was also examined; however, yields were generally low.

Aldol Condensations. The aldol reaction has often been carried out with Triton B functioning as base. In the synthesis of a cytochalasin intermediate, a seven-membered ring was closed under the action of Triton B in excellent yield (eq 5).[13] The reagent can effect the double aldol condensation of dibenzyl ketone with benzil, producing tetraphenylcyclopentadienone in excellent yield (eq 6).[14]

The action of Triton B has been used in the synthesis of 2-thiomethylindoles (eq 7).[15] Condensation of various 2-sulfonamido aldehydes with methyl methylsulfinylmethyl sulfide (55–88%) followed by treatment with H_2S and HCl gave rise to the indole products (62–80%).

Conjugate Additions. Triton B is an effective catalyst for the conjugate addition of carbon acids to Michael acceptors. Such is the case with the reaction of nitromethane and *t*-butyl acrylate mediated by a methanolic solution of (**1**) (eq 8).[16] In the given example, the triester formed was used in the preparation of dendritic macromolecules. Other bases were not as effective.[17] Nitroalkanes have also been added intramolecularly to alkynylamides[18] and intermolecularly to a variety of unsaturated systems. The addition of a benzonitrile to methyl acrylate in a Michael fashion was reported to occur in excellent yield (eq 9).[19] Heteroatomic nucleophiles have also been added in a 1,4-fashion to conjugated systems.[20]

The addition of nucleophiles to vinylsiloxycyclopropanecarboxylates catalyzed by Triton B was reported to be an efficient method for the preparation of functionalized 4-oxoalkanoates (eq 10).[21] Other catalysts were not as effective.

NuH = malonates, nitroalkanes, thiols

Benzyltrimethylammonium hydroxide also has been utilized in steroid syntheses through the Torgov reaction.[22] A cyclic 1,3-diketone can be *C*-alkylated through a formal S_N2' displacement of an ionizable tertiary hydroxy group (eq 11). Alkylation of 1,3-diketones with halides is also well known.[23]

Eq 12 shows a novel photolytic reaction in which Triton B was used to improve the conversion of an amino enone into a bridged bicyclic compound.[24]

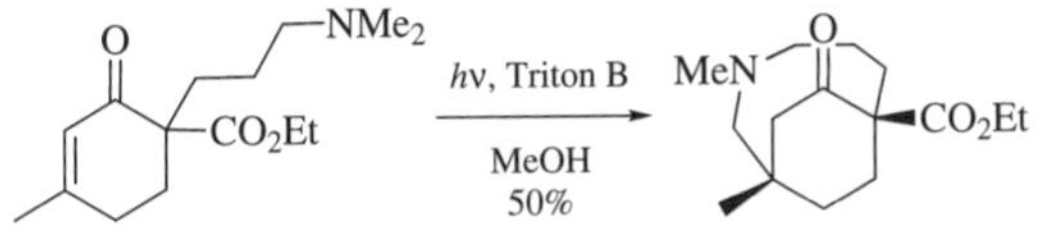

Benzyltrimethylammonium hydroxide has also been used in the conversion of an α-hydroxy-1,1-dichloride to an aldehyde, which was cyclized to generate the tetracyclic compound rhodomycinone (eq 13).[25]

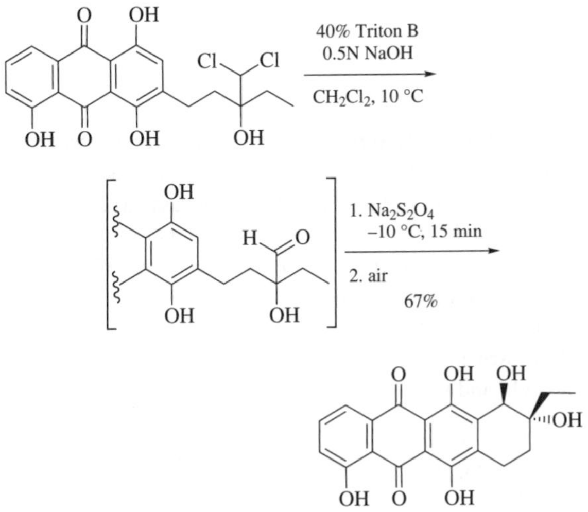

Related Reagents. Barium Hydroxide; Potassium *t*-Butoxide; Potassium Hydroxide; Potassium 2-Methyl-2-Butoxide; Sodium Hydroxide.

1. For reviews of phase transfer reactions, see (a) Keller, W. E. *Phase Transfer Reactions. Fluka Compendium*; Thieme: Stuttgart, 1986; Vols. 1 and 2. (b) Dehmlow, E. V. *Phase Transfer Catalysis*; Verlag Chemie: Deerfield Beach, FL, 1980. (c) Starks, C. M.; Liotta, C. *Phase Transfer Catalysis, Principles and Techniques*; Academic: New York, 1978. (d) Dockx, J. *S* **1973**, 441. (e) Dehmlow, E. V. *AG(E)* **1974**, *13*, 170. For a mechanistic review of hydroxide-mediated reactions under PTC conditions, see (f) Rabinovitz, M.; Cohen, Y.; Halpern, M. *AG(E)* **1986**, *25*, 960.
2. Perrin, D. D.; Armarego, W. L. F. *Purification of Laboratory Chemicals*, 3rd ed.; Pergamon: New York, 1988.
3. Collie, N.; Schryver, S. B. *JCS* **1890**, 767. See also note on stability under ***Methyltrioctylammonium Chloride***.
4. Yang, N. C.; Finnegan, R. A. *JACS* **1958**, *80*, 5845.
5. Hauser, F. M.; Chakrapani, S.; Ellenberger, W. P. *JOC* **1991**, *56*, 5248.
6. Ireland, R. E.; Wuts, P. G. M.; Ernst, B. *JACS* **1981**, *103*, 3205.
7. Asaoka, M.; Hayashibe, S.; Sonoda, S.; Takei, H. *TL* **1990**, *31*, 4761.
8. Carretero, J. C.; Ghosez, L. *TL* **1987**, *28*, 1101.
9. Kende, A. S.; Rizzi, J. P. *JACS* **1981**, *103*, 4247. Wulff, W. D.; Tang, P. C. *JACS* **1984**, *106*, 434.
10. Finger, C. *S* **1970**, 541.
11. Bestmann, H. J.; Frey, H. *LA* **1980**, 2061.
12. Frey, H.; Kaupp, G. *S* **1990**, 931.
13. Pyne, S. G.; Spellmeyer, D. C.; Chen, S.; Fuchs, P. L. *JACS* **1982**, *104*, 5728.
14. Fieser, L. S. *OSC* **1973**, *5*, 604.
15. Hewson, A. T.; Hughes, K.; Richardson, S. K.; Sharpe, D. A.; Wadsworth, A. H. *JCS(P1)* **1991**, 1565.
16. Newkome, G. R.; Behera, R. J.; Moorefield, C. N.; Baker, G. R. *JOC* **1991**, *56*, 7162.
17. Weis, C. D.; Newkome, G. R. *JOC* **1990**, *55*, 5801.
18. Patra, R.; Maiti, S. B.; Chatterjee, A.; Chakravartu, A. K. *TL* **1991**, *32*, 1363.
19. Cheng, A.; Uyeno, E.; Polgar, W.; Toll, L.; Lawson, J. A.; DeGraw, J. I.; Loew, G.; Cammerman, A.; Cammerman, N. *JMC* **1986**, *29*, 531; and references therein.
20. Examples of oxygen nucleophiles: Dasaradhi, L.; Fadnavis, N. W.; Bhalerao, U. T. *CC* **1990**, 729. Nitrogen nucleophiles: Pyne, S. G.; Chapman, S. L. *CC* **1986**, 1688. Sulfur nucleophiles: Kharasch, M. S.; Fuchs, C. F. *JOC* **1948**, *13*, 97. Stewart, J. M.; Klundt, I.; Peacock, K. *JOC* **1960**, *25*, 913.
21. Grimm, E. L.; Zschiesche, R.; Reissig, H. U. *JOC* **1985**, *50*, 5543.
22. Ananchenko, S. N.; Torgov, I. V. *TL* **1963**, 1553. Lehmann, G.; Wehlan, H.; Hilgetag, G. *CB* **1967**, *100*, 2967. Douglas, S. P.; Sawyer, J. F.; Yates, P. *TL* **1985**, *26*, 5955.
23. Mori, K.; Mori, H. *OS* **1990**, *68*, 56; *OSC* **1993**, *8*, 312.
24. Kraus, G. A.; Chen, L. *TL* **1991**, *32*, 7151.
25. Krohn, K.; Priyono, W. *T* **1984**, *40*, 4609.

Mary Ellen Bos
The R. W. Johnson Pharmaceutical Research Institute, Raritan, NJ, USA

(*R*)-1,1′-Bi-2,2′-naphthol[1]

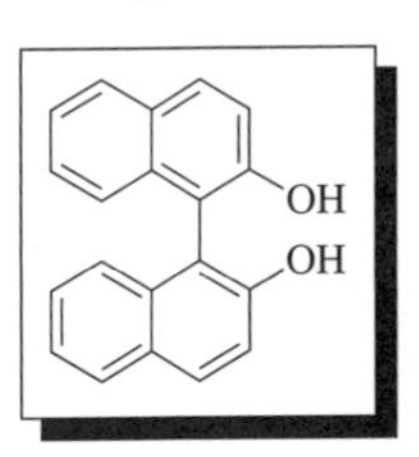

[18531-94-7] $C_{20}H_{14}O_2$ (MW 286.33)

(chiral ligand and auxiliary[1])

Alternate Name: BINOL.
Physical Data: mp 208–210 °C; $[\alpha]^{21}$ +34° ($c = 1$, THF).
Solubility: sol toluene, CH_2Cl_2, $EtNO_2$.
Form Supplied in: white solid; widely available.
Preparative Methods: racemic 1,1′-bi-2,2′-naphthol (BINOL) is most conveniently prepared by the oxidative coupling reaction of 2-naphthol in the presence of transition metal complexes (eq 1).[2] The resolution of racemic BINOL with cinchonine may be performed via the cyclic phosphate (eq 2).[3] An alternative procedure to provide directly optically active BINOL is the oxidative coupling of 2-naphthol catalyzed by Cu^{II} salt in the presence of chiral amines (eq 3).[4] The best procedure uses (+)-amphetamine as the chiral ligand and provides BINOL in 98% yield and 96% ee. Above 25 °C the Cu^{II}/(+)-amphetamine/(*S*)-BINOL complex precipitates while the more soluble Cu^{II}/(+)-amphetamine/(*R*)-BINOL complex is slowly transformed into the former complex. 9,9′-Biphenanthrene-10,10′-diol has also

been prepared in 86% yield and with 98% ee by a similar asymmetric oxidative coupling of 9-phenanthrol in the presence of (*R*)-1,2-diphenylethylamine.[5]

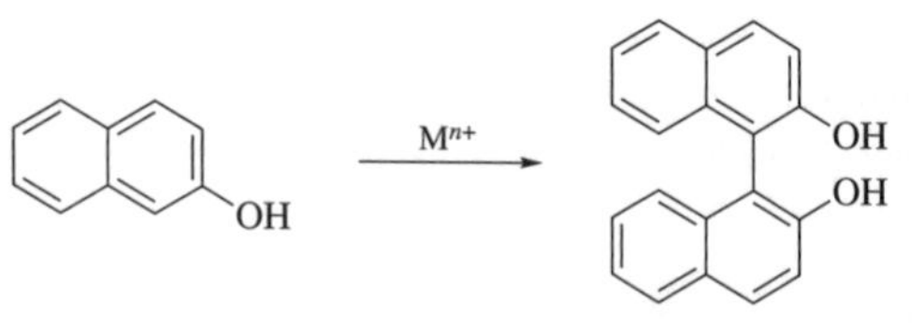

M^{n+}

OH OH OH (1)

Handling, Storage, and Precautions: keep tightly closed, store in a cool dark place; on heating in butanol at 118 °C for 24 h, BINOL lost ~1% of its optical rotation; at 100 °C for 24 h in dioxane–1.2 N HCl, BINOL lost 56% of its rotation; after 24 h at 118 °C in butanol–0.7 N KOH, BINOL lost 20% of its rotation.

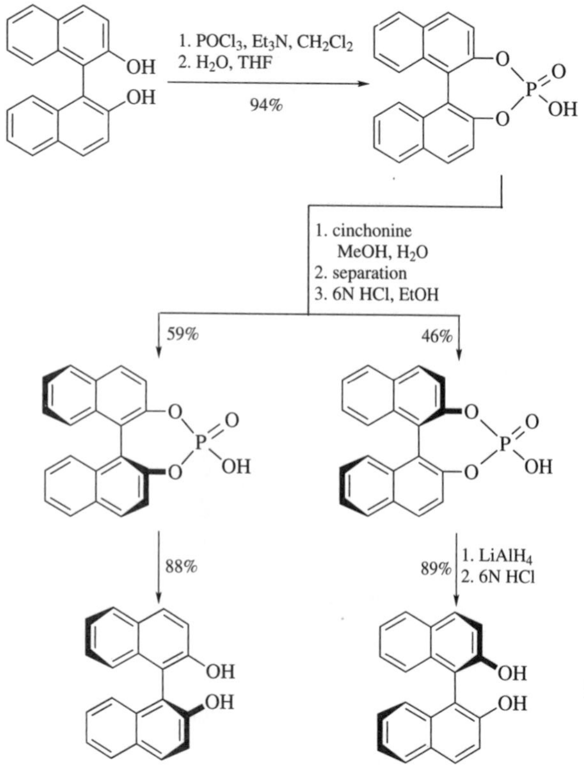

1. $POCl_3$, Et_3N, CH_2Cl_2
2. H_2O, THF

94%

1. cinchonine MeOH, H_2O
2. separation
3. 6N HCl, EtOH

59% 46%

88%

1. $LiAlH_4$
2. 6N HCl

89%

(2)

NH_2 Ph (16 equiv), $CuCl_2$ (4 equiv), MeOH, 25 °C, 20 h; 4N HCl, 98%

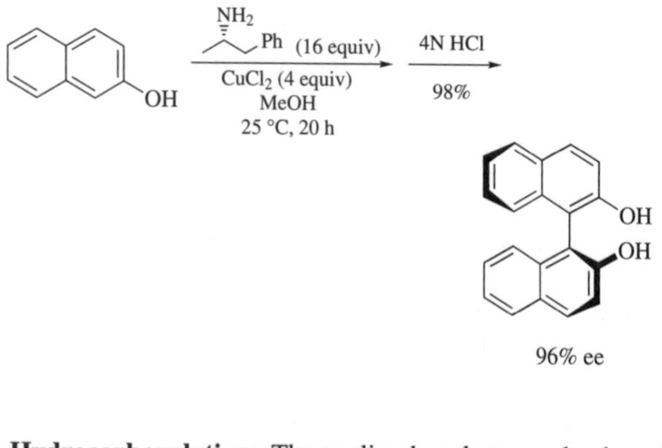

(3)

96% ee

Hydrocarboxylation. The cyclic phosphate resolved according to eq 2 can be used as the chiral ligand in the palladium(II) catalyzed asymmetric hydrocarboxylation of arylethylenes.[6] The 1-arylpropanoic acid is obtained regiospecifically with high enantioselectivity (91% ee) (eq 4).

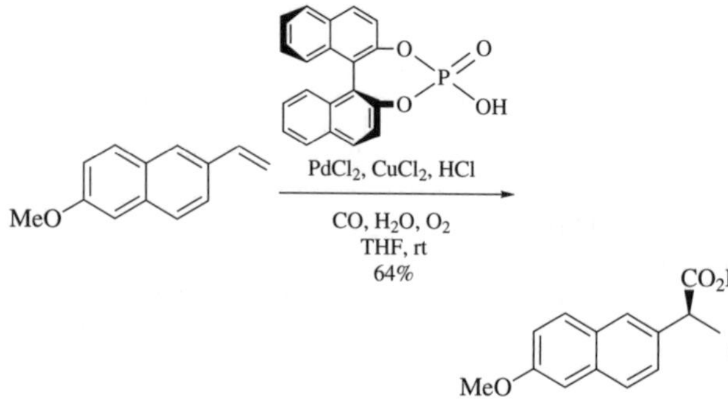

$PdCl_2$, $CuCl_2$, HCl

CO, H_2O, O_2, THF, rt, 64%

MeO CO_2H (4)

91% ee

Crown Ethers. BINOL-derived crown ethers have been reported.[7] Crown ethers containing 3,3′-disubstituted BINOL derivatives are particularly effective for asymmetric synthesis. Thus complexes of these crown ethers (e.g. ***18-Crown-6***) with ***Potassium Amide*** or ***Potassium t-Butoxide*** catalyze asymmetric Michael additions. The reaction of methyl 1-oxo-2-indancarboxylate with methyl vinyl ketone with the 3,3′-dimethyl-BINOL–crown ether/KO-*t*-Bu complex gives the Michael product in 48% yield and with 99% ee (eq 5).[8]

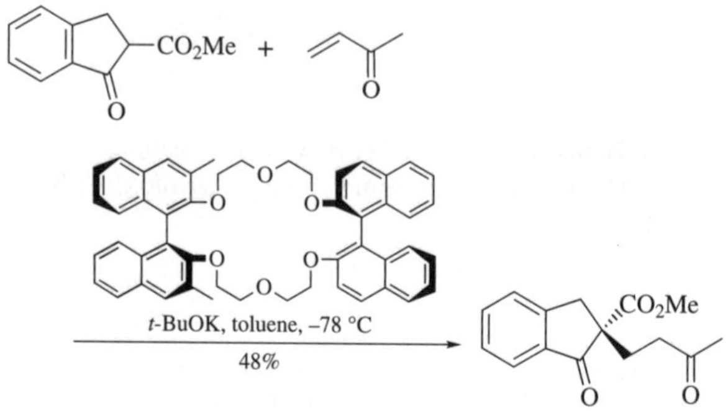

CO_2Me +

t-BuOK, toluene, –78 °C, 48%

CO_2Me (5)

99% ee

Polymerization. Complexes of BINOL-derived crown ethers with KO-*t*-Bu or BuLi have been used as initiators in the asymmetric polymerization of methacrylates.[9] Thus optically active polymers are obtained with 80–90% isotacticity. Complexes of BINOL with ***Diethylzinc*** or $CdMe_2$ also initiate the asymmetric polymerization of heterocyclic monomers.[10] The chiral initiators selectively polymerize one enantiomer to give an optically active polymer. The unreacted monomer is recovered with 92% ee at 67% conversion in the polymerization of methylthiirane with (*S*)-BINOL/Et_2Zn.

Ullmann Coupling Reaction. Axially dissymmetric biaryls have been synthesized via an intramolecular Ullmann coupling reaction of BINOL-derived aryl diesters (eq 6).[11] In the example shown, the functionalized binaphthyl is obtained with high ee after hydrolysis of the intermediate 12-membered cyclic diester.

Reduction of Prochiral Ketones. BINOL has been used as the chiral ligand of the reagent BINAL-H (see ***Lithium Aluminum***

Hydride-2,2'-Dihydroxy-1,1'-binaphthyl, Vol. B) for asymmetric reduction.[12] The reagent reduces prochiral unsaturated ketones to the corresponding secondary alcohols in up to 90% yield and >90% ee (eq 7); (*R*)-BINAL-H leads to the (*R*)-alcohols while (*S*)-BINAL-H gives the (*S*)-alcohols.

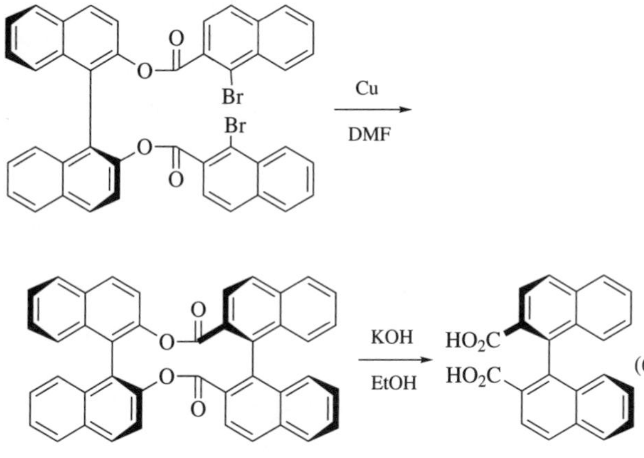

Cu, DMF

KOH, EtOH (6)

100% de

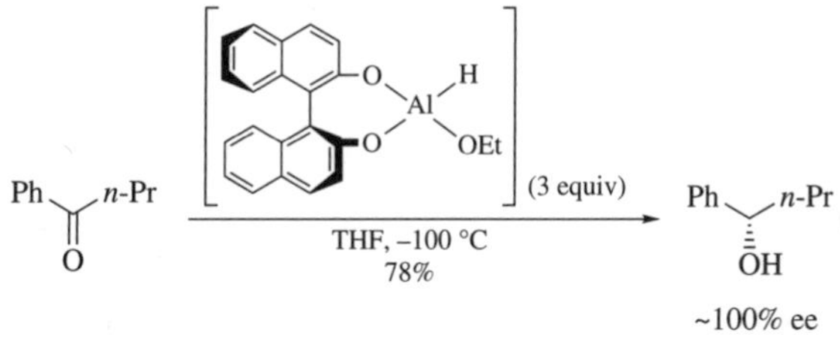

(3 equiv), THF, −100 °C, 78% (7)

~100% ee

Addition Reactions of Chiral Titanium Reagents to Aldehydes. The preparation and use of the BINOL-derived titanium complexes in the enantioselective synthesis of some benzhydrols (>90% ee) have been reported (eq 8).[13]

PhMgBr, THF, Et_2O, −78 °C to rt, 81% (8)

>90% ee

Cyanosilylation. The chiral titanium reagent, prepared from the lithium salt of BINOL with $TiCl_4$, has been used as a catalyst for the asymmetric addition of cyanotrimethylsilane to aldehydes.[14] In the example shown, the cyanohydrin is obtained with ≤82% ee (eq 9).

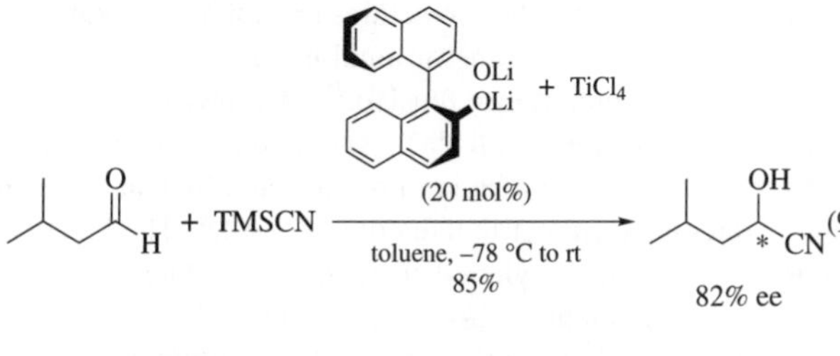

+ TMSCN, + $TiCl_4$ (20 mol%), toluene, −78 °C to rt, 85% (9)

82% ee

Diels–Alder Reactions. BINOL and its derivatives are used as the chiral ligand of chiral Lewis acid complexes for enantioselective Diels–Alder cycloadditions. BINOL-$TiCl_2$, prepared from the lithium salt of BINOL with ***Titanium(IV) Chloride***, also catalyzes the enantioselective Diels–Alder reaction of cyclopentadiene with methacrolein (eq 10)[14,15] The *exo* adduct is obtained as the major product (56% yield), but with low enantioselectivity (16% ee). More recently, BINOL-TiX_2 (X = Br or Cl) have been prepared in situ from diisopropoxytitanium dihalides ($(i\text{-PrO})_2TiX_2$, X = Br[16] or Cl[17]) with BINOL in the presence of molecular sieves (MS 4A).[16] The Diels–Alder reaction of methacrolein with 1,3-dienol derivatives can be catalyzed by BINOL-TiX_2. The *endo* adducts are obtained in high enantioselectivity (eq 11).[18] Asymmetric catalytic Diels–Alder reaction of naphthoquinone derivatives as the dienophile (eq 12)[18] can in principle provide an efficient entry to the asymmetric synthesis of anthracyclinone aglycones. The reaction of the 5-hydroxynaphthoquinone with 1-acetoxy-1,3-diene in the presence of MS-free BINOL-$TiCl_2$ (10 mol%) provides the corresponding Diels–Alder product in high chemical yield and with high enantioselectivity (76–96% ee).[18b] The Diels–Alder product is also obtained by the use of 1 equiv of 3,3′-diphenyl-BINOL/borane complex (eq 13); the structure of the intermediate has been proposed.[19]

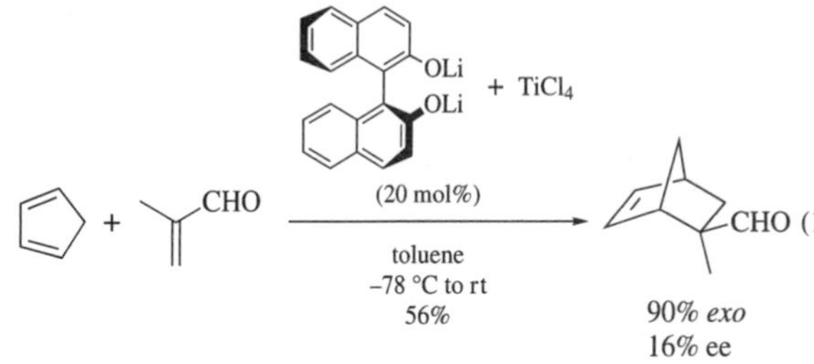

+ $TiCl_4$ (20 mol%), toluene, −78 °C to rt, 56% (10)

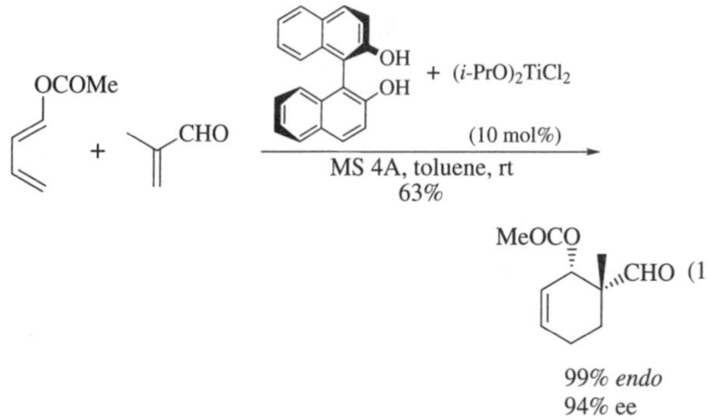

90% *exo*
16% ee

+ $(i\text{-PrO})_2TiCl_2$ (10 mol%), MS 4A, toluene, rt, 63% (11)

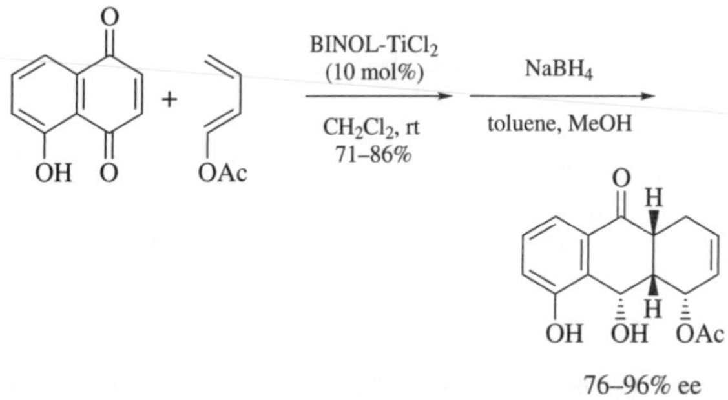

99% *endo*
94% ee

BINOL-$TiCl_2$ (10 mol%), CH_2Cl_2, rt, 71–86%; $NaBH_4$, toluene, MeOH (12)

76–96% ee

3,3′-Diphenyl-BINOL-derived chiral aluminum reagents are prepared in situ by addition of ***Ethylaluminum Dichloride*** or ***Diethylaluminum Chloride*** to 3,3′-diphenyl-BINOL. These chiral aluminum reagents promote the enantioselective Diels–Alder reaction of cyclopentadiene with the oxazolidone dienophile (eq 14).[20] *Endo* products are obtained with a high level of asymmetric induction (>90% ee); however, a stoichiometric amount of the Lewis acid is required. The preparation and use of a C_3 symmetric BINOL-derived boronate has been reported (eq 15).[21] BINOL-B(OAr)$_3$ complexes have recently been developed for the asymmetric Diels–Alder reaction with imines (eq 16).[22]

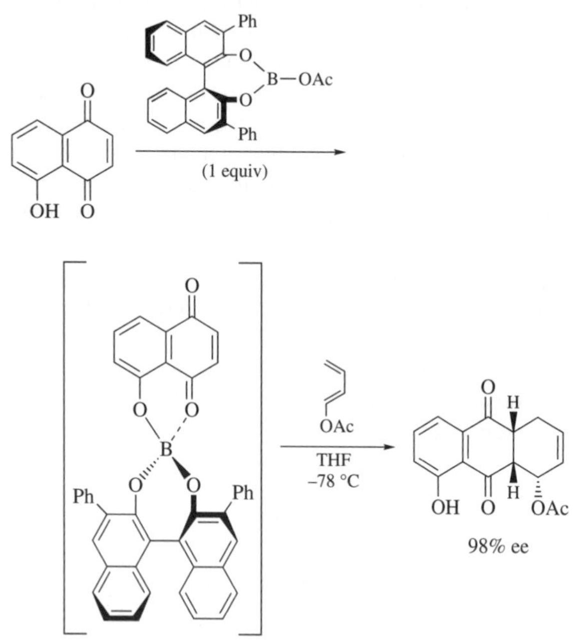

(1 equiv)

THF
−78 °C

(13)

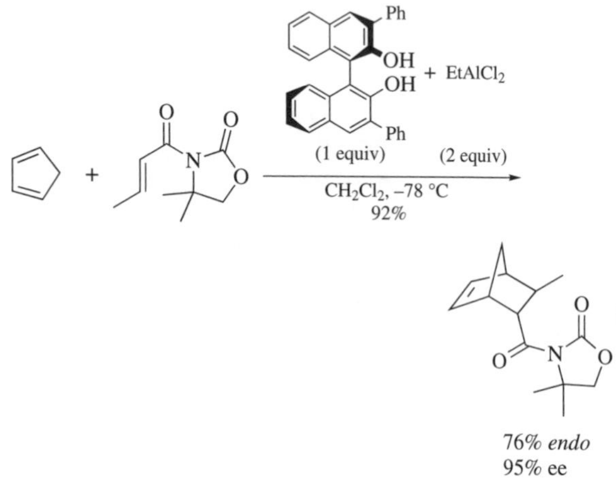

98% ee

(1 equiv) (2 equiv)

CH_2Cl_2, −78 °C
92%

(14)

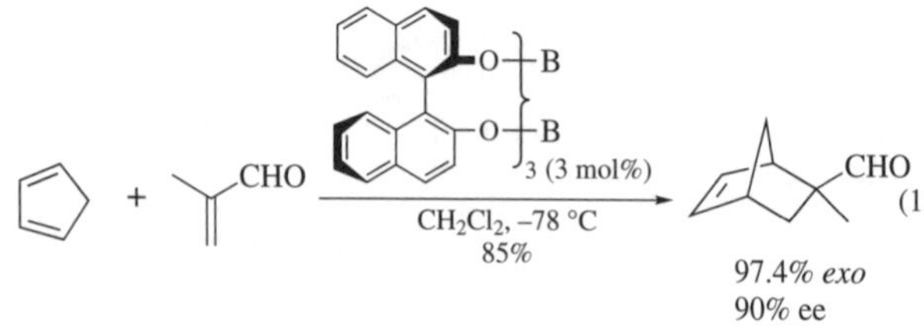

76% *endo*
95% ee

3 (3 mol%)

CH_2Cl_2, −78 °C
85%

(15)

97.4% *exo*
90% ee

Hetero Diels–Alder Reaction. Modified BINOL-derived organoaluminum reagents have been used in the asymmetric hetero Diels–Alder reaction of aldehydes (eq 17). The dihydropyrones are obtained with high *cis* diastereoselectivity and enantioselectivity.[23] The hetero Diels–Alder reaction of glyoxylates proceeds smoothly with methoxydienes using BINOL-TiCl$_2$ as a catalyst to give the *cis* product in high enantiomeric excess (eq 18).[18b,24] The hetero Diels–Alder product thus obtained can be readily converted to the lactone portion of HMG-CoA inhibitors such as mevinolin or compactin.

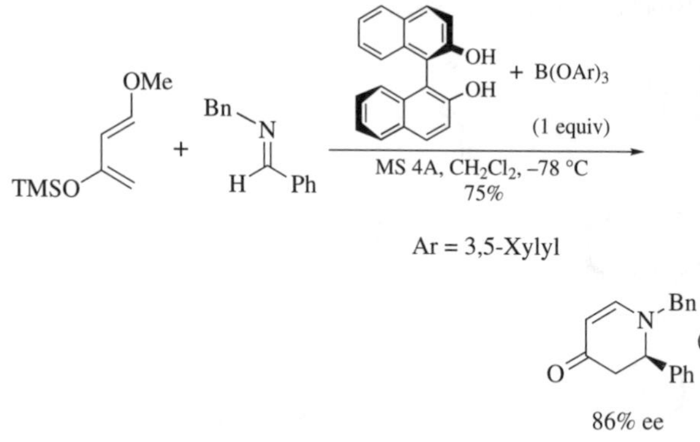

+ B(OAr)$_3$ (1 equiv)

MS 4A, CH_2Cl_2, −78 °C
75%

Ar = 3,5-Xylyl

(16)

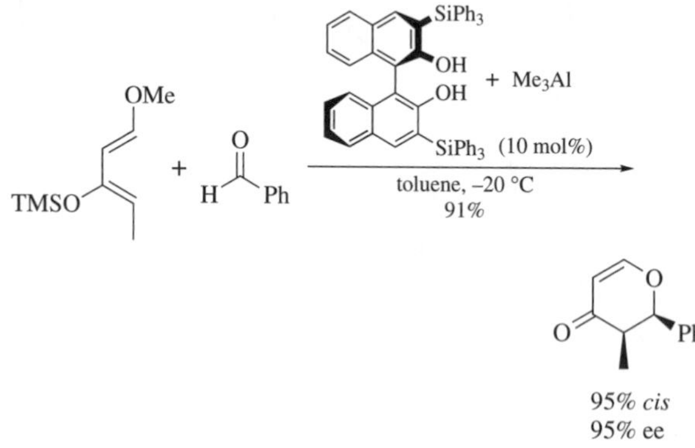

86% ee

+ Me_3Al (10 mol%)

toluene, −20 °C
91%

(17)

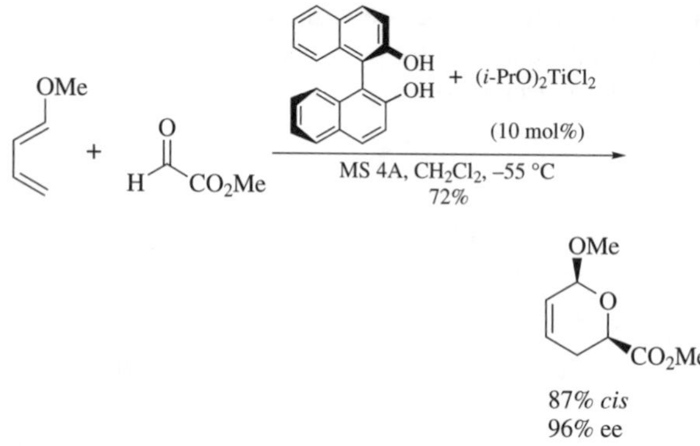

95% *cis*
95% ee

+ (*i*-PrO)$_2$TiCl$_2$ (10 mol%)

MS 4A, CH_2Cl_2, −55 °C
72%

(18)

87% *cis*
96% ee

Carbonyl–Ene Reaction. BINOL-TiX$_2$ reagent exhibits a remarkable level of asymmetric catalysis in the carbonyl–ene reaction of prochiral glyoxylates, thereby providing practical access to α-hydroxy esters.[16,25] These reactions exhibit a remarkable positive nonlinear effect (asymmetric amplification) that is of practical and mechanistic importance (eq 19).[26] The desymmetrization of prochiral ene substrates with planar symmetry by the enantiofacial selective carbonyl–ene reaction provides an efficient solution to remote internal asymmetric induction (eq 20).[27] The kinetic resolution of a racemic allylic ether by the glyoxylate–ene reaction also provides efficient access to remote but relative asymmetric induction (eq 21).[27] Both the dibromide and dichloride catalysts provide the (2*R*,5*S*)-*syn* product with 97% diastereoselectivity and ≥95% ee.

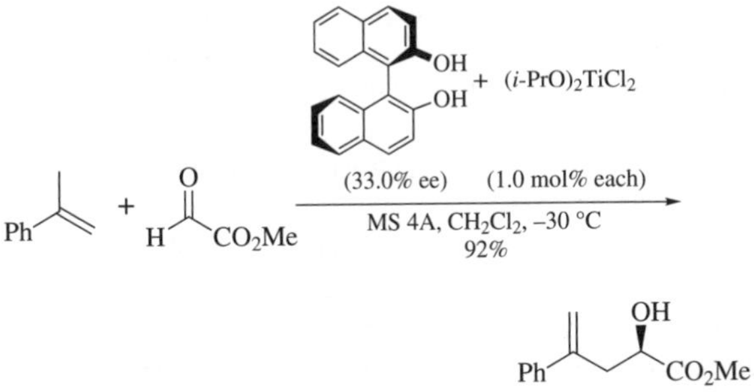

(19)

91.4% ee

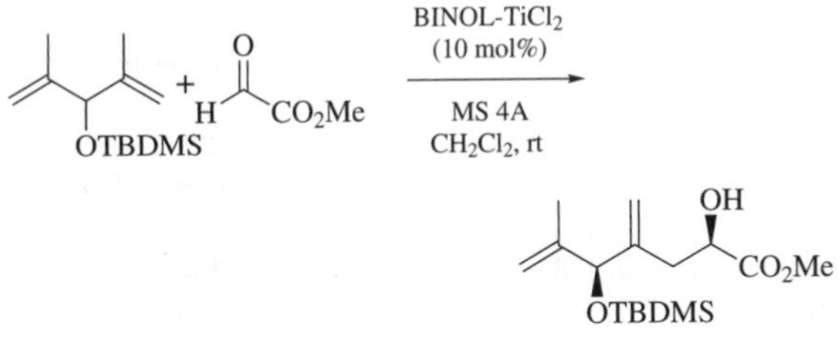

(20)

>99% *syn*
97% ee

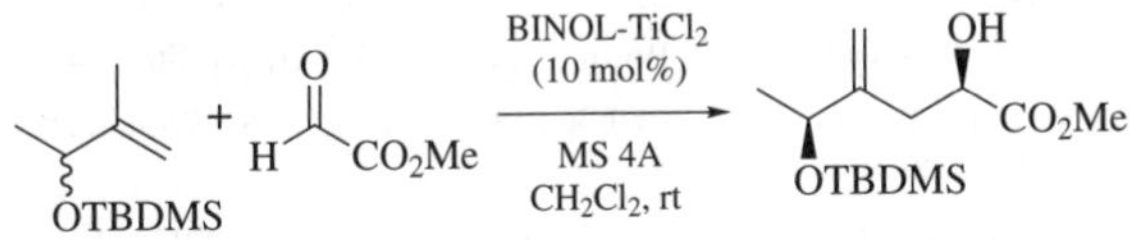

(21)

97% *syn*
>95% ee

Ene Cyclization. An intramolecular (3,4)-ene reaction of unsaturated aldehydes has been accomplished with the BINOL-derived zinc reagent.[28] Cyclization of 3-methylcitronellal with at least 3 equiv of BINOL-Zn reagent afforded the *trans*-cyclohexanol in 86% yield and 88% ee (eq 22). Asymmetric ene cyclizations of type (2,4) are also catalyzed by the BINOL-derived titanium complexes ((*R*)-BINOL-TiX_2, X = ClO_4 or OTf), modified by the perchlorate or trifluoromethanesulfonate ligand. The 7-membered cyclization of type 7-(2,4) gives the oxepane in high ee (eq 23).

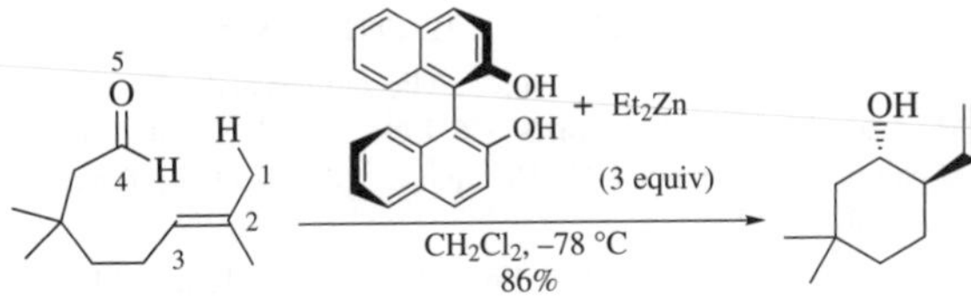

(22)

88% ee

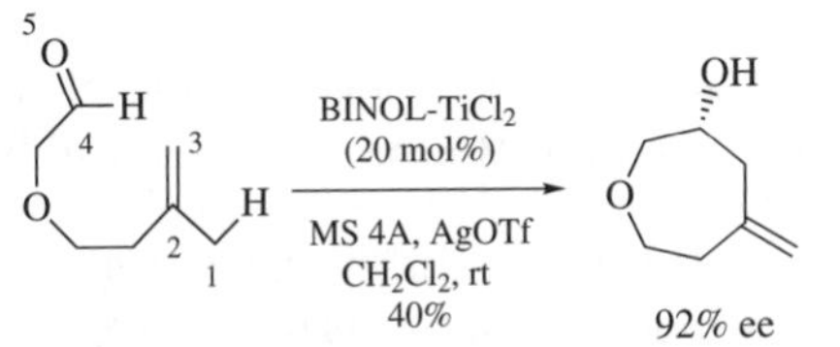

(23)

92% ee

Cationic Cyclization. A cationic cyclization of BINOL-derived neryl ether has been accomplished with an organoaluminum triflate catalyst.[29] Limonene is obtained in 54% yield and 77% ee (eq 24).

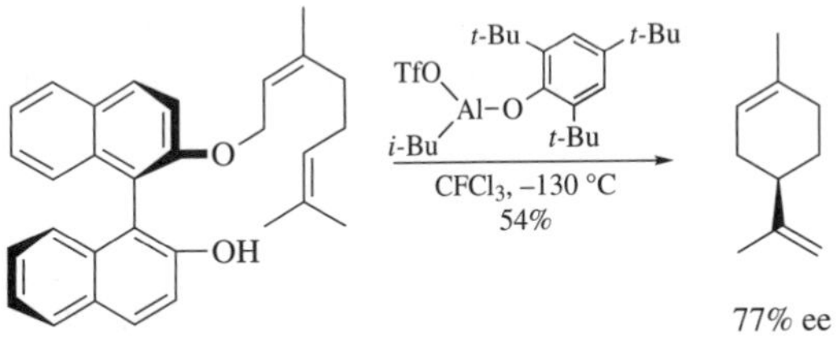

(24)

77% ee

Mukaiyama Aldol Condensation. The BINOL-derived titanium complex BINOL-$TiCl_2$ is an efficient catalyst for the Mukaiyama-type aldol reaction. Not only ketone silyl enol ether (eq 25),[30] but also ketene silyl acetals (eq 26)[31] can be used to give the aldol-type products with control of absolute and relative stereochemistry.

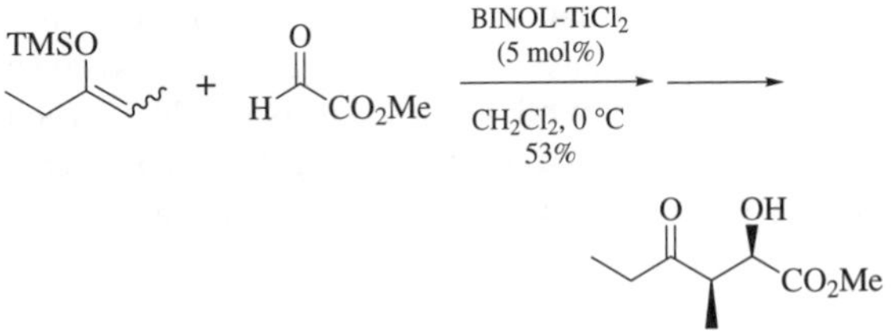

(25)

98% *syn*
99% ee

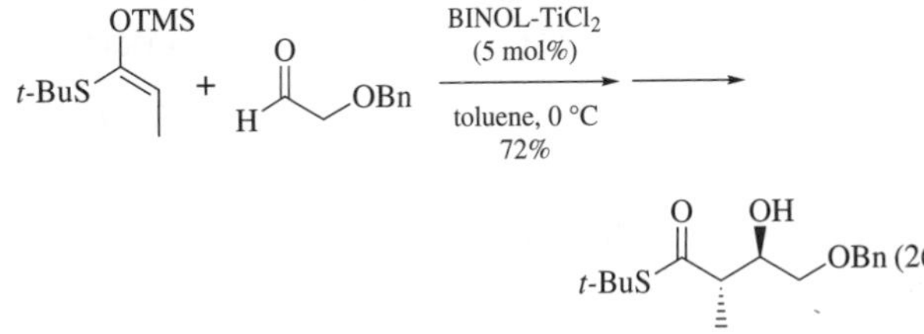

(26)

92% *anti*
90% ee

Nitro-Aldol Condensation. A BINOL-derived lanthanide complex has been used as an efficient catalyst for the nitro-aldol reaction (eq 27).[32] Interestingly enough, the presence of water and LiCl in the reaction mixture is essential to obtain the high level of asymmetric induction and chemical yield.

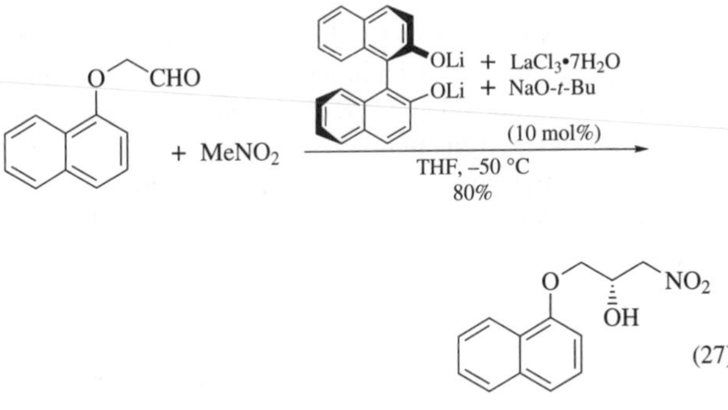

(27)

92% ee

Carbonyl Addition of Allylic Silanes and Stannanes. BINOL-$TiCl_2$ reagent also catalyzes the asymmetric carbonyl addition reaction of allylic silanes and stannanes.[33]

Thus the addition reaction of glyoxylate with (*E*)-2-butenylsilane and -stannane proceeds smoothly to give the *syn* product in high enantiomeric excess (eq 28). The reaction of aliphatic and aromatic aldehydes with allylstannane is also catalyzed by BINOL-$TiCl_2$ or BINOL-Ti(O-*i*-Pr)$_2$ to give remarkably high enantioselectivity.[34]

Bu_3Sn + $HC(O)CO_2Me$ → (BINOL-$TiCl_2$ (10 mol%), MS 4A, CH_2Cl_2, –60 °C to rt, 64%) → CO_2Me product (28)

84% *syn*
86% ee

Claisen Rearrangements. A modified BINOL-derived aluminum reagent is an effective chiral catalyst for asymmetric Claisen rearrangement of allylic vinyl ethers (eq 29).[35] The use of vinyl ethers with sterically demanding C-3 substituents is necessary for the high level of asymmetric induction.

Ph, O, $GeMe_3$ → (SiR_3, OH, OH, SiR_3 (1.1–2 equiv) + Me_3Al; CH_2Cl_2, –40 °C, 68%) → Ph, $COGeMe_3$ (29)

R_3 = *t*-$BuPh_2$

93% ee

Alkylation of BINOL-Derived Ester Enolates. The diastereoselective alkylation of BINOL-derived arylacetates affords the optically active 2-arylalkanoic acids (eq 30).[36]

OH, O, Ar → (1. *n*-BuLi, 2. MeI; THF, –78 to –30 °C, 78%) → OH, O, Ar (30)

Ar = 3-pyridyl

90% de

Related Reagents. (*R*)-1,1′-Bi-2,2′-naphtholate; (*R*)-1,1′-Bi-2,2′-naphthotitanium Dichloride; (*R*)-1,1′-Bi-2,2′-naphthotitanium Diisopropoxide; (*R,R*)-[Ethylene-1,2-bis(η^5-4,5,6,7-tetrahydro-1-indenyl)]titanium (*R*)-1,1′-Bi-2,2′-naphtholate; (–)-[Ethylene-1,2-bis(η^5-4,5,6,7-tetrahydro-1-indenyl)]-zirconium; Lithium Aluminum Hydride-2,2′-Dihydroxy-1,1′-binaphthyl.

1. (a) Rosini, C.; Franzini, L.; Raffaelli, A.; Salvadori, P. *S* **1992**, 503. (b) Miyano, S.; Hashimoto, H. *J. Synth. Org. Chem. Jpn.* **1986**, *44*, 713.
2. (a) Toda, F.; Tanaka, K.; Iwata, S. *JOC* **1989**, *54*, 3007. (b) Pirkle, W. H.; Schreiner, J. L. *JOC* **1981**, *46*, 4988. (c) McKillop, A.; Turrell, A. G.; Young, D. W.; Taylor, E. C. *JACS* **1980**, *102*, 6504. (d) Carrick, W. L.; Karapinka, G. L.; Kwiatkowski, G. T. *JOC* **1969**, *34*, 2388. (e) Dewar, M. J. S.; Nakaya, T. *JACS* **1968**, *90*, 7134. (f) Pummerer, R.; Prell, E.; Rieche, A. *CB* **1926**, *59*, 2159.
3. (a) Jacques, J.; Fouquey, C. *OS* **1988**, *67*, 1. (b) Jacques, J.; Fouquey, C.; Viterbo, R *TL* **1971**, 4617. (c) For the use of phenethylamine, see : Gong, B.; Chen, W.; Hu, B. *JOC* **1991**, *56*, 423.
4. (a) Brussee, J.; Groenendijk, J. L. G.; Koppele, J. M.; Jansen, A. C. A. *T* **1985**, *41*, 3313. (b) Smrcina, M.; Polakova, J.; Vyskocil, S.; Kocovsky, P. *JOC* **1993**, *58*, 4534.
5. Yamamoto, K.; Fukushima, H.; Nakazaki, M. *CC* **1984**, 1490.
6. Alper, H.; Hamel, N. *JACS* **1990**, *112*, 2803.
7. (a) Cram, D. J.; Cram, J. M. *ACR* **1978**, *11*, 8. (b) Helgeson, R. C.; Timko, J. M.; Moreau, P.; Peacock, S. C.; Mayer, J. M.; Cram, D. J. *JACS* **1974**, *96*, 6762.
8. Cram, D. J.; Sogah, G. D. Y. *CC* **1981**, 625.
9. Cram, D. J.; Sogah, G. D. Y. *JACS* **1985**, *107*, 8301.
10. Sepulchre, M.; Spassky, N. *Makromol. Chem. Rapid Commun.* **1981**, *2*, 261.
11. (a) Miyano, S.; Tobita, M.; Hashimoto, H. *BCJ* **1981**, *54*, 3522. (b) Miyano, S.; Fukushima, H.; Handa, S.; Ito, H.; Hashimoto, H. *BCJ* **1988**, *61*, 3249.
12. (a) Noyori, R. *CSR* **1989**, *18*, 187. (b) Noyori, R.; Tomino, I.; Yamada, M.; Nishizawa, M. *JACS* **1984**, *106*, 6717. (c) Noyori, R.; Tomino, I.; Tanimoto, Y.; Nishizawa, M. *JACS* **1984**, *106*, 6709. (d) Noyori, R.; Tomino, I.; Tanimoto, Y. *JACS* **1979**, *101*, 3129.
13. (a) Olivero, A. G.; Weidmann, B.; Seebach, D. *HCA* **1981**, *64*, 2485. (b) Seebach, D.; Beck, A. K.; Roggo, S.; Wonnacott, A. *CB* **1985**, *118*, 3673. (c) Wang, J. T.; Fan, X.; Feng, X.; Qian, Y. M. *S* **1989**, 291.
14. Reetz, M. T.; Kyung, S.-H.; Bolm, C.; Zierke, T. *CI(L)* **1986**, 824.
15. Seebach, D.; Beck, A. K.; Imwinkelried, R.; Roggo, S.; Wonnacott, A. *HCA* **1987**, *70*, 954.
16. (a) Mikami, K.; Terada, M.; Narisawa, S.; Nakai, T. *OS* **1992**, *71*, 14. (b) Mikami, K.; Terada, M.; Nakai, T. *JACS* **1990**, *112*, 3949.
17. Dijkgraff, C.; Rousseau, J. P. G. *Spectrochim. Acta* **1968**, *2*, 1213.
18. (a) Mikami, K.; Terada, M.; Motoyama, Y.; Nakai, T. *TA* **1991**, *2*, 643. (b) Mikami, K.; Motoyama, Y.; Terada, M. *JACS*, **1994**, *116*, 2812. (c) For the use of a modified BINOL-derived titanium complex, see: Maruoka, K.; Murase, N.; Yamamoto, H. *JOC* **1993**, *58*, 2938.
19. Kelly, T. R.; Whiting, A.; Chandrakumar, N. S. *JACS* **1986**, *108*, 3510.
20. Chapuis, C.; Jurczak, J. *HCA* **1987**, *70*, 436.
21. Kaufmann, D.; Boese, R. *AG(E)* **1990**, *29*, 545.
22. (a) Hattori, K.; Yamamoto, H. *T* **1993**, *49*, 1749. (b) For the use in imine-aldol reactions, see: Hattori, K.; Miyata, M.; Yamamoto, H. *JACS* **1993**, *115*, 1151.
23. (a) Maruoka, K.; Itoh, T.; Shirasaka, T.; Yamamoto, H. *JACS* **1988**, *110*, 310. (b) For the D.–A. reaction, see: Maruoka, K.; Concepcion, A. B.; Yamamoto, H. *BCJ* **1992**, *65*, 3501.
24. Terada, M.; Mikami, K.; Nakai, T. *TL* **1991**, *32*, 935.
25. (a) Glyoxylate-ene reaction with vinylic sulfides and selenides: Terada, M.; Matsukawa, S.; Mikami, K. *CC* **1993**, 327. (b) For the ene reaction of chloral, see: Maruoka, K.; Hoshino, Y.; Shirasaka, T.; Yamamoto, H. *TL* **1988**, *29*, 3967.
26. (a) Mikami, K.; Terada, M. *T* **1992**, *48*, 5671. (b) Terada, M.; Mikami, K.; Nakai, T. *CC* **1990**, 1623.
27. Mikami, K.; Narisawa, S.; Shimizu, M.; Terada, M. *JACS* **1992**, *114*, 6566.
28. Sakane, S.; Maruoka, K.; Yamamoto, H. *T* **1986**, *42*, 2203.
29. Sakane, S.; Fujiwara, J.; Maruoka, K.; Yamamoto, H. *JACS* **1983**, *105*, 6154.
30. Mikami, K.; Matsukawa, S. *JACS* **1993**, *115*, 7039.
31. Mikami, K.; Matsukawa, S. *JACS* **1994**, *116*, 4077.
32. (a) Sasai, H.; Suzuki, T.; Itoh, N.; Shibasaki, M. *TL* **1993**, *34*, 851. (b) Sasai, H.; Suzuki, T.; Arai, S.; Arai, T.; Shibasaki, M. *JACS* **1992**, *114*, 4418.
33. Aoki, S.; Mikami, K.; Terada, M.; Nakai, T. *T* **1993**, *49*, 1783.

34. (a) Costa, A. L.; Piazza, M. G.; Tagliavini, E.; Trombini, C.; Umani-Ronchi, A. *JACS* **1993**, *115*, 7001. (b) Keck, G. E.; Tarbet, K. H.; Geraci, L. S. *JACS* **1993**, *115*, 8467. (c) Keck, G. E.; Krishnamurthy, D.; Grier, M. C. *JOC* **1993**, *58*, 6543.

35. Maruoka, K.; Banno, H.; Yamamoto, H. *JACS* **1990**, *112*, 7791.

36. (a) Fuji, K.; Node, M.; Tanaka, F. *TL* **1990**, *31*, 6553. (b) Fuji, K.; Node, M.; Tanaka, F.; Hosoi, S. *TL* **1989**, *30*, 2825.

Koichi Mikami & Yukihiro Motoyama
Tokyo Institute of Technology, Japan

(*R*)-1,1′-Bi-2,2′-naphthotitanium Dichloride[1]

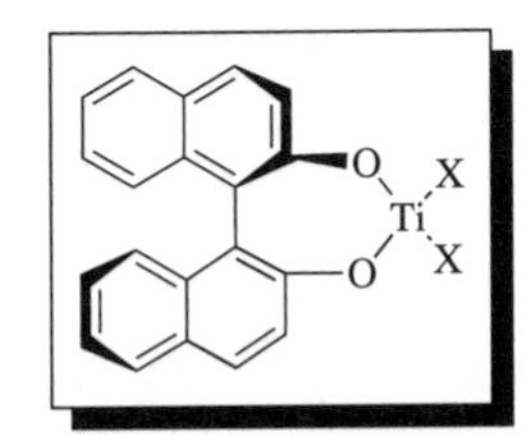

(X = Cl)		
[116051-73-1]	$C_{20}H_{12}Cl_2O_2Ti$	(MW 403.10)
(X = Br)		
[128030-80-8]	$C_{20}H_{12}Br_2O_2Ti$	(MW 492.00)
($X = ClO_4$)		
[138645-47-3]	$C_{20}H_{12}Cl_2O_{10}Ti$	(MW 531.09)
($X = OSO_2CF_3$)		
[139327-61-0]	$C_{22}H_{12}F_6O_8S_2Ti$	(MW 630.32)

(chiral Lewis acid for ene reactions,[2] Mukaiyama aldol reactions,[16] Diels–Alder reactions,[24] and cyanosilylations[27])

Alternate Name: BINOL-TiX_2.
Solubility: insol propionitrile; sol toluene, dichloromethane, and nitroethane.
Handling, Storage, and Precautions: titanium is reputed to be of low toxicity.

Introduction. The (*R*)-1,1′-bi-2,2′-naphthotitanium dihalides (BINOL-TiX_2; X = Br or Cl) are most conveniently prepared in situ from the reaction of diisopropoxytitanium dihalides ($(i\text{-PrO})_2TiX_2$; X = Br[2] or Cl[3]) with ***(R)-1,1′-Bi-2,2′-naphthol*** (BINOL) in the presence of molecular sieves (MS 4A) (eq 1).[2] When BINOL is mixed with ***Dichlorotitanium Diisopropoxide*** in the absence of MS 4A, almost no change is observed on the hydroxy-carbon signal of BINOL in the ^{13}C NMR spectrum. However, the addition of MS 4A to the solution of BINOL and $(i\text{-PrO})_2TiCl_2$ leads to a downfield shift of the hydroxy-carbon signal, indicating the formation of the BINOL-derived chiral catalyst. MS (zeolite) serves as an acid/base catalyst[4] and significantly facilitates the alkoxy ligand exchange in the in situ preparation of the chiral catalyst, BINOL-TiX_2. A 1:1 mixture of $(i\text{-PrO})_2TiX_2$ and (*R*)-BINOL in the presence of MS 4A in dichloromethane provides a red-brown solution. The molecularity of BINOL-TiX_2 in dichloromethane is ca. 2.0, depending on the concentration, particularly of homochiral (*R*)(*R*)- or (*S*)(*S*)-dimer which tends to dissociate to the monomer in lower concentration.[5]

OH, OH + $(i\text{-PrO})_2TiX_2$ —MS 4A, CH_2Cl_2, rt→ O, X, Ti, O, X (1)

X = Br or Cl

The chiral titanium complexes modified by the perchlorate or trifluoromethanesulfonate ligand such as (*R*)-1,1′-bi-2,2′-naphthotitanium diperchlorate (BINOL-$Ti(ClO_4)_2$) or (*R*)-1,1′-bi-2,2′-naphthotitanium ditriflate ((*R*)-BINOL-$Ti(OTf)_2$) can easily be prepared by the addition of ***Silver(I) Perchlorate*** or ***Silver(I) Trifluoromethanesulfonate*** (2 equiv) to BINOL-$TiCl_2$ (eq 2).[6]

O, Cl, Ti, O, Cl + AgX (2 equiv) → O, X, Ti, O, X (2)

$X = ClO_4$ or OTf

Asymmetric Catalysis of Carbonyl–Ene Reaction. (*R*)-1,1′-Bi-2,2′-naphthotitanium dihalides exhibit a remarkable level of asymmetric induction in the carbonyl–ene reaction of prochiral glyoxylate to provide practical access to α-hydroxy esters, a class of compounds of biological and synthetic importance[7] (eq 3).[2] The catalyst derived from (*R*)-BINOL leads consistently to the (*R*)-alcohol product, whereas the catalyst derived from (*S*)-BINOL affords the (*S*)-enantiomer. Generally speaking, the dibromide is superior to the dichloride in both reactivity and enantioselectivity for the reactions involving a methylene hydrogen shift in particular. On the other hand, the dichloride is lower in reactivity but superior in enantioselectivity for certain reactions involving methyl hydrogen shift. The present asymmetric catalysis is applicable to a variety of 1,1-disubstituted alkenes to provide the ene products in extremely high enantiomeric excess by judicious choice of the dibromo or dichloro catalyst. The reactions of mono- and 1,2-disubstituted alkenes afford no ene product. However, vinylic sulfides and selenides serve as alternatives to mono- and 1,2-disubstituted alkenes, giving the ene products with virtually complete enantioselectivity along with high diastereoselectivity (eq 4).[8] The synthetic advantage of vinylic sulfides and selenides is exemplified by the synthesis of enantiomerically pure (*R*)-(−)-ipsdienol, an insect aggregation pheromone.

+ H(C=O)CO_2Me —(*R*)-BINOL-TiX_2 (5 mol%), CH_2Cl_2, −30 °C→ HO, CO_2Me (3)

X = Br, 89%, 98% ee
X = Cl, 82%, 97% ee

(*R*)-BINOL-$TiCl_2$ (10 mol%), CH_2Cl_2, −30 °C, 94% (4)

X = S, R = *i*-Bu, 95% *anti*, >99% ee

Positive Nonlinear Effect[5] (Asymmetric Amplification[9]). A nonclassical phenomenon of asymmetric catalysis by the chiral BINOL-derived titanium complex is the remarkable positive nonlinear effect observed, which is of practical and mechanistic importance.[5] Convex deviation is observed from the usually assumed linear relationship between the enantiomeric purity of the BINOL ligand and the optical yield of the product. The glyoxylate–ene reaction catalyzed by the chiral titanium complex derived from a partially-resolved BINOL of 33.0% ee, for instance, provides the ene product with 91.4% ee in 92% chemical yield (eq 5). The optical yield thus obtained with a partially resolved BINOL ligand is not only much higher than the % ee of BINOL employed but is also very close to the value of 94.6% ee obtained using the enantiomerically pure BINOL. Thus the use of 35–40% ee of BINOL is sufficient to provide the equally high (>90% ee) level obtained with enantiomerically pure BINOL.

(*R*)-BINOL-$TiBr_2$ (1 mol%, 33% ee), CH_2Cl_2, −30 °C, 92% (5)

91.4% ee

Asymmetric Desymmetrization.[10] Desymmetrization of an achiral, symmetrical molecule is a potentially powerful but relatively unexplored concept for the asymmetric catalysis of carbon–carbon bond formation. While the ability of enzymes to differentiate between enantiotopic functional groups is well known,[11] little is known about the similar ability of nonenzymatic catalysts to effect carbon–carbon bond formation. The desymmetrization by the enantiofacial selective carbonyl–ene reaction of prochiral ene substrates with planar symmetry provides an efficient access to remote internal[12] asymmetric induction which is otherwise difficult to attain (eq 6).[10] The (2*R*,5*S*)-*syn* product is obtained in >99% ee along with more than 99% diastereoselectivity. The desymmetrized product thus obtained can be transformed stereoselectively by a more classical diastereoselective reaction (e.g. hydroboration).

(*R*)-BINOL-$TiCl_2$ (10 mol%); 1. 9-BBN; 2. H_2O_2, NaOH (6)

>99% *syn*, >99% ee

Kinetic Resolution.[13] On the basis of the desymmetrization concept, the kinetic resolution of a racemic substrate might be recognized as an intermolecular desymmetrization.[10] The kinetic resolution of a racemic allylic ether by the glyoxylate–ene reaction also provides an efficient access to remote relative[12] asymmetric induction. Both the dibromide and dichloride catalysts provide the (2*R*,5*S*)-*syn* product with >99% diastereoselectivity along with more than 95% ee (eq 7). The high diastereoselectivity, coupled with the high % ee, strongly suggests that the catalyst/glyoxylate complex efficiently discriminates between the two enantiomeric substrates to accomplish effective kinetic resolution. In fact, the relative rates with racemic ethers are quite large, ca. 60 and 700, respectively. As expected, the reaction of (*S*)-ene using the catalyst (*R*)-BINOL-$TiCl_2$ ('matched' catalytic system) provides complete (>99%) 1,4-*syn* diastereoselectivity in high chemical yield, whereas the reaction of (*R*)-ene using (*R*)-BINOL-$TiCl_2$ ('mismatched' catalytic system) affords a diastereomeric mixture in quite low yield (eq 8).

(*R*)-BINOL-$TiCl_2$ (10 mol%), CH_2Cl_2, rt (7)

R = *i*-Pr, >99% *syn*, 99.5% ee; 59.4% ee
R = Me, >99% *syn*, 96.2% ee; 22.0% ee

(*R*)-BINOL-$TiCl_2$ (10 mol%), CH_2Cl_2, rt, 70% (8)

>99% *syn*

Ene Cyclization.[14] The asymmetric catalysis of the intramolecular carbonyl–ene reaction not only of type (3,4) but also (2,4) employs the BINOL-derived titanium complexes ((*R*)-BINOL-TiX_2; X = ClO_4 or OTf), modified by the perchlorate and trifluoromethanesulfonate ligands.[6] The *trans*-tetrahydropyran is thus preferentially obtained in 84% ee (eq 9). The seven-membered cyclization of type 7-(2,4) gives the oxepane in high ee, where the *gem*-dimethyl groups are unnecessary (eq 10).

(*R*)-BINOL-$Ti(ClO_4)_2$ (20 mol%), CH_2Cl_2, 0 °C (9)

80% *trans*, 84% ee

(*R*)-BINOL-TiX_2 (20 mol%), CH_2Cl_2, rt (10)

X = ClO_4, 91% ee
X = OTf, 92% ee

Mukaiyama Aldol Condensation. As expected, the chiral titanium complex is also effective for a variety of carbon–carbon bond forming processes such as the aldol and the Diels–Alder reactions. The aldol process constitutes one of the most fundamental bond constructions in organic synthesis.[15] Therefore the development of chiral catalysts that promote asymmetic aldol reactions in a highly stereocontrolled and truly catalytic fashion has attracted much attention, for which the silyl enol ethers of ketones or esters have been used as a storable enolate component (Mukaiyama aldol condensation). The BINOL-derived titanium complex BINOL-$TiCl_2$ can be used as an efficient catalyst for the Mukaiyama-type aldol reaction of not only ketone silyl enol ethers but also ester silyl enol ethers with control of absolute and relative stereochemistry (eq 11).[16]

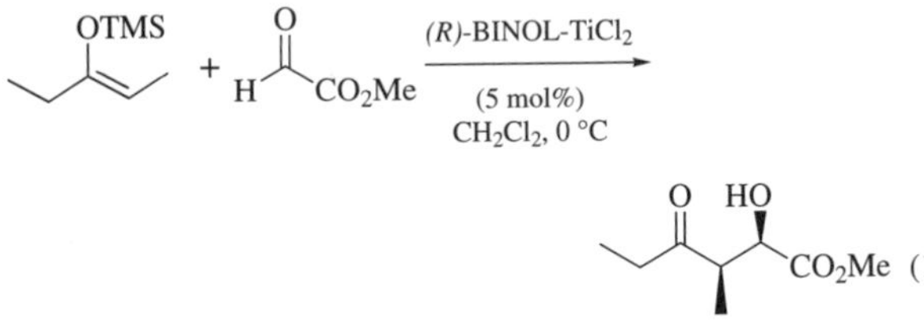

OTMS + H(C=O)CO_2Me —(*R*)-BINOL-$TiCl_2$ (5 mol%), CH_2Cl_2, 0 °C→ (11)

98% *syn*, 99% ee

Carbonyl Addition of Allylic Silanes and Stannanes.[17] The chiral titanium complex BINOL-$TiCl_2$ also catalyzes the asymmetric carbonyl addition reaction of allylic silanes and stannanes.[18] Thus the addition reaction of glyoxylate with (*E*)-2-butenylsilane and -stannane proceeds smoothly to give the *syn* product in high enantiomeric excess (eq 12). The *syn* product thus obtained can be readily converted to the lactone portion of verrucaline A. The reaction of aliphatic and aromatic aldehydes with allylstannane is also catalyzed by BINOL-$TiCl_2$ to give remarkably high enantioselectivity (eq 13).[19]

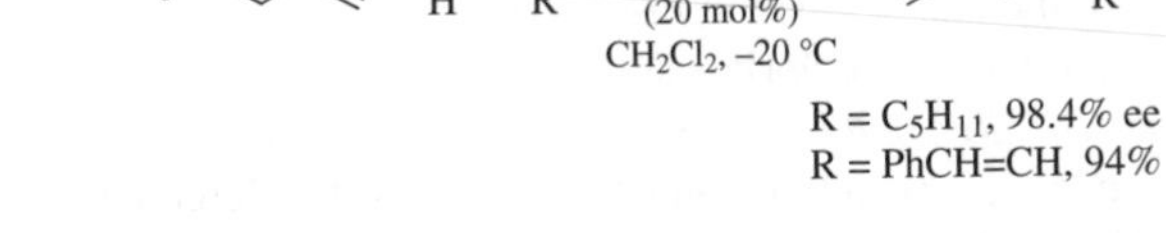

M + H(C=O)CO_2R —(*S*)-BINOL-$TiCl_2$ (10 mol%), CH_2Cl_2, −60 °C→ (12)

M = $SnBu_3$, R = *n*-Bu, 84% *syn*, 86% ee
M = $SiMe_3$, R = Me, 83% *syn*, 80% ee

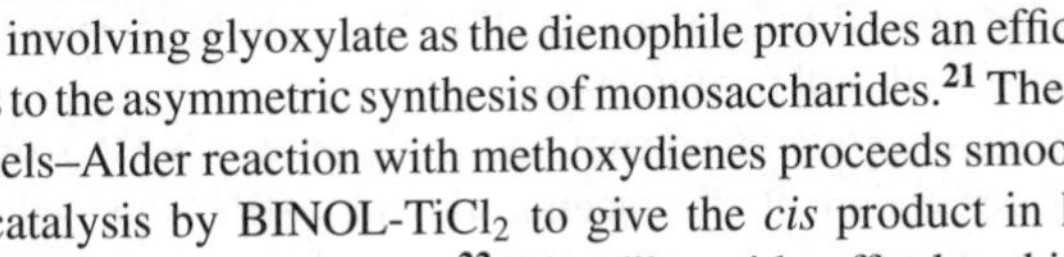

Bu_3Sn + H(C=O)R —(*S*)-BINOL-$TiCl_2$ (20 mol%), CH_2Cl_2, −20 °C→ (13)

R = C_5H_{11}, 98.4% ee
R = PhCH=CH, 94% ee

Hetero Diels–Alder Reaction.[20] The hetero-Diels–Alder reaction involving glyoxylate as the dienophile provides an efficient access to the asymmetric synthesis of monosaccharides.[21] The hetero Diels–Alder reaction with methoxydienes proceeds smoothly with catalysis by BINOL-$TiCl_2$ to give the *cis* product in high enantiomeric excess (eq 14).[22] The dibromide affords a higher *cis* selectivity, however, with a lower enantioselectivity, particularly in the *trans* adduct. The product thus obtained can be readily converted to the lactone portion of HMG-CoA inhibitors such as mevinolin or compactin.[23]

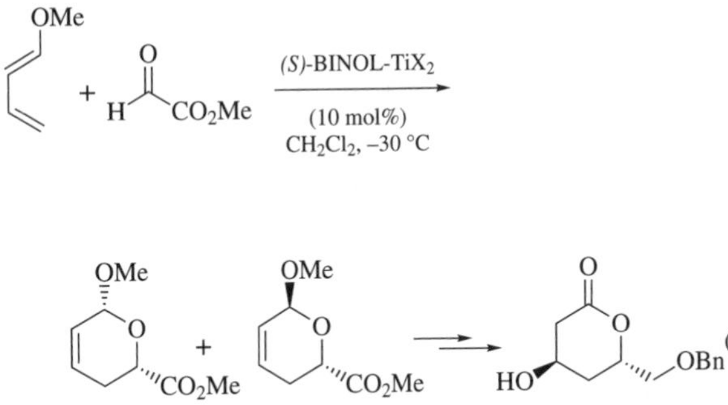

OMe diene + H(C=O)CO_2Me —(*S*)-BINOL-TiX_2 (10 mol%), CH_2Cl_2, −30 °C→ (14)

X = Cl ,78 (94% ee):22 (>90% ee)
X = Br, 84 (92% ee):16 (50% ee)

Diels–Alder Reaction.[24] The Diels–Alder reaction of methacrolein with 1,3-dienol derivatives can also be catalyzed by the chiral BINOL-derived titanium complex BINOL-$TiCl_2$. The *endo* adduct was obtained in high enantioselectivity (eq 15).[22a,25] The sense of asymmetric induction is exactly the same as observed for the asymmetric catalytic reactions shown above. Asymmetric catalytic Diels–Alder reactions with naphthoquinone derivatives as the dienophile provide an efficient entry to the asymmetric synthesis of anthracyclinone aglycones (eq 16).[26]

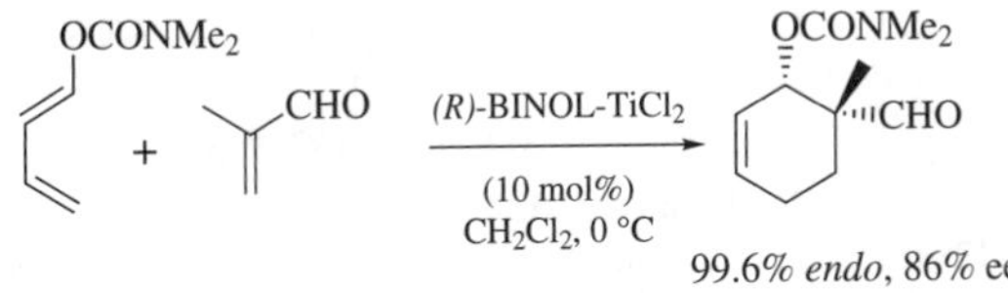

$OCONMe_2$ diene + CHO —(*R*)-BINOL-$TiCl_2$ (10 mol%), CH_2Cl_2, 0 °C→ (15)

99.6% *endo*, 86% ee

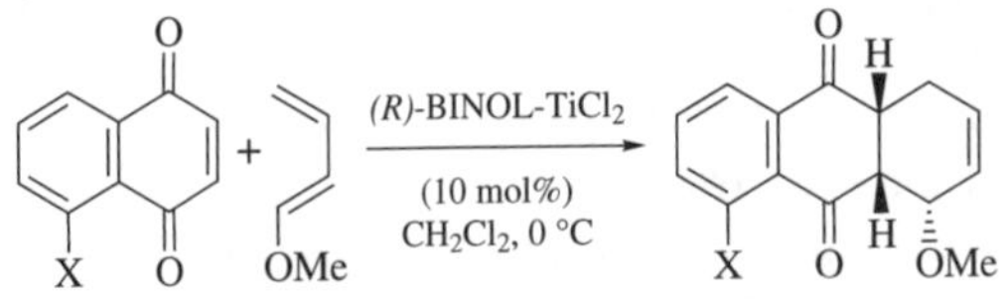

—(*R*)-BINOL-$TiCl_2$ (10 mol%), CH_2Cl_2, 0 °C→ (16)

X = H, 85% ee

Cyanosilylation.[27] Another preparative procedure of BINOL-$TiCl_2$ and the use thereof was reported in the asymmetric catalysis of the addition reaction of cyanotrimethylsilane to aldehydes.[28] The dilithium salt of BINOL in ether was treated with ***Titanium(IV) Chloride***, the red-brown mixture was warmed to room temperature, and the ether removed in vacuo. Dry benzene was added and the nondissolved solid was separated via filtration under nitrogen. Removal of the solvent delivered 50% of a sensitive red-brown solid which showed a single set of ^{13}C NMR signals (eq 17). The BINOL-$TiCl_2$ thus obtained was utilized to prepare the cyanohydrin of 3-methylbutanal in <82% ee (eq 18).

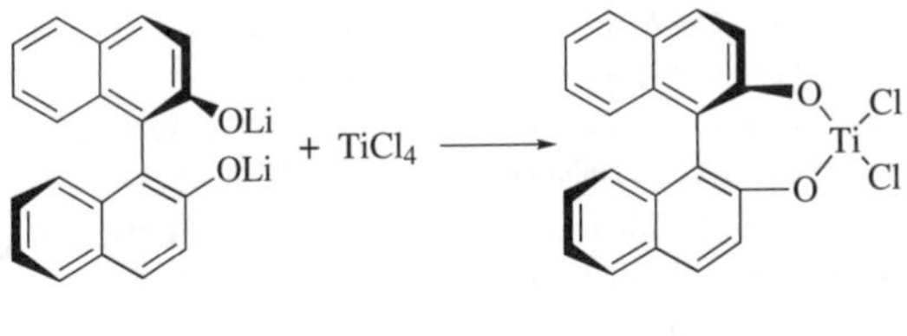

OLi OLi + $TiCl_4$ ——→ (17)

$$\text{Me}_2\text{CHCH}_2\text{CHO} + \text{Me}_3\text{SiCN} \xrightarrow[\text{toluene, –78 °C to rt; 85\%}]{(S)\text{-BINOL-TiCl}_2\ (20\ \text{mol\%})} \text{Me}_2\text{CHCH}_2\text{CH(OH)CN}\ (\leq 82\%\ \text{ee}) \quad (18)$$

Related Reagents. (*R*)-1,1′-Bi-2,2′-naphthol; (*R*)-1,1′-Bi-2,2′-naphthotitanium Diisopropoxide; Titanium(IV) Chloride.

1. (a) Mikami, K.; Shimizu, M. *CRV* **1992**, *92*, 1021. (b) Mikami, K.; Terada, M.; Narisawa, S.; Nakai, T. *SL* **1992**, 255.
2. (a) Mikami, K.; Terada, M.; Narisawa, S.; Nakai, T. *OS* **1992**, *71*, 14. (b) Mikami, K.; Terada, M.; Nakai, T. *JACS* **1990**, *112*, 3949. (c) Terada, M.; Nakai, T. *JACS* **1989**, *111*, 1940.
3. Dijkgaaf, C.; Rousseau, J. P. G. *Spectrochim. Acta* **1968**, *24A*, 1213.
4. (a) Thomas, J. M.; Theocaris, C. R. *Modern Synthetic Methods*; Springer: Berlin, 1989. (b) Onaka, M.; Izumi, Y. *Yuki Gosai Kagaku Kyokaishi* **1989**, *47*, 233. (c) Dyer, A. *An Introduction to Zeolite Molecular Sieves*; Wiley: Chichester, 1988.
5. (a) Mikami, K.; Terada, M. *T* **1992**, *48*, 5671. (b) Terada, M.; Mikami, K.; Nakai, T. *CC* **1990**, 1623.
6. (a) Mikami, K.; Sawa, E.; Terada, M. *TA* **1991**, *2*, 1403. (b) Mikami, K.; Terada, M.; Sawa, E.; Nakai, T. *TL* **1991**, *32*, 6571.
7. (a) Omura, S. *J. Synth. Org. Chem., Jpn.* **1986**, *44*, 127; (b) Hanessian, S. *Total Synthesis of Natural Products: The 'Chiron' Approach*; Pergamon: Oxford, 1983. (c) Seebach, D.; Hungerbuhler, E. *Modern Synthetic Methods*; Otto Salle: Frankfurt am Main, 1980.
8. Terada, M.; Matsukawa, S.; Mikami, K. *CC* **1993**, 327.
9. (a) Noyori, R.; Kitamura, M. *AG(E)* **1991**, *30*, 49. (b) Wynberg, H. *C* **1989**, *43*, 150. (c) Puchot, C.; Samuel, O.; Duñach, E.; Zhao, S.; Agami, C.; Kagan, H. B. *JACS* **1986**, *108*, 2353.
10. Mikami, K.; Narisawa, S.; Shimizu, M.; Terada, M. *JACS* **1992**, *114*, 6566.
11. Ward, R. S. *CSR* **1990**, *19*, 1.
12. Bartlett, P. A. *T* **1980**, *36*, 3.
13. (a) Kagan, H. B.; Fiaud, J. C. *Top. Stereochem.* **1988**, *18*, 249. (b) Brown, J. M. *CI(L)* **1988**, 612.
14. (a) Oppolzer, W.; Snieckus, V. *AG(E)* **1978**, *17*, 476. (b) Taber, D. F. *Intramolecular Diels–Alder and Alder Ene Reactions*; Springer: Berlin, 1984.
15. (a) Masamune, S.; Choy, W.; Peterson, J. S.; Sita, L. R. *AG(E)* **1985**, *24*, 1. (b) Heathcock, C. H. In *Asymmetric Synthesis*; Morrison, J. D., Ed.; Academic: New York, 1984. (c) Evans, D. A.; Nelson, J. V.; Taber, T. R. *Top. Stereochem.* **1982**, *13*, 1. (d) Mukaiyama, T. *OR* **1982**, *28*, 203.
16. Mikami, K.; Matsukawa, S. *JACS* **1993**, *115*, 7039; **1994**, *116*, 4077.
17. (a) Sakurai, H. *SL* **1989**, 1. (b) Hosomi, A. *ACR* **1988**, *21*, 200. (c) Yamamoto, Y. *ACR* **1987**, *20*, 243. (d) Hoffmann, R. W. *AG(E)* **1982**, *21*, 555.
18. Aoki, S.; Mikami, K.; Terada, M.; Nakai, T. *T* **1993**, *49*, 1783. Also see: Mikami, K.; Matsukawa, S. *TL* **1994**, *35*, 3133.
19. Costa, A. L.; Piazza, M. G.; Tagliavini, E.; Trombini, C.; Umani-Ronchi, A. *JACS* **1993**, *115*, 7001.
20. (a) Bednarski, M. D.; Lyssikatos, J. P. *COS*, **1991**, *2*, Chapter 2.5. (b) Boger, D. L.; Weinreb, S. M. *Hetero-Diels–Alder Methodology in Organic Synthesis*; Academic: New York, 1987. (c) Konowal, A.; Jurczak, J.; Zamojski, A. *T* **1976**, *32*, 2957.
21. (a) Konowal, A.; Jurczak, J.; Zamojski, A. *T* **1976**, *32*, 2957. (b) Danishefsky, S. J.; DeNinno, M. P. *AG(E)* **1987**, *26*, 15.
22. (a) Mikami, K.; Motoyama, Y.; Terada, M. *JACS* **1994**, *116*, 2812. (b) Terada, M.; Mikami, K.; Nakai, T, *TL* **1991**, *32*, 935.
23. Rosen, T.; Heathcock, C. H. *T* **1986**, *42*, 4909.
24. (a) Kagan, H. B.; Riant, O. *CRV* **1992**, *92*, 1007. (b) Oppolzer, W. *COS*, **1991**, *5*, Chapter 1.2. (c) Fringuelli, F.; Taticchi, A. *Dienes in the Diels–Alder Reaction*; Wiley: New York, 1990. (d) Taschner, M. J. *Org. Synth. Theory Appl.* **1989**, *1*, 1. (e) Paquette, L. A. In *Asymmetric Synthesis*, Morrison, J. D., Ed.; Academic: New York, 1984; Vol. 3B, Chapter 7.
25. Mikami, K.; Terada, M.; Motoyama, Y.; Nakai, T. *TA* **1991**, *2*, 643.
26. (a) Krohn, K. *T* **1990**, *46*, 291. (b) Krohn, K. *AG(E)* **1986**, *25*, 790. (c) Broadhurst, M. J.; Hassall, C. H.; Thomas, G. J. *CI(L)* **1985**, 106. (d) Arcamone, F. *Med. Res. Rev.* **1984**, *4*, 153.
27. Rasmussen, J. K.; Heilmann, S. M.; Krepski, L. R. *Adv. Silicon Chem.* **1991**, *1*, 65.
28. Reetz, M. T.; Kyung, S.-H.; Bolm, C.; Zierke, T. *CI(L)* **1986**, 824.

Koichi Mikami
Tokyo Institute of Technology, Japan

(*R*)- and (*S*)-2,2′-Bis(diphenylphosphino)-1,1′-binaphthyl[1]

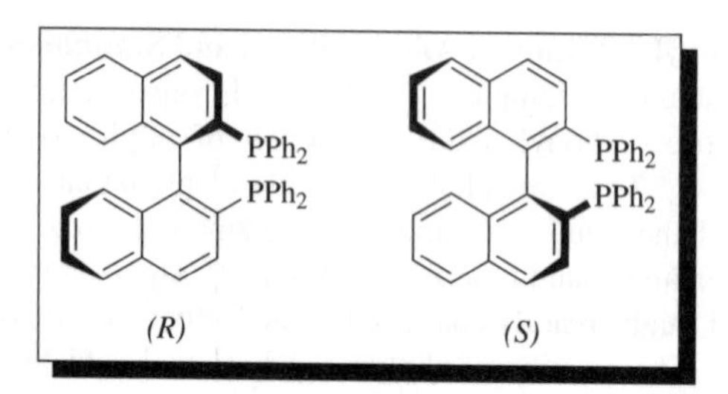

[76189-55-4] $C_{44}H_{32}P_2$ (MW 622.70)

(chiral diphosphine ligand for transition metals;[2] the complexes show high enantioselectivity band reactivity in a variety of organic reactions)

Alternate Name: BINAP.
Physical Data: mp 241–242 °C; $[\alpha]_D^{25}$ –229° (c = 0.312, benzene) for (*S*)-BINAP.[3]
Solubility: sol THF, benzene, dichloromethane; modestly sol ether, methanol, ethanol; insol water.
Form Supplied in: colorless solid.
Analysis of Reagent Purity: GLC analysis (OV-101, capillary column, 5 m, 200–280 °C) and TLC analysis (E. Merck Kieselgel 60 PF_{254}, 1:19 methanol–chloroform); R_f 0.42 (BINAPO, dioxide of BINAP), 0.67 (monoxide of BINAP), and 0.83 (BINAP). The optical purity of BINAP is analyzed after oxidizing to BINAPO by HPLC using a Pirkle column (Baker bond II) and a hexane/ethanol mixture as eluent.[3]
Preparative Methods: enantiomerically pure BINAP is obtained by resolution of the racemic dioxide, BINAPO, with camphorsulfonic acid or 2,3-di-*O*-benzoyltartaric acid followed by deoxygenation with ***Trichlorosilane*** in the presence of ***Triethylamine***.[3]
Handling, Storage, and Precautions: solid BINAP is substantially stable to air, but bottles of BINAP should be flushed with N_2

or Ar and kept tightly closed for prolonged storage. BINAP is slowly air oxidized to the monoxide in solution.

BINAP–Ru^{II} Catalyzed Asymmetric Reactions. Halogen-containing BINAP–Ru complexes are most simply prepared by reaction of $[RuCl_2(cod)]_n$ or $[RuX_2(arene)]_2$ (X = Cl, Br, or I) with BINAP.[4] Sequential treatment of $[RuCl_2(benzene)]_2$ with BINAP and sodium carboxylates affords Ru(carboxylate)$_2$(BINAP) complexes. The dicarboxylate complexes, upon treatment with strong acid HX,[5] can be converted to a series of Ru complexes empirically formulated as RuX_2(BINAP). These Ru^{II} complexes act as catalysts for asymmetric hydrogenation of various achiral and chiral unsaturated compounds.

α,β-Unsaturated carboxylic acids are hydrogenated in the presence of a small amount of $Ru(OAc)_2$(BINAP) to give the corresponding optically active saturated products in quantitative yields.[6] The reaction is carried out in methanol at ambient temperature with a substrate:catalyst (S:C) ratio of 100–600:1. The sense and degree of the enantioface differentiation are profoundly affected by hydrogen pressure and the substitution pattern of the substrates. Tiglic acid is hydrogenated quantitatively with a high enantioselectivity under a low hydrogen pressure (eq 1), whereas naproxen, a commercial anti-inflammatory agent, is obtained in 97% ee under high pressure (eq 2).[6a]

4 atm H_2, $Ru(OAc)_2$[(*R*)-BINAP], MeOH (1)

91% ee

135 atm H_2, $Ru(OAc)_2$[(*S*)-BINAP], MeOH (2)

97% ee

Enantioselective hydrogenation of certain α- and β-(acylamino)acrylic acids or esters in alcohols under 1–4 atm H_2 affords the protected α- and β-amino acids, respectively (eqs 3 and 4).[2a,7] Reaction of *N*-acylated 1-alkylidene-1,2,3,4-tetrahydroisoquinolines provides the 1*R*- or 1*S*-alkylated products. This method allows a general asymmetric synthesis of isoquinoline alkaloids (eq 5).[8]

100 atm H_2, $RuBr_2$[(*R*)-BINAP], MeOH (3)

85% ee

1 atm H_2, $Ru(OAc)_2$[(*R*)-BINAP], MeOH (4)

96% ee

1–4 atm H_2, (*R*)-BINAP–Ru, $EtOH–CH_2Cl_2$ (5)

92–100% ee

Geraniol or nerol can be converted to citronellol in 96–99% ee in quantitative yield without saturation of the C(6)–C(7) double bond (eq 6).[9] The S:C ratio approaches 50 000. The use of alcoholic solvents such as methanol or ethanol and initial H_2 pressure greater than 30 atm is required to obtain high enantioselectivity. Diastereoselective hydrogenation of the enantiomerically pure allylic alcohol with an azetidinone skeleton proceeds at atmospheric pressure in the presence of an (*R*)-BINAP–Ru complex to afford the β-methyl product, a precursor of 1β-methylcarbapenem antibiotics (eq 7).[10] Racemic allylic alcohols such as 3-methyl-2-cyclohexenol and 4-hydroxy-2-cyclopentenone can be effectively resolved by the BINAP–Ru-catalyzed hydrogenation (eq 8).[11]

100 atm H_2, $Ru(OAc)_2$[(*R*)-BINAP], MeOH (6)

99% ee

1 atm H_2, $Ru(OAc)_2$[(*R*)-TolBINAP], MeOH (7)

β:α = 99.9:0.1

4 atm H_2, $Ru(OAc)_2$[(*R*)-BINAP], MeOH (8)

46% recovery
>99% ee

Diketene is quantitatively hydrogenated to 3-methyl-3-propanolide in 92% ee (eq 9). Certain 4-methylene- and 2-alkylidene-4-butanolides as well as 2-alkylidenecyclopentanone are also hydrogenated with high enantioselectivity.[12]

100 atm H_2, [RuCl[(*S*)-BINAP](C_6H_6)]Cl, NEt_3, THF (9)

92% ee

Hydrogenation with halogen-containing BINAP–Ru complexes can convert a wide range of functionalized prochiral ketones to stereo-defined secondary alcohols with high enantiomeric purity (eq 10).[13] 3-Oxocarboxylates are among the most appropriate substrates.[13a,4d] For example, the enantioselective hydrogenation of methyl 3-oxobutanoate proceeds quantitatively in methanol with an S:C ratio of 1000–10 000 to give the hydroxy ester product in nearly 100% ee (eq 11). Halogen-containing complexes $RuX_2(BINAP)$ (X = Cl, Br, or I; polymeric form) or $[RuCl_2(BINAP)]_2NEt_3$ are used as the catalysts. Alcohols are the solvents of choice, but aprotic solvents such as dichloromethane can also be used. At room temperature the reaction requires an initial H_2 pressure of 20–100 atm, but at 80–100 °C the reaction proceeds smoothly at 4 atm H_2.[4c,4d]

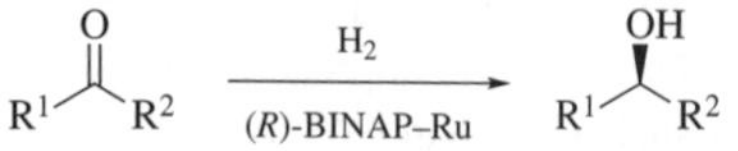

H_2, (*R*)-BINAP–Ru (10)

R^1 = alkyl, aryl; R^2 = CH_2OH, CH_2NMe_2, CH_2CH_2OH, CH_2Ac, CH_2CO_2R, CH_2COSR, CH_2CONR_2, $CH_2CH_2CO_2R$, etc.

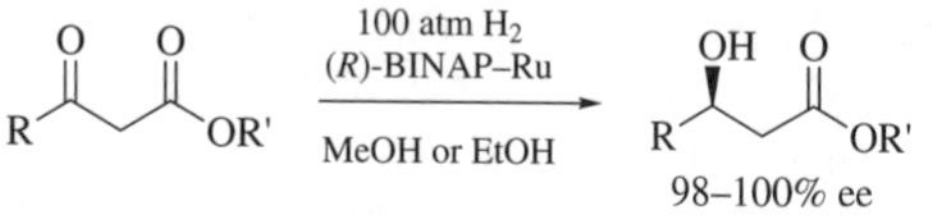

100 atm H_2, (*R*)-BINAP–Ru, MeOH or EtOH (11)

98–100% ee

R = Me, Et, Bu, *i*-Pr; R' = Me, Et, *i*-Pr, *t*-Bu

3-Oxocarboxylates possessing an additional functional group can also be hydrogenated with high enantioselectivity by choosing appropriate reaction conditions or by suitable functional group modification (eq 12).[13b,13c]

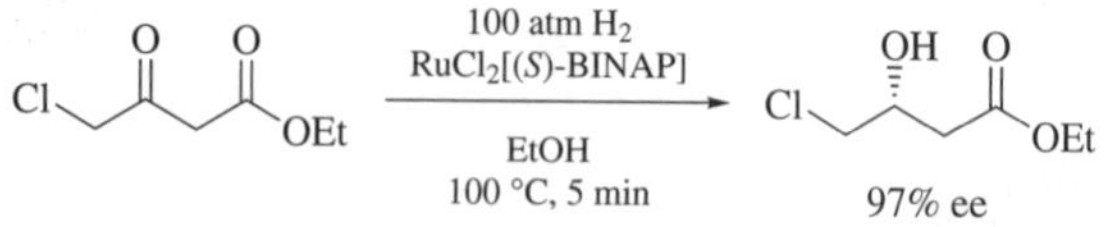

100 atm H_2, $RuCl_2[(S)\text{-BINAP}]$, EtOH, 100 °C, 5 min (12)

97% ee

The pre-existing stereogenic center in the chiral substrates profoundly affects the stereoselectivity. The (*R*)-BINAP–Ru-catalyzed reaction of (*S*)-4-(alkoxycarbonylamino)-3-oxocarboxylates give the statine series with (3*S*,4*S*) configuration almost exclusively (eq 13).[14]

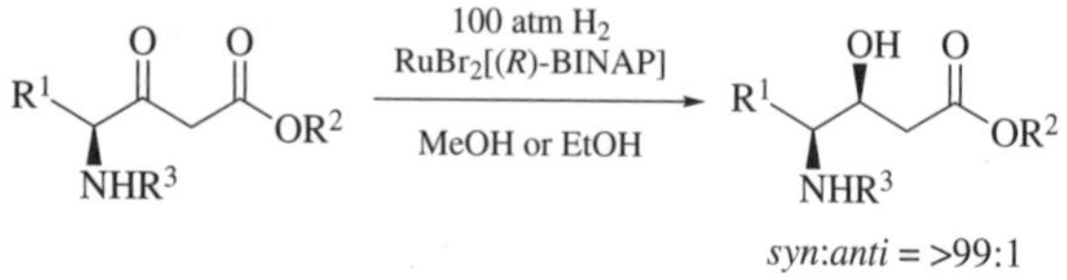

100 atm H_2, $RuBr_2[(R)\text{-BINAP}]$, MeOH or EtOH (13)

syn:*anti* = >99:1

Hydrogenation of certain racemic 2-substituted 3-oxocarboxylates occurs with high diastereo- and enantioselectivity via dynamic kinetic resolution involving in situ racemization of the substrates.[15] The (*R*)-BINAP–Ru-catalyzed reaction of 2-acylamino-3-oxocarboxylates in dichloromethane allows preparation of threonine and DOPS (anti-Parkinsonian agent) (eq 14).[16] In addition, a common intermediate for the synthesis of carbapenem antibiotics is prepared stereoselectively on an industrial scale from a 3-oxobutyric ester (**1**) with an acylaminomethyl substituent at the C(2) position.[16a] The second-order stereoselective hydrogenation of 2-ethoxycarbonylcycloalkanones gives predominantly the *trans* hydroxy esters (**2**) in high ee, whereas 2-acetyl-4-butanolide is hydrogenated to give the *syn* diastereomer (**3**).[17]

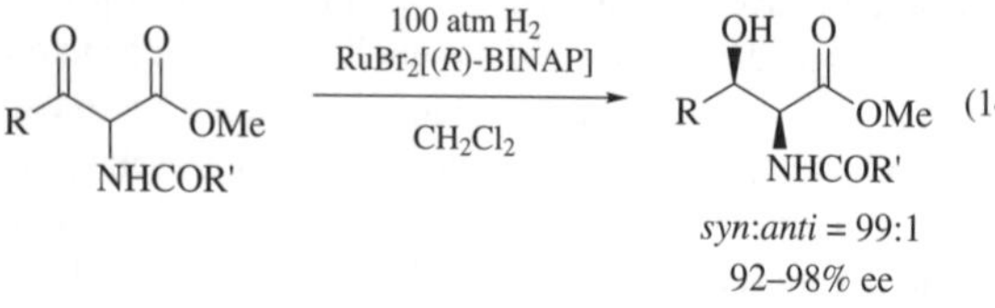

100 atm H_2, $RuBr_2[(R)\text{-BINAP}]$, CH_2Cl_2 (14)

syn:*anti* = 99:1
92–98% ee

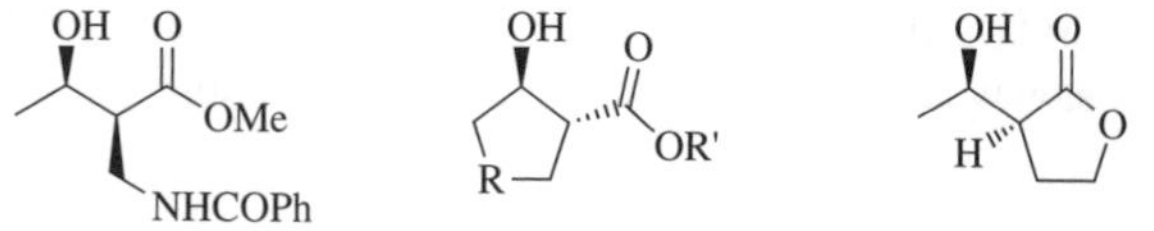

(**1**) *syn*:*anti* = 94:6, 98% ee

(**2**) R = CH_2, $(CH_2)_2$, $(CH_2)_3$; *trans*:*cis* = 93:7–99:1; 90–93% ee

(**3**) *syn*:*anti* = 98:2, 94% ee

Certain 1,2- and 1,3-diketones are doubly hydrogenated to give stereoisomeric diols. 2,4-Pentanedione, for instance, affords (*R*,*R*)- or (*S*,*S*)-2,4-pentanediol in nearly 100% ee accompanied by 1% of the *meso* diol.[13b]

A BINAP–Ru complex can hydrogenate a C=N double bond in a special cyclic sulfonimide to the sultam with >99% ee.[18]

The asymmetric transfer hydrogenation of the unsaturated carboxylic acids using formic acid or alcohols as the hydrogen source is catalyzed by $Ru(acac\text{-}F_6)(\eta^3\text{-}C_3H_5)(BINAP)$ or $[RuH(BINAP)_2]PF_6$ to produce the saturated acids in up to 97% ee (eq 15).[19]

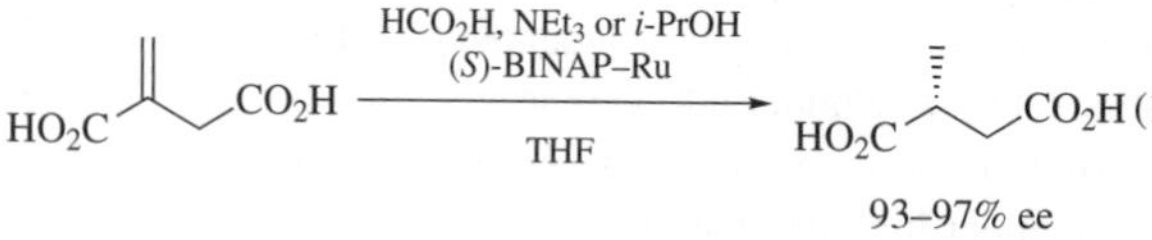

HCO_2H, NEt_3 or *i*-PrOH, (*S*)-BINAP–Ru, THF (15)

93–97% ee

BINAP–Ru complexes promote addition of arenesulfonyl chlorides to alkenes in 25–40% optical yield.[20]

BINAP–Rh^I Catalyzed Asymmetric Reactions. The rhodium(I) complexes $[Rh(BINAP)(cod)]ClO_4$, $[Rh(BINAP)(nbd)]ClO_4$, and $[Rh(BINAP)_2]ClO_4$, are prepared from $[RhCl(cod)]_2$ or ***Bis(bicyclo[2.2.1]hepta-2,5-diene)dichlorodirhodium*** and BINAP in the presence of $AgClO_4$.[21] $[Rh(BINAP)S_2]ClO_4$ is prepared by reaction of [Rh(BINAP)(cod or nbd)]ClO_4 with atmospheric pressure of hydrogen in an appropriate solvent, S.[21a] BINAP–Rh complexes catalyze a variety of asymmetric reactions.[2]

Prochiral α-(acylamino)acrylic acids or esters are hydrogenated under an initial hydrogen pressure of 3–4 atm to give the protected amino acids in up to 100% ee (eq 16).[21a] The BINAP–Rh catalyst was used for highly diastereoselective hydrogenation of a chiral homoallylic alcohol to give a fragment of the ionophore ionomycin.[22]

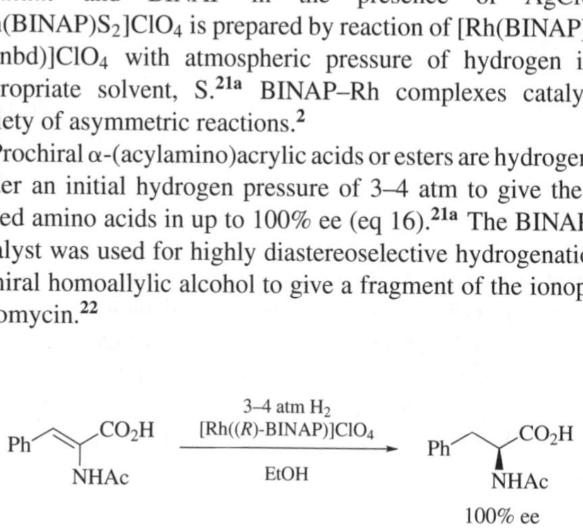

3–4 atm H_2, $[Rh((R)\text{-BINAP})]ClO_4$, EtOH (16)

100% ee

The cationic BINAP–Rh complexes catalyze asymmetric 1,3-hydrogen shifts of certain alkenes. Diethylgeranylamine can be quantitatively isomerized in THF or acetone to citronellal diethylenamine in 96–99% ee (eq 17).[23] This process is the key step in the industrial production of (−)-menthol. In the presence of a cationic (*R*)-BINAP–Rh complex, (*S*)-4-hydroxy-2-cyclopentenone is isomerized five times faster than the (*R*) enantiomer, giving a chiral intermediate of prostaglandin synthesis.[24]

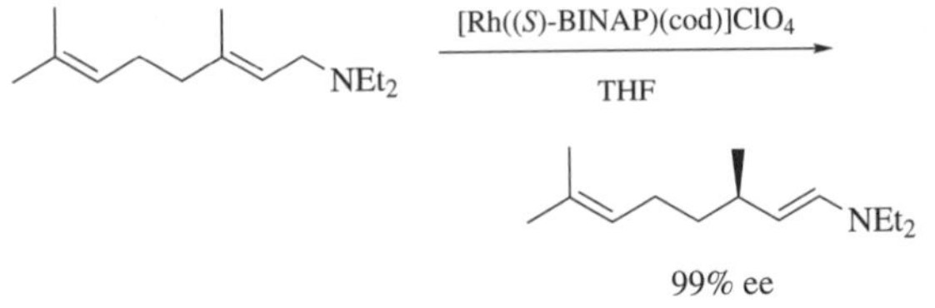

NEt_2 → [Rh((*S*)-BINAP)(cod)]ClO_4, THF → NEt_2 (17)

99% ee

Enantioselective cyclization of 4-substituted 4-pentenals to 3-substituted cyclopentanones in >99% ee is achieved with a cationic BINAP–Rh complex (eq 18).[25]

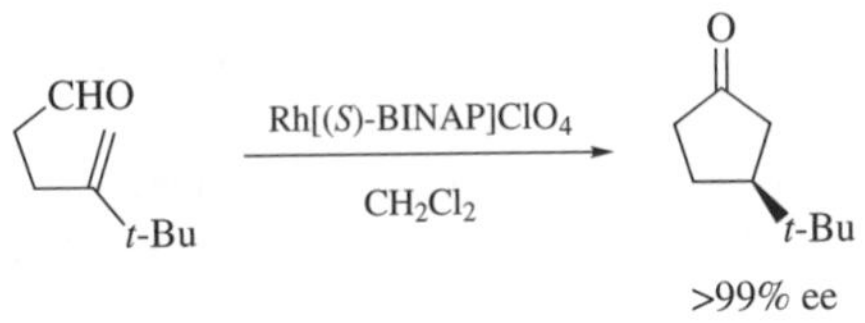

CHO, *t*-Bu → Rh[(*S*)-BINAP]ClO_4, CH_2Cl_2 → O, *t*-Bu (18)

>99% ee

Reaction of styrene and catecholborane in the presence of a BINAP–Rh complex at low temperature forms, after oxidative workup, 1-phenylethyl alcohol in 96% ee (eq 19).[26]

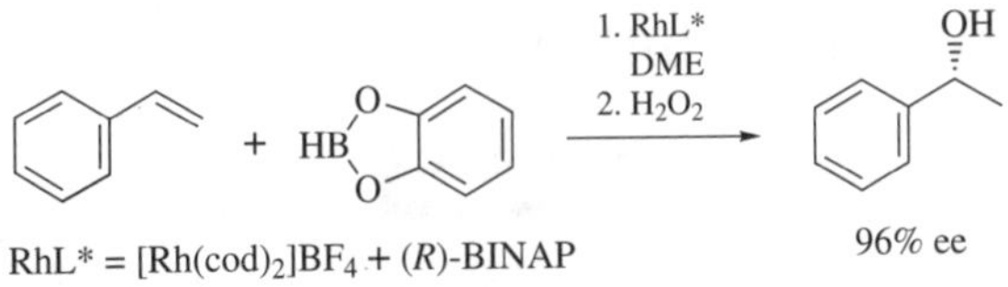

+ HB(O)(O) → 1. RhL*, DME; 2. H_2O_2 → OH (19)

RhL* = [Rh(cod)$_2$]BF_4 + (*R*)-BINAP

96% ee

Neutral BINAP–Rh complexes catalyze intramolecular hydrosilylation of alkenes. Subsequent ***Hydrogen Peroxide*** oxidation produces the optically active 1,3-diol in up to 97% ee (eq 20).[27]

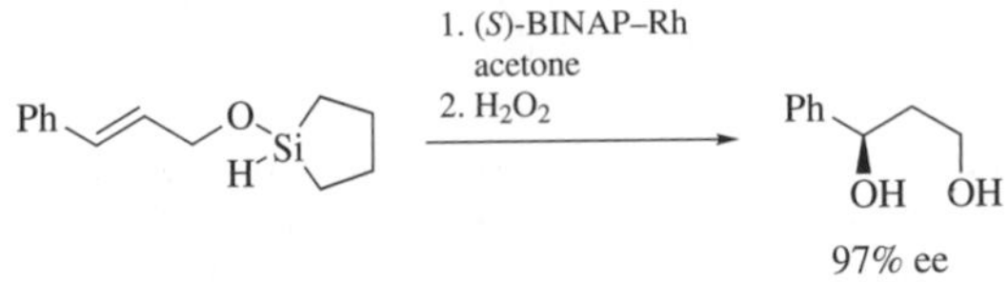

Ph, O, Si, H → 1. (*S*)-BINAP–Rh, acetone; 2. H_2O_2 → Ph, OH OH (20)

97% ee

BINAP–Pd Catalyzed Asymmetric Reactions. BINAP–Pd^0 complexes are prepared in situ from ***Bis(dibenzylideneacetone)palladium(0)*** or $Pd_2(dba)_3·CHCl_3$ and BINAP.[28] BINAP–Pd^{II} complexes are formed from ***Bis(allyl)di-μ-chlorodipalladium***, ***Palladium(II) Acetate***, or $PdCl_2(MeCN)_2$ and BINAP.[29–31]

A BINAP–Pd complex brings about enantioselective 1,4-disilylation of α,β-unsaturated ketones with chlorinated disilanes, giving enol silyl ethers in 74–92% ee (eq 21).[29]

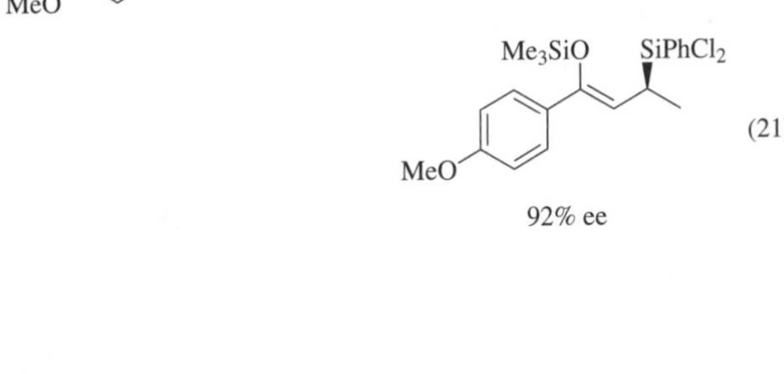

O, MeO + $Me_3Si–SiPhCl_2$ → $PdCl_2$[(*R*)-BINAP], benzene → Me_3SiO, $SiPhCl_2$, MeO (21)

92% ee

A BINAP–Pd^{II} complex catalyzes a highly enantioselective C–C bond formation between an aryl triflate and 2,3-dihydrofuran (eq 22).[30] The intramolecular version of the reaction using an alkenyl iodide in the presence of $PdCl_2$[(*R*)-BINAP] and ***Silver(I) Phosphate*** allows enantioselective formation of a bicyclic ring system (eq 23).[31]

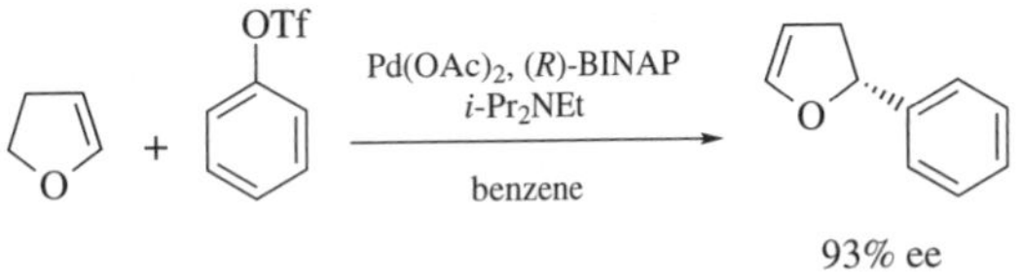

O + OTf → $Pd(OAc)_2$, (*R*)-BINAP, *i*-Pr_2NEt, benzene → O (22)

93% ee

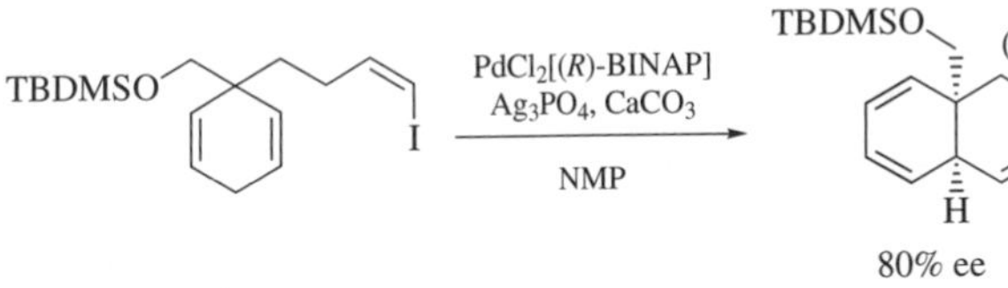

TBDMSO, I → $PdCl_2$[(*R*)-BINAP], Ag_3PO_4, $CaCO_3$, NMP → TBDMSO, H (23)

80% ee

Enantioselective electrophilic allylation of 2-acetamidomalonate esters is effected by a BINAP–Pd^0 complex (eq 24).[32]

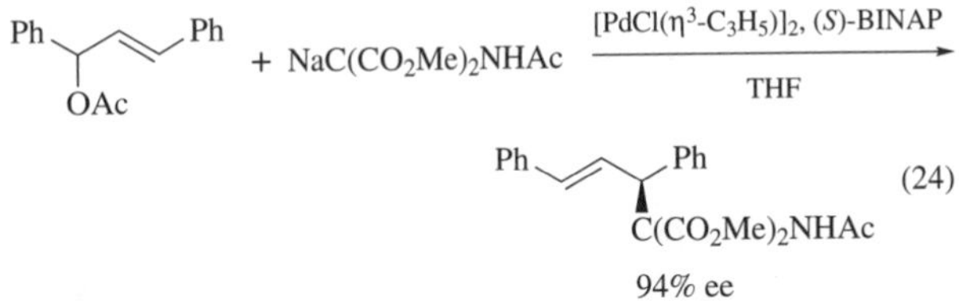

Ph, Ph, OAc + $NaC(CO_2Me)_2NHAc$ → [PdCl(η^3-C_3H_5)]$_2$, (*S*)-BINAP, THF → Ph, Ph, $C(CO_2Me)_2NHAc$ (24)

94% ee

A BINAP–Pd^0 complex catalyzes hydrocyanation of norbornene to the *exo* nitrile with up to 40% ee.[28]

BINAP–Ir^I Catalyzed Asymmetric Reactions. [Ir(BINAP)-(cod)]BF_4 is prepared from [Ir(cod)(MeCN)$_2$]-BF_4 and BINAP in THF.[33]

A combined system of the BINAP–Ir complex and bis-(*o*-dimethylaminophenyl)phenylphosphine or (*o*-dimethylaminophenyl)diphenylphosphine catalyzes hydrogenation of benzylideneacetone[33a] and cyclic aromatic ketones[33b] with modest to high enantioselectivities (eq 25).

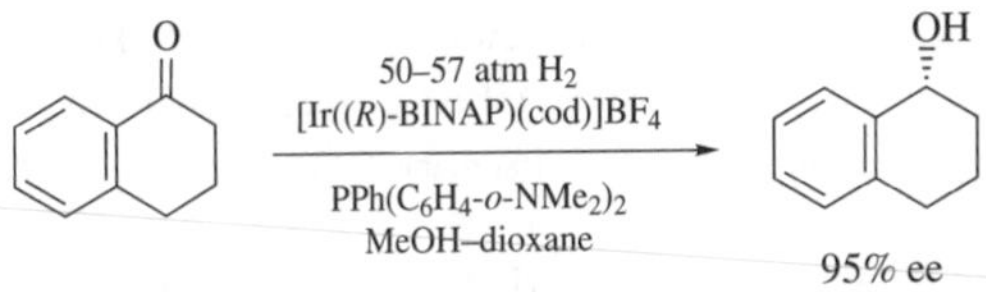

O → 50–57 atm H_2, [Ir((*R*)-BINAP)(cod)]BF_4, $PPh(C_6H_4$-*o*-$NMe_2)_2$, MeOH–dioxane → OH (25)

95% ee

Related Reagents. (Bicyclo[2.2.1]hepta-2,5-diene)-[1,4-bis(diphenylphosphino)butane]rhodium(I) Tetrafluoroborate; (Bicyclo[2.2.1]hepta-2,5-diene)[(2*S*,3*S*)bis(diphenylphosphino)-butane]rhodium Perchlorate; Bis(1,5-cyclooctadiene)rhodium Tetrafluoroborate-(*R*)-2, 2′-Bis(diphenylphosphino)-1, 1′-binaphthyl; 1, 2-Bis(2, 5-diethylphospholano)benzene; 1, 4-Bis-(diphenylphosphino)butane; (+)-*trans*-(2*S*,3*S*)-Bis(diphenylphosphino)bicyclo[2.2.1]hept-5-ene; (*R*)-*N*-[2-(*N*,*N*-Dimethylamino)ethyl]*N*-methyl-1-[(*S*)-1′, 2-bis(diphenylphosphino)-ferrocenyl]ethylamine; (2,3-*O*-Isopropylidene)-2,3-dihydroxy-1,4-bis(diphenylphosphino)butane.

---

1. (a) Miyashita, A.; Yasuda, A.; Takaya, H.; Toriumi, K.; Ito, T.; Souchi, T.; Noyori, R. *JACS* **1980**, *102*, 7932. (b) Noyori, R.; Takaya, H. *CS* **1985**, *25*, 83.
2. (a) Noyori, R.; Kitamura, M. In *Modern Synthetic Methods*; Scheffold, R., Ed.; Springer: Berlin, 1989; p 115. (b) Noyori, R. *Science* **1990**, *248*, 1194. (c) Noyori, R.; Takaya, H. *ACR* **1990**, *23*, 345. (d) Noyori, R. *Chemtech* **1992**, *22*, 360.
3. Takaya, H.; Akutagawa, S.; Noyori, R. *OS* **1988**, *67*, 20.
4. (a) Ikariya, T.; Ishii, Y.; Kawano, H.; Arai, T.; Saburi, M.; Yoshikawa, S.; Akutagawa, S. *CC* **1985**, 922. (b) Ohta, T.; Takaya, H.; Noyori, R. *IC* **1988**, *27*, 566. (c) Kitamura, M.; Tokunaga, M.; Ohkuma, T.; Noyori, R. *TL* **1991**, *32*, 4163. (d) Kitamura, M.; Tokunaga, M.; Ohkuma, T.; Noyori, R. *OS* **1992**, *71*, 1.
5. Kitamura, M.; Tokunaga, M.; Noyori, R. *JOC* **1992**, *57*, 4053.
6. (a) Ohta, T.; Takaya, H.; Kitamura, M.; Nagai, K.; Noyori, R. *JOC* **1987**, *52*, 3174. (b) Saburi, M.; Takeuchi, H.; Ogasawara, M.; Tsukahara, T.; Ishii, Y.; Ikariya, T.; Takahashi, T.; Uchida, Y. *JOM* **1992**, *428*, 155.
7. Lubell, W. D.; Kitamura, M.; Noyori, R. *TA* **1991**, *2*, 543.
8. (a) Noyori, R.; Ohta, M.; Hsiao, Y.; Kitamura, M.; Ohta, T.; Takaya, H. *JACS* **1986**, *108*, 7117. (b) Kitamura, M.; Hsiao, Y.; Noyori, R.; Takaya, H. *TL* **1987**, *28*, 4829.
9. (a) Takaya, H.; Ohta, T.; Sayo, N.; Kumobayashi, H.; Akutagawa, S.; Inoue, S.; Kasahara, I.; Noyori, R. *JACS* **1987**, *109*, 1596, 4129. (b) Takaya, H.; Ohta, T.; Inoue, S.; Tokunaga, M.; Kitamura, M.; Noyori, R. *OS* **1994**, *72*, 74.
10. Kitamura, M.; Nagai, K.; Hsiao, Y.; Noyori, R. *TL* **1990**, *31*, 549.
11. Kitamura, M.; Kasahara, I.; Manabe, K.; Noyori, R.; Takaya, H. *JOC* **1988**, *53*, 708.
12. Ohta, T.; Miyake, T.; Seido, N.; Kumobayashi, H.; Akutagawa, S.; Takaya, H. *TL* **1992**, *33*, 635.
13. (a) Noyori, R.; Ohkuma, T.; Kitamura, M.; Takaya, H.; Sayo, N.; Kumobayashi, H.; Akutagawa, S. *JACS* **1987**, *109*, 5856. (b) Kitamura, M.; Ohkuma, T.; Inoue, S.; Sayo, N.; Kumobayashi, H.; Akutagawa, S.; Ohta, T.; Takaya, H.; Noyori, R. *JACS* **1988**, *110*, 629. (c) Kitamura, M.; Ohkuma, T.; Takaya, H.; Noyori, R. *TL* **1988**, *29*, 1555. (d) Kawano, H.; Ishii, Y.; Saburi, M.; Uchida, Y. *CC* **1988**, 87. (e) Ohkuma, T.; Kitamura, M.; Noyori, R. *TL* **1990**, *31*, 5509.
14. Nishi, T.; Kitamura, M.; Ohkuma, T.; Noyori, R. *TL* **1988**, *29*, 6327.
15. (a) Kitamura, M.; Tokunaga, M.; Noyori, R. *JACS* **1993**, *115*, 144. (b) Kitamura, M.; Tokunaga, M.; Noyori, R. *T* **1993**, *49*, 1853.
16. (a) Noyori, R.; Ikeda, T.; Ohkuma, T.; Widhalm, M.; Kitamura, M.; Takaya, H.; Akutagawa, S.; Sayo, N.; Saito, T.; Taketomi, T.; Kumobayashi, H. *JACS* **1989**, *111*, 9134. (b) Genet, J. P.; Pinel, C.; Mallart, S.; Juge, S.; Thorimbert, S.; Laffitte, J. A. *TA* **1991**, *2*, 555. (c) Mashima, K.; Matsumura, Y.; Kusano, K.; Kumobayashi, H.; Sayo, N.; Hori, Y.; Ishizaki, T.; Akutagawa, S.; Takaya, H. *CC* **1991**, 609.
17. Kitamura, M.; Ohkuma, T.; Tokunaga, M.; Noyori, R. *TA* **1990**, *1*, 1.
18. Oppolzer, W.; Wills, M.; Starkemann, C.; Bernardinelli, G. *TL* **1990**, *31*, 4117.
19. (a) Brown, J. M.; Brunner, H.; Leitner, W.; Rose, M. *TA* **1991**, *2*, 331. (b) Saburi, M.; Ohnuki, M.; Ogasawara, M.; Takahashi, T.; Uchida, Y. *TL* **1992**, *33*, 5783.
20. Kameyama, M.; Kamigata, N.; Kobayashi, M. *JOC* **1987**, *52*, 3312.
21. (a) Miyashita, A.; Takaya, H.; Souchi, T.; Noyori, R. *T* **1984**, *40*, 1245. (b) Toriumi, K.; Ito, T.; Takaya, H.; Souchi, T.; Noyori, R. *Acta Crystallogr.* **1982**, *B38*, 807.
22. Evans, D. A.; Morrissey, M. M. *TL* **1984**, *25*, 4637.
23. (a) Tani, K.; Yamagata, T.; Otsuka, S.; Akutagawa, S.; Kumobayashi, H.; Taketomi, T.; Takaya, H.; Miyashita, A.; Noyori, R. *CC* **1982**, 600. (b) Inoue, S.; Takaya, H.; Tani, K.; Otsuka, S.; Sato, T.; Noyori, R. *JACS* **1990**, *112*, 4897. (c) Yamakawa, M.; Noyori, R. *OM* **1992**, *11*, 3167. (d) Tani, K.; Yamagata, T.; Tatsuno, Y.; Yamagata, Y.; Tomita, K.; Akutagawa, S.; Kumobayashi, H.; Otsuka, S. *AG(E)* **1985**, *24*, 217. (e) Otsuka, S.; Tani, K. *S* **1991**, 665.
24. Kitamura, M.; Manabe, K.; Noyori, R.; Takaya, H. *TL* **1987**, *28*, 4719.
25. Wu, X.-M.; Funakoshi, K.; Sakai, K. *TL* **1992**, *33*, 6331.
26. (a) Hayashi, T.; Matsumoto, Y.; Ito, Y. *JACS* **1989**, *111*, 3426. (b) Sato, M.; Miyaura, N.; Suzuki, A. *TL* **1990**, *31*, 231. (c) Zhang, J.; Lou, B.; Guo, G.; Dai, L. *JOC* **1991**, *56*, 1670.
27. Tamao, K.; Tohma, T.; Inui, N.; Nakayama, O.; Ito, Y. *TL* **1990**, *31*, 7333.
28. Hodgson, M.; Parker, D. *JOM* **1987**, *325*, C27.
29. Hayashi, T.; Matsumoto, Y.; Ito, Y. *JACS* **1988**, *110*, 5579.
30. Ozawa, F.; Hayashi, T. *JOM* **1992**, *428*, 267.
31. (a) Sato, Y.; Sodeoka, M.; Shibasaki, M. *CL* **1990**, 1953; Sato, Y.; Sodeoka, M.; Shibasaki, M. *JOC* **1989**, *54*, 4738. (b) Ashimori, A.; Overman, L. E. *JOC* **1992**, *57*, 4571.
32. Yamaguchi, M.; Shima, T.; Yamagishi, T.; Hida, M. *TL* **1990**, *31*, 5049.
33. (a) Mashima, K.; Akutagawa, T.; Zhang, X.; Takaya, H.; Taketomi, T.; Kumobayashi, H.; Akutagawa, S. *JOM* **1992**, *428*, 213. (b) Zhang, X.; Taketomi, T.; Yoshizumi, T.; Kumobayashi, H.; Akutagawa, S.; Mashima, K.; Takaya, H. *JACS* **1993**, *115*, 3318.

Masato Kitamura & Ryoji Noyori
Nagoya University, Japan

Bis(tri-*n*-butyltin) Oxide

$(n\text{-}Bu_3Sn)_2O$

[56-35-9] $C_{24}H_{54}OSn_2$ (MW 596.20)

promotes the oxidation of secondary alcohols and sulfides with Br_2; O- and N-activations; dehydrosulfurizations; hydrolysis catalyst)

Physical Data: bp 180 °C/2 mmHg; *d* 1.170 g cm^{-3}.
Solubility: sol ether and hexane.
Form Supplied in: colorless oil.
Handling, Storage, and Precautions: $(Bu_3Sn)_2O$ should be stored in the absence of moisture. Owing to the toxicity of organostannanes, this reagent should be handled in a well-ventilated fume hood. Contact with the eyes and skin should be avoided.

Oxidations. Benzylic, allylic, and secondary alcohols are oxidized to the corresponding carbonyl compounds by using $(Bu_3Sn)_2O$–***Bromine***.[1] This procedure is quite useful for selective oxidation of secondary alcohols in the presence of primary alcohols, which are inert under these conditions (eqs 1–3).[2] $(Bu_3Sn)_2O$–***N-Bromosuccinimide*** can also be applied to the selective oxidation of secondary alcohols (eq 4).[3]

$(Bu_3Sn)_2O$, Br_2, CH_2Cl_2, rt, 76% (1)

$(Bu_3Sn)_2O$, Br_2, CH_2Cl_2, rt, 86% (2)

$(Bu_3Sn)_2O$, Br_2, CH_2Cl_2, rt, 68% (3)

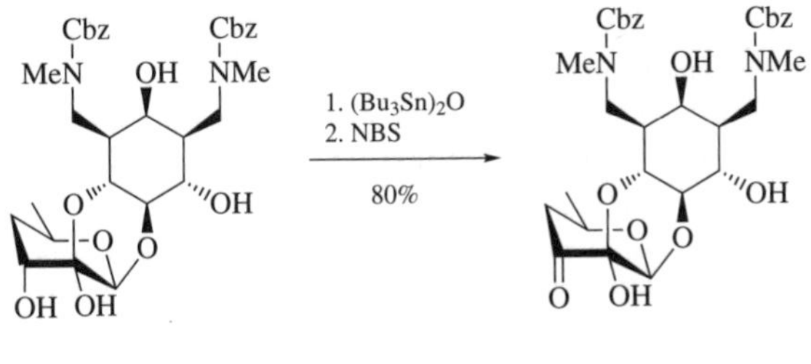

(4)

$(Bu_3Sn)_2O$–Br_2 oxidizes sulfides to sulfoxides in CH_2Cl_2 without further oxidation to sulfones, even in the presence of excess reagent (eq 5).[4] This procedure is especially useful for sulfides having long, hydrophobic alkyl chains, for which solubility problems are often encountered in the ***Sodium Periodate*** oxidation in aqueous organic solvents. Oxidation of sulfenamides to sulfinamides can be achieved without formation of sulfonamides using the reagent (eq 6).[4]

$$PhCH_2SCH_2Ph \xrightarrow[92\%]{(Bu_3Sn)_2O,\ Br_2,\ CH_2Cl_2,\ rt} PhCH_2SOCH_2Ph \quad (5)$$

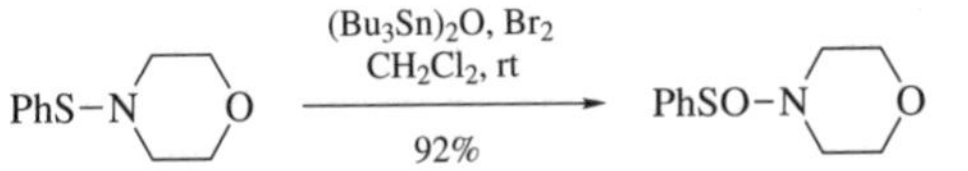

$$PhS{-}N(CH_2CH_2)_2O \xrightarrow[92\%]{(Bu_3Sn)_2O,\ Br_2,\ CH_2Cl_2,\ rt} PhSO{-}N(CH_2CH_2)_2O \quad (6)$$

$(Bu_3Sn)_2O$–Br_2–***Diphenyl Diselenide*** in refluxing $CHCl_3$ transforms alkenes into α-seleno ketones (eq 7).[5]

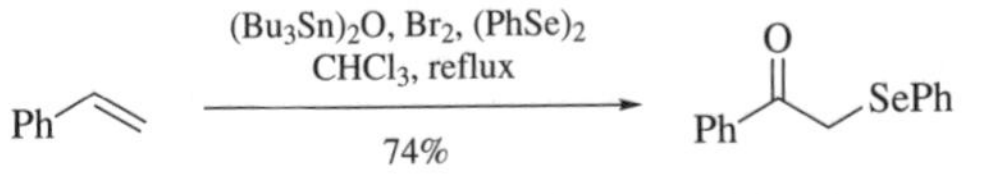

$$PhCH{=}CH_2 \xrightarrow[74\%]{(Bu_3Sn)_2O,\ Br_2,\ (PhSe)_2,\ CHCl_3,\ reflux} PhCOCH_2SePh \quad (7)$$

***O*- and *N*-activations.** $(Bu_3Sn)_2O$ has been used in the activation of hydroxy groups toward sulfamoylations, acylations, carbamoylations, and alkylations because conversion of alcohols to stannyl ethers enhances the oxygen nucleophilicity. Tributylstannyl ethers are easily prepared by heating the alcohol and $(Bu_3Sn)_2O$, with azeotropic removal of water. Sulfamoylation of alcohols can be achieved via tributyltin derivatives in high yields, whereas direct sulfamoylation gives low yields (eq 8).[6] This activation can be used for selective acylation of vicinal diols (eq 9).[7] In carbohydrate chemistry this approach is extremely useful for the regioselective acylation without the use of a blocking–deblocking technique (eq 10).[8] The order of the activation of hydroxy groups on carbohydrates has been investigated, and is shown in partial structures (**1**), (**2**), and (**3**).[9] Regioselective carbamoylation can also be accomplished by changing experimental conditions (eq 11).[10] On the other hand, alkylations of the tin derivatives are sluggish and less selective than acylations under similar conditions. Regioselective alkylation of sugar compounds, however, can be carried out in high yield by conversion to a tributyltin ether followed by addition of alkylating agent and quaternary ammonium halide catalysts (eq 12).[11]

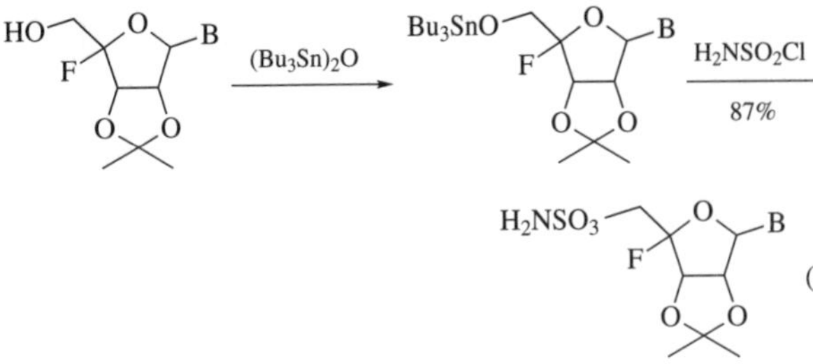

(8)

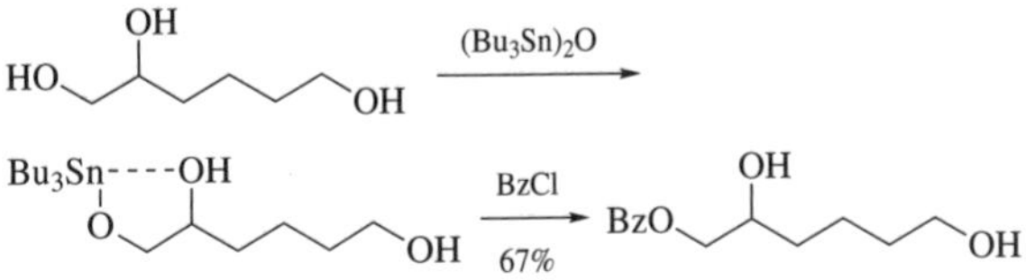

(9)

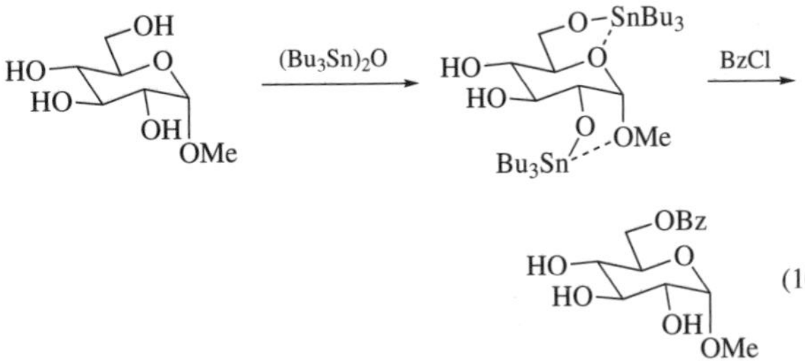

(10)

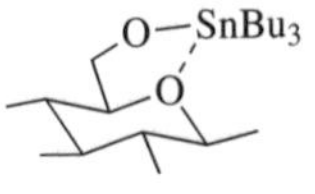

(**1**) most reactive (**2**) next most reactive (**3**) least reactive

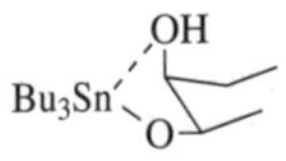

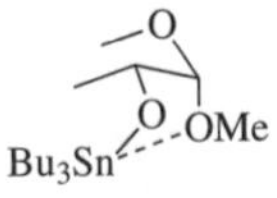

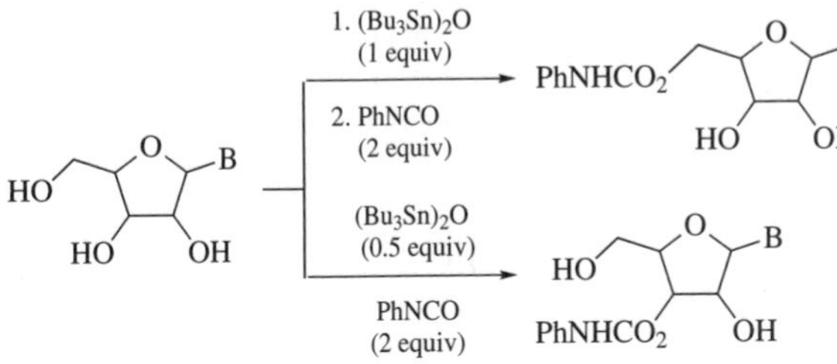

(11)

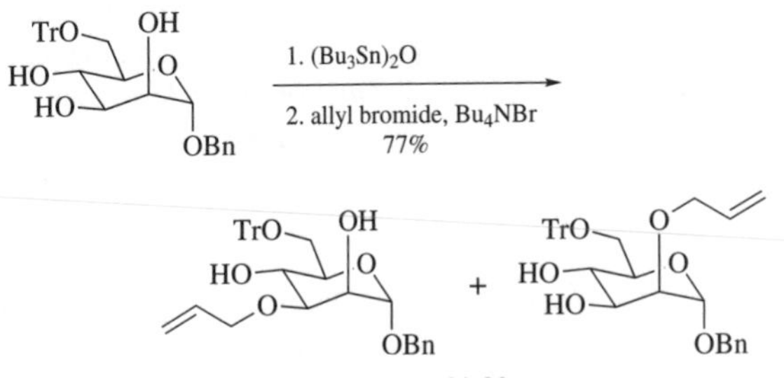

(12)

This O-activation is also effective for intramolecular alkylations such as oxetane synthesis (eq 13).[12] Similar N-activation has been used in the synthesis of pyrimidine nucleosides (eq 14).[13]

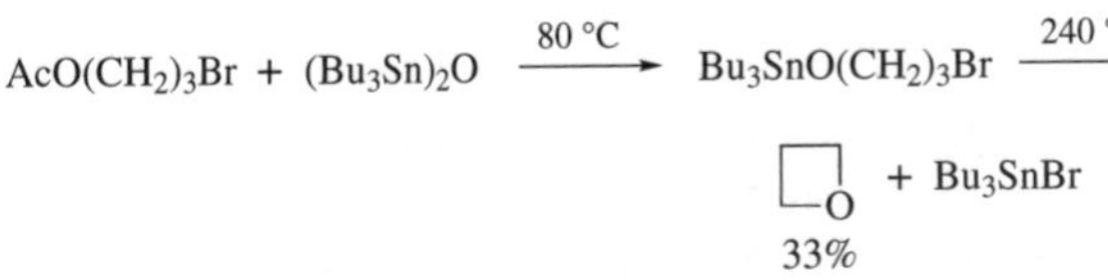

$$AcO(CH_2)_3Br + (Bu_3Sn)_2O \xrightarrow{80\ °C} Bu_3SnO(CH_2)_3Br \xrightarrow{240\ °C} \text{oxetane}\ (33\%) + Bu_3SnBr \quad (13)$$

1. $(Bu_3Sn)_2O$ 2. BzO–ribofuranose(OAc)(OBz)₂ (14)

Dehydrosulfurizations. The thiophilicity of tin compounds is often utilized in functional group transformations. Thus conversion of aromatic and aliphatic thioamides to the corresponding nitriles can be accomplished by using $(Bu_3Sn)_2O$ in boiling benzene under azeotropic conditions (eq 15).[14]

$(Bu_3Sn)_2O$, 95% (15)

Hydrolysis. Esters are efficiently hydrolyzed with $(Bu_3Sn)_2O$ under mild conditions (eq 16).[15]

$(Bu_3Sn)_2O$, 25 °C, 61% (16)

Transformation of primary alkyl bromides or iodides to the corresponding primary alcohols is achieved in good yield by using $(Bu_3Sn)_2O$–***Silver(I) Nitrate*** (or ***Silver(I) p-Toluenesulfonate***) (eq 17),[16] whereas this method is not applicable to secondary halides due to elimination.

$$MeCO_2(CH_2)_4I \xrightarrow[96\%]{(Bu_3Sn)_2O,\ AgTos,\ DMF,\ 20\,°C} MeCO_2(CH_2)_4OH \quad (17)$$

$(Bu_3Sn)_2O$ is a useful starting material for the preparation of tributyltin hydride, which is a convenient radical reducing reagent in organic synthesis. Thus ***Tri-n-butyltin Hydride*** is easily prepared by using exchange reactions of $(Bu_3Sn)_2O$ with polysiloxanes (eq 18).[17]

$$(Bu_3Sn)_2O + \left(Si(Me)(H)\text{–}O\right)_n \longrightarrow Bu_3SnH + \left[Si(Me)\text{–}O\text{–}O\text{–}Si(Me)\text{–}O\right]_n \quad (18)$$

Related Reagents. Tri-*n*-butyl(methoxy)stannane.

1. Saigo, K.; Morikawa, A.; Mukaiyama, T. *CL* **1975**, 145.
2. Ueno, Y.; Okawara, M. *TL* **1976**, 4597.
3. Hanessian, S.; Roy, R. *CJC* **1985**, *63*, 163.
4. Ueno, Y.; Inoue, T.; Okawara, M. *TL* **1977**, 2413.
5. Kuwajima, I.; Shimizu, M. *TL* **1978**, 1277.
6. Jenkins, I. D.; Verheyden, J. P. H.; Moffatt, J. G. *JACS* **1971**, *93*, 4323.
7. (a) Ogawa, T.; Matsui, M. *T* **1981**, *37*, 2363. (b) David, S.; Hanessian, S. *T* **1985**, *41*, 643.
8. (a) Crowe, A. J.; Smith, P. J. *JOM* **1976**, *110*, C57. (b) Blunden, S. J.; Smith, P. J.; Beynon, P. J.; Gillies, D. G. *Carbohydr. Res.* **1981**, *88*, 9. (c) Ogawa, T.; Matsui, M. *Carbohydr. Res.* **1977**, *56*, C1. (d) Hanessian, S.; Roy, R. *JACS* **1979**, *101*, 5839. (e) Arnarp, J.; Loenngren, J. *CC* **1980**, 1000. (f) Ogawa, T.; Nakabayashi, S.; Sasajima, K. *Carbohydr. Res.* **1981**, *96*, 29.
9. Tsuda, Y.; Haque, M. E.; Yoshimoto, K. *CPB* **1983**, *31*, 1612.
10. (a) Ishido, Y.; Hirao, I.; Sakairi, N.; Araki, Y. *H* **1979**, *13*, 181. (b) Hirao, I.; Itoh, K.; Sakairi, N.; Araki, Y.; Ishido, Y. *Carbohydr. Res.* **1982**, *109*, 181.
11. (a) Alais, J.; Veyrières, A. *JCS(P1)* **1981**, 377. (b) Veyrières, A. *JCS(P1)* **1981**, 1626.
12. Biggs, J. *TL* **1975**, 4285.
13. Ogawa, T.; Matsui, M. *JOM* **1978**, *145*, C37.
14. Lim, M.-I.; Ren, W.-Y.; Klein, R. S. *JOC* **1982**, *47*, 4594.
15. Mata, E. G.; Mascaretti, O. A. *TL* **1988**, *29*, 6893.
16. Gingras, M.; Chan, T. H. *TL* **1989**, *30*, 279.
17. Hayashi, K.; Iyoda, J.; Shiihara, I. *JOM* **1967**, *10*, 81.

Hiroshi Sano
Gunma University, Kiryu, Japan

Boron Tribromide[1]

BBr_3

[10294-33-4] BBr_3 (MW 250.52)

(Lewis acid used for deprotection of OH and NH groups; cleaves ethers or esters to alkyl bromides; bromoborates allene and alkynes)

Physical Data: mp −45 °C; bp 91.7 °C; *d* 2.650 g cm^{-3}.
Form Supplied in: colorless, fuming liquid; a 1.0 M solution in dichloromethane and hexane; $BBr_3 \cdot Me_2S$ complex is available as either a white solid or a 1.0 M solution in dichloromethane.
Purification: by distillation.
Handling, Storage, and Precautions: BBr_3 is highly moisture sensitive and decomposes in air with evolution of HBr. Store under a dry inert atmosphere and transfer by syringe or through a Teflon tube. It reacts violently with protic solvents such as water and alcohols. Ether and THF are not appropriate solvents.

Removal of Protecting Groups. BBr_3 is highly Lewis acidic. It coordinates to ethereal oxygens and promotes C–O bond cleavage to an alkyl bromide and an alkoxyborane that is hydrolyzed to an alcohol during workup (eq 1).[2]

$$R^1OR^2 \xrightarrow{BBr_3} R^1Br + Br_2BOR^2 \xrightarrow{H_2O} R^1Br + R^2OH \quad (1)$$

BBr_3 has been widely used to cleave ethers because the reaction proceeds completely under mild conditions. In a special case,

BBr_3 has been used to cleave acetals that cannot be deprotected by usual acidic conditions.[3] Because alkyl aryl ethers are cleaved at the alkyl–oxygen bond to give ArOH and alkyl bromides, BBr_3 has been most generally used for the demethylation of methyl aryl ethers,[2,4] for example as the final step of zearalenone synthesis (eq 2).[5] Problems are sometimes encountered in attempts to deprotect more than one nonadjacent methoxy group on one aromatic ring, and when stable chelates are formed.[6] The presence of a carbonyl substituent facilitates the selective deprotection of polymethoxyaryl compounds (eq 3).[7]

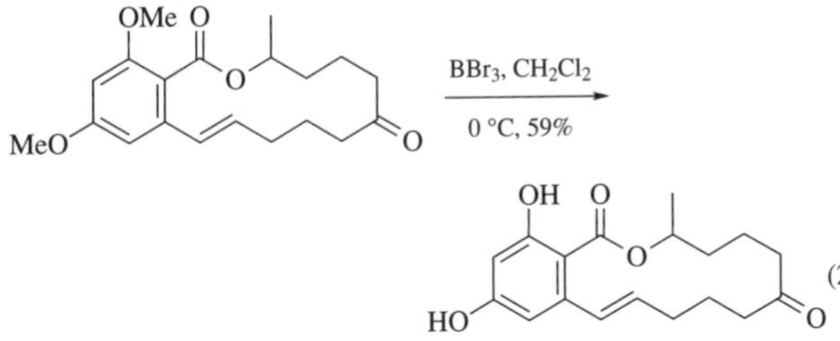

BBr3, CH2Cl2; 0 °C, 59% (2)

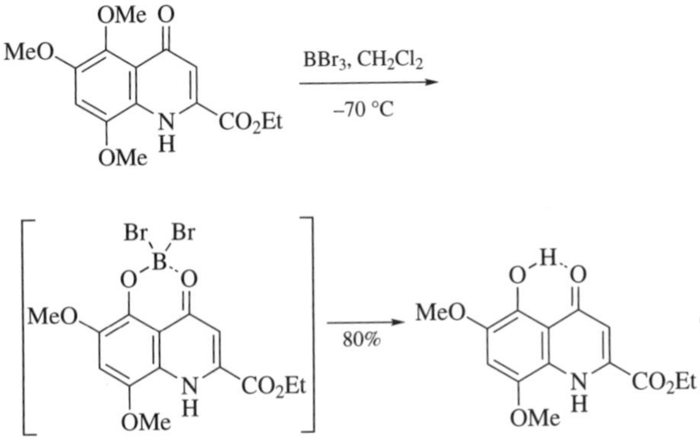

BBr3, CH2Cl2; −70 °C; 80% (3)

The cleavage of mixed dialkyl ethers occurs at the more substituted carbon–oxygen bond. Methyl ethers of secondary or tertiary alcohols give methanol and secondary or tertiary alkyl bromides selectively by the reaction with BBr_3,[8] although the addition of ***Sodium Iodide*** and ***15-Crown-5*** ether can change this selectivity (eq 4).[9] In contrast, methyl ethers of primary alcohols are generally cleaved at the Me–O bond, as demonstrated in Corey's prostaglandin synthesis (eq 5).[10]

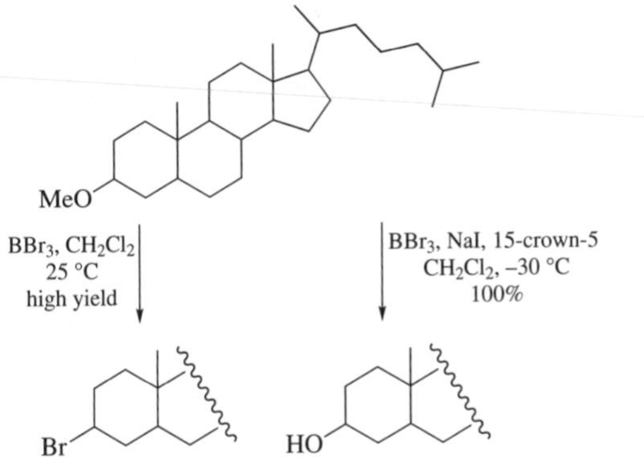

BBr3, CH2Cl2, 25 °C, high yield; BBr3, NaI, 15-crown-5, CH2Cl2, −30 °C, 100% (4)

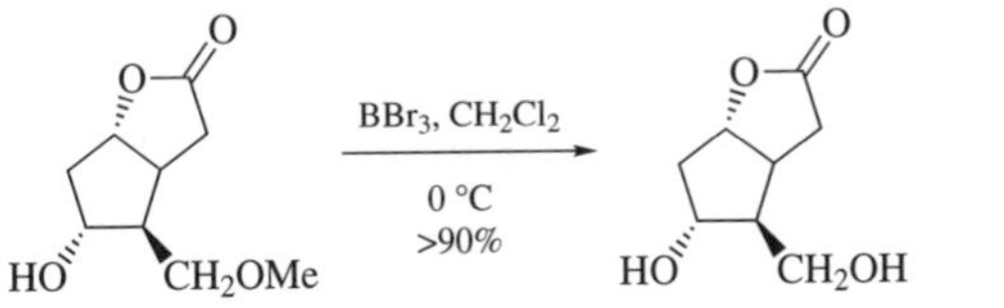

BBr3, CH2Cl2; 0 °C; >90% (5)

BBr_3 has been also used for the deprotection of carbohydrate derivatives[11] and polyoxygenated intermediates in the synthesis of deoxyvernolepin,[12] vernolepin,[13] and vernomenin.[13] Although one of the model compounds is deprotected cleanly (eq 6),[14] application of BBr_3 to more highly functionalized intermediates leads to cleavage of undesired C–O bonds competitively (eq 7).[12,13]

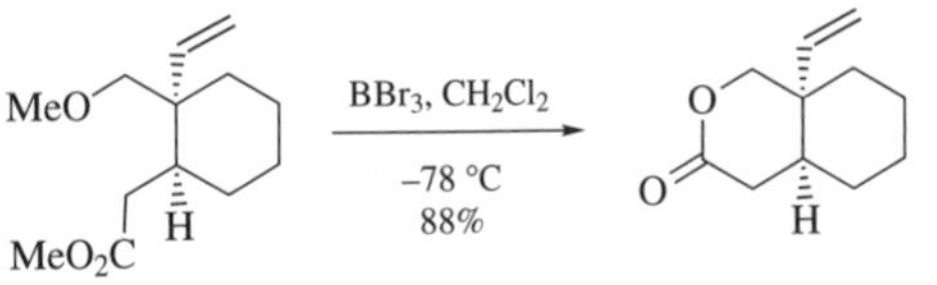

BBr3, CH2Cl2; −78 °C; 88% (6)

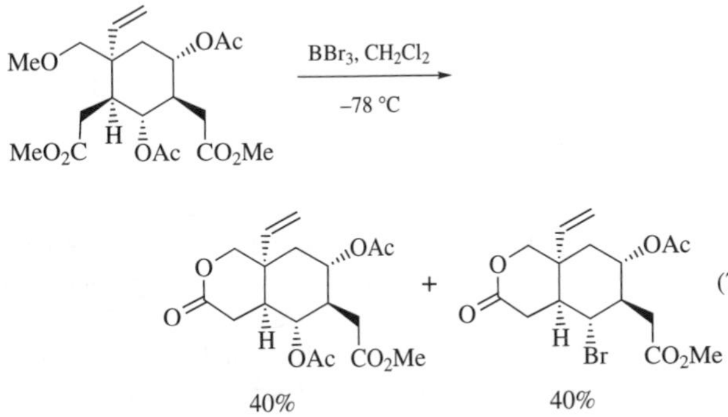

BBr3, CH2Cl2; −78 °C; 40% + 40% (7)

For the complete cleavage, 1 mol of BBr_3 is required for each ether group and other Lewis-basic functional groups. Sometimes it is difficult to find reaction conditions for the selective cleavage of the desired C–O bond. Recently, modified bromoboranes such as ***B-Bromocatecholborane***,[15] dialkylbromoboranes,[16] ***Bromobis(isopropylthio)borane***,[17] and ***9-Bromo-9-borabicyclo[3.3.1]nonane***,[18] have been introduced to cleave C–O bonds more selectively under milder conditions. $BBr_3 \cdot SMe_2$ is also effective for ether cleavage and has the advantage of being more stable than BBr_3. It can be stored for a long time and handled easily. However, a two- to fourfold excess of the reagent is necessary to complete the dealkylation of alkyl aryl ether.[19]

Amino acid protecting groups such as benzyloxycarbonyl and *t*-butoxycarbonyl groups are cleaved by BBr_3. However, the hydrolysis of the ester function also occurs under the same reaction conditions.[20] Debenzylation and debenzyloxymethylation of uracils proceed successfully in aromatic solvents, but demethylation is more sluggish and less facile (eq 8).[21]

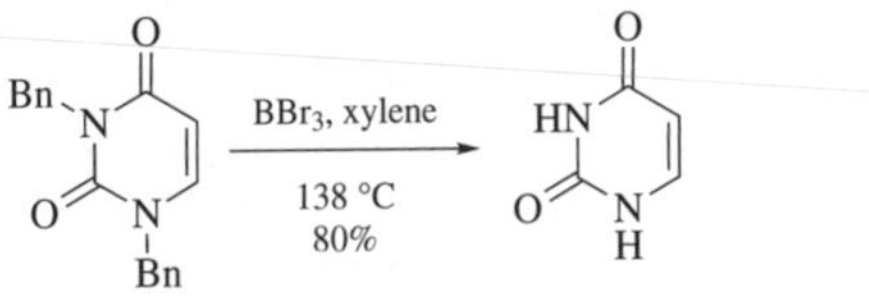

BBr3, xylene; 138 °C; 80% (8)

Substitution Reactions. BBr_3 reacts with cyclic ethers to give tris(ω-bromoalkoxy)boranes which provide ω-bromoalkanols or ω-bromoalkanals when treated with MeOH or ***Pyridinium Chlorochromate***, respectively (eq 9).[22] Unfortunately, unsymmetrically substituted ethers such as 2-methyltetrahydrofuran are cleaved nonregioselectively. Generally, ester groups survive under the reaction conditions for ether cleavage, but the ring opening of lactones occurs under mild conditions to give ω-halocarboxylic acids in good yields (eq 10).[23]

3 (tetrahydropyran) $\xrightarrow{BBr_3, CH_2Cl_2,\ 0\ °C \to \Delta}$ $(Br(CH_2)_5O)_3B$; MeOH 86% → 3 Br(CH₂)₅OH; PCC 72% → 3 Br(CH₂)₄CHO (9)

γ-butyrolactone: BBr_3, CH_2Cl_2, rt, 98% → $Br(CH_2)_3CO_2H$; BBr_3, NaI, MeCN, rt, 96% → $I(CH_2)_3CO_2H$ (10)

In the reaction with methoxybenzaldehyde, bromination of the carbonyl group takes place more rapidly than demethylation; therefore benzal bromide formation is generally observed in the reaction with aromatic aldehydes.[24] Cleavage of *t*-butyldimethylsilyl ethers or *t*-butyldiphenylsilyl ethers occurs at the C–O bond to give alkyl bromides.[25] Alcohols can be converted to alkyl bromides by this method.

In a special case, BBr_3 is used for the bromination of hydrocarbons. Adamantane is brominated by a mixture of ***Bromine***, BBr_3, and ***Aluminum Bromide*** to give 1,3-dibromoadamantane selectively.[26] Tetrachlorocyclopropene[27] and hexachlorocyclopentadiene[28] are substituted to the corresponding bromides by BBr_3 and, in the latter case, addition of $AlBr_3$ and Br_2 is effective to improve the result.[29]

Reduction of Sulfur Compounds. Alkyl and aryl sulfoxides are reduced by BBr_3 to the corresponding sulfides in good yields.[30] Addition of ***Potassium Iodide*** and a catalytic amount of ***Tetra-n-butylammonium Iodide*** is necessary for the reduction of sulfonic acids and their derivatives.[31]

Transesterification of Esters or Conversion to Amides. Transesterification reactions of carboxylic esters or conversion into the amides is promoted by a stoichiometric amount of BBr_3.[32]

Removal of Methyl Sulfide from Organoborane–Methyl Sulfide Complexes. Methyl sulfide can be removed from $BrBR_2{\cdot}SMe_2$ or $Br_2BR{\cdot}SMe_2$, which are prepared by the hydroboration reaction of alkenes or alkynes with $BrBH_2{\cdot}SMe_2$ or $Br_2BH{\cdot}SMe_2$, by using BBr_3.[33] The resulting alkenyldibromoboranes are useful for the stereoselective synthesis of bromodienes (eq 11).[34]

Bromoboration Reactions. BBr_3 does not add to isolated double bonds, but reacts with allene spontaneously even at low temperature to give (2-bromoallyl)dibromoborane,[35] which provides stable (2-bromoallyl)diphenoxyborane by the addition of anisole.[36] The diphenoxyborane derivative reacts with carbonyl compounds to give 2-bromohomoallylic alcohols in high yields (eq 12). Bromoboration of 1-alkynes provides (*Z*)-(2-bromo-1-alkenyl)dibromoboranes stereo- and regioselectively (eq 13),[37] which are applied for the synthesis of trisubstituted alkenes,[38] α,β-unsaturated esters,[39] and γ,δ-unsaturated ketones,[40] bromodienes,[41] 1,2-dihalo-1-alkenes,[42] 2-bromoalkanals,[43] and β-bromo-α,β-unsaturated amides.[44]

EtC(BBr₂SMe₂)=CHEt $\xrightarrow{BBr_3, CH_2Cl_2,\ 0\ °C\ to\ rt}$ EtC(BBr₂)=CHEt + BBr_3SMe_2; BuC≡CH, CH_2Cl_2, −78 °C to rt → [intermediate]; I_2, AcOK, THF, −78 °C to rt → bromodiene, 63%, (*Z*,*Z*) = 98% (11)

$H_2C{=}C{=}CH_2$ $\xrightarrow{1.\ BBr_3,\ -20\ °C;\ 2.\ 2\ PhOMe,\ -78\ °C\ to\ rt}$ $CH_2{=}C(Br)CH_2B(OPh)_2$ $\xrightarrow{PhCHO,\ THF,\ -78\ °C\ to\ rt}$ $CH_2{=}C(Br)CH_2CH(OH)Ph$, 89% (12)

RC≡CH $\xrightarrow{BBr_3, CH_2Cl_2,\ -78\ °C}$ R(Br)C=CH(BBr₂) (13)

Chiral Bromoborane Reagents. Complexes made from chiral 1-alkyl-2-(diphenylhydroxymethyl)pyrrolidines and BBr_3 are effective catalysts for asymmetric Diels–Alder reactions.[45] Bromoboranes prepared from chiral 1,2-diphenyl-1,2-bis(arenesulfonamido)ethanes[46,47] are used to prepare chiral allylic boranes,[47,48] allenylic borane,[49] propargylic boranes,[49] and enolates.[46,47,50] The *B*-bromodiazaborolidinene (**1**), prepared from 1,2-diphenyl-1,2-bis(*p*-toluenesulfonamido)ethane, is particularly effective in these applications. The reagents prepared from (**1**) are highly effective for the enantioselective synthesis of homoallylic alcohols (eq 14),[48] homopropargylic alcohols (eq 15),[49] propadienyl carbinols (eq 16),[49] and aldol condensation products (eq 17).[46]

(**1**) $\xrightarrow{1.\ CH_2{=}C(Cl)CH_2SnBu_3,\ CH_2Cl_2,\ 0\ °C\ to\ rt;\ 2.\ PhCHO,\ -78\ °C}$ $CH_2{=}C(Cl)CH_2CH(OH)Ph$, 84% ee (14)

(**1**) $\xrightarrow{1.\ HC{\equiv}CCH_2SnPh_3,\ CH_2Cl_2,\ 0\ °C\ to\ rt;\ 2.\ PhCHO,\ -78\ °C}$ $PhCH(OH)CH_2C{\equiv}CH$, 96% ee (15)

Ph, Ph; $TolSO_2N$–B(Br)–NSO_2Tol **(1)**; 1. $H_2C{=}C{=}CHSnPh_3$, CH_2Cl_2, 0 °C to rt; 2. PhCHO, −78 °C → Ph–CH(OH)–CH=C=CH_2 (16)
>99% ee

Ph, Ph; $TolSO_2N$–B(Br)–NSO_2Tol **(1)**; 1. 3-pentanone, *i*-Pr_2NEt, CH_2Cl_2, −78 °C; 2. EtCHO, −78 °C → (17)
98% ee, 98% *syn*

Related Reagents. Boron Trichloride; 9-Bromo-9-borabicyclo[3.3.1]nonane; *B*-Bromocatecholborane; *t*-Butyldimethylsilyl Iodide; Hydrogen Bromide; Hydrogen Iodide; Iodotrimethylsilane.

1. Bhatt, M. V.; Kulkarni, S. U. *S* **1983**, 249.
2. McOmie, J. F. W.; Watts, M. L.; West, D. E. *T* **1968**, *24*, 2289.
3. Meyers, A. I.; Nolen, R. L.; Collington, E. W.; Narwid, T. A.; Strickland, R. C. *JOC* **1973**, *38*, 1974.
4. (a) Benton, F. L.; Dillon, T. E. *JACS* **1942**, *64*, 1128. (b) Manson, D. L.; Musgrave, O. C. *JCS* **1963**, 1011. (c) McOmie, J. F. W.; Watts, M. L. *CI(L)* **1963**, 1658. (d) Blatchly, J. M.; Gardner, D. V.; McOmie, J. F. W.; Watts, M. L. *JCS(C)* **1968**, 1545.
5. (a) Vlattas, I.; Harrison, I. T.; Tökés, L.; Fried, J. H.; Cross, A. D. *JOC* **1968**, *33*, 4176. (b) Taub, D.; Girotra, N. N.; Hoffsommer, R. D.; Kuo, C. H.; Slates, H. L.; Weber, S.; Wendler, N. L. *T* **1968**, *24*, 2443.
6. (a) Stetter, H.; Wulff, C. *CB* **1960**, *93*, 1366. (b) Locksley, H. D.; Murray, I. G. *JCS(C)* **1970**, 392. (c) Bachelor, F. W.; Loman, A. A.; Snowdon, L. R. *CJC* **1970**, *48*, 1554.
7. Schäfer, W.; Franck, B. *CB* **1966**, *99*, 160.
8. Youssefyeh, R. D.; Mazur, Y. *CI(L)* **1963**, 609.
9. Niwa, H.; Hida, T.; Yamada, K. *TL* **1981**, *22*, 4239.
10. Corey, E. J.; Weinshenker, N. M.; Schaaf, T. K.; Huber, W. *JACS* **1969**, *91*, 5675.
11. Bonner, T. G.; Bourne, E. J.; McNally, S. *JCS* **1960**, 2929.
12. Grieco, P. A.; Noguez, J. A.; Masaki, Y. *JOC* **1977**, *42*, 495.
13. Grieco, P. A.; Nishizawa, M.; Burke, S. D.; Marinovic, N. *JACS* **1976**, *98*, 1612.
14. (a) Grieco, P. A.; Hiroi, K.; Reap, J. J.; Noguez, J. A. *JOC* **1975**, *40*, 1450. (b) Grieco, P. A.; Reap, J. J.; Noguez, J. A. *SC* **1975**, *5*, 155.
15. (a) Boeckman, Jr., R. K.; Potenza, J. C. *TL* **1985**, *26*, 1411. (b) King, P. F.; Stroud, S. G. *TL* **1985**, *26*, 1415.
16. (a) Guindon, Y.; Morton, H. E.; Yoakim, C. *TL* **1983**, *24*, 3969. (b) Gauthier, J. Y.; Guindon, Y. *TL* **1987**, *28*, 5985. (c) Guindon, Y.; Yoakim, C.; Morton, H. E. *TL* **1983**, *24*, 2969. (d) Guindon, Y.; Yoakim, C.; Morton, H. E. *JOC* **1984**, *49*, 3912.
17. Corey, E. J.; Hua, D. H.; Seitz, S. P. *TL* **1984**, *25*, 3.
18. Bhatt, M. V. *JOM* **1978**, *156*, 221.
19. Williard, P. G.; Fryhle, C. B. *TL* **1980**, *21*, 3731.
20. Felix, A. M. *JOC* **1974**, *39*, 1427.
21. Kundu, N. G.; Hertzberg, R. P.; Hannon, S. J. *TL* **1980**, *21*, 1109.
22. Kulkarni, S. U.; Patil, V. D. *H* **1982**, *18*, 163.
23. Olah, G. A.; Karpeles, R.; Narang, S. C. *S* **1982**, 963.
24. Lansinger, J. M.; Ronald, R. C. *SC* **1979**, *9*, 341.
25. Kim, S.; Park, J. H. *JOC* **1988**, *53*, 3111.
26. (a) Baughman, G. L. *JOC* **1964**, *29*, 238. (b) Talaty, E. R.; Cancienne, A. E.; Dupuy, A. E. *JCS(C)* **1968**, 1902.
27. Tobey, S. W.; West, R. *JACS* **1966**, *88*, 2481.
28. West, R.; Kwitowski, P. T. *JACS* **1968**, *90*, 4697.
29. Ungefug, G. A.; Roberts, C. W. *JOC* **1973**, *38*, 153.
30. Guindon, Y.; Atkinson, J. G.; Morton, H. E. *JOC* **1984**, *49*, 4538.
31. Olah, G. A.; Narang, S. C.; Field, L. D.; Karpeles, R. *JOC* **1981**, *46*, 2408.
32. Yazawa, H.; Tanaka, K.; Kariyone, K. *TL* **1974**, *15*, 3995.
33. (a) Brown, H. C.; Ravindran, N.; Kulkarni, S. U. *JOC* **1979**, *44*, 2417. (b) Brown, H. C.; Ravindran, N.; Kulkarni, S. U. *JOC* **1980**, *45*, 384. (c) Brown, H. C.; Campbell, Jr., J. B. *JOC* **1980**, *45*, 389.
34. Hyuga, S.; Takinami, S.; Hara, S.; Suzuki, A. *TL* **1986**, *27*, 977.
35. Joy, F.; Lappert, M. F.; Prokai, B. *JOM* **1966**, *5*, 506.
36. Hara, S.; Suzuki, A. *TL* **1991**, *32*, 6749.
37. (a) Lappert, M. F.; Prokai, B. *JOM* **1964**, *1*, 384. (b) Blackborow, J. R. *JOM* **1977**, *128*, 161. (c) Suzuki, A.; Hara, S. *Res. Trends Org. Chem.* **1990**, 77. (d) Suzuki, A. *PAC* **1986**, *58*, 629.
38. Satoh, Y.; Serizawa, H.; Miyaura, N.; Hara, S.; Suzuki, A. *TL* **1988**, *29*, 1811.
39. Yamashina, N.; Hyuga, S.; Hara, S.; Suzuki, A. *TL* **1989**, *30*, 6555.
40. (a) Hara, S.; Hyuga, S.; Aoyama, M.; Sato, M.; Suzuki, A. *TL* **1990**, *31*, 247. (b) Aoyama, M.; Hara, S.; Suzuki, A. *SC* **1992**, *22*, 2563.
41. Hyuga, S.; Takinami, S.; Hara, S.; Suzuki, A. *CL* **1986**, 459.
42. Hara, S.; Kato, T.; Shimizu, H.; Suzuki, A. *TL* **1985**, *26*, 1065.
43. Satoh, Y.; Tayano, T.; Koshino, H.; Hara, S.; Suzuki, A. *S* **1985**, 406.
44. Satoh, Y.; Serizawa, H.; Hara, S.; Suzuki, A. *SC* **1984**, *14*, 313.
45. Kobayashi, S.; Murakami, M.; Harada, T.; Mukaiyama, T. *CL* **1991**, 1341.
46. Corey, E. J.; Imwinkelried, R.; Pikul, S.; Xiang, Y. B. *JACS* **1989**, *111*, 5493.
47. Corey, E. J.; Kim, S. S. *TL* **1990**, *31*, 3715.
48. Corey, E. J.; Yu, C.-M.; Kim, S. S. *JACS* **1989**, *111*, 5495.
49. Corey, E. J.; Yu, C.-M.; Lee, D.-H. *JACS* **1990**, *112*, 878.
50. Corey, E. J.; Kim, S. S. *JACS* **1990**, *112*, 4976.

Akira Suzuki & Shoji Hara
Hokkaido University, Sapporo, Japan

Boron Trichloride[1]

[10294-34-5] BCl_3 (MW 117.17)

(Lewis acid capable of selective cleavage of ether and acetal protecting groups; reagent for carbonyl condensations; precursor of organoboron reagents)

Physical Data: bp 12.5 °C; *d* 1.434 g cm^{-3} (0 °C).

Solubility: sol saturated and halogenated hydrocarbon and aromatic solvents; solubility in diethyl ether is approximately 1.5 M at 0 °C; stable for several weeks in ethyl ether at 0 °C, but dec by water or alcohols.

Form Supplied in: colorless gas or fuming liquid in an ampoule; $BCl_3 \cdot SMe_2$ complex (solid) and 1 M solutions in dichloromethane, hexane, heptane, and *p*-xylene are available.

Handling, Storage, and Precautions: a poison by inhalation and an irritant to skin, eyes, and mucous membranes. Reacts exothermically with water and moist air, forming toxic and corrosive fumes. Violent reaction occurs with aniline or phosphine. All operations should be carried out in a well-ventilated fume hood without exposure to the atmosphere. The gas can be collected and measured as a liquid by condensing in a cooled centrifuge tube and then transferred to the reaction system by distillation with a slow stream of nitrogen.

Cleavage of Ethers, Acetals, and Esters. Like many other Lewis acids, BCl_3 has been extensively used as a reagent for the cleavage of a wide variety of ethers, acetals, and certain types of esters.[2] Ether cleavage procedures involve addition of BCl_3, either neat or as a solution in CH_2Cl_2, to the substrate at −80 °C. The vessel is then stoppered and allowed to warm to rt. Whereas the complexes of BCl_3 with dimethyl ether and diethyl ether are rather stable at rt, they decompose to form $ROBCl_2$ or $(RO)_2BCl$ with evolution of alkyl chloride upon heating to 56 °C.[1] Diaryl ethers are unreactive. Mixed dialkyl ethers are cleaved to give the alkyl chloride derived from C–O bond cleavage leading to the more stable carbenium ion. The transition state is predominantly S_N1 in character, as evidenced by partial racemization of chiral ethers[1,2] and the rearrangement of allyl phenyl ethers to *o*-allylphenols.[3] BCl_3 can be used for the deprotection of a variety of methoxybenzenes including hindered polymethoxybenzenes and *peri*-methoxynaphthalene.[1,2,4] When methoxy groups are *ortho* to a carbonyl group, the reaction is accelerated by the formation of a chelate between boron and the carbonyl oxygen atom (Scheme 1).[4a–c]

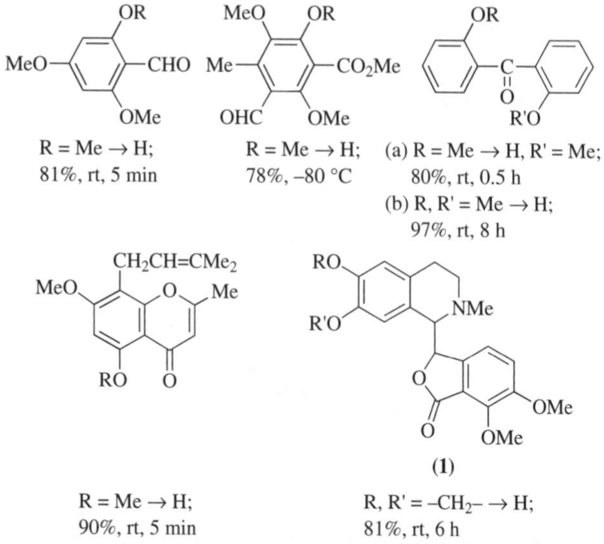

R = Me → H; 81%, rt, 5 min

R = Me → H; 78%, −80 °C

(a) R = Me → H, R' = Me; 80%, rt, 0.5 h
(b) R, R' = Me → H; 97%, rt, 8 h

R = Me → H; 90%, rt, 5 min

(**1**) R, R' = –CH2– → H; 81%, rt, 6 h

Scheme 1 Demethylation of aromatic ethers by BCl_3 in CH_2Cl_2

The reagent is less reactive than ***Boron Tribromide*** for ether cleavage; however, the type and extent of deetherification can be more easily controlled by the ratio of substrate to BCl_3 as well as the reaction temperature and time. The transformation of (−)-β-hydrastine (**1**) to (−)-cordrastine II is efficiently achieved by selective cleavage of the methylenedioxy group in preference to aromatic methoxy groups.[5] The demethylation of (−)-2-*O*-methyl-(−)-inositol in dichloromethane proceeds at −80 °C without cleavage of a tosyl ester group.[6] Methyl glycosides are converted into glycosyl chlorides at −78 °C without effecting benzyl and acetyl protecting groups.[7]

One of the difficulties with the use of BCl_3 arises from its tendency to fume profusely in air. The complex of BCl_3 with dimethyl sulfide is solid, stable in air, and handled easily. By using a two- to fourfold excess of the reagent in dichloroethane at 83 °C, aromatic methoxy and methylenedioxy groups can be cleaved in good yields.[8]

Another application of BCl_3 is for the cleavage of highly hindered esters under mild conditions. *O*-Methylpodocarpate (**2**) and methyl adamantane-1-carboxylate are cleaved at 0 °C.[9] The highly selective displacement of the acetoxy group in the presence of other potentially basic groups in 2-cephem ester (**3**) provides the corresponding allylic chloride. On the other hand, treatment of (**3**) with an excess of BCl_3 results in the cleavage of the acetoxy and *t*-butyl ester groups.[10]

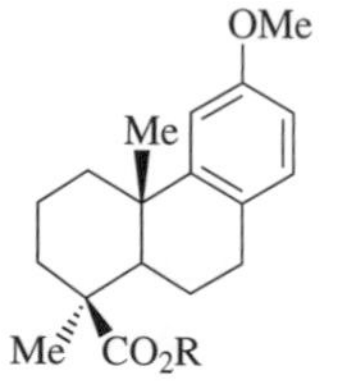

(**2**) R = Me → H; 90%
BCl_3, CH_2Cl_2, 0 °C

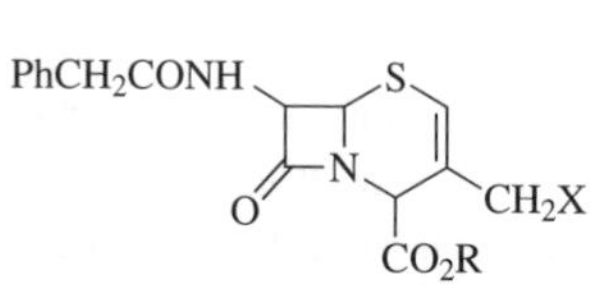

(**3**)
(1) R = *t*-Bu, X = OAc → Cl; 64%
BCl_3 (1 equiv), CH_2Cl_2, −5 °C
(2) R = *t*-Bu → H, X = OAc → OH; 68%
BCl_3 (3 equiv), CH_2Cl_2, rt

Tertiary phosphines are cleaved at the P–C bond to give diphenylphosphine oxides. Workup with ***Hydrogen Peroxide*** provides diphenylphosphinic acids (eq 1).[11]

$$\underset{X = O, S}{Ph_2PCH_2XMe} \xrightarrow[2.\ H_2O]{1.\ BCl_3,\ 0\,^\circ C \to rt} Ph_2P(O)H \xrightarrow{H_2O_2} Ph_2P(O)OH \quad (1)$$

Condensation Reactions. Boron trichloride converts ketones into (*Z*)-boron enolates at −95 °C in the presence of ***Diisopropylethylamine***. These enolates react with aldehydes with high *syn* diastereoselectivity (eq 2).[12] A similar condensation of imines with carbonyl compounds also provides crossed aldols in reasonable yields.[13] The reaction was extended to the asymmetric aldol condensation of acetophenone imine and benzaldehyde by using isobornylamine as a chiral auxiliary (48% ee).[14]

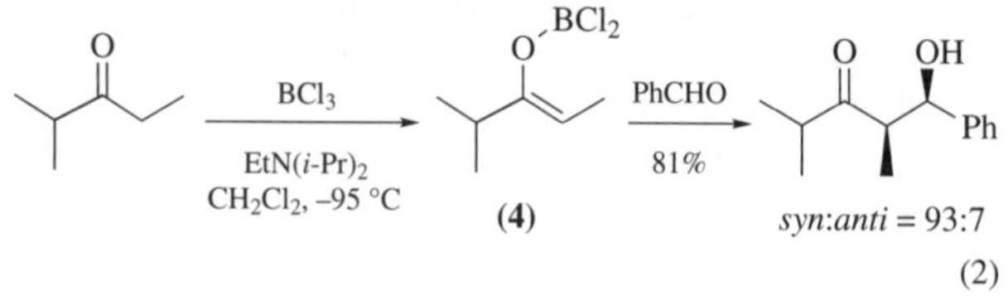

BCl_3, EtN(*i*-Pr)$_2$, CH_2Cl_2, −95 °C → (**4**); PhCHO, 81%; *syn*:*anti* = 93:7 (2)

(*N*-Alkylanilino)dichloroboranes (**5**), prepared in situ from *N*-alkylanilines and boron trichloride, are versatile intermediates for the synthesis of *ortho*-functionalized aniline derivatives (eqs 3–5).[15] The regioselective *ortho* hydroxyalkylation can be achieved with aromatic aldehydes.[16]

The reaction of (**5**) with alkyl and aryl nitriles and ***Aluminum Chloride*** catalyst provides *ortho*-acyl anilines.[16] When chloroacetonitrile is used, the products are ideal precursors for indole synthesis.[17] Use of isocyanides instead of nitriles provides *ortho*-formyl *N*-alkylanilines.[18] Although these reactions with BCl_3 are restricted to *N*-alkylanilines, the use of ***Phenylboron Dichloride*** allows the *ortho*-hydroxybenzylation of primary anilines.[19]

PhNHMe → (BCl_3, Et_3N, benzene, Δ) → PhN(Me)BCl_2 (5)

(5) → 1. ArCHO; 2. H_2O → *o*-(NHMe)C_6H_4CH(OH)Ar (3)

(5) → 1. ArCN, $AlCl_3$; 2. H_2O → *o*-(NHMe)C_6H_4C(=O)R, R = alkyl, aryl (4)

(5) → 1. ArCN; 2. AcOH, H_2O; 3. HCl, H_2O → *o*-(NHMe)C_6H_4CHO (5)

Analogously, boron trichloride induces *ortho* selective acylation of phenols at rt with nitriles, isocyanates, or acyl chlorides (eq 6).[20] The efficiency and regioselectivity of these reactions are best with BCl_3 among the representative metal halides that have been examined. In both the aniline and phenol substitutions the boron atom acts as a template to bring the reactants together, leading to cyclic intermediates and exclusively products of *ortho* substitution. A similar *ortho* selective condensation of aromatic azides with BCl_3 provides fused heterocycles containing nitrogen.[21]

3-MeOC$_6H_4$OH → 1. MeN=C=O, BCl_3, CH_2Cl_2, Δ; 2. H_3O^+ (75%) → 2-HO-4-MeO-C_6H_3C(=O)NHMe (6)

Aldehydes and ketones condense with ketene in the presence of 1 equiv of boron trichloride to give α,β-unsaturated acyl chlorides.[22] Aryl isocyanates are converted into allophanyl chlorides, which are precursors for industrially important 1,3-diazetidinediones (eq 7).[23]

PhN=C=O → (BCl_3, toluene, 0 °C, 70%) → PhNHCON(Ph)COCl → (pyridine, rt, 95%) → 1,3-diphenyl-1,3-diazetidine-2,4-dione (7)

Synthesis of Organoboron Reagents. General method of synthesis of organoboranes consists of the transmetallation reaction of organometallic compounds with BX_3.[24] Boronic acid derivatives [$RB(OH)_2$] are most conveniently synthesized by the reaction of $B(OR)_3$ with RLi or RMgX reagents, but boron trihalides are more advantageous for transmetalation reactions with less nucleophilic organometallic reagents based on Pb,[25] Hg,[26] Sn,[27] and Zr[28] (eqs 8 and 9).

$$Ph_4Sn + 2\,BCl_3 \longrightarrow 2\,PhBCl_2 + Ph_2SnCl_2 \quad (8)$$

$Cp_2Zr(Cl)CH{=}CH(CH_2)_3Me$ → (BCl_3, CH_2Cl_2, 0 °C → rt, >90%) → $Cl_2BCH{=}CH(CH_2)_3Me$ (9)

Redistribution or exchange reactions of R_3B with boron trihalides in the presence of catalytic amounts of hydride provides an efficient synthesis of RBX_2 and R_2BX.[29] Another convenient and general method for the preparation of organodichloroboranes involves treatment of alkyl, 1-alkenyl, and aryl boronates with BCl_3 in the presence of ***Iron(III) Chloride*** (3 mol %).[30] Organodichloroboranes are valuable synthetic reagents because of their high Lewis acidity, and their utility is well demonstrated in the syntheses of piperidine and pyrrolidine derivatives by the intramolecular alkylation of azides (eq 10)[31] or the synthesis of esters by the reaction with ***Ethyl Diazoacetate***.[32] The various organoborane derivatives, R_3B, R_2BCl, and $RBCl_2$, all react with organic azides and diazoacetates. However, especially facile reactions are achieved by using organodichloroboranes ($RBCl_2$).

1-(2-bromoethyl)cyclohexene → 1. $HBCl_2{\bullet}SMe_2$, BCl_3, pentane; 2. H_2O; 3. NaN_3, EtOH, Δ → *trans*-2-(2-azidoethyl)cyclohexyl-$B(OEt)_2$ → (BCl_3, CH_2Cl_2, −78 °C → rt) → 2-(2-azidoethyl)cyclohexyl-BCl_2 → ($-N_2$) → (H_2O, 40% KOH) → octahydroindole, 76% (10)

Dichloroborane and monochloroborane etherates or their methyl sulfide complexes have been prepared by the reaction of borane and boron trichloride.[33] However, hydroboration of alkenes with these borane reagents is usually very slow due to the slow dissociation of the complex. Dichloroborane prepared in pentane from boron trichloride and trimethylsilane shows unusually high reactivity with alkenes and alkynes; hydroboration is instantaneous at −78 °C (eq 11).[34]

BCl_3 → ($HSiMe_3$, pentane, −78 °C) → $HBCl_2$ → (α-pinene, 95%) → pinanyl-BCl_2 (11)

Direct boronation of benzene derivatives with BCl_3 in the presence of activated aluminum or $AlCl_3$ provides arylboronic acids after hydrolysis (eq 12).[35] Chloroboration of acetylene with boron trichloride produces dichloro(2-chloroethenyl)borane.[36] Similar reaction with phenylacetylene provides (*E*)-2-chloro-2-phenylethenylborane regio- and stereoselectively.[37]

$$PhH + BCl_3 \xrightarrow[\text{I}_2\text{ (trace), MeI (trace), 150 °C, 60–79\%}]{\text{Al}} PhBCl_2 \quad (12)$$

The syntheses of thioaldehydes, thioketones, thiolactones, and thiolactams from carbonyl compounds are readily achieved by in situ preparation of B_2S_3 from bis(tricyclohexyltin) sulfide and boron trichloride (eq 13).[38] The high sulfurating ability of this in situ prepared reagent can be attributed to its solubility in the reaction medium.

$$3\,(Cy_3Sn)=S + 2\,BCl_3 \xrightarrow[\Delta]{\text{toluene}} B_2S_3 \quad (13)$$

94%, Δ, 7 h　　92%, Δ, 3 h　　unable to isolate

Related Reagents. Bis(tricyclohexyltin); Boron Tribromide; Hydrogen Chloride; Sulfide–Boron Trichloride.

1. Gerrard, W.; Lappert, M. F. *CRV* **1958**, *58*, 1081.
2. (a) Bhatt, M. V.; Kulkarni, S. U. *S* **1983**, 249. (b) Greene, T. W. *Protective Groups in Organic Synthesis*; Wiley: New York, 1981.
3. (a) Gerrard, W.; Lappert, M. F.; Silver, H. B. *Proc. Chem. Soc.* **1957**, 19. (b) Borgulya, J.; Madeja, R.; Fahrni, P.; Hansen, H.-J.; Schmid, H.; Barner, R. *HCA* **1973**, *56*, 14.
4. (a) Dean, R. B.; Goodchild, J.; Houghton, L. E.; Martin, J. A. *TL* **1966**, 4153. (b) Arkley, V.; Attenburrow, J.; Gregory, G. I.; Walker, T. *JCS* **1962**, 1260. (c) Barton, D. H. R.; Bould, L.; Clive, D. L. J.; Magnus, P. D.; Hase, T. *JCS(C)* **1971**, 2204. (d) Carvalho, C. F.; Seargent, M. V. *CC* **1984**, 227.
5. (a) Teitel, S.; O'Brien, J.; Brossi, A. *JOC* **1972**, *37*, 3368. (b) Teitel, S.; O'Brien, J. P. *JOC* **1976**, *41*, 1657.
6. Gero, S. D. *TL* **1966**, 591.
7. Perdomo, G. R.; Krepinsky, J. J. *TL* **1987**, *28*, 5595.
8. Williard, P. G.; Fryhle, C. B. *TL* **1980**, *21*, 3731.
9. Manchand, P. S. *CC* **1971**, 667.
10. Yazawa, H.; Nakamura, H.; Tanaka, K.; Kariyone, K. *TL* **1974**, 3991.
11. Hansen, K. C.; Solleder, G. B.; Holland, C. L. *JOC* **1974**, *39*, 267.
12. Chow, H-F.; Seebach, D. *HCA* **1986**, *69*, 604.
13. Sugasawa, T.; Toyoda, T. Sasakura, K. *SC* **1979**, *9*, 515.
14. Sugasawa, T.; Toyoda, T.; *TL* **1979**, 1423.
15. Sugasawa, T. *J. Synth. Org. Chem. Jpn.* **1981**, *39*, 39.
16. Sugasawa, T.; Toyoda, T.; Adachi, M.; Sasakura, K. *JACS* **1978**, *100*, 4842.
17. Sugasawa, T.; Adachi, M.; Sasakura, K.; Kitagawa, A. *JOC* **1979**, *44*, 578.
18. Sugasawa, T.; Hamana, H.; Toyoda, T.; Adachi, M. *S* **1979**, 99.
19. Toyoda, T.; Sasakura, K.; Sugasawa, T. *TL* **1980**, *21*, 173.
20. (a) Toyoda, T.; Sasakura, K.; Sugasawa, T. *JOC* **1981**, *46*, 189. (b) Piccolo, O.; Filippini, L.; Tinucci, L.; Valoti, E.; Citterio, A. *T* **1986**, *42*, 885.
21. (a) Zanirato, P. *CC* **1983**, 1065. (b) Spagnolo, P.; Zanirato, P. *JCS(P1)* **1988**, 2615.
22. Paetzold, P. I.; Kosma, S. *CB* **1970**, *103*, 2003.
23. Helfert, H.; Fahr, E. *AG(E)* **1970**, *9*, 372.
24. (a) Nesmeyanov, A. N.; Kocheshkov, K. A. *Methods of Elemento-Organic Chemistry*; North-Holland: Amsterdam, 1967; Vol. 1, pp 20–96. (b) Mikhailov, B. M.; Bubnov, Y. N. *Organoboron Compounds in Organic Synthesis*; Harwood: Amsterdam, 1984.
25. Holliday, A. K.; Jessop, G. N. *JCS(A)* **1967**, 889.
26. Gerrard, W.; Howarth, M.; Mooney, E. F.; Pratt, D. E. *JCS* **1963**, 1582.
27. (a) Niedenzu, K.; Dawson, J. W. *JACS* **1960**, *82*, 4223. (b) Brinkman, F. E.; Stone, F. G. A. *CI(L)* **1959**, 254.
28. Cole, T. E.; Quintanilla, R.; Rodewald, S. *OM* **1991**, *10*, 3777.
29. Brown, H. C.; Levy, A. B. *JOM* **1972**, *44*, 233.
30. (a) Brindley, P. B.; Gerrard, W.; Lappert, M. F. *JCS* **1956**, 824. (b) Brown, H. C.; Salunkhe, A. M.; Argade, A. B. *OM* **1992**, *11*, 3094.
31. (a) Jego, J. M.; Carboni, B.; Vaultier, M.; Carrie', R. *CC* **1989**, 142. (b) Brown, H. C.; Salunkhe, A. M. *TL* **1993**, *34*, 1265.
32. Hooz, J.; Bridson, J. N.; Calzada, J. G.; Brown, H. C.; Midland, M. M.; Levy, A. B. *JOC* **1973**, *38*, 2574.
33. (a) Brown, H. C. *Organic Syntheses via Boranes*; Wiley: New York, 1975; pp 45–47. (b) Brown, H. C.; Kulkarni, S. U. *JOM* **1982**, *239*, 23. (c) Brown, H. C.; Ravindran, N. *IC* **1977**, *16*, 2938.
34. Soundararajan, R.; Matteson, D. S. *JOC* **1990**, *55*, 2274.
35. (a) Muetterties, E. L. *JACS* **1960**, *82*, 4163. (b) Lengyel, B.; Csakvari, B. *Z. Anorg. Allg. Chem.* **1963**, *322*, 103.
36. Lappert, M. F.; Prokai, B. *JOM* **1964**, *1*, 384.
37. Blackborow, J. R. *JCS(P2)* **1973**, 1989.
38. Steliou, K.; Mrani, M. *JACS* **1982**, *104*, 3104.

Norio Miyaura
Hokkaido University, Sapporo, Japan

Boron Trifluoride Etherate

$BF_3{\cdot}OEt_2$

($BF_3{\cdot}OEt_2$)		
[109-63-7]	$C_4H_{10}BF_3O$	(MW 141.94)
($BF_3{\cdot}MeOH$)		
[373-57-9]	CH_4BF_3O	(MW 99.85)

($BF_3{\cdot}OEt_2$: easy-to-handle and convenient source of BF_3; Lewis acid catalyst; promotes epoxide cleavage and rearrangement, control of stereoselectivity; $BF_3{\cdot}MeOH$: esterification of aliphatic and aromatic acids; cleavage of trityl ethers)

Alternate Names: boron trifluoride diethyl etherate; boron trifluoride ethyl etherate; boron trifluoride ethyl ether complex; trifluoroboron diethyl etherate.

Physical Data: $BF_3{\cdot}OEt_2$: bp 126 °C; d 1.15 g cm^{-3}; $BF_3{\cdot}MeOH$: bp 59 °C/4 mmHg; d 1.203 g cm^{-3} for 50 wt % BF_3, 0.868 g cm^{-3} for 12 wt % BF_3.

Solubility: sol benzene, chloromethanes, dioxane, ether, methanol, THF, toluene.

Form Supplied in: $BF_3{\cdot}OEt_2$: light yellow liquid, packaged under nitrogen or argon; $BF_3{\cdot}MeOH$ is available in solutions of 10–50% BF_3 in MeOH.

Preparative Methods: $BF_3{\cdot}OEt_2$ is prepared by passing BF_3 through anhydrous ether;[1a] the $BF_3{\cdot}MeOH$ complex is formed from $BF_3{\cdot}OEt_2$ and methanol.

Purification: oxidation in air darkens commercial boron trifluoride etherate; therefore the reagent should be redistilled prior to use. An excess of the etherate in ether should be distilled in an all-glass apparatus with calcium hydroxide to remove volatile acids and to reduce bumping.[1b]

Handling, Storage, and Precautions: keep away from moisture and oxidants; avoid skin contact and work in a well-ventilated fume hood.

Addition Reactions. $BF_3{\cdot}OEt_2$ facilitates the addition of moderately basic nucleophiles like alkyl-, alkenyl-, and aryllithium, imines, Grignard reagents, and enolates to a variety of electrophiles.[2]

Organolithiums undergo addition reactions with 2-isoxazolines to afford *N*-unsubstituted isoxazolidines, and to the carbon–nitrogen double bond of oxime *O*-ethers to give *O*-alkylhydroxylamines.[3] Aliphatic esters react with lithium acetylides in the presence of $BF_3{\cdot}OEt_2$ in THF at −78 °C to form alkynyl ketones in 40–80% yields.[4] Alkynylboranes, generated in situ from lithium acetylides and $BF_3{\cdot}OEt_2$, were found to react with oxiranes[5] and oxetanes[6] under mild conditions to afford β-hydroxyalkynes and γ-alkoxyalkynes, respectively. (1-Alkenyl)dialkoxyboranes react stereoselectively with α,β-unsaturated ketones[7] and esters[8] in the presence of $BF_3{\cdot}OEt_2$ to give γ,δ-unsaturated ketones and α-acyl-γ,δ-unsaturated esters, respectively.

The reaction of imines activated by $BF_3{\cdot}OEt_2$ with 4-(phenylsulfonyl)butanoic acid dianion leads to 2-piperidones in high yields.[9] (Perfluoroalkyl)lithiums, generated in situ, add to imines in the presence of $BF_3{\cdot}OEt_2$ to give perfluoroalkylated amines.[10] Enolate esters add to 3-thiazolines under mild conditions to form thiazolidines if these imines are first activated with $BF_3{\cdot}OEt_2$.[11] The carbon–nitrogen double bond of imines can be alkylated with various organometallic reagents to produce amines.[12] A solution of benzalaniline in acetone treated with $BF_3{\cdot}OEt_2$ results in the formation of β-phenyl-β-anilinoethyl methyl ketone.[13] Anilinobenzylphosphonates are synthesized in one pot using aniline, benzaldehyde, dialkyl phosphite, and $BF_3{\cdot}OEt_2$;[14] the reagent accelerates imine generation and dialkyl phosphite addition. Similarly, $BF_3{\cdot}OEt_2$ activates the nitrile group of cyanocuprates, thereby accelerating Michael reactions.[15]

The reagent activates iodobenzene for the allylation of aromatics, alcohols, and acids.[16] Allylstannanes are likewise activated for the allylation of *p*-benzoquinones, e.g. in the formation of coenzyme Q_n using polyprenylalkylstannane.[17]

Nucleophilic silanes undergo stereospecific addition to electrophilic glycols activated by Lewis acids. The glycosidation is highly stereoselective with respect to the glycosidic linkage in some cases using $BF_3{\cdot}OEt_2$. Protected pyranosides undergo stereospecific *C*-glycosidation with C-1-oxygenated allylsilanes to form α-glycosides.[18,19] α-Methoxyglycine esters react with allylsilanes and silyl enol ethers in the presence of $BF_3{\cdot}OEt_2$ to give racemic γ,δ-unsaturated α-amino acids and γ-oxo-α-amino acids, respectively.[20] β-Glucopyranosides are synthesized from an aglycon and 2,3,4,6-tetra-*O*-acetyl-β-D-glucopyranose.[21] Alcohols and silyl ethers also undergo stereoselective glycosylation with protected glycosyl fluorides to form β-glycosides.[22]

$BF_3{\cdot}OEt_2$ reverses the usual *anti* selectivity observed in the reaction of crotyl organometallic compounds (based on Cu, Cd, Hg, Sn, Tl, Ti, Zr, and V, but not on Mg, Zn, or B) with aldehydes (eq 1a) and imines (eq 1b), so that homoallyl alcohols and homoallylamines are formed, respectively.[23–28] The products show mainly *syn* diastereoselectivity. $BF_3{\cdot}OEt_2$ is the only Lewis acid which produces hydroxy- rather than halo-tetrahydropyrans from the reaction of allylstannanes with pyranosides.[29] The $BF_3{\cdot}OEt_2$ mediated condensations of γ-oxygenated allylstannanes with aldehydes (eq 1c) and with 'activated' imines (eq 1d) affords vicinal diol derivatives and 1,2-amino alcohols, respectively, with *syn* diastereoselectivity.[30,31]

The 'activated' imines are obtained from aromatic amines, aliphatic aldehydes, and α-ethoxycarbamates. The reaction of aldehydes with α-(alkoxy)-β-methylallylstannanes with aldehydes in the presence of $BF_3{\cdot}OEt_2$ gives almost exclusively *syn*-(*E*)-isomers.[31]

Y~~SnBu₃ + RC(=X)H —($BF_3{\cdot}OEt_2$)→ R–CH(X)–CH(Y)–CH=CH₂ (1)

(a) X = O, Y = Me — (a) X = OH, Y = Me
(b) X = NR², Y = Me — (b) X = NHR², Y = Me
(c) X = O, Y = OMe, OTBDMS — (c) X = OH, Y = OMe, OTBDMS
(d) X = NR², Y = OMe, OTBDMS or OCH₂OMe — (d) X = NHR², Y = OH or derivative

The reaction of α-diketones with allyltrimethylstannane in the presence of $BF_3{\cdot}OEt_2$ yields a mixture of homoallylic alcohols, with the less hindered carbonyl group being allylated predominantly.[32] The reaction between aldehydes and allylic silanes with an asymmetric ethereal functionality produces *syn*-homoallyl alcohols when ***Titanium(IV) Chloride*** is coordinated with the allylic silane and *anti* isomers with $BF_3{\cdot}OEt_2$.[33]

Chiral oxetanes can be synthesized by the $BF_3{\cdot}OEt_2$ catalyzed [2 + 2] cycloaddition reactions of 2,3-*O*-isopropylidenealdehyde-D-aldose derivatives with allylsilanes, vinyl ethers, or vinyl sulfides.[34] The regiospecificity and stereoselectivity is greater than in the photochemical reaction; *trans*-2-alkoxy- and *trans*-2-phenylthiooxetanes are the resulting products.

2-Alkylthioethyl acetates can be formed from vinyl acetates by the addition of thiols with $BF_3{\cdot}OEt_2$ as the catalyst.[35] The yield is 79%, compared to 75% when $BF_3{\cdot}OEt_2$ is used in conjunction with ***Mercury(II) Sulfate*** or ***Mercury(II) Oxide***.

α-Alkoxycarbonylallylsilanes react with acetals in the presence of $BF_3{\cdot}OEt_2$ (eq 2).[36] The products can be converted into α-methylene-γ-butyrolactones by dealkylation with ***Iodotrimethylsilane***.

Me₃Si–CH₂–C(=CH₂)CO₂Et + PhCH(OMe)₂ —($BF_3{\cdot}OEt_2$, 89%)→ CH₂=C(CO₂Et)–CH₂–CH(Ph)OMe (2)

The cuprate 1,4-conjugate addition step in the synthesis of (+)-modhephene is difficult due to the neopentyl environment of C-4 in the enone, but it can occur in the presence of $BF_3{\cdot}OEt_2$ (eq 3).[37]

—($BF_3{\cdot}OEt_2$, Me₂CuLi, 70%)→ (3)

The reagent is used as a Lewis acid catalyst for the intramolecular addition of diazo ketones to alkenes.[38] The direct synthesis of bicyclo[3.2.1]octenones from the appropriate diazo ketones using $BF_3{\cdot}OEt_2$ (eq 4) is superior to the copper-catalyzed thermal decomposition of the diazo ketone to a cyclopropyl ketone and subsequent acid-catalyzed cleavage.[38]

R, COCHN₂ —($BF_3{\cdot}OEt_2$, ClCH₂CH₂Cl, 0–27 °C, 30–51%)→ R, O (4)

$BF_3{\cdot}OEt_2$ reacts with fluorinated amines to form salts which are analogous to Vilsmeier reagents, Arnold reagents, or phosgene–immonium salts (eq 5).[39] These salts are used to acylate electron-rich aromatic compounds, introducing a fluorinated carbonyl group (eq 6).

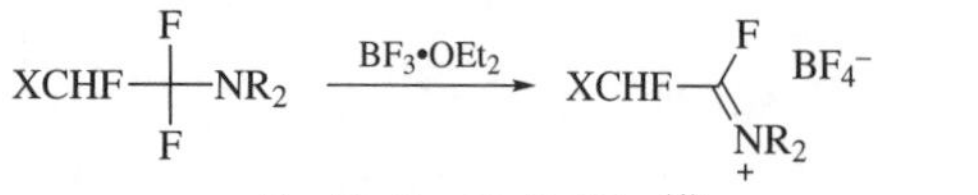

(5)

R = Et; X = Cl, F, CF_3 (1)

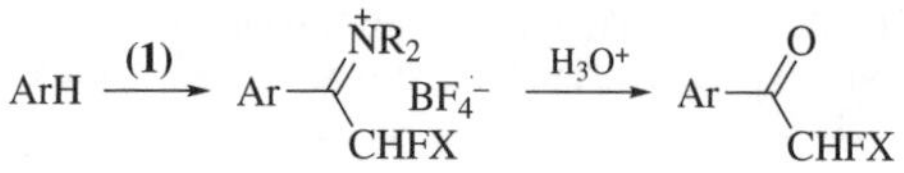

(6)

Xenon(II) Fluoride and methanol react to form ***Methyl Hypofluorite***, which reacts as a positive oxygen electrophile in the presence of BF_3 (etherate or methanol complex) to yield anti-Markovnikov fluoromethoxy products from alkenes.[40,41]

Aldol Reactions. Although ***Titanium(IV) Chloride*** is a better Lewis acid in effecting aldol reactions of aldehydes, acetals, and silyl enol ethers, $BF_3 \cdot OEt_2$ is more effective for aldol reactions with anions generated from transition metal carbenes and with tetrasubstituted enol ethers such as (*Z*)- and (*E*)-3-methyl-2-(trimethylsilyloxy)-2-pentene.[42,43] One exception involves the preparation of substituted cyclopentanediones from acetals by the aldol condensation of protected four-membered acyloin derivatives with $BF_3 \cdot OEt_2$ rather than $TiCl_4$ (eq 7).[44] The latter catalyst causes some loss of the silyl protecting group. The pinacol rearrangement is driven by the release of ring strain in the four-membered ring and controlled by an acyl group adjacent to the diol moiety.

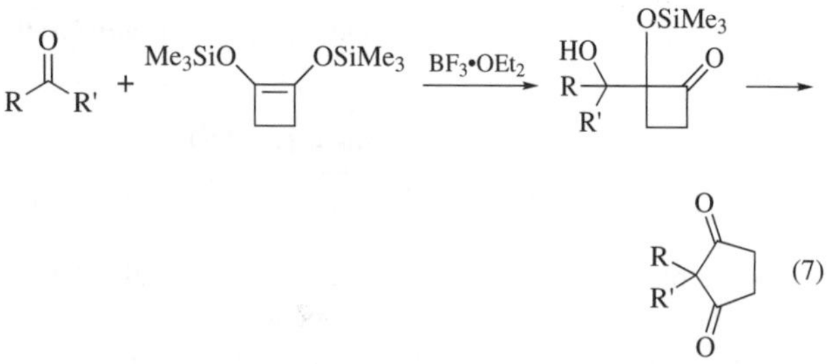

(7)

The reagent is the best promoter of the aldol reaction of 2-(trimethylsilyloxy)acrylate esters, prepared by the silylation of pyruvate esters, to afford γ-alkoxy-α-keto esters (eq 8).[45] These esters occur in a variety of important natural products.

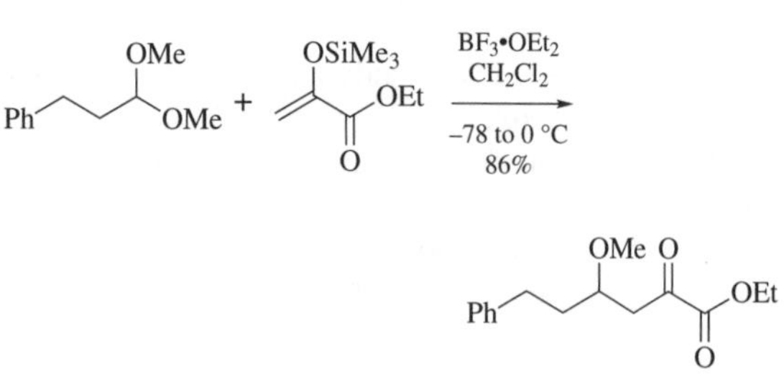

(8)

$BF_3 \cdot OEt_2$ can improve or reverse the aldehyde diastereofacial selectivity in the aldol reaction of silyl enol ethers with aldehydes, forming the *syn* adducts. For example, the reaction of the silyl enol ether of pinacolone with 2-phenylpropanal using $BF_3 \cdot OEt_2$ gives enhanced levels of Felkin selectivity relative to the addition of the corresponding lithium enolate.[46,47] In the reaction of silyl enol ethers with 3-formyl-Δ^2-isoxazolines, $BF_3 \cdot OEt_2$ gives predominantly *anti* aldol adducts, whereas other Lewis acids give *syn* aldol adducts.[48] The reagent can give high diastereofacial selectivity in the addition of silyl enol ethers or silyl ketones to chiral aldehydes.[49] In the addition of a nonstereogenic silylketene acetal to chiral, racemic α-thioaldehydes, $BF_3 \cdot OEt_2$ leads exclusively to the *anti* product.[49]

1,5-Dicarbonyl compounds are formed from the reaction of silyl enol ethers with methyl vinyl ketones in the presence of $BF_3 \cdot OEt_2$ and an alcohol (eq 9).[50] α-Methoxy ketones are formed from α-diazo ketones with $BF_3 \cdot OEt_2$ and methanol, or directly from silyl enol ethers using iodobenzene/$BF_3 \cdot OEt_2$ in methanol.[51]

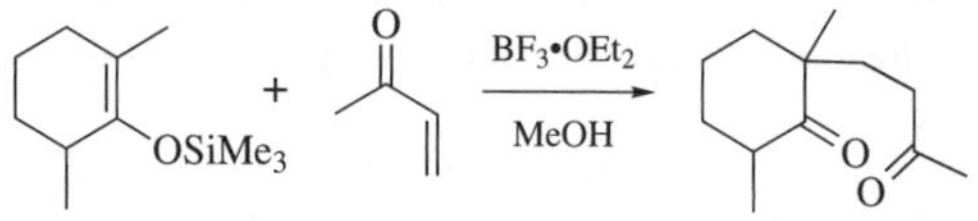

(9)

α-Mercurio ketones condense with aldehydes in the presence of $BF_3 \cdot OEt_2$ with predominant *erythro* selectivity (eq 10).[52] Enaminosilanes derived from acylic and cyclic ketones undergo *syn* selective aldol condensations in the presence of $BF_3 \cdot OEt_2$.[53]

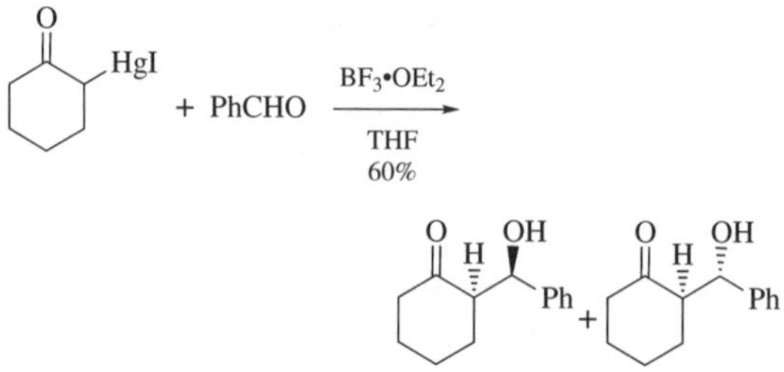

(10)

erythro 90:10 *threo*

Cyclizations. Arylamines can undergo photocyclization in the presence of $BF_3 \cdot OEt_2$ to give tricyclic products, e.g. 9-azaphenanthrene derivatives (eq 11).[54]

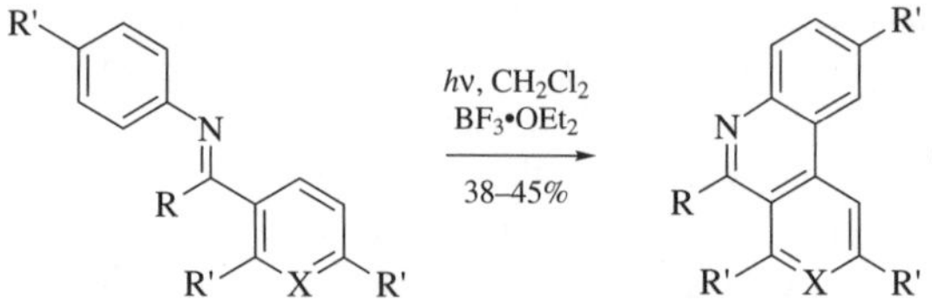

(11)

R = H, Me; R' = H, OMe; X = CH, N

Substituted phenethyl isocyanates undergo cyclization to lactams when treated with $BF_3 \cdot OEt_2$.[55] Vinyl ether epoxides (eq 12),[56] vinyl aldehydes,[57] and epoxy β-keto esters[58] all undergo cyclization with $BF_3 \cdot OEt_2$.

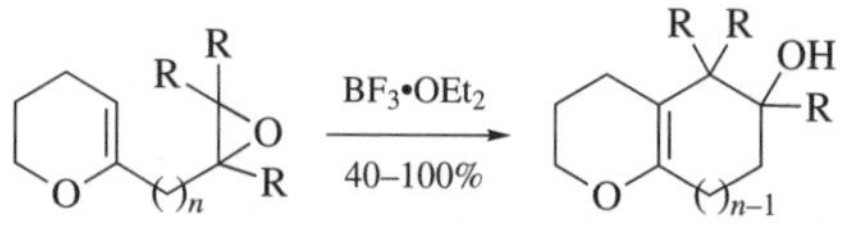

(12)

R = H, Me

β-Silyl divinyl ketones (Nazarov reagents) in the presence of $BF_3 \cdot OEt_2$ cyclize to give cyclopentenones, generally with retention of the silyl group.[59] $BF_3 \cdot OEt_2$ is used for the key step in the synthesis of the sesquiterpene trichodiene, which has adjacent quaternary centers, by catalyzing the cyclization of the dienone to the tricyclic ketone (eq 13).[60] Trifluoroacetic acid and trifluoroacetic anhydride do not catalyze this cyclization.

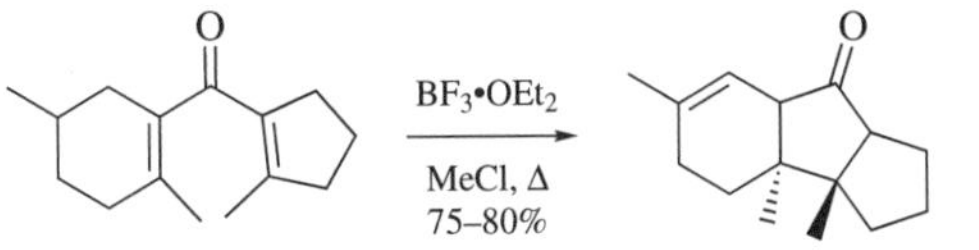

(13)

Costunolide, treated with $BF_3 \cdot OEt_2$, produces the cyclocostunolide (**2**) and a C-4 oxygenated sesquiterpene lactone (**3**), 4α-hydroxycyclocostunolide (eq 14).[61]

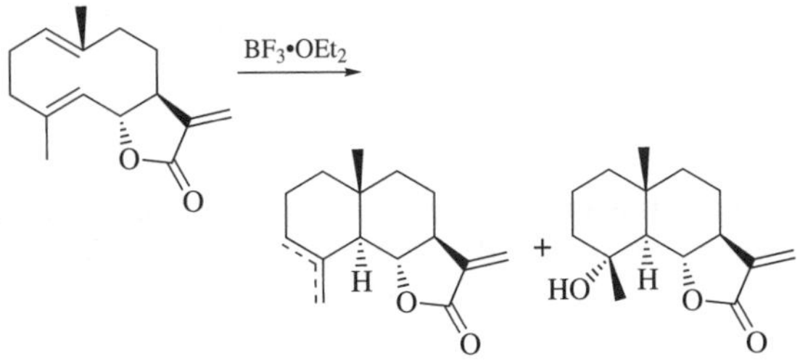

(14)

Other Condensation Reactions. $BF_3 \cdot MeOH$ and $BF_3 \cdot OEt_2$ with ethanol are widely used in the esterification of various kinds of aliphatic, aromatic, and carboxylic acids;[62] the reaction is mild, and no rearrangement of double bonds occurs. This esterification is used routinely for stable acids prior to GLC analysis. Heterocyclic carboxylic acids,[63] unsaturated organic acids,[64] biphenyl-4,4′-dicarboxylic acid,[65] 4-aminobenzoic acid,[63] and the very sensitive 1,4-dihydrobenzoic acid[65] are esterified directly.

The dianion of acetoacetate undergoes Claisen condensations with tetramethyldiamide derivatives of dicarboxylic acids to produce polyketides in the presence of $BF_3 \cdot OEt_2$ (eq 15).[66] Similarly, 3,5-dioxoalkanoates are synthesized from tertiary amides or esters with the acetoacetate dianion in the presence of $BF_3 \cdot OEt_2$ (eq 16).[66]

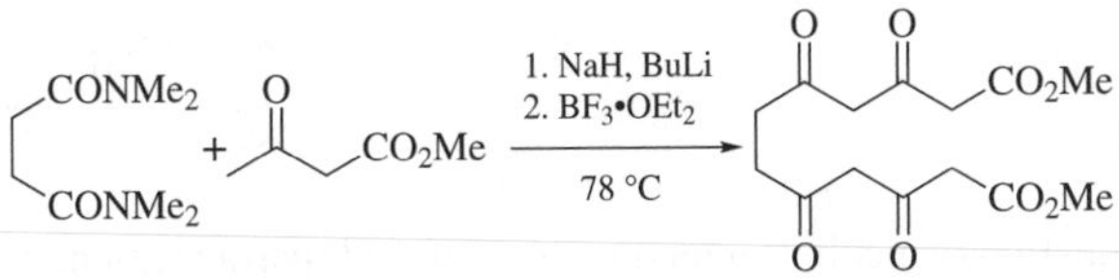

(15)

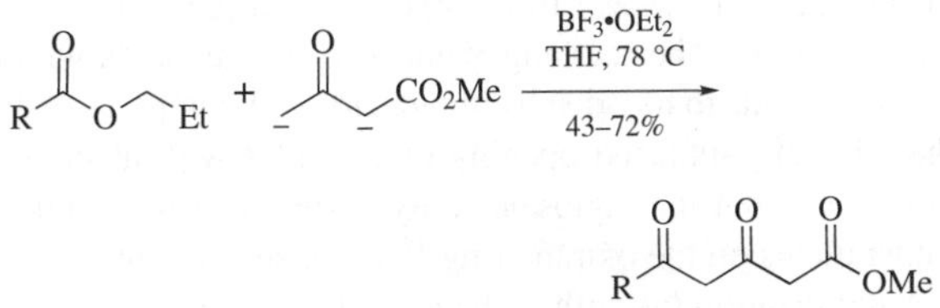

(16)

R = n-C_9H_{19}, $ClCH_2$, Ph

Aldehydes and siloxydienes undergo cyclocondensation with $BF_3 \cdot OEt_2$ to form pyrones (eq 17).[67] The stereoselectivity is influenced by the solvent.

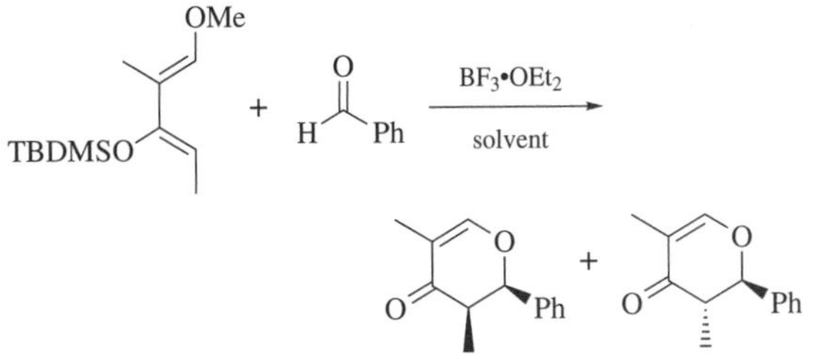

(17)

solvent: CH_2Cl_2 1:2.3
PhMe 7:1

$BF_3 \cdot OEt_2$ is effective in the direct amidation of carboxylic acids to form carboxamides (eq 18).[68] The reaction is accelerated by bases and by azeotropic removal of water.

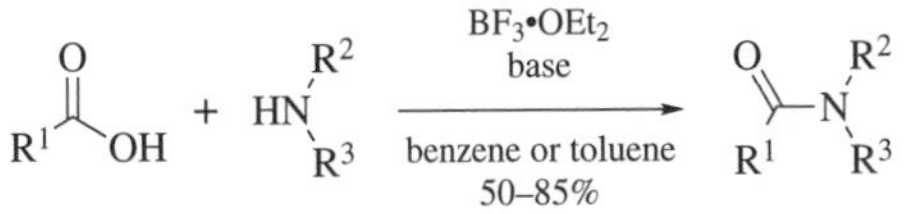

(18)

Carbamates of secondary alcohols can be prepared by a condensation reaction with the isocyanate and $BF_3 \cdot OEt_2$ or ***Aluminum Chloride***.[69] These catalysts are superior to basic catalysts such as pyridine and triethylamine. Some phenylsulfonylureas have been prepared from phenylsulfonamides and isocyanates using $BF_3 \cdot OEt_2$ as a catalyst; for example, 1-butyl-3-(p-tolylsulfonyl)urea is prepared from p-toluenesulfonamide and butyl isocyanate.[70] $BF_3 \cdot OEt_2$ is an excellent catalyst for the condensation of amines to form azomethines (eq 19).[71] The temperatures required are much lower than with ***Zinc Chloride***.

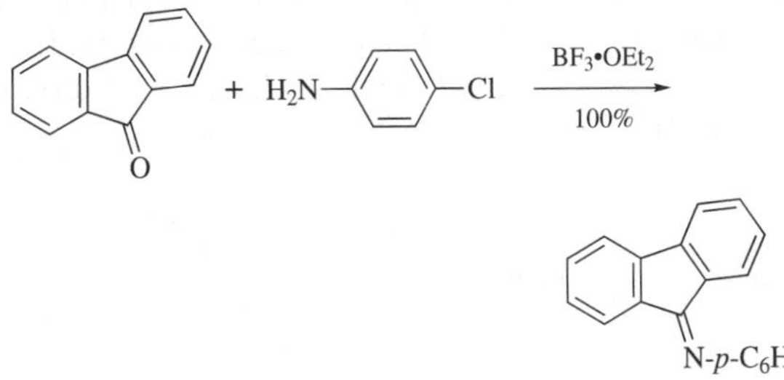

(19)

Acyltetrahydrofurans can be obtained by $BF_3 \cdot OEt_2$ catalyzed condensation of (Z)-4-hydroxy-1-alkenylcarbamates with aldehydes, with high diastereo- and enantioselectivity.[72] Pentasubstituted hydrofurans are obtained by the use of ketones.

Isobornyl ethers are obtained in high yields by the condensation of camphene with phenols at low temperatures using $BF_3 \cdot OEt_2$ as catalyst.[73] Thus camphene and 2,4-dimethylphenol react to give isobornyl 2,4-dimethylphenyl ether, which can undergo further rearrangement with $BF_3 \cdot OEt_2$ to give 2,4-dimethyl-6-isobornylphenol.[73]

The title reagent is also useful for the condensation of allylic alcohols with enols. A classic example is the reaction of phytol in dioxane with 2-methyl-1,4-naphthohydroquinone 1-monoacetate to form the dihydro monoacetate of vitamin K_1 (eq 20), which can be easily oxidized to the quinone.[74]

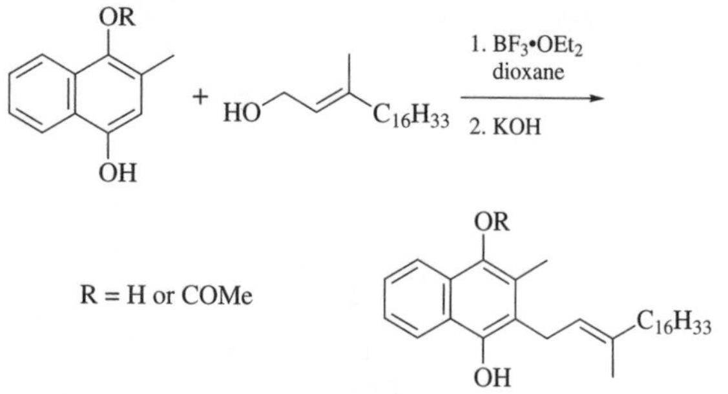

R = H or COMe (20)

$BF_3{\cdot}OEt_2$ promotes fast, mild, clean regioselective dehydration of tertiary alcohols to the thermodynamically most stable alkenes.[75] 11β-Hydroxysteroids are dehydrated by $BF_3{\cdot}OEt_2$ to give $\Delta^{9(11)}$-enes (eq 21).[76,77]

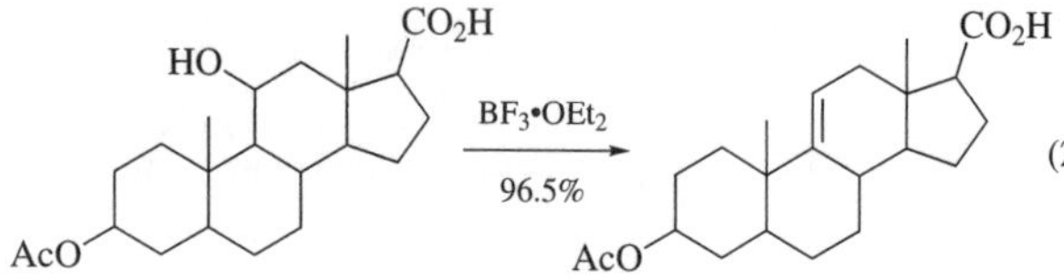

(21)

Epoxide Cleavage and Rearrangements. The treatment of epoxides with $BF_3{\cdot}OEt_2$ results in rearrangements to form aldehydes and ketones (eq 22).[78] The carbon α to the carbonyl group of an epoxy ketone migrates to give the dicarbonyl product.[79] The acyl migration in acyclic α,β-epoxy ketones proceeds through a highly concerted process, with inversion of configuration at the migration terminus.[80] With 5-substituted 2,3-epoxycyclohexanes the stereochemistry of the quaternary carbon center of the cyclopentanecarbaldehyde product is directed by the chirality of the 5-position.[81] Diketones are formed if the β-position of the α,β-epoxy ketone is unsubstituted. The 1,2-carbonyl migration of an α,β-epoxy ketone, 2-cycloheptylidenecyclopentanone oxide, occurs with $BF_3{\cdot}OEt_2$ at 25 °C to form the cyclic spiro-1,3-diketone in 1 min (eq 23).[82]

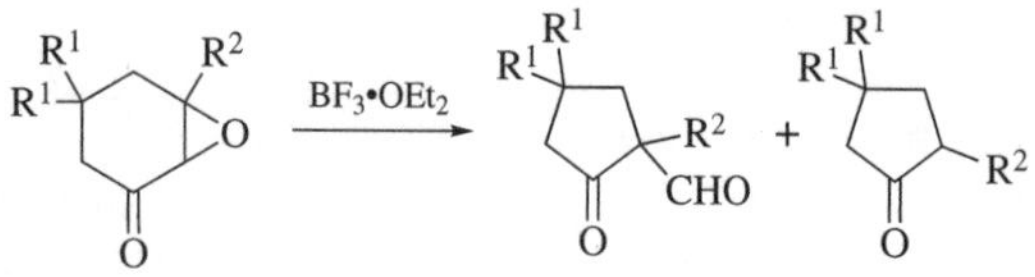

(22)

R^1 = Me, H; R^2 = Me, Ph

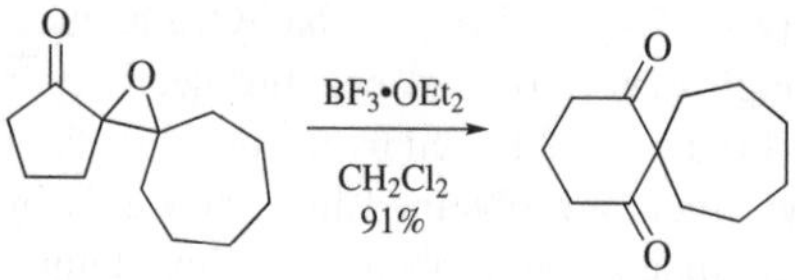

(23)

The migration of the carbonyl during epoxide cleavage is used to produce hydroxy lactones from epoxides of carboxylic acids (eq 24).[83] α-Acyl-2-indanones,[84] furans,[85] and Δ^2-oxazolines[86] (eq 25) can also be synthesized by the cleavage and rearrangement of epoxides with $BF_3{\cdot}OEt_2$. The last reaction has been conducted with sulfuric acid and with tin chloride, but the yields were lower. γ,δ-Epoxy tin compounds react with $BF_3{\cdot}OEt_2$ to give the corresponding cyclopropylcarbinyl alcohols (eq 26).[87]

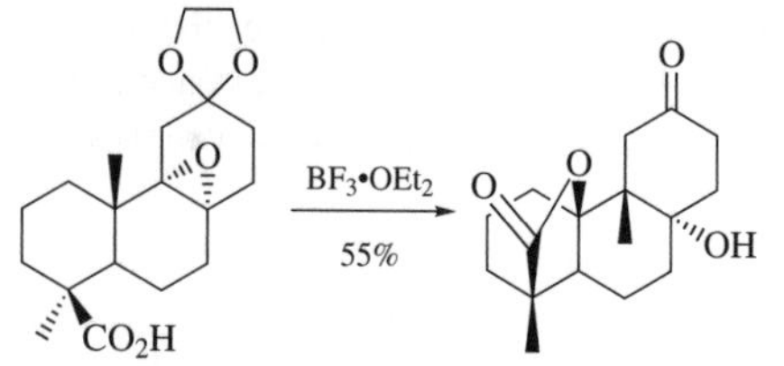

(24)

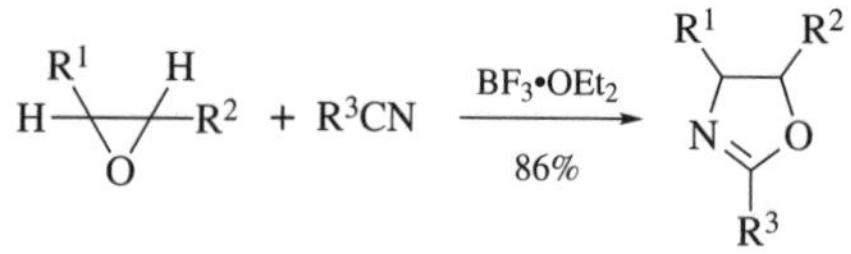

(25)

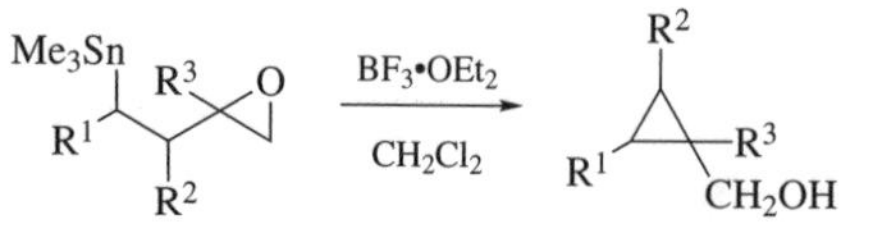

(26)

Remotely unsaturated epoxy acids undergo fission rearrangement when treated with $BF_3{\cdot}OEt_2$. Hence, *cis* and *trans* keto-cyclopropane esters are produced from the unsaturated epoxy ester methyl vernolate (eq 27).[88]

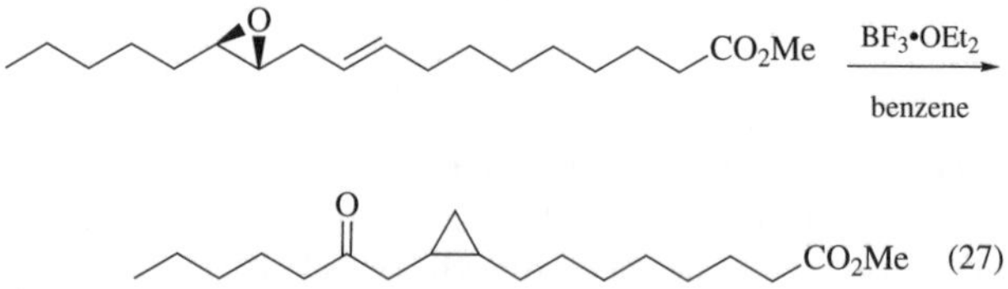

(27)

Epoxy sulfones undergo rearrangement with $BF_3{\cdot}OEt_2$ to give the corresponding aldehydes.[89] α-Epoxy sulfoxides, like other negatively substituted epoxides, undergo rearrangement in which the sulfinyl group migrates and not the hydrogen, alkyl, or aryl groups (eq 28).[89]

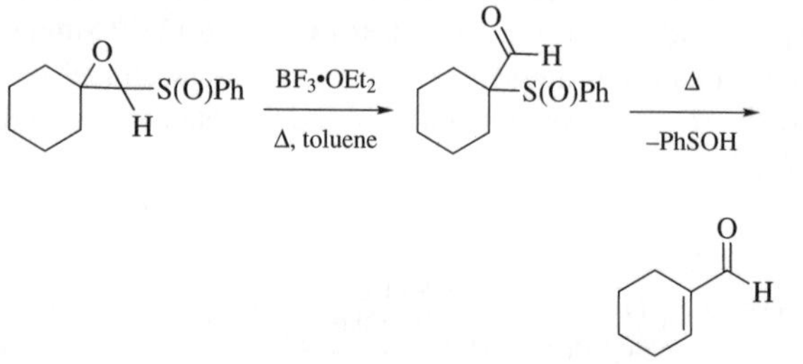

(28)

α,β-Epoxy alcohols undergo cleavage and rearrangement with $BF_3{\cdot}OEt_2$ to form β-hydroxy ketones.[90] The rearrangement is stereospecific with respect to the epoxide and generally results in *anti* migration. The rearrangement of epoxy alcohols with β-substituents leads to α,α-disubstituted carbonyl compounds.[91]

The $BF_3{\cdot}OEt_2$-induced opening of epoxides with alcohols is regioselective, but the regioselectivity varies with the nature of the substituents on the oxirane ring.[92] If the substituent provides charge stabilization (as with a phenyl ring), the internal position is attacked exclusively. On the other hand, terminal ethers are formed by the regioselective cleavage of the epoxide ring of glycidyl tosylate.[92]

A combination of cyanoborohydride and $BF_3{\cdot}OEt_2$ is used for the regio- and stereoselective cleavage of most epoxides to the less substituted alcohols resulting from *anti* ring opening.[93] The reaction rate of organocopper and cuprate reagents with slightly reactive epoxides, e.g. cyclohexene oxide, is dramatically enhanced by $BF_3 \cdot OEt_2$.[94] The Lewis acid and nucleophile work in a concerted manner so that *anti* products are formed.

Azanaphthalene *N*-oxides undergo photochemical deoxygenation reactions in benzene containing $BF_3 \cdot OEt_2$, resulting in amines in 70–80% yield;[95] these amines are important in the synthesis of heterocyclic compounds. ***Azidotrimethylsilane*** reacts with *trans*-1,2-epoxyalkylsilanes in the presence of $BF_3 \cdot OEt_2$ to produce (*Z*)-1-alkenyl azides.[96] The *cis*-1,2-epoxyalkylsilanes undergo rapid polymerization in the presence of Lewis acids.

Other Rearrangements. $BF_3 \cdot OEt_2$ is used for the regioselective rearrangement of polyprenyl aryl ethers to yield polyprenyl substituted phenols, e.g. coenzyme Q_n.[97] The reagent is used in the Fries rearrangement; for example, 5-acetyl-6-hydroxycoumaran is obtained in 96% yield from 6-acetoxycoumaran using this reagent (eq 29).[98]

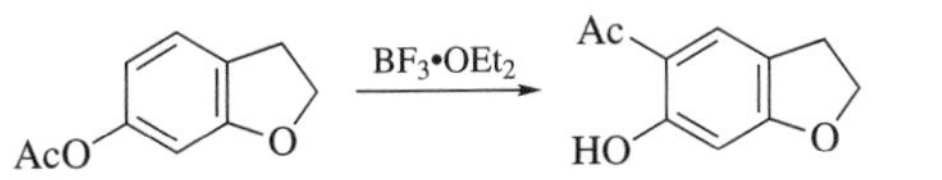

(29)

Formyl bicyclo[2.2.2]octane undergoes the retro-Claisen rearrangement to a vinyl ether in the presence of $BF_3 \cdot OEt_2$ at 0 °C (eq 30), rather than with HOAc at 110 °C.[99]

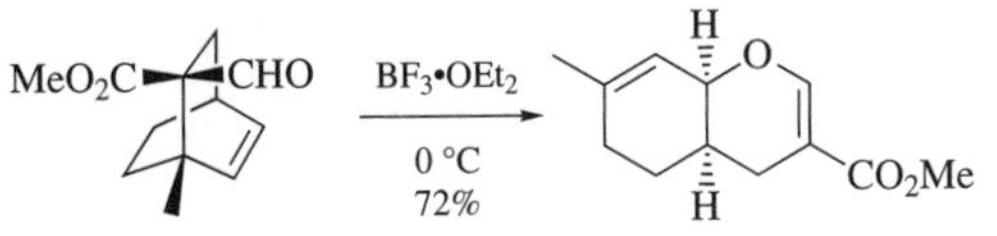

(30)

$BF_3 \cdot OEt_2$ is used for a stereospecific 1,3-alkyl migration to form *trans*-2-alkyltetrahydrofuran-3-carbaldehydes from 4,5-dihydrodioxepins (eq 31), which are obtained by the isomerization of 4,7-dihydro-1,3-dioxepins.[100] Similarly, α-alkyl-β-alkoxyaldehydes can be prepared from 1-alkenyl alkyl acetals by a 1,3-migration using $BF_3 \cdot OEt_2$ as catalyst.[101] *Syn* products are obtained from (*E*)-1-alkenyl alkyl acetals and *anti* products from the (*Z*)-acetals.

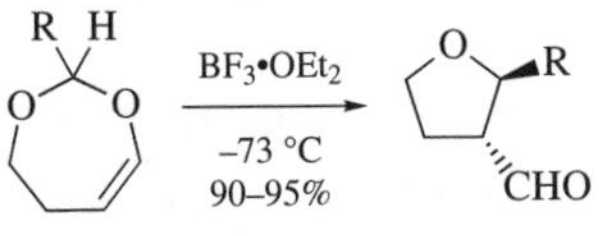

(31)

The methyl substituent, and not the cyano group, of 4-methyl-4-cyanocyclohexadienone migrates in the presence of $BF_3 \cdot OEt_2$ to give 3-methyl-4-cyanocyclohexadienone.[102] $BF_3 \cdot OEt_2$-promoted regioselective rearrangements of polyprenyl aryl ethers provide a convenient route for the preparation of polyprenyl-substituted hydroquinones (eq 32), which can be oxidized to polyprenylquinones.[103]

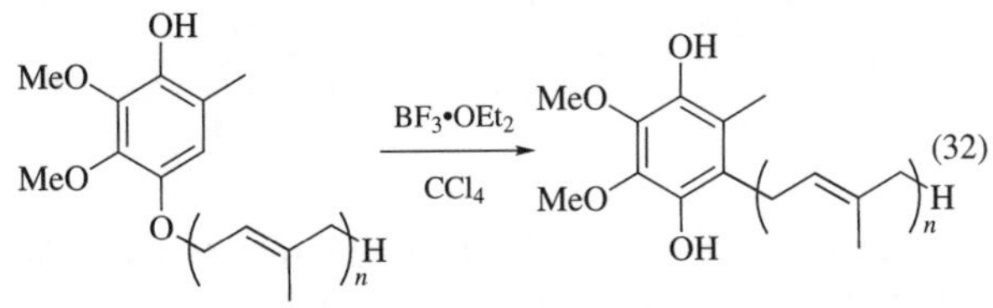

(32)

The (*E*)–(*Z*) photoisomerization of α,β-unsaturated esters,[104] cinnamic esters,[105] butenoic esters,[106] and dienoic esters[106] is catalyzed by $BF_3 \cdot OEt_2$ or ***Ethylaluminum Dichloride***. The latter two reactions also involve the photodeconjugation of α,β-unsaturated esters to β,γ-unsaturated esters. The $BF_3 \cdot MeOH$ complex is used for the isomerization of 1- and 2-butenes to form equal quantities of *cis*- and *trans*-but-2-enes;[107] the $BF_3 \cdot OEt_2$–acetic acid complex is not as effective.

The complex formed with $BF_3 \cdot OEt_2$ and ***Epichlorohydrin*** in DMF acts as a catalyst for the Beckmann rearrangement of oximes.[108] Cyclohexanone, acetaldehyde, and *syn*-benzaldehyde oximes are converted into ε-caprolactam, a mixture of *N*-methylformamide and acetamide, and *N*-phenylacetamide, respectively.

The addition of $BF_3 \cdot OEt_2$ to an α-phosphorylated imine results in the 1,3-transfer of a diphenylphosphinoyl group, with resultant migration of the C–N=C triad.[109] This method is less destructive than the thermal rearrangement. The decomposition of dimethyldioxirane in acetone to methyl acetate is accelerated with $BF_3 \cdot OEt_2$, but acetol is also formed.[110] Propene oxide undergoes polymerization with $BF_3 \cdot OEt_2$ in most solvents, but isomerizes to propionaldehyde and acetone in dioxane.[111]

Hydrolysis. $BF_3 \cdot OEt_2$ is used for stereospecific hydrolysis of methyl ethers, e.g. in the synthesis of (±)-aklavone.[112] The reagent is also used for the mild hydrolysis of dimethylhydrazones.[113] The precipitate formed by the addition of $BF_3 \cdot OEt_2$ to a dimethylhydrazone in ether is readily hydrolyzed by water to the ketone; the reaction is fast and does not affect enol acetate functionality.

Cleavage of Ethers. In aprotic, anhydrous solvents, $BF_3 \cdot MeOH$ is useful for the cleavage of trityl ethers at rt.[114] Under these conditions, *O*- and *N*-acyl groups, *O*-sulfonyl, *N*-alkoxycarbonyl, *O*-methyl, *O*-benzyl, and acetal groups are not cleaved.

$BF_3 \cdot OEt_2$ and iodide ion are extremely useful for the mild and regioselective cleavage of aliphatic ethers and for the removal of the acetal protecting group of carbonyl compounds.[115,116] Aromatic ethers are not cleaved, in contrast to other boron reagents. $BF_3 \cdot OEt_2$, in chloroform or dichloromethane, can be used for the removal of the *t*-butyldimethylsilyl (TBDMS) protecting group of hydroxyls, at 0–25 °C in 85–90% yield.[117] This is an alternative to ether cleavage with ***Tetra-n-butylammonium Fluoride*** or hydrolysis with aqueous ***Acetic Acid***.

In the presence of $BF_3 \cdot OEt_2$, dithio-substituted allylic anions react exclusively at the α-carbons of cyclic ethers, to give high yields of the corresponding alcohol products (eq 33).[118] The dithiane moiety is readily hydrolyzed with ***Mercury(II) Chloride*** to give the keto derivatives.

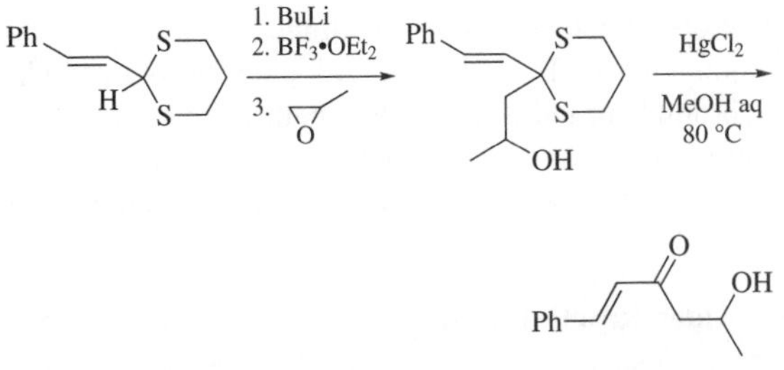

(33)

Inexpensive di-, tri-, and tetramethoxyanthraquinones can be selectively dealkylated to hydroxymethoxyanthraquinones by the formation of difluoroboron chelates with $BF_3 \cdot OEt_2$ in benzene and subsequent hydrolysis with methanol.[119] These unsymmetrically functionalized anthraquinone derivatives are useful intermediates for the synthesis of adriamycin, an antitumor agent. 2,4,6-Trimethoxytoluene reacts with cinnamic acid and $BF_3 \cdot OEt_2$, with selective demethylation, to form a boron heterocycle which can be hydrolyzed to the chalcone aurentiacin (eq 34).[120]

(34)

Reductions. In contrast to hydrosilylation reactions catalyzed by metal chlorides, aldehydes and ketones are rapidly reduced at rt by ***Triethylsilane*** and $BF_3 \cdot OEt_2$, primarily to symmetrical ethers and borate esters, respectively.[121] Aryl ketones like acetophenone and benzophenone are converted to ethylbenzene and diphenylmethane, respectively. Friedel–Crafts acylation–silane reduction reactions can also occur in one step using these reagents; thus ***Benzoyl Chloride*** reacts with benzene, triethylsilane, and $BF_3 \cdot OEt_2$ to give diphenylmethane in 30% yield.[121]

$BF_3 \cdot OEt_2$ followed by ***Diisobutylaluminum Hydride*** is used for the 1,2-reduction of γ-amino-α,β-unsaturated esters to give unsaturated amino alcohols, which are chiral building blocks for α-amino acids.[122] α,β-Unsaturated nitroalkenes can be reduced to hydroxylamines by ***Sodium Borohydride*** and $BF_3 \cdot OEt_2$ in THF;[123,124] extended reaction times result in the reduction of the hydroxylamines to alkylamines. Diphenylamine–borane is prepared from sodium borohydride, $BF_3 \cdot OEt_2$, and diphenylamine in THF at 0 °C.[125] This solid is more stable in air than $BF_3 \cdot THF$ and is almost as reactive in the reduction of aldehydes, ketones, carboxylic acids, esters, and anhydrides, as well as in the hydroboration of alkenes.

Bromination. $BF_3 \cdot OEt_2$ can catalyze the bromination of steroids that cannot be brominated in the presence of HBr or sodium acetate. Hence, 11α-bromoketones are obtained in high yields from methyl 3α,7α-diacetoxy-12-ketocholanate.[126] Bromination (at the 6α-position) and dibromination (at the 6α- and 11α-positions) of methyl 3α-acetoxy-7,12-dioxocholanate can occur, depending on the concentration of bromine.[127]

A combination of $BF_3 \cdot OEt_2$ and a halide ion (tetraethylammonium bromide or iodide in dichloromethane or chloroform, or sodium bromide or iodide in acetonitrile) is useful for the conversion of allyl, benzyl, and tertiary alcohols to the corresponding halides.[128,129]

Diels–Alder Reactions. $BF_3 \cdot OEt_2$ is used to catalyze and reverse the regiospecificity of some Diels–Alder reactions, e.g. with *peri*-hydroxylated naphthoquinones,[130] sulfur-containing compounds,[131] the reaction of 1-substituted *trans*-1,3-dienes with 2,6-dimethylbenzoquinones,[132] and the reaction of 6-methoxy-1-vinyl-3,4-dihydronaphthalene with *p*-quinones.[133] $BF_3 \cdot OEt_2$ has a drastic effect on the regioselectivity of the Diels–Alder reaction of quinoline- and isoquinoline-5,8-dione with piperylene, which produces substituted azaanthraquinones.[134] This Lewis acid is the most effective catalyst for the Diels–Alder reaction of furan with methyl acrylate, giving high *endo* selectivity in the 7-oxabicyclo[2.2.1]heptene product (eq 35).[135]

(35)

α-Vinylidenecycloalkanones, obtained by the reaction of ***Lithium Acetylide*** with epoxides and subsequent oxidation, undergo a Diels–Alder reaction at low temperature with $BF_3 \cdot OEt_2$ to form spirocyclic dienones (eq 36).[136]

(36)

Other Reactions. The 17-hydroxy group of steroids can be protected by forming the THP (*O*-tetrahydropyran-2-yl) derivative with 2,3-dihydropyran, using $BF_3 \cdot OEt_2$ as catalyst;[137] the yields are higher and the reaction times shorter than with *p*-toluenesulfonic acid monohydrate.

$BF_3 \cdot OEt_2$ catalyzes the decomposition of β,γ-unsaturated diazomethyl ketones to cyclopentenone derivatives (eq 37).[138,139] Similarly, γ,δ-unsaturated diazo ketones are decomposed to β,γ-unsaturated cyclohexenones, but in lower yields.[140]

(37)

$BF_3 \cdot OEt_2$ is an effective reagent for debenzyloxycarbonylations of methionine-containing peptides.[141] Substituted 6*H*-1,3-thiazines can be prepared in high yields from $BF_3 \cdot OEt_2$-catalyzed reactions between α,β-unsaturated aldehydes, ketones, or acetals with thioamides, thioureas, and dithiocarbamates (eq 38).[142]

(38)

α-Alkoxy ketones can be prepared from α-diazo ketones and primary, secondary, and tertiary alcohols using $BF_3 \cdot OEt_2$

in ethanol.[143] Nitrogen is released from a solution of α-***Diazoacetophenone*** and $BF_3{\cdot}OEt_2$ in ethanol to give α-ethoxyacetophenone.[143]

Anti-diols can be formed from β-hydroxy ketones using ***Tin(IV) Chloride*** or $BF_3{\cdot}OEt_2$.[144] The hydroxy ketones are silylated, treated with the Lewis acid, and then desilylated with ***Hydrogen Fluoride***. *Syn*-diols are formed if ***Zinc Chloride*** is used as the catalyst.

$BF_3{\cdot}OEt_2$ activates the formal substitution reaction of the hydroxyl group of γ- or δ-lactols with some organometallic reagents (M = Al, Zn, Sn), so that 2,5-disubstituted tetrahydrofurans or 2,6-disubstituted tetrahydropyrans are formed.[145]

A new method of nitrile synthesis from aldehydes has been discovered using *O*-(2-aminobenzoyl)hydroxylamine and $BF_3{\cdot}OEt_2$, achieving 78–94% yields (eq 39).[146]

CO_2NH_2 / NH_2 (aryl) + ArCHO —($BF_3{\cdot}OEt_2$, EtOH)→ ArCN (39)

Carbonyl compounds react predominantly at the α site of dithiocinnamyllithium if $BF_3{\cdot}OEt_2$ is present, as the hardness of the carbonyl compound is increased (eq 40).[147] The products can be hydrolyzed to α-hydroxyenones.

Ph, S, Li+, S —($BF_3{\cdot}OEt_2$, RR'CO, Et_2O)→ Ph, S, S, R, R', OH → Ph, O, R, R', OH (40)

Optically active sulfinates can be synthesized from sulfinamides and alcohols using $BF_3{\cdot}OEt_2$.[148] The reaction proceeds stereospecifically with inversion of sulfinyl configuration; the mild conditions ensure that the reaction will proceed even with alcohols with acid-labile functionality.

Related Reagents. See entries for other Lewis acids, e.g. Aluminum Chloride, Titanium(IV) Chloride, Zinc Chloride; also see entries for Boron Trifluoride (and combination reagents), and combination reagents employing boron trifluoride etherate, e.g. *n*-Butyllithium–Boron Trifluoride Etherate; Cerium(III) Acetate–Boron Trifluoride Etherate; Lithium Aluminum Hydride–Boron Trifluoride Etherate; Methylcopper–Boron Trifluoride Etherate; Tin(IV) Chloride.

---

1. (a) Hennion, G. F.; Hinton, H. D.; Nieuwland, J. A. *JACS* **1933**, *55*, 2857. (b) Zweifel, G.; Brown, H. C. *OR* **1963**, *13*, 28.
2. Eis, M. J.; Wrobel, J. E.; Ganem, B. *JACS* **1984**, *106*, 3693.
3. (a) Uno, H.; Terakawa, T.; Suzuki, H. *CL* **1989**, 1079. (b) *SL* **1991**, 559.
4. Yamaguchi, M.; Shibato, K.; Fujiwara, S.; Hirao, I. *S* **1986**, 421.
5. Yamaguchi, M.; Hirao, I. *TL* **1983**, *24*, 391.
6. Yamaguchi, M.; Nobayashi, N.; Hirao, I. *T* **1984**, *40*, 4261.
7. Hara, S.; Hyuga, S.; Aoyama, M.; Sato, M.; Suzuki, A. *TL* **1990**, *31*, 247.
8. Aoyama, M.; Hara, S.; Suzuki, A. *SC* **1992**, *22*, 2563.
9. Thompson, C. M.; Green, D. L. C.; Kubas, R. *JOC* **1988**, *53*, 5389.
10. Uno, H.; Okada, S.; Ono, T.; Shiraishi, Y.; Suzuki, H. *JOC* **1992**, *57*, 1504.
11. Volkmann, R. A.; Davies, J. T.; Meltz, C. N. *JACS* **1983**, *105*, 5946.
12. Kawate, T.; Nakagawa, M.; Yamazaki, H.; Hirayama, M.; Hino, T. *CPB* **1993**, *41*, 287.
13. Snyder, H. R.; Kornberg, H. A.; Romig, J. R. *JACS* **1939**, *61*, 3556.
14. Ha, H. J.; Nam, G. S. *SC* **1992**, *22*, 1143.
15. (a) Lipshutz, B. H.; Ellsworth, E. L.; Siahaan, T. J. *JACS* **1988**, *110*, 4834. (b) *JACS* **1989**, *111*, 1351.
16. Ochiai, M.; Fujita, E.; Arimoto, M.; Yamaguchi, H. *CPB* **1985**, *33*, 41.
17. Maruyama, K.; Naruta, Y. *JOC* **1978**, *43*, 3796.
18. Panek, J. S.; Sparks, M. A. *JOC* **1989**, *54*, 2034.
19. Giannis, A.; Sandhoff, K. *TL* **1985**, *26*, 1479.
20. Roos, E. C.; Hiemstra, H.; Speckamp, W. N.; Kaptein, B.; Kamphuis, J.; Schoemaker, H. E. *RTC* **1992**, *111*, 360.
21. Kuhn, M.; von Wartburg, A. *HCA* **1968**, *51*, 1631.
22. Kunz, H.; Sager, W. *HCA* **1985**, *68*, 283.
23. Yamamoto, Y.; Schmid, M. *CC* **1989**, 1310.
24. Yamamoto, Y.; Maruyama, K. *JOM* **1985**, *284*, C45.
25. (a) Keck, G. E.; Abbott, D. E. *TL* **1984**, *25*, 1883. (b) Keck, G. E.; Boden, E. P. *TL* **1984**, *25*, 265.
26. Keck, G. E.; Enholm, E. J. *JOC* **1985**, *50*, 146.
27. Trost, B. M.; Bonk, P. J. *JACS* **1985**, *107*, 1778.
28. Marshall, J. A.; DeHoff, B. S.; Crooks, S. L. *TL* **1987**, *28*, 527.
29. Marton, D.; Tagliavini, G.; Zordan, M.; Wardell, J. L. *JOM* **1990**, *390*, 127.
30. Ciufolini, M. A.; Spencer, G. O. *JOC* **1989**, *54*, 4739.
31. Gung, B. W.; Smith, D. T.; Wolf, M. A. *TL* **1991**, *32*, 13.
32. Takuwa, A.; Nishigaichi, Y.; Yamashita, K.; Iwamoto, H. *CL* **1990**, 1761.
33. Nishigaichi, Y.; Takuwa, A.; Jodai, A. *TL* **1991**, *32*, 2383.
34. Sugimura, H.; Osumi, K. *TL* **1989**, *30*, 1571.
35. Croxall, W. J.; Glavis, F. J.; Neher, H. T. *JACS* **1948**, *70*, 2805.
36. Hosomi, A.; Hashimoto, H.; Sakurai, H. *TL* **1980**, *21*, 951.
37. Smith, A. B. III; Jerris, P. J. *JACS* **1981**, *103*, 194.
38. Erman, W. F.; Stone, L. C. *JACS* **1971**, *93*, 2821.
39. Wakselman, C.; Tordeux, M. *CC* **1975**, 956.
40. Shellhamer, D. F.; Curtis, C. M.; Hollingsworth, D. R.; Ragains, M. L.; Richardson, R. E.; Heasley, V. L.; Shakelford, S. A.; Heasley, G. E. *JOC* **1985**, *50*, 2751.
41. Shellhamer, D. F.; Curtis, C. M.; Hollingsworth, D. R.; Ragains, M. L.; Richardson, R. E.; Heasley, V. L.; Heasley, G. E. *TL* **1982**, *23*, 2157.
42. Wulff, W. D.; Gilbertson, S. R. *JACS* **1985**, *107*, 503.
43. Yamago, S.; Machii, D.; Nakamura, E. *JOC* **1991**, *56*, 2098.
44. Nakamura, E.; Kuwajima, I. *JACS* **1977**, *99*, 961.
45. Sugimura, H.; Shigekawa, Y.; Uematsu, M. *SL* **1991**, 153.
46. Heathcock, C. H.; Flippin, L. A. *JACS* **1983**, *105*, 1667.
47. Evans, D. A.; Gage, J. R. *TL* **1990**, *31*, 6129.
48. Kamimura, A.; Marumo, S. *TL* **1990**, *31*, 5053.
49. Annunziata, R.; Cinquini, M.; Cozzi, F.; Cozzi, P. G. *TL* **1990**, *31*, 6733.
50. Duhamel, P.; Hennequin, L.; Poirier, N.; Poirier, J.-M. *TL* **1985**, *26*, 6201.
51. Moriarty, R. M.; Prakash, O.; Duncan, M. P.; Vaid, R. K. *JOC* **1987**, *52*, 150.
52. Yamamoto, Y.; Maruyama, K. *JACS* **1982**, *104*, 2323.
53. Ando, W.; Tsumaki, H. *CL* **1983**, 1409.
54. Thompson, C. M.; Docter, S. *TL* **1988**, *29*, 5213.
55. Ohta, S.; Kimoto, S. *TL* **1975**, 2279.

56. Boeckman, R. K. Jr.; Bruza, K. J.; Heinrich, G. R. *JACS* **1978**, *100*, 7101.
57. Rigby, J. H. *TL* **1982**, *23*, 1863.
58. Sum, P.-E.; Weiler, L. *CJC* **1979**, *57*, 1475.
59. Chenard, B. L.; Van Zyl, C. M.; Sanderson, D. R. *TL* **1986**, *27*, 2801.
60. Harding, K. E.; Clement, K. S. *JOC* **1984**, *49*, 3870.
61. Jain, T. C.; McCloskey, J. E. *TL* **1971**, 1415.
62. (a) Hinton, H. D.; Nieuwland, J. A. *JACS* **1932**, *54*, 2017. (b) Sowa, F. J.; Nieuwland, J. A. *JACS* **1936**, *58*, 271. (c) Hallas, G. *JCS* **1965**, 5770.
63. Kadaba, P. K. *S* **1972**, 628.
64. Kadaba, P. K. *S* **1971**, 316.
65. Marshall, J. L.; Erikson, K. C.; Folsom, T. K. *TL* **1970**, 4011.
66. Yamaguchi, M.; Shibato, K.; Nakashima, H.; Minami, T. *T* **1988**, *44*, 4767.
67. Danishefsky, S.; Chao, K.-H.; Schulte, G. *JOC* **1985**, *50*, 4650.
68. Tani, J.; Oine, T.; Inoue, I. *S* **1975**, 714.
69. Ibuka, T.; Chu, G.-N.; Aoyagi, T.; Kitada, K.; Tsukida, T.; Yoneda, F. *CPB* **1985**, *33*, 451.
70. Irie, H.; Nishimura, M.; Yoshida, M.; Ibuka, T. *JCS(P1)* **1989**, 1209.
71. Taylor, M. E.; Fletcher, T. L. *JOC* **1961**, *26*, 940.
72. Hoppe, D.; Krämer, T.; Erdbrügger, C. F.; Egert, E. *TL* **1989**, *30*, 1233.
73. Kitchen, L. J. *JACS* **1948**, *70*, 3608.
74. Hirschmann, R.; Miller, R.; Wendler, N. L. *JACS* **1954**, *76*, 4592.
75. Posner, G. H.; Shulman-Roskes, E. M.; Oh, C. H.; Carry, J.-C.; Green, J. V.; Clark, A. B.; Dai, H.; Anjeh, T. E. N. *TL* **1991**, *32*, 6489.
76. Heymann, H.; Fieser, L. F. *JACS* **1952**, *74*, 5938.
77. Clinton, R. O.; Christiansen, R. G.; Neumann, H. C.; Laskowski, S. C. *JACS* **1957**, *79*, 6475.
78. House, H. O.; Wasson, R. L. *JACS* **1957**, *79*, 1488.
79. Bird, C. W.; Yeong, Y. C.; Hudec, J. *S* **1974**, *27*.
80. Domagala, J. M.; Bach, R. D. *JACS* **1978**, *100*, 1605.
81. Obuchi, K.; Hayashibe, S.; Asaoka, M.; Takei, H. *BCJ* **1992**, *65*, 3206.
82. Bach, R. D.; Klix, R. C. *JOC* **1985**, *50*, 5438.
83. Hancock, W. S.; Mander, L. N.; Massy-Westropp, R. A. *JOC* **1973**, *38*, 4090.
84. French, L. G.; Fenlon, E. E.; Charlton, T. P. *TL* **1991**, *32*, 851.
85. Loubinoux, B.; Viriot-Villaume, M. L.; Chanot, J. J.; Caubere, P. *TL* **1975**, 843.
86. Smith, J. R. L.; Norman, R. O. C.; Stillings, M. R. *JCS(P1)* **1975**, 1200.
87. Sato, T.; Watanabe, M.; Murayama, E. *SC* **1987**, *17*, 781.
88. Conacher, H. B. S.; Gunstone, F. D. *CC* **1967**, 984.
89. Durst, T.; Tin, K.-C. *TL* **1970**, 2369.
90. Maruoka, K.; Hasegawa, M.; Yamamoto, H.; Suzuki, K.; Shimazaki, M.; Tsuchihashi, G. *JACS* **1986**, *108*, 3827.
91. Shimazaki, M.; Hara, H.; Suzuki, K.; Tsuchihashi, G. *TL* **1987**, *28*, 5891.
92. Liu, Y.; Chu, T.; Engel, R. *SC* **1992**, *22*, 2367.
93. Hutchins, R. O.; Taffer, I. M.; Burgoyne, W. *JOC* **1981**, *46*, 5214.
94. Alexakis, A.; Jachiet, D.; Normant, J. F. *T* **1986**, *42*, 5607.
95. Hata, N.; Ono, I.; Kawasaki, M. *CL* **1975**, 25.
96. Tomoda, S.; Matsumoto, Y.; Takeuchi, Y.; Nomura, Y. *BCJ* **1986**, *59*, 3283.
97. Yoshizawa, T.; Toyofuku, H.; Tachibana, K.; Kuroda, T. *CL* **1982**, 1131.
98. Davies, J. S. H.; McCrea, P. A.; Norris, W. L.; Ramage, G. R. *JCS* **1950**, 3206.
99. Boeckman, R. K. Jr.; Flann, C. J.; Poss, K. M. *JACS* **1985**, *107*, 4359.
100. Suzuki, H.; Yashima, H.; Hirose, T.; Takahashi, M., Moro-Oka, Y.; Ikawa, T. *TL* **1980**, *21*, 4927.
101. Takahashi, M.; Suzuki, H.; Moro-Oka, Y.; Ikawa, T. *TL* **1982**, *23*, 4031.
102. Marx, J. N.; Zuerker, J.; Hahn, Y. P. *TL* **1991**, *32*, 1921.
103. Yoshizawa, T.; Toyofuku, H.; Tachibana, K.; Kuroda, T. *CL* **1982**, 1131.
104. Lewis, F. D.; Oxman, J. D. *JACS* **1981**, *103*, 7345.
105. Lewis, F. D.; Oxman, J. D.; Gibson, L. L.; Hampsch, H. L.; Quillen, S. L. *JACS* **1986**, *108*, 3005.
106. Lewis, F. D.; Howard, D. K.; Barancyk, S. V.; Oxman, J. D. *JACS* **1986**, *108*, 3016.
107. Roberts, J. M.; Katovic, Z.; Eastham, A. M. *J. Polym. Sci. A1* **1970**, *8*, 3503.
108. Izumi, Y. *CL* **1990**, 2171.
109. Onys'ko, P. P.; Kim, T. V.; Kiseleva, E. I.; Sinitsa, A. D. *TL* **1992**, *33*, 691.
110. Singh, M.; Murray, R. W. *JOC* **1992**, *57*, 4263.
111. Sugiyama, S.; Ohigashi, S.; Sato, K.; Fukunaga, S.; Hayashi, H. *BCJ* **1989**, *62*, 3757.
112. Pearlman, B. A.; McNamara, J. M.; Hasan, I.; Hatakeyama, S.; Sekizaki, H.; Kishi, Y. *JACS* **1981**, *103*, 4248.
113. Gawley, R. E.; Termine, E. J. *SC* **1982**, *12*, 15.
114. Mandal, A. K.; Soni, N. R.; Ratnam, K. R. *S* **1985**, 274.
115. Mandal, A. K.; Shrotri, P. Y.; Ghogare, A. D. *S* **1986**, 221.
116. Pelter, A.; Ward, R. S.; Venkateswarlu, R.; Kamakshi, C. *T* **1992**, *48*, 7209.
117. Kelly, D. R.; Roberts, S. M.; Newton, R. F. *SC* **1979**, *9*, 295.
118. Fang, J.-M.; Chen, M.-Y. *TL* **1988**, *29*, 5939.
119. Preston, P. N.; Winwick, T.; Morley, J. O. *JCS(P1)* **1983**, 1439.
120. Schiemenz, G. P.; Schmidt, U. *LA* **1982**, 1509.
121. Doyle, M. P.; West, C. T.; Donnelly, S. J.; McOsker, C. C. *JOM* **1976**, *117*, 129.
122. Moriwake, T.; Hamano, S.; Miki, D.; Saito, S.; Torii, S. *CL* **1986**, 815.
123. Varma, R. S.; Kabalka, G. W. *OPP* **1985**, *17*, 254.
124. Varma, R. S.; Kabalka, G. W. *SC* **1985**, *15*, 843.
125. Camacho, C.; Uribe, G.; Contreras, R. *S* **1982**, 1027.
126. Yanuka, Y.; Halperin, G. *JOC* **1973**, *38*, 2587.
127. Takeda, K.; Komeno, T.; Igarashi, K. *CPB* **1956**, *4*, 343.
128. Mandal, A. K.; Mahajan, S. W. *TL* **1985**, *26*, 3863.
129. Vankar, Y. D.; Rao, C. T. *TL* **1985**, *26*, 2717.
130. Trost, B. M.; Ippen, J.; Vladuchick, W. C. *JACS* **1977**, *99*, 8116.
131. Kelly, T. R.; Montury, M. *TL* **1978**, 4311.
132. Stojanác, Z.; Dickinson, R. A.; Stojanác, N.; Woznow, R. J.; Valenta, Z. *CJC* **1975**, *53*, 616.
133. Das, J.; Kubela, R.; MacAlpine, G. A.; Stojanac, Z.; Valenta, Z. *CJC* **1979**, *57*, 3308.
134. Ohgaki, E.; Motoyoshiya, J.; Narita, S.; Kakurai, T.; Hayashi, S.; Hirakawa, K. *JCS(P1)* **1990**, 3109.
135. Kotsuki, H.; Asao, K.; Ohnishi, H. *BCJ* **1984**, *57*, 3339.
136. Gras, J.-L.; Guerin, A. *TL* **1985**, *26*, 1781.
137. Alper, H.; Dinkes, L. *S* **1972**, 81.
138. Smith, A. B. III; Branca, S. J.; Toder, B. H. *TL* **1975**, 4225.
139. Smith, A. B. III *CC* **1975**, 274.
140. Smith, A. B. III, Toder, B. H.; Branca, S. J.; Dieter, R. K. *JACS* **1981**, *103*, 1996.
141. Okamoto, M.; Kimoto, S.; Oshima, T.; Kinomura, Y.; Kawasaki, K.; Yajima, H. *CPB* **1967**, *15*, 1618.
142. Hoff, S.; Blok, A. P. *RTC* **1973**, *92*, 631.
143. Newman, M. S.; Beal, P. F. III *JACS* **1950**, *72*, 5161.
144. Anwar, S.; Davis, A. P. *T* **1988**, *44*, 3761.
145. Tomooka, K.; Matsuzawa, K.; Suzuki, K.; Tsuchihashi, G. *TL* **1987**, *28*, 6339.

146. Reddy, P. S. N.; Reddy, P. P. *SC* **1988**, *18*, 2179.
147. Fang, J.-M.; Chen, M.-Y.; Yang, W.-J. *TL* **1988**, *29*, 5937.
148. Hiroi, K.; Kitayama, R.; Sato, S. *S* **1983**, 1040.

Veronica Cornel
Emory University, Atlanta, GA, USA

Bromodimethylborane[1]

Me_2BBr

[5158-50-9] $C_2H_6BBr_2$ (MW 120.78)

(mild Lewis acid capable of selective cleavage of ethers[2,3] and acetals;[4,5] deoxygenation of sulfoxides[6])

Physical Data: mp −129 °C; bp 31–32 °C; *d* 1.238 g cm^{-3}; fp −37 °C.
Solubility: sol dichloromethane, 1,2-dichloroethane, hexane.
Form Supplied in: colorless liquid.
Preparative Method: can be conveniently prepared by treating ***Tetramethylstannane*** with ***Boron Tribromide***.[7]
Handling, Storage, and Precautions: flammable liquid, moisture sensitive; typically stored and dispensed as a 1.5–2 M solution in dichloromethane or dichloroethane. Solutions of this sort are stable for a period of months if stored at −15 °C and properly protected from moisture.

Cleavage of Ethers. Bromodimethylborane (Me_2BBr) reacts with primary, secondary, and aryl methyl ethers,[2] in addition to trityl,[8] benzyl,[2,8] and 4-methoxybenzyl[9] ethers, to regenerate the parent alcohol in good to excellent yield (e.g. eq 1). The tertiary methyl ethers examined afforded the corresponding tertiary bromides.[2] The reaction is typically carried out in dichloromethane or 1,2-dichloroethane between 0 °C and rt, in the presence of 1.3–4 equiv of Me_2BBr. The reaction is usually complete in a matter of hours. ***Triethylamine*** (0.1–0.15 equiv per equiv of Me_2BBr) is often added as an acid scavenger. 4-Methoxybenzyl ethers are more reactive and are cleaved at −78 °C, whereas aryl methyl ethers require elevated temperatures to react. Other functional groups including acetates, benzoates, alcohols, ethyl esters, and *t*-butyldiphenylsilyl ethers are recovered unchanged under the standard reaction conditions.

Me_2BBr, Et_3N; $ClCH_2CH_2Cl$, 0–25 °C, 16–18 h (1)

R = Me 93%
R = Bu 92%

Bromodimethylborane is also effective for the cleavage of cyclic ethers.[2,3] Epoxides react at −78 °C while the analogous four- to seven-membered ring heterocycles react between 0 °C and rt. In contrast to other boron-containing Lewis acids, Me_2BBr reacts via a predominantly S_N2 mechanism. Tetrahydrofuran derivatives which are substituted at the 2-position give rise to primary bromides as the major or exclusive products. The nature of the substituent has a quantitative influence on the outcome of the reaction via steric effects and/or complexation to the reagent. It is of interest to note that tetrahydrofurans can be cleaved in the presence of acyclic ethers (eq 2).[3]

Me_2BBr, Et_3N; CH_2Cl_2, 0 °C, 2 h; 87%; 4 : 1 (2)

It is also of considerable interest to note that no β-elimination of the hydroxy group was observed in the ring-opening of 2-(ethoxycarbonylmethyl)tetrahydrofurans (eq 3),[3] whereas *C*-glycosides bearing more acidic protons on the aglycon react with Me_2BBr to generate acyclic alkenes (eq 4).[10]

Me_2BBr, Et_3N; CH_2Cl_2, 0 °C, 2 h; 82% (3)

1. Me_2BBr, Et_3N, CH_2Cl_2, 0 °C, 0.5 h; 2. AcBr; 54% (4)

Bromodimethylborane has also been used in conjunction with ***Tetra-n-butylammonium Iodide*** to bring about the fragmentation of iodomethyl ether derivatives (eq 5).[11]

Me_2BBr, Bu_4NI; Et_3N, CH_2Cl_2, 0 °C, 0.5 h; 77% (5)

Cleavage of Acetals[4,5]. Cyclic and acyclic acetals react with Me_2BBr at −78 °C to generate the parent aldehydes and ketones in excellent yield (e.g. eq 6). Primary, secondary, and tertiary (2-methoxyethoxy)methyl (MEM), methoxymethyl (MOM), and (methylthio)methyl (MTM) ethers also react at −78 °C to give, after aqueous workup, the corresponding alcohol. It is interesting to note that even tertiary MEM ethers cleanly regenerate the parent alcohol without formation of the corresponding bromide or elimination products (eq 7). Treatment of an acetonide with Me_2BBr gives the parent diol in high yield (eq 8).

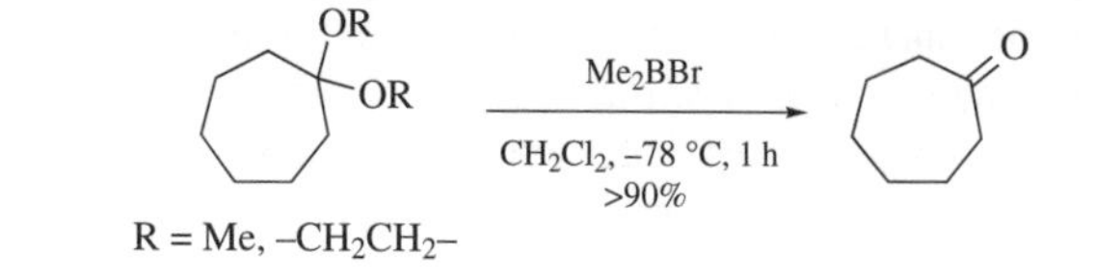

Tetrahydropyranyl (THP) and tetrahydrofuranyl (THF) ethers are converted to the corresponding alcohols by Me_2BBr at rt (eq 9), although the acetals are cleaved at $-78\,^{\circ}C$ (see below).

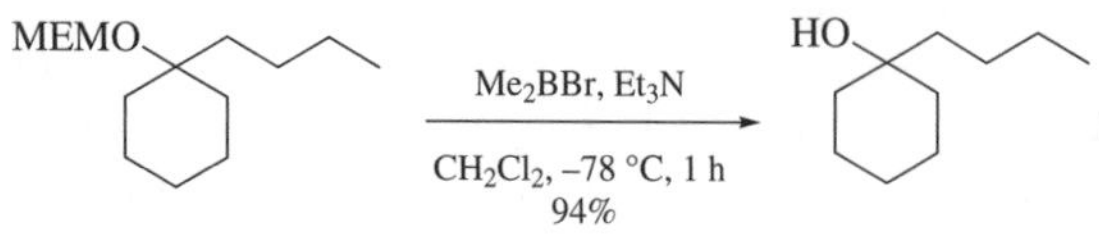

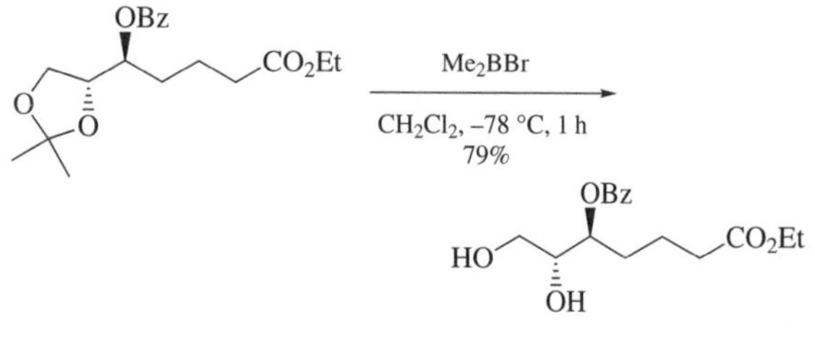

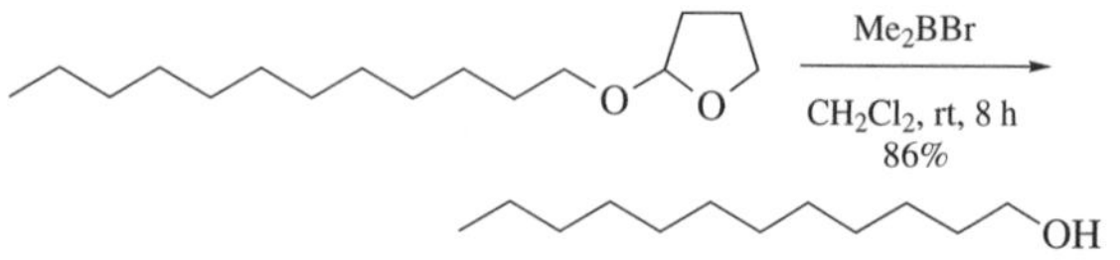

Bromodiphenylborane (Ph_2BBr) and ***9-Bromo-9-borabicyclo[3.3.1]nonane*** (Br-9-BBN) can often be used in place of Me_2BBr for the cleavage of acetals;[4,5] however, the purification of products from reactions employing Me_2BBr is facilitated by the volatility of Me_2B-containing byproducts, thus making Me_2BBr the reagent of choice in most instances.

Interconversion of Functional Groups. The reaction of Me_2BBr with MEM and MOM ethers is believed to proceed via α-bromo ether intermediates. It is possible to trap these intermediates with nucleophiles such as thiols, alcohols, and cyanide. An example of the utility of this sequence is the conversion of a readily prepared MOM ether into an MTM ether (eq 10).[12]

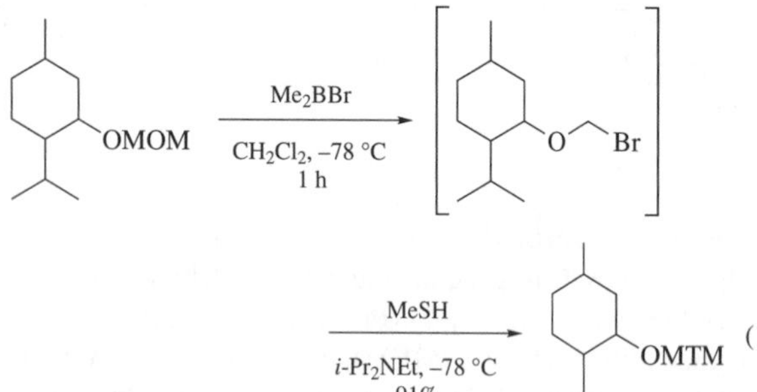

While THP and THF ethers are converted to the corresponding alcohols by Me_2BBr at rt, the acetal is cleaved at $-78\,^{\circ}C$.[13] The initial products of the reaction are acyclic α-bromo ethers. These can be trapped with a variety of nucleophiles to generate stable ring-opened products (eq 11).[13] This reaction has been extended to glycosides which, although less reactive, behave in a similar fashion.[14,15]

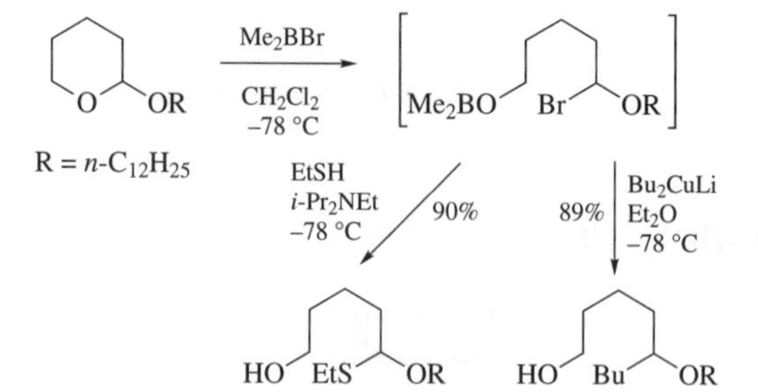

Benzylidene acetals are recovered unchanged when treated with Me_2BBr under conditions which are used to cleave other acetals.[16] It is, however, possible to cleave benzylidene acetals to generate hydroxy-*O,S*-acetals in excellent yield, by treatment with Me_2BBr at $-78\,^{\circ}C$ followed by ***Thiophenol*** (eq 12).[16] Sterically encumbered bromoboranes optimize regioselective complexation of boron to the least hindered oxygen atom and are, therefore, the reagents of choice for this process (eq 12).[16] These experiments demonstrate that benzylidene acetals do indeed react with Me_2BBr at $-78\,^{\circ}C$, like other acetals.

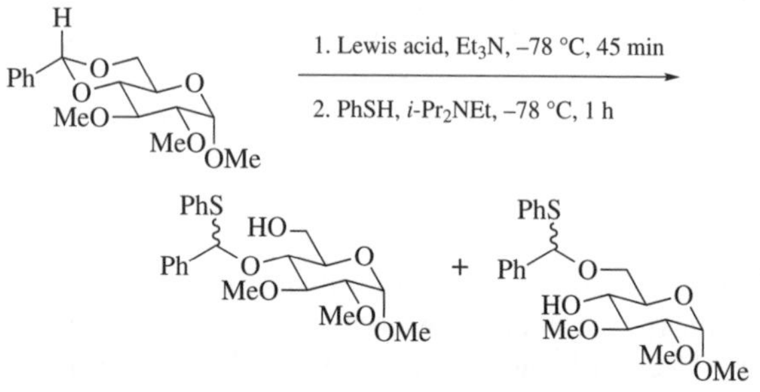

Me_2BBr	68%	17%
Ph_2BBr	96%	–
9-BrBBN	90%	–

Treatment of glycoside benzylidene acetals with a variety of disubstituted bromoboranes, followed by ***Borane-Tetrahydrofuran***, generates 4-*O*-benzyl-6-hydroxypyranosides in excellent yield (eq 13).[16]

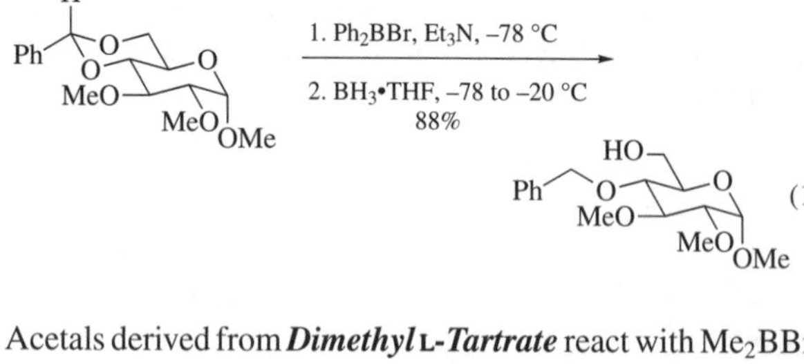

Acetals derived from ***Dimethyl L-Tartrate*** react with Me_2BBr to generate α-bromo ethers which react further with cuprate reagents to give optically active secondary alcohol derivatives (eq 14).[17] The alcohols may be liberated by treatment with ***Samarium(II) Iodide*** or by a straightforward sequence of reactions (mesylation and elimination to form an enol ether followed by exposure to methoxide in refluxing methanol). Selectivity is enhanced by the use of Ph_2BBr and by careful control of the reaction temperature at each step.

(14)

Me_2BBr, 80% 34:1
Ph_2BBr, 62% 82:1

Miscellaneous Reactions. Bromodimethylborane can also be used to convert dialkyl, aryl alkyl, and diaryl sulfoxides to the corresponding sulfides (eq 15).[6] Typically, the sulfoxides are treated with 2.5 equiv of Me_2BBr in dichloromethane at $-23\,^\circ C$ for 30 min and at $0\,^\circ C$ for 10 min. Bromine is produced in the reaction and must be removed in order to avoid possible side reactions. This is accomplished by saturating the solution with propene prior to introducing the reagent or by adding cyclohexene. Phosphine oxides and sulfones failed to react under the conditions used to deoxygenate sulfoxides.

(15)

Bromodimethylborane has also been used as a catalyst for the Pictet–Spengler reaction (eq 16)[18] and to catalyze the 1,3-transposition of an allylic lactone.[19]

(16)

Related Reagents. Boron Tribromide; 9-Bromo-9-borabicyclo[3.3.1]nonane; *B*-Bromocatecholborane; Hydrogen Bromide.

1. Guindon, Y.; Anderson, P. C.; Yoakim, C.; Girard, Y.; Berthiaume, S.; Morton, H. E. *PAC* **1988**, *60*, 1705.
2. Guindon, Y.; Yoakim, C.; Morton, H. E. *TL* **1983**, *24*, 2969.
3. Guindon, Y.; Therien, M.; Girard, Y.; Yoakim, C. *JOC* **1987**, *52*, 1680.
4. Guindon, Y.; Morton, H. E.; Yoakim, C. *TL* **1983**, *24*, 3969.
5. Guindon, Y.; Yoakim, C.; Morton, H. E. *JOC* **1984**, *49*, 3912.
6. Guindon, Y.; Atkinson, J. G.; Morton, H. E. *JOC* **1984**, *49*, 4538.
7. Nöth, H.; Vahrenkamp, H. *JOM* **1968**, *11*, 399.
8. Kodali, D. R.; Duclos Jr., R. I. *Chem. Phys. Lipids* **1992**, *61*, 169.
9. Hébert, N.; Beck, A.; Lennox, R. B.; Just, G. *JOC* **1992**, *57*, 1777.
10. Abel, S.; Linker, T.; Giese, B. *SL* **1991**, 171.
11. Gauthier, J. Y.; Guindon, Y. *TL* **1987**, *28*, 5985.
12. Morton, H. E.; Guindon, Y. *JOC* **1985**, *50*, 5379.
13. Guindon, Y.; Bernstein, M. A.; Anderson, P. C. *TL* **1987**, *28*, 2225.
14. Guindon, Y.; Anderson, P. C. *TL* **1987**, *28*, 2485.
15. Hashimoto, H.; Kawanishi, M.; Yuasa, H. *TL*, **1991**, *32*, 7087.
16. Guindon, Y.; Girard, Y.; Berthiaume, S.; Gorys, V.; Lemieux, R.; Yoakim, C. *CJC* **1990**, *68*, 897.
17. Guindon, Y.; Simoneau, B.; Yoakim, C.; Gorys, V.; Lemieux, R.; Ogilvie, W. *TL* **1991**, *32*, 5453.
18. Kawate, T.; Nakagawa, M.; Ogata, K.; Hino, T. *H* **1992**, *33*, 801.
19. Mander, L. N.; Patrick, G. L. *TL* **1990**, *31*, 423.

Yvan Guindon & Paul C. Anderson
Bio-Méga/Boehringer Ingelheim Research, Laval, Québec, Canada

Bromomagnesium Diisopropylamide

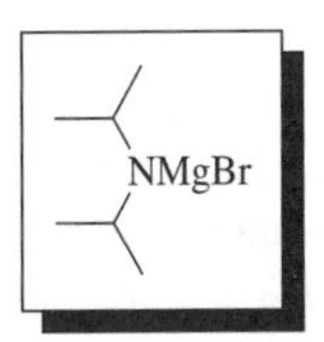

[50715-01-0] $C_6H_{14}BrMgN$ (MW 204.39)

(thermodynamic enolate generation; aldol condensation; Claisen ester condensation)

Alternate Names: diisopropylaminomagnesium bromide; BMDA.

Preparative Methods: prepared from ***Diisopropylamine*** and ***Methylmagnesium Bromide*** or ***Ethylmagnesium Bromide*** using either THF or ether as the solvent. In THFrm a 1.5 M solution was prepared at $80\,^\circ C$ and stored at $50\,^\circ C$ (presumably to prevent precipitation).[1] A refluxing 0.5 M THF solution of BMDA was reported to be stable for several hours.[2] In ether, preparations have been reported at rt; however, BMDA is practically insoluble in ether and was used as a slurry.[3,4]

Handling, Storage, and Precautions: use in a fume hood.

Ester Condensations. BMDA has been examined as a base for the Claisen ester condensation[3,5–7] and was found to be an 'excellent condensing agent for mixed condensations of methyl benzoate and methyl 2-furoate with aliphatic esters'.[5] A small study of BMDA's use in the Dieckmann cyclization of diesters seemed to show no obvious advantage over other bases.[7] Intramolecular condensations using BMDA have been reported (eqs 1 and 2).[8,9] Thiolactones have been condensed with ***Diethyl Oxalate*** (eq 3).[10]

(1)

BMDA, ether; R = H, 43%; R = Me, 35% (2)

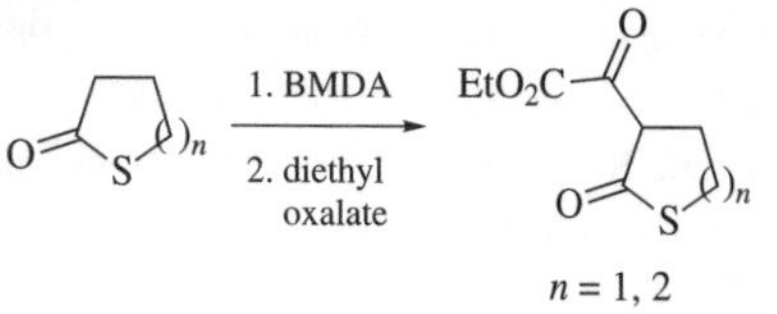

(3)

$n = 1, 2$

Nitrile Condensations. The condensation of nitriles using BMDA has been investigated. Although acetonitrile gave very poor yields, reactions of propio-, valero-, and phenylacetonitriles proceeded in 65–80% yields (eq 4). The reactions were carried out in refluxing ether. When the reactions were carried out in refluxing *n*-butyl ether, they were more complex and cyclic trimers of the pyrimidine and pyridine classes were formed.[11]

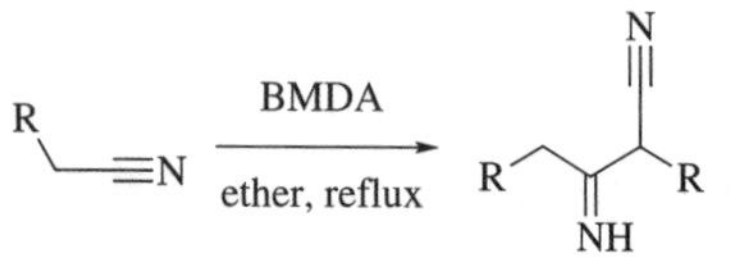

(4)

Aldol Condensations. BMDA has found utility in a number of aldol condensations where other bases have failed. In Kishi's synthesis of monensin, BMDA was used to couple the left and right halves of the molecule (eq 5).[1]

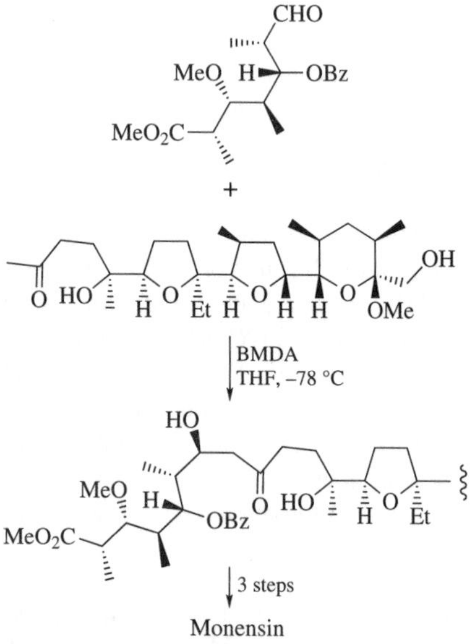

(5)

In a synthesis of the taxane ring system, Holton used BMDA to effect the intramolecular condensation to give (**2**) in 90% yield (eq 6).[12] Other reagents resulted in a retro Michael reaction.

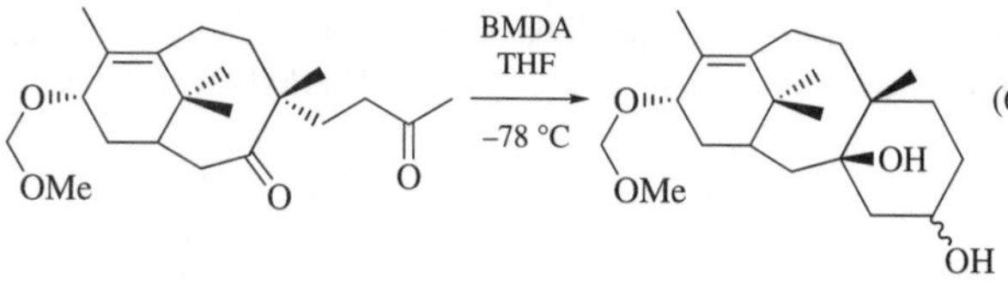

(6)

Annunziata et al. found magnesium bases such as BMDA superior in diastereoselectivity to lithium bases for the aldol condensation of the enantiomerically pure 3-*p*-tolylsulfinylmethyl-4,5-dihydroisoxazoles (eq 7).[13]

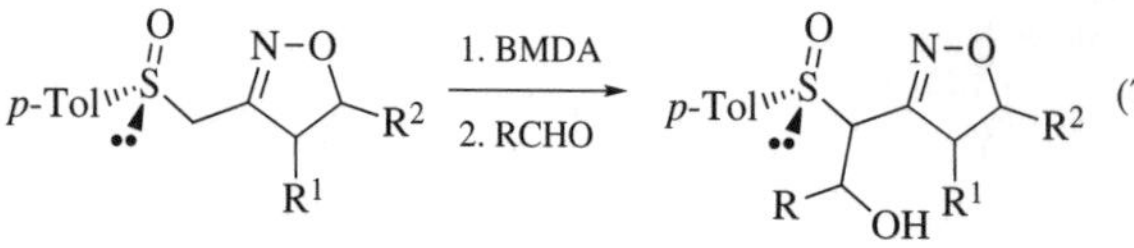

(7)

Similarly, Nokami et al. reported the condensation of another *p*-tolylsulfinyl compound with propargyl aldehyde (eq 8).[14]

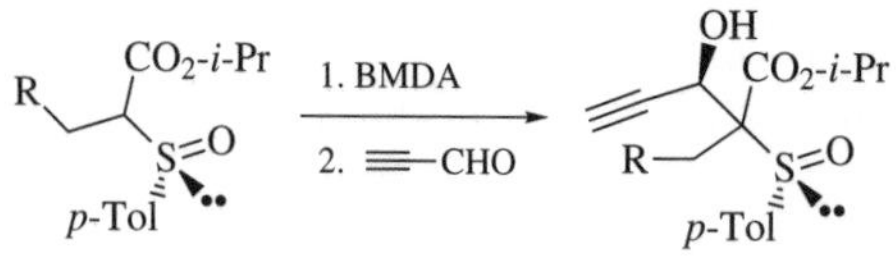

(8)

Enolate Generation. Krafft and Holton reported the examination of a number of bases for the preparation of enolates. This examination showed that BMDA was the best base for the preparation of thermodynamic enolates. The enolate was trapped with ***Chlorotrimethylsilane*** and the product obtained in 85–95% yields with ratios of thermodynamic:kinetic product (**4:5**) of ca. 97:3 (eq 9).[4] ***Lithium Diisopropylamide*** gave a 1:99 ratio of (**4:5**). Other bases examined gave predominately (**4**); however, the ratios of (**4:5**) were inferior to BMDA. This procedure has been used in the synthesis of the taxane ring system,[12] intermediates for reserpine (trapped with ***t-Butyldimethylchlorosilane***),[15] dihydropallescensin D,[16] and other systems.[17] Thermodynamic enolates have also been trapped with ***N-Phenyltrifluoromethanesulfonimide*** to give vinyl triflates.[18,19] Scott and Stille observed that the use of excess BMDA with 2-methylcyclohexanone led to lower ratios of thermodynamic:kinetic product when the enolate was trapped with either TMSCl or *N*-phenyltriflimide.[19]

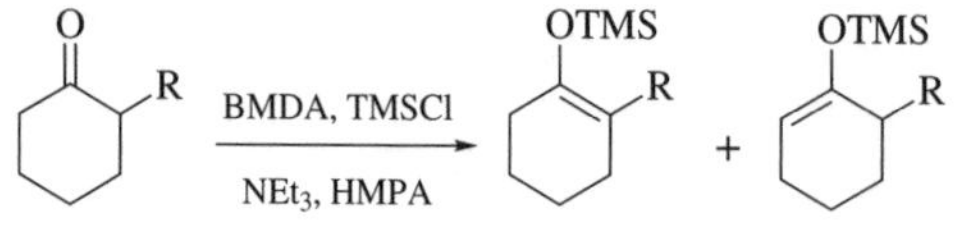

(9)

Carey and Helquist used BMDA to generate enolates in their synthesis of fused cyclopentanones (eq 10).[20]

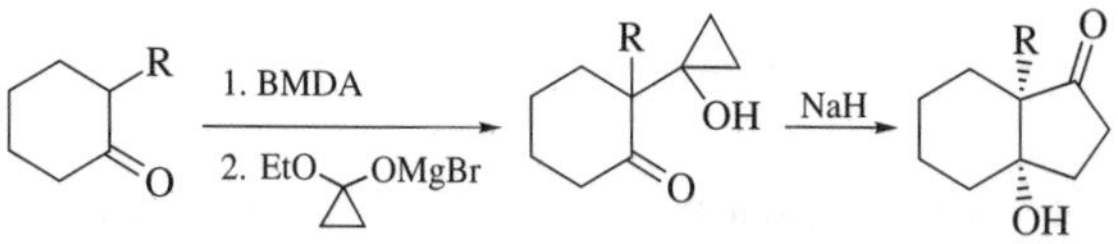

(10)

It should be noted that since BMDA can also promote aldol condensations, the use of BMDA to form enolates of ketones that are not relatively sterically hindered can lead to aldol products as the only reaction.[12,21]

Chiral Sulfoxides. One synthetic route to optically active α-sulfinyl esters is the displacement of the *O*-menthyl group of an optically active sulfinate ester (eq 11).[22–24] When R = Me the ratio of diastereomers was 1:1 and when R = Et the ratio was 3:7. A similar reaction with cyclic ketones has been reported (eq 12).[21] When $n = 1$ the ratio of diastereomers was 3.0:1 and with $n = 2$ the ratio was 1.6:1.

BMDA, RCH_2CO_2-*t*-Bu; R = H, 90%; R = Me, 68%; R = Et, 45% (11)

BMDA; *n* = 1, 70%; *n* = 2, 83% (12)

Other Reactions. BMDA has proven useful for the functionalization of cubanes. Treatment of (**6**) with BMDA in refluxing THF gave the diester (**7**) as the product,[2] whereas at rt the monoester (**8**) is formed (eq 13).[2,25]

1. BMDA; 2. CO_2; 3. CH_2N_2 (13)

(**6**)

(**7**) R = CO_2Me
(**8**) R = H

BMDA metalates (diphenylphosphinyl)ferrocene exclusively *ortho* to the diphenylphosphinyl group (eq 14).[26]

BMDA (14)

BMDA has been reported to be complementary to LDA in the acylation of 2-acyl-1,3-dithianes (**9**) (eq 15). When R = aryl, LDA gave predominately *O*-acylation (**11**), whereas BMDA gave exclusive *C*-acylation (**10**). When R = Me, LDA gave 100% *O*-acylation and BMDA gave 78% *C*-acylation.[27]

1. BMDA; 2. RCOCl (15)

(**9**) (**10**) (**11**)

Related Reagents. Bromomagnesium Diethylamide; Lithium Diisopropylamide; Lithium Hexamethyldisilazide; Lithium 2,2,6,6-Tetramethylpiperidide.

1. Fukuyama, T.; Akasaka, K.; Karanewsky, D. S.; Wang, C.-L. J.; Schmid, G.; Kishi, Y. *JACS* **1979**, *101*, 262.
2. Eaton, P. E.; Lee, C.-H.; Xiong, Y. *JACS* **1989**, *111*, 8016.
3. Frostick, F. C.; Hauser, C. R. *JACS* **1949**, *71*, 1350.
4. Krafft, M. E.; Holton, R. A. *TL* **1983**, *24*, 1345.
5. Royals, E. E.; Turpin, D. G. *JACS* **1954**, *76*, 5452.
6. Sommer, L. H.; Pioch, R. P.; Marans, N. S.; Goldberg, G. M.; Rockett, J.; Kerlin, J. *JACS* **1953**, *75*, 2932.
7. Singh, P. K.; Rajeswari, K.; Ranganayakulu, K. *IJC(B)* **1980**, *19B*, 823.
8. Haynes, L. J.; Stanners, A. H. *JCS* **1956**, 4103.
9. Eistert, v. B.; Heck, G. *LA* **1965**, *681*, 123.
10. Korte, F.; Büchel, K. H. *CB* **1960**, *93*, 1021.
11. Reynolds, G. A.; Humphlett, W. J.; Swamer, F. W.; Hauser, C. R. *JOC* **1951**, *16*, 165.
12. Holton, R. A. *JACS* **1984**, *106*, 5731.
13. Annunziata, R.; Cinquini, M.; Cozzi, F.; Restelli, A. *JCS(P1)* **1985**, 2293.
14. Nokami, J.; Ohtsuki, H.; Sakamoto, Y.; Mitsuoka, M.; Kunieda, N. *CL* **1992**, 1647.
15. Jung, M. E.; Light, L. A. *JACS* **1984**, *106*, 7614.
16. White, J. D.; Somers, T. C.; Yager, K. M. *TL* **1990**, *31*, 59.
17. Piers, E.; Friesen, R. W.; Keay, B. A. *CC* **1985**, 809.
18. Wulff, W. D.; Peterson, G. A.; Bauta, W. E.; Chan, K.-S.; Faron, K. L.; Gilbertson, S. R.; Kaesler, R. W.; Yang, D. C.; Murray, C. K. *JOC* **1986**, *51*, 277.
19. Scott, W. J.; Stille, J. K. *JACS* **1986**, *108*, 3033.
20. Carey, J. T.; Helquist, P. *TL* **1988**, *29*, 1243.
21. Carreño, M. C.; Ruano, J. L. G.; Pedregal, C.; Rubio, A. *JCS(P1)* **1989**, 1335.
22. Mioskowski, C.; Solladié, G. *TL* **1975**, 3341.
23. Solladié, G. *S* **1981**, 185.
24. Solladié, G.; Matloubi-Moghadam, F.; Luttmann, C.; Mioskowski, C. *HCA* **1982**, *65*, 1602.
25. Castaldi, G.; Colombo, R.; Allegrini, P. *TL* **1991**, *32*, 2173.
26. Sawamura, M.; Yamauchi, A.; Takegawa, T.; Ito, Y. *CC* **1991**, 874.
27. Fétizon, M.; Goulaouic, P.; Hanna, I. *SC* **1989**, *19*, 2755.

Ronald H. Erickson
Scios Nova, Baltimore, MD, USA

Brucine[1]

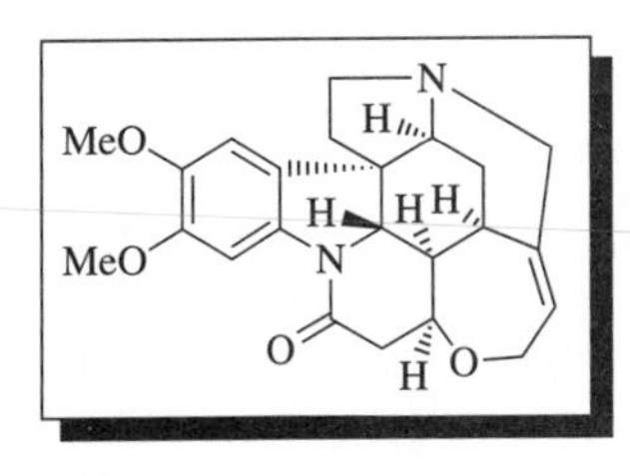

[357-57-3] $C_{23}H_{26}N_2O_4$ (MW 394.47)

(reagent for the resolution of acids, alcohols, and other neutral compounds[1])

Physical Data: colorless needles (acetone/water) mp 178 °C; $[\alpha]_D$ −79.3° (*c* 1.3, EtOH).
Solubility: very sol methanol, ethanol, and chloroform; mod sol ethyl acetate or benzene.

Form Supplied in: colorless needles or plates. The free base, which is available from multiple commercial sources, is usually hydrated. Dihydrated and tetrahydrated forms have been characterized. Anhydrous brucine can be obtained by heating at 100–120 °C in vacuo for 24 h. The hydrated forms can be used for most applications.

Purification: the commercial reagent is often used without further purification. However, the reagent can be purified by recrystallization from ethanol/water (1:1).[2] Recovered reagent[3] should be purified before reuse.

Handling, Storage, and Precautions: EXTREMELY POISONOUS. Oral LD_{50} in rats is $1\ mg\ kg^{-1}$. Handle in well-ventilated hood only.

Introduction. The alkaloid brucine has been a key resolving agent for over a century, in spite of its highly toxic nature. The group of chiral bases represented primarily by brucine, its homolog strychnine, and the cinchona alkaloids quinine, quinidine, cinchonidine, and cinchonine, has been extremely useful for the resolution of all types of acids.[1] No empirical rules have emerged from all of this work to help in predicting the optimal resolving agent for a given type of acid. Acid resolution is still primarily an empirical process that requires the evaluation of several diastereomeric salts. An inherent limitation to the use of alkaloids as resolving agents for acids is the availability of only one antipode, which sometimes allows the practical isolation of only one of the acid enantiomers in a pure form. Nevertheless, there are reports of resolutions with brucine that are so efficient that the less crystalline enantiomer can be isolated directly from the mother liquors (see below for examples). In other cases, pairs of pseudoenantiomeric cinchona alkaloids (i.e. quinine and quinidine, cinchonine and cinchonidine), or brucine and another alkaloid, display opposite selectivities for the enantiomers of a racemic acid (see below).[1a]

Resolution of Acids. The number of acids resolved with brucine is too large to attempt to list even a small portion of them in this synopsis. An excellent tabulation of all published resolutions with brucine up to 1972 is available.[1a] Only a few representative examples will be described here (eqs 1–4).[4–9] In all these cases, the resolved acids were obtained in high yield and with almost absolute enantiomeric purity. The solvents most frequently used for brucine resolutions are acetone and alcohol solvents. However, water, hexane, and others have also been used as cosolvents.

brucine, EtOH, 73%; 100% ee (1)

brucine•4H_2O, EtOH, 74%; 86% ee, absolute stereochemistry unknown (2)

brucine (anhyd), acetone, 41%; 100% ee (3)

brucine, MeOH, 45%; X = O, S; 100% ee (4)

Additional types of carboxylic acids that have been successfully resolved with brucine are represented by structures (**1**)–(**5**).[10–14]

(**1**) (**2**) (**3**) (**4**) (**5**)

As mentioned above, one of the limitations of using naturally occurring resolving agents is that only one enantiomer of the compound being resolved may be readily accessible by resolution. However, many examples have been described where brucine and some other alkaloid favor crystallization with opposite enantiomers of a given acid. For example, resolution of acid (**6**) with brucine yields the (+)-enantiomer, while cinchonidine provides material that is enriched in the (−)-enantiomer of the acid.[15] Similarly, diacid (**7**) is resolved into its (−)-enantiomer by brucine and into its (+)-enantiomer by strychnine.[16] The (+)-enantiomer of acid (**8**) can be obtained with brucine, while the (−)-enantiomer crystallizes with cinchonidine.[17] Additional examples of the same phenomenon can be found in the literature.[1a]

(**6**) (**7**) (**8**)

Resolution of Alcohols. Although not a well exploited use of brucine, a variety of secondary benzylic alcohols have been resolved by complexation and crystallization with brucine (eq 5).[18] About a dozen alcohols were obtained in close to enantiomeric purity by this procedure.[18] Also resolved by crystallization of their brucine inclusion complexes were a series of tertiary propargylic alcohols (eq 6).[19] In this case, the enantiomer that does not crystallize with brucine can be obtained in almost complete optical purity from the mother liquors.

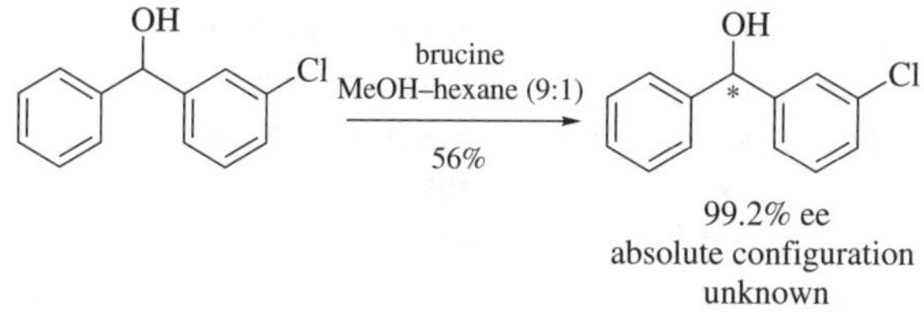

(5)

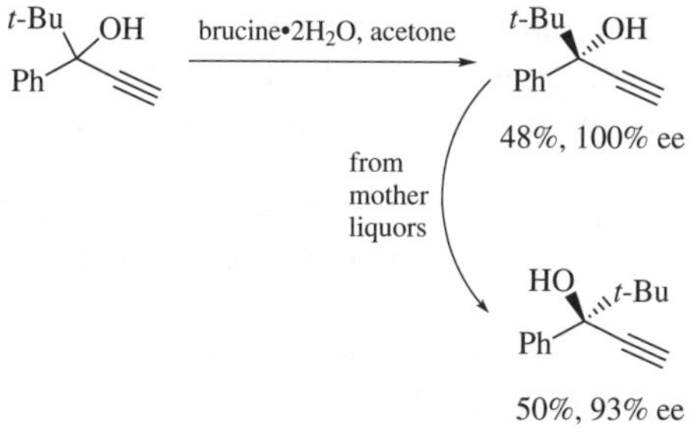

99.2% ee
absolute configuration
unknown

(6)

48%, 100% ee

from mother liquors

50%, 93% ee

A more traditional and general approach to the resolution of alcohols is the formation of the corresponding hemiphthalate or hemisuccinate esters, followed by resolution of these acidic derivatives with brucine or some other chiral base (eqs 7–9).[20–23] The resolved alcohols are liberated by alkaline hydrolysis of the esters. High enantiomeric purity is frequently achieved by this procedure, which has been applied successfully to primary, secondary, and tertiary alcohols.

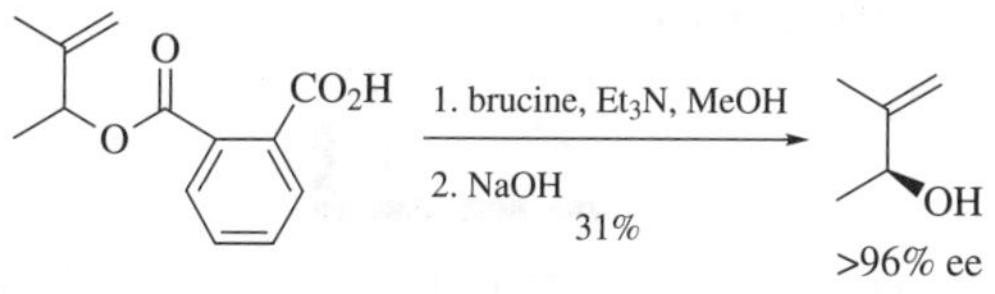

(7)

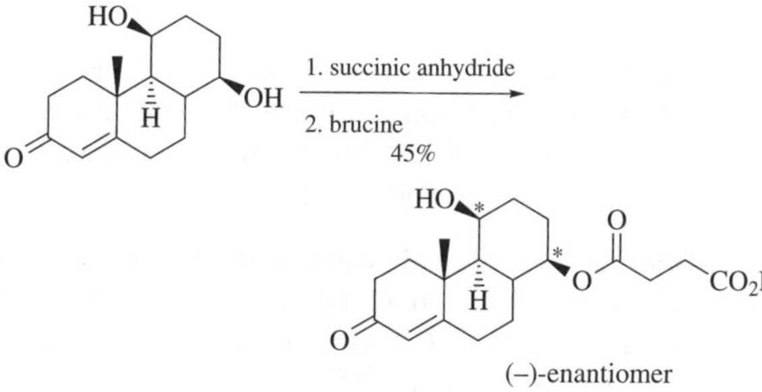

>96% ee

(8)

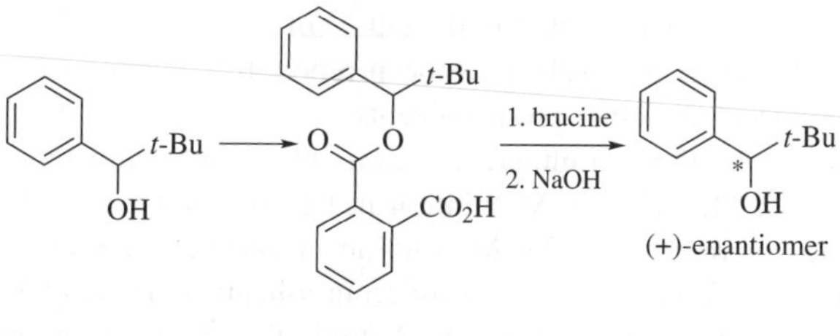

(–)-enantiomer

(9)

(+)-enantiomer

Resolution of Ketones. Brucine has not been used very extensively for the resolution of neutral compounds. However, in some cases, ketones or ketone derivatives may form diastereomeric inclusion complexes with brucine, providing an opportunity for their resolution. For example, the cyanohydrin of a bicyclic ketone has been resolved by this procedure (eq 10).[24] Following resolution of the cyanohydrin, the ketone was regenerated and determined to be of 94% ee.

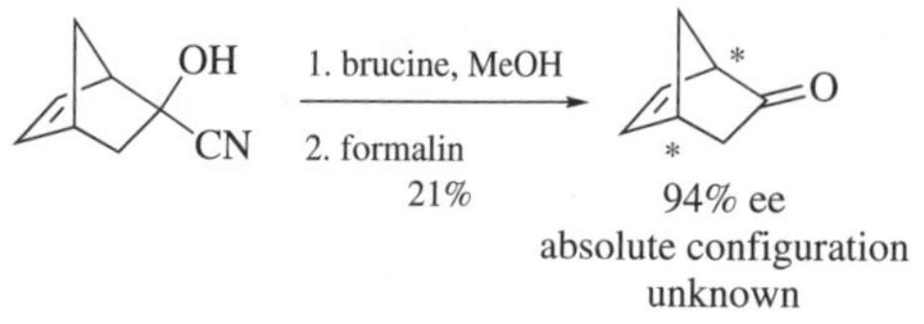

(10)

94% ee
absolute configuration
unknown

Resolution of Sulfoxides. Although it can be considered as the resolution of an unique type of carboxylic acid, some racemic sulfoxides containing carboxylic acids have been resolved via diastereomeric crystalline complexes with brucine (eq 11).[25]

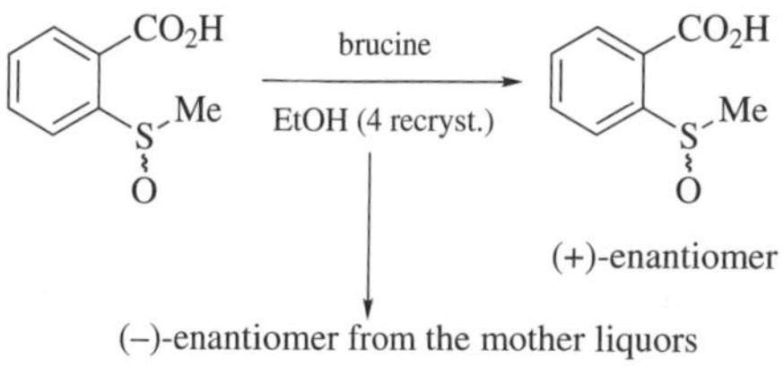

(11)

(+)-enantiomer

(–)-enantiomer from the mother liquors

Chiral Catalysis. Brucine has been utilized as chiral catalyst in a variety of reactions. For example, its incorporation into a polymer support provides a chiral catalyst for performing enantioselective benzoin condensations.[26] It has also been used as a chiral catalyst in the asymmetric synthesis of (*R*)-malic acid via the corresponding β-lactone, which results from the asymmetric cycloaddition of chloral and ketene (eq 12).[27] Though brucine yields malic acid with 68% ee, quinidine was found to be a more selective catalyst (98% ee).

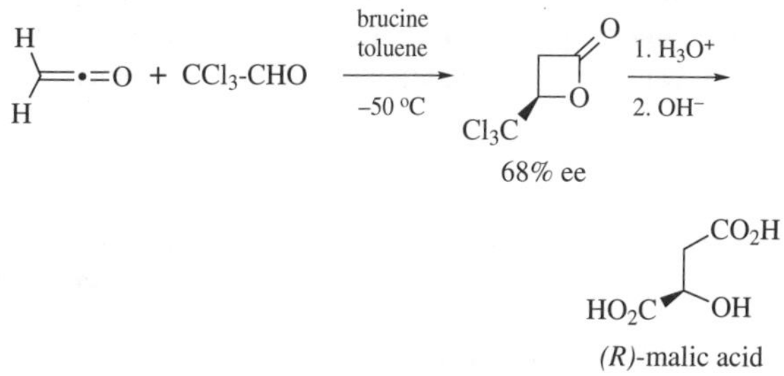

(12)

68% ee

(*R*)-malic acid

Brucine has been used as an enantioselective catalyst in the kinetic resolution of alcohols. For example, an azirinylmethanol was reacted with 0.5 equiv of ***Acetic Anhydride*** in the presence of 25 mol % brucine. The resulting acetate was found to possess 24% ee (eq 13).[28]

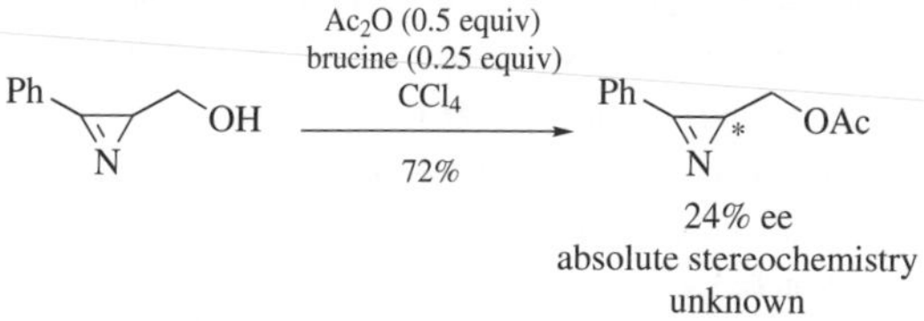

(13)

24% ee
absolute stereochemistry
unknown

Brucine has been used to produce enantiomerically enriched compounds by selective reaction with or destruction of one of the enantiomers. The optical purity of the resulting compound is usually modest, although some exceptions have been described. For example, dibromo compound (**9**) was obtained (enriched in the (−)-enantiomer) by selective destruction of the (+)-enantiomer with brucine in chloroform.[29] The resolution of (±)-2,3-dibromobutane may have also been a case of enantioselective destruction,[30] although more recent reports suggest that it is more likely a case of enantioselective entrapment in the brucine crystals (eq 14).[31]

(9)

racemic → brucine → 70% ee (14)

Miscellaneous. Brucine greatly accelerates the decarboxylation of certain β-oxo carboxylic acids at rt (eq 15),[32] as well as the decarbalkoxylation of β-oxo esters.[33] In some cases the products of these reactions possess some (modest) enantiomeric excess.[34]

brucine, acetone, 25 °C, overnight, 76% (15)

Related Reagents. (1*R*,2*S*)-Ephedrine; (2*S*,2′*S*)-2-Hydroxymethyl-1-[(1-methylpyrrolidin-2-yl)methyl]pyrrolidine; (*S*)-α-Methylbenzylamine; 1-(1-Naphthyl)ethylamine; Quinine.

1. (a) Wilen, S. H. In *Tables of Resolving Agents and Optical Resolutions*; Eliel, E. L., Ed.; University of Notre Dame Press: Notre Dame, 1972. (b) Jacques, J.; Collet, A. In *Enantiomers, Racemates and Resolutions*; Wilen, S. H., Ed.; Wiley: New York, 1981.
2. DePuy, C. H.; Breitbeil, F. W.; DeBruin, K. R. *JACS* **1966**, *88*, 3347.
3. Vogel, A. I. *Practical Organic Chemistry*; Longmans: London, 1957. p 507.
4. Allan, R. D.; Johnston, G. A. R.; Twitchin, B. *AJC* **1981**, *34*, 2231.
5. Kaifez, F.; Kovac, T.; Mihalic, M.; Belin, B.; Sunjic, V. *JHC* **1976**, *13*, 561.
6. Kanoh, S.; Hongoh, Y.; Motoi, M.; Suda, H. *BCJ* **1988**, *61*, 1032.
7. Hasaka, N.; Okigawa, M.; Kouno, I.; Kawano, N. *BCJ* **1982**, *55*, 3828.
8. Lévai, A.; Ott, J.; Snatzke, G. *M* **1992**, *123*, 919.
9. Puzicha, G.; Lévai, A.; Szilágyi, L. *M* **1988**, *119*, 933.
10. Tichy, M.; Sicher, J. *TL* **1969**, *53*, 4609.
11. Dvorken, L. V.; Smyth, R. B.; Mislow, K. *JACS* **1958**, *80*, 486.
12. McLamore, W. M.; Celmer, W. D.; Bogert, V. V.; Pennington, F. C.; Sobin, B. A.; Solomons, I. A. *JACS* **1953**, *75*, 105.
13. Sealock, R. R.; Speeter, M. E.; Schweet, R. S. *JACS* **1951**, *73*, 5386.
14. Dutta, A. S.; Morley, J. S. *CC* **1971**, 883.
15. Mislow, K.; Strinberg, I. V. *JACS* **1955**, *77*, 3807.
16. Hoffman, T. D.; Cram, D. J. *JACS* **1969**, *91*, 1000.
17. Tanabe, T.; Yajima, S.; Imaida, M. *BCJ* **1968**, *41*, 2178.
18. Toda, F.; Tanaka, K.; Koshiro, K. *TA* **1991**, *2*, 873.
19. Toda, F.; Tanaka, K. *TL* **1981**, *22*, 4669.
20. Crout, D. H. G.; Morrey, S. M. *JCS(P1)* **1983**, 2435.
21. Lukes, R. M.; Sarett, L. H. *JACS* **1954**, *76*, 1178.
22. MacLeod, R.; Welch, F. J.; Mosher, H. S. *JACS* **1960**, *82*, 876.
23. Eliel, E. L.; Kofron, J. T. *JACS* **1953**, *75*, 4585.
24. Black, K. A.; Vogel, P. *HCA* **1984**, *67*, 1612.
25. Barbieri, G.; Davoli, V.; Moretti, I.; Montanari, F.; Torre, G. *JCS(C)* **1969**, 731.
26. Castells, J.; Duñach, E. *CL* **1984**, 1859.
27. Wynberg, H.; Staring, E. G. J. *JACS* **1982**, *104*, 166.
28. Stegmann, W.; Uebelhart, P.; Heimgartner, H.; Schmid, H. *TL* **1978**, *34*, 3091.
29. Greene, F. D.; Remers, W. A.; Wilson, J. W. *JACS* **1957**, *79*, 1416.
30. Tanner, D. D.; Blackburn, E. V.; Kosugi, Y.; Ruo, T. C. S. *JACS* **1977**, *99*, 2714.
31. Pavlis, R. R.; Skell, P. S. *JOC* **1983**, *48*, 1901.
32. Hargreaves, M.; Khan, M. *M* **1978**, *109*, 799.
33. Miles, D. H.; Stagg, D. D. *JOC* **1981**, *46*, 5376.
34. Toussaint, O.; Capdevielle, P.; Maumy, M. *TL* **1987**, *28*, 539.

Juan C. Jaen
Parke-Davis Pharmaceutical Research, Ann Arbor, MI, USA

n-Butyllithium[1]

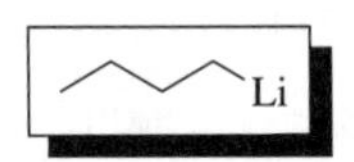

[109-72-8] C_4H_9Li (MW 64.05)

(strong base capable of lithiating carbon acids;[1] useful for heteroatom-facilitated lithiations;[2,3] useful for lithium–halogen exchange;[1,4] reagent of choice for lithium–metal transmetalation reactions[1a,c,5])

Physical Data: colorless liquid; stable at rt; eliminates LiH on heating; d^{25} 0.765; mp −76 °C; bp 80–90 °C/0.0001 mmHg; dipole moment 0.97 D.[6] ^{13}C NMR, ^{1}H NMR, ^{6}Li NMR[8] and MS studies have been reported.[7–9]

Solubility: sol hydrocarbon and ethereal solvents, but should be used at low temperature in the latter solvent type: half-lives in diethyl ether and THF have been reported;[10] reacts violently with H_2O and other protic solvents.

Form Supplied in: commercially available as approximately 1.6 M, 2.5 M, and 10.0 M solution in hexanes and in cyclohexane, approximately 2.0 M solution in pentane, and approximately 1.7 M, and 2.7 M solution in *n*-heptane. Hexameric in hydrocarbons;[1c] tetrameric in diethyl ether;[1c] dimer–tetramer equilibrium mixture in THF;[11] when used in combination with tertiary polyamines such as TMEDA and DABCO, reactivity is usually increased.[1,12]

Analysis of Reagent Purity: since the concentration of commercial solutions may vary appreciably it is necessary to standardize solutions of the reagent prior to use. A recommended method for routine analyses involves titration of the reagent with *s*-butyl alcohol using 1,10-phenanthroline or 2,2′-biquinoline as indicator.[15] Several other methods have been described.[16]

Preparative Methods: may be prepared in high yield from *n*-butyl chloride[13] or *n*-butyl bromide[14] and ***Lithium*** metal in ether or hydrocarbon solvents.

Handling, Storage, and Precautions: solutions of the reagent are pyrophoric and the reagent may catch fire if exposed to air or moisture. Handling of the reagent should be done behind a shield in a chemical fume hood. Safety goggles, chemical resistant gloves, and other protective clothing should be worn. In case of fire, a dry-powder extinguisher should be used: in no case should an extinguisher containing water or halogenated hydrocarbons be used to fight an alkyllithium fire. Bottles and reaction flasks containing the reagent should be flushed with N_2 or preferably Ar and kept tightly sealed to preclude contact with oxygen or moisture. Standard syringe/cannula techniques for air and moisture sensitive chemicals should be applied when transferring the reagent. For detailed handling techniques see Wakefield.[1b]

Lithiations. *n*-Butyllithium is a commonly used reagent for deprotonation of a variety of nitrogen (see ***Lithium Diisopropylamide***), oxygen, phosphorus and carbon acids to form lithium salts. Compared to its *s*- and *t*-butyl analogs, *n*-BuLi is less basic[17] and less reactive but it is usually the reagent of choice for deprotonation of relatively strong carbon acids. These lithiations are most favorable when the conjugated bases are stabilized by resonance or when the carbanion forms at the sp hybridized carbon of a triple bond. Thus indene,[18] triphenylmethane,[19] allylbenzene,[20] and methyl heteroaromatics (e.g. ***Pyridine***, quinoline, and isoquinoline derivatives)[21] are readily lithiated with *n*-BuLi at the benzylic position. Allenes are lithiated at C-1 or C-3 depending on the number and size of the alkyl groups at these positions,[22] and terminal alkynes react with *n*-BuLi to give lithium acetylides.[4] Propargylic hydrogens can also be removed,[23] and treatment of terminal alkynes with 2 equiv of *n*-BuLi results in the lithiation of both propargylic and acetylenic positions (eq 1).[23a]

OTHP —(2 equiv *n*-BuLi; pentane, THF)→ Li, Li, OTHP —(1. EtBr; 2. H_2O)→ OTHP (1)

82%

The metalating ability of *n*-BuLi (and other organolithiums) is greater in electron-donating solvents than in hydrocarbons. Electron-donating solvents such as diethyl ether or THF provide coordination sites for the electron deficient lithium and promote the formation of lower-order organolithium aggregates.[1a,c] These exhibit significantly higher levels of reactivity than do the higher-order oligomers present in hydrocarbon solvents. *n*-Butyllithium is often used in the presence of added lithium complexing ligands, such as ***N,N,N′,N′-Tetramethylethylenediamine*** (TMEDA) and ***1,4-Diazabicyclo[2.2.2]octane*** (DABCO), which further enhance the reactivity of this reagent.[12] Many compounds, normally unreactive toward *n*-BuLi alone (e.g. benzene), are readily lithiated[24] and even polylithiated[25] by a combination of *n*-BuLi and one of these additives. Allylic[26,27] and benzylic sites bearing no additional activating groups at the α-position (e.g. toluene)[12c] are also lithiated in the presence of lithium complexing donors (eq 2).[26]

OH —(2 equiv *n*-BuLi, TMEDA; hexane, −78 °C)→ Li, OLi —(1. Br; 2. H^+)→ OH (2)

70%

Metal alkoxides are often employed as additives to enhance the metalating ability of *n*-BuLi. Particular use has been made of the ***n-Butyllithium–Potassium t-Butoxide*** mixture,[28] which is capable of effecting rapid metalation of benzylic[29] and allylic[30] systems as well as aromatic rings.[28] Although the products of these metalations are not organolithium compounds they do, nevertheless, readily react with electrophiles. Alternatively, they may be converted into the corresponding organolithium derivatives by the addition of ***Lithium Bromide***.[30a,31] Using *t*-BuOK/*n*-BuLi for deprotonation, the one-pot synthesis of 2-(4-isobutylphenyl)propanoic acid (ibuprofen) from *p*-xylene, through a sequence of metalations and alkylations, has been achieved in 52% overall yield.[29] Dimetalation of arylalkynes with *n*-BuLi/*t*-BuOK followed by addition of electrophiles has been used as a route to *ortho*-substituted arylalkynes (eq 3).[31b]

C≡CH —(*n*-BuLi–*t*-BuOK (2:1); THF–hexane, −20 °C)→ K, C≡CLi —(1. MeI; 2. H^+)→ C≡CH (3)

88%

Facile and regioselective α-deprotonation is often effected by treatment of heteroatom-containing (e.g. oxygen, sulfur, nitrogen, and the like) compounds with *n*-BuLi, and these reactions have been extensively reviewed.[2,3,32,33] Thus sulfones,[34] certain sulfides,[35] and sulfoxides[36] can be lithiated adjacent to sulfur under various conditions, and α-heterosubstituted vinyllithium compounds are often available from the corresponding ethers,[37] thioethers,[38] chlorides,[39] and fluorides[40] via lithiation using *n*-BuLi. Isocyanides[41] and nitro compounds[42] have been lithiated adjacent to nitrogen with *n*-BuLi in THF at low temperatures (less than −60 °C) and the formation of various α-phosphorus alkyllithiums has been reported.[43] Simple ethers are also susceptible to lithiation at the α site by *n*-BuLi, especially at elevated temperatures.[44] The initial proton abstraction in these systems is generally followed by various cleavage reactions resulting from α,β- and α′,β′-eliminations as well as the Wittig rearrangement.[45] Tetrahydrofuran, for example, is rapidly lithiated by *n*-BuLi at 35 °C ($t_{1/2}$ = 10 min) to give ethylene and the lithium enolate of acetaldehyde.[46]

Due to the stabilizing effect of two sulfur atoms, 1,3-dithianes are easily lithiated at the α-position on treatment with *n*-BuLi (see ***2-Lithio-1,3-dithiane***, Vol. A).[47] The 2-lithio-1,3-dithianes constitute an important class of acyl anion equivalents, permitting electrophilic substitution to occur at the masked carbonyl carbon.[47] Hydrolysis of the 1,3-dithiane functionality into a carbonyl group is effected in the presence of mercury(II) ion (eq 4).[48]

1. *n*-BuLi, THF −20 to −75 °C; 2. $Cl(CH_2)_3Br$; *n*-BuLi, THF −75 °C; H_2O, $HgCl_2$ (4) 84% 81%

When conducted at sufficiently low temperatures (less than −78 °C), α-deprotonation can occur faster than nucleophilic addition to an electrophilic center present in the same molecule. This strategy, which involves initial lithiation followed by intramolecular nucleophilic addition of the newly generated C–Li bond to an electrophilic moiety, is useful for the construction of carbocycles including medium and large ring systems (eqs 5 and 6).[49,50]

5–10% excess *n*-BuLi, −90 to −40 °C; H^+ (5) 91%

1. *n*-BuLi, THF, DABCO, −78 °C; 2. H^+ (6) 62%

A large number of heteroaromatic compounds,[2] such as furans (see ***Furan***),[51] thiophenes (eq 7),[52] oxazoles,[53] and *N*-alkyl- and *N*-aryl substituted pyrroles, pyrazoles,[55] imidazoles,[56] triazoles,[57] and tetrazoles,[58] are lithiated under various conditions α to the ring heteroatom using *n*-BuLi. However, pyridine and other nitrogen heteroaromatics bearing the pyridine, pyrimidine, or pyrazine nucleus are generally not lithiated. Indeed, they have a tendency to undergo nucleophilic addition reactions with this reagent.[1b]

1. *n*-BuLi, Et_2O 15 °C; 2. $(CH_2O)_n$ (7) 79%

When the α-position is benzylic, propargylic, or allylic, deprotonation takes place more readily and *n*-BuLi is generally the reagent of choice for these reactions.[59,60] However, the more basic ***s-Butyllithium*** is a better reagent for deprotonation of alkyl allyl ethers[61] and certain alkyl allyl thioethers[62] which react slowly (if at all) with *n*-BuLi in THF at low temperatures (less than −65 °C).

Proton removal adjacent to a heteroatom is further facilitated if the lithium can be coordinated to proximate electron donors, such as a carbonyl oxygen, permitting the formation of 'dipole-stabilized' carbanions.[33] Thus various 2-alkenyl *N,N*-dialkylcarbamates undergo rapid α-deprotonation adjacent to oxygen on treatment with *n*-BuLi/TMEDA at −78 °C.[63,64] The resulting dipole-stabilized lithium carbanions react with ketones and aldehydes in a highly regioselective fashion providing γ-hydroxyalkylated enol esters (eq 8) which, following cleavage of the carbamoyl moiety (***Titanium(IV) Chloride***/H_2O or MeOH), afford δ-hydroxy carbonyl compounds (homoaldols) as lactols or lactol ethers.[65] Similarly, aliphatic or aromatic amides[66,67] (eq 9),[67a] phosphoramides,[68] and some formamidine derivatives (e.g. 1,2,3,6-tetrahydropyridine,[69] thiazolidine,[69] 1,3-thiazine,[69] and tetrahydroisoquinolines[70]) are selectively lithiated α to the nitrogen at the activated position. Electrophilic substitution of the intermediate organolithiums followed by hydrolytic cleavage of the amide or formamidine group provides a synthetically valuable route to α-substituted (or γ-substituted[69]) secondary amines.[33] Use of the more basic *s*-BuLi (or ***t-Butyllithium***) is generally required for deprotonation of the analogous nonbenzylic or nonallylic systems.

n-BuLi, TMEDA, Et_2O, −78 °C; Cb = CON(*i*-Pr)$_2$; 1. Me_2CHCHO; 2. HOAc 87%; (8) >97:3, >95% *E*

2 equiv *n*-BuLi, −78 °C, diglyme; *n*-BuI (9) 95%

n-Butyllithium is also used for the stereoselective α-lithiation of chiral sulfonyl compounds,[71] chiral 2-alkenyl carbamates,[64,72] various heterocyclic amine derivatives with chiral auxiliaries appended on the nitrogen (e.g. oxazoline[73] or formamidine[74,75] groups) (eq 10),[75] and chiral oxazolidinones derived from benzylamines (eq 11).[76] These elegant reactions have been applied to the asymmetric syntheses of a number of natural products.[77,78]

Ortho* Lithiations.** Heteroatom-containing substituents on aromatic rings facilitate metalation by organolithium reagents and direct the metal almost exclusively to the *ortho* position. This effect, usually referred to as *ortho* lithiation, is of considerable synthetic importance and the topic has been extensively reviewed.[2,4,79] Although *n*-BuLi (typically in THF or Et_2O with added TMEDA)[80] is capable of effecting a large number of *ortho* lithiations, the use of this reagent is somewhat limited by its tendency to undergo nucleophilic carbonyl additions with some of the most potent and useful *ortho* directors, particularly tertiary amide and carbamate functionalities (e.g. $CONEt_2$ and $OCONEt_2$).[79,81] For example, *N,N*-dimethyl- and -diethylbenzamides afford primarily aryl butyl ketones upon treatment with *n*-BuLi. The reagent of choice for these lithiations is ***s-Butyllithium, which is less nucleophilic than *n*-BuLi and hence more tolerant of electrophilic functional groups.[79] There are, however, a variety of *ortho*-directing groups that are well suited for *n*-BuLi-promoted lithiations. These include NR_2,[80] CH_2NR_2,[80] $CH_2CH_2NR_2$,[80] OMe (see ***o-Lithioanisole***),[83] OCH_2OMe,[84] SO_2NR_2,[85] C=NR,[86] 2-oxazolinyl,[85,87] F,[88] CF_3,[80] and groups that contain acidic hydrogens and themselves undergo deprotonation prior to lithiation of the aromatic ring (thus requiring the use of 2 equiv of *n*-BuLi), e.g. CONHR,[89] CH_2OH,[80] NHCO-*t*-Bu,[90] and SO_2NHR[80] (eq 12).[91] *n*-Butyllithium is also used frequently for the *ortho* lithiation of heterocyclic aromatic rings,[2] including those that contain the pyridine nucleus (eq 13).[92]

n-BuLi, THF, −100 °C; Ph(CH2)3Br, −100 °C; N2H4; α/γ-alkylation = 92:8; R* = ; i-Pr; t-BuO; Ph(CH2)3; 76%, >95% ee (10)

1. n-BuLi, THF, −78 °C; 2. MeI, 92%

1. KOH, ether; 2. Pb(OAc)4; 3. HCl, EtOH; 54%; NH2; 100% ee (11)

HO; 2 equiv n-BuLi, THF, TMEDA, −78 °C; LiO; Li; 1. CHO, MeO, NEt2, O; 2. TsOH; MeO; 77%; (12)

NHCO-t-Bu; 2 equiv n-BuLi, THF, 0 °C; t-Bu, N, OLi, Li; MeSSMe; NHCO-t-Bu, SMe; 83% (13)

Formation of Enolate Anions and Enolate Equivalents. Owing to its tendency to undergo nucleophilic addition with carbonyl groups and other electrophilic carbon–heteroatom multiple bonds (C=NR, C≡N, C=S),[1] *n*-BuLi is usually not the reagent of choice for the generation of enolate anions or enolate equivalents from active hydrogen compounds. This is done most conveniently using the less nucleophilic lithium dialkylamides (e.g. ***Lithium Diisopropylamide*** (LDA), ***Lithium 2,2,6,6-Tetramethylpiperidide*** (LiTMP), and ***Lithium Hexamethyldisilazide*** (LTSA)) prepared (often in situ) from sterically hindered secondary amines, typically by treatment with *n*-BuLi.[93] However, less reactive carbonyl compounds such as amides (eq 14)[94] and carboxylic acids[95] as well as those containing carbon–nitrogen or carbon–sulfur multiple bonds, e.g. imines,[96] oxazines,[97] nitriles,[98] some hydrazones,[99] and thioamides,[100] can be lithiated α to the electrophilic carbon with *n*-BuLi under various conditions. β-Keto esters can be alkylated at the α′-carbon using ***Sodium Hydride*** for the first deprotonation and *n*-BuLi for abstraction of the less acidic α′-proton followed by addition of alkyl halides.[101] Lithiation of unsymmetrical imines using *n*-BuLi takes place regioselectively at the most substituted α-carbon (eq 15).[96] In contrast, LDA directs metalation and subsequent alkylation predominantly to the less substituted α-position in similar systems.[102] Lithium enolates of camphor imine esters, prepared by addition of *n*-BuLi, undergo highly diastereoselective Michael additions with α,β-unsaturated esters (eq 16).[103] The tightly chelated structures of the intermediate enolates permit selective *re* face approach of the Michael acceptors, giving rise to the high degree of distereoselectivity observed.

NHPh; 2 equiv n-BuLi, THF, 0 °C; Li, Li+, NPh; 1. cyclopentanone; 2. H+; OH, NHPh; 94% (14)

1. n-BuLi, THF; 2. PhCH2Cl; 3. H3O+; 53%; Ph; 87:13 (15)

t-BuO, N, A*; 1. n-BuLi, THF, −78 °C; 2. O-t-Bu; Li; t-BuO; H; O-t-Bu; A* = ; t-BuO; 88%, syn/anti = 95:5 (16)

Metal–Halogen Interchange and Transmetalation Reactions. The metal–halogen interchange reaction which involves the exchange of halogen and lithium atoms is an important method for the preparation of organolithium compounds not readily accessible through metalation. In particular, the generation of aryl-, vinyl-, and cyclopropyllithium derivatives from the corresponding bromides or iodides on treatment with *n*-BuLi (usually in ethereal solvents at or below −78 °C) is of considerable synthetic utility (eqs 17–19).[104–106] The relative rates of exchange depend on the halide and decrease in the order I > Br > Cl > F. Fluorides and chlorides, with the exception of some polychlorinated aliphatic[107] and aromatic compounds,[108] are quite resistant to lithium–halogen interchange; instead, they tend to promote *ortho* and α lithiations.[79]

(17) 51%

(18) 85%

(19) 60%

gem-Dihalocyclopropanes react with *n*-BuLi in a highly stereoselective fashion (eq 19), although subsequent isomerization can take place.[109] Alkyl-substituted vinyllithiums can be prepared with retention of configuration[105,110] and even aryl-substituted vinyllithium compounds retain their configuration under controlled conditions (<−78 °C in THF or at rt in hydrocarbon solvents).[111] Simple alkyllithiums are generally not accessible by this route because of the unfavorable interchange equilibrium that ensues when primary or secondary halides are treated with *n*-BuLi, except in cases where the initially formed organolithium is rapidly consumed in a subsequent, irreversible reaction (eq 20).[112] Primary alkyllithiums may be prepared, however, from the corresponding alkyl iodides (but not from the bromides)[113] on treatment with the more reactive ***t*-Butyllithium** which renders the exchange operationally irreversible.[114]

n-Butyllithium is the reagent of choice for effecting a number of transmetalation reactions involving the replacement of tin, selenium, tellurium, or mercury by lithium.[1a,1c,5] These reactions are typically conducted at low temperatures (less than −60 °C) in ethereal solvents, with THF being the most commonly employed reaction medium. The tin–lithium exchange is a particularly important operation, providing a convenient route to aryl- and vinyllithiums,[115] as well as functionalized lithio derivatives which are not readily accessible by other means, such as α-alkoxylithium compounds,[116] oxiranyllithiums,[117] and amino-substituted organolithiums.[118] The Sn–Li exchange in these systems and subsequent trapping of the intermediate lithio derivatives proceed stereoselectively with retention of configuration at the tin-bearing carbon (eqs 21 and 22).[116c,118c] Consequently, enantiomerically enriched functionalized products are available through this methodology from homochiral α-hetero-substituted stannanes.[119]

(20) 95%

(21) 86%

(22) 78%

Appropriately substituted alkenic α-alkoxylithiums, derived from the corresponding tri-*n*-butyltin compounds on treatment with *n*-BuLi, have also been used to initiate 5-*exo-trig* cyclization reactions to give tetrahydrofuran derivatives (eq 23).[120] Other related ring-forming reactions have been reported.[121]

(23) *cis*:*trans* = 10:1

Organoselenium compounds, particularly selenoacetals or selenothioacetals, undergo facile lithium–selenium exchange reactions on treatment with *n*-BuLi, typically in THF at −78 °C.[122–124] The intermediate α-seleno organolithium derivatives formed in these reactions are readily trapped with electrophiles to afford a variety of synthetically useful α-functionalized selenide products[3,125] (eq 24).[126] Studies on cyclohexyl selenoacetals have shown that Li–Se exchange takes place almost exclusively at the axial position[123] and that equatorial α-lithio sulfides, derived from mixed Se,S-acetals by Li–Se exchange, epimerize within minutes at −78 °C to give the more stable axial lithio isomers.[124] *n*-Butyllithium may be used for the majority of Li–Se exchanges although selenoacetals derived from sterically hindered ketones react quite slowly with this reagent, and in these cases the use of the more reactive ***s-Butyllithium*** is warranted.[123]

A variety of organolithiums, including benzylic, vinylic, alkynic, and 1-alkoxylithium compounds, are also accessible through the lithium–tellurium exchange, which involves the treatment of diorganotellurides with *n*-BuLi at −78 °C in THF (eq 25);[127] vinyllithium derivatives have been prepared from the corresponding organomercurials by essentially the same methodology.[128]

Rearrangements. *n*-Butyllithium is used frequently to promote various anionic rearrangement reactions. 2,3-Wittig rearrangements are commonly effected by treatment of allylic and propargylic ethers with *n*-BuLi,[129] and 1,2-, 1,4-, as well as 3,4-rearrangements with similar systems are also known.[130] These reactions are most often initiated at low temperatures either by α-deprotonation adjacent to oxygen or via Li–Sn exchange, and the rearrangement occurs upon warming of the reaction mixture to 0 °C. Additional benzylic, propargylic, or allylic stabilization is necessary for deprotonation by *n*-BuLi; however, the Li–Sn method does not suffer from this limitation.[131] 2,3-Wittig rearrangements involving (*Z*)-allylic ethers proceed generally with high *syn* stereoselectivity (eq 26),[131] whereas the opposite tendency is observed with (*E*)-allylic ethers.[132] The regiochemistry of unsymmetrical bis-allylic ethers depends on the degree of substitution at the α- and γ-carbons, with the less substituted allylic position being the preferred site of deprotonation (eq 27).[133] The cyclic variant of the 2,3-Wittig rearrangement can be used for ring contraction reactions and it provides a useful method for the construction of macrocyclic ring systems (eq 28).[134]

Secondary 1,2-epoxy alcohols may be prepared through the Payne rearrangement in aprotic media by treatment of primary 2,3-epoxy alcohols with *n*-BuLi in THF with catalytic amounts of ***Lithium Chloride***,[135] and α,β-unsaturated ketones are obtained from trimethylsilyl substituted propargylic alcohols through the Brook rearrangement followed by alkylation of the intermediate allenyllithium products and acidic workup (eq 30).[136] Quaternary ammonium salts, such as benzyltrimethylammonium ion, react with *n*-BuLi to deliver nitrogen ylides which can undergo either the Stevens rearrangement to give tertiary amines or the Sommelet–Hauser rearrangement to afford *ortho* alkyl-substituted nitrogen-containing aromatics, depending on the reaction conditions (eq 31).[137] Dibenzyl thioether has been reported to undergo a Sommelet–Hauser type rearrangement on treatment with *n*-BuLi/TMEDA in HMPA,[138] and appropriately substituted sulfonic ylides, prepared from sulfonium salts on treatment with *n*-BuLi, undergo 2,3-sigmatropic rearrangements (eq 32).[139]

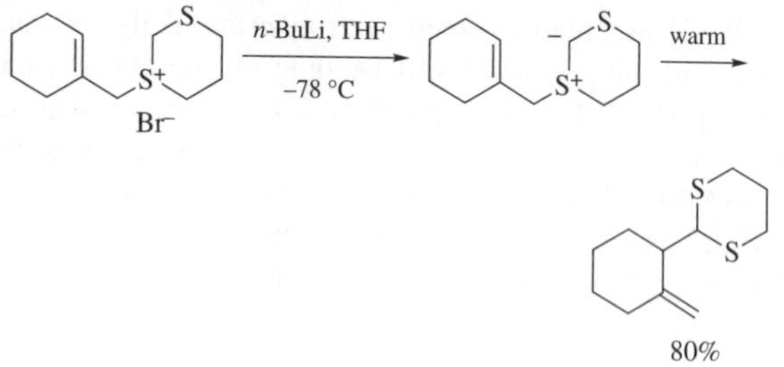

(32)

80%

Elimination Reactions. A number of eliminations can be effected using *n*-BuLi. The formation of phosphorus, nitrogen, and sulfur ylides[1a] and the generation of benzyne intermediates from aromatic halides[140] are well established processes, and α-eliminations resulting in the formation of carbenes have been used to prepare cyclopropanes,[141] oxiranes,[142] and other products via subsequent carbenoid rearrangements (eq 33).[143] Vinylidene dihalides undergo dehalogenation with concomitant formation of a triple bond on treatment with 2 equiv of *n*-BuLi.[144] The initial product of this reaction is a lithium acetylide which can be quenched with methanol to give a terminal alkyne or alkylated in situ by addition of alkyl halides to afford internal alkynes (eq 34).[145] The decomposition of (arylsulfonyl)hydrazones of aldehydes or ketones upon treatment with at least 2 equiv of *n*-BuLi (Shapiro reaction) is a useful method for the generation of vinyllithium compounds (eq 35).[146] In cases where two regioisomers can be produced, *n*-BuLi appears to promote the formation of the less substituted vinyllithium via deprotonation of the kinetically more acidic proton, whereas the use of *s*-BuLi reportedly leads to the more substituted vinyllithium product.[147]

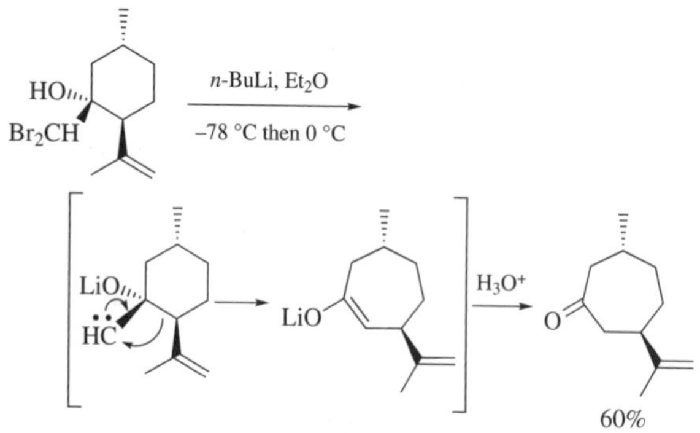

(33)

60%

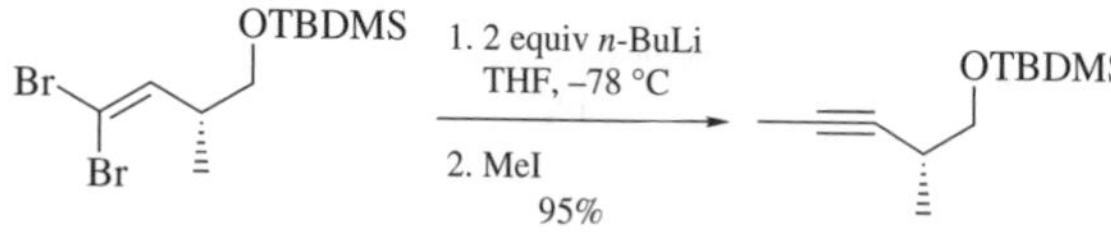

(34)

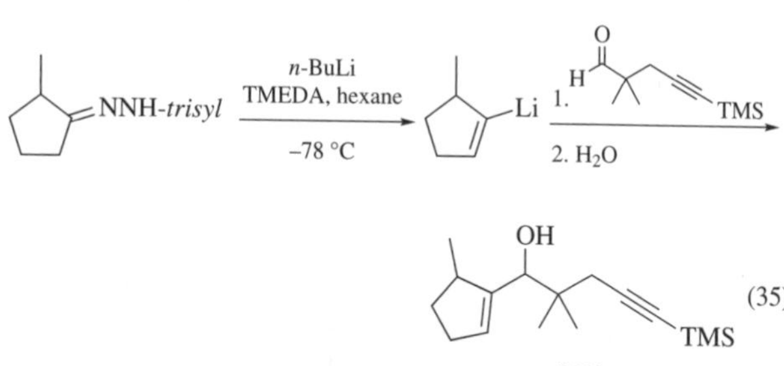

95%

(35)

83%

Related Reagents. *s*-Butyllithium; *t*-Butyllithium; *n*-Butyllithium–Boron Trifluoride Etherate; *n*-Butyllithium–Potassium *t*-Butoxide; Lithium Diisopropylamide; Methyllithium; Tungsten(VI) Chloride–*n*-Butyllithium.

1. (a) Wakefield, B. J. *The Chemistry of Organolithium Compounds*; Pergamon: Oxford, 1974. (b) Wakefield, B. J. *Organolithium Methods*; Academic: San Diego, 1990. (c) Wardell, J. L. In *Comprehensive Organometallic Chemistry*; Wilkinson, G., Ed.; Pergamon: Oxford, 1982; Chapter 2.
2. Gschwend, H. W.; Rodriguez, H. R. *OR* **1979**, *26*, 1.
3. Krief, A. *T* **1980**, *36*, 2531.
4. Wardell, J. L. In *Inorganic Reactions and Methods*; Zuckerman, J. J., Ed.; VCH: New York, 1988; Vol. 11, pp 107–129.
5. Wardell, J. L. In *Inorganic Reactions and Methods*; Zuckerman, J. J., Ed.; VCH: New York, 1988; Vol. 11, pp 31–44.
6. *Dictionary of Organometallic Compounds*; Buckingham, J., Ed.; Chapman & Hall: London, 1984; Vol. 1, p 1213.
7. (a) Seebach, D.; Hässig, R.; Gabriel, J. *HCA* **1983**, *66*, 308. (b) Heinzer, J.; Oth, J. F. M.; Seebach, D. *HCA* **1985**, *68*, 1848.
8. McGarrity, J. F. *JACS* **1985**, *107*, 1805, 1810.
9. Plavsik, D.; Srzic, D.; Klasinc, L. *JPC* **1986**, *90*, 2075.
10. Bates, R. B.; Kroposki, L. M.; Potter, P. E. *JOC* **1972**, *37*, 560.
11. Bauer, W.; Clark, T.; Schleyer, P. v. R. *JACS* **1987**, *109*, 970.
12. (a) Langer, A. W. *Adv. Chem. Ser.* **1974**, *130*, 1. (b) Smith, W. N. *Adv. Chem. Ser.* **1974**, *130*, 23. (c) West, R. *Adv. Chem. Ser.* **1974**, *130*, 211.
13. (a) Bryce-Smith, D.; Turner, E. E. *JCS* **1953**, 861. (b) Amonoo-Neizer, E. H.; Shaw, R. A.; Skovlin, D. O.; Smith, B. C. *Inorg. Synth.* **1966**, *8*, 19.
14. Jones, R. G.; Gilman, H. *OR* **1951**, *6*, 339.
15. Watson, S. C.; Eastham, J. F. *JOM* **1967**, *9*, 165.
16. See e.g. (a) Gilman, H.; Haubein, A. H. *JACS* **1944**, *66*, 1515. (b) Gilman, H.; Cartledge, F. K. *JOM* **1964**, *2*, 447. (c) Eppley, R. L.; Dixon, J. A. *JOM* **1967**, *8*, 176. (d) Collins, P. F.; Kamienski, C. W.; Esmay, D. L.; Ellestad, R. B. *AC* **1961**, *33*, 468. (e) Lipton, M. F.; Sorensen, C. M.; Sadler, A. C.; Shapiro, R. H. *JOM* **1980**, *186*, 155. (f) Bergbreiter, D. E.; Pendergrass, E. *JOC* **1981**, *46*, 219.
17. Arnett, E. M.; Moe, K. D. *JACS* **1991**, *113*, 7068.
18. Meth-Cohn, O.; Gronowitz, S. *CC* **1966**, 81.
19. Eisch, J. J. *Organomet. Synth.* **1981**, *2*, 98.
20. Herbrandson, H. F.; Mooney, D. S. *JACS* **1957**, *79*, 5809.
21. (a) Takashi, K.; Konishi, K.; Ushio, M.; Takaki, M.; Asami, R. *JOM* **1973**, *50*, 1. (b) Kaiser, E. M.; McLure, J. R. *JOM* **1979**, *175*, 11.
22. Michelot, D.; Clinet, J.-C.; Linstrumelle, G. *SC* **1982**, *12*, 739.
23. (a) Bailey, W. F.; Ovaska, T. V. *JACS* **1993**, *115*, 3080. (b) Quillinan, A. J.; Scheinmann, F. *OS* **1978**, *58*, 1.
24. Rausch, M. D.; Ciappenelli, D. J. *JOM* **1967**, *10*, 127.
25. Klein, J.; Medlik-Balan, A. *JACS* **1977**, *99*, 1473.
26. Akiyama, S.; Hooz, J. *TL* **1973**, 4115.
27. Cardillo, G.; Contento, M.; Sandri, S. *TL* **1974**, 2215.
28. Schlosser, M.; Strunk, S. *TL* **1984**, *25*, 741.
29. Faigl, F.; Schlosser, M. *TL* **1991**, *32*, 3369.
30. (a) Heus-Kloos, Y. A.; de Jong, R. L. P.; Verkruisse, H. D.; Brandsma, L.; Julia, S. *S* **1985**, 958. (b) Mordini, A.; Palio, G.; Ricci, A.; Taddei, M. *TL* **1988**, *29*, 4991.
31. (a) Ahlbrecht, H.; Dollinger, H. *TL* **1984**, *25*, 1353. (b) Hommes, H.; Verkruijsse, H. D.; Brandsma, L. *TL* **1981**, *22*, 2495.
32. (a) Ahlbrecht, H. *C* **1977**, *31*, 391. (b) Seebach, D.; Geiss, K.-H. In *New Applications of Organometallic Reagents in Organic Synthesis*; Seyferth, D., Ed.; Elsevier: Amsterdam, 1976; p 1.
33. Beak, P.; Zajdel, W. J.; Reitz, D. B. *CR* **1984**, *84*, 471.
34. Magnus, P. *T* **1977**, *33*, 2019.

35. Corey, E. J.; Seebach, D. *JOC* **1966**, *31*, 4097.
36. Durst, T. In *Comprehensive Organic Chemistry*; Barton, D. H. R.; Ollis, W. D., Eds.; Pergamon: Oxford, 1979; Vol. 3, p 171.
37. Schlosser, M.; Schaub, B.; Spahic, B.; Sleiter, G. *HCA* **1973**, *56*, 2166.
38. Schoufs, M.; Meyer, J.; Vermeer, P.; Brandsma, L *RTC* **1977**, *96*, 259.
39. Schlosser, M.; Ladenberger, V. *CB* **1967**, *100*, 3893.
40. Drakesmith, F. G.; Richardson, R. D. Stewart, O. J.; Tarrant, P. *JOC* **1968**, *33*, 286.
41. Schöllkopf, U.; Stafforst, D.; Jentsch, R. *LA* **1977**, 1167.
42. Lehr, F.; Gonnermann, J.; Seebach, D. *HCA* **1979**, *62*, 2258.
43. (a) Patterson, D. J. *JOM* **1967**, *8*, 199. (b) Appel, R.; Wander, M.; Knoll, F. *CB* **1979**, *112*, 1093. (c) Issleib, K.; Abicht, H. P. *JPR* **1970**, *312*, 456.
44. Maercker, A. *AG(E)* **1987**, *26*, 972.
45. Maercker, A.; Demuth, W. *LA* **1977**, 1909.
46. Bates, R. B.; Kroposki, L. M.; Potter, D. E. *JOC* **1972**, *37*, 560.
47. Gröbel, B. T.; Seebach, D. *S* **1977**, 357.
48. Seebach, D.; Beck, A. K. *OSC* **1988**, *6*, 316.
49. Cere, V.; Paolucci, C.; Pollicino, S.; Sandri, E.; Fava, A. *JOC* **1991**, *56*, 4513.
50. Kodama, M.; Matsuki, Y.; Ito, S. *TL* **1975**, 3065.
51. Ramanathan, V.; Levine, R. *JOC* **1962**, *27*, 1216.
52. Labaudiniere, R.; Hilboll, G.; Leon-Lomeli, A.; Lautenschläger, H.-H.; Parnham, M.; Kuhl, P.; Dereu, N. *JMC* **1992**, *35*, 3156.
53. Schroeder, R.; Schöllkopf, U.; Blume, E.; Hoppe, I. *LA* **1975**, 533.
54. Chadwick, D. J.; Cliffe, I. A.; *JCS(P1)* **1979**, 2845.
55. Butler, D. E.; Alexander, S. M. *JOC* **1972**, *37*, 215.
56. Noyce, D. S.; Stowe, G. T. *JOC* **1973**, *38*, 3762.
57. Behringer, H. Ramert, R. *LA* **1975**, 1264.
58. Raap, R. *CJC* **1971**, *49*, 2139.
59. Biellmann, J. F.; Ducep, J.-B. *OR* **1982**, *27*, 1.
60. Yamamoto, Y. *COS* **1991**, *1*, 55.
61. Evans, D. A.; Andrews, G. C.; Buckwalter, B. *JACS* **1974**, *96*, 5560.
62. Stotter, P. L.; Hornish, R. E. *JACS* **1973**, *95*, 4444.
63. Hoppe, D.; Hanko, R.; Brönneke, A.; Lichtenberg, F.; Hülsen, E. *CB* **1985**, *118*, 2822.
64. Hoppe, D. *AG(E)* **1984**, *23*, 932.
65. Hoppe, D.; Hanko, R.; Brönneke, A.; Lichtenberg, F. *AG(E)* **1981**, *20*, 1024.
66. Beak, P.; Zajdel, W. J. *JACS* **1984**, *106*, 1010.
67. (a) Tischler, A. N.; Tischler, M. H. *TL* **1978**, 3. (b) Tischler, A. N.; Tischler, M. H. *TL* **1978**, 3407.
68. Savignac, P.; Leroux, Y.; Normant, H. *T* **1975**, *31*, 877.
69. Meyers, A. I.; Edwards, P. D.; Rieker, W. F.; Bailey, T. R. *JACS* **1984**, *106*, 3270.
70. (a) Gonzalez, M. A.; Meyers, A. I. *TL* **1989**, *30*, 47. (b) Gonzalez, M. A.; Meyers, A. I. *TL* **1989**, *30*, 43.
71. Gais, H. J.; Hellmann, G. *JACS* **1992**, *114*, 4439.
72. Hoppe, D.; Krämer, T. *AG(E)* **1986**, *25*, 160.
73. Rein, K.; Goicoechea-Pappas, M.; Anklekar, T. V.; Hart, G. C.; Smith, G. A.; Gawley, R. E. *JACS* **1989**, *111*, 2211.
74. Meyers, A. I.; Elworthy, T. R. *JOC* **1992**, *57*, 4732.
75. Meyers, A. I.; Dickman, D. I.; Bailey, T. R. *JACS* **1985**, *107*, 7974.
76. Gawley, R. E.; Rein, K.; Chemburkar, S. R. *JOC* **1989**, *54*, 3002.
77. Meyers, A. I. *T* **1992**, *48*, 2589.
78. Meyers, A. I. *Aldrichim. Acta* **1985**, *18*, 59.
79. Snieckus, V. *CR* **1990**, *90*, 879.
80. Slocum, D. W.; Jennings, C. A. *JOC* **1976**, *41*, 3653.
81. Beak, P.; Snieckus, V. *ACR* **1982**, *15*, 306.
82. (a) Ludt, R. E.; Griffiths, T. S.; McGrath, K. N.; Hauser, C. R. *JOC* **1973**, *38*, 1668. (b) Beak, P.; Brown, R. A. *JOC* **1982**, *47*, 34.
83. Newman, M. S.; Kanakarajan, J. *JOC* **1980**, *45*, 2301.
84. Winkle, M. R.; Ronald, R. C. *JOC* **1982**, *47*, 2101.
85. Meyers, A. I.; Lutomski, K. *JOC* **1979**, *44*, 4464.
86. Ziegler, F. E.; Fowler, K. W. *JOC* **1976**, *41*, 1564.
87. Reuman, M.; Meyers, A. I. *T* **1985**, *41*, 837.
88. Furlano, D. C.; Calderon, S. N.; Chen, G.; Kirk, K. L. *JOC* **1988**, *53*, 3145.
89. Baldwin, J. E.; Bair, K. W. *TL* **1978**, 2559.
90. Fuhrer, W.; Gschwend, H. W. *JOC* **1979**, *44*, 1133.
91. Katsuura, K.; Snieckus, V. *CJC* **1987**, *65*, 3165.
92. Turner, J. A. *JOC* **1983**, *48*, 3401.
93. Moorhoff, C. M.; Paquette, L. A. *JOC* **1991**, *56*, 703.
94. Gay, R. L.; Hauser, C. R. *JACS* **1967**, *89*, 1647.
95. Adam, W.; Cueto, O. *JOC* **1977**, *42*, 38.
96. Hosomi, A.; Araki, Y.; Sakurai, H. *JACS* **1982**, *104*, 2081.
97. Meyers, A. I.; Malone, G. R.; Adickes, H. W. *TL* **1970**, 3715.
98. (a) Sauvetre, R.; Roux-Schmitt, M.-C.; Seyden-Penne, J. *TL* **1978**, *34*, 2135. (b) Sauvetre, R.; Seyden-Penne, J. *TL* **1976**, 3949.
99. Takano, S.; Shimazaki, Y.; Takahashi, M.; Ogasawara, K. *CC* **1988**, 1004.
100. Tamaru, Y.; Kagotini, M.; Furukawa, Y.; Amino, Y.; Yoshida, Z. *TL* **1981**, *22*, 3413.
101. Hucklin, S. N.; Weiler, L. *JACS* **1974**, *96*, 4691.
102. Fraser, R. R.; Banville, J.; Dhawan, K. C. *JACS* **1978**, *100*, 7999.
103. Kanemasa, S.; Tatsukawa, A.; Wada, E. *JOC* **1991**, *56*, 2875.
104. Laborde, E.; Kiely, J. S.; Culbertson, T. P.; Leheski, L. E. *JMC* **1993**, *36*, 1964.
105. Cahiez, G.; Bernard, D.; Normant, J. F. *S* **1976**, 245.
106. Kitatani, K.; Hiyama, T.; Nozaki, H. *JACS* **1975**, *97*, 949.
107. Hoeg, D. F.; Lusk, D. I.; Crumbliss, A. L. *JACS* **1965**, *87*, 4147.
108. Foulger, N. J.; Wakefield, B. J. *Organomet. Synth.* **1986**, *3*, 369.
109. Kitatani, K.; Hiyama, T.; Nozaki, H. *BCJ* **1977**, *50*, 3288.
110. Georgoulis, C.; Meyet, J.; Smajda, W. *JOM* **1976**, *121*, 271.
111. Panek, E. J.; Neff, B. L.; Chu, H.; Panek, M. G. *JACS* **1975**, *97*, 3996.
112. Cooke, Jr., M. P. *JOC* **1993**, *58*, 2910.
113. Bailey, W. F.; Nurmi, T. T.; Patricia, J. J.; Wang, W. *JACS* **1987**, *109*, 2442.
114. (a) Bailey, W. F.; Punzalan, E. P. *JOC* **1990**, *55*, 5404. (b) Negishi, E.; Swanson, D. R.; Rousset, C. J. *JOC* **1990**, *55*, 5406.
115. (a) Peterson, D. J.; Ward, J. F. *JOM* **1974**, *66*, 209. (b) Seyferth, D.; Mammarella, R. E. *JOM* **1979**, *77*, 53.
116. (a) Still, W. C.; Sreekumar, C. *JACS* **1980**, *102*, 1201. (b) Sawyer, J. S.; Kucerovy, A.; Macdonald, T. L.; McGarvey, G. J. *JACS* **1988**, *110*, 842. (c) Sawyer, J. S.; Macdonald, T. L.; McGarvey, G. J. *JACS* **1984**, *106*, 3376.
117. Lohse, P.; Loner, H.; Acklin, P.; Sternfeld, F.; Pfaltz, A. *TL* **1991**, *32*, 615.
118. (a) Chong, J. M.; Park, S. B. *JOC* **1992**, *57*, 2220. (b) Pearson, W. H.; Lindbeck, A. C. *JOC* **1989**, *54*, 5651. (c) Pearson, W. H.; Lindbeck, A. C. *JACS* **1991**, *113*, 8546. (d) *N*-stannylmethanimines: Pearson, W. H.; Szura, D. P.; Postich, M. J. *JACS* **1992**, *114*, 1329.
119. Chan, P. C.-M.; Chong, J. M. *TL* **1990**, *31*, 1985.
120. Broka, C. A.; Shen, T. *JACS* **1989**, *111*, 2981.
121. (a) McGarvey, G. J.; Kimura, M. *JOC* **1985**, *50*, 4652. (b) Krief, A.; Hobe, M. *TL* **1992**, *33*, 6527.
122. Krief, A.; Evrard, G.; Badaoui, E.; DeBeys, V.; Dieden, R. *TL* **1989**, *30*, 5635.
123. Krief, A.; Dumont, W.; Clarembeau, M.; Bernard, G.; Badaoui, E. *T* **1989**, *45*, 2005.

124. Reich, H. J.; Bowe, M. D. *JACS* **1990**, *112*, 8994.
125. (a) Reich, H. J. *ACR* **1979**, *12*, 22. (b) Liotta, D. *ACR* **1984**, *17*, 28. (c) Clive, D. L. *T* **1978**, *34*, 1049. (d) Davis, F. A.; Reddy, R. T. *JOC* **1992**, 2599.
126. Seebach, D.; Beck, A. K. *AG(E)* **1974**, *13*, 806.
127. Tomoki, H.; Kambe, N.; Ogawa, A.; Miyoshi, N.; Murai, S.; Sonoda, N. *AG(E)* **1987**, *26*, 1187.
128. Curtin, D. Y.; Koehl, Jr., W. J. *JACS* **1962**, *84*, 1967.
129. Marshall, J. A. *COS* **1991**, *3*, 975.
130. Crombie, L.; Darnbrough, G.; Pattenden, G. *CC* **1976**, 684.
131. Midland, M. M.; Kwon, Y. C. *TL* **1985**, *26*, 5021.
132. Tsai, D. J.-S.; Midland, M. M. *JOC* **1984**, *49*, 1842.
133. Nakai, T.; Mikami, K.; Taya, S.; Fujita, Y. *JACS* **1981**, *103*, 6492.
134. (a) Marshall, J. A.; Robinson, E. D.; Lebreton, J. *JOC* **1990**, *55*, 227. (b) Takashi, T.; Nemoto, H.; Kanda, Y.; Tsuji, J.; Fujise, Y. *JOC* **1986**, *51*, 4315.
135. Bulman Page, P. C.; Rayner, C. M.; Sutherland, I. O. *CC* **1988**, 356.
136. Kuwajima, I.; Kato, M. *TL* **1980**, *21*, 623.
137. Klein, K. P.; Van Eenam, D. N.; Hauser, C. R. *JOC* **1967**, *32*, 1155.
138. Harvey, R. G.; Cho, H. *JACS* **1974**, *96*, 2434.
139. Hunt, E.; Lythgoe, B. *CC* **1972**, 757.
140. Gilman, H.; Gorsich, R. D. *JACS* **1957**, *79*, 2625.
141. Fischer, P.; Schaefer, G. *AG(E)* **1981**, *20*, 863.
142. Cainelli, G.; Ronchi, A. U.; Bertini, F.; Grasselli, P.; Zubiani, G. *T* **1971**, *27*, 6109.
143. Bandouy, R.; Gore, J.; Ruest, L. *T* **1987**, *43*, 1099.
144. Corey, E. J.; Fuchs, P. L. *TL* **1972**, 3769.
145. Romo, D.; Johnson, D. D.; Plamondon, L.; Miwa, T.; Schreiber, S. L. *JOC* **1992**, *57*, 5060.
146. Schone, N. E.; Knudsen, M. J. *JOC* **1987**, *52*, 569.
147. Chamberlin, A. R.; Bond, F. T. *S* **1979**, 44.

Timo V. Ovaska
Connecticut College, New London, CT, USA

s-Butyllithium[1]

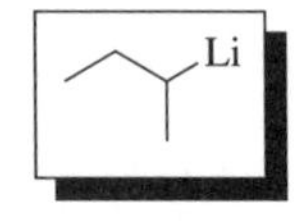

[598-30-1] C_4H_9Li (MW 64.06)

(strong base capable of lithiating weak carbon acids;[1] useful for heteroatom-facilitated lithiations;[2–4] reagent of choice for *ortho* lithiations[5])

Physical Data: colorless to pale yellow liquid; slowly eliminates LiH; d^{25} 0.783; bp 90 °C/0.05 mmHg;[6,7] ^{13}C NMR, ^{1}H NMR, ^{6}Li NMR, and MS studies have been reported.[8–11]

Solubility: sol hydrocarbon and ethereal solvents, but should be used at low temperature in the latter solvent type: half-lives in ethereal solvents have been reported;[12] reacts violently with H_2O and other protic solvents.

Form Supplied in: commercially available as approximately 1.3 M solution in cyclohexane. Mostly tetrameric in hydrocarbons;[9,13] mostly monomeric in THF;[14] when used in combination with tertiary polyamines such as ***N,N,N′,N′-Tetramethylethylenediamine*** (TMEDA) and ***1,4-Diazabicyclo[2.2.2]octane*** (DABCO), reactivity is usually increased.[1,15]

Analysis of Reagent Purity: since the concentration of commercial solutions may vary appreciably it is necessary to standardize solutions of the reagent prior to use. A recommended method for routine analyses involves titration of the reagent with *s*-butyl alcohol using 1,10-phenanthroline or 2,2′-biquinoline as indicator.[16] Several other methods have been described.[17]

Preparative Method: may be prepared in high yield from *s*-butyl chloride and ***Lithium*** metal in hydrocarbon solvents.[13]

Handling, Storage, and Precautions: solutions of the reagent are pyrophoric and the reagent may catch fire if exposed to air or moisture. Handling of the reagent should be done behind a shield in a chemical fume hood. Safety goggles, chemical resistant gloves, and other protective clothing should be worn. In case of fire, a dry-powder extinguisher should be used: in no case should an extinguisher containing water or halogenated hydrocarbons be used to fight an alkyllithium fire. Bottles and reaction flasks containing the reagent should be flushed with N_2 or preferably Ar and kept tightly sealed to preclude contact with oxygen or moisture. Standard syringe/cannula techniques for air- and moisture-sensitive chemicals should be applied when transferring the reagent.[18] For detailed handling techniques see Wakefield.[1b]

Lithiations. *s*-Butyllithium is a powerful metalating agent which is used frequently to convert a variety of organic compounds into their lithio derivatives. The resulting organolithium products may be subsequently functionalized with a large number of electrophiles, hence allowing the preparation of, for example, alcohols, carboxylic acids, ketones, esters, organosilicon compounds, and other organometallic compounds.[1] The metalations of relatively strong carbon acids, such as terminal alkynes, triarylmethanes, and methyl heteroaromatics, may be regarded as simple acid–base reactions and these are done most conveniently using the less basic ***n-Butyllithium***,[19] which is also less reactive and easier to handle than *s*-BuLi. However, the use of *s*-BuLi is warranted for deprotonation of weaker carbon acids (e.g. preparation of aryl-,[15] vinyl-,[20,21] and certain allyllithiums,[15,21,22]) and those that require a less nucleophilic lithiating agent.[5,23]

Lithiations employing *s*-BuLi are most often conducted in electron-donating solvents such as Et_2O, THF, and DME which coordinate to lithium much more strongly than do the alkyl groups of hydrocarbon solvents and thus enhance the reactivity of the organolithium species. It is generally agreed that this effect is the consequence of the depolymerization of higher order organolithium aggregates (tetramers) into smaller units (dimers and monomers).[1] The increased reactivity of organolithiums in ether solvents can, in fact, lead to α-deprotonation of the solvent molecules at elevated temperatures; in diethyl ether, for example, *s*-BuLi has a maximum lifetime of about an hour at +20 °C and THF is attacked even more readily.[12] Hence, to ensure the integrity of the organolithium reagents in these solvents, the use of low-temperature conditions (typically −78 °C) is imperative.

s-Butyllithium is often used in the presence of various lithium complexing ligands such as ***Hexamethylphosphoric Triamide***, TMEDA, and DABCO, which serve to further enhance the reactivity of this reagent.[15] The *s*-BuLi/TMEDA complex is an extremely powerful lithiating agent, effecting rapid deprotonation of many compounds, e.g. benzene, tetramethylsilane, and propene (eq 1),[15] which are unreactive toward *s*-BuLi alone. Compared to the analogous *n*-BuLi/TMEDA complex, *s*-BuLi/TMEDA is considerably more potent as a metalating agent; e.g. the lithiation of Me_4Si proceeds ~1000 times faster with *s*-BuLi/TMEDA than with *n*-BuLi/TMEDA.[15a]

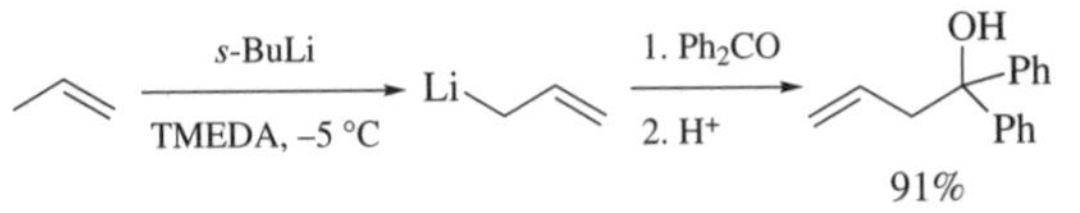

(1)

Metal alkoxides, such as ***Potassium t-Butoxide***, are also used in combination with *s*-BuLi to facilitate the metalation of weak carbon acids including aromatic compounds,[24] vinylic systems,[25] and others.[26,27] The products of these metalations are organopotassium derivatives, but they can be readily converted into the corresponding organolithium compounds by the addition of ***Lithium Bromide***.[26,27]

s-Butyllithium is often used to effect regiospecific and rapid deprotonation of heteroatom-containing compounds.[2–4,28,29] Heteroatoms such as oxygen, sulfur, nitrogen, phosphorus, and halogens enhance the acidity of α- or β-protons through either inductive or coordination effects and hence facilitate deprotonation. For example, the α-lithiation of vinyl thioethers,[30] vinyl chlorides,[31] and vinyl thioesters[32] is effected at low temperatures using *s*-BuLi in the presence of HMPA or TMEDA. The vinyllithium compounds derived from thioethers are important acyl anion equivalents which, after alkylation, are readily converted into substituted ketones (eq 2).[30]

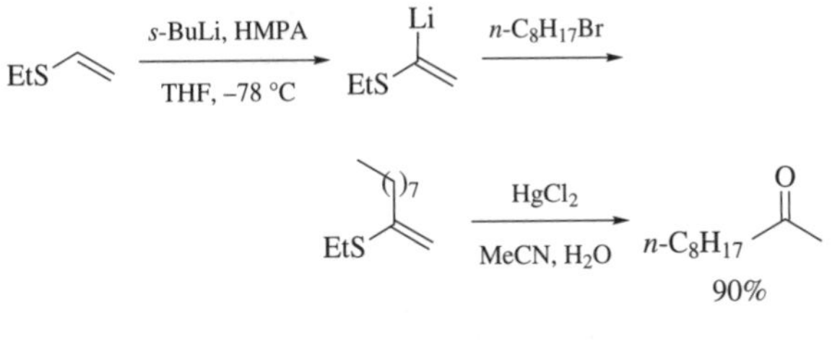

(2)

Lithiation of α-disubstituted alkenes is directed to the β-site in cases where the α-substituents are capable of activating the β-position through either inductive effects or chelation (eqs 3 and 4).[20,33] Internal coordination to the lithium by the electron-donating β-substituent in eq 4 permits deprotonation and subsequent electrophilic substitution to take place stereoselectively.[20]

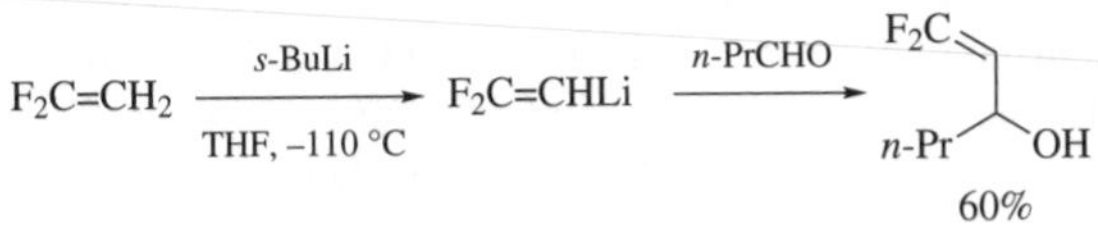

(3)

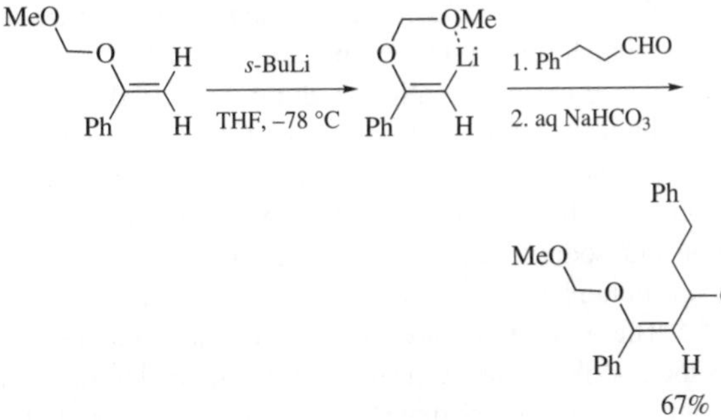

(4)

The preparation of a variety of α-heteroatom-substituted alkyllithiums is conveniently achieved using *s*-BuLi and in many cases this reagent exhibits highly original reactivity. ***(Methoxymethyl)trimethylsilane***, for example, undergoes facile lithiation on treatment with *s*-BuLi in THF at −78 °C to give $Me_3SiCHLiOMe$, which is readily hydroxyalkylated by the addition of aldehydes or ketones (eq 5).[34] With other butyllithium reagents, this reaction takes a completely different course: *n*-BuLi undergoes a nucleophilic attack on the Si atom with concomitant loss of the CH_2OMe group, and the use of ***t-Butyllithium*** gives $LiCH_2SiMe_2CH_2OMe$ through preferential deprotonation of one of the trimethylsilyl protons.[34] In a similar reaction, ***Chlorotrimethylsilane*** is lithiated α to the chlorine atom using *s*-BuLi to furnish synthetically useful α,β-epoxytrimethylsilanes after treatment with aldehydes or ketones (eq 6).[23] Again, *s*-BuLi appears to be a superior reagent for this lithiation: the use of *n*-BuLi results in loss of the chloride anion, presumably through initial nucleophilic attack of this reagent upon the silicon atom and subsequent alkyl migration, and *t*-BuLi leads to the predominant formation of ***Trimethylsilylmethyllithium***, formally derived from Li–Cl exchange.[23] Bis(trimethylsilylmethyl) selenide,[35] ***Methylenetriphenylphosphorane***,[36] phosphonic acid triamides,[37] and 1,3-oxathianes[38] are also readily lithiated on treatment with *s*-BuLi. The final ketene *O,S*-acetal product in eq 7 is formed via the Peterson alkenation process after addition of 2-methylpropanal to the lithio intermediate.[38a]

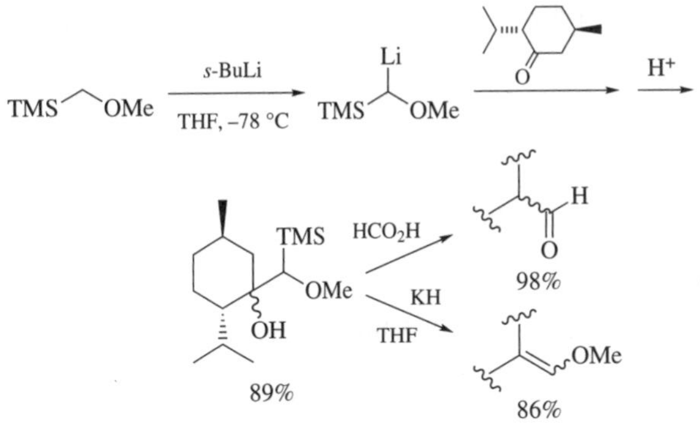

(5)

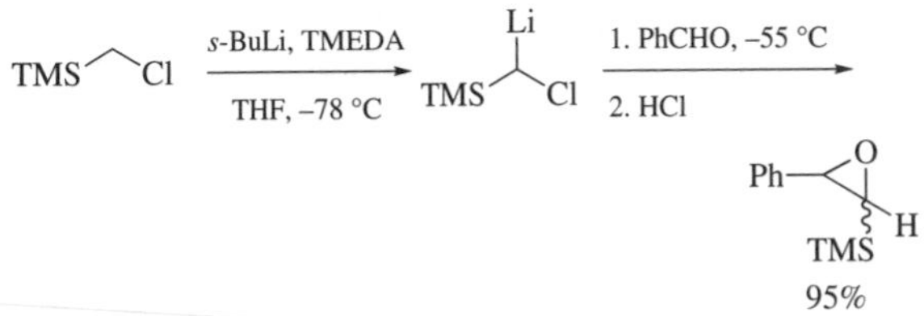

(6)

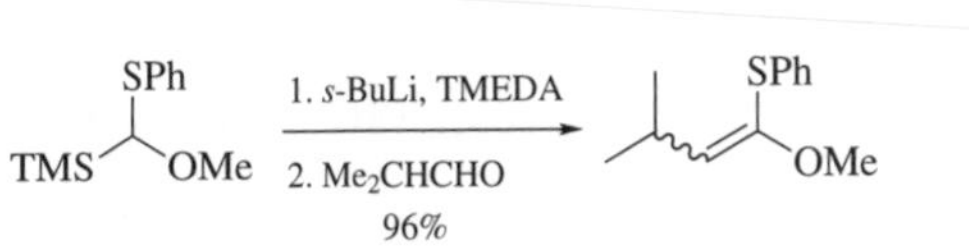

(7)

Metalation of *N*-methylpiperidine, *N*-methylpyrrolidine, and ***Trimethylamine*** takes place exclusively at the methyl group when *s*-BuLi is used in combination with *t*-BuOK (eq 8),[26] and the same mixture can be used to metalate *t*-butyl methyl ether (eq 9).[27] The intermediate organopotassium product in eq 9 is converted into the corresponding organolithium derivative by addition of LiBr to give $t\text{-}BuOCH_2Li$, the synthetic equivalent of hydroxymethyllithium ($LiCH_2OH$).[27]

Many allylic or benzylic heteroatom-containing compounds are often lithiated more conveniently using the less reactive *n*-BuLi rather than *s*-BuLi. However, the more basic *s*-BuLi is a better reagent for the deprotonation of alkyl allyl ethers[39] and certain allyl thioethers[21b] that react slowly (if at all) with *n*-BuLi in THF at low temperatures (<−65 °C). The allyllithium derivatives formed in these reactions may be alkylated or hydroxyalkylated either at the α- or the γ-position, depending on the electrophile and the structure of the allylic anion. With metalated allyl ethers, primary alkyl halides tend to attack the γ-position to afford alkylated enol ethers, but carbonyl compounds react to give products derived mainly from α-attack (eq 10).[39,40] The opposite tendency is generally observed with analogous lithiated allyl thioethers[21,41,42] which, upon addition of alkyl halides, give primarily α-alkylated products (eqs 11 and 12).[21b,42] Generally, soft electrophiles (e.g. RI) exhibit a higher propensity for γ-attack than do hard electrophiles (e.g. RCl).[42]

RX	α:γ	
n-BuCl	97:3	70%
n-BuBr	91:9	79%
n-BuI	70:30	100%

Proton removal adjacent to a heteroatom is further facilitated if the lithium can be internally coordinated to proximate electron donors, such as the carbonyl oxygen, permitting the formation of 'dipole-stabilized' carbanions.[28] Thus lithiations of various amides,[43] thioamides,[44] imides,[45] esters,[46] Boc derivatives of cyclic amines (pyrrolidines, piperidines, and hexahydroazepines),[47] thioesters,[32] *N,N*-dialkylthiocarbamates,[32] and various formamidine derivatives[48] are achieved conveniently using *s*-BuLi (eqs 13–17).[32,43b,44,47a,48a] Subsequent addition of electrophiles followed by hydrolytic removal of the activating carbonyl, carbamoyl, or formamidine moiety provides a valuable synthetic route to a variety of α-substituted amines, alcohols, and thiols.[28,46] Successful alkylation of the dipole-stabilized carbanions may require the conversion of the initial lithio carbanions into their organocuprate derivatives, e.g. by the addition of *n*-PrC≡CCu.[48a]

Ar = 2,4,6-triisopropylphenyl

Lithiation takes place more readily if the dipole-stabilized anions are further activated by an adjacent aromatic ring or a double bond and, in fact, many such compounds are most conveniently deprotonated using the less reactive *n*-BuLi.[28]

Stereoselective *s*-BuLi-promoted α-lithiations have been accomplished with various piperidine and isoquinoline derivatives bearing chiral oxazoline or formamidine substituents on the nitrogen (eqs 18 and 19).[49,50] Asymmetric deprotonations α to oxygen in carbamates and α to nitrogen in Boc-protected pyrrolidines have been effected using *s*-BuLi in the presence of ***(−)-Sparteine***, which is a homochiral lithium-complexing ligand.[51,52] These deprotonations lead to the formation of chiral dipole-stabilized carbanions which react with electrophiles under strict stereocontrol to give substituted products in high enantiomeric excess, typically >95% ee. In the case of the *N*-oxazoline derivatives of piperidine and isoquinoline, enantiomerically enriched secondary amines are obtained upon hydrolytic removal of the oxazoline moiety.[49a] These reactions are of considerable synthetic value and this methodology has been successfully applied to the asymmetric synthesis of 2-piperidines[49] and 2-pyrrolidines (eq 20),[52]

2-alkanols,[51a] 2-hydroxyalkanoic acids,[51a] alkanediols,[51b,c] and lactones (e.g. (*R*)-pantolactone) (eq 21).[51b]

s-BuLi, TMEDA
Et_2O, THF
−23 °C
Me_3SiCl
TMS
(18)
96%, ~100% de

MeO
MeO
1. s-BuLi, THF
−78 °C
2. $PhCH_2Cl$
N_2H_2
Ph
NH
(19)
88%, 86% ee
A_c =
t-Bu

1. s-BuLi, Et_2O
(−)-sparteine
−78 °C
2. Ph_2CO
3. HOAc
70%
Boc
NaOH
EtOH
reflux, 4 h
90%
Ph
OH
(20)
99.3% ee

CbxO
OCbx
s-BuLi
(−)-sparteine
Et_2O, −78 °C
Cbx =
Li
(21)
1. CO_2
2. H^+
CO_2H
5M HCl
90 °C
OH
80%, >95% ee

***ortho* Lithiations.** *s*-Butyllithium is a commonly used reagent for *ortho* lithiation of aromatic rings bearing heteroatom-containing substitutents. This methodology has been exploited extensively as a route to a wide variety of polysubstituted aromatic compounds, including natural products[53] (eq 22),[53e] and several excellent reviews have been published on the topic.[2,5,54] Included in the list of groups commonly used for *s*-BuLi-promoted *ortho* lithiations are OMe,[55] CH_2NEt_2,[55] NMe_2,[56] $CONEt_2$,[57] $OCONEt_2$,[57] SO_2NR_2,[58] 2-oxazolinyl,[57] and groups that contain acidic hydrogens and themselves undergo deprotonation prior to ring lithiation, e.g. CONHR[57,58] and SO_2NHR.[57,58] Depending on the substituent, both inductive and coordination effects can be invoked to account for the observed regiochemistry.[2,59] Although a large number of *ortho* lithiations may be conducted with *n*-BuLi,[2] the use of *s*-BuLi is generally preferred, especially when the reactions involve aromatic tertiary amides or *O*-aryl carbamates which are susceptible toward nucleophilic additions with the latter. For example, *N,N*-dimethyl- and -diethylbenzamides have been shown to afford primarily aryl butyl ketones upon treatment with *n*-BuLi.[55,60] The recommended procedure for *ortho* lithiation involves slow addition of the aromatic substrate in anhydrous THF to a slight excess of 1:1 *s*-BuLi/TMEDA in THF at −78 °C.[55] Under these conditions, the lithiation is usually complete within 5 min.

OMe
OMe
$CONEt_2$
1. s-BuLi, TMEDA
THF, −78 °C
2. MeO, MeO, N, CHO
3. TsOH, PhMe
reflux
NMe
(22)
Cordrastine I, II

When two *ortho*-directing groups are in a *meta* relationship, lithiation is generally directed to their common site. Notable exceptions to this behavior are found in cases where $CONEt_2$, CONR, or oxazoline substituents are *meta* to a dialkylamino group.[5] These systems exhibit nearly total reversal of the general trend and, as a result, almost exclusive metalation of the 6-position is observed (eq 23).[56] Considerable amounts of 6-substituted products are also obtained in cases where $OCONEt_2$ and OMe groups are in a *meta* relationship.[57] When two *ortho*-directing substituents are *ortho* or *para* to each other, the regiochemistry of the reaction depends on the relative directing abilities of these groups (eq 24).[57]

NEt_2
NMe_2
1. s-BuLi, TMEDA
THF, −78 °C
2. Me_3SiCl
93%
TMS
(23)
only isomer

$OCONEt_2$
MeO
1. s-BuLi, TMEDA
THF, −78 °C
2. DMF
88%
CHO
(24)

A large number of substituted polyaromatic and heteroaromatic compounds are also accessible through *s*-BuLi-induced *ortho* lithiation reactions. These include, for example, derivatives of naphthalene,[57] furan,[61] thiophene,[62] pyridine,[63] quinoline,[64] and N-protected imidazoles.[65] The smooth *ortho* lithiation of pyridyl (eq 25)[63] and quinolinyl systems is notable given the well-known tendency of organolithium compounds to undergo nucleophilic addition reactions with the pyridine nucleus.[1]

1. s-BuLi, TMEDA THF, −78 °C; 2. MeI; 83% (25)

Lithium–Halogen Interchange and Transmetalation Reactions. The lithium–halogen interchange involves the exchange of halogen by lithium in a reversible reaction, with the most stable organolithium species being favored at equilibrium. Hence, the most synthetically useful lithium–halogen exchanges take place between alkyllithiums and aryl halides, cyclopropyl halides, and vinyl halides. These reactions are accomplished most conveniently using *n*-BuLi rather than *s*-BuLi and, as a result, the latter has been employed only in a relatively few cases. However, the formation of select α-halo- and α-dihalolithiums,[66] aryllithium compounds,[67] and vinyllithium derivatives[68] using *s*-BuLi has been reported. The preparation of simple alkyllithium compounds by lithium–halogen exchange is usually limited to the generation of primary alkyllithiums from primary alkyl iodides by treatment with the more reactive tertiary alkyllithiums (such as *t*-BuLi).[69]

The lithium–selenium exchange has been exploited extensively for the generation of a variety of organolithiums from selenides, selenoacetals, and mixed *S*,*Se*-acetals, with *n*-BuLi being the most commonly employed reagent for these reactions.[3,70] Although *s*-BuLi can be used for all of these conversions, the primary role of this reagent is to provide a more reactive lithiating agent for the generation of synthetically useful[3,71] α-selenoalkyllithiums from selenoacetals of sterically hindered ketones.[72] For example, 2,2-bis(methylseleno)adamantane is completely unreactive toward *n*-BuLi in THF at −78 °C, but it undergoes a facile (30 min) Li–Se exchange on treatment with *s*-BuLi under identical conditions to afford the expected α-selenoorganolithium product in high yield (eq 26).[72]

SeMe, SeMe; s-BuLi, THF, −78 °C; SeMe, Li; 1. PhCHO; 2. H_3O^+; SeMe, Ph, OH; 81% (26)

The following order of reactivity of organolithium reagents toward selenoacetals has been established: *t*-BuLi/THF–hexane ≈ *s*-BuLi/THF–hexane > *n*-BuLi/THF–hexane ≈ *s*-BuLi/ether–hexane ≫ MeLi/THF–ether > *t*-BuLi/hexane ≈ *n*-BuLi/hexane.[72] As seen from this order, the reactivity of *s*-BuLi in ether–hexane is comparable to that of *n*-BuLi in THF–hexane. The α-selenoalkyllithiums generated in ether–hexane are more stable than those prepared in THF. In this medium they also exhibit higher reactivity toward certain carbonyl compounds, especially hindered[73] or highly enolizable ketones.[74] The 1-methyl-1-phenylselenoethyllithium product in eq 27 (generated in ether–hexane) undergoes nucleophilic 1,2-addition with a cyclic α,β-unsaturated ketone to afford ring-expanded ketones after treatment with TlOEt.[75]

Me, SePh, Me, SePh; 1. s-BuLi, Et_2O, −78 °C; 2. TBDMS enone; 67%

A: PhSe, OH, TBDMS + B: HO, PhSe, TBDMS; 1:2; TlOEt, $CHCl_3$, Δ

TBDMS + TBDMS (27)

14.4:1 93% from **A**
3.3:1 72% from **B**

Organolithium compounds are also accessible through the replacement of tin and tellurium by lithium, with the various organotin compounds being particularly important precursors for otherwise inaccessible organolithium derivatives (e.g. α-amino- and α-alkoxy-substituted organolithiums). Although *s*-BuLi is well suited for these transformations,[52,76] most of the synthetic applications exploiting this methodology involve the use of *n*-BuLi.

Eliminations. Arylsulfonylhydrazones undergo formal elimination reactions on treatment with 2 equiv of *s*-BuLi to give vinyllithium compounds.[77,78] Tosylhydrazones should be avoided with this reagent because they are susceptible to ring lithiation and benzylic deprotonation; instead, the use of hindered 2,4,6-triisopropylbenzenesulfonylhydrazones is recommended (eq 28).[77] In cases where two regioisomers can be produced, *s*-BuLi appears to promote the formation of the more substituted vinyllithium compounds. In contrast, the use of *n*-BuLi leads to the less substituted vinyllithium product.[78] Other notable *s*-BuLi-induced elimination reactions are the α-eliminations involving bis(phenylthio)methyllithium-containing cycloalkanol derivatives, which have been exploited as a route to ring-expanded ketones (eq 29),[79] and β-eliminations of *ortho*-lithiated 2-carboxamide-substituted furan derivatives, which have been used to prepare ring-opened enyne products.[80]

N, $NHSO_2Ar$; 2 s-BuLi, TMEDA, hexane, −55 °C; Li; n-$C_5H_{11}Br$; n-C_5H_{11}; 53% (28)

Ar = 2,4,6-triisopropylphenyl

HO SPh SPh — s-BuLi, THF, −78 to 0 °C, 98% → SPh (29)

Formation of Enolate Anions and Enolate Equivalents. *s*-Butyllithium may be used for the generation of enolate anions or enolate equivalents from active hydrogen compounds provided that nucleophilic addition of this reagent to the electrophilic carbon center is avoided. Thus the less reactive amides and imines are often suitable substrates for these reactions. For example, the bicyclic lactam in eq 30 undergoes two consecutive enolization/alkylation reactions in a highly diastereoselective fashion to afford a dialkylated product,[81] and the *N*-*t*-butylimine of 2-heptanone is deprotonated using *s*-BuLi to afford a mixture of two isomeric ketones after alkylation and hydrolysis (eq 31).[82] The less substituted ketone was obtained as the only isomer when *s*-BuLi was used with HMPA. Similar reactions have been reported for α-nitro imines,[83] and thiolactams.[84]

1. s-BuLi, THF, −78 °C; 2. allyl bromide; 3. s-BuLi, THF, −78 °C; 4. $PhCH_2Br$, −100 °C; 77% (30)

97:3 ratio of isomers

n-Bu N-t-Bu — 1. s-BuLi, THF, −78 °C; 2. $PhCH_2Cl$; 3. 2M HCl → n-Bu ... Ph + n-Bu ... Ph (31)

no additives	53:47	41%
HMPA added	100:0	61%

Rearrangements. Salicylamides are available from various aryl carbamates, including those derived from ***Pyridine*** and naphthalene,[63] through the *s*-BuLi-mediated O–C 1,3-carbamoyl migration reactions (Snieckus rearrangement).[57] This regiospecific rearrangement is the anionic equivalent of the Fries rearrangement and it involves low-temperature (−78 °C) *ortho* lithiation of aryl carbamates with *s*-BuLi/TMEDA/THF followed by warming of the reaction mixture to rt (eq 32).[85] Benzylic carbamates rearrange to give products derived from either 1,4- or 1,2-carbamoyl migration (eq 33) following treatment with *s*-BuLi in THF,[86] and lithiated phenolic esters rearrange to furnish acyl phenols even at low temperatures (eq 34).[67] *s*-Butyllithium is also capable of initiating various Wittig rearrangements involving allylic ethers, but these reactions are done more conveniently using the less reactive *n*-BuLi.

$OCONEt_2$ — 1. s-BuLi, TMEDA, THF, −78 °C; 2. warm to rt; 58% → OH $CONEt_2$ (32)

R $OCONEt_2$ — s-BuLi, THF, −78 °C, 2 h → [R Li $OCONEt_2$] — 1,2 / 1,4 (33)

1,2: R OH $CONEt_2$ — R = Me, 0%; R = H, 80%

1,4: R OH $CONEt_2$ — R = Me, 75%; R = H, 0%

OCO-t-Bu Br — 1. s-BuLi, THF, −95 °C; 2. −78 °C, 0.5 h; 71% → OH CO-t-Bu (34)

Related Reagents. *n*-Butyllithium; *t*-Butyllithium; Potassium Hydride-*s*-Butyllithium-*N*,*N*,*N*′,*N*′-Tetramethylethylenediamine.

1. (a) Wakefield, B. J. *The Chemistry of Organolithium Compounds*; Pergamon: Oxford, 1974. (b) Wakefield, B. J. *Organolithium Methods*; Academic: San Diego, 1990. (c) Wardell, J. L. In *Comprehensive Organometallic Chemistry*; Wilkinson, G., Ed.; Pergamon: Oxford, 1982; Chapter 2.
2. Gschwend, H. W.; Rodriguez, H. R. *OR* **1979**, *26*, 1.
3. Krief, A. *T* **1980**, *36*, 2531.
4. Biellmann, J. F.; Ducep, J.-B. *OR* **1982**, *27*, 1.
5. Snieckus, V. *CR* **1990**, *90*, 879.
6. Bach, R.; Wasson, J. R. In *Kirk-Othmer Encyclopedia of Chemical Technology*, 3rd ed.; Grayson, M., Ed.; Wiley: New York, 1981; Vol. 14, p 469.
7. *Dictionary of Organometallic Compounds*; Buckingham, J., Ed.; Chapman & Hall: London, 1984, Vol. 1, p 1213.
8. Bywater, S.; Lachance, P.; Worsfold, D. J. *JPC* **1975**, *79*, 2148.
9. Catala, J. M.; Clouet, G.; Brossas, J. *JOM* **1981**, *219*, 139.
10. Fraenkel, G.; Henrichs, M.; Hewitt, M.; Su, B. M. *JACS* **1984**, *106*, 255.
11. Plavsik, D.; Srzic, D.; Klasinc, L. *JPC* **1986**, *90*, 2075.
12. Gilman, H.; Haubein, A. H.; Hartzfeld, H. *JOC* **1954**, *19*, 1034.
13. Hay, D. R.; Song, Z.; Smith, S. G.; Beak, P. *JACS* **1988**, *110*, 8145.
14. Bauer, W.; Winchester, W. R.; Schleyer, P. v. R. *OM* **1987**, *6*, 2371.
15. (a) Langer, A. W. *Adv. Chem. Ser.* **1974**, *130*, 1. (b) Smith, W. N. *Adv. Chem. Ser.* **1974**, *130*, 23.
16. Watson, S. C.; Eastham, J. F. *JOM* **1967**, *9*, 165.
17. See e.g. (a) Gilman, H.; Haubein, A. H. *JACS* **1944**, *66*, 1515. (b) Gilman, H.; Cartledge, F. K. *JOM* **1964**, *2*, 447. (c) Eppley, R. L.; Dixon, J. A. *JOM* **1967**, *8*, 176. (d) Collins, P. F.; Kamienski, C. W.; Esmay, D. L.; Ellestad, R. B. *AC* **1961**, *33*, 468. (e) Lipton, M. F.; Sorensen, C. M.; Sadler, A. C.; Shapiro, R. H. *JOM* **1980**, *186*, 155. (f) Bergbreiter, D. E.; Pendergrass, E. *JOC* **1981**, *46*, 219.
18. Shriver, D. F.; Drezdon, M. A. *The Manipulation of Air Sensitive Compounds*; Wiley: New York, 1986.
19. Arnett, E. M.; Moe, K. D. *JACS* **1991**, *113*, 7068.
20. McDougal, P. G.; Rico, J. G. *TL* **1984**, *25*, 5977.
21. (a) Evans, D. A.; Andrews, G. C. *ACR* **1974**, *7*, 147. (b) Stotter, P. L.; Hornish, R. E. *JACS* **1973**, *95*, 4444.

22. (a) Evans, D. A.; Andrews, G. C.; Buckwalter, B. *JACS* **1974**, *96*, 5560. (b) Still, W. C.; Macdonald, T. L. *JOC* **1976**, *41*, 3620. (c) Still, W. C.; Macdonald, T. L. *JACS* **1974**, *96*, 5562.
23. Burford, C.; Cooke, F.; Roy, G.; Magnus, P. *T* **1983**, *39*, 867.
24. Schlosser, M.; Strunk, S. *TL* **1984**, *25*, 741.
25. Hartmann, J.; Stähle, M.; Schlosser, M. *S* **1974**, 888.
26. Ahlbrecht, H.; Dollinger, H. *TL* **1984**, *25*, 1353.
27. Corey, E. J.; Eckrich, T. M. *TL* **1983**, *24*, 3165.
28. Beak, P.; Zajdel, W. J.; Reitz, D. B. *CR* **1984**, *84*, 471.
29. Wardell, J. L. In *Inorganic Reactions and Methods*; Zuckerman, J. J., Ed.; VCH: New York, 1988; Vol. 11, p. 107–129.
30. Oshima, K.; Shimoji, K.; Takashi, H.; Yamamoto, H.; Nozaki, H. *JACS* **1973**, *95*, 2694.
31. Nelson, D. *JOC* **1984**, *49*, 2059.
32. Beak, P.; Becker, P. D. *JOC* **1982**, *47*, 3855.
33. Savetre, R.; Normant, J. F. *TL* **1981**, *22*, 957.
34. Magnus, P.; Roy, G. *OM* **1982**, *1*, 553.
35. Reich, H. J.; Shah, S. K. *JACS* **1975**, *97*, 3250.
36. Corey, E. J.; Kang, J.; Kyler, K. *TL* **1985**, *26*, 555.
37. Magnus, P.; Roy, G. *S* **1980**, 575.
38. (a) Livinghouse, T.; Hackett, S. *JOC* **1986**, *51*, 879. (b) Livinghouse, T.; Hackett, S. *TL* **1984**, *25*, 3539. (c) Fuji, K.; Ueda, M.; Sumi, K.; Fujita, E. *TL* **1981**, *22*, 2005. (d) Fuji, K.; Ueda, M.; Sumi, K.; Fujita, E. *JOC* **1985**, *50*, 662.
39. Evans, D. A.; Andrews, G. C.; Buckwalter, B. *JACS* **1974**, *96*, 5560.
40. (a) Still, W. C.; Macdonald, T. L. *JOC* **1976**, *41*, 3620. (b) Still, W. C.; Macdonald, T. L. *JACS* **1974**, *96*, 5562.
41. (a) Oshima, K.; Takasha, H.; Yamamoto, H.; Noza, H. *JACS* **1973**, *95*, 2693. (b) Oshima, K.; Yamamoto, H.; Nozaki, H. *JACS* **1973**, *95*, 4446.
42. Torii, S.; Tanaka, H.; Tomotaki, Y. *CL* **1974**, 1541.
43. (a) Reitz, D. B.; Beak, P.; Tse, A. *JOC* **1981**, *46*, 4316. (b) Beak, P.; Zajdel, W. J. *JACS* **1984**, *106*, 1010.
44. (a) Lubosch, W.; Seebach, D. *HCA* **1980**, *63*, 102. (b) Seebach, D.; Lubosch, W. *AG(E)* **1976**, *15*, 313.
45. Schlecker, R.; Seebach, D. *HCA* **1977**, *60*, 1459.
46. (a) Beak, P.; Baillargeon, M.; Carter, L. G. *JOC* **1978**, *43*, 4255. (b) Beak, P.; McKinnie, B. G. *JACS* **1977**, *99*, 5213.
47. (a) Beak, P.; Lee, W. K. *JOC* **1993**, *58*, 1109. (b) Beak, P.; Lee, W. K. *JOC* **1990**, *55*, 2578. (c) Beak, P.; Lee, W. K. *TL* **1989**, *30*, 1197.
48. (a) Meyers, A. I.; Edwards, P. D.; Rieker, W. F.; Bailey, T. R. *JACS* **1984**, *106*, 3270. (b) Meyers, A. I. *Aldrichim. Acta* **1985**, *18*, 59.
49. (a) Gawley, R. E.; Hart, G. C.; Bartolotti, L. J. *JOC* **1989**, *54*, 175. (b) Rein, K.; Goicoechea-Pappas, M.; Anklekar, T. V.; Hart, G. C.; Smith, G. A.; Gawley, R. E. *JACS* **1989**, *111*, 2211.
50. Meyers, A. I.; Gonzalez, M. A.; Struzka, V.; Akahane, A.; Guiles, J.; Warmus, J. S. *TL* **1991**, *32*, 5501.
51. (a) Hoppe, D.; Hintze, F.; Tebben, P. *AG(E)* **1990**, *29*, 1422. (b) Paetow, M.; Ahrens, H.; Hoppe, D. *TL* **1992**, *33*, 5323. (c) Ahrens, H.; Paetow, M.; Hoppe, D. *TL* **1992**, *33*, 5327. (d) Hoppe, D.; Hintze, F. *S* **1992**, 1216.
52. Kerrick, S. T.; Beak, P. *JACS* **1991**, *113*, 9708.
53. (a) Iwao, M.; Kuraishi, T. *BCJ* **1987**, *60*, 4051. (b) Mills, R. J.; Snieckus, V. *JOC* **1989**, *54*, 4386. (c) Katsuura, K.; Snieckus, V. *CJC* **1987**, *65*, 124. (d) Zani, C. L.; de Oliveira, A. B.; Snieckus, V. *TL* **1987**, *28*, 6561. (e) de Silva, S. O.; Ahmad, I.; Snieckus, V. *TL* **1978**, *19*, 5107.
54. Beak, P.; Snieckus, V. *ACR* **1982**, *15*, 306.
55. Beak, P.; Brown, R. A. *JOC* **1982**, *47*, 34.
56. Skowronska-Ptasinska, M.; Verboom, W.; Reinhoudt, D. N. *JOC* **1985**, *50*, 2690.
57. Sibi, M. A.; Snieckus, V. *JOC* **1983**, *48*, 1935.
58. Beak, P.; Tse, A.; Hawkins, J.; Chen, C.-W.; Mills, S. *T* **1983**, *39*, 1983.
59. (a) Bauer, W.; Schleyer, P. v. R. *JACS* **1989**, *111*, 7191. (b) Krizan, T. D.; Martin, J. C. *JOC* **1982**, *47*, 2681.
60. Ludt, R. E.; Griffiths, T. S.; McGrath, K. N.; Hauser, C. R. *JOC* **1973**, *38*, 1668.
61. Carpenter, A. J.; Chadwick, D. J. *JOC* **1985**, *50*, 4362.
62. Doadt, E. G.; Snieckus, V. *TL* **1985**, *26*, 1149.
63. Miah, M. A.; Snieckus, V. *JOC* **1985**, *50*, 5436.
64. Jacquelin, J. M.; Robin, Y.; Godard, A.; Queguinier, G. *CJC* **1988**, *66*, 1135.
65. Manoharan, T. S.; Brown, R. S. *JOC* **1989**, *54*, 1439.
66. Chamberlin, A. R.; Liotta, E. L.; Bond, F. T. *OS* **1983**, *61*, 141.
67. Chamberlin, A. R.; Bond, F. T. *S* **1979**, 44.
68. (a) Abraham, W. D.; Bhupathy, M.; Cohen, T. *TL* **1987**, *28*, 2203. (b) Cohen, T.; Yu, L. C. *JOC* **1984**, *49*, 605. (c) Cohen, T.; Yu, L. C. *JACS* **1983**, *105*, 2811.
69. (a) Bailey, W. F.; Punzalan, E. P. *JOC* **1990**, *55*, 5404. (b) Negishi, E.; Swanson, D. R.; Rousset, C. J. *JOC* **1990**, *55*, 5406.
70. (a) Reich, H. J. In *Organoselenium Chemistry*; Liotta, D., Ed.; Wiley: New York, 1987; p 243. (b) Seebach, D.; Peleties, N. *CB* **1972**, *105*, 511. (c) Seebach, D.; Peleties, N. *AG(E)* **1969**, *8*, 450. (d) Dumont, W.; Bayet, P.; Krief, A. *AG(E)* **1974**, *13*, 804.
71. (a) Reich, H. J. *ACR* **1979**, *12*, 22. (b) Liotta, D. *ACR* **1984**, *17*, 28. (c) Clive, D. L. *T* **1978**, *34*, 1049. (d) Davis, F. A.; Reddy, R. T. *JOC* **1992**, 2599.
72. Krief, A.; Dumont, W.; Clarembeau, M.; Berhard, G.; Badaoui, E. *T* **1989**, *45*, 2005.
73. Labar, D.; Krief, A. *CC* **1982**, 564.
74. Labar, D.; Krief, A.; Norberg, G.; Evrard, G.; Durant, F. *BSB* **1985**, *94*, 1083.
75. Paquette, L. A.; Peterson, J. R.; Ross, R. J. *JOC* **1985**, *50*, 5200.
76. (a) Reich, H. J.; Medina, M. A.; Bowe, M. D. *JACS* **1992**, *114*, 11 003. (b) Tomoki, H.; Kambe, N.; Ogawa, A.; Miyoshi, N.; Murai, S.; Sonoda, N. *AG(E)* **1987**, *26*, 1187.
77. Chamberlin, A. R.; Liotta, E. L.; Bond, F. T. *OS* **1983**, *61*, 141.
78. Chamberlin, A. R.; Bond, F. T. *S* **1979**, 44.
79. (a) Abraham, W. D.; Bhupathy, M.; Cohen, T. *TL* **1987**, *28*, 2203. (b) Cohen, T.; Yu, L. C. *JOC* **1984**, *49*, 605. (c) Cohen, T.; Yu, L. C. *JACS* **1983**, *105*, 2811.
80. Doadt, E. G.; Snieckus, V. *TL* **1985**, *26*, 1149.
81. Meyers, A. I.; Wanner, K. T. *TL* **1985**, *26*, 2047.
82. Hosomi, A.; Araki, Y.; Sakurai, H. *JACS* **1982**, *104*, 2081.
83. Denmark, S. E.; Ares, J. J. *JACS* **1988**, *110*, 4432.
84. Tamaru, Y.; Harada, T.; Yoshida, I. *JACS* **1978**, *100*, 1923.
85. Danishefsky, S.; Lee, J. Y. *JACS* **1989**, *111*, 4829.
86. Zhang, P.; Gawley, R. E. *JOC* **1993**, *58*, 3223.

Timo V. Ovaska
Connecticut College, New London, CT, USA

t-Butyllithium[1]

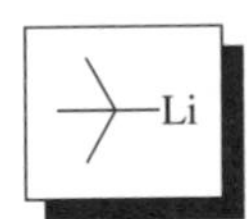

[594-19-4] C_4H_9Li (MW 64.06)

(strong base capable of lithiating weak acids;[1] useful for heteroatom-faciliated lithiations;[2–4] reagent of choice for lithium–halogen exchange;[5] can add to π-bonds[1,6])

Physical Data: colorless solid; decomposes above 140 °C with loss of LiH; sublimes at 70 °C/0.1 mmHg. X-ray structures of the solvent-free tetramer and the ether-solvated dimer[7a] and ^{13}C NMR studies in hydrocarbon[8a] and ethereal solutions[7b] have been reported.

Solubility: sol hydrocarbon solvents, diethyl ether, and THF but should be used at low temperature; half-lives in various solvents have been reported;[9] reacts violently with H_2O and other protic solvents.

Form Supplied in: commercially available as an approximately 1.7 M solution in pentane or heptane. Tetrameric in hydrocarbons,[8] dimeric in diethyl ether,[7] monomeric in THF[10a] although earlier reported as dimeric.[1,10b] In combination with tertiary polyamines such as ***N,N,N′,N′-Tetramethylethylenediamine*** (TMEDA), ***1,4-Diazabicyclo[2.2.2]octane*** (DABCO), or ***N,N,N′,N″,N″-Pentamethyldiethylenetriamine*** (PMDTA), reactivity is often increased.[1,11]

Analysis of Reagent Purity: since the concentrations of commercial solutions vary appreciably, especially after the original seal is broken, it is necessary to standardize solutions of the reagent prior to use. The classical Gilman double titration method is described in detail by Wakefield.[1b] A particularly convenient method for routine analyses involves titration of the reagent with *s*-butyl alcohol using 1,10-phenanthroline or 2,2′-biquinoline as indicator.[13] A variety of other methods have been described.[14]

Preparative Method: the reaction of *t*-butyl chloride with 1–2% Na–Li alloy in dry pentane requires particular attention to detail to achieve a reasonable yield of the reagent. Detailed procedures should be consulted.[1b,12]

Handling, Storage, and Precautions: solutions of the reagent are pyrophoric and the reagent may ignite spontaneously upon exposure to air, producing a purple flame. It is prudent to conduct all operations involving *t*-BuLi behind a shield. In case of fire, a dry-powder extinguisher should be used: in no case should an extinguisher containing water or halogenated hydrocarbons be used to fight an alkyllithium fire. Bottles and reaction flasks containing the reagent should be flushed with N_2 or preferably Ar and kept tightly sealed to preclude contact with oxygen or moisture. The reagent may be cautiously transferred under an inert gas atmosphere using standard syringe/cannula techniques.[1b] Most reactions of *t*-butyllithium are exothermic and for this reason are usually carried out at temperatures well below ambient. For detailed handling techniques see Wakefield.[1b]

Lithiations. *t*-Butyllithium is less frequently used for proton abstraction than is ***n*-Butyllithium** or ***s*-Butyllithium**. However, as it is more reactive and less nucleophilic than its isomers, *t*-BuLi may be the reagent of choice for the lithiation of relatively weak acids (eq 1). The addition of 1 equiv of a tertiary polyamine increases the metalating ability of the reagent.[1,11] Simple hydrocarbons, which are used as solvents, are generally inert to lithiation by *t*-BuLi. Although ethers and amines are often used as solvents in low temperature reactions involving *t*-BuLi, ethers,[15] THF,[16] and TMEDA[17] are readily lithiated by the reagent. Indeed, treatment of THF with *t*-BuLi has been reported to afford ethylene and the lithium enolate of acetaldehyde at temperatures as low as −78 °C.[16a]

$$t\text{-BuLi} + \text{R-H} \longrightarrow t\text{-BuH} + \text{R-Li} \quad (1)$$

Monodeprotonation of terminal alkynes is conveniently accomplished with *n*-BuLi the use of excess *t*-BuLi in hexane may lead to polylithiation.[18] For example, treatment of 1-butyne with 3 equiv of *t*-BuLi in hexane affords a mixture of allene and alkyne upon quench with an electrophile (eq 2).[19]

t-BuLi, hexanes; TMSCl, THF → TMS-substituted alkyne (30%) + tetrakis-TMS allene (50%) (2)

Highly regioselective proton abstractions may result when substrates containing such heteroatoms as oxygen, nitrogen, sulfur, and the like are treated with an organolithium. Several reviews of the topic[2–4,6,20] and a collection of detailed experimental procedures[21] are available. Two categories of heteroatom-facilitated lithiations are recognized: α-lithiation, involving removal of a proton from the carbon bearing the heteroatom, and β- (or *ortho*-) lithiation, involving abstraction of a proton from the β-position.[2,4] Lithiation of acyclic[22] (eq 3),[22a] cyclic (eq 4),[23] and conjugated enol ethers (eq 5)[24] using *t*-BuLi provides a simple route to readily functionalized α-lithio vinyl ethers (see ***1-Ethoxyvinyllithium***). Vinyl sulfides behave analogously[25] and sulfur appears to be a stronger α-director than is oxygen (eq 6).[26] Alkyl phenyl thioethers are α-lithiated by *t*-BuLi in the presence of ***Hexamethylphosphoric Triamide*** (eq 7).[27] Furans and thiophenes are readily α-lithiated by *t*-BuLi, but use of the less reactive *n*-BuLi is more convenient.[1,2,21] At temperatures below −100 °C, *t*-BuLi in THF regioselectively deprotonates enamines derived from β-acyl aldehydes in preference to 1,2- or 1,4-addition[28] (eq 8).[28a]

OMe-vinyl ether; t-BuLi, THF → α-lithio vinyl ether; cyclohexanone → HO/OMe adduct; H_2O → hydroxy ketone, 90% (3)

Dihydropyran; t-BuLi, THF, 0 °C → 2-lithio dihydropyran; methyl levulinate (CO_2Me), 68% → spiro lactone (4)

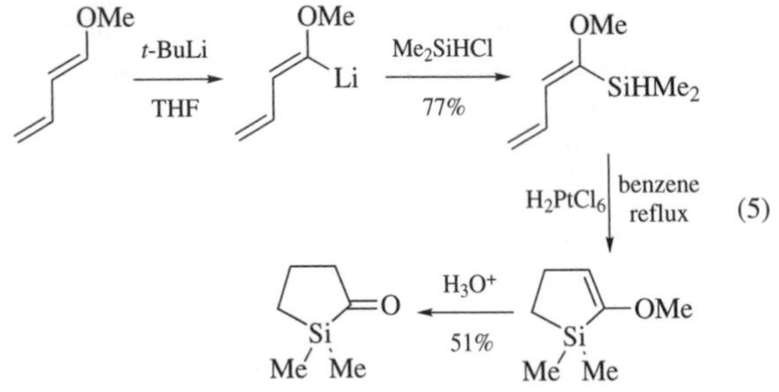

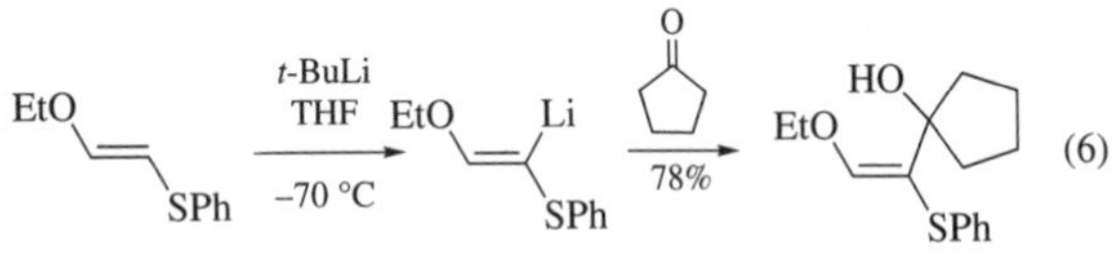

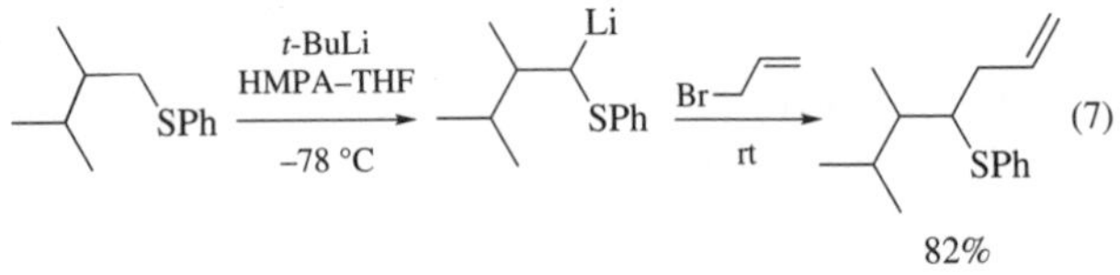

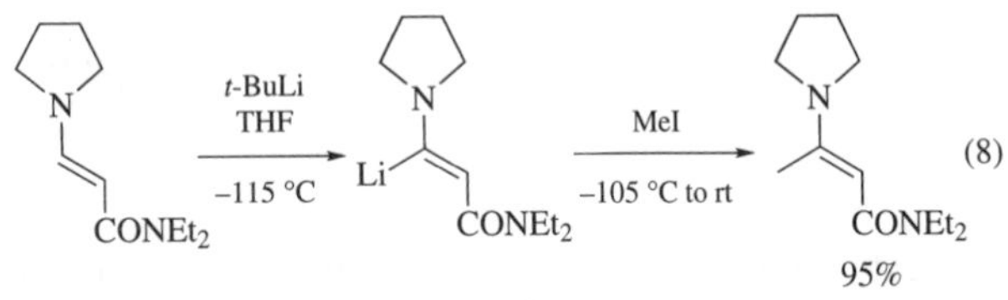

Heteroatom-directed *ortho* lithiations of aromatic compounds are most often accomplished with *n*- or *s*-BuLi, either alone or in combination with TMEDA.[4] Some benefit derives from the use of the bulkier and more basic *t*-BuLi for *ortho* lithiation of substrates bearing substituents susceptible to nucleophilic attack (eq 9)[29] or resistant to deprotonation by less reactive bases (eq 10)[9,30] The key step in a highly stereoselective asymmetric synthesis of chiral ferrocenes involves *ortho* lithiation of a ferrocenyl acetal at −78 °C using *t*-BuLi in THF (eq 11).[31]

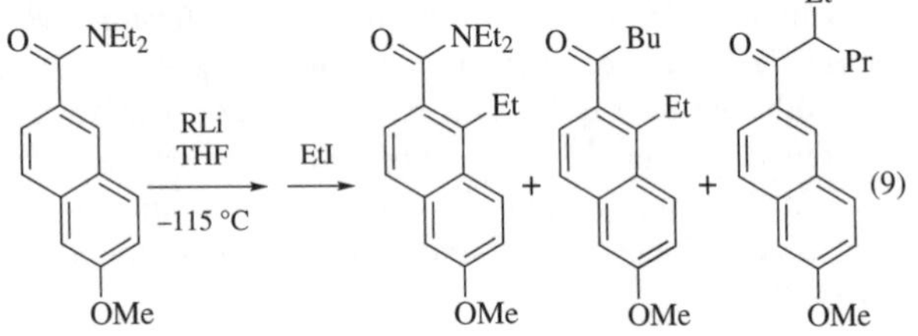

RLi			
n-BuLi	0	34%	38%
t-BuLi	68%	0	0

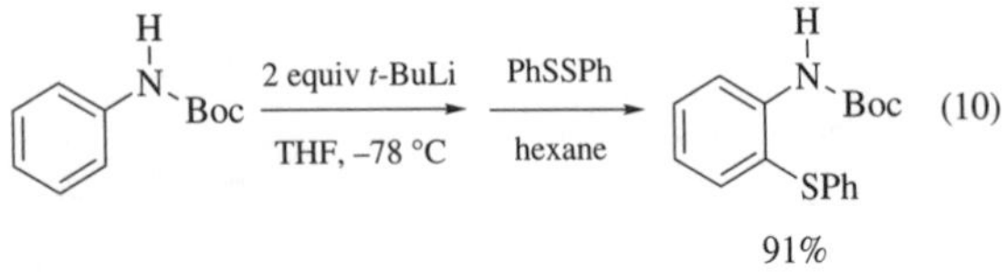

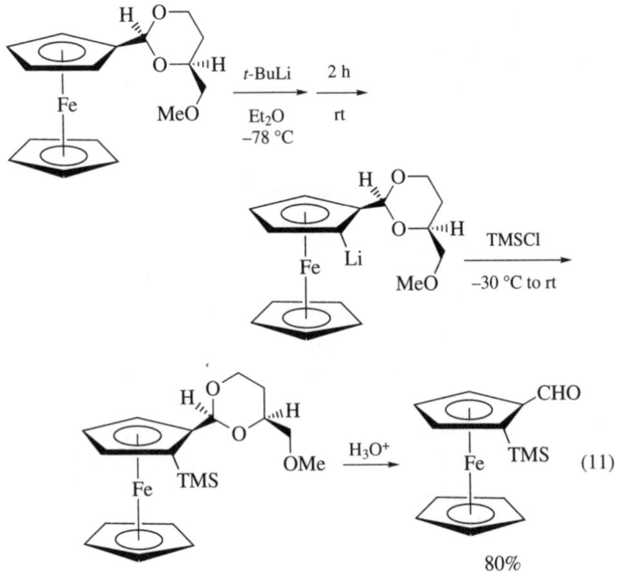

Lithium–Halogen Exchange. *t*-Butyllithium is often the reagent of choice for the preparation of an organolithium by lithium–halogen exchange. The reversible metathesis (eq 12), most readily accomplished at low temperature with bromides and iodides, leads to an equilibrium mixture favoring the more stable organolithium.[1,32] The benefits of using *t*-BuLi (rather than *n*- or *s*-BuLi) for the preparation of organolithiums by lithium–halogen exchange derives from two considerations: (1) the interchange equilibrium is favorable when an aryl, vinyl, cyclopentyl, or primary-alkyl halide is treated with a tertiary organolithium,[33] and (2) the exchange may be rendered operationally irreversible by employing 2 equiv of *t*-BuLi since 1 equiv of the reagent rapidly consumes the *t*-butyl halide generated in the reaction to give isobutane, isobutene, and lithium halide (eq 13).[34]

$$R^1\text{-X} + R^2\text{-Li} \rightleftharpoons R^1\text{-Li} + R^2\text{-X} \quad (12)$$

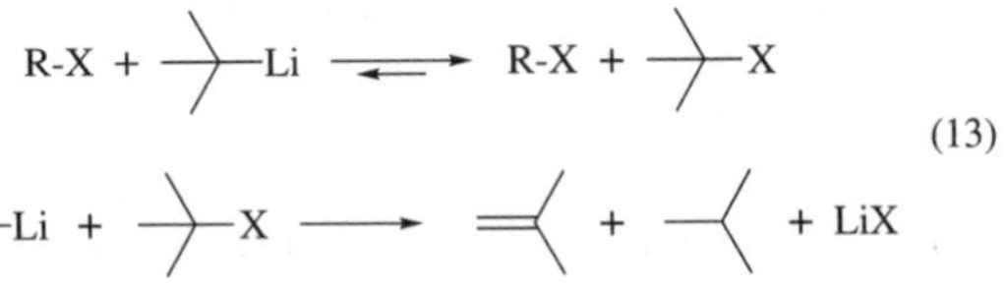

Aryl halides readily undergo lithium–halogen exchange with alkyllithiums in ethereal solvents at temperatures at or below −78 °C and the less expensive *n*-BuLi is often used for this purpose.[1,32] However, the presence of the *n*-butyl halide product can cause complications when the aryllithium is warmed since a coupling reaction may ensue[1,32] (for example, the reaction of ***Phenyllithium*** with *n*-butyl iodide in THF at −78 °C has a half-life of ~30 min). This potential problem is avoided when 2 equiv of *t*-BuLi are used for the generation of aryllithiums.

Vinyllithiums are readily prepared with retention of configuration from the corresponding vinyl bromides or iodides in a stereoselective exchange with 2 molar equiv of *t*-BuLi, provided that the reaction is conducted at temperatures below −110 °C.[35] The reactions are conducted most conveniently in the Trapp solvent,[36] a 4:4:1 mixture of THF, diethyl ether, and pentane (or hexane), by slow addition of 2 equiv of *t*-BuLi in pentane to a solution of the halide which is held at −110 to −120 °C. The preparation of (*E*)-1-hexenyllithium and its conversion to (*E*)-1-phenylthio-1-hexene is illustrative (eq 14)[35a] of the method, and detailed experimental procedures are available.[37] The stereoselective lithium–halogen exchange of vinyl halides is the key step in a highly diastereoselective metalla-Claisen rearrangement (eq 15)[38] and α-keto dianions may be prepared from the lithium enolates of α-bromo ketones by the exchange reaction with excess *t*-BuLi (eq 16).[39]

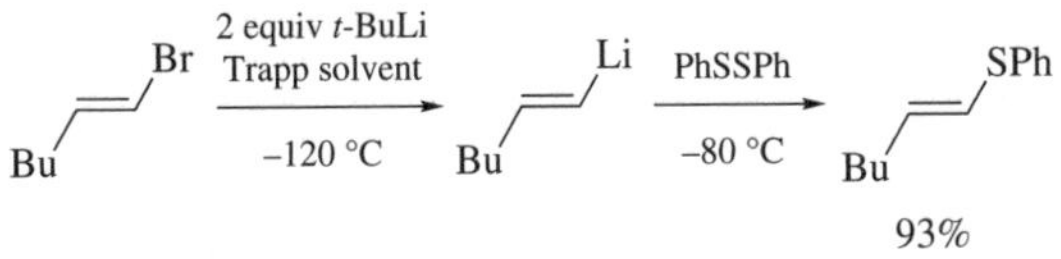

(14)

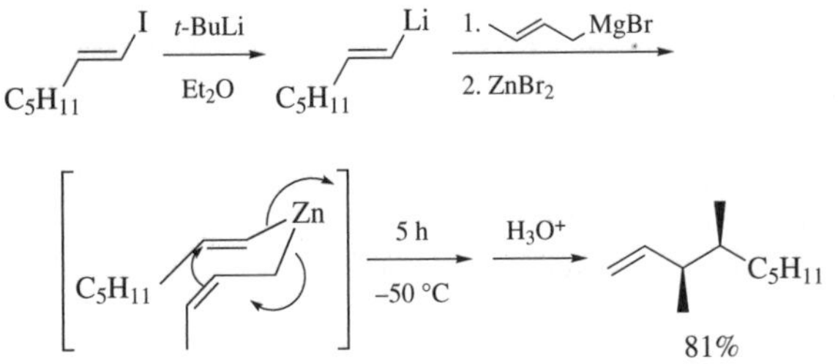

93%

(15)

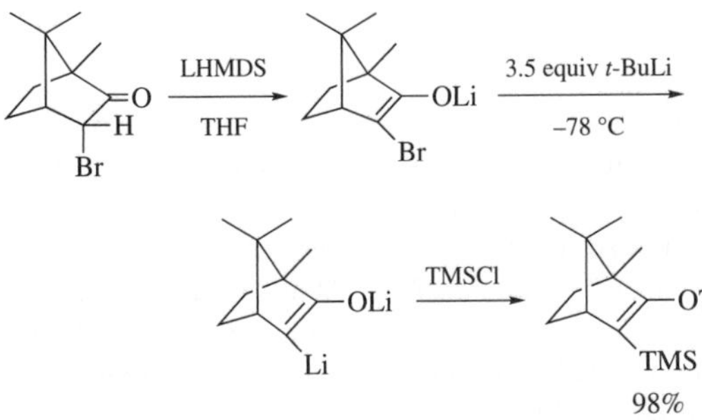

81%

(16)

98%

Cyclopropyllithiums (see ***Cyclopropyllithium***) are, in general, easily prepared from the corresponding bromides or iodides with retention of configuration by low-temperature lithium–halogen exchange using either *n*-BuLi or *t*-BuLi.[1,32] The more reactive *t*-BuLi was found to be superior for the generation of (1-ethoxycyclopropyl)lithium, a reagent useful for the preparation of cyclobutanones (eq 17).[40]

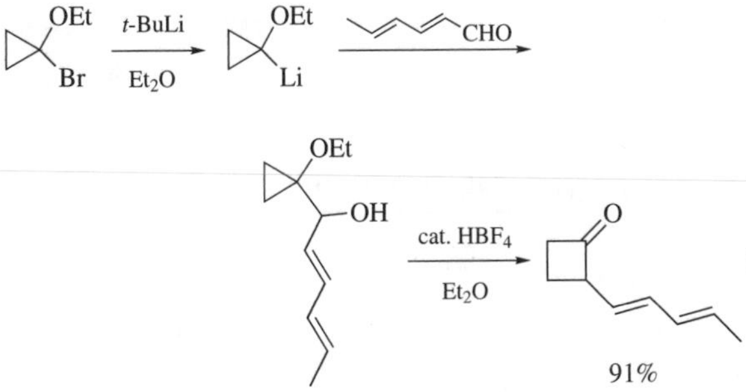

(17)

91%

The early literature suggests that metal–halogen exchange between primary alkyl halides and alkyllithiums is not, in general, a preparatively useful reaction due to unfavorable equilibria (eq 12) and competition from coupling and elimination reactions.[1,32] However, more recent mechanistic studies of the exchange reaction[41,42] have revealed that virtually all of the difficulties historically associated with use of this method for the preparation of simple alkyllithiums can be overcome by the use of *t*-BuLi and judicious choice of experimental conditions.[5] Primary alkyllithiums are conveniently prepared at −78 °C (the exchange has been conducted at temperatures as low as −131 °C)[5a] by addition of 2.1–2.2 equiv of *t*-BuLi in pentane to a solution of primary alkyl iodide in pentane–diethyl ether (3:2 by vol).[5] Under these conditions (eq 18) the exchange is exceedingly rapid, the yield of alkyllithium is excellent,[5] and the only byproduct is a small quantity of easily removed hydrocarbon derived from formal reduction of the iodide via proton abstraction from the cogenerated *t*-BuI.[43] The alkyllithium may be used at low temperature but residual *t*-BuLi remaining in solution may complicate product isolation following addition of an electrophile to the cold reaction mixture, particularly if an excess of *t*-BuLi was employed. Any residual *t*-BuLi is easily removed by simply allowing solutions containing the primary alkyllithium and *t*-BuLi to warm to room temperature: the *t*-BuLi is rapidly consumed by proton abstraction from diethyl ether, leaving clean solutions of the less reactive primary alkyllithium.[5] The success of this route to primary alkyllithiums depends critically on the appropriate choice of both halide and solvent.[5] Alkyl iodides rather than alkyl bromides or chlorides must be used in the exchange: under the conditions of the reaction (eq 18), lithium–iodine exchange most probably involves rapid attack of *t*-BuLi on the iodine atom of the halide,[42,44] while lithium–bromine exchange occurs predominantly by a single electron transfer (SET) mechanism.[42] Alkyl chlorides are essentially inert to the action of *t*-BuLi under the conditions indicated in eq 18.[42] Moreover, it is advisable to run the reaction in a solvent system that contains diethyl or a similar ether; the use of THF or other strongly coordinating Lewis bases such as TMEDA should be avoided as elimination and coupling reactions may compete with exchange in these solvents.[5a]

$$R\text{-}CH_2\text{-}I \xrightarrow[-78\,^\circ C,\ 5\ min]{t\text{-BuLi},\ n\text{-}C_5H_{12},\ Et_2O} R\text{-}CH_2\text{-}Li \xrightarrow{E^+} R\text{-}CH_2\text{-}E \quad (18)$$

Low-temperature lithium–iodine exchange with *t*-BuLi has been used to prepare 5-hexenyllithiums from 6-iodo-1-hexenes.[43] When warmed, these alkenic alkyllithiums undergo highly stereoselective and totally regiospecific 5-*exo* cyclization via a cyclohexane chair-like transition state[45] to afford high yields of (cyclopentylmethyl)lithiums that may be functionalized by reaction with an electrophile[46] (eq 19).[46b] Such isomerizations have also been conducted in a tandem fashion to give polycyclic products[47] (eq 20).[47a] Alkynic alkyllithiums, which are prepared by lithium–iodine exchange with *t*-BuLi, cyclize in a stereoselectively *syn* fashion to four-, five-, and six-membered carbocyclic rings bearing an exocyclic lithiomethylene moiety[48] (eq 21).[48a] Related chemistry initiated by lithium–selenium exchange with *t*-BuLi has also been reported.[49]

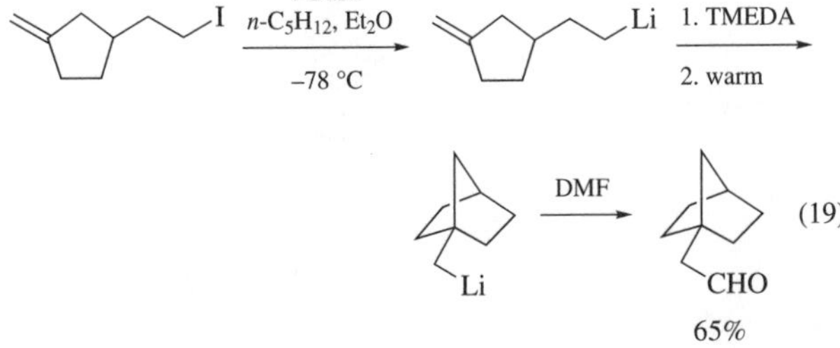

(19)

65%

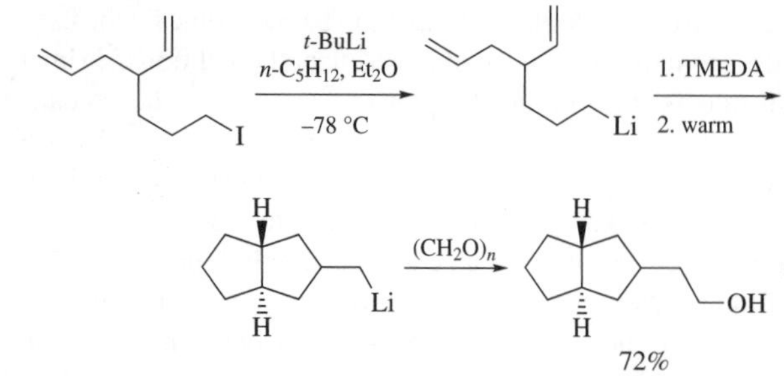

(20)

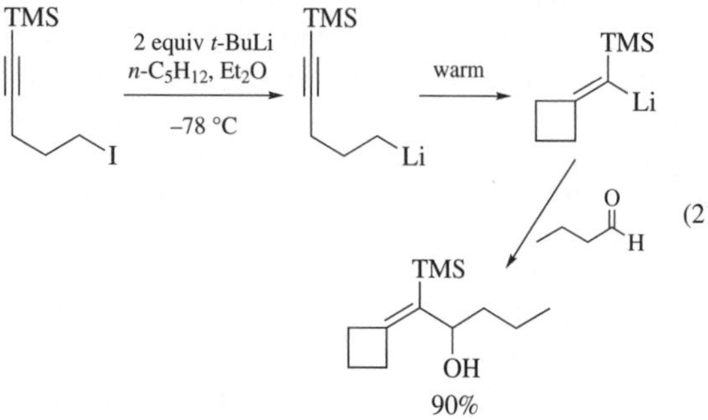

(21)

Reaction of 2 equiv of t-BuLi with α,ω-diiodoalkanes at −23 °C affords nearly quantitative yields of three- through five-membered carbocyclic rings via cyclization of an intermediate α-lithio-ω-iodoalkane (eq 22).[50] When 4 equiv of t-BuLi are used at −78 °C, α,ω-dilithioalkanes such as ***1,4-Dilithiobutane***, ***1,5-Dilithiopentane***, and 1,6-dilithiohexane result.[5b]

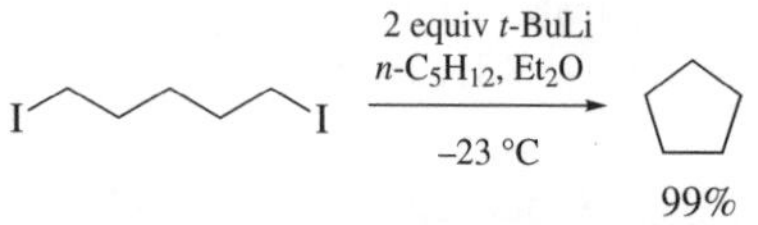

(22)

Low-temperature lithium–iodine exchange between t-BuLi and a primary alkyl iodide is more rapid than lithium–bromine exchange with aryl bromides.[51] Indeed, the iodine exchange reaction is so facile that it is sometimes possible to conduct preparatively useful lithium–iodine exchange in the presence of other reactive functional groups[1] such as ketones (eq 23).[52] Metal–halogen exchange-initiated intramolecular conjugate additions have also been reported[53] (eq 24).[53b]

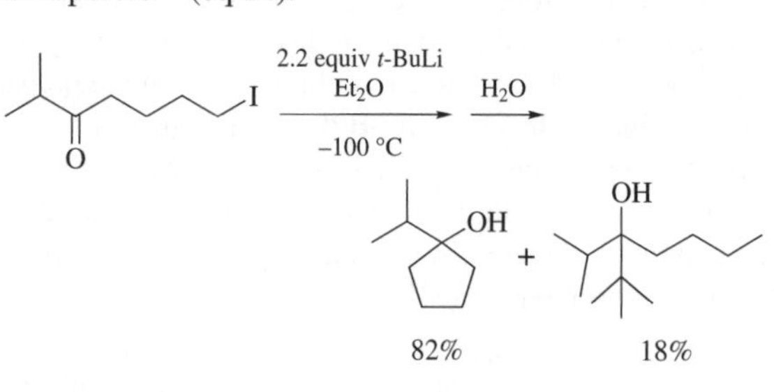

(23)

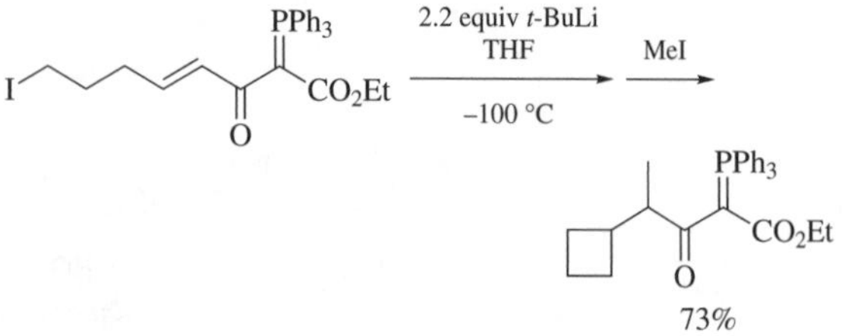

(24)

Dehalogenation of vicinal diiodides initiated by exchange with t-BuLi has been used to access highly strained alkenes[54] such as cubene (eq 25).[55]

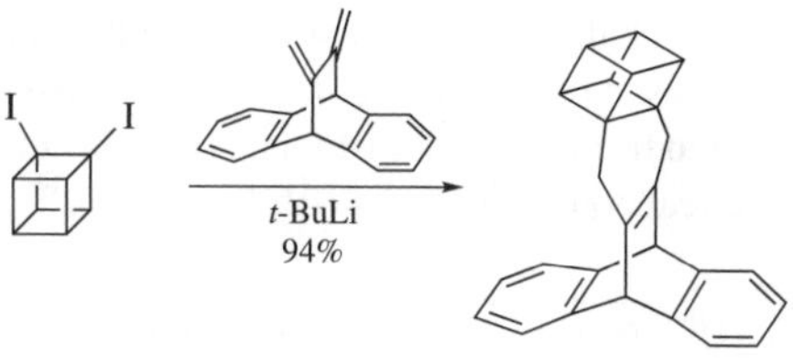

(25)

Addition of Carbon–Carbon Double Bonds. In the presence of ethers or amines, t-BuLi adds rapidly to ethylene at −25 °C at 1 atm of pressure to give neohexyllithium (3,3-dimethylbutyllithium) in quantitative yield (eq 26).[56] It should be noted that neohexyllithium may unexpectedly result from addition of t-BuLi to the ethylene generated from the decomposition of diethyl ether[15] or THF[16] when these solvents are present in reaction mixtures containing t-BuLi.[56]

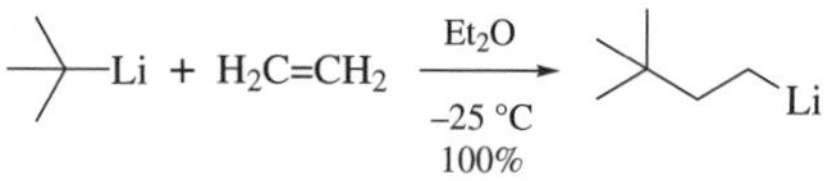

(26)

Addition of t-BuLi to other carbon–carbon double bonds is not a facile process unless: (1) the π-system is conjugated[1,57] (in which case anionic polymerization may result),[58] (2) the alkene is at least somewhat strained (eq 27),[59] (3) the resulting anion is stabilized as in vinylsilanes,[60] vinylphosphines,[61] and the like (eq 28),[62] (4) a good leaving group is present at the allylic position,[63] or (5) there is intramolecular assistance provided by a suitably placed heteroatom[6] (eq 29).[64]

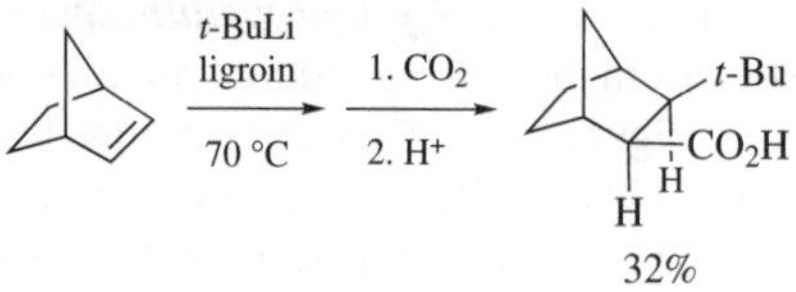

(27)

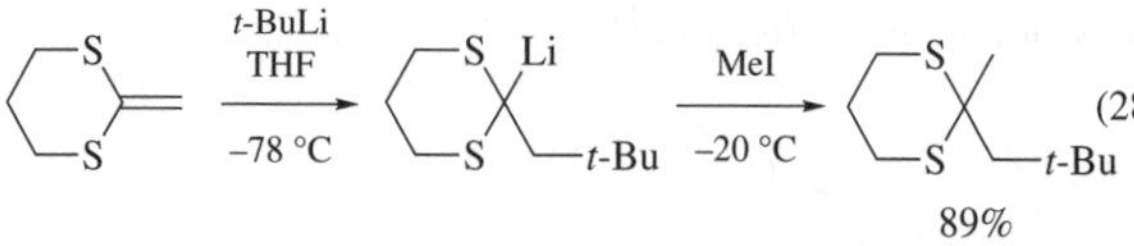

(28)

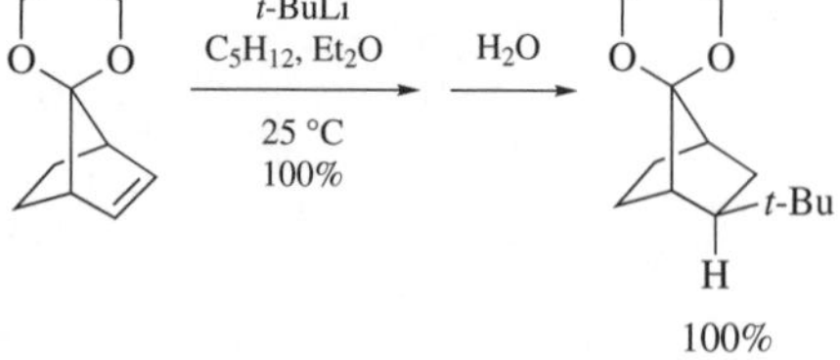

(29)

At high temperatures, alkyllithiums have been observed to alkylate aromatic rings by addition to the π-system and subsequent loss of LiH.[1,65] Of the butyllithiums, t-BuLi has been found to be more reactive than either of its isomers in the alkylation of naphthalene and other condensed aromatics.[66] Addition of t-BuLi, as well as other organolithiums, across the azomethine linkage of pyridine and similar nitrogen heterocycles is a facile process.[1]

Addition to Carbon–Heteroatom Multiple Bonds. The behavior of *t*-BuLi in reactions with carbon–heteroatom π-bonds is relatively unremarkable and parallels that of other organolithium reagents.[1] Even in cases where steric hindrance might be expected to lead to difficulties, product yields are reasonable. Thus, for example, tri-*t*-butylcarbinol (3-*t*-butyl-2,2,4,4-tetramethyl-3-pentanol) may be prepared by addition of *t*-BuLi to di-*t*-butyl ketone (2,2,4,4-tetramethyl-3-pentanone), although there is a significant amount of reduction in this case,[67] and *N*-lithio-di-*t*-butylimines may be generated by addition of *t*-BuLi to *t*-butyl cyanide.[68]

Related Reagents. *n*-Butyllithium; *s*-Butyllithium; *t*-Butylmagnesium Chloride.

---

1. (a) Wakefield, B. J. *The Chemistry of Organolithium Compounds*; Pergamon: New York, 1974. (b) Wakefield, B. J. *Organolithium Methods*; Academic: San Diego, 1988. (c) Wardell, J. L. In *Comprehensive Organometallic Chemistry*; Wilkinson, G., Ed.; Pergamon: Oxford, 1982, Vol. 1, pp 44–120. (d) Wakefield, B. J. In *Comprehensive Organic Chemistry*; Barton, D. H. R.; Ollis, W. D., Eds.; Pergamon: Oxford, 1979, Vol. 3, pp 943–967.
2. Gschwend, H. W.; Rodriguez, H. R. *OR* **1979**, *26*, 1.
3. Gilman, H.; Morton, J. W., Jr. *OR* **1954**, *8*, 258.
4. Snieckus, V. *CRV* **1990**, *90*, 879.
5. (a) Bailey, W. F.; Punzalan, E. R. *JOC* **1990**, *55*, 5404. (b) Negishi, E.; Swanson, D. R.; Rousset, C. J. *JOC* **1990**, *55*, 5406.
6. Klumpp, G. W. *RTC* **1986**, *105*, 1.
7. (a) Köttke, T.; Stälke, D. *AG(E)* **1993**, *32*, 580. (b) Bates, T. F.; Clarke, M. T.; Thomas, R. D.; *JACS* **1988**, *110*, 5109.
8. (a) Thomas, R. D.; Clarke, M. T.; Jensen, R. M.; Young, T. C. *OM* **1986**, *5*, 1851. (b) Weiner, M.; Vogel, G.; West, R. *IC* **1962**, *1*, 654.
9. Stanetty, P.; Koller, H.; Mihovilovic, M. *JOC* **1992**, *57*, 6833.
10. (a) Bauer, W.; Winchester, W. R.; Schleyer, P. v. R. *OM* **1987**, *6*, 2371. (b) Settle, F. A.; Haggerty, M.; Eastham, J. F. *JACS* **1964**, *86*, 2076.
11. Langer, A. W. *Adv. Chem. Ser.* **1974**, *130*, 1.
12. (a) Kamienski, C. W.; Esmay, D. L. *JOC* **1960**, *25*, 1807. (b) Smith, W. N.; *JOM* **1974**, *82*, 1.
13. Watson, S. C.; Eastham, J. F. *JOM* **1967**, *9*, 165.
14. (a) Collins, P. F.; Kamienski, C. W.; Esmay, D. L.; Ellestad, R. B. *Anal. Chem.* **1961**, *33*, 468. (b) Crompton, T. R. *Chemical Analysis of Organometallic Compounds*; Academic: New York, 1973. (c) Kofron, W. G.; Baclawski, L. M. *JOC* **1976**, *41*, 1879. (d) Lipton, M. F.; Sorensen, C. M.; Sadler, A. C.; Shapiro, R. H. *JOM* **1980**, *186*, 155. (e) Winkle, M. R.; Lansinger, J. M.; Ronald, R. C. *CC* **1980**, 87. (f) Bergbreiter, D. E.; Pendergrass, E. *JOC* **1981**, *46*, 219. (g) Juaristi, E.; Martinez-Richa, A.; Garcia-Rivera, A.; Cruz-Sanchez, J. S. *JOC* **1983**, *48*, 2603.
15. (a) Maercker, A.; Demuth, W. *AG(E)* **1973**, *12*, 75. (b) Maercker, A. *AG(E)* **1987**, *26*, 972.
16. (a) Jung, M. E.; Blum, R. B. *TL* **1977**, *43*, 3791. (b) Bates, R. B.; Kroposki, L. M.; Potter, D. E. *JOC* **1972**, *37*, 560.
17. Köhler, F. H.; Hertkorn, N.; Blümel, J. *CB* **1987**, *120*, 2081.
18. West, R. *Adv. Chem. Ser.* **1974**, *130*, 120.
19. West, R.; Jones, P. C. *JACS* **1969**, *91*, 6156.
20. Mallan, J. M.; Bebb, R. L. *CRV* **1969**, *69*, 693.
21. (a) Brandsma, L.; Verkruijsse, H. *Preparative Polar Organometallic Chemistry, I*; Springer: Berlin, 1987. (b) Brandsma, L. *Preparative Polar Organometallic Chemistry, II*; Springer: Berlin, 1987.
22. (a) Baldwin, J. E.; Höfle, G. A.; Lever, O. W. *JACS* **1974**, *96*, 7125. (b) Schöllkopf, U.; Hänssle, P. *LA* **1972**, *763*, 208.
23. Boeckman, R. K.; Bruza, K. J. *TL* **1977**, 4187.
24. Soderquist, J. A.; Hassner, A. *JOC* **1980**, *45*, 541.
25. Oshima, K.; Shimodi, K.; Takahashi, H.; Yamamoto, H.; Nozaki, H. *JACS* **1973**, *95*, 2694.
26. Vlattas, I.; Vecchia, L. D.; Lee, A. O. *JACS* **1976**, *98*, 2008.
27. Dolak, T. M.; Bryson, T. A. *TL* **1977**, *23*, 1961.
28. (a) Schmidt, R. R.; Talbiersky, J. *AG(E)* **1976**, *15*, 171. (b) Schmidt, R. R.; Talbiersky, J. *AG(E)* **1977**, *16*, 853. (c) Schmidt, R. R.; Talbiersky, J. *AG(E)* **1978**, *17*, 204.
29. Bindal, R. D.; Katzenellenbogen, J. A. *JOC* **1987**, *52*, 3181.
30. Muchowski, J. M.; Venuti, M. C. *JOC* **1980**, *45*, 4798.
31. Riant, O.; Samuel, O.; Kagan, H. B. *JACS* **1993**, *115*, 5835.
32. (a) Jones, R. G.; Gilman, H. *OR* **1951**, *6*, 339. (b) Jones, R. G.; Gilman, H. *CRV* **1954**, *54*, 835. (c) Schöllkopf, U. *MOC*, **1970**, *13/1*, 1.
33. Applequist, D. E.; O'Brien, D. F. *JACS* **1963**, *85*, 743.
34. Corey, E. J.; Beames, D. J. *JACS* **1972**, *94*, 7210.
35. (a) Seebach, D.; Neumann, H. *CB* **1974**, *107*, 847. (b) Neumann, H.; Seebach, D. *CB* **1978**, *111*, 2785. (c) Lee, S. H.; Schwartz, J. *JACS* **1986**, *108*, 2445.
36. (a) Köbrich, G.; Trapp, H. *CB* **1966**, *99*, 680. (b) Köbrich, G. *AG(E)* **1967**, *6*, 41.
37. Ref. 1(b), pp 29 and 137.
38. Marek, I.; Lefrancois, J.-M.; Normant, J.-F.; *SL* **1992**, 633.
39. (a) Kowalski, C. J.; O'Dowd, M. L.; Burke, M. C.; Fields, K. W. *JACS* **1980**, *102*, 5411. (b) Kowalski, C. J.; Fields, K. W. *JACS* **1982**, *104*, 1777.
40. Gadwood, R. C.; Rubino, M. R.; Nagarajan, S. C.; Michel, S. T. *JOC* **1985**, *50*, 3255.
41. Bailey, W. F.; Patricia, J. J. *JOM* **1988**, *352*, 1.
42. (a) Bailey, W. F.; Patricia, J. J.; Nurmi, T. T.; Wang, W. *TL* **1986**, *27*, 1861. (b) Bailey, W. F.; Patricia, J. J.; Nurmi, T. T. *TL* **1986**, *27*, 1865. (c) Ashby, E. C.; Pham, T. N.; Park, B. *TL* **1985**, *26*, 4691. (d) Ashby, E. C.; Pham, T. N. *JOC* **1987**, *52*, 1291.
43. Bailey, W. F.; Nurmi, T. T.; Patricia, J. J.; Wang, W. *JACS* **1987**, *109*, 2442.
44. Reich, H. J.; Green, D. P.; Phillips, N. H. *JACS* **1991**, *113*, 1414.
45. Bailey, W. F.; Khanolkar, A. D.; Gavaskar, K.; Ovaska, T. V.; Rossi, K.; Thiel, Y.; Wiberg, K. B. *JACS* **1991**, *113*, 5720.
46. (a) Bailey, W. F.; Patricia, J. J.; Del Gobbo, V. C.; Jarret, R. M.; Okarma, P. J. *JOC* **1985**, *50*, 1999. (b) Bailey, W. F.; Khanolkar, A. D. *JOC* **1990**, *55*, 6058. (c) Bailey, W. F.; Khanolkar, A. D. *T* **1991**, *47*, 7727. (d) Bailey, W. F.; Khanolkar, A. D. *OM* **1993**, *12*, 239.
47. (a) Bailey, W. F.; Khanolkar, A. D.; Gavaskar, K. V. *JACS* **1992**, *114*, 8053. (b) Bailey, W. F.; Rossi, K. *JACS* **1989**, *111*, 765.
48. (a) Bailey, W. F.; Ovaska, T. V. *JACS* **1993**, *115*, 3080. (b) Bailey, W. F.; Ovaska, T. V. *TL* **1990**, *31*, 627. (c) Wu, G.; Cederbaum, F. E.; Negishi, E. *TL* **1990**, *31*, 493. (d) Bailey, W. F.; Ovaska, T. V.; Leipert, T. K. *TL* **1989**, *30*, 3901.
49. (a) Krief, A.; Barbeaux, P. *CC* **1987**, 1214. (b) Krief, A.; Barbeaux, P. *SL* **1990**, 511. (c) Krief, A.; Barbeaux, P. *TL* **1991**, *32*, 417.
50. (a) Bailey, W. F.; Gagnier, R. P.; Patricia, J. J. *JOC* **1984**, *49*, 2098. (b) Bailey, W. F.; Gagnier, R. P. *TL* **1982**, *23*, 5123.
51. Beak, P.; Allen, D. J. *JACS* **1992**, *114*, 3420.
52. Cooke, M. P., Jr.; Houpis, I. N. *TL* **1985**, *26*, 4987.
53. (a) Cooke, M. P., Jr. *JOC* **1984**, *49*, 1144. (b) Cooke, M. P. Jr.; Widener, R. K. *JOC* **1987**, *52*, 1381. (c) Cooke, M. P. Jr. *JOC* **1992**, *57*, 1495. (d) Cooke, M. P., Jr. *JOC* **1993**, *58*, 2910.
54. Schäfer, J.; Szeimies, G. *TL* **1988**, *29*, 5253.
55. Eaton, P. E.; Maggini, M. *JACS* **1988**, *110*, 7230.
56. (a) Bartlett, P. D.; Friedman, S.; Stiles, M. *JACS* **1953**, *75*, 1771. (b) Bartlett, P. D.; Stiles, M. *JACS* **1955**, *77*, 2806. (c) Bartlett, P. D.; Tauber, S. J.; Weber, W. P. *JACS* **1969**, *91*, 6362. (d) Bartlett, P. D.; Goebel, C. V.; Weber, W. P. *JACS* **1969**, *91*, 7425.

57. (a) Glaze, W. H.; Jones, P. C. *CC* **1969**, 1434. (b) Fraenkel, G.; Estes, D. W.; Geckle, M. J. *JOM* **1980**, *185*, 147. (c) Fraenkel, G.; Geckle, J. M. *JACS* **1980**, *102*, 2869. (d) Hallden-Abberton, M.; Engelman, C.; Fraenkel, G. *JOC* **1981**, *46*, 538.
58. Morton, M. *Anionic Polymerization: Principles and Practice*; Academic: New York, 1983.
59. Mulvaney, J. E.; Gardlund, Z. G. *JOC* **1965**, *30*, 917.
60. (a) Cason, L. F.; Brooks, H. G. *JACS* **1952**, *74*, 4582. (b) Jones, P. R.; Lim, T. F. O. *JACS* **1977**, *99*, 2013. (c) Auner, N. *JOM* **1987**, *336*, 59.
61. Peterson, D. J. *JOC* **1966, 31**, 950.
62. Seebach, D.; Bürstinghaus, R.; Gröbel, B-T.; Kolb, M. *LA* **1977**, 830.
63. (a) Bailey, W. F.; Zartun, D. L. *CC* **1984**, 34. (b) Mioskowski, C.; Manna, S.; Falck, J. R. *TL* **1984**, *25*, 519.
64. Kool, M.; Klumpp, G. W. *TL* **1978**, *21*, 1873.
65. Reetz, M. T.; Schinzer, D. *AG(E)* **1977**, *16*, 44.
66. (a) Dixon, J. A.; Fishman, D. H.; *JACS* **1963**, *85*, 1356. (b) Dixon, J. A.; Fishman, D. H.; Dudinyak, R. S. *TL* **1964**, *12*, 613. (c) Eppley, R. L.; Dixon, J. A. *JACS* **1968**, *90*, 1606.
67. Bartlett, P. D.; Lefferts, E. B. *JACS* **1955**, *77*, 2804.
68. (a) Summerford, C.; Wade, K.; Wyatt, B. K. *JCS(A)* **1970**, 2016. (b) Clegg, W.; Snaith, R.; Shearer, H. M. M.; Wade, K.; Whitehead, G. *JCS(D)* **1983**, 1309.

William F. Bailey & Nanette Wachter-Jurcsak
University of Connecticut, Storrs, CT, USA

n-Butyllithium–Potassium *t*-Butoxide[1]

(*n*-BuLi)
[109-72-8] C_4H_9Li (MW 64.06)
(*t*-BuOK)
[865-47-4] C_4H_9KO (MW 112.22)

(superbase for metal–hydrogen exchange of a carbon acid R-H;[2] can generate organopotassium compounds[3])

Physical Data: see ***n*-Butyllithium** and **Potassium *t*-Butoxide**; NMR[4] and ESR[5] studies of the *n*-BuLi/*t*-BuOK mixture have been reported.
Solubility: see ***n*-Butyllithium** and **Potassium *t*-Butoxide**.
Preparative Method: prepared in situ under nitrogen by addition of equimolar potassium *t*-butoxide to a solution of *n*-butyllithium at 0 °C.
Handling, Storage, and Precautions: use in a fume hood; must be prepared and transferred under inert gas (Ar or N_2) to exclude oxygen and moisture. Solutions in THF decompose above −40 °C.

Metal–Hydrogen Exchange. Freshly prepared *n*-BuLi/*t*-BuOK is more basic than *n*-BuLi and *n*-BuK.[6] When treated with this superbase, hydrocarbons in the low acidity range of pK_a 35–50 undergo a clean metal–hydrogen exchange: benzene,[2] cyclopropane derivatives,[7,8] cumene,[7] a variety of 1- or 2-alkenes having allylic C–H bonds,[2,9–13] cyclohexene,[7,14] α-pinene,[15] conjugated dienes,[11,16–18] ene sulfides,[19] 4*H*-pyrans,[20] and 1,4-dihydropyridines.[20] *n*-BuLi/*t*-BuOK did not succeed in metalating saturated hydrocarbons such as octane and cyclohexane.

Benzene and substituted benzene, in the presence of *n*-BuLi/*t*-BuOK, undergo site selective metal–hydrogen exchange (eq 1). Best results are obtained in THF at −75 °C or −50 °C. Upon quenching with carbon dioxide, the corresponding acids are obtained in good to excellent yield.[21,22]

R
n-BuLi/*t*-BuOK, THF
K
CO_2, 60–85%
CO_2H
(1)
R = H, F, CF_3

When alkenes are treated with *n*-BuLi/*t*-BuOK, the metalation occurs regioselectively: allylic methyl groups are much more readily attacked than methylene groups, which are less reactive than methine centers.[9,12] Furthermore, *cis*-2-alkenes generally react faster than their *trans* counterparts. This holds even for 4,4-dimethyl-2-pentene (eq 2).[23]

t-Bu
n-BuLi/*t*-BuOK, pentane, 25 °C, 48 h
t-Bu, K
oxirane, pentane, 25 °C, 71%
t-Bu, OH
(2)

Above −60 °C, ethereal solvents are rapidly attacked by *n*-BuLi in the presence of *t*-BuOK.[6b] Nevertheless, acetal functions are suitable protecting groups for superbase metalations of alkenols.

The *n*-BuLi/*t*-BuOK mixture can convert 2-alkynyl ethers into allenic ethers and some α-substituted derivatives (eq 3).[24]

Bu, OMe
n-BuLi/*t*-BuOK, THF
Bu, K, OMe
Me_3SiCl, 72%
Bu, $SiMe_3$, OMe
PrCl, 74%
Bu, Pr, OMe
(3)

In the cases where metal–hydrogen exchange leads to a resonance-stabilized organometallic species, the intermediate organopotassium compounds precipitate.[3] Benzyl-, allyl-, pentadienyl-, and cyclooctadienylpotassium can be prepared in this way.[9,13,16] The accompanying lithium *t*-butoxide can be completely removed by washing with benzene or toluene. The resulting organopotassium can serve as a reagent of general applicability in preparing transition metal complexes through transmetalation (eq 4).[25]

$$\text{2-MeC}_6\text{H}_4\text{PPh}_2 \xrightarrow[\text{hexane, 70\%}]{n\text{-BuLi}/t\text{-BuOK}} \text{2-(KCH}_2)\text{C}_6\text{H}_4\text{PPh}_2 \xrightarrow{\text{M(SEt}_2)_2\text{Cl}_2} [\text{2-(Ph}_2\text{P)C}_6\text{H}_4\text{CH}_2]_2\text{M} \quad (4)$$

$M = Ni^{II}, Pd^{II}, Pt^{II}$

Related Reagents. *n*-Butyllithium.

1. Schlosser, M. *PAC* **1988**, *60*, 1627.
2. Schlosser, M. *JOM* **1967**, *8*, 9.
3. Lochmann, L.; Pospíšil, J.; Lím, D. *TL* **1966**, 257.
4. Boche, G.; Etzrodt, H. *TL* **1983**, *24*, 5477.
5. Wilhelm, D.; Clark, T.; Schleyer, P. v. R.; Courtneidge, J. L.; Davies, A. G. *JOM* **1984**, *273*, C1.
6. (a) Schlosser, M.; Strunk, S. *TL* **1984**, *25*, 741. (b) Lehmann, R.; Schlosser, M. *TL* **1984**, *25*, 745.
7. Hartmann, J.; Schlosser, M. *HCA* **1976**, *59*, 453.
8. Schlosser, M.; Schneider, P. *HCA* **1980**, *63*, 2404.
9. Schlosser, M.; Hartmann, J. *AG(E)* **1973**, *12*, 508.
10. Schlosser, M.; Hartmann, J.; David, V. *HCA* **1974**, *57*, 1567.
11. Hartmann, J.; Muthukrishnan, R.; Schlosser, M. *HCA* **1974**, *57*, 2261.
12. Schlosser, M. *AG(E)* **1974**, *13*, 701.
13. Schlosser, M.; Hartmann, J. *JACS* **1976**, *98*, 4674.
14. Hartmann, J.; Stähle, M.; Schlosser, M. *S* **1974**, 888.
15. Rauchschwalbe, G.; Schlosser, M. *HCA* **1975**, *58*, 1094.
16. Schlosser, M.; Rauchschwalbe, G. *JACS* **1978**, *100*, 3258.
17. Bosshardt, H.; Schlosser, M. *HCA* **1980**, *63*, 2393.
18. Schlosser, M.; Bosshardt, H.; Walde, A.; Stähle, M. *AG(E)* **1980**, *19*, 303.
19. Muthukrishnan, R.; Schlosser, M. *HCA* **1976**, *59*, 13.
20. Schlosser, M.; Schneider, P. *AG(E)* **1979**, *18*, 489.
21. Schlosser, M.; Katsoulos, G.; Takagishi, S. *SL* **1990**, 747.
22. Schlosser, M.; Choi, J. H.; Takagishi, S. *T* **1990**, *46*, 5633.
23. Stähle, M.; Hartmann, J.; Schlosser, M. *HCA* **1977**, *60*, 1730.
24. Verkruijsse, H. D.; Verboom, W.; Van Rijn, P. E.; Brandsma, L. *JOM* **1982**, *232*, C1.
25. Longoni, G.; Chini, P.; Canziani, F.; Fantucci, P. *G* **1974**, *104*, 249.

Xiaoyang Xia
Emory University, Atlanta, GA, USA

Calcium Carbonate

$CaCO_3$

[471-34-1] $CCaO_3$ (MW 100.09)

(relatively insoluble white solid; a useful acid scavenger and stabilizing agent for acid sensitive compounds,[1] and a coreactant in reactions producing acidic byproducts.)

Physical Data: widely available as powder (reagent grade) or ppt powder (USP). Occurs naturally as limestone, marble, and chalk. Three crystal forms are known: (1) aragonite *[14791-73-2]*; mp 825 °C (dec); *d* 2.93 g cm^{-3}; water sol 0.0015 g/100 mL (at 25 °C); (2) calcite *[13397-26-7]*; mp 1339 °C (103 atm); *d* 2.71 g cm^{-3}; water sol 0.0014 g/100 mL (at 25 °C); (3) vaterite *[13701-58-1]*; found in N. Ireland and Israel.

Handling, Storage, and Precautions: none; nontoxic; generates calcium oxide and releases carbon dioxide on heating.

Elimination. Halides (eq 1)[2] and sulfoxides (eq 2)[3] are converted to alkenes (or their tautomers) on heating with $CaCO_3$ in appropriate solvents. In the case of halo ketone eliminations (eq 3)[4] the methods of Holysz[5a] and Joly,[5b] employing lithium carbonate/lithium halide mixtures, appear to be more commonly used, and in some instances more effective.[6]

$$2\,CCl_3CH(OH)_2 + CaCO_3\ (1\ equiv) \xrightarrow[2.\ H_3O^+\quad 90\%]{1.\ H_2O,\ cat.\ NaCN,\ heat\ with\ care} 2\,CHCl_2CO_2H \quad (1)$$

(2) toluene, $CaCO_3$, reflux, 7 h (yield from cyclohexenone 76%)

(3) $CaCO_3$ (4 equiv), DMF, reflux, 82%

Halogenation. Calcium carbonate is often used with reagents such as ***N-Bromosuccinimide*** (eq 4),[7] phosphorus halides,[8] and ***Benzyltrimethylammonium Dichloroiodate*** (BTMA) (eq 5).[9]

(4) NBS (4 equiv), $CaCO_3$ (2 equiv), CCl_4, AIBN, reflux 8 h, 71%

(5) BTMA, $CaCO_3$ (1.3 equiv), CH_2Cl_2, MeOH, rt, 30 min, 96%

Substitutions. ***Silver(I) Perchlorate*** promoted substitutions of reactive halides may be effected under mild conditions, as shown in eq 6[10] and eq 7.[11] This provides a useful alternative[12] to classical Friedel–Craft conditions.

(6) $AgClO_4$ (2 equiv), $CaCO_3$ (2 equiv), 1:3 THF:ether, 76%

(7) $ArCCl_2Ph + [Me_2N\text{-}CH{=}CH\text{-}CH{=}NMe_2]^+\ ClO_4^-$; 1. $AgClO_4$ (2 equiv), $CaCO_3$ (2.5 equiv), $MeNO_2$, −35 °C; 2. hydrolysis; 90%

Pinacol Rearrangement. Rearrangement of glycol monotosylates (eq 8)[13] is effected by the procedure of Corey.[14] Solvolysis of an analogous bicyclic epoxy tosylate in an aqueous suspension of $CaCO_3$ gave a 1,5-diketone in good yield (eq 9).[15]

(8) $CaCO_3$, DMF, reflux 8 h, 100%

(9) $CaCO_3$ (1.5 equiv), 3:2 dioxane:H_2O, reflux, 24 h, 86%

Thioether and Acetal Deprotection. ***Mercury(II) Chloride*** induced cleavage of thioethers[16] and acetals (eq 10),[17] and their selenium analogs,[18] employs added $CaCO_3$.

(10) $HgCl_2$ (2.2 equiv), $CaCO_3$ (2.5 equiv), 4:1 MeCN:H_2O, 12–48 h, 84%

Cycloadditions. Cycloaddition (eq 11)[19] and related reactions (eq 12)[20] have sometimes used $CaCO_3$ as a coreactant.

CH_2=CH-OEt (10 equiv), $CaCO_3$ (5 equiv)
MeOH, 24 h
1. Amberlyst-15, THF, H_2O
2. NaOH, aq MeOH, heat
96%
(11)

$CaCO_3$ (1.1 equiv)
DMF, 105–110 °C, 24 h
70%
(12)

Other Divalent-Metal Carbonates. Other divalent-metal carbonates related to calcium carbonate include $MgCO_3$ *[546-93-0]*, ***Zinc Carbonate*** *[3486-35-9]*, $SrCO_3$ *[1633-05-2]*, and $BaCO_3$ *[513-77-9]*. In general, these carbonates have not been widely used. Barium carbonate has served a similar role in halogenation and dehydrohalogenation reactions as the calcium salt. Since barium is more toxic than calcium, its equivalent use is not recommended. Calcium, strontium, and barium carbonates have all been used as supports for noble metal catalysts in hydrogenation reactions, and in some cases appear to give better selectivity than analogous carbon- or alumina-supported catalysts.[21]

A novel coumarin synthesis (eq 13), employing zinc or magnesium carbonate,[22] avoids the acidic conditions of the Pschorr synthesis.

$ZnCO_3$ or $MgCO_3$
CH_2Cl_2
32%
(13)

Related Reagents. Calcium Hydride; Cesium Carbonate; Potassium Carbonate; Sodium Carbonate.

1. (a) Campaigne, E.; Tullar, B. F. *OSC* **1963**, *4*, 921. (b) Parish, E. J.; Luo, C.; Parish, S.; Heidepriem, R. W. *SC* **1992**, *22*, 2839.
2. Cope, A. C.; Clark, J. R.; Connor, R. *OSC* **1943**, *2*, 181.
3. Fujisawa, T.; Noda, A.; Kawara, T.; Sato, T. *CL* **1981**, 1159.
4. Green, G. F. H.; Long, A. G. *JCS* **1961**, 2532.
5. (a) Holyz, R. P. *JACS* **1953**, *75*, 4432. (b) Joly, R.; Warnant, J.; Nomine, G.; Bertin, D. *BSC* **1958**, 366.
6. Demuth, M. R.; Garrett, P.; White, J. D. *JACS* **1976**, *98*, 634.
7. Farina, G.; Zecchi, G. *S* **1977**, 755.
8. Carman, R.; Shaw, I. M. *AJC* **1976**, *29*, 133.
9. Kajigaeshi, S.; Kakinami, T.; Yamasaki, H.; Fujisaki, S.; Okamoto, T. *BCJ* **1988**, *61*, 600.
10. Shimizu, N.; Watanabe, K.; Tsuno, Y. *BCJ* **1984**, *57*, 1165 (cf. general method *JACS* **1982**, *104*, 1330).
11. Kral, V.; Arnold, Z. *S* **1982**, 823.
12. Zavada, J.; Pankom, M.; Arnold, Z. *CCC* **1976**, *41*, 1777.
13. Mazur, Y.; Nussim, M. *JACS* **1961**, *83*, 3911.
14. Corey, E. J.; Ohno, M.; Mitra, R. B.; Vatakencherry, P. A. *JACS* **1964**, *86*, 478.
15. Gray, R. W.; Dreiding, A. S. *HCA* **1977**, *60*, 1969.
16. Corey, E. J.; Bock, M. G. *TL* **1975**, 2643.
17. Langley, D. R.; Thurston, D. E. *JOC* **1987**, *52*, 91.
18. Burton, A.; Hevesi, L.; Dumont, W.; Cravader, A.; Krief, A. *S* **1979**, 877.
19. Manna, S.; Falck, J. R.; Mioskowski, C. *JOC* **1982**, *47*, 5021.
20. Tachikawa, R.; Miyadera, T.; Tamura, C; Terada, A.; Naruto, S.; Nagamatsu, E. *JCS(P1)* **1978**, 1524.
21. (a) Augustine, R. L. *Catalytic Hydrogenation; Techniques and Applications in Organic Synthesis*; Dekker: New York, 1965. (b) Rylander, P. N. *Catalytic Hydrogenation over Platinum Metals*; Academic Press: New York, 1967.
22. Buchi, G.; Weinreb, S. M. *JACS* **1971**, *93*, 746.

William Reusch
Michigan State University, East Lansing, MI, USA

Calcium Hydride[1]

[7789-78-8] H_2Ca (MW 42.10)

(used as a drying agent for reagents and solvents, as a desiccant in reactions, as a base, and as a catalyst and cocatalyst)

Physical Data: mp 816 °C, *d* 1.902 g cm^{-3}.
Form Supplied in: gray orthorhombic crystals, fine ground powder, coarse ground powder, and granules/lumps.
Handling, Storage, and Precautions: moisture sensitive and pyrophoric. It reacts vigorously with water, and should be stored and handled under dry conditions (argon or nitrogen).

Calcium hydride reacts with water to form calcium hydroxide and hydrogen gas (eq 1). It reacts much more slowly than the other metal hydrides as a base (toward organic acids), and has therefore found its most widespread use as a drying agent. Thus it has been used to dry hydrocarbons, ethers, amines (including pyridine), esters, phosphites, and alcohols. A typical procedure is to stir the liquid containing approximately 5% w/v calcium hydride for a period of time (1–2 h at least; preferably 24 h), and decanting, filtering, or distilling the liquid away from the calcium salts.[2] Traces of water and cyanide may be removed from ***N,N-Dimethylformamide*** in this manner.[3]

$$CaH_2(s) + H_2O \longrightarrow Ca(OH)_2(aq) + H_2(g) \quad (1)$$

Specific compounds dried with calcium hydride and analyzed for residual water content include benzene,[2] dioxane,[2] acetonitrile,[2,4] methanol,[5] ethanol,[5] 2-butanol[5] and *t*-butyl alcohol.[5] Benzene is dried to ≤0.2 ppm and dioxane to ≤50 ppm residual water.[5] Simply stirring acetonitrile over calcium hydride for seven days and distilling leaves 1900 ppm water.[5] A more rigorous drying protocol for acetonitrile has also been reported for electrochemical use.[4] Methanol and ethanol are dried to ~100 ppm residual water content,[5] in conflict with an early recommendation against using this drying agent.[6] Apparently, calcium hydride reacts with water considerably faster than with alcohols, making it an excellent drying agent for these compounds. On the other hand, perhaps the most convenient drying agent for the lower alcohols is 3Å molecular sieves (powdered sieves are best for all but methanol, where beads are best).[5] ***Molecular Sieves*** are as good as calcium hydride for methanol and ethanol, and superior for 2-butanol and *t*-butyl alcohol.[5]

Calcium hydride has also been used as a desiccant, removing water as it is formed in a reaction. For example, the condensation of methylboronic acid to trimethylboroxin (eq 2) proceeds in 58% yield in the presence of calcium hydride. Note that the reverse reaction is exothermic and quantitative. Other desiccants are not as efficacious.[7] Similarly, calcium hydride removes the water formed in the condensation of methylboronic acid with pinacol (eq 3).[8]

$$3\ \text{Me-B(OH)}_2 \xrightarrow[\text{Et}_2\text{O}]{\text{CaH}_2} (\text{MeBO})_3 \quad (2)$$

$$\text{Me-B(OH)}_2 + \text{pinacol} \xrightarrow[\text{pentane}]{\text{CaH}_2} \text{Me-B(pinacolate)} \quad (3)$$

Two instances have been reported where calcium hydride has apparently been used as a desiccant for residual water in reagents such as tetrabutylammonium halides. For example, benzylation of an alcohol with ***Potassium Hydroxide***, ***Benzyl Bromide***, calcium hydride, and ***Tetra-n-butylammonium Iodide*** (eq 4) was used in a ganglioside synthesis.[9] Another such example is the S_N2 ring opening of an allylic ether, which required calcium hydride, probably as a desiccant (eq 5).[10]

$$\xrightarrow[\text{BnBr, Bu}_4\text{NI, DMSO}]{\text{KOH, CaH}_2} \quad (4)$$

$$\xrightarrow[\text{CaH}_2\text{, MeCN, 94\%}]{\text{AcBr, Bu}_4\text{NBr}} \quad (5)$$

An early report describes calcium hydride as a basic catalyst for the 1,4-addition of 2-nitropropane to benzalacetophenone (eq 6).[11] In this reaction the methanol is essential (no reaction without it), and the order of addition is important (the calcium hydride must be added last).

$$\xrightarrow[\text{MeOH, 85–92\%}]{\text{CaH}_2} \quad (6)$$

Calcium hydride assists in the acylation of alcohols (acting as a base), including tertiary ones, by anhydrides and acid chlorides (eq 7).[12,13]

$$\xrightarrow[\text{CaH}_2\text{, CCl}_4\text{, ≥70\%}]{\text{ClCH}_2\text{COCl}} \quad (7)$$

The site of methylation of 2-aminoethanol is influenced by the gegenion of the hydride used: ***Sodium Hydride*** affords 96% *O*-methylation while calcium hydride yields exclusive *N*-methylation (eq 8).[14]

$$\text{H}_2\text{N(CH}_2)_2\text{OH} + \text{Me}_2\text{SO}_4 + \text{MH} \longrightarrow \text{MeNH(CH}_2)_2\text{OH} + \text{H}_2\text{N(CH}_2)_2\text{OMe} \quad (8)$$

MH = NaH 4 : 96
MH = CaH_2 100 : 0

The Sharpless asymmetric epoxidation (SAE) is an extremely valuable technique for the synthesis of epoxy alcohols from allylic alcohols.[15] Although the asymmetric epoxidation gives good yields (76–80%) and high selectivities (≥90% ee), it may require a long reaction time. Wang and co-workers studied the effect of additives on the Sharpless epoxidation of (*Z*)-2-tridecen-1-ol.[16] Addition of silica gel alone did not affect the selectivity, the yield, or the reaction time. Addition of calcium hydride shortened the reaction time but decreased the selectivity. Addition of both, however, resulted in a reaction time shortened from 96 h to 8 h, with no loss of selectivity (eq 9). The calcium hydride/silica gel-modified Sharpless reagent has since been studied on a variety of allylic alcohols,[17] and has been used by other groups as well.[18]

$$\xrightarrow[\text{silica gel, 76\%, 95\% ee}]{\text{SAE, CaH}_2} \quad (9)$$

Related Reagents. Cesium Carbonate; Sodium Hydride.

1. (a) Mackay, K. M. *Hydrogen Compounds of the Metallic Elements*; Spon: London, 1966. (b) Hurd, D. T. *An Introduction to the Chemistry of the Hydrides*; Wiley: New York, 1952. (c) Wiberg, E.; Amberger, E. *Hydrides of the Elements of Main Groups I-IV*; Elsevier: New York, 1971.
2. Burfield, D. R.; Lee, K.-H.; Smithers, R. H. *JOC* **1977**, *42*, 3060.
3. Fieser, L. F.; Fieser, M. *FF* **1975**, *5*, 247.
4. Walter, M.; Ramaley, L. *Anal. Chem.* **1973**, *45*, 165.
5. Burfield, D. R.; Smithers, R. H. *JOC* **1983**, *48*, 2420.
6. Fieser, L. F.; Fieser, M. *FF* **1967**, *1*, 105.
7. Brown, H. C.; Cole, T. E. *OM* **1985**, *4*, 816.
8. Brown, H. C.; Park, W. S.; Cha, J. S.; Cho, B. T.; Brown, C. A. *JOC* **1986**, *51*, 337.
9. Ito, Y.; Numata, M.; Sugimoto, M.; Ogawa, T. *JACS* **1989**, *111*, 8508.

10. Wakamatsu, K.; Kigoshi, H.; Niiyama, K.; Niwa, H.; Yamada, K. *T* **1986**, *42*, 5551.
11. Fishman, N.; Zuffanti, S. *JACS* **1951**, *73*, 4466.
12. Oppenauer, R. V. *M* **1966**, *97*, 62.
13. Helmchen, G.; Wegner, G. *TL* **1985**, *26*, 6051.
14. Kashima, C.; Harada, K.; Omote, Y. *CJC* **1985**, *63*, 288.
15. Katsuki, T.; Sharpless, K. B. *JACS* **1980**, *102*, 5974.
16. Wang, Z.; Zhou, W.; Lin, G. *TL* **1985**, *26*, 6221.
17. Wang, Z.; Zhou, W. *T* **1987**, *43*, 2935.
18. (a) Schwab, J. M.; Ray, T.; Ho, C. K. *JACS* **1989**, *111*, 1057. (b) Prestwich, G. D.; Graham, S. M.; Kuo, J. W.; Vogt, R. G. *JACS* **1989**, *111*, 636.

Robert E. Gawley & Arnold Davis
University of Miami, Coral Gables, FL, USA

10-Camphorsulfonic Acid

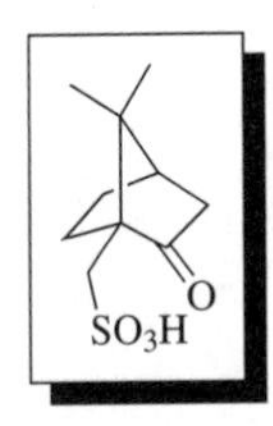

(1*S*)-(+)		
[3144-16-9]	$C_{10}H_{16}O_4S$	(MW 232.30)
(1*R*)-(−)		
[35963-20-3]		
(±)		
[5872-08-2]		

(acid catalyst,[1–8] resolving agent,[9,10] chiral auxiliary[11–20])

Physical Data: mp 203–206 °C (dec).
Solubility: sol dichloromethane, methanol, benzene; insol ether.
Form Supplied in: white crystals, racemic (±).
Analysis of Reagent Purity: melting point, NMR.
Preparative Methods: commercially available from several sources; can be prepared by sulfonation of camphor with acetic–sulfuric anhydride.[21]
Purification: recrystallize from ethyl acetate.
Handling, Storage, and Precautions: hygroscopic; corrosive.

Acid Catalyst. Camphorsulfonic acid (CSA) has been used extensively in synthetic organic chemistry as an acid catalyst. It has particularly been used in protecting group chemistry. For example, hydroxyl groups can be protected as tetrahydropyranyl (THP) ethers using dihydropyran and a catalytic amount of CSA (eq 1).[1] Both 1,2- and 1,3-diols can be selectively protected by reaction with orthoesters in the presence of camphorsulfonic acid to form the corresponding cyclic orthoester (eq 2).[2] This method of protection is particularly useful in that reduction of the orthoester with ***Diisobutylaluminum Hydride*** forms the monoacetal, which allows for preferential protection of a secondary alcohol in the presence of a primary alcohol. Ketones have also been protected using catalytic CSA (eq 3).[3]

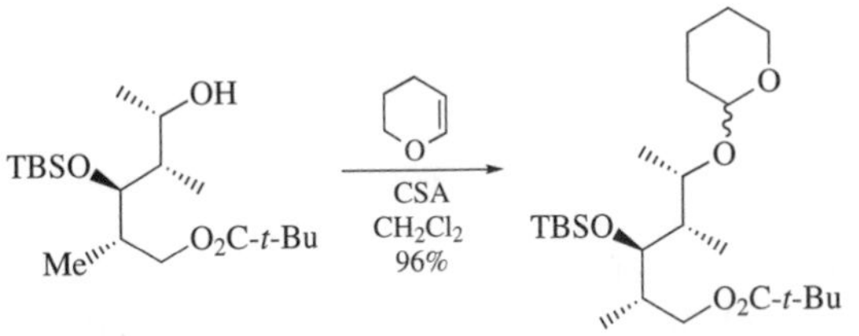

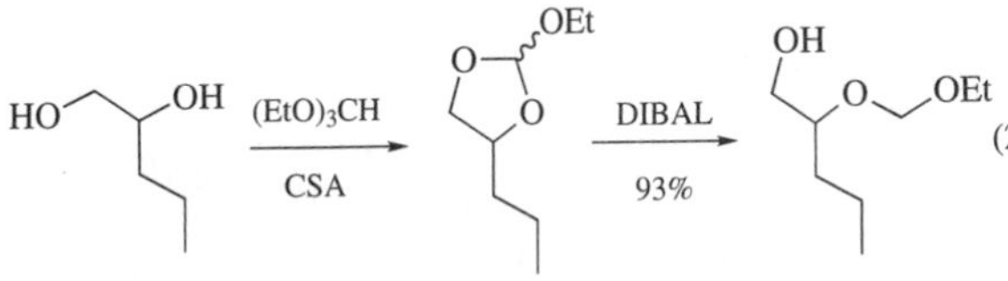

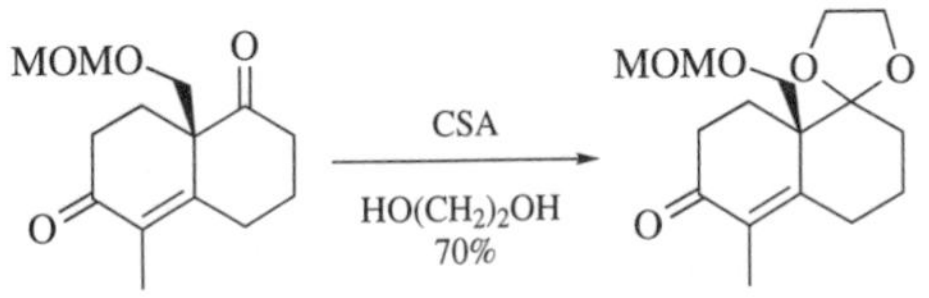

Overman has shown that camphorsulfonic acid can also be used in nucleophile-promoted alkyne–iminium cyclizations.[22] Alkylamines can react with formaldehyde and sodium iodide to yield piperidines in good yield. This methodology has been applied in the total synthesis of pumiliotoxin A (eq 4).[4]

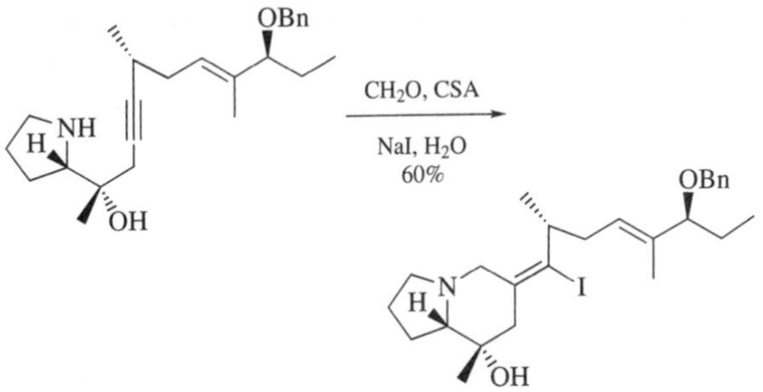

The most efficient catalyst for intramolecular opening of epoxides is CSA.[5,6] The formation of tetrahydrofurans or tetrahydropyrans is highly dependent on the structure of the hydroxy epoxide. The presence of a saturated chain at the secondary epoxide position leads to formation of tetrahydrofurans (eq 5)[5] via 5-*exo* ring closure, whereas an electron-rich double bond at this position gives tetrahydropyrans (eq 6)[6] via 6-*endo* ring closure. This methodology has also been extended to the synthesis of oxepanes (eq 7).[6]

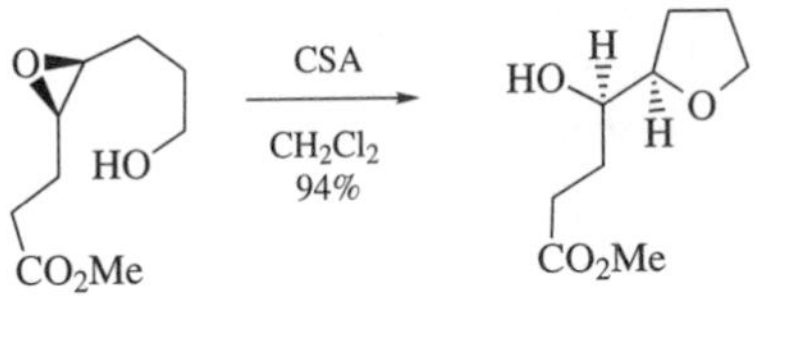

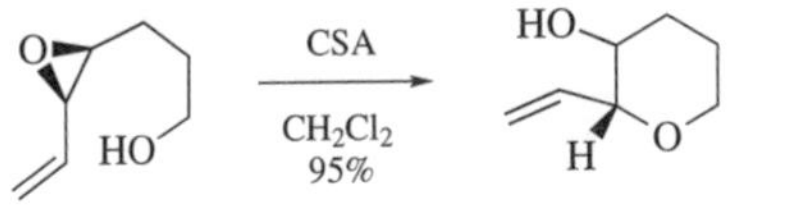

CSA; CH_2Cl_2; 75% (7)

CSA is also the acid of choice for use in phenylselenation reactions.[7] It has been used as an acid catalyst in hydroxyselenation reactions of alkenes (eq 8)[7] and organoselenium-induced cyclizations (eq 9) using ***N-Phenylselenophthalimide*** (NPSP).[7]

NSePh; H_2O, CSA; CH_2Cl_2; 89%; OH; SePh (8)

OH; NSePh; H_2O, CSA; CH_2Cl_2; 80%; SePh (9)

CSA has also been used to catalyze spiroacetalizations.[8,22] In Schreiber's approach to the talaromycins he utilized a CSA-catalyzed spiroacetalization (eq 10) and found that the use of different solvents led to varying percentages of isomeric products.[8] Other approaches to the talaromycins also utilize CSA for the required spiroacetalization.[23]

OBn; CSA; CH_2Cl_2; 69%; HO; HO; OBn (10)

Resolving Agent. Scalemic CSA has been used to resolve amines by forming diastereomeric salts which can be separated by fractional crystallization (eq 11).[9] In this instance, after obtaining the desired crystalline diastereomeric salt, the undesired diastereomer was completely transformed into the desired one by a resolution–racemization procedure (eq 12).[9] Additionally, racemic ketones can be resolved by forming enantiomeric iminium salts (eq 13).[10] Two different procedures have been devised depending on the ease of enamine formation.

Me; NH_2; Ph; (+)-CSA; NH_2; .CSA (11)

Me; NH_2; CHO; OH; Cl; Cl; H_2O; Ph; OH; Cl; Cl (12)

(+)-CSA; 88%; CSA^- (13)

Chiral Auxiliaries.

Asymmetric Diels–Alder Reactions. The commercial availability of either enantiomer of camphorsulfonic acid has made it quite useful in asymmetric Diels–Alder reactions. Reaction of the sultone (generated from CSA) with ***Lithium Diisopropylamide*** followed by esterification and β-elimination yields the crystalline acrylate (eq 14).[11] The Lewis acid-catalyzed [4 + 2] cycloaddition of 1,3-dienes with this acrylate affords the corresponding scalemic adduct which can be reduced with ***Lithium Aluminum Hydride*** to yield an enantiomerically pure alcohol (eq 15).[12]

SO_3H; 1. $NaBH_4$; 2. TsCl, py; 54%; SO_2; LDA; 58%; OH; $SO_2N(i\text{-}Pr)_2$; 1. HO(C=O)CH$_2$CH$_2$Cl, TFAA; 2. Et_3N; 75%; $SO_2N(i\text{-}Pr)_2$ (14)

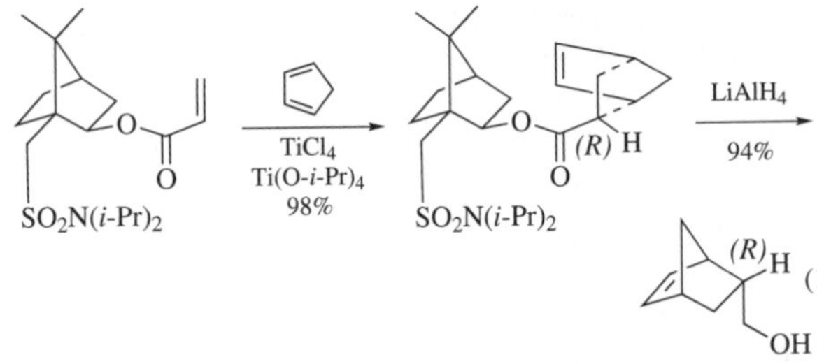

A different approach to the asymmetric Diels–Alder reaction involves the use of the sultam derived from CSA. Lewis acid-promoted reaction with dienes followed by reductive removal of the chiral auxiliary is analogous to that previously discussed for the sultone. Smith has successfully utilized this approach to synthesize the chiral acid used in the synthesis of the immunosuppressant FK-506 (eq 16).[13]

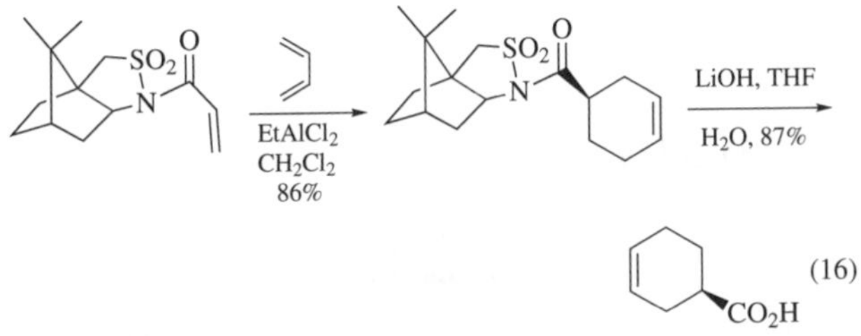

Oxaziridines. Davis has developed the use of chiral 2-sulfonyloxaziridines derived from camphorsulfonic acid as chiral auxiliaries in the asymmetric oxidation reactions.[24] Although other oxaziridines may be preferable, the camphor-derived oxaziridines can be used for the oxidation of sulfides and disulfides to sulfoxides and thiosulfinates as well as for the epoxidation of alkenes.[24] On the other hand, the camphoryloxaziridines are the preferred reagents for hydroxylation of lithium enolates of esters, amides, and ketones, as utilized in the synthesis of kjellmanianone (eq 17).[14]

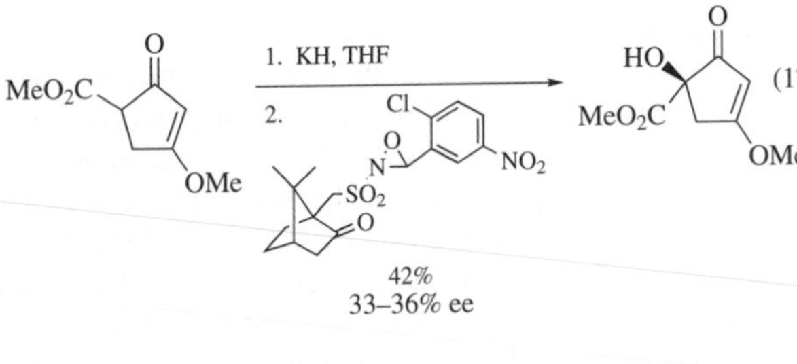

Chiral Sulfides. Optically active sulfides prepared from (+)-CSA can be used to prepare optically active 1,2-diaryloxiranes (eq 18).[15]

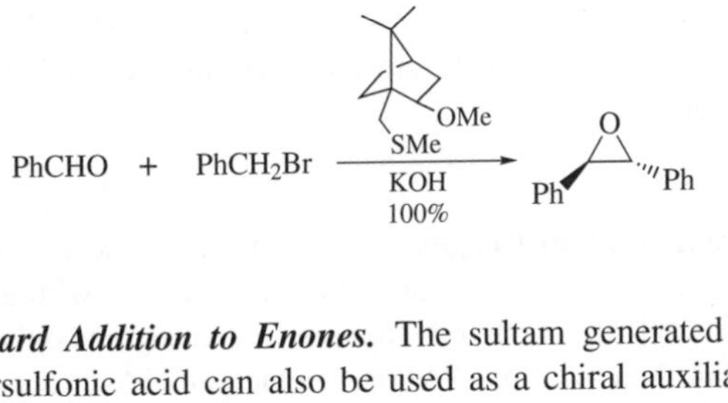

Grignard Addition to Enones. The sultam generated from camphorsulfonic acid can also be used as a chiral auxiliary in the conjugate addition of Grignard reagents to enones. Simple alkylmagnesium chlorides add in a 1,4-fashion to afford imides (eq 19).[16]

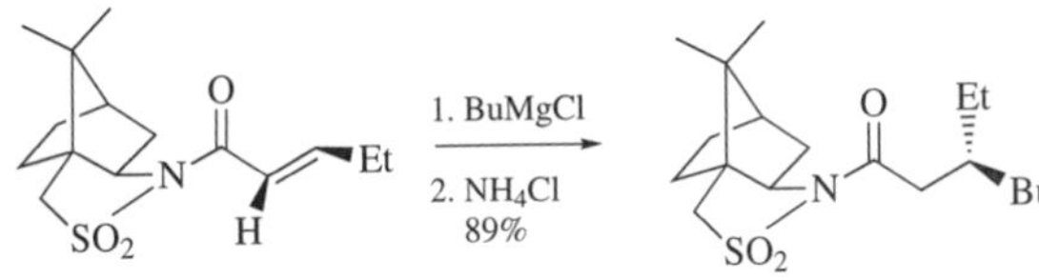

Asymmetric Hydrogenation of Camphor-Derived Sultamides. The sultamide of CSA can be used as a chiral auxiliary for synthesis of β-substituted carboxylic acids (eq 20).[17]

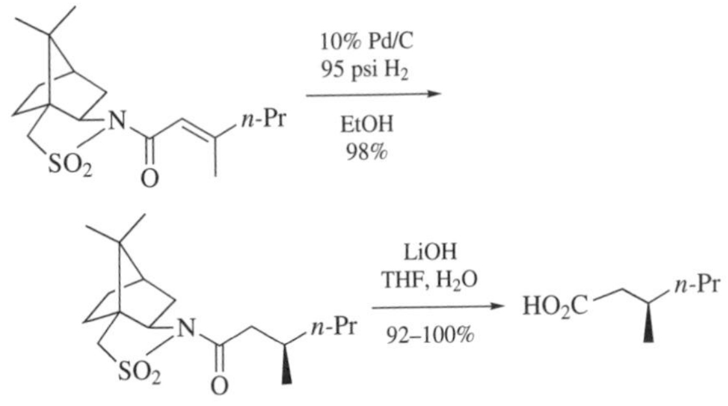

Asymmetric Acetoxylation of Esters. The silyl enol ether derived from CSA reacts with ***Lead(IV) Acetate*** to yield the α-acetoxy ester with good diastereoselectivity. Hydrolysis of the chiral auxiliary gives the α-hydroxy acid, whereas reduction affords the terminal α-glycol (eq 21).[18]

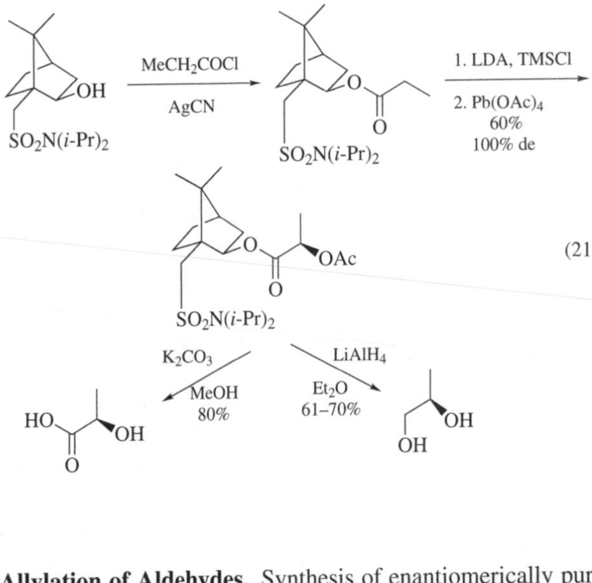

Allylation of Aldehydes. Synthesis of enantiomerically pure allyl alcohols can be accomplished by catalytic asymmetric addition of divinylzinc to aldehydes using a camphorsulfonic acid-derived catalyst (eq 22).[19]

$$H(C=O)(CH_2)_4Me + (CH_2{=}CH)_2Zn \xrightarrow[\text{90\%, >96\% ee}]{\text{amino alcohol ligand}} HO\text{–}CH(CH{=}CH_2)(CH_2)_4Me \quad (22)$$

Synthesis of Epoxides from Chiral Chlorohydrins. Asymmetric halogenation of CSA-derived esters allows for the formation of enantiomerically pure halohydrins and terminal epoxides (eq 23).[20]

$$\text{CSA ester} \xrightarrow[\text{NCS, 67\%, >96\% de}]{\text{LDA, THF, TMSCl}} \text{α-chloro ester} \quad (23)$$

$Ca(BH_4)_2$, 85% → chlorohydrin; NaOMe, 54%, >98% ee → epoxide

Related Reagents. 3-Bromocamphor-8-sulfonic Acid; (−)-(1*S*,4*R*)-Camphanic Acid; 2,4,6-Collidinium *p*-Toluenesulfonate; Pyridinium *p*-Toluenesulfonate; *p*-Toluenesulfonic Acid; Trifluoromethanesulfonic Acid.

---

1. Nicolaou, K. C.; Chakraborty, T. K.; Daines, R. A.; Simpkins, N. S. *CC* **1986**, 413.
2. Takasu, M.; Naruse, Y.; Yamamoto, H. *TL* **1988**, *29*, 1947.
3. Tamai, Y.; Hagiwara, H.; Uda, H. *JCS(P1)* **1986**, 1311.
4. Overman, L. E.; Sharp, M. J. *TL* **1988**, *29*, 901.
5. Nicolaou, K. C.; Prasad, C. V. C.; Somers, P. K.; Hwang, C.-K. *JACS* **1989**, *111*, 5330.
6. Nicolaou, K. C.; Prasad, C. V. C.; Somers, P. K.; Hwang, C.-K. *JACS* **1989**, *111*, 5335.
7. Nicolaou, K. C.; Petasis, N. A.; Claremon, D. A. *T* **1985**, *41*, 4835.
8. Schreiber, S. L.; Sommer, T. J.; Satake, K. *TL* **1985**, *26*, 17.
9. Reider, P. J.; Davis, P.; Hughes, D. L.; Grabowski, E. J. J. *JOC* **1987**, *52*, 955.
10. Adams, W. R.; Chapman, O. L.; Sieja, J. B.; Welstead, W. J., Jr. *JACS* **1966**, *88*, 162.
11. Oppolzer, W.; Chapuis, C.; Kelly, M. J. *HCA* **1983**, *66*, 2358.
12. Oppolzer, W.; Chapuis, C.; Bernardinelli, G. *TL* **1984**, *25*, 5885.
13. Smith, A. B., III; Hale, K. J.; Laakso, L. M.; Chen, K.; Riéra, A. *TL* **1989**, *30*, 6963.
14. Boschelli, D.; Smith, A. B., III; Stringer, O. D.; Jenkins, R. H., Jr. Davis, F. A. *TL* **1981**, *22*, 4385.
15. Furukawa, N.; Sugihara, Y.; Fujihara, H. *JOC* **1989**, *54*, 4222.
16. Oppolzer, W.; Poli, G.; Kingma, A. J.; Starkemann, C.; Bernardinelli, G. *HCA* **1987**, *70*, 2201.
17. Oppolzer, W.; Mills, R. J.; Réglier, M. *TL* **1986**, *27*, 183.
18. Oppolzer, W.; Dudfield, P. *HCA* **1985**, *68*, 216.
19. Oppolzer, W.; Radinov, R. N. *TL* **1988**, *29*, 5645.
20. Oppolzer, W.; Dudfield, P. *TL* **1985**, *26*, 5037.
21. Bartlett, P. D.; Knox, L. H. *OSC* **1973**, *5*, 194.
22. Overman, L. E.; Sharp, M. J. *JACS* **1988**, *110*, 612.
23. Baker, R.; Boyes, A. L.; Swain, C. J. *TL* **1989**, *30*, 985.
24. Davis, F. A.; Jenkins, R. H., Jr.; Awad, S. B.; Stringer, O. D.; Watson, W. H.; Galloy, J. *JACS* **1982**, *104*, 5412.

Ellen M. Leahy
Affymax Research Institute, Palo Alto, CA, USA

Cerium(III) Chloride[1]

$CeCl_3 \cdot 7H_2O$

[7790-86-5]	$CeCl_3$	(MW 246.47)
($\cdot 7H_2O$)		
[18618-55-8]	$H_{14}CeCl_3O_7$	(MW 372.59)

(mild Lewis acid capable of selective acetalization;[2] organocerium reagents have increased oxo-[3] and azaphilicity[4] and greatly reduced basicity;[5] in combination with $NaBH_4$ is a selective 1,2-reducing agent[6])

Alternate Name: cerous chloride; cerium trichloride.
Physical Data: mp 848 °C; bp 1727 °C; *d* 3.92 g cm^{-3}.
Solubility: insol cold H_2O; sol alcohol and acetone; slightly sol THF.
Form Supplied in: white solid; widely available.
Drying: for some applications the cerium trichloride must be strictly anhydrous. The following procedure has proven most efficacious: a one-necked base-washed flask containing the heptahydrate and a magnetic stirring bar was evacuated to 0.1 Torr and heated *slowly* to 140 °C over a 2 h period. At 70 and 100 °C, considerable amounts of water are given off, and these critical temperature zones should not be passed through too quickly. The magnetically stirred white solid is heated overnight at 140 °C, cooled, blanketed with nitrogen, treated with dry THF (10 mL g^{-1}), and agitated at rt for 3 h. To guarantee the complete removal of water, *t*-butyllithium is added dropwise until an orange color persists.
Handling, Storage, and Precautions: anhydrous $CeCl_3$ should be used as prepared for best results. Cerium is reputed to be of low toxicity.

Organocerium Reagents. Organocerates, most conveniently prepared by the reaction of lithium compounds with anhydrous $CeCl_3$ in THF, are highly oxophilic and significantly less basic than their RLi and RMgBr counterparts. As a consequence,

1,2-addition reactions involving readily enolizable ketones are not adversely affected by competing enolization.[3] Well-studied examples of this phenomenon abound.[7] Although alkyl and vinyl cerates have most often been used,[3,5,7] Cl_2CeCH_2CN,[8] Cl_2CeCH_2COOR,[9] $Cl_2CeC{\equiv}CSiMe_3$,[10] and $Cl_2CeCH_2SiMe_3$[11] are known to be equally effective. The last reagent has been utilized for the methylenation of highly enolizable ketones (eq 1)[11a] and for the conversion of carboxylic acid halides, esters, and lactones into allylsilanes (eq 2).[11b–d] Amides and nitriles also condense with organocerates with equal suppression of enolization.[3,4] In general, high levels of steric hindrance in either reaction partner can be tolerated (eq 3). In select examples, the level of double stereodifferentiation can be impressively high.[5,7,12]

Me_3SiCH_2Li, $CeCl_3$, THF, –78 °C, 83%; HF, H_2O, MeCN, 87% (1)

Me_3SiCH_2MgBr, $CeCl_3$, THF, –70 °C → rt; SiO_2, CH_2Cl_2, 92% (2)

$Ph(CH_2)_2N$ … CN; $BuCeCl_2$ (3 equiv), THF, –65 °C → rt, 90% (3)

The discovery has been made that allyl anions produced by the reductive metalation of allyl phenyl sulfides condense cleanly with α,β-unsaturated aldehydes in 1,2-fashion at the more substituted allyl terminus in the presence of ***Titanium Tetraisopropoxide***.[13] Use of the allylcerium reagent instead reverses the regioselectivity. Further, the stereochemical preference is *cis* at –78 °C (kinetic control) and *trans* when warmed to –40 °C (eq 4).[14] Grignard addition reactions to vinylogous esters are improved if promoted by $CeCl_3$ (eq 5).[15]

1. LDBB, –78 °C; 2. $CeCl_3$, –78 °C; 3. $H_2C{=}CHCHO$, –78 °C; 65%
1. LDBB, –78 °C; 2. $CeCl_3$, –78 °C → –40 °C; 3. $H_2C{=}CHCHO$, –78 °C; 48% (4)

$CeCl_3$, THF, 0 °C then H_3O^+, 75% (5)

Stereodivergent pathways relative to other organometallics have surfaced in the addition of organocerium(III) reagents to chiral α-keto amides (eq 6)[16] and 2-acyloxazolidines (eq 7).[17] Applications of this type hold considerable synthetic potential not only in additions to carbonyl compounds,[18] but also to oxime ethers,[19] oxiranes,[20] and oxazolidines.[21]

Conditions	Ratio
MeMgBr, Et_2O, –78 °C	76:24
MeMgBr, $CeCl_3$, THF, –78 °C	25:75

(6)

Conditions	Ratio
MeMgBr, Et_2O, –90 °C → –78 °C	94: 6
MeLi, $CeCl_3$, THF, –85 °C	16:84

(7)

The uniquely high reactivity of cerium reagents toward the often poorly electrophilic azomethine or C=N double bond of hydrazones has been extensively investigated. Smooth 1,2-addition occurs in good yield (67–81%) provided the air-sensitive hydrazones are acylated; when a chiral auxiliary is present, excellent diastereofacial control operates and enantiomerically enriched α-branched amines result (eq 8).[22] There appears to be a fundamental difference between 'warmed' reagents (–78 °C → 0 °C → –78 °C) and those generated and used at –78 °C. Sonication appears to facilitate conversion to the organocerate.[23] For SAMP hydrazones and related compounds the condensation is accommodating of a wide range of substitution both in the hydrazone and in the nucleophile. Subsequent reductive cleavage of the N–N bond with preservation of configuration can be performed by hydrogenolysis over W-2 ***Raney Nickel*** at 60 °C (free hydrazones; less preferred because of competing saturation of aromatic substituents if present) or by treatment with ***Lithium*** in liquid ammonia (acylated hydrazines).[22c]

R^1 = Me, CH_2CH_2OMe; 1. MeLi, $CeCl_3$, THF; 2. $ClCO_2Me$ or MeOH; 94% de
H_2, Raney Ni, 375 psi, 60 °C, 45%, R^1 = Me, R^2 = H
1. Li, NH_3, –33 °C; 2. NH_4Cl, 84%, $R^1 = CH_2CH_2OMe$, R^2 = OCOMe (8)

The preparation of (±)-1,3-diphenyl-1,3-propanediamine illustrates an alternative way in which this chemistry can be utilized (eq 9).[24] In all of the 1,2-addition reactions, the preferred reagent stoichiometry appears to be RLi:$CeCl_3$:hydrazone = 2:2:1. A detailed study of the consequences of varying the relative proportions of these constituents has indicated the optimal ratios of RLi:$CeCl_3$ to be 1:1, notwithstanding the fact that not all of the $CeCl_3$ is consumed in the transmetalation step.[25] The proposal has been made that the empirical formula of the cerate best approximates '$R_3CeCl_3Li_3$'.

Ph–CH=CH–CHO —(H_2NNH_2, EtOH, Δ)→ pyrazoline —(1. $(t\text{-BuOCO})_2O$, THF, rt; 2. PhLi, $CeCl_3$, THF, –78 °C)→ N-Boc pyrazolidine —(1. 50% CF_3CO_2H, CH_2Cl_2; 2. H_2, Raney Ni)→ H_2N/NH_2 diamine (9)

Cerium enolates, available by transmetalation of the lithium salts with $CeCl_3$ at –78 °C[26] or by reduction of α-bromo ketones with ***Cerium(III) Chloride–Tin(II) Chloride***,[27] give higher yields of crossed aldol products without altering the stereoselectivity. The $CeCl_3$–$SnCl_2$ reagent combination also acts on α,α′-dibromo ketones to give oxylallyl cations that can be captured in the usual way.[28]

Selective Reductions. Equimolar amounts of ***Sodium Borohydride***$_4$ and $CeCl_3 \cdot 7H_2O$ in methanol act on α,β-unsaturated ketones at rt or below to deliver allylic alcohols cleanly by 1,2-addition.[29] This widely used reducing agent[30] is not renowned for its diastereoselectivity[31] and asymmetric induction capabilities,[32] although exceptions are known.[33] Sometimes a stereofacial preference opposite that realized with other hydrides is encountered (eqs 10 and 11).[34]

(10)

$NaBH_4$, MeOH, –78 °C	7:93
$NaBH_4$, $CeCl_3$, MeOH, –78 °C	96: 4

(11) L-Selectride; $NaBH_4$, $CeCl_3$, MeOH, Na_2CO_3

Allylic alcohol products that are especially sensitive are known to undergo ionization and solvent capture under these mildly acidic conditions (eq 12).[35] The Lewis acidic character of $CeCl_3$ has also been used to advantage in the selective acetalization[6] of saturated aldehydes under the Luche conditions, thereby preventing their reduction.[2] Since ketones and conjugated aldehydes are less responsive, these functionalities are not transiently protected and suffer reduction (eq 13).[36] $NaBH_4$–$CeCl_3$ in MeCN transforms cinnamoyl chlorides into cinnamyl alcohols,[37] while ***Lithium Aluminum Hydride–Cerium(III) Chloride*** in hot DME or THF can effectively reduce alkyl halides and phosphine oxides.[5]

(12) $NaBH_4$, $CeCl_3$, MeOH

(13) $NaBH_4$, $CeCl_3$, EtOH, H_2O, –15 °C

Related Reagents. Lanthamum Chloride; Samarium(III) Chloride.

1. (a) Kagan, H. B.; Namy, J. L. *T* **1986**, *42*, 6573. (b) Molander, G. A. *CRV* **1992**, *92*, 29.
2. Gemal, A. L.; Luche, J.-L. *JOC* **1979**, *44*, 4187.
3. Imamoto, T.; Takiyama, N.; Nakamura, K.; Hatajima, T.; Kamiya, Y. *JACS* **1989**, *111*, 4392.
4. Ciganek, E. *JOC* **1992**, *57*, 4521.
5. (a) Imamoto, T.; Takeyama, T.; Kusumoto, T. *CL* **1985**, 1491. (b) Paquette, L. A.; Learn, K. S.; Romine, J. L.; Lin, H.-S. *JACS* **1988**, *110*, 879.
6. Luche, J.-L.; Gemal, A. L. *CC* **1978**, 976.
7. (a) Paquette, L. A.; DeRussy, D. T.; Cottrell, C. E. *JACS* **1988**, *110*, 890. (b) Paquette, L. A.; He, W.; Rogers, R. D. *JOC* **1989**, *54*, 2291.
8. Liu, H.-J.; Al-said, N. H. *TL* **1991**, *32*, 5473.
9. (a) Nagasawa, K.; Kanbara, H.; Matsushita, K.; Ito, K. *TL* **1985**, *26*, 6477. (b) Imamoto, T.; Kusumoto, T.; Tawarayama, Y.; Sugiura, Y.; Mita, T.; Hatanaka, Y.; Yokoyama, M. *JOC* **1984**, *49*, 3904.
10. (a) Suzuki, M.; Kimura, Y.; Terashima, S. *CL* **1984**, 1543. (b) Tamura, Y.; Sasho, M.; Ohe, H.; Akai, S.; Kita, Y. *TL* **1985**, *26*, 1549.
11. (a) Johnson, C. R.; Tait, B. D. *JOC* **1987**, *52*, 281. (b) Anderson, M. B.; Fuchs, P. L. *SC* **1987**, *17*, 621. (c) Narayanan, B. A.; Bunnelle, W. H. *TL* **1987**, *28*, 6261. (d) Lee, T. V.; Channon, J. A.; Cregg, C.; Porter, J. R.; Roden, F. S.; Yeoh, H. T.-L. *T* **1989**, *45*, 5877.
12. (a) Paquette, L. A.; DeRussy, D. T.; Gallucci, J. C. *JOC* **1989**, *54*, 2278. (b) Paquette, L. A.; DeRussy, D. T.; Vandenheste, T.; Rogers, R. D. *JACS* **1990**, *112*, 5562.
13. Cohen, T.; Guo, B.-S. *T* **1986**, *42*, 2803.
14. Guo, B.-S.; Doubleday, W.; Cohen, T. *JACS* **1987**, *109*, 4710.
15. Crimmins, M. T.; Dedopoulou, D. *SC* **1992**, *22*, 1953.
16. Fujisawa, T.; Ukaji, Y.; Funabora, M.; Yamashita, M.; Sato, T. *BCJ* **1990**, *63*, 1894.
17. Ukaji, Y.; Yamamoto, K.; Fukui, M.; Fujisawa, T. *TL* **1991**, *32*, 2919.
18. Kawasaki, M.; Matsuda, F.; Terashima, S. *TL* **1985**, *26*, 2693.

19. Ukaji, Y.; Kume, K.; Watai, T.; Fujisawa, T. *CL* **1991**, 173.
20. (a) Vougioukas, A. E.; Kagan, H. B. *TL* **1987**, *28*, 6065. (b) Schaumann, E.; Kirschning, A. *TL* **1988**, *29*, 4281. (c) Cohen, T.; Jeong, I.-H.; Mudryk, B.; Bhupathy, M.; Awad, M. M. A. *JOC* **1990**, *55*, 1528. (d) Marczak, S.; Wicha, J. *SC* **1990**, *20*, 1511.
21. Pridgen, L. N.; Mokhallalati, M. K.; Wu, M.-J. *JOC* **1992**, *57*, 1237.
22. (a) Denmark, S. E.; Weber, T.; Piotrowski, D. W. *JACS* **1987**, *109*, 2224. (b) Weber, T.; Edwards, J. P.; Denmark, S. E. *SL* **1989**, 20. (c) Denmark, S. E.; Nicaise, O.; Edwards, J. P. *JOC* **1990**, *55*, 6219.
23. Greeves, N.; Lyford, L. *TL* **1992**, *33*, 4759.
24. Denmark, S. E.; Kim, J.-H. *S* **1992**, 229.
25. Denmark, S. E.; Edwards, J. P.; Nicaise, O. *JOC* **1993**, *58*, 569.
26. Imamoto, T.; Kusumoto, T.; Yokoyama, M. *TL* **1983**, *24*, 5233.
27. Fukuzawa, S.; Tsuruta, T.; Fujinami, T.; Sakai, S. *JCS(P1)* **1987**, 1473.
28. Fukuzawa, S.; Fukushima, M.; Fujinami, T.; Sakai, S. *BCJ* **1989**, *62*, 2348.
29. (a) Luche, J.-L. *JACS* **1978**, *100*, 2226. (b) Luche, J.-L.; Rodriquez-Hahn, L.; Crabbé, P. *CC* **1978**, 601.
30. Examples: Godleski, S. A.; Valpey, R. S. *JOC* **1982**, *47*, 381. Block, E.; Wall, A. *JOC* **1987**, *52*, 809. Marchand, A. P.; LaRoe, W. D.; Sharma, G. V. M.; Suri, S. C.; Reddy, D. S. *JOC* **1986**, *51*, 1622. Rubin, Y.; Knobler, C. B.; Diederich, F. *JACS* **1990**, *112*, 1607.
31. (a) Danishefsky, S. J.; DeNinno, M. P.; Chen, S. *JACS* **1988**, *110*, 3929. (b) DeShong, P.; Waltermire, R. E.; Ammon, H. L. *JACS* **1988**, *110*, 1901. (c) Abelman, M. M.; Overman, L. E.; Tran, V. D. *JACS* **1990**, *112*, 6959.
32. (a) Boutin, R. H.; Rapoport, H. *JOC* **1986**, *51*, 5320. (b) Paterson, I.; Laffan, D. D. P.; Rawson, D. J. *TL* **1988**, *29*, 1461. (c) Coxon, J. M.; van Eyk, S. J.; Steel, P. J. *T* **1989**, *45*, 1029.
33. (a) Nimkar, S.; Menaldino, D.; Merrill, A. H.; Liotta, D. *TL* **1988**, *29*, 3037. (b) Rücker, G.; Hörster, H.; Gajewski, W. *SC* **1980**, *10*, 623.
34. (a) Krief, A.; Surleraux, D. *SL* **1991**, 273. (b) Kumar, V.; Amann, A.; Ourisson, G.; Luu, B. *SC* **1987**, *17*, 1279.
35. Scott, L. T.; Hashemi, M. M. *T* **1986**, *42*, 1823.
36. Replacement of MeOH by DMSO reverses this chemoselectivity: Adams, C. *SC* **1984**, *14*, 1349.
37. Lakshmy, K. V.; Mehta, P. G.; Sheth, J. P.; Triverdi, G. K. *OPP* **1985**, *17*, 251.

Leo A. Paquette
The Ohio State University, Columbus, OH, USA

Cesium Carbonate

Cs_2CO_3

[534-17-8] CCs_2O_3 (MW 325.82)

(used in the synthesis of crown ethers and macrocycles;[1] catalyzes the Horner–Emmons cyclization;[2] used as base for alkylations[3])

Physical Data: mp 610 °C (dec).
Solubility: sol water, alcohol, ether.
Form Supplied in: white powder; widely available.
Purification: crystallized from ethanol (10 mL g^{-1}) by slow evaporation.
Handling, Storage, and Precautions: irritant; handle with gloves; hygroscopic; keep away from moisture; flush eyes or skin with copious amounts of water in case of contact; possible mutagen; avoid inhalation.

Macrocyclization Reagent. Crown ethers, generally used as ionophores for various applications, can be prepared by using Cs_2CO_3. This reagent serves as both a base and a cation template in the macrocylization of dicarboxylic acids and α,ω-alkyl dihalides to generate crown ethers.[1,5] For example, a crown ether is produced in 85% yield when dicesium pyridine-3,5-dicarboxylate, generated from the acid and Cs_2CO_3, reacts with 1,11-dibromo-3,6,9-trioxoundecane in DMF upon heating the reaction mixture between 60–70 °C for 48 h (eq 1). Under similar conditions, aromatic dicarboxylic acids afford related crown ethers. This procedure also works well for crown ethers prepared from salicylic acid and bromopolyethylene glycols.[1]

Cs_2CO_3, DMF, 48 h (1)

Thia crown ethers can also be prepared with Cs_2CO_3. Yields for these reactions are consistently good (45–95%) when the cesium salts of α,ω-dithiols are treated with α,ω-dihalides in DMF (eq 2).[5] Much lower yields (1–20%) are obtained if other methods, such as high dilution techniques, are used in the absence of a cesium counterion. Similar chemistry has been used to prepare functionalized cyclophane derivatives by replacing the α,ω-dihalide with 1,3-dichloroacetone and using *o*-xylylenedithiol as the α,ω-dithiol.[6]

Cs_2CO_3, DMF, 12 h, 50 °C; $(CH_2)_6$ (2)

Chiral crown ethers have also been prepared by using Cs_2CO_3. A chiral diol and chiral ditosylate, both prepared from (*R,R*)-(+)-hydrobenzoin, have been coupled using NaH as a base and Cs_2CO_3 as a template in DMF (eq 3).[7]

NaH, Cs_2CO_3, DMF; 80 °C, 12 h; 120 °C, 48 h; 9% (3)

Tosylamides have been used to prepare aza macrocycles with Cs_2CO_3. The bicarbonate salt ($CsHCO_3$) is not basic enough to deprotonate ditosylamides.[8] Also, carboxamides and urethanes do not react as efficiently with Cs_2CO_3 as do tosylamides. Among the alkali bases, Cs_2CO_3 provided consistently higher yields than ***Lithium Carbonate***, ***Sodium Carbonate***, ***Potassium Carbonate*** and Rb_2CO_3. A typical example involved treating 1,10-bis[(*p*-tolylsulfonyl)amino]decane with two equivalents of Cs_2CO_3 in dry DMF followed by the addition of 1,6-dibromohexane and stirring the mixture at room temperature for 24 h to give the aza macrocycle in 72% yield. Reduction with sodium amalgam in buffered methanol gave the cyclic diamine (eq 4).

TsNH–[CH2]5… NHTs; 1. Cs_2CO_3, DMF; 2. $Br(CH_2)_6Br$ → N,N'-ditosyl macrocycle ($(CH_2)_8$) ; Na(Hg), Na_2HPO_4, MeOH → NH macrocycle (4)

Cs_2CO_3 has been used to prepare derivatized calix[4]arenes for use as lipophilic, water-soluble, and ionophoric receptors.[9] The calix[4]arene cone structure depends on the type of base used for alkylation of the phenolic hydroxyl groups. For example, a cone is generated when NaH is the base during alkylation of the calix[4]arene with ethyl bromoacetate in acetone. When Cs_2CO_3 is used, a 1,3-alternate cone is obtained, whereas a partial cone is formed when K_2CO_3 is used.[9]

Carcerands and carceplexes have been prepared by using Cs_2CO_3. For example, mixing a tetrathiol, a tetrachloride, and pulverized Cs_2CO_3 in DMF/THF under high dilution conditions afforded the carcerand.[10] FAB-MS of the product indicated that the cesium ion served as a template.

Lactonizations can be accomplished using Cs_2CO_3 and ω-bromocarboxylic acids at 40 °C in DMF.[4b,11] These conditions afford mixtures of lactones and diolides with their product distribution depending on the ring size. Cs_2CO_3 has been used in a macrocyclization reaction to generate steroids under anionic conditions.[12]

Horner–Emmons Base. The Horner–Wadsworth–Emmons reaction is sensitive when the carbonyl component contains acidic hydrogens, such as hydroxyl groups. This problem can be circumvented by using Cs_2CO_3 as the base. The method is somewhat dependent on the type of solvent used and the degree of hydration of Cs_2CO_3 (eq 5).[2] NaH has also been replaced with Cs_2CO_3 in a synthesis of D-*erytho*-C_{18}-sphingosine using a Wittig–Horner reaction.[13] A tandem Wittig–Michael reaction in which a lactol is treated with dimethyl phosphonoacetate in the presence of Cs_2CO_3 has been reported (eq 6).[14] An intermolecular Michael addition of a cyclic β-keto ester to an α,β-ynone has been accomplished with Cs_2CO_3 in 89% yield.[15] An intramolecular Diels–Alder reaction using Cs_2CO_3 and paraformaldehyde in THF has been used to produce the neurotoxic toxin slaframine from an amide.[16]

pyrrole-2-CHO + $(EtO)_2P(O)CH_2CO_2Et$ —(Cs_2CO_3, 1,4-dioxane, H_2O, 70 °C, 86%)→ (E)-3-(pyrrol-2-yl)acrylate ethyl ester (5)

oxabicyclic lactol (OH) —($(MeO)_2POCH_2CO_2Me$, Cs_2CO_3, 73%)→ tetrahydrofuran with CH_2CO_2Me —(500 °C)→ dihydrofuran–CH_2CO_2Me (6)

Alykylation Base. Esterification under mild conditions can be accomplished with Cs_2CO_3 or $CsHCO_3$.[3] For example, amino acid and peptide esters can be made by generating the cesium salt of the corresponding carboxylic acid followed by treating the salt with an alkyl halide in DMF (eq 7).

$$\text{Boc-Phe-Met-OH} \xrightarrow[\text{DMF, 45 °C}]{Cs_2CO_3} \text{Boc-Phe-Met-O-Cs} \xrightarrow[\text{DMF}]{\text{BnBr}} \underset{86\%}{\text{Boc-Phe-Met-OBn}} \quad (7)$$

Mono-*C*-alkylation of β-dicarbonyl compounds is complicated by competitive *O*-alkylation and dialkylation reactions.[17] *C*-Alkylation of 2,4-pentanedione has been achieved with ***Iodomethane*** in 75% yield using K_2CO_3 as a base. The use of Cs_2CO_3 by itself, or in combination with K_2CO_3, allows for improved monoalkylation reactions using higher molecular weight alkyl halides (eq 8).[18] Phenolates generated with Cs_2CO_3 are alkylated with ***Dimethyl Sulfate*** in 94% yield.[19]

$$BrCH_2CO_2Et + (MeCO)_2CH_2 \xrightarrow[CH_2Cl_2]{Cs_2CO_3} \underset{92\%}{(MeCO)_2CHCH_2CO_2Et} + \underset{8\%}{MeCOCH{=}CMeOCH_2CO_2Et} + \underset{<1\%}{MeCOC(CH_2CO_2Et){=}CMeOCH_2CO_2Et} \quad (8)$$

Dibenzyl carbonate is prepared in 63–70% yield when Cs_2CO_3 is treated with ***Benzyl Chloride*** at 100–110 °C for 12 h.[20] Cs_2CO_3 has also been used to mediate Michael additions of glycine esters. Treating a Schiff base of glycine and an enone with 10 mol % Cs_2CO_3 gives substituted prolines in 60–78% yield after hydrogenation (eq 9).[21] Replacement of NaH with 5–10 mol% Cs_2CO_3 under standard Schmidt conditions in the preparation of trichloroacetimidates allows for large scale preparation of O-α-cellobiosyl trichloroacetimidate heptacetate in good yield.[22]

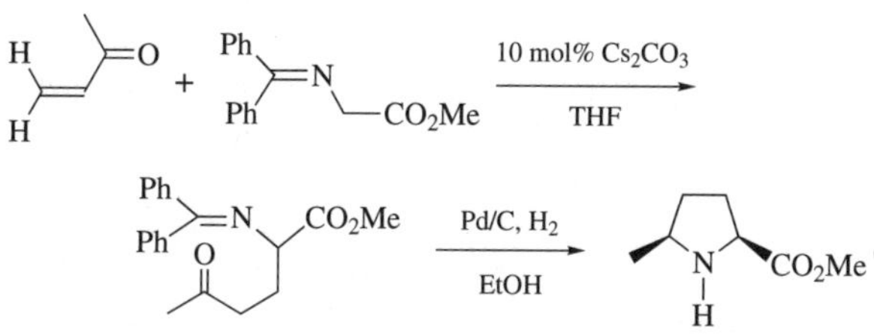

$$\text{(9)}$$

Cleavage of aryl esters has been accomplished using either Cs_2CO_3 or $CsHCO_3$. For example, refluxing resorcinol dibenzoate in DME for 24 h with 1.5 equiv Cs_2CO_3 produces the monobenzoate in greater than 95% yield.[23] Cs_2CO_3 and ***Di-t-butyl Dicarbonate*** cleave 2-oxazolidinones at 25 °C in MeOH to give Boc-amino alcohols in 70% yield (eq 10).[24]

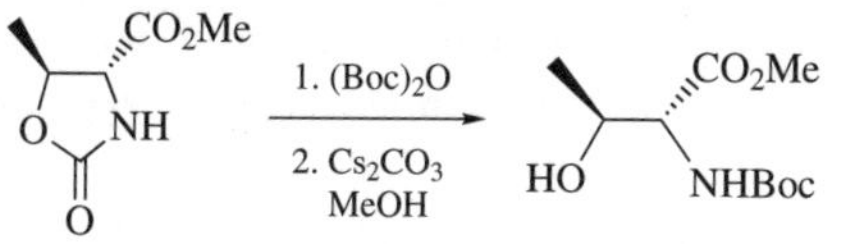

$$\text{(10)}$$

Related Reagents. Calcium Carbonate; Lithium Carbonate; Potassium Carbonate; Sodium Carbonate.

1. van Keulen, B. J.; Kellogg, R. M.; Piepers, O. *CC* **1979**, 285.
2. (a) Mouloungui, Z.; Murengezi, I.; Delmas, M.; Gaset, A. *SC* **1988**, *18*, 1241. (b) Mouloungui, Z.; Delmas, M.; Gaset, A. *JOC* **1989**, *54*, 3936.
3. Wang, S.-S.; Gisin, B. F.; Winter, D. P.; Makofske, R.; Kulesha, I. D.; Tzougraki, C.; Meienhofer, J. *JOC* **1977**, *42*, 1286.
4. (a) Piepers, O.; Kellogg, R. M. *CC* **1978**, 383. (b) Kruizinga, W. H.; Kellogg, R. M. *CC* **1979**, 286.
5. (a) Buter, J.; Kellogg, R. M. *CC* **1980**, 466. (b) Vögtle, F.; Klieser, B. *S* **1982**, 294.
6. Chiu, J.-J.; Grewal, R. S.; Hart, H.; Ward, D. L. *JOC* **1993**, *58*, 1553.
7. Crosby, J.; Stoddart, J. F.; Sun, X.; Venner, M. R. W. *S* **1993**, 141.
8. Vriesema, B. K.; Buter, J.; Kellogg, R. M. *JOC* **1984**, *49*, 110.
9. (a) Iwamoto, K.; Fujimoto, K.; Matsuda, T.; Shinkai, S. *TL* **1990**, *31*, 7169. (b) Shinkai, S.; Fujimoto, K.; Otsuka, T; Ammon, H. L. *JOC* **1992**, *57*, 1516.
10. (a) Cram, D. J.; Karbach, S.; Kim, Y. H.; Baczynskyj, L.; Marti, K.; Sampson, R. M.; Kalleymeyn, G. *JACS* **1988**, *110*, 2554. (b) Cram, D. J.; Karbach, S.; Kim, Y. H.; Baczynskyj, L.; M.; Kalleymeyn, G. W. *JACS* **1985**, *107*, 2575.
11. (a) Barbier, M. *CC* **1982**, 668. (b) Kruizinga, W. H.; Kellogg, R. M. *JACS* **1981**, *103*, 5183.
12. Lavallée, J.-F.; Deslongchamps, P. *TL* **1988**, *29*, 6033.
13. Yamanoi, T.; Akiyama, T.; Ishida, E.; Abe, H.; Amemiya, M.; Inazu, T. *CL* **1989**, 335.
14. Bloch, R.; Seck, M. *TL* **1987**, *28*, 5819.
15. (a) Lavallée, J.-F.; Berthiaume. G.; Deslongchamps, P.; *TL* **1986**, *27*, 5455. (b) Lavallée, J.-F.; Deslongchamps, P. *TL* **1987**, *28*, 3457.
16. Gobao, R. A.; Bremmer, M. L.; Weinreb, S. M. *JACS* **1982**, *104*, 7065.
17. House, H. O. *Modern Synthetic Reactions*, 2nd ed.; Benjamin: Menlo Park, CA, 1972.
18. Shrout, D. P.; Lightner, D. A. *SC* **1990**, *20*, 2075.
19. Winters, R. T.; Sercel A. D.; Showalter, H. D. H. *S* **1988**, 712.
20. Cella J. A.; Bacon, S. W. *JOC* **1984**, *49*, 1122.
21. van der Werf, A.; Kellogg, R. M. *TL* **1991**, *32*, 3727.
22. Urban F. J.; Moore, B. S.; Breitenbach, R. *TL* **1990**, *31*, 4421.
23. Zaugg, H. E. *JOC* **1976**, *41*, 3419.
24. Ishizuka, T.; Kunieda, T. *TL* **1987**, *28*, 4185.

Mark R. Sivik
Procter & Gamble, Cincinnati, OH, USA

Cesium Fluoride

[13400-13-0] CsF (MW 151.90)

(fluoride ion source for removal of silyl protecting groups; reagent for mild desilylative anion and ylide formation; base catalyst; reagent for hydrogen bond-assisted alkylation; reagent for halide exchange)

Physical Data: mp 682 °C; bp 1251 °C; d 4.115 g cm^{-3}; refractive index 1.478 (18 °C).
Solubility: sol H_2O and MeOH; insol dioxane, pyridine.
Form Supplied in: white deliquescent crystalline powder.
Purification: dry by heating at 100 °C for 2 h in vacuo.[1]
Handling, Storage, and Precautions: hygroscopic. Keep container tightly closed. Harmful if inhaled or swallowed. Causes irritation of skin, eyes, and mucous membranes and may cause allergic reaction. Incompatible with acids: store away from acids. Use in a fume hood with safety goggles and chemically resistant gloves and clothing.

Catalysis of Condensations of Carbonyl Compounds. Knoevenagel condensations of cyclohexanone and benzaldehyde with three active-methylene partners have been studied using various fluoride catalysts. Cesium and rubidium counterions lead to higher yields than the less soluble potassium, sodium, and lithium salts.[2] Silyl enol ethers react with aldehydes and ketones with CsF catalysis to form α,β-unsaturated ketones, or with unsaturated ketones by conjugate addition.[3] Trimerization of isocyanates occurs at 130 °C with CsF catalysis, forming aromatic isocyanurates in high yields.[4]

Desilylative Elimination and Carbanion Formation. Fluoride ion desilylates compounds of the type C–Si, leading to useful carbanion-like reactivity, particularly when appropriate stabilizing substituents are present. When β-substituents are reasonable leaving groups, the expected eliminations occur. These types of reactions, with which water interferes, are particularly amenable to the use of CsF, since CsF can be made anhydrous much easier than the more commonly encountered ***Tetra-n-butylammonium Fluoride***.

Cyclopropenes have been prepared by desilylation of 1,1-dichloro-2-trimethylsilylcyclopropane with CsF in diglyme at 80 °C (eq 1).[5] A similar example of an advantage of CsF (and fluoride ion sources in general) – mild elimination conditions allowing isolation of unstable species – is seen in a synthesis of allene oxides. 1-(*t*-Butyl)allene oxide can be obtained in 55% yield by treating the precursor trimethylsilyloxirane with CsF (eq 2) and trapping the volatile products in a cold trap.[6] Desilylative 1,2-elimination from cyclohexadienes has been used to obtain the interesting benzene isomers 1,2,3-cyclohexatriene and cyclohexen-3-yne, which were trapped in situ by Diel–Alder reactions.[7] Sulfine, sulfene, and alkylsulfenes can be trapped as Diels–Alder adducts after their generation by desilylative 1,2-elimination from trimethylsilylmethanesulfinyl chloride, trimethylsilylmethanesulfonyl chloride, and trimethylsilylalkanesulfonic anhydrides, respectively.[8]

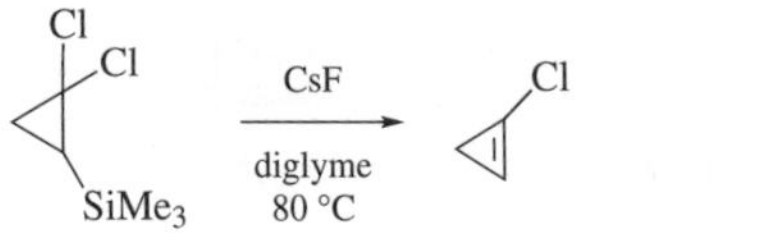

CsF, diglyme, 80 °C (1)

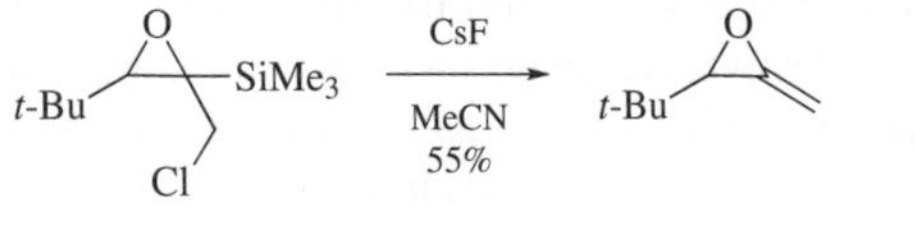

CsF, MeCN, 55% (2)

CsF-induced 1,4-elimination produces xylylene intermediates which undergo intramolecular cycloaddition with an alkene (eq 3).[9] Similar 1,4-elimination/Diels-Alder strategies have been used in the synthesis of other targets,[10] and in a cyclodimerization to give the bisthioacetal of [2,3:6,7]dibenzo-1,5-cyclooctanedione.[11]

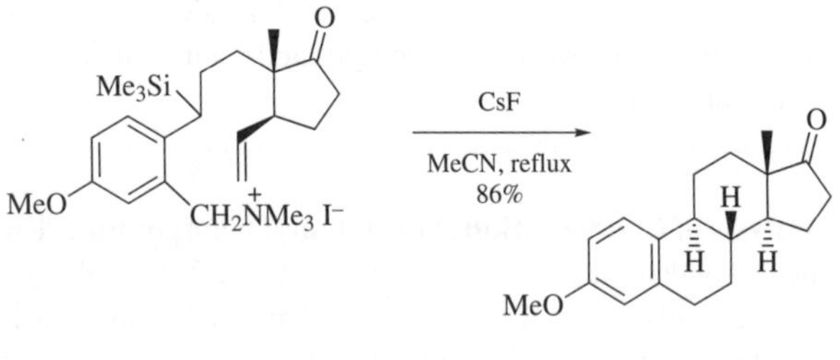

CsF, MeCN, reflux, 86% (3)

Stereoselective synthesis of 1,4-dienamine derivatives is achieved by CsF-induced desilylative ring-opening of tetrahydropyridinium salts. The diene is produced selectively as the (*Z,E*) diastereomer (eq 4). The authors contrast this behavior with that of an analogous tetrahydropyridine *N*-oxide, which undergoes spontaneous sila-Cope rearrangement to give the (*Z,Z*) diastereomer (eq 5).[12]

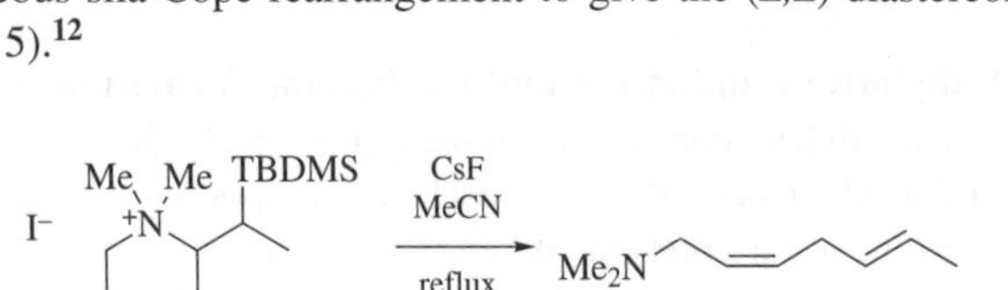

CsF, MeCN, reflux, 77% (4)

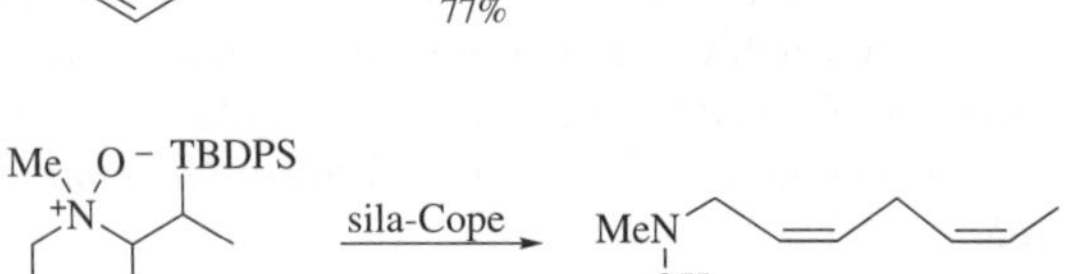

sila-Cope (5)

Numerous desilylative reactions of trimethyl(α-bromobenzyl)silanes with electrophiles are induced by CsF (eq 6).[13]

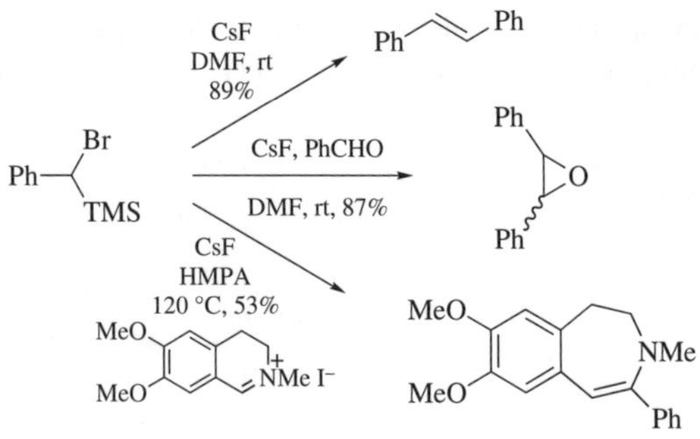

CsF, DMF, rt, 89%; CsF, PhCHO, DMF, rt, 87%; CsF, HMPA, 120 °C, 53% (6)

Acyl-, alkynyl-, benzyl-, allyl-, oxiranyl-,[14] heteroaryl-,[15] trichloromethyl-,[16] aryl-,[17] β-keto-,[18] β-sulfonyl-,[19] and cyclopropylsilanes[20] are useful carbanion precursors. When treated with CsF or other fluoride salts, they undergo desilylative addition to carbonyl compounds, including CO_2, aldehydes, ketones, and lactones; *C*-alkylation with ***Benzyl Bromide*** also can occur (eqs 7–11). Conjugate addition is also observed; nitroalkenes give 1,4-addition products, and reaction of allylsilane with cyclohexenone in the presence of CsF gives the 1,4-addition product selectively, whereas the presence of tetrabutylammonium fluoride leads to diallylated product.[21]

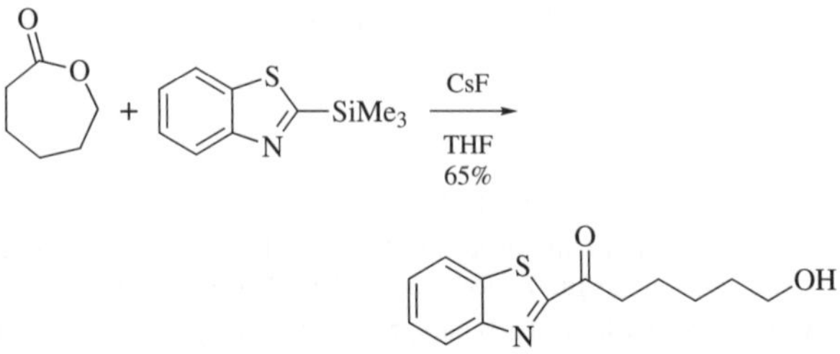

CsF, THF, 65% (7)

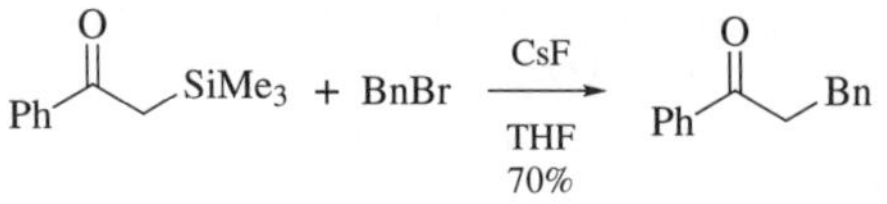

CsF, THF, 70% (8)

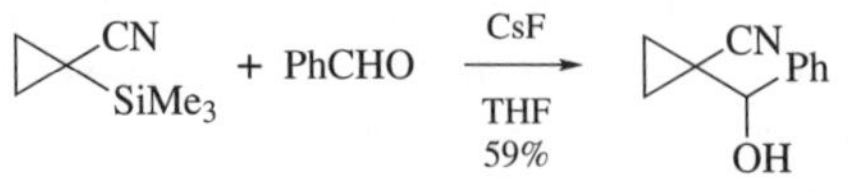

CsF, THF, 59% (9)

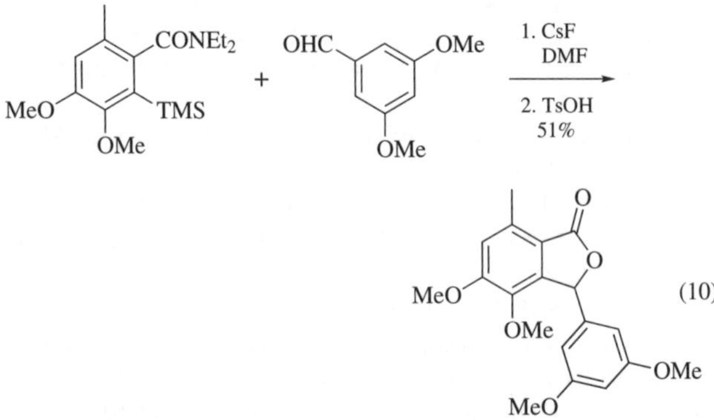

1. CsF, DMF; 2. TsOH, 51% (10)

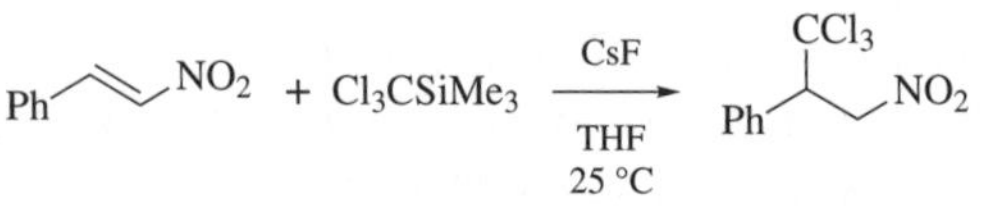

CsF, THF, 25 °C (11)

Cyclization of a carbanion produced by desilylation with CsF leads to 1,3-thiazolidines (eq 12).[22]

(12) — CsF, MeCN, rt, 25 h

Anionic reactivity remote from the silyl bond cleavage site is observed when ketene bis(trimethylsilyl) acetals or α,β-unsaturated ketene bis(trimethylsilyl) acetals are treated with CsF and an aldehyde.[23] Complementary regioselectivity occurs with ***Titanium(IV) Chloride*** (eq 13).[24]

(13) — PhCHO, CsF, THF, 78%; PhCHO, $TiCl_4$, CH_2Cl_2, 78%

A novel method for the production of thiiranes from carbonyl compounds involves desilylation with CsF (eq 14).[25]

(14) — $TMSCH_2$; RCHO, CsF, MeCN

Stannyl Anion Generation. By treatment of (trimethylsilyl)tributylstannane with CsF, a stannyl anion is formed. This method was used to achieve intramolecular reaction of a vinyl iodide with an ester, forming a spirocyclopentenone, which reacted further with the tributylstannyl anion under the reaction conditions (eq 15). Four subsequent steps completed a formal total synthesis of acorone.[26]

(15) — CO_2Me; $TMSSnBu_3$, CsF, DMF, 58%; $SnBu_3$; several steps; Acorone

Ylide Formation. Wittig reaction of a phosphonium salt with benzaldehyde is observed with CsF in DMF.[27] Similarly, CsF can be used to carry out Horner–Wadsworth–Emmons reactions.[28] Sulfonium, ammonium, and phosphonium ylides can be obtained by desilylation.[29] This ylide formation strategy has been used in a synthesis of the retronecine skeleton.[30]

Acylation and Transesterification. Esters, amides, and thioesters are obtained in high yields under neutral conditions when the carboxylic acid is treated with the appropriate protic nucleophile, 1-ethyl-2-fluoropyridinium tetrafluoroborate, and CsF in methylene chloride (eq 16). *p*-Nitrophenyl esters of optically pure *N*-phthaloylamino acids are converted to methyl esters in MeOH/CsF without the racemization observed when triethylamine is employed.[31]

(16) — CsF, CH_2Cl_2, 2-fluoro-1-ethylpyridinium, 70%

Acylations of 2,3,4,6-tetra-*O*-benzyl-D-glucopyranose in quantitative yields are achieved with acyl fluorides and CsF (eq 17). High ratios of either anomeric product can be obtained by changing the order of addition of the reagents.[32]

(17) — PhCOF, CsF, CH_2Cl_2, 98%

normal addition: α:β = 10:90
inverse addition: α:β = 75:25

Phosphate esters can be transesterified in alcohols containing excess CsF.[33] This method has been used to advantage in the regioselective protection of hydroquinones by phosphorylation in the presence of CsF.[34] Ribooligonucleotides have also been synthesized by this approach.[35]

Acylation of alcohols is very efficient with acylthiazolidine-2-thiones and CsF in warm DMF. Benzyl 3-phenylpropionate was obtained in 99% yield by this method (eq 18).[36]

(18) — BnOH, CsF, DMF, 80 °C, 99%

Hydrogen Bond-Assisted Halide Displacements. Treatment of catechols with CH_2Cl_2 in the presence of CsF results in high yields of methylenated products (eq 19) with little competition from intermolecular condensations.[37] Hydrogen bonding of the catechol with fluoride has been demonstrated, and is thought to be responsible for the facilitation of the reaction. Numerous other H-bonding substrates are also alkylated more readily under the influence of fluoride (eq 20).[38] This alkylation promotion by CsF obviates prior conversion of phthalimide to its alkali metal derivative for use in the Gabriel synthesis (eq 21).[39] CsF leads to an order of magnitude in rate enhancement relative to ***Potassium Fluoride*** in some cases, although the higher expense of CsF may negate this advantage.

(19) — + CH_2Cl_2; CsF, DMF, 110 °C, 98%

(20) — RCO_2H + $BrCH_2COPh$; CsF, DMF, 25 °C, 10 min, 91–99%

M	Time (h)	Yield (%)
K	4	73
Rb	1	82
Cs	0.2	82

1,2-Cycloalkanediols can also be methylenated under analogous conditions.[40] Glycols can be monoalkylated by treatment of their stannylene acetals with alkyl halides. CsF considerably improves the yields of these reactions.[41] Thus the dibutylstannylene acetal of dimethyl L-tartrate was monoalkylated with benzyl iodide and CsF in DMF at room temperature in 99% yield (eq 22), whereas the reaction without CsF required elevated temperatures and gave only a 13% yield. This method has recently been applied in syntheses of lepidimoide[42] and *myo*-inositol phosphates.[43] In a related reaction, glycol xanthates react with alcohols in the presence of CsF to provide monoalkylated glycols as a key step in syntheses of xylofuranosides.[44]Phosphoric acid alkylation by alkyl halides is catalyzed by CsF.[45] Several macrocyclic crown ethers have been prepared by reaction of ditosylated half-crowns with diol half-crowns in the presence of CsF.[46]

Synthesis of Acid Fluorides. Organic halides can be converted to acid fluorides by metal-catalyzed (M = Pd, Pt, Co, Rh) carbonylation in the presence of fluoride salts (eq 23). CsF is superior to fluorides with other counterions.[47]

Preparation of Alumina-Supported Fluoride Reagents. Mixing of aqueous CsF or KF with alumina followed by drying gave an effective base catalyst.[48] Its preparation was optimized by testing the activity in promoting Michael addition of nitroethane to buten-2-one and methylation of phenols with MeI.

Activation of Si–X Bonds. Silicon–hydrogen, –oxygen, and –nitrogen bonds are activated by fluoride ions under heterogeneous conditions. ***Triethoxysilane*** is a mild, efficient and highly selective reducing agent in the presence of CsF, leading to quantitative yields of 1,2-reduction products from enones (eq 24). Other bifunctional compounds also can be reduced with quantitative regioselectivity.[49]

A system of CsF and $Si(OR)_4$ effects selective 1,4-addition in condensation of ketones with activated alkenes in good yield (eq 25). A silyl enol ether is formed in situ via reaction of CsF-activated $Si(OR)_4$ with the ketone.[49] This system has also been used to effect 1,4-addition to unsaturated amides.[50] Although bis(silyl)enamines undergo sluggish reaction, if any, with various electrophiles, the presence of CsF or tributylammonium fluoride allows high-yielding reactions with carbonyl compounds to afford 2-aza-1,3-dienes (eq 26).[51]

Benzostabase Protecting Group. Aromatic and aliphatic primary amines have been protected as their benzostabase derivatives in high yield by heating with 1,2-bis(dimethylsilyl)benzene and CsF in polar aprotic solvents (eq 27).[52]

Desilylation of *O*-Silyl and *N*-Silyl Compounds. *O*-Silyl compounds serve frequently as protecting functionalities in organic synthesis. CsF is sometimes used for cleavage of these Si–O bonds, but usually has no advantage over the more commonly used tetrabutylammonium fluoride. Cleavage of O–Si bonds with CsF has been used in the preparation of cesium selenocarboxylates.[53] CsF can be used to cleave *N*-trimethylsilyl groups from amides[54] and carbodiimides.[55]

Fluorodemetalation of Organogermanium, -tin, -lead, and -boron Compounds. Cesium fluoride can be used to cleave C–M bonds; evidence for an associative mechanism in this process has been obtained.[56] Similar reactivity is observed in the reaction of organotin carboxylates and stannyl ethers with CsF and alkyl halides; the O–Sn bond is converted to O–C.[57] A borane–amine complex has been decomposed to the free amine by refluxing with CsF in Na_2CO_3/EtOH.[58]

CsF-Mediated Claisen Rearrangement. Aryl propargyl ether Claisen rearrangement leads to an intermediate allene which undergoes intramolecular capture by the concurrently formed phenol to give a pyran or a furan. A comparison of the effects of various salts on the regioselectivity revealed that CsF promotes the furan formation, despite the fact that *both* CsCl and KF promote pyran formation. The reaction also was applied in a synthesis of chelerythrine (eq 28).[59]

(28)

Chelerythrine

Alkyl Fluoride Synthesis. *n*-Octyl fluoride has been made in 77% yield (along with 15% of *n*-octane) by treatment of *n*-octyl bromide with CsF and catalytic NBu_4Br at 85 °C.[60]

Related Reagents. Diphenylsilane–Cesium Fluoride; Phenylsilane–Cesium Fluoride; Potassium Fluoride; Sodium Fluoride; Tetra-*n*-butylammonium Fluoride; (Trimethylsilyl)methanesulfonyl Chloride–Cesium Fluoride.

1. Corriu, R. J. P.; Moreau, J. J. E.; Pataud-Sat, M. *JOC* **1990**, *55*, 2878.
2. Rand, L.; Swisher, J. V.; Cronin, C. J. *JOC* **1962**, *27*, 3505.
3. Boyer, J.; Corriu, R. J. P.; Perz, R.; Reye, C. *JOM* **1980**, *184*, 157.
4. Nambu, Y.; Endo, T. *JOC* **1993**, *58*, 1932.
5. Chan, T. H.; Massuda, D. *TL* **1975**, 3383.
6. Chan, T. H.; Ong, B. S. *JOC* **1978**, *43*, 2994.
7. Shakespeare, W. C.; Johnson, R. P. *JACS* **1990**, *112*, 8578.
8. (a) Block, E.; Aslam, M. *TL* **1982**, *23*, 4203. (b) Block, E.; Wall, A. *TL* **1985**, *26*, 1425.
9. Ito, Y.; Nakatsuka, M.; Saegusa, T. *JACS* **1981**, *103*, 476.
10. (a) Djuric, S.; Sarkar, T.; Magnus, P. *JACS* **1980**, *102*, 6885. (b) Marino, J. P.; Dax, S. L. *JOC* **1984**, *49*, 3671. (c) Ito, Y.; Miyata, M.; Nakatsuka, M.; Saegusa, T. *JACS* **1981**, *103*, 5250.
11. Ito, Y.; Nakajo, E.; Sho, K.; Saegusa, T. *S* **1985**, 698.
12. Bac, N. V.; Langlois, Y. *JACS* **1982**, *104*, 7666.
13. Kessar, S. V.; Singh, P.; Kaur, N. P.; Chawla, U.; Shukla, K.; Aggarwal, P.; Venugopal, D. *JOC* **1991**, *56*, 3908.
14. Ricci, A.; Fiorenza, M.; Grifagni, M. A.; Bartolini, G.; Seconi, G. *TL* **1982**, *23*, 5079; and references therein.
15. Effenberger, F.; Spiegler, W. *CB* **1985**, *118*, 3872.
16. Cunico, R. F.; Zhang, C. *SC* **1991**, *21*, 2189.
17. Mills, R. J.; Snieckus, V. *JOC* **1989**, *54*, 4386.
18. Fiorenza, M.; Mordini, A.; Papaleo, S.; Pastorelli, S.; Ricci, A. *TL* **1985**, *26*, 787.
19. Eisch, J. J.; Behrooz, M.; Dua, S. K. *JOM* **1985**, *285*, 121.
20. (a) Ohno, M.; Tanaka, H.; Komatsu, M.; Ohshiro, Y. *SL* **1991**, 919. (b) Blankenship, C.; Wells, G. J.; Paquette, L. A. *T* **1988**, *44*, 4023. (c) Paquette, L. A.; Blankenship, C.; Wells, G. J. *JACS* **1984**, *106*, 6442.
21. Ricci, A.; Fiorenza, M.; Grifagni, M. A.; Bartolini, G.; Seconi, G. *TL* **1982**, *23*, 5079.
22. Hosomi, A.; Hayashi, S.; Hoashi, K.; Kohra, S.; Tominaga, Y. *JOC* **1987**, *52*, 4423.
23. Bellassoued, M.; Gaudemar, M. *TL* **1988**, *29*, 4551.
24. Bellassoued, M.; Ennigrou, R.; Gaudemar, M. *JOM* **1990**, *393*, 19.
25. Tominaga, Y.; Ueda, H.; Ogata, K.; Kohra, S.; Hojo, M.; Ohkuma, M.; Tomita, K.; Hosomi, A. *TL* **1992**, *33*, 85.
26. Mori, M.; Isono, N.; Kaneta, N.; Shibasaki, M. *JOC* **1993**, *58*, 2972.
27. Umemoto, T.; Gotoh, Y. *BCJ* **1991**, *64*, 2008.
28. Kawashima, T.; Ishii, T.; Inamoto, N. *BCJ* **1987**, *60*, 1831.
29. (a) Vedejs, E.; Martinez, G. R. *JACS* **1979**, *101*, 6452. (b) Machida, Y.; Shirai, N.; Sato, Y. *S* **1991**, 117.
30. Vedejs, E.; West, F. G. *JOC* **1983**, *48*, 4773.
31. Shoda, S.; Mukaiyama, T. *CL* **1980**, 391.
32. Shoda, S.; Mukaiyama, T. *CL* **1982**, 861.
33. (a) Ogilvie, K. K.; Beaucage, S. L. *CC* **1976**, 443. (b) Ogilvie, K. K.; Beaucage, S. L.; Theriault, N.; Entwistle, D. W. *JACS* **1977**, *99*, 1277.
34. Duthaler, R. O.; Lyle, P. A.; Heuberger, C. *HCA* **1984**, *67*, 1406.
35. Takaku, H.; Nomoto, T.; Murata, M.; Hata, T. *CL* **1980**, 1419.
36. Yamada, M.; Yahiro, S.; Yamano, T.; Nakatani, Y.; Ourisson, G. *BSF* **1990**, 824.
37. Clark, J. H.; Holland, H. L.; Miller, J. M. *TL* **1976**, 3361.
38. Clark, J. H.; Miller, J. M. *TL* **1977**, 599.
39. Clark, J. H. and Miller, J. M. *JACS* **1977**, *99*, 498.
40. Hagenbuch, J.-P.; Vogel, P. *C* **1977**, *31*, 136.
41. Nagashima, N.; Ohno, M. *CL* **1987**, 141.
42. Kosemura, S.; Yamamura, S.; Kakuta, H.; Mizutani, J.; Hasegawa, K. *TL* **1993**, *34*, 2653.
43. Yu, K.-L.; Fraser-Reid, B. *TL* **1988**, *29*, 979.
44. Murakami, M.; Mukaiyama, T. *CL* **1983**, 1733.
45. Takaku, H.; Kamaike, K.; Mori, H.; Ishido, Y. *CPB* **1983**, *31*, 2157.
46. deBoer, J. A. A.; Uiterwijk, J. W. H. M.; Geevers, J.; Harkema, S.; Reinhoudt, D. N. *JOC* **1983**, *48*, 4821.
47. Sakakura, T.; Chaisupakitsin, M.; Hayashi, T.; Tanaka, M. *JOM* **1987**, *334*, 205.
48. Ando, T.; Brown, S. J.; Clark, J. H.; Cork, D. G.; Hanafusa, T.; Ichihara, J.; Miller, J. M.; Robertson, M. S. *JCS(P2)* **1986**, 1133.
49. Corriu, R. J. P.; Perz, R.; Reye, C. *T* **1983**, *39*, 999.
50. (a) Chuit, C.; Corriu, R. J. P.; Perz, R.; Reye, C. *T* **1986**, *42*, 2293. (b) Corriu, R. J. P.; Perz, R. *TL* **1985**, *26*, 1311.
51. Corriu, R. J. P.; Moreau, J. J. E.; Pataud-Sat, M. *JOC* **1990**, *55*, 2878.
52. (a) Bonar-Law, R. P.; Davis, A. P.; Dorgan, B. J.; Reetz, M. T.; Wehrsig, A. *TL* **1990**, *31*, 6725. (b) Bonar-Law, R. P.; Davis, A. P.; Dorgan, B. J. *TL* **1990**, *31*, 6721.
53. Kawahara, Y.; Kato, S.; Kanda, T.; Murai, T.; Ishihara, H. *CC* **1993**, 277.
54. Kawaguchi, M.; Hamaoka, S.; Mori, M. *TL* **1993**, *34*, 6907.
55. Aumüller, A.; Hünig, S. *AG(E)* **1984**, *23*, 447.
56. Gingras, M.; Chan, T. H.; Harpp, D. N. *JOC* **1990**, *55*, 2078.
57. (a) Sato, T.; Otera, J.; Nozaki, H. *JOC* **1992**, *57*, 2166. (b) Sato, T.; Tada, T.; Otera, J.; Nozaki, H. *TL* **1989**, *30*, 1665. See also: Gingras, M.; Chan, T. H.; Harpp, D. N. *JOC* **1990**, *55*, 2078.
58. Macor, J. E.; Newman, M. E. *TL* **1991**, *32*, 3345.
59. Ishii, H.; Ishikawa, T.; Takeda, S.; Ueki, S.; Suzuki, M.; Harayama, T. *CPB* **1990**, *38*, 1775.
60. Bram, G.; Loupy, A.; Pigeon, P. *SC* **1988**, *18*, 1661.

Gregory K. Friestad & Bruce P. Branchaud
University of Oregon, Eugene, OR, USA

2,4,6-Collidine

[108-75-8] $C_8H_{11}N$ (MW 121.18)

(most useful in dehydrohalogenation reactions;[1] used for basic properties as reaction solvent or additive)

Alternate Name: 2,4,6-trimethylpyridine.

Physical Data: bp 170 °C; mp −44.5 °C; *d* 0.913 g cm^{-3} at 20 °C. Isomers of the title reagent include 2,3,4-collidine (*[2233-29-6]*; bp 192–193 °C), 2,3,5-collidine (*[695-98-7]*; bp 182–183 °C/739 mmHg), 2,3,6-collidine (*[1462-84-6]*; bp 176–178 °C), 2,4,5-collidine (*[1122-39-0]*; bp 165–168 °C), and 3,4,5-collidine (*[20579-43-5]*; bp 205–207 °C).[2–4]

Solubility: sol 3.5 g/100 mL H_2O (20 °C) and 20.8 g/100 mL H_2O (6 °C); miscible with ether; sol methanol, ethanol, chloroform, benzene, toluene, dilute acids.[3]

Form Supplied in: liquid.

Analysis of Reagent Purity: GLC.

Purification: by distillation.

Handling, Storage, and Precautions: protect from light; moisture sensitive; incompatible with strong oxidizing agents.[5]

Dehydrohalogenation Reactions. Numerous examples in the literature describe the utility of collidines to aid in dehydrohalogenation reactions. One example (eq 1) is the synthesis of 2-methyl-2-cyclohexenone (**2**) from 2-chloro-2-methylcyclohexanone (**1**).[6] Another example involves the dehydrobromination of tetrahydrofuran (**3**) to diene (**4**) en route from linalool to karahanaenone (62%, overall, eq 2).[7] Similarly, dehydrobromination of stigmasteryl acetate (**5**) affords 7-dehydrostigmasteryl acetate (**6**) in a refluxing mixture of collidine and mesitylene (eq 3).[1]

(1) → (2) [collidine] (1)
(45–49% from 2-methylcyclohexanone)

(3) → (4) [collidine, 110 °C] (2)

A key step in a synthesis of pyridocarbazoles used collidine in THF for a dehydrochlorination. A hindered base such as collidine was required to avoid cleavage of the lactone ring, the preferred reaction using a base such as *N,N*-dimethylaniline (eq 4).[8]

(5) → (6) [mesitylene, collidine, Δ; 26%] (3)

[collidine, THF, 80 °C; 82–85%] (4)

2,4,6-Collidine is used as solvent in the preparation of 2-benzylcyclopentanone (**8**) from 2-benzyl-2-methoxycarbonylcyclopentane (**7**) (eq 5).[9]

(7) → (8) [$LiI{\bullet}2H_2O$, 2,4,6-collidine, Δ; 72–76%] (5)

Cleavage of alkyl aryl ethers (72–76%) occurs in high yield with ***Lithium Iodide*** with 2,4,6-collidine as solvent.[10] In the absence of a solvent, high reaction temperatures are required.

Collidine as a Base Catalyst. Collidine is useful as a base in a variety of reactions. Sometimes it doubles as reaction base and solvent. A mixture of monobromo isomers is obtained in very high yield when 1-methylcyclopropyl-3-methyl-3-butenylcarbinol (**9**) is treated with ***Phosphorus(III) Bromide*** and collidine (eq 6).[11] The bromo compounds shown underwent further reaction with ***Zinc Bromide*** to form *trans*-1-bromo-3,7-dimethylocta-3,7-diene in 85–90% yield.

(9) [PBr_3, collidine] (6)

Collidine is effective as a proton scavenger in the Koenigs–Knorr synthesis of disaccharide glycosides. ***Silver(I) Trifluoromethanesulfonate*** and collidine are paired as a useful promoter of glycosylation reactions of alcohols with glycosyl halides. For example, the reaction of hepta-*O*-acetyl-α-D-cellobiosyl bromide (**10**) with 8-ethoxycarbonyloctanol (**11**) in the presence of silver triflate and collidine afforded a 78% yield of the corresponding 1,2-orthoacetate derivative (**12**) (eq 7).[12] When a similar reaction was carried out with *N,N,N′,N′*-tetramethylurea in place of collidine, a different reaction path was followed to give isomeric 1,2-*trans*-glycosides.

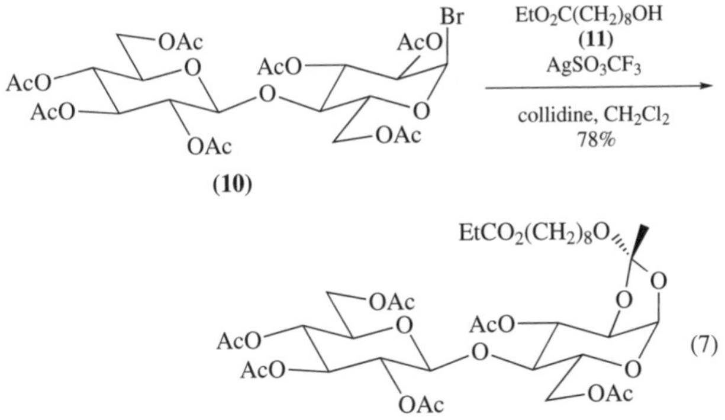

(7)

Collidine is also used as the base for glycosidation reactions promoted by ***Tin(II) Trifluoromethanesulfonate***. Starting with glucose derivative (**13**) and protected sugar derivatives, 1,2-*trans*-β-D-disaccharides have been prepared by this method (eq 8)[13]

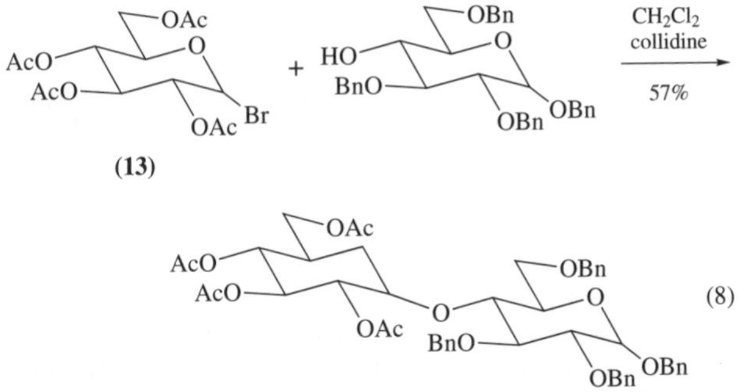

(8)

Collidine behaves as a nucleophilic catalyst in reactions such as the hydrolysis of picryl chloride. The reaction proceeds by formation of intermediate *N*-picryl-2,4,6-trimethylpyridinium ion (**14**) (eq 9).[14]

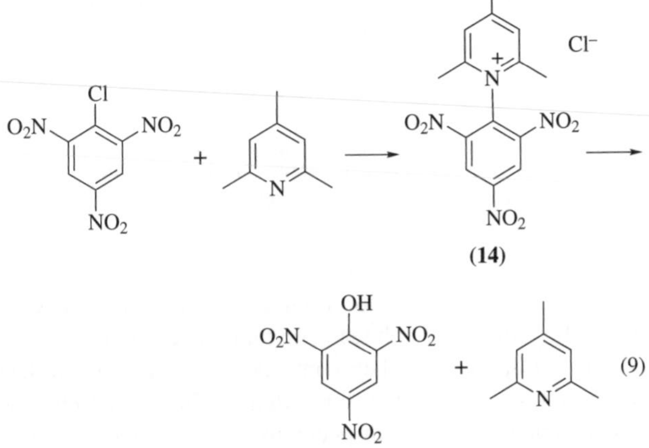

(9)

Collidine acts as a sterically hindered base in tritylation reactions of very weakly acidic compounds. The tritylation reaction proceeds in high yield for both acetone and acetonitrile, which serve as both reactant and solvent in the syntheses of 1,1,1-triphenyl-3-butanone (**15**) and 1,1,1-triphenylpropionitrile (**16**), respectively (eqs 10 and 11).[15] This tritylation is a general reaction, applicable to more strongly acidic compounds such as diethyl malonate and nitromethane. In the nitromethane reaction, a low yield of the desired product is obtained due to side reactions.

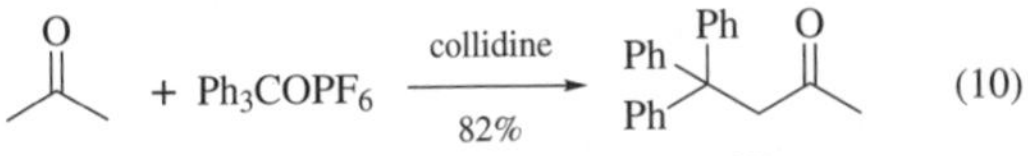

(10)

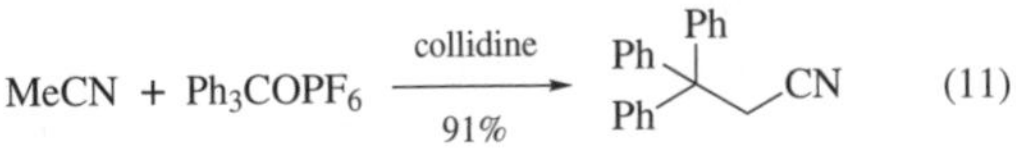

$$MeCN + Ph_3COPF_6 \xrightarrow[91\%]{\text{collidine}} Ph_3CCH_2CN \quad (11)$$

Related Reagents. 1,8-Bis(dimethylamino)naphthalene; 1,5-Diazabicyclo[4.3.0]non-5-ene; 1,8-Diazabicyclo[5.4.0]undec-7-ene; Diisopropylethylamine; 2,6-Lutidine; Phosphazene Base P_4-*t*-*Bu*; Pyridine; Quinoline.

1. Kircher, H. W.; Rosenstein, F. U. *JOC* **1973**, *38*, 2259.
2. Goe, G. L. In *Kirk-Othmer Encyclopedia of Chemical Technology*; Wiley: New York, 1982; Vol. 19, p 454.
3. *The Merck Index*, 11th ed.; Budavari, S., Ed.; Merck: Rahway, NJ, 1989; p 1529.
4. *Dictionary of Organic Compounds*, 5th ed.; Buckingham, J., Ed.; Chapman & Hall: London, 1982; p 5592.
5. *The Sigma Aldrich Library of Regulatory & Safety Data*; Aldrich: Milwaukee, WI, 1993; Vol. 2, p 2503.
6. Warnhoff, E. W.; Martin, D. G.; Johnson, W. S. *OSC* **1963**, *4*, 162.
7. Demole, E.; Enggist, P. *HCA* **1971**, *54*, 456.
8. Narasimhan, N. S.; Gokhale, S. M. *CC* **1985**, 86.
9. Elsinger, F. *OSC* **1973**, *5*, 76.
10. Harrison, I. T. *CC* **1969**, 616.
11. Brady, S. F.; Ilton, M. A.; Johnson, W. S. *JACS* **1968**, *90*, 2882.
12. Banoub, J.; Bundle, D. R. *CJC* **1979**, *57*, 2091.
13. Lubineau, A.; Malleron, A. *TL* **1985**, *26*, 1713.
14. de Rossi, R. H.; de Vargas, E. B. *JOC* **1979**, *44*, 4100.
15. Bidan, G.; Cauquis, G.; Genies, M. *T* **1979**, *35*, 177.

Angela R. Sherman
Reilly Industries, Indianapolis, IN, USA

Copper(I) Trifluoromethanesulfonate

$CuOSO_2CF_3$

[42152-44-3]	$CCuF_3O_3S$	(MW 212.62)
(2:1 benzene complex)		
[37234-97-2; 42152-46-5]	$C_8H_6Cu_2F_6O_6S_2$	(MW 503.34)

(efficient catalyst for $2\pi+2\pi$ photocycloadditions and other photoreactions of alkenes,[1] and for alkene cyclopropanation and other reactions of diazo compounds; also a selenophilic and thiophilic Lewis acid that enhances the nucleofugacity of selenide and sulfide leaving groups)

Alternate Name: copper(I) triflate.
Physical Data: moisture-sensitive white crystalline solid.
Solubility: sol MeCN, $MeCO_2H$, 2-butanone, alkenes; slightly sol benzene.
Form Supplied in: must be freshly prepared.
Preparative Methods: copper(I) trifluoromethanesulfonate (CuOTf) was first prepared as a solution in acetonitrile by synproportionation of ***Copper(II) Trifluoromethanesulfonate*** with copper(0).[2] The Cu^{I} in these solutions is strongly coordinated with acetonitrile, forming complexes analogous to ***Tetrakis(acetonitrile)copper(I) Perchlorate***.[3] A white crystalline solid benzene complex, $(CuOTf)_2 \cdot C_6H_6$, is prepared by the reaction of a suspension of ***Copper(I) Oxide*** in benzene with ***Trifluoromethanesulfonic Anhydride***.[4] Traces of ***Trifluoromethanesulfonic Acid*** apparently catalyze the reaction.[5] CuOTf is generated in situ by the reduction of $Cu(OTf)_2$ with diazo compounds.[6]
Handling, Storage, and Precautions: moisture sensitive.

Cyclopropanation with Diazo Compounds. Copper(I) triflate is a highly active catalyst for the cyclopropanation of alkenes with diazo compounds.[6] In contrast to other more extensively ligated copper catalysts, e.g. ***Copper(II) Acetylacetonate***, that favor cyclopropanation of the most highly substituted C=C bond, cyclopropanations catalyzed by CuOTf show a unique selectivity for cyclopropanation of the least alkylated C=C bond in both intermolecular (eq 1) and intramolecular (eq 2) competitions. The same selectivity is found with $Cu(OTf)_2$ as *nominal* catalyst. This is because $Cu(OTf)_2$ is reduced by the diazo compound to CuOTf, and CuOTf is the *actual* cyclopropanation catalyst in both cases.[6] Selective cyclopropanation of the least substituted C=C bond is a consequence of the alkene coordinating with the catalyst prior to interaction with the diazo compound, and the increase in stability of Cu^{I}–alkene complexes with decreasing alkyl substitution on the C=C bond. For catalysts with more strongly ligated Cu^{I}, an electrophilic carbene or carbenoid intermediate reacts with the free alkene, and the preference for cyclopropanation of the more highly substituted C=C bond arises from the enhancement of alkene nucleophilicity with increasing alkyl substitution (see ***Copper(II) Trifluoromethanesulfonate***).

N_2CHCO_2Et, CuX_n, 80 °C; products CO_2Et + EtO_2C (1)

CuX_n		Ratio	
CuOTf	17%	0.25:1	67%
$Cu(OTf)_2$	21%	0.27:1	77%
$Cu(acac)_2$	61%	1.78:1	35%

Cyclopropanecarboxylic esters are conveniently available, even from volatile alkenes, because CuOTf promotes cyclopropanations in good yields at low temperatures. Thus *trans*- and *cis*-2-butenes, boiling under reflux, react stereospecifically with ***Ethyl Diazoacetate*** to produce the corresponding ethyl 2,3-dimethylcyclopropanecarboxylates (eqs 3 and 4),[6] and cyclobutene reacts with ethyl diazoacetate at 0 °C to deliver a mixture of *exo*- and *endo*-5-ethoxycarbonylbicyclo[2.1.0]pentanes (eq 5).[7]

CH_2N_2, CuX_2, 0–10 °C (2)

X = acetylacetonate	86%	14%
X = triflate	33%	66%

+ N_2CHCO_2Et → CuOTf, 53%; CO_2Et (3)

+ N_2CHCO_2Et → CuOTf, 51%; CO_2Et (4)

+ N_2CHCO_2Et → CuOTf, cyclohexane, 0 °C; EtO_2C 51% + CO_2Et 13% (5)

CuOTf is an outstandingly effective catalyst for the synthesis of cyclopropyl phosphonates by the reaction of ***Diethyl Diazomethylphosphonate*** with alkenes (eq 6).[8] The resulting cyclopropylphosphonates are useful intermediates for the synthesis of alkylidenecyclopropanes by Wadsworth–Emmons alkenation with aromatic carbonyl compounds (eq 7).[8]

O → $N_2CHPO(OEt)_2$, CuOTf, CH_2Cl_2, 4–8 °C, 71%; O, $PO(OEt)_2$ (6)

O, $PO(OEt)_2$ → 1. BuLi, 2. Ph_2CO, 63%; O, Ph, Ph (7)

A complex of a chiral, nonracemic bis(oxazoline) with CuOTf is a highly effective catalyst for asymmetric cyclopropanation of alkenes.[9] Copper(II) triflate complexes do not catalyze the reaction unless they are first converted to Cu^{I} by reduction with a diazo compound or with phenylhydrazine. CuOTf complexes are uniquely effective. Thus the observed enantioselectivity and catalytic activity, if any, are much lower with other Cu^{I} or Cu^{II} salts including halide, cyanide, acetate, and even perchlorate. Both enantiomers of the bis(oxazoline) ligand are readily available.

Spectacularly high levels of asymmetric induction are achieved with both mono- (eq 8) and 1,1-disubstituted alkenes (eq 9).

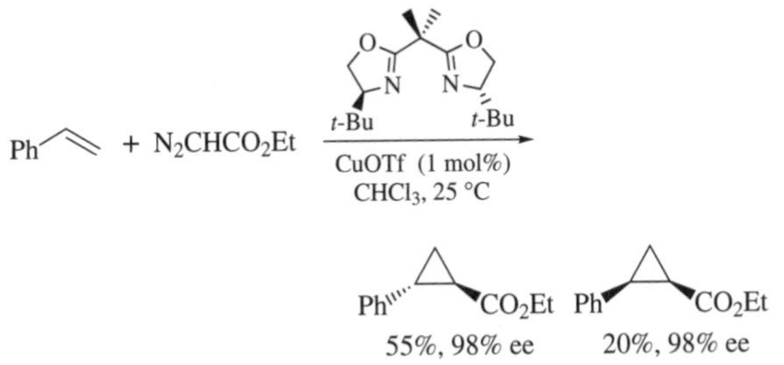

(8)

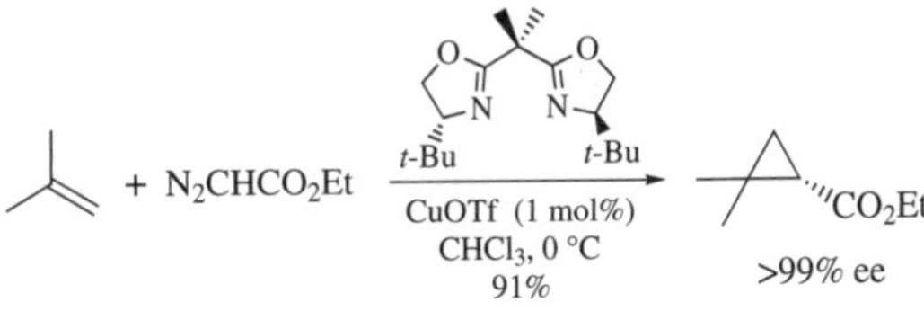

(9)

Asymmetric Aziridination. A chiral, nonracemic bis(oxazoline) complex of copper(I) triflate catalyzes asymmetric aziridination of styrene in good yield (eq 10).[9] However, enantioselectivity is not as high as the corresponding cyclopropanation (eq 8).

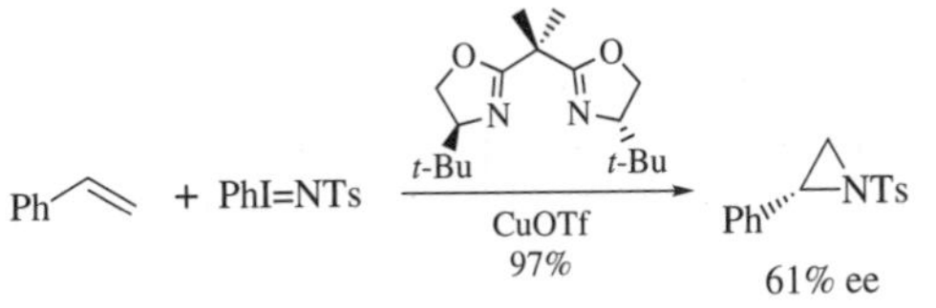

(10)

Photocycloadditions. CuOTf is an exceptionally effective catalyst for $2\pi + 2\pi$ photocycloadditions of alkenes.[1] Thus while CuBr promotes photodimerization of norbornene in only 38% yield,[10] the same reaction affords dimer in 88% yield with CuOTf as catalyst (eq 11).[11] A mechanistic study of this reaction revealed that although both 1:1 and 2:1 alkene Cu^{I} complexes are in equilibrium with free alkene and both the 1:1 and 2:1 complexes absorb UV light, only light absorbed by the 2:1 complex results in photodimerization. In other words, photodimerization requires precoordination of both C=C bonds with the Cu^{I} catalyst. Thus the exceptional ability of CuOTf, with its weakly coordinating triflate counter anion, to form π-complexes with as many as four C=C bonds[12] is of paramount importance for its effectiveness as a photodimerization catalyst.

The importance of precoordination is also evident in the CuOTf-promoted $2\pi + 2\pi$ photocycloaddition of *endo*-dicyclopentadiene. This diene forms an isolable 2:1 complex with CuOTf involving *exo*-monodentate coordination with the 8,9-C=C bond of two molecules of diene. Consequently, intermolecular $2\pi + 2\pi$ photocycloaddition involving *exo* addition to the 8,9-C=C bond is strongly favored over intramolecular reaction between the 8,9- and 3,4-C=C bonds (eq 11).[11] This contrasts with the intramolecular photocycloaddition that is promoted by high energy triplet sensitizers.[12]

Especially interesting is the *trans,anti,trans* stereochemistry of the major cyclobutane product generated in the photodimerization of cyclohexene (eq 12).[13] It was noted that the formation of this product may be the result of a preliminary CuOTf promoted *cis–trans* photoisomerization that generates a *trans*-cyclohexene intermediate (eq 13).[13] Since one face of the *trans* C=C bond is shielded by a polymethylene chain, the *trans*-cyclohexene is restricted to suprafacial additions. Although a highly strained *trans*-cyclohexene intermediate could be stabilized by coordination with Cu^{I}, such a complex has not been isolated.

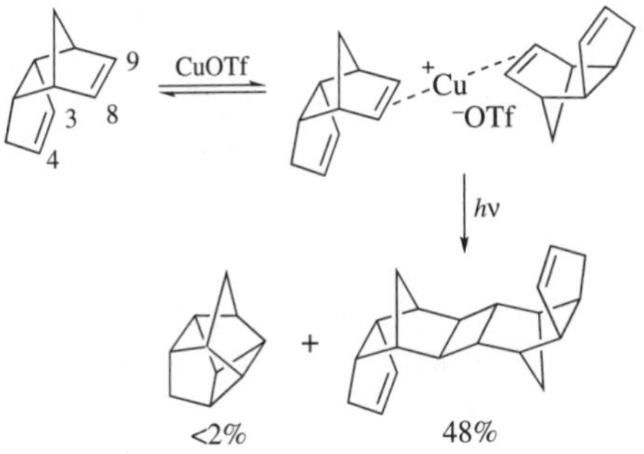

(11)

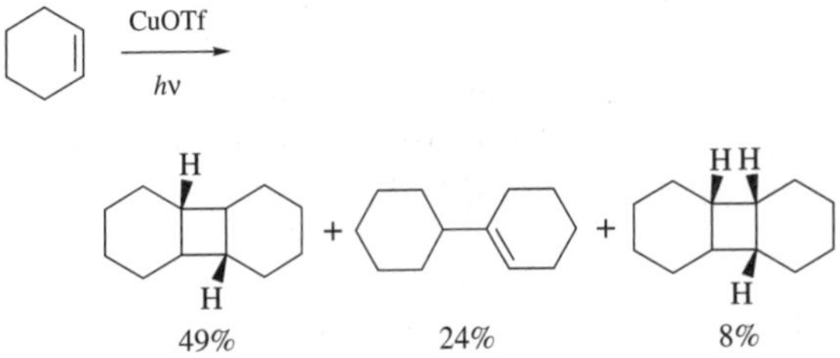

(12)

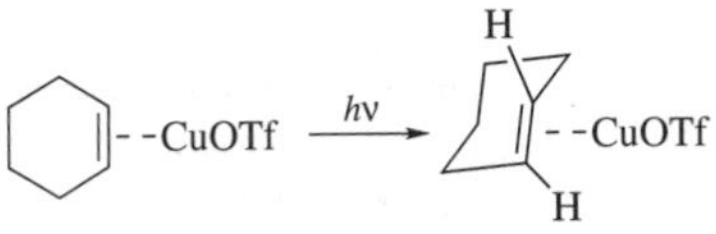

(13)

An isolable CuOTf complex of a highly strained alkene, *trans*-cycloheptene, is produced by UV irradiation of a hexane solution of *cis*-cycloheptene in the presence of CuOTf (eq 14).[14] Photocycloaddition of cycloheptene is also catalyzed by CuOTf. Surprisingly, the major product is not a *trans,anti,trans* dimer analogous to that formed from cyclohexene (eq 12) but rather a *trans,anti,trans,anti,trans* trimer (eq 15).[15]

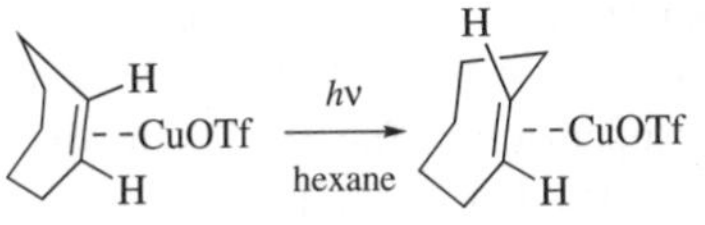

(14)

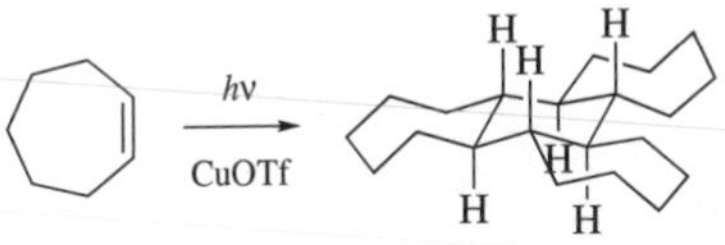

(15)

Dissolution of the *trans*-cycloheptene–CuOTf complex in cycloheptene and evaporation of the solvent delivers a tris alkene complex of CuOTf containing one *trans*-cycloheptene and two *cis*-cycloheptene ligands. Heating *trans*-cycloheptene–CuOTf in neat *cis*-cycloheptene delivers the *trans,anti,trans,anti,trans* trimer (eq 16). Experiments with *cis*-cycloheptene-d_4 show that the cyclotrimerization involves only *trans*-cycloheptene molecules, although the reaction is accelerated by the presence of *cis*-cycloheptene.[16] A likely explanation for these observations is 'concerted "template" cyclotrimerization' of a tris-*trans*-

cycloheptene–CuOTf complex formed by ligand redistribution (eq 16).[16]

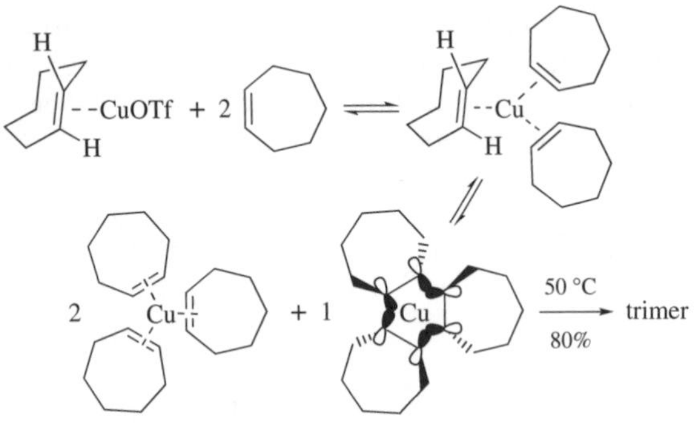

(16)

The involvement of a transient photogenerated *trans*-cyclohexene–CuOTf intermediate was also adduced to explain CuOTf catalysis of photoinduced $2\pi + 4\pi$ cycloaddition between *cis*-cyclohexene and 1,3-butadiene (eq 17).[17] In contrast to thermal Diels–Alder reactions, this reaction generates *trans*-Δ^2-octalin rather than the *cis* cycloadduct expected for a $2\pi_s + 4\pi_s$ cycloaddition. A mechanism was proposed that involves the $2\pi_s + 4\pi_s$ cycloaddition of a *trans*-cyclohexene with 1,3-butadiene in the coordination sphere of Cu^I (eq 18).[17]

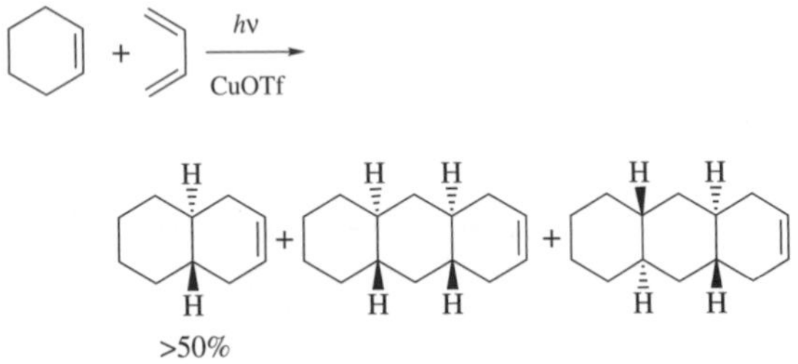

(17)

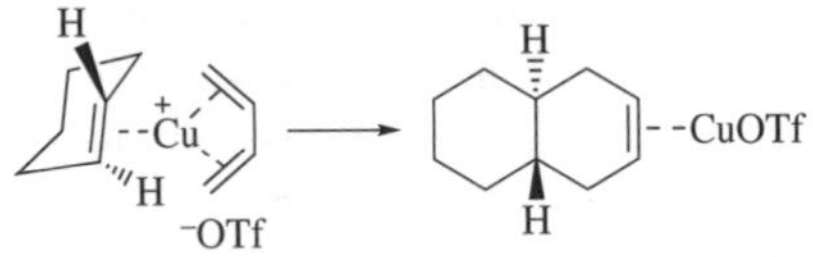

(18)

That CuOTf-catalyzed $2\pi + 2\pi$ photocycloadditions are not restricted to cyclic alkenes was first demonstrated in mixed cycloadditions involving allyl alcohol. To suppress homodimerization of *endo*-dicyclopentadiene (i.e. eq 11) the diene to Cu^I ratio is maintained at $\leq$1:1 and allyl alcohol is used as solvent. Under these conditions, a high yield of mixed cycloadduct is generated (eq 19).[18]

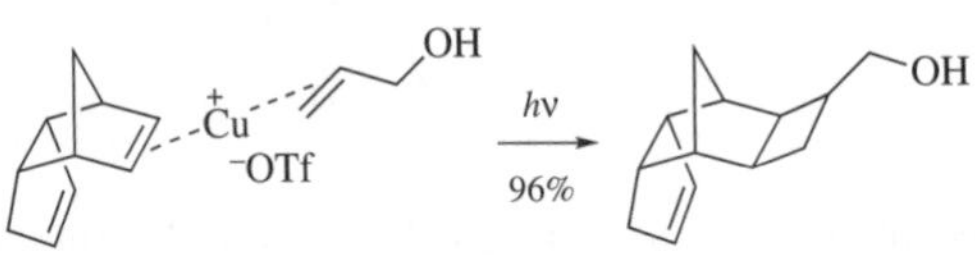

(19)

That both C=C bonds participating in $2\pi + 2\pi$ photocycloadditions can be acyclic is evident from the photobicyclization reactions of simple diallyl ethers that deliver bicyclic tetrahydrofurans (eq 20).[19,20] In conjunction with ***Ruthenium(VIII) Oxide***-catalyzed oxidation by ***Sodium Periodate***, these CuOTf-catalyzed photobicyclizations provide a synthetic route to butyrolactones from diallyl ethers (eq 20).[20] The synthetic method is applicable to the construction of multicyclic tetrahydrofurans and butyrolactones from diallyl ethers (eqs 21 and 22) as well as from homoallyl vinyl ethers (eq 23).[20]

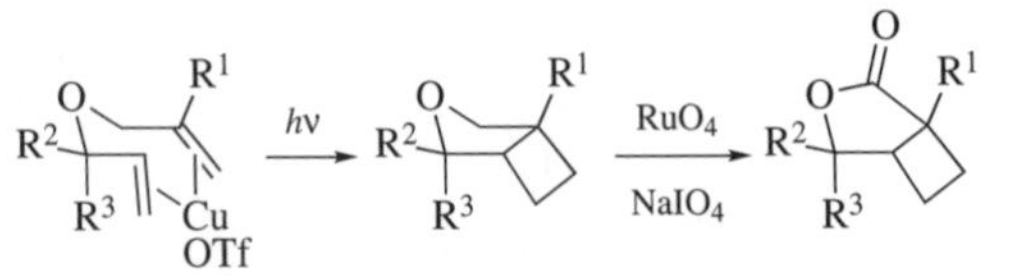

(20)

R^1	R^2	R^3	Yield (%)	Yield (%)
H	H	H	52	91
Me	Me	H	56	94
Me	Me	Me	54	44
Me	H	H	54	56
n-Bu	H	Me	83	83

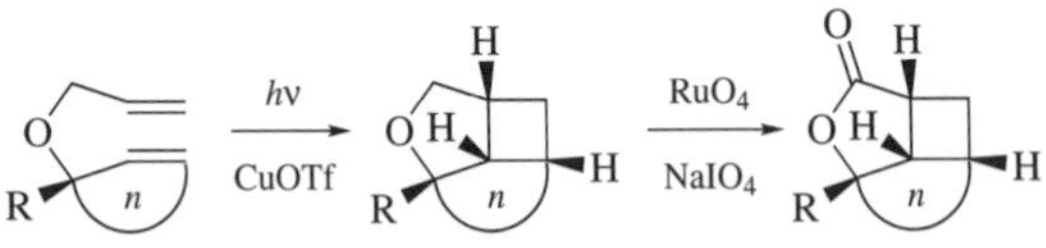

(21)

R	*n*	Yield (%)	Yield (%)
H	5	47	73
H	8	56	56
Me	5	28	87
Me	6	35	82

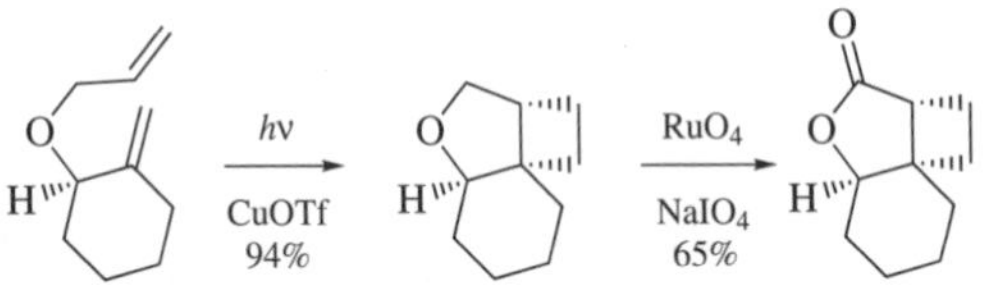

(22)

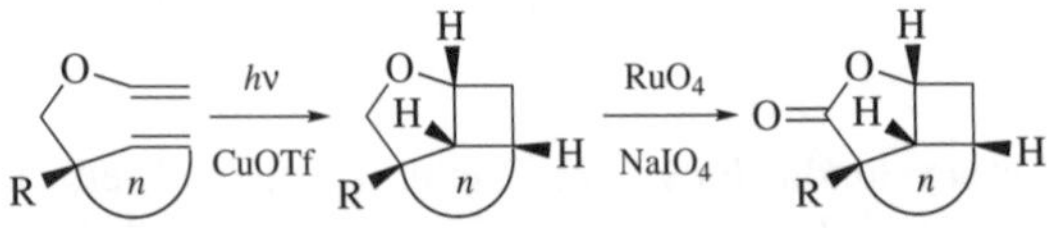

(23)

R	*n*	Yield (%)	Yield (%)
H	5	92	78
Me	5	50	71
H	6	48	85

3-Oxabicyclo[3.2.0]heptanes are also produced in the CuOTf-catalyzed photocycloadditions of allyl 2,4-hexadienyl ethers (eq 24).[21] The CuOTf-catalyzed photocycloadditions of bis-2,4-hexadienyl ethers are more complex. Thus UV irradiation of 5,5′-oxybis[(*E*)-1,3-pentadiene] in THF for 120 h produces vinylcyclohexene and tricyclo[3.3.0.0^{2,6}]octane derivatives (eq 25).[22] However, shorter irradiations reveal that these products arise by secondary CuOTf-catalyzed rearrangements of 6,7-divinyl-3-oxabicyclo[3.2.0]heptanes that are the primary photoproducts (eq 26). UV irradiation of the divinylcyclobutane intermediates in the presence of CuOTf promotes formal [1,3]- and [3,3]-sigmatropic rearrangements to produce a vinylcyclohexene and a 1,5-cyclooctadiene that is the immediate precursor of the tricyclo[3.3.0.0^{2,6}]octane.

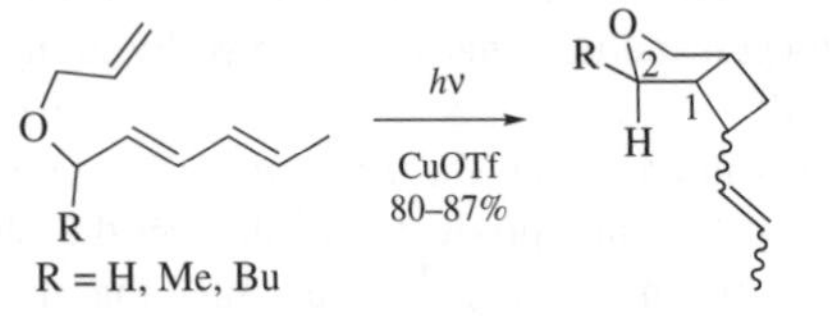

(24)

R = H, Me, Bu

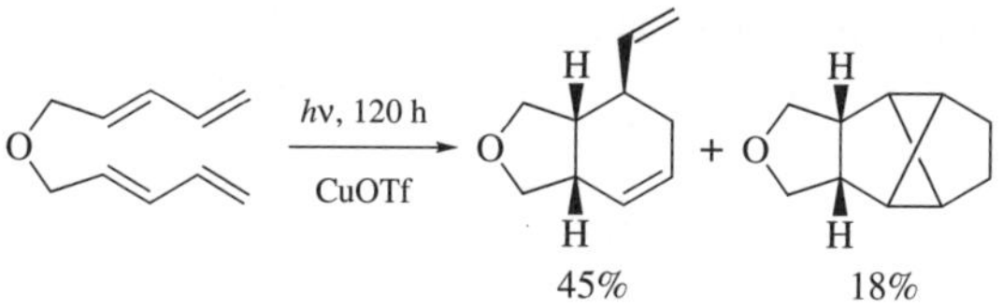

(25)

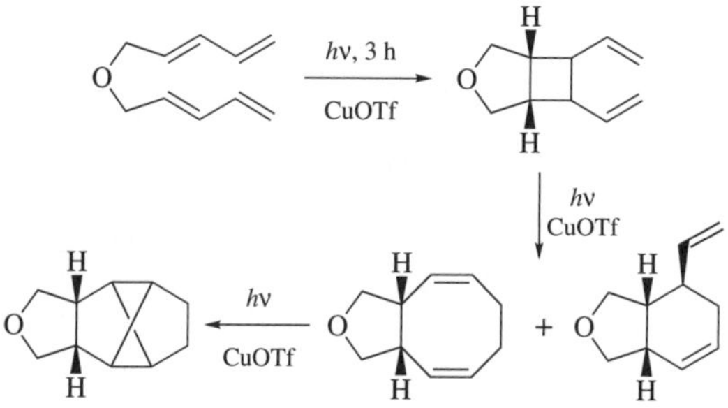

(26)

CuOTf-catalyzed photobicyclization of 1,6-heptadien-3-ols produces bicyclo[3.2.0]heptan-2-ols (eq 27).[23] In conjunction with pyrolytic fragmentation of the derived ketones, these CuOTf-catalyzed photobicyclizations provide a synthetic route to 2-cyclopenten-1-ones from 1,6-heptadien-3-ols (eq 28).[23] The derived ketones can also be converted into lactones by Baeyer–Villiger oxidation and, in conjunction with pyrolytic fragmentation, CuOTf-catalyzed photobicyclizations provide a synthetic route to enol lactones of glutaraldehydic acid from 1,6-heptadien-3-ols (eq 28).[23]

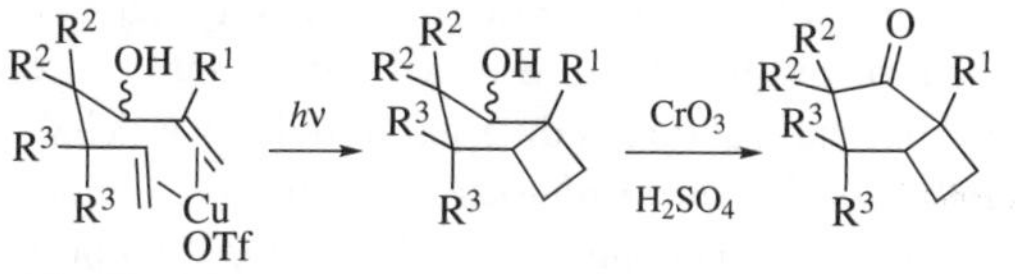

(27)

R^1	R^2	R^3	Yield (%)	Yield (%)
H	H	H	86	78
Me	H	H	81	67
H	Me	H	91	92
H	H	Me	84	92
Me	Me	H	83	93

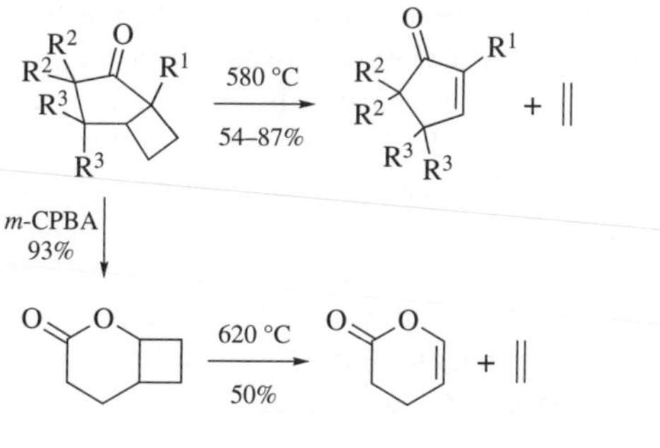

(28)

Copper(I) triflate-catalyzed photobicyclization of β- and γ-(4-pentenyl)allyl alcohols provides a synthetic route to various multicyclic carbon networks in excellent yields (eqs 29–31).[24] The reaction was exploited in a total synthesis of the panasinsene sesquiterpenes (eq 32).[25] It is especially noteworthy in this regard that attempted synthesis of a key tricyclic ketone intermediate for the panasinsenes by the well-known photocycloaddition of ***Isobutene*** to an enone failed to provide any of the requisite cyclobutyl ketone (eq 33).[25]

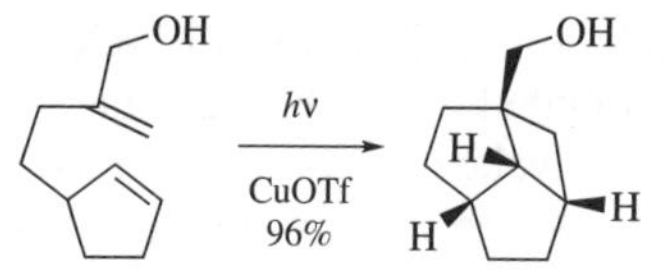

(29)

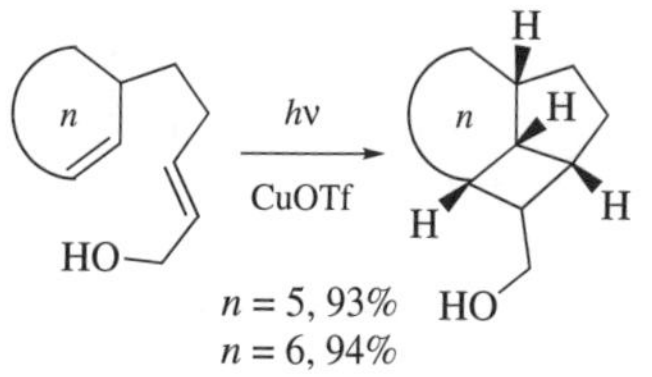

(30)

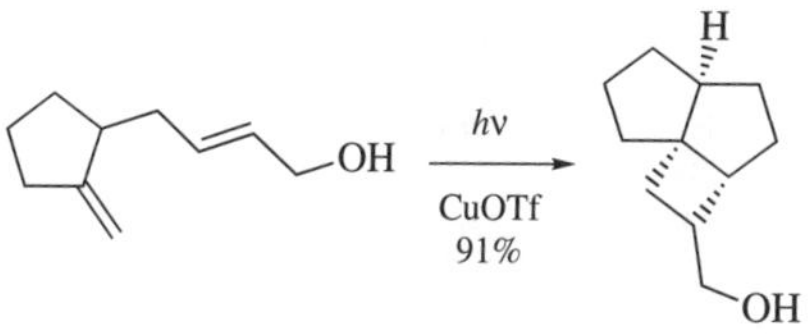

(31)

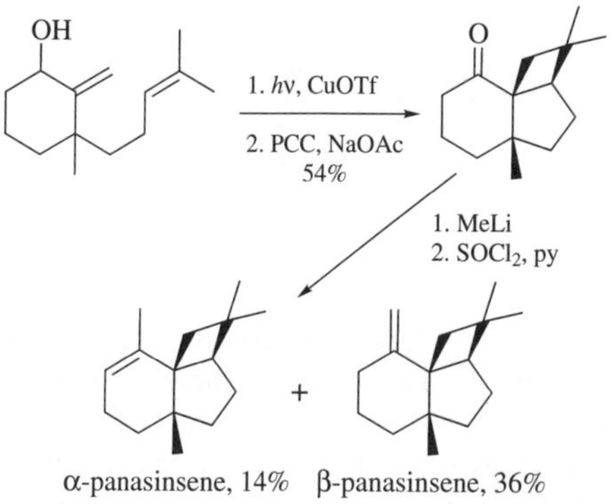

(32)

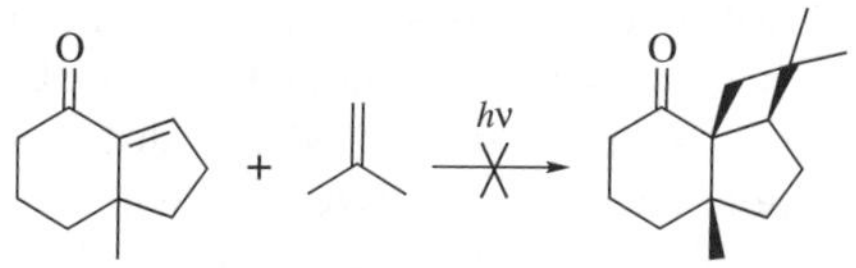

α-panasinsene, 14% β-panasinsene, 36%

(33)

In conjunction with carbocationic skeletal rearrangement, photobicyclization of 1,6-heptadien-3-ols provides a synthetic route to 7-hydroxynorbornanes (eq 34).[26] Noteworthy is the stereoselective generation of *exo*-1,2-polymethylenenorbornanes from either the *exo* or *endo* epimer of 2,3-polymethylenebicyclo[3.2.0]heptan-3-ol.

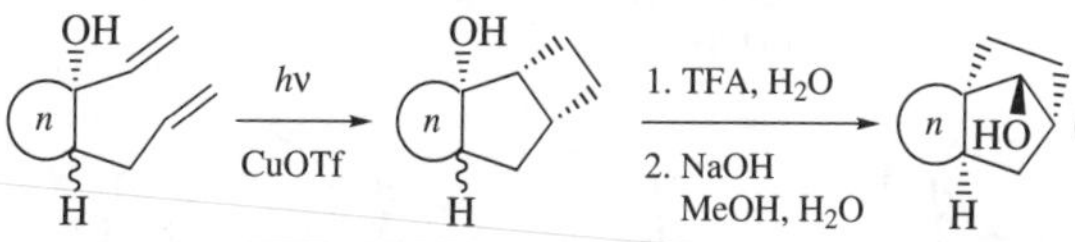

(34)

n	Yield (%)	Yield (%)
5	70–75	64–85
6	51–84	76–77
7	73	73
8	74	66

N,N-Diallylamides are recovered unchanged when irradiated in the presence of CuOTf.[27] This is because the amide chromophore interferes with photoactivation of the Cu^I–alkene complex. Thus CuOTf–alkene complexes containing one, two, three, or even four coordinated C=C bonds exhibit UV absorption at 235 +mn 5 nm (ε_{max} 2950 +mn 450).[12] The CuOTf complex of ethyl *N,N*-diallylcarbamate exhibits λ_{max} = 233.4 nm (ε_{max} 2676) but the free ligand is virtually transparent at this wavelength. Consequently, UV irradiation of ethyl *N,N*-diallylcarbamates in the presence of CuOTf delivers bicyclic (eq 35) or tricyclic (eq 36) pyrrolidines incorporating the 3-azabicyclo[3.2.0]heptane ring system.[27]

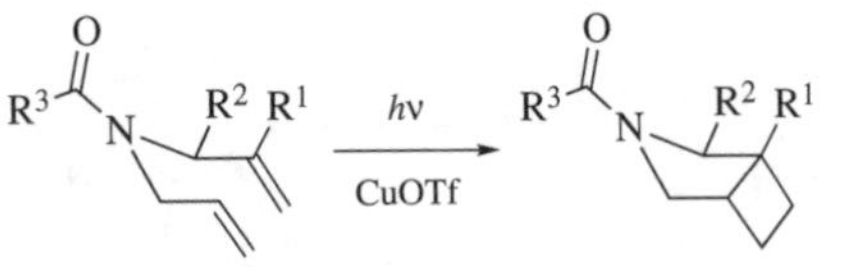

(35)

R^1	R^2	R^3	$\varepsilon_{233\ nm}$	Yield (%)
H	H	Me	192	0
H	H	H	231	0
H	H	OEt	15	74
H	Me	OEt	–	60
Me	H	OEt	–	76

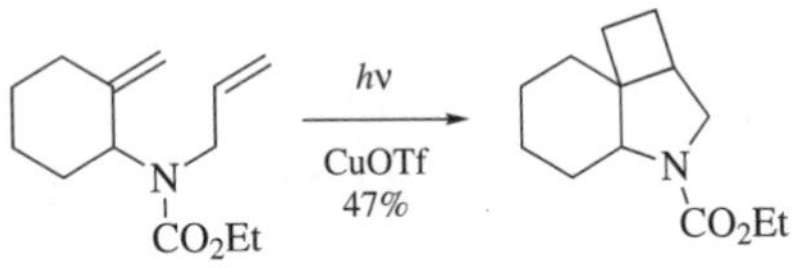

(36)

Catalyzed Diels–Alder Reactions. The uncatalyzed thermal intramolecular Diels–Alder reaction of 5,5′-oxybis[(*E*)-1,3-pentadiene] nonstereoselectively generates four isomeric 4-vinylcyclohexenes (eq 37). The major product has a *trans* ring fusion, in contrast to the single *cis* ring-fused isomer generated in the copper(I) triflate-catalyzed photoreaction of the same tetraene (eq 25). Copper(I) triflate also catalyzes a thermal Diels–Alder reaction of 5,5′-oxybis[(*E*)-1,3-pentadiene] that proceeds under milder conditions than the uncatalyzed reaction. The stereoselectivity is remarkably enhanced, generating mainly the major isomer of the uncatalyzed thermal reaction and a single *cis*-fused isomer (eq 37) that is different than the one favored in the photochemical reaction (eq 25).

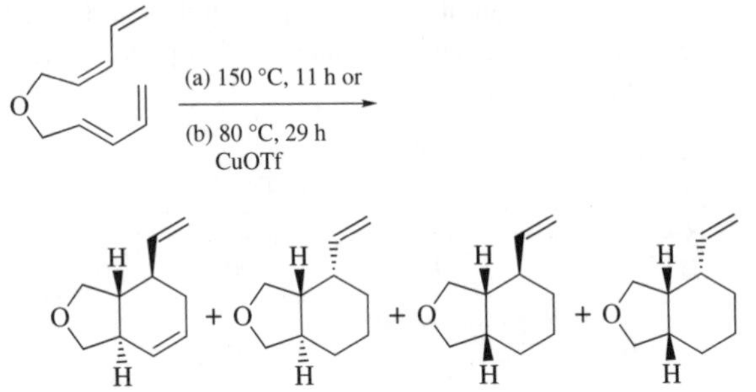

(37)

C_{sp}**–H Bond Activation.** Hydrogen–deuterium exchange between terminal alkynes and CD_3CO_2D is catalyzed by CuOTf (eq 38).[28] Proton NMR studies revealed that CuOTf and alkynes form π-complexes that rapidly exchange coordinated with free alkyne. A complex of CuOTf with 1,7-octadiene was isolated (eq 39). The complex rapidly exchanges terminal alkynic hydrogen with deuterium from CD_3CO_2D and undergoes a much slower conversion to a copper alkynide (eq 39).[28] Exchange of alkynic hydrogen and deuterium is also catalyzed by CuOTf (eq 40).[28]

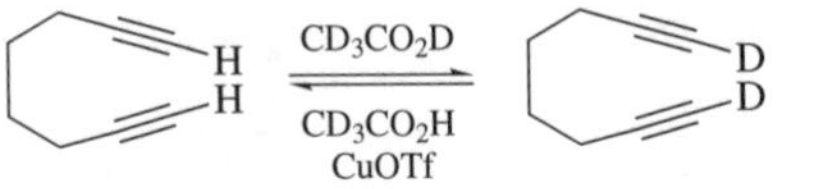

(38)

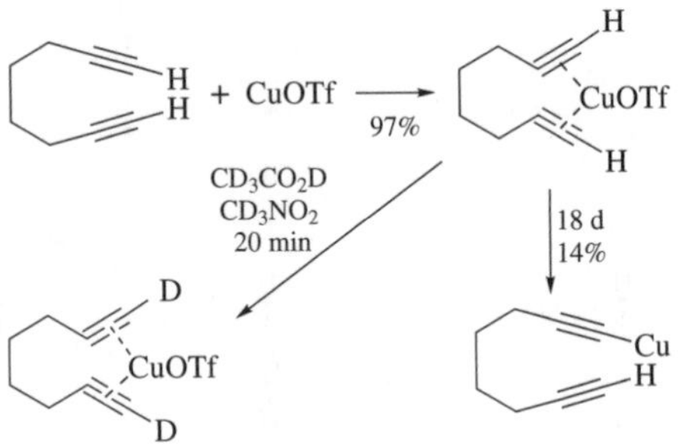

(39)

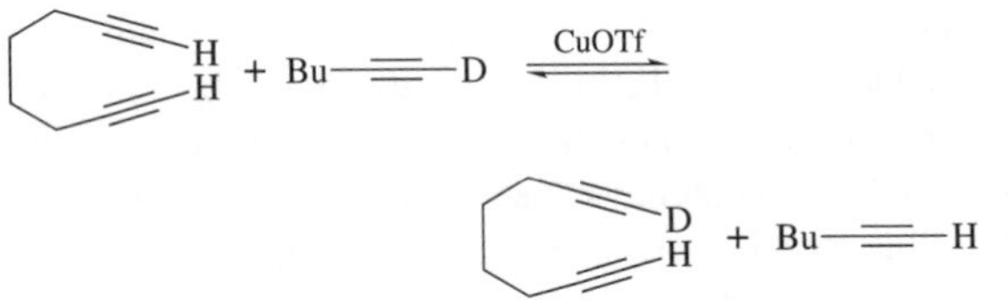

(40)

Activation of Aryl Halides. Ullmann coupling of *o*-bromonitrobenzene is accomplished under exceptionally mild conditions and in homogeneous solution by reaction with copper(I) triflate in the presence of aqueous NH_3 (eq 41).[29] Yields are enhanced by the presence of a small quantity of copper(II) triflate. That the reaction is diverted to reductive dehalogenation by ***Ammonium Tetrafluoroborate*** is presumptive evidence for an organocopper intermediate that can be captured by protonation.

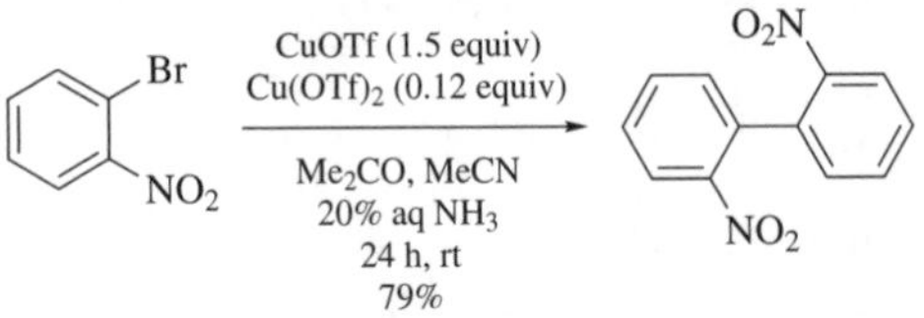

(41)

Biaryl is only a minor product from the reaction of methyl *o*-bromobenzoate with CuOTf (eq 42). The major product can result from replacement of the halide by NH_2, H, or OH, depending on reaction conditions. In the presence of 5% aqueous NH_3, methyl anthranilate is the major product.[30] More concentrated aqueous NH_3 (20%) favors the generation of methyl salicylate, and the yield of this product is enhanced by the presence of a substantial quantity of Cu^{II} ion.[29] Reductive dehalogenation is favored by the presence of ammonium ions, presumably owing to protonolysis of an arylcopper(III) intermediate.[29]

Activation of Vinyl Halides. Under the optimum conditions for reductive coupling of *o*-bromonitrobenzene (eq 41), diethyl iodofumarate gives very little coupling product; the overwhelming product was diethyl fumarate generated by hydrodehalogenation (eq 43).[29] Reductive coupling delivers *trans,trans*-1,2,3,4-tetraethoxycarbonyl-1,3-butadiene in 95% yield (GLC, or 80% of pure crystalline product) in 2 h if aqueous NH_3 is replaced by anhydrous NH_3.[29] Under the same conditions, diethyl iodomaleate undergoes 45% conversion in 20 h to deliver diethyl maleate, as well as minor amounts of *cis,cis*- and *trans,trans*-tetraethoxycarbonyl-1,3-butadiene (eq 44).[29] The stereospecificity of the reductive dehalogenations in eqs 43 and 44 is presumptive evidence for the noninvolvement of radicals in these reactions.

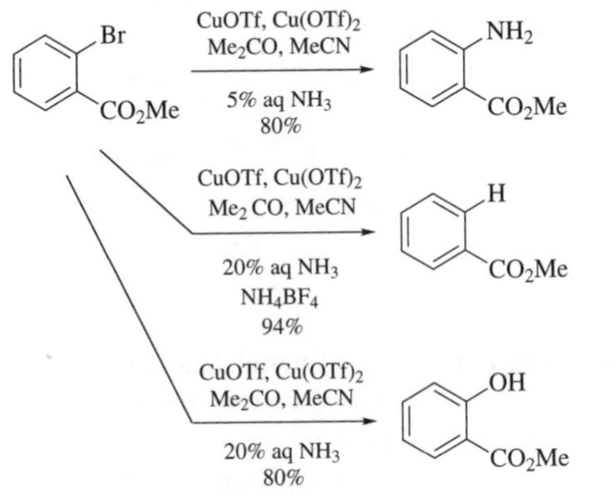

(42)

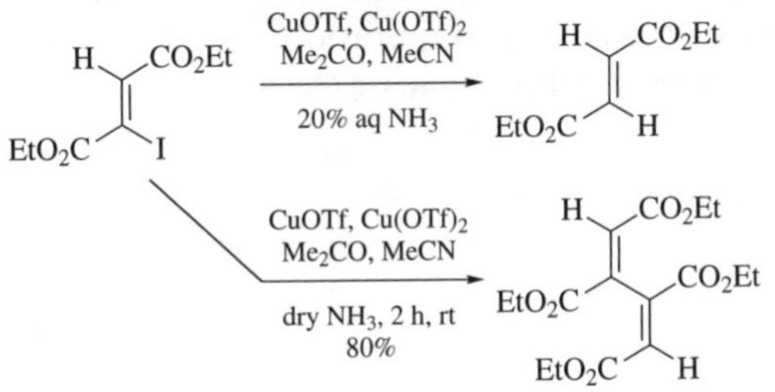

(43)

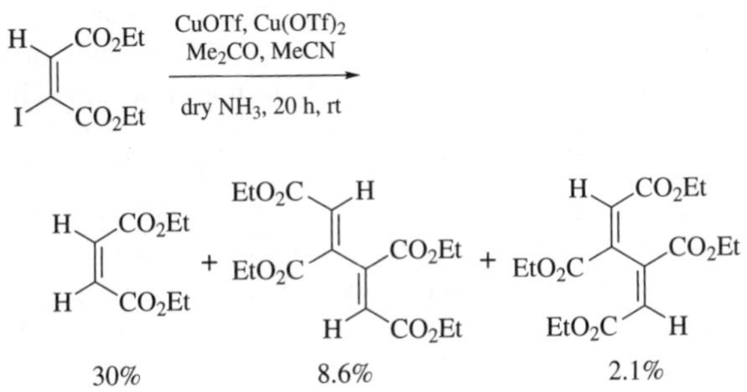

(44)

Elimination of Thiophenol from Thioacetals. Conversion of thioacetals to vinyl sulfides is accomplished under exceptionally mild conditions by treatment with $(CuOTf)_2 \cdot C_6H_6$ (eq 45).[31] The reaction involves an α-phenylthio carbocation intermediate. Three factors contribute to the effectiveness of this synthetic method: the Lewis acidity of a copper(I) cation that is unencumbered by a strongly coordinated counter anion, the solubility of the copper(I) triflate–benzene complex, and the insolubility of CuSPh in the reaction mixture. An analogous elimination reaction provides an effective route to phenylthio enol ethers from ketones (eq 46).[31]

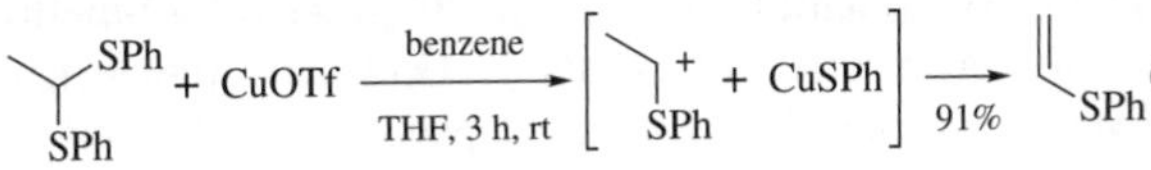

(45)

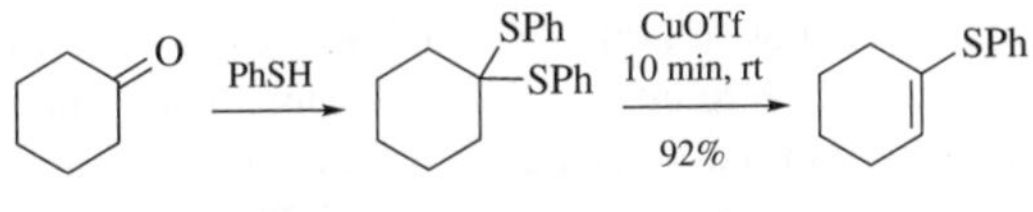

(46)

This conversion of thioacetals into vinyl sulfides was applied to a C–C connective synthesis of 2-phenylthio-1,3-butadienes from aldehydes (eq 47).[32] The key elimination step converts cyclobutanone thioacetal intermediates into 1-phenylthiocyclobutenes that undergo electrocyclic ring opening to deliver dienes.

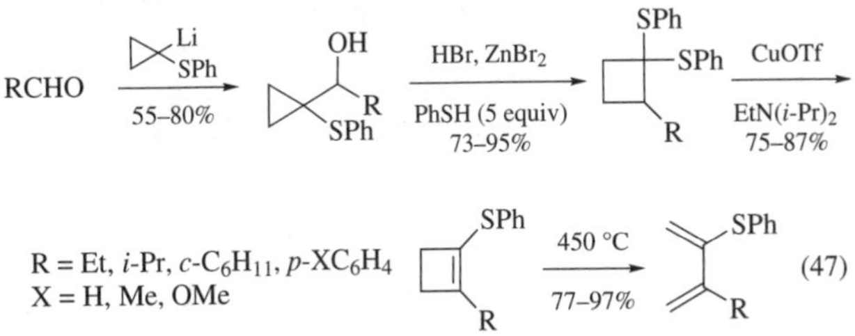

(47)

A different synthesis generates 2-phenylthio-1,3-butadienes directly by elimination of two molecules of thiophenol from β-phenylthio thioacetals that are readily available from the corresponding α,β-unsaturated ketones (eq 48).[33]

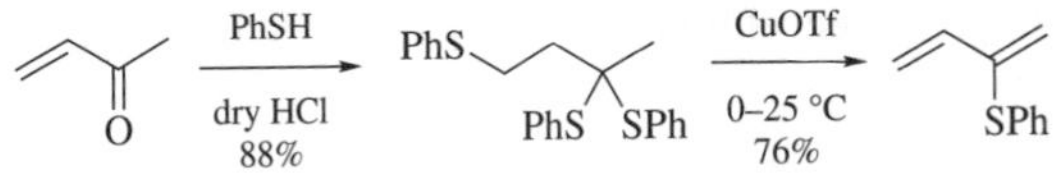

(48)

CuOTf-promoted elimination of thiophenol was exploited in two syntheses of 1-phenylthio-1,3-butadiene, one a C–C connective route from allyl bromide[31] and bis(phenylthio)-methyllithium,[34] and another from ***Crotonaldehyde*** (eq 49).[33] A topologically analogous C–C connective strategy provides 2-methoxy-1-phenylthio-1,3-butadiene from acrolein (eq 50).[5,33] That the phenylthio rather than the methoxy substituent in 2-methoxy-1-phenylthio-1,3-butadiene controls the orientation of its Diels–Alder cycloadditions is noteworthy (eq 50).

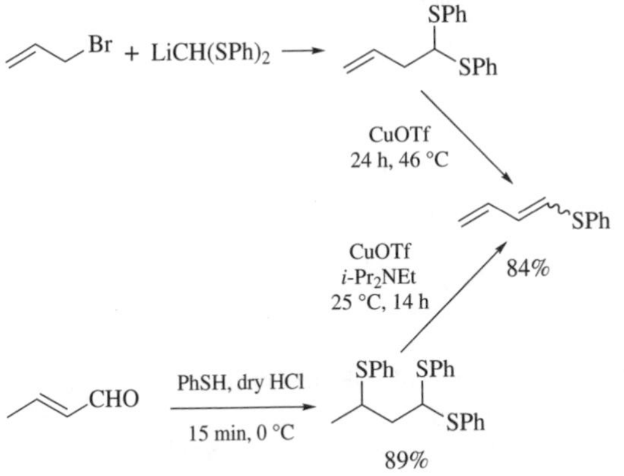

(49)

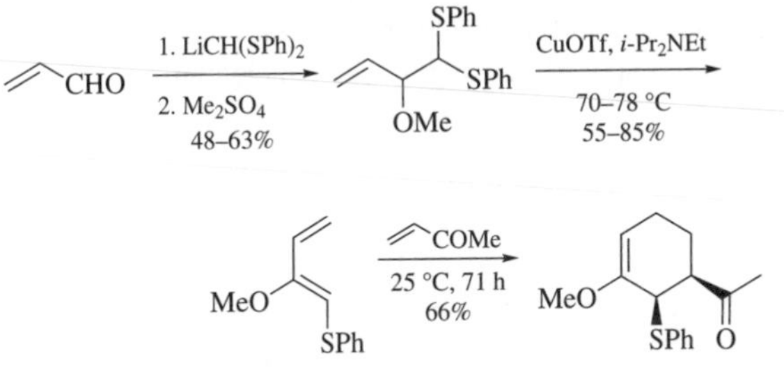

(50)

A synthesis of 4-alkyl-2-methoxy-1-phenylthio-1,3-butadienes by a simple β-elimination of thiophenol from a thioacetal is not possible owing to skeletal rearrangement that is fostered by stabilization of a cyclopropylcarbinyl carbocation intermediate by the alkyl substituent (eq 51).[35] Interconversion of an initial α-phenylthio carbocation to a more stable α-methoxy carbocation intermediate leads to the generation of a 4-alkyl-1-methoxy-2-phenylthio-1,3-butadiene instead.

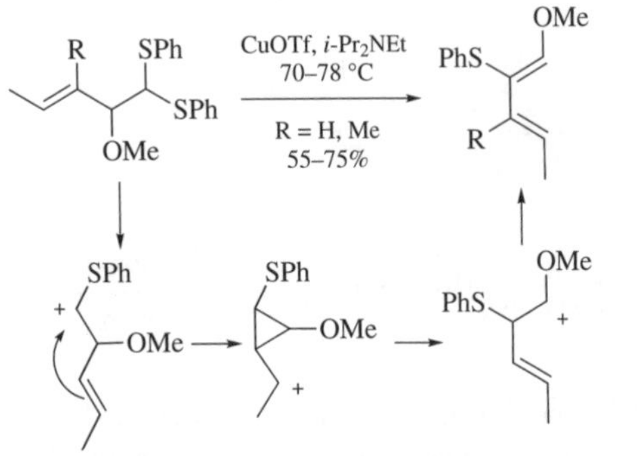

(51)

Syntheses of 1-phenylthio-1,3-butadienes from carboxylic esters (eq 52) and carboxylic acids (eq 53) are achieved by CuOTf-promoted elimination of thiophenol from intermediate thioacetals.[36]

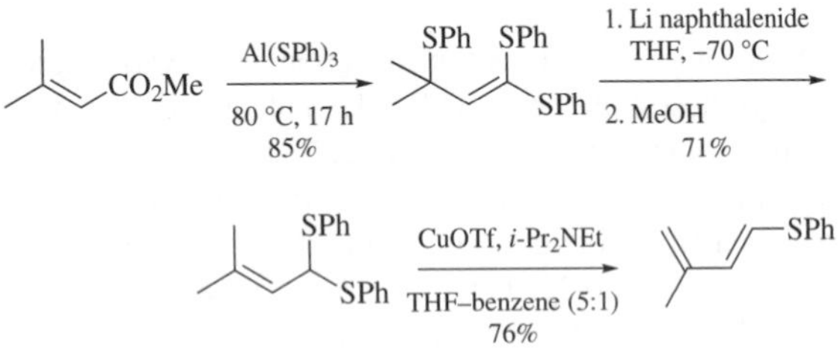

(52)

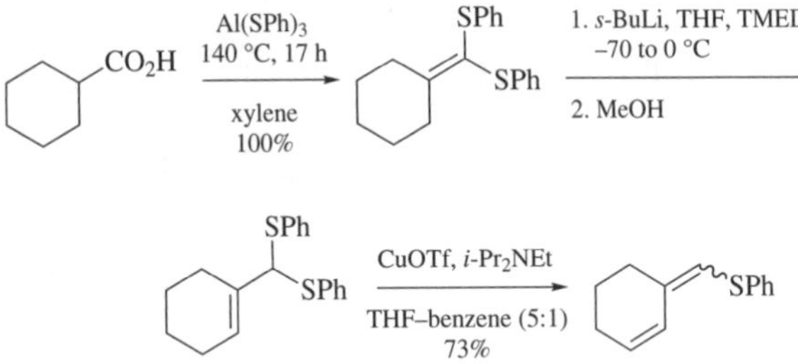

(53)

Heterocyclization of γ-Keto Dithioacetals. A C–C connective synthesis of furans is completed by a CuOTf-promoted heterocyclization of γ-keto thioacetals (eq 54).[31] Rather than simple β-elimination to generate a vinyl sulfide (eq 46), a presumed γ-keto carbocation intermediate is captured intramolecularly by an intimately juxtaposed carbonyl oxygen nucleophile.

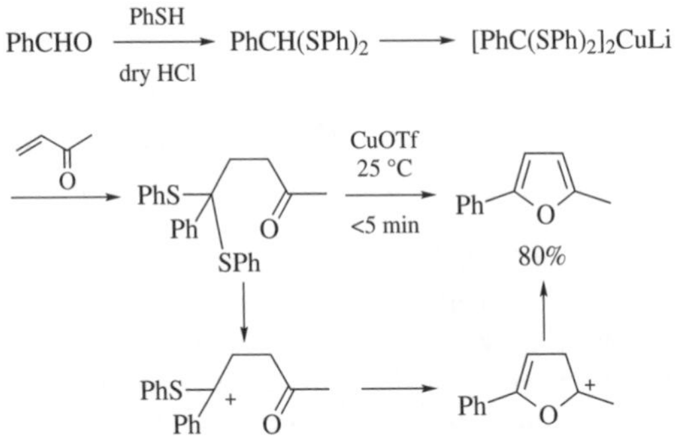

(54)

Friedel–Crafts Alkylation of Arenes with Thioacetals. $(CuOTf)_2 \cdot C_6H_6$ promotes α-thioalkylation of anisole by a dithioacetal under mild conditions (eq 55).[37]

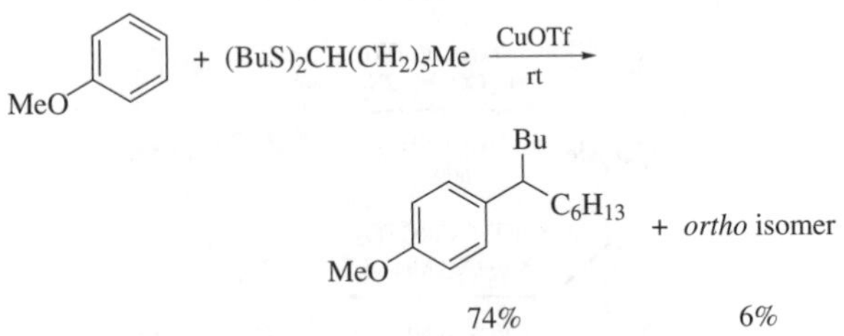

(55)

Elimination of Benzylic Phenyl Thioethers. That C–S bond activation by CuOTf is not limited to substrates that can generate sulfur-stabilized carbocation intermediates is illustrated by a C–C connective synthesis of *trans*-stilbene (eq 56).[31] The elimination of thiophenol under mild conditions is favored by benzylic stabilization of a carbocation intermediate or an E2 transition state with substantial carbocationic character.

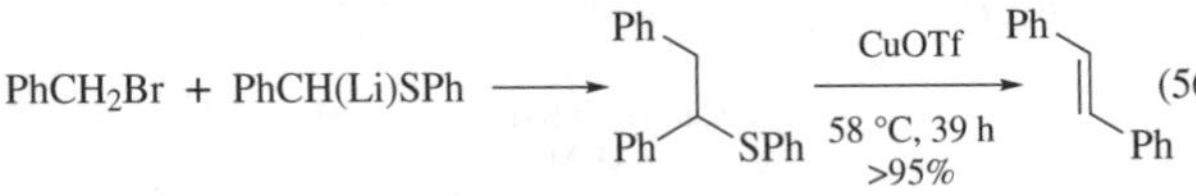

(56)

Hydrolysis of Vinylogous Thioacetals. Carbanions prepared by lithiation of γ-phenylthioallyl phenyl thioethers can serve as synthetic equivalents of β-acyl vinyl anions.[38] Umpölung of the usual electrophilic reactivity of 2-cyclohexenone is achieved by a sequence exploiting electrophilic capture of a lithiated vinylogous thioacetal and subsequent CuOTf-assisted hydrolysis (eq 57).[39] Otherwise unfunctionalized vinylogous thioacetals can be hydrolyzed to enones by ***Mercury(II) Chloride*** in wet acetonitrile.[38] However, the keto-substituted derivative in eq 57 gave only a 25% yield of enone by this method. A superior yield was obtained by CuOTf-assisted hydrolysis.[39]

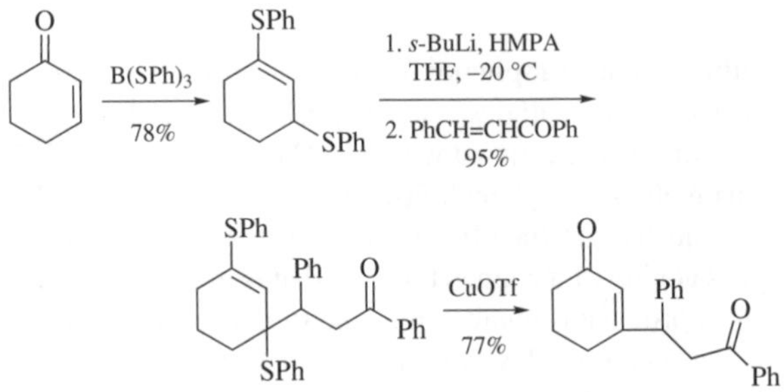

(57)

Grob Fragmentation of β-[Bis(phenylthio)-methyl]-alkoxides. A method for achieving Grob-type fragmentation of five- and six-membered rings depends upon the ability of a thiophenyl group to both stabilize a carbanion and serve as an anionic leaving group. For example, reaction of cyclohexene oxide with lithium bis(phenylthio)methide[34] produces a β-[bis(phenylthio)methyl]alkanol that undergoes fragmentation in excellent yield upon treatment with ***n-Butyllithium*** followed by CuOTf (eq 58).[40] Copper(I) trifluoroacetate is equally effective but salts of other thiophilic metals, e.g. mercury or silver, were ineffective. Treatment of the intermediate β-[bis(phenylthio)methyl]alkanol with CuOTf in the absence of added strong base leads primarily to elimination of thiophenol as expected (see eq 45). Fragmentation does not occur with only one equivalent of CuOTf. This suggests a key intermediate with at least one Cu^{I} ion to coordinate with the alkoxide and another to activate the phenylthio leaving group (eq 58).

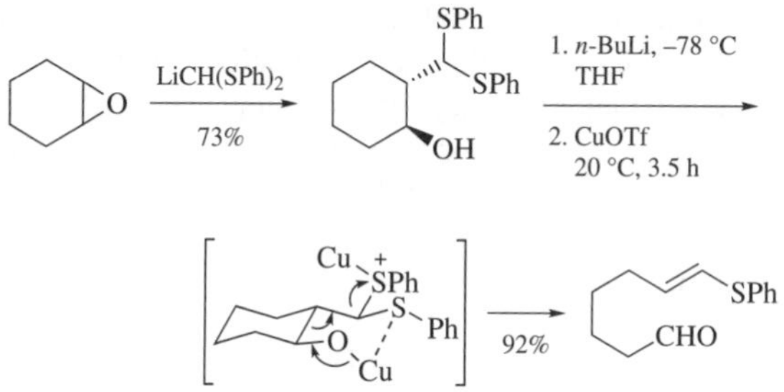

(58)

Ring-Expanding Rearrangements of α-[Bis(phenylthio)-methyl]alkanols. A one-carbon ring-expanding synthesis of α-phenylsulfenyl ketones from homologous ketones depends upon the ability of a thiophenyl group to both stabilize a carbanion and serve as an anionic leaving group. For example, reaction of cyclopentanone with lithium bis(phenylthio)methide[34] produces an α-[bis(phenylthio)methyl]alkanol that rearranges to a ring-expanded α-phenylsulfenyl ketone in good yield upon treatment with CuOTf in the presence of ***Diisopropylethylamine*** (eq 59).[41] Epoxy thioether intermediates are generated from the α-[bis(phenylthio)methyl]alkanols by intramolecular nucleophilic displacement of thiophenoxide.

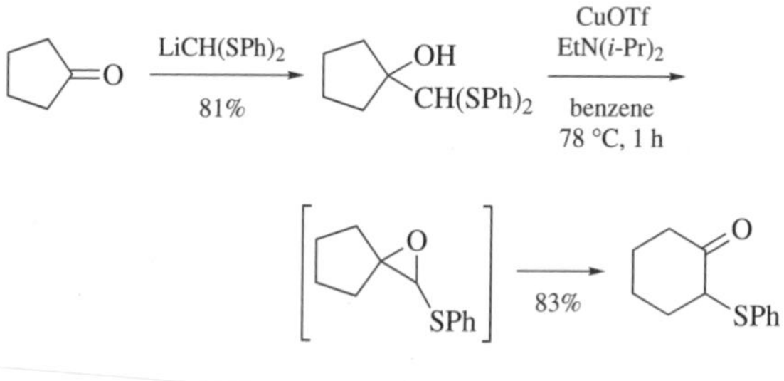

(59)

An analogous synthesis of α,α-bis(methylsulfenyl) ketones from homologous ketones by one-carbon ring expansion depends on copper(I)-promoted rearrangement of an α-tris(methylthio)methyl alkoxide intermediate (eq 60). Both ***Tetrakis(acetonitrile)copper(I) Perchlorate***[42] and ***Tetrakis(acetonitrile)copper(I) Tetrafluoroborate***[43] are effective in promoting the rearrangement but $(CuOTf)_2{\cdot}C_6H_6$, $HgCl_2$, or $Hg(TFA)_2$ are not. Apparently, the MeCN ligand is crucial. Furthermore, treatment of the intermediate α-[tris(methylthio)methyl] alcohol with CuOTf and $EtN(i\text{-}Pr)_2$ in toluene followed by aqueous workup delivers an α-hydroxy methylthio ester (eq 61),[43] in contrast to the ring-expanding rearrangement of the analogous α-[bis(phenylthio)methyl]alkanol (eq 59).[41]

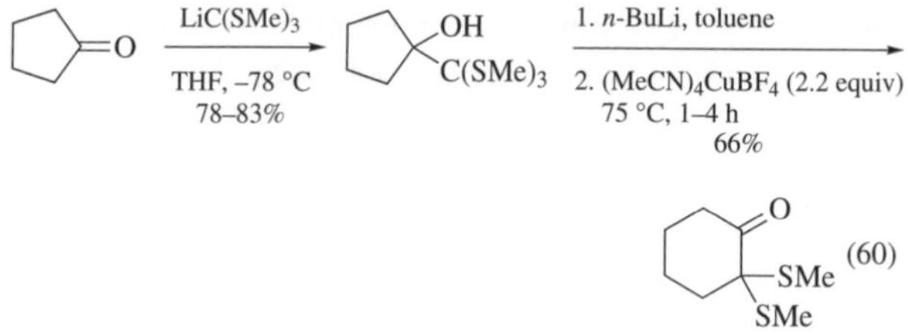

(60)

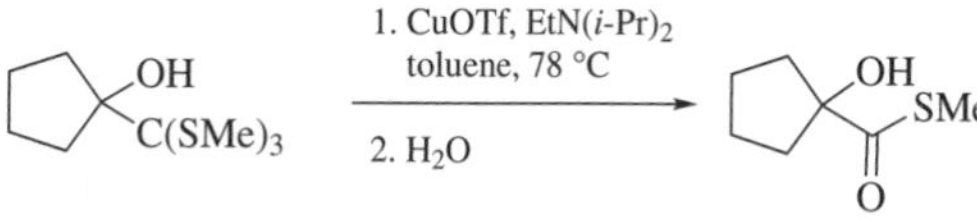

(61)

The α-[bis(phenylthio)methyl]alkanol derived from cycloheptanone does not undergo ring expansion upon treatment with CuOTf and $EtN(i\text{-}Pr)_2$ in benzene. Instead, 1,3-elimination of thiophenol delivers an epoxy thioether intermediate that undergoes a rearrangement involving 1,2-shift of a phenylsulfenyl group to produce an α-phenylsulfenyl aldehyde (eq 62).[41]

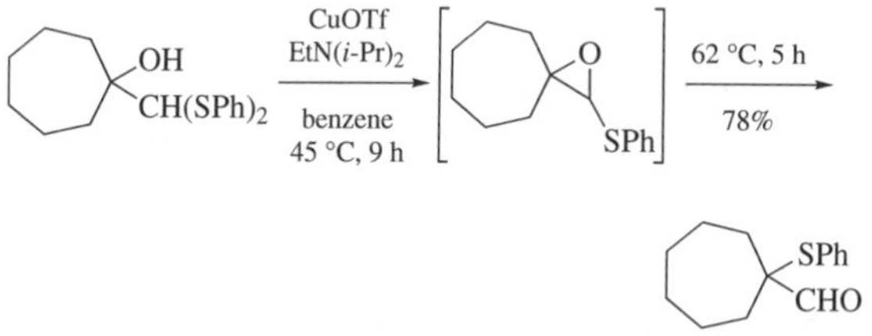

(62)

The α-[bis(phenylthio)methyl]alkanol derived from cyclohexanone, upon treatment with CuOTf and $EtN(i\text{-}Pr)_2$ in benzene, undergoes both ring-expanding rearrangement to deliver α-phenylsulfenylcycloheptanone as the major product, as well as rearrangement involving 1,2-shift of a phenylsulfenyl group to produce an α-phenylsulfenylcyclohexanecarbaldehyde (eq 63).[41] In contrast, neither ring expansion nor 1,2-shift of a methylsulfenyl group occurs upon treatment of α-[tris(methylthio)methyl]cyclohexanol with *n*-butyllithium followed by $(MeCN)_4CuBF_4$. Rather, after aqueous workup, an α-hydroxy methylthio ester is obtained (eq 64).[43]

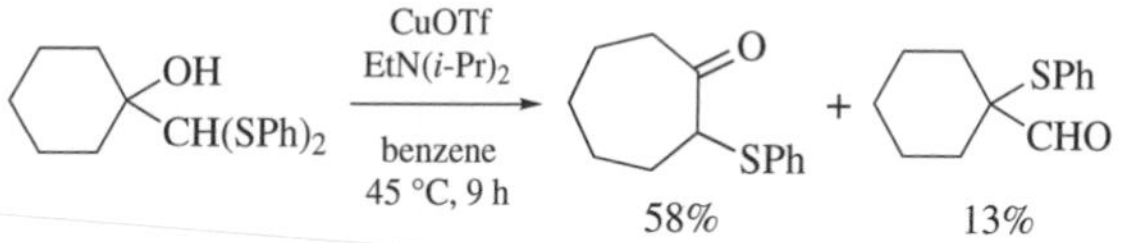

(63)

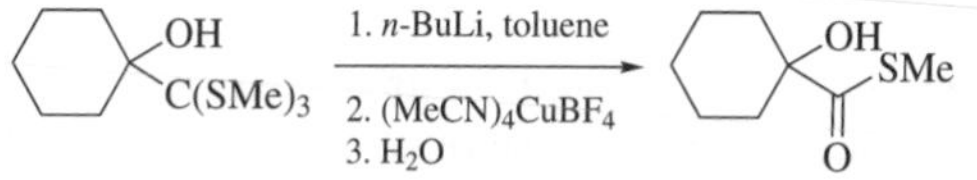

(64)

Chain-Extending Syntheses of α-Phenylsulfenyl Ketones. A C–C connective, chain-extending synthesis of α-phenylsulfenyl ketones from aldehydes (eq 65) or acyclic ketones (eq 66)[41] can be accomplished by a CuOTf-promoted activation of the α-[bis(phenylthio)methyl]alkanols generated by addition of lithium bis(phenylthio)methide.[34] Preferential migration of hydride generates phenylsulfenylmethyl ketones from aldehydes (eq 65). Regioselective insertion of a phenylsulfenylmethylene unit occurs owing to a preference for migration of the more highly substituted alkyl group of dialkyl ketones (eq 66).

Et–CHO + $LiCH(SPh)_2$ → (80%) Et–CH(OH)–CH(SPh)–SPh → (CuOTf, $EtN(i\text{-}Pr)_2$, benzene, 78 °C, 1 h, 66%) Et–CO–CH₂SPh (65)

$LiCH(SPh)_2$ + MeCO–i-Pr → (85%) Me–C(OH)(i-Pr)–$CH(SPh)_2$ → (CuOTf, $EtN(i\text{-}Pr)_2$, benzene, 78 °C, 1.5 h, 81%) MeCO–CH(SPh)–i-Pr (66)

Cyclopropanation of Enones. Conjugate addition of lithium tris(phenylthio)methide[44] to α,β-unsaturated ketones produces enolates that cyclize to bis(phenylthio)cyclopropyl ketones at −78 °C upon treatment with nearly one equivalent of $(CuOTf)_2{\cdot}C_6H_6$, i.e. 1.9 equivalents of Cu^I (eq 67).[45] The mild conditions that suffice to bring about nucleophilic displacement of thiophenoxide in the presence of CuOTf are especially noteworthy. In view of the requirement for more than one equivalent of Cu^I to achieve Grob-type fragmentation of β-[bis(phenylthio)methyl]alkoxides (see eq 58), it seems likely, although as yet unproven, that one equivalent of Cu^I coordinates strongly with the enolate oxygen and that a second equivalent of Cu^I is required to activate the thiophenoxide leaving group.

2-cyclohexenone + $LiC(SPh)_3$ → (−78 °C, THF) [lithium enolate with $C(SPh)_3$] → (CuOTf, 75%) 7,7-bis(phenylthio)bicyclo[4.1.0]heptan-2-one (67)

Vinylcyclopropanation of Enones. Conjugate addition of sulfur-stabilized allyl carbanions to α,β-unsaturated ketones produces enolates that cyclize to vinylcyclopropyl ketones upon treatment with nearly one equivalent of $(CuOTf)_2{\cdot}C_6H_6$, i.e. 1.9 equivalents of Cu^I (eq 68).[45]

Friedel–Crafts Acylation with Thio- or Selenoesters. Methylseleno esters are readily available in excellent yields by the reaction of ***Dimethylaluminum Methylselenolate*** with *O*-alkyl esters.[37] These selenoesters will acylate reactive arenes (eq 69) and heterocyclic compounds (eq 70) when activated by CuOTf, a selenophilic Lewis acid.[37] Of the potential activating metal salts tested, $(CuOTf)_2{\cdot}C_6H_6$ is uniquely effective. Mercury(II) or copper(I) trifluoroacetates that are partially organic-soluble, as well as the corresponding chlorides, silver nitrate, and copper(I) oxide that are not organic-soluble, all failed to promote any acylation. The highly reactive CuOTf–benzene complex, in dramatic contrast, was found to readily promote the acylations in benzene solution within minutes at room temperature. The presence of vinyl and keto groups is tolerated by the reaction, and while the alkyl- and vinyl-substituted derivatives afford *para* substitution only, a 2:1 mixture of *para* and *ortho* substitution occurs with the methylseleno ester of levulinic acid. Acylation of toluene is sluggish. Excellent yields of 2-acylfurans, -thiophenes, and -pyrroles are generated by this new variant of the Friedel–Crafts acylation reaction (eq 70). An intramolecular version of this reaction was shown to generate 1-tetralone from the methylseleno ester of γ-phenylbutyric acid (eq 71).[37]

2-cyclohexenone + Li–CH(SPh)–CH=C(X)–SPh → (−78 °C, THF) [lithium enolate] → (CuOTf) 2-oxobicyclo[4.1.0]heptane with CH=C(X)SPh (68)

X = H, 78%; SPh, 83%

RCOSeMe + MeO–C₆H₅ → (CuOTf (1.2 equiv), benzene, 3–40 min, rt) MeO–C₆H₄–COR (69)

R = $(CH_2)_5Me$	81%
R = $CH_2CH_2CH{=}CH_2$	63%
R = CH_2CH_2COMe	60%

$Me(CH_2)_5COSeMe$ + heteroarene (X) → (CuOTf, benzene, 15–20 min, rt) 2-acyl heteroarene–CO$(CH_2)_5Me$ (70)

X = O, 100%; S, 81%; NH, 64%

Ph(CH₂)₃COSeMe → (CuOTf (1.2 equiv), benzene, 5 min, rt, 70%) 1-tetralone (71)

Notwithstanding a prior claim that methylthio esters react only sluggishly under these conditions,[37] such a variant proved effective for a short synthesis of the 4-demethoxy-11-deoxyanthracycline skeleton (eq 72).[46] This is especially significant because methylthio esters are available by an efficient C–C connective process involving *C*-acylation of ketone lithium enolates with ***Carbon Oxysulfide*** (COS) followed by *S*-methylation with ***Iodomethane***.[46] For the deoxyanthracycline synthesis, the requisite enolate was generated by 1,4-addition of a silyl-stabilized benzyllithium derivative to 2-cyclohexenone. Treatment of the methylseleno ester with $(CuOTf)_2{\cdot}C_6H_6$ in benzene, according to the method employed with analogous seleno esters,[37] results in efficient cyclization to deliver a tetracyclic diketone in good yield.

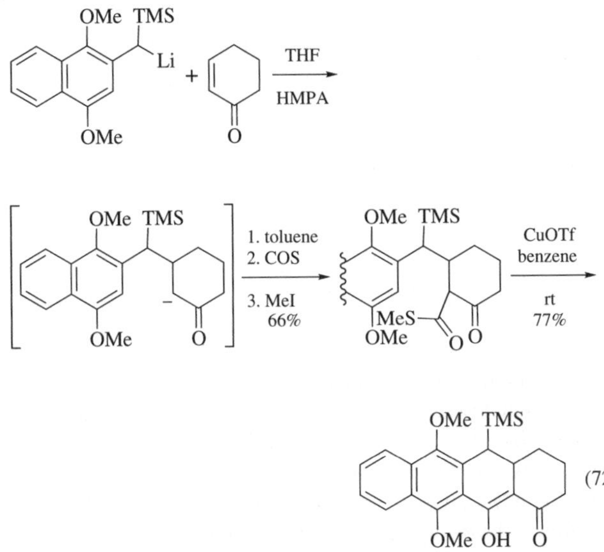

(72)

O*-Acylation with Thioesters.** Activation of a thioester with $(CuOTf)_2 \cdot C_6H_6$ was exploited as a key step in the synthesis of a macrocyclic pyrrolizidine alkaloid ester (eq 73).[47] Since thioesters are relatively unreactive acylating agents, a highly functionalized imidazolide containing acetate and *t*-butyl thioester groups selectively acylated only the primary hydroxyl in the presence of the secondary hydroxyl group in (+)-retronecine. Completion of the synthesis required activation of the *t*-butylthio ester. ***Mercury(II) Trifluoroacetate, that had proven effective for the synthesis of several natural products by lactonization,[48,49] failed to promote any lactonization in the present case.[47] Similarly, ***Mercury(II) Chloride*** and ***Cadmium Chloride***, that have proven effective for promoting lactonizations,[49] had no effect in the present case. Even copper(I) trifluoroacetate failed to induce the crucial lactonization. In contrast, CuOTf was uniquely effective for inducing the requisite macrolactonization by activating the thioester.

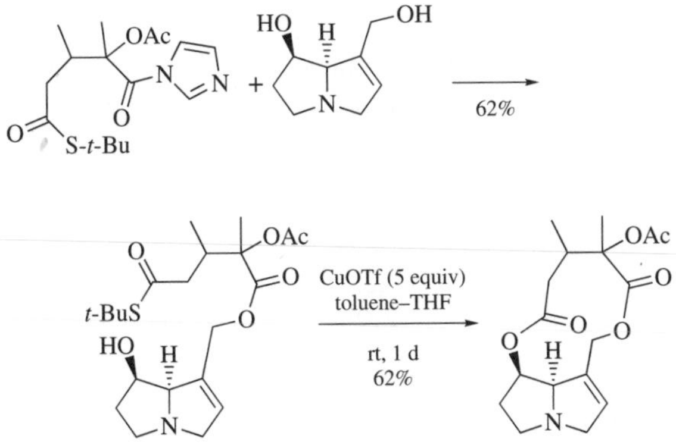

(73)

Related Reagents. Mercury(II) Acetate; Mercury(II) Chloride; Mercury(II) Oxide; Mercury(II) Trifluoroacetate; Thallium(III) Nitrate Trihydrate.

1. (a) Salomon, R. G. *Adv. Chem. Ser.* **1978**, *168*, 174. (b) Salomon, R. G. *T* **1983**, *39*, 485. (c) Salomon, R. G.; Kochi, J. K. *TL* **1973**, 2529.
2. Jenkins, C. L.; Kochi, J. K. *JACS* **1972**, *94*, 843.
3. (a) Hathaway, B. J.; Holah, D. G.; Postlethwaite, J. D. *JCS* **1961**, 3215. (b) Kubota, M.; Johnson, D. L. *J. Inorg. Nucl. Chem.* **1967**, *29*, 769.
4. (a) Salomon, R. G.; Kochi, J. K. *CC* **1972**, 559. (b) Salomon, R. G.; Kochi, J. K. *JACS* **1973**, *95*, 1889. (c) Dines, M. B. *Separ. Sci.* **1973**, *8*, 661.
5. Cohen, T.; Ruffner, R. J.; Shull, D. W.; Fogel, E. R.; Falck, J. R. *OS* **1980**, *59*, 202; *OSC* **1988**, *6*, 737.
6. Salomon, R. G.; Kochi, J. K. *JACS* **1973**, *95*, 3300.
7. Wiberg, K. B.; Kass, S. R.; Bishop, III, K. C. *JACS* **1985**, *107*, 996.
8. Lewis, R. T.; Motherwell, W. B. *TL* **1988**, *29*, 5033.
9. (a) Evans, D. A.; Woerpel, K. A.; Hinman, M. M.; Faul, M. M. *JACS* **1991**, *113*, 726. (b) Evans, D. A.; Woerpel, K. A.; Scott, M. J. *AG(E)* **1992**, *31*, 430.
10. Trecker, D. J.; Foote, R. S. In *Organic Photochemical Synthesis*; Srinivasan, R., Ed.; Wiley: New York, 1971; Vol. 1, p 81.
11. Salomon, R. G.; Kochi, J. K. *JACS* **1974**, *96*, 1137.
12. Salomon, R. G.; Kochi, J. K. *JACS* **1973**, *95*, 1889.
13. Salomon, R. G.; Folting, K.; Streib, W. E.; Kochi, J. K. *JACS* **1974**, *96*, 1145.
14. Evers, J. T. M.; Mackor, A. *RTC* **1979**, *98*, 423.
15. Evers, J. T. M.; Mackor, A. *TL* **1980**, *21*, 415.
16. Spee, T.; Mackor, A. *JACS* **1981**, *103*, 6901.
17. Evers, J. T. M.; Mackor, A. *TL* **1978**, 2317.
18. Salomon, R. G.; Sinha, A. *TL* **1978**, 1367.
19. Evers, J. T. M.; Mackor, A. *TL* **1978**, 821.
20. (a) Raychaudhuri, S. R.; Ghosh, S.; Salomon, R. G. *JACS* **1982**, *104*, 6841. (b) Ghosh, S.; Raychaudhuri, S. R.; Salomon, R. G. *JOC* **1987**, *52*, 83.
21. Avasthi, K.; Raychaudhuri, S. R.; Salomon, R. G. *JOC* **1984**, *49*, 4322.
22. Hertel, R.; Mattay, J.; Runsink, J. *JACS* **1991**, *113*, 657.
23. (a) Salomon, R. G.; Coughlin, D. J.; Easler, E. M. *JACS* **1979**, *101*, 3961. (b) Salomon, R. G.; Ghosh, S. *OS* **1984**, *62*, 125; *OSC* **1990**, *7*, 177. (c) Salomon, R. G.; Coughlin, D. J.; Ghosh, S.; Zagorski, M. G. *JACS* **1982**, *104*, 998.
24. Salomon, R. G.; Ghosh, S.; Zagorski, M. G.; Reitz, M. *JOC* **1982**, *47*, 829.
25. McMurry, J. E.; Choy, W. *TL* **1980**, *21*, 2477.
26. Avasthi, K.; Salomon, R. G. *JOC* **1986**, *51*, 2556.
27. Salomon, R. G.; Ghosh, S.; Raychaudhuri, S.; Miranti, T. S. *TL* **1984**, *25*, 3167.
28. Hefner, J. G.; Zizelman, P. M.; Durfee, L. D.; Lewandos, G. S. *JOM* **1984**, *260*, 369.
29. (a) Cohen, T.; Cristea, I. *JOC* **1975**, *40*, 3649. (b) Cohen, T.; Cristea, I. *JACS* **1976**, *98*, 748.
30. Cohen, T.; Tirpak, J. *TL* **1975**, 143.
31. Cohen, T.; Herman, G.; Falck, J. R.; Mura, Jr., A. J. *JOC* **1975**, *40*, 812.
32. Kwon, T. W.; Smith, M. B. *SC* **1992**, *22*, 2273.
33. Cohen, T.; Mura, A. J.; Shull, D. W.; Fogel, E. R.; Ruffner, R. J.; Falck, J. R. *JOC* **1976**, *41*, 3218.
34. Corey, E. J.; Seebach, D. *JOC* **1966**, *31*, 4097.
35. Cohen, T.; Kosarych, Z. *TL* **1980**, *21*, 3955.
36. Cohen, T.; Gapinski, R. E.; Hutchins, R. R. *JOC* **1979**, *44*, 3599.
37. (a) Kozikowski, A. P.; Ames, A. *JACS* **1980**, *102*, 860. (b) Kozikowski, A. P.; Ames, A. *T* **1985**, *41*, 4821.
38. (a) Corey, E. J.; Noyori, R. *TL* **1970**, 311. (b) Corey, E. J.; Erickson, B. W.; Noyori, R. *JACS* **1971**, *93*, 1724.
39. Cohen, T.; Bennett, D. A.; Mura, A. J. *JOC* **1976**, *41*, 2506.
40. Semmelhack, M. F.; Tomesch, J. C. *JOC* **1977**, *42*, 2657.
41. Cohen, T.; Kuhn, D.; Falck, J. R. *JACS* **1975**, *97*, 4749.
42. Knapp, S.; Trope, A. F.; Ornaf, R. M. *TL* **1980**, *21*, 4301.

43. Knapp, S.; Trope, A. F.; Theodore, M. S.; Hirata, N.; Barchi, J. J. *JOC* **1984**, *49*, 608.
44. Seebach, D. *AG(E)* **1967**, *6*, 442.
45. Cohen, T.; Meyers, M. *JOC* **1988**, *53*, 457.
46. Vedejs, E.; Nader, B. *JOC* **1982**, *47*, 3193.
47. Huang, J.; Meinwald, J. *JACS* **1981**, *103*, 861.
48. (a) Masamune, S. *Aldrichim. Acta* **1978**, *11*, 23. (b) Masamune, S.; Yamamoto, H.; Kamata, S.; Fukuzawa, A. *JACS* **1975**, *97*, 3513.
49. Masamune, S.; Kamata, S.; Schilling, W. *JACS* **1975**, *97*, 3515.

Robert G. Salomon
Case Western Reserve University, Cleveland, OH, USA

Copper(II) Trifluoromethanesulfonate

$Cu(OSO_2CF_3)_2$

[34946-82-2] $C_2CuF_6O_6S_2$ (MW 361.68)

(dimerization of ketone enolates and TMS enol ethers;[2a,2b] cyclization of dienolates[2c] and unsaturated silyl enol ethers;[4] allylation of ketones;[3] reactions of diazo compounds;[5–10] dehydration of alcohols[11])

Alternate Name: copper(II) triflate.
Physical Data: dec at 530 °C (no definite mp).
Solubility: sol MeOH, EtOH, DMF, MeCN, and formamide; also sol *i*-PrCN and acetone.
Form Supplied in: white powder, commercially available. Blue powder when freshly prepared (see below).
Preparative Methods: most conveniently prepared from copper(II) carbonate and triflic acid (***Trifluoromethanesulfonic Acid***) in MeCN.[1] The freshly prepared salt precipitated from Et_2O is pale blue.
Handling, Storage, and Precautions: moisture sensitive; can be handled in air for quick transfers; pure samples are only mildly corrosive. Appears to be indefinitely stable in the absence of air, moisture, and light.

Oxidative Coupling. Both intermolecular and intramolecular oxidative coupling reactions can be effected using $Cu(OTf)_2$. Examples of dimerization include one-pot syntheses of 1,4-diketones from ketone enolates or from silyl enol ethers (eqs 1 and 2),[2a,2b] and coupling of allylstannanes with TMS-enol ethers to give γ,δ-unsaturated ketones in good to moderate yields (eq 3). Other copper(II) or tin(IV) catalysts can also be used with allylstannanes. The regiochemistry depends on both the substrate and the catalyst.

2 $RC(OLi)=CR^1R^2$ + 2 $Cu(OTf)_2$ —(THF, *i*-PrCN)→ $RCOCR^1R^2CR^1R^2COR$ + 2 CuOTf + 2 LiOTf (1)

2 $R^1C(OTMS)=CHR^2$ —($Cu(OTf)_2$, Cu_2O, *i*-PrCN)→ $R^1COCHR^2CHR^2COR^1$ (2)

R^1 = Ph, R^2 = H 55%
R^1 = Ph, R^2 = Me 27%

(3) [allylstannane (R^1, R^2, $SnBu_3$, R^3) + TMS enol ether (R^4, OTMS, R^5) —$Cu(OTf)_2$→ two regioisomeric γ,δ-unsaturated ketones]

Examples of intramolecular oxidative cyclizations promoted by $Cu(OTf)_2$ include cyclization of enolates of diketones and diesters (eq 4)[2c] and oxidative cyclization of hydrolytically resistant δ,ε- and ε,ζ-unsaturated silyl enol ethers.[4] For instance, (**1**) reacts with excess $Cu_2O/Cu(OTf)_2$ in MeCN to give a 90% yield of a 20:1 mixture of the *trans*-fused and *cis*-fused tricyclic ketones (**2**) and (**3**) (eq 5).[4]

(4) [diketone (R^1, R^2) —LDA→ [dilithium dienolate] → 2,2-disubstituted cyclopentane-1,3-dione]

(5) [TBDMSO silyl enol ether (**1**) —$Cu(OTf)_2$ or CAN→ (**2**) + (**3**); H_a, H_b, H_c]

Reactions of Diazo Compounds. These include various cyclopropanations where the metal carbenoid generated from the diazoacetic acid esters/$Cu(OTf)_2$ system reacts with alkenes (eq 6) or carbodiimides (eq 7).[5,6]

(6) [alkene —N_2CHCO_2Et, $Cu(OTf)_2$→ cyclopropane-CO_2Et]

(7) [diisopropylcarbodiimide —N_2CHCO_2R, $Cu(OTf)_2$→ two aziridine products with CO_2R]

$Cu(OTf)_2$ is the reagent of choice for intramolecular cyclization of β,γ-unsaturated diazo ketones to cyclopentenones (eq 8) and for intramolecular cyclopropanation of γ,δ-unsaturated diazo ketones (eq 9).[7]

CHN2 (8)

H
CHN2 H (9)

It is a useful reagent for orthoester homologation via dialkoxycarbenium ions and for oxazole formation by reaction of ketocarbenes (via diazo esters/$Cu(OTf)_2$) with nitriles (eq 10).[8] With unsaturated nitriles, the nitrile group is selectively attacked. Kinetic and ESR evidence shows that $Cu^{II} \rightarrow Cu^{I}$ reduction is the key step.[9]

N + $N_2CHC(O)OBu$ —$Cu(OTf)_2$→ OBu (10)

The regiochemistry of the monoacetate adducts formed from *N*-methylpyrrole with the carbenoid derived from $N_2CHCO_2Et/Cu(OTf)_2$ is indicative of a reactive and less discriminating intermediate compared to carbenoids generated from N_2CHCO_2Et with other Cu^{II} reagents (eq 11).[10]

N-Me —N_2CHCO_2Et, $Cu(OTf)_2$→ N-Me (CO_2Et) + N-Me (CO_2Et) (11)

Angularly functionalized polycyclic systems may be prepared from β,γ-unsaturated diazo ketones by a vinylogous Wolff rearrangement in the presence of copper(II) triflate. Dry ***Copper(II) Acetylacetonate*** is equally suitable.[10]

Dehydration of Alcohols. Various tertiary, secondary, and primary alcohols are dehydrated with $Cu(OTf)_2$ to alkenes.[11] Preferred formation of Zaitsev orientation products and (*E*)-alkenes are indicative of a carbocationic mechanism. $Zn(OTf)_2$ and $Mg(OTf)_2$ are ineffective. In selected examples, yields are superior to those of H_2SO_4 and $POCl_3$/pyridine dehydrations.[11] See also ***Copper(II) Sulfate***.

Elimination Reactions. A $Cu(OTf)_2$/Hünig's base combination provides a useful method for converting 1,1-bis-(phenylthio)cyclobutanes to 1-(phenylthio)cyclobutenes (eq 12).[12]

SPh, SPh, R → SPh, R (12)

Reduction of Alkynyl Sulfones. The $HSiEt_2Me/Cu(OTf)_2$ system reduces alkynyl sulfones to *cis* vinylic sulfones (eq 13).[13] The yields with ***Copper(II) Tetrafluoroborate*** are higher than those with $Cu(OTf)_2$. Dimeric side products are formed in some cases.

SO_2Ph —$HSiEt_2Me$, $Cu(OTf)_2$, *i*-PrOH→ SO_2Ph (13)

Oxazoles from Ketones. Aliphatic ketones react with nitriles in the presence of $Cu(OTf)_2$ and catalytic amounts of ***p-Toluenesulfonic Acid*** in refluxing MeCN to give oxazoles.[14] The oxazole produced in this way is isomeric with that formed from a ketone and an amide (eq 14).

R^1, O, R^2 —MeCN, $Cu(OTf)_2$→ R^1, O, R^2, N; —$MeCONH_2$, Br_2→ R^1, N, R^2, O (14)

Oxidation of Alkyl Radicals. Various alkyl radicals are oxidized with Cu^{II} triflate or perchlorate to carbenium ions whose reactivities are similar to solvolytically formed cations (eq 15).[1] The synthetic utility of such $Cu(OTf)_2$-catalyzed oxidations remains to be explored.

$$R\cdot + Cu^{II}(OTf)_2 \rightleftharpoons RCu(OTf)_2 \rightleftharpoons RCuOTf^{+}\,OTf^{-} \rightleftharpoons RCu^{2+}\,2\,OTf^{-} \quad (15)$$

$$RCuOTf^{+}\,OTf^{-} \rightarrow Cu^{I}(OTf) + R^{+}$$

Related Reagents. Copper(II) Bromide; Copper(II) Chloride; Copper(II) Sulfate; Copper(II) Tetrafluoroborate; Copper(I) Trifluoromethanesulfonate.

1. Jenkins, C. L.; Kochi, J. K. *JACS* **1972**, *94*, 843.
2. (a) Kobayashi, Y.; Taguchi, T.; Tokuno, E. *TL* **1977**, 3741. (b) Kobayashi, Y.; Taguchi, T.; Morikawa, T.; Tokuno, E.; Sekiguchi, S. *CPB* **1980**, *28*, 262. (c) Kobayashi, Y.; Taguchi, T.; Morikawa, T. *TL* **1978**, 3555.
3. Takeda, T.; Ogawa, S.; Koyama, M.; Kato, T.; Fujiwara, T. *CL* **1989**, 1257.
4. Snider, B. B.; Kwon, T. *JOC* **1992**, *57*, 2399.
5. (a) Andrist, A. H.; Agnello, R. M.; Wolfe, D. C. *JOC* **1978**, *43*, 3422. (b) Doyle, M. P.; Dorow, R. L.; Buhro, W. E.; Griffin, J. H.; Tamblyn, W. H.; Trudell, M. L. *OM* **1984**, *3*, 44.
6. Hubert, A. J.; Feron, A.; Warin, R.; Teyssie, P. *TL* **1976**, 1317.
7. Doyle, M. P.; Trudell, M. L. *JOC* **1984**, *49*, 1196.
8. Moniotte, Ph. G.; Hubert, A. J.; Teyssie, Ph. *JOM* **1975**, *88*, 115.
9. Maryanoff, B. E. *JOC* **1979**, *44*, 4410.
10. Saha, B.; Bhattacharjee, G.; Ghatak, U. R. *TL* **1986**, *27*, 3913.
11. Laali, K.; Gerzina, R. J.; Flajnik, C. M.; Geric, C. M.; Dombroski, A. M. *HCA* **1987**, *70*, 607.
12. Kwon, T. W.; Smith, M. B. *SC* **1992**, *22*, 2273.

13. Ryu, I.; Kusumoto, N.; Ogawa, A.; Kambe, N.; Sonoda, N. *OM* **1989**, *8*, 2279.

14. Nagayoshi, K.; Sato, T. *CL* **1983**, 1355.

Kenneth K. Laali
Kent State University, OH, USA

18-Crown-6

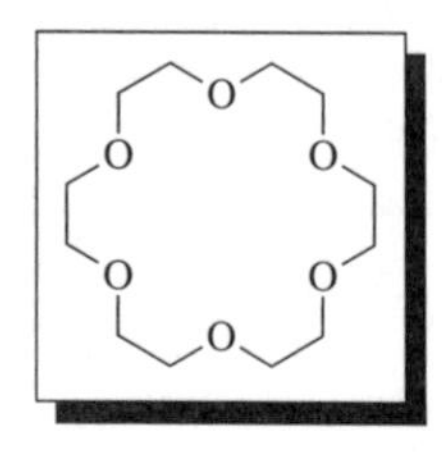

[17455-13-9] $C_{12}H_{24}O_6$ (MW 264.32)

(solubilization of a variety of metal salts, particularly potassium salts, in nonpolar solvents; activating agent for many nucleophilic substitutions and eliminations)

Physical Data: colorless, mp 36.5–38.0 °C; IR (neat) 2875, 1450, 1350, and 1120 cm^{-1}; ^1H NMR (60 MHz, CCl_4) 3.56 ppm (singlet).

Preparative Methods: commercially available. 18-Crown-6 has been synthesized by a variety of methods involving modified Williamson ether procedures.[1–6] Crude 18-crown-6 forms a crystalline complex with ***Acetonitrile*** from which the pure crown can be isolated.[4,5] 18-Crown-6 may also be isolated from the cyclic oligomerization of ***Ethylene Oxide*** in the presence of gaseous ***Boron Trifluoride*** and alkali metal cation templates such as $CsBF_4$.[7–10]

Handling, Storage, and Precautions: use in a fume hood.

Complexation with Metal Salts: Solubilities and Reactivities. 18-Crown-6 provides a simple and efficient means of solubilizing simple metal salts in nonpolar and dipolar aprotic solvents where solvation of the anionic portion of the salt should be minimal.[11] Indeed, this ligand is an effective catalyst in liquid–liquid and solid–liquid phase transfer catalysis.[12–14] Since 18-crown-6 has cavity dimensions (2.6–3.2 Å) of the same magnitude as the ionic diameter of the potassium cation (2.66 Å), it is usually more specific for potassium salts than other alkali metal salts. Nevertheless, 18-crown-6 is reasonably effective in complexing sodium and cesium salts as well. The solubilities of potassium salts in acetonitrile at 25 °C in the presence and absence of 18-crown-6 (0.15 M) are summarized in Table 1.[11]

The differences between the solubilities in the presence and the absence of crown are indications of the solubility enhancements due to the presence of 18-crown-6. The solubilities of ***Potassium Fluoride*** in benzene containing 18-crown-6 at concentrations of 1.01 M and 0.34 M are reported to be 0.052 M and 0.014 M, respectively, while the solubilities of potassium acetate in benzene containing 18-crown-6 at concentrations of 0.55 M and 1.0 M are reported to be 0.4 M and 0.8 M, respectively.[11] The solubilities of ***Lithium Fluoride***, ***Sodium Fluoride***, KF, RbF, and ***Cesium Fluoride*** in acetonitrile, ***Acetone***, THF, ***N,N-Dimethylformamide***, benzene, and cyclohexane have been determined in the presence and absence of 18-crown-6.[15]

Table 1 Solubilities of Potassium Salts in Acetonitrile at 25 °C in the Presence and Absence of 18-Crown-6

Salt	Sol. in 0.15 M 18-C-6 in MeCN	Sol. in MeCN
KF	0.0043	0.000318
KCl	0.0555	0.000243
KBr	0.135	0.00208
KI	0.202	0.105
KCN	0.129	0.00119
KOAc	0.102	0.00050
KN_3	0.138	0.00241
KSCN	0.850	0.755

The complexation and solubilization of metal salts by 18-crown-6 produces highly reactive anions. There are at least two factors that contribute the this enhanced reactivity. Since a nonpolar aprotic solvent such as benzene or a dipolar aprotic solvent such as acetonitrile should not have a great affinity for the anion, as compared to polar, protic media, the anion is not expected to be highly solvated. In addition, the increased physical separation of the cation from the anion as a result of the complexation should decrease the coulombic interaction between these oppositely charged species.[13] This latter effect is translated directly into a decrease in activation energy. As a consequence, the anion is a potent nucleophile as well as a potent base. These anionic species have been termed 'naked' anions. Studies related to the relative nucleophilicities of 'naked' anions toward benzyl tosylate in acetonitrile at 30 °C are summarized in Table 2.[11,13,16] Compared to polar, protic media, there seems to be a general leveling of nucleophilicities in acetonitrile. Several reversals of the usual order of nucleophilicities may be noted.

Table 2 Second-Order Rate Constants at 30 °C in Acetonitrile for the Reaction with Benzyl Tosylate

Nucleophile	k (M^{-1} s^{-1})	Relative rates
N_3^-	1.02	10.0
OAc^-	0.95	9.6
CN^-	0.23	2.4
Br^-	0.12	1.3
Cl^-	0.12	1.3
I^-	0.09	1.0
F^-	0.14	1.4
SCN^-	0.02	0.3

Reactions.

Halides. Potassium fluoride reacts with a variety of organic substrates under solid–liquid phase transfer catalytic conditions[17] to give both substitution and elimination products (eq 1).

(1)

The σ-anionic complex (Meisenheimer complex) resulting from the reaction of fluoride with 2,4,6-trinitrofluorobenzene (**1**) in acetonitrile has been observed using ^{1}H and ^{19}F NMR spectroscopy.[18]

(1)

Potassium fluoride or cesium fluoride solubilized in toluene with 18-crown-6 induces alkyl- and aryl-group rearrangement on chloromethyl-substituted silanes.[19]

Oxygen Anions. Acetate solubilized as the potassium salt in acetonitrile or benzene containing 18-crown-6 becomes sufficiently nucleophilic to react smoothly and quantitatively, even at room temperature, with a wide variety of organic substrates.[20,21] Substitution reactions at primary, secondary, tertiary, and benzylic positions have been demonstrated.[18] In certain cases, competing elimination processes are observed. Mixed anhydrides are formed in the reaction of the potassium or sodium salts of carboxylic acids with ***Ethyl Chloroformate***, cyanuric chloride, or ***Benzyl Chloroformate*** in acetonitrile in the presence of 18-crown-6.[22] Potassium phenylacetate has been reacted with a series of 2-bromo-substituted carbonyl compounds in the presence of 18-crown-6 to form aldehydo or keto esters which were subsequently cyclized to five-membered unsaturated lactones on further heating (eq 2).[23]

(2)

3-Bromoacetyl-7-methoxycoumarin is readily coupled with a wide variety of carboxylic acids in the presence of potassium bicarbonate and 18-crown-6[24] to form fluorescent derivatives. Reaction of alcohol mesylate with cesium acetate and 18-crown-6 in benzene is an effective method for the inversion of cyclopentyl and cyclohexyl alcohols.[25]

Reaction of 3-(bromomethyl)thiophene with the potassium salt of 4-cyano-4′-hydroxybiphenyl in THF, in the presence of 18-crown-6, produces high yields of the substitution product.[26]

Potassium Superoxide, dissolved in benzene, THF, or DMF using 18-crown-6, reacts with stoichiometric quantities of primary and secondary alkyl bromides to produce dialkyl peroxides.[27] The nucleophilic displacements by the superoxide radical anion and the intermediate alkyl peroxyanion on a chiral alkyl bromide ((*R*)-2-bromooctane) proceeds with inversion of configuration.[26] The nucleophilic reaction of alkyl halides, methanesulfonates, and tosylates in ***Dimethyl Sulfoxide***, DMF, or DME with 4 equiv of potassium superoxide in the presence of 18-crown-6 produces excellent yields of the corresponding alcohols (eq 3).[28,29]

When secondary halides were used, some elimination products accompanied the formation of the alcohol product. In general, substrate reactivity varied as follows: benzyl > primary > secondary > tertiary > aryl and I > Br > OTs > Cl. Reaction of the tosylate of (+)-(*S*)-2-octanol with superoxide produced (−)-(*R*)-2-octanol.[28] The reaction of esters of carboxylic acids with approximately a threefold excess of potassium superoxide solubilized in benzene with 18-crown-6 produces, after acidic workup, the corresponding alcohols and carboxylic acids.[30] Reaction of superoxide with the acetate ester of (−)-(*R*)-2-octanol gave only (−)-(*R*)-2-octanol, indicating that the reaction proceeds by means of an acyl–oxygen cleavage.[29] Potassium superoxide dissolved in benzene with 18-crown-6 is an effective reagent for promoting the oxidative cleavage of α-keto, α-hydroxy, and α-halo ketones, esters, and carboxylic acids to carboxylic acids.[31]

(3)

18-Crown-6 enhances the rate of anionic oxy-Cope rearrangement by ion-pair dissociation.[32]

The reaction of ethyl tosylate and ***Ethyl Iodide*** with the sodium salt of ***Ethyl Acetoacetate*** in THF has been studied in the absence and presence of 18-crown-6. In the reaction with ethyl tosylate the presence of the crown increases the amount of oxygen alkylation. In contrast, the crown has little effect on the ratio of oxygen to carbon alkylation when ethyl iodide is used.[33]

An addition reaction of 3-benzyloxymethyl-3-methyloxetane with *S*-phenyl thioacetate in the presence of 18-crown-6/potassium phenoxide gave 3-benzyloxy-2-methyl-2-phenylthiomethylpropyl acetate in excellent yields.[34]

The dibenzyl ether of optically active diethyl tartrate has been prepared on a large scale using ***Sodium Hydride***, ***Benzyl Bromide***, ***Tetra-n-butylammonium Iodide***, and a catalytic quantity of 18-crown-6.[35]

Carbon Anions. Potassium Cyanide dissolved in acetonitrile using 18-crown-6 reacts with a variety of alkyl halides to produce the corresponding alkyl cyanides.[36] As in the cases involving fluoride and acetate, some elimination processes accompanied substitution in certain cases. Under solid–liquid phase transfer catalytic conditions, primary alkyl chlorides react faster than bromides whereas the opposite is true for the corresponding secondary halides. ***Cyanotrimethylsilane*** has been synthesized from ***Chlorotrimethylsilane***.[37] *cis*-2-Chloro-4-methylcyclohexanone reacts with ‘naked’ cyanide to exclusively produce the substitution product.[11] Hydrocyanation of α,β-unsaturated nitriles and ketones with ‘naked’ cyanide in the presence of ***Acetone Cyanohydrin*** has been reported (eq 4).[38]

(4)

Either 1,4-, 2,4-, or 1,5-hexadiene has been deprotonated with Cs 18-crown-6 solutions to produce hexadienyl anions.[39]

The isomerization of 2-methylbicyclo[2.2.1]hepta-2,5-diene to 5-methylenebicyclo[2.2.1]hept-2-ene using excess ***Potassium t-Butoxide*** in DMSO in the presence and in the absence of an equivalent concentration of 18-crown-6 has been investigated (eq 5).[40] In the absence of crown, the order with respect to the base changed from zero order at high concentrations to first order at more dilute concentrations. In the presence of crown, the order with respect to base remained first order at all reported concentrations.

(5)

2,2,3-Triphenylpropylcesium rearranges in THF at 65 °C with 96% 1,2-migration of phenyl, whereas the corresponding lithium salt rearranges with at least 98% 1,2-migration of benzyl at 0 °C. The addition of 18-crown-6 as a ligand for the potassium and cesium salts significantly increases the extent of 1,2-benzyl migration.[41]

Treatment of trimethylsilyl trichloroacetate with α,α,α-trifluoroacetophenone in the presence of catalytic quantities of 18-crown-6 and ***Potassium Carbonate*** at 150 °C, followed by treatment with methanolic ***Potassium Hydroxide*** at 50 °C, produced excellent yields of Mosher's acid (α-methoxy-α-(trifluoromethyl)phenylacetic acid).[42]

18-Crown-6 is an effective catalyst in the dialkylation of *o*-nitrophenacyl derivatives.[43]

Nitrogen Anions. *N*-Propargylpyrrole was prepared by the reaction of pyrrole with powdered potassium hydroxide in toluene, catalyzed by 18-crown-6.[44]

The reaction of 'naked' nitrite with primary alkyl halides forms nitro compounds as the major product; the major byproducts are nitrite esters.[23]

Excellent yields of *N*-alkylation products were obtained in the reaction of dry alkali metal salts of 2-nitroimidazoles and methyl bromoacetate under homogeneous conditions at room temperature in the presence of 18-crown-6 in acetonitrile.[45]

Organic carbamates have been prepared in good yields from the reaction of primary amines, carbon dioxide, and an alkyl halide in the presence of 18-crown-6.[46]

Nitrogen Cations. Photolysis of 1-aminopyridinium salts, 2-aminoisoquinolinium salts, and 1-aminoquinolinium salts is reported to give aniline or a mixture of 2-, 3-, and 4-toluidines in benzene–trifluoroacetic acid or in toluene–trifluoroacetic acid.[47] An intermediate nitrenium ion is postulated. The presence of 18-crown-6 increases the yields of the above products.

Indazoles are produced in good yields from the reaction of *o*-methyl- and *o*-ethylbenzenediazonium tetrafluoroborates with two equivalents of potassium acetate in the presence of catalytic quantities of 18-crown-6 in ethanol-free chloroform.[48]

Other Anions. Reaction of potassium dihydrogenphosphide with aromatic esters in the presence of 18-crown-6 produces potassium benzoylphosphide.[49] Reaction of potassium dihydrogenphosphide with diethyl phthalate produces the potassium 18-crown-6 complex salt of the 2*H*-isophosphindoline-1,3-dione ion (**2**).[50]

Reductions. Alkoxysulfonium salts dissolved in methylene chloride are smoothly reduced with ***Sodium Cyanoborohydride*** dissolved in methanol or ethanol in the presence of 18-crown-6.[51] Even in the presence of aldehydes or ketones, only reduction of alkoxysulfonium salt was observed.

K^+/18-crown-6

(**2**)

The titanium complex generated by the reaction of ***Dichlorobis(cyclopentadienyl)titanium*** with ***Sodium Borohydride*** promotes hydroboration of alkenes and alkynes in the presence of 18-crown-6.[52,53]

Radical anions of mesitylene, toluene, and benzene are formed in the reaction with alkali metals and 18-crown-6.[54] The alkyl–oxygen bond of oxetane is cleaved with K^-/K^+ and Na^-/K^+ complexes in which the potassium cation has been complexed with 18-crown-6 to form an organometallic alkoxide.[55]

Hydrosilylation of carbonyl groups with ***Dimethyl(phenyl)silane*** proceeds in methylene chloride, benzene, or THF, in the presence of catalytic quantities of cesium, rubidium, or potassium fluoride/18-crown-6.[56]

Oxidations. The homogeneous photosensitization of oxygen by solubilizing the anionic dyes Rose Bengal and Eosin Y in methylene chloride and ***Carbon Disulfide*** using 18-crown-6 has been reported. The presence of singlet ***Oxygen*** was demonstrated by trapping with anthracene and tetramethylethylene.[57]

The carbanions from tri- and diarylmethanes are generated and oxidized to triaryl carbinols and diaryl ketones, respectively, using potassium hydroxide/DME/18-crown-6 in the presence of oxygen.[58]

Chromium(VI) Oxide, in the presence of 18-crown-6 catalyst in methylene chloride, is an efficient oxidizing system for the chemoselective oxidation of thiols to disulfides.[59]

Eliminations. The generation of 'free' carbenes was demonstrated in the reaction of ***Potassium t-Butoxide*** with ***Chloroform***, α-bromo-α-fluorotoluene, and α,α-dichlorodimethyl sulfide in the presence of 18-crown-6 from competitive reactivities toward alkenes.[60,61] ***Diazomethane*** was synthesized in 48% yield from the reaction of ***Hydrazine*** hydrate, chloroform, and potassium hydroxide in ether in the presence of catalytic quantities of 18-crown-6.[62]

Reaction of *exo*-2-norbornyl-*exo*-3*d* tosylate with the sodium salt of 2-cyclohexylcyclohexanol in triglyme produces norbornene containing no deuterium. This represents an exclusive *syn* elimination process.[63] In the presence of 18-crown-6, 27% of the product contains deuterium, thus indicating some *anti* elimination.[63] The results were interperted in terms of the effects of crown on base association and ion-pairing.

Polymerization. In the polymerization of methacrylic esters and hindered alkyl acrylates, the presence of 18-crown-6 produces living polymers in apolar solvents, such as toluene, and at temperatures as high as 0 °C.[64] The polymerization of β-lactones proceeds smoothly in the presence of ***Potassium Naphthalenide*** only after the addition of a cation complexing ligand such as 18-crown-6.[65]

The anionic polymerization of ***Propylene Oxide*** and ethylene oxide is catalyzed by the potassium salt of methoxypropanol in the presence of 18-crown-6.[66] The monomer 5-(bromomethyl)-1,3-dihydroxybenzene undergoes self-condensation in the presence of potassium carbonate and 18-crown-6 to give hyperbranched polyethers.[67] The synthesis of soluble, substituted silane high polymers by Wurtz coupling techniques is facilitated by the presence of 18-crown-6.[68] Interfacial polycondensation to produce poly(arylcarboxylate)s,[69] polyphosphates,[70] and polyphosphonates[46] has been studied in the presence of 18-crown-6. Polycondensations of 4-bromomethylbenzyl bromide were carried out with 4,4′-oxydibenzenesulfinate in the presence of 18-crown-6.[71]

Reactions on Polymers. 18-Crown-6 facilitates the alkaline hydrolysis of nitrile groups in acrylonitrile–divinylbenzene copolymer.[72] Starting from linear polycarbonate, carbonate–formal copolymers were prepared by reaction with ***Dibromomethane***, potassium hydroxide, and 18-crown-6.[73] Polystyrene has been successfully grafted onto carbon whiskers by anionic graft polymerization of styrene in which OLi groups on the surface of the carbon whiskers are activated by 18-crown-6.[74] The reaction of chloromethylated poly(styrene-*co*-divinylbenzene) with potassium superoxide in the presence of 18-crown-6 has been reported.[75]

Related Reagents. Dibenzo-18-crown-6; Dicyclohexano-18-crown-6; Potassium *t*-Butoxide-18-Crown-6; Potassium Carbonate-18-Crown-6; Potassium Hydroxide-18-Crown-6.

---

1. Pedersen, C. J. *JACS* **1967**, *89*, 7017.
2. Greene, R. N. *TL* **1972**, 1793.
3. Dale, J.; Kristiansen, P. O. *ACS* **1972**, *26*, 1471.
4. Gokel, G. W.; Cram, D. J.; Liotta, C. L.; Harris, H. P.; Cook, F. L. *JOC* **1974**, *39*, 2445.
5. Gokel, G. W.; Cram, D. J.; Liotta, C. L.; Harris, H. P.; Cook, F. L. *OS* **1977**, *57*, 30.
6. Johns, G.; Ransom, C. J.; Reese, C. B. *S* **1976**, 515.
7. Dale, J., Ger. Patent 2 523 542, 1975.
8. Dale, J.; Daasvatn, K. *CC* **1976**, 295.
9. Dale, J.; Borgen, G.; Daasvatn, J. *ACS* **1974**, *B28*, 378.
10. Dale, J.; Daasvatn, K., U.S. Patent 3 997 563, 1976.
11. Liotta, C. L. In *Synthetic Multidentate Macrocyclic Compounds*; Izatt, R. M.; Christensen, J. J., Eds.; Academic: New York, pp 111–205.
12. Dehmlow, E. V. *Phase Transfer Catalysis*; 2nd ed.; Verlag Chemie: Deerfield Beach, FL, 1983.
13. Starks, C. M.; Liotta, C. L. *Phase Transfer Catalysis: Principles and Applications*; Academic: New York, 1978.
14. Weber, W.; Gokel, G. *Phase Transfer Catalysis in Organic Synthesis*; Springer: New York, 1977.
15. Wynn, D. A.; Roth, M. M.; Pollard, B. D. *Talanta* **1984**, *31*, 1036.
16. Liotta, C. L.; Grisdale, E. E.; Hopkins, H. P. *TL* **1975**, 4205.
17. Liotta, C. L., Harris, H. P. *JACS* **1974**, *96*, 2250.
18. Terrier, F.; Ah-Kow, G.; Pouet, M.; Simonnin, M. *TL* **1976**, 227.
19. Damrauer, R.; Danahey, S. E.; Yost, V. E. *JACS* **1984**, *106*, 7633.
20. Liotta, C. L.; Harris, H. P.; McDermott, M.; Gonzalez, T.; Smith, K. *TL* **1974**, 2417.
21. Knochel, A.; Oehler, J.; Rudolph, G. *TL* **1975**, 3167.
22. Mack, M. M.; Dehm, D.; Boden, R.; Durst, H. D. Personal communication.
23. Dehm, D.; Padwa, A. *JOC* **1975**, *40*, 3139.
24. Takadate, A.; Masuda, T.; Tajima, C.; Murata, C.; Irikura, M.; Goya, S. *Anal. Sci.* **1992**, *8*, 663.
25. Torisawa, Y.; Okabe, H.; Ikegami, S. *CL* **1984**, 1555.
26. Bryce, M. R.; Chissel, A. D.; Gopal, J.; Kathirgamanathan, P.; Parker, D. *Synth. Met.* **1991**, *39*, 397.
27. Johnson, R. A.; Nidy, E. G. *JOC* **1975**, *40*, 1680.
28. Corey, E. J.; Nicolaou, K. C.; Shibasaki, M.; Machida, Y.; Shiner, C. S. *TL* **1975**, 3183.
29. San Filippo, J.; Chern, C.; Valentine, J. S. *JOC* **1975**, *40*, 1678.
30. San Filippo, J.; Chern, C.; Valentine, J. S. *JOC* **1976**, *41*, 1077.
31. San Filippo, J.; Romano, L. J.; Chern, C.; Valentine, J. S. *JOC* **1976**, *41*, 586.
32. Evans, D. A.; Golob, A. M. *JACS* **1975**, *97*, 4765.
33. Cambillau, C.; Sarthou, P.; Bram, G. *TL* **1976**, 281.
34. Nishikubo, T.; Sato, K. *CL* **1991**, 697.
35. Nemoto, H.; Takamatsu, S.; Yamamoto, Y. *JOC* **1991**, *56*, 1321.
36. Cook, F. L.; Bowers, C. W.; Liotta, C. L. *JOC* **1974**, *39*, 3416.
37. Zubrick, J. W.; Dunbar, B. I.; Durst, H. D. *TL* **1975**, 71.
38. Liotta, C. L.; Dabdoub, A.; Zalkow, L. H. *TL* **1977**, 1117.
39. Goel, S. C.; Grovenstein, E. *OM* **1992**, *11*, 1565.
40. Maskornick, M. J. *TL* **1972**, 1797.
41. Grovenstein, E.; Williamson, R. E. *JACS* **1975**, *97*, 646.
42. Goldberg, Y.; Alper, H. *JOC* **1992**, *57*, 3731.
43. Prasad, G.; Hanna, P. E. *JOC* **1991**, *56*, 7188.
44. van Eyk, S. J.; Naarmann, H.; Walker, N. P. C. *Synth. Met.* **1992**, *48*, 295.
45. Long, A.; Parrick, J.; Hodgkiss, R. J. *S* **1991**, 709.
46. Aresta, M.; Quaranta, E. *T* **1992**, *48*, 1515.
47. Takeuchi, H.; Higuchi, D.; Adachi, T. *JCS(P1)* **1991**, 1525.
48. Bartsch, R. A.; Yang, I-W. *JHC* **1984**, *21*, 1063.
49. Liotta, C. L.; McLaughlin, M. L.; O'Brien, B. A. *TL* **1984**, 1249.
50. Liotta, C. L.; McLaughlin, M. L.; Van Derveer, D. G.; O'Brien, B. A. *TL* **1984**, 1665.
51. Durst, H. D.; Subrick, J. W.; Kieczykowski, G. R. *TL* **1974**, 1777.
52. Lee, H. S.; Isagawa, K.; Otsuji, Y. *CL* **1984**, 363.
53. Lee, H. S.; Isagawa, K.; Toyoda, H.; Otsuji, Y. *CL* **1984**, 673.
54. Nelson, G. V.; von Zelewsky, A. *JACS* **1975**, *97*, 6279.
55. Jedlinski, Z.; Misiolek, A.; Jankowski, A.; Janeczek, H. *JOM* **1992**, *433*, 231.
56. Goldberg, Y.; Abele, E.; Shymanska, M.; Lukevics, E. *JOM* **1991**, *410*, 127.
57. Boden, R. M. *S* **1975**, 783.
58. Artamkina, G. A.; Grinfel'd, A. A., Beletskaya, I. P. *TL* **1984**, 4989.
59. Juaristi, M.; Aizpurua, J. M.; Lecea, B.; Palomo, C. *CJC* **1984**, *62*, 2941.
60. Moss, R. A.; Pilkiewicz, F. G. *JACS* **1974**, *96*, 5632.
61. Moss, R. A.; Joyce, M. A.; Pilkiewicz, F. G. *TL* **1975**, 2425.
62. Sepp, D. T.; Scherer, K. V.; Wever, W. P. *TL* **1974**, 2983.
63. Bartsch, R. A.; Kayser, R. H. *JACS* **1974**, *96*, 4346.
64. Varshney, S. K.; Jerome, R.; Bayard, P.; Jocobs, C.; Fayt, R.; Teyssie, P. *Macromolecules* **1992**, *25*, 4457.
65. Jedlinski, Z.; Kowalczuk, M.; Kurcok, P. *PAC* **1992**, *A29*, 1223.
66. Ding, J.; Price, C.; Booth, C. *Eur. Polym. J.* **1991**, *27*, 891, 895, 901.
67. Uhrich, K. E.; Hawker, C. J.; Frechet, J. M. J. *Macromolecules* **1992**, *25*, 4583.
68. Miller, R. D.; Thompson, D.; Sooriyakumaran, R.; Fickes, G. N. *J. Polym. Sci., Part A: Polym. Chem.* **1991**, *29*, 813.

69. Yang, C-P.; Hsiao, S-H. *J. Polym. Sci., Part A: Polym. Chem.* **1990**, *28*, 871.

70. Richards, M.; Dahiyat, B. I.; Arm, C. M.; Lin, S.; Leong, K. W. *J. Polym. Sci., Part A: Polym. Chem.* **1991**, *29*, 1157.

71. Sato, M.; Yokoyama, M. *Makromol. Chem.* **1984**, *185*, 629.

72. Trochimczuk, A. W.; Kolarz, B. N. *Eur. Polym. J.* **1992**, *28*, 1593.

73. Wang, Z. Y.; Bernard, N.; Hay, A. S. *J. Polym. Sci., Part A: Polym. Chem.* **1992**, *30*, 299.

74. Tsubokawa, N.; Yoshihara, T.; Sone, Y. *J. Polym. Sci., Part A: Polym. Chem.* **1992**, *30*, 561.

75. Kolarz, B. N.; Rapak, A. *Makromol. Chem.* **1984**, *185*, 2511.

Charles L. Liotta & Joachim Berkner
Georgia Institute of Technology, Atlanta, GA, USA

D

1,5-Diazabicyclo[4.3.0]non-5-ene[1]

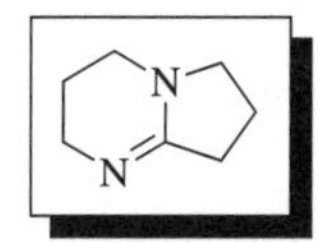

[3001-72-7] $C_7H_{12}N_2$ (MW 124.19)

(organic soluble base for elimination reactions,[2–11] isomerizations,[12–19] esterifications,[26–31] and condensations[20–25])

Alternate Name: DBN.
Physical Data: bp 95–98 °C/7.5 mmHg; *d* 1.005 g cm^{-3}.
Solubility: sol polar solvents such as water, ethanol, benzene, methylene chloride, chloroform, THF, and DMSO.
Form Supplied in: clear, colorless or light yellow liquid.
Handling, Storage, and Precautions: stable at ambient temperatures, but is hygroscopic; contact with undiluted product may cause skin irritation or burns.

Introduction. 1,5-Diazabicyclo[4.3.0]non-5-ene is an organic soluble amidine base which has been used effectively, under relatively mild conditions, for a variety of base-mediated organic transformations including eliminations, isomerizations, esterifications, and condensations. A related reagent, ***1,8-Diazabicyclo[5.4.0]undec-7-ene*** (DBU), is used for similar reactions.

Elimination Reactions. DBN has been used for dehydrohalogenations as well as for the introduction of unsaturation by elimination of sulfonic acids. Typical reactions are conducted with equimolar or excess base in solvents such as DMF, DMSO, benzene, methylene chloride, and chloroform. Products may be distilled directly from the reaction mixture or separated from the DBN salt byproduct by extraction with a nonpolar solvent. Terminal as well as internal double bonds have been introduced in this manner. Additionally, DBN has been used to prepare functionalized alkenes such as vinyl thioethers.

Oediger and Möller compared DBN with DBU in the dehydrohalogenation of bromoalkanes. Conditions were equimolar base and 80–90 °C. DBN was found to be only moderately effective, affording heptenes from bromoheptanes in 36–60% yield. Corresponding yields using DBU were 78–91%.

DBN was effective, however, for the conversion of β-substituted primary alkyl iodides into terminal alkenes.[2] Thus treatment of iodide (**1**) with 1.5 equiv of DBN in benzene at 80–90 °C for 3–4 h afforded alkene (**2**) in 91% yield (eq 1).

Both DBN and DBU have been employed in aprotic solvents for the dehydrobromination of substrates containing acetate functionality.[3–5] DBN was used in the preparation of chrysene derivative (**4**) (eq 2).[5] Diacetate (**3**) was brominated with NBS in methylene chloride, then dehydrobrominated using DBN in THF. Acetate functionality was preserved (note that DBU has been used for rt deacetylations in methanol[6]) and (**4**) was isolated in 51% yield after 16 h at −20 °C.

(**1**) —DBN, benzene, 80 °C, 3–4 h→ (**2**) (1)

(**3**) —1. NBS; 2. DBN→ (**4**) (2)

DBN promoted the elimination of methanesulfonic acid in the preparation of pyranoside (**5**) (eq 3).[7] Use of $NaHCO_3$ in place of DBN also afforded the desired pyranoside; however, the yield was low (15%) and the reaction time was longer. With NaH in refluxing DME, a mixture of regioisomeric elimination products resulted.

—DBN, DMSO, 65–70 °C, 2.5 h, 90%→ (**5**) (3)

DBN was used for the elimination of methanesulfonic acid from ester (**6**) to give *trans* α,β-unsaturated ester (**7**) in 86% yield after 4 h at rt (eq 4).[8] Interestingly, the nitrile analog of (**6**) (i.e. side chain cyano group rather than the methyl ester) afforded a 6:5 *cis*:*trans* mixture of α,β-unsaturated nitriles, probably due to the smaller steric requirements of the cyano group.

(**6**) —2 equiv DBN, benzene, rt, 4 h→ (**7**) (4)

Elimination of an intermediate thiolate in the presence of DBN was postulated to occur in a thiazolidine ring enlargement of penams (eq 5).[9,10]

—DBN, 31%→ (5)

Another rearrangement which has been postulated to occur via a DBU-mediated elimination is the conversion of 2-amino-5-imino-

4,5-dihydrothiazoles (e.g. **8**) to imidazoline thiones (e.g. **9**) and diazolidines (e.g. **10**).[11] The rearrangement is thought to proceed through a ring-opened carbodiimide intermediate which subsequently cyclizes to (**9**) or (**10**). The relative amounts of (**9**) and (**10**) formed vary with substrate structure. A representative reaction is shown in eq 6.

(8) —[1.1 equiv DBN; MeCN, reflux, 2 h; 85%]→ (9) + (10) (6)

Isomerizations. DBN has been used for base-mediated epimerizations and double bond migrations. Isomerizations generally require proton abstraction at a carbon α to a carbonyl group (or related functionality) and are thermodynamically controlled.

DBN is widely used for epimerizations of 6β-substituted penicillins to the corresponding 6α-isomers.[12–17] Epimerization of penicillins occurs rapidly and without competitive formation of 1,4-thiazepines (e.g. **12**). Methylene chloride is the most frequently used solvent. Various bases were compared in the epimerization of (**11**) (eq 7).[14] DBN and ***1,1,3,3-Tetramethylguanidine*** were equally effective. Other tertiary amine bases, including *N*-methylmorpholine, *N,N*-dimethylbenzylamine, *N*-alkylpiperidines, and ***Triethylamine***, afforded considerable amounts of thiazepine (**12**).

(11) —[cat DBN; CH_2Cl_2]→ + (12) (7)

DBN was used to achieve the equilibration of medium-ring cycloalkenones in a study of the effect of substituents on the 3-cycloalkenone/2-cycloalkenone product distribution.[18] Isomerizations were conducted in toluene or benzene from 25 to 100 °C in the presence of a catalytic amount of DBN. Generally, electron-withdrawing substituents at C-3 favored 3-cycloalkenones. Table 1 shows equilibration results for various 3-substituted cycloheptenones after treatment with a catalytic amount of DBN in toluene at 80 °C.

Table 1 Isomerization of 3-Substituted Cycloheptenones

3-Substituent	2-Cycloheptenone (%)	3-Cycloheptenone (%)
H	76.8	23.2
CO_2Me	16.8	83.2
CN	15.1	84.9
COMe	20.7	79.3

Several biomimetic methods for the conversion of amines to carbonyl compounds have used amidine bases to effect equilibration of the intermediate imine.[19–21] Corey used DBN in the oxidation of amines with mesitylglyoxal (**13**).[19] Thus cyclohexylamine and (**13**) were combined in benzene to form a Schiff base which was subsequently treated with DBN in THF/DMSO to afford cyclohexanone in 87% yield after hydrolysis (eq 8).

(13) —[1. cyclohexyl-NH_2; 2. DBN]→ cyclohexanone + mesityl-$COCH_2NH_2$ (8)

Condensations. DBN has been used to effect the reaction of active methylene compounds and other substrates containing active hydrogen. Corey used a catalytic amount of DBN to effect ring closure of the ketoaldehyde (**14**) in a prostaglandin synthesis (eq 9).[22] The cyclopentanol derivative (**15**) was obtained in 45% yield upon treatment of (**14**) with 0.1 equiv of DBU in methylene chloride at 0 °C for 24 h followed by deacetylation.

(14) —[1. cat DBN; 2. Ac_2O]→ (15) (9)

An insecticide intermediate, 1-(*p*-ethoxyphenyl)-2-nitro-1-propanol, resulted from the DBN catalyzed reaction of nitroethane and *p*-ethoxybenzaldehyde (eq 10).[23]

EtO-C_6H_4-CHO —[$EtNO_2$; cat DBN, DMSO; rt, 15 h; 86%]→ EtO-C_6H_4-CH(OH)-CH(CH_3)-NO_2 (10)

DBN has been used to generate resonance stabilized ylides in Wittig reactions.[24,25] The synthesis of dienoate (**17**) from benzaldehyde and phosphonium salt (**16**) is representative (eq 11).

(16) —[DBN; PhCHO; 72%]→ (17) (11)

Esterifications. There are few examples of the use of DBN in esterification reactions, although the related amidine, DBU, is widely used for this purpose.[26–30] Colvin reported a DBN mediated esterification in one of the final steps of the synthesis of the macrolide antibiotic pyrenophorin.[31] Thus treatment of bisimidazol-1-yl ketone (**18**) with a catalytic amount of DBN afforded bis-lactone (**19**) in 60% yield (eq 12). In the same study, Colvin and Raphael also used DBN for the quantitative selective removal of a *p*-toluenesulfonylethyl protecting group (eq 13). Stronger bases cleaved the central ester.

cat DBN (12)

(**18**) (**19**)

DBN, benzene, rt

$MeC_6H_4SO_2(CH_2)_2O$

(13)

Related Reagents. 1,8-Bis(dimethylamino)naphthalene; 1,8-Diazabicyclo[5.4.0]undec-7-ene; Diisopropylethylamine; 1,1,2,3,3-Pentaisopropylguanidine; Phosphazene Base P_4-*t*-Bu.

1. (a) Nakatani, K.; Hashimoto, S. *Yuki Gosei Kagaku Kyokai* **1975**, *33*, 925; (*CA* **1976**, *84*, 164 644y). (b) Ni, Z. *Zhongguo Yiyao Gongye Zazhi* **1991**, *22*, 180; (*CA* **1991**, *115*, 183 137k). (c) Oediger, H.; Möller, F.; Eiter, K. *S* **1972**, 591.
2. Oediger, H.; Möller, F. *AG(E)* **1967**, *6*, 76.
3. Bohlmann, F.; Haffer, G. *CB* **1968**, *101*, 2738.
4. Rao, D. R.; Lerner, L. M. *Carbohyd. Res.* **1971**, *19*, 133.
5. Utermoehlen, C. M.; Singh, M.; Lehr, R. E. *JOC* **1987**, *52*, 5574.
6. Baptistella, L. H. B.; Fernando, J.; Ballabio, K. C.; Marsaioli, A. J. *S* **1989**, 436.
7. Hanessian, S.; Plessas, N. R. *CC* **1968**, 706.
8. Tufariello, J. J.; et al. *JACS* **1979**, *101*, 2435.
9. Ramsay, B. G.; Stoodley, R. J. *JCS(C)* **1971**, 3864.
10. Ananda, G. D. S.; Steele, J.; Stoodley, R. J. *JCS(P1)* **1988**, 1765.
11. Morel, G.; Marchand, E.; Foucand, A. *JOC* **1990**, *55*, 1721.
12. Jackson, J. R.; Stoodley, R. J. *JCS(P1)* **1972**, 895.
13. Jackson, J. R.; Stoodley, R. J. *CC* **1971**, 648.
14. Ramsay, B. G.; Stoodley, R. J. *CC* **1971**, 450.
15. Vlietinck, A.; Roets, E.; Claes, P.; Vanderhaeghe, H. *TL* **1972**, 285.
16. Kovacs, Ö. K. J.; Ekström, B.; Sjöberg, B. *TL* **1969**, 1863.
17. Wolfe, S.; Lee, W. S.; Misra, R. *CC* **1970**, 1067.
18. Mease, R. C.; Hirsch, J. A. *JOC* **1984**, *49*, 2925.
19. Corey, E. J.; Achiwa, K. *JACS* **1969**, *91*, 1429.
20. Jaeger, D. A.; Broadhurst, M. D.; Cram, D. J. *JACS* **1979**, *101*, 717.
21. Buckley, T. F.; Rapoport, H. *JACS* **1982**, *104*, 4446.
22. Corey, E. J.; et al. *JACS* **1968**, *90*, 3245.
23. Holan, G. U.S. Patent 3 657 357, 1972.
24. Oediger, H.; Kobbe, H. J.; Möller, F.; Eiter, K. *CB* **1966**, *99*, 2012.
25. Flitsch, W.; Jerman, F. *LA* **1985**, 307.
26. Ono, N.; Yamada, T.; Saito, T.; Tanaka, K.; Kaji, A. *BCJ* **1978**, *51(8)*, 2401.
27. Cabré, J.; Palomo, A. L. *SC* **1984**, 413.
28. Nishikubo, T.; Iizawa, T.; Takahashi, A.; Shimokawa, T. *J. Polymer Sci., Polymer Chem. Ed.* **1990**, *28*, 105.
29. Ohta, S.; Shimabayashi, A.; Aono, M.; Okamoto, M. *S* **1982**, 833.
30. Rundel, W.; Köhler, H. *CB* **1972**, *105*, 1087.
31. Colvin, E. W.; Purcell, T. A.; Raphael, R. A. *CC* **1972**, 1031.

Ann C. Savoca
Air Products and Chemicals, Allentown, PA, USA

1,8-Diazabicyclo[5.4.0]undec-7-ene[1]

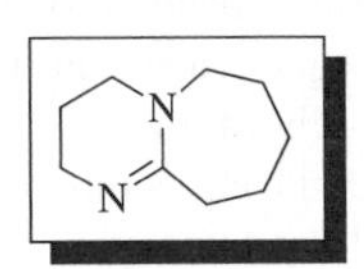

[6674-22-2] $C_9H_{16}N_2$ (MW 152.24)

(organic soluble base for elimination reactions, isomerizations, esterifications, amidations, etherifications, condensations, carboxylations/carbonylations, and halogenations)

Alternate Name: DBU.
Physical Data: bp 259–260 °C; *d* 1.0192 g cm^{-3}.
Solubility: readily sol water, ethanol, benzene, acetone, ethyl acetate, carbon tetrachloride, diethyl ether, dioxane, 1,4-butanediol, dimethyl sulfoxide; hardly sol petroleum ether.
Form Supplied in: clear, light yellow liquid.
Handling, Storage, and Precautions: stable at ambient temperatures, but is hygroscopic; overexposure to atmosphere results in water absorption which can lead to hydrolysis; rate of hydrolysis to *N*-(3-aminopropyl)-ε-caprolactam (10 wt% soln; molar ratio 1:76) is 3×10^{-4} mol%$^{-1}$ min^{-1}; half-life for a 0.657 molar soln is 33 min at 35 °C; contact with undiluted product may cause skin irritation or burns.

Introduction. 1,8-Diazabicyclo[5.4.0]undec-7-ene is an organic soluble amidine base which has been used effectively, and under relatively mild conditions, for a variety of base-mediated organic transformations including eliminations, isomerizations, esterifications, amidations, etherifications, condensations, carboxylations/carbonylations, and halogenations. A related reagent,

1,5-Diazabicyclo[4.3.0]non-5-ene (DBN), is used for similar reactions.

Elimination Reactions. DBU has been used widely for dehydrohalogenations as well as for the introduction of unsaturation by elimination of sulfonic acids. Reactions generally proceed under mild conditions and without side reactions. Typical procedures use equimolar base and elevated reaction temperatures (generally 80–100 °C). Reaction solvents such as DMF, benzene, and DMSO have been used with reaction times varying from several hours to several days. Products may be distilled directly from the reaction mixture or separated from the DBU salt byproduct by extraction with a nonpolar solvent. Terminal as well as internal double bonds can be introduced with a high degree of regioselectivity. Additionally, this method has been used to prepare functionalized alkenes such as vinyl halides and vinyl ethers. Alkynes are not typically prepared using DBU-mediated eliminations; propargyl ethers, however, are an exception.

Oediger and Möller introduced DBU in 1967 for the dehydrohalogenation of bromoalkanes. DBN was found to be less effective.[2] For example, treatment of 4-bromoheptane with equimolar DBU at 80–90 °C gave 3-heptene in 91% yield; with DBN the yield was 60%. Similar treatment of 2-bromooctane gave a 4:1 mixture of 2-octene and 1-octene in 84% yield; with DBN the yield was 40%.

DBU was used in the following one-pot procedure for converting β-disubstituted primary alkyl iodides to terminal alkenes.[3] DBN was also used effectively for this procedure. The THP-protected tosylate (**1**) was converted to the corresponding iodide with NaI in DMF; subsequent addition of 1.5 equiv of DBU and heating at 80 °C for 3–4 h afforded terminal alkene (**2**) in 82% overall yield from (**1**) (eq 1). For comparison, when (**1**) was transformed into the bromide and treated with ***Potassium t-Butoxide*** in DMSO at 50 °C, a 3:2 mixture of the terminal alkene and its rearranged trisubstituted isomer resulted.[4]

DBU, DMF 80–90 °C 3–4 h (1)

(**1**) (**2**)

DBU was found to be an effective base for converting piperidine (**3**) into 3,4-dehydropiperidine (**4**) without formation of the undesired 2,3-dehydropiperidine in a synthesis of the alkaloid sedinine (eq 2).[5]

$MeSO_3$ R' N CO_2Me DBU 70 °C, 4 d R' N CO_2Me (2)

(**3**) (**4**)

DBU has also been used to prepare (*E*)-1-iodo-1-alkenes from 1,1-diiodoalkanes.[6] In a representative procedure (eq 3), diiodobutane (**5**) was combined with equimolar DBU and heated to 100 °C until appearance of a brown solid (15–20 min). The product, (*E*)-1-iodo-1-butene (**6**), was isolated in 80% yield by distilling directly from the reaction mixture. Higher-boiling vinyl iodides required DMSO as reaction solvent and product extraction with pentane.

I I DBU 100 °C I (3)

(**5**) (**6**)

DBU is not typically used to convert vinyl halides into alkynes; in general, this conversion requires alkoxides, solid alkali, or alkali metal amides.[7] However, the (*Z*)-vinyl bromide (**7**: R = H) was nearly quantitatively converted into terminal alkynes (**8**) in 2 h with DBU in refluxing benzene; the corresponding (*E*) isomer did not react (eq 4).[8] When a (*Z/E*) mixture of (**7**) was treated with ***Potassium Carbonate*** in refluxing methyl ethyl ketone (MEK), only starting material was recovered. With β-oxygen substitution (R = CH_2OH), vinyl bromides (**7**) were quantitatively converted into alkynes in 1 h with either DBU in refluxing benzene or potassium carbonate in refluxing MEK.

MeO O Br R MeO O R (4)

(**7**) (**8**)

DBU has also been used to prepare alkenes from sulfonates. Heating tosylate (**9**) in solvent DBU for 30 min at 100 °C afforded the *cis* vinyl ether (**10**) in 72% yield (eq 5).[9] When (**9**) was treated with potassium *t*-butoxide in *t*-butanol for 30 min at 80 °C, the vinyl ether product was a 3:7 *cis*:*trans* alkene isomer mixture (72% yield).

OTs RO 13 DBU 13 OR (5)

(**9**) (**10**)

Otter used DBU to effect the elimination of methanesulfonic acid in the final step of his preparation of 1,3-dimethyl-6-propyluracil, a synthetic pyrimidine nucleoside (eq 6).[10]

Me N O OMs N Me O DBU, THF reflux, 48 h Me N O N Me O (6)

Isomerizations. 1,8-Diazabicyclo[5.4.0]undec-7-ene is used for base-mediated double bond migrations and epimerizations. Isomerizations generally require proton abstraction at a carbon α to a carbonyl group (or related functionality) and are thermodynamically controlled.

DBU was used to equilibrate a mixture of substituted pyrrolidin-2-ones in the final step of a herbicide synthesis.[11] The mixture of isomers (**11**) was allowed to stand for 1 h at rt with DBU in toluene to give the pure 3,4-*trans* isomer (**12**) in 96% yield (eq 7).

Ar O N-Ar DBU, toluene rt, 1 h Ar O N-Ar (7)

(**11**) (**12**)

Amidine bases have been used extensively for the equilibration of β-lactams. Although DBN appears to be the base of choice for such epimerizations,[12–17] the 7α-(dimethylamidino)-3-cephem ester 1α-oxide (**14**) was prepared via DBU epimerization of the corresponding 7β-isomer (**13**) (eq 8).[18]

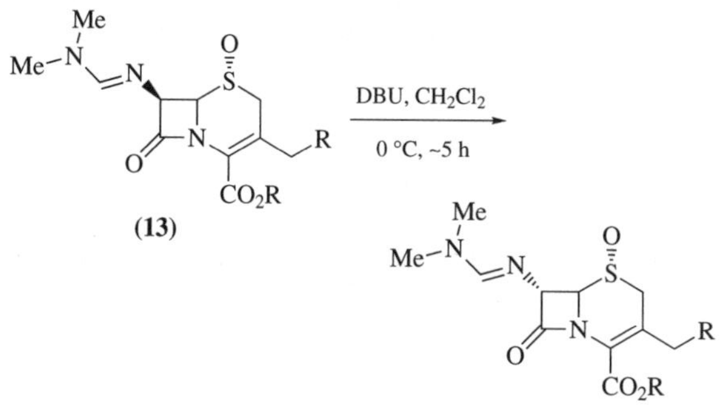

(8)

DBU has been employed to convert esters with β,γ-unsaturation into the corresponding α,β-unsaturated isomers.[19] Thus 3-pentenoate (**15**) underwent up to 60% isomerization to 2-pentenoate (**16**) in the presence of DBU at 100 °C (eq 9). Corresponding exposure of pure *cis*-2-pentenoate (**16**) to DBU at 130 °C for 4 h afforded a similar product mixture (53% *trans*-2-pentenoate, 40% of 3-pentenoate (**15**), and 7% recovered starting material), suggesting thermodynamic equilibrium.[20] For comparison, in a continuous process using ***4-Dimethylaminopyridine*** at reflux, up to 78% of the thermodynamically favored (**16**) was converted to (**15**) over the course of 30 h.

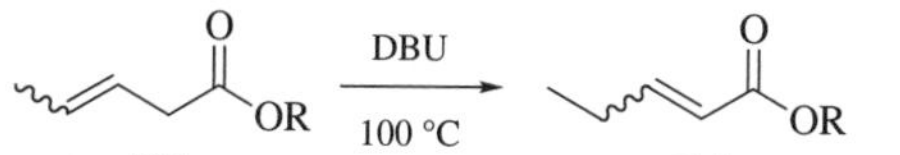

(9)

β,γ-Unsaturated nitriles (e.g. **17**) have been isomerized to the thermodynamically favored α,β-unsaturated nitriles (e.g. **18**) in the presence of catalytic DBU or DBN (eq 10).[21]

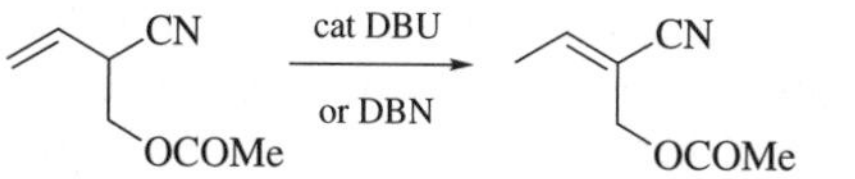

(10)

As part of a study directed toward the synthesis of dodecahedrane, DBU was used to effect the isomerization of bis-enone (**19**) into (**20**) in 90% yield (eq 11).[22] The mechanism was presumed to involve formation of the β,γ-unsaturated isomer of bis-enone (**19**).

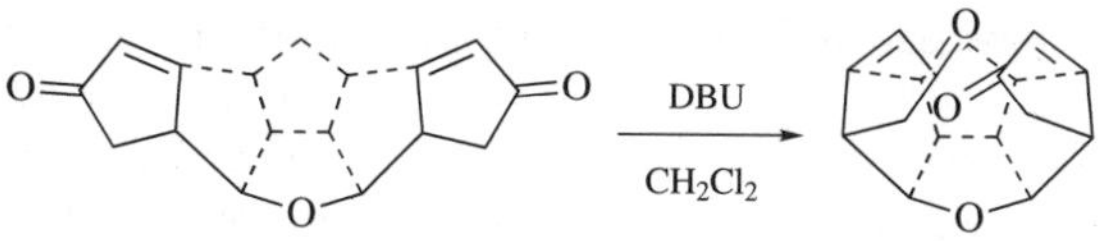

(11)

Several biomimetic methods for the conversion of amines to carbonyl compounds have used amidine bases to effect equilibration of the intermediate imine.[23–25] Rapoport used DBU in the oxidation of amines with 4-formyl-1-methylpyridinium benzenesulfonate (FMPBS).[25] As an example, phenylacetaldehyde (**22**) was obtained in 83% yield from β-phenylethylamine (**21**) after treatment with FMPBS and DBU in CH_2Cl_2/DMF (eq 12). ***Triethylamine*** was not effective as a DBU replacement except when the amine β-carbon had electron withdrawing substituents (e.g. acetophenone from α-phenylalanine).

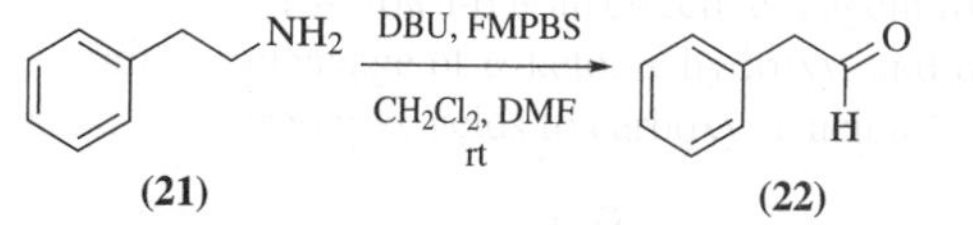

(12)

Esterifications, Amidations, and Etherifications. DBU has been used to prepare esters[26] and amides[27] from carboxylic acids as well as ethers,[28] esters,[29] and carbamates[30] from alcohols. These procedures involve proton abstraction followed by reaction of the carboxylate or alkoxide with an alkyl halide, acylating agent, or other suitable electrophile. Esterifications and amidations are generally conducted at or near rt, whereas etherifications require elevated temperatures (60–80 °C).

In 1978, Ono reported a convenient procedure for the esterification of carboxylic acids using DBU.[26] Esters were produced in high yield from acids, alkyl halides, and DBU. The advantage of this procedure is that it provides mild conditions for esterification, it is not necessary to prepare the carboxylate anion in a separate step, and side reactions, especially dehydrohalogenation, are avoided. Amino acids have been esterified without racemization using this procedure. As a representative example, benzoic acid reacted with ethyl iodide in the presence of DBU for 1 h to give ethyl benzoate in 95% yield. The same reaction using triethylamine instead of DBU afforded essentially no ethyl benzoate. With benzyl bromide, benzoic acid, and DBU in DMSO, benzyl benzoate was formed quantitatively at 30 °C in 10 min.[31] Triethylamine, when substituted for DBU, afforded benzyl benzoate in 81% after 1 h at 30 °C; ***Pyridine*** afforded a 15% yield of benzyl benzoate after 6 h.

The high yields of ester afforded by this method make it attractive for polyester synthesis.[31] Thus isophthalic acid reacted with *m*-xylylene dibromide in the presence of 2 molar equiv of DBU in DMSO at 30 °C for 3 h to afford polyester (**23**) in high yield and viscosity (eq 13). Other organic bases such as triethylamine, pyridine, *N*,*N*-dimethylaminopyridine, or a DBU–pyridine mixture did not afford any polymer.

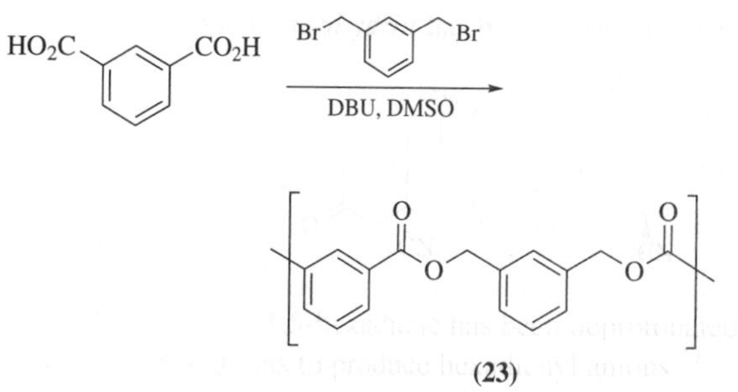

(13)

This method has also been used for the esterification of polymers.[28,32] As an example, poly(methacrylic acid) (**24**) reacted with *p*-bromomethylnitrobenzene in the presence of equimolar

DBU in DMSO at 30 °C for 3 h to afford poly(methacrylate) (**25**) at 97 mol% esterification (eq 14).

(14)

(**24**) (**25**)

The etherification of poly(4-hydroxystyrene) (**26**) is related.[32] On treatment with equimolar *p*-bromomethylnitrobenzene and DBU in HMPA at 60 °C for 24 h, (**26**) was converted into poly[4-(4-nitrobenzyloxy)styrene] (**27**) at 98 mol% etherification (eq 15). Solvents such as DMF, DMSO, and NMP were less effective for this reaction, requiring longer reaction times, higher reaction temperatures, and excess reagent. For example, a high degree of etherification (>80%) using DMF as solvent required 2 molar equivalents each of *p*-bromomethylnitrobenzene and DBU at 80 °C for more than 24 h. No reaction was observed using triethylamine or pyridine instead of DBU.

(15)

(**26**) (**27**)

Polyimides containing pendant carboxylic acids have also been esterified using 1-phenethyl bromide and DBU.[33,34]

In an alternative approach to esterification, DBU was used to accelerate the reaction of *N*-acylimidazoles with *t*-butanol in a one-pot conversion of carboxylic acids into their *t*-butyl esters. The general reaction is outlined in eq 16. Acids were treated with 1 molar equiv of ***N,N′-Carbonyldiimidazole*** in DMF under a nitrogen atmosphere at temperatures from 40 to 80 °C and reaction times of 5 to 24 h. Products were extracted from the reaction mixtures with diethyl ether. In this manner, *t*-butyl benzoate, *t*-butyl cinnamate, and *t*-butyl heptanoate were prepared in 91%, 64%, and 68% yield, respectively. With sodium *t*-butoxide rather than DBU, there was competitive formation of 3-oxoalkanoic esters with acids having one or two protons at C-2.[35] A similar limitation was noted for the conversion of *N*-acylimidazoles to *t*-butyl esters using *t*-butanol and NBS.[36]

(16)

Enolizable acyl cyanides have been converted into 1-cyano-1-alkenyl esters upon treatment with tertiary amines (e.g. DBU, pyridine, dimethylamine, and ***1,4-Diazabicyclo[2.2.2]octane***) and carboxylic acid chlorides or anhydrides.[37] With acid chlorides, equimolar base was required, whereas only a catalytic amount of base was necessary for reactions involving acid anhydrides. As an example, propionyl cyanide was treated with a stoichiometric amount of DBU and acetyl chloride in methylene chloride at rt to afford 1-cyanovinyl acetate in 61% yield.

DBU has also been used for certain deesterifications. In the case of acetates, deacetylation with DBU occurs under relatively mild conditions (rt to 80 °C; 5–45 h).[38] The method only works for esters derived from acetic acid. Methanol is the solvent of choice, although dichloromethane or benzene may be added to improve reactant solubility. It was speculated, though not confirmed, that the mechanism involves formation of the desired alcohol by elimination of ketene. As an example, acetate (**28**) was deacetylated in 93% yield to alcohol (**29**) with DBU in methanol at rt for 24 h (eq 17). The same reaction using DBU in xylene afforded only starting material.[39]

(17)

(**28**) (**29**)

Methyl esters are cleaved with DBU. High reaction temperatures and extended reaction times are required; however, the corresponding acid is generally obtained in high yield (>90%) without the use of ionic nucleophiles such as ***Lithium Iodide***, lithium thiolate, or potassium *t*-butoxide.[39] Thus a solution of methyl mesitoate, 10 equiv of DBU, and 10 equiv of *o*-xylene was heated to 165 °C for 48 h, affording mesitoic acid which was isolated in 95% yield after ether extraction (eq 18).

(18)

DBU was one of several effective bases used in an amide synthesis from *N,N*-bis(2-oxo-3-oxazolidinyl)phosphorodiamidic chloride, primary or secondary amines, and carboxylic acids.[27] Other bases included ***1-Ethylpiperidine***, triethylamine, and *N*-ethylmorpholine. Reactions were conducted at rt for 1–2 h, avoiding racemization of optically pure substrates. Thus 3,3-dimethylacrylic acid and the acid salt of (*S*)-(−)-α-methylbenzylamine in DMA were treated with 2 equiv of DBU at rt over 30 min, then allowed to react at rt for 75 min to afford amide (**30**) in 75% yield (eq 19). Other methods require higher reaction temperatures (alkyl carbamates and alkyl amines afford amides at 200 °C in the presence of tertiary amines[40]), longer reaction times (***1,3-Dicyclohexylcarbodiimide***[41] and ***Diphenylphosphinic Chloride***[42] procedures require up to 12 h and result in only modest yields for similar conversions), or excess reagents or reactants (the cyanuric chloride method[43] requires excess acid and the *o*-nitrophenyl thiocyanate/***Tri-n-butylphosphine*** method[44] uses excess reagent and amine).

(19)

(**30**)

Condensations. DBU has been used to effect condensations of active methylene compounds and other substrates containing active hydrogen. Reactions generally use equimolar DBU and aprotic solvents such as THF or benzene. Reaction times and temperatures vary.

DBU was shown to be an effective base for the Michael reaction of diethyl acetamidomalonate (**31**) with methyl acrylate in a synthesis of glutamic acid (eq 20).[45] ***1,1,3,3-Tetramethylguanidine*** and DBN were found to be equally effective. All resulted in the formation of glutamic acid derivative (**32**) quantitatively.

(20) (**31**) → (**32**) [$CH_2{=}CHCO_2Me$, DBU]

In the Knoevenagel condensation of ***Malonic Acid*** with hexanal, the β,γ-unsaturated isomer (**33**) was obtained with 94% selectivity and 56% yield after 10 h at 90 °C in the presence of equimolar DBU (eq 21).[46] Other bases selective for (**33**) included triethanolamine, triethylamine, ethylpiperidine, *N,N*-dimethylaniline, and ***2,6-Lutidine***. For comparison, pyridine, 3-methylpyridine, and 4-methylpyridine showed ∼90% selectivity for the α,β-unsaturated isomer. A β,γ-unsaturated carbonyl compound was also obtained from the condensation of ***Formaldehyde*** with pentenone in the presence of catalytic DBU or DBN.[47] These results are interesting in view of the literature reports that β,γ-unsaturated carboxylates[19,20] and nitriles[21] are isomerized to the thermodynamically favored α,β-unsaturated isomers in the presence of DBU.

(21) malonic acid → (**33**) [$Me(CH_2)_4CHO$, 90 °C, 10 h]

Diesters of homophthalic acid condensed with aromatic aldehydes in the presence of equimolar DBU in refluxing benzene for 6–10 h to give excellent yields of cinnamic esters. The reaction of dimethyl homophthalate with 3-benzyloxy-4-methoxybenzaldehyde to afford cinnamic ester (**34**) is representative (eq 22).[48] Sodium hydride, sodium alkoxides, and potassium acetate were all found to be ineffective for this reaction.

(22) → (**34**) [DBu, benzene, reflux, 6 h]

DBU catalyzed the asymmetric synthesis of δ-oxocarboxylic acids from (2*R*,3*S*)-3,4-dimethyl-5,7-dioxo-2-phenylperhydro-1,4-oxazepine (**35**).[49] The Michael reaction of (**35**) with 2-cyclopenten-1-one afforded (+)-3-cyclopentanoneacetic acid (**36**) in moderate yield (43%) and high optical purity (96%) after hydrolysis and decarboxylation (eq 23). Trityllithium (***Triphenylmethyllithium***) or potassium *t*-butoxide afforded (**36**) in similar yields (30–50%) but poorer optical purity (7–76%).

(23) (**35**) → (**36**) [1. 1.5 equiv cyclopentenone, 0.3 equiv DBU, THF, 0 °C, 3 d; 2. H_2SO_4, AcOH]

Carbonylations and Carboxylations. DBU has been used to prepare amides or imides[50–52] and esters[53] via the sequential carbonylation/alkylation of amines and alcohols, respectively. Similarly, the carboxylation of amines and alcohols affords urethanes and carbonates.[54] Reactions use stoichiometric base and are catalyzed by palladium or nickel complexes. Product yields are generally high (80–100%). Carbonylations have been conducted in DMA at elevated temperatures (115–150 °C), whereas carboxylations have been performed at rt in a variety of solvents including methylene chloride, DMSO, THF, and glyme. eqs (24) and (25) show the preparation of *N*-phenylbenzamide and allyl benzylethylcarbamate by these procedures. The efficacy of various bases was compared in the carboxylation/alkylation of benzylethylamine using a catalytic amount of tris(dibenzylideneacetone)dipalladium ($Pd_2(dba)_3$). DBU, *N*-cyclohexyl-*N′*,*N′*,*N″*,*N″*-tetramethylguanidine, and 7-methyl-1,5,7-triazabicyclo[4.4.0]dec-5-ene were preferred. DBN afforded a modest 49% yield of urethane and essentially no urethane was formed using ***Diisopropylethylamine***.

(24) [Co/Pd cat, 1.2 equiv DBU, PhI, DMAC, 115 °C, quantitative]

(25) [cat $Pd_2(dba)_3$, 1.1 equiv DBU, THF, rt, 2 h, allyl chloride, 87%]

Methods for the preparation of thiocarbamates from alcohols, carbon disulfide, and alkylating agents[55] or via the sulfur-assisted carbonylation of alcohols[56] are related; these do not require a metal catalyst. Thus *S*-benzyl *O*-*n*-butylcarbonothionate was isolated in 86% yield after the carbonylation of *n*-butanol in THF at 80 °C for 4 h in the presence of 3 equiv of powdered sulfur and 5 equiv of DBU followed by esterification with 1.2 equiv of benzyl bromide (eq 26).

(26) [1. CO, 3 equiv S, 5 equiv DBU, THF, 80 °C, 4 h; 2. 1.2 equiv BzBr]

DBU has also been used in the nickel- or palladium-catalyzed coupling of alkenes with ***Carbon Dioxide***.[57,58] As an example (eq 27), isoprene was treated with $Pd(acac)_2$, a phosphine ligand, DBU, and tributyltin ethoxide for 84 h at 80 °C under CO_2 to afford an isomeric mixture of C-10 carboxylic acids (68% after esterification and purification). It was found that DBU and tributyltin ethoxide independently promote the reaction, but not as effectively as when used as a combination.

1. $Pd(acac)_2$, ligand Bu_3SnOEt, DBU CO_2 → 2. esterification; product CO_2Me (27)

Halogenations. DBU-based brominating agents such as DBU/bromine,[59] DBU/hydrobromide perbromide,[60] and DBU/bromotrichloromethane,[61,62] have been used to brominate enolizable substrates and aromatic compounds. As an example, 3-halomethylcephems (e.g. **38**), convenient intermediates for the synthesis of 3′-substituted cephalosporins, were prepared by treatment of *exo*-methylene cephems (e.g. **37**) with DBU/bromine in THF over the temperature range −80 to 0 °C (eq 28).[59]

$PhOCH_2CONH$ ... $C_6H_4NO_2$ (**37**) —DBU, Br_2, THF→ $PhOCH_2CONH$ ... Br ... $C_6H_4NO_2$ (**38**) (28)

Miscellaneous. A synthesis of phthalocyanines and metallophthalocyanines, interesting optical and electronic materials, involves the DBU-mediated reaction of alcohols with phthalonitrile at elevated temperature.[63–67] The alkoxy-3-iminoisoindolenine (**39**) is presumed to be an intermediate (eq 29). Thus phthalonitrile was treated with equimolar DBU in refluxing ethanol to afford the corresponding phthalocyanine as a blue crystalline compound. DBN was also shown to be effective for this transformation; neither pyridine nor 1,4-diazabicyclo[2.2.2]octane promoted the formation of phthalocyanine.

CN, CN → OEt, N, NH (29)

Carboxylic acids have been phosgenated in the presence of a variety of amine bases, including DBU, to afford acid chlorides.[68] Reactions are conducted with 2 mol% base at 80–100 °C. Yields of acid chlorides are generally greater than 90%.

DBU was used to promote the in situ formation of hydrogen selenide from ***Selenium***, ***Carbon Monoxide***, and water for selective reduction of α,β-unsaturated carbonyl compounds.[69] *N*-Methylpyrrolidine was also found to be effective. Thus benzylideneacetone afforded 4-phenylbutanone in 82% yield in the presence of DBU, selenium, and carbon monoxide at 50 °C or *N*-methylpyrrolidine at 80 °C after 24 h. Isophorone underwent selective reduction of the conjugated double bond and β-ionone was reduced to 4-(2,6,6-trimethyl-1-cyclohexenyl)-2-butanone.

Related Reagents. 1,5-Diazabicyclo[4.3.0]non-5-ene.

1. (a) Nakatani, K.; Hashimoto, S. *Yuki Gosei Kagaku Kyokai* **1975**, *33*, 925 (*CA* **1976**, *84*, 164 644y). (b) Ni, Z. *Zhongguo Yiyao Gongye Zazhi* **1991**, *22*, 180 (*CA* **1991**, *115*, 183 137k). (c) Oediger, H.; Möller, F.; Eiter, K. **1972**, 591.
2. Oediger, H.; Möller, F. *AG(E)* **1967**, *6*, 76.
3. Wolff, S.; Huecas, M. E.; Agosta, W. C. *JOC* **1982**, *47*, 4358.
4. Wood, N. F.; Chang, F. C. *JOC* **1965**, *30*, 2054.
5. Ogawa, M.; Natsume, M. *H* **1985**, *23(4)*, 831.
6. Martinez, A. G. et al. *TL* **1992**, *33*, 2043.
7. (a) Kobrich, G.; Buck, P. In *Chemistry of Acetylenes* Viehe, H. G.; Ed; Dekker: New York, 1969; p 99. (b) Dehmlow, E. V.; Lissel, M. **1980**, 1 and references therein.
8. Schuda, P. F.; Heimann, M. R. *JOC* **1982**, *47*, 2484.
9. Serebrennikova, G. A.; Vtorov, I. B.; Preobrazhenskii, N. A. *ZOR* **1969**, *5*, 676; Eng. Edit., 663.
10. Otter, B. A.; Taube, A.; Fox, J. J. *JOC* **1971**, *36*, 1251.
11. Moriyasu, K. et al. **1992**, Eur. Patent Appl. 477 626-A1, 1992 (*CA* **1992**, *117*, 90 129k).
12. Jackson, J. R.; Stoodley, R. J. *JCS(P1)* **1972**, 895.
13. Jackson, J. R.; Stoodley, R. J. *CC* **1971**, 648.
14. Ramsay, B. G.; Stoodley, R. J. *CC* **1971**, 450.
15. Vlietinck, A.; Roets, E.; Claes, P.; Vanderhaeghe, H. *TL* **1972**, 285.
16. Kovacs, Ö. K. J.; Ekström, B.; Sjöberg, B. *TL* **1969**, 1863.
17. Wolfe, S.; Lee, W. S.; Misra, R. *CC* **1970**, 1067.
18. Chauvette, R. R. U.S. Patent 4 906 769, 1990.
19. Jpn. Patent 56 055 345, 1981 (*CA* **1981**, *95*, 186 653e).
20. Fischer, R.; Merger, F.; Gosch, H. J. U.S. Patent 4 906 769, 1990.
21. Disselnkötter, H.; Kurtz, P. Ger. Patent 1 941 106, 1971.
22. Mehta, G.; Nair, M. S. *JACS* **1985**, *107*, 7519.
23. Corey, E. J.; Achiwa, K. *JACS* **1969**, *91*, 1429.
24. Jaeger, D. A.; Broadhurst, M. D.; Cram, D. J. *JACS* **1979**, *101*, 717.
25. Buckley, T. F.; Rapoport, H. *JACS* **1982**, *104*, 4446.
26. Ono, N.; Yamada, T.; Saito, T.; Tanaka, K.; Kaji, A. *BCJ* **1978**, *51*, 2401.
27. Cabré, J.; Palomo, A. L. *SC* **1984**, 413.
28. Nishikubo, T.; Iizawa, T.; Takahashi, A.; Shimokawa, T. *J. Polym. Sci., Polym. Chem. Ed.* **1990**, *28*, 105.
29. Ohta, S.; Shimabayashi, A.; Aono, M.; Okamoto, M. *S* **1982**, 833.
30. Rundel, W.; Köhler, H. *CB* **1972**, *105*, 1087.
31. Nishikubo, T.; Ozaki, K. *Polym. J.* **1990**, *22*, 1043.
32. Iizawa, T.; Sato, Y. *Polym. J.* **1992**, *24*, 991.
33. Iizawa, T.; Seno, E. *Polym. J.* **1992**, *24*, 1169.
34. Seymour, R. B.; Kirshenbaum, G. S. *High Performance Polymers: Their Origin and Development*; Elsevier: New York, 1986; p 319.
35. Staab, H. A.; Mannschreck, A. *CB* **1962**, *95*, 1284.
36. Katsuki, T. *BCJ* **1976**, *49*, 2019.
37. Oku, A.; Arita, S. *BCJ* **1979**, *52*, 3337.
38. Baptistella, L. H. B.; dos Santos, J. F.; Ballabio, K. C.; Marsaioli, A. J. *S* **1989**, 436.
39. Parish, E. J.; Miles, D. H. *JOC* **1973**, *38*, 1223.
40. Falcone, S. J.; McCoy, J. J. U.S. Patent 4 336 402, 1982.
41. Bernasconi, S.; Comini, A. Corbella, A.; Gariboldi, P.; Sisti, M. *S* **1980**, 385.
42. Jackson, A. G.; Kenner, G. W.; Moore, G. A.; Ramage, R.; Thorpe, W. D. *TL* **1976**, 3627.
43. Venkataraman, K.; Wagle, D. R. *TL* **1979**, 3037.
44. Grieco, P. A.; Clark, D. S. Withers, G. P. *JOC* **1979**, *44*, 2945.

45. Potrzebowski, M. J.; Stolowich, N. J.; Scott, A. I. *J. Labelled Compd. Radiopharm.* **1990**, *28*, 355.
46. Yamanaka, H.; Yokoyama, M.; Sakamoto, T.; Shiraishi, T.; Mataichi, S.; Mizugaki, M. *H* **1983**, *20*, 1541.
47. Lantzsch, R.; Arlt, D. Ger. Patent 2 456 413, 1976.
48. Sakai, K. et al. *OPP* **1973**, *5*, 81.
49. Mukaiyama, T.; Hirako, Y.; Takeda, T. *CL* **1978**, 461.
50. Perry, R. J.; Wilson, B. D. *Macromolecules* **1993**, *26*, 1503.
51. Perry, R. J.; Turner, S. R. U.S. Patent 4 933 466, 4 933 467, and 4 933 468, 1990.
52. Yoneyama, M.; Konishi, T.; Kakimoto, M.; Imai, Y. *Macromol. Chem., Rapid Commun.* **1990**, *11*, 381.
53. Imai, Y.; Kakimoto, M.; Yoneyama, M. U.S. Patent 4 948 864, 1990.
54. McGhee, W. D.; Riley, D. P.; Christ, M. E.; Christ, K. M. *OM* **1993**, *12*, 1429.
55. Yuji, H.; Seiichi, K.; Hiroshi, T. *Chem. Express* **1989**, *4*, 805 (*CA* **1990**, *112*, 138 999v).
56. Mizuno, T.; Nishiguchi, I.; Hirashima, T.; Ogawa, A.; Kambe, N.; Sonoda, N. *TL* **1988**, *29*, 4767.
57. Hoberg, H.; Peres, Y.; Krüger, C.; Tsay, Y.-Y. *AG(E)* **1987**, *26*, 771.
58. Hoberg, H.; Minato, M. J. *JOM* **1991**, *406*, C25.
59. Koppel, G. A.; Kinnick, M. D.; Nummy, L. J. *JACS* **1977**, *99*, 2822.
60. Muathen, H. J. *JOC* **1992**, *57*, 2740.
61. Hori, Y.; Nagano, Y.; Taguchi, H.; Taniguchi, H. *Chem. Express* **1986**, *1*, 659 (*CA* **1987**, *107*, 96 302b).
62. Hori, Y.; Nagano, Y.; Uchiyama, H.; Yamada, Y.; Taniguchi, H. *Chem. Express* **1978**, 73 (*CA* **1978**, *88*, 104 177w).
63. Tomoda, H.; Saito, S.; Ogawa, S.; Shiraishi, S. *CL* **1980**, 1277.
64. Enokida, T.; Hirohashi, R. *Chem. Mater.* **1991**, *3*, 918.
65. Tomoda, H.; Saito, S.; Shiraishi, S. *CL* **1983**, 313.
66. Enokida, T.; Ehashi, S. *CL* **1988**, 179.
67. Edmonson, S. J.; Hill, J. S.; Isaacs, N. S.; Mitchell, P. C. H. *JCS(D)* **1990**, 1115.
68. Hauser, C. F.; Theiling, L. F. *JOC* **1974**, *39*, 1134.
69. Nishiyama, Y.; Makino, Y.; Hamanaka, S.; Ogawa, A.; Sonoda, N. *BCJ* **1989**, *62*, 1682.

Ann C. Savoca
Air Products and Chemicals, Allentown, PA, USA

2,6-Di-*t*-butylpyridine[1]

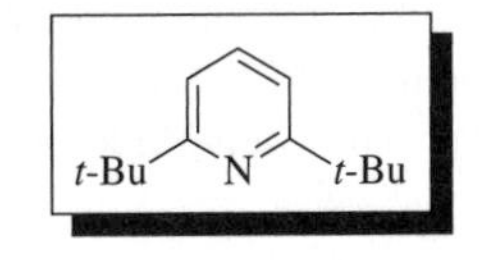

[585-48-8] $C_{13}H_{21}N$ (MW 191.32)

(weak base and proton scavenger;[2–4] prevention of acid-catalyzed side reactions;[5–7] vinyl triflate synthesis[8,9])

Alternate Names: 2,6-bis(1,1-dimethylethyl)pyridine; BDMEP.
Physical Data: mp 2.2 °C; bp 100–101 °C/23 mmHg; n_D^{20} 1.4733; pK_a 3.58 (50% EtOH);[2] pK_a 0.81 (DMSO).[3]
Solubility: insol H_2O; sol alcohol, acetone, and hexane.
Form Supplied in: colorless liquid.
Handling, Storage, and Precautions: potentially toxic; handle with care. Use in a fume hood.

Proton Scavenger. BDMEP has high basicity in the gas phase,[4] but is an abnormally weak base in solution.[2,3] The low basicity of BDMEP is due to steric hindrance of solvation.[10] Therefore BDMEP forms an HCl salt and compounds with Br_2, but does not react with MeI or BF_3.[2]

BDMEP has been used to prevent acid-catalyzed side reactions or nucleophilic reactions by the base in the synthesis of [*trans*-2,3-*trans*-5,6-$^2H_4$]-1,4-dioxane (eq 1),[5] labeled [1-^{13}C]acid chlorides (eq 2),[6] steroidal dienes,[11] and divinyl ketones. In the last case, Amberlyst/15 pretreated with BDMEP was used (eq 3).[7]

BDMEP (1)

RMgBr → 1. $^{11}CO_2$; 2. BDMEP, o-$C_6H_4(COCl)_2$; 10–30% → $R^{11}COCl$ (2)
R = Et, Pr, *c*-Bu

Amberlyst 15, BDMEP, 71% (3)

Catalysis of Enolization and Addition Reactions. It has been shown that BDMEP and its analogs (e.g. 4-methyl-BDMEP) can be used to prepare vinyl triflates from carbonyl compounds and triflic anhydride.[12] Good results were obtained with a polymer-bound BDMEP, which may be easily recycled (eq 4).[8] Intermediate vinylsilanes have been obtained from 2,3-dihydro-4*H*-pyran-4-one derivatives. These have been used in Diels–Alder reactions with alkynes to prepare silyl phenyl ethers (eq 5).[9] In the presence of BDMEP, formaldehyde acetals react with trimethylsilyl enol ethers to form β-alkoxy ketones (eq 6).[13]

Tf_2O, polymer–BDMEP, CCl_4, 25–40 °C, 24–48 h, 58–86% (4)

1. BDMEP, *i*-Pr_3SiOTf; 2. $R^3C{\equiv}CE$, $CHCl_3$, 60 °C; 70–97% (5)

R^1 = Ph, *t*-Bu, *p*-$MeOC_6H_4$; R^2 = H, CO_2Et; R^3 = H, PhCO, CO_2Me; E = PhCO, CO_2Me, CO_2Et

OTMS; Ph; $H_2C(OR^3)_2$; BDMEP; 76–92%; Ph; O; OR^3 (6)

R^3 = Me, Bn

Inhibition of Acid-Catalyzed Reactions. BDMEP was used to eliminate protic acid catalysis in aminium cation-radical initiated 'Diels–Alder' reactions of dienes,[14] and vinylcyclopropane–cyclopentene isomerizations.[15] Cationic polymerization of isobutene and α-methylstyrene by Ziegler-type catalysts in the presence of BDMEP resulted in a polymer with a higher molecular weight and noticeably narrower molecular weight distribution.[16]

Related Reagents. 1,8-Bis(dimethylamino)naphthalene; 2,4,6-Collidine; 1,5-Diazabicyclo[4.3.0]non-5-ene; 1,8-Diazabicyclo[5.4.0]undec-7-ene; Diisopropylethylamine; 2,6-Lutidine; Pyridine; Quinoline.

1. Kanner, B. *H* **1982**, *18*, 411.
2. (a) Brown, H. C.; Kanner, B. *JACS* **1953**, *75*, 3865. (b) Brown, H. C.; Kanner, B. *JACS* **1966**, *88*, 986.
3. Benoit, R. L.; Frechette, M.; Lefebvre, D. *CJC* **1988**, *66*, 1159.
4. Arnett, E. M.; Chawla, B. *JACS* **1979**, *101*, 7141.
5. Jensen, F. R.; Neese, R. A. *JOC* **1972**, *37*, 3037.
6. Luthra, S. K.; Pike, V. W.; Brady, F. *Appl. Radiat. Isot.* **1990**, *41*, 471.
7. Nakamura, E.; Kubota, K.; Isaka, M. *JOC* **1992**, *57*, 5809.
8. (a) Wright, M. E.; Pulley, S. R. *JOC* **1987**, *52*, 1623, 5036. (b) Wright, M. E.; Pulley, S. R. *JOC* **1989**, *54*, 2886.
9. Obrecht, D. *HCA* **1991**, *74*, 27.
10. (a) McDaniel, D. H.; Özcan, M. *JOC* **1968**, *33*, 1922. (b) Bernasconi, C. F.; Carre, D. J. *JACS* **1979**, *101*, 2707.
11. Prelle, A.; Winterfeldt, E. *H* **1989**, *28*, 333.
12. Stang, P. J.; Treptow, W. *S* **1980**, 283.
13. Murata, S.; Suzuki, M.; Noyori, R. *T* **1988**, *44*, 4259.
14. (a) Gassman, P. G.; Singleton, D. A. *JACS* **1984**, *106*, 7993. (b) Reynolds, D. W.; Lorenz, K. T.; Chiou, H.-S.; Bellville, D. J.; Pabon, R. A.; Bauld, N. L. *JACS* **1987**, *109*, 4960. (c) Schmittel, M.; Seggern, H. v. *AG(E)* **1991**, *30*, 999.
15. Dinnocenzo, J. P.; Conlon, D. A. *JACS* **1988**, *110*, 2324.
16. Kennedy, J. P. *J. Macromol. Sci., Chem.* **1982**, *A18*, 3.

Rafael R. Kostikov
St. Petersburg State University, Russia

Diethylaluminum Chloride[1]

[96-10-6] $C_4H_{10}AlCl$ (MW 120.56)

(strong Lewis acid that can also act as a proton scavenger; reacts with HX to give ethane and EtAlClX)

Alternate Names: chlorodiethylaluminum; diethylchloroalane.

Physical Data: mp −50 °C; bp 125 °C/50 mmHg; *d* 0.961 g cm^{-3}.
Solubility: sol most organic solvents; stable in alkanes or arenes.
Form Supplied in: commercially available neat or as solutions in hexane or toluene.
Analysis of Reagent Purity: solutions are reasonably stable but may be titrated before use by one of the standard methods.[1e]
Handling, Storage, and Precautions: must be transferred under inert gas (Ar or N_2) to exclude oxygen and water. Use in a fume hood.

Introduction. The general properties of alkylaluminum halides as Lewis acids are discussed in the entry for ***Ethylaluminum Dichloride***. Dialkylaluminum halides are less acidic than alkylaluminum dihalides. Et_2AlCl is much cheaper than ***Dimethylaluminum Chloride*** and is used more frequently than Me_2AlCl since comparable results are usually obtained. In some cases, most notably the ene reactions of carbonyl compounds, use of Me_2AlCl is preferable since its methyl groups are less nucleophilic than the ethyl groups of Et_2AlCl, which can act as a reducing agent.

Catalysis of Diels–Alder Reactions. Et_2AlCl has been extensively used as a Lewis acid catalyst for Diels–Alder reactions. *N*-Acyloxazolidinones can form both 1:1 (eq 1) and 1:2 complexes with Et_2AlCl (eq 2).[2] The 1:2 complex is ~100 times as reactive as the 1:1 complex and gives greater *endo* selectivity and higher de. Me_2AlCl gives similar selectivity with fewer byproducts.

Et_2ClAl^-; O+; N; O → COX_v (1)
endo:*exo* = 20:1; 60% de; k_{rel} = 1

$Et_2AlCl_2^-$; Et; Et; Al; +O; O+; N → COX_v (2)
endo:*exo* = 60:1; 90% de; k_{rel} = 100

Et_2AlCl has been extensively used as a Lewis acid catalyst for intermolecular[3] and intramolecular[4] Diels–Alder reactions with α,β-unsaturated ketones and esters as dienophiles. It also catalyzes inverse electron demand Diels–Alder reactions of alkenes with quinone methides[5] and Diels–Alder reactions of aldehydes as enophiles.[6]

Catalysis of Ene Reactions. Et_2AlCl has been used as a catalyst for ene reactions with ethyl propiolate as an enophile,[7] for intramolecular ene reactions of aldehydes,[8] and for intramolecular ene reactions with α,β-unsaturated esters as enophiles (eq 3).[3e,9]

EtO_2C; CO_2Et; CO_2Et; CF_3CO-N; Et_2AlCl, −78 °C, 90% → EtO_2C; CO_2Et; CO_2Et; CF_3CO-N (3)

Catalysis of Claisen and Vinylcyclopropane Rearrangements. Et_2AlCl has been used as a catalyst for Claisen rearrangement of aryl allyl ethers.[10] The rearrangement of 2-vinylcyclopropanecarboxylate esters to cyclopentenes is catalyzed by Et_2AlCl (eq 4).[11]

TBDMSO — Et_2AlCl, 0 °C, 82% → TBDMSO (4)

Generation of Electrophilic Cations. Complexation of Et_2AlCl to ketones and aldehydes activates the carbonyl group toward addition of a nucleophilic alkyl- or allylstannane or allylsilane.[12] Et_2AlCl has been used to initiate Beckmann rearrangements of oxime mesylates. The ring-expanded cation can be trapped intermolecularly by enol ethers and cyanide and intramolecularly by alkenes (eq 5).[13]

OTMS, C_6H_{13} + OMs — Et_2AlCl, −78 to 22 °C, 90% → NH, C_6H_{13} (5)

Formation and Reaction of Aluminum Enolates. Et_2AlCl has been used in modified Reformatsky reactions. Aldol adducts are obtained in good yield by reaction of an α-bromo ketone with a ketone in the presence of ***Zinc*** and Et_2AlCl in THF (eq 6).[14] Lithium enolates of esters do not react with epoxides. Reaction of lithium enolates with Et_2AlCl affords aluminum enolates that react with epoxides at the less substituted carbon (eq 7).[15]

Br + — Zn/Ag, THF, py, Et_2AlCl, 79% → OH (6)

O-*t*-Bu — 1. LDA; 2. Et_2AlCl; 3. *t*-Bu epoxide; 38% → O-*t*-Bu, OH, *t*-Bu; *syn:anti* = 95:5 (7)

Formation and Reaction of Alkynylaluminum Reagents. Lithium acetylides react with Et_2AlCl to form LiCl and diethylaluminum acetylides. The aluminum acetylides are useful reagents for carrying out S_N2 reactions on epoxides (eq 8)[15c,16] and undergo conjugate addition to enones that can adopt an *S-cis* conformation (eq 9).[17]

Et—≡CH — 1. BuLi; 2. Et_2AlCl; 3. OTBDMS epoxide; 77% → Et, OH, OTBDMS (8)

Ph—≡CH — 1. BuLi; 2. Et_2AlCl; 3. PhCH=CHC(O)Ph; 81% → Ph, Ph, O, Ph (9)

Reaction as a Nucleophile. Et_2AlCl reacts analogously to ***Ethylmagnesium Bromide*** and transfers an ethyl group to many electrophiles. Since EtMgBr and ***Ethyllithium*** are readily available, use of Et_2AlCl to deliver an ethyl group is needed only when the stereochemistry of addition is an important issue. High levels of asymmetric induction are obtained in the conjugate addition of Et_2AlCl to unsaturated acyloxazolidinones with carbohydrate-derived chiral auxiliaries (eq 10).[18] Et_2AlCl opens epoxides to chlorohydrins.[19]

OPiv, PivO, PivO, Ph — 1. Et_2AlCl, 84%; 2. LiOOH, 87% → Ph, OH, Et, O; 92% ee (10)

Related Reagents. Bis(diethylaluminum) Sulfate; Dimethylaluminum Chloride; Ethylaluminum Dichloride; Tin(IV) Chloride; Tris(acetylacetonato)cobalt–Diethylaluminum Chloride-NORPHOS; Zinc Chloride.

1. For reviews, see Ref. 1 in ***Ethylaluminum Dichloride***.
2. Evans, D. A.; Chapman, K. T.; Bisaha, J. *JACS* **1988**, *110*, 1238.
3. (a) Schlessinger, R. H.; Schultz, J. A. *JOC* **1983**, *48*, 407. (b) Cohen, T.; Kosarych, Z. *JOC* **1982**, *47*, 4005. (c) Hagiwara, H.; Okano, A.; Uda, H. *CC* **1985**, 1047. (d) Furuta, K.; Iwanaga, K.; Yamamoto, H. *TL* **1986**, *27*, 4507. (e) Oppolzer, W. *AG(E)* **1984**, *23*, 876. (f) Reetz, M. T.; Kayser, F.; Harms, K. *TL* **1992**, *33*, 3453. (g) Midland, M. M.; Koops, R. W. *JOC* **1992**, *57*, 1158.
4. (a) Roush, W. R.; Gillis, H. R. *JOC* **1982**, *47*, 4825. (b) Reich, H. J.; Eisenhart, E. K. *JOC* **1984**, *49*, 5282. (c) Shea, K. J.; Gilman, J. W. *TL* **1983**, *24*, 657. (d) Brown, P. A.; Jenkins, P. R. *JCS(P1)* **1986**, 1303. (e) Reich, H. J.; Eisenhart, E. K.; Olson, R. E.; Kelly, M. J. *JACS* **1986**, *108*, 7791. (f) Funk, R. L.; Bolton, G. L. *JACS* **1986**, *108*, 4655. (g) Taschner, M. J.; Cyr, P. T. *TL* **1990**, *31*, 5297.
5. (a) Tietze, L. F.; Brand, S.; Pfeiffer, T.; Antel, J.; Harms, K.; Sheldrick, G. M. *JACS* **1987**, *109*, 921. (b) Casiraghi, G.; Cornia, M.; Casnati, G.; Fava, G. G.; Belicchi, M. F. *CC* **1986**, 271.
6. Midland, M. M.; Afonso, M. M. *JACS* **1989**, *111*, 4368.
7. Dauben, W. G.; Brookhart, T. *JACS* **1981**, *103*, 237.
8. (a) Kamimura, A.; Yamamoto, A. *CL* **1990**, 1991. (b) Andersen, N. H.; Hadley, S. W.; Kelly, J. D.; Bacon, E. R. *JOC* **1985**, *50*, 4144.
9. (a) Oppolzer, W.; Robbiani, C.; Bättig, K. *T* **1984**, *40*, 1391. (b) Oppolzer, W.; Mirza, S. *HCA* **1984**, *67*, 730.
10. (a) Sonnenberg, F. M. *JOC* **1970**, *35*, 3166. (b) Bender, D. R.; Kanne, D.; Frazier, J. D.; Rapoport, H. *JOC* **1983**, *48*, 2709. (c) Lutz, R. P. *CR* **1984**, *84*, 205.
11. (a) Corey, E. J.; Myers, A. G. *JACS* **1985**, *107*, 5574. (b) Davies, H. M. L.; Hu, B. *TL* **1992**, *33*, 453. (c) Davies, H. M. L.; Hu, B. *JOC* **1992**, *57*, 3186.
12. (a) McDonald, T. L.; Delahunty, C. M.; Mead, K.; O'Dell, D. E. *TL* **1989**, *30*, 1473. (b) Denmark, S. E.; Weber, E. J. *JACS* **1984**, *106*, 7970. (c) Mooiweer, H. H.; Hiemstra, H.; Fortgens, H. P.; Speckamp, N. W. *TL* **1987**, *28*, 3285.

13. (a) Sakane, S.; Matsumura, Y.; Yamamura, Y.; Ishida, Y.; Maruoka, K.; Yamamoto, H. *JACS* **1983**, *105*, 672. (b) Maruoka, K.; Miyazaki, T.; Ando, M.; Matsumura, Y.; Sakane, S.; Hattori, K.; Yamamoto, H. *JACS* **1983**, *105*, 2831. (c) Matsumura, Y.; Fujiwara, J.; Maruoka, K.; Yamamoto, H. *JACS* **1983**, *105*, 6312.
14. (a) Maruoka, K.; Hashimoto, S.; Kitagawa, Y.; Yamamoto, H.; Nozaki, H. *JACS* **1977**, *99*, 7705. (b) Maruoka, K.; Hashimoto, S.; Kitigawa, Y.; Yamamoto, H.; Nozaki, H. *BCJ* **1980**, *53*, 3301. (c) Stokker, G. E.; Hoffmann, W. F.; Alberts, A. W.; Cragoe, E. J., Jr.; Deanna, A. A.; Gilfillan, J. L.; Huff, J. W.; Novello, F. C.; Prugh, J. D.; Smith, R. L.; Willard, A. K. *JMC* **1985**, *28*, 347. (d) Tsuboniwa, N.; Matsubara, S.; Morizawa, Y.; Oshima, K.; Nozaki, H. *TL* **1984**, *25*, 2569. (e) Tsuji, J.; Mandai, T. *TL* **1978**, 1817.
15. (a) Nozaki, H.; Oshima, K.; Takai, K.; Ozawa, S. *CL* **1979**, 379. (b) Sturm, T.-J.; Marolewski, A. E.; Rezenka, D. S.; Taylor, S. K. *JOC* **1989**, *54*, 2039. (c) Danishefsky, S.; Kitahara, T.; Tsai, M.; Dynak, J. *JOC* **1976**, *41*, 1669.
16. (a) Nicolaou, K. C.; Webber, S. E.; Ramphal, J.; Abe, Y. *AG(E)* **1987**, *26*, 1019. (b) Matthews, R. S.; Eickhoff, D. J. *JOC* **1985**, *50*, 3923. (c) Ishiguro, M.; Ikeda, N.; Yamamoto, H. *CL* **1982**, 1029.
17. Hooz, J.; Layton, R. B. *JACS* **1971**, *93*, 7320.
18. (a) Rück, K.; Kunz, H. *AG(E)* **1991**, *30*, 694. (b) Rück, K.; Kunz, H. *SL* **1992**, 343. (c) Rück, K.; Kunz, H. *S* **1993**, 1018.
19. Gao, L.-X.; Saitoh, H.; Feng, F.; Murai, A. *CL* **1991**, 1787.

Barry B. Snider
Brandeis University, Waltham, MA, USA

Diisopropylethylamine

i-Pr_2NEt

[7087-68-5] $C_8H_{19}N$ (MW 129.24)

(hindered non-nucleophilic amine base used in alkylations, selective generation of enolates, aldol-like reactions, and eliminations)

Alternate Names: DIPEA; DIEA; Hünig's base.
Physical Data: bp 127 °C; *d* 0.742 g cm^{-3}.
Solubility: sol most organic solvents.
Handling, Storage, and Precautions: corrosive and flammable liquid; flush containers with nitrogen or argon to prevent exposure to carbon dioxide; vapors are harmful; avoid absorption through the skin; use in a fume hood.

Metal-Catalyzed Couplings. DIPEA can be used as a base in the palladium(0)-catalyzed alkoxycarbonylation of both allyl phosphates and acetates.[1] Treatment of diethyl (*E*)-2-hexenyl phosphate with ***Tris(dibenzylideneacetone)dipalladium*** and ***Triphenylphosphine*** in the presence of 1 equiv of DIPEA in ethanol under 30 atm CO pressure leads to an 84:16 mixture of *trans*- and *cis*-ethyl heptenoates in 88% yield (eq 1). The base used neutralizes the phosphoric acid generated. Without DIPEA, the alkyl ester is not produced. Allyl acetates behave in a similar manner (eq 2). Use of other tertiary bases, such as triethylamine, leads to lower yields of products.

Pr–CH=CH–CH$_2$OP(O)(OEt)$_2$ → [CO (30 atm), Pd cat; i-Pr_2NEt, EtOH; 88%] → Pr–CH=CH–CH$_2$CO$_2$Et (1)
(*E*):(*Z*) = 84:16

1-vinylcyclohexyl acetate (OAc) → [CO (30 atm), Pd cat; i-Pr_2NEt, EtOH; 80%] → cyclohexylidene–CH$_2$CH$_2$CO$_2$Et (2)

In an asymmetric variation of the Heck reaction,[2] the sterically demanding DIPEA is used in combination with ***(R)-2,2′-Bis(diphenylphosphino)-1,1′-binaphthyl*** ((*R*)-BINAP) for generating enantiomerically enriched 2-aryl-2,3-dihydrofurans from aryl triflates.[3] However, in a comparison study, the base ***1,8-Bis(dimethylamino)naphthalene*** (proton sponge) was found to be superior in regards to enantiomeric purity for the arylation reaction (eq 3).[4,5]

p-ClC$_6$H$_4$OTf + 2,3-dihydrofuran → [Pd(OAc)$_2$, (*R*)-BINAP; base, benzene] → 2-Ar-2,3-dihydrofuran + 2-Ar-2,5-dihydrofuran (3)

base	% ee (yield)	% ee (yield)
DIPEA	82 (92)	60 (8)
Et_3N	75 (98)	9 (2)
proton sponge	96 (71)	17 (29)

Selective Enolate Formation. This base, in combination with boryl triflates, is widely applied in the enolate generation of ketones for use in directed cross-aldol reactions.[6] Reaction of 4-methyl-2-pentanone with DIPEA and ***Di-n-butylboryl Trifluoromethanesulfonate*** in ether at −78 °C produces the unisolated boron enolate. Subsequent treatment with hexanal yields the cross-aldol product in 70% yield as the sole regioisomer (eq 4).[7] Reaction takes place at the methyl group of the ketone, while none of the ketone or aldehyde self-condensation products are observed. In a complementary fashion, substitution of ***2,6-Lutidine*** for DIPEA and ***9-Borabicyclononyl Trifluoromethanesulfonate*** (9-BBN triflate) for di-*n*-butylboryl triflate yields the opposite regioisomer as the sole product (eq 4).[8]

4-methyl-2-pentanone → [1. n-Bu$_2$BOTf, i-Pr_2NEt, ether; 2. C$_5$H$_{11}$CHO; 70%] → (CH$_3$)$_2$CHCH$_2$C(O)CH$_2$CH(OH)C$_5$H$_{11}$
4-methyl-2-pentanone → [1. 9-BBN-OTf, 2,6-lutidine, ether; 2. C$_5$H$_{11}$CHO; 57%] → (CH$_3$)$_2$CHCH(CH(OH)C$_5$H$_{11}$)C(O)CH$_3$ (4)

Proper choice of boron reagent, reaction solvent, and tertiary amine base influences the enolate geometry of ketones[9] and esters.[10] The use of dialkylboron triflates or dialkylboron halides, such as *B*-chloro-9-BBN, with DIPEA in ether at −78 °C favors the (*Z*)-enol borinates of ketones, while the more sterically demanding dialkylboron halides, such as dicyclohexylboron chloride (Chx_2BCl), with ***Triethylamine*** favor the (*E*)-enol borinates (eq 5).[9a,9b]

B-Cl-9-BBN; i-Pr₂NEt, ether, −78 °C → O-9-BBN enolate, >99% Z; Chx_2BCl, Et_3N, ether, −78 °C → $OBChx_2$ enolate, >99% E (5)

Similar complementary methodologies exist for generating both the (*E*) and (*Z*) boron enolates of esters. The combination of chiral nonracemic bromoborane (**1**) and DIPEA in CH_2Cl_2 selectively converts *trans*-crotyl propionate into the (*E*)-boron enolate (**2**), which subsequently undergoes a highly enantioselective and diastereoselective Ireland–Claisen rearrangement to generate the *threo* product (**3**) in 75% yield and >97% ee. In comparison, use of triethylamine in a solvent mixture of toluene and hexane leads to a 90:10 mixture of (*Z*)- and (*E*)-boron enolates (**4**), which rearranges to generate the *erythro* product (**5**) as the major isomer in 65% yield and high enantiomeric excess (eq 6).[10a,11]

Br-BR₂*, Et₃N, toluene–hexane, −78 °C → (**4**) → (**5**) 65%, 96% ee; Br-BR₂*, i-Pr₂NEt, CH_2Cl_2, −78 °C → (**2**) → (**3**) 75%, 97% ee (6)

Br-BR₂* = ArO₂SN–B(Br)–NSO₂Ar (Ph, Ph) (**1**)

The stereoselective generation of silylketene acetals from alkyl esters and ***Triethylsilyl Perchlorate***[12] is quite effective using DIPEA as the base at −70 °C in a 1:1 solvent mixture of CH_2Cl_2 and CCl_4 (see ***2,2,6,6-Tetramethylpiperidine***).

Phenols and enols can be *O*-methylated in moderate to good yields using ***Trimethylsilyldiazomethane*** with DIPEA in methanol–acetonitrile.[13]

Thioesters[10c,10d] and oxazolidones[14] can also be selectively converted into dialkylboryl enolates using DIPEA as base and the appropriate boron triflates. The thioester enolates readily react with imines to generate β-amino thioesters (eq 7),[10e] whereas the oxazolidone boron enolates can undergo alkylation with acetoxyazetidinones (eq 8).[14] *S*-*t*-Butyl bromothioacetate undergoes a highly stereospecific Darzens condensation with substituted benzaldehydes using ***Dicyclopentylboryl Trifluoromethanesulfonate*** and DIPEA at −78 °C to afford the *trans*-glycidic ester (eq 9).[15]

1. n-Bu₂BOTf, i-Pr₂NEt, ether; 2. Ph–CH=N–CH₂Ph; 80% (7)

Et₂BOTf, i-Pr₂NEt, ether; ZnBr₂; 73%, β:α = >99:1 (8)

1. (c-C₅H₉)₂BOTf, i-Pr₂NEt, −78 °C; 2. 25 °C, 24 h (9)

Aldol-like reactions between aldehydes and nitriles,[16] alkylpyridines,[17] 2-methyloxazoles, 2-methylthiazoles,[18] glycolates, and thioglycolates[19] are possible in the presence of boryl triflates and DIPEA. Methyl-*O*-allyl glycolate tin(II) (or boron) enolates, prepared easily using DIPEA as base, undergo Wittig rearrangement to afford α-hydroxy esters with a high degree of diastereoselectivity (eq 10).[20]

Sn(OTf)₂, i-Pr₂NEt, −78 °C, 40%; >99.5:0.5 (10)

Diisopropylethylamine also finds application in the preparation of titanium enolates from esters,[21] aryl ketones,[22] and oxazolidones.[23]

Base-Promoted Alkylation. This sterically hindered amine is widely used in organic synthesis as a proton scavenger. Its lack of quaternization makes it an excellent choice of a base for use with very reactive alkylating agents.[24] In the field of protecting group chemistry, DIPEA is a particularly useful base for protection of alcohols as substituted ethers.[25] For example, the tertiary alcohol of mevalonic lactone can be protected as the *p*-

methoxybenzyloxymethyl ether using an excess of DIPEA and 3 equiv of *p*-methoxybenzyl chloromethyl ether (eq 11).[26]

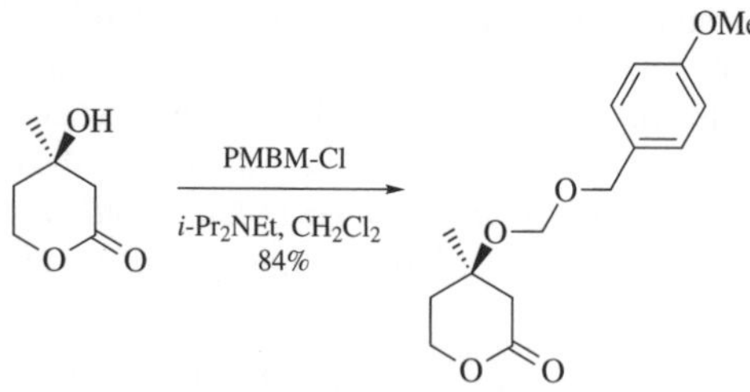

PMBM-Cl
i-Pr_2NEt, CH_2Cl_2
84%
(11)

2,4-Disubstituted oxazolones are alkylated in high yields using alkyl halides with DIPEA as a base in DMF (eq 12).[27] DIPEA can be used together with ***Triethyloxonium Tetrafluoroborate*** for the esterification of sterically hindered carboxylic acids.[28]

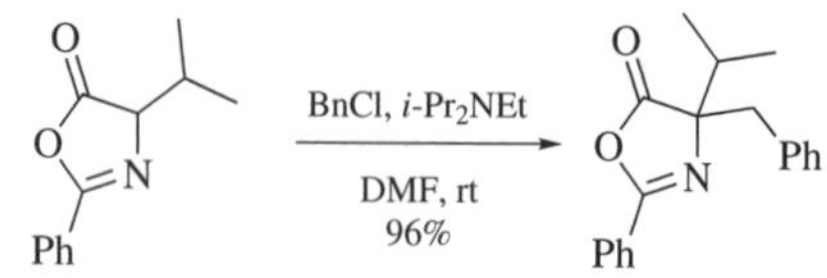

BnCl, *i*-Pr_2NEt
DMF, rt
96%
(12)

The alkylsulfination of diacetone-D-glucose with sulfinyl chlorides and DIPEA in toluene at −78 °C produces the (*S*)-sulfinates as the major products. Simply changing the base from DIPEA to pyridine and the solvent from toluene to THF results in a remarkable stereochemical reversal, affording the (*R*)-sulfinates as the major products (eq 13).[29]

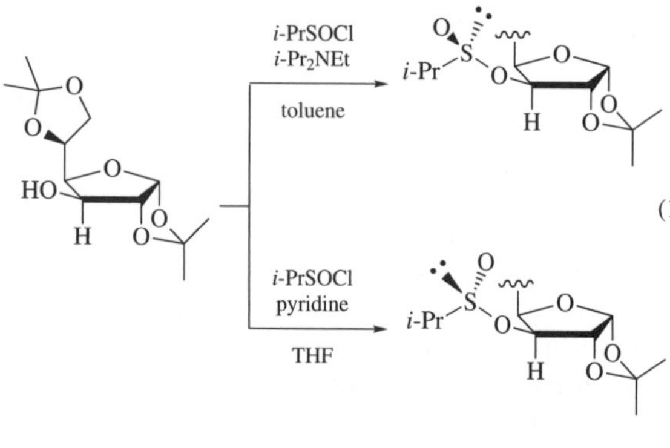

i-PrSOCl, *i*-Pr_2NEt, toluene
i-PrSOCl, pyridine, THF
(13)

The alkylation of protected uracils with alkyl halides using DIPEA in DMF or acetonitrile furnishes the alkylated uracils in good to moderate yields (eq 14). The alkylations involving ribofuranosyl bromides furnish the β-isomers as the sole products.[30]

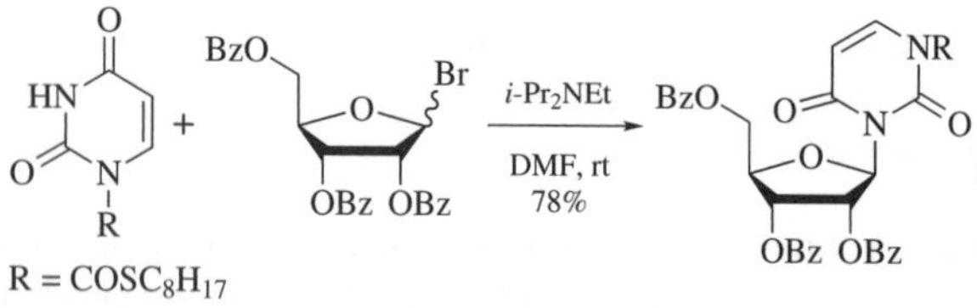

i-Pr_2NEt
DMF, rt
78%
(14)
R = $COSC_8H_{17}$

Exposure of hydroxyl vinyl ethers to ***Trifluoromethanesulfonic Anhydride*** and DIPEA in CH_2Cl_2 at −78 °C results in a stereospecific cyclization to afford cyclic hemiacetals in near quantitative yield (eq 15).[31]

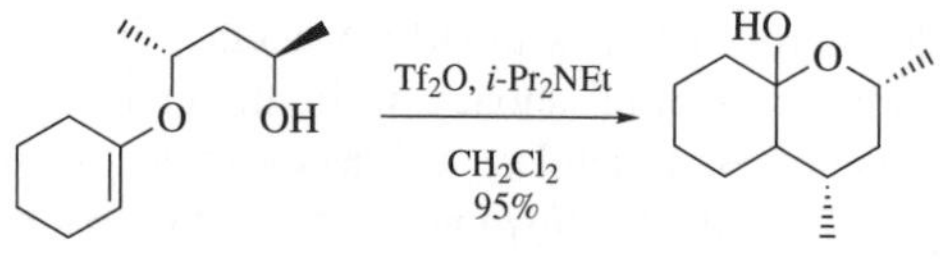

Tf_2O, *i*-Pr_2NEt
CH_2Cl_2
95%
(15)

Addition of DIPEA to a mixture of phenylsulfenyl chloride and an unsaturated alcohol or carboxylic acid results in high yields of cyclic ethers or lactones, respectively.[32] The presumed mechanism involves the intermediacy of an episulfonium ion (eq 16).

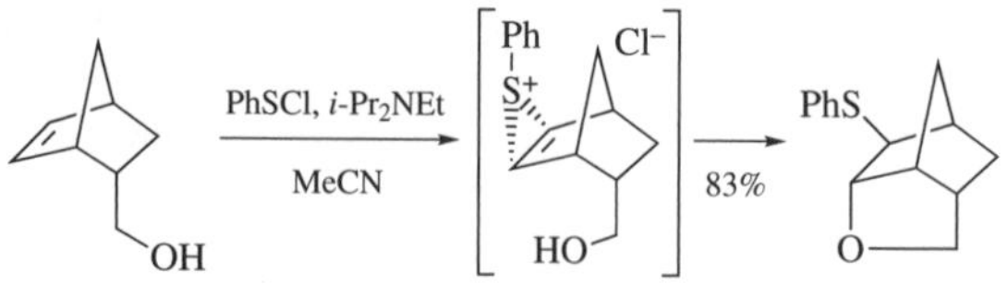

PhSCl, *i*-Pr_2NEt
MeCN
83%
(16)

Eliminations. The eliminative deoxygenation of acetals into enol ethers is accomplished using ***Trimethylsilyl Trifluoromethanesulfonate*** and a slight excess of DIPEA (eq 17);[33] similarly, treatment of 2-alkyloxazolidines with ***Chlorotrimethylsilane*** and DIPEA leads to *N*-(trimethylsilyloxyalkyl)enamines (eq 18).[34] Thioacetals follow similar chemistry to furnish vinyl sulfides.[35] Deoxygenation of sulfoxides with DIPEA and ***Iodotrimethylsilane*** produces vinyl sulfides.[36]

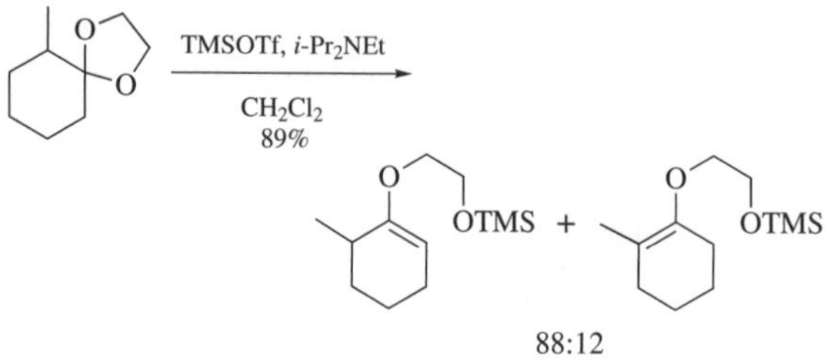

TMSOTf, *i*-Pr_2NEt
CH_2Cl_2
89%
(17) 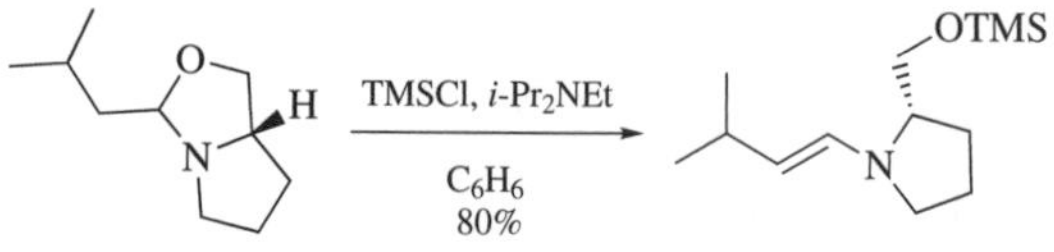
88:12

TMSCl, *i*-Pr_2NEt
C_6H_6
80%
(18)

Peptide Couplings. Diisopropylethylamine, as well as triethylamine, *N*-methylmorpholine, and other tertiary amines, find utility in the coupling of amino acids to prepare peptides.[37] The basicity and steric nature of the tertiary amine utilized during the coupling reaction influences the degree of racemization.[38]

Minor dipeptide impurities, which sometimes are difficult to remove, can form during the *N*-acylation of amino acids. Acylation of alanine using benzoyl chloride and aqueous NaOH yields *N*-benzoylalanine (60%), which contains about 1.2% of the dipeptide impurity. Use of DIPEA nearly eliminates the impurity and results in a 72% yield of *N*-acylated product.[39]

The coupling of *N*-methylated amino acids is sometimes problematic. However, the combination of 3 equiv of DIPEA and 1 equiv of bromotris(dimethylamino)phosphonium hexafluorophosphate (BroP) in CH_2Cl_2 results in good yields of desired dipeptides. Methyl *N*-methylvalinate successfully couples with other amino acids using these reaction conditions. Epimerization during the coupling is not observed (eq 19).[40]

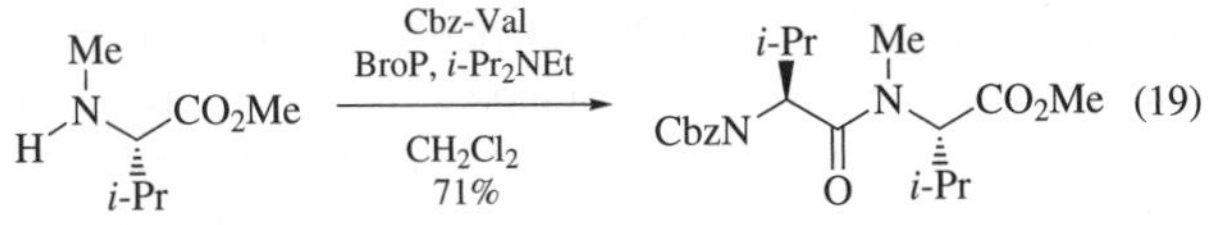

$$\xrightarrow[\text{CH}_2\text{Cl}_2,\ 71\%]{\text{Cbz-Val, BroP, } i\text{-Pr}_2\text{NEt}} \quad (19)$$

Related Reagents. 1,8-Bis(dimethylamino)naphthalene; 2,4,6-Collidine; 1,5-Diazabicyclo[4.3.0]non-5-ene; 1,8-Diazabicyclo[5.4.0]undec-7-ene; Lithium Chloride-Diisopropylethylamine; Pyridine; Triethylamine.

---

1. Murahashi, S. I.; Imada, Y.; Taniguchi, Y.; Higashiura, S. *JOC* **1993**, *58*, 1538.
2. (a) Daves, G. D., Jr.; Hallberg, A. *CRV* **1989**, *89*, 1433. (b) Heck, R. F. *Palladium Reagents in Organic Synthesis*; Academic: New York, 1985.
3. Ozawa, F.; Kubo, A.; Hayashi, T. *JACS* **1991**, *113*, 1417.
4. Ozawa, F.; Kubo, A.; Hayashi, T. *TL* **1992**, *33*, 1485.
5. For an intramolecular Heck-type reaction, see: (a) Sato, Y.; Sodeoka, M.; Shibasaki, M. *JOC* **1989**, *54*, 4738. (b) Mori, M.; Kaneta, N.; Shibasaki, M. *JOC* **1991**, *56*, 3486.
6. (a) Heathcock, C. H. In *Asymmetric Synthesis*; Morrison, J. D., Ed; Academic: New York, 1984; Vol. 3, p 111–212. (b) Evans, D. A.; Nelson, J. V.; Taber, T. R. *Top. Stereochem.* **1982**, *13*, 1.
7. (a) Mukaiyama, T.; Inoue, T. *CL* **1976**, 559. See also: (b) Mukaiyama, T.; Inomata, K.; Muraki, M. *JACS* **1973**, *95*, 967.
8. Inoue, T.; Uchimaru, T.; Mukaiyama, T. *CL* **1977**, 153.
9. (a) Brown, H. C.; Dhar, R. K.; Bakshi, R. K.; Pandiarajan, P. K.; Singaram, B. *JACS* **1989**, *111*, 3441. (b) Brown, H. C.; Dhar, R. K.; Ganesan, K.; Singaram, B. *JOC* **1992**, *57*, 499, 2716. (c) Enders, D.; Lohray, B. B. *AG(E)* **1988**, *27*, 581. (d) Evans, D. A.; Nelson, J. V.; Vogel, E.; Taber, T. R. *JACS* **1981**, *103*, 3099. (e) Van Horn, D. E.; Masamune, S. *TL* **1979**, 2229. (f) Evans, D. A.; Vogel, E.; Nelson, J. V. *JACS* **1979**, *101*, 6120. (g) Paterson, I.; Osborne, S. *TL* **1990**, *31*, 2213. (h) For sulfenylation and selenenylation of enol borinates see: Paterson, I.; Osborne, S. *SL* **1991**, 145.
10. (a) Corey, E. J.; Lee, D.-H. *JACS* **1991**, *113*, 4026. (b) Corey, E. J.; Kim, S. S. *JACS* **1990**, *112*, 4976. (c) Hirama, M.; Masamune, S. *TL* **1979**, 2225. (d) Gennari, C.; Bernardi, A.; Cardani, S.; Scolastico, C. *T* **1984**, *40*, 4059. (e) Otsuka, M.; Yoshida, M.; Kobayashi, S.; Ohno, M. *TL* **1981**, *22*, 2109.
11. See also: Paterson, I.; Lister, M. A.; McClure, C. K. *TL* **1986**, *27*, 4787.
12. Wilcox, C. S.; Babston, R. E. *TL* **1984**, *25*, 699.
13. (a) Aoyama, T.; Terasawa, S.; Sudo, K.; Shioiri, T. *CPB* **1984**, *32*, 3759 and Martin, M. *SC* **1983**, *13*, 809. (b) For *O*-ethylation of β-diketones see: Rizzardo, E. *CC* **1975**, 644. (c) For fluorosulfonation of phenols see: Roth, G. P.; Fuller, C. E. *JOC* **1991**, *56*, 3493.
14. Fuentes, L. M.; Shinkai, I.; Salzmann, T. N. *JACS* **1986**, *108*, 4675. See also: Evans, D. A.; Ennis, M. D.; Mathre, D. J. *JACS* **1982**, *104*, 1737.
15. Polniaszek, R. P.; Belmont, S. E. *SC* **1989**, *19*, 221.
16. Hamana, H.; Sugasawa, T. *CL* **1982**, 1401.
17. Hamana, H.; Sugasawa, T. *CL* **1984**, 1591.
18. Hamana, H.; Sugasawa, T. *CL* **1983**, 333.
19. (a) Sugano, Y.; Naruto, S. *CPB* **1989**, *37*, 840. (b) Sugano, Y.; Naruto, S. *CPB* **1988**, *36*, 4619.
20. Oh, T.; Wrobel, Z.; Rubenstein, S. M. *TL* **1991**, *32*, 4647.
21. Tanabe, Y.; Mukaiyama, T. *CL* **1986**, 1813.
22. Brocchini, S. J.; Eberle, M.; Lawton, R. G. *JACS* **1988**, *110*, 5211.
23. Evans, D. A.; Urpi, F.; Somers, T. C.; Clark, J. S.; Bilodeau, M. T. *JACS* **1990**, *112*, 8215.
24. (a) Guziec, F. S.; Torres, F. F. *JOC* **1993**, *58*, 1604 and references cited within. (b) Hunig, S.; Kiessel, M. *CB* **1958**, *91*, 380. (c) *FF* **1967**, *1*, 371.
25. For numerous references, see: Greene, T. W.; Wuts, P. G. M. *Protective Groups In Organic Synthesis*; Wiley: New York, 1991.
26. Kozikowski, A. P.; Wu, J.-P. *TL* **1987**, *28*, 5125.
27. Kubel, B.; Gruber, P.; Hurnaus, R.; Steglich, W. *CB* **1979**, *112*, 128.
28. Raber, D. J.; Gariano, P. *TL* **1971**, 4741.
29. Fernandez, I.; Khiar, N.; Llera, J. M.; Alcudia, F. *JOC* **1992**, *57*, 6789.
30. (a) Ozaki, S.; Watanabe, Y.; Hoshiko, T.; Fujisawa, H.; Uemura, A.; Ohrai, K. *TL* **1984**, *25*, 5061. (b) Nagase, T.; Seike, K.; Shiraishi, K.; Yamada, Y.; Ozaki, S. *CL* **1988**, 1381.
31. Kaino, M.; Naruse, Y.; Ishihara, K.; Yamamoto, H. *JOC* **1990**, *55*, 5814.
32. (a) Tuladhar, S. M.; Fallis, A. G. *TL* **1987**, *28*, 523. (b) See also: O'Malley, G. J.; Cava, M. P. *TL* **1985**, *26*, 6159.
33. Gassman, P. G.; Burns, S. J.; Pfister, K. B. *JOC* **1993**, *58*, 1449. (b) Gassman, P. G.; Burns, S. J. *JOC* **1988**, *53*, 5576.
34. Ito, Y.; Sawamura, M.; Kominami, K.; Saegusa, T. *TL* **1985**, *26*, 5303.
35. (a) Kwon, T. W.; Smith, M. B. *SC* **1992**, *22*, 2273. (b) Bartels, B.; Hunter, R.; Simon, C. D.; Tomlinson, G. D. *TL* **1987**, *28*, 2985. (c) Cohen, T.; Mura, A. J., Jr.; Shull, D. W.; Fogel, E. R.; Ruffner, R. J.; Falck, J. R. *JOC* **1976**, *41*, 3218.
36. Miller, R. D.; McKean, D. R. *TL* **1983**, *24*, 2619.
37. See: (a) Bodanszky, M.; Bodanszky, A. *The Practice of Peptide Synthesis*; Springer: Berlin, 1984. (b) Bodanszky, M.; Klausner, Y. S.; Ondetti, M. A. *Peptide Synthesis*, 2nd ed.; Wiley: New York, 1976.
38. (a) Bodanszky, M.; Bodanszky, A. *CC* **1967**, 591. (b) Williams, A. W.; Young, G. T. *JCS(P1)* **1972**, 1194. (c) Chen, F. M. F.; Lee, Y.; Steinauer, R.; Benoiton, N. L. *CJC* **1987**, *65*, 613. (d) Slebioda, M.; St-Amand, M. A.; Chen, F. M. F.; Benoiton, N. L. *CJC* **1988**, *66*, 2540.
39. Chen, F. M. F.; Benoiton, N. L. *CJC* **1987**, *65*, 1224.
40. (a) Coste, J.; Dufour, M.-N.; Pantaloni, A.; Castro, B. *TL* **1990**, *31*, 669. (b) Coste, J.; Frerot, E.; Jouin, P. *TL* **1991**, *32*, 1967.

Kirk L. Sorgi
The R. W. Johnson Pharmaceutical Research Institute, Spring House, PA, USA

Dimethylaluminum Chloride[1]

[1184-58-3] C_2H_6AlCl (MW 92.51)

(strong Lewis acid that can also act as a proton scavenger; reacts with HX to give methane and MeAlClX)

Alternate Name: chlorodimethylaluminum.
Physical Data: mp −21 °C; bp 126–127 °C; d 0.996 g cm^{-3}.
Solubility: sol most organic solvents; stable in alkanes or arenes.
Form Supplied in: commercially available neat or as solutions in hexane or toluene.
Analysis of Reagent Purity: solutions are reasonably stable but may be titrated before use by one of the standard methods.[1e]
Handling, Storage, and Precautions: must be transferred under inert gas (Ar or N_2) to exclude oxygen and water. Use in a fume hood.

Introduction. The general properties of alkylaluminum halides as Lewis acids are discussed in the entry for ***Ethylaluminum Dichloride***. Dialkylaluminum halides are less acidic than alkylaluminum dihalides. Me_2AlCl is more expensive than ***Diethylaluminum Chloride***, but the methyl group of Me_2AlCl is much less nucleophilic than the ethyl group of Et_2AlCl. Much higher yields will generally be obtained by use of Me_2AlCl in the ene reactions of carbonyl compounds. In other cases, such as the Diels–Alder reactions of α,β-unsaturated esters, comparable yields will be obtained with either Lewis acid.

Catalysis of Diels–Alder Reactions. Me_2AlCl has been used as a Lewis acid catalyst for inter- and intramolecular Diels–Alder reactions with a wide variety of dienophiles. High diastereoselectivity is obtained from chiral α,β-unsaturated *N*-acyloxazolidinones with more than 1 equiv of Me_2AlCl.[2] Use of Me_2AlCl as a catalyst affords high yields in inter- and intramolecular Diels–Alder reactions of α,β-unsaturated ketones (eq 1),[3,4] and intramolecular Diels–Alder reactions of α,β-unsaturated aldehydes.[4,5] Me_2AlCl-catalyzed Diels–Alder reactions of α,β-unsaturated *N*-acylsultams (eq 2)[6] and 1-mesityl-2,2,2-trifluoroethyl acrylate (eq 3)[7] proceed in high yield with excellent asymmetric induction. Methylaluminum sesquichloride, prepared from Me_2AlCl and ***Methylaluminum Dichloride***, catalyzes an intramolecular Diels–Alder reaction with an aldehyde as the dienophile to afford a dihydropyran.[8]

Me_2AlCl, 69% (1)

Me_2AlCl, −20 °C, 63%; TBDMSO; O_2; 100% de (2)

Me_2AlCl, −78 °C, 99%; CF_3; 100% de (3)

Catalysis of Ene Reactions[9]**.** A wide variety of Lewis acids will catalyze the ene reactions of formaldehyde with electron-rich alkenes. Electron-deficient aldehydes, such as chloral and glyoxylate esters, also undergo ene reactions with a variety of Lewis acid catalysts. Ene reactions of aliphatic and aromatic aldehydes with alkenes that can form a tertiary cation, and of formaldehyde with mono- and 1,2-disubstituted alkenes, are best carried out with 1 or more equiv of Me_2AlCl. The alcohol–Me_2AlCl complex produced in the ene reaction decomposes rapidly to give methane and a nonbasic aluminum alkoxide that does not react further (eq 4). This prevents solvolysis of the alcohol–Lewis acid complex or protonation of double bonds. Good to excellent yields of ene adducts are obtained from aliphatic and aromatic aldehydes and 1,1-di- and trisubstituted alkenes. ***Formaldehyde*** is more versatile and gives good yields of ene adducts with all classes of alkenes.[10] When less than 1 equiv of Me_2AlCl is used, γ-chloro alcohols are formed, resulting from the stereospecifically *cis* addition of the hydroxymethyl and chloride groups to the double bond (eq 5). The chloro alcohols are converted to ene-type adducts in the presence of excess Me_2AlCl. Formaldehyde undergoes Me_2AlCl-induced reactions with terminal alkynes to give a 2:3 mixture of the ene adduct allenic alcohol and the (*Z*)-3-chloro allylic alcohol in 50–75% yield (eq 6).[11]

R; $AlMe_2Cl$; $-CH_4$; AlMeCl (4)

CH_2O, Me_2AlCl → CH_2OH + Cl, CH_2OH (5)

1 equiv	20%	39%
1.5 equiv	73%	2%

CH_2O, Me_2AlCl → CH_2OH + Cl, CH_2OH, H (6)

Me_2AlCl catalyzes the ene reactions of a variety of aldehydes with (*Z*)-3β-acetoxy-5,17(20)-pregnadiene at −78 °C.[12] The stereoselectivity with aliphatic aldehydes is >10:1 in favor of the 22α-isomer, while aromatic aldehydes produce predominantly the 22β-isomer (eq 7). Me_2AlCl is also the Lewis acid of choice for ene reactions of α-halo aldehydes (eq 8).[13] Ene reactions of vinyl sulfides to produce enol silyl ethers are also catalyzed by Me_2AlCl (eq 9).[14]

AcO; RCHO, Me_2AlCl, −78 to 25 °C, 56–100%; OH, R; AcO (7)

Br, CHO + → Me_2AlCl, −78 °C, 72%; Br, OH (8)

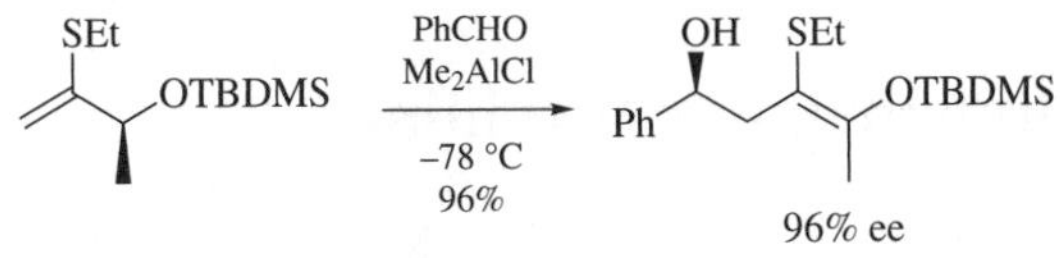

(9)

Type-I intramolecular ene reactions of aldehydes, such as citronellal, that contain electron-rich trisubstituted double bonds proceed readily thermally or with a variety of Lewis acids. Intramolecular ene reactions with less nucleophilic 1,2-disubstituted double bonds proceed efficiently with Me_2AlCl as the Lewis acid catalyst (eqs 10 and 11).[15,16]

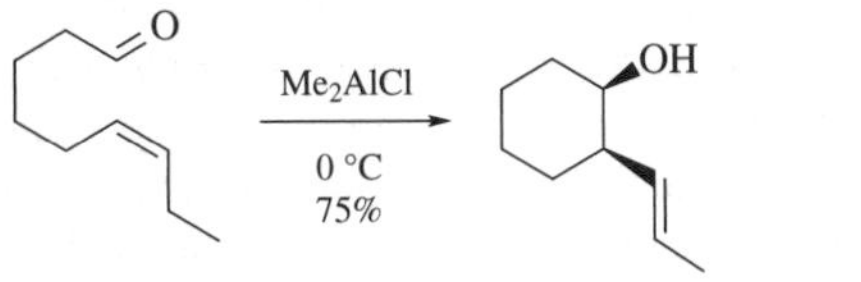

(10)

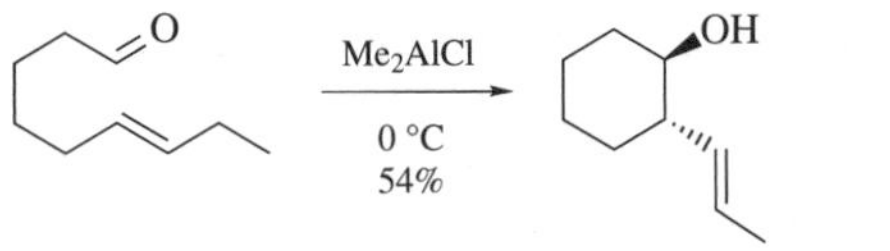

(11)

Type-II intramolecular ene reactions of aldehydes and ketones proceed readily with Me_2AlCl as the Lewis acid.[17–19] Unsaturated aldehydes and ketones can be generated in situ by Me_2AlCl-catalyzed reaction of ***Acrolein*** and ***Methyl Vinyl Ketone*** with alkylidenecycloalkanes at low temperatures (eq 12).[17] The monocyclic aldehyde reacts further under these conditions. The monocyclic ketone can be isolated at low temperature but undergoes a second ene reaction at rt to give the bicyclic alcohol. β-Keto esters form tertiary alcohols in intramolecular ene reactions. The products are stable because they are converted to the aluminum alkoxide (eq 13).[18] Intramolecular Me_2AlCl-catalyzed ene reactions have been used for the preparation of the bicyclic mevinolin ring system (eq 14).[19]

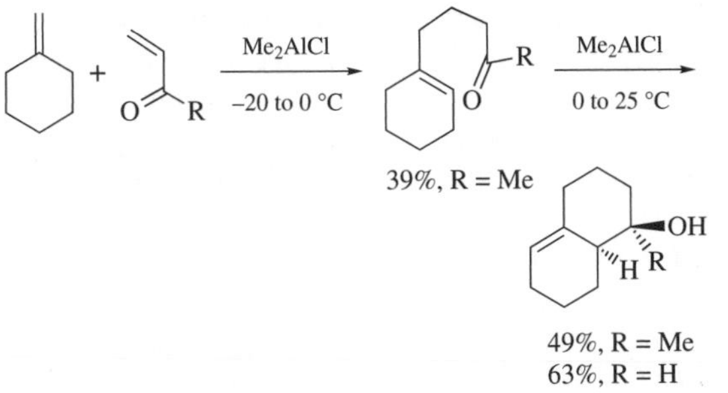

(12)

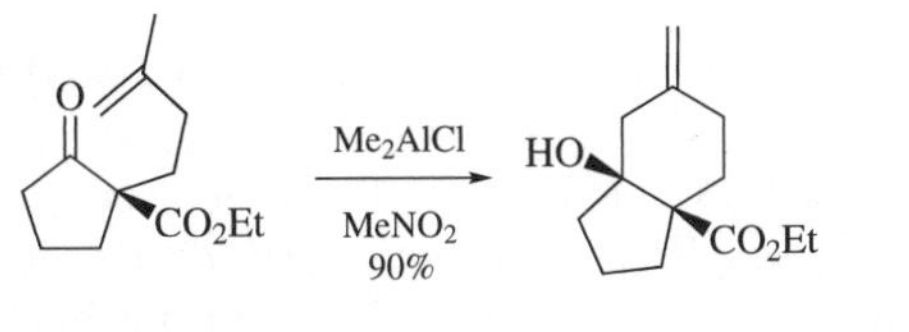

(13)

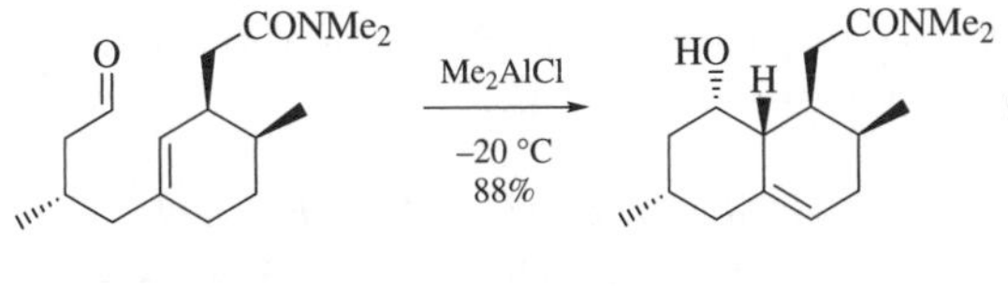

(14)

Generation of Electrophilic Cations. Me_2AlCl in dichloromethane cleaves THP ethers without deprotecting *t*-butyldimethylsilyl ethers.[20] Azido enol silyl ethers undergo Me_2AlCl-catalyzed reactions with ***Allyltributylstannane*** and enol ethers, giving conjugate addition-type products that are isolated as the silyl enol ethers (eq 15).[21] Me_2AlCl will open norbornane epoxide to the rearranged chlorohydrin.[22]

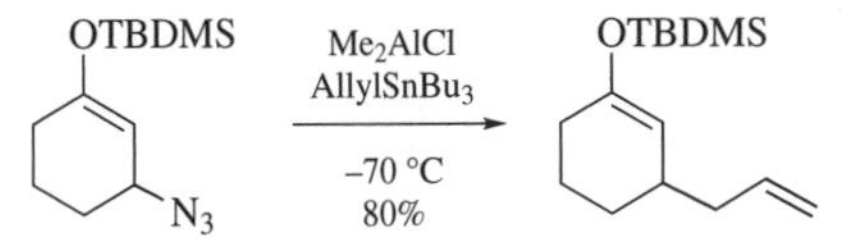

(15)

Formation and Reaction of Aluminum Enolates. Aluminum enolates prepared from esters react with imines to give a β-lactam resulting from aldol-type addition followed by ring closure (eq 16).[23]

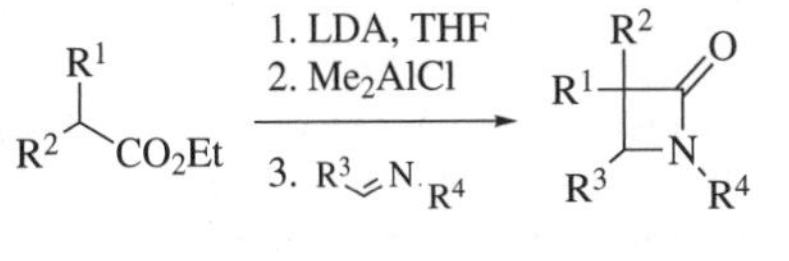

(16)

Formation and Reaction of Alkynylaluminum Reagents. Lithium acetylides react with Me_2AlCl to give dimethylaluminum acetylides that react analogously to the more commonly used diethylaluminum acetylides (see ***Diethylaluminum Ethoxyacetylide***). Addition of the aluminum acetylide to propiolactone results in an S_N2 reaction to give an alkynic acid (eq 17).[24]

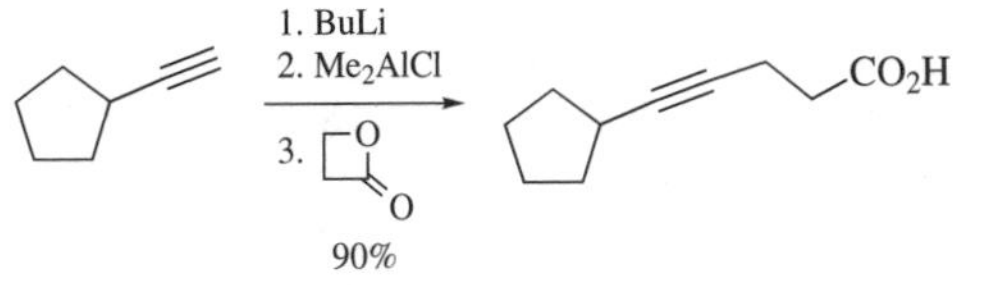

(17)

Reaction as a Nucleophile. Me_2AlCl will react analogously to MeMgBr and transfer a methyl group to many nucleophiles. Since ***Methylmagnesium Bromide*** and ***Methyllithium*** are readily available, use of Me_2AlCl to deliver a methyl group is needed only when the stereochemistry of addition is an important issue. High levels of asymmetric induction are obtained in the conjugate addition of Me_2AlCl to unsaturated acyloxazolidinones with carbohydrate-derived chiral auxiliaries (eq 18).[25] Me_2AlCl differs from higher dialkylaluminum chlorides in that methyl addition is a radical process that requires photochemical or radical initiation. Me_2AlCl will convert acid chlorides to methyl ketones.[26]

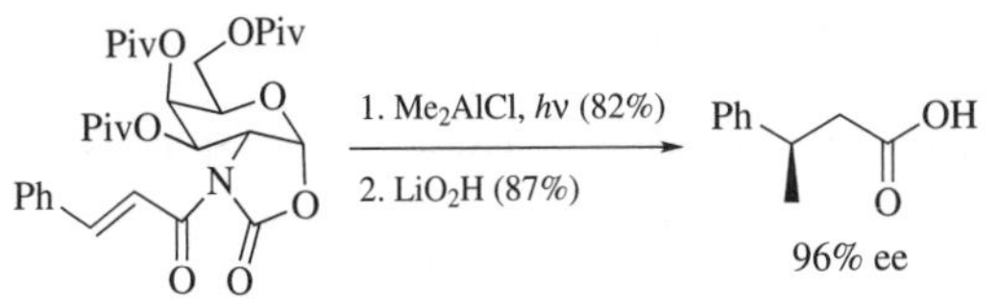

(18)

Related Reagents. 1,8-Bis(dimethylamino)naphthalene; Diethylaluminum Chloride; Dimethylaluminum Iodide; Ethylaluminum Dichloride; Methylaluminum Bis(4-bromo-2,6-di-*t*-butylphenoxide); Methylaluminum Bis(2,6-di-*t*-butyl-4-methylphenoxide); Methylaluminum Bis(2,6-di-*t*-butylphenoxide); Methylaluminum Dichloride; Trimethylsilyl Trifluoromethanesulfonate.

1. For reviews, see Ref. 1 under ***Ethylaluminum Dichloride***.
2. (a) Evans, D. A.; Chapman, K. T.; Bisaha, J. *TL* **1984**, *25*, 4071. (b) Evans, D. A.; Chapman, K. T.; Bisaha, J. *JACS* **1984**, *106*, 4261. (c) Evans, D. A.; Chapman, K. T.; Bisaha, J. *JACS* **1988**, *110*, 1238. (d) Sugahara, T.; Iwata, T.; Yamaoka, M.; Takano, S. *TL* **1989**, *30*, 1821. (e) Hauser, F. M.; Tommasi, R. A. *JOC* **1991**, *56*, 5758.
3. (a) Sakan, K.; Smith, D. A. *TL* **1984**, *25*, 2081. (b) Ireland, R. E.; Dow, W. C.; Godfrey, J. D.; Thaisrivongs, S. *JOC* **1984**, *49*, 1001.
4. Marshall, J. A.; Shearer, B. G.; Crooks, S. L. *JOC* **1987**, *52*, 1236.
5. Takeda, K.; Kobayashi, T.; Saito, K.; Yoshii, E. *JOC* **1988**, *53*, 1092.
6. Oppolzer, W.; Dupuis, D.; Poli, G.; Raynham, T. M.; Bernardinelli, G. *TL* **1988**, *29*, 5885.
7. Corey, E. J.; Cheng, X.-M.; Cimprich, K. A. *TL* **1991**, *32*, 6839.
8. Trost, B. M.; Lautens, M.; Hung, M. H.; Carmichael, C. S. *JACS* **1984**, *106*, 7641.
9. (a) Snider, B. B. *COS* **1991**, *5*, 1. (b) Snider, B. B. *COS* **1991**, *2*, 527. (c) Mikami, K.; Shimizu, M. *CRV* **1992**, *92*, 1021.
10. (a) Snider, B. B.; Rodini, D. J.; Kirk, T. C.; Cordova, R. *JACS* **1982**, *104*, 555. (b) Cartaya-Marin, C. P.; Jackson, A. C.; Snider, B. B. *JOC* **1984**, *49*, 2443. (c) Tietze, L. F.; Beifuss, U.; Antel, J.; Sheldrick, G. M. *AG(E)* **1988**, *27*, 703. (d) Metzger, J. O.; Biermann, U. *S* **1992**, 463.
11. Rodini, D. J.; Snider, B. B. *TL* **1980**, *21*, 3857.
12. (a) Mikami, K.; Loh, T.-P.; Nakai, T. *TL* **1988**, *29*, 6305. (b) Mikami, K.; Loh, T.-P.; Nakai, T. *CC* **1988**, 1430. (c) Houston, T. A.; Tanaka, Y.; Koreda, M. *JOC* **1993**, *58*, 4287.
13. Mikami, K.; Loh, T.-P.; Nakai, T. *CC* **1991**, 77.
14. Tanino, K.; Shoda, H.; Nakamura, T.; Kuwajima, I. *TL* **1992**, *33*, 1337.
15. Snider, B. B.; Karras, M.; Price, R. T.; Rodini, D. J. *JOC* **1982**, *47*, 4538.
16. (a) Smith, A. B., III; Fukui, M. *JACS* **1987**, *109*, 1269. (b) Smith, A. B., III; Fukui, M.; Vaccaro, H. A.; Empfield, J. R. *JACS* **1991**, *113*, 2071.
17. (a) Snider, B. B.; Deutsch, E. A. *JOC* **1982**, *47*, 745. (b) Snider, B. B.; Deutsch, E. A. *JOC* **1983**, *48*, 1822. (c) Snider, B. B.; Goldman, B. E. *T* **1986**, *42*, 2951.
18. Jackson, A. C.; Goldman, B. E.; Snider, B. B. *JOC* **1984**, *49*, 3988.
19. (a) Wovkulich, P. M.; Tang, P. C.; Chadha, N. K.; Batcho, A. D.; Barrish, J. C.; Uskoković, M. R. *JACS* **1989**, *111*, 2596. (b) Barrish, J. C.; Wovkulich, P. M.; Tang, P. C.; Batcho, A. D.; Uskoković, M. R. *TL* **1990**, *31*, 2235. (c) Quinkert, G.; Schmalz, H.-G.; Walzer, E.; Kowalczyk-Przewloka, T.; Dürner, G.; Bats, J. W. *AG(E)* **1987**, *26*, 61. (d) Quinkert, G.; Schmalz, H.-G.; Walzer, E.; Gross, S.; Kowalczyk-Przewloka, T.; Schierloh, C.; Dürner, G.; Bats, J. W.; Kessler, H. *LA* **1988**, 283. (e) Cohen, T.; Guo, B.-S. *T* **1986**, *42*, 2803.
20. Ogawa, Y.; Shibasaki, M. *TL* **1984**, *25*, 663.
21. Magnus, P.; Lacour, J. *JACS* **1992**, *114*, 3993.
22. Murray, T. F.; Varma, V.; Norton, J. R. *JOC* **1978**, *43*, 353.
23. Wada, M.; Aiura, H.; Akiba, K.-Y. *TL* **1987**, *28*, 3377.
24. Shinoda, M.; Iseki, K.; Oguri, T.; Hayasi, Y.; Yamada, S.-I.; Shibasaki, M. *TL* **1986**, *27*, 87.
25. (a) Rück, K.; Kunz, H. *AG(E)* **1991**, *30*, 694. (b) *SL* **1992**, 343. (c) *S* **1993**, 1018.
26. Ishibashi, H.; Takamuro, I.; Mizukami, Y.-I.; Irie, M.; Ikeda, M. *SC* **1989**, *19*, 443.

Barry B. Snider
Brandeis University, Waltham, MA, USA

4-Dimethylaminopyridine[1]

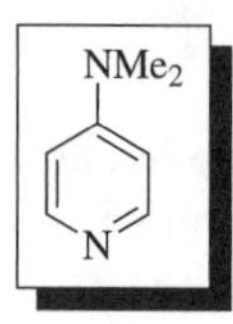

[1122-58-3] $C_7H_{10}N_2$ (MW 122.19)

(catalyst for acylation of alcohols or amines,[1–11] especially for acylations of tertiary or hindered alcohols or phenols[12] and for macrolactonizations;[13–15] catalyst for direct esterification of carboxylic acids and alcohols in the presence of dicyclohexylcarbodiimide (Steglich–Hassner esterification);[5] catalyst for silylation or tritylation of alcohols,[9,10] and for the Dakin–West reaction[20])

Alternate Name: DMAP.
Physical Data: colorless solid; mp 108–110 °C; pK_a 9.7.
Solubility: sol MeOH, $CHCl_3$, CH_2Cl_2, acetone, THF, pyridine, HOAc, EtOAc; partly sol cold hexane or water.
Form Supplied in: colorless solid; commercially available.
Preparative Methods: prepared by heating 4-pyridone with HMPA at 220 °C, or from a number of 4-substituted (Cl, OPh, SO_3H, $OSiMe_3$) pyridines by heating with DMA.[2] Prepared commercially from the 4-pyridylpyridinium salt (obtained from pyridine and $SOCl_2$) by heating with DMF at 155 °C.[1,2]
Purification: can be recrystallized from EtOAc.
Handling, Storage, and Precautions: skin irritant; corrosive, toxic solid.

Acylation of Alcohols. Several 4-aminopyridines speed up esterification of hindered alcohols with acid anhydrides by as much as 10 000 fold; of these, DMAP is the most commonly used but 4-pyrrolidinopyridine (PPY) and 4-tetramethylguanidinopyridine are somewhat more effective.[11] DMAP is usually employed in 0.05–0.2 mol equiv amounts.

DMAP catalyzes the acetylation of hindered 11β- or 12α-hydroxy steroids. The alkynic tertiary alcohol acetal in eq 1 is acetylated at rt within 20 min in the presence of excess DMAP.[3]

HO–C(Me)2–C≡C–CH(OMe)2 —(Ac_2O, DMAP; rt; 93%)→ AcO–C(Me)2–C≡C–CH(OMe)2 (1)

Esterifications mediated by ***2-Chloro-1-methylpyridinium Iodide*** also benefit from the presence of DMAP.[22]

DMAP acts as an efficient acyl transfer agent, so that alcohols resistant to acetylation by ***Acetic Anhydride–Pyridine*** usually react well in the presence of DMAP.[4a] Sterically hindered phenols can be converted into salicylaldehydes via a benzofurandione prepared by DMAP catalysis (eq 2).[4b]

Direct Esterification of Alcohols and Carboxylic Acids. Instead of using acid anhydrides for the esterification of alcohols, it is possible to carry out the reaction in one pot at rt by employing a carboxylic acid, an alcohol, ***1,3-Dicyclohexylcarbodiimide***, and DMAP.[5,6] In this manner,

N-protected amino acids and even hindered carboxylic acids can be directly esterified at rt using DCC and DMAP or 4-pyrrolidinopyridine (eq 3).[5]

1. $(ClCO)_2$, DMAP; 2. $AlCl_3$ (2)

DCC, rt; PPY, 12 h; 90% (3)

DCC–DMAP has been used in the synthesis of depsipeptides.[6b] Macrocyclic lactones have been prepared by cyclization of hydroxy carboxylic acids with DCC–DMAP. The presence of salts of DMAP,[13a] such as its trifluoroacetate, is beneficial in such cyclizations, as shown for the synthesis of a (9*S*)-dihydroerythronolide.[13b] Other macrolactonizations have been achieved using ***2,4,6-Trichlorobenzoyl Chloride***[1] and DMAP in ***Triethylamine*** at rt[14] or ***Di-2-pyridyl Carbonate*** (6 equiv) with 2 equiv of DMAP at 73 °C.[15]

Acylation of Amines. Acylation of amines is also faster in the presence of DMAP,[7] as is acylation of indoles,[8a] phosphorylation of amines or hydrazines,[2,8] and conversion of carboxylic acids into anilides by means of ***Phenyl Isocyanate***.[1] β-Lactam formation from β-amino acids has been carried out with DCC–DMAP, but epimerization occurs.[8b]

Silylation, Tritylation, and Sulfinylation of Alcohols. Tritylation, including selective tritylation of a primary alcohol in the presence of a secondary one,[9] silylation of tertiary alcohols, selective silylation to *t*-butyldimethylsilyl ethers,[6] and sulfonylation or sulfinylation[10] of alcohols proceed more readily in the presence of DMAP. Silylation of β-hydroxy ketones with ***Chlorodiisopropylsilane*** in the presence of DMAP followed by treatment with a Lewis acid gives diols (eq 4).[16]

i-Pr_2SiHCl, DMAP; $SnCl_4$ (4)

Miscellaneous Reactions. Alcohols, including tertiary ones, can be converted to their acetoacetates by reaction with ***Diketene*** in the presence of DMAP at rt.[17] Decarboxylation of β-keto esters has been carried out at pH 5–7 using 1 equiv of DMAP in refluxing wet toluene (eq 5).[18]

DMAP, toluene, H_2O, reflux, 60% (5)

Elimination of water from a *t*-alcohol in a β-hydroxy aldehyde was carried out using an excess of ***Methanesulfonyl Chloride***–DMAP–H_2O at 25 °C.[23]

Glycosidic or allylic alcohols (even when *s*-) can be converted in a 80–95% yield to alkyl chlorides by means of ***p-Toluenesulfonyl Chloride***–DMAP. Simply primary alcohols react slower and secondary ones are converted to tosylates.[24]

Aldehydes and some ketones can be converted to enol acetates by heating in the presence of TEA, Ac_2O, and DMAP.[2] DMAP catalyzes condensation of malonic acid monoesters with unsaturated aldehydes at 60 °C to afford dienoic esters (eq 6).[19]

DMAP, py, 60 °C, 86% (6)

The conversion of α-amino acids into α-amino ketones by means of acid anhydrides (Dakin–West reaction)[20] also proceeds faster in the presence of DMAP (eq 7).

1. cyclopropanecarbonyl chloride, DMAP; 2. AcOH (7)

Ketoximes can be converted to nitrimines which react with Ac_2O–DMAP to provide alkynes (eq 8).[21]

Ac_2O, DMAP; + isomeric allene (8)

For the catalysis by DMAP of the *t*-butoxylcarbonylation of alcohols, amides, carbamates, NH-pyrroles, etc., see ***Di-t-butyl Dicarbonate***.

Related Reagents. Pyridine; Tri-*n*-butylphosphine.

1. Hoefle, G.; Steglich, W.; Vorbrueggen, H. *AG(E)* **1978**, *17*, 569.
2. Scriven, E. F. V. *CSR* **1983**, *12*, 129.
3. (a) Steglich, W.; Hoefle, G. *AG(E)* **1969**, *8*, 981. (b) Hoefle, G.; Steglich, W. *S* **1972**, 619.
4. (a) Salomon, R. G., Salomon, M. F.; Zagorski, M. G.; Reuter, J. M.; Coughlin, D. J. *JACS* **1982**, *104*, 1008. (b) Zwanenburg, D. J.; Reynen, W. A. P. *S* **1976**, 624.

5. (a) Neises, B.; Steglich, W. *AG(E)* **1978**, *17*, 522. (b) Hassner, A.; Alexanian, V. *TL* **1978**, 4475.
6. (a) Ziegler, F. E.; Berger, G. D. *SC* **1979**, 539. (b) Gilon, C.; Klausner, Y.; Hassner, A. *TL* **1979**, 3811.
7. Litvinenko, L. M., Kirichenko, A. C.; *DOK* **1967**, *176*, 97. (b) Kirichenko, A. C.; Litvinenko, L. M.; Dotsenko, I. N.; Kotenko, N. G.; Nikkel'sen, E.; Berestetskaya, V. D. *DOK* **1969**, *244*, 1125 (*CA* **1979**, *90*, 157 601).
8. (a) Nickisch, K.; Klose, W.; Bohlmann, F. *CB* **1980**, *113*, 2036. (b) Kametami, T.; Nagahara, T.; Suzuki, Y.; Yokohama, S.; Huang, S.-P.; Ihara, M. *T* **1981**, *37*, 715.
9. (a) Chaudhary, S. K.; Hernandez, O. *TL* **1979**, *95*, 99. (b) Hernandez, O.; Chaudhary, S. K.; Cox, R. H.; Porter, J. *TL* **1981**, *22*, 1491.
10. Guibe-Jampel, E.; Wakselman, M.; Raulais, D. *CC* **1980**, 993.
11. Hassner, A.; Krepski, L. R.; Alexanian, V. *T* **1978**, *34*, 2069.
12. Vorbrueggen, H. (Schering AG) Ger. Offen. 2 517 774, 1976 (*CA* **1977**, *86*, 55 293).
13. (a) Boden, E. P.; Keck, G. E. *JOC* **1985**, *50*, 2394. (b) Stork, G.; Rychnovsky, S. D. *JACS* **1987**, *109*, 1565.
14. Hikota, M.; Tone, H.; Horita, K.; Yonemitsu, O. *JOC* **1990**, *55*, 7.
15. (a) Kim, S.; Lee, J. I.; Ko, Y. K. *TL* **1984**, *25*, 4943. (b) Denis, J.-N.; Greene, A. E.; Guenard, D.; Gueritte-Voegelein, F.; Mangatal, L.; Potier, P. *JACS* **1988**, *110*, 5917.
16. Anwar, S.; Davis, A. P. *T* **1988**, *44*, 3761.
17. Nudelman, A.; Kelner, R.; Broida, N.; Gottlieb, H. E. *S* **1989**, 387.
18. Taber, D. F.; Amedio J. C., Jr.; Gulino, F. *JOC* **1989**, *54*, 3474.
19. Rodriguez, J.; Waegell, B. *S* **1988**, 534.
20. (a) Buchanan, G. L. *CSR* **1988**, *17*, 91. (b) McMurry, J. *JOC* **1985**, *50*, 1112. (c) Hoefle, G., Steglich, W. *CB* **1971**, *104*, 1408.
21. Buechi, G.; Wuest, H. *JOC* **1979**, *44*, 4116.
22. Nicolaou, K. C.; Bunnage, M. E.; Koide, K. *JACS* **1994**, *116*, 8402.
23. Furukawa, J.; Morisaki, N.; Kobayashi, H.; Iwasaki, S.; Nozoe, S.; Okuda, S. *CPB* **1985**, *33*, 440.
24. Hwang, C. K.; Li, W. S.; Nicolaou, K. C. *TL* **1984**, *25*, 2295.

Alfred Hassner
Bar-Ilan University, Ramat Gan, Israel

Esterases

(enzymes of the class of hydrolases, which catalyze the hydrolysis of carboxylic acid esters[1])

Solubility: insol cold and warm H_2O.
Form Supplied in: available from various sources (microorganisms and mammalian) as powders or water suspensions.
Handling, Storage, and Precautions: stable at a pH range 6–10; can be stored at 0–4 °C for months.

Esterase-Catalyzed Hydrolysis. Hydrolytic enzymes have been accepted in organic synthesis as valuable biocatalysts, since they are commercially available at relatively low price and possess a broad substrate specificity, without necessitating use of expensive cofactors.[2] Esterases are such useful enzymes and have been widely used for the preparation of enantiomerically pure chiral compounds, by hydrolytic resolution of racemic esters or asymmetrization of prochiral substrates.[3] Well defined experimental procedures for a pig liver esterase-catalyzed saponification have been documented.[4] Generally, the enzymatic hydrolysis is carried out in an aqueous buffer, sometimes containing cosolvents,[5] at pH 7–9 and keeping the temperature at 20–25 °C. Generally, the molar equivalent NaOH for the hydrolysis is added maintaining the pH constant with an automatic titrator and, after acidification, the product is extracted with organic solvents. Esterases are commercially available from various sources, either microbial or mammalian, and in some instances also crude acetone powders can be used for the same purpose.

Pig Liver Esterase (PLE). This is the more used carboxylesterase (carboxylic-ester hydrolase, EC 3.1.1.1, *CAS 9016-18-6*) which physiologically catalyzes the hydrolysis of carboxylic acid esters to the free acid anion and alcohol.[1] PLE is a serine hydrolase which has been widely used for the preparation of chiral synthons and these applications have been fully reviewed.[6] An active-site model for interpreting and predicting the specificity of the enzyme has been published.[7] In the pioneering studies of the enzyme applications field, PLE was used for the chiral synthesis of mevalonolactone.[8] Prochiral 3-substituted glutaric acid diesters are well suited for a PLE-catalyzed asymmetrization, which leads to optically active monoesters (eq 1).[9]

R = Me, Et, Ph, CH_2Ph — MeO_2C … CO_2Me —PLE→ HO_2C … CO_2Me, >99% ee (1)

The asymmetrization of prochiral disubstituted malonates has been enantioselectively realized in the presence of PLE (eq 2).[10]

$R^1R^2C(CO_2R^3)_2$ —PLE→ $R^1R^2C(CO_2H)(CO_2R^3)$ (2)

R^1 = Ph, R^2 = Me, R^3 = Et — 86% ee
R^1 = Me, R^2 = 3,4-$MeOC_6H_3CH_2$, R^3 = Me — 93% ee
R^1 = Me, R^2 = p-MeC_6H_4, R^3 = Me — 96% ee
R^1 = Me, R^2 = CH_2O-*t*-Bu, R^3 = Me — 96% ee

The asymmetric hydrolysis of several cyclic *meso*-diesters has been accomplished and optically pure monoesters have been obtained.[11] A classical example is the hydrolysis of dimethyl *cis*-4-cyclohexene-1,2-dicarboxylate, which affords the corresponding nearly optically pure half ester, a versatile synthon for various chiral cyclohexane derivatives (eq 3).[12]

CO_2Me, CO_2Me —PLE, 98%→ CO_2H, CO_2Me, 96% ee (3)

The PLE-catalyzed asymmetric hydrolysis of *meso*-1,3-*cis*-3,5-*cis*-1,3-diacetoxy-5-benzyloxycyclohexane afforded (1*S*,3*S*,5*R*)-1-acetoxy-5-benzyloxycyclohexan-3-ol, which could be used as chiral building block for the synthesis of the compactin lactone moiety and quinic acid (eq 4).[13]

OCH_2Ph; MeOCO, OCOMe —PLE, 62%→ OCH_2Ph; HO, OCOMe, 87% ee (4)

Organometallic *meso*-diesters can be asymmetrized as well to the corresponding half ester, as shown for an (arene)tricarbonylchromium diester (eq 5).[14]

$(OC)_3Cr$; CO_2Et, CO_2Et —PLE, 85%→ $(OC)_3Cr$; CO_2H, CO_2Et, 99% ee (5)

The resolution of racemic esters is catalyzed by PLE in a highly enantioselective fashion.[3,6] Several interesting applications of this method are available. The hydrolysis of *trans*-bicyclo[2.2.1]heptane diesters has been studied to ascertain the structural requirements for the PLE hydrolysis.[15] A bulky tricyclodecadienone ester can be resolved by an highly enantioselective reaction (eq 6).[16]

CO_2Et; O —PLE→ HO_2C; O, 40%; 100% ee + CO_2Et; O, 48%; 80% ee (6)

The resolution procedure applies to racemic organometallic esters[17] and to the esters of a thianucleoside, for the preparation of pure enantiomers of an antiviral agent (2′,3′-dideoxy-5-fluoro-3′-thiacytidine) (eq 7).[18]

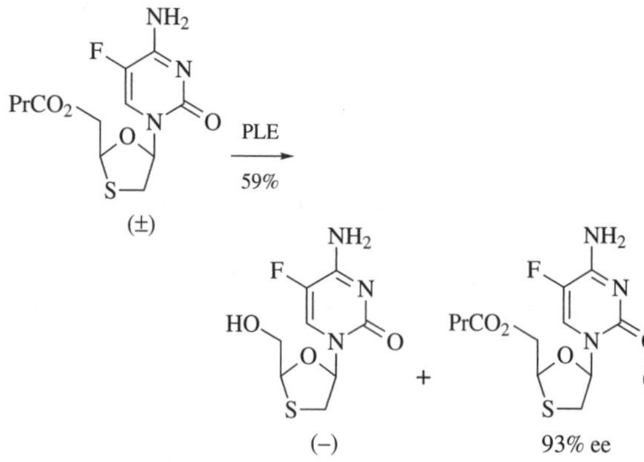

(7)

PLE has usually been applied to the enantioselective preparation of optically active compounds, but its use can be extended to chemo- or regioselective hydrolyses. A continuous process for the separation of a *cis/trans* unsaturated ester was realized using immobilized PLE (eq 8).[19]

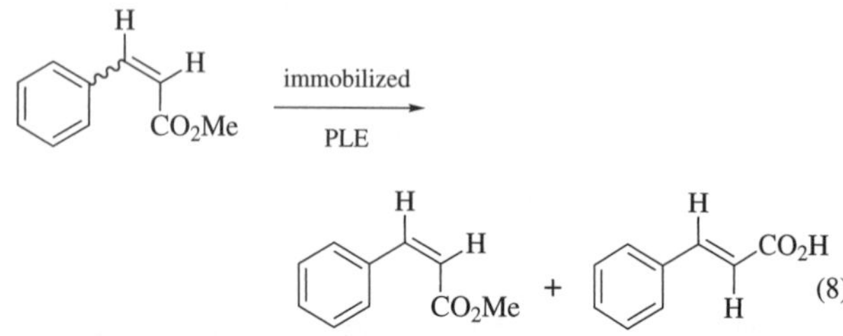

(8)

The chemoselective hydrolysis of an acetoxy group in the presence of a γ-lactone ring has been reported in the presence of PLE (eq 9).[20] In a benzylpenicillin, PLE catalyzes the chemoselective hydrolytic opening of the β-lactam ring, the methoxycarbonyl moiety remaining uneffected (eq 10).[21]

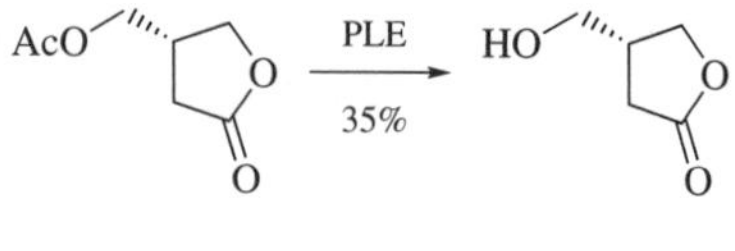

(9)

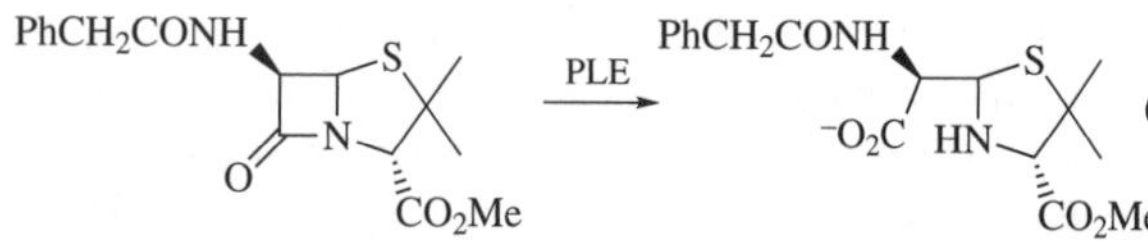

(10)

Dimethyl malate presents two ester functions α and β with respect to a hydroxy group, and PLE is able to regioselectively discriminate between these two moieties.[22]

Acetone Powder Containing Esterase Activity. The main advantage in using crude homogenates or acetone powders of organs such as liver is to have a cheap source of different enzymes. If one of these is desired for a specific substrate, the crude enzymatic mixture can be used with some advantage, compared to the purified enzymes. Pig liver acetone powder (PLAP), together with other extracts, is commercially available or can be prepared from fresh pig liver.[23] PLAP has been used for the enantioselective hydrolysis of the racemic acetate of *trans*-2-phenylcyclohexanol (eq 11).[24]

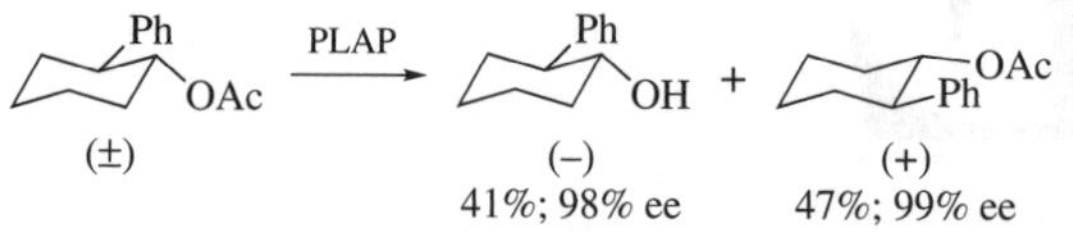

(11)

The acetates of 1-arylalkan-1-ols were successfully resolved by acetone powders (PLAP and goat liver acetone powder, GLAP) containing esterase activity (eq 12).[25]

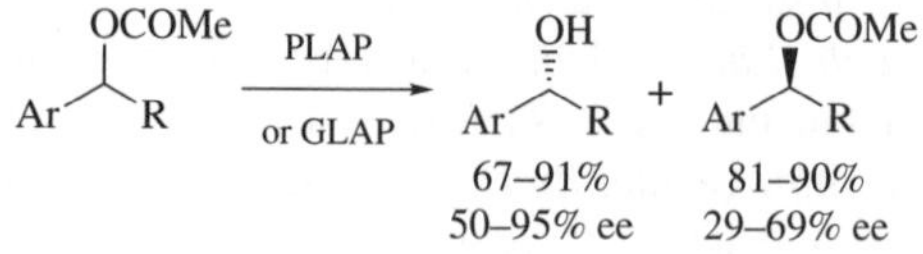

(12)

An interesting application of the esterase activity of horse liver acetone powder (HLE) has been the enantioselective hydrolysis of racemic lactones. The powder proved to be more effective than PLE in this hydrolysis, from which the unreacted lactone was recovered with high enantiomeric excess. The process seems more effective for δ and medium size lactones (eq 13).[26]

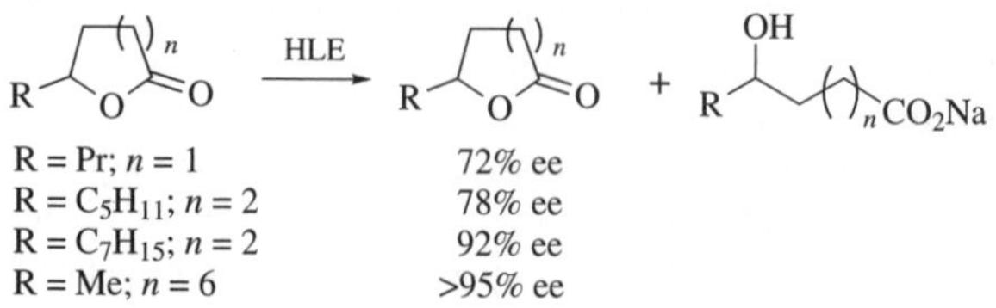

(13)

Cholesterol Esterase. This enzyme (EC 3.1.1.13; *CAS 9026-00-0*) physiologically catalyzes the hydrolysis of cholesterol esters, monoacylglycerols, and vitamin esters.[27] It has also been used for several cyclic and noncyclic substrates with variable enantioselectivity.[28] The resolution of racemic esters has been reported[29] and an interesting example is the application to the racemic acetate of an hemiacetal (eq 14).[30]

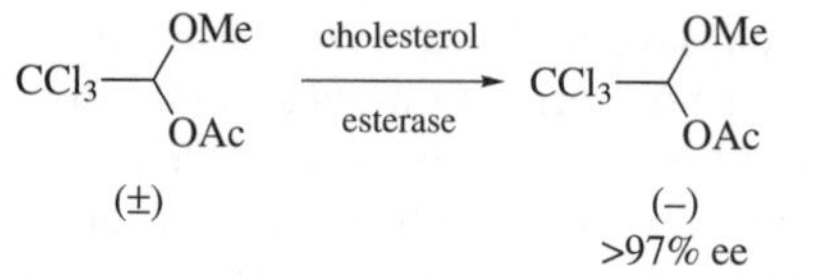

(14)

Acyl Cholinesterases. Acetylcholinesterase (AChE; EC 3.1.1.7; *CAS 9000-81-1*) is the serine esterase which catalyzes the hydrolysis of acetylcholine and possesses an esteratic site,[31] and which is responsible for unspecific hydrolyses of several substrates. Also, butyrylcholinesterase (EC 3.1.1.8; *CAS 9001-08-5*) has been sometimes used for asymmetric hydrolysis of esters.[32] Acetylcholinesterase has been used for the hydrolysis of noncyclic substrates and the results have shown satisfactory enantioselectivity.[32a,33] The enzyme from electric eel seems especially well suited to the hydrolysis of cyclic diols.[34] The asymmetrization of *cis*-3,5-diacetoxycyclopent-1-ene to (3*R*)-acetoxy-(5*S*)-hydroxycyclopent-1-ene (eq 15)[35] and the preparation of an optically active triol monoacetate starting from the triacetate of 1,3,6-trihydroxycyclohept-4-ene (eq 16)[36] are good examples of successful reactions catalyzed by acetylcholinesterase.

AcO OAc → (AChE, 94%) HO OAc; 96% ee (15)

AcO, OH, OAc → (AChE, 79%) AcO, OH, OH; >95% ee (16)

Other Esterases. Other less common esterases have been sometimes used for biocatalytic applications in organic synthesis.[37] The enzymatic approach can be the method of choice for the preparation of optically pure drugs, although sometimes special enzymes have to be prepared for this aim. By cloning a carboxylesterase into a microorganism, high level production of the esterase is made possible for the production of 2-(aryloxy)propionates and (*S*)-naproxen.[38] The esterase activity of rabbit plasma has been used for a chemoselective hydrolysis of a methylthiomethyl ester.[39] An esterase from *Candida lypolitica* has been used for the resolution of a tertiary α-substituted carboxylic acid ester.[40] Recently, a carboxyl esterase of molecular weight 30 000 ('Esterase 30 000') has been introduced for the asymmetric hydrolysis of diesters. A cyclopropyl malonate has been hydrolyzed by the esterase and the unreacted diester was recovered nearly optically pure (eq 17).[41] Diethyl 3-hydroxyglutarate, a substrate which is asymmetrized with modest enantioselectivity with PLE or other enzymes,[9e,f] has been enantioselectively hydrolyzed in the presence of Esterase 30 000 (eq 18).[42]

THPO, CO_2Me, CO_2Me → (Esterase 30,000, 42%) THPO, CO_2Me, CO_2Me; 97% ee (17)

OH, EtO_2C, CO_2Et → (Esterase 30,000, 94%) OH, EtO_2C, CO_2H; >98% ee (18)

Related Reagents. Lipases.

1. (a) Junge, W. In *Methods of Enzymatic Analysis*; Bergmeyer, H. U. Ed.; Verlag Chemie: Weinheim, 1984; Vol. IV, p 2. (b) *Enzyme Handbook*; Schomburg, D.; Salzmann, M., Eds.; Springer: Berlin, 1991; Vol. III.
2. Jones, J. B. *T* **1986**, *42*, 3351.
3. (a) Boland, W.; Frössl, C.; Lorenz, M. *S* **1991**, 1049. (b) Santaniello, E.; Ferraboschi, P.; Grisenti, P.; Manzocchi, A. *CRV* **1992**, *92*, 1071.
4. Eberle, M.; Missbach, M.; Seebach, D. *OS* **1990**, *69*, 19.
5. (a) Guanti, G.; Banfi, L.; Narisano, E.; Riva, R.; Thea, S. *TL* **1986**, *27*, 4639. (b) Björkling, F.; Boutelje, J.; Hjalmarsson, M.; Hult, K.; Norin, T. *CC* **1987**, 1041. (c) Lam, L. K. P.; Hui, R. A. H. F.; Jones, J. B. *JOC* **1986**, *51*, 2047.
6. (a) Ohno, M.; Otsuka, M. *OR* **1989**, *37*, 1. (b) Zhu, L.-M.; Tedford, M. C. *T* **1990**, *46*, 6587. (c) Jones, J. B. *PAC* **1990**, *62*, 1445.
7. Toone, E. J.; Werth, M. J.; Jones, J. B. *JACS* **1990**, *112*, 4946.
8. Huang, F.-C.; Hsu Lee, L. F.; Mittal, R. S. D.; Ravikumar, P. R.; Chan, J. A.; Sih, C. J.; Caspi, E.; Eck, C. R. *JACS* **1975**, *97*, 4144.
9. (a) Mohr, P.; Waespe-Šarčevič, N.; Tamm, C.; Gawronska, K.; Gawronski, J. K. *HCA* **1983**, *66*, 2501. (b) Brooks, D. W.; Palmer, J. T. *TL* **1983**, *24*, 3059. (c) Francis, C. J.; Jones, J. B. *CC* **1984**, 579. (d) VanMiddlesworth, F.; Wang, Y. F.; Zhou, B.-N.; DiTullio, D.; Sih, C. J. *TL* **1985**, *26*, 961. (e) Mohr, P.; Rösslein, L.; Tamm, C. *HCA* **1987**, *70*, 142. (f) Santaniello, E.; Chiari, M.; Ferraboschi, P.; Trave, S. *JOC* **1988**, *53*, 1567. (g) Andruszkiewicz, R.; Barrett, A. G. M.; Silverman, R. B. *SC* **1990**, *20*, 159. (h) Chênevert, R.; Desjardins, M. *TL* **1991**, *32*, 4249.
10. (a) Schneider, M.; Engel, N.; Boensmann, H. *AG(E)* **1984**, *23*, 66. (b) Björkling, F.; Boutelje, J.; Gatenbeck, S.; Hult, K.; Norin, T. *TL* **1985**, *26*, 4957. (c) Luyten, M.; Müller, S.; Herzog, B.; Keese, R. *HCA* **1987**, *70*, 1250. (d) De Jeso, B.; Belair, N.; Deleuze, H.; Rascle, M.-C.; Maillard, B. *TL* **1990**, *31*, 653. (e) Fadel, A.; Canet, J.-L.; Salaün, J. *SL* **1991**, 60.
11. (a) Schneider, M.; Engel, N.; Hönicke, P.; Heinemann, G.; Görisch, H. *AG(E)* **1984**, *23*, 67. (b) Sabbioni, G.; Shea, M. L.; Jones, J. B. *CC* **1984**, 236. (c) Gais, H.-J.; Lukas, K. L.; Ball, W. A.; Braun, S.; Lindner, H. J. *LA* **1986**, 687. (d) Naemura, K.; Takahashi, N.; Chikamatsu, H. *CL* **1988**, 1717. (e) Zemlicka, J.; Craine, L. E.; Heeg, M.-J.; Oliver, J. P. *JOC* **1988**, *53*, 937. (f) Brion, F.; Marie, C.; Mackiewicz, P.; Roul, J. M.; Buendia, J. *TL* **1992**, *33*, 4889. (g) Hutchinson, E. J.; Roberts, S. M.; Thorpe, A. J. *JCS(P1)* **1992**, 2245.
12. Kobayashi, S.; Kamiyama, K.; Iimori, T.; Ohno, M. *TL* **1984**, *25*, 2557.
13. Suemune, H.; Matsuno, K.; Uchida, M.; Sakai, K. *TA* **1992**, *3*, 297.
14. Malézieux, B.; Jaouen, G.; Salaün, J.; Howell, J. A. S.; Palin, M. G.; McArdle, P.; O'Gara, M.; Cunningham, D. *TA* **1992**, *3*, 375.
15. Klunder, A. J. H.; van Gastel, F. J. C.; Zwanenburg, B. *TL* **1988**, *29*, 2697.
16. Klunder, A. J. H.; Huizinga, W. B.; Hulshof, A. J. M.; Zwanenburg, B. *TL* **1986**, *27*, 2543.
17. Alcock, N. W.; Crout, D. H. G.; Henderson, C. M.; Thomas, S. E. *CC* **1988**, 746.
18. Hoong, L. K.; Strange, L. E.; Liotta, D. C.; Koszalka, G. W.; Burns, C. L.; Schinazi, R. F. *JOC* **1992**, *57*, 5563.
19. Klibanov, A. M.; Siegel, E. H. *Enzyme Microb. Technol.* **1982**, *4*, 172.
20. Wang, Y.-F.; Sih, C. J. *TL* **1984**, *25*, 4999.
21. Jones, M.; Page, M. I. *CC* **1991**, 316.
22. Papageorgiou, C.; Benezra, C. *JOC* **1985**, *50*, 1144.
23. (a) Adachi, K.; Kobayashi, S.; Ohno, M. *C* **1986**, *40*, 311. (b) Seebach, D.; Eberle, M. *C* **1986**, *40*, 315.
24. Whitesell, J. K.; Lawrence, R. M. *C* **1986**, *40*, 318.
25. Basavaiah, D.; Raju, S. B. *SC* **1991**, *21*, 1859.
26. (a) Fouque, E.; Rousseau, G. *S* **1989**, 661. (b) Guibé-Jampel, E.; Rousseau, G.; Blanco, L. *TL* **1989**, *30*, 67.
27. Rudd, E. A.; Brockman, H. L. In *Lipases*; Borgström, B.; Brockman, H. L., Eds.; Elsevier: Amsterdam, 1984; p 185.
28. Kazlauskas, R. J.; Weissfloch, A. N. E.; Rappaport, A. T.; Cuccia, L. A. *JOC* **1991**, *56*, 2656.
29. Chenault, H. K.; Kim, M.-J.; Akiyama, A.; Miyazawa, T.; Simon, E. S.; Whitesides, G. M. *JOC* **1987**, *52*, 2608.
30. Chênevert, R.; Desjardins, M.; Gagnon, R. *CL* **1990**, 33.
31. Sussman, J. L.; Harel, M.; Frolow, F.; Oefner, C.; Goldman, A.; Toker, L.; Silman, I. *Science* **1991**, *253*, 872.
32. (a) Dropsy, E. P.; Klibanov, A. M. *Biotechnol. Bioeng.* **1984**, *26*, 911. (b) Aragozzini, F.; Valenti, M.; Santaniello, E.; Ferraboschi, P; Grisenti, P. *Biocatalysis* **1992**, *5*, 325.
33. Santaniello, E.; Canevotti, R.; Casati, R.; Ceriani, L.; Ferraboschi, P; Grisenti, P. *G* **1989**, *119*, 55.
34. Danishefsky, S. J.; Cabal, M. P.; Chow, K. *JACS* **1989**, *111*, 3456.
35. Deardorff, D. R.; Matthews, A. J.; McMeekin, D. S.; Craney, C. L. *TL* **1986**, *27*, 1255.

36. Johnson, C. R.; Senanayake, C. H. *JOC* **1989**, *54*, 735.
37. Senanayake, C. H.; Bill, T. J.; Larsen, R. D.; Leazer, J.; Reider, P. J. *TL* **1992**, *33*, 5901.
38. Mutsaers, J. H. G. M.; Kooreman, H. J. *RTC* **1991**, *110*, 185.
39. Kamal, A. *SC* **1991**, *21*, 1293.
40. Yee, C.; Blythe, T. A.; McNabb, T. J.; Walts, A. E. *JOC* **1992**, *57*, 3525.
41. Fliche, C.; Braun, J.; Le Goffic, F. *SC* **1991**, *21*, 1429.
42. Monteiro, J.; Braun, J.; Le Goffic, F. *SC* **1990**, *20*, 315.

Enzo Santaniello, Patrizia Ferraboschi & Paride Grisenti
Università di Milano, Italy

Ethylaluminum Dichloride[1]

[563-43-9] $C_2H_5AlCl_2$ (MW 126.95)

(strong Lewis acid that can also act as a proton scavenger; reacts with HX to give ethane and $AlCl_2X$)

Alternate Name: dichloroethylaluminum.
Physical Data: mp 32 °C; bp 115 °C/50 mmHg; d 1.207 g cm^{-3}.
Solubility: sol most organic solvents; stable in alkanes or arenes.
Analysis of Reagent Purity: solutions are reasonably stable but may be titrated before use by one of the standard methods.[1e]
Form Supplied in: commercially available neat or as solutions in hexane or toluene.
Handling, Storage, and Precautions: must be transferred under inert gas (Ar or N_2) to exclude oxygen and water. Use in a fume hood.

Alkylaluminum Halides. Since the early 1980s, alkylaluminum halides have come into widespread use as Lewis acid catalysts. These strong Lewis acids offer many advantages over traditional metal halide Lewis acids such as ***Boron Trifluoride***, ***Aluminum Chloride***, ***Titanium(IV) Chloride***, and ***Tin(IV) Chloride***. Most importantly, the alkyl group on the aluminum will react with protons to give an alkane and a new Lewis acid. Alkylaluminum halides are therefore Brønsted bases, as well as Lewis acids. The alkyl groups are also nucleophilic, and this is the major disadvantage in the use of these compounds as Lewis acids.

Pure, anhydrous Lewis acids do not catalyze the polymerization of alkenes or the Friedel–Crafts alkylation of aromatics by alkenes. Cocatalysts, such as water or a protic acid, react with the Lewis acid to produce a very strong Brønsted acid that will protonate a double bond. Therefore, use of strictly anhydrous conditions should minimize side reactions in Lewis acid-catalyzed ene, Diels–Alder, and [2 + 2] cycloaddition reactions. Unfortunately, it is difficult to prepare anhydrous, proton-free $AlCl_3$, BF_3, etc. Alkylaluminum halides are easily prepared and stored in anhydrous form and, more importantly, scavenge any adventitious water, liberating an alkane and generating a new Lewis acid in the process.

Using alkylaluminum halides, Lewis acid-catalyzed reactions can now be carried out under aprotic conditions. This is of value when side reactions can be caused by the presence of adventitious protons; it is of special value when acidic protons are produced by the reaction. In these cases, use of the appropriate alkylaluminum halide in stoichiometric rather than catalytic amounts gives high yields of products not formed at all with other Lewis acids. The Me_2AlCl-catalyzed ene reactions of aliphatic aldehydes with alkenes, which give homoallylic alcohol–Me_2AlCl complexes that react further to give methane and the stable methylaluminum alkoxides, is an example of this type of reaction (eq 1).[1h,2] Loss of methane prevents the alcohol–Lewis acid complex from solvolyzing or protonating double bonds.

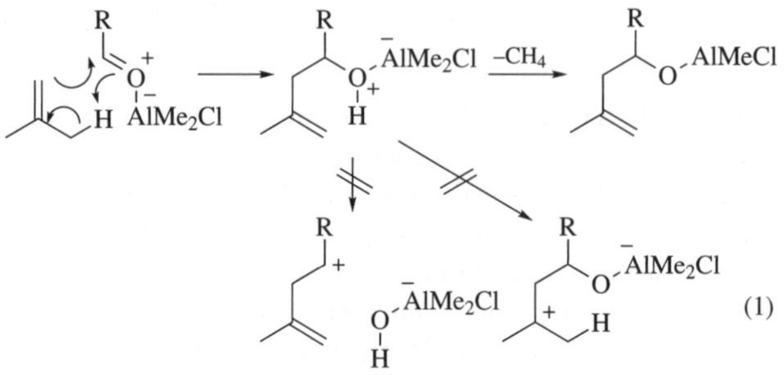

(1)

The alkylaluminum halides cover a wide range of acidity. Replacing chlorines with alkyl groups decreases Lewis acidity. $EtAlCl_2$ and ***Methylaluminum Dichloride*** are only slightly less acidic than $AlCl_3$. ***Diethylaluminum Chloride*** and ***Dimethylaluminum Chloride*** are substantially less acidic, and ***Trimethylaluminum***, Me_2AlOR, and $MeAl(OR)_2$ are even weaker Lewis acids. Alkylaluminum halides with fractional ratios of alkyl to chloride are also available. The sesquichlorides are commercially available. Other reagents can be prepared by mixing two reagents in the desired proportion. If no reaction occurs with the alkylaluminum halide, a stronger Lewis acid should be tried. If polymerization or other side reactions compete, a weaker Lewis acid should be used. The sequential ene and Prins reactions shown in eq 2 proceed cleanly with $Me_3Al_2Cl_3$.[3] Complex mixtures are obtained with the stronger Lewis acid $EtAlCl_2$ while the weaker Lewis acid Me_2AlCl reacts with ***Formaldehyde*** to give ethanol.

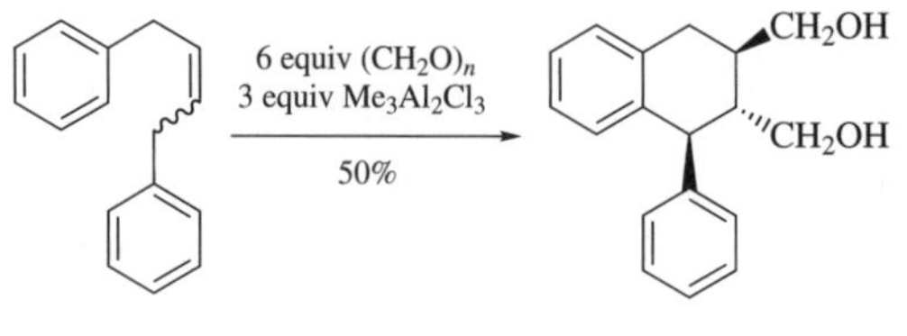

(2)

Use of more than one equivalent of Lewis acid will produce complexes that are formally 1:2 substrate Lewis acid complexes, but are more likely salts with a R_2Al–substrate cation and an aluminate anion.[4–6] This substrate in this salt is much more electrophilic and reactive than that in simple Lewis acid complexes.

These reagents are easier to use than typical inorganic Lewis acids. They are soluble in all organic solvents, including hexane and toluene, in which they are commercially available as standardized solutions. In general, alkane solvents are preferred since toluene can undergo Friedel–Crafts reactions. On a laboratory scale, these reagents are transferred by syringe like alkyllithiums and, unlike anhydrous solid Lewis acids, they do not require a glove bag or dry box for transfer.

While the Brønsted basicity of the alkyl group is advantageous, these alkyl groups are also nucleophilic. The addition of the alkyl group from the aluminum in the Lewis acid–reagent complex to the electrophilic center can be a serious side reaction. The ease of alkyl donation is $R_3Al > R_2AlCl > R_3Al_2Cl_3 > RAlCl_2$. When the nucleophilicity of the alkyl group is a problem, a Lewis acid with fewer alkyl groups should be examined. Addition of diethyl ether or another Lewis base may moderate the reaction if its greater acidity causes problems.

Ethylaluminum compounds are more nucleophilic than methylaluminum compounds and can donate a hydrogen as well as an ethyl group to the electrophilic center. Unfortunately, methylaluminum compounds must be prepared from ***Chloromethane***, while ethylaluminum compounds can be prepared much more cheaply from ***Ethylene***. Therefore ethylaluminum compounds are usually used unless the nucleophilicity of the alkyl group is a problem. Although the predominant use of alkylaluminum halides is as Lewis acids, they are occasionally used for the transfer of an alkyl group or hydride to an electrophilic center.

Eq 3 shows an unusual reaction in which the nature of the reaction depends on the amount, acidity, and alkyl group of the alkylaluminum halide.[6] Use of 1 equiv of Me_2AlCl leads to a concerted ene reaction with the side chains *cis*. Use of 2 equiv of Me_2AlCl produces a more electrophilic aldehyde complex that cyclizes to a zwitterion. Chloride transfer is the major process at −78 °C; at 0 °C, chloride transfer is reversible and a 1,5-proton transfer leads to an ene-type adduct with the side chains *trans*. Use of 2 equiv of $MeAlCl_2$ forms a cyclic zwitterion that undergoes two 1,2-hydride shifts to form the ketone. A similar zwitterion forms with $EtAlCl_2$, but β-hydride transfer leading to the saturated alcohol is faster than 1,2-hydride shifts.

Me_2AlCl (2 equiv), −78 °C, 75%; Me_2AlCl (1 equiv), −78 °C, 24%; $MeAlCl_2$ (2 equiv), −78 °C, 72%; $EtAlCl_2$ (2 equiv) −78 °C, 73%; Me_2AlCl (2 equiv), 0 °C, 34% (3)

Catalysis of Diels–Alder Reactions. $EtAlCl_2$ is a useful Lewis acid catalyst for Diels–Alder reactions. It is reported to be more efficacious for the Diels–Alder reaction of ***Acrolein*** and butadiene than either $AlCl_3$ or Et_2AlCl.[7] It is a useful catalyst for intramolecular Diels–Alder reactions with α,β-unsaturated esters (eq 4)[8] and aldehydes[9] as dienophiles. It has also proven to be a very efficient catalyst for the inter-[10] and intramolecular[11] asymmetric Diels–Alder reaction of chiral α,β-unsaturated acyl sultams (eq 5) and has been used to catalyze a wide variety of Diels–Alder reactions.[12]

MeO_2C … $EtAlCl_2$, 60% → MeO_2C, H, H (4)

$EtAlCl_2$, 53% → 99% de (5)

Catalysis of Ene Reactions. Although $AlCl_3$ can be used as a Lewis acid catalyst for ene reactions of α,β-unsaturated esters,[13a] better results are obtained more reproducibly with $EtAlCl_2$. Ene reactions of ***Methyl Propiolate*** proceed in good yield with 1,1-di-, tri-, and tetrasubstituted alkenes (eq 6).[14] A precursor to 1,25-dihydroxycholesterol can be prepared by an ene reaction with methyl propiolate. Three equiv of $EtAlCl_2$ are needed since the acetate esters are more basic than methyl propiolate (eq 7).[15] *Endo* products are obtained stereospecifically with methyl α-haloacrylates (eq 8).[14] $EtAlCl_2$ has also been used to catalyze intramolecular ene reactions (eq 9).[16]

+ CO_2Me → $EtAlCl_2$, 92% → CO_2Me (6)

AcO, AcO + CO_2Me → 3 equiv $EtAlCl_2$, 89% → CO_2Me (7)

H + Br, CO_2Me → $EtAlCl_2$, 51% → Br, CO_2Me (8)

CO_2Me → $EtAlCl_2$, 40 °C, 5 d, CH_2Cl_2, 82% → CO_2Me (9)

$EtAlCl_2$ is usually too strong a Lewis acid for ene reactions of carbonyl compounds.[13b,c] However, alkenes that contain basic sites that complex to the Lewis acid do not undergo Me_2AlCl-catalyzed ene reactions with formaldehyde. In these cases, $EtAlCl_2$ is the preferred catalyst.[17,18] The dienyl acetate shown in eq 10 reacts with excess ***Formaldehyde*** and $EtAlCl_2$ to provide the conjugated diene ene adduct that undergoes a quasi-intramolecular Diels–Alder reaction to afford a pseudomonic acid precursor. Me_2AlCl catalyzes the ene reaction of aliphatic aldehydes with 1,1-di-, tri-, and tetrasubstituted alkenes. Terminal alkenes are less nucleophilic, so the only reaction is addition of a methyl group to the aldehyde. $EtAlCl_2$, a stronger Lewis acid with a less nucleophilic alkyl group, catalyzes the reaction of aliphatic aldehydes with terminal alkenes in CH_2Cl_2 at 0 °C to give 50–60% of the ene adduct.[18] Use of $EtAlCl_2$ as a catalyst affords the best diastereoselectivity in the ene reaction of dibenzylleucinal with ***Isobutene*** (eq 11).[19] $EtAlCl_2$ has also been used to catalyze intramolecular ene reactions of trifluoromethyl ketones.[20]

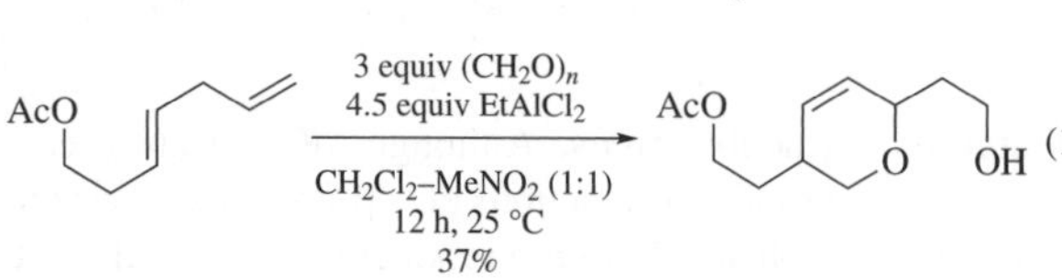

3 equiv $(CH_2O)_n$, 4.5 equiv $EtAlCl_2$, CH_2Cl_2–$MeNO_2$ (1:1), 12 h, 25 °C, 37% (10)

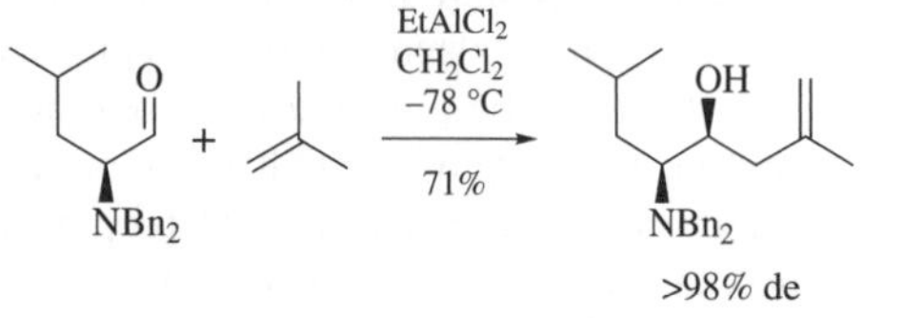

$EtAlCl_2$, CH_2Cl_2, –78 °C, 71%, >98% de (11)

Catalysis of Intramolecular Sakurai Reactions. $EtAlCl_2$ has been extensively used as a catalyst for intramolecular Sakurai additions. Enones (eqs 12 and 13)[21,22] have been most extensively explored. Different products are often obtained with fluoride or Lewis acid catalysis. $EtAlCl_2$ is the Lewis acid used most often although $TiCl_4$ and BF_3 have also been used. $EtAlCl_2$ also catalyzes intramolecular Sakurai reactions with ketones[23] and other electrophiles.[24] The cyclization of electrophilic centers onto alkylstannanes[25] and Prins-type additions to vinylsilanes[26] are also catalyzed by $EtAlCl_2$.

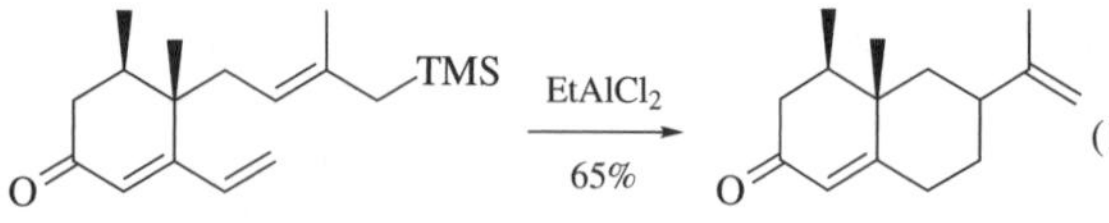

$EtAlCl_2$, 65%, nootkatone (12)

$EtAlCl_2$, 55% (13)

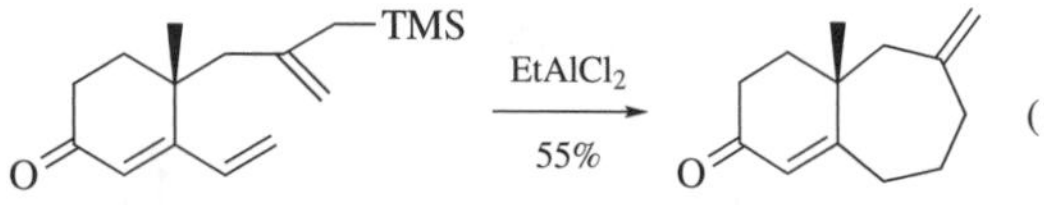

Catalysis of [2 + 2] Cycloadditions. $EtAlCl_2$ catalyzes a wide variety of [2 + 2] cycloadditions. These include the addition of alkynes or allenes to alkenes to give cyclobutenes and alkylidenecyclobutanes (eq 14),[27] the addition of electron-deficient alkenes to allenyl sulfides (eq 15),[28] the addition of propiolate esters to monosubstituted and 1,2-disubstituted alkenes to form cyclobutene carboxylates (eq 16),[14b] and the addition of allenic esters to alkenes to form cyclobutanes.[29]

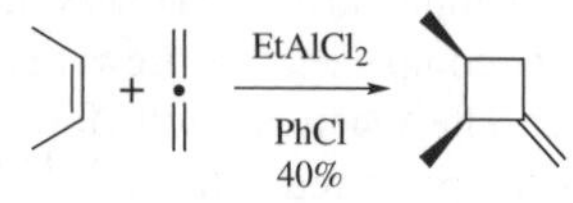

$EtAlCl_2$, PhCl, 40% (14)

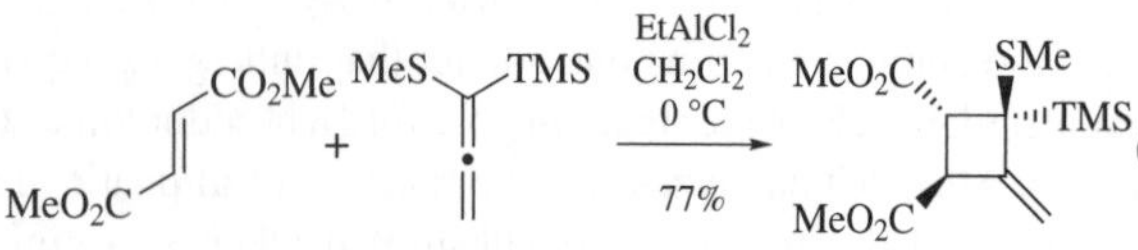

$EtAlCl_2$, CH_2Cl_2, 0 °C, 77% (15)

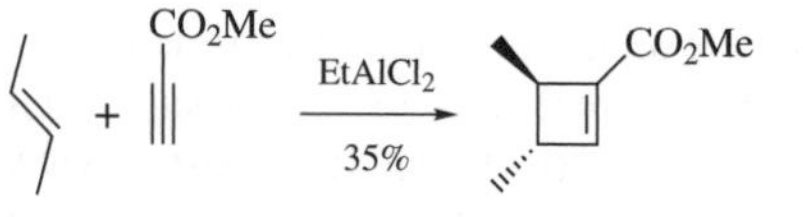

$EtAlCl_2$, 35% (16)

Generation of Electrophilic Cations. $EtAlCl_2$ has proven to be a useful Lewis acid for inducing a wide variety of electrophilic reactions. It is particularly useful since an excess of the reagent can be used so that all nucleophiles are complexed to acid. Under these conditions, intermediates tend to collapse to give cyclobutanes or undergo hydride shifts to give neutral species. Reaction of the 1:2 ***Crotonaldehyde***–$EtAlCl_2$ complex with 2-methyl-2-butene at −80 °C affords a zwitterion that collapses to give mainly the cyclobutane. Closure of the zwitterion is reversible at 0 °C. Since there are no nucleophiles in solution, two 1,2-hydride shifts take place to give the enal (eq 17).[5] Intramolecular versions of these reactions are also quite facile (eqs 18 and 19).[5,21] $EtAlCl_2$ promotes the ring enlargement of 1-acylbicyclo[4.2.0]oct-3-enes (eq 20).[30]

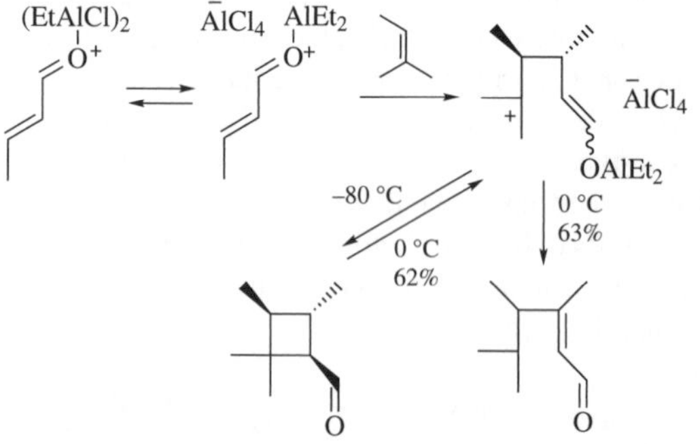

−80 °C; 0 °C, 62%; 0 °C, 63% (17)

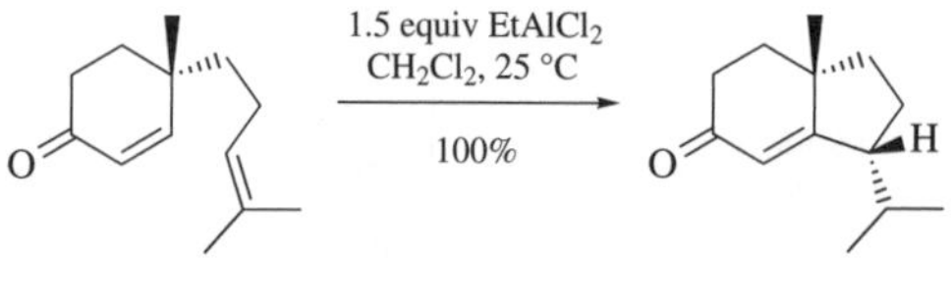

1.5 equiv $EtAlCl_2$, CH_2Cl_2, 25 °C, 100% (18)

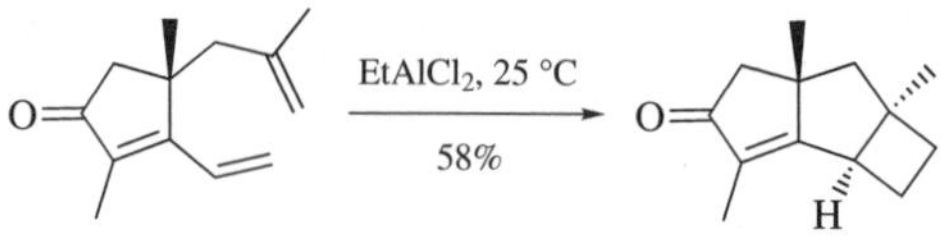

$EtAlCl_2$, 25 °C, 58% (19)

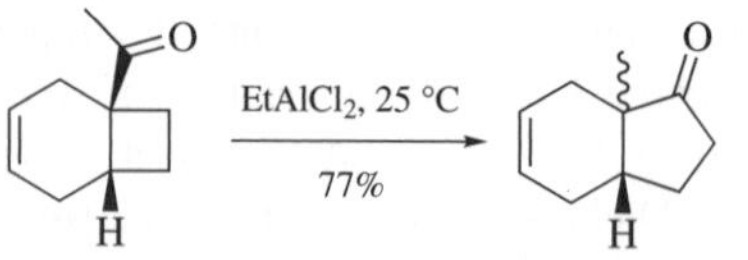

$EtAlCl_2$, 25 °C, 77%, 93:7 *cis*:*trans* (20)

$EtAlCl_2$ catalyzes the Friedel–Crafts acylation of alkenes with acid chlorides,[31] the formal [3 + 2] cycloaddition of alkenes with cyclopropane-1,1-dicarboxylates (eq 21),[32] the Friedel–Crafts alkylation of anilines and indoles with α-aminoacrylate esters,[33] and the formation of allyl sulfoxides from sulfinyl chlorides and alkenes.[34] $EtAlCl_2$ induces the Beckmann rearrangement of oxime sulfonates. The cationic intermediates can be trapped with enol silyl ethers (eq 22).[35] $EtAlCl_2$ is the preferred catalyst for addition of the cation derived from an α-chloro sulfide to an alkene to give a cation which undergoes a Friedel–Crafts alkylation (eq 23).[36]

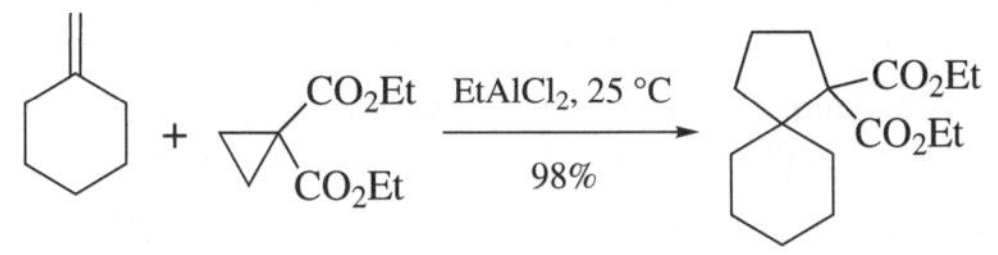

$EtAlCl_2$ reacts with aliphatic sulfones to generate an aluminum sulfinate and a cation that can be reduced to a hydrocarbon by $EtAlCl_2$,[37] trapped with a nucleophile such as ***Allyltrimethylsilane***,[38] or undergo a pinacol-type rearrangement (eq 24).[39]

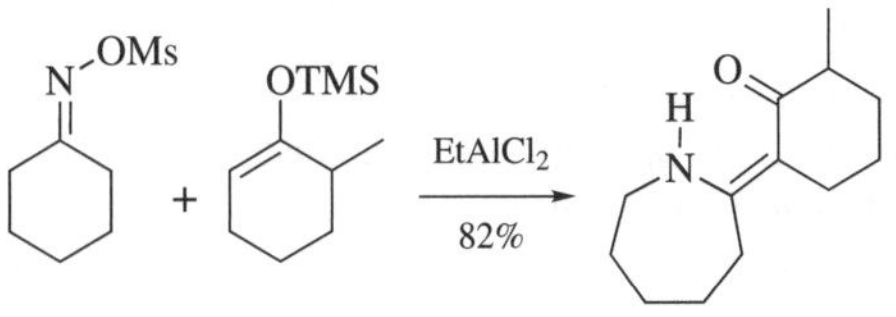

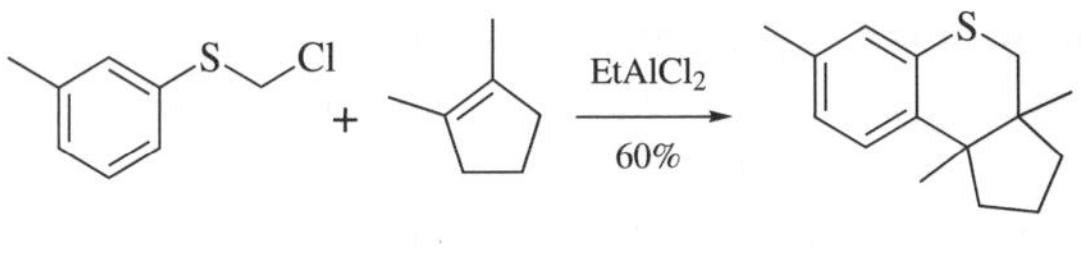

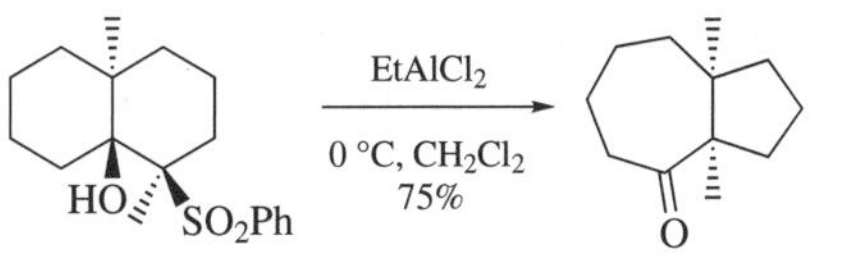

Elimination Reactions. $EtAlCl_2$ induces elimination of two molecules of HBr from the dibromide to give the dihydropyridine (eq 25).[40] The usual base-catalyzed elimination is ineffective.

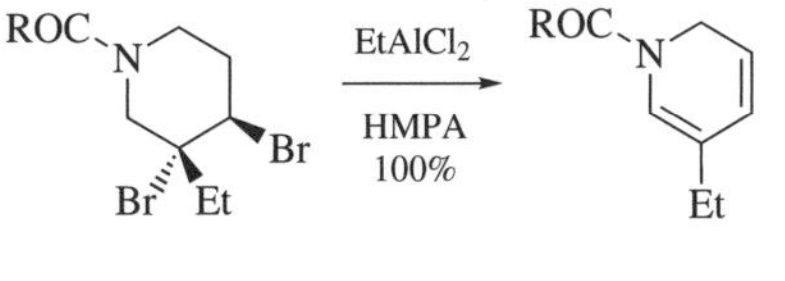

Nucleophilic Addition. $EtAlCl_2$ has been used to activate conjugated systems toward attack of an external nucleophile or to transfer a hydride or ethyl group as a nucleophile. Addition of a cuprate to the chiral amide in the presence of $EtAlCl_2$ improves the diastereoselectivity, affording a >93:7 mixture of stereoisomers (eq 26).[41] Reaction of butyrolactones with $EtAlCl_2$ reduces the lactone to a carboxylic acid by opening to the cation and hydride delivery (eq 27).[42] Sulfonimidyl chlorides react with $EtAlCl_2$ at −78 °C to provide *S*-ethyl sulfoximines in 65–95% overall yield (eq 28).[43]

Modification of Carbanions. Trimethylsilyl allylic carbanions react with aldehydes in the presence of $EtAlCl_2$ exclusively at the α-position to give *threo* adducts (eq 29).[44]

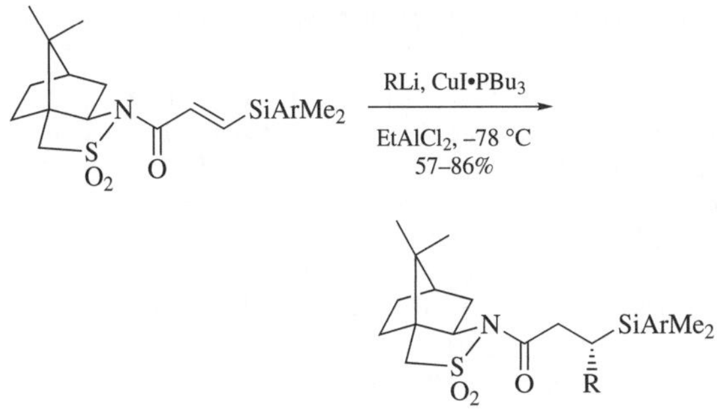

Related Reagents. Diethylaluminum Chloride; Diethylaluminum Iodide; Dimethylaluminum Chloride; Dimethylaluminum Iodide; Methylaluminum Bis(2,6-di-*t*-butyl-4-methylphenoxide); Methylaluminum Bis(2,6-di-*t*-butylphenoxide); Methylaluminum Dichloride.

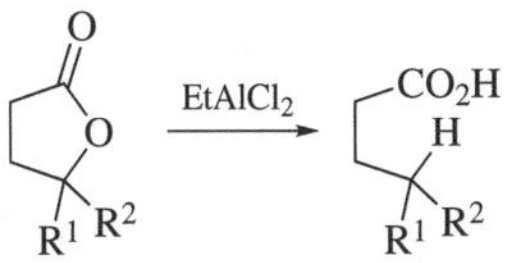

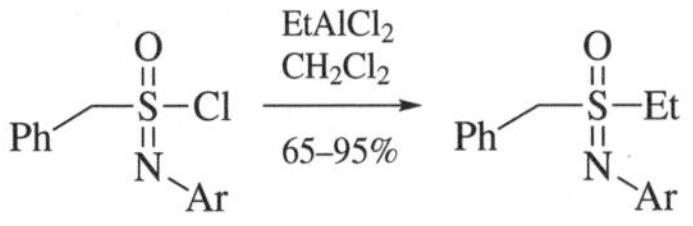

1. (a) Mole, T.; Jeffery, E. A. *Organoaluminum Compounds*; Elsevier: New York, 1972. (b) Bruno, G. *The Use of Aluminum Alkyls in Organic Synthesis*; Ethyl Corporation: Baton Rouge, LA, 1970. (c) Bruno, G. *The Use of Aluminum Alkyls in Organic Synthesis, Supplement, 1969–1972*; Ethyl Corporation: Baton Rouge, LA, 1973. (d) Honeycutt, J. B. *The Use of Aluminum Alkyls in Organic Synthesis, Supplement, 1972–1978*; Ethyl Corporation: Baton Rouge, LA, 1981. (e) *Aluminum Alkyls*; Stauffer Chemical Co.: Westport, CT, 1976. (f) Snider, B. B.; Rodini, D. J.; Karras, M.; Kirk, T. C.; Deutsch, E. A.; Cordova, R.; Price, R. T.; *T* **1981**, *37*, 3927. (g) Yamamoto, H.; Nozaki, H. *AG(E)* **1978**, *17*, 169. (h) Snider, B. B. In *Selectivities in Lewis Acid Promoted Reactions*; Schinzer, D., Ed.; Kluwer: Dordrecht, 1989; Chapter 8. (i) Maruoka, K.; Yamamoto, H. *AG(E)* **1985**, *24*, 668; *T* **1988**, *44*, 5001.
2. (a) Snider, B. B.; Rodini, D. J.; Kirk, T. C.; Cordova, R. *JACS* **1982**, *104*, 555. (b) Cartaya-Marin, C. P.; Jackson, A. C.; Snider, B. B. *JOC* **1984**, *49*, 2443.
3. Snider, B. B.; Jackson, A. C. *JOC* **1983**, *48*, 1471.
4. Evans, D. A.; Chapman, K. T.; Bisaha, J. *JACS* **1988**, *110*, 1238.
5. Snider, B. B.; Rodini, D. J.; van Straten, J. *JACS* **1980**, *102*, 5872.
6. Snider, B. B.; Karras, M.; Price, R. T.; Rodini, D. J. *JOC* **1982**, *47*, 4538.
7. Miyajima, S.; Inukai, T. *BCJ* **1972**, *45*, 1553.
8. (a) Roush, W. R.; Ko, A. I.; Gillis, H. R. *JOC* **1980**, *45*, 4264. (b) Roush, W. R.; Gillis, H. R. *JOC* **1980**, *45*, 4267; *JOC* **1982**, *47*, 4825.
9. (a) Marshall, J. A.; Audia, J. E.; Grote, J. *JOC* **1984**, *49*, 5277. (b) Marshall, J. A.; Grote, J.; Audia, J. E. *JACS* **1987**, *109*, 1186.
10. (a) Oppolzer, W.; Chapuis, C.; Bernardinelli, G. *HCA* **1984**, *67*, 1397. (b) Smith, A. B., III; Hale, K. J.; Laasko, L. M.; Chen, K.; Riéra, A. *TL* **1989**, *30*, 6963.

11. (a) Oppolzer, W.; Dupuis, D. *TL* **1985**, *26*, 5437. (b) Oppolzer, W. *AG(E)* **1984**, *23*, 876.
12. (a) Poll, T.; Metter, J. O.; Helmchen, G. *AG(E)* **1985**, *24*, 112. (b) Herndon, J. W. *JOC* **1986**, *51*, 2853. (c) Metral, J.-L.; Lauterwein, J.; Vogel, P. *HCA* **1986**, *69*, 1287. (d) Waldmann, H.; Braun, M.; Dräger, M. *AG(E)* **1990**, *29*, 1468. (e) Kametani, T.; Takeda, H.; Suzuki, Y.; Honda, T. *SC* **1985**, *15*, 499. (f) Vidari, G.; Ferrino, S.; Grieco, P. A. *JACS* **1984**, *106*, 3539. (g) Funk, R. L.; Zeller, W. E. *JOC* **1982**, *47*, 180. (h) Fringuelli, F.; Pizzo, F.; Taticchi, A.; Wenkert, E. *SC* **1986**, *16*, 245.
13. (a) Snider, B. B. *COS* **1991**, *5*, 1. (b) Snider, B. B. *COS* **1991**, *2*, 527. (c) Mikami, K.; Shimizu, M. *CRV* **1992**, *92*, 1021.
14. (a) Snider, B. B.; Rodini, D. J.; Conn, R. S. E.; Sealfon, S. *JACS* **1979**, *101*, 5283. (b) Snider, B. B.; Roush, D. M.; Rodini, D. J.; Gonzalez, D.; Spindell, D. *JOC* **1980**, *45*, 2773. (c) Snider, B. B.; Duncia, J. V. *JACS* **1980**, *102*, 5926. (d) Duncia, J. V.; Lansbury, P. T., Jr.; Miller, T.; Snider, B. B. *JACS* **1982**, *104*, 1930. (e) Snider, B. B.; Phillips, G. B. *JOC* **1983**, *48*, 3685.
15. (a) Batcho, A. D.; Berger, D. E.; Uskokovic, M. R.; Snider, B. B. *JACS* **1981**, *103*, 1293. (b) Dauben, W. G.; Brookhart, T. *JACS* **1981**, *103*, 237. (c) Batcho, A. D.; Berger, D. E.; Davoust, S. G.; Wovkulich, P. M.; Uskokovic, M. R. *HCA* **1981**, *64*, 1682. (d) Wovkulich, P. M.; Batcho, A. D.; Uskokovic, M. R. *HCA* **1984**, *67*, 612.
16. Snider, B. B.; Phillips, G. B. *JOC* **1984**, *49*, 183.
17. (a) Snider, B. B.; Phillips, G. B. *JACS* **1982**, *104*, 1113. (b) Snider, B. B.; Phillips, G. B.; Cordova, R. *JOC* **1983**, *48*, 3003.
18. Snider, B. B.; Phillips, G. B. *JOC* **1983**, *48*, 464.
19. Mikami, K.; Kaneko, M.; Loh, T.-P.; Tereda, M.; Nakai, T. *TL* **1990**, *31*, 3909.
20. Abouadellah, A.; Aubert, C.; Bégué, J.-P.; Bonnet-Delpon, D.; Guilhem, J. *JCS(P1)* **1991**, 1397.
21. (a) Majetich, G.; Behnke, M.; Hull, K. *JOC* **1985**, *50*, 3615. (b) Majetich, G.; Hull, K.; Lowery, D.; Ringold, C.; Defauw, J. In *Selectivities in Lewis Acid Promoted Reactions*; Schinzer, D., Ed.; Kluwer: Dordrecht, 1989; Chapter 9. (c) Majetich, G.; Khetani, V. *TL* **1990**, *31*, 2243. (d) Majetich, G.; Hull, K.; Casares, A. M.; Khetani, V. *JOC* **1991**, *56*, 3958. (e) Majetich, G.; Song, J.-S.; Ringold, C.; Nemeth, G. A.; Newton, M. G. *JOC* **1991**, *56*, 3973. (f) Majetich, G.; Song, J.-S.; Leigh, A. J.; Condon, S. M. *JOC* **1993**, *58*, 1030.
22. (a) Schinzer, D.; Sólyom, S.; Becker, M. *TL* **1985**, *26*, 1831. (b) Schinzer, D. *S* **1988**, 263.
23. (a) Trost, B. M.; Hiemstra, H. *JACS* **1982**, *104*, 886. (b) Trost, B. M.; Coppola, B. P. *JACS* **1982**, *104*, 6879. (c) Trost, B. M.; Fray, M. J. *TL* **1984**, *25*, 4605.
24. Wada, M.; Shigehisa, T.; Akiba, K. *TL* **1985**, *26*, 5191. (b) Pirrung, M. C.; Thomson, S. A. *TL* **1986**, *27*, 2703.
25. (a) McDonald, T. L.; Delahunty, C. M.; Mead, K.; O'Dell, D. E. *TL* **1989**, *30*, 1473. (b) Plamondon, L.; Wuest, J. D. *JOC* **1991**, *56*, 2066.
26. Casteñeda, A.; Kucera, D. J.; Overman, L. E. *JOC* **1989**, *54*, 5695.
27. (a) Lukas, J. H.; Kouwenhoven, A. P.; Baardman, F. *AG(E)* **1975**, *14*, 709. (b) Lukas, J. H.; Baardman, F.; Kouwenhoven, A. P. *AG(E)* **1976**, *15*, 369.
28. Hayashi, Y.; Niihata, S.; Narasaka, K. *CL* **1990**, 2091.
29. (a) Snider, B. B.; Spindell, D. K. *JOC* **1980**, *45*, 5017. (b) Snider, B. B.; Ron, E. *JOC* **1986**, *51*, 3643.
30. Fujiwara, T.; Tomaru, J.; Suda, A.; Takeda, T. *TL* **1992**, *33*, 2583.
31. Snider, B. B.; Jackson, A. C. *JOC* **1982**, *47*, 5393.
32. (a) Beal, R. B.; Dombroski, M. A.; Snider, B. B. *JOC* **1986**, *51*, 4391. (b) Bambal, R.; Kemmit, R. D. W. *CC* **1988**, 734.
33. Tarzia, G.; Balsamini, C.; Spadoni, G.; Duranti, E. *S* **1988**, 514.
34. Snider, B. B. *JOC* **1981**, *46*, 3155.
35. Matsumura, Y.; Fujiwara, J.; Maruoka, K.; Yamamoto, H. *JACS* **1983**, *105*, 6312.
36. (a) Ishibashi, H.; So, T. S.; Nakatani, H.; Minami, K.; Ikeda, M. *CC* **1988**, 827. (b) Ishibashi, H.; Okoda, M.; Sato, K.; Ikeda, M.; Ishiyama, K.; Tamura, Y. *CPB* **1985**, *33*, 90.
37. Trost, B. M.; Ghadiri, M. R. *JACS* **1984**, *106*, 7260.
38. Barton, D. H. R.; Boivin, J.; Sarma, J.; Da Silva, E.; Zard, S. Z. *TL* **1989**, *30*, 4237.
39. Trost, B. M.; Nielsen, J. B.; Hoogsteen, K. *JACS* **1992**, *114*, 5432.
40. (a) Raucher, S.; Lawrence, R. F. *T* **1983**, *39*, 3731. (b) *TL* **1983**, *24*, 2927.
41. Oppolzer, W.; Mills, R. J.; Pachinger, W.; Stevenson, T. *HCA* **1986**, *69*, 1542.
42. Reinheckel, H.; Sonnek, G.; Falk, F. *JPR* **1974**, *316*, 215.
43. Harmata, M. *TL* **1989**, *30*, 437.
44. Yamamoto, Y.; Saito, Y.; Maruyama, K. *CC* **1982**, 1326.

Barry B. Snider
Brandeis University, Waltham, MA, USA

Fluorosulfuric Acid–Antimony(V) Fluoride[1]

$HSO_3F–SbF_5$

[23854-38-8] F_6HO_3SSb (MW 316.83)

(a strong conjugate Brønsted–Lewis superacid system[1] widely used for the generation of stable carbocations and as catalyst and reagent for alkylation, isomerization, rearrangement, cyclization, oxyfunctionalization, formylation, sulfonation, and fluorosulfonation)

Alternate Name: Magic Acid®.

Physical Data: HSO_3F: mp −89.0 °C; bp 162.7 °C; *d* 1.743 g cm^{-3}. SbF_5: bp 149.5 °C; *d* 2.993 g cm^{-3}.

Solubility: sol SO_2ClF, liquid SO_2; solubilizes most organic compounds that are potential proton acceptors.

Form Supplied in: colorless liquid; commercially available.

Preparative Methods: $HSO_3F–SbF_5$ is prepared by mixing ***Fluorosulfuric Acid*** and ***Antimony(V) Fluoride*** at rt under dry nitrogen or argon atmosphere. The commercially available Magic Acid is a 50:50 mol % mixture of the two components. Magic Acid diluted in fluorosulfuric acid to various extents is also available. Commercial Magic Acid generally contains some HF in the form of conjugate acid.

Handling, Storage, and Precautions: Magic Acid is highly toxic, moisture sensitive, and corrosive, and should always be handled in a fume hood with proper protection. Glass is attacked by Magic Acid very slowly when moisture is excluded. Therefore, glassware may be used for handling and carrying out reactions involving Magic Acid. Teflon containers are recommended for long-term laboratory storage of Magic Acid.

Introduction. Of all the superacids, Magic Acid is probably the most widely used medium for the study of stable long-lived carbocations and other reactive cations. The general rule is that the higher the acidity of the medium used, the more stable is the carbocation generated. The acidity of the fluorosulfuric acid–antimony pentafluoride system as a function of SbF_5 content has been studied.[2,3] The increase in acidity is very sharp at low SbF_5 concentration. The H_0 value changes from −15.1 for HSO_3F to −19.8 for a mixture containing 10% SbF_5.[2] The acidity continues up to the estimated value of $H_0 = -26.5$ for 90% SbF_5 content. The H_0 value for the 1:1 molar mixture of fluorosulfuric acid and antimony pentafluoride, known as Magic Acid®, is estimated to be about −23 by a dynamic NMR study.[3] The name 'Magic Acid' originated in Olah's laboratory at Case Western Reserve University in the winter of 1966 when a piece of Christmas candle was found to dissolve readily in this acid system, giving the sharp ^1H NMR spectrum of the *t*-butyl cation, a phenomenon considered by the research student involved to be 'magic'.[1] A major reason for the wide application of Magic Acid compared with other superacid systems, besides its very high acidity, is probably the large temperature range in which it can be used. In the liquid state, it can be studied at temperatures as low as −160 °C (acid diluted with SO_2F_2 and SO_2ClF) and as high as 150 °C (neat acid).

Magic Acid has been employed as a high acidity medium for isomerization/rearrangement, alkylation, cyclization, carboxylation, formylation, oxyfunctionalization, and related reactions. It also serves as a fluorosulfonating/sulfonating agent for aromatics.

Generation of Stable Carbocations. Thanks to its high acidity, Magic Acid can be used for the generation of such reactive carbocations as the *t*-butyl cation and other alkyl cations, while fluorosulfuric acid itself is suitable only for the generation of more stable cations such as aryl- or cyclopropyl-stabilized carbocations.

Carbocations and carbodications have been generated from a variety of precursors in Magic Acid systems.[4]

Carbocation Generation from Tertiary and Secondary Alcohols.[5–11] A wide variety of aliphatic tertiary and secondary alcohols can be ionized to the corresponding alkyl cations by using Magic Acid. Formation of the *t*-butyl cation (eq 1)[5] and a cyclopropyl-stabilized dication (eq 2)[6] are representative examples. Primary (and some secondary) alcohols are protonated only at temperatures lower than −60 °C.[7] At more elevated temperatures they may cleave to give the corresponding carbocations, which, however, immediately rearrange to the more stable tertiary cations.[8,9]

t-BuOH —($HSO_3F–SbF_5$, −60 °C)→ (1)

($HSO_3F–SbF_5$, SO_2ClF, −78 °C) (2)

Similar to alcohols, aliphatic ethers,[12] thiols,[13] and sulfides are also protonated on oxygen or sulfur, respectively, at −60 °C in Magic Acid; carbocations are subsequently formed upon raising the temperature. Protonated sulfides, excluding tertiary alkyl, are resistant to cleavage up to +70 °C.[13]

Carbocation Generation from Alkyl Halides. Alkyl chlorides, fluorides, and bromides are convenient and frequently used precursors for generation of alkyl cations in $HSO_3F–SbF_5$ systems.[14] It should be noted, however, that the $HSO_3F–SbF_5$ system is less suitable than SbF_5 for the generation of alkyl, especially secondary alkyl, cations from the corresponding alkyl halides (see ***Antimony(V) Fluoride***). Ionization of cyclohexyl chloride in Magic Acid is accompanied by isomerization, yielding the 1-methyl-1-cyclopentyl cation (eq 3).[8]

—Cl —($HSO_3F–SbF_5$, −60 °C)→ (3)

Carbocation Generation from Unsaturated Hydrocarbons. Carbocations can be generated by protonation of unsaturated hydrocarbons such as alkenes and cycloalkenes,[8,11] cyclopentadienes,[15] benzenes and naphthalenes (eq 4),[16] pyrenes and cyclophanes,[17] unsaturated heterocycles,[18] and their derivatives with carbon–heteroatom multiple bonds,[1] including carbonyl and nitrile compounds and diazoalkanes (eq 5).[19] Compounds with two sites for protonation may undergo diprotonation to give dications in Magic Acid (eq 6).[15a]

HSO$_3$F–SbF$_5$, SO$_2$ClF, –83 °C (4)

$$H_2C{=}\overset{+}{N}{=}\bar{N} \xrightarrow[SO_2ClF,\ -70\ °C]{HSO_3F-SbF_5} H_2C{=}\overset{+}{N}{=}NH + Me{-}\overset{+}{N}{\equiv}N \quad (5)$$

HSO$_3$F–SbF$_5$, SO$_2$ClF, –70 °C (6)

Carbocation Generation from Saturated Hydrocarbons. Magic Acid, as a strong superacid, can abstract hydride from saturated alkanes, including straight-chain alkanes as well as branched and cyclic alkanes, at −125 to 25 °C to give alkyl cations.[6,20] For example, the 2-norbornyl cation is formed through protolytic ionization by dissolving norbornane in Magic Acid (eq 7).[20b]

HSO$_3$F–SbF$_5$, SO$_2$ClF, –70 °C (7)

Carbocation Generation by Oxidation. Oxidation of polycyclic arenes such as naphthacene and 1,2-benzanthracene (eq 8)[21] gives arene dications (see also ***Antimony(V) Fluoride***).

HSO$_3$F–SbF$_5$ or SbF$_5$–SO$_2$ClF (8)

Isomerization and Rearrangement. HSO$_3$F–SbF$_5$ is frequently used as a catalyst for the isomerization and rearrangement of terpenoids. The extremely high acidity of HSO$_3$F–SbF$_5$ allows the reaction to be carried out at temperatures as low as −100 °C and with improved selectivity.

In superacid, i.e. HSO$_3$F–SbF$_5$ or HF–SbF$_5$ systems, bicyclic phenols are isomerized to dienones in good yields (eq 9).[22] This is an unconventional isomerization process, in which the phenols lose their aromaticity to form nonaromatic dienones. Similar phenol-to-dienone isomerization of estrones occurs with HSO$_3$F–SbF$_5$ as catalyst, accompanied by major dehydrogenation products (eq 10).[23]

HSO$_3$F–SbF$_5$, 0 °C (9)

HSO$_3$F–SbF$_5$, 0 °C; 12% + 62% (10)

endo-Trimethylenenorbornane isomerizes to its *exo* isomer in quantitative yield at 0 °C under HSO$_3$F–SbF$_5$ catalysis (eq 11).[24] The isomerization will proceed further to adamantane when excess superacid is used (HSO$_3$F–SbF$_5$:substrate ratio = 1:3). For the rearrangement of trimethylenenorbornane to adamantane, HSO$_3$F–SbF$_5$ is, however, less efficient than CF$_3$SO$_3$H–SbF$_5$ or CF$_3$SO$_3$H–B(OSO$_2$CF$_3$)$_3$.[25]

HSO$_3$F–SbF$_5$, 0 °C to rt, 98% (11)

Alkylation. Benzene, alkylbenzenes, and halobenzenes undergo alkylation with 2,2,4-trimethylpentane in the presence of HSO$_3$F–SbF$_5$ at temperatures as low as −30 °C with good selectivity (eq 12).[26]

HSO$_3$F–SbF$_5$, –30 °C; 72:17:7 (12)

Cyclization. HSO$_3$F–SbF$_5$ has been used as a high-acidity catalyst for the cyclization of acyclic isoprenoids at low temperatures. The reaction course and products of the cationic cyclization depend on the acidity of the catalyst and the structural differences in the substrates. Structural changes may also lead to dramatic changes in the reaction course and products. While geranate esters (eq 13)[27] and pseudoionones (eq 14)[28] are cyclized to monocyclic derivatives, geraniol or nerol (eq 15)[29] and geranylacetone (eq 16)[30] give bicyclic ethers.

1. $HSO_3F–SbF_5$, SO_2ClF, −100 °C; 2. MeOH, 74% (13)

1. $HSO_3F–SbF_5$, SO_2ClF, −120 °C; 2. MeOH, 40% (14)

1. $HSO_3F–SbF_5$, −100 °C; 2. MeOH, 50% (15)

$HSO_3F–SbF_5$, SO_2, −65 °C, 60% (16)

Sulfonation and Fluorosulfonation. Aromatic compounds react with HSO_3F to give arenesulfonyl fluorides.[31] When the reactions are carried out in the presence of variable amounts of SbF_5, significant amounts of diaryl sulfones are obtained (eq 17).[32] In this example, the yields of arenesulfonyl fluorides decreased with increasing amounts of SbF_5. In contrast, the yields of diaryl sulfones first increased and then decreased with increasing amounts of SbF_5. The highest yield of diphenyl sulfone was obtained when the molar ratio of SbF_5:benzene was 1:1.5. Diaryl sulfones were also obtained in high yield from toluene, xylenes, and 1,2,4-trimethylbenzene under similar conditions. Sulfonation of fluoro-, chloro-, and bromobenzenes required higher molar ratios of SbF_5:arene (1:3.5) to obtain good yields of the corresponding diaryl sulfones.

$HSO_3F–SbF_5$, 0–50 °C (17)

	sulfone	SO_3F
R = H	94%	4%
R = Me	95%	1%
R = F	83%	0%

Formation of Aromatic Sulfoxides. By treatment with $HSO_3F–SbF_5$ (1:1) and ***Sulfur Dioxide***, alkylbenzenes, halobenzenes, and alkylhalobenzenes were converted to their corresponding diaryl sulfoxides along with small amounts of diaryl sulfides as minor products (eq 18).[33] In the absence of SO_2, aryl sulfone formation is the dominant process, although sulfoxide is also formed. Unsymmetrical (mixed) sulfoxides can be prepared by addition of one molar equivalent of an arene to the solution of the second arene and Magic Acid–SO_2 in Freon at low temperatures.

$HSO_3F–SbF_5$, SO_2, −30 to 0 °C (18)

R = Me, 87%; F, 55%

Formylation and Carboxylation. Formylation of aromatic compounds such as benzene, toluene, xylenes, mesitylene, indan, tetralin, and halobenzenes is achieved in $HSO_3F–SbF_5$ under atmospheric CO pressure at 0 °C (eq 19).[34] However, in the cases of alkylbenzenes, both formylation and sulfonation took place under these reaction conditions to give alkylbenzaldehydes and formylalkylbenzenesulfonyl fluorides, as well as small amounts of alkylbenzenesulfonyl fluorides and bis(alkylphenyl) sulfones. With benzene and halobenzenes, because of their lower reactivity only aldehydes were produced.

$HSO_3F–SbF_5$, CO, 0 °C to rt (19)

R = alkyl or halogen

Saturated hydrocarbons, including branched and unbranched chain alkanes as well as cycloalkanes, react with carbon monoxide in the presence of ***Copper(I) Oxide*** in $HSO_3F–SbF_5$ to afford tertiary and secondary carboxylic acids in high yield (eq 20).[35] The reaction proceeds at 0 °C under 1 atm CO. In some cases the reaction involves cleavage of C–C bonds and isomerization of the intermediate carbocations.

1. CO, Cu_2O, $HSO_3F–SbF_5$, 0 °C; 2. H_2O, 45% (20)

Oxyfunctionalization of Hydrocarbons. When treated with ***Ozone***[36] or ***Hydrogen Peroxide*** (98%)[37,38] under Magic Acid catalysis, alkanes, including methane, ethane, butanes, and higher alkanes as well as haloalkanes,[39] undergo electrophilic oxygenation followed by carbon-to-oxygen alkyl group migration giving, via alkoxycarbenium ions, ketones and alcohols (eq 21). Aliphatic alcohols, ketones, and aldehydes react with ozone in Magic Acid solution to give bifunctional oxygenated derivatives such as diketones, hydroxyl ketones, and glycols (eq 22).[40] The relative reactivity of σ-bonds in alkanes with protonated ozone was found to be $R_3C–H > R_2(H)C–H > R(H_2)C–H > C–C$.[36]

+ O_3 or H_2O_2 → 1. $HSO_3F–SbF_5$, −78 to 0 °C; 2. H_2O → =O + MeOH (21)

O_3, Magic Acid, SO_2ClF, −40 °C (22)

Aromatic compounds such as benzene, alkylbenzenes, and halobenzenes can be directly oxygenated with hydrogen peroxide in Magic Acid or other superacids, giving phenols (eq 23).[41] The phenols formed are protonated by the superacids and thus are deactivated against further electrophilic attack or oxidation. When naphthalene was treated with hydrogen peroxide in Magic Acid at −78 °C, 2-naphthol was obtained (92% regioselectivity) along with small amounts of dihydroxynaphthalenes (eq 24).[42] Unlike phenol derivatives, naphthols can be further hydroxylated with hydrogen peroxide in superacid systems to dihydroxynaphthalenes, since the unprotonated ring of the protonated naphthols can still be attacked by the electrophilic hydroxylating agent.

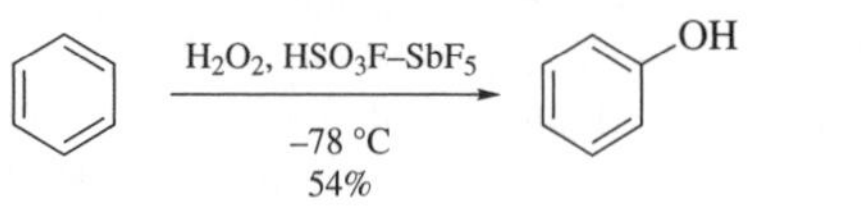

H_2O_2, $HSO_3F–SbF_5$; −78 °C; 54% (23)

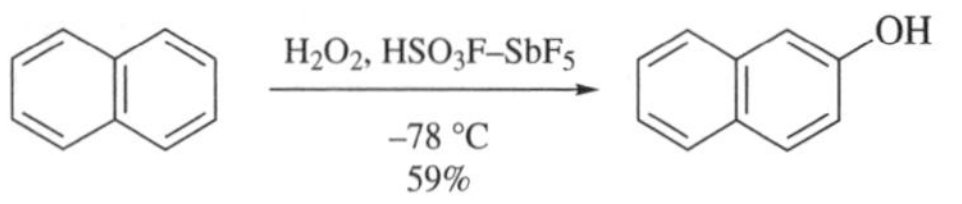

H_2O_2, $HSO_3F–SbF_5$; −78 °C; 59% (24)

Miscellaneous Reactions. In the presence of $HSO_3F–SbF_5$, ***Peroxydisulfuryl Difluoride*** ($S_2O_6F_2$) reacts smoothly with 1,1,2-trichlorotrifluoroethane at rt to give 1,2-dichlorotrifluoroethyl fluorosulfate (eq 25).[43] $S_2O_6F_2$ alone does not react with 1,1,2-trichlorotrifluoroethane even at 150 °C, while in the presence of HSO_3F the reaction occurs only at temperatures higher than 150 °C.

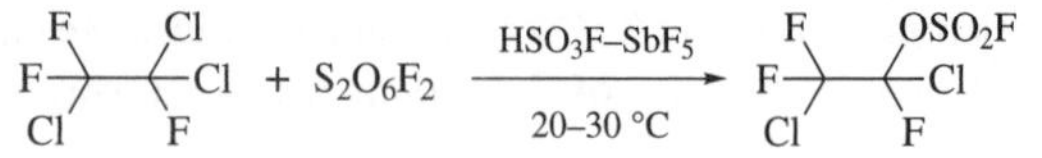

$+ S_2O_6F_2$; $HSO_3F–SbF_5$; 20–30 °C (25)

Polyfluoroarenes, such as m-$H_2C_6F_4$, m-$O_2NC_6F_4H$, m- and p-BrC_6F_4H, (p-$HC_6F_4)_2$, C_6F_5H, 1,3,5-$F_3C_6H_3$, and m-$FSO_2C_6F_4H$, have been thallated by ***Thallium(III) Trifluoroacetate*** in $HSO_3F–SbF_5$. The thallated products can be converted to polyfluoroiodoarenes by treatment with aqueous ***Sodium Iodide*** (eq 26).[44]

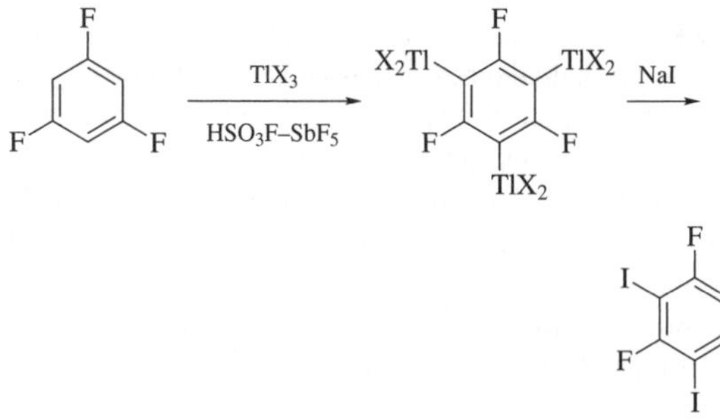

TlX_3, $HSO_3F–SbF_5$; NaI (26)

Although $HSO_3F–SbF_5$ is known to be a strong oxidizing system, it has been reported that octafluoroazoxybenzene was reduced by $HSO_3F–SbF_5$ at 20 °C to octafluoroazobenzene in quantitative yield (eq 27).[45] The mechanism of the process is still unclear.

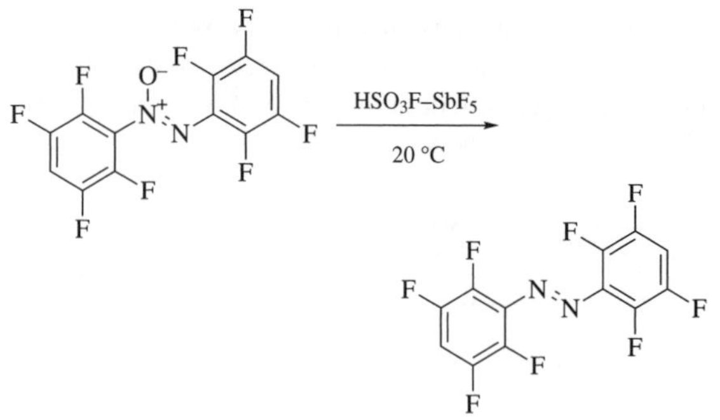

$HSO_3F–SbF_5$; 20 °C (27)

Isoalkanes are brominated by Br_2 in $HSO_3F–SbF_5$ to yield mono-, di-, and tribromoalkanes (eq 28).[46] Cleavage of C–C bonds occurs when isooctane is reacted under similar conditions, leading to butyl bromides.

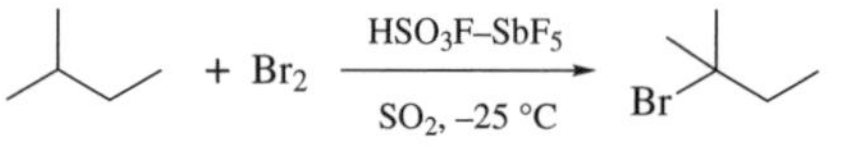

$+ Br_2$; $HSO_3F–SbF_5$; SO_2, −25 °C (28)

Fluorinative dediazoniation of an arenediazonium salt occurred at 0 °C in Magic Acid; (eq 29)[47] however, fluorosulfonation accompanied the dediazoniation.

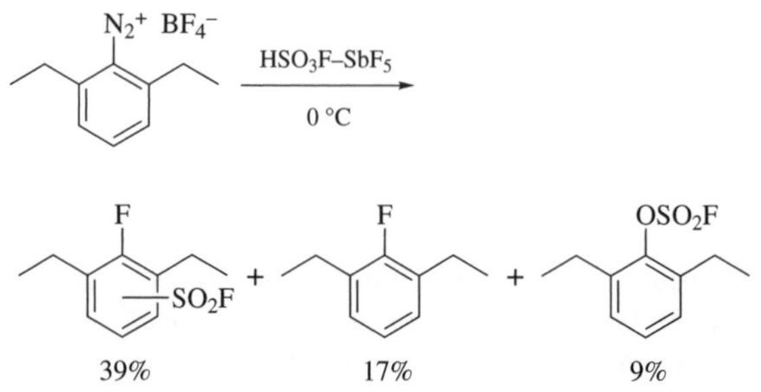

$N_2^+ BF_4^-$; $HSO_3F–SbF_5$; 0 °C; 39% + 17% + 9% (29)

Related Reagents. Antimony(V) Fluoride; Fluorosulfuric Acid; Hydrogen Fluoride; Hydrogen Fluoride-Antimony(V) Fluoride.

---

1. Olah, G. A.; Prakash, G. K. S.; Sommer, J. *Superacids*; Wiley: New York, 1985.
2. Gillespie, R. J.; Peel, T. E. *JACS* **1973**, *95*, 5173.
3. Gold, V.; Laali, K.; Morris, K. P.; Zdunek, L. Z. *CC* **1981**, 769.
4. For reviews, see (a) Hanack, M. *MOC* **1990**, *E19C*. (b) Prakash, G. K. S.; Rawdah, T. N.; Olah, G. A. *AG(E)* **1983**, *22*, 390.
5. Olah, G. A.; Comisarow, M. B.; Cupas, C. A.; Pittman, C. U., Jr. *JACS* **1965**, *87*, 2998.
6. (a) Prakash, G. K. S.; Fung, A. P.; Rawdah, T. N.; Olah, G. A. *JACS* **1985**, *107*, 2920. (b) Olah, G. A.; Reddy, V. P.; Lee, G.; Casanova, J.; Prakash, G. K. S. *JOC* **1993**, *58*, 1639.
7. Olah, G. A.; Namanworth, E. *JACS* **1966**, *88*, 5327.
8. (a) Olah, G. A.; Bollinger, J. M.; Cupas, C. A.; Lucas, J. *JACS* **1967**, *89*, 2692. (b) Olah, G. A.; Prakash, G. K. S.; Sommer, J. *Science* **1979**, *206*, 13.
9. Olah, G. A.; Sommer, J. *JACS* **1968**, *90*, 927.
10. (a) Olah, G. A.; Pittman, C. U., Jr.; Namanworth, E.; Comisarow, M. B. *JACS* **1966**, *88*, 5571. (b) Sorensen, T. S.; Rajeswari, K. *JACS* **1971**, *93*, 4222.
11. Hogeveen, H.; Kwant, P. W. *JACS* **1974**, *96*, 2208.
12. Olah, G. A.; O'Brien, D. H. *JACS* **1967**, *89*, 1725.

13. Olah, G. A.; O'Brien, D. H.; Pittman, C. U., Jr. *JACS* **1967**, *89*, 2996.
14. (a) Olah, G. A.; White, A. M. *JACS* **1969**, *91*, 5801. (b) Olah, G. A.; Donovan, D. J. *JACS* **1977**, *99*, 5026.
15. (a) Childs, R. F.; Zeya, M. *CJC* **1975**, *53*, 3425. (b) Olah, G. A.; Prakash, G. K. S.; Liang, G. *JOC* **1977**, *42*, 661. (c) Olah, G. A.; Pittman, C. U., Jr.; Sorensen, T. S. *JACS* **1966**, *88*, 2331.
16. Olah, G. A.; Mateescu, G. D.; Mo, Y. K. *JACS* **1973**, *95*, 1865.
17. (a) Prakash, G. K. S.; Rawdah, T. N.; Olah, G. A. *AG(E)* **1983**, *22*, 390. (b) Laali, K. K.; Hansen, P. E. *JOC* **1991**, *56*, 6795. (c) Laali, K. K.; Gelerinter, E; Filler, R. *JFC* **1991**, *53*, 107.
18. Yoshino, A.; Takahashi, K.; Sone, T. *BCJ* **1992**, *65*, 3228.
19. (a) McGarrity, J. F.; Cox, D. P. *JACS* **1983**, *105*, 3961. (b) Olah, G. A.; White, A. M. *JACS* **1968**, *90*, 6087. (c) Olah, G. A.; O'Brien, D. H.; Calin, M. *JACS* **1967**, *89*, 3582. (d) Olah, G. A.; White, A. M.; O'Brien, D. H. *CRV* **1970**, *70*, 561. (e) Olah, G. A.; Nakajima, T.; Prakash, G. K. S. *AG(E)* **1980**, *19*, 811.
20. (a) Olah, G. A.; Lukas, J. *JACS* **1967**, *89*, 4739. (b) Olah, G. A.; Lukas, J. *JACS* **1968**, *90*, 933.
21. (a) van der Lugt, W. T. A. M.; Buck, H. M.; Oosterhoff, L. J. *T* **1968**, *24*, 4941. (b) Olah, G. A.; Singh, B. P. *JOC* **1983**, *48*, 4830. (c) Prakash, G. K. S.; Fessner, W.-D.; Olah, G. A. *JACS* **1989**, *111*, 746.
22. Coustard, J. M.; Jacquesy, J. C. *TL* **1972**, 1341.
23. Gesson, J. P.; Jacquesy, J. C.; Jacquesy, R. *TL* **1971**, 4733.
24. Jacquesy, J. C.; Jacquesy, R.; Moreau, S.; Patoiseau, J. F. *CC* **1973**, 785.
25. Olah, G. A.; Farooq, O. *JOC* **1986**, *51*, 5410.
26. Kroeger, C. F.; Miethchen, R.; Mann, H.; Hoffmann, K.; Wiechert, K. *JPR* **1978**, *320*, 881.
27. Gavrilyuk, O. A.; Korchagina, D. V.; Bagryanskaya, I. Yu.; Gatilov, Yu. V.; Kron, A. A.; Barkhash, V. A. *ZOR* **1987**, 2124.
28. Gavrilyuk, O. A.; Korchagina, D. V.; Gatilov, Yu. V.; Mamatyuk, S. A.; Osadchii, S. A.; Fisher, E. A.; Dubovenko, Z. V.; Barkhash, V. A. *ZOR* **1987**, 457.
29. Gavrilyuk, O. A.; Korchagina, D. V.; Barkhash, V. A. *ZOR* **1989**, 2237.
30. Mustafaeva, M. T.; Smit, V. A.; Semenovskii, A. V.; Kucherov, V. F. *IZV* **1973**, 1151.
31. (a) Renoll, M. W. *JACS* **1942**, *64*, 1489. (b) Baker, W.; Coates, G. E.; Glockling, F. *JCS* **1951**, 1376. (c) Baker, B. R.; Cory, M. *JMC* **1971**, *14*, 119.
32. Tanaka, M.; Souma, Y. *JOC* **1992**, *57*, 3738.
33. Laali, K. K.; Nagvekar, D. S. *JOC* **1991**, *56*, 1867.
34. Tanaka, M.; Iyoda, J.; Souma, Y. *JOC* **1992**, *57*, 2677.
35. Souma, Y.; Sano, H. *BCJ* **1976**, 3335.
36. Olah, G. A.; Yoneda, N.; Parker, D. G. *JACS* **1976**, *98*, 5261.
37. Olah, G. A.; Yoneda, N.; Parker, D. G. *JACS* **1977**, *99*, 483.
38. Yoneda, N.; Olah, G. A. *JACS* **1977**, *99*, 3113.
39. Olah, G. A.; Yoneda, N.; Parker, D. G. *JACS* **1976**, *98*, 2251.
40. Olah, G. A.; Yoneda, N.; Ohnishi, R. *JACS* **1976**, *98*, 7341.
41. Olah, G. A.; Ohnishi, R. *JOC* **1978**, *43*, 865.
42. Olah, G. A.; Keumi, T.; Lecoq, J. C.; Fung, A. P.; Olah, J. A. *JOC* **1991**, *56*, 6148.
43. Fokin, A. V.; Studnev, Y. N.; Rapkin, A. I.; Matveenko, V. I. *IZV* **1985**, 715.
44. Deacon, G. B.; Smith, R. N. M. *AJC* **1982**, 1587.
45. Furin, G. G.; Andreevskaya, O. I.; Rezvukhin, A. I.; Yakobson, G. G. *JFC* **1985**, *28*, 1.
46. Halpern, Y. *Isr. J. Chem.* **1976**, *13*, 99.
47. Laali, K.; Szele, I.; Zollinger, H. *HCA* **1983**, 1737.

George A. Olah, G. K. Surya Prakash, Qi Wang & Xing-Ya Li
University of Southern California, Los Angeles, CA, USA

Formic Acid[1]

[64-18-6] CH_2O_2 (MW 46.03)

(formation of formate esters,[3] amides;[7] reductions;[10] transfer hydrogenation;[12] rearrangements[22])

Alternate Name: methanoic acid.

Physical Data: strongest of the simple organic acids; pK_a 3.77 (4.77 for acetic acid). Pure acid, mp 8.4 °C; bp 100.7 °C, 50 °C/120 mmHg, 25 °C/40 mmHg. Formic acid and water form an uncommon maximum boiling azeotrope, bp 107.3 °C, containing 77.5% acid. The dielectric constant of formic acid is 10 times greater than acetic acid.

Solubility: misc water in all proportions; misc EtOH, ether; mod sol C_6H_6.

Form Supplied in: commercially available as 85–95% aqueous solutions and as glacial formic acid, containing 2% water.

Analysis of Reagent Purity: formic acid is determined by titration with base. If other acids are present, formic acid content can be determined by a redox titration based on oxidation with potassium permanganate. Methods for analysis of trace organic and inorganic materials are presented.[2,23]

Purification: fractional distillation in vacuo; dehydration over $CuSO_4$ or boric anhydride.[24]

Handling, Storage, and Precautions: the strongly acidic nature of formic acid is the primary safety concern. Contact with the skin will cause immediate blistering. Immediately treat affected areas with copious amounts of water. Do not use dilute base solutions as a first treatment. Formic acid has a large heat of solution; the combined heat of neutralization and dilution will lead to thermal burns. Eye protection, gloves, and a chemical apron should be worn during all operations with concentrated formic acid. Volatile; vapors will cause intense irritation to mouth, nose, eyes, skin, and upper respiratory tract. Use of an appropriate NIOSH/MSHA respirator is recommended. Use in a fume hood.

During storage, glacial formic acid decomposes to form water and carbon monoxide. Pressure can develop in sealed containers and may result in rupture of the vessel. Ventilation should be provided to prevent the buildup of carbon monoxide in storage areas. Storage temperatures above 30 °C should be avoided.

Formic acid is incompatible with strong oxidizing reagents, bases, and finely powdered metals, furfuryl alcohol, and thallium nitrate. Contact with conc sulfuric acid will produce carbon monoxide from decomposition.

Formation of Formate Esters. Formic acid will esterify primary, secondary, and tertiary alcohols in high yield (eq 1).[3] The reaction is autocatalytic due to the high acidity of formic acid. The equilibrium position of this reaction is closer to completion than for other carboxylic acids.

Formate esters can also be produced during acid-catalyzed rearrangements (eq 2),[4] by addition to alkenes (eq 3),[5] or by ***1,3-Dicyclohexylcarbodiimide*** coupling (eq 4).[6]

Formation of Amides. Most amines react with formic acid to produce the expected amide in high yield.[7] Reaction with diamines is an important reaction for the formation of heterocyclic compounds, including benzimidazoles (eq 5)[8] and triazoles (eq 6).[9]

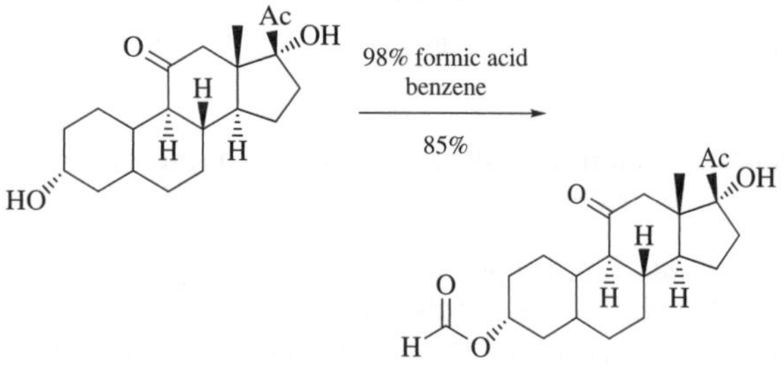

98% formic acid, benzene, 85% (1)

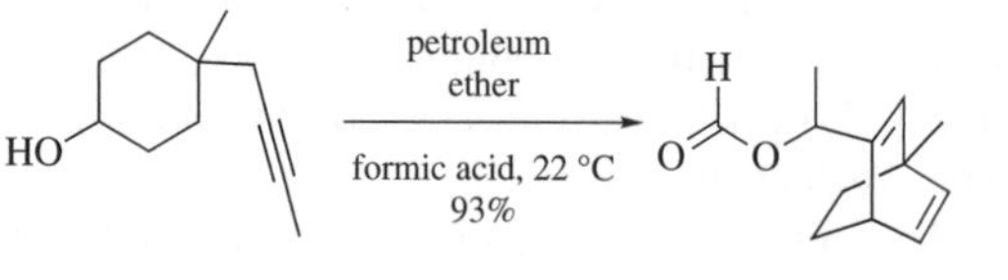

petroleum ether, formic acid, 22 °C, 93% (2)

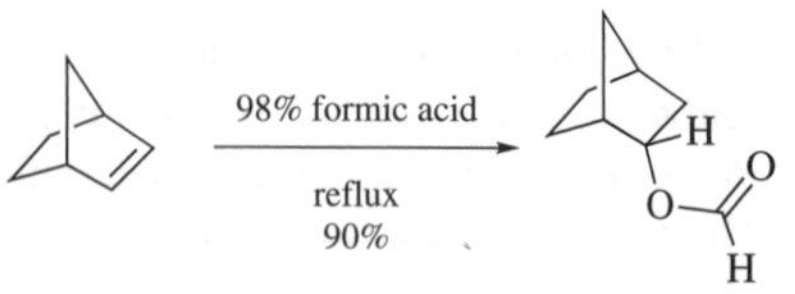

98% formic acid, reflux, 90% (3)

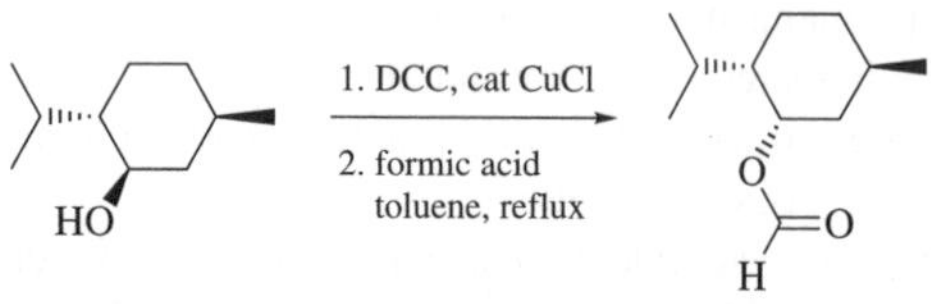

1. DCC, cat CuCl; 2. formic acid, toluene, reflux (4)

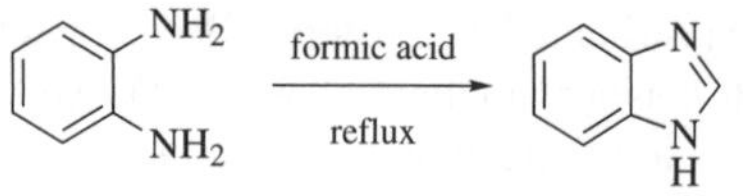

formic acid, reflux (5)

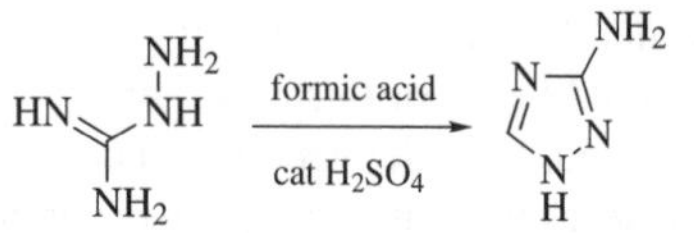

formic acid, cat H_2SO_4 (6)

Reductions with Formic Acid. Formic acid is unique among the simple organic acids in its ability to react as a reducing agent. Ketones are reduced and converted to primary amines by reaction with ammonia and formic acid (eq 7).[10] Ketones or aldehydes will react with formic acid and primary or secondary amines to produce secondary or tertiary amines (eq 8).[10a,11]

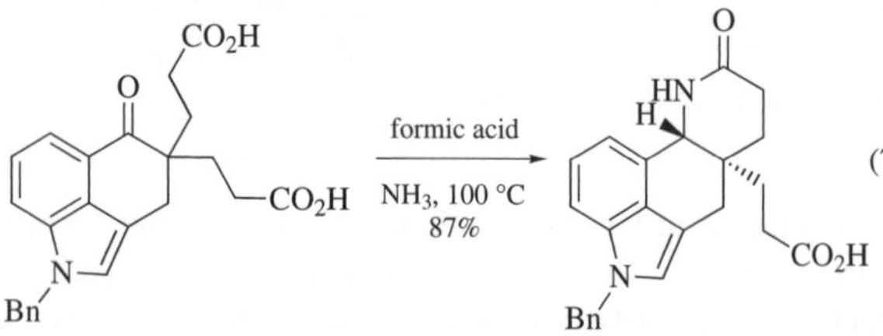

formic acid, NH_3, 100 °C, 87% (7)

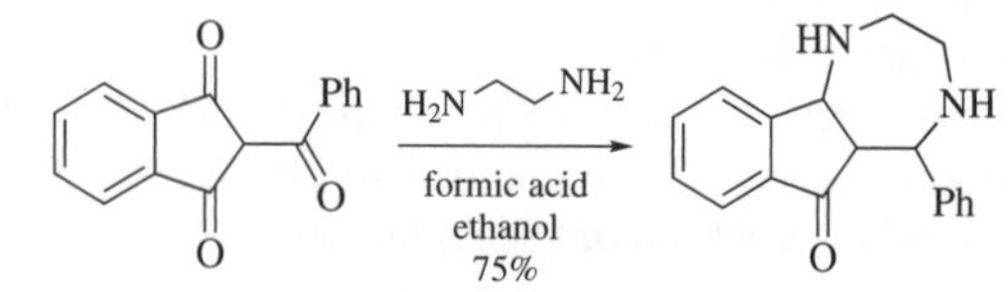

H_2N–CH_2CH_2–NH_2, formic acid, ethanol, 75% (8)

Catalytic Transfer Hydrogenation.[12] Catalytic transfer hydrogenation uses a metal catalyst and an organic hydrogen donor as a stoichiometric reducing agent. This is a useful laboratory alternative to normal catalytic reduction, as the use of flammable hydrogen gas is avoided. Formic acid and amine formates are common hydrogen donors. Formic acid has been used to reduce α,β-unsaturated aldehydes and acids[13] to the saturated compounds (eq 9). Benzyl ethers (eq 10)[14] and benzyl amines (eq 11)[15] are cleaved by transfer hydrogenation with formic acid. A variety of nitrogen functions are reduced including the nitro group,[16] azo group,[17] hydrazines,[18] and enamines.[19] The resulting amines will be formylated under the reaction conditions; use of water or alcohol as solvent will limit formylation. If the reduction results in a suitable diamine, formylation will lead to heterocycle formation (eq 12).[16a] Aromatic halides are reduced to the aromatic hydrocarbon.[20]

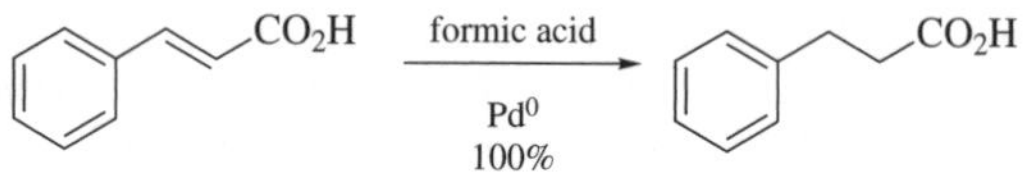

formic acid, Pd^0, 100% (9)

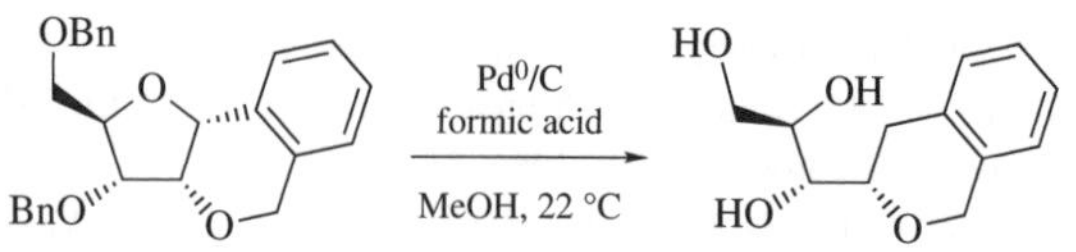

Pd^0/C, formic acid, MeOH, 22 °C (10)

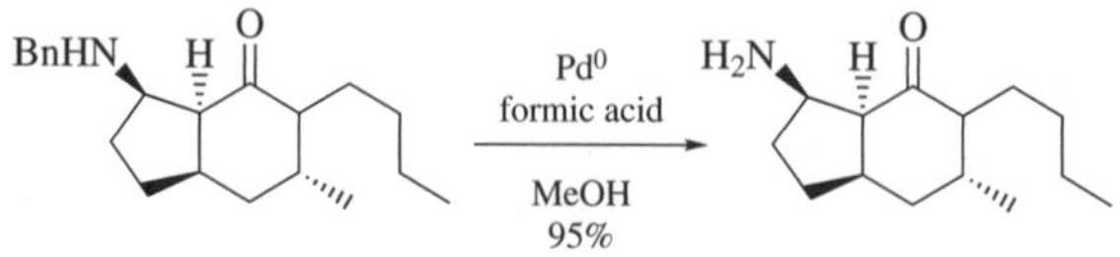

Pd^0, formic acid, MeOH, 95% (11)

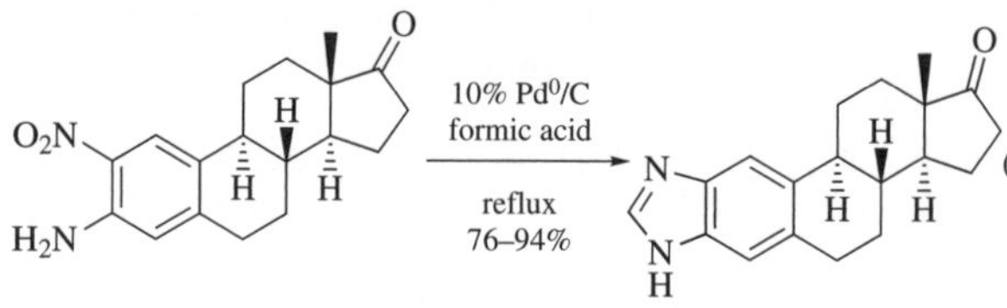

10% Pd^0/C, formic acid, reflux, 76–94% (12)

The Rupe Rearrangement. Tertiary propargyl alcohols isomerize to α,β-unsaturated ketones in the Rupe rearrangement (eq 13).[21] The reaction has been reviewed.[22] Formic acid is the most common catalyst employed.

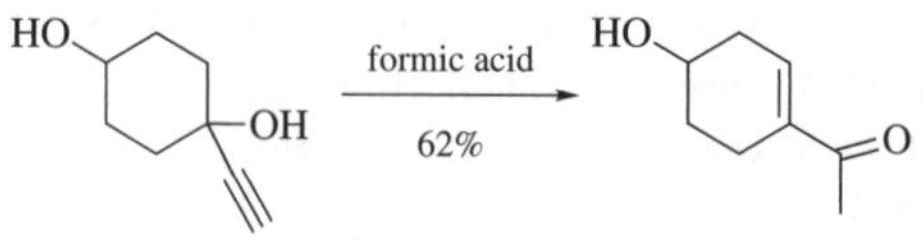

formic acid, 62% (13)

Related Reagents. For uses of other carboxylic acids in synthesis, see Acetic Acid; Acrylic Acid; Ammonium Formate; Glyoxylic Acid; Methanesulfonic Acid; Oxalic Acid and Trifluoroacetic Acid; Palladium-Triethylamine-Formic Acid.

---

1. Gibson, H. W. *CRV* **1969**, *69*, 673.
2. *Encyclopedia of Industrial Chemical Analysis*; Snell, F. D.; Etter, L. S., Eds.; Interscience: New York, 1971; Vol. 13, pp 125–131.
3. Hilscher, J.-C. *CB* **1981**, *114*, 389.
4. Kozar, L. G.; Clark, R. D.; Heathcock, C. H. *JOC* **1977**, *42*, 1386.

5. Kleinfelter, D. C.; Schleyer, P. v. R. *OS* **1962**, *42*, 79; *OSC* **1973**, *5*, 852.
6. Kaulen, J. *AG(E)* **1987**, *26*, 773.
7. Fieser, L. F.; Jones, J. E. *OSC* **1955**, *3*, 590.
8. (a) Wagner, E. C.; Millett, W. H. *OSC* **1943**, *2*, 65. (b) Mathias, L. J.; Overberger, C. G. *SC* **1975**, *5*, 461.
9. Elion, G. B.; Lange, W. H.; Hitchings, G. H. *JACS* **1956**, *78*, 2858.
10. (a) Moore, M. L. *OR* **1949**, *5*, 301. (b) Stoll, A. P.; Niklaus, P.; Troxler, F. *HCA* **1971**, *54*, 1988.
11. Mosher, W. A.; Piesch, S. *JOC* **1970**, *35*, 1026.
12. (a) Johnstone, R. A. W.; Wilby, A. H. *CRV* **1985**, *85*, 129. (b) Brieger, G.; Nestrick, T. J. *CRV* **1974**, *74*, 567.
13. (a) Elamin, B.; Park, J.-W.; Means, G. E. *TL* **1988**, *29*, 5599. (b) Cortese, N. A.; Heck, R. F. *JOC* **1978**, *43*, 3985.
14. (a) Araki, Y.; Mokubo, E.; Kobayashi, N.; Nagasawa, J. *TL* **1989**, *30*, 1115. (b) Rao, V. S.; Perlin, A. S. *Carbohydr. Res.* **1980**, *83*, 175.
15. (a) Roush, W. R.; Walts, A. E. *JACS* **1984**, *106*, 721. (b) Wang, C.-L. J.; Ripka, W. C.; Confalone, P. N. *TL* **1984**, *25*, 4613. (c) El Amin, B.; Anantharamaiah, G. M.; Royer, G. P.; Means, G. E. *JOC* **1979**, *44*, 3442.
16. (a) Leonard, N. J.; Morrice, A. G.; Sprecker, M. A. *JOC* **1975**, *40*, 356. (b) Morrice, A. G.; Sprecker, M. A.; Leonard, N. J. *JOC* **1975**, *40*, 363. (c) Entwistle, I. D.; Jackson, A. E.; Johnstone, R. A. W.; Telford, R. P. *JCS(P1)* **1977**, 443.
17. (a) Taylor, E. C.; Barton, J. W.; Osdene, T. S. *JACS* **1958**, *80*, 421. (b) Moore, J. A.; Marascia, F. J. *JACS* **1959**, *81*, 6049.
18. Schneller, S. W.; Christ, W. J. *JOC* **1981**, *46*, 1699.
19. Kikugawa, Y.; Kashimura, M. *S* **1982**, 785.
20. Pandey, P. N.; Purkayastha, M. L. *S* **1982**, 876.
21. (a) Newman, M. S.; Goble, P. H. *JACS* **1960**, *82*, 4098. (b) Takeshima, T. *JACS* **1953**, *75*, 3309.
22. Swaminathan, S.; Narayanan, K. V. *CRV* **1971**, *71*, 429.
23. *Reagent Chemicals: American Chemical Society Specifications*, 8th ed.; American Chemical Society: Washington, 1993; pp 348–350.
24. Perrin, D. D.; Armarego, W. L. F. *Purification of Laboratory Chemicals*, 3rd ed.; Pergamon: New York, 1988; p 185.

Kirk F. Eidman
Scios Nova, Baltimore, MD, USA

Hexamethyldisilazane

$Me_3Si-NH-SiMe_3$

[999-97-3] $C_6H_{19}NSi_2$ (MW 161.44)

(selective silylating reagent;[1] aminating reagent; nonnucleophilic base[2])

Alternate Name: HMDS.
Physical Data: bp 125 °C; *d* 0.765 g cm^{-3}.
Solubility: sol acetone, benzene, ethyl ether, heptane, perchloroethylene.
Form Supplied in: clear colorless liquid; widely available.
Purification: may contain trimethylsilanol or hexamethyldisiloxane; purified by distillation at ambient pressures.
Handling, Storage, and Precautions: may decompose on exposure to moist air or water, otherwise stable under normal temperatures and pressures. Harmful if swallowed, inhaled, or absorbed through skin. Fire hazard when exposed to heat, flames, or oxidizers. Use in a fume hood.

Silylation. Alcohols,[3] amines,[3] and thiols[4] can be trimethylsilylated by reaction with hexamethyldisilazane (HMDS). Ammonia is the only byproduct and is normally removed by distillation over the course of the reaction. Hydrochloride salts, which are typically encountered in silylation reactions employing chlorosilanes, are avoided, thereby obviating the need to handle large amounts of precipitates. Heating alcohols with hexamethyldisilazane to reflux is often sufficient to transfer the trimethylsilyl group (eq 1).[5] Completion of the reaction is indicated by either a change in the reflux temperature (generally a rise) or by the cessation of ammonia evolution.

$$HN(TMS)_2 + 2\,ROH \xrightarrow{\Delta} 2\,ROTMS + NH_3 \qquad (1)$$

Silylation with HMDS is most commonly carried out with acid catalysis.[5] The addition of substoichiometric amounts of ***Chlorotrimethylsilane*** (TMSCl) to the reaction mixtures has been found to be a convenient method for catalysis of the silylation reaction.[5,6] The catalytically active species is presumed to be hydrogen chloride, which is liberated upon reaction of the chlorosilane with the substrate. Alternatively, protic salts such as ammonium sulfate can be employed as the catalyst.[7] Addition of catalytic ***Lithium Iodide*** in combination with TMSCl leads to even greater reaction rates.[8] Anilines can be monosilylated by heating with excess HMDS (3 equiv) and catalytic TMSCl and catalytic LiI (eq 2). Silylation occurs without added LiI; however, the reaction is much faster in the presence of iodide, presumably due to the in situ formation of a catalytic amount of the more reactive ***Iodotrimethylsilane***.

$$\text{ArNH}_2 + 3\,\text{HMDS} \xrightarrow[\text{cat. LiI, reflux}]{\text{cat. TMSCl}} \text{ArNH(TMS)} \qquad (2)$$

R = H, alkyl, halogen

Hexamethyldisilazane is the reagent of choice for the direct trimethylsilylation of amino acids, for which TMSCl cannot be used due to the amphoteric nature of the substrate.[9] Silylation of glutamic acid with excess hexamethyldisilazane and catalytic TMSCl in either refluxing xylene or acetonitrile followed by dilution with alcohol (methanol or ethanol) yields the derived lactam in good yield (eq 3).[10]

$$HO_2C\text{–CH(NH}_2\text{)–CH}_2\text{CH}_2\text{–}CO_2H \xrightarrow[\text{2. EtOH, 93\%}]{\text{1. HMDS, cat. TMSCl}} \text{pyroglutamic acid (}HO_2C\text{-lactam)} \qquad (3)$$

The efficiency of HMDS-mediated silylations can be markedly improved by conducting reactions in polar aprotic solvents. For example, treatment of methylene chloride solutions of primary alcohols or carboxylic acids at ambient temperatures with HMDS (0.5–1 equiv) in the presence of catalytic amounts of TMSCl (0.1 equiv) gives the corresponding silyl ether and the trimethylsilyl ester, respectively (eq 4).[1] *N*-Silylation of secondary amines occurs in preference to primary alcohols when treated with 1 equiv of HMDS and 0.1 equiv TMSCl (eq 5). The silylation of secondary amines cannot be effected in the absence of solvent.[5] Secondary and tertiary alcohols can also be silylated at ambient temperatures in dichloromethane with HMDS and TMSCl mixtures; however, stoichiometric quantities of the silyl chloride are required. Catalysis by ***4-Dimethylaminopyridine*** (DMAP) is necessary for the preparation of tertiary silyl ethers.

$$\text{triol (primary OH, secondary OH, tertiary OH)} \xrightarrow[\text{TMSCl (0.1 equiv), } CH_2Cl_2, 0\,°C]{\text{HMDS (1.2 equiv)}} \text{primary OTMS, OH, OH} \qquad (4)$$

$$\text{MeNH–CH(CH}_3\text{)CH}_2\text{CH}_2\text{OH} \xrightarrow[\text{TMSCl (0.1 equiv), } CH_2Cl_2, 0\,°C]{\text{HMDS (1.2 equiv)}} \text{MeN(TMS)–CH(CH}_3\text{)CH}_2\text{CH}_2\text{OH} \qquad (5)$$

DMF is a useful solvent for HMDS-induced silylation reactions, and reaction rates 10–20 times greater than those carried out in pyridine have been reported.[11] DMSO is also an excellent solvent; however, a cosolvent such as 1,4-dioxane is required to provide miscibility with HMDS.[12]

Imidazole (ImH) catalyzes the silylation reaction of primary, secondary, and tertiary alkanethiols with hexamethyldisilazane.[12] The mechanism is proposed to involve the intermediacy of ***N-(Trimethylsilyl)imidazole*** (ImTMS), since its preparation from hexamethyldisilazane and imidazole to yield 1-(trimethylsilyl)imidazole is rapid.[13] The imidazole-catalyzed reactions of hexamethyldisilazane, however, are more efficient than the silylation reactions effected by ImTMS (eq 6 vs. eq 7) due to reversibility of the latter. Imidazole also catalyzes the reaction of HMDS with ***Hydrogen Sulfide***, which provides a convenient preparation of hexamethyldisilathiane, a reagent which has found utility in sulfur transfer reactions.[14]

HMDS (2 equiv), cat. ImH, reflux, 50% (6)

+ ImTMS ⇌ STMS + ImH (7) 15%

Silyl Enol Ethers. Silylation of 1,3-dicarbonyl compounds can be accomplished in excellent yield by heating enolizable 1,3-dicarbonyl compounds with excess HMDS (3 equiv) and catalytic imidazole (eq 8).[15]

HMDS, cat. ImH, reflux, 96% → TMSO ... O (8)

In combination with TMSI, hexamethyldisilazane is useful in the preparation of thermodynamically favored enol ethers (eq 9).[16] Reactions are carried out at rt or below and are complete within 3 h.

HMDS (1.2 equiv), TMSI (1.1 equiv), pentane → OTMS + OTMS 90:10 (9)

Related thermodynamic enolization control has been observed using metallated hexamethyldisilazide to give the more substituted bromomagnesium ketone enolates.[17] Metallation reactions of HMDS to yield Li, K, and Na derivatives are well known and the resulting nonnucleophilic bases have found extensive applications in organic synthesis.

Amination Reactions. Hexamethyldisilazane is a useful synthon for ammonia in amination reactions. Preparation of primary amides by the reaction of acyl chlorides and gaseous Ammonia, for example, is not an efficient process. Treatment of a variety of acyl halides with HMDS in dichloromethane gives, after hydrolysis, the corresponding primary amide (eq 10).[18] Omitting the hydrolysis step allows isolation of the corresponding monosilyl amide.[19]

COCl → 1. HMDS; 2. MeOH, H_3O^+ → NH_2 (10)

Reductive aminations of ketones with HMDS to yield α-branched primary amines can be effected in the presence of ***Titanium(IV) Chloride*** (eq 11).[20] The reaction is successful for sterically hindered ketones even though HMDS is a bulky amine and a poor nucleophile. The use of ammonia is precluded in these reactions since it forms an insoluble complex with $TiCl_4$.

Ar(C=O)Ar + HMDS → 1. $TiCl_4$, Et_3N; 2. $NaCNBH_3$ → Ar–CH(NH_2)–Ar (11)

The reaction of phenols with diphenylseleninic anhydride and hexamethyldisilazane gives the corresponding phenylselenoimines (eq 12).[21] The products thus obtained can be converted to the aminophenol or reductively acetylated using ***Zinc*** and ***Acetic Anhydride***. The use of ammonia or tris(trimethylsilyl)amine in place of HMDS gives only trace amounts of the selenoimines.

OH → $(PhSeO)_2O$ (1.2 equiv), HMDS (1.2 equiv), 65% → O, NSePh (12)

Related Reagents. *N,O*-Bis(trimethylsilyl)acetamide; Lithium Hexamethyldisilazide; Potassium Hexamethyldisilazide; Sodium Hexamethyldisilazide.

1. Cossy, J.; Pale, P. *TL* **1987**, *28*, 6039.
2. Colvin, E. W. *Silicon in Organic Synthesis*, Butterworths: London, 1981.
3. Speier, J. L. *JACS* **1952**, *74*, 1003.
4. Bassindale, A. R.; Walton, D. R. M. *JOM* **1970**, *25*, 389.
5. Langer, S. H.; Connell, S.; Wender, I. *JOC* **1958**, *23*, 50.
6. Sweeley, C. C.; Bentley, R.; Makita, M.; Wells, W. W. *JACS* **1963**, *85*, 2497.
7. Speier, J. L.; Zimmerman, R.; Webster, J. *JACS* **1956**, *78*, 2278.
8. Smith, A. B., III; Visnick, M.; Haseltine, J. N.; Sprengeler, P. A. *T* **1986**, *42*, 2957.
9. Birkofer, L.; Ritter, A. *AG(E)* **1965**, *4*, 417.
10. Pellegata, R.; Pinza, M.; Pifferi, G. *S* **1978**, 614.
11. Kawai, S.; Tamura, Z. *CPB* **1967**, *15*, 1493.
12. Glass, R. S. *JOM* **1973**, *61*, 83.
13. Birkofer, L.; Richter, P.; Ritter, A. *CB* **1960**, *93*, 2804.
14. Harpp, D. N.; Steliou, K. *S* **1976**, 721.
15. Torkelson, S.; Ainsworth, C. *S* **1976**, 722.
16. (a) Hoeger, C. A.; Okamura, W. H. *JACS* **1985**, *107*, 268. (b) Miller, R. D.; McKean, D. R. *S* **1979**, 730.
17. Kraft, M. E.; Holton, R. A. *TL* **1983**, *24*, 1345.
18. Pellegata R.; Italia, A.; Villa, M. *S* **1985**, 517.
19. Bowser, J. R.; Williams, P. J.; Kuvz, K. *JOC* **1983**, *48*, 4111.
20. Barney, C. L.; Huber, E. W.; McCarthy, J. R. *TL* **1990**, *31*, 5547.
21. Barton, D. H. R.; Brewster, A. G.; Ley, S. V.; Rosenfeld, M. N. *CC* **1977**, 147.

Benjamin A. Anderson
Lilly Research Laboratories, Indianapolis, IN, USA

Hexamethylphosphoric Triamide

[680-31-9] $C_6H_{18}N_3OP$ (MW 179.24)

(high Lewis acid basicity; dipolar aprotic solvent with superb ability to form cation–ligand complexes; can enhance the rates of a wide variety of main group organometallic reactions and influence regio- or stereochemistry; additive in transition metal chemistry; UV inhibitor in poly(vinyl chloride))

Alternate Names: HMPA; hexamethylphosphoramide; hexametapol; hempa; HMPT.

Physical Data: mp 7.2 °C; bp 230–232 °C/740 mmHg; *d* 1.025 g cm^{-3}; mild amine odor.

Solubility: sol water, polar and nonpolar solvents.

Form Supplied in: water-white liquid.

Drying: distilled from CaH_2 or BaO[1] at reduced pressure and stored under N_2 over molecular sieves.

Handling, Storage, and Precautions: has low to moderate acute toxicity in mammals.[2a] Inhalation exposure to HMPA has been shown to induce nasal tumors in rats,[2b] and has been classified under 'Industrial Substances Suspect of Carcinogenic Potential for Man'.[2c] Adequate precautions must be taken to avoid all forms of exposure to HMPA.

Introduction. Hexamethylphosphoric triamide has been used extensively as an additive in organolithium chemistry.[3] It is among the strongest of electron pair donors and is superior to protic solvents in that it solvates the cation much better than the anion.[4] This coordinating ability gives HMPA its unusual chemical properties.[5,6] For instance, HMPA dramatically enhances the rates of a wide variety of organolithium reactions, as well as significantly influencing regio- or stereochemistry. The reactivity or selectivity effects of HMPA are usually rationalized in terms of changes in either aggregation state or ion pair structure.[7] The breaking up of aggregates to form reactive monomers or solvent-separated ion pairs is often invoked.

Organolithium Reagent Solution Structure. The rate of metalation reactions with ***Lithium Diisopropylamide*** as base are significantly increased through the use of HMPA.[8] Treatment of LDA dimer with HMPA causes sequential solvation of the lithium cation, but no significant deaggregation;[8,9] nor does HMPA promote the break up of tetrameric unhindered phenoxides[10,11] or tetrameric MeLi.[12] A chiral bidentate lithium amide,[13] however, was converted from a dimer to a monomer by HMPA, with an increase in reactivity and enantioselectivity in deprotonation (eq 1). HMPA also converts ***Phenyllithium*** dimer into monomer.[12,14] Other aggregated lithium reagents[12] in THF (MeSLi, LiCl) or ether (Ph_2PLi, PhSeLi) are first deaggregated to monomers, and then solvent-separated ion pair species are formed.[12,15] HMPA may exert its reactivity effects by a combination of one or more of the following:

1. lowering the degree of aggregation[16,17] or forming separated ions,[18]
2. increasing reactivity through cation coordination,[17]
3. activating the aggregate through insertion into the aggregate site normally occupied by the anionic fragment,[5,19]
4. promoting triple ion (ate complex) formation.[16a,20]

HMPA; ether or toluene (1)

HMPA can have large effects on equilibria. Phenyllithium reacts with ***Diphenylmercury***, ***Diphenyl Ditelluride***, and iodobenzene in THF to form ate complexes. In THF/HMPA the ate complex formation constants are dramatically higher than in THF.[14,21]

Enolate Formation. The formation of lithium enolates is one instance where HMPA is sometimes needed.[22] The difficult generation of dimethyl tartrate acetonide enolate[23] and subsequent benzylation (eq 2), as well as the double deprotonation of methyl 3-nitropropanoate,[24] become possible (eq 3) with the addition of HMPA (or ***N,N′-Dimethylpropyleneurea***).[25]

LDA, THF, −78 °C; BnBr (2)

Cosolvent (% vol)	Yield (%)
–	<5
HMPA (17)	54
DMPU (33)	52

2 equiv LDA, THF, −78 °C (3)

Cosolvent (% vol)	Yield (%)
–	<5
HMPA (17)	50–85
DMPU (33)	50–85

Enolate Reactivity. Not only is HMPA necessary to the generation of enolates, it is often needed in the electrophilic trapping of enolates (eqs 4–6).[26,31] Studies of the electrophilic trapping of enolates have demonstrated that substantial increases in reaction rates can be achieved through the use of a polar aprotic solvent like HMPA.[4]

HMPA (4)

(5)

(6)

Solvent	Yield
THF	43%
THF–HMPA	93%

The desired [2,3]-sigmatropic rearrangement of a bis-sulfur cyano-stabilized lithium salt did not proceed (eq 7) without the addition of 25% HMPA.[27] Sometimes higher *O*/*C*-alkylation ratios are obtained in THF–HMPA.[28]

(7)

Enolate Stereochemistry. Stereochemical control of an ester enolate Claisen rearrangement was accomplished through stereoselective enolate formation.[29] The enolization of 3-pentanone with LDA afforded predominantly the (*E*)-enolate in THF and the (*Z*)-enolate in THF–HMPA, as shown by chlorotrialkylsilane trapping experiments (eq 8). Similar stereoselectivity (*Z*:*E* = 94:6) was obtained with the dipolar aprotic cosolvent DMPU.[30]

(8)

THF	77	:	23
THF–HMPA	5	:	95

In addition to altering the (*E*/*Z*) isomer ratio of enolates,[17,32] HMPA has a noticeable effect on the metalation of imines and their subsequent alkylation (eq 9).[33] When the metalation (by ***s*-Butyllithium**) of an asymmetric imine is performed in THF, a subsequent alkylation gives about a 1:1 mixture of regioisomers. In the presence of HMPA, however, only the regioisomer due to alkylation at the less-substituted site was observed. A synthetically useful solvent effect for HMPA is also observed in the asymmetric synthesis of trimethylsilyl enol ethers by chiral lithium amide bases.[34] The asymmetric induction in THF can be greatly improved by simply adding HMPA as a cosolvent.

(9)

Carbanion Formation. Often substrates that cannot be metalated by LDA or ***n*-Butyllithium** in THF can be successfully deprotonated by adding HMPA as a cosolvent. Many other weakly acidic C–H acids, e.g. (**1**)–(**6**), can be successfully metalated in the presence of HMPA.[35] HMPA also aids in the formation of dianions[36] and increases the proton abstraction efficiency of ***Sodium Hydride***.[37,38]

(**1**) (**2**) (**3**)

(**4**) (**5**) (**6**)

Carbanion Reactivity. An increase in reaction rate is observed for the reaction of alkynyllithium reagents with alkyl halides[3] and oxiranes[39] (eq 10). The strongly coordinating HMPA probably complexes the lithium cation, thereby increasing the negative charge density on the carbon and creating a much more nucleophilic alkynyl anion. A similar effect is observed for ***(Trimethylstannylmethyl)lithium***, which does not react with oxiranes in THF but in THF–HMPA the reaction proceeds readily.[40]

(10)

Cosolvent (% vol)	Yield (%)
–	<5
HMPA (17)	70
DMPU (50)	59

The decarboxylation[41] of 4-*t*-butyl-1-phenyl-1-carboxycyclohexane with ***Methyllithium*** gave a mixture of axial and equatorial products (eq 11), which was highly dependent on the nature of the solvent at the time of aqueous workup. Axial protonation was favored in ether–HMPA.

(11)

HMPA is also the only dipolar aprotic solvent to be used extensively with organomagnesium compounds.[42] Large effects are observed when HMPA is used as either a solvent or a cosolvent. As examples, HMPA accelerates addition of an allylic organomagnesium compound to aryl-substituted alkenes,[43] addition of Grignard reagents to ***Carbon Monoxide***,[44] and addition of ***Propargylmagnesium Bromide*** to allylic halides to give allene products.[45]

Carbanion Regioselectivity (1,2- vs. 1,4-Addition). The regioselectivity of addition of certain organolithium reagents to α,β-unsaturated carbonyl compounds is affected by the addition of HMPA. In the addition of 2-lithio-2-substituted-1,3-dithiane to cyclohexenone, there was a complete reversal of regioselectivity from 1,2-addition in THF to 1,4-addition with 2 equiv of HMPA present (eq 12).[46]

(12)

Additive	1,2-Addition	1,4-Addition
no HMPA	>99	<1
HMPA (2 equiv)	5	95

Lithium reagents that exhibit kinetic 1,4-addition in HMPA are shown as (7)–(14).[47] These include useful acyl anion equivalents[48] like phenylthio(trimethylsilyl)methyllithium (see ***(Phenylthiomethyl)trimethylsilane***).[49]

(7) **(8)** **(9)** **(10)**

(11) **(12)** **(13)** **(14)**

A carboxy anion equivalent was reported to undergo 1,2-addition in the absence of HMPA;[50] however, with 10 equiv of HMPA present only 1,4-addition was observed (eq 13). The addition of 1 equiv of HMPA promotes conjugate addition of alkyl and phenylthioallyl anions to cyclopentanones (eq 14) through the α-position, whereas in THF alone, irreversible 1,2-addition occurs with both α- and γ-attack.[51] The regioselectivities reported for the addition to cyclic enones of ketene dithioacetal anions[47a,52] or ***t-Butyllithium*** (eq 15)[53] are also influenced by HMPA (and counterion).

(13)

(14)

(15)

Solvent		
ether	100	–
THF	95	5
THF–5% HMPA	65	35
THF–20% HMPA	10	90

Ylide Reactivity. HMPA is used as a cosolvent in the Wittig reaction to increase reaction rate, yield, and stereoselectivity. HMPA functions as a lithium cation-complexing agent and removes LiBr salt from ether solution (LiBr/HMPA complexes form precipitates in ether).[54] Such a 'salt-free' Wittig reaction mixture[55] may be responsible for the high level of *cis*-alkene observed in reactions of nonstabilized ylides with aldehydes in THF or ether with added HMPA[25,56] (eq 16). Similarly, increased (*Z*) selectivity is observed in the Wittig alkenation of 2-oxygenated ketones (eq 17) to generate protected (*Z*)-trisubstituted allylic alcohols.[55,57]

(16)

Cosolvent (% vol)	Yield (%)	(*Z*)		(*E*)
–	46	83	:	17
HMPA (35)	44	92	:	8
DMPU (35)	39	93	:	7

(17)

	THF		THF–HMPA	
R	(*E*):(*Z*)	Yield (%)	(*E*):(*Z*)	Yield (%)
Me	2.6:1	50	8:1	76
Ph	5:1	77	10:1	84
TBS	7.6:1	61	26:1	90
Bz	1.1:1	55	36:1	86

With HMPA, Wittig reactions that give (*E*)-alkenes were also observed (eq 18),[58] as was the directed selectivity of a semistabilized arsonium ylide towards carbonyl compounds. The arsenic ylide was generated from LDA in THF or THF/HMPA solution to give exclusively epoxide (eq 19) or diene (eq 20), respectively.[59]

(18)

Ylide	Solvent	(*E*):(*Z*)	Yield (%)
Y = P	THF	65:35	91
Y = P	THF–HMPA	85:15	87
Y = As	THF	–	–
Y = As	THF–HMPA	100:0	65

(19)

(20)

Nucleophilic Cleavage of Esters and Ethers. The conversion of hindered methyl esters to carboxylic acids, and the demethylation of methyl aryl ethers, can be effectively performed by using HMPA to increase the nucleophilicity of lithium methanethiolate.[60] HMPA (or ***N,N-Dimethylformamide***) will also facilitate the cleavage of methyl aryl ethers and their methylthio analogs by sodium methaneselenolate (see ***Methaneselenol***) to give the phenol or thiophenol, respectively.[60,61] Sodium ethanethiolate (see ***Ethanethiol***) in refluxing DMF or HMPA attacks alkyl aryl selenides to give the corresponding diselenides. The most reactive combination of solvent and halide salt for the decarboxylation of β-keto esters was found to be ***Lithium Chloride***/HMPA, used as part of a stereoselective synthesis of 11-deoxyprostaglandin E_1.[62] In HMPA, the rate of ester cleavage of 2-benzyl-2-methoxycarbonyl-1-cyclopentanone with ***Sodium Cyanide*** is 30 times as fast as the more commonly used DMF.[62a]

Anion Reactivity. HMPA is one of the most potent electron pair donor solvents available for accelerating S_N2 reactions.[3,43a] The formation, for example, of an α-silyl carbanion for use in a Peterson alkenation reaction[63] can be accomplished by the displacement of silicon using ***Sodium Methoxide***[64] or ***Potassium t-Butoxide***[65] in HMPA (eq 21). The increased nucleophilicity of halide ions in the presence of HMPA is seen by the increased rate of silyl-protecting group removal with fluoride ion (***Tetra-n-butylammonium Fluoride***).[66] The substitution of aryl chlorides can be performed using sodium methoxide in HMPA[67] to give anisole derivatives, or by using sodium methanethiolate (MeSNa) (see ***Methanethiol***) in HMPA[61b,68] to generate either aryl methyl sulfides or aryl thiols, depending on the reaction conditions. The increased nucleophilicity of a magnesium alkoxide is demonstrated by the cyclization of a chloro alcohol to a 13-membered cyclic ether (eq 22) upon treatment of the compound with ***Ethylmagnesium Bromide*** in refluxing THF/HMPA.[69]

(21)

(E):(Z) = 52:1

(22)

The conversion of cyclic alkenes to 1,3-cycloalkadienes can be performed through a bromination/dehydrobromination procedure using LiCl/***Lithium Carbonate***/HMPA (eq 23).[70] The dehydrobromination of 2,3-dibromo-3-methyl-1-butanol to generate a vinyl bromide has been accomplished through the use of 2.3 equiv of LDA and 0.5 equiv of HMPA in THF at −78 °C.[71]

(23)

Low-Valent Metal Coordination (Lanthanoids). HMPA is used extensively as a solvent for dissolving-metal reductions.[72] Used as a cosolvent (5–10%), it remarkably accelerates the one-electron transfer reduction of organic halides by ***Samarium(II) Iodide***.[73] The reductions work on a variety of primary, secondary, or tertiary halides, including chlorides which could not be reduced in pure THF (eq 24). The samarium Barbier reaction, which requires hours in refluxing THF, can be performed as a titration in THF–HMPA at rt (eq 25).[74] The SmI_2/THF/HMPA system has recently been used in the deoxygenation of organoheteroatom oxides,[75] reductive dimerization of conjugated acid derivatives,[76] and selective reduction of α,β-unsaturated carbonyl compounds.[77] In addition, SmI_2 was used in a tandem radical cyclization,[78] and has been a useful reagent in Barbier-type reactions.[73b,79] The dramatic acceleration of electron transfer is also observed for ***Ytterbium(0)***.[80]

R–X → R–H (SmI_2, THF–HMPA, near quant.) (24)

X	THF Time	THF Yield	THF–HMPA Time	THF–HMPA Yield
I	6 h	95%	5 min	>95%
Br	2 days	82%	10 min	>95%
Cl	no reaction		8 h	>95%

(25)

Additive	Time	Yield
none	4 h	82%
HMPA	1 min	95%

Transition Metal Coordination. The ability of HMPA to complex to metals and alter reactivity is also expressed in palladium-catalyzed coupling reactions. For example, ethylbenzene can be formed from the Pd^0 catalyzed cross-coupling of ***Benzyl Bromide*** and ***Tetramethylstannane***, with the formation of almost no bibenzyl.[81] The reaction does not proceed in THF, but requires a highly polar solvent like HMPA or ***1-Methyl-2-pyrrolidinone***. Similarly, HMPA is necessary in the Pd^0-catalyzed coupling of acid chlorides with organotin reagents to give ketones. In highly polar solvents like HMPA the transfer of a chiral group from the tin occurs with preferential inversion of configuration (eq 26).[82] The relative rate of transfer of an alkynic or vinyl group to acid chlorides[82a] or aryl iodides[83] is also greatly accelerated by HMPA. It can increase the rate of alkylation of π-allylpalladium chloride by ester enolates,[84] and alter the chemoselectivity by leading to a cyclopropanation reaction instead of an allylic alkylation.[85]

(S)-(–) + PhCOCl → (BzPdL$_2$Cl, HMPA, 65 °C) → (R)-(–) (26)

Hydride Reductions. The reduction of organic compounds by hydride can be influenced by the choice of cosolvent. Cyanoborohydrides (e.g. ***Sodium Cyanoborohydride***) in HMPA provide a mild, effective, and selective reagent system for the reductive displacement of primary and secondary alkyl halides and sulfonate esters in a wide variety of structural types.[86] A ***Sodium Borohydride***/HMPA reagent system was used in the reduction of *N,N*-disulfonamides[87] and in the reduction of dibromides to monobromides.[37a] Similarly, ***Tri-n-butylstannane***/HMPA[88] can be utilized to chemoselectively reduce aldehydes with additional alkene or halide functionality.[89] Finally, the reduction of aldehydes and ketones with hydrosilanes proceeds in the presence of a catalytic amount of Bu_4NF in HMPA.[90] The reaction rate was much lower in DMF than in HMPA, and the reaction yields decreased considerably if less polar solvents like THF or CH_2Cl_2 were used.

Oxidation. The oxidation of acid sensitive alcohols with ***Chromium(VI) Oxide*** in HMPA is one example of the use of HMPA in oxidation reactions.[91] (CAUTION: Do not crush CrO_3 prior to reaction since violent decomposition can occur. The use of DMPU has been reported to have a similar hazardous effect). Recently, a ***Bromine***/$NaHCO_3$/HMPA system[92] was used for the oxidative esterification of alcohols with aldehydes, where the HMPA considerably accelerated the oxidation by bromine and lowered the rate of unwanted halogenation. Epoxidations of alkenes or allylic alcohols have been accomplished using MoO_5·HMPA·pyridine (***Oxodiperoxymolybdenum(pyridine)(hexamethylphosphoric triamide)***; MoOPH).[93]

Effect of HMPA on Protonation. The protonation of (9-anthryl)arylmethyllithium with various oxygen and carbon acids in THF or in THF–HMPA had a significant effect on the product ratio of C-α vs. C-10 protonation (eq 27).[94] Another study[95] found that a nitronate protonation led to mainly one diastereomeric product in a THF solution containing HMPA or DMPU (eq 28). Panek and Rodgers[96] observed stereospecific protonation of 10-*t*-butyl-9-methyl-9-lithio-9,10-dihydroanthracene: >99% *cis* protonation was observed in THF or ether with greater than >99% *trans* protonation observed in HMPA.

Inhibition by HMPA. Finally, it should be added that there are a few cases where HMPA slows the rate of a reaction. Such examples typically involved the inhibition of lithium catalysis by strong coordination of HMPA to lithium.[97] For instance, two-bond ^{13}C–^{13}C NMR coupling in organocuprates is poorly observable in ether or THF at very low temperature.[98] Exchange, however, is slowed in THF/HMPA or THF/12-crown-4, so that coupling is easily observed. The effect of HMPA suggests that Li^+ is involved in the exchange process.

(proton source, THF) (27)

Additive	Proton. at C-α	Proton. at C-10
none	57	43
HMPA (12 equiv)	>95	<5

(BuLi, THF–HMPA or THF–DMPU; PhCHO, acidic ion-exchange resin in MeOH, 75%) (28)

>95% ds

HMPA Substitutes and Analogs. Researchers have searched for an alternative to HMPA. Such a solvent must be stable to polar organometallic compounds and be comparable to HMPA in its many functions. Replacement solvents typically are useful in some applications but have limited value in others. Examples of some useful alternatives are (**15**)–(**19**).[30,42,99]

(**15**) DMPU (**16**) DEA (**17**) NEP

(**18**) (**19**) TES

Chiral analogs of HMPA, (**20**)–(**22**), have been used as ligands in transition metal complexes.[93b,100]

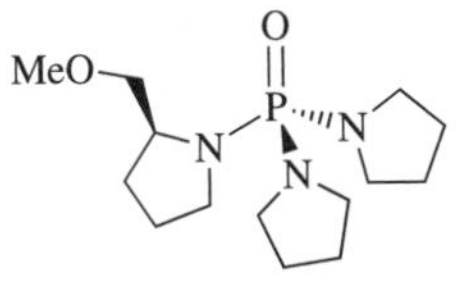

(20) 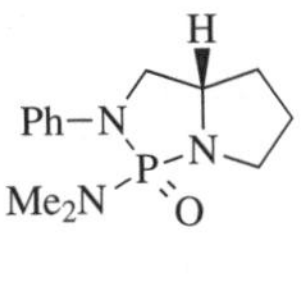 (21)

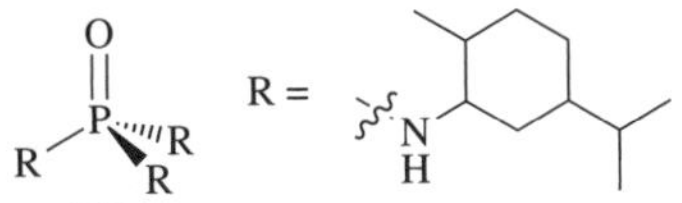

(22)

Related Reagents. N,N-Dimethylformamide; N,N'-Dimethylpropyleneurea; Dimethyl Sulfoxide; Hexamethylphosphoric Triamide–Thionyl Chloride; Lithium Chloride–Hexamethylphosphoric Triamide; 1-Methyl-2-pyrrolidinone; Potassium *t*-Butoxide–Hexamethylphosphoric Triamide; Potassium Hydride–Hexamethylphosphoric Triamide; Potassium Hydroxide–Hexamethylphosphoric Triamide.

---

1. House, H. O.; Lee, T. V. *JOC* **1978**, *43*, 4369.
2. (a) Kimbrough, R. D.; Gaines, T. B. *Nature* **1966**, *211*, 146. Shott, L. D.; Borkovec, A. B.; Knapp, W. A., Jr. *Toxicol. Appl. Pharmacol.* **1971**, *18*, 499. (b) *J. Natl. Cancer Inst.* **1982**, *68*, 157. (c) Mihal, C. P., Jr. *Am. Ind. Hyg. Assoc. J.* **1987**, *48*, 997. American Conference of Governmental Industrial Hygienists: *TLVs-Threshold Limit Values and Biological Exposure Indices for 1986–1987*; ACGIH: Cincinnati, OH; Appendix A2, p 40.
3. (a) Normant, H. *AG(E)* **1967**, *6*, 1046. (b) Normant, H. *RCR* **1970**, *39*, 457.
4. Stowell, J. C. *Carbanions in Organic Synthesis*; Wiley: New York, 1979: House, H. O. *Modern Synthetic Reactions*, 2nd ed.; Benjamin: New York, 1972.
5. Reichardt, C. *Solvent and Solvent Effects in Organic Chemistry*; VCH: Germany, 1988; p. 17. Seebach, D. *AG(E)* **1988**, *27*, 1624.
6. (a) Gutmann, V. *The Donor–Acceptor Approach to Molecular Interactions*; Plenum: New York, 1978. (b) Maria, P.-C.; Gal, J.-F. *JPC* **1985**, *89*, 1296. Maria, P.-C.; Gal, J.-F.; de Franceschi, J.; Fargin, E. *JACS* **1987**, *109*, 483. (c) Persson, I.; Sandström, M.; Goggin, P. L. *ICA* **1987**, *129*, 183.
7. Newcomb, M.; Varick, T. R.; Goh, S.-H. *JACS* **1990**, *112*, 5186. Seebach, D.; Amstutz, R.; Dunitz, J. D. *HCA* **1981**, *64*, 2622.
8. Romesburg, F. E.; Gilchrist, J. H.; Harrison, A. T.; Fuller, D. J.; Collum, D. B. *JACS* **1991**, *113*, 5751 (see also Ref 1). Romesberg, F. E.; Collum, D. B. *JACS* **1992**, *114*, 2112. Romesberg, F. E.; Bernstein, M. P.; Gilchrist, J. H.; Harrison, A. T.; Fuller, D. J.; Collum, D. B. *JACS* **1993**, *115*, 3475.
9. Galiano-Roth, A. S.; Collum, D. B. *JACS* **1989**, *111*, 6772.
10. Seebach, D. *AG(E)* **1988**, *27*, 1624.
11. Jackman, L. M.; Chen, X. *JACS* **1992**, *114*, 403.
12. Reich, H. J.; Borst, J. P.; Dykstra, R. R. *JACS* **1993**, *115*, 8728.
13. Sato, D.; Kawasaki, H.; Shimada, I.; Arata, Y.; Okamura, K.; Date, T.; Koga, K. *JACS* **1992**, *114*, 761.
14. Reich, H. J.; Green, D. P.; Phillips, N. H. *JACS* **1989**, *111*, 3444. Reich, H. J.; Green, D. P.; Phillips, N. H.; Borst, J. P. *PS* **1992**, *67*, 83.
15. Hogen-Esch, T. E.; Smid, J. *JACS* **1966**, *88*, 307. Hogen–Esch, T. E.; Smid, J. *JACS* **1966**, *88*, 318. Smid, J. *Ions and Ion Pairs in Organic Reactions*; Szwarc, M., Ed., Wiley: New York, 1972; Vol. 1, pp 85–151. O'Brien, D. H.; Russell, C. R.; Hart, A. J. *JACS* **1979**, *101*, 633. Grutzner, J. B.; Lawlor, J. M.; Jackman, L. M. *JACS* **1972**, *94*, 2306. Bartlett, P. D.; Goebel, C. V.; Weber, W. P. *JACS* **1969**, *91*, 7425. Seebach, D.; Siegel, H.; Gabriel, J.; Hässig, R. *HCA* **1980**, *63*, 2046. Seebach, D.; Hässig, R.; Gabriel, J. *HCA* **1983**, *66*, 308. Schmitt, B. J.; Schulz, G. V. *Eur. Polym. J.* **1975**, *11*, 119.
16. (a) Jackman, L. M.; Scarmoutzos, L. M.; Porter, W. *JACS* **1987**, *109*, 6524. Fraser, R. R.; Mansour, T. S. *TL* **1986**, *27*, 331. (b) House, H. O.; Prabhu, A. V.; Phillips, W. V. *JOC* **1976**, *41*, 1209.
17. Jackman, L. M.; Lange, B. C. *JACS* **1981**, *103*, 4494.
18. Corset, J.; Froment, F.; Lautie, M.-F.; Ratovelomanana, N.; Seyden-Penne, J.; Strzalko, T.; Roux-Schmitt, M.-C. *JACS* **1993**, *115*, 1684.
19. Raithby, P. R.; Reed, D.; Snaith, R.; Wright, D. S. *AG(E)* **1991**, *30*, 1011. Nudelman, N. S.; Lewkowicz, E.; Furlong, J. J. P. *JOC* **1993**, *58*, 1847.
20. Reich, H. J.; Gudmundsson, B. Ö.; Dykstra, R. R. *JACS* **1992**, *114*, 7937. Jackman, L. M.; Scarmoutzos, L. M.; Smith, B. D.; Williard, P. G. *JACS* **1988**, *110*, 6058. Barr, D.; Doyle, M. J.; Drake, S. R.; Raithby, P. R., Snaith, R.; Wright, D. S. *CC* **1988**, 1415. Fraenkel, G.; Hallden-Abberton, M. P. *JACS* **1981**, *103*, 5657.
21. Reich, H. J.; Green, D. P.; Phillips, N. H. *JACS* **1991**, *113*, 1414.
22. Tsushima, K.; Araki, K.; Murai, A. *CL* **1989**, 1313.
23. Naef, R.; Seebach, D. *AG(E)* **1981**, *20*, 1030.
24. Seebach, D.; Henning, R.; Mukhopadhyay, T. *CB* **1982**, *115*, 1705.
25. Mukhopadhyay, T.; Seebach, D. *HCA* **1982**, *65*, 385.
26. Piers, E.; Tse, H. L. A. *TL* **1984**, 3155. Cregge, R. J.; Herrman, J. L.; Lee, C. S.; Richman, J. E.; Schlessinger, R. H. *TL* **1973**, 2425. Chapdelaine, M. J.; Hulce, M. *OR* **1990**, *38*, 225. Kurth, M. J.; O'Brien, M. J. *JOC* **1985**, *50*, 3846. Odic, Y.; Pereyre, M. *JOM* **1973**, *55*, 273.
27. Snider, B. B.; Hrib, N. J.; Fuzesi, L. *JACS* **1976**, *98*, 7115.
28. Kurts, A. L.; Genkina, N. K.; Macias, I. P.; Beletskaya, I. P.; Reutov, O. A. *T* **1971**, *27*, 4777.
29. Ireland, R. E.; Mueller, R. H.; Williard, A. K. *JACS* **1976**, *98*, 2868.
30. Smrekar, O. *C* **1985**, *39*, 147.
31. Pfeffer, P. E.; Silbert, L. S. *JOC* **1970**, *35*, 262.
32. Corey, E. J.; Gross, A. W. *TL* **1984**, *25*, 495.
33. Hosomi, A.; Araki, Y.; Sakurai, H. *JACS* **1982**, *104*, 2081.
34. Shirai, R.; Tanaka, M.; Koga, K. *JACS* **1986**, *108*, 543.
35. Chan, T. H.; Chang, E. *JOC* **1974**, *39*, 3264. Chan, T. H.; Chang, E.; Vinokur, E. *TL* **1970**, 1137; Dolak, T. M.; Bryson, T. A. *TL* **1977**, 1961. Kauffmann, T. *AG(E)* **1982**, *21*, 410. Kawashima, T.; Iwama, N.; Okazaki, R. *JACS* **1993**, *115*, 2507. Grobel, B.-T.; Seebach, D. *CB* **1977**, *110*, 867. Grobel, B.-T.; Seebach, D. *AG(E)* **1974**, *13*, 83. Ager, D. J.; Cooke, G. E.; East, M. B.; Mole, S. J.; Rampersaud, A.; Webb, V. J. *OM* **1986**, *5*, 1906. Zapata, A.; Fortoul, C. R.; Acuna, C. A. *SC* **1985**, 179. Van Ende, D.; Cravador, A.; Krief, A. *JOM* **1979**, *177*, 1.
36. Bryson, T. A.; Roth, G. A.; Jing-hau, L. *TL* **1986**, *27*, 3685.
37. Caubere, P. *AG(E)* **1983**, *22*, 599.
38. Corey, E. J.; Weigel, L. O.; Chamberlin, A. R.; Lipshutz, B. *JACS* **1980**, *102*, 1439. Smith, A. B., III; Ohta, M.; Clark, W. M.; Leahy, J. W. *TL* **1993**, *34*, 3033.
39. Doolittle, R. E. *OPP* **1980**, *12*, 1. Oehlschlager, A. C.; Czyzewska, E.; Aksela, R.; Pierce, H. D., Jr. *CJC* **1986**, *64*, 1407. Merrer, Y. L.; Gravier-Pelletier, C.; Micas-Languin, D.; Mestre, F.; Dureault, A.; Depezay, J.-C. *JOC* **1989**, *54*, 2409.
40. Murayama, E.; Kikuchi, T.; Sasaki, K.; Sootome, N.; Sato, T. *CL* **1984**, 1897.
41. Gilday, J. P.; Paquette, L. A. *TL* **1988**, *29*, 4505.
42. Richey, H. G., Jr.; Farkas, J., Jr. *JOC* **1987**, *52*, 479. Marczak, S.; Wicha, J. *TL* **1993**, *34*, 6627.
43. (a) Luteri, G. F.; Fork, W. T. *JOC* **1977**, *42*, 820. (b) Luteri, G. F.; Ford, W. T. *JOM* **1976**, *105*, 139.
44. Sprangers, W. J. J. M.; Louw, R. *JCS(P2)* **1976**, 1895.
45. Harding, K. E.; Cooper, J. L.; Puckett, P. M. *JACS* **1978**, *100*, 993.
46. Brown, C. A.; Yamaichi, A. *CC* **1979**, 100.

47. (a) Zieglar, F. E.; Tam, C. C. *TL* **1979**, 4717. (b) Wartski, L.; El-Bouz, M.; Seyden-Penne, J. *JOM* **1979**, *177*, 17. Wartski, L.; El-Bouz, M.; Seyden-Penne, J.; Dumont, W.; Krief, A. *TL* **1979**, 1543. Deschamps, B. *T* **1978**, *34*, 2009. El-Bouz, M.; Nartski, L. *TL* **1980**, 2897. Luccheti, J.; Dumont, W.; Krief, A. *TL* **1979**, 2695.
48. Otera, J.; Niibo, Y.; Aikawa, H. *TL* **1987**, *28*, 2147. Zervos, M.; Wartski, L.; Seydon-Penne, J. *T* **1986**, *42*, 4963. Ager, D. J.; East, M. B. *JOC* **1986**, *51*, 3983.
49. Carey, F. A.; Court, A. S. *JOC* **1972**, *37*, 939. Ager, D. J. *TL* **1981**, *22*, 2803.
50. Hackett, S.; Livinghouse, T. *JOC* **1986**, *51*, 879. Otera, J.; Niibo, Y.; Nozaki, H. *JOC* **1989**, *54*, 5003.
51. Binns, M. R.; Haynes, R. K.; Houston, T. L.; Jackson, W. R. *TL* **1980**, *21*, 573. Binns, M. R.; Haynes, R. K.; Katsifis, A. G.; Schober, P. A.; Vonwiller, S. C. *JACS* **1988**, *110*, 5411. Binns, M. R.; Haynes, R. K. *JOC* **1981**, *46*, 3790.
52. Zieglar, F. E.; Fang, J.-M.; Tam, C. C. *JACS* **1982**, *104*, 7174.
53. Still, W. C. *JACS* **1977**, *99*, 4836; Still, W. C.; Mitra, A. *TL* **1978**, 2659.
54. Magnusson, G. *TL* **1977**, 2713. Barr, D., Doyle, M. J., Mulvey, R. E.; Raithby, P. R.; Reed, D.; Snaith, R.; Wright, D. S. *CC* **1989**, 318. Reich, H. J.; Borst, J. P.; Dykstra, R. R. *OM* **1994**, *13*, 1.
55. (a) Sreekumar, C.; Darst, K. P.; Still, W. C. *JOC* **1980**, *45*, 4260. Koreeda, M.; Patel, P. D.; Brown, L. *JOC* **1985**, *50*, 5910. (b) Vedejs, E.; Peterson, M. J. *Top. Stereochem.* **1993**, *21*, 1. Schlosser, v.-M.; Christmann, K. F. *LA* **1967**, *708*, 1.
56. Waters, R. M.; Voaden, D. J.; Warthen, J. D., Jr. *OPP* **1978**, *10*, 5. Sonnet, P. E. *OPP* **1974**, *6*, 269. Corey, E. J.; Clark, D. A.; Goto, G.; Marfat, A.; Mioskowski, C.; Samuelsson, B.; Hammarström, S. *JACS* **1980**, *102*, 1436. Wernic, D.; DiMaio, J.; Adams, J. *JOC* **1989**, *54*, 4224. Delorme, D.; Girard, Y.; Rokach, J. *JOC* **1989**, *54*, 3635. Yadagiri, P.; Lumin, S.; Falck, J. R.; Karara, A.; Capdevila, J. *TL* **1989**, *30*, 429. Vidal, J. P.; Escale, R.; Niel, G.; Rechencq, E.; Girard, J. P.; Rossi, J. C. *TL* **1989**, *30*, 5129.
57. Cereda, E.; Attolini, M.; Bellora, E.; Donetti, A. *TL* **1982**, *23*, 2219. Inoue, S.; Honda, K.; Iwase, N.; Sato, K. *BCJ* **1990**, *63*, 1629.
58. Corey, E. J.; Marfat, A.; Hoover, D. J. *TL* **1981**, *22*, 1587. Boubia, B.; Mann, A.; Bellamy, F. D.; Mioskowski, C. *AG(E)* **1990**, *29*, 1454.
59. Ousset, J. B.; Mioskowski, C.; Solladie, G. *SC* **1983**, *13*, 1193.
60. Kelly, T. R.; Dali, H. M.; Tsang, W.-G. *TL* **1977**, *44*, 3859. Evers, M. *CS* **1986**, *26*, 585.
61. (a) Evers, M.; Christiaens, L. *TL* **1983**, *24*, 377. Reich, H. J.; Cohen, M. L. *JOC* **1979**, *44*, 3148. (b) Testaferri, L.; Tiecco, M.; Tingoli, M.; Chianelli, D.; Montanucci, M. *S* **1983**, 751.
62. (a) Müller, P.; Siegfried, B. *TL* **1973**, 3565. (b) Kondo, K.; Umemota, T.; Takahatake, Y.; Tunemoto, D. *TL* **1977**, 113.
63. Ager, D. J. *OR* **1990**, *38*, 1.
64. Sakurai, H.; Nishiwaki, K.-i.; Kira, M. *TL* **1973**, 4193.
65. Bassindale, A. R.; Ellis, R. J.; Taylor, P. G. *TL* **1984**, *25*, 2705.
66. Falck, J. R.; Yadagiri, P. *JOC* **1989**, *54*, 5851.
67. Shaw, J. E.; Kunerth, D. C.; Swanson, S. B. *JOC* **1976**, *41*, 732.
68. Ashby, E. C.; Park, W. S.; Goel, A. B.; Su, W.-Y. *JOC* **1985**, *50*, 5184.
69. Marshall, J. A.; Lebreton, J.; DeHoff, B. S.; Jenson, T. M. *JOC* **1987**, *52*, 3883.
70. Normant, J. F.; Deshayes, H. *BSF(2)* **1967**, 2455. Paquette, L. A.; Meisinger, R. H.; Wingard, R. E., Jr. *JACS* **1973**, *95*, 2230. King, P. F.; Paquette, L. A. *S* **1977**, 279. Weisz, A.; Mandelbaum, A. *JOC* **1984**, *49*, 2648.
71. Roush, W. R.; Brown, B. B. *JACS* **1993**, *115*, 2268.
72. Whitesides, G. M.; Ehmann, W. J. *JOC* **1970**, *35*, 3565 and references cited therein.
73. (a) Inanaga, J.; Ishikawa, M.; Yamaguchi, M. *CL* **1987**, 1485. (b) Otsubo, K.; Kawamura, K.; Iwanaga, J.; Yamaguchi, M. *CL* **1987**, 1487.
74. Otsubo, K.; Inanaga, J.; Yamaguchi, M. *TL* **1986**, *27*, 5763.
75. Handa, Y.; Inanaga, J.; Yamaguchi, M. *CC* **1989**, 298.
76. Inanaga, J.; Handa, Y.; Tabuchi, T.; Otsubo, K.; Yamaguchi, M.; Hanamoto, T. *TL* **1991**, *32*, 6557.
77. Cabrera, A.; Alper, H. *TL* **1992**, *33*, 5007.
78. Fevig, T. L.; Elliott, R. L.; Curran, D. P. *JACS* **1988**, *110*, 5064.
79. Curran, D. P.; Wolin, R. L. *SL* **1991**, 317. Curran, D. P.; Yoo, B. *TL* **1992**, *33*, 6931.
80. Hou, Z.; Takamine, K.; Aoki, O.; Shiraishi, H.; Fujiwara, Y.; Taniguchi, H. *JOC* **1988**, *53*, 6077. Hou, Z.; Kobayashi, K.; Yamazaki, H. *CL* **1991**, 265.
81. Milstein, D.; Stille, J. K. *JACS* **1979**, *101*, 4981, 4992.
82. (a) Labadie, J. W.; Stille, J. K. *JACS* **1983**, *105*, 6129. (b) Labadie, J. W.; Stille, J. K. *JACS* **1983**, *105*, 669.
83. Bumagin, N. A.; Bumagina, I. G.; Beletskaya, I. P. *DOK* **1983**, *272*, 1384.
84. Hegedus, L. S.; Williams, R. E.; McGuire, M. A.; Hayashi, T. *JACS* **1980**, *102*, 4973.
85. Hegedus, L. S.; Darlington, W. H.; Russel, C. E. *JOC* **1980**, *45*, 5193.
86. Hutchins, R. O.; Kandasamy, D.; Maryanoff, C. A.; Masilamani, D.; Maryanoff, B. E. *JOC* **1977**, *42*, 82.
87. Hutchins, R. O.; Cistone, F.; Goldsmith, B.; Heuman, P. *JOC* **1975**, *40*, 2018.
88. Shibata, I.; Suzuki, T.; Baba, A.; Matsuda, H. *CC* **1988**, 882.
89. Shibata, I.; Yoshida, T.; Baba, A.; Matsuda, H. *CL* **1989**, 619.
90. Fujita, M.; Hiyama, T. *JOC* **1988**, *53*, 5405.
91. Cardillo, G.; Orena, M.; Sandri, S. *S* **1976**, 394.
92. Al Neirabeyeh, M.; Pujol, M. D. *TL* **1990**, *31*, 2273.
93. (a) Peyronel, J.-F.; Samuel, O.; Fiaud, J.-C. *JOC* **1987**, *52*, 5320 and references cited therein. (b) Arcoria, A.; Ballistreri, F. P.; Tomaselli, G. A.; Di Furia, F.; Modena, G. *JOC* **1986**, *51*, 2374.
94. Takagi, M.; Nojima, M.; Kusabayashi, S. *JACS* **1983**, *105*, 4676.
95. Eyer, M.; Seebach, D. *JACS* **1985**, *107*, 3601.
96. Panek, E. J.; Rodgers, T. J. *JACS* **1974**, *96*, 6921.
97. Lefour, J.-M.; Loupy, A. *T* **1978**, *34*, 2597.
98. Bertz, S. H. *JACS* **1991**, *113*, 5470.
99. Sowinski, A. F.; Whitesides, G. M. *JOC* **1979**, *44*, 2369.
100. (a) Wilson, S. R.; Price, M. F. *SC* **1982**, *12*, 657. (b) Bortolini, O.; Di Furia, F.; Modena, G.; Schionato, A. *J. Mol. Catal.* **1986**, *35*, 47.

Robert R. Dykstra
University of Wisconsin, Madison, WI, USA

Hexamethylphosphorous Triamide[1]

$(Me_2N)_3P$

[1608-26-0] $C_6H_{18}N_3P$ (MW 163.24)

(strong nucleophile;[2] used to synthesize epoxides from aldehydes[2,3] and arene oxides from aryldialdehydes;[4–7] replaces Ph_3P in the Wittig reaction;[8] with CCl_4, converts alcohols to chlorides;[9] with I_2, converts disulfides to sulfides[10] and deoxygenates sulfoxides and azoxyarenes;[11] with dialkyl azodicarboxylate and alcohol, forms mixed carbonates;[12] reduces ozonides[13])

Alternate Name: tris(dimethylamino)phosphine.
Physical Data: mp 26 °C; bp 162–164 °C/760 mmHg, 50 °C/12 mmHg; n_D^{25} 1.4636; d 0.911 g cm^{-3}.

Form Supplied in: commercially available; liquid, pure grade >97% (GC).

Preparative Methods: reaction of ***Phosphorus(III) Chloride*** with anhydrous dimethylamine; the same procedure can be used to obtain higher alkyl homologs.[14]

Purification: distillation at reduced pressure; exposure of hot liquid to air should be avoided.

Handling, Storage, and Precautions: very sensitive to air; best stored in nitrogen atmosphere; reacts with carbon dioxide; inhalation should be avoided. Use in a fume hood.

Synthesis of Epoxides. Reaction of $(Me_2N)_3P$ with aromatic aldehydes provides convenient direct synthetic access to symmetrical and unsymmetrical epoxides in generally high yields. A typical example is the reaction of *o*-chlorobenzaldehyde, which provides the corresponding stilbene oxide as a mixture of the *trans* and *cis* isomers (eq 1).[2,3]

$$2\ o\text{-}ClC_6H_4CHO + (Me_2N)_3P \xrightarrow{76\%} \text{stilbene oxide} + (Me_2N)_3P{=}O \quad (1)$$

trans:*cis* = 1.38:1

The coproduct, ***Hexamethylphosphoric Triamide***, is readily separated by taking advantage of its water solubility. A competing reaction pathway leads to formation of variable amounts of a 1:1 adduct in addition to the epoxide product (eq 2). Originally the adduct was assigned the betaine structure (**1a**). On the basis of more detailed NMR analysis, this was subsequently revised to the phosphonic diamide structure (**1b**).[15]

$$ArCHO + (Me_2N)_3P \longrightarrow \text{(1a) or (1b)} \quad (2)$$

(**1a**) (**1b**)

The ratio of products depends upon the electronegativity of the aldehyde and the mode of carrying out the reaction. Aromatic aldehydes with electronegative substituents, especially in the *ortho* position, undergo rapid exothermic reaction to yield epoxides exclusively. Conversely, aldehydes bearing electron-releasing substituents react more slowly to afford mainly 1:1 adducts. Slow addition of $(Me_2N)_3P$ to the aldehyde tends to enhance the ratio of the epoxide product. These observations are compatible with a mechanism in which an initially formed 1:1 adduct reacts with a second aldehyde molecule to form a 2:1 adduct which collapses to yield the observed products (eq 3).

$$(Me_2N)_3\overset{+}{P}{-}CH(R){-}O^- \xrightarrow{RCHO} \text{2:1 adduct} \longrightarrow (Me_2N)_3P{=}O + \text{epoxide} \quad (3)$$

$(Me_2N)_3P$ reacts also with saturated and heterocyclic aldehydes, but 1:1 adducts rather than epoxides are the predominant products. The reaction with ***Chloral*** takes a different course[2] and yields the dichlorovinyloxyphosphonium compound $Cl_2C{=}CH{-}O{-}\overset{+}{P}(NMe_2)_3\ Cl^-$.

The scope of the reaction is considerably extended by its applicability to the synthesis of mixed epoxides.[2] This is accomplished by addition of $(Me_2N)_3P$ to a mixture of aldehydes in which the less reactive aldehyde predominates. For example, addition of $(Me_2N)_3P$ to a mixture of *o*-chlorobenzaldehyde and 2-furaldehyde yields the corresponding mixed epoxide (eq 4).

$$o\text{-}ClC_6H_4CHO + OHC\text{-}(2\text{-furyl}) \xrightarrow{(Me_2N)_3P} \text{mixed epoxide} \quad (4)$$

An advantage of the method is that it allows the synthesis of epoxides unobtainable by the oxidation of alkene precursors with peroxides or peracids due to the incompatibility of functional groups with these reagents.

Synthesis of Arene Oxides. Reaction of $(Me_2N)_3P$ with aromatic dialdehydes provides arene oxides such as benz[*a*]anthracene 5,6-oxide (**2a**) (eq 5).[4–7] These compounds, also known as oxiranes, are relatively reactive, undergoing thermal and acid-catalyzed rearrangement to phenols and facile hydrolysis to dihydrodiols. Consequently, their preparation and purification requires mild reagents and conditions. The importance of this is underlined by successful synthesis of the reactive arene oxide (**2b**) in 75% yield using appropriate care,[7] despite a previous report of failure of the method.[4] While compound (**2b**) is a relatively potent mutagen, it is rapidly detoxified by mammalian cells.[6] The principal limitation of the method is the unavailability of the dialdehyde precursors, which are obtained through oxidation of the parent hydrocarbons, e.g. by ozonolysis.

$$\text{dialdehyde} + (Me_2N)_3P \longrightarrow (Me_2N)_3P{=}O + \text{arene oxide} \quad (5)$$

(**2a**) R = H
(**2b**) R = Me

Wittig and Horner–Wittig Reactions. $(Me_2N)_3P$ may be used in place of ***Triphenylphosphine*** in Wittig reactions with aldehydes and ketones (eq 6).[8] It is advantageous because the water solubility of the byproduct, hexamethylphosphoric triamide, renders it readily removable. This method has been used for the preparation of unsaturated esters as well as alkenes (eq 7).

$$PhCH_2Br + (Me_2N)_3P \longrightarrow (Me_2N)_3\overset{+}{P}CH_2Ph\ Br^- \ \textbf{(3)} \xrightarrow[Me_2CHCHO]{NaOMe} Me_2CHCH{=}CHPh \quad (6)$$

$$BrCH_2CO_2Et + (Me_2N)_3P \longrightarrow (Me_2N)_3\overset{+}{P}CH_2CO_2Et\ Br^- \ \textbf{(4)} \xrightarrow[RCHO]{NaOMe} RCH{=}CHCO_2Et \quad (7)$$

The phosphonic diamide products (**1b**) obtained from the reaction of arylaldehydes having electron-donating groups with $(Me_2N)_3P$ can be deprotonated by ***n-Butyllithium*** in DME at 0 °C.[15] These intermediates participate in Horner–Wittig-type reactions with aromatic aldehydes to give enamines in good yield (eq 8). In the examples studied the enamines have the (*E*) configuration. Mild acid hydrolysis of the reaction mixtures without isolating the intermediate enamines provides the corresponding deoxybenzoins.[15] The overall procedure represents an example of reductive nucleophilic acylation of carbonyl compounds.

$$\textbf{(1b)} \xrightarrow{BuLi} Ar^1(Me_2N)C{=}P(O)(NMe_2)_2 \xrightarrow{Ar^2CHO} Ar^1(Me_2N)C{=}CHAr^2 \longrightarrow Ar^1COCH_2Ar^2 \quad (8)$$

Conversion of Alcohols to Alkyl Chlorides and Other Derivatives. The reagent combination $(Me_2N)_3P$ and CCl_4 can be used in place of ***Triphenylphosphine–Carbon Tetrachloride*** for the conversion of alcohols to alkyl chlorides (eq 9).[9] An advantage is the ease of removal of the water-soluble coproduct $(Me_2N)_3P{=}O$. The mechanism entails initial rapid formation of a quasiphosphonium ion, followed by reaction with an alcohol with displacement of chloride, and nucleophilic attack by the chloride ion on the carbon atom of the alcohol in a final rate-determining step to yield an alkyl chloride.

$$(Me_2N)_3P{:}\ Cl{-}CCl_3 \xrightarrow{fast} (Me_2N)_3\overset{+}{P}{-}Cl\ Cl_3C^- \xrightarrow[fast]{ROH}$$

$$Cl_3CH + (Me_2N)_3\overset{+}{P}{-}O{-}R\ Cl^- \xrightarrow{slow} (Me_2N)_3P{=}O + RCl \quad (9)$$

This reagent reacts more rapidly with primary than with secondary alcohols. This property has been made use of to transform the primary hydroxy groups of sugars to salts, which then may be converted to halides (Cl, Br, I), azides, amines, thiols, thiocyanates, etc. by reaction with appropriate nucleophiles (eq 10).[16] Arylalkyl ethers and thioethers may also be prepared by appropriate modification of this method.[17] These reactions generally proceed with high stereoselectivity. Thus reaction of chiral 2-octanol with this reagent afforded 2-chlorooctane with complete inversion of configuration. Also, conversion of the salt prepared from reaction of (*R*)-(−)-2-octanol with $(Me_2N)_3P/CCl_4$ at low temperature to the corresponding hexafluorophosphate salt, followed by reaction of this with potassium phenolate in DMF, gave optically pure (*S*)-(+)-2-phenoxyoctane in 93% yield (eq 11).[17]

$$\text{methyl glycoside (6-}CH_2OH) \xrightarrow[CCl_4,\ THF]{(Me_2N)_3P} \text{6-}CH_2O\overset{+}{P}(NMe_2)_3\ Cl^- \xrightarrow{Nu^-} \text{6-}CH_2{-}Nu \quad (10)$$

Nu = Cl, Br, I, N_3, NH_2, SCN, H

$$(Me_2N)_3\overset{+}{P}{-}O{-}CH(CH_3)CH_2CH_2Bu\ PF_6^- + PhOK \longrightarrow PhO{-}CH(CH_3)CH_2CH_2Bu + (R_2N)_3P{=}O \quad (11)$$

$(Me_2N)_3P/CCl_4$ may also be employed for the selective functionalization of primary long-chain diols.[17] Reactions of diols of this type with $(Me_2N)_3P$ and CCl_4 in THF followed by addition of KPF_5 gives mono salts in high yield (eq 12).[18] THF serves to precipitate the mono salts as they are formed, thereby blocking their conversion to bis salts. Reactions of the mono salts with various nucleophiles provides the corresponding monosubstituted primary alcohols.

$$HOCH_2(CH_2)_nCH_2OH \xrightarrow[CCl_4,\ -40\ °C;\ KPF_5]{(Me_2N)_3P} (Me_2N)_3\overset{+}{P}OCH_2(CH_2)_nCH_2OH\ PF_5^- \xrightarrow{Nu^-} NuCH_2(CH_2)_nCH_2OH \quad (12)$$

n = 3–9 Nu = I, N_3, CN, SCN, MeO

Conversion of Disulfides to Sulfides. Alkyl, aralkyl, and alicyclic disulfides undergo facile desulfurization to the corresponding sulfides on treatment with $(Me_2N)_3P$ or $(Et_2N)_3P$ (eq 13).[10] For example, reaction of methyl phenyl disulfide with $(Et_2N)_3P$ in benzene at rt for ≈1 min furnishes methyl phenyl sulfide in 86% yield. The desulfurization process is stereospecific, in that inversion of configuration occurs at one of the carbon atoms α to the disulfide group. Thus desulfurization of *cis*-3,6-dimethoxycarbonyl-1,2-dithiane affords a quantitative yield of *trans*-2,5-dimethoxycarbonylthiolane (eq 14). It is worthy of note that the rates of these reactions are markedly enhanced by solvents of high polarity.

$$PhSSMe + (R_2N)_3P \longrightarrow PhSMe + (R_2N)_3P{=}S \quad (13)$$

$$\textit{cis}\text{-}MeO_2C\text{-dithiane-}CO_2Me \xrightarrow{(R_2N)_3P} \textit{trans}\text{-}MeO_2C\text{-thiolane-}CO_2Me + (R_2N)_3P{=}S \quad (14)$$

Deoxygenation of Sulfoxides and Azoxyarenes. Sulfoxides are deoxygenated to sulfides under mild conditions with $(Me_2N)_3P$ activated with ***Iodine*** in acetonitrile (eq 15).[11] Equimolar ratios of the sulfoxide, $(Me_2N)_3P$, and I_2 are generally employed. Yields are superior to those obtained with either $(Me_2N)_3P/CCl_4$ or Ph_3P/I_2. Reaction time is reduced by addition of ***Sodium Iodide***. Azoxyarenes, such as azoxybenzene, are converted to azoarenes with this reagent combination under similar mild conditions (eq 16).[11]

$$Ph_2S{=}O + (Me_2N)_3P \xrightarrow{I_2} Ph_2S + (Me_2N)_3P{=}O \quad (15)$$

$$PhN^{+}(O^{-}){=}NPh + (Me_2N)_3P \xrightarrow{I_2} PhN{=}NPh + (Me_2N)_3P{=}O \quad (16)$$

Preparation of Mixed Carbonates. The reaction of $(Me_2N)_3P$ with alcohols and dialkyl azodicarboxylates proceeds smoothly at rt to provide mixed dialkyl carbonate esters in moderate to good yields (eq 17).[12] An advantage of the method over the chloroformate method is the neutrality of the conditions employed. It should be noted that the related system ***Triphenylphosphine–Diethyl Azodicarboxylate*** converts alcohols into amines.

$$R^1OH + R^2O_2C{-}N{=}N{-}CO_2R^2 \xrightarrow{(R^3{}_2N)_3P} R^1O{-}CO{-}OR^2 + HCO_2R^2 + N_2 \quad (17)$$

Reduction of Ozonides. In the synthesis of ecdysone from ergosterol, the ozonide product produced from (**5**) was reduced with $(Me_2N)_3P$ under mild conditions to the aldehyde (**6**) without isomerization (eq 18).[13] The scope of this method has not been investigated.

AcO, AcO, OH, O, *i*-Pr (**5**) $\xrightarrow[2.\ (Me_2N)_3P]{1.\ O_3}$ AcO, AcO, OH, O, CHO (**6**) (18)

Related Reagents. Diphosphorus Tetraiodide; Raney Nickel; Tri-*n*-butylphosphine; Triphenylphosphine.

1. Wurziger, H. *Kontakte (Darmstadt)* **1990**, 13.
2. Mark, V. *JACS* **1963**, *85*, 1884.
3. Mark, V. *OSC* **1973**, *5*, 358.
4. Newman, M. S.; Blum, S. *JACS* **1964**, *86*, 5598.
5. Harvey, R. G. *S* **1986**, 605.
6. Harvey, R. G. *Polycyclic Aromatic Hydrocarbons: Chemistry and Carcinogenesis*; Cambridge University Press: Cambridge, 1991; Chapter 12.
7. Harvey, R. G.; Goh, S. H.; Cortez, C. *JACS* **1975**, *97*, 3468.
8. Oediger, H.; Eiter, K. *LA* **1965**, *682*, 58.
9. Downie, I. M.; Lee, J. B.; Matough, M. F. S. *CC* **1968**, 1350.
10. Harpp, D. N.; Gleason, J. G. *JACS* **1971**, *93*, 2437.
11. Olah, G. A.; Gupta, B. G. B.; Narang, S. C. *JOC* **1978**, *43*, 4503.
12. Grynkiewicz, G.; Jurczak, J.; Zamojski, A. *T* **1975**, *31*, 1411.
13. Furlenmeier, A.; Fürst, A.; Langemann, A.; Waldvogel, G.; Hocks, P.; Kerb, U.; Wiechert, R. *HCA* **1967**, *50*, 2387.
14. Mark, V. *OSC* **1973**, *5*, 602.
15. Babudri, F.; Fiandanese, V.; Musio, R.; Naso, F.; Sciavovelli, O.; Scilimati, A. *S* **1991**, 225.
16. Castro, B.; Chapleur, Y.; Gross, B. *BSF(2)* **1973**, 3034.
17. Downie, I. M.; Heaney, H.; Kemp, G. *AG(E)* **1975**, *14*, 370.
18. Boigegrain, R.; Castro, B.; Selve, C. *TL* **1975**, 2529.

Ronald G. Harvey
University of Chicago, IL, USA

Hydrazine[1]

N_2H_4

(N_2H_4)		
[302-01-2]	H_4N_2	(MW 32.06)
(hydrate)		
[10217-52-4]		
(monohydrate)		
[7803-57-8]	H_5N_2O	(MW 49.07)
(monohydrochloride)		
[2644-70-4]	ClH_5N_2	(MW 68.52)
(dihydrochloride)		
[5341-61-7]	$Cl_2H_6N_2$	(MW 104.98)
(sulfate)		
[10034-93-2]	$H_6N_2O_4S$	(MW 130.15)

(reducing agent used in the conversion of carbonyls to methylene compounds;[1] reduces alkenes,[9] alkynes,[9] and nitro groups;[14] converts α,β-epoxy ketones to allylic alcohols;[32] synthesis of hydrazides;[35] synthesis of dinitrogen containing heterocycles[42–46])

Physical Data: mp 1.4 °C; bp 113.5 °C; *d* 1.021 g cm^{-3}.

Solubility: sol water, ethanol, methanol, propyl and isobutyl alcohols.

Form Supplied in: anhydrate, colorless oil that fumes in air; hydrate and monohydrate, colorless oils; monohydrochloride, dihydrochloride, sulfate, white solids; all widely available.

Analysis of Reagent Purity: titration.[1]

Purification: anhydrous hydrazine can be prepared by treating hydrazine hydrate with BaO, $Ba(OH)_2$, CaO, NaOH, or Na. Treatment with sodamide has been attempted but this yields diimide, NaOH, and ammonia. An excess of sodamide led to an explosion at 70 °C. The hydrate can be treated with boric acid to give the hydrazinium borate, which is dehydrated by heating. Further heating gives diimide.[1]

Handling, Storage, and Precautions: caution must be taken to avoid prolonged exposure to vapors as this can cause serious damage to the eyes and lungs. In cases of skin contact, wash the affected area immediately as burns similar to alkali contact can occur. Standard protective clothing including an ammonia gas mask are recommended. The vapors of hydrazine are flammable (ignition temperature 270 °C in presence of air). There have been reports of hydrazine, in contact with organic material such as wool or rags, burning spontaneously. Metal oxides can also initiate combustion of hydrazine. Hydrazine and its solutions should be stored in glass containers under nitrogen for extended periods. There are no significant precautions for reaction vessel type with hydrazine; however, there have been reports that stainless steel vessels must be checked for significant oxide formation prior to use. Use in a fume hood.

Reductions. The use of hydrazine in the reduction of carbonyl compounds to their corresponding methylene groups via the Wolff–Kishner reduction has been covered extensively in the literature.[1] The procedure involves the reaction of a carbonyl-containing compound with hydrazine at high temperatures in the presence of a base (usually ***Sodium Hydroxide*** or ***Potassium Hydroxide***). The intermediate hydrazone is converted directly to the fully reduced species. A modification of the original conditions was used by Paquette in the synthesis of (±)-isocomene (eq 1).[2]

$H_2NNH_2{\bullet}H_2O$, K_2CO_3, 200 °C, 80% (1)

Unfortunately, the original procedure suffers from the drawback of high temperatures, which makes large-scale runs impractical. The Huang–Minlon modification[3] of this procedure revolutionized the reaction, making it usable on large scales. This procedure involves direct reduction of the carbonyl compound with hydrazine hydrate in the presence of sodium or potassium hydroxide in diethylene glycol. The procedure is widely applicable to a variety of acid-labile substrates but caution must be taken where base-sensitive functionalities are present. This reaction has seen widespread use in the preparation of a variety of compounds. Other modifications[4] have allowed widespread application of this useful transformation. Barton and co-workers further elaborated the Huang–Minlon modifications by using anhydrous hydrazine and ***Sodium*** metal to ensure totally anhydrous conditions. This protocol allowed the reduction of sterically hindered ketones, such as in the deoxygenation of 11-keto steroids (eq 2).[4a] Cram utilized dry DMSO and ***Potassium t-Butoxide*** in the reduction of hydrazones. This procedure is limited in that the hydrazones must be prepared and isolated prior to reduction.[4b] The Henbest modification[4c] involves the utilization of dry toluene and potassium *t*-butoxide. The advantage of this procedure is the low temperatures needed (110 °C) but it suffers from the drawback that, again, preformed hydrazones must be used. Utilizing modified Wolff–Kishner conditions, 2,4-dehydroadamantanone is converted to 8,9-dehydroadamantane (eq 3).[5]

AcO; H_2NNH_2, RONa, Ac_2O, 69%; AcO (2)

(3)

Hindered aldehydes have been reduced using this procedure.[6] This example is particularly noteworthy in that the aldehyde is sterically hindered and resistant to other methods for conversion to the methyl group.[6a] Note also that the acetal survives the manipulation (eq 4). The reaction is equally useful in the reduction of semicarbazones or azines.

OHC; H_2NNH_2, KOH, $(HOCH_2CH_2)_2O$ (4)

In a similar reaction, hydrazine has been shown to desulfurize thioacetals, cyclic and acyclic, to methylene groups (eq 5). The reaction is run in diethylene glycol in the presence of potassium hydroxide, conditions similar to the Huang–Minlon protocol. Yields are generally good (60–95%). In situations where base sensitivity is a concern, the potassium hydroxide may be omitted. Higher temperatures are then required.[7]

R^2, R^1; H_2NNH_2, KOH, diethylene glycol; R^2, R^1 (5)

Hydrazine, via in situ copper(II)-catalyzed conversion to ***Diimide***, is a useful reagent in the reduction of carbon and nitrogen multiple bonds. The reagent is more reactive to symmetrical rather than polar multiple bonds (C=N, C=O, N=O, S=O, etc.)[8] and reviews of diimide reductions are available.[9] The generation of diimide from hydrazine has been well documented and a wide variety of oxidizing agents can be employed: oxygen (air),[10] ***Hydrogen Peroxide***,[10] ***Potassium Ferricyanide***,[11] ***Mercury(II) Oxide***,[11] ***Sodium Periodate***,[12] and hypervalent ***Iodine***[13] have all been reported. The reductions are stereospecific, with addition occurring *cis* on the less sterically hindered face of the substrate.

Other functional groups have been reduced using hydrazine. Nitroarenes are converted to anilines[14] in the presence of a variety of catalysts such as ***Raney Nickel***,[14a,15] platinum,[14a] ruthenium,[14a] ***Palladium on Carbon***,[16] β-iron(III) oxide,[17] and iron(III) chloride with activated carbon.[18] Graphite/hydrazine reduces aliphatic and aromatic nitro compounds in excellent yields.[19] Halonitrobenzenes generally give excellent yields of haloanilines. In experiments where palladium catalysts are used, significant dehalogenation occurs to an extent that this can be considered a general dehalogenation method.[20] Oximes have also been reduced.[21]

Hydrazones. Reaction of hydrazine with aldehydes and ketones is not generally useful due to competing azine formation or competing Wolff–Kishner reduction. Exceptions have been documented. Recommended conditions for hydrazone preparation are to reflux equimolar amounts of the carbonyl component and hydrazine in *n*-butanol.[22,23] A more useful method for simple hydrazone synthesis involves reaction of the carbonyl compound with dimethylhydrazine followed by an exchange reaction with hydrazine.[24] For substrates where an azine is formed, the hydrazone can be prepared by refluxing the azine with anhydrous hydrazine.[25] *gem*-Dibromo compounds have been converted to hydrazones by reaction with hydrazine (eq 6).[26]

H_2NNH_2, Et_2O, MeOH (6)

Hydrazones are useful synthetic intermediates and have been converted to vinyl iodides[27] and vinyl selenides (eq 7) (see also ***p*-Toluenesulfonylhydrazide**).[28]

RX, base (7)

R = I, PhSe; RX = I_2, PhSeBr
base = pentaalkylguanidine, triethylamine

Diazomalonates have been prepared from dialkyl mesoxylates via the ***Silver(I) Oxide***-catalyzed decomposition of the intermediate hydrazones.[29] Monohydrazones of 1,2-diketones yield ketenes after mercury(II) oxide oxidation followed by heating.[30] Dihydrazones of the same compounds give alkynes under similar conditions.[31]

Wharton Reaction. α,β-Epoxy ketones and aldehydes rearrange in the presence of hydrazine, via the epoxy hydrazone, to give the corresponding allylic alcohols. This reaction has been successful in the steroid field but, due to low yields, has seen limited use as a general synthetic tool. Some general reaction conditions have been set. If the intermediate epoxy hydrazone is isolable, treatment with a strong base (potassium *t*-butoxide or ***Potassium Diisopropylamide***) gives good yields, whereas ***Triethylamine*** can be used with nonisolable epoxy hydrazones (eq 8).[32]

(8)

Some deviations from expected Wharton reaction products have been reported in the literature. Investigators found that in some specific cases, treatment of α,β-epoxy ketones under Wharton conditions gives cyclized allylic alcohols (eq 9). No mechanistic interpretation of these observations has been offered. Related compounds have given the expected products, and it therefore appears this phenomenon is case-specific.[33]

H_2NNH_2 (9)

Cyclic α,β-epoxy ketones have been fragmented upon treatment with hydrazine to give alkynic aldehydes.[34]

Hydrazides. Acyl halides,[35] esters, and amides react with hydrazine to form hydrazides which are themselves useful synthetic intermediates. Treatment of the hydrazide with nitrous acid yields the acyl azide which, upon heating, gives isocyanates (Curtius rearrangement).[36] Di- or trichlorides are obtained upon reaction with ***Phosphorus(V) Chloride***.[37] Crotonate and other esters have been cleaved with hydrazine to liberate the free alcohol (eq 10).[38]

H_2NNH_2 (10)

Hydrazine deacylates amides (Gabriel amine synthesis) via the Ing–Manske protocol.[39] This procedure has its limitations, as shown in the synthesis of penicillins and cephalosporins where it was observed that hydrazine reacts with the azetidinone ring. In this case, ***Sodium Sulfide*** was used.[40]

Heterocycle Synthesis. The reaction of hydrazine with α,β-unsaturated ketones yields pyrazoles.[41,42] Although the products can be isolated as such, they are useful intermediates in the synthesis of cyclopropanes upon pyrolysis of cyclopropyl acetates after treatment with ***Lead(IV) Acetate*** (eq 11).

H_2NNH_2; $Pb(OAc)_4$, heat; 60% (11)

3,5-Diaminopyrazoles were prepared by the addition of hydrazine (eq 12), in refluxing ethanol, to benzylmalononitriles (42–73%).[43] Likewise, hydrazine reacted with 1,1-diacetylcyclopropyl ketones to give β-ethyl-1,2-azole derivatives. The reaction mixture must have a nucleophilic component (usually the solvent, i.e. methanol) to facilitate the opening of the cyclopropane ring. Without this, no identifiable products are obtained (eq 13).[44]

H_2NNH_2, ethanol, reflux, 42–73% (12)

Ar = Ph, 4-MeC_6H_4, 3-$NO_2C_6H_4$

R^1, R^2 = Me, Ph
X = Cl, Br, OMe, OEt, OPh, OAc, CN, etc.

(13)

In an attempt to reduce the nitro group of nitroimidazoles, an unexpected triazole product was obtained in 66% yield. The suggested mechanism involves addition of the hydrazine to the ring, followed by fragmentation and recombination to give the observed product (eq 14).[45]

(14)

Finally, hydrazine dihydrochloride reacted with 2-alkoxynaphthaldehydes to give a product which resulted from an intramolecular $[3^+ + 2]$ criss-cross cycloaddition (42–87%) (eq 15).[46]

1. 2 equiv H_2NNH_2 ethanol reflux
2. Et_3N

(15)

Peptide Synthesis. Treatment of acyl hydrazides with nitrous acid leads to the formation of acid azides which react with amines to form amides in good yield. This procedure has been used in peptide synthesis, but is largely superseded by coupling reagents such as ***1,3-Dicyclohexylcarbodiimide***.[47]

1. (a) Todd, D. *OR* **1948**, *4*, 378. (b) Szmant, H. H. *AG(E)* **1968**, *7*, 120. (c) Reusch, W. *Reduction*; Dekker: New York, 1968, pp 171–185. (d) Clark, C. *Hydrazine*; Mathieson Chemical Corp.: Baltimore, MD, 1953.
2. Paquette, L. A.; Han, Y. K. *JOC* **1979**, *44*, 4014.
3. (a) Huang-Minlon *JACS* **1946**, *68*, 2487; **1949**, *71*, 3301. (b) Durham, L. J.; McLeod, D. J.; Cason, J. *OSC* **1963**, *4*, 510. (c) Hunig, S.; Lucke, E.; Brenninger, W. *OS* **1963**, *43*, 34.
4. (a) Barton, D. H. R.; Ives, D. A. J.; Thomas, B. R. *JCS* **1955**, 2056. (b) Cram, D. J.; Sahyun, M. R. V.; Knox, G. R. *JACS* **1962**, *90*, 7287. (c) Grundon, M. F.; Henbest, H. B.; Scott, M. D. *JCS* **1963**, 1855. (d) Moffett, R. B.; Hunter, J. H. *JACS* **1951**, *73*, 1973. (e) Nagata, W.; Itazaki, H. *CI(L)* **1964**, 1194.
5. Murray, R. K., Jr.; Babiak, K. A. *JOC* **1973**, *38*, 2556.
6. (a) Zalkow, L. H.; Girotra, N. N. *JOC* **1964**, *29*, 1299. (b) Aquila, H. *Ann. Chim.* **1968**, *721*, 117.
7. van Tamelen, E. E.; Dewey, R. S.; Lease, M. F.; Pirkle, W. H. *JACS* **1961**, *83*, 4302.
8. Georgian, V.; Harrisson, R.; Gubisch, N. *JACS* **1959**, *81*, 5834.
9. (a) Miller, C. E. *J. Chem. Educ.* **1965**, *42*, 254. (b) Hunig, S.; Muller, H. R.; Thier, W. *AG(E)* **1965**, *4*, 271. (c) Hammersma, J. W.; Snyder, E. I. *JOC* **1965**, *30*, 3985.
10. Buyle, R.; Van Overstraeten, A. *CI(L)* **1964**, 839.
11. Ohno, M.; Okamoto, M. *TL* **1964**, 2423.
12. Hoffman, J. M., Jr.; Schlessinger, R. H. *CC* **1971**, 1245.
13. Moriarty, R. M.; Vaid, R. K.; Duncan, M. P. *SC* **1987**, *17*, 703.
14. (a) Furst, A.; Berlo, R. C.; Hooton, S. *CRV* **1965**, *65*, 51. (b) Miyata, T.; Ishino, Y.; Hirashima, T. *S* **1978**, 834.
15. Ayynger, N. R.; Lugada, A. C.; Nikrad, P. V.; Sharma, V. K. *S* **1981**, 640.
16. (a) Pietra, S. *AC(R)* **1955**, *45*, 850. (b) Rondestvedt, C. S., Jr.; Johnson, T. A. *Chem. Eng. News* **1977**, 38. (c) Bavin, P. M. G. *OS* **1960**, *40*, 5.
17. Weiser, H. B.; Milligan, W. O.; Cook, E. L. *Inorg. Synth.* **1946**, 215.
18. Hirashima, T.; Manabe, O. *CL* **1975**, 259.
19. Han, B. H.; Shin, D. H.; Cho, S. Y. *TL* **1985**, *26*, 6233.
20. Mosby, W. L. *CI(L)* **1959**, 1348.
21. Lloyd, D.; McDougall, R. H.; Wasson, F. I. *JCS* **1965**, 822.
22. Schonberg, A.; Fateen, A. E. K.; Sammour, A. E. M. A. *JACS* **1957**, *79*, 6020.
23. Baltzly, R.; Mehta, N. B.; Russell, P. B.; Brooks, R. E.; Grivsky, E. M.; Steinberg, A. M. *JOC* **1961**, *26*, 3669.
24. Newkome, G. R.; Fishel, D. L. *JOC* **1966**, *31*, 677.
25. Day, A. C.; Whiting, M. C. *OS* **1970**, *50*, 3.
26. McBee, E. T.; Sienkowski, K. J. *JOC* **1973**, *38*, 1340.
27. Barton, D. H. R.; Basiardes, G.; Fourrey, J.-L. *TL* **1983**, *24*, 1605.
28. Barton, D. H. R.; Basiardes, G.; Fourrey, J.-L. *TL* **1984**, *25*, 1287.
29. Ciganek, E. *JOC* **1965**, *30*, 4366.
30. (a) Nenitzescu, C. D.; Solomonica, E. *OSC* **1943**, *2*, 496. (b) Smith, L. I.; Hoehn, H. H. *OSC* **1955**, *3*, 356.
31. Cope, A. C.; Smith, D. S.; Cotter, R. J. *OSC* **1963**, *4*, 377.
32. Dupuy, C.; Luche, J. L. *T* **1989**, *45*, 3437.
33. (a) Ohloff, G.; Unde, G. *HCA* **1970**, *53*, 531. (b) Schulte-Elte, K. N.; Rautenstrauch, V.; Ohloff, G. *HCA* **1971**, *54*, 1805. (c) Stork, G.; Williard, P. G. *JACS* **1977**, *99*, 7067.
34. (a) Felix, D.; Wintner, C.; Eschenmoser, A. *OS* **1976**, *55*, 52. (b) Felix, D.; Muller, R. K.; Joos, R.; Schreiber, J.; Eschenmoser, A. *HCA* **1972**, *55*, 1276.
35. ans Stoye, P. In *The Chemistry of Amides (The Chemistry of Functional Groups)*; Zabicky, J., Ed.; Interscience: New York, 1970; pp 515–600.
36. (a) *The Chemistry of the Azido Group*; Interscience: New York, 1971. (b) Pfister, J. R.; Wymann, W. E. *S* **1983**, 38.
37. (a) Mikhailov, Matyushecheva, Derkach, Yagupol'skii *ZOR* **1970**, *6*, 147. (b) Mikhailov, Matyushecheva, Yagupol'skii *ZOR* **1973**, *9*, 1847.
38. Arentzen, R.; Reese, C. B. *CC* **1977**, 270.
39. Ing, H. R.; Manske, R. H. F. *JCS* **1926**, 2348.
40. Kukolja, S.; Lammert, S. R. *JACS* **1975**, *97*, 5582 and 5583.
41. Freeman, J. P. *JOC* **1964**, *29*, 1379.
42. Reimlinger, H.; Vandewalle, J. J. M. *ANY* **1968**, *720*, 117.
43. Vequero, J. J.; Fuentes, L.; Del Castillo, J. C.; Pérez, M. I.; Garcia, J. L.; Soto, J. L. *S* **1987**, 33.
44. Kefirov, N. S.; Kozhushkov, S. I.; Kuzetsova, T. S. *T* **1986**, *42*, 709.
45. Goldman, P.; Ramos, S. M.; Wuest, J. D. *JOC* **1984**, *49*, 932.

46. Shimizu, T.; Hayashi, Y.; Miki, M.; Teramura, K. *JOC* **1987**, *52*, 2277.
47. Bodanszky, M. *The Principles of Peptide Synthesis*; Springer: New York, 1984; p 16.

Brian A. Roden
Abbott Laboratories, North Chicago, IL, USA

Hydrobromic Acid[1]

[10035-10-6] BrH (MW 80.91)

(preparation of alkyl and vinyl bromides; preparation of phenols from alkyl aryl ethers; in combination with hydrogen peroxide is an in situ source of bromine for the preparation of alkyl and aryl bromides)

Physical Data: aqueous solution forms a constant boiling azeotrope containing ca. 48% HBr at 760 mmHg; $d = 1.49\ \text{g cm}^{-3}$. Anhydrous gas, d 2.71 g L^{-1}; mp −86.9 °C; bp −66.8 °C.
Solubility: very sol water and protic solvents.
Form Supplied in: as anhydrous gas in cylinders; as aqueous solutions of various concentrations; widely available in all forms.
Analysis of Reagent Purity: titration.
Handling, Storage, and Precautions: hydrogen bromide is a corrosive, colorless, nonflammable gas which forms a white cloud when exposed to air; as concentrated solutions, hydrobromic acid is a colorless to light yellow corrosive liquid which fumes when exposed to air; the acid can cause severe skin burns, damage to the respiratory and digestive tract, and/or visual damage; repeated exposure may cause dermatitis and photosensitization; the gas and solutions of hydrobromic acid should be handled with adequate ventilation and proper skin and eye protection. Use in a fume hood.

Acid Catalysis. ***Hydrogen Bromide*** is completely ionized in all but the most concentrated aqueous solutions, making it a strong Lewis acid. However, the expense of hydrobromic acid relative to ***Hydrochloric Acid*** and other mineral acids, as well as the greater nucleophilicity of bromide, has limited its use as an acid catalyst.

Bromomethylation. Concentrated hydrobromic acid or anhydrous hydrogen bromide have been used with ***Paraformaldehyde*** or 1,3,5-trioxane for the bromomethylation of aromatic compounds (eq 1). Formation of bis(bromomethyl) ether, a carcinogenic compound, under the reaction conditions is problematic. This side reaction has limited the use of the bromomethylation process. Phenols are so reactive under the reaction conditions that they are frequently deactivated through preparation of their acyl derivatives prior to bromomethylation. The reaction may be run in the presence of Brønsted acid catalysts.[2] Treatment of dibenzyl diselenide with ***Zinc*** and hydrobromic acid followed by paraformaldehyde and hydrogen bromide produced high yields of bromomethyl benzyl selenide (eq 2).[3] Analogous chemistry is observed with benzyl sulfide (eq 3). Treatment of an aryl alkyl ketone with ***Bis(dimethylamino)methane*** and hydrogen bromide has produced moderate yields of the ketone, which was bromomethylated α to the carbonyl (eq 4).[4]

$$\text{ArH} + \text{CH}_2\text{O} + \text{HBr} \longrightarrow \text{ArCH}_2\text{Br} \quad (1)$$

$$\text{RSeSeR} + \text{CH}_2\text{O} + \text{HBr} \longrightarrow \text{RSeCH}_2\text{Br} \quad (2)$$

$$\text{RSH} + \text{CH}_2\text{O} + \text{HBr} \longrightarrow \text{RSCH}_2\text{Br} \quad (3)$$

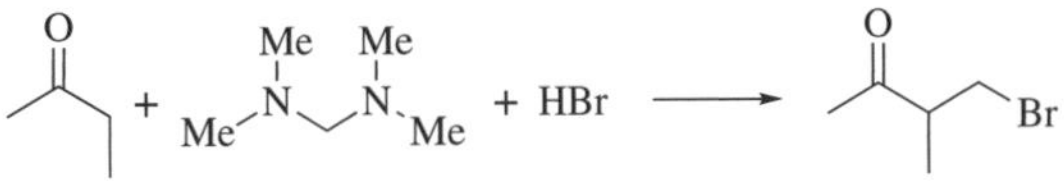

(4)

Addition to Single Bonds in Three-Membered Rings. Hydrogen bromide is less commonly used for the preparation of derivatives of terpenes containing three-membered rings than hydrochloric acid. Cleavage of cyclopropanes produces addition products which are expected to arise from bromide addition to the most stable carbonium ion (analogous to Markovnikov addition), some rearrangements do occur, and steric factors do play a role in the reaction (eq 5). Simple alkyl derivatives of the parent cyclopropane can give polymeric materials as the main product. As an example, treatment of *cis*-carane with hydrogen bromide results in cleavage of the cyclopropane ring with isolation of the tertiary bromide.[5]

(5)

Cyclopropylcarbinols react with concentrated hydrobromic acid under mild conditions to yield 1-bromo-3-butenes (eq 6).[6] The reaction is regiospecific with secondary and some tertiary alcohols. With alkynic cyclopropylcarbinols, stereospecificity of the product double bond can be controlled through the use of ***Octacarbonyldicobalt***.[7]

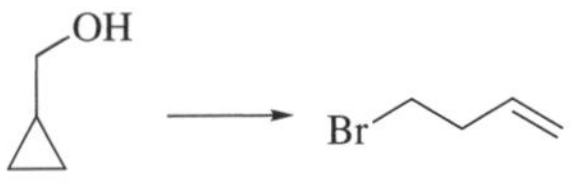

(6)

In acetic acid, acyl cyclopropane derivatives add anhydrous hydrogen bromide to yield 4-bromobutanone derivatives (eq 7). This addition is stereo- as well as regioselective.[8]

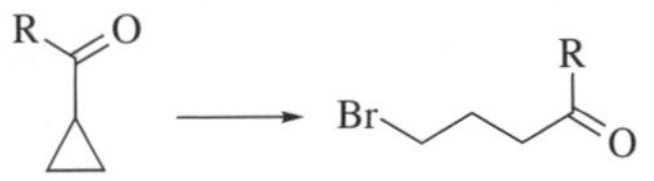

(7)

Oxiranes react readily with hydrobromic acid to yield addition products (eq 8). The reaction proceeds with inversion at carbon. The expected product is that in which the bromide adds to the least-hindered carbon. Addition to epoxycyclohexanes gives products in which attack of bromide is axial.[9] Analogous chemistry is observed with aziridines (eq 9).[10]

(8)

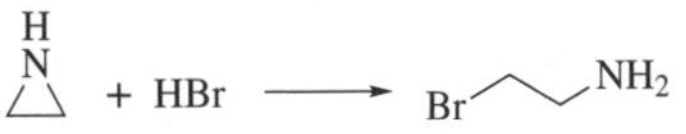

(9)

Addition to Single Bonds in Four-, Five-, and Six-Membered Rings. Although the parent acylcyclobutanes are stable to hydrogen bromide, [3.2.2]propellanes, [4.2.2]propellanes, and cubanes react to give addition products (eqs 10 and 11) or rearrangement products (eq 12), as shown.[11]

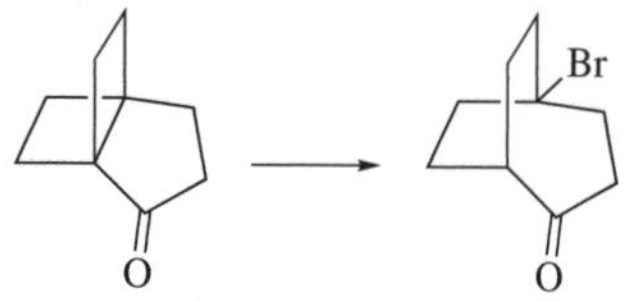

(10)

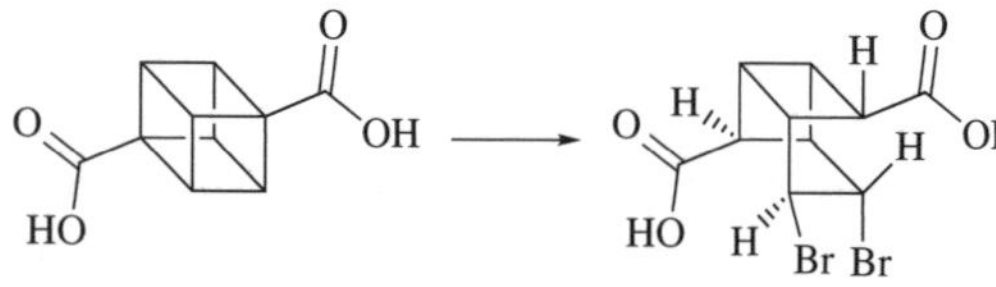

(11)

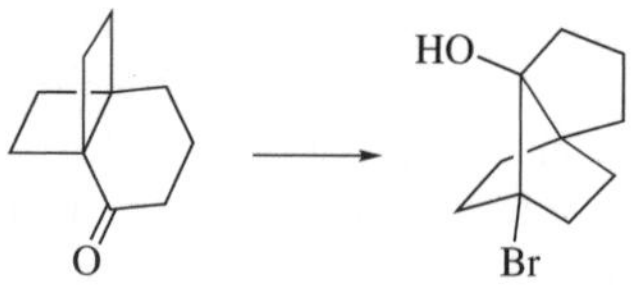

(12)

Oxetanes (eq 13) and tetrahydrofurans (eq 14) can be opened by anhydrous hydrogen bromide under a variety of conditions.[12]

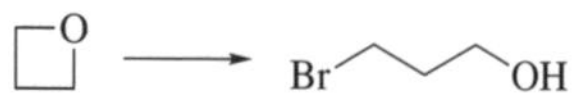

(13)

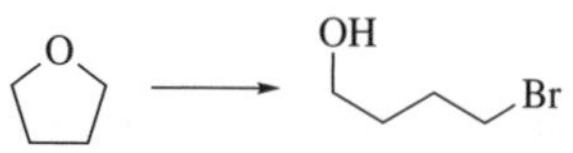

(14)

Four-, five-, and six-membered lactones add anhydrous hydrogen bromide to yield the acyclic bromide (eq 15). When run in glacial acetic acid, the free carboxylic acid is isolated; in alcohols, the ester is produced.[13]

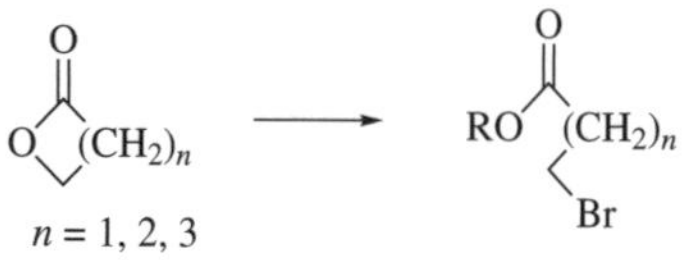

$n = 1, 2, 3$ (15)

Reaction with Ethers. Ethers react with hydrogen bromide under a variety of conditions to yield the corresponding bromide and alcohol (eq 16). Acetic acid is often used as solvent, but other carboxylic acids have also been used. Frequently, under the reaction conditions, the alcohol is also converted to its bromide.[14] Reports of explosive reactions of ethers and hydrogen bromide have been reviewed.[15]

$$R^2OR^1 + HBr \longrightarrow R^2OH + R^2Br + R^1OH + R^1Br \quad (16)$$

Aryl methyl ethers can be cleaved to phenols with hydrogen bromide in glacial acetic acid or by concentrated hydrobromic acid.[16] This transformation is commonly accomplished using boron trihalides, ***Pyridinium Chloride***, or ***Iodotrimethylsilane***.[17]

Addition to Carbon–Carbon Multiple Bonds. Hydrogen bromide reacts with alkenes more rapidly than hydrogen chloride. When care is taken to avoid radical conditions, the products which are obtained are those expected from Markovnikov addition (eq 17). ***Iron(III) Chloride***, iron(III) bromide, or ***Aluminum Bromide*** are the most commonly used Lewis acids to activate unreactive double bonds. Hydrobromination of double bonds under radical conditions can lead to mixtures of products. When an electron-withdrawing group is attached directly to the double bond, the bromide typically adds β to that group.[18]

$$R^2HC{=}CHR^1 + HBr \longrightarrow R^2H_2C{-}CHBrR^1 \quad (17)$$

Treatment of allene with anhydrous hydrogen bromide results in formation of the expected Markovnikov addition product and 1,3-dimethyl-1,3-dibromocyclobutane (eq 18). 1,3-Disubstituted allenes, when treated with anhydrous hydrogen bromide, produce mixtures of HBr addition products resulting from addition of the proton to either the central ketene carbon or to a terminal ketene carbon (eq 19). 1,1-Disubstituted allenes add the proton to the central allene carbon and bromide to the terminal carbon to produce 1,1-dialkyl-3-bromo-1-propenes (eq 20).[19]

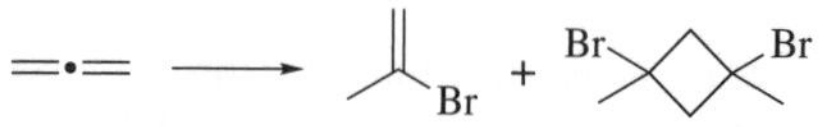

(18)

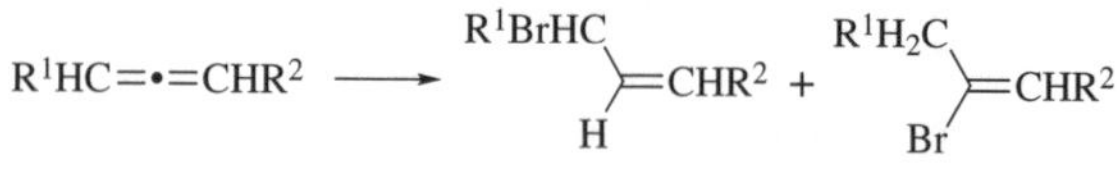

$$R^1HC{=}{\bullet}{=}CHR^2 \longrightarrow R^1BrHC(H)C{=}CHR^2 + R^1H_2C(Br)C{=}CHR^2 \quad (19)$$

$$={\bullet}{=}CR^1R^2 \longrightarrow BrH_2C(H)C{=}CR^1R^2 \quad (20)$$

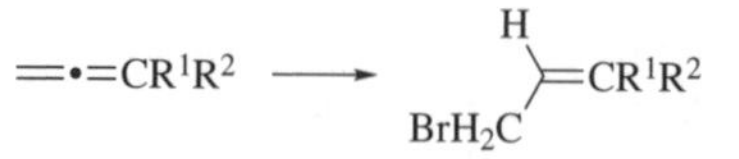

Addition of hydrogen bromide to alkynes is typically slow. Addition of ammonium bromide salts or ***Copper(I) Bromide*** produces a dramatic acceleration in the rate of the addition (eqs 21 and 22). In the absence of radicals, the isolated products are typically those expected from Markovnikov addition,[20] although mixtures of products have been reported when the carbon α to the triple bond bears an amine.[21] Under radical conditions, hydrogen bromide adds to alkynes to yield anti-Markovnikov products.[22] In the presence of ***Copper(II) Bromide*** and ammonium bromide, aqueous hydrobromic acid adds to vinylacetylene to yield 2-bromobutadiene (eq 23).[23]

$$R^2C{\equiv}CR^1 + HBr \longrightarrow R^2HC{=}CBrR^1 + R^2H_2C{-}CBr_2R^1 \quad (21)$$

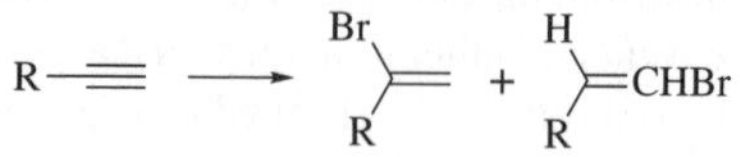

(22)

(23)

Hydrobromic acid is superior to hydrochloric acid for the conversion of 3-acylprop-2-ynal diethyl acetals to 3-acylprop-2-enoic acids (eq 24).[24]

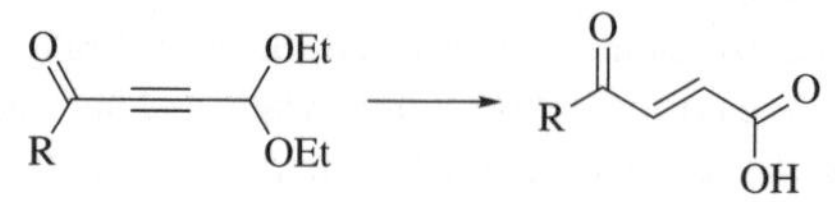

(24)

Reactions with Alcohols. Hydrobromic acid reacts with primary, secondary, tertiary, allylic, benzylic, and propargylic alcohols to give the corresponding bromides (eq 25). As is the case with the conversion of alcohols to chlorides, a large number of alternative reagents are available. Although hydrobromic acid is readily available, greater selectivity is often achieved using reagents such as ***Zinc Bromide/Triphenylphosphine/Diethyl Azodicarboxylate***, ***Phosphorus(III) Bromide***, or PPh_3/***Carbon Tetrabromide***. Treatment of alkyl phosphites, phosphonates, and diphenyl phosphinites with hydrogen bromide produces the alkyl bromide in which inversion at carbon has occurred. Yields are higher and conditions milder than the corresponding reaction with hydrogen chloride.[25]

$$\mathrm{ROH} \longrightarrow \mathrm{RBr} \qquad (25)$$

The reactions of carbohydrates and their derivatives with hydrogen bromide at the anomeric hydroxy are especially facile examples of the conversion of alcohols to bromides.[26]

When treated with hydrogen bromide, cyclobutylcarbinol rearranges to cyclopentyl bromide (eq 26).[27] The analogous [2.1.1]bicyclocarbinol produces the primary bromide with hydrogen bromide (eq 27).[28]

(26)

(27)

Concentrated hydrobromic acid reacts with 1,1-dialkylpropargyl alcohols to give products which are dependent upon the reaction conditions. Products isolated when 3-methylbut-1-yn-3-ol was the starting alcohol include 1-bromo-3,3-dimethylallene, 3-bromo-3-methyl-1-butyne, 1-bromo-3-methyl-1,3-butadiene, 1,3-dibromo-3-methylbutene, and 1,2,3-tribromo-3-methylbutane (eq 28). Tertiary propargyl alcohols, in the presence of copper(I) bromide, ammonium bromide, and 45–48% hydrobromic acid, rapidly produce 1-bromoallenes. Secondary propargyl alcohols produce 1-bromoallenes using 60% hydrobromic acid, copper(I) bromide, and ammonium bromide.[29]

(28)

Reactions with Diazo Compounds. Arylamines can be converted to aryl bromides by treatment with ***Sodium Nitrite***/hydrobromic acid/***Copper*** or copper(I) bromide (eq 29).[30] Hydrobromic acid converts α-diazo ketones to α-bromo ketones in good yield (eq 30).[31] Pure enantiomers of serine and threonine give good yields and high enantiomeric purity of α-bromo acids when treated with nitrite and hydrobromic acid.[32]

$$\mathrm{ArNH_2} \longrightarrow [\mathrm{ArN_3}] \longrightarrow \mathrm{ArBr} \qquad (29)$$

(30)

Reactions with Nitriles. Addition of anhydrous hydrogen bromide to nitriles produces imidoyl bromides (eq 31).[33] Treatment of *N*-alkylimidoyl bromides with hydrogen bromide results in isolation of the corresponding iminium bromides (eq 32).[34] Methylene bis(thiocyanate) reacts with hydrogen bromide to produce the cyclic imidoyl bromide (eq 33).[35] Methyl and phenyl thiocyanate react with 2 equiv of hydrogen bromide to produce 1-bromothioformimidate salts (eq 34).[36] Cyanogen di-*N*-oxide reacts with hydrobromic acid to produce the hydroxamoyl bromide analog of oxalyl bromide (eq 35).[37]

$$\mathrm{R{-}C{\equiv}N} \longrightarrow \mathrm{RC(=NH)Br} \qquad (31)$$

$$\mathrm{R^2C(=NR^1)Br} \longrightarrow \mathrm{R^2C(=N^+HR^1)Br\ Br^-} \qquad (32)$$

$$\mathrm{CH_2(SCN)_2} \longrightarrow \text{(cyclic imidoyl bromide, NH•HBr)} \qquad (33)$$

$$\mathrm{RSCN} \longrightarrow \mathrm{RSC(=NH_2^+)Br\ Br^-} \qquad (34)$$

$$\mathrm{ONC{-}CNO} \longrightarrow \mathrm{BrC(=NOH)C(=NOH)Br} \qquad (35)$$

Reactions with Sulfur Compounds. Thiols react with paraformaldehyde and hydrobromic acid to yield bromomethyl thioethers (see bromomethylation above). Benzenesulfonamides, benzenesulfonohydrazides, and benzenesulfinic acids can all react with hydrobromic acid to yield disulfides or sulfenyl bromides, depending upon the reaction conditions (eqs 36–38). Hydrogen bromide appears to be better than hydrogen chloride for the preparation of disulfides from benzenesulfonamides, but less satisfactory than hydrogen chloride for the conversion of benzenesulfonohydrazides to disulfides.[38] Sulfoxides are converted into bromosulfonium bromides or sulfides (eq 39).[39] Chloro(trifluoromethyl)sulfine reacts with anhydrous hydrogen bromide to produce 1-bromo-1-chloro-2,2,2-trifluoroethylsulfenyl bromide in high yield (eq 40).[40] Additional information is included under the section on the in situ generation of ***Bromine*** (see below).

$$\mathrm{ArSO_2NR_2} \longrightarrow \mathrm{ArSBr} + \mathrm{ArSSAr} \qquad (36)$$

$$\mathrm{ArSO_2NHNR_2} \longrightarrow \mathrm{ArSBr} + \mathrm{ArSSAr} \qquad (37)$$

$$\mathrm{ArSO_2H} \longrightarrow \mathrm{ArSBr} + \mathrm{ArSSAr} \qquad (38)$$

$$\mathrm{Me_2S{=}O} \longrightarrow \mathrm{Me_2S^+Br\ Br^-} \longrightarrow \mathrm{MeSMe} \qquad (39)$$

$$\mathrm{CF_3C(=S{=}O)Cl} \longrightarrow \mathrm{CF_3CBrCl{-}SBr} \qquad (40)$$

Reactions with Silicon Compounds. Phenylsilanes react with hydrogen bromide to yield benzene and the bromosilane (eq 41). The reaction is more facile than the reaction with hydrogen chloride. Increasing the electronegativity of substituents on silicon decreases the ease with which the aryl–silicon bond is broken.[41] *t*-Butyldimethylsilyl ethers of phenols are cleaved to phenols at rt using a mixture of hydrobromic acid and ***Potassium Fluoride*** (eq 42).[42] Triethylaminosilanes are converted to the corresponding bromosilanes in the presence of hydrobromic acid/***Sulfuric Acid*** (eq 43).[43]

$$\mathrm{Ph{-}SiR_3} \longrightarrow \mathrm{Br{-}SiR_3} \qquad (41)$$

$$\mathrm{PhO{-}SiR_3} \longrightarrow \mathrm{PhOH} \qquad (42)$$

$$\mathrm{X(CH_3)CH{-}Si(NH_2)Et_2} \longrightarrow \mathrm{X(CH_3)CH{-}SiBrEt_2} \qquad (43)$$

Transhalogenation Reactions. Alkyl chlorides can be converted to alkyl bromides using hydrogen bromide in the presence of iron(III) bromide (eq 44).[44] This conversion can also be effected under neutral conditions by heating the chloride with a metal bromide in acetone, an alcohol, or ethyl bromide.[45]

$$\mathrm{RCl} \longrightarrow \mathrm{RBr} \qquad (44)$$

Acid chlorides react with anhydrous hydrogen bromide to yield the corresponding acid bromides (eq 45).[46] This conversion may also be effected using ***Bromotrimethylsilane***.[47]

$$\mathrm{RC(O)Cl} \longrightarrow \mathrm{RC(O)Br} \qquad (45)$$

Trichloromethylsulfenyl chloride reacts with concentrated hydrobromic acid to yield trichloromethylsulfenyl bromide (eq 46).[48] Dichloromethyl methyl sulfide produces dibromomethyl methyl sulfide when treated with anhydrous hydrogen bromide (eq 47).[49]

$$\mathrm{Cl_3CSCl} \longrightarrow \mathrm{Cl_3CSBr} \qquad (46)$$

$$\mathrm{Cl_2HCSMe} \longrightarrow \mathrm{Br_2HCSMe} \qquad (47)$$

Bromide Isomerization. α,α-Dibromo ketones equilibrate to α,α′-dibromo ketones in the presence of dilute hydrobromic acid (eq 48).[50]

$$\mathrm{Br_2CHC(O)CH_3} \rightleftharpoons \mathrm{BrCH_2C(O)CH_2Br} \qquad (48)$$

Organoselenium, Organogermanium, and Organorhenium Chemistry. Selenols react with paraformaldehyde and hydrogen bromide to produce bromomethyl selenides (see bromomethylation above). Methyl phenyl selenides are cleaved by hydrogen bromide in acetic acid to yield phenyl selenols (eq 49).[51] Alkyl phenyl selenoxides react with hydrogen bromide to yield the alkyl bromides (eq 50).[52] Selenonium nitroylides give bromonitromethane derivatives when reacted with hydrogen bromide in ether (eq 51).[53]

$$\mathrm{PhSeMe} \longrightarrow \mathrm{PhSeH} \qquad (49)$$

$$\mathrm{PhSe(O){-}CHR^1R^2} \longrightarrow \mathrm{Br{-}CHR^1R^2} \qquad (50)$$

$$\mathrm{Ph_2Se{=}C(NO_2)R} \longrightarrow \mathrm{Br{-}CH(NO_2)R} \qquad (51)$$

1,1-Dihydroxy-2,3-diphenylgermirene is converted to the dibromide with anhydrous hydrogen bromide in benzene (eq 52).[54]

$$\mathrm{Ph_2C_2Ge(OH)_2} \longrightarrow \mathrm{Ph_2C_2GeBr_2} \qquad (52)$$

The pure enantiomer of the pseudotetrahedral rhenium alkyl complex in eq 53 reacts with concentrated hydrobromic acid to produce the pure reduced enantiomer. Cleavage of the rhenium–carbon bond occurs with retention at both carbon and rhenium.[55]

$$\mathrm{Ar{-}C(T)(D){-}Re(NO)(PPh_3)(Cp)} \longrightarrow \mathrm{Ar{-}C(H)(D)(T)} \qquad (53)$$

In Situ Generation of Bromine. ***Dimethyl Sulfoxide*** reacts with hydrobromic acid at about 80 °C to produce dimethyl sulfide, water, and bromine (eq 54). The DMSO/HBr reagent has been used to oxidize 1,3-diketones to 1,2,3-triketones, acetophenones to phenylglyoxals, benzylamines to imines, and 4,5-dihydropyridazin-3(2*H*)-ones to pyridazin-3(2*H*)-ones.[56] The combination is also effective in converting stilbenes, 1,2-dibromo-1,2-diarylethanes, and 2-bromo-1,2-diarylethanols to benzils.[57] It is possible to use a catalytic amount of hydrogen bromide in some of these reactions.

$$\mathrm{Me_2S{=}O} + \mathrm{HBr} \longrightarrow \mathrm{Me_2S} + \mathrm{H_2O} + \mathrm{Br_2} \qquad (54)$$

Hydrobromic acid and ***Hydrogen Peroxide*** are an effective combination for the in situ generation of bromine (eq 55). This combination of reagents can be used for the bromination of alkenes and aromatics, for the preparation of bromohydrins from alkenes, and for the preparation of benzylic bromides.[58]

$$2\,\mathrm{HBr} + \mathrm{H_2O_2} \longrightarrow 2\,\mathrm{H_2O} + \mathrm{Br_2} \qquad (55)$$

Hydrobromic acid is a source of bromine when irradiated in the presence of air or oxygen (eq 56). Ethylbenzene, when photooxidized, yields a mixture of acetophenone, 1-phenylethanol, and 1-phenylbromoethane.[59]

$$4\,\mathrm{HBr} + \mathrm{O_2} \xrightarrow{h\nu} 2\,\mathrm{Br_2} + 2\,\mathrm{H_2O} \qquad (56)$$

Related Reagents. Boron Tribromide; Formaldehyde–Hydrogen Bromide; Hydrogen Bromide; Phosphorus(III) Bromide.

---

1. (a) Brasted, R. C. In *Comprehensive Inorganic Chemistry*; Sneed, M. C.; Maynard, J. L.; Brasted, R. C., Eds.; Van Nostrand: New York, 1954; Vol. III, p 118. (b) Downs, A. J.; Adams, C. J. In *Comprehensive Inorganic Chemistry*; Bailar, J. C., Jr., Ed.; Pergamon: Oxford, 1973; Vol. 2, p 1280.
2. (a) Fields, D. L.; Miller, J. B.; Reynolds, D. D. *JOC* **1964**, *29*, 2640. (b) Schetty, G. *HCA* **1948**, *31*, 1229 (*CA* **1949**, *43*, 203d). (c) Bailey, P. S.; Bath, S. S.; Thomsen, W. F.; Nelson, H. H.; Kawas, E. E. *JOC* **1956**, *21*, 297. (d) Böhmer, V.; Marschollek, F.; Zetta, L. *JOC* **1987**, *52*, 3200. (e) Mitchell, R. H.; Iyer, V. S. *SL* **1989**, 55.
3. Reich, H. J.; Jasperse, C. P.; Renga, J. M. *JOC* **1986**, *51*, 2981.
4. Moussavi, Z.; Depreux, P.; Lesieur, D. *SC* **1991**, *21*, 271.
5. Bardyshev, I. I.; Buinova, É. F.; Protashchik, I. V. *JOU* **1971**, *7*, 2398 (*CA* **1972**, *76*, 46 311z).
6. (a) Julia, M.; Julia, S.; Guegan, R. *BSF* **1960**, 1072 (*CA* **1961**, *55*, 5567b). (b) Julia, M.; Julia, S.; Tchen, S.-Y. *BSF* **1961**, 1849 (*CA* **1962**, *57*, 4535f). (c) Julia, M.; Julia, S.; Amaudric du Chaffaut, J. *BSF* **1960**, 1735 (*CA* **1961**, *55*, 17 525d). (d) Gualtieri, F.; Teodori, E.; Bellucci, C.; Pesce, E.; Piacenza, G. *JMC* **1985**, *28*, 1621. (e) Julia, M.; Julia, S.; Stalla-Bourdillon, B.; Descoins, C. *BSF* **1964**, 2533 (*CA* **1965**, *62*, 5182d). (f) Julia, M.; Descoins, C. *BSF* **1962**, 1933 (*CA* **1963**, *58*, 12 414h). (g) Hatakeyama, S.; Numata, H.; Osanai, K.; Takano, S. *CC* **1989**, 1893.
7. (a) McCormick, J. P.; Barton, D. L. *CC* **1975**, 303. (b) Descoins, C.; Samain, D. *TL* **1976**, 745.
8. (a) Takano, S.; Iwata, H.; Ogasawara, K. *H* **1978**, *9*, 1249. (b) Murray, R. K., Jr.; Morgan, T. K., Jr. *JOC* **1975**, *40*, 2642. (c) Grieco, P. A.; Masaki, Y. *JOC* **1975**, *40*, 150.
9. (a) For a discussion of factors influencing oxirane cleavage, see Buchanan, J. G.; Sable, H. Z. In *Selective Organic Transformations*; Thyagarajan, B. S., Ed.; Wiley: New York, 1972; Vol. 2, p 1. (b) For a review on the mechanism of epoxide cleavage, see Parker, R. E.; Isaacs, N. S. *CRV* **1959**, *59*, 737. (c) Bordwell, F. G.; Frame, R. R.; Strong, J. G. *JOC* **1968**, *33*, 3385. (d) Layachi, K.; Ariès-Gautron, I.; Guerro, M.; Robert, A. *T* **1992**, *48*, 1585. (e) Hudrlik, P. F.; Hudrlik, A. M.; Rona, R. J.; Misra, R. N.; Withers, G. P. *JACS* **1977**, *99*, 1993.
10. (a) Jenkins, T. C.; Naylor, M. A.; O'Neill, P.; Threadgill, M. D.; Cole, S.; Stratford, I. J.; Adams, G. E.; Fielden, E. M.; Suto, M. J.; Stier, M. A. *JMC* **1990**, *33*, 2603. (b) Heine, H. W.; Proctor, Z. *JOC* **1958**, *23*, 1554. (c) Buss, D. H.; Hough, L.; Richardson, A. C. *JCS* **1965**, 2736. (d) Gensler, W. J. *JACS* **1948**, *70*, 1843.
11. (a) Eaton, P. E.; Jobe, P. G.; Reingold, I. D. *JACS* **1984**, *106*, 6437. (b) Eaton, P. E.; Jobe, P. G.; Nyi, K. *JACS* **1980**, *102*, 6636. (c) Eaton, P. E.; Millikan, R.; Engel, P. *JOC* **1990**, *55*, 2823.
12. (a) Wilson, E. R.; Frankel, M. B. *JOC* **1985**, *50*, 3211. (b) Paul, R.; Tchelitcheff, S. *BSF* **1953**, 1014. (c) Ogawa, S.; Suzuki, M.; Tonegawa, T. *BCJ* **1988**, *61*, 1824.
13. (a) Zaugg, H. E. *JACS* **1950**, *72*, 2998. (b) Plieninger, H. *CB* **1950**, *83*, 268. (c) Stork, G.; Hill, R. K. *JACS* **1957**, *79*, 495. (d) Cottrell, I. F.; Hands, D.; Kennedy, D. J.; Paul, K. J.; Wright, S. H. B.; Hoogsteen, K. *JCS(P1)* **1991**, 1091. (e) Orlek, B. S.; Wadsworth, H.; Wyman, P.; Hadley, M. S. *TL* **1991**, *32*, 1241.
14. (a) Landini, D.; Montanari, F.; Rolla, F. *S* **1978**, 771. (b) Newkome, G. R.; Gupta, V. K.; Griffin, R. W.; Arai, S. *JOC* **1987**, *52*, 5480. (c) Grubbs, R. H.; Pancoast, T. A.; Grey, R. A. *TL* **1974**, 2425.
15. Leleu, M. J. *Cah. Notes Doc.* **1976**, *82*, 127 (*CA* **1978**, *88*, 26 952d).
16. (a) Doxsee, K. M.; Feigel, M.; Stewart, K. D.; Canary, J. W.; Knobler, C. B.; Cram, D. J. *JACS* **1987**, *109*, 3098. (b) Ramesh, D.; Kar, G. K.; Chatterjee, B. G.; Ray, J. K. *JOC* **1988**, *53*, 212.
17. For a compilation of specific reagents used to convert aryl ethers to phenols, see Larock, R. C. *Comprehensive Organic Transformations*; VCH: New York, 1989; p 502.
18. (a) Roedig, A. *MOC* **1960**, *V/4*, 102. (b) Larock, R. C.; Leong, W. W. *COS* **1991**, *4*, 279. (c) Traynham, J. G.; Pascual, O. S. *JOC* **1956**, *21*, 1362. (d) Dowd, P.; Shapiro, M.; Kang, J. *T* **1984**, *40*, 3069.
19. (a) Taylor, D. R. *CRV* **1967**, *67*, 338. (b) Caserio, M. C. In *Selective Organic Transformations*; Thyagarajan, B. S., Ed.; Wiley: New York, 1970; Vol. 1, p 239.
20. (a) Hiyama, T.; Wakasa, N.; Ueda, T.; Kusumoto, T. *BCJ* **1990**, *63*, 640. (b) Baird, M. S.; Dale, C. M.; Lytollis, W.; Simpson, M. J. *TL* **1992**, *33*, 1521. (c) Grob, C. A.; Cseh, G. *HCA* **1964**, *47*, 194 (*CA* **1964**, *60*, 10 492g). (d) Cousseau, J. *S* **1980**, 805.
21. Mori, M.; Higuchi, Y.; Kagechika, K.; Shibasaki, M. *H* **1989**, *29*, 853.
22. Skell, P. S.; Allen, R. G. *JACS* **1958**, *80*, 5997.
23. Keegstra, M. A.; Verkruijsse, H. D.; Andringa, H.; Brandsma, L. *SC* **1991**, *21*, 721.
24. Obrecht, D.; Weiss, B. *HCA* **1989**, *72*, 117.
25. (a) Hudson, H. R. *S* **1969**, 112. (b) For a compilation of specific reagents used to convert alcohols to bromides, see Larock, R. C. *Comprehensive Organic Transformations*; VCH: New York, 1989; pp 354 and 361.
26. (a) Fernez, A.; Stoffyn, P. J. *T* **1959**, *6*, 139 (*CA* **1959**, *53*, 21 675h). (b) Wysocki, R. J.; Siddiqui, M. A.; Barchi, J. J.; Driscoll, J. S.; Marquez, V. E. *S* **1991**, 1005.
27. Demjanow, N. J. *CB* **1907**, *40*, 4959.
28. Wiberg, K. B.; Lowry, B. R.; Colby, T. H. *JACS* **1961**, *83*, 3998.
29. (a) Landor, S. R.; Patel, A. N.; Whiter, P. F.; Greaves, P. M. *JCS(C)*, **1966**, 1223. (b) Moulin, F. *HCA* **1951**, *34*, 2416 (*CA* **1952**, *46*, 7036h). (c) Favorskaya, T. A. *ZOB* **1940**, *10*, 461 (*CA* **1940**, *34*, 7845[1]).
30. (a) Rinehart, K. L., Jr.; Kobayashi, J.; Harbour, G. C.; Gilmore, J.; Mascal, M.; Holt, T. G.; Shield, L. S.; Lafargue, F. *JACS* **1987**, *109*, 3378. (b) Bigelow, L. A. *OSC* **1941**, *1*, 135. (c) Hartwell, J. L. *OSC* **1955**, *3*, 185.
31. (a) Dauben, W. G.; Hiskey, C. F.; Muhs, M. A. *JACS* **1952**, *74*, 2082. (b) Pettit, G. R.; Green, B.; Das Gupta, A. K.; Whitehouse, P. A.; Yardley, J. P. *JOC* **1970**, *35*, 1381.
32. (a) Larchevêque, M.; Petit, Y. *TL* **1987**, *28*, 1993. (b) Larchevêque, M.; Mambu, L.; Petit, Y. *SC* **1991**, *21*, 2295.
33. Ulrich, H. *The Chemistry of Imidoyl Halides*; Plenum: New York, 1968; pp 66–68.
34. Ulrich, H. *The Chemistry of Imidoyl Halides*; Plenum: New York, 1968; p 100.
35. Johnson, F.; Madroñero, R. *Adv. Heterocycl. Chem.* **1966**, *6*, 131.
36. Allenstein, E.; Quis, P. *CB* **1964**, *97*, 3162 (*CA* **1965**, *62*, 1562g).
37. Grundmann, C.; Mini, V.; Dean, J. M.; Frommeld, H. D. *LA* **1965**, *687*, 191.
38. (a) Burawoy, A.; Vellins, C. E. *JCS* **1954**, 90. (b) Burawoy, A.; Turner, C. *JCS* **1950**, 469. (c) Yung, D. K.; Forrest, T. P.; Manzer, A. R.; Gilroy, M. L. *JPS* **1977**, *66*, 1009. (d) Searles, S.; Nukina, S. *CRV* **1959**, *59*, 1095.
39. Iselin, B. *HCA* **1961**, *44*, 61 (*CA* **1961**, *55*, 17 522d).
40. Fritz, H.; Sundermeyer, W. *CB* **1989**, *122*, 1757 (*CA* **1989**, *111*, 153 179a).
41. (a) Mitter, F. K.; Pollhammer, G. I.; Hengge, E. *JOM* **1986**, *314*, 1. (b) Hager, R.; Steigelmann, O.; Müller, G.; Schmidbaur, H. *CB* **1989**, *122*, 2115. (c) Schmidbaur, H.; Zech, J.; Rankin, D. W. H.; Robertson, H. E. *CB* **1991**, *124*, 1953. (d) Matsumoto, H.; Yokoyama, N.; Sakamoto, A.; Aramaki, Y.; Endo, R.; Nagai, Y. *CL* **1986**, 1643. (e) Fritz, G.; Kummer, D. *Z. Anorg. Allg. Chem.* **1961**, *308*, 105 (*CA* **1961**, *55*, 18 412a).
42. Sinhababu, A. K.; Kawase, M.; Borchardt, R. T. *S* **1988**, 710.
43. Bailey, D. L.; Sommer, L. H.; Whitmore, F. C. *JACS* **1948**, *70*, 435.
44. (a) Yoon, K. B.; Kochi, J. K. *CC* **1987**, 1013. (b) Roedig, A. *MOC* **1960**, *V/4*, 354.
45. (a) Bowers, S. D., Jr.; Sturtevant, J. M. *JACS* **1955**, *77*, 4903. (b) Bailey, W. J.; Fujiwara, E. *JACS* **1955**, *77*, 165. (c) Willy, W. E.; McKean, D. R.; Garcia, B. A. *BCJ* **1976**, *49*, 1989.

46. Staudinger, H.; Anthes, E. *CB* **1913**, *46*, 1417 (*CA* **1913**, *7*, 2576).
47. Schmidt, A. H.; Russ, M.; Grosse, D. *S* **1981**, 216 (*CA* **1981**, *95*, 42 299w).
48. Dear, R. E. A.; Gilbert, E. E. *S* **1972**, 310.
49. Boberg, F.; Winter, G.; Schultze, G. R. *CB* **1956**, *89*, 1160 (*CA* **1957**, *51*, 3434f).
50. (a) Djerassi, C.; Scholz, C. R. *JACS* **1947**, *69*, 2404. (b) Szabó, L.; Tóth, I.; Tke, L.; Kolonits, P.; Szantay, C. *CB* **1976**, *109*, 3390 (*CA* **1977**, *86*, 55 266x). (c) Djerassi, C.; Scholz, C. R. *JOC* **1948**, *13*, 697.
51. Christiaens, L.; Renson, M. *BSB* **1970**, *79*, 235 (*CA* **1970**, *73*, 3726j).
52. Hevesi, L.; Sevrin, M.; Krief, A. *TL* **1976**, 2651.
53. Semenov, V. V.; Mel'nikova, L. G.; Shevelev, S. A.; Fainzil'berg, A. A. *IZV* **1980**, 138 (*CA* **1980**, *92*, 215 200a).
54. Volpin, M. E.; Koreshkov, Yu. D.; Dulova, V. G.; Kursanov, D. N. *T* **1962**, *18*, 107.
55. O'Connor, E. J.; Kobayashi, M.; Floss, H. G.; Gladysz, J. A. *JACS* **1987**, *109*, 4837.
56. (a) Schipper, E.; Cinnamon, M.; Rascher, L.; Chiang, Y. H.; Oroshnik, W. *TL* **1968**, 6201. (b) Fletcher, T. L.; Pan, H.-L. *JCS* **1965**, 4588. (c) Gilman, H.; Eisch, J. *JACS* **1955**, *77*, 3862. (d) Desmond, R.; Mills, S.; Volante, R. P.; Shinkai, I. *SC* **1989**, *19*, 379. (e) Nakao, T.; Obata, M.; Yamaguchi, Y.; Tahara, T. *CPB* **1991**, *39*, 524.
57. (a) Yusubov, M. S.; Filimonov, V. D. *JOU* **1989**, *25*, 199. (b) Yusubov, M. S.; Filimonov, V. D. *JOU* **1989**, *25*, 1410. (c) Yusubov, M. S.; Filimonov, V. D.; Ogorodnikov, V. D. *IZV* **1991**, 868 (*CA* **1991**, *115*, 28 773w).
58. (a) Johnson, R.; Reeve, K. *Spec. Chem.* **1992**, *12*, 292. (b) Brandsma, L.; de Jong, R. L. P. *SC* **1990**, *20*, 1697.
59. (a) Fuchs, B.; Mayer, W. J. W.; Abramson, S. *CC* **1985**, 1711. (b) Nakada, M.; Fukushi, S.; Hirota, M. *BCJ* **1990**, *63*, 944.

John E. Mills
R. W. Johnson Pharmaceutical Research Institute, Spring House, PA, USA

Hydrochloric Acid[1]

ClH

[7647-01-0] ClH (MW 36.46)

(Strong Brønsted acid used for general acid catalysis; preparation of salts of amines; preparation of alkyl, allyl, aryl, benzyl, and vinyl chlorides; protodemetalation reagent; in combination with hydrogen peroxide is an in situ source of chlorine for the preparation of alkyl and aryl chlorides)

Physical Data: aqueous solution: freezing point of ca. 31% solution −46 °C; forms constant boiling azeotrope containing 20.22% HCl with water (bp 108.58 °C/760 mmHg; *d* 1.096 g cm^{-3}). Anhydrous gas: *d* 1.639 g L^{-1}; mp −114.22 °C; bp −85.05 °C/760 mmHg.

Solubility: very sol water, protic solvents, ethers.

Form Supplied in: as anhydrous gas in cylinders; as aqueous solutions of various concentrations; widely available in all forms.

Analysis of Reagent Purity: titration.

Handling, Storage, and Precautions: hydrogen chloride is a corrosive, colorless, nonflammable gas which forms a white cloud when exposed to air; as concentrated solutions, hydrochloric acid is a colorless to light-yellow corrosive liquid which fumes when exposed to air; the acid can cause severe skin burns, damage to the respiratory and digestive tract, and/or visual damage; repeated exposure may cause dermatitis and photosensitization; the gas and solutions of hydrochloric acid should be handled with adequate ventilation and proper skin and eye protection. Use in a fume hood.

Acid Catalysis. ***Hydrogen Chloride*** is completely ionized in all but the most concentrated aqueous solutions. In addition, the nucleophilicity[2] of the halide ions follows the general order $I^- > Br^- > Cl^- > F^-$. Consequently, in many of the reactions in which it is employed, it is used as an acid catalyst. Reactions employing hydrochloric acid as a catalyst include the hydrolysis of esters to acids, the hydrolysis of nitriles and imides to amides, acids, and amines,[3] the hydrolysis of amides to acids and amines,[4] the hydrolysis of imines and enamines to ketones and amines, the hydrolysis of nitroso compounds to ketones, the hydrolysis of ketals, aminals, and enol ethers to ketones, the hydrolysis of acetals to aldehydes, and the hydrolysis of ethers to alcohols.[5] Anhydrous hydrogen chloride in an alcohol is frequently used in Fischer esterifications and in acetalization reactions. Hydrochloric acid has been used to effect numerous molecular rearrangements. Carbon–carbon bond-forming reactions catalyzed by hydrochloric acid include the aldol condensation and the Mannich reaction.[6] The Bergius–Willstatter saccharification process utilizes hydrochloric acid to convert cellulose to fermentable sugar.

Amine Salt Formation. Many amines form solid hydrochloride salts. The low toxicity of chloride ion and favorable physical properties of many of these salts has resulted in the use of such salts for a large number of pharmaceuticals.[7] Hydrochloride salts are also used for characterization of amines.

Chloroalkylation. Concentrated hydrochloric acid and anhydrous hydrogen chloride have been used with ***Paraformaldehyde*** for the chloromethylation of aromatic compounds (eq 1). Formation of bis(chloromethyl) ether, a carcinogenic compound, under the reaction conditions is problematic. This side reaction has limited the use of the chloromethylation process.[8] In the presence of thiols, acetaldehyde will react with hydrogen chloride to yield alkyl chloroethyl thioethers (eq 2).[9]

$$ArH + CH_2O + HCl \longrightarrow ArCH_2Cl \quad (1)$$

$$RSH + RCHO + HCl \longrightarrow RSCHClR \quad (2)$$

Amidomethylation. *N*-Acylhemiaminals can react with pyrroles in ethanol saturated with hydrogen chloride to produce the amidomethylated product (eq 3). Such amidomethylations are more commonly performed using ***Sulfuric Acid*** as catalyst.[10]

(3)

Addition to Single Bonds in Three-Membered Rings. Hydrogen chloride will react with some cyclopropanes to yield ring-opened products (eq 4). These addition reactions typically yield products expected from Markovnikov addition. Δ^3-Carene (**1**) and Δ^4-carene (**2**) both produce a mixture of sylvestrene dihydrochloride (**3**) and dipentene dihydrochloride (**4**).[11]

$$\text{cyclopropane} + HCl \longrightarrow Cl\text{–}CH_2CH_2CH_3 \quad (4)$$

(**1**) (**2**) (**3**) (**4**)

Anhydrous hydrogen chloride can react with oxiranes and aziridines to yield addition products (eqs 5 and 6). Oxiranes and aziridines derived from cyclohexenes open preferentially through axial attack by chloride.[12] Other Lewis acids, including ***Aluminum Chloride***, ***Tin(IV) Chloride***, and ***Iron(III) Chloride***, also produce 2-chloroethanols when reacted with oxiranes. In the presence of sulfur nucleophiles, good yields of the sulfur addition product may be obtained.[13]

$$\text{oxirane} + HCl \longrightarrow Cl\text{–}CH_2CH_2\text{–}OH \quad (5)$$

$$\text{aziridine} + HCl \longrightarrow Cl\text{–}CH_2CH_2\text{–}NH_2 \quad (6)$$

Addition to Single Bonds in Five-Membered Rings. Treatment of γ-butyrolactone with hydrochloric acid produces 4-chlorobutyric acid in good yield (eq 7).[14] Ring opening of 1,4-dihydro-1,4-epoxybenzene derivatives with hydrochloric acid results in isolation of the corresponding phenols (eq 8). This aromatization may be very regiospecific.[15,5a]

(7)

(8)

Reaction with Ethers. In addition to the reactions with cyclic ethers cited above, hydrogen chloride reacts with acyclic ethers to produce the corresponding alcohols and chlorides (eq 9). The rate of the reaction is dependent upon the structure of the ether.[16] In the presence of ***Zinc Chloride***, hydrogen chloride reacts with propargyl ethers to produce propargyl chlorides (eq 10).[17]

$$R^2OR^1 + HCl \longrightarrow R^2OH + R^2Cl + R^1OH + R^1Cl \quad (9)$$

(10)

Addition to Carbon–Carbon Multiple Bonds[1b]. Hydrochloric acid reacts with alkenes to produce either alcohols or alkyl chlorides. Anhydrous hydrogen chloride is typically the reagent used for the preparation of alkyl chlorides (eq 11). The product of the reaction is dependent upon the substrate and reaction conditions; Markovnikov addition is typically observed, but addition may be either *syn* or *anti*. The kinetic product formed from addition of HCl to alkenes may be unstable under the reaction conditions and may rearrange to yield thermodynamically more stable products. Addition of hydrogen chloride to α-pinene (**5**) leads initially to pinene hydrochloride (**6**), which isomerizes mainly to bornyl chloride (**7**) containing some fenchyl chloride (**8**).[18] Analogous rearrangements are observed with camphenes.[19] Addition to some alkenes and alkynes may require elevated temperatures, elevated pressures, or addition of a Lewis acid. The reaction with alkynes produces vinyl chlorides and dichloroalkanes (eq 12).[20] The addition to conjugated double bonds frequently leads to products which are a mixture of 1,2- and 1,4-addition products. When the diene is conjugated to an electron-withdrawing group, anti-Markovnikov addition is common.[21] Selectivity may be observed in the reaction of butadiene derivatives.[22] Allenes frequently give mixtures of products due to acid-catalyzed migration of double bonds. The major product at low temperature is frequently the product formed from protonation at the central allene carbon.[23]

$$R^2HC{=}CHR^1 + HCl \longrightarrow R^2H_2C\text{–}CHClR^1 \quad (11)$$

(**5**) (**6**) (**7**) (**8**)

$$R^2C{\equiv}CR^1 + HCl \longrightarrow R^2HC{=}CClR^1 + R^2H_2C\text{–}CCl_2R^1 \quad (12)$$

Addition of anhydrous hydrogen chloride to 1-nitro-1-alkenes can produce 1,2-dichloroaldoximes (eq 13).[24]

(13)

Reactions with Alcohols. Concentrated hydrochloric acid and anhydrous hydrogen chloride react with alcohols to produce alkyl chlorides (eq 14).[25] Allylic,[26] benzylic,[27] or tertiary alcohols[28] typically are most useful as substrates for conversion to chlorides. Rearrangement may occur. On a laboratory scale, hydrochloric acid has been largely replaced by the use of phosphorus reagents such as ***Triphenylphosphine***/CCl_4, ***Triphenylphosphine***

Dichloride, or Ph_3P/***Diethyl Azodicarboxylate***/Cl^- to achieve the conversion of an alcohol to a chloride.[29] These reagents frequently give less rearrangement and more stereospecificity than hydrochloric acid, ***Phosphorus(V) Chloride***, ***Phosphorus Oxychloride***, ***Thionyl Chloride***, ***Phosphorus(III) Chloride***, or ***Dimethylchloromethyleneammonium Chloride***. Treatment of alkyl phosphites, phosphonates, and diphenyl phosphinites derived from alcohols with hydrogen chloride produces the alkyl chloride in which inversion at carbon has occurred. Yields are lower and conditions are more rigorous than the corresponding reaction with ***Hydrogen Bromide***.[30]

$$\text{ROH} \longrightarrow \text{RCl} \quad (14)$$

The reactions of carbohydrates and their derivatives with hydrogen chloride at the anomeric hydroxyl are examples of the facile conversion of alcohols to chlorides.[31] Protected and unprotected sugars react with methanol in the presence of dilute hydrochloric acid to produce the methyl ether at the anomeric center.[32]

Reactions with Diazo Compounds. The reaction of diazo compounds derived from α-amino acids or α-amino ketones with hydrochloric acid results in isolation of the corresponding racemic α-chloro carbonyl compound (eq 15).[33]

$$R^1C(=O)CH(R^2)N{=}N^+{=}N^- \longrightarrow R^1C(=O)CH(R^2)Cl \quad (15)$$

Reactions with Nitriles and Their Derivatives. Addition of anhydrous hydrogen chloride to nitriles produces imidoyl chlorides (eq 16).[34] Treatment of *N*-alkylimidoyl chlorides with hydrogen chloride results in isolation of the corresponding iminium halides (eq 17).[35] Aryl cyanates react with hydrogen chloride to produce haloformimidinium halides (eq 18).[36] Molecules containing two nitrile groups frequently give cyclic amidinium products.[37] Hydrogen chloride reacts with nitrile oxides to yield hydroxamoyl chlorides (eq 19). Hydroxamoyl chlorides are also produced in the reaction of phenylnitromethane with hydrogen chloride and in the reaction of acetophenones with isopropyl nitrite and hydrogen chloride.[38]

$$R{-}{\equiv}N \longrightarrow RC(=NH)Cl \quad (16)$$

$$R^2C(=NR^1)Cl \longrightarrow R^2C(=\overset{+}{N}HR^1)Cl\ Cl^- \quad (17)$$

$$RO{-}{\equiv}N \longrightarrow ROC(=NH)Cl \quad (18)$$

$$R{-}{\equiv}\overset{+}{N}{-}O^- \longrightarrow RC(=NOH)Cl \quad (19)$$

Reaction with Nitrogen–Sulfur Bonds. Hydrogen chloride has been used to cleave the nitrogen–sulfur bond of 2,2,2-trifluoro-1,1-diphenylethylsulfenyl-protected amines in high yield (eq 20).[39]

$$R^1NHSR^2 + HCl \longrightarrow R^1NH_2 + R^2SCl \quad (20)$$

Hydrolysis of 1,1-Dichloroethylenes. Hydrochloric acid in glacial acetic acid or alcohols has been used to convert 1,1-dichloroethylenes to carboxylic acids (eq 21).[40] This reaction appears to require additional conjugation or the presence of an allylic leaving group to proceed smoothly. The use of sulfuric acid appears to be more general since additional conjugation is not required.[41] ***Potassium Hydroxide*** has also been used to effect this hydrolysis.[42]

$$R^1R^2C{=}CCl_2 \longrightarrow R^1R^2CHC(=O)OR^3 \quad (21)$$

Silicon and Sulfur Chemistry. Protodesilylation with hydrochloric acid is most useful for substrates in which the silicon–carbon bond to be cleaved is aryl (eq 22), vinylic (eq 23), benzylic, or allylic (eq 24).[43] Phenylsilanes react with hydrogen chloride to yield benzene and the chlorosilane. The reaction is less facile than the reaction with hydrogen bromide.[44] Increasing the electronegativity of substituents on silicon decreases the ease with which the aryl–silicon bond is broken. Hindered aryl silanes may require fluoride ion.[45] Allylic protodesilylation normally occurs with double bond migration (eq 24).[46] Protodesilylation of a silicon–carbon bond in which the carbon is sp hybridized may require the addition of a fluoride source (eq 25).[47]

$$ArSiR^1R^2R^3 + HCl \longrightarrow ArH \quad (22)$$

$$CH_2{=}CH{-}SiR^1R^2R^3 \longrightarrow CH_2{=}CH{-}H \quad (23)$$

$$CH_2{=}CHCH_2SiR^1R^2R^3 \longrightarrow CH_3CH{=}CH_2 \quad (24)$$

$$R^4{-}{\equiv}{-}SiR^1R^2R^3 \xrightarrow[\text{DMF}]{\text{KF}} R^4{-}{\equiv} \quad (25)$$

Triethylaminosilanes are converted to the corresponding chlorosilanes in the presence of hydrochloric acid/sulfuric acid (eq 26).[48]

$$X{-}CH(CH_3){-}Si(NH_2)Et_2 \longrightarrow X{-}CH(CH_3){-}Si(Cl)Et_2 \quad (26)$$

Silyl ethers may be cleaved using hydrochloric acid (eq 27).[49] Fluoride ion is an alternative which is frequently used when the substrate is acid sensitive.

$$R^4OSiR^1R^2R^3 \longrightarrow R^4OH \quad (27)$$

Acylsilanes have been obtained from 1-silyl-1-enol ethers (eq 28), 1-silyl-1-aminoethylenes (eq 29), and mixed *O*-alkyl-*O*-silyl acetals of acylsilanes (eq 30) upon treatment with dilute hydrochloric acid.[50] Analogous chemistry has been observed with germanium and tin compounds.[51]

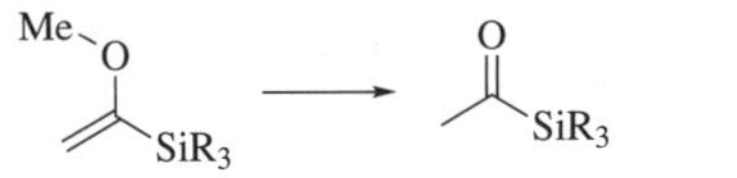

(28)

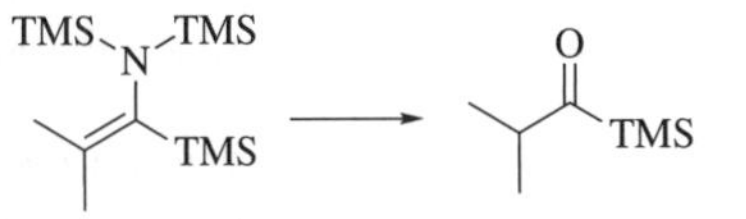

(29)

(30)

Phenylsilane reacts with anhydrous hydrogen chloride in ether in the presence of ***Aluminum Chloride*** to yield chlorophenylsilane (eq 31).[52]

$$PhSiH_3 \longrightarrow PhSiH_2Cl \quad (31)$$

Diphenylsilanediol yields the cyclic siloxane in moderate yield when treated with concentrated hydrochloric acid in ether at reflux (eq 32).[53] This reaction is also effected using amines.

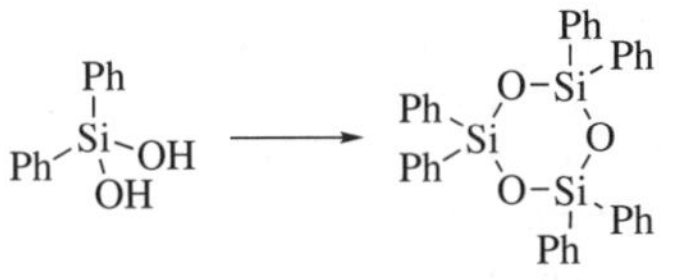

(32)

Thioacyl chlorides have been prepared from thioketenes and anhydrous hydrogen chloride at low temperatures (eq 33).[54]

(33)

Transhalogenation Reactions. 3-Bromo-4,5-dihydroisoxazole derivatives can be converted to the corresponding chlorides by treatment with hydrochloric acid and ***Lithium Chloride*** (eq 34).[55]

(34)

Organometallic Chemistry. Carbon–metal bonds are typically cleaved by hydrochloric acid to yield the metal chloride and the hydrocarbon (eq 35).[56] Oxygen–tin bonds are also cleaved by treatment with hydrochloric acid.[57]

$$RM + HCl \longrightarrow RH \quad (35)$$

Alkyl phenyl selenoxides react with hydrogen chloride to yield the alkyl chlorides (eq 36). Conversion to the bromide with hydrogen bromide proceeds faster and in higher yield.[58]

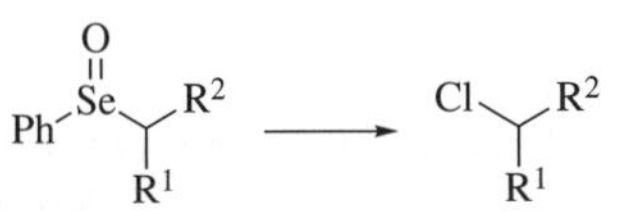

(36)

Treatment of the zirconium metallacycle shown with hydrochloric acid led to formation of the indole in high yield (eq 37).[59]

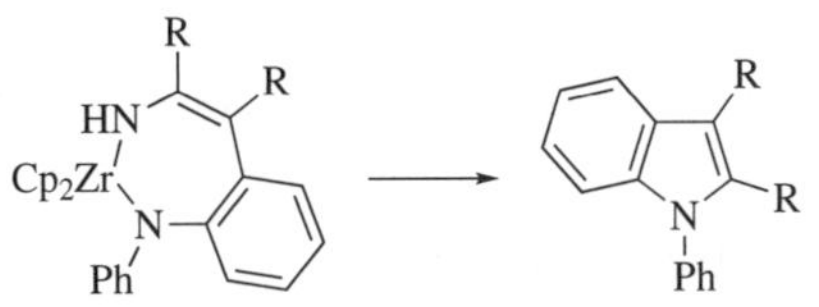

(37)

Treatment of a tungsten complex with 2 equiv of hydrochloric acid in ether led to formation of a new chlorotungsten compound (eq 38).[60]

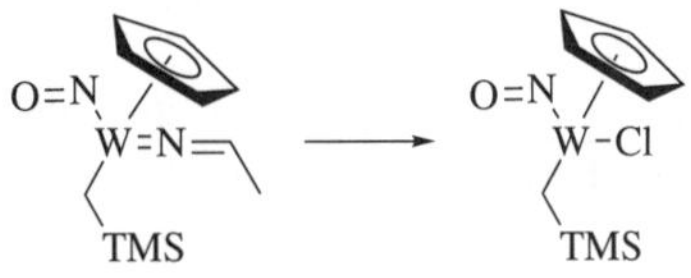

(38)

1,1-Dihydroxy-2,3-diphenylgermirene is converted to the dichloride with anhydrous hydrogen chloride in benzene (eq 39).[61]

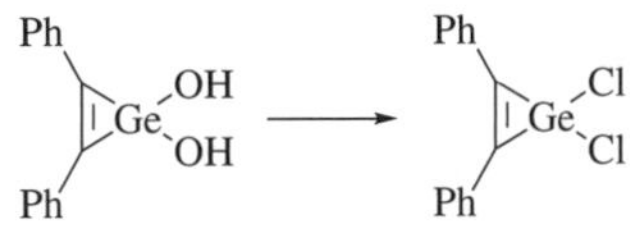

(39)

In Situ Generation of Chlorine and Hypochlorite. Hydrochloric acid and ***Hydrogen Peroxide*** are an effective combination for the in situ generation of ***Chlorine*** (eq 40). This combination of reagents can be used for the halogenation of alkenes and aromatics.[62]

$$HCl + [O] \longrightarrow Cl_2 \quad (40)$$

Treatment of 1,2-diphenylacetylene with HCl and ***Iodosylbenzene*** on silica gel appears to proceed through formation of chlorine to yield a mixture of (*E*)- and (*Z*)-1,2-diphenyl-1,2-dichloroethylene. However, under similar conditions, 1,1,2-triphenylethylene produces 2-chloro-1,1,2-triphenylethanol.[63] Chlorides are obtained in higher yields than bromides, which are obtained in higher yields than iodides when HBr or HI are substituted for HCl in the reaction. 1-Phenylpropyne reacts with oxone (***Potassium Monoperoxysulfate***) and hydrochloric acid in DMF to produce 2,2-dichloropropiophenone in high yield.[64] Replacement of oxone with ***m-Chloroperbenzoic Acid*** results in lower yields.

Related Reagents. Aluminum Chloride; Formaldehyde–Hydrogen Chloride; Hydrogen Chloride; Tin(IV) Chloride; Zinc Chloride.

---

1. (a) Stroh, R. *MOC* **1962**, *V/3*, 811. (b) Larock, R. C.; Leong, W. W. *COS* **1991**, *4*, 272. (c) Bohlmann, R. *COS* **1991**, *6*, 206. (d) Kantlehner, W. *COS* **1991**, *6*, 497. (e) Rosenberg, D. S. In *Kirk-Othmer Encyclopedia of Chemical Technology*; Wiley: New York, 1980; Vol. 12, p 983.
2. (a) Wells, P. R. *CRV* **1963**, *63*, 212. (b) Bunnett, J. F. *Ann. Rev. Phys. Chem.* **1963**, *14*, 271. (c) Bunnett, J. F. *JACS* **1961**, *83*, 4956.

3. (a) Paris, G.; Berlinguet, L.; Gaudry, R. *OSC* **1963**, *4*, 496. (b) Horning, E. C.; Finelli, A. F. *OSC* **1963**, *4*, 790. (c) Allen, C. F. H.; Johnson, H. B. *OSC* **1963**, *4*, 804. (d) Dunn, M. S.; Smart, B. W. *OSC* **1963**, *4*, 55.
4. (a) Searles, S.; Nukina, S. *CRV* **1959**, *59*, 1078. (b) Huntress, E. H.; Walter, H. C. *JACS* **1948**, *70*, 3702.
5. (a) Batt, D. G.; Jones, D. G.; La Greca, S. *JOC* **1991**, *56*, 6704. (b) Axtell, H. C.; Howell, W. M.; Schmid, L. G.; Cann, M. C.; *JOC* **1991**, *56*, 3906. (c) Schwartz, A.; Madan, P. B.; Mohacsi, E.; O'Brien, J. P.; Todaro, L. J.; Coffen, D. L. *JOC* **1992**, *57*, 851.
6. Yi, L.; Zou, J. H.; Lei, H. S.; Lin, X. M.; Zhang, M. X. *OPP* **1991**, *23*, 673.
7. (a) Berge, S. M.; Bighley, L. D.; Monkhouse, D. C. *JPS* **1977**, *66*, 1. (b) Wells, J. I. *Pharmaceutical Preformulation: The Physicochemical Properties of Drug Substances*; Halsted: New York, 1988; p 29.
8. (a) Fuson, R. C.; McKeever, C. H. *OR* **1942**, *1*, 63. (b) Zahn, H.; Dietrich, R.; Gerstner, W. *CB* **1955**, *88*, 1737.
9. Holland, H. L.; Contreras, L.; Ratemi, E. S. *SC* **1992**, *22*, 1473.
10. Zaugg, H. E. *S* **1984**, 85.
11. Simonsen, J. L. *The Terpenes*; 2nd ed.; Cambridge University Press: London, 1949; Vol. II, pp 64–77.
12. (a) Layachi, K; Ariès-Gautron, I; Guerro, M; Robert, A. *T* **1992**, *48*, 1585 (*CA* **1992**, *116*, 193 825t). (b) Legters, J.; Willems, J. G. H.; Thijs, L.; Zwanenburg, B. *RTC* **1992**, *111*, 59. (c) Addy, J. K.; Parker, R. E. *JCS* **1965**, 644. (d) Addy, J. K.; Parker, R. E. *JCS* **1963**, 915. (e) Buss, D. H.; Hough, L.; Richardson, A. C. *JCS* **1965**, 2736.
13. Schwartz, A.; Madan, P. B.; Mohacsi, E.; O'Brien, J. P.; Todaro, L. J.; Coffen, D. L. *JOC* **1992**, *57*, 851.
14. (a) Hardt, P; Stravs, A.; Abgottspon, P; U.S. Patent 5 087 745. (b) Knobler, Y.; Frankel, M. *JCS* **1958**, 1629. (c) Hoffmann, M. G.; Zeiss, H.-J. *TL* **1992**, *33*, 2669.
15. Axtell, H. C.; Howell, W. M.; Schmid, L. G.; Cann, M. C. *JOC* **1991**, *56*, 3906.
16. Norris, J. F.; Rigby, G. W. *JACS* **1932**, *54*, 2088.
17. (a) Scott, L. T.; DeCicco, G. J.; Hyun, J. L.; Reinhardt, G. *JACS* **1985**, *107*, 6546. (b) Hennion, G. F.; Sheehan, J. J.; Maloney, D. E. *JACS* **1950**, *72*, 3542.
18. Simonsen, J. L. *The Terpenes*, 2nd ed.; Cambridge University Press: London, 1949; Vol. II, pp 156, 171.
19. (a) Meerwein, H.; van Emster, K. *CB* **1922**, *55*, 2500. (b) Brecknell, D. J.; Carman, R. M.; Greenfield, K. L. *AJC* **1984**, *37*, 1075.
20. Fahey, R. C.; Payne, M. T.; Lee, D.-J. *JOC* **1974**, *39*, 1124.
21. Scherkenbeck, J.; Böttger, D.; Welzel, P. *T* **1988**, *44*, 2439.
22. (a) Kharasch, M. S.; Kritchevsky, J.; Mayo, F. R. *JOC* **1938**, *2*, 489. (b) Hatch, L. F.; Nesbitt, S. S. *JACS* **1950**, *72*, 727. (c) Hatch, L. F.; Journeay, G. E. *JACS* **1953**, *75*, 3712.
23. Caserio, M. C. In *Selective Organic Transformations*; Thyagarajan, B. S., Ed.; Wiley: New York, 1970; Vol. 1, p 239.
24. Heath, R. L.; Rose, J. D. *JCS* **1947**, 1485.
25. (a) Conant, J. B.; Quayle, O. R. *OSC* **1941**, *1*, 292, 294. (b) Marvel, C. S.; Calvery, H. O. *OSC* **1941**, *1*, 533. (c) Landini, D.; Montanari, F.; Rolla, F. *S* **1974**, 37. (d) Hennion, G. F.; Boisselle, A. P. *JOC* **1961**, *26*, 725. (e) DeWolfe, R. H.; Young, W. G. *CRV* **1956**, *56*, 801.
26. Meléndez, E.; Pardo, M. del C. *BSF* **1974**, 632 (*CA* **1974**, *81*, 90 965t).
27. (a) Pourahmady, N.; Vickery, E. H.; Eisenbraun, E. J. *JOC* **1982**, *47*, 2590. (b) Boekelheide, V.; Vick, G. K. *JACS* **1956**, *78*, 653.
28. Norris, J. F.; Olmsted, A. W. *OSC* **1941**, *1*, 144.
29. Bohlmann, R. *COS* **1991**, *6*, 204.
30. (a) Hudson, H. R. *S* **1969**, 112. (b) For a compilation of specific reagents used to convert alcohols to chlorides, see Larock, R. C. *Comprehensive Organic Transformations*; VCH: New York, 1989; p 354.
31. (a) Fox, J. J.; Goodman, I. *JACS* **1951**, *73*, 3256. (b) Kālè, V. N.; Clive, D. L. J. *JOC* **1984**, *49*, 1554.
32. Smith, F.; Van Cleve, J. W. *JACS* **1955**, *77*, 3159.
33. (a) Van Atta, R. E.; Zook, H. D.; Elving, P. J. *JACS* **1954**, *76*, 1185. (b) McPhee, W. D.; Klingsberg, E. *OSC* **1955**, *3*, 119.
34. Ulrich, H. *The Chemistry of Imidoyl Halides*; Plenum: New York, 1968; pp 66–68.
35. Ulrich, H. *The Chemistry of Imidoyl Halides*; Plenum: New York, 1968; p 100.
36. Martin, D.; Weise, A. *CB* **1967**, *100*, 3736 (*CA* **1968**, *68*, 21 357r).
37. Duquette, L. G.; Johnson, F. *T* **1967**, *23*, 4517, 4539.
38. Ulrich, H. *The Chemistry of Imidoyl Halides*; Plenum: New York, 1968; pp 159–161.
39. Netscher, T.; Weller, T. *T* **1991**, *47*, 8145.
40. (a) Nesmeyanov, A. N.; Freidlina, R. Kh.; Zakharkin, L. I. *DOK* **1954**, *99*, 781 (*CA* **1955**, *49*, 15 797). (b) Zakharkin, L. I.; Sorokina, L. P. *IZV* **1959**, 936 (*CA* **1960**, *54*, 1402f).
41. (a) Nesmeyanov, A. N.; Freidlina, R. Kh.; Semenov, N. A. *IZV* **1960**, 1969 (*CA* **1961**, *55*, 13 363f). (b) Randriamahefa, S.; Deschamps, P.; Gallo, R. *S* **1985**, 493. (c) Kruper, W. J.; Emmons, A. H. *JOC* **1991**, *56*, 3323.
42. Grummitt, O.; Buck, A.; Egan, R. *OSC* **1955**, *3*, 270.
43. Hillard, R. L., III; Vollhardt, K. P. C. *JACS* **1977**, *99*, 4058.
44. Fritz, G.; Kummer, D. *Z. Anorg. Allg. Chem.* **1961**, *308*, 105 (*CA* **1961**, *55*, 18 412a).
45. Corey, E. J.; Xiang, Y. B. *TL* **1987**, *28*, 5403.
46. Salomon, R. G.; Salomon, M. F.; Zagorski, M. G.; Reuter, J. M.; Coughlin, D. J. *JACS* **1982**, *104*, 1008.
47. Andreini, B. P.; Benetti, M.; Carpita, A.; Rossi, R. *T* **1987**, *43*, 4591.
48. Bailey, D. L.; Sommer, L. H.; Whitmore, F. C. *JACS* **1948**, *70*, 435.
49. Le Roux, C.; Maraval, M.; Borredon, M. E.; Gaspard-Iloughmane, H.; Dubac, J. *TL* **1992**, *33*, 1053 (*CA* **1992**, *116*, 214 265c).
50. (a) Leroux, Y.; Mantione, R. *JOM* **1971**, *30*, 295 (*CA* **1971**, *75*, 88 252r). (b) Soderquist, J. A.; Rivera, I.; Negron, A. *JOC* **1989**, *54*, 4051. (c) Soderquist, J. A. *OS* **1990**, *68*, 25. (d) Picard, J. P.; Aizpurua, J. M.; Elyusufi, A.; Kowalski, P. *JOM* **1990**, *391*, 13. (e) Picard, J. P.; Calas, R.; Dunoguès, J.; Duffaut, N.; Gerval, J.; Lapouyade, P. *JOC* **1979**, *44*, 420.
51. Soderquist, J. A.; Hassner, A. *JACS* **1980**, *102*, 1577.
52. Schmidbaur, H.; Zech, J.; Rankin, D. W. H.; Robertson, H. E. *CB* **1991**, *124*, 1953.
53. (a) Takiguchi, T. *JOC* **1959**, *24*, 989. (b) Kohama, S. *NKZ* **1963**, *84*, 422 (*CA* **1963**, *59*, 14 016d).
54. Seybold, G. *AG(E)* **1975**, *14*, 703.
55. Rohloff, J. C.; Robinson, J., III; Gardner, J. O. *TL* **1992**, *33*, 3113.
56. (a) Gilman, H.; Marshall, F. J. *JACS* **1949**, *71*, 2066. (b) Kharasch, M. S.; Flenner, A. L. *JACS* **1932**, *54*, 674. (c) Gilman, H.; Towne, E. B. *JACS* **1939**, *61*, 739.
57. Hatem, J.; Henriet-Bernard, C.; Grimaldi, J.; Maurin, R. *TL* **1992**, *33*, 1057.
58. Hevesi, L.; Sevrin, M.; Krief, A. *TL* **1976**, 2651.
59. Walsh, P. J.; Carney, M. J.; Bergman, R. G. *JACS* **1991**, *113*, 6343.
60. Debad, J. D.; Legzdins, P.; Batchelor, R. J.; Einstein, F. W. B. *OM* **1992**, *11*, 6.
61. Volpin, M. E.; Koreshkov, Yu. D.; Dulova, V. G.; Kursanov, D. N. *T* **1962**, *18*, 107.
62. Johnson, R.; Reeve, K. *Spec. Chem.* **1992**, *12*, 292.
63. Sohmiya, H.; Kimura, T.; Bauchat, P.; Fujita, M.; Ando, T. *CL* **1991**, 1391.
64. Kim, K. K.; Kim, J. N.; Kim, K. M.; Kim, H. R.; Ryu, E. K. *CL* **1992**, 603.

John E. Mills
R. W. Johnson Pharmaceutical Research Institute, Spring House, PA, USA

Hydrogen Bromide[1]

[10035-10-6] BrH (MW 80.91)

(reagent for electrophilic[5] and radical-mediated[1b] hydrobromination of alkenes and alkynes; cleaves epoxides[1c] and ethers;[20] converts alcohols[13] and chlorides[32] to bromides; converts α-diazo ketones to α-bromomethyl ketones,[28a] and 1,4-cyclohexanediontles to phenols[29])

Alternate Name: ***Hydrobromic Acid***
Solubility: sol most organic solvents.[2]
Form Supplied in: widely available as compressed gas; 30% solution in AcOH or $EtCO_2H$; 48% aqueous solution.
Preparative Methods: can be generated in situ by treatment of refluxing tetrahydronaphthalene with Br_2;[3] treatment of Ph_3PHBr with refluxing xylene.[4]
Handling, Storage, and Precautions: highly toxic and corrosive. This reagent should be handled in a fume hood.

Hydrobromination of Alkenes and Alkynes. HBr undergoes addition readily to most alkenes and alkynes.[1a] However, radical and ionic addition usually compete, affording a mixture of products unless the system is symmetrically substituted (eq 1).[5]

(1) [scheme: C_5H_{11}-CH=CH₂ + Br• ⇌ radical C_5H_{11}CH•CH₂Br; HBr → $C_5H_{11}CH_2CH_2Br$ 83%; HBr ⇌ cation $C_5H_{11}CH^+CH_3$; Br⁻ → $C_5H_{11}CHBrCH_3$ 11%]

Radical addition can be facilitated by light, peroxides, and other radical-generating systems.[1b] Limiting reaction to ionic addition is more challenging, requiring the rigorous absence of light, oxygen, and peroxide impurities, along with the use of a radical inhibitor.[6] Ionic addition to alkenes without competing radical reaction can also be effected with aqueous HBr at 115 °C and a phase-transfer catalyst (eq 2).[7]

(2) C_5H_{11}CH=CH₂ → (48% HBr, $C_{16}H_{33}(Bu)_3PBr$, 115 °C, 2 h; 94%) C_5H_{11}CHBrCH₃

More conveniently, competing radical reaction can be avoided by conducting the addition in the presence of appropriately prepared silica gel or alumina, which also accelerates the rate of ionic addition.[5,8] Addition is rendered even more convenient by the use of various inorganic and organic acid bromides that undergo reaction with silica gel or alumina to afford HBr in situ (eq 3).[5]

(3) C_5H_{11}CH=CH₂ → ($(COBr)_2$, Al_2O_3, 25 °C, 20 min; 99%) C_5H_{11}CHBrCH₃

Surface-mediated hydrobromination of phenylalkynes initially affords the *syn* adducts, which undergo subsequent equilibration with the thermodynamically more stable (*Z*) isomers (eq 4).[5] Thus either isomer can be obtained in high yield by the appropriate choice of reaction conditions. Surface-mediated hydrobromination of terminal alkylalkynes affords the corresponding 2-bromo-1-alkenes in good yield (eq 5).[5] Several other methods had previously been developed for this difficult transformation.[9]

(4) Ph–C≡C–R → (PBr_3, SiO_2 or Al_2O_3) Ph–CBr=CH–R (syn adduct) ⇌ Ph–CBr=CH–R (*Z*)

R = Me, Ph or *t*-Bu

(5) C_5H_{11}–C≡CH → (AcBr, Al_2O_3, 25 °C, 24 h; 88%) C_5H_{11}–CBr=CH₂

Cleavage of Epoxides to Bromohydrins. The addition of HBr to epoxides to give bromohydrins proceeds readily with either 48% aqueous HBr or anhydrous HBr in a variety of organic solvents.[1c,10] The stereoselectivity generally parallels that of HCl additions (see ***Hydrogen Chloride***), and is similarly affected by changes in solvent and substitution. Typically, ring opening of unsymmetrical epoxides favors formation of the bromohydrin in which bromine is at the less highly substituted position. This preference is greater for HBr than HCl additions because of the greater size and nucleophilicity of the bromide ion.[1c,11]

Reaction with Alcohols. The reaction of HBr with alcohols to form alkyl bromides is a general, high-yield reaction that can be effected under a variety of conditions. The conversion of primary and secondary alkyl alcohols to the corresponding bromides with anhydrous HBr requires elevated temperatures (100 °C),[12] but benzylic and allylic alcohols are reactive in $CHCl_3$ solution at 25 °C.[13] Alcohols are also converted to alkyl bromides at elevated temperatures with 48% aqueous HBr.[14] Alcohols resistant to substitution under these conditions are often reactive in the presence of H_2SO_4 as a catalyst.[12,15] The use of phase-transfer catalysts has little effect on the rate of the reaction, but has been found to reduce significantly the extent of skeletal rearrangement that often accompanies substitution (eq 6).[16] Tertiary bromides can be synthesized from the corresponding alcohols under especially mild conditions in the presence of ***Lithium Bromide***.[17]

(6) Bu–CH(CH₂CH₃)–CH₂OH → (48% HBr, 100 °C) Bu–CH(Et)–CH₂Br + Bu–CBr(Me)–Et + Bu–CH₂–CBr(Me)₂

no catalyst	82%	16%	2%
phase transfer catalyst	99%		

Vicinal diols, when treated with 48% aqueous HBr in AcOH, give 2-bromo acetates with good regio- and stereoselectivity (eq 7).[18] In the absence of AcOH, some diols are dehydrated to give dienes.[19]

(7)

(S)-(+)-propane-1,2-diol → (HBr, AcOH, 45 min, 78–85%) → (S)-(–)-2-acetoxy-1-bromopropane

A series of bis(benzylic) diols has been converted to 2,3-disubstituted indenes with HBr in good yields (eq 8), whereas reactions with other hydrogen halides followed a different course.[20] This method failed for the synthesis of monoalkyl indenes but was satisfactory for the synthesis of monoaryl systems.

(8) 48% HBr, Δ, 6 h, 84%

Cleavage Reactions. The cleavage of ethers by HBr is a synthetically useful method for the synthesis of alkyl bromides and for the deprotection of phenols.[21] Dialkyl ethers are cleaved with 48% aqueous HBr at elevated temperatures in the presence of H_2SO_4[22] or a phase-transfer catalyst.[23] Under these conditions, most dialkyl ethers give high yields (90–95%) of alkyl bromides, while alkyl aryl ethers give phenols (90%) and alkyl bromides (90%). Aryl methyl ethers are readily cleaved with the use of 48% aqueous HBr to give the phenol and MeBr.[24] The cleavage of cyclic lactones by HBr, in a manner analogous to the cleavage of ethers, has also been reported.[25] When triphenylphosphonium bromide is used as an in situ source of HBr, methyl, ethyl, and benzyl ethers are cleaved at elevated temperatures to give alkene products in good yield along with a tetrasubstituted phosphonium salt (eq 9).[26] Similarly, aryl alkyl ethers are converted to phenols, and benzyl and methyl esters to carboxylic acids.

(9) OEt (cyclohexyl) → ($Ph_3P{\bullet}HBr$, 180 °C, 67%) → cyclohexene + $Ph_3EtP^+\ Br^-$

AcOH and CF_3CO_2H solutions of HBr efficiently cleave benzyloxycarbonyl groups from protected peptides.[27] Anhydrous HBr selectively cleaves alkyl groups from aryl amines.[28] Thus dialkyl-substituted anilines were converted to monoalkyl-substituted anilines at 150 °C, and monoalkyl anilines converted to aniline at 195 °C. This method is not useful for cleavage of trialkyl amines.

Synthesis of α-Bromo Ketones. α-Diazo ketones are converted to α-bromomethyl ketones by either anhydrous HBr in Et_2O or 48% aqueous HBr in various solvents (eq 10).[29] This is a useful, high-yield, method for the regiospecific conversion of a carboxylic acid to the corresponding bromomethyl ketone, since the required α-diazo ketone is readily synthesized from reaction of the corresponding acid chloride with ***Diazomethane***.

(10) $PhCH_2COCHN_2$ → (HBr, Et_2O, 0 °C, 0.5 h, 55%) → $PhCH_2COCH_2Br$

Reductions with HBr. HBr has been used as a reagent for the reduction of 1,4-cyclohexanediones to phenols,[30] as well as for the cyclization and aromatization of a variety of carbonyl-substituted arenes (eq 11).[31]

(11) CHO → (48% HBr, AcOH, Δ, 2 min, 93%)

Although HBr generally reduces aryl sulfoxides, its uses are limited due to ensuing bromination of the aryl ring.[32]

Miscellaneous Reactions. Alkyl chlorides, which are generally more readily available than alkyl bromides, are converted into the corresponding bromides by anhydrous HBr in the presence of catalytic amounts of $FeBr_3$.[33] Secondary, tertiary, and allylic bromides are formed in high yields from the corresponding chlorides, while primary systems undergo extensive cationic rearrangement. Alkyl iodides have been converted into the corresponding bromides with 42% HBr in the presence of HNO_3.[34]

Several arylamines have been brominated at the *ortho* position by HBr in DMSO.[35] Since removal of the amine functionality is possible in good yield, this provides a regiospecific method of arene bromination.

HBr reacts with aldehydes in the presence of arenes to give bromoalkylated products[36] and in the presence of alcohols to give α-bromomethyl ethers in good yields (eq 12).[37]

(12) $PhCH_2OH + HCHO$ → (HBr, 2 h, 97%) → $PhCH_2OCH_2Br$

Addition of anhydrous HBr to α,β-unsaturated aldehydes or ketones, followed by acetalization, readily affords β-bromoacetals (eq 13).[38]

(13) 1. HBr, CH_2Cl_2; 2. $(EtO)_3CH$, HO OH (diol), 54%; Br

Related Reagents. Boron Tribromide; Formaldehyde–Hydrogen Bromide; Hydrobromic Acid; Phosphorus(III) Bromide.

1. (a) Larock, R. C.; Leong, W. W. *COS* **1991**, *4*, 269. (b) Stacey, F. W.; Harris, J. F., Jr. *OR* **1963**, *13*, 150. (c) Parker, R. E.; Isaacs, N. S. *CRV* **1959**, *59*, 737.
2. Fogg, P. G. T.; Gerrard, W.; Clever, H. L. In *Solubility Data Series*; Lorimer, J. W., Ed.; Pergamon: Oxford, 1990; Vol. 42.
3. Maxson, R. N. *Inorg. Synth.* **1939**, *1*, 149.
4. Hercouet, A.; Le Corre, M. *S* **1988**, 157.
5. (a) Kropp, P. J.; Daus, K. A.; Crawford, S. D.; Tubergen, M. W.; Kepler, K. D.; Craig, S. L.; Wilson, V. P. *JACS* **1990**, *112*, 7433. (b) Kropp, P.

J.; Daus, K. A.; Tubergen, M. W.; Kepler, K. D.; Wilson, V. P.; Craig, S. L.; Baillargeon, M. M.; Breton, G. W. *JACS* **1993**, *115*, 3071. (c) Kropp, P. J.; Crawford, S. D. *JOC* **1994**, *59*, 3102.

6. Walling, C. *Free Radicals in Solution*; Wiley: New York, 1957; pp 291–296.
7. Landini, D.; Rolla, F. *JOC* **1980**, *45*, 3527.
8. Walborsky, H. M.; Topolski, M. *JACS* **1992**, *114*, 3455.
9. (a) Boeckman, R. K., Jr.; Blum, D. M. *JOC* **1974**, *39*, 3307. (b) Cousseau, J. S. **1980**, 805. (c) Hara, S.; Dojo, H.; Takinami, S.; Suzuki, A. *TL* **1983**, *24*, 731.
10. (a) Buchanan, J. G.; Sable, H. Z. In *Selective Organic Transformations*; Thyagarajan, B. S., Ed.; Wiley: New York, 1972; Vol. 2, pp 1–92. (b) Armarego, W. L. F. In *Stereochemistry of Heterocyclic Compounds*; Taylor, E. C.; Weissberger, A., Eds.; Wiley: New York, 1977; pp 23–25. (c) Bartok, M.; Lang, K. L. In *The Chemistry of Ethers, Crown Ethers, Hydroxyl Groups and Their Sulfur Analogues*; Patai, S., Ed.; Wiley: New York, 1980; Part 2, pp 655–659.
11. Stewart, C. A.; VanderWerf, C. A. *JACS* **1954**, *76*, 1259.
12. (a) Reid, E. E.; Ruhoff, J. R.; Burnett, R. E. *OSC* **1943**, *2*, 246. (b) McEwen, W. L. *OSC* **1955**, *3*, 227.
13. Doxsee, K. M.; Feigel, M.; Stewart, K. D.; Canary, J. W.; Knobler, C. B.; Cram, D. J. *JACS* **1987**, *109*, 3098.
14. Vogel, A. I. *JCS* **1943**, 636.
15. Kamm, O.; Marvel, C. S. *OSC* **1941**, *1*, 25.
16. Dakka, G.; Sasson, Y. *TL* **1987**, *28*, 1223.
17. Masada, H.; Murotani, Y. *BCJ* **1980**, *53*, 1181.
18. (a) Golding, B. T.; Hall, D. R.; Sakrikar, S. *JCS(P1)* **1973**, 1214. (b) Ellis, M. K.; Golding, B. T. *OSC* **1990**, *7*, 356.
19. Allen, C. F. H.; Bell, A. *OSC* **1955**, *3*, 312.
20. Parham, W. E.; Sayed, Y. A. *S* **1976**, 116.
21. Bhatt, M. V.; Kulkarni, S. U. *S* **1983**, 249.
22. (a) Andrus, D. W. *OSC* **1941**, *3*, 692. (b) Newkome, G. R.; Gupta, V. K.; Griffin, R. W.; Arai, S. *JOC* **1987**, *52*, 5480.
23. Landini, D.; Montanari, F; Rolla, F. *S* **1978**, 771.
24. (a) Clarke, H. T.; Taylor, E. R. *OSC* **1941**, *1*, 150. (b) Surrey, A. R. *OSC* **1955**, *3*, 753.
25. (a) Lavety, J.; Proctor, G. R. *OSC* **1973**, *5*, 545. (b) ApSimon, J.; Seguin, R. *SC* **1980**, *10*, 897.
26. Bestmann, H. J.; Mott, L.; Lienert, J. *LA* **1967**, *709*, 105.
27. Bodansky, M.; Bodansky, A. In *The Practice of Peptide Synthesis*; Springer: New York, 1984; pp 165–168.
28. Chambers, R. A.; Pearson, D. E. *JOC* **1963**, *28*, 3144.
29. (a) Catch, J. R.; Elliott, D. F.; Hey, D. H.; Jones, E. R. H. *JCS* **1948**, 278. (b) Balenovic, K.; Cerar, D.; Filipovic, L. *JOC* **1953**, *18*, 868.
30. Rao, C. G.; Rengaraju, S.; Bhatt, M. V. *CC* **1974**, 584.
31. (a) Bradsher, C. K.; Winston, J. J., Jr. *JACS* **1954**, *76*, 734. (b) Canonne, P.; Holm, P.; Leitch, L. C. *CJC* **1967**, *45*, 2151.
32. Madesclaire, M. *T* **1988**, *44*, 6537.
33. Yoon, K. B.; Kochi, J. K. *JOC* **1989**, 3028.
34. Svetlakov, N. V.; Moisak, I. E.; Averko-Antonovich, I. G. *JOU* **1969**, *5*, 971.
35. (a) Pan, H.-L.; Fletcher, T. L. *S* **1973**, 610. (b) Fletcher, T. L.; Pan, H.-L. *JCS* **1965**, 4588.
36. Olah, G. A.; Tolgyesi, W. S. In *Friedel-Crafts and Related Reactions*; Olah, G. A., Ed.; Interscience: New York, 1964; Vol. 2, Part 2, pp 1–92.
37. Connor, D. S.; Klein, G. N.; Taylor, G. N.; Boeckman, Jr., R. K.; Medwid, J. B. *OSC* **1988**, *6*, 101.
38. Stowell, J. C.; Keith, D. R.; King, B. T. *OS* **1984**, *62*, 140.

Gary W. Breton & Paul J. Kropp
University of North Carolina, Chapel Hill, NC, USA

Hydrogen Chloride[1]

[7647-01-0] ClH (MW 36.46)

(reagent for hydrochlorination of alkenes and alkynes;[4] cleaves epoxides[1b] and ethers;[21a] converts alcohols to chlorides[12b] and diols to cyclic ethers;[17] chloroalkylates arenes;[22] converts aldehydes to α-chloro ethers[23b])

Alternate Name: ***Hydrochloric Acid***
Solubility: sol most organic solvents.[2]
Form Supplied in: widely available; compressed gas; 1 M solution in AcOH, Et_2O, or Me_2S; 4 M solution in dioxane; 37% aqueous solution.
Preparative Methods: addition of H_2SO_4 to NaCl or 37% aqueous HCl.[3]
Handling, Storage, and Precautions: highly toxic and corrosive; handle only in a fume hood.

Hydrochlorination of Alkenes and Alkynes. HCl undergoes solution-phase addition readily to C=C double bonds that are strained or from which the resulting carbocation is benzylic or tertiary.[1a] However, other alkenes do not undergo addition at preparatively useful rates.[4] Although addition can be facilitated by Lewis acid catalysis,[5] mono- and 1,2-disubstituted alkenes undergo polymerization under these conditions.[5a] The rate of addition is inversely proportional to the electron donor strength of the solvent, following the order heptane ≈ $CHCl_3$ > xylene > nitrobenzene ≫ MeOH > dioxane > Et_2O.[6,7] In the strongly donating solvent Et_2O, even highly reactive alkenes undergo slow addition unless one of the reactants is present in high concentration. Additions conducted in solutions saturated with HCl exhibit an inverse temperature coefficient because of the increased solubility of HCl at lower temperatures.[3b]

Alkynes undergo addition more slowly than alkenes, requiring extended reaction times, elevated temperatures, and, usually, Lewis acid catalysis.[1a] However, dialkylalkynes afford the (*Z*)-vinyl chloride on treatment with refluxing aqueous HCl (eq 1).[8]

Pr—≡—Pr —(37% HCl, 80 °C, 18 h; 81%)→ Pr–C(Cl)=CH–Pr (1)

Addition to alkenes and alkynes is greatly facilitated by the presence of appropriately prepared silica gel or alumina.[4] Alkenes and alkynes that exhibit little or no reaction with HCl in solution readily undergo addition under these conditions. The reaction is rendered even more convenient by the use of various inorganic and organic acid chlorides that afford HCl in situ in the presence of silica gel or alumina. Surface-mediated hydrochlorination of 1,2-dimethylcyclohexene in CH_2Cl_2 gives initially the *syn* adduct, which undergoes equilibration with the thermodynamically more stable *trans* isomer under the reaction conditions (eq 2).[4] Thus either isomer can be obtained in high yield through the proper choice of reaction conditions. Similarly, phenylalkynes initially afford *syn* adducts, which undergo subsequent equilibration with the thermodynamically more stable (*Z*) isomers (eq 3).[4] Again, either isomer can be obtained in high yield.

(2)

$$\text{Ph—≡—R} \xrightarrow[\text{SiO}_2 \text{ or Al}_2\text{O}_3]{\text{SOCl}_2} \text{Ph(Cl)C=CHR} \rightleftharpoons \text{Ph(Cl)C=CHR} \quad (3)$$

R = Me or Ph

Cleavage of Epoxides to Chlorohydrins. The addition of HCl to epoxides to form chlorohydrins proceeds readily with either 37% aqueous HCl or solutions of anhydrous HCl in a variety of organic solvents.[1b,9] For simple alkyl-substituted oxiranes, addition typically occurs through backside attack of chloride ion on the protonated epoxide, resulting in net inversion of the carbon center (eq 4).[1b,9] For aryl- or vinyl-substituted epoxides (in which more carbocationic character is involved in the transition state during ring opening), the stereochemical outcome may range from complete retention to predominant inversion and is highly solvent dependent.[10] Anhydrous conditions and solvents of low dielectric strength favor *syn* cleavage, while *anti* cleavage is favored in the presence of water or in hydroxylic solvents.[10]

37% HCl, 2 h, 77% (4)

Cleavage of simple alkyl-substituted epoxides under anhydrous conditions typically favors formation of the chlorohydrin in which chlorine is at the less highly substituted position (eqs 5 and 6).[11] More highly substituted epoxides, particularly aryl-substituted, give increasing amounts of the opposite regioisomer. Regioselectivity is also very sensitive to the solvent system employed for the reaction (eqs 5 and 6).

HCl (5)

THF	84%	16%
THF/H_2O	40%	60%

HCl (6)

THF	62%	38%
THF/H_2O	25%	75%

Reaction with Alcohols. The reaction of HCl with alcohols to form alkyl chlorides is a general reaction, giving good to high yields of products. Primary and secondary aliphatic alcohols are most easily converted to the corresponding chlorides with either 37% aqueous HCl or anhydrous HCl at elevated temperatures in the presence of ***Zinc Chloride***.[12] Phase-transfer catalysis has also been employed in the synthesis of primary chlorides from alcohols.[13] The need for a catalyst can be avoided by using the highly polar solvent HMPA.[14] Tertiary,[7,15a] benzylic,[15b] and allylic[15c] alcohols are readily converted to chlorides at 25 °C, or lower, without the need for catalysts. Glycerol can be selectively mono- or dichlorinated by controlled addition to HCl to AcOH solutions.[16] Bis(benzylic) diols have been converted in good yields to substituted cyclic ethers with HCl, whereas reaction with HBr or HI followed a completely different course (eq 7).[17]

37% HCl, Δ, 6 h, 80% (7)

Reductions with HCl. HCl has been used to reduce a series of 1,4-cyclohexanediones to the corresponding phenols in good yield (eq 8).[18]

37% HCl, Δ, 15 h, 70% (8)

α-Diazo ketones are reduced to α-chloromethyl ketones by either anhydrous HCl in organic solvents or 37% aqueous HCl in Et_2O.[19] Generally, good to high yields are obtained. Chloroacetone was synthesized in this manner without the complicating formation of dichlorides (eq 9).[19c]

HCl, Et_2O, 68% (9)

Although aryl sulfoxides are reduced to sulfides by HCl, accompanying ring chlorination limits the usefulness of the reaction.[20]

Cleavage of Ethers. Allyl, *t*-butyl, trityl, benzhydryl, and benzyl ethers are cleaved by HCl in AcOH (eq 10).[21a] In some cases, aryl methyl ethers have been successfully cleaved (eq 11).[21b]

37% HCl, AcOH, 80 °C, 1.5 h, 89% (10)

20% HCl, Δ, 13 h, 57% (11)

Reaction with Aldehydes. Arenes react readily with mixtures of HCl and formaldehyde in the presence of a Lewis acid, usually $ZnCl_2$, to give the chloromethylated derivative.[22] Yields are good and the reaction conditions can be controlled to afford predominantly mono- or disubstituted products. Chloroalkylations can be effected with other aldehydes such as propanal and butanal. In the presence of alcohols, HCl and aldehydes give high conversions to α-chloro ethers (eq 12).[23]

Ph⌒OH + HCHO —(HCl, 83%)→ Ph⌒O⌒Cl (12)

Related Reagents. Boron Trichloride; Formaldehyde–Hydrogen Chloride; Hydrochloric Acid; Zinc Chloride.

---

1. (a) Larock, R. C.; Leong, W. W. *COS* **1991**, *4*, 269. (b) Parker, R. E.; Isaacs, N. S. *CRV* **1959**, *59*, 737.
2. Fogg, P. G. T.; Gerrard, W.; Clever, H. L. In *Solubility Data Series*; Lorimer, J. W.; Ed.; Pergamon: Oxford, 1990; Vol. 42.
3. (a) Maxson, R. N. *Inorg. Synth.* **1939**, *1*, 147. (b) Brown, H. C.; Rei, M.-H. *JOC* **1966**, *31*, 1090.
4. (a) Kropp, P. J.; Daus, K. A.; Crawford, S. D.; Tubergen, M. W.; Kepler, K. D.; Craig, S. L.; Wilson, V. P. *JACS* **1990**, *112*, 7433. (b) Kropp, P. J.; Daus, K. A.; Tubergen, M. W.; Kepler, K. D.; Wilson, V. P.; Craig, S. L.; Baillargeon, M. M.; Breton, G. W. *JACS* **1993**, *115*, 3071. (c) Kropp, P. J.; Crawford, S. D. *JOC* **1994**, *59*, 3102.
5. (a) Shields, T. C. *CJC* **1971**, *49*, 1142. (b) Hassner, A.; Fibiger, R. F. *S* **1984**, 960.
6. (a) O'Connor, S. F.; Baldinger, L. H.; Vogt, R. R.; Hennion, G. F. *JACS* **1939**, *61*, 1454. (b) Hennion, G. F.; Irwin, C. F. *JACS* **1941**, *63*, 860.
7. For a different order, see: Brown, H. C.; Liu, K.-T. *JACS* **1975**, *97*, 600.
8. Hudrlik, P. F.; Kulkarni, A. K.; Jain, S.; Hudrlik, A. M. *T* **1983**, *39*, 877.
9. (a) Lucas, H. J.; Gould, C. W., Jr. *JACS* **1941**, *63*, 2541. (b) Buchanan, J. G.; Sable, H. Z. In *Selective Organic Transformations*; Thyagarajan, B. S., Ed.; Wiley: New York, 1972; Vol. 2, pp 1–92. (c) Armarego, W. L. F. In *Stereochemistry of Heterocyclic Compounds*; Taylor, E. C.; Weissberger, A., Eds.; Wiley: New York, 1977; pp 23–25. (d) Bartok, M.; Lang, K. L. In *The Chemistry of Ethers, Crown Ethers, Hydroxyl Groups and Their Sulfur Analogues*; Patai, S., Ed.; Wiley: New York, 1980; Part 2, pp 655–657.
10. Berti, G.; Macchia, B.; Macchia, F. *T* **1972**, *28*, 1299.
11. Lamaty, G.; Maloq, R.; Selve, C.; Sivade, A.; Wylde, J. *JCS(P2)* **1975**, 1119.
12. (a) Copenhaver, J. E.; Whaley, A. M. *OSC* **1941**, *1*, 142. (b) Vogel, A. I. *JCS* **1943**, 636. (c) Atwood, M. T. *J. Am. Oil Chem. Soc.* **1963**, *40*, 64.
13. Landini, D.; Montanari, F.; Rolla, F. *S* **1974**, 37.
14. Fuchs, R.; Cole, L. L. *CJC* **1975**, *53*, 3620.
15. (a) Norris, J. F.; Olmsted, A. W. *OSC* **1941**, *1*, 144. (b) Pourahmady, N.; Vickery, E. H.; Eisenbraun, E. J. *JOC* **1982**, *47*, 2590. (c) Melendez, E.; Pardo, M. C. *BSF* **1974**, 632.
16. Conant, J. B.; Quayle, O. R. *OSC* **1941**, *1*, 292, 294.
17. Parham, W. E.; Sayed, Y. A. *S* **1976**, 116.
18. Rao, C. G.; Rengaraju, S.; Bhatt, M. V. *CC* **1974**, 584.
19. (a) McPhee, W. D.; Klingsberg, E. *OSC* **1955**, *3*, 119. (b) Dauben, W. G.; Hiskey, C. F.; Muhs, M. A. *JACS* **1952**, *74*, 2082. (c) Van Atta, R. E.; Zook, H. D.; Elving, P. J. *JACS* **1954**, *76*, 1185.
20. Madesclaire, M. *T* **1988**, *44*, 6537.
21. (a) Bhatt, M. V.; Kulkarni, S. U. *S* **1983**, 249. (b) Brossi, A.; Blount, J. F.; O'Brien, J.; Teitel, S. *JACS* **1971**, *93*, 6248.
22. Olah, G. A., Tolgyesi, W. S. In *Friedel–Crafts and Related Reactions*; Olah, G. A., Ed.; Interscience: New York, 1964; Vol. 2, Part 2, pp 1–92.
23. (a) Marvel, C. S.; Porter, P. K. *OSC* **1932**, *1*, 377. (b) Grummitt, O.; Budewitz, E. P.; Chudd, C. C. *OSC* **1963**, *4*, 748. (c) Connor, D. S.; Klein, G. W.; Taylor, G. N.; Boeckman, R. K.; Medwid, J. B. *OSC* **1988**, *6*, 101.

Gary W. Breton & Paul J. Kropp
University of North Carolina, Chapel Hill, NC, USA

Hydrogen Fluoride[1]

[7664-39-3] FH (MW 20.01)

(strong Brønsted–Lowry acid[2] capable of fluorinating numerous organic substrates;[1] cleaves silyl[3] and peptide[4] protecting groups; effects lignocellulose solvolysis;[5] catalyzes a number of electrophilic aromatic substitution reactions[6])

Alternate Name: ***Hydrofluoric Acid***
Physical Data: mp −83.37 °C; bp 19.54 °C; *d* 1.015 g cm^{-3} (0 °C), 0.958 g cm^{-3} (25 °C).
Solubility: sol water (52.7 wt %, 6 theoretical plates); strong proton donor to alcohols, carboxylic acids, ethers, and ketones (unstable); insol aliphatic hydrocarbons; very slightly sol aromatic hydrocarbons.
Form Supplied in: anhydrous liquid; 48–50% aqueous solution; most common impurity is fluorosilicic acid (<100 ppm).
Purification: for virtually all synthetic purposes, anhydrous hydrogen fluoride (AHF) is supplied in sufficiently high purity from commercial sources. Ultrapure AHF can be obtained by either distillation[7] of commercially available AHF or thermal decomposition[8] of potassium acid fluoride.
Handling, Storage, and Precautions: HF is an extremely toxic material. HF can cause severe damage to the respiratory system and will cause severe burns to tissue, e.g. skin, fingernails, mouth, eyes, etc.; penetration through tissue into bone is possible. If contact with HF occurs, the affected area must be flushed immediately with copious amounts of water for at least 15 min. **Immediate medical attention must be sought.** Recommended personnel protective equipment includes the use of gloves (neoprene/nitrile/rubber composite), goggles, a faceshield, and an apron while working in a well-ventilated hood, preferably equipped with a HF monitor.

AHF is supplied either in lecture bottles or 3 lb metal cylinders; aq HF is supplied in polyethylene bottles. The use of glassware and stainless steel vessels should be avoided when handling either AHF or aq HF.[9] For reactions run at ambient temperature and pressure, polyethylene vessels can be employed; copper and iron vessels are also suitable. However, the use of an autoclave constructed of high metallurgy such as Monel or nickel, employing inert atmosphere techniques, is highly recommended.

Fluorinations. AHF is an effective reagent for the fluorination of a wide variety of organic substrates.[1] However, AHF in combination with organic bases,[10] for example Olah's reagent (HF–pyridine),[11] and nucleophilic fluoride transfer agents[12] are more frequently employed than AHF, particularly when monofluorination of an organic substrate is desired.

Addition of AHF to unsaturated hydrocarbons occurs by an electrophilic mechanism and, for alkenes, in a Markovnikov fashion. Addition of AHF to alkynes[13] proceeds with poor stereochemical control to give mixtures of (*E*)- and (*Z*)-vinyl halides and is often accompanied by the formation of *gem*-difluoroalkanes and fluoropolymers.

AHF reacts with carboxylic acid halides,[14] esters,[15] and anhydrides[14] to give carboxylic acid fluorides. Acetoacetyl fluoride can be prepared in near quantitative yield by adding AHF to diketene.[16] At temperatures $\leq 0\,^\circ C$, AHF is known to add to isocyanates[17] to yield the corresponding carbamoyl fluorides.

Epoxides and aziridines undergo ring opening in the presence of AHF. For example, stereospecific ring opening of an epoxide[18] can be accomplished as a means of introducing fluorine into the B ring of a steroid (eq 1).

HF (anhyd), THF
CHCl$_3$
74%

(1)

AHF addition to *trans*-2,3-diphenylaziridine[19] occurs stereospecifically to afford the *erythro* isomer exclusively, whereas the *threo* isomer predominates upon treatment with Olah's reagent (eq 2).

HF–py 92%
HF–py
AHF 100%

(2)

Halofluorination of alkenes with ***t-Butyl Hypochlorite***, ***N-Bromosuccinimide***, and ***N-Iodosuccinimide*** in the presence of AHF[20] proceeds in good yield, although superior yields again can be obtained with Olah's reagent.[21] In the presence of AHF, addition of ***N-Bromoacetamide*** and *N*-iodosuccinimide to cyclohexene occurs with a high degree of stereoselectivity to produce *trans* isomers in 43% and 72% yield, respectively.[22] Halofluorination occurs in a Markovnikov fashion.[23] Synthesis of *gem*-fluorohalides can be achieved by adding AHF to vinyl halides.[24]

Commercially,[25] the use of AHF for the synthesis of fluoroorganic compounds via either halogen exchange (HALEX) or decomposition of diazonium salts is carried out by this route for economic reasons. For synthetic purposes, inorganic fluorides,[1f] tetraalkylammonium fluorides,[12] and HF·base complexes[10] are preferred fluorinating agents. High yields of α,α,α-trifluoromethoxybenzene[26] via halogen exchange and fluoropyridines via decomposition of the corresponding diazonium salt[27] in the presence of AHF can be obtained.

Bond-Cleaving Reactions. Desilylation of silyl ethers[28] and silyl enol ethers[29] is most frequently accomplished using HF/MeCN.[3] Compounds containing both alcoholic and phenolic *t*-butyldimethylsilyl ethers can be desilylated with HF/MeCN chemoselectively, leaving the silyl ether of the phenol unscathed (eq 3).[30]

2 equiv 40% HF
92%

(3)

Cleavage of silyl enol ethers result in the formation of α-alkylidene-β-lactams[31] and migration of C=C bonds, leading to γ-substituted cyclohexenals[32] and β-substituted cyclopentenones.[33]

Concomitant ring formation during desilylation of *t*-butyldimethylsilyl ethers occurs with a high degree of stereoselectivity to give lactones,[34] and is a facile route to butenolides,[35] β-methylene-γ-butyrolactones,[36] bis-β-ketomacrolides,[37] spiroacetals (eq 4),[38] and dispiroacetals.[39]

40% HF, MeCN
39% + 8%

(4)

Similar treatment of *t*-butyldimethylsilyl enol ethers leads to the formation of cyclopentanecarboxylic acid[40] and *cis/trans* carbocycles.[41]

Desilylation of *erythro*-α-silyloxyalkylboranes with concomitant protiodeboronation gives alkenes (eq 5) with (*E*:*Z*) ratios as high as 95:5 in 73% isolated yield.[42]

Cleavage of Si–C bonds also is an entry into alkenes,[43] aldehydes,[44] and α,β-unsaturated aldehydes.[45] Cleavage of the Si–N bond gives amines.[46]

Hydrogen fluoride is a versatile reagent in peptide chemistry, cleaving *N*-benzyloxycarbonyl, *S*-benzyl, and *S*-*p*-methoxybenzyl protecting groups of peptides.[5] Removal of protecting groups of aspartine residues (eq 6),[47] *N*-nitroarginine residues,[48] *O*-dimethylphosphinyltyrosine,[49] and the *O*-phosphate ester of tyrosine[50] is accomplished in HF/anisole at $0\,^\circ C$ in ≤ 1 h. However, under similar conditions, side reactions of glutamyl peptides[51] lead to Friedel–Crafts acylation of anisole.

aq HF, MeCN

90.2% 2.8% (5)

anisole, HF acetone

0.5 h, –78 to 0 °C 70%

R = resin support

(6)

Peptide protecting groups such as phosphonamides[52] and sulfonamides[53] are cleaved in the presence of HF/anisole. In the case of the tripeptide Gly–Lys–Gly, the sulfonyl protecting group of the Lys residue can be chemoselectively removed in the presence of a *t*-Boc protecting group.[54] Polysaccharide solvolysis and glycoprotein deglycosylation is efficiently performed with HF.[55]

Catalyst in Electrophilic Aromatic Substitutions. In many instances, AHF offers distinct advantages over traditional Lewis acids for the catalysis of electrophilic aromatic substitution reactions. Unlike most Lewis acids, AHF acts as both catalyst and solvent,[56] and can be removed by distillation.[57]

Alkylation of arenes with alkenes,[6a,58] alkyl halides,[6b,59] and methylcyclopropane[60] proceeds in moderate to good yield. Alkylation of arenes with concomitant ring closure leads to cyclic[61] and heterocyclic compounds.[62] Alkylation of benzene with sodium nitronate in the presence of AHF gives benzaldehyde oxime in 78% yield.[63]

The distribution of *o*-, *m*-, and *p*-isomers is known to be a function of temperature. For example, alkylation of phenol with ***Isobutene*** at −40 °C gives 2-*t*-butylphenol exclusively, while at 0 °C, 3-*t*-butylphenol is obtained in 88% yield; above 30 °C, 4-*t*-butylphenol begins to predominate.[64]

AHF is an excellent catalyst for Friedel–Crafts acylation of substrates as diverse as phenols,[65] thiophenes,[66] and ferrocene.[67] Acylation of phenols is known to proceed with a high degree of regioselectivity to yield *p*-isomers almost exclusively (eq 7).[68]

(7)

R^1 = H, Me, Ph, OMe, SMe
R^2 = Me, Et, Ph, CH_2Cl
X = F, Cl, Br, OH, OAc

Similarly, Fries rearrangement of phenyl esters leads to formation of the *p*-isomer[69] unless of course the *para* position is blocked; then *ortho* substitution occurs.[70] Although thiophenol does not undergo acylation in the presence of AHF, thioanisole does, giving the *p*-isomer.[71] As opposed to Lewis acids, AHF-catalyzed acylation of 2-methoxynaphthalene proceeds with high *para* regioselectivity to yield 6-methoxy-2-acetonaphthone.[72]

Intramolecular cyclizations of 3-phenylpropionic acid and 4-phenylbutanoic acid afford α-tetralone and α-hydrindone in 92% and 73% yield, respectively.[73] More sophisticated polycyclic compounds, such as anthracyclinones,[74] can be constructed, as can benzo[*a*]fluoranthene[75] via reductive cyclization.

Other electrophilic aromatic substitutions do not proceed generally as well in AHF, as is the case for the sulfonation of benzene[76] and rearrangement of *p*-cresyl benzenesulfonate[77] to 2-hydroxy-4-methyl diphenyl sulfone. However, nitration,[78] amidomethylation,[79] and thioamidation[80] of aromatic substrates can be achieved in good to excellent yield.

Related Reagents. *N*-Bromosuccinimide–Hydrogen Fluoride; Hydrofluoric Acid; Hydrogen Fluoride–Antimony(V) Fluoride; Potassium Fluoride; Pyridinium Poly(hydrogen fluoride); Sodium Fluoride; Tetra-*n*-butylammonium Fluoride.

1. (a) *Fluorine Chemistry*; Simons, J. H., Ed.; Academic: New York, 1950; Vol. 1, Chapters 6–7. (b) *Fluorine Chemistry*; Simons, J. H.; Ed.; Academic: New York, 1954; Vol. 2, Chapter 4. (c) Hudlicky, M. *Organic Fluorine Chemistry*; Plenum: New York, 1971; Chapters 1, 2, 4. (d) *Kirk-Othmer Encyclopedia of Chemical Technology*; Grayson, M., Ed.; Wiley: New York, 1980; Vol. 10, pp 733–753. (e) *Gmelin Handbook of Inorganic Chemistry*; Koschel D., Ed.; Springer: New York, 1982; Suppl. Vol. 3. (f) *Syntheses of Fluoroorganic Compounds*; Knunyants, I. L.; Yakobson, G. G., Eds.; Springer: New York, 1985. (g) *Fluorine: The First Hundred Years (1886–1986)*; Banks, R. E.; Sharp, D. W. A.; Tatlow, J. C., Eds.; Elsevier: New York, 1986; Chapters 1, 4. (h) *Ullmann's Encyclopedia of Industrial Chemistry*; Gerhartz, W.; Ed.; VCH: New York, 1988; Vol. A11, pp 308–316. (i) Jache, A. W. *Fluorine-Containing Molecules*; VCH: New York, 1988; Chapter 9. (j) Hudlicky, M. *Chemistry of Organic Fluorine Compounds*, 2nd ed.; Prentice Hall: New York, 1992.
2. The Hammett acidity function, $H_0 = -10.2$ (25 °C): see ref. 1(e), p 183.
3. Newton, R. F.; Reynolds, D. P.; Finch, M. A. W.; Kelly, D. R.; Roberts, S. M. *TL* **1979**, 3981.
4. (a) Sakakibara, S.; Shimonishi, Y. K.; Kishida, Y.; Okada, M.; Sugihara, H. *BCJ* **1967**, *40*, 2164. (b) Sakakibara, S.; Kishida, Y.; Nishizawa, R.; Shimonishi, Y. *BCJ* **1968**, *41*, 438.
5. Hawley, M. C.; Selke, S. M.; Lamport, D. T. A. *Energy Agric.* **1983**, *2*, 219.
6. (a) Simons, J. H.; Archer, S. *JACS* **1938**, *60*, 2952. (b) Simons, J. H.; Archer, S. *JACS* **1938**, *60*, 2953. (c) Simons, J. H.; Randall, D. I.; Archer, S. *JACS* **1939**, *61*, 1795.
7. For purification methods, see ref. 1(e), pp 2–5.

8. Kilpatrick, M.; Luborsky, F. E. *JACS* **1953**, *75*, 577.
9. Degnan, T. F. *CA* **1976**, *84*, 137 860j.
10. (a) Yoneda, N. *T* **1991**, *47*, 5329. (b) *Synthetic Fluorine Chemistry*; Olah, G. A.; Chambers, R. D.; Prakash, G. K., Eds.; Wiley: New York, 1992; Chapter 8.
11. (a) Olah, G. A.; Nojima, M.; Kerekes, I. *S* **1973**, 779. (b) Olah, G. A.; Welch, J. T.; Vankar, Y. D.; Nojima, M.; Kerekes, I.; Olah, J. A. *JOC* **1979**, *44*, 3872.
12. Mascaretti, O. A. *Aldrichim. Acta* **1993**, *26*, 47.
13. Newkirk, A. E. *JACS* **1946**, *68*, 2467.
14. Olah, G. A.; Kuhn, S. J. *JOC* **1961**, *26*, 237.
15. Rothman, E. S.; Moore, G. G.; Serota, S. *JOC* **1969**, *34*, 2486.
16. Olah, G. A.; Kuhn, S. J. *JOC* **1961**, *26*, 225.
17. (a) Buckley, G. D.; Piggott, H. A.; Welch, A. J. E. *JCS* **1945**, 864. (b) Durden, J. A. U.S. Patent 4 304 735, 1981.
18. Fried, J.; Sabo, E. F. *JACS* **1957**, *79*, 1130.
19. (a) Alvernhe, G.; Kozolowska-Gramsz, E.; Lacombe-Bar, S.; Laurent, A. *TL* **1978**, 5203. (b) Alvernhe, G. M.; Ennakoua, C. M.; Lacombe, S. M.; Laurent, A. J. *JOC* **1981**, *46*, 4938.
20. Olah, G. A.; Bollinger, J. M. *JACS* **1967**, *89*, 4744.
21. Olah, G. A.; Nojima, M.; Kerekes, I. *S* **1973**, 780.
22. Bowers, A.; Ibáñez, L. C.; Denot, E.; Becerra, R. *JACS* **1960**, *82*, 4001.
23. (a) Pattison, F. L. M.; Peters, D. A. V.; Dean, F. H. *CJC* **1965**, *43*, 1689. (b) Pattison, F. L. M.; Buchanan, R. L.; Dean, F. H. *CJC* **1965**, *43*, 1700. (c) Dean, F. H.; Pattison, F. L. M. *CJC* **1965**, *43*, 2415.
24. (a) Hopff, H.; Valkanas, G. *HCA* **1963**, *46*, 1818. (b) Webb, J. L.; Corn, J. E. *JOC* **1973**, *38*, 2091.
25. *Organofluorine Chemicals and Their Industrial Application*; Banks, R. E., Ed.; Horwood: Chichester, 1979.
26. Olah, G. A.; Yamato, T.; Hashimoto, T.; Shih, J. G.; Trivedi, N.; Singh, B. P.; Piteau, M.; Olah, J. A. *JACS* **1987**, *109*, 3708.
27. Fukuharaa, T.; Yoneda, N.; Suzuki, A. *JFC* **1988**, *38*, 435.
28. (a) Anwar, S.; Davis, A. P. *CC* **1986**, 831. (b) Hauser, F. M.; Hewawasam, P.; Mal, D. *JACS* **1988**, *110*, 2919.
29. Ried, W.; Reiher, U. *CB* **1987**, *120*, 1597.
30. Collington, E. W.; Finch, H.; Smith, I. J. *TL* **1985**, *26*, 681.
31. Palomo, C.; Aizpurua, J. M.; López, M. C.; Aurrekoetxea, N.; Oiarbide, M. *TL* **1990**, *31*, 6425.
32. Jones, T. K.; Denmark, S. E. *JOC* **1985**, *50*, 4037.
33. Kozikowski, A. P.; Jung, S. H. *JOC* **1986**, *51*, 3400.
34. DeShong, P.; Simpson, D. M.; Lin, M.-T. *TL* **1989**, *30*, 2885.
35. Larson, G. L.; Prieto, J. A.; Gonzalez, P. *SC* **1989**, *19*, 2779.
36. Greene, A. E.; Coelho, F.; Depres, J.-P. *JOC* **1985**, *50*, 1973.
37. Fox, C. M. J.; Ley, S. V.; Slawin, A. M. Z.; Williams, D. J. *CC* **1985**, 1805.
38. (a) Amouroux, R. *H* **1984**, *22*, 1489. (b) Paterson, I.; Craw, P. A. *TL* **1989**, *30*, 5799.
39. Kocieński, P.; Fall, Y.; Whitby, R. *JCS(P1)* **1989**, 841.
40. Khan, K. M.; Knight, D. W. *CC* **1991**, 1699.
41. Cameron, A. G.; Knight, D. W. *TL* **1982**, *23*, 5455.
42. Pelter, A.; Buss, D.; Colclough, E. *CC* **1987**, 297.
43. (a) Ochiai, M.; Tada, S.-I.; Sumi, K.; Fujita, E. *CC* **1982**, 281. (b) Johnson, C. R.; Tait, B. D. *JOC* **1987**, *52*, 281.
44. Schonauer, K.; Zbiral, E. *TL* **1983**, *24*, 573.
45. DeShong, P.; Leginus, J. M. *JOC* **1984**, *49*, 3421.
46. Schwartz, E.; Shanzer, A. *TL* **1982**, *23*, 979.
47. Blake, J. *Int. J. Peptide Protein Res.* **1979**, *13*, 418.
48. (a) Lenard, J. *JOC* **1967**, *32*, 250. (b) Inouye, K.; Sasaki, A.; Yoshida, N. *BCJ* **1974**, *47*, 202.
49. Ueki, M.; Sano, Y.; Sori, I.; Shinozaki, K. *TL* **1986**, *27*, 4181.
50. Kitas, E. A.; Perich, J. W.; Johns, R. B.; Tregear, G. W. *TL* **1988**, *29*, 3591.
51. Feinberg, R. S.; Merrifield, R. B. *CA* **1977**, *86*, 5785v.
52. Greenhalgh, R.; Blanchfield, J. R. *CJC* **1966**, *44*, 501.
53. (a) Rodricks, J. V.; Rapoport, H. *JOC* **1971**, *36*, 46. (b) Bosin, T. R.; Hanson, R. N.; Rodricks, J. V.; Simpson, R. A.; Rapoport, H. *JOC* **1973**, *38*, 1591.
54. Fukuda, T.; Kitada, C.; Fujino, M. *CC* **1978**, 220.
55. (a) Rorrer, G. L.; Hawley, M. C.; Selke, S. M.; Lamport, D. T. A.; Dey, P. M. *CA* **1991**, *114*, 3818u. (b) Defaye, J.; Pedersen, C. *CA* **1992**, *116*, 216 678p.
56. (a) Zingaro, R. A. *CA* **1977**, *85*, 201 343t. (b) Baasner, B.; Klauke, E. *JFC* **1982**, *19*, 553.
57. Reid, E. B.; Yost, J. F. *JACS* **1950**, *72*, 5232.
58. (a) Simons, J. H.; Archer, S. *JACS* **1938**, *60*, 986. (b) Calcott, W. S.; Tinker, J. M.; Weinmayr, V. *JACS* **1939**, *61*, 949.
59. Simons, J. H.; Archer, S. *JACS* **1939**, *61*, 1521.
60. Pines, H.; Huntsman, W. D.; Ipatieff, V. N. *JACS* **1951**, *73*, 4343.
61. Renfrow, W. B.; Renfrow, A.; Shoun, E.; Sears, C. A. *JACS* **1951**, *73*, 317.
62. (a) Mondon; Aumann, G.; Oelrich, E. *CB* **1972**, *105*, 2025. (b) Michne, W. F. *JOC* **1976**, *41*, 894.
63. Berrier, C.; Brahmi, R.; Carreyre, H.; Coustard, J. M.; Jacquesy, J. C. *TL* **1989**, *30*, 5763.
64. Norell, J. R. *JOC* **1973**, *38*, 1929.
65. Bailey, D. *CI(L)* **1971**, 682.
66. (a) Hartough, H. D.; Kosak, A. I. *JACS* **1947**, *69*, 3093. (b) Dann, O.; Kokorudz, M.; Gropper, R. *CB* **1954**, *87*, 140.
67. Weinmayr, V. *JACS* **1955**, *77*, 3009.
68. Mott, G. N. U.S. Patent 4 607 125, 1986.
69. Dann, O.; Mylius, G. *LA* **1954**, *587*, 1.
70. Norell, J. R. *JOC* **1973**, *38*, 1924.
71. Aslam, M.; Davenport, K. G.; Stansbury, W. F. *JOC* **1991**, *56*, 5955.
72. Davenport, K. G.; Linstid, H. C. U.S. Patent 4 593 125, 1986.
73. Fieser, L. F.; Hershberg, E. B. *JACS* **1939**, *61*, 1272.
74. Braun, M. *T* **1984**, *40*, 4585.
75. Ray, J. K.; Harvey, R. G. *JOC* **1982**, *47*, 3335.
76. Simons, J. H.; Passino, H. J.; Archer, S. *JACS* **1941**, *63*, 608.
77. Simons, J. H.; Archer, S.; Randall, D. I. *JACS* **1940**, *62*, 485.
78. Finger, G. C.; Reed, F. H.; Maynert, E. W.; Weiner, A. M. *JACS* **1951**, *73*, 149.
79. Desbois, M. *Actual. Chim.* **1987**, 147.
80. Feiring, A. E. *JOC* **1976**, *41*, 148.

Kenneth G. Davenport
Hoechst Celanese Corporation, Corpus Christi, TX, USA

Hydrogen Iodide[1]

[10034-85-2] HI (MW 127.91)

(electrophilic hydriodination of alkenes and alkynes;[3–10] cleavage of epoxides,[1b,11,12] ethers, and acetals;[13] conversion of alcohols to iodides;[14–17] reducing agent for many groups including quinones,[24,25] α-diketones,[23] α-ketols,[23] α-halo ketones,[19] α-diazo ketones,[22] and sulfoxides;[30] reductive cyclization of keto acids[27,28])

Alternate Name: hydriodic acid.
Physical Data: 57% aqueous solution: bp 127 °C; *d* 1.70 g cm^{-3}.
Solubility: sol most common organic solvents.
Form Supplied in: compressed gas; colorless 57% aq solution; widely available.
Preparative Methods: from the reaction of tetrahydronaphthalene with I_2;[2] can be generated in situ (a) from Me_3SiCl and NaI in the presence of water,[3] (b) from I_2 and activated alumina,[4] (c) from KI and H_3PO_4,[5] and (d) from $Et_2PhN{\cdot}BI_3$ and AcOH.[6]
Purification: distillation of the aqueous azeotrope; concentrated solutions can be regenerated after long storage by treatment with hypophosphorous acid.
Handling, Storage, and Precautions: store protected from air and light at or below rt. Highly corrosive and toxic. This reagent should be handled in a fume hood.

Hydriodination of Alkenes and Alkynes. Being a stronger acid, HI undergoes addition more readily than ***Hydrogen Chloride*** or ***Hydrogen Bromide*** to most alkenes and alkynes.[1a] Moreover, there is no competing radical addition as with HBr. However, because of the difficulty in generating and transferring anhydrous HI, addition of HI has received less attention than addition of HCl and HBr. As mentioned above, several techniques have been developed for generating HI in situ. These include the use of KI and H_3PO_4 (eq 1);[5] Me_3SiCl and NaI in the presence of water (eq 2);[3] I_2 and activated Al_2O_3 (eq 2);[4] and the $Et_2PhN{\cdot}BI_3$ complex and AcOH (eq 3).[6] Alternatively, $I(py)_2BF_4$ has been used with the hydride donor Et_3SiH (eq 1).[7]

(1)

KI, H_3PO_4, 80 °C, 3 h — 88–90%
$I(py)_2BF_4$, HBF_4, Et_3SiH, 20 °C, 1 h — 50%

C_5H_{11} → C_5H_{11} (I) (2)

TMSCl, NaI, H_2O, 25 °C, 1 h — 98%
I_2, Al_2O_3, 36 °C, 2 h — 83%
57% HI, $C_{16}H_{33}(Bu)_3PBr$, 115 °C, 0.25 h — 97%
TMSI, SiO_2, 25 °C, 1 h — 98%

$Et_2PhN{\bullet}BI_3$, AcOH, 82% — I (3)

Aqueous HI has been used with a phase-transfer catalyst to hydriodinate alkenes (eq 2).[8] Similarly, it has been used to convert dialkylalkynes to the corresponding (*Z*)-vinyl iodides (eq 4).[9]

Pr—≡—Pr → (57% HI, 80 °C, 4 h; 92%) Pr(I)C=CH–Pr (4)

A particularly convenient method for generating HI in situ involves the use of various inorganic and organic iodides in the presence of appropriately prepared silica gel or alumina (eq 2).[10] These adsorbents also facilitate the addition process. Surface-mediated hydriodination of phenylalkynes affords the (*E*) isomers, resulting from *syn* addition (eq 5).[10] The regiochemical course of these hydroiodinations follows Markovnikov's rule.

Ph—≡—R → (PI_3, CH_2Cl_2, Al_2O_3, 25 °C, 0.3–3 h; 76–85%) Ph(I)C=CH–R (5)
R = Me, Ph or *t*-Bu

Cleavage of Epoxides to Iodohydrins. The addition of HI to epoxides to give iodohydrins proceeds readily using either aqueous HI or anhydrous HI in organic solvents.[1b,11] Because of the difficulty of preparing anhydrous HI, aqueous solutions have most often been used for this transformation. The stereo- and regioselectivity of the addition process is similar to the general trends discussed for the corresponding additions with HCl (see ***Hydrogen Chloride***). Trimethylsilyl-substituted epoxides give the corresponding iodohydrins with particularly high stereo- and regiospecificity (eq 6).[12]

Hex, Hex, TMS epoxide → (57% HI, Et_2O, 0 °C; 100%) Hex, I, OH, TMS, Hex (6)

Cleavage of Ethers and Acetals. HI readily cleaves ethers to alcohols and/or iodides[13] and is an attractive reagent for this transformation from the standpoint of economy and convenience. Primary and secondary alkyl methyl ethers are cleaved to afford alcohols (or derivatives), while benzyl and tertiary alkyl ethers often yield iodides. Acetals react in a similar fashion to produce ketones, although this deprotection method rarely offers advantages over more common procedures.

Conversion of Alcohols and Chlorides to Iodides. The reaction of 57% aqueous HI with saturated primary and secondary alcohols at elevated temperatures leads in fair to high yield to the corresponding iodides.[14] Tertiary iodides have been synthesized in good to high yields from the corresponding alcohols under especially mild conditions by 55% aqueous HI in the presence of ***Lithium Iodide***.[15] Allylic alcohols are transformed to allylic iodides by HI generated in situ from Me_3SiCl/NaI.[16] Benzylic alcohols are subject to conversion to the saturated system, presumably via iodine substitution and ensuing reduction (eq 7).[17]

OH, OH → (48% HI, Δ, 6 h; 73%) (7)

Alkyl iodides can also be synthesized via treatment of secondary and tertiary alkyl chlorides with anhydrous HI in the presence of catalytic amounts of FeI_3.[18]

Reduction of α-Substituted Ketones. Treatment of various α-substituted ketones with HI leads to reductive scission of the α-substituent. Reductive dehalogenation of α-halo ketones can thus be accomplished with HI to furnish the corresponding ketones in high yield (eq 8).[19] Reaction occurs readily even with sterically hindered substrates. Related procedures employing cat. NaI or 57% aq HI and phosphorous acid in acetonitrile[20] or NaI in concd

H_2SO_4[21] require long reaction times, high temperatures, and/or reactive substrates and are less satisfactory.

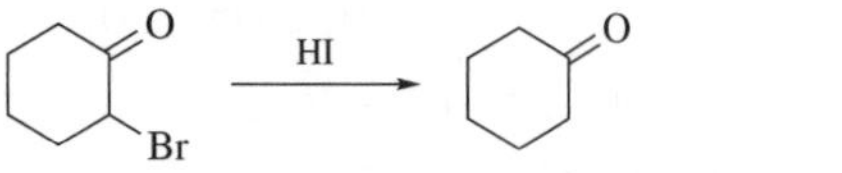

(8)

α-Diazo ketones are reduced to methyl ketones by 47% aqueous HI in $CHCl_3$ (eq 9).[22] The reaction with HI differs from that of HBr and HCl, which give halomethyl ketones as products. Presumably, the initially formed iodomethyl ketone is reduced to the saturated ketone under the reaction conditions.

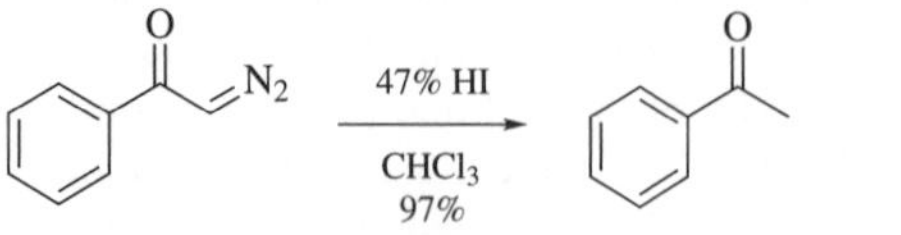

(9)

α-Diketones and α-ketols are reduced to the corresponding saturated ketones in good yields by aqueous HI in acetic acid at reflux (eq 10).[23]

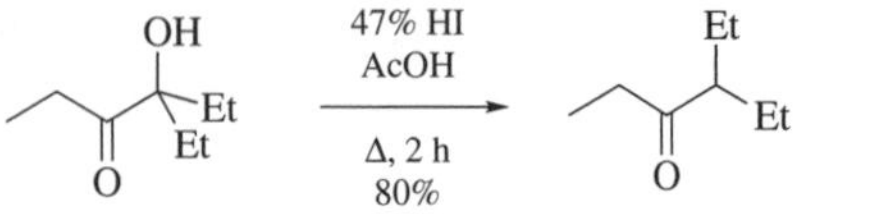

(10)

Reduction of Quinones and Phenols to Arenes. Polycyclic quinones may be reduced to polyarenes by HI in HOAc at reflux (eq 11).[24,25] In resistant cases, concentrated aqueous HI may be employed; addition of phosphorus often results in cleaner reaction by removing the I_2 formed. Large excess of HI or prolonged reaction time may lead to overreduction. Since hydroquinones and phenols are intermediates in these reactions, they are also readily reducible with this reagent. Reductive methylation of quinones can be accomplished in high yield by reaction of polycyclic quinones with excess ***Methyllithium*** followed by reduction with HI (eq 12).[26]

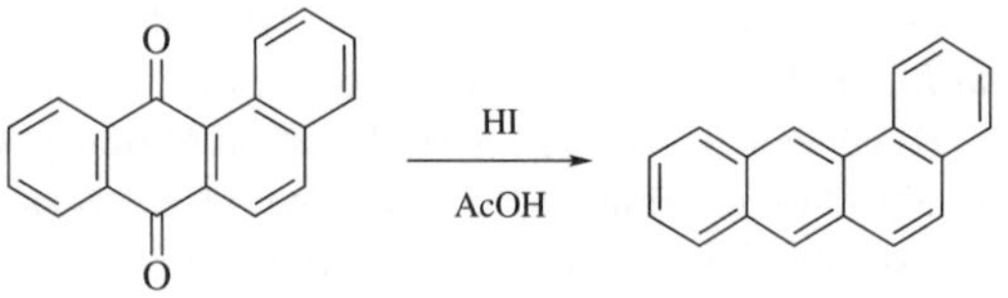

(11)

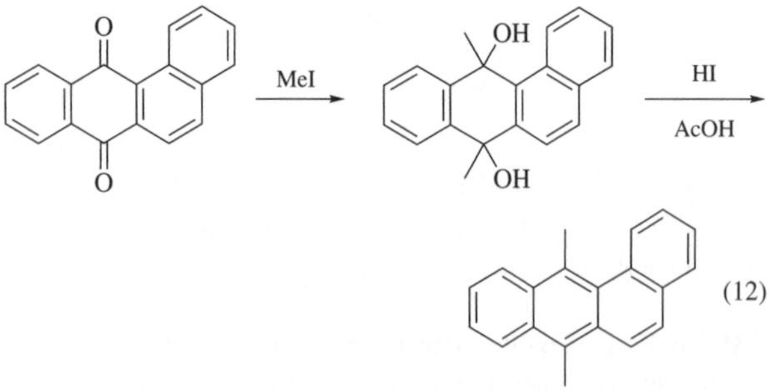

(12)

A method for the construction of fused polyarenes entails reaction of a smaller aromatic ring system with phthalic anhydride followed by reductive cyclization of the keto acid product with HI in acetic acid to form a polyarene with two additional rings (eq 13).[27,28] This method conveniently combines three steps (reduction of the carbonyl group, cyclodehydration, and reduction) into one step.

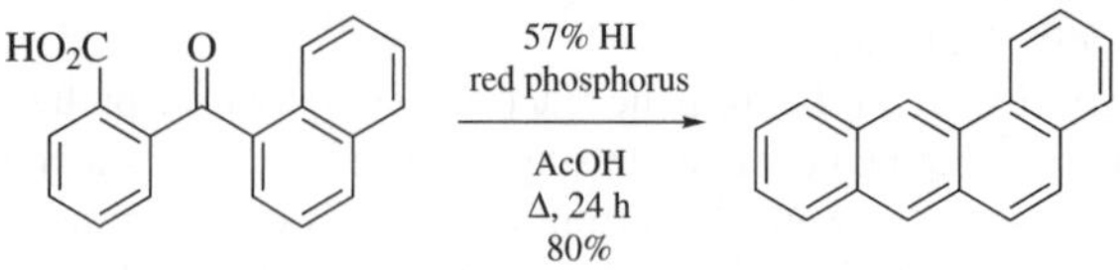

(13)

Reductive Deoxygenation of Aryl Ketones. The combination HI/P/HOAc effectively deoxygenates aryl ketones (eq 14).[27–29] While this method is utilized infrequently, it represents a useful alternative to the better known Wolff–Kishner and Clemmensen reduction methods.

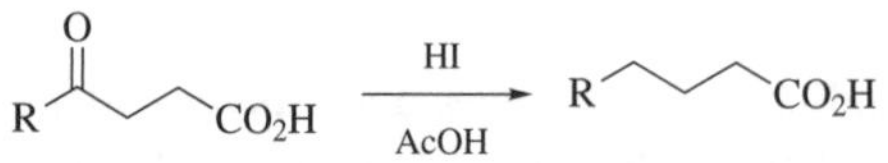

(14)

Reduction of Sulfoxides to Sulfides. Sulfoxides are readily deoxygenated by HI without the complicating halogenation that often accompanies reduction using HBr or HCl.[30]

Reduction of Alkenylsilanes. Hydriodic acid reacts with vinylsilanes with replacement of R_3Si by hydrogen (eq 15).[31] A small amount of I_2 and water (or D_2O) is also effective. These reactions usually occur with retention of configuration.

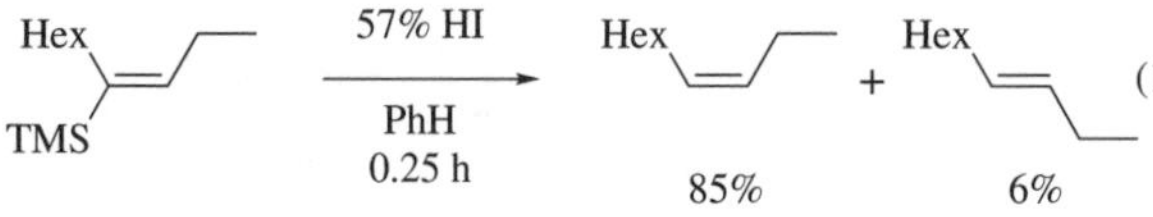

(15)

Related Reagents. Aluminum Iodide; Diphosphorus Tetraiodide; Iodotrimethylsilane; Lithium Iodide; Phosphorus(III) Iodide; Potassium Iodide; Tetra-*n*-butylammonium Iodide; Triphenylphosphine-Iodine.

1. (a) Larock, R. C.; LeLong, W. W. *COS* **1991**, *4*, 269. (b) Parker, R. E.; Isaacs, N. S. *CRV* **1959**, *59*, 737.
2. Hoffman, C. J. *Inorg. Synth.* **1963**, *7*, 180.
3. Irifune, S.; Kibayashi, T.; Ishii, Y.; Ogawa, M. *S* **1988**, 366.
4. Pagni, R. M.; Kabalka, G. W.; Boothe, R.; Gaetano, K.; Stewart, L. J.; Conaway, R.; Dial, C.; Gray, D.; Larson, S.; Luidhardt, T. *JOC* **1988**, *53*, 4477.
5. (a) Stone, H.; Shechter, H. *OSC* **1963**, *4*, 543. (b) Kropp, P. J.; Adkins, R. *JACS* **1991**, *113*, 2709.
6. Reddy, C. K.; Periasamy, M. *TL* **1990**, *31*, 1919.
7. Barluenga, J.; Gonzalez, J. M.; Campos, P. J.; Asensio, G. *AG(E)* **1985**, *24*, 319.
8. Landini, D.; Rolla, F. *JOC* **1980**, *45*, 3527.
9. Hudrlik, P. F.; Kulkarni, A. K.; Jain, S.; Hudrlik, A. M. *T* **1983**, *39*, 877.
10. (a) Kropp, P. J.; Daus, K. A.; Crawford, S. D.; Tubergen, M. W.; Kepler, K. D.; Craig, S. L.; Wilson, V. P. *JACS* **1990**, *112*, 7433. (b) Kropp, P. J.; Daus, K. A.; Tubergen, M. W.; Kepler, K. D.; Wilson, V. P.; Craig, S. L.; Baillargeon, M. M.; Breton, G. W. *JACS* **1993**, *115*, 3071. (c) Kropp, P. J., Crawford, S. D. *JOC*, **1994**, *59*, 3102.

11. (a) Buchanan, J. G.; Sable, H. Z. In *Selective Organic Transformations*; Thyagarajan, B. S., Ed.; Wiley: New York, 1972; Vol. 2, pp 1–92. (b) Armarego, W. L. F. In *Stereochemistry of Heterocyclic Compounds*; Taylor E. C.; Weissberger, A., Eds.; Wiley: New York, 1977; Vol. 2, pp 23–25. (c) Bártok, M.; Láng, K. L. In *The Chemistry of Functional Groups. Supplement E: The Chemistry of Ethers, Crown Ethers, Hydroxyl Groups and Their Sulfur Analogues*, Patai, S., Ed.; Wiley: New York, 1980; Part 2, pp 655–657. (d) Owen, L. N.; Saharia, G. S. *JCS* **1953**, 2582.
12. Obayashi, M.; Utimoto, K.; Nozaki, H. *TL* **1978**, 1383.
13. (a) Bhatt, M. V.; Kulkarni, S. U. *S* **1983**, 249. (b) Deulofeu, V.; Guerrero, T. J. *OSC* **1955**, *3*, 586.
14. Vogel, A. I. *JCS* **1943**, 636.
15. Masada, H.; Murotani, Y. *BCJ* **1980**, *53*, 1181.
16. (a) Kanai, T.; Irifune, S.; Ishii, Y.; Ogawa, M. *S* **1989**, 283. (b) Kanai, T.; Kanagawa, Y.; Ishii, Y. *JOC* **1990**, *55*, 3274.
17. Parham, W. E.; Sayed, Y. A. *S* **1976**, 116.
18. Yoon, K. B.; Kochi, J. K. *JOC* **1989**, *54*, 3028.
19. Penso, M.; Mottadelli, S.; Albanese, D. *SC* **1993**, *23*, 1385.
20. Mandal, A. K.; Nijasure, A. M. *SL* **1990**, 554.
21. Gemal, A. L.; Luche, J. L. *TL* **1980**, *21*, 3195.
22. (a) Wolfrom, M. L.; Brown, R. L. *JACS* **1943**, *65*, 1516. (b) Pojer, P. M.; Ritchie, E.; Taylor, W. C. *AJC* **1968**, *21*, 1375.
23. (a) Reusch, W.; LaMahieu, R. *JACS* **1964**, *86*, 3068. (b) Hoeger, C. A.; Johnston, A. D.; Okamura, W. H. *JACS* **1987**, *109*, 4690.
24. Konieczny, M.; Harvey, R. G. *JOC* **1979**, *44*, 4813.
25. Konieczny, M.; Harvey, R. G. *OSC* **1990**, *7*, 18.
26. Konieczny, M.; Harvey, R. G. *JOC* **1980**, *45*, 1308.
27. Platt, K. L.; Oesch, F. *JOC* **1981**, *46*, 2601.
28. Harvey, R. G.; Leyba, C.; Konieczny, M.; Fu, P. P.; Sukumaran, K. B. *JOC* **1978**, *43*, 3423.
29. Ansell, L. L.; Rangarajan, T.; Burgess, W. M.; Eisenbraun, E. J.; Keen, G. W.; Hamming, M. C. *OPP* **1976**, *8*, 133.
30. (a) Madesclaire, M. *T* **1988**, *44*, 6537. (b) Ookuni, I.; Fry, A. *JOC* **1971**, *36*, 4097.
31. Utimoto, K.; Kitai, M.; Nozaki, H. *TL* **1975**, 2825.

Gary W. Breton & Paul J. Kropp
University of North Carolina, Chapel Hill, NC, USA

Ronald G. Harvey
University of Chicago, IL, USA

Imidazole

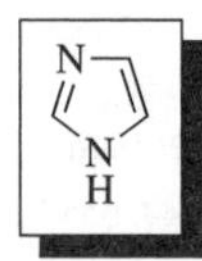

[288-32-4] $C_3H_4N_2$ (MW 68.09)

(nucleophilic catalyst for silylations and acylations; buffer; weak base; iodination methods)

Alternate Names: Im; iminazole; 1,3-diazole; glyoxaline.
Physical Data: mp 90–91 °C; bp 255 °C, 138 °C/12 mmHg.
Solubility: sol water, alcohols, ether, acetone, chloroform.
Form Supplied in: colorless crystalline solid; widely available.
Drying: 40 °C in vacuo over P_2O_5.
Purification: can be crystallized from C_6H_6, CCl_4, CH_2Cl_2, EtOH, petroleum ether, acetone–petroleum ether, or water; can also be purified by vacuum distillation, sublimation, or by zone refining.

Introduction. Protonated Imidazole has a pK_a of 7.1 and is thus a stronger base than thiazole (pK_a 2.5), oxazole (pK_a 0.8), and pyridine (pK_a 5.2). It is both a good acceptor and donor of hydrogen bonds. The pK_a for loss of the N–H is ~14.2, i.e. imidazole is a very weak acid.

Silylations. Imidazole is a standard component in silylations of alcohols as well as carboxylic acids, amines, and a variety of other functions, typically in combination with a silyl chloride in DMF (eq 1).[1] A very widely used procedure for alcohol protection is by conversion into the corresponding *t*-butyldimethylsilyl (TBDMS or TBS) ether using the method;[2] in other solvents such as pyridine or THF the reactions are much slower, probably because the primary silylating reagent is *t*-BuMe$_2$Si–Im. In this and many other aspects, imidazole resembles another very useful transfer catalyst, ***4-Dimethylaminopyridine*** (DMAP). Similarly, bulkier and hence more stable silyl groups, such as *t*-butyldiphenylsilyl (TBDPS)[3] and triisopropylsilyl (TIPS),[4] can be introduced. Times for completion of reaction at 20 °C vary (0.5–20 h); these silylating agents, originally developed for nucleoside protection,[5] usually react faster with primary alcohols and with certain secondary alcohols, thus allowing selective protection of polyols to be achieved efficiently. (A faster alternative involves the use of ***1,8-Diazabicyclo[5.4.0]undec-7-ene*** (DBU) in place of imidazole, in a variety of solvents such as CH_2Cl_2, C_6H_6, or MeCN, in combination with R_3SiCl).[6] 1,3-Diones can be efficiently *O*-silylated using TBDMSCl–Im[7] or ***Hexamethyldisilazane*** (HMDS) and imidazole;[8] other reagents are not as suitable, even in cases where the enol content is high. The products are useful as *trans*-silylating agents.[7] The HMDS–Im combination is also useful for the silylation of thiols.[9]

$$R^1\text{CH}_2\text{OH} + R^2{}_3\text{SiCl} \xrightarrow[\text{DMF}]{\text{imidazole}} R^1\text{CH}_2\text{OSiR}^2{}_3 \quad (1)$$

Ester Hydrolysis. Inspired by evidence that the imidazole ring of histidine residues present in various hydrolytic enzymes is responsible for their proteolytic activities, imidazole itself has been shown to be an excellent catalyst of ester hydrolysis (e.g. eq 2).[10] In intramolecular transesterifications and hydrolyses of 2-hydroxymethylbenzoic acid derivatives, the accelerating role of imidazole is due to its ability to act as a proton transfer catalyst rather than as a nucleophile.[11]

(2) [triacetoxyflavone → imidazole, 60% aq EtOH, Δ → trihydroxyflavone; labels: AcO, OAc, OAc, O; HO, OH, OH, O]

Peptide Coupling. Peptide couplings involving *p*-nitrophenyl and related esters are dramatically accelerated by the addition of imidazole.[12] However, such reactions, which probably proceed by way of an acylimidazole, can be prone to racemization, in which case ***1,2,4-Triazole*** can be a superior activator.[13] Imidazole also catalyzes peptide coupling using the ***Triphenyl Phosphite*** method, with negligible racemization when the reactions are carried out in dioxane or DMF,[14] and is useful for the activation of phosphomonoester groups in nucleotide coupling, in combination with an arylsulfonyl chloride.[15]

Acylimidazoles and Nucleophiles. Acylimidazoles are readily prepared from the parent carboxylic acids by reaction of the derived acid chloride with imidazole or directly using ***N,N′-Carbonyldiimidazole***. These intermediates react smoothly with a variety of nucleophiles including Grignard reagents (eq 3),[16] ***Lithium Aluminium Hydride*** (eq 4),[16] and nitronates (eq 5).[17] At −20 °C, aroylimidazoles can be reduced to the corresponding aldehydes in the presence of an ester function.[16]

$$\text{RCO–Im} \xrightarrow{\text{PhMgBr}} \text{RCOPh} \quad (3)$$

$$\text{RCO–Im} \xrightarrow{\text{LiAlH}_4} \text{RCHO} \quad (4)$$

$$\text{RCO–Im} \xrightarrow{\text{CH}_2\text{=N}^+(\text{O}^-)\text{O}^-} \text{RCOCH}_2\text{NO}_2 \quad (5)$$

The activation provided by an imidazole substituent is further illustrated in a route to 1,3-oxathiole-2-thiones from sodium 1-imidazolecarbodithioate, derived from the sodium salt of imidazole and CS_2 (eq 6).[18]

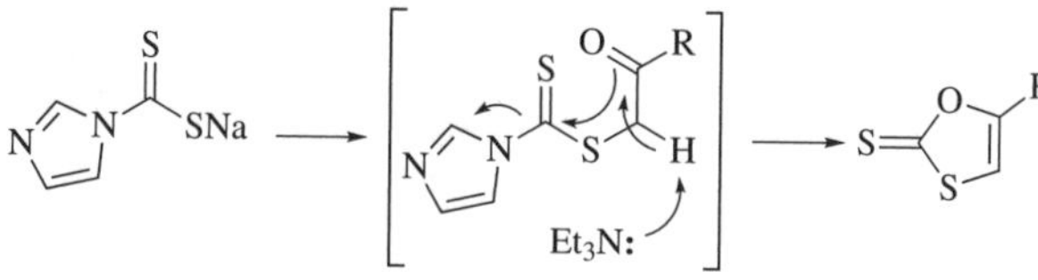

Other Uses. Imidazole is one of the best catalysts for the preparation of acid chlorides from the corresponding carboxylic acids and phosgene.[19] Aryl triflates can be obtained from phenols, or better phenolates, using ***Trifluoromethanesulfonic Anhydride*** in combination with imidazole; *N*-triflylimidazole is the reactive species.[20] Photochemical deconjugations of enones can be erratic but are promoted by the presence of a weak base such as imidazole or pyridine in polar solvents.[21]

Iodination of Alcohols. Imidazole, ***Triphenylphosphine***, and ***Iodine*** in hot toluene,[22] or preferably toluene–acetonitrile mixtures,[23] is an excellent combination for the conversion of alcohols into iodides, ROH → RI. Secondary alcohols react with inversion (but see below), although the method can be used for the selective iodination of primary hydroxyls. Applications in the area of natural product synthesis[24] emphasize the mildness and generality of the method as well as providing alternative recipes; sometimes, 2,4,5-triiodoimidazole can be a superior reagent.[22] Similarly, $Ph_3P–Cl_2–Im$ can be used for the preparation of alkyl chlorides, ROH → RCl, and the addition of imidazole in the ***Triphenylphosphine–Carbon Tetrachloride*** method for alcohol chlorination has a beneficial effect.[25]

Diol Deoxygenation. The $Ph_3P–Im–I_2$ combination can also be used to convert *vic*-diols into the corresponding alkenes, although 2,4,5-triiodoimidazole is more effective than imidazole itself.[26] Alternative combinations are ***Triphenylphosphine–Iodoform–Imidazole***, which can deoxygenate *cis*-diols but which is better suited to *trans*-isomers,[27] and ***Chlorodiphenylphosphine***–I_2–Im, which can be used to deoxygenate *vic*-diols when both are secondary or when one is secondary and one is primary.[28]

Epoxidation. ***t-Butyl Hydroperoxide*** in combination with $MoO_2(acac)_2$ can be used to oxidatively cleave alkenes, but will epoxidize such functions in the presence of a metalloporphyrin or a simple amine, the choice of which depends upon the substrate structure. Imidazole is the most suitable for 1-phenylpropene, ***Pyridine*** for stilbene, and *N,N*-dimethylethylenediamine for 1-alkenes.[29]

Related Reagents. 1,5-Diazabicyclo[4.3.0]non-5-ene; 1,8-Diazabicyclo[5.4.0]undec-7-ene; 4-Dimethylaminopyridine; Pyridine; Tri-*n*-butylphosphine.

1. Lalonde, M.; Chan, T. H. *S* **1985**, 817. Greene, T. W.; Wuts, P. G. M. *Protecting Groups in Organic Synthesis*, 2nd ed.; Wiley: New York, 1991. Kocieński, P. J. *Protecting Groups*; Thieme: Stuttgart, 1994.
2. Corey, E. J.; Venkateswarlu, A. *JACS* **1972**, *94*, 6190.
3. Hanessian, S.; Lavallee, P. *CJC* **1975**, *53*, 2975.
4. Cunico, R. F.; Bedell, L. *JOC* **1980**, *45*, 4797.
5. Ogilvie, K. K. *CJC* **1973**, *51*, 3799. Ogilvie, K. K.; Iwacha, D. J. *TL* **1973**, 317. Ogilvie, K. K.; Sadana, K. L.; Thompson, E. A.; Quilliam, M. A.; Westmore, J. B. *TL* **1974**, 2861. Ogilvie, K. K.; Thompson, E. A.; Quilliam, M. A.; Westmore, J. B. *TL* **1974**, 2865.
6. Aizpurua, J. M.; Palomo, C. *TL* **1985**, *26*, 475.
7. Veysoglu, T.; Mitscher, L. A. *TL* **1981**, *22*, 1299, 1303.
8. Torkelson, S.; Ainsworth, C. *S* **1976**, 722.
9. Glass, R. S. *JOM* **1973**, *61*, 83.
10. Bender, M. L.; Turnquest, B. W. *JACS* **1957**, *79*, 1652. Bruice, T. C.; Schmir, G. L. *JACS* **1957**, *79*, 1663. Bender, M. L. *CRV* **1960**, *60*, 82. Looker, J. H.; Holm, M. J.; Minor, J. L.; Kagal, S. A. *JHC* **1964**, *1*, 253.
11. Fife, T. H.; Benjamin, B. M. *JACS* **1973**, *95*, 2059. Kirby, A. J.; Lloyd, G. J. *JCS(P2)* **1974**, 637. Chiong, K. N. G.; Lewis, S. D.; Shafer, J. A. *JACS* **1975**, *97*, 418. Pollack, R. M.; Dumsha, T. C. *JACS* **1975**, *97*, 377. Belke, C. J.; Su, S. C. K.; Shafer, J. A. *JACS* **1971**, *93*, 4552.
12. Mazur, R. H. *JOC* **1963**, *28*, 2498. Wieland, T.; Vogeler, K. *AG(E)* **1963**, *2*, 42; *LA* **1964**, *680*, 125. McGahren, W. J.; Goodman, M. *T* **1967**, *23*, 2017. Stewart, F. H. C. *CI(L)* **1967**, 1960.
13. Beyerman, H. C.; van der Brink, W. M.; Weygand, F.; Prox, A.; Konig, W.; Schmidhammer, L.; Nintz, E. *RTC* **1965**, *84*, 213.
14. Mitin, Y. V.; Glinskaya, O. V. *TL* **1969**, 5267.
15. Berlin, Yu. A.; Chakhmakhcheva, O. G.; Efimov, V. A.; Kolosov, M. N.; Korobko, V. G. *TL* **1973**, 1353.
16. Staab, H. A.; Braunling, H. *LA* **1962**, *654*, 119. Staab, H. A.; Jost, E. *LA* **1962**, *655*, 90. Staab, H. A. *AG(E)* **1962**, *1*, 351.
17. Baker, D. C.; Putt, S. R. *S* **1978**, 478.
18. Ishida, M.; Sugiura, K.; Takagi, K.; Hiraoka, H.; Kato, S. *CL* **1988**, 1705.
19. Hauser, C. F.; Theiling, L. F. *JOC* **1974**, *39*, 1134.
20. Effenberger, F.; Mack, K. E. *TL* **1970**, 3947.
21. Eng, S. L.; Ricard, R.; Wan, C. S. K.; Weedon, A. C. *CC* **1983**, 236.
22. Garegg, P. J.; Samuelsson, B. *CC* **1979**, 978; *JCS(P1)* **1980**, 2866.
23. Garegg, P. J.; Johansson, R.; Ortega, C.; Samuelsson, B. *JCS(P1)* **1982**, 681.
24. Corey, E. J.; Pyne, S. G.; Su, W. *TL* **1983**, *24*, 4883. Berlage, U.; Schmidt, J.; Peters, U.; Welzel, P. *TL* **1987**, *28*, 3091. Corey, E. J.; Nagata, R. *TL* **1987**, *28*, 5391. Soll, R. M.; Seitz, S. P. *TL* **1987**, *28*, 5457.
25. Garegg, P. J.; Johansson, R.; Samuelsson, B. *S* **1984**, 168.
26. Garegg, P. J.; Samuelsson, B. *S* **1979**, *469*, 813.
27. Bessodes, M.; Abushanab, E.; Panzica, R. P. *CC* **1981**, 26.
28. Liu, Z.; Classon, B.; Samuelsson, B. *JOC* **1990**, *55*, 4273.
29. Kato, J.; Ota, H.; Matsukawa, K.; Endo, T. *TL* **1988**, *29*, 2843.

David W. Knight
Nottingham University, UK

Iron(III) Chloride[1]

[7705-08-0] Cl_3Fe (MW 162.20)

(mild oxidant capable of phenolic coupling,[1] dimerizing aryllithiums[6] and ketone enolates;[7,8] mild Lewis acid: catalyzes ene reactions,[21] Nazarov cyclizations,[18–20] Michael additions,[24] and acetonations[29])

Alternate Name: ferric chloride.
Physical Data: mp 306 °C; *d* 2.898 g cm^{-3}.[25]
Solubility: 74.4 g/100 mL cold water, 535.7 g mL^{-1} boiling water; v sol alcohol, MeOH, ether, 63 g mL^{-1} in acetone (18 °C).
Form Supplied in: black crystalline powder; widely available.
Preparative Methods: anhydrous $FeCl_3$ available commercially is adequate for most purposes. However, the anhydrous material can be obtained from the hydrate by drying with thionyl chloride[7] or azeotropic distillation with benzene.[12]
Handling, Storage, and Precautions: is hygroscopic and corrosive; inhalation or ingestion may be fatal. It causes eye and skin irritation. It should be stored and handled under an inert dry atmosphere.[36] Use in a fume hood.

Oxidative Properties.[1] $FeCl_3$ oxidizes a wide array of functionalities, such as certain phenols to quinones (eq 1), dithiols to disulfides (eq 2), and 2-hydroxycyclohexanone to 1,2-cyclohexanedione.[1] Inter- and intramolecular oxidative dimerization of aromatics gives rise to such products as magnolol, metacyclophanes,[1] and crinine alkaloids (eq 3).[2] Phenolic ethylamines and *N*-acetyloxyamides can be cyclized to indoles (eq 4)[3] and oxindoles (eq 5),[4] respectively. Dimerization of aryllithium or Grignard reagents yields intermediates for cyclophane[5] and perylenequinone[6] synthesis (eq 6). Inter-[7] and intramolecular[8] ketone enolates can be converted to 1,4-diketones (eq 7), and lithium salts of allylic sulfones afford 1,6-disulfones.[9]

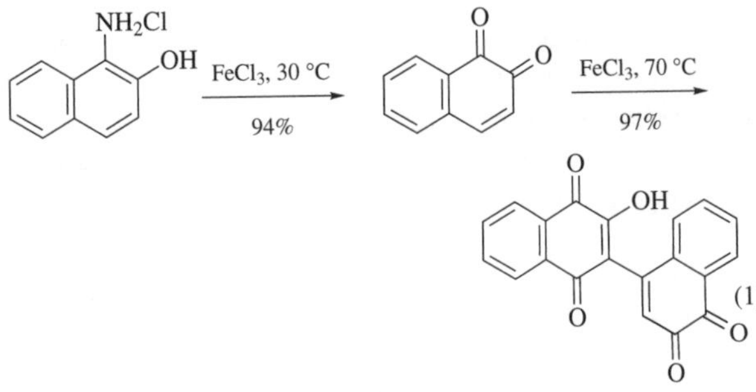

(1)

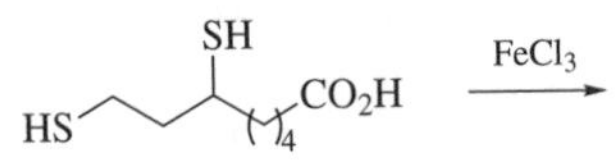

(2)

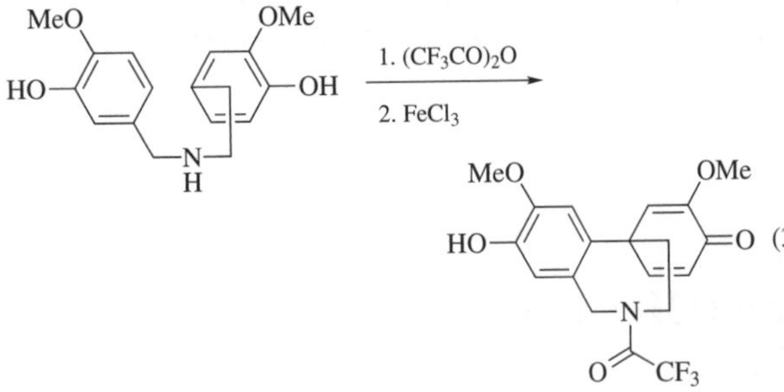

(3)

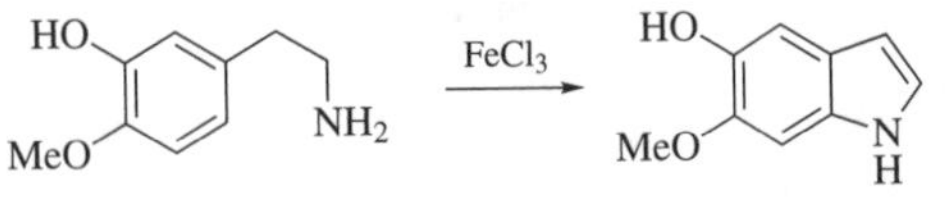

(4)

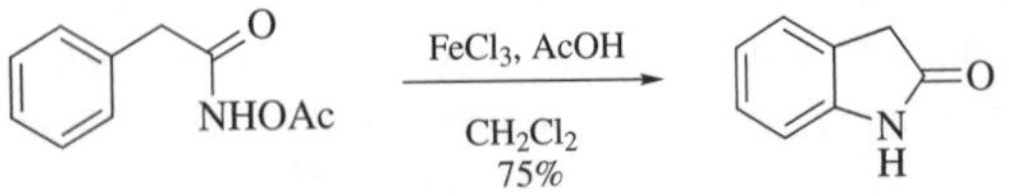

(5)

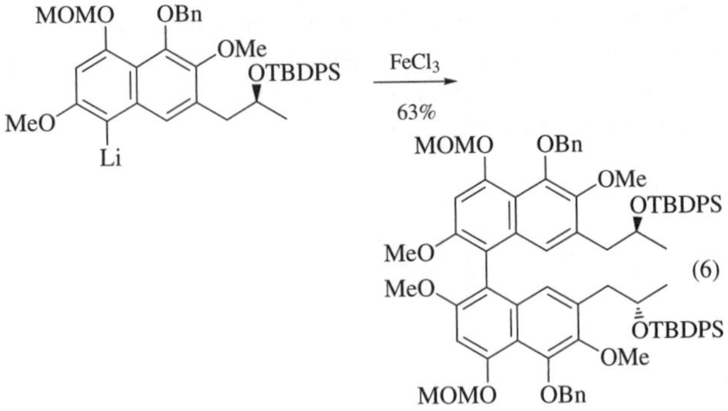

(6)

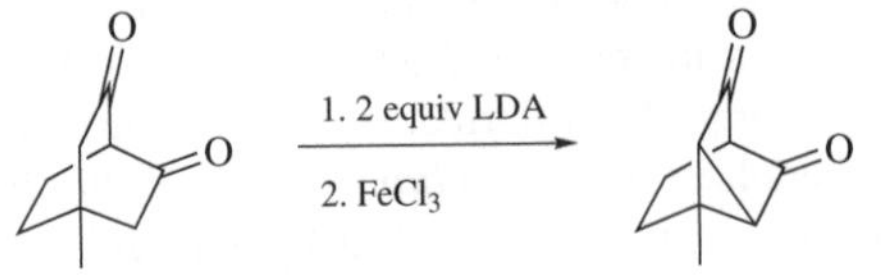

(7)

Stereoselective cross-coupling of alkenyl halides with Grignard reagents is catalyzed by $FeCl_3$ (45–83%) (eqs 8 and 9).[10] Propargyl halides also react to afford allenes.[11] A study of Fe^{III} catalysts revealed that ***Tris(dibenzoylmethide)iron(III)*** was the most useful.[12]

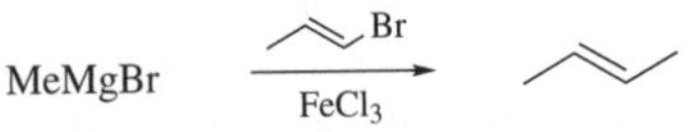

(8)

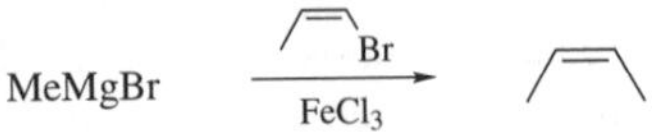

(9)

Alkylcyclopentanones can be dehydrogenated to cyclopentenones, but ***Copper(I) Chloride*** is a better catalyst.[13] Trimethylsilyloxybicyclo[*n*.1.0]alkanes can be oxidatively cleaved, providing a three-step method of ring expansion (eq 10).[14] Cycloalkanones are cleaved with $FeCl_3$/MeOH under O_2 to ω-oxo esters; this reaction works best with flanking methyl groups (eq 11).[15] Photooxidation of alkenes with $FeCl_3$ can yield a variety of useful chloroketones depending on the starting material,[16] and photoreaction of carbohydrates in pyridine induces a selective C(1)–C(2) bond cleavage, in contrast to ***Titanium(IV) Chloride*** (C(5)–C(6) cleavage) (eq 12).[17] $FeCl_3$/EtOH can also be used to disengage tricarbonyliron complex ligands.

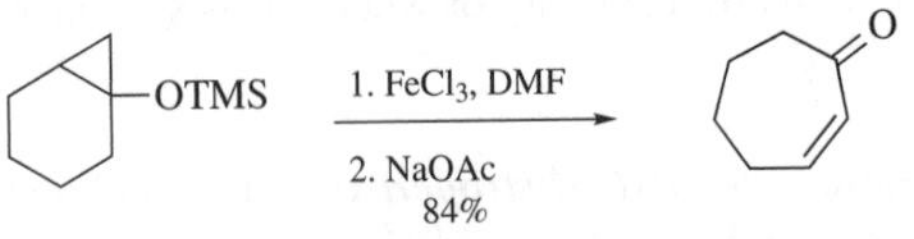

(10)

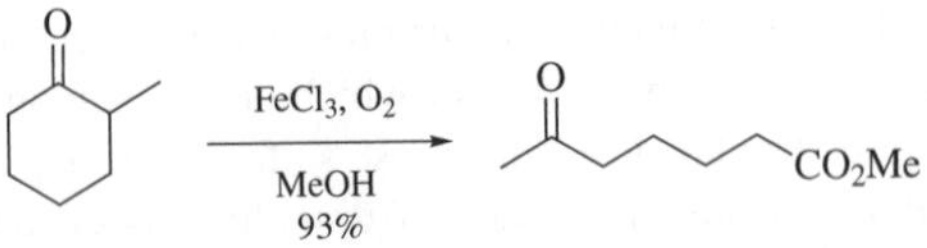

(11)

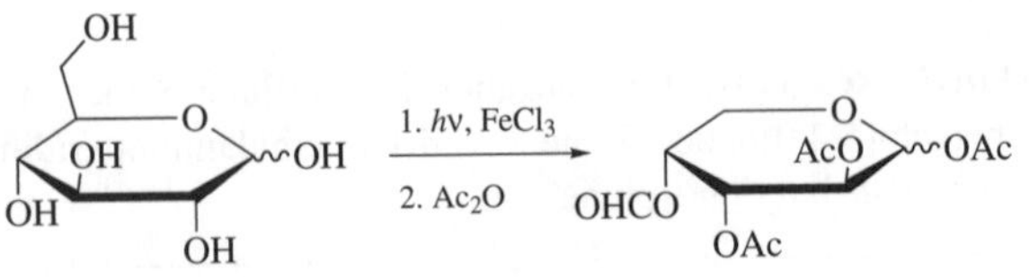

(12)

Lewis Acid Mediated Reactions. Silicon-directed Nazarov cyclizations occur readily in dichloromethane catalyzed by $FeCl_3$, utilizing the cation-stabilizing effect of silicon.[18] Cyclohexenyl systems afford only *cis*-fused ring products. The reaction has been elaborated to the preparation of linear tricycles with β-silyldivinyl ketones at low temperature (eq 13).[19] Optically active β′-silyl divinyl ketones have been used to demonstrate that cyclization occurs with essentially complete control by silicon in the *anti* S'_E sense.[20] $FeCl_3$ is the best Lewis acid catalyst for the intramolecular ene reaction of the Knoevenagel adduct from citronellal and dimethyl malonate at low temperature (eq 14).[21] However, the basic alumina supported catalyst can give more reliable results. The ene reaction of an unsaturated ester of an allylic alcohol yields a chlorolactone cleanly at 25 °C.[22] This reaction produces only one of four possible diastereomers, with clean *trans* addition to the double bond occurring (eq 15). 1-Silyloxycycloalkanecarbaldehydes undergo ring expansion to 2-silyloxycycloalkanones (82–89%) (eq 16). $FeCl_3$ catalysis provides the best selectivity derived from rearrangement of the more substituted α-carbon atom.[23] $FeCl_3$-catalyzed addition of primary and secondary amines to acrylates occurs exclusively 1,4 with no polymerization (79–97%) (eq 17).[24]

FeCl3, –50 °C; 79% (13)

FeCl3, CH2Cl2; 94% (14)

FeCl3, CH2Cl2; 85% (15)

FeCl3, –23 °C; 82% (16)

CO2Et + Et2NH → FeCl3, 96% → Et2N CO2Et (17)

In the field of protecting group chemistry $FeCl_3$ will cleave benzyl[25] and silyl ethers,[26] convert MEM ethers to carboxylic esters,[27] and when dispersed on 3Å molecular sieves catalyzes the formation of MOM ethers.[28] In the area of carbohydrate chemistry, $FeCl_3$ is proving a versatile reagent for acetylation, acetonation, acetolysis, transesterification, *O*-glycosidation of β-per-*O*-acetates, formation of oxazolines, direct conversion of 1,3,4,6-tetra-*O*-acetyl-2-deoxy-2-acylamido-β-D-glucopyranoses into their *O*-glycosides, preparation of 1-thioalkyl(aryl)-β-D-hexopyranosides from the peracetylated hexopyranoses having a 1,2-*trans* configuration,[29] and as an anomerization catalyst for the preparation of alkyl-α-glycopyranosides (eq 18).[30]

1. FeCl3, CH2Cl2; 2. RCl, AgOTf; 90%; R = 4-MeOCinn (18)

Substituted amidines have been prepared from a nitrile compound, an alkyl halide, an amine, and $FeCl_3$ in a one-pot synthesis (40–80%) (eq 19).[31] $FeCl_3$ in ether converts epoxides into chlorohydrins. Fused bicyclic epoxides yield *trans*-chlorohydrins (eq 20).[32] Friedel–Crafts acylation of activated (Me, OMe substituents) aromatics occurs readily with optically active *N*-phthaloyl-α-amino acid chlorides catalyzed by $FeCl_3$ (1–5 mol%).[33] Trialkylboranes react with $FeCl_3$ in THF/H_2O to afford alkyl chlorides in excellent yield.[34] *t*-Alkyl and benzylic chlorides can be converted to the iodides on reaction with ***Sodium Iodide*** in benzene catalyzed by $FeCl_3$.[35]

$R^1CN + R^2Cl$ + piperidine → FeCl3 → $R^1C(=NR^2)N$(piperidinyl) (19)

1. FeCl3, Et2O; 2. H2O; 78% (20)

Related Reagents. Copper(I) Chloride; Ethylaluminum Dichloride; Iron(III) Chloride–Acetic Anhydride; Iron(III) Chloride–Alumina; Iron(III) Chloride–Dimethylformamide; Iron(III) Chloride–Silica Gel; Iron(III) Chloride–Sodium Hydride; Tin(IV) Chloride; Zinc Chloride.

1. *FF* **1967**, *1*, 390.
2. Franck, B.; Lubs, H. J. *AG(E)* **1968**, *7*, 223.
3. Kametani, T.; Noguchi, I.; Nyu, K.; Takano, S. *TL* **1970**, 723.
4. Cherest, M.; Lusinchi, X. *TL* **1989**, **30**, 715.
5. Schirch, P. F. T.; Boekelheide, V. *JACS* **1979**, *101*, 3125.
6. Broka, C. A. *TL* **1991**, *32*, 859.
7. Frazier, R. H., Jr.; Harlow, R. L. *JOC* **1980**, *45*, 5408.
8. Poupart, M.-A.; Paquette, L. A. *TL* **1988**, *29*, 269.
9. Buchi, G.; Freidinger, R. M. *TL* **1985**, *26*, 5923.
10. Tamura, M.; Kochi, J. *S* **1971**, 303.
11. Pasto, D. J.; Hennion, G. F.; Shults, R. H.; Waterhouse, A.; Chou, S.-K. *JOC* **1976**, *41*, 3496.
12. Neumann, S. M.; Kochi, J. K. *JOC* **1975**, *40*, 599.
13. Cardinale, G.; Laan, J. A. M.; Russell, S. W.; Ward, J. P. *RTC* **1982**, *101*, 199.
14. Ito, Y.; Fujii, S.; Saegusa, T. *JOC* **1976**, *41*, 2073.
15. Ito, S.; Matsumoto, M. *JOC* **1983**, *48*, 1133.
16. Kohda, A.; Nagayoshi, K.; Maemoto, K.; Sato, T. *JOC* **1983**, *48*, 425.
17. Ichikawa, S.; Tomita, I.; Hosaka, A.; Sato, T. *BCJ* **1988**, *61*, 513.
18. Denmark, S. E.; Habermas, K. L.; Hite, G. A.; Jones, T. K. *T* **1986**, *42*, 2821.
19. Denmark, S. E.; Klix, R. C. *T* **1988**, *44*, 4043.
20. Denmark, S. E.; Wallace, M. A.; Walker, C. B., Jr. *JOC* **1990**, *55*, 5543.
21. Tietze, L. F.; Beifuss, U. *S* **1988**, 359.
22. Snider, B. B.; Roush, D. M. *JOC* **1979**, *44*, 4229.

23. Matsuda, T.; Tanino, K.; Kuwajima, I. *TL* **1989**, *30*, 4267.
24. Cabral, J.; Laszlo, P.; Mahe, L. *TL* **1989**, *30*, 3969.
25. Park, M. H.; Takeda, R.; Nakanishi, K. *TL* **1987**, *28*, 3823.
26. Dalla Cort, A. *SC* **1990**, *20*, 757.
27. Gross, R. S.; Watt, D. S. *SC* **1987**, *17*, 1749.
28. Patney, H. K. *SL* **1992**, 567.
29. Dasgupta, F.; Garegg, P. J. *ACS* **1989**, *43*, 471 and references therein.
30. Ikemoto, N.; Kim, O. K.; Lo, L.-C.; Satyanarayana, V.; Chang, M.; Nakanishi, K. *TL* **1992**, *33*, 4295.
31. Fuks, R. *T* **1973**, *29*, 2147.
32. Kagan, J.; Firth, B. E.; Shih, N. Y.; Boyajian, C. G. *JOC* **1977**, *42*, 343.
33. Effenberger, F.; Steegmuller, D. *CB* **1988**, *121*, 117 (*CA* **1988**, *108*, 75 799z).
34. Arase, A.; Masuda, Y.; Suzuki, A. *BCJ* **1974**, *47*, 2511.
35. Miller, J. A.; Nunn, M. J. *JCS(P1)* **1976**, 416.
36. *Sigma-Aldrich Library of Chemical Safety Data*, 2nd ed.; Lenga, R. E., Ed.; Sigma-Aldrich: Milwaukee, WI, 1988; p 1680A.

Andrew D. White
Parke-Davis Pharmaceutical Research, Ann Arbor, MI, USA

Lipases[1]

[9001-62-1] (MW 30000–100000)

(catalyst for asymmetric transformations of chiral or prochiral alcohols or acids by hydrolysis or esterification reactions[1])

Solubility: powder sol aqueous solutions; suspension in organic media.

Form Supplied in: usually a white or brownish powder, but also immobilized on an appropriate support. Lipases from microbial sources are virtually homogeneous in terms of hydrolytic activity, while mammalian and plant lipase preparations contain several interfering enzymes including proteases and esterases.

Analysis of Reagent Purity: assay by titrimetry.[2]

Handling, Storage, and Precautions: must be stored in a refrigerator at 0–5 °C. Avoid breathing or inhaling dust. Avoid too vigorous stirring.

Lipase General Aspects. Enzymes are now widely recognized as practical catalysts for asymmetric synthesis.[1b,c] Lipases are among the most widely applied and versatile biocatalysts in organic synthesis as can be witnessed by a number of recent reviews.[1] There are several reasons for this. They are readily available, do not require cofactors, are inexpensive and highly stable, exhibit broad substrate specificity, do not require water-soluble substrates, mechanistically are relatively well understood and, finally, are splendidly suited to retain a high degree of activity in organic media.

More than 20 lipases are now commercially available, either free or immobilized, from animal, plant, and microbial sources.[2] Amongst the lipases, the pig pancreatic lipase (PPL), the yeast lipase from *Candida cylindracea* (*rugosa*) (CCL), and the bacteria lipases from *Pseudomonas fluorescens* (*cepecia*) (PFL) and other unclassified *Pseudomonas* species (PSL) have been most widely used. The experimental methods are very straightforward and little different in their execution from conventional chemical reactions. Hydrolysis reactions are conducted on the soluble lipase in buffered aqueous solutions, commonly in the presence of an organic cosolvent. In organic media the enzyme is added as a powder or in an immobilized form and the resulting suspension stirred or (better) shaken at approximately 40 °C. The enzyme is removed by filtration.

Since their action toward substrates in terms of chemo-, regio-, and enantioselectivity varies considerably, it is important to have a large selection of lipases to find the right enzyme for a specific reaction by traditional biocatalyst screening. Alternative strategies for improving enantioselectivity of the already existing commercial lipases have been developed,[1a,3] including product recycling,[4] solvent screening,[5] water content control,[6] and immobilization. In addition to this, several active site models have been proposed to predict the enantiopreference of certain lipases.[1a,7]

Lipases have been used in three main types of asymmetric transformations: kinetic resolution of racemic carboxylic acids or alcohols, enantioselective group differentiations of *meso* dicarboxylic acids or diols, and enantiotopic group differentiation of prochiral dicarboxylic acid and diol derivatives. Hydrolysis has been the most widely used technique, but complementary esterification or transesterification procedures are increasingly coming into use. Lipases are used most frequently in transformations involving chiral alcohols rather than acids, unlike the pig liver esterase (PLE),[1a,b,c] which is most frequently employed on esters of chiral carboxylic acids. Lipases are also gaining increasing importance in solving problems of regioselectivity of various polyol and carbohydrate compounds.[1c,e,f,8] They have found application in stereoselective transformations involving lactonization and oligomerization of hydroxy acids and esters.[1e,f] Finally, a minor but useful advantage of the lipases is their mildness, which is particularly important in transformations involving labile compounds.[1a]

The range of nucleophiles that lipases accept is not confined to water or alcohols. There are numerous examples of amines,[9] hydrazine,[10] phenols,[11] and hydrogen peroxide.[12] Proteases have frequently been used in biocatalytic transformations involving ester hydrolysis and esterification reactions and their different stereoselection often provides a useful complement to the lipases.[1a,c,f]

Kinetic resolution of racemic compounds is by far the most common transformation catalyzed by lipases, in which the enzyme discriminates between the two enantiomeric constituents of a racemic mixture. It is important to note that the maximum yield of a kinetic resolution is restricted to 50% for each enantiomer based on the starting material. The prochiral route and transformations involving *meso* compounds, 'the *meso*-trick', have the advantage of potentially obtaining a 100% yield of pure enantiomer. A theoretical quantitative analysis of the kinetics involved in the biocatalytic processes described above has been developed.[1a,d,e] The enantiomeric ratio (E), an index of enantioselectivity, can be calculated from the extent of conversion and the corresponding enantiomeric excess (ee) values of either the product or the remaining substrate. The results reveal that for an irreversible process, such as hydrolysis, the optimum in both chemical and optical yield for the faster hydrolyzed enantiomer is to be expected near 40% conversion, and for the remaining slower hydrolyzed enantiomer around 60% conversion. For a high enantiomeric ratio (>100), high enantioselectivity is expected for both enantiomers at 50% conversion.

Under almost anhydrous conditions in organic medium,[1e,f] lipases can be used in the reverse mode for direct ester synthesis from carboxylic acids and alcohols, as well as transesterifications (acyl transfer reactions) which can be divided into alcoholysis (ester and alcohol), acidolysis (ester and acid), and interesterification (ester–ester interchange). The direct esterification and alcoholysis in particular have been most frequently used in asymmetric transformations involving lipases. The parameters that influence enzymatic catalysis in organic solvents have been intensively studied and discussed.[1a,e,f]

Besides ester synthesis being favored over hydrolysis, there are several major advantages of undertaking biocatalytic reactions in anhydrous media: increased solubility of nonpolar substrates, ease of product and enzyme recovery, enhanced thermal stability of enzymes and substrate specificity, and enantioselectivity regulation by the solvent. The main disadvantages include lower catalytic activity in organic media and reversibility, which limits the yield and works against the kinetic resolution, lowering the enantioselectivity of such processes. There are several strategies available to overcome these problems.[1a,e,f] Enol esters, such as vinyl or isopropenyl esters, are by far the most commonly used acyl transfer agents to ensure irreversibility by tautomerization of the enol leaving group.[13,14] Anhydrides,[15] *S*-phenyl thioacetate,[16] acyloxypyridines,[17] and oximes[18] have also been applied in a similar manner as acyl donors. Active trifluoro-[19] and trichloroethyl[20] esters have similarly been used to suppress the reversibility by speeding the acyl–enzyme formation and generating the weakly nucleophilic trifluoro- or trichloroethanol. Primary alcohols have also been used as acyl acceptors in transesterifications (deacylations) involving esters of more bulky and less nucleophilic secondary alcohols.[21]

Kinetic Resolution by Hydrolysis. Until very recently, kinetic resolution of racemic alcohols as ester derivatives was by far the most common type of asymmetric transformations involving lipases.[1a] There are number of examples involving acyclic secondary alcohols, such as the glyceraldehyde derivative in eq 1[22] and various related alkyl- and aryloxy substituted chloride and tosylate glycerol derivatives.[22,23]

PSL, H_2O (1)

46% >98% ee; 47% 97% ee

A wide variety of other alcohol substrates has been resolved,[1a] including aryl substituted secondary alcohols,[20,24] α-alkyl-β-hydroxy esters,[25] β-hydroxy nitriles,[26] and fluoroorganic compounds.[27] Active chloroacetate esters are commonly used to speed up the hydrolysis reactions, as exemplified in eq 2.[28] Primary acyclic alcohols possessing a stereogenic center that have been resolved include 2,3-epoxy alcohols,[29,30] 2-amino alcohols,[31] and crown ethers.[32]

PSL, H_2O (2)

99% ee; 99% ee

Lipase-catalyzed asymmetric hydrolysis has also been conducted on numerous monocyclic, variously substituted five-, six-, and seven-membered cycloalkane and cycloalkene secondary alcohols and diols.[1a] More recent reports include *cis*-4-acetoxyflavan,[33] substituted cyclopentenones,[34] and the 1,2-bis(hydroxymethyl)cyclobutanol derivative exemplified in eq 3.[35]

PFL, H_2O (3)

racemic; 48% >99% ee; 50% >99% ee

Various bicyclic racemic alcohols have been resolved by asymmetric hydrolysis of their corresponding esters. Generally, the *exo* isomers appear to be far inferior substrates compared with the *endo* substrates.[1a] Eq 4 illustrates the resolution of a bicyclic derivative of the Corey lactone type.[36]

PFL, H_2O (4)

(±); 50% >99% ee; 50% >99% ee

There are also several reports on the enantioselective hydrolysis of bicyclic secondary alcohols possessing the bicyclo[2.2.1]heptane and bicyclo[2.2.2]octane framework.[37] Again, with this type of substrate the lipases appear to exhibit strong preference for the *endo* isomers with the (*R*)-configured esters preferentially hydrolyzed.

Various chiral acids have also been resolved by lipase-catalyzed asymmetric hydrolysis.[1a] The reports include variously α-substituted acids[3,38] as well as the tertiary α-benzyloxy ester exemplified in eq 5.[39] Remethylation and repeated hydrolysis afforded the (*S*)-enantiomer in eq 5 optically pure. More recent examples include esters of glycidic acid,[40] β-aryl-β-hydroxy acid,[41] and sulfinyl alkanoates.[42]

CCL, H_2O (5)

40% >99% ee; 52% 82% ee

Kinetic Resolution by Transesterification. Asymmetric transformation involving acylation of chiral alcohols is by far the most common example of kinetic resolution by lipase-catalyzed transesterification, most commonly with irreversible vinyl esters.[1a,15] This field is now becoming the most widely applied technique involving lipases. Recent reports of the numerous secondary alcohol substrates include various monocyclic (eq 6)[43] and acyclic[44] compounds, cyanohydrins,[45] sulfones,[46] and glycals,[47] to name a few.

PFL, vinyl acetate, t-BuOMe (6): 50%, 100% ee; 48%, 100% ee

There are also several reports of enantioselective transesterification involving primary alcohols possessing stereogenic centers by similar acylation procedures, such as 2,3-epoxy alcohols (eq 7),[48] norbornene-derived iodolactones,[49] and 1,3-propanediols.[50]

PFL, vinyl acetate, $CHCl_3$ (7): 34%, 98% ee; 32%, 98% ee

Enantioselective lipase-catalyzed transesterification involving deacylation of esters of racemic primary or secondary alcohols with primary alcohols, most frequently *n*-butanol, serving as an acyl acceptor, is fairly common.[1a] Recent examples include esters of amino alcohols,[51] isoserine,[52] chlorohydrins,[53] and various tosyloxybutanoate esters (eq 8).[54]

PFL, BuOH, hexane (8): 47%, 97% ee; 47%, 95% ee

Kinetic resolution involving acidolysis of esters of racemic secondary alcohols and acids or transesterification of chiral acids does not have many examples in the literature.[1a]

Kinetic Resolution by Direct Esterification. This is the least common strategy for kinetic resolution and is most commonly executed on racemic alcohols with carboxylic acids in organic solvents.[1a] Reports include several alicyclic secondary alcohols such as menthol[55] and various aliphatic secondary alcohols.[56] Kinetic resolution of a variety of racemic saturated, unsaturated, and α-substituted carboxylic acids has also been effected by direct esterification with various alcohols.[20,57]

In addition to this, there are several reports of asymmetric esterification of racemic alcohols with anhydrides as acyl donors. Examples include various primary and secondary alcohols,[15] bicyclic secondary alcohols of the norbornane type,[58] amino alcohols,[59] and ferrocenes.[60] This is exemplified in eq 9 for 1-phenylethanol.[15]

PFL, benzene, propionic anhydride (9): 40%, >95% ee; 46%, 94% ee

Prochiral Compounds. The enantiodifferentiation of prochiral compounds by lipase-catalyzed hydrolysis and transesterification reactions is fairly common, with prochiral 1,3-diols most frequently employed as substrates.[1a] Recent reports of asymmetric hydrolysis include diesters of 2-substituted 1,3-propanediols[61] and 2-*O*-protected glycerol derivatives.[8] The asymmetric transesterification of prochiral diols such as 2-*O*-benzylglycerol[8,13a] and various other 2-substituted 1,3-propanediol derivatives[13b,62] is also fairly common, most frequently with ***Vinyl Acetate*** as an irreversible acyl transfer agent.

There are also recent reports of the lipase-catalyzed enantioselective hydrolysis of prochiral diacid derivatives such as 2-substituted malonates,[63] barbiturates,[64] and highly substituted, sterically hindered 1,4-dihydropyridine derivatives using acyloxymethyl groups to enhance the reaction rate.[65] An example of a prochiral diester hydrolysis is illustrated in eq 10.[66]

PSL, H_2O, 93% (10): 99% ee

***Meso* Compounds.** Although pig liver esterase is by far the most suitable enzyme for asymmetric transformations involving *meso* compounds, especially diacids, there are several reports on the lipase-catalyzed hydrolysis and transesterification reactions of cyclic diol derivatives.[1a] The former includes variously substituted cycloalkene diacetates, cyclohexylidene protected erythritol diacetate,[67] piperidine derivatives,[68] and the *exo*-acetonide in eq 11.[69] Complementary results are clearly demonstrated in eqs 11 and 12 for the hydrolysis and esterification processes.

PFL, H_2O, 71% (11): >99% ee

PFL, CH_2=CHOAc, 81% (12): >99% ee

The asymmetric transesterification of cyclic *meso*-diols, usually with vinyl acetate as an irreversible acyl transfer agent, includes monocyclic cycloalkene diol derivatives,[70] bicyclic diols,[71] such as the *exo*-acetonide in eq 12,[69] bicyclic diols of the norbornyl type,[72] and organometallic 1,2-bis(hydroxymethyl)ferrocene possessing planar chirality.[73]

Regioselective Biotransformations with Lipases. Lipases are gaining increasing importance in solving problems of regioselectivity of various polyol and carbohydrate compounds.[1a,c,e,f,8] A variety of diols or the corresponding acetates as well as polyhydric phenol acetates[74] have been acylated or deacylated in a highly regioselective manner in high yields by lipase-catalyzed transesterification reactions. Regioselective direct esterification of aliphatic 1,2-diols[75] and inositol derivatives[76] using anhydrides as acylating agents has recently been reported. Primary hydroxyl groups are exclusively transformed, as would be anticipated on steric grounds. One example of a highly regioselective and at the same time highly enantioselective hydrolysis of a racemic diester is demonstrated in eq 13.[77]

PPL, H_2O (13)

47%, >95% ee   38%, 73% ee

There are also several reports on highly regioselective transesterification of various steroid derivatives, one example being displayed in eq 14 in which butyration occurred exclusively at the 3β-hydroxyl group by *Chromobacterium viscosum* lipase (CVL).[78] Opposite regioselectivity toward the 17β-hydroxyl group was observed with subtilisin protease.[78]

CVL, $PrCO_2CH_2CF_3$, acetone, 83% (14)

There are numerous examples of highly regioselective lipase-catalyzed hydrolysis and acylation/deacylation processes involving monosaccharide and carbohydrate derivatives.[1a,f,8] Usually, the biotransformation processes occur preferentially and in many cases exclusively on the primary hydroxyl group (eq 15),[79] but highly regioselective transformations have also been described on secondary alcoholic groups for various carbohydrate derivatives possessing an acyl or alkyl protection on the primary hydroxyl moiety. Recent reports include highly regioselective acetylation of pyranosidic and furanosidic monosaccharide derivatives[80] and alkoxycarbonylation of nucleosides with oxime carbonates.[18]

CCL, H_2O, 10% acetone, 90% (15)

Lactonization and Polycondensation. The lipase-catalyzed intramolecular transesterification of a range of ω-hydroxy esters has been investigated extensively[1a,e,f] and was observed to be very dependent on the chain length of the substrate (eq 16).

lipase (16)

For longer-chain hydroxy esters ($n = 13$, 14) the corresponding macrolide was accomplished in high yield with very little diolide formed (diolide increased considerably with lower n). With medium-sized hydroxy esters the product profile became considerably more complex, consisting of a complex mixture of di-, tri-, tetra-, and pentalactones.[28,81] Shorter-chain unsubstituted β-, δ-, and ϵ-hydroxy esters almost exclusively underwent intermolecular transesterification to afford the corresponding oligomers. δ-Substituted δ-hydroxy esters[82] and γ-hydroxy esters[83] underwent lactonization with a high degree of enantioselectivity.

Prochiral γ-hydroxy diesters underwent enantioselective lactonization with PPL to afford the (*S*)-lactone in a highly enantioselective fashion (eq 17).[83a] Formation of macrocyclic lactones by the condensation of diacids or diesters with diols, leading to mono- and dilactones,[84] linear oligomeric esters, or high molecular weight optically active polymers,[85] depending upon type of substrates as well as reaction conditions, has also been described.

PPL, hexane, 80% (17)

>98% ee

Mildness and Miscellaneous Reactions. The mildness of the lipases has been particularly well suited in transformations involving labile compounds that are likely to undergo decomposition when conventional chemical methods are applied,[1a] such as the long-chain polyunsaturated ω-3-type fatty acids[86] and highly labile prostaglandin precursor derivatives.[87] Under mild conditions, lipase was exploited to hydrolyze the peracetal protected hydroperoxy derivative in eq 18 to afford the corresponding acid without affecting the peracetal protection moiety.[88]

PFL, H_2O, 99% (18)

Various miscellaneous lipase-catalyzed reactions have been reported,[1a] including lipase-mediated epoxidation of alkenes,[12] transamidation,[89] thiotransesterification of thioesters for the preparation of optically active thiols,[90] regio- and chemoselective peptide acylation,[91] lactamization,[92] and highly enantioselective hydrolysis of racemic oxazolin-5-ones which undergo a rapid keto–enol tautomerism to afford optically pure amino acids, thus exceeding the 50% yield limit.[93]

Finally, lipases are able to differentiate enantiotopic faces of appropriately substituted enol esters to afford optically active ketones,[94] indicating that simultaneously upon hydrolysis of the acyl group, protonation occurs from one specified side of the double bond of the enol ester without formation of an enol intermediate (eq 19).[94a]

lipase, whole cells, 98% (19)

94% ee

Related Reagents. Esterases.

1. (a) Haraldsson, G. G. In *The Chemistry of the Functional Groups, Supplement B2: The Chemistry of Acid Derivatives*; Patai, S., Ed.; Wiley: Chichester, 1992; Vol. 2, Part 2, pp 1395–1473. (b) Jones, J. B. *T* **1986**, *42*, 3351. (c) Crout, D. H. G.; Christen, M. In *Modern Synthetic Methods*; Scheffold, R., Ed.; Springer: Berlin, 1989; Vol. 5, pp 1–114. (d) Sih, C. J.; Wu, S.-H. In *Topics in Stereochemistry*; Eliel, E. L.; Wilen, S. H., Eds.; Wiley: New York, 1989; Vol. 19, pp 63–125. (e) Chen, C.-S.; Sih, C. J. *AG(E)* **1989**, *28*, 695. (f) Klibanov, A. M. *ACR* **1990**, *23*, 114. (g) Xie, Z.-F. *TA* **1991**, *2*, 733. (h) Boland, W.; Frössl, C.; Lorenz, M. *S* **1991**, 1049.
2. Eigtved, P. In *Advances in Applied Lipid Research*; JAI: Greenwich, CT, 1992; Vol. 1, pp 1–64.
3. Wu, S.-H.; Guo, Z.-W.; Sih, C. J. *JACS* **1990**, *112*, 1990.
4. Chen, C. S.; Fujimoto, Y.; Girdaukas, G.; Sih, C. J. *JACS* **1982**, *104*, 7294.
5. Tawaki, S.; Klibanov, A. M. *JACS* **1992**, *114*, 1882.
6. Secundo, F.; Riva, S.; Carrea, G. *TA* **1992**, *3*, 267.
7. Kazlauskas, R. J.; Weissfloch, A. N. E.; Rappaport, A. T.; Cuccia, L. A. *JOC* **1991**, *56*, 2656.
8. Drueckhammer, D. G.; Hennen, W. J.; Pederson, R. L.; Barbas, III, C. F.; Gautheron, C. M.; Krach, T.; Wong, C.-H. *S* **1991**, 499.
9. Garcia, M. J.; Rebolledo, F.; Gotor, V. *TA* **1992**, *3*, 1519.
10. Astorga, C.; Rebolledo, F.; Gotor, V. *S* **1991**, 350.
11. Nicolosi, G.; Piattelli, M.; Sanfilippo, C. *T* **1992**, *48*, 2477.
12. Björkling, F.; Frykman, H.; Godtfredsen, S. E.; Kirk, O. *T* **1992**, *48*, 4587.
13. (a) Wang, Y.-F.; Wong, C.-H. *JOC* **1988**, *53*, 3127. (b) Wang, Y.-F.; Lalonde, J. J.; Momongan, M.; Bergbreiter, D. E.; Wong, C.-H. *JACS* **1988**, *110*, 7200.
14. Faber, K.; Riva, S. *S* **1992**, 895.
15. Bianchi, D.; Cesti, P.; Battistel, E. *JOC* **1988**, *53*, 5531.
16. Akita, H.; Umezawa, I.; Takano, M.; Matsukura, H.; Oishi, T. *CPB* **1991**, *39*, 3094.
17. Keumi, T.; Hiraoka, Y.; Ban, T.; Takahashi, I.; Kitajima, H. *CL* **1991**, 1989.
18. Morís, F.; Gotor, V. *T* **1992**, *48*, 9869.
19. Stokes, T. M.; Oehlschlager, A. C. *TL* **1987**, *28*, 2091.
20. Kirchner, G.; Scollar, M. P.; Klibanov, A. M. *JACS* **1985**, *107*, 7072.
21. Bevinakatti, H. S.; Banerji, A. A.; Newadkar, R. V. *JOC* **1989**, *54*, 2453.
22. von der Osten, C. H.; Sinskey, A. J.; Barbas, III, C. F.; Pederson, R. L.; Wang, Y.-F.; Wong, C.-H. *JACS* **1989**, *111*, 3924.
23. (a) Pederson, R. L.; Liu, K. K.-C.; Rutan, J. F.; Chen, L.; Wong, C.-H. *JOC* **1990**, *55*, 4897. (b) Ader, U.; Schneider, M. P. *TA* **1992**, *3*, 201. (c) Ader, U.; Schneider, M. P. *TA* **1992**, *3*, 521.
24. Hiratake, J.; Inagaki, M.; Nishioka, T.; Oda, J. *JOC* **1988**, *53*, 6130.
25. Itoh, T.; Kuroda, K.; Tomasada, M.; Takagi, Y. *JOC* **1991**, *56*, 797.
26. Itoh, T.; Takagi, Y.; Nishiyama, S. *JOC* **1991**, *56*, 1521.
27. Bravo, P.; Resnati, G. *TA* **1990**, *1*, 661.
28. Ngooi, T. K.; Scilimati, A.; Guo, Z.-w.; Sih, C. J. *JOC* **1989**, *54*, 911.
29. Ladner, W. E.; Whitesides, G. M. *JACS* **1984**, *106*, 7250.
30. Pawlak, J. L.; Berchtold, G. A. *JOC* **1987**, *52*, 1765.
31. Francalanci, F.; Cesti, P.; Cabri, W.; Bianchi, D.; Martinengo, T.; Foa, M. *JOC* **1987**, *52*, 5079.
32. Tsukube, H.; Betchaku, A.; Hiyama, Y.; Itoh, T. *CC* **1992**, 1751.
33. Izumi, T.; Hino, T.; Kasahara, A. *JCS(P1)* **1992**, 1265.
34. Danda, H.; Nagatomi, T.; Maehara, A.; Umemura, T. *T* **1991**, *47*, 8701.
35. Chen, X.; Siddiqi, S. M.; Schneller, S. W. *TL* **1992**, *33*, 2249.
36. Sugahara, T.; Satoh, I.; Yamada, O.; Takano, S. *CPB* **1991**, *39*, 2758.
37. Oberhauser, T.; Faber, K.; Griengl, H. *T* **1989**, *45*, 1679.
38. Kalaritis, P.; Regenye, R. W.; Partridge, J. J.; Coffen, D. L. *JOC* **1990**, *55*, 812.
39. Sugai, T.; Kakeya, H.; Ohta, H. *JOC* **1990**, *55*, 4643.
40. Gentile, A.; Giordano, C.; Fuganti, C.; Ghirotto, L.; Servi, S. *JOC* **1992**, *57*, 6635.
41. Boaz, N. W. *JOC* **1992**, *57*, 4289.
42. Burgess, K.; Henderson, I.; Ho, K.-K. *JOC* **1992**, *57*, 1290.
43. (a) Carrea, G.; Danieli, B.; Palmisano, G.; Riva, S.; Santagostino, M. *TA* **1992**, *3*, 775. (b) Takano, S.; Yamane, T.; Takahashi, M.; Ogasawara, K. *TA* **1992**, *3*, 837.
44. Morgan, B.; Oehlschlager, A. C.; Stokes, T. M. *JOC* **1992**, *57*, 3231.
45. Inagaki, M.; Hiratake, J.; Nishioka, T.; Oda, J. *JOC* **1992**, *57*, 5643.
46. Carretero, J. C.; Dominguez, E. *JOC* **1992**, *57*, 3867.
47. Berkowitz, D. B.; Danishefsky, S. J.; Schulte, G. K. *JACS* **1992**, *114*, 4518.
48. Ferraboschi, P.; Brembilla, D.; Grisenti, P.; Santaniello, E. *JOC* **1991**, *56*, 5478.
49. Janssen, A. J. M.; Klunder, A. J. H.; Zwanenburg, B. *T* **1991**, *47*, 5513.
50. Grisenti, P.; Ferraboschi, P.; Manzocchi, A.; Santaniello, E. *T* **1992**, *48*, 3827.
51. Kanerva, L. T.; Rahiala, K.; Vänttinen, E. *JCS(P1)* **1992**, 1759.
52. Lu, Y.; Miet, C.; Kunesch, N.; Poisson, J. E. *TA* **1991**, *2*, 871.
53. Bevinakatti, H. S.; Banerji, A. A. *JOC* **1991**, *56*, 5372.
54. Chen, C.-S.; Liu Y.-C.; Marsella, M. *JCS(P1)* **1990**, 2559.
55. Langrand, G.; Baratti, J.; Buono, G.; Triantaphylides, C. *TL* **1986**, *27*, 29.
56. (a) Sonnet, P. E. *JOC* **1987**, *52*, 3477. (b) Lutz, D.; Guldner, A.; Thums, R.; Schreier, P. *TA* **1990**, *1*, 783.
57. Engel, K.-H. *TA* **1991**, *2*, 165.
58. Berger, B.; Rabiller, C. G.; Königsberger, K.; Faber, K.; Griengl, H. *TA* **1990**, *1*, 541.
59. Kamal, A.; Rao, M. V. *TA* **1991**, *2*, 751.
60. Izumi, T.; Tamura, F.; Sasaki, K. *BCJ* **1992**, *65*, 2784.
61. Guanti, G.; Banfi, L.; Narisano, E. *JOC* **1992**, *57*, 1540.
62. Didier, E.; Loubinoux, B.; Ramos Tombo, G. M.; Rihs, G. *T* **1991**, *47*, 4941.
63. Gutman, A. L.; Shkolnik, E.; Shapira, M. *T* **1992**, *48*, 8775.
64. Murata, M.; Achiwa, K. *TL* **1991**, *32*, 6763.
65. Holdgrün, X. K.; Sih, C. J. *TL* **1991**, *32*, 3465.
66. Hughes, D. L.; Bergan, J. J.; Amato, J. S.; Bhupathy, M.; Leazer, J. L.; McNamara, J. M.; Sidler, D. R.; Reider, P. J.; Grabowski, E. J. J. *JOC* **1990**, *55*, 6252.
67. Gais, H.-J.; Hemmerle, H.; Kossek, S. *S* **1992**, 169.
68. (a) Chênevert, R.; Dickman, M. *TA* **1992**, *3*, 1021. (b) Momose, T.; Toyooka, N.; Jin, M. *TL* **1992**, *33*, 5389.
69. Tanaka, M.; Yoshioka, M.; Sakai, K. *CC* **1992**, 1454.
70. (a) Mekrami, M.; Sicsic, S. *TA* **1992**, *3*, 431. (b) Harris, K. J.; Gu, Q.-M.; Shih, Y.-E.; Girdaukas, G.; Sih, C. J. *TL* **1991**, *32*, 3941.
71. Theil, F.; Schick, H.; Winter, G.; Reck, G. *T* **1991**, *47*, 7569.
72. (a) Andreu, C.; Marco, J. A.; Asensio, G. *JCS(P1)* **1990**, 3209. (b) Murata, M.; Uchida, H.; Achiwa, K. *CPB* **1992**, *40*, 2610.
73. Nicolosi, G.; Morrone, R.; Patti, A.; Piattelli, M. *TA* **1992**, *3*, 753.
74. Natoli, M.; Nicolosi, G.; Piattelli, M. *JOC* **1992**, *57*, 5776.
75. Bosetti, A.; Bianchi, D.; Cesti, P.; Golini, P.; Spezia, S. *JCS(P1)* **1992**, 2395.
76. Ling, L.; Watanabe, Y.; Akiyama, T.; Ozaki, S. *TL* **1992**, *33*, 1911.
77. Guibé-Jampel, E.; Rousseau, G.; Salaün, J. *CC* **1987**, 1080.
78. Riva, S.; Klibanov, A. M. *JACS* **1988**, *110*, 3291.
79. Sweers, H. M.; Wong, C.-H. *JACS* **1986**, *108*, 6421.
80. (a) Theil, F.; Schick, H. *S* **1991**, 533. (b) Chinn, M. J.; Iacazio, G.; Spackman, D. G.; Turner, N. J.; Roberts, S. M. *JCS(P1)* **1992**, 661.
81. Guo, Z.-W.; Ngooi, T. K.; Scilimati, A.; Fülling, G.; Sih, C. J. *TL* **1988**, *29*, 5583.

82. (a) Bonini, C.; Pucci, P.; Viggiani, L. *JOC* **1991**, *56*, 4050. (b) Henkel, B.; Kunath, A.; Schick, H. *LA* **1992**, 809.
83. (a) Gutman, A. L.; Zuobi, K.; Bravdo, T. *JOC* **1990**, *55*, 3546. (b) Huffer, M.; Schreier, P. *TA* **1991**, *2*, 1157.
84. Guo, Z.-W.; Sih, C. J. *JACS* **1988**, *110*, 1999.
85. (a) Margolin, A. L.; Crenne, J.-Y.; Klibanov, A. M. *TL* **1987**, *28*, 1607. (b) Margolin, A. L.; Fitzpatrick, P. A.; Dubin, P. L.; Klibanov, A. M. *JACS* **1991**, *113*, 4693.
86. (a) Haraldsson, G. G.; Höskuldsson, P. A.; Sigurdsson, S. Th.; Thorsteinsson, F.; Gudbjarnason, S. *TL* **1989**, *30*, 1671. (b) Haraldsson, G. G.; Almarsson, Ö. *ACS* **1991**, *45*, 723.
87. (a) Porter, N. A.; Byers, J. D.; Holden, K. M.; Menzel, D. B. *JACS* **1979**, *101*, 4319. (b) Lin, C.-H.; Alexander, D. L.; Chidester, C. G.; Gorman, R. R.; Johnson, R. A. *JACS* **1982**, *104*, 1621.
88. Baba, N.; Yoneda, K.; Tahara, S.; Iwasa, J.; Kaneko, T.; Matsuo, M. *CC* **1990**, 1281.
89. Gotor, V.; Brieva, R.; González, C.; Rebolledo, F. *T* **1991**, *47*, 9207.
90. Bianchi, D.; Cesti, P. *JOC* **1990**, *55*, 5657.
91. Gardossi, L.; Bianchi, D.; Klibanov, A. M. *JACS* **1991**, *113*, 6328.
92. Gutman, A. L.; Meyer, E.; Yue, X.; Abell, C. *TL* **1992**, *33*, 3943.
93. Gu, R.-L.; Lee, I.-S.; Sih, C. J. *TL* **1992**, *33*, 1953.
94. (a) Ohta, H.; Matsumoto, K.; Tsutsumi, S.; Ihori, T. *CC* **1989**, 485. (b) Sugai, T.; Kakeya, H.; Ohta, H.; Morooka, M.; Ohba, S. *T* **1989**, *45*, 6135.

Gudmundur G. Haraldsson
University of Iceland, Reykjavik, Iceland

Lithium Amide[1]

$LiNH_2$

[7782-89-0] H_2LiN (MW 22.97)

(strong base; used in *N*-alkylation of aromatic amines,[5,6] Claisen condensations,[7] α-alkylations of carbonyl compounds,[13] eliminations[21] and isomerizations,[26] synthesis of ethynyl compounds[16] and alkynyl carbinols[30])

Physical Data: mp 380–400 °C; *d* 1.178 g cm^{-3}.
Solubility: sl sol liq NH_3, ethanol; insol ether, benzene, toluene.
Form Supplied in: gray-white powder; widely available.
Analysis of Reagent Purity: several titration procedures are available.[2]
Preparative Method: in a typical procedure[3] for the preparation of lithium amide in ammonia, a small piece of ***Lithium*** metal is added to commercial anhyd liquid NH_3 with stirring; after the almost immediate appearance of a blue color, a few crystals of iron(III) nitrate are added, followed by small portions of lithium metal; after about 20 min the blue color disappears and a gray suspension of lithium amide is formed.
Handling, Storage, and Precautions: the dry solid is flammable, air- and moisture-sensitive, and must be stored in tightly stoppered bottles; in water, it decomposes slowly when in lumps, faster in smaller particle sizes; like other alkali metal amides, the reagent must be guarded against air oxidation to prevent the formation of potentially explosive substances;[4] samples which develop a yellow or green or darker color should be properly disposed of; if left open for minimal periods during weighing, the titer remains fairly constant; contact with skin causes burns; use in a fume hood.

N-Alkylation of Amines. Lithium amide has been employed as a powerful base in the *N*-alkylation of heterocyclic aromatic amines.[5] A variety of *N*-substituted 2-aminopyridines, 2-aminopyrimidines, and 2-aminolepidines have been prepared by treating the respective amine with 2 equiv $LiNH_2$ and the alkyl halide (eq 1).[6] Lithium amide is more easily handled than ***Sodium Amide*** and is therefore the reagent of choice in these reactions.

(2-amino-4-methylquinoline) —[$LiNH_2$ / RCH_2Cl]→ (2-NHCH$_2$R-4-methylquinoline) (1)

Condensations of Esters with Carbonyl Compounds. Hauser and Puterbaugh found $LiNH_2$ superior to sodium amide in condensations of *t*-butyl acetate with ketones (eq 2).[7] Condensations with acetophenone, *p*-nitrocaprophenone, benzaldehyde, acetone, and cyclohexanone proceeded in good yields (53–76%).

$MeCO_2$-*t*-Bu —[1. $LiNH_2$; 2. PhCOMe]→ Ph–C(OH)(Me)–CH_2CO_2-*t*-Bu (2)

This condensation method is a useful alternative to the Reformatsky reaction, and in certain cases compares favorably in terms of yields and practicality. Later studies simulated a Reformatsky-type reaction employing *t*-butyl acetate instead of an α-halo ester.[8] It was found that this type of aldol condensation may be effected more conveniently and in higher yields by means of $LiNH_2$, and that the use of ***Zinc Chloride*** was unnecessary in this case. Use of $NaNH_2$ under similar conditions failed to give β-hydroxy esters. Hauser and Lindsay further reported[9] that when ethyl acetate was employed instead of *t*-butyl acetate, self-condensation of the ester could be circumvented by using 2 equiv $LiNH_2$. Under these conditions, even benzophenone could be condensed with ethyl acetate. It is suggested that the extra equivalent of $LiNH_2$ coordinates with the monolithio salt of ethyl acetate, as in (**1**), or in a dimer or trimer. Some sort of coordination is indicated, since an extra equivalent of $LiNH_2$ retards the self-condensation of EtOAc; $NaNH_2$, which should coordinate to a smaller degree, fails to exhibit such a retarding effect under similar conditions.[3,10]

(**1**)

Ethyl 3-alkoxy-2-butenoates have been condensed with aldehydes in the presence of $LiNH_2$ at the γ-position to give dihydropyrones which, under the basic conditions employed, give rise to 3-alkoxy-*cis*-2-*trans*-4-unsaturated acids (eq 3).[11]

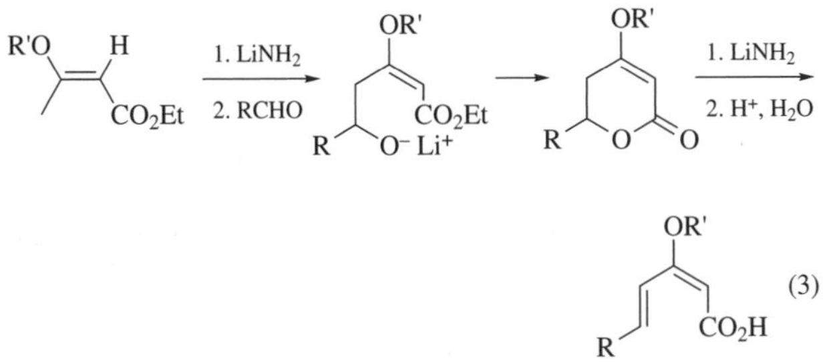

(3)

Along similar lines, enol silyl ethers are alkylated in high yields with lithium amide and alkyl halides.[12] When alkylation of (2) is attempted with ***Methyllithium*** and 3-bromopropionitrile, E2 elimination predominates, yielding polyacrylonitrile and cyclohexanone; the same reaction using $LiNH_2$ instead results in a 48% yield of the alkylated product (eq 4).

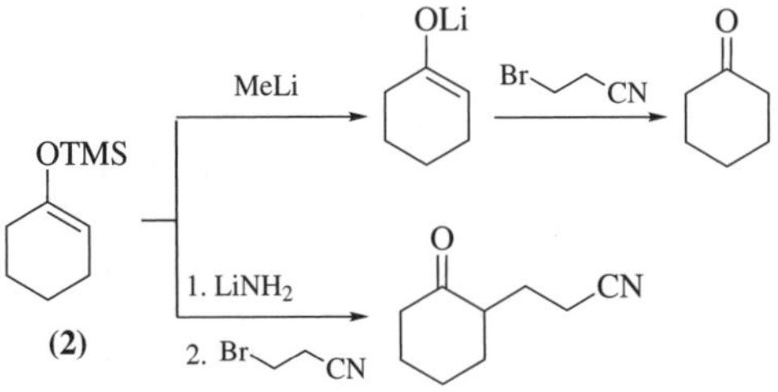

(4)

Alkylation of *t*-Butyl Esters with Organic Halides. The direct alkylation of *t*-butyl acetate is a valuable alternative to the malonic ester method for preparing mono- and dibasic carboxylic acids. Monoalkylation is observed with various alkyl or alkenyl halides in very good yields using $LiNH_2$ (eqs 5 and 6);[13] in contrast, $NaNH_2$ gives rise to dialkylated products under the same conditions. *t*-Butyl propionate and *n*-butyrate have also been alkylated in excellent yields using $LiNH_2$. With 1,4-dibromobutane, suberic acid was obtained in quantitative yield after saponification. 1,2-Dibromoethane gave succinic acid in moderate yield, whereas use of $NaNH_2$ in the latter case did not provide any product.

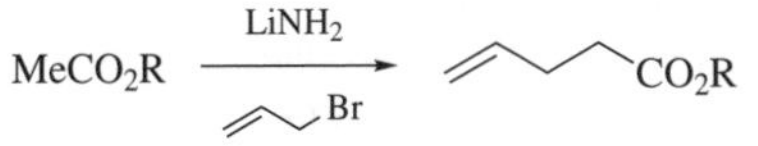

$$MeCO_2R \xrightarrow[\text{allyl bromide}]{LiNH_2} CH_2{=}CHCH_2CH_2CO_2R \quad (5)$$

$$MeCO_2R \xrightarrow[Br(CH_2)_nBr]{LiNH_2} RO_2C(CH_2)_{n+2}CO_2R \quad (6)$$

Treatment of *t*-butyl acetate with $LiNH_2$ and 1,4-dibromo-2-butene gave the diester of 4-octenedioic acid (eq 7). α,β-Unsaturated esters furnished α-mono- and α,α-dialkylation products with $LiNH_2$ (eq 8); no γ-alkylated products were formed in these reactions.[14]

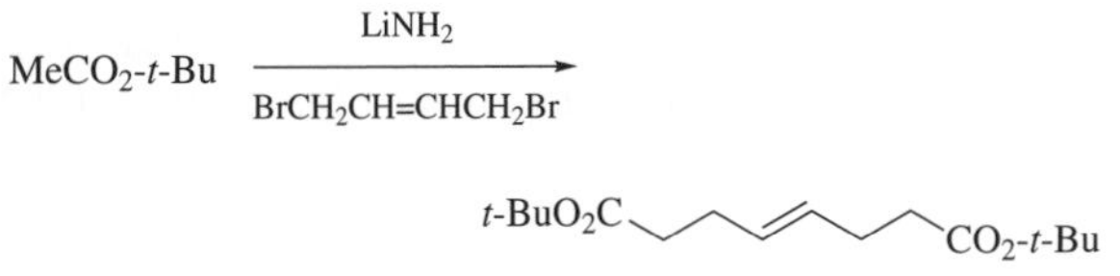

$$MeCO_2\text{-}t\text{-}Bu \xrightarrow[BrCH_2CH{=}CHCH_2Br]{LiNH_2} t\text{-}BuO_2C(CH_2)_2CH{=}CH(CH_2)_2CO_2\text{-}t\text{-}Bu \quad (7)$$

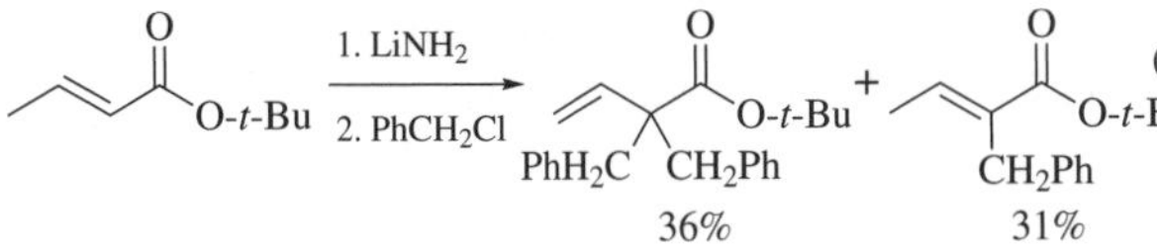

(8) 36% 31%

Alkylation and Condensation of Substituted Succinic Acid Derivatives. Both monoester isomers of 2-alkyl- or arylsuccinic acids were alkylated exclusively on the carbon adjacent to the ester function (see **3**) using 2 equiv $LiNH_2$.[15] In contrast, the anion obtained from diethyl 2-methylsuccinate gave on methylation a mixture of alkylated esters, including the diesters of 2,2-dimethylsuccinic and 2,3-dimethylsuccinic acids.

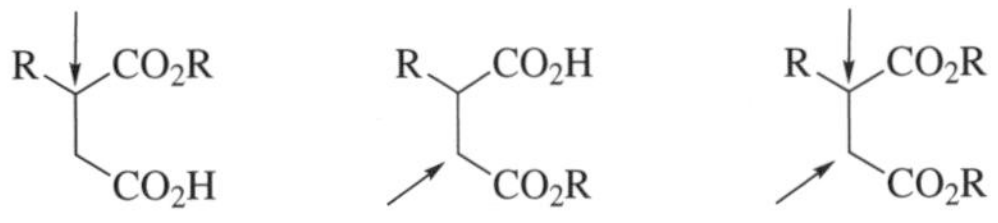

(3) (4) (5)

Condensation of monoester dianions with ketones or benzaldehyde similarly took place at the carbon α to the ester group, but only when the methylene group was not substituted (i.e. **4**). Condensation products were not obtained from the isomer (**3**).

Reactions with Alkyne Derivatives. Alkynes carrying functional groups such as NHR or OH can be alkylated in high yields in the presence of lithium amide (eqs 9 and 10).[16–20] 1-Alkyn-ω-ols are alkylated by primary or secondary alkyl halides in 50–80% yields, and it is not necessary to protect the alcohols.

$$HO(CH_2)_nC{\equiv}CH \xrightarrow[RX]{2\text{ equiv }LiNH_2} HO(CH_2)_nC{\equiv}CR \quad (9)$$

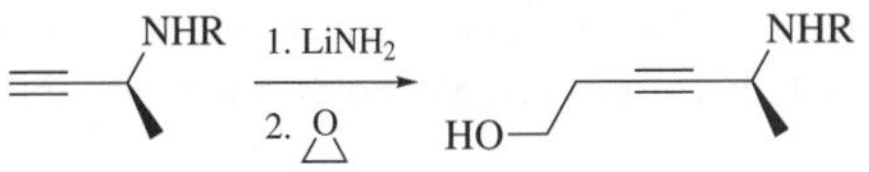

(10)

Eliminations and Isomerizations. $LiNH_2$ has been found superior to sodium amide in dehydrohalogenations of γ-methallyl chloride and α-methallyl chloride to give 3-methylcyclopropene and 1-methylcyclopropene, respectively (eqs 11 and 12).[21–23] ***Potassium Amide***, on the other hand, isomerizes the α-isomer to methylenecyclopropane.[24]

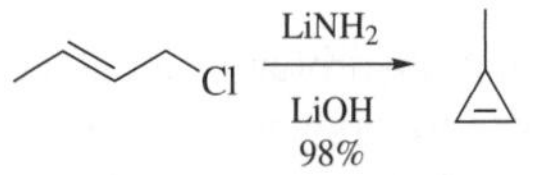

(11) $LiNH_2$, LiOH, 98%

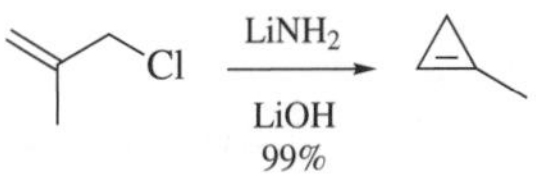

(12) $LiNH_2$, LiOH, 99%

In the dimerization of cyclopropene in the presence of alkali amides (eq 13), two opposing factors influence the course of the reaction.[25] On the one hand, the rate of cyclopropenylcyclopropane formation is retarded in the presence of stronger base (KNH_2) due to lower concentrations of unmetalated cyclopropene. On the other hand, increasing base strength favors the equilibrium between (**6**) and (**7**) in favor of (**7**), thus securing a greater yield of bicyclopropylidene (**8**). The best yields of (**8**) were obtained using $LiNH_2$ in liquid NH_3, albeit with much longer reaction times (4 weeks at −50 °C) than with $NaNH_2$.

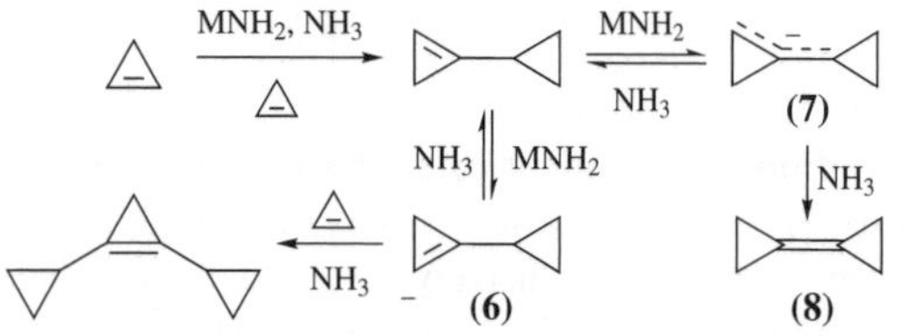

(13)

The propargyl sulfides (**9**) and (**10**) have been isomerized in good yields to the corresponding allenes with $LiNH_2$ in liquid NH_3 (eqs 14 and 15).[26–28]

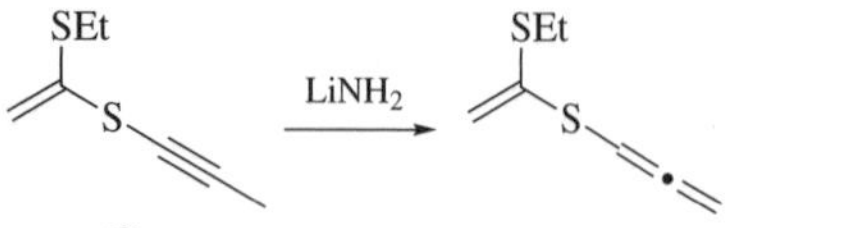

(14)

RC≡CSCH₂C≡CH (**10**) —$LiNH_2$→ RC≡CSCH=C=CH₂ (15)

Chiral propargyl alcohols can be prepared[29] from allylic alcohols by Sharpless asymmetric epoxidation,[30] conversion of the alcohol product into the corresponding chloride, and treatment with $LiNH_2$ in liquid NH_3 (eq 16). Use of ***Lithium Diisopropylamide*** (LDA) gives comparable yields. The same reaction with ***n-Butyllithium*** in THF at −33 °C results in mixtures of the chlorovinyl alcohol, propargyl alcohol, and starting material. It appears that *n*-BuLi reacts indiscriminately with both the epoxy chloride and the chlorovinyl alcohol formed during the reaction.

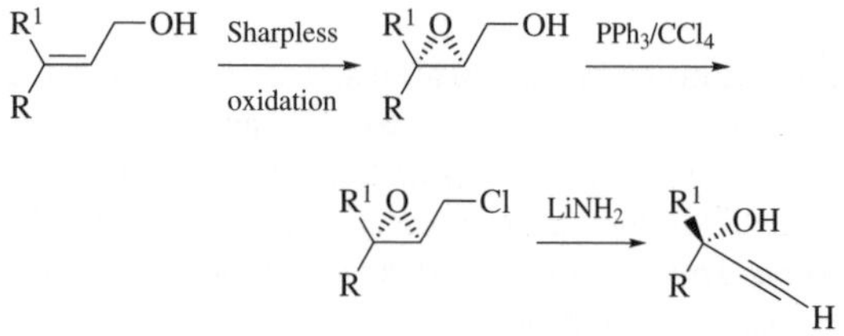

(16)

2-Chloromethyltetrahydrofuran undergoes ring opening followed by dehydrochlorination with 3 equiv of $LiNH_2$.[31] The lithio acetylide (**11**) formed in situ can be alkylated to give 4-alkynyl alcohols (eq 17).

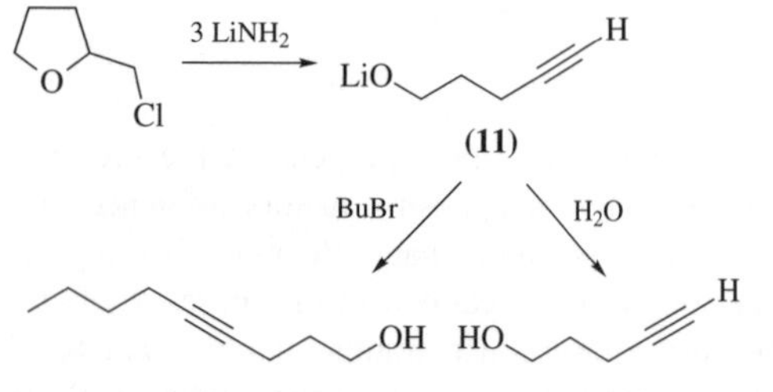

(17)

This reaction has been employed in the synthesis of chiral alkynyl compounds from simple carbohydrate precursors (eq 18).[32] The elimination reaction here is chemoselective, since the other isopropylidene group in the substrate remains unaffected.

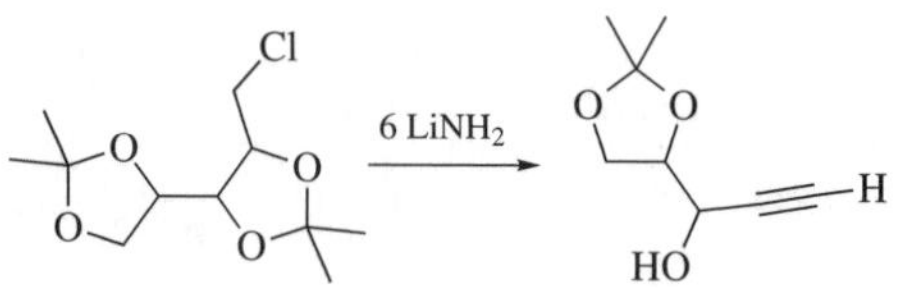

(18)

Related Reagents. Lithium Diethylamide; Lithium Diisopropylamide; Lithium Hexamethyldisilazide; Lithium Piperidide; Lithium Pyrrolidide; Lithium 2,2,6,6-Tetramethylpiperidide; Potassium Amide; Sodium Amide.

1. (a) Bergstrom, F. W.; Fernelius, W. C. *CRV* **1933**, *12*, 43; (b) Bergstrom, F. W.; Fernelius, W. C. *CRV* **1937**, *20*, 413.
2. See references cited in: Duhamel, L.; Plaquevent, J.-C. *JOM* **1993**, *448*, 1.
3. Dunnavant, W. R.; Hauser, C. R. *JOC* **1960**, *25*, 503.
4. Leffler, M. T. *OR* **1942**, *1*, 91.
5. Kaye, I. A. *JACS* **1949**, *71*, 2322.
6. Kaye, I. A.; Kogon, I. C. *JACS* **1951**, *73*, 5891.
7. Hauser, C. R.; Puterbaugh, W. H. *JACS* **1953**, *75*, 1068.
8. Hauser, C. R.; Puterbaugh, W. H. *JACS* **1951**, *73*, 2972.
9. Hauser, C. R.; Lindsay, J. K. *JACS* **1955**, *77*, 1050.
10. (a) Dunnavant, W. R.; Hauser, C. R. *OS* **1964**, *44*, 56. (b) Dunnavant, W. R.; Hauser, C. R. *OSC* **1973**, *5*, 564.
11. Smissman, E. E.; Voldeng, A. N. *JOC* **1964**, *29*, 3161.
12. (a) Binkley, E. S.; Heathcock, C. H. *JOC* **1975**, *40*, 2156. (b) Patterson, J. W., Jr.; Fried, J. H. *JOC* **1974**, *39*, 2506.
13. Sisido, K.; Kazama, Y.; Kodama, H.; Nozaki, H. *JACS* **1959**, *81*, 5817.
14. Sisido, K.; Sei, K.; Nozaki, H. *JOC* **1962**, *27*, 2681.
15. Kofron, W. G.; Wideman, L. G. *JOC* **1972**, *37*, 555.
16. Flahaut, J.; Miginiac, P. *HCA* **1978**, *61*, 2275.
17. Rao, A. V. R.; Reddy, E. R. *TL* **1986**, *27*, 2279.
18. Rao, A. V. R.; Reddy, E. R.; Sharma, G. V. M.; Yadagiri, P.; Yadav, J. S. *TL* **1985**, *26*, 465.
19. Claesson, A.; Olsson, L. I.; Sullivan, G. R.; Mosher, H. S. *JACS* **1975**, *97*, 2919.
20. (a) Landor, S. R.; Punja, N. *TL* **1966**, 4905. (b) Cowie, J. S.; Landor, P. D.; Landor, S. R.; Punja, N. *JCS(P1)* **1972**, 2197.
21. Wawzonek, S.; Studnicka, B. J.; Zigman, A. R. *JOC* **1969**, *34*, 1316.
22. (a) Köster, R.; Arora, S.; Binger, P. *AG(E)* **1970**, *9*, 810; (b) Köster, R.; Arora, S.; Binger, P. *LA* **1973**, 1219.
23. Fisher, F.; Applequist, D. E. *JOC* **1965**, *30*, 2089.
24. Le Perchec, P.; Conia, J. M. *TL* **1970**, 1587.
25. (a) Schipperijn, A. J. *RTC* **1971**, *90*, 1110. (b) Schipperijn, A. J.; Smael, P. *RTC* **1973**, *92*, 1121.
26. Schuijl, P. J. W.; Brandsma, L. *RTC* **1969**, *88*, 1201.
27. Meijer, J.; Brandsma, L. *RTC* **1972**, *91*, 578.
28. Brandsma, L.; Jonker, C.; Berg, M. H. *RTC* **1965**, *84*, 560.
29. Yadav, J. S.; Deshpande, P. K.; Sharma, G. V. M. *T* **1990**, *46*, 7033.
30. Katsuki, T.; Sharpless, K. B. *JACS* **1980**, *102*, 5974.
31. Ohloff, G.; Vial, C.; Näf, F.; Pawlak, M. *HCA* **1977**, *60*, 1161.
32. (a) Yadav, J. S.; Chander, M. C.; Rao, C. S. *TL* **1989**, *30*, 5455. (b) Yadav, J. S.; Krishna, P. R.; Gurjar, M. K. *T* **1989**, *45*, 6263.

Ihsan Erden
San Francisco State University, CA, USA

Lithium Bromide[1]

LiBr

[7550-35-8] BrLi (MW 86.85)

(source of nucleophilic bromide;z[2] mild Lewis acid;[1] salt effects in organometallic reactions;[1] epoxide opening[1])

Physical Data: mp 550 °C; bp 1265 °C; *d* 3.464 g cm^{-3}.
Solubility: 145 g/100 mL H_2O (4 °C); 254 g/100 mL H_2O (90 °C); 73 g/100 mL EtOH (40 °C); 8 g/100 mL MeOH; sol ether, glycol, pentanol, acetone; slightly sol pyridine.
Form Supplied in: anhyd white solid, or as hydrate.
Purification: dry for 1 h at 120 °C/0.1 mmHg before use; or dry by heating in vacuo at 70 °C (oil bath) for 24 h, then store at 110 °C until use.
Handling, Storage, and Precautions: for best results, dry before use in anhyd reactions.

Alkyl and Alkenyl Bromides. LiBr has been extensively used as a source of bromide in nucleophilic substitution and addition reactions. Interconversion of halides[2] and transformation of alcohols to alkyl bromides via the corresponding sulfonate[3] or trifluoroacetate[4] have been widely used in organic synthesis. Primary and secondary alcohols have been directly converted to alkyl bromides upon treatment with a mixture of ***Triphenylphosphine***, ***Diethyl Azodicarboxylate***, and LiBr.[5]

(*Z*)-3-Bromopropenoates and -propenoic acids have been synthesized stereoselectively by the reaction of LiBr and propiolates or propiolic acid (eq 1).[6]

≡—CO_2Et —(LiBr, AcOH; 70 °C, 15 h; 91%)→ Br_/=_CO_2Et (1)

Heterolytic Cleavage of C–X Bonds. In the presence of a Lewis acid, LiBr acts as a nucleophile in the opening of 1,2-oxiranes to produce bromohydrins (eq 2).[7] In the absence of an external Lewis acid or nucleophile, epoxides generally give rise to products resulting from ring-contraction reactions (eq 3).

(cyclohexene oxide) —(LiBr, AcOH; THF, rt; 90%)→ (trans-2-bromocyclohexanol: OH, Br) (2)

(cyclohexene oxide) —(LiBr, alumina; toluene, reflux; 77%)→ (cyclopentyl)–CHO (3)

LiBr-mediated decomposition of dioxaphospholanes results in the exclusive formation of the epoxide, whereas the thermal decomposition produces a mixture of products (eq 4).[8]

(diol: OH, Ph, OH, Ph) —(1. $(EtO)_2PPh_3$; 2. LiBr, rt; 97%)→ (epoxide: Ph, O, Ph) (4)

Protection of alcohols as their MOM ethers can be achieved using a mixture of ***Dimethoxymethane***, LiBr, and ***p-Toluenesulfonic Acid***.[9]

Bifunctional Reagents. Activated α-bromo ketones are smoothly converted into the corresponding silyl enol ethers when treated with a mixture of LiBr/R_3N/***Chlorotrimethylsilane***.[10] Aldehydes are converted into the corresponding α,β-unsaturated esters using ***Triethyl Phosphonoacetate*** and ***Triethylamine*** in the presence of LiBr (eq 5).[11,12] Similar conditions were extensively used in the asymmetric cycloaddition and Michael addition reactions of *N*-lithiated azomethine ylides (eq 6).[13]

PhCHO + $(EtO)_2P(O)CH_2CO_2Et$ —(Et_3N, MX; MeCN; 25 °C, 3 h)→ Ph–CH=CH–CO_2Et (5)

MX = LiCl, 77%; LiBr, 93%; $MgCl_2$, 15%; $MgBr_2$, 71%

t-Bu–CH=N–CH(CH_3)–CO_2Me —(1. LiBr, DBU; 2. CH_2=CH–CO_2Me; 61%)→ *t*-Bu–CH=N–C(CH_3)(CO_2Me)–$CH_2CH_2CO_2Me$ (6)

Additive for Organometallic Transformations. The addition of LiBr and ***Lithium Iodide*** was shown to enhance the rate of organozinc formation from primary alkyl chlorides, sulfonates, and phosphonates, and ***Zinc*** dust.[14] Beneficent effects of LiBr addition have also been reported for the Heck-type coupling reactions[15] and for the nickel-catalyzed cross-couplings of alkenyl and α-metalated alkenyl sulfoximines with organozinc reagents.[16] The addition of 2 equiv of LiBr significantly enhances the yield of the conjugate addition products in reactions of certain organocopper reagents (eq 7).[17]

(cyclohexenone) —($MeCu(PCy_2)Li$)→ (3-methylcyclohexanone) (7)

LiBr (0 equiv), 61%
LiBr (2 equiv), 96%

Finally, concentrated solutions of LiBr are also known to alter significantly the solubility and the reactivity of amino acids and peptides in organic solvents.[18]

Related Reagents. Phosphorus(III) Bromide; Sodium Bromide; Tetra-*n*-butylammonium Bromide; Tetramethylammonium Bromide; Triphenylphosphine Dibromide.

---

1. Loupy, A.; Tchoubar, B. *Salt effects in Organic and Organometallic Chemistry*; VCH: Weinheim, 1992.
2. Sasson, Y.; Weiss, M.; Loupy, A.; Bram, G.; Pardo, C. *CC* **1986**, 1250.
3. (a) Ingold, K. U.; Walton, J. C. *JACS* **1987**, *109*, 6937. (b) McMurry, J. E.; Erion, M. D. *JACS* **1985**, *107*, 2712.
4. Camps, F.; Gasol, V.; Guerrero, A. *S* **1987**, 511.
5. Manna, S.; Falck, J. R. Mioskowski, C. *SC* **1985**, *15*, 663.
6. (a) Ma, S.; Lu, X. *TL* **1990**, *31*, 7653. (b) Ma, S.; Lu, X. *CC* **1990**, 1643.

7. (a) Bonini, C.; Giuliano, C.; Righi, G.; Rossi, L. *SC* **1992**, *22*, 1863. (b) Shimizu, M.; Yoshida, A.; Fujisawa, T. *SL* **1992**, 204. (c) Bajwa, J. S.; Anderson, R. C. *TL* **1991**, *32*, 3021.
8. (a) Murray, W. T.; Evans Jr., S. A. *NJC* **1989**, *13*, 329. (b) Murray, W. T.; Evans, S. A., Jr. *JOC* **1989**, *54*, 2440.
9. Gras, J.-L.; Chang, Y.-Y. K. W.; Guérin, A. *S* **1985**, 74.
10. Duhamel, L.; Tombret, F.; Poirier, J. M. *OPP* **1985**, *17*, 99.
11. Rathke, M. W.; Nowak, M. *JOC* **1985**, *50*, 2624.
12. Seyden-Penne, J. *BSF* **1988**, 238.
13. (a) Kanemasa, S.; Tatsukawa, A.; Wada, E. *JOC* **1991**, *56*, 2875. (b) Kanemasa, S.; Uchida, O.; Wada, E. *JOC* **1990**, *55*, 4411. (c) Kanemasa, S.; Yoshioka, M.; Tsuge, O. *BCJ* **1989**, *62*, 869. (d) Kanemasa, S.; Yamamoto, H.; Wada, E.; Sakurai, T.; Urushido, K. *BCJ* **1990**, *63*, 2857.
14. Jubert, C.; Knochel, P. *JOC* **1992**, *57*, 5425.
15. (a) Cabri, W.; Candiani, I.; DeBernardinis, S.; Francalanci, F.; Penco, S. *JOC* **1991**, *56*, 5796. (b) Karabelas, K.; Hallberg, A. *JOC* **1989**, *54*, 1773.
16. Erdelmeier, I.; Gais, H.-J. *JACS* **1989**, *111*, 1125.
17. Bertz, S. H.; Dabbagh, G. *JOC* **1984**, *49*, 1119.
18. Seebach, D. *Aldrichim. Acta* **1992**, *25*, 59.

André B. Charette
Université de Montréal, Québec, Canada

Lithium *t*-Butoxide

t-BuOLi

[1907-33-1] C_4H_9LiO (MW 80.07)

(weakly basic and nucleophilic metal alkoxide)[1]

Physical Data: crystalline solid, subl over 170–205 °C/760 mmHg; dec above 250 °C.[2]

Form Supplied in: commercially available as a white powder.

Preparative Methods: the crystalline material is prepared using a Schlenk apparatus by reaction of *t*-BuOH with a small excess of ***Lithium*** sand in PhMe at rt for 24 h;[2] the excess metal is removed by filtration and the solvent is removed in vacuo; the remaining solid is sublimed at 110–120 °C/0.01 mmHg. In situ preparation is accomplished by adding a solution of anhyd *t*-BuOH in THF to 1.6 M ***n-Butyllithium*** in hexane,[1a] or by slowly adding a 1.55 M solution of *n*-BuLi in hexane to an excess of the anhyd alcohol under N_2;[1b] the reaction is exothermic, and a water bath should be used to keep the temperature of the mixture near rt.

Handling, Storage, and Precautions: use same procedure as for ***Potassium t-Butoxide***, i.e. handle and conduct reactions in a fume hood under an inert atmosphere; for critical experiments, it is recommended that the reagent be freshly prepared.

Preparation of *t*-Butyl Esters. *t*-BuOLi reacts with acid chlorides of hindered carboxylic acids to give the corresponding *t*-butyl esters in high yields (eq 1).[1] The ester product shown in eq 1 is not obtained by the more conventional method of reacting the acid chloride with *t*-BuOH in the presence of $PhNMe_2$.[3]

Cl + *t*-BuOLi → (Et_2O, *t*-BuOH; reflux; 79–82%) O-*t*-Bu (1)

Oxidation of Alcohols. Alkoxymagnesium bromides of secondary alcohols and allylic and benzylic alcohols are oxidized to the corresponding ketones or aldehydes with ***N-Chlorosuccinimide*** in the presence of *t*-BuOLi.[4] The method is ineffective for primary saturated and some unsaturated alcohols, but oxidations of the former substrates to aldehydes occur readily if ***t-Butoxymagnesium Bromide*** is substituted for *t*-BuOLi.[5] Secondary alcohols (eq 2) and primary benzylic (eq 3) and allylic alcohols are oxidized to the corresponding carbonyl compounds in good yields using 2–3 equiv of a 1:1 mixture of ***Copper(II) Bromide***–*t*-BuOLi in THF at rt.[6] However, the reagent is not effective for the conversion of primary aliphatic alcohols to aldehydes.

Ph–CH(OH) → (3 equiv $CuBr_2$–*t*-BuOLi; THF, rt; 87%) Ph ketone (2)

MeO–C6H4–CH2OH → (2.2 equiv $CuBr_2$–*t*-BuOLi; THF, rt; 97%) MeO–C6H4–CHO (3)

Condensation Reactions. In the presence of dipolar aprotic additives, e.g. ***Hexamethylphosphoric Triamide*** (eq 4) or ***12-Crown-4***, which reduce the degree of aggregation, *t*-BuOLi catalyzes the Michael addition of methyl cyanoacetate to benzylideneacetophenone.[7] A similar yield of the Michael adduct is obtained if *t*-BuOK is used as the base, and no additive is required.

PhCH=CHCOPh + $NCCH_2CO_2Me$ → (0.1 equiv *t*-BuOLi, 0.2 equiv HMPA; THF, rt; 71%) Ph, Ph, O, NC, CO_2Me (4)

Additions to Hindered Ketones. In the absence of various additive salts, the reaction of $i\text{-}Pr_2Mg$ with $t\text{-}Bu_2CO$ leads mainly to simple reduction of the carbonyl group.[8] In the presence of *t*-BuOLi, addition to the carbonyl group occurs in good yield (eq 5).[8] *t*-BuOLi, MeOK, and $(Me_2CH)_2CHOK$ are more effective than MeONa or quaternary ammonium salts in preventing the reduction process. The exact role of the additive in favoring addition to the hindered ketone over reduction is unknown.

i-Pr$_2$Mg, t-BuOLi additive, Et$_2$O, 25 °C (5)

	HO, i-Pr product	OH product
t-BuOLi (1.0–2.0 equiv)	81–84%	0%
no additive	1.3%	88%

β-Elimination Reactions. 1-Methoxyacenaphthene undergoes loss of MeOH via an E1cB mechanism.[9] t-BuOLi/t-BuOH favors exchange and elimination of the proton *syn* to the methoxy group much more than do other alkali metal alkoxides. It has been suggested that the strong tendency of the lithium cation of the base ion pair (or an aggregate) to coordinate with the ether oxygen of the substrate accounts for this.[9]

Related Reagents. t-Butoxymagnesium Bromide; Potassium t-Butoxide; Potassium t-Heptoxide; Potassium Methoxide–Dimethyl Sulfoxide; Potassium 2-Methyl-2-butoxide.

1. (a) Kaiser, E. M.; Woodruff, R. A. *JOC* **1970**, *35*, 1198. (b) Crowther, G. P.; Kaiser, E. M.; Woodruff, R. A.; Hauser, C. R. *OSC* **1988**, *6*, 259.
2. Chisholm, M. H.; Drake, S. R.; Naiini, A. A.; Streib, W. E. *Polyhedron* **1991**, *10*, 805.
3. Hauser, C. R.; Hudson, B. E.; Abramovitch, B.; Shivers, J. C. *OSC* **1955**, *3*, 142.
4. Mukaiyama, T.; Tsunoda, M.; Saigo, K. *CL* **1975**, 691.
5. Narasaka, K.; Morikawa, A.; Saigo, K.; Mukaiyama, T. *BCJ* **1977**, *50*, 2773.
6. Yamaguchi, J.; Yamamoto, S.; Takeda, T. *CL* **1992**, 1185.
7. Cossentini, M.; Strzalko, T.; Seyden-Penne, J. *BSF(2)* **1987**, 531.
8. Richey, H. G., Jr.; DeStephano, J. P. *JOC* **1990**, *55*, 3281.
9. Hunter, D. H.; Shearing, D. J. *JACS* **1971**, *93*, 2348.

Drury Caine
University of Alabama, Tuscaloosa, AL, USA

Lithium Carbonate

Li_2CO_3

[554-13-2] CLi_2O_3 (MW 73.89)

(base for dehydrohalogenation reactions and halide displacements; reagent for regioselective alkylations)

Physical Data: mp 723 °C; bp 1310 °C; d 2.11 g cm^{-3}.
Solubility: sol cold (1 g/78 mL), hot H_2O (1 g/140 mL); insol alcohol, acetone.
Form Supplied in: white powder; widely available.
Drying: details are available.[1]

Dehydrohalogenation. Lithium carbonate in refluxing DMF is an excellent base for dehydrohalogenation reactions. LiCl in DMF caused rearrangement of the alkene to the more substituted α,β-position.[2] Treatment with a stronger base, lithium carbonate, led to good yields of the less substituted enone (eq 1). Substituted tropones have been prepared in a similar manner from alkyl-substituted tribrominated cycloheptanones (eq 2).[3]

Li_2CO_3, DMF, 140 °C, 38% (1)

Li_2CO_3, DMF, reflux, 75% (2)

Halide Displacement. A preparation of steroidal oxetanones has been reported using Li_2CO_3 in refluxing DMF. It was found that either the *cis* or the *trans* bromides can give the α-oxetanone in good yield (eq 3).[4]

Li_2CO_3, DMF, reflux, 69% (3)

Alkylations. Lithium carbonate used in conjunction with DMF and various alkylating agents has been found to regioselectively alkylate bisphenolic, 1,3-dicarbonyl, and thiol substrates that contain hydrogens that are more acidic than phenol (pK_a 8).[5] It has been determined that phenolic compounds containing electron-withdrawing groups alkylate at either the *ortho* or *para* positions exclusively (eq 4). Attempts to use other bases such as ***Potassium Carbonate*** or ***Sodium Carbonate*** showed a loss of selectivity. Using more reactive alkylating agents such as propargyl bromide (see under ***Propargyl Chloride***, Vol. A) or ***Benzyl Bromide*** also reduces the selectivity of alkylation. This sequence has been used to differentiate between the two hydroxyl groups present in alizarin (eq 5).[5]

MeI, Li_2CO_3, DMF, 18 h, 55 °C, 90% (4)

MeI, Li_2CO_3, DMF, 16 h, 50 °C, 82% (5)

1,3-Dicarbonyl compounds such as β-keto esters or β-diketones are sufficiently acidic to alkylate as well (eq 6).[5]

$C_5H_{11}I$, Li_2CO_3, DMF, 22 h, 80 °C, 76% (6)

Thiophenoxyphenols have shown selectivity of alkylation with lithium carbonate. The more acidic S–H can be preferentially alkylated in deference to a phenolic O–H bond.[5] Using potassium carbonate in the same alkylation scheme gives only 4% of the desired product.

Base Assistance. Spirocyclic ethers can be generated via intramolecular nucleophilic attack of hydroxide upon π-(allyl)palladium complexes (eq 7). The attack of an external nucleophile can then be directed for either *cis* or *trans* addition, depending on the concentration of nucleophile present.[6] In the presence of 3 equiv of Li_2CO_3, *cis* migration of acetate was the only reaction observed. Conversely, the addition of 1.8 equiv of chloride ion causes chloride to be added from the face opposite the metal.

5% $Pd(OAc)_2$, 3.0 equiv Li_2CO_3, 24 h, 86% → AcO... >98% *trans* (7)

5% $Pd(OAc)_2$, 1.8 equiv LiCl, 24 h, 73% → Cl... >99% *cis*

Related Reagents. Lithium Carbonate–Lithium Bromide; Potassium Carbonate; Sodium Carbonate.

1. *Inorg. Synth.* **1939**, *1*, 1.
2. House, H. O.; Bashe, R. W. II. *JOC* **1965**, *30*, 2942.
3. (a) Jones, G. *JCS(C)* **1970**, 1230. (b) Collington, E. W.; Jones, G. *JCS(C)* **1969**, 2656. (c) Collington, E. W.; Jones, G. *CC* **1968**, 958.
4. (a) Hanna, R.; Maalouf, G.; Muckensturm, B. *T* **1973**, *29*, 2297. (b) Rowland, A. T.; Bennett, R. J.; Shoupe, T. S. *JOC* **1968**, *33*, 2426.
5. Wymann, W. E.; Davis, R.; Patterson, J. W.; Pfister, J. *SC* **1988**, *18*, 1379.
6. Bäckvall, J. E.; Andersson, P. G. *JOC* **1991**, *56*, 2274.

Dennis Wright & Mark C. McMills
Ohio University, Athens, OH, USA

Lithium Chloride

[7447-41-8] ClLi (MW 42.39)

(source of Cl^- as nucleophile and ligand; weak Lewis acid that modifies the reactivity of enolates, lithium dialkylamides, and other Lewis bases)

Physical Data: mp 605 °C; bp 1325–1360 °C; *d* 2.068 g cm^{-3}.
Solubility: very sol H_2O; sol methanol, ethanol, acetone, acetonitrile, THF, DMF, DMSO, HMPA.
Form Supplied in: white solid, widely available.
Drying: deliquescent; for most applications, drying at 150 °C for 3 h is sufficient; for higher purity, recrystallization from methanol, followed by drying at 140 °C/0.5 mmHg overnight, is recommended.
Handling, Storage, and Precautions: of low toxicity; take directly from the oven when dryness is required.

Source of Chloride Nucleophile. The solubility of LiCl in many organic dipolar solvents renders it an effective source of nucleophilic chloride anion. Lithium chloride converts alcohols to alkyl chlorides[1] under Mitsunobu conditions,[2] or by way of the corresponding sulfonates[3] or other leaving groups.[4] This salt cleanly and regioselectively opens epoxides to chlorohydrins in the presence of acids and Lewis acids such as ***Acetic Acid***,[5] Amberlyst 15 resin,[6] and ***Titanium(IV) Chloride***.[7] In the presence of acetic acid, LiCl regio- and stereoselectively hydrochlorinates 2-propynoic acid and its derivatives to form the corresponding derivatives of (*Z*)-3-chloropropenoic acid.[8] Oxidative decarboxylation of carboxylic acids by ***Lead(IV) Acetate*** in the presence of 1 equiv of LiCl generates the corresponding chlorides.[9]

In wet DMSO, LiCl dealkoxycarbonylates various activated esters (eq 1).[10,11] If the reaction is performed in anhyd solvent the reaction generates a carbanion intermediate, which can undergo inter- or intramolecular alkylation or elimination. Other inorganic salts (NaCN, NaCl, ***Lithium Iodide***) and other dipolar aprotic solvents (HMPA, DMF) can also be employed. Under similar conditions, lithium chloride cleaves alkyl aryl ethers having electron-withdrawing substituents at the *ortho* or *para* positions.[12]

$$XC(R^1)(R^2)CO_2R^3 \xrightarrow[\text{wet DMSO}]{\text{LiCl}} XC(R^1)(R^2)H \quad (1)$$

$X = CO_2R$, COR, CN, SO_2R; R^3 = Me, Et

Source of Chloride Ligand. In palladium-catalyzed reactions, LiCl is often the reagent of choice as a source of chloride ligand. Lithium chloride is a necessary component in palladium-catalyzed coupling and carbonylative coupling reactions of organostannanes and vinyl triflates.[13,14] Lithium chloride has a dramatic effect on the stereochemical course of palladium-catalyzed 1,4-additions to 1,3-dienes.[15] Treatment of 1,3-cyclohexadiene with ***Palladium(II) Acetate*** and LiOAc and the oxidizing agents ***1,4-Benzoquinone*** and ***Manganese Dioxide*** affords 1,4-*trans*-diacetoxy-2-cyclohexene (eq 2). In the presence of a catalytic quantity of LiCl, the *cis* isomer is formed (eq 3). If 2 equiv LiCl are added, the *cis*-acetoxychloro compound forms (eq 4). These methods are general for both cyclic and acyclic dienes, and have recently been extended to the stereospecific formation of fused heterocycles.[16] Lithium chloride is also used in the preparation of ***Dilithium Tetrachloropalladate(II)***[17] and zinc organocuprate reagents.[18]

$Pd(OAc)_2$, LiOAc, no added LiCl; 1,4-benzoquinone, MnO_2, 93% → AcO...OAc >91% *trans* (2)

$Pd(OAc)_2$, LiOAc, 0.2 equiv LiCl; 1,4-benzoquinone, MnO_2, 85% → AcO...OAc >96% *cis* (3)

$Pd(OAc)_2$, LiOAc, 2 equiv LiCl; 1,4-benzoquinone, MnO_2; 89% → Cl–(cyclohexene)–OAc, >98% *cis* (4)

Weak Lewis Acid. Lithium chloride is a weak Lewis acid that forms mixed aggregates with lithium dialkylamides, enolates, alkoxides, peptides, and related 'hard' Lewis bases.[19] Thus LiCl often has a dramatic effect on reactions involving these species. In the deprotonation of 3-pentanone by ***Lithium 2,2,6,6-Tetramethylpiperidide*** (LTMP), addition of 0.3 equiv LiCl increases the (*E*)/(*Z*) selectivity from 9:1 to 52:1 (eq 5).[20] Enhancement in the enantioselectivity of deprotonation of prochiral ketones by a chiral lithium amide has also been reported.[21] Lithium chloride stabilizes anions derived from α-phosphonoacetates, permitting amine and amidine bases to be used to perform Horner–Wadsworth–Emmons reactions on base-sensitive aldehydes under exceptionally mild conditions.[22] Lithium chloride and other lithium salts disrupt peptide aggregation and increase the solubilities of peptides in THF and other ethereal solvents, often by 100-fold or greater.[23] These effects render LiCl a useful additive in the chemical modification of peptides (e.g. by the formation and alkylation of peptide enolates).[19,24] Lithium chloride has also shown promise as an additive in solid-phase peptide synthesis, increasing resin swelling and improving the efficiencies of difficult coupling steps.[25]

1. LiTMP (LiCl); 2. TMSCl, THF → OTMS + OTMS (5)

0.0 equiv LiCl 9:1
0.3 equiv LiCl 52:1

Related Reagents. Lithium Chloride–Diisopropylethylamine; Lithium Chloride–Hexamethylphosphoric Triamide; Zinc Chloride.

1. Magid, R. M. *T* **1980**, *36*, 1901.
2. Manna, S.; Falck, J. R.; Mioskowski, C. *SC* **1985**, *15*, 663.
3. (a) Owen, L. N.; Robins, P. A. *JCS* **1949**, 320. (b) Owen, L. N.; Robins, P. A. *JCS* **1949**, 326. (c) Eglinton, G.; Whiting, M. C. *JCS* **1950**, 3650. (d) Collington, E. W.; Meyers, A. I. *JOC* **1971**, *36*, 3044. (e) Stork, G.; Grieco, P. A.; Gregson, M. *TL* **1969**, *18*, 1393.
4. (a) Czernecki, S.; Georgoulis, C. *BSF* **1975**, 405. (b) Camps, F.; Gasol, V.; Guerrero, A. *S* **1987**, 511.
5. Bajwa, J. S.; Anderson, R. C. *TL* **1991**, *32*, 3021.
6. Bonini, C.; Giuliano, C.; Righi, G.; Rossi, L. *SC* **1992**, *22*, 1863.
7. Shimizu, M.; Yoshida, A.; Fujisawa, T. *SL* **1992**, 204.
8. (a) Ma, S.; Lu, X. *TL* **1990**, *31*, 7653. (b) Ma, S.; Lu, X.; Li, Z. *JOC* **1992**, *57*, 709.
9. (a) Kochi, J. K. *JACS* **1965**, *87*, 2500. (b) Review: Sheldon, R. A., Kochi, J. K. *OR* **1972**, *19*, 279.
10. Krapcho, A. P.; Weimaster, J. F.; Eldridge, J. M.; Jahngen, E. G. E., Jr.; Lovey, A. J.; Stephens, W. P. *JOC* **1978**, *43*, 138.
11. Reviews: (a) Krapcho, A. P. *S* **1982**, 805. (b) Krapcho, A. P. *S* **1982**, 893.
12. Bernard, A. M.; Ghiani, M. R.; Piras, P. P.; Rivoldini, A. *S* **1989**, 287.
13. (a) Scott, W. J.; Crisp, G. T.; Stille, J. K. *JACS* **1984**, *106*, 4630. (b) Crisp, G. T.; Scott, W. L.; Stille, J. K. *JACS* **1984**, *106*, 7500.
14. Reviews: (a) Stille, J. K. *AG(E)* **1986**, *25*, 508. (b) Scott, W. J.; McMurry, J. E. *ACR* **1988**, *21*, 47.
15. (a) Bäckvall, J. E.; Byström, S. E.; Nordberg, R. E. *JOC* **1984**, *49*, 4619. (b) Bäckvall, J. E.; Nyström, J. E.; Nordberg, R. E. *JACS* **1985**, *107*, 3676.
16. (a) Bäckvall, J. E.; Andersson, P. G. *JACS* **1992**, *114*, 6374. (b) Review: Bäckvall, J. E. *PAC* **1992**, *64*, 429.
17. Lipshutz, B. H.; Sengupta, S. *OR* **1992**, *41*, 135.
18. (a) Knochel, P.; Yeh, M. C. P.; Berk, S. C.; Talbert, J. *JOC* **1988**, *53*, 2390. (b) Jubert, C.; Knochel, P. *JOC* **1992**, *57*, 5431. (c) Ibuka, T.; Yoshizawa, H.; Habashita, H.; Fujii, N.; Chounan, Y.; Tanaka, M.; Yamamoto, Y. *TL* **1992**, *33*, 3783. (d) Yamamoto, Y.; Chounan, Y.; Tanaka, M.; Ibuka, T. *JOC* **1992**, *57*, 1024. (e) Knochel, P.; Rozema, M. J.; Tucker, C. E.; Retherford, C.; Furlong, M.; AchyuthaRao, S. *PAC* **1992**, *64*, 361.
19. Seebach, D. *AG(E)* **1988**, *27*, 1624.
20. (a) Hall, P. L.; Gilchrist, J. H.; Collum, D. B. *JACS* **1991**, *113*, 9571. (b) Hall, P. L.; Gilchrist, J. H.; Harrison, A. T.; Fuller, D. J.; Collum, D. B. *JACS* **1991**, *113*, 9575.
21. Bunn, B. J.; Simpkins, N. S. *JOC* **1993**, *58*, 533.
22. (a) Blanchette, M. A.; Choy, W.; Davis, J. T.; Essenfeld, A. P.; Masamune, S.; Roush, W. R.; Sakai, T. *TL* **1984**, *25*, 2183. (b) Rathke, M. W.; Nowak, M. *JOC* **1985**, *50*, 2624.
23. Seebach, D.; Thaler, A.; Beck, A. K. *HCA* **1989**, *72*, 857.
24. Seebach, D.; Bossler, H.; Gründler, H.; Shoda, S.-i.; Wenger, R. *HCA* **1991**, *74*, 197.
25. (a) Thaler, A.; Seebach, D.; Cardinaux, F. *HCA* **1991**, *74*, 617. (b) Thaler, A.; Seebach, D.; Cardinaux, F. *HCA* **1991**, *74*, 628.

James S. Nowick
University of California, Irvine, CA, USA

Guido Lutterbach
Johannes Gutenberg University, Mainz, Germany

Lithium Diethylamide

[816-43-3] $C_4H_{10}LiN$ (MW 79.09)

(strong base; nitrogen nucleophile)

Physical Data: $pK_a = 31.7$.[1]
Solubility: sol THF, Et_2O, Et_2O–hexane mixtures, Et_2O–benzene mixtures; insol hydrocarbon solvents.
Form Supplied in: white solid.
Preparative Methods: by treating diethylamine with ***Phenyllithium*** or ***n-Butyllithium*** or, alternatively, by treating diethylamine with ***Lithium*** metal in the presence of ***Hexamethylphosphoric Triamide***[2] or an electron acceptor.[3]
Handling, Storage, and Precautions: reacts with water vigorously and ignites spontaneously in air. When heated to decomposition, it emits toxic fumes of NO_x. Should be kept and handled under argon. Use in a fume hood.

Lithium Diethylamide as Base. Lithium diethylamide is a strong base whose use may be complementary to ***Lithium Diisopropylamide***. For example, $LiNEt_2$ serves to abstract the ω-proton of sorbic acid to generate a lithium trienolate which undergoes regioselective condensation with aldehydes and ketones (eq 1),[4] and to prepare carbocyclic nitriles by a sequential inter- and intramolecular alkylation route (eq 2).[5] $LiNEt_2$ has been employed with a sterically hindered lactam for an α-deprotonation–cyclization reaction as part of the synthesis of (−)-δ-*N*-normethylskytanthine, an alkaloid from *Tecoma arequipensis* (eq 3).[6] $LiNEt_2$ may be a superior base to LDA for enolate formation.[7]

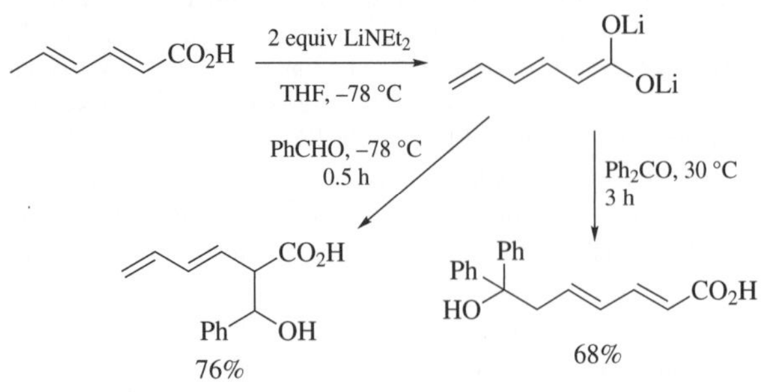

(1)

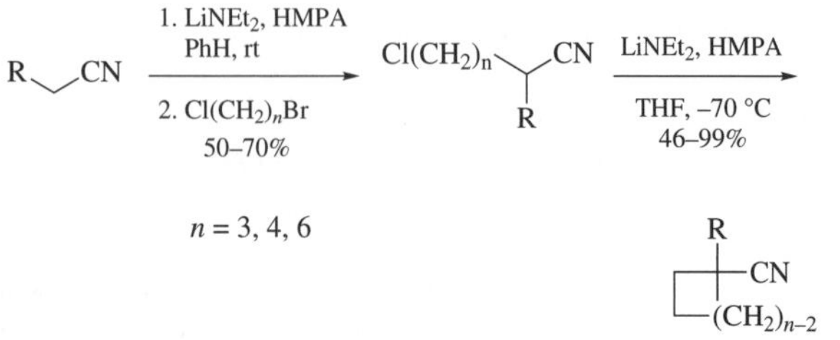

(2)

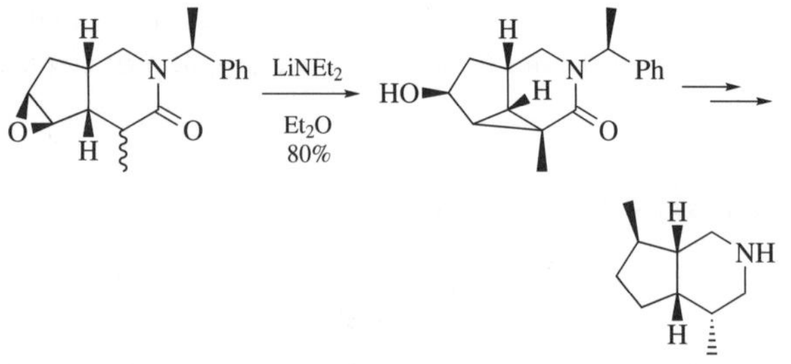

(3)

Lithium diethylamide is widely used for base-induced epoxide opening to allylic alcohols and rearrangement via carbenoids.[8] Two reaction pathways are observed: β-*syn*-elimination[8c] to afford an allylic alcohol (eq 4) and α-elimination to generate a carbenoid followed by C–H insertion (eqs 5 and 6). Allylic deprotonation followed by intramolecular epoxide ring opening has been observed (eq 6). The β-elimination process is often highly stereoselective and therefore has synthetic value.[8h]

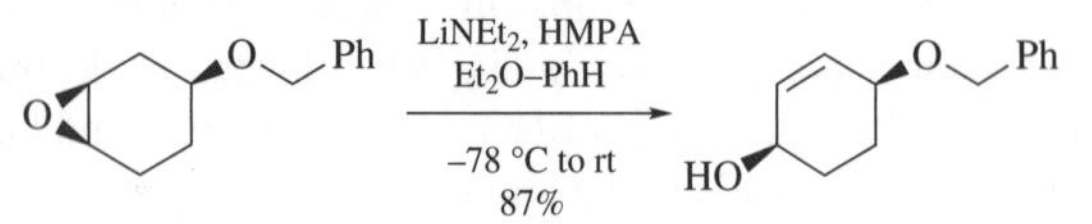

(4)

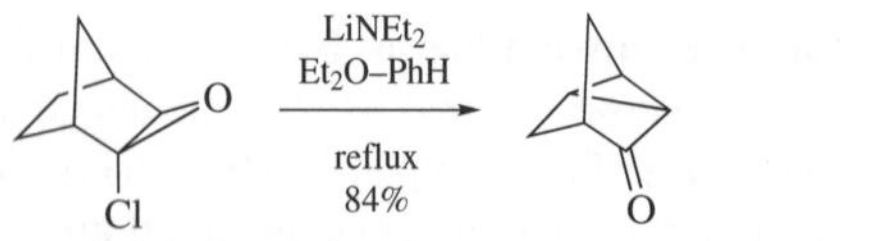

(5)

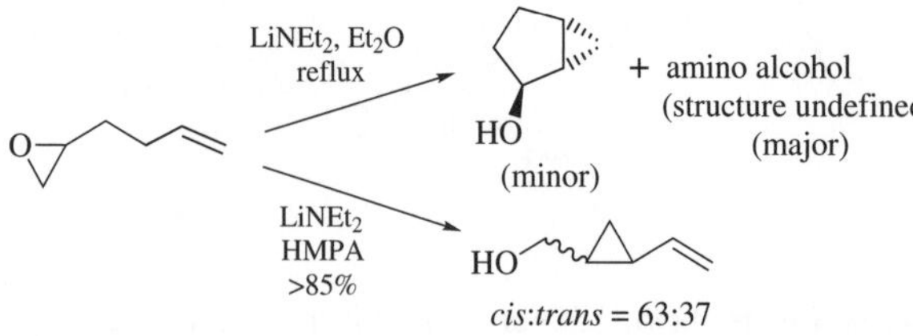

(6)

$LiNEt_2$ is sufficiently basic to effect α-deprotonation of allenic ethers. Subsequent reaction with ketones followed by acid treatment affords 3-furanones (eq 7).[9]

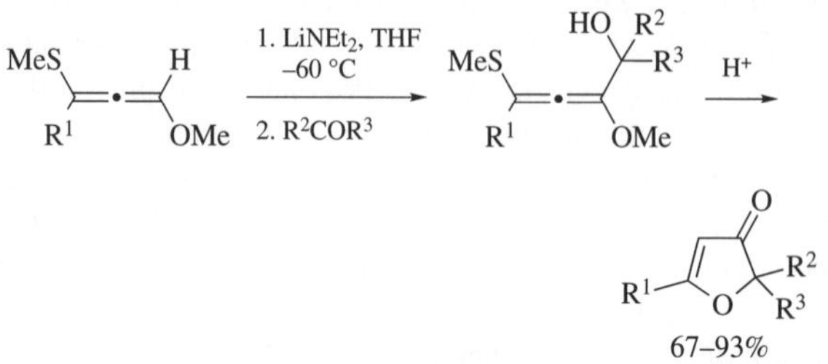

(7)

Treatment of tetraphenylphosphonium bromide with $LiNEt_2$ leads to 9-phenyldibenzophosphole[10] which may be further transformed into Wittig–Horner reagents (eq 8), useful for highly stereoselective synthesis of alkenes.

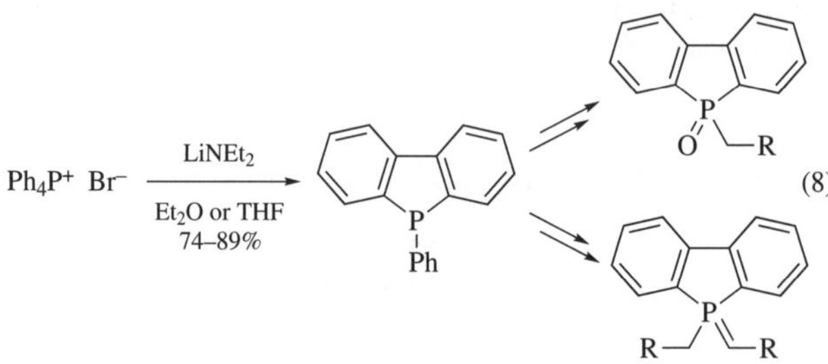

(8)

Lithium Diethylamide as Nucleophile. Treatment of aldehydes with a $LiNEt_2$–***Titanium Tetraisopropoxide*** complex generates α-amino alkoxides which undergo reaction with Grignard reagents to afford tertiary amines (eq 9).[11] $LiNEt_2$ attacks THP ethers of 2,4-alkadien-1-ols regioselectively at C-5 to furnish amino dienes with predominately (*E,E*) stereochemistry (eq 10).[12] Treatment of perfluorinated alkanes with $LiNEt_2$ results in β-elimination to provide terminal alkenes (eq 11).[13] Further treatment with $LiNEt_2$ leads to addition and fluoride elimination to give enamines which, after hydrolysis, produce perfluorinated amides. Similarly, chloroalkynes lead to ynamines (eq 12).[14]

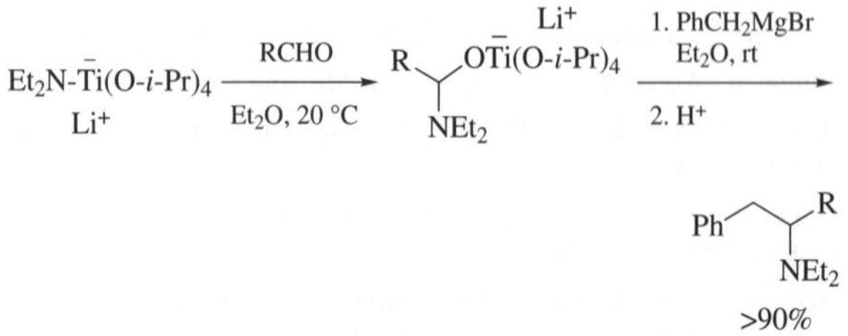

(9)

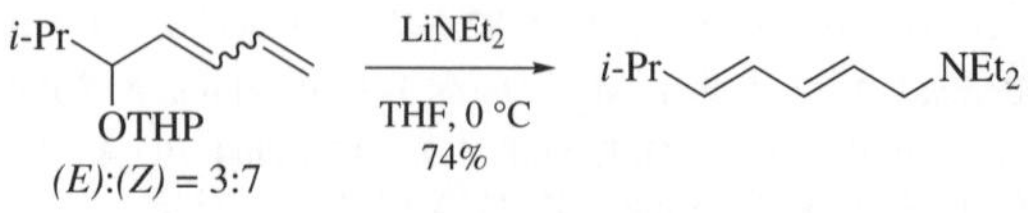

(10)

$$R_FCF_2CHF_2 \xrightarrow[Et_2O,\ -10\ ^\circ C]{LiNEt_2} [R_FCF{=}CF_2] \xrightarrow{LiNEt_2}$$

$$R_FCF{=}CFNEt_2 \xrightarrow{H^+} R_FCHFCONEt_2 \quad (11)$$

40–60%

$R_F = CF_3(CF_2)_3$-, $PhCH_2O(CF_2)_4$-, etc.

$$\text{EtO(cyclopropyl)H–C≡C–Cl} \xrightarrow[Et_2O,\ -20\ ^\circ C;\ 38\%]{LiNEt_2} \text{EtO(cyclopropyl)H–C≡C–NEt}_2 \quad (12)$$

Related Reagents. Lithium Amide; Lithium Diisopropylamide; Lithium Hexamethyldisilazide; Lithium *N*-Methylpiperazide; Lithium Morpholide; Lithium Piperidide; Lithium Pyrrolidide; Lithium 2,2,6,6-Tetramethylpiperidide.

---

1. Ahlbrecht, H.; Schneider, G. *T* **1986**, *42*, 4729.
2. Normant, H.; Cuvigny, T.; Reisdorf, D. *CR(C)* **1969**, *268*, 521.
3. Gaudemar-Bardone, F.; Gaudemar, M. *S* **1979**, 463.
4. Ballester, P.; Costa, A.; Garcia-Raso, A.; Gomez-Solivellas, A.; Mestres, R. *TL* **1985**, *26*, 3625.
5. Larcheveque, M.; Mulot, P.; Cuvigny, T. *JOM* **1973**, *57*, C33.
6. Cid, M. M.; Eggnauer, U.; Weber, H. P.; Pombo-Villar, E. *TL* **1991**, *32*, 7233.
7. Seebach, D.; Wasmuth, D. *AG(E)* **1981**, *20*, 971.
8. (a) Crandall, J. K.; Chang, L.-H. *JOC* **1967**, *32*, 435. (b) Rickborn, B.; Thummel, R. P. *JOC* **1969**, *34*, 3583. (c) Thummel, R. P.; Rickborn, B. *JACS* **1970**, *92*, 2064. (d) Thummel, R. P.; Rickborn, B. *JOC* **1972**, *37*, 3919. (e) McDonald, R. N.; Steppel, R. N.; Cousins, R. C. *JOC* **1975**, *40*, 1694. (f) Apparu, M.; Barrelle, M. *TL* **1976**, 2837. (g) Williams, D. R.; Grote, J. *JOC* **1983**, *48*, 134. (h) Nemoto, H.; Morizumi, M.; Nagai, M.; Fukumoto, K.; Kametani, T. *JCS(P1)* **1988**, 885.
9. Carlson, R. M.; Jones, R. W.; Hatcher, A. S. *TL* **1975**, 1741.
10. (a) Hoffmann, H. *CB* **1962**, *95*, 2563. (b) Cornforth, J.; Cornforth, R. H.; Gray, R. T. *JCS(P1)* **1982**, 2289. (c) Roberts, T. G.; Whitham, G. H. *JCS(P1)* **1985**, 1953. (d) Elliott, J.; Warren, S. *TL* **1986**, *27*, 645. (e) Vedejs, E.; Marth, C. *TL* **1987**, *28*, 3445.
11. Takahashi, H.; Tsubuki, T.; Higashiyama, K. *S* **1988**, 238.
12. Ishii, T.; Kawamura, N.; Matsubara, S.; Utimoto, K.; Kozima, S.; Hitomi, T. *JOC* **1987**, *52*, 4416.
13. Wakselman, C.; Nguyen, T. *JOC* **1977**, *42*, 565.
14. Keyaniyan, S.; Apel, M.; Richmond, J. P.; De Meijere, A. *AG(E)* **1985**, *24*, 770.

Masao Tsukazaki & Victor Snieckus
University of Waterloo, Ontario, Canada

Lithium Diisopropylamide[1,2]

[4111-54-0] $C_6H_{14}LiN$ (MW 107.15)

(hindered nonnucleophilic strong base used for carbanion generation, especially kinetic enolates,[4] α-heteroatom, allylic,[5,6] and aromatic and heteroaromatic carbanions;[7–9] unavoidable in organic synthesis)

Alternate Name: LDA.
Physical Data: powder, melts with decomposition; $pK_a = 35.7$ in THF.[3]
Solubility: sol Et_2O, THF, DME, HMPA; unstable above 0 °C in these solvents; stable in hexane or pentane (0.5–0.6 M) at rt for weeks, when not cooled or concentrated;[10] the complex with one molecule of THF is soluble in alkanes like cyclohexane and heptane.
Form Supplied in: solid; 2.0 M solution in heptane/THF/ethylbenzene stabilized with $Mg(i\text{-}Pr_2N)_2$; 10 wt % suspension in hexanes; LDA·THF complex, 1.5 M solution in cyclohexane and 2.0 M in heptane.
Analysis of Reagent Purity: several titration methods have been described.[120]
Preparative Methods: generally prepared directly before use from anhydrous ***Diisopropylamine*** and commercially available solutions of ***n-Butyllithium***. Another process, especially useful for the preparation of large quantities, is the reaction of styrene with 2 equiv of ***Lithium*** and 2 equiv of $i\text{-}Pr_2NH$ in Et_2O.[11]
Handling, Storage, and Precautions: very moisture- and air-sensitive, and should always be kept in an inert atmosphere; irritating to the skin and mucous membranes; therefore contact with skin, eyes, and other tissues and organs should be avoided; proper protection is necessary. Use in a fume hood.

Enolates. LDA is undeniably the first-choice base for the quantitative formation of enolates in general and kinetic (usually less substituted) enolates in particular. It was first used for this purpose by Hamell and Levine;[12] now it is a rare day (or night) that a chemist does not use or a graduate student does not propose LDA for a carbanionic transformation. Its advantages (fast, complete, and regiospecific enolate generation, unreactivity to alkyl halides, lack of interfering products) over earlier developed bases (e.g. ***Sodium Hydride***, ***Triphenylmethyllithium***)[13] are no longer noted. Except for enolates of aldehydes, which are very reactive and undergo self-aldol condensation reactions,[14] the derived lithio enolates can be used for a broad spectrum of reactions including *O*- and *C*-alkylations and acylations, transmetalation to other synthetically useful metallo enolates,[15] and a multitude of carbanion-based condensations and rearrangements. The trapping of kinetically derived lithio enolates with ***Chlorotrimethylsilane*** as silyl enol ethers[16] (e.g. eqs 1 and 2),[17] originally used for separation from minor regioisomers and regeneration (with ***Methyllithium*** or ***Lithium Amide***) for the purpose of regiospecific alkylation, has been superseded in general by methods of their direct alkylation under Lewis acid catalysis.[18]

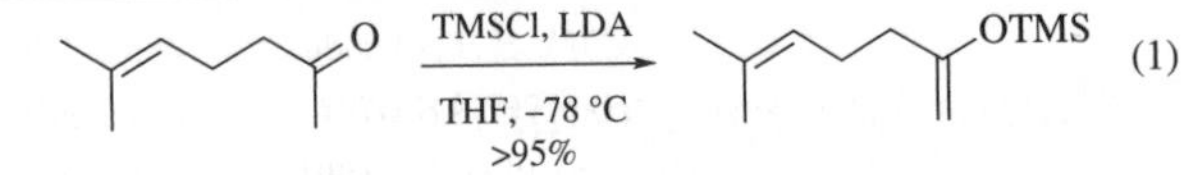

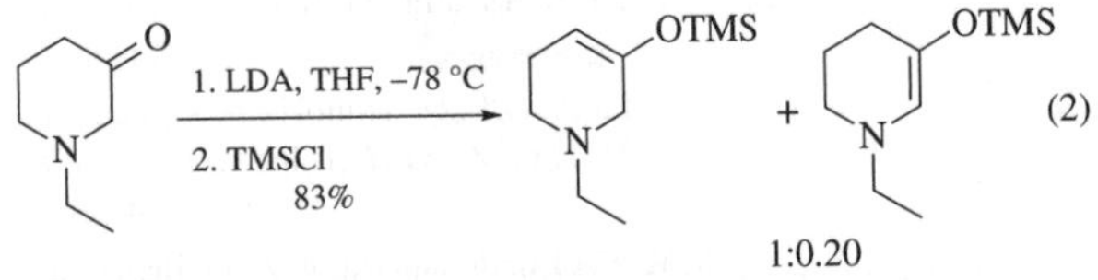

Consonant with its pK_a, LDA is used to derive a variety of α-stabilized carbanions, e.g. ketones (eqs 1 and 3),[19,20] α-amino ketones,[21] ω-bromo ketones (eq 4),[10] imines (eqs 6 and 7),[22–24] imino ethers (eq 8),[25] carboxylic acids (eq 9),[26] carboxylic esters,[27] lactones,[28] amides,[29] lactams,[30] imides,[31] and nitriles[32] which may be alkylated or used in aldol condensations. Enolates of methyl ketones may be *O*-alkylated with ***Diethyl Phosphorochloridate*** to give enol phosphates which, upon β-elimination, afford terminal alkynes in high yields (eq 5).[33]

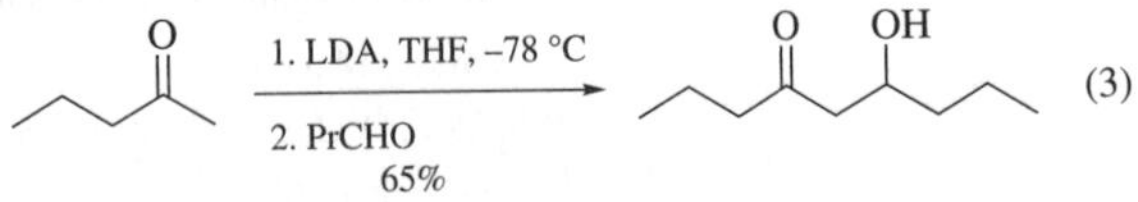

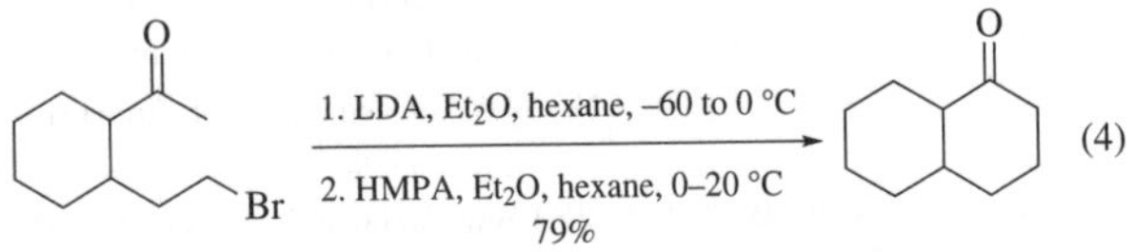

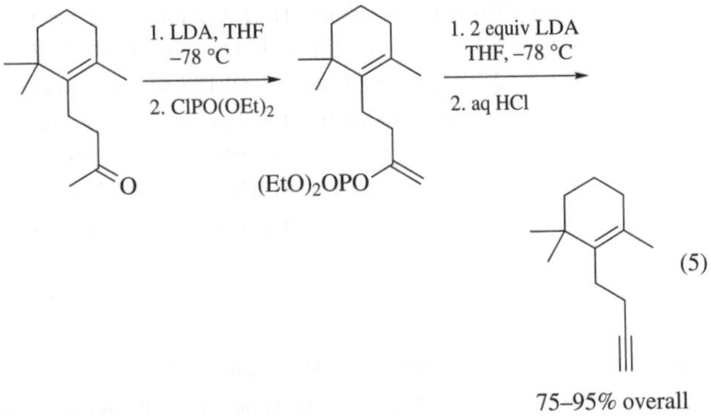

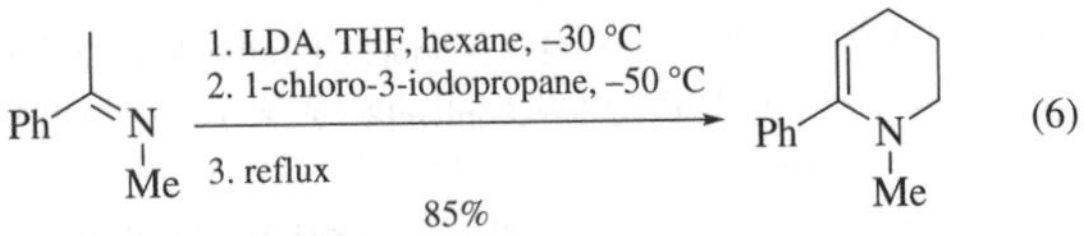

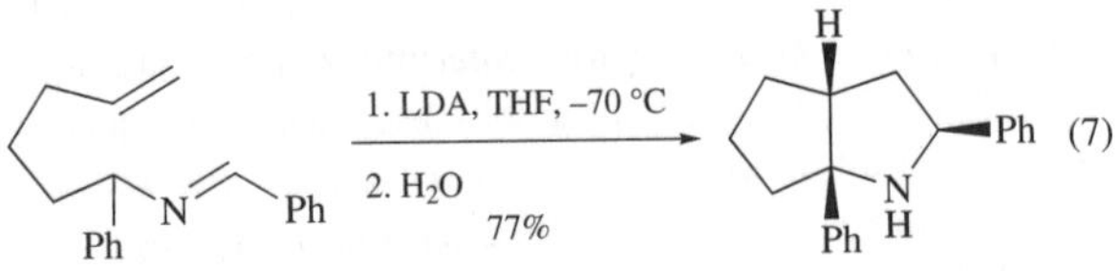

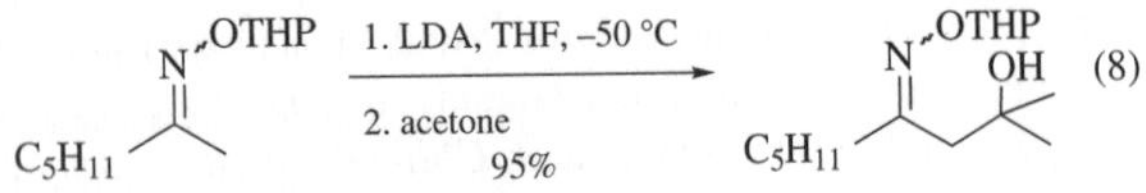

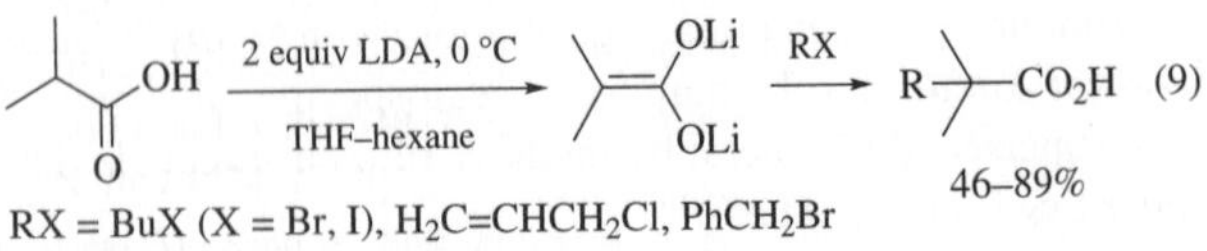

RX = BuX (X = Br, I), $H_2C{=}CHCH_2Cl$, $PhCH_2Br$

The (*Z*)/(*E*) stereoselectivity of enolate formation is dictated by the structure of the starting carbonyl compound and the base used for deprotonation. Compared to LDA, ***Lithium 2,2,6,6-Tetramethylpiperidide*** usually favors (*E*)-enolates whereas ***Lithium Hexamethyldisilazide*** preferentially leads to (*Z*)-enolates (eq 10).[34] With a *caveat* for any generalization, enolate configuration usually determines the stereochemical result in the product; for example, using a hindered ester and a bulky aldehyde combination, excellent stereoselectivities in aldol reactions are observed (eq 11).[27]

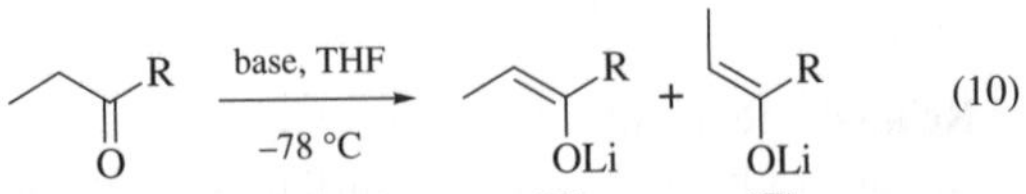

R = alkyl, alkoxy, NR'$_2$ (*Z*) (*E*)

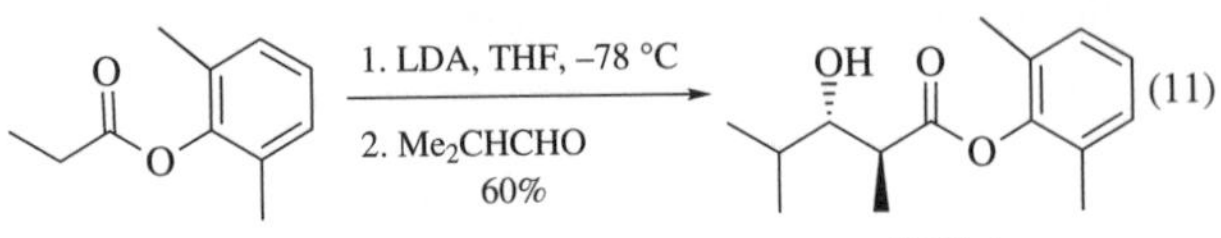

>96% de

A useful reaction of ketone enolates is their oxidative coupling,[35] e.g. in the formation of a tricyclic intermediate towards the synthesis of the diterpene cerorubenic acid-III (eq 12).[36]

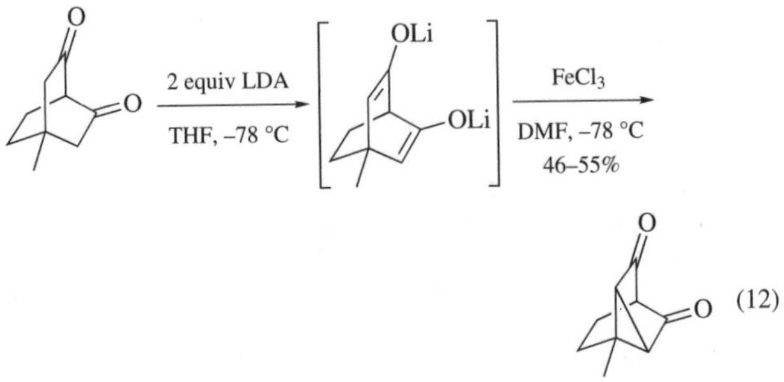

Alkylation of β-lactone enolates proceeds with high stereoselectivity dictated by strong stereocontrol from the C-4 R substituent (eq 13).[28] A similar result is obtained with the lithium enolates of diarylazetidinones which, on reaction with aldehydes, alkylate with high *cis* diastereoselectivity (eq 14).[30]

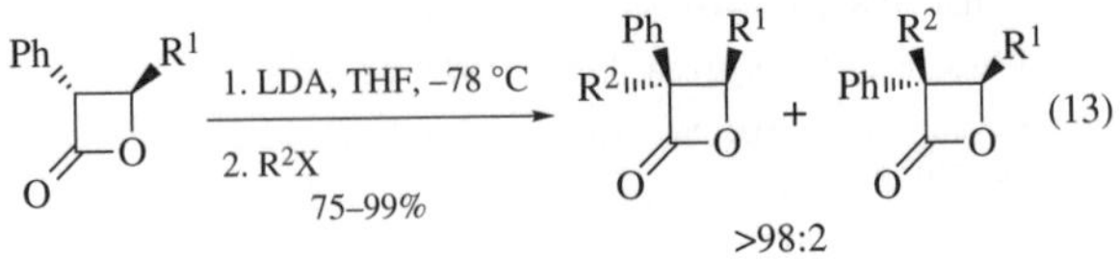

>98:2

R^1 = *t*-Bu, R^2 = Me, Et, CH_2Ph, CH_2CO_2Me
R^1 = *i*-Pr, R^2 = Me, $CH_2CH{=}CH_2$, CO_2Me, $CH_2C{\equiv}CH$, COPh

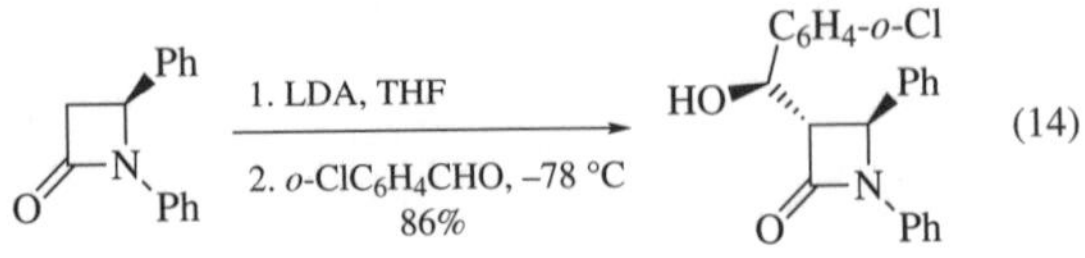

Sarcosine (*N*-methylglycine)-containing tripeptides and hexapeptides are poly-deprotonated in the presence of excess ***Lithium Chloride*** to give amide enolates which give *C*-alkylated sarcosine products with ***Iodomethane***, ***Allyl Bromide***, and ***Benzyl Bromide*** (eq 15).[29] With (*S*) configuration amino acids, the newly formed stereogenic center tends to have the (*R*) configuration.

α-Nitrile carbanions are generally useful in alkylation and other reactions with electrophiles, e.g. eq 16.[32]

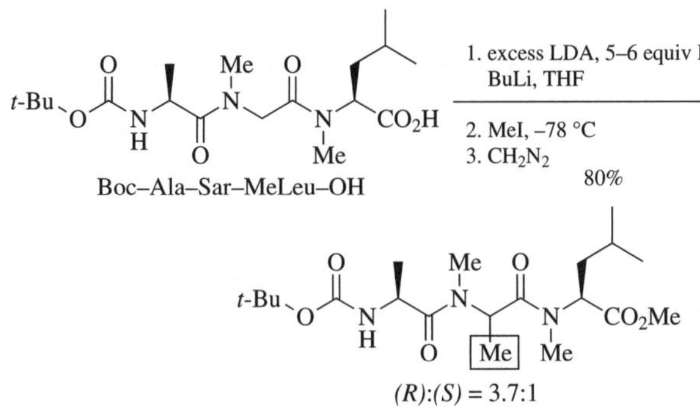

1. excess LDA, 5–6 equiv LiCl BuLi, THF
2. MeI, –78 °C
3. CH_2N_2
80%

Boc–Ala–Sar–MeLeu–OH (15)

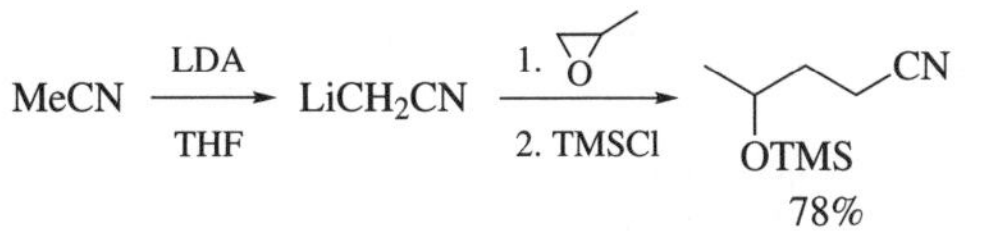

$(R):(S) = 3.7:1$

MeCN —LDA, THF→ $LiCH_2CN$ —1. epoxide; 2. TMSCl→ 78% (16)

α,β-Unsaturated Carbonyl Compounds. Of particular synthetic value is the generation of kinetic enolates of cyclic enones which may be alkylated (e.g. eq 17)[37] or undergo more demanding processes, e.g. double Michael addition (eq 18).[38] Imines of α,β-unsaturated aldehydes and ketones with δ-acidic hydrogens undergo clean deconjugative α-alkylation (eq 19).[23,39]

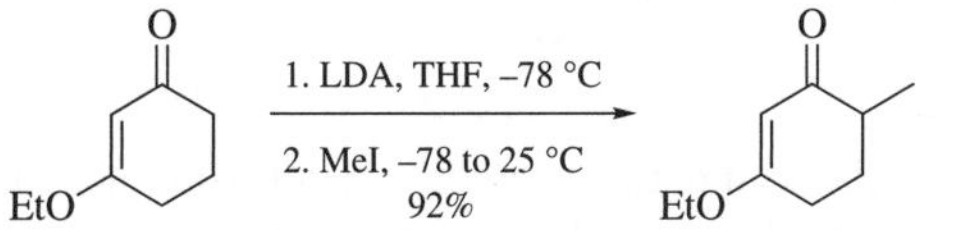

1. LDA, THF, –78 °C
2. MeI, –78 to 25 °C
92% (17)

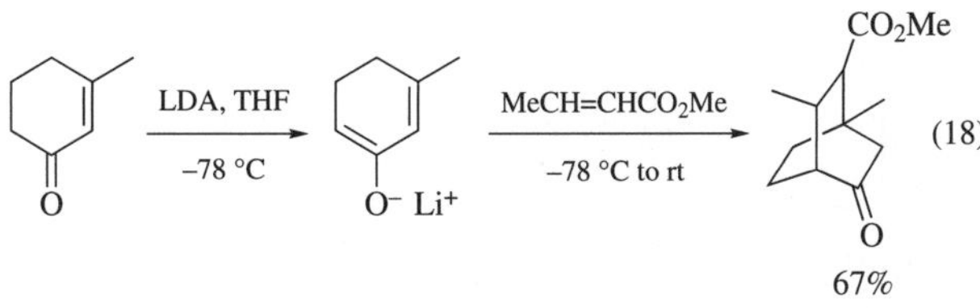

LDA, THF, –78 °C; $MeCH{=}CHCO_2Me$, –78 °C to rt; 67% (18)

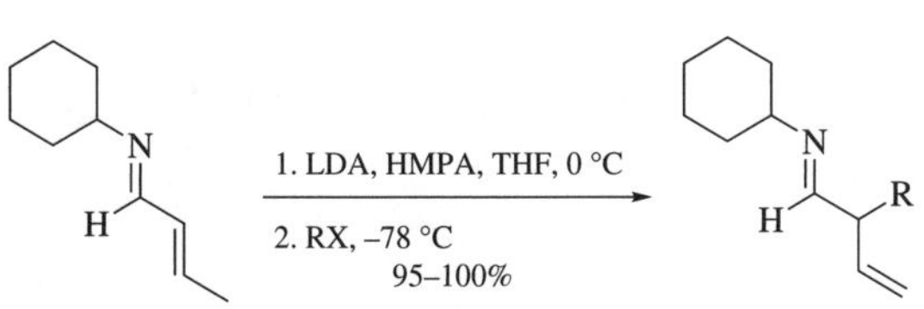

1. LDA, HMPA, THF, 0 °C
2. RX, –78 °C
95–100% (19)

R = Me, $H_2C{=}CHCH_2$, $HC{\equiv}CCH_2$, $PhCH_2$, *i*-Pr, Bu
X = I, Br

Copper dienolates of α,β-unsaturated acids, prepared by ***Copper(I) Iodide*** transmetalation, can be selectively γ-alkylated with allylic halides; the use of nonallylic electrophiles leads mainly to α-alkylation (eq 20).[40] Copper enolates of α,β-unsaturated esters[41] and amides[42] show similar behavior.

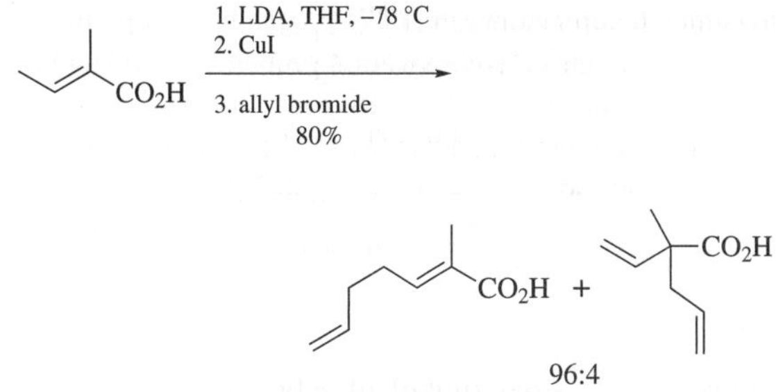

1. LDA, THF, –78 °C
2. CuI
3. allyl bromide
80%

96:4 (20)

Enolates with Chiral Auxiliaries. Enantioselective alkylation of carbonyl derivatives encompassing chiral auxiliaries constitutes an important synthetic process. The anions derived from aldehydes,[43] acyclic ketones,[44,45] and cyclic ketones with ***(S)-1-Amino-2-methoxymethylpyrrolidine*** (SAMP)[46] are used to obtain alkylated products in good to excellent yields and high enantioselectivity (e.g. eq 21).[46]

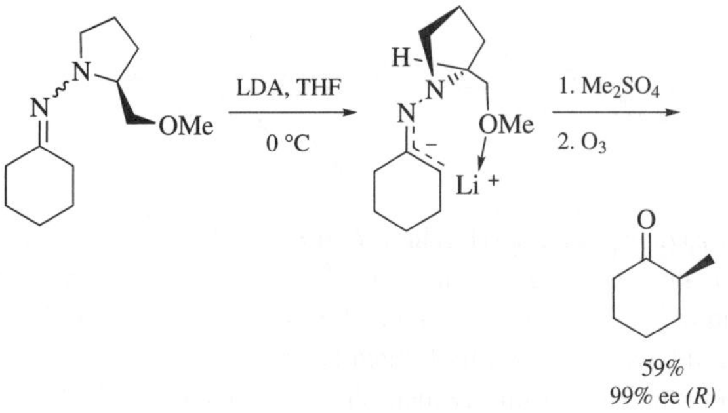

LDA, THF, 0 °C; 1. Me_2SO_4; 2. O_3

59%
99% ee (*R*) (21)

Metalation[47,48] of chiral oxazolines, derived from (1*S*,2*S*)-1-phenyl-2-amino-1,3-propanediol, followed by alkylation and hydrolysis, leads to optically active dialkylacetic acids, e.g. eq 22,[49] 2-substituted butyrolactones and valerolactones,[50] β-hydroxy and β-methoxy acids,[51] 2-hydroxy carboxylic acids,[52] and 3-substituted alkanoic acids (eq 23).[53]

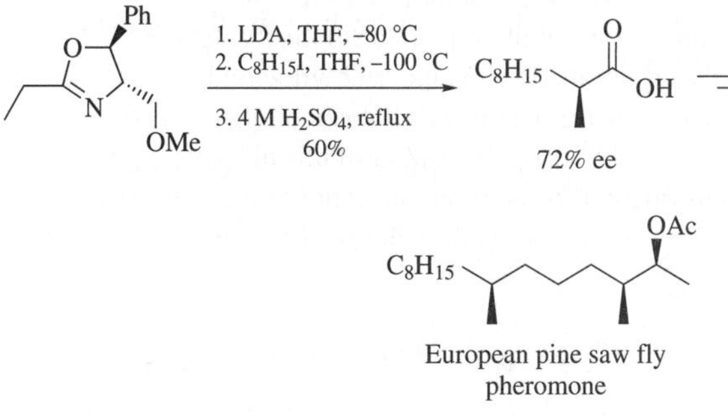

1. LDA, THF, –80 °C
2. $C_8H_{15}I$, THF, –100 °C
3. 4 M H_2SO_4, reflux
60%

72% ee

European pine saw fly pheromone (22)

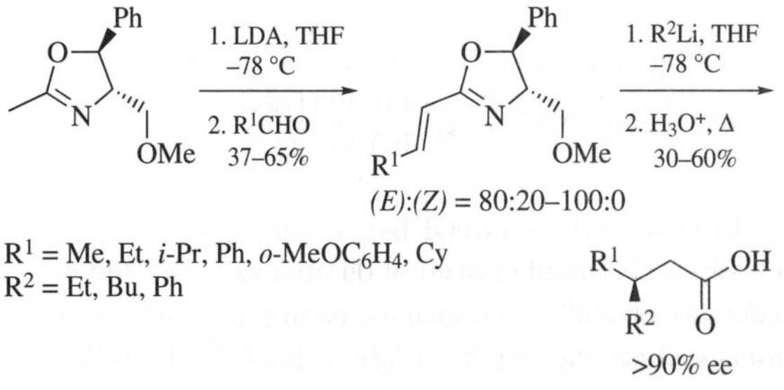

1. LDA, THF –78 °C
2. R^1CHO
37–65%

$(E):(Z) = 80:20–100:0$

1. R^2Li, THF –78 °C
2. H_3O^+, Δ
30–60%

R^1 = Me, Et, *i*-Pr, Ph, *o*-$MeOC_6H_4$, Cy
R^2 = Et, Bu, Ph

>90% ee (23)

Imide enolates derived from (*S*)-valinol and (1*S*,2*R*)-norephedrine and obtained by either LDA or ***Sodium Hexa-***

methyldisilazide deprotonation (eq 24)[54] exhibit complementary and highly diastereoselective alkylation properties. Mild and nondestructive removal of the chiral auxiliary to yield carboxylic acids, esters, or alcohols contributes to the significance of this protocol in small- and large-scale synthesis.[55,56]

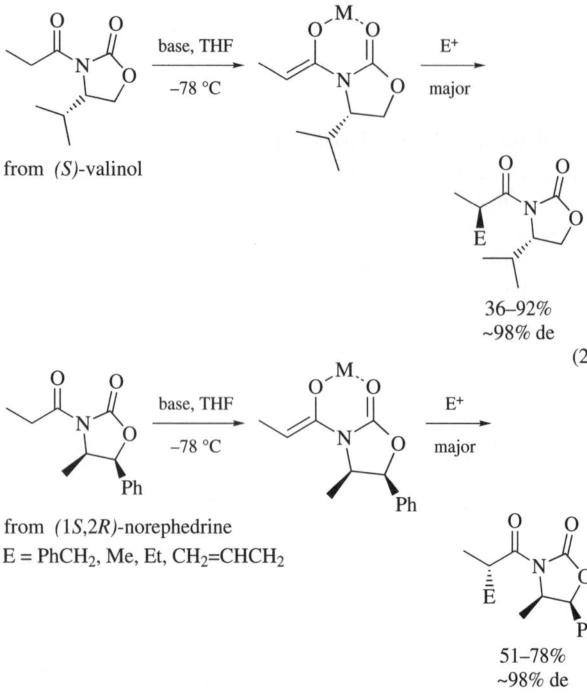

Metalated *t*-butyl 2-*t*-butyl-2,5-dihydro-4-methoxyimidazole-1-carboxylate is alkylated in good yields and with high *trans* diastereoselectivity (>99:1); hydrolysis of the resulting adducts liberates the α-amino acid methyl esters in high yields (eq 25).[57] Using this method, the (*S*)-*t*-butyl 2-*t*-butylimidazole derivative gave, upon isopropylation and hydrolysis, the L-valine methyl ester with 81% ee.

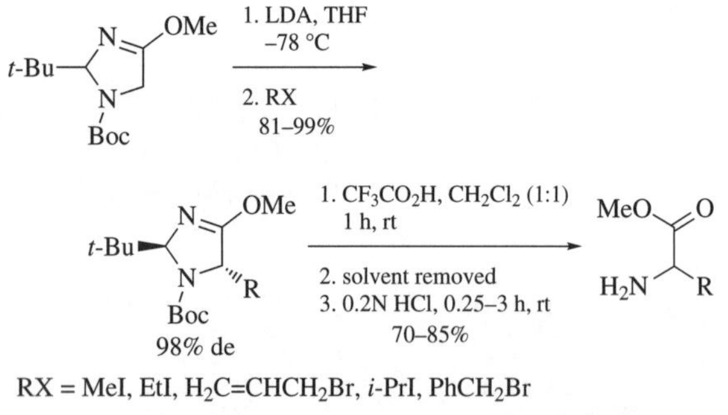

Alkylations and aldol condensations of aldehydes and ketones with enolates of chiral dioxolanes proceed generally with high diastereoselection, e.g. eq 26[58,59] and eq 27.[60] The magnesium enolate of *S*-(+)-2-acetoxy-1,1,2-triphenylethanol generated by transmetalation with ***Magnesium Bromide*** has enjoyed considerable success in aldol condensations, e.g. eq 28.[61]

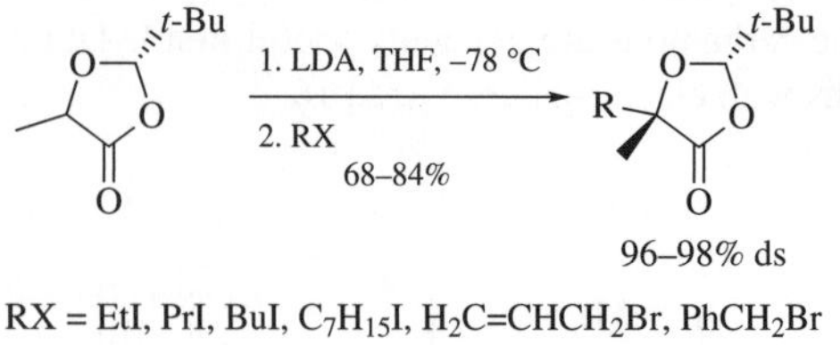

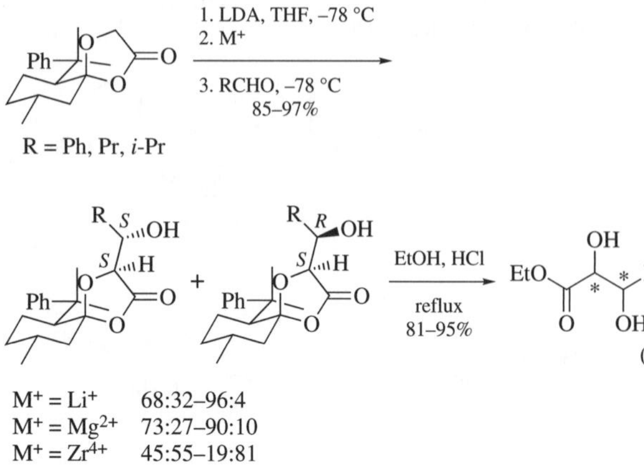

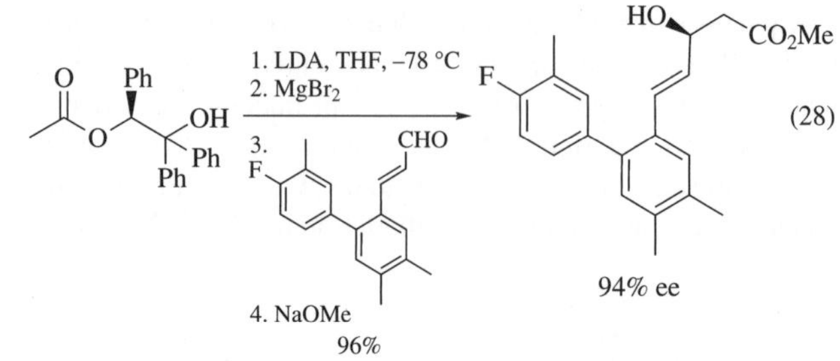

β-Lactams with high ee values[62] result from the condensation of lithium enolates of 10-diisopropylsulfonamide isobornyl esters with azadienes (eq 29).[63,64]

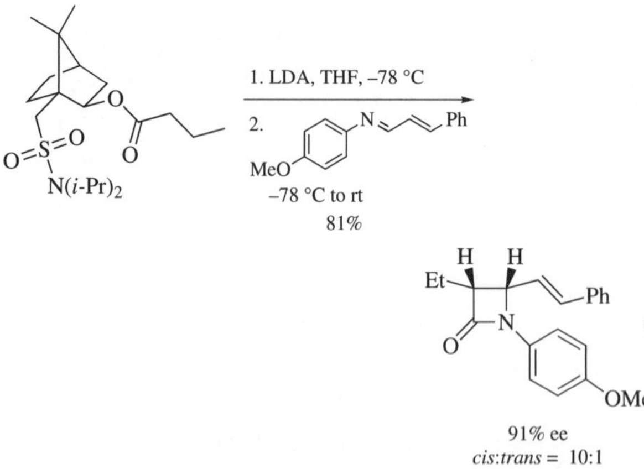

Enolate Rearrangements.

Ireland–Claisen Rearrangement.[65,66] Silyl enol ethers of allyl esters undergo highly stereoselective Ireland–Claisen rearrangement to afford 4-pentenoic acids.[67] The structure of the ester and the reaction conditions dictate the stereoselectivity. It has seen wide synthetic application, e.g. the construction of unnatural (−)-trichodiene (eq 30).[68]

1. 2.2 equiv LDA, TMSCl Et_3N, THF, −110 °C
2. reflux, 12 h
3. H_3O^+
75%

92:8 (30)

Wittig Rearrangements.[69] Enolates derived from chiral propargyloxyacetic acids lead to chiral allenic esters, via a rapid diastereoselective [2,3]-Wittig rearrangement followed by esterification (eq 31).[70] Analogous rearrangements are feasible with carbohydrate derivatives[71] and steroids (eq 32).[72] In some cases, improved enantioselectivities resulted when the lithium enolate was transmetalated with ***Dichlorobis(cyclopentadienyl)zirconium***.[73] Benzylic, allylic, propargylic, and related lithio derivatives of cyanohydrin ethers undergo the [2,3]-Wittig rearrangement to afford ketones, e.g. eq 33.[74,75] Lithiated *N*-benzyl- and *N*-allylazetidinones lead, via [1,2]-Wittig rearrangement exhibiting radical character, to pyrrolidones (eq 34).[76]

1. LDA, THF, −78 °C
2. CH_2N_2
80%

90% ee (*R*) → (*S*), (*R*) 86% de (31)

1. 2.3 equiv LDA THF, −56 °C
2. CH_2N_2
85%

>95% de (32)

LDA, −78 °C
THF–HMPA
60% (33)

LDA, THF
−78 °C
93–97% (34)

R = Ph, CH=CH_2

Stevens Rearrangement. Ammonium ylides undergo a stereoselective Stevens [2,3]-rearrangement when treated with LDA to afford *N*-methyl-2,3-disubstituted piperidines (eq 35).[77]

LDA, THF
65 °C
62–75%

R = Ph, OEt

only *cis* (35)

Heteroatom-Stabilized Carbanions. Heteroatom-stabilized and allylic carbanions serve as homoenolate anions and acyl anion equivalents,[5,78] e.g. α-anions of protected cyanohydrins of aldehydes and α,β-unsaturated aldehydes are intermediates in general syntheses of ketones and α,β-unsaturated ketones (eq 36).[79] Allylic anions of cyanohydrin ethers may be α-alkylated (eq 37)[80,81] or, if warmed to −25 °C, may undergo 1,3-silyl migration to cyanoenolates which may be trapped with TMSCl.[82] Metalated α-aminonitriles of aldehydes are used for the synthesis of ketones and enamines (eq 38).[83] Similarly, allylic anions from 2-morpholino-3-alkenenitriles undergo predominantly α-*C*-alkylation to give, after hydrolysis, α,β-unsaturated ketones (eq 39).[84]

1. LDA, THF, −78 °C
2. HMPA
3. RBr

1. 5% H_2SO_4 in MeOH
2. 0.5 N NaOH

(36)

R = Bu, Hex, Dec	80–85%
R = *i*-Pr, *c*-Pent, *c*-Hex	41–80%
R = H_2C=$CHCH_2$	76%

1. LDA, THF, −78 °C
2. BuBr, −78 °C to rt
73%

1. Et_3N•HF
2. NaOH
75%

(37)

α-Sulfoxide- and sulfone-stabilized carbanions are highly useful synthetic intermediates, e.g. α-lithiated (+)-methyl *p*-tolyl (*R*)-sulfoxide reacts with α,β-unsaturated aldehydes to give, after dehydration, enantiomerically pure 1-[*p*-tolylsulfinyl]-1,3-butadienes (eq 40),[85] metalation of phenyl 3-tosyloxybutyl sulfone leads to efficient cyclopropane ring formation (eq 41),[86] and precursors for the Ramberg–Bäcklund rearrangement are prepared via α-lithiated cyclic sulfones (eq 42).[87]

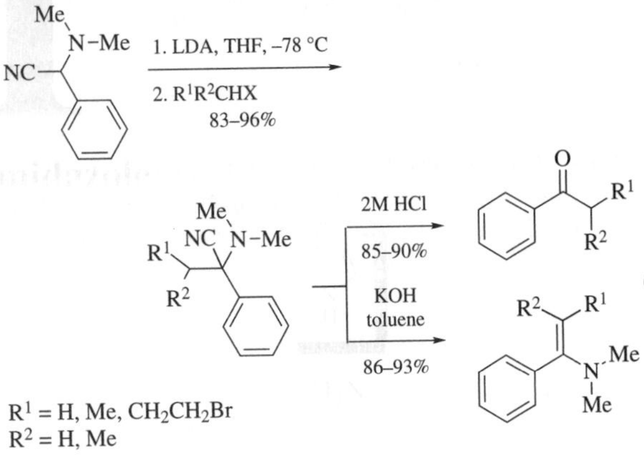

R^1 = H, Me, CH_2CH_2Br
R^2 = H, Me

(38)

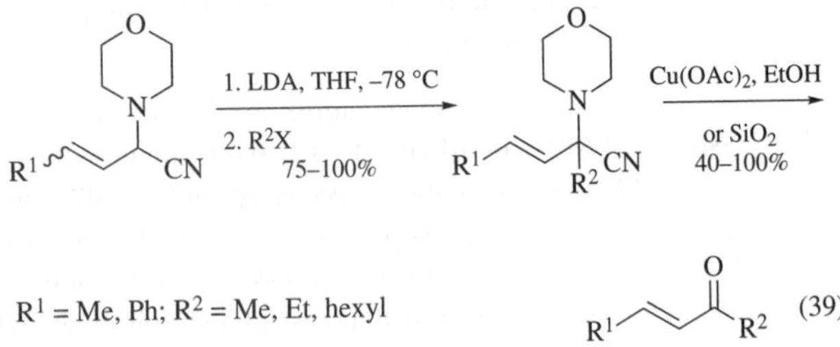

R^1 = Me, Ph; R^2 = Me, Et, hexyl

(39)

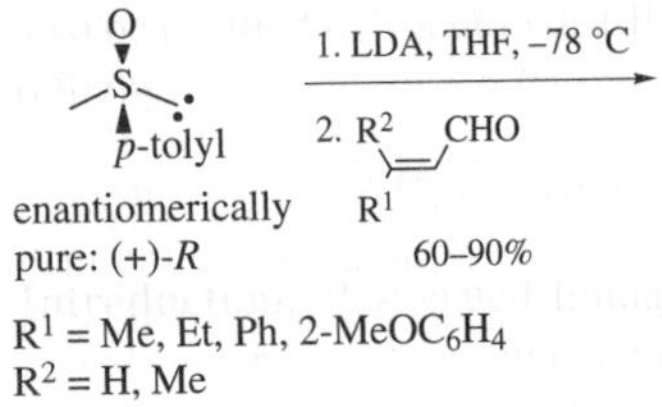

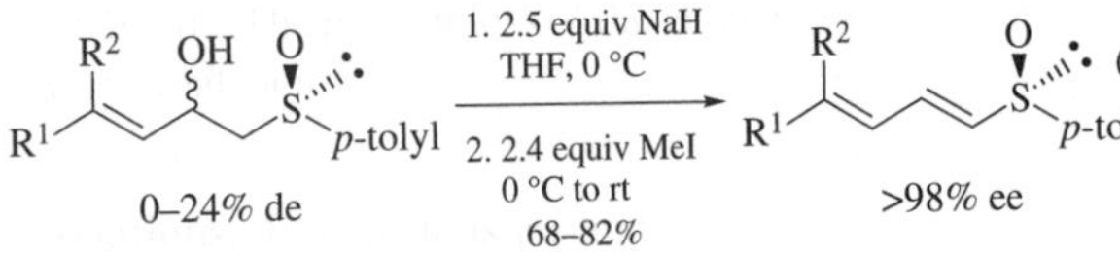

enantiomerically pure: (+)-*R*
R^1 = Me, Et, Ph, 2-MeOC$_6$H$_4$
R^2 = H, Me

(40)

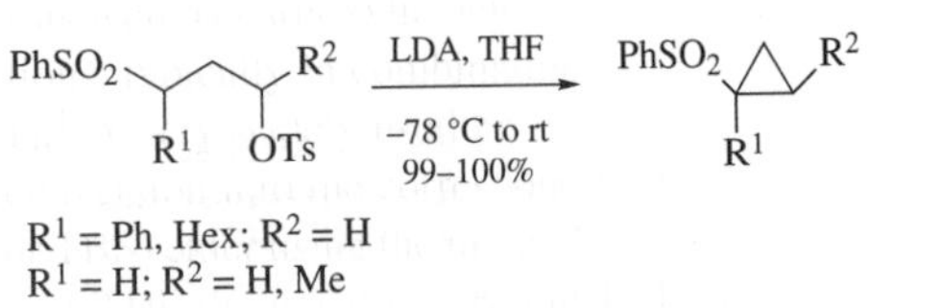

(41)

R^1 = Ph, Hex; R^2 = H
R^1 = H; R^2 = H, Me

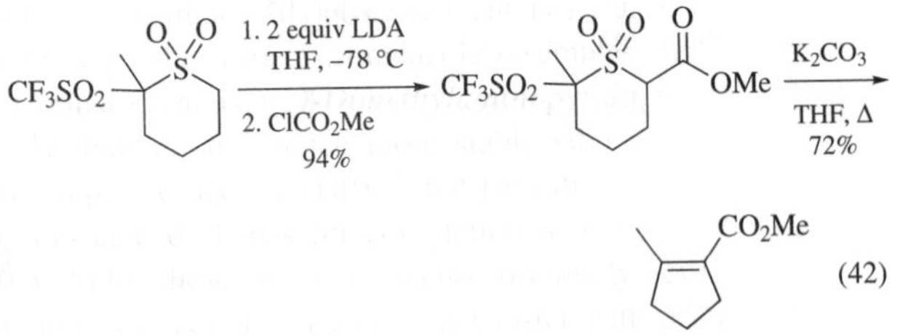

(42)

Lithiated allylic sulfoxides may be α-alkylated and the resulting products subjected to [2,3]-sigmatropic rearrangement induced by a thiophile to give allylic alcohols (eq 43).[6,88] In contrast, alkenyl aryl sulfoxides produce α-lithiated species which are alkylated with MeI or PhCHO in good yields (eq 44).[89,90] LDA has also been used to metalate allylic[91] and propargylic selenides[92,93] as well as aryl vinyl selenides.[94]

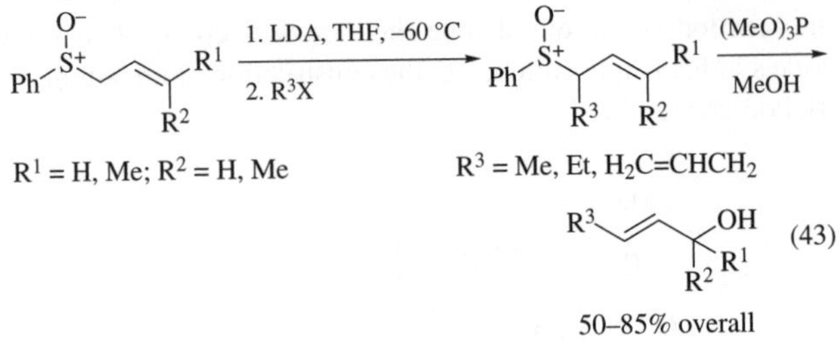

R^1 = H, Me; R^2 = H, Me
R^3 = Me, Et, $H_2C{=}CHCH_2$

(43)

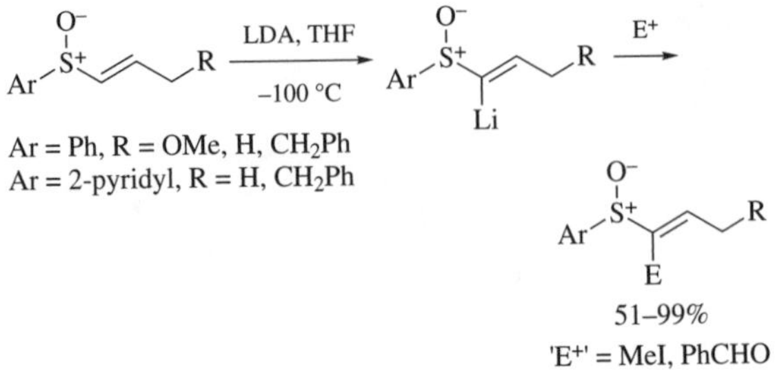

50–85% overall

Ar = Ph, R = OMe, H, CH_2Ph
Ar = 2-pyridyl, R = H, CH_2Ph

(44)

51–99%
'E$^+$' = MeI, PhCHO

Aromatic and Heteroaromatic Metalations.

Aromatics. The utility of LDA for kinetic deprotonation of aromatic substrates is compromised by its insufficient basicity compared to the alkyllithiums.[8] Selective lateral deprotonation may be achieved which complements the *ortho*-metalation process (eq 45).[95]

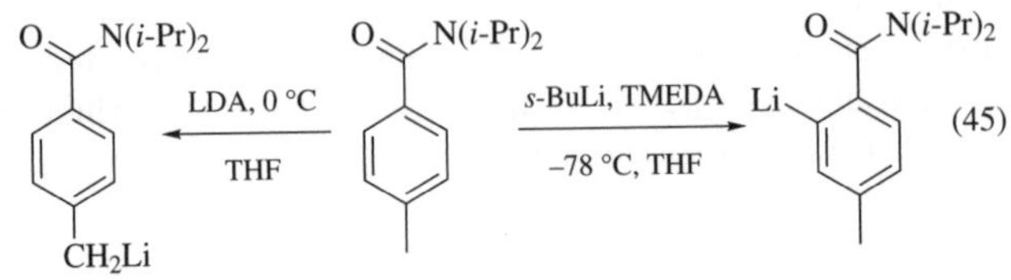

(45)

Of considerable synthetic utility is the LDA-induced conversion of bromobenzenes into benzynes. Thus low-temperature deprotonation of *m*-alkoxyaryl bromides followed by warming to rt in the presence of furan gives cycloadducts (eq 46);[96] the lithio species may also be trapped with CO_2 at −78 °C to give unusual aromatic substitution patterns. Treatment of aryl triflates with LDA leads to *N,N*-isopropylanilines in good yield (67–98%).[97]

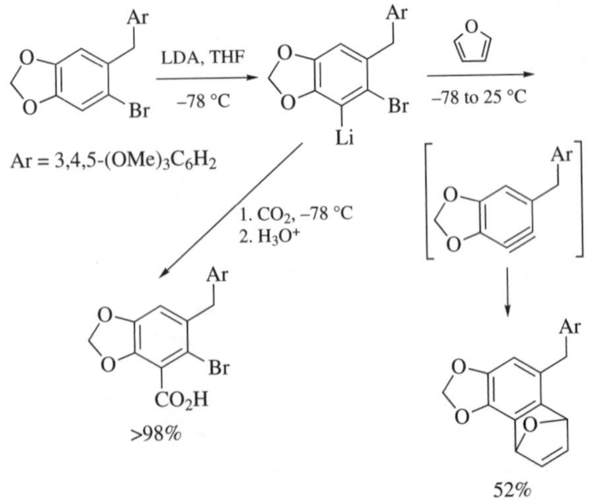

Ar = 3,4,5-(OMe)$_3$C$_6$H$_2$

(46)

Remote metalation of biaryls and *m*-teraryls provides a general synthesis of substituted and condensed fluorenones (eq 47).[98] Similarly, biaryl *O*-carbamates undergo remote metalation and anionic Fries rearrangement to 2-hydroxy-2′-carboxamidobiaryls, which are efficiently transformed into dibenzo[*b,d*]pyranones.[99]

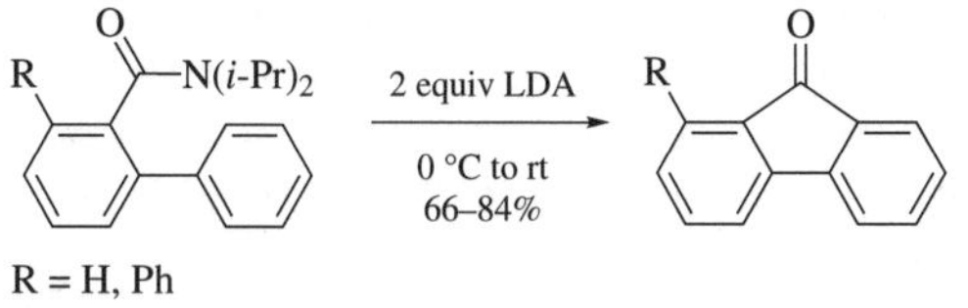

(47) R = H, Ph; 2 equiv LDA, 0 °C to rt, 66–84%

Lateral Lithiation.[7] Laterally metalated *o*-toluic acids (eq 48),[100] esters (eq 49),[101] and amides (eq 50)[102] are important intermediates for chain extension and carbo- and hetero-ring annulation. Remote lateral lithiation of 2-methyl-2′-carboxaminobiaryls constitutes a general regiospecific synthesis of 9-phenanthrols (eq 51).[103–105]

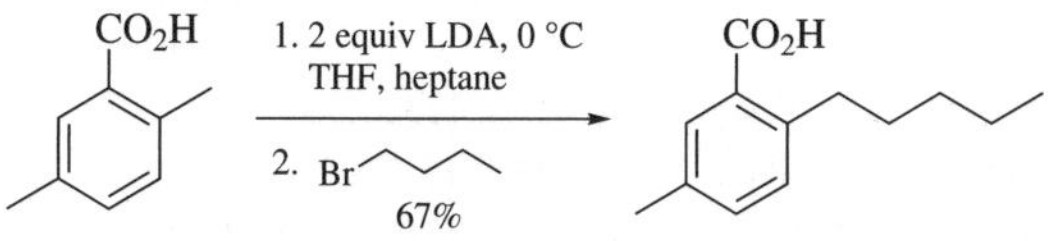

(48) 1. 2 equiv LDA, 0 °C, THF, heptane; 2. Br⁀⁀; 67%

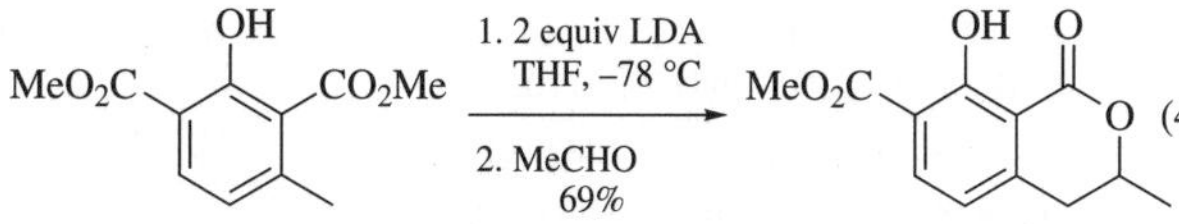

(49) 1. 2 equiv LDA, THF, –78 °C; 2. MeCHO; 69%

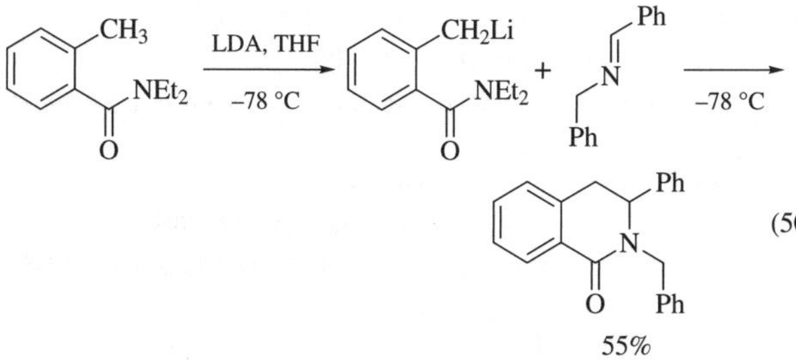

(50) LDA, THF, –78 °C; –78 °C; 55%

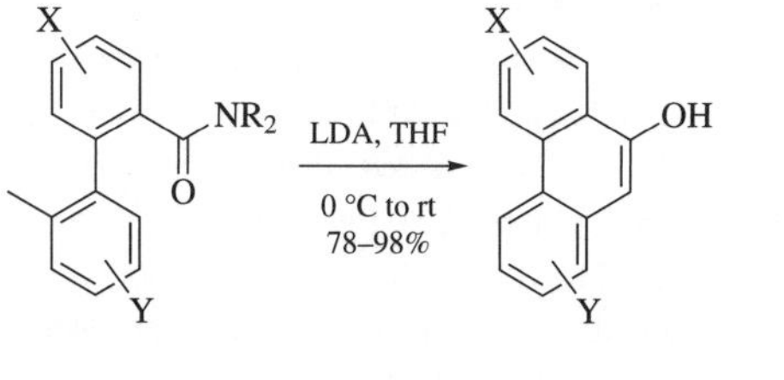

(51) LDA, THF, 0 °C to rt, 78–98%

α-Lithiated *o*-tolyl isocyanides may be alkylated and, following a second metalation with LiTMP, converted into 3-substituted indoles, e.g. eq 52.[106] Michael addition of α-lithiated 3-sulfonyl- and 3-cyanophthalides to α,β-unsaturated ketones and esters initiates a regioselective aromatic ring annulation process to give naphthalenes, e.g. eq 53.[107] Lithiated 3-cyanophthalides also undergo reaction with in situ-generated arynes to give anthraquinones (eq 54).[108] Similarly, aza-anthraquinones are obtained using 3-bromopyridine as the 3,4-pyridyne precursor.[109]

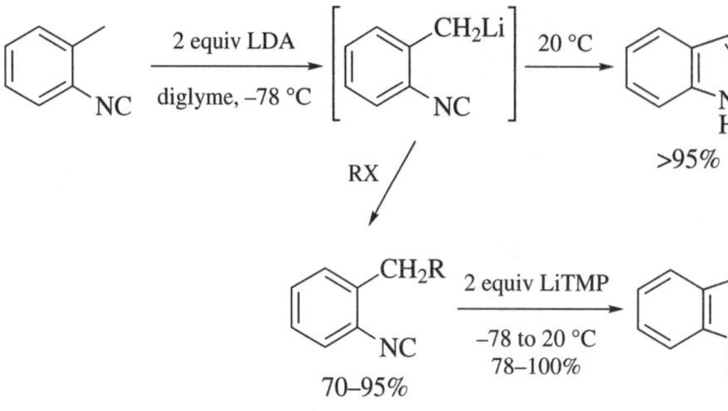

(52) 2 equiv LDA, diglyme, –78 °C; 20 °C; >95%; RX; 70–95%; 2 equiv LiTMP, –78 to 20 °C, 78–100%

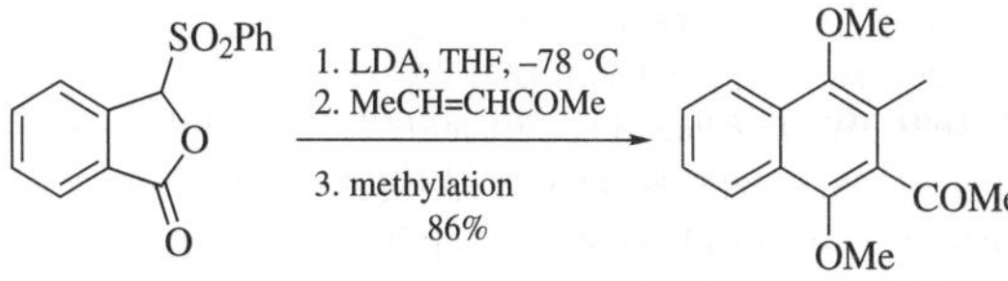

(53) 1. LDA, THF, –78 °C; 2. MeCH=CHCOMe; 3. methylation; 86%

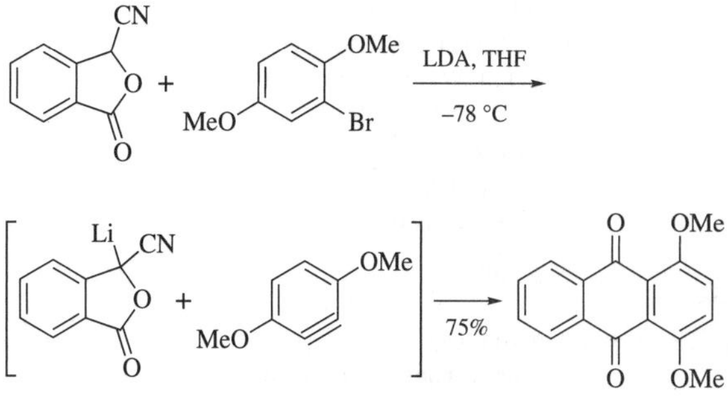

(54) LDA, THF, –78 °C; 75%

Heteroaromatics.[9] LDA (and LiTMP) are advantageous bases for the directed *ortho*-metalation chemistry of pyridines[110,111] and, to a much lesser extent, pyrimidines.[112] Several halopyridines can be lithiated *ortho* to the halo substituent and trapped with various electrophiles (eq 55).[113]

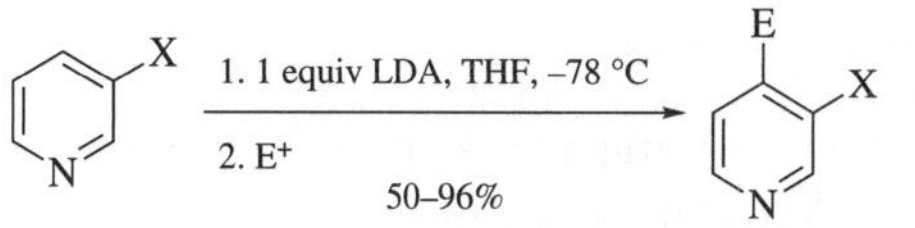

(55) 1. 1 equiv LDA, THF, –78 °C; 2. E^+; 50–96%

X = F, Cl, Br; 'E^+' = TMSCl, I_2, $(PhS)_2$, PhCHO, Ph_2CO

Lithiated furan- and thiophenecarboxylic acids (eq 56),[114,115] indoles (eq 57),[116] indole-3-carboxylic acids (eq 58),[117] benzofurancarboxylic acids,[118] and thiazole- and oxazolecarboxylic acids[119] serve well in heteroaromatic synthesis.

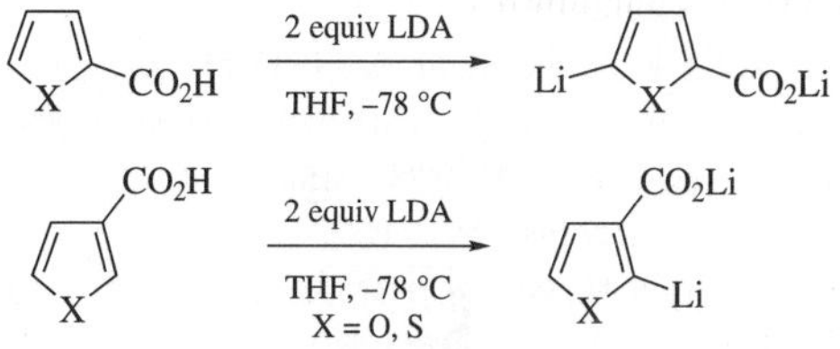

(56) 2 equiv LDA, THF, –78 °C; 2 equiv LDA, THF, –78 °C, X = O, S

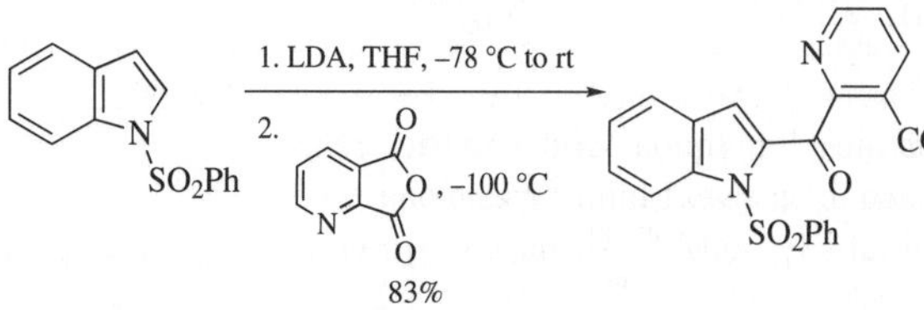

(57) 1. LDA, THF, –78 °C to rt; 2. –100 °C; 83%

$$\text{1-methylindole-3-CO}_2\text{H} \xrightarrow[\text{THF, -78 °C}]{\text{2 equiv LDA}} \text{2-lithio-1-methylindole-3-CO}_2\text{Li} \xrightarrow{E^+} \text{2-E-1-methylindole-3-CO}_2\text{H} \quad (58)$$

40–90%

E = TMS, Me, PhCH(OH), PrCH(OH), Ph_2COH
PhC(OH)Me, Me_2COH

Related Reagents. Lithium Amide; Lithium Diethylamide; Lithium Hexamethyldisilazide; Lithium Piperidide; Lithium Pyrrolidide; Lithium 2,2,6,6-Tetramethylpiperidide; Potassium Diisopropylamide.

1. Reed, F.; Rathman, T. L. *Spec. Chem.* **1989**, *9*, 174.
2. Rathman, T. L. *Spec. Chem.* **1989**, *9*, 300.
3. Fraser, R. R.; Mansour, T. S. *JOC* **1984**, *49*, 3442.
4. Caine, D. *COS* **1991**, *3*, 1.
5. Yamamoto, Y. *COS* **1991**, *2*, 55.
6. Evans, D. A.; Andrews, G. C. *ACR* **1974**, *7*, 147.
7. Clark, R. D.; Jahangir, A. *OR* **1995**, submitted.
8. Snieckus, V. *CRV* **1990**, *90*, 879.
9. Queguiner, G.; Marsais, F.; Snieckus, V.; Epsztajn, J. *Adv. Heterocyclic Chem.* **1991**, *52*, 187.
10. House, H. O.; Phillips, W. V.; Sayer, T. S. B.; Yau, C. C. *JOC* **1978**, *43*, 700.
11. Reetz, M. T.; Maier, W. F. *LA* **1980**, 1471.
12. Hamell, M.; Levine, R. *JOC* **1950**, *15*, 162.
13. House, H. O.; Czuba, L. J.; Gall, M.; Olmstead, H. D. *JOC* **1969**, *34*, 2324.
14. Mekelburger, H. B.; Wilcox, C. S. *COS* **1991**, *2*, 99.
15. Evans, D. A. In *Asymmetric Synthesis*; Morrison, J. D., Ed.; Academic: New York, 1984; Vol. 3, p 1.
16. Brownbridge, P. *S* **1983**, 1.
17. Corey, E. J.; Gross, A. W. *TL* **1984**, *25*, 495.
18. Mukaiyama, T. *OR* **1982**, *28*, 203.
19. d'Angelo, J. *T* **1976**, *32*, 2979.
20. Stork, G.; Kraus, G. A.; Garcia, G. A. *JOC* **1974**, *39*, 3459.
21. Garst, M. E.; Bonfiglio, J. N.; Grudoski, D. A.; Marks, J. *JOC* **1980**, *45*, 2307.
22. Evans, D. A. *JACS* **1970**, *92*, 7593.
23. Whitesell, J. K.; Whitesell, M. A. *S* **1983**, 517.
24. Pearson, W. H.; Walters, M. A.; Oswell, K. D. *JACS* **1986**, *108*, 2769.
25. Ensley, H. E.; Lohr, R. *TL* **1978**, 1415.
26. Creger, P. L. *JACS* **1967**, *89*, 2500.
27. Montgomery, S. H.; Pirrung, M. C.; Heathcock, C. H. *OS* **1985**, *63*, 99; *OSC* **1990**, *7*, 190.
28. Mulzer, J.; Kerkmann, T. *JACS* **1980**, *102*, 3620.
29. Seebach, D.; Bossler, H.; Gründler, H.; Shoda, S. *HCA* **1991**, *74*, 197.
30. Otto, H. H.; Mayrhofer, R.; Bergmann, H. J. *LA* **1983**, 1152.
31. Garratt, P. J.; Hollowood, F. *JOC* **1982**, *47*, 68.
32. Murata, S.; Matsuda, I. *S* **1978**, 221.
33. Negishi, E.; King, A. O.; Klima, W. L.; Patterson, W.; Silveira, A. *JOC* **1980**, *45*, 2526.
34. Heathcock, C. H.; Buse, C. T.; Kleschick, W. A.; Pirrung, M. C.; Sohn, J. E.; Lampe, J. *JOC* **1980**, *45*, 1066.
35. Frazier, R. H., Jr.; Harlow, R. L. *JOC* **1980**, *45*, 5408.
36. Paquette, L. A.; Poupart, M.-A. *JOC* **1993**, *58*, 4245.
37. Kende, A. S.; Fludzinski, P. *OS* **1986**, *64*, 68; *OSC* **1990**, *7*, 208.
38. Spitzner, D.; Engler, A. *OS* **1988**, *66*, 37; *OSC* **1993**, *8*, 219.
39. Kieczykowski, G. R.; Schlessinger, R. H.; Sulsky, R. B. *TL* **1976**, 597.
40. Savu, P. M.; Katzenellenbogen, J. A. *JOC* **1981**, *46*, 239.
41. Katzenellenbogen, J. A.; Crumrine, A. L. *JACS* **1974**, *96*, 5662.
42. Majewski, M.; Mpango, G. B.; Thomas, M. T.; Wu, A.; Snieckus, V. *JOC* **1981**, *46*, 2029.
43. Enders, D. In *Current Trends in Organic Synthesis*; Nozaki, H., Ed.; Pergamon: New York, 1983; p 151.
44. Enders, D.; Eichenauer, H. *AG(E)* **1979**, *18*, 397.
45. Enders, D.; Baus, U. *LA* **1983**, 1439.
46. Enders, D. In *Asymmetric Synthesis*; Morrison, J. D., Ed.; Academic: New York, 1984; Vol. 3, p 275.
47. Lutomski, K. A.; Meyers, A. I. In *Asymmetric Synthesis*; Morrison, J. D., Ed.; Academic: New York, 1984; Vol. 3, p 213.
48. Meyers, A. I.; Knaus, G.; Kamata, K. *JACS* **1974**, *96*, 268.
49. Byström, S.; Högberg, H.-E.; Norin, T. *T* **1981**, *37*, 2249.
50. Meyers, A. I.; Yamamoto, Y.; Mihelich, E. D.; Bell, R. A. *JOC* **1980**, *45*, 2792.
51. Meyers, A. I.; Knaus, G. *TL* **1974**, 1333.
52. Meyers, A. I.; Slade, J. *JOC* **1980**, *45*, 2785.
53. Meyers, A. I.; Smith, R. K.; Whitten, C. E. *JOC* **1979**, *44*, 2250.
54. Evans, D. A.; Ennis, M. D.; Mathre, D. J. *JACS* **1982**, *104*, 1737.
55. Evans, D. A.; Takacs, J. M.; McGee, L. R.; Ennis, M. D.; Mathre, D. J.; Bartroli, J. *PAC* **1981**, *53*, 1109.
56. Evans, D. A. *Aldrichim. Acta* **1982**, *15*, 23.
57. Blank, S.; Seebach, D. *AG(E)* **1993**, *32*, 1765.
58. Seebach, D.; Naef, R. *HCA* **1981**, *64*, 2704.
59. Seebach, D.; Naef, R.; Calderari, G. *T* **1984**, *40*, 1313.
60. Pearson, W. H.; Cheng, M.-C. *JOC* **1987**, *52*, 3176.
61. Lynch, J. E.; Volante, R. P.; Wattley, R. V.; Shinkai, I. *TL* **1987**, *28*, 1385.
62. Hart, D. J.; Lee, C.-S. *JACS* **1986**, *108*, 6054.
63. Oppolzer, W.; Chapuis, C.; Bernardinelli, G. *TL* **1984**, *25*, 5885.
64. Oppolzer, W.; Dudfield, P.; Stevenson, T.; Godel, T. *HCA* **1985**, *68*, 212.
65. Wipf, P. *COS* **1991**, *5*, 827.
66. Ireland, R. E.; Mueller, R. H. *JACS* **1972**, *94*, 5897.
67. Ireland, R. E.; Willard, A. K. *TL* **1975**, 3975.
68. Gilbert, J. C.; Selliah, R. D. *JOC* **1993**, *58*, 6255.
69. Mikami, K.; Nakai, T. *S* **1991**, 594.
70. Marshall, J. A.; Wang, X. *JOC* **1990**, *55*, 2995.
71. Kakinuma, K.; Li, H.-Y. *TL* **1989**, *30*, 4157.
72. Koreeda, M.; Ricca, D. J. *JOC* **1986**, *51*, 4090.
73. Uchikawa, M.; Katsuki, T.; Yamaguchi, M. *TL* **1986**, *27*, 4581.
74. Cazes, B.; Julia, S. *SC* **1977**, *7*, 273.
75. Cazes, B.; Julia, S. *SC* **1977**, *7*, 113.
76. Durst, T.; Van Den Elzen, R.; LeBelle, M. J. *JACS* **1972**, *94*, 9261.
77. Neeson, S. J.; Stevenson, P. J. *TL* **1988**, *29*, 3993.
78. Krief, A. *COS* **1991**, *3*, 85.
79. Stork, G.; Maldonado, L. *JACS* **1971**, *93*, 5286.
80. Jacobson, R. M.; Lahm, G. P.; Clader, J. W. *JOC* **1980**, *45*, 395.
81. Hertenstein, U.; Hünig, S.; Öller, M. *CB* **1980**, *113*, 3783.
82. Hertenstein, U.; Hünig, S.; Reichelt, H.; Schaller, R. *CB* **1982**, *115*, 261.

83. Ahlbrecht, H.; Raab, W.; Vonderheid, C. *S* **1979**, 127.
84. Takahashi, K.; Honma, A.; Ogura, K.; Iida, H. *CL* **1982**, 1263.
85. Solladié, G.; Ruiz, P.; Colobert, F.; Carreño, M. C.; Garcia-Ruano, J. L. *S* **1991**, 1011.
86. Chang, Y. H.; Pinnick, H. W. *JOC* **1978**, *43*, 373.
87. Hendrickson, J. B.; Boudreaux, G. J.; Palumbo, P. S. *JACS* **1986**, *108*, 2358.
88. Evans, D. A.; Andrews, G. C.; Fujimoto, T. T.; Wells, D. *TL* **1973**, 1385.
89. Okamura, H.; Mitsuhira, Y.; Miura, M.; Takei, H. *CL* **1978**, 517.
90. Posner, G. H.; Tang, P.; Mallamo, J. P. *TL* **1978**, 3995.
91. Reich, H. J.; Clark, M. C.; Willis, W. W., Jr. *JOC* **1982**, *47*, 1618.
92. Reich, H. J.; Shah, S. K. *JACS* **1977**, *99*, 263.
93. Reich, H. J.; Shah, S. K.; Gold, P. M.; Olson, R. E. *JACS* **1981**, *103*, 3112.
94. Reich, H. J.; Willis, W. W., Jr.; Clark, P. D. *JOC* **1981**, *46*, 2775.
95. Beak, P.; Brown, R. A. *JOC* **1982**, *47*, 34.
96. Jung, M. E.; Lowen, G. T. *TL* **1986**, *27*, 5319.
97. Wickham, P. P.; Hazen, K. H.; Guo, H.; Jones, G.; Hardee Reuter, K.; Scott, W. J. *JOC* **1991**, *56*, 2045.
98. Fu, J. M.; Zhao, B. P.; Sharp, M. J.; Snieckus, V. *JOC* **1991**, *56*, 1683.
99. Wang, W.; Snieckus, V. *JOC* **1992**, *57*, 424.
100. Creger, P. L. *JACS* **1970**, *92*, 1396.
101. Kraus, G. A. *JOC* **1981**, *46*, 201.
102. Clark, R. D.; Jahangir *JOC* **1987**, *52*, 5378.
103. Sharp, M. J.; Cheng, W.; Snieckus, V. *TL* **1987**, *28*, 5093.
104. Fu, J.; Sharp, M. J.; Snieckus, V. *TL* **1988**, *29*, 5459.
105. Fu, J.; Zhao, B.; Sharp, M. J.; Snieckus, V. *JOC* **1991**, *56*, 1683.
106. Ito, Y.; Kobayashi, K.; Saegusa, T. *JACS* **1977**, *99*, 3532.
107. Hauser, F. M.; Rhee, R. P. *JOC* **1978**, *43*, 178.
108. Khanapure, S. P.; Reddy, R. T.; Biehl, E. R. *JOC* **1987**, *52*, 5685.
109. Khanapure, S. P.; Biehl, E. R. *H* **1988**, *27*, 2643.
110. Marsais, F.; Trécourt, F.; Bréant, P.; Quéguiner, G. *JHC* **1988**, *25*, 81.
111. Rocca, P.; Cochennec, C.; Marsais, F.; Thomas-dit-Dumont, L.; Mallet, M.; Godard, A.; Quéguiner, G. *JOC* **1993**, *58*, 7832.
112. Plé, N.; Turck, A.; Martin, P.; Barbey, S.; Quéguiner, G. *TL* **1993**, *34*, 1605.
113. Gribble, G. W.; Saulnier, M. G. *TL* **1980**, *21*, 4137.
114. Knight, D. W.; Nott, A. P. *JCS(P1)* **1981**, 1125.
115. Knight, D. W.; Nott, A. P. *JCS(P1)* **1983**, 791.
116. Gribble, G. W.; Fletcher, G. L.; Ketcha, D. M.; Rajopadhye, M. *JOC* **1989**, *54*, 3264.
117. Buttery, C. D.; Jones, R. G.; Knight, D. W. *SL* **1991**, 315.
118. Buttery, C. D.; Knight, D. W.; Nott, A. P. *TL* **1982**, *23*, 4127.
119. Cornwall, P.; Dell, C. P.; Knight, D. W. *TL* **1987**, *28*, 3585.
120. Duhamel, L.; Plaquevent, J.-C. *JOM* **1993**, *448*, 1 and references cited therein.

Wouter I. Iwema Bakker, Poh Lee Wong & Victor Snieckus
University of Waterloo, Ontario, Canada

Lithium Hexamethyldisilazide

$(Me_3Si)_2NLi$

[4039-32-1] $C_6H_{18}LiNSi_2$ (MW 167.37)

(strong nonnucleophilic base)

Alternate Names: LHMDS; lithium bis(trimethylsilyl)amide.

Physical Data: distillable low-melting solid; mp 70–72 °C, bp 115 °C/1 mmHg.[2] LHMDS is a cyclic trimer in the solid state,[3] whereas in benzene solution it exists in a monomer–dimer equilibrium.[4] LHMDS is less soluble, less basic, more stable, and much less sensitive to air compared to ***Lithium Diisopropylamide***. pK_a 29.5 (THF, 27 °C).[1]

Solubility: sol most nonpolar solvents, e.g. aromatic hydrocarbons, hexanes, THF.

Form Supplied in: colorless crystalline solid, 1 M solution in THF or hexanes, 1.3 M solution in THF, 1 M solution in THF/cyclohexane.

Preparative Methods: conveniently prepared by the reaction of ***Hexamethyldisilazane*** with ***n-Butyllithium*** in hexane. For most uses the hexane is then evaporated and replaced with THF.[5]

Handling, Storage, and Precautions: a flammable, moisture sensitive solid; stable in a nitrogen atmosphere. Use in a fume hood.

Ketone Enolates. A high yielding synthesis of 6-aryl-4,6-dioxohexanoic acids, precursors to antiinflammatory agents, is achieved using LHMDS (eq 1).[6] This process is applicable to large scale and involves relatively high reaction temperatures. The use of LDA gives reduced yields and small amounts of a diisopropylamide byproduct.

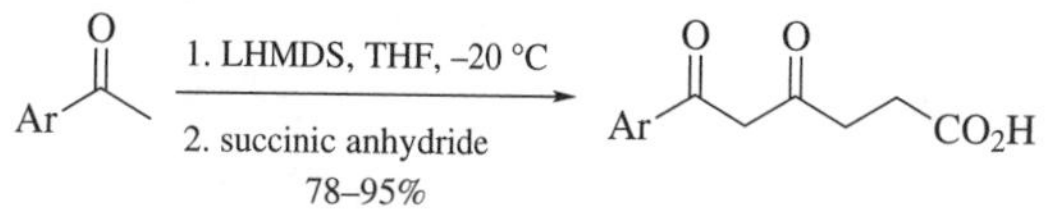

1. LHMDS, THF, –20 °C
2. succinic anhydride
78–95%
(1)

Ar = 4-ClC_6H_4, 2-thienyl, 3-pyridyl, 4-MeC_6H_4

Ester Enolates. Enantiomerically pure amino acids may ultimately be prepared via stereospecific ester enolate generation using an oxazolidine chiral auxiliary (eq 2).[7] Moderate diastereoselectivity is observed using ***Potassium Hexamethyldisilazide***.

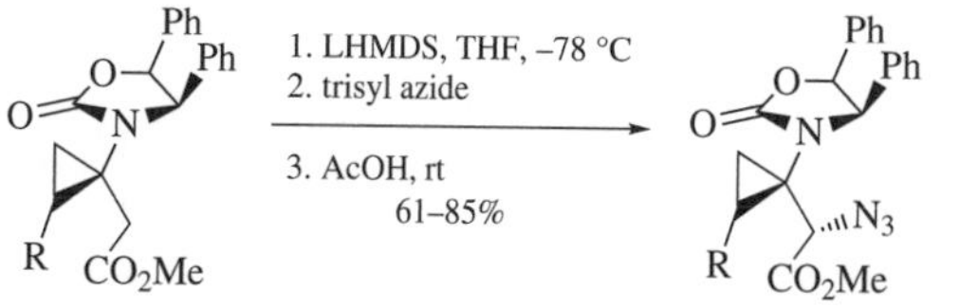

1. LHMDS, THF, –78 °C
2. trisyl azide
3. AcOH, rt
61–85%
(2)

trisyl = 2,4,6-triisopropylbenzenesulfonyl

Lithio ethyl acetate is prepared in quantitative yield by reaction of LHMDS with ethyl acetate in THF at −78 °C.[8a] Reaction with carbonyl compounds leads to condensation products in high yield (eqs 3 and 4).[8] No racemization of the α-silyloxy esters occurs (eq 4).

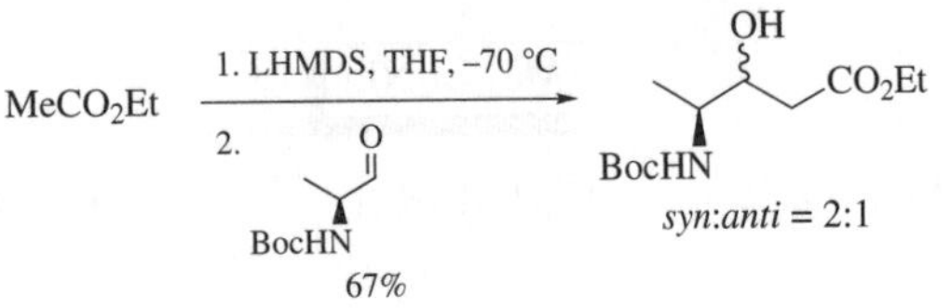

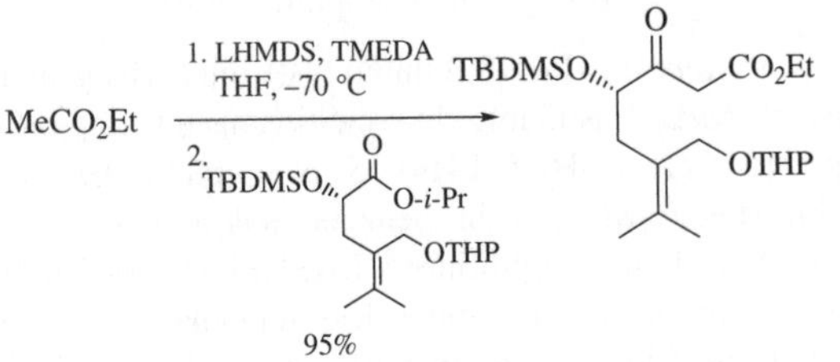

Kinetic Enolates. LHMDS is the recommended base for the generation of kinetic enolates. The resulting enolates are more regiostable than those generated with the corresponding sodium base, ***Sodium Hexamethyldisilazide***. Thus reaction of Δ^4-3-keto steroids with LHMDS yields 2,4-dienolate ions which can be methylated at C-2 or trapped as 2,4-dienolsilyl ethers (eq 5).[9] Use of ***Potassium t-Butoxide***/*t*-BuOH produces the thermodynamically more stable 3,5-dienolate. Acid-catalyzed conditions yields the 3,5-enol ether. Enolates generated with LHMDS may serve as ketone protecting groups during metal hydride reductions (eq 6).[10]

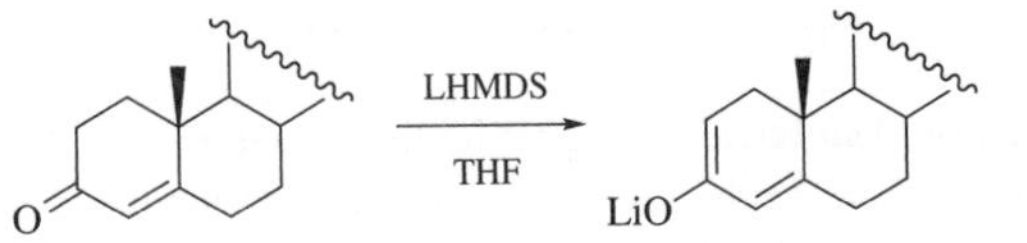

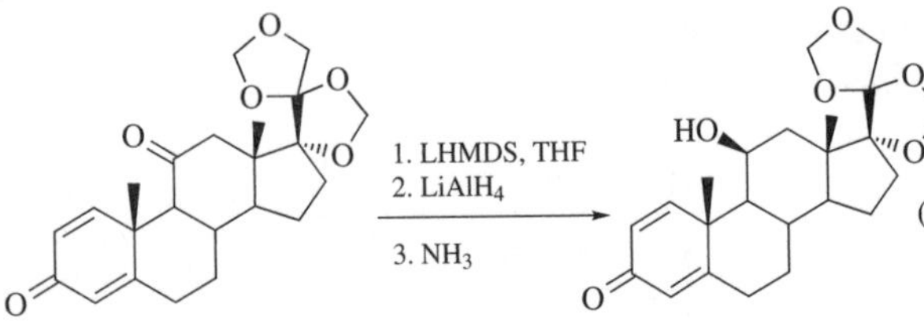

LHMDS has also been used in directed aldol condensations. The compatibility of the base with a silyl ether moiety is of note in the synthesis of (±)-[6]-gingerol (eq 7).[11]

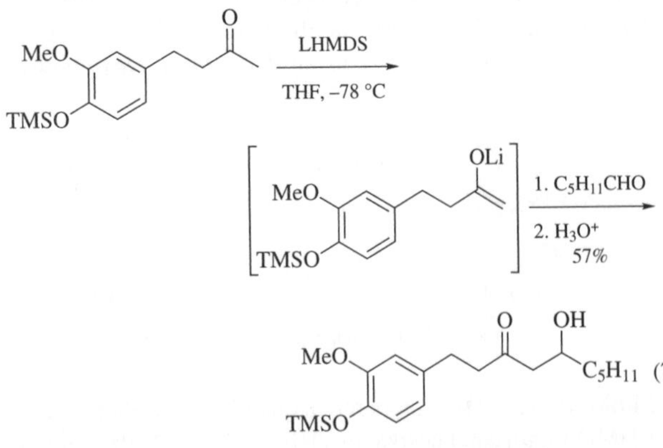

Darzens Condensation. The Darzens reaction invariably fails with aldehydes due to competing base-catalyzed self-condensation reactions.[12] With LHMDS as base, even acetaldehyde provides the desired glycidic ester products in high yield (eq 8).[13]

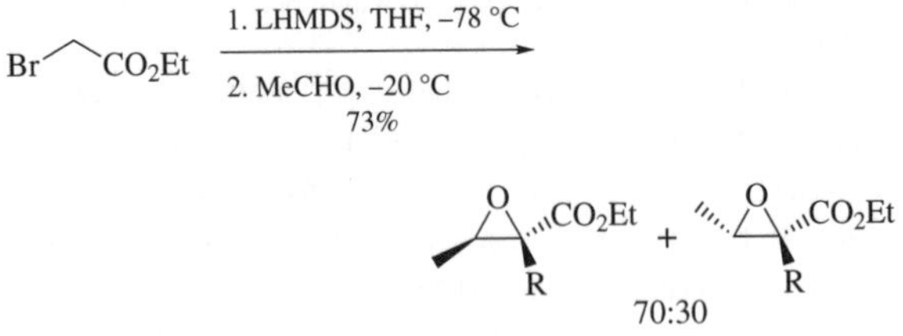

Intramolecular Cyclizations. LHMDS-mediated intramolecular cyclizations have been demonstrated (eq 9).[14] The choice of counter cation has a dramatic effect on the stereochemistry of the cyclization.

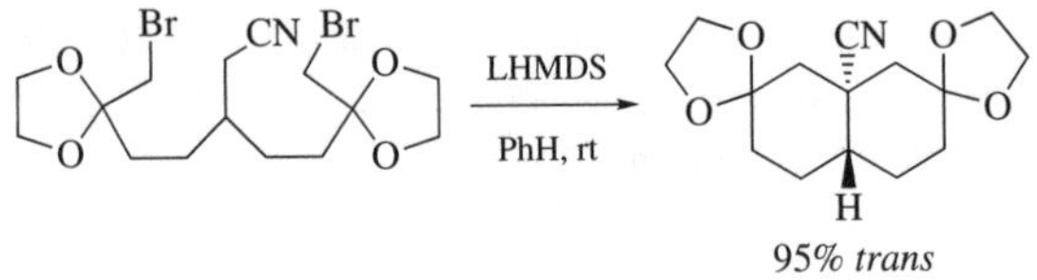

Ester Enolate Claisen Rearrangement. LHMDS is comparable to LDA for the stereoselective Ireland–Claisen rearrangement of ester enolates (eq 10).[15]

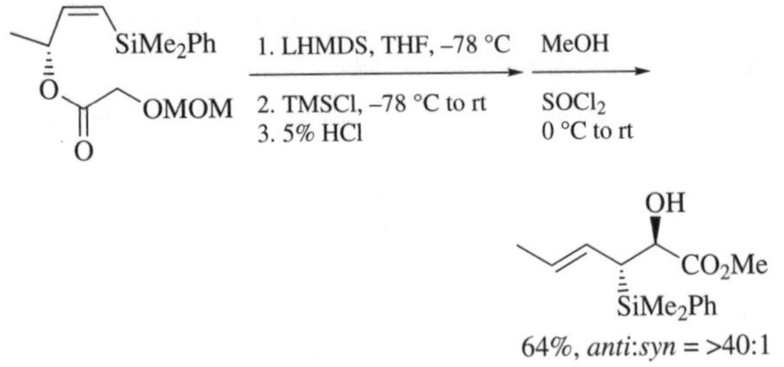

Intramolecular Double Michael Addition. LHMDS-mediated sequential Michael reactions constitute the key component of a total synthesis of the diterpene alkaloid atisine (eq 11).[16]

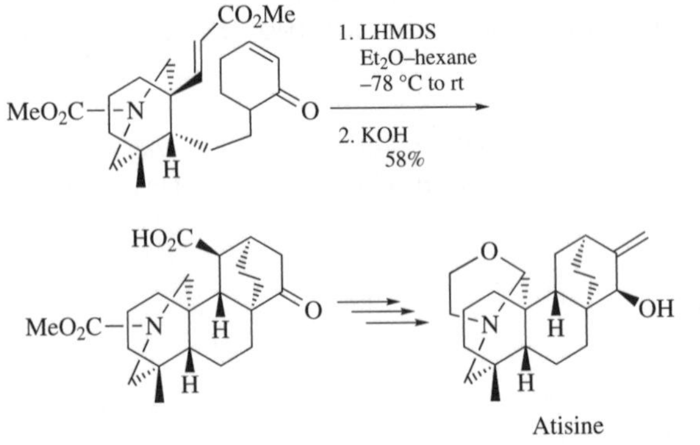

Synthesis of Primary Amines. *N,N*-Bis(trimethylsilyl)-methoxymethylamine, formally a $^{+}CH_2NH_2$ equivalent, is obtained in high yield by treating chloromethyl methyl ether with LHMDS. Treatment of the bis-silylamine with organometallic reagents followed by mild solvolysis gives primary amines in good to excellent yield (eq 12).[17]

$$ClCH_2OMe \xrightarrow[\text{THF, 0 °C; 86\%}]{\text{LHMDS}} (TMS)_2NCH_2OMe \xrightarrow[\text{rt or reflux}]{\text{RMgX, Et}_2\text{O}}$$

$$(TMS)_2NCH_2R \xrightarrow{HO^-} RCH_2NH_2 \quad \text{50–92\%} \qquad (12)$$

***N*-Trimethylsilylaldimines.** Aldehydes, even enolizable ones, undergo Peterson reactions with LHDMS to give *N*-trimethylsilylaldimines (eq 13),[18] which are valuable intermediates for a variety of systems, including primary amines,[18a,19] β-lactams,[18] and α-methylene-γ-lactams (eq 14).[20] Extension of this chemistry to include α-keto ester substrates allows for the preparation of α-amino esters.[21]

$$RCH{=}O + LHMDS \xrightarrow{THF} RCH{=}NTMS \qquad (13)$$

RCH=NTMS: 1. Zn, $CH_2{=}C(CH_2Br)CO_2Et$; 2. H_3O^+ → 5-R-3-methylene-pyrrolidin-2-one (14); $R^1R^2C(Li)CO_2Et$ → β-lactam (R^1, R^2, R, NH, O); $LiAlH_4$ → RCH_2NH_2; 1. R^1M, 2. H_3O^+ → $RCH(NH_2)R^1$

***N,N*-Bis(trimethylsilyl)aminomethyl Acetylide.** The reaction of LHMDS with propargyl bromide constitutes a straightforward route to a γ-amino lithium acetylide, a useful precursor to a wide variety of unsaturated protected primary amines (eq 15).[22]

$$2\ LHMDS + HC{\equiv}CCH_2Br \xrightarrow[-20\ °C]{Et_2O} Li{-}C{\equiv}CCH_2N(TMS)_2 \qquad (15)$$

For example, reaction with aromatic aldehydes gives α-alkynyl amino alcohols, which can be trapped as their silyl ethers. Base-catalyzed isomerization to an allenic isomer followed by hydrolysis and concomitant cyclization affords 2-substituted pyrroles (eq 16).[22]

$$Li{-}C{\equiv}CCH_2N(TMS)_2 \xrightarrow[\text{2. TMSCl, rt; 68\%}]{\text{1. PhCHO, THF, –78 °C to rt}} PhCH(OTMS)C{\equiv}CCH_2N(TMS)_2 \xrightarrow{LHMDS}$$

$$[Ph(TMSO)C{=}C{=}CH{-}CH_2N(TMS)_2] \xrightarrow[64\%]{H_2O} \text{2-phenylpyrrole} \qquad (16)$$

β-Ketosilanes. β-Ketosilanes may be prepared from α-bromo ketones using LHMDS to generate intermediate silyl enol ethers followed by metal–halogen exchange (eq 17).[23] They undergo facile rearrangement to silyl enol ethers and are also substrates, after carbonyl reduction, for overall Peterson alkenation.

$$RC(O)CH(R^1)Br \xrightarrow[\text{2. TMSCl}]{\text{1. LHMDS}} RC(OTMS){=}C(R^1)Br \xrightarrow[\text{2. NH}_4\text{Cl}]{\text{1. 2 equiv BuLi}} RC(O)CH(R^1)TMS \qquad (17)$$

Related Reagents. Lithium Amide; Lithium Diethylamide; Lithium Diisopropylamide; Lithium Piperidide; Lithium Pyrrolidide; Lithium 2,2,6,6-Tetramethylpiperidide; Potassium Hexamethyldisilazide; Sodium Hexamethyldisilazide.

1. Wannagat, U.; Nierderprüm, H. *CB* **1961**, *94*, 1540.
2. (a) Mootz, D.; Zinnius, A.; Böttcher, B. *AG(E)* **1969**, *8*, 378. (b) Rogers, R. D.; Atwood, J. L.; Grüning, R. *JOM* **1978**, *157*, 229. For further structural information see (c) Lappert, M. F.; Power, P. P.; Sanger, A. R.; Srivastava, R. C. *Metal and Metalloid Amides*; Wiley: New York, 1980.
3. Kimura, B. Y.; Brown, T. L. *JOM* **1971**, *26*, 57.
4. Fraser, R. R.; Mansour, T. S. *JOC* **1984**, *49*, 3442.
5. (a) Rathke, M. W. *OSC* **1988**, *6*, 598. (b) Amonoo-Neizer, E. H.; Shaw, R. A.; Skovlin, D. O.; Smith, B. C. *Inorg. Synth.* **1966**, *8*, 19.
6. Murray, W.; Wachter, M.; Barton, D.; Forero-Kelly, Y. *S* **1991**, 18.
7. Es-Sayed, M.; Gratkowski, C.; Krass, N.; Meyers, A. I.; de Meijere, A. *SL* **1992**, 962.
8. (a) Rathke, M. W. *JACS* **1970**, *92*, 3222. (b) Mori, K.; Matsuda, H. *LA* **1992**, 131. (c) Pettersson, L.; Magnusson, G.; Frejd, T. *ACS* **1993**, *47*, 196.
9. Tanabe, M.; Crowe, D. F. *CC* **1973**, 564.
10. Barton, D. H. R.; Hesse, R. H.; Pechet, M. M.; Wiltshire, C. *CC* **1972**, 1017.
11. Denniff, P.; Whiting, D. A. *CC* **1976**, 712.
12. (a) Newman, M. S.; Magerlein, B. J. *OR* **1949**, *5*, 413. (b) Morrison, J. D.; Mosher, H. S. *Asymmetric Organic Reactions*; Prentice–Hall: New York, 1971.
13. Borch, R. F. *TL* **1972**, 3761.
14. (a) Stork, G.; Gardner, J. O.; Boeckman, Jr., R. K.; Parker, K. A. *JACS* **1973**, *95*, 2014. (b) Stork, G.; Boeckman, Jr., R. K. *JACS* **1973**, *95*, 2016. (c) Stork, G.; Cohen, J. F. *JACS* **1974**, *96*, 5270.
15. (a) Ireland, R. E; Daub, J. P. *JOC* **1981**, *46*, 479. (b) Fujisawa, T.; Maehata, E.; Kohama, H.; Sato, T. *CL* **1985**, 1457. (c) Sato, T.; Tsunekawa, H.; Kohama, H.; Fujisawa, T. *CL* **1986**, 1553. (d) Ireland, R. E.; Wipf, P.; Armstrong, III, J. D. *JOC* **1991**, *56*, 650. (e) Panek, J. S.; Clark, T. D. *JOC* **1992**, *57*, 4323.
16. Ihara, M.; Suzuki, M.; Fukumoto, K.; Kabuto, C. *JACS* **1990**, *112*, 1164.
17. (a) Morimoto, T.; Takahashi, T.; Sekiya, M. *CC* **1984**, 794. For other closely related examples of amine synthesis see: (b) King, F. D.; Walton, D. R. M. *CC* **1974**, 256. (c) Murai, T.; Yamamoto, M.; Kondo, S.; Kato, S. *JOC* **1993**, *58*, 7440.
18. (a) Hart, D. J.; Kanai, K.-i.; Thomas, D. G.; Yang, T.-K. *JOC* **1983**, *48*, 289. (b) Cainelli, G.; Giacomini, D.; Panunzio, M.; Martelli, G.; Spunta, G. *TL* **1987**, *28*, 5369. (c) Andreoli, P.; Billi, L.; Cainelli, G.; Panunzio, M.; Bandini, E.; Martelli, G.; Spunta, G. *T* **1991**, *47*, 9061 and references cited therein. (d) Colvin, E. W. *Silicon Reagents in Organic Synthesis*; Academic: London, 1988; p 73 and references cited therein.
19. Leboutet, L.; Courtois, G.; Miginiac, L. *JOM* **1991**, *420*, 155.
20. El Alami, N.; Belaud, C.; Villieras, J. *SC* **1988**, *18*, 2073.
21. Matsuda, Y.; Tanimoto, S.; Okamoto, T.; Ali, S. M. *JCS(P1)* **1989**, 279.

22. Corriu, R. J. P.; Huynh, V.; Iqbal, J.; Moreau, J. J. E.; Vernhet, C. *T* **1992**, *48*, 6231.
23. (a) Sampson, P.; Hammond, G. B.; Weimer, D. F. *JOC* **1986**, *51*, 4342. (b) Kowalski, C. J.; O'Dowd, M. L.; Burke, M. C.; Fields, K. W. *JACS* **1980**, *102*, 5411.

Matthew Gray & Victor Snieckus
University of Waterloo, Ontario, Canada

Lithium Hydroxide

LiOH

[1310-65-2] HLiO (MW 23.95)
($\cdot H_2O$)
[1310-66-3] H_3LiO_2 (MW 41.97)

(very strong alkali; reacts readily with acids and is used as a base as well as a nucleophile in organic reactions, particularly in the hydrolysis of esters and amides)

Physical Data: *d* (anhyd.) 2.54 g cm^{-3}; (monohyd.) 1.51 g cm^{-3}; mp (anhyd.) 471 °C.
Solubility: sol water: (w/w) 10.7% at 0 °C, 10.9% at 20 °C, 14.8% at 100 °C; pH of 1.0 N soln. about 14; slightly sol ethanol.
Form Supplied in: the commercially available material is the monohydrate: small, white monoclinic crystals.
Handling, Storage, and Precautions: corrosive. Harmful if swallowed, inhaled, or absorbed through skin. Toxicity in rat: LD_{50} 365 mg kg^{-1} (oral dose). Material is extremely destructive to tissue of the mucous membranes and upper respiratory tract, eyes, and skin. Inhalation may be fatal; do not breathe dust. Avoid contact with eyes, skin, and clothing. Incompatible with strong oxidizing agents and strong acids. Absorbs moisture and CO_2 from air. Hygroscopic; store in a cool dry place. Wash thoroughly after handling.

Hydrolysis of Esters. The methyl ester of the 9,11-azo analog of prostaglandin endoperoxide PGH_2 (**1**) was hydrolyzed under mild conditions using 0.15 N LiOH in 2.5:1 THF–H_2O at 0 °C for 2.5 h to give the 9,11-azo analog of PGH_2 (**2**) in 99% isolated yield (eq 1).[1]

CO_2Me / C_5H_{11} / OH — 0.15 N LiOH; THF, H_2O; 99% — CO_2H / C_5H_{11} / OH (1)

(**1**) (**2**)

Hydrolysis of Urethanes to Alcohols. The *p*-phenylphenylcarbamoyl group served as a protecting group for the C-11 alcohol and also as a directing group to control the stereochemistry of the carbonyl reduction at C-15 in the synthesis of the prostaglandin synthon (**3**). This protecting group was removed following the ketone reduction by treating (**3**) with 1 M aq LiOH at 120 °C for 72 h to afford (**4**) in >90% yield.[2]

(**3**) R = *p*-PhC_6H_4NHCO-
(**4**) R = H

Hydrolysis of Bile Acid Methyl Esters. Bile acid methyl esters were hydrolyzed under mild conditions with LiOH in aq. methanol at 25 °C to the bile acids. This procedure gave higher yields (80–98%) compared to conventional ***Sodium Hydroxide*** or ***Potassium Hydroxide*** hydrolysis reactions (48–55%). The procedure caused no racemization or elimination side reactions, e.g. (23*R*)-methyl 3α,7α,23α-trihydrox y-5β-cholan-24-oate (**5**) gave the corresponding acid (**6**) in 96% isolated yield (eq 2).[3]

(**5**) — LiOH, MeOH–H_2O, rt, 20 min, 96% — (**6**) (2)

Hydrolysis of *N*-Boc Lactams and *N*-Boc Secondary Amides. *N*-Boc derivatives of lactams were regioselectively hydrolyzed with LiOH in aq. THF to the corresponding ω-amino acids. For example, *N*-Boc valerolactam was hydrolyzed with 3.0 equiv of LiOH in aq. THF at rt to give the *N*-Boc-δ-aminovaleric acid in 90% yield (eq 3). The reaction conditions are very mild and provide the ω-amino group in a protected form, thus permitting further elaboration of the carboxylic acid residue. The procedure was applied successfully to the hydrolysis of *N*-Boc secondary amides to the corresponding acids.[4]

1. LiOH (3 equiv), aq THF, rt; 2. H^+; 90% — HO_2C~~~NHBoc (3)

Cleavage of 2-Oxazolidinone Chiral Auxiliaries. The hydrolysis of the *N*-(α-azidoacyl)- and *N*-[α-((+)-MTPA-amino)acyl]-2-oxazolidinone derivatives (**7**) with LiOH (2 equiv) in THF–H_2O (3:1) for 30 min at 0 °C afforded the corresponding α-amino acid synthons (**8**) in 95–100% yields (eq 4). No detectable racemization was observed when R = Bn and about 1% racemization occurred with the phenylglycine synthons (R = Ph).[5]

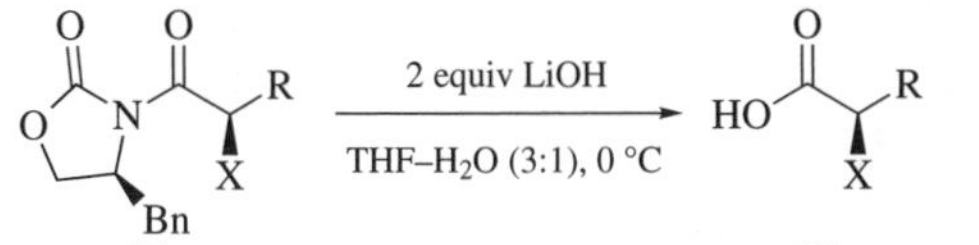

2 equiv LiOH, THF–H_2O (3:1), 0 °C (4)

(**7**) (**8**)

X = N_3, NH((+)-MTPA) 95–100%
R = Bn, Ph

Regioselective Hydrolysis of Carboximide Derivatives. The regioselective hydrolysis of the carboximide derivative (**9**) was achieved with excess 2 N aq LiOH in dioxane at rt overnight to give the desired β-hydroxy-α-amino acid (**10**) and a byproduct (**11**) (eq 5) in isolated yields of 83% and 17% respectively (a 5:1 ratio). The products (**10**) and (**11**) are the result of endocyclic and exocyclic carbonyl attack by the hydroxide ion, respectively. The success of this procedure was based on the selection of the sterically demanding *N*-protecting Boc group, which largely directed the regioselectivity of the attack by the hydroxide ion to the less hindered oxazolidinone carbonyl.[6]

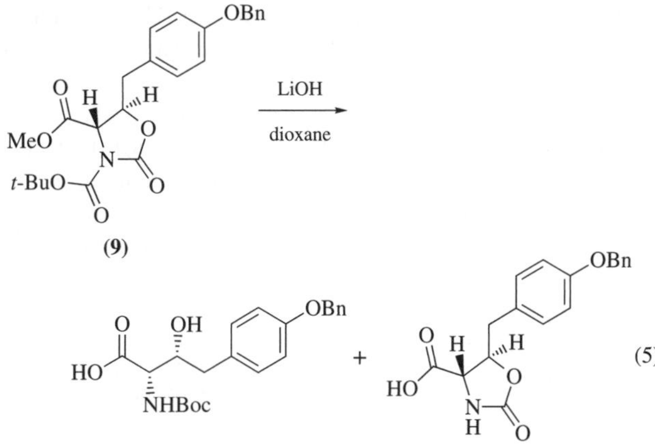

LiOH, dioxane

(**9**)

(**10**) 83% + (**11**) 17% (5)

If, on the other hand, the hydrolysis was to be directed towards the exocyclic carbonyl of a highly branched *N*-acyl oxazolidinone derivative, a complementary procedure was developed employing lithium hydroperoxide to achieve that regioselectivity.[7]

Hydrolysis of *N*-Monosubstituted Amides via Acetoxypivalimides. Hydrolysis of *N*-monosubstituted amides under mild conditions without epimerization at the α-position was accomplished in two steps. The amides (**12**) were first converted into the *N*-acetoxypivaloyl derivatives (**13**) (eq 6). These derivatives were then treated with LiOH in THF to give the carboxylic acids (**15**) in good yields. The procedure takes advantage of the easier hydrolysis of imides than amides and the intramolecular nucleophilic attack by the acetoxy oxygen on the imide carbonyl (eq 7) causing an *N–O* acyl migration to form the intermediate acyloxypivalamide (**14**). Further hydrolysis of (**14**) gave the desired carboxylic acids (**15**) and *N*-alkylhydroxypivalamides (**16**).[8]

The method is mild enough to use with *N*-butyl-(2*S*,3*S*)-dimethyl-4-pentenamide to give (2*S*,3*S*)-dimethyl-4-pentenoic acid in 76% yield with practically no epimerization (eq 8).[8]

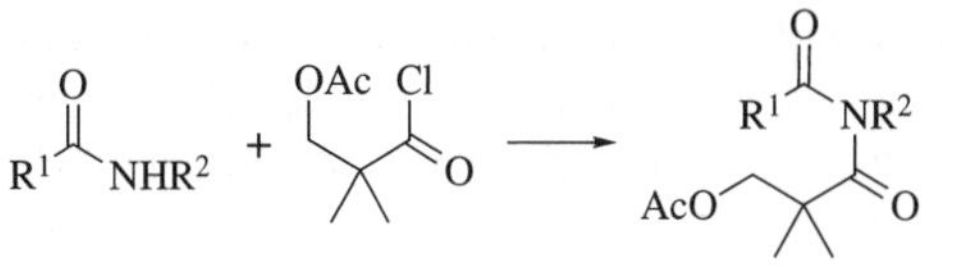

(**12**) + OAc Cl → (**13**) (6)

(**13**) LiOH, THF → [... (**14**)] → R^1CO_2H + (**16**) (7)

(**15**) (**16**)

syn:*anti* = 99.5:0.5 — 1. OAc Cl; 2. LiOH, 76% → *syn*:*anti* = 99.2:0.8 (8)

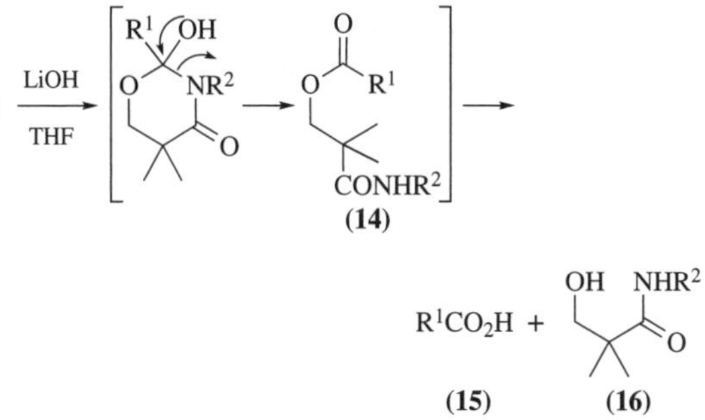

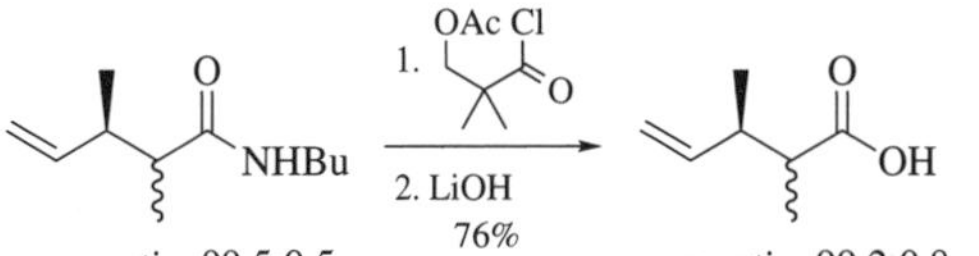

Selective *O*-Acetyl Hydrolysis of *N,O*-Diacetylhydroxamates. Treatment of *N*-acetoxy-*N*-1-[4-(phenylmethoxy)-phenyl]ethylacetamide (**17**) with LiOH in 2:1 *i*-PrOH–H_2O selectively hydrolyzed the *O*-acetyl to give the hydroxamic acid (**18**) (eq 9). The procedure was described as part of a two-step conversion of hydroxylamines into hydroxamic acids.[9]

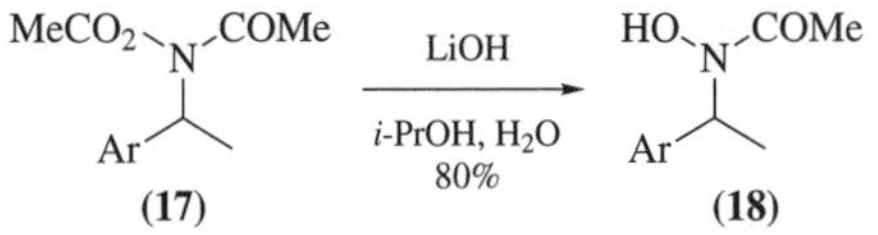

LiOH, *i*-PrOH, H_2O, 80% (9)

(**17**) (**18**)

Ar = *p*-$PhCH_2OC_6H_4$-

Selective *O*-Deacetylation in Carbohydrates. The *O*-acetyl groups in the penta-*O*-acetyl tri-*N*-acyl disaccharide (**19**) (eq 10) were selectively hydrolyzed with 0.6 M LiOH for 1 h at rt to give, after purification and recrystallization, the pentahydroxy tri-*N*-acyl disaccharide (**20**) in 92% yield.[10]

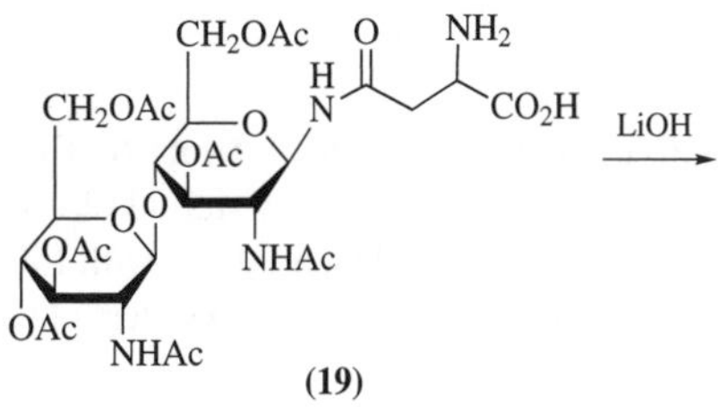

LiOH (10)

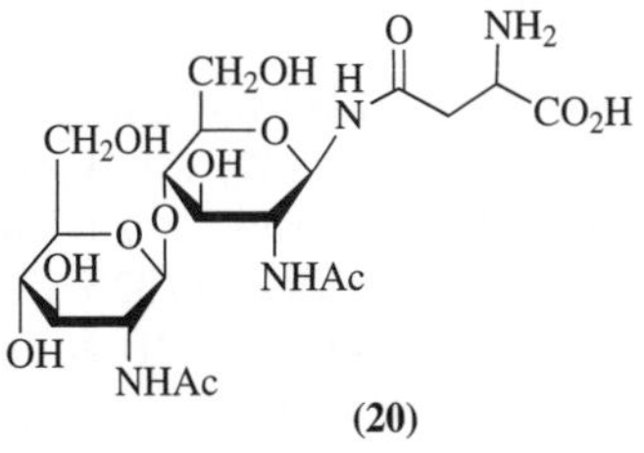

(**19**) (**20**)

Dihydroquinoxalones from Ethyl Alkylglycidates. The reactions of *o*-phenylenediamine with ethyl 2,3-epoxycrotonate (**21a**), ethyl 2,3-epoxypentenoate (**21b**), or ethyl 2,3-epoxy-3,3-dimethylacrylate (**21c**) were catalyzed by LiOH and gave the dihydroquinoxalones (**22a–c**) (eq 11). The initial nucleophilic attack occurred on the α-carbon of (**21**) followed by intramolecular aminolysis to give the cyclized product. When unsubstituted epoxyacrylates and cinnamates were used, the initial attack occurred on the β-carbon and gave the uncyclized hydroxyamino esters.[11]

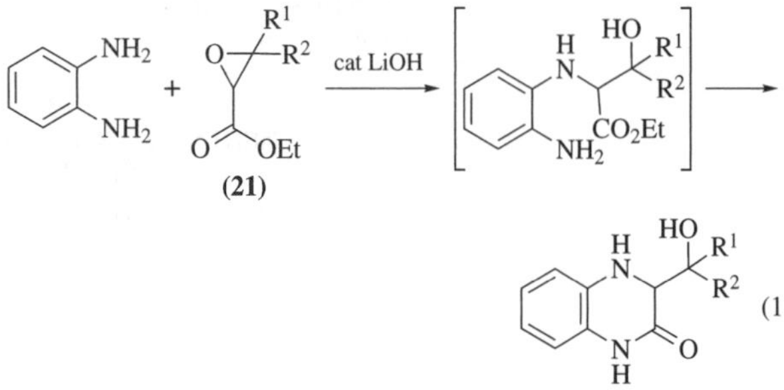

(**21**) —cat LiOH→ (11)

(**22a**) R^1 = Me, R^2 = H
(**22b**) R^1 = Et, R^2 = H
(**22c**) R^1 = R^2 = Me

Synthesis of 3-Hydroxybenzo[*b*]thiophene-2-carboxylates. Substituted methyl 3-hydroxybenzo[*b*]thiophene-2-carboxylates (**24**) were prepared in yields ranging from 50 to 85% by the LiOH-catalyzed reaction of substituted methyl *o*-nitrobenzoates (**23**) and methyl thioglycolate in dry DMF (eq 12). Under the reaction conditions the activated nitro group was displaced by the thiolate anion, followed by cyclization.[12]

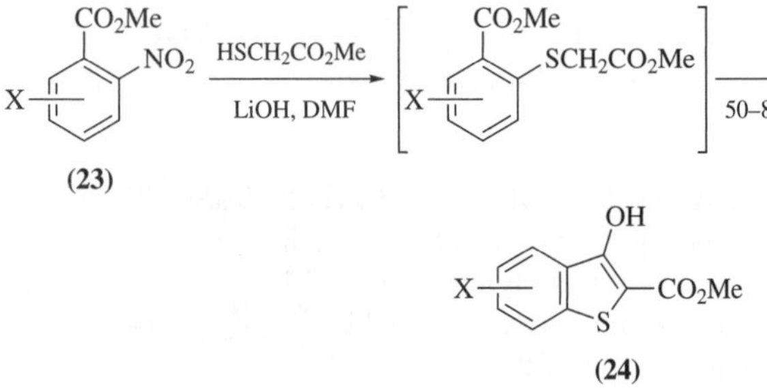

(**23**) —$HSCH_2CO_2Me$, LiOH, DMF→ —50–85%→ (**24**) (12)

Stereochemical Control in Conversion of Diols to Epoxides. The intermediate 22-tosyloxy-23-pivaloxy steroid (**25**) was prepared from the corresponding diol and then treated with LiOH in dioxane–H_2O at rt for 8 h to give the 22,23-epoxy steroid (**26**) in 92% yield (eq 13).[13]

Oxetane Formation from 3,4-Epoxy Alcohols. Treatment of the 3,4-epoxy alcohol (**27**) with LiOH, NaOH, or KOH in aq DMSO at 140–150 °C gave a mixture of the oxetane (**28**) in 49% yield and the 1,2,4-triol (**29**) in 32% yield (eq 14). Reaction times were 15, 45, and 90 min for LiOH, NaOH, and KOH, respectively. The procedure was presented as a nonphotochemical synthesis of oxetanes.[14]

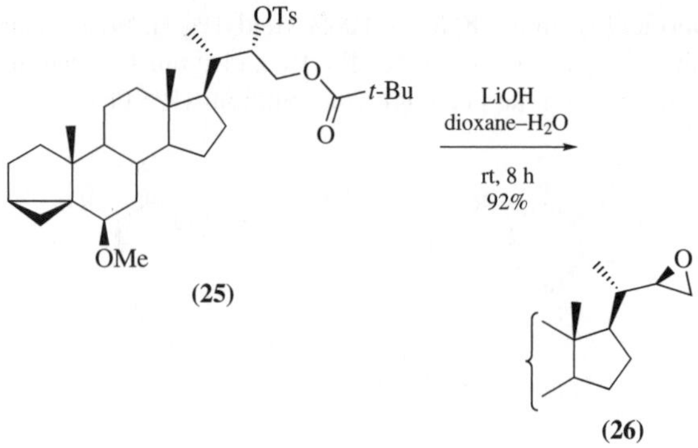

(**25**) —LiOH, dioxane–H_2O, rt, 8 h, 92%→ (**26**) (13)

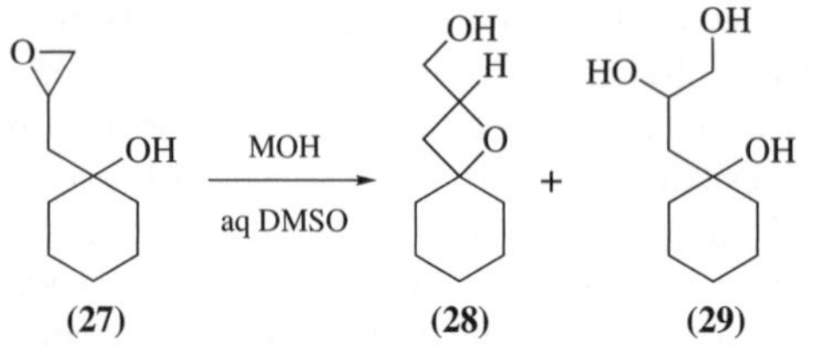

(**27**) —MOH, aq DMSO→ (**28**) + (**29**) (14)

M = Li, Na or K

Formation of *trans*-Hydrindanones. Treatment of 8-oxo-2-methyl-6-nonenal (**30**) with LiOH in MeOH gave a 4:1 mixture of *trans*- and *cis*-hydroxyhydrindanones (**32**). The reaction proceeded via consecutive intramolecular Michael and aldol addition reactions. The stereochemical control was greatly improved by using the better chelating Zr^{IV} *n*-propoxide in benzene to effect the initial Michael addition, yielding the monocyclic ketoaldehyde (**31**), and then adding LiOH/MeOH to effect the aldol addition step (eq 15). This modification gave the desired hydroxyhydrindanone (**32**) in 90% yield as a 40:1 mixture of the *trans* and *cis* isomers.[15]

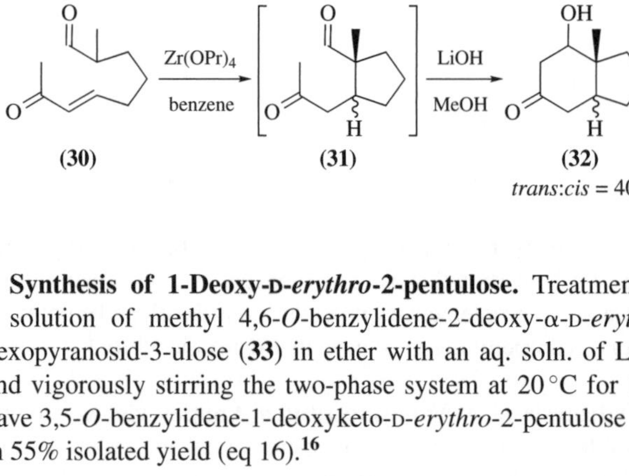

(**30**) —$Zr(OPr)_4$, benzene→ (**31**) —LiOH, MeOH→ (**32**) (15)

trans:*cis* = 40:1

Synthesis of 1-Deoxy-D-*erythro*-2-pentulose. Treatment of a solution of methyl 4,6-*O*-benzylidene-2-deoxy-α-D-*erythro*-hexopyranosid-3-ulose (**33**) in ether with an aq. soln. of LiOH and vigorously stirring the two-phase system at 20 °C for 16 h gave 3,5-*O*-benzylidene-1-deoxyketo-D-*erythro*-2-pentulose (**34**) in 55% isolated yield (eq 16).[16]

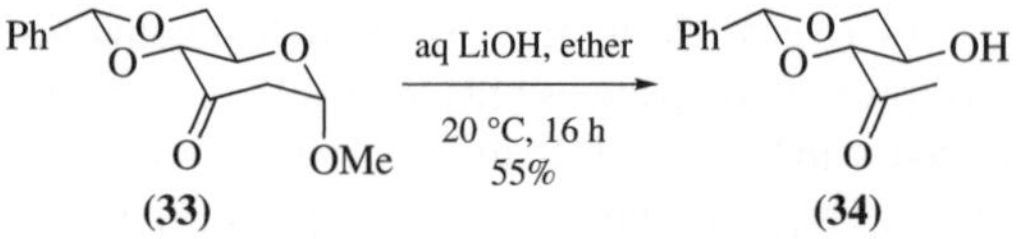

(**33**) —aq LiOH, ether, 20 °C, 16 h, 55%→ (**34**) (16)

Conversion of Mixed Phosphoric Acid Triesters to Mixed Phosphoric Acid Diesters. In a three-step sequence for the conversion of bis(*p*-nitrophenyl) hydrogen phosphate to dialkyl phosphates (**36**), the intermediate dialkyl *p*-nitrophenyl phosphates (**35**) were selectively hydrolyzed with 1N LiOH in acetonitrile at rt to give the desired dialkyl phosphates under mild conditions (eq 17). Various mixed dialkyl phosphates were prepared in good overall yields without isolation of any synthetic intermediate.[17]

$$R^1O(R^2O)P(=O)O\text{-}C_6H_4\text{-}NO_2 \; \mathbf{(35)} \xrightarrow[2.\ H^+]{1.\ 1\ N\ LiOH,\ MeCN} R^1O\text{-}P(=O)(OR^2)\text{-}OH \; \mathbf{(36)} \quad (17)$$

R^1	R^2	overall yield
$Ph(CH_2)_2$	$MeO(CH_2)_2$	80%
$Ph(CH_2)_2$	$HC{\equiv}C\text{-}CH_2$	76%
$MeO(CH_2)_2$	$HC{\equiv}C\text{-}CH_2$	75%

Related Reagents. Barium Hydroxide; Potassium Hydroxide; Sodium Hydroxide; Tetra-butylammonium Hydroxide; Tetramethylammonium Hydroxide.

1. Corey, E. J.; Narasaka, K.; Shibasaki, M. *JACS* **1976**, *98*, 6417.
2. Corey, E. J.; Becker, K. B.; Varma, R. K. *JACS* **1972**, *94*, 8616.
3. Dayal, B.; Salen, G.; Toome, B.; Tint, G. S.; Shefer, S.; Padia, J. *Steroids* **1990**, *55*, 233.
4. Flynn, D. L.; Zelle, R. E.; Grieco, P. A. *JOC* **1983**, *48*, 2424.
5. Evans, D. A.; Ellman, J. A.; Dorow, R. L. *TL* **1987**, *28*, 1123.
6. Evans, D. A.; Weber, A. E. *JACS* **1987**, *109*, 7151.
7. Evans, D. A.; Britton, T. C.; Ellman, J. A. *TL* **1987**, *28*, 6141.
8. Tsunoda, T.; Sasaki, O.; Ito, S. *TL* **1990**, *31*, 731.
9. Summers, J. B.; Gunn, B. P.; Martin, J. G.; Martin, M. B.; Mazdiyasni, H.; Stewart, A. O.; Young, P. R.; Bouska, J. B.; Goetze, A. M.; Dyer, R. D.; Brooks, D. W.; Carter, G. W. *JMC* **1988**, *31*, 1960.
10. Spinola, M.; Jeanloz, R. W. *JBC* **1970**, *245*, 4158.
11. Murata, S.; Sugimoto, T.; Matsuura, S. *H* **1987**, *26*, 883.
12. Beck, J. R. *JOC* **1973**, *38*, 4086.
13. Koreeda, M.; Ricca, D. J. *JOC* **1986**, *51*, 4090.
14. Murai, A.; Ono, M.; Masamune, T. *CC* **1976**, 864.
15. Stork, G.; Shiner, C. S.; Winkler, J. D. *JACS* **1982**, *104*, 310.
16. Fischer, J.-C; Horton, D.; Weckerle, W. *CJC* **1977**, *55*, 4078.
17. Mukaiyama, T.; Morito, N.; Watanabe, Y. *CL* **1979**, 531.

Ahmed F. Abdel-Magid
The R. W. Johnson Pharmaceutical Research Institute, Spring House, PA, USA

Lithium Iodide[1]

LiI

[10377-51-2] ILi (MW 133.84)

(ester cleavage and decarboxylation;[2] source of nucleophilic iodide;[3] mild Lewis acid;[1] salt effects in organometallic reactions;[1] epoxide opening[4])

Physical Data: mp 449 °C; bp 1180 °C; *d* 4.076 g cm^{-3}.
Solubility: 165 g/100 mL H_2O (20 °C); 433 g/100 mL H_2O (80 °C); 251 g/100 mL EtOH (20 °C); 343 g/100 mL MeOH (20 °C); 43 g/100 mL acetone (18 °C); very sol NH_4OH.
Form Supplied in: anhydrous white solid or as the hydrate.
Preparative Methods: the anhydrous salt of high purity can be prepared from lithium hydride and iodine in ether.[5]
Purification: crystallized from hot H_2O (0.5 mL g^{-1}) by cooling in $CaCl_2$–ice or from acetone. LiI is dried for 2 h at 120 °C (0.1 mmHg, P_2O_5) before use.
Handling, Storage, and Precautions: for best results, LiI should be dried prior to use in anhydrous reactions.

Heterolytic C–X Bond Cleaving Reactions. In the presence of amine bases, LiI has been extensively used as a mild reagent for the chemoselective cleavage of methyl esters (eq 1).[6] Decarboxylation of methyl esters usually occurs when an electron-withdrawing group is present at the α-position of the ester (eq 2).[7] Ester-type glycosyl linkages of acidic tri- and diterpenes can also be selectively cleaved under these conditions.[8] Aryl methyl ethers can be demethylated to afford the corresponding phenols upon heating with LiI and *s*-collidine.[9]

$$\text{EtO}_2\text{C-substrate-CH}_2\text{CO}_2\text{Me} \xrightarrow[91\%]{\text{LiI, py, reflux}} \text{EtO}_2\text{C-substrate-CH}_2\text{CO}_2\text{H} \quad (1)$$

$$\text{substrate(CO}_2\text{Me)}_2 \xrightarrow[\text{reflux, } 71\%]{\text{LiI}\bullet 2\text{H}_2\text{O, } s\text{-collidine}} \text{product-CO}_2\text{Me} \quad (2)$$

1,2-Oxiranes are readily opened by LiI and a Lewis acid to produce iodohydrins (eq 3).[4] Conversely, 1-oxaspiro[2.2]pentanes and 1-oxaspiro[3.2]hexanes give rise to bond migration products.[10] β-Vinyl-β-propiolactone is efficiently opened by LiI to produce the corresponding substituted allyl iodide (eq 4).[11]

$$\text{BnO-epoxide-OBn} \xrightarrow[\text{THF, rt, } 100\%]{\text{LiI, AcOH}} \text{BnO-CH}_2\text{CH(OH)CH(I)CH}_2\text{-OBn} \quad (3)$$

$$\beta\text{-vinyl-}\beta\text{-propiolactone} \xrightarrow[\text{rt, 1 h, } 85\%]{\text{LiI, THF}} \text{I-CH}_2\text{CH=CHCH}_2\text{-CO}_2\text{H} \quad (4)$$

Alkyl and Alkenyl Iodides. LiI has been used as a source of iodide in nucleophilic substitution and addition reactions. Primary alcohols have been directly converted to alkyl iodides upon treatment with a mixture of ***Triphenylphosphine***, ***Diethyl Azodicarboxylate***, and LiI.[3] Tertiary alcohols can be converted into tertiary alkyl iodides upon treatment with ***Hydrogen Iodide*** in the presence of LiI.[12]

(*Z*)-3-Iodopropenoates and -propenoic acids have been synthesized stereoselectively by the reaction of LiI and propiolates or propiolic acid.[13]

C–C Bond Forming Reactions. LiI was shown to efficiently catalyze the Michael addition of β-dicarbonyl compounds,[14] and the intramolecular allylsilane addition to imines to produce 4-methylenepiperidine derivatives (eq 5).[15]

TMS; t-Bu, N, Ph → LiI, THF, 40 °C, 12 h, 68% → t-Bu, N, H, Ph (5)

LiI as an Additive for Organometallic-Mediated Transformations[16]. The *syn*/*anti* selectivity in the reduction of β-alkoxy ketones is drastically increased by the addition of LiI (eq 6).[17]

O, O, O, OH, OTBDPS → $LiAlH_4$, Et_2O, –78 °C, 96%

O, O, OH, OH, OTBDPS + O, O, OH, OH, OTBDPS (6)

with LiI *syn*:*anti* = 89:11
without LiI *syn*:*anti* = 79:21

The addition of ***Lithium Bromide*** and LiI was shown to enhance the rate of organozinc formation from primary alkyl chlorides, sulfonates, and phosphonates, and zinc dust.[18] Beneficial effects of LiI addition have also been reported for Heck-type coupling reactions[19] and in conjugate addition to chiral vinyl sulfoximines.[20]

The (*E*)/(*Z*) alkenic ratio in Wittig-type alkenations was shown to be dependent on the amount of Li salt present.[21]

Reduction of α-Alkoxycarbonyl Derivatives. α-Halo ketones are reduced to the corresponding ketones upon treatment with a mixture of LiI and ***Boron Trifluoride Etherate***.[22]

Related Reagents. Aluminum Iodide; Iodotrimethylsilane; Lithium Chloride; Triphenylphosphine–Iodine.

---

1. Loupy, A.; Tchoubar, B. *Salt Effects in Organic and Organometallic Chemistry*; VCH: Weinheim, 1992.
2. (a) McMurry, J. *OR* **1976**, *24*, 187. (b) Krapcho, A. P. *S* **1982**, 805. (c) Krapcho, A. P. *S* **1982**, 893.
3. Manna, S.; Falck, J. R.; Mioskowski, C. *SC* **1985**, *15*, 663.
4. (a) Bonini, C.; Giuliano, C.; Righi, G.; Rossi, L. *SC* **1992**, *22*, 1863. (b) Shimizu, M.; Yoshida, A.; Fujisawa, T. *SL* **1992**, 204. (c) Bajwa, J. S.; Anderson, R. C. *TL* **1991**, *32*, 3021.
5. Taylor, M. D.; Grant, L. R. *JACS* **1955**, *77*, 1507.
6. Magnus, P.; Gallagher, T. *CC* **1984**, 389.
7. Johnson, F.; Paul, K. G.; Favara, D. *JOC* **1982**, *47*, 4254.
8. Ohtani, K.; Mizutani, K.; Kasai, R.; Tanaka, O. *TL* **1984**, *25*, 4537.
9. (a) Kende, A. S.; Rizzi, J. P. *JACS* **1981**, *103*, 4247. (b) Harrison, I. T. *CC* **1969**, 616.
10. (a) Salaün, J.; Conia, J. M. *CC* **1971**, 1579. (b) Aue, D. H.; Meshishnek, M. J.; Shellhamer, D. F. *TL* **1973**, 4799.
11. Fujisawa, T.; Sato, T.; Takeuchi, M. *CL* **1982**, 71.
12. Masada, H.; Murotani, Y. *BCJ* **1980**, *53*, 1181.
13. (a) Ma, S.; Lu, X. *TL* **1990**, *31*, 7653. (b) Ma, S.; Lu, X. *CC* **1990**, 1643.
14. Antonioletti, R.; Bonadies, F.; Monteagudo, E. S.; Scettri, A. *TL* **1991**, *32*, 5373.
15. Bell, T. W.; Hu, L.-Y. *TL* **1988**, *29*, 4819.
16. For the effect of LiI on organocopper reagents see: Lipshutz, B. H.; Kayser, F.; Siegmann, K. *TL* **1993**, *34*, 6693.
17. (a) Mori, Y.; Kuhara, M.; Takeuchi, A.; Suzuki, M. *TL* **1988**, *29*, 5419. (b) Mori, Y.; Takeuchi, A.; Kageyama, H.; Suzuki, M. *TL* **1988**, *29*, 5423.
18. Jubert, C.; Knochel, P. *JOC* **1992**, *57*, 5425.
19. Cabri, W.; Candiani, I.; DeBernardinis, S.; Francalanci, F.; Penco, S.; Santi, R. *JOC* **1991**, *56*, 5796.
20. (a) Pyne, S. G. *JOC* **1986**, *51*, 81. (b) Pyne, S. G. *TL* **1986**, *27*, 1691.
21. (a) Soderquist, J. A.; Anderson, C. L. *TL* **1988**, *29*, 2425. (b) Soderquist, J. A.; Anderson, C. L. *TL* **1988**, *29*, 2777. (c) Buss, A. D.; Warren, S.; Leake, J. S.; Whitham, G. H. *JCS(P1)* **1983**, 2215. (d) Buss, A. D.; Warren, S. *JCS(P1)* **1985**, 2307.
22. Townsend, J. M.; Spencer, T. A. *TL* **1971**, 137.

André B. Charette
Université de Montréal, Québec, Canada

Lithium *N*-Methylpiperazide

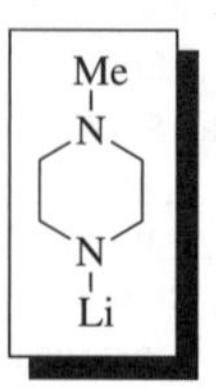

[105563-31-3] $C_5H_{11}LiN_2$ (MW 106.12)

(in situ protection of aryl aldehydes via α-amino alkoxide formation; α-amino alkoxides can direct or block *ortho*-lithiation[1])

Alternate Name: LNMP.
Preparative Method: prepared in situ by adding ***n-Butyllithium*** to *N*-methylpiperazine in anhydrmetous solvents.
Handling, Storage, and Precautions: highly flammable; should be kept under a nitrogen or argon atmosphere.

Lithium *N*-methylpiperazide is conveniently used as a reagent for in situ protection of aryl aldehydes through α-amino alkoxide formation.[1] Stability of LNMP-derived α-amino alkoxides to alkyllithium reagents and their solubility in organic solvents are reasons why LNMP is preferred over similar reagents.[1] The α-amino alkoxides obtained in situ from substituted benzaldehydes are not easily *ortho*-lithiated; however, refluxing a benzene solution of the α-amino alkoxide and excess ***n-Butyllithium*** for 3–12 h effects *ortho*-metalation. Addition of excess electrophile followed by aqueous workup gives moderate to good yields of the desired *ortho*-substituted aryl aldehydes (eq 1).[2]

CHO, R → 1. LNMP; 2. 3 equiv BuLi; 3. E^+ → CHO, E, R (1)

Regioselective substitution of bis-activated benzaldehydes can be achieved by lithiation/alkylation of an α-amino alkoxide obtained from LNMP. *p*-Anisaldehyde, on treatment with LNMP and ***s*-Butyllithium** in the presence of ***N,N,N′,N′*-Tetramethylethylenediamine** and subsequent quenching with **Iodomethane**, gives 4-methoxy-3-methylbenzaldehyde (eq 2).[3] This is in sharp contrast to the analogous reaction using ***N*-Lithio-*N,N′,N′*-trimethylethylenediamine** (LTMDA) in place of LNMP, where 4-methoxy-2-methylbenzaldehyde is produced in good yield.[1] Under the same reaction conditions, *o*- and *m*-anisaldehydes give C-3 and C-4 lithiated products, respectively. Both 3,5- and 2,4-dimethoxybenzaldehydes can be metalated between the methoxy substituents in excellent yield using LNMP, *s*-BuLi, and TMEDA.

1. LNMP
2. 2 equiv *s*-BuLi, TMEDA
3. MeI
73% (2)

Heterocyclic aldehydes can be metalated regioselectively using similar methodology to that described above.[4] For example, 2-thiophenecarbaldehyde on reaction with LNMP, *n*-BuLi, and TMEDA and subsequent quenching with iodomethane, gives 5-methyl-2-thiophenecarbaldehyde in 77% yield (eq 3).

1. LNMP
2. 3 equiv BuLi
3. MeI
77% (3)

When LTMDA is used instead of LNMP, a mixture of 3-and 5-methyl-2-thiophenecarbaldehydes is obtained. However, 3-thiophenecarbaldehyde and LNMP under similar reaction conditions give a mixture of C-5 and C-2 substituted products in a ratio of 83:17. Alkylation of 2-furaldehyde occurs exclusively at the 5-position when treated with LNMP, *n*-butyllithium, and iodomethane.[4] In a similar manner, 3-furaldehyde gives mainly C-2 alkylation, but 2-methyl-3-furaldehyde, on treatment with LNMP, *n*-BuLi, and iodomethane, gives 2,5-dimethyl-3-furaldehyde in 50% yield. Using similar reaction conditions, *N*-methylpyrrole-2-carbaldehyde and *N*-methylindole-2-carbaldehyde give the corresponding C-5 and C-3 substituted products on reaction with LNMP, *n*-BuLi, TMEDA, and iodomethane.[4] Interestingly, methoxypyridinecarbaldehydes, on treatment with LNMP, alkyllithium, and TMEDA, give metalation *ortho* to the methoxy group (Scheme 1).[5] This is in sharp contrast to LTMDA-assisted metalations, where substitution occurs *ortho* to the aldehyde group.

A regioselective lithiation/alkylation of a 1-Boc-3-formyl-1,4-dihydropyridine has been effected using LNMP and ***Mesityllithium*** as the base.[6] Dealkylation of several *o*-alkoxyaryl aldehydes can be achieved via the corresponding α-amino alkoxides prepared with LNMP.[7]

Scheme 1

Related Reagents. Lithium Diethylamide; Lithium Diisopropylamide; *N*-Lithio-*N,N′,N′*-trimethylethylenediamine; Lithium Morpholide.

1. Comins, D. L. *SL* **1992**, 615.
2. Comins, D. L.; Brown, J. D. *TL* **1981**, *22*, 4213.
3. Comins, D. L.; Brown, J. D. *JOC* **1984**, *49*, 1078.
4. Comins, D. L.; Killpack, M. O. *JOC* **1987**, *52*, 104.
5. Comins, D. L.; Killpack, M. O. *JOC* **1990**, *55*, 69.
6. Comins, D. L.; Weglarz, M. A. *JOC* **1988**, *53*, 4437.
7. Gillies, B.; Loft, M. S. *SC* **1988**, *18*, 191.

Daniel L. Comins & Sajan P. Joseph
North Carolina State University, Raleigh, NC, USA

Lithium Perchlorate[1]

$LiClO_4$

[7791-03-9] $ClLiO_4$ (MW 106.39)

(mild Lewis acid[1] for cycloaddition reactions,[2] conjugate additions,[3] ring opening of epoxides,[4] and ring expansion of cyclopropanes[5])

Physical Data: mp 236 °C; bp 430 °C (dec); *d* 2.428 g cm^{-3}.

Solubility: 60 g/100 mL H_2O (25 °C); 150 g/100 mL H_2O (89 °C); 152 g/100 mL EtOH (25 °C); 137 g/100 mL acetone (25 °C); 182 g/100 mL MeOH (25 °C); 114 g/100 mL ether.

Form Supplied in: white solid; widely available in anhydrous form or as the trihydrate; it is usually used in solution in ether or MeOH.

Purification: anhydrous lithium perchlorate is prepared by heating the commercially available anhydrous material or its trihydrate at 160 °C for 48 h under high vacuum (P_2O_5 trap).

Handling, Storage, and Precautions: the anhydrous material should be used as prepared for best results. The decomposition of lithium perchlorate starts at about 400 °C and becomes rapid at 430 °C, yielding lithium chloride and oxygen. Perchlorates are potentially explosive and should be handled with caution.

Lewis Acid Catalyst for Cycloaddition Reactions and Carbonyl Addition Reactions. This reagent, usually prepared as a

5.0 M solution in diethyl ether, produces a dramatic rate acceleration of Diels–Alder reactions (eq 1). Evidence shows that this rate acceleration, which was initially thought to be a result of high internal solvent pressure, is due to the Lewis acid character of the lithium ion.[6]

(1)

5.0M $LiClO_4$, Et_2O, 5 h, rt, 80% (*endo*:*exo* = 2.9:1)
benzene, 72 h, 60 °C, 74% (*endo*:*exo* = 1:11.5)

Under these conditions, reasonable levels of diastereoselectivity have been observed in the reaction between a chiral diene and ***N*-Phenylmaleimide** (eq 2).[7] An interesting protecting group dependence of diastereoselectivities has also been observed in the hetero-Diels–Alder reaction of *N*-protected α-amino aldehydes with 1-methoxy-3-*t*-butyldimethylsilyloxybutadiene to produce dihydropyrones (eqs 3 and 4).[8]

5.0 M $LiClO_4$, Et_2O, −30 °C, 12 h, 74%

(2) 7.4:1

1. 3.0 M $LiClO_4$, Et_2O; 2. TFA, CH_2Cl_2, 66%

(3) single diastereomer

1. 3.0 M $LiClO_4$, Et_2O; 2. TFA, CH_2Cl_2, 79%

(4) 8:1

O-Silylated ketene acetals undergo 1,4-conjugate addition to hindered α,β-unsaturated carbonyl systems[3] and quinones[9] in the presence of $LiClO_4$.

[1,3]-Sigmatropic Rearrangements. In contrast to the [3,3]-sigmatropic rearrangement observed under thermal conditions, allyl vinyl ethers undergo [1,3]-sigmatropic rearrangements at rt when submitted to 1.5–3.0 M $LiClO_4$ in Et_2O (eq 5).[10]

3.0M $LiClO_4$, Et_2O, 90%

(5) 5:1

Epoxide Opening. $LiClO_4$ is an efficient promotor for the regioselective nucleophilic opening of oxiranes with amines,[11] cyanide,[12] azide,[13] thiols,[14] halides,[15] and lithium acetylides.[16] The regioselective opening of oxiranes with lithium enolates derived from ketones has also been observed in the presence of $LiClO_4$ (eq 6).[17]

$LiClO_4$, THF, 25 °C, 24 h, 90%

(6) without $LiClO_4$, 30%

Ring Expansion. The condensation of aldehydes and ketones with diphenylsulfonium cyclopropylide produces oxaspiropentanes which undergo ring expansion to produce cyclobutanones upon treatment with lithium perchlorate.[18]

Related Reagents. Diethylaluminum Chloride; Dimethylaluminum Chloride; Lithium Tetrafluoroborate; Magnesium Bromide.

1. (a) Loupy, A.; Tchoubar, B. *Salts Effects in Organic and Organometallic Chemistry*; VCH: Weinheim, 1992. (b) Grieco, P. A. *Aldrichim. Acta* **1991**, *24*, 59. (c) Waldmann, H. *AG(E)* **1991**, *30*, 1306.
2. (a) Grieco, P. A.; Nunes, J. J.; Gaul, M. D. *JACS* **1990**, *112*, 4595. (b) Braun, R.; Sauer, J. *CB* **1986**, *119*, 1269.
3. Grieco, P. A.; Cooke, R. J.; Henry, K. J.; VanderRoest, J. M. *TL* **1991**, *32*, 4665.
4. Chini, M.; Crotti, P.; Flippin, L. A.; Macchia, F. *JOC* **1990**, *55*, 4265.
5. Rickborn, B.; Gerkin, R. M. *JACS* **1971**, *93*, 1693.
6. (a) Forman, M. A.; Dailey, W. P. *JACS* **1991**, *113*, 2761. (b) Desimoni, G.; Faita, G.; Righetti, P. P.; Tacconi, G. *T* **1991**, *47*, 8399.
7. Hatakeyama, S.; Sugawara, K.; Takano, S. *CC* **1992**, 953.
8. Grieco, P. A.; Moher, E. D. *TL* **1993**, *34*, 5567.
9. Ipaktschi, J.; Heydari, A. *AG(E)* **1992**, *31*, 313.
10. Grieco, P. A.; Clark, J. D.; Jagoe, C. T. *JACS* **1991**, *113*, 5488.
11. Chini, M.; Crotti, P.; Macchia, F. *JOC* **1991**, *56*, 5939.
12. Chini, M.; Crotti, P.; Favero, L.; Macchia, F. *TL* **1991**, *32*, 4775.

13. Chini, M.; Crotti, P.; Macchia, F. *TL* **1990**, *31*, 5641.
14. Chini, M.; Crotti, P.; Giovani, E.; Macchia, F.; Pineschi, M. *SL* **1992**, 303.
15. (a) Chini, M.; Crotti, P.; Gardelli, C.; Macchia, F. *T* **1992**, *48*, 3805. (b) Chini, M.; Crotti, P.; Flippin, L. A.; Macchia, F. *JOC* **1990**, *55*, 4265.
16. Chini, M.; Crotti, P.; Favero, L.; Macchia, F. *TL* **1991**, *32*, 6617.
17. Chini, M.; Crotti, P.; Favero, L.; Pineschi, M. *TL* **1991**, *32*, 7583.
18. Trost, B. M.; Bogdanowicz, M. J. *JACS* **1973**, *95*, 5321.

André B. Charette
Université de Montréal, Québec, Canada

Lithium Tetrafluoroborate[1]

$LiBF_4$

[14283-07-9] BF_4Li (MW 93.75)

(fluoride source for cleavage of silyl ethers;[2] mild Lewis acid capable of promoting acetal hydrolysis,[3] rearrangement of oxaspiropentanes,[4] facilitating the addition of arenesulfenyl chlorides to alkenes;[5] catalysis of Diels–Alder-9 cycloadditions;[6] used as an electrolyte in oxidative cleavage of 2-oxo-1-cyclohexanones[7])

Physical Data: mp 5 °C.
Solubility: sol methanol, acetonitrile.
Form Supplied in: anhydrous white solid; also available as a 1 M solution in acetonitrile.
Drying: anhydrous material is obtained by overnight drying in an oven at 80–90 °C.[1]
Handling, Storage, and Precautions: extremely hygroscopic; should be handled and stored under an inert atmosphere; contact irritant.

Cleavage of Ethers. $LiBF_4$ is an effective reagent for the removal of silyl ethers.[2,8,9] An excess of the salt in acetonitrile at ambient temperature effects cleavage of *t*-butyldimethylsilyl ethers (eqs 1–3).

3 equiv $LiBF_4$, CH_2Cl_2, MeCN, 60 h, rt, 86% (1)

$LiBF_4$, CH_2Cl_2, MeCN, 5 h, rt, 82% (2)

$LiBF_4$, THF, MeCN, reflux, 3 d, 91% (3)

Use of this reagent under more vigorous conditions results in removal of (trimethylsilyl)ethoxymethyl (SEM) ethers and methoxymethyl (MOM) ethers (eqs 4, 5).[11]

$LiBF_4$, MeCN, H_2O, 72 °C, 100% (4)

$LiBF_4$, MeCN, 50 °C, 24 h, 89% (5)

$LiBF_4$ is particularly effective in the deprotection of β-trimethylsilylethyl protected glycosides (eq 6).[12] Notably, the β-anomers undergo hydrolysis more rapidly than the corresponding α-anomers. Benzylidene acetals are cleaved under these conditions (eq 7).

$LiBF_4$, MeCN, 70 °C, 82% (6)

$LiBF_4$, MeCN, 70 °C, 95% (7)

Hydrolysis of Acetals. Treatment of acetals with $LiBF_4$ in wet acetonitrile provides carbonyl compounds in moderate yield.[3] Aliphatic acetals (eq 8) give the corresponding carbonyl products in near quantitative yield. Cyclic dioxolane derivatives are hydrolyzed at a slower rate (eq 9). Substitution of the dioxolane further hinders the reaction (eq 10). Aliphatic aldehydes protected as 1,3-dioxanes are inert to these hydrolysis conditions.

(8) $LiBF_4$, 2% H_2O, MeCN, 10 h, rt, 95%

(9) $LiBF_4$, 2% H_2O, MeCN, 5 h, rt, 40%

(10) $LiBF_4$, 2% H_2O, MeCN, 5 h, rt, <5%

Lewis Acid Promoted Reactions. $LiBF_4$ promotes a high yielding rearrangement of oxaspiropentanes to cyclobutanones (eq 11).[4] In this regard, $LiBF_4$ is reported to be superior to ***Lithium Perchlorate***.

(11) 1. cyclopropylidene=SPh_2; 2. $LiBF_4$, 99%

$LiBF_4$ is an effective catalyst for the intramolecular Diels–Alder reaction. A single *cis*-fused cycloadduct was obtained in quantitative yield from the triene (eq 12).[6] The use of other Lewis acids gives poor selectivity and lower yields of the bicyclic product.

(12) 1.1 equiv $LiBF_4$, PhH, MeCN, 72 h, rt, 100%

$LiBF_4$ has been employed as an electrolyte in the oxidative cleavage of 2-oxy-1-cyclohexanones, which gives good yields of the ring-opened dicarbonyl compound (eq 13).[7]

(13) $-2e^-$, MeOH, $LiBF_4$, (Pt) electrode, 80–88%

The Lewis acid character of $LiBF_4$ enhances the electrophilicity of arenesulfenyl chlorides in their additions to carbon–carbon double bonds.[5] In appropriately disposed bicyclic systems (eq 14), transannular bond formation is observed.

(14) 2-nitrobenzenesulfenyl chloride, 4 equiv $LiBF_4$, MeCN, 20 °C

Related Reagents. Lithium Fluoride; Lithium Perchlorate; Tetra-butylammonium Fluoride.

1. Shapiro, I.; Weiss, H. O. *JACS* **1953**, *75*, 1753.
2. Metcalf, B. W.; Burkhart, J. P.; Jund, K. *TL* **1980**, *21*, 35.
3. Lipshutz, B. H.; Harvey, P. F. *SC* **1982**, *12*, 267.
4. Trost, B. M.; Preckel, M. *JACS* **1973**, *95*, 7862.
5. Zefirov, N. S.; Koz'min, A. S.; Kirin, V. N.; Zhdankin, V. V.; Caple, R. *JOC* **1981**, *46*, 5265.
6. Smith, D. A.; Houk, K. N. *TL* **1991**, *32*, 1549.
7. Torii, S.; Inokuchi, T.; Oi, R. *JOC* **1982**, *47*, 47.
8. Grisenti, P.; Ferraboschi, P.; Mazocchi, A.; Santaniello, E. *TL* **1992**, *48*, 3827.
9. Magnus, P.; Mugrage, B.; DeLuca, M. R.; Cain, G. A. *JACS* **1990**, *112*, 5220.
10. Ireland, R. E.; Varney, M. D. *JOC* **1986**, *51*, 635.
11. Watanabe, H.; Mori, K. *JCS(P1)* **1991**, *12*, 2919.
12. Lipshutz, B. H.; Pegram, J. J.; Money, M. C. *TL* **1981**, *22*, 4603.

Paul J. Coleman
Indiana University, Bloomington, IN, USA

Lithium 2,2,6,6-Tetramethylpiperidide

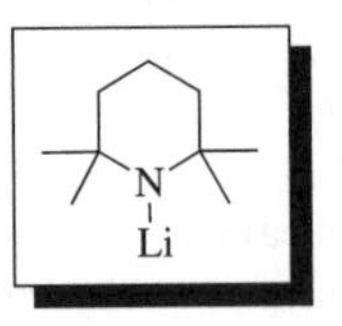

[38227-87-1] $C_9H_{18}LiN$ (MW 147.22)

(strong, highly hindered, nonnucleophilic base ($pK_a = 37.3$)[1] capable of selective deprotonation of aromatics, heteroaromatics, and aliphatic C–H acidic sites in the presence of a variety of functional groups; also compatible with several electrophiles for in situ quenching of kinetically derived lithiated species[2–5])

Alternate Names: LiTMP; LTMP.

Physical Data: exists in THF solution as a dimer–monomer equilibrium mixture; additives such as HMPA increase monomer concentration.[6,7] X-ray structure determination shows that LiTMP crystallizes as a tetramer from hexane/pentane mixtures.[8]

Solubility: sol most organic solvents including THF, Et_2O, hexane.

Preparative Method: prepared in Et_2O or THF solutions immediately before use by treatment of commercially available dry ***2,2,6,6-Tetramethylpiperidine*** with ***n-Butyllithium*** (1:1). Tetramethylpiperidine is dried by heating a mixture of the base with CaH_2 at reflux for 4 h in a preflamed flask, followed by distillation at atmospheric pressure (bp 152 °C). It should be stored in a septum sealed bottle.

Handling, Storage, and Precautions: LiTMP solutions are pyrophoric, can cause severe burns, and should always be handled and transferred under an inert atmosphere. Solutions of LiTMP show a loss of activity (50% in THF; 60% in Et_2O) after 12 h at 24 °C.[9] Use in a fume hood.

Benzyne Formation. The earliest uses of LiTMP as a base involved the deprotonation of benzyl chloride to give phenylcarbene and of 2-chloroanisole to give 3-methoxybenzyne.[10] LiTMP may also be used for heteroaryne generation; for example, treatment of 1,3-bis(TMS)isobenzofuran with 3-bromopyridine in the presence of LiTMP gives the corresponding cycloaddition product (eq 1).[11]

(1)

In situ Compatibility with Electrophiles. Martin first demonstrated that the low nucleophilicity of LiTMP makes it compatible, at low temperatures, with certain electrophiles, e.g. ***Chlorotrimethylsilane***, ***Trimethyl Borate***.[2a] This allows the preparation of 2- and 2,6-silylated benzoates and benzonitriles by equilibrium controlled *ortho*-lithiation processes (eq 2). The distribution of mono- and disilylated products may be controlled by the number of equivalents of LiTMP; these products may be converted into corresponding bromo and iodo derivatives by *ipso*-halodesilylation.[2a,12] Similar ***Lithium Diisopropylamide***–TMSCl compatibility has been demonstrated.[5,13]

(2)

R = C≡N, CO_2-*i*-Pr

Analogous use of the LiTMP–***Mercury(II) Chloride*** combination allows selective functionalization of cubanecarboxamides (eq 3) and cyclopropanecarboxamides (eq 4),[14,15] presumably as a consequence of the high s character of the C–H bonds in these systems.

(3)

(4)

Am = CON(*i*-Pr)$_2$

Application of Martin's conditions to *N*,*N*-dimethylbenzamide leads, surprisingly, to α,α′-disilylation in good yield (eq 5).[16] Further treatment with LiTMP/TMSCl gives the *ortho*-silylated product. The bis-TMS derivative serves as a useful directed metalation group and may be readily converted to benzoic acid, benzaldehyde, and benzyl alcohol derivatives.

(5)

Remote Metalation. LiTMP may be used interchangeably with LDA to effect conversion of diaryl amides into fluorenones by virtue of a complex induced proximity effect (eq 6).[17] This general route complements Friedel–Crafts chemistry for the preparation of fluorenones.

(6)

Directed *ortho* Metalation. *N*-*t*-Boc-pyrrole undergoes clean deprotonation with LiTMP.[18] Low-temperature LiTMP lithiation of pyrazinecarboxamide followed by deuteration leads to isomeric products reflecting the temperature at which the reaction is quenched (eq 7).[3] This is rationalized by the thermodynamic stability of the *ortho*-amido lithiated species at the higher temperature.

(7)

Am = CONH-*t*-Bu

	3-D product	6-D product
0 °C	0%	75%
−70 °C	40%	25%

The use of LiTMP in heterocyclic directed *ortho*-metalation chemistry is advantageous. For example, pyridine amide and oxazoline derivatives are smoothly deprotonated and lead, after electrophile quench, to 4-substituted products (e.g. eq 8),[19] which may be used for the synthesis of natural products (e.g. eq 9) and pharmaceuticals.[20–22] In contrast to LDA and, of course, alkyllithiums, no nucleophilic attack of the pyridine ring is observed.[19] The considerable C–H acidity of methylpyridines allows deprotonation–condensation reactions which are of general synthetic value (eq 9).[20]

(8)

E^+ = D_2O, 70%; PhCHO, 50%; Et_2CO, 52%; MeI, 80%; BuI, 9%

(9)

Sesbanine

In some cases, LiTMP complements the metalation regioselectivity of LDA. For example, LDA metalation of 2,4-

dichloropyrimidine gives the 5-substituted product whereas LiTMP affords mainly the product of 6-substitution (eq 10).[23]

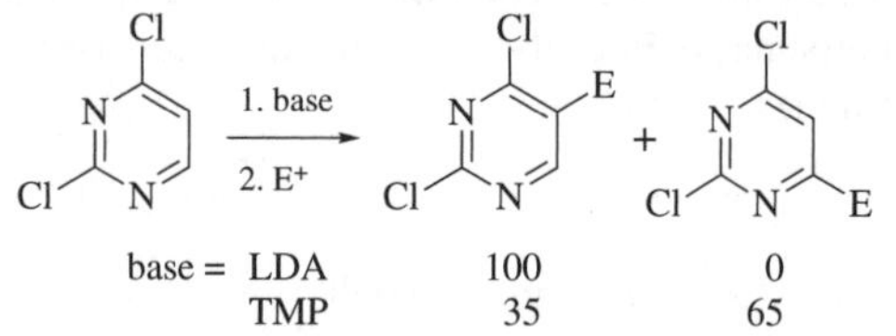

base =	5-E	6-E
LDA	100	0
TMP	35	65

(10)

Enolate Formation. LiTMP is widely used to generate α-stabilized carbanions. For example, in the synthesis of paniculide A, the use of LiTMP is critical; LDA fails to give the required product (eq 11).[24]

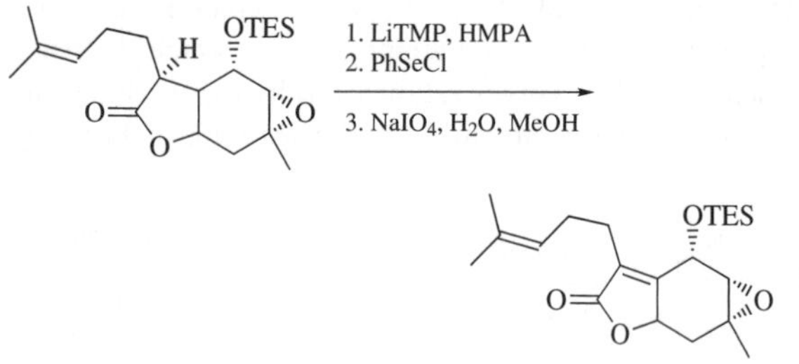

(11)

LiTMP is especially useful in cases where the reactants are sensitive to nucleophilic attack (e.g. eq 12) (cf. 66% yield with LDA).[10]

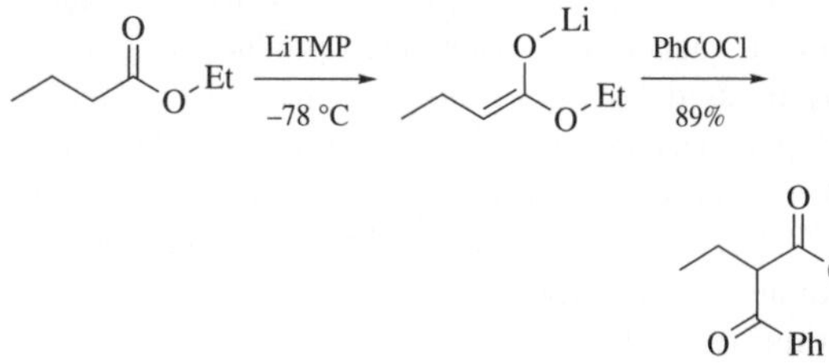

(12)

LiTMP is also used to deprotonate nonenolizable sites. For example, bridgehead deprotonation of (−)-camphenilone is followed by self-condensation to give a high yield of dimeric product (eq 13).[25]

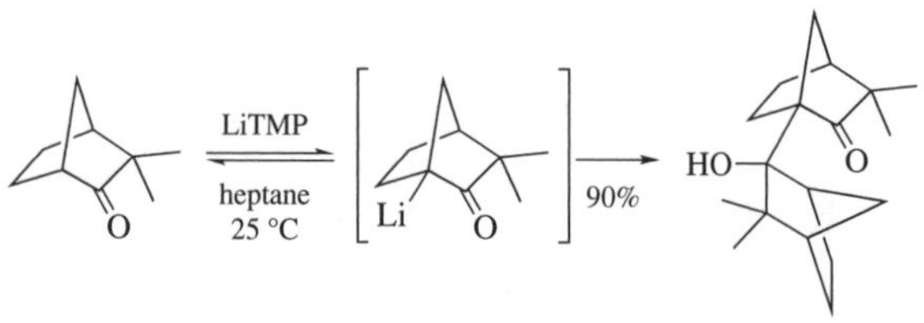

(13)

Under similar conditions, LiTMP generates the acyllithium of a nonenolizable aldehyde which undergoes self-condensation to give an acyloin (eq 14).[25]

The primary reservation concerning the use of LiTMP in synthesis, particularly on a large scale, is the cost of tetramethylpiperidine relative to other amine bases. However, an aqueous acid extraction of the reaction mixture allows the recovery of tetramethylpiperidinium salt from which the free base can be obtained by neutralization and distillation. This recovery method is dependent on the acid stability of the reaction products.

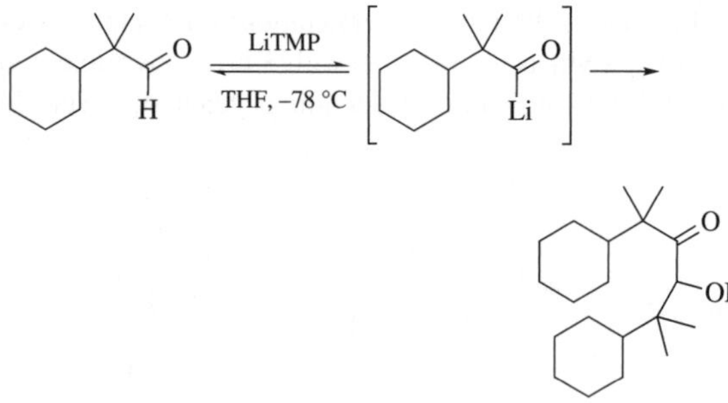

(14)

Related Reagents. Lithium Diethylamide; Lithium Diisopropylamide; Lithium Hexamethyldisilazide; Lithium Piperidide; Lithium Pyrrolidide; Potassium Hexamethyldisilazide; Sodium Hexamethyldisilazide.

1. Fraser, R. R.; Mansour, T. S. *JOC* **1984**, *49*, 3443.
2. (a) Krizan, T.; Martin, J. C. *JACS* **1983**, *105*, 6155. (b) Krizan, T. D.; Martin, J. C. *JOC* **1982**, *47*, 2681. (c) Fraser, R. R.; Savard, S. *CJC* **1986**, *64*, 621.
3. Turck, A.; Plé, N.; Trohay, D.; Ndzi, B.; Quéguiner, G. *JHC* **1992**, *29*, 699.
4. Beak, P.; Lee, B. *JOC* **1989**, *54*, 458.
5. Eaton, P. E.; Martin, R. M. *JOC* **1988**, *53*, 2728.
6. Renaud, P.; Fox, M. A. *JACS* **1988**, *110*, 5705.
7. Romesberg, F. E.; Gilchrist, J. H.; Harrison, A. T.; Fuller, D. J.; Collum, D. B. *JACS* **1991**, *113*, 5751.
8. Lappert, M. F.; Slade, M. J.; Singh, A. *JACS* **1983**, *105*, 302.
9. Kopka, I. E.; Fataftah, A.; Rathke, M. W. *JOC* **1987**, *52*, 448.
10. Olofson, R. A.; Dougherty, C. M. *JACS* **1973**, *95*, 582.
11. Crump, S. L.; Netka, J.; Rickborn, B. *JOC* **1985**, *50*, 2746.
12. Unrau, C. M.; Campbell, M. G.; Snieckus, V. *TL* **1992**, *23*, 2773.
13. Corey, E. J.; Gross, A. W. *TL* **1984**, *25*, 495.
14. Eaton, P. E.; Castaldi, G. *JACS* **1985**, *107*, 724.
15. Eaton, P. E.; Daniels, R. G.; Casucci, D.; Cunkle, G. T.; Engel, P. *JOC* **1987**, *52*, 2100.
16. Cuevas, J.-C.; Patil, P.; Snieckus, V. *TL* **1989**, *30*, 5841.
17. Zhao, B.; Snieckus, V., unpublished results.
18. Hasan, I.; Marinelli, E. R.; Lin, L.-C. C.; Fowler, F. W.; Levy, A. B. *JOC* **1981**, *46*, 157.
19. Meyers, A. I.; Gabel, R. A. *JOC* **1982**, *47*, 2633.
20. Iwao, M.; Kuraishi, T. *TL* **1983**, *24*, 2649.
21. Watanabe, M.; Shinoda, E.; Shimizu, Y.; Furukawa, S.; Iwao, M.; Kuraishi, T. *T* **1987**, *43*, 5281.
22. Dunbar, P. G.; Martin, A. R. *H* **1987**, *26*, 3165.
23. Plé, N.; Turck, A.; Martin, P.; Barbey, S.; Quéguiner, G. *TL* **1993**, *34*, 1605.
24. Smith, A. B., III; Richard, R. E. *JOC* **1981**, *46*, 4814.
25. Shiner, C. S.; Berks, A. H.; Fisher, A. M. *JACS* **1988**, *110*, 957.

Mike Campbell & Victor Snieckus
University of Waterloo, Ontario, Canada

2,6-Lutidine

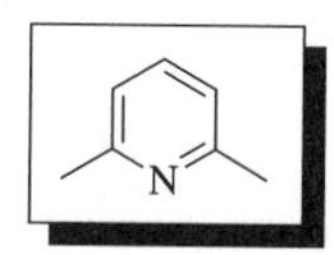

[108-48-5] C_7H_9N (MW 107.17)

(basic reagent/catalyst for silylation of alcohols,[2–10] for Rosenmund reduction,[14] and for preparation of boron enolates[15])

Physical Data: colorless oily liquid with an odor of pyridine and peppermint; bp 144.0 °C/760 mmHg, 79 °C/87 mmHg; mp −5.9 °C; *dE* (20 °C) 0.9252 g cm^{-3}; n_D^{20} 1.49797. The major synthetic applications reported for 2,6-lutidine exploit its weakly nucleophilic nature (a result of steric crowding at the ring nitrogen) but moderately basic character (pK_a of its conjugate acid is 6.7).

Solubility: sol common organic solvents (e.g. ether, THF, DMF, alcohol); shows considerable water solubility (ca. 27% (w/w) at 45 °C).

Form Supplied in: a widely available article of commerce. Redistilled (99+ % purity) is available commercially.

Drying: commonly dried by treatment with KOH or sodium followed by distillation, or by refluxing with, followed by distillation from, BaO[1] or CaH_2.[2]

Purification: both 3- and 4-picolines are common contaminants which can be removed by distillation from $AlCl_3$ (14 g/100 mL 2,6-lutidine, which also removes traces of water), or by addition of BF_3 (4 mL/100 mL of 2,6-lutidine) to anhydrous, fractionally distilled 2,6-lutidine, followed by redistillation.[1]

Handling, Storage, and Precautions: toxic flammable liquid; should be handled with caution in a fume hood.

Alcohol Protection Reactions. The protection of alcohols as hindered silyl ethers proceeds in >90% yield using the appropriate trialkylsilyl triflate in the presence of anhydrous 2,6-lutidine as an acid scavenger. ***Triisopropylsilyl Trifluoromethanesulfonate***/2,6-lutidine in CH_2Cl_2 (−78 °C to 0 °C) allows for the protection of primary and secondary (but not tertiary) alcohols (eq 1).[2,3] Tertiary alcohols can be protected as the somewhat less crowded *i*-$PrEt_2Si$ ether (*i*-$PrEt_2SiOTf$/2,6-lutidine, CH_2Cl_2, rt).[4] The silylation of tertiary and unreactive secondary alcohols can also be achieved using ***t-Butyldimethylsilyl Trifluoromethanesulfonate***/2,6-lutidine in CH_2Cl_2 (0–25 °C);[2,5] silylation at −78 °C allows for the selective protection of a secondary alcohol in the presence of a secondary allylic alcohol (eq 2).[6] The silylation of secondary allylic alcohols as the bulky tribenzylsilyl ether or tri-*p*-xylylsilyl ether can be accomplished using the appropriate tris(arylmethyl)chlorosilane/2,6-lutidine in DMF at −20 °C.[7] The bis-protection of 1,2-, 1,3-, and 1,4-diols as the corresponding dialkylsilylene derivatives can be achieved in high yield using ***Diisopropylsilyl Bis(trifluoromethanesulfonate)*** or ***Di-t-butylsilyl Bis(trifluoromethanesulfonate)*** at rt in $CDCl_3$ in the presence of 2,6-lutidine (eq 3);[8] interestingly, 2,6-lutidine was found to retard the 3′,5′-protection of nucleosides in DMF solution.[9] Protection of a 1,3-diol as its methylene acetal can be achieved using ***Trimethylsilyl Trifluoromethanesulfonate***/2,6-lutidine in $MeOCH_2OMe$ (0 °C, 15 min, 79%).[10]

(1) *i*-Pr_3SiOTf, 2,6-lutidine, CH_2Cl_2, 0 °C, 2 h, 91%

(2) TBDMSOTf, 2,6-lutidine, CH_2Cl_2, −78 °C, 30 min, 91%

(3) *t*-$Bu_2Si(OTf)_2$, 2,6-lutidine (3 equiv), $CDCl_3$, 25 °C, 96%

Other Protection/Deprotection Reactions. 2,6-Lutidine is more effective than ***Silver(I) Carbonate*** or other amine bases at mediating the conversion of a 2-acetylpyranosyl bromide to an orthoester (eq 4).[11] Side reactions are minimized due to the lower acidity of the conjugate acid of 2,6-lutidine (pK_a 6.7) when compared to the other substituted pyridines examined. *Exo* stereoselectivity is maximized when using 2,6-lutidine rather than silver carbonate. 2,6-Lutidine considerably accelerates bond cleavage during the triarylamine radical cation-mediated oxidative cleavage of primary and secondary benzyl[12] and *p*-methoxybenzyl[13] ethers to the corresponding alcohols.

(4) I–C6H4–CH_2OH, 2,6-lutidine, $CHCl_3$, 69%

Rosenmund Reduction. 2,6-Lutidine is usefully employed as an HCl scavenger in the catalytic hydrogenation of aliphatic acyl chlorides to aldehydes (Rosenmund reduction) (eq 5).[14] These reactions are run in THF at <0.25 M in substrate to prevent poisoning of the Pd catalyst. 2,6-Lutidine is far superior to other basic additives (e.g. ***Diisopropylethylamine***, *N,N*-dimethylaniline, NaOAc). This method can tolerate ketone and ester functionality, proceeds on both unhindered and crowded acyl chlorides, and is especially suitable for the preparation of sensitive aldehydes. These reactions are faster and superior to those performed under classical Rosenmund reduction conditions (no added base), where high reaction temperatures are required and where the HCl byproduct can mediate side reactions.

(5) < 0.25 M; 1 atm H_2, 10% Pd/C, 2,6-lutidine, THF, rt, 1–2 h, 89%

Preparation of Boron Enolates. Dialkylboryl triflates react with ketones in the presence of a hindered tertiary amine base (typically 2,6-lutidine or ***Diisopropylethylamine***) to afford boron enolates that are useful in stereoselective directed aldol condensation reactions.[15] 2,6-Lutidine is superior to other tertiary amines for the regioselective formation of the more stable 9-BBN boron enolate from alkyl methyl ketones using ***9-Borabicyclononyl Trifluoromethanesulfonate*** (eq 6).[15b,16,17] The 2,6-lutidinium triflate byproduct is presumably sufficiently acidic to facilitate enolate equilibration, resulting in thermodynamically controlled enolate formation.[15b,17] Interestingly, alkyl ethyl ketones are converted under apparently identical conditions to the less substituted boron enolate[18] with relatively high (>10:1) (*Z*) stereoselectivity (eq 7);[17,18] these reactions presumably proceed under kinetic control.[17] An opposite sense of regiocontrol has been observed when preparing dibutylboron enolates rather than 9-BBN enolates from alkyl methyl ketones using ***Di-n-butylboryl Trifluoromethanesulfonate***/2,6-lutidine,[17] although use of the more hindered and basic diisopropylethylamine in place of 2,6-lutidine affords much higher levels of regiocontrol.[17] Poor (*Z*)-dibutylboron enolate stereoselectivity is observed with alkyl ethyl ketones using 2,6-lutidine under kinetic control;[19,20] somewhat improved (>3:1) (*Z*) selectivity is achieved under thermodynamic control.[20] Much higher (*Z*) stereoselectivity is achieved under strictly kinetic control using diisopropylethylamine as base.[17,19,20]

9-BBNOTf, 2,6-lutidine; ether, −78 °C, 16 h — only (6)

9-BBNOTf, 2,6-lutidine; ether, −78 °C — (*Z*):(*E*) = 10:1 (7)

Radical Cyclopentannulation Reactions. ***Zinc/Chlorotrimethylsilane***-mediated cyclopentannulation reactions, involving the addition of an α-siloxy radical to an alkene or alkyne, are best performed in the presence of 2,6-lutidine, which prevents proton and $ZnCl_2$-catalyzed elimination of the tertiary siloxy group in the immediate product (eq 8).[21]

1. Zn, TMSCl, 2,6-lutidine, THF, reflux, 12–18 h; 2. aq work-up — CO_2Me 68% + CO_2Me 14% (8)

Oxidation of Allylic Alcohols. 2,6-Lutidine is added to prevent HCl-mediated side reactions during the ruthenium(II)-catalyzed oxidation of allylic alcohols (including labile allylic alcohols such as retinol) to α,β-unsaturated aldehydes using molecular oxygen (eq 9).[22] Other less hindered pyridine derivatives poison the catalyst and are ineffective.

Bromination of Sensitive Allylic Alcohols. Preparation of highly labile allyl bromides can be achieved using ***Thionyl Bromide***/2,6-lutidine (eq 10).[23] Virtually no bromide is formed in the absence of 2,6-lutidine.

3 mol % $RuCl_2(PPh_3)_3$, 3–5 mol % 2,6-lutidine; 1 atm O_2, $ClCH_2CH_2Cl$, rt, 48 h — 57% (9)

HO–(…)$_5$–CO_2Et → $SOBr_2$, 2,6-lutidine; −20 °C, 25 min, >70% → Br–(…)$_5$–CO_2Et (10)

Related Reagents. 2,4,6-Collidine; 1,5-Diazabicyclo-[4.3.0]non-5-ene; 1,8-Diazabicyclo[5.4.0]undec-7-ene; Diisopropylethylamine; Pyridine; Triethylamine; Quinoline.

1. Perrin, D. D.; Armarego, W. L. F. *Purification of Laboratory Chemicals*, 3rd ed.; Pergamon: Oxford, 1988; pp 212–213.
2. Corey, E. J.; Cho, H.; Rucker, C.; Hua, D. H. *TL* **1981**, *22*, 3455.
3. Tanaka, K.; Yoda, H.; Isobe, Y.; Kaji, A. *JOC* **1986**, *51*, 1856.
4. Toshima, K.; Mukaiyama, S.; Kinoshita, M.; Tatsuta, K. *TL* **1989**, *30*, 6413.
5. Andrews, R. C.; Teague, S. J.; Meyers, A. I. *JACS* **1988**, *110*, 7854.
6. Askin, D.; Angst, C.; Danishefsky, S. *JOC* **1987**, *52*, 622.
7. Corey, E. J.; Ensley, H. E. *JOC* **1973**, *38*, 3187.
8. Corey, E. J.; Hopkins, P. B. *TL* **1982**, *23*, 4871.
9. Furusawa, K.; Ueno, K.; Katsura, T. *CL* **1990**, 97.
10. Matsuda, F.; Kawasaki, M.; Terashima, S. *TL* **1985**, *26*, 4639.
11. Mazurek, M.; Perlin, A. S. *CJC* **1965**, *43*, 1918.
12. Schmidt, W.; Steckhan, E. *AG(E)* **1979**, *18*, 801.
13. Schmidt, W.; Steckhan, E. *AG(E)* **1978**, *17*, 673.
14. Burgstahler, A. W.; Weigel, L. O.; Shaefer, C. G. *S* **1976**, 767.
15. (a) Braun, M. In *Advances in Carbanion Chemistry*; Snieckus, V., Ed.; Jai: Greenwich, CT, 1992; Vol. 1, pp 177–247. (b) Evans, D. A.; Nelson, J. V.; Taber, T. R. *Top. Stereochem.* **1982**, *13*, 1.
16. Inoue, T.; Uchimaru, T.; Mukaiyama, T. *CL* **1977**, 153.
17. Inoue, T.; Mukaiyama, T. *BCJ* **1980**, *53*, 174.
18. Van Horn, D. E.; Masamune, S. *TL* **1979**, 2229.
19. Evans, D. A.; Vogel, E.; Nelson, J. V. *JACS* **1979**, *101*, 6120.
20. Evans, D. A.; Nelson, J. V.; Vogel, E.; Taber, T. R. *JACS* **1981**, *103*, 3099.
21. Corey, E. J.; Pyne, S. G. *TL* **1983**, *24*, 2821.
22. Matsumoto, M.; Ito, S. *CC* **1981**, 907.
23. Boschelli, D.; Takemasa, T.; Nishitani, Y.; Masamune, S. *TL* **1985**, *26*, 5239.

Paul Sampson & Thomas E. Janini
Kent State University, OH, USA

Magnesium Bromide

$MgBr_2$

($MgBr_2$)		
[2923-28-6]	Br_2Mg	(MW 184.11)
($MgBr_2 \cdot 6H_2O$)		
[13446-53-2]	$H_{12}Br_2MgO_6$	(MW 292.23)
($MgBr_2 \cdot OEt_2$)		
[29858-07-9]	$C_4H_{10}Br_2MgO$	(MW 258.25)

(Lewis acid capable of catalyzing selective nucleophilic additions,[1] cycloadditions,[2] rearrangements,[3] coupling reactions;[4] effective brominating agent[5])

Physical Data: mp 165 °C (dec) (etherate >300 °C; fp 35 °C).
Solubility: sol alcohol, H_2O; etherate sol common organic solvents.
Form Supplied in: white solid, widely available; etherate gray solid.
Preparative Method: the etherate is easily prepared from reacting a slight excess of ***Magnesium*** turnings with ***1,2-Dibromoethane*** in anhydrous diethyl ether.[13]
Handling, Storage, and Precautions: etherate is flammable and moisture sensitive; freshly prepared material is most reactive and anhydrous. Irritant.

Nucleophilic Additions. $MgBr_2$ has been shown to form discrete bidentate chelates with various species,[6] particularly α- and/or α,β-alkoxy carbonyl compounds,[7] and thus functions as a diastereofacial control element in many nucleophilic addition reactions. In many cases, its inclusion completely reverses the nonchelation-controlled stereochemistry observed with nonchelating Lewis acids such as ***Boron Trifluoride Etherate***. Highest diastereoselectivity is observed with α-substituted aldehydes (eq 1).[8] High selectivities are observed for β-alkoxy aldehydes as well, including cases where three contiguous chiral centers are defined during the reaction (eq 2).[9]

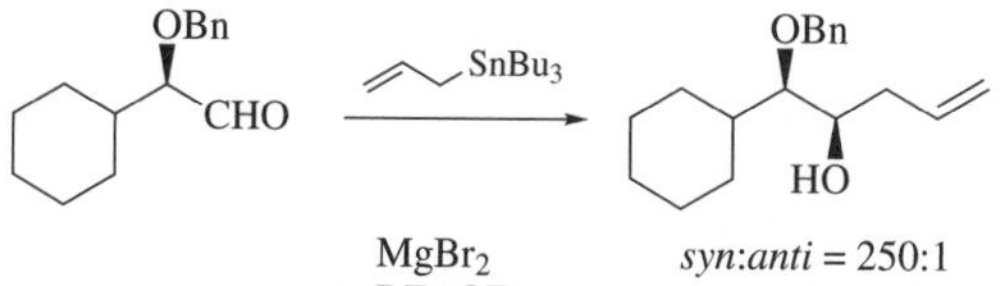

(1)
MgBr2 — *syn:anti* = 250:1
$BF_3 \cdot OEt_2$ — *syn:anti* = 39:61

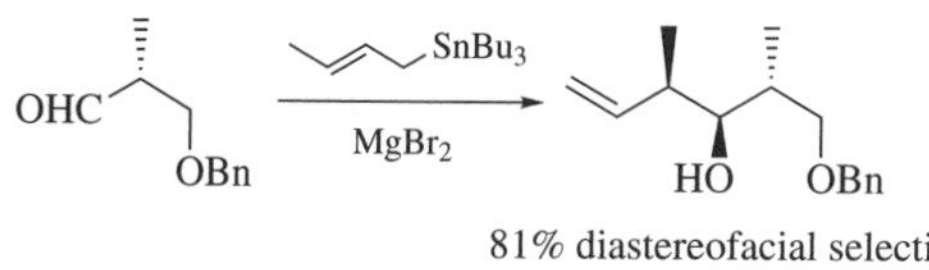

(2)
81% diastereofacial selectivity

Nucleophilic additions to α,β-dialkoxy aldehydes via allyl silanes (eq 3)[10] or silyl ketene acetals (Mukaiyama reaction) (eq 4)[11] exhibit similarly high selectivities. α-Thio aldehydes also react under $MgBr_2$-catalyzed Mukaiyama conditions with efficient stereocontrol (eq 5).[1]

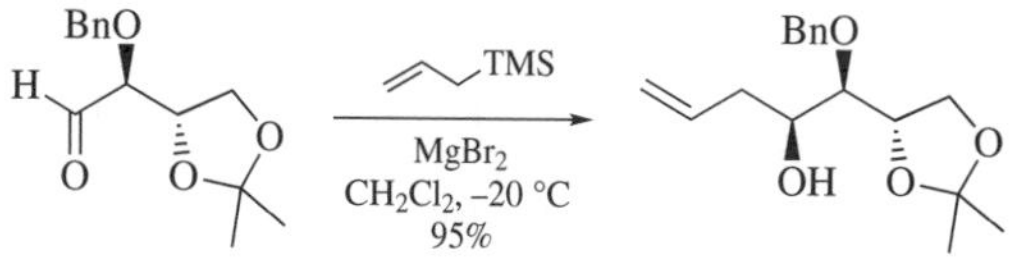

(3)
$MgBr_2$, CH_2Cl_2, –20 °C, 95%
syn:anti >98:2

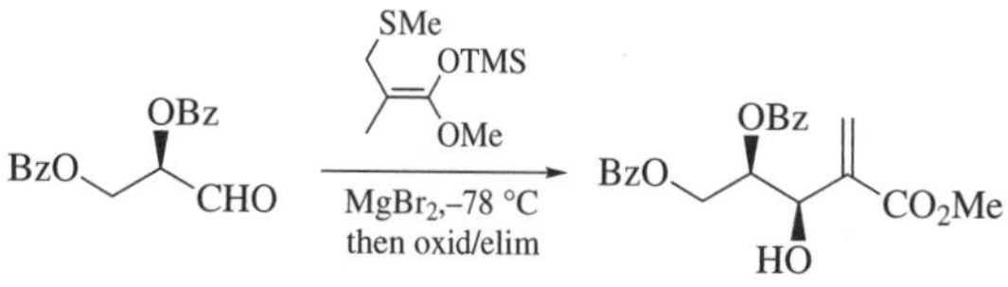

(4)
$MgBr_2$, –78 °C, then oxid/elim
syn:anti >55:1

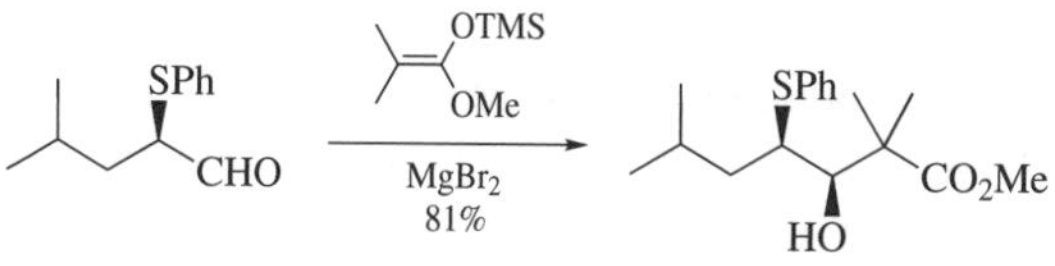

(5)
$MgBr_2$, 81%
syn:anti = 87:13

Rearrangements. The oxophilic nature of $MgBr_2$ renders it effective in mediating many rearrangements wherein polarization of a C–O bond initiates the process. A classic application involves the conversion of an epoxide to an aldehyde (eq 6).[12]

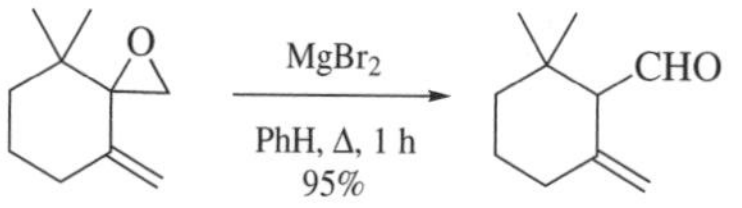

(6)
$MgBr_2$, PhH, Δ, 1 h, 95%

β-Lactones, when treated with $MgBr_2$, undergo ionization with further rearrangement to afford either butyrolactones or β,γ-unsaturated carboxylic acids (eq 7).[3] The reaction course is dependent upon whether the cation resulting from lactone ionization can rearrange to a more or equivalently stable cation, in which case ring expansion is observed. If the β-lactone bears an α-chloro substituent, ring expansion is accompanied by elimination of HCl to afford butenolides (eq 8).[13]

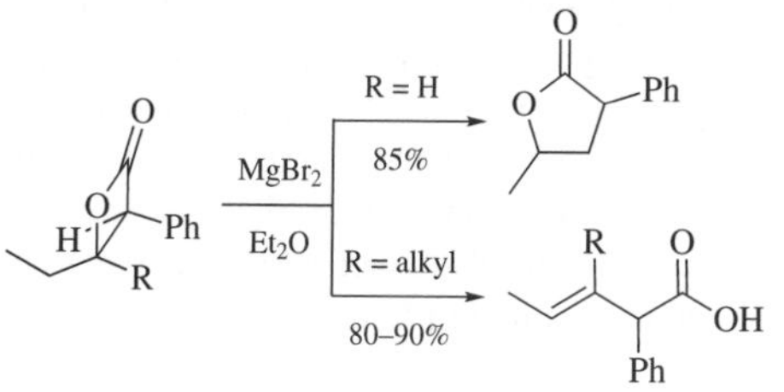

(7)
$MgBr_2$, Et_2O; R = H, 85%; R = alkyl, 80–90%

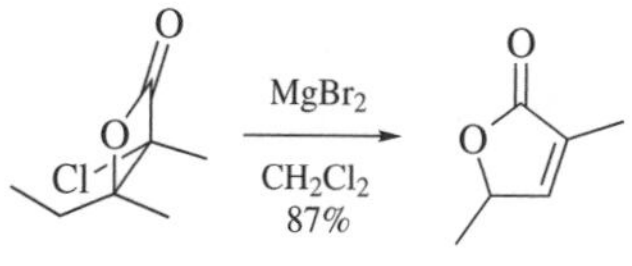

(8)
$MgBr_2$, CH_2Cl_2, 87%

Cycloadditions. The $MgBr_2$-mediated cyclocondensation of a Danishefsky-type diene with chiral α-alkoxy aldehydes affords a single diastereomer, which reflects a reacting conformer in which the alkoxy group is *syn* to the carbonyl, which is then attacked from its less-hindered face (eq 9).[2] The cycloaddition of ynamines with cycloalkenones occurred selectively at the carbonyl, while in the absence of $MgBr_2$ reaction at the alkene C=C bonds was observed (eq 10).[14]

OHC OBn, Et; MgBr$_2$, THF, 0 °C then TFA, CH$_2$Cl$_2$ 78%; only isomer detected (9)

MgBr$_2$ 60–70% (10)

Organometallic Reactions. $MgBr_2$ often increases the yields of Grignard reactions, as in the synthesis of cyclopropanols from 1,3-dichloroacetone (eq 11).[15] It also serves to form Grignard reagents in situ via first lithiation followed by transmetalation with $MgBr_2$. This technique enables the formation of vinyl Grignards from vinyl sulfones (eq 12),[16] and of α-silyl Grignard reagents from allyl silanes (eq 13).[17] In the latter case, the presence of $MgBr_2$ provided regioselectivity at the α-position; without it, substitution at the γ-position was predominant.

1. RMgBr, MgBr$_2$ Et$_2$O; 2. Li then H$_3$O$^+$ (11)

1. MeLi, THF, –95 °C; 2. MgBr$_2$, R^{2}R^{3}CO, H$^+$; ca. 75% (12)

1. *t*-BuLi, HMPA; 2. MgBr$_2$; 3. R^{1}R^{2}CO (13)

Bromination. Magnesium bromide serves as a source of bromide ion, for displacement of sulfonate esters under mild conditions and in high yield. The reaction is known to proceed with complete inversion at the reacting carbon center when backside attack is possible (eq 14).[5] Even bromination of highly congested bridgehead positions is possible via triflate displacement, although the reaction requires high temperatures, long reaction times, and activation via ultrasound (eq 15).[18]

MgBr$_2$ (1.3 equiv) Et$_2$O, CH$_2$Cl$_2$; 0 °C, 30 min; 95%; $[\alpha]_D$ –9.5°; $[\alpha]_D$ +9° (14)

MgBr$_2$, CS$_2$)))); 140 °C, 168 h; 48% (15)

Epoxides are also converted regiospecifically to bromohydrins by $MgBr_2$.[19]

Carbonyl Condensations. Bis- or tris-TMS ketenimines condense with ketones, mediated by $MgBr_2$, to produce an intermediate that loses hexamethylsiloxane, affording 2-alkenenitriles in high yield with high (*E*) selectivity (eq 16).[20] $MgBr_2$ also enables the use of very mild bases like ***Triethylamine*** in Horner–Wadsworth–Emmons reactions, enabling the synthesis of unsaturated esters from aldehydes or ketones without the need for strongly basic conditions (eq 17).[21]

MgBr$_2$, C$_6$H$_6$, Et$_2$O; highly (*E*) selective (16)

(EtO)$_2$POCH$_2$CO$_2$Et, MgBr$_2$, TEA; THF, rt, 24 h; 81% (17)

Related Reagents. Diethylaluminum Chloride; Dimethylaluminum Chloride; Lithium Perchlorate; Tin(IV) Chloride; Titanium Tetraisopropoxide; Zinc Bromide; Zinc Chloride; Zinc Iodide.

1. Annunziata, R.; Cinquini, M.; Cozzi, F.; Cozzi, P. G.; Consolandi, E. *JOC* **1992**, *57*, 456.
2. Danishefsky, S.; Pearson, W. H.; Harvey, D. F.; Maring, C. J.; Springer, J. P. *JACS* **1985**, *107*, 1256.
3. (a) Black, T. H.; Hall, J. A.; Sheu, R. G. *JOC* **1988**, *53*, 2371. (b) Black, T. H.; Eisenbeis, S. A.; McDermott, T. S.; Maluleka, S. L. *T* **1990**, *46*, 2307.
4. Cai, D.; Still, W. C. *JOC* **1988**, *53*, 464.
5. Hannesian, S.; Kagotani, M.; Komaglou, K. *H* **1989**, *28*, 1115.
6. Keck, G. E.; Castellino, S. *JACS* **1986**, *108*, 3847.
7. Chen, X.; Hortelano, E. R.; Eliel, E. L.; Frye, S. V. *JACS* **1992**, *114*, 1778.
8. Keck, G. E.; Boden, E. P. *TL* **1984**, *25*, 265.
9. Keck, G. E.; Abbott, D. E. *TL* **1984**, *25*, 1883.
10. Williams, D. R.; Klingler, F. D. *TL* **1987**, *28*, 869.
11. Bernardi, A.; Cardani, S.; Colombo, L.; Poli, G.; Schimperna, G.; Scolastico, C. *JOC* **1987**, *52*, 888.
12. Serramedan, D.; Marc, F.; Pereyre, M.; Filliatre, C.; Chabardes, P.; Delmond, B. *TL* **1992**, *33*, 4457.
13. Black, T. H.; McDermott, T. S.; Brown, G. A. *TL* **1991**, *32*, 6501.
14. Ficini, J.; Krief, A.; Guingant, A.; Desmaele, D. *TL* **1981**, *22*, 725.
15. Barluenga, J.; Florez, J.; Yus, M. *S* **1983**, 647.
16. Eisch, J. J.; Galle, J. E. *JOC* **1979**, *44*, 3279.
17. Lau, P. W. K.; Chan, T. H. *TL* **1978**, 2383.
18. Martinez, A. G.; Vilar, E. T.; Lopez, J. C.; Alonso, J. M.; Hanack, M.; Subramanian, L. R. *S* **1991**, 353.
19. Ueda, Y.; Maynard, S. C. *TL* **1988**, *29*, 5197.
20. Matsuda, I.; Okada, H.; Izumi, Y. *BCJ* **1983**, *56*, 528.
21. Rathke, M. W.; Nowak, M. *JOC* **1985**, *50*, 2624.

T. Howard Black
Eastern Illinois University, Charleston, IL, USA

Magnesium Methoxide

$(MeO)_2Mg$

[109-88-6] $C_2H_6MgO_2$ (MW 86.39)

(weakly basic metal alkoxide; forms stable chelates with enolizable carbonyl compounds[1])

Solubility: slightly sol MeOH, EtOH, DMF.

Form Supplied in: commercially available as neat white powder or as 8 wt % solution in MeOH.

Preparative Method: by addition of small batches of ***Magnesium*** metal to anhydrous MeOH.[1]

Handling, Storage, and Precautions: use same precautions as for other metal alkoxides (see ***Potassium t-Butoxide***); avoid contact with eyes and clothing; handle and conduct reactions under an inert atmosphere. For critical experiments, it is recommended that the reagent be freshly prepared. Use in a fume hood.

Preparation of Magnesium Methyl Carbonate. The reaction of $(MeO)_2Mg$ with CO_2 in DMF solution leads to the formation of $MeOMgOCO_2Me$ (eq 1).[2] This reagent is very useful for the carboxylation of α-methylene ketones, lactones, nitro compounds, etc.

$$(MeO)_2Mg + CO_2 \underset{DMF}{\rightleftharpoons} MeOMgOCO_2Me \quad (1)$$

Carbonyl Condensation Reactions. Because of the ability of the magnesium cation to form stable chelates with carbonyl compounds, $(MeO)_2Mg$ is a useful base for the promotion of a variety of carbonyl condensation reactions. The reagent catalyzes the condensation of triacetic acid lactone methyl ether with aromatic aldehydes to yield styryl monopyrone methyl ethers (eq 2).[3] The condensation of 2,2-dimethoxy-3-butanone with benzaldehyde in the presence of $(MeO)_2Mg$ leads to a magnesium chelate of the initially formed α-diketone condensation product, which undergoes intramolecular Michael addition to form 4-phenylcyclopentane-1,2-dione (eq 3).[4] Dimethyl oxalate reacts with 2,2-dimethoxy-3-butanone under similar conditions to give 2,5-dimethoxy-*p*-benzoquinone.[4] The use of other bases such as sodium alkoxides, tertiary amines, or quaternary ammonium salts in these reactions leads to unidentified products. Treatment of cyclohexenone with dimethyl 2-methoxycarbonylglutarate in the presence of $(MeO)_2Mg$ in MeOH results in a Michael addition followed by an aldol condensation to yield a decalin-1,8-dione derivative which exists primarily in its tautomeric enol forms (eq 4).[5]

OMe + PhCHO —$(MeO)_2Mg$, MeOH, reflux, 33%→ OMe, Ph (2)

MeO OMe + PhCHO —$(MeO)_2Mg$, MeOH, reflux, 64%→ Ph (3)

+ MeO_2C CO_2Me CO_2Me —$(MeO)_2Mg$, MeOH, reflux, 75%→ MeO_2C CO_2Me (4)

A mixture of diastereomers of 6-isopropyl-9-methylbicyclo[4.3.0]nonen-3-one is obtained by a tandem Michael–aldol reaction of an appropriate acyclic unsaturated keto aldehyde using $(MeO)_2Mg$ as the base (eq 5).[6] The reaction is reasonably stereoselective for the *trans*-fused ring system, but it is somewhat more stereoselective and the overall yield is considerably higher when ***Zirconium Tetraisopropoxide*** is used as the base. The ability of the zirconium cation to form even stronger chelates than the magnesium cation presumably accounts for this.[6b] A good yield of the cyclization products is also obtained with ***Sodium Methoxide***, but the stereoselectivity is low.

CHO, O —base→ —$(F_3CCO)_2O$, DBU→ (5)

$(MeO)_2Mg$	10:1.5:1	51%
$(i\text{-PrO})_4Zr$	10:1:1	80%
MeONa	3.4:1.2:1	80%

The methyl ester of phenylsulfinylacetic acid condenses with aliphatic aldehydes to give γ-hydroxy-α,β-unsaturated esters directly (eq 6).[7] The α-phenylsulfinyl-α,β -unsaturated ester, which is obtained by dehydration of the aldol product, undergoes deconjugation of the double bond to the β,γ-position. Then, a [2,3]-sigmatropic rearrangement of the sulfoxide group occurs, followed by cleavage of the sulfenate. A two-step process which involves the use of ***Sodium Hydride/Zinc Chloride*** catalyst to effect the aldol condensation and dehydration followed by treatment with an amine base yields the same types of products.

Ph–S(O)–CO_2Me + $(\ )_4$CHO —$(MeO)_2Mg$, MeOH, rt, 58%→ OH, $(\ )_3$, CO_2Me (6)

The Darzens condensation of methyl chloroacetate with benzaldehyde occurs relatively slowly in the presence of $(MeO)_2Mg$ in MeOH.[8] These conditions allow the isolation of methyl 2-chloro-3-hydroxy-3-phenylpropanoate and methyl 2-hydroxy-3-methoxy-3-phenylpropanoate, as well as the products of the Cannizzaro reaction of benzaldehyde, in addition to the expected methyl 2,3-epoxy-3-phenylpropanoate (glycidic ester).

Since the magnesium cation has a high affinity to form stable chelates with β-dicarbonyl compounds, the course of cyclizations of polyketonic–polyenolic systems may change significantly in going from alkali metal alkoxide bases to $(MeO)_2Mg$.[3b,9,10] For example, dimethyl 2,4-diacetylglutaconate cyclizes to an acyl pyrone in relatively high yield upon treatment with ≥2.0 equiv of MeONa in MeOH/PhH (eq 7).[9a] The same pyrone is formed if the diester is treated with 1.0 equiv of $(MeO)_2Mg$, but when ≥2.0 equiv of this reagent is used, a phenol derivative and a resorcinol derivative are also formed. Apparently, the chelate of the initially formed acyl pyrone undergoes ring opening with the excess base to form the chelate of a glutaconic ester dianion, which undergoes an intramolecular Michael or aldol reaction to form a phenol derivative or an intramolecular Claisen condensation to form a resorcinol derivative.[9a] Also, bispyrones undergo ring openings to triketo esters followed by aldol cyclizations, sometimes accompanied by decarboxylations to mixtures of resorcinol derivatives upon treatment with KOH in MeOH (eq 8),[10] but phloroglucinol derivatives, which apparently arise via Claisen condensation of open chain magnesium bischelates, are obtained in low yields with excess $(MeO)_2Mg$ (eq 9).[3b,9a]

(7)

	acyl pyrone	phenol	resorcinol
2.0 equiv $(MeO)_2Mg$	–	55%	26%
2.0 equiv MeONa	81%	–	–

(8)

5.4% 18% 10%

(9)

(MeO)₂Mg, MeOH, reflux, 8.5%

Triacetic acid methyl ester is conveniently prepared by treating dehydroacetic acid with $(MeO)_2Mg$ in MeOH (eq 10).[11] The ring opening reaction occurs more slowly when MeOLi or MeONa are employed as the bases.

(10)

$(MeO)_2Mg$, MeOH, reflux, 85%

Related Reagents. Aluminum *t*-Butoxide; Methyl Magnesium Carbonate; Potassium *t*-Butoxide; Sodium Ethoxide; Sodium Methoxide; Zirconium Tetraisopropoxide.

1. *FF* **1969**, *2*, 255.
2. (a) Stiles, M.; Finkbeiner, H. L. *JACS* **1959**, *81*, 505. (b) Finkbeiner, H. L.; Stiles, M. *JACS* **1963**, *85*, 616.
3. (a) Bu'Lock, J. D.; Smith, H. G. *JCS* **1960**, 502. (b) Douglas, J. L.; Money, T. *T* **1967**, *23*, 3545.
4. Muxfeldt, H.; Weigele, M.; Van Rheenen, V. *JOC* **1965**, *30*, 3573.
5. Schank, K.; Moell, N. *CB* **1969**, *102*, 71.
6. (a) Attah-Poku, S. K.; Chau, F.; Yadav, V. K.; Fallis, A. G. *JOC* **1985**, *50*, 3418. (b) Stork, G.; Shiner, C. S.; Winkler, J. D. *JACS* **1982**, *104*, 310.
7. Cass, Q. B.; Jaxa-Chamiec, A. A.; Sammes, P. C. *CC* **1981**, 1248.
8. Svoboda, J.; Nič, M.; Paleček, J. *CCC* **1992**, *57*, 119.
9. (a) Crombie, L.; James, A. W. G. *CC* **1966**, 357. (b) Crombie, L.; Games, D. E.; Knight, M. H. *CC* **1966**, 355. (c) Crombie, L.; Games, D. E.; Knight, M. H. *TL* **1964**, 2313.
10. Money, T.; Comer, F. W.; Webster, G. R. B.; Wright, I. G.; Scott, A. I. *T* **1967**, *23*, 3435.
11. Batelaan, J. G. *SC* **1976**, *6*, 81.

Drury Caine

The University of Alabama, Tuscaloosa, AL, USA

Mercury(II) Acetate

$Hg(OAc)_2$

[1600-27-7] $C_4H_6HgO_4$ (MW 318.69)

(inter- and intramolecular oxy-,[1–3] amino-, and amidomercuration[1,40,43,47] of alkenes and alkynes; cleavage of cyclopropanes;[64,65] mercuration of aromatics;[75] *o*-mercuration of phenols;[76] mild oxidation[101,102])

Alternate Name: mercuric acetate.
Physical Data: mp 179–182 °C; bp dec; *d* 3.28 g cm^{-3}.
Solubility: sol CH_2Cl_2, AcOH; slightly sol alcohols; insol benzene, hexane, etc.; sol H_2O (slowly hydrolyzed to HgO).
Form Supplied in: white, hygroscopic crystals.
Drying: recrystallization from AcOH followed by storage in a desiccator.
Handling, Storage, and Precautions: acute poison. $Hg(OAc)_2$ is easily absorbed through the skin and is corrosive. Exposure to all mercury compounds is to be strictly avoided. Releases toxic Hg fumes when heated to decomposition. Protect from light.

Oxymercuration.[1,2] Electrophilic Hg^{2+} is capable of attacking alkenic double bonds to effect Markovnikov addition,[3] typically with *anti* stereochemistry (eq 1).[4,5] However, *syn* attack can be enforced if the *anti*-face is severely hindered;[4] thus norbornene,[5] *trans*-cyclooctene,[6] and *trans*-cyclononene[6] give *syn* addition products.[6] Solvomercuration occurs in hydroxylic solvents (water or alcohols)[1,7,8] except when *t*-BuOH is utilized.[7] Using Hg^{II} salts with less nucleophilic anions (e.g. $CF_3CO_2^-$) improves the yields of the solvomercuration products.[7] In the presence of Cl_2 or Br_2, the HgX group in the original adduct is replaced by halogen.[7] Peroxymercuration of alkenes occurs in the presence of peroxides.[9] Methoxymercuration of alkynes[10] or allenes[11] in methanol gives mercurated vinyl ethers (eq 2),[10] whereas vinyl acetates are formed in AcOH.[12] Mercury(II) acetate also catalyzes the perborate or MoO_5·HMPA-mediated oxidation of terminal alkynes to α-acetoxy ketones (eq 3).[13]

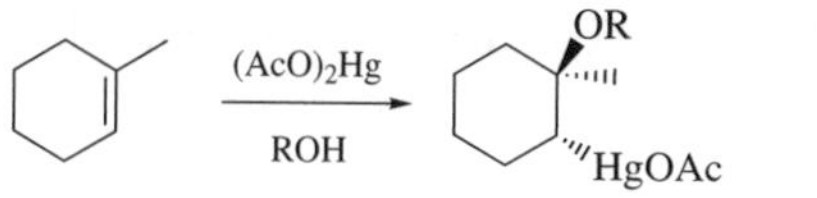

(1)

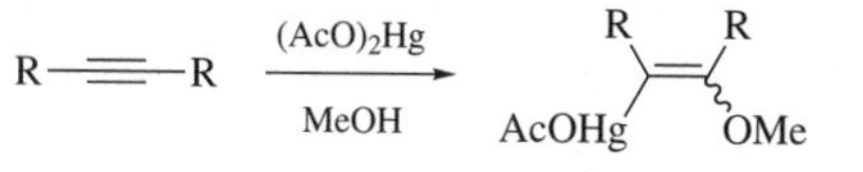

(2)

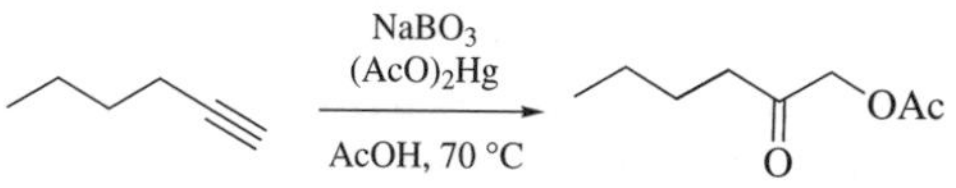

(3)

Although mercuration exhibits features common to most electrophilic additions,[14] it differs in details. Thus the rate-limiting step for mercuration is the cleavage of the intermediate mercuronium ion by a nucleophile; in contrast, bromination proceeds by the rate-limiting formation of the bromonium ion.[15]

Mercuration is very sensitive to steric factors in which it parallels hydroboration.[16] The following reactivity of alkenes has been observed: terminal disubstituted > terminal monosubstituted > internal disubstituted trisubstituted > tetrasubstituted (eq 4);[17] *cis*-alkenes are more reactive than their *trans* counterparts[17] and enol ethers are much more reactive than any other alkene.[18] Attempted asymmetric mercuration in the presence of cyclodextrin or in chiral micelles has only been partly successful, giving 3–55% ee; in one case the ee was 96%.[19]

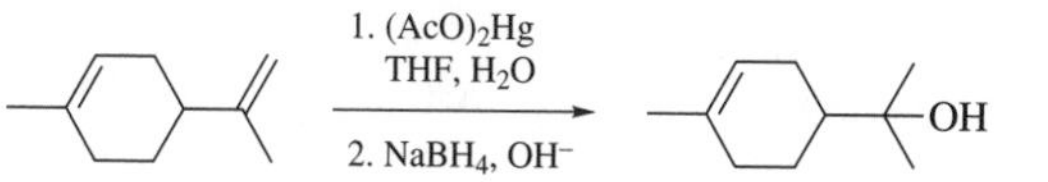

(4)

Neighboring group participation can be employed to construct cyclic structures such as tetrahydrofurans and tetrahydropyrans[3c,20–22] (eqs 5 and 6),[3c,20,21] spiroacetals (eq 7),[23] lactones,[24] and cyclic peroxides,[25] or to introduce a hydroxy group stereoselectively.[26] 4-Pentynoic acid reacts with $(AcO)_2Hg$ to produce α′-angelicalactone (eq 8).[27] A neighboring ether or hydroxy group can steer the approach of Hg^{2+} to the double bond by coordination and thus exercise a relatively high degree of stereocontrol (up to 93% de).[28]

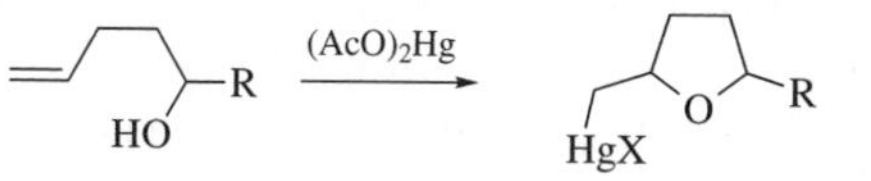

(5)

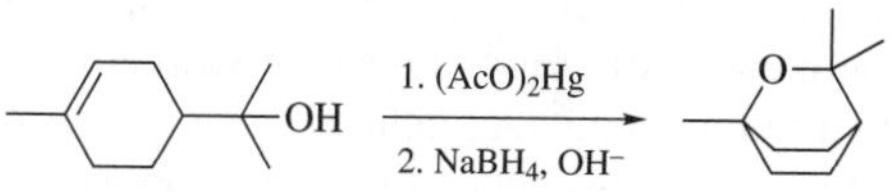

(6)

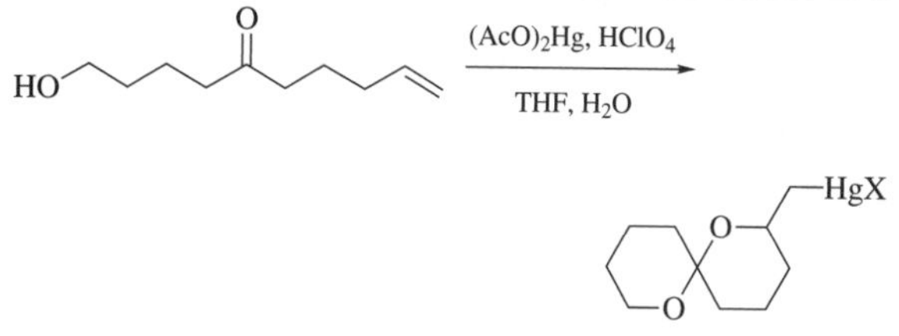

(7)

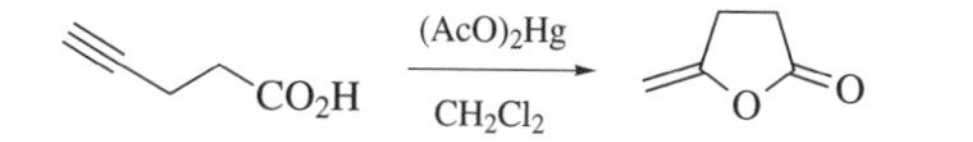

(8)

The adducts can be demercurated by hydrides (e.g. ***Sodium Borohydride***)[1] or sulfur compounds (***1,3-Propanedithiol*** or ***Sodium Trithiocarbonate***)[29,30] so that the method can be used as a formal addition of ROH across a double bond. While the hydride reduction may result in inversion (4:1), predominant retention of configuration (20:1) has been reported for the latter method.[29,30] The hydride reduction proceeds via a C-centered radical which can be trapped in a number of ways, e.g. by addition across an electron-deficient double bond[31–34] (eqs 9 and 10),[31,32] or by oxygen.[35] The latter method ($O_2/NaBH_4$) replaces HgX with OH.[1,36] Transmetalation of organomercurials by palladium is another option.[1,36]

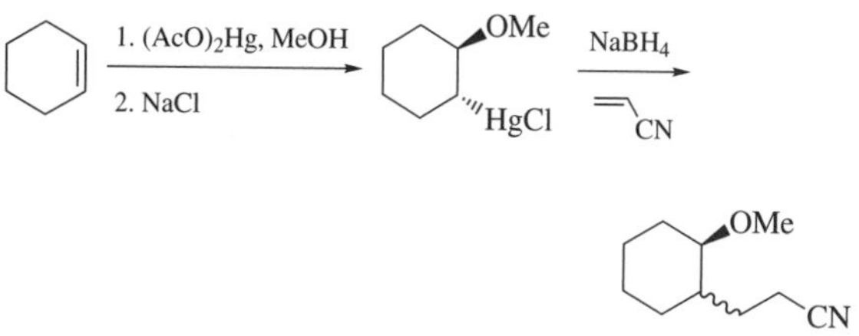

(9)

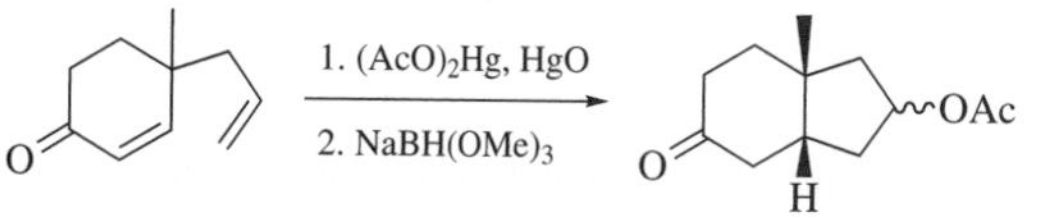

(10)

On treatment with bromine, NBS, NBA, or KBr_3, organomercurials are converted into the corresponding halides[37] (in a non-stereoselective fashion[38]). If applied to α,β-unsaturated acids,[37a,b] this reaction can be employed as a key step for the synthesis of serine or threonine (eq 11).[39]

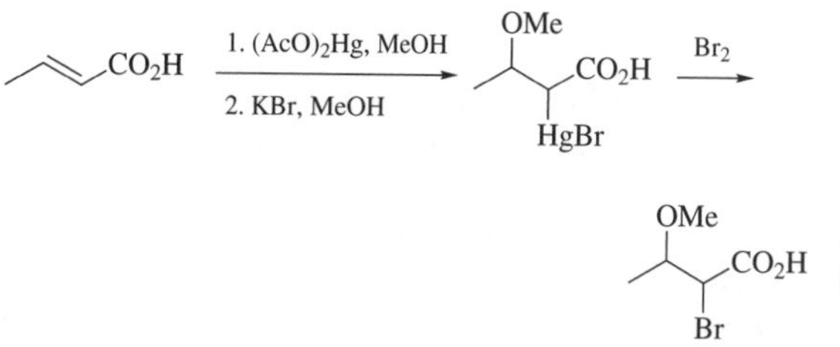

(11)

Amino- and Amidomercuration. Amino alkenes can similarly be cyclized by $(AcO)_2Hg$ to afford various nitrogen heterocycles[22b,d,e,40] via a 5- or 6-$(N)^n$-*exo-trig* pathway (eq 12).[41] This method has also been applied to the synthesis of indoles.[42] An interesting intra/intermolecular version has been developed for the stereoselective synthesis of piperidines (eq 13).[43,44]

1. $(AcO)_2Hg$, MeOH
2. $NaBH_4$
58% (12)

1. $(AcO)_2Hg$, $ArNH_2$
2. KBr, MeOH, H_2O
3. $NaBH_4$ (13)

Intramolecular amidomercuration is another variation on the same theme[32–34,35d,45] which works even with β-lactams (eq 14).[46] Intermolecular amidomercuration can be accomplished in acetonitrile through a Ritter-type reaction.[47]

1. $(AcO)_2Hg$
2. $NaBH_4$
75% (14)

Unsaturated oximes also undergo a ring-closure reaction on treatment with $(AcO)_2Hg$ to produce nitrones which can be trapped via a [3 + 2] cycloaddition (eq 15).[48]

$(AcO)_2Hg$, THF, 25 °C (15)

3-Alken-1-ynes undergo catalytic aminomercuration in the presence of $(AcO)_2Hg$ or HgCl at elevated temperatures to produce enamines.[49] By contrast, propargylic alcohols ($HC{\equiv}CCH_2OH$) undergo oxidative aminomercuration to afford bis-aminated aldehydes, e.g. (*Z*)-PhNHCH=C(NHPh)CH=O.[50] Catalytic aminomercuration of protected propargylic alcohols gives enamines, $CH_2{=}C(NR^1_2)CH_2OR^2$, as the result of a regioselective Markovnikov addition–demercuration.[51] Dipropargyl ethers produce aminofurans.[51]

Polyene Cyclization. Treatment of dienes or trienes with $(AcO)_2Hg$ triggers cyclization analogous to the Johnson polyene cyclizations.[52,53] The HgX group in the resulting cyclic organomercurial can best be eliminated to afford the corresponding alkene by photochemically induced transmetalation with ***Diphenyl Diselenide*** followed by standard selenoxide elimination.[54] Allenes with another alkenic bond can also be employed in this cyclization.[55] By contrast, 1,4-cyclohexadiene gives a mixture of mono- and bis-methoxymercurated products with no transannular C–C bond formation.[56]

Addition of Alcohols to Vinyl Ethers. Vinyl ethers such as $CH_2{=}CHOEt$ undergo a formal exchange reaction with another alcohol in the presence of $(AcO)_2Hg$, which occurs via elimination of EtOH from the intermediate.[1,57] Acetals can also be isolated from this reaction.[18,58] Similar reactions occur with vinyl esters (e.g. vinyl acetate)[59] and vinyl thioethers.[60] Ketone trimethylsilyl enolates give α-mercurio ketones[61] which are capable of the ***Boron Trifluoride Etherate***-catalyzed aldol condensation with aldehydes;[62] the reaction exhibits a high degree of *syn* stereoselectivity, comparable with that obtained from the Zr or Ti enolates.[61]

Mercuration of Enamines. Enamines react with $(AcO)_2Hg$ in DMF to give intermediate iminium salts ($R^1_2N^+{=}CHC(HgOAc)R^2R^3$), which are reduced with $NaBH_4$ to afford the corresponding tertiary amines $R^1_2NCH_2CHR^2R^3$ (50–90%).[63]

Ring Opening of Cyclopropanes. The cleavage of cyclopropanes with Hg^{II} occurs[34,64] predominantly with inversion of configuration at the carbon attacked by Hg^{2+} (corner opening)[65,66] and nucleophiles such as AcO^- or MeOH are incorporated in the product (eq 16).[65] Peroxides can also be employed as nucleophiles to produce the corresponding organic peroxides.[67] Cyclopropenes are either opened on treatment with $(AcO)_2Hg$ to produce allylic acetates[68] or undergo addition across the C=C bond (if the cleavage is suppressed by an electron-withdrawing group).[69]

HOMe, Hg^{2+} —$(AcO)_2Hg$, MeOH→ OMe, HgOAc (16)

Allylic Oxidation. At higher temperatures (typically 70–150 °C), alkenes are oxidized with $(AcO)_2Hg$ to give allylic acetates (eq 17; the Treibs reaction)[70,71] or dienes;[72] the yields vary in the range 20–70%.[73] Similarly, enolizable ketones and carboxylic acids undergo α-acetoxylation, but the yield seldom surpasses other methods.[70] Certain alkenes are oxidized to α,β-unsaturated ketones.[74]

H, Hg-OAc, AcO → AcO, Hg → OAc (17)

Mercuration of Aromatics. Being an electrophile, $(AcO)_2Hg$ effects aromatic electrophilic substitution of reactive aromatics.[75] A strong *o*-directing effect of groups capable of coordination with Hg^{2+} has been observed. Thus, for instance, on treatment with $(AcO)_2Hg$ in boiling water followed by NaCl quenching, phenol affords exclusively *o*-chloromercuriophenol (eq 18).[76] Mercuration of the following aromatics and heteroaromatics[37c,d] has been described: acetanilide[77] and its derivatives,[78] aromatic Schiff bases,[79] 3,4,5-trimethoxybenzoic acid,[80] estradiol,[81] α-phenylpyridine,[82] furan,[83] furfuryl diacetate,[84] pyrrole,[83,85] thiophene,[86] 1,3-thiazole,[87] indole,[88] pyrimidines,[89] pyridazines,[90] diazines,[91] trithiadiazepine,[92] ferrocene,[93] tricarbonyl(η-cyclobutadiene)iron,[94] pentamethylruthenocene,[95] Ni^{II}, Pd^{II}, and Pt^{II} porphyrins,[96] and others.[1] Since organomercurials RHgX are readily converted to the corresponding bromide RBr on reaction with Br_2, Hg^{II} salts can serve as effective catalysts in aromatic bromination.[97] An indirect route to mercurioaromatics ArHgOAc is by transmetalation of the corresponding silyl[98] or stannyl[99] derivatives ($ArSiMe_3$ or $ArSnBu_3$) with $(AcO)_2Hg$.[98] The resulting organomercurials can be further transmetalated with borane.[98,99] Reversed transmetalation of organoboranes with $(AcO)_2Hg$ ($R^1BR^2_2 \rightarrow R^1HgX$) has also been described.[100]

$$\text{PhOH} \xrightarrow{(AcO)_2Hg} \text{2-(HgOAc)C}_6\text{H}_4\text{OH} \xrightarrow[44\%]{NaCl} \text{2-(HgCl)C}_6\text{H}_4\text{OH} \quad (18)$$

Mercuriaromatics are coupled to give bisaryl derivatives on catalytic reaction with $PdCl_2$[101] or $[Rh(CO)_2Cl_2]_2LiCl$[102,103] in a process similar to Ullmann coupling.

Dehydrogenation of Tertiary Amines. Tertiary amines react with Hg^{II} to generate iminium salts (eq 19), which on treatment with base are converted into enamines.[104] Alternatively, the iminium intermediate can be trapped by an internal nucleophile (eq 19).[105] The reaction is improved by adding ethylenediaminetetraacetic acid (EDTA).[106]

(19)

Miscellaneous Reactions. $(AcO)_2Hg$ has been found to oxidize α-amino ketones to α-dicarbonyl compounds (50–95%),[107] to effect *cis* → *trans* isomerization of alkene,[108] to induce hydrolysis of vinyl chlorides to the corresponding ketones (a modification of the Wichterle reaction),[109] to mediate Koenigs–Knorr-type reactions (in analogy to Ag^+),[110] to oxidize β-aminocyclohexenones to *m*-aminophenols,[111] to hydrate nitriles (to give primary amides),[112] to effect oxidative degradation of penicillins,[113] and to mediate Claisen rearrangement.[114] Trialkylboranes R_3B, obtained by hydroboration of alkenes, react with $(AcO)_2Hg$ and I_2 to afford terminal acetates ROAc; this is an alternative to the usual H_2O_2/OH^- oxidation.[115] This B → Hg → OAc transformation has also been accomplished with vinylic substrates.[100b] On treatment with $(AcO)_2Hg$ in MeOH followed by demercuration with KSCN, cinnamyl esters $PhCH{=}CHCH_2OCOR$ undergo cleavage to furnish $PhCH(OMe)CH{=}CH_2$ (eq 20). The reaction is catalyzed by HNO_3 (apparently, mercury(II) nitrate is the reactive species) and can be used for protection–deprotection of carboxylic acids.[116]

$$RCO_2CH_2CH{=}CHPh \xrightarrow[\text{HNO}_3\text{ cat, MeOH}]{(AcO)_2Hg} RCO_2CH_2CH(HgOAc)CH(OMe)Ph \xrightarrow[H_2O]{KSCN} RCO_2H + CH_2{=}CHCH(OMe)Ph \quad (20)$$

Related Reagents. Mercury(II) Acetate–Ethylene diaminetetraacetic Acid; Mercury(II) Acetate–Sodium Trimethoxyborohydride; Mercury(II) Chloride; Mercury(II) Nitrate; Mercury(II) Perchlorate; Mercury(II) Trifluoroacetate; Thallium(III) Acetate; Thallium(III) Trifluoroacetate.

1. (a) Larock, R. C. *AG(E)* **1978**, *17*, 27. (b) Larock, R. C. *T* **1982**, *38*, 1713. (c) Larock, R. C. *Organomercury Compounds in Organic Synthesis*; Springer: Berlin, 1985. (d) Larock, R. C. *Solvomercuration/Demercuration Reactions in Organic Synthesis*; Springer: Berlin, 1986. (e) Jerkunica, J. M.; Traylor, T. G. *OSC* **1988**, *6*, 766.
2. Kitching, W. *Organomet. Chem. Rev.* **1968**, *3*, 61.
3. (a) Brown, H. C.; Geoghegan, P., Jr. *JACS* **1967**, *89*, 1522. (b) Brown, H. C.; Hammar, W. J. *JACS* **1967**, *89*, 1524. (c) Brown, H. C.; Kawakami, J. H.; Ikegami, S. *JACS* **1967**, *89*, 1525. (d) Brown, H. C.; Geoghegan, P. Jr., *JOC* **1970**, *35*, 1844. (e) Brown, H. C.; Geoghegan, P. J., Jr.; Kurek, J. T.; Lynch, G. J. *Organomet. Chem. Synth.* **1970**, *1*, 7. (f) Brown, H. C.; Kurek, J. T.; Rei, M.-H.; Thompson, K. L. *JOC* **1984**, *49*, 2551. (g) Shul'man, A. I.; Mikhailova, N. P.; Sarkisov, Yu. S.; V'yunov, K. A. *ZOR* **1988**, *24*, 2342. (h) Skorobogatova, E. V.; Sokolova, T. N.; Malisova, N. V.; Kartashov, V. R.; Zefirov, N. S. *ZOR* **1986**, *22*, 2150.
4. (a) Senda, Y.; Kamiyama, S.; Imaizumi, S. *JCS(P1)* **1978**, 530. (b) Chatt, J. *CRV* **1951**, *48*, 7. (c) Kartashov, V. R.; Sokolova, T. N.; Skorobogatova, E. V.; Bazhenov, D. V.; Roznyatovskii, V. A. *ZOR* **1987**, *23*, 2245. (d) Kartashov, V. R.; Sokolova, T. N.; Skorobogatova, E. V.; Grishin, Yu. K.; Bazhenov, D. V.; Roznyatovskii, V. A.; Koz'min, A. S.; Zefirov, N. S. *IZV* **1987**, 1914. (e) Kartashov, V. R.; Sokolova, T. N.; Vasil'eva, O. V.; Timofeev, I. V.; Grishin, Yu. K.; Bazhenov, D. V.; Zefirov, N. S. *ZOR* **1990**, *26*, 1800.
5. (a) Traylor, T. G.; Baker, A. W. *JACS* **1963**, *85*, 2746. (b) Traylor, T. G. *JACS* **1964**, *86*, 244. (c) Michael, J. P.; Blom, N. F.; Glintenkamp, L.-A. *JCS(P1)* **1991**, 1855. (d) Kartashov, V. R.; Sokolova, T. N.; Timofeev, I. V.; Grishin, Yu. K.; Bazhenov, D. V.; Zefirov, N. S. *ZOR* **1991**, *27*, 2077. (e) Lambert, J. B.; Emblidge, R. W.; Zhao, Y. *JOC* **1994**, *59*, 5397.
6. Waters, W. L.; Traylor, T. G.; Factor, A. *JOC* **1973**, *38*, 2306.
7. (a) Brown, H. C.; Rei, M.-H. *JACS* **1969**, *91*, 5646. (b) Barluenga, J.; Martínez-Gallo, J. M.; Nájera, C.; Yus, M. *CC* **1985**, 1422. (c) Barluenga, J.; Martínez-Gallo, J. M.; Nájera, C.; Yus, M. *JCR(S)* **1986**, 274.
8. (a) Khomutov, A. R.; Khurs, E. N.; Khomutov, R. M. *Bioorg. Khim.* **1988**, *14*, 385 (*CA* **1989**, *110*, 212973j). (b) Kuznetsova, T. S.; Kozhushkov, S. I.; Lukin, K. A.; Grishin, Yu. K.; Bazhenov, D. V.; Koz'min, A. S.; Zefirov, N. S. *ZOR* **1991**, *27*, 78.
9. (a) Bloodworth, A. J.; Bowyer, K. J.; Mitchell, J. C. *JOC* **1987**, *52*, 1124. (b) Bloodworth, A. J.; Spencer, M. D. *JOM* **1990**, *386*, 299. (c) Bloodworth, A. J.; Chan, K. H.; Cooksey, C. J.; Hargreaves, N. *JCS(P1)* **1991**, 1923. (d) Courtneidge, J. L.; Bush, M. *JOM* **1992**, *437*, 57.
10. (a) Bassetti, M.; Floris, B. *JOC* **1986**, *51*, 4140. (b) Bassetti, M.; Floris, B.; Spadafora, G. *JOC* **1989**, *54*, 5934. (c) Bassetti, M.; Bocelli, G. *CC* **1990**, 257. (d) Kartashov, V. R.; Sokolova, T. N.; Grishin, Yu. K.; Bazhenov, D. V.; Zefirov, N. S. *Metalloorg. Khim.* **1992**, *5*, 969 (*CA* **1993**, *118*, 39103r).
11. Pasto, D. J.; Sugi, K. D. *JOC* **1991**, *56*, 4157.
12. (a) Bach, R. D.; Woodard, R. A.; Anderson, T. J.; Glick, M. D. *JOC* **1982**, *47*, 3707. (b) Kartashov, V. R.; Sokolova, T. N.; Malisova, N. V.; Skorobogatova, E. V.; Grishin, Yu. K.; Roznyatovskii, V. A.; Bazhenov, D. V.; Zefirov, N. S. *ZOR* **1986**, *22*, 2232. (c) Kartashov, V. R.; Sokolova, T. N.; Skorobogatova, E. V.; Chernov, A. N.; Bazhenov, D. V.; Grishin, Yu. K.; Ustynyuk, Yu. A.; Zefirov, N. S. *ZOR* **1989**, *25*, 1846. (d) Kartashov, V. R.; Sokolova, T. N.; Grishin, Yu. K.; Bazhenov, D. V.; Zefirov, N. S. *ZOR* **1990**, *26*, 1126.

13. (a) Reed, K. L.; Gupton, J. T.; McFarlane, K. L. *SC* **1989**, *19*, 2595. (b) Ballistreri, F. P.; Failla, S.; Tomaselli, G. A.; Curci, R. *TL* **1986**, *27*, 5139.
14. (a) Lewis, A.; Azoro, J. *TL* **1979**, 3627. (b) Cabaleiro, M. C.; Ayala, A. D.; Johnson, M. D. *JCS(P2)* **1973**, 1207.
15. Vardhan, H. B.; Bach, R. D. *JOC* **1992**, *57*, 4948.
16. (a) Nelson, D. J.; Cooper, P. J.; Coerver, J. M. *TL* **1987**, *28*, 943. (b) Broughton, H. B.; Green, S. M.; Rzepa, H. S. *CC* **1992**, 998. (c) For an earlier work, see: Takaishi, N.; Fujikura, Y.; Inamoto, Y. *JOC* **1975**, *40*, 3767.
17. (a) Brown, H. C.; Geoghegan, P. J., Jr. *JOC* **1972**, *37*, 1937. (b) Brown, H. C.; Geoghegan, P. J., Jr.; Lynch, G. J.; Kurek, J. T. *JOC* **1972**, *37*, 1941. (c) Sakai, M. *TL* **1973**, 347. (d) Morisaki, M.; Rubio-Lightbourn, J.; Ikekawa, N. *CPB* **1973**, *21*, 457. (e) Link, C. M.; Jansen, D. K.; Sukenik, C. H. *JACS* **1980**, *102*, 7798.
18. (a) Carnevale, G.; Davini, E.; Iavarone, C.; Trogolo, C. *JOC* **1988**, *53*, 5343. (b) Carnevale, G.; Davini, E.; Iavarone, C.; Trogolo, C. *JCS(P1)* **1990**, 989.
19. (a) Rao, K. R.; Sampathkumar, H. M. *SC* **1993**, *23*, 1877. (b) Zhang, Y.; Bao, W.; Dong, H. *SC* **1993**, *23*, 3029. (c) Micelles also increase nucleophile selectivity: Livneh, M.; Sutter, J. K.; Sukenik, C. N. *JOC* **1987**, *52*, 5039.
20. Coxon, J. M.; Hartshorn, M. P.; Mitchell, J. W.; Richards, K. E. *CI(L)* **1968**, 652.
21. Larock, R. C.; Harrison, L. W. *JACS* **1984**, *106*, 4218.
22. (a) Hill, C. L.; Whitesides, G. M. *JACS* **1974**, *96*, 870. (b) Spéziale, V.; Roussel, J.; Lattes, A. *JHC* **1974**, *11*, 771. (c) Grundon, M. F.; Stewart, D.; Watts, W. E. *CC* **1975**, 772. (d) Spéziale, V.; Amat, M.; Lattes, A. *JHC* **1976**, *13*, 349. (e) Benhamou, M.-C.; Etemad-Moghadam, G.; Spéziale, V.; Lattes, A. *JHC* **1978**, *15*, 1313. (f) Pougny, J.-R.; Nassr, M. A. M.; Sinaÿ, P. *CC* **1981**, 375. (g) Hanessian, S.; Kloss, J.; Sugawara, T. *JACS* **1986**, *108*, 2758. (h) Amate, Y.; García-Granados, A.; López, F. A.; de Buruaga, A. S. *S* **1991**, 371. (i) Barrero, A. F.; Sánchez, J. F.; Altarejos, J.; Perales, A.; Torres, R. *JCS(P1)* **1991**, 2513. (j) Boschetti, A.; Nicotra, F.; Panza, L.; Russo, G. *JOC* **1988**, *53*, 4181. (k) Kraus, G. A.; Li, J.; Gordon, M. S.; Jensen, J. H. *JOC* **1995**, *60*, 1154.
23. (a) Kitching, W.; Lewis, J. A.; Fletcher, M. T.; DeVoss, J. J.; Drew, R. A. I.; Moore, C. J. *CC* **1986**, 855. (b) DeVoss, J. J.; Jamie, J. F.; Blanchfield, J. T.; Fletcher, M. T.; O'Shea, M. G.; Kitching, W. *T* **1991**, *47*, 1985.
24. Tureček, F. *CCC* **1982**, *47*, 858.
25. Bloodworth, A. J.; Tallant, N. A. *CC* **1992**, 428.
26. Ward, J.; Börner, A.; Kagan, H. B. *TA* **1992**, *3*, 849.
27. (a) Amos, R. A.; Katzenellenbogen, J. A. *JOC* **1978**, *43*, 560. For general hydration of alkynes to produce ketones, see: (b) Myddleton, W. W.; Barrett, A. W.; Seager, J. H. *JACS* **1930**, *52*, 4405. (c) Koulkes, M. *BSF(2)* **1953**, 402. (d) Kagan, H. B.; Marquet, A.; Jacques, J. *BSF(2)* **1960**, 1079.
28. (a) Thaisrivongs, S.; Seebach, D. *JACS* **1983**, *105*, 7407. (b) Giese, B.; Bartman, D. *TL* **1985**, *26*, 1197.
29. (a) Gouzoules, F. H.; Whitney, R. A. *TL* **1985**, *26*, 3441. (b) Gouzoules, F. H.; Whitney, R. A. *JOC* **1986**, *51*, 2024.
30. Bartlett, P. A.; Adams, J. L. *JACS* **1980**, *102*, 337.
31. (a) Giese, B.; Heuck, K. *CB* **1979**, *112*, 3759. (b) Giese, B.; Heuck, K. *TL* **1980**, *21*, 1829. (c) Giese, B.; Heuck, K. *CB* **1981**, *114*, 1572. (d) Giese, B.; Heuck, K.; Lüning, U. *TL* **1981**, *22*, 2155. (e) Kozikowski, A. P.; Nieduzak, T. R.; Scripko, J. *OM* **1982**, *1*, 675. (f) Giese, B.; Horler, H.; Zwick, W. *TL* **1982**, *23*, 931. (g) Giese, B.; Horler, H. *T* **1985**, *41*, 4025. (h) Giese, B. *Radicals in Organic Synthesis*; Pergamon: Oxford, 1986.
32. (a) Danishefsky, S.; Chackalamannil, S.; Uang, B.-J. *JOC* **1982**, *47*, 2231. (b) Danishefsky, S.; Taniyama, E.; Webb, R. R. *TL* **1983**, *24*, 11. (c) Danishefsky, S.; Taniyama, E. *TL* **1983**, *24*, 15.
33. (a) Carruthers, W.; Williams, M. J.; Cox, M. T. *CC* **1984**, 1235. (b) Kahn, M.; Devens, B. *TL* **1986**, *27*, 4841. (c) Dasaradhi, L.; Bhalerao, U. T. *SC* **1987**, *17*, 1845. (d) Weinges, K.; Sipos, W. *CB* **1988**, *121*, 363.
34. (a) Giese, B.; Horler, H.; Zwick, W. *TL* **1982**, *23*, 931. (b) Giese, B.; Horler, H. *TL* **1983**, *24*, 3221. (c) Giese, B.; Zwick, W. *CB* **1982**, *115*, 2526. (d) Giese, B.; Zwick, W. *CB* **1983**, *116*, 1264. (e) Giese, B.; Hasskerl, T.; Lüning, U. *CB* **1984**, *117*, 859. (f) Giese, B.; Gröninger, K. *TL* **1984**, *25*, 2743. (g) Giese, B. *AG(E)* **1985**, *24*, 553. (h) Barluenga, J.; Yus, M. *CRV* **1988**, *88*, 487. (i) See ref. 31(h).
35. (a) Quirk, R. P. *JOC* **1972**, *37*, 3554. (b) Quirk, R. P.; Lea, R. E. *TL* **1974**, 1925. (c) Quirk, R. P.; Lea, R. E. *JACS* **1976**, *98*, 5973. (d) Takahata, H.; Bandoh, H.; Momose, T. *JOC* **1992**, *57*, 4401 and references cited therein.
36. (a) Rodeheaver, G. T.; Hunt, D. F. *CC* **1971**, 818. (b) Hunt, D. F.; Rodeheaver, G. T. *TL* **1972**, 3595. (c) Heumann, A.; Bäckvall, J.-E. *AG(E)* **1985**, *24*, 207. (d) Kočovský, P.; Šrogl, J.; Gogoll, A.; Hanuš, V.; Polášek, M. *CC* **1992**, 1086. (e) Walkup, R. D.; Kim, S. W.; Wagy, S. D. *JOC* **1993**, *58*, 6486. See also Ref. 38.
37. (a) Abderhalden, E.; Heyns, K. *CB* **1934**, *67*, 530. (b) Schiltz, L. R.; Carter, H. E. *JBC* **1936**, *116*, 793. (c) Perlmutter, M.; Satyamurthy, N.; Luxen, A.; Phelps, M. E.; Barrio, J. R. *Appl. Radiat. Isot.* **1990**, *41*, 801 (*CA* **1991**, *114*, 143945v). (d) Voronkov, M. G.; Trofimova, O. M.; Chernov, N. F. *IZV* **1985**, 2148. (e) Brady, J. H.; Redhouse, A. D.; Wakefield, B. J. *JCR(S)* **1982**, 137.
38. Kočovský, P.; Šrogl, J.; Pour, M.; Gogoll, A. *JACS* **1994**, *116*, 186.
39. (a) Carter, H. E.; West, H. D. *OSC* **1955**, *3*, 774. (b) Carter, H. E.; West, H. D. *OSC* **1955**, *3*, 813.
40. (a) Kanne, D. B.; Ashworth, D. J.; Cheng, M. T.; Mutter, L. C. *JACS* **1986**, *108*, 7864. (b) Barluenga, J.; Aznar, F.; Fraiz, S., Jr.; Pinto, A. C. *TL* **1991**, *32*, 3205. (c) Esser, F. *S* **1987**, 460. (d) Le Moigne, F.; Mercier, A.; Tordo, P. *TL* **1991**, *32*, 3841.
41. Saitoh, Y.; Moriyama, Y.; Hirota, H.; Takahashi, T.; Khoung-Huu, Q. *BCJ* **1981**, *54*, 488.
42. Majumdar, K. C.; De, R. N.; Saha, S. *TL* **1990**, *31*, 1207.
43. (a) Barluenga, J.; Nájera, C.; Yus, M. *S* **1979**, 896. (b) Barluenga, J.; Villamaña, J.; Fañanás, F. J.; Yus, M. *CC* **1982**, 355.
44. For aminomercuration of vinyl or allylsilanes, see: (a) Barluenga, J.; Jiménez, C.; Nájera, C.; Yus, M. *S* **1982**, 414. (b) Voronkov, M. G.; Kirpichenko, S. V.; Abrosimova, A. T.; Albanov, A. I.; Keiko, V. V.; Lavrent'yev, V. I. *JOM* **1987**, *326*, 159.
45. (a) Harding, K. E.; Burks, S. R. *JOC* **1984**, *49*, 40. (b) Harding, K. E.; Marman, T. H. *JOC* **1984**, *49*, 2838. (c) Liu, P. S. *JOC* **1987**, *52*, 4717. (d) Takacs, J. M.; Helle, M. A.; Yang, L. *TL* **1989**, *30*, 1777. (e) Takacs, J. M.; Helle, M. A.; Sanyal, B. J.; Eberspracher, T. A. *TL* **1990**, *31*, 6765. (f) Amoroso, R.; Cardillo, G.; Tomasini, C. *H* **1992**, *34*, 349.
46. Aida, T.; Legault, R.; Dugat, D.; Durst, T. *TL* **1979**, 4993.
47. Brown, H. C.; Kurek, J. T. *JACS* **1969**, *91*, 5647.
48. Grigg, R.; Hadjisoteriou, M.; Kennewell, P.; Markandu, J.; Thornton-Pett, M. *CC* **1992**, 1388.
49. (a) Davtyan, S. Zh.; Chobanyan, Zh. A.; Badanyan, Sh. O. *Arm. Khim. Zh.* **1983**, *36*, 508 (*CA* **1984**, *100*, 67447c). (b) Barluenga, J.; Aznar, F.; Liz, R.; Cabal, M.-P. *CC* **1985**, 1375. (c) Barluenga, J.; Aznar, F.; Valdés, C.; Cabal, M.-P. *JOC* **1991**, *56*, 6166. (d) Barluenga, J.; Aznar, F.; Liz, R.; Cabal, M.-P. *S* **1986**, 960.
50. Barluenga, J.; Aznar, F.; Liz, R. *CC* **1986**, 1180.
51. (a) Barluenga, J.; Aznar, F.; Liz, R.; Postigo, C. *CC* **1986**, 1465. (b) Barluenga, J.; Aznar, F.; Bayod, M. *TL* **1988**, *29*, 5029.
52. (a) Julia, M.; Fourneron, J.-D. *TL* **1973**, 3429. (b) Julia, M.; Gasquez, E. C. *BSF(2)* **1973**, 1796. (c) Julia, M.; Colomer, E.; Julia, S. *BSF(2)* **1966**, 2397.
53. (a) Tkachev, A. V.; Mamatyuk, V. I.; Dubovenko, Zh. V. *ZOR* **1987**, *23*, 526. (b) Nishizawa, M.; Takao, H.; Kanoh, N.; Asoh, K.; Hatakeyama, S.; Yamada, H. *TL* **1994**, *35*, 5693.
54. Erion, M. D.; McMurry, J. E. *TL* **1985**, *26*, 559.
55. Delbecq, F.; Goré, J. *TL* **1976**, 3459.
56. Beger, J.; Thomas, B.; Vogel, T.; Lang, R. *JPR* **1991**, *333*, 447.

57. (a) Rhoads, S. J.; Raulins, N. R. *OR* **1975**, *22*, 1. (b) Ziegler, F. E. *ACR* **1977**, *10*, 227. (c) Carpino, L. A. *JACS* **1958**, *80*, 599. (d) Daub, G. W.; Sanchez, M. G.; Cromer, R. A.; Gibson, L. L. *JOC* **1982**, *47*, 743. (e) McMurry, J. E.; Kočovský, P. *TL* **1985**, *26*, 2171.
58. Boeckman, R. K., Jr.; Flann, C. J. *TL* **1983**, *24*, 4923.
59. (a) Swern, D.; Jordan, E. F., Jr. *OSC* **1963**, *4*, 977. (b) Watanabe, W. H.; Conlon, L. E. *JACS* **1957**, *79*, 2828. (c) Burgstahler, A. W.; Nordin, I. C. *JACS* **1961**, *83*, 198. (d) Ogiso, A.; Iwai, I. *CPB* **1964**, *12*, 820. (e) Büchi, G.; White, J. D. *JACS* **1964**, *86*, 2884. (f) Fukuda, W.; Sato, H.; Kakiuchi, H. *BCJ* **1986**, *59*, 751.
60. Davidson, A. H.; Floyd, C. D.; Lewis, C. N.; Myers, P. L. *CC* **1988**, 1417.
61. (a) House, H. O.; Auerbach, R. A.; Gall, M.; Peet, N. P. *JOC* **1973**, *38*, 514. For the formation of the α-mercuriocarbonyl compound via a [3,3]-sigmatropic rearrangement, see: (b) Bluthe, N.; Goré, J.; Malacria, M. *T* **1986**, *42*, 1333. (c) Bluthe, N.; Malacria, M.; Goré, J. *T* **1984**, *40*, 3277.
62. Yamamoto, Y.; Maruyama, K. *JACS* **1982**, *104*, 2323.
63. Bach, R. D.; Mitra, D. K. *CC* **1971**, 1433.
64. (a) Blossey, E. C. *Steroids* **1969**, *14*, 725. (b) Salomon, R. G.; Gleim, R. D. *JOC* **1976**, *41*, 1529. (c) Collum, D. B.; Mohamadi, F.; Hallock, J. H. *JACS* **1983**, *105*, 6882. (d) Langbein, G.; Siemann, H.-J.; Gruner, I.; Müller, C. *T* **1986**, *42*, 937. (e) Newman, M. S.; Van der Zwan, M. C. *JOC* **1974**, *39*, 1186. (f) Bandaev, S. G.; Eshnazarov, Yu. Kh.; Nasyrov, I. M.; Mochalov, S. S.; Shabarov, Yu. S. *ZOR* **1988**, *24*, 733. (g) Bandaev, S. G.; Eshnazarov, Yu. Kh.; Mochalov, S. S.; Shabarov, Yu. S.; Zefirov, N. S. *Metalloorg. Khim.* **1992**, *5*, 690 (*CA* **1992**, *117*, 251 458j). (h) Razin, V. V.; Zadonskaya, N. Yu. *ZOR* **1990**, *26*, 2342. (i) Razin, V. V.; Genaev, A. M.; Dobronravov, A. N. *ZOR* **1992**, *28*, 104. (j) Bandaev, S. G.; Mochalov, S. S.; Shabarov, Yu. S.; Zefirov, N. S. *Metalloorg. Khim.* **1992**, *5*, 604 (*CA* **1993**, *118*, 39 101p).
65. (a) Coxon, J. M.; Steel, P. J.; Whittington, B. J.; Battiste, M. A. *JACS* **1988**, *110*, 2988. (b) Coxon, J. M.; Steel, P. J.; Whittington, B. J.; Battiste, M. A. *JOC* **1989**, *54*, 1383. (c) Coxon, J. M.; Steel, P. J.; Whittington, B. I. *JOC* **1990**, *55*, 4136.
66. (a) DePuy, C. H.; McGirk, R. H. *JACS* **1973**, *95*, 2366. (b) Lambert, J. B.; Chelius, E. C.; Bible, R. H., Jr.; Hajdu, E. *JACS* **1991**, *113*, 1331. (c) For the most recent discussion, see Ref. 38.
67. (a) Bloodworth, A. J.; Lampman, G. M. *JOC* **1988**, *53*, 2668. (b) Bloodworth, A. J.; Chan, K. H.; Cooksey, C. J.; Hargreaves, N. *JCS(P1)* **1991**, 1923. (c) Bloodworth, A. J.; Korkodilos, D. *TL* **1991**, *32*, 6953.
68. Shirafuji, T.; Nozaki, H. *T* **1973**, *29*, 77.
69. (a) Kartashov, V. R.; Skorobogatova, E. V.; Sokolova, T. N.; Vasil'eva, O. V.; Malisova, N. V.; Grishin, Yu. K.; Bazhenov, D. V.; Zefirov, N. S. *ZOR* **1991**, *27*, 1240. (b) Kartashov, V. R.; Skorobogatova, E. V.; Malisova, N. V.; Grishin, Yu. K.; Bazhenov, D. V.; Zefirov, N. S. *ZOR* **1991**, *27*, 2490.
70. (a) Treibs, W.; Bast, H. *LA* **1949**, *561*, 165. (b) Boev, V. I.; Dombrovskii, A. V. *ZOB* **1981**, *51*, 2241. (c) Grdenić, D.; Kordar-Čolig, B.; Matković-Čalogović, D.; Skirica, M.; Popović, Z. *JOM* **1991**, *411*, 19. (d) House, H. O.; Thompson, H. W. *JOC* **1961**, *26*, 3729. (e) Jefferies, P. R.; Macbeth, A. K.; Milligan, B. *JCS* **1954**, 705. (f) Brenner, A.; Schinz, H. *HCA* **1952**, *35*, 1615. (g) Zalkow, L. H.; Ellis, J. W.; Brennan, M. R. *JOC* **1963**, *28*, 1705. For a review, see: (h) Arzoumanian, H.; Metzger, J. *S* **1971**, 527. (i) Rawlinson, D. J.; Sosnovsky, G. *S* **1973**, 567. (j) House, H. O. *Modern Synthetic Reactions*, 2nd ed.; Benjamin: Menlo Park, 1972; Chapter 7.
71. (a) Treibs, W.; Lucius, G.; Kögler, H.; Breslauer, H. *LA* **1953**, *581*, 59. (b) Treibs, W.; Weissenfels, M. *CB* **1960**, *93*, 1374. (c) Alkonyi, I. *CB* **1962**, *95*, 279. (d) Kergomard, A.; Tardivat, J. C.; Vuillerme, J. P. *BSF* **1974**, 2572. (e) Sleezer, P. D.; Winstein, S.; Young, W. G. *JACS* **1963**, *85*, 1890. (f) Anderson, C. B.; Winstein, S. *JOC* **1963**, *28*, 605. (g) Rappoport, Z.; Sleezer, P. D.; Winstein, S.; Young, W. G. *TL* **1965**, 3719. (h) Rappoport, Z.; Dyall, L. K.; Winstein, S.; Young, W. G. *TL* **1970**, 3483. (i) Rappoport, Z.; Winstein, S.; Young, W. G. *JACS* **1972**, *94*, 2320. (j) Mrozik, H.; Eskola, P.; Fisher, M. H. *JOC* **1986**, *51*, 3058.
72. (a) Windaus, A.; Brunken, J. *LA* **1928**, *460*, 225. (b) Windaus, A.; Dithmar, K.; Mürke, H.; Suckfüll, F. *LA* **1931**, *488*, 91. (c) Windaus, A.; Riemann, U.; Zühlsdorff, G. *LA* **1942**, *552*, 135. (d) Windaus, A.; Riemann, U.; Rüggeberg, H. H.; Zühlsdorff, G. *LA* **1942**, *552*, 142. (e) Ruyle, W. V.; Jacob, T. A.; Chemerda, J. M.; Chamberlin, E. M.; Rosenburg, D. W.; Sita, G. E.; Erickson, R. L.; Aliminosa, L. M.; Tishler, M. *JACS* **1953**, *75*, 2604. (f) Saucy, G.; Geistlich, P.; Helbling, R.; Heusser, H. *HCA* **1954**, *37*, 250. (g) Barton, D. H. R.; Rosenfelder, W. J. *JCS* **1951**, 2381.
73. (a) Mercury(II) trifluoroacetate seems to be a better choice: Reischl, W.; Kalchhauser, H. *TL* **1992**, *33*, 2451. Analogous oxidation is known for Pd^{II} and $(AcO)_4Pb$. Pd^{II}: (b) McMurry, J. E.; Kočovský, P. *TL* **1984**, *25*, 4187. (c) Heumann, A.; Åkermark, B. *AG(E)* **1984**, *23*, 453. (d) Hansson, S.; Heumann, A.; Rein, T.; Åkermark, B. *JOC* **1990**, *55*, 975. Pb^{IV}: (e) Whitham, G. H. *JCS* **1961**, 2232. (f) Cooper, M. A.; Salmon, J. R.; Whittaker, D.; Scheidegger, U. *JCS(B)* **1967**, 1259. (g) Mori, K. *ABC* **1976**, *40*, 415.
74. Blossey, E. C.; Kucinski, P. *CC* **1973**, 56.
75. (a) Gleghorn, J. *CR(C)* **1986**, *303*, 1425. (b) Courtneidge, J. L.; Davies, A. G.; Gregory, R. S.; McGuchan, D. C.; Yazdi, S. N. *CC* **1987**, 1192. (c) Davies, A. G.; McGuchan, D. C. *OM* **1991**, *10*, 329.
76. Whitmore, F. C.; Hanson, E. R. *OSC* **1941**, *1*, 161.
77. Bernardi, A. *G* **1926**, *56*, 337.
78. (a) Ragno, M. *G* **1938**, *68*, 738. (b) Ragno, M. *G* **1940**, *70*, 423.
79. Ding, K.; Wu, Y.; Hu, H.; Shen, L.; Wang, X. *OM* **1992**, *11*, 3849.
80. Vicente, J.; Abad, J.-A.; Sandoval, A.; Jones, P. G. *JOM* **1992**, *434*, 1.
81. Kirk, D. N.; Slade, C. J. *CC* **1982**, 563.
82. Constable, E. C.; Leese, T. A.; Tocher, D. A. *CC* **1989**, 570.
83. Ciusa, R.; Grilla, G. *G* **1927**, *57*, 323.
84. Scheibler, H.; Jeschke, J.; Beiser, W. *JPR* **1933**, *136*, 232.
85. (a) O'Connor, G. N.; Crawford, J. V.; Wang, C.-H. *JOC* **1965**, *30*, 4090. (b) Ganske, J. A.; Pandey, R. K.; Postich, M. J.; Snow, K. M.; Smith, K. M. *JOC* **1989**, *54*, 4801.
86. Buu-Hoï, N. P. *BSF* **1958**, 1407.
87. Travagli, G. *G* **1955**, *85*, 926.
88. Banerji, A.; Sarkar, M. *Proc. Indian Acad. Sci., Ser. Chem. Sci.* **1982**, *91*, 247 (*CA* **1983**, *98*, 54 069w).
89. (a) Skulski, L.; Kujawa, A.; Wroczynski, P. *KGS* **1989**, *249* (*CA* **1989**, *111*, 232 442j). (b) Skulski, L.; Kujawa, A.; Kujawa, T. M. *Bull. Pol. Acad. Sci., Chem.* **1988**, *35*, 499 (*CA* **1989**, *110*, 115 000g). (c) Skulski, L.; Baranowski, A.; Lempke, T. *Bull. Pol. Acad. Sci., Chem.* **1991**, *39*, 459 (*CA* **1992**, *117*, 171 072p).
90. Kandile, N. G.; Soliman, A. A.; El Sawi, E. A. *Synth. React. Inorg. Metal-Org. Chem.* **1989**, *19*, 779 (*CA* **1990**, *112*, 235 463y).
91. El-Sawi, E. A.; Ahmed, M. A. *Acta Chim. Hung.* **1987**, *124*, 657 (*CA* **1989**, *110*, 57 783b).
92. Rees, C. W.; Surtees, J. R. J. *JCS(P1)* **1991**, 2945.
93. Neto, A. F.; Miller, J.; Kiyan, N. Z.; Miyata, Y. *An. Acad. Bras. Cienc.* **1989**, *61*, 419 (*CA* **1991**, *115*, 29 549q).
94. Amiet, G.; Nicholas, K.; Pettit, R. *CC* **1970**, 161.
95. Winter, C. H.; Han, Y.-H.; Heeg, M. J. *OM* **1992**, *11*, 3169.
96. Buchler, J. W.; Herget, G. *ZN(B)* **1987**, *42*, 1003 (*CA* **1988**, *108*, 94 761x).
97. Walker, T.; Warburton, W. K.; Webb, G. B. *JCS* **1962**, 1277.
98. (a) Moody, C. J.; Shah, P. *JCS(P1)* **1989**, 2463. (b) Jackson, P. M.; Moody, C. J. *SL* **1990**, 521.
99. Hylarides, M. D.; Willbur, D. S.; Hadley, S. W.; Fritzberg, A. R. *JOM* **1989**, *367*, 259.
100. (a) Hall, L. D.; Neeser, J.-R. *CC* **1982**, 887. (b) Brown, H. C.; Larock, R. C.; Gupta, S. K.; Rajagopalan, S.; Bhat, N. G. *JOC* **1989**, *54*, 6079.
101. Larock, R. C. *JOC* **1976**, *41*, 2241.
102. (a) Larock, R. C.; Bernhardt, J. C. *JOC* **1977**, *42*, 1680. (b) Takagi, K.; Hayama, N.; Okamoto, N.; Sakakibara, Y.; Oka, S. *BCJ* **1977**, *50*, 2741.

103. For a review, see: (a) Fanta, P. E. *CRV* **1946**, *38*, 139. (b) Fanta, P. E. *CRV* **1964**, *64*, 613. (c) Bacon, R. G. R.; Hill, H. A. O. *QR* **1965**, *19*, 95. (d) Fanta, P. E. *S* **1974**, 9. (e) Sainsbury, M. *T* **1980**, *36*, 3327.
104. (a) Leonard, N. J.; Hay, A. S.; Fulmer, R. W.; Gash, V. W. *JACS* **1955**, *77*, 439. (b) Leonard, N. J.; Fulmer, R. W.; Hay, A. S. *JACS* **1956**, *78*, 3457. (c) Leonard, N. J.; Miller, L. A.; Thomas, P. D. *JACS* **1956**, *78*, 3463. (d) Chevolot, L.; Husson, H.-P.; Potier, P. *T* **1975**, *31*, 2491. (e) Aimi, N.; Yamanaka, E.; Endo, J.; Sakai, S.; Haginiwa, J. *TL* **1972**, 1081. (f) For a review and comparison with other isoelectronic cations (Tl^{3+} and Pb^{4+}), see: Butler, R. N. *CRV* **1984**, *84*, 249.
105. (a) Wenkert, E. *JACS* **1962**, *84*, 98. (b) Kutney, J. P.; Brown, R. T.; Piers, E. *JACS* **1964**, *86*, 2286. (c) Kutney, J. P.; Brown, R. T.; Piers, E. *JACS* **1964**, *86*, 2287. (d) Kutney, J. P.; Piers, E.; Brown, R. T. *JACS* **1970**, *92*, 1700. (e) Kutney, J. P.; Piers, E.; Brown, R. T.; Hadfield, J. R. *JACS* **1970**, *92*, 1708.
106. (a) Knabe, J. *AP* **1959**, *292*, 416. (b) Knabe, J. *AP* **1960**, *293*, 121. (c) Knabe, J.; Grund, G. *AP* **1963**, *296*, 854. (d) Knabe, J.; Roloff, H. *CB* **1964**, *97*, 3452.
107. Möhrle, H.; Schittenhelm, D. *CB* **1971**, *104*, 2475.
108. Mills, J. S. *JCS(C)* **1967**, 2514.
109. (a) Julia, M.; Blasioli, C. *BSF* **1976**, 1941. (b) Martin, S. F.; Chou, T. *TL* **1978**, 1943. (c) Martin, S. F.; Chou, T. *JOC* **1978**, *43*, 1027. For reviews on the Wichterle reaction, see: (d) Jung, M. E. *T* **1976**, *32*, 3. (e) Gawley, R. E. *S* **1976**, 777.
110. (a) Thompson, A.; Wolform, M. L.; Inatome, M. *JACS* **1955**, *77*, 3160. (b) Wolform, M. L.; Groebke, W. *JOC* **1963**, *28*, 2986. (c) Kočovský, P.; Pour, M. *JOC* **1990**, *55*, 5580.
111. (a) Iida, H.; Yuasa, Y.; Kibayashi, C. *S* **1982**, 471. (b) Iida, H.; Yuasa, Y.; Kibayashi, C. *TL* **1982**, *23*, 3591.
112. Plummer, B. F.; Menendez, B. F.; Songster, M. *JOC* **1989**, *54*, 718.
113. Stoodley, R. J.; Whitehouse, N. R. *JCS(P1)* **1973**, 32.
114. (a) Barluenga, J.; Aznar, F.; Liz, R. *S* **1984**, 304. (b) Barluenga, J.; Aznar, F.; Liz, R.; Bayod, M. *CC* **1984**, 1427. (c) Barluenga, J.; Aznar, F.; Liz, R.; Bayod, M. *JOC* **1987**, *52*, 5190. (d) Barluenga, J.; Aznar, F.; Fraiz, S., Jr.; Pinto, A. C. *TL* **1991**, *32*, 3205.
115. Larock, R. C. *JOC* **1974**, *39*, 834.
116. Corey, E. J.; Tius, M. A. *TL* **1977**, 2081.

Pavel Kočovský
University of Leicester, UK

Mercury(II) Chloride[1]

[7487-94-7] Cl_2Hg (MW 271.49)

(electrophilic mercuration of multiple bonds;[1] cleavage of vinyl sulfides and thioacetals;[17] transmetalation;[1] preparation of amalgams[30–33])

Alternate Name: mercuric chloride.
Physical Data: mp 277 °C; bp 302 °C; *d* 5.440 g cm^{-3}.
Solubility: sol H_2O, alcohol, ether, glycerol, acetic acid, acetone, ethyl acetate; slightly sol benzene, pyridine, CS_2.
Form Supplied in: white rhombic crystals.
Handling, Storage, and Precautions: violent poison; may be fatal if swallowed in 0.2–0.4 g doses. Exposure to any mercury reagent is to be avoided. Teratogen; mutagen; irritant. Reacts violently with K, Na. Releases toxic Hg vapor when heated to decomposition. Handle in a fume hood.

Electrophilic Attack on Multiple Bonds. Although less electrophilic than other Hg^{II} reagents, $HgCl_2$ has been successfully employed in electrophilic cyclization of various dienes[1,2] (see also ***Mercury(II) Acetate***) (eq 1);[3] an allylic hydroxyl controls the diastereoselectivity of the latter reaction.[3] Aromatization of certain conjugated systems has also been observed on treatment with $HgCl_2$.[4] Similar to Tl^{I} salts,[5] $HgCl_2$ promotes iodocyclization of alkenic alcohols.[6] In the presence of a halogen (Cl_2 or Br_2), $HgCl_2$ facilitates halogenation of a C=C bond.[2]

1. $HgCl_2$, *t*-BuOH; 2. $NaBH_4$; 55% (1)

Intramolecular aminomercuration of δ,ε-unsaturated amines has also been accomplished with $HgCl_2$[7,8] (eq 2).[8] The stereochemistry of the reaction is solvent dependent[8] and may be reversible.[9]

1. $HgCl_2$; 2. $NaBH_4$ (2)

THF	10:90	36%
THF, H_2O	87:13	64%

Terminal alkynes (RC≡CH) add MeOH in the presence of ***Triethylamine*** and a catalytic amount of $HgCl_2$ to give enol ethers of the corresponding ketones (RC(OMe)=CH_2).[10] This reaction parallels the well-known $HgSO_4$-catalyzed hydration of alkynes, producing ketones. 3-Alken-1-ynes undergo catalytic aminomercuration in the presence of $HgCl_2$ at 70 °C over 3–6 h to produce enamines.[11] By contrast, propargylic alcohols (HC≡CCH_2OH) undergo oxidative aminomercuration to afford bis-aminated aldehydes, e.g. (*Z*)-PhNHCH=C(NHPh)CH=O.[12] Propargyl amines (HC≡C–CH_2NR_2) add $HgCl_2$ in aqueous HCl to give ClCH=C(HgCl)–CH_2NR_2.[13]

Treatment of silyl enol ethers of ε-alkynic ketones or aldehydes with $HgCl_2$ (1.1 equiv) and ***Hexamethyldisilazane*** (0.2 equiv; acid scavenger) induces cyclization (eq 3).[14]

$HgCl_2$, $(TMS)_2NH$, CH_2Cl_2, 30 °C (3)

Enol ethers derived from carbohydrates can be readily converted into carbocycles via a $HgCl_2$-mediated reaction which involves an electrophilic attack at the C=C bond to generate the corresponding ketoaldehyde, which cyclizes spontaneously via an intramolecular aldol condensation (eq 4).[15]

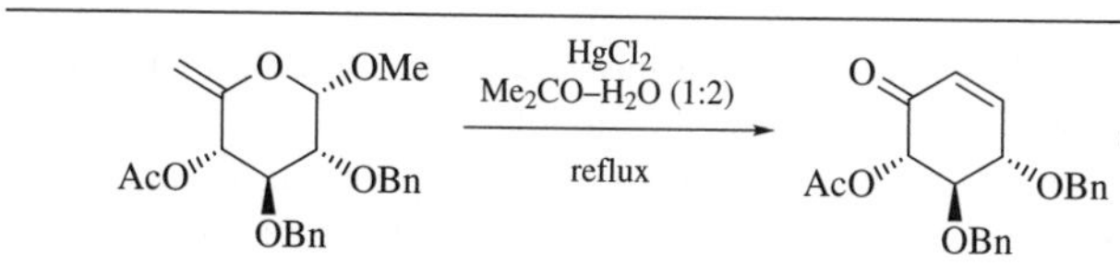

(4)

Aldehydes $RCH_2CH{=}O$ (R = Me, Et) afford α,α-bischloromercurated products on treatment with excess $HgCl_2$.[16]

Hydrolysis of Vinyl Sulfides and Thioacetals to Carbonyl Compounds[17]. Whereas the hydrolysis of vinyl sulfides to ketones works well with a mixture of $HgCl_2$ and an additive (HgO, $CaCO_3$, or $CdCO_3$), the reaction leading to aldehydes often gives unsatisfactory results. In this case, yields can be dramatically improved if HCl is first added across the double bond of the vinyl sulfide (RCH=CHSPh) to generate $R{-}CH_2CH(Cl)SPh$. The latter intermediate is then quantitatively hydrolyzed by $HgCl_2$ and water to the aldehyde $RCH_2CH{=}O$.[18]

Thioacetals[19,20] and *O,S*-acetals[21] are hydrolyzed by means of $HgCl_2$ to the corresponding carbonyl compounds; addition of ***Calcium Carbonate*** usually improves the yields. This method, involving spontaneous spirocyclization of the resulting keto group, has been employed in the synthesis of talaromycin B (eq 5).[20]

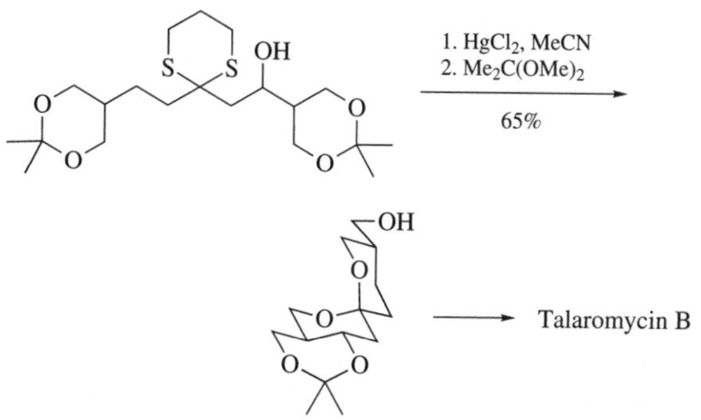

Talaromycin B (5)

Methylthiomethyl (MTM) ethers can be converted into 2-methoxyethoxy (MEM), methoxymethyl (MOM), or ethoxymethyl (EOM) ethers on reaction with $HgCl_2$ and $MeOCH_2CH_2OH$, MeOH, or EtOH, respectively, in 70–80% yields.[22]

Addition of $HgCl_2$ to boronate ate complexes derived from *O,S*-acetals induces B → C migration. This sequence has been used to obtain optically pure aldehydes (eq 6).[23] Selenoacetals are similarly hydrolyzed by $HgCl_2/CaCO_3$ in acetonitrile.[24]

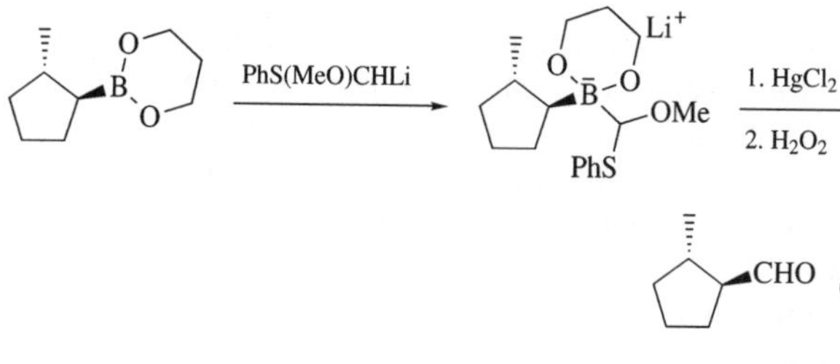

(6)

Preparation of Organomercurials by Exchange Reactions. Among the methods developed for the synthesis of organomercurials is the transmetalation of other organometallics with $HgCl_2$ (e.g. ArLi → ArHgCl or RMgCl → RHgCl),[1,25] and reactions of aromatic diazonium salts with $HgCl_2$ and copper ($ArN_2^+ Cl^- \rightarrow ArHgCl$).[26] The yields in the latter methods do not exceed 50%.[25] Sodium *p*-toluenesulfinate is also converted into the corresponding organomercurial (MeC_6H_4HgCl) on reaction with $HgCl_2$.[27] Vinylmercury chlorides (RCH=CH–HgCl) can be prepared by transmetalation of the corresponding vinylalanes, which, in turn, are available from terminal alkynes; the transmetalation occurs with >98% retention of configuration.[28] A stable metalated cubane derivative has been obtained by lithiation of the diisopropylamide of cubanecarboxylic acid with ***Lithium 2,2,6,6-Tetramethylpiperidide*** followed by transmetalation with $HgCl_2$.[29] Reversed transmetalation (cubane–HgCl → cubane–Li) has also been described.[29]

Amalgams. Mercury(II) chloride has been extensively utilized for the preparation of a variety of amalgams (e.g. Zn,[30] Mg,[31] and Al)[32,33] to be employed in reductive processes such as Clemmensen reduction (with Zn)[30] or pinacol coupling (Mg),[31] and to prepare, for example, aluminum ethoxide[32] and *t*-butoxide.[33]

Miscellaneous. Penam derivatives result from the $HgCl_2$-promoted ring closure of azetidin-2-one.[34] Mercury(II) chloride seems to be a reagent of choice for isolation of histidine from a mixture of amino acids in the form of an insoluble complex.[35] In combination with iodine, $HgCl_2$ facilitates α-iodination of enolizable ketones and aldehydes.[36]

Related Reagents. Mercury(II) Acetate; Mercury(II) Chloride–Cadmium Carbonate; Mercury(II) Chloride–Silver(I) Nitrite; Mercury(II) Trifluoroacetate.

1. (a) Larock, R. C. *AG(E)* **1978**, *17*, 27. (b) Larock, R. C. *T* **1982**, *38*, 1713. (c) Larock, R. C. *Organomercury Compounds in Organic Synthesis*; Springer: Berlin, 1985. (d) Larock, R. C. *Solvomercuration/Demercuration Reactions in Organic Synthesis*; Springer: Berlin, 1986.
2. (a) Vardhan, H. B.; Bach, R. D. *JOC* **1992**, *57*, 4948. (b) Barluenga, J.; Martínez-Gallo, J. M.; Nájera, C.; Yus, M. *CC* **1985**, 1422.
3. (a) Henbest, H. B.; Nicholls, B. *JCS* **1959**, 227. (b) Henbest, H. B.; McElkinney, R. S. *JCS* **1959**, 1834. (c) Matsuki, Y.; Kodama, M.; Itô, S. *TL* **1979**, 2901.
4. Rozenberg, V. I.; Gavrilova, G. V.; Ginzburg, B. I.; Nikanorov, V. A.; Reutov, O. A. *IZV* **1982**, 1916; *BAU* **1982**, *31*, 1707.
5. Kočovský, P.; Pour, M. *JOC* **1990**, *55*, 5580.
6. Forsyth, C. J.; Clardy, J. *JACS* **1990**, *112*, 3497.
7. Périé, J. J.; Laval, J. P.; Roussel, J.; Lattes, A. *TL* **1971**, 4399.
8. Tokuda, M.; Yamada, Y.; Suginome, H. *CL* **1988**, 1289.
9. Barluenga, J.; Perez-Prieto, J.; Bayon, A. M. *T* **1984**, *40*, 1199.
10. Barluenga, J.; Aznar, F.; Bayod, M. *S* **1988**, 144.
11. (a) Barluenga, J.; Aznar, F.; Liz, R.; Cabal, M. P. *CC* **1985**, 1375. Similar reaction occurs with $(AcO)_2Hg$: (b) Davtyan, S. Zh.; Chobanyan, Zh. A.; Badanyan, Sh. O. *Arm. Khim. Zh.* **1983**, *36*, 508 (*CA* **1984**, *100*, 67 447c). (c) Barluenga, J.; Aznar, F.; Valdez, C.; Cabal, M. P. *JOC* **1991**, *56*, 6166. (d) Barluenga, J.; Aznar, F.; Liz, R.; Cabal, M. P. *S* **1986**, 960.
12. Barluenga, J.; Aznar, F.; Liz, R. *CC* **1986**, 1180.
13. Larock, R. C.; Burns, L. D.; Varaprath, S.; Russell, C. E.; Richardson, J. W., Jr.; Janakiraman, M. N.; Jacobson, R. A. *OM* **1987**, *6*, 1780.
14. (a) Drouin, J.; Bonaventura, M.-A.; Coia, J.-M. *JACS* **1985**, *107*, 1726. (b) Conia, J. M.; LePerchec, P. *S* **1975**, 1. (c) Forsyth, C. J.; Clardy, J. *JACS* **1990**, *112*, 3497.
15. Chida, N.; Ohtsuka, M.; Nakazawa, K.; Ogawa, S. *CC* **1989**, 436.

16. Korpar-Colig, B.; Popovic, Z.; Sikirica, M. *Croat. Chem. Acta* **1984**, *57*, 689 (*CA* **1985**, *102*, 220 968m).
17. Stachel, D. P. N. *CSR* **1977**, *6*, 345.
18. Mura, A. J., Jr.; Majetich, G.; Grieco, P. A.; Cohen, T. *TL* **1975**, 4437.
19. Seebach, D.; Beck, A. K. *OSC* **1988**, *6*, 316.
20. (a) Schreiber, S. L.; Sommer, T. J. *TL* **1983**, *24*, 4781. (b) Kozikowski, A. P.; Scripko, J. G. *JACS* **1984**, *106*, 353.
21. Jensen, J. L.; Maynard, D. F.; Shaw, G. R.; Smith, T. W., Jr. *JOC* **1992**, *57*, 1982.
22. Chowdhury, P. K.; Sharma, D. N.; Sharma, R. R. *CI(L)* **1984**, 803.
23. (a) Brown, H. C.; Imai, T. *JACS* **1983**, *105*, 6285. (b) Brown, H. C.; Imai, T.; Desai, M. C.; Singaram, B. *JACS* **1985**, *107*, 4980.
24. Burton, A.; Hevesi, L.; Dumont, W.; Cravador, A.; Krief, A. *S* **1979**, 877.
25. (a) Eaton, P. E.; Martin, R. M. *JOC* **1988**, *53*, 2728. (b) Wells, A. P.; Kitching, W. *JCS(P1)* **1995**, 527.
26. Nesmeyanov, A. N. *OSC* **1943**, *2*, 432.
27. Whitmore, F. C.; Hamilton, F. H.; Thurman, N. *OSC* **1941**, *1*, 519.
28. Negishi, E.; Jadhav, K. P.; Daotien, N. *TL* **1982**, *23*, 2085.
29. (a) Eaton, P.; Castaldi, G. U.S. Patent Appl. 613 708 (*CA* **1986**, *105*, 172 705n) (b) Eaton, P. E.; Cunkle, G. T.; Marchioro, G.; Martin, R. M. *JACS* **1987**, *109*, 948.
30. (a) Martin, E. L. *OSC* **1943**, *2*, 499. (b) Schwarz, R.; Hering, H. *OSC* **1963**, *4*, 203. (c) Shriner, R. L.; Berger, A. *OSC* **1955**, *3*, 786.
31. Adams, R.; Adams, E. W. *OSC* **1941**, *1*, 459.
32. Chalmers, W. *OSC* **1943**, *2*, 598.
33. Wayne, W.; Adkins, H. *OSC* **1955**, *3*, 367.
34. Sheehan, J. C.; Piper, J. V. *JOC* **1973**, *38*, 3492.
35. Foster, G. L.; Shemin, D. *OSC* **1943**, *2*, 330.
36. Barluenga, J.; Martinez-Gallo, J. M.; Najera, C.; Yus, M. *S* **1986**, 678.

Pavel Kočovský
University of Leicester, UK

Mercury(II) Trifluoroacetate[1]

$(CF_3CO_2)_2Hg$

[13257-51-7] $C_4F_6HgO_4$ (MW 426.63)

(oxymercuration[3] and aminomercuration[17] of alkenes; polyene cyclization;[21] allylic oxidation;[32] macrolide synthesis[37])

Alternate Name: mercuric trifluoroacetate.
Physical Data: mp 171–173 °C.
Solubility: sol THF, DME, dioxane; insol hexane.
Preparative Methods: can be prepared from ***Mercury(II) Oxide*** in ***Trifluoroacetic Acid***[2–4] or generated in situ from ***Mercury(II) Acetate*** and CF_3CO_2H.[8b]
Form Supplied in: white, hygroscopic crystals.
Handling, Storage, and Precautions: acute poison; is easily absorbed through the skin and is corrosive; exposure to all mercury compounds is to be strictly avoided. Releases toxic Hg fumes when heated to decomposition. Protect from light. Use in a fume hood.

Oxymercuration.[3–5] The reactivity of mercury(II) trifluoroacetate is similar to that of ***Mercury(II) Acetate*** but exhibits higher electrophilicity.[3] For instance, while cholesterol and other alkenes with a trisubstituted double bond are normally reluctant to react with $(AcO)_2Hg$ (unless neighboring group participation boosts the reaction), $(CF_3CO_2)_2Hg$ reacts readily with cholesteryl acetate (eq 1).[6] Even α,β-unsaturated esters react with $(CF_3CO_2)_2Hg$; the corresponding adducts are readily transformed into α-bromo derivatives (eq 2).[7] Generally, in the presence of Cl_2 or Br_2, the HgX group in the original adduct is replaced by halogen.[1,5]

$(CF_3CO_2)_2Hg$ (1)

1. $(CF_3CO_2)_2Hg$, R^2OH; 2. KBr; Br_2 (2)

Increased selectivity in oxymercuration has been observed when $(CF_3CO_2)_2Hg$ and other Hg^{II} salts were generated in situ from HgO and RCO_2H in an inert solvent (CH_2Cl_2, hexane, or aq THF) on sonication.[4]

Whereas oxymercuration normally occurs with *anti* stereoselectivity,[3,4] products of *syn* addition have been observed for norbornene derivatives.[8]

Intramolecular oxymercuration[9] has been extensively utilized for the construction of various oxygen heterocycles, usually with high stereoselectivity (eq 3).[10] Both isolated[9,10] and conjugated[11] double bonds react, and 5-*exo-trig*, 5-*endo-trig*, and 6-*exo-trig* reactions have all been observed.[9–11] Similarly, cyclic peroxides are formed from the unsaturated peroxides.[12]

1. $(CF_3CO_2)_2Hg$; 2. NaCl (3)

The relatively stable trichloroacetals derived from allylic alcohols can be utilized as intermediates to accomplish stereo- and regiocontrolled mercuration which gives *cis*-1,2-diols (eq 4).[13] However, this method has the following limitations: (1) the original hydroxy group must be equatorial, (2) the double bond cannot be trisubstituted, (3) the reaction is not regioselective for acyclic allylic alcohols, and (4) the reaction is unsuccessful with homoallylic alcohols.[13]

1. $(CF_3CO_2)_2Hg$; 2. $NaBH_4$; Na or Zn (4)

Hydroxy allenes seem to be particularly suitable substrates as the sp^2-bonded mercury can be transmetalated with palladium salts, which expands the versatility of this methodology (eq 5).[14] Alkynic alcohols also produce vinylmercurials (with some preference for (*Z*) isomers)[15] which can be further elaborated, e.g. by conversion into vinyl halides on reaction with sources of electrophilic halogen (***N-Iodosuccinimide***).[16]

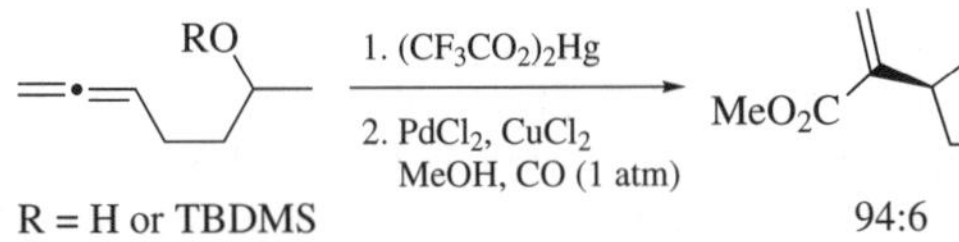

Aminomercuration. Intramolecular aminomercuration followed by demercuration is a useful method of construction of a variety of nitrogen heterocycles such as aza sugars.[17] Intramolecular amidomercuration, employing $(CF_3CO_2)_2Hg$, has also been reported.[18] It has been shown that the steric course of mercuration can be controlled by a neighboring nitrogen group (eq 6).[19]

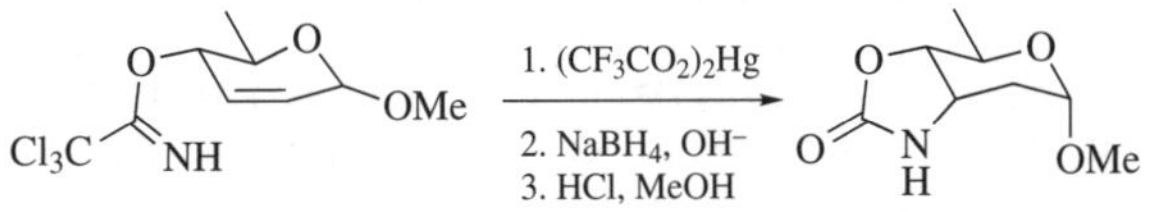

Azidomercuration of an alkenic double bond can be accomplished in quantitative yield (in the case of a monosubstituted double bond) by the reaction carried out with $(CF_3CO_2)_2Hg$ and ***Sodium Azide*** (3:1) in THF–H_2O (1:1) at 60 °C for 48 h.[20] This method has been utilized for syntheses of amino sugars. By contrast, $(AcO)_2Hg$ gives only 20% in the same case.[20]

Polyene Cyclization. Acyclic isoprenoids are stereospecifically cyclized on reaction with $(CF_3CO_2)_2Hg$ to afford polycyclic structures (eq 7).[21] An aromatic ring can be used in place of the nucleophilic double bond in a similar type of cyclization.[22]

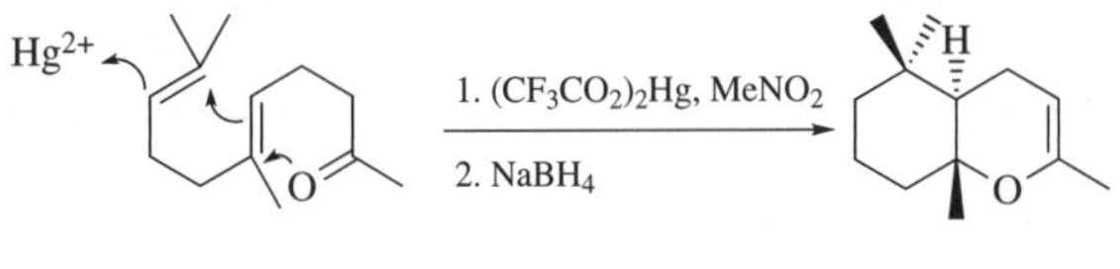

Cyclopropane Ring Opening. Mercury(II) trifluoroacetate readily opens up cyclopropane rings.[23] The reaction is usually faster than that with $(AcO)_2Hg$ and slower than for ***Mercury(II) Nitrate***.

Mercuration of Aromatics. Being a good electrophile, $(CF_3CO_2)_2Hg$ and other Hg^{II} salts also attack aromatic rings.[24,25] Combined with the subsequent reaction with iodine, this method has long served to accomplish iodination in cases where the arene is inert to direct iodination. Polyhaloaromatics can also be synthesized in this way.[26]

[3,3]-Sigmatropic Rearrangements. Oxy-Cope rearrangement can be initiated at rt by $(CF_3CO_2)_2Hg$ (1 equiv); the intermediate α-mercurio ketone is then demercurated with ***Sodium Borohydride***.[27] The reaction can be carried out with a catalytic amount of Hg^{II} and an excess of lithium triflate or trifluoroacetate.[28] The Claisen rearrangement of allylic carbamates can be best achieved via a suprafacial [3,3]-migration mediated by $(CF_3CO_2)_2Hg$ (eq 8); the reaction is thermodynamically controlled.[29] Using a carbamate derived from an optically active amine, formation one of the diastereoisomers may be preferred; however, this method seems to have a limited value due to the low de.[30]

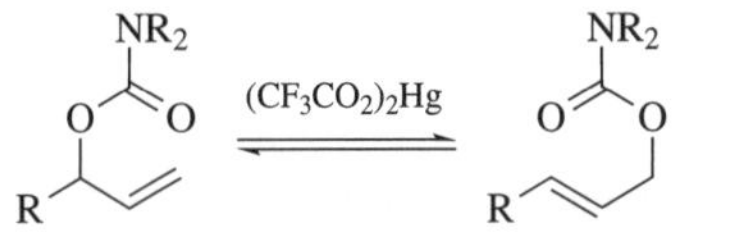

Allylic Oxidation (Treibs Reaction[31]).[32] More effective than the acetate, $(CF_3CO_2)_2Hg$ facilitates allylic oxidation[33,34] of alkenes which are generally reluctant to be oxymercurated (namely those with trisubstituted double bonds). Thus, cholesterol and its esters are converted, after hydrolysis, to the 3β,6β-dihydroxy-4-ene at 0 °C in 40–70% yields.[33]

Hydrolysis of Vinyl Chlorides. The original Wichterle reaction[35] relied on the hydrolysis of vinyl chlorides with H_2SO_4. The improved version utilizing $(CF_3CO_2)_2Hg$ allows the control of regioselectivity by the solvent.[36] Thus, hydrolysis in $MeNO_2$, CH_2Cl_2, or AcOH gives only 1,5-diketone (to be cyclized) in up to 97% yield, whereas hydrolysis in MeOH gives mainly the 1,4-diketone (eq 9).[36]

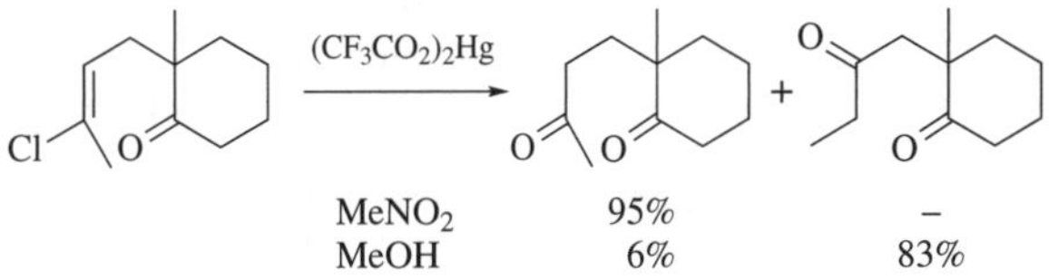

Synthesis of Macrolides. *t*-Butyl thioesters are converted to esters on reaction with $(CF_3CO_2)_2Hg$ or $(MeSO_3)_2Hg$. This method is general for synthesis of macrolides (eq 10) but cannot be used for substrates containing electron-rich centers such as double bonds (which undergo oxymercuration); other less electrophilic cations can be employed in those instances (Cu^+, Cu^{2+}, or Ag^+).[37]

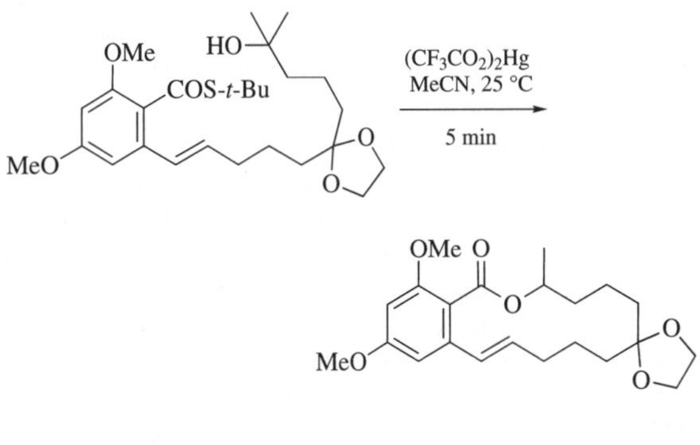

Miscellaneous. Like HgO or Tl^{III}, mercury(II) trifluoroacetate oxidizes phenols and hydroquinones to quinones in the presence of an acid scavenger.[38] Monohydrazones of α-diketones are oxidized with $(CF_3CO_2)_2Hg$ to give α-diazo ketones.[2] Enol esters, obtained by the reaction of 4-dimethylamino-3-butyn-2-one with ω-hydroxy carboxylic acids, react with $(CF_3CO_2)_2Hg$ or ***10-Camphorsulfonic Acid*** to give macrocyclic lactones.[39] 1,3-Dioxolanium salts are converted to α-mercurio derivatives of carboxylic acids on reaction with $(CF_3CO_2)_2Hg$ or $(AcO)_2Hg$.[40] TMS derivatives of tertiary vinyl alcohols undergo Wagner–Meerwein migration on treatment with $(CF_3CO_2)_2Hg$; this can be used as a ring-expansion methodology in cyclic systems.[41]

Related Reagents. *t*-Butyldimethylsilyl Hydroperoxide–Mercury(II) Trifluoroacetate; Mercury(II) Acetate; Mercury(II) Chloride.

---

1. (a) Larock, R. C. *AG(E)* **1978**, *17*, 27. (b) Larock, R. C. *T* **1982**, *38*, 1713. (c) Larock, R. C. *Organomercury Compounds in Organic Synthesis*; Springer: Berlin, 1985. (d) Larock, R. C. *Solvomercuration/Demercuration Reactions in Organic Synthesis*; Springer: Berlin, 1986.
2. Newman M. S.; Arkell, A. *JOC* **1959**, *24*, 385.
3. (a) Brown, H. C.; Rei, M.-H. *JACS* **1969**, *91*, 5646. (b) Brown, H. C.; Kurek, J. T.; Rei, M.-H.; Thompson, K. L. *JOC* **1984**, *49*, 2551.
4. Eihorn, J.; Eihorn, C.; Luche, J. L. *JOC* **1989**, *54*, 4479.
5. (a) Holtmeier, W.; Hobert, K.; Welzel, P. *TL* **1976**, 4709. (b) Barluenga, J.; Martínez-Gallo, J. M.; Nájera, C.; Yus, M. *CC* **1985**, 1422. (c) Barluenga, J.; Martínez-Gallo, J. M.; Nájera, C.; Yus, M. *JCR(S)* 1986, 274.
6. Torrini, I.; Romeo, A. *TL* **1975**, 2605.
7. Anelli, P. L.; Beltrami, A.; Lolli, M.; Uggeri, F. *SC* **1993**, *23*, 2639.
8. (a) Michael, J. P.; Blom, N. F.; Glintenkamp, L.-A. *JCS(P1)* **1991**, 1855. (b) Traylor, T. G. *ACR* **1969**, *2*, 152. (c) Grishin, Yu. K.; Bazhenov, D. V.; Ustynyuk, Yu. A.; Skorobogatova, E. V.; Malisova, N. V.; Kartashov, V. R.; Zefirov, N. S. *DOK* **1986**, *289*, 620. (d) Chernov, A. N.; Furmanova, N. G.; Kartashov, V. R.; Skorobogatova, E. V.; Malisova, N. V.; Zefirov, N. S. *DOK* **1986**, *289*, 1391.
9. (a) Welzel, P.; Holtmeier, W.; Wessling, B. *LA* **1978**, 1327. (b) Kočovský, P.; Pour, M. *JOC* **1990**, *55*, 5580. (c) Salomon, R. G.; Roy, S.; Salomon, M. F. *TL* **1988**, *29*, 769. (d) Kozikowski, A. P.; Lee, J. *JOC* **1990**, *55*, 863.
10. Arnone, A.; Bravo, P.; Resnati, G.; Viani, F. *JCS(P1)* **1991**, 1315.
11. Cardillo, G.; Hashem, Md. A.; Tomasini, C. *JCS(P1)* **1990**, 1487.
12. (a) Adam, W.; Sakanishi, K. *JACS* **1978**, *100*, 3935. (b) Bloodworth, A. J.; Johnson, K. A. *TL* **1994**, *35*, 8057.
13. Overman, L. E.; Campbell, C. B. *JOC* **1974**, *39*, 1474.
14. (a) Walkup, R. D.; Park G. *TL* **1987**, *28*, 1023. (b) Walkup, R. D.; Park G. *TL* **1988**, *29*, 5505. (c) Walkup, R. D.; Park G. *JACS* **1990**, *112*, 1597. (d) Walkup, R. D.; Kim, S. W.; Wagy, S. D. *JOC* **1993**, *58*, 6486.
15. (a) Bassetti, M.; Trovato, M. P.; Bocelli, G. *OM* **1990**, *9*, 2292. (b) Barluenga, J.; Aznar, F.; Liz, R.; Bayod, M. *CC* **1988**, 121.
16. (a) Riediker, M.; Schwartz, J. *JACS* **1982**, *104*, 5842. (b) Rollinson, S. W.; Amos, R. A.; Katzenellenbogen, J. A. *JACS* **1981**, *103*, 4114.
17. (a) Bernotas, R. C.; Ganem, B. *TL* **1985**, *26*, 1123. (b) Tong, M. K.; Blumenthal, E. M.; Ganem, B. *TL* **1990**, *31*, 1683.
18. (a) Takacs, J. M.; Helle, M. A.; Takusagawa, F. *TL* **1989**, *30*, 7321. (b) Takacs, J. M.; Helle, M. A.; Sanyal, B. J.; Eberspacher, T. A. *TL* **1990**, *31*, 6765. (c) Amoroso, R.; Cardillo, G.; Tomasini, C.; Tortoreto, P. *JOC* **1992**, *57*, 1082.
19. Giuliano, R. M.; Deisenroth, T. W.; Frank, W. C. *JOC* **1986**, *51*, 2304.
20. Czernecki, S.; Georgoulis, C.; Provelenghiou, C. *TL* **1979**, 4841.
21. (a) Kurbanov, M.; Semenovsky, A. V.; Smit, W. A.; Shmelev, L. V.; Kucherov, V. F. *TL* **1972**, 2175. (b) Mustafaeva, M. T.; Smit, V. A.; Semenovskii, A. V.; Kucherov, V. F. *IZV* **1973**, 1111. (c) Hoye, T. R.; Kurth, M. J. *JOC* **1978**, *43*, 3693. (d) Hoye, T. R.; Kurth, M. J. *JOC* **1979**, *44*, 3461. (e) Corey, E. J.; Tius, M. A.; Das, J. *JACS* **1980**, *102*, 1742. (f) Erion, M. D.; McMurry, J. E. *TL* **1985**, *26*, 559. (g) Nishizawa, M.; Takenaka, H.; Hayashi, Y. *JOC* **1986**, *51*, 806. (h) Lavey, B. J.; Westkaemper, J. G.; Spencer, T. A. *JOC* **1994**, *59*, 5492. (i) Nishizawa, M.; Takao, H.; Kanoh, N.; Asoh, K.; Hatakeyama, S.; Yamada, H. *TL* **1994**, *35*, 5693. (j) Nishizawa, M.; Morikuni, E.; Asoh, K.; Kan, Y.; Uenoyama, K.; Imagawa, H. *SL* **1995**, 169.
22. Gopalsamy, A.; Balasubramanian, K. K. *CC* **1988**, 34.
23. (a) Collum, D. B.; Mohamadi, F.; Hallock, J. S. *JACS* **1983**, *105*, 6882. (b) Collum, D. B.; Still, W. C.; Mohamadi, F. *JACS* **1986**, *108*, 2094. (c) Kočovský, P.; Grech, J. M.; Mitchell, W. L. *JOC* **1995**, *60*, 482.
24. (a) Dessau, R. M.; Shih, S.; Heiba, E. I. *JACS* **1970**, *92*, 412. (b) McKillop, A.; Turrell, A. G.; Young, D. W.; Taylor, E. C. *JACS* **1980**, *102*, 6504. (c) Lau, W.; Kochi, J. K. *JACS* **1984**, *106*, 7100. (d) Courtneidge, J. L.; Davies, A. G.; Gregory, P. S.; McGuchan, D. C.; Yazdi, S. N. *CC* **1987**, 1192.
25. (a) Deacon, G. B.; O'Conor, M. J.; Stretton, G. N. *AJC* **1986**, *39*, 953. (b) Humphries, R. E.; Massey, A. G. *Phosphorus Sulfur* **1988**, *36*, 135. (c) Büchler, J. W.; Herget, G. *ZN(B)* **1987**, *42*, 1003 (*CA* **1988**, *108*, 94 761x). (d) Baranowski, A.; Skulski, L. *Bull. Pol. Acad. Sci., Chem.* **1991**, *39*, 29 (*CA* **1991**, *115*, 91 948x).
26. (a) Deacon, G. B.; Farquharson, G. J. *AJC* **1976**, *29*, 627. (b) Deacon, G. B.; Farquharson, G. J. *AJC* **1977**, *30*, 293. (c) Deacon, G. B.; Farquharson, G. J. *AJC* **1977**, *30*, 1701. (d) Brady, J. H.; Redhouse, A. D.; Wakefield, B. J. *JCR(S)* **1982**, 137. (e) Perlmutter, M.; Satyamurthy, N.; Luxen, A.; Phelps, M. E.; Barrio, J. R. *Appl. Radiat. Isot.* **1990**, *41*, 801 (*CA* **1991**, *114*, 143 945v).
27. Bluthe, N.; Malacria, M.; Gore, J. *TL* **1982**, *23*, 4263.
28. (a) Bluthe, N.; Malacria, M.; Gore, J. *T* **1984**, *40*, 3277. (b) Bluthe, N.; Gore, J.; Malacria, M. *T* **1986**, *42*, 1333.
29. Overman, L. E.; Campbell, C. B.; Knoll, F. M. *JACS* **1978**, *100*, 4822.
30. Trost, B. M.; Timko, J. M.; Stanton, J. L. *CC* **1978**, 436.
31. Treibs, W. *N* **1948**, *35*, 125.
32. (a) Arzoumanian, H.; Metzger, J. *S* **1971**, 527. (b) Rawlinson, D. J.; Sosnovsky, G. *S* **1973**, 567.
33. Massiot, G.; Husson, H.-P.; Potier, P. *S* **1974**, 722.
34. Reischl, W.; Kalchhauser, H. *TL* **1992**, *33*, 2451.
35. (a) Wichterle, O. *CCC* **1947**, *12*, 93. (b) Wichterle, O.; Procházka, J.; Hofman, J. *CCC* **1948**, *13*, 300. For reviews on Wichterle reaction, see: (c) Jung, M. E. *T* **1976**, *32*, 3. (d) Gawley, R. E. *S* **1976**, 777.
36. Yoshioka, H.; Takasaki, K.; Kobayashi, M.; Matsumoto, T. *TL* **1979**, 3489.
37. (a) Masamune, S.; Kamata, S.; Schilling, W. *JACS* **1975**, *97*, 3515. (b) Masamune, S.; Yamamoto, H.; Kamata, S.; Fukuzawa, A. *JACS* **1975**, *97*, 3513. (c) Masamune, S.; Hayase, Y.; Schilling, W.; Chan, W. K.; Bates, G. S. *JACS* **1977**, *99*, 6756. (d) Masamune, S. *Aldrichim. Acta* **1978**, *11*, 23.
38. McKillop, A.; Young, D. W. *SC* **1977**, *7*, 467.
39. Gais, H.-J. *TL* **1984**, *25*, 273.
40. Boev, V. I.; Dombrovskii, A. V. *ZOB* **1981**, *51*, 1927.
41. Kim, S.; Uh, K. H. *TL* **1992**, *33*, 4325.

Pavel Kočovský
University of Leicester, UK

Mesityllithium[1,2]

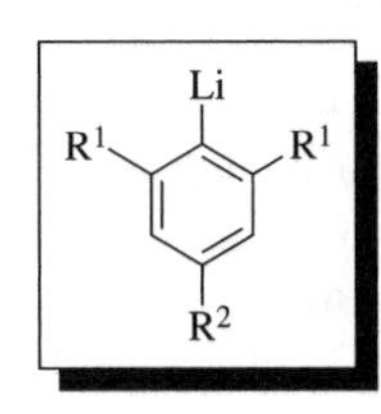

(**1**; $R^1 = R^2 = Me$)		
[5806-59-7]	$C_9H_{11}Li$	(MW 126.14)
(**2**; $R^1 = R^2 = i$-Pr)		
[74226-59-8]	$C_{15}H_{23}Li$	(MW 210.32)
(**3**; $R^1 = R^2 = t$-Bu)		
[35383-91-6]	$C_{18}H_{29}Li$	(MW 252.41)
(**4**; $R^1 = R^2 = t\text{-}BuCH_2$)		
[76804-36-9]	$C_{21}H_{35}Li$	(MW 294.50)
(**5**; $R^1 = R^2 = CF_3$)		
[444-40-6]	$C_9H_2F_9Li$	(MW 288.05)
(**6**; $R^1 = R^2 = (SiMe_3)_2CH$)		
[125444-10-2]	$C_{27}H_{59}LiSi_6$	(MW 559.34)
(**7**; $R^1 = Me$, $R^2 = t$-Bu)		
[126062-41-7]	$C_{12}H_{17}Li$	(MW 168.23)
(**8**; $R^1 = Et$, $R^2 = t$-Bu)		
[149266-65-9]	$C_{14}H_{21}Li$	(MW 196.29)
(**9**; $R^1 = R^2 = Me$, $\cdot Et_2O$)		
[119088-34-5]	$C_{13}H_{21}LiO$	(MW 200.28)
(**10**; $R^1 = R^2 = i$-Pr, $\cdot Et_2O$)		
[107742-58-5]	$C_{19}H_{33}LiO$	(MW 284.46)
(**11**; $R^1 = R^2 = t$-Bu, ·3THF)		
[149764-17-0]	$C_{30}H_{53}LiO_3$	(MW 468.77)
(**12**; $R^1 = R^2 = Et$, $\cdot Et_2O$)		
[107742-59-6]	$C_{16}H_{22}LiO$	(MW 237.32)
(**13**; $R^1 = R^2 = Ph$, $\cdot 2Et_2O$)		
[133777-60-3]	$C_{32}H_{37}LiO_2$	(MW 460.63)

(used for the generation of other amine-free organolithium compounds; derivatives are also synthetic building blocks for the preparation of compounds containing low-coordinate main Group element functional groups which are otherwise not stable, i.e. steric protection)

Alternate Name: 2,4,6-trimethylphenyllithium.

Preparative Methods: the corresponding bromobenzene derivative is treated with 2 equiv ***t*-Butyllithium**[3] in THF solution,[4,5] a method which leads to a 1:1 mixture of the desired Li derivatives (**1**)–(**8**) and ***Lithium Bromide*** (no alkyl halide is present under these conditions; eq 1).

$$\text{ArBr}(R^1,R^1,R^2) + 2\ t\text{-BuLi} \xrightarrow[-CH_2=CMe_2]{-CHMe_3} \text{ArLi}(R^1,R^1,R^2) + LiBr \quad (1)$$

According to X-ray structure determinations and molecular weight measurements in solution, the highly hindered organolithium compounds in the series (**1**)–(**13**) are monomeric,[6,7] i.e. not aggregated as are most other Li derivatives.

Handling, Storage, and Precautions: solutions of (**1**)–(**8**) may be kept under Ar or N_2 in a refrigerator; normally they are used right away; with (**1**), (**2**), and (**4**), a transmetalation from the nuclear to a benzylic position would lead to thermodynamically more stable Li compounds (such rearrangements[8] are catalyzed by tertiary amines, for instance ***N,N,N′,N′*-Tetramethylethylenediamine**).

Use of *o,o′*-Disubstituted Compounds as Bases. Two examples are shown in eqs 2 and 3.[3,4] The 1:1 acylation of Li enolates shown in the first reaction gives best yields of diketones such as (**14**) at −75 °C when the enolate solutions are amine-free The enolate formation from benzyl methyl ketone produces the thermodynamically less stable isomer (**15**) preferentially over the more stable one (**16**).[4,9]

Cycloheptanone → 1. (**1**); 2. $ClOCCH_2CH_2NO_2$, −78 °C; 3. H_3O^+ → (**14**) (2)

Benzyl methyl ketone → 1. (**3**); 2. TMSCl → (**15**) + (**16**), 74:26 (3)

Preparation of Low-Coordinate Main Group Element Derivatives. Due to the steric hindrance, especially in the 2,4,6-tri-*t*-butyl aryl compounds (**3**) and (**11**), called supermesityllithium, it is possible to prepare derivatives of compounds which are otherwise not stable because they would be too reactive to be isolated. Three examples are the selenoaldehyde (**17**),[10] the dibora-allene anion derivative (**18**),[11] and the diphosphene (**19**).[12] A literature search reveals that this technique of steric protection was used all over the Periodic Table, including transition metals (consider also the principle of sterically protected but electronically effective functional groups,[13] and compare also the stabilization of tetrahedrane by *t*-butyl substitution[14]).

(**17**) (**18**) (**19**)

Related Reagents. Lithium Diisopropylamide; Lithium 2,2,6,6-Tetramethylpiperidide; Phenyllithium.

1. Brandsma, L.; Verkruijsse, H. *Preparative Polar Organometallic Chemistry*; Springer: Berlin, 1987; vol 1.
2. Wakefield, B. J. *Organolithium Methods*; Academic: London, 1988.
3. (a) Seebach, D.; Neumann, H. *CB* **1974**, *107*, 847 (*CA* **1974**, *80*, 145 527p). (b) Neumann, H.; Seebach, D. *TL* **1976**, 4839. (c) Neumann, H.; Seebach, D. *CB* **1978**, *111*, 2785 (*CA* **1978**, *89*, 162 560x).
4. (a) Beck, A. K.; Hoekstra, M. S.; Seebach, D. *TL* **1977**, 1187. (b) Seebach, D.; Weller, T.; Protschuk, G.; Beck, A. K.; Hoekstra, M. S. *HCA* **1981**, *64*, 716 (*CA* **1981**, *95*, 186 681n).
5. Yoshifuji, M.; Nakamura, T.; Inamoto, N. *TL* **1987**, *28*, 6325.
6. Maetzke, T.; Seebach, D. *HCA* **1989**, *72*, 624 (*CA* **1990**, *112*, 55 649e).
7. (a) Olmstead, M. M.; Power, P. P. *JOM* **1991**, *408*, 1. (b) Girolami, G. S.; Riehl, M. E.; Suslick, K. S.; Wilson, S. R. *OM* **1992**, *11*, 3907. (c) Ruhlandt-Senge, K.; Ellison, J. J.; Wehmschulte, R. J.; Pauer, F.; Power, P. P. *JACS* **1993**, *115*, 11353.
8. See for instance: Meyer, N.; Seebach, D. *CB* **1980**, *113*, 1304 (*CA* **1980**, *93*, 71 166q).
9. Another method of obtaining thermodynamically less stable Li enolates is the trapping of in situ generated ketenes by RLi: (a) Häner, R.; Laube, T.; Seebach, D. *JACS* **1985**, *107*, 5396. (b) Seebach, D.; Amstutz, R.; Laube, T.; Schweizer, W. B.; Dunitz, J. D. *JACS* **1985**, *107*, 5403.
10. (a) Okazaki, R.; Ishii, A.; Inamoto, N. *CC* **1986**, 71. (b) Ishii, A.; Okazaki, R.; Inamoto, N. *BCJ* **1986**, *59*, 2529. (c) Ishii, A.; Okazaki, R.; Inamoto, N. *BCJ* **1987**, *60*, 1037.
11. Hunold, R.; Allwohn, J.; Baum, G.; Massa, W.; Berndt, A. *AG(E)* **1988**, *27*, 961.
12. Yoshifuji, M.; Shima, I.; Inamoto, N. *JACS* **1981**, *103*, 4587.
13. (a) Schlecker, R.; Seebach, D.; Lubosch, W. *HCA* **1978**, *61*, 512 (*CA* **1978**, *89*, 59 583h). (b) Seebach, D.; Hassel, T. *AG(E)* **1978**, *17*, 274. (c) Hassel, T.; Seebach, D. *AG(E)* **1979**, *18*, 399. (d) Seebach, D.; Locher, R. *AG(E)* **1979**, *18*, 957. (e) Seebach, D.; Ertas, M.; Locher, R.; Schweizer, W. B. *HCA* **1985**, *68*, 264 (*CA* **1985**, *103*, 123 099s). (f) Seebach, D.; Häner, R.; Vettiger, T. *HCA* **1987**, *70*, 1507 (*CA* **1988**, *108*, 150 938z). (g) Vettiger, T.; Seebach, D. *LA* **1990**, 195 (*CA* **1990**, *112*, 139 774e). (h) Suzuki, K.; Seebach, D. *LA* **1992**, 51. (i) Seebach, D.; Pfammatter, E.; Gramlich, V.; Bremi, T.; Kühnle, F.; Portmann, S.; Tironi, I. *LA* **1992**, 1145.
14. Maier, G. *AG(E)* **1988**, *27*, 309.

Albert K. Beck, Robert Dahinden & Dieter Seebach
Eidgenössische Technische Hochschule, Zürich, Switzerland

Methanesulfonic Acid

$MeSO_3H$

[75-75-2] CH_4O_3S (MW 96.12)

(cyclocondensation reagent; precursor for methanesulfonyl chloride and anhydride; catalyst for polymerization, alkylation, and esterification reactions)

Alternate Name: MsOH.
Physical Data: mp 20 °C; bp 167 °C/10 mmHg, 122 °C/1 mmHg; d 1.481 g cm^{-3}; n_{20}^{D} 1.3210.
Solubility: sol water, ethanol, ether; insol hexane; very sparingly sol benzene, toluene.
Form Supplied in: technical quality is 95% pure containing 2% water.
Analysis of Reagent Purity: ^{17}O NMR.[1]
Preparative Method: by oxidation of ***Dimethyl Sulfide*** with ***Dimethyl Sulfoxide*** in the presence of water and a catalytic amount of ***Hydrogen Bromide***.[17]
Purification: stir with P_2O_5 (20 g for 500 mL of the acid) at 100 °C for 0.5 h and then distill under vacuum.
Handling, Storage, and Precautions: irritant and highly corrosive liquid; should be stored in glass containers. Use in a fume hood.

Cyclization Reactions. MsOH is a weaker acid than ***Trifluoromethanesulfonic Acid*** and hence only few reports exist on the use of pure MsOH in cyclocondensation reactions.[2] MsOH in dichloromethane effects a symmetry allowed cyclization of the precursor diene to afford 1,2,3,4-tetramethyl-5-(trifluoromethyl)cyclopentadiene, which is used as a ligand in organometallic chemistry (eq 1).[3]

HO CF3 — $MeSO_3H$, CH_2Cl_2, rt, 82% → CF3 (1)

A 1:10 solution by weight of ***Phosphorus(V) Oxide*** in MsOH[4] is a convenient alternative to ***Polyphosphoric Acid*** for cyclization reactions. For example, the classical preparation of cyclopentenones via intramolecular acylation of alkenoic acids or their lactones (eq 2) and Beckmann rearrangement using polyphosphoric acid (eq 3) give comparable yields when performed with $MsOH/P_2O_5$.[4]

O O — $MeSO_3H$, P_2O_5, 92% → O (2)

NOH — $MeSO_3H$, P_2O_5, 96% → O NH (3)

MsOH in conjunction with P_2O_5 has been used in the rearrangement of 2-vinylcyclobutanones to spiro and fused cyclopentenones (eqs 4 and 5).[5]

O — $MeSO_3H$, P_2O_5 → O 51% + O 13% (4)

O — $MeSO_3H$, P_2O_5, 48% → O (5)

3-Unsubstituted indoles are formed regioselectively by treatment of precursor hydrazones with $MsOH/P_2O_5$.[6] In the example given in eq 6,[6] only 2–3% of the undesired 3-isomer is formed. Because decomposition sometimes occurs, it is advisable to dilute the reagent with a suitable polar, nonbasic solvent like sulfolane or dichloromethane.

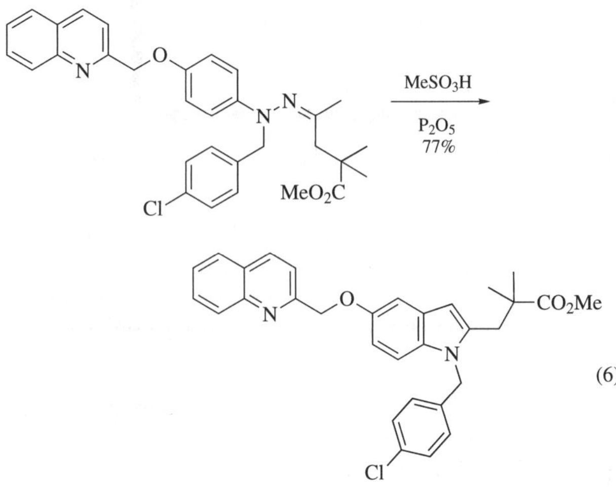

$MeSO_3H$, P_2O_5, 77% (6)

MsOH itself is a better cyclizing agent than an admixture with P_2O_5 for the cyclization of 3-arylpropanoic and 4-arylbutanoic acids to 1-indanones (eq 7) and 1-tetralones (eq 8).[7]

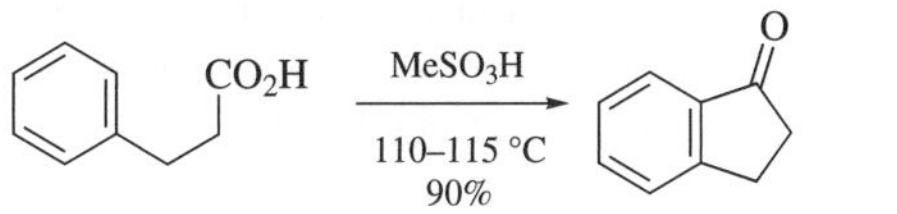

$MeSO_3H$, 110–115 °C, 90% (7)

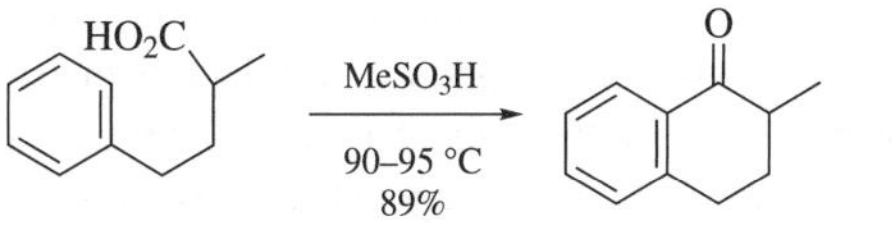

$MeSO_3H$, 90–95 °C, 89% (8)

Cyclization with neat MsOH is also observed in the formation of cyclopentenones from 2-vinylcyclobutanones (eq 9).[5]

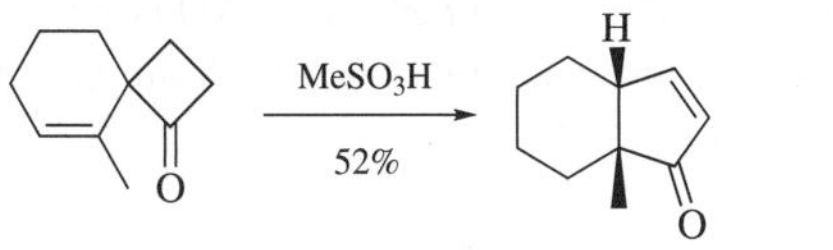

$MeSO_3H$, 52% (9)

MsOH has also been employed successfully in cyclocondensation reactions in the field of heterocycles (eq 10).[8,18] Thus the hexahydroimidazo[1,2-*a*]pyrimidine-5,7-dione shown in eq 10 gives the corresponding cyclized product in 81% yield.[8]

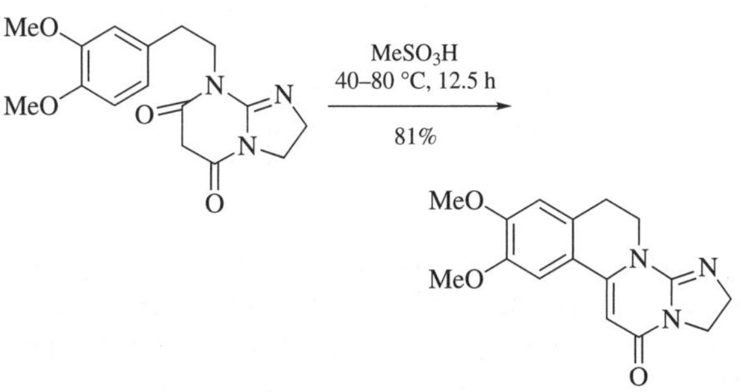

$MeSO_3H$, 40–80 °C, 12.5 h, 81% (10)

Other Applications. MsOH is superior to ***Sulfuric Acid*** as the solvent and catalyst for the conversion of benzoic acid to ***Perbenzoic Acid*** (eq 11).[9]

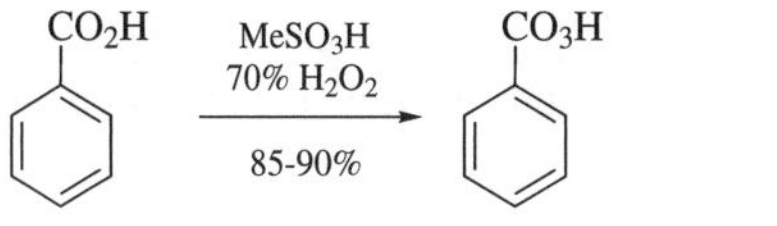

$MeSO_3H$, 70% H_2O_2, 85-90% (11)

MsOH in the presence of methionine is the reagent of choice and an excellent substitute for ***Boron Tribromide*** for *O*-demethylation of opioid derivatives (eq 12).[10] Among the other reagents tested, only TfOH is as effective as MsOH/methionine.

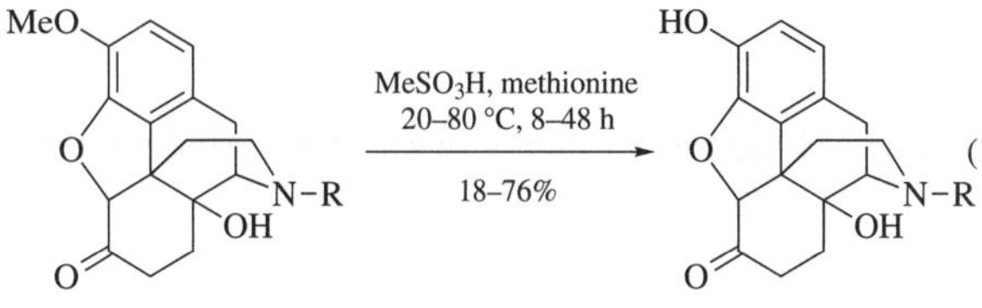

$MeSO_3H$, methionine, 20–80 °C, 8–48 h, 18–76% (12)

R = Me, CO_2Et

Schmidt rearrangement of optically active cyclic β-keto esters with retention of configuration is effectively carried out with MsOH in the presence of ***Sodium Azide*** (eq 13).[11]

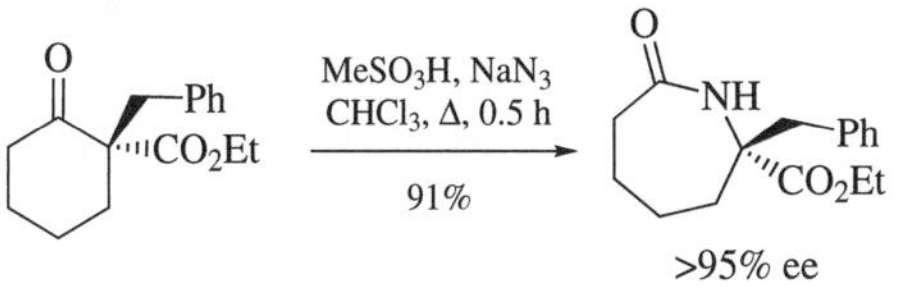

$MeSO_3H$, NaN_3, $CHCl_3$, Δ, 0.5 h, 91% (13)

>95% ee

Attempted acid-induced cyclization with MsOH of an intermediate diazo ketone involved in the synthesis of tricyclo[5.2.1.$0^{4,10}$]decane-2,5,8-trione affords the corresponding methylsulfonyloxy derivative via the protonated diazonium salt (eq 14).[12]

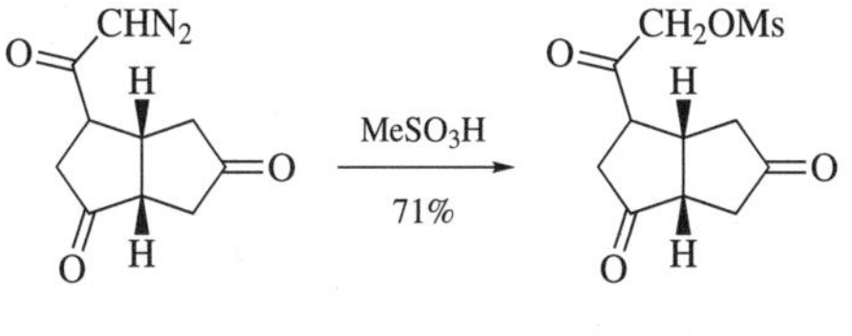

$MeSO_3H$, 71% (14)

MsOH is a useful reagent for the condensation of 2-(hydroxymethylene)cyclohexanone with sulfonamides in the presence of molecular sieves to afford products in the *cis-u* diastereoisomeric forms and with >90% stereoselectivity (eq 15).[13]

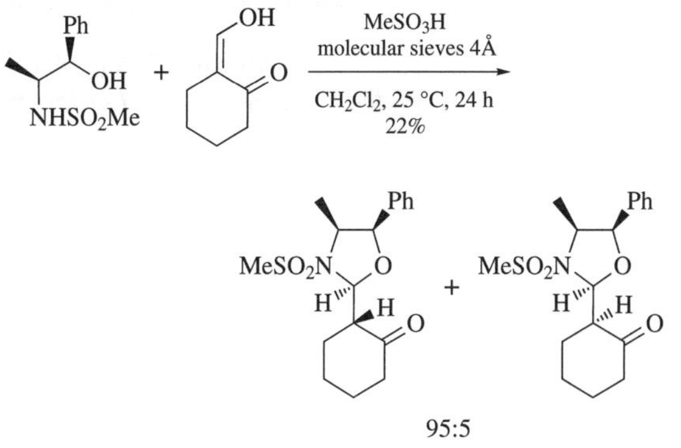

$MeSO_3H$, molecular sieves 4Å, CH_2Cl_2, 25 °C, 24 h, 22% (15)

95:5

The reaction of trimethylphosphine–borane with MsOH in dichloromethane gives the methanesulfonate derivative of the borane (eq 16).[14] This compound can be condensed with diphenylphosphine–borane in the presence of ***Sodium Hydride*** to give the corresponding dimer. By repeating the sequences of mesylation and condensation, a tetramer containing a linear P–B bond has been synthesized.[14]

$$Me-\overset{Me}{\underset{Me}{P^{+}}}-\bar{B}H_3 \xrightarrow[CH_2Cl_2,\ rt\ 89\%]{MeSO_3H} Me-\overset{Me}{\underset{Me}{P^{+}}}-\bar{B}H_2OMs \quad (16)$$

MsOH has also been used to deblock the benzyl protecting group[15] and to carry out the acidic hydrolysis of esters.[16]

Related Reagents. Phosphorus(V) Oxide–Methanesulfonic Acid; Polyphosphoric Acid; Trifluoroacetic Acid; Trifluoromethanesulfonic Acid.

---

1. Ilczyszyn, M. *JPC* **1991**, *95*, 7621.
2. Newman, M. S.; Davis, C. D. *JOC* **1967**, *32*, 66.
3. Gassman, P. G.; Mickelson, J. W.; Sowa, J. R., Jr. *JACS* **1992**, *114*, 6942.
4. Eaton, P. E.; Carlson, G. R.; Lee, J. T. *JOC* **1973**, *38*, 4071.
5. Matz, J. R.; Cohen, T. *TL* **1981**, *22*, 2459.
6. Zhao, D.; Hughes, D. L.; Bender, D. R.; De Marco, A. M.; Reider, P. J. *JOC* **1991**, *56*, 3001.
7. Premasagar, V.; Palaniswamy, V. A.; Eisenbraun, E. *JOC* **1981**, *46*, 2974.
8. Esser, F.; Pook, K.-H.; Carpy, A. *S* **1990**, 72.
9. (a) Silbert, L. S.; Siegel, E.; Swern, D. *JOC* **1962**, *27*, 1336. (b) Silbert, L. S.; Siegel, E.; Swern, D. *OS* **1963**, *43*, 93; *OSC* **1973**, *5*, 904.
10. Andre, J.-D.; Dormoy, J.-R.; Heymes, A. *SC* **1992**, *22*, 2313.
11. Georg, G. I.; Guan, X.; Kant, J. *BML* **1991**, *1*, 125.
12. Almansa, C.; Carceller, E.; Moyano, A.; Serratosa, F. *T* **1986**, *42*, 3637.
13. Hoppe, I.; Hoffmann, H.; Gärtner, I.; Krettek, T.; Hoppe, D. *S* **1991**, 1157.
14. Imamoto, T.; Oshiki, T. *TL* **1989**, *30*, 383.
15. Loev, B.; Haas, M. A.; Dowalo, F. *CI(L)* **1968**, 973.
16. Loev, B. *CI(L)* **1964**, 193.
17. Lowe, O. G. *JOC* **1976**, *41*, 2061.
18. Esser, F.; Pook, K.-H.; Carpy, A.; Leger, J. M. *S* **1994**, 77.

Lakshminarayanapuram R. Subramanian & Michael Hanack
Universität Tübingen, Germany

Antonio García Martínez
Universidad Complutense de Madrid, Spain

Methylaluminum Bis(2,6-di-*t*-butyl-4-methylphenoxide)

[65260-44-8] $C_{31}H_{49}AlO_2$ (MW 480.78)

(designer Lewis acid for amphiphilic alkylation,[1,4,5] amphiphilic reduction,[6] conjugate addition,[7–10] discrimination of two different substrates (ethers and ketones),[12–14] stereocontrolled cycloadditions,[15–17] and polymerization[18])

Alternate Name: MAD.
Physical Data: a crystal structure of a MAD–benzophenone complex has been reported.[3]
Solubility: sol CH_2Cl_2, toluene, hexane.
Form Supplied in: prepared from commercially available reagents and used in situ.
Preparative Method: prepared by reaction of a 1–2 M hexane solution of ***Trimethylaluminum*** with 2 equiv of 2,6-di-*t*-butyl-4-methylphenol in toluene or CH_2Cl_2 at rt for 1 h.[1,2]
Handling, Storage, and Precautions: the dry solid and solutions are highly flammable and must be handled in the absence of oxygen and moisture. The solution should be used as prepared for best results. Use in a fume hood.

Amphiphilic Alkylation. The exceptionally bulky, oxygenophilic organoaluminum reagent methylaluminum bis(2,6-di-*t*-butyl-4-methylphenoxide) has been developed for the stereoselective activation of carbonyl groups. Combination of MAD with Grignard reagents or organolithium reagents generates the amphiphilic alkylation systems, where substituted cyclohexanones and cyclopentanones afford equatorial alcohols with high stereoselectivity (eqs 1–3).[1] The X-ray crystallographic determination of a benzophenone–MAD complex shows that the ketone coordinates such that the aluminum is in the nodal plane of the C=O π-bond.[3]

1. AlR_3; 2. MeLi, –78 °C → axial alcohol + equatorial alcohol (1)

Al reagent		axial:equatorial
none		79:21
Me_3Al		76:24
i-Bu_3Al		64:36
Me_2AlOPh		72:28
MAD	84%	1:99

This methodology has been used in the final step of the stereocontrolled synthesis of a defense substance of the termite *Nastitermes princeps*, a secotrinervitane diterpene. Whereas ***Methyllithium*** adds exclusively to the β-face of the ketone to furnish the α-alcohol as the sole isolable product, attack of methyllithium occurs preferably at the α-face of the carbonyl in the amphiphilic alkylation system using MAD/MeLi to give the desired substance as a major product (eq 4).[4]

1. MAD; 2. MeLi; CH_2Cl_2, ether, −78 °C; 69%; 9:91 (2)

(3)

MeLi	77%	75:25
MAD, MeLi	82%	1:99

toluene, ether, −78 °C (4)

MeLi	91%	0:100
MAD, MeLi	30%	75:25

Alkylation of α-substituted aldehydes with the MAD/RMgX system provides *anti*-alcohols preferentially (eq 5).[1] This amphiphilic system allows the chemoselective carbonyl alkylation and reduction of aldehydes in the presence of ketones (eq 6).[5]

1. MAD; 2. EtMgBr; 90%; 25:75 (5)

1. MAD; 2. MeLi; 72% (6)

Amphiphilic Reduction. In contrast to the facile MAD-mediated alkylation of cyclic ketones with primary organolithiums or Grignard reagents, both alkylation and reduction take place with *s*-alkylmagnesium halides. When a bulky nucleophile such as *t*-BuMgCl is employed for the amphiphilic reaction system, substituted cyclohexanones afford equatorial alcohols exclusively as reduction products with high selectivity (eq 7).[6]

1. MAD; 2. *t*-BuMgBr, 0 °C, 2.5 h; 88%; 1:99 (7)

Conjugate Addition. The MAD/RLi system can also be used for conjugate alkylation and reduction of α,β-unsaturated ketones with organolithiums and ***Lithium n-Butyl(diisobutyl)aluminum Hydride***, respectively (eqs 8 and 9).[7,8] This method is particularly effective for the conjugate alkylation and reduction of quinone monoacetals and their derivatives (eqs 10 and 11).[9,10]

1. MAD; 2. nucleophile (8)

MeLi	68%	29:71
BuLi	59%	17:83
PhLi	71%	33:67
$H_2C{=}C(O\text{-}t\text{-}Bu)OLi$	87%	10:90

1. MAD; 2. $LiBu(i\text{-}Bu)_2AlH$; 83% (9)

1. MAD; 2. PhLi; 80% (10)

1. MAD; 2. $H_2C{=}C(Me)Li$; 75% (11)

Discrimination of Two Different Oxygen-Containing Substrates. MAD can be used as a Lewis acidic receptor for the discrimination of two different oxygen-containing substrates based on selective Lewis acid–base complex formation.[11] Indeed, in an equimolar mixture of methyl and ethyl ether substrates, MAD coordinates to the methyl ether exclusively (eq 12). This chemistry has been applied to a new type of coordination chromatography using polymeric, bulky organoaluminum compounds as a stationary phase that allows a surprisingly clean separation of structurally similar methyl ethers from ethyl ethers.[12,13]

$Ph(CH_2)_3OMe + Ph(CH_2)_3OEt \xrightarrow[(MAD)]{AlR_3} Ph(CH_2)_3O(Me){\cdot}AlR_3 + Ph(CH_2)_3OEt$ (12)

Selective reduction of sterically more hindered ketones has been effected with the MAD/***Diisobutylaluminum Hydride*** system by selective complexation of the less hindered ketones with MAD and subsequent attack of *i*-Bu_2AlH at the free, more hindered ketones (eq 13).[14]

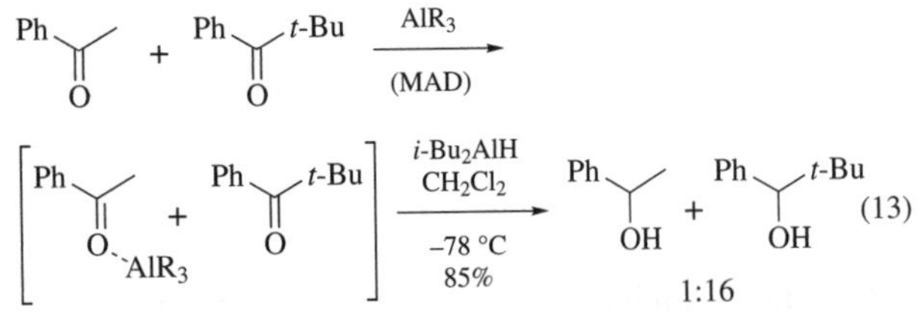

The selective binding behavior of Lewis acidic MAD for two different ester groups allows the regio- and stereocontrolled Diels–Alder reaction of unsymmetrical fumarates (eq 14).[15]

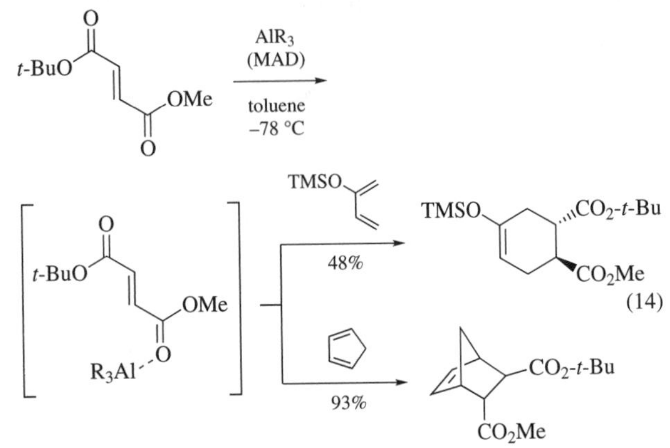

Cycloadditions. The steric effects of MAD on stereoselectivity in the Diels–Alder reactions of cyclic dienes and α,β-unsaturated aldehydes has been studied (eq 15).[16] MAD promotes very mild and highly stereocontrolled [2+4] cycloaddition between a pyrone sulfone and an enantiomerically pure vinyl ether. After 12 h at −45 °C in 4:1 toluene/CH_2Cl_2 with 0.5 equiv of MAD and 2.5 equiv of the vinyl ether, cycloadducts are isolated in 93% yield as a 98:2 ratio of the *endo* diastereomers (eq 16).[17]

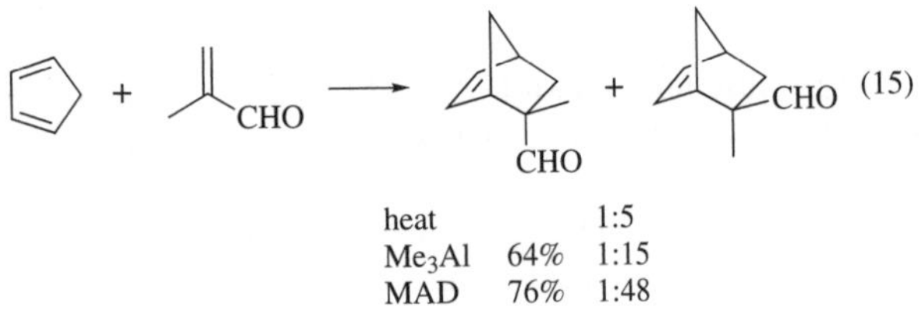

heat		1:5
Me_3Al	64%	1:15
MAD	76%	1:48

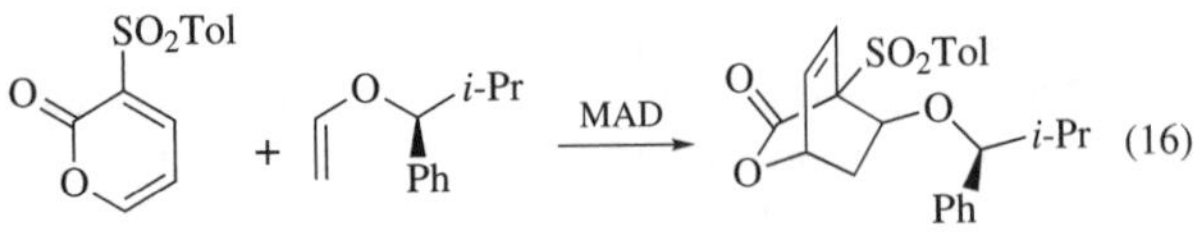

Polymerization. A novel, Lewis acid-induced Michael addition of an enolate complex to an α,β-unsaturated ester has been developed, where the unfavorable reaction between the nucleophile and Lewis acid is sterically suppressed.[18] For example, the addition polymerization of methyl methacrylate (MMA) via an enolate complex of aluminum tetraphenylporphyrin (**1**) as the nucleophilic growing species has been effected in the presence of MAD, producing polymer molecules cleanly with a narrow molecular weight distribution (eq 17).

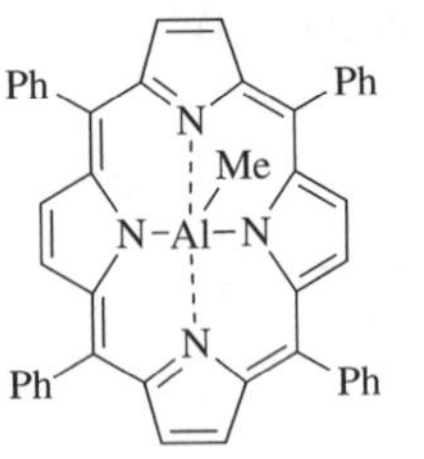

(**1**) (TPP)AlMe

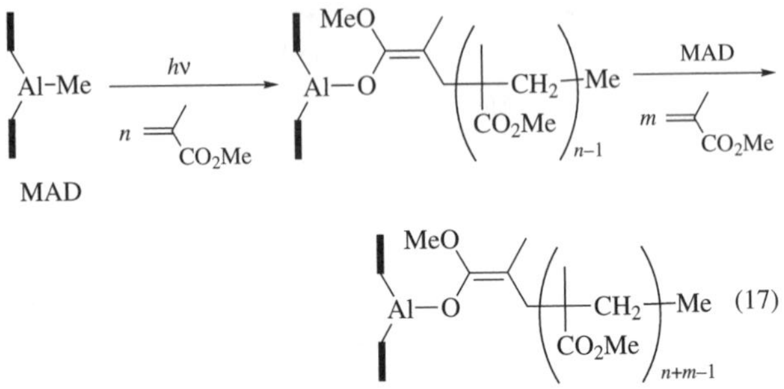

Related Reagents. Diethylaluminum Chloride; Dimethylaluminum Chloride; Ethylaluminum Dichloride; Methylaluminum Bis(2,6-di-*t*-butyl-4-methylphenoxide); Methylaluminum Bis-(2,6-di-*t*-butylphenoxide); Methylaluminum Dichloride; Tetrahydro-1-methyl-3,3-diphenyl-1*H*,3*H*-pyrrolo[1,2-c][1,3,2]oxazaborole.

1. (a) Maruoka, K.; Itoh, T.; Yamamoto, H. *JACS* **1985**, *107*, 4573. (b) Maruoka, K.; Itoh, T.; Sakurai, M.; Nonoshita, K.; Yamamoto, H. *JACS* **1988**, *110*, 3588.
2. (a) Starowieyski, K. B.; Pasynkiewicz, S.; Skowronska-Ptasinska, M. *JOM* **1975**, *90*, C43. (b) Skowronska-Ptasinska, M.; Starowieyski, K. B.; Pasynkiewicz, S. *JOM* **1977**, *141*, 149. (c) Skowronska-Ptasinska, M.; Starowieyski, K. B.; Pasynkiewicz, S.; Carewska, M. *JOM* **1978**, *160*, 403.
3. (a) Power, M. B.; Bott, S. G.; Atwood, J. L.; Barron, A. R. *JACS* **1990**, *112*, 3446. (b) Healy, M. D.; Ziller, J. W.; Barron, A. R. *JACS* **1990**, *112*, 2949.
4. (a) Kato, T.; Hirukawa, T.; Uyehara, T.; Yamamoto, Y. *TL* **1987**, *28*, 1439. (b) Hirukawa, T.; Shudo, T.; Kato, T. *JCS(P1)* **1993**, 217.
5. Maruoka, K.; Araki, Y.; Yamamoto, H. *TL* **1988**, *29*, 3101.
6. Maruoka, K.; Sakurai, M.; Yamamoto, H. *TL* **1985**, *26*, 3853.
7. Maruoka, K.; Nonoshita, K.; Yamamoto, H. *TL* **1987**, *28*, 5723.
8. Nonoshita, K.; Maruoka, K.; Yamamoto, H. *BCJ* **1988**, *61*, 2241.
9. Stern, A. J.; Rohde, J. J.; Swenton, J. S. *JOC* **1989**, *54*, 4413.
10. Doty, B. J.; Morrow, G. W. *TL* **1990**, *31*, 6125.
11. Maruoka, K.; Nagahara, S.; Yamamoto, H. *TL* **1990**, *31*, 5475.
12. Maruoka, K.; Nagahara, S.; Yamamoto, H. *JACS* **1990**, *112*, 6115.
13. (a) Maruoka, K.; Nagahara, S.; Yamamoto, H. *BCJ* **1990**, *63*, 3354. (b) Healy, M. D.; Power, M. B.; Barron, A. R. *J. Coord. Chem.* **1990**, *21*, 363.
14. Maruoka, K.; Araki, Y.; Yamamoto, H. *JACS* **1988**, *110*, 2650.
15. Maruoka, K.; Saito, S.; Yamamoto, H. *JACS* **1992**, *114*, 1089.
16. Maruoka, K.; Nonoshita, K.; Yamamoto, H. *SC* **1988**, *18*, 1453.
17. Posner, G. H.; Kinter, C. M. *JOC* **1990**, *55*, 3967.

18. (a) Kuroki, M.; Watanabe, T.; Aida, T.; Inoue, S. *JACS* **1991**, *113*, 5903. (b) Adachi, T.; Sugimoto, H.; Aida, T.; Inoue, S. *Macromolecules* **1992**, *25*, 2280. (c) Adachi, T.; Sugimoto, H.; Aida, T.; Inoue, S. *Macromolecules* **1993**, *26*, 1238.

Keiji Maruoka & Hisashi Yamamoto
Nagoya University, Japan

Methylaluminum Bis(2,6-di-*t*-butylphenoxide)

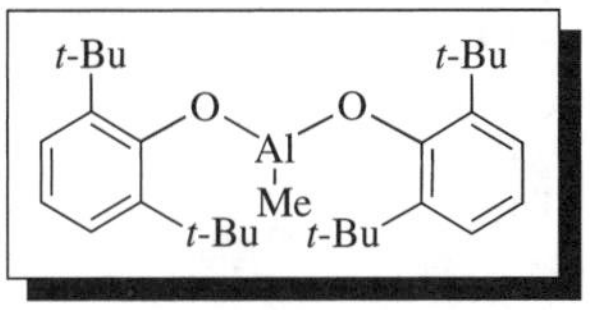

[128353-30-0] $C_{29}H_{45}AlO_2$ (MW 452.72)

(designer Lewis acid for amphiphilic alkylation and conjugate addition)

Solubility: sol CH_2Cl_2, toluene, hexane.

Form Supplied in: prepared from commercially available reagents and used in situ.

Preparative Method: prepared by reaction of a 1–2 M hexane solution of ***Trimethylaluminum*** with 2 equiv of 2,6-di-*t*-butylphenol in toluene or CH_2Cl_2 at rt for 1 h in a similar manner to that described in the preparation of MAD or MAT.[1]

Handling, Storage, and Precautions: the dry solid and solutions are highly flammable and must be handled in the absence of oxygen and moisture. The solution should be used as prepared for best results. Use in a fume hood.

Amphiphilic Alkylation. The exceptionally bulky, oxygenophilic organoaluminum reagent methylaluminum bis(2,6-di-*t*-butylphenoxide), like ***Methylaluminum Bis(2,6-di-t-butyl-4-methylphenoxide)*** (MAD) and ***Methylaluminum Bis(2,4,6-tri-t-butylphenoxide)*** (MAT), has been used for stereoselective activation of carbonyl groups. Combination of this modified organoaluminum compound, as a bulky Lewis acid, with Grignard reagents or organolithium reagents generates the amphiphilic alkylation system, in which substituted cyclohexanones afford equatorial alcohols with high selectivity (eq 1).[2] This approach is quite useful in the stereoselective alkylation of steroidal ketones. Thus reaction of cholestan-3-one with ***Methyllithium*** gives predominantly 3β-methylcholestan-3α-ol (axial alcohol), whereas the amphiphilic methylation of the ketone gives 3α-methylcholestan-3β-ol (equatorial alcohol) exclusively (eq 2).[1,2]

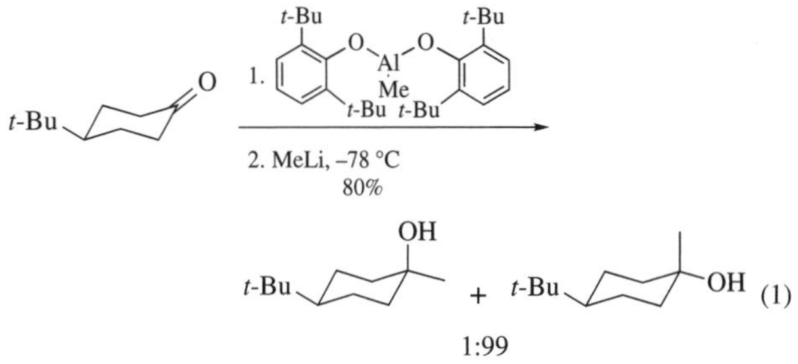

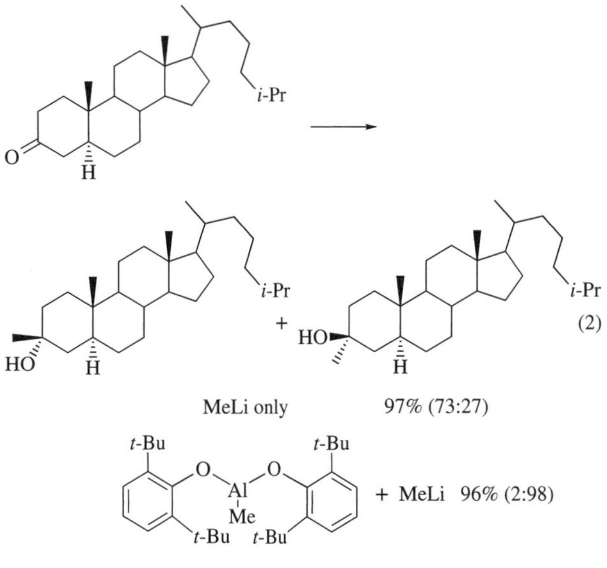

The amphiphilic alkylation of α-substituted aldehydes using the title aluminum reagent and Grignard reagents affords *anti*-alcohols preferentially (eq 3), as also observed with MAD/RMgX or MAT/RMgX.[2]

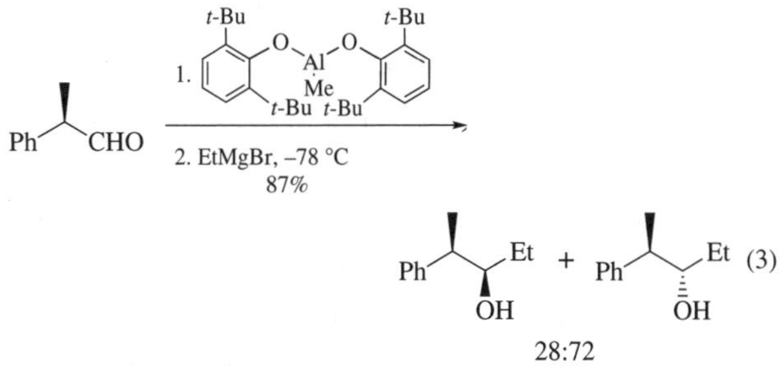

1. reagent; 2. MeLi, –78 °C; 80%; 1:99 (1)

MeLi only 97% (73:27); reagent + MeLi 96% (2:98) (2)

1. reagent; 2. EtMgBr, –78 °C; 87%; 28:72 (3)

Conjugate Addition. The title reagent has also been used for conjugate alkylation and reduction of α,β-unsaturated ketones with organolithiums and ***Lithium n-Butyl(diisobutyl)aluminum Hydride***, respectively (eqs 4 and 5).[3] The 1,4 vs. 1,2 selectivity is not much different from that observed with MAD or MAT.[4]

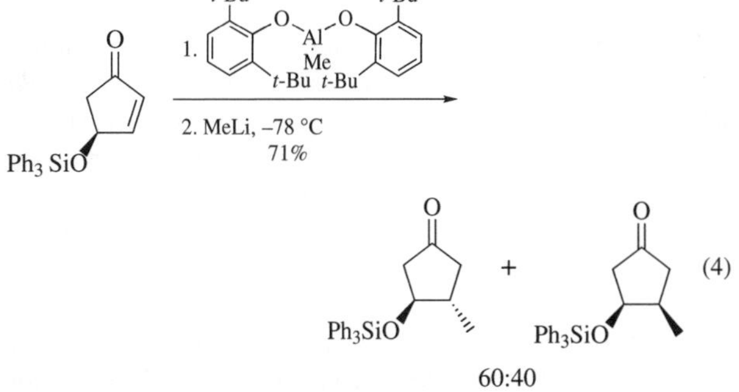

1. reagent; 2. MeLi, –78 °C; 71%; 60:40 (4)

1. MAD (t-Bu, t-Bu, Me, t-Bu t-Bu)
2. LiBu(i-Bu)$_2$AlH, –78 °C

Ph, Ph (5)

58% 25%

Related Reagents. Diethylaluminum Chloride; Ethylaluminum Dichloride; Methylaluminum Bis(2,6-di-t-butyl-4-methylphenoxide); Methylaluminum Bis(2,6-di-t-butylphenoxide).

1. (a) Maruoka, K.; Itoh, T.; Yamamoto, H. *JACS* **1985**, *107*, 4573. (b) Maruoka, K.; Itoh, T.; Sakurai, M.; Nonoshita, K.; Yamamoto, H. *JACS* **1988**, *110*, 3588.
2. Itoh, T.; Maruoka, K.; Yamamoto, H., unpublished results.
3. Nonoshita, K.; Maruoka, K.; Yamamoto, H., unpublished results.
4. (a) Maruoka, K.; Nonoshita, K.; Yamamoto, H. *TL* **1987**, *28*, 5723. (b) Nonoshita, K.; Maruoka, K.; Yamamoto, H. *BCJ* **1988**, *61*, 2241.

Keiji Maruoka & Hisashi Yamamoto
Nagoya University, Japan

Methylaluminum Dichloride[1]

MeAlCl$_2$

[917-65-7] CH_3AlCl_2 (MW 112.92)

(strong Lewis acid that can also act as a proton scavenger; reacts with HX to give methane and $AlCl_2X$)

Alternate Name: dichloro(methyl)aluminum.
Physical Data: mp 72.7 °C; bp 94–95 °C/100 mmHg.
Solubility: sol most organic solvents; stable in alkanes or arenes.
Form Supplied in: commercially available as solutions in hexane or toluene.
Analysis of Reagent Purity: solutions are reasonably stable but may be titrated before use by one of the standard methods.[1b]
Handling, Storage, and Precautions: must be transferred under inert gas (Ar or N_2) to exclude oxygen and water. Use in a fume hood.

Introduction. The general properties of alkylaluminum halides as Lewis acids are discussed in the entry for ***Ethylaluminum Dichloride***. MeAlCl$_2$ is used less frequently than EtAlCl$_2$, since EtAlCl$_2$ is much cheaper and comparable results are usually obtained. In some cases, use of MeAlCl$_2$ is preferable, since the methyl group of MeAlCl$_2$ is less nucleophilic than the ethyl group of EtAlCl$_2$ and EtAlCl$_2$ can transfer a hydride.

Catalysis of Diels–Alder Reactions. MeAlCl$_2$ has been used as a catalyst for Diels–Alder[2–4] and retro-Diels–Alder reactions.[5] MeAlCl$_2$ is the catalyst of choice for intramolecular Diels–Alder reactions of furans (eq 1) due to the ease of handling and reaction workup.[2]

MeAlCl$_2$, CH_2Cl_2, –78 °C, 2 h, 63% (1)

Catalysis of Ene Reactions. MeAlCl$_2$ has been used as a catalyst for ene reactions of trifluoroacetaldehyde[6] and for intramolecular ene reactions of trifluoromethyl ketones (eq 2).[7] Use of EtAlCl$_2$ leads to the ene adduct in lower yield accompanied by 25% of the saturated analog resulting from hydride delivery to the zwitterion.

MeAlCl$_2$, –78 °C, 1 h, 90% (2)

Generation of Electrophilic Cations. 1:2 Aldehyde– or ketone–MeAlCl$_2$ complexes add intramolecularly to alkenes to give zwitterions that undergo 1,2-hydride and alkyl shifts to regenerate a ketone (eqs 3 and 4).[8–11] EtAlCl$_2$ can transfer a hydride to the zwitterion to give the reduced product (eq 3).[8]

2 equiv MeAlCl$_2$, –78 °C, 72%; 2 equiv EtAlCl$_2$, –78 °C, 73% (3)

MeAlCl$_2$, CH_2Cl_2, 90 °C, 2 h, 53% (4)

MeAlCl$_2$ has been used as the catalyst for epoxide-initiated cation–alkene cyclizations.[12] A tertiary alcohol has been converted to a hydrazide by reaction with MeAlCl$_2$ in the presence of ***Mesitylenesulfonylhydrazide***.[13] α-Trimethylsilyl enones can be prepared by isomerization of 1-(trimethylsilyl)-2-propynyl trimethylsilyl ethers with MeAlCl$_2$.[14]

Nucleophilic Additions. MeAlCl$_2$ has occasionally been used to introduce a methyl group. γ-Lactones react with MeAlCl$_2$ to give carboxylic acids with a methyl group at the γ-position.[15] The reagent prepared from MeAlCl$_2$ and ***Dichlorobis(cyclopentadienyl)titanium*** adds to trimethylsilylalkynes in a *syn* fashion.[16] Optimal stereoselectivity is obtained using MeAlCl$_2$ to transfer a methyl group to a formyl amide (eq 5).[17]

MeAlCl$_2$, CH_2Cl_2, 0 °C, 2 h, 66%; ds = 97:3 (5)

Related Reagents. Diethylaluminum Chloride; Dimethylaluminum Chloride; Ethylaluminum Dichloride.

---

1. (a) For reviews, see ref. 1 in ***Ethylaluminum Dichloride***. (b) *Aluminum Alkyls*; Stauffer Chemical Co.: Westport, CT, 1976.
2. (a) Rogers, C.; Keay, B. A. *TL* **1991**, *32*, 6477. (b) Rogers, C.; Keay, B. A. *CJC* **1992**, *70*, 2929.
3. Boeckman, R. K., Jr.; Nelson, S. G.; Gaul, M. D. *JACS* **1992**, *114*, 2258.
4. Trost, B. M.; Lautens, M.; Hung, M. H.; Carmichael, C. S. *JACS* **1984**, *106*, 7641.
5. Grieco, P. A.; Abood, N. *JOC* **1989**, *54*, 6008.
6. Ogawa, K.; Nagai, T.; Nonomura, M.; Takagi, T.; Koyama, M.; Ando, A.; Miki, T.; Kumadaki, I. *CPB* **1991**, *39*, 1707.
7. Abouadellah, A.; Aubert, C.; Bégué, J.-P.; Bonnet-Delpon, D.; Guilhem, J. *JCS(P1)* **1991**, 1397.
8. Snider, B. B.; Karras, M.; Price, R. T.; Rodini, D. J. *JOC* **1982**, *47*, 4538.
9. Snider, B. B.; Rodini, D. J.; van Straten, J. *JACS* **1980**, *102*, 5872.
10. Snider, B. B.; Kirk, T. C. *JACS* **1983**, *105*, 2364.
11. Snider, B. B.; Cartaya-Marin, C. P. *JOC* **1984**, *49*, 153.
12. Corey, E. J.; Sodeoka, M. *TL* **1991**, *32*, 7005.
13. Wood, J. L.; Porco, J. A., Jr.; Taunton, J.; Lee, A. Y.; Clardy, J.; Schreiber, S. L. *JACS* **1992**, *114*, 5898.
14. Enda, J.; Kuwajima, I. *CC* **1984**, 1589.
15. Reinheckel, H.; Sonnek, G.; Falk, F. *JPR* **1974**, *316*, 215.
16. Eisch, J. J.; Piotrowski, A. M.; Brownstein, S. K.; Gabe, E. J.; Lee, F. L. *JACS* **1985**, *107*, 7219.
17. Fujii, H.; Taniguchi, M.; Oshima, K.; Utimoto, K. *TL* **1992**, *33*, 4579.

Barry B. Snider
Brandeis University, Waltham, MA, USA

Molecular Sieves[1]

(electrocyclic reaction cocatalyst; mild acid catalyst; desiccant)

Form Supplied in: the most common forms are bead, pellet, and powdered solids with cavity sizes of 3, 4, 5 and 10 Å. The bead and pellet forms are adequate for drying solvents, while the powdered form is preferred for use in most reactions.

Preparative Methods: sieves are most effective if activated prior to use by drying under vacuum (<1 mmHg) at 300 °C for at least 15 h.

Handling, Storage, and Precautions: all forms of sieves readily absorb water upon exposure to air and are therefore best stored in a desiccator. Sieves can be recycled by (a) washing well with an organic solvent, (b) drying at 100 °C for several hours, and (c) reactivation at ≥200 °C. Skin contact should be avoided as the desiccant properties of the sieves cause irritation.

General Information. Molecular sieves are metal aluminosilicates of the general formula $M_{2/n}O{\cdot}Al_2O_3{\cdot}xSiO_2{\cdot}yH_2O$ (where n is the valence of the metal, M) characterized by a regular (zeolite) structure and cavity size which is retained even with loss of hydration.[2,3] Although they occur in Nature, most sieves are manufactured commercially as they can be designed with properties specific to their application. Many variations have been synthesized and the field of zeolite design is one of intense investigation.[4] The Linde Division of the Union Carbide Corporation is a major supplier of molecular sieves for synthetic organic applications. Of their products, the 3, 4, and 5 Å as well as 13X sieves are the most commonly employed; these differ both in pore size (3, 4, 5, and 10 Å, respectively) and cation constitution (K, Na, Ca, and Na, respectively).[5] Thus a sieve appropriate to a specific application can be selected.

Diels–Alder Catalysis. Whether combined with a Lewis acid or used alone, molecular sieves are powerful accelerators of this concerted process. In the former case, great progress has been made toward combining this reagent with a chiral Lewis acid to induce asymmetry during reaction between two achiral molecules.[6–12] This has been achieved through the use of the chiral titanate (**1**) and achiral auxiliary (**2**) which aids in coordination of the dienophile (eqs 1 and 2). Not only is this a highly successful method for the enantioselective generation of the stereocenters, but it is also catalytic. The power of this methodology is elegantly demonstrated in eq 2, taken from Narasaka's synthesis of the hydronaphthalene portions of mevinic acids, where this catalyst system is employed to introduce four contiguous chiral centers with complete diastereoselectivity and very high enantioselectivity from an achiral precursor. Although a less dramatic example, eq 3 displays the ability of sieves to function independently as cycloaddition catalysts,[13] as attested to by the mild temperature and short reaction time.

(**1**) (**2**) = X

5% (**1**), MS 4A; 0 °C, toluene–PE; 94% (1)
94% ee

30% (**1**), MS 4A; rt, toluene–PE, 150 h; 70% (2)
single diastereoisomer
>95% ee

Ene Reactions. Utilization of the reagent in ene reactions parallels its use in Diels–Alder cycloadditions, although the choice of the Lewis acid cocatalyst is substrate dependent. The three most effective chiral metal complexes employed to date are either the tartrate-based titanate (**1**), titanium-complexed commercially available 1,1′-binaphthol, or the more bulky binaphthol-derived system (**3**) (eqs 4–6).[14–16] Performance of these catalytic systems is comparable to those employed in the Diels–Alder reaction in both enantioselectivity and mildness of reaction conditions.

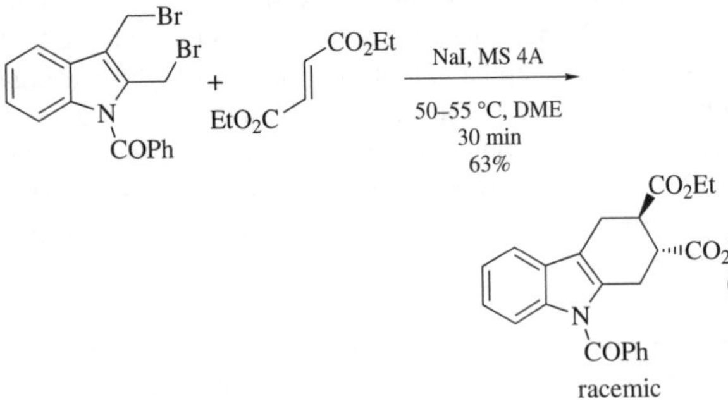

Other Electrocyclic Reactions. Finally, molecular sieves have been successful in facilitating asymmetric [2 + 2] cycloadditions when (**1**) is present, and in promoting [3 + 2] dipolar cycloadditions. Again, in the former case, the optical yields are extremely good as can be seen from eqs 7–9.[17–20] In the latter case, the sieves generate the nitrile oxide in situ and this species then reacts with the acrylate acceptor to yield the isoxazole (eq 10). Although this method requires a relatively long reaction time, the mild conditions have the advantage of suppressing side reactions, including dimerization of the nitrile oxide, resulting in high yields of very pure material.[21]

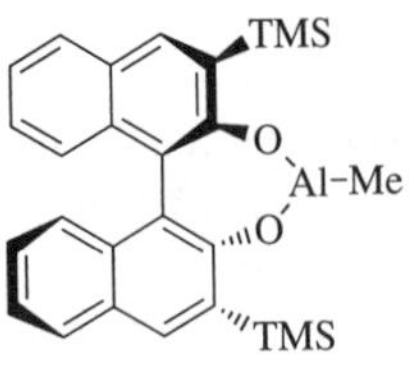

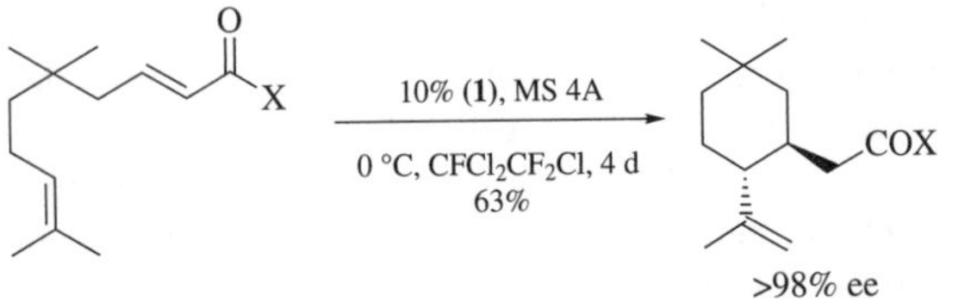

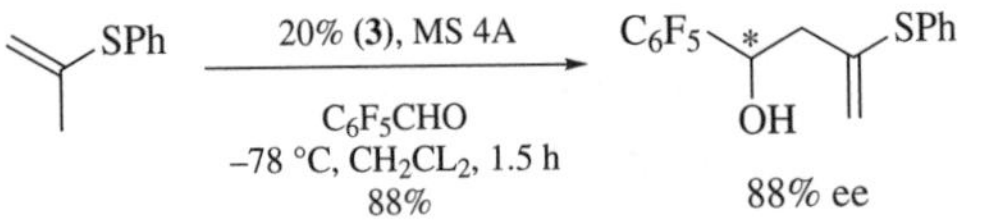

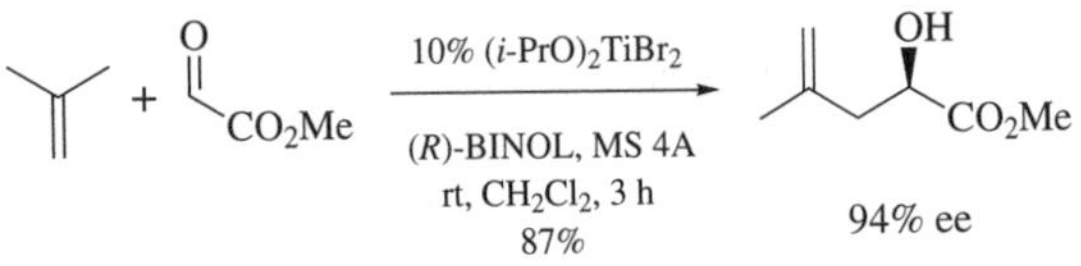

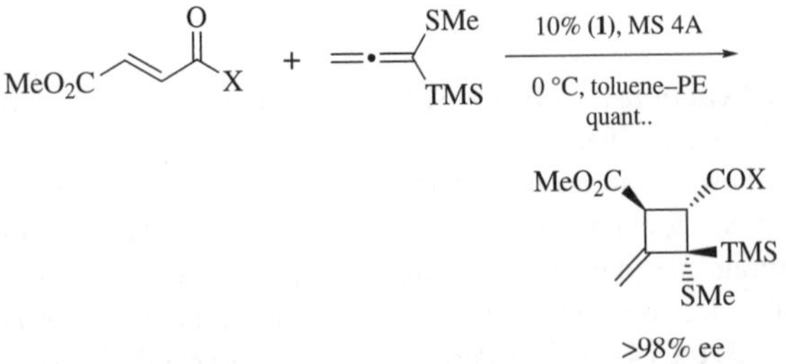

Acid Scavenging. In addition to their use as cocatalysts, molecular sieves also function as acid scavengers, making them especially suited to the suppression of acid-catalyzed side reactions such as polymerization. For example, they are employed in the preparation of high-purity methacrylic acid esters from methacryloyl chloride and various alcohols (primary, secondary, tertiary, and benzylic).[22] This acid-scavenging ability has also proven useful in the direct acylation of acid-sensitive, unreactive tertiary hydroxyl groups and of acid- and base-sensitive amides with acyl chlorides.[23,24]

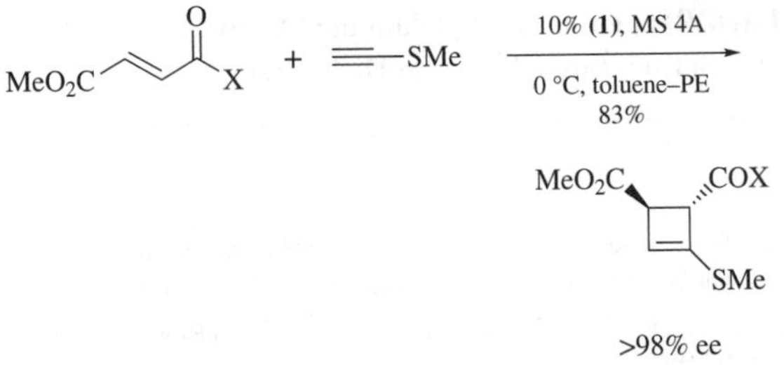

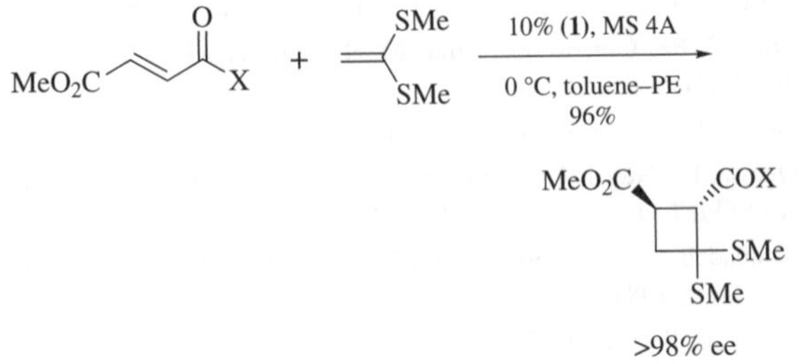

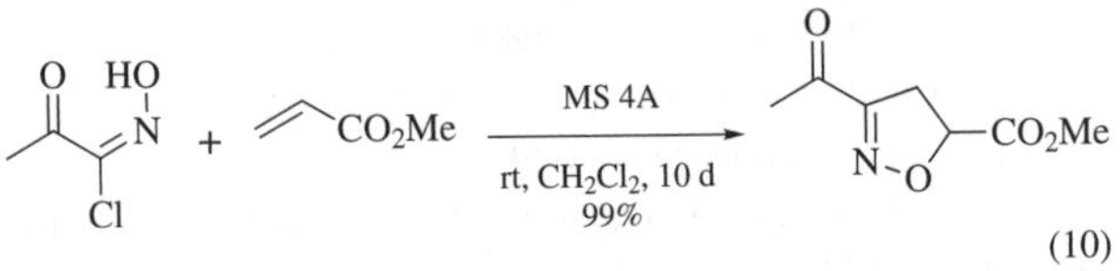

Sorbtion. The reagent's sorbtion ability has been exploited in a wide range of carbonyl and carboxylate transformations. Key to their utility is the inclusion of molecules such as water and small alcohols into the sieve cavities while excluding the larger compounds, thus allowing greater control of equilibria. By this method, ketimines and enamines derived from sterically encumbered precursors are more accessible.[25,26] Likewise, triphenylphosphazenes can be obtained by reaction of *N*-aminotriphenylphosphinime with various aldehydes and ketones.[27] Reductive amination of carbonyls proceeds in better yield with molecular sieves present to absorb water.[28] When coupled with tertiary amines such as ***1,5-Diazabicyclo[4.3.0]non-5-ene*** or ***1,8-Diazabicyclo[5.4.0]undec-7-ene***, 3 or 4 Å sieves can effect the alkenation of δ-alkoxy-α,β-unsaturated aldehydes efficiently.[29,30] Amide synthesis with molecular sieves provides a general, high yielding, and chemoselective route to secondary amides free of byproducts and impurities.[31] Transesterification of methyl esters with branched primary, secondary, and tertiary alcohols has been reported with the 5 Å sieve.[3] Zeolites have been utilized in the preparation of an asymmetric hydrocyanating agent by reaction with titanate (**1**) followed by treatment with 2 equiv of ***Cyanotrimethylsilane*** at ambient temperature.[32,33] Addition of this reagent to aldehydes at −78 °C in toluene provides the corresponding cyanohydrins in yields of 67–92% and optical purities ranging from 61% to 93%.

Acid Catalysis. The Lewis acid reactivity of this reagent can be applied to Michael-type reactions as shown in eqs 11 and 12.[34,35] The resulting ring systems can be further transformed into a variety of useful synthetic building blocks. Molecular sieves can also be coupled with Lewis acids to promote acetal formation and exchange. When used with ***p-Toluenesulfonic Acid*** as a cocatalyst, sieves provide a facile synthesis of acetals from carbonyls, not only when primary but also when secondary alcohols are involved.[36,37] Employment with ***Boron Trifluoride Etherate*** catalyzes exchange between (tributylstannyl)methanol and ***Dimethoxymethane*** to produce the useful hydroxymethyl anion equivalent ***Tri-n-butyl[(methoxymethoxy)methyl]stannane*** in high yield.[38,39]

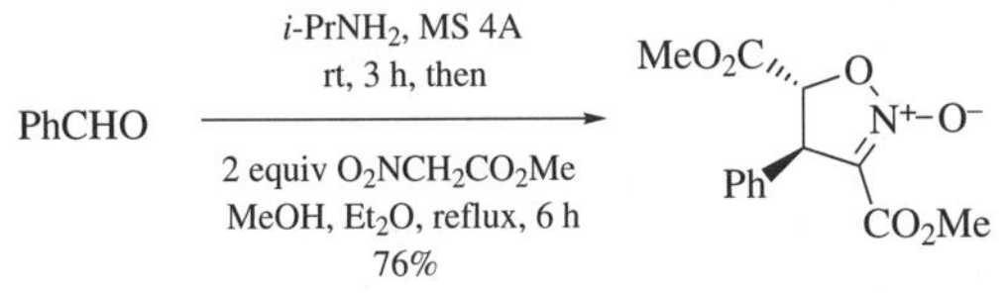

PhCHO → (i-$PrNH_2$, MS 4A, rt, 3 h, then 2 equiv $O_2NCH_2CO_2Me$, MeOH, Et_2O, reflux, 6 h; 76%) (11)

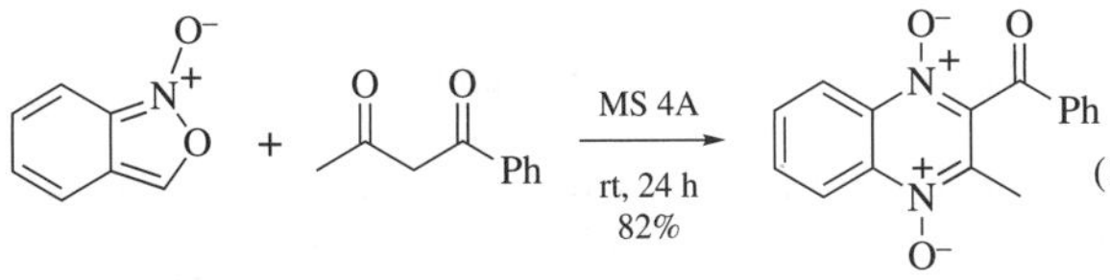

MS 4A, rt, 24 h, 82% (12)

Oxidations. Titanium silicate molecular sieves have served as catalysts in the selective oxidation of thioethers to sulfoxides.[40] They effect this transformation under mild conditions (1 equiv of H_2O_2 in refluxing acetone) with little over-oxidation to the sulfone. When used as a promoter in ***Pyridinium Chlorochromate*** and ***Pyridinium Dichromate*** oxidations of nucleoside derivatives, sieves work remarkably well, in contrast to other additives such as ***Alumina***, Celite, or silica gel which fail to accelerate these reactions.[41] As drying agents, sieves are a crucial component of the Sharpless catalytic asymmetric epoxidation.[42]

Miscellaneous Reactions. The shape selectivity of zeolites has been exploited to selectively brominate either a hindered double bond in the presence of an unhindered double bond or vice versa, depending on the reaction conditions.[43] If the hindered and unhindered alkene mixture is allowed to equilibrate prior to bromine addition, bromination of the hindered alkene is greatly favored (up to a 95:5 preference). Conversely, inclusion of the bromine in the sieves followed by addition of the alkene mixture shows opposite selectivity.

Molecular sieves have been employed as acid scavengers in the transition metal-catalyzed synthesis of carboxylic acids and esters from iodides under base-free conditions, as represented in eq 13.[44] Another transition metal-catalyzed reaction (eq 14) applies sieves as a cocatalyst with ***Iron(III) Chloride*** to form a nonreducing disaccharide in an alternative to the conventional Koenigs–Knorr method.[45] In concert with ***Tin(II) Trifluoromethanesulfonate*** and ***2,4,6-Collidine***, molecular sieves promote the coupling of acetobromoglucose with various protected sugar derivatives to form exclusively *trans*-β-D-glucosides with glucose as the reducing unit.[46] Finally, the use of zeolites as cocatalysts in palladium-catalyzed oxidative cyclizations has led to substantial improvements in the diastereoselectivity of these reactions.[47,48] This is exemplified in eq 15, which shows only a 17% de in the absence of molecular sieves.

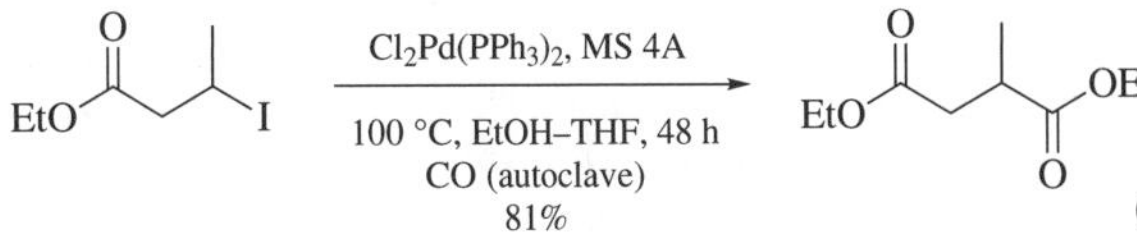

$Cl_2Pd(PPh_3)_2$, MS 4A, 100 °C, EtOH–THF, 48 h, CO (autoclave), 81% (13)

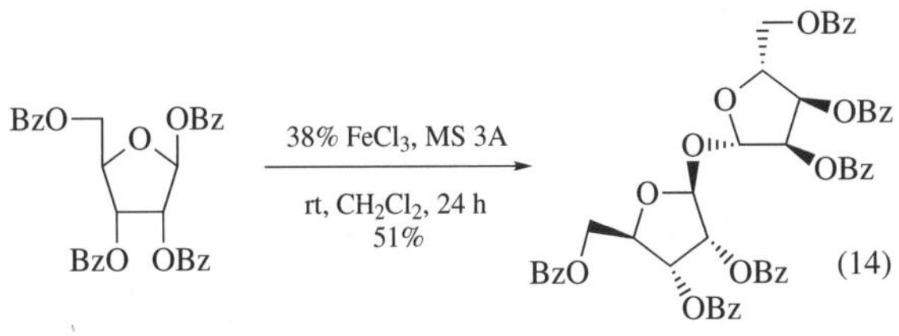

38% $FeCl_3$, MS 3A, rt, CH_2Cl_2, 24 h, 51% (14)

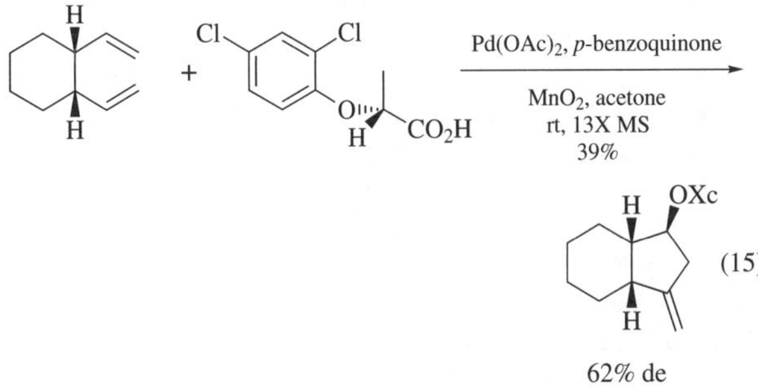

$Pd(OAc)_2$, *p*-benzoquinone, MnO_2, acetone, rt, 13X MS, 39%; 62% de (15)

Related Reagents. Alumina; 1,8-Diazabicyclo[5.4.0]undec-7-ene; *p*-Toluenesulfonic Acid.

1. (a) Hölderich, W.; Hesse, M.; Näumann, F. *AG(E)* **1988**, *27*, 226. (b) Davis, M. *ACR* **1993**, *26*, 111. (c) Dyer, A. *An Introduction to Zeolite Molecular Sieves*; Wiley: New York, 1988. (d) *Studies in Surface Science and Catalysis*; Bekkum, H. V.; Flanigen, E. M.; Jansen, J. C., Eds.; Elsevier: New York, 1991; Vol. 58. (e) *Molecular Sieve Catalysts*; Michiels, P.; De Herdt, O. C. E., Eds.; Pergamon: Oxford, 1987.
2. Hersh, C. K. *Molecular Sieves*; Reinhold: New York, 1961.
3. *FF* **1967**, *1*, 703.
4. Szostak, R. *Handbook of Molecular Sieves*; Van Nostrand Reinhold: New York, 1992.
5. Linde Division *Specialty Gasses*; Union Carbide, 1985; Section 9.
6. Narasaka, K.; Inoue, M.; Okada, N. *CL* **1986**, 1109.
7. Narasaka, K.; Inoue, M.; Yamada, T.; Sugimuri, J.; Iwasawa, N. *CL* **1987**, 2409.
8. Iwasawa, N.; Hayashi, Y.; Sakurai, H.; Narasaka, K. *CL* **1989**, 1581.
9. Iwasawa, N.; Sugimori, J.; Kawase, Y.; Narasaka, K. *CL* **1989**, 1947.
10. Narasaka, K.; Tanaka, H.; Kanai, F. *BCJ* **1991**, *64*, 387.
11. Narasaka, K.; Iwasawa, N.; Inoue, M.; Yamada, T.; Nakashima, M.; Sugimori, J. *JACS* **1989**, *111*, 5340.
12. Narasaka, K.; Saitou, M.; Iwasawa, N. *TA* **1991**, *2*, 1305.
13. Pindur, U.; Haber, M. *H* **1991**, *32*, 1463.
14. Narasaka, K.; Hayashi, Y.; Shimada, S. *CL* **1988**, 1609.
15. Maruoka, K.; Hoshino, Y.; Shirasaka, T.; Yamamoto, H. *TL* **1988**, *29*, 3967.
16. Mikami, K.; Terada, M.; Nakai, T. *JACS* **1990**, *112*, 3949.
17. Ichikawa, Y.; Narita, A.; Shiozawa, A.; Hayashi, Y.; Narasaka, K. *CC* **1989**, 1919.
18. Hayashi, Y.; Narasaka, K. *CL* **1989**, 793.

19. Hayashi, Y.; Narasaka, K. *CL* **1990**, 1295.
20. Hayashi, Y.; Niihata, S.; Narasaka, K. *CL* **1990**, 2091.
21. Kim, J. N.; Ryu, E. K. *H* **1990**, *31*, 1693.
22. Banks, A. R.; Fibiger, R. F.; Jones, T. *JOC* **1977**, *42*, 3965.
23. Nakamura, T.; Fukatsu, S.; Seki, S.; Niida, T. *CL* **1978**, 1293.
24. Weinstock, L. M.; Karady, S.; Roberts, F. E.; Hoinowski, A. M.; Brenner, G. S.; Lee, T. B. K.; Lumma, W. C.; Sletzinger, M. *TL* **1975**, 3979.
25. Taguchi, K.; Westheimer, F. H. *JOC* **1971**, *36*, 1570.
26. Bonnett, R.; Emerson, T. R. *JCS* **1965**, 4508.
27. Walker, C. C.; Shechter, H. *TL* **1965**, 1447.
28. Borch, R. F.; Bernstein, M. D.; Durst, H. D. *JACS* **1971**, *93*, 2897.
29. Ishida, A.; Mukaiyama, T. *CL* **1975**, 1167.
30. Mukaiyama, T.; Ishida, A. *CL* **1975**, 1201.
31. Cossy, J.; Pale-Grosdemange, C. *TL* **1989**, *30*, 2771.
32. Narasaka, K.; Yamada, T.; Minamikawa, H. *CL* **1987**, 2073.
33. Minamikawa, H.; Hayakawa, S.; Yamada, T.; Iwasawa, N.; Narasaka, K. *BCJ* **1988**, *61*, 4379.
34. Mélot, J. M.; Texier-Boullet, F.; Foucard, A. *S* **1988**, 558.
35. Takabatake, T.; Hasegawa, M. *JHC* **1987**, *24*, 529.
36. Roelofsen, D. P.; Wils, E. R. J.; van Bekkum, H. *RTC* **1971**, *90*, 1141.
37. Roelofsen, D. P.; van Bekkum, H. *S* **1972**, 419.
38. Danheiser, R. L.; Gee, S. K.; Perez, J. J. *JACS* **1986**, *108*, 806.
39. Danheiser, R. L.; Romines, K. R.; Koyama, H.; Gee, S. K.; Johnson, C. R.; Medich, J. R. *OS* **1993**, *71*, 133.
40. Reddy, R. S.; Reddy, J. S.; Kumar, R.; Kumar, P. *CC* **1992**, 84.
41. Herscovici, J.; Antonakis, K. *CC* **1980**, 561.
42. Gao, Y.; Hanson, R. M.; Klunder, J. M.; Ko, S. Y.; Masamune, H.; Sharpless, K. B. *JACS* **1987**, *109*, 5765.
43. Smith, K.; Fry, K. B. *CC* **1992**, 187.
44. Urata, H.; Hu, N.; Maekawa, H.; Fuchikami, T. *TL* **1991**, *32*, 4733.
45. Lerner, L. M. *Carbohydr. Res.* **1990**, *207*, 138.
46. Lubineau, A.; Malleron, A. *TL* **1985**, *26*, 1713.
47. Heumann, A.; Tottie, L.; Moberg, C. *CC* **1991**, 218.
48. Tottie, L.; Baeckström, P.; Moberg, C.; Tegenfeldt, J.; Heumann, A. *JOC* **1992**, *57*, 6579.

James C. Lanter
The Ohio State University, Columbus, OH, USA

Montmorillonite K10

[1318-93-0]

(catalyzes protection reactions of carbonyl and hydroxy groups; promotes ene,[22] condensation,[25] and alkene addition[27] reactions)

Physical Data: the surface acidity of dry K10 corresponds to a Hammett acidity function $H_0 = -6$ to -8.[13]

Form Supplied in: yellowish-grey dusty powder. Forms with water a mud that is difficult to filter, more easily separated by centrifugation; with most organic solvents, forms a well-settling, easy-to-filter suspension.[6]

Handling, Storage, and Precautions: avoid breathing dust; keep in closed containers sheltered from exposure to volatile compounds and moisture.

General. Montmorillonite clays are layered silicates and are among the numerous inorganic supports for reagents used in organic synthesis.[1,2] The interlayer cations are exchangeable, thus allowing alteration of the acidic nature of the material by simple ion-exchange procedures.[3,4] Presently, in fine organic synthesis, the most frequently used montmorillonite is K10, an acidic catalyst, manufactured by alteration of montmorillonite (by calcination and washing with mineral acid; this is probably a proprietary process).

The first part of this article specifically deals with representative laboratory applications to fine chemistry of clearly identified, unaltered K10, excluding its modified forms (cation-exchanged, doped by salt deposition, pillared, etc.) and industrial uses in bulk. This illustrative medley shows the prowess of K10 as a strong Brønsted acidic catalyst. The second part deals with cation-exchanged (mainly Fe^{III}) montmorillonite. Clayfen and claycop, versatile stoichiometric reagents obtained by metal nitrate deposition on K10,[5] are used in oxidation and nitration reactions. They are treated under ***Iron(III) Nitrate–K10 Montmorillonite Clay*** and ***Copper(II) Nitrate–K10 Bentonite Clay***.

K10 is often confused, both in name and in use, with other clay-based acidic catalysts (KSF, K10F, Girdler catalyst, 'acid treated' or 'H^+-exchanged' montmorillonite or clay, etc.) that can be effectively interchanged for K10 in some applications. Between the 1930s and the 1960s, such acid-treated montmorillonites were common industrial catalysts, especially in petroleum processing, but have now been superseded by zeolites.

Activation.[6] K10 clay may be used crude, or after simple thermal activation. Its acidic properties are boosted by cation exchange (i.e. by iron(III)[7] or zinc(II)[8]) or by deposition of Lewis acids, such as zinc(II)[9,10] or iron(III)[11] chloride (i.e. 'clayzic' and 'clayfec'). In addition, K10 is a support of choice for reacting salts, for example nitrates of thallium(III),[12] iron(III) ('clayfen'),[5] or copper(II) ('claycop').[5] Multifarious modifications (with a commensurate number of brand names) result in a surprisingly wide range of applications; coupled with the frequent imprecise identification of the clay (K10 or one of its possible substitutes mentioned above), they turn K10 into a Proteus impossible to grab and to trace exhaustively in the literature.

Preparation of Acetals. Trimethyl orthoformate (see ***Triethyl Orthoformate***) impregnated on K10 affords easy preparation of dimethyl acetals,[14] complete within a few minutes at room temperature in inert solvents such as carbon tetrachloride or hexane (eq 1). The recovered clay can be reused.

HC(OMe)$_3$, K10; 5 min, 100%; MeO OMe (1)

Cyclic diacetals of glutaraldehyde are prepared in fair yields by K10-promoted reaction of 2-ethoxy-2,3-dihydro-4*H*-pyran with diols, under benzene azeotropic dehydration (eq 2).[15]

OH; OH; +; O; OEt; K10; benzene, Dean–Stark; 2–3 h; 80%; O; O; –(CH$_2$)$_3$–; O; O (2)

Diastereoisomeric acetal formation catalyzed by K10 has been applied to the resolution of racemic ketones, with diethyl (+)-(*R*,*R*)-tartrate as an optically active vicinal diol.[16]

1,3-Dioxolanes are also prepared by K10-catalyzed reaction of 1-chloro-2,3-epoxypropane (***Epichlorohydrin***) with aldehydes or ketones, in carbon tetrachloride at reflux (eq 3).[17] In the reaction of acetone with the epichlorohydrin, the efficiency of catalysts varies in the order: K10 (70%) > ***Tin(IV) Chloride*** (65%) > ***Boron Trifluoride*** (60%) = ***Hydrochloric Acid*** (60%) > ***Phosphorus(V) Oxide*** (57%).

K10, CCl_4, reflux, 8 h, 70% (3)

Preparation of Enamines. Ketones and amines form enamines in the presence of K10 at reflux in benzene or toluene, with azeotropic elimination of water (eq 4). Typical reactions are over within 3–4 h. With cyclohexanone, the efficiency depends on the nature of the secondary amine: ***Pyrrolidine*** (75%) > ***Morpholine*** (71%) > ***Piperidine*** (55%) > ***Dibutylamine*** (34%).[18] Acetophenone requires longer heating.[19]

K10, toluene, reflux, 70 h, 54% (4)

The K10-catalyzed reaction of aniline with β-keto esters gives enamines chemoselectively, avoiding the competing formation of anilide observed with other acidic catalysts.[20,21]

Synthesis of γ-Lactones via the Ene Reaction. K10 catalyzes the ene reaction of diethyl oxomalonate and methyl-substituted alkenes at a rather low temperature for this reaction (80 °C), followed by lactonization (eq 5).[22] When alkene isomerization precedes the ene step, it results in a mixture of lactones. Using kaolinite instead of K10 stops the reaction at the ene intermediate, before lactonization.

CO_2Et, CO_2Et; K10, 80 °C, 72 h; OH, CO_2Et + OH, CO_2Et (5)

50:50 mixture of diastereoisomers 44:49 78%

Synthesis of Enol Thioethers. Using a Dean–Stark water separator, K10 catalyzes formation of alkyl- and arylthioalkenes from cyclic ketones and thiols or thiophenols, in refluxing toluene (eq 6). A similar catalysis is effected by KSF (in a faster reaction) and K10F.[23] The isomer distribution is under thermodynamic control.

SH; K10, toluene, reflux, 6 h, 81% (6)

Preparation of Monoethers of 3-Chloro-1,2-propanediol. Alcohols react regioselectively with 1-chloro-2,3-epoxypropane to form 1-alkoxy-2-hydroxy-3-chloropropanes. The K10-catalyzed process is carried out in refluxing carbon tetrachloride for 2.5 h (eq 7).[24] Yields are similar to those obtained by ***Sulfuric Acid*** catalysis.

K10, CCl_4, reflux, 2.5 h, 60% (7)

α,β-Unsaturated Aldehydes via Condensation of Acetals with Vinyl Ethers. K10-catalyzed reaction of diethyl acetals with ***Ethyl Vinyl Ether*** leads to 1,1,3-trialkoxyalkanes. Hydrolysis turns these into *trans*-α,β-unsaturated aldehydes.[25] The reaction is performed close to ambient temperatures (eq 8). K10 is superior to previously reported catalysts, such as ***Boron Trifluoride*** or ***Iron(III) Chloride***. The addition is almost instantaneous and needs no solvent. Cyclohexanone diethyl acetal gives an analogous reaction.

OEt, OEt + OEt; K10, rt, 80%; OEt OEt, OEt; H^+, 83% (8)

Protective Tetrahydropyranylation of Alcohols and Phenols. With an excess of ***3,4-Dihydro-2H-pyran***, in the presence of K10 at room temperature, alcohols are transformed quantitatively into their tetrahydropyranyl derivatives. Run in dichloromethane at room temperature, the reaction is complete within 5–30 min (eq 9). The procedure is applicable to primary, secondary, tertiary, and polyfunctional alcohols as well as to phenols.[26]

K10, CH_2Cl_2, rt, 30 min, 83% (9)

Markovnikov Addition of Hydrochloric Acid to Alkenes. 1-Chloro-1-methylcyclohexane, the formal Markovnikov adduct of hydrochloric acid and 1-methylcyclohexene, becomes largely predominant when ***Sulfuryl Chloride*** is the chlorine source and K10 the solid acid.[27] The reaction at 0 °C, in dry methylene chloride, is complete within 2 h (eq 10).

K10, SO_2Cl_2, CH_2Cl_2, 0 °C, 2 h, 98% (10)

1,1:1,2 = 91:9

Porphyrin Synthesis. *Meso*-tetraalkylporphyrins are formed in good yields from condensation of aliphatic aldehydes with pyrrole; thermally activated K10 catalyzes the polymerization–cyclization to porphyrinogen, followed by ***p-Chloranil*** oxidation (eq 11).[28]

1. K10, CH_2Cl_2; 2. *p*-chloranil, 46% (11)

Meso-tetraarylporphyrins, with four identical or with tuneable ratios of different aryl substituents, are made by taking advantage of modified K10 ('clayfen' or Fe^{III}-exchanged) properties.[29,30]

Iron(III)-Doped Montmorillonite.

General Considerations. The acid strength of some cation-exchanged montmorillonites is between ***Methanesulfonic Acid*** (a strong acid) and ***Trifluoromethanesulfonic Acid*** (a superacid) and, in some instances, their catalytic activity is greater than that of a superacid.[31] Iron montmorillonite is prepared by mixing the clay with various Fe^{III} compounds in water.[8,32] The resulting material is filtered and dehydrated to afford the active solid-acid catalyst. These solid-acid catalysts are relatively inexpensive and are generally used in very small quantities to catalyze a wide variety of reactions, including Friedel–Crafts alkylation and acylation, Diels–Alder reactions, and aldol condensations.[1,5]

Diels–Alder Reactions.[33] Stereoselective Diels–Alder reactions involving an oxygen-containing dienophile are accelerated in the presence of Fe^{III}-doped montmorillonite in organic solvents (eq 12).[34] Furans also undergo Diels–Alder reactions with ***Acrolein*** and ***Methyl Vinyl Ketone*** in CH_2Cl_2 to give the corresponding cycloadducts in moderate yield (eq 13).[35] The iron-doped clay also catalyzes the radical ion-initiated self-Diels–Alder cycloaddition of unactivated dienophiles such as 1,3-cyclohexadiene and 2,4-dimethyl-1,3-pentadiene (eq 14).[36]

CHO; K10–Fe^{3+}, CH_2Cl_2, −24 °C, 60%; CHO (12)

K10–Fe^{3+}, CH_2Cl_2, −24 °C, 98% (13)

endo:*exo* = 13.5:1

K10–Fe^{3+}, 4-*t*-butylphenol (10%), CH_2Cl_2, 0 °C (14)

endo:*exo* = 4:1

The role of Fe^{III}-impregnated montmorillonite, and other cation-exchanged montmorillonites, in asymmetric Diels–Alder reactions was found to be limited to the use of small chiral auxiliaries; the results obtained from these reactions are similar to those of homogeneous aluminum catalysts (eq 15).[33]

endo:*exo* = 98:2, 39% de

(*S*) (*S*) (*S*) CO_2R^* + (*R*) (*R*) (*R*) CO_2R^* + (*R*) CO_2R^* (*S*) (*R*) + (*S*) CO_2R^* (*S*) (*R*) (15)

Friedel–Crafts Acylation and Alkylation.[37,38] The Friedel–Crafts acylation of aromatic substrates with various acyclic carboxylic acids in the presence of cation-exchanged (H^+, Al^{3+}, Ni^{2+}, Zr^{2+}, Ce^{3+}, Cu^{2+}, La^{3+}) montmorillonites has been reported.[39] Curiously, the use of iron-doped montmorillonite was not included in the report; however, some catalysis is expected. Under these conditions, the yield of the desired ketones was found to be dependent on acid chain length and the nature of the interlayer cation.

The direct arylation of a saturated hydrocarbon, namely adamantane, in benzene using $FeCl_3$-impregnated K10 was recently reported.[11] Additionally, Friedel–Crafts chlorination of adamantane in CCl_4 using the same catalyst was also reported. The alkylation of aromatic substrates with halides under clay catalysis gave much higher yields than conventional Friedel–Crafts reactions employing ***Titanium(IV) Chloride*** or ***Aluminum Chloride*** as catalyst.[8] Higher levels of dialkylation were observed in some cases. The alkylation of aromatic compounds with alcohols and alkenes was also found to be catalyzed with very low levels of cation-exchanged montmorillonites, as compared to standard Lewis acid catalysis; however, iron-doped clays performed poorly compared to other metal-doped clays.

Aldol Condensations. Cation-exchanged montmorillonites accelerate the aldol condensation of silyl enol ethers with acetals and aldehydes.[40] Similarly, the aldol reaction of silyl ketene acetals with electrophiles is catalyzed by solid-acid catalysts. Neither report discussed the use of iron montmorillonite for these reactions; however, some reactivity is anticipated.

Miscellaneous Reactions. The coupling of silyl ketene acetals (enolsilanes) with pyridine derivatives bearing an electron-withdrawing substituent, namely cyano, in the *meta* position is catalyzed by iron montmorillonite and other similar solid-acid catalysts (eq 16).[41]

R^1 = Me, Et; R^2 = H, Me; R^3 = H, Me (16)

The resulting *N*-silyldihydropyridines easily undergo desilylation by treatment with ***Cerium(IV) Ammonium Nitrate*** to afford the desired dihydropyridine derivative. The reactivity was found to be dependent on the montmorillonite counterion and to follow the order: $Fe^{3+} > Co^{2+} > Cu^{2+} \approx Zn^{2+} > Al^{3+} \approx Ni^{2+} \approx Sn^{4+}$.

1. Cornélis, A.; Laszlo, P. *SL* **1994**, 155.
2. McKillop, A.; Young, D. W. *S* **1979**, 401.
3. Theng, B. K. G. *The Chemistry of Clay–Organic Reactions*; Hilger: London, 1974.
4. Thomas, J. M. In *Intercalation Chemistry*; Whittingham, M. S.; Jacobson, J. A., Eds.; Academic: New York, 1982; p 55.
5. Cornélis, A.; Laszlo, P. *S* **1985**, 909.
6. Cornélis, A. In *Preparative Chemistry Using Supported Reagents*; Laszlo, P., Ed.; Academic: New York, 1987; pp 99–111.
7. Cornélis, A.; Gerstmans, A.; Laszlo, P.; Mathy, A.; Zieba, I. *Catal. Lett.* **1990**, *6*, 103.
8. Laszlo, P.; Mathy, A. *HCA* **1987**, *70*, 577.
9. Clark, J. A.; Kybett, A. B.; Macquarrie, D. J.; Barlow, S. J.; Landon, P. *CC* **1989**, 1353.
10. Cornélis, A.; Laszlo, P.; Wang, S. *TL* **1993**, *34*, 3849.
11. Chalais, S.; Cornélis, A.; Gerstmans, A.; Kolodziejski, W.; Laszlo, P.; Mathy, A.; Métra, P. *HCA* **1985**, *68*, 1196.
12. Taylor, E. C.; Chiang, C.-S.; McKillop, A.; White, J. F. *JACS* **1976**, *98*, 6750.
13. Pennetreau, P. PhD Thesis, University of Liège (Belgium), 1986.
14. Taylor, E. C.; Chiang, C.-S. *S* **1977**, 467.
15. Vu Moc Thuy; Maitte, P. *BSF(2)* **1979**, 264.
16. Conan, J. Y.; Natat, A.; Guinot, F.; Lamaty, G. *BSF(2)* **1974**, 1400.
17. Vu Moc Thuy; Petit, H.; Maitte, P. *BSB* **1980**, *89*, 759.
18. Hünig, S.; Benzing, E.; Lücke, E. *CB* **1957**, *90*, 2833.
19. Hünig, S.; Hübner, K.; Benzing, E. *CB* **1962**, *95*, 926.
20. Werner, W. *T* **1969**, *25*, 255.
21. Werner, W. *T* **1971**, *27*, 1755.
22. Roudier, J.-F.; Foucaud, A. *TL* **1984**, *25*, 4375.
23. Labiad, B.; Villemin, D. *S* **1989**, 143.
24. Vu Moc Thuy; Petit, H.; Maitte, P. *BSB* **1982**, *91*, 261.
25. Fishman, D.; Klug, J. T.; Shani, A. *S* **1981**, 137.
26. Hoyer, S.; Laszlo, P.; Orlovic, M.; Polla, E. *S* **1986**, 655.
27. Delaude, L.; Laszlo, P. *TL* **1991**, *32*, 3705.
28. Onaka, M.; Shinoda, T.; Izumi, Y.; Nolen, R. *CL* **1993**, 117.
29. Cornélis, A.; Laszlo, P.; Pennetreau, P. *Clay Minerals* **1983**, *18*, 437.
30. Laszlo, P.; Luchetti, J. *CL* **1993**, 449.
31. Kawai, M.; Onaka, M.; Isumi, Y. *BCJ* **1988**, *61*, 1237.
32. Tennakoon, D. T. B.; Thomas, J. M.; Tricker, M. J.; Williams, J. O. *JCS(D)* **1974**, 2207.
33. Cativiela, C.; Figueras, F.; Fraile, J. M.; Garcia, J. I.; Mayoral, J. A. *TA* **1993**, *4*, 223 and references therein.
34. Laszlo, P.; Lucchetti, J. *TL* **1984**, *25*, 2147.
35. Laszlo, P.; Lucchetti, J. *TL* **1984**, *25*, 4387.
36. Laszlo, P.; Lucchetti, J. *TL* **1984**, *25*, 1567.
37. Olah, G. A. *Friedel–Crafts Chemistry*; Wiley: New York, 1973.
38. Olah, G. A.; Reddy, V. P.; Prakash, G. K. S. In *Kirk-Othmer Encyclopedia of Chemical Technology*, 4th ed.; Wiley: New York, 1994; Vol. 11, p 1042
39. Chiche, B.; Finiels, A.; Gauthier, C.; Geneste, P.; Graille, J.; Piock, D. *J. Mol. Catal.* **1987**, *42*, 229.
40. Onaka, M.; Ohno, R.; Kawai, M.; Isumi, Y. *BCJ* **1987**, *60*, 2689.
41. Onaka, M.; Ohno, R.; Izumi, Y. *TL* **1989**, *30*, 747.

André Cornélis & Pierre Laszlo
Université de Liège, Belgium

Mark W. Zettler
The Dow Chemical Company, Midland, MI, USA

Nitric Acid

HNO3

[7697-37-2] HNO_3 (MW 63.02)
(fuming)
[52583-42-3]

(nitration and oxidation of organic molecules)

Physical Data: mp −41.6 °C; bp 83 °C; *d* 1.50 g cm^{-3}.[1]
Solubility: sol H_2O.
Form Supplied in: clear colorless liquid (69–71% in H_2O); fuming nitric acid is a colorless to pale yellow liquid, HNO_3 content >90% widely available.
Preparative Methods: anhydrous nitric acid can be prepared by distilling fuming nitric acid from an equal volume of concentrated sulfuric acid.
Handling, Storage, and Precautions: strong acid; oxidizing agent. Colorless acid may discolor on exposure to light. Anhydrous nitric acid decomposes above the freezing point to give NO_2, H_2O, and O_2. Emits toxic fumes of nitrogen oxides. Hygroscopic. Contact with other material may cause fire. Poison. Corrosive. Avoid contact and inhalation. May be fatal if inhaled, swallowed, or absorbed through skin. Material is extremely destructive to tissue of the mucous membranes and upper respiratory tract, eyes, and skin.[2] Many of the reactions mentioned in this article require special care in order to avoid uncontrollable reactions and the possibility of explosions. Many nitrated compounds are unstable. The reader is therefore strongly urged to refer to original literature procedures.

Nitric acid holds an important place in the history of organic synthesis. It is used primarily for the nitration of organic molecules and to effect a wide variety of oxidative transformations. The advantage of nitric acid as a reagent is that it allows simple and straightforward isolation of products. However, it is not a very selective oxidant.

Nitration of Simple Aromatic Systems.[3] Nitration of aromatics has been studied extensively. The mechanism by which nitration occurs is believed to involve electrophilic attack by NO_2^+. The concentration of the active species increases in the presence of the more acidic ***Sulfuric Acid***. With this mixed reagent, the nitration of simple benzene derivatives[4] and polyaromatic ring systems[5] has been accomplished. For example, treatment of methyl benzoate with concentrated nitric acid gives methyl 3-nitrobenzoate in 81–85% yield.[4a] In polyaromatic ring systems, nitration usually occurs selectively on the more electron rich aromatic ring (eq 1).[6] Under the reaction conditions, anilines are protonated and the aniline ring is relatively unreactive (eq 2).[5e]

Nitration of Aromatic Heterocycles. Several types of aromatic heterocycles have been nitrated with nitric acid.[7] For example, when thiophene is treated with nitric acid in acetic anhydride, 2-nitrothiophene is formed in 70–85% yield. In order to prepare 3-nitrothiophene, a deactivating strategy is required (eq 3). 2,4-Dinitrothiophene can be obtained by nitration of either 2-nitro- or 3-nitrothiophene.[7a] Nitration of the *N*-oxide of 2-methylpyridine allows the introduction of a nitro group at the 4 position in good yield (eq 4).[7b]

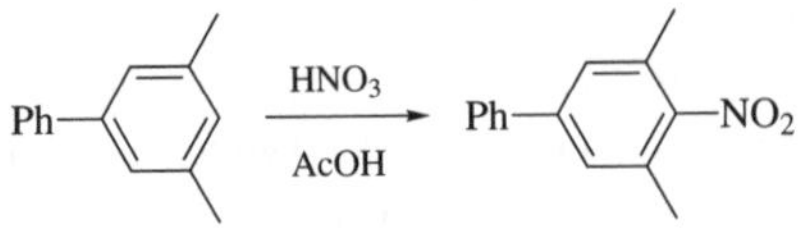

(1) HNO3, AcOH

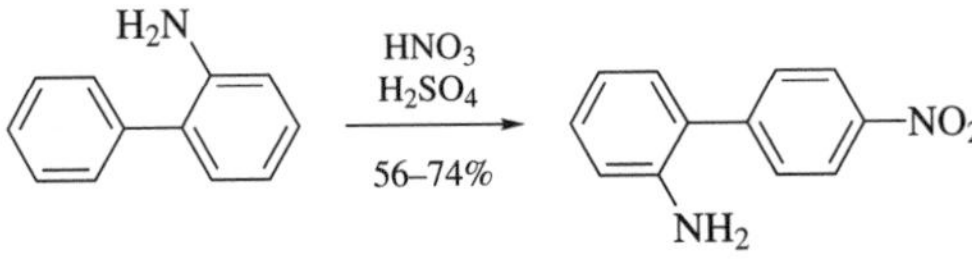

(2) HNO3, H2SO4, 56–74%

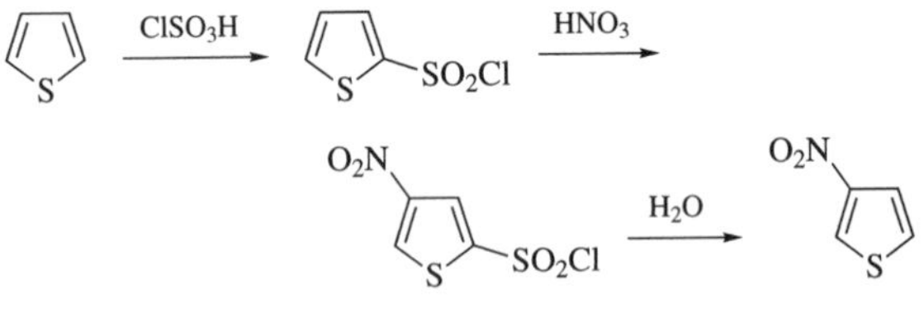

(3) ClSO3H; HNO3; H2O

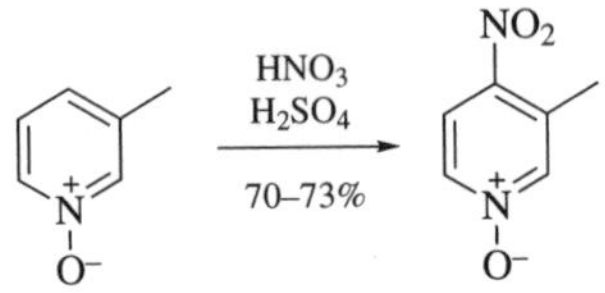

(4) HNO3, H2SO4, 70–73%

Nitration of Alkenes.[8] Alkenes may also be nitrated by nitric acid. This reaction has been exploited in the synthesis of a number of steroid derivatives.[9] Fuming nitric acid converts cholesteryl acetate to 6-nitrocholesteryl acetate in good yield (eq 5). The nitration of the more highly functionalized dienyl acetate (eq 6) provides the corresponding nitro steroid. Treatment of 1,1-dichloro-2-fluoroethylene with nitric acid in concentrated sulfuric acid provides fluoronitroacetyl chloride in 16% yield.[10] 2-Sulfolenes have also been nitrated.[11]

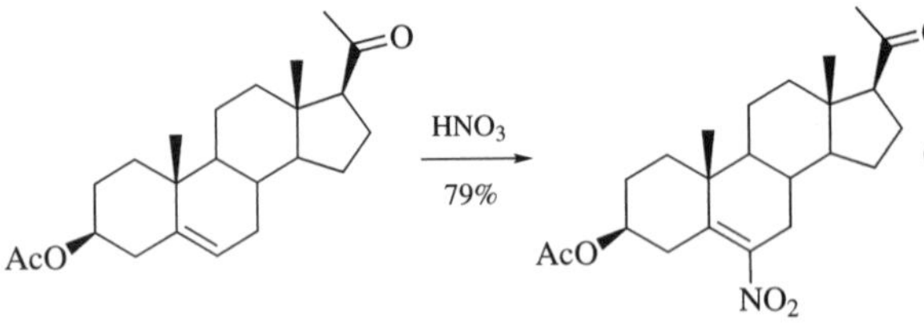

(5) HNO3, 79%

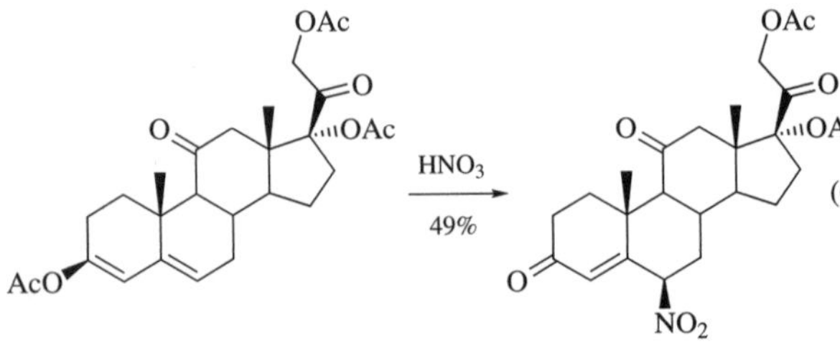

(6) HNO3, 49%

Nitration of Active Methylene Carbons.[8,12] Active methylene carbons are nitrated by a number of reagents.[12,13] With nitric acid the nitration of β-diketones can be achieved (eq 7). Nitration of diethyl malonate provides diethyl nitromalonate in 92% yield. α-Nitro ketones are obtained by nitration of ketones or enol acetates with the reagent derived from nitric acid and ***Acetic Anhydride*** in moderate to good yields (eq 8).[14]

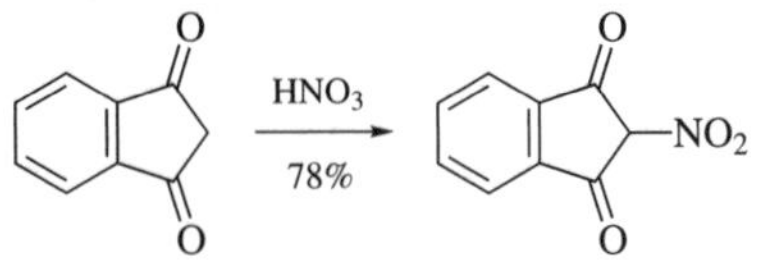

(7)

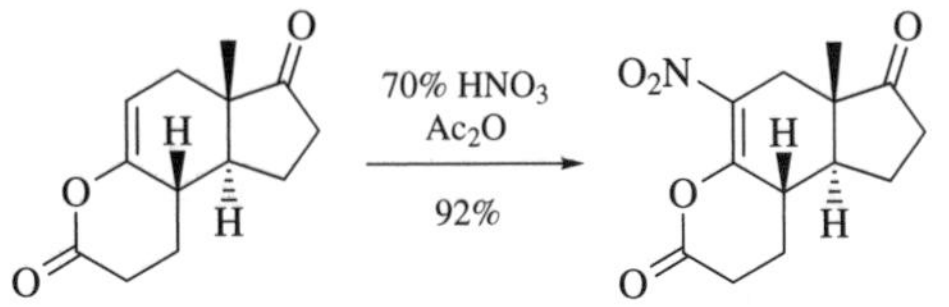

(8)

Nitration of Heteroatoms. Secondary amines and primary and secondary amides can be converted to *N*-nitro compounds by direct nitration with nitric acid.[15,16] Although most primary amines cannot be nitrated directly, it is possible to obtain primary nitramines by nitration of the corresponding dichloroamines (eq 9).[17] Treatment of pyrrolidone with nitric acid and acetic acid in the presence of copper provides *N*-nitroso-2-pyrrolidone in 70% yield;[18] note that this transformation represents *N*-nitrosation.

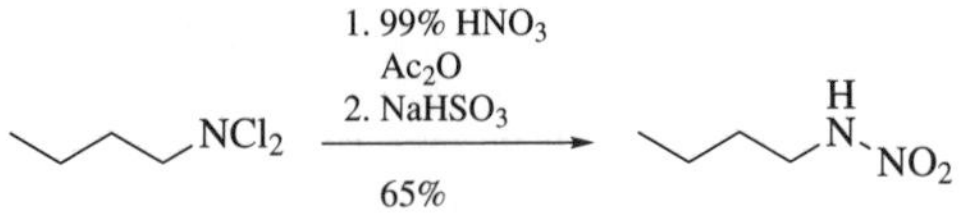

(9)

Nitrolysis of dialkyl *t*-butylamines with nitric acid and sulfuric acid or acetic anhydride provides secondary nitramines (eq 10).[19] More recently, Suri has shown that the reagent derived from ammonium nitrate and trifluoroacetic acid is effective for *N*-nitration.[20]

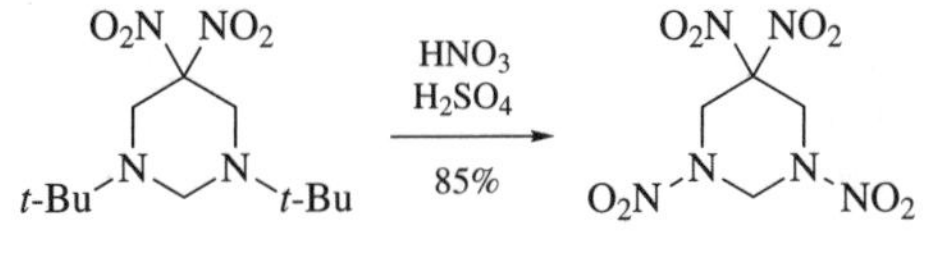

(10)

Acyl nitrates can be prepared conveniently on a laboratory scale by treatment of 90% nitric acid with a tenfold excess of the corresponding acid anhydrides.[21]

Oxidation of Alcohols, Aldehydes, and Esters. Nitric acid oxidizes alcohols and aldehydes to the corresponding carboxylic acids. For example, 1-chloro-3-propanol is oxidized to 3-chloropropanoic acid in 78–79% yield.[22] 3-Chloropropionaldehyde affords the same product.[23] Primary alcohols protected as esters are oxidized to the carboxylic acids (eq 11).[24] In a two-phase solvent system (for example dimethyl ether and water), the oxidation of benzyl alcohols can be controlled so that aldehydes are obtained.[25]

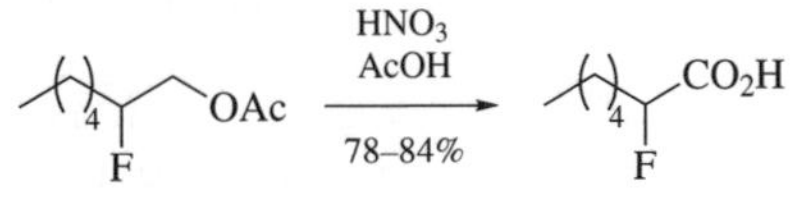

(11)

Oxidative Cleavage. Treatment of cyclohexanone with 33% nitric acid gives adipic acid in quantitative yield;[26] oxidation of cyclohexanol affords the same product (eq 12).[27] Glutaric acid is obtained in 70–75% yield by oxidative cleavage of 3,4-dihydro-2*H*-pyran.[27]

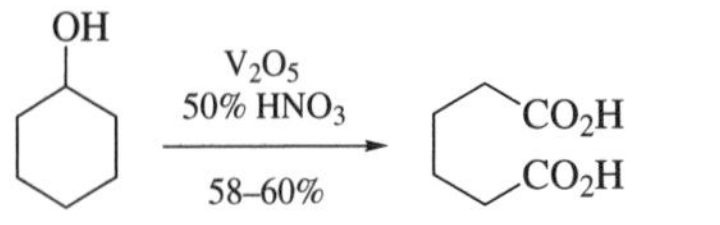

(12)

Oxidation at Benzylic Position. Nitric acid oxidizes many aromatic alkyl substituents to the carboxylic acid group. Thus toluene is oxidized to benzoic acid in 85–90% yield.[28] Oxidation of ethylbenzene with 15% nitric acid also gives benzoic acid in 80% yield. The reaction is general and has also been applied to the oxidation of pyridine derivatives. When 4-methylpyridine is treated with 10% nitric acid in phosphoric acid at elevated temperature and pressure, 4-pyridinecarboxylic acid is obtained in 93% yield.[29] The reaction of *p*-isopropyltoluene can be controlled to give the partially oxidized product, *p*-methylbenzoic acid, in 56–59% yield.[27] Additional examples of selective benzylic oxidations are shown in eqs 13 and 14.[30,31]

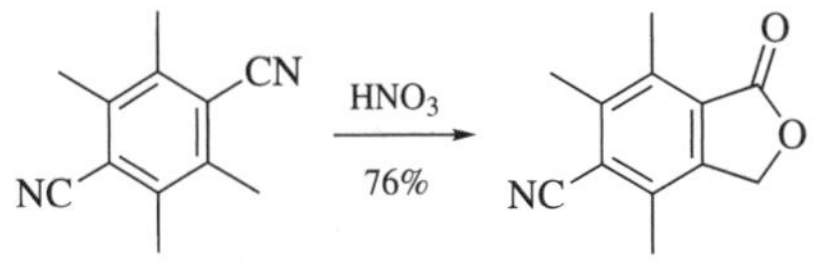

(13)

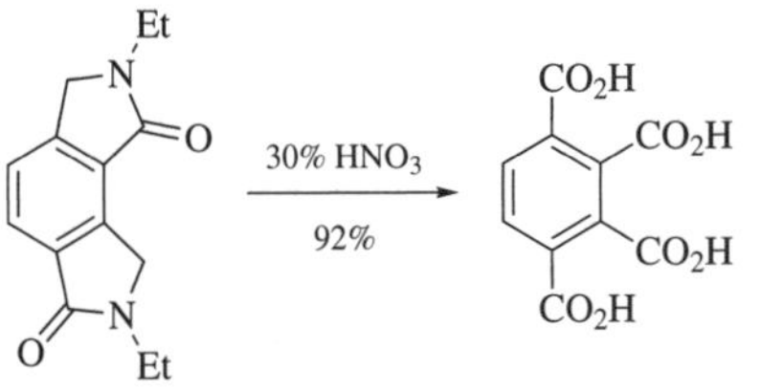

(14)

Oxidation to Quinones. Nitric acid oxidizes a wide variety of hydroquinone derivatives to quinones (eq 15).[32] Aminonaphthols can be converted to naphthoquinones by treatment with nitric acid (eq 16).[27] Perhalogenated aromatic systems have also been oxidized to quinones (eq 17).[33]

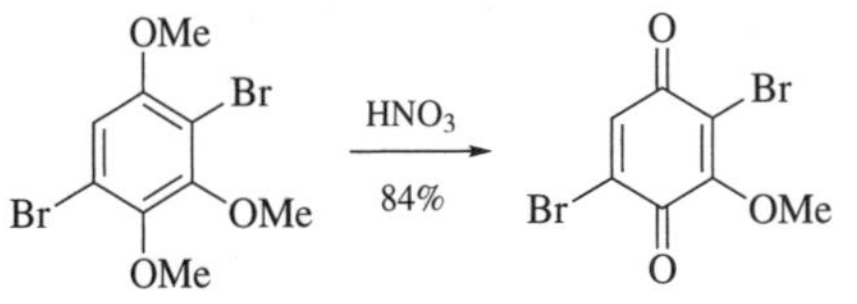

(15)

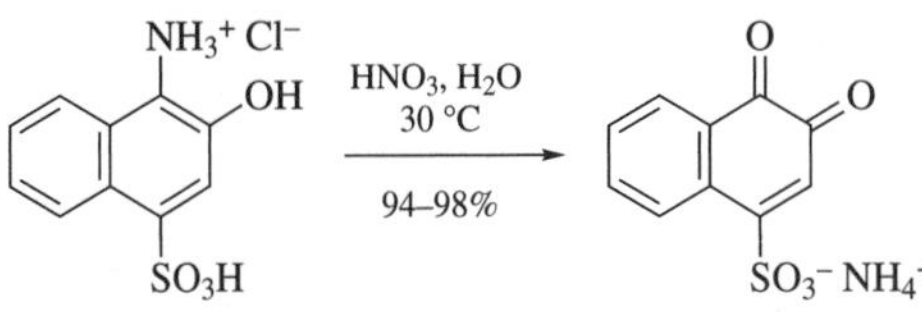

(16)

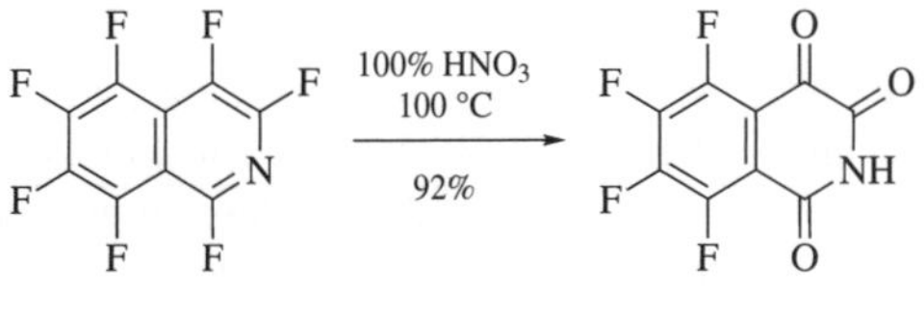

(17)

Dehydrogenation and Aromatization. Dihydropyridines can be aromatized by dilute nitric and sulfuric acid (eq 18).[34] Diethyl hydrazodicarboxylate is dehydrogenated by fuming nitric acid to diethyl azodicarboxylate in 70–80% yield.[35]

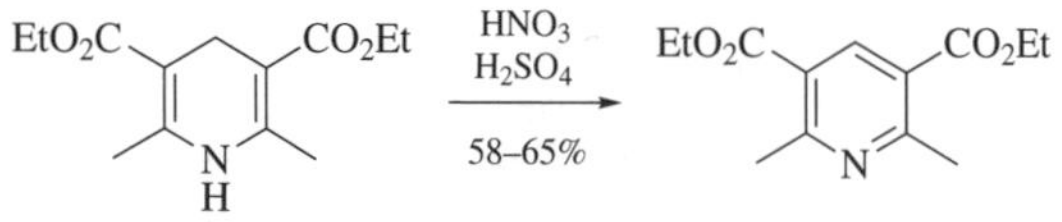

HNO_3, H_2SO_4, 58–65% (18)

Oxidation of Heteroatoms. The nitroso group is oxidized efficiently to the nitro group by nitric acid.[36] For example, 2,4-dinitrosoresorcinol is converted efficiently to 2,4,6-trinitroresorcinol with concentrated nitric acid (eq 19).[37]

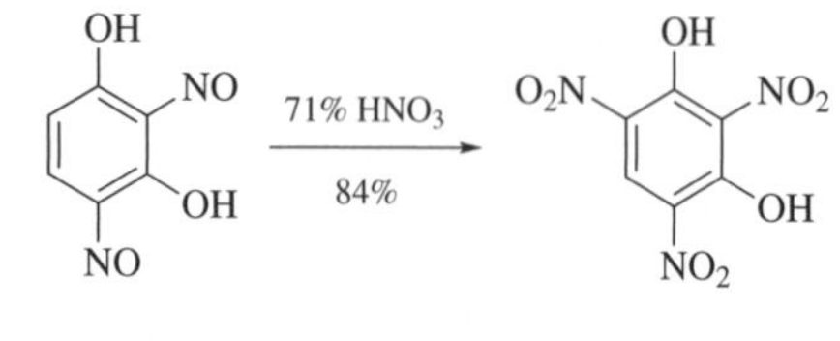

71% HNO_3, 84% (19)

Azoxycyclohexane can be obtained by oxidation of azocyclohexane.[38] Dialkyl sulfides have been oxidized to the corresponding sulfones[39] and sulfoxides;[40] thiols provide sulfonic acids (eq 20).[41] Iodoso compounds have been obtained from oxidation of aryl iodides (eq 21).[42] Nitric acid converts 2-amino-6-nitrobenzonitriles to substituted 1,2,3-benzotriazin-4(3*H*)-one N^2-oxides (eq 22).[43]

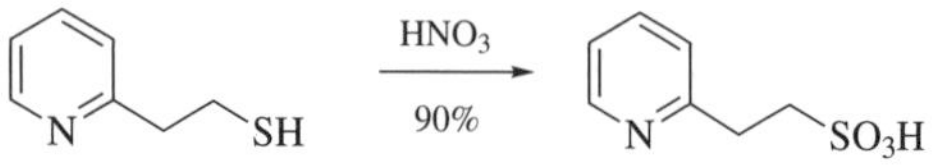

HNO_3, 90% (20)

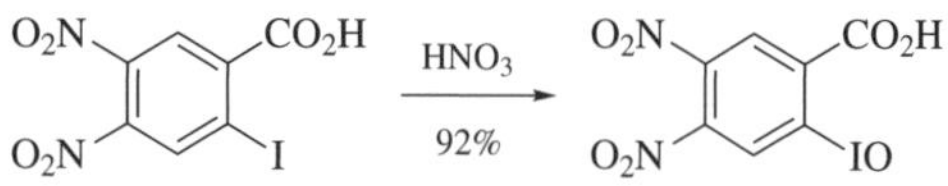

HNO_3, 92% (21)

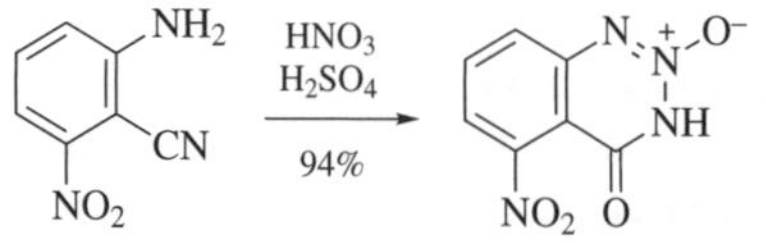

HNO_3, H_2SO_4, 94% (22)

Other Uses. Dichloromaleic anhydride has been obtained in 81% yield by treatment of hexachlorobutadiene with fuming nitric acid followed by concentrated sulfuric acid.[25] Pyrroles react rapidly with nitric acid to give pyrrolinones (eq 23).[44] Desulfurization of 1,2,4-triazole-3-thiol with nitric acid presumably involves the formation of the sulfonic acid, which is then hydrolyzed to triazole in 52–58% yield.[27,45]

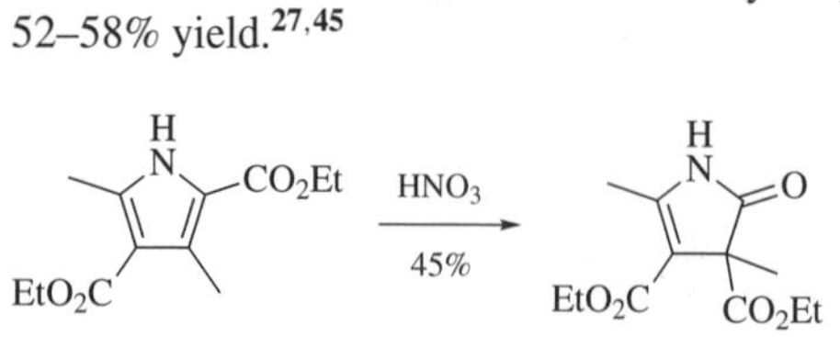

HNO_3, 45% (23)

Primary alkyl halides have also been oxidized to provide the corresponding carboxylic acids. On treatment with concentrated nitric acid, *trans*-2,3-bis(iodomethyl)-*p*-dioxane provides *trans*-*p*-dioxane-2,3-dicarboxylic acid in 73% yield.[46]

Related Reagents. Acetyl Nitrate; Nitronium Tetrafluoroborate; Nitrosylsulfuric Acid.

1. Stern, S. A.; Mullhaupt, J. T.; Kay, W. B. *CRV* **1960**, *60*, 185.
2. (a) *The Merck Index*, 11th ed.; Budavari, S., Ed.; Merck: Rahway, NJ, 1989; pp 6495–6497. (b) *The Sigma Aldrich Library of Chemical Safety*, 2nd ed.; Leng, R. E., Ed.; Sigma-Aldrich Corporation: Milwaukee, WI, 1988; pp 2546B–C.
3. (a) Hoggett, J. G.; Moodie, R. B.; Penton, J. R.; Schofield, K. *Nitration and Aromatic Reactivity*; Cambridge University Press: Cambridge, 1971. (b) Schofield, K. *Aromatic Nitration*; Cambridge University Press: Cambridge, 1980.
4. (a) Kamm, O.; Segur, J. B. *OSC* **1941**, *1*, 372. (b) Robertson, G. B. *OSC* **1941**, *1*, 396. (c) Culhane, P. J.; Woodward, G. E. *OSC* **1941**, *1*, 408. (d) Smith, L. I. *OSC* **1943**, *2*, 254. (e) Corson, B. B.; Hazen, R. K. *OSC* **1943**, *2*, 434. (f) Powell, G.; Johnson, F. R. *OSC* **1943**, *2*, 449. (g) Huntress, E. H.; Shriner, R. L. *OSC* **1943**, *2*, 459. (h) Brewster, R. Q.; Williams, B.; Phillips, R. *OSC* **1955**, *3*, 337. (i) Icke, R. N.; Redemann, C. E.; Wisegarver, B. B.; Alles, G. A. *OSC* **1955**, *3*, 644. (j) Kobe, K. A.; Doumani, T. F. *OSC* **1955**, *3*, 653. (k) Fitch, H. M. *OSC* **1955**, *3*, 658. (l) Fanta, P. E.; Tarbell, D. S. *OSC* **1955**, *3*, 661. (m) Howard, J. C. *OSC* **1963**, *4*, 42. (n) Schultz, H. P. *OSC* **1963**, *4*, 364. (o) Buckles, R. E.; Bellis, M. P. *OSC* **1963**, *4*, 722. (p) Fetscher, C. A. *OSC* **1963**, *4*, 735. (q) Boyer, J. H.; Buriks, R. S. *OSC* **1973**, *5*, 1067.
5. (a) Hartman, W. W.; Smith, L. A. *OSC* **1943**, *2*, 438. (b) Kuhn, W. E. *OSC* **1943**, *2*, 447. (c) Woolfolk, E. O.; Orchin, M. *OSC* **1955**, *3*, 837. (d) Braun, C. E.; Cook, C. D.; Merritt, C., Jr.; Rousseau, J. E. *OSC* **1963**, *4*, 711. (e) Mendenhall, G. D.; Smith, P. A. S. *OSC* **1975**, *5*, 829. (f) Newman, M. S.; Boden, H. *OSC* **1975**, *5*, 1029. (g) Vouros, P.; Petersen, B.; Dafeldecker, W. P.; Neumeyer, J. L. *JOC* **1977**, *42*, 744. (h) Keumi, T.; Tomioka, N.; Hamanaka, K.; Kakihara, H.; Fukushima, M.; Morita, T.; Kitajima, H. *JOC* **1991**, *56*, 4671.
6. (a) Grieve, W. S. M.; Hey, D. H. *JCS* **1932**, 2245. (b) Hey, D. H. *JCS* **1932**, 2636. (c) Hey, D. H.; Buckley Jackson, E. R. *JCS* **1934**, 645.
7. (a) Babasnian, V. S. *OSC* **1943**, *2*, 466. (b) Taylor, E. C., Jr.; Crovetti, A. J. *OSC* **1963**, *4*, 654. (c) Fox, B. A.; Threlfall, T. L. *OSC* **1973**, *5*, 346. (d) Kolb, V. M.; Darling, S. D.; Koster, D. F.; Meyers, C. Y. *JOC* **1984**, *49*, 1636. (e) Szabo, K. J.; Hörnfeldt, A.-B.; Gronowitz, S. *JOC* **1991**, *56*, 1590. (f) Einhorn, J.; Demerseman, P.; Royer, R. *CJC* **1983**, *61*, 2287.
8. For a review on the synthesis of aliphatic and alicyclic nitro compounds, see: Kornblum, N. *OR* **1960**, *12*, 101.
9. Laron, H. O. In *The Chemistry of the Nitro and Nitroso Groups*; Feuer, H., Ed.; Wiley: New York, 1969; Part 1, pp. 323–324.
10. Martinov, I. V.; Kruglyak, Y. L. *JGU* **1965**, *35*, 974.
11. Titova, M. V.; Berestovitskaya, V. M.; Perekalin, V. V. *JOU* **1981**, *17*, 1172.
12. See Ref. 9; Part 1, pp 310–316.
13. Feuer, H. In *The Chemistry of Amino, Nitroso and Nitro Compounds and Their Derivatives*; Patai, S., Ed.; Wiley: New York, 1982; Part 2, p 805.
14. (a) Dampawan, P.; Zajac, W. W. *JOC* **1982**, *47*, 1176. (b) Stork, G.; Clark, G.; Weller, T. *TL* **1984**, *25*, 5367.
15. For a review on the formation of the nitroamine group, see: Wright, G. F. In *The Chemistry of the Nitro and Nitroso Groups*; Feuer, H., Ed.; Wiley: New York, 1969; Part 1, pp 613–684.
16. (a) Willer, R. L.; Atkins, R. L. *JOC* **1984**, *49*, 5147. (b) Rowlands, D. A. *Synthetic Reagents*; Pizey, J. S., Ed.; Wiley: New York, 1985; Vol. 6, pp 359–360.
17. Smart, G. N. R.; Wright, G. F. *Can. J. Res.* **1948**, *26B*, 284 (*CA* **1948**, *42*, 5844a).
18. McQuinn, R. L.; Cheng, Y.-C.; Digenis, G. A. *SC* **1979**, *9*, 25.
19. Cichra, D. A.; Adolph, H. G. *JOC* **1982**, *47*, 2474.

20. Suri, S. C.; Chapman, R. D. *S* **1988**, 743.
21. Bachman, G. B.; Biermann, T. F. *JOC* **1970**, *35*, 4229.
22. Hudlicky, M. *Oxidations in Organic Chemistry*; American Chemical Society: Washington, 1990; p 127.
23. Haines, A. H. *Methods for the Oxidation of Organic Compounds*; Academic: San Diego, 1988; p 247.
24. See Ref. 22; p 224.
25. Fieser, M.; Fieser, L. F. *FF* **1975**, *5*, 474.
26. See Ref. 22; p 211.
27. Fieser, M.; Fieser, L. F. *FF* **1967**, *1*, 733.
28. See Ref. 22; pp 105–106.
29. See Ref. 22; pp 108–109.
30. Suzuki, H.; Hanafusa, T. *S* **1974**, 432.
31. Kajimoto, T.; Tsuji, J. *JOC* **1983**, *48*, 1685.
32. For an extensive review, see: Musgrave, O. C. *CRV* **1969**, *69*, 499.
33. (a) See Ref. 22; pp 113–114. (b) Suzuki, H.; Ishizaka, K.; Maruyama, S.; Hanafusa, T. *CC* **1975**, 51.
34. See Ref. 22; pp 52, 241.
35. See Ref. 22; p 233.
36. Iffland, D. C.; Yen, T.-F. *JACS* **1954**, *76*, 4083.
37. Fieser, M.; Fieser, L. F. *FF* **1972**, *3*, 212.
38. Langley, B. W.; Lythgoe, B.; Riggs, N. V. *JCS* **1951**, 2309.
39. See Ref. 22; p 257.
40. Goheen, D. W.; Bennett, C. F. *JOC* **1961**, *26*, 1331.
41. See Ref. 22; p 252.
42. See Ref. 22; p 266.
43. Mitschker, A.; Wedemeyer, K. *S* **1988**, 517.
44. Moon, M. W. *JOC* **1977**, *42*, 2219.
45. Whitehead, C. W.; Traverso, J. J. *JACS* **1956**, *78*, 5294.
46. Summerbell, R. K.; Lestina, G. J. *JACS* **1957**, *79*, 3878.

Kathlyn A. Parker & Mark W. Ledeboer
Brown University, Providence, RI, USA

P

Phosphazene Base P_4-*t*-Bu

[111324-04-0] $C_{22}H_{63}N_{13}P_4$ (MW 633.86)

(extremely strong, highly hindered, kinetically highly active base;[1] generates exceptionally nucleophilic 'naked' carbanions from a wide range of carbon acids;[2] the very low Lewis acidity *of the cation suppresses side reactions commonly observed with metal organyls, e.g. aldol and ester condensations*[3] *and β-eliminations*[4])

Alternate Name: 3-*t*-butylimino-1,1,1,5,5,5-hexakis(dimethylamino)-3-[tris(dimethylamino)phosphor anylidene]amino-1λ^5, 3λ^5,5λ^5-1,4-triphosphazadiene.

Physical Data: mp ca. 207 °C (dec); pK_{BH}+ 30.1.[5]

Solubility: very sol THF, ether, hexane, benzene, toluene; sol with protonation in protic solvents, MeCN; reacts rapidly with all types of haloalkanes except fluoroalkanes.

Form Supplied in: 1 M solution in hexane; commercially available.

Analysis of Reagent Purity: NMR in benzene-d_6. ^{1}H NMR: δ 1.83 (s, 9 H), 2.73 (d, J=10 Hz, 54 H); ^{13}C NMR: δ 35.6 (br d, J 13 Hz), 38.08 (d, J=4 Hz), 51.31 (d, J=5.5 Hz); ^{31}P NMR: δ −24.44 (q, J=20 Hz), 5.74 (br m).

Purification: all water-soluble salts of P_4-*t*-Bu can be converted to the HBF_4 salt by precipitation from aqueous solution with $NaBF_4$. Water-insoluble salts are first converted to the chloride by means of a column charged with strongly basic anion exchange resin (Cl^- form, MeOH). P_4-*t*-Bu·HBF_4 is recrystallized from aqueous ethylamine and dried in vacuo at 60 °C. 2.35 g (60 mmol) of potassium metal and 5 mg of $Fe(NO_3)_3{\cdot}9H_2O$ are added to 50 mL of anhydrous NH_3(l) with stirring (glass-covered stirring bar) under N_2 and the solution kept at ca. −40 °C until the color turns to gray and evolution of H_2 ceases. A solution of 21.6 g (30.0 mmol) of P_4-*t*-Bu·HBF_4 in 50 mL of THF is added and the mixture stirred at −40 °C for another 15 min. After evaporation of solvents in vacuo, the residue is extracted with 60 mL of hexane (caution! KNH_2 is pyrophoric), rigorously protecting from moisture. The solvent is removed at reduced pressure, affording up to 18.3 g (96%) of the crystalline base. P_4-*t*-Bu can be sublimed at 160 °C/10^{-3} mmHg, but protic impurities are not removed thereby.

For most applications, P_4-*t*-Bu must be strictly anhydrous, but water content is not readily detected by NMR. The following procedure has proven to be effective in the elimination of small amounts of protic impurities: to a 0.5 M solution of the base in hexane or heptane, ethyl bromide (at least 3 mol per mol of H_2O) is added. After 0.5 h at rt, diethyl ether and excess ethyl bromide (bp 31 °C) are removed by evaporating part of the solvent in vacuo, the precipitated hydrobromide salt of the base is filtered off, the solvent is removed completely in vacuo, and the residue is dried in a high vacuum.

Handling, Storage, and Precautions: P_4-*t*-Bu is extremely hygroscopic and must be stored and handled so as to rigorously exclude moisture. It is thermally stable up to ca. 120 °C, extremely resistant towards (basic) hydrolysis, and likewise insensitive to dry oxygen.

Alkylations of Carbanions. P_4-*t*-Bu is a member of a novel class of kinetically highly active uncharged bases,[1,2,6–8] with pK_{BH}+ values ranging from 13[5,6] to ca. 34;[1,2,5,7,8] among the strongest of these phosphazene bases, it is the most readily available.[1] In the presence of alkylating agents, in situ alkylation of low acidic substrates in concentrated (ca. 0.5 M) THF solution is generally extremely rapid on (gradual) addition of P_4-*t*-Bu at −100 °C to −78 °C. Due to the high solubilizing power of phosphazene bases, solubility problems are scarce. Separation of products from salts of the base is easily achieved, e.g. by direct precipitation of its halide salts with diethyl ether or benzene, by extraction (CH_2Cl_2) or precipitation ($NaBF_4$) of salts from aqueous solution, or by filtration over silica gel. The very low Lewis acidity of the huge cation contrasts sharply with the characteristics of lithium amide bases. Thus Lewis acid-catalyzed side reactions, e.g. aldol or ester condensations in alkylations of enolates, are effectively suppressed; even β-lactones are easily mono-[2] or peralkylated (eq 1).[8] In cases where the corresponding lithium organyls decompose entirely via β-alkoxide elimination,[4,9] 'naked' enolates of β-alkoxy esters undergo clean alkylation.

+ 4 EtI —(2 equiv P_4-*t*-Bu, THF, −100 °C; 68%)→ Et, Et product; —(LDA or $KN(TMS)_2$)→ only decomposition (1)

Using more hindered bases like P_4-*t*-Oct[2,7,10] (*t*-Oct = 1,1,3,3-tetramethylbutyl) enhances the selectivity for monoalkylation considerably (eq 2).[2] This tendency also holds for selective monoalkylation of secondary carbon centers.[8] Even sterically congested quaternary centers are formed with great ease (eq 3).[4]

CO_2Me / CO_2Me —(EtI, THF, −78 °C)→ starting material + Et, CO_2Me, CO_2Me + Et, CO_2Me, CO_2Me, Et (2)

	starting material		
P_4-*t*-Bu	28%	41%, 97% de	14%
P_4-*t*-Oct	9%	71%, 98% de	8%
LDA, HMPA		no alkylation product	
$K[N(TMS)_2]$		no alkylation product	

(3)

Alkylation of nitriles is not complicated by Thorpe condensation, as observed with ***Lithium Diisopropylamide*** as base. Alkylation of 1,2-dinitriles with no elimination of hydrocyanic acid (eq 4)[8] is achieved in high yield.

NC–CH2CH2–CN + EtI → (base, THF, –78 °C to rt) NC–C(Et)2–C(Et)2–CN (4)

base	yield
P_4-*t*-Bu	98%
LDA	0.5%
LDA, HMPA	15%
$KN(TMS)_2$	2%

Alkylation of Nitrogen Acids. *N*-Perbenzylation of rather insoluble *N*-Boc-protected or cyclic peptides leads to easily soluble, fully protected derivatives.[11] Occasionally, regioselective *C*-alkylation of glycine or sarcosine subunits is also observed (eq 5).

P_4-*t*-Bu, BnBr, THF, –100 °C, 80% (5)

Anionic Polymerization. Polymerization of methyl methacrylate occurs in the presence of P_4-*t*-Bu. Using ethyl acetate as initiator in THF at elevated temperatures, high molecular weights and narrow molecular weight distributions are achieved,[3] 'backbiting' by ester condensation[12] being effectively suppressed. Presumably due to lack of 'chelate control' by the cation, the stereochemistry differs from that of polymerizations with metal organyls, syndiotactic diads being favored (eq 6).

EtOAc (1 equiv) + P_4-*t*-Bu (1.1 equiv) → (126 equiv methyl methacrylate, THF, 60 °C) $EtOCO(CH_2C(CH_3)E)_nH$ (6)

$M_n = 15200$; $D = 1.11$

$E = CO_2Me$ 78:22

Other Applications. The high steric hindrance of P_4-*t*-Bu enables the formation of isomerically pure 1-alkenes from primary halides at rt in almost quantitative yield.[1] The simplicity of the NMR spectra, the UV transparency (end absorption below 230 nm), and the simplicity of handling recommends P_4-*t*-Bu and other phosphazene bases for obtaining spectroscopic data of highly basic or unstable 'naked' anions.[13]

Related Commercially Available Bases. A number of other phosphazene bases are commercially available. P_1-*t*-Oct (**1**), P_1-*t*-Bu (**2**), BEMP (**3**), BEMP bound to a Merrifield polymer, and BTPP (**4**)[10] are hindered bases, suitable for *O*-alkylation of carbohydrates[6,14] and cyanohydrins,[6] *N*-alkylation of nucleosides[6] and carbamates,[15] *C*-alkylation of alkyl malonic esters,[6] *O*-tosylation of amino alcohols,[16] and aldol condensations;[17] reactions with these bases are often considerably more selective than with metal bases.

$(Me_2N)_3P{=}NCMe_2CH_2$-*t*-Bu

(1) P_1-*t*-Oct, $pK_{BH^+} = 14.0$

$(Me_2N)_3P{=}N$-*t*-Bu

(2) P_1-*t*-Bu, $pK_{BH^+} = 14.4$

(3) R = Me: BEMP, $pK_{BH^+} = 15.1$; R = H_2C-(P)

(4) BTPP, $pK_{BH^+} = 15.8$

N-Alkylation of phthalimide (Gabriel synthesis) occurs in homogeneous acetonitrile solution at rt.[18] In contrast to DBU or guanidines, all phosphazene bases are highly resistant towards basic hydrolysis, thus representing easy-to-recover catalysts for the basic hydrolysis of sensitive esters in a relatively apolar medium (eq 7).[19]

BEMP polymer, $THF–H_2O$ (5:1), 55 °C, 5 h, 90% (7)

P_2-Et (**5**)[10,20] is only a moderately hindered base ($pK_{BH^+} = 20.2$),[5] suitable for E2 elimination reactions of secondary halides, which is ca. 4 orders of magnitude more reactive and less easily alkylated than DBU ($pK_{BH^+} = 11.8$).[5]

$(Me_2N)_3P{=}N{-}P(NMe_2)_2{=}NEt$

(5)

Related Reagents. Collidine; 1,5-Diazabicyclo[4.3.0]-nonene-5; 1,8-Diazabicyclo[5.4.0]undecene-7.

1. Schwesinger, R.; Schlemper, H. *AG(E)* **1987**, *26*, 1167.
2. Schwesinger, R. *Nachr. Chem. Tech. Lab.* **1990**, *38*, 1214 (*CA* **1991**, *114*, 23 099a).
3. Pietzonka, T.; Seebach, D. *AG(E)* **1993**, *32*, 716.
4. Pietzonka, T.; Seebach, D. *CB* **1990**, *124*, 1837.
5. The relative pK_a value of the conjugate cation in MeCN, based on 9-phenylfluorene = 18.49; values beyond 22 are extrapolated from a THF scale.
6. Schwesinger, R. *C* **1985**, *39*, 269.
7. Schwesinger, R.; Hasenfratz, C.; Schlemper, H.; Walz, L.; Peters, E.-M.; Peters, K.; von Schnering, H. G. *AG(E)* **1993**, *32*, 1361.
8. Schwesinger, R.; Hasenfratz, C. Unpublished results.
9. Seebach, D.; Aebi, J. D.; Gander-Coquoz, M.; Naef, R. *H* **1987**, *70*, 1194.
10. Schwesinger, R.; Willaredt, J.; Schlemper, H.; Keller, M.; Schmitt, D.; Fritz, H. *CB* **1994**, *127*, 2435.
11. Pietzonka, T.; Seebach, D. *AG(E)* **1992**, *31*, 1481.
12. Posner, G. H.; Shulman-Roskes, E. M. *JOC* **1989**, *54*, 3514.
13. Fletschinger, M.; Zipperer, B.; Fritz, H.; Prinzbach, H. *TL* **1987**, *28*, 2517; Gais, H.-J.; Vollhardt, J.; Krüger, C. *AG(E)* **1988**, *27*, 1108; Braun, J.; Hasenfratz, C.; Schwesinger, R.; Limbach, H.-H. *AG(E)* **1994**, *33*, 2215.
14. Netscher, T.; Schwesinger, R.; Trupp, B.; Prinzbach, H. *TL* **1987**, *28*, 2115; Sproat, B. S.; Beijer, B.; Iribarren, A. *Nucleic Acid Res.* **1990**, *18*, 41.
15. Schubert, J.; Schwesinger, R.; Knothe, L.; Prinzbach, H. *LA* **1986**, 2009; Kühlmeyer, R.; Seitz, B.; Weller, T.; Fritz, H.; Schwesinger, R.; Prinzbach, H. *CB* **1989**, *122*, 1729; Falk-Heppner, M.; Keller, M.; Prinzbach, H. *AG(E)* **1989**, *28*, 1253.
16. Prinzbach, H.; Lutz, G. Unpublished results.
17. Montforts, F. P.; Schwartz, U. M. *AG(E)* **1985**, *24*, 775.
18. Schwesinger, R. Unpublished results.
19. Schwesinger, R.; Willaredt, J. Unpublished results.
20. Schwesinger, R.; Schlemper, H. Unpublished results.

Reinhard Schwesinger
University of Freiburg in Breisgau, Germany

Phosphoric Acid

H_3PO_4

[7664-38-2] H_3O_4P (MW 98.00)

(acid catalyst; dehydrating agent; phosphorylating agent[1])

Physical Data: mp 41 °C; bp 158 °C.
Solubility: sol water, formic acid, acetic acid.
Form Supplied in: commercially available; low melting white solid or colorless liquid.
Analysis of Reagent Purity: titration.
Preparative Method: anhydrous H_3PO_4 can be prepared by dissolving P_2O_5 in 85% H_3PO_4.
Purification: dry at 150 °C. At temperatures above 200 °C, monomeric H_3PO_4 changes to oligomeric metaphosphoric acid.
Handling, Storage, and Precautions: corrosive.

Introduction. Phosphoric acid is a strong, nonoxidizing acid that is available in a number of different forms. Anhydrous, monomeric H_3PO_4 (orthophosphoric acid) is a low melting solid that can be purchased commercially or prepared by adding P_2O_5 to commercial 85% aqueous H_3PO_4. This combination has been used to iodinate alcohols, cleave ethers, and hydroiodinate alkenes, in addition to a variety of uses listed below.[2] Metaphosphoric acid *[37267-86-0]* is an oligomeric form, also commercially available.[3]

Hydrolysis of Imines, Amides, and Nitriles. The hydrolysis of imines,[4] amides, and nitriles to carboxylic acids (eq 1) using H_3PO_4 is a time-honored technique.[5] There are numerous methods for accomplishing this transformation.[6]

$$\text{Ph-CN} \xrightarrow[\text{70–90\%}]{\text{100\% H}_3\text{PO}_4,\ 155\ ^\circ\text{C, 4 h}} \text{Ph-CO}_2\text{H} \quad (1)$$

Preparation of Phosphates.[1] Phosphate mono- and dialkyl esters have been prepared from phosphoric acid (see also ***Phosphorus Oxychloride***). Treatment of (**1**) with H_3PO_4 at rt provides the corresponding phosphate in good yield (eq 2).[7] At high temperatures, diesterification can occur (eq 3).[8]

(eq 2): (**1**) (NH₂•HCl, HO, OH, N, HCl) → 100% H_3PO_4, 25 °C, 70% → phosphate (NH₂•HCl, HO, O–P(O)(OH)₂, N, HCl)

(eq 3): HO-(tricyclic alcohol) → H_3PO_4, K_2CO_3, xylene, 195 °C, 76% → O=P(OH)[O-(tricyclic)]₂

An effective, large-scale synthesis of diammonium acetylphosphate has been developed, and involves direct acylation of H_3PO_4. Either ketene[9] or acetic anhydride[10] (eq 4) can be used in this process.

(eq 4): Ac_2O → 1. H_3PO_4, EtOAc, 0 °C; 2. NH_3, MeOH, −30 °C; 86% → $CH_3C(O)OP(O)(ONH_4)_2$

Acid Cyclization Catalyst. A common use for H_3PO_4, as a solution in water, a liquid acid, or in anhydrous form, is as an acid cyclization catalyst. Cyclization of cross-conjugated ketones in H_3PO_4/formic acid leads to 2,3-dialkylcyclopentenones (eq 5) rather than the 3,4-dialkyl products expected of the Nazarov cyclization.[11] This result is obtained when either the ketone or the corresponding ethylene acetal is used as starting material. Similar results are obtained using HBr/HOAc, although, in these examples, some of the 3,4-dialkyl products are obtained as well.

H_3PO_4, HCO_2H, 90 °C, 3 h, 77% (5)

An intermolecular example of an alkene-cation addition reaction provides access to chromenes through isoprenylation of a phenol (eq 6).[12]

H_3PO_4, xylene, 78% (6)

Cyclization of (**2**) in phosphoric acid results in an azabicyclononane (eq 7).[13]

85% H_3PO_4, 73% (7)

(**2**)

Phosphoric acid-mediated condensation of indole derivatives with ketones provides access to 3-substituted indoles (eq 8).[14]

2N H_3PO_4, 88% (8)

Cyclization of enol acetates (eq 9), δ,ϵ-unsaturated aldehydes,[15] and ketene dithioacetals (eq 10) can also be accomplished by heating in H_3PO_4. The cyclization of (**3**) results in dithiopyridines by way of an intramolecular Ritter reaction followed by a 1,3-methylthio shift. Lewis acids, ***Boron Trifluoride Etherate*** in particular, lead to simple dehydration.[16]

85% aq. H_3PO_4, toluene, 100 °C, 80% (9)

trans:*cis* = 8:1

88% H_3PO_4, 130 °C, 3 h, 82% (10)

(**3**)

Debenzylation. Treatment of (**4**), an *N*-benzyl derivative of biotin, with anhydrous H_3PO_4 and phenol at elevated temperature leads to debenzylation (eq 11).[17] This reaction provides a nonreductive route to deprotection of the urea while not attacking the unsaturated ester.

H_3PO_4, phenol, 150 °C, 3 h, 53% (11)

(**4**)

Related Reagents. Phosphorus(V) Oxide–Phosphoric Acid; Phosphorus Oxychloride; Polyphosphoric Acid; Sulfuric Acid.

1. *MOC* **1982**, *E2*, 491.
2. *FF* **1967**, *1*, 872.
3. *Kirk-Othmer Encyclopedia of Chemical Technology*; Wiley: New York, 1978; Vol. 17, pp 428, 448–452.
4. Mislow, K.; McGinn, F. A. *JACS* **1958**, *80*, 6036.
5. Berger, G.; Olivier, S. C. J. *RTC* **1927**, *46*, 600.
6. Haslam, E. *T* **1980**, *36*, 2409.
7. Wilson, A. N.; Harris, S. A. *JACS* **1951**, *73*, 4693.
8. Inamoto, Y.; Aigami, K.; Kadono, T.; Nakayama, H.; Takatsuki, A.; Tamura, G. *JMC* **1977**, *20*, 1371.
9. Whitesides, G. M.; Siegel, M.; Garrett, P. *JOC* **1975**, *40*, 2516.
10. Lewis, J. M.; Haynie, S. L.; Whitesides, G. M. *JOC* **1979**, *44*, 864.
11. Hirano, S.; Hiyama, T.; Nozaki, H. *TL* **1974**, 1429.
12. Ahluwalia, V. K.; Arora, K. K. *T* **1981**, *37*, 1437.
13. Beretta, M. G.; Rindone, B.; Scolastico, C. *S* **1975**, 440.
14. Freter, K. *JOC* **1975**, *40*, 2525.
15. Saucy, G.; Ireland, R. E.; Bordner, J.; Dickerson, R. E. *JOC* **1971**, *36*, 1195.
16. Gupta, A. K.; Ila, H.; Junjappa, H. *TL* **1988**, *29*, 6633.
17. Field, G. F.; Zally, W. J.; Sternbach, L. H.; Blount, J. F. *JOC* **1976**, *41*, 3853.

Mark S. Meier
University of Kentucky, Lexington, KY, USA

Phosphorus(V) Oxide–Methanesulfonic Acid[1]

P_2O_5–$MeSO_3H$

[39394-84-8]	$CH_4O_8P_2S$	(MW 238.06)
(P_2O_5)		
[1314-56-3]	O_5P_2	(MW 141.94)
($MeSO_3H$)		
[75-75-2]	CH_4O_3S	(MW 96.12)

(acidic dehydrating agent used in cycloalkenone synthesis,[1] Friedel–Crafts reactions,[1] the Fischer indole synthesis,[2] the

Beckmann rearrangement,[1] and other dehydrations; an alternative to polyphosphoric acid[1])

Alternate Name: Eaton's reagent.

Physical Data: 7.5 wt% solution: bp 122 °C/1 mmHg; *d* 1.500 g cm^{-3}.

Solubility: sol ether, alcohol, MeCN, CH_2Cl_2; insol toluene, hexane.[2,3]

Form Supplied in: 7.5 wt% solution is commercially available.

Preparative Method: prepared[1] by adding ***Phosphorus(V) Oxide*** (P_2O_5, 36 g) in one portion to ***Methanesulfonic Acid*** (360 g) and stirring at rt[3] until the P_2O_5 dissolves.[4] Although Eaton recommends the use of freshly distilled methanesulfonic acid to allow for a clean workup and good yields,[1] others report using the acid as purchased.[5,6]

Handling, Storage, and Precautions: Eaton's reagent is toxic and corrosive. Direct contact with this reagent should be avoided. The solution begins to yellow upon standing for long periods of time; however, this does not appear to affect the viability of the reagent.[1] Use in a fume hood.

Reagent Description. The reagent was conceived as an alternative to the widely used, but often inconvenient, ***Polyphosphoric Acid*** (PPA) (see also ***Polyphosphate Ester***, PPE).[1] Eaton's reagent successfully addresses the drawbacks of PPA's physical properties. It is much less viscous, and is, therefore, easier to stir. Organic compounds are generally soluble in Eaton's reagent, and the hydrolytic workup is less tedious.[1] Reactions are run at ambient or slightly elevated temperatures. Standard aqueous workup is easy and clean. Eaton recommends quenching the reaction with water; quenching in ice may cause methanesulfonic anhydride to precipitate and be extracted into the organic layer; quenching in aqueous base may cause extensive foaming.[1] In addition to its ease of handling, yields obtained with Eaton's reagent compare favorably with those obtained with PPA.[7] Few modifications of Eaton's original procedure have appeared. A 1:5 by weight ratio has been reported to be as effective as a 1:10 ratio.[8] It has been noted that, to avoid polymer formation, only the minimum amount of reagent needed to effect condensation should be used.[9] The nature of the reagent has not been rigorously determined. It appears that the reactive or catalytic species may vary by reaction. In certain acid-catalyzed reactions, P_2O_5 has been found to be superfluous.[10]

Cycloalkenones. $P_2O_5/MeSO_3H$ is used as a reagent in several reactions leading to cycloalkenones. First described in Eaton's original paper, the lactone-to-cyclopentenone rearrangement (eq 1)[1] has since found wide use.[11]

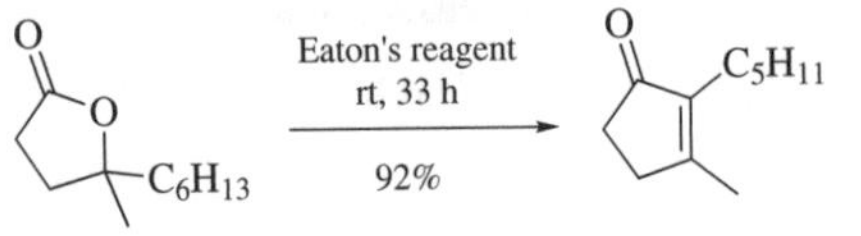

Eaton's reagent, rt, 33 h; 92% (1)

In a related reaction, readily available nitroalkanoic acids cyclize to form cyclopentenones (eq 2).[12] As illustrated in eq 3,[13a] vinylcyclobutanones undergo acyl migration to produce either cyclopentenones or cyclohexenones.[13]

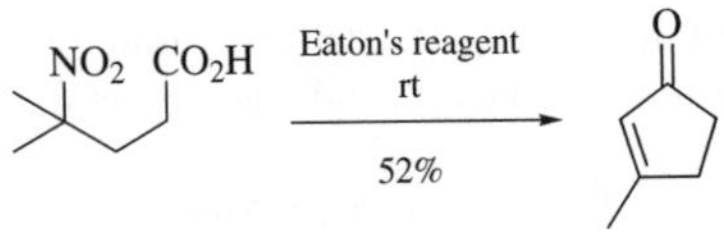

Eaton's reagent, rt; 52% (2)

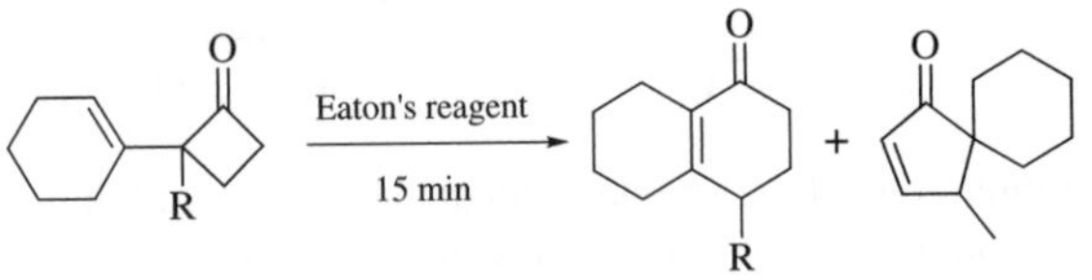

Eaton's reagent, 15 min (3)

R = H	65%	0%
R = Me	13%	51%

Vinylcyclopentenones have undergone the Nazarov cyclization in good yield in the presence of Eaton's reagent (eq 4).[14] However, other reagents may be more generally useful, since there are reports of Eaton's reagent not providing optimal results in this reaction.[15]

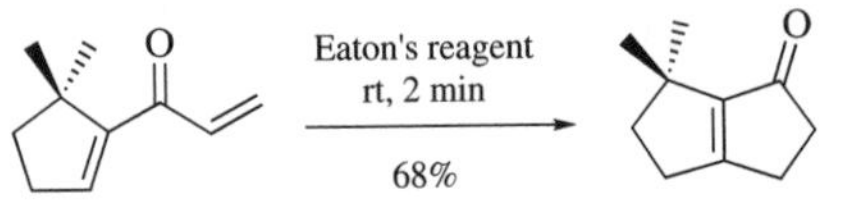

Eaton's reagent, rt, 2 min; 68% (4)

Friedel–Crafts Acylations of Aromatic Rings. Eaton's reagent has been used widely and very effectively[16a] to catalyze Friedel–Crafts acylations.[16] One of the few potential drawbacks is the deprotection of an aryl ether. The examples shown below compare the utility of Eaton's reagent vs. PPA with regard to this deprotection problem. In eq 5, although the cyclization proceeds with an undesired protecting group exchange, the cyclization fails in PPA.[17] Deprotection of an aryl methyl ether is avoided by using Eaton's reagent in place of PPA in one case (eq 6),[18] but not in another.[19] Intramolecular Friedel–Crafts acylations have been observed to occur without the addition of P_2O_5.[10] A comparative study found this observation to be generally applicable to intramolecular acylations, but not intermolecular acylations.[10b,20]

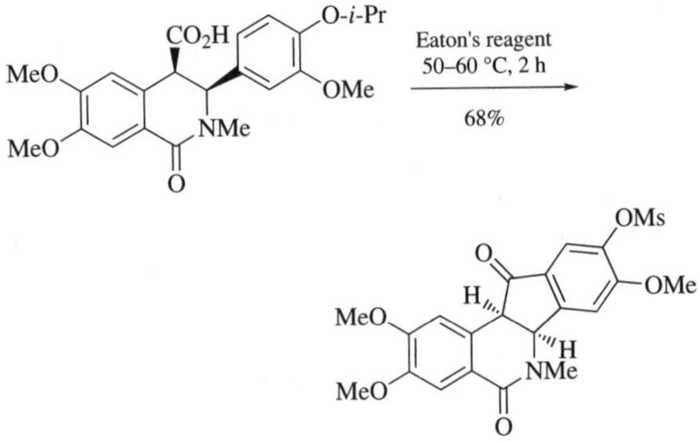

Eaton's reagent, 50–60 °C, 2 h; 68% (5)

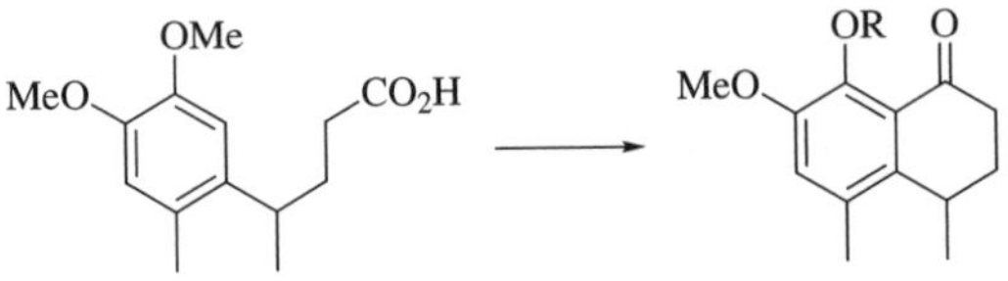

(6)

Eaton's reagent	R = Me, 70%
PPA	R = H, 28%

Friedel–Crafts Alkylations. $P_2O_5/MeSO_3H$ compares favorably with other reagents in the Friedel–Crafts alkylation reaction.[21] Mechanistic aspects of this reaction have been discussed.[21b] eq 7 shows an alkene-initiated alkylation that provides (+)-*O*-methylpodocarpate selectively.[22]

Eaton's reagent, rt, 15 min, 92% (7)

Dehydration. Alcohols have been dehydrated to alkenes with Eaton's reagent (eq 8).[23] In a formal dehydration, a cyclopentenone has been transformed into a diene (eq 9).[24]

Eaton's reagent, 25 °C, 1 h (8)

91:9

Eaton's reagent, 60 °C, 6 h, 30–50% (9)

Fischer Indole Synthesis. The use of Eaton's reagent as the acid catalyst in the Fischer indole reaction results in unprecedented regiocontrol favoring 2-substituted indoles (eq 10).[2] In cases where the harshness of the reagent results in low yields of indoles, dilution of the reaction mixture in sulfolane or CH_2Cl_2 attenuates the problem. Mechanistic studies indicate that the catalytic species, in this reaction, is $MeSO_3H$. The role of P_2O_5 is to act as a drying agent. Further experiments indicate that for the Friedel–Crafts acylation this is not the case; a mixed anhydride is the catalytic species.[2]

NHNH₂ + ... H^+ (10)

AcOH	100: 1	
PPA	50:50	
Eaton's reagent	78:22	95%

Heterocycle Preparation. Various heterocycles have been prepared through $P_2O_5/MeSO_3H$-mediated cyclizations. Condensation, and subsequent dehydration, of aminothiophenol and the appropriate acid provides benzothiazoles (eq 11).[25]

NH_2, SH + AcOH — Eaton's reagent, 25 °C, 1 h, 70 °C, 10 h, 93% (11)

Oxadiazoles can be prepared from diacylhydrazines (eq 12).[26] Furans are formed from the cyclodehydration of a phenolic ketone (eq 13).[27]

Eaton's reagent, 75 °C, 4 h, 84% (12)

Eaton's reagent, rt, 1.5 h, 94% (13)

Butenolides have been prepared by cyclization of keto esters (eq 14)[28] or by elimination of H_2O from a preformed hydroxy butenolide (eq 15).[29]

Eaton's reagent, 50 °C, 5.2 h, 74% (14)

Eaton's reagent, rt, 1.5 h, 95% (15)

Yields in the synthesis of thiadiazolo[3,2-*a*]pyrimidin-5-ones have been greatly improved by using Eaton's reagent in place of PPA (eq 16).[8b]

(16)

P_2O_5, $MeSO_3H$; 1:5	87%
PPA	18%

Eaton's reagent is superior to PPA in the addition of an amide across a double bond (eq 17).[30] In another synthesis of lactams, 3-alkenamides reacted stereoselectively with benzaldehyde to provide lactams containing three contiguous stereogenic centers (eq 18).[31]

Eaton's reagent, 80–90 °C, 6 h, 60% (17)

PhCHO + [H$_2$N–C(=O)–CH$_2$–CH=CH–CH$_3$] —(Eaton's reagent, 35 °C, 18 h, 63%)→ [MsO–CH(CH$_3$)-substituted 5-phenylpyrrolidin-2-one] (18)

Beckmann Rearrangement. Eaton's disclosure of $P_2O_5/MeSO_3H$ as an alternative to PPA compared the two reagents' ability to effect the Beckmann rearrangement.[1] Eaton's reagent has been reported to be superior to other reagents at inducing stereospecific rearrangement of the (*E*)- and (*Z*)-oximes of phenylacetone.[4] However, this is not a general finding. Rearrangement of the oxime in eq 19 does not provide the product expected from an *anti*-migration process.[32]

[oxime, HO–N] —(Eaton's reagent, 100 °C, 1 h, 72%)→ [lactam, HN, O] (19)

Related Reagents. Camphorsulfonic Acid; Methanesulfonic Acid; Phosphoric Acid; Phosphorus pentoxide; Polyphosphate Ester; Polyphosphoric Acid.

1. Eaton, P. E.; Carlson, G. R.; Lee, J. T. *JOC* **1973**, *38*, 4071.
2. This paper reports that in the preparation of 2-substituted indoles Eaton's reagent is superior to PPA, H_2SO_4, and PPSE: Zhao, D.; Hughes, D. L.; Bender, D. R.; DeMarco, A. M.; Reider, P. J. *JOC* **1991**, *56*, 3001.
3. These authors report the presence of a finely-divided solid after stirring for 6 h. It was removed by filtration under nitrogen.
4. Alternatively, the solution may be heated during dissolution of the P_2O_5. See: Stradling, S. S.; Hornick, D.; Lee, J.; Riley, J. *J. Chem. Educ.* **1983**, *60*, 502.
5. Akhtar, S. R.; Crivello, J. V.; Lee, J. L. *JOC* **1990**, *55*, 4222.
6. Corey, E. J.; Boger, D. L. *TL* **1978**, 5.
7. Examples of exceptions: (a) PPA is superior to either Eaton's reagent or H_2SO_4 in a Friedel–Crafts acylation: Hormi, O. E. O.; Moisio, M. R.; Sund, B. C. *JOC* **1987**, *52*, 5272. (b) PPA is superior to Eaton's reagent, $BF_3{\cdot}OEt_2$, HCO_2H, $ZnCl_2$, CF_3CO_2H, *p*-TsOH, and H_2SO_4 in a Friedel–Crafts alkylation: Maskill, H. *JCS(P1)* **1987**, 1739. (c) PPA is superior to Eaton's reagent or $CF_3CO_2H/(CF_3CO)_2O/BF_3{\cdot}OEt_2$ in a Friedel–Crafts acylation: Hands, D.; Marley, H.; Skittrall, S. J.; Wright, S. H. B.; Verhoeven, T. R. *JHC* **1986**, *23*, 1333. (d) PPA is superior to Eaton's reagent in a Friedel–Crafts acylation: Bosch, J.; Rubiralta, M.; Domingo, A.; Bolos, J.; Linares, A.; Minguillon, C.; Amat, M.; Bonjoch, J. *JOC* **1985**, *50*, 1516. (e) PPA is superior to Eaton's reagent or PPE in a Friedel–Crafts acylation: Jilek, J.; Holubek, J.; Svatek, E.; Schlanger, J.; Pomykacek, J.; Protiva, M. *CCC* **1985**, *50*, 519. (f) In a Friedel–Crafts acylation, where PPA or Eaton's reagent fails to give satisfactory results, the corresponding acid chloride is cyclized using $AlCl_3$: Barco, A.; Benetti, S.; Pollini, G. P. *OPP* **1976**, *8*, 7.
8. (a) Eaton, P. E.; Mueller, R. H.; Carlson, G. R.; Cullison, D. A.; Cooper, G. F.; Chou, T.-C.; Krebs, E.-P. *JACS* **1977**, *99*, 2751. (b) Tsuji, T.; Takenaka, K. *BCJ* **1982**, *55*, 637.
9. Parish, W. W.; Stott, P. E.; McCausland, C. W.; Bradshaw, J. S. *JOC* **1978**, *43*, 4577.
10. (a) Leon, A.; Daub, G.; Silverman, I. R. *JOC* **1984**, *49*, 4544. (b) Premasagar, V.; Palaniswamy, V. A.; Eisenbraun, E. J. *JOC* **1981**, *46*, 2974.
11. (a) Jacobson, R. M.; Lahm, G. P.; Clader, J. W. *JOC* **1980**, *45*, 395. (b) Inouye, Y.; Fukaya, C.; Kakisawa, H. *BCJ* **1981**, *54*, 1117. (c) Murthy, Y. V. S.; Pillai, C. N. *T* **1992**, *48*, 5331. (d) Eaton, P. E.; Srikrishna, A.; Uggeri, F. *JOC* **1984**, *49*, 1728. (e) Pohmakotr, M.; Reutrakul, V.; Phongpradit, T.; Chansri, A. *CL* **1982**, 687. (f) Baldwin, J. E.; Beckwith, P. L. M. *CC* **1983**, 279. (g) Mundy, B. P.; Wilkening, D.; Lipkowitz, K. B. *JOC* **1985**, *50*, 5727. (h) Mehta, G.; Karra, S. R. *TL* **1991**, *32*, 3215. (i) Ho, T.-L.; Yeh, W.-L; Yule, J.; Liu, H.-J. *CJC* **1992**, *70*, 1375.
12. Ho, T.-L. *CC* **1980**, 1149.
13. (a) Matz, J. R.; Cohen, T. *TL* **1981**, *22*, 2459. (b) For a related ring expansion of 1-alkenylcyclopropanols to cyclopentenones, see: Barnier, J.-P.; Karkour, B.; Salaun, J. *CC* **1985**, 1270.
14. Paquette, L. A.; Stevens, K. E. *CJC* **1984**, *62*, 2415.
15. (a) This paper reports obtaining Nazarov cyclization products in 8–10% yield with either Eaton's reagent or $FeCl_3$. A silicon assisted Nazarov was also explored: Cheney, D. L.; Paquette, L. A. *JOC* **1989**, *54*, 3334. (b) PPA is superior to Eaton's reagent or methanesulfonic acid in effecting cyclization of 1,1′-dicyclopentenyl ketone: Eaton, P. E.; Giordano, C.; Schloemer, G.; Vogel, U. *JOC* **1976**, *41*, 2238. (c) Many other reagents including HCO_2H/H_3PO_4, HCl, H_2SO_4, $SnCl_4$, and TsOH have been used in this type of Nazarov cyclization. For a review of the Nazarov cyclization, see: Santelli-Rouvier, C.; Santelli, M. *S* **1983**, 429.
16. Examples: (a) McGarry, L. W.; Detty, M. R. *JOC* **1990**, *55*, 4349. (b) Grunewald, G. L.; Sall, D. J.; Monn, J. A. *JMC* **1988**, *31*, 433. (c) Russell, R. K.; Rampulla, R. A.; van Nievelt, C. E.; Klaubert, D. H. *JHC* **1990**, *27*, 1761. (d) Ye, Q.; Grunewald, G. L. *JMC* **1989**, *32*, 478. (e) Kelly, T. R.; Ghoshal, M. *JACS* **1985**, *107*, 3879. (f) Eck, G.; Julia, M.; Pfeiffer, B.; Rolando, C. *TL* **1985**, *26*, 4723. (g) Kitazawa, S.; Kimura, K.; Yano, H.; Shono, T. *JACS* **1984**, *106*, 6978. (h) Stott, P. E.; Bradshaw, J. S.; Parish, W. W.; Copper, J. W. *JOC* **1980**, *45*, 4716. (i) Cushman, M.; Abbaspour, A.; Gupta, Y. P. *JACS* **1983**, *105*, 2873. (j) Acton, D.; Hill, G.; Tait, B. S. *JMC* **1983**, *26*, 1131. (k) Miller, S. J.; Proctor, G. R.; Scopes, D. I. C. *JCS(P1)* **1982**, 2927.
17. Cushman, M.; Mohan, P. *JMC* **1985**, *28*, 1031.
18. Inouye, Y.; Uchida, Y.; Kakisawa, H. *BCJ* **1977**, *50*, 961.
19. Falling, S. N.; Rapoport, H. *JOC* **1980**, *45*, 1260.
20. For an example of an intermolecular acylation of cyclohexenone, see: Cargill, R. L.; Jackson, T. E. *JOC* **1973**, *38*, 2125.
21. (a) Fox, J. L.; Chen, C. H.; Stenberg, J. F. *OPP* **1985**, *17*, 169. (b) Davis, B. R.; Hinds, M. G.; Johnson, S. J. *AJC* **1985**, *38*, 1815.
22. Hao, X.-J.; Node, M.; Fuji, K. *JCS(P1)* **1992**, 1505.
23. Ziegler, F. E.; Fang, J.-M.; Tam, C. C. *JACS* **1982**, *104*, 7174.
24. Scott, L. T.; Minton, M. A.; Kirms, M. A. *JACS* **1980**, *102*, 6311.
25. Boger, D. L. *JOC* **1978**, *43*, 2296.
26. Rigo, B.; Couturier, D. *JHC* **1986**, *23*, 253.
27. Cambie, R. C.; Howe, T. A.; Pausler, M. G.; Rutledge, P. S.; Woodgate, P. D. *AJC* **1987**, *40*, 1063.
28. Schultz, A. G.; Yee, Y. K. *JOC* **1976**, *41*, 561.
29. Schultz, A. G.; Godfrey, J. D. *JACS* **1980**, *102*, 2414.
30. Tilley, J. W.; Clader, J. W.; Wirkus, M.; Blount, J. F. *JOC* **1985**, *50*, 2220.
31. Marson, C. M.; Grabowska, U.; Walsgrove, T.; Eggleston, D. S.; Baures, P. W. *JOC* **1991**, *56*, 2603.
32. Jeffs, P. W.; Molina, G.; Cortese, N. A.; Hauck, P. R.; Wolfram, J. *JOC* **1982**, *47*, 3876.

Lisa A. Dixon
The R. W. Johnson Pharmaceutical Research Institute, Raritan, NJ, USA

Polyphosphate Ester

[–] $C_8H_{20}O_{12}P_4$ (MW 432.14)

(dehydrating agent; used in the Bischler–Napieralski reaction;[1] Friedel–Crafts acylation;[2] conversion of carboxylic acids to esters,[3] thiol esters,[4] amides,[5,6] and nitriles;[6,7] Beckmann rearrangement;[7] and heterocycle preparation; an alternative to PPA in dehydration reactions)

Alternate Name: PPE.
Physical Data: d 1.463 g cm^{-3}; n 1.440.[8]
Solubility: sol $CHCl_3$; reacts vigorously with H_2O.
Form Supplied in: not commercially available. May be prepared as described below. It is a colorless to yellowish substance which forms a stiff gel below 0 °C.
Analysis of Reagent Purity: PPE (3% solution in $CHCl_3$) shows a characteristic band in its IR spectrum at 1330 cm^{-1}. This band disappears and a new, broad band appears at 1200–1260 cm^{-1} when PPE is treated with a trace amount of H_2O.[5] ^{1}H and ^{31}P NMR techniques may be employed to determine composition of the mixture and structural features of each component.[9] Elemental analysis is not reliable.[9]
Preparative Methods: was originally prepared by Langheld,[12] and has been referred to as 'Langheld esters'.[9,10] The preparation reported by Cava et al.[13] is a compilation of three previous procedures.[8,14,15] ***Phosphorus(V) Oxide*** (150 g) is added to a solution of anhydrous ether (300 mL) and alcohol-free ***Chloroform*** (150 mL). The reaction mixture is refluxed under dry nitrogen for 4 d and the resulting clear solution decanted from a small amount of residue. The solution is concentrated to a colorless syrup in a rotary evaporator; residual traces of solvent are removed by heating the syrup for 36 h at 40 °C in vacuo.[13] More harsh conditions result in unusable material.[11] While this preparation is sufficient for most synthetic applications, Van Wazer et al. reported that PPE prepared from triethyl orthophosphate and phosphorus pentoxide was superior for biochemical studies.[9]
Handling, Storage, and Precautions: Usually prepared fresh, but may be stored for at least 1 month.[4] PPE is unstable above 110–120 °C. A gradual decomposition occurs which becomes vigorous with evolution of gas at 150–160 °C.[11] Direct contact with PPE should be avoided. Use in a fume hood.

Reagent Description. PPE is another of the phosphate-based reagents that are related to ***Polyphosphoric Acid*** (PPA) (see also ***Phosphorus(V) Oxide–Methanesulfonic Acid***). It differs from PPA in that it is aprotic and soluble in organic media. It is often compared to the related polyphosphoric acid trimethylsilyl ester (PPSE). In general, the advantages of PPE include its solubility in organic solvents, the mild conditions under which it is used, and its relatively nonhazardous, nonnoxious nature. A disadvantage is its time-consuming preparation.[16] The reagent is composed of a mixture of polymeric phosphoric acid esters.[8–10] Variations in the reactants and the reaction conditions may result in differences in the reagent's composition.[5,9,10] It is presumed to activate carboxylic acids through formation of mixed anhydrides.[17]

Bischler–Napieralski Reaction[1,18]. One of the most cited uses of PPE is as a reagent in the Bischler–Napieralski reaction (eq 1).[18a] The utility of PPE compares favorably to other reagents, such as ***Phosphorus Oxychloride***, ***Phosphorus(V) Oxide***, and PPA.[18a,b,h,i]

NHCOPh → N, N H, Ph (1)

PPE, 68%; P_2O_5, 36%

Friedel–Crafts Acylation.[2,19] Intramolecular Friedel–Crafts acylations have been carried out by treatment of aromatic acids with PPE (eq 2).[19a] Depending on the substrate, PPE may be as effective[20] or more effective[21] than other reagents.

CO_2H —PPE, 20 °C, 2 h, 92%→ (2)

Carboxylic Acid Derivatives. Esters,[3] lactones,[22] thiol esters,[4] thiolactones,[23] amides,[5] and acylureas[24] have all been prepared from a PPE-mediated reaction of a carboxylic acid and the appropriate coupling partner. There are many excellent methods for effecting the above transformations;[25] PPE is distinguished from these mainly by its low cost. eq 3 describes the synthesis of phenyl esters,[3a] which were shown to be stable in PPE. This is in contrast to PPA, which can effect the Fries rearrangement of phenyl esters.[3a] Aryl and alkyl thiols have been condensed with alkyl, alkenyl, and aryl carboxylic acids,[4] and malonic acid.[26] Hindered thiol esters can be prepared by this method (eq 4).[4] Sensitive substrates such as penicillin G could be reacted at low temperature with the addition of pyridine.[4]

$$PhCO_2H + PhOH \xrightarrow[90\%]{\text{PPE, rt, 24 h}} PhCO_2Ph \quad (3)$$

2,6-dichlorobenzoic acid (CO_2H, Cl, Cl) + t-BuSH —PPE, rt, 15 h, 85%→ COS-t-Bu (Cl, Cl) (4)

Conversion of Carboxylic Acids and Amides to Nitriles. PPE has been used to dehydrate amides to nitriles.[27] More recently, a one-pot conversion of carboxylic acids to nitriles has been reported.[6] Treatment of an acid with PPE under an atmosphere of ammonia results in the formation of an intermediate amide, which upon further treatment with PPE undergoes dehydration to provide the nitrile (eq 5).[6] The amide may be obtained upon quenching the reaction after the initial condensation. Advantages of this method include use of a nonnoxious reagent, and reaction conditions that are relatively mild.[5,6,16] In an example of this reaction in a more complex substrate, this methodology has been applied to a synthesis of the canthine alkaloid skeleton.[28]

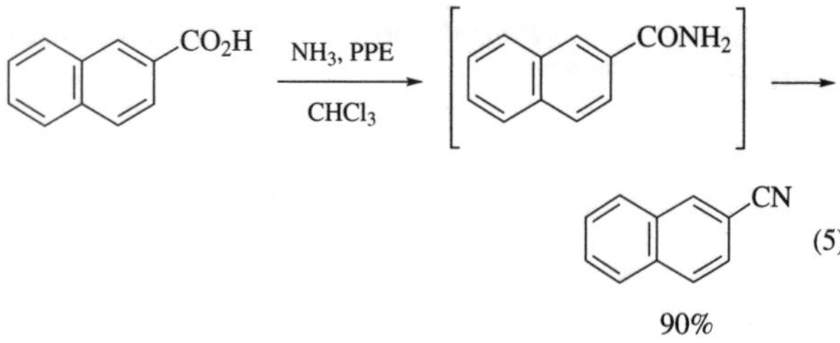

(5)

Beckmann Rearrangement. Upon treatment with PPA, ketoximes have undergone the Beckmann rearrangement (eq 6).[7] Under more forcing conditions, ketoximes provide amidines (eq 6). Submission of the oxime of benzaldehyde to similar conditions failed to produce benzamide; the only product obtained was benzonitrile. Treatment of benzamide under the same conditions did not lead to the nitrile. This suggests that dehydration of the aldoxime did not proceed through the intermediacy of the amide.[7] PPE was found to be equivalent to PPSE in terms of yields obtained in the Beckmann rearrangement of oximes.[29]

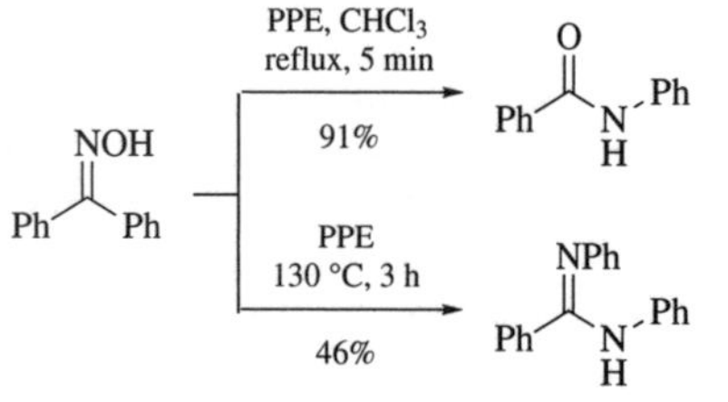

(6)

Fischer Indole Synthesis. PPE has been utilized in the Fischer indole synthesis. Yields obtained ranged from 21 to 86% (eq 7).[30] A side product was the C-3 alkylated indole (see the section on ethylation).[30]

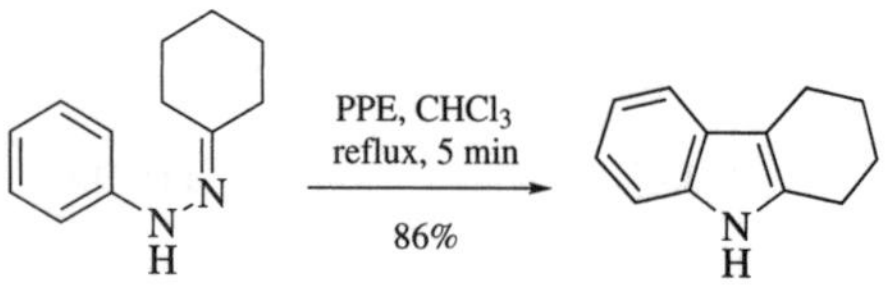

(7)

Heterocycle Preparation. Various heterocycles have been produced through PPE-mediated cyclization reactions. Benzimidazoles (eq 8),[14] benzoxazoles,[31] and benzothiazoles[31] have been prepared through condensation of an *ortho*-substituted aniline and an acid (eq 8).[14]

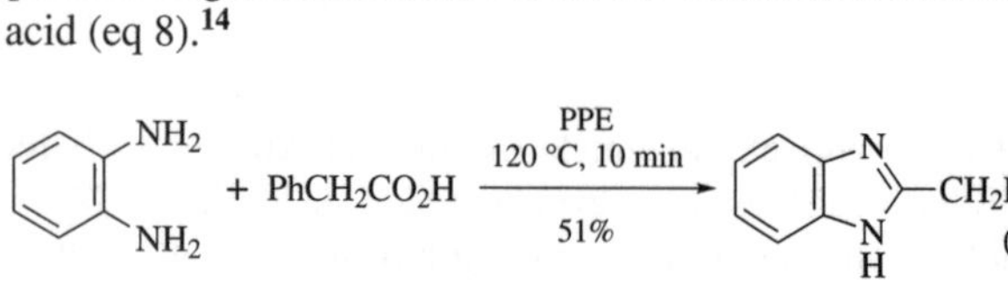

(8)

In related work, it was reported that PPE was utilized in the synthesis of benzothiazoles.[32] However, in the preparation of benzimidazoles, 6 N HCl was reported to be the preferred reagent; in the preparation of benzoxazoles, PPA was a better reagent.[32a] Other heterocycles prepared using PPE as a reagent include pyrrolo[2,3-*b*]pyridines,[33] pyrrolo[3,4-*b*]pyridines,[34] pyrrolo[3,2-*b*]pyridines,[35] benzimidazo[1,2-*c*]indazolo[2,3-*a*]-quinazolines,[36] 1,3-diazepines,[37] 1,3-diazocines,[37] 1,4-dihydro-4-oxoquinolines,[38] and 1,3-thiazin-4-ones.[39]

Miscellaneous Transformations. 5,6-Dihydro-2(1*H*)-pyridinones have been prepared stereoselectively by condensing 3-alkenamides with aryl aldehydes (eq 9).[40] PPA is also effective in this reaction; however, the yields are lower.

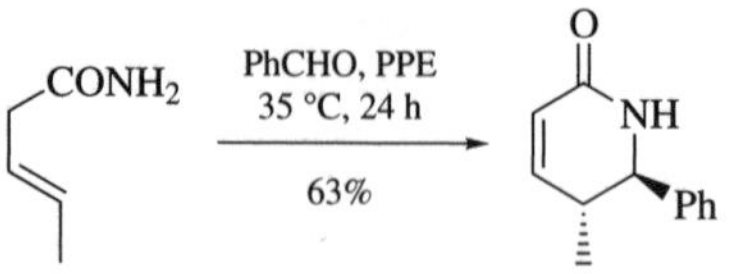

(9)

A one-carbon homologation of benzyl alcohols to amides or esters is shown in eq 10. Alkyl (and aryl) alcohols and secondary amines react with a cobalt intermediate to provide products with yields ranging from 20–82%.[41]

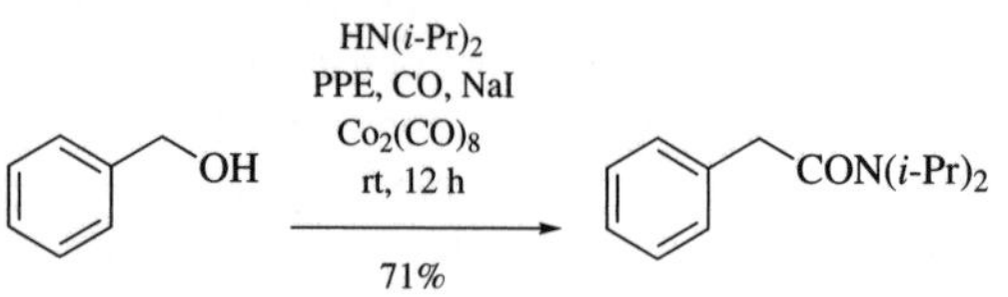

(10)

Ethylation. During a Fischer indole synthesis, PPE was observed to alkylate indoles.[30] Further studies revealed that at 160 °C the ethyl indolenine was obtained in moderate yield. The diethyl and *N*-ethyl compounds were obtained in minor amounts (eq 11).[11]

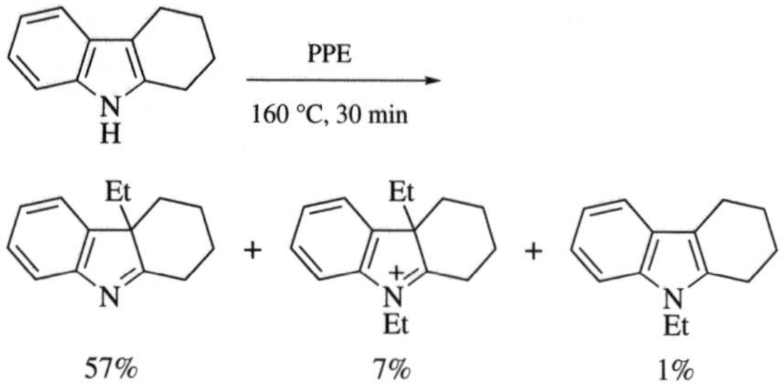

(11)

N-Alkylation of various amines has been reported. A tautomeric imidazole was alkylated in 74% yield (eq 12).[42] Yields for other methylation (polyphosphate methyl ester) or ethylation (polyphosphate ethyl ester) reactions varied from 44–74%.[42]

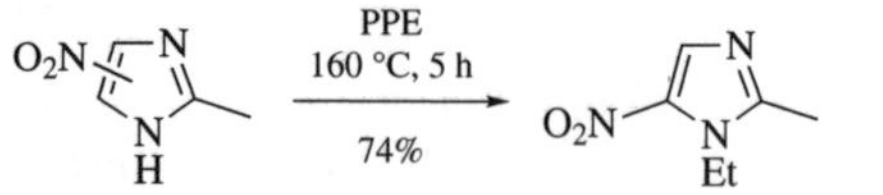

(12)

Related Reagents. Phosphoric Acid; Phosphorus(V) Oxide–Methanesulfonic Acid; Polyphosphoric Acid.

---

1. For a general discussion of the Bischler–Napieralski reaction, see: (a) Whaley, W. M.; Govindachari, T. R. *OR* **1951**, *6*, 74. (b) Fodor, G.; Nagubandi, S. *T* **1980**, *36*, 1279. (c) Kametani, T.; Fukumoto, K. In *The Chemistry of Heterocyclic Compounds*; Grethe, G., Ed.; Wiley: New York, 1981; Vol. 38, Part 1, pp 139–274.
2. For general references, see: March, J. *Advanced Organic Chemistry*, 3rd ed.; Wiley: New York, 1985; pp 484–487.

3. (a) Kanaoka, Y.; Tanizawa, K.; Sato, E.; Yonemitsu, O.; Ban, Y. *CPB* **1967**, *15*, 593. (b) El Seoud, O. A.; Pivetta, F.; El Seoud, M. I.; Farah, J. P. S.; Martins, A. *JOC* **1979**, *44*, 4832.
4. Imamoto, T.; Kodera, M.; Yokoyama, M. *S* **1982**, 134.
5. Kanaoka, Y.; Machida, M.; Yonemitsu, O.; Ban, Y. *CPB* **1965**, *13*, 1065.
6. Imamoto, T.; Takaoka, T.; Yokoyama, M. *S* **1983**, 142.
7. Kanaoka, Y.; Yonemitsu, O.; Sato, E.; Ban, Y. *CPB* **1968**, *16*, 280.
8. Pollmann, W.; Schramm, G. *BBA* **1964**, *80*, 1.
9. Van Wazer, J. R.; Norval, S. *JACS* **1966**, *88*, 4415.
10. Burkhardt, G.; Klein, M. P.; Calvin, M. *JACS* **1965**, *87*, 591.
11. Yonemitsu, O.; Miyashita, K.; Ban, Y.; Kanaoka, Y. *T* **1969**, *25*, 95.
12. Langheld, K. *CB* **1910**, *43*, 1857.
13. Cava, M. P.; Lakshmikantham, M. V.; Mitchell, M. J. *JOC* **1969**, *34*, 2665.
14. Kanaoka, Y.; Yonemitsu, O.; Tanizawa, K.; Ban, Y. *CPB* **1964**, *12*, 773.
15. Schramm, G.; Grotsch, H.; Pollmann, W. *AG(E)* **1962**, *1*, 1.
16. Yokoyama, M.; Yoshida, S.; Imamoto, T. *S* **1982**, 591.
17. Cheng, K.-F.; Wong, T.-T.; Wong, W.-T.; Lai, T.-F. *JCS(P1)* **1990**, 2487.
18. (a) Kanaoka, Y.; Sato, E.; Ban, Y. *CPB* **1967**, *15*, 101. (b) Doskotch, R. W.; Phillipson, J. D.; Ray, A. B.; Beal, J. L. *JOC* **1971**, *36*, 2409. (c) Matsuo, K.; Okumura, M.; Tanaka, K. *CPB* **1982**, *30*, 4170. (d) Matsuo, K.; Okumura, M.; Tanaka, K. *CL* **1982**, 1339. (e) Pandit, U. K.; Das, B.; Chatterjee, A. *T* **1987**, *43*, 4235. (f) Fujii, T.; Yamada, K.; Minami, S.; Yoshifuji, S.; Ohba, M. *CPB* **1983**, *31*, 2583. (g) Lenz, G. R.; Woo, C.-M. *JHC* **1981**, *18*, 691. (h) Ishida, A.; Nakamura, T.; Irie, K.; Oh-ishi, T. *CPB* **1985**, *33*, 3237. (i) Sano, T.; Toda, J.; Maehara, N.; Tsuda, Y. *CJC* **1987**, *65*, 94. (j) Kanaoka, Y.; Sato, E.; Yonemitsu, O.; Ban, Y. *TL* **1964**, 2419.
19. (a) Zjawiony, J.; Peterson, J. R. *OPP* **1991**, *23*, 163. (b) Girard, Y.; Atkinson, J. G.; Belanger, P. C.; Fuentes, J. J.; Rokach, J.; Rooney, C. S.; Remy, D. C.; Hunt, C. A. *JOC* **1983**, *48*, 3220.
20. Feliz, M.; Bosch, J.; Mauleón, D.; Amat, M.; Domingo, A. *JOC* **1982**, *47*, 2435.
21. (a) Kelly, T. R.; Chandrakumar, N. S.; Saha, J. K. *JOC* **1989**, *54*, 980. (b) Imanishi, T.; Nakai, A.; Yagi, N.; Hanaoka, M. *CPB* **1981**, *29*, 901. (c) Begley, W. J.; Grimshaw, J. *JCS(P1)* **1977**, 2324.
22. Lele, S. R.; Hosangadi, B. D. *IJC(B)* **1979**, *18*, 533.
23. Vegh, D.; Morel, J.; Decroix, B.; Zalupsky, P. *SC* **1992**, *22*, 2057.
24. Heinicke, G.; Hung, T. V.; Prager, R. H.; Ward, A. D. *AJC* **1984**, *37*, 831.
25. (a) For other methods of ester and lactone formation, see: March, J. *Advanced Organic Chemistry*, 3rd ed.; Wiley: New York, 1985; pp 348–351, 353–354. (b) For esterification of carboxylic acids using alkylphosphoric esters (APEs), see: Balasubramaniyan, V.; Bhatia, V. G.; Wagh, S. B. *T* **1983**, *39*, 1475. (c) For other methods of thiol ester formation, see Ref. 5 and March, J. *Advanced Organic Chemistry*, 3rd ed.; Wiley: New York, 1985; pp 362–363. (d) For other methods of amide formation, see March, J. *Advanced Organic Chemistry*, 3rd ed.; Wiley: New York, 1985; pp 370–377.
26. Imamoto, T.; Kodera, M.; Yokoyama, M. *BCJ* **1982**, *55*, 2303.
27. (a) Kanaoka, Y.; Kuga, T.; Tanizawa, K. *CPB* **1970**, *18*, 397. (b) For a list of other reagents used in the amide to nitrile conversion, see Ref. 16.
28. Benson, S. C.; Li, J.-H.; Snyder, J. K. *JOC* **1992**, *57*, 5285.
29. Imamoto, T.; Yokoyama, H.; Yokoyama, M. *TL* **1981**, *22*, 1803.
30. Kanaoka, Y.; Ban, Y.; Miyashita, K.; Irie, K.; Yonemitsu, O. *CPB* **1966**, *14*, 934.
31. Kanaoka, Y.; Hamada, T.; Yonemitsu, O. *CPB* **1970**, *18*, 587.
32. (a) Yalcin, I.; Oren, I.; Sener, E.; Akin, A.; Ucarturk, N. *Eur. J. Med. Chem.* **1992**, *27*, 395. (b) Yoshino, K.; Kohno, T.; Uno, T.; Morita, T.; Tsukamoto, G. *JMC* **1986**, *29*, 820.
33. Vishwakarma, L. C.; Sowell, J. W. *JHC* **1985**, *22*, 1429.
34. Bayomi, S. M.; Price, K. E.; Sowell, J. W. *JHC* **1985**, *22*, 729.
35. Bayomi, S. M.; Price, K. E.; Sowell, J. W. *JHC* **1985**, *22*, 83.
36. Reddy, V. R. K.; Reddy, P. S. N.; Ratnam, C. V. *SC* **1991**, *21*, 49.
37. Perillo, I.; Fernández, B.; Lamdan, S. *JCS(P2)* **1977**, 2068.
38. Okumura, K.; Adachi, T.; Tomie, M.; Kondo, K.; Inoue, I. *JCS(P1)* **1972**, 173.
39. (a) Yokoyama, M.; Sato, K.; Tateno, H.; Hatanaka, H. *JCS(P1)* **1987**, 623. (b) Yokoyama, M.; Kodera, M.; Imamoto, T. *JOC* **1984**, *49*, 74.
40. Marson, C. M.; Grabowska, U.; Walsgrove, T. *JOC* **1992**, *57*, 5045.
41. Imamoto, T.; Kusumoto, T.; Yokoyama, M. *BCJ* **1982**, *55*, 643.
42. Oklobdzija, M.; Sunjic, V.; Kajfez, F.; Caplar, V.; Kolbah, D. *S* **1975**, 596.

Lisa A. Dixon
The R. W. Johnson Pharmaceutical Research Institute, Raritan, NJ, USA

Polyphosphoric Acid[1]

[8017-16-1]

(moderately strong mineral acid with powerful dehydrating properties; used for intramolecular and intermolecular acylations, heterocyclic synthesis, and acid-catalyzed rearrangements)

Alternate Name: PPA.

Physical Data: hygroscopic, highly viscous, clear, colorless, or light amber; specific gravity 2.060 at 83% phosphorus pentoxide content.

Solubility: dissolution in any protic solvent will result in solvolysis of the reagent; dissolution in polar aprotic solvents could result in dehydration or destruction of the solvent; polyphosphoric acid is neither soluble in nor reacts with nonpolar organics such as toluene or hexane.

Form Supplied in: inexpensive and commercially available from most major suppliers.

Preparative Methods: by mixing x mL of ***Phosphoric Acid*** (85%, d 1.7 g mL^{-1}) with 2.2 x g of ***Phosphorus(V) Oxide*** (P_2O_5) followed by heating to 200 °C for 30 min.

Handling, Storage, and Precautions: normally used as the solvent so that a 10–50 fold excess is routinely employed. Due to high viscosity, PPA is difficult to pour and stir at rt, but is much easier to work with at temperatures above 60 °C. Addition of cosolvents, such as xylene, has facilitated the difficult workup usually associated with PPA.[2] Eaton's reagent (see ***Phosphorus(V) Oxide–Methanesulfonic Acid***) has been found to perform similar chemistry at lower temperatures without the viscosity problems. When diluting PPA or working up a reaction, ice is normally used to moderate the exothermic reaction that occurs with water. PPA has the ability to burn mucous membranes immediately and unprotected skin with time. Other than the corrosive nature of this reagent it has low inherent toxicity. Use in a fume hood.

Description. Polyphosphoric acid is a mixture of orthophosphoric acid and linear phosphoric acids. In order to simplify discussion of this reagent, the complex mixture is described empirically as a wt % of P_2O_5 in water. The distribution of phosphoric acids that are found in PPA is dependent upon the wt % of P_2O_5. Commercially available PPA contains 82–85% P_2O_5, with 83%

P_2O_5 considered to be the standard. At this concentration there is no free water and the distribution of phosphoric acids is approximately 6% orthophosphoric acid, 19% pyrophosphoric acid, and 11% triphosphoric acid, while the remaining material is linear phosphoric acids up to a chain length of approximately 14 phosphoric acid units.[3] Only at wt % of P_2O_5 over 84% do appreciable high weight polymeric species occur. Neutralization of the most acidic protons in PPA is accomplished at pH 3.8–4.2 and corresponds to one strongly acidic proton for each phosphorus atom.

The powerful dehydrating properties of PPA, low nucleophilicity of the phosphoric acid media, and moderate acidity explain why this reagent is so widely used. Unlike ***Sulfuric Acid***, PPA has a low propensity to cause oxidation of the substrate and is also able to dissolve organic compounds. PPA has demonstrated rates of dehydration equal to that of 100% sulfuric acid even though it is a much weaker acid.

Cyclization of Acids, Esters, Ketones, Aldehydes, Acetals, Alcohols, and Alkenes onto Aromatic Rings. Polyphosphoric acid is the reagent of choice to cyclize aromatic carboxylic acids to indanones (eq 1),[4] tetralones (eq 2),[5] and benzosuberones (eq 3).[6] Anomalous results for the cyclization of 3-(2-methoxyphenyl)propionic acid led researchers to discover a method for synthesis of metacyclophanes (eq 4).[7] Another interesting reaction that demonstrates the utility of PPA in forming cyclic aromatic ketones is the double cyclization of biscarboxylic acids (eq 5).[8] Carboxylic esters often demonstrate the ability to be cyclized as readily as the acids (eq 6).[9]

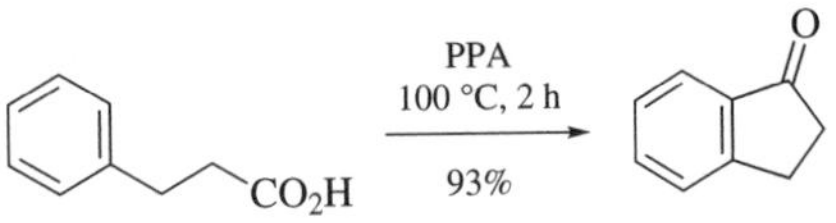

CO_2H —PPA, 100 °C, 2 h, 93%→ (1)

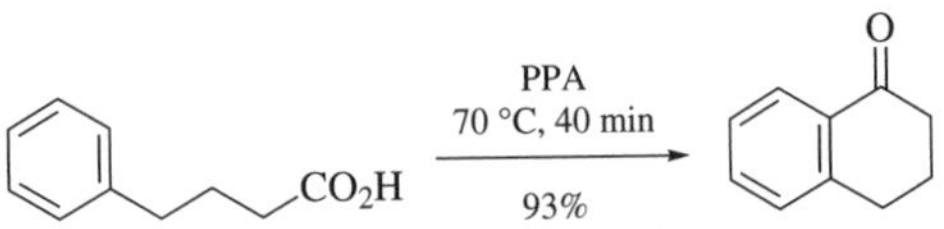

CO_2H —PPA, 70 °C, 40 min, 93%→ (2)

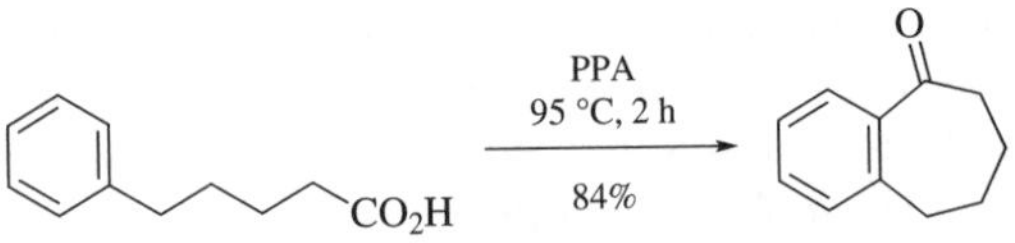

CO_2H —PPA, 95 °C, 2 h, 84%→ (3)

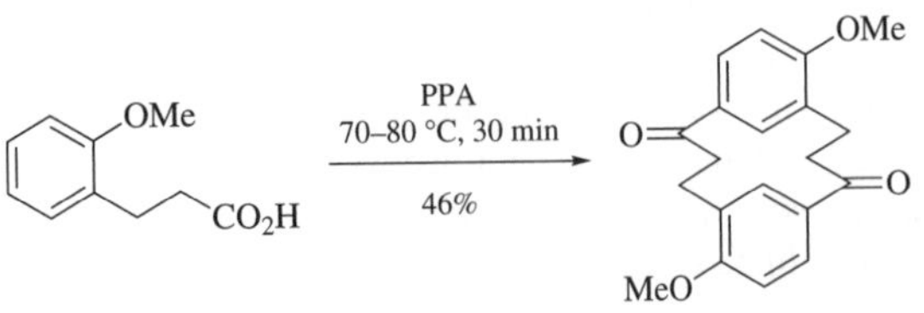

OMe, CO_2H —PPA, 70–80 °C, 30 min, 46%→ OMe, MeO (4)

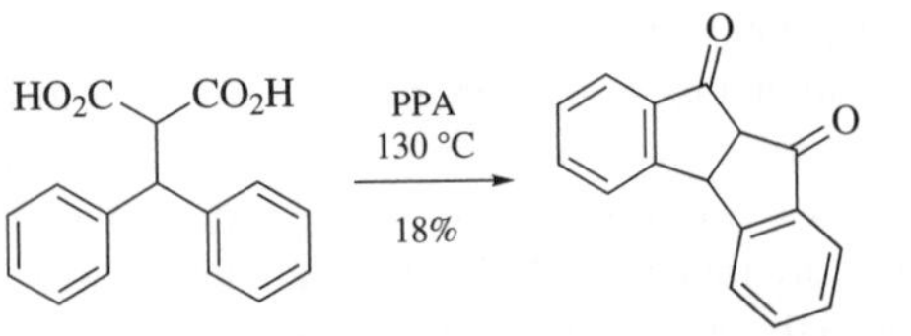

HO_2C, CO_2H —PPA, 130 °C, 18%→ (5)

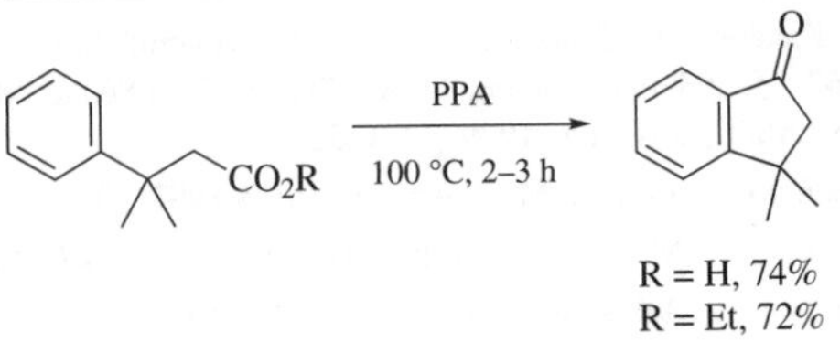

CO_2R —PPA, 100 °C, 2–3 h→ (6)

R = H, 74%
R = Et, 72%

Methoxy or alkyl substitution of the aromatic ring has also been found to allow shorter reaction times and lower reaction temperatures.[10] Cyclization of ketones (eq 7),[11] aldehydes (eq 8),[12] and acetals (eq 9)[13] occurs with dehydration to give cyclic alkenes. Tertiary or benzylic alcohols are usually the only alcohols which give straightforward cyclization products (eq 10).[14] Secondary and primary alcohols usually rearrange (Wagner–Meerwein rearrangements) before cyclization can occur. Just as in alcohols, alkenes are also prone to rearrangement unless the carbenium ion formed upon protonation is tertiary or benzylic. Low to moderate yields have been reported for cyclization of alkenes onto aromatic rings (eq 11).[15]

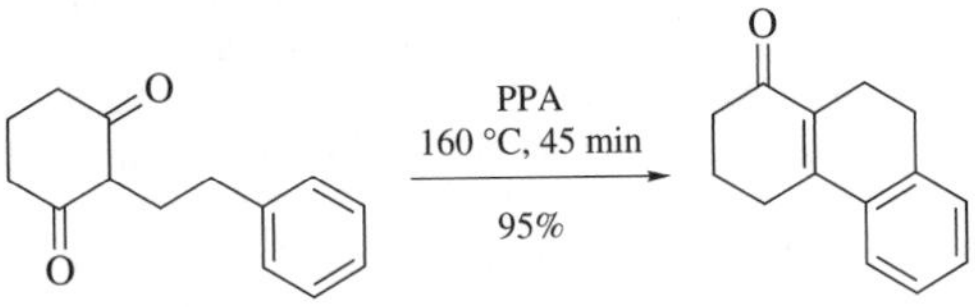

—PPA, 160 °C, 45 min, 95%→ (7)

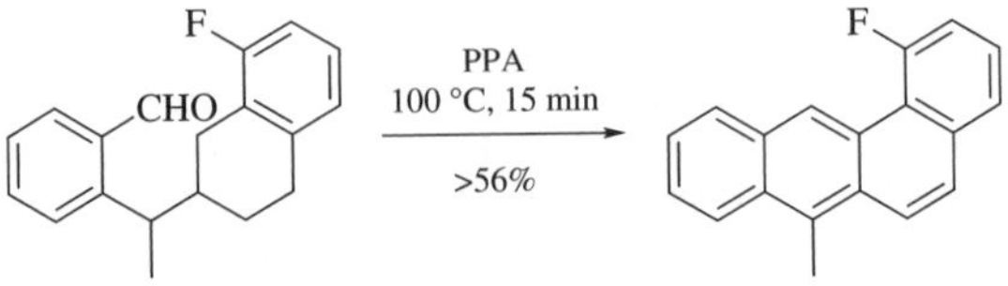

F, CHO —PPA, 100 °C, 15 min, >56%→ F (8)

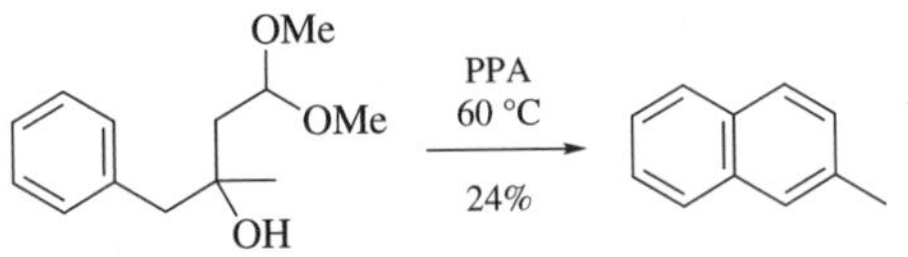

OMe, OMe, OH —PPA, 60 °C, 24%→ (9)

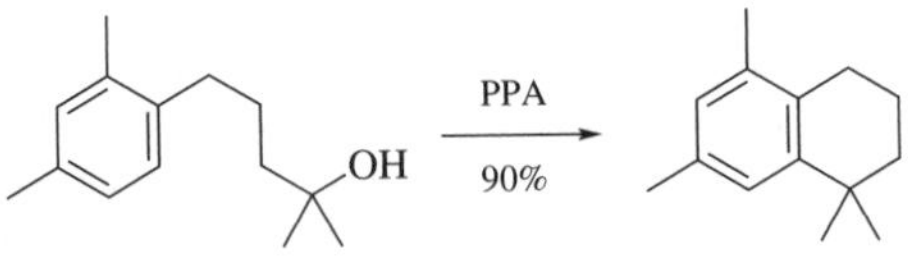

OH —PPA, 90%→ (10)

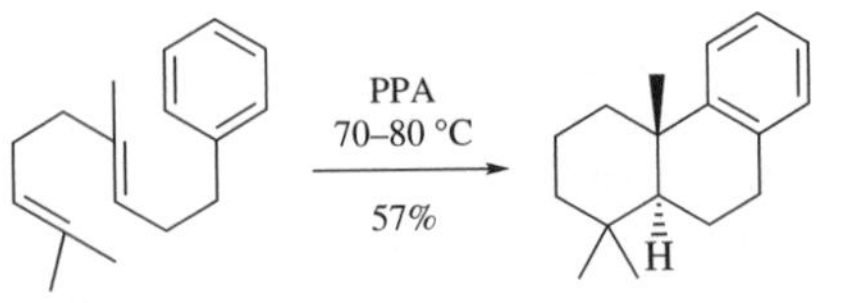

—PPA, 70–80 °C, 57%→ H (11)

Cyclization onto Nonaromatic Moieties. Cyclopentenones have been synthesized from carboxylic acids in good yield (eqs 12–14).[16] Compared with other protic acids, PPA has demonstrated the ability to favor carbon–carbon bond formation over lactone formation.[17]

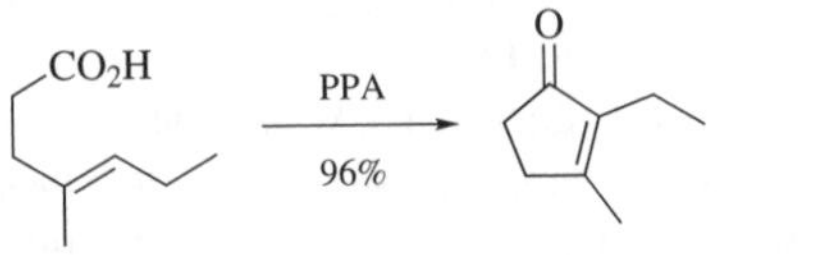

CO_2H —PPA, 96%→ (12)

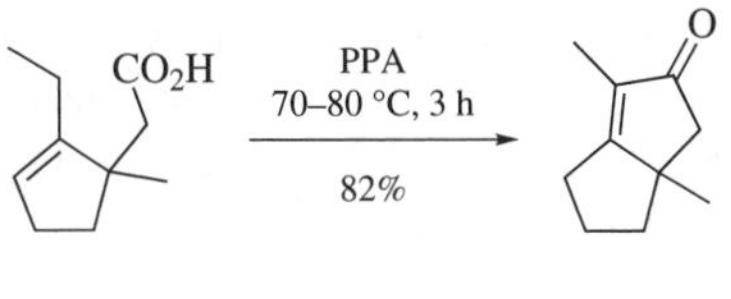

(13)

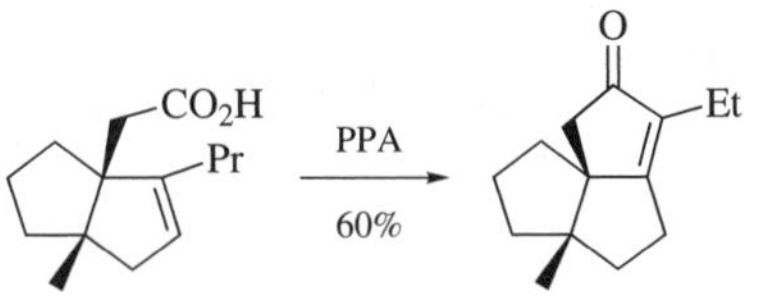

(14)

Synthesis of cyclohexenones from alkenyl acids has been demonstrated (eq 15);[18] however, formation of methylcyclopentenones and lactones may occur when possible (eq 16).[19]

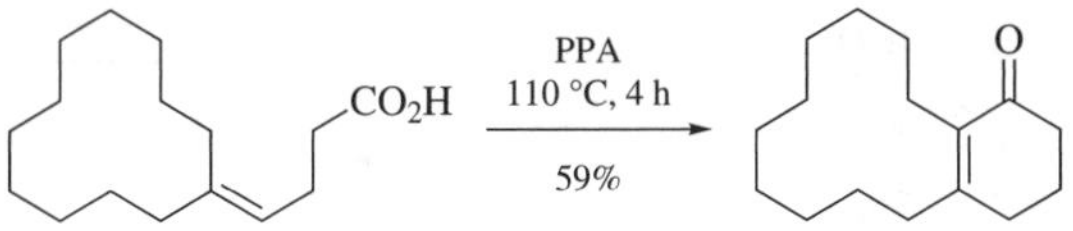

(15)

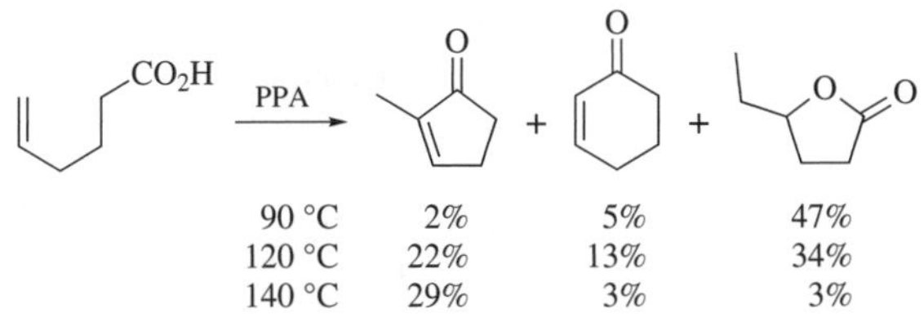

(16)

90 °C	2%	5%	47%
120 °C	22%	13%	34%
140 °C	29%	3%	3%

Cyclization Reactions which Form Heterocycles. Use of PPA as an acid catalyst to form heterocyclic compounds has been exhaustively reviewed in the literature.[1] The ability of PPA to be used in place of more acidic or more nucleophilic reagents has led to applications in a variety of heterocyclic systems.

Nitrogen Heterocycles. Although ***Zinc Chloride*** is normally used as catalyst, indoles substituted in the 2-position can be obtained from hydrazones by ring closure with PPA (Fisher indole synthesis) (eq 17).[20]

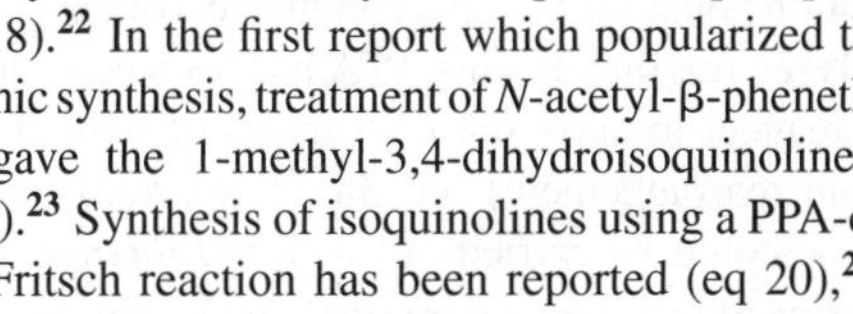

(17)

Use of PPA in the Bischler–Napieralski reaction has shown superior results to other reagents for construction of the isoquinoline ring system.[21] For example, dihydroisoquinolines are obtained from phenethylformamides in yields superior to phosphorus pentoxide (eq 18).[22] In the first report which popularized the use of PPA in organic synthesis, treatment of *N*-acetyl-β-phenethylamine with PPA gave the 1-methyl-3,4-dihydroisoquinoline in 23% yield (eq 19).[23] Synthesis of isoquinolines using a PPA-catalyzed Pomeranz–Fritsch reaction has been reported (eq 20),[24] but the low yields and poor reproducibility of this reaction have been overcome by the use of ***Hydrogen Chloride***/dioxane to cyclize the *N*-tosyl derivative.[25]

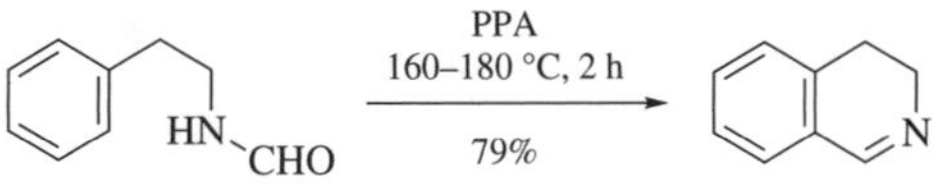

(18)

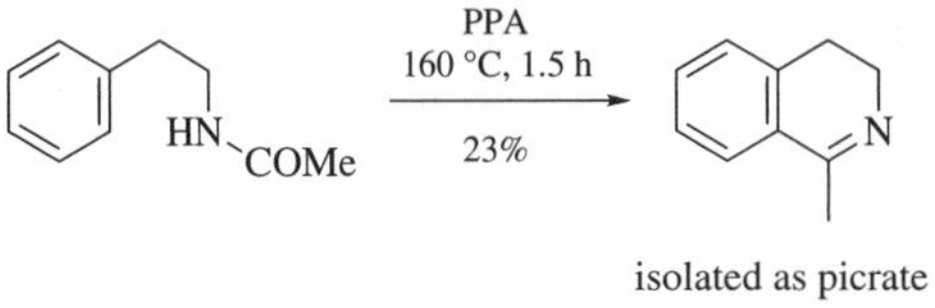

(19)

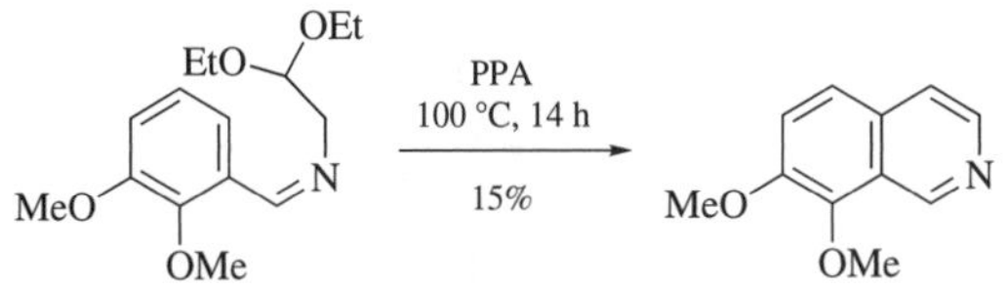

isolated as picrate

(20)

The quinoline carbon framework can be assembled by ring closure of an aromatic acid to give a keto quinoline (eq 21).[26] Using PPA as catalyst, phenylquinolinones can be prepared from 3-aryl-3-hydroxypropionanilides (eq 22),[27] or from amido ketones (eq 23).[28]

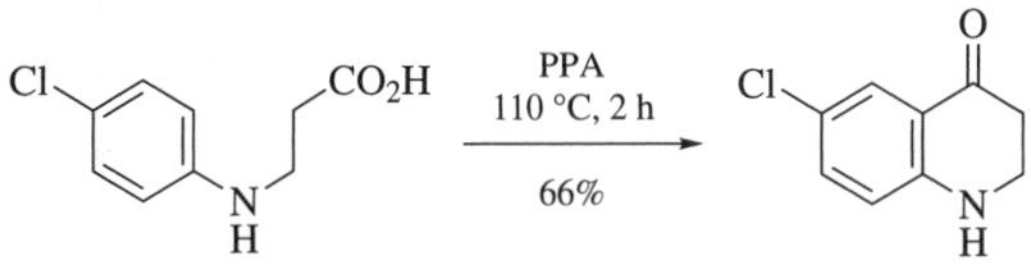

(21)

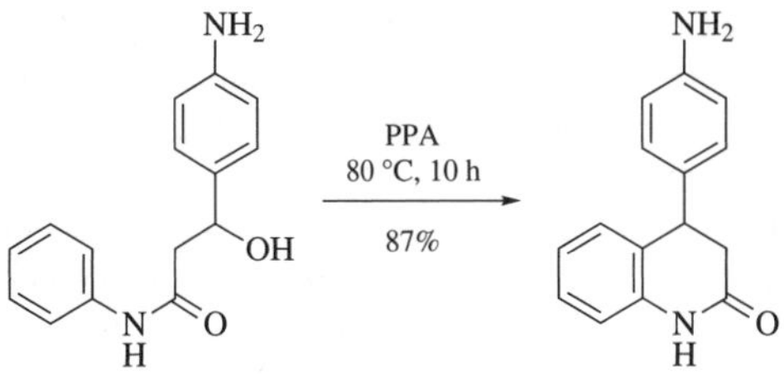

(22)

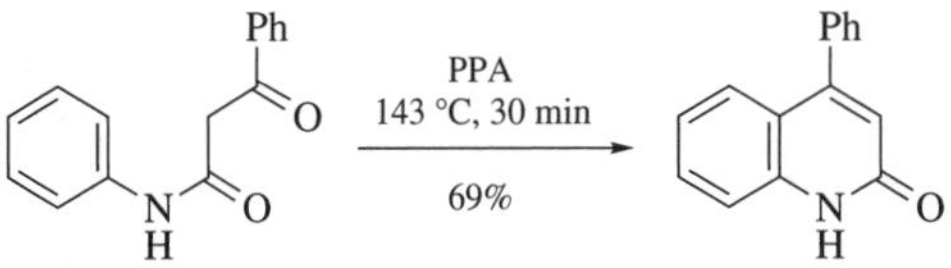

(23)

When attempting to effect a Beckmann rearrangement it was found that the oxime of a hexahydrobenzindolizine did not give the expected amide but instead dehydrated. This was followed by ring opening then ring closure to give a dihydrobenzonaphthyridinone (eq 24). When the hexahydrobenzindolizine itself was treated with PPA the compound simply dehydrated without rearrangement to give the dihydrobenzindolizone (eq 25). Sulfuric acid completely failed to effect this dehydration and Eaton's reagent was reported to give lower yields of product.[29]

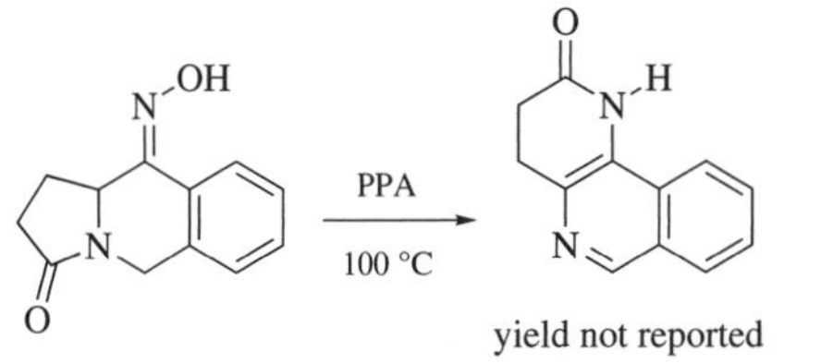

(24)

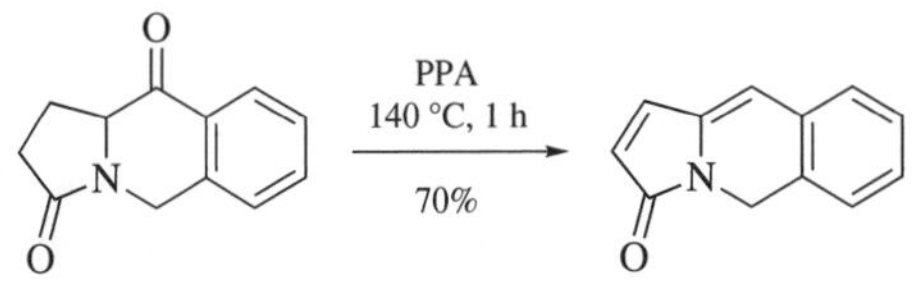

yield not reported

(25)

Alkyl- or chloro-substituted isatins were obtained more conveniently with PPA than with sulfuric acid (eq 26).[30] Synthesis of the bacterial coenzyme methoxatin was facilitated by use of PPA to synthesize the pivotal isatin intermediate (eq 27).[31]

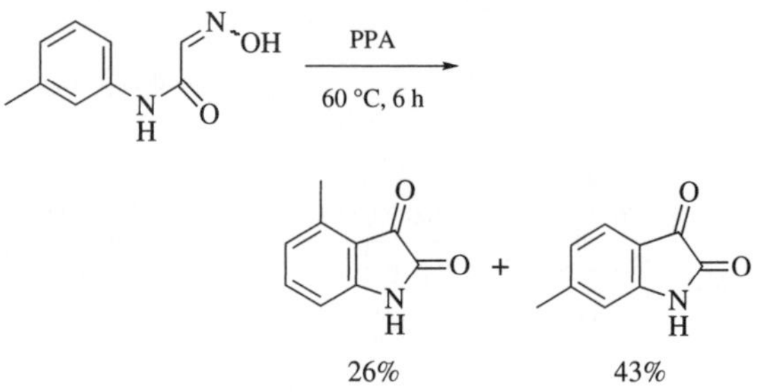

PPA, 60 °C, 6 h; 26%; 43% (26)

PPA, 100 °C, 20 min, 70% (27)

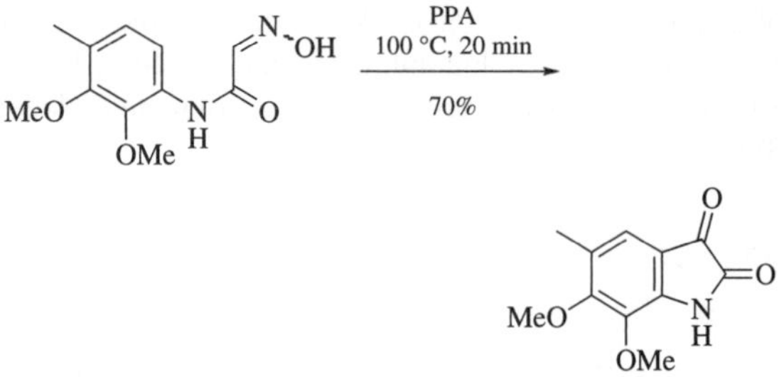

Complex lactams were stereoselectively assembled in a beautifully simple reaction between 3-alkenamides and benzaldehyde (eq 28).[32] Oxazolinones can be made utilizing the Erlenmeyer azlactone synthesis. Use of PPA cleanly affords the (*E*) isomer whereas other methods provide only the (*Z*) isomers or mixtures of (*E*) and (*Z*) isomers (eq 29).[33] Other nitrogen-containing heterocyclic systems which can be obtained using PPA include benzimidazoles (eq 30)[34] and triazoles (eq 31).[35]

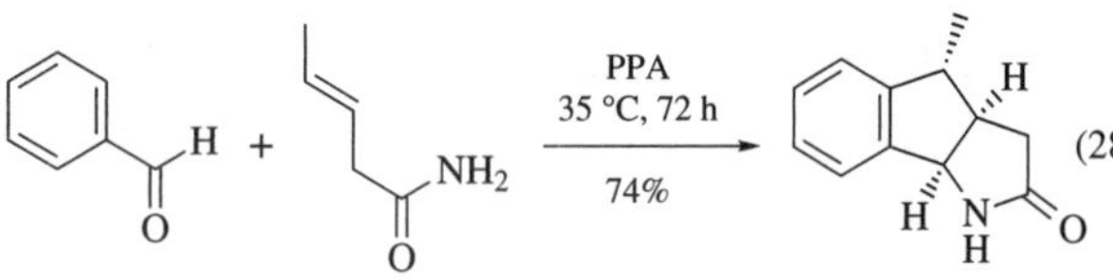

PPA, 35 °C, 72 h, 74% (28)

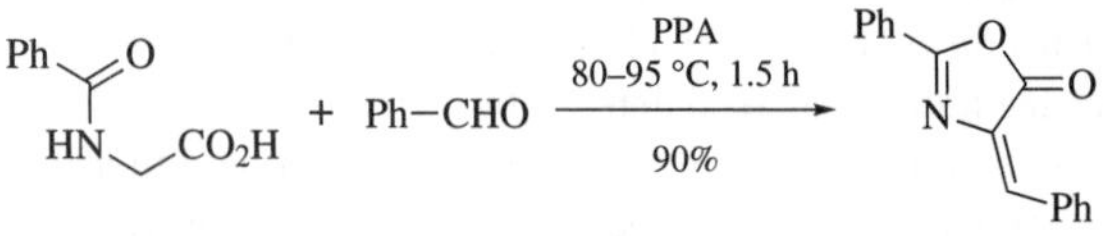

Ph-CHO, PPA, 80–95 °C, 1.5 h, 90% (29)

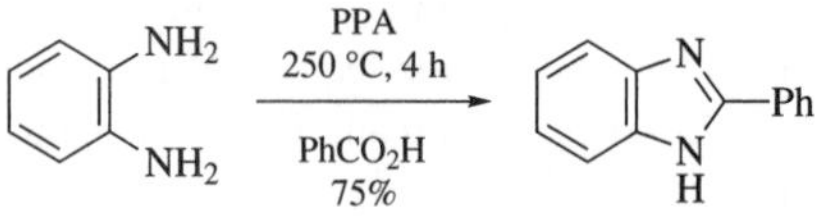

PPA, 250 °C, 4 h, $PhCO_2H$, 75% (30)

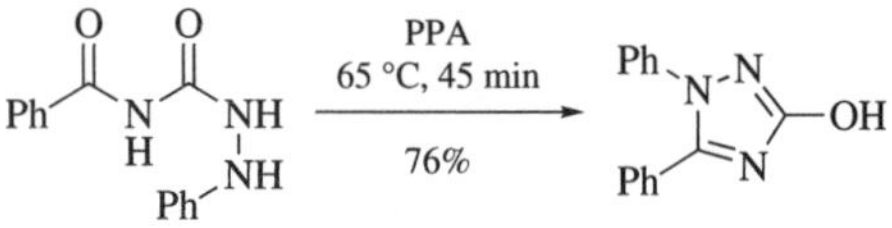

PPA, 65 °C, 45 min, 76% (31)

Oxygen and Sulfur Heterocycles. Diphenylfurans are formed in higher yields with PPA than with sulfuric acid, ***Acetic Anhydride***, or phosphorus pentoxide as the dehydrating/cyclization agent (eq 32).[36] Similar to cyclization of aromatic acids to form aromatic cyclic ketones, the replacement of the benzylic methylene group with an oxygen affords chromanones in good yield (eq 33).[37] The structurally related flavones can be prepared by an intramolecular 1,4-addition catalyzed by PPA (eq 34).[38] Seven-membered rings which contain oxygen, such as benzoxepinones, can be made in good yield from 4-phenoxybutyric acids (eq 35).[39] Benzofurans are formed in good yield by cyclization of α-phenoxy ketones (eq 36).[40] Thienopyrroles are obtained by ring closure of a pyrrolecarboxylic acid (eq 37).[41]

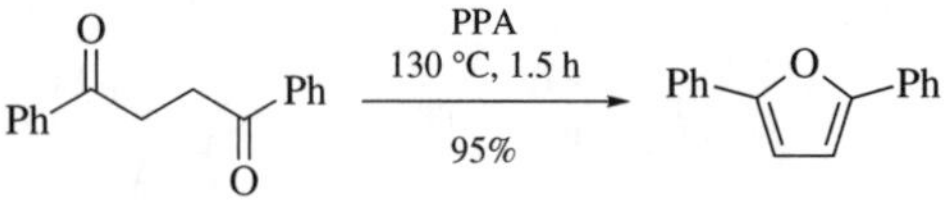

PPA, 130 °C, 1.5 h, 95% (32)

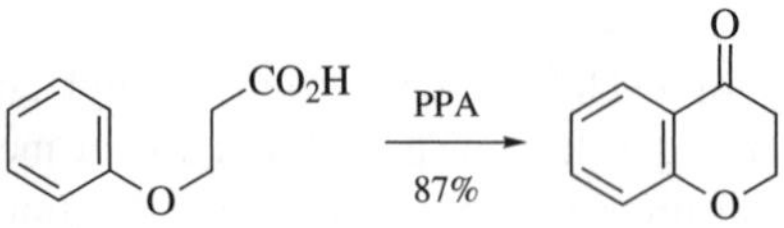

PPA, 87% (33)

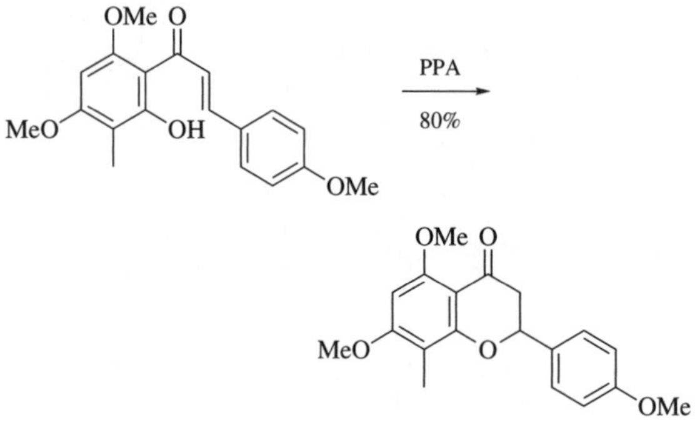

PPA, 80% (34)

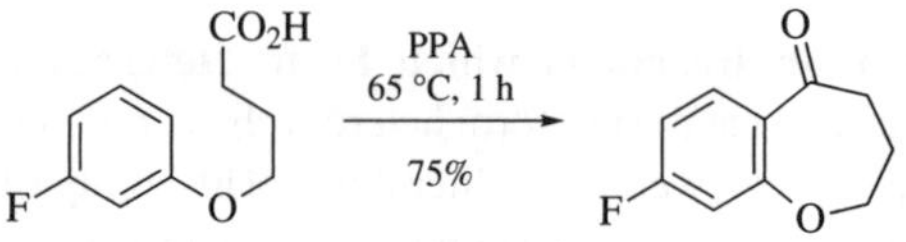

PPA, 65 °C, 1 h, 75% (35)

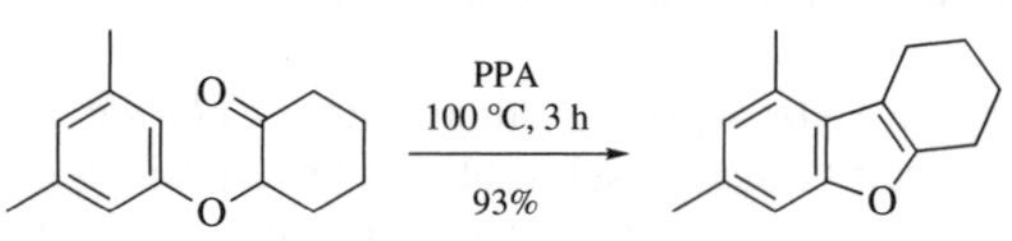

PPA, 100 °C, 3 h, 93% (36)

PPA, 125 °C, 36% 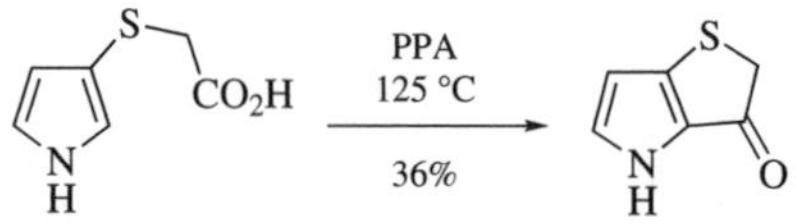 (37)

Rearrangements and Isomerizations. The same characteristics which facilitate cyclizations, such as media with low nucleophilicity, good solvation power, relatively mild acidity, and low oxidation potential, are also conducive to clean, high-yielding acid-catalyzed rearrangements. PPA has been considered to be an effective reagent to carry out conversion of oximes into amides (Beckmann rearrangement) (eq 38).[42] Rearrangement of a decalin oxime could be carried out with ***p-Toluenesulfonyl Chloride/Pyridine*** to maintain the original *cis*-decalin stereochemistry or PPA could be used to allow formation of the *trans*-decalin system through an alternative mechanism (eq 39).[43]

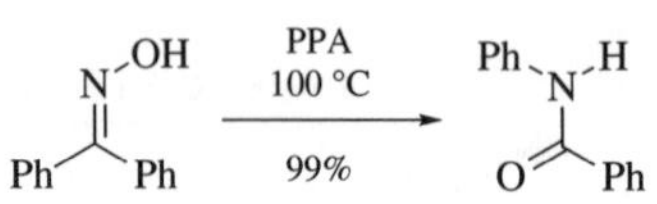

PPA, 100 °C, 99% (38)

(39) yield not specified

Treatment of aromatic carboxylic acids with ***Nitromethane*** and PPA (Lossen rearrangement) gives high yields of anilines (eq 40).[44] Although PPA can be used to catalyze the reaction of ***Hydrazoic Acid*** with carboxylic acids, ketones, and aldehydes (Schmidt rearrangement), sulfuric acid is usually the reagent of choice. One example of PPA being equivalent to sulfuric acid in the Schmidt rearrangement is when it is applied to acetophenone (eq 41).[45]

(40) + $MeNO_2$ —PPA, 115 °C, 1.5 h, 60%→

(41) + NaN_3 —PPA, 50 °C, 7 h, 98%→

Wagner–Meerwein rearrangements and, more generally, carbenium ion-mediated carbon skeleton rearrangements can be effected with PPA. Lewis acids are generally preferred due to the ease of use and workup. An example of a high yield PPA induced Wagner–Meerwein rearrangement is formation of the phenanthrene skeleton from substituted fluorenes (eq 42).[46] Ring contraction was found to occur when a thiochromene was treated with PPA (eq 43).[47]

(42) PPA, 160 °C, 0.5 h, 89%

(43) PPA, 100 °C, 3 h, moderate yield

Acid-catalyzed isomerizations which can be affected with PPA include conversion of *trans*-cinnamic acid to *cis*-cinnamic acid (eq 44), *trans,trans*-diene conversion to *trans,cis*-dienes (eq 45),[48] and azlactone isomerizations (eq 46).[33]

(44) PPA, 80–95 °C, 1.5 h, 100%

(45) PPA, 80–95 °C, 1.5 h, yield not specified

(46) PPA, 100 °C, 1.5 h, 90%

Intermolecular Reactions. PPA is generally used as an intramolecular catalyst but has demonstrated limited utility for intermolecular alkylations. Due to the elevated temperatures necessary to achieve catalysis (and reduce PPA viscosity), most reactions give mixtures of products favoring multiple substitution. Consequently, most Friedel–Crafts alkylations are carried out with ***Aluminum Chloride*** or another Lewis acid which can be more readily controlled. In the case of activated aromatics, such as phenol, alkylation can proceed under moderate conditions (eq 47).[49]

(47) PPA, 85 °C, 40 min, 58% combined yield

Acylations with PPA are much more prevalent. In the seven years between the first popularized use of PPA in 1950[23] and the Popp and McEwen review,[1b] over 200 intermolecular acylations were reported. One of the first acylations was the reaction between cyclohexene and ***Acetic Acid*** (eq 48).[50] Acylation of activated aromatics proceeds in high yield (eq 49).[51] Acylation of phenols is problematic due to competing ester formation (eq 50).[52]

(48) + $MeCO_2H$ —PPA, 55 °C, 45 min, 60%→

(49) PPA, 75 °C, 5 h, 85%

(50) + $Ph-CO_2H$ —PPA, 100 °C, 20 min→ 16% + 68%

Miscellaneous Uses of Polyphosphoric Acid. Nitrile hydrolysis to amides is commonly carried out using PPA. At 100–110 °C, nitriles are routinely converted to amides (eq 51).[53]

$$Ph-CN \xrightarrow[96\%]{\text{PPA, 110 °C, 1 h}} Ph-CONH_2 \quad (51)$$

When the temperature is raised to 200 °C, decyanation of aromatic nitriles often results.[54] Reagents with unique abilities have been developed by mixing PPA with other agents. The reduced acidity of PPA compared to other mineral acids proved beneficial when a PPA/*Ethanethiol* mixture was used to open one epoxide selectively in a bis-epoxy steroid (eq 52).[55] PPA/$POCl_3$ affords a medium which is used to convert tertiary alcohols into chlorides (eq 53).[56]

PPA, EtSH; 90 °C (52)

PPA, $POCl_3$; Δ; 75–80% (53)

In an attempt to generate HCl in situ during cyclization of a tetraamine, a PPA/NaCl mixture was superior to PPA alone (eq 54).[57] PPA/*Acetic Acid* was used to synthesize nonenolizable β-diketones (eq 55).[58] Although Wagner–Meerwein rearrangements predominate when primary carbonium ions are formed with PPA, a mixture of PPA/***Potassium Iodide*** allowed conversion of a primary methyl ether to a primary alkyl iodide in high yield (eq 56).[59] It has been reported that nitrations carried out using PPA/***Nitric Acid*** are less hazardous than Ac_2O/HNO_3 mixtures (eq 57).[60]

PPA, NaCl; 165 °C, 4 h; 33% (54)

PPA, AcOH; 100 °C, 7 h; 75% (55)

PPA, KI; 135–140 °C, 5 h; 99% (56)

PPA, HNO_3; 60 °C, 1 h; 78% (57)

Related Reagents. Phosphoric Acid; Phosphorus(V) Oxide; Phosphorus(V) Oxide–Methanesulfonic Acid; Phosphorus Oxychloride; Polyphosphate Ester.

1. The most recent (literature up to 1981) comprehensive synthetic review is: (a) Pizey, J. S. *Synthetic Reagents*; Ellis Horwood: Chichester, 1985; Vol. 6. Two very useful comprehensive reviews which cover the synthetic applications of PPA up to 1960 are: (b) Popp, F. D.; McEwen, W. E. *CRV* **1958**, *58*, 321; and (c) Uhlig, F.; Snyder, H. R. In *Advances in Organic Chemistry: Methods and Results*; Raphael, R. A.; Taylor, E. C.; Wynberg, H., Eds.; Interscience: New York, 1960; Vol. 1, p 35. Additional reviews: (d) Marthe, J.- P.; Munavalli, S. *BSF* **1963**, 2679. (e) Krongauz, E. S.; Rusanov, A. L.; Renard, T. L. *RCR* **1970**, *39*, 747. (f) Verhe, R.; Schamp, N. *Ind. Chim. Belg.* **1973**, *38*, 945.
2. Guy, A.; Guetté, J. P. *S* **1980**, 222.
3. Jameson, R. F. *JCS* **1959**, 752.
4. Gilmore, Jr., J. C. *JACS* **1951**, *73*, 5879.
5. Koo, J. *JACS* **1953**, *75*, 1891.
6. Horton, W. J.; Walker, F. E. *JACS* **1952**, *74*, 758.
7. Zhang, J.; Hertzler, R. L.; Holt, E. M.; Vickstrom, T.; Eisenbraun, E. J. *JOC* **1993**, *58*, 556.
8. Allen, J. M.; Johnston, K. M.; Shotter, R. G. *CI(L)* **1976**, 108.
9. Horii, Z.; Ninomiya, K.; Tamura, Y. *YZ* **1956**, *76*, 163.
10. Pizey, J. S. *Synthetic Reagents*; Ellis Horwood: Chichester, 1985; Vol. 6, p 245.
11. Birch, A. J.; Smith, H. *JCS* **1951**, 1882.
12. Newman, M. S.; Seshadri, S. *JOC* **1962**, *27*, 76.
13. Kotchetkov, N. K.; Nifant′ev, E. J.; Nesmeyanov, A. N. *DOK* **1955**, *104*, 422.
14. Nasipuri, D.; De Dalal, I. *IJC* **1973**, *11*, 823.
15. Nasipuri, D.; Chaundhury, S. R. R.; Mitra, A.; Ghosh, C. K. *IJC* **1972**, *10*, 136.
16. Dorsch, M.; Jager, V.; Sponlein, W. *AG(E)* **1984**, *23*, 798.
17. Ansell, M. F.; Palmer, M. H. *QR* **1964**, *18*, 211.
18. Rae, I. D.; Umbrasas, B. N. *AJC* **1975**, *28*, 2669.
19. Ansell, M. F.; Emmett, J. C.; Coombs, R. V. *JCS* **1968**, 217.
20. Kissman, H. M.; Farnsworth, D. W.; Witkop, B. *JACS* **1952**, *74*, 3948.
21. Cannon, J. G.; Webster, G. L. *J. Am. Pharm. Assoc.* **1958**, *47*, 353.
22. Pratt, E. F.; Rice, R. G.; Luckenbaugh, R. W. *JACS* **1957**, *79*, 1212.
23. Snyder, H. R.; Werber, F. X. *JACS* **1950**, *72*, 2962.
24. Djerassi, C.; Markley, F. X.; Ehrlich, R. *JOC* **1956**, *21*, 975.
25. Boger, D. L.; Brotherton, C. E.; Kelley, M. D. *T* **1981**, *37*, 3977.
26. (a) Koo, J. *JOC* **1961**, *26*, 2440. (b) Koo, J. *JOC* **1963**, *28*, 1134.
27. Hazai, L.; Deak, G.; Sohar, P.; Toth, G.; Tamas, J. *JHC* **1991**, *28*, 919.
28. Stephenson, E. F. M. *JCS* **1956**, 2557.
29. Barbry, D.; Couturier, D. *JHC* **1990**, *27*, 1383.
30. Grimshaw, J.; Begley, W. J. *S* **1974**, 496.
31. Gainor, J. A.; Weinreb, S. M. *JOC* **1982**, *47*, 2833.
32. Marson, C. M.; Grabowska, U.; Walsgrove, T.; Eggleston, D. S.; Baures, P. W. *JOC* **1991**, *56*, 2603.
33. Rao, Y. S. *JOC* **1976**, *41*, 722.
34. Hein, D. W.; Alheim, R. J.; Leavitt, J. J. *JACS* **1957**, *79*, 427.
35. Arcus, C. L.; Prydal, B. S. *JCS* **1957**, 1091.
36. Nowlin, G. *JACS* **1950**, *72*, 5754.
37. Loudon, J. D.; Razdan, R. K. *JCS* **1954**, 4299.
38. Nakazawa, K.; Matsuura, S. *YZ* **1955**, *75*, 469.
39. Freedman, J.; Stewart, K. T. *JHC* **1989**, *26*, 1547.
40. Trippett, S. *JCS* **1957**, 419.

41. Matteson, D. S.; Snyder, H. R. *JACS* **1957**, *79*, 3610.
42. Horning, E. C.; Stromberg, V. L. *JACS* **1952**, *74*, 2680.
43. Hill, R. K.; Chortyk, O. T. *JACS* **1962**, *84*, 1064.
44. Bachman, G. B.; Goldwater, J. E. *JOC* **1964**, *29*, 2576.
45. Conley, R. T. *JOC* **1958**, *23*, 1330.
46. Bavin, P. M. G.; Dewar, M. J. S. *JCS* **1955**, 4477.
47. MacNicol, D. D.; McKendrick, J. J. *TL* **1973**, 2593.
48. Rao, Y. S.; Filler, R. *CC* **1976**, 471.
49. Gardner, P. D. *JACS* **1954**, *76*, 4550.
50. Dev, S. *JIC* **1956**, *33*, 703.
51. Barrio, J. R.; Barrio, M. D. C. G.; Vernengo, M. J. *JMC* **1971**, *14*, 898.
52. Nakazawa, K.; Baba, S. *YZ* **1955**, *75*, 378.
53. Snyder, H. R.; Elston, C. T. *JACS* **1954**, *76*, 3039.
54. Ceder, O.; Vernmark, K. *ACS* **1973**, *27*, 3259.
55. Tomoeda, M.; Furuta, T.; Koga, T. *CPB* **1967**, *15*, 887.
56. Kopecky, J.; Smejkal, J. *TL* **1967**, 1931.
57. Snyder, H. R.; Konecky, M. S. *JACS* **1958**, *80*, 4388.
58. Gerlach, H.; Muller, W. *AG(E)* **1972**, *11*, 1030.
59. (a) Cope, A. C.; Burrows, E. P.; Derieg, M. E.; Moon, S.; Wirth, W. D. *JACS* **1965**, *87*, 5452. (b) Stone, H.; Shechter, H. *JOC* **1950**, *15*, 491.
60. Kispersky, J. P.; Klager, K. *JACS* **1955**, *77*, 5433.

John H. Dodd
The R. W. Johnson Pharmaceutical Research Institute, Raritan, NJ, USA

Potassium Amide

KNH_2

[17242-52-3] H_2KN (MW 55.13)

(strong base and nucleophile; used for the generation and trapping of arynes; has been used extensively to study the reactivity of heterocyclic systems)

Solubility: 1.7 M in liquid ammonia;[1] 1.3×10^{-4} M in THF.[2]

Preparative Methods: a solution of potassium amide in liquid Ammonia is prepared by adding pieces of ***Potassium*** to liquid ***Ammonia*** in an ordinary three-necked flask equipped with a mechanical stirrer.[3] A piece of potassium is added to liquid ammonia and after the appearance of a blue color, a few crystals of iron(III) nitrate hydrate are added as catalyst. The remaining pieces of potassium are added at a rate which maintains active hydrogen evolution. Discharge of the deep blue color indicates complete conversion to potassium amide. External cooling is not required since the evaporation of ammonia will provide ample cooling.[4] If the reaction must be maintained at −78 °C then a dry ice–acetone condenser is necessary, otherwise an air condenser is sufficient. The resulting opaque mixture contains potassium amide which is mostly in solution. A more elaborate two-flask assembly for the generation and transfer of a potassium amide solution has also been described.[5]

Handling, Storage, and Precautions: potassium amide is flammable and ignites on contact with moisture. Excess material is destroyed by careful treatment with ethanol or isopropanol. In the preparation of potassium amide the following precautions should be noted. Potassium is a silvery gray metal but it can form an explosive peroxide coating. If it acquires an orange or red color or an appreciable oxide coating it should be considered extremely hazardous. Extreme caution should be exercised in any attempt to isolate potassium amide as it is suspected to be shock sensitive following partial oxidation. An explosion has been reported during the isolation of dry potassium amide.[6] Reactions should be performed in a fume hood to prevent exposure to ammonia. Hydrogen is evolved during the generation of potassium amide. No ignition source should be present.

Introduction. Potassium amide is both a strong base and a strong nucleophile and thus it has been used most effectively in reactions which exploit both of these properties, such as the generation and trapping of arynes, the amination of aromatic systems, and the rearrangement of various heterocyclic systems. Examples of these types of transformations are described below. The reagent has also been used simply as a strong base to induce deprotonation or elimination reactions, provided there are no competing nucleophilic reaction pathways available. This is currently a less important feature of the reagent given the ready availability of strong, nonnucleophilic bases, but a few examples are described at the end of this section.

Aryne Formation. Potassium amide in liquid ammonia has been used extensively in the generation and trapping of benzynes via dehydrohalogenation of phenyl halides.[7] One of the definitive experiments providing evidence for the existence of benzyne involved the treatment of [1-^{14}C]chlorobenzene with potassium amide in liquid ammonia to provide equal amounts of [1-^{14}C]aniline and [2-^{14}C]aniline (eq 1).[8] The potassium amide–ammonia system has since been used for the preparation of various substituted anilines.[9]

Cl; KNH_2, NH_3; NH_3; 43%; NH_2 + NH_2 (1)

Potassium amide has also been utilized extensively for the generation and trapping of various six-membered hetarynes including those generated from halogenated pyridines, diazines, isoquinolines, naphthyridines, and certain multicyclic systems.[10] Of the various hetarynes, the evidence supporting the existence of 3,4-pyridyne is recognized as the most convincing. Treatment of either 3- or 4-chloropyridine with potassium amide in liquid ammonia provides a constant ratio of the isomeric amine products (eq 2).[11]

Cl; N; or; Cl; N; KNH_2, NH_3; N; 70%; 1/3; NH_2; N; +; NH_2; 2/3; N (2)

Although mixtures of amines are usually formed in potassium amide-induced aryne formations, in some cases selectivity for one isomer can be achieved. For example, it was found that treatment of a tricyclic bromobenzo[*f*]quinoline with potassium amide in liquid ammonia provided only one product via its postulated hetaryne intermediate (eq 3).[12] Steric interference by the angular ring is presumed to block formation of the other isomer.

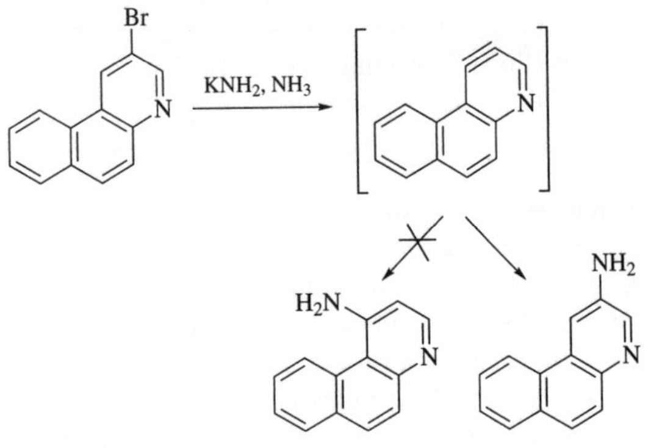

KNH_2, NH_3 (3)

Even in the presence of potassium amide, an intramolecular nucleophile can often compete effectively with amide ion to trap an aryne intermediate, resulting in ring closure. For example, 2-phenylbenzothiazole has been prepared in 90% yield utilizing this strategy (eq 4).[13] Intramolecular cyclization onto potassium amide-generated benzyne derivatives has been achieved with carbon, nitrogen, oxygen, and sulfur nucleophiles.[14] The strategy has also been successfully used with certain hetarynes, particularly 5-substituted 3,4-pyridynes (eq 5).[15] In some cases, however, when potassium amide is used as the aryne-generating base, competitive amination can occur to a significant extent (eq 6).

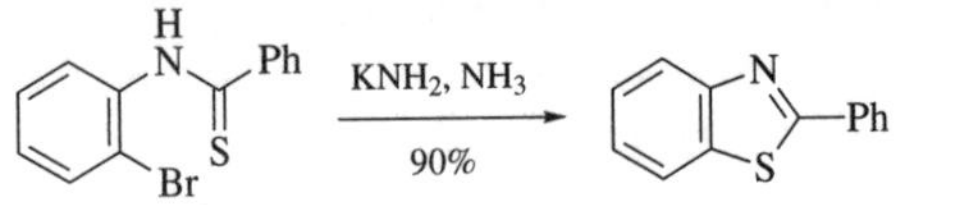

KNH_2, NH_3 90% (4)

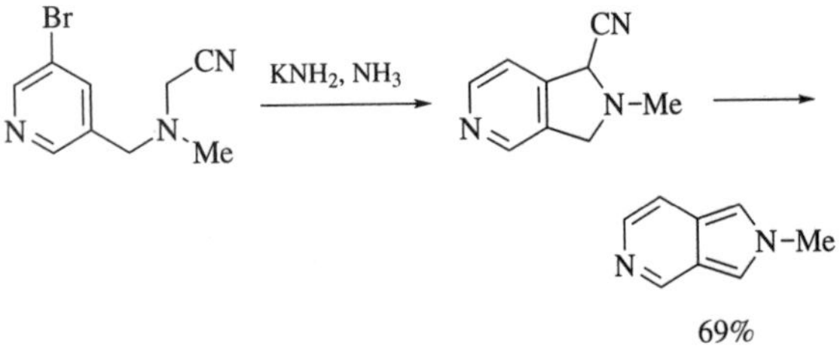

KNH_2, NH_3 (5) 69%

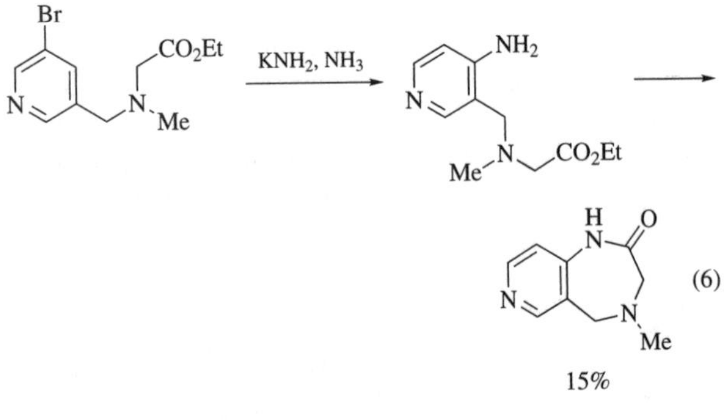

KNH_2, NH_3 (6) 15%

Potassium amide-generated arynes can also be trapped intramolecularly by a sufficiently nucleophilic phenyl ring. For example, appendage of a negatively charged atom to an aromatic ring can confer sufficient nucleophilicity to the *ortho* and *para* positions for this type of reaction.[16] An example of this strategy is shown in the synthesis of phenanthridines from haloanils (eq 7).[17] In this case, potassium amide is used not only to generate the requisite benzyne but also to activate the system to ring closure via addition of amide ion to the azomethine linkage. For the reaction to succeed, the nucleophilic addition of the amide ion must be very fast relative to the formation of the benzyne. This phenanthridine synthesis tolerates a variety of substituents in the aniline ring, with the exception of hydroxy and nitro groups which are thought to slow the amide addition to the azomethine group significantly.[18]

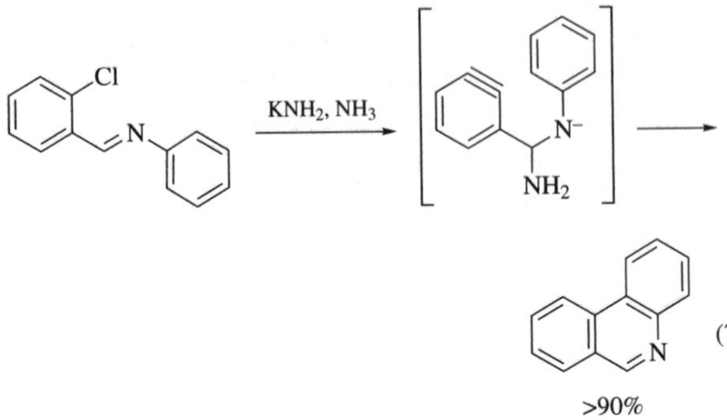

KNH_2, NH_3 (7) >90%

Rearrangement of Heterocyclic Systems. The generation of pyridyne utilizing potassium amide prompted the investigation of the reaction of potassium amide with other heterocyclic azine systems in a search for other hetarynes. However, the halogenated precursors to hetarynes are often more reactive toward nonaryne reactions and in some cases alternative mechanisms for amination are involved. Although it was initially believed that 4-bromo-6-phenylpyrimidine reacted with potassium amide via a 6-substituted 4,5-didehydro intermediate, extensive examinations of the reaction of potassium amide with halopyrimidines led to the elucidation of another mechanism for nucleophilic substitution which proceeds through a ring opened intermediate (eq 8).[19] This mechanism is referred to as S_N(ANRORC) for addition of the nucleophile, ring opening, and ring closure. The evidence accumulated to support its existence has been reviewed.[20] In particular, labeling studies have been used to demonstrate that the amide anion nitrogen becomes incorporated into the ring system (eq 9). This mechanism has been shown to be operative (to a greater or lesser extent) in the reaction of potassium amide with a variety of halogen-substituted heterocyclic systems in addition to pyrimidines, including quinazolines, triazines, and purines. A review of the extensive literature in the area, categorized by ring type, is available.[21]

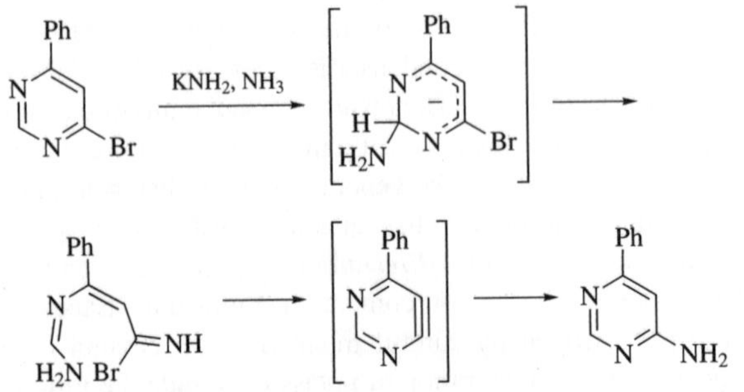

KNH_2, NH_3 (8)

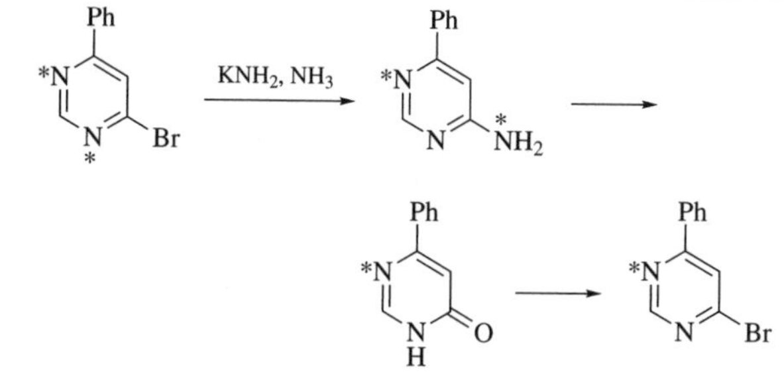

(9)

Under the conditions of potassium amide in liquid ammonia, many other examples of skeletal rearrangements of heterocyclic systems are known.[22] For example, ring contractions have been documented, such as that of 2-chloropyrazine to 2-cyanoimidazole (eq 10).[23] In this system, as in many cases, multiple rearrangement pathways lead to multiple products. The details of these potassium amide-induced heterocyclic ring transformation studies are available in a monograph.[22a]

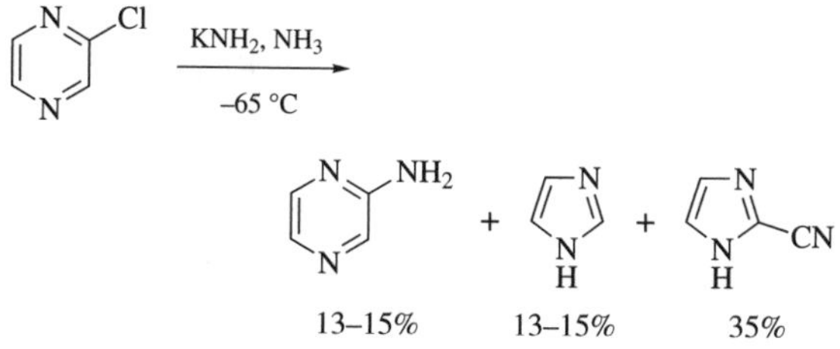

KNH_2, NH_3, –65 °C

13–15% + 13–15% + 35% (10)

A particularly useful transformation is the rearrangement of 2-substituted 4-halopyrimidines into *s*-triazines via a ring-opened intermediate (eq 11).[24] This reaction allows for the preparation of unsymmetrically substituted *s*-triazines which are difficult to obtain by other methods. Competing formation of the 4-aminopyrimidine is minimized by utilizing the chloro derivatives (X = Cl).

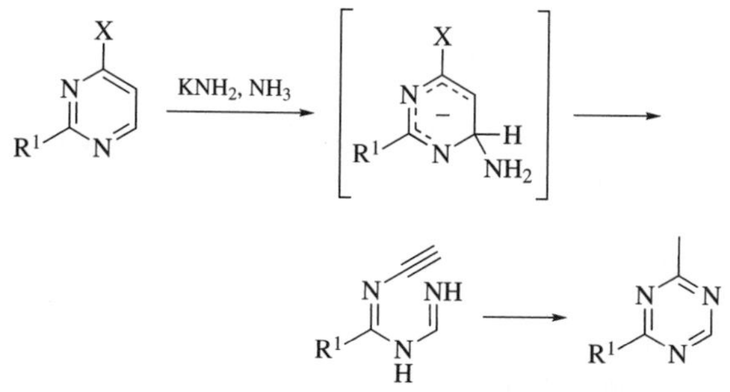

(11)

R^1 = alkyl, aryl <40%
$R^1 = R^2{}_2N$ 80–90%

Direct Amination of Aromatic Rings. Potassium amide can be used to induce a direct nucleophilic substitution of amide anion for hydrogen in certain azines.[25] Potassium amide in liquid ammonia adds readily to pyrazine, pyrimidine, and pyridazine to give anionic σ-adducts which can be oxidized by ***Potassium Permanganate*** to give the corresponding aminoaza heterocycle in good yield (eq 12).[26] More highly electron deficient systems, like pteridines and nitroaza aromatics, are able to add ammonia itself to give neutral σ-adducts which are also oxidized by $KMnO_4$ to heteroarylamines.[27] This modified Chichibabin reaction has been reviewed elsewhere.[28]

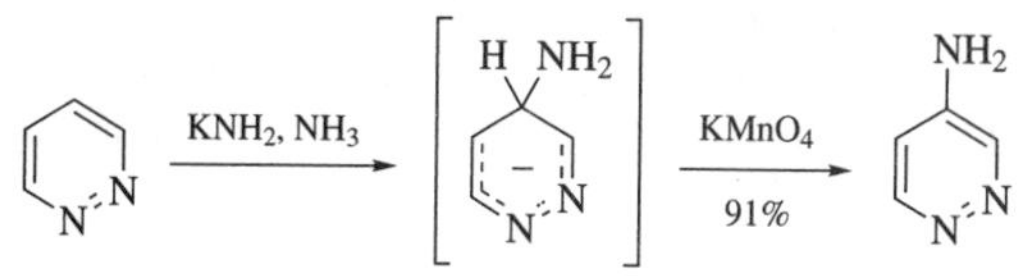

KNH_2, NH_3; $KMnO_4$, 91% (12)

Potassium amide can also displace a suitably situated halide on a heteroaromatic ring to provide the aromatic amine, but this reaction is often accompanied by other products from competing rearrangement pathways (eq 10). Simple phenols can be converted to the corresponding aniline in a two-step process involving treatment of the aryl diethyl phosphate ester with potassium amide and potassium metal in liquid ammonia (eq 13).[29]

$$ArOH + NaOH + (EtO)_2POCl \xrightarrow{80\text{–}90\%} ArOPO(OEt)_2 \xrightarrow[56\text{–}78\%]{KNH_2,\ K\ in\ NH_3} ArNH_2 \quad (13)$$

Anion Generation. Potassium amide is a strong base which can be used in simple deprotonation reactions for the generation of various anions. In enolate chemistry, potassium amide in liquid ammonia has been used to generate the dianions of β-diketones and β-ketoaldehydes.[30] These species can then be regioselectively alkylated at the γ-position. In unsymmetrical β-diketones the second deprotonation occurs at the less substituted γ-position (eq 14). The scope of this reaction, including a tabular survey of known examples, has been carefully reviewed.[30] Potassium amide in liquid ammonia has also been used for the preparation of 1,3-dinitro-2-keto derivatives from the reaction of cycloalkanones with alkyl nitrates.[31]

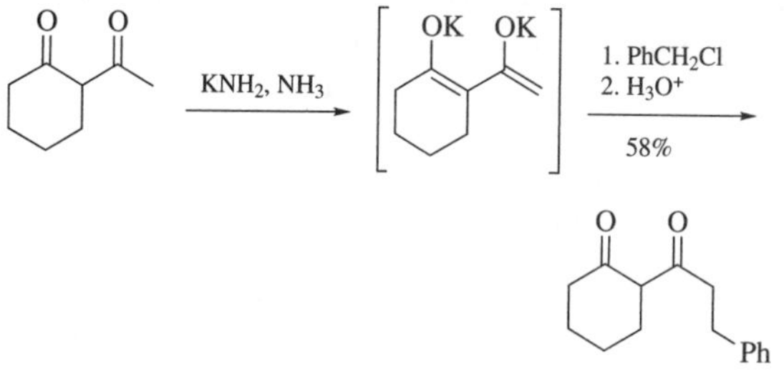

KNH_2, NH_3; 1. $PhCH_2Cl$ 2. H_3O^+, 58% (14)

Deprotonation of benzyl ethers by potassium amide in liquid ammonia has been used to effect the Wittig rearrangement.[32] In a preparation of phenanthrene, use of potassium amide accomplished the rearrangement to the carbinol in 90% yield after 1 h, whereas ***Phenyllithium*** required 1 week (eq 15).[33]

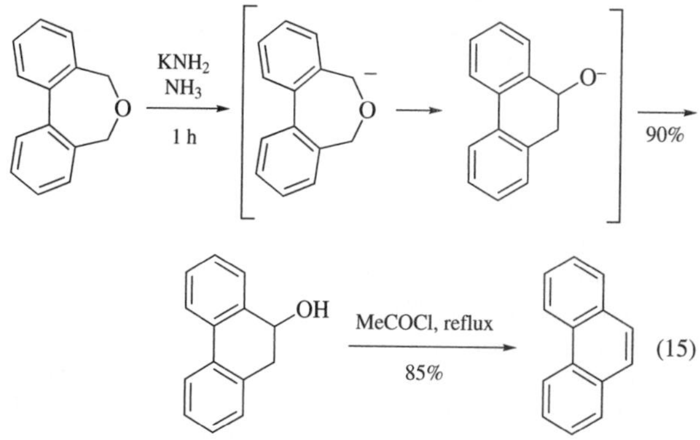

KNH_2, NH_3, 1 h; 90%; MeCOCl, reflux, 85% (15)

Examples of large-scale preparations utilizing potassium amide in liquid ammonia for the benzylic deprotonation of lutidine and diphenylacetonitrile have been published.[3,34] Allylic deprotonation of 2-chloromethyl-1-butene with potassium amide in THF at 65 °C has been used in an effective preparation of vinylcyclopropane (eq 16).[35]

Et, Cl; KNH_2, THF, 65 °C; ca. 60% (16)

Elimination Reactions. Potassium amide in liquid ammonia can be used as a base to induce elimination reactions.[36] For example, this reagent has been used in a preparation of dimethoxycyclopropene via an intramolecular alkylation followed by elimination (eq 17).[37] Potassium amide-induced elimination followed by an additional deprotonation has also been used to generate 8,8-dimethylcyclooctatrienyl anion (eq 18).[38]

MeO OMe, Br, Cl; KNH_2, NH_3; 40–50%; MeO OMe (17)

Cl + Cl; KNH_2, NH_3; − (18)

Related Reagents. Lithium Amide; Lithium Diisopropylamide; Potassium 3-Aminopropylamide; Potassium Diisopropylamide; Potassium Hexamethyldisilazide; Sodium Amide.

1. Biehl, E. R.; Stewart, W.; Marks, A.; Reeves, P. C. *JOC* **1979**, *44*, 3674.
2. Buncel, E.; Menon, B. *JOM* **1977**, *141*, 1.
3. Hauser, C. R.; Dunnavant, W. R. *OSC* **1963**, *4*, 962.
4. *FF* **1967**, *1*, 907.
5. Bunnett, J. F.; Hrutfiord, B. F.; Williamson, S. M. *OSC* **1973**, *5*, 12.
6. Sanders, D. R. *Chem. Eng. News* **1986**, *64* (21), 2.
7. Hoffmann, R. W. *Dehydrobenzene and Cycloalkynes*; Academic: New York, 1967; Chapter 1 and references therein.
8. Roberts, J. D.; Simmons, H. E., Jr.; Carlsmith, L. A.; Vaughan, C. W. *JACS* **1953**, *75*, 3290.
9. Hoffmann, R. W. *Dehydrobenzene and Cycloalkynes*; Academic: New York, 1967; pp 115–119.
10. Reinecke, M. G. *T* **1982**, *38*, 427.
11. Pieterse, M. J.; den Hertog, H. J. *RTC* **1961**, *80*, 1376.
12. Reinecke, M. G. *T* **1982**, *38*, 485.
13. Hrutford, B. F.; Bunnett, J. F. *JACS* **1958**, *80*, 2021.
14. Hoffmann, R. W. *Dehydrobenzene and Cycloalkynes*; Academic: New York, 1967; pp 150–164.
15. Ahmed, I.; Cheeseman, G. W. H.; Jaques, B. *T* **1979**, *35*, 1145.
16. Kessar, S. V. *ACR* **1978**, *11*, 283.
17. Kessar, S. V.; Gopal, R.; Singh, M. *T* **1973**, *29*, 167.
18. Kessar, S. V.; Pal, D.; Singh, M. *T* **1973**, *29*, 177.
19. de Valk, J.; van der Plas, H. C. *RTC* **1971**, *90*, 1239.
20. van der Plas, H. C. *ACR* **1978**, *11*, 462.
21. van der Plas, H. C. *T* **1985**, *41*, 237.
22. For examples see: (a) van der Plas, H. C. *Ring Transformations of Heterocycles*; Academic: New York, 1973; Vol. 2. (b) Rykowski, A.; van der Plas, H. C. *JOC* **1987**, *52*, 71. (c) Nagel, A.; van der Plas, H. C.; Geurtsen, G.; van der Kuilen, A. *JHC* **1979**, *16*, 305.
23. Lont, P. J.; van der Plas, H. C.; Koudijs, A. *RTC* **1971**, *90*, 207.
24. van der Plas, H. C. *Ring Transformations of Heterocycles*; Academic: New York, 1973; Vol. 2, pp 135–141.
25. The general area of nucleophilic substitution (including KNH_2) of hydrogen in azines has been reviewed: Chupakhin, O. N.; Charushin, V. N.; van der Plas, H. C. *T* **1988**, *44*, 1.
26. Hara, H.; van der Plas, H. C. *JHC* **1982**, *19*, 1285.
27. (a) Hara, H.; van der Plas, H. C. *JHC* **1982**, *19*, 1527. (b) Wozniak, M.; van der Plas, H. C.; van Veldhuizen, B. *JHC* **1983**, *20*, 9.
28. van der Plas, H. C.; Wozniak, M. *Croat. Chem. Acta* **1986**, *59*, 33.
29. Rossi, R. A.; Bunnett, J. F. *JOC* **1972**, *37*, 3570.
30. Harris, T. M.; Harris, C. M. *OR* **1969**, *17*, 155.
31. Feuer, H.; Hall, A. M.; Golden, S.; Reitz, R. L. *JOC* **1968**, *33*, 3622.
32. Hauser, C. R.; Kantor, S. W. *JACS* **1951**, *73*, 1437.
33. Weinheimer, A. J.; Kantor, S. W.; Hauser, C. R. *JOC* **1953**, *18*, 801.
34. Kofron, W. G.; Baclawski, L. M. *OSC* **1988**, *6*, 611.
35. Arora, S.; Binger, P.; Köster, R. *S* **1973**, 146.
36. Hauser, C. R.; Skell, P. S.; Bright, R. D.; Renfrow, W. B. *JACS* **1947**, *69*, 589.
37. Baucom, K. B.; Butler, G. B. *JOC* **1972**, *37*, 1730.
38. Staley, S. W.; Pearl, N. J. *JACS* **1973**, *95*, 2731.

Katherine S. Takaki
Bristol-Myers Squibb Co., Wallingford, CT, USA

Potassium 3-Aminopropylamide

$KNH(CH_2)_3NH_2$

[56038-00-7] $C_3H_9KN_2$ (MW 112.24)

(strong base[1] used primarily for the isomerization of internal alkynes to terminal alkynes;[2] can also be used to isomerize alkenes;[3] also used to promote intramolecular transamidation reactions[4])

Alternate Name: KAPA.
Solubility: highly sol ($\geq$1.5 M) in excess 1,3-diaminopropane
Preparative Methods: usually prepared as a solution in excess, dry 1,3-diaminopropane.[5] The most common preparation involves the direct, quantitative reaction of ***Potassium Hydride*** with excess amine at 20–25 °C.[6] Alternative preparations have been developed to avoid the handling and purification of KH. The reagent has been prepared by the reaction of ***Potassium Amide*** and 1,3-diaminopropane followed by evaporation of excess ***Ammonia*** in vacuo.[7] The reagent has also been prepared by heating ***Potassium*** and 1,3-diaminopropane under sonication.[8] A trace of iron(III) nitrate increases the rate of reagent formation, although it is not essential. Addition of ***Potassium t-Butoxide*** to lithium 3-aminopropylamide (from ***Lithium*** and 1,3-diaminopropane) probably generates KAPA in situ via cation exchange.[9]
Handling, Storage, and Precautions: moisture sensitive. A solution of KAPA in 1,3-diaminopropane appears stable for at least 8 h at rt and possibly even for 2–3 days.[2,10] Use in a fume hood.

Isomerization of Alkynes. Potassium 3-aminopropylamide is best known for its use in the 'acetylene zipper' reaction.[2] This reagent induces a rapid migration of triple bonds from the interior of a carbon chain to the terminus. Previously reported isomerizations of alkynes required very strong bases (e.g. ***Sodium***, ***Sodium Amide***) and elevated temperatures to effect a one-position isomerization of a 2-alkyne to a 1-alkyne. Migrations over more than one position were reported to be generally unsatisfactory or very slow. By contrast, KAPA produces rapid, multiposition alkyne migrations to provide the terminal alkyne within seconds at 0 °C (eq 1). Although internal alkynes are favored thermodynamically, this isomerization is driven forward by the formation (possibly aided by precipitation) of the stable metal acetylide. Quenching of the acetylide with water provides the terminal alkyne in excellent yield. In nonlinear systems, migration is blocked by branching in the chain.

$$Me(CH_2)_5{-}{\equiv}{-}(CH_2)_5Me \xrightarrow{\text{KAPA}} Me(CH_2)_{11}{-}{\equiv}{-}K \xrightarrow{H_2O} Me(CH_2)_{11}{-}{\equiv} \quad (1)$$

89%

The reaction has also been studied in functionalized systems. In alkynic alcohols, KAPA-induced migration of the triple bond proceeds to the chain terminus which is remote from the OH function (eq 2).[11] It is believed that the anionic intermediate in the migration of the triple bond is inhibited from migrating toward the similarly charged alkoxide.[11a] If the hydroxy group is attached to a chiral carbon, this phenomenon protects the configuration of that carbon during the migration process. Isomerization of chiral propargylic alcohols proceeds to give the terminal alkynic alcohols with no significant loss of enantiomeric purity (eq 3).[12]

$$HO(CH_2)_7{-}{\equiv}{-}(CH_2)_7Me \xrightarrow[\text{90\%}]{\text{KAPA, 20 °C, 1 h}} HO(CH_2)_{15}{-}{\equiv} \quad (2)$$

$$Me(CH_2)_9CH(OH){-}{\equiv}{-}(CH_2)_4Me \xrightarrow{\text{KAPA}} Me(CH_2)_9CH(OH)(CH_2)_5{-}{\equiv} \quad (3)$$

>99% ee

Successful isomerizations have also been carried out with ω-alkynylorganoboranes and propynyl ethers.[13] Isomerizations of alkynic carboxylic acids generally lead to a mixture of products and poor yields of the terminal alkyne.[14] However, if the carboxylic acid group is separated from the alkyne by an intervening carbinol, the carboxylate can be carried through the migration reaction without difficulty.[15]

KAPA can also be used for the in situ generation of an internal alkyne which then rapidly isomerizes to the terminal position. Allenes, vinyl sulfides, and cyclic ethers have been used as alkyne equivalents to provide terminal alkynes (eq 4).[16]

$$\text{2-tetrahydrofuranyl–CH(SPh)(CH}_2)_6\text{Me} \xrightarrow{\text{KAPA}} [\text{KO(CH}_2)_3\text{CH=C(SPh)(CH}_2)_6\text{Me}] \longrightarrow HO(CH_2)_{10}{-}{\equiv} \quad (4)$$

84%

Isomerization of Exocyclic Alkenes. Trace quantities of KAPA can be used in a two-phase reaction to effect the rapid, low-temperature isomerization of neat, exocyclic alkenes on a large scale.[3] In particular, KAPA has been used for the isomerization of (−)-β-pinene to the less accessible (−)-α-pinene in high yield and without racemization (eq 5).[17]

$$\text{(−)-β-pinene (92.1\% ee)} \xrightarrow[\text{93\%}]{\text{KAPA}} \text{(−)-α-pinene (92\% ee)} \quad (5)$$

Intramolecular Transamidation Reactions. KAPA is recommended for use in the 'zip' reaction for the ring enlargement of aminolactams (eq 6).[4,18] It is believed that the reaction is driven in part by formation of the resonance-stabilized amide anion.

$$\text{13-membered lactam N-(CH}_2)_3\text{NH(CH}_2)_3\text{NH}_2 \xrightarrow[\text{90\%}]{\text{KAPA}} \text{21-membered macrocyclic lactam (NH, NH, NH)} \quad (6)$$

Related Reagents. Lithium Amide; Lithium Diisopropylamide; Potassium Diisopropylamide; Potassium Hexamethyldisilazide.

1. KAPA has been reported to be over a million times more basic than the potassium salt of DMSO: Arnett, E. M.; Venkatasubramaniam, K. G. *TL* **1981**, *22*, 987.
2. Brown, C. A.; Yamashita, A. *JACS* **1975**, *97*, 891.
3. Brown, C. A. *S* **1978**, 754.
4. Kramer, U.; Guggisberg, A.; Hesse, M.; Schmid, H. *AG(E)* **1977**, *16*, 861.
5. Burfield, D. R.; Smithers, R. H.; Tan, A. S.-C. *JOC* **1981**, *46*, 629.
6. Brown, C. A. *CC* **1975**, 222.
7. Hommes, H.; Brandsma, L. *RTC* **1977**, *96*, 160.
8. (a) Kimmel, T.; Becker, D. *JOC* **1984**, *49*, 2494. (b) Theobald, P. G.; Okamura, W. H. *JOC* **1990**, *55*, 741.
9. Abrams, S. R. *CJC* **1984**, *62*, 1333.
10. No change in the NMR spectrum noted after 2–3 days: Arnett, E. M.; Venkatasubramaniam, K. G. *JOC* **1983**, *48*, 1569.
11. (a) Brown, C. A.; Yamashita, A. *CC* **1976**, 959. (b) Lindhoudt, J. C.; van Mourik, G. L.; Pabon, H. J. J. *TL* **1976**, 2565. (c) Abrams, S. R.; Shaw, A. C. *OS* **1988**, *66*, 127. (d) Abrams, S. R.; Shaw, A. C. *OSC* **1993**, *8*, 146.
12. (a) Midland, M. M.; Halterman, R. L; Brown, C. A.; Yamaichi, A. *TL* **1981**, *22*, 4171. (b) Maas, R. L.; Ingram, C. D.; Porter, A. T.; Oates, J. A.; Taber, D. F.; Brash, A. R. *JBC* **1985**, *260*, 4217. (c) Heathcock, C. H.; Stafford, J. A. *JOC* **1992**, *57*, 2566.
13. (a) Brown, C. A.; Negishi, E.-I. *CC* **1977**, 318. (b) Almansa, C.; Moyano, A.; Pericàs, M. A.; Serratosa, F. *S* **1988**, 707.
14. Conditions have been reported for the isomerization of the triple bond of stearolic acid to the terminal position in moderate yield: Augustin, K. E.; Schäfer, H. J. *LA* **1991**, 1037.

15. Burger, A.; Clark, J. E.; Nishimoto, M.; Muerhoff, A. S.; Masters, B. S. S.; de Montellano, O. *JMC* **1993**, *36*, 1418.
16. (a) Brown, C. A. *JOC* **1978**, *43*, 3083. (b) Grattan, T. J.; Whitehurst, J. S. *JCS(P1)* **1990**, 11. (c) Barajas, L.; Hernández, J. E.; Torres, S. *SC* **1990**, *20*, 2733.
17. (a) Brown, C. A.; Jadhav, P. K. *OS* **1987**, *65*, 224. (b) Brown, C. A.; Jadhav, P. K. *OSC* **1993**, *8*, 553.
18. Kramer, U.; Guggisberg, A.; Hesse, M.; Schmid, H. *AG(E)* **1978**, *17*, 200.

Katherine S. Takaki
Bristol-Myers Squibb Co., Wallingford, CT, USA

Potassium *t*-Butoxide

t-BuOK

[865-47-4] C_4H_9KO (MW 112.23)

(strong alkoxide base capable of deprotonating many carbon and other Brønsted acids; relatively poor nucleophile[1])

Physical Data: mp 256–258 °C (dec).
Solubility: sol/100g solvent at 25–26 °C: hexane 0.27 g, toluene 2.27 g, ether 4.34 g, *t*-BuOH 17.80 g, THF 25.00 g.
Form Supplied in: white, hygroscopic powder; widely available commercially; also available as a 1.0 M solution in THF.
Preparative Methods: sublimation (220 °C/1 mmHg; 180 °C/0.05 mmHg) of the commercial material prior to use is recommended. For critical experiments, the reagent should be freshly prepared prior to use. Solutions of the reagent in *t*-BuOH may be prepared by reaction of the anhydrous alcohol with ***Potassium*** under nitrogen,[2] or the solvent may be removed (finally at 150 °C/0.1–1.0 mmHg for 2 h) and solutions prepared using other solvents.[3]
Handling, Storage, and Precautions: do not breathe dust; avoid contact with eyes, skin, and clothing. Reacts with water, oxygen, and carbon dioxide, and may ignite on exposure to air at elevated temperatures. Handle in an inert atmosphere box or bag and conduct reactions under an inert atmosphere in a fume hood. Store in small lots in sealed containers under nitrogen.

Introduction. Potassium *t*-butoxide is intermediate in power among the bases which are commonly employed in modern organic synthesis. It is a stronger base than the alkali metal hydroxides and primary and secondary alkali metal alkoxides,[1] but it is a weaker base than the alkali metal amides and their alkyl derivatives, e.g. the versatile strong base ***Lithium Diisopropylamide***.[4] The continued popularity of *t*-BuOK results from its commercial availability and the fact that its base strength is highly dependent on the choice of reaction solvent. It is strongly basic in DMSO, where it exists primarily as ligand-separated ion pairs and dissociated ions, but its strength is significantly decreased in solvents such as benzene, THF, and DME, where its state of aggregation is largely tetrameric. DMSO is able to enhance the basicity of the *t*-butoxide anion by selectively complexing with the potassium cation. Other additives, such as the dipolar aprotic solvent ***Hexamethylphosphoric Triamide*** (HMPA) and ***18-Crown-6***, have a similar effect. This section will provide cursory coverage of the reactions of *t*-BuOK in *t*-BuOH and in relatively nonpolar aprotic solvents. The unique features of ***Potassium t-Butoxide-Dimethyl Sulfoxide***, ***Potassium t-Butoxide-Hexamethylphosphoric Triamide***, and ***Potassium t-Butoxide-18-Crown-6***, as well as those of the ***Potassium t-Butoxide-t-Butyl Alcohol Complex (1:1)***, are described in the Encyclopedia of Reagents for Organic Synthesis.

Alkylations. Many bases which are weaker than *t*-BuOK are capable of essentially quantitative conversion of active methylene compounds into the corresponding enolates or other anions.[5] However, the alkylation of diethyl malonate with a bicyclic secondary tosylate (eq 1)[6] and the alkylation of ethyl *n*-butylacetoacetate with *n*-BuI (eq 2)[7] provide examples of cases where the use of *t*-BuOK in *t*-BuOH is very effective. In the latter reaction, cleavage of the product via a retro-Claisen reaction is minimized with the sterically hindered base and yields obtained are higher than when ***Sodium Ethoxide*** or EtOK in EtOH, ***Sodium*** in dioxane or toluene, or ***Sodium Hydride*** in toluene are used for the enolate formation.

OTs, H + $EtO_2C\diagdown\diagup CO_2Et$ — *t*-BuOK, *t*-BuOH, reflux, 65% → H, CO_2Et, CO_2Et (1)

Bu, COMe, CO_2Et — *t*-BuOK, *t*-BuOH, reflux → BuI, *t*-BuOH → Bu, Bu, COMe, CO_2Et, 80% (2)

Potassium *t*-butoxide in *t*-BuOH or ethereal solvents is not capable of effecting quantitative formation of enolates of unactivated saturated ketones;[8] also, because potassium enolates are subject to rapid proton transfer reactions, their intermolecular alkylations are complicated by equilibration of structurally isomeric enolates, polyalkylation, and aldol condensation reactions.[9] Thus reactions of preformed lithium enolates (generated by deprotonation of ketones with LDA or by indirect procedures) with alkylating agents provide the method of choice for regioselective alkylation of unsymmetrical ketones.[9,10] However, as illustrated in eqs 3 and 4,[11,12] ketones capable of forming only a single enolate, such as symmetrical cyclic ketones and those containing α-methylene blocking groups, are readily alkylated via *t*-BuOK-promoted reactions. Also, *t*-BuOK in *t*-BuOH or benzene frequently has been employed for the alkylation of α,α-disubstituted aldehydes.[9]

O + $Cl\diagup\diagdown SMe_2^+ I^-$ — *t*-BuOK, *t*-BuOH, 70% → O (3)

1. *t*-BuOK, *t*-BuOH; 2. (methallyl chloride); 3. KOH, H_2O, diethylene glycol, heat; 84%; 4:1 (4)

As illustrated in eq 5, α,β-unsaturated ketones undergo α,α-dialkylation when treated with excess *t*-BuOK/*t*-BuOH and an alkylating agent.[13] The α,α-dimethylated β,γ-unsaturated ketone is formed by conversion of the initially produced α-methylated β,γ-unsaturated ketone to its dienolate, which undergoes a second methylation faster than the β,γ-double bond is isomerized to the α,β-position.[14]

1. *t*-BuOK (3 equiv), *t*-BuOH; 2. MeI (excess); 77% (5)

In contrast to intermolecular processes, intramolecular alkylations are frequently performed with *t*-BuOK in various solvents.[19] In eqs 6 and 7 are shown examples of *endo*- and *exo*-cycloalkylations of cyclic saturated ketones which lead to new five-membered rings.[15,16]

t-BuOK, *t*-BuOH; 65% (6)

t-BuOK, PhH; 70% (7)

An interesting example of how a change in the base can influence the course of a cycloalkylation reaction is shown in eq 8.[17] Since the reaction with *t*-BuOK involves equilibrating conditions, *exo*-cycloalkylation occurs via the more-substituted enolate, which is more thermodynamically stable. On the other hand, when LDA is used as the base, *endo*-cycloalkylation occurs via the kinetically formed terminal enolate.

t-BuOK, *t*-BuOH, pentane, 25 °C; 86–94%; LDA, THF, −72 to 65 °C; 77–84% (8)

Intramolecular alkylations of α,β-unsaturated ketones may occur at the α-, α′-, or γ-positions depending upon the nature of the base, the leaving group, and other structural features. A recent example involving α-cycloalkylation using *t*-BuOK is shown in eq 9.[18]

t-BuOK, *t*-BuOH, THF, 25 °C; 75% (9)

The reaction of α-halo esters, ketones, nitriles, and related compounds with appropriate organoboranes in the presence of *t*-BuOK can lead to replacement of the halogen with an alkyl or an aryl group.[19] An example of this reaction using an α-bromo ester is shown in eq 10.[20] THF is a more effective solvent than *t*-BuOH for alkylations of α-bromo ketones using this methodology.[21] Potassium 2,6-di-*t*-butylphenoxide, a mild, sterically hindered base, is much more effective than *t*-BuOK for the alkylation of highly reactive α-halo ketones, e.g. bromoacetone, and α-halo nitriles.[22]

t-BuOK, *t*-BuOH, THF; 75% (10)

Condensation Reactions. Traditionally, intermolecular aldol condensation reactions have been performed under equilibrating conditions using weaker bases than *t*-BuOK in protic solvents.[23] Since the mid-1970s, new methodology has focused on directed aldol condensations which involve the use of preformed ***Acetaldehyde*** and Group 2 enolates,[24] Group 13 enolates,[25] and transition metal enolates.[26] Although examples of the use of *t*-BuOK in intramolecular aldol condensations are limited, complex diketones (eq 11)[27] and keto aldehydes (eq 12)[28] have been cyclized with this base.

t-BuOK, *t*-BuOH, rt; 79% (11)

t-BuOK is the most commonly used base for the Darzens condensation of an α-halo ester with a ketone or an aromatic aldehyde to yield an α,β-epoxy or glycidic ester.[29] In the example shown in eq 13,[30] the aldol step is reversible and the epoxide ring closure step is rate limiting. This leads to the product with the ester group and the bulky β-substituent *trans*. However, in other systems the opposite stereochemical result often occurs because the aldol condensation step is rate limiting. In addition to esters, a wide variety of α-halo compounds that contain electron-withdrawing groups may participate in these types of reaction.[29c]

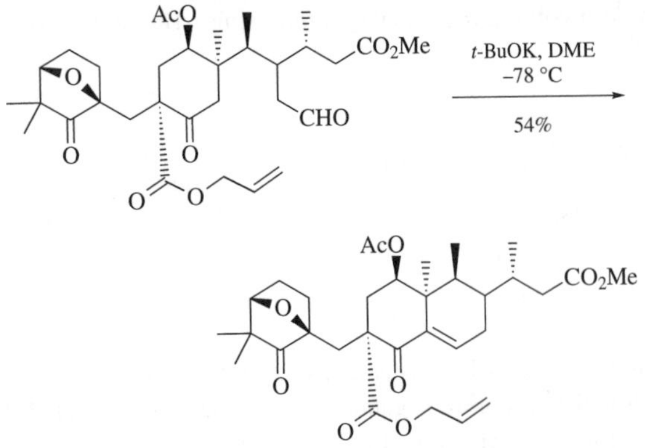

t-BuOK, DME, –78 °C, 54% (12)

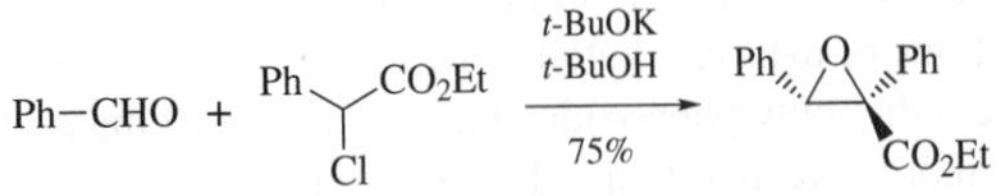

t-BuOK, *t*-BuOH, 75% (13)

The Dieckmann cyclization of diesters and related reactions has found an enormous amount of use in the synthesis of five-, six-, seven-, and even larger-membered rings.[31] In unsymmetrical systems, steric effects and the stability of the product enolates determine the regiochemistry of the reaction. As shown in eqs 14 and 15,[32,33] *t*-BuOK is an effective base for these reactions when used in *t*-BuOH or other solvents such as benzene. In the former example, 40% of unchanged starting material is recovered when MeONa/PhH is used to effect the cyclization.

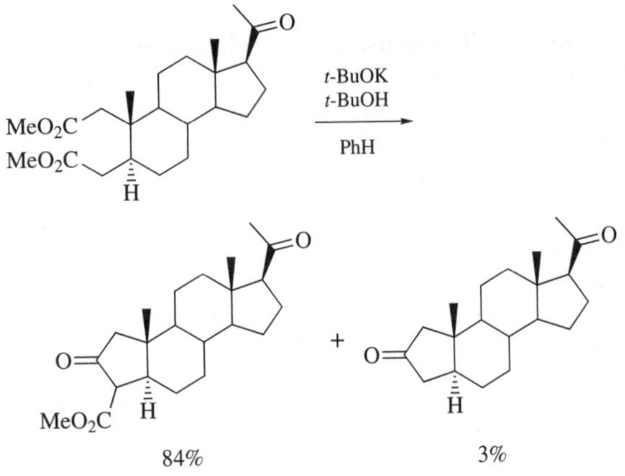

t-BuOK, *t*-BuOH, PhH; 84% + 3% (14)

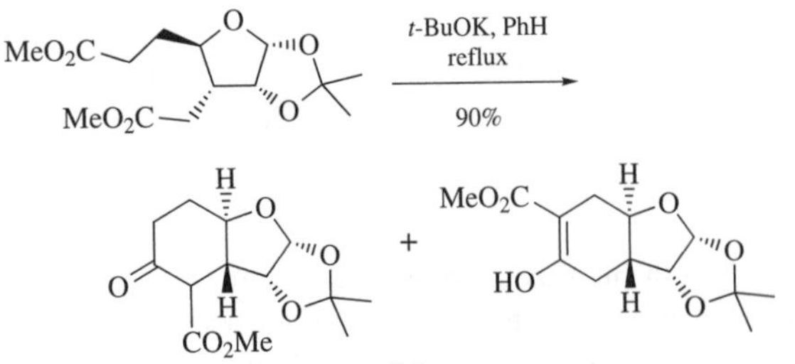

t-BuOK, PhH, reflux, 90%; 8:1 (15)

The nature of the base can profoundly influence the regiochemistry of the reaction. *t*-BuOK favors kinetic control in the reaction shown in eq 16 and the product derived from cyclization of the enolate having a β-amino group is obtained. However, when EtONa/EtOH is employed, the more stable β-keto ester enolate resulting from thermodynamic control is obtained.[34] In addition to diesters, dinitriles, ε-keto esters, ε-cyano esters, ε-sulfinyl esters, and ε-phosphonium esters may participate in these reactions.[31]

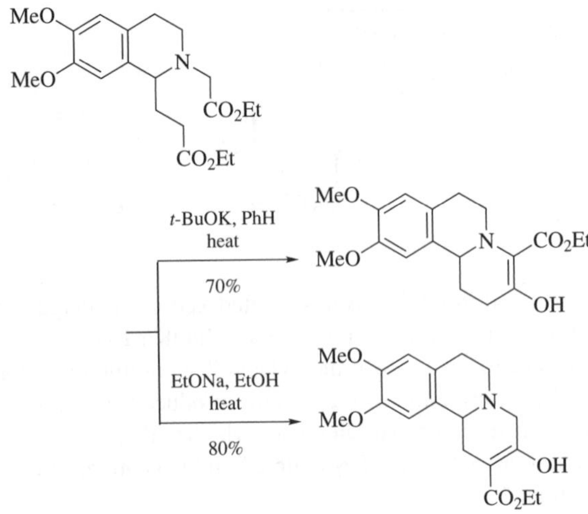

t-BuOK, PhH, heat, 70%; EtONa, EtOH, heat, 80% (16)

t-BuOK is a better base than EtONa for the Stobbe condensation (eq 17) because yields are higher, reaction times are shorter, and frequent side reactions of ketone or aldehyde reduction are avoided.[35]

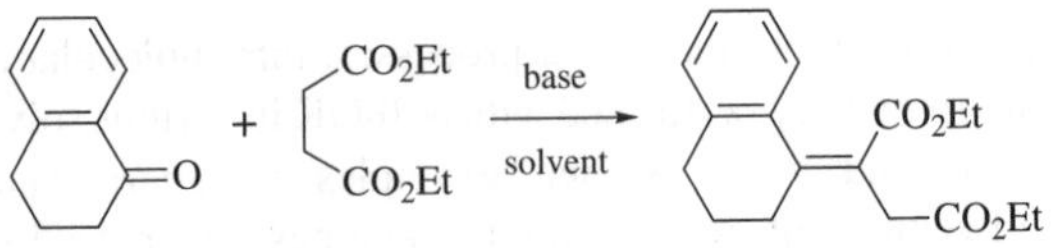

base, solvent (17)

Base, solvent	Yield (%)
EtONa (2 equiv), EtOH, reflux, 6 h	79
t-BuOK (1.1 equiv), *t*-BuOH, reflux, 40 min	89–94

Although the use of *t*-BuOK for the generation of unstabilized ylides from phosphonium salts has been rare, it is the base of choice for the generation of ***Methylenetriphenylphosphorane*** for the Wittig reaction of hindered ketones (eq 18).[36] No 1,1-di-*t*-butylethylene is obtained from di-*t*-butyl ketone when NaH/DMSO is used to generate the ylide.[37]

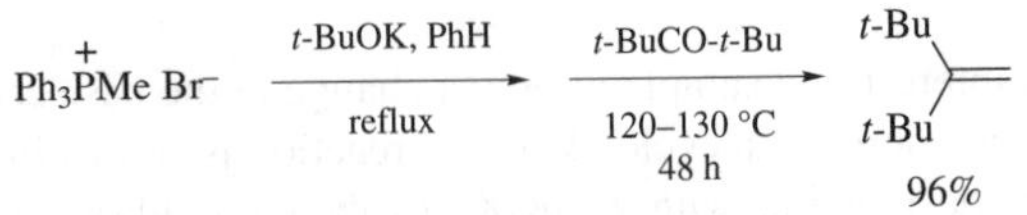

$Ph_3\overset{+}{P}Me\ Br^-$ — *t*-BuOK, PhH, reflux → *t*-BuCO-*t*-Bu, 120–130 °C, 48 h → 96% (18)

Elimination Reactions. *t*-BuOK is a widely used base for both α- and β-elimination reactions. It is the most effective base in the conventional alkoxide–haloform reaction for the generation of dihalocarbenes.[38] This procedure still finds general use (eq 19),[39] but since it requires anhydrous conditions, it has been replaced to a degree by use of phase-transfer catalysts.[40] Vinylidene carbenes have also been produced from the reaction of α-halo allenes with *t*-BuOK.[41]

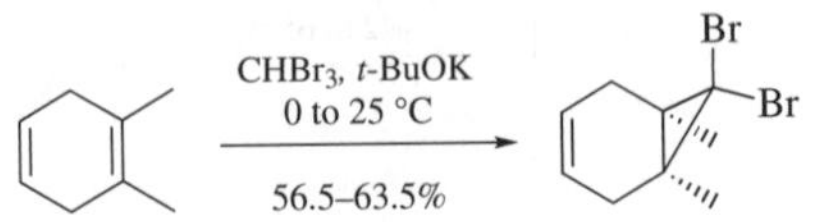

$CHBr_3$, *t*-BuOK, 0 to 25 °C, 56.5–63.5% (19)

Substrates containing a host of leaving groups, such as alkyl chlorides (eqs 20 and 21),[42,43] bromides (eqs 22–24),[44–46] tosylates, mesylates (eq 25),[47] and even sulfenates and sulfinates (eq 26),[48] undergo β-eliminations with *t*-BuOK in the solid phase or in *t*-BuOH or various nonpolar solvents.

t-BuOK, THF, −78 to 25 °C, 80–88% (20)

t-BuOK, THF, 80% (21)

t-BuOK, xylene, 75% (22)

t-BuOK (solid), 67%, *anti*-elimination (23)

t-BuOK, ether, 0 to 5 °C; 52–67% (24)

t-BuOK, t-BuOH; X = O, 32%; X = S, 65% (25)

t-BuOK, THF, 68% (26)

The regiochemical and stereochemical results of β-eliminations of 2-substituted acyclic alkanes with *t*-BuOK in various solvents have been extensively investigated.[49] Space does not permit the details of these investigations to be presented, but a few brief generalizations may be noted. (1) *t*-BuOK normally gives more of the terminal alkene than primary alkoxide bases but less than extremely bulky bases such as potassium tricyclohexylmethoxide. (2) A greater proportion of the terminal alkene is produced in solvents such as *t*-BuOH, where the base is highly aggregated, than in DMSO where it is substantially monomeric and partially dissociated. (3) The lower the state of aggregation of the base, the higher the *trans*:*cis* ratio of the disubstituted 2-alkene. (4) As base aggregation increases, the *syn*:*anti* elimination ratio increases. As far as the *trans*:*cis* ratio is concerned, cyclic halides give the opposite results from open-chain systems, i.e. this ratio is higher when base aggregation is greater (eq 27).[50]

t-BuOK, solvent (27)

Solvent	*trans* (%)	*cis* (%)
PhH	83	1.5
PhH, dicyclo-hexyl-18-crown-6	9	76
DMF	7	72

Fragmentation Reactions. Because it is a relatively strong base and a relatively weak nucleophile, *t*-BuOK has been the most popular base for effecting Grob-type fragmentations of cyclic 1,3-diol derivatives.[51] In *t*-BuOH or nonpolar solvents, sufficiently high concentrations of alkoxide ions of 1,3-diol derivatives are produced for fragmentations to proceed smoothly if relatively good leaving groups are present (eq 28).[52] When relatively poor leaving groups, e.g. sulfinates, are involved, it is necessary to increase the strength of *t*-BuOK by the addition of a dipolar aprotic solvent (eq 29).[53] Other base–solvent combinations which may offer advantages over *t*-BuOK/*t*-BuOH for certain fragmentation substrates include NaH/THF, $DMSO^-$ Na^+/DMSO, and LAH/ether.[51]

t-BuOK, t-BuOH, 40 °C; t-BuOK, t-BuOH, 47 °C; 88–91% (28)

t-BuOK, PhMe, HMPA, 120 °C, 72% (29)

Isomerizations of Unsaturated Compounds. *t*-BuOK is an effective base for bringing about migrations of double bonds in alkenes and alkynes via carbanion intermediates,[1] but since the base promotes these reactions most effectively in DMSO, they will be described in more detail under ***Potassium t-Butoxide-Dimethyl Sulfoxide***. Important examples of enone deconjugations with *t*-BuOK/*t*-BuOH which proceed via di- and trienolate intermediates are shown in eqs 30 and 31.[54,55] Potassium *t*-pentoxide is effective in promoting the latter reaction, but various lithium amide bases are not, apparently because they deprotonate the enone at the α′-position regioselectively. The isomerization of α,β-unsaturated imines to alkenyl imines (eq 32) is an important step in an alternative method for reduction–alkylation of α,β-unsaturated ketones.[56]

Ketone Cleavage Reactions. *t*-BuOK in ether containing water is the medium of choice for the cleavage of nonenolizable ketones such as nortricyclanone (eq 33).[57] Optically active tertiary α-phenyl phenyl ketones are cleaved in fair yields with high retention of configuration with the base in *t*-BuOH or PhH (eq 34).[58] *t*-BuOK is more effective than *t*-BuONa or ***Lithium t-Butoxide*** in these reactions.

Michael Additions. *t*-BuOK is one of an arsenal of bases that can be used in Michael addition reactions.[59] A catalyst prepared by impregnation of xonotlite with the base promotes the Michael addition of the β-diketone dimedone to 2 mol of MVK (eq 35).[60] Unsymmetrical ketones like 2-methylcyclohexanone yield Michael adducts primarily at the more substituted α-carbon atom (eq 36).[8a] Active methylene compounds undergo double Michael additions to enynes in the presence of *t*-BuOK (eq 37).[61] This type of reaction has been used in the total synthesis of griseofulvin. A tricyclo[5.3.1.$0^{1,5}$]undecanedione, a precursor of (±)-cedrene, is available by a highly regioselective intramolecular Michael addition using *t*-BuOK (eq 38).[62]

Oxidation Reactions. The use of *t*-BuOK to convert organic substrates to carbanionic species which react with molecular oxygen via a radical process has been reviewed.[1] Ketones and esters are the most common substrates for these reactions. Oxidations of unsymmetrical ketones occur via the more thermodynamically stable potassium enolates; tetrasubstituted enolates yield stable hydroperoxides, while hydroperoxides derived from trisubstituted enolates are further oxidized to α-diketones or their enol forms.[9a]

Rearrangement Reactions. Benzil is converted into the ester of benzylic acid in high yield upon treatment with *t*-BuOK in *t*-BuOH/PhH (eq 39).[63] Lower yields are obtained if the individual solvents alone are used or if the base is replaced by MeONa or EtONa.

Monotosylates of *cis*-1,2-diols, such as that derived from α-pinene, undergo pinacol-type rearrangements upon treatment with *t*-BuOK (eq 40).[64] The *trans* isomer is converted to the epoxide under the same conditions (eq 41).[64] 4-Benzoyloxycyclohexanone undergoes a mechanistically interesting rearrangement to benzoylcyclopropanepropionic acid when treated with *t*-BuOK/*t*-BuOH (eq 42).[65]

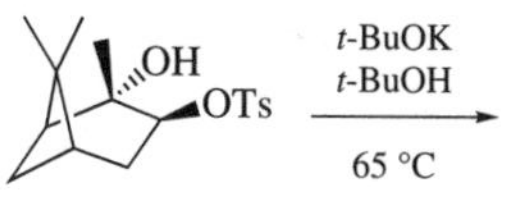

(41)

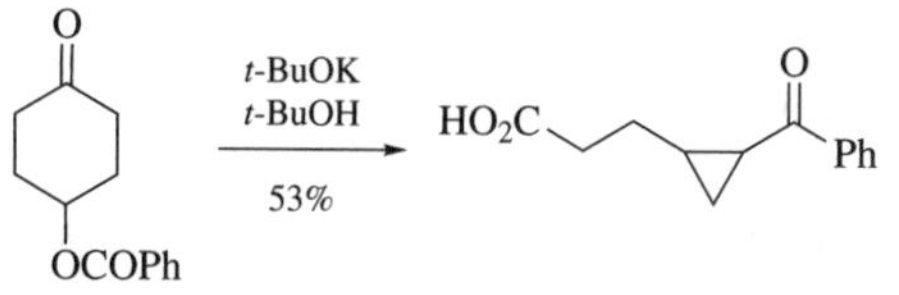

(42)

(46)

The Ramberg–Bäcklund rearrangement of α-halo sulfones is frequently carried out with t-BuOK.[66] This reaction provides a useful route to deuterium-labeled alkenes (eq 43).[67] t-BuOK is presumably the active base in a modification which involves the direct conversion of a sulfone into an alkene with KOH and a mixture of t-BuOH/CCl_4.[68] The base converts the sulfone to the α-sulfonyl carbanion, which undergoes chlorination with CCl_4 by a single-electron transfer process. Proton abstraction from the α′-position of the α-chloro ketone by the base yields a thiirane 1,1-dioxide which loses SO_2 to yield the alkene. t-BuOK is frequently used directly in this modified procedure (eq 44).[69] Cyclic enediynes are available from the corresponding α-chloro sulfones using t-BuOK to effect the Ramberg–Bäcklund rearrangement.[70]

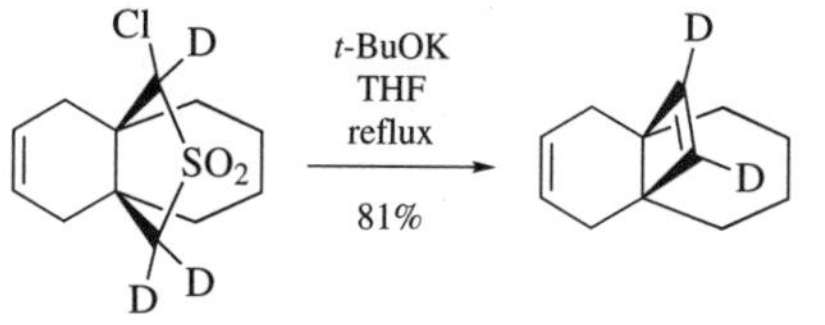

(43)

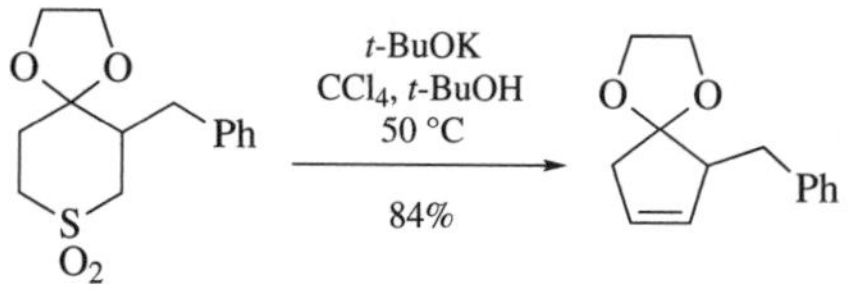

(44)

Bromomethylenecyclobutanes undergo ring enlargement reactions to 1-bromocyclopentenes when heated in the presence of solid t-BuOK (eq 45).[71]

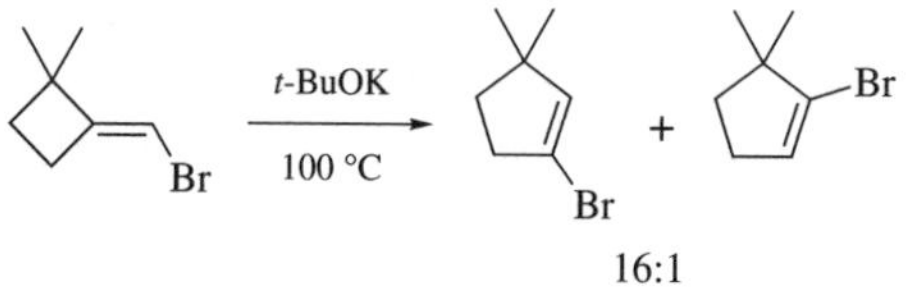

(45)

Reduction Reactions. A modification of the Wolff–Kishner reduction, which is particularly useful for the reduction of α,β-unsaturated ketones, involves the reaction of carbonyl hydrazones and semicarbazones with t-BuOK in PhMe at reflux.[72] The reduction of (±)-3-oxo-α-cadinol to (±)-α-cadinol and its isomer (eq 46) provides an example of this method.[73]

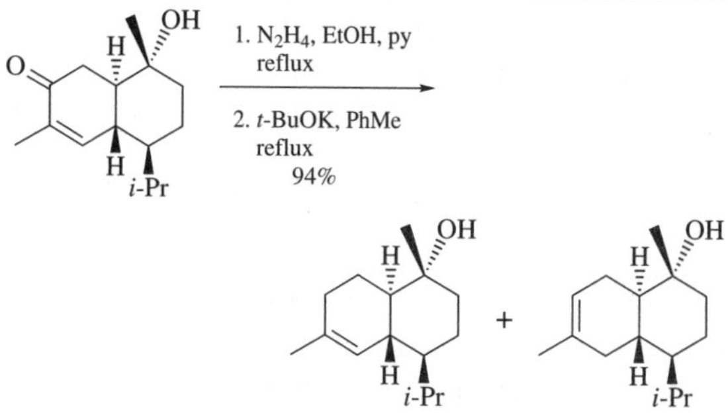

Related Reagents. n-Butyllithium-Potassium t-Butoxide; Potassium Amide; Potassium t-Butoxide-Benzophenone; Potassium t-Butoxide-t-Butyl Alcohol Complex; Potassium t-Butoxide-18-Crown-6; Potassium t-Butoxide-Dimethyl Sulfoxide; Potassium t-Butoxide-Hexamethylphosphoric Triamide; Potassium Diisopropylamide; Potassium t-Heptoxide; Potassium Hexamethyldisilazide; Potassium Hydroxide; Potassium 2-Methyl-2-butoxide.

---

1. Pearson, D. E.; Buehler, C. A. *CRV* **1974**, *74*, 45.
2. Johnson, W. S.; Schneider, W. P. *OSC* **1963**, *4*, 132.
3. (a) Skattebøl, L.; Solomon, S. *OS* **1969**, *49*, 35. (b) Skattebøl, L.; Solomon, S. *OSC* **1973**, *5*, 306.
4. House, H. O.; Czuba, L. J.; Gall, M.; Olmstead, H. D. *JOC* **1969**, *34*, 2324.
5. (a) Carey, F. A.; Sundberg, R. J. *Advanced Organic Chemistry*, 3rd ed.; Plenum: New York, 1990; Part B, p 3. (b) House, H. O. *Modern Synthetic Reactions*, 2nd ed.; Benjamin: Menlo Park, CA, 1972; pp 510–546.
6. Marshall, J. A.; Carroll, R. D. *JOC* **1965**, *30*, 2748.
7. Renfrow, W. B.; Renfrow, A. *JACS* **1946**, *68*, 1801.
8. (a) House, H. O.; Roelofs, W. L.; Trost, B. M. *JOC* **1966**, *31*, 646. (b) Malhotra, S. K.; Johnson, F. *JACS* **1965**, *87*, 5513. (c) Brown, C. A. *CC* **1974**, 680.
9. (a) Caine, D. In *Carbon–Carbon Bond Formation*; Augustine, R. L., Ed.; Dekker: New York, 1979; Vol. 1, Chapter 2. (b) Caine, D. *COS* **1991**, *3*, 1.
10. Evans, D. A. In *Asymmetric Synthesis*; Morrison, J. D., Ed.; Academic: New York, 1984; Vol. 3, Chapter 1.
11. Ruber, S. M.; Ronald, R. C. *TL* **1984**, *25*, 5501.
12. Piers, E.; Britton, R. W.; deWaal, W. *CJC* **1969**, *47*, 831.
13. Dauben, W. G.; Ashcraft, A. C. *JACS* **1963**, *85*, 3673.
14. Ringold, H. J.; Malhotra, S. K. *JACS* **1962**, *84*, 3402.
15. Cargill, R. L.; Bushey, D. F.; Ellis, P. D.; Wolff, S.; Agosta, W. C. *JOC* **1974**, *39*, 573.
16. Christol, H.; Mousserson, M.; Plenat, M. F. *BSF* **1959**, 543.
17. House, H. O.; Sayer, T. S. B.; Yau, C. C. *JOC* **1978**, *43*, 2153.
18. Srikrishna, A.; Hemamalini, P.; Sharma, G. V. R. *JOC* **1993**, *58*, 2509.
19. (a) Brown, H. C.; Rogić, M. M. *Organomet. Chem. Synth.* **1972**, *1*, 305. (b) Rogić, M. M. *Intra-Sci. Chem. Rep.* **1973**, *7*, 155.
20. (a) Brown, H. C.; Rogić, M. M.; Rathke, M. W.; Kabalka, G. W. *JACS* **1968**, *90*, 818, 1911; (b) Brown, H. C.; Rogić, M. M.; Rathke, M. W.; Kabalka, G. W. *JACS* **1969**, *91*, 2150.
21. Brown, H. C.; Rogić, M. M.; Rathke, M. W. *JACS* **1968**, *90*, 6218.
22. Brown, H. C.; Nambu, H.; Rogić, M. M. *JACS* **1969**, *91*, 6852, 6854, 6855.
23. Heathcock, C. H. *COS* **1991**, *2*, 133.
24. Heathcock, C. H. *COS* **1991**, *2*, 181.

25. Kim, B. M.; Williams, S. F.; Masamune, S. *COS* **1991**, *2*, 239.
26. Paterson, I. *COS* **1991**, *2*, 301.
27. Trost, B. M.; Shuey, C. D.; DiNinno, F., Jr.; McElvain, S. S. *JACS* **1979**, *101*, 1284.
28. Murai, A.; Tanimoto, N.; Sakamoto, N.; Masamune, T. *JACS* **1988**, *110*, 1985.
29. (a) Newman, M. S.; Magerlein, B. J. *OR* **1949**, *5*, 413. (b) Ballester, M. *CRV* **1955**, *55*, 283. (c) Rosen, T. *COS* **1991**, *2*, 409.
30. Zimmerman, H. E.; Ahramjiam, L. *JACS* **1960**, *82*, 5459.
31. (a) Schaefer, J. P.; Bloomfield, J. J. *OR* **1967**, *15*, 1. (b) Davis, B. R.; Garratt, P. J. *COS* **1991**, *2*, 795.
32. Nace, H. R.; Smith, A. H. *JOC* **1973**, *38*, 1941.
33. Georges, M.; Tam, T.-F.; Fraser-Reid, B. *JOC* **1985**, *50*, 5747.
34. Blasko, G.; Kardos, J.; Baitz-Gács, E.; Simonyi, M.; Szantay, C. *H* **1986**, *24*, 2887.
35. Johnson, W. S.; Daub, G. H. *OR* **1951**, *6*, 1.
36. Fitjer, L.; Quabeck, U. *SC* **1985**, *15*, 855.
37. Abruscato, G. J.; Binder, R. G.; Tidwell, T. T. *JOC* **1972**, *37*, 1787.
38. Parham, W. E.; Schweizer, E. E. *OR* **1963**, *13*, 55.
39. Taylor, R. T.; Paquette, L. A. *OSC* **1990**, *7*, 200.
40. (a) Starks, C. M.; Liotta, C. *Phase Transfer Catalysis: Principles and Techniques*; Academic: New York, 1978; Chapter 6. (b) Starks, C. M.; Liotta, C. L.; Halpern, M. *Phase-Transfer Catalysis: Fundamentals, Applications, and Industrial Perspectives*; Chapman & Hall: New York, 1994; Chapter 8.
41. Hartzler, H. D. *JOC* **1964**, *29*, 1311.
42. Pulwer, M. J.; Blacklock, T. J. *OSC* **1990**, *7*, 203.
43. Halton, B.; Milsom, P. J. *CC* **1971**, 814.
44. Johnson, W. S.; Johns, W. F. *JACS* **1957**, *79*, 2005.
45. Tremelling, M. J.; Hopper, S. P.; Mendelowitz, P. C. *JOC* **1978**, *43*, 3076.
46. (a) Paquette, L. A.; Barrett, J. H. *OS* **1969**, *49*, 62. (b) Paquette, L. A.; Barrett, J. H. *OSC* **1973**, *5*, 467.
47. Quinn, C. B.; Wiseman, J. R. *JACS* **1973**, *95*, 1342, 6120.
48. Mandai, T.; Hara, K.; Nakajima, T.; Kawada, M.; Otera, J. *TL* **1983**, *24*, 4993.
49. (a) Bartsch, R. A.; Zavada, J. *CRV* **1980**, *80*, 453. (b) Krebs, A.; Swienty-Busch, J. *COS* **1991**, *6*, 949.
50. Swoboda, M.; Hapala, J.; Zavada, J. *TL* **1972**, 265.
51. (a) Becker, K. B.; Grob, C. A. In *The Chemistry of Double-Bonded Functional Groups*; Patai, S., Ed.; Wiley: New York, 1977; Part 2, p 653. (b) Caine, D. *OPP* **1988**, *20*, 3. (c) Weyerstahl, P.; Marschall, H. *COS* **1991**, *6*, 1041.
52. Heathcock, C. H.; Badger, R. A. *JOC* **1972**, *37*, 234.
53. Fischli, A.; Branca, Q.; Daly, J. *HCA* **1976**, *59*, 2443.
54. Ringold, H. J.; Malhotra, S. K. *TL* **1962**, 669.
55. Corey, E. J.; Cyr, C. R. *TL* **1974**, 1761.
56. Wender, P. A.; Eissenstat, M. A. *JACS* **1978**, *100*, 292.
57. (a) Gassman, P. G.; Lumb, J. T.; Zalar, F. V. *JACS* **1967**, *89*, 946. (b) Gassman, P. G.; Zalar, F. V. *JACS* **1966**, *88*, 2252.
58. (a) Paquette, L. A.; Gilday, J. P. *JOC* **1988**, *53*, 4972. (b) Paquette, L. A.; Ra, C. S. *JOC* **1988**, *53*, 4978.
59. Jung, M. E. *COS* **1991**, *4*, 1.
60. Houbreckts, Y.; Laszlo, P.; Pennetreau, P. *TL* **1986**, *27*, 705.
61. Stork, G.; Tomasz, M. *JACS* **1964**, *86*, 471.
62. Horton, M.; Pattenden, G. *TL* **1983**, *24*, 2125.
63. Doering, W. von E.; Urban, R. S. *JACS* **1956**, *78*, 5938.
64. Carlson, R. G.; Pierce, J. K. *TL* **1968**, 6213.
65. Yates, P.; Anderson, C. D. *JACS* **1963**, *85*, 2937.
66. (a) Paquette, L. A. *OR* **1977**, *25*, 1. (b) Clough, J. M. *COS* **1991**, *3*, 861.
67. Paquette, L. A.; Wingard, R. E., Jr.; Photis, J. M. *JACS* **1974**, *96*, 5801.
68. (a) Meyers, C. Y.; Matthews, W. S.; Ho, L. L.; Kolb, V. M.; Parady, T. E. In *Catalysis in Organic Synthesis*; Smith, G. V., Ed.; Academic: New York, 1977; p 197. (b) Meyers, C. Y. In *Topics in Organic Sulfur Chemistry*; Tisler, M., Ed.; University Press: Ljubljana, 1978; p 207.
69. Matsuyama, H.; Miyazawa, Y.; Takei, Y.; Kobayashi, M. *JOC* **1987**, *52*, 1703.
70. Nicolaou, K. C.; Zuccarello, G.; Ogawa, Y.; Schweiger, E. J.; Kumazawa, T. *JACS* **1988**, *110*, 4866.
71. Erickson, K. L. *JOC* **1971**, *36*, 1031.
72. Grundon, M. F.; Henbest, H. B.; Scott, M. D. *JCS* **1963**, 1855.
73. Caine, D.; Frobese, A. S. *TL* **1977**, 3107.

Drury Caine

The University of Alabama, Tuscaloosa, AL, USA

Potassium Carbonate[1]

K_2CO_3

[584-08-7] CK_2O_3 (MW 138.21)

(inorganic base used in the alkylation of phenols[2] and 1,3-dicarbonyl compounds;[7] generation of sulfur[12] and phosphorus ylides;[13] also used to dry organic solvents[16])

Physical Data: mp 891 °C; *d* 2.43 g cm^{-3}; hygroscopic until 16.4% water is absorbed; pH of saturated aqueous solution is 11.6.
Solubility: 1.12 g mL^{-1} in water at 20 °C; practically insol alcohol.
Form Supplied in: white solid; widely available. Use in form supplied for best results.
Purification: crystallize from water 70 °C; dry over H_2SO_4.
Handling, Storage, and Precautions: keep in a tightly sealed container; avoid inhalation.

Alkylation of Phenols. The mildly basic nature of this reagent makes it ideal for the selective deprotonation of organic acids. This property makes K_2CO_3 a useful reagent for the alkylation of phenols. The procedure[2] for phenol alkylation involves dissolving the phenol and alkylating agent in acetone with excess K_2CO_3 added as an insoluble component (eq 1). The reaction is run at reflux. Workup consists of removal of solvent and addition of water. The product is isolated by either extraction or filtration. Popular alternative solvents are DMF[3] and MeCN.[4] If the aqueous solution needs to be neutralized, it is convenient to remove excess K_2CO_3 via filtration before solvent removal and subsequent water addition.

OH NO2 + Br —K_2CO_3, acetone, 80%→ O NO2 (1)

Sulfides can also be alkylated using K_2CO_3 in DMF, as demonstrated in the shell closure reaction used to capture guests within linked anisyl moieties.[5]

Hydrolysis Reactions. Ester hydrolysis using K_2CO_3 in MeOH was found to give clean conversion to the allylic alcohol (eq 2).[6]

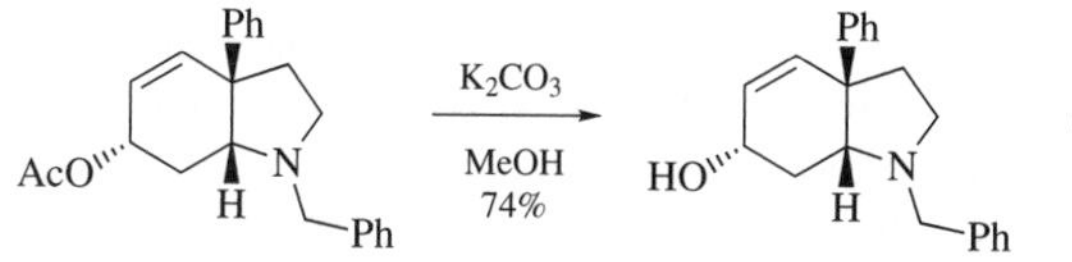

K_2CO_3, MeOH, 74% (2)

When amides that contain α,β-amino alcohols undergo diazotization followed by treatment with aqueous K_2CO_3,[7] they are cleanly converted to the epoxide (eq 3), while the β-amino ester gives the corresponding alcohol (eq 4).

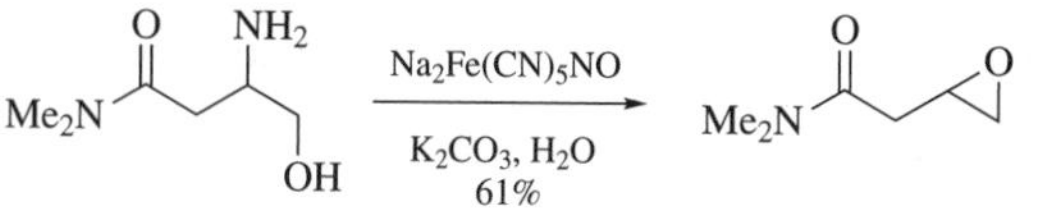

$Na_2Fe(CN)_5NO$; K_2CO_3, H_2O, 61% (3)

$Na_2Fe(CN)_5NO$; K_2CO_3, H_2O, 58% (4)

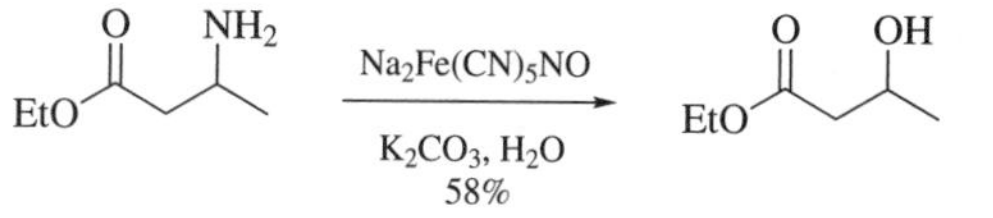

Reactions of 1,3-Dicarbonyl Compounds. The enolate of β-keto esters can be alkylated in a procedure developed by Claisen[8] that is analogous to the alkylation of phenols. The β-keto ester and alkylating agent are combined with K_2CO_3 in acetone and the reaction heated to reflux. The reaction is useful for the *C*-alkylation of substrates. In the case of ethyl 2-oxocyclopentanecarboxylate[9] the reaction gave good yields and high selectivity for *C*-alkylation with unobstructed alkylating agents (eq 5).

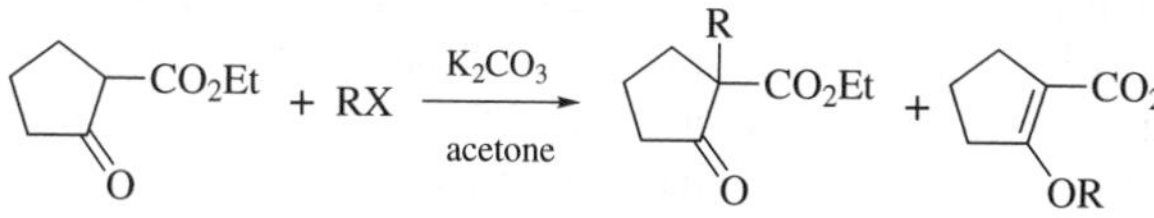

+ RX → (K_2CO_3, acetone) (5)

RX	*C*-alkylation (%)	*O*-alkylation (%)
MeI	100	<1
$(MeO)_2SO_2$	100	<1
H_2C=$CHCH_2Br$	91	<1
$BrCH_2CO_2Et$	92	<1
i-PrI	78	18
$PhCH_2OCH_2Cl$	49	21

The reagent also shows good selectivity in the monoalkylation of β-dicarbonyl compounds. The reaction of 2-chloro-2-methylpropanal with malonic esters in THF or ether gives good yield of the monoalkylated malonic ester (eq 6).[10]

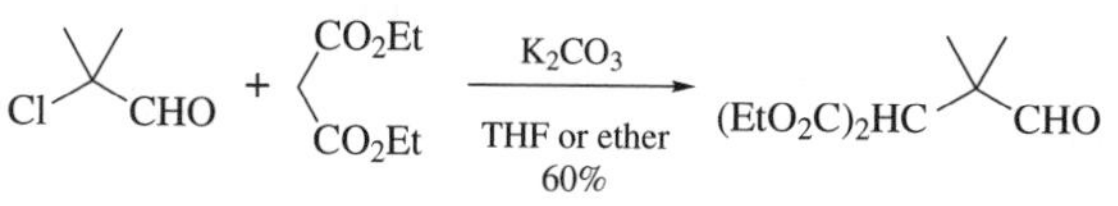

K_2CO_3; THF or ether, 60% → $(EtO_2C)_2HC$–C(Me)$_2$–CHO (6)

Aldehydes can be reacted with 1,3-dicarbonyl compounds in THF to give α,β-unsaturated carbonyl compounds with good (*Z*/*E*) selectivity (eq 7).[11]

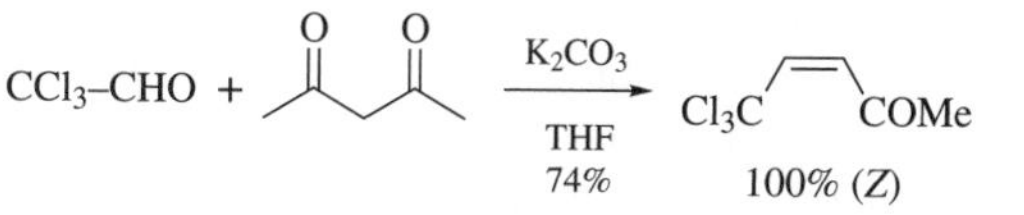

CCl_3–CHO + acetylacetone → (K_2CO_3, THF, 74%) Cl_3C–CH=CH–COMe, 100% (*Z*) (7)

The reagent is also useful for the generation of intramolecular Michael addition products with excellent stereoselectivity (eq 8).[12]

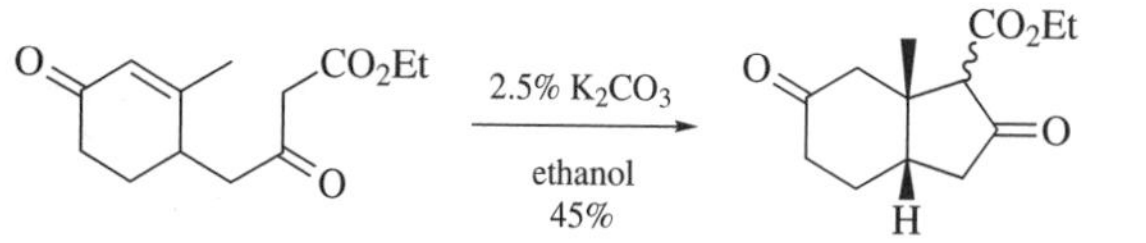

2.5% K_2CO_3, ethanol, 45% (8)

Generation of Phosphorus Ylides. Wittig reactions can be done either under heterogeneous (water–1,4-dioxane) conditions[13] or more conveniently using 18-crown-6 as a phase-transfer catalyst (see ***Potassium Carbonate-18-Crown-6***).[14] The phase-transfer conditions give alkenes in good yields with a preference for the (*E*)-isomer (eq 9). The procedure involves the combination of alkylphosphonium salt, K_2CO_3, aldehyde, and ***18-Crown-6*** in either CH_2Cl_2 or THF. The solvent is removed, product extracted with petroleum ether, and filtered through silica gel.

$$R^1CHO + Ph_3\overset{+}{P}CH_2R^2\ X^- \xrightarrow[CH_2Cl_2\text{ or THF}]{K_2CO_3,\ 18\text{-crown-6}} R^1CH{=}CHR^2 \quad (9)$$

R^1	R^2	(*E*) %
Ph	Ph	70
Ph	Me	78
Et	Ph	75

Generation of Sulfur Ylides. β-Ketosulfonium salts react with K_2CO_3 to generate the corresponding ylide. α-Methylthio ketones are reacted with allylic halides to afford sulfonium salts that are reacted with aqueous K_2CO_3 to generate a sulfur ylide[15] that can undergo [2,3]-sigmatropic rearrangement (eq 10).

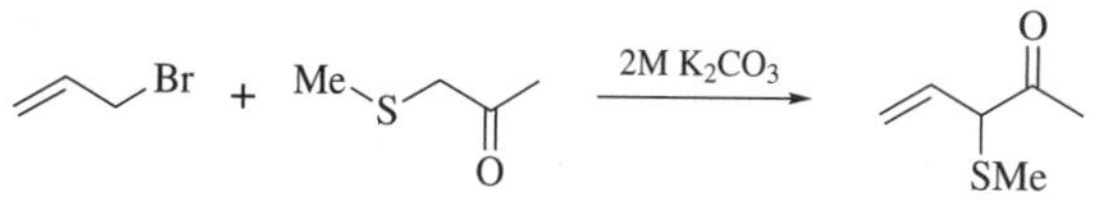

2M K_2CO_3 (10)

Related Reagents. Lithium Hydroxide; Potassium Carbonate-18-Crown-6; Potassium Hydride; Sodium Carbonate; Sodium Hydroxide.

1. *FF* **1975**, *5*, 552.
2. Allen, C. F.; Gates, J. W. Jr. *OSC* **1955**, *3*, 140.
3. Sherman, J. C.; Knobler, C. B.; Cram D. J. *JACS* **1991**, *113*, 2194.
4. Davis, R.; Muchowski, J. M. *S* **1982**, 987.
5. Bryant, J. A.; Blanda, M. T.; Vincenti, M.; Cram, D. J. *JACS* **1991**, *113*, 2167.
6. Martin, S. F.; Campbell, C. L. *TL* **1987**, *28*, 503.
7. McGarvey, G. J.; Kimura, M. *JOC* **1986**, *51*, 3913.
8. *MOC* **1952**, *8*, 602.
9. Barco, A.; Benetti, S.; Pollini, G. P. *S* **1973**, 316.
10. Takeda, A.; Sadao, T.; Oota, Y. *JOC* **1973**, *38*, 4148.
11. Tsuboi, S.; Uno, T.; Takeda, A. *CL* **1978**, 1325.
12. Stork, G.; Taber, D. F.; Marx, M. *TL* **1978**, 2445.
13. Le Bigot, Y.; Delmas, M.; Gaset, A. *SC* **1982**, *12*, 107.
14. Boden, R. M. *S* **1975**, 783.

15. Ogura, K.; Furukawa, S.; Tsuchihashi, G. *JACS* **1980**, *102*, 2125.
16. Gordon, A. J.; Ford, R. A. *The Chemist's Companion: A Handbook of Practical Data, Techniques, and References*, Wiley: New York, 1972; p 445.

Kurt D. Deshayes
Bowling Green State University, OH, USA

Potassium Diisopropylamide

KN(*i*-Pr)$_2$

[*67459-71-6*] $C_6H_{14}KN$ (MW 139.31)

(non-nucleophilic strong base)

Alternate Name: KDA.
Solubility: sol THF; insol benzene.
Analysis of Reagent Purity: amide bases can be titrated utilizing 1,6-dihydro-6-butyl-2,2′-bipyridine.[1] This indicator is able to determine the concentration of amide base solutions even in the presence of alkoxide.
Preparative Methods: KDA is prepared in situ immediately prior to use. In the original paper, the reagent was prepared either by the addition of diisopropylamine to a mixture of *n*-butylpotassium–lithium *t*-butoxide in hexane at 0 °C, or by the addition of ***n-Butyllithium*** to a solution of ***Potassium t-Butoxide*** and ***Diisopropylamine*** in THF at −78 °C.[2] It is also common to treat a solution of ***Lithium Diisopropylamide*** with potassium *t*-butoxide.[3] This reagent has been referred to in the literature as KDA, KDA-lithium *t*-butoxide, and LIDAKOR (lithium diorganylamide potassium alcoholate), but an examination of the experimental procedures indicates that all of the authors are, for practical purposes, referring to the same reagent. The exchange of lithium for potassium between organolithium compounds and potassium alkoxides is well established and has been reviewed.[4]
Handling, Storage, and Precautions: moisture sensitive. Use in a fume hood.

Deprotonation Reactions. KDA is used to effect deprotonations when a strong, non-nucleophilic base is required. It is considerably more reactive than its lithium counterpart and thus it is often used in situations where ***Lithium Diisopropylamide*** (LDA) has been tried and shown to be ineffective. Even in cases where LDA is effective for deprotonation, KDA can be used advantageously to promote anion generation more rapidly and at lower temperatures and to provide a more reactive derivative. KDA is also used when the potassium counterion is important for the outcome of a reaction. For example, it is well recognized that, in contrast to lithium enolates, when potassium enolates are employed, enolate equilibration is normally faster than alkylation, resulting in predominant alkylation of the thermodynamically more stable enolate.[5] The examples outlined below have been selected to illustrate these types of situations in which the unique properties of KDA are exploited.

KDA is used periodically for the generation of enolates. In a recent example, an oxindole system was successfully alkylated with an aziridine via the dianion formed by treatment with 2 equiv of KDA in THF at −78 °C (eq 1).[6] KDA can also be used to generate the dianions of carboxylic acids which then add predominantly 1,2 to α,β-unsaturated carbonyl compounds.[7] The counterion in this reaction has a pronounced effect on the ratio of 1,2- to 1,4-addition. The heightened reactivity of KDA has proved useful when enolate stability is a concern. In one example it was determined that the enolate of a particular pyrrolidinone was unstable above −70 °C, but the compound was successfully deprotonated at −90 °C within 20 min using KDA.[8] KDA has also been used when greater reactivity is required in the generated enolate. Potassium enolates have been shown to undergo a conjugate addition to a vinyl sulfone in cases where the lithium enolate was completely unreactive.[9]

1. 2 equiv KDA, THF, −78 °C
2. N$^+$ $^-$OTf
3. −78 °C to rt
53%

(1)

In one of the first reports of this reagent, it was shown that KDA can be used for the metalation of nitrosamines (eq 2).[10] The reaction takes place more rapidly than with LDA and gives a more reactive intermediate toward alkylation. KDA is also preferred over LDA for the generation of alkyl phosphonate anions prior to mono- or difluorination using *N*-fluorobenzenesulfonimide.[11] In the deprotonation of alkoxyalkyl phenyl sulfones (acyl anion equivalents), KDA must be used if the anion is to be added to a ketone rather than simply alkylated.[12]

1. base
2. PrI
(2)

Base	Time (min)	Yield
LDA	210	37%
KDA	30	41%
LDA, HMPA	90	49%
KDA, HMPA	30	60%

KDA has been used for the deprotonation of dimethylhydrazones and oxime ethers.[13] In most cases, deprotonation is complete in THF at −78 °C within 15 min. By contrast, the conditions for successful deprotonations using *n*-butyllithium or LDA are substrate dependent, requiring more time for optimization on a case by case basis. The anion of acetone dimethylhydrazone, generated by treatment with KDA, has been added to a vinyl sulfone and the resulting α-sulfonyl anion further alkylated (eq 3).[14]

KDA appears to be the base of choice for the generation of selenium-stabilized carbanions. It has been used to effect the deprotonation of 1-(phenylseleno)alkenes and bis(phenylseleno)acetals to give carbanions which react rapidly with a variety of electrophiles (eq 4).[15] Carbonyl compounds which contain an α-hydrogen undergo carbonyl addition rather than enolate formation. By comparison, LDA does not deprotonate these systems at an appreciable rate. In the case of α-aryl bis(phenylseleno)-acetals, the use of LDA results in the recovery of a significant amount of starting material while the use of KDA produces only the desired product (eq 4).[16] KDA will also successfully deprotonate simple benzyl selenides and 1,1,1-tris(trifluoromethylseleno)methane.[16,17] However, when dibenzyl diselenide is treated with KDA, cleavage of the Se–Se bond is the predominant reaction in a rare example of KDA acting as a nucleophile.[18]

1. KDA, THF, –78 °C; 2. PhO_2S-...; 3. allyl bromide; 92% (3)

1. base; 2. E^+ (4)

R^1	Base	E	R^2	Yield
Me	LDA	$PhCH_2Br$	H	0%
Me	KDA	$PhCH_2Br$	H	88%
Ph	LDA	PhCHO	OH	26%
Ph	KDA	PhCHO	OH	79%

KDA has been used for the deprotonation of phenylthiotributylstannyltrimethylsilylmethane (eq 5).[19] The potassium anion is formed in 15 min and readily alkylated, whereas the analogous lithium anion is formed in only 80% after 1 h and requires a complexing agent like TMEDA to achieve good yields of alkylated product. Transmetalation of the tributylstannyl group with *n*-butyllithium followed by alkylation provides α-silyl sulfides which can be subsequently converted to ketones. If, instead of an alkyl halide, the potassium anion is reacted with a nonenolizable carbonyl compound, a vinylstannane is obtained via a Peterson reaction (eq 5).[20] No control over alkene geometry was observed in this case. However, cation dependent geometric control has been shown to occur in the Peterson reaction between bistrimethylsilyl acetate enolates and aldehydes (eq 6).[21] Employing the potassium enolate generated with KDA gives significant selectivity for the (*E*)-isomer, suggesting that elimination occurs largely from a nonchelated intermediate.

KDA, –78 °C; MeI, –78 °C to rt, 89%; PhCHO, –78 °C to rt, 63% (5)

THF, –78 °C; RCHO (6)

M	R	(*E*):(*Z*)
K	*i*-Pr	>100:1
Li	*i*-Pr	6.5:1
K	Ph	17:1
Li	Ph	9.4:1

KDA has been used for the metalation of imidazoles in cases where *n*-butyllithium and LDA were ineffective.[22] In a synthesis of a dinucleating hexaimidazole ligand, both rings of a bis-imidazole system were metalated and derivatized using KDA.[23]

The high reactivity and kinetic basicity of KDA allow it to deprotonate even simple aromatic methyl groups. For this reason, toluene is not recommended as a solvent for KDA reactions. The deprotonation of dimethylnaphthalene, dimethylanthracene, and dimethyltriptycene systems have been studied using KDA.[24] KDA has also been used for the direct metalation of isoprene.[25]

Elimination Reactions. KDA has been used for the generation of dienols from homoallylic cyclic ethers (eq 7).[26] KDA promotes the ring-opening elimination without reacting with the resulting diene. By contrast, the use of LDA decreases yields dramatically. In this system, KDA is generated from LDA in the presence of a catalytic amount (10 mol %) of potassium *t*-butoxide. Presumably the product potassium alkoxide that is formed reacts with remaining LDA to generate a constant, low level of KDA which helps to minimize undesirable side reactions.

'catalytic KDA'; 72% (7)

KDA has also been used to promote the smooth ring opening of epoxides to generate allylic alcohols in good yields.[27] It is felt that the combination of lithium and potassium bases used in the generation of KDA both play important roles in facilitating this reaction: the electrophilic lithium may help to open the epoxide while the heavier alkali metal potassium confers maximum basicity to its anion to facilitate deprotonation. Where possible, *trans*-alkenols are the preferred or exclusive product, probably arising via a *syn*-periplanar elimination route.

KDA has been cited in an improved procedure for the Wharton transposition.[28] This multistep 1,3-transposition of oxygen proceeds through a base-induced rearrangement of an α,β-epoxyhydrazone. For stable epoxyhydrazones, yields are improved considerably by the use of KDA or potassium *t*-butoxide (eq 8).

KDA, –100 °C; 68% (8)

Related Reagents. Lithium Diisopropylamide; Lithium Hexamethyldisilazide; Potassium 3-Aminopropylamide; Potassium Hexamethyldisilazide.

1. Ireland, R. E.; Meissner, R. S. *JOC* **1991**, *56*, 4566.
2. Raucher, S.; Koolpe, G. A. *JOC* **1978**, *43*, 3794.
3. Lochmann, L.; Trekoval, J. *JOM* **1979**, *179*, 123.
4. Lochmann, L.; Trekoval, J. *CCC* **1988**, *53*, 76.
5. Caine, D. *COS* **1991**, *3*, 1.
6. Carroll, W. A.; Grieco, P. A. *JACS* **1993**, *115*, 1164.
7. Mulzer, J.; Brüntrup, G.; Hartz, G.; Kühl, U.; Blaschek, U.; Böhrer, G. *CB* **1981**, *114*, 3701.
8. Fray, M. J.; Bull, D. J.; James, K. *SL* **1992**, 709.
9. Hamann, P. R.; Fuchs, P. L. *JOC* **1983**, *48*, 914.
10. Renger, B.; Hügel, H.; Wykypiel, W.; Seebach, D. *CB* **1978**, *111*, 2630.
11. Differding, E.; Duthaler, R. O.; Krieger, A.; Rüegg, G. M.; Schmit, C. *SL* **1991**, 395.
12. Tanaka, K.; Matsui, S.; Kaji, A. *BCJ* **1980**, *53*, 3619.
13. Gawley, R. E.; Termine, E. J.; Aube, J. *TL* **1980**, *21*, 3115.
14. Pyne, S. G.; Spellmeyer, D. C.; Chen, S.; Fuchs, P. L. *JACS* **1982**, *104*, 5728.
15. Raucher, S.; Koolpe, G. A. *JOC* **1978**, *43*, 3794.
16. Clarembeau, M.; Krief, A. *TL* **1986**, *27*, 1723.
17. Haas, A.; Kempf, K. W. *T* **1984**, *40*, 4963.
18. Krief, A.; Trabelsi, M.; Dumont, W. *SL* **1992**, 638.
19. Ager, D. *JCS(P1)* **1986**, 195.
20. Ager, D. J.; Cooke, G. E.; East, M. B.; Mole, S. J.; Rampersaud, A.; Webb, V. J. *OM* **1986**, *5*, 1906.
21. Boeckman, R. K. Jr.; Chinn, R. L. *TL* **1985**, *26*, 5005.
22. Iddon, B.; Lim, B. L. *CC* **1981**, 1095.
23. Tolman, W. B.; Rardin, R. L.; Lippard, S. J. *JACS* **1989**, *111*, 4532.
24. (a) Inagaki, S.; Imai, T.; Mori, Y. *BCJ* **1989**, *62*, 79. (b) Inagaki, S.; Imai, T.; Kawata, H. *CL* **1985**, 1191.
25. Klusener, P. A. A.; Tip, L.; Brandsma, L. *T* **1991**, *47*, 2041.
26. Margot, C.; Schlosser, M. *TL* **1985**, *26*, 1035.
27. Mordini, A.; Ben Rayama, E.; Margot, C.; Schlosser, M. *T* **1990**, *46*, 2401.
28. Dupuy, C.; Luche, J. L. *T* **1989**, *45*, 3437.

Katherine S. Takaki
Bristol-Myers Squibb Co., Wallingford, CT, USA

Potassium Fluoride[1]

[7789-23-3] FK (MW 58.10)

(fluorinating reagent;[1] desilylating reagent;[11] base[29])

Physical Data: mp 846 °C; bp 1505 °C; *d* 2.48 g mL^{-1}; vapor pressure 1 mmHg at 885 °C.

Solubility: sol water, HF, NH_3; slightly sol alcohol; insol most organic solvents.

Form Supplied in: white to colcolorless crystals and powder.

Handling, Storage, and Precautions: anhydrous KF should be handled and used under dry N_2 for best results. KF reacts with acid to form the toxic gas hydrogen fluoride.

Fluorination Reagent. Potassium fluoride is widely used as a fluorinating reagent to prepare various organic fluorides. Aromatic[1] or vinyl fluorides[2] activated by electron-withdrawing groups are readily prepared from an exchange reaction of KF with aromatic or vinyl chlorides in polar aprotic solvents such as DMF, DMSO, or tetramethylene sulfone. In the presence of Ph_4PBr[3] or phthaloyl difluoride,[4] the fluoride ion replaces not only chlorine but also the nitro group in activated nitrobenzenes in DMSO. Reaction of polychlorobenzenes with KF gives mono-, di-, and trifluorobenzenes, depending on the reaction conditions.[1f,5] Perfluorobenzene and perfluoropyridine have been available from chlorine–fluorine exchange reactions.[6] The best results are obtained from heating a mixture of perchloroaromatic substrates and KF in the absence of solvent at an elevated temperature. Displacement of halide by fluoride ion also occurs with alkyl halides,[7] particularly in the presence of CaF_2,[8] but it becomes more difficult with polychloroalkanes.

The chlorine–fluorine exchange reaction has been applied to prepare acyl or sulfonyl fluorides.[9] Aryl acyl fluorides, alkyl acyl fluorides and sulfonyl fluorides are readily obtained from the reaction of acyl or sulfonyl chlorides with KF in the presence of CaF_2 in MeCN at rt, although the reaction is slow with KF alone.[8] Perfluorinated analogs are more active and the exchange reaction with KF is completed within a few hours in MeCN at rt.[10]

Desilylating Reagent. Due to formation of a strong silicon–fluorine bond, KF readily reacts with trialkylsilyl ethers[11] or acyl,[12] alkyl,[13] vinyl,[14] and alkynyl[15] silanes to remove the trialkylsilyl group. In the presence of a leaving group in the β-position, the silanes undergo smooth β-elimination reactions with KF in DMSO. Alkenes are obtained from saturated silanes (eq 1),[16] and alkynes[17] and allenes (eq 2)[18] are produced from vinyl silanes. Silylvinyl triflates are decomposed with KF and a crown ether at −20 °C to give vinylidene carbenes in essentially quantitative yield and which can be trapped by alkenes to give cyclopropanes (eq 3).[19]

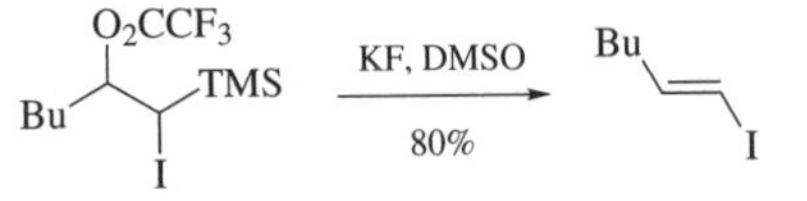

KF, DMSO; 80% (1)

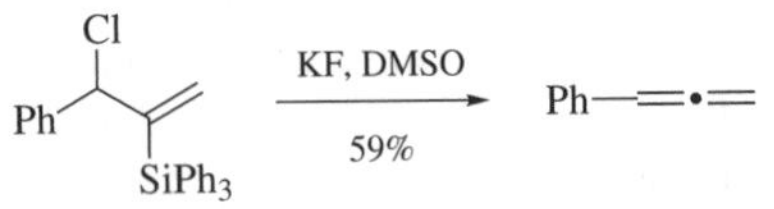

KF, DMSO; 59% (2)

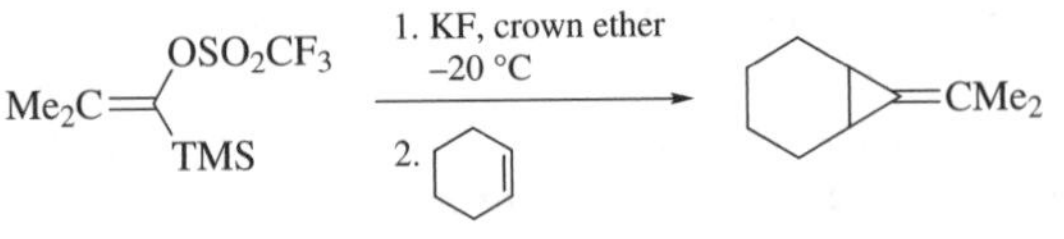

1. KF, crown ether, −20 °C; 2. cyclohexene (3)

Vinyl trialkylsilyl ethers liberate ketones or aldehydes with KF and a proton source.[20] In the presence of a carbonyl group, aldol condensation occurs to give β-hydroxy ketones (eq 4).[21] The intramolecular aldol reaction also goes smoothly to produce a cyclic product (eq 5).[22]

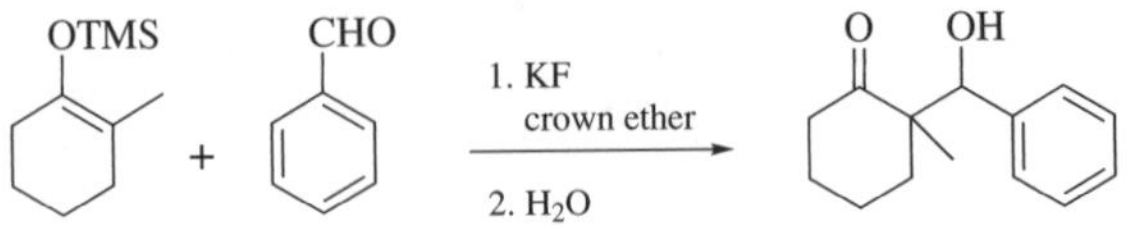

1. KF, crown ether; 2. H_2O (4)

In the presence of KF, perfluoroalkyl-[23] or perfluorophenyltrimethylsilanes[24] react with aldehydes, ketones, and acyl fluorides to give fluoro alcohols (eq 6). Sulfones were obtained in good yields when sulfonyl fluorides were used as substrates. These reactions usually proceed in aprotic solvents, such as MeCN and PhCN, at or near rt.

$$(CF_3)(CF_2Cl)C{=}O + Me_3SiCF_3 \xrightarrow[\text{2. } H_2SO_4,\ 88\%]{\text{1. KF}} ClCF_2(CF_3)_2C(OH) \quad (6)$$

Certain carbon–silicon bonds can be oxidized with 30% ***Hydrogen Peroxide*** and KF to give the corresponding alcohols in good yields.[25] This method is applicable to one-pot synthesis of anti-Markovnikov alcohols from terminal alkenes by hydrosilylation and oxidation (eq 7).[26] The oxidizing ability of 90% H_2O_2 is only slightly greater than that of 30% H_2O_2; 70% ***t-Butyl Hydroperoxide*** is less active. Although ***m-Chloroperbenzoic Acid*** also works well, 30% H_2O_2 is more economical and the reaction is cleaner. The stereochemistry of the reaction center is retained under these conditions (eq 8).[27] Various chiral diols are available by asymmetric catalytic intramolecular hydrosilylation of internally substituted alkenes using Rh complexes as catalysts, followed by oxidation with H_2O_2 and KF (eq 9).[28] Vinylsilanes are also oxidized with H_2O_2 and KF to give ketones.[26]

$$MeO_2C(CH_2)_8CH{=}CH_2 \xrightarrow[\text{2. 30\% } H_2O_2,\ KF,\ 79\%]{\text{1. } HSi(OEt)_3,\ RhCl(PPh_3)_3} MeO_2C(CH_2)_{10}OH \quad (7)$$

KF promotes carbonylative coupling reactions of organofluorosilanes with aryl iodides in the presence of a palladium catalyst (eq 10).[52]

Use as a Base.[29] KF in DMSO is a good dehydrohalogenation reagent for vinyl and alkyl halides to form alkynes,[30] allenes,[30] alkenes,[31] and dienes.[32] The presence of crown ethers facilitates the elimination reaction, particularly when MeCN is used as a solvent.[30]

Alkylation of alcohols,[33] phenols,[34] thiols,[35] and amines[36] with alkyl halides is promoted by KF. This method has been applied to synthesis of crown ethers.[37] Heterocycles are also readily alkylated with allyl halides (eq 11).[38] Arylation of phenols with activated aryl fluorides has been accomplished by using KF as a catalyst.[39]

The conversion of carboxylic acids and alkyl halides to esters (eq 12) has proven especially useful since employment of KF gives high yields of products.[39] This reaction may be conducted by using the reactant acid as the bulk solvent; DMF can be used as a solvent when starting acids are solids.[40] The alkyl iodides are the most reactive substrates.[35]

$$PhCOCH_2Br + RCO_2H \xrightarrow[\text{91–99\%}]{\text{KF, DMF}} RCO_2CH_2COPh \quad (12)$$

KF is widely used as a catalyst in Michael additions and aldol and Knoevenagel condensations. The use of KF has several advantages: a strong base is not required; catalytic amounts of KF are sufficient; separation of the catalyst is easy; yields are high with high selectivity; and the reactions are often successful when other strong bases are ineffective. Nitro compounds,[41] nitriles,[42] esters,[43] 1,3-diketones,[44] and heterocycles such as imidazoles (eq 13)[45] are usually used as Michael addition donors, and α,β-unsaturated aliphatic ketones, esters, nitro compounds, and nitriles are employed as the acceptors. Protic solvents such as ethanol are the most common, presumably due to the good solubility of KF in these solvents.[46] MeCN and benzene are also employed, but the presence of a crown ether such as ***18-Crown-6*** is necessary for this reaction.[47]

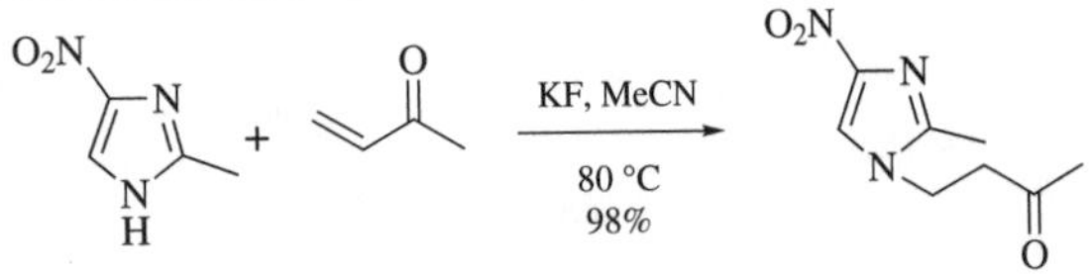

(13)

Various nitro alcohols are obtained by aldol condensation of nitroalkanes with aldehydes.[48] At elevated temperatures, the Knoevenagel reaction occurs to give unsaturated products (eq 14).[49] The cyclization of 1,4-diketones to cyclopentenones is also achieved with KF and a crown ether in refluxing xylene (eq 15).[50] Aromatic compounds are, however, usually obtained upon treatment of 1,3-diketones with KF in DMF (eq 16).[51] Reaction of 1,2-diketones with KF/DMF gives 1,4-benzoquinone derivatives (eq 17).[51]

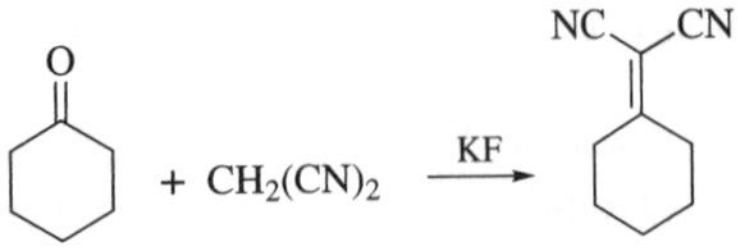

(14)

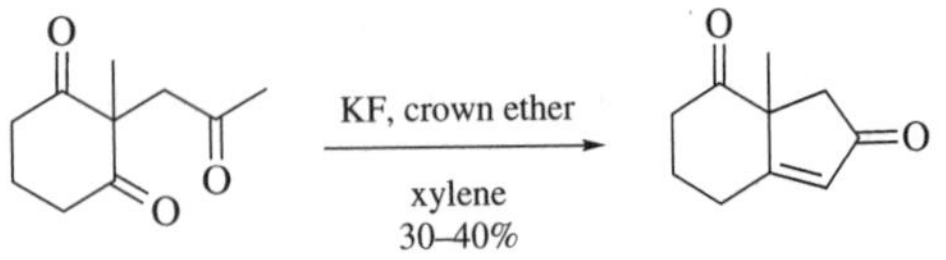

(15)

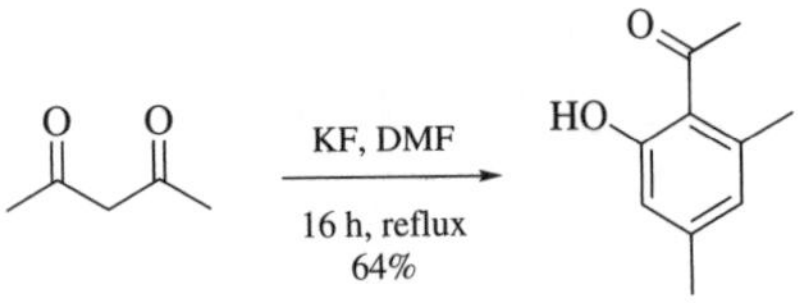

(16)

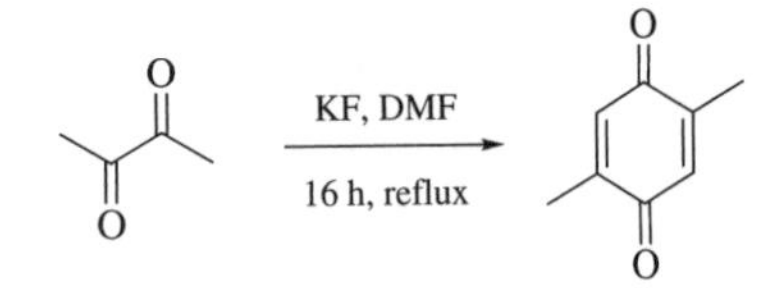

(17)

Related Reagents. Lithium Fluoride; Potassium Fluoride–Alumina; Potassium Fluoride–Celite.

1. (a) Hudlicky, M. *Chemistry of Organic Fluorine Compounds*, 2nd ed.; Horwood: New York, 1992; pp 119–127. (b) Finger, G. C.; Kruse, C. W. *JACS* **1956**, *78*, 6034. (c) Starr, L. D.; Finger, G. C. *CI(L)* **1962**, 1328. (d) Finger, G. C.; Dickerson, D. R.; Adl, T.; Hodgins, T. *CC* **1965**, 430. (e) Ishikawa, N.; Kitazume, T.; Yamazaki, T.; Mochida, Y.; Tatsuno, T. *CL* **1981**, 761. (f) Pews, R. G.; Gall, J. A. *JFC* **1990**, *50*, 371. (g) Suzuki, H.; Kimura, Y. *JFC* **1991**, *52*, 341. (h) Kimura, Y.; Suzuki, H. *TL* **1989**, *30*, 1271. (i) *Reagent Chemicals: American Chemical Society Specifications*, 8th ed.; ACS: Washington, 1993; pp 563–565.
2. (a) See Ref. 1(a), pp 117–118. (b) Wallenfels, K.; Witzler, F. *T* **1967**, *23*, 1359. (c) Bertram, H.-J.; Böhm, S.; Born, L. *S* **1991**, 937.
3. (a) Clark, J. H.; Boechat, N. *CI(L)* **1991**, 436. (b) Beaumont, A. J.; Clark, J. H. *JFC* **1991**, *52*, 295.
4. Maggini, M.; Passudetti, M.; Gonzales-Trueba, G.; Prato, M.; Quintily, U.; Scorrano, G. *JOC* **1991**, *56*, 6406.
5. (a) Pews, R. G.; Gall, J. A. *JFC* **1991**, *52*, 307. (b) Porwisiak, J.; Dmowski, W. *JFC* **1991**, *51*, 131.
6. (a) Fielding, H. C.; Gallimore, L. P.; Roberts, H. L.; Tittle, B. *JCS(C)* **1966**, 2142. (b) Chambers, R. D.; Hutchinson, J.; Musgrave, W. K. R. *JCS* **1964**, 3573.
7. See Ref. 1(a), pp 114–117. (b) Shahak, I.; Bergmann, E. D. *JCS* **1967**, 319. (c) Stadlbauer, W.; Laschober, R.; Lutschouning, H.; Schindler, G.; Kappe, T. *M* **1992**, *123*, 617.
8. (a) Clark, J. H.; Hyde, A. J.; Smith, D. K. *CC* **1986**, 791. (b) Ichihara, J.; Matsuo, T.; Hanafusa, T.; Ando, T. *CC* **1986**, 793.
9. (a) Nesmejanov, A. N.; Kahn, E. J. *CB* **1934**, *67*, 370. (b) Davis, W.; Dick, J. H. *JCS* **1931**, 2104. (c) Dear, R. E. A.; Gilbert, E. E. *JOC* **1968**, *33*, 1690.
10. Hu, L. Q.; DesMarteau, D. D. *IC* **1993**, *32*, 5007.
11. (a) Carpino, L. A.; Sau, A. C. *CC* **1979**, 514. (b) Sinhababu, A. K.; Kawase, M.; Borchardt, R. T. *S* **1988**, 710. (c) Rosini, G.; Marotta, E.; Righi, P.; Seerden, J. P. *JOC* **1991**, *56*, 6258.
12. (a) Degl'Innocenti, A.; Pike, S.; Walton, D. R. M.; Seconi, G.; Ricci, A.; Fiorenza, M. *CC* **1980**, 1201. (b) Schinzer, D.; Heathcock, C. H. *TL* **1981**, *22*, 1881.
13. Alcaraz, C.; Carretero, J. C.; Dominguez, E. *TL* **1991**, *32*, 1385.
14. Chan, T. H.; Mychajlowskij, W. *TL* **1974**, 3479.
15. (a) Semmelhack, M. F.; Neu, T.; Foubelo, F. *TL* **1992**, *33*, 3277. (b) Xu, Z.; Byun, H. S.; Bittman, R. *JOC* **1991**, *56*, 7183. (c) Courtemanche, G.; Normant, J. F. *TL* **1991**, *32*, 5317.
16. Miller, R. B.; Reichenbach, T. *TL* **1974**, 543.
17. Cunico, R. F.; Dexheimer, E. M. *JACS* **1972**, *94*, 2868.
18. Chan, T. H.; Mychajlowskij, W.; Ong, B. S.; Harpp, D. N. *JOC* **1978**, *43*, 1526.
19. Stang, P. J.; Fox, D. P. *JOC* **1977**, *42*, 1667.
20. Chuit, C.; Foulon, J. P.; Normant, J. F. *T* **1981**, *37*, 1385.
21. Noyori, R.; Yokoyama, K.; Sakata, J.; Kuwajima, I.; Nakamura, E.; Shimizu, M. *JACS* **1977**, *99*, 1265.
22. Sano, T.; Toda, J.; Tsuda, Y. *CPB* **1992**, *40*, 36.
23. (a) Kotun, S. P.; Anderson, J. D. O.; DesMarteau, D. D. *JOC* **1992**, *57*, 1124. (b) Anderson, J. D. O.; Pennington, W. T.; DesMarteau, D. D. *IC* **1993**, *32*, 5079.
24. Patel, N. R.; Kirchmeier, R. L. *IC* **1992**, *31*, 2537.
25. (a) Tamao, K.; Ishida, N. *JOM* **1984**, *269*, C37. (b) Xi, Z.; Agback, P.; Plavec, J.; Sandström, A.; Chattopadhyaya, J. *T* **1992**, *48*, 349. (c) Tamao, K.; Kawachi, A.; Ito, Y. *JACS* **1992**, *114*, 3989. (d) Barrett, A. G. M.; Malecha, J. W. *JOC* **1991**, *56*, 5243.
26. Tamao, K.; Ishida, N.; Tanaka, T.; Kumada, M. *OM* **1983**, *2*, 1694.
27. (a) Uozumi, Y.; Hayashi, T. *JACS* **1991**, *113*, 9887. (b) Tamao, K.; Nakajo, E.; Ito, Y. *JOC* **1987**, *52*, 957. (c) Roush, W. R.; Grover, P. T. *T* **1992**, *48*, 1981.
28. Bergens, S. H.; Noheda, P.; Whelan, J.; Bosnich, B. *JACS* **1992**, *114*, 2121.
29. Clark, J. H. *CRV* **1980**, *80*, 429.
30. Naso, F.; Ronzini, L. *JCS(P1)* **1974**, 340.
31. Clark, J. H.; Emsley, J. *JCS(D)* **1975**, 2129.
32. Chollet, A.; Hagenbuch, J. P.; Vogel, P. *HCA* **1979**, *62*, 511.
33. Lundt, I.; Pedersen, C. *S* **1992**, 669.
34. (a) Wu, W. L.; Chen, S. E.; Chang, W. L.; Chen, C. F.; Lee, A. R. *Eur. J. Med. Chem.* **1992**, *27*, 353. (b) González, A. G.; Barrera, J. B.; Hernández, C. Y. *H* **1992**, *34*, 1311.
35. Clark, J. H.; Miller, J. M. *JACS* **1977**, *99*, 498.
36. Clark, J. H.; Miller, J. M. *CC* **1976**, 229.
37. Reinhoudt, D. N.; Jong, F.; Tomassen, H. P. M. *TL* **1979**, 2067.
38. Halazy, S.; Gross-Bergès, V. *CC* **1992**, 743.
39. Emsley, J.; Hoyte, O. P. A.; Overill, R. E. *JACS* **1978**, *100*, 3303.
40. Clark, J. H.; Miller, J. M. *TL* **1977**, 599.
41. Thomas, A.; Manjunatha, S. G.; Rajappa, S. *HCA* **1992**, *75*, 715.
42. Apsimon, J. W.; Hooper, J. W.; Laishes, B. A. *CJC* **1970**, *48*, 3064.
43. Lawrence, R. W.; Perlmutter, P. *CL* **1992**, 305.
44. Yanami, T.; Kato, M.; Yoshikoshi, A. *CC* **1975**, 726.

45. Rao, A. K. S. B.; Rao, C. G.; Singh, B. B. *JCR(S)* **1991**, 350.
46. Kambe, S.; Yasuda, H. *BCJ* **1966**, *39*, 2549.
47. Belsky, I. *CC* **1977**, 237.
48. (a) Kambe, S.; Yasuda, H. *BCJ* **1968**, *41*, 1444. (b) Beck, A. K.; Seebach, D. *CB* **1991**, *124*, 2897.
49. Rand, L.; Swisher, J. V.; Cronin, C. J. *JOC* **1962**, *27*, 3505.
50. Dauben, W. G.; Hart, D. J. *JOC* **1977**, *42*, 3787.
51. Clark, J. H.; Miller, J. M. *JCS(P1)* **1977**, 2063.
52. Hatanaka, Y.; Fukushima, S.; Hiyama, T. *T* **1992**, *48*, 2113.

Qi Han & Hui-Yin Li
DuPont Merck, Wilmington, DE, USA

Potassium Hexamethyldisilazide

$KN(SiMe_3)_2$

[40949-94-8] $C_6H_{18}KNSi_2$ (MW 199.53)

(sterically hindered base)

Alternate Names: KHMDS; potassium bis(trimethylsilyl)amide.
Solubility: sol THF, ether, benzene, toluene.[1]
Form Supplied in: commercially available as moisture-sensitive, tan powder, 95% pure, and 0.5 M solution in toluene.
Analysis of Reagent Purity: solid state structures of $[KN(SiMe_3)_2]_2$[3] and $[KN(SiMe_3)_2 \cdot 2\ \text{toluene}]_2$[4] have been determined by X-ray diffraction; solutions may be titrated using fluorene,[2] 2,2′-bipyridine,[5] and 4-phenylbenzylidene benzylamine[6] as indicators.
Preparative Methods: prepared and isolated by the procedure of Wannagat and Niederpruem.[1] A more convenient in situ generation from ***Potassium Hydride*** and ***Hexamethyldisilazane*** is described by Brown.[2]
Handling, Storage, and Precautions: the dry solid and solutions are flammable and must be stored in the absence of moisture. These should be handled and stored under a nitrogen atmosphere. Use in a fume hood.

Use as a Sterically Hindered Base for Enolate Generation. Potassium bis(trimethysilyl)amide, $KN(TMS)_2$, has been shown to be a good base for the formation of kinetic enolates from carbonyl groups bearing α-hydrogens.[7] For example, treatment of 2-methylcyclohexanone with $KN(TMS)_2$ at low temperature followed by trapping with ***Triethylborane*** and ***Iodomethane*** gave good selectivity for 2,6-dimethylcyclohexanone (eq 1). In comparison, the use of ***Potassium Hydride*** for this transformation gave good selectivity for 2,2-dimethylcyclohexanone, which is the product derived from the thermodynamic enolate (eq 2).[8]

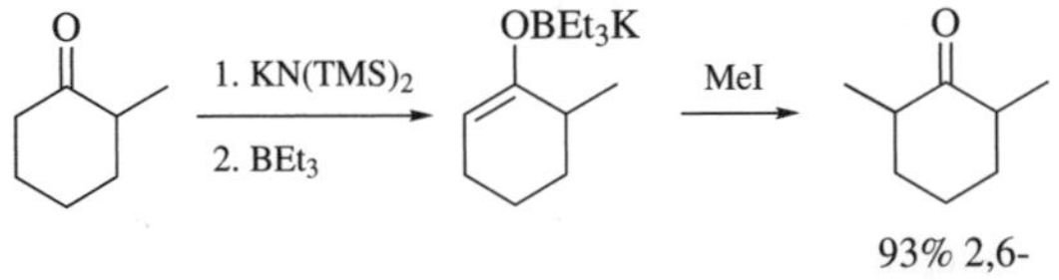

(1)

93% 2,6-
86% yield

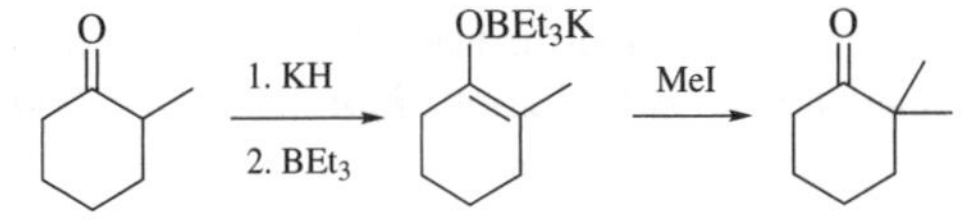

(2)

90% 2,2-
79% yield

This reagent has been shown to be a good base for the generation of highly reactive potassium enolates;[9] for example, treatment of various ketones and esters bearing α-hydrogens with $KN(TMS)_2$ followed by 2 equiv of N-F-saccharinsultam allowed isolation of the difluorinated product (eq 3).

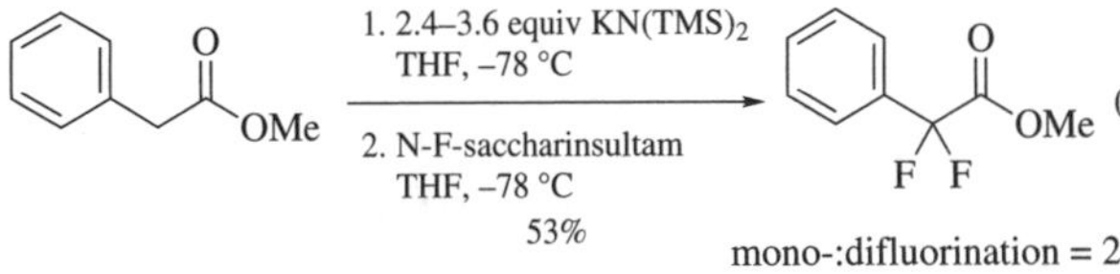

(3)

mono-:difluorination = 2:98

In a study on the electrophilic azide transfer to chiral enolates, Evans[10] found that the use of potassium bis(trimethylsilyl)amide was crucial for this process. The $KN(TMS)_2$ played a dual role in the reaction; as a base, it was used for the stereoselective generation of the (*Z*)-enolate (**1**). Reaction of this enolate with trisyl azide gave an intermediate triazene species (**2**) (eq 4). The potassium counterion from the $KN(TMS)_2$ used for enolate formation was important for the decomposition of the triazene to the desired azide. Use of other hindered bases such as ***Lithium Hexamethyldisilazide*** allowed preparation of the intermediate triazene; however, the lithium ion did not catalyze the decomposition of the triazene to the azide.[10a] This methodology has been utilized in the synthesis of cyclic tripeptides.[10b]

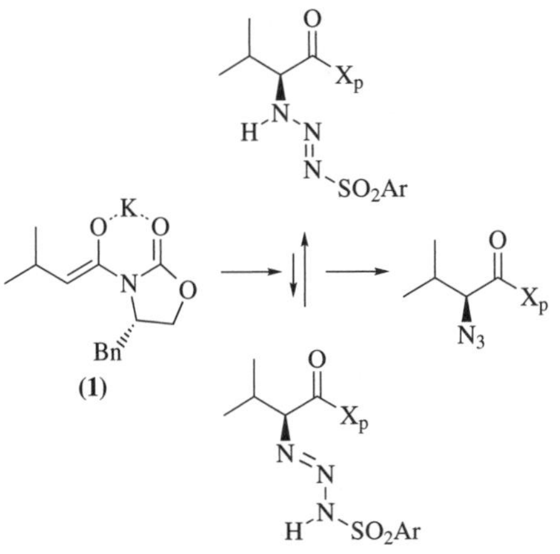

(1) (2) (4)

Treatment of carbonyl species bearing acidic α-hydrogens with potassium bis(trimethylsilyl)amide has also been shown to generate anions which, due to the larger, less coordinating potassium cation, allow the negative charge to be stabilized by other features in the molecule rather than as the potassium enolate. Treatment of 9-acetyl-*cis,cis,cis,cis*-cyclonona-1,3,5,7-tetraene with this reagent gave an anionic species which was characterized by spectroscopic methods to be more like the [9]annulene anion than the nonafulvene enolate. In this case the negative charge is more fully stabilized by delocalization into the ring to form the aromatic species rather than as the potassium enolate. Use of the bis(trimethylsilyl)amide bearing the more strongly coordinating lithium cation led to an intermediate which appeared to be lithium nonafulvene enolate. Addition of ***Chlorotrimethylsilane*** to each of these intermediates gave the same nonafulvenesilyl enol ether (eq 5).[11]

KN(TMS)$_2$ LiN(TMS)$_2$ K$^+$ OLi TMSCl TMSCl OTMS (5)

Selective Formation of Linear Conjugated Dienolates. Potassium bis(trimethylsilyl)amide has been shown to be an efficient base for the selective generation of linear-conjugated dienolates from α,β-unsaturated ketones.[12] As shown in eqs 6 and 7, treatment of both cyclic and acyclic α,β-unsaturated enones with KN(TMS)$_2$ in a solvent mixture of DMF/THF (2:1) followed by quenching with ***Methyl Chloroformate*** gave excellent selectivities for the products derived from the linear dienolate anion. In comparison, the use of lithium bases for this reaction gave products derived from the cross-conjugated dienolate anions. This methodology, however, did not work for 1-cyclohexenyl methyl ketone, in which case the product from the cross-conjugated dienolate anion was isolated exclusively (eq 8).

1. base; 2. ClCO$_2$Me → OCO$_2$Me + OCO$_2$Me (6)

Base	Linear	Cross	Yield (%)
KN(TMS)$_2$	>99	–	34
LiN(TMS)$_2$	75	25	68
LDA	–	99	44

1. base; 2. ClCO$_2$Me → OCO$_2$Me + OCO$_2$Me (7)

Base	Linear	Cross	Yield (%)
KN(TMS)$_2$	99	–	34
LDA	16	84	50

Stereoselective Generation of Alkyl (*Z*)-3-Alkenoates. Deconjugative isomerization of 2-alkenoates to 3-alkenoates occurs via γ-deprotonation of the α,β-unsaturated ester to form an intermediate dienolate anion. In most cases, the α-carbon is more reactive to protonation[13] and allows for the isolation of the 3-alkenoate. If the C-4 position bears a methyl group, this transformation is usually stereospecific, leading to the (*Z*)-3-alkenoate; however, when groups larger than a methyl occupy the C-4 position, the reaction becomes increasingly stereorandom.[13d] Potassium bis(trimethylsilyl)amide, however, was shown to be a good base for the stereoselective isomerization of 2,4-dimethyl-3-pentyl (*E*)-2-dodecenoate, which bears a long C-4 substituent, to the corresponding (*Z*)-3-dodecenoate (eq 9).[14a]

1. KN(TMS)$_2$; 2. ClCO$_2$Me; 25% → OCO$_2$Me (not detected) + OCO$_2$Me (>99) (8)

CO$_2$CH(*i*-Pr)$_2$ —base, THF, –78 °C→ CO$_2$CH(*i*-Pr)$_2$ (9)

Base	(*Z*):(*E*)	Yield (%)
LDA/HMPA	84:16	85
KN(TMS)$_2$	97:3	64

This reagent has been used to stereoselectively prepare (*Z*)-3-alkenoate moieties for use in the syntheses of insect pheromones.[14]

Generation of α-Keto Acid Equivalents (Dianions of Glycolic Acid Thioacetals). Potassium bis(trimethylsilyl)amide was found to be the optimal reagent for the generation of the dianion of glycolic acid thioacetals. This reagent may be used to effect a nucleophilic α-keto acid homologation. Treatment of the starting bis(ethylthio)acetic acid with KN(TMS)$_2$ proceeded to give the corresponding soluble dianionic species. This underwent alkylation with a variety of halides and tosylates (eq 10) and subsequent hydrolysis allowed isolation of the desired α-keto acids.[15]

EtS, SEt, OH —1. 2 equiv KN(TMS)$_2$; 2. RX; 3. H$_3$O$^+$→ EtS, EtS, R, OH (10)

RX	Yield (%)
MeI	100
EtOTs	100
i-PrOTs	72
CyOTs	64
PhCH$_2$Cl	100

The dianion was also shown to undergo ring-opening reactions with epoxides and aziridines (eq 11).

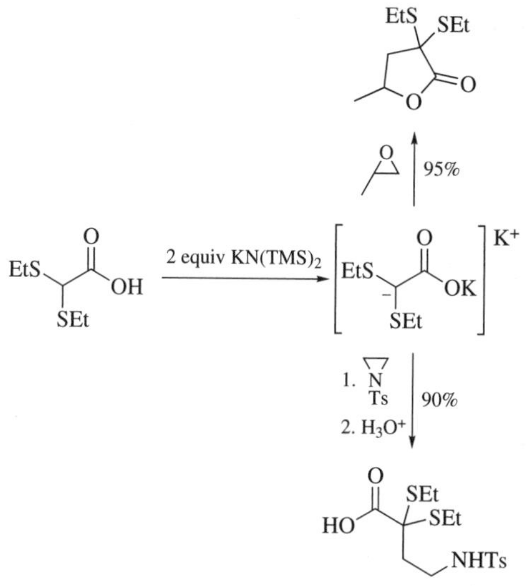

EtS SEt O O 95% O O K+ EtS OH 2 equiv KN(TMS)$_2$ EtS OK SEt SEt (11) 1. N Ts 90% 2. H_3O^+ O SEt HO SEt NHTs

Generation of Ylides and Phosphonate Anions.

Ylides. In the Wittig reaction, lithium salt-free conditions have been shown to improve (*Z*/*E*) ratios of the alkenes which are prepared;[16] ***Sodium Hexamethyldisilazide*** has been shown to be a good base for generating these conditions. In a Wittig-based synthesis of (*Z*)-trisubstituted allylic alcohols, potassium bis(trimethylsilyl)amide was shown to be the reagent of choice for preparing the starting ylides.[17] These were allowed to react with protected α-hydroxy ketones and depending upon the substitution pattern of the ylide and/or the ketone, stereoselectivities ranging from good to excellent were achieved (eqs 12–14).

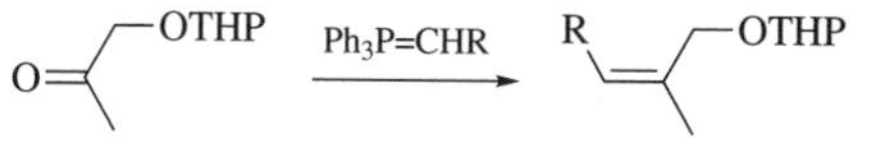

OTHP Ph$_3$P=CHR R OTHP (12)

R	(*Z*):(*E*)	Yield (%)
n-Pr	60:1	87
i-Pr	6:1	45

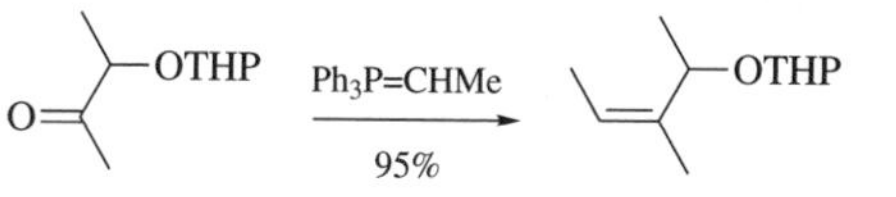

OTHP Ph$_3$P=CHMe 95% OTHP (13)

(*Z*):(*E*) = 200:1

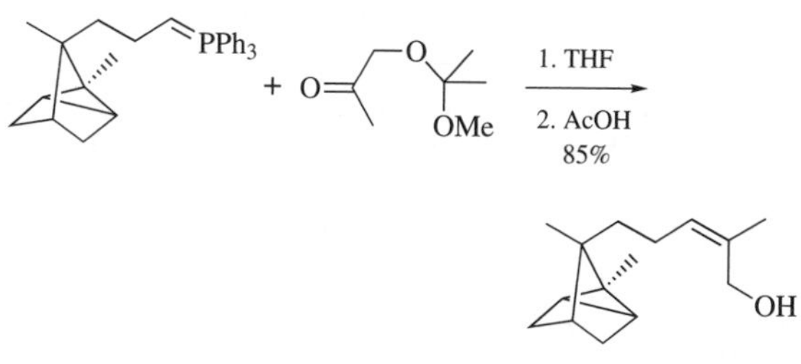

PPh$_3$ + O OMe 1. THF 2. AcOH 85%

OH (14)

>99% stereoisomeric purity

Phosphonates. In a Horner–Emmons-based synthesis of di- and trisubstituted (*Z*)-α,β-unsaturated esters, the strongly dissociated base system of potassium bis(trimethylsilyl)amide/***18-Crown-6*** was used to prepare the desired phosphonate anions. This base system, coupled with highly electrophilic bis(trifluoroethyl)phosphono esters, gave phosphonate anions which, when allowed to react with aldehydes, gave excellent selectivity for the (*Z*)-α,β-unsaturated esters (eq 15).[18]

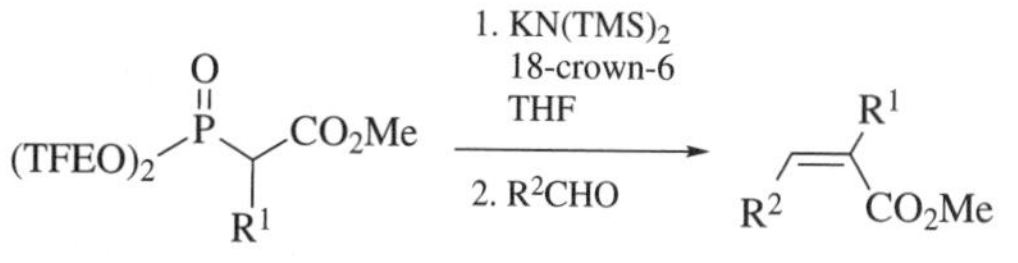

(TFEO)$_2$P(O)CH(R^1)CO$_2$Me → 1. KN(TMS)$_2$, 18-crown-6, THF; 2. R^{2}CHO → R^{2}CH=C(R^1)CO$_2$Me (15)

R^{2}CHO	R^1	(*Z*):(*E*)	Yield (%)
Me(CH$_2$)$_6$CHO	H	12:1	90
Me(CH$_2$)$_6$CHO	Me	46:1	88
Me(CH$_2$)$_2$CH=CHCHO	H	>50:1	87
Me(CH$_2$)$_2$CH=CHCHO	Me	>50:1	79
CyCHO	H	4:1	71
CyCHO	Me	>50:1	80
PhCHO	H	>50:1	95
PhCHO	Me	30:1	95

Intramolecular Cyclizations.

Haloacetal Cyclizations. Intramolecular closure of a carbanion onto an α-haloacetal has been shown to be a valuable method for the formation of carbocycles.[19a] Potassium bis(trimethylsilyl)amide was found to be the most useful base for the formation of the necessary carbanions. This methodology may be used for the formation of single carbocycles (eq 16), for annulation onto existing ring systems (eq 17), and for the formation of multiple ring systems in a single step (eq 18). In the case of annulations forming decalin or hydrindan systems, this ring closure proceeded to give largely the *cis*-fused bicycles (eq 17). For the reaction shown in eq 18, in which two rings are being formed, the stereochemistry of the ring closure was found to be dependent upon the counter ion of the bis(trimethylsilyl)amide; use of the potassium base allowed isolation of the *cis*-decalin system as the major product (95%), whereas use of lithium bis(trimethylsilyl)amide led to the isolation of the *trans*-decalin (95%).[19]

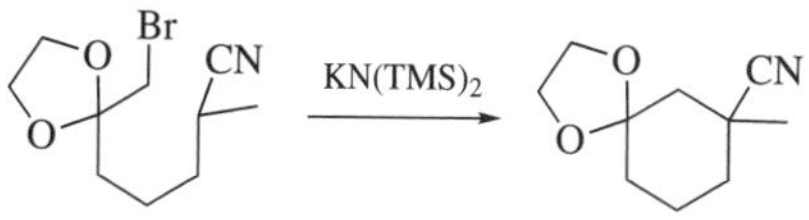

Br O CN O KN(TMS)$_2$ O CN O (16)

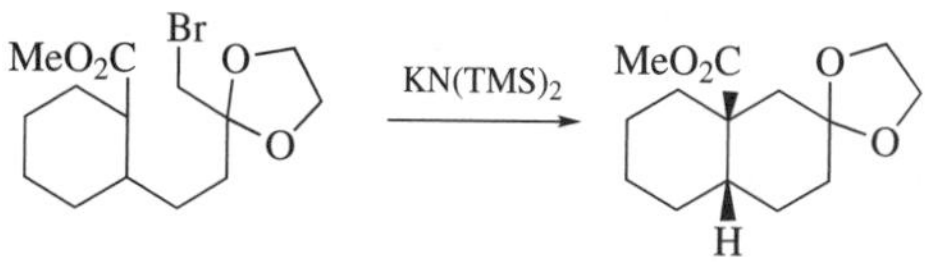

MeO$_2$C Br O O KN(TMS)$_2$ MeO$_2$C O O H (17)

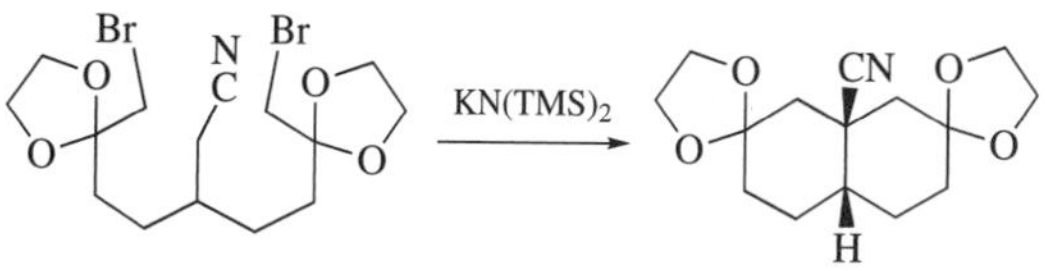

Br N C Br O O O O KN(TMS)$_2$ O CN O O O H (18)

95% *cis* + 5% *trans*

Intramolecular Lactonization. In a general method for the formation of 14- and 16-membered lactones via intramolecular alkylation,[20] potassium bis(trimethylsilyl) amide was shown to be a useful base for this transformation (eq 19).

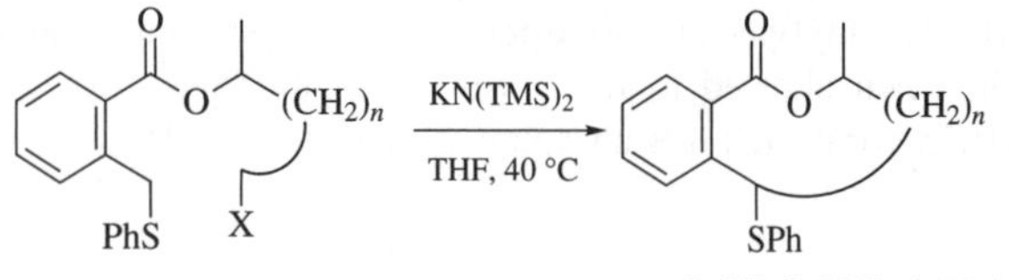

n = 5, 0%; 7, 75%; 8, 71% (19)

Intramolecular Rearrangement. Potassium bis(tri-methylsilyl)amide was shown to be a good base for the generation of a diallylic anion which underwent a biogenetically inspired intramolecular cyclization, forming (±)-dicytopterene B (eq 20).[21]

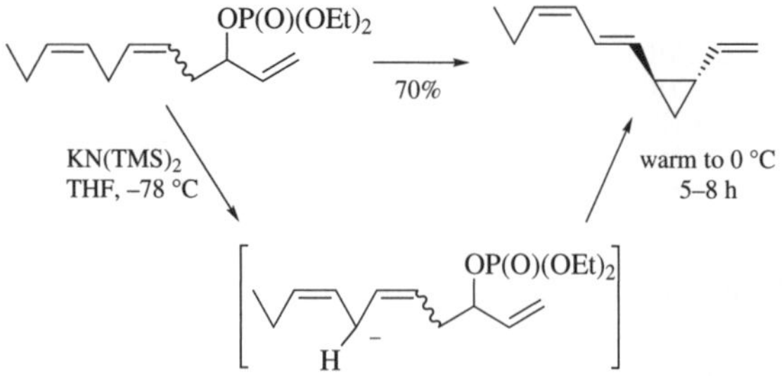

(20)

Synthesis of Vinyl Fluorides. Addition of potassium bis(trimethylsilyl)amide to β-fluoro-β-silyl alcohols was shown to selectively effect a Peterson-type alkenation reaction to form vinyl fluorides (eq 21).[22] Treatment of a primary β-fluoro-β-silyl alcohol with $KN(TMS)_2$ led cleanly to the terminal alkene. Use of a *syn*-substituted secondary alcohol led to the stereoselective formation of the (*Z*)-substituted alkene (eq 22); reaction of the *anti*-isomer, however, demonstrated no (*Z*:*E*) selectivity.

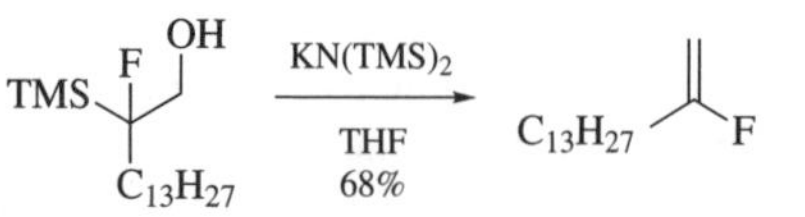

(21)

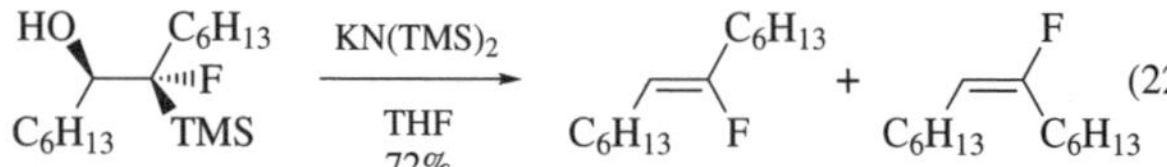

(22)

Oxyanionic Cope Rearrangement. Potassium bis(trimethylsilyl)amide/18-crown-6 was shown to be a convenient alternative to potassium hydride for the generation of anions for oxyanionic Cope rearrangements (eq 23).[23]

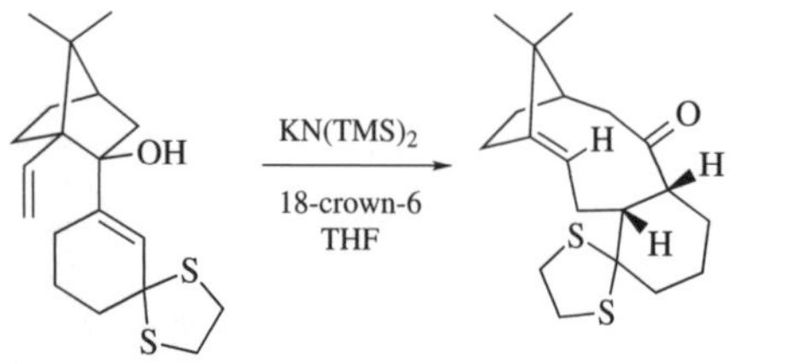

(23)

Stereoselective Synthesis of Functionalized Cyclopentenes. Potassium bis(trimethylsilyl)amide was shown to be an effective base for the base-induced ring contraction of thiocarbonyl Diels–Alder adducts (eq 24).[24] ***Lithium Diisopropylamide*** has also been shown to be equally effective for this transformation.

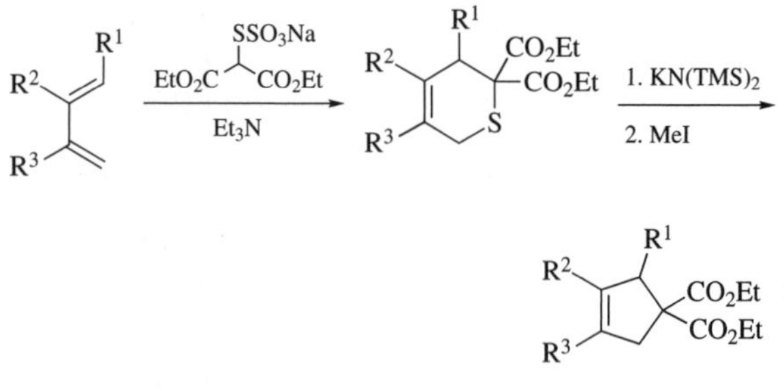

(24)

Related Reagents. Lithium Diisopropylamine; Lithium Hexamethyldisilazide; Potassium Diisopropylamide; Sodium Hexamethyldisilazide.

---

1. Wannagat, U.; Niederpruem, H. *CB* **1961**, *94*, 1540.
2. (a) Brown, C. A. *S* **1974**, 427. (b) Brown, C. A. *JOC* **1974**, *39*, 3913.
3. Tesh, K. F.; Hanusa, T. P.; Huffman, J. C. *IC* **1990**, *29*, 1584.
4. Williard, P. G. *Acta Crystallogr.* **1988**, *C44*, 270.
5. Ireland, R. E.; Meissner, R. S. *JOC* **1991**, *56*, 4566.
6. Duhamel, L.; Plaquevent, J.-C. *JOM* **1993**, *448*, 1.
7. (a) Evans, D. A. In *Asymmetric Synthesis*; Morrison, J. D., Ed.; Academic: New York, 1984; Vol. 3, p 1. (b) Brown, C. A. *JOC* **1974**, *39*, 3913.
8. Negishi, E.; Chatterjee, S. *TL* **1983**, *24*, 1341.
9. Differding, E.; Rueegg, G.; Lang, R. W. *TL* **1991**, *32*, 1779.
10. (a) Evans, D. A.; Britton, T. C. *JACS* **1987**, *109*, 6881. (b) Evans, D. A.; Ellman, J. A. *JACS* **1989**, *111*, 1063.
11. (a) Boche, G.; Heidenhain, F. *AG(E)* **1978**, *17*, 283. (b) Boche, G.; Heidenhain, F.; Thiel, W.; Eiben, R. *CB* **1982**, *115*, 3167.
12. Kawanisi, M.; Itoh, Y.; Hieda, T.; Kozima, S.; Hitomi, T.; Kobayashi, K. *CL* **1985**, 647.
13. (a) Rathke, M. W.; Sullivan, D. *TL* **1972**, 4249. (b) Herrman, J. L.; Kieczykowski, G. R.; Schlessinger, R. H. *TL* **1973**, 2433. (c) Krebs, E. P. *HCA* **1981**, *64*, 1023. (d) Kende, A. S.; Toder, B. H. *JOC* **1982**, *47*, 163. (e) Ikeda, Y.; Yamamoto, H. *TL* **1984**, *25*, 5181.
14. (a) Ikeda, Y.; Ukai, J.; Ikeda, N.; Yamamoto, H. *T* **1987**, *43*, 743. (b) Chattopadhyay, A.; Mamdapur, V. R. *SC* **1990**, *20*, 2225.
15. (a) Bates, G. S. *CC* **1979**, 161. (b) Bates, G. S.; Ramaswamy, S. *CJC* **1980**, *58*, 716.
16. (a) Schlosser, M.; Christmann, K. F. *LA* **1967**, *708*, 1. (b) Schlosser, M. *Top. Stereochem.* **1970**, *5*, 1. (c) Schlosser, M.; Schaub, B.; de Oliveira-Neto, J.; Jeganathan, S. *C* **1986**, *40*, 244. (d) Schaub, B.; Jeganathan, S.; Schlosser, M. *C* **1986**, *40*, 246.
17. Sreekumar, C.; Darst, K. P.; Still, W. C. *JOC* **1980**, *45*, 4260.
18. Still, W. C.; Gennari, C. *TL* **1983**, *24*, 4405.
19. (a) Stork, G.; Gardner, J. O.; Boeckman, R. K., Jr.; Parker, K. A. *JACS* **1973**, *95*, 2014. (b) Stork, G.; Boeckman, R. K., Jr. *JACS* **1973**, *95*, 2016.
20. Takahashi, T.; Kazuyuki, K.; Tsuji, J. *TL* **1978**, 4917.
21. Abraham, W. D.; Cohen, T. *JACS* **1991**, *113*, 2313.
22. Shimizu, M.; Yoshioka, H. *TL* **1989**, *30*, 967.
23. Paquette, L. A.; Pegg, N. A.; Toops, D.; Maynard, G. D.; Rogers, R. D. *JACS* **1990**, *112*, 277.
24. Larsen, S. D. *JACS* **1988**, *110*, 5932.

Brett T. Watson

Bristol-Myers Squibb Pharmaceutical Research Institute, Wallingford, CT, USA

Potassium Hydride[1]

KH

[7693-26-7] HK (MW 40.11)

(base and hydride donor to Lewis acids such as boranes and borates; used for deprotonation, cyclization–condensation, elimination, rearrangement reactions, and as a reducing agent).

Physical Data: solid; dec on heating; *d* 1.47 g mL^{-1}; n_D 1.453.
Solubility: dec in cold and hot water; insol CS_2, ether, benzene.
Form Supplied in: as a dispersion in mineral oil, 20–35% by weight. A standardization procedure has been published.[1]
Purification: commercial KH dispersion is made by reduction of metallic potassium,[2] and so KH is often contaminated with traces of potassium or potassium superoxide. These can be removed by pretreatment with iodine to produce a reagent with superior consistency in several reactions that are sensitive to such impurities.[3]
Handling, Storage, and Precautions: dispersion is a liquid, and tends to settle upon standing. Prolonged storage produces a compacted solid that must be broken up to achieve a homogeneous dispersion. Brown suggests[1] using a long-handled screwdriver to break up the compacted material and leaving a Teflon-covered stir bar in the container to aid dispersion. Clamping the sealed polyethylene bottle containing the dispersion to a Parr shaker for 15 min also works well.

The mineral oil may be removed by adding pentane to small quantities of the dispersion, stirring the slurry, then allowing the hydride to settle. The pentane/mineral oil supernatant may be pipetted off, but care should be exercised to quench carefully any hydride present in the supernatant with a small amount of methanol or ethanol before disposal. Two or three such rinses are sufficient to remove all traces of mineral oil.

Irritant. Great care must be taken in handling, and all operations involving the manipulation of the dry solid material should be conducted under an inert atmosphere in a fume hood.

Introduction. The following is divided into two major classes of reactions: KH as a base, and KH as a reducing agent. The acid–base reactions are further divided into the type of acid (OH, NH, CH, or Sn/GeH). Secondary effects, such as rate acceleration in oxy-Cope reactions, are listed under the appropriate acid–base reaction.

The reactions of saline hydrides occur at the crystal surface. The crystal lattice energies decrease from LiH to CsH; KH appears to have the optimum lattice energy and hydride radius for surface reactions. It is thus usually superior to LiH or NaH in the reactions discussed below.

Acid–Base Reactions.

Oxygen Acids. Potassium hydride reacts rapidly and quantitatively with acids such as carboxylic acids, phenols, and alcohols.[1] Of particular note is its ability to rapidly deprotonate tertiary alcohols and hindered phenols, in instances where ***Sodium Hydride*** or elemental ***Potassium*** react sluggishly or not at all. For example, triethylcarbinol and 2,6-di-*t*-butylphenol are quantitatively deprotonated in less than 5 min by KH in either ether or diglyme at 20 °C (eqs 1 and 2). For triethylcarbinol, NaH is ineffective, and K is sluggish in comparison.[4]

$$Et_3C{-}OH \xrightarrow[THF]{KH} Et_3C{-}OK \quad (1)$$

$$\text{2,6-di-}t\text{-Bu-}C_6H_3OH \xrightarrow[THF]{KH} \text{2,6-di-}t\text{-Bu-}C_6H_3OK \quad (2)$$

An interesting conformational effect is seen when *p*-*t*-butylcalix[4]arene is tetraethylated. When KH is used as the base, the partial cone conformation predominates, whereas with NaH, the cone is produced exclusively (eq 3).[5]

(3) EtI, KH, DMF or THF; partial cone, 64% + cone, 2%

An unusual cyclization of hydroxyallenes to dihydrofurans is mediated by KH (eq 4).[6] The reaction only proceeds when the K is complexed to ***18-Crown-6***, or when the base is ***Potassium t-Butoxide*** in refluxing *t*-butanol. Interestingly, no products of [1,3]- or [3,3]-rearrangement were detected.

(4) KH, 18-crown-6, THF, 74%

Cyclization of a KH-generated alkoxide is a method for the 100% stereoselective formation of spiroacetals by an intramolecular 1,4-addition to sulfoxides (eqs 5 and 6).[7] Lower selectivity resulted when either NaH or ***n-Butyllithium*** was used in the first step.

(5) 1. KH, THF (90%); 2. Raney Ni (79%)

(6) 1. KH, THF (90%); 2. Raney Ni (88%)

The elimination of trimethylsilanol from β-hydroxysilanes is a highly selective *syn* elimination when mediated by KH (eq 7).[8] In contrast, acid-catalyzed elimination is *anti* selective. The KH reaction is complete in 1 h at rt, whereas NaH requires 20 h in HMPA and results in a lower yield of product.

The *syn* elimination (eq 7) can be coupled with an intramolecular epoxide ring-opening to effect a stereoselective synthesis of α-alkylidenetetrahydrofurans (eq 8).[9]

Rate enhancements on the order of 10^{10} to 10^{17} are observed for a [3,3]-sigmatropic rearrangement (oxy-Cope) of an *endo*-vinylbicyclo[2.2.2]octene alkoxide (generated with KH in the presence of HMPA or crown ether) vs. the alcohol (eq 9).[10] Certain substrates for this reaction are quite sensitive to impurities in the KH, but pretreatment with ***Iodine*** eliminates the problem.[3]

Rate accelerations of $\geq 10^6$ by alkoxides have been observed for a [4+2] cycloreversion (Alder–Rickert reaction).[11] Eq 10 illustrates an example; replacement of the methyls with either BOM or acetonide protecting groups is also possible.[12]

Nitrogen Acids. Amines such as ***Diisopropylamine*** are not deprotonated by KH, although ***1,2-Diaminoethane***, diisobutylamine, and ***Pyrrolidine*** may be kaliated in excess amine solvent.[4a] *N*-Isopropylaniline and ***Hexamethyldisilazane*** are kaliated effectively and quantitatively in THF (eqs 11 and 12).[1,4b]

Deprotonation of indoles is also effective, and has been used to 'protect' the N–H bond prior to a lithium–halogen exchange.[13] Subsequent reaction with an electrophile occurs selectively at the lithiated position (eq 13).

E = CHO, MeCO, $CONH_2$, TMS, $SnMe_3$

Carbon Acids. Dimethyl Sulfoxide is deprotonated in ≤10 min with KH in THF at rt (eq 14).[1] Under similar conditions, NaH is essentially unreactive. Cyclopentadiene (eq 15) and fluorene are deprotonated quantitatively as well.[1]

Triphenylmethane is not deprotonated directly by KH, unless a catalytic amount of DMSO[1] (via in situ formation of dimsylpotassium, which in turn deprotonates triphenylmethane) or a crown ether is present (eq 16).[14]

Potassium hydride deprotonates ketones such as acetone, cyclohexanone, and isobutyrophenone with little or no self-condensation or reduction.[1] For unsymmetrical ketones such as 2-methylcyclohexanone, a mixture of regioisomers is produced. *O*-Acylation[15] and silylation[16] are thus facilitated. Permethylation of cyclopentanone can be achieved by addition of the ketone to a THF suspension of KH, followed by ***Iodomethane*** (eq 17).[17]

Monoalkylation can be achieved by treating the potassium enolate with ***Triethylborane*** prior to alkylation (eq 18).[18]

α,β-Unsaturated ketones give γ-alkylation, although polymerization can be a problem.[1] The enamines of β-diketones, however, can be alkylated in good yield (eq 19).[19] KH also mediates the Claisen condensation of esters (eq 20).[20]

1. KH, 2. MeI, 84% (19)

KH, THF, 82% (20)

Potassium hydride facilitated a tandem intramolecular Michael reaction–Claisen condensation in the synthesis of aklavinones.[21] In the absence of any additive, the 'unnatural' *C*-10 isomer was the only product observed in the NMR, but in the presence of 2.2.2-cryptand, the desired isomer was produced in 53% isolated yield (eq 21).

TBDMSO; CO_2Me; OMe; KH, HMPA, 2,2,2-cryptand, THF, 53%; TBDMSO, CO_2Me, H, 10, OH, OMe, H (21)

Ester enolate formation has been used to eliminate an acylamino group in a synthesis of condensed heterocycles, such as the indanone shown in eq 22.[22] The azabicycloheptene starting material is available by a Diels–Alder reaction, and the *N*-acyl group is removed in situ.

COMe, N, CO_2Et, $CHCO_2Et$; KH, THF, 60–75%; CO_2Et, O, N, H (22)

Silicon, Tin, and Germanium Acids. Trimethylsilane, tributylstannane, and tributylgermane are efficiently metalated by KH (eq 23).[23] These reactions are sensitive to impurities in the KH,[24] but pretreatment of the KH with iodine alleviates the difficulties.[3]

$$R_3XH \xrightarrow[60\text{–}80\%]{\text{KH, THF}} R_3X^- \ K^+ \quad (23)$$

X = Si, Sn, Ge
R = Me, Bu, Ph

Reductions. Potassium salts of selenanes,[25] silanes, and stannanes are also produced by reduction of Se–Se, Si–Si, and Sn–Sn bonds by KH (eq 24).[23]

$$R_nXXR_n \xrightarrow{\text{KH, THF}} R_nX^- \ K^+ \quad (24)$$

X = Se (n = 1), Si, Sn (n = 3)
R = Me, Ph

Potassium hydride reduces hindered boranes and borates to trialkyl (or trialkoxy) borohydrides.[1,26] For example, tri-*s*-butylborane is reduced in 93% yield (eq 25).

$$(s\text{-Bu})_3B \xrightarrow{\text{KH, THF}} (s\text{-Bu})_3BH^- \ K^+ \quad (25)$$

Reduction of butylpotassium with hydrogen produces a 'super-active' form of KH that reduces ketones and alkyl halides in high yields (eq 26).[27] This active hydride also deprotonates aldehydes and ketones at low temperature, but reduction is often a side reaction.

$$C_9H_{19}CH_2Br \xrightarrow[95\%]{\text{KH}^*\text{, THF}} C_9H_{19}Me \quad (26)$$

Related Reagents. Calcium Hydride; Potassium Hydride–*s*-Butyllithium-*N*,*N*,*N*′,*N*′-Tetramethylethylenediamine; Potassium Hydride–Hexamethylphosphoric Triamide; Sodium Hydride.

---

1. Brown, C. A. *JOC* **1974**, *39*, 3913.
2. Wiberg, E.; Amberger, E. *Hydrides of the Elements of Main Groups I–IV*; Elsevier: New York, 1971; pp 34–35.
3. Macdonald, T. L.; Natalie, K. J., Jr.; Prasad, G.; Sawyer, J. S. *JOC* **1986**, *51*, 1124.
4. (a) Brown, C. A. *JACS* **1973**, *95*, 982. (b) Brown, C. A. *S* **1974**, 427.
5. Groenen, L. C.; Ruël, B. H. M.; Casnati, A.; Timmerman, P.; Verboom, W.; Harkema, S.; Pochini, A.; Ungaro, R.; Reinhoudt, D. N. *TL* **1991**, *32*, 2675.
6. Gange, D.; Magnus, P. *JACS* **1978**, *100*, 7746.
7. Iwata, C.; Hattori, K.; Uchida, S.; Imanishi, T. *TL* **1984**, *25*, 2995.
8. Hudrlik, P. F.; Peterson, D. *JACS* **1975**, *97*, 1464.
9. Luo, F.-T.; Negishi, E.-I. *JOC* **1983**, *48*, 5144.
10. Evans, D. A.; Golob, A. M. *JACS* **1975**, *97*, 4765.
11. Papies, O.; Grimme, W. *TL* **1980**, *21*, 2799.
12. Knapp, S.; Ornaf, R. M.; Rodriques, K. E. *JACS* **1983**, *105*, 5494.
13. Yang, Y.-H.; Martin, A. R.; Nelson, D. L.; Regan, J. *H* **1992**, *34*, 1169.
14. Buncel, E.; Menon, B. *CC* **1976**, 648.
15. Jung, F.; Ladjama, D.; Riehl, J. J. *S* **1979**, 507.
16. Baigrie, L. M.: Lenoir, D.; Seikaly, H. R.; Tidwell, T. T. *JOC* **1985**, *50*, 2105.
17. Millard, A. A.; Rathke, M. W. *JOC* **1978**, *43*, 1834.
18. Negishi, E.-I.; Idacavage, M. J. *TL* **1979**, 845.
19. Gammill, R. B.; Bryson, T. A. *S* **1976**, 401.
20. Brown, C. A. *S* **1975**, 326.
21. Uno, H.; Naruta, Y.; Maruyama, K. *T* **1984**, *40*, 4725.
22. Kozikowski, A. P.; Kuniak, M. P. *JOC* **1978**, *43*, 2083.
23. (a) Corriu, R. J. P.; Guerin, C. *CC* **1980**, 168. (b) Corriu, R. J. P.; Guerin, C. *JOM* **1980**, *197*, C19.
24. Newcomb, M.; Smith, M. G. *JOM* **1982**, *228*, 61.

25. Krief, A.; Trabelsi, M.; Dumont, W. *S* **1992**, 933.
26. Brown, C. A. *JACS* **1973**, *95*, 4100.
27. Pi, R.; Friedl, T.; Schleyer, P. v. R.; Klusener, P.; Brandsma, L. *JOC* **1987**, *52*, 4299.

Robert E. Gawley & Xiaojie Zhang
University of Miami, Coral Gables, FL, USA

Potassium Hydroxide

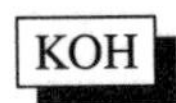

[1310-58-3] HKO (MW 56.11)

(very strong alkali; reacts readily with acids; used in nucleophilic substitution reactions, addition reactions, and basic hydrolysis reactions; occasionally employed as a catalyst for aldol-type reactions; ethanolic solution (alcoholic KOH) is traditionally used in dehydrohalogenation of halides)

Physical Data: mp 361 °C; *d* 2.044 g cm^{-3}.
Solubility: sol 0.9 part water, 0.6 part boiling water, or 3 parts ethanol; its dissolution in water is exothermic.
Form Supplied in: white lumps, rods, or pellets; commercially available pellets contain 85–88% KOH and 10–15% water.
Handling, Storage, and Precautions: toxic (oral-rat: LD_{50}: 365 mg kg^{-1}). Corrosive. Harmful if swallowed, inhaled, or absorbed through skin. Material is extremely destructive to tissue of the mucous membranes and upper respiratory tract. Inhalation may be fatal. Do not breathe dust. Avoid contact with eyes, skin, and clothing. Store in a cool dry place. Extremely hygroscopic. Wash thoroughly after handling.[1]
Incompatible with acids, acid chlorides, acid anhydrides, and aluminum metal. Absorbs moisture and CO_2 from air.

Elimination Reactions. Alcoholic KOH is the most frequently used base for dehydrohalogenation of alkyl halides in the synthesis of alkenes.[2] Some representative examples showing the use of KOH in elimination reactions are listed here.

Preparation of Diphenylacetylene. Diphenylacetylene was prepared via the dehydrohalogenation of stilbene dibromide. The dibromide was heated at reflux temperature with KOH in absolute ethanol for 24 h. After workup and recrystallization, a 66–69% yield of pure diphenylacetylene was obtained (eq 1).[3]

$$\text{PhCH(Br)CH(Br)Ph} \xrightarrow[\text{reflux, 24 h; 66–69\%}]{\text{KOH, EtOH}} \text{Ph—≡—Ph} \quad (1)$$

An alternative procedure for the synthesis of diphenylacetylene used the quaternary enammonium salt of deoxybenzoin via a Hofmann elimination reaction. Thus treatment of the methylated pyrrolidine enamine of deoxybenzoin (**1**) with 40% aq KOH at reflux afforded an 86% yield of diphenylacetylene (eq 2). The reaction was not as successful with other enamines.[4]

$$(\mathbf{1}) \xrightarrow[\text{reflux; 86\%}]{\text{40\% aq KOH}} \text{Ph—≡—Ph} \quad (2)$$

Preparation of Muconic Acid. Muconic acid (1,3-butadiene-1,4-dicarboxylic acid) (**3**) was prepared by dehydrohalogenation and hydrolysis of diethyl α,δ-dibromoadipate (**2**) upon heating with KOH in MeOH at reflux (eq 3). The acid was obtained in 37–43% yield.[5]

$$\text{EtO}_2\text{C–CH(Br)CH}_2\text{CH}_2\text{CH(Br)–CO}_2\text{Et}\ (\mathbf{2}) \xrightarrow[\text{2. H}^+]{\text{1. KOH, MeOH}} \text{HO}_2\text{C–CH=CH–CH=CH–CO}_2\text{H}\ (\mathbf{3}) \quad (3)$$

Synthesis of Coumarilic Acid. 2-Benzofurancarboxylic acid (coumarilic acid) (**5**) was prepared from coumarin dibromide (**4**) by the action of alcoholic KOH (eq 4). The product was isolated, after acidification, in 82–88% yield.[6]

$$(\mathbf{4}) \xrightarrow[\text{2. HCl}]{\text{1. KOH, EtOH}} (\mathbf{5}) \quad (4)$$

Hydrolysis of Hindered Esters. A solid–liquid phase-transfer catalysis system without organic solvents was developed for the hydrolysis of hindered esters such as mesitoic esters. The system consists of powdered KOH and 2% Aliquat 336 (methyltrioctylammonium chloride). The best results were obtained with 5 mol equiv of powdered KOH + 2% Aliquat 336 at 85 °C for 5 h (eq 5).[7]

$$\text{MesCO}_2\text{R} \xrightarrow[\text{2. HCl}]{\text{1. KOH, Aliquat 336, 85 °C}} \text{MesCO}_2\text{H} \quad (5)$$

R = Me 93%
R = C_8H_{17} 87%

Hydrolysis of Nitriles to Amides. Several nitriles were chemoselectively hydrolyzed to the corresponding amides when heated at reflux in *t*-butyl alcohol containing powdered solid KOH. Thus benzonitrile was converted to benzamide in 94% yield when refluxed in *t*-BuOH containing powdered KOH for 20 min (eq 6). The amides were not further hydrolyzed to the carboxylic acids under these conditions; this was explained by the formation of an insoluble K salt which thus precluded further nucleophilic attack. The use of MeOH as a solvent resulted in a lower yield and more hydrolysis to the carboxylic acid.[8]

$$\text{PhCN} \xrightarrow[\text{reflux, 20 min; 94\%}]{\text{solid KOH, } t\text{-BuOH}} \text{PhCONH}_2 \quad (6)$$

Formation of Hydroxamic Acids from Ethyl Benzoate. Potassium benzohydroxamate was prepared in 60% yield by treatment of ethyl benzoate with hydroxylamine in the presence of KOH in MeOH. The K salt was converted to benzohydroxamic acid with dil AcOH (eq 7).[9]

$$PhCO_2Et + H_2NOH \xrightarrow[\text{2. AcOH}\quad 60\%]{\text{1. KOH, MeOH}} PhCONHOH \quad (7)$$

Reduction of Aromatic Aldehydes via Crossed Cannizzaro Reaction. Certain benzyl alcohols were prepared from the corresponding aromatic aldehydes via the crossed Cannizzaro reaction with formaldehyde and KOH in 80–90% yield.[10] Thus the reaction of *p*-tolualdehyde with formalin and KOH in methanol at 60–70 °C gave about 80% yield of *p*-methylbenzyl alcohol (eq 8).[11]

$$p\text{-}MeC_6H_4CHO \xrightarrow[\text{MeOH, 60–70 °C}]{\text{HCHO, KOH}} p\text{-}MeC_6H_4CH_2OH + HCO_2K \quad (8)$$

Intramolecular Aldol Reactions. Treatment of several substituted 5-keto-2-hexenals (**6**) with KOH (2–3 equiv) in anhyd MeOH at rt for 2–3 h resulted in intramolecular aldol condensation reactions which led to the formation of the substituted phenols (**7**) in 74–100% yields (eq 9). Other catalysts including pyrrolidinium acetate, pyridinium tosylate, and $BF_3 \cdot Et_2O$ were ineffective.[12]

(**6**) $\xrightarrow[\text{2–3 h, rt}]{\text{KOH, MeOH}}$ (**7**) (9)

R^1	$R^2 + R^3$	Yield
H	$-CH_2CH_2CH(t\text{-Bu})CH_2-$	92%
Me	$-CH_2CH_2CH(t\text{-Bu})CH_2-$	96%
Me	$-CH_2CH_2CH_2-$	80%
Me	$-CH_2CH_2CH_2CH_2CH_2-$	74%

Hydroxydeamination of Primary Amines. Treatment of 1-substituted 1-tosylhydrazines (**8**) with KOH in refluxing ethanol in the presence of atmospheric oxygen afforded the corresponding alcohols (**9**) in high yields (eq 10). The reaction is believed to proceed via the hydroperoxide (**10**), since it was the major isolated product at rt. When the hydroperoxide was subjected to the typical reaction conditions, the corresponding alcohol (**9**) was obtained in excellent yield, with ethanol presumably acting as the reductant. The procedure provides a convenient route for replacement of the nitrogen of a primary amine by a hydroxy group. The sequence includes the conversion of the primary amine into the corresponding *p*-toluenesulfonamide followed by *N*-amination using ***Chloramine*** or ***O-(Mesitylsulfonyl)hydroxylamine*** to form the tosylhydrazine.[13]

Reactions with Alkyl Halides. The alkali hydroxides including KOH are very effective in nucleophilic substitution reactions of alkyl halides. Aq KOH is traditionally used to convert alkyl halides to alcohols.[14] Occasionally, it is used to convert *gem*-dihalides to ketones and trihalomethyl compounds to carboxylic acids.[15]

(**8**) $\xrightarrow[\text{O}_2\text{, rt}]{\text{KOH, EtOH}}$ (**10**); (**8**) $\xrightarrow[\text{O}_2\text{, reflux}]{\text{KOH, EtOH}}$ (**9**); (**10**) $\xrightarrow[\text{reflux}]{\text{KOH, EtOH}}$ (**9**) (10)

yields from tosylhydrazine: $R = p\text{-}MeOC_6H_4$ 98%
$R = p\text{-}ClC_6H_4$ 95%
$R = C_{14}H_{29}$ 91%

Reactions with Trichloromethyl Derivatives. Treatment of a trichloromethyl derivative with KOH usually gives the corresponding carboxylic acid, e.g. 1,1-bis(*p*-chlorophenyl)-2,2,2-trichloroethane (**11**) was converted to bis(*p*-chlorophenyl)acetic acid (**12**) in 69–73% yield when reacted with KOH in aq diethylene glycol at 134–137 °C (eq 11).[15]

$$\underset{(\mathbf{11})}{Ar_2CHCCl_3} \xrightarrow[\text{2. } H_2SO_4\quad 69\text{–}73\%]{\text{1. KOH, } H_2O\text{, diethylene glycol}} \underset{(\mathbf{12})}{Ar_2CHCO_2H} \quad (11)$$

$Ar = p\text{-}ClC_6H_4$

Substituted 1,1,1-trichloromethyl-2-ols [(trichloromethyl)-carbinols] react with methanolic KOH to form substituted α-methoxyacetic acids (**15b**) in excellent yields if the substituent group is aryl, alkyl or dialkyl, or vinyl. The reaction fails with ethynylcarbinols. The reaction works equally well if methanol is replaced by ethanol or other alcohols. In a proposed mechanism, the (trichloromethyl)carbinol (**13**) is converted in situ to the dichloro epoxide (**14**). The epoxide is opened by methanol to give (**15a**) which is converted to the acid (**15b**) under the reaction conditions and workup (eq 12).[16]

(**13**) $\xrightarrow{\text{KOH}}$ [(**14**)] $\xrightarrow{\text{MeOH}}$ (**15a**) X = Cl; (**15b**) X = OH (12)

Preparation of (Trichloromethyl)carbinols. Several (trichloromethyl)carbinols were synthesized in very good yields from aldehydes or ketones upon treatment with $CHCl_3$ in DMF in the presence of methanolic KOH (eq 13). The MeOH/DMF mixture provided a homogeneous reaction that permitted facile product formation at low temperature and was superior to using either solvent alone.[17]

$$R^1COR^2 + CHCl_3 \xrightarrow[\text{DMF, -5 °C}]{\text{KOH, MeOH}} R^1R^2C(OH)CCl_3 \quad (13)$$

$R^1 = Ph, R^2 = H$ 99%
$R^1 = p\text{-}NO_2C_6H_4, R^2 = H$ 90%
$R^1, R^2 = (CH_2)_5$ 68%

Oxidation of Aromatic Compounds. 2,3-Dimethyltetrahydroanthraquinone (**16**) was dehydrogenated to 2,3-dimethylanthraquinone (**17**) upon treatment with ethanolic KOH and air for 24 h (eq 14). The product was isolated in 94–96%.[18]

$$(16) \xrightarrow[\text{EtOH}]{O_2,\ \text{KOH}} (17) \quad (14)$$

Benzylation of Aromatic Primary Amines. Heating a mixture of a primary aromatic amine and benzyl alcohol in the presence of KOH at a temperature between 250–280 °C (eq 15) gave the corresponding *N*-benzyl aromatic amine in high yield.[19] For example, heating a mixture of 2-aminopyridine (1 equiv) and benzyl alcohol (1.4 equiv) to 250 °C in the presence of a catalytic amount of KOH (0.14 equiv) afforded 2-(benzylamino)pyridine in 98–99% yield.[20]

$$\text{ArNH}_2 + \text{PhCH}_2\text{OH} \xrightarrow[250\text{–}280\ °\text{C}]{\text{KOH}} \text{ArNHCH}_2\text{Ph} \quad (15)$$

Ar = 2-pyridyl, 98–99%

Reductive Benzylation of Aromatic Nitro Compounds. Treatment of various aromatic nitro compounds with excess benzylamine and KOH at 225–260 °C, while removing the water generated, resulted in the formation of *N*-benzyl aromatic amines in low yields (11–51%). Under these conditions the nitro compound was first reduced to the aniline, which was benzylated according to the previous reference (eq 16).[21]

$$\text{ArNO}_2 \xrightarrow[\text{KOH}]{\text{PhCH}_2\text{OH}} [\text{ArNH}_2] \xrightarrow[\text{KOH}]{\text{PhCH}_2\text{OH}} \text{ArNHCH}_2\text{Ph} \quad (16)$$

Conversion of Lactones into Benzyloxy Carboxylic Acids. Treatment of lactones with benzyl chlorides and powdered KOH in dry toluene at reflux temperature resulted in their conversion to the corresponding benzyloxy carboxylic acids.[22] The reaction was used to prepare γ-, δ-, and ϵ-benzyloxy or *p*-methoxybenzyloxy carboxylic acids from the appropriate lactones (eq 17).[23]

1. ArCH_2Cl, KOH, toluene, Δ; 2. esterification (17)

n = 1, 2, or 3 — CO_2Me — Ar = Ph, $p\text{-MeOC}_6\text{H}_4$

Rearrangement. The substituted 4-hydroxycyclohexanone (**18**) rearranged partially to the isomeric compound (**19**) when heated with KOH in 35:1 *t*-BuOH–H_2O at reflux (eq 18). The reaction was reversible and the ratio of (**18**):(**19**) at equilibrium was 65:35. The rearrangement was explained by a base-induced transannular 1,4-hydride shift.[24]

$$(18) \underset{}{\overset{\text{KOH},\ t\text{-BuOH, H}_2\text{O, reflux}}{\rightleftharpoons}} (19) \quad (18)$$

Epimerization of *meso*-Hydrobenzoins. (±)-Hydrobenzoin (**21**) was obtained in 64% yield by heating either *erythro*- or *threo*-3-phenylglyceric acid (**20a**) with KOH to 160 °C under reduced pressure. Under these conditions the (±)-diastereomer of hydrobenzoin is the thermodynamically more stable isomer. Thus applying the reaction to *meso*-hydrobenzoin (**20b**) gave the pure (±)-isomer in 90% yield in a few minutes (eq 19). Lower conversion (70%) was obtained with ***Potassium*** metal in refluxing toluene, and an attempt to use NaOH was unsuccessful. The KOH epimerization was applied successfully to certain other substituted *meso*- and *erythro*-hydrobenzoins.[25]

$$(20) \xrightarrow[15\ \text{mmHg}]{\text{KOH, 160 °C}} (21) \quad (19)$$

(**20a**) R = CO_2H
(**20b**) R = Ph

Huang-Minlon Modification of the Wolff–Kishner Reduction. The use of KOH was introduced as a substitute for metallic ***Sodium*** or ***Sodium Methoxide*** in Wolff–Kishner reductions of ketones. This modification allowed the use of ***Hydrazine*** hydrate and made the procedure simpler and economical.[26]

Removal of 2-Hydroxypropyl Group in *p*-Ethynylbenzoic Acid Synthesis. Potassium *p*-ethynylbenzoate (**23**) was obtained from 4-[4-(methoxycarbonyl)phenyl]-2-methyl-3-butyn-2-ol (**22**) in 98.5% yield and 99% purity upon heating at reflux in BuOH and 4 equiv of KOH (or NaOH) for 10 min (eq 20). The salt was precipitated from solution as it was formed.[27]

$$(22) \xrightarrow[\text{10 min, 98.5\%}]{\text{KOH (4 equiv), BuOH, reflux}} (23) \quad (20)$$

Related Reagents. Lithium Hydroxide; Potassium Hydroxide–Alumina; Potassium Hydroxide–Carbon Tetrachloride; Potassium Hydroxide-18-Crown-6; Potassium Hydroxide-Dimethyl Sulfoxide; Potassium Hydroxide-Hexamethylphosphoric Triamide; Sodium Hydroxide.

1. For complete safety data on KOH, see: *The Sigma-Aldrich Library of Chemical Safety Data*; Lenga, R. E., Ed.; Sigma-Aldrich: Milwaukee, WI, 1985; p 1535C.
2. March, J. In *Advanced Organic Chemistry. Reactions, Mechanisms, and Structure*, 4th ed.; Wiley: New York, 1992; p 1023.

3. Smith, L. I.; Falkof, M. M. *OSC* **1955**, *3*, 350.
4. Hendrickson, J. B.; Sufrin, J. R. *TL* **1973**, 1513.
5. Guha, P. C.; Sankaran, D. K. *OSC* **1955**, *3*, 623.
6. Fuson, R. C.; Kneisley, J. W.; Kaiser, E. W. *OSC* **1955**, *3*, 209.
7. Loupy, A.; Pedoussaut, M.; Sansoulet, J. *JOC* **1986**, *51*, 740.
8. Hall, J. H.; Gisler, M. *JOC* **1976**, *41*, 3769.
9. (a) Hauser, C. R.; Renfrow, Jr., W. B. *OSC* **1943**, *2*, 67. (b) Renfrow, Jr., W. B.; Hauser, C. R. *JACS* **1937**, *59*, 2308.
10. Davidson, D.; Weiss, M. *OSC* **1943**, *2*, 590.
11. Davidson, D.; Bogert, M. T. *JACS* **1935**, *57*, 905.
12. Tius, M. A.; Thurkauf, A.; Truesdell, J. W. *TL* **1982**, *23*, 2823.
13. Guziec, Jr., F. S.; Wei, D. *TL* **1992**, *33*, 7465.
14. Ref. 2, p 370.
15. Grummitt, O.; Buck, A.; Egan, R. *OSC* **1955**, *3*, 270.
16. Reeve, W.; Steckel, T. F. *CJC* **1980**, *58*, 2784 and references therein.
17. Wyvratt, J. M.; Hazen, G. G.; Wienstock, L. M. *JOC* **1987**, *52*, 944.
18. Allen, C. F. H.; Bell, A. *OSC* **1955**, *3*, 310.
19. Sprinzak, Y. *JACS* **1956**, *78*, 3207.
20. Sprinzak, Y. *OSC* **1963**, *4*, 91.
21. Miyano, S.; Abe, N.; Uno, A. *CPB* **1966**, *14*, 731.
22. Eyre, D. H.; Harrison, J. W.; Lythgoe, B. *JCS(C)* **1967**, 452.
23. Hoye, T. R.; Kurth, M. J.; Lo, V. *TL* **1981**, *22*, 815.
24. Warnhoff, E. W. *CC* **1976**, 517.
25. Collet, A. *S* **1973**, 664.
26. Huang-Minlon *JACS* **1946**, *68*, 2487.
27. Melissaris, A. P.; Litt, M. H. *JOC* **1992**, *57*, 6998.

Ahmed F. Abdel-Magid
The R. W. Johnson Pharmaceutical Research Institute, Spring House, PA, USA

Propionic Acid

CO_2H

[79-09-4] $C_3H_6O_2$ (MW 74.09)

(catalyst for orthoester Claisen rearrangement;[1] solvent for malonate diester decarboxylation[13] and porphyrin synthesis[14])

Alternate Name: propanoic acid.
Physical Data: clear liquid, bp 141 °C, 40 °C/10 mmHg; mp −23 °C; *d* 0.993 g cm^{-3}.[1]
Solubility: miscible with water, ethanol, ether, chloroform.
Form Supplied in: neat liquid; commercially available.
Analysis of Reagent Purity: by titrimetric assay; carbonyl compound impurity tests are known.[15]
Purification: usually by repeated fractional distillation.[16]
Handling, Storage, and Precautions: corrosive organic acid with an acrid odor. It is especially destructive to tissue of mucous membranes and upper respiratory tract, eyes, and skin. Inhalation is a major hazard and may be fatal. Use of a NIOSH/MSHA approved respirator is recommended. Use gloves, eye protection, and protective clothing to avoid contact. The acid should never be mixed with basic solvents, oxidizers, and reducing agents. Use in a fume hood.

Orthoester Claisen Rearrangement. Propionic acid is the classic acid catalyst for the Johnson orthoester Claisen rearrangement of an allylic alcohol and trialkyl orthoester (see ***Triethyl Orthoacetate***). Examples have been reported for all allylic bond substitutions, including 1,1-disubstituted (eq 1),[2] (*E*),[3] (*Z*),[4] trisubstituted,[5] and tetrasubstituted.[6]

HO — cat. propionic acid, $(EtO)_3CMe$, 1 h, 138 °C → EtO(C=O) (1)

The reaction of a propargyl alcohol and triethyl orthoacetate produces an allene ester (eq 2).[7] Allenic alcohols react in a similar manner to give diene esters in moderate yield (eq 3).[8]

OH — cat. propionic acid, $(EtO)_3CMe$, 1 h, 100 °C, 4 h, 180 °C → OEt, O (2)

OH, O, N — $(EtO)_3CMe$, cat. propionic acid, 20 h, 140 °C → O, N, O, OMe (3)

The reaction of 2-butynediol with excess trimethyl orthopropionate and catalytic propionic acid yields a symmetrical diene diester through the novel double Claisen orthoester rearrangement (eq 4).[9]

HO, OH — $(MeO)_3CEt$, cat. propionic acid, 3 h, 50–110 °C, distill methanol, 88% → MeO, O, OMe, O (4)

Other Variations on the Orthoester Claisen Rearrangement. The condensation of a dialkyl acetal and a primary allylic alcohol to form an unsaturated ketone has been catalyzed with propionic acid (eq 5).[10] The condensation of an allylic alcohol with ***Triethyl Methanetricarboxylate*** yields, after decarboxylation, an unsaturated ester (eq 6).[11]

OH + MeO OMe — cat. propionic acid, 24 h, 120 °C, 69% → O, OMe, O (5)

$$HC(CO_2Et)_3 + \text{HO-CH}_2\text{CH=CHCH}_3 \xrightarrow[\text{3 h, 140 °C; 74\%}]{(EtO)_3CMe,\ \text{cat. propionic acid}} \text{product (OEt ester)} \quad (6)$$

Decarboxylation of Malonic Esters. The decarboxylation of malonic ester derivatives to esters is general in propionic acid (eq 7).[12] This reaction is complementary to the basic hydrolysis of malonate esters to carboxylic acids.

$$PhCH_2CH(CO_2Et)_2 \xrightarrow[\text{48 h, reflux; 85\%}]{\text{propionic acid}} PhCH_2CH_2CO_2Et \quad (7)$$

Synthesis of Porphyrins. The condensation of an aldehyde and pyrrole in propionic acid leads to the formation of a tetrasubstituted porphyrin.[13] In an improved porphyrin synthesis, 2-acylpyrroles are reduced to the corresponding alcohols and converted to porphyrins by condensation and oxidation in refluxing propionic acid (eq 8).[14]

$$\text{2-(CH(OH)R)pyrrole} \xrightarrow[\text{0.25 equiv } Zn(OAc)_2]{\text{propionic acid, reflux, } O_2} \text{Zn tetra-R-porphyrin} \quad (8)$$

Related Reagents. Acetic Acid; Pivalic Acid; *p*-Toluenesulfonic Acid; Trifluoroacetic Acid.

---

1. *The Sigma–Aldrich Library of Chemical Safety Data*, 2nd ed.; Lenga, R. E., Ed.; Sigma–Aldrich: Milwaukee, 1988; vol. 2, p 2957.
2. (a) Johnson, W. S.; Werthemann, L.; Bartlett, W. R.; Brocksom, T. J.; Li, T.; Faulkner, D. J.; Petersen, M. R. *JACS* **1970**, *92*, 741. (b) Henrick, C. A.; Schaub, F.; Siddall, J. B. *JACS* **1972**, *94*, 5374.
3. Tadano, K.; Shimada, K.; Miyake, A.; Ishihara, J.; Ogawa, S. *BCJ* **1989**, *62*, 3978.
4. Nishikimi, Y.; Iimori, T.; Sodeoka, M.; Shibasaki, M. *JOC* **1989**, *54*, 3354.
5. (a) Tadano, K. I.; Idogaki, Y.; Yamada, H.; Suami, T. *JOC* **1987**, *52*, 1201. (b) Tadano, K. I.; Ishihara, J.; Yamada, H.; Ogawa, S. *JOC* **1989**, *54*, 1223.
6. Taguchi, T.; Morikawa, T.; Kitagawa, O.; Mishima, T.; Kobayashi, Y. *CPB* **1985**, *33*, 5137.
7. (a) Dauben, W. G.; Shapiro, G. *JOC* **1984**, *49*, 4252. (b) Crandall, J. K.; Tindell, G. L. *CC* **1970**, 1411. (c) Henderson, M. A.; Heathcock, C. H. *JOC* **1988**, *53*, 4736.
8. (a) Behrens, U.; Wolff, C.; Hoppe, D. *S* **1991**, 644. (b) Sleeman, M. J.; Meehan, G. V. *TL* **1989**, *30*, 3345.
9. Ishino, Y.; Nishiguchi, I.; Kim, M.; Hirashima, T. *S* **1982**, *9*, 740.
10. (a) Daub, G. W.; Lunt, S. R. *TL* **1983**, *24*, 4397. (b) Daub, G. W.; Sanchez, M. G.; Cromer, R. A.; Gibson, L. L. *JOC* **1982**, *47*, 743.
11. Kulkarni, M. G.; Sebastian, M. T. *SC* **1991**, *21*, 581.
12. Brown, R. T.; Jones, M. F. *JCR(S)* **1984**, 332.
13. Datta-Gupta, N.; Malakar, D.; Jenkins, C.; Strange, C. *BCJ* **1988**, *61*, 2274.
14. Kuroda, Y.; Murase, H.; Suzuki, Y.; Ogoshi, H. *TL* **1989**, *30*, 2411.
15. *Reagent Chemicals: American Chemical Society Specifications*, 8th ed.; American Chemical Society: Washington, 1993; pp 606–608.
16. Perrin, D. D.; Armarego, W. L. F. *Purification of Laboratory Chemicals*, 3rd. ed.; Pergamon: New York, 1988; p 264.

Kirk F. Eidman
Scios Nova, Baltimore, MD, USA

Pyridine

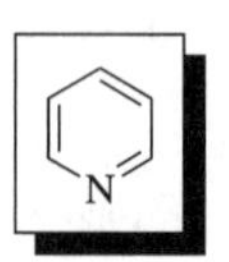

[110-86-1] C_5H_5N (MW 79.11)

(weak base useful as acid scavenging solvent or catalyst, especially for condensation,[1] dehalogenation,[2] halogenation,[3] and acylation[4] reactions)

Physical Data: [5] bp 115.3 °C; mp −41.6 °C; forms azeotropic mixture with water, bp 93.6 °C (41.3 wt % H_2O); steam volatile; *d* 0.9830 g cm^{-3} at 20 °C; pK_a 5.22 (in H_2O at 20 °C).
Solubility: miscible water, alcohol, ether, petroleum ether, and numerous other organic solvents.
Form Supplied in: colorless liquid; widely available.
Analysis of Reagent Purity: titration, GLC.
Purification: distillation.
Handling, Storage, and Precautions: flammable solvent with flash point 20 °C; hygroscopic; LD_{50} rat (oral) 891 mg kg^{-1};[5] minimum detectable odor 0.012 ppm;[5] incompatible with acids, acid chlorides, oxidizing agents, and chloroformates.[6]

Use as a Base in Condensation Reactions. Pyridine can be used as a base in cyclocondensation reactions. When the reaction of 2-acylphenoxyacetic acids in acetic anhydride is carried out in the presence of pyridine, formation of benzofuran-2-carboxylic acids is preferred. When weaker bases such as sodium acetate or sodium formate are used for the reaction, mixtures of benzofurans and benzofuran-2-carboxylic acids are produced, with the benzofuran derivatives as the major products.[7] For example, 2-acetyl-4-nitrophenoxyacetic acid (**1**) gives 65% 3-methyl-5-nitrobenzofuran-2-carboxylic acid (**2**) and 25% 3-methyl-5-nitrobenzofuran (**3**) when treated with ***Acetic Anhydride*** in pyridine. The same reaction in the presence of sodium acetate instead of pyridine gives a reversal of products with (**2**) as the minor product (33%) and (**3**) as the major product (60%) (eq 1).

$$\mathbf{(1)} \xrightarrow[Ac_2O,\ 110\ °C]{\text{pyridine}} \mathbf{(2)}\ 65\% + \mathbf{(3)}\ 25\% \quad (1)$$

Pyridine has been used effectively as a catalyst in the Knoevenagel condensation reaction.[8] Depending upon the nature of the base employed, the product selectivity can be altered. Different stereochemistry has been observed for aromatic heterocyclic bases such as pyridine and aliphatic tertiary amines such as ***Triethylamine***. Reaction of hexanal (**4**) with ***Malonic Acid*** (**5**) in the presence of pyridine as the base gives the α,β-unsaturated acid (**6**) as the major product (91:9 α,β:β,γ). When bases such as triethylamine are used in the reaction, a higher overall yield is obtained (76%); however, the corresponding β,γ-unsaturated acid (**7**) is the predominant product[1] (2:98 α,β:β,γ) (eq 2). One factor that appears important in determining the product ratio is the steric hindrance of the base. Pyridine, without the steric hindrance, shows a definite preference for the α,β-unsaturated carboxylic acid.

(4) + (5) —[$-H_2O$, $-CO_2$; 90 °C, pyridine]→ (6) α,β + (7) β,γ (2)

α,β:β,γ = 91:9, 64%

Condensation of furfural (**8**) with malonic acid (**5**) in the presence of pyridine affords an excellent yield of furylacrylic acid (**9**) (eq 3).[9] Similarly, condensation of *m*-nitrobenzaldehyde with malonic acid proceeds in the presence of pyridine as catalyst.[10]

(8) + $CH_2(CO_2H)_2$ (5) —[pyridine, Δ, 91–92%]→ (9) (3)

Yield improvements in the Knoevenagel condensation reaction have been obtained when the reaction is carried out in the presence of ***Titanium(IV) Chloride*** and pyridine in either tetrahydrofuran or dioxane solvent.[11] For example, the condensation of ***Acetaldehyde*** (**10**) with ***Diethyl Malonate***, (**11**) affords an 86% yield of product (**12**), as compared to a 25% yield without the promoter (eq 4).[8,12]

(10) + (11) —[dioxane, 70 h, 0 °C; $TiCl_4$, pyridine, 86%]→ (12) (4)

Condensation of carboxylic acids with amines in pyridine in the presence of a phosphorous(III) acid/iodine complex is an effective method for the synthesis of amides (eq 5).[13] Pyridine acts as a base and a solvent in this reaction.

—[$P(OH)_3$, pyridine, I_2; 75–85 °C, 5–6 h, 91%]→ (5)

Organoaluminum compounds such as ***Diisobutylaluminum Phenoxide*** with pyridine are good reagents for the regioselective aldol condensation of methyl ketones.[14] This system works effectively with pyridine acting as a base to inhibit undesirable proton transfer reactions (eq 6). Commonly used methods such as formation of the silyl enolate are not practicable here since they do not undergo regioselective attack on the methyl side of the ketone. Cross-aldol condensation reactions can also be carried out regiospecifically in the presence of a tertiary amine such as pyridine or ***2,6-Lutidine***.[15]

—[*i*-Bu_2AlOPh, pyridine, THF; 0 °C to reflux, 86%]→ (6)

1:1

Oxidation Reactions. An extensive amount of research has been carried out on the Gif system, and related systems, for the selective oxidation of saturated hydrocarbons. Pyridine is commonly used as a solvent in these reactions which involve a metal-catalyzed oxidation using ***Hydrogen Peroxide*** in a pyridine–acetic acid solvent system (eq 7). Copper, zinc, and iron are commonly used in the reactions.[16–19] Conversions are typically in the range of 10–25%; however, the selectivities are nearly quantitative.

—[$FeCl_2{\cdot}4H_2O$; Zn^0, pyridine, AcOH]→ (7)

1:1

Oxidation reactions of alcohols by ***Lead(IV) Acetate*** can be accelerated in the presence of pyridine. A kinetic study of this reaction has been carried out by Banerji using benzyl alcohol (**13**) as substrate (eq 8).[20]

(13) —[$Pb(OAc)_4$; pyridine]→ (8)

Oxidation of alcohols can also be effectively carried out using ***Chlorine*** in pyridine to give aldehydes and ketones. Secondary alcohols can be selectively oxidized in the presence of primary alcohols.[21] For example, treatment of 5β-cholestane-3β,19-diol (**14**) with chlorine and pyridine affords selective oxidation to 19-hydroxy-: 5β-cholestan-3-one (**15**) in almost quantitative yield (eq 9).

Aqueous pyridine is a very effective solvent for the oxidation of sulfides to sulfoxides with ***Phenyliodine(III) Dichloride***. High yields of the sulfoxides are typically obtained without contamination by sulfones (eq 10).[22]

Halogenation Reactions. Treatment of enones with ***Iodine*** and pyridine in carbon tetrachloride results in direct iodination. Good yields of 2-iodo enones are obtained when a solution of iodine (1.2–1.4 equiv) dissolved in a 1:1 (v/v) mixture of pyridine/carbon tetrachloride is added to cycloalkenones (eq 11).[23]

(14) $\xrightarrow[\text{pyridine, ~100\%}]{Cl_2, CCl_4}$ (15) (9)

$\xrightarrow[\text{20 °C, 85\%}]{PhICl_2, \text{ pyridine, } H_2O}$ (10)

$\xrightarrow[\text{0 °C, 57\%}]{I_2, \ CCl_4, \text{ pyridine}}$ (11)

Pyridine is employed as a base in the conversion of tetrahydrofurfuryl alcohol (**16**) to the corresponding bromide (**17**). A fair yield of the bromide is obtained (eq 12).[24] Addition of pyridine greatly increases the yield of product.[3]

(16) $\xrightarrow[\text{–5 °C to rt, 53–61\%}]{PBr_3, \text{ pyridine}}$ (17) (12)

In the bromination of 1-methylaminoanthraquinone (**18**), a good yield of 1-methylamino-4-bromoanthraquinone (**19**) is obtained using pyridine as acid scavenger and solvent (eq 13).[25]

(18) $\xrightarrow[\Delta, \text{ 70–74\%}]{Br_2, \text{ pyridine}}$ (19) (13)

Pyridine and 2,6-lutidine are useful as acid scavengers in the chlorination of α,β-unsaturated ketones and esters. For the ketones, the pyridine bases typically favor the Markovnikov isomer; however, for esters, pyridine shows little effect on the Markovnikov (alkoxy adjacent to carbonyl)/anti-Markovnikov isomer ratios (eq 14).[26]

$\xrightarrow[\text{MeOH}]{Cl_2, \text{ pyridine}}$ (14) 56:28:14

Dehalogenation Reactions. Dehalogenation of α-halo ketones can be effected by a number of reagents including pyridine and ***Tin(II) Chloride*** (eqs 15 and 16).[2] In the absence of the pyridine, the yield of acetophenone (**20**) drops to 75% and the yield of cyclohexanone (**21**) to 10%. Sulfur compounds such as sodium sulfite and ***Sodium Sulfide***, as well as benzene and aniline, are capable of dehalogenation in the presence of a promoter such as tin(II) chloride. An advantage of this reaction is that both aliphatic and aromatic α-halo ketones can be dehalogenated in good to excellent yields.

$\xrightarrow[H_2O, \text{ THF, reflux, 91\%}]{SnCl_2, \text{ pyridine}}$ (20) (15)

$\xrightarrow[H_2O, \text{ THF, reflux, 65\%}]{SnCl_2, \text{ pyridine}}$ (21) (16)

Dehydrochlorination of 2-(1-chloroethyl)thiophene (**22**) in the presence of pyridine gives 2-vinylthiophene (**23**) in 50–55% overall yield starting from thiophene (eq 17).[27] Didehydrochlorination of 2,2-dichloro-4-alkylbutanolides affords 4-alkylidene-2-butenolides upon reflux in pyridine (eq 18).[28]

(22) $\xrightarrow[\Delta]{\text{pyridine}}$ (23) 50–55% from thiophene (17)

$\xrightarrow[\Delta, \text{ 58\%}]{\text{pyridine}}$ (18) (Z):(E) = 84:16

Epoxidations. Taking advantage of its utility as a base, pyridine has been used for epoxidations in anhydrous organic systems with α-azo hydroperoxides as the epoxidizing reagents (eq 19).[29] ***Sodium Hydroxide*** works equally well or better with some of the epoxidizing agents; however, with the hydroperoxide (**24**), sodium hydroxide is a poor catalyst. Sodium hydroxide-catalyzed dehydration of the azo hydroperoxide, which affords *N*-(4-bromophenyl)-*N*′-benzyldiazene, is a competing side reaction.[30]

+ (24) $\xrightarrow[\text{reflux, 79\%}]{\text{pyridine}}$ (19)

Coupling Reactions. Dimerization of aromatic nitrile oxides can be catalyzed by pyridine, affording 3,6-diaryl-1,4,2,5-dioxadiazines in good yields (eq 20).[31] Other nucleophiles such as

4-phenylpyridine, 4-methylpyridine, ***4-Dimethylaminopyridine***, and ***N-Methylimidazole*** are also suitable catalysts for the reaction.

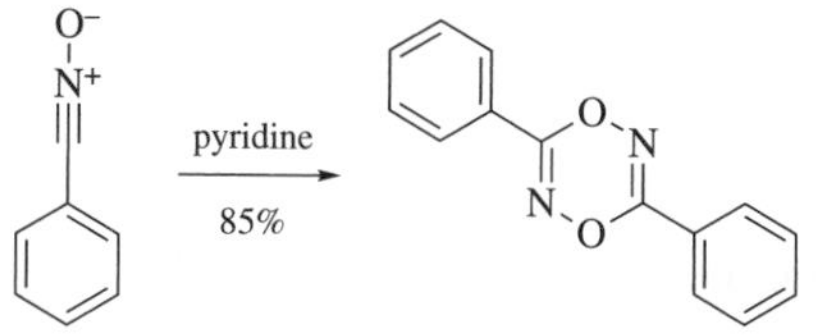

(20)

Esterifications. Pyridine is a widely used catalyst for acylation reactions. Acetylation reactions are effectively carried out in the presence of hexachlorocyclophosphazatriene and pyridine.[32] Acetylation of phenols by ***Acetic Anhydride*** in carbon tetrachloride is also catalyzed by pyridine. In the absence of the pyridine catalyst, at 0 °C and 25 °C, no acylation of the phenols is observed (eq 21).[4]

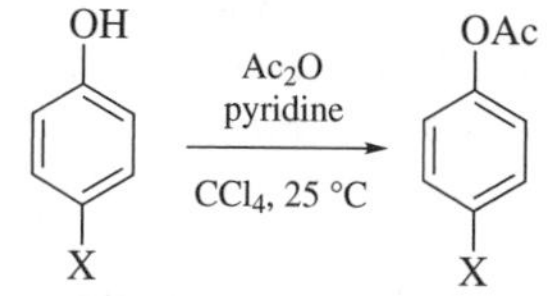

(21)

X = Cl, Me

Acylation of 6,7-dimethoxy-1-methyl-3,4-dihydroisoquinoline (**25**) with acetic anhydride in pyridine affords 2-acetyl-6,7-dimethoxy-1-methylene-1,2,3,4-tetrahydroisoquinoline (**26**) in 72–77% yield (eq 22).[33]

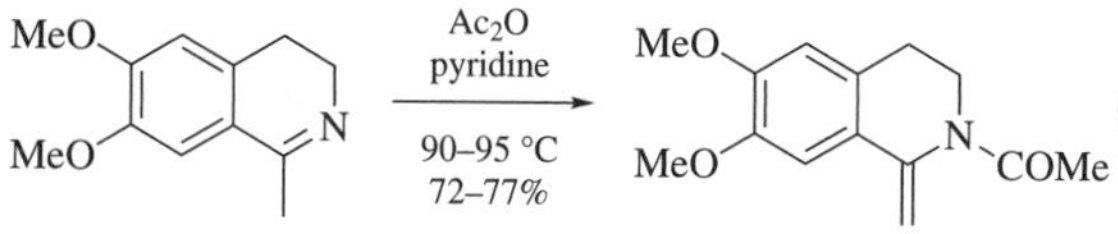

(22)

Treatment of *o*-hydroxyacetophenone (**27**) with ***Benzoyl Chloride*** in pyridine affords *o*-benzoyloxyacetophenone (**28**) in excellent yield (eq 23).[34]

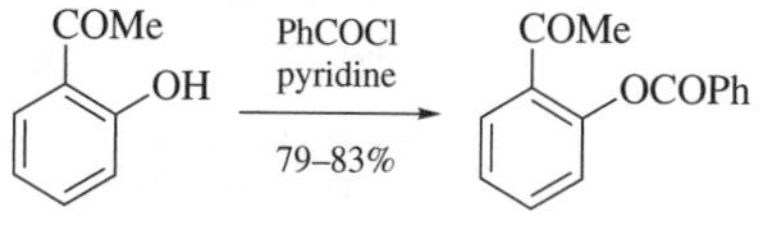

(23)

Synthesis of Benzonitriles. Arylthallium(III) salts with ***Copper(I) Cyanide*** or copper(II) cyanide afford fair to excellent yields of benzonitriles when heated at reflux in pyridine solvent (eq 24).[35] The use of acetonitrile as solvent gives lower product yields.

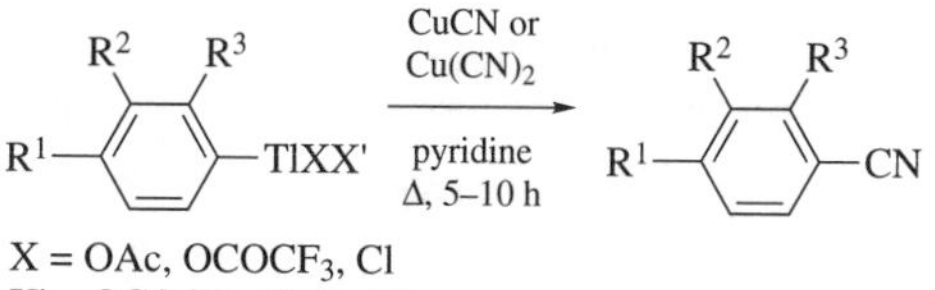

(24)

$X = OAc, OCOCF_3, Cl$
$X' = OCOCF_3, ClO_4, Cl$

Catalyst Poison. Pyridine and ***Quinoline*** have been used in conjunction with ***Palladium on Barium Sulfate*** as a catalyst system for the reduction of alkynes to *cis*-alkenes.[36]

Related Reagents. Borane–Pyridine; Chlorine–Pyridine; 2,4,6-Collidine; Copper(II) Sulfate-Pyridine; Di-*t*-butyl Chromate Pyridine; 2,6-Di-*t*-butylpyridine; 4-Dimethylaminopyridine; Dimethyl Sulfoxide-Sulfur Trioxide/Pyridine; 2,6-Lutidine; Quinoline; Sulfur Trioxide–Pyridine.

---

1. Yamanaka, H.; Yokoyama, M.; Sakamoto, T.; Shiraishi, T.; Sagi, M.; Mizugaki, M. *H* **1983**, *20*, 1541.
2. Ono, A.; Maruyama, T.; Kamimura, J. *S* **1987**, 1093.
3. Dox, A. W.; Jones, E. G. *JACS* **1928**, *50*, 2033.
4. Bonner, T. G.; McNamara, P. *JCS(B)* **1968**, 795.
5. Goe, G. L. In *Kirk-Othmer Encyclopedia of Chemical Technology*, 3rd ed.; Wiley: New York, 1982; Vol. 19, p 454.
6. *The Sigma-Aldrich Library of Regulatory & Safety Data*; Sigma-Aldrich: Milwaukee, 1993; Vol. 2, p 2489.
7. Horaguchi, T.; Matsuda, S.; Tanemura, K. *JHC* **1987**, *24*, 965.
8. Jones, G. *OR* **1967**, *15*, 204.
9. Rajagopalan, S.; Raman, P. V. A. *OSC* **1955**, *3*, 425.
10. Wiley, R. H.; Smith, N. R. *OSC* **1963**, *4*, 731.
11. Lehnert, W. *TL* **1970**, 4723.
12. Kon, G. A. R.; Speight, E. A. *JCS* **1926**, 2727.
13. Chiriac, C. I. *RRC* **1985**, *30*, 799.
14. Tsuji, J.; Yamada, T.; Kaito, M.; Mandai, T. *BCJ* **1980**, *53*, 1417.
15. Inoue, T.; Uchimaru, T.; Mukaiyama, T. *CL* **1977**, 153.
16. Barton, D. H. R.; Beviere, S. D.; Chavasiri, W.; Csuhai, E.; Doller, D. *T* **1992**, *48*, 2895.
17. Tung, H.-C.; Kang, C.; Sawyer, D. T. *JACS* **1992**, *114*, 3445.
18. Barton, D. H. R.; Beviere, S. D.; Chavasiri, W.; Csuhai, E.; Doller, D.; Liu, W.-G. *JACS* **1992**, *114*, 2147.
19. Barton, D. H. R.; Doller, D. *ACR* **1992**, *25*, 504.
20. Banerji, K. K.; Banerjee, S. K.; Shanker, R. *IJC(A)* **1977**, *15A*, 702.
21. Wicha, J.; Zarecki, A. *TL* **1974**, 3059.
22. Barbieri, G.; Cinquini, M.; Colonna, S.; Montanari, F. *JCS(C)* **1968**, 659.
23. Johnson, C. R.; Adams, J. P.; Braun, M. P.; Senanyake, C. B. W.; Wovkulich, P. M.; Uskokovic, M. R. *TL* **1992**, *33*, 917.
24. Smith, L. H. *OSC* **1955**, *3*, 793.
25. Wilson, C. V. *OSC* **1955**, *3*, 575.
26. Heasley, V. L. *JCS(P2)* **1991**, 393.
27. Emerson, W. S.; Patrick, T. M., Jr. *OSC* **1963**, *4*, 980.
28. Nakano, T.; Nagai, Y. *CC* **1981**, 815.
29. Tezuka, T.; Iwaki, M. *H* **1984**, *22*, 725.
30. Tezuka, T.; Iwaki, M. *TL* **1983**, *24*, 3109.
31. De Sarlo, F.; Guarna, A. *JCS(P2)* **1976**, 626.
32. Shumeiko, A. E.; Vapirov, V. V.; Titskii, G. D.; Kurchenko, L. P. *ZOB* **1990**, *60*, 2666 (*CA* **1991**, *114*, 246 557z).
33. Brossi, A.; Dolan, L. A.; Teitel, S. *OSC* **1988**, *6*, 1.
34. Wheeler, T. S. *OSC* **1963**, *4*, 478.
35. Uemura, S.; Ikeda, Y.; Ichikawa, K. *T* **1972**, *28*, 3025.
36. Danben, W. G.; Hart, D. J. *JOC* **1977**, *42*, 3787.

Angela R. Sherman
Reilly Industries, Indianapolis, IN, USA

Pyridinium *p*-Toluenesulfonate

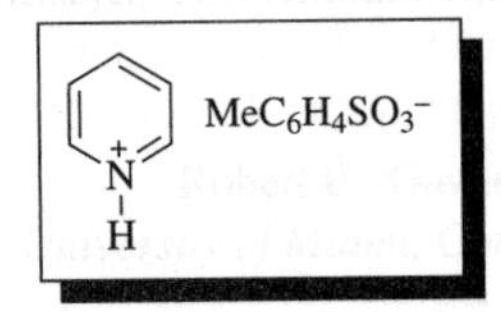

[24057-28-1] $C_{12}H_{13}NO_3S$ (MW 251.33)

(used as an acid catalyst for the protection and deprotection of alcohols and acetals)

Alternate Name: PPTS.
Physical Data: mp 120–121 °C.[1,2]
Solubility: sol benzene, toluene, dichloromethane, chloroform, ethanol, and acetone.
Form Supplied in: white, moisture-sensitive crystalline solid.
Preparative Method: prepared in a 95% yield by the addition of ***p*-Toluenesulfonic Acid** to ***Pyridine*** at 22–24 °C.[1,2]
Purification: recrystallized from acetone.[1,2]
Handling, Storage, and Precautions: the title reagent is moisture sensitive, requiring storage under nitrogen in a desiccator over a suitable drying agent. Stored in this manner, the reagent has an excellent shelf life. Irritant; handle with gloves.

Properties and General Considerations. PPTS (**1**) is a weakly acidic salt frequently used in catalytic quantity for the protection of alcohols as tetrahydropyranyl (eq 1)[1] and trimethylsilyl ethers (eq 2).[3]

ROH —(1), dihydropyran; benzene, reflux→ [THP ether, OR] (1)

ROH —(1), (TMS)$_2$O; benzene, reflux→ ROTMS (2)

PPTS is particularly favored for use with acid-sensitive functionalities such as allylic alcohols, epoxides, and acetals. *N*-Tosyl-L-norephedrine has been converted to its oxazolidine using a catalytic amount of (**1**) (25 mol%) and ***Triethyl Orthoformate*** (eq 3).[3]

TsNH, HO [norephedrine] —(1), (MeO)$_3$CH; benzene, reflux; 96%→ [oxazolidine: Ts, N, MeO, O] (3)

p-Methoxyphenol has been protected as its THP ether (80% yield) using 4 equiv of (**1**).[4] All these reactions are run in dry aprotic solvent (benzene, toluene, or dichloromethane) and water formed during the course of the reaction is removed by azeotropic distillation or by 4 Å molecular sieves. Yields are good to excellent.

Cleavage of *t*-Butyldimethylsilyl Ethers. A catalytic quantity of (**1**) (30 mol %) in ethanol selectively cleaves *t*-butyldimethylsilyl ethers in the presence of *t*-butyldiphenylsilyl ethers (eq 4).[5]

TBDPSO~~~OTBDMS —(1) (30 mol %); ethanol; 92%→ TBDPSO~~~OH (4)

Acetalization. PPTS is an excellent catalyst for the preparation of 1,3-dioxolane acetals from ketones and ethanediol (90–95% yield).[6] The weak acidic nature of (**1**) makes it ideal for acid-sensitive compounds.

$R^1C(O)R^2$ —(1), ethanediol; benzene, reflux→ [1,3-dioxolane of R^1, R^2] (5)

The reaction is carried out using 10–30 mol % of (**1**) in refluxing benzene, accompanied by the azeotropic removal of water (Dean–Stark apparatus). Similar conditions have been used for the preparation of acetals using (*R,R*)-hydrobenzoin and a tricyclic benzaldehyde derivative.[7]

Deprotection of Acid-Sensitive Protecting Groups via Transacetalization. PPTS is a versatile catalyst for the deprotection of acid-sensitive acetal protecting groups. For example, 10–30 mol % of (**1**) efficiently cleaves tetrahydropyran ethers (THP) (eq 6),[1] 1,3-dioxolanes (eq 7),[6] methoxymethyl ethers (MOM), and methoxyethoxymethyl ethers (MEM) (eq 8).[8]

[THP ether, OR] —(1), ethanol; rt→ ROH (6)

[1,3-dioxolane of R^1, R^2] —(1), acetone; water, reflux→ $R^1C(O)R^2$ (7)

[allylic OR] —(1), *t*-BuOH or 2-butanone→ [allylic OH] (8)

R = CH_2OMe (MOM), 94%
R = $CH_2OCH_2CH_2OMe$ (MEM), 93%

2-Methoxybutadienes. PPTS catalyzes the isomerization of allenyl ethers to 1-alkyl-2-methoxybutadienes. Freshly prepared crude allenyl ethers are treated with (**1**) (1 mol %) in dichloromethane at room temperature (eq 9).[9]

[allenyl ether: OMe, R] —(1), CH_2Cl_2; rt; 40–50%→ [MeO, R diene] (9)

2-Alkyl-3-acylfurans. PPTS converts 5-acetoxy-4,5-dihydrofurans to the corresponding furans (eq 10).[10] The reaction is carried out in refluxing toluene for 3–6 h. Yields are poor to modest.

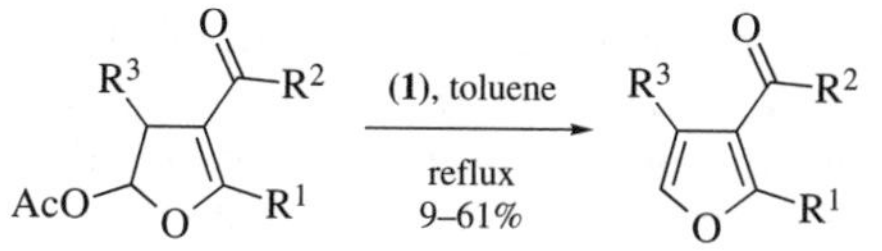

(10)

α-Alkoxyhydrazines. Reaction of α-sulfinylhydrazones with alcohols in the presence of (**1**) (10–30 mol %) results in α-oxy substituted hydrazines (eq 11).[11]

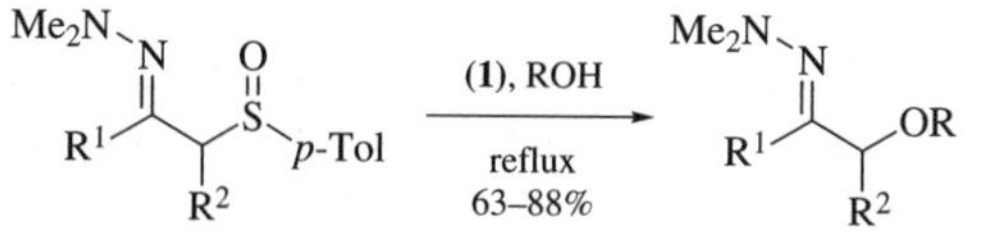

(11)

Related Reagents. Propionic Acid; *p*-Toluenesulfonic Acid; Trifluoroacetic Acid.

1. Miyashita, N.; Yoshikoshi, A.; Grieco, P. A. *JOC* **1977**, *42*, 3772.
2. Freeman, F.; Kim, D. S. H. L.; Rodriguez, E. *JOC* **1992**, *57*, 1722.
3. Pinnick, H. W.; Bal, B. S.; Lajis, N. H. *TL* **1979**, 4261.
4. Cottet, F.; Cottier, L.; Descotes, G.; Srivastara, R. M. *JHC* **1988**, *25*, 1481.
5. Prakash, C.; Saleh, S.; Blair, I. A. *TL* **1989**, *30*, 19.
6. Sterzycki, R. *S* **1979**, 724.
7. Halterman, R. L.; Jan, S-T. *JOC* **1991**, *56*, 5253.
8. Monti, H.; Leandri, G.; Ringuet, M. K.; Corriol, C. *SC* **1983**, *13*, 1021.
9. Kucerovy, A.; Neunschwander, K.; Weinreb, S. M. *SC* **1983**, *13*, 875.
10. Baciocchi, E.; Ruzziconi, R. *SC* **1988**, *18*, 1841.
11. Pfieger, P.; Mioskowski, C.; Salaun, J. P.; Weissbart, D.; Durst, F. *TL* **1989**, *30*, 2791.

Adam A. Galan
Parke Davis Pharmaceutical Research Division, Ann Arbor, MI, USA

Pyrrolidine

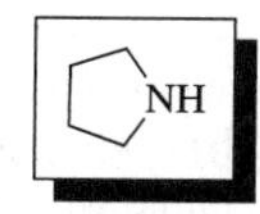

[123-75-1] C_4H_9N (MW 71.14)

(secondary amine used for the formation and utilization of enamines;[2] catalyst for aldol cyclization[17] and Mannich condensation;[23] formation of secondary amides[28])

Physical Data: bp 88.5–89 °C; *d* 0.8618 g cm^{-3}; pK_a 11.1.
Solubility: miscible with water; sol alsolcohol, chloroform, and ether.
Form Supplied in: liquid, 99%; widely available.
Purification: distilled under nitrogen after drying with sodium or BaO.[1]
Handling, Storage, and Precautions: flammable; fumes in air; strong base; irritant. Use in a fume hood.

Enamine Preparation and Utilization[2]. The principal use of pyrrolidine in synthesis is in the formation and utilization of ketone and aldehyde enamines. For preparation of the enamine a benzene or toluene solution of the carbonyl compound and pyrrolidine is heated at reflux with azeotropic removal of water. Acid catalysts or molecular sieves as water trapping agents are frequently employed. Cyclopentanone[3–5] and cyclohexanone,[3–6] for example, react in benzene to form enamines without added acid catalyst (eqs 1 and 2), as does β-tetralone (eq 3).[7]

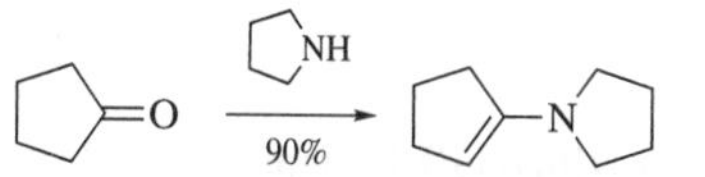

(1)

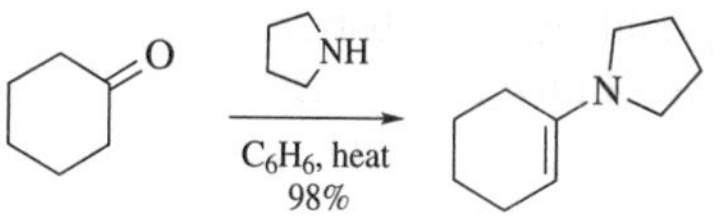

(2)

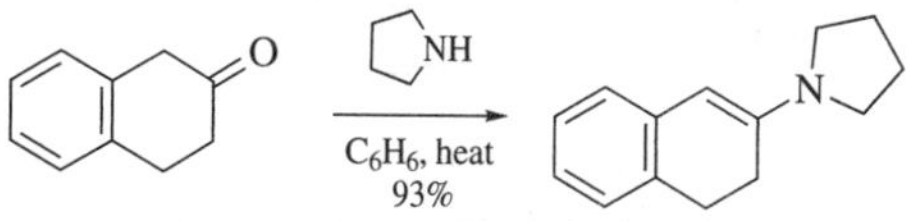

(3)

In toluene solution, cyclododecanone is converted to the pyrrolidine enamine in the presence of ***Boron Trifluoride Etherate*** (eq 4).[8]

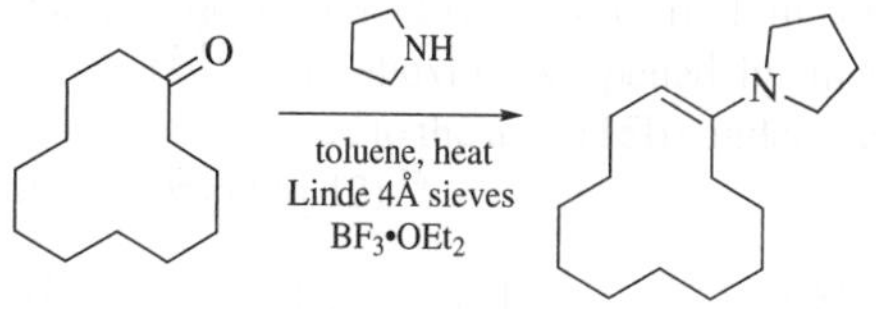

(4)

Acyclic ketones form pyrrolidine enamines sluggishly or with significant self-condensation. The enamine of diethyl ketone may be prepared in modest yield in the presence of 4 Å molecular sieves (eq 5).[5]

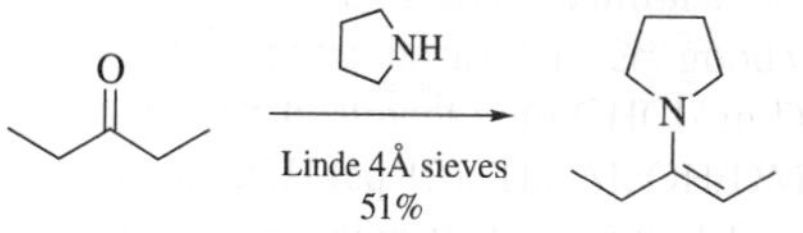

(5)

The pyrrolidine enamine of a relatively low-boiling aldehyde, isobutyraldehyde,[9] may be prepared by heating the base and the carbonyl component without additional solvent (eq 6).

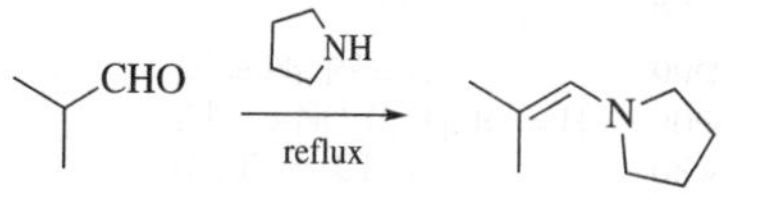

(6)

The reaction of ketones with pyrrolidine, ***Dimethyl Diazomethylphosphonate***, and base yields the pyrrolidine enamine of the homologous aldehyde (eq 7).[10]

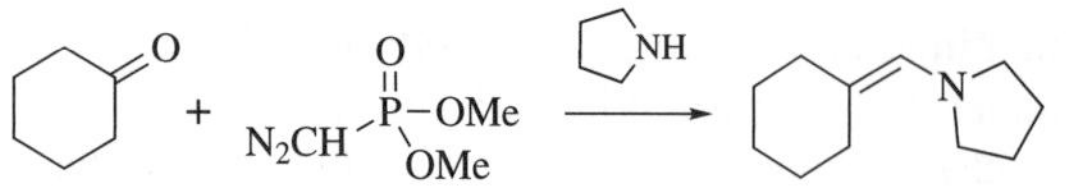

(7)

Enamino ketones have been prepared from β-chlorovinyl ketones (eq 8)[11] or directly from the corresponding β-dicarbonyl compound (eq 9).[12]

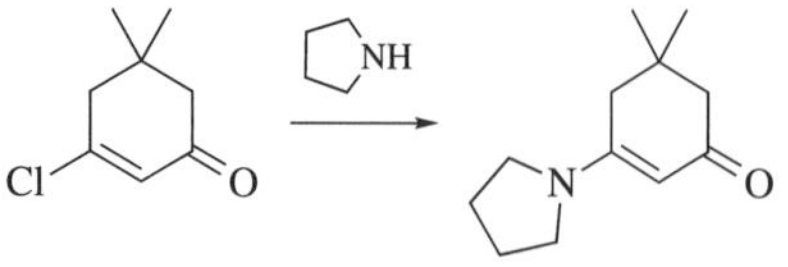

(8)

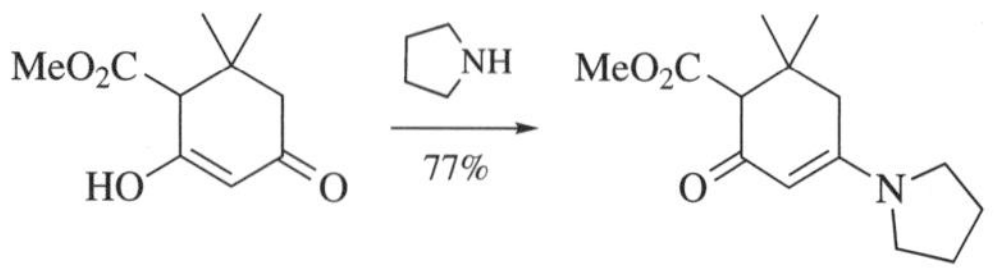

77%

(9)

The reaction of pyrrolidine with monoalkylated cyclohexanones results in the formation of the less substituted enamine as the major product (eq 10).[5]

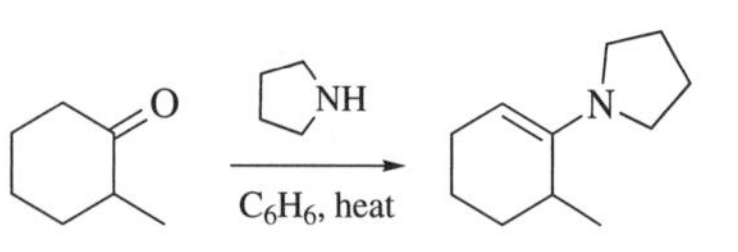

C_6H_6, heat

(10)

Regioselectivity is also observed in the formation of an enamine from 3-tri-*n*-butylstannylcyclohexanone and pyrrolidine (eq 11).[13]

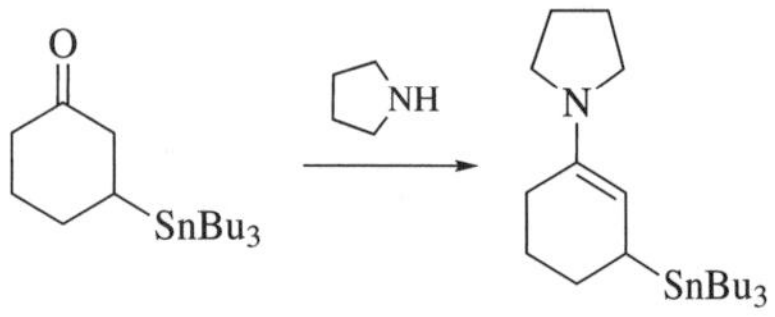

(11)

Pyrrolidine enamines react with both alkyl halides and electrophilic alkenes[2–5,7,14] to provide α-alkylated ketones and aldehydes (eqs 12 and 13). The great advantage of enamines over simple enolates in these reactions is that they do not overalkylate. Monosubstitution products are generally obtained. Acylation at carbon to afford β-dicarbonyl compounds occurs upon exposure of pyrrolidine enamines to acyl halides, but less effectively than with the corresponding morpholine enamine.[5]

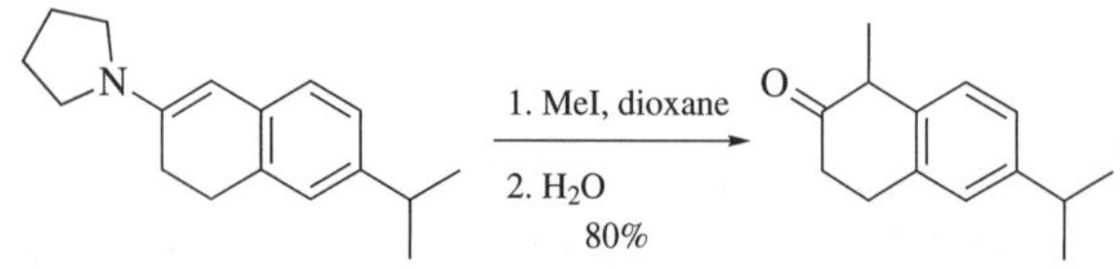

1. MeI, dioxane
2. H_2O
80%

(12)

64%

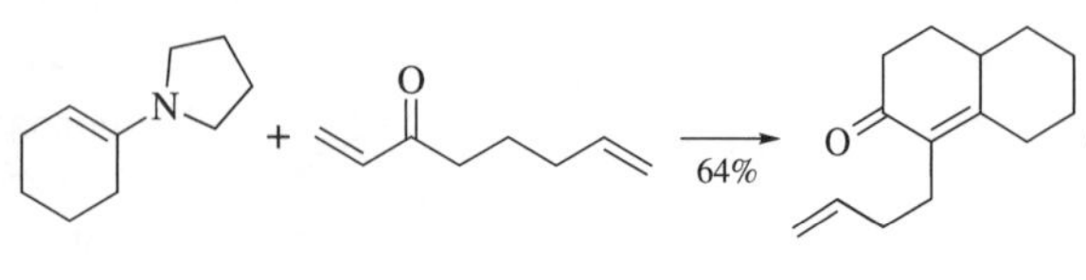

(13)

Pyrrolidine enamines also undergo [2+4] cycloaddition reactions with ***1,2,4-Triazine*** to provide, following loss of dinitrogen and aromatization, substituted pyridines (eq 14) and isoquinolines.[15,16]

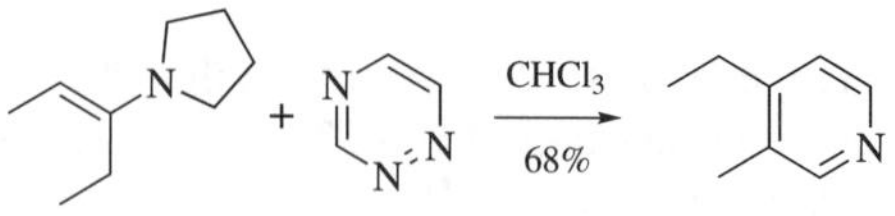

$CHCl_3$
68%

(14)

Condensation Reactions. Pyrrolidine is a common catalyst for the aldol cyclization phase of the Robinson annulation process for both diketones[14,17–20] and keto aldehydes (eqs 15 and 16).[21]

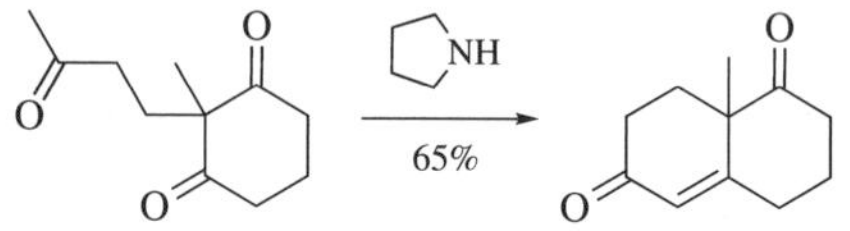

65%

(15)

HOAc

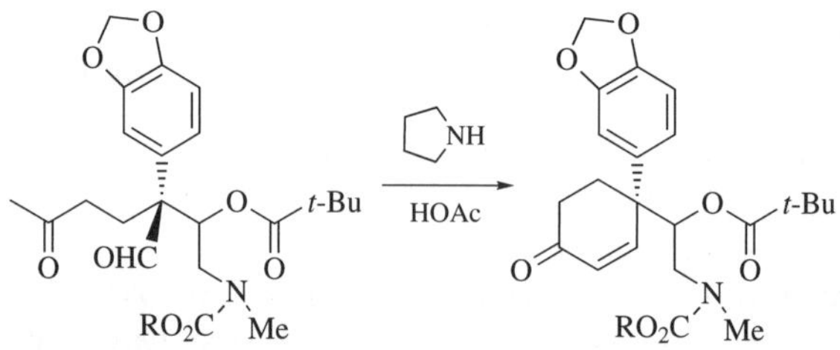

(16)

R = $CH_2CH{=}CH_2$

Spirocyclization has been shown to be favored over hydronaphthalenone closure in the reactions of substituted formylcyclohexanedione ethers (eq 17).[22]

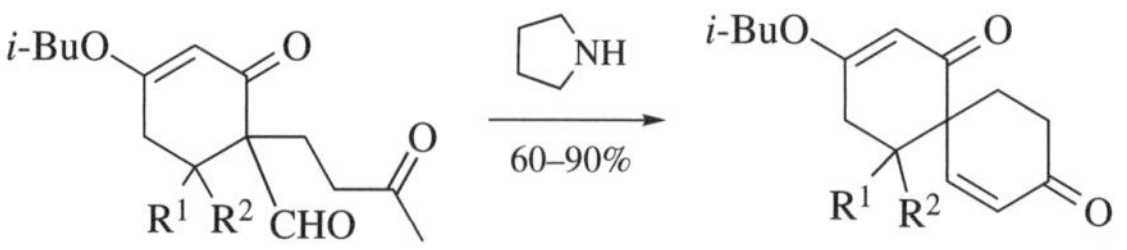

60–90%

(17)

R^1, R^2 = H; R^1 = H, R^2 = Me; R^1, R^2 = Me

In similar fashion to aldol cyclizations, Mannich-type condensations are frequently carried out with pyrrolidine as the secondary amine component. The enamine of cyclohexanone reacts with an *o*-hydroxyacetophenone to afford a spirochromanone product (eq 18).[23]

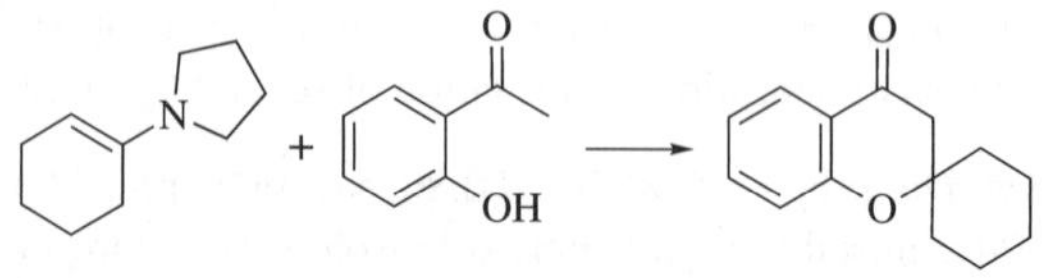

(18)

Phenols acting as the enol component of the Mannich process react with aldehydes in the presence of pyrrolidine (eq 19).[24]

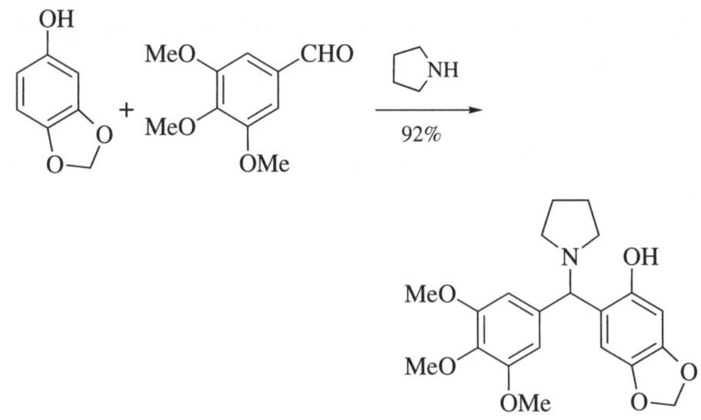

(19)

Pyrrolidine is also the most effective catalyst for the condensation of both aldehydes and ketones with cyclopentadiene to provide fulvenes (eqs 20 and 21).[25–27] In these cases the pyrrolidine has been shown to play the roles of both base catalyst and enamine-forming amine. The rates of such condensations are greater with pyrrolidine than with other secondary amine catalysts, a result attributable to the minimum steric hindrance exhibited by the cyclic amine.

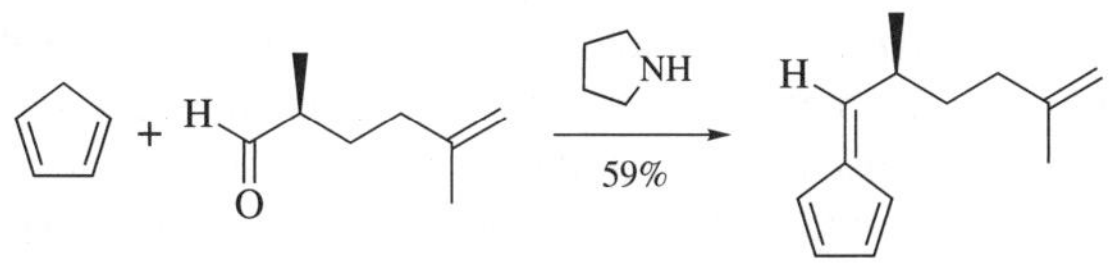

(20)

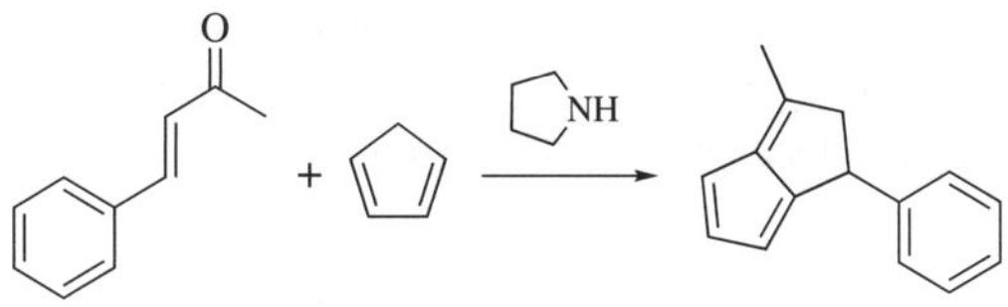

(21)

Amide Formation. Pyrrolidine is an effective reagent for the formation of secondary amides from esters (eq 22) and urethanes from thiocarbonates (eq 23).[28,29]

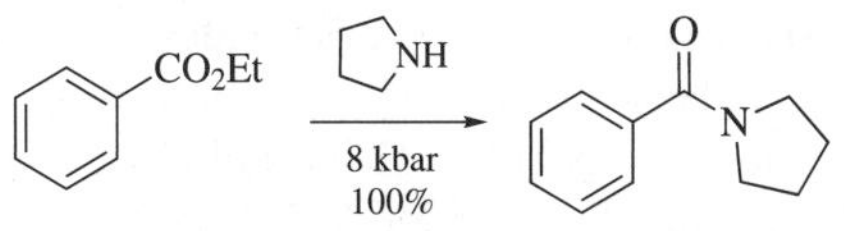

(22)

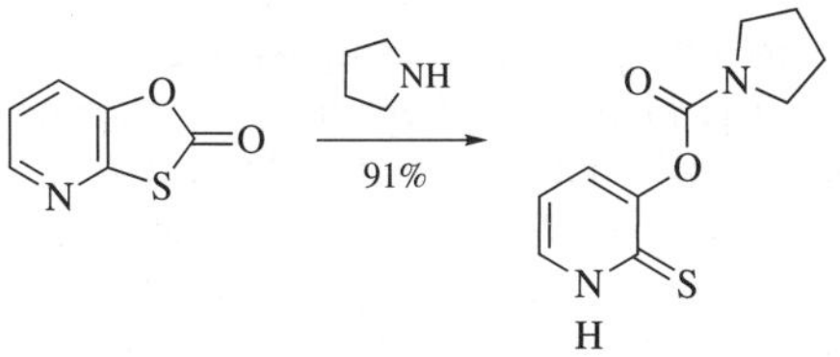

(23)

Related Reagents. (*S*)-1-Amino-2-methoxymethylpyrrolidine; Diisopropylamine; *trans*-2,5-Dimethylpyrrolidine; (*S*)-2-Methoxymethylpyrrolidine; Morpholine; Pyridine; Quinoline.

1. Perrin, D. D.; Armarego, W. L. F. *Purification of Laboratory Chemicals*, 3rd ed.; Pergamon: Oxford, 1988; p 542.
2. Cook, A. G. *Enamines: Synthesis, Structure, and Reactions*, 2nd ed.; Dekker: New York, 1988; p 717.
3. Stork, G.; Terrell, R.; Szmuszkovicz, J. *JACS* **1954**, *76*, 2029.
4. Stork, G.; Landesman, H. K. *JACS* **1956**, *78*, 5128.
5. Stork, G.; Brizzolara, A.; Landesman, H.; Szmuszkovicz, J.; Terrell, R. *JACS* **1963**, *85*, 207.
6. Woodward, R. B.; Pachter, I. J.; Scheinbaum, M. L. *OSC* **1988**, *6*, 1014.
7. Stork, G.; Schulenberg, J. W. *JACS* **1962**, *84*, 284.
8. (a) Hamada, Y.; Shioiri, T. *OS* **1984**, *62*, 191. (b) Hamada, Y.; Shioiri, T. *OSC* **1990**, *7*, 135.
9. Chan, Y.; Epstein, W. W. *OSC* **1988**, *6*, 496.
10. Gilbert, J. C.; Weerasooriya, U. *TL* **1980**, *21*, 2041.
11. Leonard, N. J.; Adamcik, J. A. *JACS* **1959**, *81*, 595.
12. Greenhill, J. V.; Chaaban, I.; Steel, P. J. *JHC* **1992**, *29*, 1375.
13. Ahlbrecht, H.; Weber, P. *S* **1989**, 117.
14. Tsuji, J. *S* **1984**, 369.
15. Boger, D. L.; Schumacher, J.; Mullican, M. D.; Patel, M.; Panek, J. S. *JOC* **1982**, *47*, 2673.
16. Boger, D. *T* **1983**, *39*, 2869.
17. Ramachandran, S.; Newman, M. S. *OSC* **1973**, *5*, 486.
18. Spencer, T. A.; Schmiegel, K. K.; Williamson, K. L. *JACS* **1963**, *85*, 3785.
19. Scanio, C. J. V.; Hill, L. P. *S* **1970**, 651.
20. Begbie, A. L.; Golding, B. T. *JCS(P1)* **1972**, 602.
21. Martin, S. F.; Davidsen, S. K. *JACS* **1984**, *106*, 6431.
22. de Groot, A.; Jansen, B. J. M. *TL* **1976**, 2709.
23. Kabbe, H. J. *S* **1978**, 886.
24. Jurd, L. *JHC* **1988**, *25*, 89.
25. Stone, K. J.; Little, R. D. *JOC* **1984**, *49*, 1849.
26. Stone, K. J.; Little, R. D. *JACS* **1985**, *107*, 2495.
27. Griesbeck, A. G. *JOC* **1989**, *54*, 4981.
28. Matsumoto, K.; Hashimoto, S.; Otani, S. *AG(E)* **1986**, *25*, 565.
29. Matsumoto, K.; Hashimoto, S.; Uchida, T.; Okamoto, T.; Otani, S. *CB* **1989**, *122*, 1357.

David Goldsmith
Emory University, Atlanta, GA, USA

Quinine

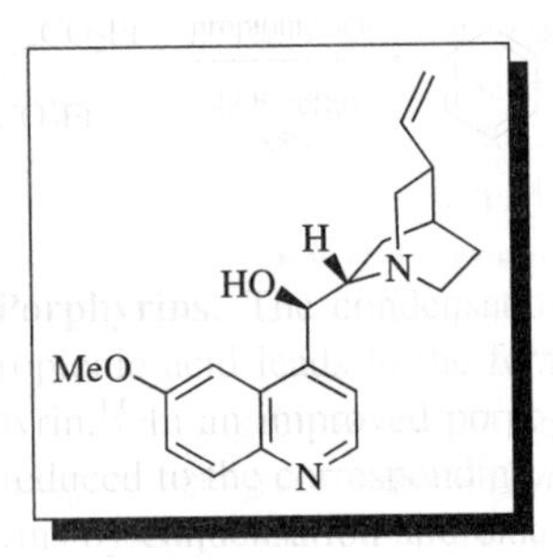

[130-95-0] $C_{20}H_{23}N_2O_2$ (MW 324.45)

(chiral catalyst[1–14])

Physical Data: mp 173–175 °C.
Solubility: sol hot water, methanol, benzene, chloroform, ether, glycerol; insol pet ether.
Form Supplied in: crystalline solid; 90% purity.
Analysis of Reagent Purity: NMR, mp.
Preparative Methods: commercially available from several sources.
Purification: recrystallize from absolute ethanol.
Handling, Storage, and Precautions: toxic; irritant.

Asymmetric Diels–Alder Reactions. Chiral bases, including quinine, have been used as catalysts in Diels–Alder reactions (eq 1).[1] The reactions take place at room temperature or below and require 1–10% equiv of the alkaloid. The asymmetric induction that is observed can be attributed to complex formation between the achiral dienolate and the chiral amine.[1]

(1) quinine, $CHCl_3$, 0 °C, 96%; 32% ee

Preparation of Chiral Sulfinates. Optically active sulfinates can be prepared by reaction of a symmetrical sulfite with ***t*-Butylmagnesium Chloride** in the presence of an optically active amino alcohol. The best enantioselectivity has been observed using quinine as the optically active amine (eq 2).[2] An alternative approach to this new enantioselective asymmetric synthesis of alkyl *t*-butylsulfinates would be reaction of a racemic sulfinate with *t*-butylmagnesium chloride complexed by optically active alkaloids (eq 3).[2] In this case, kinetic resolution of the racemic sulfinate leads to an optically active sulfinate and an optically active sulfoxide.

$$\text{EtO-S(O)-OEt} + t\text{-BuMgCl} \xrightarrow[\text{ether, 62\%}]{\text{quinine}} t\text{-Bu-S(O)-OEt} \quad (R)\text{, 69\% ee} \tag{2}$$

$$p\text{-Tol-S(O)-O-}i\text{-Pr} + t\text{-BuMgCl} \xrightarrow[\text{ether}]{\text{quinine}} p\text{-Tol-S(O)-O-}i\text{-Pr}\ ((S)\text{, 33\% yield, 33\% ee}) + t\text{-Bu-S(O)-}p\text{-Tol}\ ((S)\text{, 66\% yield, 13\% ee}) \tag{3}$$

Stereoselective Addition of Diethylzinc to Aldehydes. Wynberg has found that the cinchona alkaloids catalyze the reaction of ***Diethylzinc*** and aldehydes to form optically active alcohols (eq 4).[3] The highest enantiomeric excess obtained was from reactions which used quinine as the catalyst. Results show that the hydroxyl group of the catalyst hydrogen bonds with the aldehyde and that the diethylzinc interacts with the vinyl group of the catalyst as well, but it has not been determined if one or two catalyst molecules are involved in the transition state. Similar results have been obtained using a furan aldehyde.[4]

$$\text{PhCHO} + ZnEt_2 \xrightarrow[\text{2. HCl, } H_2O;\ 92\%]{\text{1. quinine, toluene}} \text{PhCH(OH)Et} \quad \text{68\% ee} \tag{4}$$

Synthesis of Optically Active Epoxides. Alkaloids and alkaloid salts have been successfully used as catalysts for the asymmetric synthesis of epoxides. The use of chiral catalysts such as quinine or quinium benzylchloride (QUIBEC) have allowed access to optically active epoxides through a variety of reaction conditions, including oxidation using ***Hydrogen Peroxide*** (eq 5),[5] Darzens condensations (eq 6),[6] epoxidation of ketones by ***Sodium Hypochlorite*** (eq 7),[6] halohydrin ring closure (eq 8),[6] and cyanide addition to α-halo ketones (eq 9).[6] Although the relative stereochemistry of most of the products has not been determined, enantiomerically enriched materials have been isolated. A more recent example has been published in which optically active 2,3-epoxycyclohexanone has been synthesized by oxidation with ***t*-Butyl Hydroperoxide** in the presence of QUIBEC and the absolute stereochemistry of the product established (eq 10).[7]

(5) chalcone; QUIBEC, 30% H_2O_2, 90%; (optical purity not determined)

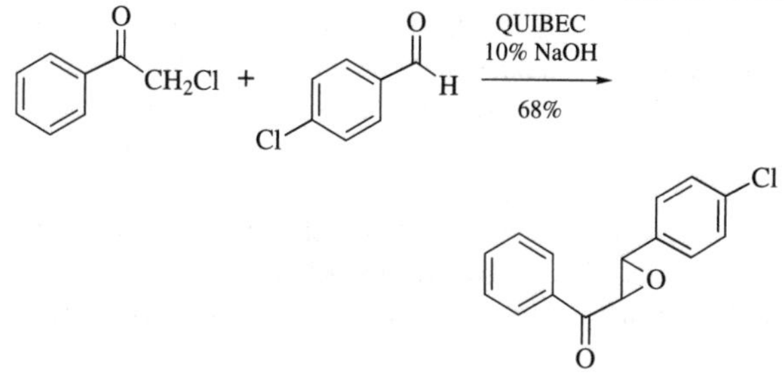

QUIBEC, 10% NaOH, 68%

8% ee (6)

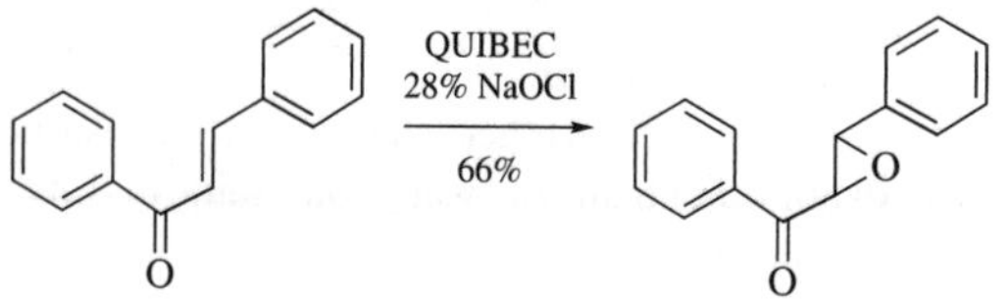

QUIBEC, 28% NaOCl, 66%

25–30% ee (7)

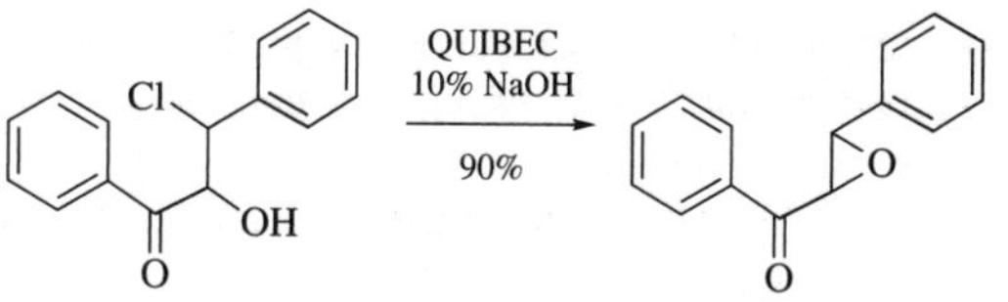

QUIBEC, 10% NaOH, 90%

6% ee (8)

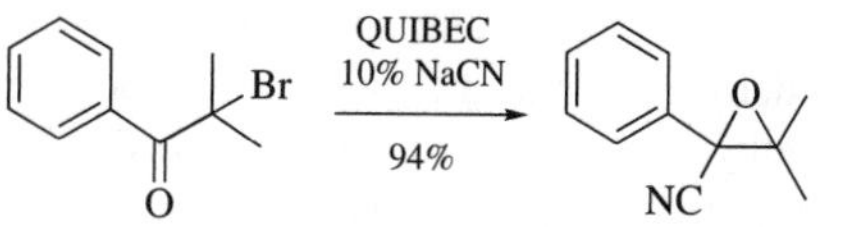

QUIBEC, 10% NaCN, 94%

(optical purity not determined) (9)

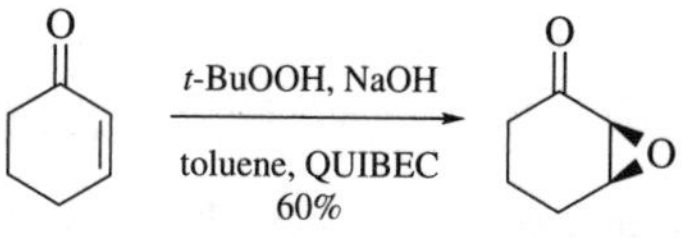

t-BuOOH, NaOH; toluene, QUIBEC; 60%

20–23% ee (10)

Asymmetric Michael Reactions. Asymmetric induction has been observed in Michael-type addition reactions that are catalyzed by chiral amines.[8] The *N*-benzyl fluoride salt of quinine has been particularly successful since the fluoride ion serves as a base and the aminium ion as a source of chirality.[9] Drastic improvements in optical purity (1–23%) have resulted by changing from quinine to the *N*-benzyl fluoride salt (eq 11).[9]

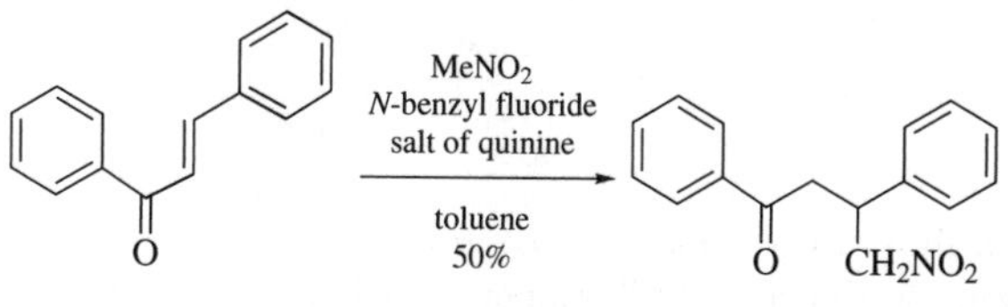

$MeNO_2$, *N*-benzyl fluoride salt of quinine; toluene; 50%

CH_2NO_2

23% ee (11)

Asymmetric Synthesis of β-Keto Sulfides. Quinine can be used to catalyze asymmetrically the addition of thiols to cyclohexenone, thus forming β-keto sulfides (eq 12).[10] The absolute stereochemistry of the products has not been determined.

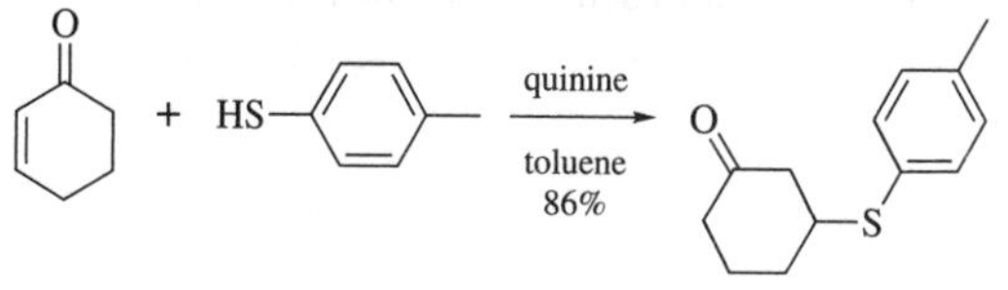

quinine; toluene; 86%

46% ee (12)

Asymmetric Reduction of Ketones. Alkyl phenyl ketones can be asymmetrically reduced to the corresponding alcohol using ***Sodium Borohydride*** under phase-transfer conditions in the presence of a catalytic amount of QUIBEC (eq 13).[11] The results indicate that the asymmetric reduction is due to the rigidity of the catalyst as well as the β-position of the hydroxyl group on the quinine molecule. The asymmetric induction is much lower with a γ-hydroxyl group.[11]

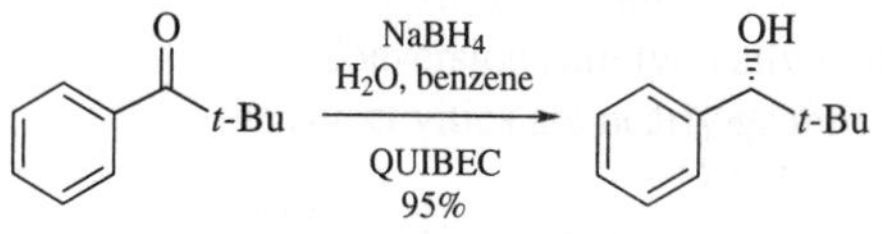

$NaBH_4$, H_2O, benzene; QUIBEC; 95%

32% ee (13)

Synthesis of Optically Active β-Hydroxy Esters. Chiral amino alcohols such as quinine have been used in the enantioselective synthesis of β-hydroxy esters via an indium-induced Reformatsky reaction (eq 14).[12] Although the enantioselectivities are not particularly high, aromatic aldehydes have produced the best results to date. The absolute stereochemistry of the products has not yet been assigned.

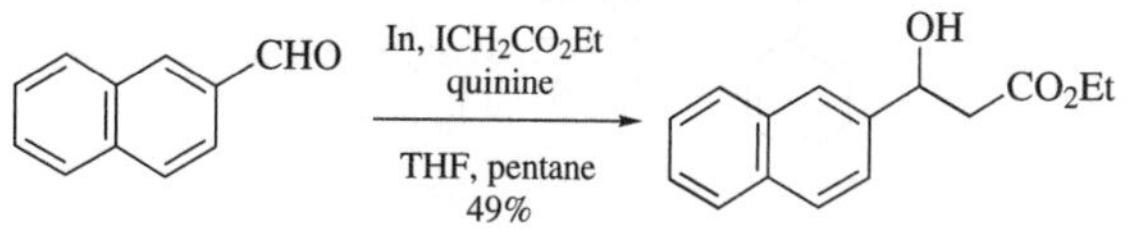

In, ICH_2CO_2Et, quinine; THF, pentane; 49%

49% ee (14)

Preparation of Polymeric Catalyst. A quinine/***Acrylonitrile*** copolymer has been successfully synthesized via radical polymerization using ***Azobisisobutyronitrile*** (AIBN) as initiator (eq 15).[13] The polymer can be prepared such that the vinyl group is the connecting site and the amino alcohol portion can either be free or protected. These copolymers are thermally stable and are soluble in polar aprotic solvents such as DMF and DMSO, but insoluble in common organic solvents. Preliminary experiments have shown that these copolymers can be used as asymmetric catalysts.[13]

Asymmetric Addition of Thioglycolic Acid to Nitro Alkenes. Quinine has been used to catalyze the addition of thioglycolic acid to nitro alkenes (eq 16).[14] Enantiomerically enriched materials have been isolated, although the absolute stereochemistry of the products has not been assigned. The direction and extent of asymmetric induction seems to be dependent on the catalyst/acid ratio, thereby pointing to interaction between the carbonyl of the acid and the alkaloid nitrogen as being responsible for the asymmetric induction.[14]

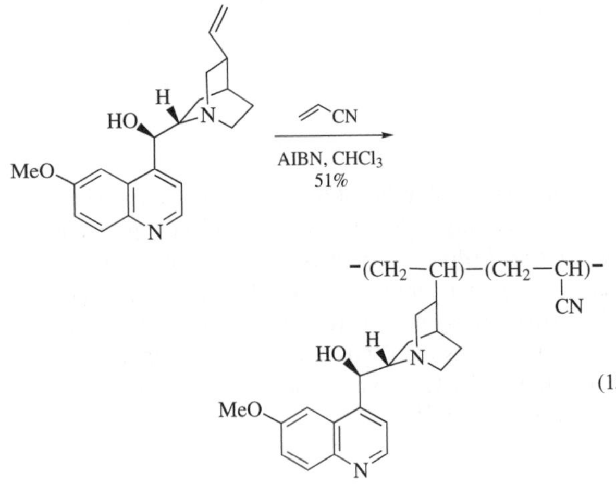

(15)

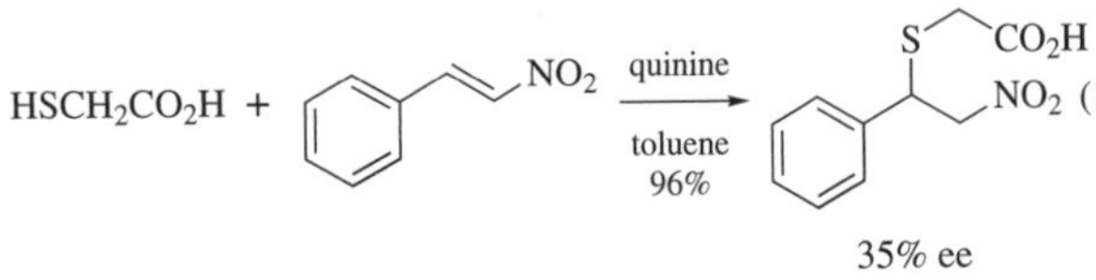

$$HSCH_2CO_2H + PhCH{=}CHNO_2 \xrightarrow[\text{toluene, 96\%}]{\text{quinine}} PhCH(SCH_2CO_2H)CH_2NO_2 \quad (16)$$

35% ee

Related Reagents. (−)-Sparteine.

1. Riant, O.; Kagan, H. B. *TL* **1989**, *30*, 7403.
2. Drabowicz, J.; Legedź, S.; Mikolajczyk, M. *T* **1988**, *44*, 5243.
3. Smaardijk, A. A.; Wynberg, H. *JOC* **1987**, *52*, 135.
4. van Oeveren, A.; Menge, W.; Feringa, B. L. *TL* **1989**, *30*, 6427.
5. Helder, R.; Hummelen, J. C.; Laane, R. W. P. M.; Wiering, J. S.; Wynberg, H. *TL* **1976**, 1831.
6. Hummelen, J. C.; Wynberg, H. *TL* **1978**, 1089.
7. Wynberg, H.; Marsman, B. *JOC* **1980**, *45*, 158.
8. Wynberg, H.; Helder, R. *TL* **1975**, 4057.
9. Colonna, S.; Hiemstra, H.; Wynberg, H. *CC* **1978**, 238.
10. Helder, R.; Arends, R.; Bolt, W.; Hiemstra, H.; Wynberg, H. *TL* **1977**, 2181.
11. Colonna, S.; Fornasier, R. *JCS(P1)* **1978**, 371.
12. Johar, P. S.; Araki, S.; Butsugan, Y. *JCS(P1)* **1992**, 711.
13. Kobayashi, N.; Iwai, K. *JACS* **1978**, *100*, 7071.
14. Kobayashi, N.; Iwai, K. *JOC* **1981**, *46*, 1823.

Ellen M. Leahy
Affymax Research Institute, Palo Alto, CA, USA

Quinoline

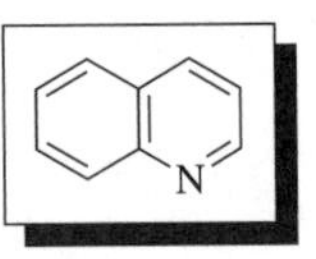

[91-22-5] C_8H_7N (MW 129.16)

(useful as a base, solvent and/or catalyst, especially for decarboxylation reactions[1] and the Rosenmund reaction[2,3])

Physical Data: mp −15 °C; bp 237.6 °C;[4] steam volatile; *d* 1.0858 g cm^{-3} at 30 °C; pK_a 9.5.[5]
Solubility: sol ethanol, ethyl ether, acetone, carbon disulfide; more readily sol hot water than cold water; quinoline dissolves sulfur, phosphorus, and arsenic trioxide.
Form Supplied in: colorless liquid.
Analysis of Reagent Purity: GLC.
Purification: distillation.
Handling, Storage, and Precautions: very hygroscopic; darkens on storage in light; package with protection from light and moisture; LD_{50} rat (oral) 460 mg kg^{-1}; incompatible with strong oxidizing agents and strong acids.[5,6]

Dehydrohalogenation Reactions. Quinoline is sometimes used as a base or solvent for dehydrohalogenation reactions because of its basic properties. A wide variety of substrates, ranging from very simple to quite complex compounds, have been effectively dehydrohalogenated with quinoline. For example, 3-bromo-3-methyl-2-butanone (**1**) underwent reaction with quinoline to give a mixture of products, 3-methyl-2-butanone (**2**) and 2-methyl-1-buten-3-one (**3**) (eq 1).[7]

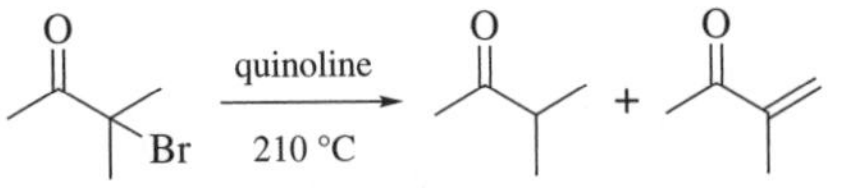

(1) —quinoline, 210 °C→ (2) 11% + (3) 43% (1)

Trans-3-penten-2-one (**5**) has been prepared from 4-chloropentan-2-one (**4**) using quinoline as the acid scavenger (eq 2).[8]

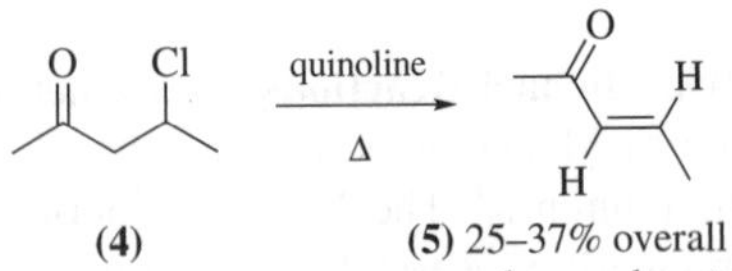

(4) —quinoline, Δ→ (5) 25–37% overall from propylene and acetyl chloride (2)

Dehydrochlorination of 2-oxa-7,7-dichloronorcarane (**6**) in quinoline gives an 83% yield of 2,3-dihydro-6-chlorooxepine (**7**) (eq 3).[9]

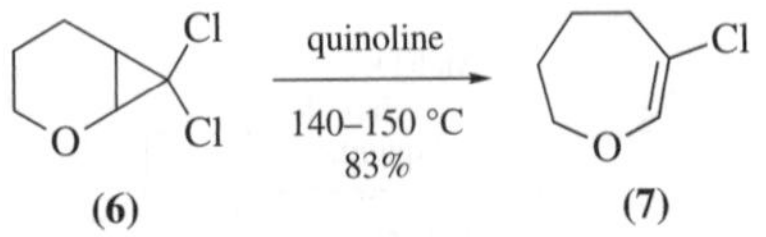

(6) —quinoline, 140–150 °C, 83%→ (7) (3)

Similarly, when 7,7-dibromonorcarane (**8**) is treated with quinoline, by manipulation of reaction conditions the bromocyclohepta-1,3-diene derivatives can be isolated or total dehydrobromination to ***1,3,5-Cycloheptatriene*** (**9**) can be effected (eq 4).[10]

(8) → quinoline, 200 °C, 66% → (9) (4)

Selective dehydrobromination of (α-bromovinyl)chlorosilanes can be carried out in quinoline, affording fair to good yields of the corresponding ethynylchlorosilanes (eq 5).[11]

quinoline, 180–200 °C, 77% (5)

Decarboxylation Reactions. Taking advantage of its basic properties, quinoline is generally useful as a solvent for decarboxylation reactions. It is especially suitable because its relatively high boiling point facilitates the decarboxylation. Examples include the decarboxylation of 3-methyl-2-furoic acid (**10**) to 3-methylfuran (**11**) (eq 6),[12] *m*-nitrocinnamic acid (**12**) to *m*-nitrostyrene (**13**) (eq 7),[1] and α-phenylcinnamic acid (**14**) to *cis*-stilbene (**15**) (eq 8).[13] Copper catalysts such as ***Copper*** or ***Copper Chromite*** are used to effect the decarboxylations.

(10) → Cu^0, quinoline, 250 °C, 83–89% → (11) (6)

(12) → Cu^0, quinoline, 185–195 °C, 56–60% → (13) (7)

(14) → Cu chromite, quinoline, 210–220 °C, 62–65% → (15) (8)

Preparation of Isothiocyanates. ***Ethoxycarbonyl Isothiocyanate*** (**18**), useful as a synthetic reagent for heterocyclic syntheses,[14] can be prepared in excellent yield from the reaction of ***Ethyl Chloroformate*** (**16**) with ***Sodium Thiocyanate*** (**17**) using quinoline as a base catalyst (eq 9).[15] Only trace amounts of the isomeric thiocyanates are formed in the reaction. In the absence of the base catalyst, only a moderate (65%) yield of the desired product is obtained, along with significant (10%) contamination by the isomeric ethoxycarbonyl thiocyanate. ***Pyridine*** can also be used as the catalyst here; however, it reacts faster, which sometimes leads to undesired byproduct formation.

EtO(C=O)Cl (16) + NaSCN (17) → quinoline, 80% → EtO(C=O)NCS (18) (9)

Quinoline as a Solvent. When quinoline is used as a solvent for the reaction of ***Phenylcopper*** with iodoarenes, metal–halogen exchange and unsymmetrical coupling reactions occur as shown in eq 10.[16] The products suggest that basic solvents such as quinoline solvate the phenylcopper, leading to a complex mixture of products. Similar activity is observed in pyridine solvent.

quinoline, 50 °C, 25 h; 4:3:1 (10)

Acylation Reactions. Many basic catalysts have been used to carry out acylation reactions. The stereoselectivity of the reaction of diacetoxy or dibenzoyloxysuccinic anhydrides with racemic alcohols is affected by the choice of base catalyst. Quinoline is one of the most effective base catalysts for enhancing the stereoselectivity of the reaction of (2*R*,3*R*)-2,3-diacetoxysuccinic anhydride with 1-phenylethanol in comparison with catalysts such as pyridine, 3- and 4-methylpyridines, and isoquinoline. Other effective basic catalysts are pyridine derivatives substituted in the 2-position of the ring such as 2-methylpyridine and 2,6-dimethylpyridine.[17]

Formation of Allenes. Quinoline can be used as an acid scrubber in the preparation of allenes via vinyl triflate intermediates. The elimination of ***Trifluoromethanesulfonic Acid*** in quinoline proceeds in good yield (eqs 11 and 12).[18]

OTf → quinoline, 100 °C, 79% (11)

Et, OTf → quinoline, 100 °C, 85% → Et (12)

Rosenmund Reaction. A catalyst poison prepared from quinoline and ***Sulfur*** is useful for controlling the reaction of β-naphthoyl chloride (**19**) with hydrogen gas and ***Palladium on Barium Sulfate*** catalyst.[2] If control of the reaction is not maintained by catalyst poisoning to reduce activity, further reduction beyond the desired β-naphthaldehyde product (**20**) is often observed (eq 13).[3]

(19) → Pd, $BaSO_4$, H_2, quinoline, S, 74–81% → (20) (13)

Elimination Reactions. Quinoline in DMSO facilitates the elimination of dimethyl sulfide from sulfonium salt (**21**), affording the very reactive cyclobutenone derivative (**22**) (eq 14).[19]

quinoline, DMSO (14)

$Me_2\overset{+}{S}$ FSO_3^-

(**21**) (**22**)

Related Reagents. 2,4,6-Collidine; 2,6-Lutidine; Pyridine.

---

1. Wiley, R. H.; Smith, N. R. *OSC* **1963**, *4*, 731.
2. Rosenmund, K. W.; Zetzsche, F. *CB* **1921**, *54*, 425.
3. Hershberg, E. B.; Cason, J. *OSC* **1955**, *3*, 627.
4. Holter, S. N. In *Kirk-Othmer Encyclopedia of Chemical Technology*; Wiley: New York, 1982; Vol. 19, p 532.
5. *The Merck Index*, 11th edn.; Budavari, S., Ed.; Merck: Rahway, NJ, 1989; p 1285.
6. *The Sigma-Aldrich Library of Regulatory & Safety Data*; Sigma-Aldrich: Milwaukee, WI, 1993; Vol. 2, p 2611.
7. Griesbaum, K.; Kibar, R. *CB* **1973**, *106*, 1041.
8. Odom, H. C.; Pinder, A. R. *OSC* **1988**, *6*, 883.
9. Schweizer, E. E.; Parham, W. E. *JACS* **1960**, *82*, 4085.
10. Lindsay, D. G.; Reese, C. B. *T* **1965**, *21*, 1673.
11. Matsumoto, H.; Kato, T.; Matsubara, I.; Hoshino, Y.; Nagai, Y. *CL* **1979**, 1287.
12. Burness, D. M. *OSC* **1963**, *4*, 628.
13. Buckles, R. E.; Wheeler, N. G. *OSC* **1963**, *4*, 857.
14. George, B.; Papadopoulos, E. P. *JHC* **1983**, *20*, 1127.
15. Lewellyn, M. E.; Wang, S. S.; Strydom, P. J. *JOC* **1990**, *55*, 5230.
16. Nilsson, M.; Wennerstrom, O. *TL* **1968**, 3307.
17. Bell, K. H. *AJC* **1981**, *34*, 671.
18. Stang, P. J.; Hargrove, R. J. *JOC* **1975**, *40*, 657.
19. Kelly, T. R.; McNutt, R. W. *TL* **1975**, 285.

Angela R. Sherman
Reilly Industries, Indianapolis, IN, USA

Silver(I) Tetrafluoroborate

[14104-20-2] $AgBF_4$ (MW 194.68)

(mild Lewis acid with a high affinity for organic halides)

Physical Data: mp 200 °C (dec).

Solubility: sol benzene, toluene, nitromethane, diethyl ether, water.

Form Supplied in: white solid; widely available.

Analysis of Reagent Purity: contents of Ag can be assayed conveniently by volumetric titration of Ag^I.

Preparative Method: can be prepared by reacting ***Silver(I) Fluoride*** with ***Boron Trifluoride*** in nitromethane.[1]

Handling, Storage, and Precautions: should be protected from light and moisture; very hygroscopic.

Introduction. This reagent has replaced ***Silver(I) Perchlorate*** to a large extent because of the sensitivity of perchlorates.

Activation of Acyl Chlorides. In several cases, $AgBF_4$ has been used to increase the reactivity of acyl chlorides towards nucleophiles.[2] For example, *N*-acylammonium salts were prepared for the first time by the reaction of a tertiary amine and an acyl chloride in the presence of $AgBF_4$ (eq 1).[3]

Nucleophilic Substitution on Alkyl Halides by Heteroatoms. A number of more or less activated alkyl halides, such as benzyl halides[4] and allyl halides,[5] undergo substitution reactions mediated by $AgBF_4$ in the presence of a heteroatom nucleophile. For example, treatment of pentamethylcyclopentadienyl bromide with $AgBF_4$ in the presence of a nucleophile gives the corresponding substituted product (eq 2). Thiols, amines, and alcohols have been used as nucleophiles.[6]

Adenine analogs are prepared stereoselectively from cyclopentene derivatives using a two-step procedure (eq 3). The reaction probably involves a seleniranium salt as an intermediate.[7]

Intramolecular substitutions mediated by Ag^+ do not seem to require activated halides.[8] For example, ω-chloro amides react with $AgBF_4$, giving products from intramolecular attack of the amide oxygen. Depending on the structure of the amide, imino lactones,[9] imino lactonium salts,[10] or lactone hydrazones (eq 4)[11] are obtained as products. Fluorination of α-bromo ketones using $AgBF_4$ has also been reported.[12]

Nucleophilic Aromatic Substitution. In one case, it has been reported that $AgBF_4$ promotes the nucleophilic substitution of an aromatic chloride (eq 5).[13] This is not due to activation of the halide, but apparently to suppression of halide-promoted decomplexation of the arene–manganese derivative.

Carbon–Carbon Bond Formation via Cationic Intermediates. In analogy with the heteroatom substitutions described above, certain aliphatic halides undergo substitution reactions with carbon nucleophiles promoted by $AgBF_4$. For example, Eschenmoser and co-workers used $AgBF_4$ in order to transform α-chloro nitrones into 1,3-dipoles which react with ordinary alkenes in a cycloaddition manner (eq 6).[14]

Livinghouse and co-workers have shown that acylnitrilium ions, prepared from isocyanides and an acid chloride followed by treatment with $AgBF_4$, are useful intermediates in the synthesis of nitrogen-containing heterocycles (eq 7).[15]

OTBS; 1. (isobutyryl chloride); 2. $AgBF_4$, CH_2Cl_2, –78 °C, 78% (7)

In certain cases, allylsilanes[16] and trimethylsilyl enol ethers[17] react with alkyl halides with the formation of a new carbon–carbon bond. α-Bromo imidates[18] (eq 8) and β-chloro imines[19] have been reported to undergo electrophilic aromatic substitution on relatively electron-rich aromatics in the presence of $AgBF_4$.

$AgBF_4$, $(ClCH_2)_2$, rt, 60% (8)

Synthesis via Iminium Ions. α-Cyano amines react with $AgBF_4$ with the formation of an intermediate iminium ion.[20] This has been used synthetically as a method for removal of the cyano group either by a consecutive reduction[21,22] to the amine (eq 9) or by elimination to the imine[23] or enamine.[24]

1. $AgBF_4$, THF, rt; 2. $Zn(BH_4)_2$; 92% (9)

Rearrangements. A number of strained alkyl and/or reactive halides, such as cyclopropyl[25] and bicyclic[26,27] chlorides, rearrange on treatment with $AgBF_4$. For example, β-bromotetrahydropyrans rearrange to tetrahydrofurans stereoselectively on treatment with $AgBF_4$ (eq 10).[28] Other examples include the rearrangement of α-haloalkyl aryl ketones into arylacetic acid derivatives,[29] and the rearrangement of α-haloalkylsilanes upon treatment with $AgBF_4$.[30]

$AgBF_4$, MeOH, rt, 88% (10)

In the presence of strained hydrocarbons, $AgBF_4$ functions as a mild Lewis acid and causes rearrangements.[31–33] For example, the tricyclic hydrocarbon (**1**) rearranges upon treatment with a catalytic amount of $AgBF_4$ to the less strained hydrocarbon (**2**) (eq 11).[34]

$AgBF_4$, $CDCl_3$, rt; (**1**) → (**2**) (11)

Numerous examples include the rearrangement of propargyl esters into allenyl esters (see also ***Silver(I) Trifluoromethanesulfonate***)[35] or to dihydrofurans,[36] the Claisen rearrangement of aryl allenylmethyl ethers,[37] and the rearrangement of silyloxycyclopropanes (eq 12, also effected by ***Copper(II) Tetrafluoroborate***).[38]

TMSO; $AgBF_4$, Et_2O, –20 °C, 70% (12)

Activation of Thiol Esters. Pyridyl thiol esters are converted into esters on treatment with $AgBF_4$ and an alcohol.[39] Acylation of alkynylsilanes can also be carried out using thiol esters in the presence of $AgBF_4$.[40]

Alkylation of Thioethers. Thioethers can be methylated by ***Iodomethane*** in the presence of $AgBF_4$.[41] Benzylation of thioethers in the presence of $AgBF_4$ has also been reported.[42]

Electrophilic Aromatic Substitution. Electrophilic nitration using a combination of NO_2Cl and $AgBF_4$ has been reported.[43] Conversion of arylsilanes into iodides and bromides has been achieved using a combination of the halogen and $AgBF_4$ (eq 13).[44]

TMS–(thiophene $S O_2$)–TMS; $AgBF_4$, I_2, CH_2Cl_2, 0 °C, 97%; I–(thiophene $S O_2$)–I (13)

Catalysis of Cycloadditions. Addition of catalytic amounts of $AgBF_4$ greatly increases the selectivity of [2 + 4] cycloadditions of benzyne.[45]

Related Reagents. Dimethyl Sulfoxide–Silver Tetrafluoroborate; Lithium Perchlorate; Lithium Tetrafluoroborate; Silver Trifluoroacetate; Silver(I) Trifluoromethanesulfonate.

1. Olah, G. A.; Quinn, H. W. *J. Inorg. Nucl. Chem.* **1960**, *14*, 295.
2. Schegolev, A. A.; Smit, W. A.; Roitburd, G. V.; Kucherov, V. F. *TL* **1974**, 3373.
3. King, J. A., Jr.; Bryant, G. L., Jr. *JOC* **1992**, *57*, 5136.
4. Zimmerman, H. E.; Paskovich, D. H. *JACS* **1964**, *86*, 2149.
5. Bloodworth, A. J.; Tallant, N. A. *CC* **1992**, 428.
6. Jutzi, P.; Mix, A. *CB* **1992**, *125*, 951.
7. Wolff-Kugel, D.; Halazy, S. *TL* **1991**, *32*, 6341.
8. Lucchini, V.; Modena, G.; Pasquato, L. *CC* **1992**, 293.
9. Peter, H.; Brugger, M.; Schreiber, J.; Eschenmoser, A. *HCA* **1963**, *46*, 577.
10. Nader, R. B.; Kaloustain, M. K. *TL* **1979**, 1477.
11. Enders, D.; Brauer-Scheib, S.; Fey, P. *S* **1985**, 393.
12. Fry, A. J.; Migron, Y. *TL* **1979**, 3357.
13. Pearson, A. J.; Shin, H. *T* **1992**, *48*, 7527.
14. Kempe, H. M.; Das Gupta, T. K.; Blatt, K.; Gygax, P.; Felix, D.; Eschenmoser, A. *HCA* **1972**, *55*, 2187.
15. Lee, C. H.; Westling, M.; Livinghouse, T.; Williams, A. C. *JACS* **1992**, *114*, 4089; Luedtke, G.; Westling, M.; Livinghouse, T. *T* **1992**, *48*, 2209.
16. Nishiyama, H.; Naritomi, T.; Sakuta, T.; Itoh, K. *JOC* **1983**, *48*, 1557.
17. Padwa, A.; Ishida, M. *TL* **1991**, *41*, 5673; Padwa, A.; Austin, D. J.; Ishida, M.; Muller, C. M.; Murphree, S. S.; Yeske, P. E. *JOC* **1992**, *57*, 1161.

18. Shatzmiller, S.; Bercovici, S. *LA* **1992**, 997.
19. Kuehne, M.; Matson, P. A.; Bornmann, W. G. *JOC* **1991**, *56*, 513.
20. Grierson, D. S.; Bettiol, J. L.; Buck, I.; Husson, H. P. *JOC* **1992**, *57*, 6414.
21. Bettiol, J. L.; Buck, I.; Husson, H. P.; Grierson, D. S. *TL* **1991**, *32*, 5413.
22. Theodorakis, E.; Royer, J.; Husson, H. P. *SC* **1991**, 521.
23. Belattar, A.; Saxton, J. E. *JCS(P1)* **1992**, 1583.
24. Agami, C.; Couty, F.; Lin, J. *H* **1993**, *36*, 25.
25. Birch, A. J.; Keeton, R. *JCS(C)* **1968**, 109.
26. Yamada, Y.; Kimura, M.; Nagaoka, H.; Ohnishi, K. *TL* **1977**, 2379.
27. Kraus, G. A.; Zheng, D. *SL* **1993**, 71.
28. Ting, P. C.; Bartlett, P. A. *JACS* **1984**, *106*, 2668.
29. Giordano, C.; Castaldi, G.; Casagrande, F.; Belli, A. *JCS(P1)* **1982**, 2575.
30. Eaborn, C.; Lickiss, P. D.; Najim, S. T.; Stanczyk, W. A. *JCS(P2)* **1993**, 59; Eaborn, C.; Lickiss, P. D.; Najim, S. T. *JCS(P2)* **1993**, 391.
31. Paquette, L. A. *S* **1975**, 349.
32. Paquette, L. A. *JACS* **1970**, *92*, 5765.
33. Fitjer, L.; Justus, K.; Puder, P.; Dittmer, M.; Hassler, C.; Noltemeyer, M. *AG(E)* **1991**, *30*, 436.
34. Paquette, L. A.; Leichter, L. M. *JACS* **1972**, *94*, 3653.
35. Koch-Pomeranz, U.; Hansen, H. J.; Schmid, H. *HCA* **1973**, *56*, 2981.
36. Shigemans, Y.; Yasui, M.; Ohrai, S.; Sasaki, M.; Sashiwa, H.; Saimoto, H. *JOC* **1991**, *56*, 910.
37. Dikshit, D. K.; Singh, S.; Panday, S. K. *JCR(S)* **1991**, 298.
38. Ruy, I.; Ando, M.; Ogawa, A.; Murai, S.; Sonoda, N. *JACS* **1983**, *105*, 7192.
39. Gerlach, H.; Thalmann, A. *HCA* **1974**, *57*, 2661.
40. Kawanami, Y.; Katsuki, T.; Yamaguchi, M. *TL* **1983**, *24*, 5131.
41. Ishibashi, H.; Tabata, T.; Kobayashi, T.; Takamuro, I.; Ikeda, M. *CPB* **1991**, *39*, 2878.
42. Beerli, R.; Borschberg, H. J. *HCA* **1991**, *74*, 110.
43. Olah, G. A.; Pavláth, A.; Kuhn, S. *CI(L)* **1957**, *50*; Kuhn, S. J.; Olah, G. A. *JACS* **1961**, *83*, 4564.
44. Furukawa, N.; Hoshiai, H.; Shibutani, T.; Higaki, M.; Iwasaki, F.; Fujihara, H. *H* **1992**, *34*, 1085.
45. Crews, P.; Beard, J. *JOC* **1973**, *38*, 529.

Lars-G. Wistrand
Nycomed Innovation, Malmö, Sweden

Sodium Amide

$NaNH_2$

[7782-92-5] H_2NNa (MW 39.02)

(strong base;[3] strong nucleophile[38])

Alternate Name: sodamide.
Physical Data: mp 210 °C; bp 400 °C/760 mmHg.
Solubility: sol liq ammonia (~1 mol L^{-1} at −33 °C).[1]
Form Supplied in: commercially available as a powder; easily prepared in the laboratory.
Preparative Methods: combination of ***Ammonia***, small quantities of an iron(III) salt, and ***Sodium*** leads to formation of a black catalyst, whereupon the remainder of the sodium is added. Published procedures differ in details.[2]
Handling, Storage, and Precautions: flammable; corrosive; when opened to air, decomposes and forms a potentially explosive yellow byproduct.[1]

Reaction as a Base. Sodamide often serves as a base to generate reactive anions.[3] In DMSO in the presence of various bases, including sodamide, carbohydrates are benzylated in good yield with ***Benzyl Chloride***.[3a] Reaction of (**1**) and (**2**) in the presence[3b] of sodamide gives (**3**) (eq 1). Sodamide is effective in generating the acetonitrile anion for reaction with sulfines.[4] Deprotonation of phenylacetic esters in the presence of sodamide allows aldol reaction with benzaldehyde derivatives to afford 2,3-diaryl-3-hydroxypropionic acids.[5] Similarly, reaction of acetophenone and ethyl chloroacetate (eq 2) gives the Darzens' product (**4**).[6] Treatment of primary anilines and cyanopyridines with sodamide leads to good yields of carboxamidines.[7] Oxygenation of hindered 4-alkylphenols in the presence of sodamide provides a convenient source of quinols.[8]

(**1**) + (**2**) —$NaNH_2$, 69%→ (**3**) (1)

Ph(C=O)Me —$ClCH_2CO_2Et$, $NaNH_2$, 62–64%→ (**4**) (2)

Sodamide in THF with boric acid neutralization has proven effective for the deconjugation of conjugated unsaturated steroids.[9] The presence of sodamide in liquid ammonia at low temperature facilitates interconversion of 1,4- and 1,3-cyclohexadienes.[10] Deprotonation of 2-bromothiophenes and 2-halothianaphthalenes affords the 3-halo isomers via a series of complex equilibria.[11] Cyclopropenes, which possess an acidity comparable to alkynes, are rapidly metalated by sodamide (and other alkali amides) to produce reactive intermediates for alkylation.[12] Selective deprotonation occurs with a wide variety of acidic methyl, methylene, and methine hydrogens adjacent to carbonyls or attached to heterocycles. For example, 2,4-lutidine (**5**) undergoes deprotonation (eq 3) to (**6**) followed by reaction with ethyl benzoate to yield (**7**).[13a] Deprotonation followed by reaction with electrophiles is a powerful method for generating complex carbon skeletons.[13] Examination of the role of bases, including sodamide, on the stereochemistry (including isomerization) of products formed in the Michael reaction has been reported.[14] In the racemization of the single stereogenic center in nicotine, sodamide was inferior to ***Potassium t-Butoxide***.[15]

(**5**) —$NaNH_2$→ [(**6**)] —$PhCO_2Et$, 85%→ (**7**) (3)

Dianion Generation. Numerous early investigations into dianion chemistry[16] employed sodamide as the base. Conversion of the simple heterocycle (**8**) into the corresponding dianion with sodamide in liquid ammonia followed by reaction with benzonitrile (eq 4) led to an interesting rearrangement product (**9**).[17] β-Dicarbonyl dianions are routinely prepared by reaction with sodamide. These strongly nucleophilic species undergo regioselective alkylation (eq 5) by reaction of disodioacetylacetone (**10**) (much more soluble in liquid ammonia than its dipotassium counterpart[16a]) with 11-bromoundecanoic acid to give (**11**)[18] and reaction of (**10**) (eq 6) with diphenyliodonium chloride to yield (**12**).[19]

(**8**) —PhCN, $NaNH_2$, 58%→ (**9**) (4)

(**10**) + $Br(CH_2)_{10}CO_2Li$ —H^+, 82%→ (**11**) (5)

(**10**) + Ph_2ICl —H^+, 60–64%→ (**12**) (6)

Elimination Reactions. Sodamide's utility as a reagent for elimination reactions is illustrated by the following selected examples. Methiodide (**13**) undergoes facile loss of HI and diethylmethylamine to generate methyl vinyl ketone.[20] Five isomeric alkenes and a cyclopropane result from treatment of 2-benzyl-3-phenylpropyltrimethylammonium iodide with sodamide.[21] Upon reaction with sodamide, various thioamides eliminate hydrogen sulfide to form ynamines in fair yield.[22] In the presence of sodamide, *cis*-1,4-dichloro-2-butene (**14**) yields mainly *trans*-1-chloro-1,3-butadiene (**15**) (eq 7) while *trans*-1,4-dichloro-2-butene gives a preponderence of *cis*-1-chloro-1,3-butadiene (**16**) (eq 8).[23] Upon warming a mixture of methallyl chloride (**17**) and sodamide (eq 9), there is formed methylenecyclopropane (**18**) and 1-methylcyclopropene (**19**).[24] Sodamide, ***Sodium Hydride***, and ***Sodium Methoxide*** all have utility in the Bamford–Stevens reaction for the conversion of tosylhydrazones into alkenes.[25]

(**13**)

(**14**) —$NaNH_2$, 52%→ (**15**) (7)

—$NaNH_2$, 72%→ (**16**) (8)

(**17**) —$NaNH_2$, 72%→ (**18**) + (**19**) (9)

Preparation of Alkynes. Sodamide-mediated elimination of one or two moles of HX from a suitable substrate is a classical method for the synthesis of alkynes. For example, β-bromostyrene with sodamide in liquid ammonia provides an excellent source of phenylacetylene.[26] Cyclohexylpropyne (**21**) can be generated by reaction (eq 10) of vinyl bromide (**20**) with 3 equiv (excess) of sodamide.[27] Oleic acid (**22**) can be transformed into stearolic acid (**23**) by a straightforward sequence (eq 11) involving bromination followed by reaction with excess sodamide.[28] Similar methodology has been employed to synthesize many other alkynes.[29] Dehydrohalogenation with concomitant ether cleavage provides an efficient route to complex alkynes. For example, reaction of (**24**) with sodamide (eq 12) provides the hydroxylic terminal pentyne (**25**).[29j] Alkyne–allene isomerization has been accomplished with sodamide.[30]

(**20**) —3 equiv $NaNH_2$, 66%→ (**21**) (10)

(**22**) —1. Br_2; 2. 3 equiv $NaNH_2$, H^+, 42–52%→ (**23**) (11)

(**24**) —3.5 equiv $NaNH_2$; NH_3, NH_4Cl, 75–85%→ HO (**25**) (12)

Aryne Chemistry. Among the many existing methods for the generation of arynes,[31] reaction of a halobenzene derivative with sodamide (as in the example (eq 13) of (**26**) going to (**27**)[32a]) is a commonly employed procedure.[32] The highly reactive intermediate arynes can be made to undergo reaction with nucleophiles other than amide anion. Thus bromobenzene (**28**) is converted (eq 14) into aryl sulfide (**29**).[33a] Sodamide-generated arynes have also been reacted with more complex species,[34] as illustrated by the transformation (eq 15) of (**30**) into (**31**) followed by cyclization to (**32**).[34a] Intramolecular benzyne reactions involving sodamide have been used successfully in the synthesis of aporphine alkaloids.[35]

(**26**) —$NaNH_2$→ [aryne] —RNH_2, 68–85%→ (**27**) (13)

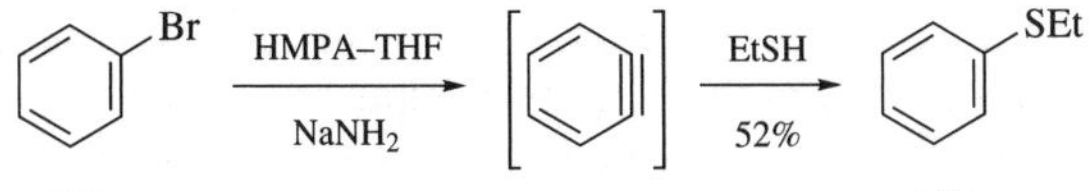

(14)

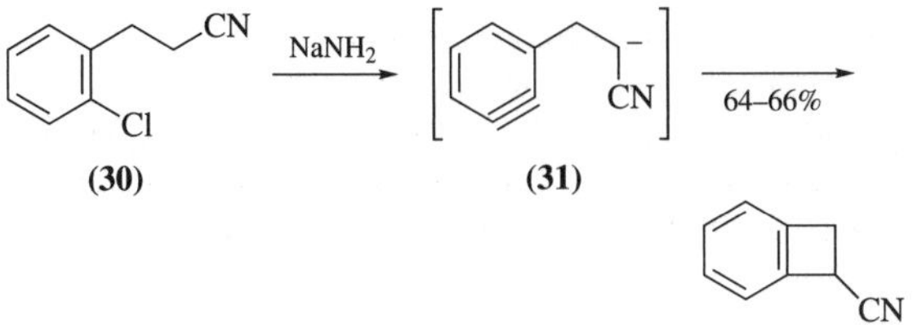

(15)

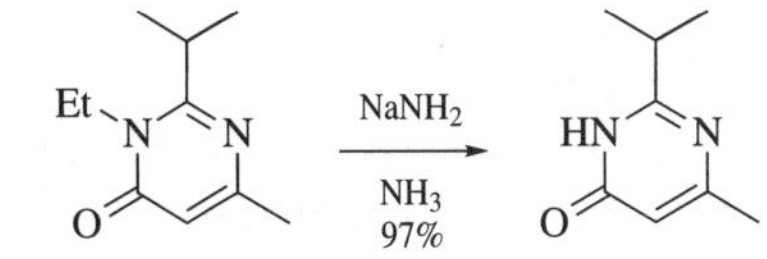

(19)

Cleavage and Rearrangement. Sodamide is involved in many cleavage and rearrangement reactions. Cleavage reactions,[46] with specific reference to the Haller–Bauer reaction,[47] exemplified by (**35**) going to (**36**) (eq 20),[47d] are a convenient synthetic transform. It is significant that the addition of ***1,4-Diazabicyclo[2.2.2]octane*** (DABCO) permits the Haller–Bauer reaction to be performed with commercial sodamide.[47d] Rearrangement reactions involving sodamide are well-known,[48] with several being common name reactions such as the Truce–Stiles,[49] the Sommelet–Hauser,[50a] and the Stevens[50a] reactions. A typical Sommelet–Hauser rearrangement is illustrated by (**37**) going to (**38**) (eq 21).[50b] Vinylpyridines undergo polymerization in sodamide/liquid ammonia.[51]

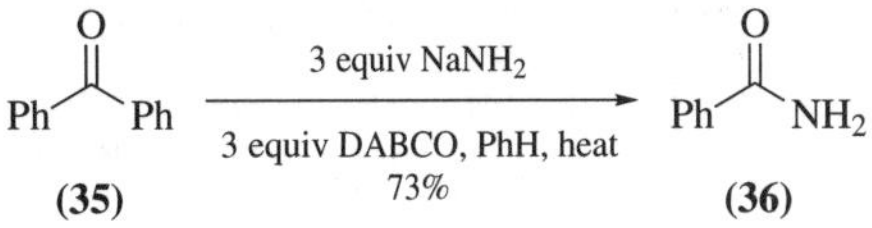

(20)

(21)

Generation of Ylides. Sodamide is a common base for the generation of ylides in the Wittig reaction.[36] The commercially available instant ylide consists[37a] of a 1:1 stoichiometric mixture of ***Methyltriphenylphosphonium Bromide*** and sodium amide (eq 16).[37b]

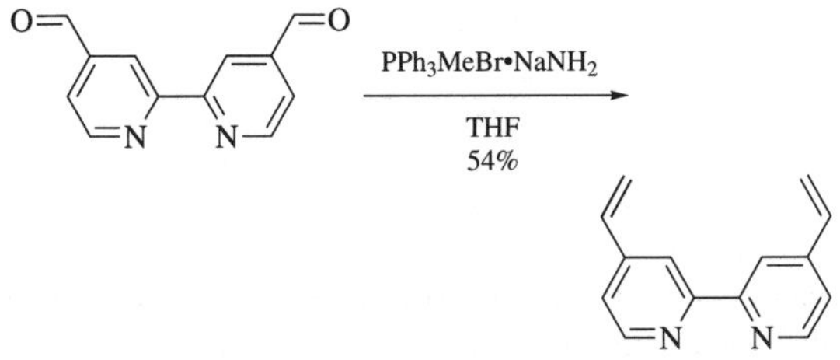

(16)

Reaction as a Nucleophile. Nucleophilic addition reactions are a major feature of sodamide chemistry. Addition followed by intramolecular attack provides a convenient methodology for the construction of unusual adducts.[38] Sodamide, sodamide/potassamide mixtures, and other alkali metal amides have been found to catalyze the amination of alkenes.[39] The Chichibabin reaction and its variants[40] provide a useful route to numerous substituted heterocycles. The addition–elimination reaction of sodamide on a heterocyclic substrate is nicely illustrated by the transformation (eq 17) of (**33**) into 6-methylisocytosine (**34**).[41] Nucleophilic addition reactions to nitro-substituted aromatic substrates have been observed.[42] Also intriguing are the various reaction pathways observed for heterocycles containing an appended trifluoromethyl group.[43] Photochemically assisted additions of sodamide have been reported (eq 18).[44] Sodamide is also an effective reagent for accomplishing *N*-dealkylations (eq 19)[45a] and *N*-deacylations.[45b]

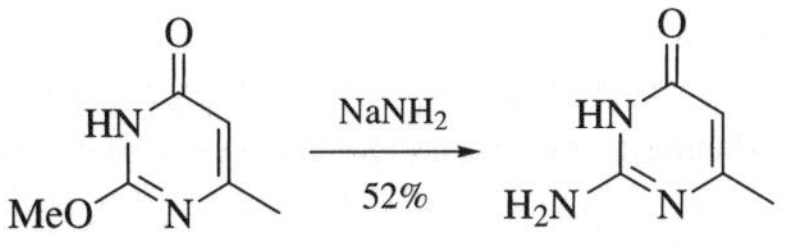

(17)

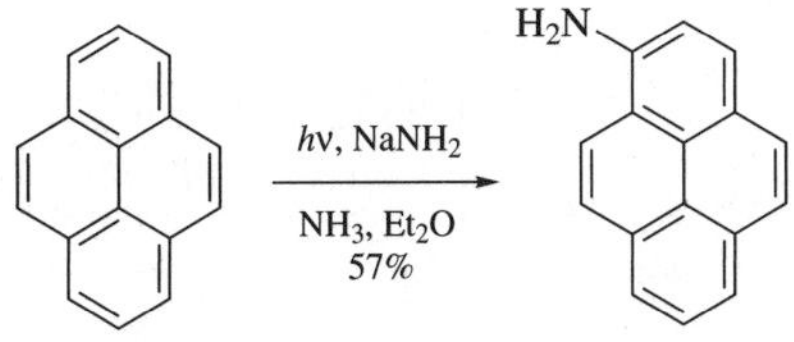

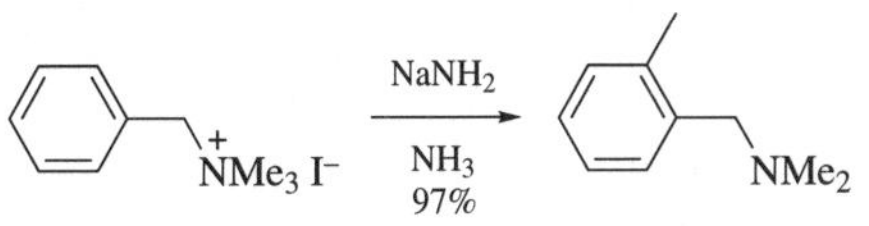

(18)

In recent years, sodamide has been combined with other bases (especially with alkali metal *t*-butoxides) to create a whole family of so-called complex bases with exceptional properties (see ***Sodium Amide–Sodium t-Butoxide***).[52] Typical applications of these bases are in the *syn* elimination depicted[52d] by (**39**) going to (**40**) and (**41**) (eq 22) and the carbanion alkylation involving the conversion of (**42**) to (**43**) (eq 23).[52f]

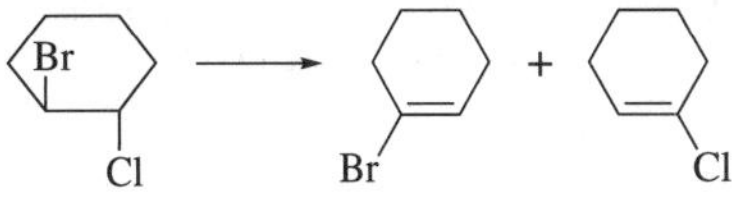

(22)

	(40):(41)
$NaNH_2$, *t*-BuONa, 87%	65:35
$NaNH_2$, *t*-BuONa, 15-crown-5, 76%	3:97

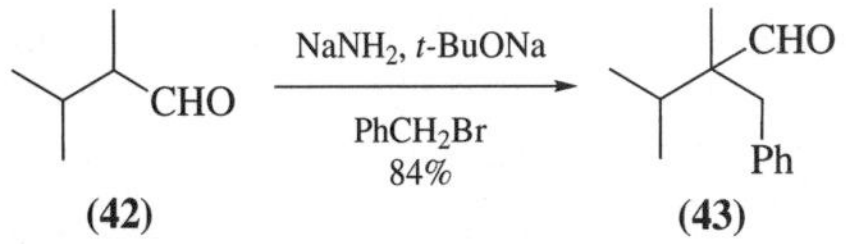

(23)

Related Reagents. Lithium Amide; Potassium Amide; Potassium *t*-Butoxide; Sodium Amide–Sodium *t*-Butoxide; Sodium–Ammonia; Sodium Hydride.

1. *FF* **1967**, *1*, 1034.
2. (a) Vaughn, T. H.; Vogt, R. R.; Nieuwland, J. A. *JACS* **1934**, *56*, 2120. (b) Hauser, C. R.; Adams, J. T.; Levine, R. *OSC* **1955**, *3*, 291. (c) Hauser, C. R.; Dunnavant, W. R. *OS* **1960**, *40*, 38. (d) Jones, E. R. H.; Eglinton, G.; Whiting, M. C.; Shaw, B. L *OSC* **1963**, *4*, 404. (e) Khan, N. A.; Deatherage, F. E.; Brown, J. B. *OSC* **1963**, *4*, 851. (f) Greenlee, K. W.; Henne, A. L. *Inorg. Synth.* **1946**, *2*, 128.

3. (a) Iwashige, T.; Saeki, H. *CPB* **1967**, *15*, 1803. (b) Ireland, R. E.; Kierstead, R. C. *JOC* **1966**, *31*, 2543.
4. Loontjes, J. A.; van der Leij, M.; Zwanenberg, B. *RTC* **1980**, *99*, 39.
5. Kratchanov, C. G.; Kirtchev, N. A. *S* **1971**, 317.
6. Allen, C. F. H.; VanAllan, J. *OSC* **1955**, *3*, 727.
7. Hisano, T.; Tasaki, M.; Tsumoto, K.; Matsuoka, T.; Ichikawa, M. *CPB* **1983**, *31*, 2484.
8. Nishinaga, A.; Itahara, T.; Matsuura, T. *BCJ* **1975**, *48*, 1683.
9. Shapiro, E. L.; Leggatt, T.; Weber, L.; Olivetto, E. P.; Tanabe, M.; Crowe, D. F. *Steroids* **1964**, *3*, 183.
10. Rabideau, P. W.; Huser, D. L. *JOC* **1983**, *48*, 4266.
11. (a) Reinecke, M. G.; Hollingworth, T. A. *JOC* **1972**, *37*, 4257. (b) Brandsma, L.; de Jong, R. L. P. *SC* **1990**, *20*, 1697.
12. (a) Schipperijn, A. J.; Smael, P. *RTC* **1973**, *92*, 1121. (b) Schipperijn, A. J.; Smael, P. *RTC* **1973**, *92*, 1159.
13. (a) Levine, R.; Dimmig, D. A.; Kadunce, W. M. *JOC* **1974**, *39*, 3834. (b) Yamamoto, M.; Sugiyama, N. *BCJ* **1975**, *48*, 508. (c) Kaiser, E. M.; Bartling, G. J.; Thomas, W. R.; Nichols, S. B.; Nash, D. R. *JOC* **1973**, *38*, 71. (d) Harris, T. M.; Harris, C. M.; Wachter, M. P. *T* **1968**, *24*, 6897. (e) Vanderwerf, C. A.; Lemmermann, L. V. *OSC* **1955**, *3*, 44. (f) Coffman, D. D. *OSC* **1955**, *3*, 320. (g) Hauser, C. R.; Adams, J. T.; Levine, R. *OSC* **1955**, *3*, 291. (h) Potts, K. T.; Saxton, J. E. *OS* **1960**, *40*, 68. (i) Kaiser, E. M.; Bartling, G. J. *JOC* **1972**, *37*, 490. (j) Rash, F. H.; Boatman, S.; Hauser, C. R. *JOC* **1967**, *32*, 372.
14. (a) Gospodova, T. S.; Stefanovsky, Y. N. *M* **1990**, *121*, 275. (b) Viteva, L. Z.; Stefanovsky, Y. N. *M* **1982**, *113*, 181.
15. Tsujino, Y.; Shibata, S.; Katsuyama, A.; Kisaki, T.; Kaneko, H. *H* **1982**, *19*, 2151.
16. (a) Harris, T. M.; Harris, C. M. *OR* **1969**, *17*, 155. (b) Harris, T. M.; Harris, C. M. *JOC* **1966**, *31*, 1032.
17. Kashima, C.; Yammamoto, M.; Kobayashi, S.; Sugiyama, N. *BCJ* **1974**, *47*, 1805.
18. Pendarvis, R. O.; Hampton, K. G. *JOC* **1974**, *39*, 2289.
19. Hampton, K. G.; Harris, T. M.; Hauser, C. R. *OS* **1971**, *51*, 128.
20. (a) duFeu, E. C.; McQuillin, F. J.; Robinson, R. *JCS* **1937**, 53. (b) Cornforth, J. W.; Robinson, R. *JCS* **1949**, 1855.
21. Bumgardner, C. L.; Iwerks, H. *JACS* **1966**, *88*, 5518.
22. Halleux, A.; Reimlinger, H.; Viehe, H. G. *TL* **1970**, 3141.
23. Heasley, V. L.; Lais, B. R. *JOC* **1968**, *33*, 2571.
24. (a) Fisher, F.; Applequist, D. E. *JOC* **1965**, *30*, 2089. (b) Salaun, J. R.; Conia, J. M. *CC* **1971**, 1579. (c) Koster, R.; Arora, S.; Binger, P. *S* **1971**, 322. (d) Arora, S.; Binger, P.; Koster, R. *S* **1973**, 146. (e) Fitjer, L.; Conia, J.-M. *AG(E)* **1973**, *12*, 332.
25. Kirmse, W.; von Bullow, B.-G.; Schepp, H. *LA* **1966**, *691*, 41.
26. Vaughan, T. H.; Vogt, R. R.; Nieuwland, J. A. *JACS* **1934**, *56*, 2120.
27. Lespieau, R.; Bourguel, M. *OSC* **1941**, *1*, 191.
28. Khan, N. A.; Deatherage, F. E.; Brown, J. E. *OSC* **1963**, *4*, 851.
29. (a) Khan, N. A. *OSC* **1963**, *4*, 969. (b) Ashworth, P. J.; Mansfield, G. H.; Whiting, M. C. *OSC* **1963**, *4*, 128. (c) Messeguer, A.; Serratosa, F.; Rivera, J. *TL*. **1973**, 2895. (d) Armitage, J. B.; Jones, E. R. H.; Whiting, M. C. *JCS* **1953**, 3317. (e) Bohlmann, F. *CB* **1951**, 84, 545. (f) Jones, E. R. H.; Eglinton, G.; Whiting, M. C. Shaw, B. L. *OSC* **1963**, *4*, 404. (g) Wasserman, H. H.; Wharton, P. S. *JACS* **1960**, *82*, 661. (h) Newman, M. S.; Geib, J. R.; Stalick, W. M. *OPP* **1972**, *4*, 89. (i) Brandsma, L.; Harryvan, E.; Arens, J. F. *RTC* **1968**, *87*, 1238. (j) Jones, E. R. H.; Eglinton, G.; Whiting, M. C. *OSC* **1963**, *4*, 755.
30. (a) Carr, M. D.; Gan, L. H.; Reid, I. *JCS(P2)* **1973**, 672. (b) Montijn, P. P.; Kupecz, A.; Brandsma, L.; Arens, J. F. *RTC* **1969**, 88, 958.
31. Hoffmann, R. W., *Dehydrobenzene and Cycloalkynes*; Academic: New York, 1967.
32. (a) Biehl, E. R.; Patrizi, R.; Reeves, P. C. *JOC* **1971**, *36*, 3252. (b) Biehl, E. R.; Stewart, W.; Marks, A.; Reeves, P. C. *JOC* **1979**, *44*, 3674. (c) Levine, R.; Biehl, E. R. *JOC* **1975**, *40*, 1835. (d) Biehl, E. R.; Smith, S. M.; Reeves, P. C. *JOC* **1971**, *36*, 1841. (e) Biehl, E. R.; Nieh, E.; Hsu, K. C. *JOC* **1969**, *34*, 3595. (f) Biehl, E. R.; Hsu, K. C.; Nieh, E. *JOC* **1970**, *35*, 2454. (g) Kraakman, P. A.; Valk, J.-M.; Niederländer, H. A. G.; Brower, D. B. E.; Bickelhaupt, F. M.; de Wolf, W. H.; Bickelhaupt, F.; Stam, C. H. *JACS* **1990**, *112*, 6638. (h) Apeloig, Y.; Arad, D.; Halton, B.; Randall, C. J. *JACS* **1986**, *108*, 4932.
33. (a) Caubere, P. *BSF* **1967**, 3446, 3451. (b) Carre, M. C.; Ezzinadi, A. S.; Zouaoui, M. A.; Geoffroy, P.; Caubere, P. *SC* **1989**, *19*, 3323.
34. (a) Skorcz, J. A.; Kaminski, F. E. *OS* **1968**, *48*, 53. (b) Carre, M.-C.; Gregoire, B.; Caubere, P. *JOC* **1984**, *49*, 2050. (c) Loubinoux, B.; Caubere, P. *S* **1974**, 201. (d) Buske, G. R.; Ford, W. T. *JOC* **1976**, *41*, 1995.
35. Kametani, T.; Fukumoto, K.; Nakano, T. *JHC* **1972**, *9*, 1363.
36. (a) Moiseenkov, A. M.; Schaub, B.; Margot, C.; Schlosser, M. *TL* **1985**, *26*, 305. (b) Schaub, B.; Blaser, G.; Schlosser, M. *TL* **1985**, *26*, 307. (c) Schlosser, M.; Schaub, B.; de Oliveira-Neto, J.; Jeganathan, S. *C* **1986**, *40*, 244. (d) Schaub, B.; Jeganathan, S.; Schlosser, M. *C* **1986**, *40*, 246. (e) Dauphin, G.; David, L.; Duprat, P.; Kergomard, A.; Veshambre, H. *S* **1973**, 149. (f) Takahashi, H.; Fujiwara, K.; Ohta, M. *BCJ* **1962**, *35*, 1498. (g) Yamamoto, Y.; Schimidbaur, H. *CC* **1975**, 668. (h) Quast, H.; Jakobi, H. *CB* **1991**, *124*, 1619.
37. (a) Schlosser, M.; Schaub, B. *C* **1982**, *36*, 396. (b) Ciana, L. D.; Dressick, W. J.; von Zelewsky, A. *JHC* **1990**, *27*, 163.
38. (a) Barnard, I. F.; Elvidge, J. A. *JCS(P1)* **1983**, 1813. (b) Yamagouchi, K. *BCJ* **1976**, *49*, 1366.
39. Pez, G. P.; Galle, J. E. *PAC* **1985**, *57*, 1917.
40. Vorbruggen, H. *Adv. Heterocycl. Chem.* **1990**, *49*, 117.
41. Botta, M.; De Angelis, F.; Finizia, G.; Gambacorta, A.; Nicoletti, R. *SC* **1985**, *15*, 27.
42. Gandhi, S. S.; Gibson, M. S.; Kaldas, M. L.; Vines, S. M. *JOC* **1979**, *44*, 4705.
43. (a) Kobayashi, Y.; Kumadaki, I.; Taguchi, S.; Hanzawa, Y. *TL* **1970**, 3901. (b) Kobayashi, Y.; Kumadaki, I.; Hanzawa, Y.; Minura, M. *CPB* **1975**, *23*, 2044. (c) Kobayashi, Y.; Kumadaki, I.; Hanzawa, Y.; Mimura, M. *CPB* **1975**, *23*, 636. (d) Kobayashi, Y.; Kumudaki, I.; Taguchi, S.; Hanzawa, Y. *CPB* **1972**, *20*, 1047.
44. Tintel, C.; Rietmeyer, F. J.; Cornelisse, J. *RTC* **1983**, *102*, 224.
45. (a) Hirai, Y.; Egawa, H.; Yamada, S.; Yamazaki, T. *H* **1983**, *20*, 1243. (b) Fraenkel, G.; Cooper, J. W. *JACS* **1971**, *93*, 7228.
46. (a) Furukawa, N.; Tanaka, H.; Oae, S. *BCJ* **1968**, *41*, 1463. (b) Shiotani, S.; Kometani, T. *CPB* **1973**, *21*, 1160.
47. (a) Hamlin, K. E.; Weston, A. W. *OR* **1957**, *9*, 1. (b) Alexander, E. C.; Tom, T. *TL* **1978**, 1741. (c) Paquette, L. A.; Maynard, G. D. *JOC* **1989**, *54*, 5054. (d) Kaiser, E. M.; Warner, C. D. *S* **1975**, 395.
48. (a) Mason, J. G.; Youssef, A. K.; Ogliaruso, M. A. *JOC* **1975**, *40*, 3015. (b) Youssef, A. K.; Ogliaruso, M. A. *JOC* **1973**, *38*, 3998. (c) Sarel, S.; Klug, J. T.; Taube, A. *JOC* **1970**, *35*, 1850. (d) Klein, K. P.; Hauser, C. R. *JOC* **1966**, *31*, 4275.
49. Crowther, G. P.; Hauser, C. R. *JOC* **1968**, *33*, 2228.
50. (a) Pine, S. H. *OR* **1970**, *18*, 403. (b) Kantor, S. W.; Hauser, C. R. *JACS* **1951**, *73*, 4122. (c) Giumanini, A. G.; Trombini, C.; Lercker, G.; Lepley, A. R. *JOC* **1976**, *41*, 2187.
51. Laurin, D.; Parravano, G. *J. Polym. Sci. Part A-1, Polym. Chem. Ed.* **1968**, *6*, 1047.
52. (a) Caubere, P. *ACR* **1974**, *7*, 301. (b) Caubere, P. *Top. Curr. Chem.* **1978**, *73*, 49. (c) Ndebeka, G.; Raynal, S.; Caubere, P. *JOC* **1980**, *45*, 5394. (d) Croft, A. P.; Bartsch, R. A. *JOC* **1983**, *48*, 876. (e) Croft, A. P.; Bartsch, R. A. *TL* **1983**, *24*, 2737. (f) Carre, M. C.; Ndebeka, G.; Riondel, A.; Bourgasser, P.; Caubere, P. *TL* **1984**, *25*, 1551. (g) Raynal, S. *Eur. Polym. J.* **1986**, *22*, 559.

John L. Belletire & R. Jeffery Rauh
The University of Cincinnati, OH, USA

Sodium Carbonate

Na_2CO_3

[497-19-8] CNa_2O_3 (MW 105.99)

(widely used base)

Physical Data: mp 851 °C; *d* 2.532 g cm^{-3}.
Solubility: sol glycerol, water.
Form Supplied in: white powder.
Handling, Storage, and Precautions: moderately toxic; severe irritant; noncombustible; stable; decomposition occurs slowly above 400 °C to give carbon dioxide and sodium oxide; incompatible with strong acids and aluminum; hygroscopic.

Selective Hydrolysis. Cyanoethylated β-keto esters can be hydrolyzed with aqueous sodium carbonate to the corresponding δ-keto nitrile (eq 1).[1] The use of sodium hydroxide in this reaction leads to hydrolysis of the nitrile, affording the keto acid.

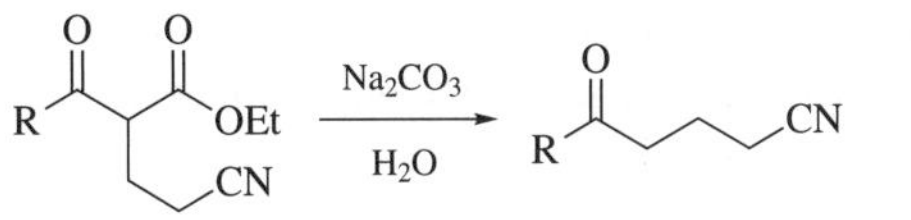

(1)

The selective hydrolysis of esters in the presence of acetyl-protected amines can be achieved with aqueous sodium carbonate (eq 2).[2]

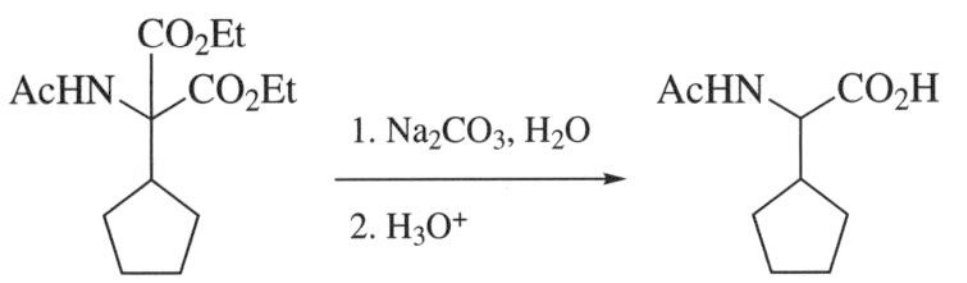

(2)

McFadyen–Stevens Reduction. The decomposition of sulfonylhydrazines to aldehydes was achieved using sodium carbonate as base. The sulfonylhydrazine was heated in ethylene glycol to 150 °C, and 4–6 equiv of sodium carbonate added. The reaction is complete in approximately 30 s. This reaction has been applied to the formation of aromatic,[3] heterocyclic,[3] and aliphatic aldehydes (eq 3).[4]

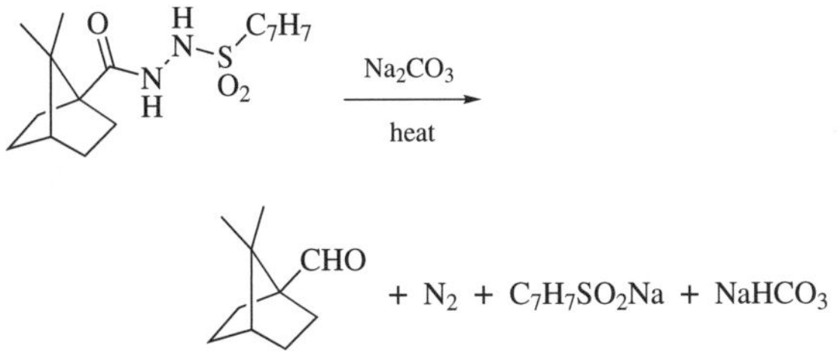

$$+ N_2 + C_7H_7SO_2Na + NaHCO_3 \quad (3)$$

Carbon–Carbon Bond Forming Reactions. The reaction of β-keto esters with α,β-unsaturated aldehydes in the presence of sodium carbonate and a phase-transfer catalyst afforded the Michael addition product (eq 4).[5] The use of sodium hydroxide in this reaction led to substantial decomposition.

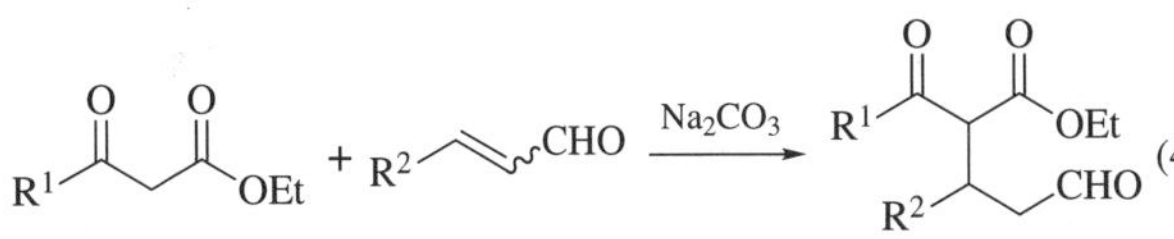

(4)

This methodology was used to prepare 3-oxo-5-phenyl-1-cyclohexene from ethyl acetoacetate and cinnamic acid (eq 5).[5] Under the same conditions, β-cyano esters and α,β-unsaturated aldehydes undergo Knoevenagel condensation (eq 6).[5]

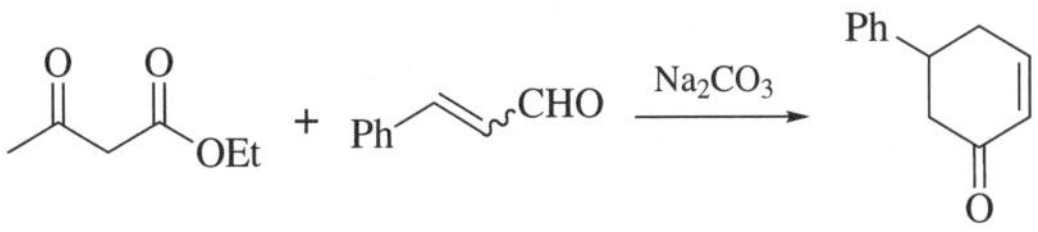

(5)

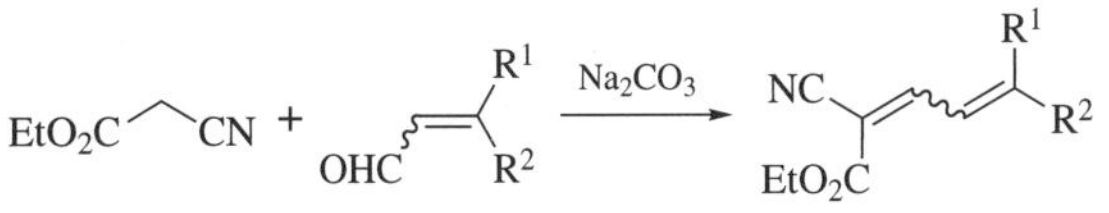

(6)

Hydrogenation. Sodium carbonate is an effective acid scavenger for hydrogenations. The hydrogenation of *m*-nitro-*O*-benzoylmandelate with ***Raney Nickel*** and sodium carbonate in ethanol at 180 °C afforded the ethyl ester of *m*-aminophenylacetic acid (eq 7).[6] Other bases were found to be less effective in removing the benzoic acid, resulting in slow and incomplete reactions. Sodium carbonate is also used in the sodium hydrophosphite hydrogenations of many functional groups.[7]

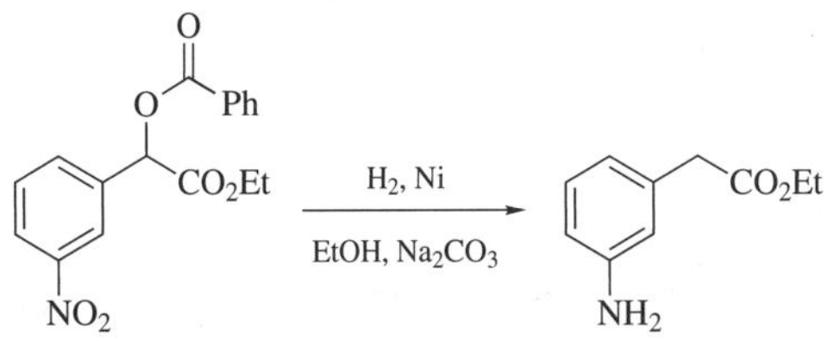

(7)

Synthesis of Phenyliodonium β-Diketonates. β-Diketones react with ***(Diacetoxyiodo)benzene*** in the presence of sodium carbonate to afford phenyliodonium β-diketonates (eq 8).[8] Ozonolysis of the phenyliodonium β-diketonates provides an efficient synthesis of triketones.

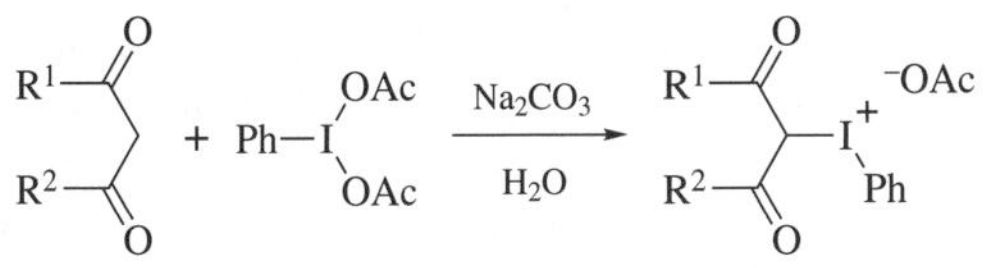

(8)

Other Applications. Sodium carbonate is used in the alkylation of amines,[9] debromination,[10] hydrolysis,[11] imidazoline synthesis,[12] as a buffer,[13] and a neutralization agent.[14]

Related Reagents. Barium Hydroxide; Cadmium Carbonate; Calcium Carbonate; Cesium Carbonate; Lithium Carbonate; Potassium Carbonate; Potassium Hydroxide; Sodium Hydroxide.

---

1. Levine, R.; Yoho, C. *JACS* **1952**, *74*, 5597.
2. Hill, J. T.; Dunn, F. W. *JOC* **1965**, *30*, 1321.
3. Mosettig, E. *OR* **1954**, *8*, 232.
4. Sprecher, M.; Feldkimel, M.; Wilchek, M. *JOC* **1961**, *26*, 3664.

5. Kryshtal, G. V.; Kulganek, V. V.; Kucherov, V. F.; Yanovskaya, L. A. *S* **1979**, 107.
6. Cronyn, M. W. *JOC* **1949**, *14*, 1013.
7. For examples, see: Boyer, S. K.; Bach, J.; McKenna, J.; Jagdmann, E. *JOC* **1985**, *50*, 3408.
8. Lick, C.; Schank, K. *S* **1983**, 392.
9. Hu, M. W.; Singh, P.; Ullman, E. F. *JOC* **1980**, *45*, 1711.
10. Wolinsky, J.; Erickson, K. L. *JOC* **1965**, *30*, 2208.
11. Brown, J.; Brown, R. *CJC* **1955**, *33*, 1819.
12. Jung, S.-H.; Kohn, H. *JACS* **1985**, *107*, 2931.
13. Emmons, W. D.; Pagano, A. S. *JACS* **1955**, *77*, 89.
14. For examples, see: Izumi, Y.; Okada, H.; Matsuda, I. *CL* **1983**, *97*; Obrecht, R.; Hermann, R.; Ugi, I. *S* **1985**, 400.

Roger Harrington
Giba-Geigy, Summit, NJ, USA

Sodium Ethoxide[1]

NaOEt

[141-52-6] C_2H_5NaO (MW 68.06)

(used as a base for the α-deprotonation of carbonyl-containing compounds for subsequent intermolecular[2] or intramolecular[1f] condensations, displacements,[3] or skeletal rearrangements,[1d,4] arylacetonitriles,[5] nitro-containing aliphatic compounds,[6] sulfonium salts,[3j,7] for the dehydration of carbinolamines[8] and for dehydrohalogenation,[1b,9] for the *N*-deprotonation of amides,[10a–d] tosylamines,[10e] amine hydrochlorides[1c] or toluenesulfonates,[10f] cyanamide,[10g] and the *S*-deprotonations of sulfides,[1c,11] often followed by cyclizations;[1c,10b,10g,11] can be used as a nucleophile in *ipso* substitution reactions of vinyl sulfides,[12a] aromatic halides,[12b–i] sometimes catalyzed by copper[12c] or palladium,[12h,i] aryl sulfones,[12j] and aromatic nitro compounds,[12k] in the Williamson ether synthesis,[13a–c] in displacements of halo,[13d–h] nitro,[13e] thiooxy,[13g,h] and phenoxy[13g] groups from dichloromethane and chloroform analogs, in a novel transesterification–conjugate addition protocol of acrylic esters,[14] with α-nitro epoxides to form α-ethoxy ketones,[15] in the nucleophilic attack on nitriles[1e,16] and polyhaloalkenes,[17] in reaction with chlorodiphenylphosphine to form Arbuzov precursors,[18] and with Grignard reagents to form organomagnesium ethoxides[19])

Physical Data: mp >300 °C.
Solubility: sol ethanol, diethyl ether.
Form Supplied in: white or yellowish powder or as 21 wt % solution in ethanol; widely available.
Preparative Methods: it is often necessary to prepare sodium ethoxide immediately prior to its use. Preparation of a 6–10% solution in ethanol: to commercial absolute ethanol is added the required amount of ***Sodium*** in lumps or slices with or without stirring and with or without a nitrogen atmosphere and the solution is cooled or heated as required until the metal is dissolved. The evolved hydrogen should be vented into a hood. Addition of ethanol to sodium is also reported. Alcohol-free reagent can also be prepared.[1a]
Handling, Storage, and Precautions: is a hygroscopic, flammable, corrosive, and toxic solid which will decompose upon exposure to air. The solid or solution reagent should be tightly sealed under an inert atmosphere in a dark bottle.

General Discussion. Kinetic studies of the proton transfer reaction between bis(4-nitrophenyl)methane and alkoxide base systems reveal that sodium ethoxide is a kinetically faster ($k_H = 2.16$ M^{-1} s^{-1}) base than either sodium isopropoxide ($k_H = 0.280$ M^{-1} s^{-1}) or sodium *t*-butoxide ($k_H = 1.05$ M^{-1} s^{-1}) (eq 1).[20a] Reaction between alkali alkoxides and *p*-nitrophenyl methanesulfonate have revealed the following order of reactivity: LiOEt< NaOEt < CsOEt ≈ KOEt ≈ KOEt + 18-crown-6 < KOEt + [2.2.2]cryptand.[20b]

$$Ar_2CH_2 + RONa \rightleftharpoons Ar_2CHNa + ROH \quad (1)$$

Condensations of esters with the enolate of acetone can be effected using sodium ethoxide as a base to yield products such as acetylacetone[2a] and the precursor to chelidonic acid depicted in eq 2.[2b] The corresponding enolates of methyl aryl ketones can similarly be utilized.[2c,1f]

NaOEt, EtOH; H_2O, HCl (2)
76–79%

Condensation of the sodium ethoxide-generated enolate of cyclohexanone with ethyl formate yields 2-hydroxymethylenecyclohexanone which, upon treatment with hydrazine monohydrate, forms indazole in excellent yield (eq 3).[2d]

+ HCO_2Et; NaOEt, EtOH, 0 °C to rt; $N_2H_4 \cdot H_2O$ (3)
95–98%

Sodium ethoxide is often the base of choice in the α-acylation[2e–i] and alkylation[3h] of esters. Interesting intramolecular cyclizations sometimes follow in good yield (eq 4).

Alkylation of malonate esters through the intermediacy of enolates generated using sodium ethoxide is quite common.[3] Subsequent decarboxylation to the carboxylic acid can then be induced, as depicted in eq 5.[3d]

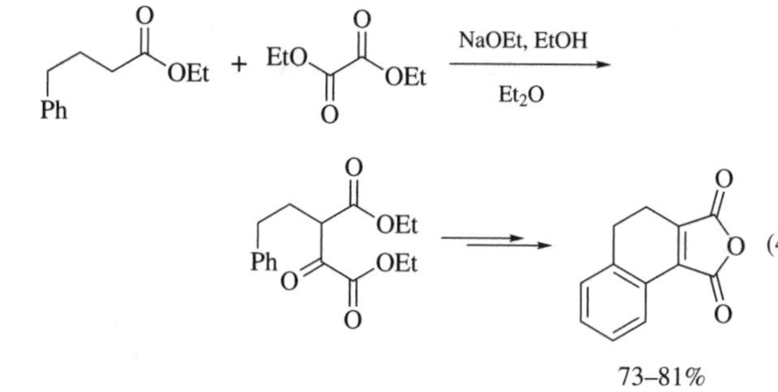

Ph, OEt + EtO, OEt — NaOEt, EtOH; Et_2O →

Ph, OEt, OEt → (4)

73–81%

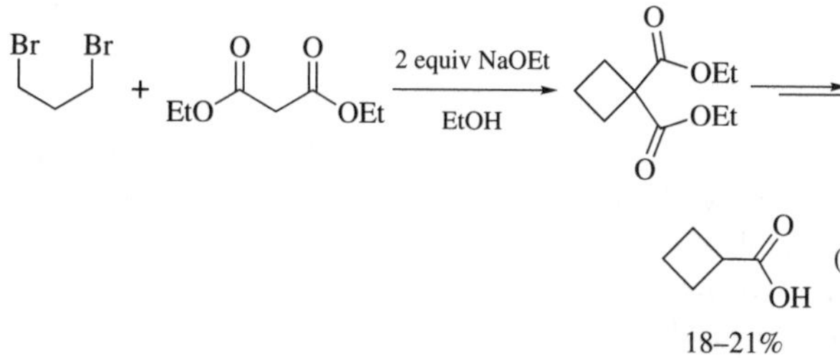

Br Br + EtO, OEt — 2 equiv NaOEt; EtOH → OEt, OEt →

OH (5)

18–21%

Sodium ethoxide and ***Sodium Hydroxide*** have both proven effective in the cyclopropanation of various γ,δ-epoxy ketones, as illustrated in eq 6. Adducts are generated in good to excellent yields.[4]

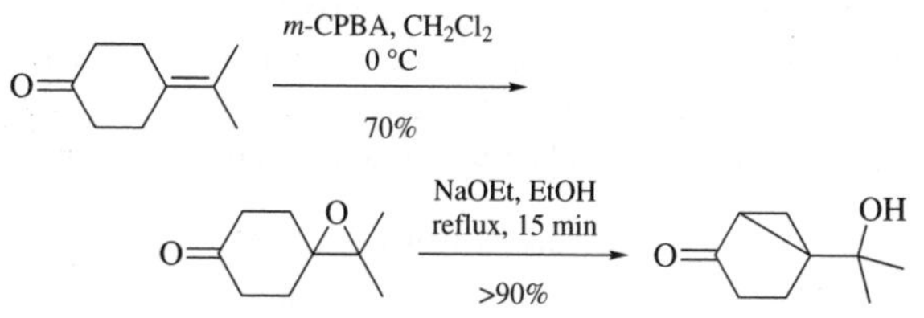

O= — *m*-CPBA, CH_2Cl_2; 0 °C; 70% →

— NaOEt, EtOH; reflux, 15 min; >90% → OH (6)

Deprotonation α to nitriles has been found to be strongly influenced by the purity of the sodium ethoxide used[5a] and can be done selectively over α-deprotonation of an ester.[5b] Reactions with aromatic aldehydes lead to condensation adducts (eq 7).[5c]

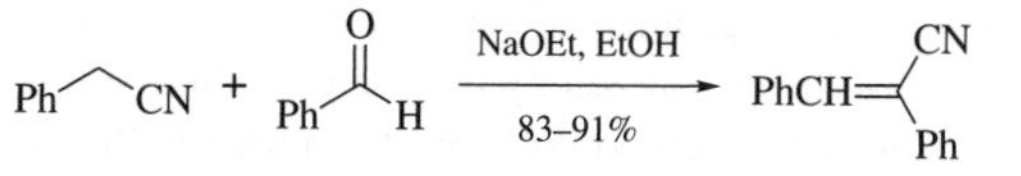

Ph CN + Ph H — NaOEt, EtOH; 83–91% → PhCH=C(CN)Ph (7)

Deprotonations α to nitro functionalities have been found to proceed with greater ease using sodium ethoxide rather than ***Sodium Methoxide***.[6] The in situ generated anion of 2-nitropropane transforms a benzylic bromide to its aromatic aldehyde derivative in good yield (eq 8).[7a]

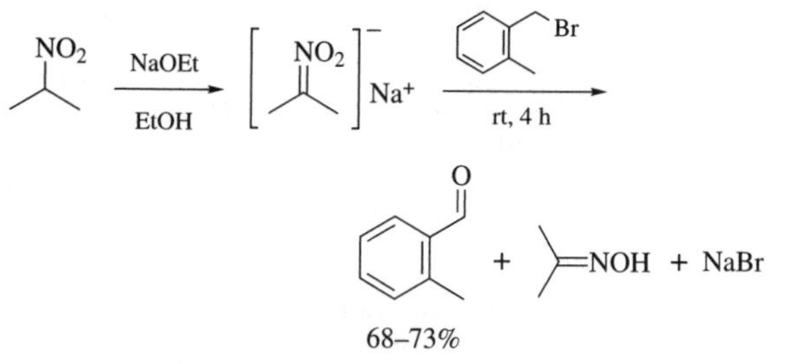

NO_2 — NaOEt; EtOH → [NO_2]$^-$ Na^+ — Br; rt, 4 h →

+ =NOH + NaBr (8)

68–73%

Sodium ethoxide serves as a multipurpose base to generate both the ylide of dimethyl-2-propynylsulfonium bromide and the anion of acetylacetone in the synthesis of 3-acetyl-2,4-dimethylfuran illustrated in eq 9.[3j] The intermediacy of an allene has been suggested.[7b]

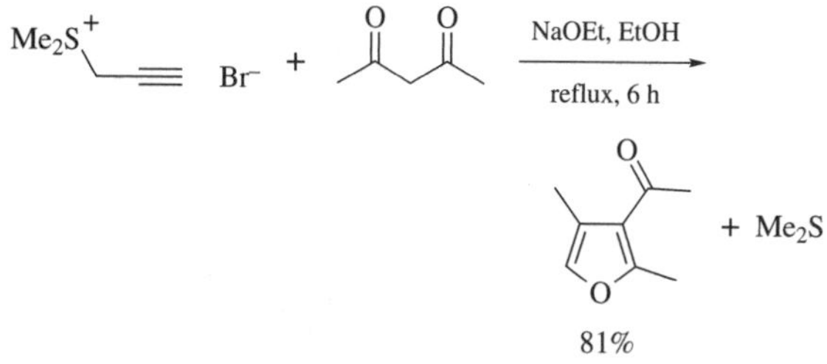

Me_2S^+ Br^- + — NaOEt, EtOH; reflux, 6 h →

+ Me_2S (9)

81%

Sodium ethoxide both dehydrates and nucleophilically assists in the deacylation of the carbinolamine intermediate depicted in eq 10.[8] Furthermore, sodium ethoxide-mediated dehydrohalogenation in DMSO solvent has been achieved in excellent yield (eq 11).[1b,9]

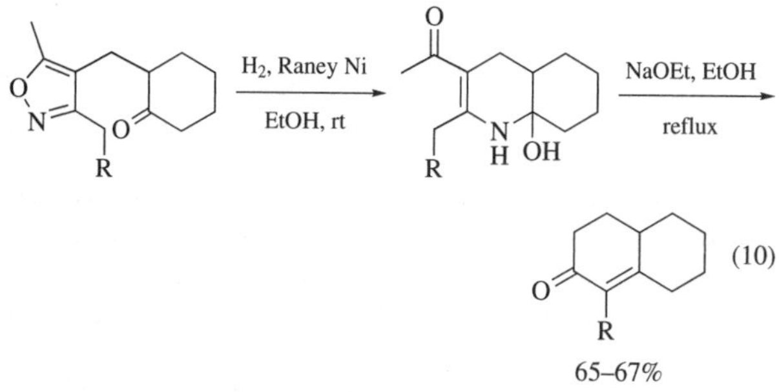

R — H_2, Raney Ni; EtOH, rt → R, N H OH — NaOEt, EtOH; reflux →

R (10)

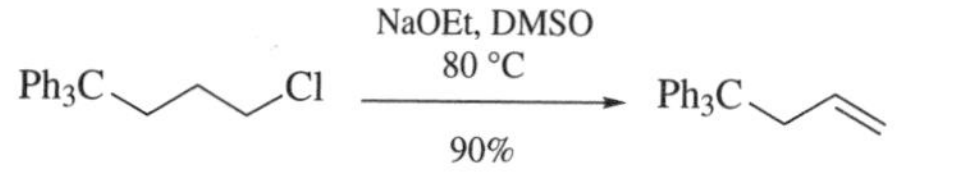

65–67%

Ph_3C Cl — NaOEt, DMSO; 80 °C; 90% → Ph_3C (11)

Heteroatom deprotonations using sodium ethoxide as the base are also quite common. Sodium ethoxide serves as an effective base when the nitrogen of cyanoacetamide must be condensed on an ester. Subsequent intramolecular cyclization proceeding via α-deprotonation to the nitrile leads to excellent yield of the dicyanoglutarimide shown in eq 12.[10a]

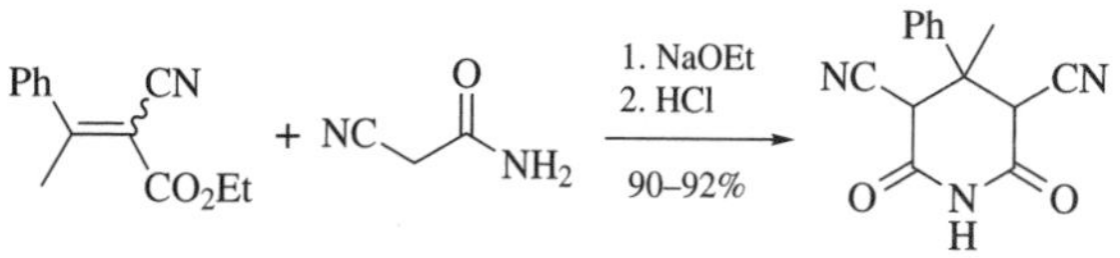

Ph CN CO_2Et + NC NH_2 — 1. NaOEt; 2. HCl; 90–92% → Ph, NC, CN, N H (12)

Other *N*-deprotonations include that of urea,[10b] thiourea,[10c] succinimide,[10d] tosylamines,[10e] cysteamine hydrochloride (eq 13),[1c] ribofuranosylamine toluenesulfonate,[10f] and cyanamide (eq 14).[10g] Similarly, *S*-deprotonations of sulfides (eq 13)[1c,11] have been reported. Often, adducts can be used in subsequent cyclizations.[1c,10b,g,11]

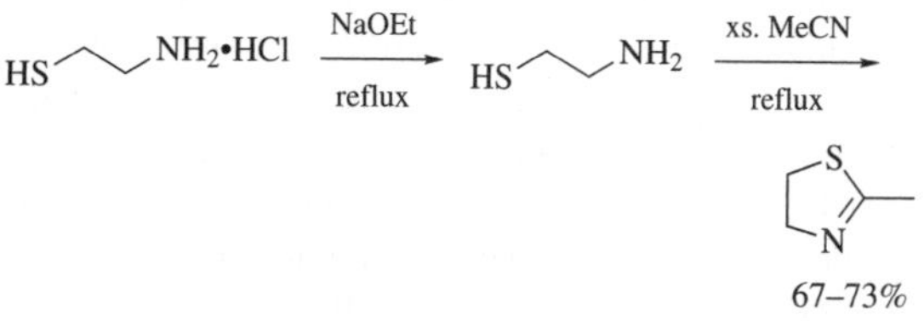

HS NH_2•HCl — NaOEt; reflux → HS NH_2 — xs. MeCN; reflux →

S N (13)

67–73%

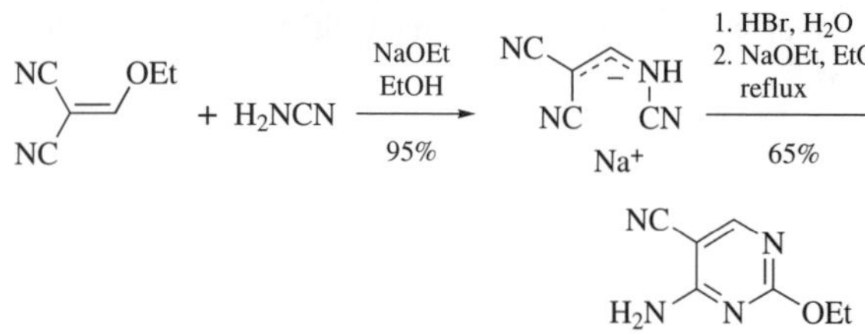

(14)

The synthesis of 6,6-dialkoxyfulvenes from 6,6-bis(methylthio)fulvene has been reported in good yield via treatment with sodium ethoxide or related nucleophiles, followed by thermolysis (eq 15).[12a]

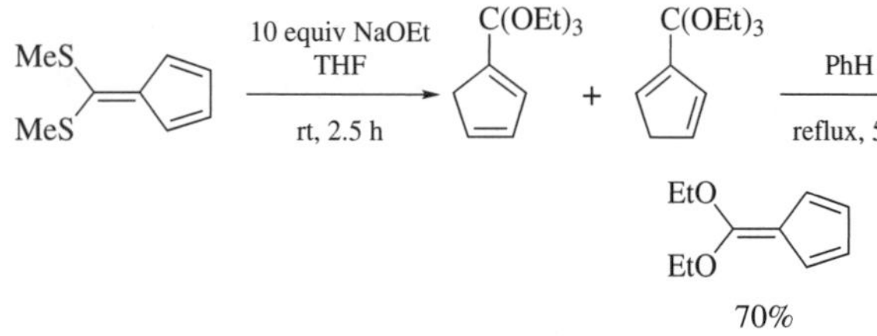

(15)

Addition of a large excess of sodium ethoxide or methoxide to 11-chloro[5]metacyclophanes leads to good yields of *ipso* alkoxy-substituted adducts.[12b] Sodium ethoxide proves to be the fastest of a series of nucleophiles in *ipso* substitution reactions of aromatic bromides employing a ***Copper(I) Bromide*** catalyst (eq 16). Once again, purity of alkoxide seems to be important.[12c] Substitution of aryl iodides is also known.[12d]

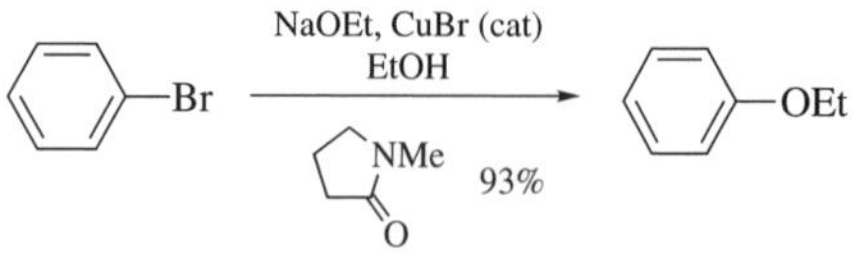

(16)

Ipso ethoxy substitutions of halo-pyrimidines,[10g] -purines (eq 17),[12e] -thiophenes,[12c,f] -furans,[12c] and -triazines[12g] using sodium ethoxide have also been reported in good yield.

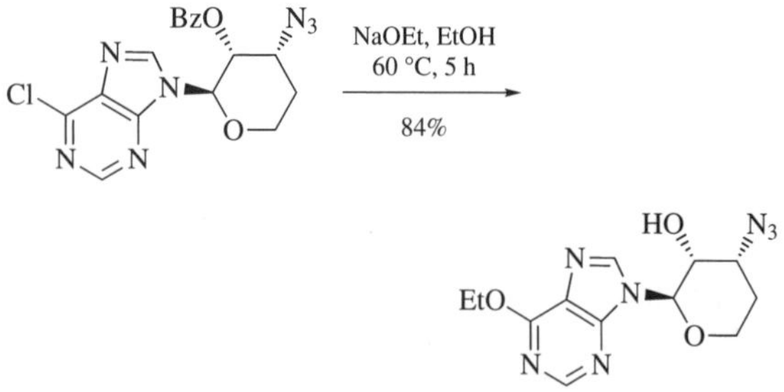

(17)

Aryl iodides, vinyl bromides, and tricarbonyl(chloroarene)chromium complexes have been found to react under mild conditions with sodium alkoxides in the presence of catalytic dichlorobis(triphenylphosphine)palladium(II) (see ***Palladium(II) Chloride***) to afford the corresponding benzoate esters as the major products (eq 18).[12h,i] It has been rationalized that the higher base strength of sodium ethoxide compared to that of sodium methoxide allows the former to successfully decarbonylate ethyl formate, whereas the latter proves more sluggish, yielding poorer results.

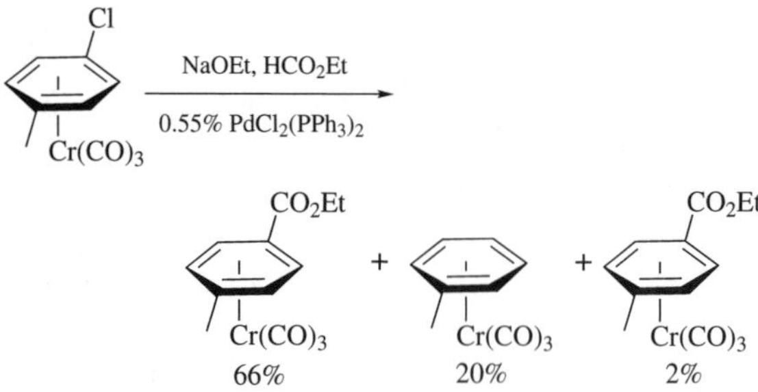

(18)

Ipso substitution of 2- and 4-pyridyl sulfones is also possible using a variety of nucleophiles, including sodium ethoxide. The sulfone moiety is preferentially displaced over a chlorine substituent also present on the aromatic ring (eq 19). The same reaction utilizing sulfoxides gives mixed results, while sulfides fail to react.[12j]

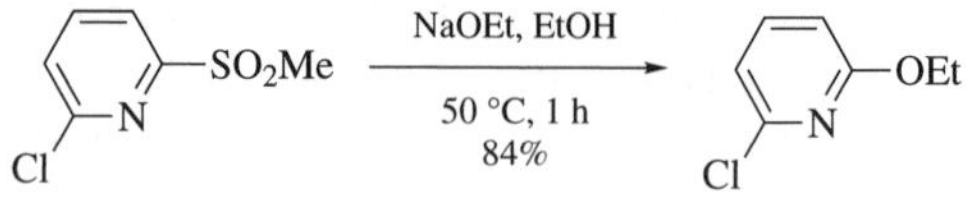

(19)

It has been suggested that displacement of the nitro group of the radical anion of *p*-nitrobenzophenone by sodium ethoxide is followed by chain transfer from the resulting *p*-ethoxybenzophenone radical, resulting in the adduct pictured in eq 20.[12k]

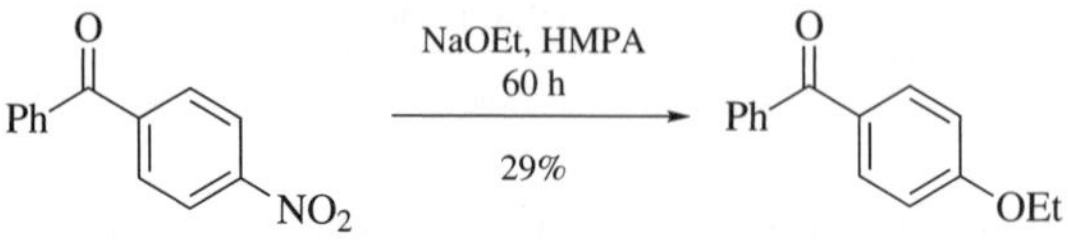

(20)

Sodium ethoxide is frequently employed as the nucleophile in the Williamson ether synthesis,[13a–c] as illustrated in eq 21. Furthermore, polydisplacements of halo,[13d–h] nitro,[13e] thiooxy,[13g,h] and phenoxy[13g] groups of dichloromethane and chloroform analogs have been realized by the use of sodium ethoxide as the nucleophile. This methodology has been used in the syntheses of ethyl orthocarbonate (eq 22) and ethyl diethoxyacetate (eq 23) in moderate to good yields.

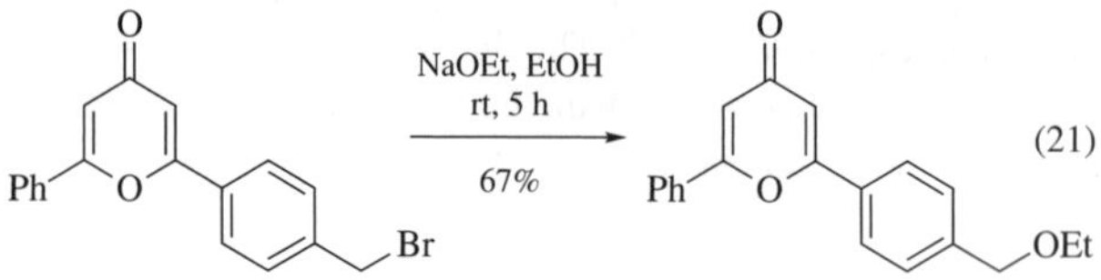

(21)

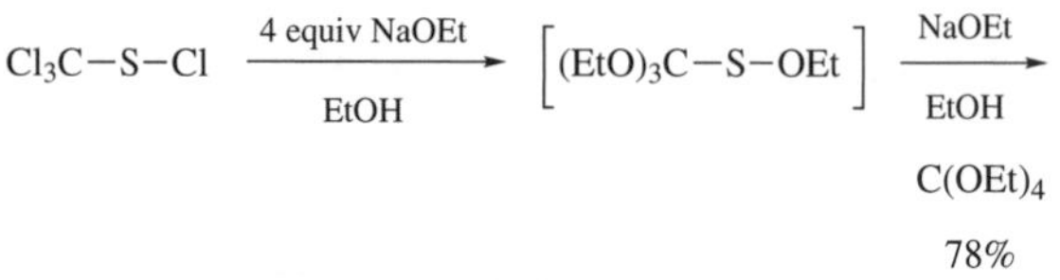

$$Cl_3C{-}S{-}Cl \xrightarrow[\text{EtOH}]{\text{4 equiv NaOEt}} [(EtO)_3C{-}S{-}OEt] \xrightarrow[\text{EtOH}]{\text{NaOEt}} C(OEt)_4 \quad 78\% \qquad (22)$$

$$Cl_2HC{-}CO_2H \xrightarrow[\text{45–50\%}]{\text{1. 3 equiv NaOEt; 2. HCl, EtOH}} (EtO)_2HC{-}CO_2Et \qquad (23)$$

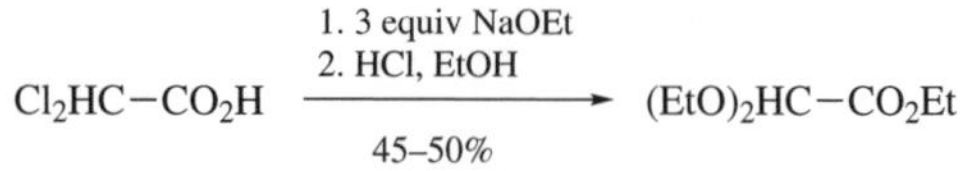

Sodium ethoxide has been employed for a transesterification–conjugate addition protocol, which takes advantage of the *syn*-selective addition of sodium alkoxides to acrylic esters such as the one depicted in eq 24.[14]

NaOEt, EtOH, −50 °C, >85% (24)

syn:*anti* = 86:14

An interesting synthesis of α-ethoxy ketones employs nucleophilic reaction of sodium ethoxide with α-nitro epoxides. The reaction appears to be quite general and a variety of nucleophiles may be employed (eq 25).[15]

NaOEt, EtOH, rt, 2 min, 64% (25)

Nucleophilic attack of sodium ethoxide on 3,4- or 2,3-cyanomethyl cyanopyridines followed by cyclization provides access to amino alkoxy naphthyridines (eq 26).[1e,16] The reaction is believed to proceed through an imidate intermediate.[16a] Unfortunately, treatment of the analogous carbocyclic 2-cyanobenzyl cyanides under similar conditions leads to dimerization of starting material, and cyclization must be achieved under acidic conditions.[16b]

NaOEt, EtOH, reflux to rt, 72% (26)

Sodium ethoxide is also known to nucleophilically add to polyhaloalkenes with preferential attack on the methylene with the highest degree of fluorine substitution (eqs 27 and 28).[17] Excess nucleophile unveils an ethyl ester (eq 28).

NaOEt, EtOH, 92–97% (27)

1 equiv NaOEt, EtOH, 80 °C, 40 h; 3 equiv NaOEt (28)

80–85%

Reaction of sodium ethoxide with chlorodiphenylphosphine provides access to Arbuzov precursors, which can be further reacted with suitable electrophiles to form phosphine oxides (eq 29).[18]

Ph_2P-Cl → (NaOEt, EtOH, Et_2O, 0 °C, 0.5 h, 96%) Ph_2P-OEt → (DMF, reflux, 15 h, 82%) (29)

Treatment of Grignard reagents (RMgBr) with sodium ethoxide offers a synthetic route to organomagnesium ethoxides (RMgOEt).[19] No yields are reported for this reaction.

Related Reagents. Potassium *t*-Butoxide; Potassium 2-Methyl-2-butoxide; Sodium Hydroxide; Sodium Methoxide.

1. (a) *FF* **1967**, *1*, 1065. (b) *FF* **1969**, *2*, 157. (c) *FF* **1972**, *3*, 265. (d) *FF* **1974**, *4*, 451. (e) *FF* **1977**, *6*, 540. (f) *FF* **1986**, *12*, 402.
2. (a) Denoon, Jr., C. E. *OSC* **1955**, *3*, 16. (b) Riegel, E. R.; Zwilgmeyer, F. *OSC* **1943**, *2*, 126. (c) Magnani, A.; McElvain, S. M. *OSC* **1955**, *3*, 251. (d) Tishler, M.; Gal, G.; Stein, G. A. *OSC* **1963**, *4*, 536. (e) Briese, R. R.; McElvain, S. M. *JACS* **1933**, *55*, 1697. (f) Hershberg, E. B.; Fieser, L. F. *OSC* **1943**, *2*, 194. (g) Floyd, D. E.; Miller, S. E. *OSC* **1963**, *4*, 141. (h) Holmes, H. L.; Trevoy, L. W. *OSC* **1955**, *3*, 301. (i) Friedman, L.; Kosower, E. *OSC* **1955**, *3*, 510.
3. (a) Allen, F.; Kalm, M. J. *OSC* **1963**, *4*, 616. (b) Adams, R.; Kamm, R. M. *OSC* **1941**, *1*, 245. (c) Cox, R. F. B.; McElvain, S. M. *OSC* **1943**, *2*, 279. (d) Callen, J. E.; Dornfield, C. A.; Coleman, G. H. *OSC* **1955**, *3*, 212. (e) Moffett, R. B. *OSC* **1963**, *4*, 291. (f) Mariella, R. P.; Raube, R. *OSC* **1963**, *4*, 288. (g) Andruzzi, F.; Hvilsted, S. *Polymer* **1991**, *32*, 2294. (h) Marvel, C. S.; King, W. B. *OSC* **1941**, *1*, 246. (i) Shriner, R. L.; Todd, H. R. *OSC* **1943**, *2*, 200. (j) Howes, P. D.; Stirling, C. J. M. *OSC* **1988**, *6*, 31.
4. (a) Gaoni, Y. *T* **1972**, *28*, 5525. (b) Gaoni, Y. *T* **1972**, *28*, 5533.
5. (a) Horning, E. C.; Finell, A. F. *OSC* **1963**, *4*, 461. (b) Coan, S. B.; Becker, E. I. *OSC* **1963**, *4*, 174. (c) Womack, E. B.; McWhirter, J. *OSC* **1955**, *3*, 714.
6. Dauben, Jr., H. J.; Ringold, H. J.; Wade, R. H.; Pearson, D. L.; Anderson, Jr.; A. G. *OSC* **1963**, *4*, 221.
7. (a) Hass, W. B.; Bender, M. L. *OSC* **1963**, *4*, 932. (b) Batty, J. W.; Howes, P. D.; Stirling, C. J. M. *JCS(P1)* **1973**, 65.
8. McMurry, J. E. *OSC* **1988**, *6*, 781.
9. Norman, R. O. C.; Thomas, C. B. *JCS(C)* **1967**, 1115.
10. (a) McElvain, S. M.; Clemens, D. H. *OSC* **1963**, *4*, 662. (b) Sherman, W. R.; Taylor, Jr., E. C. *OSC* **1963**, *4*, 247. (c) Ulbricht, T. L. V.; Okuda, T.; Price, C. C. *OSC* **1963**, *4*, 566. (d) Crockett, G. C.; Koch, T. D. *OSC* **1988**, *6*, 226. (e) Atkins, T. J.; Richman, J. E.; Oettle, W. F. *OSC* **1988**, *6*, 652. (f) Espie, J. C.; Lhomme, M. F.; Morat, C.; Lhomme, J. *TL* **1990**, *31*, 1423. (g) Schmidt, H.-W.; Koitz, G.; Junek, H. *JHC* **1987**, *24*, 1305.
11. Gillis, R. G.; Lacey, A. B. *OSC* **1963**, *4*, 396.
12. (a) Gupta, I.; Yates, P. *SC* **1982**, *12*, 1007. (b) Kraakman, P. A.; Valk, J.-M.; Niederländer, H. A. G.; Brouwer, D. B. E.; Bickelhaupt, F. M.; de Wolf, W. H.; Bickelhaupt, F.; Stam, C. H. *JACS* **1990**, *112*, 6638. (c) Keegstra, M. A.; Peters, T. H. A.; Brandsma, L. *T* **1992**, *48*, 3633. (d) Somei, M.; Yamada, F.; Kunimoto, M.; Kaneko, C. *H* **1984**, *22*, 797. (e) Carret, G.; Grouiller, A.; Chabannes, B.; Pacheco, H. *Nucleosides Nucleotides* **1986**, *3*, 331. (f) Puschmann, I.; Erker, T. *H* **1993**, *36*, 1323. (g) Konno, S.; Ohba, S.; Agata, M.; Aizawa, Y.; Sagi, M.; Yamanaka, H. *H* **1987**, *26*, 3259. (h) Carpentier, J.-F.; Castanet, Y.; Brocard, J.; Mortreux, A.; Petit, F. *TL* **1991**, *32*, 4705. (i) Carpentier, J.-F.; Castanet, Y.; Brocard, J.; Mortreux, A.; Petit, F. *TL* **1992**, *33*, 2001. (j) Furukawa, N.; Ogawa, S.; Kawai, T. *JCS(P1)* **1984**, 1839. (k) Denney, D. B.; Denney, D. Z.; Perez, A. J. *T* **1993**, *49*, 4463.
13. (a) Slomkowski, S.; Winnik, M. A.; Furlong, P.; Reynolds, W. F. *Macromolecules* **1989**, *22*, 503. (b) Marei, M. G.; Mishrikey, M. M.; El-Kholy, I. El-S. *Acta Chim. Hung.* **1987**, *124*, 733. (c) Marei, M. G.; Mishrikey, M. M.; El-Kholy, I. El-S. *IJC(B)* **1987**, *26B*, 163. (d) Moffett, R. B. *OSC* **1963**, *4*, 427. (e) Roberts, J. D.; McMahon, R. E. *OSC* **1963**, *4*, 457. (f) Kaufmann, W. E.; Dreger, E. E. *OSC* **1941**, *1*, 253. (g) Connolly, J. M.; Dyson, G. M. *JCS* **1936**, 827. (h) Tieckelmann, H.; Post, H. W. *JOC* **1948**, *13*, 265.
14. Mulzer, J.; Kappert, M.; Huttner, G.; Jibril, I. *AG(E)* **1984**, *23*, 704.

15. Vankar, Y. B.; Shah, K.; Bawa, A.; Singh, S. P. *T* **1991**, *47*, 8883.
16. (a) Alhaique, F.; Riccieri, F. M.; Santucci, E. *TL* **1975**, 173. (b) Johnson, F.; Nasutavicus, W. W. *JOC* **1962**, *27*, 3953.
17. (a) Shainyan, B. A.; Rappoport, Z. *JOC* **1993**, *58*, 3421. (b) Englund, B. *OSC* **1963**, *4*, 184.
18. Yang, Z.; Geise, H. J.; Nouwen, J.; Adriaensens, P.; Franco, D.; Vanderzande, D.; Martens, H.; Gelan, J.; Mehbod, M. *Synth. Met.* **1992**, *47*, 111.
19. Gupta, S.; Sharma, S.; Narula, A. K. *JOM* **1993**, *452*, 1.
20. (a) Schroeder, G. *React. Kinet. Catal. Lett.* **1992**, *46*, 51. (b) Pregel, M. J.; Buncel, E. *JACS* **1993**, *115*, 10.

K. Sinclair Whitaker
Wayne State University, Detroit, MI, USA

D. Todd Whitaker
Detroit Country Day School, Beverly Hills, MI, USA

Sodium Hexamethyldisilazide[1]

$NaN(SiMe_3)_2$

[1070-89-9] $C_6H_{18}NNaSi_2$ (MW 183.42)

(useful as a sterically hindered base and as a nucleophile)

Alternate Names: NaHMDS; sodium bis(trimethylsilyl)amide.
Physical Data: mp 171–175 °C; bp 170 °C/2 mmHg.
Solubility: sol THF, ether, benzene, toluene.[1]
Form Supplied in: (a) off-white powder (95%); (b) solution in THF (1.0 M); (c) solution in toluene (0.6 M).
Analysis of Reagent Purity: THF solutions of the reagent may be titrated using 4-phenylbenzylidenebenzylamine as an indicator.[2]
Handling, Storage, and Precautions: the dry solid and solutions are flammable and must be stored in the absence of moisture. These should be handled and stored under a nitrogen atmosphere. Use in a fume hood.

Introduction. Sodium bis(trimethylsilyl)amide is a synthetically useful reagent in that it combines both high basicity[3] and nucleophilicity,[4] each of which may be exploited for useful organic transformations such as selective formation of enolates,[5] preparation of Wittig reagents,[6] formation of acyl anion equivalents,[7] and the generation of carbenoid species.[8] As a nucleophile, it has been used as a nitrogen source for the preparation of primary amines.[9,10]

Sterically Hindered Base for Enolate Formation. Like other metal dialkylamide bases, sodium bis(trimethylsilyl)amide is sufficiently basic to deprotonate carbonyl-activated carbon acids[5] and is sterically hindered, allowing good initial kinetic vs. thermodynamic deprotonation ratios.[11] The presence of the sodium counterion also allows for subsequent equilibration to the thermodynamically more stable enolate.[5f] More recently, this base has been used in the stereoselective generation of enolates for subsequent alkylation or oxidation in asymmetric syntheses.[12] As shown in eq 1, NaHMDS was used to selectively generate a (*Z*)-enolate; alkylation with ***Iodomethane*** proceeded with excellent diastereoselectivity.[12a] In this case, use of the sodium enolate was preferred as it was more reactive than the corresponding lithium enolate at lower temperatures.

NaHMDS → MeI (1)

79%
99:1 diastereoselectivity

The reagent has been used for the enolization of carbonyl compounds in a number of syntheses.[13] For ketones and aldehydes which do not have enolizable protons, NaHMDS may be used to prepare the corresponding TMS-imine.[14]

Generation of Ylides for Wittig Reactions. In the Wittig reaction, salt-free conditions have been shown to improve (*Z*):(*E*) ratios of the alkenes which are prepared.[15] NaHMDS has been shown to be a good base for generating ylides under lithium-salt-free conditions.[6] It has been used in a number of syntheses to selectively prepare (*Z*)-alkenes.[16] Ylides generated under these conditions have been shown to undergo other ylide reactions such as *C*-acylations of thiolesters and inter- and intramolecular cyclization.[6] Although Wittig-based syntheses of vinyl halides exist,[17] NaHMDS has been shown to be the base of choice for the generation of iodomethylenetriphenylphosphorane for the stereoselective synthesis of (*Z*)-1-iodoalkenes from aldehydes and ketones (eq 2).[18]

$[Ph_3PCH_2I]^+ I^-$ —NaHMDS, THF→ $[Ph_3PCHI]$; PhCHO → 96%, (*Z*):(*E*) = 62:1; cyclohexanone → 61% (2)

NaHMDS has been shown to be the necessary base for the generation of the ylide anion of sodium cyanotriphenylphosphoranylidenemethanide, which may be alkylated with various electrophiles and in turn used as an ylide to react with carbonyl compounds.[19] NaHMDS was used as the base of choice in a Horner–Emmons–Wadsworth-based synthesis of terminal conjugated enynes.[20]

Intramolecular Alkylation via Protected Cyanohydrins (Acyl Anion Equivalents). Although NaHMDS was not the base of choice for the generation of protected cyanohydrin acyl carbanion equivalents in the original references,[21] it has been shown to be an important reagent for intramolecular alkylation using this strategy (eqs 3 and 4).[7,22] The advantages of this reagent are (a) that it allows high yields of intramolecularly cyclized products with little intermolecular alkylation and (b) the carbanion produced in this manner acts only as a nucleophile without isomerization of double bonds α,β to the anion or other existing double bonds in the molecule. Small and medium rings as well as macrocycles[22a] have been reported using this methodology (eqs 3 and 4).

(3)

(4)

Generation of Carbenoid Species. Metal bis(trimethylsilyl)amides may be used to effect α-eliminations.[23] It is proposed that these nucleophilic agents undergo a hydrogen–metal exchange reaction with polyhalomethanes to give stable carbenoid species.[23b] NaHMDS has been used to generate carbenoid species which have been used in a one-step synthesis of monobromocyclopropanes (eqs 5 and 6).[23c,d] NaHMDS has been shown to give better yields than the corresponding lithium or potassium amides in this reaction.

(5)

cis:trans = 1.5:1

(6)

A similar study which evaluated the use of NaHMDS versus ***n*-Butyllithium** for the generation of the active carbenoid species from 1,1-dichloroethane and subsequent reaction with alkenes, forming 1-chloro-1-methylcyclopropanes, suggested that the amide gave very similar results to those with *n*-butyllithium.[24]

In an initial report, the carbenoid species formed by the treatment of diiodomethane with NaHMDS was shown to react as a nucleophile, displacing primary halides and leading to a synthesis of 1,1-diiodoalkanes; this is formally a 1,1-diiodomethylene homologation (eq 7).[25] This methodology is limited in that electrophiles which contain functionality that allows facile E2 elimination (i.e. allyl) form a mixture of the desired 1,1-diiodo compound and the iododiene. In the case of ***Allyl Bromide***, addition of 2 equiv of the sodium reagent allows isolation of the iododiene as the major product.

CH_2I_2 → (NaHMDS, THF, −90 °C) → $[NaCHI_2]$ → (RX, THF, 61–66%) → $RCHI_2$, R = Et, Bu, heptyl; → (1/2 allyl bromide, 83%) → iododiene, (E):(Z) = 40:60 (7)

Synthesis of Primary Amines. The nucleophilic properties of this reagent may be utilized in the S_N2 displacement of primary alkyl bromides, iodides, and tosylates to form bis(trimethylsilyl)amines (**1**) (eq 8).[9a] HCl hydrolysis of (**1**) allows isolation of the corresponding hydrochloride salt of the amine, which may be readily separated from the byproduct, bis(trimethylsilyl) ether. In one example a secondary allylic bromide also underwent the conversion with good yield.

$RX + NaHMDS \longrightarrow R{-}N(TMS)_2 \longrightarrow RNH_3Cl + (TMS)_2O$ (8)

R	X	(1)	(2)
Me	I	75%	99%
Et	OTs	77%	98%
Br	Br	73%	100%
cyclohex-2-enyl	Br	66%	97%

Aminomethylation. NaHMDS may be used as the nitrogen source in a general method for the addition of an aminomethyl group (eq 9).[10] The reagent is allowed to react with chloromethyl methyl ether, forming the intermediate aminoether. Addition of Grignard reagents to this compound allows the displacement of the methoxy group, leaving the bis(trimethylsilyl)-protected amines. Acidic hydrolysis of these allows isolation of the hydrochloride salt of the corresponding amine in good yields.

$MeOCH_2Cl$ → (NaHMDS, THF) → $MeOCH_2N(TMS)_2$ → (1. RMgX, Et_2O; 2. HCl, H_2O) → $RCH_2\overset{+}{N}H_3\ Cl^-$ (9)

R = Me, 89%; allyl, 76%; Cy, 75%; Ph, 78%; propargyl, 66%

Related Reagents. Lithium Diisopropylamide; Lithium Hexamethyldisilazide; Potassium Hexamethyldisilazide.

1. Wannagat, U.; Niederpruem, H. *CB* **1961**, *94*, 1540.
2. Duhamel, L.; Plaquevent, J. C. *JOM* **1993**, *448*, 1.
3. Barletta, G.; Chung, A. C.; Rios, C. B.; Jordan, F.; Schlegel, J. M. *JACS* **1990**, *112*, 8144.
4. (a) Capozzi, G.; Gori, L.; Menichetti, S. *TL* **1990**, *31*, 6213. (b) Capozzi, G.; Gori, L.; Menichetti, S.; Nativi, C. *JCS(P1)* **1992**, 1923.
5. (a) Evans, D. A. In *Asymmetric Synthesis*; Morrison, J. D., Ed.; Academic: New York, 1984; Vol. 3, p 1. (b) Tanabe, M.; Crowe, D. F. *CC* **1969**, 1498. (c) Barton, D. H. R.; Hesse, R. H.; Pechet, M. M.; Wiltshire, C. *CC* **1972**, 1017. (d) Krüger, C. R.; Rochow, E. *JOM* **1964**, *1*, 476. (e) Krüger, C. R.; Rochow, E. G. *AG(E)* **1963**, *2*, 617. (f) Gaudemar, M.; Bellassoued, M. *TL* **1989**, *30*, 2779.
6. Bestmann, H. J.; Stransky, W.; Vostrowsky, O. *CB* **1976**, *109*, 1694.
7. Stork, G.; Depezay, J. C.; d'Angelo, J. *TL* **1975**, 389.

8. Martel, B.; Hiriart, J. M. *S* **1972**, 201.
9. (a) Bestmann, H. J.; Woelfel, G. *CB* **1984**, *117*, 1250. (b) Anteunis, M. J. O.; Callens, R. De Witte M.; Reyniers, M. F.; Spiessens, L. *BSB* **1987**, *96*, 545.
10. Bestmann, H. J.; Woelfel, G.; Mederer, K. *S* **1987**, 848.
11. Barton, D. H. R.; Hesse, R. H.; Tarzia, G.; Pechet, M. M. *CC* **1969**, 1497.
12. (a) Evans, D. A.; Ennis, M. D.; Mathre, D. J. *JACS* **1982**, *104*, 1737. (b) Evans, D. A.; Morrissey, M. M.; Dorow, R. L. *JACS* **1985**, *107*, 4346. (c) Davis, F. A.; Haque, M. S. Przeslawski, R. M. *JOC* **1989**, *54*, 2021.
13. (a) Schmidt, U.; Riedl, B. *CC* **1992**, 1186. (b) Glazer, E. A.; Koss, D. A.; Olson, J. A.; Ricketts, A. P.; Schaaf, T. K.; Wiscount, R. J. Jr. *JMC* **1992**, *35*, 1839.
14. Krueger, C.; Rochow, E. G.; Wannagat, U. *CB* **1963**, *96*, 2132.
15. (a) Schlosser, M.; Christmann, K. F. *LA* **1967**, *708*, 1. (b) Schlosser, M. *Top. Stereochem.* **1970**, *5*, 1. (c) Schlosser, M.; Schaub, B.; de Oliveira-Neto, J.; Jeganathan, S. *C* **1986**, *40*, 244. (d) Schaub, B.; Jeganathan, S.; Schlosser, M. *C* **1986**, *40*, 246.
16. (a) Corey, E. J.; Su, W. *TL* **1990**, *31*, 3833. (b) Niwa, H.; Inagaki, H.; Yamada, K. *TL* **1991**, *32*, 5127. (c) Chattopadhyay, A.; Mamdapur, V. R. *SC* **1990**, *20*, 2225. (d) Mueller, S.; Schmidt, R. R. *HCA* **1993**, *76*, 616.
17. (a) Miyano, S.; Izumi, Y.; Fuji, K.; Ohno, Y.; Hashimoto, H. *BCJ* **1979**, *52*, 1197. (b) Smithers, R. H. *JOC* **1978**, *43*, 2833.
18. Stork, G.; Zhao, K. *TL* **1989**, *30*, 2173.
19. Bestmann, H. J.; Schmidt, M. *AG(E)* **1987**, *26*, 79.
20. Gibson, A. W.; Humphrey, G. R.; Kennedy, D. J.; Wright, S. H. B. *S* **1991**, 414.
21. (a) Stork, G.; Maldonado, L. *JACS* **1971**, *93*, 5286. (b) Stork, G.; Maldonado, L. *JACS* **1974**, *96*, 5272.
22. (a) Takahashi, T.; Nagashima, T. Tsuji, J. *TL* **1981**, 1359; (b) Takahashi, T.; Nemoto, H.; Tsuji, J. *TL* **1983**, 2005.
23. (a) Martel, B.; Aly, E. *JOM* **1971**, *29*, 61; (b) Martel, B.; Hiriart, J. M. *TL* **1971**, 2737. (c) Martel, B.; Hiriart, J. M. *S* **1972**, 201. (d) Martel, B.; Hiriart, J. M. *AG(E)* **1972**, *11*, 326.
24. Arora, S.; Binger, P. *S* **1974**, 801.
25. Charreau, P.; Julia, M.; Verpeaux, J. N. *BSF(2)* **1990**, *127*, 275.

Brett T. Watson
Bristol-Myers Squibb Pharmaceutical Research Institute, Wallingford, CT, USA

Sodium Hydride[1]

[7646-69-7] HNa (MW 24.00)

(used as a base for the deprotonation of alcohols, phenols, amides (NH), ketones, esters, and stannanes; used as a reducing agent for disulfides, disilanes, azides, and isoquinolines)

Physical Data: mp 800 °C (dec); *d* 1.396 g cm^{-3}.
Solubility: decomposes in water; insol all organic solvents; insol liq $NlsH_3$; sol molten sodium.
Form Supplied in: free-flowing gray powder (95% dry hydride); gray powder dispersed in mineral oil.
Handling, Storage, and Precautions: the dispersion is a solid and may be handled in the air. The mineral oil may be removed from the dispersion by stirring with pentane, then allowing the hydride to settle. The pentane/mineral oil supernatant may be pipetted off, but care should be exercised to quench carefully any hydride in the supernatant with a small amount of an alcohol before disposal. The dry powder should only be handled in an inert atmosphere.

Sodium hydride dust is a severe irritant and all operations should be done in a fume hood, under a dry atmosphere. Sodium hydride is stable in dry air at temperatures of up to 230 °C before ignition occurs; in moist air, however, the hydride rapidly decomposes, and if the material is a very fine powder, spontaneous ignition can occur as a result of the heat evolved from the hydrolysis reaction. Sodium hydride reacts more violently with water than sodium metal (eq 1); the heat of reaction usually causes hydrogen ignition.

$$NaH + H_2O \longrightarrow NaOH + H_2 \qquad (1)$$

Introduction. The following is arranged by reaction type: NaH acting as a base on oxygen, nitrogen, germanium/silicon, and carbon acids, and as a reducing agent.

Oxygen Acids (Alcohol Deprotonation). Sodium hydride may be used as a base in the Williamson ether synthesis in neat benzyl chloride,[2] in DMSO,[3] or in THF (eq 2).[4] Phenols may also be deprotonated and alkylated in THF.[4b]

$$Ph_3COH + NaH + MeI \xrightarrow[85\%]{THF} Ph_3COMe \qquad (2)$$

Curiously, tertiary propargylic alcohols may be alkylated in preference to either axial or equatorial secondary alcohols, using sodium hydride in DMF (eq 3).[5]

1. 2 equiv NaH, DMF; 2. MeI (excess) (3)

C-5β (C-3 axial): 91%
C-5α (C-3 equatorial): 93%

Sodium hydride in DMF is also used to deprotonate carbohydrate derivatives, for methylation or benzylation (eq 4).[6]

1. NaH, DMF; 2. MeI; 79% (4)

Unstable benzyl tosylates may be made by deprotonation of benzyl alcohols and acylation with ***p*-Toluenesulfonyl Chloride** (eq 5).[7]

$$PhCH_2OH \xrightarrow[\text{2. } p\text{-TsCl, } -70 \text{ to } 25\,^{\circ}C,\ 80\%]{\text{1. NaH, ether, reflux, 15 h}} PhCH_2OTs \qquad (5)$$

An interesting conformational effect is seen when *p-t*-butylcalix[4]arene is tetraethylated. When ***Potassium Hydride*** is used as the base, the partial cone conformation predominates (i.e. one of the aryl groups is inverted), whereas with sodium hydride, the cone is produced exclusively (eq 6).[8]

(6) [scheme: p-t-butylcalix[4]arene (OH) → EtI, NaH, DMF or MeCN → tetraethyl ether (OEt), cone]

Deprotonation of vinylsilane-allylic alcohols using sodium hydride in HMPA is followed by an 'essentially quantitative' C → O silicon migration (eq 7).[9]

(7) [scheme: TMS-vinyl allylic alcohol → NaH (catalytic), HMPA, >95% → OTMS product]

Nitrogen Acids. Sodium hydride in DMSO, HMPA, NMP, or DMA assists the transamination of esters (eq 8).[10]

(8) [scheme: ethyl nicotinate (CO_2Et) + cyclohexylamine (NH_2) → NaH, DMSO, 85% → N-cyclohexylnicotinamide]

Acyl amino acids and peptides may be alkylated on nitrogen using sodium hydride, with no racemization (eq 9).[11] A slight change of reaction conditions allows simultaneous esterification.[12]

(9) [scheme: Cbz-leucine (BnO, CO_2H) → 1. 3 equiv NaH; 2. 8 equiv MeI; THF–DMF (10:1) → N-methyl derivative]

A similar intramolecular alkylation has been used to make β-lactams (eq 10).[13]

(10) [scheme: chloro precursor (Ph, S, Cl, CO_2Me) → NaH, CH_2Cl_2, DMF, 82% → β-lactam (MeO_2C)]

Germanium/Silicon Acids. Germanium–hydrogen and silicon–hydrogen bonds are quantitatively cleaved with sodium hydride in ethereal solvents (eq 11).[14]

$$R_3XH + NaH \xrightarrow[>95\%]{DME} R_3XNa \quad (11)$$

R = Bu, Ph; X = Si, Ge

Carbon Acids. Active methylene compounds, such as malonates and β-keto esters, can be deprotonated with sodium hydride and alkylated on carbon (eqs 12 and 13).[15] Alkylation of Reissert anions is also facile with sodium hydride (eq 14).[16]

(12) [scheme: ethyl acetoacetate → 1. NaH, DMF; 2. EtBr; 74% → α-ethyl product]

(13) [scheme: diethyl malonate derivative (Ph, Ph) → 1. NaH, DMF; 2. BuBr; 75% → butylated product]

(14) [scheme: Reissert compound (N-COPh, CN) → 1. NaH, DMF; 2. $PhCH_2Cl$; 3. H_2O; 84% → benzylated product]

Normally, sodium enolates of ketones alkylate on oxygen. A 'superactive' form of sodium hydride is formed when butylsodium is reduced with hydrogen; superactive sodium hydride is an excellent base for essentially quantitative deprotonation of ketones, trapped as their silyl ethers.[17] For example, cyclododecanone is converted to its enol ether (containing a 'hyperstable' double bond) in 92% yield (eq 15).

(15) [scheme: cyclododecanone → NaH, TMSCl, hexane, 0 °C, 92% → OTMS enol ether]

More commonly, sodium hydride is used as a base for carbonyl condensation reactions. For example, Claisen condensations of ethyl acetate[18] and ethyl isovalerate[19] are effected by sodium hydride. Condensations of cyclohexanone with methyl benzoate[19] and ethyl formate (eq 16)[20] are also facile.[21] Sodium hydride can also doubly deprotonate β-diketones, allowing acylation at the less acidic site (eq 17).[22]

(16) [scheme: cyclohexanone → NaH, HCO_2Et, 70–74% → 2-(hydroxymethylene)cyclohexanone]

(17) [scheme: benzoylacetone → 1. 4 equiv NaH; 2. $PhCO_2Et$; 87% → triketone]

An undergraduate experiment using sodium hydride involves the crossed condensation of ethyl acetate and dimethyl phthalate (eq 18).[23] Sodium hydride is also effective as a base in the Stobbe condensation[24] and the Darzens condensation.[25] It is also effective in a stereoselective intramolecular Michael reaction (eq 19).[26]

(18) [scheme: dimethyl phthalate + $MeCO_2Et$ → NaH, 54–83% → ONa indenone CO_2Me]

(19) [scheme: MeO_2C keto enone → NaH (catalytic), benzene, 90% → cyclopentanone product]

The Dieckmann condensation of esters[27] and thioesters[28] is mediated by sodium hydride (eq 20). Conditions for the latter are significantly more mild than for the former, and the yields are higher.

$$\text{(CH}_2)_4(\text{COXEt})_2 \xrightarrow[\text{X = S, NaH, DME, EtSH, 20 °C, 90\%}]{\text{X = O, NaH, toluene, 45 °C, 65–80\%}} \text{2-oxocyclopentane-COXEt} \quad (20)$$

Sodium hydride may be used to cleave formate esters and formanilides (eqs 21 and 22).[29] The mechanism apparently involves removal of the formyl proton and loss of carbon monoxide.

$$\text{BuOCHO} + \text{NaH} \xrightarrow[68\%]{\text{DME}} \text{BuONa} + \text{CO} + \text{H}_2 \quad (21)$$

$$\text{Ph(Me)NCHO} + \text{NaH} \xrightarrow[70\%]{\text{DME}} \text{Ph(Me)NNa} + \text{CO} + \text{H}_2 \quad (22)$$

Dehydrohalogenation with sodium hydride is a means of making methylenecyclopropanes (eq 23).[30]

$$\text{Br-cyclopropane-CO}_2\text{Et} \xrightarrow[\text{EtOH, 60\%}]{\text{NaH}} \text{methylenecyclopropane-CO}_2\text{Et} \quad (23)$$

Enolate formation apparently accelerates the Diels–Alder cycloaddition/cycloreversion, shown in eq 24, which occurs at room temperature.[31]

$$\text{homophthalic anhydride} + \text{1,4-naphthoquinone} \xrightarrow[\text{THF, 88\%}]{\text{NaH}} \text{hydroxyanthraquinone} \quad (24)$$

An unusual cyclization of *N*-allyl-α,β-unsaturated amides is mediated by sodium hydride in refluxing xylene (eq 25).[32] The reaction is thought to proceed by intramolecular 1,4-addition of the dianion shown.

$$\text{N-allyl-3-methylbut-2-enamide} \xrightarrow[\text{xylene}]{\text{NaH}} [\text{dianion}] \xrightarrow{70\%} \text{azepinone} \quad (25)$$

Reductions. The sodium salt of trimethylsilane is produced quantitatively by reduction of hexamethyldisilane with sodium hydride (eq 26).[33]

$$\text{TMS–TMS} + \text{NaH} \xrightarrow[>95\%]{} \text{TMSNa} + \text{TMSH} \quad (26)$$

Sodium hydride in DMSO is an effective medium for the reduction of disulfide bonds in proteins under aprotic conditions.[34] When the molar ratio of hydride to 1/2 cystine residues exceeds 2:1, essentially complete reduction of the disulfide bonds of bovine serum albumin is achieved.

Azides are reduced to amines by sodium hydride, although the yields are moderate (eq 27).[35] Sodium hydride also reduces isoquinoline to 1,2-dihydroisoquinoline in good yield (eq 28).[36]

$$\text{BuN}_3 + \text{NaH} \xrightarrow[39\%]{} \text{BuNH}_2 \quad (27)$$

$$\text{isoquinoline} \xrightarrow[\text{2. Ac}_2\text{O, benzene; 77\%}]{\text{1. 3–5 equiv NaH, HMPA}} \text{N-Ac-1,2-dihydroisoquinoline} \quad (28)$$

Related Reagents. Calcium Hydride; Iron(III) Chloride–Sodium Hydride; Lithium Aluminum Hydride; Potassium Hydride; Potassium Hydride-*s*-Butyllithium-*N*,*N*,*N*′,*N*′-Tetramethylethylenediamine; Potassium Hydride–Hexamethylphosphoric Triamide; Sodium Borohydride; Sodium Hydride–Copper(II) Acetate–Sodium *t*-Pentoxide; Sodium Hydride–Nickel(II) Acetate–Sodium *t*-Pentoxide; Sodium Hydride–Palladium(II) Acetate–Sodium *t*-Pentoxide; Tris(cyclopentadienyl)lanthanum–Sodium Hydride.

1. (a) Mackay, K. M. *Hydrogen Compounds of the Metallic Elements*; Spon: London, 1966. (b) Hurd, D. T. *An Introduction to the Chemistry of the Hydrides*; Wiley: New York, 1952. (c) Wiberg, E.; Amberger, E. *Hydrides of the Elements of Main Groups I–IV*; Elsevier: New York, 1971.
2. Tate, M. E.; Bishop, C. T. *CJC* **1963**, *41*, 1801.
3. Doornbos, T.; Strating, J. *SC* **1971**, *1*, 175.
4. (a) Stoochnoff, B. A.; Benoiton, N. L. *TL* **1973**, 21. (b) Brown, C. A.; Barton, D.; Sivaram, S. *S* **1974**, 434.
5. Hajos, Z. G.; Duncan, G. R. *CJC* **1975**, *53*, 2971.
6. (a) Brimacombe, J. S.; Jones, B. D.; Stacey, M.; Willard, J. J. *Carbohydr. Res.* **1966**, *2*, 167. (b) Brimacombe, J. S.; Ching, O. A.; Stacey, M. *JCS(C)* **1969**, 197.
7. Kochi, J. K.; Hammond, G. S. *JACS* **1953**, *75*, 3443.
8. Groenen, L. C.; Ruël, B. H. M.; Casnati, A.; Timmerman, P.; Verboom, W.; Harkema, S.; Pochini, A.; Ungaro, R.; Reinhoudt, D. N. *TL* **1991**, *32*, 2675.
9. Sato, F.; Tanaka, Y.; Sato, M. *CC* **1983**, 165.
10. Singh, B. *TL* **1971**, 321.
11. Coggins, J. R.; Benoiton, N. L. *CJC* **1971**, *49*, 1968.
12. McDermott, J. R.; Benoiton, N. L. *CJC* **1973**, *51*, 1915.
13. (a) Baldwin, J. E.; Christie, M. A.; Haber, S. B.; Kruse, L. I. *JACS* **1976**, *98*, 3045. (b) Wasserman, H. H.; Hlasta, D. J.; Tremper, A. W.; Wu, J. S. *TL* **1979**, 549. (c) Wasserman, H. H.; Hlasta, D. J. *JACS* **1978**, *100*, 6780.
14. Corriu, R. J. P.; Guerin, C. *JOM* **1980**, *197*, C19.
15. Zaugg, H. E.; Dunnigan, D. A.; Michaels, R. J.; Swett, L. R.; Wang, T. S.; Sommers, A. H.; DeNet, R. W. *JOC* **1961**, *26*, 644.
16. (a) Kershaw, J. R.; Uff, B. C. *CC* **1966**, 331. (b) Uff, B. C.; Kershaw, J. R. *JCS(C)* **1969**, 666.
17. Pi, R.; Friedl, T.; Schleyer, P. v. R.; Klusener, P.; Brandsma, L. *JOC* **1987**, *52*, 4299.
18. Hinckley, A. A. *Sodium Hydride Dispersions*; Metal Hydrides, Inc., 1964 (quoted in: *FF* **1967**, *1*, 1075).
19. Swamer, F. W.; Hauser, C. R. *JACS* **1946**, *68*, 2647.
20. Ainsworth, C. *OSC* **1963**, *4*, 536.
21. (a) Bloomfield, J. J. *JOC* **1961**, *26*, 4112. (b) Bloomfield, J. J. *JOC* **1962**, *27*, 2742. (c) Anselme, J. P. *JOC* **1967**, *32*, 3716.
22. Miles, M. L.; Harris, T. M.; Hauser, C. R. *JOC* **1965**, *30*, 1007.
23. Gruen, H.; Norcross, B. E. *J. Chem. Educ.* **1965**, *42*, 268.
24. (a) Ref. 18. (b) Daub, G. H.; Johnson, W. S. *JACS* **1948**, *70*, 418. (c) Daub, G. H.; Johnson, W. S. *JACS* **1950**, *72*, 501.
25. (a) Ref. 18. (b) Della Pergola, R.; DiBattista, P. *SC* **1984**, *14*, 121.
26. Stork, G.; Winkler, J. D.; Saccomano, N. A. *TL* **1983**, *24*, 465.
27. Pinkney, P. S. *OSC* **1943**, *2*, 116.

28. Liu, H.-J.; Lai, H. K. *TL* **1979**, 1193.
29. Powers, J. C.; Seidner, R.; Parsons, T. G. *TL* **1965**, 1713.
30. Carbon, J. A.; Martin, W. B.; Swett, L. R. *JACS* **1958**, *80*, 1002.
31. Tamura, Y.; Sasho, M.; Nakagawa, K.; Tsugoshi, T.; Kita, Y. *JOC* **1984**, *49*, 473.
32. Bortolussi, M.; Bloch, R.; Conia, J. M. *TL* **1977**, 2289.
33. Corriu, R. J. P.; Guérin, C. *CC* **1980**, 168.
34. Krull, L. H.; Friedman, M. *Biochem. Biophys. Res. Commun.* **1967**, *29*, 373.
35. Lee, Y.-J.; Closson, W. D. *TL* **1974**, 381.
36. Natsume, M.; Kumadaki, S.; Kanda, Y.; Kiuchi, K. *TL* **1973**, 2335.

Robert E. Gawley
University of Miami, Coral Gables, FL, USA

Sodium Hydroxide[1]

NaOH

[1310-73-2] HNaO (MW 40.00)

(inorganic source of hydroxide ion for the saponification of carboxylic acid derivatives;[3,4] alkylation of phenols,[7] alcohols,[13] aldehydes,[18] and ketones;[19] transformation of amides to amines,[11] generation of dichlorocarbenes;[22] also useful for removing water from amines[27])

Physical Data: mp 318.4 °C; *d* 2.13 g cm^{-3}.
Solubility: in water: 0.42 g mL^{-1} at 0 °C; 3.47 g mL^{-1} at 100 °C.
Form Supplied in: pellets or standard aqueous solutions of variable concentrations.
Purification: dissolve in dry ethanol, 100 g L^{-1}, filter solution through a fine frit. Concentrate the solution under vacuum until a thick slurry is formed. Place slurry on a coarse sintered-glass disk, remove mother liquor and wash several times with dry ethanol. Vacuum dry crystals with mild heating for 30 h to give fine white powder.
Handling, Storage, and Precautions: hygroscopic; keep in a sealed container in a dry environment. Keep away from flame. Avoid contact and ingestion.

Hydrolysis of Carboxylic Acid Derivatives. The hydrolysis of esters, amides, nitriles, and ureas to the corresponding carboxylic acid can be accomplished with NaOH as hydroxide source, although milder methodology is available.[2] Ester hydrolysis is typically carried out in combinations of 1 N NaOH and either ethanol or methanol.[3] The hydrolysis of amides, nitriles and ureas is usually accomplished using excess NaOH in refluxing ethylene glycol with trace amounts of water.[4]

Dehydration of Sulfoxides and Selenoxides. Thiophenes and selenophenes can be generated by the dehydration of the corresponding sulfoxides[5] and selenoxides[6] using 40% and 50% NaOH (aq), respectively.

Alkylation Reactions. The NaOH-promoted conversion of phenols and carboxylic acids to their corresponding ethers[7] and esters[8] is accomplished in almost quantitative yield when excess ***Hexamethylphosphoric Triamide*** (HMPA) is added. The procedure calls for substrate, alkylating agent, 25% NaOH (aq), and HMPA to be combined at room temperature (eq 1).

CO_2H + I⁀ —(25% NaOH, HMPA; 99%)→ CO_2Et (1)

Phenols can be methylated with dimethyl sulfate,[9] and epoxides can be derived from halohydrins[10] using NaOH to generate the phenolate and alcoholate anions.

Conversion of Amides to Amines. The conversion of amides to amines containing one less carbon, the Hoffman reaction, is accomplished by the combination of bromine (or chlorine) with NaOH followed by heating (eq 2).[11] The reaction works well with aliphatic, aryl, and heterocyclic amides. The reaction can also be carried out in alcohol solutions, and the resulting urethane hydrolyzed in a subsequent step.

$$\text{(3-pyridyl)CONH}_2 + Br_2 + 4\,NaOH \xrightarrow{87\%} \text{(3-pyridyl)NH}_2 \quad (2)$$

Phase-Transfer-Catalyzed Alkylation Reactions. The use of phase-transfer-catalyzed two-phase reactions introduces hydroxide ion into organic solvent.[12] Under these conditions, hydroxide acts as a powerful base that efficiently promotes many transformations. Phase-transfer methodology is superior to the traditional Williamson synthesis of ethers (eq 3).[13]

$$BuOH + PhCH_2Cl \xrightarrow[92\%]{50\%\ NaOH,\ 5\%\ TBAB} PhCH_2OBu \quad (3)$$

Reaction conditions consist of a fivefold excess of 50% NaOH (aq), alcohol, excess alkyl chloride, and 3–5 mol % tetrabutylammonium bisulfate (TBAB). This method does not work with secondary alkyl halides but was successful with secondary alcohols. Alcohols can also be methylated under phase-transfer conditions using ***Dimethyl Sulfate***, 50% NaOH (aq), alcohol, and ***Tetra-n-butylammonium Iodide*** (TBAI) with methylene chloride or petroleum ether as the second phase.[14] This method works well for primary alcohols, is sluggish for secondary alcohols, and works not at all for tertiary alcohols. The use of a chiral phase-transfer catalyst gave alkylation of racemic alcohols[15] and ring closure of racemic chlorohydrins[16] with only moderate enantiomeric excess. Diphenylphosphinic hydrazide was alkylated in high yield using a mixture of solid K_2CO_3, NaOH, and tetra-*n*-butylammonium sulfate in boiling benzene.[17]

The combination of TBAI and 50% NaOH (aq) promotes aldehyde alkylation in modest yield.[18] The use of *N*-benzylcinchonine salts as phase-transfer catalysts results in excellent yield and good enantioselectivity in the alkylation of indonanes (eq 4).[19]

N-(p-trifluoromethylbenzyl)cinconium bromide, 50% NaOH, toluene, 92%; 99% ee (4)

Phase-Transfer-Catalyzed Ylide Chemistry. Wittig reactions can be done by combining 50% NaOH (aq), TBAI, a phosphonium salt, and a ketone in CH_2Cl_2. Benzylic phosphonium salts are converted to alkenes in good yields, but no stereoselectivity is observed (eq 5).[20]

50% NaOH, CH_2Cl_2, TBAI; 40% 40% (5)

Similarly, trimethylsulfonium iodide (eq 6) and trimethyloxosulfonium iodide (eq 7) are converted to the corresponding sulfur ylides.[21]

CH_2Cl_2, NaOH, TBAI, 50 °C, 48 h, >90% (6)

CH_2Cl_2, NaOH, TBAI, 50 °C, 20 h, 86% (7)

Phase-Transfer-Catalyzed Carbene Chemistry. Dichlorocarbene can be generated from chloroform using 50% NaOH (aq) with 1–5% phase-transfer catalyst. This methodology can be used for dichlorocyclopropanation (eq 8),[22] C–H insertion (eq 9)[23] and the conversion of primary amines into isocyanides.[24]

50% NaOH, $CHCl_3$, $Me_3\overset{+}{N}$–Bn Cl^-, 100% (8)

Micelles have been shown to give increased yields compared to tetraalkylammonium salt phase-transfer catalysts.[25] Similiar reactions can be done using bromoform, but the ease with which dibromocarbene undergoes hydrolysis makes this methodology less useful than the generation of dichlorocarbene.[26]

Related Reagents. Potassium Carbonate; Potassium Hydroxide; Sodium Carbonate.

1. Fyfe, C. A. *The Chemistry of the Hydroxy Group, Part 1*; Wiley: New York, 1971.
2. Haslam, E. *T* **1980**, *36*, 2409.
3. Baser, W. R.; Hauser, C. R. *OSC* **1963**, *4*, 582, 628.
4. (a) Newman, M. S.; Wise R. M. *JACS* **1956**, *78*, 450. (b) Tsai, L.; Miwa, T.; Newman, M. S. *JACS* **1957**, *79*, 450. (c) Pearson, D. E.; Baxter, J. F.; Carter, K. N. *OSC* **1955**, *3*, 154.
5. Horner, C. J.; Saris, L. E.; Lakshmikantham, M. V.; Cava, M. P. *TL* **1976**, 2581.
6. Saris, L. E.; Cava, M. P. *JACS* **1976**, *98*, 867.
7. Shaw, J. E.; Kunerth, D. C.; Sherry, J. J. *TL* **1973**, 689.
8. Shaw, J. E.; Kunerth, D. C. *JOC* **1974**, *39*, 1968.
9. Hiers, G. S.; Hager, F. D. *OSC* **1941**, *1*, 58.
10. Elderfield, R. C. *Heterocyclic Compounds*; Wiley: New York, 1950; Vol. 1, Chapter 1.
11. Wallis, E. S.; Lane, J. F. *OR* **1946**, *3*, 267.
12. Weber, W. P.; Gokel, G. W. *Phase Transfer Catalysis in Organic Synthesis*; Springer: Berlin, 1977.
13. Freedman, H. H.; Dubois, R. A. *TL* **1975**, 3251.
14. Merz, A. *AG(E)* **1973**, *12*, 846.
15. Verbicky, J. W. Jr.; O'Neil E. A. *JOC* **1985**, *50*, 1787.
16. Hummelen, J. C.; Wynberg, H.; *TL* **1978**, 1089.
17. Mlotkowska, B.; Zwierzak, A. *TL* **1978**, 4731.
18. Dietl, H. K.; Brannock, K. C. *TL* **1973**, 1273.
19. Bhattacharya, A.; Dolling, U-H.; Grabowski, E. J. J.; Karady, S.; Ryan K. M.; Weinstock, L. M. *AG(E)* **1986**, *25*, 476.
20. Märkl, G.; Merz, A. *S* **1973**, 295.
21. Merz, A.; Märkl, G. *AG(E)* **1973**, *12*, 815.
22. (a) Starks, C. M. *JACS* **1971**, *93*, 195. (b) Moss, R. A.; Smudlin, D. J. *JOC* **1976**, *41*, 611.
23. Tabushi, I.; Yoshida, Z-I.; Takahashi, N. *JACS* **1970**, *92*, 6670.
24. Weber, W. P.; Gokel, G. W.; Ugi, I. K. *AG(E)* **1972**, *11*, 530.
25. Joshi, G. C.; Singh, N.; Pande, L. M. *TL* **1972**, 1461.
26. Skattebøl, L.; Abiskaroun, G.; Greibrokk, T. *TL* **1973**, 1367.
27. Gordon, A. J.; Ford, R. A. *The Chemist's Companion: A Handbook of Practical Data, Techniques, and References*; Wiley: New York, 1972; 446.

Kurt D. Deshayes
Bowling Green State University, OH, USA

Sodium Methylsulfinylmethylide[1]

$CH_3-S(=O)-CH_2Na$

[15590-23-5] C_2H_5NaOS (MW 100.13)

(strong base and nucleophile; very useful for the introduction of the methylsulfinylmethyl group[1])

Alternate Names: sodium dimsylate; dimsylsodium; NaDMSO.
Solubility: sol DMSO.

Form Supplied in: not commercially available.

Analysis of Reagent Purity: titration with formanilide using triphenylmamethane as indicator.[2]

Preparative Method: prepared by the reaction of ***Sodium Hydride*** with ***Dimethyl Sulfoxide*** for 1 h at 70 °C. ***Sodium Amide*** may also be used as base.[3]

Handling, Storage, and Precautions: this reagent is not exceptionally stable over long periods of time. Modifications in preparation and storage have led to an increased shelf life.[4] However, it is probably best prepared as needed and used quickly. Hydrogen is emitted during the preparation of this reagent and decomposition occurs at elevated temperatures.[5] Due caution should be exercised in the preparation, particularly on a large scale. Reports have been made of explosions during the large scale preparation of this reagent or its attempted isolation.[1,6]

Introduction. The pK_a of DMSO is 35. Consequently, one might expect its conjugate base to be a potent Brønsted base and this is indeed the case. In addition to the sodium salt, both the lithium and potassium salts of DMSO have found widespread use in synthesis. After its introduction by Corey,[2] explorations of the chemistry of sodium methylsulfinylmethylide (NaDMSO) blossomed and are summarized nicely in two reviews.[1]

Sodium Methylsulfinylmethylide as Base.

Generation of Ylides. NaDMSO as a solution in DMSO has been used extensively in the preparation of phosphorus and sulfur ylides.[1,7,8] This medium is often the one of choice for the generation and reaction of Wittig reagents (eq 1).[7,9] A less common application of Wittig reagents generated with NaDMSO is in an oxidative coupling of a bis-ylide to form a cyclic alkene (eq 2).[10]

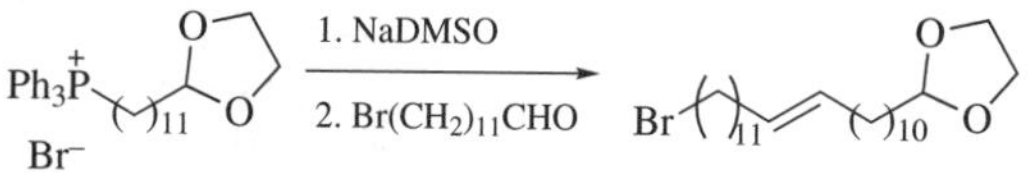

(1)

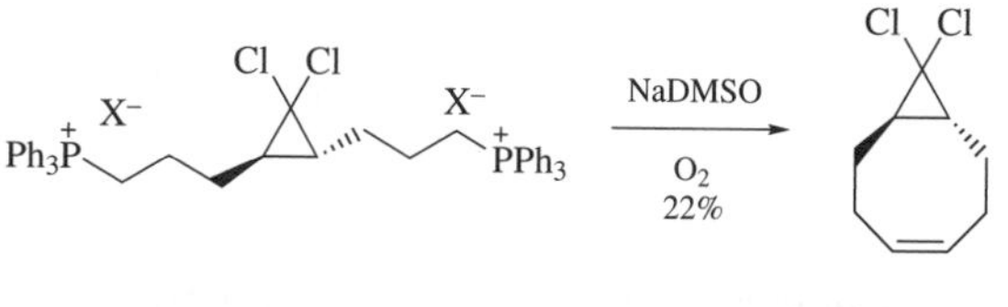

(2)

The methylene transfer reagents ***Dimethylsulfonium Methylide*** and ***Dimethylsulfoxonium Methylide*** are conveniently generated with NaDMSO in DMSO.[11] These and other sulfonium ylides are useful in the synthesis of epoxides and cyclopropanes (eqs 3 and 4).[8,11] Also intriguing is the use of dimethyloxosulfoxonium methylide to give dienes (eq 5).[12]

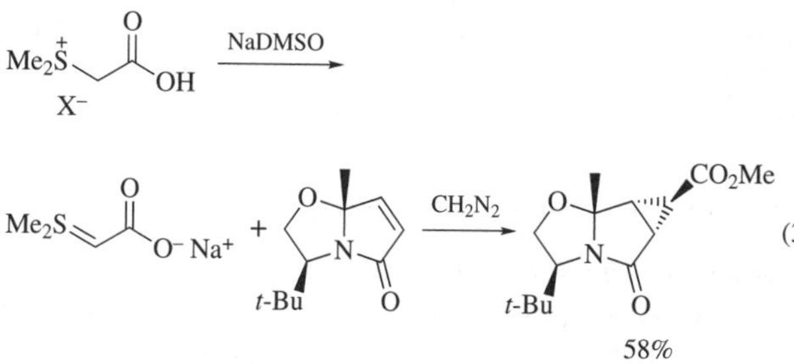

(3)

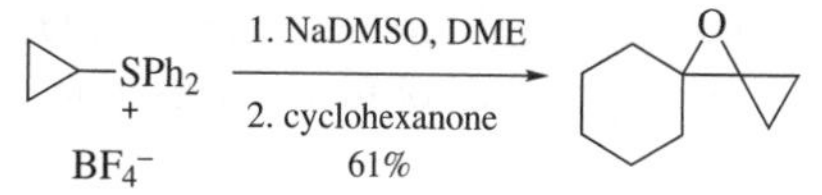

(4)

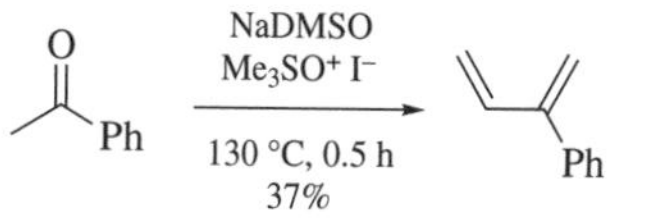

(5)

Diaminosulfonium salts can be deprotonated with NaDMSO to give rearrangement products or epoxides upon reaction with aldehydes (eq 6).[13] A variety of ammonium ylides have been generated using NaDMSO.[14,15] These species are disposed to undergo Wittig, Stevens, or Sommelet–Hauser rearrangements, depending on their specific constitution (eqs 7 and 8). Changes in base and other reaction conditions can change product distributions.[14a,b]

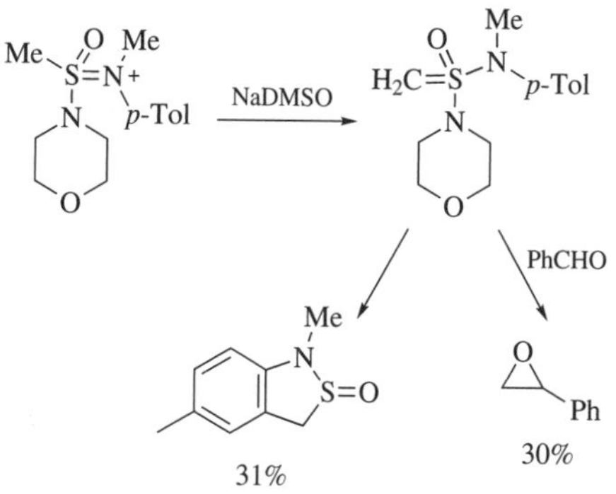

(6)

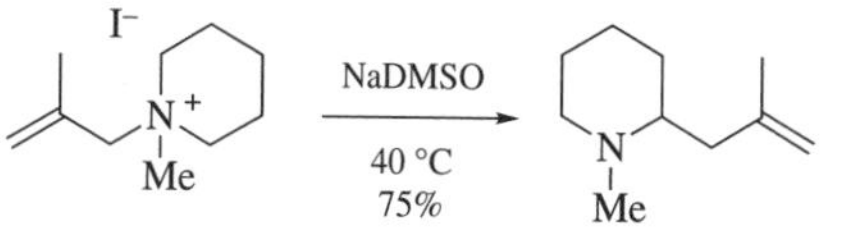

(7)

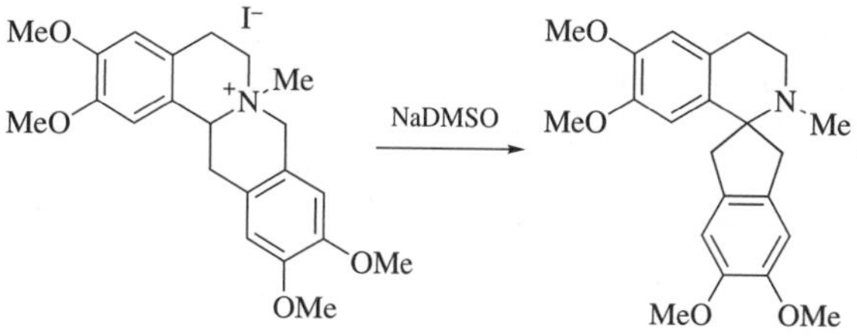

(8)

Eliminations. Elimination reactions using NaDMSO have apparently not been extensively investigated, although the use of metal alkoxides in DMSO has received considerable attention.[16] Whether the active agent in such mixtures is NaDMSO is not clear. Eliminations clearly mediated by NaDMSO are known.[17] For example, treatment of (**1**) with NaDMSO at 25 °C gives the elimination product (**2**) in 91% yield (eq 9).[17a] Longer reaction times result in addition of DMSO to the newly formed double bond. Heating results in isolation of a dealkylation product (**4**), a clear example of the diverse reactivity associated with NaDMSO. Alkynes are generated by the reaction of 1,2-dibromoalkanes with excess NaDMSO.[18] Aryl halides can give benzynes in the presence of NaDMSO (eq 10).[19] Further, addition of DMSO is possible, leading to a unique mode of functionalization (eq 11).[19c]

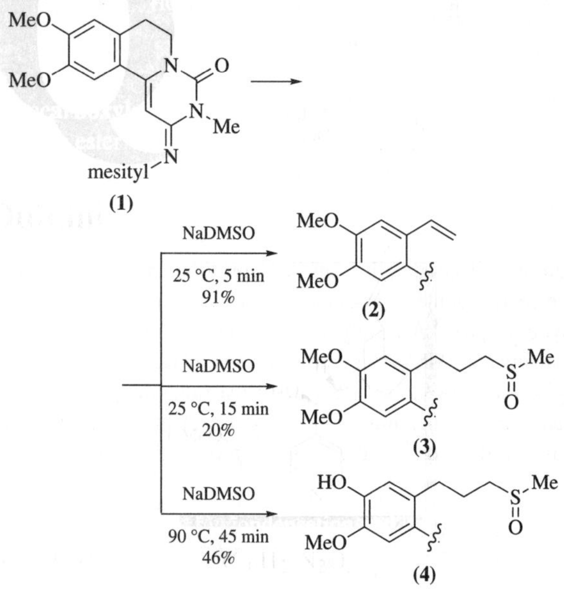

(9)

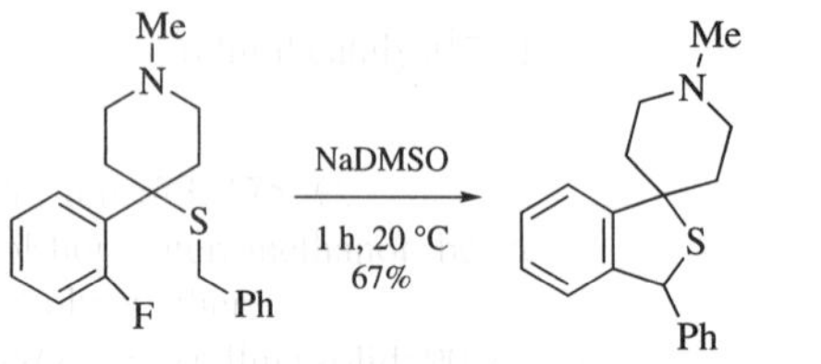

(10)

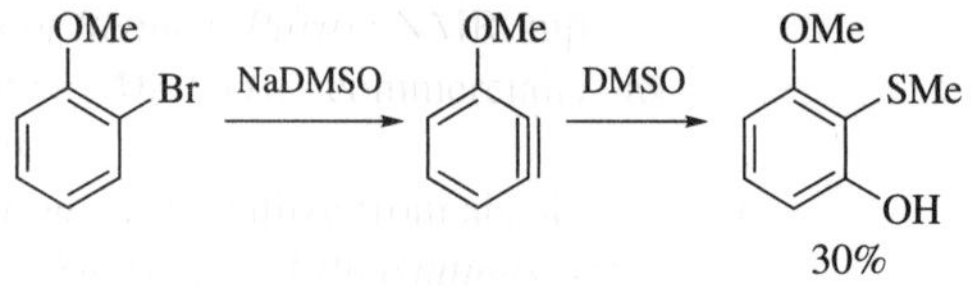

(11)

Rearrangements. NaDMSO has been used to promote double bond isomerization, leading to aromatization in the case shown (eq 12).[20] An anionotropic rearrangement of a cyclohexadienone to a substituted hydroquinone has been reported (eq 13).[21] The use of NaDMSO in DMSO has been described as an optimal choice for executing the carbanion accelerated Claisen rearrangement (eq 14).[22] Grob-type fragmentations mediated by NaDMSO have been used in total synthesis (eq 15).[23]

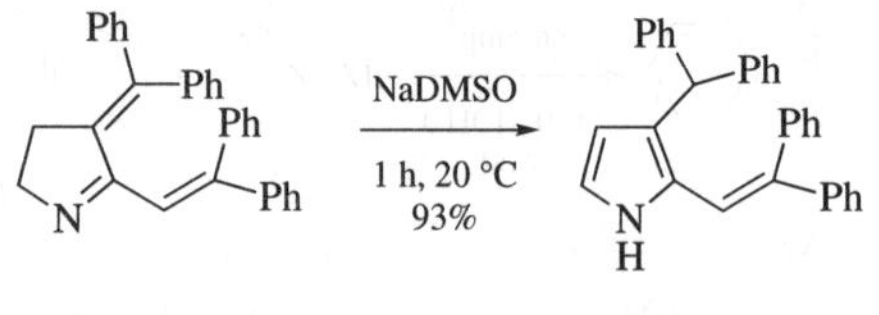

(12)

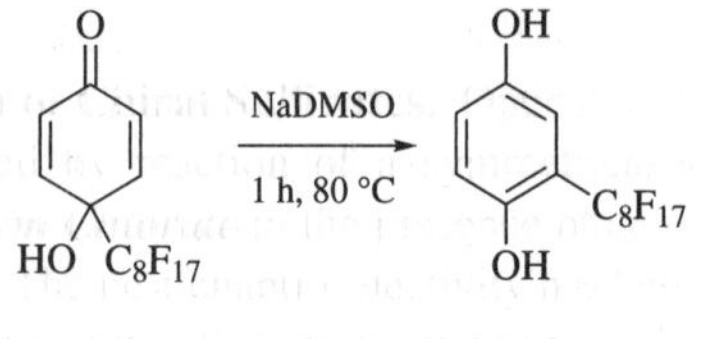

(13)

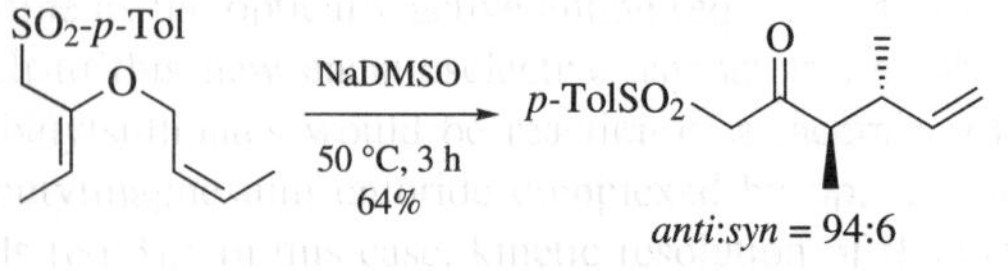

(14)

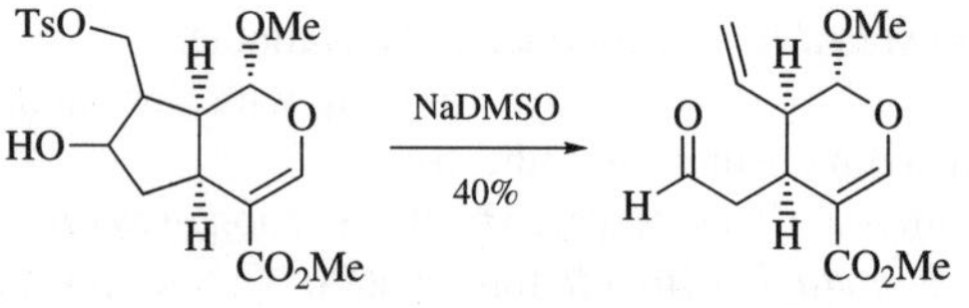

(15)

Anion Alkylation and Acylation. NaDMSO has been used as a base to create new anions or carbanions, which then can be functionalized via normal alkylation or acylation procedures. Intramolecular aminolysis of an ester mediated by NaDMSO has been reported (eq 16).[24] A tandem double Michael–Dieckmann condensation leading to a complex tricyclic structure has been developed (eq 17).[25] Generation of a ketone enolate with NaDMSO followed by an intramolecular alkylation has been a part of a number of total syntheses.[26] Alkylations of sulfoximine and sulfone (Ramberg–Bäcklund rearrangement) carbanions derived from NaDMSO are known.[27] The intramolecular oxidative coupling of nitro-stabilized carbanions can lead to highly functionalized cyclopropanes which are potentially useful as high energy materials (eq 18).[28]

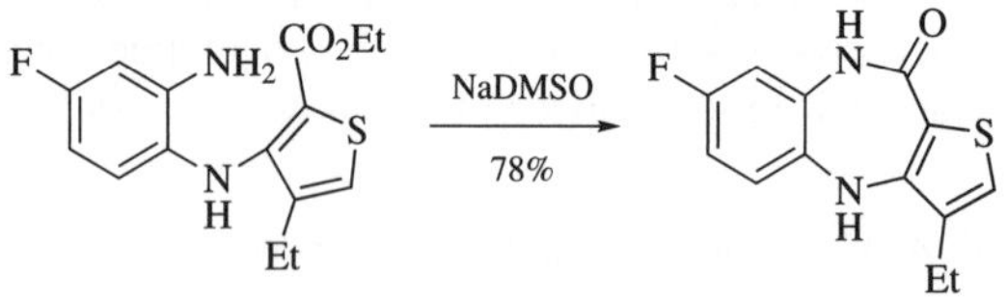

(16)

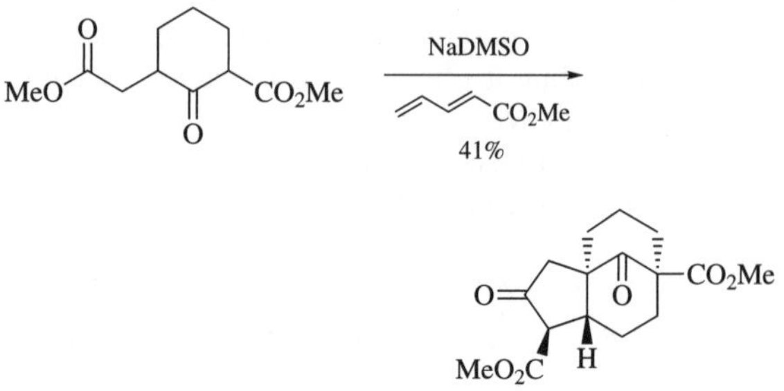

(17)

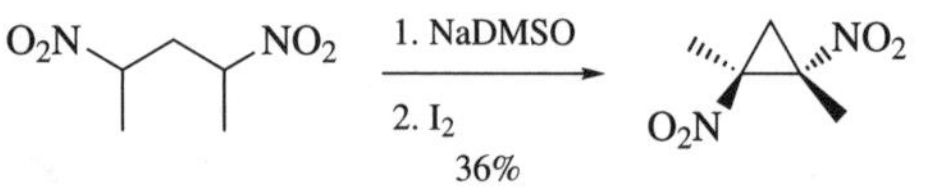

(18)

Ether Synthesis. The application of NaDMSO to the Williamson ether synthesis has been documented.[29] The potential for alkoxide fragmentation exists and some discretion must be exercised in using this base for ether formation.[30] An intramolecular version of the Williamson reaction mediated by NaDMSO leads to an oxetane in high yield.[31] The preparation of otherwise difficulty accessible xanthates has been realized using NaDMSO.[32] Very common is the modification of oligo- and polysaccharides via Williamson ether synthesis (Hokomori reaction) to facilitate handling and analysis of these compounds.[33] Other hydroxylic polymers can also be functionalized in this way.[34]

Sodium Methylsulfinylmethylide as Nucleophile.

Reactions with Esters. Perhaps one of the most useful reactions of NaDMSO is its condensation with esters to produce β-keto sulfoxides. The rich chemistry of these difunctional compounds makes accessible a wide variety of other organic compounds.[1,35] One of the most straightforward applications of NaDMSO is the synthesis of methyl ketones from esters (eq 19).[36] Thermal elimination of methylsulfenic acid after alkylation leads to α,β-unsaturated ketones (eq 20).[37] Among the wide variety of possible transformations of β-keto sulfoxides, another which stands out is the Pummerer reaction, which allows for the formation of carbon–carbon bonds via an 'umpolung' of the reactivity adjacent to the carbonyl group (eq 21).[38] A synthesis of ninhydrin was developed based on this type of chemistry.[38c] Methacrylate polymers have been modified by reaction with NaDMSO.[39]

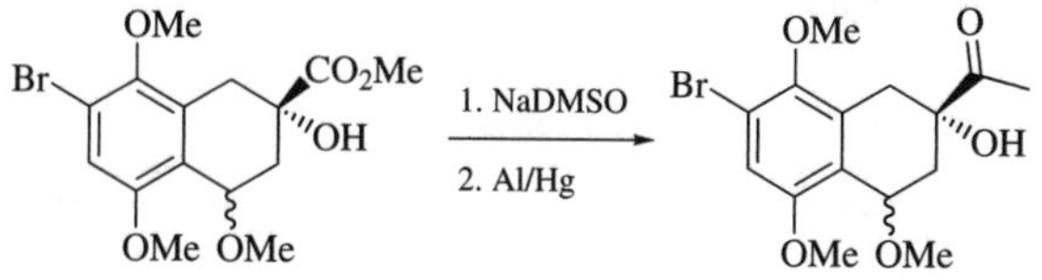

1. NaDMSO; 2. Al/Hg (19)

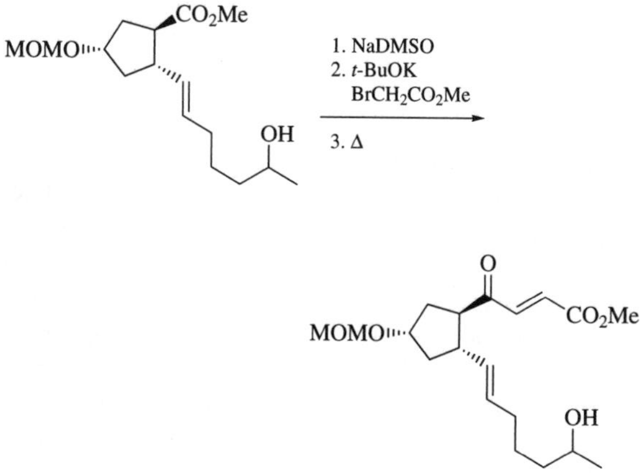

1. NaDMSO; 2. t-BuOK, $BrCH_2CO_2Me$; 3. Δ (20)

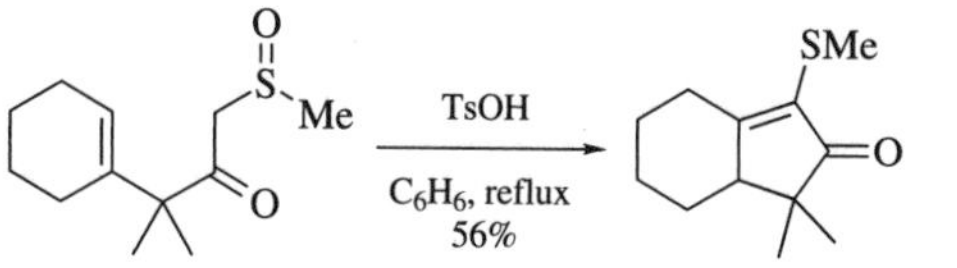

TsOH, C_6H_6, reflux, 56% (21)

Reactions with Aldehydes and Ketones. The reaction of nonenolizable ketones and aldehydes with NaDMSO generally proceeds smoothly to give β-hydroxy sulfoxides.[1] The chemistry of the latter is not as rich as that of β-keto sulfoxides, to which they can be converted. However, dehydration leads efficiently to α,β-unsaturated sulfoxides which can be transformed to the corresponding sulfides or sulfones.[40] A one-pot procedure using ***Sodium*** metal to produce NaDMSO and serve as a reductant leads to α,β-unsaturated sulfides directly.[41] Interestingly, a dianion of DMSO can be prepared from the reaction of DMSO with 2.2 equiv of $NaNH_2$. Reaction with benzophenone gives the expected adduct, albeit in only modest yield (eq 22).[3] A unique synthesis of 3-phenylindole based upon the reaction of NaDMSO with 2-aminobenzophenone has been reported (eq 23).[42]

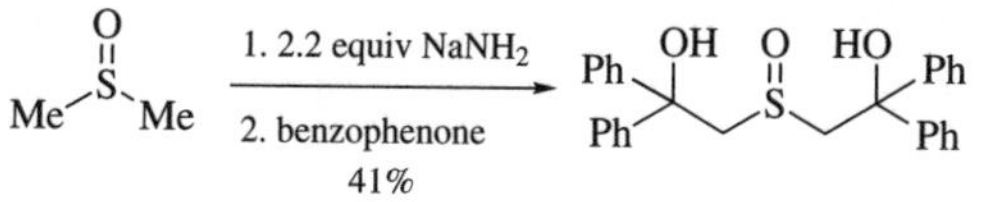

1. 2.2 equiv $NaNH_2$; 2. benzophenone, 41% (22)

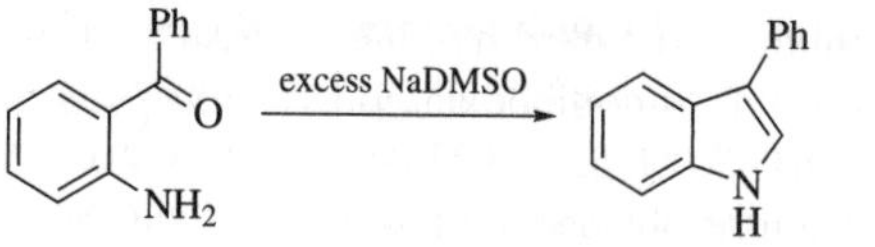

excess NaDMSO (23)

Enolizable ketones and aldehydes often give enolates upon reaction with NaDMSO as well as the expected addition products, limiting the utility of the reagent with these systems. Several unusual reactions of NaDMSO with enolizable ketones have been reported. For example, treatment of 4-heptanone with NaDMSO at elevated temperatures results in diene formation (eq 24).[43] Similarly, the reaction of cyclopentanone with NaDMSO gives a diene resulting from condensation followed by nucleophilic addition of NaDMSO and fragmentation (eq 25).[44]

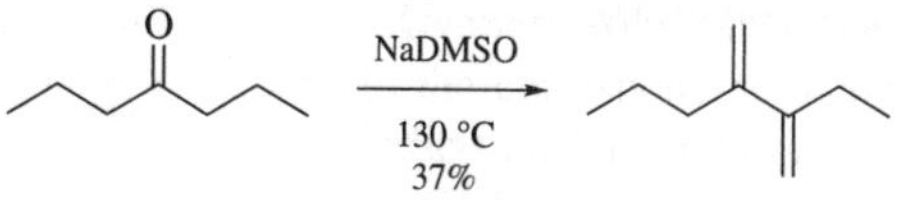

NaDMSO, 130 °C, 37% (24)

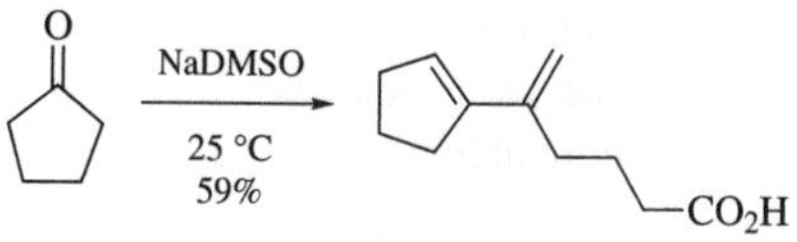

NaDMSO, 25 °C, 59% (25)

Reactions with Imines and Related Compounds. The reaction of imines with NaDMSO has not been extensively studied.[2] However, several interesting applications with heterocycles at least formally possessing imine functional groups have been reported. Treatment of 1-methylquinoline with excess NaDMSO results in a tandem 1,4–1,2 addition to give a unique β-amino sulfoxide (eq 26).[45] A mechanistically intriguing synthesis of phenanthrene is the result of the reaction of a benzoquinoline *N*-oxide with NaDMSO (eq 27).[46] Finally, a synthesis of dibenzo[*a,f*]quinolizines has been developed based on the addition of NaDMSO to an imine followed by trapping of the resulting amide with a pendant benzyne (eq 28).[47]

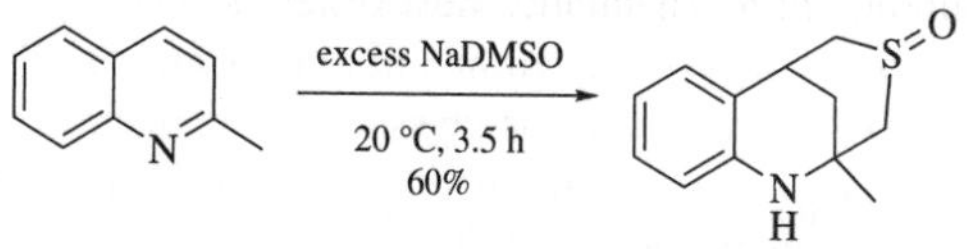

excess NaDMSO, 20 °C, 3.5 h, 60% (26)

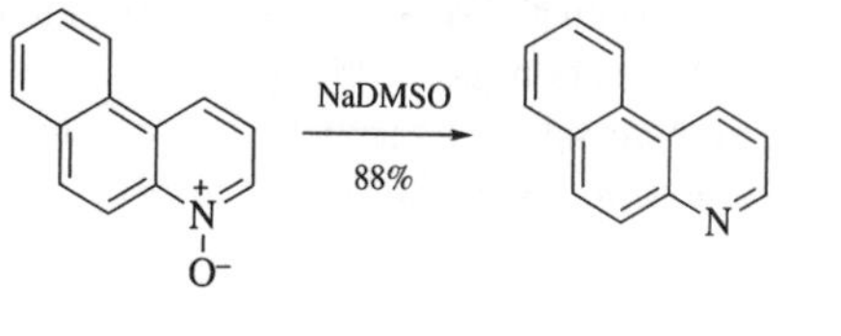

NaDMSO, 88% (27)

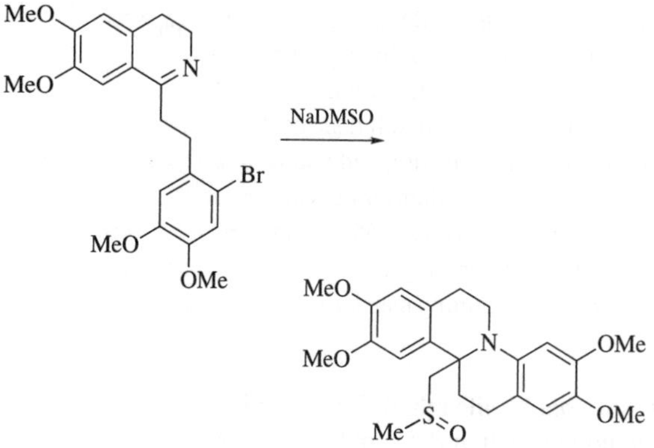

NaDMSO (28)

Reactions with Alkenes and Alkynes. The reaction of NaDMSO with alkenes or alkynes conjugated to an aryl ring or alkene is well known.[1] 1,1-Diphenylethene gives a 1:1 addition product in quantitative yield upon reaction with NaDMSO.[48] From a synthetic perspective, one of the most useful of these reactions is the alkylation, especially the methylation, of stilbenes (eq 29).[49] This occurs by initial attack of NaDMSO on the unsaturated system followed by the elimination of methanesulfenic acid and isomerization. Dienes and other polyenes are subject to the same type of chemistry, but yields are modest and isomer formation can be a problem (eq 30).[50] Reaction of NaDMSO with diphenylacetylene can lead to either simple addition products or those based on an addition–elimination sequence, depending on the reaction conditions (eq 31).[51] Addition of NaDMSO to unsaturated systems has been used in polymer synthesis.[52]

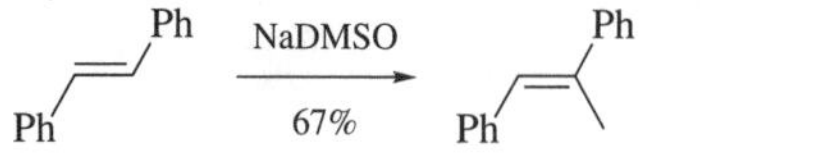

NaDMSO, 67% (29)

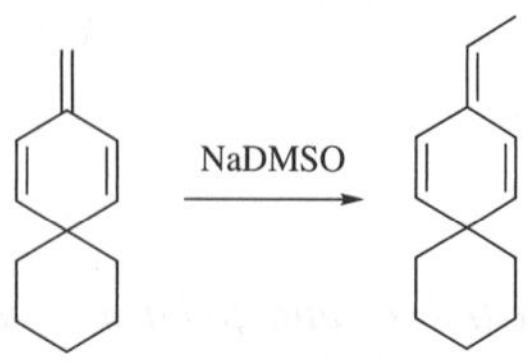

NaDMSO (30)

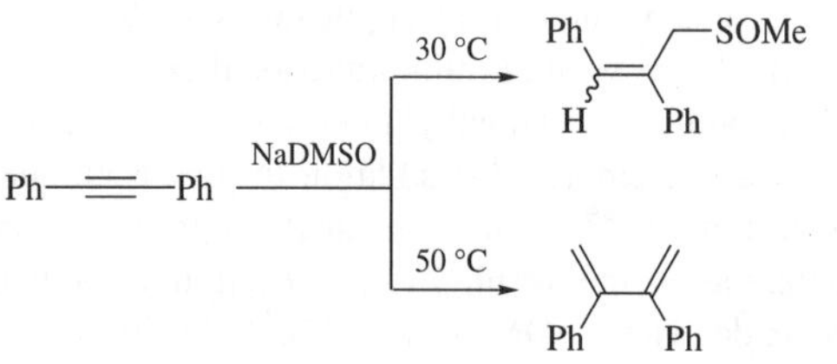

NaDMSO; 30 °C; 50 °C; SOMe (31)

Reactions with Aromatics. The reactions of NaDMSO with aromatic electrophiles can be categorized as occurring through either benzyne or addition–elimination mechanisms. Reaction of NaDMSO with chlorobenzene gives a mixture of sulfoxides, presumably via a benzyne intermediate (eq 32).[53] This chemistry has been used in the modification of polystyrenes.[54] Various fluoroaromatics react with NaDMSO to produce substitution products via addition–elimination (eq 33).[55] A variety of condensed aromatics (e.g. anthracene) undergo methylation analogous to that of stilbene upon reaction with NaDMSO (eq 34).[56]

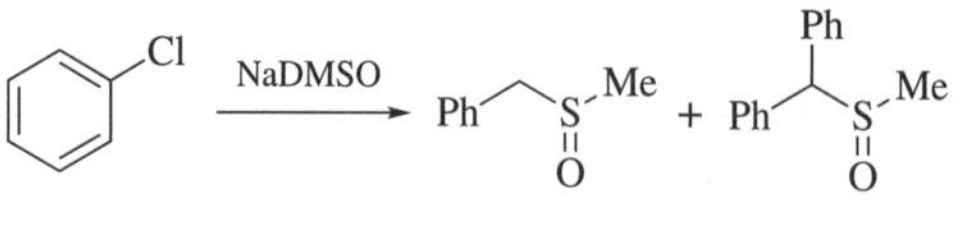

NaDMSO (32)

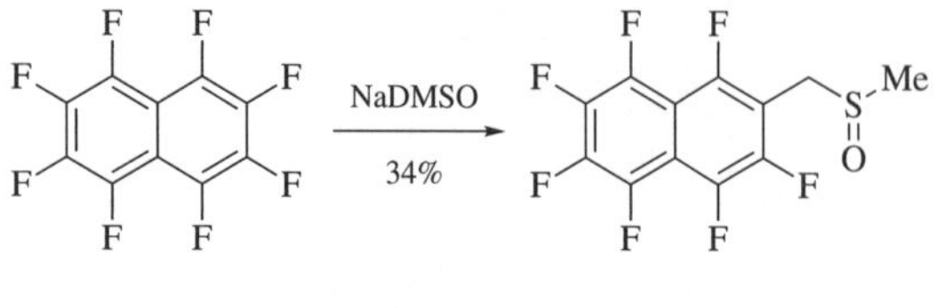

NaDMSO, 34% (33)

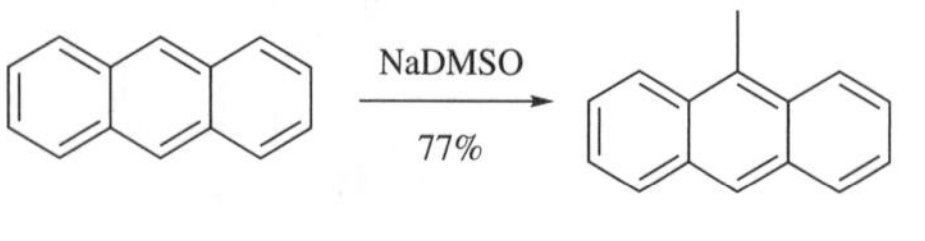

NaDMSO, 77% (34)

Reactions with Halides and Related Compounds. Alkylation of NaDMSO with primary halides and tosylates results in the formation of the expected sulfoxides.[57] However, reaction with benzyl chloride gives stilbene as the major product, suggesting that the basicity of NaDMSO must be considered even with reactive electrophiles.[2] More hindered systems favor elimination. 1,2,5,6-Tetrabromocyclooctane debrominates to 1,5-cyclooctadiene upon reaction with NaDMSO (eq 35).[58] With ***Potassium t-Butoxide*** in DMSO the major product is one of elimination, namely cyclooctatetraene. Monodebromination also occurs with *gem*-dibromocyclopropanes.[59] In a process which presumably proceeds via S_N2 substitution, reaction of imidates with NaDMSO leads to amides (eq 36).[60]

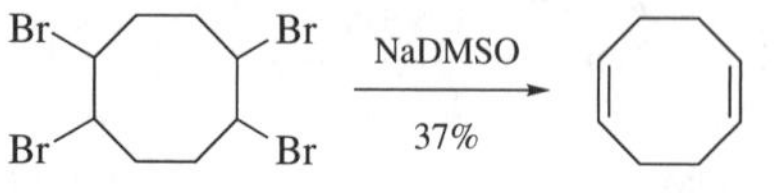

NaDMSO, 37% (35)

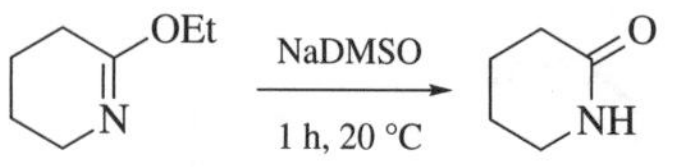

NaDMSO, 1 h, 20 °C (36)

Reactions with Epoxides. The reaction of NaDMSO with epoxides does not appear to have been widely investigated. Nevertheless, some synthetically useful transformations have been documented. Ring opening with NaDMSO followed by thermal alkene formation was developed as a route to an optically pure secondary allylic alcohol (eq 37).[61] A related transformation involves the reaction of trimethylsilyl-substituted epoxides (eq 38).[62] Ring opening at the TMS-substituted carbon, followed by desilylation and sulfenate elimination, leads to allyl alcohols in good yield in a one-pot process.

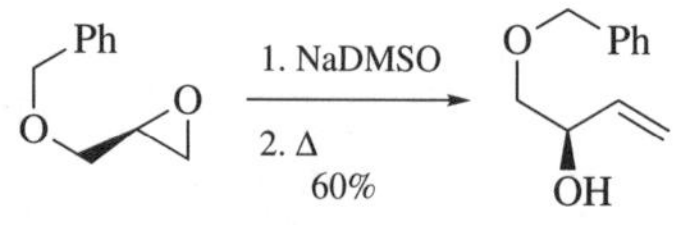

1. NaDMSO, 2. Δ, 60% (37)

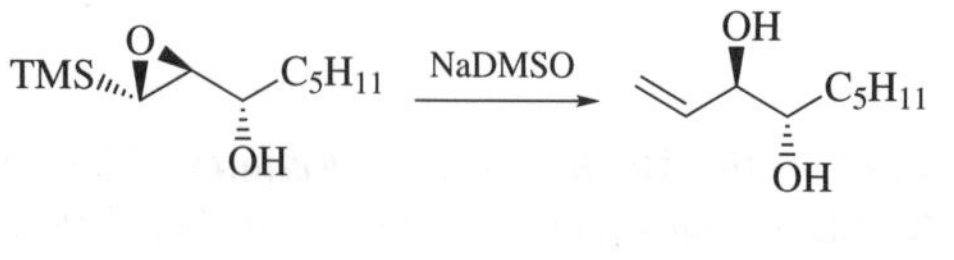

NaDMSO (38)

Reactions with Phosphorus and Sulfur Electrophiles. A one-pot synthesis of α,β-unsaturated sulfoxides begins with the reaction of NaDMSO with ***Diethyl Phosphorochloridate*** to give a Horner–Emmons reagent which reacts with aldehydes in an efficient manner (eq 39).[63] Bis-sulfoxides are prepared by the reaction of NaDMSO and a diastereomerically pure methyl sulfinate ester.[64] Some kinetic resolution is observed in this reaction. The thiophilic addition of NaDMSO to sulfines also leads to bis-sulfoxides (eq 40).[65]

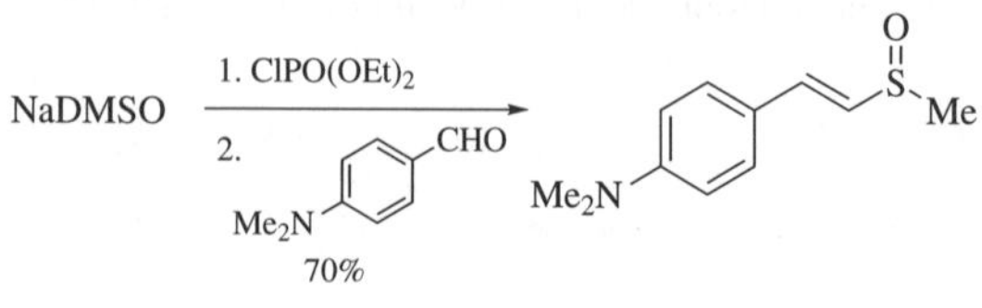

NaDMSO; 1. $ClPO(OEt)_2$; 2. Me_2N-C6H4-CHO; 70% (39)

$$(p\text{-Tol})_2C{=}S{=}O \xrightarrow[80\%]{\text{NaDMSO}} p\text{-Tol}_2CH\text{-}S(O)\text{-}CH_2\text{-}S(O)\text{-}Me \quad (40)$$

Related Reagents. Lithium Methylsulfinylmethylide; Potassium Methylsulfinylmethylide; Sodium Diisopropylamide; Sodium Hexamethyldisilazide.

---

1. (a) Durst, T. *Adv. Org. Chem.* **1969**, *6*, 285. (b) Hauthal, H. G.; Lorenz, D. In *Dimethyl Sulphoxide*; Martin, D.; Hauthal, H. G., Eds.; Wiley: New York, 1971; pp 349–374.
2. Corey, E. J.; Chaykovsky, M. *JACS* **1965**, *87*, 1345.
3. Kaiser, E. M.; Beard, R. D.; Hauser, C. R. *JOM* **1973**, *59*, 53.
4. Sjöberg, S. *TL* **1966**, 6383.
5. Price, C. C.; Yukuta, T. *JOC* **1969**, *34*, 2503.
6. (a) Leleu, J. *Cah. Notes. Doc.* **1976**, *85*, 583 (*CA* **1978**, *88*, 26 914t). (b) Itoh, M.; Morisaki, S.; Muranaga, K.; Matsunaga, T.; Tohyama, K.; Tamura, M.; Yoshida, T. *Anzen Kogaku* **1984**, *23*, 269 (*CA* **1985**, *102*, 100 117).
7. Gosney, I.; Rowley, A. G. In *Organophosphorus Reagents in Organic Synthesis*; Cadogan, J. I. G., Ed.; Academic: London, 1979; pp 17–153.
8. (a) Romo, D.; Meyers, A. I. *JOC* **1992**, *57*, 6265. (b) Trost, B. M.; Bogdanowicz, M. J. *JACS* **1973**, *95*, 5298. (c) Trost, B. M.; Bogdanowicz, M. J. *JACS* **1973**, *95*, 5321.
9. (a) Paynter, O. I.; Simmonds, D. J.; Whiting, M. C. *CC* **1982**, 1165. (b) Hall, D. R.; Beevor, P. S.; Lester, R.; Poppi, R. G.; Nesbitt, B. F. *CI(L)* **1975**, 216.
10. Deyrup, J. A.; Betkouski, M. F. *JOC* **1975**, *40*, 284.
11. (a) Corey, E. J.; Chaykovsky, M. *JACS* **1965**, *87*, 1353. (b) Trost, B. M.; Melvin, L. S., Jr. *Sulfur Ylides: Emerging Synthetic Intermediates*; Academic: New York, 1975.
12. Yurchenko, A. G.; Kyrij, A. B.; Likhotvorik, I. R.; Melnik, N. N.; Zaharh, P.; Bzhezovski, V. V.; Kushko, A. O. *S* **1991**, 393.
13. Okuma, K.; Higuchi, N.; Kaji, S.; Takeuchi, H.; Ohta, H.; Matsuyama, H.; Kamigata, N.; Kobayashi, M. *BCJ* **1990**, *63*, 3223.
14. (a) Dietrich, V. W.; Schulze, K.; Mühlstädt, M. *JPR* **1977**, *319*, 799. (b) Dietrich, W.; Schulze, K.; Mühlstädt, M. *JPR* **1977**, *319*, 667.
15. (a) Kano, S.; Yokomatsu, T.; Komiyama, E.; Tokita, S.; Takahagi, Y.; Shibuya, S. *CPB* **1975**, *23*, 1171. (b) Kano, S.; Yokomatsu, T.; Ono, T.; Takahagi, Y.; Shibuya, S. *CPB* **1977**, *25*, 2510.
16. See Ref. 1b, pp 174–197.
17. (a) Lal, B.; Gidwani, R. M.; de Souza, N. J. *JOC* **1990**, *55*, 5117. (b) Kano, S.; Komiyama, E.; Nawa, K.; Shibuya, S. *CPB* **1976**, *24*, 310. (c) Kano, S.; Yokomatsu, T.; Shibuya, S. *CPB* **1977**, *25*, 2401.
18. Klein, J.; Gurfinkel, E. *T* **1970**, *26*, 2127.
19. (a) Ong, H. H.; Profitt, J. A.; Anderson, V. B.; Kruse, H.; Wilker, J. C.; Geyer, H. M., III *JMC* **1981**, *24*, 74. (b) Kano, S.; Ogawa, T.; Yokomatsu, T.; Takahagi, Y.; Komiyama, E.; Shibuya, S. *H* **1975**, *3*, 129. (c) Birch, A. J.; Chamberlain, K. B.; Oloyede, S. S. *AJC* **1971**, *24*, 2179.
20. Wittig, G.; Hesse, A. *LA* **1975**, 1831.
21. Uno, H.; Yayama, A.; Suzuki, H. *CL* **1991**, 1165.
22. Denmark, S. E.; Harmata, M. A.; White, K. S. *JACS* **1989**, *111*, 8878.
23. (a) Kinast, G.; Tietze, L.-F. *CB* **1976**, *109*, 3626. (b) Corey, E. J.; Mitra, R. B.; Uda, H. *JACS* **1964**, *86*, 485.
24. Chakrabarti, J. K.; Hicks, T. A.; Hotten, T. M.; Tupper, D. E. *JCS(P1)* **1978**, 937.
25. Danishefsky, S.; Hatch, W. E.; Sax, M.; Abola, E.; Pletcher, J. *JACS* **1973**, *95*, 2410.
26. (a) Kelly, R. B.; Eber, J.; Hung, I.-K. *CC* **1973**, 689. (b) Corey, E. J.; Watt, D. S. *JACS* **1973**, *95*, 2303. (c) Heathcock, C. H. *JACS* **1966**, *88*, 4110.
27. (a) Morton, D. R., Jr.; Brokaw, F. C. *JOC* **1979**, *44*, 2880. (b) Scholz, D.; Burtscher, D. *LA* **1985**, 517. (c) Paquette, L. A. *OR* **1977**, *25*, 1. (d) Johnson, C. R. *Aldrichim. Acta* **1985**, *18*, 3.
28. Wade, P. A.; Dailey, W. P.; Carroll, P. J. *JACS* **1987**, *109*, 5452.
29. Sjöberg, B.; Sjöberg, K. *ACS* **1972**, *26*, 275.
30. Partington, S. M.; Watt, C. I. F. *JCS(P2)* **1988**, 983.
31. Corey, E. J.; Mitra, R. B.; Uda, H. *JACS* **1964**, *86*, 485.
32. de Groot, A.; Evanhius, B.; Wynberg, H. *JOC* **1968**, *33*, 2214.
33. Zähringer, U.; Rietschel, E. T. *Carbohydr. Res.* **1986**, *152*, 81. (b) Lee, K.-S.; Gilbert, R. D. *Carbohydr. Res.* **1981**, *88*, 162.
34. Galin, J. C. *J. Appl. Polym. Sci.* **1971**, *15*, 213.
35. For leading references, see: Ibarra, C. A.; Rogríguez, R. C.; Monreal, M. C. F.; Navarrao, F. J. G.; Tesorero, J. M. *JOC* **1989**, *54*, 5620.
36. Swenton, J. S.; Anderson, D. K.; Jackson, D. K.; Narasimhan, L. *JOC* **1981**, *46*, 4825.
37. (a) Bartlett, P. A.; Green, F. R., III *JACS* **1978**, *100*, 4858. (b) Dal Pozzo, A.; Acquasaliente, M.; Buraschi, M.; Anderson, B. M. *S* **1984**, 926.
38. (a) Isibashi, H.; Okada, M.; Komatsu, H.; Ikeda, M. *S* **1985**, 643. (b) Oikawa, Y.; Yonemitsu, O. *JOC* **1976**, *41*, 1118. (c) Becker, H.-D.; Russell, G. A. *JOC* **1963**, *28*, 1896. (d) De Lucchi, O.; Miotti, U.; Modena, G. *OR* **1991**, *40*, 157.
39. (a) Katsutoshi, N.; Harada, A.; Oyamada, M. *NKK* **1983**, 713. (b) Arranz, F.; Galin, J. C. *Makromol. Chem.* **1972**, *152*, 185.
40. (a) Fillion, H.; Boucherle, A. *BSF* **1971**, 3674. (b) Fillion, H.; Duc, C. L.; Agnius-Delord, C. *BSF* **1974**, 2923.
41. Kojima, T.; Fujisawa, T. *CL* **1978**, 1425.
42. Bravo, P.; Gavdiano, G.; Ponti, P. P. *CI(L)* **1971**, 253.
43. Yurchenko, A. G.; Kirii, A. V.; Mel'nik, N. N.; Likhotvorik, I. R. *ZOR* **1990**, *26*, 2230.
44. Comer, W. T.; Temple, D. L. *JOC* **1973**, *38*, 2121.
45. Kato, H.; Takeuchi, I.; Hamada, Y.; Ono, M.; Hirota, M. *TL* **1978**, 135.
46. Hamada, Y.; Takeuchi, I. *JOC* **1977**, *42*, 4209.
47. (a) Kano, S.; Yokomatsu, T.; Shibuya, S. *CPB* **1975**, *23*, 1098. (b) Kano, S.; Yokomatsu, T. *TL* **1978**, 1209.
48. Walling, C.; Bollyky, L. *JOC* **1964**, *29*, 2699.
49. (a) James, B. G.; Pattenden, G. *CC* **1973**, 145. (b) Feldman, M.; Danishefsky, S.; Levine, R. *JOC* **1966**, *31*, 4322.
50. Murray, D. F. *JOC* **1983**, *48*, 4860.
51. (a) Iwai, I.; Ide, J. *CPB* **1965**, *13*, 663. (b) Iwai, I.; Ide, J. *OS* **1970**, *50*, 62.
52. (a) Priola, A.; Trossarelli, L. *Makromol. Chem.* **1970**, *139*, 281. (b) Kriz, J.; Benes, M. J.; Peska, J. *CCC* **1967**, *32*, 4043.
53. Corey, E. J.; Chaykovsky, M. *JACS* **1962**, *84*, 866.
54. Janout, M.; Kahovec, J.; Hrudkova, H.; Svec, F.; Cefelin, P. *Polym. Bull. (Berlin)* **1984**, *11*, 215.
55. (a) Brooke, G. M.; Ferguson, J. A. K. J. *JFC* **1988**, *41*, 263. (b) Brooke, G. M.; Mawson, S. D. *JCS(P1)* **1990**, 1919.
56. Nozaki, H.; Yamamoto, Y. Noyori, R. *TL* **1966**, 1123.
57. Entwistle, I. D.; Johnstone, R. A. W. *CC* **1965**, 29.
58. Cardenas, C. G.; Khafaji, A. N.; Osborn, C. L.; Gardner, P. D. *CI(L)* **1965**, 345.
59. Osborn, C. L.; Shields, T. C.; Shoulders, B. A.; Cardenas, C. G.; Gardner, P. D. *CI(L)* **1965**, 766.
60. Kano, S.; Yokomatsu, T.; Hibino, S.; Imamura, K.; Shibuya, S. *H* **1977**, *6*, 1319.
61. Takano, S.; Tomita, S.; Iwabuchi, Y.; Ogasawara, R. *S* **1988**, 610.
62. Kobayashi, Y.; Ito, Y. I.; Urabe, H.; Sato, F. *SL* **1991**, 813.
63. Almog, J.; Weissman, B. A. *S* **1973**, 164.

64. Kunieda, N.; Nokami, J.; Kinoshita, M. *BCJ* **1976**, *49*, 256.
65. Loontjes, J. A.; van der Leij, M.; Zwanenberg, B. *RTC* **1980**, *99*, 39.

Michael Harmata
University of Missouri-Columbia, MO, USA

(–)-Sparteine[1]

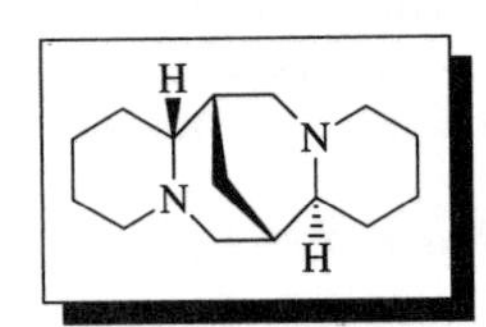

[90-39-1]	$C_{15}H_{26}N_2$	(MW 234.43)
(sulfate pentahydrate)		
[6160-12-9]	$C_{15}H_{38}N_2O_9S$	(MW 422.62)

(reagent for chiral modification of organo-lithium, -magnesium, and -zinc reagents[1,2])

Alternate Name: [(7*S*)-(7α,7aα,14α,14aβ)]-dodecahydro-7,14-methano-2*H*,6*H*-dipyrido[1,2-*a*:1′,2′-*e*][1,5]diazocine.
Physical Data: bp 137–138 °C/1 mm Hg; *d* 1.02 g cm^{-3}; $[\alpha]_D^{20}$ −17.5° (*c* = 2, EtOH). X-Ray structures of several complexes of metal salts,[3] alkyllithium derivatives,[4] and of allylpalladium[5] and studies on the conformation in solution[6] and a NMR study on the structure of the 2-propyllithium–ether–(–)-sparteine complex[7] have been reported.
Solubility: 0.3 g/100 ml H_2O at 20 °C; sol ether, hexane.
Form Supplied in: free base: colorless viscous fluid.
Handling, Storage, and Precautions: highly toxic in the digestive tract. Keep in refrigerator at 0 °C. Moderately hygroscopic; dehydration by drying an ethereal solution over ***Calcium Hydride***. Is easily recovered by extraction of alkaline aqueous solutions. Use in a fume hood.

Chiral Modification of Achiral Organometallic Reagents. The addition of ***n-Butyllithium*** or ***Ethylmagnesium Bromide*** to aldehydes or ketones in the presence of (–)- sparteine resulted in the formation of optically active secondary or tertiary alcohols with 20% ee or lower.[8] Optically active acyl sulfoxides (≤15% ee) were obtained by acylation of ***p-Tolylsulfinylmethyllithium***.[9] The asymmetric Reformatsky reaction of ethyl bromoacetate with benzaldehyde proceeds with 95% ee,[10] in an exceptional case (eq 1).[11]

Br‿CO₂Et + PhCHO —[Zn, (MeO)₂CH₂, (–)-sparteine; 40%]→ PhCH(OH)CH₂CO₂Et (1)
95% ee

Equilibration of Configurationally Labile Organolithium Reagents. The equilibration of diastereomeric pairs of alkyllithium–(–)-sparteine complexes and trapping by achiral electrophiles gives enantioenriched products. Examples are α-(*N*,*N*-diisopropylcarbamoyloxy)benzyllithium in ether,[12] not in THF,[13] 1-phenylethyllithium,[8a] and the dilithium salt of *N*-methyl-3-phenylpropanoic acid amide (eq 2).[14]

Ph‿C(O)NHMe —[2 equiv *s*-BuLi, (–)-sparteine]→ [Ph*‿C(OLi)=NMe] Li⁺•(–)-sparteine —[1. BuI; 2. H_2O]→ PhCH(Bu)CH₂C(O)NHMe (2)
77%, 88% ee

The deprotonation[15] of (*E*)-2-butenyl *N*,*N*-diisopropylcarbamate leads to (1*S*,2*E*)-1-(*N*,*N*-diisopropylcarbamoyloxy)-2-butenyllithium–(–)-sparteine[16] with ≥90% de after crystallization, combined with a second-order asymmetric transformation (eq 3).[4d] It has been applied in the enantioselective synthesis of γ-lactones,[16] such as (+)-eldanolide (eq 3),[17] dihydroavermectin B_{1b},[18] and doubly branched sugar analogs.[19]

s-BuLi, (–)-sparteine, pentane, cyclohexane, −70 °C; crystallization; 3 equiv Ti(O-*i*-Pr)₄; Ti(O-*i*-Pr)₃, OCb; 3 steps (3)
62%, >97% ds, 92% ee

Generation of Enantioenriched, Configurationally Stable Organolithium Reagents.[15,20] (1*S*,2*E*)-1-(*N*,*N*-Diisopropylcarbamoyloxy)-1-methyl-2-butenyllithium–(–)-sparteine is configurationally stable in solution and is obtained by kinetic resolution of the racemic 2-alkenyl carbamate by *n*-butyllithium–(–)-sparteine with ≥80% de (eq 4).[21] The enantioenriched allylstannane, obtained on γ-stannylation, was used as chiral homoenolate reagent.[21a] The methoxycarbonylation (α, inversion) yields enantioenriched 3-alkenoates.[21b]

Alkyl carbamates, derived from 2,2,4,4-tetramethyl-1,3-oxazolidine (R–CH₂–OCby), are deprotonated by ***s-Butyllithium***–(–)-sparteine with differentiation between the enantiotopic protons (eq 5).[22,20] The pro-*S* proton is removed with high stereoselectivity and reliability, and, subsequently, stereospecifically substituted by electrophiles with stereoretention to give enantiomerically enriched secondary alcohols (≥95% ee) after deprotection.[22b]

BuLi, (–)-sparteine / pentane

CbO Li•(–)-sparteine + OCb (4)

41%, 80% ee

Bu_3SnCl

Bu_3Sn OCb + Bu_3Sn OCb

48%, >80% ee, 87:13

s-BuLi, (–)-sparteine / Et_2O, –78 °C

1. CO_2 2. CH_2N_2

OCby

1. $MeSO_3H$, MeOH 2. $Ba(OH)_2$, MeOH (5)

70–90%, >95% ee

R = alkyl, Bn_2NCH_2, $Bn_2N(CH_2)_2$, $CbyOCH_2CH_2$

The ee values in the enantioselective deprotonation are independent of the size of the attached alkyl residue. The method tolerates several substituents, e.g. 2-[23] or 3-dibenzylamino,[24] 3- or 4-(*N,N*-dialkylcarbamoyloxy),[25] or 4-TBDMSO.[25a] Essentially enantiopure 2-hydroxy acids,[22a] β-amino alkanols,[24] γ-amino alkanols,[23] cyclopropyl carbamates,[25a] and 2-hydroxy-4-butanolides[25a] were obtained. Extraordinary high (>70) kinetic H/D isotope effects were observed in the deprotonation of chiral 1-deuteroalkyl carbamates.[26] Kinetic resolution of racemic alkyl carbamates was achieved.[27]

N-Boc-pyrrolidines are similarly deprotonated and furnish enantioenriched 2-substituted pyrrolidines (eq 6).[28]

1. s-BuLi, (–)-sparteine 2. TMSCl 65% (6)

CO_2-*t*-Bu → TMS, CO_2-*t*-Bu, 95% ee

Further Applications. Chiral 1,1-diaryl-2-propynols are resolved by mutual crystallization with (−)-sparteine.[29] Low ee values were achieved in Pd-mediated alkylations.[30] Numerous attempts at enantioselective, alkyllithium-catalyzed polymerizations of alkenes in the presence of (−)-sparteine have been reported.[31]

Related Reagents. (+)-Sparteine (pachycarpine[32]) is best prepared by resolution of (±)-sparteine, obtained from *rac*-lupanine[33] or by total synthesis[34] with (−)-10-camphorsulfonic acid.[35] (*S*)-α-Methylbenzylamine; (*S*)-Proline; Quinine; *N,N,N′,N′*-Tetramethylethylenediamine.

---

1. Boczon, W. *H* **1992**, *33*, 1101.
2. Review: Tomioka, K. *S* **1990**, 541.
3. For leading references see: Review: Kuroda, R.; Mason, S. F. *JCS(D)* **1977**, 371.
4. (a) Engelhardt, L. M.; Leung, W.-P.; Raston, C. L.; Salem, G.; Twiss, P.; White, A. H. *JCS(D)* **1988**, 2403. (b) Byrne, L. T.; Engelhardt, L. M.; Jacobsen, G. E.; Leung, W.-P.; Papasergio, R. I.; Raston, C. L.; Skelton, B. W.; Twiss, P.; White, A. H. *JCS(D)* **1989**, 105. (c) Marsch, M.; Harms, K.; Zschage, O.; Hoppe, D.; Boche, G. *AG* **1991**, *103*, 338; *AG(E)* **1991**, *30*, 321. (d) Ledig, B.; Marsch, M.; Harms, K.; Boche, G. *AG* **1992**, *104*, 80; *AG(E)* **1992**, *31*, 79.
5. Togni, A.; Rihs, G.; Pregosin, P. S.; Ammann, C. *HCA* **1990**, *73*, 723.
6. (a) Bohlmann, F.; Schumann, D.; Arndt, C. *TL* **1965**, 2705. (b) Wiewiorowski, M.; Edwards, O. E.; Bratek-Wiewiorowska, M. D. *CJC* **1967**, *45*, 1447.
7. Gallagher, D. J.; Kerrick, S. T.; Beak, P. *JACS* **1992**, *114*, 5872.
8. (a) Nozaki, H.; Aratani, T.; Noyori, R. *T* **1971**, *27*, 905. (b) Nozaki, H.; Aratani, T.; Toraya, T. *TL* **1968**, 4097. (c) Aratani, T.; Gonda, T.; Nozaki, H. *T* **1970**, *26*, 5453.
9. Kunieda, N.; Kinoshita, M. *PS* **1981**, *10*, 383.
10. Guetté, M.; Capillon, J.; Guetté, J.-P. *T* **1973**, *29*, 3659.
11. Hansen, M. M.; Bartlett, P. A.; Heathcock, C. H. *OM* **1987**, *6*, 2069.
12. Hoppe, D.; Retzow, S. unpublished.
13. Zhang, P.; Gawley, R. E. *JOC* **1993**, *58*, 3223.
14. Beak, P.; Du, H. *JACS* **1993**, *115*, 2516.
15. Reviews: (a) Hoppe, D.; Krämer, T.; Schwark, J.-R.; Zschage, O. *PAC* **1990**, *62*, 1999. (b) Kunz, H.; Waldmann, H. *Chemtracts Org. Chem.* **1990**, *3*, 421.
16. (a) Zschage, O.; Hoppe, D. *T* **1992**, *48*, 5657. (b) Hoppe, D.; Zschage, O. *AG* **1989**, *101*, 67; *AG(E)* **1989**, *28*, 69.
17. Paulsen, H.; Hoppe, D. *T* **1992**, *48*, 5667.
18. Férézou, J. P.; Julia, M.; Khourzom, R.; Pancrazi, A.; Robert, P. *SL* **1991**, 611.
19. Peschke, B.; Lüssmann, J.; Dyrbusch, M.; Hoppe, D. *CB* **1992**, *125*, 1421.
20. Review: Knochel, P. *AG* **1992**, *104*, 1486; *AG(E)* **1992**, *31*, 1459.
21. (a) Zschage, O.; Schwark, J.-R.; Krämer, T.; Hoppe, D. *T* **1992**, *48*, 8377. (b) Zschage, O.; Hoppe, D. *T* **1992**, *48*, 8389. (c) Zschage, O.; Schwark, J.-R.; Hoppe, D. *AG* **1990**, *102*, 336; *AG(E)* **1990**, 29, 296.
22. (a) Hoppe, D.; Hintze, F.; Tebben, P. *AG* **1990**, *102*, 1457; *AG(E)* **1990**, *29*, 1422. (b) Hintze, F.; Hoppe, D. *S* **1992**, 1216.
23. Schwerdtfeger, J.; Hoppe, D. *AG* **1992**, *104*, 1547; *AG(E)* **1992**, *31*, 1505.
24. Sommerfeld, P.; Hoppe, D. *SL* **1992**, 764.
25. (a) Paetow, M.; Ahrens, H.; Hoppe, D. *TL* **1992**, *33*, 5323. (b) Ahrens, H.; Paetow, M.; Hoppe, D. *TL* **1992**, *33*, 5327.
26. Hoppe, D.; Paetow, M.; Hintze, F. *AG* **1993**, *105*, 430; *AG(E)* **1993**, *32*, 394.
27. Haller, J.; Hense, T.; Hoppe, D. *SL* **1993**, 726.
28. Kerrick, S. T.; Beak, P. *JACS* **1991**, *113*, 9708.
29. (a) Toda, F.; Tanaka, K.; Ueda, H.; Oshima, T. *CC* **1983**, 743. (b) Toda, F.; Tanaka, K.; Ueda, H.; Oshima, T. *Isr. J. Chem.* **1985**, *25*, 338.
30. Trost, B. M.; Dietsche, T. J. *JACS* **1973**, *95*, 8200.
31. For leading references see: Nakano, T.; Okamoto, Y.; Hatada, K. *JACS* **1992**, *114*, 1318.
32. Orechoff, A.; Rabinowitch, M.; Konowalowa, R. *CB* **1933**, *66*, 621.

33. Clemo, G. R.; Raper, R.; Short, W. S. *JCS* **1949**, 663.
34. van Tamelen, E. E.; Foltz, R. L. *JACS* **1960**, *82*, 1960.
35. Ebner, T.; Eichelbaum, M.; Fischer, P.; Meese, C. O. *AP* **1989**, *322*, 399.

Dieter Hoppe
University of Münster, Germany

Sulfur Dioxide

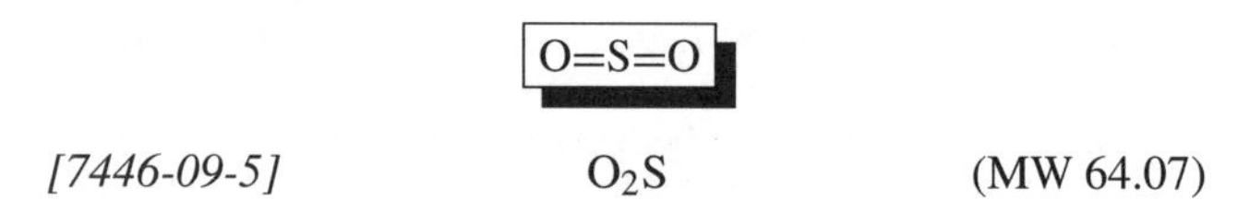

[7446-09-5] O_2S (MW 64.07)

(reacts with 1,3-dienes in reversible cheletropic addition;[2,3] SO_2 extrusion from 2,5-dihydrothiophene 1,1-dioxides generates 1,3-dienes,[9,10] *o*-quinodimethanes;[4,5] inserts into carbon–metal bonds;[14–16] reacts with ylides to give heterocycles;[17,181] electrophilic promoter of Mannich reactions;[20,21] catalyst for isomerization of alkenes,[24] dienes,[22,23] and nitrile oxides;[25] involved in sulfonic acid[26] and sulfonyl chloride synthesis;[27] reacts with Wittig reagents or α-silyl carbanions to give sulfines;[28,29] reacts with polyenes,[30,31] cyclopropanes;[32] reduces pyridine *N*-oxides[36] and ozonides;[37] solvent for superacid-promoted reactions[41])

Physical Data: [1]mp −75.5 °C; bp −10.0 °C; *d* (liquid) 1.434 g cm^{-3} at 0 °C; *d* (vapor, 0 °C) 2.264 g cm^{-3}.

Solubility: sol H_2O, alcohol, acetic acid, sulfuric acid, benzene, acetonitrile.

Form Supplied in: colorless liquid or pungent, nonflammable gas; supplied in small cylinders; widely available.

Handling, Storage, and Precautions: corrosive irritant to respiratory system, skin, eyes; mildly toxic by inhalation; teratogenic; reacts violently with numerous substances; use only in well-ventilated fume hood.[1]

Reversible Cheletropic Addition of SO_2 to 1,3-Dienes and Analogs. Much of the chemistry of SO_2 involves its reversible cheletropic addition to 1,3-dienes to give 2,5-dihydrothiophene 1,1-dioxides (sulfolenes) (eq 1).[2] Much less commonly observed is the alternative, hetero-Diels–Alder cycloaddition mode, also reversible, leading to the sultine (sulfinic ester) structure.[3] At low temperature this latter mode can be kinetically favored with reactive 1,3-dienes, but the cheletropic addition is generally favored thermodynamically.[3]

(1)

Trapping and regeneration of reactive *o*-quinodimethanes via cheletropic addition and extrusion of SO_2 has proven useful in heterocyclic (eqs 2 and 3)[4] and carbocyclic systems (eq 4).[5] Thermal extrusion of SO_2 from the cyclic sulfones yielded the reactive *o*-quinodimethanes, which were trapped with various dienophiles.

(2)

(3)

(4)

Linear cheletropic addition of SO_2 to conjugated diallenes affords 2,5-bis(alkylidene)-2,5-dihydrothiophene 1,1-dioxides via the disrotatory process shown (eq 5).[6]

(5)

Similarly, addition of SO_2 to 3-methyl-1,2,4-pentatrienyl-1-phosphonate (eq 6)[7] or 5-methyl-1,3,4-hexatrienyl-3-phosphonate esters (eq 7)[8] affords the corresponding phosphorylated alkylidene-2,5-dihydrothiophene 1,1-dioxides.

(6)

(7)

Chemical separation of a mixture of 1,3-diene geometric isomers was accomplished by selective cheletropic addition of SO_2 to the *trans* diastereomer (eq 8).[9] Chromatographic purification of the dihydrothiophene and thermolysis cleanly afforded *trans*-9,11-dodecadien-1-yl acetate, a sex pheromone of the female red bollworm moth.

An elaboration of this tactic is shown in eq 9 for the synthesis of (*E*)-1-substituted 1,3-dienes and (*E,E*)-1,4-disubstituted 1,3-dienes.[10] The tricyclic sulfone (**1**) can be metalated and alkylated once or twice, giving (**2**) or (**3**), respectively. Thermolysis in the gas phase at 650 °C resulted in sequential loss of cyclopentadiene and SO_2 to provide the dienes in 91–98% isomeric purity.

Acyclic Sulfone Synthesis. Addition of Grignard and organolithium reagents to SO_2 is known to produce salts of sulfinic acids.[11] An example extending this to the enolate derived from camphor is shown in eq 10.[12] The intermediate β-keto sulfinate was alkylated to afford the β-keto sulfone shown in modest yield.

A high yield synthesis of diphenyl sulfone results from the phenylation of SO_2 with pentaphenylbismuth (eq 11),[13a] which proceeds more readily than the corresponding reaction with ***Triphenylbismuthine***.[13b]

$$Ph_5Bi + SO_2 \xrightarrow{95\%} PhSO_2Ph \qquad (11)$$

Insertion of SO_2 into Carbon–Metal Bonds. Sulfur dioxide inserts into cyclohex-2-enylstannanes in chloroform via a *syn* S_E' process (eqs 12 and 13).[14] The γ-*syn* stereospecificity is lost upon changing the solvent to methanol.

The bis(η^1,η^3-allyl)palladium phosphine complex in eq 14 inserts SO_2 to provide the *S*-sulfinate derivative, which undergoes reductive elimination to a 1:1 mixture of diallyl and allyl 2-propenyl sulfone upon treatment with ***Carbon Monoxide***.[15]

Triethylaluminum etherate reacts with SO_2 to afford the aluminum salt of ethylsulfinic acid (eq 15).[16]

$$Et_3Al{\bullet}OEt_2 + SO_2 \longrightarrow (EtSO_2)_3Al \qquad (15)$$

Heterocycle Synthesis. Combination of a dihaloketene, a Schiff base and SO_2 to give 5,5-dihalo-4-oxo-1,3-thiazolidine 1,1-dioxides is exemplified in eq 16.[17] These products can be dehalogenated or heated to extrude SO_2, yielding β-lactams.

X = Cl, 81%; Br, 67%

Generation of a thiocarbonyl ylide in liquid SO_2 from a Δ^3-1,3,4-thiadiazoline results in a regioselective 1,3-dipolar cycloaddition, giving 1,2,4-oxadithiolane 2-oxides (eq 17).[18]

Reaction of the functionalized formamidine in eq 18 with SO_2 in the presence of ***Triethylamine*** affords the 1,2,4-oxathiazole 2-oxide shown.[19]

Use of SO_2 in Mannich Reactions. Treatment of 2,5-dimethylphenol with SO_2 apparently allows the formation of a half-sulfite ester (eq 19). Addition of an aminol ether affords selective *o*-aminoalkylation.[20] Without SO_2, the major pathway of Mannich reaction with 2,5-dimethylphenol leads to the *p*-aminoalkylation product.

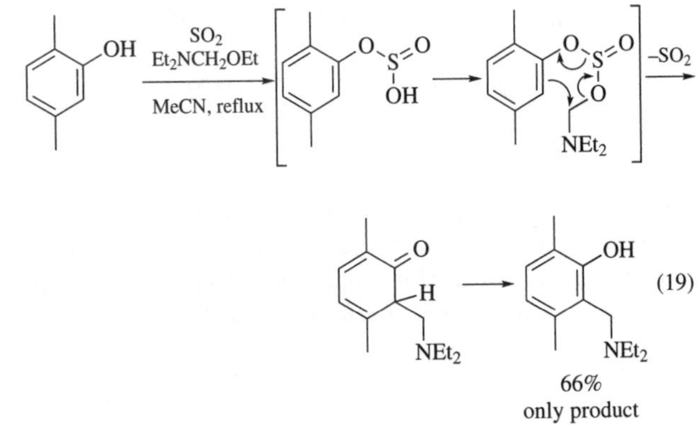

SO_2, Et_2NCH_2OEt, MeCN, reflux; $-SO_2$

66% only product (19)

Sulfur dioxide serves as an electrophilic reagent for activating bis(dialkylamino)methanes or alkoxy(dialkylamino)methanes for Mannich reactions with 2-methylfuran (eq 20), *N*-methylpyrrole (eq 21), and indole (eq 22).[21] Dipolar adducts of SO_2 and the bis(amino)methane or alkoxyaminomethane reagents are proposed as intermediates, rather than free iminium salts.

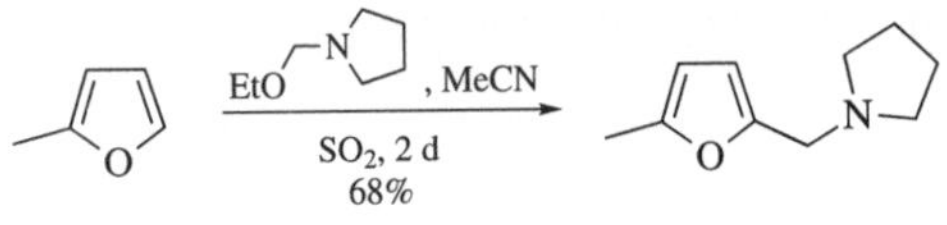

EtO, MeCN; SO_2, 2 d; 68% (20)

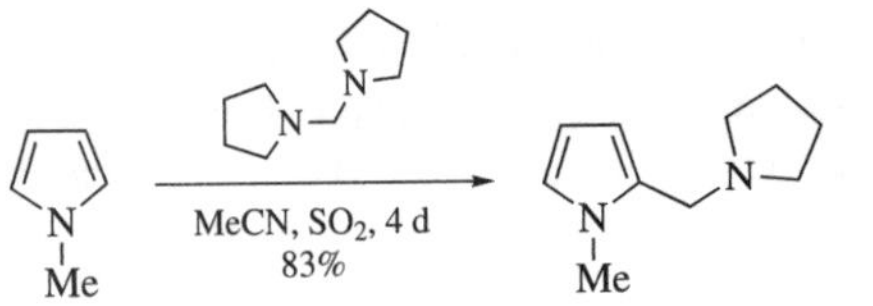

MeCN, SO_2, 4 d; 83% (21)

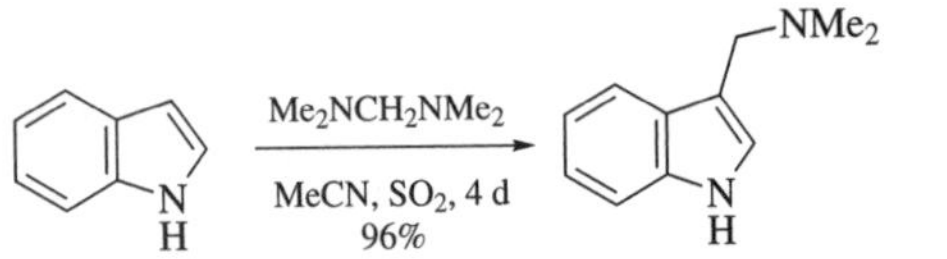

$Me_2NCH_2NMe_2$; MeCN, SO_2, 4 d; 96% (22)

Isomerization of Dienes, Alkenes, and Nitrile Oxides by SO_2. Treatment of ergosterol acetate with SO_2 in pyridine causes rearrangement of the 5,7-diene to the 6,8(14)-diene isomer (eq 23).[22] This result is fundamentally different from related sterol diene rearrangements catalyzed by acid. A consistent rationale for this SO_2-induced isomerization is shown.

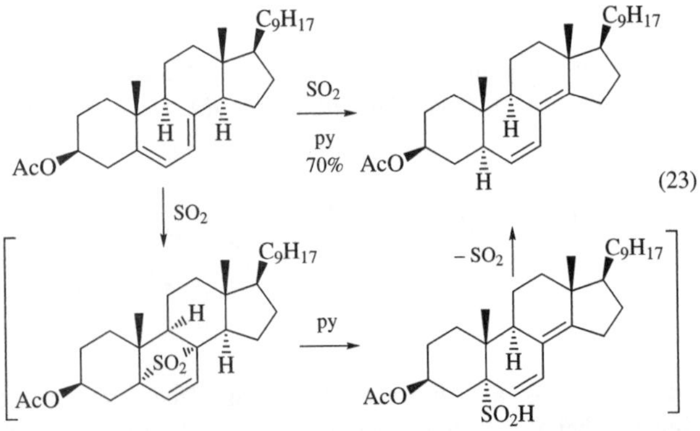

SO_2, py, 70% (23); SO_2; py; $-SO_2$

Addition of SO_2 to (5*Z*)- and (5*E*)-vitamin D_3 (**4**) yields the sulfolene derivatives (**5**) (eq 24).[23] Thermal extrusion of SO_2 from (**5**) affords isotachysterol$_3$ (**6**), alone or mixed with isovitamin D_3 (**7**). Treatment of sulfone (**5**) with methanolic ***Potassium Hydroxide*** or ***Alumina*** results in the formation of (5*E*)-vitamin D_3.

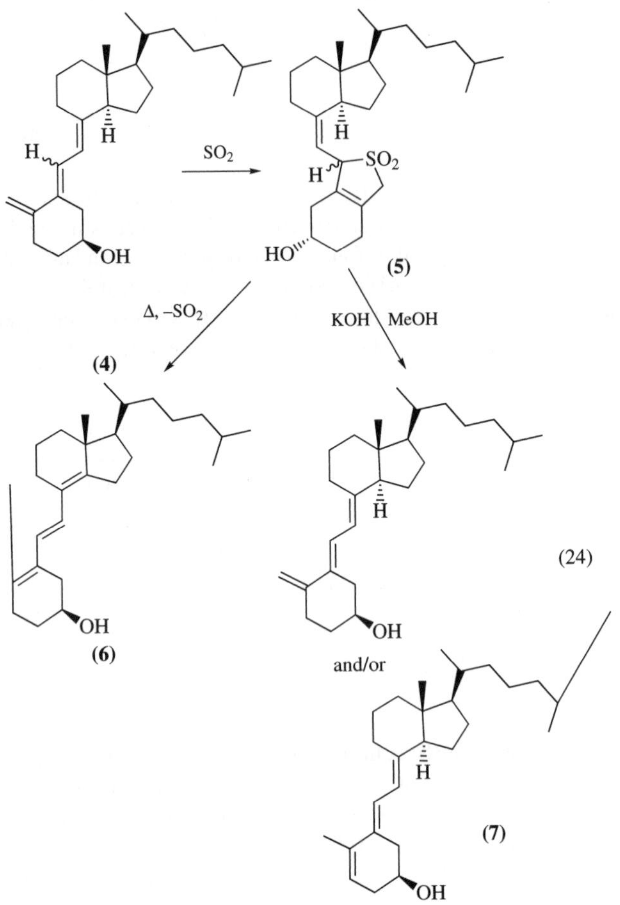

(**4**) SO_2 → (**5**); Δ, $-SO_2$ → (**6**) and/or (**7**); KOH MeOH (24)

Regiospecific alkene isomerization has been effected by SO_2 with a variety of acyclic and cyclic alkenes, including β-pinene (eq 25) and methylenecyclohexane (eq 26).[24a] The sequence of steps presented in eq 26 has been advanced to rationalize the process. Interestingly, if the reaction is attempted in the presence of D_2O, alkene migration is completely suppressed, and all allylic hydrogens in the substrate alkene are exchanged for deuterium.[24b]

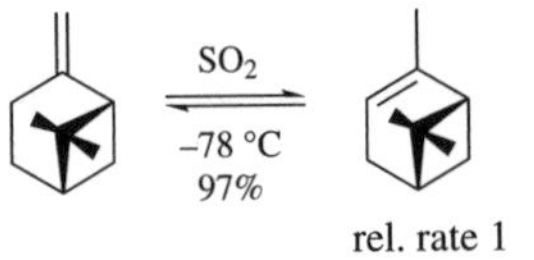

SO_2, $-78\ °C$, 97%; rel. rate 1 (25)

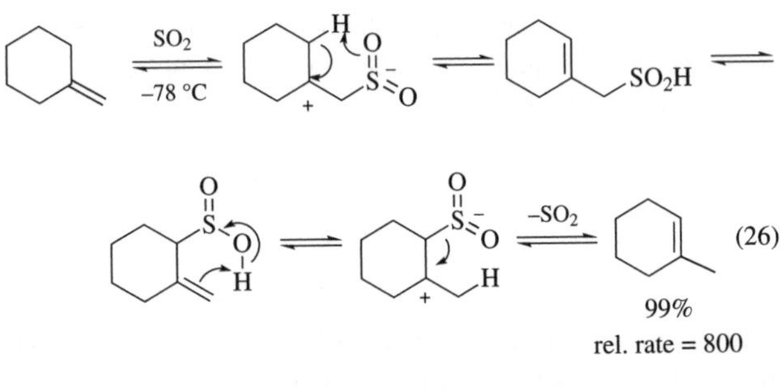

SO_2, $-78\ °C$; SO_2H; $-SO_2$; 99%; rel. rate = 800 (26)

SO_2 catalyzes the conversion of nitrile oxides to isocyanates via a dipolar cycloaddition followed by regioisomeric loss of SO_2 and rearrangement (eq 27).[25] The intermediate 1,3,2,4-dioxathiazole 2-oxides can be isolated, or converted to hydroxamic acids upon hydrolysis.

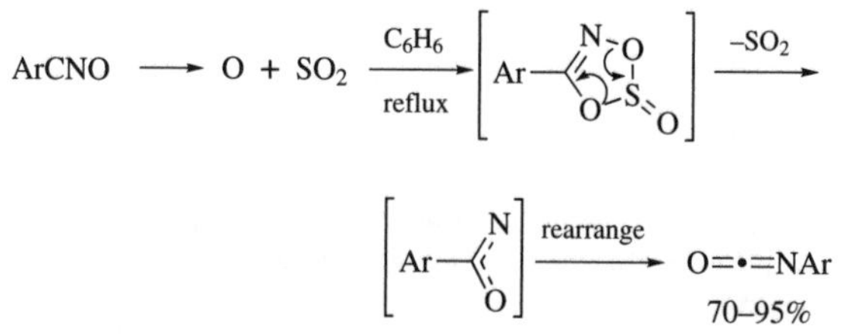

$$ArCNO \longrightarrow O + SO_2 \xrightarrow[\text{reflux}]{C_6H_6} [\ldots] \xrightarrow{-SO_2} [\ldots] \xrightarrow{\text{rearrange}} O{=}\bullet{=}NAr \quad (27)$$

70–95%

Sulfonic Acid and Sulfonyl Chloride Synthesis. Condensation of benzimidazole with SO_2 and an aldehyde or ketone results in the formation of benzimidazoylalkylsulfonic acids (eq 28).[26]

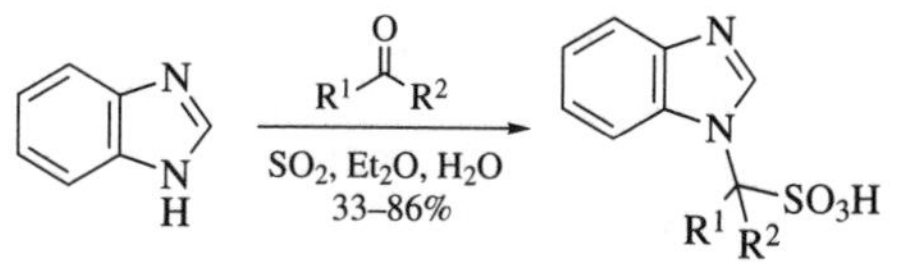

$$\xrightarrow[SO_2,\ Et_2O,\ H_2O\ \ 33\text{–}86\%]{R^1COR^2} \quad (28)$$

A wide variety of substituted anilines, especially those with electron-withdrawing substituents, can be converted to substituted arenesulfonyl chlorides via their diazonium salts (eq 29).[27]

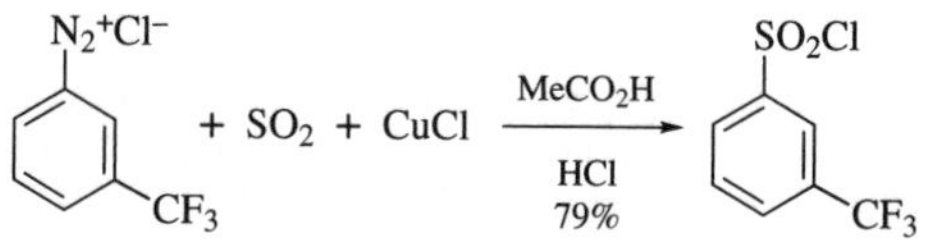

$$N_2^+Cl^- \ (CF_3) + SO_2 + CuCl \xrightarrow[HCl\ \ 79\%]{MeCO_2H} SO_2Cl\ (CF_3) \quad (29)$$

Synthesis of Sulfines. Reaction of SO_2 with fluorenylidenetriphenylphosphorane in benzene gave the corresponding sulfine (eq 30).[28] In a related fashion, reaction of SO_2 with an α-silyl carbanion leads to the sulfine product (eq 31).[29] Thus alkylidenation of SO_2 is possible with Wittig or Peterson reagents.

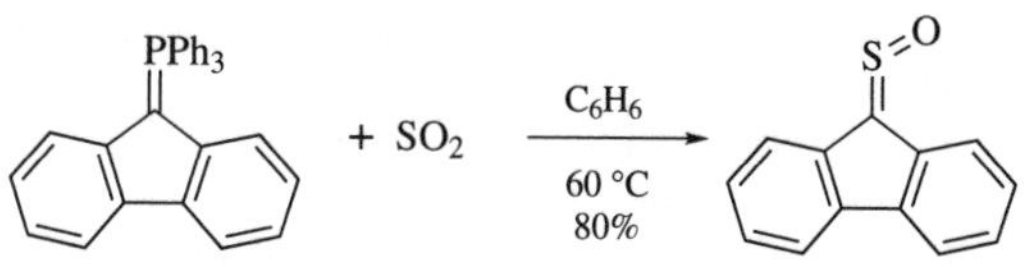

$$PPh_3 + SO_2 \xrightarrow[60\ ^\circ C\ \ 80\%]{C_6H_6} S{=}O \quad (30)$$

$$TMS \xrightarrow[2.\ SO_2,\ -78\ \text{to}\ 20\ ^\circ C\ \ 80\%]{1.\ BuLi,\ THF\ \ -78\ \text{to}\ 20\ ^\circ C} S{=}O \quad (31)$$

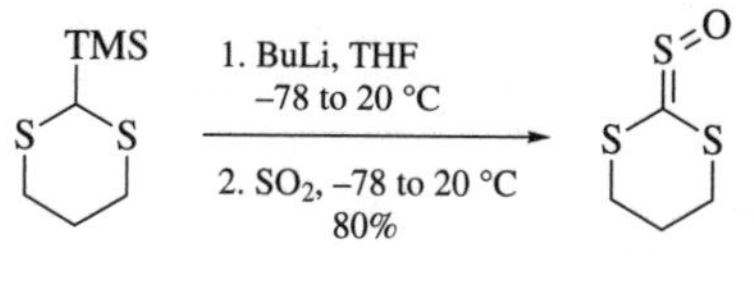

Miscellaneous Addition Reactions of SO_2. A 1,6-addition of SO_2 to *cis*-3-hexatriene gives 2,7-dihydrothiepin 1,1-dioxide in high yield (eq 32).[30] Fully unsaturated thiepin dioxides can be elaborated from this and analogous triene–SO_2 adducts, or the triene can be regenerated by thermolysis.

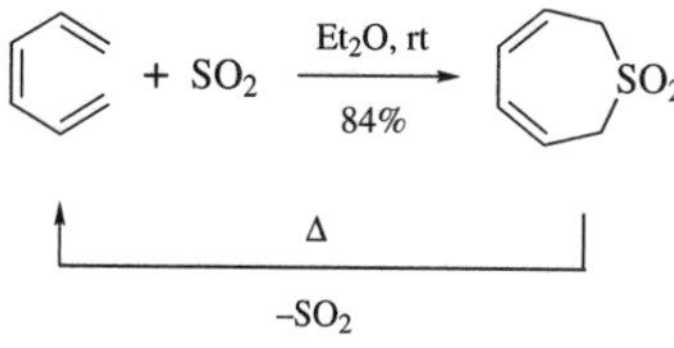

$$+ SO_2 \xrightarrow[84\%]{Et_2O,\ rt} SO_2 \quad (32)$$

Δ, $-SO_2$

A cyclo(copolymerization) between dimethyldiallylammonium chloride and SO_2 afforded a 3:1 mixture of the 10- and 20-membered macrocycles containing pyrrolidinium and sulfonyl groups (eq 33).[31]

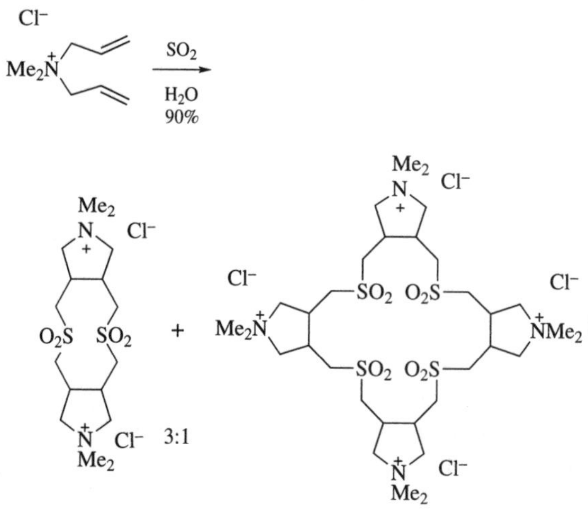

$$Cl^-\ Me_2\overset{+}{N} \xrightarrow[H_2O\ \ 90\%]{SO_2} \quad (33)$$

3:1

Aryl-substituted cyclopropanes undergo γ-sultine formation when treated with SO_2 in protic acid. The ring opening is regioselective, but the substituted 1,2-oxathiolane 2-oxides are formed as mixtures of diastereomers (eq 34).[32]

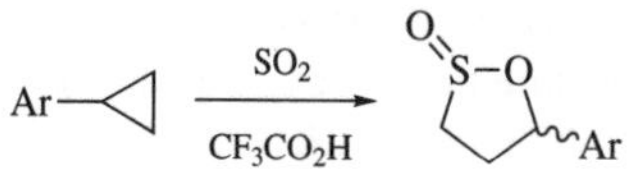

$$Ar \xrightarrow[CF_3CO_2H]{SO_2} Ar \quad (34)$$

Bicyclo[5.1.0]octa-2,5-diene reacts with dry SO_2 at 150 °C in toluene to afford a moderate yield of the bicyclic sulfone shown (eq 35).[33]

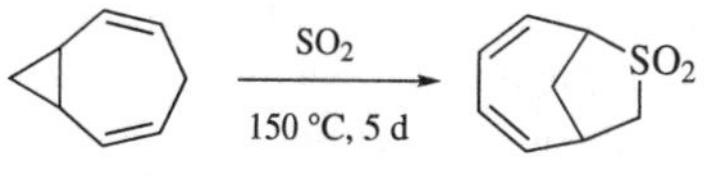

$$\xrightarrow[150\ ^\circ C,\ 5\ d]{SO_2} SO_2 \quad (35)$$

Photoaddition of 4,5-pyrenedione and SO_2 proceeds to give the cyclic sulfate (eq 36) in high yield.[34]

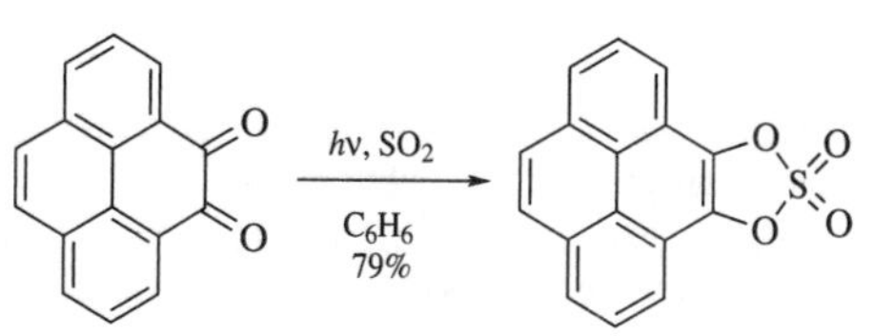

$$\xrightarrow[C_6H_6\ \ 79\%]{h\nu,\ SO_2} \quad (36)$$

Cheletropic $[_\pi 2_s + _\pi 2_s + _\omega 2_s]$ addition of SO_2 to norbornadiene provides a nearly quantitative yield of the tetracyclic sulfone in eq 37.[35] Of several homoconjugated dienes examined, only norbornadiene exhibited this mode of reactivity.

rt, 36 h → O_2S almost quantitative (37)

Miscellaneous Reactions of SO_2. Two examples of SO_2 serving as a reductant are illustrated in eqs 38 and 39. Pyridine *N*-oxides are deoxygenated under mild conditions in moderate yields.[36] The ozonide derived from sulfolene (itself a product of SO_2 plus butadiene) is reduced by SO_2 to afford the acetal. Treatment of this with ammonium chloride gives 4*H*-1,4-thiazine 1,1-dioxide (eq 39).[37]

R^2, R^1 pyridine N-oxide + SO_2 —dioxane, reflux, 3 h, 21–78%→ pyridine + SO_3 (38)

1. O_3, EtOH, CH_2Cl_2, –78 °C; 2. SO_2 → EtO, O, OEt acetal; NH_4Cl, $MeCO_2H$, reflux → H-N thiazine dioxide (39)

Vinyldiazomethane reacts with SO_2 leading, via the presumed *cis*-divinyl episulfone intermediate, to the dihydrothiepin 1,1-dioxide through a Cope rearrangement (eq 40).[38]

CHN_2 —SO_2→ [episulfone] → dihydrothiepin dioxide —1. NBS; 2. Et_3N→ thiepin dioxide (40)

Sulfur dioxide also serves as an electrophilic agent for the conversion of orthoesters to esters and dialkyl sulfites. For example, ***Triethyl Orthoacetate*** gives only ethyl acetate and diethyl sulfite as products upon treatment with SO_2 (eq 41).[39]

$$MeC(OEt)_3 + SO_2 \xrightarrow[24\ h]{-70\ °C\ to\ rt} MeC(O)OEt + EtOS(O)OEt \quad (41)$$

Formation of dithioacetals and *O,S*-acetals from ketones and aldehydes can be catalyzed by SO_2, with yields comparable or superior to protic catalysts.[40] Finally, liquid SO_2 has served as a solvent for a variety of superacid-mediated cyclization, rearrangement, and sulfinylation reactions.[41]

Related Reagents. Copper(I) Chloride–Sulfur Dioxide.

---

1. Mahn, W. J. *Academic Laboratory Chemical Hazards Guidebook*; Van Nostrand Reinhold: New York, 1991; p 264.
2. (a) de Bruin, G. *Proc. K. Ned. Akad. Wet.* **1914**, *17*, 585. (b) Backer, H. J.; Strating, J. *RTC* **1943**, *62*, 815.
3. (a) Jung, F.; Molin, M.; Van Den Elzen, R.; Durst, T. *JACS* **1974**, *96*, 935. (b) Heldeweg, R. F.; Hogeveen, H. *JACS* **1976**, *98*, 2341. (c) Vogel, P.; Deguin, B. *JACS* **1992**, *114*, 9210. (d) Sordo, J. A.; Sordo, T. L.; Suárez, D. *JACS* **1994**, *116*, 763.
4. Chaloner, L. M.; Crew, A. P. A.; O'Neill, P. M.; Storr, R. C.; Yelland, M. *T* **1992**, *48*, 8101.
5. (a) Jones, D. W.; Pomfret, A. *JCS(P1)* **1991**, 263. See also: (b) Charlton, J. L.; Plourde, G. L.; Koh, K.; Secco, A. S. *CJC* **1990**, *68*, 2022.
6. Kleveland, K.; Skattebol, L. *ACS* **1975**, *29*, 27.
7. (a) Tancheva, T. N.; Angelov, Ch. M.; Mondeshka, D. M. *H* **1985**, *23*, 843. (b) Angelov, Ch. M.; Vachkov, K. V. *TL* **1981**, *22*, 2517. (c) Angelov, Ch. M.; Kirilov, M.; Vachkov, K. V.; Spassov, S. L. *TL* **1980**, *21*, 3507.
8. Enchev, D. D.; Angelov, Ch. M.; Kirilov, M. *CS* **1987**, *27*, 295.
9. Nesbitt, B. F.; Beevor, P. S.; Cole, R. A.; Lester, R.; Poppi, R. G. *TL* **1973**, 4669.
10. (a) Bloch, R.; Abecassis, J. *TL* **1982**, *23*, 3277. (b) Bloch, R.; Abecassis, J. *TL* **1983**, *24*, 1247. See also: (c) Schmitthenner, H. F.; Weinreb, S. M. *JOC* **1980**, *45*, 3372.
11. (a) Rosenheim, A.; Singer, L. *CB* **1904**, *37*, 2152. (b) Kitching, W.; Fong, C. W. *Organomet. Chem. Rev. A* **1970**, *5*, 281. (c) Truce, E.; Lyons, J. F. *JACS* **1951**, *73*, 126. (d) Truce, E.; Wellisch, E. *JACS* **1952**, *74*, 5177.
12. Singh, N. P.; Biellmann, J.-F. *SC* **1988**, *18*, 1061.
13. (a) Sharutin, V. V.; Ermoshkin, A. E. *BAU* **1987**, 2414. (b) Sheikh, S. L. A.; Smith, B. C. *CC* **1968**, 1474.
14. Young, D.; Kitching, W. *OM* **1988**, *7*, 1196.
15. Hung, T.; Jolly, P. W.; Wilke, G. *JOM* **1980**, *190*, C5.
16. (a) Baker, E. B.; Sisler, H. H. *JACS* **1953**, *75*, 5193. (b) Ziegler, K.; Krupp, F.; Weyer, K.; Larbig, W. *LA* **1960**, *629*, 251.
17. Bellus, D. *HCA* **1975**, *58*, 2509.
18. Mloston, G. *BSB* **1990**, *99*, 265.
19. Grützmacher, H.; Roesky, H. W. *CB* **1987**, *120*, 995 (*CA* **1987**, *107*, 23 296t).
20. Fairhurst, R. A.; Heaney, H.; Papageorgiou, G.; Wilkins, R. F. *TL* **1988**, *29*, 5801.
21. Eyley, S. C.; Heaney, H.; Papageorgiou, G.; Wilkins, R. F. *TL* **1988**, *29*, 2997.
22. (a) Laubach, G. D.; Schreiber, E. C.; Agnello, E. J.; Brunigs, K. J. *JACS* **1956**, *78*, 4743. (b) Laubach, G. D.; Schreiber, E. C.; Agnello, E. J.; Lightfoot, E. N.; Brunigs, K. J. *JACS* **1953**, *75*, 1514.
23. Reischl, W.; Zbiral, E. *HCA* **1979**, *62*, 1763.
24. (a) Rogic, M. M.; Masilamani, D. *JACS* **1977**, *99*, 5219. (b) Masilamani, D.; Rogic, M. M. *JACS* **1978**, *100*, 4634.
25. (a) Trickes, G.; Meier, H. *AG(E)* **1977**, *16*, 555. (b) Grundmann, C.; Kochs, P.; Boal, J. R. *LA* **1972**, *761*, 162.
26. Badeev, Yu. V.; Pozdeev, O. K.; Ivanov, V. B.; Korobkova, V. D.; Andreev, S. V.; Ergorova, E. V.; Batyeva, E. S. *Khim.-Farm. Zh.* **1991**, *25*, 33 (*CA* **1991**, *115*, 114 426s).
27. Hoffman, R. V. *OS* **1981**, *60*, 121.
28. Zwanenburg, B.; Venier, C. G.; Porskamp, P. A. T. W.; van der Leij, M. *TL* **1978**, 807.
29. van der Leij, M.; Porskamp, P. A. T. W.; Lammerink, B. H. M.; Zwanenburg, B. *TL* **1978**, 811.
30. (a) Mock, W. L.; McCausland, J. H. *JOC* **1976**, *41*, 242. (b) Mock, W. L. *JACS* **1967**, *89*, 1281.
31. Leplyanin, G. V.; Vorob'eva, A. I.; Sysoeva, L. B.; Khalilov, L. M.; Shamaeva, Z. G.; Kozlov, V. G.; Tolstikov, G. A. *Dokl. Chem. (Engl. Transl.)* **1988**, *303*, 355.
32. (a) Bondarenko, O. B.; Saginova, L. G.; Voevodskaya, T. I.; Shabarov, Yu. S. *JOU* **1990**, *26*, 473. (b) Bondarenko, O. B.; Saginova, L. G.; Voevodskaya, T. I.; Buevich, A. V.; Yufit, D. S.; Struchkov, Yu. T.; Shabarov, Yu. S. *JOU* **1990**, *26*, 236.
33. Dalling, J.; Gall, J. H.; MacNicol, D. D. *TL* **1979**, 4789.
34. Tintel, C.; Terheijden, J.; Lugtenburg, J.; Cornelisse, J. *TL* **1987**, *28*, 2057.
35. De Lucchi, O.; Lucchini, V. *CC* **1982**, 1105.
36. Daniher, F. A.; Hackley, B. E., Jr. *JOC* **1966**, *31*, 4267.

37. Noland, W. E.; DeMaster, R. D. *OS* **1972**, *52*, 135.
38. Paquette, L. A.; Maiorana, S. *CC* **1971**, 313.
39. Rogic, M. M.; Klein, K. P.; Balquist, J. M.; Oxenrider, B. C. *JOC* **1976**, *41*, 482.
40. Burczyk, B.; Kortylewicz, Z. *S* **1982**, 831.
41. (a) Dean, C.; Whittaker, D. *JCS(P2)* **1990**, 1275. (b) Carr, G.; Dean, C.; Whittaker, D. *JCS(P2)* **1989**, 71. (c) Baig, M. A.; Banthorpe, D. V.; Carr, G.; Whittaker, D. *JCS(P2)* **1990**, 163. (d) Laali, K. K.; Nagvekar, D. S. *JOC* **1991**, *56*, 1867.

Steven D. Burke
University of Wisconsin, Madison, WI, USA

Sulfuric Acid[1]

H_2SO_4

[7664-93-9] H_2O_4S (MW 98.09)

(widely used protic acid solvent and catalyst;[1] can oxidize aliphatic and aromatic hydrocarbons;[1d] can sulfonate aromatic rings[1b])

Physical Data: mp 3 °C (98% sulfuric acid); bp 290+ °C; *d* 1.841 g cm^{-3} (96–98 sulfuric acid).
Solubility: sol water.
Form Supplied in: liquid sold in aqueous solutions of concentrations 78, 93, 95–98, 99, 100 wt %. Sulfuric acid (fuming) contains 18–24% free ***Sulfur Trioxide***.
Preparative Method: 100% sulfuric acid can be prepared by adding 95–98% sulfuric acid (concentrated sulfuric acid) to fuming sulfuric acid.[2]

Protic Acid Solvent and Catalyst.[3] Sulfuric acid is an inexpensive, easily handled protic acid, solvent, and catalyst. Typical workup procedures for reactions in H_2SO_4 involve aqueous dilution prior to product separation.

Hydrolyses. Nitroalkanes are readily available and their hydrolysis is an important synthetic tool. Hydrolysis of primary nitroalkanes with H_2SO_4, the most effective catalyst, gives carboxylic acids.[4] The salts of primary or secondary nitroalkanes, when hydrolyzed with H_2SO_4, form aldehydes or ketones (also see ***Titanium(III) Chloride***).[5] This reaction has been applied to β,γ-unsaturated nitroalkenes as a mild route to α,β-unsaturated aldehydes.[5c] An improved two-layer method treats the nitronate anion with H_2SO_4 in pentane; the product aldehyde dissolves in pentane and avoids contact with acid.[5d]

Vinyl halides are hydrolyzed by H_2SO_4 in the Wichterle reaction, a route to 1,5-diketones in which 1,3-dichloro-*cis*-2-butene serves as a methyl vinyl ketone equivalent.[6] The hydrolysis can be controlled to avoid acid-catalyzed aldol condensation (see condensations, below).[7]

Sulfuric acid is a useful catalyst for cleavage of protecting groups,[8] and has been used to cleave TBDMS protecting groups in the presence of TBDPS groups.[9] A useful method for resolution of chiral ketones involves formation and separation of chiral hydrazones followed by hydrolysis with 10% H_2SO_4.[10]

Hydrations. Alkyne hydration generally involves mercury(II) ion catalysts.[1b] However, H_2SO_4 hydrates the alkyne (**1**) and also catalyzes a subsequent regio- and stereoselective cyclopentanone annulation via the Nazarov cyclization (eq 1).[11] Nitriles can be selectively hydrated to amides using strong H_2SO_4.[12]

i-Pr, OH, OH; H_2SO_4, MeOH; (**1**); i-Pr, O, 64%; + i-Pr, O, 10% (1)

Additions.[1a,b] The Ritter reaction, in which nitriles add to alkenes in conc H_2SO_4, is a useful procedure for preparation of amides of *t*-alkylcarbinamines.[1b] As applied to the *threo* α-halo alcohol (**2**), retention of stereochemistry is observed (eq 2).[13] H_2SO_4 catalyzes carbonylation of α,β-unsaturated aldehydes in a general synthesis of 3,4-dialkyl-2(5*H*)-furanones (eq 3).[14] Michael additions to conjugated ketones are catalyzed by H_2SO_4 (see condensations, below).[7]

HO, Br (**2**) + MeCN —H_2SO_4, 81%→ AcHN, Br (2)

Et, O + CO (6–9 MPa) —96% H_2SO_4, 44%→ Et, O, O (3)

Sulfuric acid catalyzes the regioselective methoxybromination of α,β-unsaturated carbonyl compounds with ***N-Bromosuccinimide*** (eq 4); ***Boron Trifluoride***, ***Acetic Acid***, and ***Phosphoric Acid*** were unsatisfactory catalysts.[15] Sensitivity to acid catalyst was also noted in the intramolecular diazo ketone cyclization of β,γ-unsaturated diazomethyl ketone (**3**) (eq 5); H_2SO_4 gave a rearrangement product (**4**).[16]

O, Ph —conc H_2SO_4 (cat.), NBS, MeOH, 91%→ MeO, O, Ph, Br (4)

Dehydrations and β-Eliminations.[1b] Useful stereospecific H_2SO_4-catalyzed *anti* elimination of *threo*-β-hydroxyalkylsilanes is stereoselective for the *cis*-alkene, while ***Potassium Hydride*** mediated *syn* elimination affords selectively *trans*-alkenes (eq 6).[17]

98% H_2SO_4, $MeNO_2$, 25 °C, 79% → (4); BF_3 etherate, $MeNO_2$, 25 °C, 70% — from (3) (5)

H_2SO_4, THF	92%	100: 0
KH, THF	96%	5:95

(6)

Electrophilic Substitutions.[1a] Sulfuric acid catalyzes nitration of aromatic carbocycles[1a,b] and heterocycles.[18] Benzenes with *meta*-directing groups can be alkylated by primary and secondary alcohols in H_2SO_4, ***Polyphosphoric Acid***, or 85% phosphoric acid; even nitrobenzene can be alkylated by ethanol.[19] H_2SO_4 catalyzes the α-amidoalkylation reaction[20] and is especially useful for Friedel–Craft ketone synthesis using anhydrides.[1a] For acylations with acids, PPA avoids charring, sulfonation, and ester cleavage and is generally a preferred reagent (***Hydrogen Fluoride*** and ***Trifluoromethanesulfonic Acid***).[21] Keto acids and phenol react selectively to give phenolic esters with PPA, but ring substitution products with H_2SO_4.[22]

The utility of H_2SO_4 as a catalyst for the substitution of alkanes is evidenced in the formation of carboxylic acids by *trans* carboxylation (eq 7);[23] the hydrocarbon must have a tertiary hydrogen and the acid source for CO must be a tertiary alkyl acid.

+ $Et_2(Me)CCO_2H$ —(98% H_2SO_4, 30 °C, 84%)→ CO_2H (7)

Carbonyl Reactions.[1a,b] H_2SO_4 is considerably more effective than ***p-Toluenesulfonic Acid*** for conversion of anthrone to 9-alkoxyanthracenes.[24] A practical procedure for regioselective formation of pyridoxine dimethyl acetal which replaces anhydrous TsOH utilizes 96% H_2SO_4 (eq 8).[25]

MeCOMe, H_2SO_4 (10 equiv), 74% (8)

H_2SO_4 catalyzes esterification of highly hindered aromatic acids,[1b] and it catalyzes the formation of *N*-acylamides from acid anhydrides and amides.[26] A rapid esterification procedure involves reaction of primary, secondary, or tertiary alcohols with acids in H_2SO_4 using ultrasound.[27]

Condensations. Sulfuric acid is a useful reagent for the synthesis of heterocycles by dehydrative cyclization.[18,28] Yields in the Skraup quinoline synthesis, which utilizes sulfuric acid as the condensing agent, are remarkably sensitive to H_2SO_4 concentration.[29]

Sulfuric acid-catalyzed aldol condensations of 1,5-diketones in H_2SO_4 are under thermodynamical control[30] and products may differ from those of base-catalyzed reactions (eqs 9 and 10).[7,31,32]

conc H_2SO_4, benzene, heat → major; NaOEt, EtOH, heat, 87% (9)

3N H_2SO_4; 81%	95: 5	
LDA, ether; 82%	21:79	

(10)

Rearrangements. The choice of acid catalyst can influence skeletal rearrangements; for example, H_2SO_4 and H_3PO_4 can afford different products in polyene cyclizations (eq 11).[33] The reductive rearrangement of alcohol (**5**) gave different products in H_2SO_4 and H_3PO_4 (eq 12).[34]

95% H_2SO_4, 25 °C, 97%; 85% H_3PO_4, 100 °C, 72% (11)

conc H_2SO_4, C_5H_{12}, 97%; 85% H_3PO_4, C_7H_{16}, 72% — from (5) (12)

For functional group isomerizations, H_2SO_4 is useful in the Beckmann rearrangement of oximes and the Hofmann–Loffler–Freytag reaction of *N*-haloamines and amides.[1b] H_2SO_4 is superior to BF_3 in the isomerization of α-epoxycyclopentanones to α-hydroxycyclopentenones (eq 13).[35] The formation of 2-oxoadamantane from bicyclo[3.3.1]nonane-

2,6-diol is highly sensitive to acid concentration and requires 95% H_2SO_4 for optimum yield.[36]

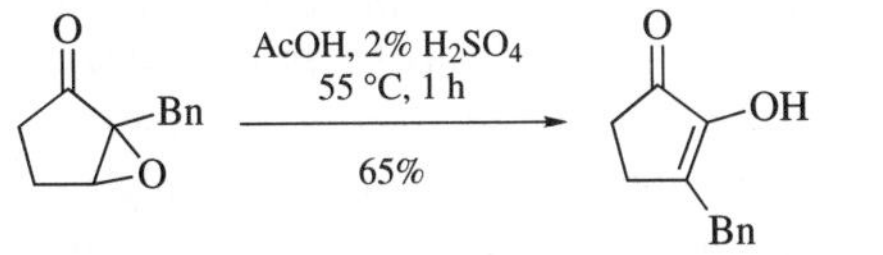

(13)

Choice and strength of acid catalyst can have regiochemical consequences. H_2SO_4-catalyzed Wallach rearrangement of azoxybenzenes provides mainly *p*-hydroxyazobenzenes, while ***Antimony(V) Chloride*** gives mainly *ortho* products.[37] Regiochemistry in the Schmidt reaction of ketones can be dependent upon H_2SO_4 concentration (eq 14).[38]

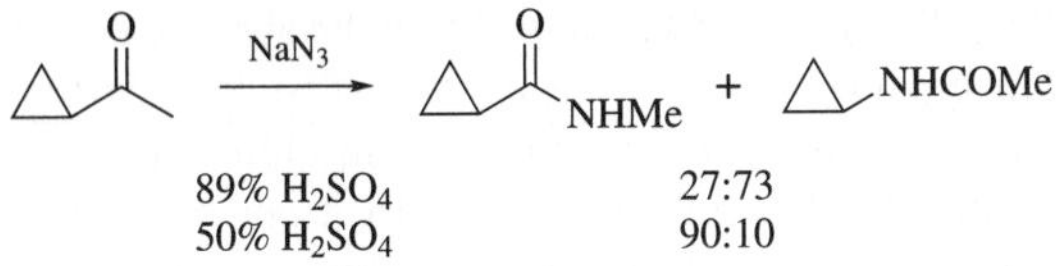

(14)

89% H_2SO_4	27:73
50% H_2SO_4	90:10

H_2SO_4 can catalyze stereochemical isomerizations (eq 15); thermodynamic conditions afforded *cis*-lactone (**6**); the diastereomeric *cis*-lactone is formed under kinetic control with ***Formic Acid*** or ***Tin(IV) Chloride***.[39] If a chiral center is present, enantioselective allylic alcohol rearrangement[40] and enantioselective alkene cyclization can be catalyzed by H_2SO_4 (eq 16).[41]

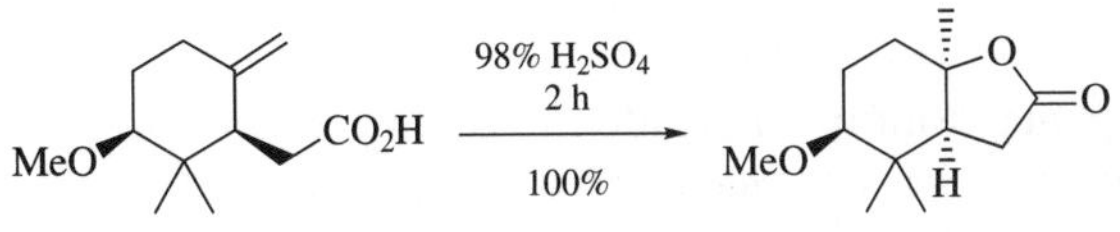

(15)

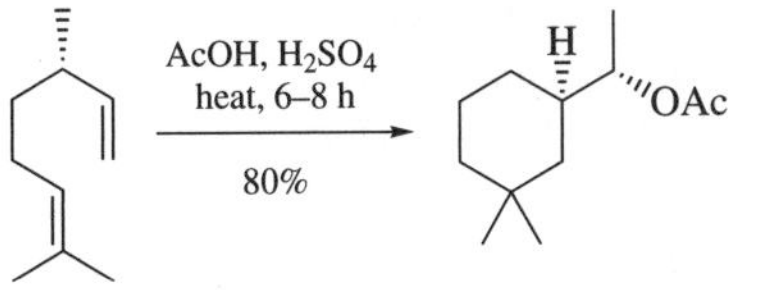

(16)

Catalyzed Oxidations.[1b] Sulfuric acid is the catalyst of choice for ***m-Chloroperbenzoic Acid*** oxidation of unreactive 11-keto steroids,[42] and it catalyzes the $NaBO_3$ oxidation of alkenes in Ac_2O to *trans*-diols.[43]

Dehydrogenation. H_2SO_4 acts as solvent, catalyst, and selective oxidizing agent in the formation of 2-pyridones from cyclic cyano ketones (eq 17).[44] Aromatization of unsaturated carbocyclic rings can be effected using H_2SO_4 and heat,[1d] and α-alkylcyclohexanones can be converted to *o*-alkylphenyl acetates by reaction using H_2SO_4 (eq 18).[45] Intermolecular dehydrogenation of an aminonaphthalene to a biphenyl occurs with 66% H_2SO_4 (eq 19).[46]

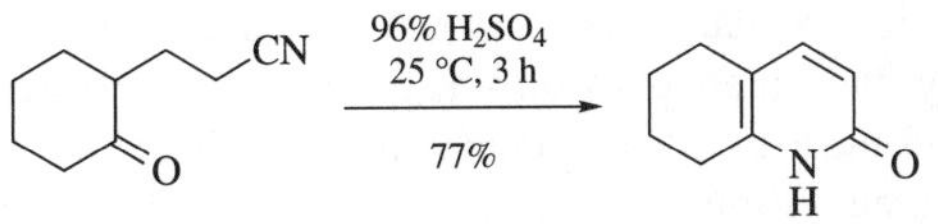

(17)

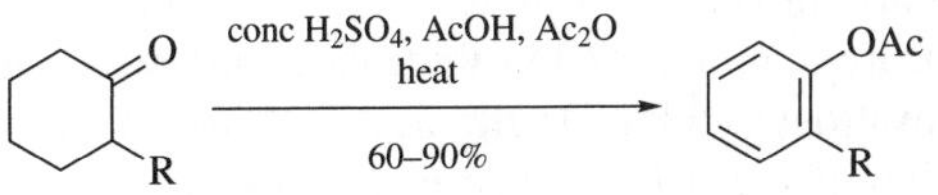

(18)

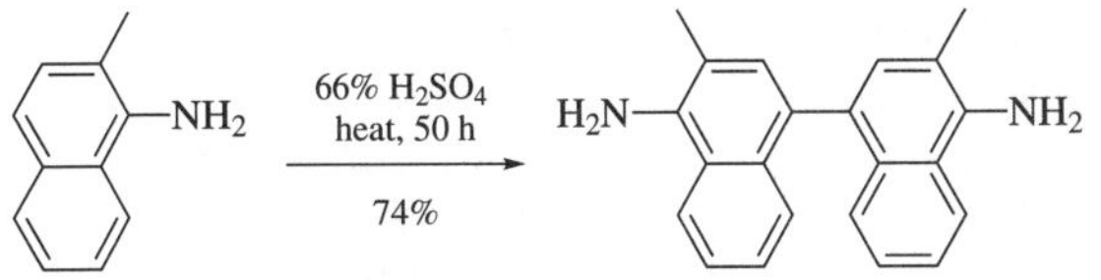

(19)

Hydroxylation. Hydroxylation of nitro and hydroxyl-substituted fused aromatic rings can be effected with sulfuric acid under forcing conditions.[1d] Adamantanone has been prepared from adamantane using 98% H_2SO_4 as oxidant.[47]

Sulfonation. Sulfonation of β-carbolines occurs in concentrated H_2SO_4; pyrrole, indole, and carbazole do not sulfonate under the conditions.[48]

Formation of Reducing Agents. An easily scaled-up method to convert amino acids to amino alcohols without affecting *N*-tosyl or *N*-Cbz groups uses H_2SO_4/***Sodium Borohydride*** (see also ***Aluminum Hydride***).[49]

Related Reagents. 10-Camphorsulfonic Acid; Fluorosulfuric Acid–Antimony(V) Fluoride; Methanesulfonic Acid; *p*-Toluenesulfonic Acid.

---

1. (a) Olah, G. A. *Friedel-Craft and Related Reactions*; Interscience: New York, 1963–65; Vols. I–IV. (b) March, J. *Advanced Organic Chemistry*, 4th ed.; Wiley: New York, 1992. (c) *Kirk-Othmer Encyclopedia of Chemical Technology*, 3rd ed.; Wiley: New York, 1983; Vol. 22, pp 190–232. (d) *MOC* **1981**, *IV/1a*, 323.
2. *FF* **1981**, *9*, 441.
3. Cox, R. A. *ACR* **1987**, *20*, 27.
4. Crandall, R. B., Locke, A. W. *JCS(B)* **1968**, 98.
5. (a) Noland, W. E. *CRV* **1955**, *55*, 137. (b) Pinnick, H. W. *OR* **1990**, *38*, 655. (c) Lou, J.-D.; Lou, W.-X. *S* **1987**, 179. (d) Chikashita, H.; Morita, Y.; Itoh, K. *SC* **1987**, *17*, 677.
6. House, H. O. *Modern Synthetic Reactions*, 2nd ed.; Benjamin: New York, 1972; p 611.
7. Steen, R. v. d.; Biescheuvel, P. L.; Erkelens, C.; Mathies, R. A.; Lugtenburg, J. *RTC* **1989**, *108*, 83.
8. (a) Kunz, H.; Waldmann, H. *COS* **1991**, *6*, Chapter 3.1. (b) Greene, T. W.; Wuts, P. G. M. *Protective Groups in Organic Synthesis*, 2nd ed.; Wiley: New York, 1991.
9. Franke, F.; Guthrie, R. D. *AJC* **1978**, *31*, 1285.
10. Fernandez, F.; Perez, C. *H* **1987**, *26*, 2411.
11. Hiyama, T.; Shinoda, M.; Nozaki, H. *JACS* **1979**, *101*, 1599.
12. Zabricky, J. *The Chemistry of Amides*; Interscience: New York, 1970; p 119.
13. Wohl, R. A. *JOC* **1973**, *38*, 3099.
14. Woo, E. P.; Cheng, F. C. W. *JOC* **1986**, *51*, 3706.
15. Heasley, V. L.; Wade, K. E.; Aucoin, T. G.; Gipe, D. E.; Shellhamer, D. F. *JOC* **1983**, *48*, 1377.
16. Satyanarayana, G. O. S. V.; Roy, S. C.; Ghatak, U. R. *JOC* **1982**, *47*, 5353.
17. Hudrlik, P. F.; Peterson, D. *JACS* **1975**, *97*, 1464.
18. Newkome, G. R.; Paudler, W. W. *Contemporary Heterocyclic Chemistry*; Wiley: New York, 1982, p 104.
19. Shen, Y.-S.; Liu, H.-X.; Wu, M.; Du, W.-Q., Chen, Y.-Q.; Li, N.-P. *JOC* **1991**, *56*, 7160.
20. Zaugg, H. E.; Martin, W. B. *OR* **1965**, *14*, 52.

21. Popp, F. D.; McEwen, W. E. *CRV* **1958**, *58*, 321.
22. Bader, A. R.; Kontowicz, A. D. *JACS* **1954**, *76*, 4465.
23. Lazzeri, V.; Jalal, R.; Poinas, R.; Gallo, R. *NJC* **1992**, *16*, 521.
24. Pirkle, W. H.; Finn, J. M. *JOC* **1983**, *48*, 2779.
25. Wu, Y., Ahlberg, P. *ACS* **1989**, *43*, 1009.
26. Challis, B. C.; Challis, J. A. In *Comprehensive Organic Chemistry*; Barton, D. H. R.; Ollis, W. D., Eds.; Pergamon: Oxford, 1979; Vol. 2, p 982.
27. Khurana, J. M.; Sahoo, P. K.; Maikap, G. C. *SC* **1990**, *20*, 2267.
28. Paquette, L. A. *Principles of Modern Heterocyclic Chemistry*; Benjamin: New York, 1968.
29. Manske, R. H. F.; Kulka, M. *OR* **1953**, *7*, 59.
30. Nielsen, A. T.; Houlihan, W. J. *OR* **1968**, *16*, 1.
31. Larcheveque, M.; Valette, G.; Cuvigny, T. *S* **1977**, 424.
32. Still, W. C.; Middlesworth, F. L. v. *JOC* **1977**, *42*, 1258.
33. Johnson, W. S. *ACR* **1968**, *1*, 1.
34. Takaishi, N.; Inamoto, Y.; Tsuchihashi, K.; Aigami, K.; Fujikura, Y. *JOC* **1976**, *41*, 771.
35. Barco, A.; Benetti, S.; Pollini, G. P.; Taddia, R. *S* **1975**, 104.
36. Averina, N. V.; Zefirov, N. S. *CC* **1973**, 197.
37. Yamamoto, J.; Nishigaki, Y.; Imagawa, M.; Umezu, M.; Matsuura, T. *CL* **1976**, 261.
38. Fikes, L. E.; Shechter, H. *JOC* **1979**, *44*, 741.
39. Rouessac, F.; Zamarlik, H. *TL* **1979**, *20*, 3421.
40. Fehr, T.; Stadler, P. A. *HCA* **1975**, *58*, 2484.
41. Ansari, H. R. *T* **1973**, *29*, 1559.
42. Suginome, H.; Yamada, S.; Wang, J. B. *JOC* **1990**, *55*, 2170.
43. Xie, G.; Xu, L.; Hu, J.; Ma, S.; Hou, W.; Tao, F. *TL* **1988**, *29*, 2967.
44. Meyers, A. I.; Garcia-Munoz, G. *JOC* **1964**, *29*, 1435.
45. Kablaoui, M. S. *JOC* **1974**, *39*, 2126.
46. Fierz-David, H. E.; Blangey, L.; Dubendorfer, H. *HCA* **1946**, *29*, 1661.
47. Geluk, H. W.; Keiser, V. G. *OSC* **1988**, *6*, 48.
48. Munoz, M. A.; Balon, M.; Carmona, C.; Hidalgo, J.; Poveda, M. L. *H* **1988**, *27*, 2067.
49. Abiko, A.; Masamune, S. *TL* **1992**, *33*, 5517.

Grant R. Krow
Temple University, Philadelphia, PA, USA

Sulfur Trioxide[1]

[7446-11-9] O_3S (MW 80.07)

(highly reactive electrophilic agent for replacing (i) a hydrogen atom or other substituent bonded to carbon by an SO_3H or derived sulfo group, forming a carbon–sulfur bond (called sulfonation), (ii) a hydrogen atom bonded to oxygen by an SO_3H group, forming an oxygen–sulfur bond (called sulfation or *O*-sulfonation, and (iii) a hydrogen atom bonded to nitrogen by an SO_3H group, forming a nitrogen–sulfur bond (called sulfamation or *N*-sulfonation);[1a,1b] reagent for cycloaddition to carbon–carbon double bonds;[1a,2] moderate oxidizing agent[3])

Physical Data: mp 16.8 °C; bp 44.7 °C; *d* 1.970 g cm^{-3}.

Solubility: miscible in all proportions with liquid SO_2; sol dichloromethane, chloroform, carbon tetrachloride, 1,1,1-trichloroethane (some react slowly with SO_3, e.g. CCl_4 yields phosgene);[4] SO_3 in dilute solution ($\leq$10 mol %) in SO_2, CCl_3F, and CCl_4 is present as monomer; at higher concentrations the cyclic trimer is also present;[5] sol nitromethane (reacts at 0–15 °C slowly, and at 25–40 °C sometimes violently to give nitromethanesulfonic acid and other compounds);[6] modestly sol 1,4-dioxane; the resulting complex is unstable;[7,8] the solid adduct can decompose violently on standing for some time at rt;[9] reacts violently with water.

Form Supplied in: liquid/solid, containing 1% of stabilizer to prevent polymerization, in sealed glass container.

Handling, Storage, and Precautions: in the absence of a suitable stabilizer, SO_3 shows a strong tendency to polymerize when exposed even to traces of moisture. Use in a fume hood. Keep the container of SO_3 tightly closed and dry. Upon handling open to air, the liquid gives off fumes of a sulfur trioxide–sulfuric acid spray! Reacts violently with water. Causes burns to the skin; is very toxic by inhalation and if swallowed. Contact with combustible material may cause fire. Wear eye and face protection and suitable protective clothing. In case of contact with eyes, rinse immediately with plenty of water, immediately remove all contaminated clothing and seek medical advice.

Organic Sulfur Trioxide Reagents. Apart from SO_3 and its less reactive addition compounds, formed with suitable Lewis bases[10] such as ***Pyridine***,[11] ***Trimethylamine***,[12] ***Dimethyl Sulfide***,[13] sulfolane,[14] ***Triphenylphosphine***,[15,16] triphenylbismuthine,[16] triphenylphosphine oxide,[17] tricylohexyl-[16] and trimethylphosphine oxide,[17] triethyl phosphate,[18] ***Dimethyl Sulfoxide***,[19] ***N,N-Dimethylformamide***,[20] ***Nitromethane***,[21] 1,4-dioxane,[1a–d] and 1,4-oxathiane,[22] there is a class of sulfonating sulfate ester reagents, e.g. dimethyl polysulfate ($MeOSO_2(OSO_2)_nOMe$)[23] and trimethylsilyl chlorosulfate.[24,25] There are also the protic sulfonating reagents, concentrated aqueous and fuming ***Sulfuric Acid***, and ***Fluorosulfuric Acid***, ***Chlorosulfonic Acid***, and acetylsulfuric acid.[1d] The reactivity order of the Lewis base complexes of SO_3 varies strongly. On the basis of direct experimental evidence[26] and of relative $pK_{B\text{-}SO_3}$[27] and pK_{BH^+}[28] values, the sulfonation reactivity is suggested to increase in the order trimethylamine–SO_3 < ***Sulfur Trioxide–Pyridine***, DMSO–SO_3 < *N,N*-dimethylformamide–SO_3, ***Trimethyl Phosphate***–SO_3 < ***Sulfur Trioxide-1,4-Dioxane*** < sulfolane–SO_3, nitromethane–SO_3.

Aromatic Sulfonation. Reaction of an arene (ArH) with SO_3 leads to sulfonation with formation of arenesulfonic acids ($ArSO_3H$). Mechanistically two different stages can be recognized, viz. 'primary' and 'secondary' sulfonation.[1f,29,30] In the primary stage, arenepyrosulfonic acids (ArS_2O_6H) are formed. A pyrosulfonic acid is a mixed anhydride, liable to disproportionate to give arenesulfonic anhydride and $H_2S_2O_7$.[30b,31–34] Working up the reaction mixture with water and heating the resulting aqueous mixture to reflux for 15 min leads to complete hydrolysis of the pyrosulfonic acids and sulfonic anhydrides to the corresponding arenesulfonic acids.[31] Sulfonylation to give diaryl sulfones is another possible complication in reaction of arenes with SO_3, especially when using deactivated arenes at high concentrations.[1f]

The steric requirements of a sulfonic acid group are very similar to those of a *t*-butyl group and prevent sulfonation *ortho* to a *t*-butyl group. Sulfonation of *t*-butylbenzene with SO_3 gives 98% of the 4-sulfonic acid (4-S), and that of *m*-di-*t*-butylbenzene gives 98% of the 5-S.[35] Reaction of *p*-di-*t*-butylbenzene with SO_3 in CCl_3F yields 58% of *p*-*t*-butylbenzenesulfonic acid by direct sulfo-de-*t*-butylation.[35]

Sulfonic acid isomer distribution data for the SO_3 sulfonation are available for alkylbenzenes[36] and their halogeno derivatives;[37] phenol and anisole,[31,38] and their methyl,[38,39] halogeno,[38,40] and hydroxy and methoxy derivatives;[32,41] and naphthalene,[36] and its methyl,[42] and hydroxy and methoxy derivatives.[34,43] SO_3 sulfonation isomer distribution data are also available for a number of polycyclic aromatic hydrocarbons and 1,6-methano[10]annulenes, including some alkyl derivatives.[36]

Deviating Sulfonation Behavior of 9-Alkylanthracenes. Reaction of 9-methylanthracene with SO_3 in dioxane leads exclusively to methyl sulfonation to give sulfonic acid (**1**) quantitatively.[44] Under comparable conditions, 9-*t*-butylanthracene gives δ-sultone (**2**).[45a] Sulfonation of 9-pentylanthracene gives predominantly sulfonic acid (**3**) with some (**4**).[45b] The α-alkenyl-γ-sulfonic acid (**3**) is formed via the corresponding 9-alkenylanthracene as an intermediate.[45c]

(**1**) R = H
(**4**) R = Bu

(**2**)

(**3**)

Positional Selectivity. Judging from the data for toluene,[46] *o*-xylene,[47] 1-methylnaphthalene,[42a,48] phenol,[38,49] anisole,[38,49] and 2,3-dihydrobenzofuran[50] and -pyran,[50] the positional selectivity is significantly greater for sulfonation with SO_3 than with sulfuric acid containing 90 wt % H_2SO_4.

For the sulfonation of an (alkyl)arene with SO_3 in differing solvents, the variation in the isomer distribution is limited.[14,51] With phenol and anisole the *ortho*:*para* sulfonation ratio is significantly larger when using CH_2Cl_2 than the complex-forming solvents nitromethane and dioxane.[31] With CH_2Cl_2 the SO_3 forms instead a complex with the C(sp^2) oxygen of the substrate which, as result of intramolecular transfer of SO_3, leads to enhanced *ortho* sulfonation. With substituted phenols and anilines, subject to the positions of the other substituents, the positional selectivity changes very significantly on varying the [SO_3]:[substrate] from $\leq$1.0 to $\geq$4.0 (Table 1).

As for the phenol derivatives, this illustrates the importance of the initial sulfation equilibrium (k_1, k_{-1}) (eq 1). Applying $\leq$1.0 equiv SO_3 the effective substrate species being sulfonated is the hydroxyarene H–Ar–OH (k_2), the OH substituent of which is strongly activating and *para* directing, whereas on using a large excess of SO_3 the entity undergoing sulfonation is the corresponding aryl hydrogen sulfate H–Ar–OSO_3H (k_3), the OSO_3H substituent being deactivating and *para* (+ *ortho*) directing. This may accordingly lead to a different substitution pattern. The sulfation equilibrium (k_1, k_{-1}) constant is temperature dependent. Phenol with 1.0 equiv SO_3 in nitromethane at −35 °C is rapidly sulfated to give phenyl hydrogen sulfate quantitatively, which at $\geq$0 °C isomerizes to phenol-4-S as the only eventual product.[38] The increase of both the 2- to 3-S ratio in the reaction of *p*-methoxyphenol with 1.5 equiv SO_3,[41] and the 2- to 4-S ratio in the reaction of 1-naphthol with 1.0 equiv SO_3[43a] with increasing reaction temperature, were ascribed to the increase in the k_{-1}/k_1 ratio.

Table 1 Sulfonation of Phenol and Aniline Derivatives in Nitromethane at 0 °C

Substrate[a]	SO_3 (equiv)	Product mixture composition (%)[a]
2-MeO-P[41]	1.0	4-S (76) 5-S (24)
	4.0	(19) (81)
	4.0[b]	(<2) (>98)
3,5-Me_2-P[39]	0.9	2-S (89) 4-S (11)
	4.0	(25) (75)
2,6-Me_2-P[40b]	0.9	3-S (<2) 4-S (>98)
	6.0	(77) (23)
2,6-Me_2-A[40b]	1.0	3-S (<2) 4-S (>98)
	8.0	(87) (13)

[a] P, A, and S stand for phenol, aniline, and SO_3H, respectively.
[b] Reversed addition

$$\begin{array}{ccc} H_2Ar\text{–}XH + SO_3 & \underset{k_{-1}}{\overset{k_1}{\rightleftharpoons}} & H_2Ar\text{–}XSO_3H \\ \downarrow k_2\ SO_3 & & \downarrow k_3\ SO_3 \\ HO_3S\text{–}HAr\text{–}XH & & HO_3S\text{–}HAr'\text{–}XSO_3H \\ & & SO_3 \uparrow\downarrow -SO_3 \\ & & HO_3S\text{–}HAr'\text{–}XH \end{array} \quad (1)$$

X = O, NH
Ar = C_6H_3, C_6H_2X, $C_{10}H_5$, $C_{10}H_4X$

Sulfonation of Unsaturated Aliphatic Compounds. Sulfur trioxide reacts vigorously with linear[52] and branched[53] alkenes to give rise to β-sultones as the primary sulfonation products.[1a,1b,2] Since neat SO_3 is too reactive, complexes of SO_3 with dioxane[8] or pyridine[11] are used to moderate the sulfonation reaction. The formation of β-sultone is stereo- (*syn*) and regioselective, obeying Markovnikov's rule. However, these β-sultones are thermally unstable at rt and their rearrangement leads to complex mixtures of alkenesulfonic acids and γ- and δ-sultones. Yields of the isolated alkenesulfonic acids (eq 2)[54] and sultones (eq 3)[53] vary considerably with the alkene structure and reaction conditions. Halogenated,[1h] in particular fluorinated,[1i] ethylenes react with SO_3 to give relatively stable halogenated β-sultones. This cycloaddition of SO_3 at the double bond proceeds in a regioselective way (eq 4).[55]

SO_3, 97% (2)

t-Bu; SO_3, 83%; O–S, O_2 (3)

$$F_2C{=}CFCl \xrightarrow[95\%]{SO_3} \text{F}_2\text{C–CFCl (O–SO}_2\text{ ring)} \quad (4)$$

Reaction of an alkene with an excess of SO_3 gives a cyclic sulfonate–sulfate anhydride, also referred to as carbyl sulfate or pyrosulfate (eq 5).[56] This carbyl sulfate is formed by a slow insertion of SO_3 into the intermediate β-sultone.[52a] The complex of sulfur trioxide with dimethyl sulfide reacts with alkenes and alkynes to afford sulfobetaines in good yields (eq 6).[13,57] These sulfobetaines are produced by nucleophilic attack of the dimethyl sulfide on the initially formed β-sultones.

(5)

R = H, 94%; Me, 94%; 2,6-$Cl_2C_6H_3$, 96%

(6) $Me_2S–SO_3$, 78%

Conjugated dienes are sulfonated by sulfur trioxide reagents to give β-unsaturated δ-sultones (eq 7). The yields vary strongly; they increase with the number of alkyl substituents at the 2- and 3-positions of the alkadiene.[58]

(7) SO_3, 80%

Functionalized alkenes containing a phenyl or carboxylic acid group at appropriate distance from the double bond undergo intramolecular cyclization during the sulfonation.[59] On reaction of (*E*)-4-hexenoic acid[59a] with SO_3 a sulfo-δ-lactone is formed (eq 8), and a Friedel–Crafts type cyclization is observed on sulfonation of (*E*)-5-phenyl-2-pentene[59b] (eq 9); both cyclizations proceed quantitatively and stereospecifically.

(8)

(9)

Sulfonation of Saturated Aliphatic Compounds. Sulfur trioxide reacts with aliphatic acids[60] or esters[61] to give, initially, insertion of SO_3 into the carboxylic acid or ester group, followed by sulfonation at the α-carbon (eq 10). Reactions of aldehydes and ketones with SO_3 also afford the α-sulfonated products.[62]

(10)

R^1 = H, alkyl; R^2 = H, alkyl

It should be noted that sulfonation of linear alkylbenzenes, linear long-chain α-alkenes, and fatty esters and the sulfation of fatty alcohols by SO_3–air mixtures are widely applied processes in industry for the production of surfactants.[63]

Related Reagents. Dimethyl Sulfoxide–Sulfur Trioxide/ Pyridine; Sulfur Trioxide–1,4-Dioxane; Sulfur Trioxide– Pyridine.

1. (a) Gilbert, E. E. *CRV* **1962**, *62*, 549. (b) Gilbert, E. E. *Sulfonation and Related Reactions*; Interscience: New York, 1965; Chapters I and II. (c) Gilbert, E. E. *S* **1969**, 3. (d) Cerfontain, H. *Mechanistic Aspects in Aromatic Sulfonation and Desulfonation*; Interscience: New York, 1968; Chapters 1 and 2. (e) Cerfontain, H. *JOC* **1982**, *47*, 4680, (f) Cerfontain, H. *RTC* **1985**, *104*, 153. (g) *The Chemistry of Sulphonic Acids, Esters and Derivatives*; Patai, S.; Rappoport, Z., Eds.; Wiley: New York, 1991; Chapters 10, 19, 22. (h) Knunjanz, I. L.; Sokolski, G. A. *AG* **1972**, *11*, 623. (i) Mohtasham, J.; Gard, G. L. *Coord. Chem. Rev.* **1992**, *112*, 47.
2. Roberts, D. W.; Williams, D. L. *T* **1987**, *43*, 1027.
3. Mark, V.; Zengierski, L.; Pattison, V. A.; Walker L. E. *JACS* **1971**, *93*, 3538.
4. Oddo, G.; Sconzo, A. *G* **1927**, *57*, 83 (*CA* **1927**, *21*, 1771).
5. Gillespie, R. J.; Robinson, E. A. *CJC* **1961**, *39*, 2189.
6. Terent'ev, A. P.; Yanovskaya, L. A.; Berlin, A. M.; Borisov, E. A. *Vestn. Mosk. Univ. Fiz.-Math. Estestven. Nauk* **1953**, 117 (*CA* **1955**, *49*, 8092f).
7. Bordwell, F. G.; Crosby, G. W. *JACS* **1956**, *78*, 5367.
8. Suter, C. M.; Evans, P. B.; Kiefer, J. M. *JACS* **1938**, *60*, 538.
9. Sisler, H. H.; Audrieth, L. F. *Inorg. Synth.* **1946**, *2*, 173.
10. Baumgarten, P. *CB* **1926**, *59*, 1166.
11. (a) Scully, J. F.; Brown, E. V. *JOC* **1954**, *19*, 894. (b) Terent'ev, A. P.; Dombrovskii, A. V. *JGU* **1949**, *19*, 1467.
12. Burg, A. B. *JACS* **1943**, *65*, 1629.
13. Nagayama, M.; Okumura, O.; Yaguchi, K.; Mori, A. *BCJ* **1974**, *47*, 2473.
14. Hayashi, T.; Iida, H.; Ogata, I. *J. Appl. Chem. Biotechnol.* **1976**, *26*, 513 (*CA* **1977**, *86*, 50 108n).
15. Galpin, I. J.; Kenner, G. W.; Marston, A.; Mills, O. S. *CC* **1981**, 789.
16. Becke-Goehring, M.; Thielemann, H. *Z. Anorg. Allgem. Chem.* **1961**, *308*, 33 (*CA* **1961**, *55*, 21 937h).
17. Burg, A. B.; McKee, W. E. *JACS* **1951**, *73*, 4590.
18. Sakota, N.; Kohora, K. *Nippon Kagaku Zasshi* **1965**, *86*, 747 (*CA* **1966**, *65*, 667h).
19. Whistler, R. L.; King, A. H.; Ruffini, G.; Lucas, F. A. *Arch. Biochem. Biophys.* **1967**, *121*, 358 (*CA* **1967**, *67*, 79 018b).
20. (a) Kenner, G. W.; Stedman, R. J. *JCS* **1952**, 2069. (b) Wolfram M. L.; Han, T. M. S. *JACS* **1959**, *81*, 1764.
21. Cerfontain, H.; Koeberg-Telder, A. *RTC* **1970**, *89*, 569.
22. Quaedvlieg, M. *MOC* **1955**, *9*, 364.
23. Van Wazer, J. R.; Grant, D.; Dungan, C. H. *JACS* **1965**, *87*, 3333.
24. (a) Birot, M.; Dunogues, J.; Duffaut, N.; Calas, R.; Lefort, M. *BSF* **1978**, *1*, 442. (b) Bourgeois, P.; Duffaut, N. *BSF* **1980**, 2, 195.
25. (a) Hofmann, K.; Simchen, G. *LA* **1982**, *282*. (b) Hofmann, K.; Simchen, G. *LA* **1984**, 39.

26. De Wit, P.; Kruk, C.; Cerfontain, H. *RTC* **1986**, *105*, 266.
27. Auger, Y.; Delesalle, G.; Fischer, J. C.; Wartel, M. *J. Electroanal. Chem.* **1980**, *106*, 149 (*CA* **1980**, *92*, 153 823v).
28. (a) Lowry, T. H.; Schueller-Richardson, K. *Mechanism and Theory in Organic Chemistry*, 3rd ed.; Harper and Row: New York, 1987; p 297. (b) Ege, S. N. *Organic Chemistry*, 2nd ed.; Heath: Lexington, MA, 1989.
29. (a) Christensen, N. H. *ACS* **1963**, *17*, 2253. (b) Christensen, N. H. *ACS* **1964**, *18*, 954. (c) Guyer, P.; Fleury, R.; Reich, H. U. *C* **1968**, *22*, 40.
30. (a) Bosscher, J. K.; Cerfontain, H. *RTC* **1968**, *87*, 873. (b) *T* **1968**, *24*, 6543.
31. Ansink, H. R. W.; Cerfontain, H. *RTC* **1992**, *111*, 183.
32. Cerfontain, H.; Ansink, H. R. W.; Coenjaarts, N. J.; De Graaf, E. J.; Koeberg-Telder, A. *RTC* **1989**, *108*, 445.
33. Ansink, H. R. W.; De Graaf, E. J.; Zelvelder, E.; Cerfontain, H. *CJC* **1993**, *71*, 210.
34. Ansink, H. R. W.; Zelvelder, E.; De Graaf, E. J.; Cerfontain, H. *JCS(P2)* **1993**, 721.
35. Ris, C.; Cerfontain, H. *JCS(P2)* **1975**, 1438.
36. Cerfontain, H. *RTC* **1985**, *104*, 153 (Tables III–V).
37. Cerfontain, H.; Koeberg-Telder, A.; Laali, K.; Lambrechts, H. J. A.; De Wit, P. *RTC* **1982**, *101*, 390.
38. Cerfontain, H.; Koeberg-Telder, A.; Lambrechts, H. J. A.; De Wit, P. *JOC* **1984**, *49*, 4917.
39. Goossens, H. D.; Lambrechts, H. J. A.; Cerfontain, H.; De Wit, P. *RTC* **1988**, *107*, 426.
40. (a) De Wit, P.; Cerfontain, H. *RTC* **1988**, *107*, 121, 418. (b) De Wit, P.; Woldhuis, A. F.; Cerfontain, H. *RTC* **1988**, *107*, 668.
41. Cerfontain, H.; Coenjaarts, N. J.; Koeberg-Telder, A. *RTC* **1989**, *108*, 7.
42. (a) Lammertsma, K.; Cerfontain, H. *JCS(P2)* **1979**, 673. (b) Lambrechts, H. J. A.; Cerfontain, H. *T* **1982**, *38*, 1667.
43. (a) Ansink, H. R. W.; Zelvelder, E.; Cerfontain, H. *RTC* **1993**, *112*, 210. (b) Ansink, H. R. W.; Zelvelder, E.; Cerfontain, H. *RTC* **1993**, *112*, 216. (c) Ansink, H. R. W., De Graaf, E. J.; Zelvelder, E.; Cerfontain, H. *RTC* **1992**, *111*, 499.
44. Cerfontain, H.; Koeberg-Telder, A.; Ris, C.; Schenk, C. *JCS(P2)* **1975**, 966.
45. (a) Van de Griendt, F.; Cerfontain, H. *TL* **1978**, 3263. (b) Van de Griendt, F.; Cerfontain, H. *JCS(P2)* **1980**, 13. (c) Van de Griendt, F.; Cerfontain, H. *JCS(P2)* **1980**, 19.
46. (a) Van Albada, M. P.; Cerfontain, H.; Koeberg-Telder, A. *RTC* **1972**, 91, 33. (b) Cerfontain, H.; Sixma, F. L. J.; Vollbracht, L. *RTC* **1963**, *82*, 659.
47. (a) Cerfontain, H.; Koeberg-Telder, A.; Van Kuipers, E. *JCS(P2)* **1972**, 2091. (b) Prinsen, A. J.; Cerfontain, H. *RTC* **1969**, *88*, 833.
48. Lammertsma, K.; Verlaan, C. J.; Cerfontain, H. *JCS(P2)* **1978**, 719.
49. Cerfontain, H.; Lambrechts, H. J. A.; Schaasberg-Nienhuis, Z. R. H.; Coombes, R. G.; Hadjigeorgiou, P.; Tucker, G. P. *JCS(P2)* **1985**, 659.
50. Ansink, H. R. W.; Cerfontain, H. *RTC* **1989**, *108*, 395.
51. El Homsi, A.; Gilot, B.; Canselier, J-P. *J. Appl. Chem. Biotechnol.* **1978**, *28*, 405 (*CA* **1979**, *90*, 121 114g).
52. (a) Bordwell, F. G.; Peterson, M. L. *JACS* **1954**, *76*, 3952. (b) Püschel, F.; Kaiser, C. *CB* **1965**, *98*, 735.
53. (a) Bordwell, F. G.; Chapman, R. D.; Osborne, C. E.; *JACS* **1959**, *81*, 2002. (b) Robbins, M. D.; Broaddus, C. D. *JOC* **1974**, *39*, 2459.
54. Arnold, R. T.; Amidon, R. W.; Dodson, R. M. *JACS* **1950**, *72*, 2871.
55. Dmitriev, M. A.; Sokol'skii, G. A.; Knunyants, I. L. *IZV* **1960**, 792.
56. (a) Breslow, D. S.; Hough, R. R. *JACS* **1957**, *79*, 5000. (b) Weinreich, G. H.; Jufresa, M. *BSF* **1965**, 787. (c) Sheehan, J. C.; Zoller, U. *JOC* **1975**, *40*, 1179.
57. Shastin, A. V.; Popkova, T. V.; Balenkova, E. S.; Bel'skii, V. K.; Lazhko, E. I. *JOU* **1987**, *23*, 2039.
58. (a) Schonk, R. M.; Bakker, B. H.; Cerfontain, H. *RTC* **1993**, *112*, 201. (b) Akiyama, T.; Sugihara, M.; Imagawa, T.; Kawanisi, M. *BCJ* **1978**, *51*, 1251.
59. (a) Schonk, R. M.; Bakker, B. H.; Cerfontain, H. *RTC* **1992**, *111*, 389. (b) *RTC* **1992**, *111*, 478.
60. Truce, W. E.; Olson, C. E. *JACS* **1953**, *75*, 1651.
61. Smith, F. D.; Stirton, A. J. *J. Am. Oil Chem. Soc.* **1967**, *44*, 405 (*CA* **1967**, *67*, 90 351a).
62. Truce, W. E.; Alfieri, C. C. *JACS* **1950**, *72*, 2740.
63. Surfactant Science Series *Anionic Surfactants*; Linfield, W. M., Ed.; Dekker: New York, 1976; Vol. 7.

Hans Cerfontain & Bert H. Bakker
University of Amsterdam, The Netherlands

Tetra-*n*-butylammonium Fluoride[1]

n-Bu_4NF

(TBAF)		
[429-41-4]	$C_{16}H_{36}FN$	(MW 261.53)
(TBAF·$3H_2O$)		
[87749-50-6]	$C_{16}H_{42}FNO_3$	(MW 315.59)
(TBAF·xH_2O)		
[22206-57-1]		

(can be used for most fluoride-assisted reactions; deprotection of silyl groups;[1e] desilylation;[2,3] fluorination;[4] used as a base[5,6])

Alternate Name: TBAF.
Physical Data: TBAF·xH_2O: mp 62–63 °C.
Solubility: sol H_2O, THF, MeCN.
Form Supplied in: trihydrate, 1.0 M solution in THF, and 75 wt % solution in water.
Preparative Method: aqueous ***Hydrofluoric Acid*** is passed through an Amberlite IRA 410 OH column, followed by an aqueous solution of ***Tetra-n-butylammonium Bromide***. After the resin is washed with water, the combined water fractions are repeatedly evaporated until no water is present. Tetrabutylammonium fluoride is collected as an oil in quantitative yield.
Handling, Storage, and Precautions: use in a fume hood.

Deprotection of Silyl Groups. Tetrabutylammonium fluoride has been used widely as a reagent for the efficient cleavage of various silyl protecting groups such as *O*-silyls of nucleosides,[7,8] pyrophosphate,[9] *N*-silyls,[10,11] CO_2 -silyl, and *S*-silyl derivatives.[1e] These reactions are often carried out under very mild conditions in excellent yields. Thus it has been used in the synthesis of base-sensitive chlorohydrins (eq 1)[12] and β-lactams.[10,13] 2-(Trimethylsilyl)ethoxymethyl groups can also be effectively removed from various substrates (eq 2).[14–18] Silyl ethers can be converted to esters in one pot when they are treated with TBAF, followed by exposure to acyl chlorides[19,20] or anhydride[21] in the presence of base (eq 3). Treatment of triisopropylsilyl enol ethers with ***Iodosylbenzene/Azidotrimethylsilane***, followed by desilylation and elimination with TBAF, gives good yields of the α,β-unsaturated ketones (eq 4).[22,23]

(1) [scheme: OTBDMS chlorohydrin → OH; Bu_4NF, THF, 25 °C, 30 min, 50%]

Cyclobutanone alkyl silyl acetals, obtained from [2+2] cycloadditions, can be deprotected with 1 equiv of TBAF in THF to give the open-chain cyano esters in excellent yields (eq 5).[24] When 4-chloro-2-cyanocyclobutane alkyl silyl acetals are used as substrates for this reaction, (*E*/*Z*) mixtures of 2-cyanocyclopropanecarboxylates are obtained by an intramolecular cyclization (eq 6).

(2) [scheme: TBAF, THF, reflux, 60%]

(3) [scheme: 1. TBAF, THF; 2. Et_3N, 80 °C]

(4) [scheme: TBAF, THF, 0 °C]

(5) [scheme: TBAF, THF, 90%]

(6) [scheme: TBAF, THF, 83%]

11-Membered pyrrolizidine dilactones have been synthesized by treating a trimethylsilylethyl ester with TBAF in MeCN to form an anion, which then undergoes cyclization by displacement of the mesylate.

Desilylation Reagent. Cleavage of carbon–silicon bonds with fluoride has been studied very extensively. TBAF is a very powerful reagent for desilylation of a wide range of silicon-containing compounds, such as vinylsilanes,[2,25,26] alkynylsilanes,[23,27] arylsilanes,[28,29] acylsilanes,[30] β-silyl sulfones,[31–33] and other silane derivatives.[3,34–37] It appears that cleavage of sp-C–Si bonds is more facile than that of sp^2-C–Si and sp^3-C–Si bonds and that substituted groups, such as phenyl and alkoxyl, can often facilitate cleavage. A dimethylphenylsilyl group can be removed from a vinyl carbon by TBAF with retention of the alkene stereochemistry (eq 7).[2] This method has been applied to the synthesis of terminal conjugated trienes (eqs 8 and 9).[38] The five-membered siloxanes can be desilylated with 3 equiv of TBAF in DMF and this protodesilylation is very sensitive to subtle structure changes (eq 10).[39]

(7) [scheme: TBAF, 80 °C, THF–DMSO (2:1), 99%]

(8) [scheme: 1. AcONa, AcOH, 40 °C, 99%; 2. TBAF, DMSO–THF, 80 °C, 99%; 98%]

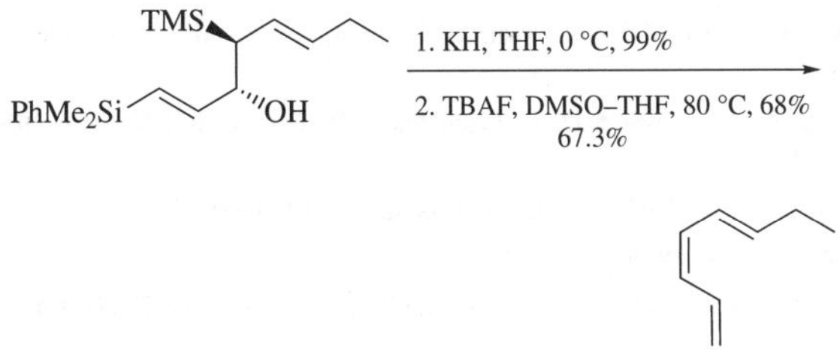

(9)

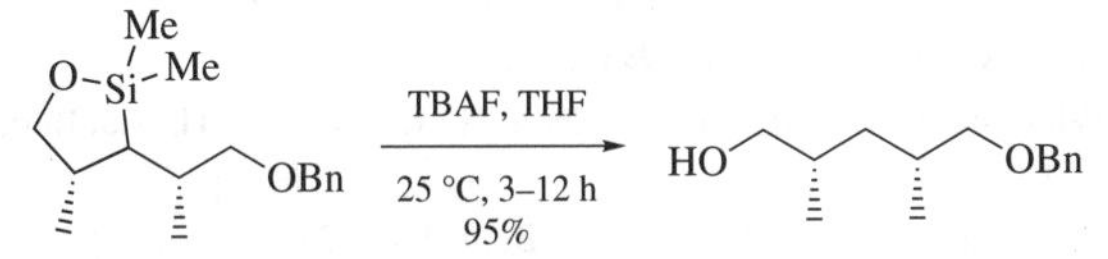

(10)

The anions, generated in situ by desilylation of silylacetylenes,[40,41] allylsilanes,[42–44] propargylsilanes,[45] α-silyloxetanones,[46] bis(trimethylsilylmethyl) sulfides,[47] and other silane derivatives,[48–51] can undergo nucleophilic addition to ketones and aldehydes (eq 11).[52] *N*-(*C,C*-bis(trimethylsilyl)methyl) amido derivatives can add to aldehydes followed by Peterson alkenation to form acyl enamines.[48,53] Treatment of 2-trimethylsilyl-1,3-dithianes can generate dithianyl anions, which are capable of carbocyclization via direct addition to carbonyl or Michael addition (eq 12). The fluoride-catalyzed Michael additions are more general than Lewis acid-catalyzed reactions and proceed well even for those compounds with enolizable protons and/or severe steric hindrance (eq 13).[54,55]

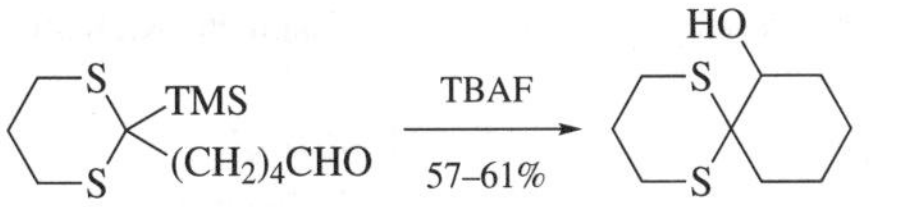

(11)

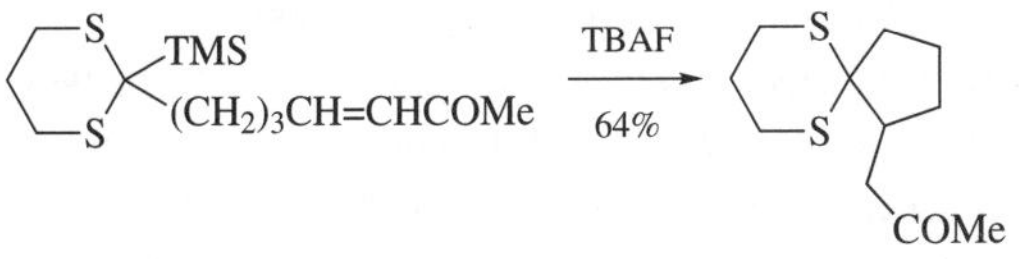

(12)

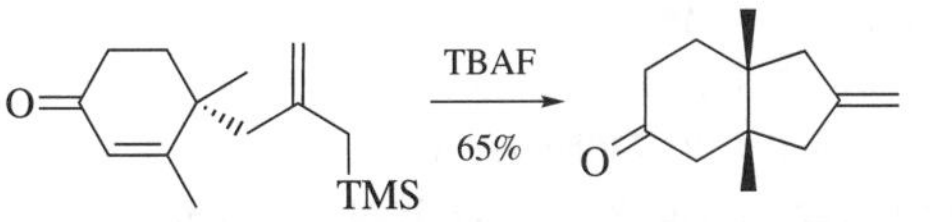

(13)

Direct fluoride-induced trifluoromethylation of α-keto esters (eq 14),[56] ketones,[57] aldehydes,[58,59] and sulfoxides[59] have been reported using ***Trifluoromethyltrimethylsilane*** with TBAF in THF.

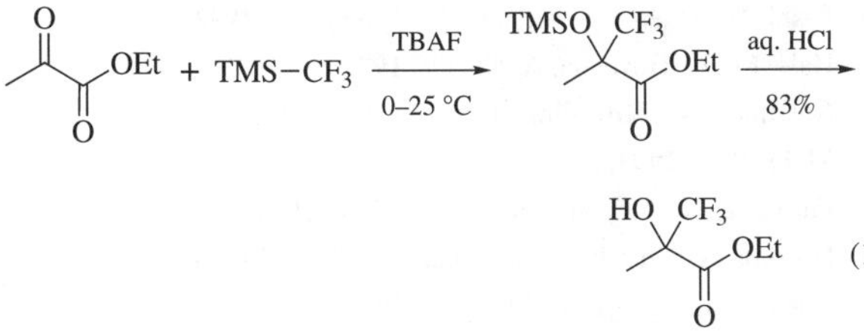

(14)

Desilylation of some compounds can generate very reactive species such as benzynes,[60] pyridynes,[61] xylylenes,[62,63] and benzofuran-2,3-xylylenes.[64] 1,4-Elimination of *o*-(α-trimethylsilylalkyl)benzyltrimethylammonium halides with TBAF in acetonitrile generates *o*-xylylenes, which undergo intermolecular and intramolecular cycloadditions (eq 15).[62–64] Treatment of α-silyl disulfides with ***Cesium Fluoride*** or TBAF forms thioaldehydes, which have been trapped by cycloaddition with cyclopentadiene (eq 16).[65]

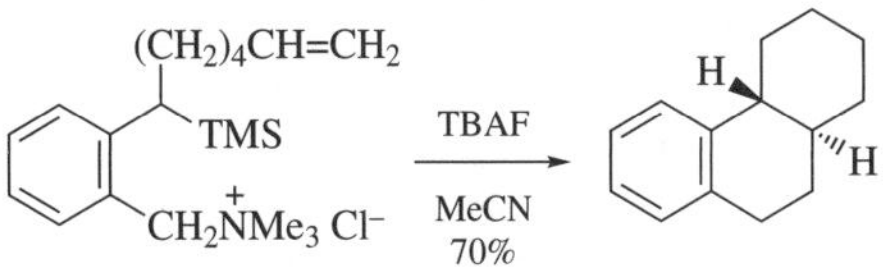

(15)

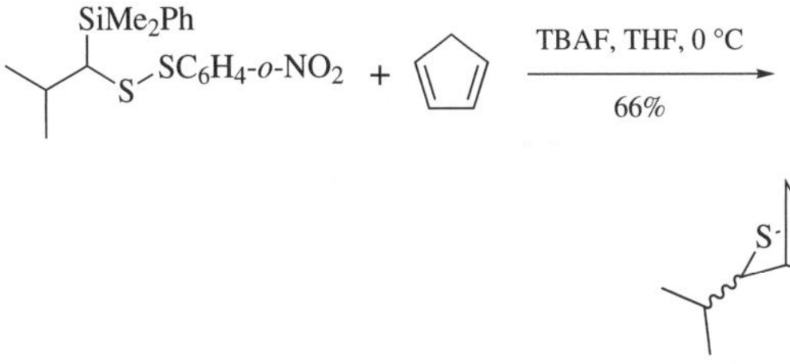

(16)

exo:*endo* = 1:7

Use as a Base. TBAF has been widely used for a variety of base-catalyzed reactions such as alkylation,[66] elimination,[67] halogenation,[68] Michael addition,[69–71] aldol condensation, and intramolecular cyclizations.[5,72–74] It is especially useful when other inorganic bases face solubility problems in organic solvents. The reactions are usually carried out below 100 °C due to the low thermal stability of TBAF.[1e]

TBAF is very useful for alkylation of nucleic acid derivatives. Methylation[75] or benzylation[66] of uracil gives almost quantitative yields of alkylated product when using alkyl bromides, dialkyl sulfates (eq 17), trialkyl phosphates, or alkyl chlorides with TBAF. Alkylation of the thiol anions generated from deprotection by 1,2-dibromoethane produces interesting tetrachalcogenofulvalenes.[14] Under phase-transfer conditions, selective mono- and dialkylations of malononitrile have been achieved by using neat TBAF with ***Potassium Carbonate*** or ***Potassium t-Butoxide*** and controlling the amount of alkyl bromides or iodides used (eqs 18 and 19).[76]

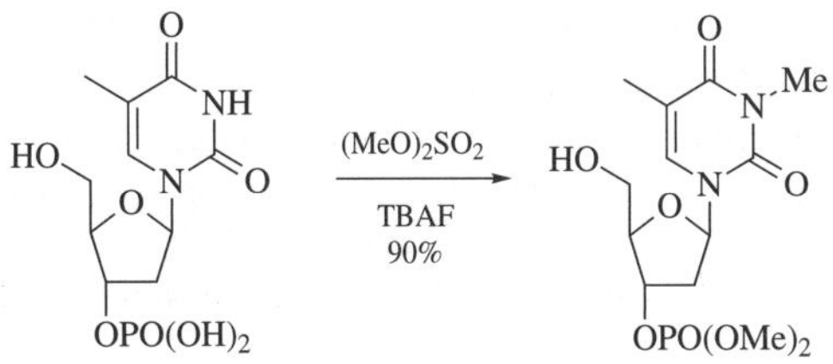

(17)

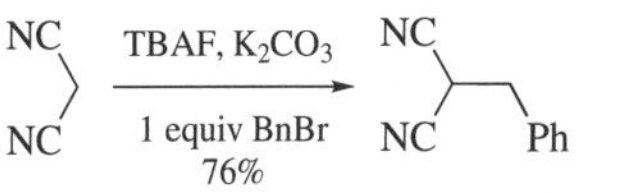

(18)

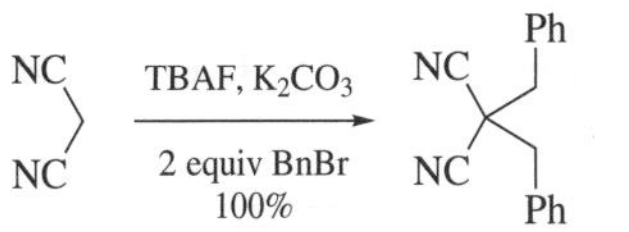

(19)

Enol silyl ethers react with aldehydes with a catalytic amount of TBAF to give the aldol silyl ethers in good yields. These reactions generally proceed under very mild conditions and within shorter periods of time than conventional strong acidic or basic conditions. The products from 4-*t*-butyl-1-methyl-2-(trimethylsilyloxy)cyclohexene and benzaldehyde show very good axial selectivity and a little *anti–syn* selectivity (eq 20).[77] The aldol condensation of ketones and aldehydes can be achieved in one pot when ethyl (trimethylsilyl)acetate is used as a silylation agent with TBAF (eq 21).

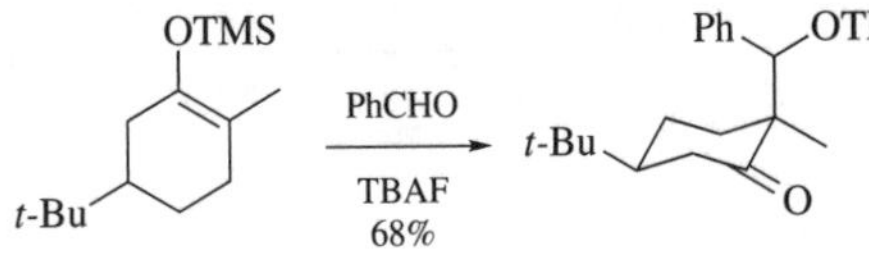

(20)

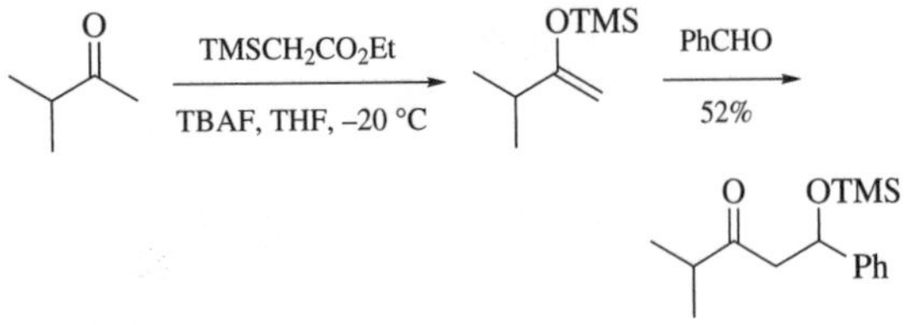

(21)

Silyl nitronates undergo aldol condensation with aldehydes in the presence of a catalytic amount of anhydrous TBAF to form highly diastereoselective *erythro* products, which can be elaborated to give synthetically useful 1,2-amino alcohols (eq 22).[6,78] A one-pot procedure has been developed for direct aldol condensation of nitroalkanes with aldehydes by using TBAF trihydrate with ***Triethylamine*** and ***t-Butyldimethylchlorosilane***.[79] It appears that silyl nitronates are not reactive intermediates in this case, and the reactions proceed by a different mechanism.

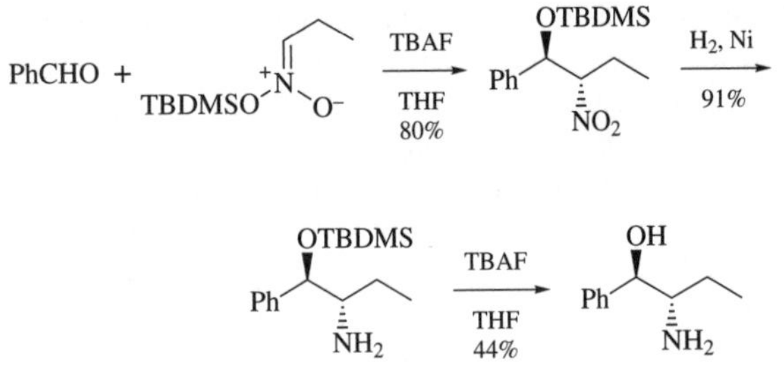

(22)

Miscellaneous. Fluoride ion from anhydrous TBAF undergoes nucleophilic displacement of tosylates,[4,80] halides,[80] and aryl nitro compounds[81] to give fluorinated products. When used with ***N-Bromosuccinimide***, bromofluorination products are obtained.[82]

Several important peptide-protecting groups such as 9-fluorenylmethyloxycarbonyl,[83] benzyl,[84] 4-nitrobenzyl,[85] 2,2,2-trichloroethyl,[85] and acetonyl (eq 23)[86] can be removed by TBAF under mild conditions.

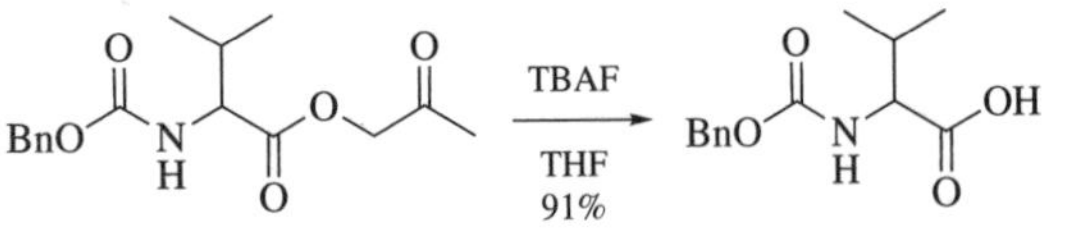

(23)

Related Reagents. Lithium Fluoride; Potassium Fluoride; Sodium Fluoride.

---

1. (a) Corey, E. J.; Snider, B. B. *JACS* **1972**, *94*, 2549. (b) Hudlicky, M. *Chemistry of Organic Fluorine Compounds*, 2nd ed.; Horwood: New York, 1992. (c) Umemoto, T. *Yuki Gosei Kagaku Kyokaishi* **1992**, *50*, 338. (d) Clark, J. H. *CRV* **1980**, *80*, 429. (e) Greene, T. W.; Wuts, P. G. M. *Protective Groups in Organic Synthesis*, 2nd ed.; Wiley: New York, 1991. (f) Sharma, R. K.; Fry, J. L. *JOC* **1983**, *48*, 2112. (g) Cox, D. P.; Terpinski, J.; Lawrynowicz, W. *JOC* **1984**, *49*, 3216.
2. Oda, H.; Sato, M.; Morizawa, Y.; Oshima, K.; Nozaki, H. *T* **1985**, *41*, 3257.
3. Dhar, R. K.; Clawson, D. K.; Fronczek, F. R.; Rabideau, P. W. *JOC* **1992**, *57*, 2917.
4. Gerdes, J. M.; Bishop, J. E.; Mathis, C. A. *JFC* **1991**, *51*, 149.
5. Pless, J. *JOC* **1974**, *39*, 2644.
6. Seebach, D.; Beck, A. K.; Mukhopadhyay, T.; Thomas, E. *HCA* **1982**, *65*, 1101.
7. Krawczyk, S. H.; Townsend, L. B. *TL* **1991**, *32*, 5693.
8. Meier, C.; Tam, H.-D. *SL* **1991**, 227.
9. Valentijn, A. R. P. M.; van der Marel, G. A.; Cohen, L. H.; van Boom, J. *SL* **1991**, 663.
10. Hanessian, S.; Sumi, K.; Vanasse, B. *SL* **1992**, 33.
11. Kita, Y.; Shibata, N.; Tamura, O.; Miki, T. *CPB* **1991**, *39*, 2225.
12. Solladié-Cavallo, A.; Quazzotti, S.; Fischer, J.; DeCian, A. *JOC* **1992**, *57*, 174.
13. Konosu, T.; Oida, S. *CPB* **1991**, *39*, 2212.
14. Zambounis, J. S.; Mayer, C. W. *TL* **1991**, *32*, 2737.
15. Kita, H.; Tohma, H.; Inagaki, M.; Hatanaka, K. *H* **1992**, *33*, 503.
16. Stephenson, G. R.; Owen, D. A.; Finch, H.; Swanson, S. *TL* **1991**, *32*, 1291.
17. Fugina, N.; Holzer, W.; Wasicky, M. *H* **1992**, *34*, 303.
18. Shakya, S.; Durst, T. *H* **1992**, *34*, 67.
19. Beaucage, S. L.; Ogilvie, K. K. *TL* **1977**, 1691.
20. Ma, C.; Miller, M. J. *TL* **1991**, *32*, 2577.
21. Mandai, T.; Murakami, T.; Kawada, M.; Tsuji, J. *TL* **1991**, *32*, 3399.
22. Magnus, P.; Evans, A.; Lacour, J. *TL* **1992**, *33*, 2933.
23. Ihara, M.; Suzuki, S.; Taniguchi, N.; Fukumoto, K.; Kabuto, C. *CC* **1991**, 1168.
24. Rousseau, G.; Quendo, A. *T* **1992**, *48*, 6361.
25. Fleming, I.; Newton, T. W.; Sabin, V.; Zammattio, F. *T* **1992**, *48*, 7793.
26. Ito, T.; Okamoto, S.; Sato, F. *TL* **1990**, *31*, 6399.
27. Lopp, M.; Kanger, T.; Müraus, A.; Pehk, T.; Lille, Ü. *TA* **1991**, *2*, 943.
28. Yu, S.; Keay, B. A. *JCS(P1)* **1991**, 2600.
29. Mukai, C.; Kim, I. J.; Hanaoka, M. *TA* **1992**, *3*, 1007.
30. Degl'Innocenti, A.; Stucchi, E.; Capperucci, A.; Mordini, A.; Reginato, G.; Ricci, A. *SL* **1992**, 329.
31. Kocienski, P. J. *TL* **1979**, 2649.
32. Kocienski, P. J. *JOC* **1980**, *45*, 2037.
33. Hsiao, C. N.; Hannick, S. M. *TL* **1990**, *31*, 6609.
34. Bonini, B. F.; Masiero, S.; Mazzanti, G.; Zani, P. *TL* **1991**, *32*, 2971.
35. Nativi, C.; Palio, G.; Taddei, M. *TL* **1991**, *32*, 1583.
36. Okamoto, S.; Yoshino, T.; Tsujiyama, H.; Sato, F. *TL* **1991**, *32*, 5793.
37. Kobayashi, Y.; Ito, T.; Yamakawa, I.; Urabe, H.; Sato, F. *SL* **1991**, 813.
38. Kishi, N.; Maeda, T.; Mikami, K.; Nakai, T. *T* **1992**, *48*, 4087.
39. Hale, M. R.; Hoveyda, A. H. *JOC* **1992**, *57*, 1643.
40. Nakamura, E.; Kuwajima, I. *AG(E)* **1976**, *15*, 498.
41. Mohr, P. *TL* **1991**, *32*, 2223.
42. Furuta, K.; Mouri, M.; Yamamoto, H. *SL* **1991**, 561.
43. Hosomi, A.; Shirahata, A.; Sakurai, H. *TL* **1978**, 3043.
44. Nakamura, H.; Oya, T.; Murai, A. *BCJ* **1992**, *65*, 929.
45. Pornet, J. *TL* **1981**, *22*, 455.
46. Mead, K. T.; Park, M. *JOC* **1992**, *57*, 2511.
47. Hosomi, A.; Ogata, K.; Ohkuma, M.; Hojo, M. *SL* **1991**, 557.
48. Lasarte, J.; Palomo, C.; Picard, J. P.; Dunogues, J.; Aizpurua, J. M. *CC* **1989**, 72.
49. Watanabe, Y.; Takeda, T.; Anbo, K.; Ueno, Y.; Toru, T. *CL* **1992**, 159.
50. Paquette, L. A.; Blankenship, C.; Wells, G. J. *JACS* **1984**, *106*, 6442.
51. Seitz, D. E.; Milius, R. A.; Quick, J. *TL* **1982**, *23*, 1439.

52. Grotjahn, D. B.; Andersen, N. H. *CC* **1981**, 306.
53. Palomo, C.; Aizpurua, J. M.; Legido, M.; Picard, J. P.; Dunogues, J.; Constantieux, T. *TL* **1992**, *33*, 3903.
54. Majetich, G.; Casares, A.; Chapman, D.; Behnke, M. *JOC* **1986**, *51*, 1745.
55. Majetich, G.; Desmond, R. W.; Soria, J. J. *JOC* **1986**, *51*, 1753.
56. Ramaiah, P.; Prakash, G. K. S. *SL* **1991**, 643.
57. Coombs, M. M.; Zepik, H. H. *CC* **1992**, 1376.
58. Bansal, R. C.; Dean, B.; Hakomori, S.; Toyokuni, T. *CC* **1991**, 796.
59. Patel, N. R.; Kirchmeier, R. L. *IC* **1992**, *31*, 2537.
60. Himeshima, Y.; Sonoda, T.; Kobayashi, H. *CL* **1983**, 1211.
61. Tsukazaki, M.; Snieckus, V. *H* **1992**, *33*, 533.
62. Ito, Y.; Nakatsuka, M.; Saegusa, T. *JACS* **1980**, *102*, 863.
63. Ito, Y.; Miyata, S.; Nakatsuka, M.; Saegusa, T. *JOC* **1981**, *46*, 1043.
64. Bedford, S. B.; Begley, M. J.; Cornwall, P.; Knight, D. W. *SL* **1991**, 627.
65. Krafft, G. A.; Meinke, P. T. *TL* **1985**, *26*, 1947.
66. Botta, M.; Summa, V.; Saladino, R.; Nicoletti, R. *SC* **1991**, *21*, 2181.
67. Ben Ayed, T.; Amri, H.; El Gaied, M. M. *T* **1991**, *47*, 9621.
68. Sasson, Y.; Webster, O. W. *CC* **1992**, 1200.
69. Kuwajima, I.; Murofushi, T.; Nakamura, E. *S* **1976**, 602.
70. Yamamoto, Y.; Okano, H.; Yamada, J. *TL* **1991**, *32*, 4749.
71. Arya, P.; Wayner, D. D. M. *TL* **1991**, *32*, 6265.
72. Taguchi, T.; Suda, Y.; Hamochi, M.; Fujino, Y.; Iitaka, Y. *CL* **1991**, 1425.
73. Ley, S. V.; Smith, S. C.; Woodward, P. R. *T* **1992**, *48*, 1145, 3203.
74. White, J. D.; Ohira, S. *JOC* **1986**, *51*, 5492.
75. Ogilvie, K. K.; Beaucage, S. L.; Gillen, M. F. *TL* **1978**, 1663.
76. Díez-Barra, E.; De La Hoz, A.; Moreno, A.; Sánchez-Verdú, P. *JCS(P1)* **1991**, 2589.
77. Nakamura, E.; Shimizu, M.; Kuwajima, I.; Sakata, J.; Yokoyama, K.; Noyori, R. *JOC* **1983**, *48*, 932.
78. Colvin, E. W.; Seebach, D. *CC* **1978**, 689.
79. Fernández, R.; Gasch, C.; Gómez-Sánchez, A.; Vílchez, J. E. *TL* **1991**, *32*, 3225.
80. Cox, D. P.; Terpinski, J.; Lawrynowicz, W. *JOC* **1984**, *49*, 3216.
81. Clark, J. H.; Smith, D. K. *TL* **1985**, *26*, 2233.
82. Maeda, M.; Abe, M.; Kojima, M. *JFC* **1987**, *34*, 337.
83. Ueki, M.; Amemiya, M. *TL* **1987**, *28*, 6617.
84. Ueki, M.; Aoki, H.; Katoh, T. *TL* **1993**, *34*, 2783.
85. Namikoshi, M.; Kundu, B.; Rinehart, K. L. *JOC* **1991**, *56*, 5464.
86. Kundu, B. *TL* **1992**, *33*, 3193.

Hui-Yin Li
Du Pont Merck Pharmaceutical Company, Wilmington, DE, USA

Tetra-*n*-butylammonium Hydroxide

n-Bu_4NOH

[2052-49-5] $C_{16}H_{37}NO$ (MW 259.54)

(quaternary ammonium salt; strong base; phase-transfer catalyst[1])

Alternate Name: TBAH.
Solubility: sol most organic solvents (alcohols, hydrocarbons, aromatics, halogenated solvents), H_2O.
Form Supplied in: 40 wt % solution in H_2O; 1.0 M solution in MeOH. A typical impurity is tributylamine.
Analysis of Reagent Purity: aqueous titration with HCl versus phenolphthalein;[2] nonaqueous titration with benzoic acid in pyridine and thymol blue indicator.[3]
Preparative Method: prepared in situ from ammonium halides.[2]
Handling, Storage, and Precautions: methanol solution is hygroscopic; highly toxic; may decompose on heating.[4]

Alkylations. Tetrabutylammonium hydroxide (**1**) has been used to effect a variety of organic transformations. A strongly basic catalyst, (**1**) was able to promote alkylation (eq 1)[5] and arylation[6] of nitroalkanes under phase-transfer conditions. The intermediate tetrabutylammonium *aci*-nitronates generated were more reactive than the corresponding lithium salts.

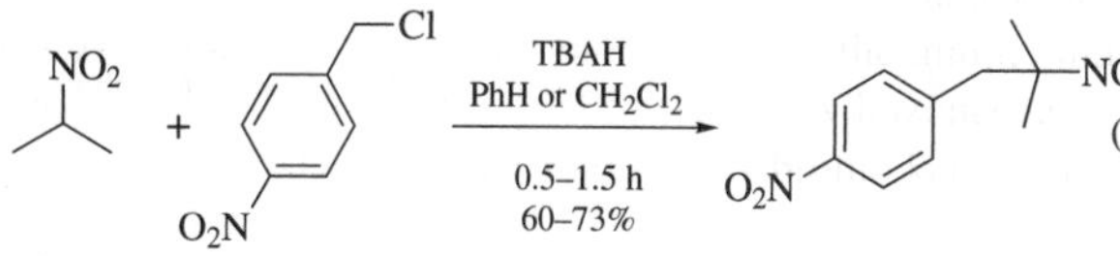

TBAH, PhH or CH_2Cl_2, 0.5–1.5 h, 60–73% (1)

The monoalkylation of cyanomethanephosphonic diamides was reported to occur with high selectivity and good yield (eq 2).[7] Strong amide bases or ***Sodium Hydride*** led to formation of mixtures of mono- and dialkylation products. Alkylations of other active hydrogen compounds have been studied.[8]

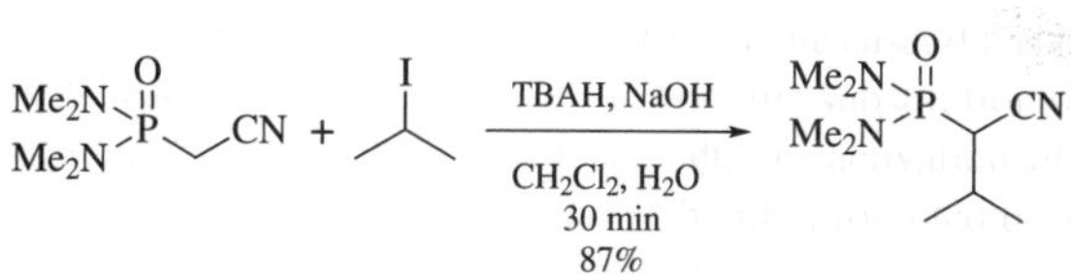

TBAH, NaOH, CH_2Cl_2, H_2O, 30 min, 87% (2)

The selective monobenzylation of an indole nitrogen under phase-transfer conditions proceeds in better than 95% yield for a variety of substrates (eq 3).[9] The reaction is procedurally simple when compared to the alkylation of the dianion normally generated under anhydrous conditions with ***n-Butyllithium*** (***Hexamethylphosphoric Triamide***, THF, 0 °C), which produces only a 60% yield of product.

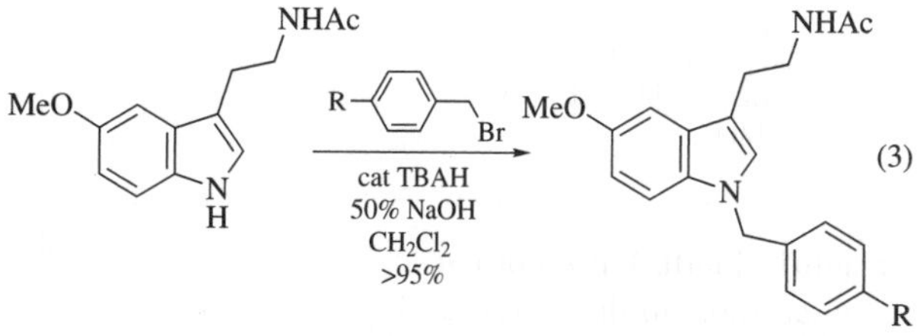

cat TBAH, 50% NaOH, CH_2Cl_2, >95% (3)

R = CO_2Me

An extension of this chemistry is seen in the intramolecular alkylations of nitrogen which have been reported in the preparation of dihydropyridones (eq 4).[10]

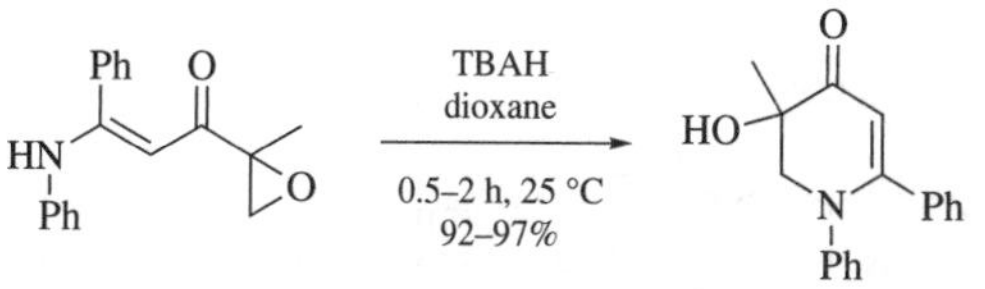

TBAH, dioxane, 0.5–2 h, 25 °C, 92–97% (4)

Tetrabutylammonium hydroxide has also been used in the Robinson annulation of dihydrocarvone with ***Methyl Vinyl Ketone*** in 63% yield, which is reported to improve upon other methods.[11] Dehydration does not occur in the reaction and workup is simplified (eq 5). Similarly, intramolecular Horner–Emmons closure of an activated phosphono ester was used in the synthesis of (±)-silphinene (eq 6).[12]

TBAH, MVK; EtOH, DMSO, 0 °C (5)

TBAH; PhH, H_2O, 70% (6)

Dehydrations. Vicinal dihydro diols of polycyclic aromatic hydrocarbons such as the benzo[*a*]pyrene derivative (**2**) were dehydrated under mild conditions by treatment with TBAH in methanol (eq 7).[13] Alkali hydroxides did not effect the reaction in these cases without a catalyst present (crown ether or ammonium salt), and methanol was necessary as solvent. The regioselectivity for the dehydration (**3**:**4** = 13:87) was opposite to that from a ***Phosphoric Acid***-catalyzed dehydration (**3**:**4** = 97:3). The base-catalyzed reaction did not proceed for substrates which lacked aromatization capability. A patent has been obtained for the monodehydration of a fluorinated dihydroxybenzene under similar conditions.[14]

TBAH; MeOH, 60 °C, 89% (7)

(**2**)

(**3**) $R^1 = H$, $R^2 = OH$
(**4**) $R^1 = OH$, $R^2 = H$

Carbenes. The reaction of (**1**) with ***Chloroform*** under phase-transfer conditions is a good method for generating dichlorocarbene.[1c] This reaction was used to cleanly generate aldehyde (**6**) from dehydronuciferene (**5**), which is nucleophilic enough to attack dichlorocarbene (eq 8).[15] In contrast, compound (**5**) could not be alkylated with ***Iodomethane***.

50% NaOH, cat TBAH; $CHCl_3$, 60 °C, 4 h (8)

(**5**) (**6**)

A phase-transfer reaction mediated by tetrabutylammonium hydroxide was found to be most suitable for the halo lactonization of alkynoic acids to prepare halo enol lactones (eq 9).[16] The reaction was carried out with halosuccinimides in the presence of catalytic TBAH, and yields of 70–93% were obtained.

halosuccinimide, 2 mol % TBAH; $KHCO_3$, CH_2Cl_2, H_2O (9)

X = Br, 89%
X = I, 81%

Related Reagents. Benzyltrimethylammonium Hydroxide; Lithium Hydroxide; Potassium Hydroxide; Sodium Hydroxide; Tetramethylammonium Hydroxide.

1. For reviews of phase transfer reactions, see (a) Keller, W. E. *Phase Transfer Reactions. Fluka Compendium*; Thieme Verlag: Stuttgart, 1986; Vols. 1 and 2. (b) Dehmlow, E. V. *Phase Transfer Catalysis*; Verlag Chemie: Deerfield Beach, FL, 1980. (c) Starks, C. M.; Liotta, C. *Phase Transfer Catalysis, Principles and Techniques*; Academic: New York, 1978. (d) Dockx, J. *S* **1973**, 441. (e) Dehmlow, E. V. *AG(E)* **1974**, *13*, 170. For a mechanistic review of hydroxide-mediated reactions under PTC conditions, see: (f) Rabinovitz, M.; Cohen, Y.; Halpern, M. *AG(E)* **1986**, *25*, 960.
2. (a) In alcohol using silver oxide and Bu_4I: Cluett, M. L. *Anal. Chem.* **1959**, *31*, 610. (b) In alcohol using ion exchange resin and Bu_4I: Harlow, G. A.; Noble, C. M.; Wyld, G. E. A. *Anal. Chem.* **1956**, *28*, 787.
3. Gandrud, B. W.; Lazrus, A. L. *Anal. Chem.* **1983**, *55*, 988.
4. Smith, P. A. S.; Frank, S. *JACS* **1952**, *74*, 509. See also Dehmlow, E. V.; Knufinke, V. *JCR(S)* **1989**, 224 for a study on the stability and performance of various phase transfer catalysts.
5. Burt, B. L.; Freeman, D. J.; Gray, P. G.; Norris, R. K.; Randles, D. *TL* **1977**, 3063.
6. Norris, R. K.; Randles, D. *AJC* **1979**, *32*, 2413.
7. Blanchard, J.; Collignon, N.; Savignac, P.; Normant, H. *S* **1975**, 655.
8. For malonates, benzyl nitrile, and benzyl methyl ketone, see, for example: Brändström, A.; Junggren, U. *TL* **1972**, 473. For cyclic diketones, see: Kagechika, K.; Shibasaki, M. *JOC* **1991**, *56*, 4093.
9. de Silva, S. O.; Snieckus, V. *CJC* **1978**, *56*, 1621.
10. Zvonok, A. M.; Kuz'menok, N. M.; Stanishevskii, L. S. *KGS* **1988**, 1022.
11. Chen, E. Y. *SC* **1983**, *13*, 927.
12. Rao, Y. K.; Nagarajan, M. *TL* **1988**, *29*, 107.
13. McCourt, D. W.; Roller, P. P.; Gelboin, H. V. *JOC* **1981**, *46*, 4157.
14. Ryback, G. U.S. Patent 4 855 512, 1989 (*CA* **1989**, *110*, 134 878r).
15. Saá, J. M.; Cava, M. P. *JOC* **1977**, *42*, 347.
16. Krafft, G. A.; Katzenellenbogen, J. A. *JACS* **1981**, *103*, 5459.

Mary Ellen Bos
The R. W. Johnson Pharmaceutical Research Institute, Raritan, NJ, USA

Tetrafluoroboric Acid

BF_4H

[16872-11-0]	BF_4H	(MW 87.82)
($HBF_4 \cdot OMe_2$) *[67969-83-9]*	$C_2H_7BF_4O$	(MW 133.90)
($HBF_4 \cdot OEt_2$) *[67969-82-8]*	$C_4H_{11}BF_4O$	(MW 161.96)

(strong acid with $pK_a = -0.44$[1] and a noncoordinating counterion[2])

Solubility: sol H_2O, alcohols and ethers.

Form Supplied in: 48% aq solution; diethyl ether complex; dimethyl ether complex.

Purification: commercial 50% solutions can be concentrated by evaporation at 50–60 °C/5 mmHg to give a residue of 11 N total acidity. Water can also be removed by slow addition to ice-cold acetic anhydride (dangerously exothermic reaction: caution is advised!).[3]

Handling, Storage, and Precautions: storage in glass containers is not recommended, although dilute solutions can be used in glass containers over short periods. Poisonous: causes burns to eyes, skin, and mucous membranes and may be fatal if ingested. Use in a fume hood with safety goggles and chemically resistant gloves and clothing. Incompatible with cyanides and strong bases: decomposes to form HF.

General Considerations. Tetrafluoroboric acid is a useful strong Brønsted acid with a nonnucleophilic counterion. Its solubility in polar organic solvents makes it a useful strong acid under nonaqueous conditions.

Preparation of Arenediazonium Tetrafluoroborates. Arenediazonium salts can be precipitated as their tetrafluoroborate salts by addition of a cold solution of HBF_4 to a solution of the initial arenediazonium chloride. 4-Methoxybenzenediazonium tetrafluoroborate is thus obtained in 94–98% yield. The arylamine can be diazotized with ***Sodium Nitrite*** in the presence of HBF_4, and the diazonium fluoroborate then precipitates directly. 3-Nitroaniline, when treated in this manner, provides a 90–97% yield of 3-nitrobenzenediazonium tetrafluoroborate (eq 1).[4] Diazotizations of arylamines and heteroarylamines in organic solvents can be conveniently conducted using $HBF_4{\cdot}OEt_2$ and alkyl nitrites.[5]

NH_2 / NO_2 — $NaNO_2$, aq. HBF_4, 0 °C, 90–97% → N_2BF_4 / NO_2 (1)

Protecting-Group Manipulations. A preparation of monoesters of glutamic and aspartic acids using $HBF_4{\cdot}OEt_2$ as a catalyst has been developed. The diesters are generally side products when other acids are used, but $HBF_4{\cdot}OEt_2$ appears to suppress the diesterification. Thus L-glutamic acid γ-benzyl ester was obtained in 94% yield by treatment of the carboxylic acid in BnOH with the $HBF_4{\cdot}OEt_2$ catalyst and a drying agent.[6]

When di-*t*-butyl carbonate is treated with DMAP·HBF_4, water-soluble Boc-pyridinium tetrafluoroborate is formed.[36] This reagent installs the Boc protecting group onto amino acids in aqueous NaOH. Thus L-proline is converted to its *N*-Boc derivative in nearly quantitative yield in 10 min.

Carbohydrate protection and deprotection reactions are amenable to HBF_4 catalysis; aqueous HBF_4 and $HBF_4{\cdot}OEt_2$ are complementary in these applications. Transacetalization with benzaldehyde dimethyl acetal and ethereal HBF_4 gives a monobenzylidene product without disturbing the isopropylidene and trityl groups of the substrate (eq 2). Aqueous HBF_4 is useful in selective deprotection reactions, cleaving a trityl group in the presence of other acid-sensitive functionality (eq 3).[7]

TrO, HO, OH — $PhCH(OMe)_2$, $HBF_4{\cdot}OEt_2$, DMF, 25 °C, 83% → TrO, Ph (2)

OTr, OAc, AcO, OBn, NHBz — aq HBF_4, 95% → OH, OAc, AcO, OBn, NHBz (3)

Recently developed in solid-phase peptide synthesis is the use of HBF_4/thioanisole/*m*-cresol as a general deprotection reagent. Most commonly used groups are cleaved from termini or side chain functionality under mild conditions (1 M HBF_4, 4 °C, 30–60 min) in very high yields. Human glucagon (a 29 residue peptide) has been synthesized as a demonstration of the effectiveness of the method,[8] which has been successful in both solid-phase and solution-phase syntheses.[9]

General Acid Catalysis. HBF_4 is a versatile acid catalyst which is applicable to many typical acid-catalyzed reactions. Some adducts obtained upon reaction of cyclopropyl phenyl sulfide anion with carbonyl compounds can be rearranged to cyclobutanones under the catalytic influence of HBF_4 (eq 4), although different acids have been required for other substrates.[10] Acid-catalyzed rearrangement of 2-cyclopropylphenyl phenyl ether or sulfide with HBF_4 has been reported (eq 5).[11]

PhS, OH — HBF_4, ether, 20 °C, 84% → O (4)

ZPh, Z = S, O — HBF_4, 92% → Et, Z (5)

Acid-catalyzed oxidation of epoxides with $HBF_4{\cdot}OMe_2$/DMSO results in the formation of α-hydroxy ketones (eq 6).[12] This procedure in an acidic medium complements the α-hydroxylation of ketone enolates under strongly basic conditions.

O — $HBF_4{\cdot}OMe_2$, DMSO, 4Å mol. sieves, 88% → O, OH (6)

An intramolecular alkylation of a diazomethyl ketone was achieved with catalysis by HBF_4, providing an angularly fused cyclopentanone hydrofluorene (eq 7).[13]

O, N_2 — 48% HBF_4, $MeNO_2$, 84% → O (7)

Other catalytic applications of HBF_4 include alkene isomerizations,[14] alkylation of alcohols with diazoalkanes,[15] preparations of substituted pyridines,[16] hydrolysis of α-hydroxyketene or α-(methylthio)ketene thioacetals to α,β-unsaturated thioesters,[17] and terpene formation from isoprenic precursors.[18]

Preparation of Annulated Triazolones. Cyclization of the isocyanate in eq 8 in the presence of HBF_4 affords an oxotriazolium tetrafluoroborate, which then rearranges to form a 1,5-heteroannulated 1,2-dihydro-2-phenyl-3*H*-1,2,4-triazol-3-one (eq 8).[19]

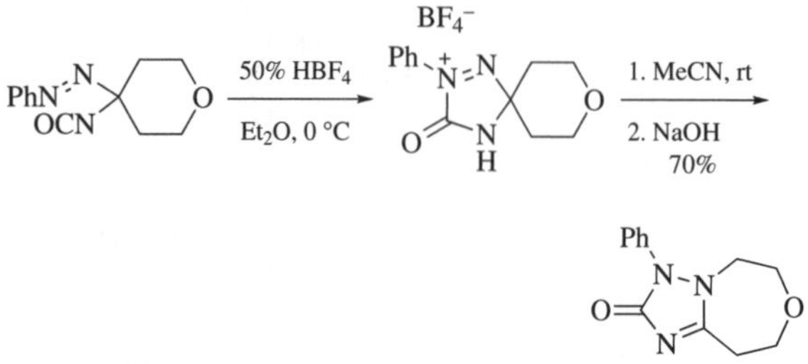

(8)

Carbenium Tetrafluoroborate Preparation. 4,6,8-Trimethylazulene is converted by ethereal HBF_4 into 4,6,8-trimethylazulenium tetrafluoroborate.[20] A convenient preparation of tropylium tetrafluoroborate employs HBF_4 to precipitate the product from a solution of the double salt $[C_7H_7]PCl_6\cdot[C_7H_7]Cl$ in ethanol (eq 9). An indefinitely stable, nonhygroscopic, and nonexplosive white solid is obtained, a distinct advantage of a fluoroborate salt.[21]

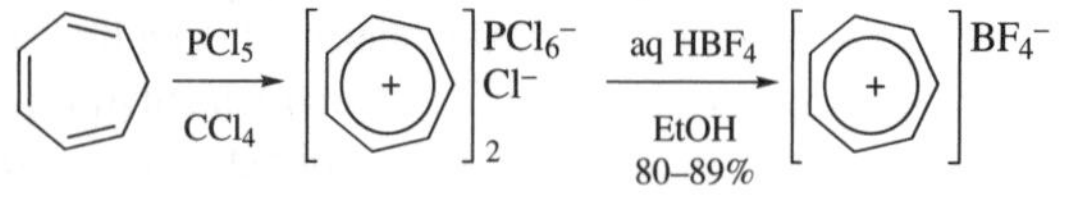

(9)

The formyl cation equivalent 1,3-benzodithiolylium tetrafluoroborate can be made in 94% yield by treatment of a dithioorthoformate with HBF_4/Ac_2O (eq 10).[22] 2-Deuterio-1,3-benzothiolylium tetrafluoroborate prepared by this method has been used to produce 1-deuterioaldehydes by homologation.[23] An analogous preparation of a similar formyl cation equivalent, 1,3-dithiolan-2-ylium tetrafluoroborate, employs $HBF_4\cdot OEt_2$ (eq 11) to provide the intermediate salt. Subsequent reaction with a silyl enol ether forms a masked 2-formylcyclohexanone in high yield (eq 11).

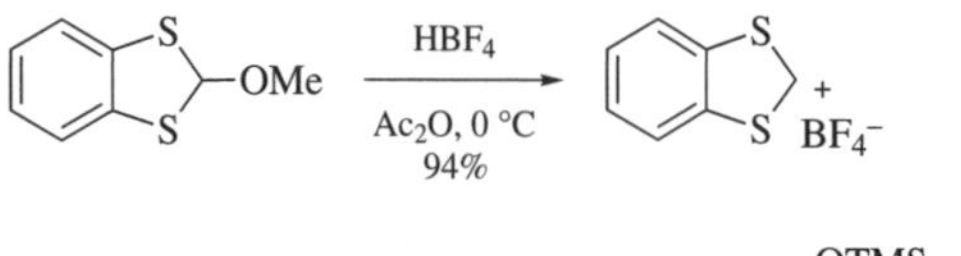

(10)

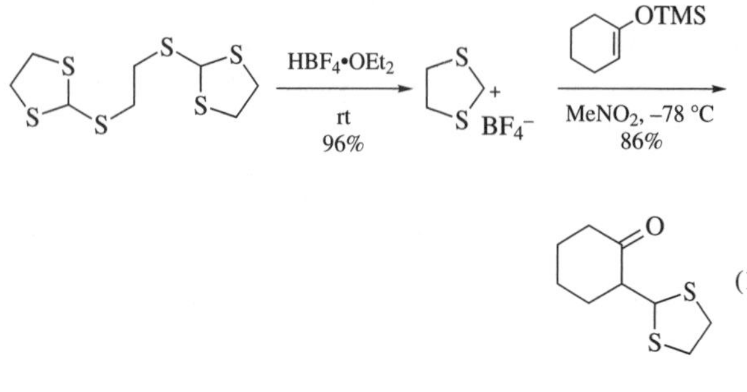

(11)

Ethynylfluorenylium dyes can be obtained by treatment of appropriate tertiary alcohols with HBF_4 (eq 12).[24]

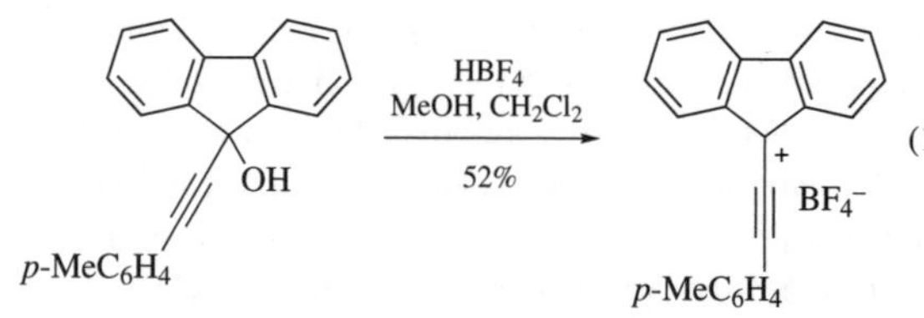

(12)

In addition to the aforementioned carbenium ions and diazonium ions, numerous other organic cations can be obtained as their fluoroborate salts by treatment of appropriate precursors with HBF_4.[25]

Dimerization of Carbodiimides. Treatment of alkyl carbodiimides with anhydrous $HBF_4\cdot OEt_2$ in CH_2Cl_2 results in rapid dimerization to tetrafluoroborate salts in 95% yield (eq 13). Basification converts the salts to diazetidines.[26] In the same work, aryl carbodiimides undergo a similar reaction, but substituted quinazolines are obtained.

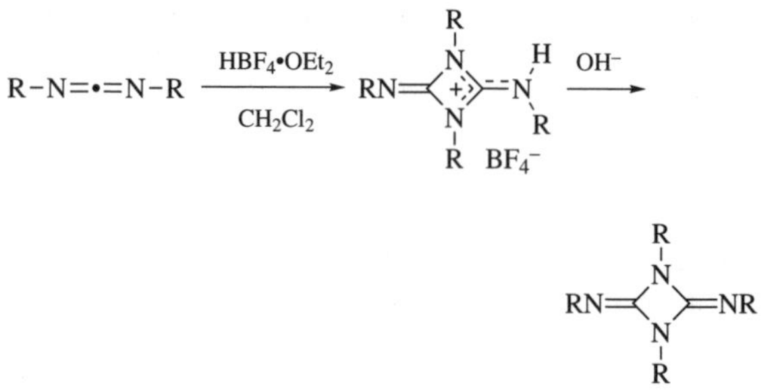

(13)

Mercury(II) Oxide/Tetrafluoroboric Acid. Yellow ***Mercury(II) Oxide*** is added to 48% aqueous HBF_4 to yield, upon solvent removal, $HgO\cdot 2HBF_4$ as a hygroscopic white solid.[27] This reagent is useful in applications involving mercuration of alkenes, including diamination of alkenes and preparation of *trans*-cinnamyl ethers from allylbenzene.[28] Alkylations of carboxylic acids[29] and alcohols[30] with alkyl halides are also facilitated by $HgO\cdot 2HBF_4$. Mercury(II) oxide and HBF_4 in alcohol effected mild solvolysis of 2-hydroxytrithioorthoesters to yield α-hydroxycarboxylic esters in high yield.[31] See also ***Mercury(II) Oxide-Tetrafluoroboric Acid***.

Preparation of a Useful Hypervalent Iodine Reagent. When treated with $HBF_4\cdot OMe_2$ at low temperatures, ***Iodosylbenzene*** reacts with silyl enol ethers to form a hypervalent iodine adduct capable of useful carbon–carbon bond formation reactions with alkenes (eq 14).[32]

Synthesis of Cationic Organometallic Complexes. Tetrafluoroborate is frequently encountered as the counterion in cationic organometallic compounds; its lack of nucleophilic reactivity makes HBF_4 and its etherates ideal reagents for delivery of protons without side reactions. The poorly coordinating conjugate base of HBF_4 allows substrates greater opportunity to bind to metals in organometallic reactions requiring the presence of acids.[33]

$$PhIO + HBF_4{\cdot}OMe_2 \xrightarrow[-50\ to\ -20\ °C]{CH_2Cl_2} PhIO{\cdot}HBF_4$$

OTMS, Ph

$[PhICH_2COPh]^+BF_4^-$ (14)

cyclohexene; OTMS; $Me_2C{=}CMe_2$

Ph, O, 90%; Ph, O, O, 50%; Ph, O, 80%

Propargylium complexes of cobalt, obtained by treatment of propargylic alcohols with $HBF_4{\cdot}OEt_2$, have been studied with regard to their selectivity as alkylating agents. *N*-Acetyl-3,4-dimethoxyphenethylamine undergoes selective aromatic substitution, whereas the unprotected amine undergoes *N,N*-dialkylation but not aromatic substitution (eq 15).[34]

HO, $Co(CO)_3$, $Co(CO)_3$ $\xrightarrow{HBF_4{\cdot}OEt_2}$

^+H_2C—≡, BF_4^- $Co_2(CO)_6$; MeO, MeO, NHAc → MeO, MeO, NHAc, $(OC)_6Co_2$ (15)

Oxidation. In a reaction proposed as a model for substrate reactions at metal–sulfur centers of enzymes, HBF_4 apparently functions as an oxidizing agent in a two-electron oxidation of a ruthenium benzenedithiolate complex (eq 16).[35]

NBu_4^+ [PMe_3, S, Ru, S, S, S, PMe_3]$^-$ $\xrightarrow[20\ °C]{HBF_4,\ MeOH}$ [PMe_3, S, Ru, S, S, S, PMe_3] $^+BF_4^-$ (16)

Related Reagents. Ammonium Tetrafluoroborate; Methanesulfonic Acid; Trifluoroacetic Acid.

1. (a) Sudakova, T. N.; Krasnoshchekov, V. V. *Zh. Neorg. Khim.* **1978**, *23*, 1506. (b) Acidity relative to other strong acids: Bessiere, J. *BSF* **1969**, *9*, 3356.
2. Ellis, R.; Henderson, R. A.; Hills, A.; Hughes, D. L. *JOM* **1987**, *333*, C6.
3. (a) Lichtenberg, D. W.; Wojcicki, A. *JOM* **1975**, *94*, 311. (b) Wudl, F.; Kaplan, M. L. *Inorg. Synth.* **1979**, *19*, 27.
4. Roe, A. *OR* **1949**, *5*, 193. See also: (a) Starkey, E. B. *OSC* **1943**, *2*, 225. (b) Curtin, D. Y.; Ursprung, J. A. *JOC* **1956**, *21*, 1221. (c) Schiemann, G.; Winkelmuller, W. *OSC* **1943**, *2*, 299.
5. (a) Cohen, T.; Dietz, A. G., Jr.; Miser, J. R. *JOC* **1977**, *42*, 2053. (b) Allmann, R.; Debaerdemaeker, T.; Grehn, W. *CB* **1974**, *107*, 1555.
6. Albert, R.; Danklmaier, J.; Honig, H.; Kandolf, H. *S* **1987**, 635.
7. Albert, R.; Dax, K.; Pleschko, R.; Stutz, A. E. *Carbohydr. Res.* **1985**, *137*, 282.
8. Akaji, K.; Yoshida, M.; Tatsumi, T.; Kimura, T.; Fujiwara, Y.; Kiso, Y. *CC* **1990**, 288.
9. Kiso, Y.; Yoshida, M.; Tatsumi, T.; Kimura, T.; Fujiwara, Y.; Akaji, K. *CPB* **1989**, *37*, 3432.
10. Trost, B. M.; Keeley, D. E.; Arndt, H. C.; Bogdanowicz, M. J. *JACS* **1977**, *99*, 3088.
11. Shabarov, Y. S.; Pisanova, E. V.; Saginova, L. G. *ZOR* **1980**, *16*, 418 (*CA* **1981**, *94*, 3819a).
12. Tsuji, T. *BCJ* **1989**, *62*, 645.
13. Ray, C.; Saha, B.; Ghatak, U. R. *SC* **1991**, *21*, 1223.
14. Powell, J. W.; Whiting, M. C. *Proc. Chem. Soc.* **1960**, 412.
15. (a) Neeman, M.; Johnson, W. S. *OS* **1961**, *41*, 9. (b) Brückner, R.; Peiseler, B. *TL* **1988**, *29*, 5233.
16. (a) Schulz, W.; Pracejus, H.; Oehme, G. *J. Mol. Catal.* **1991**, *66*, 29. (b) Kanemasa, S.; Asai, Y.; Tanaka, J. *BCJ* **1991**, *64*, 375.
17. Dieter, R. K.; Lin, Y. J.; Dieter, J. W. *JOC* **1984**, *49*, 3183.
18. Babin, D.; Fourneron, J.-D.; Julia, M. *BSF(2)* **1980**, 588.
19. Gstasch, H.; Seil, P. *S* **1990**, 1048.
20. (a) Hafner, K.; Pelster, H.; Schneider, J. *LA* **1961**, *650*, 62. (b) Hafner, K.; Pelster, H.; Patzelt, H. *LA* **1961**, *650*, 80.
21. Conrow, K. *OS* **1963**, *43*, 101.
22. Nakayama, J.; Fujiwara, K.; Hoshino, M. *CL* **1975**, 1099.
23. Nakayama, J. *BCJ* **1982**, *55*, 2289.
24. Nakatsuji, S.; Nakazumi, H.; Fukuma, H.; Yahiro, T.; Nakashima, K.; Iyoda, M.; Akiyama, S. *JCS(P1)* **1991**, 1881.
25. A few examples: (a) oxotriazolium: Gstasch, H.; Seil, P. *S* **1990**, 1048. (b) pyridinium: Paley, M. S.; Meehan, E. J.; Smith, C. D.; Rosenberger, F. E.; Howard, S. C.; Harris, J. M. *JOC* **1989**, *54*, 3432. (c) pyridinium: Guibe-Jampel, E.; Wakselman, M. *S* **1977**, 772. (d) tetrameric dication from 2-aminobenzaldehyde: Skuratowicz, J. S.; Madden, I. L.; Busch, D. H. *IC* **1977**, *16*, 1721. (e) tetrathiafulvenium: Wudl, F. *JACS* **1975**, *97*, 1962. Wudl, F.; Kaplan, M. L. *Inorg. Synth.* **1979**, *19*, 27. (f) sulfonium: LaRochelle, R. W.; Trost, B. M. *JACS* **1971**, *93*, 6077. (g) diazetidinium: Hartke, K.; Rossbach, F. *AG(E)* **1968**, *7*, 72.
26. Hartke, K.; Rossbach, F. *AG(E)* **1968**, *7*, 72.
27. Barluenga, J.; Alonso-Cires, L.; Asensio, G. *S* **1979**, 962.
28. Barluenga, J.; Alonso-Cires, L.; Asensio, G. *TL* **1981**, *22*, 2239.
29. Barluenga, J.; Alonso-Cires, L.; Campos, P. J.; Asenio, G. *S* **1983**, 649.
30. Barluenga, J.; Alonso-Cires, L.; Campos, P. J.; Asensio, G. *T* **1984**, *40*, 2563.
31. Scholz, D. *SC* **1982**, *12*, 527.
32. Zhdankin, V. V.; Tykwinski, R.; Caple, R.; Berglund, B.; Koz'min, A. S.; Zefirov, N. S. *TL* **1988**, *29*, 3703.
33. Ellis, R.; Henderson, R. A.; Hills, A.; Hughes, D. L. *JOM* **1987**, *333*, C6. Some other recent examples of the use of $HBF_4{\cdot}OEt_2$ in synthesis and/or reactions of organometallics: (a) Field, J. S.; Haines, R. J.; Stewart, M. W.; Sundermeyer, J.; Woollam, S. F. *JCS(D)* **1993**, 947. (b) Dawson, D. M.; Henderson, R. A.; Hills, A.; Hughes, D. L. *JCS(D)* **1992**, 973. (c) Lemos, M. A. N. D. A.; Pombeiro, A. J. L.; Hughes, D. L.; Richards, R. L. *JOM* **1992**, *434*, C6. (d) Arliguie, T.; Chaudret, B.; Jalon, F. A.; Otero, A.; Lopez, J. A.; Lahoz, F. J. *OM* **1991**, *10*, 1888. (e) Bassner, S. L.; Sheridan, J. B.; Kelley, C.; Geoffroy, G. L. *OM* **1989**, *8*, 2121. For use of $HBF_4{\cdot}OMe_2$: (a) Schrock, R. R.; Liu, A. H.; O'Regan, M. B.; Finch, W. C.; Payack, J. F. *IC* **1988**, *27*, 3574. (b) Blagg, J.; Davies, S. G.; Goodfellow, C. L.; Sutton, K. H. *JCS(P1)* **1987**, 1805.

34. Gruselle, M.; Philomin, V.; Chaminant, F.; Jaouen, G.; Nicholas, K. M. *JOM* **1990**, *399*, 317.

35. Sellmann, D.; Binker, G.; Knoch, F. *ZN(B)* **1987**, *42*, 1298.

36. Guibe-Jampel, E.; Wakselman, M. *S* **1977**, 772.

Gregory K. Friestad & Bruce P. Branchaud
University of Oregon, Eugene, OR, USA

Tetrahydro-1-methyl-3,3-diphenyl-1*H*,3*H*-pyrrolo[1,2-*c*][1,3,2]oxazaborole[1]

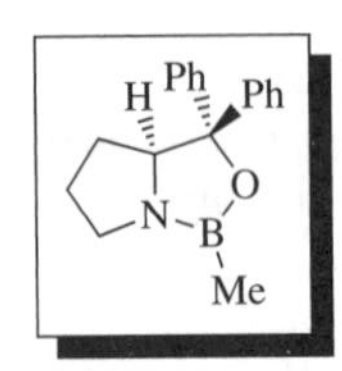

(*S*)
[112022-81-8] $C_{18}H_{20}BNO$ (MW 277.20)
($\cdot BH_3$)
[112022-90-9]
(*R*)112022-83-0

(one of many chiral oxazaborolidines/chiral Lewis acids useful as enantioselective catalysts for the reduction of prochiral ketones,[1–3] imines,[4] and oximes,[2e,f,5] and the reduction of 2-pyranones to afford chiral biaryls;[6] other chiral oxazaborolidines have been used for the addition of diethylzinc to aldehydes,[7] asymmetric hydroboration,[8a,b] the Diels–Alder reaction,[9–11] and the aldol reaction[12,13])

Physical Data: mp 79–81 °C.
Solubility: very sol THF, CH_2Cl_2, toluene.
Preparative Methods: see text.
Purification: Kugelrohr distillation (50 °C/0.001 mbar)
Handling, Storage, and Precautions: the free oxazaborolidine must be rigorously protected from exposure to moisture. The crystalline borane complex is more stable, and is the preferred form to handle and store this catalyst.

Enantioselective Ketone Reduction. The major application of chiral oxazaborolidines has been the stoichiometric (as the oxazaborolidine–borane complex) (eq 1) and catalytic (in the presence of a stoichiometric borane source) (eq 2) enantioselective reduction of prochiral ketones.[1] These asymmetric catalysts work best for the reduction of aryl alkyl ketones, often providing very high (>95% ee) levels of enantioselectivity.

(1)

(2)

Following from the work of Itsuno[2] and Corey,[3] over 75 chiral oxazaborolidine catalysts have been reported for the reduction of prochiral ketones [(**1**),[2,3a,14,15a,e,f,16d–f,17b] (**2**),[16d,18b] (**3**),[3,6,19b–e,20,21,26c] (**4**),[16a] (**5**),[1b,16c,22] (**6**),[22b] (**7**),[3d,18a] (**8**),[16b] (**9**),[23] (**10**),[24] (**11**),[24] (**12**),[19a]]. Oxazaborolidines derived from proline (**3**) (see ***α,α-Diphenyl-2-pyrrolidinem ethanol***) and valine (**1**; R^4 = *i*-Pr) (see ***2-Amino-3-methyl-1,1-diphenyl-1-butanol***) have received the most attention.

(1) (2) (3)

(4) (5) (6)

(7) (8) (9)

(10) (11) (12)

Unsubstituted (B–H) oxazaborolidines (**16**) are prepared from a chiral β-amino alcohol (**13**) and a source of borane (***Diborane***, ***Borane–Tetrahydrofuran***, ***Borane–Dimethyl Sulfide***, or $H_3B \cdot NMe_3$) via a multistep process (eq 3). Formation of the initial amine–borane complex (**14**) is generally exothermic, and this intermediate can often be isolated. Gentle heating with the loss of one mole of hydrogen results in the formation of (**15**). Continued heating with the loss of a second mole of hydrogen then affords oxazaborolidine (**16**). When R^4 and R^5 are connected, forming a four- or five-membered ring, more forcing conditions (70–75 °C, 1.7 bar, 48–72 h) are required to effect this conversion due to the additional ring strain. [*Caution*: under these conditions, borane or diborane in the vapor phase can begin to decompose.[25]] Finally, additional borane is added to afford the oxazaborolidine–borane complex (**17**).

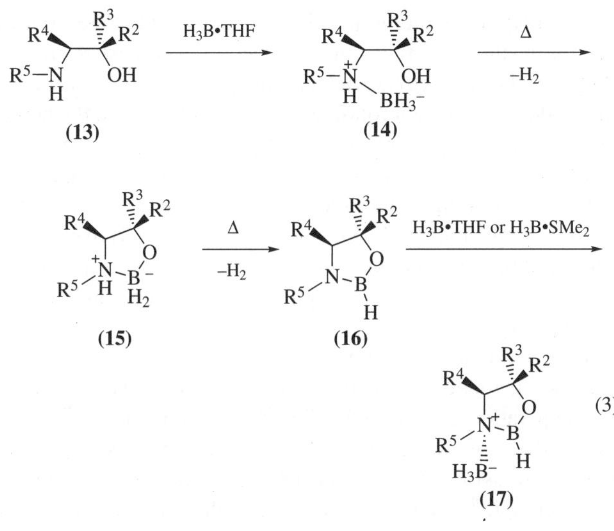

(3)

Free oxazaborolidine (**16**), by itself, will not reduce ketones. Furthermore, (**16**) is not particularly stable, reacting with moisture (H_2O), air (O_2), unreacted amino alcohol, other alcohols,[8c] or, depending on the substituents, with itself to form various dimers.[3a,8c,d,15d,26,27a] This instability is due to the strain of a partial double bond between nitrogen and boron (eq 4). Formation of the oxazaborolidine–borane complex (**17**) tends to release some of this strain. As such, (**16**) and (**17**) are generally prepared and used in situ without isolation; in many cases, they have not been fully characterized.[17c]

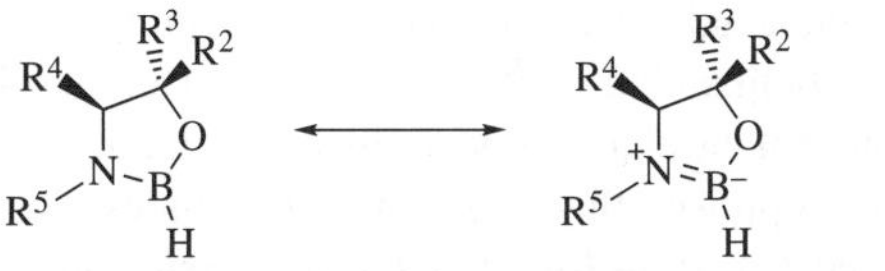

(4)

Oxazaborolidines substituted at boron (**1**; R^1 = alkyl, aryl) are prepared from a chiral β-amino alcohol and the corresponding boronic acid in a two-step process (eq 5).[3b,9] Heat and an efficient method of water removal (i.e. azeotropic distillation, molecular sieves) are required to drive the second step. When R^4 and R^5 are connected, more forcing conditions are necessary, both to complete the second step and to prevent the intermediate from proceeding to an alternate disproportionation product.[21] Alternative procedures using bis(diethylamino)phenylborane (eq 6),[26a,b] trisubstituted boroxines (eq 7),[21,27] and ethyl or butyl bis(trifluoroethyl)boronate esters (eq 8)[19e] have been developed to circumvent these problems. The substituted oxazaborolidines are more stable than unsubstituted (B–H) oxazaborolidines (i.e. they can be handled in the presence of air, and do not form dimers), but are still prone to decomposition by moisture (H_2O).[21] In many cases the substituted oxazaborolidines have been isolated, purified, and characterized.

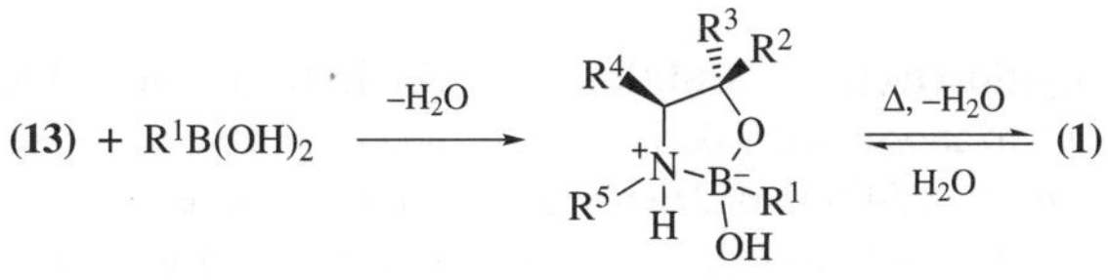

(**13**) + $R^1B(OH)_2$ $\xrightarrow{-H_2O}$ ⇌ (**1**) (5)

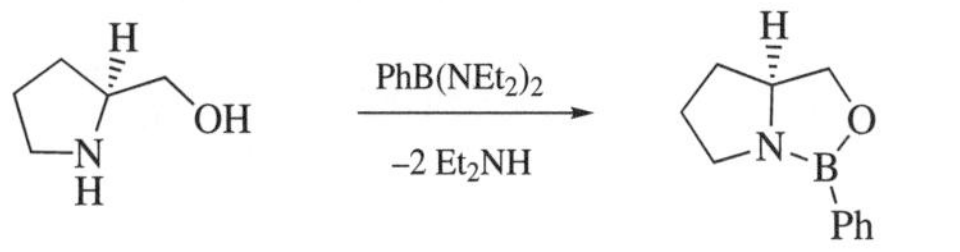

(6)

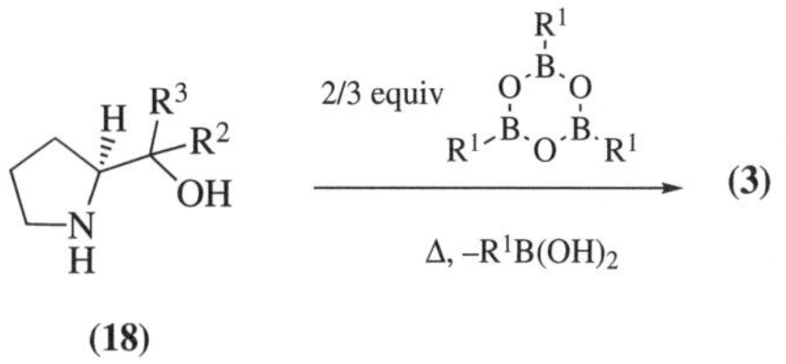

(7)

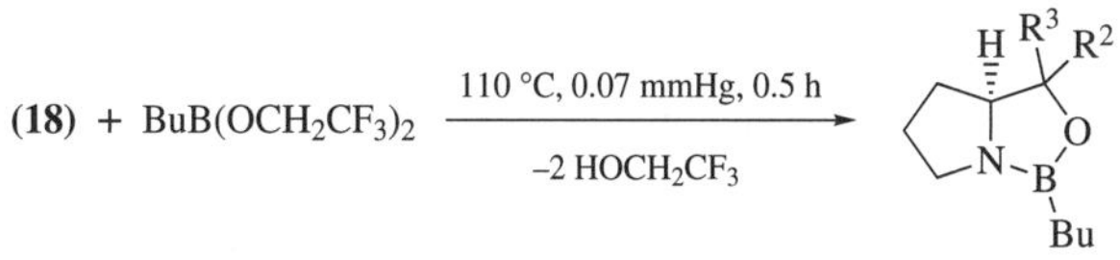

(**18**) + $BuB(OCH_2CF_3)_2$ $\xrightarrow[-2\ HOCH_2CF_3]{110\ °C,\ 0.07\ mmHg,\ 0.5\ h}$ (8)

Substituted oxazaborolidines also react with borane (B_2H_6, $H_3B{\cdot}THF$, or $H_3B{\cdot}SMe_2$) to form an oxazaborolidine–borane complex (**19**) (eq 9).[3b,27] The oxazaborolidine–borane complex, by releasing the strain of the partial double bond between the ring boron and nitrogen, is more stable than the free oxazaborolidine, and in many cases exists as a stable crystalline solid.[21c,27,28]

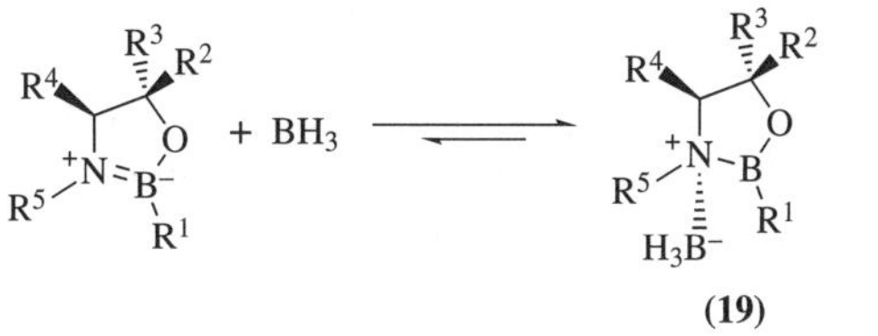

(9)

The oxazaborolidine–borane complex (**19**) can be used stoichiometrically (eq 1) or catalytically (eq 10) for the enantioselective reduction of prochiral ketones.[27a] When used catalytically, the oxazaborolidine–borane complex (**19**) is the second intermediate in the catalytic cycle (eq 10) proposed to explain the behavior of the oxazaborolidine catalyst.[3a,29] Subsequent coordination between the Lewis acidic ring boron and the carbonyl oxygen activates the ketone toward reduction. Intramolecular hydride transfer from the BH_3 coordinated to the ring nitrogen then occurs via a six-membered ring chair transition state.[17b,27a,30] Following hydride transfer, the alkoxy–BH_2 dissociates, and oxazaborolidine (**1**) is free to begin the cycle again. The diastereomeric transition state model (**20**), leading to the enantiomeric carbinol product, is disfavored due to unfavorable 1,3-diaxial steric interactions between R_L and R^1. Additional work will be required to better understand the catalytic cycle and the intermediates involved to further improve the oxazaborolidine catalysts. The behavior of the catalysts has been the subject of molecular orbital calculations in a series of 12 papers.[31] It should be noted, however, that not all of the results and conclusions are supported by experimental observations.

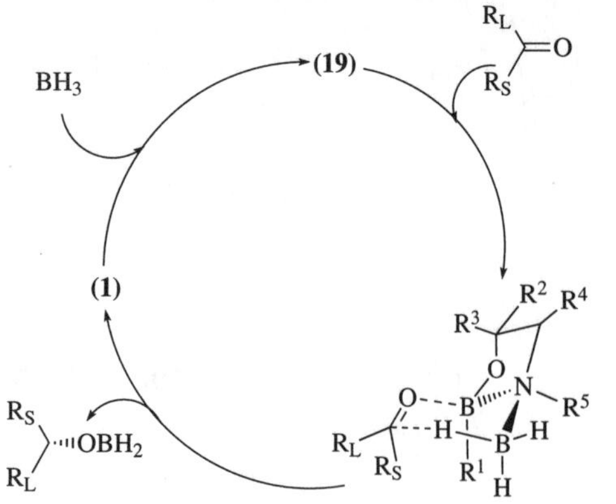

(10)

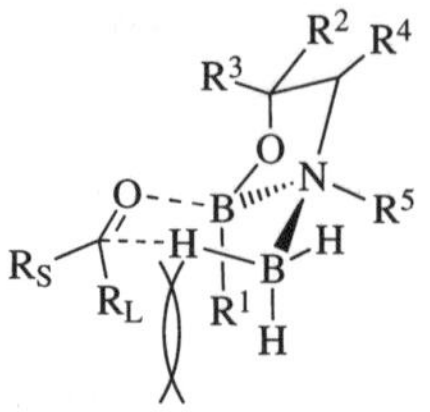

(20)

The enantioselectivities reported for the reduction of acetophenone and 1-tetralone using several representative chiral (4*S*)-oxazaborolidine catalysts are summarized in Table 1. The oxazaborolidines derived from (*S*)-azetidinecarboxylic acid and (*S*)-proline provide the best results. It is interesting to note the reversal in enantioselectivity going from catalyst (**5a**) to (**6a**).

Oxazaborolidine catalyzed reductions are generally performed in an aprotic solvent, such as dichloromethane, THF, or toluene. When the reactions are run in a Lewis basic solvent, such as THF, the solvent competes with the oxazaborolidine to complex with the borane, which can have an effect on the enantioselectivity and/or rate of the reaction.[27a] The solubility of the oxazaborolidine–borane complex can be the limiting factor for reactions run in toluene, although this problem has been circumvented by using oxazaborolidines with more lipophilic substituents (R^1 = *n*-Bu; R^2, R^3 = 2-naphthyl).[19b–d] We have found dichloromethane to be the best overall solvent for these reactions.[27a]

The reactions are typically performed using $H_3B \cdot THF$, $H_3B \cdot SMe_2$, or ***Catecholborane***[19d] as the hydride source. When using $H_3B \cdot THF$ or $H_3B \cdot SMe_2$, two of the three hydrides are effectively utilized.[27a] This is only true for reactions run at temperatures greater than −40 °C. At lower temperatures, only one hydride is transferred at a reasonable rate. When two hydrides are used, there is some evidence that the enantioselectivity for transfer of the second hydride is different, and may in fact be lower.[27a] Whether this implies that an alternative catalytic cycle operates, whereby the alkoxy–BH_2 intermediate generated during the first hydride transfer remains coordinated to the oxazaborolidine, and then transfers the second hydride (with a different degree of enantioselectivity), or that some other intermediate present is active, but not as an enantioselective reducing agent, will require further investigation. In any event, the amount of BH_3 used should be at least 0.5 mole per mole of ketone plus an amount equal to the oxazaborolidine catalyst, with the possibility that 1 mole per mole provides slightly higher enantioselectivity. When catecholborane is used as the hydride source, a 50–100% excess of this reagent is used.

The mode of addition and the reaction temperature both affect the enantioselectivity of the reaction. The best results are obtained when the ketone is added slowly to a solution of the oxazaborolidine (or oxazaborolidine–borane complex) and the borane source, at as low a temperature that provides a reasonable reaction rate.[27a] This is in contrast to a previous report that indicated that oxazaborolidine-catalyzed reductions 'lose stereoselectivity at lower temperatures'.[19d] With unsubstituted (R^1 = H) oxazaborolidines, higher temperatures may be required due to incomplete formation of the catalyst, the presence of dimers, and/or other intermediates.[26c]

In their role as enantioselective catalysts for the reduction of prochiral ketones, chiral oxazaborolidines have been used for the preparation of prostaglandins,[3a] PAF antagonists,[3a] a key intermediate of ginkgolide B,[32a] bilobalide,[32b] a key intermediate of forskolin,[32c] (*R*)- and (*S*)-fluoxetine,[32d] (*R*)- and (*S*)-isopreterenol,[19c] vitamin D analogs,[33] the carbonic anhydrase inhibitor MK-0417,[21b] the dopamine D1 agonist A-77636,[20b] taxol,[34] the LTD_4 antagonists L-695,499 and L-699,392,[35] the β-adrenergic agonist CL 316,243,[36] and the antiarrhythmic MK-0499.[37] They have also been used for the synthesis of chiral amines,[38,39] α-hydroxy acids,[19d,40a] benzylic thiols,[40c] the enantioselective reduction of trihalomethyl ketones,[40a,b,d] and ketones containing various heteroatoms.[17a,21b,27a,35,37]

Enantioselective Reduction of Imines and Ketoxime *O*-Ethers. In addition to the reduction of prochiral ketones, chiral oxazaborolidines have been employed as enantioselective reagents and catalysts for the reduction of imines (eq 11)[4,23] and ketoxime *O*-ethers (eq 12)[2e,f,5] to give chiral amines. It is interesting to note that the enantioselectivity for the reduction of ketoxime *O*-ethers is opposite that of ketones and imines. For more information, see ***2-Amino-3-methyl-1,1-diphenyl-1-butanol***.

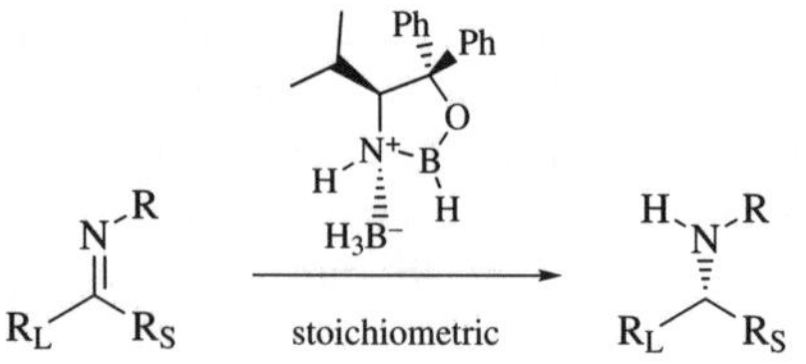

stoichiometric (11)

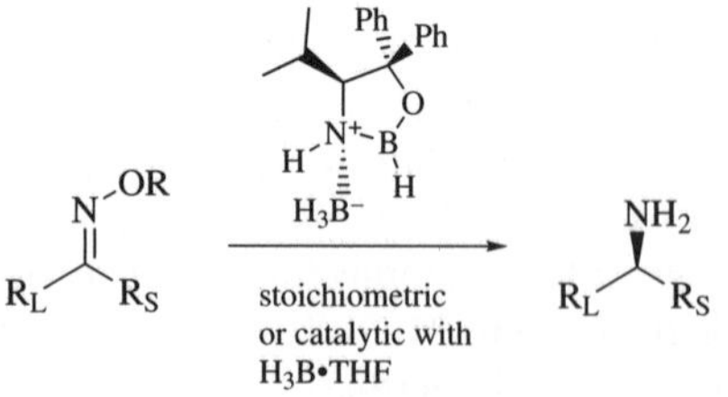

stoichiometric or catalytic with $H_3B \cdot THF$ (12)

Enantioselective Addition of Diethylzinc to Aldehydes. Oxazaborolidines derived from ephedrine have been used to catalyze the addition of ***Diethylzinc*** to aldehydes (eq 13).[7] Both the rate and enantioselectivity are optimized when R^1 = H. Aro-

Table 1 Chiral Oxazaborolidine Catalyzed Reduction of Acetophenone and 1-Tetralone

Catalyst	R^1	R^2,R^3	R^4 (mol %)	Catalyst (ee %)	Acetophenone (ee %)	1-Tetralone
(**1a**)[3a]	H	Ph	*i*-Pr	10	94.7 (*R*)	–
(**2a**)[16d]	H	Ph	–	10	98 (*R*)	–
(**3a**)[3a]	H	Ph	–	10	97 (*R*)	89 (*R*)
(**3b**)[21b]	Me	Ph	–	10	98 (*R*)	94 (*R*)
(**3b**)·BH_3[27a]	Me	Ph	–	5	97.6 (*R*)	99.0 (*R*)
(**3b**)·BH_3[27a]	Me	Ph	–	100	99.8 (*R*)	99.2 (*R*)
(**5a**)[22b]	H	Ph	–	10	96 (*R*)	79 (*R*)
(**6a**)[22b]	H	Ph	–	10	90 (*S*)	79 (*S*)
(**7a**)[18a]	H	Ph	–	10	87 (*R*)	–
(**8a**)[16b]	H	Ph	–	10	71 (*R*)	44 (*R*)
(**9a**)[23]	H	H	–	110	88 (*R*)	–
(**12a**)[19a]	Me	Ph	–	10	97.5 (*R*)	95.3 (*R*)

matic aldehydes generally react faster than aliphatic aldehydes, and the enantioselectivity for aromatic aldehydes is good to excellent (86–96% ee).

RCHO + Et_2Zn → (5 mol% catalyst) → R–CH(OH)–Et (13)

Other Applications. Chiral oxazaborolidines derived from ephedrine have also been used in asymmetric hydroborations,[8a,b] and as reagents to determine the enantiomeric purity of secondary alcohols.[8c] Chiral 1,3,2-oxazaborolidin-5-ones derived from amino acids have been used as asymmetric catalysts for the Diels–Alder reaction,[9–11] and the aldol reaction.[12,13]

Related Reagents. 2-Amino-3-methyl-1,1-diphenyl-1-butanol; α,α-Diphenyl-2-pyrrolidinemethanol; Ephedrine-borane; Norephedrine-Borane.

---

1. (a) Wallbaum, S.; Martens, J. *TA* **1992**, *3*, 1475. (b) Singh, V. K. *S* **1992**, 605. (c) Deloux, L.; Srebnik M. *CRV* **1993**, *93*, 763.
2. (a) Hirao, A.; Itsuno, S.; Nakahama, S.; Yamazaki, N. *CC* **1981**, 315. (b) Itsuno, S.; Hirao, A.; Nakahama, S.; Yamazaki, N. *JCS(P1)* **1983**, 1673. (c) Itsuno, S.; Ito, K.; Hirao, A.; Nakahama, S. *CC* **1983**, 469. (d) Itsuno, S.; Ito, K.; Hirao, A.; Nakahama, S. *JOC* **1984**, *49*, 555. (e) Itsuno, S.; Nakano, M.; Miyazaki, K.; Masuda, H.; Ito, K.; Hirao, A.; Nakahama, S. *JCS(P1)* **1985**, 2039. (f) Itsuno, S.; Nakano, M.; Ito, K.; Hirao, A.; Owa, M.; Kanda, N.; Nakahama, S. *JCS(P1)* **1985**, 2615.
3. (a) Corey, E. J.; Bakshi, R. K.; Shibata, S. *JACS* **1987**, *109*, 5551. (b) Corey, E. J.; Bakshi, R. K.; Shibata, S.; Chen, C. P.; Singh, V. K. *JACS* **1987**, *109*, 7925. (c) Corey, E. J.; Shibata, S.; Bakshi, R. K. *JOC* **1988**, *53*, 2861. (d) Corey, E. J. U.S. Patent 4 943 635, 1990.
4. (a) Cho, B. T.; Chun, Y. S. *JCS(P1)* **1990**, 3200. (b) Cho, B. T.; Chun, Y. S. *TA* **1992**, *3*, 337.
5. (a) Itsuno, S.; Sakurai, Y.; Ito, K.; Hirao, A.; Nakahama, S. *BCJ* **1987**, *60*, 395. (b) Itsuno, S.; Sakurai, Y.; Shimizu, K.; Ito, K. *JCS(P1)* **1989**, 1548. (c) Itsuno, S.; Sakurai, Y.; Shimizu, K.; Ito, K. *JCS(P1)* **1990**, 1859.
6. Bringmann, G.; Hartung, T. *AG(E)* **1992**, *31*, 761.
7. Joshi, N. N.; Srebnik, M.; Brown, H. C. *TL* **1989**, *30*, 5551.
8. (a) Brown, J. M.; Lloyd-Jones, G. C. *TA* **1990**, *1*, 869. (b) Brown, J. M.; Lloyd-Jones, G. C. *CC* **1992**, 710. (c) Brown, J. M.; Leppard, S. W.; Lloyd-Jones, G. C. *TA* **1992**, *3*, 261. (d) Brown, J. M.; Lloyd-Jones, G. C.; Layzell, T. P. *TA* **1993**, *4*, 2151.
9. Takasu, M.; Yamamoto, H. *SL* **1990**, 194.
10. (a) Sartor, D.; Saffrich, J.; Helmchen, G. *SL* **1990**, 197. (b) Sartor, D.; Saffrich, J.; Helmchen, G.; Richards, C. J.; Lambert, H. *TA* **1991**, *2*, 639.
11. (a) Corey, E. J.; Loh, T.-P. *JACS* **1991**, *113*, 8966. (b) Corey, E. J.; Loh, T.-P.; Roper, T. D.; Azimioara, M. D.; Noe, M. C. *JACS* **1992**, *114*, 8290.
12. Kiyooka, S.; Kaneko, Y.; Komura, M.; Matsuo, H.; Nakano, M. *JOC* **1991**, *56*, 2276.
13. Parmee, E. R.; Tempkin, O.; Masamune, S.; Abiko, A. *JACS* **1991**, *113*, 9365.
14. Mandal, A. K.; Kasar, T. G.; Mahajan, S. W.; Jawalkar, D. G. *SC* **1987**, *17*, 563.
15. (a) Grundon, M. F.; McCleery, D. G.; Wilson, J. W. *JCS(P1)* **1981**, 231. (b) Mancilla, T.; Santiesteban, F.; Contreras, R.; Klaebe, A. *TL* **1982**, *23*, 1561. (c) Tlahuext, H.; Contreras, R. *TA* **1992**, *3*, 727. (d) Tlahuext, H. Contreras, R. *TA* **1992**, *3*, 1145 (e) Cho, B. T.; Chun, Y. S. *TA* **1992**, *3*, 1539 (f) Berenguer, R.; Garcia, J.; Gonzalez, M.; Vilarrasa, J. *TA* **1993**, *4*, 13.
16. (a) Wallbaum, S.; Martens, J. *TA* **1991**, *2*, 1093. (b) Stingl, K.; Martens, J.; Wallbaum, S. *TA* **1992**, *3*, 223. (c) Martens, J.; Dauelsberg, C.; Behnen, W.; Wallbaum, S. *TA* **1992**, *3*, 347. (d) Behnen, W.; Dauelsberg, C.; Wallbaum, S.; Martens, J. *SC* **1992**, *22*, 2143. (e) Mehler, T.; Martens, J. *TA* **1993**, *4*, 1983. (f) Mehler, T.; Martens, J. *TA* **1993**, *4*, 2299.
17. (a) Quallich, G. J.; Woodall, T. M. *TL* **1993**, *34*, 785. (b) Quallich, G. J.; Woodall, T. M. *TL* **1993**, *34*, 4145. (c) Quallich, G. J.; Woodall, T. M. *SL* **1993**, 929.
18. (a) Rao, A. V. R.; Gurjar, M. K.; Sharma, P. A.; Kaiwar, V. *TL* **1990**, *31*, 2341. (b) Rao, A. V. R.; Gurjar, M. K.; Kaiwar, V. *TA* **1992**, *3*, 859.
19. (a) Corey, E. J.; Chen, C. P.; Reichard, G. A. *TL* **1989**, *30*, 5547. (b) Corey, E. J.; Link, J. O. *TL* **1989**, *30*, 6275. (c) Corey, E. J.; Link, J. O. *TL* **1990**, *31*, 601. (d) Corey, E. J.; Bakshi, R. K. *TL* **1990**, *31*, 611. (e) Corey, E. J.; Link, J. O. *TL* **1992**, *33*, 4141.
20. (a) DeNinno, M. P.; Perner, R. J.; Lijewski, L. *TL* **1990**, *31*, 7415. (b) DeNinno, M. P.; Perner, R. J.; Morton, H. E.; DiDomenico, Jr., S. *JOC* **1992**, *57*, 7115.
21. (a) Mathre, D. J.; Jones, T. K.; Xavier, L. C.; Blacklock, T. J.; Reamer, R. A.; Mohan, J. J.; Jones, E. T. T.; Hoogsteen, K.; Baum, M. W.; Grabowski, E. J. J. *JOC* **1991**, *56*, 751. (b) Jones, T. K.; Mohan, J. J.; Xavier, L. C.; Blacklock, T. J.; Mathre, D. J.; Sohar, P.; Jones, E. T. T.; Reamer, R. A.;

Roberts, F. E.; Grabowski, E. J. J. *JOC* **1991**, *56*, 763. (c) Blacklock, T. J.; Jones, T. K.; Mathre, D. J.; Xavier, L. C. U.S. Patent 5 039 802, 1991. (d) Blacklock, T. J.; Jones, T. K.; Mathre, D. J.; Xavier, L. C. U.S. Patent 5 264 585, 1993. (e) Shinkai, I. *JHC* **1992**, *29*, 627.

22. (a) Youn, I. K.; Lee, S. W.; Pak, C. S. *TL* **1988**, *29*, 4453. (b) Kim, Y. H.; Park, D. H.; Byun, I. S.; Yoon, I. K.; Park, C. S. *JOC* **1993**, *58*, 4511.
23. Nakagawa, M.; Kawate, T.; Kikikawa, T.; Yamada, H.; Matsui, T.; Hino, T. *T* **1993**, *49*, 1739.
24. Tanaka, K.; Matsui, J.; Suzuki, H. *CC* **1991**, 1311.
25. (a) Long, L. H. *J. Inorg. Nucl. Chem.* **1970**, *32*, 1097. (b) Fernandez, H.; Grotewold, J.; Previtali, C. M. *JCS(D)* **1973**, 2090. (c) Gibb, T. C.; Greenwood, N. N.; Spalding, T. R.; Taylorson, D. *JCS(D)* **1979**, 1398.
26. (a) Bielawski, J.; Niedenzu, K. *Synth. React. Inorg. Met.-Org. Chem.* **1980**, *10*, 479. (b) Cragg, R. H.; Miller, T. J. *JOM* **1985**, *294*, 1. (c) Brunel, J. M.; Maffei, M.; Buono, G. *TA* **1993**, *4*, 2255.
27. (a) Mathre, D. J.; Thompson, A. S.; Douglas, A. W.; Hoogsteen, K.; Carroll, J. D.; Corley, E. G.; Grabowski, E. J. J. *JOC* **1993**, *58*, 2880. (b) Blacklock, T. J.; Jones, T. K.; Mathre, D. J.; Xavier, L. C. U.S. Patent 5 189 177, 1993. (c) Carroll, J. D.; Mathre, D. J.; Corley, E. G.; Thompson, A. S. U.S. Patent 5 264 574, 1993.
28. Corey, E. J.; Azimioara, M.; Sarshar, S. *TL* **1992**, *24*, 3429.
29. Evans, D. A. *Science* **1988**, *240*, 420.
30. Jones, D. K.; Liotta, D. C.; Shinkai, I.; Mathre, D. J. *JOC* **1993**, *58*, 799.
31. Nevalainen, V. *TA* **1993**, *4*, 2001; and references contained therein.
32. (a) Corey, E. J.; Gavai, A. V. *TL* **1988**, *29*, 3201. (b) Corey, E. J.; Su, W.-G. *TL* **1988**, *29*, 3423. (c) Corey, E. J.; Jardine, P. D. S.; Mohri, T. *TL* **1988**, *29*, 6409. (d) Corey, E. J.; Reichard, G. A. *TL* **1989**, *30*, 5207.
33. (a) Kabat, M.; Kiegiel, J.; Cohen, N.; Toth, K.; Wovkulich, P. M.; Uskokovic, M. R. *TL* **1991**, *32*, 2343. (b) Lee, A. S.; Norman, A. W.; Okamura, W. H. *JOC* **1992**, *57*, 3846.
34. Nicolaou, K. C.; Hwang, C.-K.; Sorensen, E. J.; Clairborne, C. F. *CC* **1992**, 1117.
35. (a) Labelle, M.; Prasit, P.; Belley, M.; Blouin, M.; Champion, E.; Charette, L.; DeLuca, J. G.; Dufresne, C.; Frenette, R.; Gauthier, J. Y.; Grimm, E.; Grossman, S. J.; Guay, D.; Herold, E. G.; Jones, T. R.; Lau, Y.; Leblanc, Y.; Leger, S.; Lord, A.; McAuliffe, M.; McFarlane, C.; Masson, P.; Metters, K. M.; Ouimet, N.; Patrick, D. H.; Perrier, H.; Piechuta, H.; Roy, P.; Williams, H.; Wang, Z.; Xiang, Y. B.; Zamboni, R. J.; Ford-Hutchinson, A. W.; Young, R. N. *BML* **1992**, *2*, 1141. (b) King, A. O.; Corley, E. G.; Anderson, R. K.; Larsen, R. D.; Verhoeven, T. R.; Reider, P. J.; Xiang, Y. B.; Belley, M.; Leblanc, Y.; Labelle, M.; Prasit, P.; Zamboni, R. J. *JOC* **1993**, *58*, 3731.
36. Bloom, J. D.; Dutia, M. D.; Johnson, B. D.; Wissner, A.; Burns, M. G.; Largis, E. E.; Dolan, J. A.; Claus, T. H. *JMC* **1992**, *35*, 3081.
37. Cai, D.; Tschaen, D.; Shi, Y.-J.; Verhoeven, T. R.; Reamer, R. A.; Douglas, A. W. *TL* **1993**, *34*, 3243.
38. Chen, C.-P.; Prasad, K.; Repic, O. *TL* **1991**, *32*, 7175.
39. Thompson, A. S.; Humphrey, G. R.; DeMarco, A. M.; Mathre, D. J.; Grabowski, E. J. J. *JOC* **1993**, *58*, 5886.
40. (a) Corey, E. J.; Cheng, X. M.; Cimprich, K. A.; Sarshar, S. *TL* **1991**, *32*, 6835. (b) Corey, E. J.; Link, J. O. *TL* **1992**, *33*, 3431. (c) Corey, E. J.; Cimprich, K. A. *TL* **1992**, *33*, 4099. (d) Corey, E. J.; Link, J. O.; Bakshi, R. K. *TL* **1992**, *33*, 7107.

David J. Mathre & Ichiro Shinkai
Merck Research Laboratories, Rahway, NJ, USA

N,N,N′,N′-Tetramethylethylenediamine[1]

Me_2N–CH_2CH_2–NMe_2

[110-18-9] $C_6H_{16}N_2$ (MW 116.24)

(bidentate tertiary amine Lewis base with good solvating properties; used as an additive to stabilize and activate organometallic reagents and inorganic salts; enhances the rate of metalation of a variety of aromatic and unsaturated systems as well as influencing the regiochemical outcome of these reactions; effective as a neutral amine in base catalyzed reactions)

Alternate Name: TMEDA.
Physical Data: bp 121 °C; *d* 0.781 g cm^{-3}.
Solubility: very sol water, most organic solvents.
Form Supplied in: colorless liquid, typically of 99% purity as obtained commercially.
Drying: for uses in conjunction with organometallic reagents, moisture exclusion is necessary. Removal of water is best achieved by refluxing over lithium aluminum hydride or calcium hydride for 2 h under nitrogen and distilling immediately prior to use.
Handling, Storage, and Precautions: TMEDA should be used directly after distilling. However, it may be stored under nitrogen and transferred by using a syringe and septum cap as required. For most applications the amine is removed during aqueous workup simply by washing with water owing to its high water solubility. Use in a fume hood.

Lithiation of Difficult Substrates. TMEDA, through an erstwhile perception of enhanced chelating ability,[1b] but more likely through presentation of a more labile environment, activates organolithium reagents.[1j] ***n*-Butyllithium** forms hexamers in hexane but in the presence of TMEDA exists as a solvated tetramer.[2] Thus the use of the *n*-butyllithium/TMEDA complex in hexane[3] effects the dilithiation of ***Furan*** and thiophene,[4] and lithiation of benzene,[5] in high yields. The allylic deprotonation of unactivated alkenes is normally difficult to achieve with BuLi alone. However, in the presence of TMEDA, propene is monolithiated or dilithiated in the allylic position,[6] and limonene is selectively lithiated at C-10.[7] The resulting allylic carbanion and electrophiles such as ***Paraformaldehyde*** give functionalized products (eq 1).[8]

7 … 10 —BuLi, TMEDA→ Li+ —$(HCHO)_n$→ HO (1)

However, when vinylic metalation is desired, competing allylic deprotonation may occur. In general, thermodynamic acidity and the kinetic preference for vinylic deprotonation of cyclic alkenes decrease with increasing ring size.[1h] The stable alkane-soluble reagent ***n-Butyllithium–Potassium t-Butoxide***–TMEDA in hexane[9] metalates ***Ethylene*** with potassium[10] and effects selective vinylic deprotonation of cyclopentene (eq 2),[11] cyclobutene,[12] norbornene, and norbornadiene.[10]

BuLi, t-BuOK, TMEDA; hexane, –20 °C; 9:1 (2)

N-Alkyl- and N-arylpyrroles are readily α-lithiated.[1c] For example, N-methylpyrrole is deprotonated with ***Ethyllithium***/TMEDA and the lithiated product has been treated with ***Carbon Dioxide*** to give the corresponding carboxylic acid in 70% yield (eq 3).[13]

1. EtLi, TMEDA; 2. CO_2; 70%; CO_2H (3)

Numerous examples exist in which TMEDA not only facilitates the lithiation of aromatic and heteroaromatic substrates but also controls the regioselectivity of lithiation.[1c] While tertiary benzamides are susceptible to nucleophilic attack by n-butyllithium to give aryl butyl ketones, the use of ***s-Butyllithium***/TMEDA in THF at −78 °C provides the synthetically useful *ortho* metalated tertiary benzamide which may be treated with a large variety of electrophiles (eq 4).[1d,14] Even with compounds having a second more acidic site the above conditions allow *ortho* lithiation to take place under kinetic control. Thus a p-toluamide is *ortho* lithiated with s-butyllithium/TMEDA in THF at −78 °C, but when ***Lithium Diisopropylamide*** is used as the base in THF at 0 °C the thermodynamically favored benzyllithium species is obtained (eq 5).[15] The very marked influence of TMEDA on the lithiation of naphthyl methyl ether in hydrocarbon solvents is dramatically illustrated in the example in eq 6.[16]

NEt_2; s-BuLi, TMEDA; THF, –78 °C; EX (4)

EX = D_2O, MeI, DMF, CO_2, $(CO_2Et)_2$, ArCHO, I_2, TMSCl, $B(OMe)_3/H_2O_2$, PhNCO, Ph_2CO

$N(i\text{-}Pr)_2$; s-BuLi, TMEDA; THF, –78 °C; LDA; THF, 0 °C (5)

1. RLi; 2. CO_2, CH_2N_2 (6)

R = Bu, hexane/TMEDA >99.3:<0.3 60%
R = t-Bu, cyclohexane 1:99 35%

The use of the s-butyllithium/TMEDA system in THF at −78 °C is widely employed for the lithiation of amide and thioamide derivatives adjacent to nitrogen.[1f] The resulting lithiated species undergo reaction with a variety of electrophiles such as aldehydes and ketones[17a] and dihalides (eq 7).[17b]

s-BuLi, TMEDA; –78 °C (7)

Ligand for Crystallographic Studies. TMEDA markedly facilitates the isolation of otherwise inaccessible crystalline organolithium reagents suitable for structural determination by X-ray crystallographic means.[18] In most cases, bidentate binding of the TMEDA ligand to the lithium ion is observed.

Control of Regioselectivity and Stereoselectivity. The recognition by Ireland and co-workers[19] that ***Hexamethylphosphoric Triamide*** has a profound effect on the stereochemistry of lithium enolates has led to the examination of the effects of other additives, as the ability to control enolate stereochemistry is of utmost importance for the stereochemical outcome of aldol reactions. Kinetic deprotonation of 3-pentanone with ***Lithium 2,2,6,6-Tetramethylpiperidide*** at 0 °C in THF containing varying amounts of HMPA or TMEDA was found to give predominantly the (Z)-enolate at a base:ketone:additive ratio of ca. 1:1:1, whereas with a base:ketone:additive ratio 1:0.25:1, formation of the (E)-enolate was favored (Table 1).[20] This remarkable result contrasts with those cases where HMPA:base ratios were varied towards larger amounts of HMPA, which favored formation of the (Z)-enolate.[21]

However, TMEDA unlike HMPA, does not cause flow over from a carbonyl to conjugate addition manifold for many lithiated systems. For example, lithiated allylic sulfides undergo conjugate (or '1,4') addition to cyclopent-2-enone in the presence of HMPA (see ***Allyl Phenyl Sulfide***), but in the presence of TMEDA, carbonyl addition only is observed.[22] The perception that TMEDA is unable to form solvent-separated ion pairs required for conjugate addition in this case now requires reevaluation.[1j,23] In the reaction of lithio α-trimethylsilylmethyl phenyl sulfide with

Table 1 Deprotonation of 3-Pentanone with LiTMP (1.0 mmol) in THF at 0 C in the Presence of TMEDA or HMPA[19]

3-Pentanone (mmol)	TMEDA (mmol)	HMPA (mmol)	(*E*)-Enolate	(*Z*)-Enolate	Total % yield of silyl enol ether derivatives
0.9	1.0	–	17	83	70
0.45	1.0	–	91	9	90
0.25	1.0	–	95	5	70
0.9	–	1.0	8	92	89
0.45	–	1.0	65	35	75
0.25	–	1.0	66	34	80

cyclohexenone, HMPA promotes predominant conjugate addition, whereas TMEDA has little effect on the normal carbonyl addition pathway taken in THF alone (eq 8).[24]

PhS⌒TMS —BuLi→ PhS(Li)TMS; + cyclohexenone, HMPA/THF, rt, 80% → 3-substituted cyclohexanone (CH(SPh)TMS); + cyclohexenone, TMEDA/THF, rt → [alkoxide] —87%→ vinyl sulfide (SPh) (8)

Likewise, TMEDA in THF has little effect on the stereochemical outcome of the Horner–Wittig reaction of lithiated ethyldiphenylphosphine oxide with benzaldehyde in THF at low temperature compared to the reaction in THF alone: a very slight enhancement in favor of the *erythro* (*anti*) hydroxy phosphine oxide intermediate (*erythro*:*threo* from 85:15 to 88:12), thus leading to slightly enhanced (*Z*)-alkene formation, is observed. By contrast, a reaction in ether alone provides less of the *erythro* product (*erythro*:*threo* 60:40).[25] These examples serve to emphasize current thought that TMEDA is not a 'good' chelating agent for lithium in relation to THF itself.[1j] It is noteworthy that the substitution of methyl in TMEDA by chiral binaphthylmethyl ligands generates a reagent which efficiently catalyzes asymmetric addition of butyllithium to benzaldehyde in ether at low temperature (eq 9).[26]

1. BuLi; 2. PhCHO, 73% → Bu(Ph)CHOH, 95% ee (9)

Butylmagnesium bromide in THF in the presence of excess TMEDA undergoes addition to a chiral crotonamide derivative to give the conjugate adduct in modest diastereomeric excess (67%) compared with 16% in the absence of TMEDA.[27] On the other hand, diastereoselection in the alkylation of enolates of chiral diamides derived from piperazines in THF containing TMEDA was minimal, with better results being provided by HMPA.[28]

Stabilization and Activation of Organometallic Reagents. The development and use of organocopper reagents as nucleophiles in the conjugate addition to enones is an important area of organic synthesis. It has been found that the reactivity of aryl- and alkylcopper reagents is dramatically improved through the use of ***Chlorotrimethylsilane*** and TMEDA.[29] The TMEDA not only stabilizes and solubilizes the organocopper reagent but also facilitates the trapping of the resulting enolates, thereby affording silyl enol ethers in excellent yields.[29a] Such a role is also played by HMPA[30] and ***4-Dimethylaminopyridine***.[30b] However, the low toxicity and cost of TMEDA makes it an attractive alternative. Conjugate addition of ***Lithium Di-n-butylcuprate*** to methyl 2-butynoate in diethyl ether provides a 74:26 mixture of (*E*)- and (*Z*)-alkylated enoates, whereas in the presence of TMEDA this ratio increases to 97:3. Stereoselectivity of the conjugate addition of the copper reagent derived from butylmagnesium bromide and ***Copper(I) Iodide*** to ethyl pentynoate in diethyl ether is also enhanced when TMEDA or pyrrolidone are used as additives, to give the (*E*)- and (*Z*)-enoates in a ratio of 99:1. In contrast, the use of HMPA affords a selectivity of only 78:22 for the (*E*)-isomer.[31]

The preparation of a trifluoromethylvinyl anion equivalent has been described in which the vinyl bromide is converted into the zinc reagent with ***Zinc/Silver Couple*** in the presence of TMEDA. The conversion proceeds very cleanly to afford a thermally stable vinylzinc bromide–TMEDA complex which can undergo reactions with electrophiles. TMEDA is essential for the conversion (eq 10).[32]

$CH_2{=}C(CF_3)Br$ —Zn(Ag), TMEDA/THF→ $CH_2{=}C(CF_3)ZnBr{\cdot}TMEDA$ (10)

The intramolecular insertion of unactivated alkenes into carbon–lithium bonds to give cycloalkylmethyllithium compounds provides a high-yielding alternative to analogous radical cyclizations.[1g,33] In many cases the addition of Lewis bases such as TMEDA increases the rate of cyclization[1g,33] and dramatically improves those cyclizations which are otherwise sluggish at room temperature (eq 11).[34]

Open chain allylic alcohols add organolithium reagents in the presence of TMEDA. The reactions are regio- and stereoselective;

the suggestion is made that the TMEDA complexes the alkoxide lithium counterion, allowing the alkoxide to orientate the incoming organolithium reagent and stabilize the resulting intermediate (eq 12).[35]

1. *t*-BuLi, pentane–Et_2O (3:2)
2. H^+, H_2O

(11)

TMEDA (2 equiv)	68%	31%
no TMEDA	6%	93%

PrLi, TMEDA
68%

(12)

threo:*erytho* = >50:1

Inorganic Complexes useful in Organometallic Reactions and Organic Synthesis. The complexing properties of TMEDA have made it possible to prepare and handle salts which are otherwise air and moisture sensitive. Thus ***Zinc Chloride*** in the presence of one equivalent of TMEDA forms a crystalline air stable solid, $ZnCl_2{\cdot}$TMEDA,[36] which with three equivalents of an alkyllithium reagent is converted into trialkylzinclithium. Likewise, CuI and TMEDA react to form the CuI·TMEDA complex,[29] a stable solid which is used for the preparation of stabilized organocopper reagents. The rate of Cu^I-catalyzed oxidative coupling of terminal alkynes in the presence of oxygen to form diynes is considerably increased by using TMEDA as a solubilizing agent for ***Copper(I) Chloride***.[37] Magnesium hydride, rapidly acquiring widespread recognition as a versatile reducing agent (see ***Magnesium Hydride–Copper(I) Iodide***), is prepared from phenylsilane and ***Dibutylmagnesium*** in the presence of TMEDA to give a very active THF-soluble complex free of halide or impurities derived from the usual reducing agents such as ***Lithium Aluminium Hydride***.[38] TMEDA (see also ***1,4-Diazabicyclo[2.2.2]octane***) forms insoluble adducts with boranes and alanes. In particular, the formation of air-stable adducts with monoalkyl boranes is of synthetic usefulness. The free monoalkyl borane may be regenerated by treating the adduct with ***Boron Trifluoride Etherate*** in THF and filtering off the newly formed TMEDA·$2BF_3$ precipitate.[1e]

Base Catalyzed Reactions. TMEDA can be monoprotonated (pK_a 8.97) and diprotonated (pK_a 5.85).[39] Titanium enolate formation from ketones and acid derivatives has been achieved by using ***Titanium(IV) Chloride*** and tertiary amines including TMEDA in dichloromethane at 0 °C.[40] The reactive species, which is likely to be a complex with the tertiary amine, undergoes aldol reaction with aldehydes to form *syn* adducts with high stereoselectivity (eq 13).

In the case of TMEDA, stereoselection in favor of the *syn* product (98:2) is enhanced over that achieved with ***Diisopropylethylamine*** (94:6).[40] Along with bases such as ***Triethylamine*** and ethylisopropylamine, TMEDA facilitates the preparation of cyanohydrin trimethylsilyl ethers from aldehydes and ***Cyanotrimethylsilane***.[41] It has been suggested that coordination by nitrogen induces formation of an active hypervalent cyanation intermediate from cyanotrimethylsilane. The conjugate addition of thiols to enones has been successfully catalyzed by using TMEDA in methanol at room temperature, as exemplified by the reaction of 10-mercaptoisoborneol and 4-*t*-butoxycyclopentenone (eq 14).[42] In this case the relative mildness of the reaction conditions prevents subsequent elimination of *t*-butoxide from occurring to give the unwanted enone.

$TiCl_4$, TMEDA
CH_2Cl_2, 0 °C

$OTiCl_3$

CHO

(13)

TMEDA	98:2
i-Pr_2NEt	94:6

TMEDA, MeOH
93–95%

(14)

Related Reagents. Hexamethylphosphoric Triamide; *N,N,N′,N′,N′*-Pentamethyldiethylenetriamine; Potassium Hydride-*s*-Butyllithium-*N,N,N′,N′*-Tetramethylethylenediamine; (−)-Sparteine.

1. (a) Agami, C. *BSF(2)* **1970**, 1619. (b) Wakefield, B. J. *The Chemistry of Organolithium Compounds*; Pergamon: Oxford, 1974. (c) Gschwend, H. W.; Rodriguez, H. R. *OR* **1979**, *26*, 1. (d) Beak, P.; Snieckus, V. *ACR* **1982**, *15*, 306. (e) Singaram, B.; Pai, G. G. *H* **1982**, *18*, 387. (f) Beak, P.; Zajdel, W. J.; Reitz, D. B. *CRV* **1984**, *84*, 471. (g) Klumpp, G. W. *RTC* **1986**, *105*, 1. (h) Brandsma, L.; Verkruijsse, H. *Preparative Polar Organometallic Chemistry 1*; Springer: Berlin, 1987. (i) *Advances in Carbanion Chemistry*; Snieckus, V., Ed.; JAI: Greenwich, CT, 1992; Vol. 1. (j) For a review and critical analysis of TMEDA complexation to lithium, see: Collum, D. B. *ACR* **1992**, *25*, 448.
2. Lewis, H. L.; Brown, T. L. *JACS* **1970**, *92*, 4664.
3. Peterson, D. J. *JOC* **1967**, *32*, 1717.
4. Chadwick, D. J.; Willbe, C. *JCS(P1)* **1977**, 887.
5. Eberhardt, G. G.; Butte, W. A. *JOC* **1964**, *29*, 2928. Rausch, M. D.; Ciappenelli, D. J. *JOM* **1967**, *10*, 127.
6. Klein, J.; Medlik-Balan, A. *CC* **1975**, 877.
7. Crawford, R. J.; Erman, W. F.; Broaddus, C. D. *JACS* **1972**, *94*, 4298.
8. Crawford, R. J. *JOC* **1972**, *37*, 3543.
9. Schade, C.; Bauer, W.; Schleyer, P. v. R. *JOM* **1985**, *295*, C25.
10. Brandsma, L.; Verkruijsse, H. D.; Schade, C.; Schleyer, P. v. R. *CC* **1986**, 260.
11. Broaddus, C. D.; Muck, D. L. *JACS* **1967**, *89*, 6533.
12. Stähle, M.; Lehmann, R.; Kramar, J.; Schlosser, M. *C* **1985**, *39*, 229.
13. Gjøs, N.; Gronowitz, S. *ACS* **1971**, *25*, 2596.

14. Beak, P.; Brown, R. A. *JOC* **1977**, *42*, 1823.
15. Beak, P.; Brown, R. A. *JOC* **1982**, *47*, 34.
16. Shirley, D. A.; Cheng, C. F. *JOM* **1969**, *20*, 251.
17. (a) Reitz, D. B.; Beak, P.; Tse, A. *JOC* **1981**, *46*, 4316. Beak, P.; Zajdel, W. J. *JACS* **1984**, *106*, 1010. (b) Lubosch, W.; Seebach, D. *HCA* **1980**, *63*, 102.
18. Seebach, D. *AG(E)* **1988**, *27*, 1624. Boche, G. *AG(E)* **1989**, *28*, 277; Zarges, W.; Marsch, M.; Harms, K.; Koch, W.; Frenking, G.; Boche, G. *CB* **1991**, *124*, 543.
19. Ireland, R. E.; Mueller, R. H.; Willard, A. K. *JACS* **1976**, *98*, 2868.
20. Fataftah, Z. A.; Kopka, I. E.; Rathke, M. W. *JACS* **1980**, *102*, 3959.
21. Romesberg, F. E.; Gilchrist, J. H.; Harrison, A. T.; Fuller, D. J.; Collum, D. B. *JACS* **1991**, *113*, 5751.
22. Binns, M. R.; Haynes, R. K.; Houston, T. L.; Jackson, W. R. *TL* **1980**, *21*, 573.
23. Cohen, T.; Abraham, W. D.; Myers, M. *JACS* **1987**, *109*, 7923. Binns, M. R.; Haynes, R. K.; Katsifis, A. G.; Schober, P. A.; Vonwiller, S. C. *JOC* **1989**, *54*, 1960.
24. Ager, D. J.; East, M. B. *JOC* **1986**, *51*, 3983.
25. Buss, A. D.; Warren, S. *TL* **1983**, *24*, 3931.
26. Mazaleyrat, J.-P.; Cram, D. J. *JACS* **1981**, *103*, 4585; Maigrot, N.; Mazaleyrat, J.-P. *CC* **1985**, 508.
27. Soai, K.; Machida, H.; Yokota, N. *JCS(P1)* **1987**, 1909.
28. Soai, K.; Hayashi, H.; Shinozaki, A.; Umebayashi, H.; Yamada, Y. *BCJ* **1987**, *60*, 3450.
29. (a) Johnson, C. R.; Marren, T. J. *TL* **1987**, *28*, 27. (b) Van Heerden, P. S.; Bezuidenhoudt, B. C. B.; Steenkamp, J. A.; Ferreira, D. *TL* **1992**, *33*, 2383 and cited references.
30. (a) Horiguchi, Y.; Matsuzawa, S.; Nakamura, E.; Kuwajima, I. *TL* **1986**, *27*, 4025. (b). Nakamura, E.; Matsuzawa, S.; Horiguchi, Y.; Kuwajima, I. *TL* **1986**, *27*, 4029.
31. Anderson, R. J.; Corbin, V. L.; Cotterrell, G.; Cox, G. R.; Henrick, C. A.; Schaub, F.; Siddall, J. B. *JACS* **1975**, *97*, 1197.
32. Jiang, B.; Xu, Y. *JOC* **1991**, *56*, 7336.
33. Bailey, W. F.; Khanolkar, A. D.; Gavaskar, K.; Ovaska, T. V.; Rossi, K.; Thiel, Y.; Wiberg, K. B. *JACS* **1991**, *113*, 5720.
34. Bailey, W. F.; Nurmi, T. T.; Patricia, J. T.; Wang, W. *JACS* **1987**, *109*, 2442.
35. Felkin, H.; Swierczewski, G.; Tambuté, A. *TL* **1969**, 707.
36. Watson, R. A.; Kjonaas, R. A. *TL* **1986**, *27*, 1437. Isobe, M.; Kondo, S.; Nagasawa, N.; Goto, T. *CL* **1977**, 679.
37. Hay, A. S. *JOC* **1962**, *27*, 3320. Jones, G. E.; Kendrick, D. A.; Holmes, A. B. *OS* **1987**, *65*, 52.
38. Michalczyk, M. J. *OM* **1992**, *11*, 2307.
39. Spialter, L.; Moshier, R. W. *JACS* **1957**, *79*, 5955.
40. Evans, D. A.; Rieger, D. L.; Bilodeau, M. T.; Urpi, F. *JACS* **1991**, *113*, 1047.
41. Kobayashi, S.; Tsuchiya, Y.; Mukaiyama, T. *CL* **1991**, 537.
42. Eschler, B. M.; Haynes, R. K.; Ironside, M. D.; Kremmydas, S.; Ridley, D. D.; Hambley, T. W. *JOC* **1991**, *56*, 4760.

Richard K. Haynes
Hong Kong University of Science and Technology, Hong Kong

Simone C. Vonwiller
The University of Sydney, NSW, Australia

1,1,3,3-Tetramethylguanidine

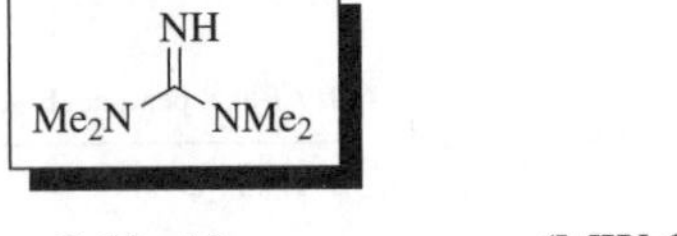

[80-70-6] $C_5H_{13}N_3$ (MW 115.21)

(strong base used to generate diazo compounds, protected 3β-substituted steroids, amino acid derivatives without racemization, α-nitroalkyl anions, β-glycosylated phenols, amides from acid chlorides; for the selective cleavage of Cbz groups; for cleavage of peptides from a resin; for the catalysis of conjugate addition reactions, as an azide counterion in asymmetric synthesis; for the catalysis of silylation reactions; for the conversion of ketone hydrazones to vinyl iodides)

Alternate Name: TMG.
Physical Data: bp 165 °C, 52–54 °C/11 mmHg; mp 60 °C; *d* 0.918 g cm^{-3}.
Solubility: freely sol most organic solvents; sol water.
Purification: distillation in vacuo.
Handling, Storage, and Precautions: this liquid is corrosive. It is harmful if swallowed, inhaled, or absorbed through the skin. It should be handled in a fume hood; the handler should wear chemical safety goggles, rubber gloves, and a respirator. Bottles of the liquid should be flushed with an inert atmosphere such as nitrogen or argon to minimize exposure to carbon dioxide.

Diazodiphenylmethane Preparation. 1,1,3,3-Tetramethylguanidine is the base of choice for use in the preparation of ***Diphenyldiazomethane*** from the corresponding ketone hydrazone (eq 1).[1]

$$Ph_2C{=}N{-}NH_2 \xrightarrow[\text{trace } I_2,\ \text{base}]{CH_3C(O)OOH} Ph_2C{=}\overset{+}{N}{=}N^- \quad (1)$$

Base	Yield (%)
N H	34
TMG	96
$NaHCO_3$ PTC, Bu_4NCl	66

Protected 3β-Substituted Steroids. Protection of 3β-substituted steroids by conversion to 6-oxo-3α,5-cyclo-5α-steroid derivatives is facile with TMG (eq 2).[2]

Amino Acid Derivatives without Racemization. TMG readily forms soluble salts with numerous amino acids, thus facilitating many reactions. The variety of amino acid derivatives prepared without racemization include *N*-Boc from ***t-Butyl Azidoformate***[3] or from *t*-butyl phenyl carbonate,[4] *N*-trifluoroacetyl,[5] and peptide coupling products (eq 3).[6,7]

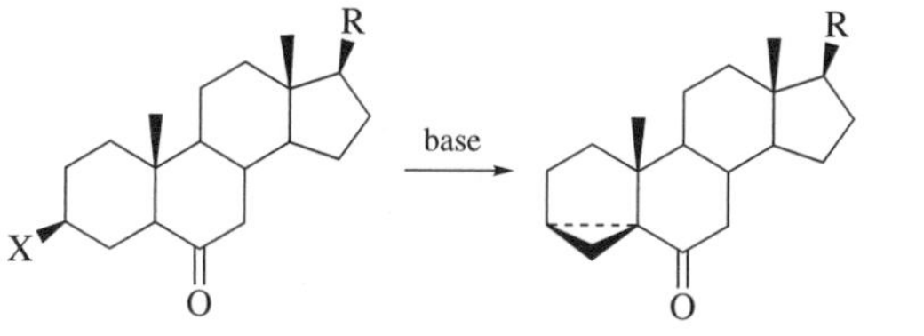

(2)

R	X	Base	Yield (%)
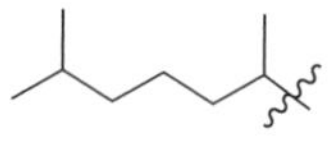 (side chain A)	TsO	TMG	98
	Cl	TMG	95
	TsO	NaOEt, EtOH	85
	Cl	NaOEt, EtOH	95
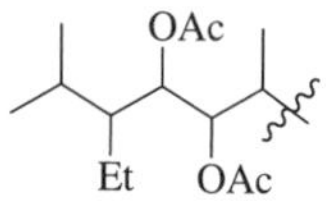 (side chain B: OAc, Et, OAc)	Cl	TMG	90
	Cl	NaOEt, EtOH	NA*

*Transesterification byproducts

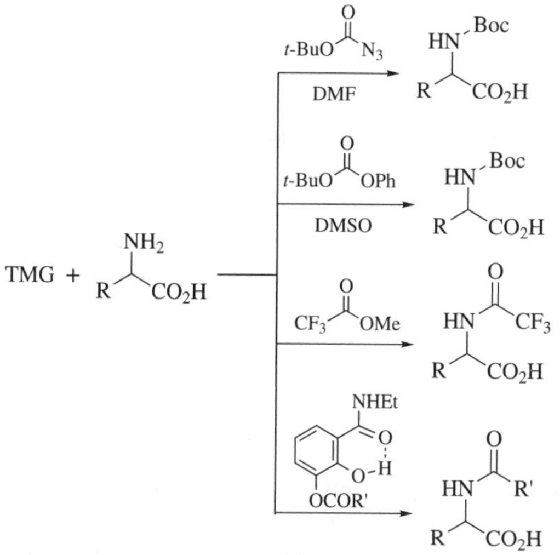

(3)

Insoluble amino acids such as taurine are converted to γ-L-glutamyl taurine with TMG as base (eq 4).[8]

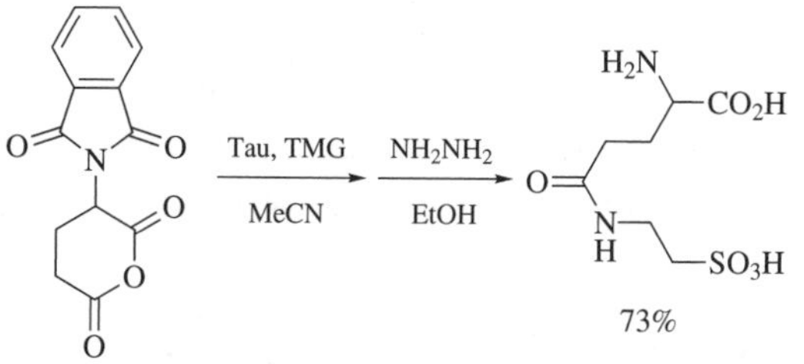

(4)

α-Nitroalkyl Anion Formation. TMG is often chosen as the basic catalyst for conjugate additions of nitromethane. An early report[9] cited isolation of monomeric addition products to unsaturated esters as a key result (eq 5). Highly selective additions of nitromethane to levoglucosenone afford excellent yields of condensation products (eq 6).[10]

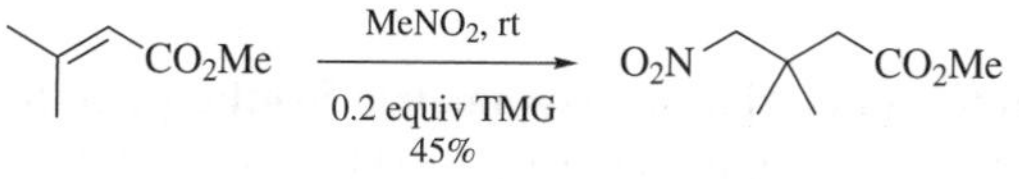

(5)

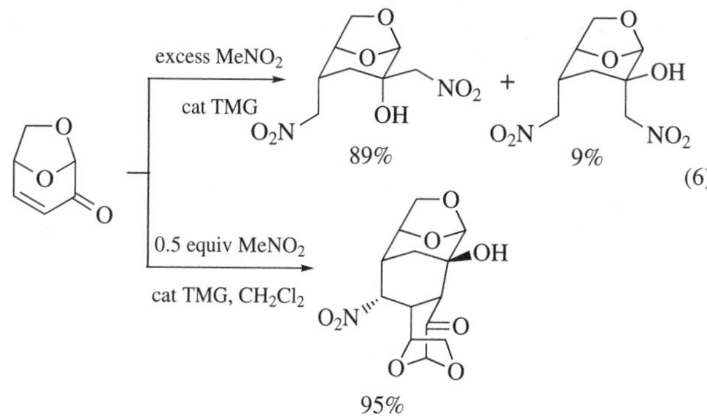

(6)

Addition of nitromethane to 4-oxygenated 2-substituted cyclopent-2-enones provides a facile entry to prostaglandin intermediates (eq 7).[11] Precursors of γ-aminobutyric acid analogs are readily available in good yield from nitromethane additions to 2-alkenoic esters (eq 8).[12]

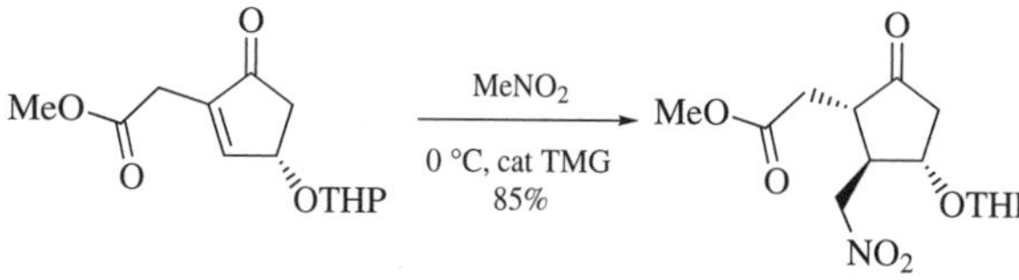

(7)

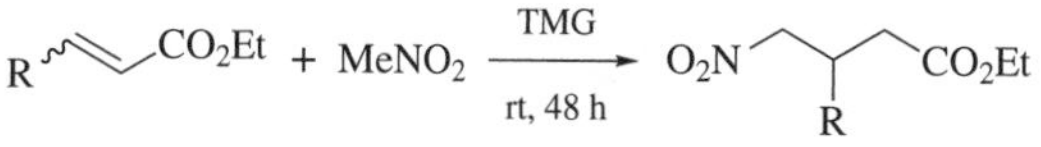

(8)

R = Me, 74%; *i*-Pr, 72%; Bu, 65%; *s*-Bu, 74%; *i*-Bu, 66%

With more hindered nitroalkanes, ***1,8-Diazabicyclo[5.4.0]-undec-7-ene*** (DBU) is required as base for high yields of Michael products (eq 9).[13]

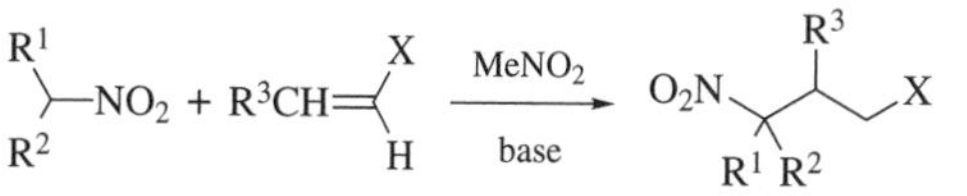

(9)

R^1	R^2	R^3	X	Base	% Product
Me	C_6H_{13}	H	-COMe	TMG	80
Me	*i*-Bu	H	$-CO_2Et$	TMG	80
Me	C_6H_{13}	H	-CN	TMG	82
Me	Me	H	$-SO_2Ph$	TMG	93
Me	Me	Ph	$-CO_2Me$	DBU	90
Bn	Me	Me	-CN	DBU	74
Me	Me	Ph	$-SO_2Ph$	DBU	90

Nitroalkane addition to 3-methylene-2,3-dihydrothiophene *S,S*-dioxide furnishes a new route to 1,3-dienes, including a precursor to ipsenol (eq 10).[14]

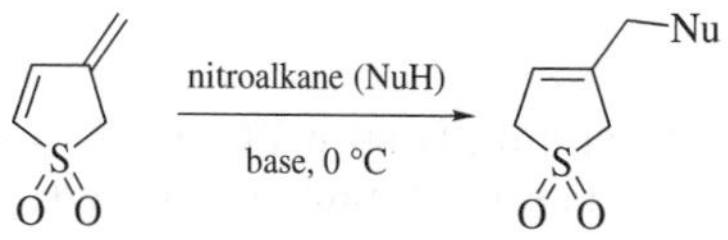

(10)

NuH	base	% Product
$MeNO_2$	DBU	42
$PrNO_2$	TMG	42
i-$PrNO_2$	TMG	78

β-Glycosylated Phenol Formation. While inorganic bases (CsF, K_2CO_3) or weaker bases such as 2,6-lutidine inhibit the reaction, TMG is especially effective at enhancing β-selectivity in glycosidation of phenols (eq 11).[15]

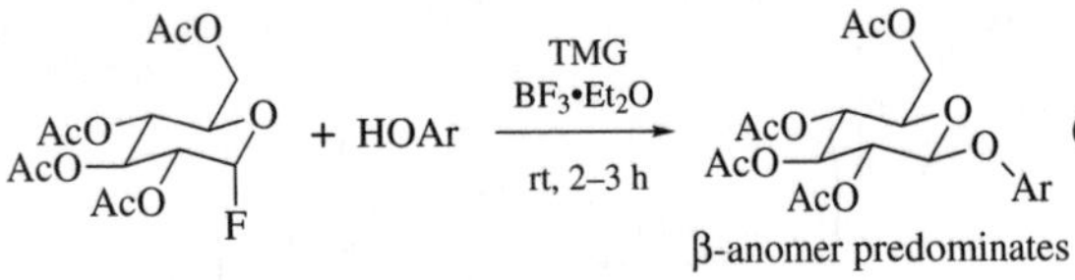

β-anomer predominates (11)

	α:β ratio	
Ar	Presence TMG	Absence TMG
p-$MeOC_6H_4$	β only	51:49
2,6-$(MeO)_2C_6H_3$	β only	NA
2,6-$Me_2C_6H_3$	β only	NA
o-ClC_6H_4	10:90	80:20
1-Naphthyl	3:97	40:60
p-$O_2NC_6H_4$	32:68	82:18

Amides from Acid Chlorides. For the preparation of chiral, nonracemic C_2-symmetrical pyrrolidine-derived auxiliaries, TMG is the base of choice (eq 12).[16]

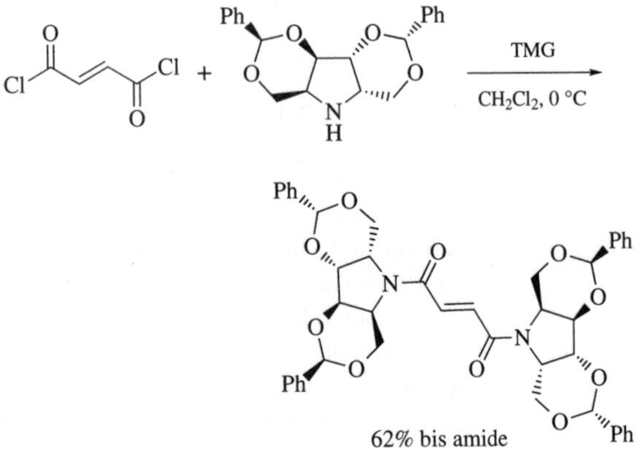

(12)

62% bis amide
no isomerization noted

Selective Cleavage of Z Groups. Selective removal of a primary benzyloxycarbonyl (Cbz, Z) group in the presence of both primary t-butoxycarbonyl (Boc) and secondary Z groups gives primary amines by TMG-catalyzed methanolysis (eq 13).[17,18]

$$\text{BocN(Z)-(CH}_2)_3\text{-N(Z)-(CH}_2)_4\text{-N(Z)Boc} \xrightarrow[\text{MeOH, rt}]{\text{TMG}} \text{BocNH-(CH}_2)_3\text{-N(Z)-(CH}_2)_4\text{-NHBoc} \quad (13)$$

Cleavage of Peptides from a Resin. TMG-catalyzed elimination effects removal of protected peptides from a 2-[4-(hydroxymethyl)phenylacetoxy]propionyl resin (eq 14).[19]

Catalysis of Conjugate Addition Reactions. Thiol addition to α,β-unsaturated nitroalkenes, in the presence of ***Formaldehyde***, leads to precursors of allyl alcohols in high yields (88–97%) (eq 15).[20] Conjugate addition of hydroxylamine to β-alkoxy acrylonitrile derivatives provides an entry to 4-thiocarbamoyl-5-aminoisoxazoles (eq 16).[21] Conjugate addition of benzyl alcohol catalyzed by TMG provides an adduct with no racemization noted at the initial stereocenter (eq 17).[22]

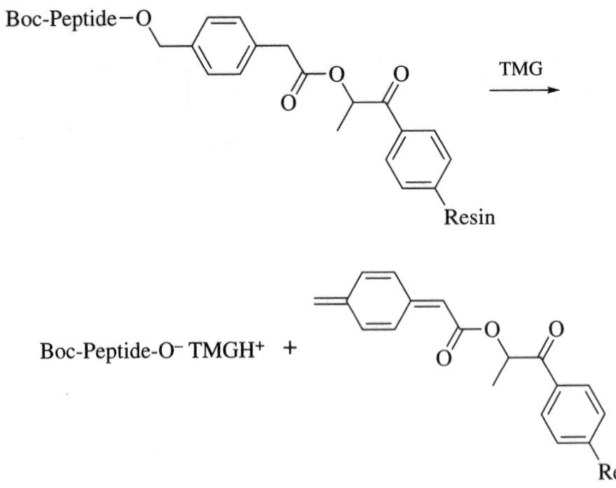

(14)

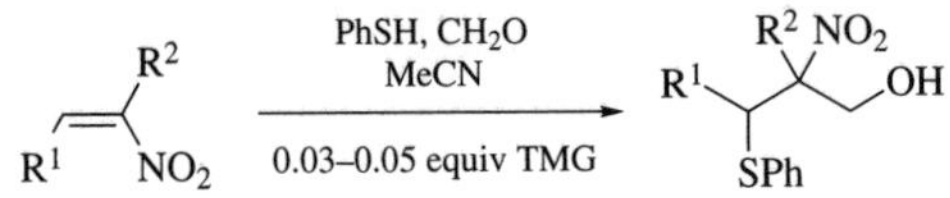

(15)

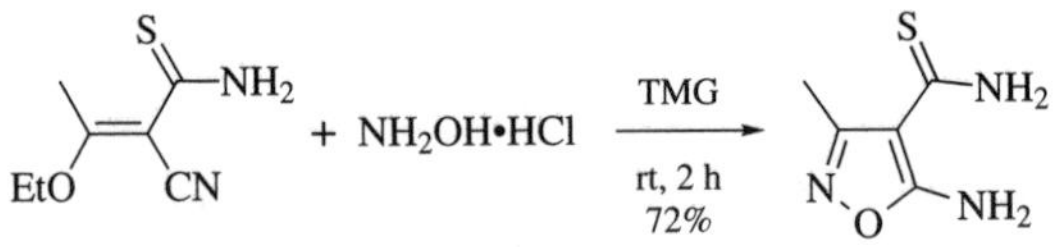

(16)

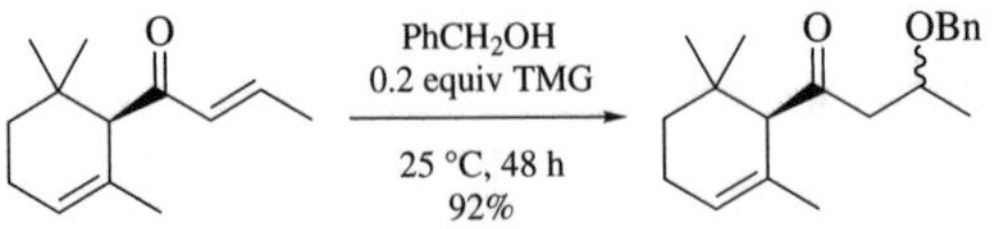

(17)

1:1 mix of diastereomers

A practical synthesis of (−)-huperzine A utilizes TMG as the base in a key step involving the assemblage of the tricyclic intermediate (eq 18).[23,24] Enantioselective synthesis of α-allokainoic acid uses tandem conjugate additions catalyzed by TMG (eq 19).[25]

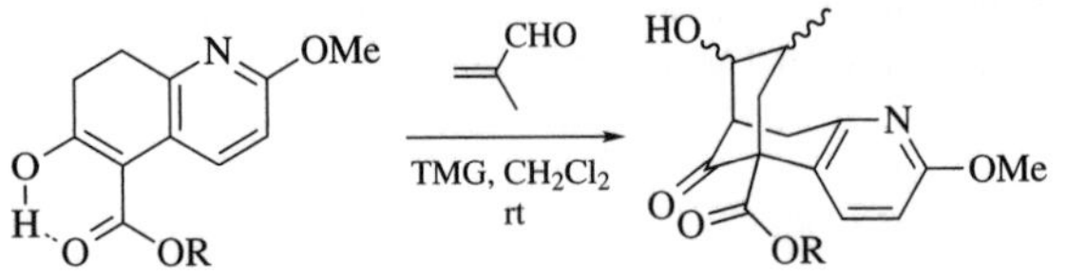

(18)

R = Me, 93%; (–)-8-phenylmenthyl, 91%*

*Major diastereomer of dehydration leads to natural (–)-huperzine

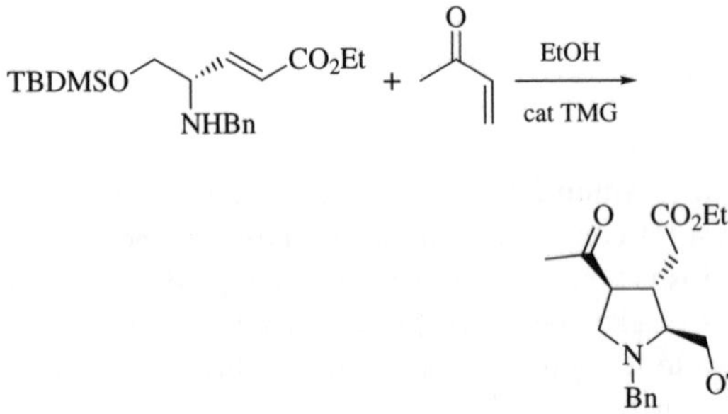

(19)

Azide Counterion in Asymmetric Synthesis. In the asymmetric synthesis of α-amino acids via the electrophilic azidation of chiral nonracemic imide enolates, stereospecific bromide displacement by tetramethylguanidinium azide was a high yielding step (eq 20).[26,27]

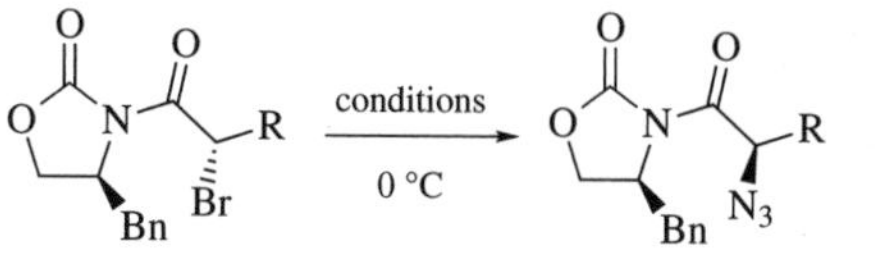

conditions, 0 °C (20)

Conditions	R	Ratio of diastereomers bromide (R):(S)[a]	azide (R):(S)[a]
NaN_3, DMSO	Bn	95:5	14:86
TMGA,[b] CH_2Cl_2	Bn	95:5	4:96
TMGA,[b] CH_2Cl_2	Ph	78:22	22:78
TMGA,[b] CH_2Cl_2	*i*-Pr	96:4	4:96

[a] Absolute stereochemistry of the carbon bearing the bromide or azide
[b] Tetramethylguanidium azide

Catalysis of Silylation Reactions. While ***Triethylamine*** is typically used as the HCl scavenger in the formation of *t*-butyldimethylsilyl ethers, TMG as a catalyst tends to accelerate the reaction (eq 21).[28]

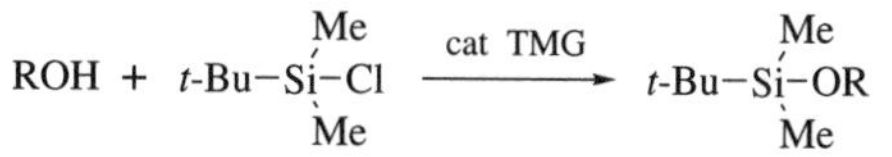

$$\text{ROH} + t\text{-Bu-Si(Me)}_2\text{-Cl} \xrightarrow{\text{cat TMG}} t\text{-Bu-Si(Me)}_2\text{-OR} \quad (21)$$

ROH	Solvent	Time (h)	Yield (%)
$PhCH_2CH_2OH$	DMF	0.3	95
	MeCN	0.3	96
PhCH(OH)CH_3	DMF	1.0	93
	MeCN	12.0	96

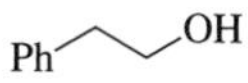

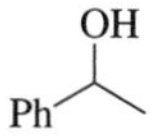

Conversion of Ketone Hydrazones to Vinyl Iodides. Oxidation of ketone hydrazones by iodine in the presence of TMG occurs in high yields (eq 22).[29]

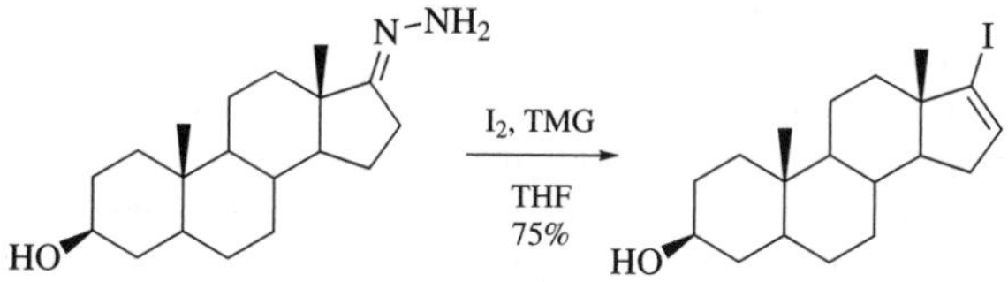

I_2, TMG; THF; 75% (22)

Related Reagents. *t*-Butyltetramethylguanidine; 1,1,2,3,3-Pentaisopropylguanidine.

1. Adamson, J. R.; Bywood, R.; Eastlick, D. T.; Gallagher, G.; Walker, D.; Wilson, E. M. *JCS(P1)* **1975**, 2030.
2. Anastasia, M.; Allevi, P.; Ciuffreda, P.; Fiecchi, A. *S* **1983**, 123.
3. Ali, A.; Fahrenholz, F.; Weinstein, B. *AG(E)* **1972**, *11*, 289.
4. Ragnarsson, U.; Karlsson, S. M.; Sandberg, B. E.; Larsson, L.-E. *OS* **1973**, *53*, 25; *OSC* **1988**, *6*, 203.
5. Steglich, W.; Hinze, S. *S* **1976**, 399.
6. Kemp, D. S.; Wrobel, S. J., Jr.; Wang, S.-W.; Bernstein, Z.; Rebek, J., Jr. *T* **1974**, *30*, 3969.
7. Kemp, D. S.; Wang, S.-W.; Rebek, J., Jr.; Mollan, R. C.; Banquer, C.; Subramanyam, G. *T* **1974**, *30*, 3955.
8. Gulyas, J.; Sebestyen, F.; Hercsel-Szepespataky, J.; Furka, A. *OPP* **1987**, *19*, 64.
9. Pollini, G. P.; Barco, A.; De Giuli, G. *S* **1972**, 44.
10. Forsyth, A. C.; Paton, R. M.; Watt, I. *TL* **1989**, *30*, 993.
11. Baraldi, P. G.; Barco, A.; Benetti, S.; Pollini, G. P.; Simoni, D.; Zanirato, V. *T* **1987**, *43*, 4669.
12. Andruszkiewicz, R.; Silverman, R. B. *S* **1989**, 953.
13. Ono, N.; Kamimura, A.; Miyake, H.; Hamamoto, I.; Kaji, A. *JOC* **1985**, *50*, 3692.
14. Nomoto, T.; Takayama, H. *CC* **1989**, 295.
15. Yamaguchi, M.; Horiguchi, A.; Fukuda, A.; Minami, T. *JCS(P1)* **1990**, 1079.
16. Veit, A.; Lenz, R.; Seiler, M. E.; Neuburger, M.; Zehnder, M.; Giese, B. *HCA* **1993**, *76*, 441.
17. Almeida, L. M. S.; Grehn, L.; Ragnarsson, U. *CC* **1987**, 1250.
18. Almeida, L. M. S.; Grehn, L.; Ragnarsson, U. *JCS(P1)* **1988**, 1905.
19. Whitney, D. B.; Tam, J. P.; Merrifield, R. B. *T* **1984**, *40*, 4237.
20. Ono, N.; Kamimura, A.; Kaji, A. *TL* **1984**, *25*, 5319.
21. Vicentini, C. B.; Veronese, A. C.; Poli, T.; Guarneri, M.; Giori, P.; Ferretti, V. *JHC* **1990**, *27*, 1481.
22. Fehr, C.; Guntern, O. *HCA* **1992**, *75*, 1023.
23. Yamada, F.; Kozikowski, A. P.; Reddy, E. R.; Pang, Y.-P.; Miller, J. H.; McKinney, M. *JACS* **1991**, *113*, 4695.
24. Xia, Y.; Kozikowski, A. P. *JACS* **1989**, *111*, 4116.
25. Barco, A.; Benetti, S.; Casolari, A.; Pollini, G. P.; Spalluto, G. *TL* **1990**, *31*, 4917.
26. Evans, D. A.; Ellman, J. A.; Dorow, R. L. *TL* **1987**, *28*, 1123.
27. Evans, D. A.; Britton, T. C.; Ellman, J. A.; Dorow, R. L. *JACS* **1990**, *112*, 4011.
28. Kim, S.; Chang, H. *SC* **1984**, *14*, 899.
29. Barton, D. H. R.; Bashiardes, G.; Fourrey, J.-L. *TL* **1983**, *24*, 1605.

Cynthia A. Maryanoff
The R. W. Johnson Pharmaceutical Research Institute, Spring House, PA, USA

Tin(IV) Chloride

[7646-78-8] Cl_4Sn (MW 260.51)

(strong Lewis acid used to promote nucleophilic additions, pericyclic reactions, and cationic rearrangements; chlorination reagent)

Alternate Name: stannic chloride.
Physical Data: colorless liquid; mp −33 °C; bp 114.1 °C; *d* 2.226 g cm^{-3}.
Solubility: reacts violently with water; sol cold H_2O; dec hot H_2O; sol alcohol, Et_2O, CCl_4, benzene, toluene, acetone.
Form Supplied in: colorless liquid; 1 M soln in CH_2Cl_2 or heptane; widely available.
Purification: reflux with mercury or P_2O_5 for several hours, then distill under reduced nitrogen pressure into receiver with P_2O_5. Redistill. Typical impurities: hydrates.

Handling, Storage, and Precautions: hygroscopic; should be stored in a glove box or over P_2O_5 to minimize exposure to moisture. Containers should be flushed with N_2 or Ar and tightly sealed. Perform all manipulations under N_2 or Ar. Solvating with H_2O liberates much heat. Use in a fume hood.

Introduction. $SnCl_4$ is used extensively in organic synthesis as a Lewis acid for enhancing a variety of reactions. $SnCl_4$ is classified as a strong Lewis acid according to HSAB theory, and therefore interacts preferentially with hard oxygen and nitrogen bases. Six-coordinate 1:2 species and 1:1 chelates are the most stable coordination complexes, although 1:1 five-coordinate species are also possible.[1] $SnCl_4$ can be used in stoichiometric amounts, in which case it is considered a 'promoter', or in substoichiometric amounts as a catalyst, depending upon the nature of the reaction. $SnCl_4$ is an attractive alternative to boron, aluminum, and titanium Lewis acids because it is monomeric, highly soluble in organic solvents, and relatively easy to handle. $SnCl_4$ and $TiCl_4$ are among the most common Lewis acids employed in 'chelation control' strategies for asymmetric induction. However, $SnCl_4$ is not often the Lewis acid of choice for optimum selectivities and yields.

$SnCl_4$ is also the principal source for alkyltin chlorides, R_nSnCl_{4-n}.[2] Allyltrialkyltin reagents react with $SnCl_4$ to produce allyltrichlorotin species through an S_E2' pathway (eq 1).[3] Silyl enol ethers react with $SnCl_4$ to give α-trichlorotin ketones (eq 2).[4] Transmetalation or metathesis reactions of this type are competing pathways to nucleophilic addition reactions where $SnCl_4$ is present as an external Lewis acid. As a consequence, four important experimental variables must be considered when using $SnCl_4$ as a promoter: (1) the stoichiometry between the substrate and the Lewis acid; (2) the reaction temperature; (3) the nature of the Lewis base site(s) in the substrate; and (4) the order of addition. These variables influence the reaction pathway and product distribution.[5]

SnR3 + SnCl4 ⟶ SnCl3 + SnBu3Cl (1)

OTMS, R + SnCl4 ⟶ O, R, SnCl3 + TMSCl (2)

Nucleophilic Additions to Aldehydes. $SnCl_4$ is effective in promoting the addition of nucleophiles to simple aldehydes. Among the most synthetically useful additions are allylstannane and -silane additions. The product distribution in the stannane reactions can be influenced by the order of addition, stoichiometry, and reaction temperature. The *anti* geometry of the tin–aldehyde complex is favored due to steric interactions. Furthermore, the six-coordinate 2:1 complex is most likely the reactive intermediate in these systems. The use of crotylstannanes provides evidence for competing transmetalation reaction pathways (eq 3).[6] Superior selectivities are provided by ***Titanium(IV) Chloride***.

The presence of additional Lewis base sites within the molecule can result in the formation of chelates with $SnCl_4$ or $TiCl_4$, which can lead to 1,2- or 1,3-asymmetric induction with the appropriate substitution at the C-2 or C-3 centers. NMR studies have provided a basis for explaining the levels of diastereofacial selectivity observed in nucleophilic additions to Lewis acid chelates of β-alkoxy aldehydes with substitution at the C-2 or C-3 positions.[7] These studies reveal that $SnCl_4$ chelates are dynamically unstable when substrates are sterically crowded at the alkoxy center, thus enhancing the formation of 2:1 complexes and/or competing metathesis pathways. Furthermore, for β-siloxy aldehydes, the 2:1 $SnCl_4$ complex is formed preferentially over the corresponding chelate.[8]

O, H; SnBu3; Lewis acid ⟶ OH, R

normal addition	1.3 equiv $SnCl_4$	22.8	26.0	36.4	14.8
inverse addition	1.3 equiv $SnCl_4$	21.8	74.9	1.2	2.2
normal addition	1.05 equiv $TiCl_4$	90.5	7.0	2.1	0.5
inverse addition	2.1 equiv $TiCl_4$	4.4	90.8	–	4.9

(3)

Mukaiyama Aldol Additions. Lewis acid-promoted additions of a chiral aldehyde to a silyl enol ether or silyl ketene acetal (the Mukaiyama[9] aldol addition) occurs with good diastereofacial selectivity.[10] The reaction has been investigated with nonheterosubstituted aldehydes, α- and β-alkoxy aldehydes,[11] α- and β-amino aldehydes,[12] and thio-substituted aldehydes.[13] High diastereoselectivity is observed in the $SnCl_4$- or $TiCl_4$-promoted aldol addition of silyl enol ethers to α- and β-alkoxy aldehydes. Prior chelation of the aldehyde before addition of the enol silane is important because certain enol silanes interact with $SnCl_4$ to produce α-trichlorostannyl ketones, which provide lower selectivity.[14] Simple diastereoselectivity is independent of the geometry of the enol silane, and the reaction does not proceed through prior Si–Ti or Si–Sn exchange. Good *anti* selectivities (up to 98:2) are obtained in the $SnCl_4$-promoted reactions of chiral α-thio-substituted aldehydes only with α-phenylthio-substituted aldehydes (eq 4). Stereorandom results are obtained with $SnCl_4$ when other alkylthio-substituted aldehydes, such as α-isopropylthio-substituted aldehydes, are used. ***Boron Trifluoride Etherate*** catalysis gives better *anti* selectivities than $SnCl_4$ for aldehydes with smaller alkylthio substituents. Excellent *syn* selectivities are obtained for α-thio-substituted aldehydes with $TiCl_4$.

R¹S, R, O + OMe, OTMS; Lewis acid ⟶ R¹S, R, CO2Me, OH (*syn*) + R¹S, R, CO2Me, OH (*anti*) (4)

Additions to Nitriles. $SnCl_4$-promoted addition of malonates and bromomalonates to simple nitriles (not electron deficient) gives α,β-dehydro-β-amino acid derivatives (eq 5).[15] $SnCl_4$ is the Lewis acid of choice for the condensation of aroyl chlorides with sodium isocyanate, affording aroyl isocyanates in 70–85% yields.[16] Nonaromatic acyl chlorides react under more variable reaction conditions.

1. $SnCl_4$; 2. Na_2CO_3; 55% (5)

Hydrochlorination of Allenic Ketones. $SnCl_4$ is also a source for generating chloride anions which form new carbon–chlorine bonds. This occurs through a ligand exchange pathway which has been exploited in the formation of β-chloro enones from conjugated allenic ketones (eq 6).[17] Yields range from 36–82% with complete selectivity for the *trans* geometry. A variety of substituents (R^1, R^2) can be tolerated including aryl, rings, and alkoxymethyl groups (R^1).

$SnCl_4$, 20 °C; benzene (6)

Glycosylations. The reaction of glycofuranosides having a free hydroxyl group at C-2 with functionalized organosilanes, in the presence of $SnCl_4$, provides *C*-glycosyl compounds in high stereoselectivity (eq 7).[18] Organosilanes such as 4-(chlorodimethylsilyl)toluene, chlorodimethylvinylsilane, ***Allyltrimethylsilane***, and allylchlorodimethylsilane are effective reagents. The presence of a leaving group on the silane is essential for good selectivity since the reaction proceeds intramolecularly through a 2-*O*-organosilyl glycoside. The availability of furanosides in the ribo, xylo, and arabino series make this reaction valuable for the stereoselective synthesis of *C*-furanosides. Regioselective glycosylation of nitrogen-containing heterocycles is also effectively promoted by $SnCl_4$, and has been used in the synthesis of pentostatin-like nucleosides, such as (**1**).[19]

$SnCl_4$ (7)

R = H: pentostatin
R = OH: coformycin

(**1**)

Selective De-*O*-benzylation. Regioselective de-*O*-benzylation of polyols and perbenzylated sugars is achieved with organotin reagents or other Lewis acids.[20,21] The equatorial *O*-benzyl group of 1,6-anhydro-2,3,4-tri-*O*-benzyl-β-D-mannopyranose is selectively cleaved by $SnCl_4$ or $TiCl_4$ (eq 8).[2] The equatorial *O*-benzyl group is also selectively cleaved when one of the axial *O*-benzyl groups is replaced by an *O*-methyl group. The 2-*O*-benzyl group of 1,2,3-tris(benzyloxy)propane is selectively cleaved (eq 9), but no debenzylation is observed with 1,2-bis(benzyloxy)ethane.

0.5 h (8)

$SnCl_4$	92%	5%
$TiCl_4$	77%	19%

$SnCl_4$; 86% (9)

Rearrangement of Allylic Acetals. Lewis acid-promoted ($SnCl_4$ or ***Diethylaluminum Chloride***) rearrangements of allylic acetals provide substituted tetrahydrofurans.[22] Upon addition of Lewis acid, (**2**) rearranges to the *all-cis* furan (**3**) (eq 10). No racemization is observed with optically active allylic acetals; however, addition of KOH completely epimerizes the furan–carbonyl bond, as does quenching at room temperature. Acetals successfully undergo similar rearrangement provided the alkene is substituted. Completely substituted tetrahydrofurans are synthesized stereoselectively (>97% ee) by the rearrangement of disubstituted allyl acetals (eq 11). This reaction is related to the acid-catalyzed rearrangements of 5-methyl-5-vinyloxazolidines to 3-acetylpyrrolidines, which involves an aza-Cope rearrangement and Mannich cyclization.[23]

$SnCl_4$, –70 °C; –10 °C, 2 h; 58% (10)

(**2**) (**3**)

$SnCl_4$, –70 °C; –10 °C, 2 h (11)

R^1 = Me, R^2 = H; 90%
R^1 = H, R^2 = Me; 73%

The rearrangement is also useful for furan annulations, through enlargement of the starting carbocycle.[24] Thus addition of $SnCl_4$ to either diastereomer of the allylic acetal (**4**) produces the *cis*-fused cycloheptatetrahydrofuran (**5**) in 48–76% yield (eq 12). Acetals derived from *trans*-diols rearrange to the same *cis*-fused bicyclics in higher yield. The stereochemistry of a terminal alkene is transmitted to the C-3 carbon of the bicyclic products (eq 13). Rearrangements of acetals require substitution at the internal alkene carbon.

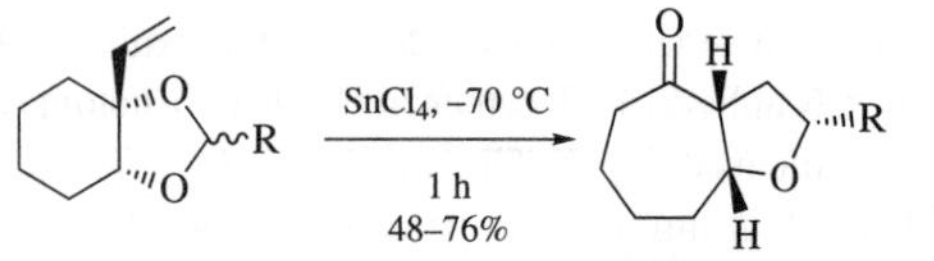

(12)

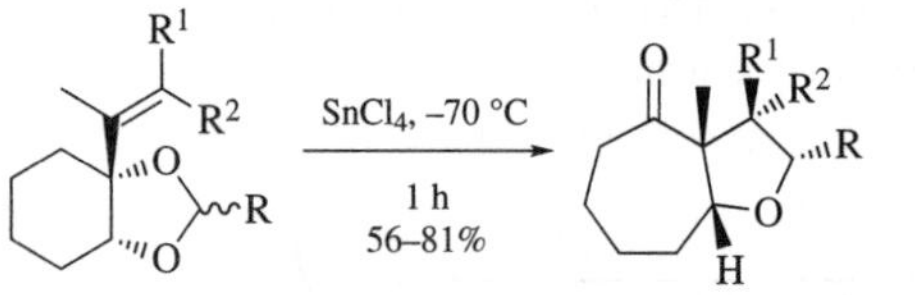

(13)

(17)

(18)

(19)

α-*t*-Alkylations. $SnCl_4$-promoted α-*t*-alkylations of alkenyl β-dicarbonyl compounds is a particularly useful cyclization reaction.[25] Cyclization occurs through initial formation of a stannyl enol ether, followed by protonation of the alkene to form a carbocation which undergoes subsequent closure (eq 14). The analogous α-*s*-alkylation reactions are best catalyzed by other Lewis acids.

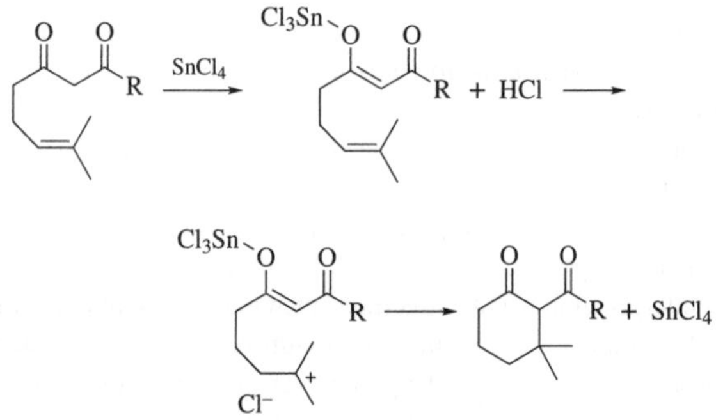

(14)

This reaction is useful for cyclizations involving 6-*endo*-trigonal (eq 14) and 'allowed' 7-*endo* trigonal processes (eq 15), but not for those involving 5-*endo* trigonal processes (eq 16). These observations are consistent with the Baldwin rules.

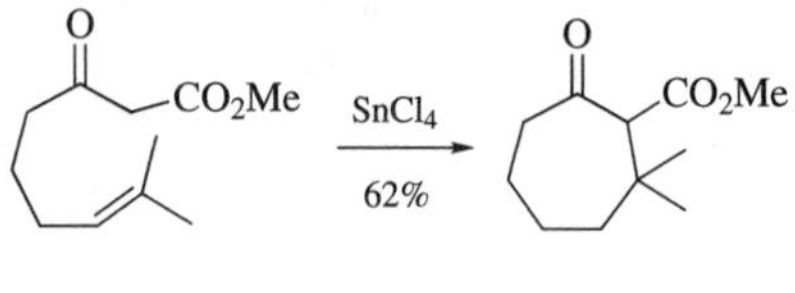

(15)

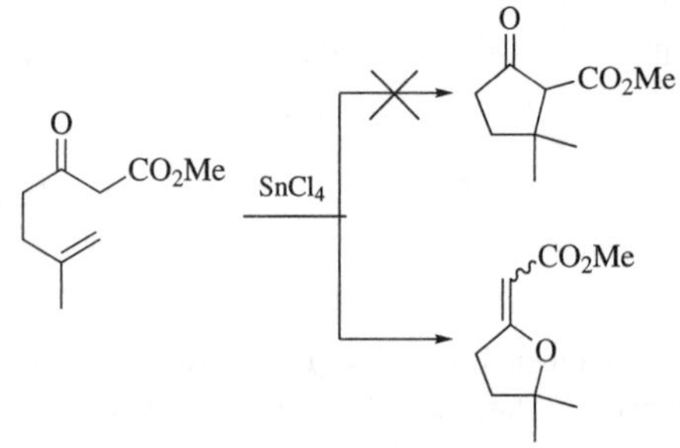

(16)

Reactions involving 4- and 6-*exo* trigonal cyclizations result in poor yields or undesired products, while those involving 5-*exo* trigonal cyclizations produce higher yields (eq 17). This synthetic strategy can also be used to form bicyclic and spiro compounds (eqs 18 and 19).

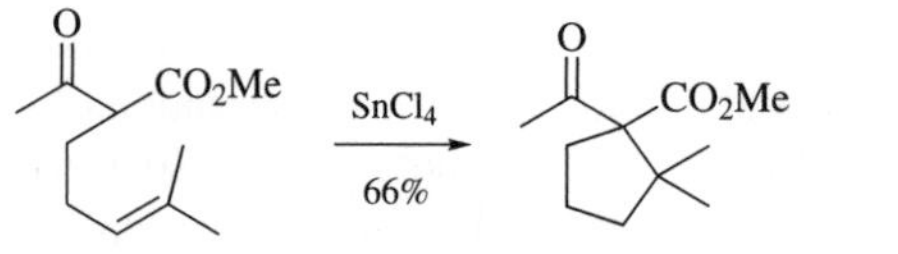

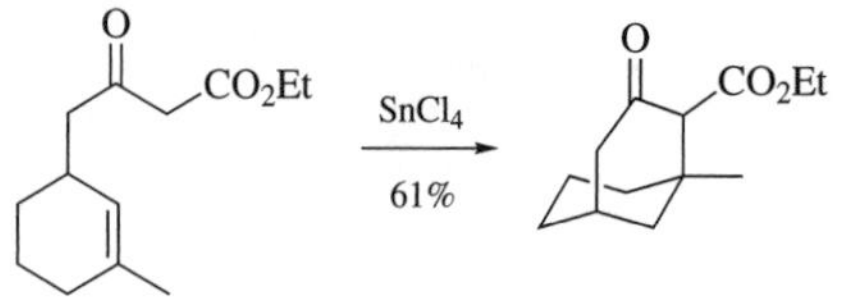

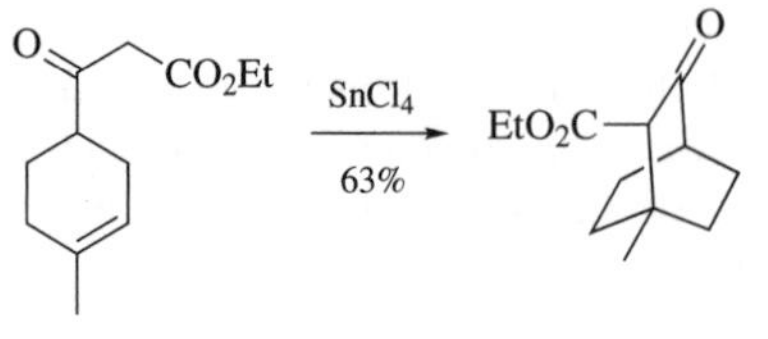

Alkene Cyclizations. Cationic cyclizations of polyenes, containing initiating groups such as cyclic acetals, are promoted by $SnCl_4$ and have been utilized in the synthesis of *cis*- and *trans*-decalins, *cis*- and *trans*-octalins, and tri- and tetracyclic terpenoids and steroids.[26] In most instances, *all-trans*-alkenes yield products with *trans,anti,trans* stereochemistry (eq 20), while *cis*-alkenes lead to *syn* stereochemistry at the newly formed ring junctions. The stereoselectivity of polyene cyclizations are often greatly diminished when the terminating alkene is a vinyl group rather than an isopropenyl group. Acyclic compounds which contain terminal acyclic acetals and alkenes or vinylsilanes can be cyclized in a similar fashion to yield eight- and nine-membered cyclic ethers (eq 21).[27]

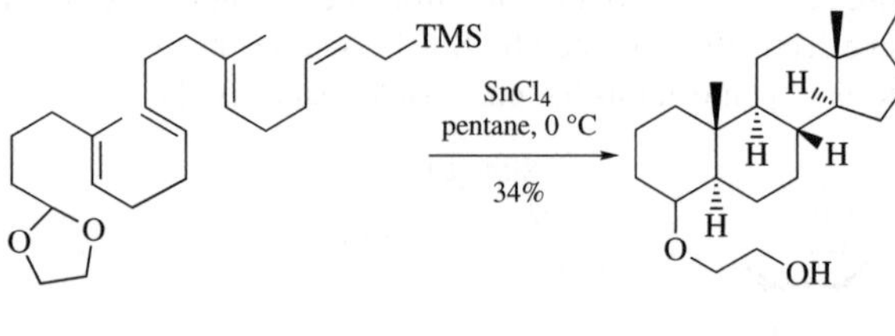

(20)

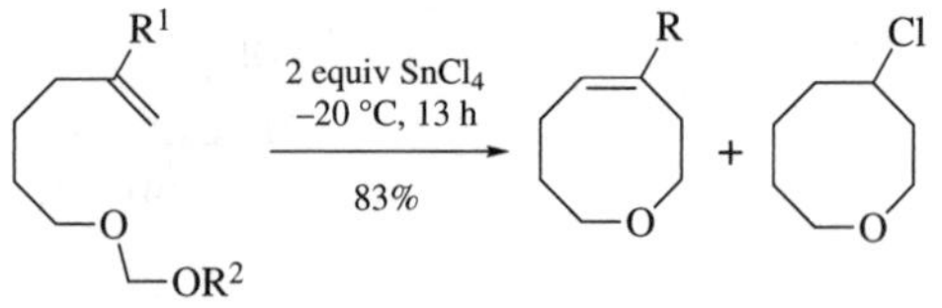

(21)

R^1 = H, TMS; R^2 = $(CH_2)_2OMe$

The analogous cyclization of chiral imines occurs in high yields (75–85%) with good asymmetric induction (36–65% ee).[28] For example, the cyclization of aldimine (**6**), derived from methyl citronellal, using $SnCl_4$ affords only the *trans*-substituted aminocyclohexane (**7**) in high yield (eq 22). *Exo* products are formed exclusively or preferentially over the thermodynamically favored *endo* products.

$$\textbf{(6)} \xrightarrow[84\%]{SnCl_4} \textbf{(7)} \quad (22)$$

$SnCl_4$-induced cyclizations between alkenes and enol acetates result in cycloalkanes or bicycloalkanes in high yield (eq 23). It is interesting to note that the TMSOTf-catalyzed reaction can yield fused products rather than bicyclo products. Alkenic carboxylic esters, allylic alcohols, sulfones, and sulfonate esters are also cyclized in the presence of $SnCl_4$; however, alkenic oxiranes often cyclize in poor yield.[26a]

(23) [$SnCl_4$, 98%; TMSOTf, 65%]

$SnCl_4$ is also effective in the opening of cyclopropane rings to produce cationic intermediates useful in cyclization reactions. For example, the cyclization of aryl cyclopropyl ketones to form aryl tetralones, precursors of aryl lignan lactones and aryl naphthalene lignans, is mediated by $SnCl_4$ (eq 24).[29] The reaction is successful in nitromethane, but not in benzene or methylene chloride. Analogous cyclizations with epoxides result in very low yields (2–5%).

(24) [$SnCl_4$, 63%]

Polymerization. Cationic polymerizations are catalyzed by $SnCl_4$ and other Lewis acids (eq 25). Propagation is based upon the formation of a cationic species upon complexation with $SnCl_4$.[30] Radical pathways are also possible for polymer propagation.[31]

(25) [HB, $SnCl_4$]

Diels–Alder Reactions. Diels–Alder reactions are enhanced through the complexation of dienophiles or dienes by Lewis acids.[32] Furthermore, Lewis acids have been successfully employed in asymmetric Diels–Alder additions.[33] Although $SnCl_4$ is a useful Lewis acid in Diels–Alder reactions, in most instances titanium or aluminum Lewis acids provide higher yields and/or selectivities. The stereoselectivity in Lewis acid-promoted Diels–Alder reactions between chiral α,β-unsaturated *N*-acyloxazolidinones shows unexpected selectivities as a function of the Lewis acid (eq 26).[34] Optimum selectivity is expected for chelated intermediates, yet both $SnCl_4$ and $TiCl_4$ perform poorly relative to Et_2AlCl (1.4 equiv). The formation of the $SnCl_4$–*N*-acyloxazolidinone chelate has been confirmed by solution NMR studies.[35] These data suggest that other factors such as the steric bulk associated with complexes may contribute to stereoselectivity.

(26) [Lewis acid; *exo* I, *exo* II, *endo* I, *endo* II]

Lewis acid	conv %	Σ*endo*:Σ*exo*	*endo* I:*endo* II
1.1 equiv $SnCl_4$	70	14.9	3.1
1.1 equiv $TiCl_4$	100	9.9	2.7
1.4 equiv Et_2AlCl	100	50.0	17

In Lewis acid-promoted Diels–Alder reactions of cyclopentadiene with the acrylate of (*S*)-ethyl lactate, good diastereofacial and *endo*/*exo* selectivity are obtained with $SnCl_4$ (84:16; *endo*/*exo* = 18:1) and $TiCl_4$ (85:15; *endo*/*exo* = 16:1).[36] It is interesting to note that boron, aluminum, and zirconium Lewis acids give the opposite diastereofacial selectivity (33:67 to 48:52). Competing polymerization of the diene is observed in methylene chloride, particularly with $TiCl_4$, but not in solvent mixtures containing *n*-hexane.

Cycloalkenones generally perform poorly as dienophiles in Diels–Alder reactions but their reactivity can be enhanced by Lewis acids.[37] $SnCl_4$ is effective in promoting the Diels–Alder reaction between simple 1,3-butadienes, such as isoprene and piperylene, and cyclopentenone esters. For example, the $SnCl_4$-promoted cycloaddition between (**8**) and isoprene is completely regioselective, providing the substituted indene in 86% yield (eq 27).[38] However, cycloaddition does not occur in the presence of $SnCl_4$ when the diene contains an oxygen-bearing substituent such as an alkoxy or siloxy group. In these cases, as is generally true for the Diels–Alder reactions of cycloalkenones, other Lewis acids are more effective. For example, $SnCl_4$-promotion of the cycloaddition between (**8**) and 3-methyl-2-(*t*-butyldimethylsiloxy)butadiene yields 37% of the desired product, while ***Zinc Chloride*** provides a 90% yield. When furan or 2-methyl-1-alkylsiloxybutadiene are utilized as dienes, only decomposition of the starting material is observed with $SnCl_4$.

(27) [(**8**) + isoprene, $SnCl_4$]

The Lewis acid-promoted Diels–Alder reaction has been employed in the assembly of steroid skeletons.[39] The cycloaddition reaction between a substituted bicyclic diene and 2,6-dimethylbenzoquinone produces two stereoisomers in a 1:5

ratio with a yield of 83% when $SnCl_4$ is used in acetonitrile. $TiCl_4$ provides slightly higher selectivities (1:8) but lower yield (70%) (eq 28).

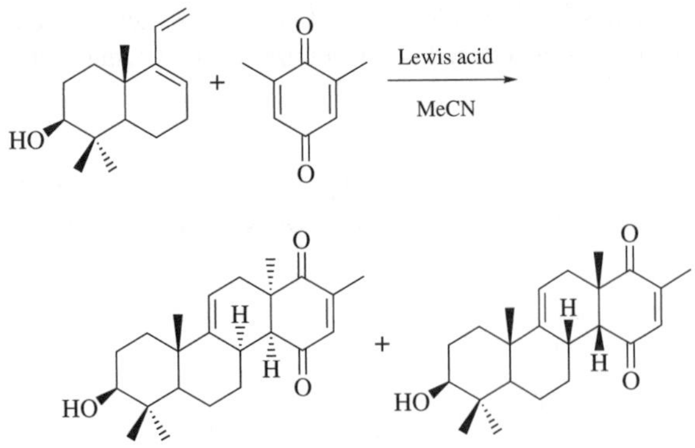

(28)

$SnCl_4$	1:5
$TiCl_4$	1:8

When the dienophile *N*-α-methylbenzylmaleimide (**9**) is reacted with 2-*t*-butyl-1,3-butadiene in the presence of Lewis acids, cycloadducts (**10**) and (**11**) are formed (eq 29).[40] While $SnCl_4$ provides (**10**) and (**11**) in a 5:1 ratio, $TiCl_4$ and $EtAlCl_2$ both provide a 15:1 ratio. Polymerization of the diene competes with adduct formation under all conditions.

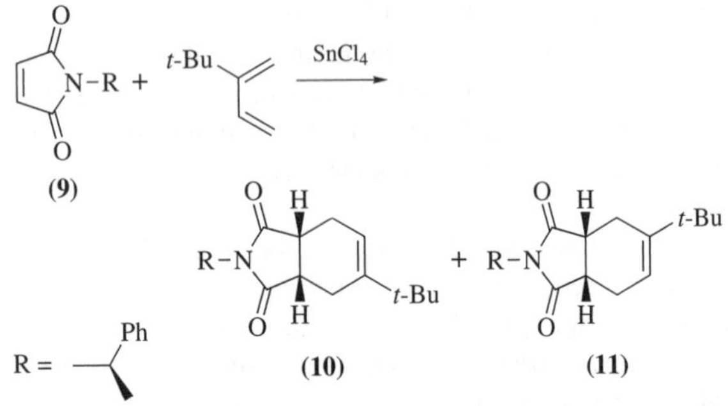

(29)

[4 + 3] Cycloadditions. Oxyallyl cations,[41] which react as C_3 rather than C_2 components in cyclization reactions, are generated by the addition of $SnCl_4$ to substrates which contain silyl enol ethers which are conjugated with a carbonyl moiety. Thus 2-(trimethylsiloxy)propenal undergoes cyclization with cyclopentadiene or furan (eq 30).[42] Substituted 1,1-dimethoxyacetones also form these intermediates and undergo subsequent cyclizations (eq 31).[43] This method complements the usual synthesis of oxyallyl cations involving reductive elimination of halogens from halogenated ketones or electronically equivalent structures.[44]

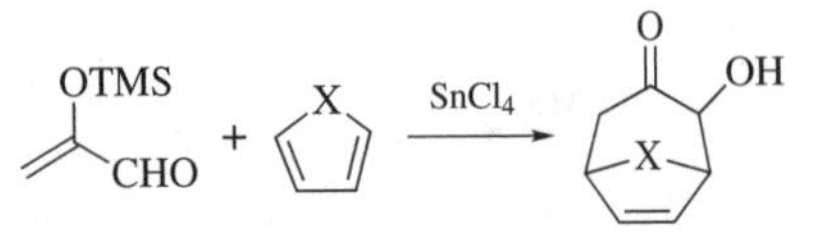

(30)

X = O, CH_2

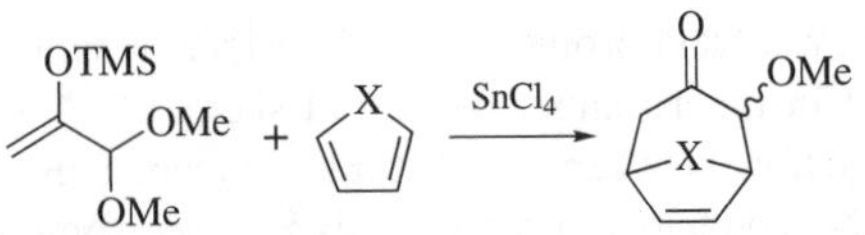

(31)

[3 + 2] Cycloadditions. Lewis acid-mediated [3 + 2] cycloadditions of oxazoles and aldehydes or diethyl ketomalonate have been observed using organoaluminum and Sn^{IV} Lewis acids.[45] The reactions are highly regioselective, with stereoselectivity extremely dependent upon Lewis acid (eq 32). For example, the (BINOL)AlMe-promoted reaction between benzaldehyde and the oxazole (**12**) provides the oxazoline with a *cis*/*trans* ratio of 98:2. The selectivity is reversed with $SnCl_4$ which provides a *cis*/*trans* ratio of 15:85. *trans*-5-Substituted 4-alkoxycarbonyl-2-oxazolines are synthesized under thermodynamic conditions in the aldol reaction of isocyanoacetates with aldehydes.[46]

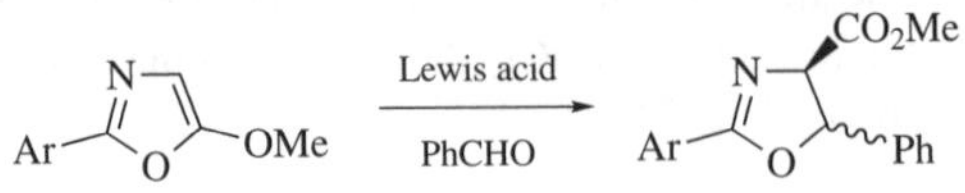

(**12**)

(32)

[2 + 2] Cycloadditions. The regioselectivity in the cycloaddition reactions of 2-alkoxy-5-allyl-1,4-benzoquinones with styrenes is controlled by the choice of Ti^{IV} or $SnCl_4$ Lewis acids (eq 33).[47] The use of an excess of $TiCl_4$ or mixtures of $TiCl_4$ and $Ti(O\text{-}i\text{-}Pr)_4$ produces cyclobutane (**13**) as the major or exclusive product, while $SnCl_4$ promotion with one equivalent of Lewis acid results in the formation of (**14**) only. These reactions represent a classic example of the mechanistic variability often associated with seemingly modest changes in Lewis acid.

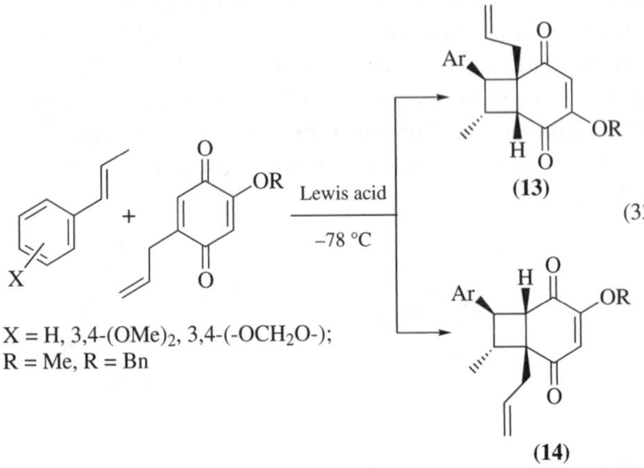

X = H, 3,4-$(OMe)_2$, 3,4-(-OCH_2O-);
R = Me, R = Bn

(33)

Ene Reactions. The Lewis acid-catalyzed ene reaction is synthetically useful methodology for forming new carbon–carbon bonds.[48] Ene reactions utilizing reactive enophiles such as formaldehyde and chloral can be promoted by $SnCl_4$. $SnCl_4$ also enhances intramolecular ene reactions, such as the cyclization of (**15**) which produces the α-hydroxy δ-lactone in 85% yield (eq 34).[49] The ene cyclization of citronellal to give isopulegol has also been reported.[50] Proton scavenging aluminum Lewis acids such as $RAlCl_2$ are most often used in ene reactions to eliminate proton-induced side reactions.

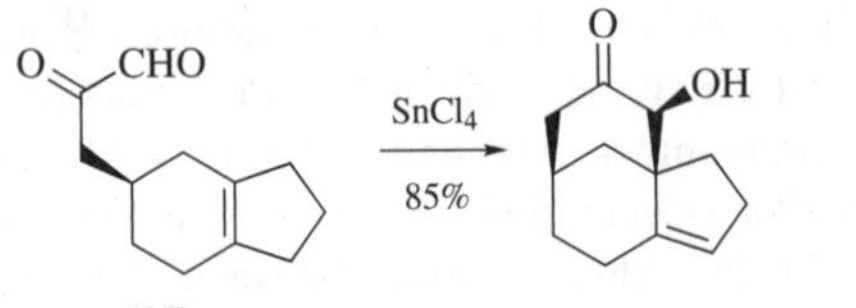

(**15**)

(34)

Related Reagents. Boron Trifluoride etherate; Tin(IV) Chloride–Zinc Chloride; Titanium Tetrachloride; Zinc Chloride.

1. (a) Shambayati, S.; Crowe, W. E.; Schreiber, S. L. *AG(E)* **1990**, *29*, 256. (b) Reetz, M. T. In *Selectivities in Lewis Acid Promoted Reactions*; Schinzer, D., Ed.; Kluwer: Dordrecht, 1989; pp 107–125. (c) Denmark, S. E.; Almstead, N. G. *JACS* **1993**, *115*, 3133.
2. Davies, G. A.; Smith, P. J. In *Comprehensive Organometallic Chemistry*; Wilkinson, G.; Stone, F. G. A.; Abel, E. W., Eds.; Pergamon: New York, 1982; Vol. 2, p 519.
3. Naruta, Y.; Nishigaichi, Y.; Maruyama, K. *T* **1989**, *45*, 1067.
4. (a) Nakamura, E.; Kuwajima, I. *CL* **1983**, 59. (b) Yamaguchi, M.; Hayashi, A.; Hirama, M. *JACS* **1993**, *115*, 3362. (c) Yamaguchi, M.; Hayashi, A.; Hirama, M. *CL* **1992**, 2479.
5. (a) Keck, G. E.; Castellino, S.; Andrus, M. B. In *Selectivities in Lewis Acid Promoted Reactions*; Schinzer, D., Ed.; Kluwer: Dordrecht, 1989; pp 73–105. (b) Keck, G. E.; Andrus, M. B.; Castellino, S. *JACS* **1989**, *111*, 8136. (c) Denmark, S. E.; Wilson, T.; Wilson, T. M. *JACS* **1988**, *110*, 984. (d) Boaretto, A.; Marton, D.; Tagliavini, G.; Ganis, P. *JOM* **1987**, *321*, 199. (e) Yamamoto, T.; Maeda, N.; Maruyama, K. *CC* **1983**, 742. (f) Quintard, J. P.; Elissondo, B.; Pereyre, M. *JOC* **1983**, *48*, 1559.
6. Keck, G. E.; Abbott, D. E.; Boden, E. P.; Enholm, E. J. *TL* **1984**, *25*, 3927.
7. (a) Keck, G. E.; Castellino, S. *JACS* **1986**, *108*, 3847. (b) Keck, G. E.; Castellino, S.; Wiley, M. R. *JOC* **1986**, *51*, 5478.
8. Keck, G. E.; Castellino, S. *TL* **1987**, *28*, 281.
9. Mukaiyama, T.; Banno, K.; Narasaka, K. *JACS* **1974**, *96*, 7503.
10. Review of Mukaiyama aldol reaction: Gennan, C. *COS* 1991, Vol. 2.
11. Reetz, M. T.; Kesseler, K.; Jung, A. *T* **1984**, *40*, 4327.
12. (a) Reetz, M. T. *AG(E)* **1984**, *23*, 556. (b) ref 11.
13. (a) Annunziata, R.; Cinquini, M.; Cozzi, F.; Cozzi, P. G.; Consolandi, E. *JOC* **1992**, *57*, 456. (b) Annunziata, R.; Cinquini, M.; Cozzi, F.; Cozzi, P. G. *TL* **1990**, *31*, 6733.
14. Nakamura, E.; Kawajima, I. *TL* **1983**, *24*, 3343.
15. Scavo, F.; Helquist, P. *TL* **1985**, *26*, 2603.
16. Deng, M. Z.; Caubere, P.; Senet, J. P.; Lecolier, S. *T* **1988**, *44*, 6079.
17. Gras, J. L.; Galledou, B. S. *BSF(2)* **1982**, 89.
18. Martin, O. R.; Rao, S. P.; Kurz, K. G.; El-Shenawy, H. A. *JACS* **1988**, *110*, 8698.
19. Showalter, H. D. H.; Putt, S. R. *TL* **1981**, *22*, 3155.
20. Wagner, D.; Verheyden, J. P. H.; Moffat, J. G. *JOC* **1974**, *39*, 24.
21. Hori, H.; Nishida, Y.; Ohrui, H.; Meguro, H. *JOC* **1989**, *54*, 1346.
22. Hopkins, M. H.; Overman, L. E. *JACS* **1987**, *109*, 4748.
23. Overman, L. E.; Kakimoto, M. E.; Okazaki, M. E.; Meier, G. P. *JACS* **1983**, *105*, 6622.
24. Herrington, P. M.; Hopkins, M. H.; Mishra, P.; Brown, M. J.; Overman, L. E. *JOC* **1987**, *52*, 3711.
25. Review of α-alkylations to carbonyl compounds: Reetz, M. T. *AG(E)* **1982**, *21*, 96.
26. (a) Review of asymmetric alkene cyclization: Bartlett, P. A. *Asymmetric Synthesis*; Morrison, J. D., Ed.; Academic: New York, 1984; Vol. 3, Part B, pp 341–409. (b) Review of thermal cycloadditions: Fallis, A. G.; Lu, Y.-F. *Advances in Cycloaddition*; Curran, D. P., Ed.; JAI: Greenwich, CT, 1993; Vol. 3, pp 1–66.
27. (a) Overman, L. E.; Blumenkopf, T. A.; Castaneda, A.; Thompson, A. S. *JACS* **1986**, *108*, 3516. (b) Overman, L. E.; Castaneda, A.; Blumenkopf, T. A. *JACS* **1986**, *108*, 1303.
28. Demailly, G.; Solladie, G. *JOC* **1981**, *46*, 3102.
29. Murphy, W. S.; Waltanansin, S. *JCS(P1)* **1982**, 1029.
30. (a) Kamigaito, M.; Madea, Y.; Sawamota, M.; Higashimura, T. *Macromolecules* **1993**, *26*, 1643. (b) Takahashi, T.; Yokozawa, T.; Endo, T. *Makromol. Chem.* **1991**, *192*, 1207. (c) Ran, R. C.; Mao, G. P. J. *Macromol. Sci. Chem.* **1990**, *A27*, 125. (d) Kurita, K.; Inoue, S.; Yamamura, K.; Yoshino, H.; Ishii, S.; Nishimura, S. I. *Macromolecules* **1992**, *25*, 3791. (e) Yokozawa, T.; Hayashi, R.; Endo, T. *Macromolecules* **1993**, *26*, 3313.
31. (a) Tanaka, H.; Kato, H.; Sakai, I.; Sato, T.; Ota, T. *Makromol. Chem. Rapid Commun.* **1987**, *8*, 223. (b) Yuan, Y.; Song, H.; Xu, G. *Polym. Int.* **1993**, *31*, 397.
32. Birney, D. M.; Houk, K. N. *JACS* **1990**, *112*, 4127.
33. For leading references on asymmetric Diels–Alder reactions, see: (a) Paquette, L. A. In *Asymmetric Synthesis*; Morrison, J. D., Ed.; Academic: New York, 1984; Vol. 3, pp 455–483. (b) Oppolzer, W. *AG(E)* **1984**, *23*, 876. (c) Carruthers, W. *Cycloaddition Reactions in Organic Synthesis*; Pergamon: New York, 1990; pp 61–72.
34. Evans, D. A.; Chapman, K. T.; Bisaha, J. *JACS* **1988**, *110*, 1238.
35. Castellino, S. *JOC* **1990**, *55*, 5197.
36. Poll, T.; Helmchen, G.; Bauer, B. *TL* **1984**, *25*, 2191.
37. (a) Fringuelli, F.; Pizzo, F.; Taticchi, A.; Wenkert, E. *JOC* **1983**, *48*, 2802. (b) Fringuelli, F.; Pizzo, F.; Taticchi, A.; Halls, T. D. J.; Wenkert, E. *JOC* **1982**, *47*, 5056.
38. Liu, H. J.; Ulibarri, G.; Browne, E. N. C. *CJC* **1992**, *70*, 1545.
39. Arseniyadis, A.; Rodriguez, R.; Spanevello, J. C.; Thompson, A.; Guittet, E.; Ourisson, G. *T* **1992**, *48*, 1255.
40. Baldwin, S. W.; Greenspan, P.; Alaimo, C.; McPhail, A. T. *TL* **1991**, *42*, 5877.
41. For a recent review of oxyallyl cations, see: Mann, J. *T* **1986**, *42*, 4611.
42. Masatomi, O.; Kohki, M.; Tatsuya, H.; Shoji, E. *JOC* **1990**, *55*, 6086.
43. Murray, D. H.; Albizati, K. F. *TL* **1990**, *31*, 4109.
44. Hoffman, H. M. R. *AG(E)* **1973**, *12*, 819; **1984**, *23*, 1.
45. Suga, H.; Shi, X.; Fujieda, H.; Ibata, T. *TL* **1991**, *32*, 6911.
46. For examples of enantioselective synthesis of *trans*-4-alkoxy-2-oxazolines, see: Ito. Y; Sawamura, M.; Shirakawa, E.; Hayashizaki, K.; Hayashi, T. *TL* **1988**, *29*, 235; *T* **1988**, *44*, 5253.
47. Engler, T. A.; Wei, D.; Latavic, M. A. *TL* **1993**, *34*, 1429.
48. Reviews of ene reactions: (a) Hoffman, H. M. R. *AG(E)* **1969**, *8*, 556. (b) Oppolzer, W.; Sniekus, V. *AG(E)* **1978**, *17*, 476. (c) Snyder, B. B. *ACR* **1980**, *13*, 426.
49. Lindner, D. L.; Doherty, J. B.; Shoham, G.; Woodward, R. B. *TL* **1982**, *23*, 5111.
50. Nakatani, Y.; Kawashima, K. *S* **1978**, 147.

Stephen Castellino
Rhône-Poulenc Ag. Co., Research Triangle Park, NC, USA

David E. Volk
North Dakota State University, Fargo, ND, USA

Tin(II) Trifluoromethanesulfonate

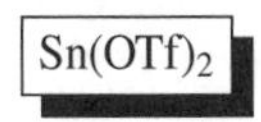

[62086-04-8] $C_2F_6O_6S_2Sn$ (MW 416.87)

(mild Lewis acid used primarily for the generation of tin(II) enolates for stereoselective aldol and Michael reactions;[1] [2,3]-Wittig and Ireland–Claisen rearrangements;[2] addition of tin(II) acetylides to aldehydes;[3] asymmetric allylation of aldehydes[4])

Physical Data: mp >300 °C.
Form Supplied in: fine white powder.[5]
Purification: can be purified by washing with diethyl ether.[6]
Handling, Storage, and Precautions: moisture and air sensitive; all handling and storage should be done under nitrogen.

Aldol Reaction. Mukaiyama has developed the use of tin(II) triflate in diastereoselective and enantioselective aldol-type reactions.[1,7] Initially, the stereoselective aldol reactions were demonstrated using a stoichiometric amount of tin(II) triflate.[8] The reaction between 3-acylthiazolidine-2-thione and 3-phenylpropionaldehyde is a representative example of a diastereoselective *syn*-aldol synthesis (eq 1).

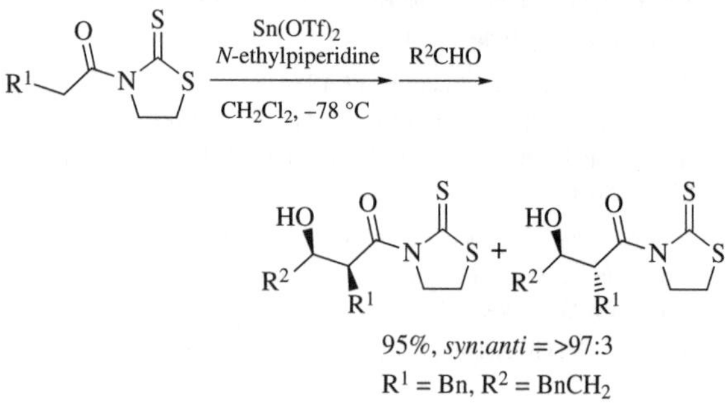

$Sn(OTf)_2$, *N*-ethylpiperidine, CH_2Cl_2, −78 °C; R^2CHO (1)

95%, *syn:anti* = >97:3
R^1 = Bn, R^2 = $BnCH_2$

Enantioselective aldol-type reactions have been achieved by addition of a chiral diamine to the reaction mixture.[9] Specifically, addition of (*S*)-1-methyl-2-[(piperidin-1-ylmethyl]pyrrolidine to the reaction of 3-acetylthiazolidine-2-thione and 3-phenylpropionaldehyde provides the *syn*-aldol with greater than 90% ee (eq 2).

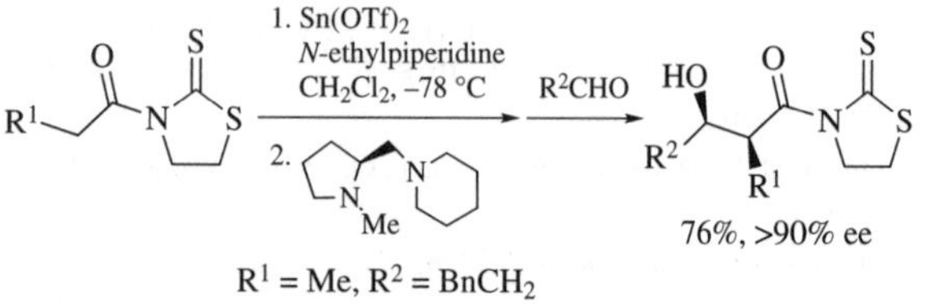

1. $Sn(OTf)_2$, *N*-ethylpiperidine, CH_2Cl_2, −78 °C; 2. chiral diamine; R^2CHO (2)

76%, >90% ee
R^1 = Me, R^2 = $BnCH_2$

Reactions of ketene silyl acetals with various aldehydes in the presence of a stoichiometric amount of a tin(II) triflate–tin(IV) additive–chiral diamine complex have also been performed with excellent enantioselectivity (eq 3).[10]

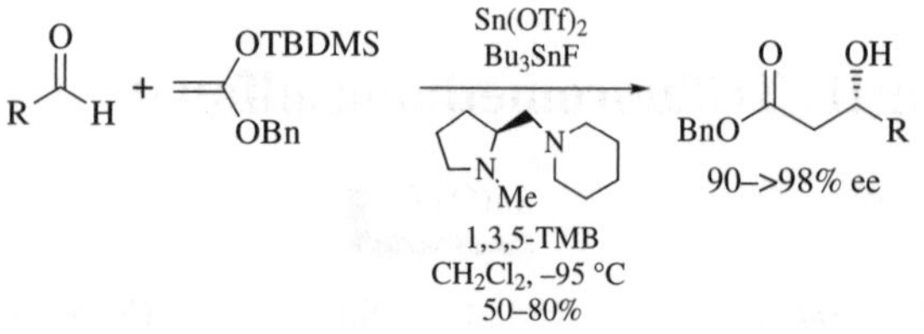

RCHO + OTBDMS/OBn ketene acetal; $Sn(OTf)_2$, Bu_3SnF, chiral diamine, 1,3,5-TMB, CH_2Cl_2, −95 °C, 50–80% (3)

90–>98% ee

Stereoselective aldol reactions using a catalytic amount of tin(II) triflate have been developed. The diastereoselective reaction between ***Methyl Vinyl Ketone*** and benzaldehyde first demonstrated the usefulness of tin(II) triflate as a catalyst (eq 4).[1]

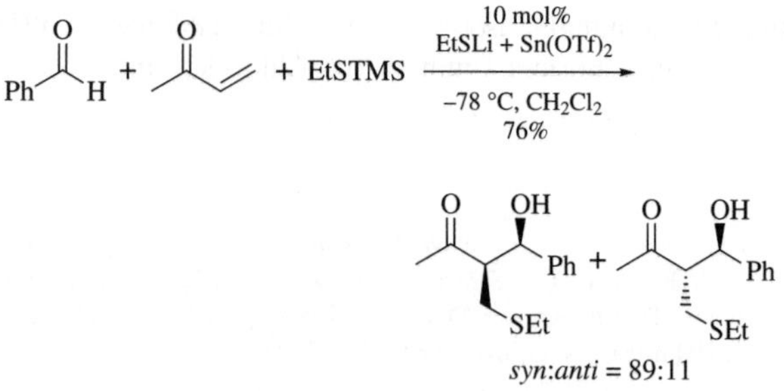

PhCHO + methyl vinyl ketone + EtSTMS; 10 mol% EtSLi + $Sn(OTf)_2$, −78 °C, CH_2Cl_2, 76% (4)

syn:anti = 89:11

The development of catalytic enantioselective reactions of silyl enol ethers with aliphatic, aromatic, alkynic, and α,β-unsaturated aldehydes has broadened the scope of this catalytic process (eqs 5–7).[7a,b]

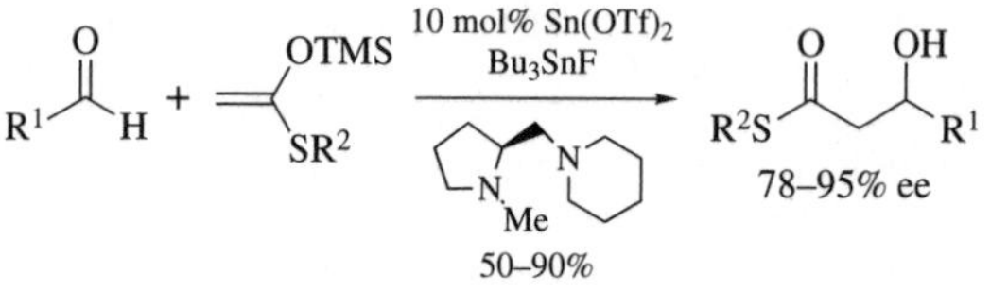

R^1CHO + OTMS/SR^2 enol ether; 10 mol% $Sn(OTf)_2$, Bu_3SnF, chiral diamine, 50–90% (5)

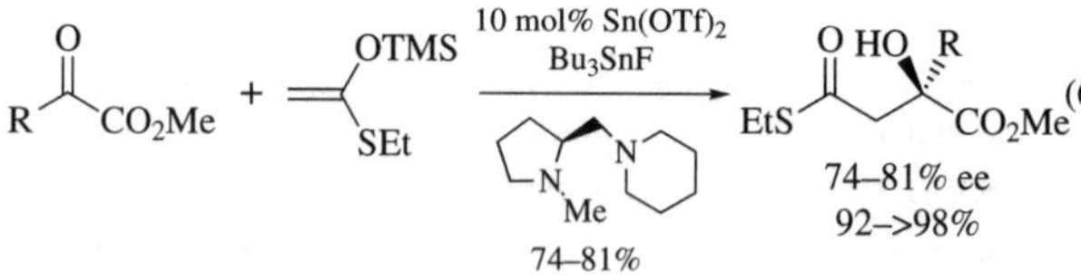

78–95% ee

$RCOCO_2Me$ + OTMS/SEt enol ether; 10 mol% $Sn(OTf)_2$, Bu_3SnF, chiral diamine, 74–81% (6)

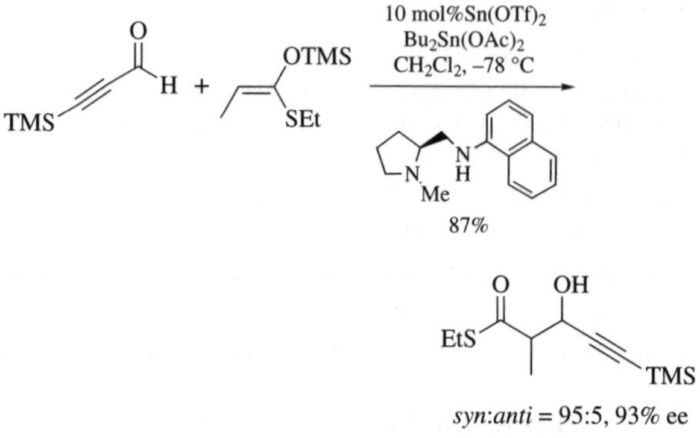

74–81% ee
92–>98%

TMS-alkynal + OTMS/SEt enol ether; 10 mol% $Sn(OTf)_2$, $Bu_2Sn(OAc)_2$, CH_2Cl_2, −78 °C, chiral diamine, 87% (7)

syn:anti = 95:5, 93% ee

In addition, Mukaiyama has shown that the stereochemistry of these reactions is controlled by the tin(II) triflate–tin(IV) additive–chiral diamine complex and is therefore independent of the chirality of the aldehyde (eqs 8 and 9).[7b,11]

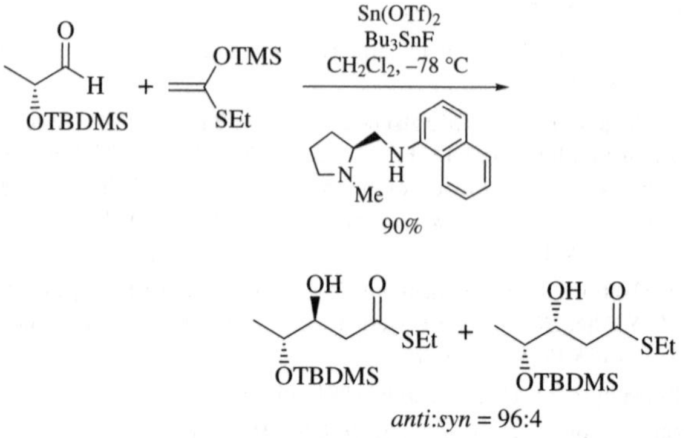

OTBDMS aldehyde + OTMS/SEt enol ether; $Sn(OTf)_2$, Bu_3SnF, CH_2Cl_2, −78 °C, chiral diamine, 90% (8)

anti:syn = 96:4

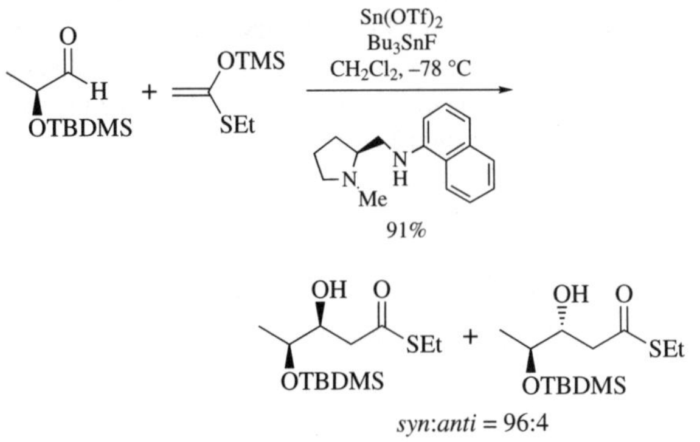

syn:anti = 96:4 (9)

Michael Reaction. The development of catalytic aldol-type reactions led to the discovery of a tin(II) triflate-catalyzed asymmetric Michael reaction.[1] Specifically, tin(II) enolates, prepared in situ from silyl enethiolates with catalytic amounts of tin(II) triflate and a chiral diamine, react with various α,β-unsaturated ketones to give chiral Michael adducts with 40–70% ee (eq 10).[12] Previously, asymmetric Michael reactions of tin(II) enolates with enones had been demonstrated in stoichiometric systems (eq 11).[13]

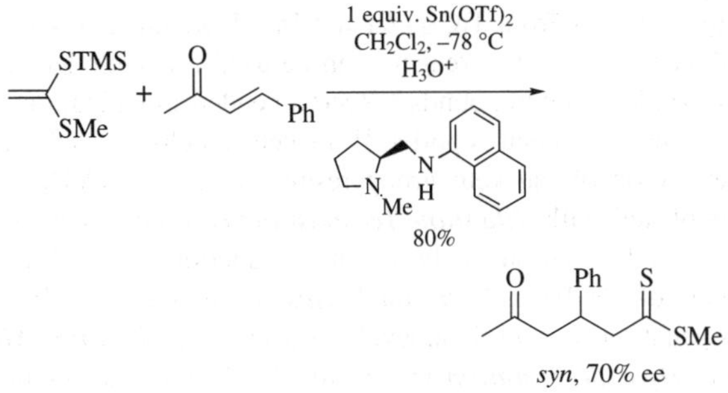

syn, 70% ee (10)

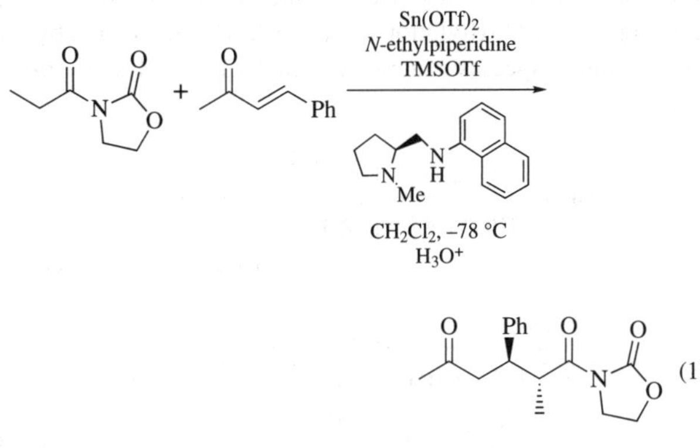

(11)

[2,3]-Wittig Rearrangements. Tin(II) enolates, generated from allylic glycolate esters and tin(II) triflate in the presence of ***Diisopropylethylamine***, undergo [2,3]-Wittig rearrangements upon warming to room temperature (eq 12).[2a] The corresponding boron enolates of the esters rearrange in a similar manner but with less stereoselectivity.

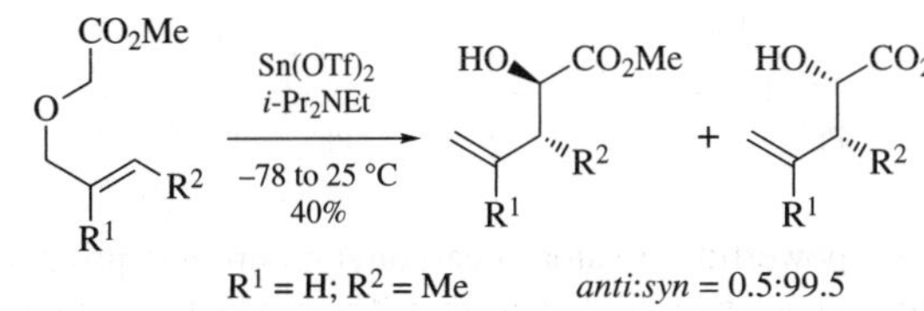

R^1 = H; R^2 = Me anti:syn = 0.5:99.5 (12)

Ireland–Claisen Rearrangement. An isolated case of an α-alkoxy ester rearranging via a tin(II) enolate has been reported (eq 13).[2b]

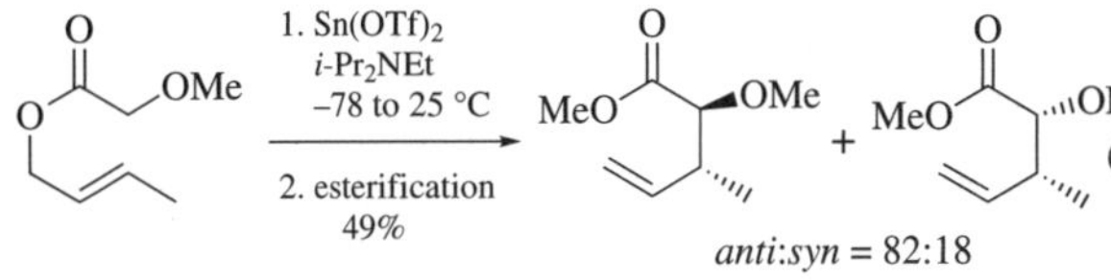

anti:syn = 82:18 (13)

Addition of 1-Alkynes to Aldehydes. In the presence of a stoichiometric amount of tin(II) triflate, 1-alkynes react with aldehydes and ketones to give alkynic alcohols in 57–91% yield (eq 14).[3] The mechanism of this reaction is, at present, unclear. It has been proposed that the tin(II) triflate reacts with the alkyne to form a tin acetylide which adds to the aldehyde.

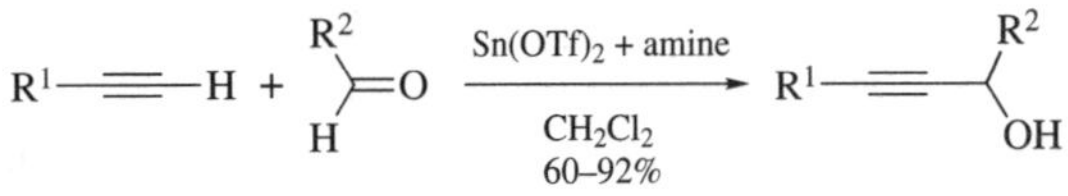

(14)

Asymmetric Allylation. Mukaiyama has designed a 'chiral allylating agent' consisting of an allyldialkylaluminum and a chiral diamine chelated to tin(II) triflate. The reaction of this agent with various aldehydes gives corresponding homoallylic alcohols with good to excellent enantioselectivity (eq 15).[4] Other tin(II) compounds ($SnCl_2$, $SnBr_2$, SnF_2, $Sn(OAc)_2$) react in this system to give the homoallylic alcohols in good yield but with virtually no enantioselectivity.

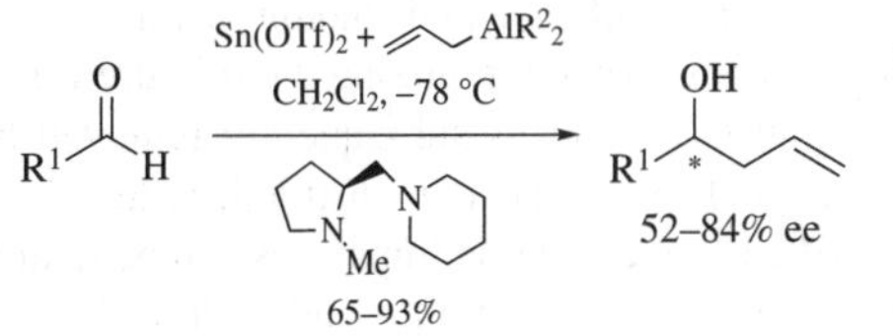

52–84% ee (15)

Related Reagents. Tin(II) Bromide; Tin(II) Chloride; Tri-*n*-butyl(iodoacetoxy)stannane.

1. Iwasawa, N.; Yura, T.; Mukaiyama, T. *T* **1989**, *45*, 1197.
2. (a) Oh, T.; Wrobel, Z.; Devine, P. *SL* **1992**, 81. (b) Oh, T.; Wrobel, Z.; Rubenstein, S. M. *TL* **1991**, *32*, 4647.
3. Yamaguchi, M.; Hayashi, A.; Minami, T. *JOC* **1991**, *56*, 4091.
4. Mukaiyama, T.; Minowa, N. *BCJ* **1987**, *60*, 3697.
5. Tin(II) triflate is commercially available in >97% purity.
6. Evans, D.; Weber, A. *JACS* **1986**, *108*, 6757.
7. (a) Kobayashi, S.; Furuya, M.; Ohtsubo, A.; Mukaiyama, T. *TA* **1991**, *2*, 635. (b) Kobayashi, S.; Furuya, M.; Ohtsubo, A.; Mukaiyama, T. *CL* **1991**, 989. (c) Kobayashi, S.; Ohtsubo, A.; Mukaiyama, T. *CL* **1991**, 831.
8. Mukaiyama, T.; Iwasawa, N.; Stevens, R. W.; Haga, T. *T* **1984**, *40*, 1381.
9. (a) Mukaiyama, T.; Asanuma, H.; Hachiya, I.; Harada, T.; Kobayashi, S. *CL* **1991**, 1209. (b) Kobayashi, S.; Uchiro, H.; Fujishita, Y.; Shiina, I.; Mukaiyama, T. *JACS* **1991**, *113*, 4247.
10. Mukaiyama, T.; Kobayashi, S.; Sano, T. *T* **1990**, *46*, 4653.

11. For more details on the tin(II) triflate–tin(IV) additive–chiral diamine complex see: Mukaiyama, T.; Uchiro, H.; Kobayashi, S. *CL* **1989**, 1757.
12. Roush, W. *Chemtracts–Org. Chem.* **1988**, 439.
13. Yura, T.; Iwasawa, N.; Mukaiyama, T. *CL* **1988**, 1021.

Samantha Janisse
Eli Lilly and Company, Indianapolis, IN, USA

Titanium(IV) Chloride

[7550-45-0] Cl_4Ti (MW 189.68)

(Lewis acid catalyst;[52] affects stereochemical course in cycloaddition[38,39] and aldol reactions;[4–17] electrophilic substitutions;[55] powerful dehydrating agent; when reduced to low-valent state, effects C–C bond formation by reductive coupling;[7–9] reduction of functional groups[81])

Physical Data: mp −24 °C, bp 136.4 °C, *d* 1.726 g cm^{-3}.
Solubility: sol THF, toluene, CH_2Cl_2.
Form Supplied in: neat as colorless liquid; solution in CH_2Cl_2; solution in toluene; $TiCl_4 \cdot 2THF$ *[31011-57-1]*.
Preparative Methods: see Perrin et al.[1]
Purification: reflux with mercury or a small amount of pure copper turnings and distill under N_2 in an all-glass system. Organic material can be removed by adding aluminum chloride hexahydrate as a slurry with an equal amount of water (ca. 2% weight of the amount of $TiCl_4$), refluxing the mixture for 2–6 h while bubbling in Cl_2 which is subsequently removed by a stream of dry air, before the $TiCl_4$ is distilled, refluxed with copper, and distilled again. Volatile impurities can be removed using a technique of freezing, pumping, and melting.[2]
Handling, Storage, and Precautions: moisture sensitive; reacts violently, almost explosively, with water; highly flammable; toxic if inhaled or swallowed; causes burns on contact with skin; use in a fume hood.

General Discussion. $TiCl_4$ is a strong Lewis acid and is used as such in organic reactions. It has high affinity for oxygenated organic molecules and possesses a powerful dehydrating ability. Low-valent titanium, from the reduction of $TiCl_4$ by metal or metal hydrides, is used for reductive coupling reactions and for reduction of functional groups.

Carbon–carbon bond formation by reductive coupling of ketones by low-valent titanium leads to vicinal diols and alkenes, as does coupling using low-valent reagents from ***Titanium(III) Chloride***. Reviews are available on the preparation and reactions of low-valent titanium.[3–6] The reagent from ***Titanium(IV) Chloride–Zinc*** in THF or dioxane in the presence of pyridine transforms ketones into tetrasubstituted alkenes. Unsymmetrical ketones yield (*E*)/(*Z*)-isomer mixtures (eq 1). Strongly hindered ketones react slowly and may preferentially be reduced to alcohols.[7] Reductive coupling of diketo sulfides yields 2,5-dihydrothiophenes (eq 2).[8] The macrocyclic porphycene has been obtained, albeit in low yield, by McMurry coupling of diformylbipyrrole (eq 3).[9] Aldimines are reductively coupled in a reaction analogous to the reaction of carbonyl compounds. The product is a 1:1 mixture of *meso*- and (±)-isomers (eq 4).[10]

$TiCl_4$–Zn, py, THF (1)

$TiCl_4$–Zn, CH_2Cl_2, 0 °C, 70–90% (2)

$TiCl_4$–Zn, THF, py, 2–3% (3)

$TiCl_4$–Mg, 92% (4)

Methylenation of aldehydes and ketones results from reactions with the complex from $TiCl_4$–Zn and ***Diiodomethane*** or ***Dibromomethane*** (eq 5). The reagent can be used for methylenation of enolizable oxo compounds.[6] Recommended modifications to the reagent have been reported.[11] In keto aldehydes, selective methenylation of the keto group results when the aldehyde is precomplexed with ***Titanium Tetrakis(diethylamide)***. Chemoselective methenylation of the aldehyde function is possible by direct use of CH_2I_2/Zn/***Titanium Tetraisopropoxide*** (eq 6).[12] Cyclopropanation to *gem*-dihalocyclopropanes uses ***Lithium Aluminum Hydride-Titanium(IV) Chloride***. The exclusion of a strong base, as frequently used in alternative procedures, is an advantage (eq 7).[13] Allylation of imines has been effected by low-valent titanium species generated from $TiCl_4$ and ***Aluminum*** foil (eq 8).[14]

CH_2Br_2 → ($TiCl_4$–Zn, THF, CH_2Cl_2) → [ZnBr / ZnBr] → (R^1R^2CO) (5)
R^1, R^2 = H, aryl, alkyl

Ti(O-*i*-Pr)$_4$–Zn, CH_2I_2, 83%; $Ti(NEt_2)_4$, CH_2Cl_2; $TiCl_4$–Zn, CH_2I_2, 76% (6)

+ CCl_4, $TiCl_4$–LAH, THF, 0 °C (7)

$TiCl_4$–Al, 83% (8)

$TiCl_4$ is a powerful activator of carbonyl groups and promotes nucleophilic attack by a silyl enol ether. The product is a titanium salt of an aldol which, on hydrolysis, yields a β-hydroxy ketone. $TiCl_4$ is generally the best catalyst for this reaction. The temperature range for reactions with ketones is normally 0–20 °C; aldehydes react even at −78 °C, which allows for chemoselectivity (eq 9).[15] In α- or β-alkoxy aldehydes, the aldol reaction can proceed with high 1,2- or 1,3-asymmetric induction. With the nonchelating Lewis acid ***Boron Trifluoride Etherate***, the diastereoselectivity may be opposite to that obtained for the chelating $TiCl_4$ or ***Tin(IV) Chloride*** (eq 10).[16]

(9)

(10)

$TiCl_4$, 85%	>95:5	
BF_3		10:90

Titanium enolates are generally prepared by transmetalation of alkali-metal enolates but may also arise as structural parts of intermediates in $TiCl_4$-promoted reactions. This is illustrated in a stereoselective alkylation using an oxazolidinone as a chiral auxiliary (eq 11). The enolate, or its ate complex, may be the intermediate in the reaction.[17]

(11)

99%, >100:1

The stereoselectivity in $TiCl_4$-promoted reaction of silyl ketene acetals with aldehydes may be improved by addition of ***Triphenylphosphine*** (eq 12).[18] Enol ethers, as well as enol acetates, can be the nucleophile (eqs 13 and 14).[19] 2-Acetoxyfuran, in analogy to vinyl acetates, reacts with aldehydes to furnish 4-substituted butenolides under the influence of $TiCl_4$ (eq 15).[20]

(12)

anti:*syn* = 10.5:1
no PPh_3, 4:1

(13)

(14)

(15)

Silyl enol ethers react with acetals at −78 °C to form β-alkoxy ketones.[21] In intramolecular reactions, six-, seven-, and eight-membered rings are formed.[22] With 1,3-dioxolanes (acetals), 1–2 equiv of $TiCl_4$ leads to pyranone formation (eq 16), whereas no cyclization products are obtained with $SnCl_4$ or $ZnCl_2$.[23]

(16)

Alkynyltributyltin compounds react with steroidal aldehydes. In the presence of $TiCl_4$ the reaction gives 9:1 diastereoselectivity (eq 17). Reactions of alkynylmetals with chiral aldehydes generally show only slight diastereoselectivity.[24] With silyloxyacetylenes and aldehydes, α,β-unsaturated carboxylic acid esters are formed with high (*E*) selectivity (eq 18).[25]

(17)

9:1

(18)

Allylsilanes are regiospecific in their Lewis acid-catalyzed reactions, the electrophile bonding to the terminus of the allyl unit remote from the silyl group (eq 19).[26,27] Allylstannanes are less reliable in this respect.[26,28] The example in eq 19 illustrates regioselectivity.[29] On extended conjugation the reaction takes place at the terminus of the extended system (eq 20).[30]

(19)

(20)

Intramolecular reactions work well, both for allylstannanes and allylsilanes. An epoxide can be the electrophile. Opening of the epoxide with allylic attack gives the cyclic product (eq 21). $TiCl_4$ is superior to $Ti(O\text{-}i\text{-}Pr)_4$, $SnCl_4$, and ***Aluminum Chloride***, the other Lewis acids tested in the cyclization reaction. The reaction is stereospecific.[31]

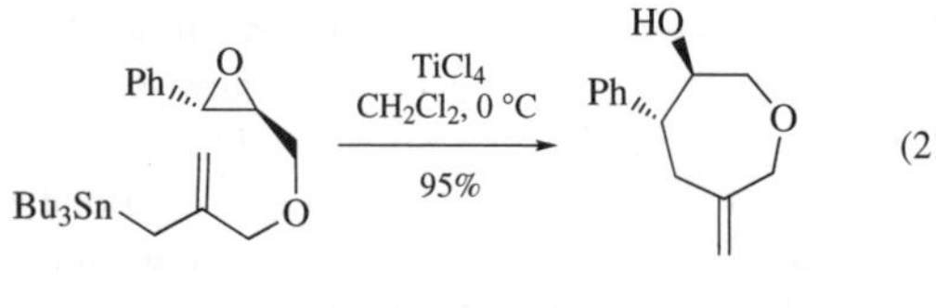

Ph, O, Bu_3Sn, O — $TiCl_4$, CH_2Cl_2, 0 °C, 95% → HO, Ph, O (21)

Stereochemical aspects in these $TiCl_4$-promoted reactions have been covered in reviews.[26,27] Acetals are very good electrophiles for allylsilanes, and may be better than the parent oxo compounds because the products are less prone to further reaction; intramolecular reactions are facile.[26,32] In the unsymmetrical acetal in eq 22, the methoxyethoxy group departs selectively because of chelation to the Lewis acid.[33]

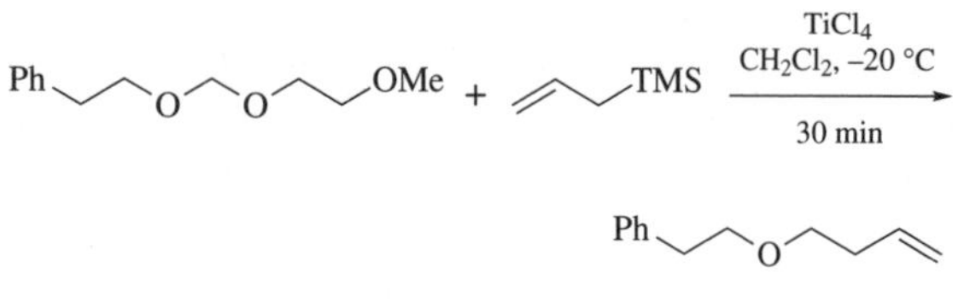

Ph O O OMe + TMS — $TiCl_4$, CH_2Cl_2, −20 °C, 30 min → Ph O (22)

The $TiCl_4$-mediated addition of allenylsilanes to aldehydes and ketones provides a general, regiocontrolled route to a wide variety of substituted homopropargylic alcohols. With acetals, corresponding ethers are formed (eq 23).[34] Allenylation of acetals results from the reaction of propargylsilanes with acetals. The products are α-allenyl ethers (eq 24).[35]

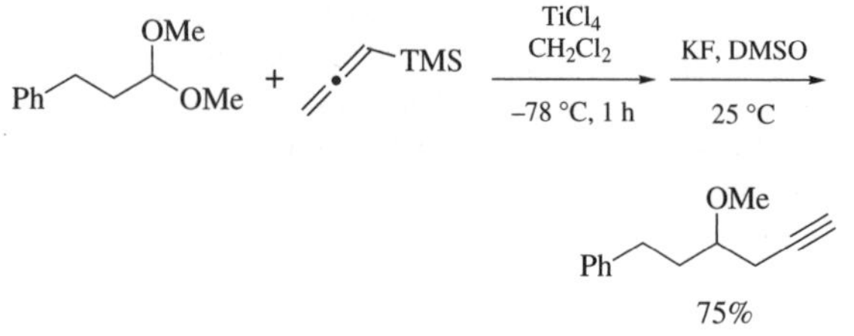

OMe, Ph, OMe + TMS — $TiCl_4$, CH_2Cl_2, −78 °C, 1 h; KF, DMSO, 25 °C → OMe, Ph, 75% (23)

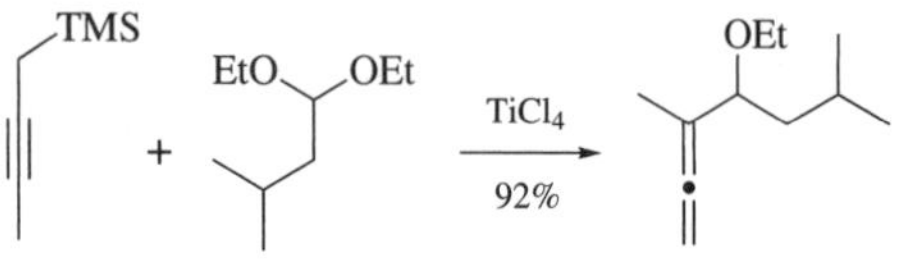

TMS + EtO OEt — $TiCl_4$, 92% → OEt (24)

Homoallylamines are formed in $TiCl_4$- or $BF_3{\cdot}OEt_2$-mediated reactions between allylstannanes and aldimines. Crotyltributyltin gives mainly *syn*-β-methyl homoallylamines under optimal conditions (eq 25).[36]

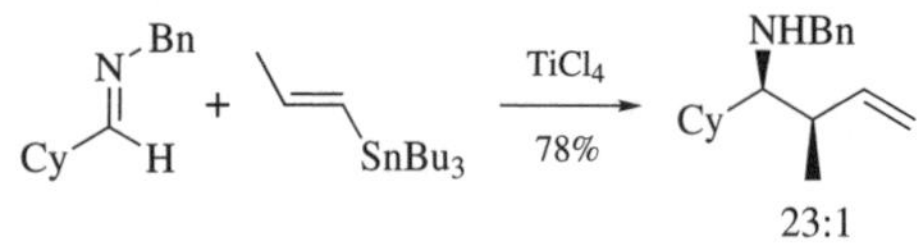

N-Bn, Cy, H + $SnBu_3$ — $TiCl_4$, 78% → NHBn, Cy, 23:1 (25)

Diastereoselective Mannich reactions may result with $TiCl_4$ as adjuvant (eq 26); transmetalation of the initial lithium alkoxide adduct and displacement by a lithium enolate gives a diastereoselectivity of 78%.[37]

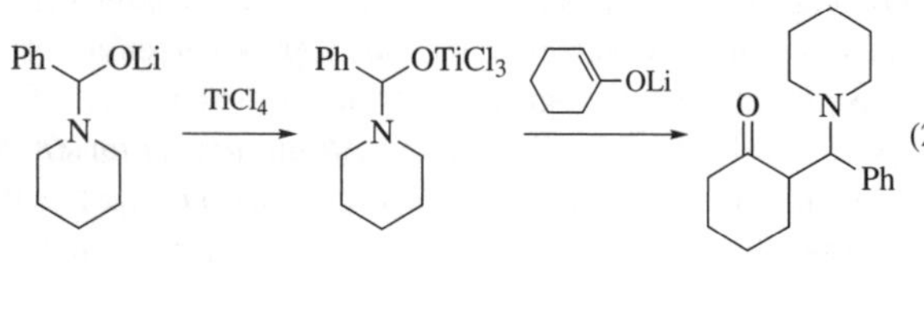

Ph OLi, N — $TiCl_4$ → Ph $OTiCl_3$, N — OLi → O, N, Ph (26)

Lewis acid catalysis increases the reactivity of dienophiles in Diels–Alder reactions by complexing to basic sites on the dienophile. In aldehydes, complexation takes place via the lone pair on the carbonyl oxygen. The stereochemistry is strongly influenced by the Lewis acids.[38] Under chelating conditions, when α-alkoxy aldehydes are used, the prevalent products from $TiCl_4$ catalysis have *cis* configuration (eqs 27 and 28). Stereochemical aspects of pericyclic pathways or passage through aldol type intermediates have been summarized and discussed.[38,39]

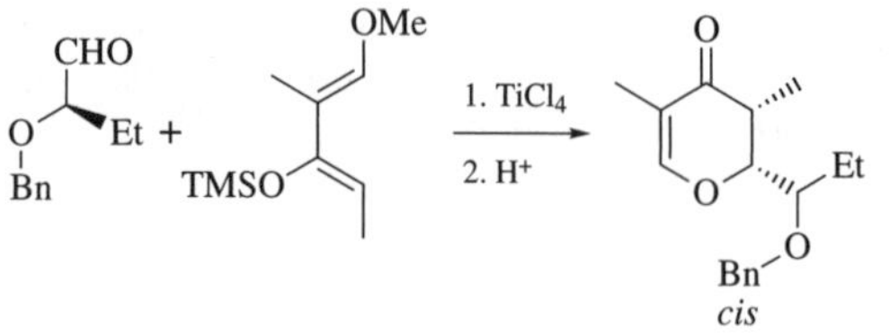

CHO, O, Bn, Et + OMe, TMSO — 1. $TiCl_4$, 2. H^+ → O, O, Et, Bn, O, *cis* (27)

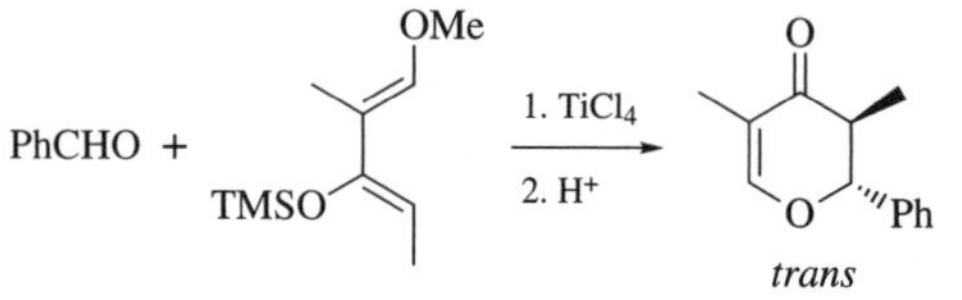

PhCHO + OMe, TMSO — 1. $TiCl_4$, 2. H^+ → O, O, Ph, *trans* (28)

Ketene silyl acetals can be dimerized to succinates on treatment with $TiCl_4$ in CH_2Cl_2 (eq 29). No reaction occurs with $TiCl_3$, nor are other metal salts efficient.[40] β-Amino esters are formed in the presence of Schiff bases (eq 30).[41]

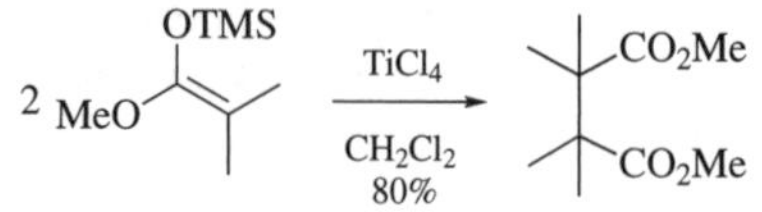

2 OTMS, MeO — $TiCl_4$, CH_2Cl_2, 80% → CO_2Me, CO_2Me (29)

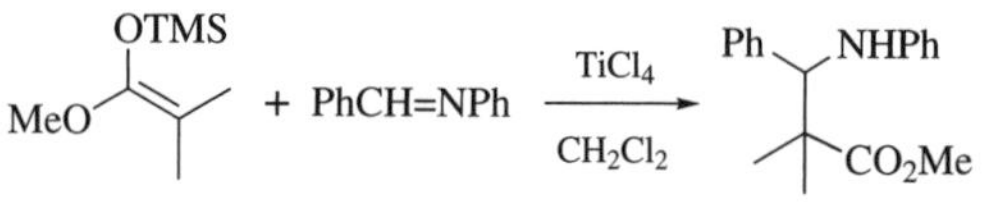

OTMS, MeO + PhCH=NPh — $TiCl_4$, CH_2Cl_2 → Ph NHPh, CO_2Me (30)

Cycloaddition of alkenes to quinones is effected by Ti^{IV} derivatives. The composition of the Ti^{IV} adjuvant largely controls the type of cycloadduct. $TiCl_4$ favors [3+2] cycloaddition; [2+2] cycloaddition requires a mixed $TiCl_4$–$Ti(O\text{-}i\text{-}Pr)_4$ catalyst (eq 31).[42]

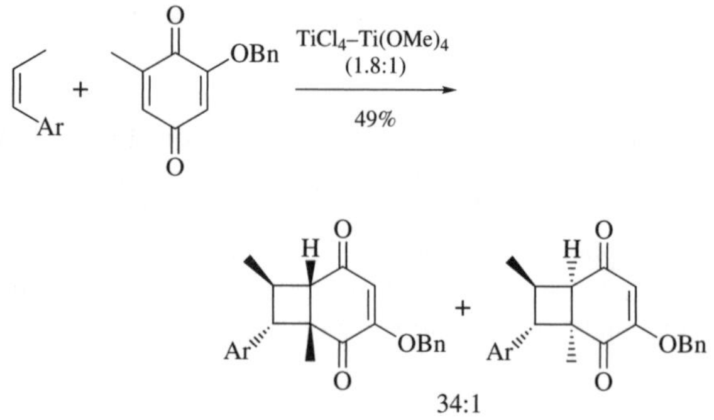

Ar + O, OBn, O — $TiCl_4$–$Ti(OMe)_4$ (1.8:1), 49% → H O, Ar, OBn, O + H O, Ar, OBn, O; 34:1 (31)

The $TiCl_4$-promoted Michael reaction proceeds under very mild conditions (−78 °C); this suppresses side-reactions and 1,5-dicarbonyl compounds are formed in good yields.[3] For $TiCl_4$-sensitive compounds, a mixture of $TiCl_4$ and $Ti(O\text{-}i\text{-}Pr)_4$ is used. From silyl enol ethers and α,β-unsaturated ketones, 1,5-dicarbonyl compounds are formed (eq 32).[3] The reaction also proceeds for α,β-unsaturated acetals.[43] Silylketene acetals react with α,β-unsaturated ketones or their acetals to form δ-oxo esters (eq 33).[44]

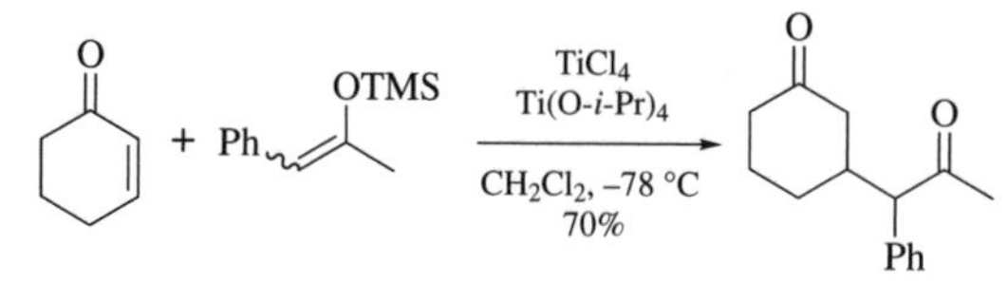

(32)

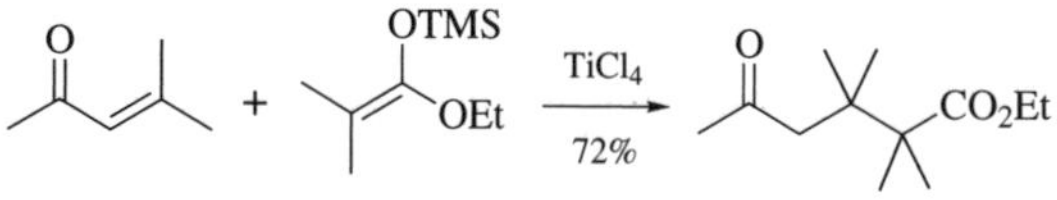

(33)

Conjugate addition of allylsilanes to enones results in regiospecific introduction of the allyl group (eq 34).[45] The reaction can be intramolecular (eq 35).[46] $TiCl_4$ or $SnCl_4$ activates nitro alkenes for Michael addition with silyl enol ethers or ketene silyl acetals. The silyl nitronate product is hydrolyzed to a 1,4-diketone or γ-keto ester (eq 36).[47]

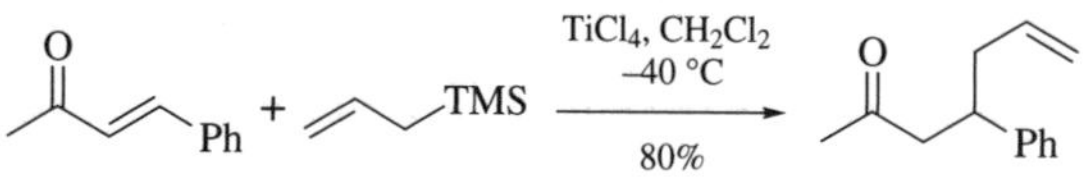

(34)

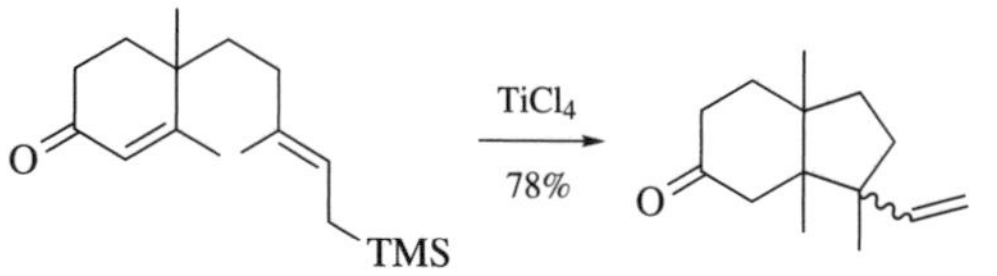

(35)

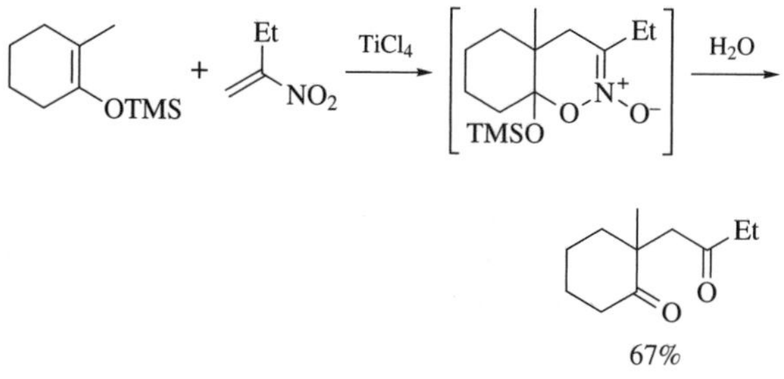

(36)

Conjugate propynylation of enones results from $TiCl_4$-mediated addition of allenylstannanes. Other Lewis acids are ineffective in this reaction (eq 37).[48]

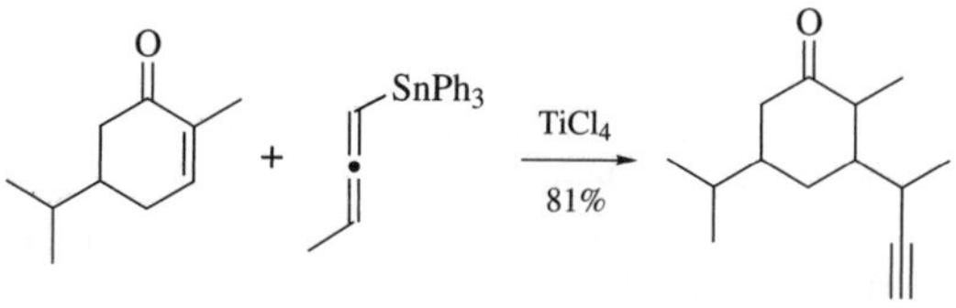

(37)

The 3,4-dichloro derivative of squaric acid, on $TiCl_4$-catalysis, reacts with silyl enol ethers or allylsilanes to form 1,2- or 1,4-adducts depending on the substitution. The 3,4-diethoxy derivative, however, adds silyl enol ethers in a 1,4-fashion; the adduct subsequently eliminates the ethoxy group (eq 38).[49]

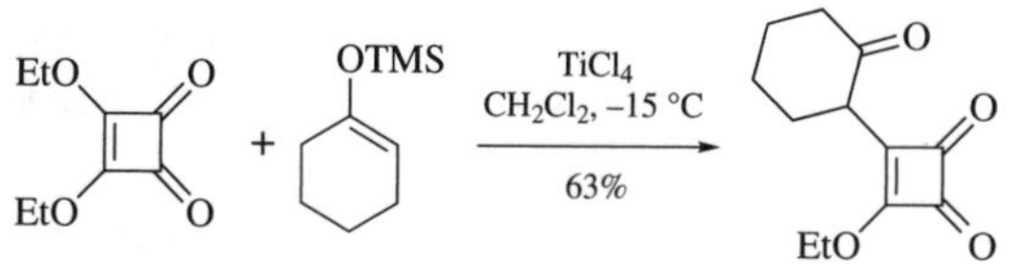

(38)

The Knoevenagel reaction with $TiCl_4$ and a tertiary base (eq 39) is recommended over methods which rely on strongly basic conditions.[50] The $TiCl_4$-procedure at low temperature is suitable for base-sensitive substrates.[51]

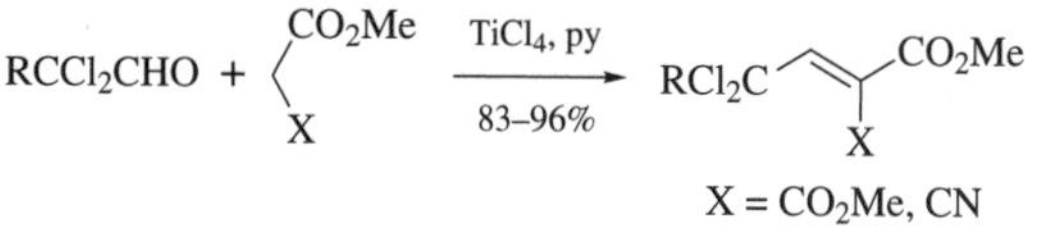

(39)

$TiCl_4$, as a Lewis acid, is used as a catalyst in Friedel–Crafts reactions. $AlCl_3$, $SnCl_4$, and $BF_3 \cdot OEt_2$ are more commonly used Friedel–Crafts catalysts for reactions with arenes.[52] Use of $TiCl_4$ as catalyst in the preparation of aromatic aldehydes is shown in eq 40.[53]

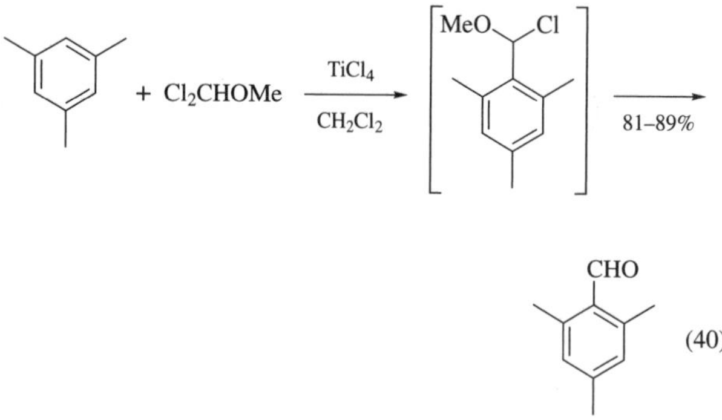

(40)

The formylation reaction can be used to prepare (*E*)-α,β-unsaturated aldehydes from vinylsilanes by *ipso* substitution (eq 41).[54] Regioselectivity in alkylation reactions may depend on the Lewis catalyst. In a fluorene, regioselective 1,8-chloromethylation results with $TiCl_4$ (eq 42).[55] $TiCl_4$ activates nitroalkenes for electrophilic substitution into arenes. The intermediate is hydrolyzed to the oxoalkylated product (eq 43).[56]

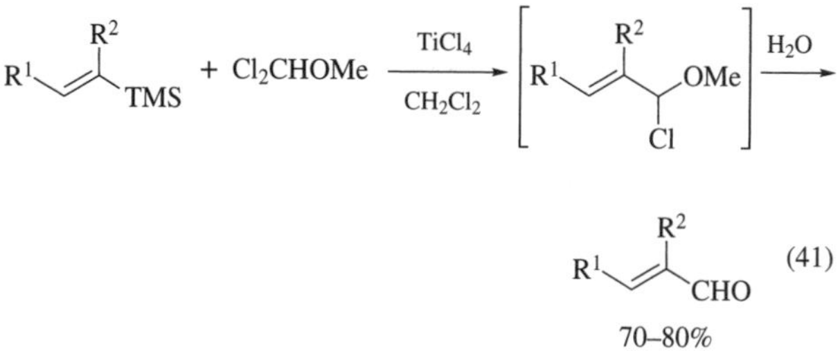

(41)

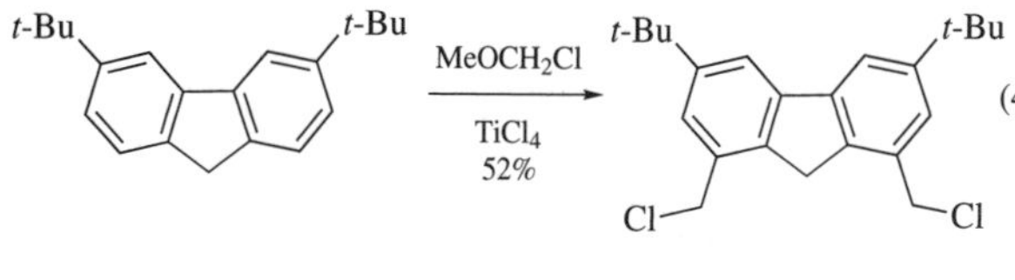

(42)

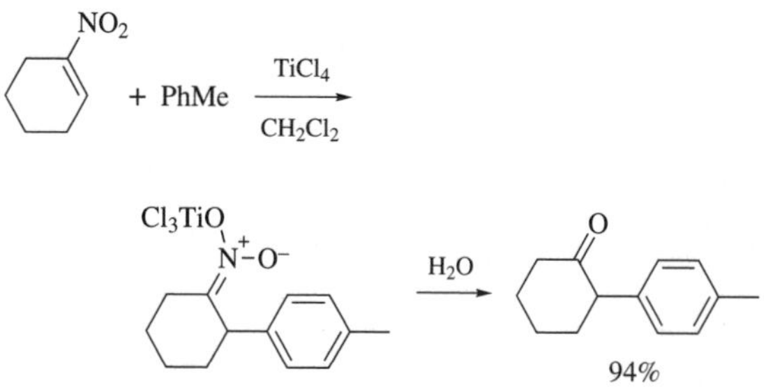

(43)

$TiCl_4$ is generally a good catalyst for Friedel–Crafts acylation of activated alkenes. In the acylation of pyrrolidine-2,4-diones, particularly with unsaturated acyl substrates, $TiCl_4$ is superior to $SnCl_4$ and $BF_3 \cdot OEt_2$ (eq 44).[57]

(44) [scheme: pyrrolidinedione + MeCH=CHCOCl, $TiCl_4$, $PhNO_2$, 52%]

The ready *ipso* substitution of a silyl group favors substitution rather than addition of electrophiles in alkenylsilanes.[58] Direct acylation of isobutene is not satisfactory, but the acylation is successful on silyl derivatives (eq 45).[59] Intramolecular acylation of alkenylsilanes leads to cyclic products. Even (*E*)-silylalkenes have been cyclized to enones (eq 46).[60]

(45) [scheme: $TiCl_4$, 65%]

(46) [scheme: $TiCl_4$, CH_2Cl_2]

In analogy to the acylation reactions, vinylsilanes can be alkylated by *ipso* substitution. In the example in eq 47, the MEM group is activated as leaving group by metal complexation[61] The silyl group in allylsilanes directs the incoming electrophile to the allylic γ-carbon (eq 48). $TiCl_4$ is one of several Lewis acids used for catalysis in acylations.[58] The same reaction is seen in allylsilanes with extended conjugation.[62] Eq 49 shows the $TiCl_4$-catalyzed alkylation of an allylsilane.[63]

(47) [scheme: $TiCl_4$, CH_2Cl_2, −5 °C, 78%]

(48) [scheme: + RCOCl, $TiCl_4$, CH_2Cl_2, −78 °C]

(49) [scheme: $TiCl_4$, 88%]

$TiCl_4$ is a useful catalyst for the Fries rearrangement of phenol esters to *o*- or *p*-hydroxy ketones (eq 50). $TiCl_4$ is a cleaner catalyst than the more frequently used $AlCl_3$, which may cause alkyl migrations.[64]

(50) [scheme: $TiCl_4$, 140 °C, 2 h; 75–82% + 5–10%]

$TiCl_4$ is both a strong Lewis acid and a powerful dehydrating agent, and hence useful as a water scavenger in the synthesis of enamines. It is particularly useful for the preparation of enamines of acyclic ketones. It is recommended that $TiCl_4$ is complexed with the amine before addition of the ketone.[65] Highly sterically hindered enamines are available by this method (eq 51).[66] Primary amines generally react very slowly with aromatic ketones to form imines. $TiCl_4$ is a good catalyst (eq 52).[67]

(51) [scheme: + Me_2NH, $TiCl_4$, 72%]

(52) $Ar_2C=O + H_2NR \xrightarrow[\text{PhH, 20 °C, 75–98\%}]{TiCl_4} Ar_2C=NR$

$TiCl_4$ with a tertiary base provides mild conditions for dehydration of both aldoximes and primary acid amides to form nitriles.[68] Vinyl sulfides are formed from oxo compounds using $TiCl_4$ and a tertiary amine.[3] $TiCl_4$ activates carbon–carbon double bonds for thiol addition (eq 53).[3]

(53) $CH_2=CHCO_2Me + EtSH \xrightarrow[\text{CH}_2\text{Cl}_2\text{, 20 °C, 94\%}]{TiCl_4} EtSCH_2CH_2CO_2Me$

$TiCl_4$ mediates thioacetalization of aldehydes and ketones with alkanethiols or alkanedithiols in yields >90%. The reaction is satisfactory also for readily enolizable oxo compounds.[69] γ-Lactols, which are generally more stable than acyclic analogs, are amenable to dithiol cleavage (eq 54).[70]

(54) [scheme: + ethanedithiol, $TiCl_4$, 85%]

α-Hydroxy amides are formed in a reaction involving an isocyanide, $TiCl_4$, and an aldehyde or a ketone (eq 55).[71] Vinyl chlorides undergo ready hydrolysis on combined use of $TiCl_4$ and $MeOH–H_2O$ (eq 56).[72] $TiCl_4$-mediated hydrolysis of vinyl sulfides is a good preparative route to ketones.[3,73]

(55) MeNC $\xrightarrow[CH_2Cl_2]{TiCl_4}$ Me–N=C(Cl)TiCl$_3$ $\xrightarrow{R^1COR^2}$ $R^1R^2C(OTiCl_3)C(Cl)=NMe$ $\xrightarrow{H_2O}$ $R^1R^2C(OH)C(O)NHMe$ 65–95%

$$\xrightarrow[\text{CH}_2\text{Cl}_2,\ \text{Me}_2\text{CO},\ 20\ ^\circ\text{C}\ \ 96\%]{\text{TiCl}_4,\ \text{H}_2\text{O–MeOH}} \qquad (56)$$

Low-valent titanium can be used to reduce sulfides and haloarenes to the corresponding hydrocarbons (eq 57).[3,74] Low-valent titanium, prepared from $TiCl_4$ and ***Magnesium Amalgam***, will reduce nitroarenes to amines in THF/*t*-BuOH at 0 °C without affecting halo, cyano, and ester groups.[75] $SnCl_2{\cdot}2H_2O$ is an alternative reagent for nitro group reduction.[76]

$$\xrightarrow[\text{THF},\ \Delta\ \ 83\%]{\text{TiCl}_4\text{–LAH (1:2)}} \qquad (57)$$

Deoxygenation of sulfoxides is a rapid reaction using $[TiCl_2]$ formed in situ by reduction of $TiCl_4$ with Zn dust in CH_2Cl_2 or Et_2O at rt, with yields in the range 85–90%.[77] For deoxygenation of *N*-oxides of pyridine-based heterocycles, a reagent prepared from $TiCl_4$–$NaBH_4$ (1:2) in DME has been described.[78] Carboxylic acids can be reduced to primary alcohols by $TiCl_4$–$NaBH_4$ (ratio 1:3). For reduction of amides and lactams the optimum molar ratio is 1:2.[79]

Several low-valent metal species have been found active in reductive elimination of vicinal dibromides to the corresponding alkenes. In most cases, e.g. with $TiCl_4$–LAH and $TiCl_4$–Zn, the reactions proceed through predominant *anti* elimination to yield alkenes with high isomeric purity.[80] The low-valent titanium reagent obtained from $TiCl_4$–LAH (ca. 2:1) will saturate the double bond of enedicarboxylates in the presence of triethylamine (eq 58).[81]

$$\xrightarrow[\text{THF},\ 100\ ^\circ\text{C}]{\text{TiCl}_4\text{–LAH, NEt}_3} \qquad (58)$$

Related Reagents. Dibromomethane–Zinc–Titanium(IV) Chloride; Diiodomethane–Zinc–Titanium(IV) Chloride; (4*R*,5*R*)-2,2-Dimethyl-4,5-bis(hydroxydiphenylmethyl)-1,3-di-oxolane-Titanium(IV) Chloride; Lithium Aluminum Hydride–Titanium(IV) Chloride; Titanium(IV) Chloride–Diazabicyclo-[5.4.0]undec-7-ene; Titanium(IV) Chloride–2,2,6,6-Tetramethylpiperidine; Titanium(IV) Chloride–Triethylaluminum; Titanium-(IV) Chloride–Zinc.

1. Perrin, D. D.; Armarego, W. L. F.; Perrin, D. R. *Purification of Laboratory Chemicals*, 2nd ed.; Pergamon: Oxford, 1980; p 542.
2. *Gmelin Handbuch der Anorganischen Chemie*; Verlag Chemie: Weinheim, 1951; *Titan*, p 92, 299.
3. Mukaiyama, T. *AG(E)* **1977**, *16*, 817.
4. Betschart, C.; Seebach, D. *C* **1989**, *43*, 39.
5. Pons, J.-M.; Santelli, M. *T* **1988**, *44*, 4295.
6. Reetz, M. T. *Organotitanium Reagents in Organic Synthesis*; Springer: Berlin, 1986; p 223.
7. Lenoir, D. *S* **1977**, 553.
8. (a) Nakayama, J.; Machida, H.; Hoshino, M. *TL* **1985**, *26*, 1981. (b) Nakayama, J.; Machida, H.; Satio, R.; Hoshino, M. *TL* **1985**, *26*, 1983.
9. Vogel, E.; Köcher, M.; Schmickler, H.; Lex, J. *AG(E)* **1986**, *25*, 257.
10. Betschart, C.; Schmidt, B.; Seebach, D. *HCA* **1988**, *71*, 1999.
11. (a) Lombardo, L. *TL* **1982**, *23*, 4293. (b) Lombardo, L. *OS* **1987**, *65*, 81.
12. Hibino, J.; Okazoe, T.; Takai, K.; Nozaki, H. *TL* **1985**, *26*, 5579.
13. Mukaiyama, T.; Shiono, M.; Watanabe, K.; Onaka, M. *CL* **1975**, 711.
14. Tanaka, H.; Inoue, K.; Pokorski, U.; Taniguchi, M.; Torii, S. *TL* **1990**, *31*, 3023.
15. Mukaiyama, T.; Narasaka, K.; Banno, K. *JACS* **1974**, *96*, 7503.
16. (a) Reetz, M. T. *AG(E)* **1984**, *23*, 556. (b) Reetz, M. T.; Kesseler, K. *CC* **1984**, 1079.
17. Evans, D. A.; Urpí, F.; Somers, T. C.; Clark, J. S.; Bilodeau, M. T. *JACS* **1990**, *112*, 8215.
18. Palazzi, C.; Colombo, L.; Gennari, C. *TL* **1986**, *27*, 1735.
19. (a) Kitazawa, E.; Imamura, T.; Saigo, K.; Mukaiyama, T. *CL* **1975**, 569. (b) Mukaiyama, T.; Izawa, T.; Saigo, K. *CL* **1974**, 323.
20. Shono, T.; Matsumura, Y.; Yamane, S. *TL* **1981**, *22*, 3269.
21. Mukaiyama, T.; Banno, K. *CL* **1976**, 279.
22. Mukaiyama, T. *OR* **1982**, *28*, 203.
23. Cockerill, G. S.; Kocienski, P. *CC* **1983**, 705.
24. Yamamoto, Y.; Nishii, S.; Maruyama, K. *CC* **1986**, 102.
25. Kowalski, C. J.; Sakdarat, S. *JOC* **1990**, *55*, 1977.
26. Fleming, I. *COS* **1991**, *2*, 563.
27. Fleming, I.; Dunoguės, J.; Smithers, R. H. *OR* **1989**, *37*, 57.
28. Yamamoto, Y. *ACR* **1987**, *20*, 243.
29. Hosomi, A.; Shirahata, A.; Sakurai, H. *TL* **1978**, 3043.
30. Seyferth, D.; Pornet, J.; Weinstein, R. M. *OM* **1982**, *1*, 1651.
31. (a) Molander, G. A.; Shubert, D. C. *JACS* **1987**, *109*, 576. (b) Molander, G. A.; Andrews, S. W. *JOC* **1989**, *54*, 3114.
32. Mukaiyama, T.; Murakami, M. *S* **1987**, 1043.
33. Nishiyama, H.; Itoh, K. *JOC* **1982**, *47*, 2496.
34. (a) Danheiser, R. L.; Carini, D. J.; Kwasigroch, C. A. *JOC* **1986**, *51*, 3870. (b) Danheiser, R. L.; Carini, D. J. *JOC* **1980**, *45*, 3925.
35. Pornet, J.; Miginiac, L.; Jaworski, K.; Randrianoelina, B. *OM* **1985**, *4*, 333.
36. Keck, G. E.; Enholm, E. J. *JOC* **1985**, *50*, 146.
37. Seebach, D.; Betschart, C.; Schiess, M. *HCA* **1984**, *67*, 1593.
38. Bednarski, M. D.; Lyssikatos, J. P. *COS* **1991**, *2*, 661.
39. Danishefsky, S. J.; Pearson, W. H.; Harvey, D. F. *JACS* **1984**, *106*, 2456.
40. Inaba, S.; Ojima, I. *TL* **1977**, 2009.
41. Ojima, I.; Inaba, S.; Yoshida, K. *TL* **1977**, 3643.
42. Engler, T. A.; Combrink, K. D.; Ray, J. E. *JACS* **1988**, *110*, 7931.
43. Narasaka, K.; Soai, K.; Aikawa, Y.; Mukaiyama, T. *BCJ* **1976**, *49*, 779.
44. Saigo, K.; Osaki, M.; Mukaiyama, T. *CL* **1976**, 163.
45. Hosomi, A.; Sakurai, H. *JACS* **1977**, *99*, 1673.
46. Majetich, G.; Hull, K.; Defauw, J.; Shawe, T. *TL* **1985**, *26*, 2755.
47. Miyashita, M.; Yanami, T.; Kumazawa, T.; Yoshikoshi, A. *JACS* **1984**, *106*, 2149.
48. Haruta, J.; Nishi, K.; Matsuda, S.; Tamura, Y.; Kita, Y. *CC* **1989**, 1065.
49. Ohno, M.; Yamamoto, Y.; Shirasaki, Y.; Eguchi, S. *JCS(P1)* **1993**, 263.
50. (a) Lehnert, W. *S* **1974**, 667. (b) Campaigne, E.; Beckman, J. C. *S* **1978**, 385.
51. Courtheyn, D.; Verhe, R.; De Kimpe, N.; De Buyck, L.; Schamp, N. *JOC* **1981**, *46*, 3226.

52. (a) Olah, G. A.; Krishnamurti, R.; Prakash, G. K. S. *COS* **1991**, *3*, 293. (b) Heaney, H. *COS* **1991**, *2*, 733.
53. (a) Rieche, A.; Gross, H.; Höft, E. *OS* **1967**, *47*, 1. (b) Rieche, A.; Gross, H.; Höft, E.; Beyer, E. *OS* **1967**, *47*, 47.
54. (a) Yamamoto, K.; Nunokawa, O.; Tsuji, J. *S* **1977**, 721. (b) Yamamoto, K.; Yoshitake, J.; Qui, N. T., Tsuji, J. *CL* **1978**, 859.
55. Tsuge, A.; Yamasaki, T.; Moriguchi, T.; Matsuda, T.; Nagano, Y.; Nago, H.; Mataka, S.; Kajigaeshi, S.; Tashiro, M. *S* **1993**, 205.
56. Lee, K.; Oh, D. Y. *TL* **1988**, *29*, 2977.
57. Jones, R. C. F.; Sumaria, S. *TL* **1978**, 3173.
58. Eyley, S. *COS* **1991**, *2*, 707.
59. Pillot, J.-P.; Bennetau, B.; Dunogues, J.; Calas, R. *TL* **1980**, *21*, 4717.
60. (a) Burke, S. D.; Murtiashaw, C. W.; Dike, M. S.; Strickland, S. M. S.; Saunders, J. O. *JOC* **1981**, *46*, 2400. (b) Nakamura, E.; Fukuzaki, K.; Kuwajima, I. *CC* **1983**, 499.
61. Overman, L. E.; Castañeda, A.; Blumenkopf, T. A. *JACS* **1986**, *108*, 1303.
62. Hosomi, A.; Saito, M.; Sakurai, H. *TL* **1979**, 429.
63. (a) Albaugh-Robertson, P.; Katzenellenbogen, J. A. *TL* **1982**, *23*, 723. (b) Morizawa, Y.; Kanemoto, S.; Oshima, K.; Nozaki, H. *TL* **1982**, *23*, 2953.
64. Martin, R.; Demerseman, P. *S* **1992**, 738.
65. Carlson, R.; Nilsson, Å.; Strömqvist, M. *ACS* **1983**, *B37*, 7.
66. (a) White, W. A.; Weingarten, H. *JOC* **1967**, *32*, 213. (b) White, W. A.; Chupp, J. P.; Weingarten, H. *JOC* **1967**, *32*, 3246.
67. Moretti, I.; Torre, G. *S* **1970**, 141.
68. (a) Lehnert, W. *TL* **1971**, 559. (b) Lehnert, W. *TL* **1971**, 1501.
69. Kumar, V.; Dev, S. *TL* **1983**, *24*, 1289.
70. Bulman-Page, P. C.; Roberts, R. A.; Paquette, L. A. *TL* **1983**, *24*, 3555.
71. Schiess, M.; Seebach, D. *HCA* **1983**, *66*, 1618.
72. (a) Mukaiyama, T.; Imamoto, T.; Kobayashi, S. *CL* **1973**, 261. (b) Mukaiyama, T.; Imamoto, T.; Kobayashi, S. *CL* **1973**, 715.
73. Seebach, D.; Neumann, H. *CB* **1974**, *107*, 847.
74. Mukaiyama, T.; Hayashi, M.; Narasaka, K. *CL* **1973**, 291.
75. George, J.; Chandrasekaran, S. *SC* **1983**, *13*, 495.
76. (a) Ballamy, F. D.; Ou, K. *TL* **1984**, *25*, 839. (b) Varma, R. S.; Kabalka, G. W. *CL* **1985**, 243.
77. Drabowicz, J.; Mikolajczyk, M. *S* **1978**, 138.
78. Kano, S.; Tanaka, Y.; Hibino, S. *H* **1980**, *14*, 39.
79. Kano, S.; Tanaka, Y.; Sugino, E.; Hibino, S. *S* **1980**, 695.
80. Imamato, T. *COS* **1991**, *1*, 231.
81. Hung, C. W.; Wong, H. N. C. *TL* **1987**, *28*, 2393.

Lise-Lotte Gundersen
Norwegian College of Pharmacy, Oslo, Norway

Frode Rise & Kjell Undheim
University of Oslo, Norway

Titanium Tetraisopropoxide[1]

$Ti(O\text{-}i\text{-}Pr)_4$

[546-68-9] $C_{12}H_{28}O_4Ti$ (MW 284.28)

(mild Lewis acid used as a catalyst in transesterification reactions, nucleophilic cleavages of 2,3-epoxy alcohols and isomerization reactions; additive in the Sharpless epoxidation reaction and in various reactions involving nucleophilic additions to carbonyl and α,β-unsaturated carbonyl compounds)

Physical Data: mp 18–20 °C; bp 218 °C/10 mmHg; *d* 0.955 g cm^{-3}.
Solubility: sol a wide range of solvents including ethers, organohalides, alcohols, benzene.
Form Supplied in: low-melting solid; widely available.
Handling, Storage, and Precautions: flammable and moisture sensitive; acts as an irritant.

Transesterification and Lactamization Reactions.[2–5] $Ti(O\text{-}i\text{-}Pr)_4$, as well as other titanium(IV) alkoxides, has been recommended as an exceptionally mild and efficient transesterification catalyst which can be used with many acid-sensitive substrates (eq 1).[2,3] Thus the acetonide as well as C=O, OH, OTBDMS, and lactam functional groups (eq 2) are unaffected by these conditions, although acetates are hydrolyzed to the parent alcohol. The alcohol solvent employed in such processes need not be anhydrous, nor need it be identical with the OR group in the titanate, because exchange of these moieties is generally slow compared to the transesterification reaction.

MeO_2C / MeO_2C acetonide → [$Ti(O\text{-}i\text{-}Pr)_4$, *i*-PrOH, 70 °C, 91%] → *i*-PrO_2C / *i*-PrO_2C acetonide (1)

β-lactam methyl ester (OMe) → [$Ti(O\text{-}i\text{-}Pr)_4$, *i*-PrOH, 70 °C, 91%] → β-lactam isopropyl ester (O-*i*-Pr) (2)

N-Protected esters of dipeptides can be transesterified using $Ti(O\text{-}i\text{-}Pr)_4$ in the presence of 4 Å molecular sieves and this procedure provides a good way of converting methyl esters into their benzyl counterparts. Such reactions (eq 3) proceed without racemization in 70–85% yield.[4] In a related process,[5] β-, γ-, and δ-amino acids undergo lactamization on treatment with $Ti(O\text{-}i\text{-}Pr)_4$ in refluxing 1,2-dichloroethane (eq 4).

BocHN–Leu–NH–Phe–OMe + HOBn → [$Ti(O\text{-}i\text{-}Pr)_4$, THF, 65 °C, 16 h, 73%] → BocHN–Leu–NH–Phe–OBn (3)

HO_2C–$CH(R^1)$–$(CH_2)_n$–NHR → [$Ti(O\text{-}i\text{-}Pr)_4$, $Cl(CH_2)_2Cl$, 83 °C, 3–26 h, 35–93%] → lactam (4)

R, R^1 = H, Me
n = 1–3

Nucleophilic Cleavage of 2,3-Epoxy Alcohols and Related Compounds.[6–10] Titanium alkoxides are weak Lewis acids which generally have no effect on simple epoxides. However, reaction of 2,3-epoxy alcohols (available, for example, from the Sharpless-type epoxidation of allylic alcohols) with nucleophiles in the presence of $Ti(O\text{-}i\text{-}Pr)_4$ results in highly regioselective ring-cleavage reactions involving preferential nucleophilic attack at C-3 (eq 5).[6] In the absence of the titanium alkoxide, no reaction is observed under otherwise identical conditions, except in the case of PhSNa.

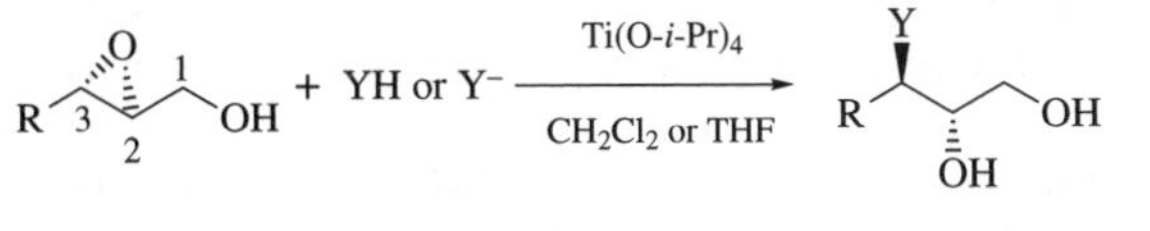

(5)

$R = Me(CH_2)_2$	$YH = CH_2{=}CHCH_2OH$	90%; C-3:C-2 > 100:1
$R = Me(CH_2)_2$	$YH = Et_2NH$	90%; C-3:C-2 = 20:1
$R = Me(CH_2)_2$	$YH = PhCO_2H$	74%; C-3:C-2 > 100:1
$R = Me(CH_2)_2$	YH = TsOH (with 2,6-lutidine)	64%; C-3:C-2 > 100:1
$R = Me(CH_2)_2$	YH = PhSH	95%; C-3:C-2 = 6.4:1
$R = Me(CH_2)_6$	YH = AcSH	91%; C-3:C-2 > 100:1
$R = Me(CH_2)_6$	$Y^- = NO_3^-$	86%; C-3:C-2 > 100:1

These types of conversion can be extended[7] to 2,3-epoxy carboxylic acids (glycidic acids) and the related amides. The former compounds are readily available through ***Ruthenium(VIII) Oxide***-mediated oxidation of the appropriate 2,3-epoxy alcohols (eq 6).

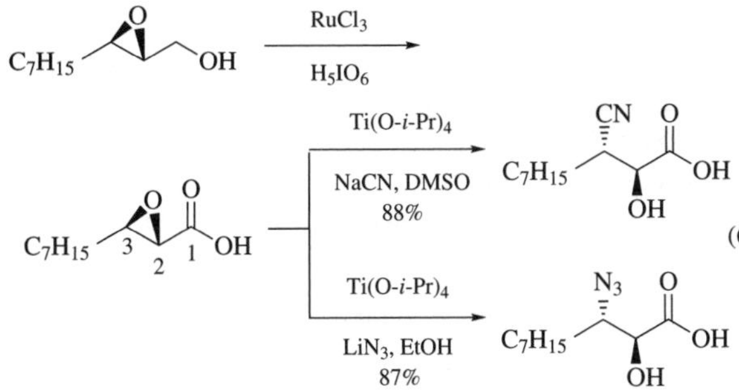

(6)

The C-3:C-2 selectivity is, in the cases shown above, greater than 100:1, although lower selectivities (e.g. 20:1) are observed with other nucleophiles such as Et_2NH and PhSH. It should be noted that glycidic acids and amides react preferentially with dialkylamines at C-2, but one equivalent or greater of $Ti(O\text{-}i\text{-}Pr)_4$ ensures reaction occurs with high selectivity at C-3. Combining this type of protocol with the Sharpless asymmetric epoxidation reaction has permitted the development of stereoselective syntheses (eqs 7 and 8)[6,7] of all four stereoisomers of the unusual *N*-terminal amino acid of amastatin, a tripeptide competitive inhibitor of aminopeptidases.[8]

The reaction of both open-chain and cyclic 2,3-epoxy alcohols with molecular ***Bromine*** or ***Iodine*** in the presence of $Ti(O\text{-}i\text{-}Pr)_4$ at 0 °C leads to the regioselective formation of halo diols (eq 9).[9a] Interestingly, if these reactions are conducted at 25 °C a 1:1 mixture of the C-2 and C-3 cleavage products is obtained, and the same outcome is observed, even at 0 °C, when the acetate derivative of the 2,3-epoxy alcohol is involved as substrate. Dialkylamine hydrochlorides can be used as sources of halide nucleophiles in these types of epoxide ring-cleavage reactions.[9b]

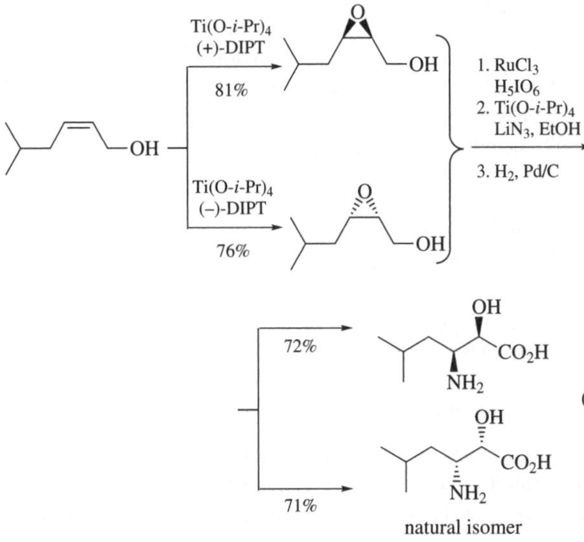

(7)

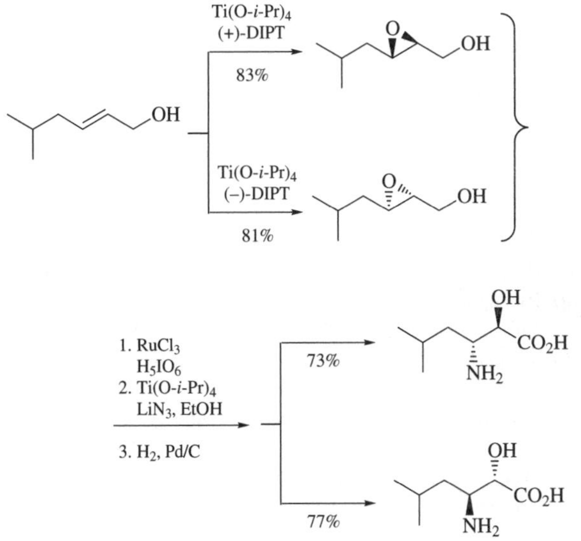

(8)

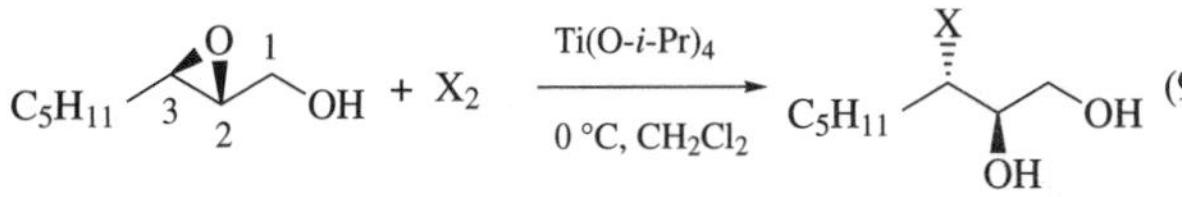

(9)

87% (X = I); C-3:C-2 = 7:1
78% (X = Br); C-3:C-2 = 15:1

2,3-Epithio alcohols have been obtained by reacting 2,3-epoxy alcohols with ***Thiourea*** at room temperature or 0 °C in the presence of $Ti(O\text{-}i\text{-}Pr)_4$ and using THF as solvent (eq 10).[10] The reactions proceed with high regio- and stereoselectivity, *trans*-substituted 2,3-epoxy alcohols giving only *trans*-2,3-epithio alcohols with complete inversion of configuration at both stereogenic centers. However, when *cis*-2,3-epoxy alcohols are used as starting materials the yields of epithio alcohols were low and thiodiols were also formed. Epithiocinnamyl alcohols could also be prepared from the corresponding epoxycinnamyl alcohols at 0 °C. However, these products were found to decompose to cinnamyl alcohol and sulfur on standing. Without Ti(O-*i*-Pr)$_4$, thiourea was insoluble in THF and the reaction did not proceed. One equivalent of Ti(O-*i*-Pr)$_4$ was required to achieve complete reaction, and THF was the best solvent (no reaction was observed in ether, CH_2Cl_2, or benzene under similar conditions).

1. H_2NCSNH_2,Ti(O-*i*-Pr)$_4$, 25 °C, THF, 2–3 h; 2. $NaHCO_3$ (satd); 77% (10)

Isomerization Reactions.[11–14] The reaction of certain 2,3-epoxy alcohols with Ti(O-*i*-Pr)$_4$ can result in isomerization. For example (eq 11), reaction of the illustrated substrate in CH_2Cl_2 results in rearrangement to the isomeric enediol, and this conversion represents a key step in a synthesis of the marine natural product pleraplysillin.[11]

Ti(O-*i*-Pr)$_4$, 0 °C to rt, CH_2Cl_2, 77% (11)

Under similar conditions, the pendant double bond attached to a 2,3-epoxy alcohol acts as an internal nucleophile attacking at C-3, resulting, after proton loss, in a mixture of cyclized products (eq 12).[12] The cyclopropane-containing products are believed to arise via a retro-homo Prins reaction. Pendant triple bonds can also participate in related cyclization reactions and cyclic allenes result (eq 13).[12] The observation that the *threo* isomer of the substrate shown in eq 12 is stable to Ti(O-*i*-Pr)$_4$ has led to the suggestion that an intramolecular metal alkoxide is the active catalyst in successful cyclization reactions.

Ti(O-*i*-Pr)$_4$, rt, CH_2Cl_2 (12)

20–30%; R = CHO, R = CH_2OH 30–40%

Ti(O-*i*-Pr)$_4$, 25 °C, CH_2Cl_2, 50% (13)

Ti(O-*i*-Pr)$_4$ has also played a key role in the synthesis of taxanes (eq 14).[12b,c]

Ti(O-*i*-Pr)$_4$, 40 °C, CH_2Cl_2, 45 min, 90% (14)

Allylic hydroperoxides, which are readily obtained by reaction of the corresponding alkene with ***Singlet Oxygen***, have been shown to isomerize to the corresponding 2,3-epoxy alcohol when treated with catalytic amounts of Ti(O-*i*-Pr)$_4$ (eq 15).[13] The title reagent is the one of choice when converting di-, tri-, and tetrasubstituted alkenes (both cyclic and acyclic) into the corresponding 2,3-epoxy alcohols by this protocol. The reactions are generally highly stereoselective and deoxygenation of the allylic hydroperoxide (to give the corresponding allylic alcohol) is not normally a process which competes significantly with the isomerization reaction.

$^1O_2$, TPP, 0 °C, CH_2Cl_2; Ti(O-*i*-Pr)$_4$ (16 mol%), 0 °C, CH_2Cl_2, 91% (15)

cis:*trans* = 98:2

This type of chemistry has been extended to the preparation of epoxy diols from chiral allylic alcohols (eq 16).[13,14] The methodology is impressive in that three successive chiral centers are constructed with predictable configuration. Furthermore, the rapid rate of the isomerization process is remarkable, given that α,β-unsaturated diols are generally poor substrates for titanium metal-mediated epoxidations. This rate enhancement is attributed to the tridentate nature of the intermediate hydroperoxides.

$^1O_2$, TPP; Ti(O-*i*-Pr)$_4$ (0.05 equiv), –25 °C, CH_2Cl_2, 1 h (16)

80% of a 86:14 mixture

Asymmetric Epoxidation Reactions.[15] While Ti(O-*i*-Pr)$_4$ clearly has the capacity to bring about the nucleophilic ring-cleavage of 2,3-epoxy alcohols (see above), it remains the preferred species for the preparation of the titanium tartrate complex central to the Sharpless asymmetric epoxidation process (see, for example, eq 7). Since *t*-butoxide-mediated ring-opening of 2-substituted 2,3-epoxy alcohols (a subclass of epoxy alcohols particularly sensitive to nucleophilic ring-cleavage) is much slower than by isopropoxide, the use of Ti(O-*t*-Bu)$_4$ is sometimes recommended in place of Ti(O-*i*-Pr)$_4$. However, with the reduced amount of catalyst that is now needed for all asymmetric epoxidations, this precaution appears unnecessary in most instances.

Nucleophilic Additions to Carbonyl Compounds.[16–28] A wide range of organometallic compounds react with Ti(O-*i*-Pr)$_4$ to produce organotitanium/titanium-'ate' species which may exhibit reactivities that differ significantly from those of their precursor.[1] Thus Ti(O-*i*-Pr)$_4$ forms '-ate' complexes with, amongst others, Grignard reagents and the resulting species show useful selectivities in their reaction with carbonyl compounds. For example, the complex with allylmagnesium chloride is highly selective in its reaction with aldehydes in the presence of ketones, or ketones in the presence of esters (eq 17).[16] Interestingly, the corresponding amino titanium-'ate' complexes react selectively with ketones in the presence of aldehydes.

OEt + MgCl; Ti(O-*i*-Pr)$_4$, –78 °C, THF, 81%; OH, OEt (17)

Reaction of certain sulfur-substituted allylic anions with Ti(O-*i*-Pr)$_4$ produces a 3-(alkylthio)allyltitanium reagent that condenses, through its α-terminus, with aldehydes to give *anti*-β-hydroxy sulfides in a highly stereo- and regioselective manner (eq 18). These latter compounds can be transformed, stereoselectively, into *trans*-vinyloxiranes or 1,3-dienes.[17]

Li+, RS, α, γ; Ti(O-*i*-Pr)$_4$, –78 °C, THF; [RS, α, γ, TiL$_n$]; R^{1}CHO, –78 to 0 °C, 70–99%

SR, R^1, OH; *erythro*:*threo* = >6:1; Me$_3$OBF$_4$ then NaOH (aq.) 62–96%; R^1, O; 1. BuLi, MsCl 2. *t*-BuLi R = Ph; SPh, R^1 (18)

The titanium species derived from sequential treatment of α-alkoxy-substituted allylsilanes with ***s*-Butyllithium** then Ti(O-*i*-Pr)$_4$ engages in a Peterson alkenation reaction with aldehydes to give, via electrophilic attack at the α-terminus of the allyl anion, 2-oxygenated 1,3-butadienes which can be hydrolyzed to the corresponding vinyl ketone (eq 19).[18a]

OEt, O, TMS, α, γ; 1. *s*-BuLi, THF, –78 °C 2. Ti(O-*i*-Pr)$_4$ 3. R^{1}CHO 45–80%

OEt, O, R^1; HCl, THF, 100%; O, R^1 (19)

The titanium-'ate' complexes of α-methoxy allylic phosphine oxides, generated in situ by reaction of the corresponding lithium anion and Ti(O-*i*-Pr)$_4$, condense with aldehydes exclusively at the α-position to produce homoallylic alcohols in a diastereoselective fashion.[18b] The overall result is the three-carbon homologation of the original aldehyde, and this protocol has been used in a synthesis of (−)-aplysin-20 from nerolidol.[19] The titanium-'ate' complex produced by reaction of the chiral lithium anion of an (*E*)-crotyl carbamate with Ti(O-*i*-Pr)$_4$ affords γ-condensation products (homoaldols) on reaction with aldehydes.[18c,d] Allyl anions produced by the reductive metalation of allyl phenyl sulfides condense with α,β-unsaturated aldehydes in a 1,2-manner at the more substituted (α) allyl terminus in the presence of Ti(O-*i*-Pr)$_4$.[20] 1,2-Addition of dialkylzincs to α,β-unsaturated aldehydes can be achieved with useful levels of enantiocontrol when the reaction is conducted using a chiral titanium(IV) catalyst in the presence of Ti(O-*i*-Pr)$_4$ (eq 20).[21] Higher ee values are observed when an α-substituent (e.g. bromine) is attached to the substrate aldehyde, but a β-substituent *cis*-related to the carbonyl group has the opposite effect.

TIPSO, O; R$_2$Zn, Ti(O-*i*-Pr)$_4$, –60 to –55 °C, 8–12 h; Tf, N, Ti(O-*i*-Pr)$_2$, N, Tf; TIPSO, OH, R (20)

R = Et, 68%; 93% ee
R = C$_5$H$_{11}$, 75%; 85% ee
R = (CH$_2$)$_5$OPiv, 90%; 92% ee

Highly enantioselective trimethylsilylcyanation of various aldehydes can be achieved by using ***Cyanotrimethylsilane*** in the presence of a modified Sharpless catalyst consisting of Ti(O-*i*-Pr)$_4$ and chiral diisopropyl tartrate.[22] Best results are obtained using dichloromethane as solvent and isopropanol as additive and running the reaction at 0 °C. The same type of catalyst system also effects the asymmetric ring-opening of symmetrical cycloalkene oxides with ***Azidotrimethylsilane***. *trans*-2-Azidocycloalkanols are obtained in up to 63% optical yield.[22] The titanium amide complexes derived from the reaction of lithium dialkylamides with Ti(O-*i*-Pr)$_4$ condense with alkyl and aryl aldehydes and the resulting aminal derivatives undergo C–O bond displacement by benzylmagnesium chloride, thereby generating α-substituted β-phenethylamines (eq 21).[23]

LiNR$^1_2$; 1. Ti(O-*i*-Pr)$_4$, Et$_2$O, –20 °C, 20 min 2. R^{2}CHO, 20 °C, 3–5 h 3. PhCH$_2$MgCl, Et$_2$O, rt, 1 h 4. H$_2$O 73–93%; Ph, R^2, NR$^1_2$ (21)

R^1 = alkyl; R^2 = alkyl, aryl

The aldol-type condensation of aldehydes and ketones with ketenimines[24] and ketones[25] can be catalyzed by titanium alkoxides and, in appropriate cases, useful levels of stereocontrol can be achieved (eq 22).

TMS, =C=N, TMS + C$_8$H$_{17}$CHO; TiCl$_4$, Ti(O-*i*-Pr)$_4$, CH$_2$Cl$_2$, –78 °C

HO, C$_8$H$_{17}$, TMS, CN; BF$_3$•Et$_2$O, 86%; C$_8$H$_{17}$, CN (22)

E:*Z* = 8:92

A variety of useful reducing agents have been generated by combining hydrides with $Ti(O\text{-}i\text{-}Pr)_4$.[26–28] For example, the combination of 5 mol % $Ti(O\text{-}i\text{-}Pr)_4$ with 2.5–3.0 equiv of ***Triethoxysilane*** cleanly hydrosilylates esters to silyl ethers at 40–55 °C, and these latter compounds can be converted into the corresponding primary alcohols via aqueous alkaline hydrolysis (eq 23).[26] The actual reducing agent is presumed to be a titanium hydride species which is produced by a σ-bond metathesis process involving $Ti(O\text{-}i\text{-}Pr)_4$ and the silane. The procedure has considerable merit in that no added solvent is required and the active reagent can be generated and used in air. Halides, epoxides, alcohols, and an alkyne all survive the reduction process. ***Lithium Borohydride*** reduction of 2,3-epoxy alcohols yields 1,2-diols highly regioselectively when used in the presence of $Ti(O\text{-}i\text{-}Pr)_4$,[27] while the combination of the reagent with ***Sodium Cyanoborohydride*** is reported[28] to offer superior results in reductive amination processes with difficult carbonyls and those sensitive to acidic conditions.

$$RC(O)OR^1 \xrightarrow[\text{40–55 °C, 4–22 h}]{\text{5\% Ti(O-}i\text{-Pr)}_4\text{, 2.3–3 equiv (EtO)}_3\text{SiH}} \xrightarrow[\text{rt, 2–4.5 h}]{\text{1 M NaOH, THF}} RCH_2OH \quad \text{70–95\%} \tag{23}$$

Miscellaneous Applications.[29–31] The chemoselective oxidation of alcohols and diols using $Ti(O\text{-}i\text{-}Pr)_4$/***t-Butyl Hydroperoxide*** has been reported.[29] The title reagent has also been employed as a catalyst in Diels–Alder reactions[30] and as an additive in the palladium-catalyzed reaction of aryl-substituted allylic alcohols with zinc enolates of β-dicarbonyl compounds (eq 24).[31] The latter reaction is presumed to generate C-allylated β-dicarbonyl compounds as the primary products of reaction, but these compounds suffer deacylation in the presence of $Ti(O\text{-}i\text{-}Pr)_4$.

$$PhCH{=}CHCH_2OH + CH_3C(O)CH_2C(O)CH_3 \xrightarrow[\text{MeO(CH}_2)_2\text{OEt, 120 °C}]{\text{ZnCl}_2\text{, Et}_3\text{N, PdCl}_2\text{(PPh}_3)_2\text{, Ti(O-}i\text{-Pr)}_4} PhCH{=}CHCH_2CH_2C(O)CH_3 \tag{24}$$

Related Reagents. Aluminum Isopropoxide; Zirconium Tetraisopropoxide.

1. (a) Shiihara, I.; Schwartz, W. T., Jr.; Post, H. W. *CRV* **1961**, *61*, 1. (b) Reetz, M. T. *Top. Curr. Chem.* **1982**, *106*, 3. (c) Seebach, D.; Weidmann, B; Widler, L. *Mod. Synth. Methods* **1983**, *3*, 217. (d) Reetz, M. T. *Organotitanium Reagents in Organic Synthesis*; Springer: Berlin, 1986. (e) Hoppe, D.; Krämer, T.; Schwark, J.-R.; Zschage, O. *PAC* **1990**, *62*, 1999.
2. (a) Imwinkelried, R.; Schiess, M.; Seebach, D. *OS* **1987**, *65*, 230. (b) Seebach, D.; Hungerbühler, E.; Schnurrenberger, P.; Weidmann, B.; Züger, M. *S* **1982**, 138.
3. (a) Schnurrenberger, P.; Züger, M. F.; Seebach, D. *HCA* **1982**, *65*, 1197. (b) Férézou, J. P.; Julia, M.; Liu, L. W.; Pancrazi, A. *SL* **1991**, 618.
4. Rehwinkel, H.; Steglich, W. *S* **1982**, 826.
5. Mader, M.; Helquist, P. *TL* **1988**, *29*, 3049.
6. (a) Caron, M.; Sharpless, K. B. *JOC* **1985**, *50*, 1557. (b) *Aldrichim. Acta* **1985**, *18*, 53.
7. Chong, J. M.; Sharpless, K. B. *JOC* **1985**, *50*, 1560.
8. Tobe, H.; Morishima, H.; Aoyagi, T.; Umezawa, H.; Ishiki, K.; Nakamura, K.; Yoshioka, T.; Shimauchi, Y.; Inui, T. *ABC* **1982**, *46*, 1865.
9. (a) Alvarez, E.; Nuñez, T.; Martin, V. S. *JOC* **1990**, *55*, 3429. (b) Gao, L.; Murai, A. *CL* **1989**, 357.
10. Gao, Y.; Sharpless, K. B. *JOC* **1988**, *53*, 4114.
11. Masaki, Y.; Hashimoto, K.; Serizawa, Y.; Kaji, K. *BCJ* **1984**, *57*, 3476.
12. (a) Morgans, D. J., Jr.; Sharpless, K. B.; Traynor, S. G. *JACS* **1981**, *103*, 462. (b) Holton, R. A.; Juo, R. R.; Kim, H. B.; Williams, A. D.; Harusawa, S.; Lowenthal, R. E.; Yogai, S. *JACS* **1988**, *110*, 6558. (c) Wender, P. A.; Mucciaro, T. P. *JACS* **1992**, *114*, 5878.
13. (a) Mihelich, E. D. US Patent 4 345 984 (*CA* **1983**, *98*, 125 739c). (b) Adam, W.; Braun, M.; Griesbeck, A.; Lucchini, V.; Staab, E.; Will, B. *JACS* **1989**, *111*, 203. (c) Adam, W.; Nestler, B. *JACS* **1993**, *115*, 7226.
14. Adam, W.; Nestler, B. *AG(E)* **1993**, *32*, 733.
15. Johnson, R. A.; Sharpless, K. B. *COS* **1991**, *7*, 389.
16. Reetz, M. T.; Wenderoth, B. *TL* **1982**, *23*, 5259.
17. Furuta, K.; Ikeda, Y.; Meguriya, N.; Ikeda, N.; Yamamoto, H. *BCJ* **1984**, *57*, 2781.
18. (a) Murai, A.; Abiko, A.; Shimada, N.; Masamune, T. *TL* **1984**, *25*, 4951. (b) Birse, E. F.; McKenzie, A.; Murray, A. W. *JCS(P1)* **1988**, 1039. (c) Férézou, J. P.; Julia, M.; Khourzom, R.; Pancrazi, A.; Robert, P. *SL* **1991**, 611. (d) Hoppe, D.; Zschage, O. *AG(E)* **1989**, *28*, 69.
19. Murai, A.; Abiko, A.; Masamune, T. *TL* **1984**, *25*, 4955.
20. Cohen, T.; Guo, B.-S. *T* **1986**, *42*, 2803.
21. Rozema, M. J.; Eisenberg, C.; Lütjens, H.; Ostwald, R.; Belyk, K.; Knochel, P. *TL* **1993**, *34*, 3115.
22. (a) Hayashi, M.; Matsuda, T.; Oguni, N. *JCS(P1)* **1992**, 3135. (b) Hayashi, M.; Kohmura, K.; Oguni, N. *SL* **1991**, 774.
23. Takahashi, H.; Tsubuki, T.; Higashiyama, K. *S* **1988**, 238.
24. Okada, H.; Matsuda, I.; Izumi, Y. *CL* **1983**, 97.
25. Vuitel, L.; Jacot-Guillarmod, A. *HCA* **1974**, *57*, 1703.
26. Berk, S. C.; Buchwald, S. L. *JOC* **1992**, *57*, 3751.
27. Dai, L.; Lou, B.; Zhang, Y.; Guo, G. *TL* **1986**, *27*, 4343.
28. Mattson, R. J.; Pham, K. M.; Leuck, D. J.; Cowen, K. A. *JOC* **1990**, *55*, 2552.
29. Yamawaki, K.; Ishii, Y.; Ogawa, M. *Chem. Express* **1986**, *1*, 95.
30. McFarlane, A. K.; Thomas, G.; Whiting, A. *TL* **1993**, *34*, 2379.
31. Itoh, K.; Hamaguchi, N.; Miura, M.; Nomura, M. *JCS(P1)* **1992**, 2833.

Martin G. Banwell
University of Melbourne, Parkville, Victoria, Australia

p-Toluenesulfonic Acid

$$p\text{-}CH_3C_6H_4SO_3H$$

[104-15-4]	$C_7H_8O_3S$	(MW 172.22)
(monohydrate)		
[6192-52-5]	$C_7H_{10}O_4S$	(MW 190.24)
(Na salt)		
[657-84-1]	$C_7H_7NaO_3S$	(MW 194.20)

(acid catalyst frequently used in nonpolar media; effective in carbonyl protection–deprotection; selectively cleaves N-Boc, other amine protecting groups; superior for enol ether, acetate preparation; used in esterifications, dehydrations, isomerizations, rearrangements)

Alternate Name: tosic acid.

Physical Data: the anhydrous acid exists as monoclinic leaflets or prisms, mp 106–107 °C; pK_a −6.62 (H_2SO_4).[1] There is also a metastable form, mp 38 °C. The monohydrate is a white crystalline powder, mp 103–106 °C.

Solubility: sol water (67 g/100 mL), ethanol, ethyl ether. The sodium salt is very sol water.

Form Supplied in: widely available as the monohydrate, which is commonly used. Various metal salts are also commercially available.

Analysis of Reagent Purity: by acid–base titrimetry.[46]

Purification: precipitated or crystallized from HCl soln, aq EtOH; the free acid has been crystallized from several organic solvents.[47]

Handling, Storage, and Precautions: highly toxic, oxidizing agent. Extremely irritating to the skin and mucous membranes. Use of gloves and protective clothing is recommended.[2] Use in a fume hood.

Acid Catalyst. Tosic acid is one of the most widely used organic acid catalysts, particularly in nonpolar solvents. It is utilized frequently in many of the common acid-catalyzed reactions and transformations in organic chemistry, including esterification, formation of acetals, dehydration processes, preparation of enol ethers and acetates, and rearrangement and isomerization processes. A comprehensive literature review of even its recent uses is beyond the scope of this publication; representative examples of each of the above-mentioned classes are presented.

Formation and Cleavage of Acetals. Tosic acid is perhaps the most common acid catalyst for the protection of ketones as acetals. The reaction is conventionally carried out in refluxing toluene or benzene with removal of water by a Dean–Stark trap (eq 1).[3] Although ***Boron Trifluoride Etherate*** is the customary catalyst for the condensation of ketones with alkyldithiols, tosic acid is a milder reagent and has been used for this purpose (eq 2).[4] *p*-Toluenesulfonic acid may also be used in lieu of $BF_3 \cdot Et_2O$ in the tetrahydropyranylation and methoxytetrahydropyranylation of alcohols (3 examples).[5]

HO⌒OH; p-TsOH, benzene, reflux, 95% (1)

HS⌒⌒SH; p-TsOH, toluene, reflux, 98% (2)

Treatment of acetals (eqs 3 and 4)[6,7] with aqueous tosic acid effects cleavage to the carbonyl compounds.

p-TsOH; H_2O, CH_2Cl_2 (3)

p-TsOH; EtOH, H_2O, 97% (4)

In a recently described synthesis of (+)-phyllanthocin,[8] spiroacetalization of (**1**) to form key intermediate (**2**) was studied (eq 5). Acid-catalyzed cyclization by tosic acid was found to be superior to ***10-Camphorsulfonic Acid***- or base-catalyzed cyclization using ***1,8-Diazabicyclo[5.4.0]undec-7-ene***.

(1) p-TsOH; benzene, 72% (2) (5)

Esterification and Lactone Formation. Tosic acid has been used in lieu of mineral acid in Fischer esterifications of carboxylic acids (eq 6).[9] Hydroxy acids (eq 7)[10,11] and aldehyde carboxylic acids (eq 8)[10] may be cyclized to lactones and enol lactones, respectively, by treatment with tosic acid in organic solvent.

p-TsOH; EtOH (6)

p-TsOH; CH_2Cl_2, 62% (7)

p-TsOH; cyclohexane, 52% (8)

Tosic acid has also been used to catalyze internal translactonization processes (eq 9).[12]

p-TsOH; 93% (9)

Dehydration Processes. Dehydration of ketols to α,β-unsaturated compounds is effectively catalyzed by tosic acid;[13] addition of calcium chloride to the reaction mixture has been noted to give superior results in some instances.[14,15] Other acid-catalyzed processes can occur concomitantly, such as cleavage of silyl ethers.[16] Tosic acid adsorbed on silica gel was found to be an effective catalyst for the dehydration of secondary and tertiary alcohols (16 examples),[17] including a number of steroid alcohols which are resistant to most methods of catalyzed dehydration (eq 10).

(10) p-TsOH, silica gel; 45 °C, 1 h; 87%

p-Toluenesulfonic acid has been used to catalyze the formation of enamines; water is removed via azeotropic distillation.[18] The dehydration of primary nitro compounds by tosic acid provides access to nitrile oxides, which can subsequently engage in 1,3-dipolar cycloaddition reactions.[19] Oximes may be dehydrated to nitriles by heating with tosic acid in DMF.[20] Reaction of cyclohexanone oximes with ketene in the presence of tosic acid results in aromatization to aryl amines.[21] This constitutes a milder method for Semmler–Wolff aromatization than those previously reported.

Cationic Rearrangements and Isomerizations. In a study of Wagner–Meerwein rearrangements of tricyclo[4.3.2.0]undecanones catalyzed by tosic acid,[22] (**3**) was converted to (**4**) in refluxing benzene in 82% yield (eq 11). A similar rearrangement has been utilized in the synthesis of cyclopentanoid sesquiterpenes (eq 12).[23] Similar processes in which the migrating atom is sulfur have been reported;[24] sulfur-containing bicyclo[10.5.0]alkenes and related compounds have been obtained via tosic acid-catalyzed ring expansions of sulfoxides (eq 13).[25]

(11) (**3**) → (**4**); p-TsOH; benzene, reflux; 82%

(12) p-TsOH; benzene, reflux; 98%

(13) p-TsOH; benzene, reflux; 75%

Treatment of tertiary vinyl alcohols with tosic acid in a mixture of ***Acetic Acid*** and ***Acetic Anhydride*** effects their conversion to allylic acetates (eq 14).[26] Treatment of β-hydroxyalkyl phenyl sulfides with tosic acid in refluxing benzene results in migration of the phenylthio group to generate allylic sulfides.[27]

(14) p-TsOH; Ac_2O, AcOH; 91%

Synthesis of Enol Ethers and Acetates. Tosic acid is, in general, superior to other common catalysts (***Sulfuric Acid***, ***Phosphoric Acid***, potassium acetate) for enolization in the conversion of ketones to their enol acetates using acetic anhydride.[28,29] This methodology has been used extensively in steroid synthesis.[30–33] Isopropenyl acetate may also be used with removal of acetone by slow distillation (eq 15).[34] Enol ethers may similarly be obtained by treatment with tosic acid and alcohol in refluxing benzene or toluene followed by azeotropic removal of water.[35]

(15) OAc; p-TsOH; 89%

Synthesis of Steroid Acetates. Tosic acid has been used to good effect in the acetylation of steroid substrates, replacing the commonly used ***Pyridine*** catalyst. Treatment of cholestane-3β,5α,6β-triol with pyridine and acetic anhydride afforded the 3,6-diacetate; heating the triol with tosic acid in acetic anhydride provided the desired triacetate.[36] Acetylation of the 17α-hydroxyl group of progesterone has been accomplished by this method.[37,38]

Addition of Alcohols to Nitriles (Ritter Reaction). Esters may be prepared by the acid-catalyzed addition of alcohols to nitriles. Tosic acid is the preferred catalyst for this reaction.[39]

Cleavage of Amine Protecting Groups. *t*-Butyloxycarbonyl (Boc) groups may be cleaved from protected amines in the presence of *t*-butyl- and *p*-methoxybenzyl esters by the action of tosic acid in a mixture of ethanol and ether (eq 16) (seven examples).[40] *p*-Methoxybenzyloxycarbonyl groups may be removed with tosic acid in acetonitrile.[41]

Boc-D-Ala–D-Ala–OBu → (p-TsOH, EtOH; 3 h, Et_2O; 93%) D-Ala–D-Ala–OBu (16)

Synthesis of Substituted Methylbenzenes. Tosic acid is lithiated at the 2-position with 2 equiv of ***n-Butyllithium***; the resulting anion may be reacted with various electrophiles. Desulfonylation of the substituted product constitutes a synthesis of *meta*-substituted toluenes.[42]

Synthesis of Sulfones. Alkyl *p*-tolyl sulfones may be prepared by reaction of the ***p-Toluenesulfonyl Chloride*** with Grignard reagents. *p*-Tolyl aryl sulfones can be synthesized by the condensation of tosic acid with aromatic compounds in ***Polyphosphoric Acid***;[43] alternatively, the sulfonyl chloride may be reacted with aromatic substrates under Friedel–Crafts conditions.[43] Milder conditions using ***Phosphorus(V) Oxide–Methanesulfonic Acid*** have been reported.[44]

Amine Salts. Amines are frequently converted to their tosylate salts for characterization.[45]

Related Reagents. Camphorsulfonic Acid; Hydrochloric Acid; Methanesulfonic Acid; Sulfuric Acid; Trifluoroacetic Acid; Trifluoromethanesulfonic Anhydride.

---

1. (a) *Dictionary of Organic Compounds*, 5th ed.; Chapman and Hall: New York, 1982; Vol. 4, p 3749. (b) *The Merck Index*, 11th ed.; Budavari, S., Ed.; Merck: Rahway, NJ, 1989; p 1501.
2. *The Sigma-Aldrich Library of Chemical Safety Data*, 2nd ed.; Lenga, R. E., Ed.; Sigma-Aldrich: Milwaukee, 1988; Vol. 2, p 3366.
3. Belmont, D. T.; Paquette, L. A. *JOC* **1985**, *50*, 4102.
4. Takano, S.; Yonaga, M.; Morimoto, M.; Ogasawara, K. *JCS(P1)* **1985**, 305.
5. van Boom, J. H.; Herschied, J. D. M. *S* **1973**, 169.
6. Oikawa, Y.; Nishi, T.; Yonemitsu, O. *JCS(P1)* **1985**, 7.
7. Hagiwara, H.; Uda, H. *JCS(P1)* **1985**, 1157.
8. Smith, A. B., III; Empfield, J. R.; Vaccaro, H. A. *TL* **1989**, *30*, 7325.
9. Cope, A. C.; Herrick, E. C. *OSC* **1963**, *4*, 304.
10. Suemune, H.; Oda, K.; Saeki, S.; Sakai, K. *CPB* **1988**, *36*, 172.
11. Johnson, W. S.; Bauer, V. J.; Margrave, J. L.; Frisch, M. A.; Dreger, L. H.; Hubbard, W. N. *JACS* **1961**, *83*, 606.
12. Corey, E. J.; Brunelle, D. J.; Nicolaou, K. C. *JACS* **1977**, *99*, 7359.
13. Sondheimer, F.; Mechoulam, R.; Sprecher, M. *T* **1964**, *20*, 2473.
14. Wenkert, E.; Stevens, T. E. *JACS* **1956**, *78*, 2318.
15. Spencer, T. A.; Schmiegel, K. K.; Schmiegel, W. W. *JOC* **1965**, *30*, 1626.
16. Zhang, W.-Y.; Jakiela, D. J.; Maul, A.; Knors, C.; Lauher, J. W.; Helquist, P.; Enders, D. *JACS* **1988**, *110*, 4652.
17. D'Onofrio, F.; Scettri, A. *S* **1985**, 1159.
18. Hünig, S.; Lücke, E.; Brenninger, W. *OS* **1961**, *41*, 65; *OSC* **1973**, *5*, 808.
19. Shimizu, T.; Hayashi, Y.; Teramura, K. *BCJ* **1984**, *57*, 2531.
20. Antonowa, A.; Hauptmann, S. *ZC* **1976**, *16*, 17.
21. Tamura, Y.; Yoshimoto, Y.; Sakai, K.; Kita, Y. *S* **1980**, 483.
22. Peet, N. P.; Cargill, R. L.; Bushey, D. F. *JOC* **1973**, *38*, 1218.
23. Pirrung, M. C. *JACS* **1979**, *101*, 7130.
24. Chen, C. H.; Donatelli, B. A. *JOC* **1976**, *41*, 3053.
25. Nickon, A.; Rodriguez, A. D.; Shirhatti, V.; Ganguly, R. *JOC* **1985**, *50*, 4218.
26. Babler, J. H.; Olsen, D. O. *TL* **1974**, 351.
27. (a) Brownbridge, P.; Warren, S. *CC* **1975**, 820. (b) Brownbridge, P.; Fleming, I.; Pearce, A.; Warren, S. *CC* **1976**, 751.
28. Bedoukian, P. Z. *JACS* **1945**, *67*, 1430.
29. (a) Mao, C.-L.; Hauser, C. R. *OS* **1971**, *51*, 90. (b) Mao, C.-L.; Hauser, C. R. *OSC* **1988**, *6*, 245.
30. Kritchevsky, T. H.; Gallagher, T. F. *JACS* **1951**, *73*, 184.
31. Koechlin, B. A.; Kritchevsky, T. H.; Gallagher, T. F. *JACS* **1951**, *73*, 189.
32. Kritchevsky, T. H.; Garmaise, D. L.; Gallagher, T. F. *JACS* **1952**, *74*, 483.
33. Marshall, C. W.; Kritchevsky, T. H.; Lieberman, S.; Gallagher, T. F. *JACS* **1948**, *70*, 1837.
34. Evans, E. H.; Hewson, A. T.; March, L. A.; Nowell, I. W.; Wadsworth, A. H. *JCS(P1)* **1987**, 137.
35. (a) Gannon, W. F.; House, H. O. *OS* **1960**, *40*, 41; (b) Gannon, W. F.; House, H. O. *OSC* **1973**, *5*, 539.
36. Davis, M.; Petrow, V. *JCS* **1949**, 2536.
37. Turner, R. B. *JACS* **1952**, *74*, 4220.
38. Minlon, H.; Wilson, E.; Wendler, N. L.; Tishler, M. *JACS* **1952**, *74*, 5394.
39. James, F. L.; Bryan, W. H. *JOC* **1958**, *23*, 1225.
40. Goodacre, J.; Ponsford, R. J.; Stirling, I. *TL* **1975**, 3609.
41. Yamada, H.; Tobiki, H.; Tanno, N.; Suzuki, H.; Jimpo, K.; Ueda, S.; Nakagome, T. *BCJ* **1984**, *57*, 3333.
42. Figuly, G. D.; Martin, J. C. *JOC* **1980**, *45*, 3728.
43. (a) Sandler, S. R.; Karo, W. *Organic Functional Group Preparations*; Academic: New York, 1968; Vol. 1, p 500. (b) Sandler, S. R.; Karo, W. *Organic Functional Group Preparations*, 2nd ed.; Academic: New York, 1983; Vol. 1, p 619.
44. Ueda, M.; Uchiyama, K.; Kano, T. *S* **1984**, 323.
45. Bargar, T. M.; Broersma, R. J.; Creemer, L. C.; McCarthy, J. R.; Hornsperger, J.-M.; Attwood, P. V.; Jung, M. J. *JACS* **1988**, *110*, 2975.
46. *Reagent Chemicals: American Chemical Society Specifications*, 8th ed.; American Chemical Society: Washington, 1993; pp 762–763.
47. Perrin, D. D.; Armarego, W. L. F. *Purification of Laboratory Chemicals*, 3rd ed.; Pergamon: New York, 1988; p 291.

Gregory S. Hamilton
Scios Nova, Baltimore, MD, USA

Tri-*n*-butyl(methoxy)stannane[1]

n-Bu_3SnOMe

[1067-52-3] $C_{13}H_{30}OSn$ (MW 321.14)

(mild methoxide source with strong nucleophilic properties but with reduced basicity, making it less likely to induce elimination reactions; precursor to other alkoxytributylstannanes)

Alternate Names: tributyltin methoxtride; methoxytributylstannane.
Physical Data: bp 90 °C/0.1 mmHg; *d* 1.129 g cm^{-3}; ^{1}H NMR ($CDCl_3$) 3.54 ppm (OMe); ^{119}Sn NMR 83 ppm.
Solubility: sol common organic solvents.
Form Supplied in: colorless liquid; widely available.
Handling, Storage, and Precautions: toxic, as are most organotin reagents; should be manipulated in a fume hood and handled with gloves. Readily hydrolyzable, it must be stored in the absence of moisture and CO_2. Exposure to moist air leads to hexabutyldistannoxane and tributylstannyl carbonate.

Introduction. This reagent can be used in substitution, oxidation, or addition reactions. It is also employed as a catalyst for urethane, polylactone, polylactide, and polycarbonate synthesis.

Substitution Reactions. Methoxytributylstannane and related compounds react with acyl chlorides or anhydrides to give esters under very mild conditions (eq 1).[2] This smooth reaction is particularly useful in carbohydrate chemistry.[1d]

$$\mathrm{Bu_3SnOMe + RCOX \xrightarrow{-Bu_3SnX} RCO_2Me} \quad (1)$$
$$\mathrm{X = Cl, OCOR}$$

Methoxytributylstannane is also useful for the preparation of peresters, as reactive stannyl peroxide intermediates are quantitatively obtained from the corresponding hydroperoxides.[3] Methoxytriethylstannane can be used under palladium catalysis to give aryl esters from aryl iodides and carbon monoxide (eq 2).[4]

$$\mathrm{ArI + CO + Et_3SnOMe \xrightarrow[-Et_3SnI]{PhPdI(PPh_3)_2} ArCO_2Me} \quad (2)$$

These esterifications have been generalized to organosulfur and organophosphorus chemistry. With sulfenyl chlorides,[5] Bu_3SnOMe gives access to sulfenates in better yields than does ***Sodium Methoxide*** (eq 3). *O*-Alkyl *S,S*-diaryl phosphorodithioates are converted to *O,O*-dialkyl *S*-aryl phosphorothioates (eq 4)[6] and dinucleoside *S*-aryl phosphorothioates give dinucleoside *O*-methyl phosphates (eq 5).[7]

$$PhSCl + Bu_3SnOMe \xrightarrow[80\%]{-Bu_3SnCl} PhSOMe \quad (3)$$

$$(ArS)_2P(O)OR + Bu_3SnOMe \xrightarrow{-Bu_3SnSAr} ArSP(O)(OR)OMe \quad (4)$$

DMTrO, Th^bz, ArS–P(O), BzO; Bu3SnOMe, –Bu3SnSAr, 80% → MeO–P(O) (5)

In the presence of ***Lithium Chloride*** as an activating agent, methoxytributylstannane promotes mild, chemoselective debenzoylations. Thus debenzoylation of 7,13-diacetylbaccatin III (a precursor of taxol) occurs only at C-2 (eq 6). In a highly polar solvent, NMP, none of the four other acyl groups is affected to any appreciable extent.[8] *O*-Deacetylation of sugars is readily accomplished with methoxytributylstannane. Anomeric acetates are more reactive than primary or secondary acetates, thereby enabling selective removal.[9]

Bu3SnOMe, NMP, 67% (6)

Dioxybis(tributylstannane), prepared from methoxytributylstannane and ***Hydrogen Peroxide***, allows the preparation of primary, secondary, and tertiary dialkyl peroxides from the corresponding triflates in good yields.[10]

The modulation of the nucleophilicity of an alkoxy group by the tin atom is particularly useful in methoxylation reactions,[11] where the absence of base even allows the use of base sensitive halides (eq 7).

$$Bu_3SnOMe + RX \longrightarrow ROMe + Bu_3SnX \quad (7)$$

Alkyl glycosides can be prepared by using alkoxytributylstannanes. Per-*O*-acetyl-α-D-glucopyranosyl bromide reacts with methoxytributylstannane in the presence of catalytic amounts of ***Tin(IV) Chloride*** to give methyl per-*O*-acetyl-β-D-glycoside (eq 8). Replacement of tin(IV) chloride by 0.5 equiv of triethylammonium bromide completely changes the course of the reaction; epimerization to the more reactive β-anomeric configuration occurs during the reaction, affording an *exo*-orthoester (eq 9).[12]

Bu3SnOMe, cat. SnCl4, –Bu3SnBr, 86% (8)

Bu3SnOMe, 0.5 equiv Et4NBr, 78% (9)

When 2-, 3-, 4-, or 5-halogenated alcohols are used, methoxytributylstannane first reacts with the hydroxyl in an exchange reaction. Then the alkoxide intermediates undergo intramolecular alkoxylation to give oxiranes, oxetanes, tetrahydrofurans, or tetrahydropyrans, respectively (eq 10).[13] The displacement is stereospecific, as *erythro*-3-bromo-2-butanol leads mainly to (*E*)-butene oxide and *threo*-3-bromo-2-butanol to the (*Z*)-isomer ($n = 1$).

Bu3SnOMe, –MeOH; –Bu3SnX (10)

Methoxylation with methoxytributylstannane is not limited to halogenated derivatives. Under palladium catalysis, allyl acetates are transformed into allyl methyl ethers. This chemoselective approach to allyl etherification is often stereoselective and tolerates other electrophilic functional groups such as primary halides. It has been applied to carbohydrate chemistry (eq 11).[14]

1. Bu3SnOMe; 2. PhCH=CHCH2OAc, Pd(PPh3)4, 51% (11)

β-Lactams are stereospecifically methoxylated at the 4-position by methoxytributylstannane when a phenylsulfinyl group is used as leaving group (eq 12). Thus 3-substituted or unsubstituted 4-phenylsulfinyl-2-azetidinones lead to the corresponding 4-methoxy-2-azetidinones in the presence of catalytic amounts of ***Trimethylsilyl Trifluoromethanesulfonate*** at rt.[15]

+ Bu3SnOMe, cat. TMSOTf (12)

Enoxytributylstannanes (tin enolates) can be regiospecifically prepared by transesterification of enol acetates derived from aldehydes[16] or ketones[17] with methoxytributylstannane. The formation of new carbon–carbon bonds can be accomplished with primary alkyl halides, allyl halides, or functional halides. This coupling is regiospecific (eqs 13 and 14).

$$\text{OAc (cyclohexenyl)} \xrightarrow{Bu_3SnOMe} \text{OSnBu}_3 \xrightarrow{MeI} \text{2,2-dimethylcyclohexanone (81\%)} \quad (13)$$

$$\text{OAc} \xrightarrow{Bu_3SnOMe} \text{OSnBu}_3 \xrightarrow{MeI} \text{product (82\%)} \quad (14)$$

Under palladium catalysis, aryl[18], vinyl,[19] or heteroaryl[20] halides, or allyl acetates[21] may be used instead of allyl or alkyl halides (eqs 15 and 16). This reaction is also regiospecific.[22] It allows arylation of chroman-4-ones into isoflavanones.[23] In contrast, a different reaction takes place in acetonitrile, which leads to α-enones. This process has been applied to the synthesis of α,β-unsaturated large-ring ketones from enol esters (eq 17).[24]

$$\text{OAc} \xrightarrow[\text{2. PhBr, cat. Pd; 67\%}]{\text{1. } Bu_3SnOMe} \text{PhCH}_2\text{COEt} \quad (15)$$

$$\text{OAc} \xrightarrow[\text{2. PhBr, cat. Pd; 60\%}]{\text{1. } Bu_3SnOMe} \text{MeCOCH(Me)Ph} \quad (16)$$

$$\text{OAc} \xrightarrow[\text{2. cat. Pd, MeCN; 85\%}]{\text{1. } Bu_3SnOMe} \text{cyclododecenone} \quad (17)$$

Metalation of alkynes by methoxytributylstannane gives alkynyltributylstannanes, which are able to undergo further reactions with acyl chlorides, leading to alkynyl ketones (eq 18).[25] With 1,3-dienes, [4 + 2] cycloadditions give cyclic vinylic organotins which are easily transformed into functional cyclohexadienes (eq 19) through vinyllithium intermediates.[26]

$$R^1\text{−≡} \xrightarrow[-MeOH]{Bu_3SnOMe} R^1\text{−≡−SnBu}_3 \xrightarrow[-Bu_3SnCl]{R^2COCl} R^1\text{−≡−COR}^2 \quad (18)$$

$$R\text{−≡} \xrightarrow[\text{2. butadiene}]{\text{1. } Bu_3SnOMe, -MeOH} \text{(R, SnBu}_3\text{)cyclohexadiene} \xrightarrow[-Bu_4Sn]{BuLi} \text{(R, Li)cyclohexadiene} \quad (19)$$

Oxidation Reactions. Alkoxystannanes, conveniently prepared from methoxytributylstannane and alcohols, are oxidized under very mild conditions by bromotrichloromethane,[27] ***1,1-Di-t-butyl Peroxide***,[28] ***Bromine***,[29] ***N-Bromosuccinimide***,[30] or ***Nitronium Tetrafluoroborate*** (eq 20).[31] Secondary alcohols are selectively transformed into ketones, even in the presence of primary alcohols.[32]

$$R^1R^2CHOH \xrightarrow[\text{2. } \cdot CX_3 \text{ or } Br_2 \text{ or NBS...}]{\text{1. } Bu_3SnOMe} R^1R^2C{=}O \quad (20)$$

Disulfides are obtained when thiols are treated with methoxytributylstannane and anhydrous ***Iron(III) Chloride***, through an oxidation reaction involving thiostannane intermediates (eq 21). Tertiary, as well as primary, secondary, or aryl thiols may also be employed. Various functionalities, such as hydroxy, amino, amido, or ester groups, remain intact under the reaction conditions.[33]

$$RSH + Bu_3SnOMe \xrightarrow{FeCl_3} 1/2\ RSSR + Bu_3SnCl \quad (21)$$

Addition Reactions. Enoxytributylstannanes, prepared from methoxytributylstannane, add to aldehydes[34] to give predominately the *anti*-aldols at low temperature, whereas *syn* selectivity is observed at higher temperature (eq 22).[35]

$$\text{AcO(Ph)C=CHMe} \xrightarrow[\text{2. BnCHO}]{\text{1. } Bu_3SnOMe} \text{PhCOCH(Me)CH(OH)Bn} + \text{PhCOCH(Me)CH(OH)Bn} \quad (22)$$

−78 °C	95:5
45 °C	10:90

Intramolecular opening of oxiranes by tin alkoxides leads to cyclic ethers. After treatment by methoxytributylstannane, substituted 3-epoxybutanols stereospecifically give either oxetanemethanols (eq 23) or 3-oxolanols (eq 24) on heating, depending on the substituents.[36]

$$\text{HO–epoxide} \xrightarrow[\text{2. }\Delta;\ 84\%]{\text{1. } Bu_3SnOMe} \text{oxetane–OH} \quad (23)$$

$$\text{HO–epoxide} \xrightarrow[\text{2. }\Delta;\ 80\%]{\text{1. } Bu_3SnOMe} \text{oxolan-3-ol} \quad (24)$$

Nucleophilic additions of methoxytributylstannane to isocyanates or ketenes lead to reactive intermediates giving heterocycles after further condensations. Mixed isocyanurates are obtained when different isocyanates are involved in the reaction (eq 25).[37]

$$EtNCO + Bu_3SnOMe \longrightarrow EtN(SnBu_3)CO_2Me \xrightarrow{PhNCO} \text{1-Et-3,5-Ph}_2\text{-isocyanurate (67\%)} \quad (25)$$

Alkoxides resulting from the opening of halogenated lactones by methoxytributylstannane add to isocyanates to give 2-oxazolidones (eq 26),[38] or 2-oxazinones,[39] depending on the ring size of the starting lactones. Other heterocumulenes, such as diimides, isocyanates or carbon dioxide, react equally well (eq 27).

1. Bu_3SnOMe; 2. RNCO → MeO_2C-oxazolidinone + Bu_3SnBr (26)

1. Bu_3SnOMe; 2. PhNCS — 1. Bu_3SnOMe; 2. PhNCNPh — 1. Bu_3SnOMe; 2. RNCO (27)

A Darzens reaction can be performed under mild, neutral conditions with the stannyl carbamate resulting from the addition of methoxytributylstannane to ethyl isocyanate (eq 28). This adduct selectively generates organotin enolates from α-halo ketones even when enolizable α′-hydrogens are present in the halo ketone.[40]

$$\text{EtNCO} + \text{Bu}_3\text{SnOMe} \longrightarrow \text{Bu}_3\text{Sn-N(Et)-CO}_2\text{Me} \xrightarrow[\text{2. R}^2\text{CHO}]{\text{1. R}^1\text{COCH}_2\text{X}} \text{R}^1\text{CO-epoxide-R}^2 \quad (28)$$

Methoxytributylstannane adds to silylketenes under very mild conditions (eq 29). Condensation of the adduct with aldehydes in the presence of titanium tetrachloride gives β-hydroxy-α-silyl esters with very high *syn* selectivity. A stereocontrolled elimination in these esters subsequently leads to (*E*)- or (*Z*)-α,β-unsaturated esters, depending on the conditions used.[41]

TBDMS ketene: 1. Bu_3SnOMe; 2. RCHO; then $BF_3{\bullet}OEt_2$ or NaH (29)

Stannyl quinones are formed by the thermal isomerization of 4-stannyloxy-4-alkynylcyclobutenones (eq 30) via a ketene intermediate. Once this intramolecular rearrangement is accomplished, the stannyl group can be substituted under palladium catalysis by various organic halides to give substituted quinones.[42]

Bu_3SnOMe, $-MeOH$; then allyl bromide, cat. Pd (30)

Iminocarbonylation of aryl bromides, involving the palladium-catalyzed condensation of *t*-butyl isocyanide, methoxytributylstannane, and an aryl bromide, leads to imidates (eq 31).[43]

$$\text{PhBr} + t\text{-BuNC} + \text{Bu}_3\text{SnOMe} \xrightarrow[-\text{Bu}_3\text{SnBr},\ 63\%]{\text{cat. Pd}} \text{Ph(MeO)C=N-}t\text{-Bu} \quad (31)$$

Catalysis. Methoxytributylstannane is used as a catalyst for polymerization of lactones,[44] lactides,[45] and 1,3-dioxan-2-one.[46] The reaction is suggested to proceed by a coordinative insertion mechanism involving the cleavage of the acyl–oxygen bond of the monomer, and not by an ionic mechanism.

Related Reagents. Di-*n*-butyltin Oxide; Sodium Methoxide; Tri-*n*-butyltin Trifluoromethanesulfonate.

---

1. (a) Bloodworth, A. J.; Davies, A. *Organotin Compounds*; Sawyer. A., Ed.; Dekker: New York, 1971; Vol. 1, p 153. (b) Pereyre, M.; Quintard, J. P.; Rahm, A. *Tin in Organic Synthesis*; Butterworths: London, 1986. (c) Davies, A.; Smith, P. J. *Comprehensive Organometallic Chemistry*; Wilkinson, G.; Stone, F. G. A.; Abel, E. W., Eds.; Pergamon: Oxford, 1982; Vol. 2, p 536. (d) David, S.; Hanessian, S. *T* **1985**, *41*, 643.
2. Valade, J.; Pereyre, M. *CR* **1962**, *254*, 3693.
3. Haynes, R. K.; Vonwiller, S. C. *CC* **1990**, 448.
4. Bumagin, N. A.; Gulevich, Y. V.; Beletskaya, I. P. *JOM* **1985**, *285*, 415.
5. Armitage, D. A. *S* **1984**, 1042.
6. Watanabe, Y.; Mukaiyama, T. *CL* **1979**, 389.
7. Ohuchi, S.; Ayukawa, H.; Hata, T. *CL* **1992**, 1501.
8. Farina, V.; Huang, S. *TL* **1992**, *33*, 3979.
9. Herzig, J.; Nudelman, A.; Gottlieb, H. E. *Carbohydr. Res.* **1988**, *177*, 21.
10. (a) Salomon, M. F.; Salomon, R. G. *JACS* **1977**, *99*, 3500. (b) Salomon, M. F.; Salomon, R. G. *JACS* **1977**, *99*, 3501.
11. Pommier, J. C.; Valade, J. *CR(C)* **1965**, *260*, 4549.
12. (a) Ogawa, T.; Matsui, M. *Carbohydr. Res.* **1976**, *51*, C13. (b) Ogawa, T.; Katano, K.; Sasajima, K.; Matsui, M. *T* **1981**, *37*, 2779.
13. (a) Delmond, B.; Pommier, J. C.; Valade, J. *JOM* **1973**, *47*, 337. (b) Delmond, B.; Pommier, J. C.; Valade, J. *JOM* **1972**, *35*, 91.
14. Keinan, E.; Sahai, M.; Roth, Z.; Nudelman, A.; Herzig, J. *JOC* **1985**, *50*, 3558.
15. Kita, Y.; Shibata, N.; Yoshida, N.; Tohjo, T. *CPB* **1992**, *40*, 1044.

16. Pereyre, M.; Bellegarde, B.; Mendelsohn, J.; Valade, J. *JOM* **1968**, *11*, 97.
17. Jung, M. E.; Blum, R. B. *TL* **1977**, 3791.
18. Kosugi, M.; Suzuki, M.; Hagiwara, I.; Goto, K.; Saitoh, K.; Migita, T. *CL* **1982**, 939.
19. Kosugi, M.; Hagiwara, I.; Migita, T. *CL* **1983**, 839.
20. Nair, V.; Turner, G. A.; Buenger, G. S.; Chamberlain, S. D. *JOC* **1988**, *53*, 3051.
21. Tsuji, J.; Minami, I.; Shimizu, I. *TL* **1983**, *24*, 4713.
22. Kosugi, M.; Hagiwara, I.; Sumiya, Y.; Migita, T. *CC* **1983**, 344.
23. Donnelly, D. M. X.; Finet, J. P.; Stenson, P. H. *H* **1989**, *28*, 15.
24. (a) Tsuji, J.; Minami, I.; Shimizu, I. *TL* **1983**, *24*, 5639. (b) Minami, I.; Takahashi, K.; Shimizu, I.; Kimura, T.; Tsuji, J. *T* **1986**, *42*, 2971.
25. Logue, M. W.; Teng, K. *JOC* **1982**, *47*, 2549.
26. (a) Jousseaume, B.; Villeneuve, P. *T* **1989**, *45*, 1145. (b) Jousseaume, B.; Villeneuve, P. *J. Labelled Compd. Radiopharm.* **1988**, *25*, 717.
27. Pommier, J. C.; Chevolleau, D. *JOM* **1974**, *74*, 405.
28. Godet, J. Y.; Pereyre, M.; Pommier, J. C.; Chevolleau, D. *JOM* **1973**, *55*, C15.
29. (a) Saigo, K.; Morikawa, A.; Mukaiyama, T. *CL* **1975**, 145. (b) Saigo, K.; Morikawa, A.; Mukaiyama, T. *BCJ* **1976**, *49*, 1656.
30. Ogawa, T.; Matsui, M. *JACS* **1976**, *98*, 1629.
31. Olah, G. A.; Ho, T. L. *S* **1976**, 609.
32. Ueno, Y.; Okawara, M. *TL* **1976**, 4597.
33. Sato, T.; Otera, J.; Nozaki, H. *TL* **1990**, *31*, 3591.
34. Noltes, J. G.; Creemers, H. M. J. C.; Van der Kerk, G. J. M. *JOM* **1968**, *11*, P21.
35. Labadie, S. S.; Stille, J. K. *T* **1984**, *40*, 2329.
36. (a) Bats, J. P.; Moulines, J.; Pommier, J. C. *TL* **1976**, 2249. (b) Bats, J. P.; Moulines, J.; Picard, P.; Leclercq, D. *TL* **1980**, *21*, 3051. (c) Bats, J. P.; Moulines, J.; Picard, P.; Leclercq, D. *T* **1982**, *38*, 2139.
37. (a) Davies, A. G. *S* **1969**, 56. (b) Bloodworth, A. J.; Davies, A. G. *JCS* **1965**, 6858.
38. Shibata, I.; Toyota, M.; Baba, A.; Matsuda, H. *JOC* **1990**, *55*, 2487.
39. Shibata, I.; Toyota, M.; Baba, A.; Matsuda, H. *S* **1988**, 486.
40. Shibata, I.; Yamasaki, H.; Baba, A.; Matsuda, H. *JOC* **1992**, 6909.
41. Akai. S.; Tsuzuki, Y.; Matsuda., S.; Kitagaki, S.; Kita. Y. *SL* **1991**, 911.
42. Liebeskind, L. S.; Foster, B. S. *JACS* **1990**, *112*, 8612.
43. Kosugi, M.; Ogata, T.; Tamura, H.; Sano, H.; Migita, T. *CL* **1986**, 1197.
44. Kricheldorf, H. R.; Scharnagl, N. *J. Macromol. Sci.* **1989**, *A26*, 951.
45. Kricheldorf, H. R.; Boettcher, C.; Toennes, K. U. *Polymer* **1992**, *33*, 2817.
46. (a) Kricheldorf, H. R.; Jenssen, J.; Kreiser-Saunders, I. *Makromol. Chem.* **1991**, *192*, 2391. (b) Kricheldorf, H. R.; Sumbel, M. V.; Kreiser-Saunders, I. *Macromolecules* **1991**, *24*, 1944.

Bernard Jousseaume
Université Bordeaux I, Talence, France

Tri-*n*-butyltin Trifluoromethanesulfonate[1]

$n\text{-}Bu_3SnOSO_2CF_3$

[68725-14-4] $C_{13}H_{27}F_3O_3SSn$ (MW 439.18)

(powerful stannylating reagent;[1] readily stannylates vinylcuprates and vinylaluminum compounds;[1–3] catalyzes hydrostannation of aldehydes and ketones by tin hydrides[4])

Physical Data: mp 41–43 °C; bp 155–167 °C/0.08 mmHg.
Solubility: freely sol polar and nonpolar organic solvents.
Preparative Methods: by analogy to the preparation of trialkyltin carboxylates,[5] tributyltin trifluoromethanesulfonate is prepared from ***Bis(tri-n-butyltin) Oxide*** and either ***Trifluoromethanesulfonic Acid*** (with subsequent removal of water) or ***Trifluoromethanesulfonic Anhydride***,[2] as in eq 1.

$$(CF_3SO_2)O + (Bu_3Sn)_2O \xrightarrow[93\%]{} 2\ Bu_3SnOSO_2CF_3 \quad (1)$$

The following procedure is quite convenient: freshly distilled trifluoromethanesulfonic anhydride[6] is caused to react with an equal molar amount of tributyltin oxide cooled at 20 °C under nitrogen. When the exothermic reaction completely subsides, the flask is set up for a short-path distillation and the resulting red oil is distilled at about 0.1 mmHg. The distillate is collected without cooling to prevent crystallization of the product in the condenser. Alternatively, this reagent can be prepared in situ by the action of trifluoromethanesulfonic acid on ***Tri-n-butylstannane*** in benzene or dichloroethane.[4]

Handling, Storage, and Precautions: this reagent is deliquescent. Therefore it is best handled as a solution in benzene, ether, or hexane, stored at room temperature under nitrogen. Solutions can be dried with 4Åmolecular sieves. As with certain other trialkyltin electrophiles, tributyltin trifluoromethanesulfonate is foul smelling. The toxicity of this class of compounds has been described.[7] Use in a fume hood.

Transmetalation Reactions. The powerful electrophilic property of this reagent makes it particularly attractive for the conversion of vinylcuprates and vinylalanes into vinylstannanes. For example, the reaction of vinyl heterocuprates[8] with this reagent occurs readily to give the corresponding (*Z*)-vinylstannane, even at −78 °C (eq 2). Stannylation of vinylalanes[9] occurs readily to give the corresponding vinylstannane (eq 3). Complementary methods are available for the preparation of both (*Z*)-[2] and (*E*)-vinylstannanes.[1,3] The configuration of the vinylstannane is controlled at the vinylaluminum stage: treatment of ***Propargyl Alcohol*** with ***Lithium Aluminium Hydride*** followed by tributyltin trifluoromethanesulfonate affords the (*Z*)-allyl alcohol, whereas if the addition of $LiAlH_4$ is followed by NaOMe and then the reagent, the (*E*)-allyl alcohol results (eq 4).[1]

$$RMgBr + HC{\equiv}CH \xrightarrow[\text{2. } Bu_3SnOSO_2CF_3,\ -78\ ^\circ C]{\text{1. CuI, THF, }-30\ ^\circ C} R\text{–CH=CH–}SnBu_3 \quad (2)$$

R = Et, 79%; C_5H_{11}, 90%

1. Red-Al, DME, 23 °C
2. $Bu_3SnOSO_2CF_3$, 62 °C
89%

OH, SnBu3 (3)

The (*E*)-vinylstannane also results when the alkyne is treated sequentially with ***Diisobutylaluminum Hydride***, ***Methyllithium***, and then the title reagent.[3] For example, treatment of 1-octyne with DIBAL in hexane followed by MeLi and then the reagent afforded (*E*)-1-tributylstannyl-1-octene in good yield (eq 5).

0.5 equiv $LiAlH_4$, THF; $Bu_3SnOSO_2CF_3$; 1. NaOMe 2. $Bu_3SnOSO_2CF_3$ (4)

1. DIBAL, hexane, 85 °C 2. MeLi, −78 °C 3. $Bu_3SnOSO_2CF_3$ 60% (5)

The title reagent is also suitable for the preparation of allylstannanes from allyllithium reagents,[10] and vinylstannanes from vinyllithium reagents.[11]

Hydrostannation Catalyst. This triflate is a catalyst in the hydrostannation of aldehydes and ketones by tributyltin hydride.[4] Treatment of aldehydes and ketones with one equiv of Bu_3SnH in benzene or dichloroethane in the presence of 20% of the reagent afforded the corresponding primary or secondary alcohol in high yield. Using this system, aldehydes are preferentially reduced in the presence of methyl ketones (ca. 99:1). In addition, the reagent can be employed as a cocatalyst with $PdCl_2(PPh_3)_2$ in the 1,4-reduction of enone systems (eq 6).

1.4 equiv Bu_3SnH, 0.2 equiv $Bu_3SnOSO_2CF_3$, 0.01 equiv $PdCl_2(PPh_3)_2$, 10 min, rt (6)

Stereospecific Arylation of Arylsulfanyl Lactones. The title reagent catalyzes the rapid and stereospecific cyclization of certain chiral arylsulfanyl lactones to the corresponding optically pure γ-butyrolactones (eq 7).[12]

$Bu_3SnOSO_2CF_3$, toluene, 40–50 °C, 30 min, 85% (7)

Related Reagents. Tri-*n*-butylchlorostannane.

1. Eckrich, T. M. Ph. D. Thesis, Harvard University, Cambridge, MA, 1984.
2. Corey, E. J.; Eckrich, T. M. *TL* **1984**, *25*, 2419.
3. Groh, B. L.; Kreager, A. F.; Schneider, J. B. *SC* **1991**, *21*, 2065.
4. Yang, T. X.; Four, P.; Guibé, F.; Balavoine, G. *NJC* **1984**, *8*, 611.
5. Poller, R. C. *The Chemistry of Organotin Compounds*; Academic: New York, 1970; p 173.
6. Burdon, J.; Farazmund, I.; Stacey, M.; Tathen, J. C. *JCS* **1957**, 2574.
7. Poller, R. C. *The Chemistry of Organotin Compounds*; Academic: New York, 1970; p 271.
8. Westmijze, H.; Meijer, J.; Bos, H. J. T.; Vermeer, P. *RTC* **1976**, *95*, 299, 304.
9. Negishi, E. *Organometallics in Organic Synthesis*; Wiley: New York, 1980; Vol. 1, pp 357–362.
10. Corey, E. J.; Walker, J. C. *JACS* **1987**, *109*, 8108.
11. Corey, E. J.; Yu, C.-M.; Kim, S. S. *JACS* **1989**, *111*, 5495.
12. Marino, J. P.; Laborde, E.; Paley, R. S. *JACS* **1988**, *110*, 966.

Thomas M. Eckrich
Eli Lilly and Company, Lafayette, IN, USA

Triethylamine

Et_3N

[121-44-8] $C_6H_{15}N$ (MW 101.22)

(tertiary amine base used in oxidation, dehydrohalogenation, and substitution reactions)

Alternate Name: TEA.
Physical Data: bp 88.8 °C; mp −115 °C; *d* 0.726 g cm^{-3}.
Solubility: sol most organic solvents.
Purification: dried over $CaSO_4$, $LiAlH_4$, 4Å sieves, CaH_2, KOH, or K_2CO_3, then distilled from BaO, sodium, P_2O_5 or CaH_2.[1]
Handling, Storage, and Precautions: is a corrosive and flammable liquid. Bottles of triethylamine should be flushed with nitrogen or argon to prevent exposure to carbon dioxide. The vapors are harmful and care should be taken to avoid absorption through the skin. Use in a fume hood.

Introduction. The most widely used organic amine base in synthetic organic chemistry is probably triethylamine. Its popularity stems from availability and low cost, along with ease of removal by distillation due to a mid-range boiling point (88.8 °C). Also, the hydrochloride and hydrobromide salts are somewhat insoluble in organic solvents, such as diethyl ether, and may be removed by simple filtration. Triethylamine finds wide use in oxidations, reductions, eliminations, substitutions, and addition reactions. There follows a brief compilation of the uses of triethylamine.

Oxidations. The addition of triethylamine and ***Acetic Anhydride*** during the workup of the ozonolysis of cycloalkenes allows for the selective differentiation of the oxidized termini.[2] Ozonolysis of cycloalkenes in MeOH buffered with $NaHCO_3$ generates the intermediate α-methoxy hydroperoxide aldehyde which dehydrates upon exposure to acetic anhydride and triethylamine to produce aldehyde esters (eq 1). Intermediate peroxy acetals are produced with the addition of ***p-Toluenesulfonic Acid*** to the ozonolysis reaction and similarly dehydrate with triethylamine and acetic anhydride to afford acetal esters. Omitting the acetic anhydride and triethylamine leads to acetal aldehydes under reduction conditions (eq 2).

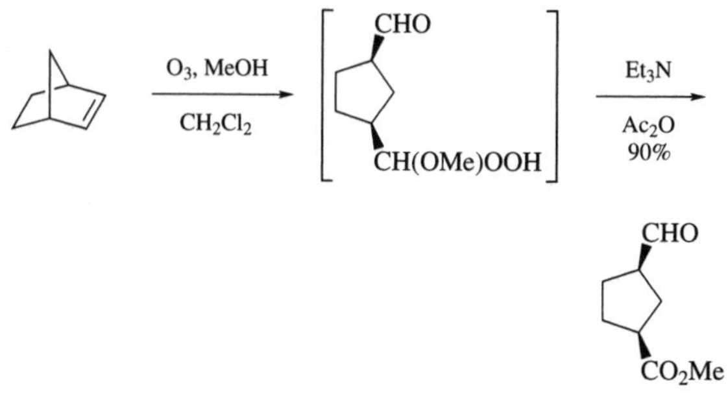

(1)

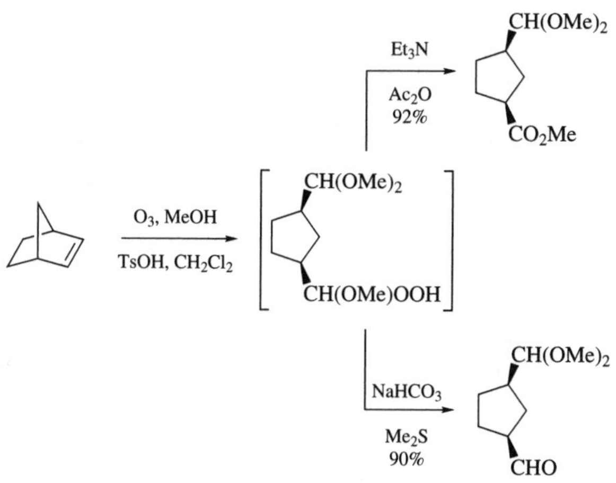

(2)

In the Swern oxidation[3] of alcohols to ketones and aldehydes, employing ***Dimethyl Sulfoxide-Oxalyl Chloride***, triethylamine is usually the amine base utilized in the basification step of the reaction. Mechanistically, the basification step includes deprotonation of the methyl group of the alkoxysulfonium salt intermediate to form an ylide which then yields the carbonyl compound via an intramolecular proton transfer (eq 3).[3b] A similar mechanism is encountered with ***N-Chlorosuccinimide-Dimethyl Sulfide*** and TEA.[4]

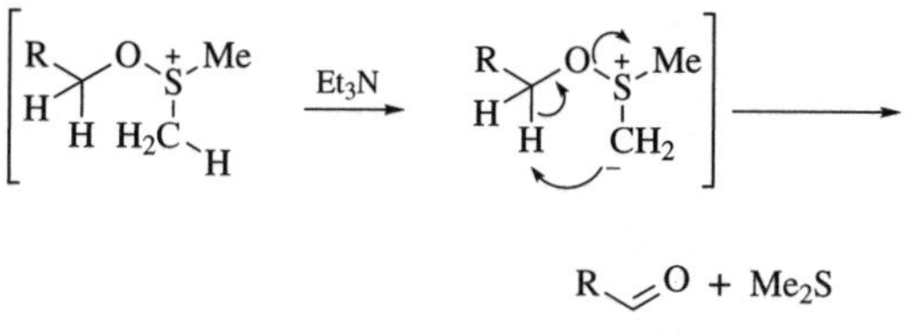

(3)

The steric nature of the amine affects the overall performance of the oxidations. In comparative studies involving oxidations with ***Dimethyl Sulfoxide–Trifluoroacetic Anhydride***[5] and ***Dimethyl Sulfoxide–Methanesulfonic Anhydride***[6] the sterically hindered ***Diisopropylethylamine*** was found to be superior to TEA (eq 4).[3b]

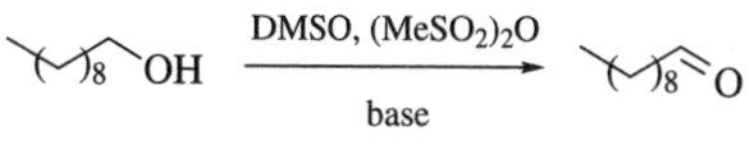

(4)

Base	Yield (%)
Et_3N	68
DIPEA	94

The treatment of vicinal diols with an excess of DMSO/trifluoroacetic anhydride followed by basification with TEA results in complete oxidation to α-diketones in good yields (eq 5).[7]

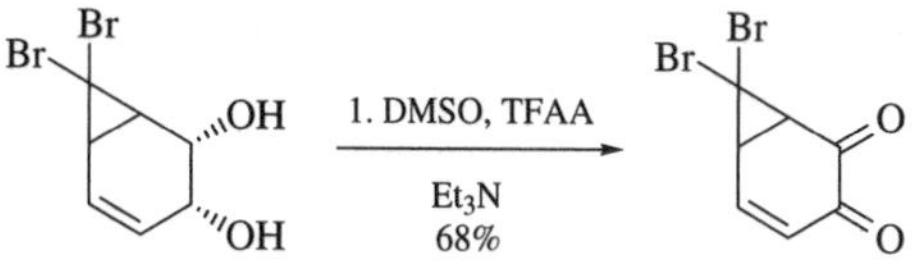

(5)

Eliminations. Dehydrohalogenation of alkyl halides to generate alkenes is normally carried out using alcoholic ***Potassium Hydroxide***; however triethylamine has some utility.[8] For example, bridgehead enones are produced efficiently by treatment of β-bromo ketones with 2 equiv of TEA at 0 °C. The highly reactive enones are not isolated (due to their instability) but can be subsequently trapped with a variety of dienes in a Diels–Alder reaction (eq 6).[9]

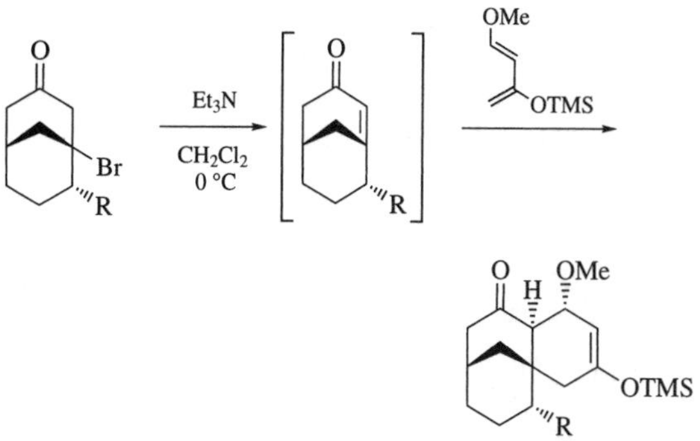

(6)

Triethylamine accomplishes the dehydrobromination of α,α′-dibromo sulfones using TEA in CH_2Cl_2 to afford α,β-unsaturated bromomethyl sulfones in good yields. Further treatment of the sulfones with ***Potassium t-Butoxide*** induces a Ramberg–Bäcklund-like reaction and leads to 1,3-dienes in moderate to good yields (eq 7).[10]

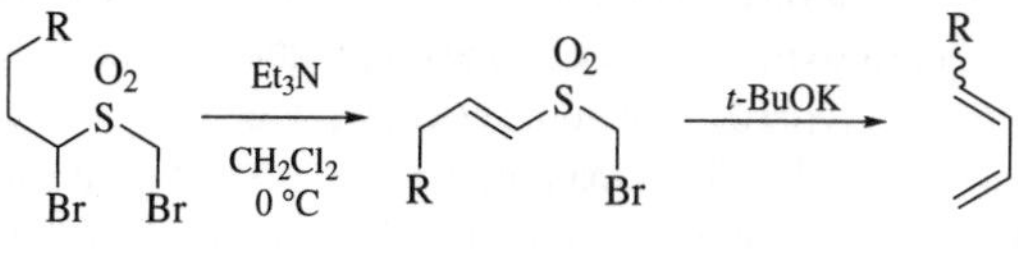

(7)

Chloroalkylidene malonates are converted via a decarboxylation–elimination reaction to alkynic esters by treatment with triethylamine at 90 °C (eq 8).[11] Likewise, triethylamine is the preferred base for the conversion of *erythro*-2,3-dibromobutanoic acid to *cis*-1-bromopropene (eq 9). Pyridine, Na_2CO_3, or $NaHCO_3$ in DMF lead to poor yields of the bromopropene.[12]

Exposure of 2,3-dibromo-3,3-difluoropropionyl chloride to ca. 1 equiv of triethylamine in CH_2Cl_2 at 0 °C induces dehydrobromination and leads to 2-bromo-3,3-difluoroacryloyl chloride in 63% yield (eq 10).[13] Under these reaction conditions, the ketene product of dehydrochlorination is not observed.

Dehydrohalogenation of acid chlorides with a tertiary amine is the typical method for the preparation of ketenes,[14] most likely involving an acylammonium intermediate.[15] Other methods are available which involve carboxyl group-activating agents and tertiary amine bases. TEA-generated ketenes find wide application in organic synthesis, in particular in the synthesis of β-lactams[16] using stereo- and enantioselective ketene–imine cycloaddition methodology. For example, the diastereoselective Staudinger [2 + 2] ketene–imine cycloaddition reaction[17] between chiral nonracemic oxazolidinone-*N*-acetyl chlorides and imines[18] generates β-lactams in high yields and with good stereocontrol (eq 11).[19]

The dehydration of primary nitroalkanes and the dehydrochlorination of chloroximes allows for the preparation of highly reactive nitrile oxides (eq 12). The nitrile oxides, which are not isolated, undergo facile [3 + 2] cycloaddition reactions[20] with a variety of trapping agents. Subjecting nitroalkanes to 2 equiv of ***Phenyl Isocyanate*** and a catalytic amount of triethylamine[21] effects dehydration to form nitrile oxides (eq 13).[22] Alternatively, treatment of oximes with aqueous ***Sodium Hypochlorite*** and triethylamine in CH_2Cl_2 efficiently produces the nitrile oxides (eq 14).[23]

Substitutions. Triethylamine finds use as a proton scavenger in palladium-catalyzed coupling reactions involving aryl, alkenyl, allyl, and alkyl derivatives.[24] The palladium-catalyzed coupling of vinyl halides with alkenes (Heck reaction) is an invaluable method for forming carbon–carbon bonds.[25] The elements of an intramolecular Heck reaction and the coupling of allylic alcohols with vinyl halides to afford aldehydes or ketones have been combined to generate moderate to excellent yields of cyclized products (eq 15).[26]

In an asymmetric variation of the Heck reaction, triethylamine is used in combination with ***(R)-2,2′-Bis(diphenylphosphino)-1,1′-binaphthyl*** ((*R*)-BINAP) for preparing enantiomerically enriched 2-aryl-2,3-dihydrofurans from aryl triflates.[27] Though TEA resulted in the highest degree of diastereoselectivity, the base ***1,8-Bis(dimethylamino)naphthalene*** (proton sponge) was found to be superior in regards to enantiomeric purity (eq 16).[28,29]

Base	% ee (yield)	% ee (yield)
DIPEA	82 (92)	60 (8)
Et_3N	75 (98)	9 (2)
proton sponge	96 (71)	17 (29)

Triethylamine also finds use in the enolboronation of various carbonyl compounds using dialkylboron halides.[30] These boron enolates are valuable in the stereocontrolled aldol reaction.

Proper choice of boron reagent, reaction solvent, reaction temperature, and tertiary amine base influences the enolate geometry of ketones[31] and esters.[32] The use of the sterically demanding dialkylboron halides, such as dicyclohexylboron chloride (Chx_2BCl), with triethylamine favors formation of the (*E*)-enol borinates, while dialkylboron triflates or dialkylboron halides, such as *B*-chloro-9-BBN, with diisopropylethylamine in ether at −78 °C favor the (*Z*)-enol borinates (eq 17).[31a,b] For similar complementary methodologies for generating both the (*E*)- and (*Z*)-boron enolates of esters, see ***Diisopropylethylamine***.

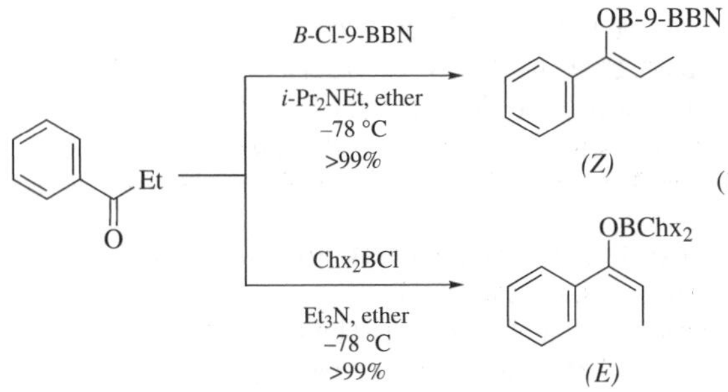

(17)

The asymmetric aldol addition involving *N*-acyloxazolidinones and aldehydes can be carried out with high stereoselectivity utilizing ***Tin(II) Trifluoromethanesulfonate*** and triethylamine. Treatment of *N*-(isothiocyanoacetyl)-2-oxazolidinones and aldehydes with tin(II) triflate and TEA at −78 °C in THF leads to high yields of aldol products with high diastereoselectivity (eq 18).[33]

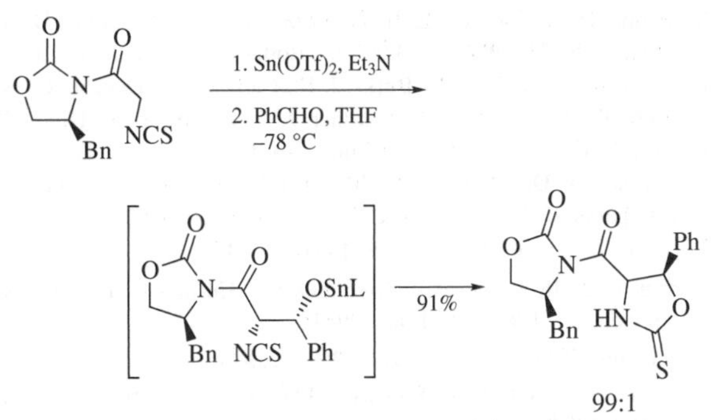

(18)

The regioselective synthesis of silyl enol ethers involves the trapping of the enolate anion of ketones and esters under kinetic or thermodynamic conditions.[34] Treatment of the unsymmetrical ketone 2-methylcyclohexanone with TEA and ***Chlorotrimethylsilane*** in DMF affords a 22:78 mixture of the kinetic and thermodynamic trimethylsilyl enol ethers. Using ***Lithium Diisopropylamide*** in DME under kinetic control leads to a >99:1 mixture of the kinetic and thermodynamic silyl enol ethers in 74% yield (eq 19).[35] Likewise, treatment of acetone with chlorotrimethylsilane, TEA, and anhydrous ***Sodium Iodide*** in acetonitrile leads to acetone trimethylsilyl enol ether in good yield.[36]

The mixture of TEA and ***Zinc Chloride*** is effective for converting α,β-unsaturated ketones and aldehydes into the corresponding silyl enol ethers. The Danishefsky–Kitahara diene is prepared in 68% yield by treatment of 4-methoxy-3-buten-2-one with an excess of TEA and TMSCl in the presence of a catalytic amount of anhydrous $ZnCl_2$ in benzene (eq 20).[37] Under similar reaction conditions, crotonaldehyde and 3-methylcrotonaldehyde are also converted into their corresponding enol ethers. 3-Methylcrotonaldehyde leads to an 80/20 mixture of (*E*/*Z*) dienes, while crotonaldehyde yields the (*E*)-silyl enol ether exclusively (eq 21).[38]

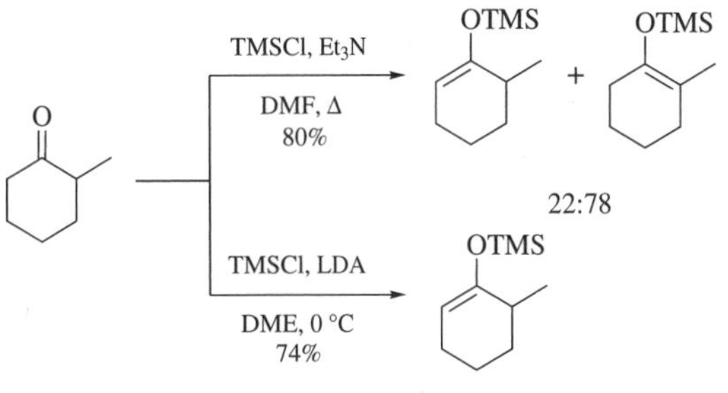

(19)

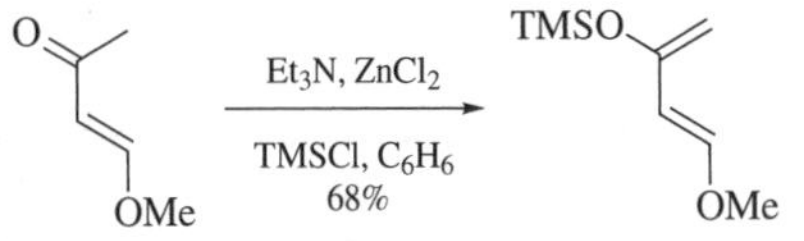

(20)

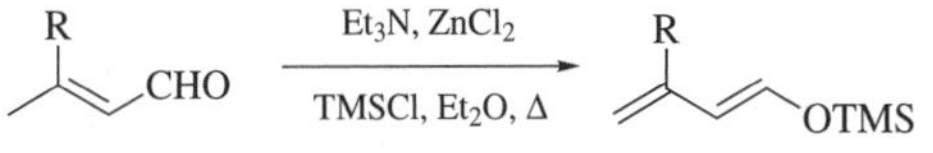

(21)

Triethylamine in combination with ***Triethylsilyl Perchlorate*** is somewhat selective in the generation of the (*Z*)-silylketene acetal of isopropyl propionate (eq 22)[39] (see ***2,2,6,6-Tetramethylpiperidine*** for a comparison of other bases). Triethylamine and ***t-Butyldimethylsilyl Trifluoromethanesulfonate*** (TBDMSOTf) are quite effective in preparing silyl enol ethers from sterically hindered ketones and lactones (eq 23).[40]

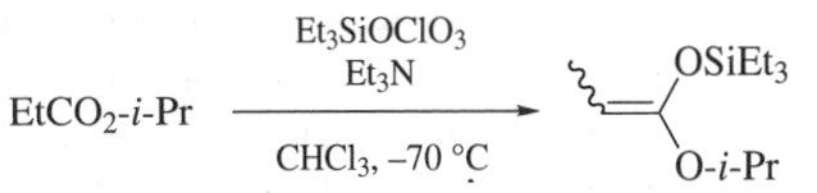

(22)

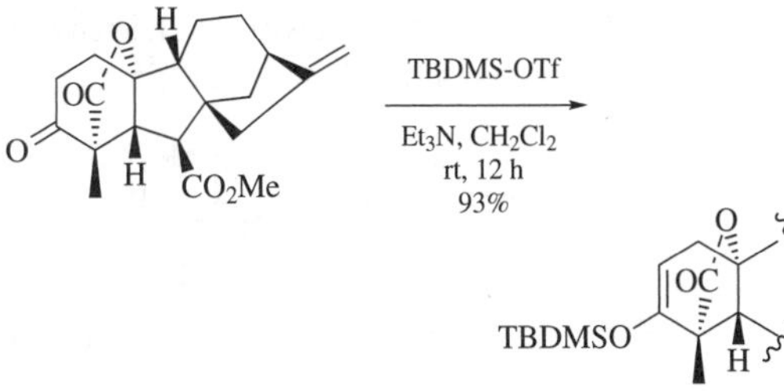

(23)

Triethylamine is the base of choice in neutralizing the acids liberated in preparing (1) diazo ketones from acid chlorides and ***Diazomethane***, (2) mixed anhydrides from carboxylic acids and alkyl haloformates, and (3) esters from carboxylic acids and alkyl halides, such as phenacyl bromide.[41]

Triethylamine is particularly useful as a proton scavenger in the field of protective group chemistry.[42] For just a few examples: alcohols have been protected as substituted methyl ethers with such alkylating agents as ***Chloromethyl Methyl Sulfide***[43] and ***t-Butyl Chloromethyl Ether***[44] using TEA as a base. Primary alcohols can be selectively silylated in the presence of secondary and tertiary alcohols using ***t-Butyldiphenylchlorosilane*** with TEA and a catalytic amount of ***4-Dimethylaminopyridine*** (eq 24).[45] The selective benzoylation of diols can be achieved using 1-(benzoyloxy)benzotriazole (BOBT) and TEA in CH_2Cl_2 at room temperature (eq 25).[46] Diols, in particular 1,2- and 1,3-diols, react with ***Di-t-butyldichlorosilane*** in the presence of TEA and a variety of silyl transfer agents to afford the di-*t*-butylsilylene derivatives in good yields (eq 26).[47]

HO– (Ur) ... OH OH; Ur = uridine —[TBDPSCl; Et_3N, cat. DMAP; DMF, rt; 68%]→ TBDPSO– (Ur) ... OH OH (24)

Ph ... OH, OH, OMe —[BOBT, Et_3N; CH_2Cl_2, rt; 90%]→ Ph ... OBz, OH, OMe (25)

O, H ... OH OH H OAc —[*t*-Bu_2SiCl_2; HOBT, Et_3N; AcCN; 84%]→ O, H ... H, OAc, O–Si(*t*-Bu)(*t*-Bu)–O (26)

Carbamates are cleaved easily to alcohols upon exposure to ***Trichlorosilane*** and TEA[48] and carboxylic acids are conveniently esterified using equimolar amounts of alkyl chloroformates with triethylamine and a catalytic amount of ***4-Dimethylaminopyridine***.[49] TEA is an effective base for the alkylation (protection) of carboxylic acids with ***Chloroacetonitrile*** to produce cyanomethyl esters.[50] Primary amines with pK_a values of 10–11 react with 1,1,4,4-tetramethyl-1,4-dichlorodisilethylene in the presence of TEA at rt to afford the disilylazacyclopentane derivatives in high yields (eq 27).[51] Amines with lower pK_a values require ***n-Butyllithium*** as the base or can be protected via a ***Zinc Iodide***-catalyzed trans-silylation with 1,1,4,4-tetramethyl-1,4-bis(*N,N*-dimethylamino)disilethylene (eq 28).[52]

H_2N, H ... N, CO_2Me —[$Me_2Si(Cl)CH_2CH_2Si(Cl)Me_2$; Et_3N, CH_2Cl_2; 75%]→ $SiMe_2$, Si–N, Me_2, H ... N, CO_2Me (27)

Br–C_6H_4–NH_2 —[$Me_2Si(NMe_2)CH_2CH_2Si(NMe_2)Me_2$; ZnI_2, 140 °C; 93%]→ Br–C_6H_4–N($SiMe_2$)($SiMe_2$) (28)

Triethylamine and other tertiary amines find utility in the derivatization of amino acids as well as coupling reactions to prepare peptides.[53] The basicity and steric nature of the tertiary amine utilized during the coupling reaction influences the degree of racemization.[54]

Related Reagents. Collidine; 1,5-Diazabicyclo[4.3.0]-nonene-5; 1,8-Diazabicyclo[5.4.0]-undecene; Diisopropylethylamine; Palladium–Triethylamine–Formic Acid.

---

1. Perrin, D. D.; Armarego, W. L. F. *Purification of Laboratory Chemicals*, 3rd ed.; Pergamon: Oxford, 1988; p 296.
2. (a) Schreiber, S. L.; Claus, R. E.; Reagan, J. *TL* **1982**, *23*, 3867. (b) Claus, R. E.; Schreiber, S. L. *OSC* **1990**, *7*, 168. (c) See also: Bailey, P. S. *Ozonation in Organic Chemistry*; Academic: New York, 1978; Vol. 1.
3. (a) Mancuso, A. J.; Huang, S.-L.; Swern, D. *JOC* **1978**, *43*, 2480. (b) Omura, K.; Swern, D. *T* **1978**, *34*, 1651. (c) Mancuso, A. J.; Swern, D. *S* **1981**, 165. (d) For other examples of 'Swern-like' oxidations, see: Hudlicky, M. *Oxidations in Organic Chemistry*; American Chemical Society: Washington, 1990.
4. Corey, E. J.; Kim, C. U. *JACS* **1972**, *94*, 7586.
5. Huang, S. L.; Omura, K.; Swern, D. *S* **1978**, 297.
6. Albright, J. D. *JOC* **1974**, *39*, 1977.
7. Amon, C. M.; Banwell, M. G.; Gravatt, G. L. *JOC* **1987**, *52*, 4851.
8. For examples, see: Fieser, L. F.; Fieser, M. *FF* **1967**, *1*, 1201.
9. See (a) Kraus, G. A.; Hon, Y.-S. *JOC* **1986**, *51*, 116. (b) Kraus, G. A.; Hon, Y.-S.; Sy, J.; Raggon, J. *JOC* **1988**, *53*, 1397, and references cited therein.
10. Block, E.; Aslam, M. *JACS* **1983**, *105*, 6164 and 6165.
11. Hormi, O. *OSC* **1993**, *8*, 247.
12. Fuller, C. E.; Walker, D. G. *JOC* **1991**, *56*, 4066.
13. Brahms, J. C.; Dailey, W. P. *JOC* **1991**, *56*, 900.
14. Koppel, G. A. In *Small Ring Heterocycles*; Hassner, A., Ed.; Wiley: New York, 1983.
15. Wasserman, H. H.; Piper, J. U.; Dehmlow, E. V. *JOC* **1973**, *38*, 1451 and references cited therein.
16. Georg, G. I.; Ravikumar, V. T. In *The Organic Chemistry of β-Lactams*; Georg, G. I. Ed.; VCH: New York, 1992; pp 295–368.
17. (a) Thomas, R. C. *TL* **1989**, *30*, 5239. (b) Cooper, R. D. G.; Daugherty, B. W.; Boyd, D. B. *PAC* **1987**, *59*, 485.
18. (a) Evans, D. A.; Sjogren, E. B. *TL* **1985**, *26*, 3783. (b) Evans, D. A.; Sjogren, E. B. *TL* **1985**, *26*, 3787. (c) Bodurow, C. C.; Boyer, B. D.; Brennan, J.; Bunnell, C. A.; Burks, J. E.; Carr, M. A.; Doecke, C. W.; Eckrich, T. M.; Fisher, J. W.; Gardner, J. P.; Graves, B. J.; Hines, P.; Hoying, R. C.; Jackson, B. G.; Kinnick, M. D.; Kochert, C. D.; Lewis, J. S.; Luke, W. D.; Moore, L. L.; Morin, J. M., Jr.; Nist, R. L.; Prather, D. E.; Sparks, D. L.; Vladuchick, W. C. *TL* **1989**, *30*, 2321.
19. Boger, D. L.; Myers, J. B., Jr. *JOC* **1991**, *56*, 5385.
20. Curran, D. P. In *Advances in Cycloaddition*; Curran, D. P., Ed.; JAI: Greenwich, CT, 1988; Vol. 1, pp 129–189.
21. Mukaiyama, T.; Hoshino, T. *JACS* **1960**, *82*, 5339.
22. (a) Kozikowski, A. P.; Stein, P. D. *JACS* **1982**, *104*, 4023. (b) Kozikowski, A. P. *ACR* **1984**, *17*, 410.
23. Lee, G. A. *S* **1982**, 508.
24. (a) Daves, G. D., Jr.; Hallberg, A. *CRV* **1989**, *89*, 1433. (b) Heck, R. F. *Palladium Reagents in Organic Syntheses*; Academic: New York, 1985.
25. Heck, R. F. *OR* **1982**, *27*, 345.
26. (a) Gaudin, J.-M. *TL* **1991**, *32*, 6113, and references cited therein. (b) See also: Shi, L.; Narula, C. K.; Mak, K. T.; Kao, L.; Xu, Y.; Heck, R. F. *JOC* **1983**, *48*, 3894.
27. Ozawa, F.; Kubo, A.; Hayashi, T. *JACS* **1991**, *113*, 1417.
28. Ozawa, F.; Kubo, A.; Hayashi, T. *TL* **1992**, *33*, 1485.
29. For an intramolecular Heck-type reaction, see: (a) Sato, Y.; Sodeoka, M.; Shibasaki, M. *JOC* **1989**, *54*, 4738. (b) Mori, M.; Kaneta, N.; Shibasaki, M. *JOC* **1991**, *56*, 3486.
30. Brown, H. C.; Ganesan, K.; Dhar, R. K. *JOC* **1992**, *57*, 3767.
31. (a) Brown, H. C.; Dhar, R. K.; Bakshi, R. K.; Pandiarajan, P. K.; Singaram, B. *JACS* **1989**, *111*, 3441. (b) Brown, H. C.; Dhar, R. K.; Ganesan, K.; Singaram, B. *JOC* **1992**, *57*, 499 and 2716. (c) Enders, D.; Lohray, B. B. *AG(E)* **1988**, *27*, 581. (d) Evans, D. A.; Nelson, J. V.; Vogel, E.; Taber, T. R. *JACS* **1981**, *103*, 3099. (e) Van Horn, D. E.; Masamune, S. *TL* **1979**, 2229. (f) Evans, D. A.; Vogel, E.; Nelson, J. V. *JACS* **1979**, *101*, 6120. (g) Paterson, I.; Osborne, S. *TL* **1990**, *31*, 2213. (h) For sulfenylation and selenenylation of enol borinates, see: Paterson, I.; Osborne, S. *SL* **1991**, 145.

32. (a) Corey, E. J.; Lee, D.-H. *JACS* **1991**, *113*, 4026. (b) Corey, E. J.; Kim, S. S. *JACS* **1990**, *112*, 4976. (c) Hirama, M.; Masamune, S. *TL* **1979**, 2225. (d) Gennari, C.; Bernardi, A.; Cardani, S.; Scolastico, C. *T* **1984**, *40*, 4059. (e) Otsuka, M.; Yoshida, M.; Kobayashi, S.; Ohno, M. *TL* **1981**, *22*, 2109.
33. (a) Lago, M. A.; Samanen, J.; Elliott, J. D. *JOC* **1992**, *57*, 3493. (b) Evans, D. A.; Weber, A. E. *JACS* **1986**, *108*, 6757. (c) See also: Evans, D. A.; Sjogren, E. B.; Weber, A. E.; Conn, R. E. *TL* **1987**, *28*, 39 and Iseki, K.; Oishi, S.; Taguchi, T.; Kobayashi, Y. *TL* **1993**, *34*, 8147.
34. (a) Colvin, E. W. *Silicon in Organic Synthesis*; Butterworths: London, 1981; Chapter 17. (b) Rasmussen, J. K. *S* **1977**, 91. (c) Fleming, I. *C* **1980**, *34*, 265. (d) Brownbridge, P. *S* **1983**, 1 and 85. (e) Taylor, R. J. K. *S* **1985**, 364.
35. (a) House, H. O.; Czuba, L. J.; Gall, M.; Olmstead, H. D. *JOC* **1969**, *34*, 2324. (b) Fleming, I.; Paterson, I. *S* **1979**, 736. (c) Reetz, M. T.; Chatzhosifidis, I.; Hubner, F.; Heimbach, H. *OSC* **1990**, *7*, 424. (d) Jung, M. E.; McCombs, C. A. *OS* **1978**, *58*, 163; *OSC* **1988**, *6*, 445.
36. Walshe, N. D. A.; Goodwin, G. B. T.; Smith, G. C.; Woodward, F. E. *OS* **1987**, *65*, 1; *OSC* **1993**, *8*, 1.
37. (a) Danishefsky, S.; Kitahara, T. *JACS* **1974**, *96*, 7807. (b) Danishefsky, S.; Kitahara, T.; Schuda, P. F. *OSC* **1990**, *7*, 312.
38. Gaonac'h, O.; Maddaluno, J.; Chauvin, J.; Duhamel, L. *JOC* **1991**, *56*, 4045.
39. Wilcox, C. S.; Babston, R. E. *TL* **1984**, *25*, 699.
40. Mander, L. N.; Sethi, S. P. *TL* **1984**, *25*, 5953.
41. For examples, see: Fieser, L. F.; Fieser, M. *FF* **1967**, *1*, 1198.
42. For numerous references, see: Greene, T. W.; Wuts, P. G. M. *Protective Groups In Organic Synthesis,* 2nd ed.; Wiley: New York, 1991.
43. Suzuki, K.; Inanaga, J.; Yamaguchi, M. *CL* **1979**, 1277.
44. Pinnick, H. W.; Lajis, N. H. *JOC* **1978**, *43*, 3964.
45. (a) Chaudhary, S. K.; Hernandez, O. *TL* **1979**, 99. (b) See also: Guindon, Y.; Yoakim, C.; Bernstein, M. A.; Morton, H. E. *TL* **1985**, *26*, 1185. (c) Hanessian, S.; Lavallee, P. *CJC* **1975**, *53*, 2975.
46. (a) Soll, R. M.; Seitz, S. P. *TL* **1987**, *28*, 5457. (b) See also: Kim, S.; Chang, H.; Kim, W. J. *JOC* **1985**, *50*, 1751.
47. Trost, B. M.; Caldwell, C. G.; Murayama, E.; Heissler, D. *JOC* **1983**, *48*, 3252.
48. Pirkle, W. H.; Hauske, J. R. *JOC* **1977**, *42*, 2781.
49. Kim, S.; Kim, Y. C.; Lee, J. I. *TL* **1983**, *24*, 3365.
50. Hugel, H. M.; Bhaskar, K. V.; Longmore, R. W. *SC* **1992**, *22*, 693.
51. Djuric, S.; Venit, J.; Magnus, P. *TL* **1981**, *22*, 1787.
52. Guggenheim, T. L. *TL* **1984**, *25*, 1253.
53. See: (a) Bodanszky, M.; Bodanszky, A. *The Practice of Peptide Synthesis,* 2nd ed.; Springer: Berlin, 1994. (b) Bodanszky, M.; Klausner, Y. S.; Ondetti, M. A. *Peptide Synthesis*, 2nd ed.; Wiley: New York, 1976.
54. (a) Bodanszky, M.; Bodanszky, A. *CC* **1967**, 591. (b) Williams, A. W.; Young, G. T. *JCS(P1)* **1972**, 1194. (c) Chen, F. M. F.; Lee, Y.; Steinauer, R.; Benoiton, N. L. *CJC* **1987**, *65*, 613. (d) Slebioda, M.; St-Amand, M. A.; Chen, F. M. F.; Benoiton, N. L. *CJC* **1988**, *66*, 2540.

Kirk L. Sorgi
The R. W. Johnson Pharmaceutical Research Institute, Spring House, PA, USA

Triethyl Phosphite[1]

$(EtO)_3P$

[122-52-1] $C_6H_{15}O_3P$ (MW 166.18)

(can react with electrophiles to form phosphonates or phosphates; can function as a reducing agent for a variety of functional groups; forms a stable complex with copper(I) iodide)

Physical Data: bp 155–157 °C/760 mmHg; *d* 0.958 g cm^{-3}.
Solubility: sol most organic solvents.
Form Supplied in: clear, free flowing liquid in >97% purity.
Handling, Storage, and Precautions: should be stored in a dry place, preferably in a fume hood due to its pungent odor. The reagent is purified[1b] by treating with sodium (to remove water and dialkyl phosphonate), followed by decanting and distillation. The purified material may then be stored over activated molecular sieves.

Transesterification. Hoffman reported that triethyl phosphite is easily transesterified when heated in the presence of aliphatic alcohols.[2] The uncatalyzed transformation occurs in three distinct steps which proceed at approximately equal rates. Transesterification using trimethanolethane in the presence of triethylamine affords the corresponding crystalline phosphite (eq 1).[3]

(EtO)3P, Et3N, Δ; 90% (1)

Organophosphonates. Triethyl phosphite[4] has been used extensively in the synthesis of phosphonates via the Arbuzov reaction.[5,6] This general protocol works well with a variety of electrophiles and is carried out by simply heating the reactants between 100 °C and 150 °C using phosphite as solvent. Phosphonates bearing electron withdrawing groups β to phosphorus represent a valuable class of reagents which have been creatively applied to the stereoselective synthesis of alkenes (eqs 2–6).[6g–k] Acyl phosphonates, prepared from the corresponding acid halides, have been utilized as intermediates in the overall reduction of acids to aldehydes (eq 7),[7] and as intermediates in the preparation of vinylphosphonates (eq 8).[8]

Cl; (EtO)3P, Δ; P(OEt)2; NaH, DME, benzaldehyde, 64% (2)

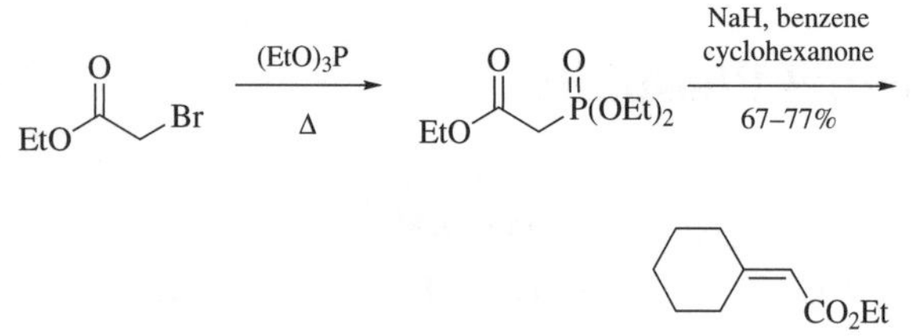

EtO(C=O)CH₂Br → $(EtO)_3P$, Δ → EtO(C=O)CH₂P(O)(OEt)₂ → NaH, benzene, cyclohexanone, 67–77% → cyclohexylidene–CH–CO_2Et (3)

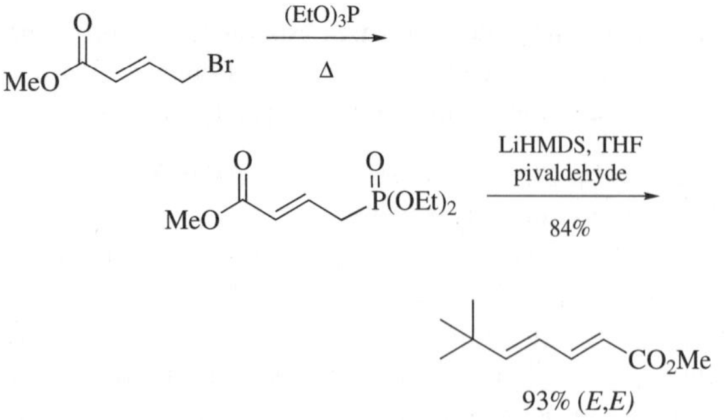

MeO(C=O)CH=CHCH₂Br → $(EtO)_3P$, Δ → MeO(C=O)CH=CHCH₂P(O)(OEt)₂ → LiHMDS, THF, pivaldehyde, 84% → CO_2Me, 93% (*E,E*) (4)

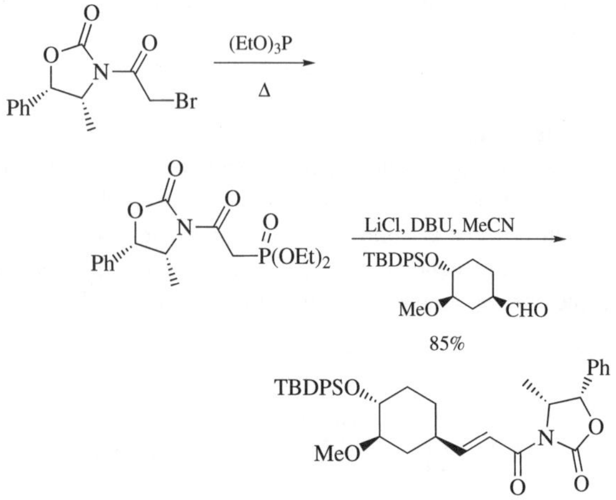

Ph, oxazolidinone N–C(O)CH₂Br → $(EtO)_3P$, Δ → N–C(O)CH₂P(O)(OEt)₂ → LiCl, DBU, MeCN, TBDPSO, MeO, CHO, 85% → TBDPSO, MeO, Ph, N, O (5)

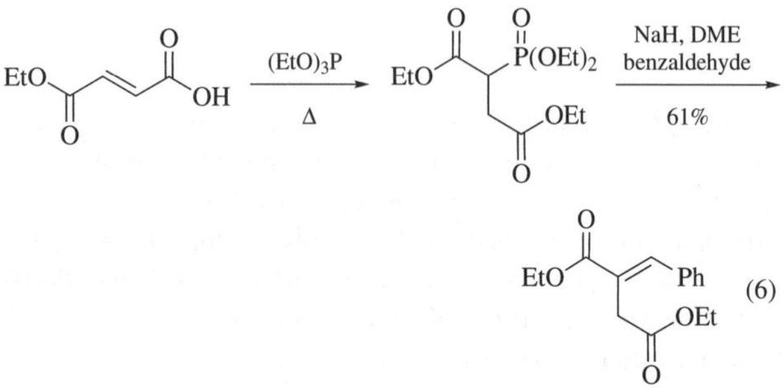

EtO(C=O)CH=CH(C=O)OH → $(EtO)_3P$, Δ → EtO(C=O)CH(P(O)(OEt)₂)CH₂(C=O)OEt → NaH, DME, benzaldehyde, 61% → EtO(C=O)C(=CHPh)CH₂(C=O)OEt (6)

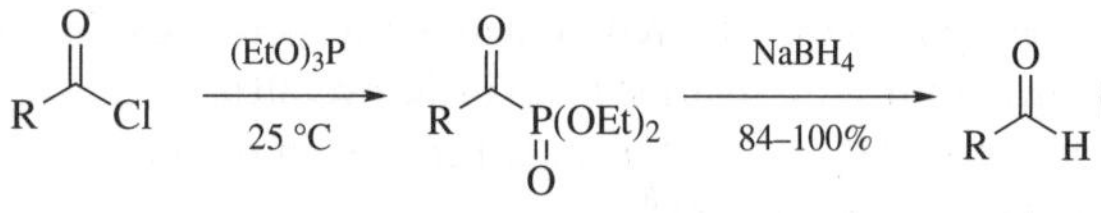

R(C=O)Cl → $(EtO)_3P$, 25 °C → R(C=O)P(O)(OEt)₂ → $NaBH_4$, 84–100% → R(C=O)H (7)

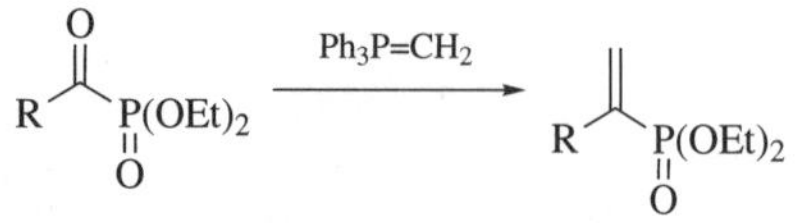

R(C=O)P(O)(OEt)₂ → $Ph_3P{=}CH_2$ → R(C=CH₂)P(O)(OEt)₂ (8)

A nonclassical Arbuzov approach to α-methoxybenzylphosphonates employs the treatment of acetals of aromatic aldehydes with triethyl phosphite in the presence of ***Boron Trifluoride Etherate*** (eq 9).[9] In contrast, direct arene phosphorylation has been achieved through treatment of aromatic substrates with triethyl phosphite and ***Cerium(IV) Ammonium Nitrate***.[10] In this case the active species is believed to be a phosphite radical cation (eq 10).

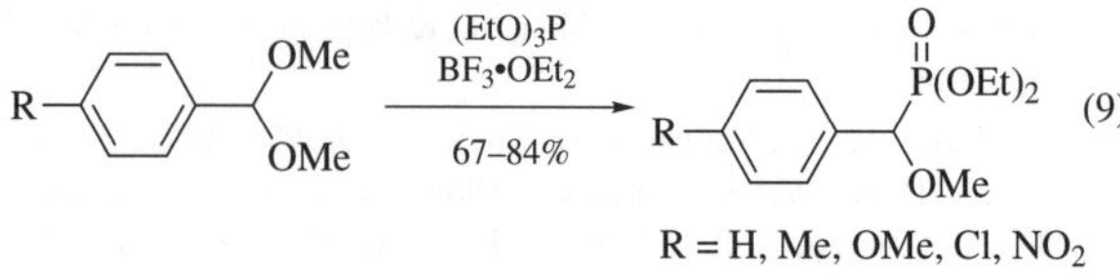

R–C₆H₄–CH(OMe)₂ → $(EtO)_3P$, $BF_3{\bullet}OEt_2$, 67–84% → R–C₆H₄–CH(OMe)P(O)(OEt)₂ (9)

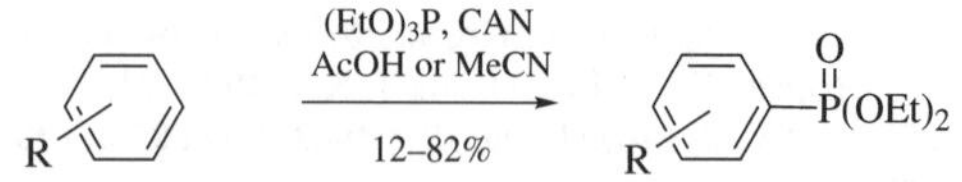

R = H, Me, OMe, Cl, NO_2

R–C₆H₅ → $(EtO)_3P$, CAN, AcOH or MeCN, 12–82% → R–C₆H₄–P(O)(OEt)₂ (10)

Vinyl Phosphates. Perkow[11,12] discovered that α-halo aldehydes and ketones are converted to vinyl phosphates upon treatment with triethyl phosphite (eq 11). These intermediates are easily reduced to alkenes under dissolving metal conditions (eq 12).[13] In cases where the vinyl phosphate is acyclic, treatment with a strong base generally provides good yields of the corresponding alkyne (eq 13).[14]

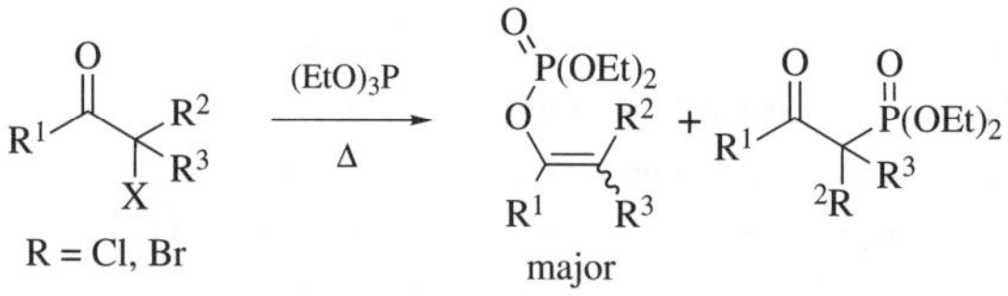

$R^1C(O)CXR^2R^3$ → $(EtO)_3P$, Δ → $(EtO)_2P(O)O–C(R^1){=}CR^2R^3$ (major) + $R^1C(O)C(R^2R^3)P(O)(OEt)_2$ (11)

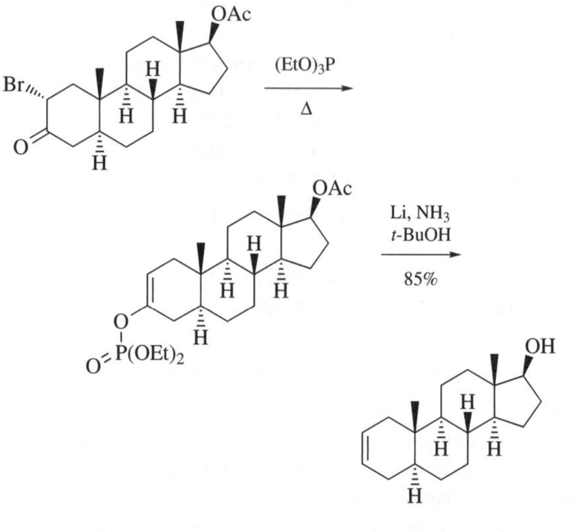

R = Cl, Br

Br, O, H, OAc (steroid) → $(EtO)_3P$, Δ → O=P(OEt)₂–O, H, OAc (enol phosphate) → Li, NH_3, *t*-BuOH, 85% → H, OH (12)

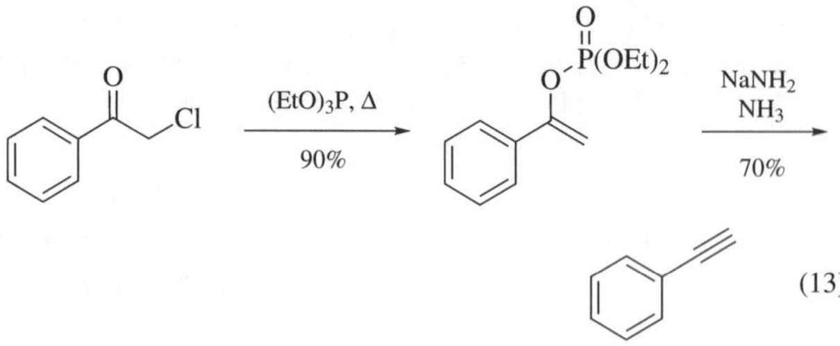

PhC(O)CH₂Cl → $(EtO)_3P$, Δ, 90% → PhC(=CH₂)OP(O)(OEt)₂ → $NaNH_2$, NH_3, 70% → PhC≡CH (13)

Deoxygenation. Triethyl phosphite is capable of deoxygenating a variety of organic substrates. Hydroperoxides, for example, are rapidly converted to alcohols after treatment with one equivalent of the reagent at ambient temperature (eq 14).[15] The reduction of peroxides generated in situ via enolate oxidation is a powerful technique for achieving regio and stereoselective α-hydroxylation (eq 15).[16] In addition to reducing hydroperoxides, triethyl phosphite has been utilized for the deoxygenation of *endo*-peroxides (eq 16),[17] diaroyl peroxides (eq 17)[18] and the reductive coupling of phthalic anhydride (eq 18).[19]

(14): $(EtO)_3P$, toluene, rt; 100%

(15): NaH, *t*-BuOH, DMF; $(EtO)_3P$, then O_2; 64%

(16): $(EtO)_3P$; 160–170 °C

(17): $(EtO)_3P$

(18): $(EtO)_3P$

Scott reported that terminal epoxides can be reduced using forcing conditions while internal epoxides are generally unaffected, even at temperatures approaching 200 °C.[20] In contrast, Saegusa and co-workers discovered that α-keto acids are easily reduced to α-hydroxy acids under mild conditions in good to excellent yields (eq 19).[21]

(19): $(EtO)_3P$, rt; 67–94%

The deoxygenation of nitrogenous functionalities using triethyl phosphite has also received attention.[22] Reductive cyclization of aromatic nitro-containing compounds has provided entry into a variety of heterocycles (eqs 20 and 21).[23] It has been suggested that a highly reactive nitrene intermediate is involved in the cyclization process.[24] Aryl nitroso compounds subjected to these conditions experience a similar fate, although the reactive intermediate in this case is believed to be polar in nature (eq 22).[25]

(20): $(EtO)_3P$, 160 °C; 83%

(21): $(EtO)_3P$, 160 °C; 85%

(22): $(EtO)_3P$, 160 °C; no yield given; R = Me, OMe

Azides are also subject to reduction by triethyl phosphite via a modified Staudinger process (eq 23).[26]

(23): $(EtO)_3P$, rt; HCl; 0–80%

R^1 = alkyl; $R^1 = R^2$ = H
$R^1 = R^2$ = alkyl; R^3 = H
$R^1 = R^2 = R^3$ = alkyl

Desulfurization. A frequently encountered application of triethyl phosphite is the conversion of cyclic thiocarbonates to the corresponding alkenes (Corey–Winter synthesis).[27] The reaction proceeds stereospecifically via a carbene intermediate and often results in good yields of desired alkene (eqs 24 and 25).[28]

(24): $(EtO)_3P$, Δ

(25): $(EtO)_3P$, Δ

The conversion of thiiranes to alkenes proceeds readily, in contrast to epoxide reduction.[29] For example, the episulfides of *cis*- and *trans*-2-butene are reduced smoothly and with high stereospecificity. Thiols[30] and disulfides[31] are reduced efficiently as well. In some cases, reductive desulfurization can be coupled with carbon–carbon bond formation. For example, reaction of di-*n*-butyl disulfide with triethyl phosphite in the presence of carbon monoxide gives a nearly quantitative yield of the corresponding homologated thioester (eq 26).[32]

(26): $(EtO)_3P$, CO; 99%

In a completely unrelated example, [2,2]paracyclophane was synthesized from the corresponding bis-thioether via photochemical extrusion of sulfur mediated by triethyl phosphite (eq 27).[33] The concept of linking proximal carbon atoms with the extrusion of sulfur was used successfully by Eschenmoser in his synthesis of a vitamin B_{12} intermediate (eq 28),[34] while a modification of Barton's method allowed the synthesis of the highly strained diquadricyclanylidene (eq 29).[35]

(27) $(EtO)_3P$, *hν*, 85%

(28) $(EtO)_3P$, Δ, 50%

(29) $(EtO)_3P$, Δ, 20%

A bimolecular coupling mediated by triethyl phosphite was applied to the synthesis of an interesting unsymmetrical tetrathiafulvalene, although yields were low due to competing homo coupling (eq 30).[36]

(30) $(EtO)_3P$, Δ, 3%

Carbonyl Adducts. Triethyl phosphite reacts with α-diketones to form isolable cyclic phosphate esters. Upon additional heating, the corresponding alkynes are obtained (eq 31).[37]

(31) $(EtO)_3P$, Δ; R = H, Cl, Me, OMe; Δ, 20–60%

Use in Cuprate Chemistry. When added to solutions of *cis*-divinylcuprates, trimethyl phosphite facilitates the transfer of both vinyl groups via stabilization of the vinylcopper intermediate (eq 32).[38]

(32) $(EtO)_3P$, R^1X; $Cu{\bullet}P(OEt)_3$; R^1X

Copper(I) Iodide–Triethyl Phosphite is formed in 89% yield by the reaction of triethyl phosphite and ***Copper(I) Iodide*** in refluxing benzene.[39] The salt has been used to form arylcopper intermediates which participate in a modified Ullmann coupling sequence (eq 33).[40]

(33) BuLi, $CuI{\bullet}(EtO)_3P$; H_3O^+, 58%

Related Reagents. Trimethyl Phosphite.

1. (a) Preparation: Schuetz, R. D.; Jacobs, R. L. *JOC* **1961**, *26*, 3467. (b) Perrin, D. D.; Armarego, W. L. F. *Purification of Laboratory Chemicals*; Pergamon: New York, 1988; p 297.
2. Hoffmann, F. W.; Ess, R. J., Usinger, R. P., Jr. *JACS* **1956**, *78*, 5817.
3. Wadsworth, W. S., Jr.; Emmons, W. D. *JACS* **1962**, *84*, 610.
4. (a) Schuetz, R. D.; Jacobs, R. L. *JOC* **1961**, *26*, 3467 (b) Ford-Moore, A. H.; Perry, B. J. *OSC* **1963**, *4*, 955.
5. For reviews of the Arbuzov, reaction, see: (a) Arbuzov, B. A. *PAC* **1964**, *9*, 307. (b) Kosolapoff, G. M. *OR* **1951**, *6*, 273. (c) Redmore, D. *CR* **1971**, *71*, 315. (d) Bhattacharya, A. K.; Thyagarajan, G. *CRV* **1981**, *81*, 415. (e) Cadogan, J. I. G. *Organophosphorus Reagents in Organic Synthesis*; Academic: New York, 1979.
6. (a) Horner, L.; Hoffmann, H.; Wipple, H. G. *CB* **1958**, *91*, 61. (b) Horner, L.; Hoffmann, H.; Wipple, H. G.; Klahre, G. *CB* **1959**, *92*, 2499. (c) Wadsworth, W. S., Jr.; Emmons, W. D. *JACS* **1961**, *83*, 1733. For reviews, see: (d) Wadsworth, W. S. *OR* **1977**, *25*, 73. (e) Boutagy J.; Thomas, R. *CRV* **1974**, *74*, 87. (f) Carey, F. A.; Sundberg, R. J. *Advanced Organic Chemistry*, 3rd ed.; Plenum: New York, 1990; Part B, pp 100–102. (g) See Ref. 3c. (h) Wadsworth, W. S., Jr.; Emmons, W. D. *OS* **1965**, *45*, 44. (i) Rousch, W. R. *JACS* **1978**, *100*, 3599. (j) Nicolaou, K. C.; Bertinato, P.; Piscopio, A. D.; Chakraborty, T. K.; Minowa, N. *CC* **1993**, 619. (k) See Ref. 5c.
7. Horner, L.; Roder, H. *CB* **1970**, *103*, 2984.
8. Kojima, M.; Yamashita, M.; Yoshida, H.; Ogata, T. *S* **1979**, 147.
9. Burkhouse, D.; Zimmer, H. *S* **1984**, 330.

10. Kottmann, H; Skarzewski, J.; Effenberger, F. *S* **1987**, 797.
11. Perkow, N.; Ullerich, K.; Meyer, F. *N* **1952**, *39*, 353.
12. For reviews of the Perkow reaction, see: (a) Lichtenthaler, F. W. *CR* **1961**, *61*, 608. (b) See Ref 5a.
13. (a) Fetizon, M.; Jurion, M.; Anh, N. T. *CC* **1969**, 112. (b) Ireland, R. E.; Pfister, G. *TL* **1969**, 2145. For reduction of enol phosphates to the corresponding saturated hydrocarbons, see: (c) Coates, R. M.; Shah, S. K.; Mason, R. W. *JACS* **1982**, *104*, 2198 and (d) Heathcock, C. H.; Davidsen, S. K.; Mills, S.; Sanner, M. A. *JACS* **1986**, *108*, 5650.
14. (a) Craig, J. C.; Moyle, M. *JCS* **1963**, 3712. (b) Negishi, E.; King, A. O.; Klima, W. L. *JOC* **1980**, *45*, 2526. (c) McMurry, J. E.; Bosch, G. K. *JOC* **1987**, *52*, 4885.
15. Karasch, M. S.; Mosher, R. A.; Bengelsdorf, I. S. *JOC* **1960**, *25*, 1000.
16. (a) Kido, F.; Kitahara, H.; Yoshikoshi, A. *JOC* **1986**, *51*, 1478. (b) Hartwig, W.; Born, L. *JOC* **1987**, *52*, 4352. (c) Gardner, J. N.; Carlon, F. E.; Gnoj, O. *JOC* **1968**, *33*, 3294. (d) Gardner, J. N.; Popper, T. L.; Carlon, F. E.; Gnoj, O.; Herzog, H. L. *JOC* **1968**, *33*, 3695.
17. (a) Kametani, T.; Ogasawara, K. *CI(L)* **1968**, 1772. (b) Horner, L.; Jurgeleit, W. *LA* **1955**, *591*, 138.
18. Burn, A. J.; Cadogan, J. I. G.; Bunyan, P. J. *JCS* **1963**, 1527.
19. Ramirez, F.; Yamanaka, H.; Basedow, O. H. *JACS* **1961**, *83*, 173.
20. (a) Scott, C. B. *JOC* **1957**, *22*, 1118. (b) Neureiter, N. P.; Bordwell, F. G. *JACS* **1959**, *81*, 578.
21. Saegusa, T.; Kobayashi, S.; Kimura, Y.; Yokoyama, T. *JOC* **1977**, *42*, 2797.
22. Cadogan, J. I. G. *S* **1969**, 11.
23. (a) Cadogan, J. I. G.; Cameron-Wood, M.; Mackie, R. K.; Searle, R. J. G. *JCS* **1965**, 4831. (b) Cadogan, J. I. G.; Searle, R. J. G. *CI(L)* **1963**, 1282. (c) Sundberg, R. J. *JOC* **1965**, *30*, 3604. (d) Cadogan, J. I. G.; Mackie, R. K.; Todd, M. J. *CC* **1966**, 491. (e) Grundmann, C. *CB* **1964**, *97*, 575.
24. Brooke, P. K.; Herbert, R. B.; Holliman, E. G. *TL* **1973**, 761.
25. Smolinsky, G.; Feuer, B. I. *JOC* **1966**, *31*, 3882.
26. (a) Koziara, A.; Osowska-Pacewicka, K.; Zawadzki, S.; Zwierzak, A. *S* **1985**, 202. (b) Koziara, A.; Zwierzak, A. *TL* **1987**, *28*, 6513.
27. (a) Corey, E. J.; Winter, R. A. E. *JACS* **1963**, *85*, 2677. (b) Corey, E. J.; Carey, F. A.; Winter, R. A. E. *JACS* **1965**, *87*, 934.
28. For a review, see: Block, E. *OR* **1983**, *30*, 457.
29. (a) Davis, R. E. *JOC* **1958**, *23*, 1767. (b) Schuetz, R. D.; Jacobs, R. L. *JOC* **1958**, *23*, 1799.
30. (a) Hoffman, F. W.; Ess, R. J.; Simmons, T. C.; Hanzel, R. S. *JACS* **1956**, *78*, 6414. (b) Walling, C.; Rabinowitz, R. *JACS* **1959**, *81*, 1243.
31. Jacobson, H. I.; Harvey, R. G.; Jensen, E. V. *JACS* **1955**, *77*, 6064.
32. Walling, C.; Basedow, O. H.; Savas, E. S. *JACS* **1960**, *82*, 2181.
33. Brink, M. *S* **1975**, 807.
34. Eschenmoser, A. *QR* **1970**, *24*, 366.
35. Sauter, H.; Horster, H. G.; Prinzbach, H. *AG(E)* **1973**, *12*, 991.
36. Spencer, H. K.; Cava, M. P.; Garito, A. F. *CC* **1976**, 966.
37. (a) Mukaiyama, T.; Nambu, H.; Kumamoto, T. *JOC* **1964**, 2243. (b) Ramirez, F.; Desai, N. B. *JACS* **1963**, *85*, 3252.
38. Alexakis, A.; Cahiez, G.; Normant, J. F. *S* **1979**, 826.
39. Nishizawa, Y. *BCJ* **1961**, *34*, 1170.
40. (a) Ziegler, F. E.; Fowler, K. W.; Kanfer, S. *JACS* **1976**, *98*, 8282. (b) Ziegler, F. E.; Fowler, K. W.; Sinha, N. D. *TL* **1978**, 2767. (c) Ziegler, F. E.; Fowler, K. W.; Rodgers, W. B.; Wester, R. T. *OS* **1987**, *65*, 108.

Anthony D. Piscopio
Pfizer, Groton, CT, USA

Trifluoroacetic Acid

CF_3CO_2H

[76-05-1] $C_2HF_3O_2$ (MW 114.03)

(solvent; acid catalyst for diverse organic transformations, including solvolysis,[9] rearrangements,[3] reductions,[17] oxidations,[22] and trifluoromethylation[8,16])

Alternate Name: TFA.
Physical Data: [1] freely flowing clear liquid, mp −15.4 °C, bp 72 °C; pK_a 0.23 (25 °C, H_2O); *d* 1.480 g cm^{-3}. Trifluoroacetic acid and water form an azeotrope, bp 105.5 °C, 20.8% water.
Solubility: miscible with water and most organic solvents, but has limited solubility in alkanes (with more than six carbons) and carbon disulfide.
Form Supplied in: neat liquid; commercially available.
Purification: distilled from traces of $(CF_3CO)_2O$ or P_2O_5; $KMnO_4$ has caused *serious explosions*.[33]
Handling, Storage, and Precautions: [2] is a strong organic acid. It is extremely corrosive and especially destructive to tissue of mucous membranes. Inhalation is a major hazard. Use of a NIOSH/MSHA approved respirator is recommended. Use in a fume hood with gloves and protective clothing to avoid contact. TFA is hygroscopic. Listed incompatibilities include oxidizing and reducing agents; however, literature references report the use of both types of reagents with TFA. TFA should never be mixed with basic solvents or acid-sensitive materials. The acute toxicity of TFA is low.

Acid-Catalyzed Rearrangements. There are many examples of rearrangements catalyzed by TFA, including acid-catalyzed epoxide ring opening (eq 1),[3] biomimetic cyclizations (eq 2),[4] Cope rearrangements (eq 3),[5] and natural product synthesis (eq 4).[6] The cited references are given as examples; a comprehensive listing is beyond the scope of this publication. Often these reactions are initiated through protonation and dehydration to provide a cationic intermediate for cyclization. TFA is a general catalyst for most acid-catalyzed rearrangements. The physical properties of TFA may provide benefits over alternative acids. The volatility of this catalyst will allow product isolation by simple solvent evaporation. Less volatile alternatives, such as ***Sulfuric Acid*** or ***p-Toluenesulfonic Acid***, may require neutralization or an extractive workup. Owing to the low nucleophilicity of trifluoroacetate anion, TFA has been used as a solvent for basic research into solvolysis mechanisms.[7]

O, O, OMe — TFA, $CHCl_3$, 70% → O, O (1)

Synthesis of Trifluoromethyl Organic and Organometallic Compounds. The reaction of TFA with Grignard reagents is general for the formation of trifluoromethyl ketones (eq 5).[8] The reaction is useful only with readily available Grignard reagents, as 1 equiv of organometallic reagent is consumed deprotonating TFA. The best yields are obtained using 2.5–3.0 mol of Grignard reagent per equiv of TFA. Phenyl, alkynyl, and normal alkyl Grignard reagents give superior results. Strongly reducing Grignard reagents, such as ***Isopropylmagnesium Bromide***, tend to produce trifluoromethyl alcohols from reduction of the initially formed ketone.

0.5% TFA, CH_2Cl_2, –35 °C, 1 h (2)

TFA, CH_2Cl_2, 25 °C, 90 min (3)

TFA, ether, 0 °C (4)

CF_3CO_2H (0.33 equiv), 58% (5)

The condensation of ***Mercury(II) Oxide*** with 2 mol of TFA produces ***Mercury(II) Trifluoroacetate***. Thermal decarboxylation in the presence of carbonate followed by sublimation yields bis(trifluoromethyl)mercury.[9] Treatment of bis(trifluoromethyl)mercury with Cu^0 in NMP or DMA produces a stable trifluoromethylcopper reagent.[10] Addition of an aromatic or benzylic halide leads to displacement and incorporation of the trifluoromethyl group (eq 6). Alternatives for the introduction of a trifluoromethyl group as a nucleophile are limited. The trifluoromethyl derivatives of lithium and magnesium are unknown, probably due to fluoride elimination to produce difluorocarbene.[11] Bis(trifluoromethyl)mercury has been used to synthesize trifluoromethyl organometallic derivatives of germanium,[12] tin,[13] zinc,[14] and cadmium.[15]

$$Hg(CF_3)_2 + Cu^0 \xrightarrow[140\ °C]{\text{NMP or DMA}} CF_3Cu + Hg^0$$

$$CF_3Cu + \text{4-}IC_6H_4NO_2 \xrightarrow[88\%]{150\ °C,\ 3\ h} \text{4-}CF_3C_6H_4NO_2 \quad (6)$$

Direct trifluoromethylation of electron-poor aromatic and heterocyclic systems can be accomplished by treatment with TFA and ***Xenon(II) Fluoride***.[16] The initial reaction is thought to produce $Xe(O_2CCF_3)_2$ which undergoes decomposition to yield carbon dioxide, xenon, and trifluoromethyl radicals. Aromatic and heterocyclic compounds can react with the trifluoromethyl radicals, often with a high degree of selectivity. Yields of trifluoromethylated products vary between good to moderate, but the reaction allows entry into structures that would otherwise be difficult to synthesize. The chemistry has been extended to heterocyclic systems, as demonstrated with an intermediate reaction in the production of 5-(trifluoromethyl)-2′-deoxyuridine, a known antiviral compound (eq 7).[16]

CF_3CO_2H, XeF_2, 25 °C, 33% (7)

Reductions with Boron and Silicon Hydrides. ***Sodium Borohydride***,[17] ***Sodium Cyanoborohydride***,[18] ***Borane-Tetrahydrofuran***,[19] and ***Triethylsilane***[20] have been used in conjunction with TFA to reduce a variety of functional groups. These reactions generally proceed by protonation of a functional group followed by delivery of hydride.

Of special note is the reduction of cobalt-complexed secondary α-alkynic alcohols with sodium borohydride and TFA.[21] Oxidative decomposition of the resulting cobalt complex produces secondary alkynes in good to moderate yields. Production of the identical diastereomer from either epimeric alkynic alcohol is consistent with formation of a common intermediate by protonation and dehydration, followed by stereoselective hydride addition (eq 8).

1. $Co_2(CO)_8$ 2. TFA, $NaBH_4$ 3. $Fe(NO_3)_3$ (8)

Baeyer–Villiger Oxidations in TFA. Trifluoroacetic acid has been reported to catalyze the action of ***m-Chloroperbenzoic Acid*** in the Baeyer–Villiger oxidation of cyclic and acyclic ketones.[22]

Trifluoroperacetic Acid (TFPAA) is remarkably efficient for the oxidation of ketones in the Baeyer–Villiger reaction.[23] This reagent has been prepared from ***Trifluoroacetic Anhydride*** and concentrated ***Hydrogen Peroxide***. Since 90% hydrogen peroxide is no longer available as a commercial reagent, alternatives are needed for classical oxidation procedures. A combination of TFA and sodium percarbonate has been used as a replacement for TFPAA in the Baeyer–Villiger reaction (eq 9).[24] Yields vary between good to excellent. The procedure is not applicable to aliphatic ketones, as TFA esters are produced from transesterification with the solvent.

Cleavage of Nitrogen- and Oxygen-Protecting Groups. Trifluoroacetic acid has found many applications in the removal of protecting groups. Examples include solvolysis under aqueous and anhydrous conditions. Groups that have been cleaved with TFA include *N*-Boc,[25] *N*-benzyloxymethyl,[26] benzyl ether,[27] *p*-methoxybenzyl ether,[28] *t*-butyl ether,[29] *t*-butyloxymethyl ether,[30] triphenylmethyl ether,[31] and dimethyl acetals.[32]

Na_2CO_4, TFA, 0 °C, 40 min; 25 °C, 14 h; 76% (9)

Related Reagents. Acetic Acid; Fluorosulfuric Acid; Formic Acid; Hydrochloric Acid; Hydrofluoric Acid; Oxalic Acid; Sulfuric Acid; *p*-Toluenesulfonic Acid; Trichloroacetic Acid; Triethylsilane-Trifluoroacetic Acid.

1. Astrologes, G. In *Kirk-Othmer Encyclopedia of Chemical Technology*, 3rd ed.; Wiley: New York, 1979; Vol. 10, pp 891–896.
2. *The Sigma-Aldrich Library of Chemical Safety Data*, 2nd ed.; Lenga, R. E., Ed.; Sigma-Aldrich: Milwaukee, 1988; Vol. 2, p 3439.
3. Cory, R. M.; Ritchie, B. M.; Shrier, A. M. *TL* **1990**, *31*, 6789.
4. (a) Schmid, R.; Huesmann, P. L.; Johnson, W. S. *JACS* **1980**, *102*, 5122. (b) Amupitan, J.; Sutherland, J. K. *CC* **1980**, 398. (c) Johnson, W. S.; Lindell, S. D.; Steele, J. *JACS* **1987**, *109*, 5852.
5. Dauben, W. G.; Chollet, A. *TL* **1981**, *22*, 1583.
6. Volkmann, R. A.; Andrews, G. C.; Johnson, W. S. *JACS* **1975**, *97*, 4777.
7. (a) Peterson, P. E.; Casey, C.; Tao, E. V. P.; Agtarap, A.; Thompson, G. *JACS* **1965**, *87*, 5163. (b) Peterson, P. E.; Bopp, R. J.; Chevli, D. M.; Curran, E. L.; Dillard, D. E.; Kamat, R. J. *JACS* **1967**, *89*, 5902.
8. (a) Sykes, A.; Tatlow, J. C.; Thomas, C. R. *CI(L)* **1955**, 630. (b) Dishart, K. T.; Levine, R. *JACS* **1956**, *78*, 2268. (c) Sykes, A.; Tatlow, J. C.; Thomas, C. R. *JCS* **1956**, 835. (d) Margaretha, P.; Schröder, C.; Wolff, S.; Agosta, W. C. *JOC* **1983**, *48*, 1925.
9. Eujen, R. *Inorg. Synth.* **1986**, *24*, 52.
10. Kondratenko, N. V.; Vechirko, E. P.; Yagupolskii, L. M. *S* **1980**, 932.
11. Mareda, J.; Rondan, N. G.; Houk, K. N. *JACS* **1983**, *105*, 6997.
12. Lagow, R. J.; Eujen, R.; Gerchman, L. L.; Morrison, J. A. *JACS* **1978**, *100*, 1722.
13. (a) Krause, L. J.; Morrison, J. A. *IC* **1980**, *19*, 604. (b) Eujen, R.; Lagow, R. J. *JCS(D)* **1978**, 541.
14. Lui, E. K. S. *IC* **1980**, *19*, 266.
15. (a) Eujen, R.; Thurmann, U. *JOM* **1992**, *433*, 63. (b) Ontiveros, C. D.; Morrison, J. A. *Inorg. Synth.* **1986**, *24*, 55.
16. (a) Tanabe, Y.; Matsuo, N.; Ohno, N. *JOC* **1988**, *53*, 4582. (b) Frohn, H. J.; Jakobs, S. *CC* **1989**, 625.
17. (a) Ketcha, D. M.; Lieurance, B. A.; Homan, D. F. J.; Gribble, G. W. *JOC* **1989**, *54*, 4350. (b) Gribble, G. W.; Leese, R. M.; Evans, B. E. *S* **1977**, 172.
18. (a) Hegedus, L. S.; McKearin, J. M. *JACS* **1982**, *104*, 2444. (b) Dailey, O. D., Jr.; Fuchs, P. L. *JOC* **1980**, *45*, 216.
19. Maryanoff, B. E.; McComsey, D. F. *JOC* **1978**, *43*, 2733.
20. (a) West, C. T.; Donnelly, S. J.; Kooistra, D. A.; Doyle, M. P. *JOC* **1973**, *38*, 2675. (b) Hamada, A.; Chang, Y. A.; Uretsky, N.; Miller, D. D. *JMC* **1984**, *27*, 675. (c) Magnus, P.; Gallagher, T.; Schultz, J.; Or, Y.; Ananthanarayan, T. P. *JACS* **1987**, *109*, 2706.
21. Nicholas, K. M.; Siegel, J. *JACS* **1985**, *107*, 4999.
22. (a) Koch, S. S. C.; Chamberlin, A. R. *SC* **1989**, *19*, 829. (b) Takano, S.; Ohashi, K.; Sugihara, T.; Ogasawara, K. *CL* **1991**, 203.
23. Emmons, W. D.; Lucas, G. B. *JACS* **1955**, *77*, 2287.
24. Olah, G. A.; Wang, Q.; Trivedi, N. J.; Prakash, G. K. S. *S* **1991**, 739.
25. Lundt, B. F.; Johansen, N. L.; Voelund, A., Markussen, J. *Int. J. Pept. Protein Res.* **1978**, *12*, 258.
26. Defrees, S. A.; Reddy, K. S.; Cassady, J. M. *SC* **1988**, *18*, 213.
27. Kālè, V. N.; Clive, D. L. J. *JOC* **1984**, *49*, 1554.
28. (a) Kende, A. S.; Veits, J. E.; Lorah, D. P.; Ebetino, F. H. *TL* **1984**, *25*, 2423. (b) Kende, A. S.; Lorah, D.; Boatman, R. J. *JACS* **1981**, *103*, 1271.
29. (a) Potman, R. P.; Janssen, N. J. M. L.; Scheeren, J. W.; Nivard, R. J. F. *JOC* **1984**, *49*, 3628. (b) Aben, R. W.; Scheeren, H. W. *JCS(P1)* **1979**, 3132.
30. Pinnick, H. W.; Lajis, N. H. *JOC* **1978**, *43*, 3964.
31. Prasad, K.; Repič, O. *TL* **1984**, *25*, 3391.
32. Zamboni, R.; Rokach, J. *TL* **1982**, *23*, 2631.
33. Perrin, D. D.; Armarego, W. L. F. *Purification of Laboratory Chemicals*, 3rd ed.; Pergamon: New York, 1988; p 297.

Kirk F. Eidman
Scios Nova, Baltimore, MD, USA

Trifluoromethanesulfonic Acid[1]

CF_3SO_2OH

[1493-13-6] CHF_3O_3S (MW 150.09)

(one of the strongest organic acids; catalyst for oligomerization/polymerization of alkenes and ethers; precursor for triflic anhydride and several metal triflates; acid catalyst in various reactions)

Alternate Name: triflic acid.
Physical Data: bp 162 °C/760 mmHg, 84 °C/43 mmHg, 54 °C/8 mmHg; d 1.696 g cm^{-3}.
Solubility: sol water and in many polar organic solvents such as DMF, sulfolane, DMSO, dimethyl sulfone, acetonitrile; sol alcohols, ketones, ethers, and esters, but these generally are not suitable inert solvents (see below).
Analysis of Reagent Purity: IR;[2] ^{19}F NMR.[3]
Preparative Methods: best prepared by basic hydrolysis of CF_3SO_2F followed by acidification.[2]
Purification: distilled with a small amount of Tf_2O.[4]
Handling, Storage, and Precautions: is a stable, hygroscopic liquid which fumes copiously on exposure to moist air. Transfer under dry nitrogen is recommended. Contact with cork, rubber, and plasticized materials will cause rapid discoloration of the acid and deterioration of the materials. Samples are best stored in sealed glass ampules or glass bottles with Kel-F™ or PTFE plastic screw cap linings. Use in a fume hood.

Reaction with P_2O_5. Trifluoromethanesulfonic acid (TfOH) reacts with an excess of ***Phosphorus(V) Oxide*** to give ***Trifluoromethanesulfonic Anhydride*** (eq 1),[5] while treatment with a smaller amount of P_2O_5 ($TfOH:P_2O_5 = 6:1$) and slower distillation leads to trifluoromethyl triflate (eq 2).[6]

$$CF_3SO_3H + P_2O_5\ (\text{excess}) \xrightarrow[-H_2O]{\Delta} (CF_3SO_2)_2O \qquad (1)$$

$$6\,CF_3SO_3H + P_2O_5 \xrightarrow[70\%]{\Delta} 3\,CF_3SO_2OCF_3 + 3\,SO_2 + 2\,H_3PO_4 \quad (2)$$

The synthetic utility of trifluoromethyl triflate as a trifluoromethanesulfonylating agent is severely limited, because the reagent is rapidly destroyed by a fluoride-ion chain reaction in the presence of other nucleophiles.[7]

Dehydration of a 2:1 mixture of CF_3CO_2H and TfOH with P_2O_5 affords trifluoroacetyl triflate (eq 3),[8] which is a very reactive agent for trifluoroacetylations at O, N, C, or halogen centers (eq 3).[8a]

$$CF_3SO_3H \xrightarrow[P_2O_5,\ \Delta\ 75\%]{CF_3CO_2H} CF_3CO_2OSO_2CF_3 \xrightarrow[-MeOTf\ 86\%]{anisole,\ 65\ °C,\ 28\ h} PhOCOCF_3 \quad (3)$$

Protonation and Related Reactions. TfOH is one of the strongest monoprotic organic acids known. The acid, and its conjugate base ($CF_3SO_3^-$), have extreme thermal stability, are resistant to oxidation and reduction, and are not a source of fluoride ions, even in the presence of strong nucleophiles. They do not lead to sulfonation as do ***Sulfuric Acid***, ***Fluorosulfuric Acid***, and ***Chlorosulfonic Acid*** in some reactions. TfOH is therefore effectively employed in protonation reactions.

The strong protonating property of TfOH is used to generate allyl cations from suitable precursors in low-temperature ionic Diels–Alder reactions. 3,3-Diethoxypropene and 2-vinyl-1,3-dioxolane add to cyclohexa-1,3-diene in the presence of TfOH to give the corresponding Diels–Alder adducts, the latter in high yield (eq 4).[9]

CF_3SO_3H, CH_2Cl_2; $H_2C{=}CHCH(OEt)_2$ 57%; 2-vinyl-1,3-dioxolane 94% (4)

An intramolecular Diels–Alder reaction with high stereoselectivity occurs involving allyl cations by protonation of allyl alcohols (eq 5).[10]

CF_3SO_3H, CH_2Cl_2, −23 °C (5)

Alkynes and allenes are protonated with TfOH to give vinyl triflates (eqs 6 and 7),[11] which are precursors to vinyl cations.

CF_3SO_3H, pentane, −30 °C, 70%; OTf + OTf, 65:35 (6)

CF_3SO_3H, −78 °C, 60–80%; OTf (7)

A convenient synthesis of pyrimidines is developed by protonation of alkynes with TfOH in the presence of nitriles (eq 8).[12]

$R^1{-}{\equiv} + R^2CN \xrightarrow[68–94\%]{CF_3SO_3H,\ 0\ °C\ to\ rt,\ 24\ h}$ pyrimidine (R^1, R^2, R^2) (8)

$R^1 = R^2$ = alkyl, aryl

Triflic acid catalyzes the transformation of α-hydroxy carbonyl compounds to ketones (eq 9).[13]

OH, COPh; CF_3SO_3H, CH_2Cl_2, 0 °C to rt; −PhCHO, 97%; O (9)

Oximes undergo Beckmann rearrangement with TfOH in the presence of Bu_4NReO_4 to give amides in high yield (eq 10).[14]

$Ph_2C{=}NOH \xrightarrow[98\%]{CF_3SO_3H,\ Bu_4NReO_4,\ MeNO_2,\ 20\ min} PhNHC(O)Ph$ (10)

TfOH protonates nitroalkenes, even nitroethylene, to give *N*,*N*-dihydroxyiminium carbenium ions, which react with arenes to give arylated oximes. This overall process provides a route to α-aryl methyl ketones from 2-nitropropene (eq 11)[15] and constitutes a versatile synthetic method for the preparation of α-arylated ketones, otherwise difficult to synthesize by the conventional Friedel–Crafts reaction.

NO_2; CF_3SO_3H; PhH; H_2O, 85%; Ph, O (11)

TfOH catalyzes the removal of *N*-*t*-butyl groups from *N*-substituted *N*-*t*-butylcarbamates to give carbamate-protected primary amines (eq 12).[16]

t-Bu, O, Ph, O, N(*t*-Bu), OBn; 1 mol % CF_3SO_3H; 95%; *t*-Bu, O, Ph, O, NH, OBn (12)

The methyl group attached to the phenolic oxygen of tyrosine is smoothly cleaved by TfOH in the presence of ***Thioanisole*** (eq 13).[17] This deblocking method was successfully applied to the synthesis of a new potent enkephalin derivative.

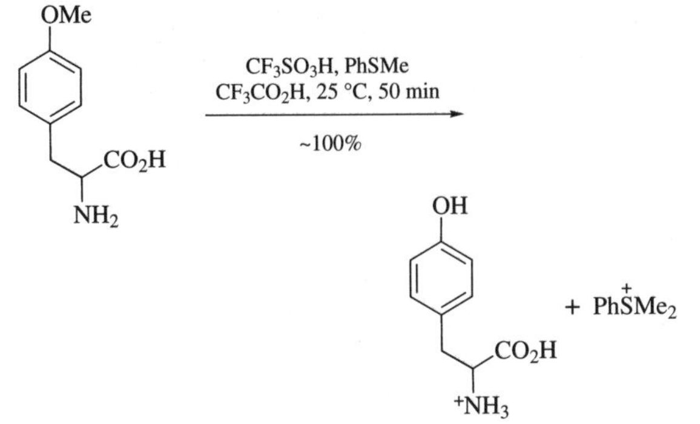

CF_3SO_3H, PhSMe
CF_3CO_2H, 25 °C, 50 min
~100%

+ PhS^+Me_2 (13)

1,3,4-Oxadiazoles are prepared in good yields from silylated diacylhydrazines (formed in situ) by acid-catalyzed cyclization using TfOH (eq 14).[18]

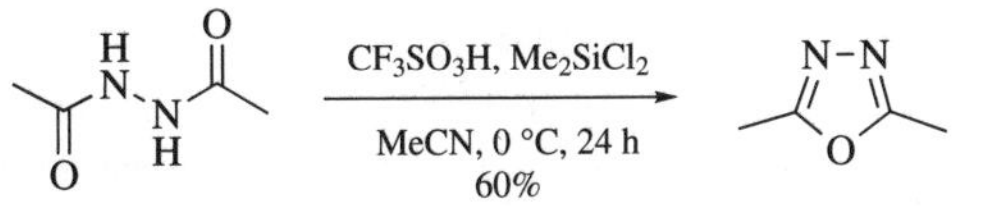

CF_3SO_3H, Me_2SiCl_2
MeCN, 0 °C, 24 h
60%
(14)

TfOH protonates naphthalene at room temperature to give a complex mixture of products.[19] TfOH promotes aldol reaction of silyl enol ethers with aldehydes and acetals, leading to new C–C bond formation (eq 15).[20] TfOH competes well with other reagents employed for the aldol reaction, while ***Methanesulfonic Acid*** does not afford any product.

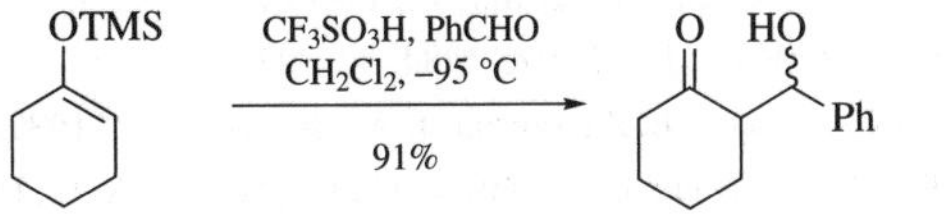

CF_3SO_3H, PhCHO
CH_2Cl_2, –95 °C
91%
(15)

Cyclization of 3- and 4-arylalkanoic acids to bicyclic ketones is effected by TfOH via the corresponding acid chlorides (eq 16).[21]

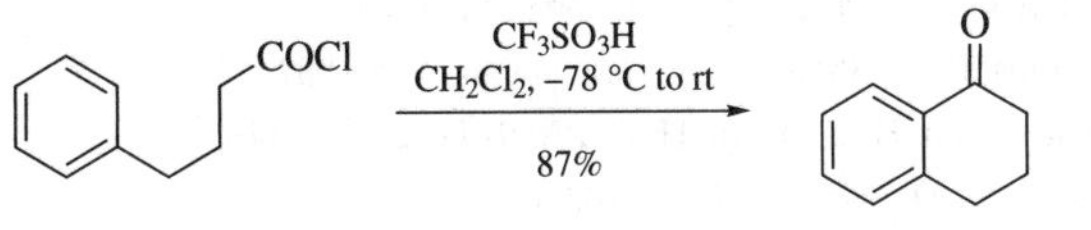

CF_3SO_3H
CH_2Cl_2, –78 °C to rt
87%
(16)

Allylic *O*-methylisoureas are cyclized with TfOH containing ***Benzeneselenenyl Trifluoromethanesulfonate*** to 5,6-dihydro-1,3-oxazines (eq 17).[22]

CF_3SO_3H
PhSeOTf
66%
(17)

Tscherniac amidomethylation of aromatics with *N*-hydroxymethylphthalimide in TfOH proceeds smoothly at room temperature to give the corresponding α-amido-methylated products (eq 18).[23]

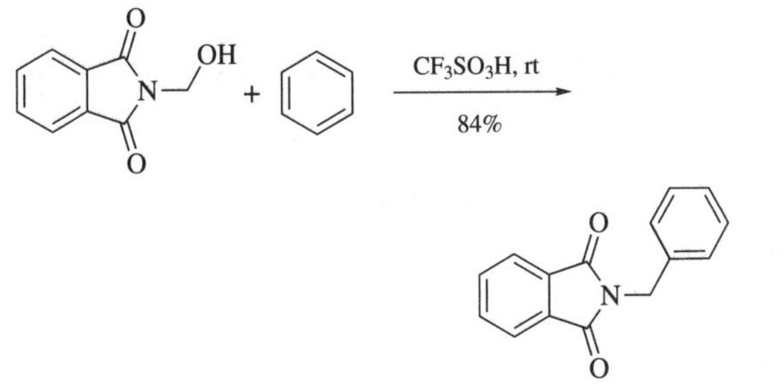

CF_3SO_3H, rt
84%
(18)

TfOH catalyzes the amination[24] and phenylamination[25] of aromatics via the corresponding aminodiazonium ion generated from ***Azidotrimethylsilane*** and ***Phenyl Azide*** respectively (eq 19).

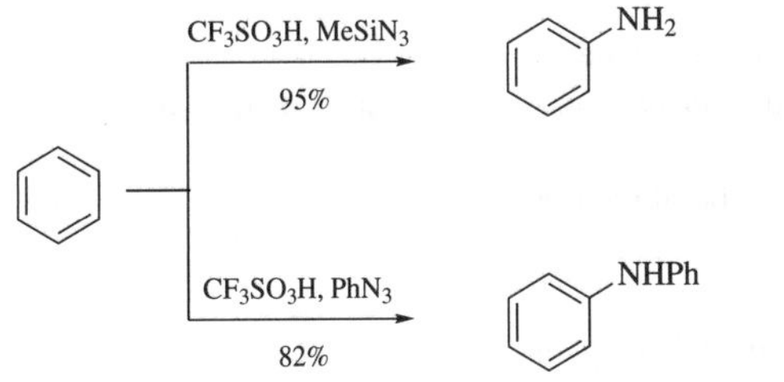

CF_3SO_3H, $MeSiN_3$
95%

CF_3SO_3H, PhN_3
82%
(19)

Electrophilic hydroxylation of aromatics is carried out by protonation of ***Bis(trimethylsilyl) Peroxide*** with TfOH in the presence of the substrate (eq 20).[26]

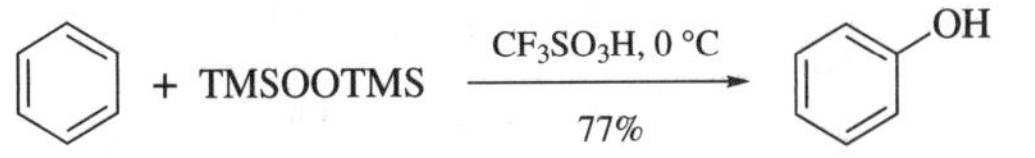

+ TMSOOTMS
CF_3SO_3H, 0 °C
77%
(20)

Phenol and 2,3,5,6-tetramethylphenol are protonated with TfOH under irradiation to afford rearranged products (eqs 21 and 22).[27]

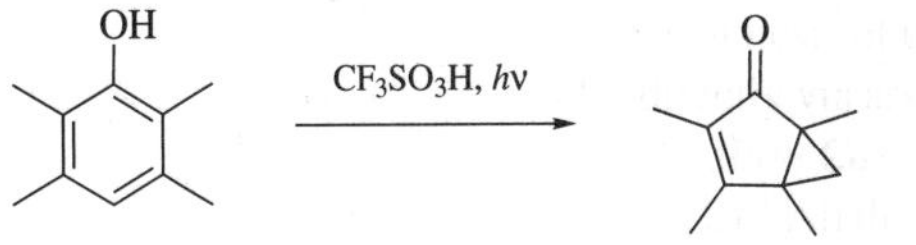

CF_3SO_3H, *hν*
(21)

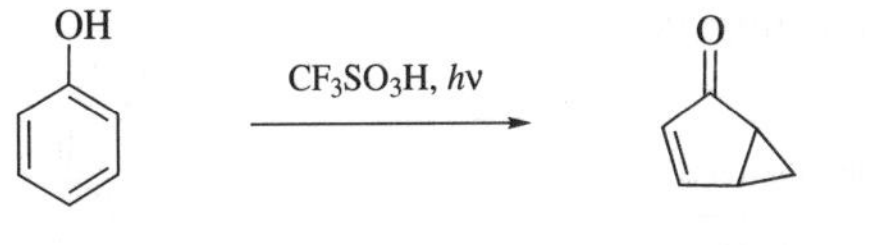

CF_3SO_3H, *hν*
(22)

Other Applications. TfOH is the starting material for the preparation of the electrophilic reagent ***Trimethylsilyl Trifluoromethanesulfonate***. The latter is prepared by reacting TfOH with ***Chlorotrimethylsilane***[28] or more conveniently with Me_4Si (eq 23).[29]

$$CF_3SO_3H \xrightarrow[12\text{–}99\%]{\text{TMSCl or } Me_4Si} CF_3SO_3TMS \quad (23)$$

Functionalized silyl triflates can also be prepared using TfOH (eq 24).[30]

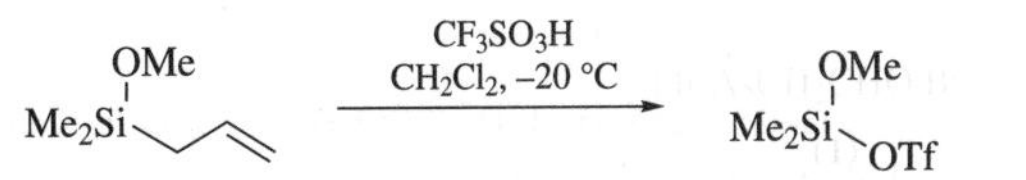

CF_3SO_3H
CH_2Cl_2, –20 °C
(24)

Reaction of aromatic compounds with ***Bis(pyridine)-iodonium(I) Tetrafluoroborate*** in the presence of TfOH is an effective method to form the monoiodo compounds regioselectively (eq 25).[31]

IPy_2BF_4 + CHO; CF_3SO_3H, CH_2Cl_2, rt; 80%; CHO, I (25)

Ionic hydrogenation of alkenes with trialkylsilanes is possible in the presence of the strong acid TfOH, even at −75 °C (eq 26).[32]

Et; CF_3SO_3H, Et_3SiH, CH_2Cl_2, −75 °C; 98%; Et (26)

Hydroxycarbonyl compounds can be selectively reduced to carbonyl compounds by means of TfOH in the presence of trialkylboranes (eq 27).[33]

CHO, OH; CF_3SO_3H, Et_3B; 81%; CHO (27)

The triphenylmethyl cation is nitrated with ***Nitronium Tetrafluoroborate*** in the presence of TfOH (eq 28).[34]

Ph_3CBF_4; 1. CF_3SO_3H, NO_2BF_4; 2. ionic hydrogenation; 64%; Ph, Ph, NO_2 (28)

Sterically hindered azidophenyltriazines decompose in TfOH at 0 °C to give isomeric triflates (eq 29).[35]

NH_2, N, N, Cl, N_3, H_2N, N, H; CF_3SO_3H, CF_3CO_2H, 0 °C; 98%; TfO, NH_2, Cl, N, N, NH_2, H_2N, N, H (29)

Benzoyl triflate prepared from TfOH and ***Benzoyl Chloride*** is a mild and effective benzoylating agent for sterically hindered alcohols[36] and acylative ring expansion reactions.[37] The applications of TfOH in Koch–Haaf carboxylation,[38] Fries rearrangement,[39] and sequential chain extension in carbohydrates[40] are also documented. Recent applications of TfOH in cyclization reactions have been published.[41–43]

Related Reagents. 10-Camphorsulfonic Acid; Methanesulfonic Acid; Sulfuric Acid; Trifluoroacetic Acid.

1. (a) Howells, R. D.; McCown, J. D. *CRV* **1977**, *77*, 69. (b) Stang, P. J.; White, M. R. *Aldrichim. Acta* **1983**, *16*, 15.
2. (a) Haszeldine, R. N.; Kidd, J. M. *JCS* **1954**, 4228. (b) Burdon, J.; Farazmand, I.; Stacey, M.; Tatlow, J. C. *JCS* **1957**, 2574.
3. Matjaszewski, K.; Sigwalt, P. *M* **1986**, *187*, 2299 (*CA* **1987**, *106*, 5495).
4. (a) Sagl, D.; Martin, J. C. *JACS* **1988**, *110*, 5827. (b) Saito, S.; Sato, Y.; Ohwada, T.; Shudo, K. *CPB* **1991**, *39*, 2718.
5. Stang, P. J.; Hanack, M.; Subramanian, L. R. *S* **1982**, 85.
6. Hassani, M. O.; Germain, A.; Brunel, D.; Commeyras, A. *TL* **1981**, *22*, 65.
7. Taylor, S. L.; Martin, J. C. *JOC* **1987**, *52*, 4147.
8. (a) Forbus, T. R., Jr.; Taylor, S. L.; Martin, J. C. *JOC* **1987**, *52*, 4156. (b) Taylor, S. L.; Forbus, T. R., Jr.; Martin, J. C. *OSC* **1990**, *7*, 506.
9. Gassman, P. G.; Singleton, D. A.; Wilwerding, J. J.; Chavan, S. P. *JACS* **1987**, *109*, 2182.
10. (a) Gassman, P. G.; Singleton, D. A. *JOC* **1986**, *51*, 3075. (b) Gorman, D. B.; Gassman, P. G. *JOC* **1995**, *60*, 977.
11. (a) Stang, P. J.; Summerville, R. H. *JACS* **1969**, *91*, 4600. (b) Summerville, R. H.; Senkler, C. A.; Schleyer, P. v. R.; Dueber, T. E.; Stang, P. J. *JACS* **1974**, *96*, 1100.
12. García Martínez, A.; Herrera Fernandez, A.; Martínez Alvarez, R.; Silva Losada, M. C.; Molero Vilchez, D.; Subramanian, L. R.; Hanack, M. *S* **1990**, 881.
13. Olah, G. A.; Wu, A. *JOC* **1991**, *56*, 2531.
14. Narasaka, K.; Kusama, H.; Yamashita, Y.; Sato, H. *CL* **1993**, 489.
15. Okabe, K.; Ohwada, T.; Ohta, T.; Shudo, K. *JOC* **1989**, *54*, 733.
16. Earle, M. J.; Fairhurst, R. A.; Heaney, H.; Papageorgiou, G. *SL* **1990**, 621.
17. Kiso, Y.; Nakamura, S.; Ito, K.; Ukawa, K.; Kitagawa, K.; Akita, T.; Moritoki, H. *CC* **1979**, 971.
18. Rigo, B.; Cauliez, P.; Fasseur, D.; Couturier, D. *SC* **1988**, *18*, 1247.
19. Launikonis, A.; Sasse, W. H. F.; Willing, I. R. *AJC* **1993**, *46*, 427.
20. Kawai, M.; Onaka, M.; Izumi, Y. *BCJ* **1988**, *61*, 1237.
21. Hulin, B.; Koreeda, M. *JOC* **1984**, *49*, 207.
22. Freire, R.; León, E. Z.; Salazar, J. A.; Suárez, E. *CC* **1989**, 452.
23. Olah, G. A.; Wang, Q.; Sandford, G.; Oxyzoglou, A. B.; Prakash, G. K. S. *S* **1993**, 1077.
24. Olah, G. A.; Ernst, T. D. *JOC* **1989**, *54*, 1203.
25. Olah, G. A.; Ramaiah, P.; Wang, Q.; Prakash, G. S. K. *JOC* **1993**, *58*, 6900.
26. Olah, G. A.; Ernst, T. D. *JOC* **1989**, *54*, 1204.
27. Childs, R. F.; Shaw, G. S.; Varadarajan, A. *S* **1982**, 198.
28. Marsmann, H. C.; Horn, H. G. *ZN(B)* **1972**, *27*, 1448.
29. Demuth, M.; Mikhail, G. *S* **1982**, 827.
30. Uhlig, W. *JOM* **1993**, *452*, 29.
31. Barluenga, J.; González, J. M.; García-Martín, M. A.; Campos, P. J.; Asensio, G. *JOC* **1993**, *58*, 2058.
32. Bullock, R. M.; Rappoli, B. J. *CC* **1989**, 1447.
33. Olah, G. A.; Wu, A.-H. *S* **1991**, 407.
34. Olah, G. A.; Wang, Q.; Orlinkov, A.; Ramaiah, P. *JOC* **1993**, *58*, 5017.
35. Stevens, M. F. G.; Chui, W. K.; Castro, M. A. *JHC* **1993**, *30*, 849.
36. Brown, L.; Koreeda, M. *JOC* **1984**, *49*, 3875.
37. Takeuchi, K.; Ohga, Y.; Munakata, M.; Kitagawa, T.; Kinoshita, T. *TL* **1992**, *33*, 3335.
38. Booth, B. L.; El-Fekky, T. A. *JCS(P1)* **1979**, 2441.
39. Effenberger, F.; Klenk, H.; Reiter, P. L. *AG(E)* **1973**, *12*, 775.
40. Auzanneau, F.-I.; Bundle, D. R. *CJC* **1993**, *71*, 534.
41. Marson, C. M.; Fallah, A. *TL* **1994**, *35*, 293.

42. Saito, S.; Sato, Y.; Ohwada, T.; Shudo, K. *JACS* **1994**, *116*, 2312.
43. Pearson, W. H.; Fang, W.; Kamp, J. W. *JOC* **1994**, *59*, 2682.

Lakshminarayanapuram R. Subramanian, Antonio García Martínez, & Michael Hanack
Universität Tübingen, Germany

Trimethyl Phosphite[1]

$(MeO)_3P$

[121-45-9] $C_3H_9O_3P$ (MW 124.09)

(reducing agent for various functional groups; can function as a potent thiophile for sulfur extrusion; forms a stable complex with copper(I) iodide and methylcopper)

Physical Data: bp 111–112 °C/760 mmHg; *d* 1.052 g cm^{-3}.
Solubility: sol m cost organic solvents.
Form Supplied in: clear, free flowing liquid; commercially available in >99% purity.
Purification: treatment with sodium (to remove water and dialkyl phosphonate), followed by decanting and distillation.[2]
Handling, Storage, and Precautions: store in a dry place, preferably in a fume hood due to its pungent odor; purified material should be stored over activated molecular sieves.

Use in the Arbuzov and Perkow Reactions. Although used less frequently than ***Triethyl Phosphite***, trimethyl phosphite has seen significant application in the synthesis of organophosphonates (eq 1)[3] and phosphates (eq 2)[4] via Arbuzov and Perkow reactions, respectively. For example, an interesting α-diazophosphonate reagent developed by Seyferth has been applied to the one-step conversion of aldehydes to terminal alkynes (eq 3).[5] Since the potentially hazardous diazophosphonate is purified by distillation, use of the relatively volatile dimethyl derivative is warranted.

(MeO)3P, Δ; X = Cl, Br, I (1)

(MeO)3P, Δ; X = Cl, Br (2)

(MeO)3P, Δ; 1. N2H4 2. NaNO2, HCl

t-BuOK, THF; RCHO −78 °C to rt excellent yields; R—≡—H (3)

Carbonyl Adducts. Trimethyl phosphite reacts readily with 1,4- and 1,2-benzoquinones to give methyl esters of hydroquinone monophosphates (eq 4)[6] and cyclic oxyphosphoranes (eq 5),[6] respectively. Ramirez and co-workers reported that trimethyl phosphite promotes the dimerization of methyl pyruvate to give, after basic hydrolysis, a mixture of diastereomeric tartrates (eq 6).[7]

(MeO)3P, Δ; 100% (4)

(MeO)3P, Δ; 100% (5)

(MeO)3P, Δ; NaOH 90% (6)

Deoxygenation. The reductive decomposition of ozonides can be conveniently accomplished using trimethyl phosphite (eq 7).[8] The innocuous byproduct, trimethyl phosphate, is easily removed from the crude reaction mixture either by extraction or evaporation.

O3, CH2Cl2–MeOH (MeO)3P, −78 °C; 95% (7)

The reagent has been equally efficacious in the conversion of nitrile oxides to nitriles (eq 8).[9] Additionally, Seebach and coworkers have reported the synthesis of a cyclic ϵ-tetrazine via reduction of the corresponding *N*-oxide (eq 9).[10]

$(MeO)_3P$, 93% (8)

$(MeO)_3P$, 100% (9)

Desulfurization. Trimethyl phosphite (in addition to triethyl phosphite) has been applied successfully to the Corey–Winter alkene synthesis.[11] The key step of the sequence involves phosphite-mediated decomposition of a thiocarbonate derivative which proceeds stereospecifically and often results in good yields of desired alkene (eqs 10 and 11).[12] Alkynes have also been synthesized using this method, albeit in lower overall yields (eq 12).[13]

$(MeO)_3P$, Δ, good yields (10)

$(MeO)_3P$, Δ, good yields (11)

TMSO, OTMS; 1. MeLi, THF 2. CS_2, MeI; $(MeO)_3P$, Δ, 25–35%; R—≡—R (12)

Treatment of 1,3-dithiacyclohexane-2-thione[14] with trimethyl phosphite affords the corresponding ylide quantitatively. The phosphorane has been applied the one-carbon homologation of aldehydes to carboxylic acids (eq 13).[15]

$(MeO)_3P$, Δ, 100%; RCHO; H_2O (13)

In addition, trimethyl phosphite has been used to capture intermediates formed during [2,3]-sigmatropic rearrangement of allylic sulfoxides (eq 14)[16] and in the conversion of penicillin derivatives to azetidinones (eq 15).[17]

Δ; $(MeO)_3P$, 51% (14)

$(MeO)_3P$, RCOOH, 65–86% (15)

Dehalogenation. Dershowitz[18] has reported that vicinal dibromides are smoothly converted to alkenes by heating in the presence of trimethyl phosphite. The reagent was successfully applied to systems where other reagents, such as ***Sodium Iodide*** or ***Zinc*** dust proved unsatisfactory (eq 16). The dehydrohalogenation of a steroidal allylic bromide has also been reported (eq 17).[19]

$(MeO)_3P$, Δ, 92% (16)

$(MeO)_3P$, Δ, 56% (17)

Copper Complexes. ***Copper(I) Iodide-Trimethyl Phosphite*** is formed in 84% yield by the reaction of trimethyl phosphite and ***Copper(I) Iodide*** in refluxing benzene.[20] The salt has been used as a catalyst in the decomposition of diethyl diazomalonate to give ethers (eq 18)[21] and cyclopropanes (eq 19).[20] Furanones have been prepared using related methodology (eq 20).[22]

$(MeO)_3P$, CuI, EtOH, 82% (18)

1,3-cyclooctadiene, $(MeO)_3P$•CuI, 100% (19)

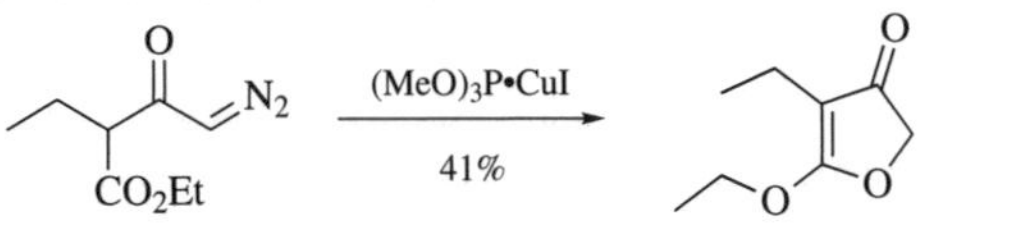

(20)

Trimethyl phosphite–methylcopper is a relatively stable complex of ***Methylcopper***. The reagent adds readily, in a conjugate sense, to cyclohexenones, with a strong preference for axial attack (eq 21).[23]

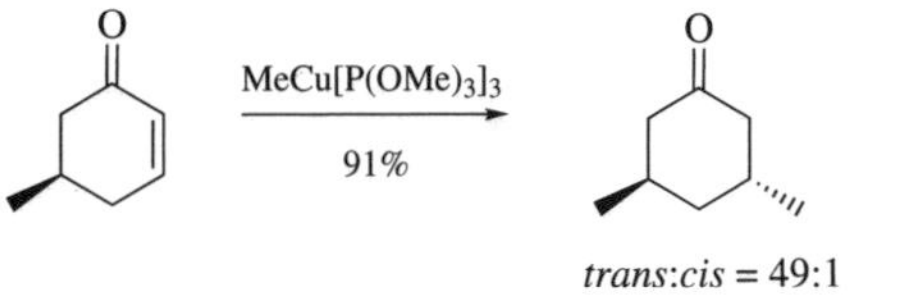

(21)

trans:*cis* = 49:1

Esterification. Trimethyl phosphite has been used in the conversion of a sensitive indolecarboxylic acid to the corresponding methyl ester under neutral conditions (eq 22).[24]

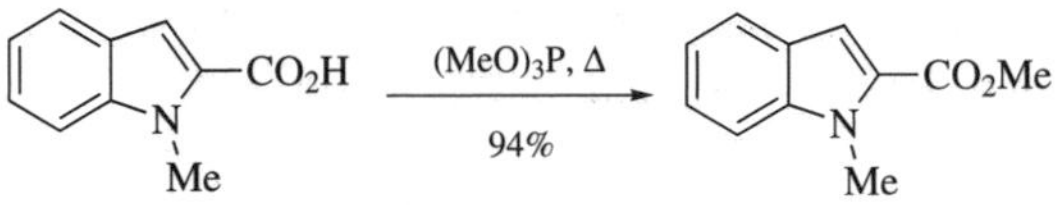

(22)

Related Reagents. Triethyl Phosphite.

1. Schuetz, R. D.; Jacobs, R. L. *JOC* **1961**, *26*, 3467.
2. Perrin, D. D.; Armarego, W. L. F. *Purification of Laboratory Chemicals*, 3rd ed.; Pergamon: New York, 1988; p 297.
3. For reviews of the Arbuzov reaction, see: (a) Arbuzov, B. A. *PAC* **1964**, *9*, 307. (b) Kosolapoff, G. M. *OR* **1951**, *6*, 273. (c) Redmore, D. *CRV* **1971**, *71*, 315. (d) Bhattacharya, A. K.; Thyagarajan, G. *CRV* **1981**, *81*, 415. (e) Cadogen, J. I. G. *Organophosphorus Reagents in Organic Synthesis*; Academic: New York, 1979. For application to the synthesis of alkenes, see: (f) Horner, L.; Hoffman, H.; Wipple, H. G. *CB* **1958**, *91*, 61. (g) Horner, L.; Hoffman, H.; Wipple, H. G. *CB* **1959**, *92*, 2499. (h) Wadsworth, W. S., Jr.; Emmons, W. D. *JACS* **1961**, *83*, 1733. For reviews see: (i) Wadsworth W. S. *OR* **1977**, *25*, 73. (j) Boutagy, J.; Thomas, R. *CRV* **1974**, *74*, 87. (k) Carey, F. A.; Sundberg, R. J. *Advanced Organic Chemistry*, 3rd ed.; Plenum: New York, 1990; Part B, pp 100–102. (l) Wadsworth, W. S., Jr.; Emmons, W. D. *OS* **1965**, *45*, 44. (m) Roush, W. R. *JACS* **1978**, *100*, 3599. (n) Nicolaou, K. C.; Bertinato, P.; Piscopio, A. D.; Chakraborty, T. K.; Minowa, N. *CC* **1993**, 619 and references cited therein.
4. (a) Perkov, N.; Ullerich, K.; Meyer, F. *N* **1952**, *39*, 353. For reviews of the Perkow reaction, see: (b) Lichtenthaler, F. W. *CRV* **1961**, *61*, 607. (c) Arbuzov, B. A. *PAC* **1964**, *9*, 307.
5. (a) Seyferth, D.; Marmor, R. S.; Hilbert, P. J. *JOC* **1971**, *36*, 1379. (b) Colvin, E. W.; Hamill, B. J. *CC* **1973**, 151. (c) Colvin, E. W.; Hamill, B. J. *JCS(P1)* **1977**, 869. (d) Gilbert, J. C.; Weerasooriya, U. *JOC* **1979**, *44*, 4997. (e) Nakatsuka, M.; Ragan, J. A.; Sammakia, T.; Smith, D. B.; Uehling, D. E.; Schreiber, S. L. *JACS* **1990**, *112*, 5583.
6. Ramirez, F.; Desai, N. B. *JACS* **1963**, *85*, 3252.
7. Ramirez, F.; Desai, N. B.; Ramanathan, N. *TL* **1963**, 323.
8. (a) Knowles, W. S.; Thompson, Q. E. *JOC* **1960**, *25*, 1031. (b) Stille, J. K.; Foster, R. T. *JOC* **1963**, *28*, 2703. (c) Stevens, R. V.; Beaulieu, N.; Chan, W. H.; Daniewski, A. R.; Takishi, T.; Waldner, A.; Willard, P. G.; Zutter, U. *JACS* **1986**, *108*, 1039. (d) Murray, R. W. *ACR* **1968**, *1*, 313.
9. Grundmann, C.; Frommeld, H. D. *JOC* **1965**, *30*, 2077.
10. (a) Seebach, D.; Enders, D.; Renger, B.; Brugel, W. *AG(E)* **1973**, *12*, 495. (b) Seebach, D.; Enders, D. *AG(E)* **1972**, *11*, 301.
11. (a) Corey, E. J.; Winter, R. A. E. *JACS* **1963**, *85*, 2677. (b) Corey, E. J.; Carey, F. A.; Winter, R. A. E. *JACS* **1965**, *87*, 934.
12. For a review see: Block, E. *OR* **1983**, *30*, 457.
13. Bauer, D. P.; Macomber, R. S. *JOC* **1976**, *41*, 2640.
14. Mills, W. H.; Saunders, B. C. *JCS* **1931**, 537.
15. Corey, E. J.; Markl, G. *TL* **1967**, 3201.
16. (a) Brown, W. L.; Fallis, A. G. *TL* **1985**, *26*, 607. (b) Bickart, P.; Carson, F. W.; Jacobus, J.; Miller, E. G.; Mislow, K. *JACS* **1968**, *90*, 4869. (c) Tang, R.; Mislow, K. *JACS* **1970**, *92*, 2100. (d) Grieco, P. *CC* **1972**, 702. (e) Evans, D. A.; Andrews, G. C. *ACR* **1974**, *7*, 147. (f) Hoffman, R. W.; Goldman, S.; Maak, N.; Gerlach, R.; Frickel, F.; Steinbach, G. *CB* **1980**, *113*, 819. (g) Isobe, M.; Iio, H.; Kitamura, M.; Goto, T. *CL* **1978**, 541.
17. Suarato, A.; Lombardi, P.; Galliani, C.; Franceschi, G. *TL* **1978**, 4059.
18. Dershowitz, S.; Proskauer, S. *JOC* **1961**, *26*, 3595.
19. Hunziker, F.; Mullner, F. X. *H* **1958**, *41*, 70.
20. Peace, B. W.; Carman, F.; Wulfman, D. F. *S* **1971**, 658.
21. Pelliciari, R.; Cogolli, P. *S* **1975**, 269.
22. Bien, S.; Gillon, A. *TL* **1974**, 3073.
23. House, H. O.; Fischer, W. F., Jr. *JOC* **1968**, *33*, 949.
24. (a) Szmuskovicz, J. *OPP* **1972**, *4*, 51. (b) Kamai, G.; Kukhtin, V. A.; Strogova, O. A. *CA* **1957**, *51*, 11 994b.

Anthony D. Piscopio
Pfizer, Groton, CT, USA

Trimethylsilyl Trifluoromethanesulfonate[1]

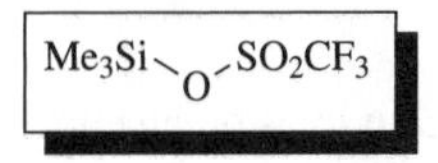

[88248-68-4; 27607-77-8] $C_4H_9F_3O_3SSi$ (MW 222.29)

Alternate Name: TMSOTf.
Physical Data: bp 45–47 °C/17 mmHg, 39–40 °C/12mmHg; *d* 1.225 g cm^{-3}.
Solubility: sol aliphatic and aromatic hydrocarbons, haloalkanes, ethers.
Form Supplied in: colorless liquid; commercially available.
Preparative Methods: may be prepared by a variety of methods.[2]
Handling, Storage, and Precautions: flammable; corrosive; very hygroscopic.

Silylation. TMSOTf is widely used in the conversion of carbonyl compounds to their enol ethers. The conversion is some 10^9 faster with TMSOTf/***Triethylamine*** than with ***Chlorotrimethylsilane*** (eqs 1–3).[3–5]

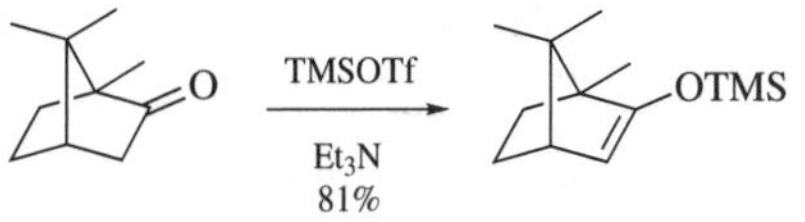

(1)

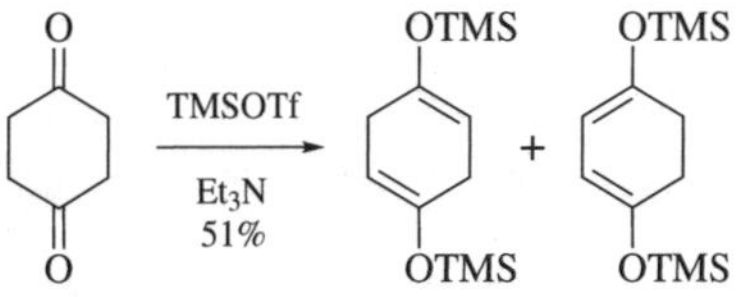

(2)

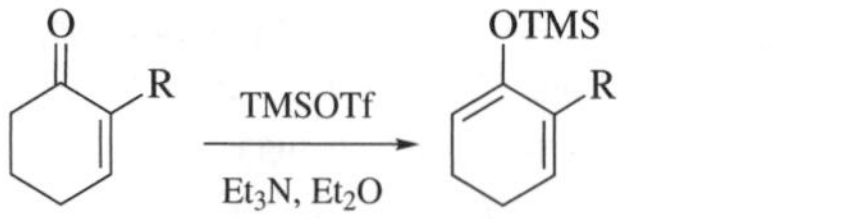

(3)

Dicarbonyl compounds are converted to the corresponding bis-enol ethers; this method is an improvement over the previous two-step method (eq 4).[6]

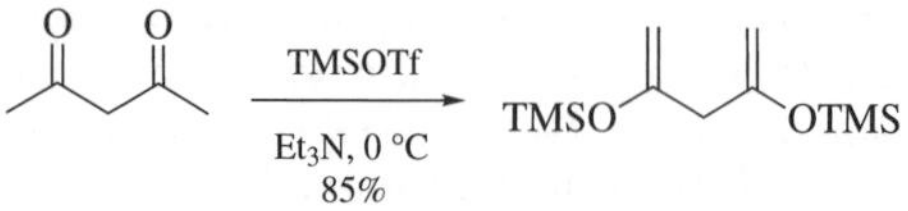

(4)

In general, TMSOTf has a tendency to *C*-silylation which is seen most clearly in the reaction of esters, where *C*-silylation dominates over *O*-silylation. The exact ratio of products obtained depends on the ester structure[7] (eq 5).[8] Nitriles undergo *C*-silylation; primary nitriles may undergo *C,C*-disilylation.[9]

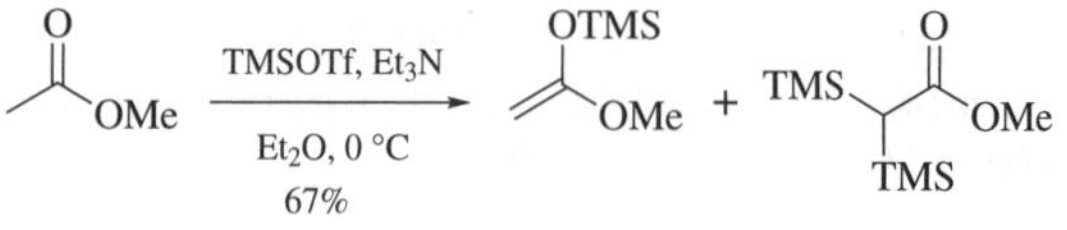

(5)

84:16

TMS enol ethers may be prepared by rearrangement of α-ketosilanes in the presence of catalytic TMSOTf (eq 6).[10,11]

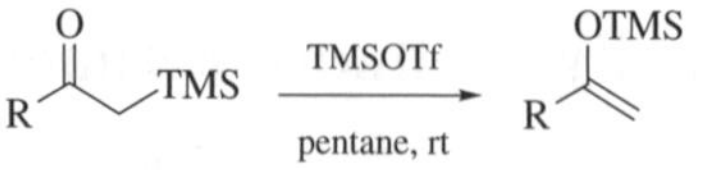

(6)

Enhanced regioselectivity is obtained when trimethylsilyl enol ethers are prepared by treatment of α-trimethylsilyl ketones with catalytic TMSOTf (eq 7).[12]

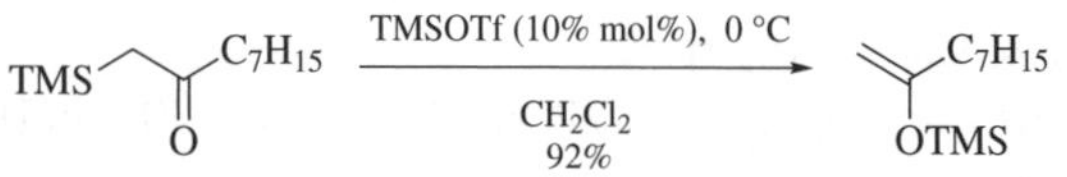

(7)

The reaction of imines with TMSOTf in the presence of Et_3N gives *N*-silylenamines.[13]

Ethers do not react, but epoxides are cleaved to give silyl ethers of allylic alcohols in the presence of TMSOTf and ***1,8-Diazabicyclo[5.4.0]undec-7-ene***; The regiochemistry of the reaction is dependent on the structure of the epoxide (eq 8).[14]

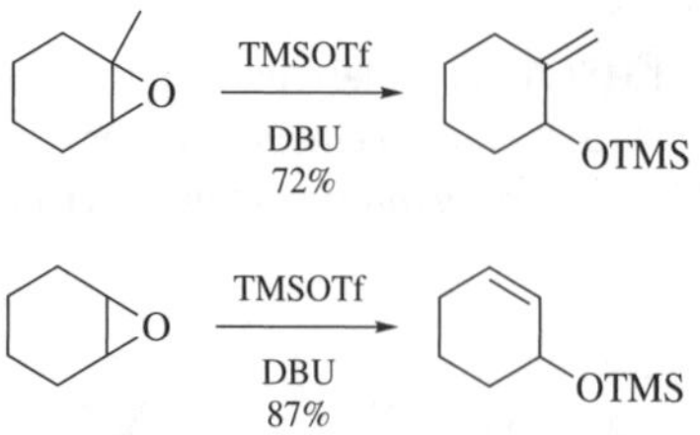

(8)

Indoles and pyrroles undergo efficient *C*-silylation with TMSOTf (eq 9).[15]

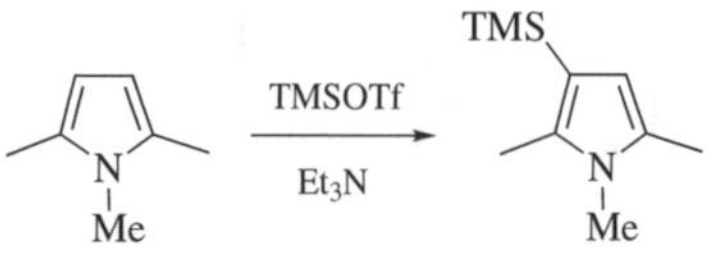

(9)

t-Butyl esters are dealkylatively silylated to give TMS esters by TMSOTf; benzyl esters are inert under the same conditions.[16]

Imines formed from unsaturated amines and α-carbonyl esters undergo ene reactions in the presence of TMSOTf to form cyclic amino acids.[17]

Carbonyl Activation. 1,3-Dioxolanation of conjugated enals is facilitated by TMSOTf in the presence of 1,2-bis(trimethylsilyloxy)ethane. In particular, highly selective protection of sterically differentiated ketones is possible (eq 10).[18] Selective protection of ketones in the presence of enals is also facilitated (eq 11).[19]

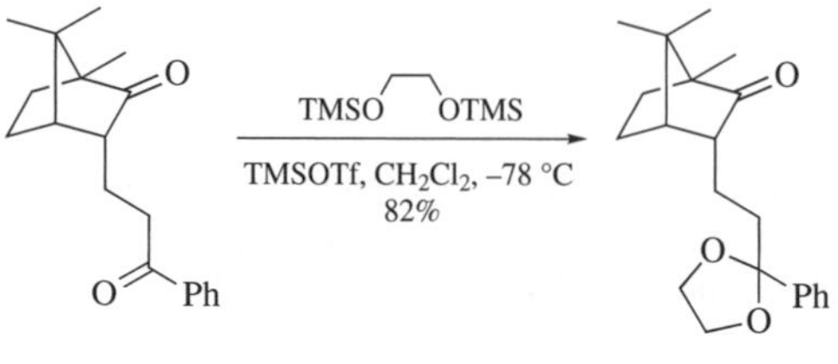

(10)

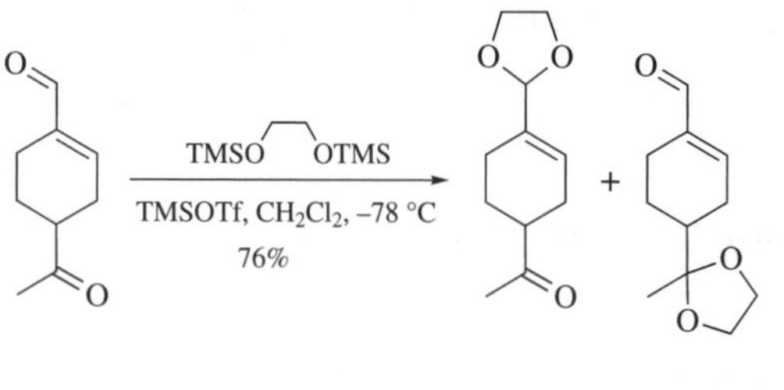

(11)

1:27

The similar reaction of 2-alkyl-1,3-disilyloxypropanes with chiral ketones is highly selective and has been used to prepare spiroacetal starting materials for an asymmetric synthesis of α-tocopherol subunits (eq 12).[20]

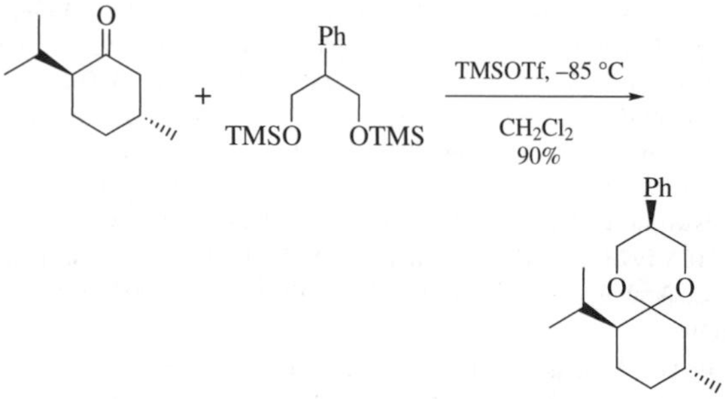

(12)

The preparation of spiro-fused dioxolanes (useful as chiral glycolic enolate equivalents) also employs TMSOTf (eq 13).[21]

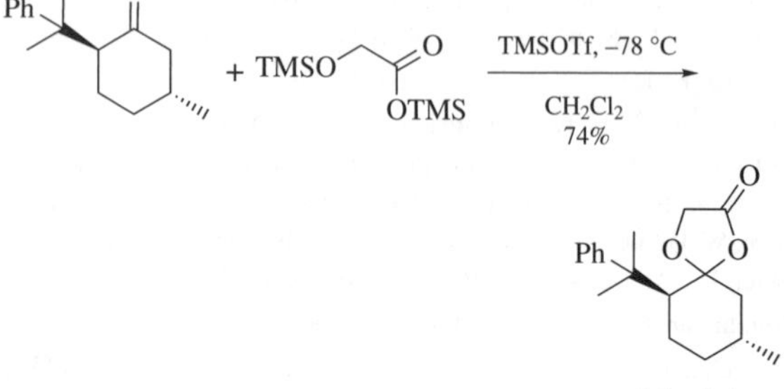

(13)

~ 1:1 mixture

TMSOTf mediates a stereoselective aldol-type condensation of silyl enol ethers and acetals (or orthoesters). The nonbasic reaction conditions are extremely mild. TMSOTf catalyzes many aldol-type reactions; in particular, the reaction of relatively nonnucleophilic enol derivatives with carbonyl compounds is facile in the presence of the silyl triflate. The activation of acetals was first reported by Noyori and has since been widely employed (eq 14).[22,23]

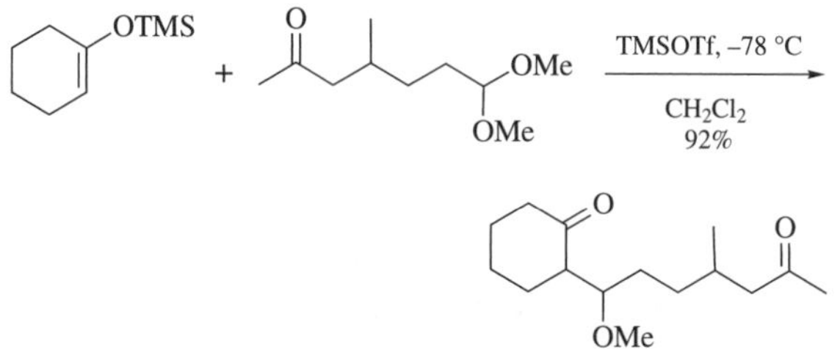

TMSOTf, –78 °C; CH_2Cl_2; 92% (14)

In an extension to this work, TMSOTf catalyzes the first step of a [3 + 2] annulation sequence which allows facile synthesis of fused cyclopentanes possessing bridgehead hydroxy groups (eq 15).[24]

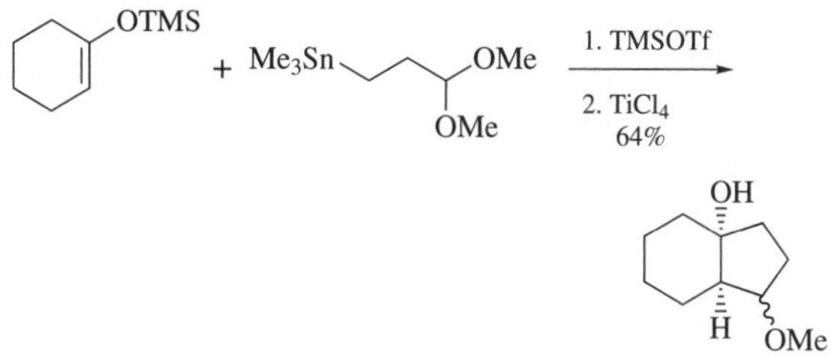

1. TMSOTf; 2. $TiCl_4$; 64% (15)

The use of TMSOTf in aldol reactions of silyl enol ethers and ketene acetals with aldehydes is ubiquitous. Many refinements of the basic reaction have appeared. An example is shown in eq 16.[25]

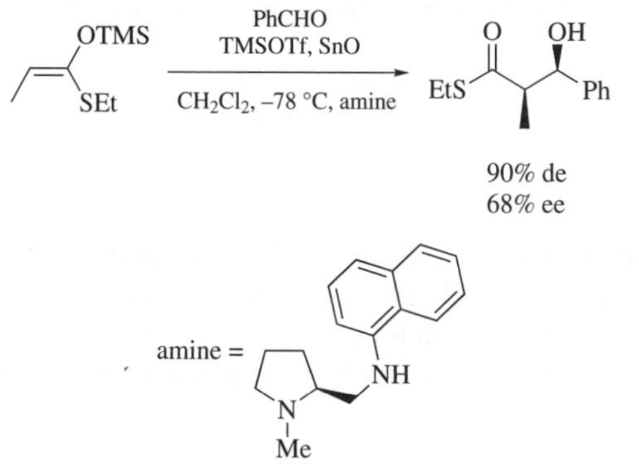

PhCHO, TMSOTf, SnO; CH_2Cl_2, –78 °C, amine (16)

90% de
68% ee

The use of TMSOTf in the reaction of silyl ketene acetals with imines offers an improvement over other methods (such as Ti^{IV}- or Zn^{II}-mediated processes) in that truly catalytic amounts of activator may be used (eq 17);[26] this reaction may be used as the crucial step in a general synthesis of 3-(1′-hydroxyethyl)-2-azetidinones (eq 18).[27]

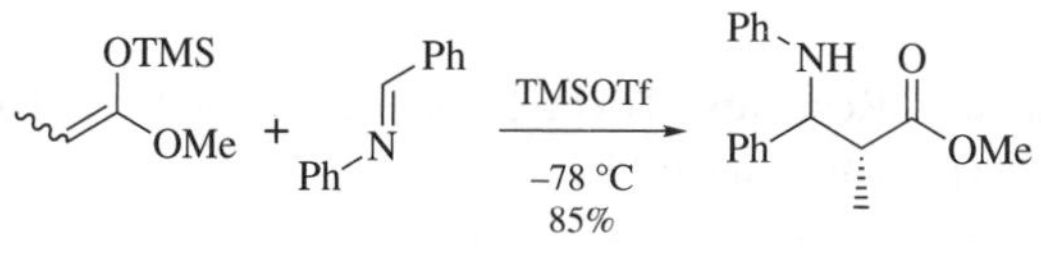

TMSOTf; –78 °C; 85% (17)

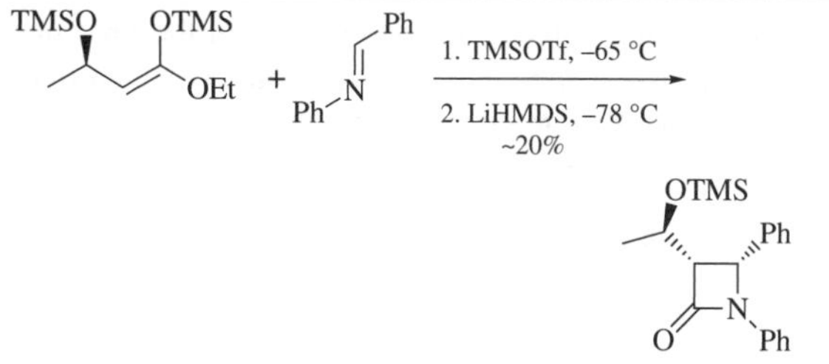

1. TMSOTf, –65 °C; 2. LiHMDS, –78 °C; ~20% (18)

Stereoselective cyclization of α,β-unsaturated enamide esters is induced by TMSOTf and has been used as a route to quinolizidines and indolizidines (eq 19).[28]

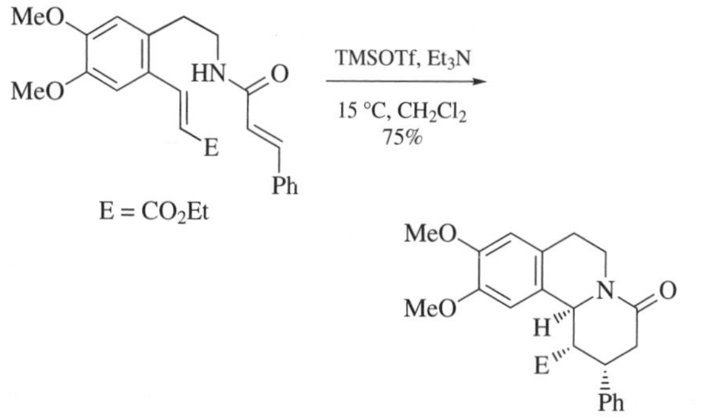

TMSOTf, Et_3N; 15 °C, CH_2Cl_2; 75%; E = CO_2Et (19)

The formation of nitrones by reaction of aldehydes and ketones with ***N*-Methyl-*N*,*O*-bis(trimethylsilyl)hydroxylamine** is accelerated when TMSOTf is used as a catalyst; the acceleration is particularly pronounced when the carbonyl group is under a strong electronic influence (eq 20).[29]

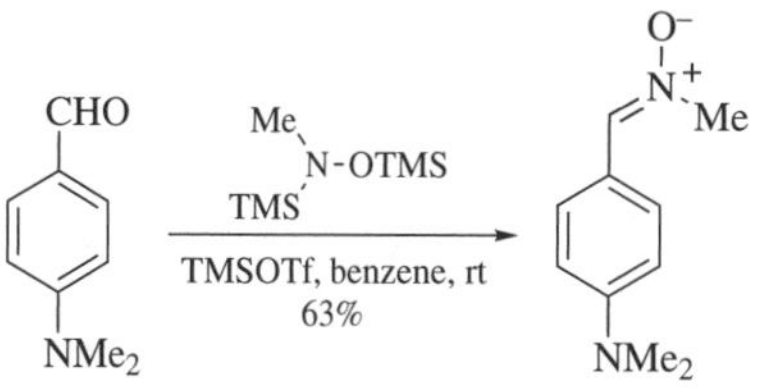

TMSOTf, benzene, rt; 63% (20)

β-Stannylcyclohexanones undergo a stereoselective ring contraction when treated with TMSOTf at low temperature. When other Lewis acids were employed, a mixture of ring-contracted and protiodestannylated products was obtained (eq 21).[30]

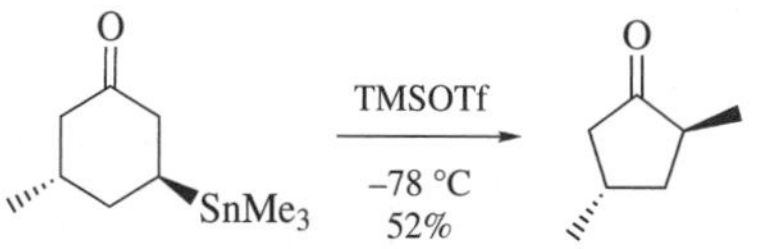

TMSOTf; –78 °C; 52% (21)

The often difficult conjugate addition of alkynyl organometallic reagents to enones is greatly facilitated by TMSOTf. In particular, alkynyl zinc reagents (normally unreactive with α,β-unsaturated carbonyl compounds) add in good yield (eq 22).[31] The proportion of 1,4-addition depends on the substitution pattern of the substrate.

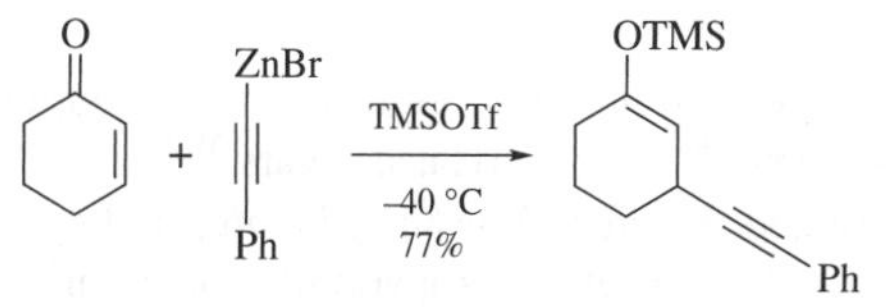

TMSOTf; –40 °C; 77% (22)

The 1,4-addition of phosphines to enones in the presence of TMSOTf gives β-phosphonium silyl enol ethers, which may be deprotonated and alkylated in situ (eq 23).[32]

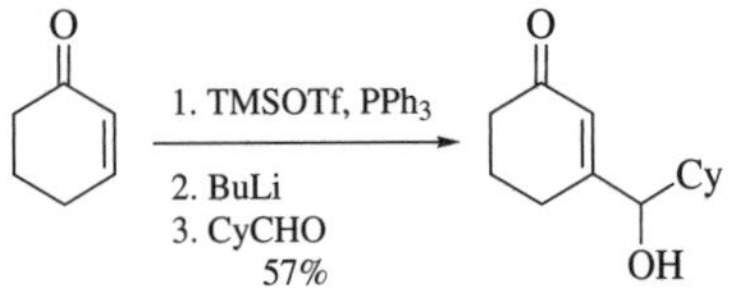

(23)

Miscellaneous. Methyl glucopyranosides and glycopyranosyl chlorides undergo allylation with allylsilanes under TMSOTf catalysis to give predominantly α-allylated carbohydrate analogs (eq 24).[33]

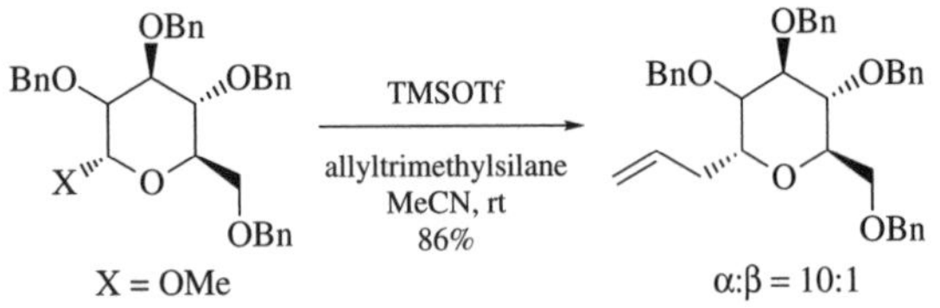

(24)

Glycosidation is a reaction of massive importance and widespread employment. TMSOTf activates many selective glycosidation reactions (eq 25).[34]

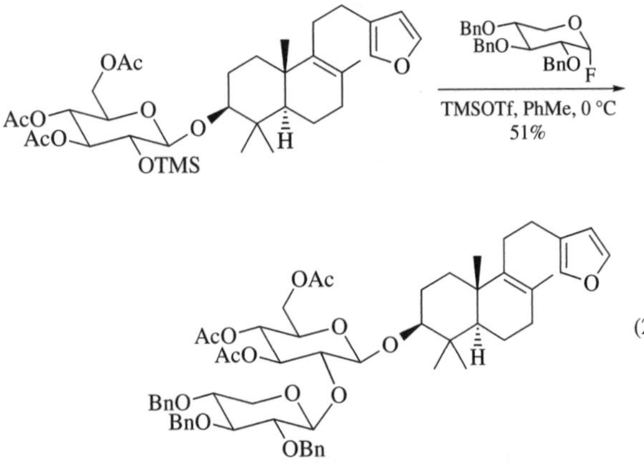

(25)

TMSOTf activation for coupling of 1-*O*-acylated glycosyl donors has been employed in a synthesis of avermectin disaccharides (eq 26).[35]

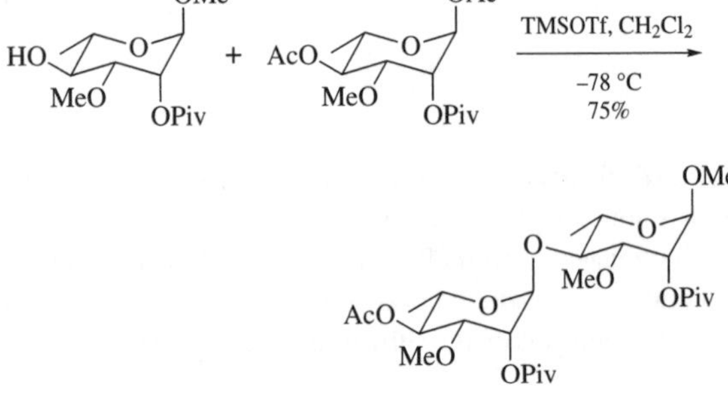

(26)

Similar activation is efficient in couplings with trichloroimidates[36] and *O*-silylated sugars.[37,38]

2-Substituted Δ^3-piperidines may be prepared by the reaction of 4-hydroxy-1,2,3,4-tetrahydropyridines with a variety of carbon and heteronucleophiles in the presence of TMSOTf (eqs 27 and 28).[39]

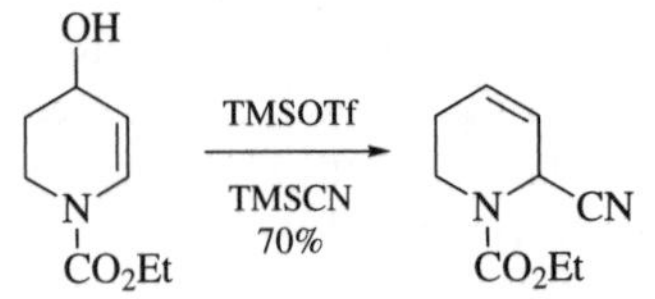

(27)

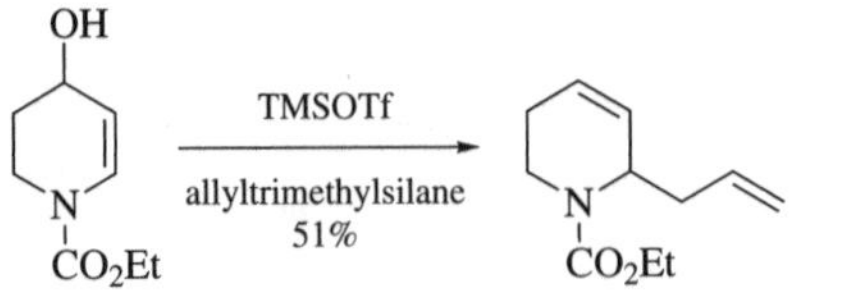

(28)

Iodolactamization is facilitated by the sequential reaction of unsaturated amides with TMSOTf and ***Iodine*** (eq 29).[40]

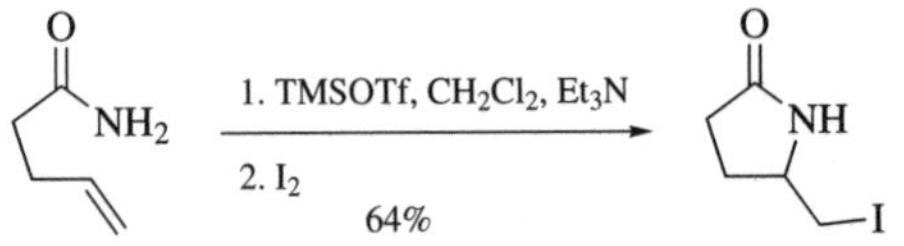

(29)

By use of a silicon-directed Beckmann fragmentation, cyclic (*E*)-β-trimethylsilylketoxime acetates are cleaved in high yield in the presence of catalytic TMSOTf to give the corresponding unsaturated nitriles. Regio- and stereocontrol are complete (eq 30).[41]

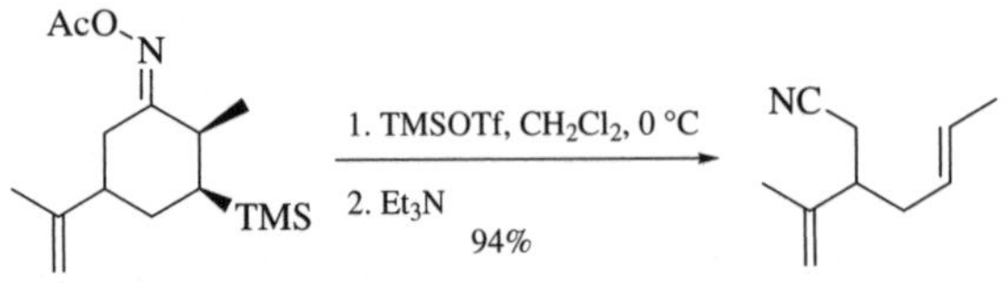

(30)

A general route to enol ethers is provided by the reaction of acetals with TMSOTf in the presence of a hindered base (eq 31).[42] The method is efficient for dioxolanes and noncyclic acetals.

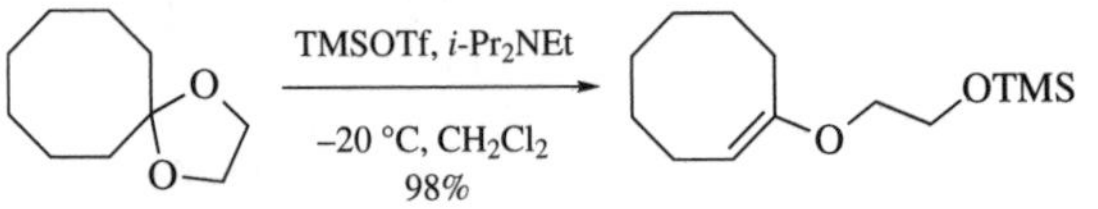

(31)

α-Halo sulfoxides are converted to α-halovinyl sulfides by reaction with excess TMSOTf (eq 32),[43] while α-cyano- and α-alkoxycarbonyl sulfoxides undergo a similar reaction (eq 33).[44] TMSOTf is reported as much superior to ***Iodotrimethylsilane*** in these reactions.

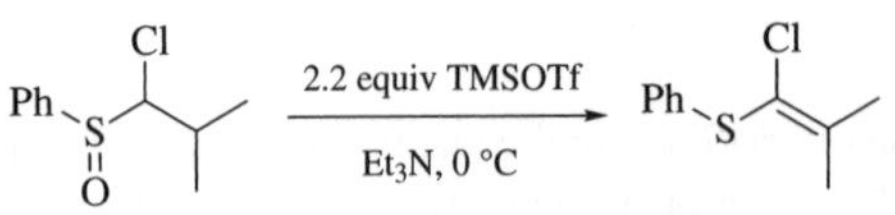

(32)

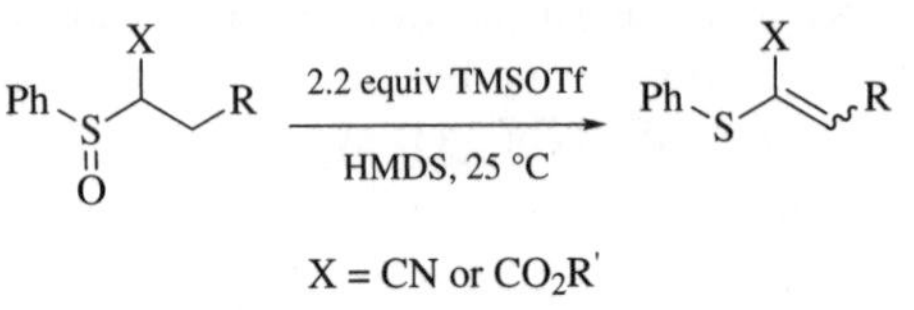

(33)

X = CN or CO_2R'

Related Reagents. Triethylsilyl Perchlorate; Trifluoromethanesulfonic Anhydride.

1. Reviews: (a) Emde, H.; Domsch, D.; Feger, H.; Frick, U.; Götz, H. H.; Hofmann, K.; Kober, W.; Krägeloh, K.; Oesterle, T.; Steppan, W.; West, W.; Simchen, G. *S* **1982**, 1. (b) Noyori, R.; Murata, S.; Suzuki, M. *T* **1981**, *37*, 3899. (c) Stang, P. J.; White, M. R. *Aldrichim. Acta* **1983**, *16*, 15. Preparation: (d) Olah, G. H.; Husain, A.; Gupta, B. G. B.; Salem, G. F.; Narang, S. C. *JOC* **1981**, *46*, 5212. (e) Morita, T.; Okamoto, Y.; Sakurai, H. *S* **1981**, 745. (f) Demuth, M.; Mikhail, G. *S* **1982**, 827. (g) Ballester, M.; Palomo, A. L. *S* **1983**, 571. (h) Demuth, M.; Mikhail, G. *T* **1983**, *39*, 991. (i) Aizpurua, J. M.; Palomo, C. *S* **1985**, 206.
2. Simchen, G.; Kober, W. *S* **1976**, 259.
3. Hergott, H. H.; Simchen, G. *LA* **1980**, 1718.
4. Simchen, G.; Kober, W. *S* **1976**, 259.
5. Emde, H.; Götz, A.; Hofmann, K.; Simchen, G. *LA* **1981**, 1643.
6. Krägeloh, K.; Simchen, G. *S* **1981**, 30.
7. Emde, H.; Simchen, G. *LA* **1983**, 816.
8. Emde, H.; Simchen, G. *S* **1977**, 636.
9. Emde, H.; Simchen, G. *S* **1977**, 867.
10. Yamamoto, Y.; Ohdoi, K.; Nakatani, M.; Akiba, K. *CL* **1984**, 1967.
11. Emde, H.; Götz, A.; Hofmann, K.; Simchen, G. *LA* **1981**, 1643.
12. Matsuda, I.; Sato, S.; Hattori, M.; Izumi, Y. *TL* **1985**, *26*, 3215.
13. Ahlbrecht, H.; Düber, E. O. *S* **1980**, 630.
14. Murata, S.; Suzuki, M.; Noyori, R. *JACS* **1980**, *102*, 2738.
15. Frick, U.; Simchen, G. *S* **1984**, 929.
16. Borgulya, J.; Bernauer, K. *S* **1980**, 545.
17. Tietze, L. F.; Bratz, M. *S* **1989**, 439.
18. Hwu, J. R.; Wetzel, J. M. *JOC* **1985**, *50*, 3946.
19. Hwu, J. R.; Robl, J. A. *JOC* **1987**, *52*, 188.
20. Harada, T.; Hayashiya, T.; Wada, I.; Iwa-ake, N.; Oku, A. *JACS* **1987**, *109*, 527.
21. Pearson, W. H.; Cheng, M-C. *JACS* **1986**, *51*, 3746.
22. Murata, S.; Suzuki, M.; Noyori, R. *JACS* **1980**, *102*, 3248.
23. Murata, S.; Suzuki, M.; Noyori, R. *T* **1988**, *44*, 4259.
24. Lee, T. V.; Richardson, K. A. *TL* **1985**, *26*, 3629.
25. Mukaiyama, T.; Uchiro, H.; Kobayashi, S. *CL* **1990**, 1147.
26. Guanti, G.; Narisano, E.; Banfi, L. *TL* **1987**, *28*, 4331.
27. Guanti, G.; Narisano, E.; Banfi, L. *TL* **1987**, *28*, 4335.
28. Ihara, M.; Tsuruta, M.; Fukumoto, K.; Kametani, T. *CC* **1985**, 1159.
29. Robl, J. A.; Hwu, J. R. *JOC* **1985**, *50*, 5913.
30. Sato, T.; Watanabe, T.; Hayata, T.; Tsukui, T. *CC* **1989**, 153.
31. Kim, S.; Lee, J. M. *TL* **1990**, *31*, 7627.
32. Kim, S.; Lee, P. H. *TL* **1988**, *29*, 5413.
33. Hosomi, A.; Sakata, Y.; Sakurai, H. *TL* **1984**, *25*, 2383.
34. Yamada, H.; Nishizawa, M *T* **1992**, 3021.
35. Rainer, H.; Scharf, H.-D.; Runsink, J. *LA* **1992**, 103.
36. Schmidt, R. R. *AG(E)* **1986**, *25*, 212.
37. Tietze, L.-F.; Fischer, R.; Guder, H.-J. *TL* **1982**, *23*, 4661.
38. Mukaiyama, T.; Matsubara, K. *CL* **1992**, 1041.
39. Kozikowski, A. P.; Park, P. *JOC* **1984**, *49*, 1674.
40. Knapp, S.; Rodriques, K. E. *TL* **1985**, *26*, 1803.
41. Nishiyama, H.; Sakuta, K.; Osaka, N.; Itoh, K. *TL* **1983**, *24*, 4021.
42. Gassman, P. G.; Burns, S. J. *JOC* **1988**, *53*, 5574.
43. Miller, R. D.; Hässig, R., *SC* **1984**, *14*, 1285.
44. Miller, R. D.; Hässig, R., *TL* **1985**, *26*, 2395.

Joseph Sweeney & Gemma Perkins
University of Bristol, UK

Triphenylarsine[1]

Ph_3As

[603-32-7] $C_{18}H_{15}As$ (MW 306.25)

(nucleophilic agent for the synthesis of arsonium salts, which undergo epoxidation or alkenation reactions; can be used as a ligand)

Physical Data: plates (EtOH), mp 60.5 °C; bp 360 °C, 232–234 °C/14 mmHg.
Solubility: insol H_2O, ethanol; sol ether, THF, acetonitrile, etc.
Form Supplied in: commercially available.
Preparative Method: prepared from arsenic trichloride, chlorobenzene, and powdered sodium in benzene under reflux.[2]
Handling, Storage, and Precautions: use in a fume hood.

Introduction. Arsonium ylides are more reactive than the corresponding phosphonium ylides since the 'covalent' canonical form (ylene form) makes a smaller contribution to the overall structure of arsonium ylides (eq 1) than to that of phosphonium ylides.[1] This has been supported by X-ray crystallography.[3]

$$Ph_3As{=}CR^1R^2 \longleftrightarrow Ph_3\overset{+}{As}{-}\overset{-}{C}R^1R^2 \quad (1)$$

Arsonium Salts. Reaction of triphenylarsine with halo compounds forms arsonium salts which are converted to ylides on treatment with base. These perform alkenation reactions under PTC conditions.[4] A general procedure for the synthesis of unsaturated aldehydes, ketones, esters, and amides directly via arsonium salts in the presence of a weak base (solid ***Potassium Carbonate***) under PTC conditions at rt has been devised (eq 2).[5] All the arsonium salts are stable and can be stored for a long time. The Ph_3AsO byproduct can be easily reconverted to Ph_3As by reduction.[6]

$$R(CH{=}CH)_mCHO + Ph_3\overset{+}{As}(CH{=}CH)_nXBr^- \xrightarrow[\text{trace } H_2O\text{, rt} \; -Ph_3AsO]{K_2CO_3\text{ (s), solvent}} R(CH{=}CH)_{m+n+1}X \quad (2)$$

$m = 0, 1, 2$; $n = 0, 1$
X = CHO, COMe, CO_2R, $CONR^1R^2$

Formylmethyltriphenylarsonium bromide (**1**) (eq 3) reacts with aldehydes to give (*E*)-α,β-enals exclusively (eq 4).[7,8]

$$MeCHO + O\langle\;\rangle O{\bullet}Br_2 \longrightarrow BrCH_2CHO \xrightarrow{Ph_3As} Ph_3\overset{+}{As}CH_2CHO\ Br^- \quad (3)$$

(**1**)

$$\text{RCHO} + \text{Ph}_3\overset{+}{\text{As}}\text{CH}_2\text{CHO Br}^- \ (\mathbf{1}) \xrightarrow[\text{rt, }-\text{Ph}_3\text{AsO}\ \ 81\text{–}98\%]{\text{K}_2\text{CO}_3\text{ (s), THF–Et}_2\text{O (trace H}_2\text{O)}} \text{R–CH=CH–CHO}\ (\mathbf{2}) \quad (4)$$

Formylallyltriphenylarsonium bromide (**3**) reacts with aldehydes to give mixtures comprising mostly (2*E*,4*E*)- with some (2*E*,4*Z*)-dienals. The latter can be isomerized to the former by treating with a catalytic amount of ***Iodine*** in daylight (eq 5).[9]

$$\text{RCHO} + \text{Ph}_3\overset{+}{\text{As}}\text{CH}_2\text{CH=CHCHO Br}^- \ (\mathbf{3}) \xrightarrow[\text{rt, }-\text{Ph}_3\text{AsO}]{\text{K}_2\text{CO}_3\text{ (s), THF–Et}_2\text{O (trace H}_2\text{O)}}$$

$$\text{R–CH=CH–CH=CH–CHO} \xrightarrow{\text{I}_2,\ h\nu} \text{R–CH=CH–CH=CH–CHO}\ (\mathbf{4}) \quad (5)$$

2-(Oxoamido)triphenylarsonium bromides (**5**)[10] react with saturated and unsaturated aldehydes at rt in the presence of K_2CO_3(s) to afford (2*E*)-unsaturated amides (**6**) or (2*E*,4*E*)-dienamides in excellent yields. No (*Z*) stereoisomer is detected (eq 6).[10] A vinylog (**7**)[11] reacts with aromatic aldehydes to give exclusively the (2*E*,4*E*)-products in 80–98% yield (eq 7); with aliphatic aldehydes, (2*E*,4*E*/2*E*,4*Z*) products are formed in a ratio of 85/15. The (2*E*,4*Z*) product can be isomerized to the (2*E*,4*E*) product.[11] The reactions of aldehydes with arsonium salts are listed in Table 1.

$$\text{RCHO} + \text{Ph}_3\overset{+}{\text{As}}\text{CH}_2\text{CONR}^1\text{R}^2\text{ Br}^- \ (\mathbf{5}) \xrightarrow[\text{or MeCN–HCONH}_2\text{, K}_2\text{CO}_3\text{, rt}]{\text{THF or MeCN (trace H}_2\text{O)}} \text{R–CH=CH–C(O)–NR}^1\text{R}^2\ (\mathbf{6}) \quad (6)$$

$$\text{RCHO} + \text{Ph}_3\overset{+}{\text{As}}\text{CH}_2\text{CH=CHC(O)NH}i\text{-Bu Br}^- \ (\mathbf{7}) \xrightarrow{\text{as eq (6)}} \xrightarrow[\text{CH}_2\text{Cl}_2,\ h\nu]{\text{cat. I}_2} \text{R–CH=CH–CH=CH–C(O)–NH}i\text{-Bu} \quad (7)$$

Triphenylarsonium salts have been used to synthesize a variety of natural products under very mild conditions: (*E*,*Z*)-,diene sex pheromones,[12] pellitorine,[10b] trichonine,[13] *Achillea* amide,[10b] *Otanthus maritima* amide,[10a] (+)-yingzhaosu A,[14] LTA_4 methyl ester,[15] 19-hydroxy-LTB_4,[16] and lipoxins A_4 and B_4[17] have all been successfully synthesized.

Since tributylarsine is more reactive towards halides than triphenylarsine, Wittig-type alkenation of carbonyl compounds can be performed catalytically. Thus the reaction of various aldehydes with ***Methyl Bromoacetate*** (or ***Bromoacetone***) in the presence of ***Triphenyl Phosphite*** and Bu_3As (0.2 equiv) provides α,β-unsaturated esters (or α,β-unsaturated ketones) in 60–87% yields and with (*E*)/(*Z*) ratios of 97:3–99:1 (eq 8).[18]

$$\text{RCHO} + \text{BrCH}_2\text{CO}_2\text{Me} + (\text{PhO})_3\text{P} + \text{K}_2\text{CO}_3 \xrightarrow[\text{THF–MeCN, rt}]{\text{cat. Bu}_3\text{As}} \text{RCH=CHCO}_2\text{Me} + (\text{PhO})_3\text{PO} \quad (8)$$

An efficient one-pot synthesis of α-iodo- α,β-unsaturated esters, ketones, and nitriles via arsonium salts has been reported (eq 9).[19] A one-pot synthesis of *trans*-fluorovinylic epoxides has also been achieved (eq 10).[20]

$$\text{Ph}_3\overset{+}{\text{As}}\text{CH}_2\text{X Br}^- \xrightarrow[\text{MeCN, 10 °C}]{\text{I}_2\text{, K}_2\text{CO}_3\text{ (s)}} \text{Ph}_3\overset{+}{\text{As}}\text{CHIX I}^- \xrightarrow[\text{trace H}_2\text{O, 25 °C}\ \ 60\text{–}97\%]{\text{RCHO, K}_2\text{CO}_3\text{ (s)}} \text{RCH=CHIX} \quad (9)$$

X = CO_2Me, CN, COMe (*Z*):(*E*) = 61:39–95:5

$$\text{Ph}_3\overset{+}{\text{As}}\text{–}\overset{-}{\text{C}}(\text{CF}_3)\text{–CH=CMe}_2 \xrightarrow{\text{RCHO}} \text{R–epoxide–C(F}_3\text{C)=CMe}_2 \quad (10)$$

Silylated enynyl carboxylic esters can be synthesized with high stereoselectivity by the reaction of trimethylsilyl-2-propynylidenetriphenylarsorane, prepared in situ from the corresponding arsonium salt with ***n-Butyllithium***, and $BrCH_2CO_2R$ (eq 11).[21] Triphenylarsoranylideneketene reacts with 2-benzoylpyrrole in methylene chloride to give 1-phenylpyrrolizin-3-one in 85% yield (eq 12).[22]

$$\text{Ph}_3\overset{+}{\text{As}}\text{–}\overset{-}{\text{C}}(\text{COR})\text{–C≡C–TMS} \xrightarrow[46\text{–}60\%]{-\text{Ph}_3\text{As}} \text{TMS–C≡C–CH=CH–CO}_2\text{R} \quad (11)$$

(*E*):(*Z*) ≈ 50:50

$$\text{Ph}_3\text{As=C=C=O} + \text{2-benzoylpyrrole} \longrightarrow [\text{Ph}_3\text{As=CH–C(O)–N-pyrrolyl–C(O)Ph}] \xrightarrow[85\%]{-\text{Ph}_3\text{AsO}} \text{1-phenylpyrrolizin-3-one} \quad (12)$$

Methylenetriphenylarsorane ($Ph_3As{=}CH_2$), which is thermally unstable both in the solid state and in solution, has been isolated and characterized by analytical and spectroscopic methods.[23] Triphenylarsonium ethylide ($Ph_3As{=}CHCH_3$), prepared from triphenylethylarsonium tetrafluoroborate with ***Potassium Hexamethyldisilazide*** in THF/HMPT at −40 °C, reacts with aliphatic aldehydes to give *trans*-epoxides with high selectivity (eq 13). Stereoselection is lower with aromatic aldehydes (83% (*E*) for benzaldehyde). The reagent also reacts with ketones to form trisubstituted epoxides.[24]

$$\text{Me(CH}_2)_6\text{CHO} \xrightarrow[-78\text{ to }25\text{ °C}\ \ 80\%]{\text{Ph}_3\text{AsCHMe}} \textit{trans}\text{-Me(CH}_2)_6\text{-epoxide-Me} \quad (13)$$

99% (*E*)

The synthesis and reactivity of (3,3-diisopropoxypropyl)triphenylarsonium ylide have been reported. This

Table 1 Reaction of Aldehydes (RCHO) with Arsonium Salts

Reagent		Mp (°C)	Product	Yield (%)	Ratio of isomer
$Ph_3\overset{+}{A}sCH_2CHO\ Br^-$	**(1)**	160–161	RCH=CHCHO	81–98	only (*E*)
$Ph_3\overset{+}{A}sCH_2CH{=}CHCHO\ Br^-$	**(2)**	153–154	RCH=CHCH=CHCHO	79–98	(2*E*,4*E*):(2*E*,4*Z*) = 1:1–4:1
$Ph_3\overset{+}{A}sCH_2CO$-pyrrolidinyl Br^-	**(5a)**	191–192	RCH=CHCO-pyrrolidinyl	81–98	only (*E*)
$Ph_3\overset{+}{A}sCH_2CO$-piperidinyl Br^-	**(5b)**	167–168	RCH=CHCO-piperidinyl	88–99	only (*E*)
$Ph_3As^+CH_2CONH$-*i*-Bu Br^-	**(5c)**	174–175	RCH=CHCONH-*i*-Bu	80–98	only (*E*)

reagent can be considered as a β-formylvinyl anion equivalent (eq 14), as shown by the conversion of aldehydes to 4-hydroxy-2(*E*)-enals under very mild conditions (eq 15).[25] This route has been successfully applied to the total synthesis of (±)-hepoxilin A_3.[26]

$$Ph_3\overset{+}{A}s\overset{-}{C}HCH_2CH(O\text{-}i\text{-}Pr)_2 \equiv \overset{-}{C}H{=}CHCHO \quad (14)$$

$$Ph_3\overset{+}{A}s{-}\overset{-}{C}H_2 + ICH_2CH(O\text{-}i\text{-}Pr)_2 \xrightarrow[80\%]{THF\text{–}HMPA} Ph_3\overset{+}{A}sCH_2CH_2CH(O\text{-}i\text{-}Pr)_2\ I^- \xrightarrow[\text{2. RCHO, –78 °C; 56–70\%}]{\text{1. LDA, THF, –40 °C}}$$

$$\text{R-epoxide-CH}_2CH(O\text{-}i\text{-}Pr)_2 \xrightarrow{TFA,\ CHCl_3} \text{R-epoxide-CH}_2CHO \xrightarrow[65\text{–}86\%]{Et_3N,\ Et_2O} RCH(OH)CH{=}CHCHO \quad (15)$$

Homologation of aldehydes using (phenylthiomethylene)triphenylarsorane has been reported. Reaction with aldehydes gives exclusively α-phenylthio epoxides in THF and enol phenol thioethers in THF/HMPA. The former adducts are readily transformed to α-thiophenoxy carbonyl compounds and the latter to one-carbon homologated aldehydes (eq 16).[27]

$$\text{R(R')(SPh)-epoxide} \xleftarrow[THF]{Ph_3As{=}CR'SPh} RCHO \xrightarrow[THF\text{–}HMPA]{Ph_3As{=}CR'SPh} RCH{=}CR'SPh \quad (16)$$

$$\downarrow \qquad\qquad\qquad\qquad \downarrow$$

$$RCOCHR'SPh \longrightarrow RCOCH_2R' \qquad\qquad RCH_2COR'$$

Large rate enhancement in Stille cross-coupling reactions is observed with triphenylarsine (a factor of 70 over the triphenyl phosphine-based catalyst) (eq 17).[28]

$$4\text{-}AcC_6H_4OTf + CH_2{=}CHSnBu_3 \xrightarrow[Ph_3As,\ 40\ °C;\ 95\%]{Pd_2(dba)_3,\ NMP} 4\text{-}AcC_6H_4CH{=}CH_2 \quad (17)$$

Related Reagents. Dibutyl Telluride; Tri-*n*-butylstibine; Triphenylphosphine.

1. (a) Huang, Y. Z.; Shen, Y. C. *Adv. Organomet. Chem.* **1982**, *20*, 115. (b) Huang, Y. Z.; Xu, Y.; Li, Z. *OPP* **1982**, *14*, 373. (c) Lloyd, D.; Gosney, I.; Ormiston, R. A. *CSR* **1987**, *16*, 45.
2. Shriner, R. L.; Wolf, C. N. *OSC* **1963**, *4*, 910.
3. Shao, M. C.; Jin, X. L.; Tang, Y. Q.; Huang, Q. C.; Huang, Y. Z. *TL* **1982**, *23*, 5343.
4. Shi, L. L.; Xiao, W.; Ge, Y.; Huang, Y. Z. *Acta Chim. Sin.* **1986**, *44*, 421.
5. (a) Huang, Y. Z.; Shi, L. L.; Yang, J. H.; Xiao, W. J. *Youji Huaxue* **1988**, 10. (b) Huang, Y. Z.; Shi, L. L.; Yang, J. H.; Xiao, W. J.; Li, S. W.; Wang, W. B. In *Heteroatom Chemistry*; Block, E., Ed.; VCH: New York 1990; pp 189–206.
6. (a) Xing, Y. D.; Hou, X. L.; Huang, N. Z. *TL* **1981**, *22*, 4727. (b) Lu, X.; Wang, Q. W.; Tao, X. C.; Sun, J. H.; Lei, G. X. *Acta Chim. Sin.* **1985**, *43*, 450.
7. Huang, Y. Z.; Shi, L. L.; Yang, J. H. *TL* **1985**, *26*, 6447.
8. Billimoria, J. D.; Maclagan, N. F. *JCS* **1954**, 3257.
9. Yang, J. H.; Shi, L. L.; Xiao, W. J.; Wen, X. Q.; Huang, Y. Z. *HC* **1990**, *1*, 75.
10. (a) Huang, Y. Z.; Shi, L. L.; Yang, J. H.; Zhang, J. T. *TL* **1987**, *28*, 2159. (b) Shi, L. L.; Yang, J. H.; Wen, X. Q.; Huang, Y. Z. *TL* **1988**, *29*, 3949.
11. Yang, J. H. Doctoral Dissertation, Shanghai Institute of Organic Chemistry, 1988.
12. Huang, Y. Z.; Shi, L. L.; Yang, J. H.; Cai, Z. W. *JOC* **1987**, *52*, 3558.
13. Shi, L. L.; Yang, J. H.; Li, M.; Huang, Y. Z. *LA* **1988**, 377.
14. Xu, X. X.; Zhu, J.; Huang, D. Z.; Zhou, W. S. *TL* **1991**, *32*, 5785.

15. Wang, Y. F.; Li, J. C.; Wu, Y. L.; Huang, Y. Z.; Shi, L. L.; Yang, J. H. *TL* **1986**, *27*, 4583.
16. Le Merrer, Y.; Bonnet, A.; Depezay, J. C. *TL* **1988**, *29*, 2647.
17. Gravier-Pelletier, C.; Dumas, J.; LeMerrer, Y.; Depezay, J. C. *TL* **1991**, *32*, 1165.
18. Shi, L. L.; Wang, W. B.; Wang, Y. C.; Huang, Y. Z. *JOC* **1989**, *54*, 2027. See also *CHEMTRACTS – Org. Chem.* **1989**, *2*, 300.
19. Huang, Y. Z.; Shi, L. L.; Li, S. W.; Huang, R. *SC* **1989**, *19*, 2639.
20. Shen, Y.; Liao, Q.; Qiu, W. *CC* **1988**, 1309.
21. Shen, Y.; Xiang, Y. *HC* **1992**, *3*, 547.
22. Bestmann, H. J.; Bansal, R. K. *TL* **1981**, *22*, 3839.
23. Yamamoto, Y.; Schmidbaur, H. *CC* **1975**, 668.
24. Still, W. C.; Novack, V. J. *JACS* **1981**, *103*, 1283.
25. Chabert, P.; Ousset, J. B.; Mioskowski, C. *TL* **1989**, *30*, 179.
26. Chabert, P.; Mioskowski, C.; Falck, J. R. *TL* **1989**, *30*, 2545.
27. Boubia, B.; Mioskowski, C.; Manna, S.; Falck, J. R. *TL* **1989**, *30*, 6023.
28. Farina, V.; Krishnan, B. *JACS* **1991**, *113*, 9585.

Yao-Zeng Huang, Li-Lan Shi & Zhang-Lin Zhou
Shanghai Institute of Organic Chemistry, Academia Sinica, China

Triphenylcarbenium Tetrafluoroborate

$Ph_3C^+ BF_4^-$

[341-02-6] $C_{19}H_{15}BF_4$ (MW 330.15)

(easily prepared[1] hydride abstractor used for conversion of dihydroaromatics to aromatics,[2–4] and the preparation of aromatic and benzylic cations;[5–8] oxidative hydrolysis of ketals[9] and thioketals;[10] conversion of acetonides to α-hydroxy ketones;[9] oxidation of acetals[11] and thioacetals;[12] selective oxidation of alcohols and ethers to ketones;[9,13–15] oxidation of silyl enol ethers to enones;[16] hydrolysis of TBS and MTM ethers;[17] oxidation of amines and amides to iminium salts;[18–20] oxidation of organometallics to give alkenes;[21–23] sensitizer for photooxidation using molecular oxygen;[24] Lewis acid catalyst for various reactions;[25] polymerization catalyst;[26] other reactions[27–30])

Alternate Name: trityl fluoroborate.
Physical Data: mp ~200 °C (dec).
Solubility: sol most standard organic solvents; reacts with some nucleophilic solvents.
Form Supplied in: yellow solid; commercially available.
Preparative Methods: the most convenient procedure involves the reaction of Ph_3CCl with ***Silver(I) Tetrafluoroborate*** in ethanol.[1b] The most economical route employs the reaction of Ph_3CCl with the anhydrous ***Tetrafluoroboric Acid***–Et_2O complex.[1c]
Purification: recrystallization of commercial samples from a minimal amount of dry MeCN provides material of improved purity, but the recovery is poor.[1a]
Handling, Storage, and Precautions: moisture-sensitive and corrosive. Recrystallized reagent can be stored at rt for several months in a desiccator without significant decomposition. This compound is much less light-sensitive than other trityl salts such as the perchlorate.[1a]

Preparation of Aromatic Compounds via Dehydrogenation. Dihydroaromatic compounds are easily converted into the corresponding aromatic compound by treatment with triphenylcarbenium tetrafluoroborate followed by base.[2] Certain α,α-disubstituted dihydroaromatics are converted to the 1,4-dialkylaromatic compounds with rearrangement (eq 1).[3] Nonbenzenoid aromatic systems, e.g. benzazulene[4a] or dibenzosesquifulvalene,[4b] are readily prepared from their dihydro counterparts. Aromatic cations are also easily prepared by hydride abstraction, for example, tropylium ion (e.g. in the synthesis of heptalene (eq 2)),[5] cyclopropenyl cation,[6] and others, including heterocyclic systems.[7] Some benzylic cations, especially ferrocenyl cations,[8] can also be formed by either hydride abstraction or trityl addition.

Bn Bn — $Ph_3C^+BF_4^-$, AcOH, Δ, 0.5 h, 67% → Bn, Bn (1)

1. $Ph_3C^+ BF_4^-$ 88%, 2. Me_3N, 41% (2)

Oxidation by Hydride Abstraction. In the early 1970s, Barton developed a method for the oxidative hydrolysis of ketals to ketones, e.g. in the tetracycline series (eq 3).[9] The same conditions can also be used to hydrolyze thioketals.[10] Acetonides of 1,2-diols are oxidized to the α-hydroxy ketones in good yield by this reagent (eq 4).[9] The hydrogen of acetals is easily abstracted (eq 5), providing a method for the conversion of benzylidene units in sugars to the hydroxy benzoates.[11] The hydrogen of dithioacetals is also abstracted to give the salts.[12] Since benzylic hydrogens are readily abstracted, this is also a method for deprotection of benzyl ethers.[9,13] Trimethylsilyl, *t*-butyl, and trityl ethers of simple alcohols are oxidized to the corresponding ketones and aldehydes in good yield. Primary–secondary diols are selectively oxidized at the secondary center to give hydroxy ketones by this method (eq 6).[14] 2,2-Disubstituted 1,4-diols are oxidized only at the 4-position to give the corresponding lactones.[15] Trimethylsilyl enol ethers are oxidized to α,β-unsaturated ketones, thereby providing a method for ketone to enone conversion (eq 7).[16] *t*-Butyldimethylsilyl (TBDMS) ethers are not oxidized but rather hydrolyzed to the alcohols, as are methylthiomethyl (MTM) ethers.[17] Benzylic amines and amides can be oxidized to the iminium salts,[18] allylic amines and enamines afford eniminium salts,[19] and orthoamides give triaminocarbocations.[20]

O H ... HO O O ... O Ph — $Ph_3C^+ BF_4^-$, 65% → O H ... H O ... O Ph (3)

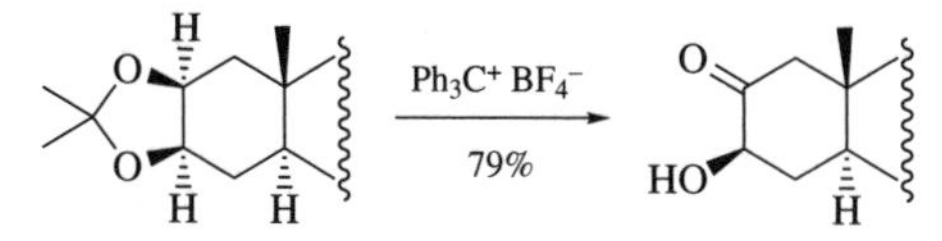

(4)

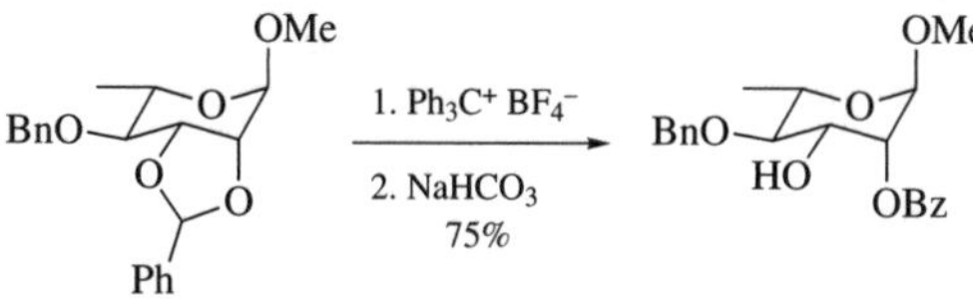

(5)

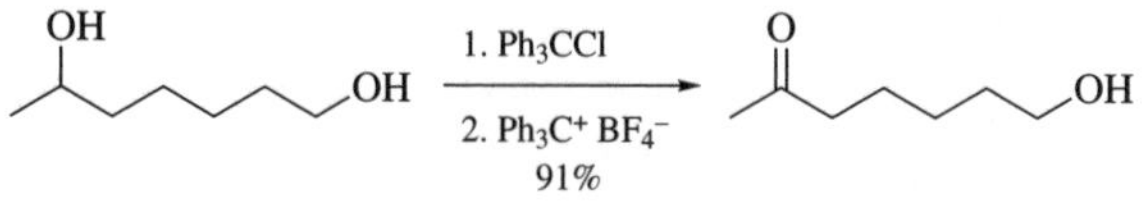

(6)

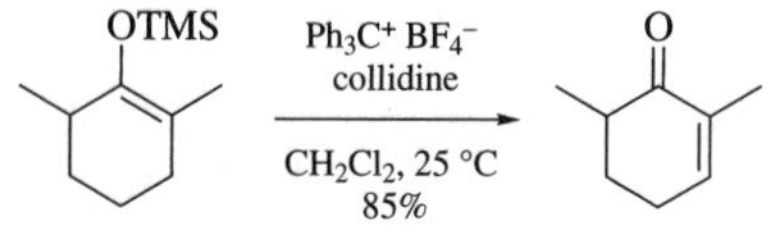

(7)

Generation of Alkenes from Organometallics. Various β-metalloalkanes can be oxidized by trityl fluoroborate to the corresponding alkenes.[21–23] The highest yields are obtained for the β-iron derivatives (eq 8), which are easily prepared from the corresponding halides or tosylates.[21] Grignard reagents and organolithiums also undergo this reaction (eq 9),[22] as do Group 14 organometallics (silanes, stannanes, etc.).[23]

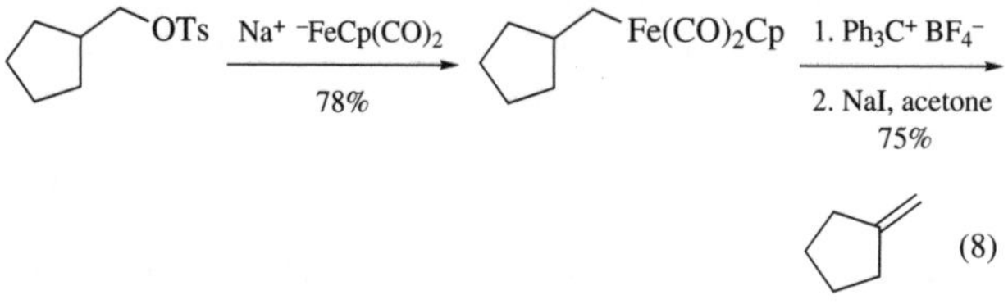

(8)

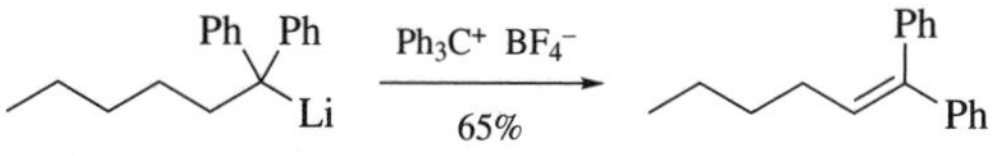

(9)

Sensitizer of Photooxygenation. Barton showed that oxygen, in the presence of trityl fluoroborate and ordinary light, adds to cisoid dienes at −78 °C in very high yields.[24] For example, the peroxide of ergosterol acetate is formed in quantitative yields under these conditions (eq 10),[24a,b] which have been used also for photocycloreversions of cyclobutanes.[24c]

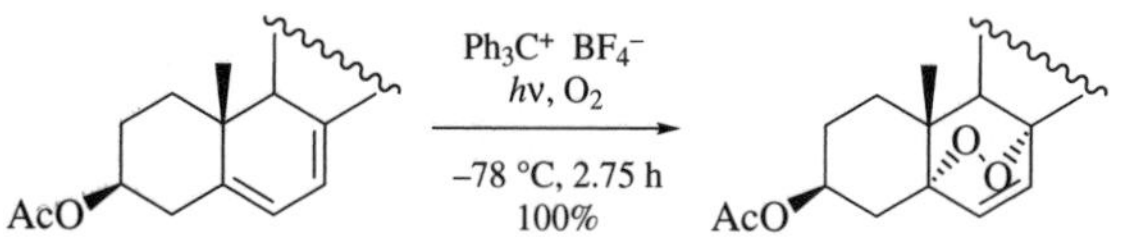

(10)

Lewis Acid Catalysis. Trityl fluoroborate is a good Lewis acid for various transformations,[25] e.g. the Mukaiyama-type aldol reaction using a dithioacetal and silyl enol ether (eq 11).[25a] It has also been used as the catalyst for the formation of glycosides from alcohols and sugar dimethylthiophosphinates (eq 12)[25b] and for the formation of disaccharides from a protected α-cyanoacetal of glucose and a 6-*O*-trityl hexose.[25c] Michael additions of various silyl nucleophiles to conjugated dithiolenium cations also proceed well (eq 13).[25d,e] Finally, the [4 + 2] cycloaddition of cyclic dienes and oxygenated allyl cations has been effected with trityl fluoroborate.[25f]

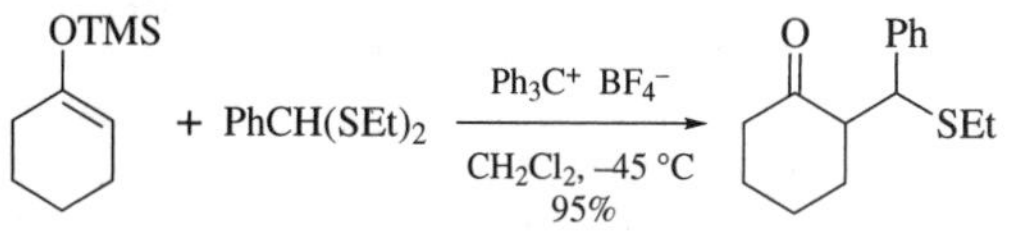

(11)

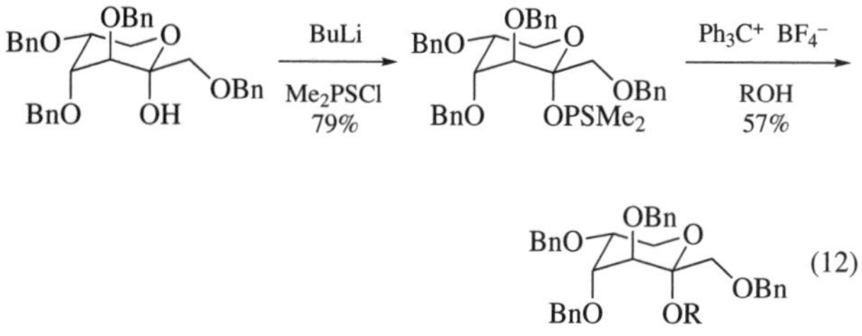

(12)

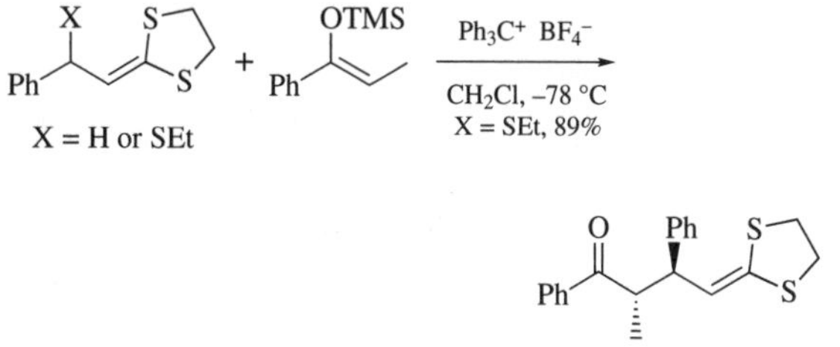

(13)

Polymerization Catalyst. Several types of polymerization[26] have been promoted by trityl fluoroborate, including reactions of orthocarbonates[26a] and orthoesters,[26b–d] vinyl ethers,[26e–g] epoxides,[26h,i] and lactones.[26j,k]

Other Reactions. Trityl fluoroborate has been used often to prepare cationic organometallic complexes, as in the conversion of dienyl complexes of iron, ruthenium, and osmium into their cationic derivatives.[27] It alkylates pyridines on the nitrogen atom in a preparation of dihydropyridines[28a] and acts as a tritylating agent.[28b] It has also been used in attempts to form silyl cations and silyl fluorides from silanes.[29] Finally, it has been reported to be a useful desiccant.[30]

Related Reagents. Trimethylsilyl Trifluoromethanesulfonate; Triphenylmethyl Perchlorate.

1. (a) Dauben, H. J., Jr.; Honnen, L. R.; Harmon, K. M. *JOC* **1960**, *25*, 1442. (b) Fukui, K.; Ohkubo, K.; Yamabe, T. *BCJ* **1969**, *42*, 312. (c) Olah, G. A.; Svoboda, J. J.; Olah, J. A. *S* **1972**, 544.
2. (a) Müller, P. *HCA* **1973**, *56*, 1243. (b) Giese, G.; Heesing, A. *CB* **1990**, *123*, 2373.
3. (a) Karger, M. H.; Mazur, Y. *JOC* **1971**, *36*, 540. (b) Acheson, R. M.; Flowerday, R. F. *JCS(P1)* **1975**, 2065.
4. (a) O'Leary, M. A.; Richardson, G. W.; Wege, D. *T* **1981**, *37*, 813. (b) Prinzbach, H.; Seip, D.; Knothe, L.; Faisst, W. *LA* **1966**, *698*, 34.
5. (a) Dauben, H. J., Jr.; Gadecki, F. A.; Harmon, K. M.; Pearson, D. L. *JACS* **1957**, *79*, 4557. (b) Dauben, H. J., Jr.; Bertelli, D. J. *JACS* **1961**, *83*, 4657, 4659. (c) Peter-Katalinic, J.; Zsindely, J.; Schmid, H. *HCA* **1973**, *56*, 2796. (d) Vogel, E.; Ippen, J. *AG(E)* **1974**, *13*, 734. (e) Beeby, J.; Garratt, P. J. *JOC* **1973**, *38*, 3051. (f) Murata, I.; Yamamoto, K.; Kayane, Y. *AG(E)* **1974**, *13*, 807, 808. (g) Kuroda, S.; Asao, T. *TL* **1977**, 285. (h) Komatsu, K.; Takeuchi, K.; Arima, M.; Waki, Y.; Shirai, S.; Okamoto, K. *BCJ* **1982**, *55*, 3257. (i) Müller, J.; Mertschenk, B. *CB* **1972**, *105*, 3346. (j) Schweikert, O.; Netscher, T.; Knothe, L.; Prinzbach, H. *CB*

1984, *117*, 2045. (k) Bindl, J.; Seitz, P.; Seitz, U.; Salbeck, E.; Salbeck, J.; Daub, J. *CB* **1987**, *120*, 1747.

6. (a) Zimmerman, H. E.; Aasen, S. M. *JOC* **1978**, *43*, 1493. (b) Komatsu, K.; Tomioka, I.; Okamoto, K. *BCJ* **1979**, *52*, 856.
7. (a) Yamamura, K.; Miyake, H.; Murata, I. *JOC* **1986**, *51*, 251. (b) Matsumoto, S.; Masuda, H.; Iwata, K.; Mitsunobu, O. *TL* **1973**, 1733. (c) Yano, S.; Nishino, K.; Nakasuji, K.; Murata, I. *CL* **1978**, 723. (d) Kedik, L. M.; Freger, A. A.; Viktorova, E. A. *KGS* **1976**, *12*, 328 (*Chem. Heterocycl. Compd. (Engl. Transl.)* **1976**, *12*, 279). (e) Reichardt, C.; Schäfer, G.; Milart, P. *CCC* **1990**, *55*, 97.
8. (a) Müller, P. *HCA* **1973**, *56*, 500. (b) Boev, V. I.; Dombrovskii, A. V. *ZOB* **1987**, *57*, 938, 633. (c) Klimova, E. I.; Pushin, A. N.; Sazonova, V. A. *ZOB* **1987**, *57*, 2336. (d) Abram, T. S.; Watts, W. E. *JCS(P1)* **1975**, 113; *JOM* **1975**, *87*, C39. (e) Barua, P.; Barua, N. C.; Sharma, R. P. *TL* **1983**, *24*, 5801. (f) Akgun, E.; Tunali, M. *AP* **1988**, *321*, 921.
9. (a) Barton, D. H. R.; Magnus, P. D.; Smith, G.; Strecker, G.; Zurr, D. *JCS(P1)* **1972**, 542. (b) Barton, D. H. R.; Magnus, P. D.; Smith, G.; Zurr, D. *CC* **1971**, 861.
10. Ohshima, M.; Murakami, M.; Mukaiyama, T. *CL* **1986**, 1593.
11. (a) Hanessian, S.; Staub, A. P. A. *TL* **1973**, 3551. (b) Jacobsen, S.; Pedersen, C. *ACS* **1974**, *28B*, 1024, 866. (c) Wessel, H.-P.; Bundle, D. R. *JCS(P1)* **1985**, 2251.
12. (a) Nakayama, J.; Fujiwara, K.; Hoshino, M. *CL* **1975**, 1099; *BCJ* **1976**, *49*, 3567. (b) Nakayama, J.; Imura, M.; Hoshino, M. *BCJ* **1980**, *53*, 1661. (c) Nakayama, J. *BCJ* **1982**, *55*, 2289. (d) Bock, H.; Brähler, G.; Henkel, U.; Schlecker, R.; Seebach, D. *CB* **1980**, *113*, 289. (e) Neidlein, R.; Droste-Tran-Viet, D.; Gieren, A.; Kokkinidis, M.; Wilckens, R.; Geserich, H.-P.; Ruppel, W. *HCA* **1984**, *67*, 574. (f) However, azide abstraction is seen with azidodithioacetals: Nakayama, J.; Fujiwara, K.; Hoshino, M. *JOC* **1980**, *45*, 2024.
13. (a) Barton, D. H. R.; Magnus, P. D.; Streckert, G.; Zurr, D. *CC* **1971**, 1109. (b) Doyle, M. P.; Siegfried, B. *JACS* **1976**, *98*, 163. (c) Hoye, T. R.; Kurth, M. J. *JACS* **1979**, *101*, 5065. (d) For simple ethers, see: Deno, N. C.; Potter, N. H. *JACS* **1967**, *89*, 3550.
14. (a) Jung, M. E. *JOC* **1976**, *41*, 1479. (b) Jung, M. E.; Speltz, L. M. *JACS* **1976**, *98*, 7882. (c) Jung, M. E.; Brown, R. W. *TL* **1978**, 2771.
15. Doyle, M. P.; Dow, R. L.; Bagheri, V.; Patrie, W. J. *JOC* **1983**, *48*, 476; *TL* **1980**, *21*, 2795.
16. (a) Jung, M. E.; Pan, Y.-G.; Rathke, M. W.; Sullivan, D. F.; Woodbury, R. P. *JOC* **1977**, *42*, 3961. (b) Reetz, M. T.; Stephan, W. *LA* **1980**, 533.
17. (a) Metcalf, B. W.; Burkhardt, J. P.; Jund, K. *TL* **1980**, *21*, 35. (b) Chowdhury, P. K.; Sharma, R. P.; Baruah, J. N. *TL* **1983**, *24*, 4485. (c) Niwa, H.; Miyachi, Y. *BCJ* **1991**, *64*, 716.
18. (a) Damico, R.; Broaddus, C. D. *JOC* **1966**, *31*, 1607. (b) Barton, D. H. R.; Bracho, R. D.; Gunatilaka, A. A. L.; Widdowson, D. A. *JCS(P1)* **1975**, 579. (c) Wanner, K. T.; Praschak, I.; Nagel, U. *AP* **1990**, *322*, 335; *H* **1989**, *29*, 29.
19. Reetz, M. T.; Stephan, W.; Maier, W. F. *SC* **1980**, *10*, 867.
20. Erhardt, J. M.; Grover, E. R.; Wuest, J. D. *JACS* **1980**, *102*, 6365.
21. (a) Laycock, D. E.; Hartgerink, J.; Baird, M. C. *JOC* **1980**, *45*, 291. (b) Laycock, D. E.; Baird, M. C. *TL* **1978**, 3307. (c) Slack, D.; Baird, M. C. *CC* **1974**, 701. (d) Bly, R. S.; Bly, R. K.; Hossain, M. M.; Silverman, G. S.; Wallace, E. *T* **1986**, *42*, 1093. (e) Bly, R. S.; Silverman, G. S.; Bly, R. K. *OM* **1985**, *4*, 374.
22. Reetz, M. T.; Schinzer, D. *AG(E)* **1977**, *16*, 44.
23. (a) Traylor, T. G.; Berwin, H. J.; Jerkunica, J.; Hall, M. L. *PAC* **1972**, *30*, 597. (b) Jerkunica, J. M.; Traylor, T. G. *JACS* **1971**, *93*, 6278. (c) Washburne, S. S.; Szendroi, R. *JOC* **1981**, *46*, 691. (d) Washburne, S. S.; Simolike, J. B. *JOM* **1974**, *81*, 41. (e) However, organostannanes lacking a β-hydrogen afford alkyltriphenylmethanes in good yield. Kashin, A. N.; Bumagin, N. A.; Beletskaya, I. P.; Reutov, O. A. *JOM* **1979**, *171*, 321.
24. (a) Barton, D. H. R.; Haynes, R. K.; Leclerc, G.; Magnus, P. D.; Menzies, I. D. *JCS(P1)* **1975**, 2055. (b) Barton, D. H. R.; Leclerc, G.; Magnus, P. D.; Menzies, I. D. *CC* **1972**, 447. (c) Okada, K.; Hisamitsu, K.; Mukai, T. *TL* **1981**, *22*, 1251. (d) Futamura, S.; Kamiya, Y. *CL* **1989**, 1703.
25. (a) Ohshima, M.; Murakami, M.; Mukaiyama, T. *CL* **1985**, 1871. (b) Inazu, T.; Yamanoi, T. Jpn. Patent 02 240 093, 02 255 693 (*CA* **1991**, *114*, 143 907j, 143 908k); Jpn. Patent 01 233 295 (*CA* **1990**, *112*, 198 972r). (c) Bochkov, A. F.; Kochetkov, N. K. *Carbohydr. Res.* **1975**, *39*, 355; for polymerizations of carbohydrate cyclic orthoesters, see: Bochkov, A. F.; Chernetskii, V. N.; Kochetkov, N. K. *Carbohydr. Res.* **1975**, *43*, 35; *BAU* **1975**, *24*, 396. (d) Hashimoto, Y.; Mukaiyama, T. *CL* **1986**, 1623, 755. (e) Hashimoto, Y.; Sugumi, H.; Okauchi, T.; Mukaiyama, T. *CL* **1987**, 1691. (f) Murray, D. H.; Albizati, K. F. *TL* **1990**, *31*, 4109.
26. (a) Endo, T.; Sato, H.; Takata, T. *Macromolecules* **1987**, *20*, 1416. (b) Uno, H.; Endo, T.; Okawara, M. *J. Polym. Sci., Polym. Chem. Ed.* **1985**, *23*, 63. (c) Nishida, H.; Ogata, T. Jpn. Patent 62 295 920 (*CA* **1988**, *109*, 57 030h). (d) See also Ref. 25c. (e) Kunitake, T. *J. Macromol. Sci., Chem.* **1975**, *A9*, 797. (f) Kunitake, T.; Takarabe, K.; Tsugawa, S. *Polym. J.* **1976**, *8*, 363. (g) Spange, S.; Dreier, R.; Opitz, G.; Heublein, G. *Acta Polym.* **1989**, *40*, 55. (h) Mijangos, F.; León, L. M. *J. Polym. Sci., Polym. Lett. Ed.* **1983**, *21*, 885; *Eur. Polym. J.* **1983**, *19*, 29. (i) Bruzga, P.; Grazulevicius, J.; Kavaliunas, R.; Kublickas, R. *Polym. Bull. (Berlin)* **1991**, *26*, 193. (j) Khomyakov, A. K.; Gorelikov, A. T.; Shapet'ko, N. N.; Lyudvig, E. B. *Vysokomol. Soedin., Ser. A* **1976**, *18*, 1699, 1053; *DOK* **1975**, *222*, 1111.
27. (a) For a review, see any basic organometallic text, e.g. Coates, G. E.; Green, M. L. H.; Wade, K. *Organometallic Compounds*; Methuen: London, 1968; Vol. 2, pp 136ff. (b) Birch, A. J.; Cross, P. E.; Lewis, J.; White, D. A. *CI(L)* **1964**, 838. (c) Cotton, F. A.; Deeming, A. J.; Josty, P. L.; Ullah, S. S.; Domingos, A. J. P.; Johnson, B. F. G.; Lewis, J. *JACS* **1971**, *93*, 4624.
28. (a) Lyle, R. E.; Boyce, C. B. *JOC* **1974**, *39*, 3708. (b) Hanessian, S.; Staub, A. P. A. *TL* **1973**, 3555.
29. (a) Sommer, L. H.; Bauman, D. L. *JACS* **1969**, *91*, 7076. (b) Bulkowski, J. E.; Stacy, R.; Van Dyke, C. H. *JOM* **1975**, *87*, 137. (c) Chojnowski, J.; Fortuniak, W.; Stanczyk, W. *JACS* **1987**, *109*, 7776.
30. Burfield, D. R.; Lee, K.-H.; Smithers, R. H. *JOC* **1977**, *42*, 3060.

Michael E. Jung
University of California, Los Angeles, CA, USA

Z

Zinc Bromide[1]

$ZnBr_2$

[7699-45-8] Br_2Zn (MW 225.19)

(used in the preparation of organozinc reagents via transmetalation;[1] a mild Lewis acid useful for promoting addition[2] and substitution reactions[3])

Physical Data: mp 394 °C; bp 697 °C (dec); d 4.201 g cm^{-3}.
Solubility: sol Et_2O, H_2O (1 g/ 3;25 mL), 90% EtOH (1 g/ 0.5 mL).
Form Supplied in: granular white powder; principal impurity is H_2O.
Analysis of Reagent Purity: melting point.
Purification: heat to 300 °C under vacuum (2×10^{-2} mmHg) for 1 h, then sublime.
Handling, Storage, and Precautions: very hygroscopic; store under anhydrous conditions. Irritant.

Organozinc Reagents. The transmetalation of organomagnesium, organolithium, and organocopper reagents by anhydrous $ZnBr_2$ in ethereal solvents offers a convenient method of preparing organozinc bromides and diorganozinc reagents.[1a] Alternatively, anhydrous $ZnBr_2$ may be reduced by potassium metal to result in highly activated Zn^0, which is useful for the preparation of zinc reagents through oxidative addition to organic halides.[4] Alkyl, allylic, and propargylic zinc reagents derived by these methods have shown considerable value in their stereoselective and regioselective addition reactions with aldehydes, ketones, imines, and iminium salts.[1a,5] Zinc enolates used in the Reformatsky reaction may also be prepared through transmetalation using $ZnBr_2$.[1b] Organozinc species are especially useful in palladium- and nickel-catalyzed coupling reactions of sp^2 carbon centers. In this fashion, sp^2–sp^3 (eq 1)[6] and sp^2–sp^2 (eqs 2 and 3)[7,8] carbon–carbon bonds are formed selectively in high yields. The enantioselective cross coupling of secondary Grignard reagents with vinyl bromide is strongly affected by the presence of $ZnBr_2$, which accelerates the reaction and inverts its enantioselectivity (eq 4).[9]

PhCH$_2$Li → (1. $ZnBr_2$, THF; 2. 4-I-C$_6$H$_4$-CO$_2$Me, Ni(PPh$_3$)$_4$, THF–Et$_2$O, 25 °C; 88%) → Ph-CH$_2$-C$_6$H$_4$-CO$_2$Me (1)

Organozinc intermediates formed via transmetalation using $ZnBr_2$ have been used to effect carbozincation of alkenes and alkynes through metallo-ene and metallo-Claisen reactions. Both intermolecular and intramolecular variants of these reactions have been described, often proceeding with high levels of stereoselectivity and affording organometallic products that may be used in subsequent transformations (eqs 5 and 6),[10] including alkenation (eq 6).[10b,c] Bimetallic zinc–zirconium reagents have also been developed that offer a method for the alkenation of carbonyl compounds (eq 7).[11]

OCONEt$_2$ / Li → (1. $ZnBr_2$, THF; 2. Br-CH=CH-Ph, Pd(PPh$_3$)$_4$; 80%) → OCONEt$_2$ / Ph (2)

Et(CH=CH)$_2$CuLi → ($ZnBr_2$, Pd(PPh$_3$)$_4$, THF–Et$_2$O; I-CH=CH-C$_5$H$_{11}$; 96%) → Et / C$_5$H$_{11}$ diene
Et(CH=CH)$_2$CuLi → ($ZnBr_2$, THF, Pd(PPh$_3$)$_4$; 4-I-C$_6$H$_4$-CO$_2$Me; 81%) → Et-CH=CH-C$_6$H$_4$-CO$_2$Me (3)

MgCl → (vinyl bromide, NiCl$_2$, Et$_2$O, 0 °C; MeS–CH$_2$CH$_2$–CH(NMe$_2$)–PPh$_2$) → R^1 R^2 (4)

without $ZnBr_2$ R^1 = Me, R^2 = H; >95% (52% ee)
with $ZnBr_2$ R^1 = H, R^2 = Me; >95% (49% ee)

TMS / OMe → (1. BuLi, THF; 2. $ZnBr_2$; 80%) → X / OMe / TMS (5)

X = ZnBr → (H_2O) → X = H

t-BuO / I / Pr → (1. t-BuLi; 2. CH$_2$=CHCH$_2$CH$_2$MgBr; 3. $ZnBr_2$) → [Pr / t-Bu / O / Zn] →

M^1 / M^2 / Pr / O-t-Bu → (1. PhCHO; 2. H_3O^+; 68%) → Ph / Pr / O-t-Bu, 95% de (6)

M^1, M^2 = metal → (H_3O^+) → M^1, M^2 = H, 75%, 95% de

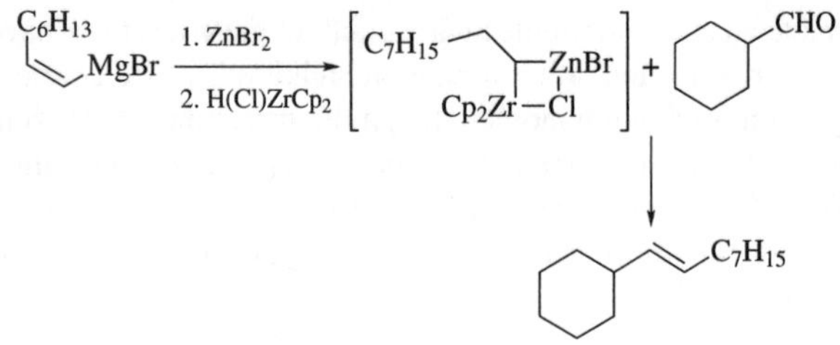

(7)

83%, 100% *(E)*

Concerted Ring-Forming Reactions. The mild Lewis acid character of $ZnBr_2$ sometime imparts a catalytic effect on thermally allowed pericyclic reactions. The rate and stereoselectivity of cycloaddition reactions (eq 8),[12] including dipolar cycloadditions (eq 9),[13] are significantly improved by the presence of this zinc salt.

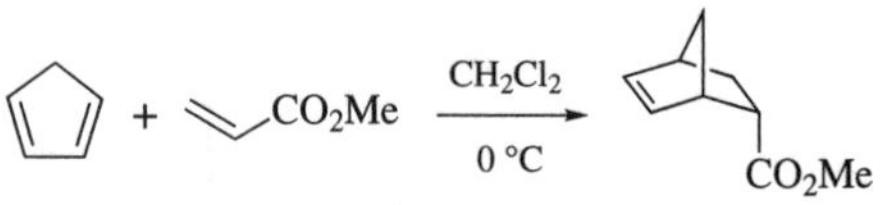

(8)

no catalyst, 5 h — 70%, 78% *endo*
with $ZnBr_2$, 3 h — 80%, 95% *endo*

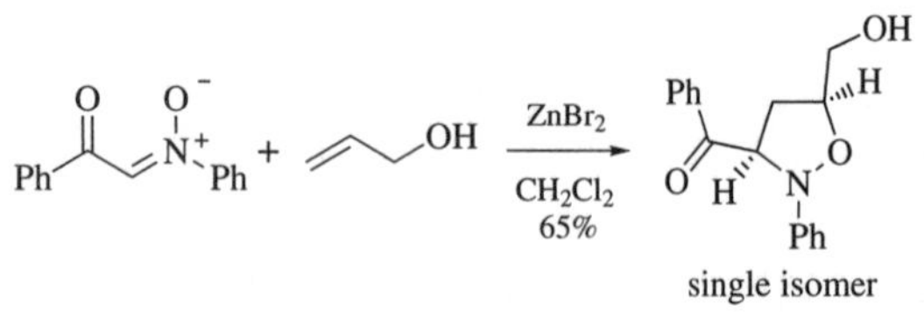

(9)

single isomer

Some intramolecular ene reactions benefit from $ZnBr_2$ catalysis to afford the cyclic products under milder conditions, in higher yields and selectivities (eqs 10 and 11).[14,15] Generally, the use of $ZnBr_2$ is preferred over ***Zinc Chloride*** or ***Zinc Iodide*** in this type of reaction.[15]

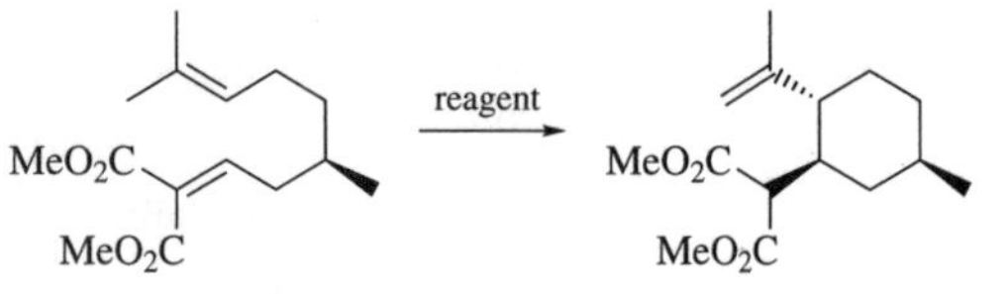

(10)

180 °C, *o*-dichlorobenzene — 66% (83% de)
$ZnBr_2$, CH_2Cl_2, 25 °C — 79% (95% de)

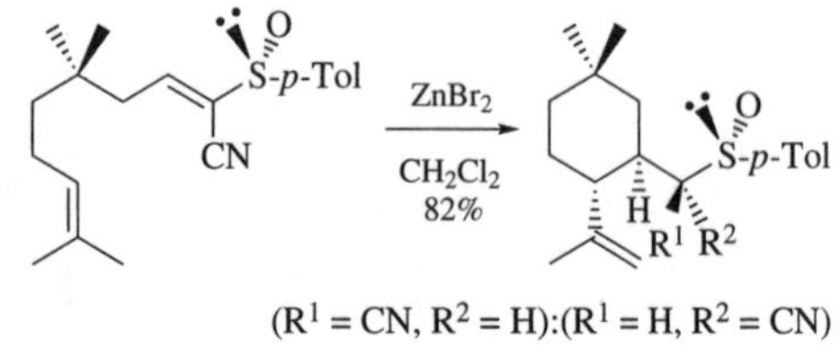

(11)

$(R^1 = CN, R^2 = H):(R^1 = H, R^2 = CN) = 88:12$

Activation of C=X Bonds. Lewis acid activation of carbonyl compounds by $ZnBr_2$ promotes the addition of allylsilanes and silyl ketene acetals.[16] Addition to imines has also been reported.[17]

In general, other Lewis acids have been found to be more useful, though in some instances $ZnBr_2$ has proven to be advantageous (eq 12).[2]

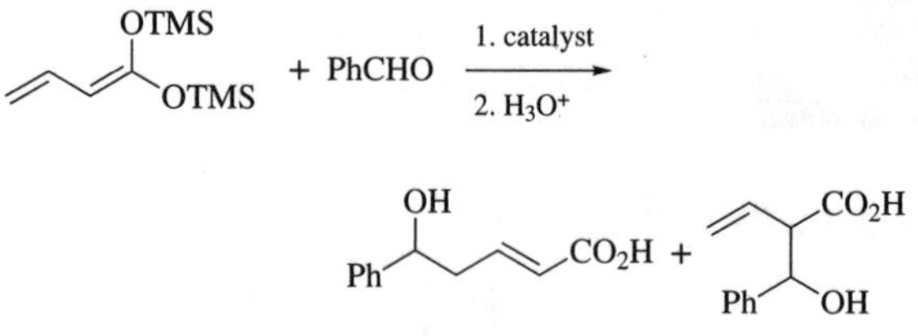

(12)

$ZnBr_2$, THF, 20 °C, 6 h — 100:0
CsF, CH_2Cl_2. 20 °C, 14 h — 5:95

Activation of C–X Bonds. Even more important than carbonyl activation, $ZnBr_2$ promotes substitution reactions with suitably active organic halides with a variety of nucleophiles. Alkylation of silyl enol ethers and silyl ketene acetals using benzyl and allyl halides proceeds smoothly (eq 13).[3] Especially useful electrophiles are α-thio halides which afford products that may be desulfurized or oxidatively eliminated to result in α,β-unsaturated ketones, esters, and lactones (eq 14).[18] Other electrophiles that have been used with these alkenic nucleophiles include ***Chloromethyl Methyl Ether***, $HC(OMe)_3$, and ***Acetyl Chloride***.[3,19]

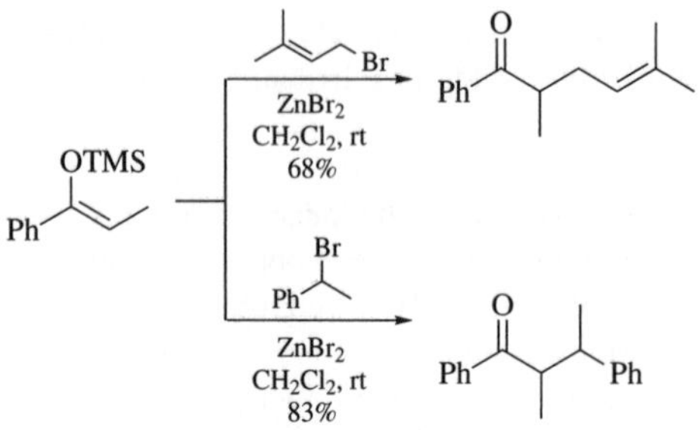

(13)

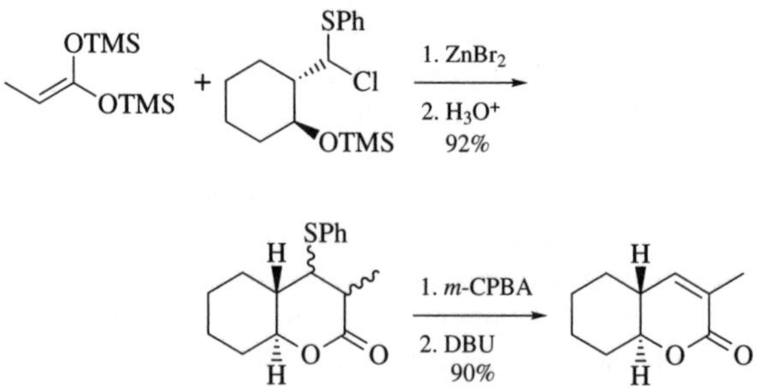

(14)

Enol ethers and allylic silanes and stannanes will engage cyclic α-seleno sulfoxides,[20] ω-acetoxy lactams,[21] and acyl glycosides (eq 15)[22] in the presence of $ZnBr_2$ catalysis. Along these lines, it has been found that $ZnBr_2$ is superior to ***Boron Trifluoride Etherate*** in promoting glycoside bond formation using trichloroimidate-activated glycosides (eq 16).[23] Imidazole carbamates are also effective activating groups for $ZnBr_2$-mediated glycosylation (eq 17).[24]

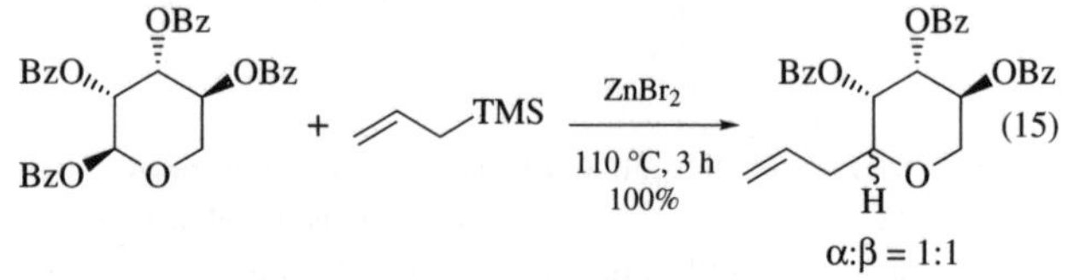

(15)

α:β = 1:1

60% $ZnBr_2$, CH_2Cl_2, 3Å mol sieves, Δ (16)

Ph-CH(OH)-CO_2Me, $ZnBr_2$, THF, Et_2O, Δ, 88% (17)

10:1 de

Cyclic acetals also undergo highly selective, Lewis acid-dependent ring opening substitution with ***Cyanotrimethylsilane*** (eq 18).[25]

TMSCN (18)

$ZnBr_2$, CH_2Cl_2, 25 °C, 20 h	1:250
$TiCl_4$, CH_2Cl_2, 25 °C, 20 h	250:1

Reduction. Complexation with $ZnBr_2$ has been shown to markedly improve stereoselectivity in the reduction of certain heteroatom-substituted ketones (eqs 19 and 20).[26,27] Furthermore, the *anti* selectivity observed in $BF_3 \cdot OEt_2$-mediated intramolecular hydrosilylation of ketones is reversed when $ZnBr_2$ is used instead (eq 21).[28]

1. $ZnBr_2$, MeOH; 2. $NaBH_4$; 100% (19)

DIBAL, $ZnBr_2$, THF, 91% (20)

66:34

(21)

$BF_3 \cdot OEt_2$, CH_2Cl_2, –80 °C, 2 h	23:1
$ZnBr_2$, CH_2Cl_2, –80 °C, 8 h	1:6

Deprotection. $ZnBr_2$ is a very mild reagent for several deprotection protocols, including the detritylation of nucleotides[29] and deoxynucleotides,[30] *N*-deacylation of *N,O*-peracylated nucleotides,[31] and the selective removal of Boc groups from secondary amines in the presence of Boc-protected primary amines.[32] Perhaps the most widespread use of $ZnBr_2$ for deprotection is in the mild removal of MEM ethers to afford free alcohols (eq 22).[33]

$ZnBr_2$, CH_2Cl_2, 25 °C, 10 h, 93% (22)

Miscellaneous. An important method for the synthesis of stereodefined trisubstituted double bonds involves the treatment of cyclopropyl bromides with $ZnBr_2$. The (*E*) isomer is obtained almost exclusively by this method (eq 23).[34]

1. PBr_3, LiBr, collidine, Et_2O; 2. $ZnBr_2$, Et_2O; >95% (23)

(*E*)

The rearrangement of a variety of terpene oxides has been examined (eq 24).[35] While $ZnBr_2$ is generally a satisfactory catalyst for this purpose, other Lewis acids, including $ZnCl_2$[36] and ***Magnesium Bromide***,[37] are advantageous in some instances.

(–)-α-Pinene → ; $ZnBr_2$, PhH, 80 °C; 88% (24)

In the presence of $ZnBr_2$/48% ***Hydrobromic Acid***, suitably functionalized cyclopropanes undergo ring expansion to afford cyclobutane (eq 25)[38] and α-methylene butyrolactone products (eq 26).[39] One-carbon ring expansion has been reported when certain trimethylsilyl dimethyl acetals are exposed to $ZnBr_2$ with warming (eq 27).[40]

48% HBr; $ZnBr_2$, PhSH, 95% — 1. $Cu(OTf)_2$, i-Pr_2NEt; 2. 450 °C, 81% (25)

48% HBr, $ZnBr_2$; EtOH, 100 °C, 6 h, 50% (26)

$ZnBr_2$, CH_2Cl_2; 40 °C, 30 min, 100% (27)

Related Reagents. Zinc Chloride; Zinc Iodide; Zinc p-Toluenesulfonate; Zinc Trifluoromethanesulfonate.

---

1. (a) Knochel, P. *COS* **1991**, *1*, Chapter 1.7. (b) Rathke, M. W.; Weipert, P. *COS* **1991**, *2*, Chapter 1.8.
2. For an example: Bellassoued, M.; Ennigrou, R.; Gaudemar, M. *JOM* **1988**, *338*, 149.
3. For examples: (a) Reetz, M. T.; Maier, W. F. *AG(E)* **1978**, *17*, 48. (b) Reetz, M. T.; Chatziiosifidis, I.; Löwe, W. F.; Maier, W. F. *TL* **1979**, 1427. (c) Paterson, I. *TL* **1979**, 1519.
4. Riecke, R. D.; Uhm, S. J.; Hudnall, P. M. *CC* **1973**, 269.
5. For representative examples of allylic and propargylic zinc reagents: (a) Yamamoto, Y.; Nishii, S.; Maruyama, K.; Komatsu, T.; Ito, W. *JACS* **1986**, *108*, 7778. (b) Yamamoto, Y.; Ito, W. *T* **1988**, *44*, 5414. (c) Yamamoto, Y.; Ito, W.; Maruyama, K. *CC* **1985**, 1131. (d) Yamanoto, Y.; Komatsu, T.; Maruyama, K. *CC* **1985**, 814. (e) Fronza, G.; Fuganti, C.; Grasselli, P.; Pedrocchi-Fantoni, G.; Zirotti, C. *TL* **1982**, *23*, 4143. (f) Fujisawa, T.; Kojima, E.; Itoh, T.; Sato, T. *TL* **1985**, *26*, 6089. (g) Pornet, J.; Miginiac, L. *BSF* **1975**, 841. (h) Yamamoto, Y.; Komatsu, T.; Maruyama, K. *JOM* **1985**, *285*, 31. (i) Bouchoule, C.; Miginiac, P. *CR(C)* **1968**, *266*, 1614. (j) Miginiac, L.; Mauzé, B. *BSF* **1968**, 3832. (k) Arous-Chtara, R.; Gaudemar, M.; Moreau, J.-L. *CR(C)* **1976**, *282*, 687. (l) Moreau, J.-L.; Gaudemar, M. *BSF* **1971**, 3071. (m) Miginiac, L.; Mauzé, B. *BSF* **1968**, 2544.
6. Negishi, E.; King, A. O.; Okudado, N. *JOC* **1977**, *42*, 1821.
7. Sengupta, S.; Snieckus, V. *JOC* **1990**, *55*, 5680. See also: Gilchrist, T. L.; Summersell, R. J. *TL* **1987**, *28*, 1469.
8. (a) Jabri, N.; Alexakis, A.; Normant, J. F. *BSF(2)* **1983**, 321. (b) Jabri, N.; Alexakis, A.; Normant, J. F. *TL* **1982**, *23*, 1589. (c) Jabri, N.; Alexakis, A.; Normant, J. F. *TL* **1981**, *22*, 959. (d) Jabri, N.; Alexakis, A.; Normant, J. F. *TL* **1981**, *22*, 3851.
9. Cross, G.; Vriesema, B. K.; Boven, G.; Kellogg, R. M.; van Bolhuis, F. *JOM* **1989**, *370*, 357.
10. (a) Courtemanche, G.; Normant, J.-F. *TL* **1991**, *32*, 5317. (b) Marek, I.; Normant, J.-F. *TL* **1991**, *32*, 5973. (c) Marek, I.; Lefrançois, J.-M.; Normant, J.-F. *SL* **1992**, 633.
11. Tucker, C. E.; Knochel, P. *JACS* **1991**, *113*, 9888.
12. Narayana Murthy, Y. V. S.; Pillai, C. N. *SC* **1991**, *21*, 783. See also: López, R.; Carretero, J. C. *TA* **1991**, *2*, 93.
13. Kanemasa, S.; Tsuruoka, T.; Wada, E. *TL* **1993**, *34*, 87.
14. Tietze, L. F.; Biefuss, U.; Ruther, M. *JOC* **1989**, *54*, 3120. See also: (a) Tietze, L. F.; Ruther, M. *CB* **1990**, *123*, 1387. (b) Nakatani, Y.; Kawashima, K. *S* **1978**, 147.
15. Hiroi, K.; Umemura, M. *TL* **1992**, *33*, 3343.
16. (a) Mikami, K.; Kawamoto, K.; Loh, T.-P.; Nakai, T. *CC* **1990**, 1161. (b) Bellassoued, M.; Gaudemar, M. *TL* **1988**, *29*, 4551.
17. Gaudemar, M.; Bellassoued, M. *TL* **1990**, *31*, 349.
18. Khan, H. A.; Paterson, I. *TL* **1982**, *23*, 5083. See also: (a) Paterson, I. *T* **1988**, *44*, 4207. (b) Khan, H. A.; Paterson, I. *TL* **1982**, *23*, 4811. (c) Paterson, I.; Fleming, I. *TL* **1979**, *20*, 993, 995, 2179.
19. Fleming, I.; Goldhill, J.; Paterson, I. *TL* **1979**, 3209.
20. Ren, P.; Ribezzo, M. *JACS* **1991**, *113*, 7803.
21. Ohta, T.; Shiokawa, S.; Iwashita, E.; Nozoe, S. *H* **1992**, *34*, 895.
22. Kozikowski, A. P.; Sorgi, K. L. *TL* **1982**, *23*, 2281.
23. Urban, F. J.; Moore, B. S.; Breitenbach, R. *TL* **1990**, *31*, 4421.
24. Ford, M. J.; Ley, S. V. *SL* **1990**, 255.
25. Corcoran, R. C. *TL* **1990**, *31*, 2101.
26. Bartnik, R.; Lesniak, S.; Laurent, A. *TL* **1981**, *22*, 4811.
27. Barros, D.; Carreño, M. C.; Ruano, J. L. G.; Maestro, M. C. *TL* **1992**, *33*, 2733.
28. Anwar, S.; Davis, A. P. *T* **1988**, *44*, 3761.
29. Waldemeier, F.; De Bernardini, S.; Leach, C. A.; Tamm, C. *HCA* **1982**, *65*, 2472.
30. (a) Kohli, V.; Blöcker, H.; Köster, H. *TL* **1980**, *21*, 2683. (b) Matteuci, M. D.; Caruthers, M. H. *TL* **1980**, *21*, 3243.
31. Kierzek, R.; Ito, H.; Bhatt, R.; Itakura, K. *TL* **1981**, *22*, 3761.
32. Nigam, S. C.; Mann, A.; Taddei, M.; Wermuth, C.-G. *SC* **1989**, *19*, 3139.
33. Corey, E. J.; Gras, J.-L.; Ulrich, P. *TL* **1976**, 809.
34. Johnson, W. S.; Li, T.; Faulkner, D. J.; Campbell, S. F. *JACS* **1968**, *90*, 6225. See also: (a) Brady, S. F.; Ilton, M. A.; Johnson, W. S. *JACS* **1968**, *90*, 2882. (b) Nakamura, H.; Yamamoto, H.; Nozaki, H. *TL* **1973**, 111.
35. Lewis, J. B.; Hendrick, G. W. *JOC* **1965**, *30*, 4271. See also: (a) Settine, R. L.; Parks, G. L.; Hunter, G. L. K. *JOC* **1964**, *29*, 616. (b) Bessière-Chréieu, Y.; Bras, J. P. *CR(C)* **1970**, *271*, 200. (c) Clark, Jr., B. C.; Chafin, T. C.; Lee, P. L.; Hunter, G. L. K. *JOC* **1978**, *43*, 519. (d) Watanabe, H.; Katsuhara, J.; Yamamoto, N. *BCJ* **1971**, *44*, 1328.
36. Kaminski, J.; Schwegler, M. A.; Hoefnagel, A. J.; van Bekkum, H. *RTC* **1992**, *111*, 432.
37. Serramedan, D.; Marc, F.; Pereyre, M.; Filliatre, C.; Chabardès, P.; Delmond, B. *TL* **1992**, *33*, 4457.
38. Kwan, T. W.; Smith, M. B. *SC* **1992**, *22*, 2273.
39. Hudrlik, P. F.; Rudnick, L. R.; Korzeniowski, S. H. *JACS* **1973**, *95*, 6848.
40. (a) Tanino, K.; Katoh, T.; Kuwajima, I. *TL* **1988**, *29*, 1815. (b) Tanino, K.; Katoh, T.; Kuwajima, I. *TL* **1988**, *29*, 1819. See also: Tanino, K.; Sato, K.; Kuwajima, I. *TL* **1989**, *30*, 6551.

Glenn J. McGarvey
University of Virginia, Charlottesville, VA, USA

Zinc Chloride[1]

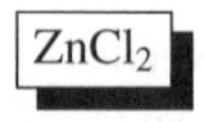

[7646-85-7] Cl_2Zn (MW 136.29)

(used in the preparation of organozinc reagents via transmetalation;[1] a mild Lewis acid useful for promoting cycloaddition,[2] substitution,[3] and addition reactions,[4] including electrophilic aromatic additions;[5] has found use in selective reductions[6])

Physical Data: mp 293 °C; bp 732 °C; *d* 2.907 g cm^{-3}.

Solubility: sol H_2O (432 g/100 g at 25 °C), EtOH (1 g/1.3 mL), glycerol (1 g/2 mL).

Form Supplied in: white, odorless, very deliquescent granules; principal impurities are H_2O and zinc oxychloride.

Analysis of Reagent Purity: melting point.

Purification: reflux (50 g) in dioxane (400 mL) in the presence of Zn^0 dust, then filter hot and allow to cool to precipitate purified $ZnCl_2$. Also, anhydrous material may be sublimed under a stream of dry HCl, followed by heating to 400 °C in a stream of dry N_2.

Handling, Storage, and Precautions: very hygroscopic; store under anhydrous conditions; moderately irritating to skin and mucous membranes.

Organozinc Reagents. The transmetalation of organomagnesium, organolithium, and organocopper reagents is an important and versatile method of preparing useful zinc reagents.[1a] Alternatively, $ZnCl_2$ may be employed for direct insertion of Zn^0 into carbon–halogen bonds using Mg^0/ultrasound[7] or prior reduction with K^0.[8] The addition of the resulting allylic and propargylic or allenic zinc reagents to carbonyl compounds, imines, and iminium salts represents an important method of selective carbon–carbon bond formation.[9,11] These reactions are generally guided by chelation and the Zimmerman–Traxler transition state[10] to control the relative stereochemistry about the new carbon–carbon bond (eq 1),[12] as well as with respect to preexisting stereocenters (eq 2).[13] Anions derived from propargylic deprotonation add to carbonyl compounds with a high degree of regiochemical integrity in the presence of $ZnCl_2$, i.e. allenic organozinc intermediates cleanly afford the propargylic product (eq 3).[11b]

1. *s*-BuLi, THF, –78 °C; 2. Lewis acid; 3. PhCHO, THF (1)

Lewis acid		
none	63%	53:47
0.5 equiv $ZnCl_2$	68%	89:11

1. then H_3O^+; 2. PhCOCl; 60% (2) >98:2

1. *s*-BuLi, hexane, –78 °C; 2. $ZnCl_2$; 3. (cyclohexenone); 95% (3)

Alkylzinc reagents, often in the presence of copper salts, effectively participate in conjugate addition reactions[14] and clean S_N2' reactions (eq 4).[15]

MeLi: 1. $ZnCl_2$, THF; 2. CuCl; 3. OTBDMS/MsO substrate (CO_2Me), 99%; product OTBDMS, CO_2Me, >99:1 (4)

Organozinc reagents are superior species for palladium- and nickel-catalyzed coupling reactions.[16] This offers an exceptional method for the selective formation of sp^2–sp^3 (eq 5),[17] sp^2–sp^2 (eqs 6 and 7),[18,19] and sp^2–sp (eq 8)[20] carbon–carbon bonds. In addition, the palladium-catalyzed coupling reactions of sp and sp^2 halides with vinylalanes (eq 9),[21] vinylcuprates,[22] vinylzirconium (eq 10)[23] and acyliron species[24] often proceed more effectively in the presence of $ZnCl_2$. This effect has been noted in other metal-mediated carbon bond formations,[25] though the role of this $ZnCl_2$ catalysis is not understood at present.

+ ClZn–CH_2–SiMe(O-*i*-Pr)$_2$; $PdCl_2$(dppf), THF, Δ, 95% (5)

1. *t*-BuLi, THF, –78 °C; 2. $ZnCl_2$; 3. [Pd(dba)$_2$]•2PPh_3, Ph/Br; 68% (6)

1. *t*-BuLi, Et_2O, –78 °C; 2. $ZnCl_2$, THF; 3. $PdCl_2(PPh_3)_2$, *i*-$BuAlH_4$, THF, *t*-Bu/I/OTMS/*t*-Bu; 60% (7)

1. BuLi; 2. $ZnCl_2$; 3. $Pd(PPh_3)_4$, 2-iodothiophene; 92% (8)

C_8H_{17}–CH=CH–Al(*i*-Bu)$_2$ + I/Bu; $Pd(PPh_3)_4$, $ZnCl_2$, THF, rt, 66% (9)

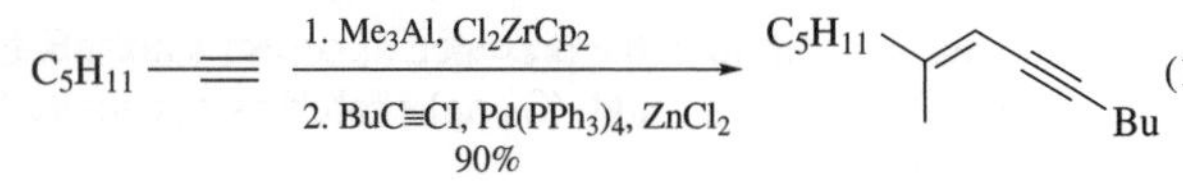

Organozinc reagents have been successfully exploited in asymmetric carbon bond formation.[26] Chiral glyoxylate esters engage organozinc species derived from Grignard reagents in selective addition reactions to afford enantiomerically enriched α-hydroxy acids (eq 11).[27] Enantioselective addition reactions of Grignard reagents have been achieved by sequential addition of $ZnCl_2$ and a chiral catalyst to afford secondary alcohols (eq 12).[28]

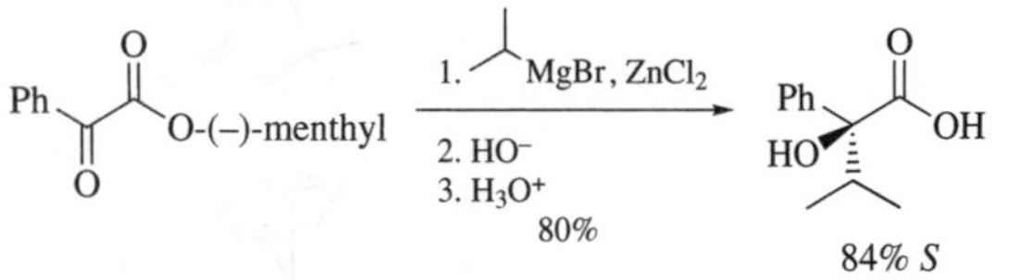

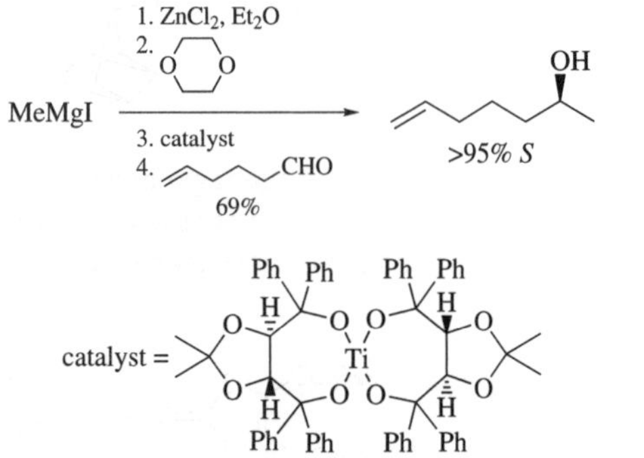

Zinc Enolates. Zinc enolates can be prepared by deprotonation of carbonyl compounds using standard bases, followed by transmetalation with $ZnCl_2$.[1b] These enolates offer important opportunities for stereoselective aldol condensations, including higher yields and stereochemical selection for *threo* crossed-aldol products (eq 13).[29] Both of these consequences are the result of the reversible formation of a six-membered zinc chelate intermediate (**1**) which favors the *anti* disposition of its substituents. Though their use has been mostly replaced by kinetically controlled aldol methodology, zinc enolates sometimes present advantages in specific cases.[30] For example, chelated zinc enolates derived from α-amino acids have been shown to be of value in the stereoselective synthesis of β-lactams (eq 14).[31] Controlled monoalkylation of unsymmetrical ketones has been accomplished via treatment of zinc enolates with α-chlorothio ethers (eq 15),[32] and mild allylation of β-dicarbonyl compounds may be realized through exposure of the corresponding zinc enolate to an allyl alcohol in the presence of a palladium catalyst (eq 16).[33]

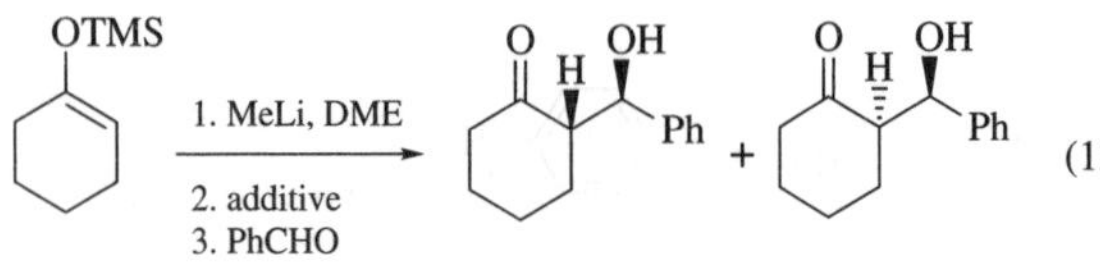

Homoenolates. Readily available mixed Me_3Si/alkyl acetals are converted into zinc homoenolates in high yield through exposure to $ZnCl_2$ in Et_2O (eq 17).[34] These mildly reactive species, which tolerate asymmetry α to the carbonyl, are useful for a variety of bond-forming reactions. In the presence of Me_3Si activation they undergo addition to aldehydes and, when admixed with CuBr·DMS (see ***Copper(I) Bromide***), can be allylated or conjugatively introduced to α,β-unsaturated ketones (eq 18).[35] Copper salts are not required for conjugate addition to propargylic esters (eq 19).[36] These zinc species also participate in useful palladium-catalyzed bond-forming processes (eq 20).[37]

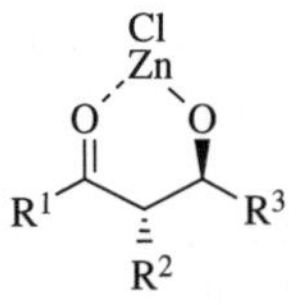

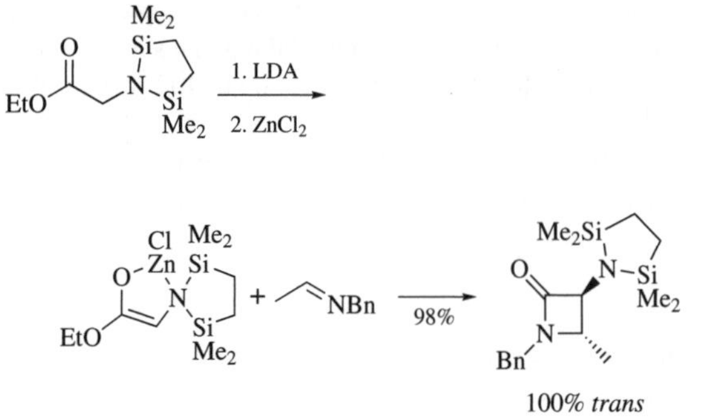

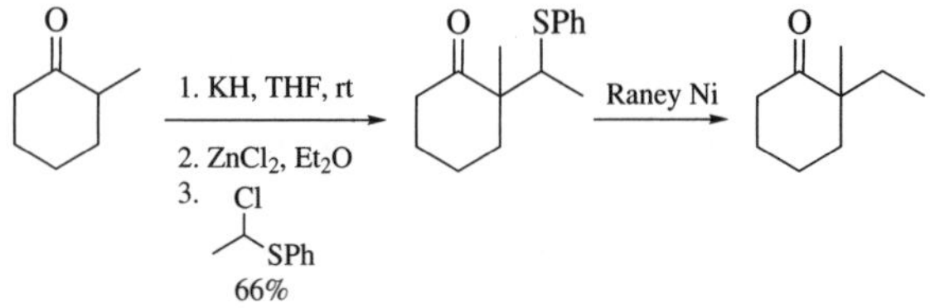

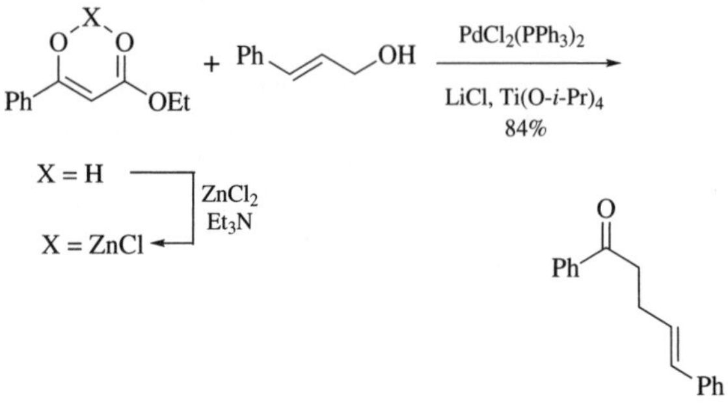

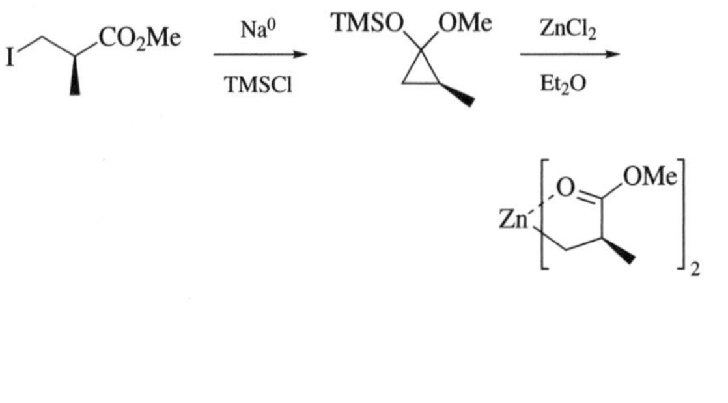

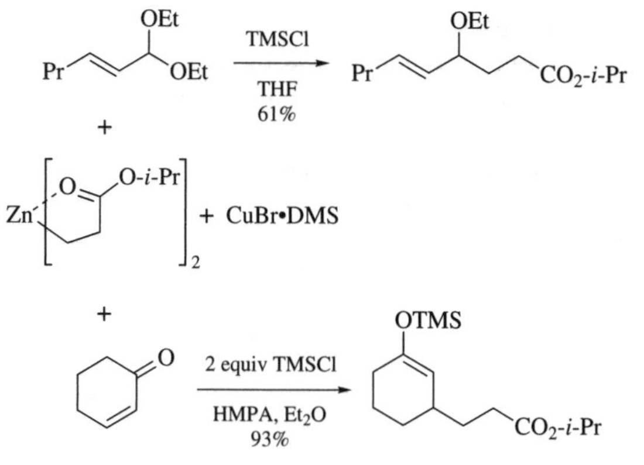

(18)

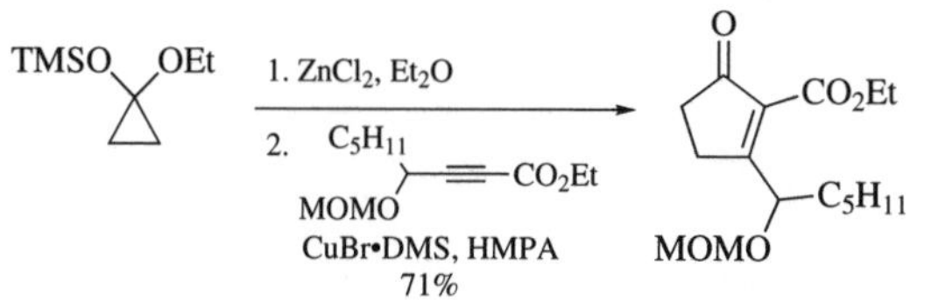

(19)

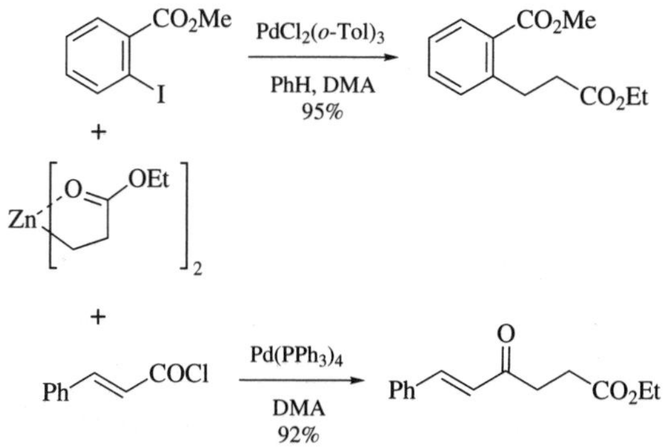

(20)

Cycloaddition Reactions. Catalysis by $ZnCl_2$ is often a powerful influence on cycloaddition reactions.[2] In addition to improving the rate of Diels–Alder reactions, enhanced control over regioselectivity (eq 21)[2c] and stereochemistry (eq 22)[2e] may be observed. An exceedingly important application of $ZnCl_2$ catalysis is found in the reaction of electron-rich dienes with carbonyl compounds.[2a,b] Intensive mechanistic studies on these reactions has uncovered two mechanistic pathways that afford stereochemically contrasting products, depending upon which Lewis acid is employed (eq 23).[38] It was concluded that cycloaddition reactions catalyzed by $ZnCl_2$ proceed via a classical [4 + 2] concerted process, whereas Lewis acids such as ***Boron Trifluoride Etherate*** and ***Titanium(IV) Chloride*** afford products through sequential aldol/cyclization processes. Examples abound wherein advantage has been taken of the predictable Cram–Felkin selectivity of this $ZnCl_2$ catalysis to exploit asymmetric variants of this reaction to synthesize a variety of natural products (eq 24).[39] The selective cycloaddition of electron-rich dienes with imines has also been catalyzed by $ZnCl_2$ (eq 25).[40]

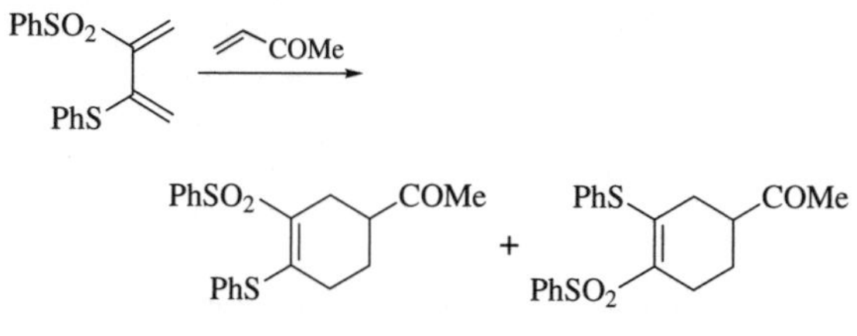

(21)

190 °C, 7 h	81%	3:1
$ZnCl_2$, rt, 24 h	90%	100:0

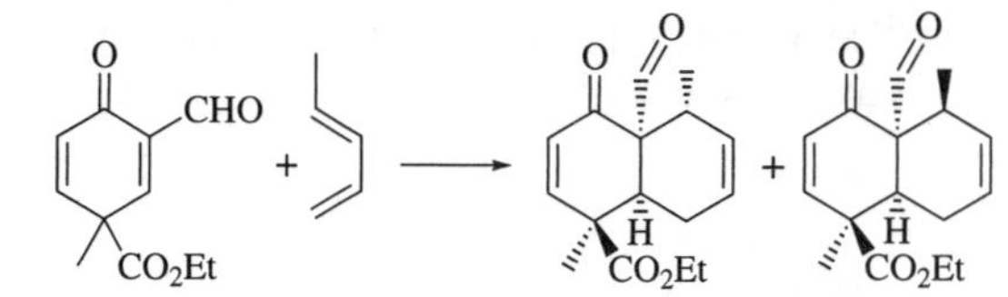

(22)

78 °C, 24 h		4.5:1
$ZnCl_2$, 0 °C, 5 h, 95%		100:0

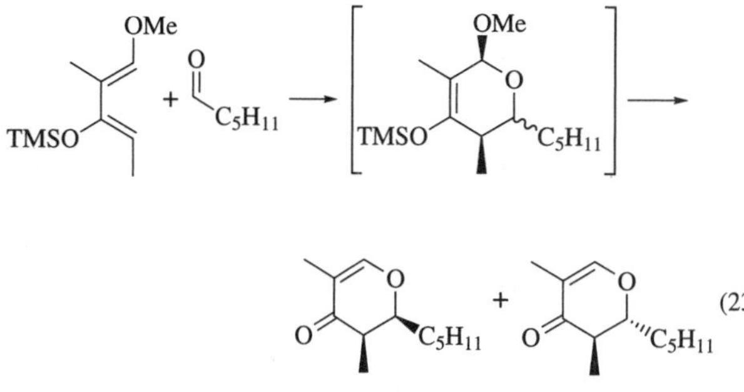

(23)

BF_3•OEt_2, CH_2Cl_2, −78 °C	21%	69%
$ZnCl_2$, THF, 25 °C	91%	2%

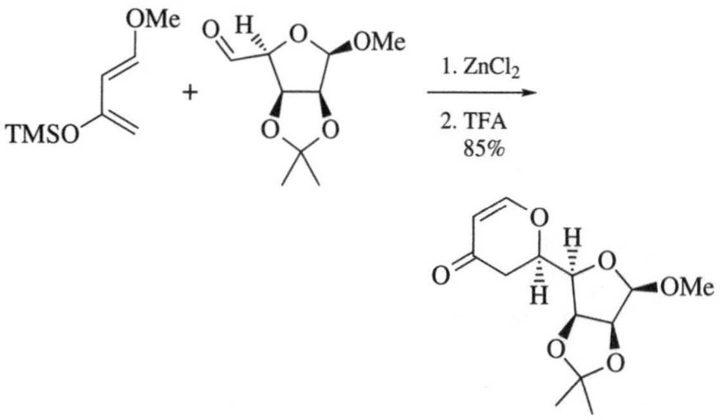

(24)

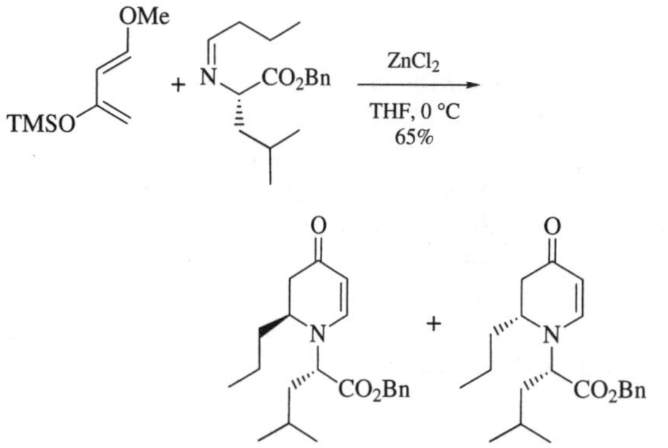

(25)

>9:1

Activation of C=X Bonds. The mild Lewis acid character of $ZnCl_2$ is frequently exploited to promote the addition of various nucleophiles to carbon–heteroatom double bonds.[4] The well-established Knoevenagel condensation and related reactions have been effectively catalyzed by $ZnCl_2$ (eq 26).[41] The addition of enol ethers and ketene acetals to aldehydes and ketones has been noted (eq 27),[42] though $ZnCl_2$ has been used less widely in these aldol-type condensations than other Lewis acids, including $TiCl_4$, ***Magnesium Bromide***, and ***Tin(IV) Chloride***, to name a few.[4] In some instances, excellent levels of stereocontrol have been observed (eq 28).[43] Analogous additions to imines have been noted as well.[4e–g,44] Some conjugate addition reactions may also benefit from $ZnCl_2$ catalysis (eq 29).[45]

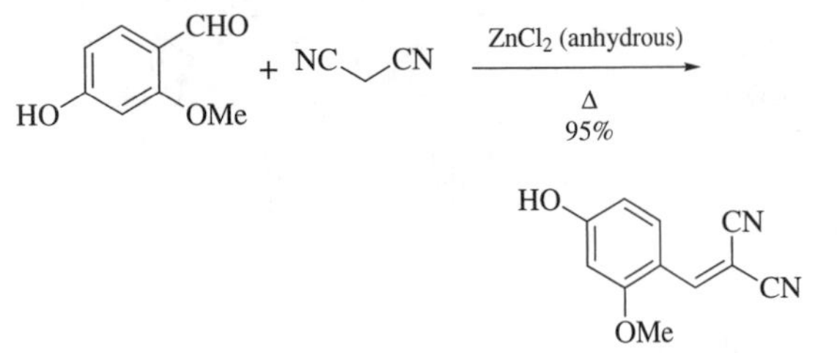

(26)

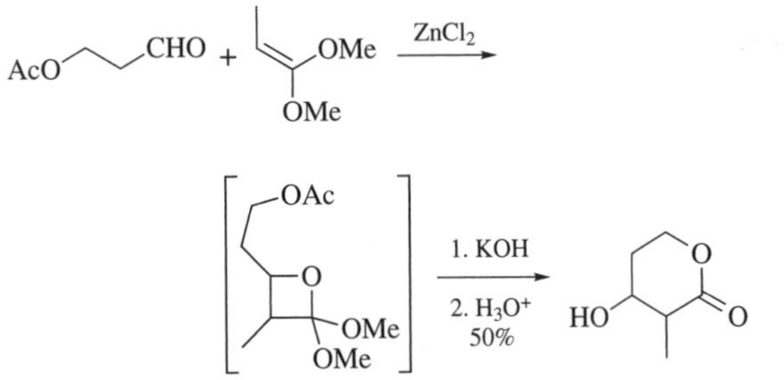

(27)

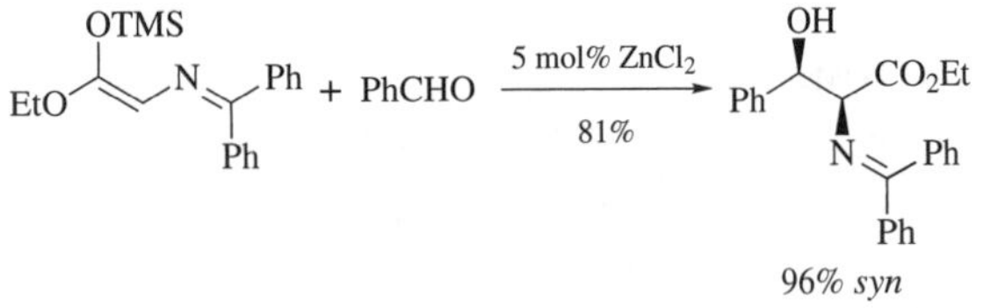

(28)

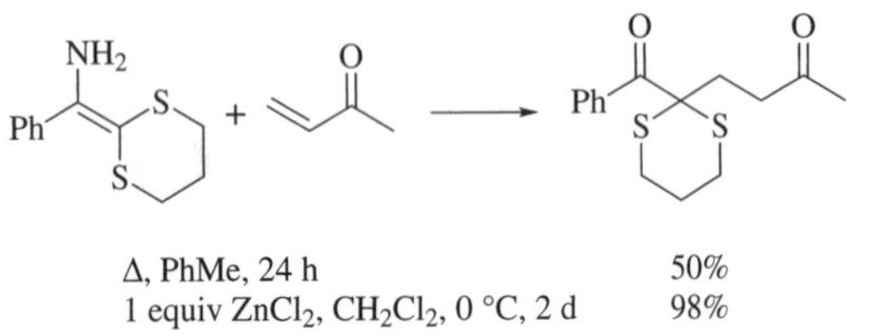

(29)

Δ, PhMe, 24 h — 50%
1 equiv $ZnCl_2$, CH_2Cl_2, 0 °C, 2 d — 98%

The formation of cyanohydrins using ***Cyanotrimethylsilane*** and ***Isoselenocyanatotrimethylsilane*** has been effectively catalyzed by $ZnCl_2$ (eq 30),[46] as has Strecker amino acid synthesis via the treatment of imines with $Me_3SiCN/ZnCl_2$.[47] The combination of carbonyl compounds with ***Acetyl Chloride*** or ***Acetyl Bromide*** may be promoted by $ZnCl_2$ to afford protected vicinal halohydrins (eq 31).[48]

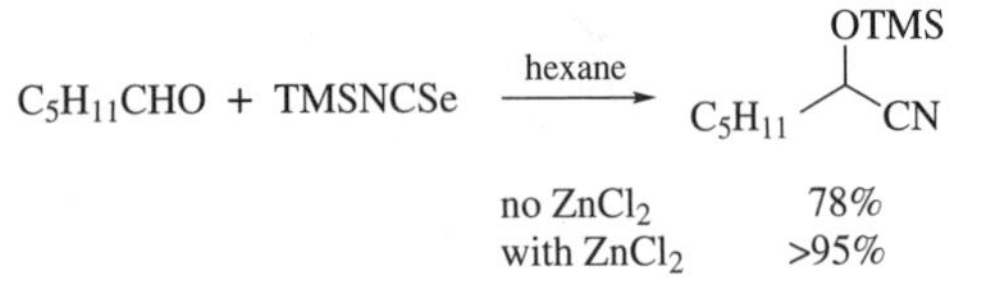

$C_5H_{11}CHO$ + TMSNCSe → (hexane) (30)

no $ZnCl_2$ — 78%
with $ZnCl_2$ — >95%

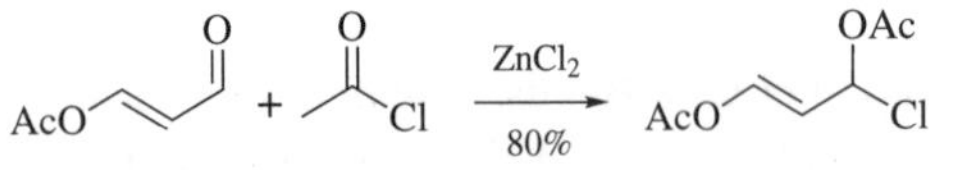

(31)

It is noteworthy that the treatment of carbonyl compounds with ***Chlorotrimethylsilane***/$ZnCl_2$ results in a useful synthesis of Me_3Si enol ethers which, in turn, may be useful for other carbon bond-forming processes (eq 32).[49]

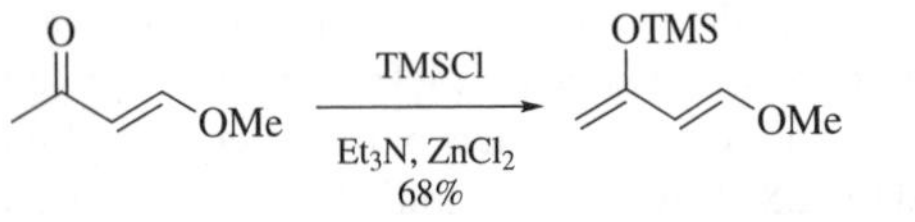

(32)

Activation of C–X Bonds. The activation of C–X single bonds toward nucleophilic substitution is also mediated by the Lewis acidic character of $ZnCl_2$.[3] Benzylic (eq 33),[50] allylic (eq 34),[3d,51] propargylic,[52] and tertiary halides (eq 35)[53] undergo substitution with mild carbon and heteroatom[3e] nucleophiles.

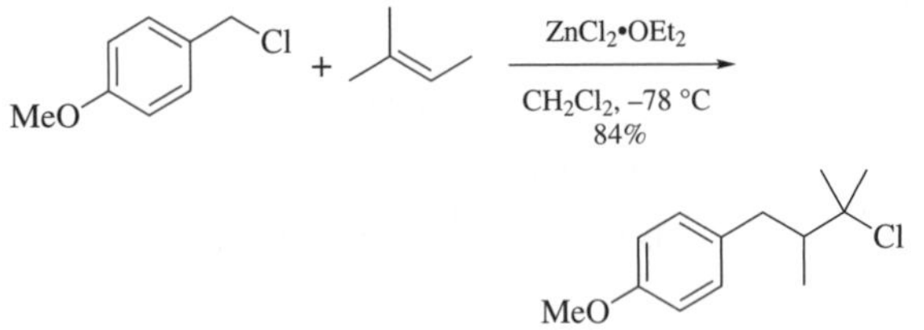

(33)

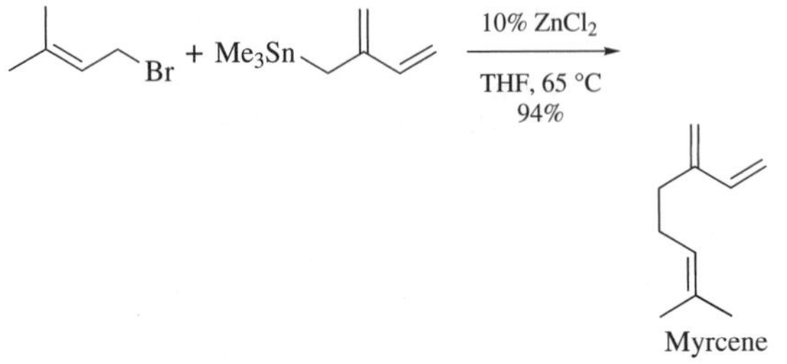

(34)

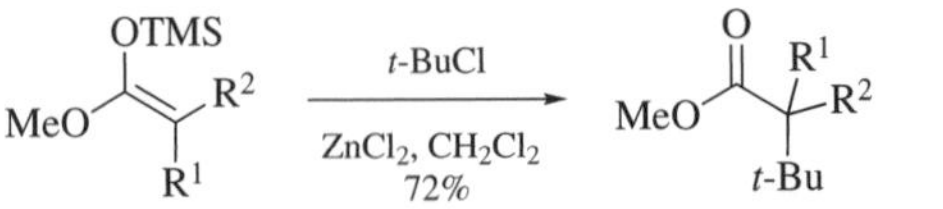

(35)

In a similar fashion, acetals (eq 36)[54] and orthoesters (eq 37)[55] may be used as electrophiles in substitution reactions with electron-rich alkenic nucleophiles. The combination of $ZnCl_2$ with co-catalysts has sometimes proven advantageous in these reactions (eq 38).[56]

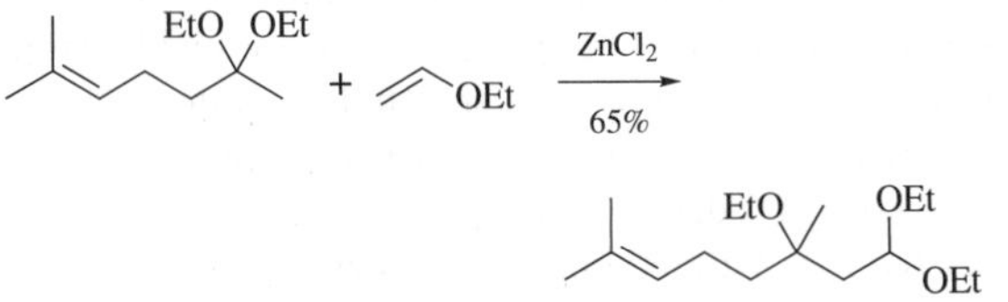

(36)

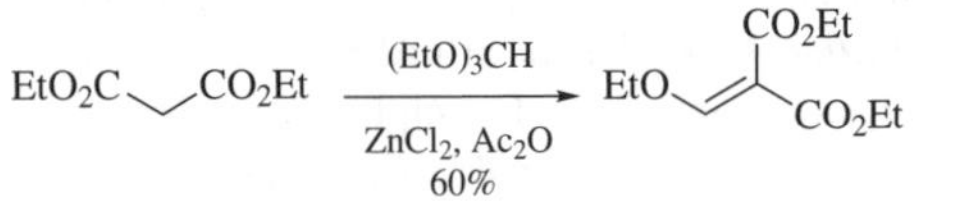

(37)

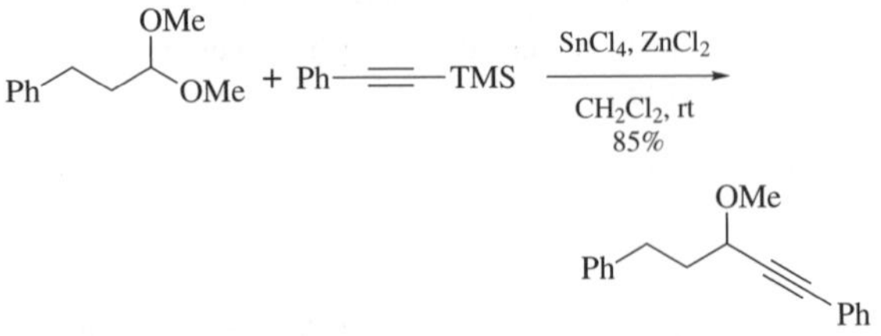

(38)

The regioselective ring opening reactions of epoxides (eq 39),[57,58] oxetanes,[58] and tetrahydrofurans (eq 40)[59] has been promoted by $ZnCl_2$ to afford adducts with suitable nucleophiles.

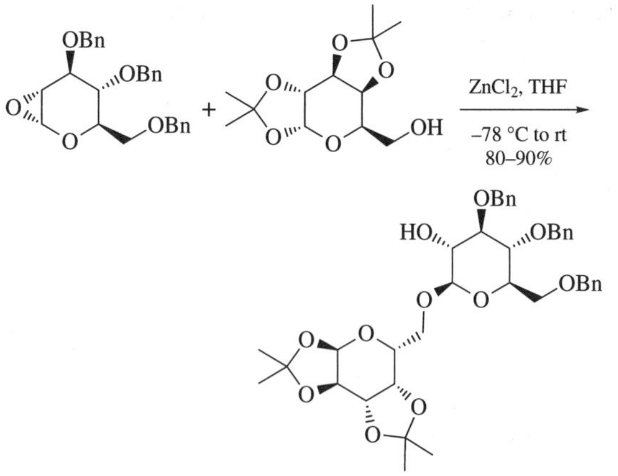

(39)

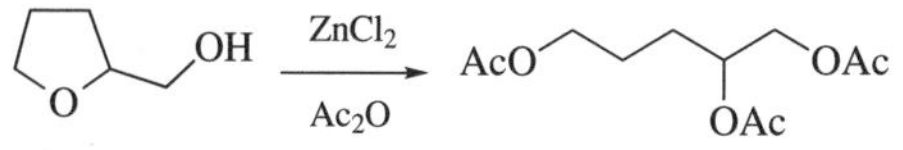

(40)

Activation by $ZnCl_2$ of allylic (eq 41)[60] and propargylic chlorides (eq 42),[61] as well as α-chloroenamines (eq 43),[62] in the presence of simple alkenes has been shown to yield four-membered and five-membered cycloadducts.

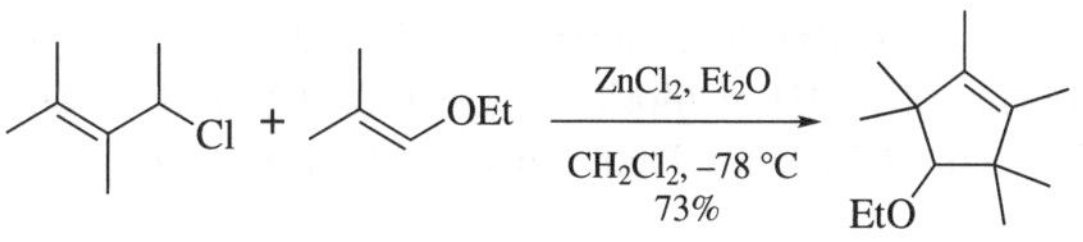

(41)

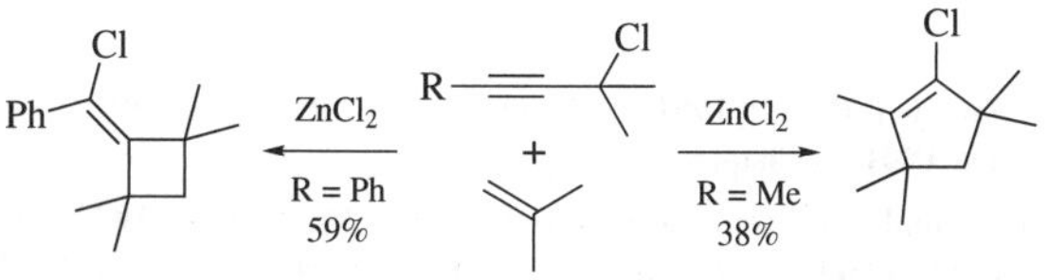

(42)

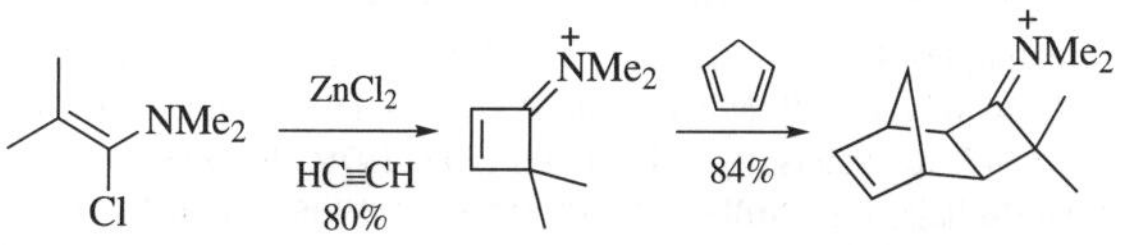

(43)

Chlorination of alcohols by ***Thionyl Chloride***,[63] the preparation of acyl chlorides from lactones and anhydrides,[64] and the bromination and iodination of aromatic rings by ***Benzyltrimethylammonium Tribromide*** and ***Benzyltrimethylammonium Dichloroiodate***, respectively,[65] are all effectively catalyzed by the presence of $ZnCl_2$. In addition, $ZnCl_2$ acts as a source of chloride for the halogenation of primary, secondary, and allylic alcohols using ***Triphenylphosphine–Diethyl Azodicarboxylate***.[66]

Reduction. The very useful reducing agent ***Zinc Borohydride*** is prepared by exposure of $NaBH_4$ to $ZnCl_2$ in ether solvents.[67] Its use in selective reductions is described elsewhere.[6a] A complex reducing agent resulting from the mixture of NaH/*t*-pentyl-OH/$ZnCl_2$, given the acronym ZnCRA (***Zinc Complex Reducing Agents***), is found to open epoxides in a highly regioselective fashion, favoring hydride delivery at the least hindered position.[68] A related reagent, ZnCRASi, which includes Me_3SiCl in the mixture, selectively reduces ketones in high yield, though the levels of selectivity do not compete with other selective reagents.[69]

The reduction of ketones and aldehydes by silicon and tin hydrides in the presence of $ZnCl_2$ has been documented.[70] In the presence of Pd^0 catalysts and $ZnCl_2$, these hydrides selectively reduce α,β-unsaturated aldehydes and ketones to the corresponding saturated products (eq 44).[71] The presence of $ZnCl_2$ has been shown to modify the reactivity of several common reducing agents. For example, the mixture of ***Sodium Cyanoborohydride*** and $ZnCl_2$ selectively reduces tertiary, allylic, and benzylic halides (eq 45),[72] ***Sodium Borohydride***, in the presence of $ZnCl_2$ and $PhNMe_2$ will reduce aryl esters to primary alcohols,[73] and ***Lithium Aluminum Hydride*** with $ZnCl_2/CuCl_2$ desulfurizes dithianes (eq 46).[74]

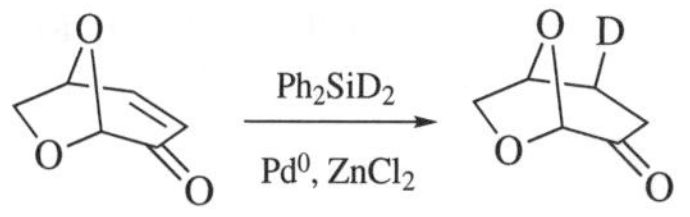

(44)

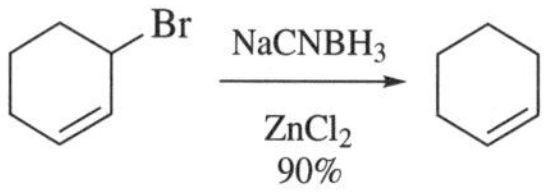

(45)

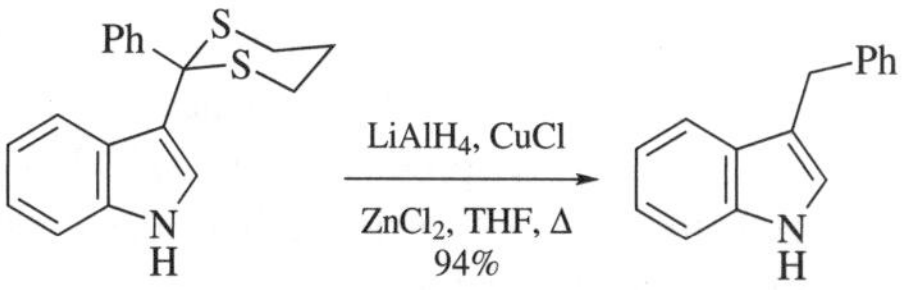

(46)

Stereoselectivity is also modified by the presence of $ZnCl_2$ (eq 47).[75] Enantioselective reduction of aryl ketones has been observed with ***Diisobutylaluminum Hydride*** (DIBAL) modified by $ZnCl_2$ and chiral diamine ligands (eq 48).[76]

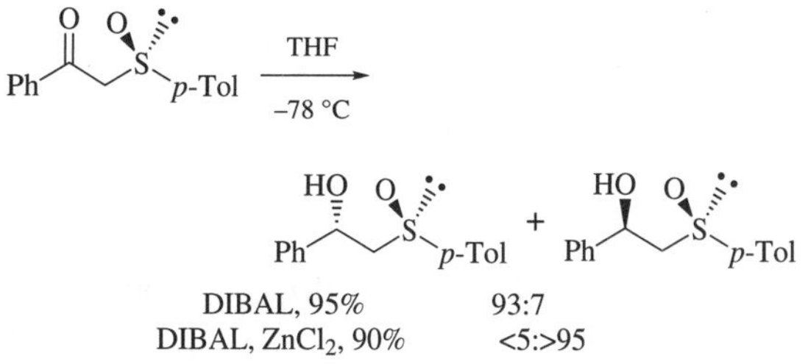

(47)

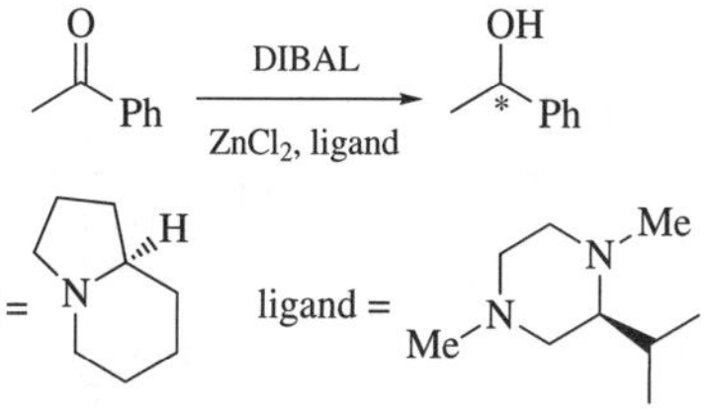

(48)

Protection/Deprotection. The acetylation of carbohydrates and other alcohols has been realized using ***Acetic Anhydride***/$ZnCl_2$ (eq 49).[77] It is found that $ZnCl_2$ imparts selectivity to both acetylation[78] and acetonide formation[79] of polyols, which is useful in the synthetic manipulation of carbohydrates. Synthetically useful selective deprotection of acetates (eq 50)[80] and dimethyl acetals (eq 51)[81] has been reported to be mediated by $ZnCl_2$.

(49) Ac_2O, $ZnCl_2$; H_2O, Δ; 79%

(50) $ZnCl_2$, EtOH: R^1 = H, R^2 = Ac, 100%; $SnCl_2$: R^1 = Ac, R^2 = H, 100%

(51) AcCl; $ZnCl_2$; 89%

Acylation. Unsaturated esters are obtained through acylation of alkenes by anhydrides using activation by $ZnCl_2$ (eq 52).[82] The mixture of $RCOCl/ZnCl_2$ is effective in the acylation of silyl enol ethers to afford β-dicarbonyl products (eq 53).[83] Friedel–Crafts acylation is catalyzed by $ZnCl_2$, using anhydrides or acyl halides as the electrophiles (eqs 54 and 55).[84,85]

(52) $(EtCO)_2O$; $ZnCl_2$, H_2O; 79%

(53) t-BuCOCl; $ZnCl_2$; 90%

(54) Ac_2O; $ZnCl_2$, Δ; 65%

(55) $ZnCl_2$; $POCl_3$; 68%

Aromatic Substitution. Several important classes of aromatic substitutions are mediated by $ZnCl_2$, including the Hoesch reaction (eq 56)[86] and the Fischer indole synthesis (eq 57).[87] Haloalkylation of aromatic rings using ***Formaldehyde*** or ***Chloromethyl Methyl Ether*** is readily accomplished through the agency of $ZnCl_2$ and warming (eq 58).[88]

(56) MeCN, $ZnCl_2$; HCl, Et_2O; H_2O; 87%

(57) $ZnCl_2$; cumene, Δ; 86%

(58) $ClCH_2OMe$; $ZnCl_2$, Δ; 86%

Related Reagents. Aluminum Chloride; Diphenylsilane–Tetrakis(triphenylphosphine)palladium(0)–Zinc Chloride; Phosphorus(III) Chloride–Zinc(II) Chloride; Phosphorus Oxychloride–Zinc(II) Chloride; Tin(IV) Chloride; Tin(IV) Chloride–Zinc Chloride; Titanium(IV) Chloride; Zinc Chloride Etherate.

---

1. (a) Knochel, P. *COS* **1991**, *1*, Chapter 1.7. (b) Rathke, M. W.; Weipert, P. *COS* **1991**, *2*, Chapter 1.8.
2. For examples: (a) Danishefsky, S.; DeNinno, M. P. *AG(E)* **1987**, *26*, 15. (b) Danishefsky, S. *Aldrichim. Acta* **1986**, *19*, 59. (c) Chou, S.-S.; Sun, D.-J. *CC* **1988**, 1176. (d) Liu, H.-J.; Ulibarri, G.; Browne, E. N. C. *CJC* **1992**, *70*, 1545. (e) Liu, H.-J.; Han, Y. *TL* **1993**, *34*, 423.
3. For examples: (a) Zhai, D.; Zhai, W.; Williams, R. M. *JACS* **1988**, *110*, 2501. (b) Williams, R. M.; Sinclair, P. J.; Zhai, D.; Chen, D. *JACS* **1988**, *110*, 1547. (c) Paterson, I.; Fleming, I. *TL* **1979**, *20*, 993, 995, 2175. (d) Godschalz, J. P.; Stille, J. K. *TL* **1983**, *24*, 1905. (e) Miller, J. A. *TL* **1975**, 2050. (f) Shikhmamedbekova, A. Z.; Sultanov, R. A. *JGU* **1970**, *40*, 72. (g) Ishibashi, H.; Nakatani, H.; Umei, Y.; Yamamoto, W.; Ikeda, M. *JCS(P1)* **1987**, 589. (h) Mori, I.; Bartlett, P. A.; Heathcock, C. H. *JACS* **1987**, *109*, 7199.
4. For examples: (a) Eliel, E. L.; Hutchins, Sr., R. O.; Knoeber, M. *OS* **1970**, *50*, 38. (b) Angle, S. R.; Turnbull, K. D. *JACS* **1989**, *111*, 1136. (c) Takai, K.; Heathcock, C. H. *JOC* **1985**, *50*, 3247. (d) Chiba, T.; Nakai, T. *CL* **1987**, 2187. (e) Taguchi, T.; Kitagawa, O.; Suda, Y.; Ohkawa, W.; Hashimoto, A.; Iitaka, Y.; Kobayashi, Y. *TL* **1988**, *29*, 5291. (f) Kunz, H.; Pfrengle, W. *AG(E)* **1989**, *28*, 1067. (g) Kunz, H.; Pfrengle, W. *JOC* **1989**, *54*, 4261.
5. For examples: (a) Gulati, K. C.; Seth, S. R.; Venkataraman, K. *OSC* **1943**, *2*, 522. (b) Böhmer, V.; Deveaux, J. *OPP* **1972**, *4*, 283. (c) Chapman, N. B.; Clarke, K.; Hughes, H. *JCS* **1965**, 1424.
6. For examples: (a) Oishi, T.; Nakata, T. *ACR* **1983**, *24*, 2653. (b) Fort, Y.; Feghouli, A.; Vanderesse, R.; Caubère, P. *JOC* **1990**, *55*, 5911. (c) Solladié, G.; Demailly, G.; Greck, C. *OS* **1977**, *56*, 8.
7. Boerma, J. In *Comprehensive Organometallic Chemistry*; Wilkinson, G., Ed.; Pergamon: Oxford, 1982, *2*, Chapter 16.
8. Rieke, R. D.; Uhm, S. J. *S* **1975**, 452.
9. Allylic organozinc chlorides: (a) Masuyama, Y.; Kinugawa, N.; Kurusu, Y. *JOC* **1987**, *52*, 3702. (b) Tamao, K.; Nakajo, E.; Ito, Y. *JOC* **1987**,

52, 957. (c) Pétrier, C.; Luche, J.-L. *JOC* **1985**, *50*, 910. (d) Tamao, K.; Nakajo, E.; Ito, Y. *T* **1988**, *44*, 3997. (e) Jacobson, R. N.; Clader, J. W. *TL* **1980**, *21*, 1205. (f) Hua, D. H.; Chon-Yu-King, R.; McKie, J. A.; Myer, L. *JACS* **1987**, *109*, 5026. (g) Courtois, G.; Harama, M.; Miginiac, P. *JOM* **1981**, *218*, 275. (h) Evans, D. A.; Sjogren, E. B. *TL* **1986**, *27*, 4961. (i) Fujisawa, T.; Kofima, E.; Itoh, T.; Sato, T. *TL* **1985**, *26*, 6089. (j) Fang, J.-M.; Hong, B.-C. *JOC* **1987**, *52*, 3162. (k) Auvray, P.; Knochel, P.; Normant, J.-F. *TL* **1986**, *27*, 5091. For general reviews on the reactions of allylic organometallic reagents, see ref. 10.

10. (a) Yamamoto, Y. *ACR* **1987**, *20*, 243. (b) Hofmann, R. W. *AG(E)* **1982**, *21*, 555. (c) Yamamoto, Y.; Maruyama, K. *H* **1982**, *18*, 357. (d) Courtois, G.; Miginiac, L. *JOM* **1974**, *69*, 1.
11. (a) Zweifel, G.; Hahn, G. *JOC* **1984**, *49*, 4565. (b) Evans, D. A.; Nelson, J. V. *JACS* **1980**, *102*, 774.
12. Verlhac, J.-B.; Pereyre, M. *JOM* **1990**, *391*, 283.
13. (a) Fuganti, C.; Grasselli, P.; Pedrocchi-Fantoni, G. *JOC* **1983**, *48*, 909. (b) Fronza, G.; Fuganti, C.; Grasselli, P.; Pedrocchi-Fantoni, G. *J. Carbohydr. Chem.* **1983**, *2*, 225.
14. Watson, R. A.; Kjonaas, R. A. *TL* **1986**, *27*, 1437.
15. Yamamoto, Y.; Chounan, Y.; Tanaka, M.; Ibuka, T. *JOC* **1992**, *57*, 1024. See also: Arai, M.; Kawasuji, T.; Nakamura, E. *CL* **1993**, 357.
16. For examples: (a) Negishi, E.; Takahashi, T.; King, A. O. *OS* **1988**, *66*, 67. (b) Murahashi, S.-I.; Yamamura, M.; Yanagisawa, K.; Mita, N.; Kondo, K.; Negishi, E. *JOC* **1983**, *48*, 1560. (c) Tius, M. A.; Trehan, S. *JOC* **1986**, *51*, 765. (d) Rossi, R.; Carpita, A.; Cossi, P. *T* **1992**, *48*, 8801. (e) Shiragami, H.; Kawamoto, T.; Imi, K.; Matsubara, S.; Utimoto, K.; Nozaki, H. *T* **1988**, *44*, 4009. (f) Matsushita, H.; Negishi, E. *JACS* **1981**, *103*, 2882. (g) Pelter, A.; Rowlands, M.; Jenkins, I. H. *TL* **1987**, *28*, 5213. (h) Andreini, B. P.; Carpita, A.; Rossi, R. *TL* **1988**, *29*, 2239. (i) Negishi, E.; Okukado, N.; Lovich, S. F.; Luo, T.-T. *JOC* **1984**, *49*, 2629.
17. Tamao, K.; Ishida, M.; Kumada, M. *JOC* **1983**, *48*, 2120.
18. Russell, C. E.; Hegedus, L. S. *JACS* **1983**, *105*, 943.
19. Takahashi, K.; Ogiyama, M. *CC* **1990**, 1196.
20. (a) King, A. O.; Okukado, N.; Negishi, E. *CC* **1977**, 683. (b) King, A. O.; Negishi, E.; Villiani, Jr., F. J.; Silveira, A. *JOC* **1978**, *43*, 358.
21. Negishi, E.; Takahashi, T.; Baba, S. *OS* **1988**, *66*, 60. See also: Zweifel, G.; Miller, J. A. *OR* **1984**, *32*, 375.
22. (a) Jabri, N.; Alexakis, A.; Normant, J.-F. *BSF(2)* **1983**, *321*, 332. (b) Jabri, N.; Alexakis, A.; Normant, J.-F. *TL* **1981**, *22*, 959, 3851. (c) Nunomoto, S.; Kawakami, Y.; Yamashita, Y. *JOC* **1983**, *48*, 1912.
23. Negishi, E.; Okukado, N.; King, A. O.; Van Horn, D. E.; Spiegel, B. I. *JACS* **1978**, *100*, 2254. See also: Van Horn, D. E.; Valente, L. F.; Idacavage, M. J.; Negishi, E. *JOM* **1978**, *156*, C20.
24. Koga, T.; Makinouchi, S.; Okukado, N. *CL* **1988**, 1141.
25. (a) Godschalx, J.; Stille, J. K. *TL* **1980**, *21*, 2599. (b) Erdelmeier, I.; Gais, H.-J. *JACS* **1989**, *111*, 1125. (c) Cahiez, G.; Chavant, P.-Y. *TL* **1989**, *30*, 7373.
26. For examples: (a) Jansen, J. F. G. A.; Feringa, B. L. *JOC* **1990**, *55*, 4168. (b) Soai, K.; Kawase, Y.; Oshio, A. *JCS(P1)* **1991**, 1613.
27. Boireau, G.; Deberly, A.; Abeuhaim, D. *T* **1989**, *45*, 5837. See also: Fujisawa, T.; Ukaji, Y.; Funabora, M.; Yamashita, M.; Sato, T. *BCJ* **1990**, *63*, 1894.
28. Seebach, D.; Behrendt, L.; Felix, D. *AG(E)* **1991**, *30*, 1008.
29. House, H. O.; Crumrine, D. S.; Teranishi, A. Y.; Olmstead, H. D. *JACS* **1973**, *95*, 3310.
30. For a general review on stereoselective aldol condensations, see: Evans, D. A.; Nelson, J. V.; Taber, T. R. *Top. Stereochem.* **1982**, *13*, 1.
31. van der Steen, F. H.; Jastrzebski, J. T. B. H.; van Koten, G. *TL* **1988**, *29*, 765, 2467. See also: van der Steen, F. H.; Boersma, J.; Spek, A. L.; van Koten, G. *OM* **1991**, *10*, 2467.
32. Groth, U.; Huhn, T.; Richter, N. *LA* **1993**, 49.
33. Itoh, K.; Hamaguchi, N.; Miura, M.; Nomura, M. *JCS(P1)* **1992**, 2833.
34. Nakamura, E.; Sekiya, K.; Kuwajima, I. *TL* **1987**, *28*, 337.
35. Nakamura, E.; Aoki, S.; Sekiya, K.; Oshino, H.; Kuwajima, I. *JACS* **1987**, *109*, 8056.
36. Crimmins, M. T.; Nantermet, P. G. *JOC* **1990**, *55*, 4235.
37. (a) Tamaru, Y.; Ochiai, H.; Nakamura, T.; Yoshida, Z. *TL* **1986**, *27*, 955. (b) Tamaru, Y.; Ochiai, H.; Nakamura, T.; Tsubaki, K.; Yoshida, Z.-i. *TL* **1985**, *26*, 5559.
38. (a) Danishefsky, S. J.; Kerwin, Jr., J. F.; Kobayashi, S. *JACS* **1982**, *104*, 358. (b) Danishefsky, S. J.; Larson, E. R.; Askin, D. *JACS* **1982**, *104*, 6437. (c) Larson, E. R.; Danishefsky, S. J. *JACS* **1982**, *104*, 6458. (d) Danishefsky, S. J.; Maring, C. J. *JACS* **1985**, *107*, 1269. (e) Danishefsky, S. J.; Larson, E. R.; Askin, D.; Kato, N. *JACS* **1985**, *107*, 1246. (f) Danishefsky, S. J.; Kato, N.; Askin, D.; Kerwin, Jr., J. F. *JACS* **1982**, *104*, 360. (g) Midland, M. M.; Graham, R. S. *JACS* **1984**, *106*, 4294. (h) Garner, P. *TL* **1984**, *25*, 5855.
39. Danishefsky, S. J.; Hungate, R. *JACS* **1986**, *108*, 2486.
40. Waldmann, H.; Braun, M. *JOC* **1992**, *57*, 4444.
41. Rao, P. S.; Venkataratnam, R. V. *TL* **1991**, *32*, 5821. See also Ref. 4a.
42. Hofstraat, R. G.; Lange, J.; Scheeren, H. W.; Nivard, R. J. F. *JCS(P1)* **1988**, 2315.
43. van der Werf, A. W.; Kellogg, R. M.; van Bolhuis, F. *CC* **1991**, 682.
44. Beslin, P.; Marion, P. *TL* **1992**, *33*, 5339.
45. Page, P. C. B.; Harkin, S. A.; Marchington, A. P. *SC* **1989**, *19*, 1655. See also: Yamauchi, M.; Shirota, M.; Watanabe, T. *H* **1990**, *31*, 1699.
46. Sukata, K. *JOC* **1989**, *54*, 2015. See also: (a) Deuchert, K.; Hertenstein, W.; Wehner, G. *CB* **1979**, *112*, 2045. (b) Ruano, J. L. G.; Castro, A. M. M.; Rodriguez, J. H. *TL* **1991**, *32*, 3195.
47. Kunz, H.; Sager, W.; Pfrengle, W.; Schanzenback, D. *TL* **1988**, *29*, 4397.
48. (a) Neuenschwander, M.; Bigler, P.; Christen, K.; Iseli, R.; Kyburz, R.; Mühle, H. *HCA* **1978**, *61*, 2047. (b) Bigler, P.; Schonholzer, S.; Neuenschwander, M. *HCA* **1978**, *61*, 2059. (c) Bigler, P.; Neuenschwander, M. *HCA* **1978**, *61*, 2165.
49. Danishefsky, S. J.; Kitahara, T. *JACS* **1974**, *96*, 7807.
50. Bäuml, E.; Tschemschlok, K.; Pock, R.; Mayr, H. *TL* **1988**, *29*, 6925. See also: (a) Clark, J. H.; Kybett, A. P.; Macquarrie, D. C.; Barlow, S. J.; Landon, P. *CC* **1989**, 1353. (b) Reetz, M. T.; Sauerwald, M. *TL* **1983**, *24*, 2837.
51. Other examples: (a) Koschinsky, R.; Köhli, T.-P.; Mayr, H. *TL* **1988**, *29*, 5641. (b) Alonso, F.; Yus, M. *T* **1991**, *47*, 9119.
52. Mayr, H.; Klein, H. *JOC* **1981**, *46*, 4097.
53. Reetz, M. T.; Schwellus, K. *TL* **1978**, 1455.
54. Isler, O.; Schudel, P. *Adv. Org. Chem.* **1963**, *4*, 128. See also: Oriyama, T.; Iwanami, K.; Miyauchi, Y.; Koga, G. *BCJ* **1990**, 3716.
55. Parham, W. E.; Reed, L. J. *OSC* **1955**, *3*, 395. See also: (a) Grégoire, de Bellemont, E. *BSF* **1901**, *25*, 18. (b) Hatanaka, K.; Tanimoto, S.; Sugimoto, T.; Okano, M. *TL* **1981**, *22*, 3243.
56. Hayashi, M.; Inubushi, A.; Mukaiyama, T. *BCJ* **1988**, *61*, 4037.
57. Halcomb, R. L.; Danishefsky, S. J. *JACS* **1989**, *111*, 6661.
58. Scheeren, H. W.; Dahman, F. J. M.; Bakker, C. G. *TL* **1979**, 2925. See also: (a) Sukata, K. *BCJ* **1990**, *63*, 825. (b) Hug, E.; Mellor, M.; Scovell, E. G.; Sutherland, J. K. *CC* **1978**, 526.
59. Grummett, O.; Stearns, J. A.; Arters, A. A. *OSC* **1955**, *3*, 833.
60. Klein, H.; Mayr, H. *AG(E)* **1981**, *20*, 1027.
61. Mayr, H.; Seitz, B.; Halberstat-Kausch, I.-K. *JOC* **1981**, *46*, 1041.
62. Hoornaert, C.; Hesbain-Frisque, A. M.; Ghosez, L. *AG(E)* **1975**, *14*, 569. See also: Sidani, A.; Marchand-Brynaert, J.; Ghosez, L. *AG(E)* **1974**, *13*, 267.
63. Squires, T. G.; Schmidt, W. W.; McCandlish, Jr., C. S. *JOC* **1975**, *40*, 134.
64. (a) Goel, O. P.; Seamans, R. E. *S* **1973**, 538. (b) Kyrides, L. P. *OSC* **1943**, *2*, 528.
65. (a) Kajigaeshi, S.; Kakinami, T.; Moriwaki, M.; Tanaka, T.; Fujisaki, S.; Okamoto, T. *BCJ* **1989**, *62*, 439. (b) Kajigaeshi, S.; Kakinami, T.; Watanabe, F.; Odanoto, T. *BCJ* **1989**, *62*, 1349.

66. Ho, P.-T.; Davies, N. *JOC* **1984**, *49*, 3027.
67. Gensler, W. J.; Johnson, F.; Sloan, A. D. B. *JACS* **1960**, *82*, 6074.
68. Fort, Y.; Vanderesse, R.; Caubère, P. *TL* **1985**, *26*, 3111.
69. See Ref. 6b and Caubère, P.; Vanderesse, R.; Fort, Y. *ACS* **1991**, *45*, 742.
70. (a) Anwar, S.; Davies, A. P. *T* **1988**, *44*, 3761. (b) Laurent, A. J.; Lesniak, S. *TL* **1992**, *33*, 3311.
71. Keinan, E.; Greenspoon, N. *JACS* **1986**, *108*, 7314.
72. Kim, S.; Kim, Y. J.; Ahn, K. H. *TL* **1983**, *24*, 3369.
73. Yamakawa, T.; Masaki, M.; Nohira, H. *BCJ* **1991**, *64*, 2730.
74. (a) Mukaiyama, T. *IJS(B)* **1972**, *7*, 173. (b) Stütz, P.; Stadler, P. A. *OS* **1977**, *56*, 8.
75. Solladié, G.; Demailly, G.; Greck, C. *TL* **1985**, *26*, 435. See also: Hanamoto, T.; Fuchikama, T. *JOC* **1990**, *55*, 4969 and Ref. 6c.
76. Falorni, M.; Giacomelli, G.; Lardicci, L. *G* **1990**, *120*, 765.
77. Braun, C. E.; Cook, C. D. *OS* **1961**, *41*, 79.
78. Hanessian, S.; Kagotani, M. *Carbohydr. Res.* **1990**, *202*, 67.
79. Angyal, S. J.; Gilham, P. T.; Macdonald, C. G. *JCS* **1957**, 1417.
80. Griffen, R. J.; Lowe, P. R. *JSC(P1)* **1992**, 1811.
81. Chang, C.; Chu, K. C.; Yue, S. *SC* **1992**, *22*, 1217.
82. Groves, J. K.; Jones, N. *JCS(C)* **1968**, 2898. See also: (a) Marshall, J. A.; Andersen, N. H.; Schlicher, J. W. *JOC* **1979**, *35*, 858. (b) Groves, J. K.; Jones, N. *JCS(C)* **1969**, 2350. (c) House, H. O.; Gilmore, W. F. *JACS* **1961**, *83*, 3980.
83. Tirpak, R. E.; Rathke, M. W. *JOC* **1982**, *47*, 5099. See also: Reetz, M. T.; Kyung, S.-H. *TL* **1985**, *26*, 6333.
84. Cooper, S. R. *OSC* **1955**, *3*, 761. See also: Baddeley, G.; Williamson, R. *JCS* **1956**, 4647. See also: Zani, C. L.; de Oliveira, A. B.; Snieckus, V. *TL* **1987**, *28*, 6561.
85. Dike, S. Y.; Merchant, J. R.; Sapre, N. Y. *T* **1991**, *47*, 4775. See also: (a) Shah, V. R.; Bose, J. L.; Shah, R. C. *JOC* **1960**, *25*, 677. (b) Dallacker, F.; Kratzer, P.; Lipp, M. *LA* **1961**, *643*, 97.
86. Ref. 5a and Ruske, W. In *Friedel–Crafts and Related Reactions*; Olah, G. A., Ed.; Interscience: New York, 1964, Vol. 3, p 383.
87. Ref. 5c and (a) Shriner, R. L.; Ashley, W. C.; Welch, E. *OSC* **1955**, *3*, 725. (b) Prochazki, M. P.; Cartson, R. *ACS* **1989**, *43*, 651.
88. Ref. 5b and Olah, G. A.; Tolgyesi, W. S. In *Friedel–Crafts and Related Reactions*; Olah, G. A., Ed.; Interscience: New York, 1964, Vol. 2, p 659.

Glenn J. McGarvey
University of Virginia, Charlottesville, VA, USA

Zinc Iodide[1]

ZnI_2

[10139-47-6] I_2Zn (MW 319.19)

(used in the preparation of organozinc reagents via transmetalation;[1] a mild Lewis acid useful for promoting addition[2] and substitution reactions[3])

Physical Data: mp 446 °C; bp 625 °C (dec); *d* 4.740 g cm^{-3}.
Solubility: sol H_2O (1 g/0.3 mL), glycerol (1 g/2 mL); freely sol EtOH, Et_2O.
Form Supplied in: white, odorless, granular solid; principal impurities are H_2O and iodine.
Analysis of Reagent Purity: mp.
Purification: heat to 300 °C under vacuum for 1 h, then sublime.
Handling, Storage, and Precautions: very hygroscopic and light sensitive; store under anhydrous conditions in the absence of light.

Organozinc Reagents. Organozinc reagents may be prepared through transmetalation of organolithium, organomagnesium, and organocopper species with ZnI_2, although ***Zinc Chloride*** and ***Zinc Bromide*** are used far more frequently.[1a] It is more usually the case that organozinc iodides are prepared by zinc insertion into alkyl iodides using ***Zinc/Copper Couple***.[1a] These zinc reagents possess reactivity analogous to the other organozinc halides, finding particular use in palladium-catalyzed coupling reactions (eq 1).[4] An alternative method for the preparation of the Simmons–Smith reagent (***Iodomethylzinc Iodide***) involving the treatment of ***Diazomethane*** with ZnI_2 has been reported (eq 2).[5]

+ IZn CO2Et; $PdCl_2(PPh_3)_2$; PhH, DMA, rt; 78%; N; I; CO2Et (1)

$$CH_2N_2 + ZnI_2 \longrightarrow ICH_2ZnI + N_2 \quad (2)$$

Cycloaddition Reactions. The catalytic effect of ZnI_2 on the Diels–Alder reaction has been noted (eq 3),[6] but its use in such cycloaddition reactions is rare compared with $ZnCl_2$ and $ZnBr_2$.

O; + CO2Me; ZnI_2; 40 °C; 55%; O; CO2Me; H (3)
endo:*exo* = 67:33

Activation of C=X Bonds. Catalysis of aldol condensation reactions using silyl ketene acetals and ZnI_2 has been the subject of several studies.[7] It is observed that ZnI_2 favors the activation of functionalized carbonyl compounds via β-chelates to impart useful stereoselectivity (eq 4).[7a] In analogous fashion, diastereoselective additions to imines and nitrones have been reported which offer useful access to β-lactams (eq 5).[8] α,β-Unsaturated esters are subject to conjugate addition reactions by silyl ketene acetals, also through the agency of ZnI_2 activation (eq 6).[9]

OTBDMS; MeO; + O O CHO; ZnI_2; MeCN; 67%; MeO; O OH; O O (4)
96% *anti*

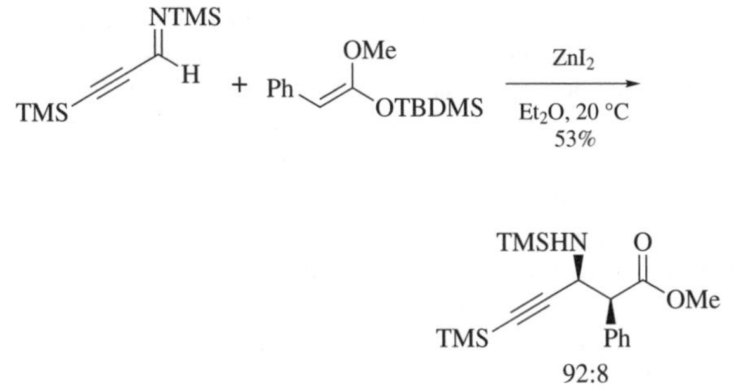

(5)

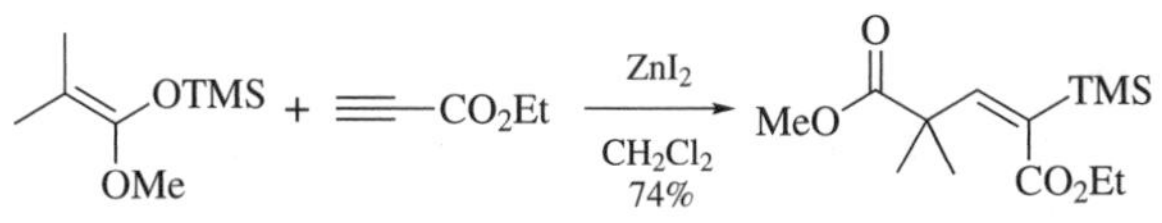

(6)

An important role for ZnI_2 has been found in the catalysis of R_3SiCN addition to ketones and aldehydes to afford silyl protected cyanohydrins.[10] This is a very general reaction that is effective even with very hindered carbonyl compounds (eq 7).[11] Diastereoselective cyanohydrin formation has been reported when these reaction conditions are applied to asymmetric carbonyl substrates (eq 8).[2a]

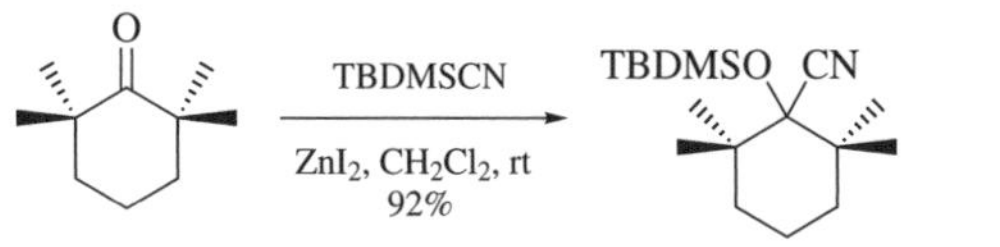

(7)

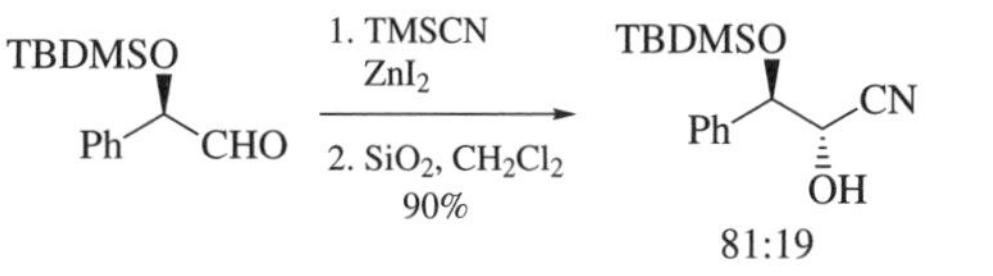

(8)

Activation of C–X Bonds. The Lewis acidity of ZnI_2 may be exploited to activate various carbon–heteroatom bonds to nucleophilic substitution. Treatment of epoxides and oxetanes with ***Cyanotrimethylsilane***/ZnI_2 results in selective C–N bond formation with ring opening (eq 9).[12] Similarly, C–S and C–Se bonds may be formed by treating cyclic ethers with ZnI_2 and $RSSiMe_3$ and $RSeSiMe_3$, respectively (eq 10).[13]

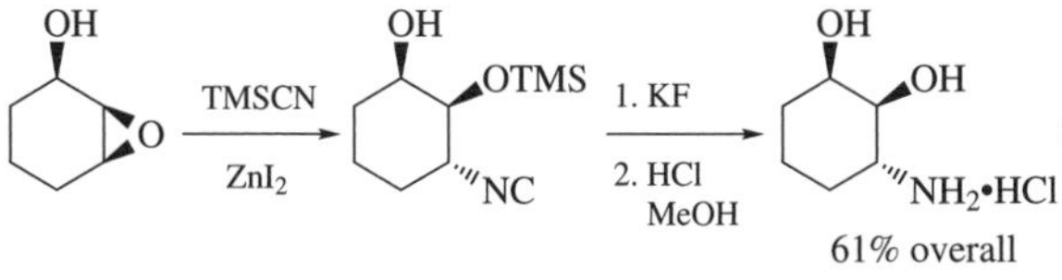

(9)

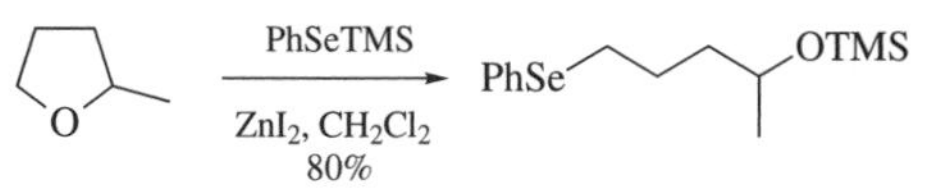

(10)

Treatment of 4-acetoxy- and 4-sulfoxyazetidin-2-ones with ZnI_2 results in the formation of the corresponding imine or iminium species which subsequently suffers silyl-mediated addition. These reactions result in the formation of *trans* substitution (eqs 11 and 12).[14,15]

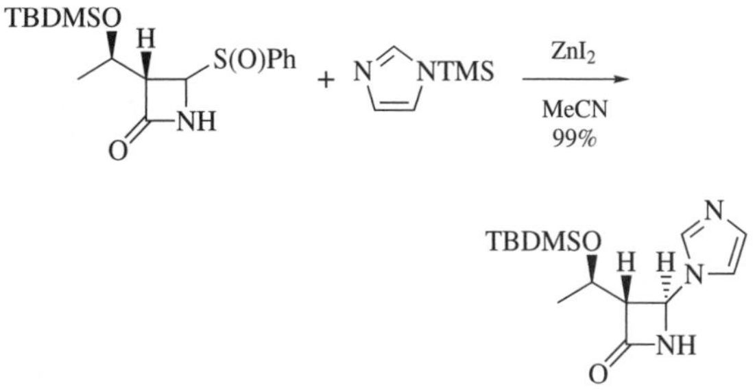

(11)

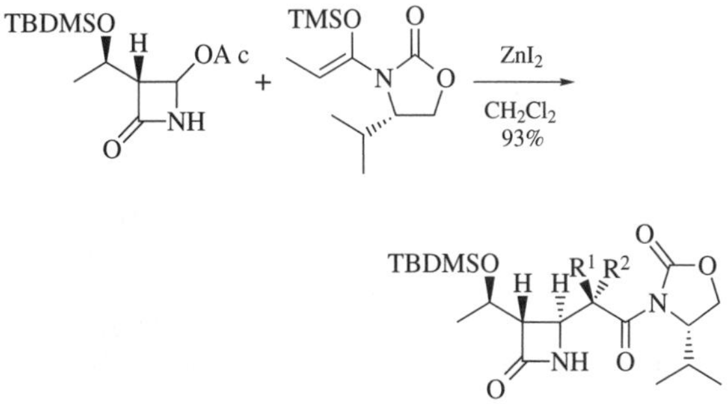

(12)

(R^1 = Me, R^2 = H):(R^1 = H, R^2 = Me) = 3:1

Substitution reactions of orthoesters[3b] and acetals,[3c] including anomeric bond formation in carbohydrates (eq 13),[16] have been catalyzed by ZnI_2.

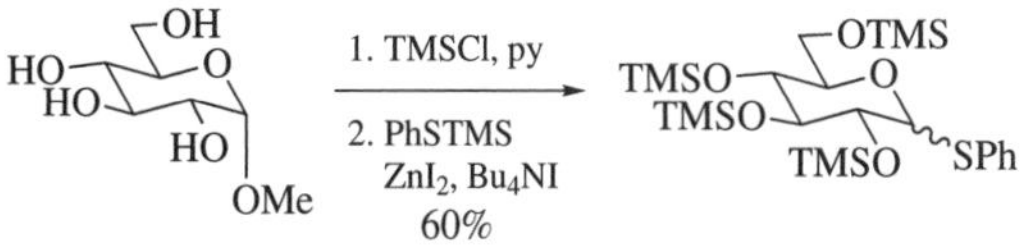

(13)

Reduction. Allyl and aryl ketones, aldehydes, and alcohols are reduced to the corresponding hydrocarbon by ***Sodium Cyanoborohydride***/ZnI_2 (eq 14).[17] This is a reasonably reactive reducing mixture which will also attack nitro and ester groups.

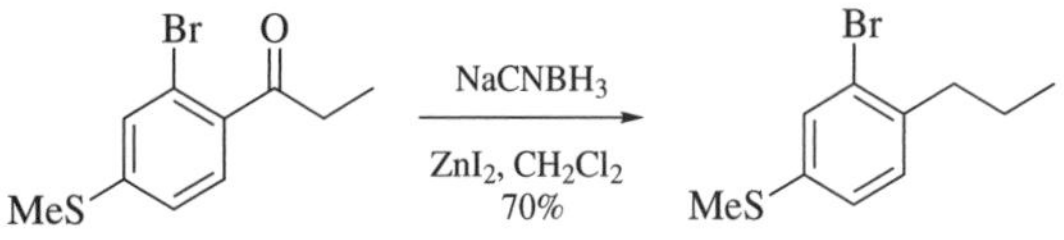

(14)

Deprotection. Methyl and benzyl ethers may be cleaved by the combination of ***(Phenylthio)trimethylsilane***/ZnI_2 (eq 15).[18] Similar cleavage of alkyl and benzyl ethers takes place using ***Acetic Anhydride***/ZnI_2 to afford the corresponding acetate (eq 16).[19]

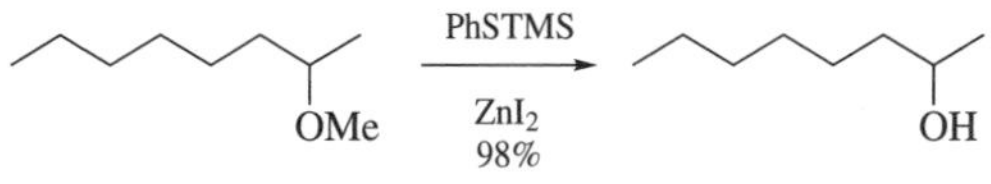

(15)

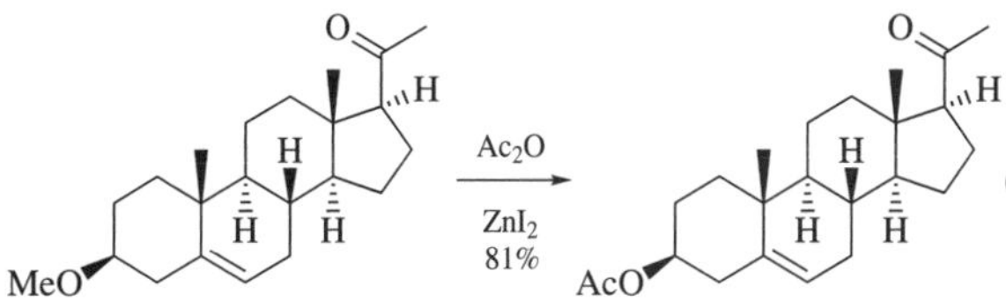

(16)

Related Reagents. Zinc Bromide; Zinc Chloride; Zinc/Copper Couple.

1. (a) Knochel, P. *COS* **1991**, *1*, Chapter 1.7. (b) Rathke, M. W.; Weipert, P. *COS* **1991**, *2*, Chapter 1.8.
2. For examples: (a) Effenberger, T.; Hopf, M.; Ziegler, T.; Hudelmayer, J. *CB* **1991**, *124*, 1651. (b) Klimba, P. G.; Singleton, D. A. *JOC* **1992**, *57*, 1733. (c) Colvin, E. W.; McGarry, D. G. *CC* **1985**, 539.
3. For examples: (a) Paquette, L. A.; Lagerwall, D. R.; King, J. L.; Niwayama, S.; Skerlj, R. *TL* **1991**, *32*, 5259. (b) Howk, B. W.; Sauer, J. C. *OSC* **1963**, *4*, 801. (c) Kubota, T.; Iijima, M.; Tanaka, T. *TL* **1992**, *33*, 1351.
4. (a) Sakamoto, T.; Nishimura, S.; Kondo, Y.; Yamanaka, H. *S* **1988**, 485. (b) See also: Yamanaka, H.; An-naka, M.; Kondo, Y.; Sakamoto, T. *CPB* **1985**, *38*, 4309.
5. (a) Wittig, G.; Schwarzenback, K. *LA* **1961**, *650*, 1. (b) Wittig, G.; Wingher, F. *LA* **1962**, *656*, 18.
6. Brion, F. *TL* **1982**, *23*, 5239.
7. (a) Kita, Y.; Yasuda, H.; Tamura, O.; Itoh, F.; Yuan Ke, Y.; Tamura, Y. *TL* **1985**, *26*, 5777. (b) Kita, Y.; Tamura, O.; Itoh, F.; Yasuda, H.; Kishino, H.; Yuan Ke, Y.; Tamura, Y. *JOC* **1988**, *53*, 554. (c) Annunziata, R.; Cinquini, M.; Cozzi, P. G.; Consolandi, E. *JOC* **1992**, *57*, 456.
8. (a) Colvin, E. W.; McGarry, D.; Nugent, M. J. *T* **1988**, *44*, 4157, and reference 2c. See also: (b) Kita, Y.; Itoh, F.; Tamura, O.; Yuan Ke, Y.; Tamura, Y. *TL* **1987**, *28*, 1431. (c) Reider, P. J.; Grabowski, E. J. J. *TL* **1982**, *23*, 2293. (d) Chiba, T.; Nakai, T. *CL* **1987**, 2187. (e) Chiba, T.; Nagatsuma, M.; Nakai, T. *CL* **1985**, 1343.
9. Quendo, A.; Rousseau, G. *TL* **1988**, *29*, 6443.
10. (a) Rassmussen, J. K.; Heihmann, S. M. *S* **1978**, 219. (b) Gassman, P.; Talley, J. J. *TL* **1978**, 3773. (c) Foley, L. H. *SC* **1984**, *14*, 1291. (d) Higuchi, K.; Onaka, M.; Izumi, Y. *CC* **1991**, 1035. (e) Quast, H.; Carlson, J.; Klaubert, C. A.; Peters, E.-M.; Peters, K.; von Schnering, H. G. *LA* **1992**, 759. (f) Batra, M. S.; Brunet, E. *TL* **1993**, *34*, 711.
11. Golinski, M.; Brock, C. P.; Watt, D. S. *JOC* **1993**, *58*, 159.
12. (a) Gassman, P. G.; Gremban, R. S. *TL* **1984**, *25*, 3259. (b) See also: Gassman, P. G.; Haberman, L. M. *TL* **1985**, *26*, 4971.
13. (a) Miyoshi, N.; Hatayama, Y.; Ryu, I.; Kambe, N.; Murai, T.; Murai, S.; Sonoda *S* **1988**, 175. (b) See also: Guidon, Y.; Young, R. N.; Frenette, R. *SC* **1981**, *11*, 391.
14. (a) Kita, Y.; Shibata, N.; Yoshida, N.; Tohjo, T. *TL* **1991**, *32*, 2375. (b) See also: Kita, Y.; Shibata, N.; Miki, T.; Takemura, Y.; Tamura, O. *CC* **1990**, 727.
15. Fuentes, L. M.; Shinkai, I.; Salzmann, T. N. *JACS* **1986**, *108*, 4675.
16. (a) Hanessian, S.; Guidon, Y. *Carbohydr. Res.* **1980**, *86*, C3. (b) Chu, S.-H. L.; Anderson, L. *Carbohydr. Res.* **1976**, *50*, 227.
17. Lau, C. K.; Dufresne, C.; Bélanger, P. C.; Piétré, S.; Scheigetz, J. *JOC* **1986**, *51*, 3038.
18. Hanessian, S.; Guidon, Y. *TL* **1980**, *21*, 2305.
19. Benedetti, M. O. V.; Monteagudo, E. S.; Burton, G. *JCS(S)* **1990**, 248.

Glenn J. McGarvey
University of Virginia, Charlottesville, VA, USA

List of Contributors

Ahmed F. Abdel-Magid — *The R. W. Johnson Pharmaceutical Research Institute, Spring House, PA, USA*
- Barium Hydroxide 24
- Lithium Hydroxide 224
- Potassium Hydroxide 310

Benjamin A. Anderson — *Lilly Research Laboratories, Indianapolis, IN, USA*
- Hexamethyldisilazane 158

Paul C. Anderson — *Bio-Méga/Boehringer Ingelheim Research, Laval, Québec, Canada*
Bromodimethylborane 59

William F. Bailey — *University of Connecticut, Storrs, CT, USA*
- *t*-Butyllithium 81

Bert H. Bakker — *University of Amsterdam, The Netherlands*
- Sulfur Trioxide 360

Wouter I. Iwema Bakker — *University of Waterloo, Ontario, Canada*
- Lithium Diisopropylamide 213

Martin G. Banwell — *University of Melbourne, Parkville, Victoria, Australia*
- Titanium Tetraisopropoxide 398

Albert K. Beck — *Eidgenössische Technische Hochschule, Zürich, Switzerland*
- Mesityllithium 251

John L. Belletire — *The University of Cincinnati, OH, USA*
- Sodium Amide 329

Joachim Berkner — *Georgia Institute of Technology, Atlanta, GA, USA*
- 18-Crown-6 118

T. Howard Black — *Eastern Illinois University, Charleston, IL, USA*
- Magnesium Bromide 237

Mary Ellen Bos — *The R. W. Johnson Pharmaceutical Research Institute, Raritan, NJ, USA*
- Benzyltrimethylammonium Hydroxide 27
- Tetra-*n*-butylammonium Hydroxide 367

Bruce P. Branchaud — *University of Oregon, Eugene, OR, USA*
- Cesium Fluoride 99
- Tetrafluoroboric Acid 368

Gary W. Breton — *University of North Carolina, Chapel Hill, NC, USA*
- Hydrogen Bromide 183
- Hydrogen Chloride 185

Steven D. Burke — *University of Wisconsin, Madison, WI, USA*
- Sulfur Dioxide 352

Drury Caine — *The University of Alabama, Tuscaloosa, AL, USA*
- Magnesium Methoxide 239
- Potassium *t*-Butoxide 290
- Lithium *t*-Butoxide 208

Mike Campbell — *University of Waterloo, Ontario, Canada*
- Lithium 2,2,6,6-Tetramethylpiperidide 232

Hans Cerfontain — *University of Amsterdam, The Netherlands*
- Sulfur Trioxide 360

André B. Charette — *Université de Montréal, Québec, Canada*
- Lithium Bromide 207
- Lithium Iodide 227
- Lithium Perchlorate 229

Paul J. Coleman — *Indiana University, Bloomington, IN, USA*
- Lithium Tetrafluoroborate 231

Daniel L. Comins — *North Carolina State University, Raleigh, NC, USA*
- Lithium *N*-Methylpiperazide 228

Veronica Cornel — *Emory University, Atlanta, GA, USA*
- Boron Trifluoride Etherate 50

Robert Dahinden — *Eidgenössische Technische Hochschule, Zürich, Switzerland*
- Mesityllithium 251

Kenneth G. Davenport — *Hoechst Celanese Corporation, Corpus Christi, TX, USA*
- Hydrogen Fluoride 187

Arnold Davis — *University of Miami, Coral Gables, FL, USA*
- Calcium Hydride 89

Kurt D. Deshayes — *Bowling Green State University, OH, USA*
- Potassium Carbonate 296
- Sodium Hydroxide 343

Lisa A. Dixon — *The R. W. Johnson Pharmaceutical Research Institute, Raritan, NJ, USA*
- Phosphorus(V) Oxide–Methanesulfonic Acid 273
- Polyphosphate Ester 277

John H. Dodd — *The R. W. Johnson Pharmaceutical Research Institute, Raritan, NJ, USA*
- Polyphosphoric Acid 279

Robert R. Dykstra — *University of Wisconsin, Madison, WI, USA*
- Hexamethylphosphoric Triamide 160

Thomas M. Eckrich — *Eli Lilly and Company, Lafayette, IN, USA*
- Tri-*n*-butyltin Trifluoromethanesulfonate 409

Kirk F. Eidman — *Scios Nova, Baltimore, MD, USA*
- Formic Acid 155
- Propionic Acid 313
- Trifluoroacetic Acid 419

Ihsan Erden — *San Francisco State University, CA, USA*
- Lithium Amide 204

Ronald H. Erickson — *Scios Nova, Baltimore, MD, USA*
- Bromomagnesium Diisopropylamide 61

Agnes Fabre — *The Ohio State University, Columbus, OH, USA*
- Alumina 9

Patrizia Ferraboschi — *Università di Milano, Italy*
- Esterases 143

Alan S. Florjancic — *The Ohio State University, Columbus, OH, USA*
- Alumina 9

Gregory K. Friestad — *University of Oregon, Eugene, OR, USA*
- Cesium Fluoride 99
- Tetrafluoroboric Acid 368

Adam A. Galan — *Parke Davis Pharmaceutical Research Division, Ann Arbor, MI, USA*
- Pyridinium *p*-Toluenesulfonate 318

Paul Galatsis — *University of Guelph, Ontario, Canada*
- Aluminum Chloride 12

Robert E. Gawley — *University of Miami, Coral Gables, FL, USA*
- Calcium Hydride 89
- Potassium Hydride 307
- Sodium Hydride 340

David Goldsmith — *Emory University, Atlanta, GA, USA*
- Pyrrolidine 319

Matthew Gray — *University of Waterloo, Ontario, Canada*
- Lithium Hexamethyldisilazide 221

Paride Grisenti — *Università di Milano, Italy*
- Esterases 143

Yvan Guindon — *Bio-Méga/Boehringer Ingelheim Research, Laval, Québec, Canada*
- Bromodimethylborane 59

Gregory S. Hamilton — *Scios Nova, Baltimore, MD, USA*
- *p*-Toluenesulfonic Acid 402

Michael Hanack — *Universität Tübingen, Germany*
- Trifluoromethanesulfonic Acid 421

Qi Han — *DuPont Merck, Wilmington, DE, USA*
- Potassium Fluoride 300

Gudmundur G. Haraldsson — *University of Iceland, Reykjavik, Iceland*
- Lipases 199

Shoji Hara — *Hokkaido University, Sapporo, Japan*
- Boron Tribromide 44

Michael Harmata — *University of Missouri-Columbia, MO, USA*
- Sodium Methylsulfinylmethylide 344

Roger Harrington — *Giba-Geigy, Summit, NJ, USA*
- Sodium Carbonate 333

Ronald G. Harvey — *University of Chicago, IL, USA*
- Hexamethylphosphorous Triamide 166
- Hydrogen Iodide 190

Alfred Hassner *Bar-Ilan University, Ramat Gan, Israel*
- 4-Dimethylaminopyridine 140

Dieter Hoppe *University of Münster, Germany*
- (−)-Sparteine 350

Yao-Zeng Huang *Shanghai Institute of Organic Chemistry, Academia Sinica, China*
- Triphenylarsine 431

Kazuaki Ishihara *Nagoya University, Japan*
- Aluminum Isopropoxide 16

Seiji Iwasa *The Ohio State University, Columbus, OH, USA*
- Alumina 9

Juan C. Jaen *Parke-Davis Pharmaceutical Research, Ann Arbor, MI, USA*
- Brucine 63

Thomas E. Janini *Kent State University, OH, USA*
- 2,6-Lutidine 235

Samantha Janisse *Eli Lilly and Company, Indianapolis, IN, USA*
- Tin(II) Trifluoromethanesulfonate 389

Sajan P. Joseph *North Carolina State University, Raleigh, NC, USA*
- Lithium *N*-Methylpiperazide 228

Bernard Jousseaume *Université Bordeaux I, Talence, France*
- Tri-*n*-butyl(methoxy)stannane 405

Michael E. Jung *University of California, Los Angeles, CA, USA*
- Triphenylcarbenium Tetrafluoroborate 434

Masato Kitamura *Nagoya University, Japan*
- (*R*)- and (*S*)-2,2′-Bis(diphenylphosphino)-1,1′-binaphthyl 38

David W. Knight *Nottingham University, UK*
- Imidazole 194

Pavel Kočovský *University of Leicester, UK*
- Mercury(II) Acetate 240
- Mercury(II) Chloride 246
- Mercury(II) Trifluoroacetate 248

Rafael R. Kostikov *St. Petersburg State University, Russia*
- 2,6-Di-*t*-butylpyridine 131

Paul J. Kropp *University of North Carolina, Chapel Hill, NC, USA*
- Hydrogen Bromide 183
- Hydrogen Chloride 185

Grant R. Krow *Temple University, Philadelphia, PA, USA*
- Sulfuric Acid 357

Kenneth K. Laali *Kent State University, OH, USA*
- Copper(II) Trifluoromethanesulfonate 116

James C. Lanter *The Ohio State University, Columbus, OH, USA*
- Molecular Sieves 259

Reinhard Schwesinger — *University of Freiburg in Breisgau, Germany*
- Phosphazene Base P_4-*t*-Bu 270

Dieter Seebach — *Eidgenössische Technische Hochschule, Zürich, Switzerland*
- Mesityllithium 251

Angela R. Sherman — *Reilly Industries, Indianapolis, IN, USA*
- 2,4,6-Collidine 104
- Pyridine 314
- Quinoline 324

Ichiro Shinkai — *Merck Research Laboratories, Rahway, NJ, USA*
- Tetrahydro-1-methyl-3,3-diphenyl-1*H*,3*H*-pyrrolo[1,2-*c*][1,3,2]oxazaborole 372

Li-Lan Shi — *Shanghai Institute of Organic Chemistry, Academia Sinica, China*
- Triphenylarsine 431

Mark R. Sivik — *Procter & Gamble, Cincinnati, OH, USA*
- Cesium Carbonate 97

Barry B. Snider — *Brandeis University, Waltham, MA, USA*
- Diethylaluminum Chloride 132
- Dimethylaluminum Chloride 137
- Ethylaluminum Dichloride 146
- Methylaluminum Dichloride 258

Victor Snieckus — *University of Waterloo, Ontario, Canada*
- Lithium 2,2,6,6-Tetramethylpiperidide 232
- Lithium Diethylamide 211
- Lithium Diisopropylamide 213
- Lithium Hexamethyldisilazide 221

Kirk L. Sorgi — *The R. W. Johnson Pharmaceutical Research Institute, Spring House, PA, USA*
- Diisopropylethylamine 134
- Triethylamine 410

Lakshminarayanapuram R. Subramanian — *Universität Tübingen, Germany*
- Trifluoromethanesulfonic Acid 421

Akira Suzuki — *Hokkaido University, Sapporo, Japan*
- Boron Tribromide 44

Joseph Sweeney — *University of Bristol, UK*
- Trimethylsilyl Trifluoromethanesulfonate 427

Katherine S. Takaki — *Bristol-Myers Squibb Co., Wallingford, CT, USA*
- Potassium 3-Aminopropylamide 288
- Potassium Amide 285
- Potassium Diisopropylamide 298

Masao Tsukazaki — *University of Waterloo, Ontario, Canada*
- Lithium Diethylamide 211

Kjell Undheim — *University of Oslo, Norway*
- Titanium(IV) Chloride 392

David E. Volk — *North Dakota State University, Fargo, ND, USA*
- Tin(IV) Chloride 383

Simone C. Vonwiller	*The University of Sydney, NSW, Australia*	
	• *N,N,N′,N′*-Tetramethylethylenediamine	376
Nanette Wachter-Jurcsak	*University of Connecticut, Storrs, CT, USA*	
	• *t*-Butyllithium	81
Qi Wang	*University of Southern California, Los Angeles, CA, USA*	
	• Antimony(V) Fluoride	19
	• Fluorosulfuric Acid–Antimony(V) Fluoride	151
Brett T. Watson	*Bristol-Myers Squibb Pharmaceutical Research Institute, Wallingford, CT, USA*	
	• Potassium Hexamethyldisilazide	303
	• Sodium Hexamethyldisilazide	338
D. Todd Whitaker	*Detroit Country Day School, Beverly Hills, MI, USA*	
	• Sodium Ethoxide	334
Andrew D. White	*Parke-Davis Pharmaceutical Research, Ann Arbor, MI, USA*	
	• Iron(III) Chloride	195
Lars-G. Wistrand	*Nycomed Innovation, Malmö, Sweden*	
	• Silver(I) Tetrafluoroborate	327
Poh Lee Wong	*University of Waterloo, Ontario, Canada*	
	• Lithium Diisopropylamide	213
Dennis Wright	*Ohio University, Athens, OH, USA*	
	• Lithium Carbonate	209
Xiaoyang Xia	*Emory University, Atlanta, GA, USA*	
	• *n*-Butyllithium–Potassium *t*-Butoxide	86
Hisashi Yamamoto	*Nagoya University, Japan*	
	• Methylaluminum Bis(2,6-di-*t*-butyl-4-methylphenoxide)	254
Hisashi Yamamoto	*Nagoya University, Japan*	
	• Aluminum Isopropoxide	16
	• Methylaluminum Bis(2,6-di-*t*-butylphenoxide)	257
Mark W. Zettler	*The Dow Chemical Company, Midland, MI, USA*	
	• Montmorillonite K10	262
Xiaojie Zhang	*University of Miami, Coral Gables, FL, USA*	
	• Potassium Hydride	307
Zhang-Lin Zhou	*Shanghai Institute of Organic Chemistry, Academia Sinica, China*	
	• Triphenylarsine	431

Reagent Formula Index

Subject Index

Reference Abbreviations

ABC	Agric. Biol. Chem.
AC(R)	Ann. Chim. (Rome)
ACR	Acc. Chem. Res.
ACS	Acta Chem. Scand.
AF	Arzneim.-Forsch.
AG	Angew. Chem.
AG(E)	Angew. Chem., Int. Ed. Engl.
AJC	Aust. J. Chem.
AK	Ark. Kemi
ANY	Ann. N. Y. Acad. Sci.
AP	Arch. Pharm. (Weinheim, Ger.)
B	Biochemistry
BAU	Bull. Acad. Sci. USSR, Div. Chem. Sci.
BBA	Biochim. Biophys. Acta
BCJ	Bull. Chem. Soc. Jpn.
BJ	Biochem. J.
BML	Bioorg. Med. Chem. Lett.
BSB	Bull. Soc. Chim. Belg.
BSF(2)	Bull. Soc. Chem. Fr. Part 2
C	Chimia
CA	Chem. Abstr.
CB	Ber. Dtsch. Chem. Ges./Chem. Ber.
CC	Chem. Commun./J. Chem. Soc., Chem. Commun.
CCC	Collect. Czech. Chem. Commun.
CED	J. Chem. Eng. Data
CI(L)	Chem. Ind. (London)
CJC	Can. J. Chem.
CL	Chem. Lett.
COS	Comprehensive Organic Synthesis
CPB	Chem. Pharm. Bull.
CR(C)	C. R. Hebd. Seances Acad. Sci., Ser. C
CRV	Chem. Rev.
CS	Chem. Ser.
CSR	Chem. Soc. Rev.
CZ	Chem.-Ztg.
DOK	Dokl. Akad. Nauk SSSR
E	Experientia
FES	Farmaco Ed. Sci.
FF	Fieser & Fieser
G	Gazz. Chim. Ital.
H	Heterocycles
HC	Heteroatom Chem.
HCA	Helv. Chim. Acta
IC	Inorg. Chem.
ICA	Inorg. Chim. Acta
IJC(B)	Indian J. Chem., Sect. B
IJS(B)	Int. J. Sulfur Chem., Part B
IZV	Izv. Akad. Nauk SSSR, Ser. Khim.
JACS	J. Am. Chem. Soc.
JBC	J. Biol. Chem.
JCP	J. Chem. Phys.
JCR(M)	J. Chem. Res. (M)
JCR(S)	J. Chem. Res. (S)
JCS	J. Chem. Soc.
JCS(C)	J. Chem. Soc. (C)
JCS(D)	J. Chem. Soc., Dalton Trans.
JCS(F)	J. Chem. Soc., Faraday Trans.
JCS(P1)	J. Chem. Soc., Perkin Trans. 1
JCS(P2)	J. Chem. Soc., Perkin Trans. 2
JFC	J. Fluorine Chem.
JGU	J. Gen. Chem. USSR (Engl. Transl.)
JHC	J. Heterocycl. Chem.
JIC	J. Indian Chem., Soc.
JMC	J. Med. Chem.
JMR	J. Magn. Reson.
JOC	J. Org. Chem.
JOM	J. Organomet. Chem.
JOU	J. Org. Chem. USSR (Engl. Transl.)
JPOC	J. Phys. Org. Chem.
JPP	J. Photochem. Photobiol.
JPR	J. Prakt. Chem.
JPS	J. Pharm. Sci.
KGS	Khim. Geterotsikl. Soedin.
LA	Justus Liebigs Ann. Chem./Liebigs Ann. Chem.
M	Monatsh. Chem.
MOC	Methoden Org. Chem. (Houben-Weyl)
MRC	Magn. Reson. Chem.
N	Naturwissenschaften
NJC	Nouv. J. Chim.
NKK	Nippon Kagaku Kaishi
OM	Organometallics
OMR	Org. Magn. Reson.
OPP	Org. Prep. Proced. Int.
OR	Org. React.
OS	Org. Synth.
OSC	Org. Synth., Coll. Vol.
P	Phytochemistry
PAC	Pure Appl. Chem.
PIA(A)	Proc. Indian Acad. Sci., Sect. A